Nonlinear Optimal and Flatness-based Control Methods and Applications for Complex Dynamical Systems

Other related titles:

You may also like

- PBCE133 | Rigatos | Control and Estimation of Dynamical Nonlinear and PDE Systems: Theory and applications | 2022
- PBCE126 | Monteriu | Fault Diagnosis and Fault-Tolerant Control of Robotic Systems | 2020
- PBCE1130 | Petkov | Design of Embedded Robust Control Systems Using MATLAB®/Simulink® | 2018
- PBCE1040 | Guvenc | Control of Mechatronic Systems | 2017
- PBCE0840 | Ding | Nonlinear and Adaptive Control Systems | 2013

We also publish a wide range of books on the following topics:
Computing and Networks
Control, Robotics and Sensors
Electrical Regulations
Electromagnetics and Radar
Energy Engineering
Healthcare Technologies
History and Management of Technology
IET Codes and Guidance
Materials, Circuits and Devices
Model Forms
Nanomaterials and Nanotechnologies
Optics, Photonics and Lasers
Production, Design and Manufacturing
Security
Telecommunications
Transportation

All books are available in print via https://shop.theiet.org or as eBooks via our Digital Library https://digital-library.theiet.org.

IET CONTROL, SENSORS AND ROBOTICS SERIES 136

Nonlinear Optimal and Flatness-based Control Methods and Applications for Complex Dynamical Systems

Gerasimos Rigatos, Masoud Abbaszadeh, and Pierluigi Siano

The Institution of Engineering and Technology

About the IET

This book is published by the Institution of Engineering and Technology (The IET).

We inspire, inform, and influence the global engineering community to engineer a better world. As a diverse home across engineering and technology, we share knowledge that helps make better sense of the world, to accelerate innovation and solve the global challenges that matter.

The IET is a not-for-profit organization. The surplus we make from our books is used to support activities and products for the engineering community and promote the positive role of science, engineering and technology in the world. This includes education resources and outreach, scholarships and awards, events and courses, publications, professional development and mentoring, and advocacy to governments.

To discover more about the IET please visit https://www.theiet.org/.

About IET books

The IET publishes books across many engineering and technology disciplines. Our authors and editors offer fresh perspectives from universities and industry. Within our subject areas, we have several book series steered by editorial boards made up of leading subject experts.

We peer review each book at the proposal stage to ensure the quality and relevance of our publications.

Get involved

If you are interested in becoming an author, editor, series advisor, or peer reviewer, please visit https://www.theiet.org/publishing/publishing-with-iet-books/ or contact author_support@theiet.org.

Discovering our electronic content

All of our books are available online via the IET's Digital Library. Our Digital Library is the home of technical documents, eBooks, conference publications, real-life case studies, and journal articles. To find out more, please visit https://digital-library.theiet.org.

In collaboration with the United Nations and the International Publishers Association, the IET is a Signatory member of the SDG Publishers Compact. The Compact aims to accelerate progress to achieve the Sustainable Development Goals (SDGs) by 2030. Signatories aspire to develop sustainable practices and act as champions of the SDGs during the Decade of Action (2020-30), publishing books and journals that will help inform, develop, and inspire action in that direction.

In line with our sustainable goals, our UK printing partner has FSC accreditation, which is reducing our environmental impact on the planet. We use a print-on-demand model to further reduce our carbon footprint.

Published by The Institution of Engineering and Technology, London, United Kingdom

The Institution of Engineering and Technology (the "**Publisher**") is registered as a Charity in England & Wales (no. 211014) and Scotland (no. SC038698).

First published 2025

The Institution of Engineering and Technology
Futures Place,
Kings Way, Stevenage,
Herts, SG1 2UA,
United Kingdom

www.theiet.org

British Library Cataloguing in Publication Data

A catalogue record for this product is available from the British Library

ISBN 978-1-83953-819-3 (hardback)
ISBN 978-1-83953-820-9 (PDF)

Typeset in India by MPS Limited

Cover Image: Stocktrek Images via *Getty Images*

In memory of my mother Diamantina Rigatou (1939–2018)
Gerasimos Rigatos

To my wife Elham, and my sons, Arman and Ario
Masoud Abbaszadeh

To my family
Pierluigi Siano

Contents

Acknowledgement

The authors of the monograph appreciate the review remarks of colleagues working in the research area of nonlinear control for complex dynamical systems. These comments have enabled to arrive at a concise, meaningful, and clear presentation of the analyzed nonlinear optimal and flatness-based control methods for non-trivial dynamical systems with nonlinear dynamics and have contributed to developing a research document that will be useful for the engineering and academic community.

Overview

Control of nonlinear dynamical systems has been often based on the concept of diffeomorphisms. This requires state-variables and state-space description transformations that bring the system into an equivalent linear form. In the new state-space description, a solution for both the control and state estimation problem is enabled. However, far from being an ideal solution, the use of such transformations is not always a straightforward procedure since the controlled system should previously satisfy feedback linearizability conditions. Besides, in this approach, the control inputs are computed for the linearized equivalent model, and it is necessary to apply inverse transformations so as to find the control signals that should be finally used in the initial nonlinear state-space description of the system. Thus, one may come against singularity issues which signify the existence of state-space regions where the inverse transformations cannot be performed because of generating non-bounded (infinite) control inputs.

Taking into account the above, in the present monograph, novel solutions to the control problem of complex nonlinear dynamical systems are developed and tested. These are (i) a nonlinear optimal (H-infinity) control approach and (ii) a flatness-based control approach implemented in successive loops. The new control methods are not constrained by the aforementioned shortcomings of global linearization-based control schemes (complicated changes of state variables, forward and backward state-space transformations, singularities). It is shown that such methods can be used in a wide class of nonlinear dynamical systems without needing to transform the systems' state-space model into equivalent linearized forms. It is also shown that the new control methods can be implemented in a computationally simple manner and are also followed by global stability proofs.

The monograph's nonlinear optimal (H-infinity) control scheme is a novel contribution to the area of nonlinear control. The method is based on an approximate linearization of the dynamics of the controlled system taking place at each sampling instance around a temporary operating point which is defined by the present value of the system's state vector and by the last sampled value of the control inputs vector. The linearization process relies on first-order Taylor series expansion and on the computation of the associated Jacobian matrices. For the approximately linearized model of the system, a stabilizing H-infinity feedback controller is designed. The controller's feedback gains are computed by solving repetitively (at each time-step of the control algorithm) an algebraic Riccati equation. The global stability properties of the nonlinear optimal control scheme are proven through Lyapunov analysis. The method of nonlinear optimal (H-infinity) control is of proven global stability and remains computationally tractable. It retains the advantages of linear optimal control, that is fast and accurate tracking of reference setpoints under moderate variations of the control inputs. The nonlinear optimal control method is applicable to a wider class of dynamical systems than approaches based on the solution of state-dependent Riccati equations (SDRE). The SDRE approaches can be applied only to dynamical systems which can be transformed to the linear parameter varying (LPV) form. Besides, the novel nonlinear optimal (H-infinity) control method performs better than optimal control approaches which are based on approximations of the solution of the Hamilton–Jacobi–Bellman equation by Galerkin series expansion (the stability properties of the Galerkin series-based optimal control schemes are still unproven).

The monograph introduces also a second solution to the problem of nonlinear control of complex dynamical systems without the need to apply changes in state variables (diffeomorphisms) and complicated state-space transformations. The new solution is a flatness-based control approach implemented in successive loops. In this method, the dynamic model of the nonlinear system is separated into subsystems which are connected in a cascading manner. This control method is directly applicable (but not limited) to dynamical systems of the triangular form and to nonlinear systems which can be transformed into such a form. The state-space model of the initial nonlinear system is decomposed into cascading subsystems which satisfy differential flatness properties. For each subsystem of the state-space model, a virtual control input is computed, capable of inverting the subsystem's dynamics and of eliminating the subsystem's tracking error. The control input which is actually applied to the initial nonlinear system is computed from the last row of the state-space description. This control input incorporates in a recursive manner all virtual control inputs which were computed from the individual subsystems included in the initial state-space equation. The control input that should be applied to the nonlinear system so as to assure that all state vector elements will converge to the desirable setpoints is obtained at each iteration of the control algorithm by tracing backward the subsystems of the state-space model.

Concisely, in the present monograph, novel solutions to the control problem of complex nonlinear dynamical systems are developed and tested. These are (i) a nonlinear optimal (H-infinity) control approach and (ii) a flatness-based control approach implemented in successive loops. The new control methods are free of shortcomings met in control schemes which are based on diffeomorphisms and global linearization (complicated changes of state variables, forward and backward state-space transformations, singularities). It is shown that such methods can be used in a wide class of nonlinear dynamical systems without needing to transform the systems' state-space model into equivalent linearized forms. It is also shown that the new control methods can be implemented in a computationally simple manner and are also followed by global stability proofs. Potential application domains of the nonlinear optimal and flatness-based control concepts for complex dynamical systems are: (i) robotic systems; (ii) robotic cranes and pendulums; (iii) space robots, aerospace systems, and satellites; (iv) mechatronic systems; (v) power electronics; (vi) biosystems; (vii) financial systems.

Gerasimos Rigatos
Athens, Greece
December 2024

Masoud Abbaszadeh
Clifton Park, NY, USA
December 2024

Pierluigi Siano
Fisciano, Salerno, Italy
December 2024

Glossary

AC/DC: alternating current/direct current
ADCS: attitude determination and control subsystem
AMB: active magnetic bearing
ARE: algebraic Riccati equation
CDPR: cable-driven parallel robot
CLT: central limit theorem
CNC: computer numerical control
DC/DC: direct current/direct current
DFIG: doubly fed induction generator
DOF: degrees of freedom
DKF: derivative-free nonlinear Kalman filter
EKF: extended Kalman filter
EMF: electromagnetic force
FSRM: free-floating space robotic manipulator
HKF: H-infinity Kalman filter
H_∞ control: H-infinity control
H_∞ Kalman filter: H-infinity Kalman filter
HVDC: high voltage direct current
IM: induction motor
KF: Kalman filter
LMI: linear matrix inequality
LPV: linear parameter varying system
LQR: linear quadratic regulator
LQG: linear quadratic Gaussian
MIMO: multi-input multi-output
MPC: model predictive control
NES: normalized error square
NMPC: nonlinear model predictive control
ODE: ordinary differential equation
PDE: partial differential equation
PID: proportional derivative integral
PMLSM: permanent magnet linear synchronous motor
PMLSG: permanent magnet linear synchronous generator
PMSG: permanent magnet synchronous generator
PMSM: permanent magnet synchronous motor
PWM: pulse width modulation
RSV: reentry space vehicle
RMSE: root mean square error
SCARA: selective compliance articulated robot arm
SDRE: state-dependent Riccati equation
SISO: single-input single-output

SMC: sliding mode control
VSC: voltage source converter
VSI: voltage source inverter
WEC: wave energy converter

Preface

A main approach in the control of nonlinear dynamical systems has been so far the concept of diffeomorphisms, that is of state-variable and state-space description transformations that allow for bringing the system into an equivalent linear form where a solution for both the control and state estimation problem becomes feasible. However, the application of such transformations is not always a straightforward procedure since the controlled system should previously satisfy feedback linearizability conditions. Besides, once the control inputs have been computed for the linearized equivalent model, it is necessary to apply inverse transformations so as to find the control signals that should be finally used in the initial nonlinear state-space description of the system. This process often comes against singularity issues which signify that there may exist certain state-space regions where the inverse transformations cannot be performed because of generating non-bounded (infinite) control inputs.

Taking into account the above, in the present monograph, novel solutions to the control problem of complex nonlinear dynamical systems are developed and tested. These are (i) a nonlinear optimal (H-infinity) control approach and (ii) a flatness-based control approach implemented in successive loops. The new control methods are not constrained by the aforementioned shortcomings of global linearization-based control schemes (complicated changes of state variables, forward and backward state-space transformations, singularities). It is shown that such methods can be used in a wide class of nonlinear dynamical systems without needing to transform the systems' state-space model into equivalent linearized forms. It is also shown that the new control methods can be implemented in a computationally simple manner and are also followed by global stability proofs.

The monograph's nonlinear optimal (H-infinity) control scheme is a novel contribution to the area of nonlinear control. The method is based on an approximate linearization of the dynamics of the controlled system taking place at each sampling instance around a temporary operating point which is defined by the present value of the system's state vector and by the last sampled value of the control inputs vector. The linearization process relies on first-order Taylor series expansion and on the computation of the associated Jacobian matrices. For the approximately linearized model of the system, a stabilizing H-infinity feedback controller is designed. The controller's feedback gains are computed by solving repetitively (at each time-step of the control algorithm) an algebraic Riccati equation. The global stability properties of the nonlinear optimal control scheme are proven through Lyapunov analysis. The method of nonlinear optimal (H-infinity) control is of proven global stability and remains computationally tractable. It retains the advantages of linear optimal control, that is fast and accurate tracking of reference setpoints under moderate variations of the control inputs. The nonlinear optimal control method is applicable to a wider class of dynamical systems than approaches based on the solution of state-dependent Riccati equations (SDRE). The SDRE approaches can be applied only to dynamical systems which can be transformed to the linear parameter varying (LPV) form. Besides, the novel nonlinear optimal (H-infinity) control method performs better than optimal control approaches which are based on approximations of the solution of the Hamilton–Jacobi–Bellman equation by Galerkin series expansion (the stability properties of the Galerkin series-based optimal control schemes are still unproven).

A comparison of the nonlinear optimal (H-infinity) control method against other linear and nonlinear control schemes for complex dynamical systems has shown the following: (1) unlike Lie algebra-based control, the new nonlinear optimal control method does not rely on complicated transformations (diffeomorphisms) of the system's state variables. The control inputs that the nonlinear optimal control method computes can be applied directly to the initial nonlinear dynamics of the system and are not used on its transformed equivalent description. The inverse transformations which are met in global linearization-based control are avoided and consequently one does not come against singularity issues. (2) Unlike model predictive control and nonlinear model predictive control, the nonlinear optimal control method is of proven global stability. It is known that model predictive control is a linear control method which if applied to systems with complex nonlinear dynamics the stability of the control loop will be lost. Besides, in nonlinear model predictive control, the convergence of the iterative search for an optimum depends on initialization and parameter values selection, and consequently, the global stability properties of the NMPC method cannot be ensured either. (3) Unlike sliding-mode and backstepping control, the nonlinear optimal control method does not require the state-space description of the system to be found in a specific form. About sliding-mode control, it is known that when the controlled system is not found in the input–output linearized form, the definition of the sliding surface can be an intuitive procedure. About backstepping control, it is known that it cannot be directly applied to a dynamical system if the related state-space model is not found in the strict-feedback (backstepping integral) form. (4) Unlike PID control, the nonlinear optimal control method is of proven global stability, the selection of the controller's parameters does not rely on a heuristics-based tuning procedure, and the stability of the control loop is ensured in case of changes of operating points. (5) Unlike multiple local models-based control, the nonlinear optimal control method uses only one linearization point and needs the solution of only one Riccati equation so as to compute the stabilizing feedback gains of the controller. Equivalently, the nonlinear optimal control method does not require the solution of complicated linear matrix inequalities. Consequently, in terms of computation load, the nonlinear optimal control method is much more efficient.

The monograph introduces also a second solution to the problem of nonlinear control of complex dynamical systems without the need to apply changes of state variables (diffeomorphisms) and complicated state-space transformations. The new solution is a flatness-based control approach implemented in successive loops. In this method, the dynamic model of the nonlinear system is separated into subsystems which are connected in a cascading manner. This control method is directly applicable (but not limited) to dynamical systems of the triangular form and to nonlinear systems which can be transformed into such a form. The state-space model of the initial nonlinear system is decomposed into cascading subsystems which satisfy differential flatness properties. For each subsystem of the state-space model, a virtual control input is computed, capable of inverting the subsystem's dynamics and of eliminating the subsystem's tracking error. The control input which is actually applied to the initial nonlinear system is computed from the last row of the state-space description. This control input incorporates in a recursive manner all virtual control inputs which were computed from the individual subsystems included in the initial state-space equation. The control input that should be applied to the nonlinear system so as to assure that all state vector elements will converge to the desirable setpoints is obtained at each iteration of the control algorithm by tracing backward the subsystems of the state-space model.

Concisely, in the present monograph, novel solutions to the control problem of complex nonlinear dynamical systems are developed and tested. These are (i) a nonlinear optimal (H-infinity) control approach and (ii) a flatness-based control approach implemented in successive loops. The new control methods are free of shortcomings met in control schemes which are based on diffeomorphisms and global linearization (complicated changes of state variables, forward and backward state-space transformations, singularities). It is shown that such methods can be used in a wide class of nonlinear dynamical systems without needing to transform the systems' state-space model into equivalent linearized forms. It is also shown that the new control methods can be implemented in a computationally simple manner and are also followed by global stability proofs. Potential application domains of the nonlinear optimal and flatness-based control concepts for complex dynamical systems are: (i) robotic systems; (ii) robotic cranes and pendulums; (iii) space robots, aerospace systems, and satellites; (iv) mechatronic systems; (v) power electronics; (vi) biosystems; (vii) financial systems.

The monograph is structured in the following chapters:

Chapter 1 analyzes nonlinear optimal and flatness-based control methods for robotic systems. First, the chapter treats the nonlinear optimal control problem of the 6-DOF parallel Stewart robotic manipulators. Next, the control problem of 4-DOF SCARA robots is treated. Additionally, the problem of nonlinear control of dual-arm robotic manipulators is solved, with the use of a nonlinear optimal control method. Finally, flatness-based control in successive loops is applied to the dynamic model of electropneumatic robots and particularly to exoskeletons with electropneumatic actuation.

Chapter 2 analyzes further nonlinear optimal and flatness-based control methods for robotic systems. First, the problem of multi-loop flatness-based control for dual-arm robotic manipulators is solved. Next, a multi-loop flatness-based control method is developed for the dynamic model of an electrohydraulic robotic manipulator. Additionally, a multi-loop flatness-based control approach is proposed for the dynamic model of a multi-DOF flexible-joint robotic manipulator. Moreover, a multi-loop flatness-based control scheme is applied to the model of a three-link redundant robotic manipulator. Finally, the chapter analyzes a method for flatness-based disturbance observer and control with application to robotic mining excavators

Chapter 3 analyzes nonlinear optimal and flatness-based control methods for robotic cranes and underactuated pendulums. First, the problem of nonlinear optimal control of the 4-DOF robotic tower crane is solved. Next, the problem of nonlinear optimal control of the dual PMLSM-driven gantry crane is treated. Moreover, a nonlinear optimal control scheme is developed for the 2-cable-driven parallel robot which functions as a crane. Finally, nonlinear optimal control is applied to the dynamic model of the 4-cable-driven parallel robot which functions as a crane.

Chapter 4 analyzes further nonlinear optimal and flatness-based control methods for robotic cranes and underactuated pendulums. First, a multi-loop flatness-based control method is proposed for the control and stabilization problem of the dual PMLSM-driven gantry crane. Next, the problem of nonlinear optimal control of parallel inverted pendulums is treated. Additionally, a nonlinear optimal control method is proposed for the control and stabilization problem of the rotary double inverted pendulum (double Furuta pendulum). Moreover, a nonlinear optimal control scheme is developed for the dynamic model of the three-link biped robot.

Chapter 5 analyzes nonlinear optimal and flatness-based control for aerospace systems. First, it solves the nonlinear optimal control problem for the 4-DOF dynamic model of a reentry space vehicle. Next, it applies nonlinear optimal control and multi-loop flatness-based control to the 6-DOF attitude dynamics of a reentry space vehicles. Furthermore, it applies multi-loop flatness-based control to the 3-DOF dynamic model of the attitude of microsatellites. Finally, it applies multi-loop flatness-based control to the dynamic model of the 3-DOF free-floating space robotic manipulator.

Chapter 6 analyzes further nonlinear optimal and flatness-based control methods for aerospace systems. First, it solves the nonlinear optimal control for the dynamic model of the 3-DOF free-floating space robotic manipulator. Besides, it analyzes nonlinear optimal control for the dynamic model of the underactuated tethered satellite. Next, it generalizes toward nonlinear optimal control for the dynamic model of a tethered multi-satellite system. Finally, it presents results on nonlinear optimal control for the dynamic model of the wing-rock effect that appears in aircrafts and reentry space vehicles.

Chapter 7 analyzes nonlinear optimal control and multi-loop flatness-based control for mechatronic systems and for power electronics. First, it solves the nonlinear optimal and flatness-based control problem for the dynamic model of an induction motor-driven desalination unit. Moreover, it develops a multi-loop flatness-based control method for centrifugal gas compressors driven by induction motors and permanent magnet synchronous motors. Besides, it introduces nonlinear optimal control for wave energy conversion units which consist of a permanent magnet linear synchronous generator connected to a three-phase AC/DC converter. Additionally, it analyzes nonlinear optimal control and multi-loop flatness-based control for the dynamic model of 5-DOF active magnetic bearings. Next, a differential flatness and Kalman filter-based disturbance observer is developed for the fault diagnosis problem of 5-DOF active magnetic bearings. Finally, a differential flatness and Kalman filter-based disturbance observer is proposed for the fault diagnosis problem of the VSC-HVDC transmission system in the electric power grid.

Chapter 8 analyzes nonlinear optimal control and flatness-based control for dynamical models of biological systems and financial systems. First, nonlinear optimal control is applied for the treatment of bacterial infections with resistance to antibiotics. Additionally, nonlinear optimal control is proposed for the optimized infusion of chemotherapy and the treatment of leukemia. In the same context, nonlinear optimal control and flatness-based control are proposed for treatment of tumor growth. In general, nonlinear optimal control and multi-loop flatness-based control can be used for the stabilization of competing population dynamics among which cell populations are a specific case. Finally, in this chapter nonlinear optimal control is used for the optimized management of multi-echelon supply chain networks under time-delays.

The monograph is completed with two appendices which elaborate on the theoretical background of the nonlinear optimal and flatness-based control methods for complex dynamical systems:

Appendix A analyzes nonlinear optimal control and Lie algebra-based control. In the first section of this appendix control and state estimation based on approximate linearization of the dynamic models of nonlinear systems is explained. Linearization is performed sequentially through Taylor series expansion and the solution of a Riccati equation. The H-infinity Kalman filter is developed as a robust state estimator. In the second section of the chapter, global linearization-based control and state estimation is developed for nonlinear systems. To this end, input–output and input–state linearization conditions are analyzed. Results from Lie algebra-based control and Lie algebra-based state observers are explained. The concept of dynamic extension is also outlined.

Appendix B analyzes differential flatness theory and flatness-based control methods. In the first section of the appendix, global linearization-based control with the use of differential flatness theory is analyzed, comprising issues such as differential flatness for finite dimensional systems, equivalence and differential flatness, feedback control and equivalence flatness-based control, and state feedback for systems with model uncertainties and classification of types of differentially flat systems. In the second section of this appendix, the concept of flatness-based control of nonlinear dynamical systems in cascading loops is analyzed. This comprises issues such as decomposition of the state-space model into cascading differentially flat subsystems, tracking error dynamics for flatness-based control in successive loops, and comparison to backstepping control for nonlinear systems.

By providing solutions to the control problem of complex dynamical systems, the monograph has contributed to the advancement of the subject area. Based on these findings and by further exploiting the new nonlinear control methods that the monograph has analyzed, it is expected that additional progress in this technological domain will be achieved in the forthcoming years.

Gerasimos Rigatos
Athens. Greece
December 2024

Masoud Abbaszadeh
Clifton Park, NY, USA
December 2024

Pierluigi Siano
Fisciano, Salerno, Italy
December 2024

About the authors

Gerasimos Rigatos is a research director at the Industrial Systems Institute in Greece. He leads research cooperation projects in the areas of nonlinear control, nonlinear filtering, and control of distributed parameter systems. He has been a visiting professor at Université Paris XI, France; Ecole Centrale de Nantes, France, Harper-Adams University College, UK; University of Northumbria, UK; and University of Salerno, Italy. He is a senior member of IEEE and a member and CEng of the IET. He holds an editor's position for the *Journal of Information Sciences*, the *Journal of Advanced Robotic Systems*, and the *SAE Journal of Electrified Vehicles*. He received his PhD degree from the Department of Electrical and Computer Engineering of the National Technical University of Athens (NTUA), Greece.

Masoud Abbaszadeh is a principal research engineer and technical leader of the Cyber-Physical Security & Resilience portfolio at GE Vernova Research Center, NY, USA. He is also an adjunct professor at Rensselaer Polytechnic Institute, NY, USA. His research interests include estimation and detection theory, robust and nonlinear control, and machine learning with applications in cyber-physical resilience and autonomy. He has published 4 books, over 170 peer-reviewed articles, and over 80 issued or pending patents. He is an associate editor of *IEEE Transactions on Control Systems Technology*, and a member of IEEE CSS Conference Editorial Board. He received his BSc, MSc, and PhD degrees in Electrical and Computer Engineering from Amirkabir University of Technology (Iran), Sharif University of Technology (Iran), and University of Alberta (Canada), respectively.

Pierluigi Siano (M'09–SM'14) received an MSc degree in electronic engineering and a PhD degree in information and electrical engineering from the University of Salerno, Salerno, Italy, in 2001 and 2006, respectively. He is a full professor of electrical power systems and scientific director of the Smart Grids and Smart Cities Laboratory with the Department of Management & Innovation Systems, University of Salerno. His research activities are centered on demand response, energy management, the integration of distributed energy resources in smart grids, and electricity markets, as well as the planning and management of power systems. In these research fields, he has co-authored more than 800 articles, including more than 400 international journals that received in Scopus more than 24,000 citations with an H-index equal to 75. Since 2019, he has been awarded as a Highly Cited Researcher in Engineering by the Web of Science Group. He has been the Chair of the IES TC on Smart Grids. He is an editor for the Power & Energy Society Section of *IEEE Access*, *IEEE Transactions on Power Systems*, *IEEE Transactions on Industrial Informatics*, *IEEE Transactions on Industrial Electronics*, and *IEEE Systems*.

Chapter 1

Nonlinear optimal and flatness-based control for robotic systems I

Abstract

The chapter analyzes nonlinear optimal and flatness-based control methods for robotic systems. First, the chapter treats the nonlinear optimal control problem of the 6-DOF parallel Stewart robotic manipulators. Next, the control problem of 4-DOF SCARA robots is treated. Additionally, the problem of nonlinear control of dual-arm robotic manipulators is solved, with the use of a nonlinear optimal control method. Finally, flatness-based control in successive loops is applied to the dynamic model of electropneumatic robots and particularly to exoskeletons with electropneumatic actuation.

1.1 Nonlinear optimal control of the 6-DOF parallel Stewart robotic manipulator

1.1.1 Outline

The present monograph begins by analyzing nonlinear optimal and flatness-based control methods for advanced robotic systems. For instance, one can consider parallel Stewart robotic manipulators which are widely used in industry and for assisting humans in several tasks [461,1080]. Among applications of parallel Stewart platforms, one can distinguish additive manufacturing tasks, 6-DOF flight simulators, orientation of LEO satellite antennas, and heave compensation in offshore installations [968,853,113,114]. They have 6-DOF and consist of a fixed and a moving platform which are connected with the use of extensible links (legs) [472,471]. The legs of such parallel robots change their length with the use of electric, hydraulic, or pneumatic actuators [874,875]. The moving platform can perform both translational and rotational motion [852,460]. The translational motion is described by the cartesian coordinates of the center of the moving platform expressed in the coordinates system of the fixed platform, as well as of the associated linear velocities [957,886]. The rotational motion consists of roll, pitch, and yaw turn angles, as well as of the associated angular velocities [593,793]. There are three Euler angles that denote the turn motion of the moving plate around the axes of the fixed coordinates frame. So far, several nonlinear control schemes have been applied to the 6-DOF Stewart parallel robotic platform [149,307]. One can distinguish optimal and nonlinear model predictive control (NMPC) methods [885,925,630,134]. Besides, there are results on sliding-mode control and backstepping control [341,362]. The compensation of model uncertainties has been attempted with the use of adaptive and neural control methods [509,185]. Robustness to parametric changes and external perturbations is among the primary objectives in the design of feedback control for Stewart platforms [819,593].

In this section, a novel nonlinear optimal control method is proposed for the 6-DOF dynamic model of the parallel Stewart robotic platform [725,721,741]. It is shown that the Stewart robotic platform is a differentially flat system. Next, the dynamic model of the Stewart parallel robotic platform undergoes approximate linearization at each sampling instance around a temporary operating point which is defined by the present value of the robot's state vector and by the last sampled value of the control inputs vector. The linearization process is based on first-order Taylor series expansion and on the computation of the associated Jacobian matrices [705,60,706]. The modeling error which is due to the truncation of higher-order terms from the Taylor series expansion is considered to be a perturbation which is asymptotically compensated by the robustness of the control algorithm. For the approximately linearized model of the system, an H-infinity feedback controller is designed. The H-infinity controller provides a solution to the optimal control problem of the Stewart parallel robotic platform under model uncertainty and perturbations.

Actually, the H-infinity feedback controller represents a min–max differential game which takes place between (i) the control inputs of the parallel robotic platform which try to minimize a cost function that contains a quadratic term of the state vector's tracking error, and (ii) the model uncertainty and external disturbance terms which try to maximize this cost function [725,721,741]. To select the gains of the H-infinity feedback controller an algebraic Riccati equation is solved repetitively at each sampling period based on the locally linearized model of the robot [711,732,876]. The global stability properties of the control method are proven through Lyapunov analysis/ First, it is proven that the H-infinity tracking performance criterion is satisfied for the robot. This demonstrates also robustness to parametric variations and model invalidation. Under moderate conditions, it is also proven that the parallel robot's control loop is globally asymptotically stable. To apply state estimation-based control without the need to measure the entire state vector of the parallel robot the H-infinity Kalman filter is used as a robust state estimator. The H-infinity control and estimation scheme achieves fast and accurate tracking of setpoints by the state variables of the parallel robot under moderate variation of the control inputs.

The proposed nonlinear optimal control method goes beyond past approaches for treating the optimal control problem for nonlinear dynamical systems [721–732]. In past concepts for nonlinear optimal control, the state-space model of the system is linearized around points of the desirable trajectory. On the contrary, in the new nonlinear optimal control method, linearization is performed at each sampling instance around the present value of the system's state vector. The span of the Taylor expansion is kept small and the induced linearization error is negligible. Besides, in the new nonlinear optimal control method, the Riccati equation which is used for computing the feedback gains of the controller is novel. Compared to NMPC which is a popular scheme for treating the optimal control problem in nonlinear dynamical systems, the new nonlinear optimal control method is of proven global stability and the convergence of its iterative search for the optimum does not depend on initialization and ad hoc parametrization of the controller. Compared to Linear Parameter Varying (LPV) control schemes which are based on the solution of State Dependent Riccati Equations, the new nonlinear optimal control method does need to transform the controlled system into the LPV form and thus it is applicable to a wider class of dynamical systems. Finally, compared to nonlinear optimal control methods which are based on Pontryagin's maximum principle, the new nonlinear optimal control method does not need to redefine a Hamiltonian function on an ad hoc basis [308,170,855]. Moreover, the computation of the stabilizing feedback control signals at each sampling instance does not need the solution of a system of differential equations so as to compute the so-called "adjoint coefficients."

1.1.2 Dynamic model of the 6-DOF Stewart parallel robot

The diagram of the 6-DOF parallel robotic Stewart platform is shown in Figure 1.1. It comprises a fixed basis with a reference frame at its center which is defined as $O_BX_BY_BZ_B$ and a moving platform with a reference frame at its center which is defined as $O_PX_PY_PZ_P$. The coordinates of the center O_P of the moving platform's reference frame are expressed in the coordinates system of the fixed basis and are denoted as $[x_r, y_r, z_r]^T$. Moreover, the Euler rotation angles of the moving platform around the axes of the reference frame of the fixed basis are denoted as $[a, \beta, \gamma]^T$. The complete state vector of the moving platform is defined as $x = [x_r, y_r, z_r, a, \beta, \gamma]^T$ [461,1080].

The fixed basis is connected to the moving basis through extensible links (legs). Each link emanates from a point B_i, $i = 1, 2, \cdots, 6$ on the fixed basis and terminates at a point P_i, $i = 1, 2, \cdots, 6$ on the moving basis. Due to the extension and contraction of the variable-length legs which connect the two bases, the moving base can perform (i) a translational motion defined by the position of its center of gravity, expressed in the reference frame of the fixed base with coordinates which are denoted by vector $[x_r, y_r, z_r]^T$, (ii) a rotational motion around the axes of the fixed frame which is described by the Euler angles' vector $[a, \beta, \gamma]^T$.

The state vector of the 6-DOF parallel Stewart robotic platform is denoted as [461,1080]:

$$q = [x_r, y_r, z_r, a, \beta, \gamma]^T \tag{1.1}$$

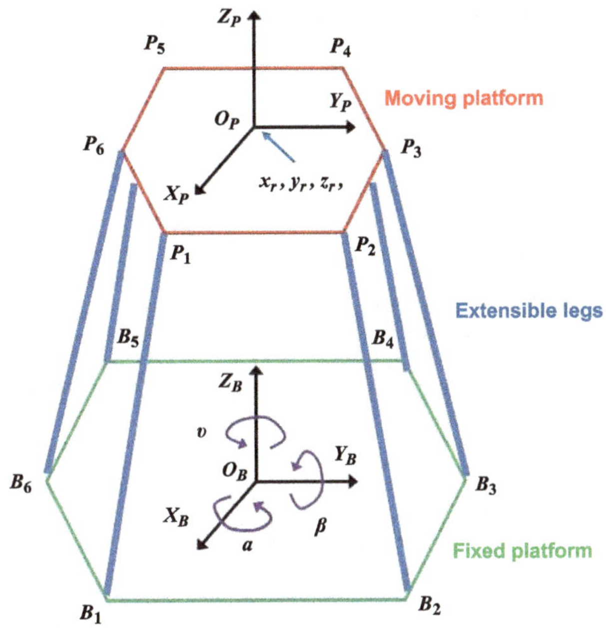

Figure 1.1 Diagram of the 6-DOF Stewart parallel robotic platform

where x_r represents longitudinal (surge) motion, y_r denotes lateral (sway) motion, and z_r represents vertical (heave) motion. Furthermore, a denotes roll turn motion, β represents pitch turn motion, and γ denotes yaw turn motion.

The dynamic model of the Stewart platform has the form [461, 1080]:

$$M(q)\ddot{q} + V(q,\dot{q})\dot{q} + G(q) = \tau + \tau_d \tag{1.2}$$

where $M(q)$ is the inertia matrix, $V(q,\dot{q})$ is the Coriolis and centrifugal forces matrix and $C(q)$ is the gravitational forces vector. The generalized torques vector is denoted as τ while the disturbances' vector is written as τ_d. Using that the control forces from the actuators are generated at the side of the fixed basis, while the torques which are exerted on the moving platform are transferred to the other end of the extensible legs, it holds that

$$\tau = J^T f \tag{1.3}$$

where the Jacobian matrix J connects the variation of the length of the extensible legs $\dot{l}_i$ to the vector of the linear and angular velocity of the moving platform, that is

$$\dot{l} = J\dot{q} \tag{1.4}$$

The inertia matrix M of the robotic platform is defined as

$$M = \begin{pmatrix} m & 0 & 0 & 0 & 0 & 0 \\ 0 & m & 0 & 0 & 0 & 0 \\ 0 & 0 & m & 0 & 0 & 0 \\ 0 & 0 & 0 & m_{44} & m_{45} & m_{46} \\ 0 & 0 & 0 & m_{54} & m_{55} & m_{56} \\ 0 & 0 & 0 & m_{64} & m_{65} & m_{66} \end{pmatrix} \tag{1.5}$$

where $m_{44} = I_x cos^2(\beta)cos^2(\gamma) + I_y cos^2(\beta)sin^2(\gamma) + I_z sin^2(\beta)$, $m_{45} = m_{54} = (I_x - I_y)cos(\beta)cos(\gamma)sin(\gamma)$, $m_{46} = m_{64} = I_z sin(\beta)$, $m_{55} = I_x sin^2(\gamma) + I_y cos^2(\gamma)$, $m_{56} = m_{65} = 0$, and $m_{66} = I_z$.

The elements of the Coriolis and centrifugal forces matrix $V(q,\dot{q})$ are

$$V(q,\dot{q}) = \begin{pmatrix} 0 & 0 & 0 & 0 & 0 & 0 \\ 0 & 0 & 0 & 0 & 0 & 0 \\ 0 & 0 & 0 & 0 & 0 & 0 \\ 0 & 0 & 0 & V_{44} & V_{45} & V_{46} \\ 0 & 0 & 0 & V_{54} & V_{55} & V_{56} \\ 0 & 0 & 0 & V_{64} & V_{65} & V_{66} \end{pmatrix} \tag{1.6}$$

where $V_{44} = K_1\dot{\beta} + K_2\dot{\gamma}$, $V_{45} = K_1\dot{a} + K_2\dot{\beta} + K_3\dot{\gamma}$, $V_{46} = K_2\dot{a} + K_3\dot{\beta}$, $V_{54} = -K_1\dot{a} + K_3\dot{\gamma}$, $V_{55} = K_4\dot{\gamma}$, $V_{56} = K_3\dot{a} + K_4\dot{\beta}$, $V_{64} = -K_2\dot{a} - K_3\dot{\beta}$, $V_{65} = -K_3\dot{a} - K_4\dot{\beta}$, $V_{66} = 0$.
and $K_1 = cos(\beta)sin(\beta)(cos^2(\gamma)I_x sin^2(\gamma)I_y - I_z)$, $K_2 = -cos^2(\beta)cos(\gamma)sin(\gamma)(I_x - I_y)$, $K_3 = \frac{1}{2}(cos(\gamma) - sin(\gamma))(cos(\gamma) + sin(\gamma))(I_x - I_y)$, $K_4 = cos(\gamma)sin(\gamma)(I_x - I_y)$, $K_5 = -cos(\gamma)sin(\gamma)sin(\beta)(I_x - I_y)$.

The gravitational forces vector is

$$G = \begin{pmatrix} 0 & 0 & -mg & 0 & 0 & 0 \end{pmatrix}^T \tag{1.7}$$

The following rotation matrix $R_{a\beta\gamma}$ is defined for transforming points initially expressed in the coordinates frame of the moving platform into the coordinates frame of the fixed platform:

$$R_{a\beta\gamma} = \begin{pmatrix} cos(a)cos(\beta) & cos(a)sin(\beta)sin(\gamma) - sin(a)cos(\gamma) & cos(\beta)sin(\beta)cos(\gamma) + sin(a)sin(\gamma) \\ sin(a)cos(\beta) & sin(a)sin(\beta)sin(\gamma) - cos(a)cos(\gamma) & sin(a)sin(\beta)cos(\gamma) - cos(a)sin(\gamma) \\ sin(\beta) & cos(\beta)sin(\gamma) & cos(\beta)cos(\gamma) \end{pmatrix} \tag{1.8}$$

The inverse of the inertia matrix M is given by

$$M^{-1} = \frac{1}{detM} \begin{pmatrix} M_{11} & -M_{21} & M_{31} & -M_{41} & M_{51} & -M_{61} \\ -M_{12} & M_{22} & -M_{32} & M_{42} & -M_{52} & M_{62} \\ M_{13} & -M_{23} & M_{33} & -M_{43} & M_{53} & -M_{63} \\ -M_{14} & M_{24} & -M_{34} & M_{44} & -M_{54} & M_{64} \\ M_{15} & -M_{25} & M_{35} & -M_{45} & M_{55} & -M_{65} \\ -M_{16} & M_{26} & -M_{36} & M_{46} & -M_{56} & M_{66} \end{pmatrix} \tag{1.9}$$

The determinant of the inertia matrix M is given by

$$\begin{gathered} detM = m^3\left[m_{44}(m_{55}m_{66} - m_{65}m_{56}) - m_{45}(m_{54}m_{66} - m_{64}m_{56}) + m_{46}(m_{54}m_{65} - m_{64}m_{55})\right] \\ \Rightarrow detM = m^3 d\bar{e}tM \end{gathered} \tag{1.10}$$

Besides, the following subdeterminants are computed:

$$M_{11} = m^3\left[m_{44}(m_{55}m_{66} - m_{65}m_{56}) - m_{45}(m_{54}m_{66} - m_{64}m_{56}) + m_{46}(m_{54}m_{65} - m_{64}m_{55})\right] = \frac{detM}{m} \tag{1.11}$$

$$M_{12} = 0 \quad M_{13} = 0 \quad M_{14} = 0 \quad M_{15} = 0 \quad M_{16} = 0 \tag{1.12}$$

$$M_{22} = m^3\left[m_{44}(m_{55}m_{66} - m_{65}m_{56}) - m_{45}(m_{54}m_{66} - m_{64}m_{56}) + m_{46}(m_{54}m_{65} - m_{64}m_{55})\right] = \frac{detM}{m} \tag{1.13}$$

$$M_{21} = 0 \quad M_{23} = 0 \quad M_{24} = 0 \quad M_{25} = 0 \quad M_{26} = 0 \tag{1.14}$$

$$M_{33} = m^3\left[m_{44}(m_{55}m_{66} - m_{65}m_{56}) - m_{45}(m_{54}m_{66} - m_{64}m_{56}) + m_{46}(m_{54}m_{65} - m_{64}m_{55})\right] = \frac{detM}{m} \tag{1.15}$$

$$M_{31} = 0 \quad M_{32} = 0 \quad M_{34} = 0 \quad M_{35} = 0 \quad M_{36} = 0 \tag{1.16}$$

$$M_{41} = 0 \quad M_{42} = 0 \quad M_{43} = 0 \tag{1.17}$$

$$\begin{aligned} M_{44} &= m^3(m_{55}m_{66} - m_{65}m_{56}) = m^3\bar{M}_{44} \\ M_{45} &= m^3(m_{54}m_{66} - m_{64}m_{56}) = m^3\bar{M}_{45} \\ M_{46} &= m^3(m_{54}m_{65} - m_{64}m_{55}) = m^3\bar{M}_{46} \end{aligned} \tag{1.18}$$

$$M_{51}=0 \quad M_{52}=0 \quad M_{53}=0 \tag{1.19}$$

$$\begin{aligned} M_{54}&=m^3(m_{45}m_{66}-m_{65}m_{46})=m^3\bar{M}_{54}\\ M_{55}&=m^3(m_{44}m_{66}-m_{64}m_{46})=m^3\bar{M}_{55}\\ M_{56}&=m^3(m_{44}m_{65}-m_{64}m_{45})=m^3\bar{M}_{56} \end{aligned} \tag{1.20}$$

$$M_{61}=0 \quad M_{62}=0 \quad M_{63}=0 \tag{1.21}$$

$$\begin{aligned} M_{64}&=m^3(m_{45}m_{56}-m_{55}m_{46})=m^3\bar{M}_{64}\\ M_{65}&=m^3(m_{44}m_{56}-m_{54}m_{46})=m^3\bar{M}_{65}\\ M_{66}&=m^3(m_{44}m_{55}-m_{54}m_{45})=m^3\bar{M}_{66} \end{aligned} \tag{1.22}$$

Thus, the inverse of the inertia matrix M can be also written in the form:

$$M^{-1}=\begin{pmatrix} \frac{1}{m} & 0 & 0 & 0 & 0 & 0\\ 0 & \frac{1}{m} & 0 & 0 & 0 & 0\\ 0 & 0 & \frac{1}{m} & 0 & 0 & 0\\ 0 & 0 & 0 & \frac{\bar{M}_{44}}{detM} & \frac{\bar{M}_{45}}{detM} & \frac{\bar{M}_{46}}{detM}\\ 0 & 0 & 0 & \frac{\bar{M}_{54}}{detM} & \frac{\bar{M}_{55}}{detM} & \frac{\bar{M}_{56}}{detM}\\ 0 & 0 & 0 & \frac{\bar{M}_{64}}{detM} & \frac{\bar{M}_{65}}{detM} & \frac{\bar{M}_{66}}{detM} \end{pmatrix} \tag{1.23}$$

Next, the dynamic model of the Stewart platform is written as

$$\ddot{q}=-M^{-1}(q)[V(q.\dot{q})\dot{q}+G(q)]+M^{-1}(q)\tau \tag{1.24}$$

The term $M^{-1}(q)[V(q.\dot{q})\dot{q}+G(q)]$ is given by

$$M^{-1}(q)[V(q.\dot{q})\dot{q}+G(q)]=$$

$$\begin{pmatrix} 0\\ 0\\ -g\\ \frac{\bar{M}_{44}}{detM}(V_{44}\dot{a}+V_{45}\dot{\beta}+V_{46}\dot{\gamma})+\frac{\bar{M}_{45}}{detM}(V_{54}\dot{a}+V_{55}\dot{\beta}+V_{56}\dot{\gamma})+\frac{\bar{M}_{46}}{detM}(V_{64}\dot{a}+V_{65}\dot{\beta}+V_{66}\dot{\gamma})\\ \frac{\bar{M}_{54}}{detM}(V_{44}\dot{a}+V_{45}\dot{\beta}+V_{46}\dot{\gamma})+\frac{\bar{M}_{55}}{detM}(V_{54}\dot{a}+V_{55}\dot{\beta}+V_{56}\dot{\gamma})+\frac{\bar{M}_{56}}{detM}(V_{64}\dot{a}+V_{65}\dot{\beta}+V_{66}\dot{\gamma})\\ \frac{\bar{M}_{64}}{detM}(V_{44}\dot{a}+V_{45}\dot{\beta}+V_{46}\dot{\gamma})+\frac{\bar{M}_{65}}{detM}(V_{54}\dot{a}+V_{55}\dot{\beta}+V_{56}\dot{\gamma})+\frac{\bar{M}_{66}}{detM}(V_{64}\dot{a}+V_{65}\dot{\beta}+V_{66}\dot{\gamma}) \end{pmatrix} \tag{1.25}$$

Next, the state vector of the parallel robot is defined as

$$\begin{aligned} x&=[x_1,x_2,x_3,x_4,x_5,x_6,x_7,x_8,x_8,x_{10},x_{11},x_{12}]^T \text{ or}\\ x&=[x_r,\dot{x}_r,y_r,\dot{y}_r,z_r,\dot{z}_r,a,\dot{a},\beta,\dot{\beta},\gamma,\dot{\gamma}]^T \end{aligned} \tag{1.26}$$

The control inputs vector of the robotic platform is written as

$$\begin{aligned} u&=[u_1,u_2,u_3,u_4,u_5,u_6]^T \text{ or}\\ u&=[\tau_1,\tau_2,\tau_3,\tau_4,\tau_5,\tau_6]^T \end{aligned} \tag{1.27}$$

Thus, the dynamic model of the 6-DOF parallel Stewart robotic platform is described as

$$\begin{pmatrix} \dot{x}_1 \\ \dot{x}_2 \\ \dot{x}_3 \\ \dot{x}_4 \\ \dot{x}_5 \\ \dot{x}_6 \\ \dot{x}_7 \\ \dot{x}_8 \\ \dot{x}_9 \\ \dot{x}_{10} \\ \dot{x}_{11} \\ \dot{x}_{12} \end{pmatrix} = \begin{pmatrix} x_2 \\ 0 \\ x_4 \\ 0 \\ x_6 \\ -g \\ x_8 \\ \frac{\bar{M}_{44}}{detM}(V_{44}\dot{a} + V_{45}\dot{\beta} + V_{46}\dot{\gamma}) + \frac{\bar{M}_{45}}{detM}(V_{54}\dot{a} + V_{55}\dot{\beta} + V_{56}\dot{\gamma}) + \frac{\bar{M}_{46}}{detM}(V_{64}\dot{a} + V_{65}\dot{\beta} + V_{66}\dot{\gamma}) \\ x_{10} \\ \frac{\bar{M}_{54}}{detM}(V_{44}\dot{a} + V_{45}\dot{\beta} + V_{46}\dot{\gamma}) + \frac{\bar{M}_{55}}{detM}(V_{54}\dot{a} + V_{55}\dot{\beta} + V_{56}\dot{\gamma}) + \frac{\bar{M}_{56}}{detM}(V_{64}\dot{a} + V_{65}\dot{\beta} + V_{66}\dot{\gamma}) \\ x_{12} \\ \frac{\bar{M}_{64}}{detM}(V_{44}\dot{a} + V_{45}\dot{\beta} + V_{46}\dot{\gamma}) + \frac{\bar{M}_{65}}{detM}(V_{54}\dot{a} + V_{55}\dot{\beta} + V_{56}\dot{\gamma}) + \frac{\bar{M}_{66}}{detM}(V_{64}\dot{a} + V_{65}\dot{\beta} + V_{66}\dot{\gamma}) \end{pmatrix} + \tag{1.28}$$

$$+ \begin{pmatrix} 0 & 0 & 0 & 0 & 0 & 0 \\ \frac{1}{m} & 0 & 0 & 0 & 0 & 0 \\ 0 & 0 & 0 & 0 & 0 & 0 \\ 0 & \frac{1}{m} & 0 & 0 & 0 & 0 \\ 0 & 0 & 0 & 0 & 0 & 0 \\ 0 & 0 & \frac{1}{m} & 0 & 0 & 0 \\ 0 & 0 & 0 & 0 & 0 & 0 \\ 0 & 0 & 0 & \frac{\bar{M}_{44}}{detM} & \frac{\bar{M}_{45}}{detM} & \frac{\bar{M}_{46}}{detM} \\ 0 & 0 & 0 & 0 & 0 & 0 \\ 0 & 0 & 0 & \frac{\bar{M}_{54}}{detM} & \frac{\bar{M}_{55}}{detM} & \frac{\bar{M}_{56}}{detM} \\ 0 & 0 & 0 & 0 & 0 & 0 \\ 0 & 0 & 0 & \frac{\bar{M}_{64}}{detM} & \frac{\bar{M}_{65}}{detM} & \frac{\bar{M}_{66}}{detM} \end{pmatrix} \begin{pmatrix} u_1 \\ u_2 \\ u_3 \\ u_4 \\ u_5 \\ u_6 \end{pmatrix}$$

Concisely, the dynamic model of the robot can be written in the nonlinear affine-in-the-input state-space form:

$$\dot{x} = f(x) + g(x)u \tag{1.29}$$

where $x \in R^{12\times1}$, $f(x) \in R^{12\times1}$, $g(x) \in R^{12\times6}$, and $u \in R^{6\times1}$.

It can be also demonstrated that the dynamic model of the 6-DOF Stewart robotic platform is a differentially flat system, with a flat outputs vector that consists of its odd-numbered state variables $Y = [x_1, x_3, x_5, x_7, x_9, x_{11}]^T$. Indeed, it holds that the even-numbered state variables of the system are differential functions of the flat outputs:

$$\dot{x}_2 = x_1 \quad \dot{x}_4 = x_4 \quad \dot{x}_6 = x_5 \quad \dot{x}_8 = x_7 \quad \dot{x}_{10} = x_9 \quad \dot{x}_{11} = x_{12} \tag{1.30}$$

while from Eq. (1.29), one can solve for the control inputs vector u thus arriving at a relation of the form:

$$u = h_u(Y, \dot{Y}) \tag{1.31}$$

Consequently, all state variables and the control inputs vector of the 6-DOF Stewart parallel robot can be written as differential functions of the robot's flat outputs, and the system is differentially flat. The differential flatness property is an implicit proof of the controllability of this parallel robotic system.

1.1.3 Approximate linearization of the dynamic model of the Stewart platform

The dynamic model of the Stewart platform being initially in the nonlinear state-space form $\dot{x} = f(x) + g(x)u$ undergoes approximate linearization through first-order Taylor series expansion and through the computation of the associated Jacobian matrices. The linearization process takes place at each sampling instance around the temporary operating point (x^*, u^*) which is defined by the present value x^* of the robot's state vector and by the last sampled value u^* of the control inputs vector. The modeling error which is due to the truncation of higher-order terms from the Taylor series expansion is considered to be a perturbation which is asymptotically compensated by the robustness of the control algorithm.

The linearization process brings the dynamics of the robot to the linearized form:

$$\begin{aligned} \dot{x} &= Ax + Bu + \tilde{d} \\ y &= Cx \end{aligned} \tag{1.32}$$

Using that the first three columns of the control inputs gains matrix $g_i(x)$ $i = 1, 2, 3$ are zero vectors, the Jacobian matrices A, B are computed as

$$\begin{aligned} A = \nabla_x[f(x) + g(x)u] \mid_{(x^*,u^*)} \Rightarrow A = \nabla_x[f(x)] \mid_{(x^*,u^*)} + \nabla_x g_4(x)u_4 \mid_{(x^*,u^*)} + \\ + \nabla_x g_5(x)u_5 \mid_{(x^*,u^*)} \nabla_x g_6(x)u_6 \mid_{(x^*,u^*)} \end{aligned} \tag{1.33}$$

$$B = \nabla_u[f(x) + g(x)u] \mid_{(x^*,u^*)} \Rightarrow B = g(x) \mid_{(x^*,u^*)} \tag{1.34}$$

The disturbance vector $\tilde{d}$ may comprise (i) the modeling error due to truncation of higher-order terms from the Taylor series expansion, (ii) exogenous perturbations, and (iii) sensor measurement noise of any distribution.

- The analytical computation of the Jacobian matrix $\nabla_x f(x) \mid_{(x^*,u^*)}$ is given below:

First row of the Jacobian matrix $\nabla_x f(x) \mid_{(x^*,u^*)}$: $\frac{\partial f_1}{\partial x_i} = 0$ for $i = 1, \cdots, 12$ and $i \neq 2$ and $\frac{\partial f_1}{\partial x_2} = 1$.

Second row of the Jacobian matrix $\nabla_x f(x) \mid_{(x^*,u^*)}$: $\frac{\partial f_2}{\partial x_i} = 0$ for $i = 1, \cdots, 12$.

Third row of the Jacobian matrix $\nabla_x f(x) \mid_{(x^*,u^*)}$: $\frac{\partial f_3}{\partial x_i} = 0$ for $i = 1, \cdots, 12$ and $i \neq 4$ and $\frac{\partial f_3}{\partial x_4} = 1$.

Fourth row of the Jacobian matrix $\nabla_x f(x) \mid_{(x^*,u^*)}$: $\frac{\partial f_4}{\partial x_i} = 0$ for $i = 1, \cdots, 12$.

Fifth row of the Jacobian matrix $\nabla_x f(x) \mid_{(x^*,u^*)}$: $\frac{\partial f_5}{\partial x_i} = 0$ for $i = 1, \cdots, 12$ and $i \neq 6$ and $\frac{\partial f_5}{\partial x_6} = 1$.

Sixth row of the Jacobian matrix $\nabla_x f(x) \mid_{(x^*,u^*)}$: $\frac{\partial f_6}{\partial x_i} = 0$ for $i = 1, \cdots, 12$.

Seventh row of the Jacobian matrix $\nabla_x f(x) \mid_{(x^*,u^*)}$: $\frac{\partial f_7}{\partial x_i} = 0$ for $i = 1, \cdots, 12$ and $i \neq 8$ and $\frac{\partial f_7}{\partial x_8} = 1$.

Eighth row of the Jacobian matrix $\nabla_x f(x) \mid_{(x^*,u^*)}$:

It holds that $f_8(x) = \frac{f_{8,num}}{f_{8,den}}$ where $f_{8,num} = \bar{M}_{44}(V_{44}\dot{a} + V_{45}\dot{\beta} + V_{46}\dot{\gamma}) + \bar{M}_{45}(V_{54}\dot{a} + V_{55}\dot{\beta} + V_{56}\dot{\gamma}) + \bar{M}_{46}(V_{64}\dot{a} + V_{65}\dot{\beta} + V_{66}\dot{\gamma})$ and $f_{8,den} = det\bar{M}$.
It holds that

$$\frac{\partial f_8}{\partial x_i} = \frac{\frac{\partial f_{8,num}}{\partial x_i} det\bar{M} - f_{8,num}\frac{\partial det\bar{M}}{\partial x_i}}{det\bar{M}^2} \tag{1.35}$$

where

$$\begin{aligned} \frac{\partial f_{8,num}}{\partial x_i} = & \frac{\partial \bar{M}_{44}}{\partial x_i}(V_{44}\dot{a} + V_{45}\dot{\beta} + V_{46}\dot{\gamma}) + \bar{M}_{44}\left(\frac{\partial V_{44}}{\partial x_i}\dot{a} + V_{44}\frac{\partial \dot{a}}{\partial x_i} + \frac{\partial V_{45}}{\partial x_i}\dot{\beta} + V_{45}\frac{\partial \dot{\beta}}{\partial x_i} + \frac{\partial V_{46}}{\partial x_i}\dot{\gamma} + V_{46}\frac{\partial \dot{\gamma}}{\partial x_i}\right) + \\ & \frac{\partial \bar{M}_{45}}{\partial x_i}(V_{54}\dot{a} + V_{55}\dot{\beta} + V_{56}\dot{\gamma}) + \bar{M}_{45}\left(\frac{\partial V_{54}}{\partial x_i}\dot{a} + V_{54}\frac{\partial \dot{a}}{\partial x_i} + \frac{\partial V_{55}}{\partial x_i}\dot{\beta} + V_{55}\frac{\partial \dot{\beta}}{\partial x_i} + \frac{\partial V_{56}}{\partial x_i}\dot{\gamma} + V_{56}\frac{\partial \dot{\gamma}}{\partial x_i}\right) + \\ & \frac{\partial \bar{M}_{46}}{\partial x_i}(V_{64}\dot{a} + V_{65}\dot{\beta} + V_{66}\dot{\gamma}) + \bar{M}_{46}\left(\frac{\partial V_{64}}{\partial x_i}\dot{a} + V_{64}\frac{\partial \dot{a}}{\partial x_i} + \frac{\partial V_{65}}{\partial x_i}\dot{\beta} + V_{65}\frac{\partial \dot{\beta}}{\partial x_i} + \frac{\partial V_{66}}{\partial x_i}\dot{\gamma} + V_{66}\frac{\partial \dot{\gamma}}{\partial x_i}\right) \end{aligned} \tag{1.36}$$

and

$$\begin{aligned}\frac{\partial detM}{\partial x_i} = m^3 \Big\{ & \frac{\partial m_{44}}{\partial x_i}(m_{55}m_{66} - m_{65}m_{56}) + m_{44}\left(\frac{\partial m_{55}}{\partial x_i}m_{66} + m_{55}\frac{\partial m_{66}}{\partial x_i} - \frac{\partial m_{65}}{\partial x_i}m_{56} - m_{56}\frac{\partial m_{56}}{\partial x_i}\right) - \\ & \frac{\partial m_{45}}{\partial x_i}(m_{54}m_{66} - m_{64}m_{56}) + m_{45}\left(\frac{\partial m_{54}}{\partial x_i}m_{66} + m_{54}\frac{\partial m_{66}}{\partial x_i} - \frac{\partial m_{64}}{\partial x_i}m_{56} - m_{64}\frac{\partial m_{56}}{\partial x_i}\right) + \\ & \frac{\partial m_{46}}{\partial x_i}(m_{54}m_{65} - m_{64}m_{55}) + m_{46}\left(\frac{\partial m_{54}}{\partial x_i}m_{65} + m_{54}\frac{\partial m_{65}}{\partial x_i} - \frac{\partial m_{64}}{\partial x_i}m_{55} - m_{64}\frac{\partial m_{55}}{\partial x_i}\right)\Big\}\end{aligned} \tag{1.37}$$

Ninth row of the Jacobian matrix $\nabla_x f(x) \mid_{(x^*,u^*)}$: $\frac{\partial f_9}{\partial x_i} = 0$ for $i = 1, \cdots, 12$ and $i \neq 10$ and $\frac{\partial f_9}{\partial x_{10}} = 1$.

Tenth row of the Jacobian matrix $\nabla_x f(x) \mid_{(x^*,u^*)}$:

It holds that $f_{10}(x) = \frac{f_{10,num}}{f_{10,den}}$ where $f_{10,num} = \bar{M}_{54}(V_{44}a + V_{45}\beta + V_{46}\gamma) + \bar{M}_{55}(V_{54}a + V_{55}\beta + V_{56}\gamma) + \bar{M}_{56}(V_{64}a + V_{65}\beta + V_{66}\gamma)$ and $f_{10,den} = de\bar{t}M$.

It holds that

$$\frac{\partial f_{10}}{\partial x_i} = \frac{\frac{\partial f_{10,num}}{\partial x_i}de\bar{t}M - f_{10,num}\frac{\partial de\bar{t}M}{\partial x_i}}{de\bar{t}M^2} \tag{1.38}$$

where

$$\begin{aligned}\frac{\partial f_{10,num}}{\partial x_i} = & \frac{\partial \bar{M}_{54}}{\partial x_i}(V_{44}\dot{a} + V_{45}\dot{\beta} + V_{46}\dot{\gamma}) + \bar{M}_{44}\left(\frac{\partial V_{44}}{\partial x_i}\dot{a} + V_{44}\frac{\partial \dot{a}}{\partial x_i} + \frac{\partial V_{45}}{\partial x_i}\dot{\beta} + V_{45}\frac{\partial \dot{\beta}}{\partial x_i} + \frac{\partial V_{46}}{\partial x_i}\dot{\gamma} + V_{46}\frac{\partial \dot{\gamma}}{\partial x_i}\right) + \\ & \frac{\partial \bar{M}_{55}}{\partial x_i}(V_{54}\dot{a} + V_{55}\dot{\beta} + V_{56}\dot{\gamma}) + \bar{M}_{45}\left(\frac{\partial V_{54}}{\partial x_i}\dot{a} + V_{54}\frac{\partial \dot{a}}{\partial x_i} + \frac{\partial V_{55}}{\partial x_i}\dot{\beta} + V_{55}\frac{\partial \dot{\beta}}{\partial x_i} + \frac{\partial V_{56}}{\partial x_i}\dot{\gamma} + V_{56}\frac{\partial \dot{\gamma}}{\partial x_i}\right) + \\ & \frac{\partial \bar{M}_{56}}{\partial x_i}(V_{64}\dot{a} + V_{65}\dot{\beta} + V_{66}\dot{\gamma}) + \bar{M}_{46}\left(\frac{\partial V_{64}}{\partial x_i}\dot{a} + V_{64}\frac{\partial \dot{a}}{\partial x_i} + \frac{\partial V_{65}}{\partial x_i}\dot{\beta} + V_{65}\frac{\partial \dot{\beta}}{\partial x_i} + \frac{\partial V_{66}}{\partial x_i}\dot{\gamma} + V_{66}\frac{\partial \dot{\gamma}}{\partial x_i}\right)\end{aligned} \tag{1.39}$$

Eleventh row of the Jacobian matrix $\nabla_x f(x) \mid_{(x^*,u^*)}$: $\frac{\partial f_{11}}{\partial x_i} = 0$ for $i = 1, \cdots, 12$ and $i \neq 12$ and $\frac{\partial f_{11}}{\partial x_{12}} = 1$.

Twelfth row of the Jacobian matrix $\nabla_x f(x) \mid_{(x^*,u^*)}$.

It holds that $f_{12}(x) = \frac{f_{12,num}}{f_{12,den}}$ where $f_{12,num} = \bar{M}_{64}(V_{44}a + V_{45}\beta + V_{46}\gamma) + \bar{M}_{55}(V_{54}a + V_{65}\beta + V_{56}\gamma) + \bar{M}_{66}(V_{64}a + V_{65}\beta + V_{66}\gamma)$ and $f_{12,den} = de\bar{t}M$.

It holds that

$$\frac{\partial f_{12}}{\partial x_i} = \frac{\frac{\partial f_{12,num}}{\partial x_i}de\bar{t}M - f_{10,num}\frac{\partial de\bar{t}M}{\partial x_i}}{de\bar{t}M^2} \tag{1.40}$$

where

$$\begin{aligned}\frac{\partial f_{12,num}}{\partial x_i} = & \frac{\partial \bar{M}_{64}}{\partial x_i}(V_{44}\dot{a} + V_{45}\dot{\beta} + V_{46}\dot{\gamma}) + \bar{M}_{44}\left(\frac{\partial V_{44}}{\partial x_i}\dot{a} + V_{44}\frac{\partial \dot{a}}{\partial x_i} + \frac{\partial V_{45}}{\partial x_i}\dot{\beta} + V_{45}\frac{\partial \dot{\beta}}{\partial x_i} + \frac{\partial V_{46}}{\partial x_i}\dot{\gamma} + V_{46}\frac{\partial \dot{\gamma}}{\partial x_i}\right) + \\ & \frac{\partial \bar{M}_{65}}{\partial x_i}(V_{54}\dot{a} + V_{55}\dot{\beta} + V_{56}\dot{\gamma}) + \bar{M}_{45}\left(\frac{\partial V_{54}}{\partial x_i}\dot{a} + V_{54}\frac{\partial \dot{a}}{\partial x_i} + \frac{\partial V_{55}}{\partial x_i}\dot{\beta} + V_{55}\frac{\partial \dot{\beta}}{\partial x_i} + \frac{\partial V_{56}}{\partial x_i}\dot{\gamma} + V_{56}\frac{\partial \dot{\gamma}}{\partial x_i}\right) + \\ & \frac{\partial \bar{M}_{46}}{\partial x_i}(V_{64}\dot{a} + V_{65}\dot{\beta} + V_{66}\dot{\gamma}) + \bar{M}_{66}\left(\frac{\partial V_{64}}{\partial x_i}\dot{a} + V_{64}\frac{\partial \dot{a}}{\partial x_i} + \frac{\partial V_{65}}{\partial x_i}\dot{\beta} + V_{65}\frac{\partial \dot{\beta}}{\partial x_i} + \frac{\partial V_{66}}{\partial x_i}\dot{\gamma} + V_{66}\frac{\partial \dot{\gamma}}{\partial x_i}\right)\end{aligned} \tag{1.41}$$

About the first three columns of the control inputs gain matrix $g(x)$ it holds that:

$$\nabla_x g_1(x) \mid_{(x^*,u^*)} = 0_{12\times 12} \quad \nabla_x g_2(x) \mid_{(x^*,u^*)} = 0_{12\times 12} \quad \nabla_x g_3(x) \mid_{(x^*,u^*)} = 0_{12\times 12} \tag{1.42}$$

- The analytical computation of the Jacobian matrix $\nabla_x g_4(x) \mid_{(x^*,u^*)}$ is given below:

First row of the Jacobian matrix $\nabla_x g_4(x) \mid_{(x^*,u^*)}$: $\frac{\partial g_{4,1}}{\partial x_i} = 0$ for $i = 1. \cdots, 12$.

Second row of the Jacobian matrix $\nabla_x g_4(x) \mid_{(x^*,u^*)}$: $\frac{\partial g_{4,2}}{\partial x_i} = 0$ for $i = 1. \cdots, 12$.

Third row of the Jacobian matrix $\nabla_x g_4(x) \mid_{(x^*,u^*)}$: $\frac{\partial g_{4,3}}{\partial x_i} = 0$ for $i = 1.\cdots, 12$.

Fourth row of the Jacobian matrix $\nabla_x g_4(x) \mid_{(x^*,u^*)}$: $\frac{\partial g_{4,4}}{\partial x_i} = 0$ for $i = 1.\cdots, 12$.

Fifth row of the Jacobian matrix $\nabla_x g_4(x) \mid_{(x^*,u^*)}$: $\frac{\partial g_{4,5}}{\partial x_i} = 0$ for $i = 1.\cdots, 12$.

Sixth row of the Jacobian matrix $\nabla_x g_4(x) \mid_{(x^*,u^*)}$: $\frac{\partial g_{4,6}}{\partial x_i} = 0$ for $i = 1.\cdots, 12$.

Seventh row of the Jacobian matrix $\nabla_x g_4(x) \mid_{(x^*,u^*)}$: $\frac{\partial g_{4,7}}{\partial x_i} = 0$ for $i = 1.\cdots, 12$.

Eighth row of the Jacobian matrix $\nabla_x g_4(x) \mid_{(x^*,u^*)}$:

$$\frac{\partial g_{4,8}}{\partial x_i} = \frac{\frac{\partial M_{44}}{\partial x_i}\bar{M} - M_{44}\frac{\partial de\bar{t}M}{\partial x_i}}{de\bar{t}M^2} \tag{1.43}$$

Ninth row of the Jacobian matrix $\nabla_x g_4(x) \mid_{(x^*,u^*)}$: $\frac{\partial g_{4,9}}{\partial x_i} = 0$ for $i = 1.\cdots, 12$.

Tenth row of the Jacobian matrix $\nabla_x g_4(x) \mid_{(x^*,u^*)}$:

$$\frac{\partial g_{4,10}}{\partial x_i} = \frac{\frac{\partial M_{54}}{\partial x_i}\bar{M} - M_{54}\frac{\partial de\bar{t}M}{\partial x_i}}{de\bar{t}M^2} \tag{1.44}$$

Eleventh row of the Jacobian matrix $\nabla_x g_4(x) \mid_{(x^*,u^*)}$: $\frac{\partial g_{4,11}}{\partial x_i} = 0$ for $i = 1.\cdots, 12$.

Twelfth row of the Jacobian matrix $\nabla_x g_4(x) \mid_{(x^*,u^*)}$:

$$\frac{\partial g_{4,12}}{\partial x_i} = \frac{\frac{\partial M_{64}}{\partial x_i}\bar{M} - M_{64}\frac{\partial de\bar{t}M}{\partial x_i}}{de\bar{t}M^2} \tag{1.45}$$

• The analytical computation of the Jacobian matrix $\nabla_x g_5(x) \mid_{(x^*,u^*)}$ is given below:

First row of the Jacobian matrix $\nabla_x g_5(x) \mid_{(x^*,u^*)}$: $\frac{\partial g_{5,1}}{\partial x_i} = 0$ for $i = 1.\cdots, 12$.

Second row of the Jacobian matrix $\nabla_x g_5(x) \mid_{(x^*,u^*)}$: $\frac{\partial g_{5,2}}{\partial x_i} = 0$ for $i = 1.\cdots, 12$.

Third row of the Jacobian matrix $\nabla_x g_5(x) \mid_{(x^*,u^*)}$: $\frac{\partial g_{5,3}}{\partial x_i} = 0$ for $i = 1.\cdots, 12$.

Fourth row of the Jacobian matrix $\nabla_x g_5(x) \mid_{(x^*,u^*)}$: $\frac{\partial g_{5,4}}{\partial x_i} = 0$ for $i = 1.\cdots, 12$.

Fifth row of the Jacobian matrix $\nabla_x g_5(x) \mid_{(x^*,u^*)}$: $\frac{\partial g_{5,5}}{\partial x_i} = 0$ for $i = 1.\cdots, 12$.

Sixth row of the Jacobian matrix $\nabla_x g_5(x) \mid_{(x^*,u^*)}$: $\frac{\partial g_{5,6}}{\partial x_i} = 0$ for $i = 1.\cdots, 12$.

Seventh row of the Jacobian matrix $\nabla_x g_4(x) \mid_{(x^*,u^*)}$: $\frac{\partial g_{5,7}}{\partial x_i} = 0$ for $i = 1.\cdots, 12$.

Eighth row of the Jacobian matrix $\nabla_x g_5(x) \mid_{(x^*,u^*)}$:

$$\frac{\partial g_{5,8}}{\partial x_i} = \frac{\frac{\partial M_{45}}{\partial x_i}\bar{M} - M_{45}\frac{\partial de\bar{t}M}{\partial x_i}}{de\bar{t}M^2} \tag{1.46}$$

Ninth row of the Jacobian matrix $\nabla_x g_5(x) \mid_{(x^*,u^*)}$: $\frac{\partial g_{5,9}}{\partial x_i} = 0$ for $i = 1.\cdots, 12$.

Tenth row of the Jacobian matrix $\nabla_x g_4(x) \mid_{(x^*,u^*)}$:

$$\frac{\partial g_{5,10}}{\partial x_i} = \frac{\frac{\partial M_{55}}{\partial x_i}\bar{M} - M_{55}\frac{\partial det\bar{M}}{\partial x_i}}{det\bar{M}^2} \quad (1.47)$$

Eleventh row of the Jacobian matrix $\nabla_x g_5(x) \mid_{(x^*,u^*)}$: $\frac{\partial g_{5,11}}{\partial x_i} = 0$ for $i = 1.\cdots,12$.

Twelfth row of the Jacobian matrix $\nabla_x g_5(x) \mid_{(x^*,u^*)}$:

$$\frac{\partial g_{5,12}}{\partial x_i} = \frac{\frac{\partial M_{65}}{\partial x_i}\bar{M} - M_{65}\frac{\partial det\bar{M}}{\partial x_i}}{det\bar{M}^2} \quad (1.48)$$

• The analytical computation of the Jacobian matrix $\nabla_x g_6(x) \mid_{(x^*,u^*)}$ is given below:

First row of the Jacobian matrix $\nabla_x g_6(x) \mid_{(x^*,u^*)}$: $\frac{\partial g_{6,1}}{\partial x_i} = 0$ for $i = 1.\cdots,12$.

Second row of the Jacobian matrix $\nabla_x g_6(x) \mid_{(x^*,u^*)}$: $\frac{\partial g_{6,2}}{\partial x_i} = 0$ for $i = 1.\cdots,12$.

Third row of the Jacobian matrix $\nabla_x g_6(x) \mid_{(x^*,u^*)}$: $\frac{\partial g_{6,3}}{\partial x_i} = 0$ for $i = 1.\cdots,12$.

Fourth row of the Jacobian matrix $\nabla_x g_6(x) \mid_{(x^*,u^*)}$: $\frac{\partial g_{6,4}}{\partial x_i} = 0$ for $i = 1.\cdots,12$.

Fifth row of the Jacobian matrix $\nabla_x g_6(x) \mid_{(x^*,u^*)}$: $\frac{\partial g_{6,5}}{\partial x_i} = 0$ for $i = 1.\cdots,12$.

Sixth row of the Jacobian matrix $\nabla_x g_6(x) \mid_{(x^*,u^*)}$: $\frac{\partial g_{6,6}}{\partial x_i} = 0$ for $i = 1.\cdots,12$.

Seventh row of the Jacobian matrix $\nabla_x g_6(x) \mid_{(x^*,u^*)}$: $\frac{\partial g_{6,7}}{\partial x_i} = 0$ for $i = 1.\cdots,12$.

Eighth row of the Jacobian matrix $\nabla_x g_6(x) \mid_{(x^*,u^*)}$:

$$\frac{\partial g_{6,8}}{\partial x_i} = \frac{\frac{\partial M_{46}}{\partial x_i}\bar{M} - M_{46}\frac{\partial det\bar{M}}{\partial x_i}}{det\bar{M}^2} \quad (1.49)$$

Ninth row of the Jacobian matrix $\nabla_x g_6(x) \mid_{(x^*,u^*)}$: $\frac{\partial g_{6,9}}{\partial x_i} = 0$ for $i = 1.\cdots,12$.

Tenth row of the Jacobian matrix $\nabla_x g_6(x) \mid_{(x^*,u^*)}$:

$$\frac{\partial g_{6,10}}{\partial x_i} = \frac{\frac{\partial M_{56}}{\partial x_i}\bar{M} - M_{56}\frac{\partial det\bar{M}}{\partial x_i}}{det\bar{M}^2} \quad (1.50)$$

Eleventh row of the Jacobian matrix $\nabla_x g_6(x) \mid_{(x^*,u^*)}$: $\frac{\partial g_{6,11}}{\partial x_i} = 0$ for $i = 1.\cdots,12$.

Twelfth row of the Jacobian matrix $\nabla_x g_6(x) \mid_{(x^*,u^*)}$:

$$\frac{\partial g_{6,12}}{\partial x_i} = \frac{\frac{\partial M_{66}}{\partial x_i}\bar{M} - M_{66}\frac{\partial det\bar{M}}{\partial x_i}}{det\bar{M}^2} \quad (1.51)$$

• Computation of the partial derivatives of the subdeterminants $\bar{M}_{ij}$, $i = 1,\cdots,6, j = 1,\cdots,6$.

It holds that for $i = 1, \cdots, 12$:

$$\begin{aligned}\frac{\partial \bar{M}_{11}}{\partial x_i} &= \frac{\partial m_{44}}{\partial x_i}(m_{55}m_{66} - m_{65}m_{56}) + \\ &+ m_{14}\left(\frac{\partial m_{55}}{\partial x_i}m_{66} + m_{55}\frac{\partial m_{66}}{\partial x_i} - \frac{\partial m_{65}}{\partial x_i}m_{56} - m_{65}\frac{\partial m_{56}}{\partial x_i}\right) \\ &- \frac{\partial m_{45}}{\partial x_i}(m_{54}m_{66} - m_{64}m_{56}) \\ &- m_{45}\left(\frac{\partial m_{45}}{\partial x_i}m_{66} + m_{54}\frac{\partial m_{66}}{\partial x_i} - \frac{\partial m_{64}}{\partial x_i}m_{56} - m_{64}\frac{\partial m_{56}}{\partial x_i}\right) \\ &+ \frac{\partial m_{46}}{\partial x_i}(m_{54}m_{65} - m_{64}m_{55}) \\ &+ m_{46}\left(\frac{\partial m_{54}}{\partial x_i}m_{65} + m_{54}\frac{\partial m_{65}}{\partial x_i} - \frac{\partial m_{64}}{\partial x_i}m_{55} - m_{64}\frac{\partial m_{55}}{\partial x_i}\right)\end{aligned} \tag{1.52}$$

Besides, for $i = 1, \cdots, 12$, one has that $\frac{\partial \bar{M}_{12}}{\partial x_i} = 0$, $\frac{\partial \bar{M}_{13}}{\partial x_i} = 0$, $\frac{\partial \bar{M}_{14}}{\partial x_i} = 0$, $\frac{\partial \bar{M}_{15}}{\partial x_i} = 0$, and $\frac{\partial \bar{M}_{16}}{\partial x_i} = 0$.

Additionally, for $i = 1, \cdots, 12$ it holds that $\frac{\partial \bar{M}_{21}}{\partial x_i} = 0$, $\frac{\partial \bar{M}_{22}}{\partial x_i} = \frac{\partial \bar{M}_{11}}{\partial x_i}$, $\frac{\partial \bar{M}_{23}}{\partial x_i} = 0$, $\frac{\partial \bar{M}_{24}}{\partial x_i} = 0$, $\frac{\partial \bar{M}_{25}}{\partial x_i} = 0$, and $\frac{\partial \bar{M}_{26}}{\partial x_i} = 0$.

Moreover, for $i = 1, \cdots, 12$ it holds that $\frac{\partial \bar{M}_{31}}{\partial x_i} = 0$, $\frac{\partial \bar{M}_{31}}{\partial x_i} = 0$, $\frac{\partial \bar{M}_{33}}{\partial x_i} = \frac{\partial \bar{M}_{11}}{\partial x_i}$, $\frac{\partial \bar{M}_{34}}{\partial x_i} = 0$, $\frac{\partial \bar{M}_{35}}{\partial x_i} = 0$, and $\frac{\partial \bar{M}_{36}}{\partial x_i} = 0$.

It also holds that $\frac{\partial \bar{M}_{41}}{\partial x_i} = 0$, $\frac{\partial \bar{M}_{42}}{\partial x_i} = 0$, and $\frac{\partial \bar{M}_{43}}{\partial x_i} = 0$.
Additionally, using that $\bar{M}_{44} = (m_{55}m_{66} - m_{65}m_{56})$, one gets that

$$\frac{\partial \bar{M}_{44}}{\partial x_i} = \frac{\partial m_{55}}{\partial x_i}m_{66} + m_{55}\frac{\partial m_{66}}{\partial x_i} - \frac{\partial m_{65}}{\partial x_i}m_{56} - m_{65}\frac{\partial m_{56}}{\partial x_i} \tag{1.53}$$

Using that $\bar{M}_{45} = (m_{54}m_{66} - m_{64}m_{56})$, one gets

$$\frac{\partial \bar{M}_{45}}{\partial x_i} = \frac{\partial m_{54}}{\partial x_i}m_{66} + m_{54}\frac{\partial m_{66}}{\partial x_i} - \frac{\partial m_{64}}{\partial x_i}m_{56} - m_{64}\frac{\partial m_{56}}{\partial x_i} \tag{1.54}$$

Using that $\bar{M}_{46} = (m_{44}m_{65} - m_{64}m_{45})$, one gets

$$\frac{\partial \bar{M}_{46}}{\partial x_i} = \frac{\partial m_{54}}{\partial x_i}m_{65} + m_{54}\frac{\partial m_{65}}{\partial x_i} - \frac{\partial m_{64}}{\partial x_i}m_{55} - m_{64}\frac{\partial m_{55}}{\partial x_i} \tag{1.55}$$

It also holds that $\frac{\partial \bar{M}_{51}}{\partial x_i} = 0$, $\frac{\partial \bar{M}_{52}}{\partial x_i} = 0$, and $\frac{\partial \bar{M}_{53}}{\partial x_i} = 0$.
Additionally, using that $\bar{M}_{54} = (m_{45}m_{66} - m_{65}m_{46})$, one gets that

$$\frac{\partial \bar{M}_{54}}{\partial x_i} = \frac{\partial m_{45}}{\partial x_i}m_{66} + m_{45}\frac{\partial m_{66}}{\partial x_i} - \frac{\partial m_{65}}{\partial x_i}m_{46} - m_{65}\frac{\partial m_{46}}{\partial x_i} \tag{1.56}$$

Using that $\bar{M}_{55} = (m_{44}m_{66} - m_{64}m_{46})$, one gets

$$\frac{\partial \bar{M}_{55}}{\partial x_i} = \frac{\partial m_{44}}{\partial x_i}m_{66} + m_{44}\frac{\partial m_{66}}{\partial x_i} - \frac{\partial m_{64}}{\partial x_i}m_{46} - m_{64}\frac{\partial m_{46}}{\partial x_i} \tag{1.57}$$

Using that $\bar{M}_{56} = (m_{44}m_{65} - m_{64}m_{45})$, one gets

$$\frac{\partial \bar{M}_{56}}{\partial x_i} = \frac{\partial m_{44}}{\partial x_i}m_{65} + m_{44}\frac{\partial m_{65}}{\partial x_i} - \frac{\partial m_{64}}{\partial x_i}m_{45} - m_{64}\frac{\partial m_{45}}{\partial x_i} \tag{1.58}$$

It also holds that $\frac{\partial \bar{M}_{61}}{\partial x_i} = 0$, $\frac{\partial \bar{M}_{62}}{\partial x_i} = 0$, and $\frac{\partial \bar{M}_{63}}{\partial x_i} = 0$.
Additionally, using that $\bar{M}_{64} = (m_{45}m_{56} - m_{55}m_{46})$, one gets that

$$\frac{\partial \bar{M}_{64}}{\partial x_i} = \frac{\partial m_{45}}{\partial x_i}m_{56} + m_{45}\frac{\partial m_{56}}{\partial x_i} - \frac{\partial m_{55}}{\partial x_i}m_{46} - m_{55}\frac{\partial m_{46}}{\partial x_i} \tag{1.59}$$

Using that $\bar{M}_{65} = (m_{44}m_{56} - m_{54}m_{46})$, one gets

$$\frac{\partial \bar{M}_{65}}{\partial x_i} = \frac{\partial m_{44}}{\partial x_i}m_{56} + m_{44}\frac{\partial m_{56}}{\partial x_i} - \frac{\partial m_{54}}{\partial x_i}m_{46} - m_{54}\frac{\partial m_{46}}{\partial x_i} \tag{1.60}$$

Using that $\bar{M}_{66} = (m_{44}m_{55} - m_{54}m_{45})$, one gets

$$\frac{\partial \bar{M}_{66}}{\partial x_i} = \frac{\partial m_{44}}{\partial x_i}m_{55} + m_{44}\frac{\partial m_{55}}{\partial x_i} - \frac{\partial m_{54}}{\partial x_i}m_{45} - m_{54}\frac{\partial m_{45}}{\partial x_i} \tag{1.61}$$

• Computation of the partial derivatives of the elements of the Coriolis matrix V_{ij}, $i=1,\cdots,6$, and $=1,\cdots,6$

For $i=1,2,\cdots,12$ it holds that $\frac{\partial V_{11}}{\partial x_i}=0$, $\frac{\partial V_{12}}{\partial x_i}=0$, $\frac{\partial V_{13}}{\partial x_i}=0$, $\frac{\partial V_{14}}{\partial x_i}=0$, $\frac{\partial V_{15}}{\partial x_i}=0$, $\frac{\partial V_{16}}{\partial x_i}=0$

For $i=1,2,\cdots,12$ it holds that $\frac{\partial V_{21}}{\partial x_i}=0$, $\frac{\partial V_{22}}{\partial x_i}=0$, $\frac{\partial V_{23}}{\partial x_i}=0$, $\frac{\partial V_{24}}{\partial x_i}=0$, $\frac{\partial V_{25}}{\partial x_i}=0$, $\frac{\partial V_{26}}{\partial x_i}=0$

For $i=1,2,\cdots,12$ it holds that $\frac{\partial V_{31}}{\partial x_i}=0$, $\frac{\partial V_{32}}{\partial x_i}=0$, $\frac{\partial V_{33}}{\partial x_i}=0$, $\frac{\partial V_{34}}{\partial x_i}=0$, $\frac{\partial V_{35}}{\partial x_i}=0$, $\frac{\partial V_{36}}{\partial x_i}=0$

For $i=1,2,\cdots,12$ it holds that $\frac{\partial V_{41}}{\partial x_i}=0$, $\frac{\partial V_{42}}{\partial x_i}=0$, $\frac{\partial V_{43}}{\partial x_i}=0$. Besides, one has that

$$\frac{\partial V_{44}}{\partial x_i}=\frac{\partial K_1}{\partial x_i}\dot{\beta}+K_1\frac{\partial\dot{\beta}}{\partial x_i}+\frac{\partial K_2}{\partial x_i}\dot{\gamma}+K_2\frac{\partial\dot{\gamma}}{\partial x_i} \tag{1.62}$$

$$\frac{\partial V_{45}}{\partial x_i}=\frac{\partial K_1}{\partial x_i}\dot{a}+K_1\frac{\partial\dot{a}}{\partial x_i}+\frac{\partial K_5}{\partial x_i}\dot{\beta}+K_5\frac{\partial\dot{\beta}}{\partial x_i}+\frac{\partial K_3}{\partial x_i}\dot{\gamma}+K_3\frac{\partial\dot{\gamma}}{\partial x_i} \tag{1.63}$$

$$\frac{\partial V_{46}}{\partial x_i}=\frac{\partial K_2}{\partial x_i}\dot{a}+K_2\frac{\partial\dot{a}}{\partial x_i}+\frac{\partial K_3}{\partial x_i}\dot{\beta}+K_3\frac{\partial\dot{\beta}}{\partial x_i} \tag{1.64}$$

For $i=1,2,\cdots,12$ it holds that $\frac{\partial V_{51}}{\partial x_i}=0$, $\frac{\partial V_{52}}{\partial x_i}=0$, $\frac{\partial V_{53}}{\partial x_i}=0$. Besides, one has that

$$\frac{\partial V_{54}}{\partial x_i}=-\frac{\partial K_1}{\partial x_i}\dot{\beta}-K_1\frac{\partial\dot{\beta}}{\partial x_i}+\frac{\partial K_3}{\partial x_i}\dot{\gamma}+K_3\frac{\partial\dot{\gamma}}{\partial x_i} \tag{1.65}$$

$$\frac{\partial V_{55}}{\partial x_i}=\frac{\partial K_4}{\partial x_i}\dot{\gamma}+K_4\frac{\partial\dot{\gamma}}{\partial x_i} \tag{1.66}$$

$$\frac{\partial V_{56}}{\partial x_i}=\frac{\partial K_3}{\partial x_i}\dot{a}+K_3\frac{\partial\dot{a}}{\partial x_i}+\frac{\partial K_4}{\partial x_i}\dot{\beta}+K_4\frac{\partial\dot{\beta}}{\partial x_i} \tag{1.67}$$

For $i=1,2,\cdots,12$ it holds that $\frac{\partial V_{61}}{\partial x_i}=0$, $\frac{\partial V_{62}}{\partial x_i}=0$, $\frac{\partial V_{63}}{\partial x_i}=0$. Besides, one has that

$$\frac{\partial V_{64}}{\partial x_i}=-\frac{\partial K_2}{\partial x_i}\dot{a}-K_2\frac{\partial\dot{a}}{\partial x_i}-\frac{\partial K_3}{\partial x_i}\dot{\beta}+K_3\frac{\partial\dot{\beta}}{\partial x_i} \tag{1.68}$$

$$\frac{\partial V_{65}}{\partial x_i}=-\frac{\partial K_3}{\partial x_i}\dot{a}-K_3\frac{\partial\dot{a}}{\partial x_i}-\frac{\partial K_4}{\partial x_i}\dot{\beta}+K_4\frac{\partial\dot{\beta}}{\partial x_i} \tag{1.69}$$

$$\frac{\partial V_{66}}{\partial x_i}=0 \tag{1.70}$$

• Computation of the partial derivatives of the terms K_i, $i=1,\cdots,5$:

It holds that $\frac{\partial K_1}{\partial x_i}=0$ for $i=1,2\cdots,12$ and $i{\neq}9,11$, while $\frac{\partial K_1}{\partial x_9}=[cos^2(x_9)-sin^2(x_9)](cos^2(x_{11})I_x+sin^2(x_{11}I_y)-I_z)$ and $\frac{\partial K_1}{\partial x_{11}}=cos(x_9)sin(x_9)[-2cos(x_{11})sin(x_{11})I_x+2sin(x_{11})cos(x_{11})I_y-I_z]$.

Besides, it holds that $\frac{\partial K_2}{\partial x_i}=0$ for $i=1,2\cdots,12$ and $i{\neq}9,11$, while $\frac{\partial K_2}{\partial x_9}=2cos(x_9)sin(x_9)cos(x_{11})sin(x_{11})(I_x-I_y)$ and $\frac{\partial K_2}{\partial x_{11}}=-cos^2(x_9)[cos^2(x_{11})-sin^2(x_{11})](I_x-I_y)$.

Additionally, it holds that $\frac{\partial K_3}{\partial x_i}=0$ for $i=1,2\cdots,12$ and $i{\neq}9,11$, while $\frac{\partial K_3}{\partial x_9}=-\frac{1}{2}sin(x_9)(cos(x_{11})-sin(x_{11}))(cos(x_{11})+sin(x_{11}))(I_x-I_y)$ and $\frac{\partial K_3}{\partial x_{11}}=\frac{1}{2}cos(x_9)[(-sin(x_{11})-cos(x_{11}))(cos(x_{11})+sin(x_{11}))+(cos(x_{11})-sin(x_{11}))(-sin(x_{11})+cos(x_{11}))](I_x-I_y)$.

Furthermore, it holds that $\frac{\partial K_4}{\partial x_i}=0$ for $i=1,2\cdots,12$ and $i{\neq}11$, while $\frac{\partial K_4}{\partial x_{11}}=[cos^2(x_{11})-sin^2(x_{11})](I_x-I_y)$.

Finally, it holds that $\frac{\partial K_5}{\partial x_i}=0$ for $i=1,2\cdots,12$ and $i{\neq}9,11$, while $\frac{\partial K_5}{\partial x_9}=cos(x_{11})sin(x_{11})sin(x_9)(I_x-I_y)$ and $\frac{\partial K_5}{\partial x_{11}}=[cos^2(x_{11})-sin^2(x_{11})]sin(x_9)(I_x-I_y)$.

• Computation of the partial derivatives of the elements of the inertia matrix m_{ij}, $i=1,\cdots,6$ and $j=1,\cdots,6$.

It holds that $\frac{\partial m_{1i}}{\partial x_i}$, $\frac{\partial m_{2i}}{\partial x_i}$, $\frac{\partial m_{13i}}{\partial x_i}$ for $i=1,\cdots,6$.

It also holds that $\frac{\partial m_{41}}{\partial x_i}=0$, $\frac{\partial m_{42}}{\partial x_i}=0$, $\frac{\partial m_{43}}{\partial x_i}=0$.

Additionally, it holds that $\frac{\partial m_{44}}{\partial x_i}=0$ for $i=1,2,\cdots,12$ and $i{\neq}9,11$ while $\frac{\partial m_{44}}{\partial x_9}=I_x(-2cos(x_9)sin(x_9)cos^2(x_{11}))+I_y(-2cos(x_9)sin(x_9)sin^2(x_{11}))+I_z(2sin(x_9)cos(x_9))$, and $\frac{\partial m_{44}}{\partial x_{11}}=[-2I_xcos^2(x_9)+2I_ysin^2(x_9)]sin(x_{11})cos(x_{11})$.

Moreover, it holds that $\frac{\partial m_{45}}{\partial x_i}=0$ for $i=1,2,\cdots,12$ and $i{\neq}9,11$ while $\frac{\partial m_{45}}{\partial x_9}=-(I_x-I_y)sin(x_9)cos(x_{11})sin(x_{11})$, and $\frac{\partial m_{45}}{\partial x_{11}}=(I_x-I_y)cos(x_9)[cos^2(x_{11})-sin^2(x_{11})]$.

Furthermore, it holds that $\frac{\partial m_{46}}{\partial x_i}=0$ for $i=1,2,\cdots,12$ and $i{\neq}9$ while $\frac{\partial m_{46}}{\partial x_9}=I_zcos(x_9)$.

It also holds that $\frac{\partial m_{51}}{\partial x_i}=0$, $\frac{\partial m_{52}}{\partial x_i}=0$, $\frac{\partial m_{53}}{\partial x_i}=0$.

Additionally, it holds that $\frac{\partial m_{54}}{\partial x_i}=0$ for $i=1,2,\cdots,12$ and $i{\neq}9,11$ while $\frac{\partial m_{54}}{\partial x_9}=-(I_x-I_y)sin(x_9)cos(x_{11})sin(x_{11})$, and $\frac{\partial m_{54}}{\partial x_{11}}=(I_x-I_y)cos(x_9)[cos^2(x_{11})-sin^2(x_{11})]$.

Moreover, it holds that $\frac{\partial m_{55}}{\partial x_i}=0$ for $i=1,2,\cdots,12$ and $i{\neq}11$ while $\frac{\partial m_{55}}{\partial x_{11}}=2(I_x-I_y)sin(x_{11})cos(x_{11})$.

Furthermore, it holds that $\frac{\partial m_{56}}{\partial x_i}=0$ for $i=1,2,\cdots,12$.

It also holds that $\frac{\partial m_{61}}{\partial x_i}=0$, $\frac{\partial m_{62}}{\partial x_i}=0$, $\frac{\partial m_{63}}{\partial x_i}=0$.

Additionally, it holds that $\frac{\partial m_{64}}{\partial x_i}=0$ for $i=1,2,\cdots,12$ and $i{\neq}9$ while $\frac{\partial m_{64}}{\partial x_9}=I_zcos(x_9)$.

Moreover, it holds that $\frac{\partial m_{65}}{\partial x_i}=0$ for $i=1,2,\cdots,12$.

Finally, it holds that $\frac{\partial m_{66}}{\partial x_i}=0$ for $i=1,2,\cdots,12$.

• Computation of the partial derivatives of the functions that coincide with Euler angles of the robot and their time derivatives:

It holds that $\frac{\partial a}{\partial x_i}=0$ for $i=1,2,\cdots,12$ and $i{\neq}7$ while $\frac{\partial a}{\partial x_7}=1$.

Besides, $\frac{\partial \dot{a}}{\partial x_i}=0$ for $i=1,2,\cdots,12$ and $i{\neq}8$ while $\frac{\partial \dot{a}}{\partial x_8}=1$.

It holds that $\frac{\partial \beta}{\partial x_i}=0$ for $i=1,2,\cdots,12$ and $i{\neq}9$ while $\frac{\partial \beta}{\partial x_9}=1$.

Furthermore $\frac{\partial \dot{\beta}}{\partial x_i}=0$ for $i=1,2,\cdots,12$ and $i{\neq}10$ while $\frac{\partial \dot{\beta}}{\partial x_{10}}=1$.

It holds that $\frac{\partial \gamma}{\partial x_i}=0$ for $i=1,2,\cdots,12$ and $i{\neq}11$ while $\frac{\partial \gamma}{\partial x_{11}}=1$.

Additionally, $\frac{\partial \dot{\gamma}}{\partial x_i}=0$ for $i=1,2,\cdots,11$ while $\frac{\partial \dot{\gamma}}{\partial x_{12}}=1$.

This above-given linearization approach which has been followed for implementing the nonlinear optimal control scheme results in a quite accurate model of the system's dynamics. Consider, for instance, the following affine-in-the-input state-space model:

$$\begin{aligned}
&\dot{x}=f(x)+g(x)u\Rightarrow\\
&\dot{x}=[f(x^*)+\nabla_xf(x)\mid_{x^*}(x-x^*)]+[g(x^*)+\nabla_xg(x)\mid_{x^*}(x-x^*)]u^*+g(x^*)u^*+g(x^*)(u-u^*)+\tilde{d}_1\Rightarrow\\
&\dot{x}=[\nabla_xf(x)\mid_{x^*}+\nabla_xg(x)\mid_{x^*}u^*]x+g(x^*)u-[\nabla_xf(x)\mid_{x^*}+\nabla_xg(x)\mid_{x^*}u^*]x^*+f(x^*)+g(x^*)u^*+\tilde{d}_1
\end{aligned} \tag{1.71}$$

where $\tilde{d}_1$ is the modeling error due to truncation of higher-order terms in the Taylor series expansion of $f(x)$ and $g(x)$. Next, by defining $A = [\nabla_x f(x) \mid_{x^*} + \nabla_x g(x) \mid_{x^*} u^*]$, $B = g(x^*)$, one obtains

$$\dot{x} = Ax + Bu - Ax^* + f(x^*) + g(x^*)u^* + \tilde{d}_1 \tag{1.72}$$

Moreover, by denoting $\tilde{d} = -Ax^* + f(x^*) + g(x^*)u^* + \tilde{d}_1$ about the cumulative modeling error term in the Taylor series expansion procedure, one has

$$\dot{x} = Ax + Bu + \tilde{d} \tag{1.73}$$

which is the approximately linearized model of the dynamics of the system of Eq. (1.32). The term $f(x^*) + g(x^*)u^*$ is the derivative of the state vector at (x^*, u^*) which is almost annihilated by $-Ax^*$.

1.1.4 An H-infinity controller for the Stewart parallel robotic platform

1.1.4.1 Min–max control and disturbance rejection

In the H_∞ control approach, a feedback control scheme is designed for setpoint tracking by the state vector of the Stewart parallel robotic platform and for simultaneous disturbance rejection, considering that the disturbance affects the system in the worst possible manner. For the approximately linearized model of Eq. (1.32), the disturbances' effects are incorporated in the following quadratic cost function [725]:

$$J(t) = \frac{1}{2}\int_0^T [y^T(t)y(t) + ru^T(t)u(t) - \rho^2\tilde{d}^T(t)\tilde{d}(t)]dt, \quad r, \rho > 0 \tag{1.74}$$

The significance of the negative sign in the cost function's term that is associated with the perturbation variable $\tilde{d}(t)$ is that the disturbance tries to maximize the cost function $J(t)$ while the control signal $u(t)$ tries to minimize it.

The physical meaning of the relation given above is that the control signal and the disturbances compete with each other within a min–max differential game. This problem of min–max optimization can be written as $min_u max_{\tilde{d}} J(u, \tilde{d})$. The objective of the optimization procedure is to compute a control signal $u(t)$ which can compensate for the worst possible disturbance that affects the system. However, the solution to the min–max optimization problem is directly related to the value of parameter ρ. This means that there is an upper bound in the disturbances' magnitude that can be annihilated by the control signal.

1.1.4.2 Design of the stabilizing feedback controller

After linearization around its current operating point (x^*, u^*), the dynamic model of the 6-DOF Stewart parallel robotic platform is written as

$$\dot{x} = Ax + Bu + d_1 \tag{1.75}$$

Parameter d_1 stands for the linearization error in the 6-DOF Stewart parallel robotic platform's model appearing above in Eq. (1.75). The reference setpoints for the state vector of the 6-DOF Stewart parallel robotic platform are denoted by $\mathbf{x_d} = [x_1^d, \cdots, x_{12}^d]$. Tracking of this trajectory is achieved after applying the control input u_d. At every time instant the control input u_d is assumed to differ from the control input u appearing in Eq. (1.75) by an amount equal to Δu, that is $u_d = u + \Delta u$:

$$\dot{x}_d = Ax_d + Bu_d + d_2 \tag{1.76}$$

The dynamics of the controlled system described in Eq. (1.75) can be also written as

$$\dot{x} = Ax + Bu + Bu_d - Bu_d + d_1 \tag{1.77}$$

and by denoting $d_3 = -Bu_d + d_1$ as an aggregate disturbance term, one obtains

$$\dot{x} = Ax + Bu + Bu_d + d_3 \tag{1.78}$$

By subtracting Eq. (1.76) from Eq. (1.78), one has

$$\dot{x} - \dot{x}_d = A(x - x_d) + Bu + d_3 - d_2 \tag{1.79}$$

By denoting the tracking error as $e = x - x_d$ and the aggregate disturbance term as $L\tilde{d} = d_3 - d_2$, the tracking error dynamics becomes

$$\dot{e} = Ae + Bu + L\tilde{d} \tag{1.80}$$

where L is a disturbance inputs gain matrix. For the linearized system, the cost function of Eq. (1.74) is defined, where coefficient r determines the penalization of the control input and the weight coefficient ρ determines the reward of the disturbances' effects. Then, as it will be also shown in subsection 1.1.5, the optimal feedback control law is given by

$$u(t) = -Ke(t) \tag{1.81}$$

with $K = \frac{1}{r}B^TP$ where P is a positive definite symmetric matrix which is obtained from the solution of the Riccati equation:

$$A^TP + PA + Q - P(\tfrac{2}{r}BB^T - \tfrac{1}{\rho^2}LL^T)P = 0 \tag{1.82}$$

where Q is also a semi-positive definite symmetric matrix. The worst-case disturbance is given by $\tilde{d}(t) = \frac{1}{\rho^2}L^TPx(t)$. The computation of the worst-case disturbance that this controller can sustain, comes from superposition of Bellman's optimality principle when considering that the 6-DOF Stewart parallel robotic platform is affected by two separate inputs, that is the control input u and the cumulative disturbance input $\tilde{d}$. From the previous relation about the worst-case disturbance, it can be seen that the smallest value of the attenuation coefficient ρ which results into an admissible solution for the methods's algebraic Riccati equation, is the one that provides the control loop with maximum robustness. The diagram of the considered control loop is depicted in Figure 1.2.

Figure 1.2 gives also a block diagram about the nonlinear optimal control scheme. The implementation stages of the control method are outlined as follows: (1) At each sampling instance there is linearization of the state-space model of the 6-DOF Stewart parallel robotic platform around a time-varying operating point which is defined by the present value of the system's state vector and by the last sampled value of the control inputs vector. The linearization is performed through first-order Taylor series expansion and the computation of the associated Jacobian matrices, (2) for the approximately linearized model of the 6-DOF Stewart parallel robotic platform an H-infinity feedback controller is designed. To compute the controller's feedback gains an algebraic Riccati equation is also repetitively solved at each

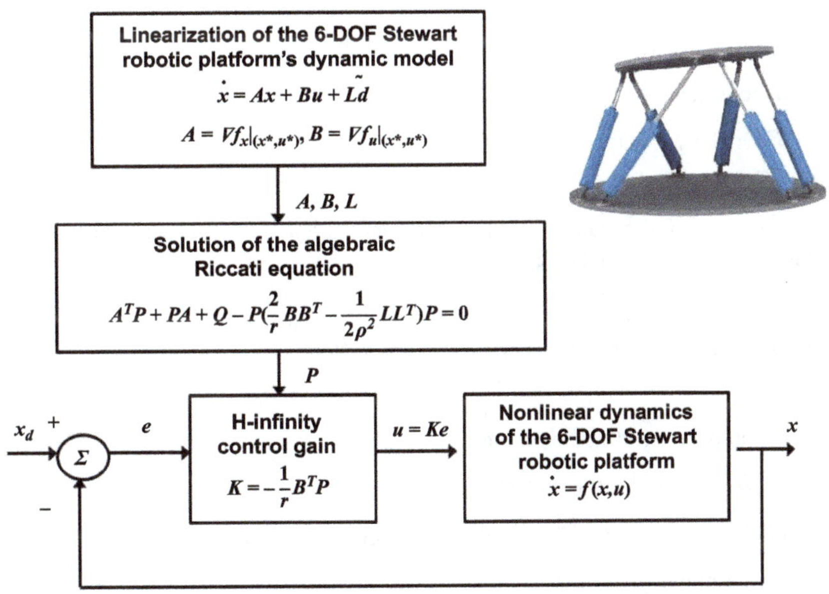

Figure 1.2 Diagram of the nonlinear optimal control loop for the 6-DOF Stewart parallel robotic platform

time-step of the control algorithm. Main parameters of this Riccati equation are gains r and ρ as well as gain matrices Q and L, (3) the computed optimal control inputs are applied directly to the initial nonlinear state-space model of the system. The global stability of the control method and the elimination of the state vector's tracking error are confirmed through Lyapunov analysis. The computational effort for the implementation of this control method is moderate and the solution of the method's Riccati equation is achieved in a time-interval which is smaller than the sampling period.

1.1.5 Lyapunov stability analysis

1.1.5.1 Stability proof

Through Lyapunov stability analysis, it will be shown that the proposed nonlinear control scheme assures H_∞ tracking performance for the 6-DOF Stewart parallel robotic platform and that in case of bounded disturbance terms, asymptotic convergence to the reference setpoints is achieved. The tracking error dynamics for the Stewart parallel robotic platform is written in the form [721,725]:

$$\dot{e} = Ae + Bu + L\tilde{d} \tag{1.83}$$

where in the Stewart parallel robotic platform's case $L \in R^{12\times 12}$ is the disturbance inputs gain matrix. Variable $\tilde{d}$ denotes model uncertainties and external disturbances of the Stewart parallel robotic platform's model. The following Lyapunov equation is considered:

$$V = \frac{1}{2}e^T P e \tag{1.84}$$

where $e = x - x_d$ is the tracking error. By differentiating with respect to time, one obtains

$$\begin{aligned} \dot{V} &= \tfrac{1}{2}\dot{e}^T P e + \tfrac{1}{2}e^T P \dot{e} \Rightarrow \\ \dot{V} &= \tfrac{1}{2}[Ae + Bu + L\tilde{d}]^T P e + \tfrac{1}{2}e^T P [Ae + Bu + L\tilde{d}] \Rightarrow \end{aligned} \tag{1.85}$$

$$\begin{aligned} \dot{V} = &\tfrac{1}{2}[e^T A^T + u^T B^T + \tilde{d}^T L^T] P e + \\ &+ \tfrac{1}{2}e^T P [Ae + Bu + L\tilde{d}] \Rightarrow \end{aligned} \tag{1.86}$$

$$\begin{aligned} \dot{V} = &\tfrac{1}{2}e^T A^T P e + \tfrac{1}{2}u^T B^T P e + \tfrac{1}{2}\tilde{d}^T L^T P e + \\ &\tfrac{1}{2}e^T P A e + \tfrac{1}{2}e^T P B u + \tfrac{1}{2}e^T P L \tilde{d} \end{aligned} \tag{1.87}$$

The previous equation is rewritten as

$$\begin{aligned} \dot{V} = &\tfrac{1}{2}e^T (A^T P + PA) e + (\tfrac{1}{2}u^T B^T P e + \tfrac{1}{2}e^T P B u) + \\ &+ (\tfrac{1}{2}\tilde{d}^T L^T P e + \tfrac{1}{2}e^T P L \tilde{d}) \end{aligned} \tag{1.88}$$

Assumption: For given positive definite matrix Q and coefficients r and ρ, there exists a positive definite matrix P, which is the solution of the following matrix equation:

$$A^T P + PA = -Q + P(\tfrac{2}{r}BB^T - \tfrac{1}{\rho^2}LL^T)P \tag{1.89}$$

Moreover, the following feedback control law is applied to the system:

$$u = -\tfrac{1}{r}B^T P e \tag{1.90}$$

By substituting Eq. (1.89) and Eq. (1.90), one obtains

$$\dot{V} = \tfrac{1}{2}e^T[-Q + P(\tfrac{2}{r}BB^T - \tfrac{1}{\rho^2}LL^T)P]e + e^T PB(-\tfrac{1}{r}B^T P e) + e^T P L \tilde{d} \Rightarrow \tag{1.91}$$

$$\dot{V} = -\tfrac{1}{2}e^T Q e + (\tfrac{1}{r}e^T P B B^T P e - \tfrac{1}{2\rho^2}e^T P L L^T P e) - \tfrac{1}{r}e^T P B B^T P e + e^T P L \tilde{d} \tag{1.92}$$

which after intermediate operations gives

$$\dot{V} = -\tfrac{1}{2}e^T Q e - \tfrac{1}{2\rho^2}e^T P L L^T P e + e^T P L \tilde{d} \tag{1.93}$$

or, equivalently,

$$\begin{aligned}\dot{V} = &-\tfrac{1}{2}e^TQe - \tfrac{1}{2\rho^2}e^TPLL^TPe+ \\ &+\tfrac{1}{2}e^TPL\tilde{d} + \tfrac{1}{2}\tilde{d}^TL^TPe\end{aligned} \tag{1.94}$$

Lemma: The following inequality holds [721,725]:

$$\tfrac{1}{2}e^TL\tilde{d} + \tfrac{1}{2}\tilde{d}L^TPe - \tfrac{1}{2\rho^2}e^TPLL^TPe \leq \tfrac{1}{2}\rho^2\tilde{d}^T\tilde{d} \tag{1.95}$$

Proof: The binomial $(\rho\alpha - \frac{1}{\rho}b)^2$ is considered. Expanding the left part of the above inequality, one gets

$$\begin{aligned}&\rho^2a^2 + \tfrac{1}{\rho^2}b^2 - 2ab \geq 0 \Rightarrow \tfrac{1}{2}\rho^2a^2 + \tfrac{1}{2\rho^2}b^2 - ab \geq 0 \Rightarrow \\ &ab - \tfrac{1}{2\rho^2}b^2 \leq \tfrac{1}{2}\rho^2a^2 \Rightarrow \tfrac{1}{2}ab + \tfrac{1}{2}ab - \tfrac{1}{2\rho^2}b^2 \leq \tfrac{1}{2}\rho^2a^2\end{aligned} \tag{1.96}$$

The following substitutions are carried out: $a = \tilde{d}$ and $b = e^TPL$ and the previous relation becomes

$$\tfrac{1}{2}\tilde{d}^TL^TPe + \tfrac{1}{2}e^TPL\tilde{d} - \tfrac{1}{2\rho^2}e^TPLL^TPe \leq \tfrac{1}{2}\rho^2\tilde{d}^T\tilde{d} \tag{1.97}$$

Eq. (1.97) is substituted in Eq. (1.94) and the inequality is enforced, thus giving [721,725]

$$\dot{V} \leq -\tfrac{1}{2}e^TQe + \tfrac{1}{2}\rho^2\tilde{d}^T\tilde{d} \tag{1.98}$$

Eq. (1.98) shows that the H_∞ tracking performance criterion is satisfied. The integration of $\dot{V}$ from 0 to T gives

$$\begin{aligned}&\int_0^T\dot{V}(t)dt \leq -\tfrac{1}{2}\int_0^T||e||_Q^2dt + \tfrac{1}{2}\rho^2\int_0^T||\tilde{d}||^2dt \Rightarrow \\ &2V(T) + \int_0^T||e||_Q^2dt \leq 2V(0) + \rho^2\int_0^T||\tilde{d}||^2dt\end{aligned} \tag{1.99}$$

Moreover, if there exists a positive constant $M_d > 0$ such that

$$\int_0^\infty ||\tilde{d}||^2dt \leq M_d \tag{1.100}$$

then, one gets

$$\int_0^\infty ||e||_Q^2dt \leq 2V(0) + \rho^2M_d \tag{1.101}$$

Thus, the integral $\int_0^\infty ||e||_Q^2dt$ is bounded. Moreover, $V(T)$ is bounded and from the definition of the Lyapunov function V in Eq. (1.84) it becomes clear that $e(t)$ will be also bounded since $e(t) \in \Omega_e = \{e|e^TPe \leq 2V(0) + \rho^2M_d\}$. According to the above and with the use of Barbalat's Lemma, one obtains $lim_{t\to\infty}e(t) = 0$.

Through the stages of the stability proof, one arrives at Eq. (1.98), which shows that the H-infinity tracking performance criterion holds. By selecting the attenuation coefficient ρ to be sufficiently small and in particular to satisfy $\rho^2 < ||e||_Q^2/||\tilde{d}||^2$, one has that the first derivative of the Lyapunov function is upper bounded by 0. This condition holds at each sampling instance and consequently global stability for the control loop can be concluded.

1.1.5.2 Robust state estimation with the use of the H_∞ Kalman filter

The control loop has to be implemented with the use of information provided by a small number of sensors and by processing only a small number of state variables. Actually, one can implement feedback control by measuring only the state variables of the 6-DOF Stewart parallel robotic platform which coincide with the flat outputs. To reconstruct the missing information about the state vector of the parallel robotic platform it is proposed to use a filtering scheme and based on it to apply state estimation-based control [721,725]. By denoting as $A(k)$, $B(k)$, and $C(k)$ the discrete-time equivalents of matrices A, B, and C of the linearized state-space model of the system, the recursion of the H_∞ Kalman filter, for the model of the Stewart parallel robotic platform, can be formulated in terms of a *measurement update* and a *time update* part.

Measurement update:

$$\begin{aligned} D(k) &= [I - \theta W(k)P^-(k) + C^T(k)R(k)^{-1}C(k)P^-(k)]^{-1} \\ K(k) &= P^-(k)D(k)C^T(k)R(k)^{-1} \\ \hat{x}(k) &= \hat{x}^-(k) + K(k)[y(k) - C\hat{x}^-(k)] \end{aligned} \tag{1.102}$$

Time update:

$$\begin{aligned} \hat{x}^-(k+1) &= A(k)x(k) + B(k)u(k) \\ P^-(k+1) &= A(k)P^-(k)D(k)A^T(k) + Q(k) \end{aligned} \tag{1.103}$$

where it is assumed that parameter θ is sufficiently small to assure that the covariance matrix $P^-(k)^{-1} - \theta W(k) + C^T(k)R(k)^{-1}C(k)$ will be positive definite. When $\theta = 0$ the H_∞ Kalman filter becomes equivalent to the standard Kalman filter. One can measure only a part of the state vector of the Stewart parallel robotic platform consisting of the odd-numbered state variables $Y = [x_1, x_3, x_5, x_7, x_9, x_{11}]^T$, and can estimate through filtering the rest of the state vector elements, that is the even-numbered state variables (linear and angular velocities) $\dot{Y} = [x_2, x_4, x_6, x_8, x_{10}, x_{12}]^T$.

1.1.6 Simulation tests

The performance of the nonlinear optimal control scheme for the Stewart parallel robotic platform is also confirmed through simulation experiments. To apply the nonlinear optimal control method, the dynamic model of parallel robot undergoes linearization with Jacobian matrices computation at each sampling instance while the algebraic Riccati equation of Eq. (1.89) is also solved at each sampling interval using the MATLAB® function "aresolv()". Indicative values about the parameters of the parallel robotic system have been $m = 20kg$, $I_x = 1.45kg{\cdot}m^2$, $I_y = 1.45kg{\cdot}m^2$, and $I_z = 2.5kg{\cdot}m^2$. The obtained results are depicted in Figures 1.3–1.18. The real values of the state variables of the parallel robot are shown in blue color, the associated setpoints are plotted in red color while the estimated state variables which are provided by the H-infinity Kalman filter are given in green color. Through all simulation diagrams, it can be noticed that the nonlinear optimal control method achieves fast and accurate tracking of setpoints under moderate variations of the control inputs.

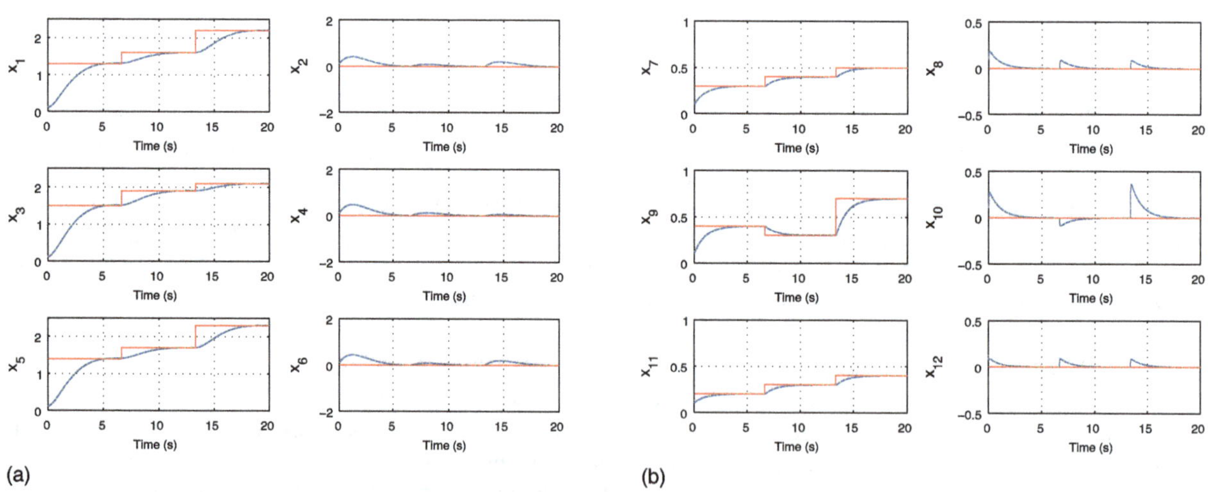

Figure 1.3 Tracking of setpoint 1 by the 6-DOF Stewart parallel robotic platform with the use of nonlinear optimal control: (a) convergence of state variables x_1 to x_6 (blue lines) to the associated setpoints (red lines) and estimated values provided by Kalman filtering; (b) convergence of state variables x_7 to x_{12} (blue lines) to the associated setpoints (red lines) and estimated values provided by Kalman filtering

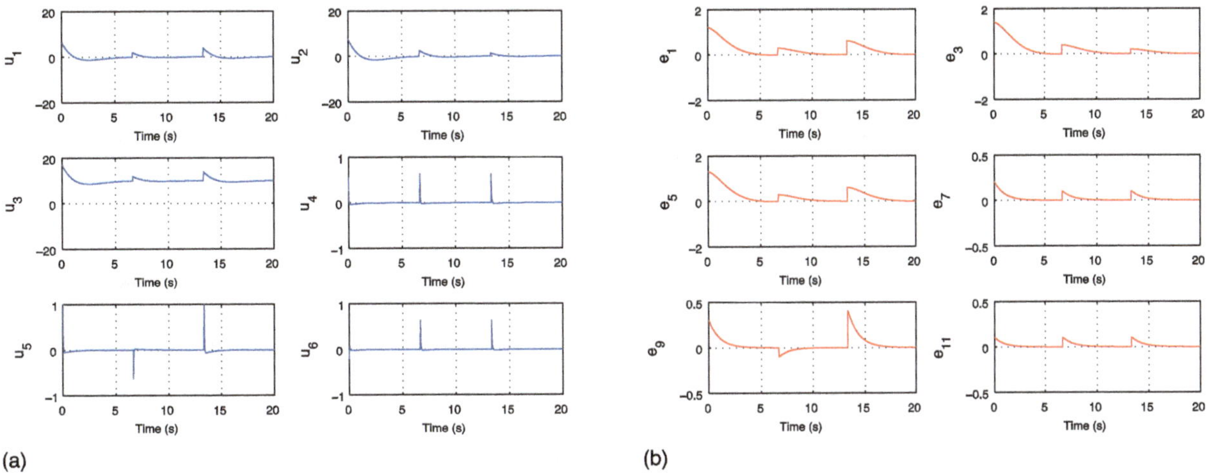

Figure 1.4 Tracking of setpoint 1 by the 6-DOF Stewart parallel robotic platform with the use of nonlinear optimal control: (a) variations of the control inputs u_1 to u_6 (blue lines); (b) variation of the tracking error variables e_i, $i = 1,3,5,7,9,11$ associated with the state variables x_i, $i = 1,3,5,7,9,11$

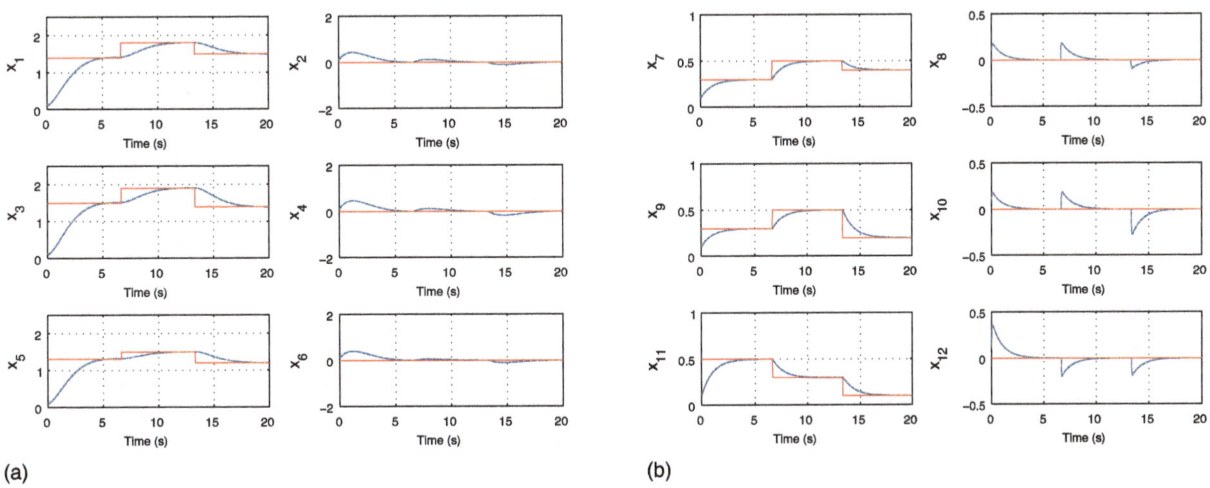

Figure 1.5 Tracking of setpoint 2 by the 6-DOF Stewart parallel robotic platform with the use of nonlinear optimal control: (a) convergence of state variables x_1 to x_6 (blue lines) to the associated setpoints (red lines) and estimated values provided by Kalman filtering; (b) convergence of state variables x_7 to x_{12} (blue lines) to the associated setpoints (red lines) and estimated values provided by Kalman filtering

The transient performance of the nonlinear optimal control method depends on the selection of values for the coefficients r, ρ, and Q of the above-noted Riccati equation. By assigning small values to gain r, one ensures elimination of the state vector's tracking error. Moreover, for relatively large values of the diagonal elements of the diagonal matrix Q, one achieves faster convergence of the state variables to setpoints. Furthermore, the smallest values of the attenuation coefficient ρ for which one still obtains a valid solution of the Riccati equation in the form of a positive definite and symmetric matrix P is the one that provides maximum robustness to the control loop. Indicative values about the above-noted gains have been $r = 0.005$, $\rho = 0.1$ while the diagonal elements of matrix Q were $q_{ii} = 0.2, i = 1, \cdots, 12$.

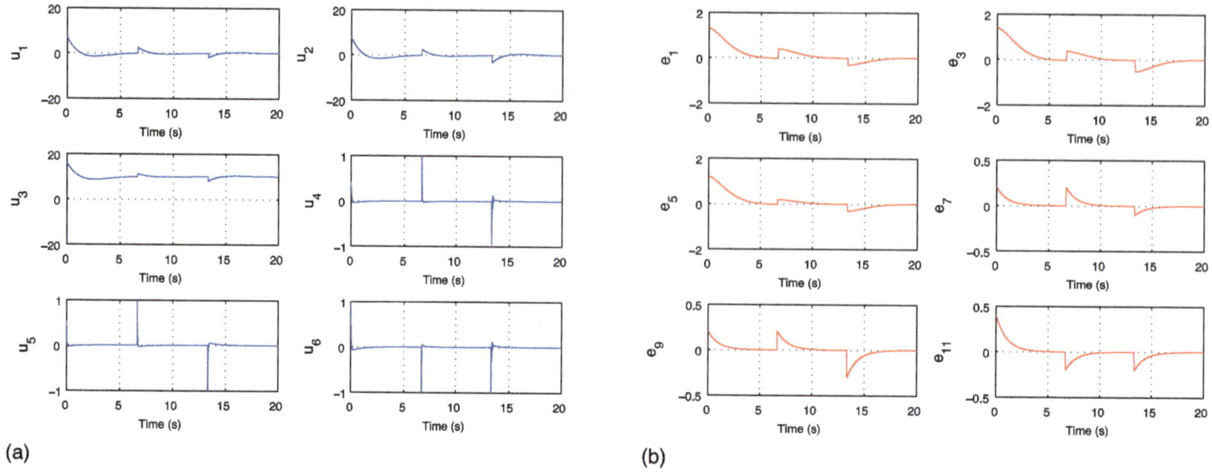

Figure 1.6 Tracking of setpoint 2 by the 6-DOF Stewart parallel robotic platform with the use of nonlinear optimal control: (a) variations of the control inputs u_1 to u_6 (blue lines); (b) variation of the tracking error variables e_i, $i = 1,3,5,7,9,11$ associated with the state variables x_i, $i = 1,3,5,7,9,11$

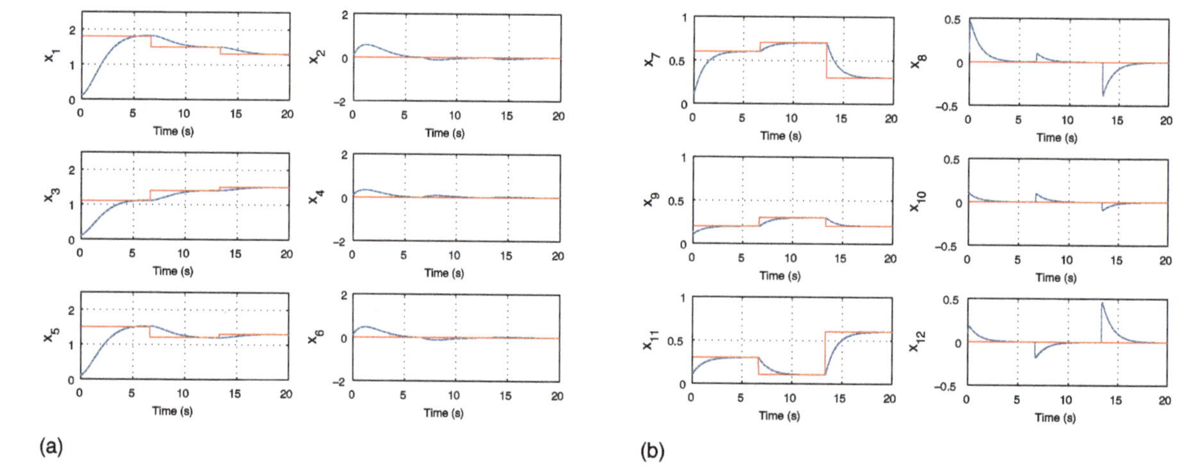

Figure 1.7 Tracking of setpoint 3 by the 6-DOF Stewart parallel robotic platform with the use of nonlinear optimal control: (a) convergence of state variables x_1 to x_6 (blue lines) to the associated setpoints (red lines) and estimated values provided by Kalman filtering; (b) convergence of state variables x_7 to x_{12} (blue lines) to the associated setpoints (red lines) and estimated values provided by Kalman filtering

The nonlinear optimal control method is "energy optimal" because it achieves all control objectives under minimum dispersion of energy. Moreover, it is noteworthy that the nonlinear optimal control method is applied directly on the initial nonlinear state-space model of the parallel robot without the need for changes of state variables, and forward or inverse transformations.

A comparison against other nonlinear control methods one could have considered for the 6-DOF parallel Stewart robotic platform shows that the proposed nonlinear optimal control scheme exhibits specific advantages. Unlike global linearization control approaches (such as Lie algebra-based control), the nonlinear optimal control method

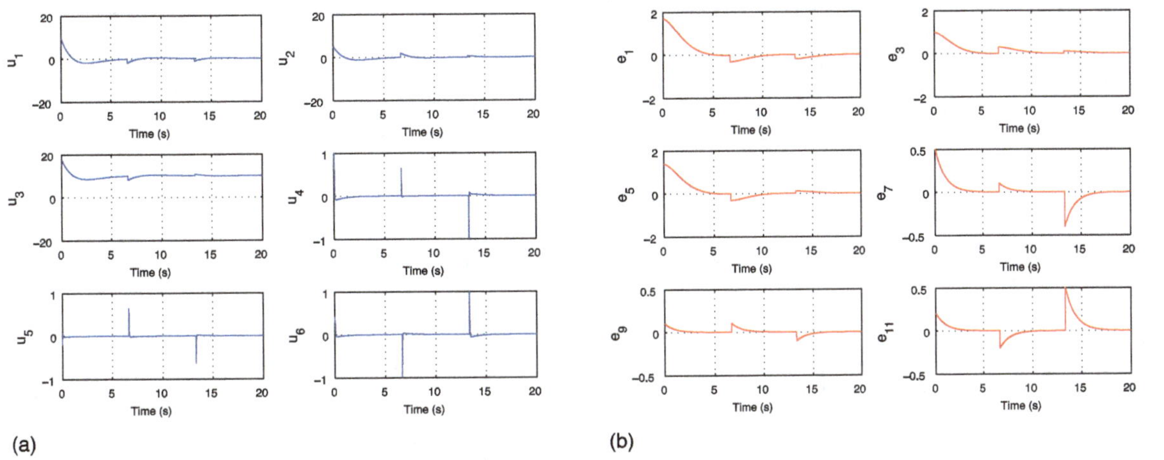

Figure 1.8 Tracking of setpoint 3 by the 6-DOF Stewart parallel robotic platform with the use of nonlinear optimal control: (a) variations of the control inputs u_1 to u_6 (blue lines); (b) variation of the tracking error variables e_i, $i = 1,3,5,7,9,11$ associated with the state variables x_i, $i = 1,3,5,7,9,11$

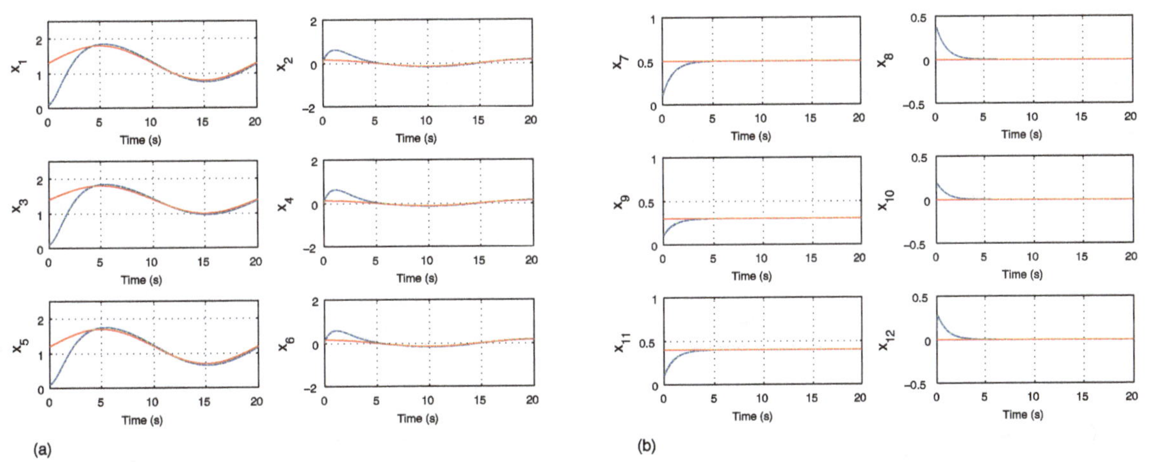

Figure 1.9 Tracking of setpoint 4 by the 6-DOF Stewart parallel robotic platform with the use of nonlinear optimal control: (a) convergence of state variables x_1 to x_6 (blue lines) to the associated setpoints (red lines) and estimated values provided by Kalman filtering; (b) convergence of state variables x_7 to x_{12} (blue lines) to the associated setpoints (red lines) and estimated values provided by Kalman filtering

does not need any changes of state variables or any transformation of the state-space model of the controlled system. Unlike NMPC, the new nonlinear optimal control method is of proven global stability, does not use any empirical parametrization and the convergence of its iterative search for the optimum does not depend on initial conditions. Unlike sliding-mode control and backstepping control, the use of the new nonlinear optimal control method does not require the controlled system to be in a specific state-space form (such as the input–output linearized form, the canonical form, or the strict-feedback form). For instance, unless the controlled system is in the input–output linearized form the selection of sliding surfaces which is needed in sliding-mode control becomes an intuitive procedure. Furthermore, if the controlled system is not found in the strict-feedback (backstepping integral) form the use of backstepping control is

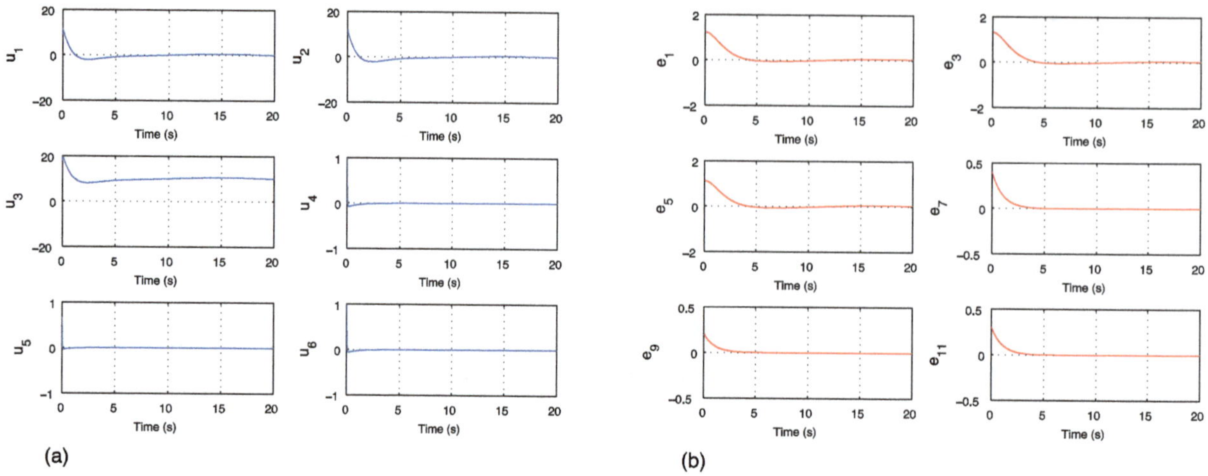

Figure 1.10 Tracking of setpoint 4 by the 6-DOF Stewart parallel robotic platform with the use of nonlinear optimal control: (a) variations of the control inputs u_1 to u_6 (blue lines); (b) variation of the tracking error variables e_i, $i = 1,3,5,7,9,11$ associated with the state variables x_i, $i = 1,3,5,7,9,11$

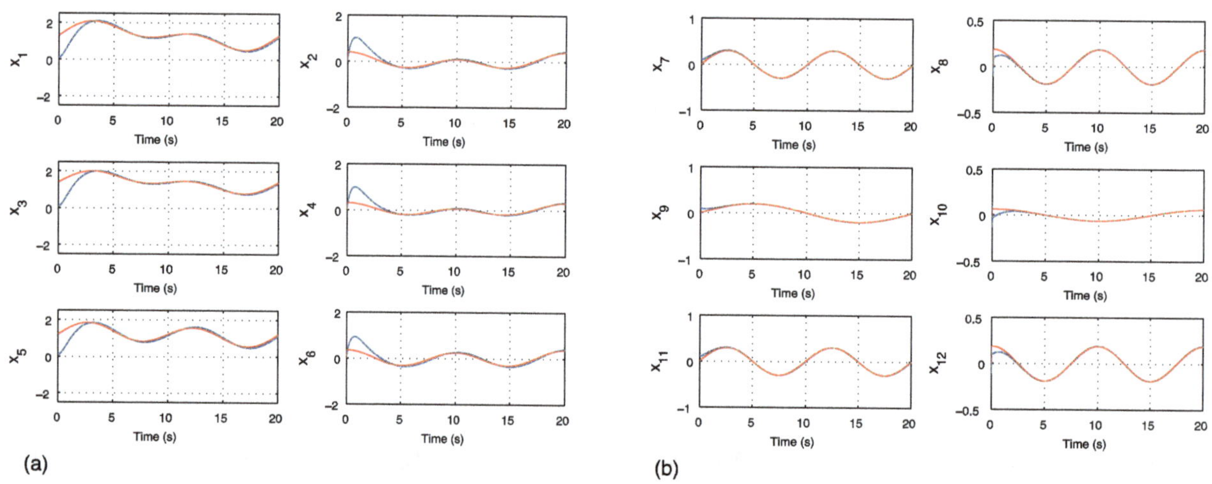

Figure 1.11 Tracking of setpoint 5 by the 6-DOF Stewart parallel robotic platform with the use of nonlinear optimal control: (a) convergence of state variables x_1 to x_6 (blue lines) to the associated setpoints (red lines) and estimated values provided by Kalman filtering; (b) convergence of state variables x_7 to x_{12} (blue lines) to the associated setpoints (red lines) and estimated values provided by Kalman filtering

not feasible. Unlike proportional derivative integral control, the selection of controller parameters in the new nonlinear optimal control method is not an empirical process while there is proof of the global stability of the associated control loop, as well as reliable functioning of the controlled system in changes of operating points. Finally, unlike local multi-model feedback control systems, the new nonlinear optimal control method uses one single linearization point and does need the solution of multiple Riccati equations or LMIs, thus keeping also moderate the computational effort for its implementation.

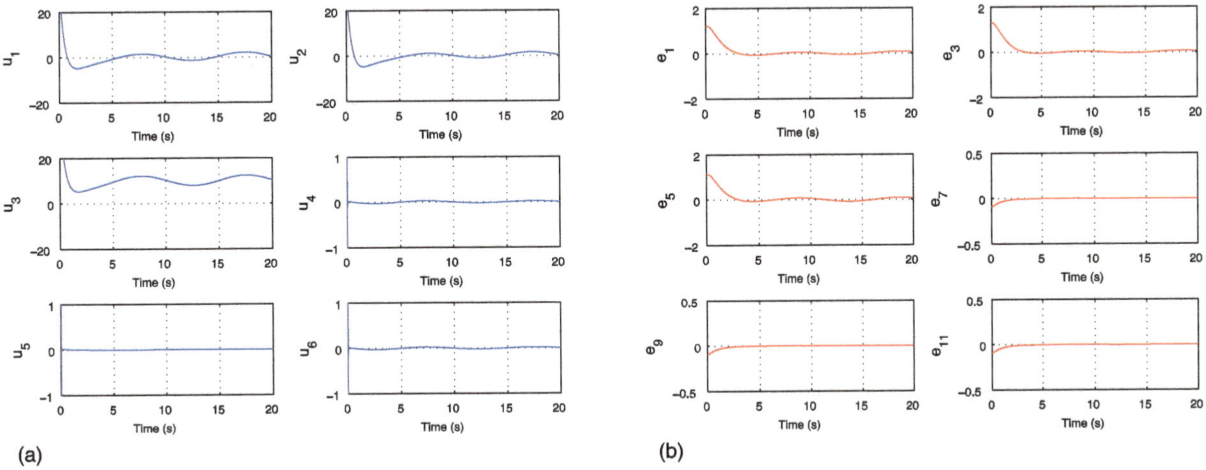

Figure 1.12 *Tracking of setpoint 5 by the 6-DOF Stewart parallel robotic platform with the use of nonlinear optimal control: (a) variations of the control inputs u_1 to u_6 (blue lines); (b) variation of the tracking error variables e_i, $i = 1,3,5,7,9,11$ associated with the state variables x_i, $i = 1,3,5,7,9,11$*

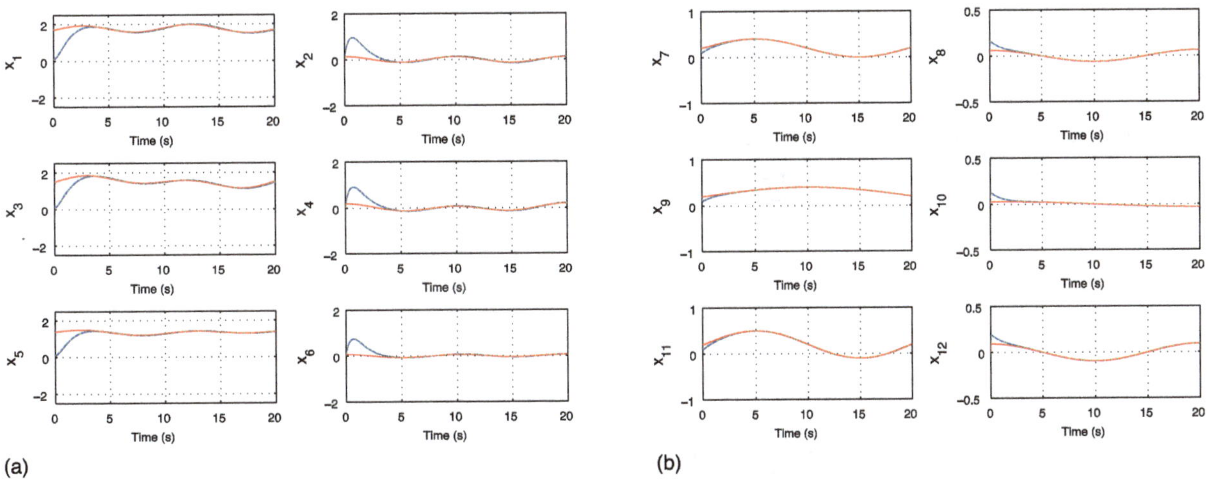

Figure 1.13 *Tracking of setpoint 6 by the 6-DOF Stewart parallel robotic platform with the use of nonlinear optimal control: (a) convergence of state variables x_1 to x_6 (blue lines) to the associated setpoints (red lines) and estimated values provided by Kalman filtering; (b) convergence of state variables x_7 to x_{12} (blue lines) to the associated setpoints (red lines) and estimated values provided by Kalman filtering*

1.2 Nonlinear optimal control for a 4-DOF SCARA robotic manipulator

1.2.1 Outline

Selective compliance articulated robot arm (SCARA) robots are widely used in industrial tasks as well as in the teaching of robotics and in the related testing of new robot control algorithms [721,900,210,1073,338]. The rapid development of the Computer, Communication and Consumer Electronics Industry (3C industry) has led also to the

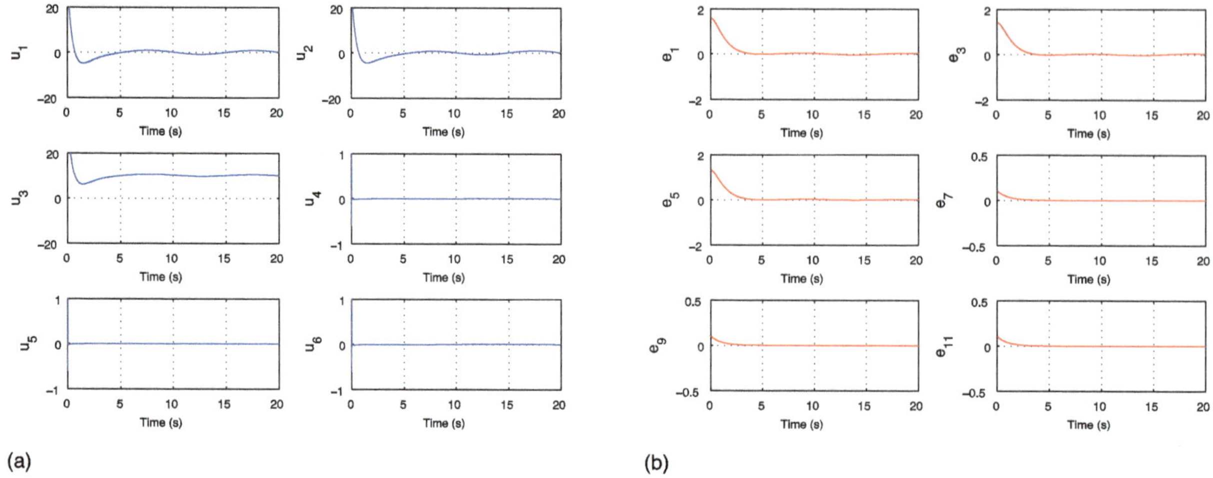

Figure 1.14 Tracking of setpoint 6 by the 6-DOF Stewart parallel robotic platform with the use of nonlinear optimal control: (a) variations of the control inputs u_1 to u_6 (blue lines); (b) variation of the tracking error variables e_i, $i = 1,3,5,7,9,11$ associated with the state variables x_i, $i = 1,3,5,7,9,11$

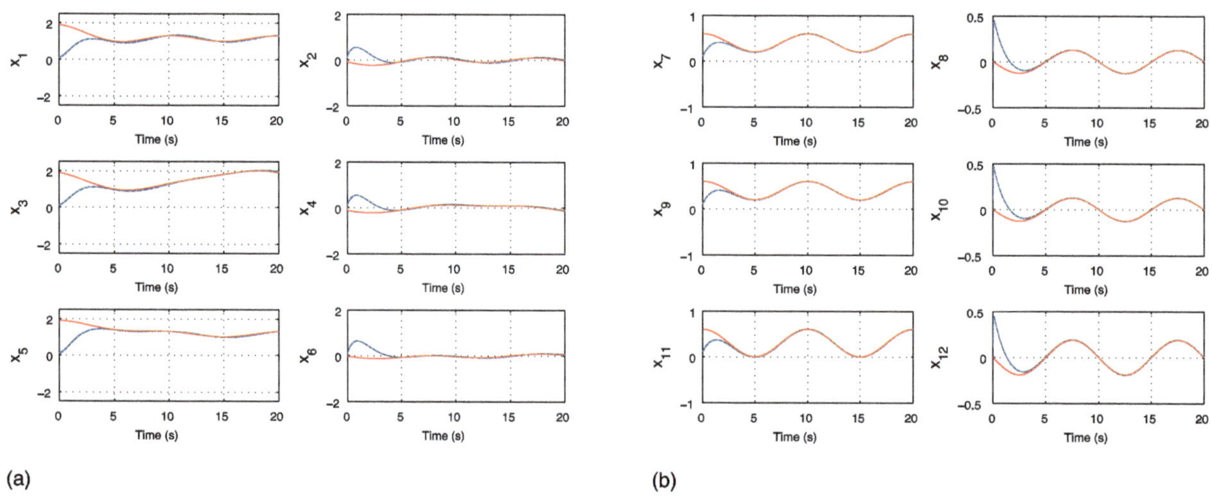

Figure 1.15 Tracking of setpoint 7 by the 6-DOF Stewart parallel robotic platform with the use of nonlinear optimal control: (a) convergence of state variables x_1 to x_6 (blue lines) to the associated setpoints (red lines) and estimated values provided by Kalman filtering; (b) convergence of state variables x_7 to x_{12} (blue lines) to the associated setpoints (red lines) and estimated values provided by Kalman filtering

spread of the use of SCARA robots [583,190,240,651,456]. SCARA robots exhibit agility in assembly tasks for the 3C industry and particularly in the fabrication of electronic devices, as well as in welding, handling of objects, and pick and place tasks with high speed, short time-cycle, accurate path following, and in general much flexible operation [889,887,888,157,77]. Of course, to achieve the precise execution of such tasks, SCARA robots have to be

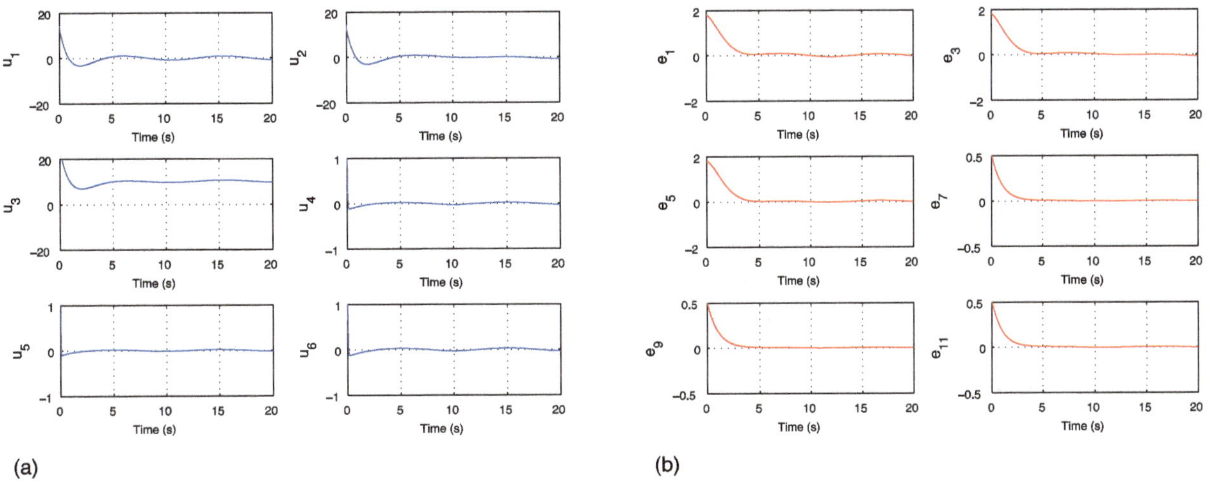

Figure 1.16 Tracking of setpoint 7 by the 6-DOF Stewart parallel robotic platform with the use of nonlinear optimal control: (a) variations of the control inputs u_1 to u_6 (blue lines); (b) variation of the tracking error variables e_i, $i = 1,3,5,7,9,11$ associated with the state variables x_i, $i = 1,3,5,7,9,11$

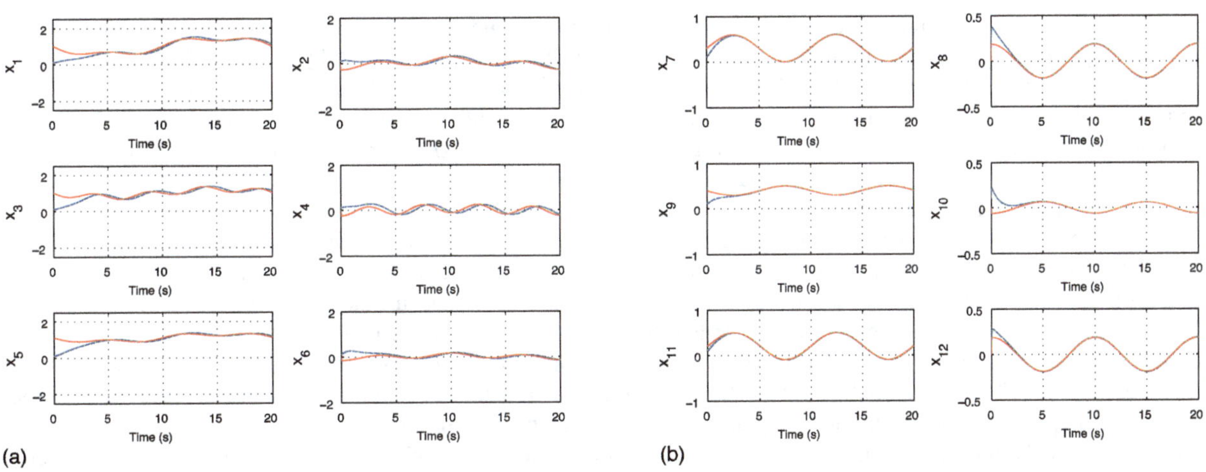

Figure 1.17 Tracking of setpoint 8 by the 6-DOF Stewart parallel robotic platform with the use of nonlinear optimal control: (a) convergence of state variables x_1 to x_6 (blue lines) to the associated setpoints (red lines) and estimated values provided by Kalman filtering; (b) convergence of state variables x_7 to x_{12} (blue lines) to the associated setpoints (red lines) and estimated values provided by Kalman filtering

equipped with computationally powerful microprocessors and have to be also supplied with elaborated nonlinear control algorithms [851,899,200,168,71].

SCARA robots are high-performance robotic manipulators with relatively simple structure. With three revolute joints (named as shoulder, elbow, and wrist, respectively) a SCARA robot can move its end-effector horizontally, while with a prismatic joint, it can move the end-effector vertically [124,801,585,30,385,813]. The configuration of the SCARA robot (Figure 1.19) is outlined as follows: First, it comprises a revolute joint about the vertical axis. This

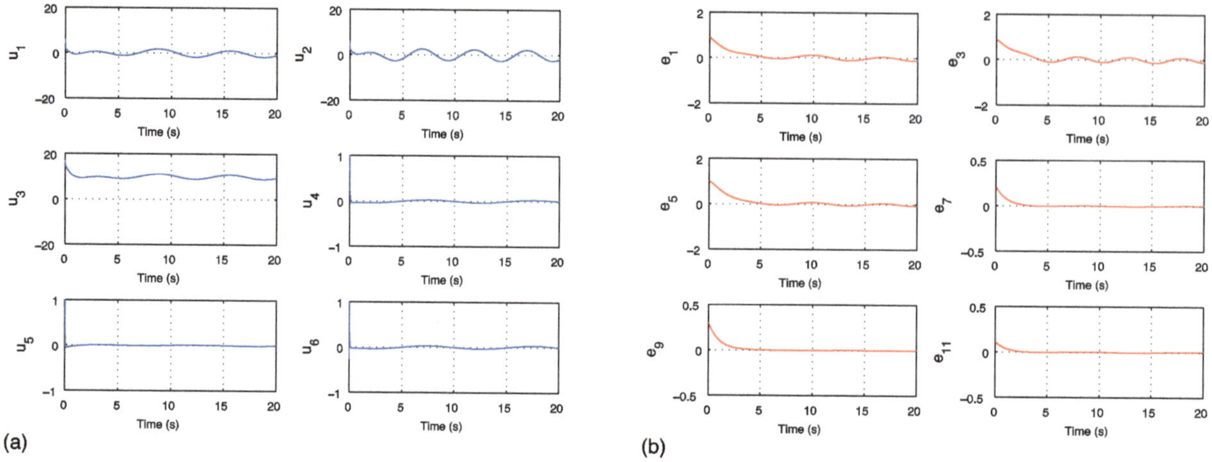

Figure 1.18 Tracking of setpoint 8 by the 6-DOF Stewart parallel robotic platform with the use of nonlinear optimal control: (a) variations of the control inputs u_1 to u_6 (blue lines); (b) variation of the tracking error variables e_i, $i = 1,3,5,7,9,11$ associated with the state variables x_i, $i = 1,3,5,7,9,11$

joint swings a rigid arm and at the end of this arm there is a second revolute joint which swings the second arm again about the vertical axis. The first two revolute joints enable to move horizontally a load picked by the robot's end-effector. A prismatic joint (tool) is mounted at the end of the second arm. This can move straight up and down. Finally, at the end of the tool, there is a third revolute joint which allows for the precise positioning and orientation of the load. Indicative recent results showing the application of SCARA robots in industrial tasks can be found in [776,1009,376,890,340]. Potential application areas for this section's developments, always related to industrial robotics, are described in [934,805,1095,443].

In the present section, a nonlinear optimal control method is proposed for the nonlinear model of a 4-DOF SCARA robot [719,725]. The dynamic model of the 4-DOF SCARA robot undergoes first approximate linearization around a temporary operating point which is updated at each sampling instance. This operating point is defined by the present value of the robot's state vector and by the last sampled value of the control inputs vector. The linearization process relies on first-order Taylor series expansion and on the computation of the associated Jacobian matrices [705,60,706]. The modeling error which is due to the truncation of higher-order terms in the Taylor series expansion, is proven to be small and is asymptotically compensated by the robustness of the control algorithm. For the approximately linearized state-space description of the system, a stabilizing H-infinity feedback controller is defined.

The proposed H-infinity controller achieves the solution of the optimal control problem for the SCARA robot under model uncertainty and external perturbations. Actually, it represents a min–max differential game which takes place between (i) the control inputs of the system that try to minimize a cost function comprising a quadratic term of the state vector's tracking error (ii) the model uncertainty and exogenous perturbation terms which try to maximize this cost function. To compute the stabilizing feedback gains of this controller an algebraic Riccati equation has to be also solved at each time-step of the control method [721,711]. The global stability properties of the control scheme are proven through Lyapunov analysis. First, it is proven that the control loop satisfies the H-infinity tracking performance criterion [721,876]. Next, it is proven that under moderate conditions, global asymptotic stability of the control loop is ensured. To implement state estimation-based control without need to measure the entire state vector of the system the H-infinity Kalman filter is used as a robust state estimator [721]. The nonlinear optimal control method retains the advantages of linear optimal control, that is fast and accurate tracking of reference setpoints under moderate variations of the control inputs.

1.2.2 Dynamic model of the 4-DOF SCARA robotic manipulator

1.2.2.1 State-space model of the SCARA robot

The diagram of the 4-DOF SCARA robot is shown in Figure 1.19. The associated state-space model of the robot's dynamics takes the form [721,900]

$$M(\theta)\ddot{\theta}+\tilde{C}(\theta,\dot{\theta})\dot{\theta}+G(\theta)+\tilde{d}=\tau(t) \tag{1.104}$$

where $\theta=[\theta_1,\theta_2,\theta_3,\theta_4]^T$ is the robot's vector of state variables, θ_i, $i=1,2,4$ are the joints' turn angle, $\dot{\theta}_i$, $i=1,2,4$ are the joints' angular speeds, θ_3 is the position of the prismatic joint, $\dot{\theta}_3$ is the velocity of the prismatic joint, $\tilde{d}$ is the disturbances' vector, $M(\theta)$ is the inertia matrix, $\tilde{C}(\theta,\dot{\theta})$ is the Coriolis and centrifugal forces matrix, and $G(\theta)$ is the gravitational forces vector. These parameters of the robotic model are defined as follows [900]: ,

$$M(\theta)=\begin{pmatrix} p_1+p_2cos(\theta_2) & p_3+0.5p_2cos(\theta_2) & 0 & -p_5 \\ p_2+0.5p_2\cos(\theta_2) & p_2 & 0 & -p_5 \\ 0 & 0 & p_4 & 0 \\ -p_5 & -p_5 0 & p_5 & \end{pmatrix} \tag{1.105}$$

$$\tilde{C}(\theta,\dot{\theta})=\begin{pmatrix} -p_2cos(\theta_1)\dot{\theta}_2 & -0.5p_2sin(\theta_2)\dot{\theta}_2 & 0 & 0 \\ 0.5p_2sin(\theta_2)\dot{\theta}_1 & 0 & 0 & 0 \\ 0 & 0 & 0 & 0 \\ 0 & 0 & 0 & 0 \end{pmatrix} \quad G(\theta)=\begin{pmatrix} 0 \\ 0 \\ p_4\bar{g} \\ 0 \end{pmatrix} \quad \tilde{d}=\begin{pmatrix} b_1\dot{\theta}_1 \\ b_2\dot{\theta}_2 \\ b_3\dot{\theta}_3 \\ b_4\dot{\theta}_4 \end{pmatrix} \tag{1.106}$$

In the previous model τ_i are the control inputs, I_i are the moments of inertia around the centroid (the center of rotation of each link is at distance κ from the end of the link), m_i is the mass of the i-th link, d_i is the center of mass of the i-th link, l_i is the length of the i-th link, and θ_i is the angle (position) i-th joint, while $\bar{g}$ is the acceleration of gravity. It holds that

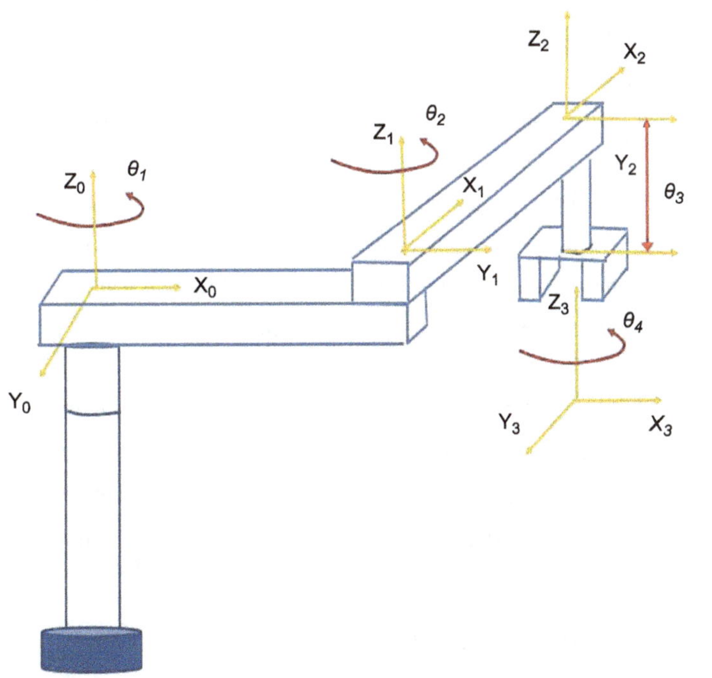

Figure 1.19 The 4-DOF SCARA robotic manipulator and the reference frames (coordinate systems) associated with the robot's links and joints

$I_i = m_i\kappa_i^2$, $p_1 = \sum_{i=1}^{4} I_i + m_1 d_1^2 + m_2(d_1^2 + l_1^2) + (m_3 + m_4)(l_1^2 + l_2^2)$, $p_2 = 2[l_1 d_2 m_2 + l_1 l_2(m_3 + m_4)]$, $p_3 = m_3 + m_4$, $p_4 = m_3 + m_4$ and $p_5 = I_4$.

About the elements of the inertia matrix, one has: $m_{11} = p_1 + p_2 cos(\theta_2)$, $m_{12} = p_3 + 0.5p_2 cos(\theta_2)$, $m_{13} = 0$, $m_{14} = -p_5$, $m_{21} = p_2 + 0.5p_2\cos(\theta_2)$, $m_{22} = p_2$, $m_{23} = 0$, $m_{34} = -p_5$, $m_{31} = 0$, $m_{32} = 0$, $m_{33} = p_4$, $m_{34} = 0$, $m_{41} = -p_5$, $m_{42} = -p_5$, $m_{43} = 0$, $m_{44} = p_5$.

About the elements of the Coriolis matrix, one has: $c_{11} = -p_2 cos(\theta_1)\dot{\theta}_2$, $c_{12} = -0.5p_2 sin(\theta_2)\dot{\theta}_2$, $c_{13} = 0$ $c_{14} = 0$, $c_{21} = 0.5p_2 sin(\theta_2)\dot{\theta}_1$, $c_{22} = 0$, $c_{23} = 0$, $c_{24} = 0$, $c_{31} = 0$, $c_{32} = 0$, $c_{33} = 0$, $c_{34} = 0$, $c_{41} = 0$, $c_{42} = 0$, $c_{43} = 0$, $c_{44} = 0$.

About the elements of the Gravitational forces vector, one has: $g_1 = 0$, $g_2 = 0$, $g_3 = p_4\bar{g}$, and $g_4 = 0$.

About the elements of the disturbances (friction) vector, one has: $d_1 = b_1\dot{\theta}_1$, $d_2 = b_2\dot{\theta}_2$, $d_3 = b_3\dot{\theta}_3$, and $d_4 = b_4\dot{\theta}_4$.

Next, the inverse of the inertia matrix M is defined as

$$N^{-1} = \frac{1}{detM}\begin{pmatrix} M_{11} & -M_{21} & M_{31} & -M_{41} \\ -M_{12} & M_{22} & -M_{32} & M_{42} \\ M_{13} & -M_{23} & M_{33} & -M_{43} \\ -M_{14} & M_{24} & -M_{34} & M_{44} \end{pmatrix} \tag{1.107}$$

where the above-noted subdeterminants M_{ij} $i = 1, \cdots, 4$ and $j = 1, \cdots, 4$ are defined as

$$\begin{aligned}
M_{11} &= m_{22}(m_{33}m_{44} - m_{43}m_{34}) - m_{23}(m_{32}m_{44} - m_{43}m_{34} + m_{24})(m_{32}m_{43} - m_{42}m_{33}) \\
M_{12} &= m_{21}(m_{33}m_{44} - m_{43}m_{34}) - m_{23}(m_{31}m_{44} - m_{41}m_{34} + m_{24})(m_{31}m_{43} - m_{41}m_{33}) \\
M_{13} &= m_{21}(m_{32}m_{44} - m_{42}m_{34}) - m_{22}(m_{31}m_{44} - m_{41}m_{34} + m_{24})(m_{31}m_{43} - m_{41}m_{33}) \\
M_{14} &= m_{21}(m_{32}m_{43} - m_{42}m_{33}) - m_{22}(m_{31}m_{43} - m_{41}m_{33} + m_{23})(m_{31}m_{42} - m_{41}m_{32})
\end{aligned}$$

$$\begin{aligned}
M_{21} &= m_{12}(m_{33}m_{44} - m_{43}m_{34}) - m_{13}(m_{32}m_{44} - m_{42}m_{34} + m_{14})(m_{32}m_{43} - m_{42}m_{33}) \\
M_{22} &= m_{11}(m_{23}m_{24} - m_{43}m_{44}) - m_{13}(m_{31}m_{44} - m_{41}m_{14} + m_{14})(m_{31}m_{43} - m_{41}m_{23}) \\
M_{23} &= m_{11}(m_{32}m_{44} - m_{42}m_{44}) - m_{12}(m_{31}m_{44} - m_{41}m_{34} + m_{14})(m_{31}m_{42} - m_{41}m_{32}) \\
M_{24} &= m_{11}(m_{32}m_{43} - m_{42}m_{33}) - m_{12}(m_{31}m_{43} - m_{41}m_{33} + m_{13})(m_{31}m_{42} - m_{41}m_{32})
\end{aligned}$$

$$\begin{aligned}
M_{31} &= m_{12}(m_{23}m_{44} - m_{43}m_{24}) - m_{13}(m_{22}m_{44} - m_{42}m_{24} + m_{14})(m_{22}m_{43} - m_{42}m_{23}) \\
M_{32} &= m_{11}(m_{23}m_{44} - m_{42}m_{24}) - m_{13}(m_{12}m_{44} - m_{41}m_{24} + m_{14})(m_{12}m_{43} - m_{41}m_{23}) \\
M_{33} &= m_{11}(m_{22}m_{44} - m_{42}m_{24}) - m_{12}(m_{21}m_{44} - m_{41}m_{24} + m_{14})(m_{41}m_{22} - m_{21}m_{42}) \\
M_{34} &= m_{11}(m_{22}m_{43} - m_{42}m_{23}) - m_{12}(m_{21}m_{43} - m_{41}m_{23} + m_{13})(m_{21}m_{42} - m_{41}m_{22})
\end{aligned}$$

$$\begin{aligned}
M_{41} &= m_{12}(m_{23}m_{34} - m_{33}m_{24}) - m_{13}(m_{22}m_{34} - m_{32}m_{23} + m_{14})(m_{22}m_{33} - m_{32}m_{23}) \\
M_{42} &= m_{11}(m_{22}m_{43} - m_{33}m_{42}) - m_{31}(m_{12}m_{43} - m_{13}m_{44} + m_{41})(m_{12}m_{33} - m_{31}m_{32}) \\
M_{43} &= m_{11}(m_{22}m_{34} - m_{32}m_{24}) - m_{12}(m_{21}m_{34} - m_{31}m_{24} + m_{14})(m_{21}m_{32} - m_{31}m_{22}) \\
M_{44} &= m_{11}(m_{22}m_{33} - m_{32}m_{23}) - m_{12}(m_{21}m_{33} - m_{31}m_{23} + m_{13})(m_{21}m_{32} - m_{31}m_{22})
\end{aligned}$$

The determinant of matrix M is

$$detM = m_{11}M_{11} - m_{12}M_{12} + m_{13}M_{13} - m_{14}M_{14}$$

For the dynamic model of the SCARA robot that was initially written in the form

$$M(\theta)\ddot{\theta} + \tilde{C}(\theta,\dot{\theta})\dot{\theta} + G(\theta) + \tilde{d} = \tau \tag{1.108}$$

it holds that

$$\tilde{C}(\theta,\dot{\theta})\dot{\theta} = \begin{pmatrix} c_{11} & c_{12} & c_{13} & c_{14} \\ c_{21} & c_{22} & c_{23} & c_{24} \\ c_{31} & c_{32} & c_{33} & c_{34} \\ c_{41} & c_{42} & c_{43} & c_{44} \end{pmatrix} \begin{pmatrix} \dot{\theta}_1 \\ \dot{\theta}_2 \\ \dot{\theta}_3 \\ \dot{\theta}_4 \end{pmatrix} \tag{1.109}$$

or equivalently,

$$C(\theta,\dot{\theta}) = \begin{pmatrix} c_1 \\ c_2 \\ c_3 \\ c_4 \end{pmatrix} = \begin{pmatrix} c_{11}\dot{\theta}_1 & c_{12}\dot{\theta}_2 & c_{13}\dot{\theta}_3 & c_{14}\dot{\theta}_4 \\ c_{21}\dot{\theta}_2 & c_{22}\dot{\theta}_2 & c_{23}\dot{\theta}_3 & c_{24}\dot{\theta}_4 \\ c_{31}\dot{\theta}_3 & c_{32}\dot{\theta}_2 & c_{33}\dot{\theta}_3 & c_{34}\dot{\theta}_4 \\ c_{41}\dot{\theta}_4 & c_{42}\dot{\theta}_2 & c_{43}\dot{\theta}_3 & c_{44}\dot{\theta}_4 \end{pmatrix} \tag{1.110}$$

Consequently, the dynamic model of the robot can be written as

$$\begin{gathered} M(\theta)\ddot{\theta} + C(\theta,\dot{\theta}) + G(\theta) + \tilde{d}(\dot{\theta}) = \tau \Rightarrow \\ \ddot{\theta} + M^{-1}(\theta)C(\theta,\dot{\theta}) + M^{-1}(\theta)G(\theta) + M^{-1}(\theta))\tilde{d}(\dot{\theta}) = M^{-1}(\theta)\tau \Rightarrow \\ \ddot{\theta} = -M^{-1}(\theta)C(\theta,\dot{\theta}) - M^{-1}(\theta)G(\theta) - M^{-1}(\theta))\tilde{d}(\dot{\theta}) + M^{-1}(\theta)\tau \Rightarrow \\ \ddot{\theta} = -M^{-1}(\theta)[C(\theta,\dot{\theta}) + G(\theta) + \tilde{d}(\dot{\theta})] + M^{-1}(\theta)\tau \end{gathered} \tag{1.111}$$

Therefore, the dynamic model of the SCARA robot is written as

$$\begin{aligned} \begin{pmatrix} \ddot{\theta}_1 \\ \ddot{\theta}_2 \\ \ddot{\theta}_3 \\ \ddot{\theta}_4 \end{pmatrix} = -\tfrac{1}{detM} & \begin{pmatrix} M_{11} & -M_{21} & M_{31} & -M_{41} \\ -M_{12} & M_{22} & -M_{32} & M_{42} \\ M_{13} & -M_{23} & M_{33} & -M_{43} \\ -M_{14} & M_{24} & -M_{34} & M_{44} \end{pmatrix} \begin{pmatrix} c_1 + g_1 + d_1 \\ c_2 + g_2 + d_2 \\ c_3 + g_3 + d_3 \\ c_4 + g_4 + d_4 \end{pmatrix} + \\ & + \tfrac{1}{detM} \begin{pmatrix} M_{11} & -M_{21} & M_{31} & -M_{41} \\ -M_{12} & M_{22} & -M_{32} & M_{42} \\ M_{13} & -M_{23} & M_{33} & -M_{43} \\ -M_{14} & M_{24} & -M_{34} & M_{44} \end{pmatrix} \begin{pmatrix} \tau_1 \\ \tau_2 \\ \tau_3 \\ \tau_4 \end{pmatrix} \end{aligned} \tag{1.112}$$

Equivalently, using that the torques vector $\tau=[\tau_1,\tau_2,\tau_3,\tau_4]^T$ is the control inputs vector $u=[u_1,u_2,u_3,u_4]^T$, the dynamic model of the SCARA robot is written as

$$\begin{pmatrix}\ddot{\theta}_1\\ \ddot{\theta}_2\\ \ddot{\theta}_3\\ \ddot{\theta}_4\end{pmatrix}=\begin{pmatrix}\frac{-M_{11}(c_1+g_1+d_1)+M_{21}(c_2+g_2+d_2)-M_{31}(c_3+g_3+d_3)+M_{41}(c_4+g_4+d_4)}{detM}\\ \frac{M_{12}(c_1+g_1+d_1)-M_{22}(c_2+g_2+d_2)+M_{32}(c_3+g_3+d_3)-M_{42}(c_4+g_4+d_4)}{detM}\\ \frac{-M_{13}(c_1+g_1+d_1)+M_{23}(c_2+g_2+d_2)-M_{33}(c_3+g_3+d_3)+M_{43}(c_4+g_4+d_4)}{detM}\\ \frac{M_{14}(c_1+g_1+d_1)-M_{24}(c_2+g_2+d_2)+M_{34}(c_3+g_3+d_3)-M_{44}(c_4+g_4+d_4)}{detM}\end{pmatrix}+$$
$$+\begin{pmatrix}\frac{M_{11}}{detM} & -\frac{M_{21}}{detM} & \frac{M_{31}}{detM} & -\frac{M_{41}}{detM}\\ -\frac{M_{12}}{detM} & \frac{M_{22}}{detM} & -\frac{M_{32}}{detM} & \frac{M_{42}}{detM}\\ \frac{M_{13}}{detM} & -\frac{M_{23}}{detM} & \frac{M_{33}}{detM} & -\frac{M_{34}}{detM}\\ -\frac{M_{14}}{detM} & \frac{M_{24}}{detM} & -\frac{M_{34}}{detM} & \frac{M_{44}}{detM}\end{pmatrix}\begin{pmatrix}u_1\\ u_2\\ u_3\\ u_4\end{pmatrix} \tag{1.113}$$

The state vector of the SCARA robot is

$$\begin{matrix}x=[x_1,x_2,x_3,x_4,x_5,x_6,x_7,x_8]^T\Rightarrow\\ x=[\theta_1,\dot{\theta}_1,\theta_2,\dot{\theta}_2,\theta_3,\dot{\theta}_3,\theta_4,\dot{\theta}_4]^T\end{matrix} \tag{1.114}$$

Moreover, the following functions are defined:

$$\begin{matrix}f_1(x)=x_2 & f_2(x)=\frac{-M_{11}(c_1+g_1+d_1)+M_{21}(c_2+g_2+d_2)-M_{31}(c_3+g_3+d_3)+M_{41}(c_4+g_4+d_4)}{detM}\\ f_3(x)=x_4 & f_4(x)=\frac{M_{12}(c_1+g_1+d_1)-M_{22}(c_2+g_2+d_2)+M_{32}(c_3+g_3+d_3)-M_{42}(c_4+g_4+d_4)}{detM}\\ f_5(x)=x_6 & f_6(x)=\frac{-M_{13}(c_1+g_1+d_1)+M_{23}(c_2+g_2+d_2)-M_{33}(c_3+g_3+d_3)+M_{43}(c_4+g_4+d_4)}{detM}\\ f_7(x)=x_8 & f_8(x)=\frac{M_{14}(c_1+g_1+d_1)-M_{24}(c_2+g_2+d_2)+M_{34}(c_3+g_3+d_3)-M_{44}(c_4+g_4+d_4)}{detM}\end{matrix}$$

$$\begin{matrix}g_{11}(x)=0 & g_{12}(x)=0 & g_{13}=0 & g_{14}=0\\ g_{21}(x)=\frac{M_{11}}{detM} & g_{22}(x)=-\frac{M_{21}}{detM} & g_{23}(x)=\frac{M_{31}}{detM} & g_{24}(x)=-\frac{M_{41}}{detM}\\ g_{31}(x)=0 & g_{32}(x)=0 & g_{33}(x)=0 & g_{34}(x)=0\\ g_{41}(x)=-\frac{M_{12}}{detM} & g_{42}(x)=\frac{M_{22}}{detM} & g_{43}(x)=-\frac{M_{32}}{detM} & g_{44}(x)=\frac{M_{42}}{detM}\\ g_{51}(x)=0 & g_{52}(x)=0 & g_{53}(x)=0 & g_{54}(x)=0\\ g_{61}(x)=\frac{M_{13}}{detM} & g_{62}(x)=-\frac{M_{23}}{detM} & g_{63}(x)=\frac{M_{33}}{detM} & g_{64}(x)=-\frac{M_{43}}{detM}\\ g_{71}(x)=0 & g_{72}(x)=0 & g_{73}(x)=0 & g_{74}(x)=0\\ g_{81}(x)=-\frac{M_{14}}{detM} & g_{82}(x)=\frac{M_{24}}{detM} & g_{83}(x)=-\frac{M_{34}}{detM} & g_{84}(x)=\frac{M_{44}}{detM}\end{matrix}$$

Thus, the state-space model of the 4-DOF SCARA robot is written as

$$\begin{pmatrix}\dot{x}_1\\ \dot{x}_2\\ \dot{x}_3\\ \dot{x}_4\\ \dot{x}_5\\ \dot{x}_6\\ \dot{x}_7\\ \dot{x}_8\end{pmatrix}=\begin{pmatrix}f_1(x)\\ f_2(x)\\ f_3(x)\\ f_4(x)\\ f_5(x)\\ f_6(x)\\ f_7(x)\\ f_8(x)\end{pmatrix}+\begin{pmatrix}g_{11}(x) & g_{12}(x) & g_{13}(x) & g_{14}(x)\\ g_{21}(x) & g_{22}(x) & g_{23}(x) & g_{24}(x)\\ g_{31}(x) & g_{32}(x) & g_{33}(x) & g_{34}(x)\\ g_{41}(x) & g_{42}(x) & g_{43}(x) & g_{44}(x)\\ g_{51}(x) & g_{52}(x) & g_{53}(x) & g_{54}(x)\\ g_{61}(x) & g_{62}(x) & g_{63}(x) & g_{64}(x)\\ g_{71}(x) & g_{72}(x) & g_{73}(x) & g_{74}(x)\\ g_{81}(x) & g_{82}(x) & g_{83}(x) & g_{84}(x)\end{pmatrix}\begin{pmatrix}\tau_1\\ \tau_2\\ \tau_3\\ \tau_4\end{pmatrix} \tag{1.115}$$

or in concise form one has the affine-in-the-input nonlinear state-space model

$$\dot{x}=f(x)+g(x)u \tag{1.116}$$

where $x\in R^{8\times 1}, f(x)\in R^{8\times 1}$, $g(x)\in R^{8\times 4}$, and $u\in R^{4\times 1}$.

1.2.3 Approximate linearization of the dynamic model of the SCARA robot

1.2.3.1 Linearization of the robot's dynamics

The dynamic model of the 4-DOF SCARA robot being initially expressed in the state-space form

$$\dot{x}=f(x)+g(x)u \tag{1.117}$$

undergoes approximate linearization at each sampling instance around the temporary operating point (x^*,u^*), where x^* is the present value of the system's state vector and u^* is the last sampled value of the control inputs vector. The linearization process is based on Taylor series expansion and gives

$$\dot{x}=Ax+Bu+\tilde{d} \tag{1.118}$$

where $\tilde{d}$ is the cumulative disturbances' vector which may be due to truncation of higher-order terms from the Taylor series expansion, (b) exogenous perturbations, (c) sensor measurements noise of any distribution. Matrices A and B are Jacobian matrices of the Taylor series expansion which are defined as

$$\begin{aligned}A&=\nabla_x[f(x)+g(x)u]\mid_{(x^*,u^*)}\Rightarrow\\ A&=\nabla_x f(x)\mid_{(x^*,u^*)}+\nabla_x g_1(x)u\mid_{(x^*,u_1^*)}+\nabla_x g_2(x)u_2\mid_{(x^*,u^*)}\\ &\quad+\nabla_x g_3(x)u_3\mid_{(x^*,u^*)}+\nabla_x g_4(x)u_4\mid_{(x^*,u^*)}\end{aligned} \tag{1.119}$$

$$B=\nabla_u[f(x)+g(x)u]\mid_{(x^*,u^*)}\Rightarrow B=g(x)\mid_{(x^*,u^*)} \tag{1.120}$$

where $g_i(x),\ i=1,\cdots,4$ are the columns of the control inputs gain matrix $g(x)$.

This linearization approach which has been followed for implementing the nonlinear optimal control scheme results into a quite accurate model of the system's dynamics. Consider again the affine-in-the-input state-space model:

$$\begin{aligned}\dot{x}&=f(x)+g(x)u\Rightarrow\\ \dot{x}&=[f(x^*)+\nabla_x f(x)\mid_{x^*}(x-x^*)]+[g(x^*)+\nabla_x g(x)\mid_{x^*}(x-x^*)]u^*+g(x^*)u^*+g(x^*)(u-u^*)+\tilde{d}_1\Rightarrow\\ \dot{x}&=[\nabla_x f(x)\mid_{x^*}+\nabla_x g(x)\mid_{x^*}u^*]x+g(x^*)u-[\nabla_x f(x)\mid_{x^*}+\nabla_x g(x)\mid_{x^*}u^*]x^*+f(x^*)+g(x^*)u^*+\tilde{d}_1\end{aligned} \tag{1.121}$$

where $\tilde{d}_1$ is the modeling error due to truncation of higher-order terms in the Taylor series expansion of $f(x)$ and $g(x)$. Next, by defining $A=[\nabla_x f(x)\mid_{x^*}+\nabla_x g(x)\mid_{x^*}u^*]$, $B=g(x^*)$, one obtains

$$\dot{x}=Ax+Bu-Ax^*+f(x^*)+g(x^*)u^*+\tilde{d}_1 \tag{1.122}$$

Moreover, by denoting $\tilde{d} = -Ax^* + f(x^*) + g(x^*)u^* + \tilde{d}_1$ about the cumulative modeling error term in the Taylor series expansion procedure, one has

$$\dot{x} = Ax + Bu + \tilde{d} \tag{1.123}$$

which is the approximately linearized model of the dynamics of the system of Eq. (1.118). The term $f(x^*) + g(x^*)u^*$ is the derivative of the state vector at (x^*, u^*) which is almost annihilated by $-Ax^*$.

1.2.3.2 Computation of the Jacobian matrices

The computation of the Jacobian matrices A and B proceeds as follows:

Computation of the Jacobian matrix $\nabla_x f(x) \mid_{(x^*,u^*)}$:

First row of the Jacobian matrix $\nabla_x f(x) \mid_{(x^*,u^*)}$: $\frac{\partial f_1}{\partial x_1} = 0$, $\frac{\partial f_1}{\partial x_2} = 1$, $\frac{\partial f_1}{\partial x_3} = 0$, $\frac{\partial f_1}{\partial x_4} = 0$, $\frac{\partial f_1}{\partial x_5} = 0$, $\frac{\partial f_1}{\partial x_6} = 0$, $\frac{\partial f_1}{\partial x_7} = 0$, and $\frac{\partial f_1}{\partial x_8} = 0$.

Second row of the Jacobian matrix $\nabla_x f(x) \mid_{(x^*,u^*)}$: It holds that $f_2(x) = \frac{f_{2,num}}{f_{2,den}}$ with $f_{2,num} = -M_{11}(c_1 + g_1 + d + 1) + M_{21}(c_2 + g_2 + d_2) - M_{31}(c_3 + g_3 + d_3) + M_{41}(c_4 + g_4 + d_4)$ and $f_{2,den} = detM$. Thus, for $i = 1, 2, \cdots, 8$, one has

$$\begin{aligned}\frac{\partial f_{2,num}}{\partial x_1} &= -\frac{\partial M_{11}}{\partial x_i}(c_1 + g_1 + d_1) - M_{11}\left(\frac{\partial c_1}{\partial x_i} + \frac{\partial g_1}{\partial x_i} + \frac{\partial d_1}{\partial x_i}\right) + \\ &+ \frac{\partial M_{21}}{\partial x_i}(c_2 + g_2 + d_2) + M_{21}\left(\frac{\partial c_2}{\partial x_i} + \frac{\partial g_2}{\partial x_i} + \frac{\partial d_2}{\partial x_i}\right) - \\ &- \frac{\partial M_{31}}{\partial x_i}(c_3 + g_3 + d_3) - M_{31}\left(\frac{\partial c_3}{\partial x_i} + \frac{\partial g_3}{\partial x_i} + \frac{\partial d_3}{\partial x_i}\right) + \\ &+ \frac{\partial M_{41}}{\partial x_i}(c_4 + g_4 + d_4) + M_{41}\left(\frac{\partial c_4}{\partial x_i} + \frac{\partial g_4}{\partial x_i} + \frac{\partial d_4}{\partial x_i}\right)\end{aligned} \tag{1.124}$$

and also

$$\frac{\partial f_{2,den}}{\partial x_i} = \frac{\partial detM}{\partial x_i} \tag{1.125}$$

and finally

$$\frac{\partial f_2}{\partial x_i} = \frac{\frac{\partial f_{2,num}}{\partial x_i} f_{2,den} - f_{2,num}\frac{\partial f_{2,den}}{\partial x_i}}{detM^2} \tag{1.126}$$

Third row of the Jacobian matrix $\nabla_x f(x) \mid_{(x^*,u^*)}$: $\frac{\partial f_3}{\partial x_1} = 0$, $\frac{\partial f_3}{\partial x_2} = 0$, $\frac{\partial f_3}{\partial x_3} = 0$, $\frac{\partial f_3}{\partial x_4} = 1$, $\frac{\partial f_3}{\partial x_5} = 0$, $\frac{\partial f_3}{\partial x_6} = 0$, $\frac{\partial f_3}{\partial x_7} = 0$, and $\frac{\partial f_3}{\partial x_8} = 0$.

Fourth row of the Jacobian matrix $\nabla_x f(x) \mid_{(x^*,u^*)}$: It holds that $f_4(x) = \frac{f_{4,num}}{f_{4,den}}$ with $f_{4,num} = M_{12}(c_1 + g_1 + d + 1) - M_{22}(c_2 + g_2 + d_2) + M_{32}(c_3 + g_3 + d_3) - M_{42}(c_4 + g_4 + d_4)$ and $f_{4,den} = detM$. Thus, for $i = 1, 2, \cdots, 8$, one has

$$\begin{aligned}\frac{\partial f_{4,num}}{\partial x_1} &= \frac{\partial M_{12}}{\partial x_i}(c_1 + g_1 + d_1) + M_{12}\left(\frac{\partial c_1}{\partial x_i} + \frac{\partial g_1}{\partial x_i} + \frac{\partial d_1}{\partial x_i}\right) \\ &- \frac{\partial M_{22}}{\partial x_i}(c_2 + g_2 + d_2) - M_{22}\left(\frac{\partial c_2}{\partial x_i} + \frac{\partial g_2}{\partial x_i} + \frac{\partial d_2}{\partial x_i}\right) \\ &\frac{\partial M_{32}}{\partial x_i}(c_3 + g_3 + d_3) + M_{32}\left(\frac{\partial c_3}{\partial x_i} + \frac{\partial g_3}{\partial x_i} + \frac{\partial d_3}{\partial x_i}\right) - \\ &- \frac{\partial M_{42}}{\partial x_i}(c_4 + g_4 + d_4) - M_{42}\left(\frac{\partial c_4}{\partial x_i} + \frac{\partial g_4}{\partial x_i} + \frac{\partial d_4}{\partial x_i}\right)\end{aligned} \tag{1.127}$$

and also

$$\frac{\partial f_{4,den}}{\partial x_i} = \frac{\partial detM}{\partial x_i} \tag{1.128}$$

and finally

$$\frac{\partial f_4}{\partial x_i} = \frac{\frac{\partial f_{4,num}}{\partial x_i} f_{4,den} - f_{4,num}\frac{\partial f_{4,den}}{\partial x_i}}{detM^2} \tag{1.129}$$

Fifth row of the Jacobian matrix $\nabla_x f(x) \mid_{(x^*,u^*)}$: $\frac{\partial f_5}{\partial x_1}=0$, $\frac{\partial f_5}{\partial x_2}=0$, $\frac{\partial f_5}{\partial x_3}=0$, $\frac{\partial f_5}{\partial x_4}=0$, $\frac{\partial f_5}{\partial x_5}=0$, $\frac{\partial f_5}{\partial x_6}=1$, $\frac{\partial f_5}{\partial x_7}=0$, and $\frac{\partial f_5}{\partial x_8}=0$.

Sixth row of the Jacobian matrix $\nabla_x f(x) \mid_{(x^*,u^*)}$: It holds that $f_6(x)=\frac{f_{6,num}}{f_{6,den}}$ with $f_{6,num}=-M_{13}(c_1+g_1+d+1)+M_{23}(c_2+g_2+d_2)-M_{33}(c_3+g_3+d_3)+M_{43}(c_4+g_4+d_4)$ and $f_{6,den}=detM$. Thus, for $i=1,2,\cdots,8$, one has

$$\begin{aligned}\frac{\partial f_{6,num}}{\partial x_1} = & -\frac{\partial M_{13}}{\partial x_i}(c_1+g_1+d_1)-M_{13}\left(\frac{\partial c_1}{\partial x_i}+\frac{\partial g_1}{\partial x_i}+\frac{\partial d_1}{\partial x_i}\right)+ \\ & +\frac{\partial M_{23}}{\partial x_i}(c_2+g_2+d_2)+M_{23}\left(\frac{\partial c_2}{\partial x_i}+\frac{\partial g_2}{\partial x_i}+\frac{\partial d_2}{\partial x_i}\right)- \\ & -\frac{\partial M_{33}}{\partial x_i}(c_3+g_3+d_3)-M_{33}\left(\frac{\partial c_3}{\partial x_i}+\frac{\partial g_3}{\partial x_i}+\frac{\partial d_3}{\partial x_i}\right)+ \\ & +\frac{\partial M_{43}}{\partial x_i}(c_4+g_4+d_4)+M_{43}\left(\frac{\partial c_4}{\partial x_i}+\frac{\partial g_4}{\partial x_i}+\frac{\partial d_4}{\partial x_i}\right)\end{aligned} \tag{1.130}$$

and also

$$\frac{\partial f_{6,den}}{\partial x_i}=\frac{\partial detM}{\partial x_i} \tag{1.131}$$

and finally

$$\frac{\partial f_6}{\partial x_i}=\frac{\frac{\partial f_{6,num}}{\partial x_i}f_{6,den}-f_{6,num}\frac{\partial f_{6,den}}{\partial x_i}}{detM^2} \tag{1.132}$$

Seventh row of the Jacobian matrix $\nabla_x f(x) \mid_{(x^*,u^*)}$: $\frac{\partial f_7}{\partial x_1}=0$, $\frac{\partial f_7}{\partial x_2}=0$, $\frac{\partial f_7}{\partial x_3}=0$, $\frac{\partial f_7}{\partial x_4}=0$, $\frac{\partial f_7}{\partial x_5}=0$, $\frac{\partial f_7}{\partial x_6}=0$, $\frac{\partial f_7}{\partial x_7}=0$, and $\frac{\partial f_7}{\partial x_8}=1$.

Eighth row of the Jacobian matrix $\nabla_x f(x) \mid_{(x^*,u^*)}$: It holds that $f_4(x)=\frac{f_{8,num}}{f_{8,den}}$ with $f_{8,num}=M_{14}(c_1+g_1+d+1)-M_{24}(c_2+g_2+d_2)+M_{34}(c_3+g_3+d_3)-M_{44}(c_4+g_4+d_4)$ and $f_{8,den}=detM$. Thus, for $i=1,2,\cdots,8$, one has

$$\begin{aligned}\frac{\partial f_{8,num}}{\partial x_1} = & \frac{\partial M_{14}}{\partial x_i}(c_1+g_1+d_1)+M_{14}\left(\frac{\partial c_1}{\partial x_i}+\frac{\partial g_1}{\partial x_i}+\frac{\partial d_1}{\partial x_i}\right)+ \\ & -\frac{\partial M_{24}}{\partial x_i}(c_2+g_2+d_2)-M_{24}\left(\frac{\partial c_2}{\partial x_i}+\frac{\partial g_2}{\partial x_i}+\frac{\partial d_2}{\partial x_i}\right)- \\ & \frac{\partial M_{34}}{\partial x_i}(c_3+g_3+d_3)+M_{34}\left(\frac{\partial c_3}{\partial x_i}+\frac{\partial g_3}{\partial x_i}+\frac{\partial d_3}{\partial x_i}\right)- \\ & -\frac{\partial M_{44}}{\partial x_i}(c_4+g_4+d_4)-M_{44}\left(\frac{\partial c_4}{\partial x_i}+\frac{\partial g_4}{\partial x_i}+\frac{\partial d_4}{\partial x_i}\right)\end{aligned} \tag{1.133}$$

and also

$$\frac{\partial f_{8,den}}{\partial x_i}=\frac{\partial detM}{\partial x_i} \tag{1.134}$$

and finally

$$\frac{\partial f_8}{\partial x_i}=\frac{\frac{\partial f_{8,num}}{\partial x_i}f_{8,den}-f_{8,num}\frac{\partial f_{8,den}}{\partial x_i}}{detM^2} \tag{1.135}$$

Computation of the Jacobian matrix $\nabla_x g_1(x) \mid_{(x^*,u^*)}$.

First row of the Jacobian matrix $\nabla_x g_1(x) \mid_{(x^*,u^*)}$: $\frac{\partial g_{11}(x)}{\partial x_1}=0$ for $i=1,2,\cdots,8$.

Second row of the Jacobian matrix $\nabla_x g_1(x) \mid_{(x^*,u^*)}$: $\frac{\partial g_{21}(x)}{\partial x_i}=\frac{\frac{\partial M_{11}}{\partial x_i}detM-M_{11}\frac{\partial detM}{\partial x_i}}{detM^2}$, for $i=1,2,\cdots,8$.

Third row of the Jacobian matrix $\nabla_x g_1(x) \mid_{(x^*,u^*)}$: $\frac{\partial g_{31}(x)}{\partial x_i}=0$, for $i=1,2,\cdots,8$.

Fourth row of the Jacobian matrix $\nabla_x g_1(x) \mid_{(x^*,u^*)}$: $\frac{\partial g_{41}(x)}{\partial x_i}=\frac{-\frac{\partial M_{12}}{\partial x_i}detM+M_{12}\frac{\partial detM}{\partial x_i}}{detM^2}$, for $i=1,2,\cdots,8$.

Fifth row of the Jacobian matrix $\nabla_x g_1(x) \mid_{(x^*,u^*)}$: $\frac{\partial g_{51}(x)}{\partial x_i} = 0$, for $i = 1, 2, \cdots, 8$.

Sixth row of the Jacobian matrix $\nabla_x g_1(x) \mid_{(x^*,u^*)}$: $\frac{\partial g_{61}(x)}{\partial x_i} = \frac{\frac{\partial M_{13}}{\partial x_i} detM - M_{13}\frac{\partial detM}{\partial x_i}}{detM^2}$, for $i = 1, 2, \cdots, 8$.

Seventh row of the Jacobian matrix $\nabla_x g_1(x) \mid_{(x^*,u^*)}$: $\frac{\partial g_{71}(x)}{\partial x_i} = 0$, for $i = 1, 2, \cdots, 8$.

Eighth row of the Jacobian matrix $\nabla_x g_1(x) \mid_{(x^*,u^*)}$: $\frac{\partial g_{81}(x)}{\partial x_i} = \frac{-\frac{\partial M_{14}}{\partial x_i} detM + M_{14}\frac{\partial detM}{\partial x_i}}{detM^2}$, for $i = 1, 2, \cdots, 8$.

Computation of the Jacobian matrix $\nabla_x g_2(x) \mid_{(x^*,u^*)}$.

First row of the Jacobian matrix $\nabla_x g_2(x) \mid_{(x^*,u^*)}$: $\frac{\partial g_{12}(x)}{\partial x_1} = 0$ for $i = 1, 2, \cdots, 8$.

Second row of the Jacobian matrix $\nabla_x g_2(x) \mid_{(x^*,u^*)}$: $\frac{\partial g_{22}(x)}{\partial x_i} = \frac{-\frac{\partial M_{21}}{\partial x_i} detM + M_{21}\frac{\partial detM}{\partial x_i}}{detM^2}$, for $i = 1, 2, \cdots, 8$.

Third row of the Jacobian matrix $\nabla_x g_2(x) \mid_{(x^*,u^*)}$: $\frac{\partial g_{32}(x)}{\partial x_i} = 0$, for $i = 1, 2, \cdots, 8$.

Fourth row of the Jacobian matrix $\nabla_x g_2(x) \mid_{(x^*,u^*)}$: $\frac{\partial g_{42}(x)}{\partial x_i} = \frac{\frac{\partial M_{22}}{\partial x_i} detM - M_{22}\frac{\partial detM}{\partial x_i}}{detM^2}$, for $i = 1, 2, \cdots, 8$.

Fifth row of the Jacobian matrix $\nabla_x g_2(x) \mid_{(x^*,u^*)}$: $\frac{\partial g_{52}(x)}{\partial x_i} = 0$, for $i = 1, 2, \cdots, 8$.

Sixth row of the Jacobian matrix $\nabla_x g_2(x) \mid_{(x^*,u^*)}$: $\frac{\partial g_{62}(x)}{\partial x_i} = \frac{-\frac{\partial M_{23}}{\partial x_i} detM + M_{23}\frac{\partial detM}{\partial x_i}}{detM^2}$, for $i = 1, 2, \cdots, 8$.

Seventh row of the Jacobian matrix $\nabla_x g_2(x) \mid_{(x^*,u^*)}$: $\frac{\partial g_{72}(x)}{\partial x_i} = 0$, for $i = 1, 2, \cdots, 8$.

Eighth row of the Jacobian matrix $\nabla_x g_2(x) \mid_{(x^*,u^*)}$: $\frac{\partial g_{82}(x)}{\partial x_i} = \frac{\frac{\partial M_{24}}{\partial x_i} detM - M_{24}\frac{\partial detM}{\partial x_i}}{detM^2}$, for $i = 1, 2, \cdots, 8$.

Computation of the Jacobian matrix $\nabla_x g_3(x) \mid_{(x^*,u^*)}$.

First row of the Jacobian matrix $\nabla_x g_3(x) \mid_{(x^*,u^*)}$: $\frac{\partial g_{13}(x)}{\partial x_1} = 0$ for $i = 1, 2, \cdots, 8$.

Second row of the Jacobian matrix $\nabla_x g_3(x) \mid_{(x^*,u^*)}$: $\frac{\partial g_{23}(x)}{\partial x_i} = \frac{\frac{\partial M_{31}}{\partial x_i} detM - M_{31}\frac{\partial detM}{\partial x_i}}{detM^2}$, for $i = 1, 2, \cdots, 8$.

Third row of the Jacobian matrix $\nabla_x g_3(x) \mid_{(x^*,u^*)}$: $\frac{\partial g_{33}(x)}{\partial x_i} = 0$, for $i = 1, 2, \cdots, 8$.

Fourth row of the Jacobian matrix $\nabla_x g_3(x) \mid_{(x^*,u^*)}$: $\frac{\partial g_{43}(x)}{\partial x_i} = \frac{-\frac{\partial M_{32}}{\partial x_i} detM + M_{32}\frac{\partial detM}{\partial x_i}}{detM^2}$, for $i = 1, 2, \cdots, 8$.

Fifth row of the Jacobian matrix $\nabla_x g_3(x) \mid_{(x^*,u^*)}$: $\frac{\partial g_{53}(x)}{\partial x_i} = 0$, for $i = 1, 2, \cdots, 8$.

Sixth row of the Jacobian matrix $\nabla_x g_3(x) \mid_{(x^*,u^*)}$: $\frac{\partial g_{63}(x)}{\partial x_i} = \frac{\frac{\partial M_{33}}{\partial x_i} detM - M_{33}\frac{\partial detM}{\partial x_i}}{detM^2}$, for $i = 1, 2, \cdots, 8$.

Seventh row of the Jacobian matrix $\nabla_x g_3(x) \mid_{(x^*,u^*)}$: $\frac{\partial g_{73}(x)}{\partial x_i} = 0$, for $i = 1, 2, \cdots, 8$.

Eighth row of the Jacobian matrix $\nabla_x g_3(x) \mid_{(x^*,u^*)}$: $\frac{\partial g_{83}(x)}{\partial x_i} = \frac{-\frac{\partial M_{34}}{\partial x_i} detM + M_{34}\frac{\partial detM}{\partial x_i}}{detM^2}$, for $i = 1, 2, \cdots, 8$.

Computation of the Jacobian matrix $\nabla_x g_4(x) \mid_{(x^*,u^*)}$.

First row of the Jacobian matrix $\nabla_x g_4(x)\mid_{(x^*,u^*)}$: $\frac{\partial g_{14}(x)}{\partial x_1}=0$ for $i=1,2,\cdots,8$.

Second row of the Jacobian matrix $\nabla_x g_4(x)\mid_{(x^*,u^*)}$: $\frac{\partial g_{24}(x)}{\partial x_i}=\frac{-\frac{\partial M_{41}}{\partial x_i}detM+M_{41}\frac{\partial detM}{\partial x_i}}{detM^2}$, for $i=1,2,\cdots,8$.

Third row of the Jacobian matrix $\nabla_x g_4(x)\mid_{(x^*,u^*)}$: $\frac{\partial g_{34}(x)}{\partial x_i}=0$, for $i=1,2,\cdots,8$.

Fourth row of the Jacobian matrix $\nabla_x g_4(x)\mid_{(x^*,u^*)}$: $\frac{\partial g_{44}(x)}{\partial x_i}=\frac{\frac{\partial M_{42}}{\partial x_i}detM-M_{42}\frac{\partial detM}{\partial x_i}}{detM^2}$, for $i=1,2,\cdots,8$.

Fifth row of the Jacobian matrix $\nabla_x g_4(x)\mid_{(x^*,u^*)}$: $\frac{\partial g_{54}(x)}{\partial x_i}=0$, for $i=1,2,\cdots,8$.

Sixth row of the Jacobian matrix $\nabla_x g_4(x)\mid_{(x^*,u^*)}$: $\frac{\partial g_{64}(x)}{\partial x_i}=\frac{-\frac{\partial M_{33}}{\partial x_i}detM+M_{33}\frac{\partial detM}{\partial x_i}}{detM^2}$, for $i=1,2,\cdots,8$.

Seventh row of the Jacobian matrix $\nabla_x g_4(x)\mid_{(x^*,u^*)}$: $\frac{\partial g_{74}(x)}{\partial x_i}=0$, for $i=1,2,\cdots,8$.

Eighth row of the Jacobian matrix $\nabla_x g_4(x)\mid_{(x^*,u^*)}$: $\frac{\partial g_{84}(x)}{\partial x_i}=\frac{\frac{\partial M_{44}}{\partial x_i}detM-M_{44}\frac{\partial detM}{\partial x_i}}{detM^2}$, for $i=1,2,\cdots,8$.

Next, one computes the partial derivatives of the subdeterminants M_{ij} and of the determinant $detM$:

$$\begin{aligned}\frac{\partial M_{11}}{\partial x_i}&=\frac{\partial m_{22}}{\partial x_i}(m_{33}m_{44}-m_{43}m_{34})+\\ +m_{22}&\left(\frac{\partial m_{33}}{\partial x_i}m_{44}+m_{33}\frac{\partial m_{44}}{\partial x_i}-\frac{\partial m_{43}}{\partial x_i}m_{34}-m_{43}\frac{\partial m_{34}}{\partial x_i}\right)-\\ &-\frac{\partial m_{23}}{\partial x_i}(m_{32}m_{44}-m_{42}m_{34}-)\\ -m_{23}&\left(\frac{\partial m_{33}}{\partial x_i}m_{44}+m_{32}\frac{\partial m_{44}}{\partial x_i}-\frac{\partial m_{42}}{\partial x_i}m_{34}-m_{42}\frac{\partial m_{34}}{\partial x_i}\right)+\\ &+\frac{\partial m_{24}}{\partial x_i}(m_{32}m_{43}-m_{42}m_{33})+\\ +m_{24}&\left(\frac{\partial m_{32}}{\partial x_i}m_{43}+m_{32}\frac{\partial m_{43}}{\partial x_i}-\frac{\partial m_{42}}{\partial x_i}m_{33}-m_{42}\frac{\partial m_{33}}{\partial x_i}\right)\end{aligned}\tag{1.136}$$

Equivalently, one has

$$\begin{aligned}\frac{\partial M_{12}}{\partial x_i}&=\frac{\partial m_{21}}{\partial x_i}(m_{33}m_{44}-m_{43}m_{34})+\\ +m_{21}&(\frac{\partial m_{13}}{\partial x_i}m_{44}+m_{13}\frac{\partial m_{44}}{\partial x_i}-\frac{\partial m_{43}}{\partial x_i}m_{34}-m_{43}\frac{\partial m_{34}}{\partial x_i})-\\ &-\frac{\partial m_{23}}{\partial x_i}(m_{31}m_{44}-m_{41}m_{34}-)\\ -m_{23}&(\frac{\partial m_{31}}{\partial x_i}m_{44}+m_{31}\frac{\partial m_{44}}{\partial x_i}-\frac{\partial m_{41}}{\partial x_i}m_{34}-m_{41}\frac{\partial m_{34}}{\partial x_i})+\\ &+\frac{\partial m_{24}}{\partial x_i}(m_{31}m_{43}-m_{41}m_{33})+\\ +m_{24}&\left(\frac{\partial m_{31}}{\partial x_i}m_{43}+m_{31}\frac{\partial m_{43}}{\partial x_i}-\frac{\partial m_{41}}{\partial x_i}m_{33}-m_{41}\frac{\partial m_{33}}{\partial x_i}\right)\end{aligned}\tag{1.137}$$

Moreover, it holds that

$$\begin{aligned}\frac{\partial M_{13}}{\partial x_i}&=\frac{\partial m_{21}}{\partial x_i}(m_{32}m_{44}-m_{42}m_{34})+\\ +m_{21}&\left(\frac{\partial m_{32}}{\partial x_i}m_{44}+m_{32}\frac{\partial m_{44}}{\partial x_i}-\frac{\partial m_{43}}{\partial x_i}m_{34}-m_{43}\frac{\partial m_{34}}{\partial x_i}\right)-\\ &-\frac{\partial m_{23}}{\partial x_i}(m_{31}m_{44}-m_{41}m_{34}-)\\ -m_{23}&\left(\frac{\partial m_{31}}{\partial x_i}m_{44}+m_{31}\frac{\partial m_{44}}{\partial x_i}-\frac{\partial m_{41}}{\partial x_i}m_{34}-m_{41}\frac{\partial m_{34}}{\partial x_i}\right)+\\ &+\frac{\partial m_{24}}{\partial x_i}(m_{31}m_{42}-m_{41}m_{32})+\\ +m_{24}&\left(\frac{\partial m_{31}}{\partial x_i}m_{42}+m_{31}\frac{\partial m_{42}}{\partial x_i}-\frac{\partial m_{41}}{\partial x_i}m_{32}-m_{41}\frac{\partial m_{32}}{\partial x_i}\right)\end{aligned}\tag{1.138}$$

Additionally, it holds that

$$\begin{array}{c}\frac{\partial M_{14}}{\partial x_i}=\frac{\partial m_{21}}{\partial x_i}(m_{32}m_{43}-m_{42}m_{33})+\\ +m_{21}\left(\frac{\partial m_{32}}{\partial x_i}m_{43}+m_{32}\frac{\partial m_{43}}{\partial x_i}-\frac{\partial m_{44}}{\partial x_i}m_{33}-m_{42}\frac{\partial m_{32}}{\partial x_i}\right)-\\ -\frac{\partial m_{22}}{\partial x_i}(m_{31}m_{43}-m_{41}m_{33}-)\\ -m_{22}\left(\frac{\partial m_{31}}{\partial x_i}m_{43}+m_{31}\frac{\partial m_{43}}{\partial x_i}-\frac{\partial m_{33}}{\partial x_i}m_{33}-m_{41}\frac{\partial m_{33}}{\partial x_i}\right)+\\ +\frac{\partial m_{23}}{\partial x_i}(m_{31}m_{42}-m_{41}m_{32})+\\ +m_{23}\left(\frac{\partial m_{31}}{\partial x_i}m_{42}+m_{31}\frac{\partial m_{42}}{\partial x_i}-\frac{\partial m_{41}}{\partial x_i}m_{32}-m_{41}\frac{\partial m_{32}}{\partial x_i}\right)\end{array}\tag{1.139}$$

In a similar manner, one obtains

$$\begin{array}{c}\frac{\partial M_{21}}{\partial x_i}=\frac{\partial m_{12}}{\partial x_i}(m_{33}m_{44}-m_{43}m_{34})+\\ +m_{12}\left(\frac{\partial m_{33}}{\partial x_i}m_{44}+m_{33}\frac{\partial m_{44}}{\partial x_i}-\frac{\partial m_{43}}{\partial x_i}m_{33}-m_{43}\frac{\partial m_{34}}{\partial x_i}\right)-\\ -\frac{\partial m_{13}}{\partial x_i}(m_{32}m_{44}-m_{42}m_{34}-)\\ -m_{13}\left(\frac{\partial m_{32}}{\partial x_i}m_{44}+m_{32}\frac{\partial m_{44}}{\partial x_i}-\frac{\partial m_{42}}{\partial x_i}m_{34}-m_{42}\frac{\partial m_{34}}{\partial x_i}\right)+\\ +\frac{\partial m_{14}}{\partial x_i}(m_{32}m_{43}-m_{42}m_{33})+\\ +m_{14}\left(\frac{\partial m_{32}}{\partial x_i}m_{43}+m_{32}\frac{\partial m_{43}}{\partial x_i}-\frac{\partial m_{42}}{\partial x_i}m_{33}-m_{42}\frac{\partial m_{33}}{\partial x_i}\right)\end{array}\tag{1.140}$$

Equivalently, one has

$$\begin{array}{c}\frac{\partial M_{22}}{\partial x_i}=\frac{\partial m_{11}}{\partial x_i}(m_{23}m_{24}-m_{43}m_{44})+\\ +m_{11}\left(\frac{\partial m_{23}}{\partial x_i}m_{24}+m_{23}\frac{\partial m_{24}}{\partial x_i}-\frac{\partial m_{43}}{\partial x_i}m_{44}-m_{43}\frac{\partial m_{44}}{\partial x_i}\right)-\\ -\frac{\partial m_{13}}{\partial x_i}(m_{31}m_{44}-m_{41}m_{24}-)\\ -m_{13}\left(\frac{\partial m_{31}}{\partial x_i}m_{44}+m_{31}\frac{\partial m_{44}}{\partial x_i}-\frac{\partial m_{41}}{\partial x_i}m_{24}-m_{41}\frac{\partial m_{24}}{\partial x_i}\right)+\\ +\frac{\partial m_{14}}{\partial x_i}(m_{31}m_{43}-m_{41}m_{23})+\\ +m_{14}\left(\frac{\partial m_{31}}{\partial x_i}m_{43}+m_{31}\frac{\partial m_{43}}{\partial x_i}-\frac{\partial m_{41}}{\partial x_i}m_{23}-m_{41}\frac{\partial m_{23}}{\partial x_i}\right)\end{array}\tag{1.141}$$

Following this procedure, one gets

$$\begin{array}{c}\frac{\partial M_{23}}{\partial x_i}=\frac{\partial m_{11}}{\partial x_i}(m_{32}m_{44}-m_{42}m_{34})+\\ +m_{11}\left(\frac{\partial m_{32}}{\partial x_i}m_{44}+m_{32}\frac{\partial m_{44}}{\partial x_i}-\frac{\partial m_{42}}{\partial x_i}m_{34}-m_{42}\frac{\partial m_{34}}{\partial x_i}\right)-\\ -\frac{\partial m_{12}}{\partial x_i}(m_{31}m_{43}-m_{41}m_{33}-)\\ -m_{12}\left(\frac{\partial m_{31}}{\partial x_i}m_{43}+m_{31}\frac{\partial m_{43}}{\partial x_i}-\frac{\partial m_{41}}{\partial x_i}m_{33}-m_{41}\frac{\partial m_{33}}{\partial x_i}\right)+\\ +\frac{\partial m_{14}}{\partial x_i}(m_{31}m_{42}-m_{41}m_{32})+\\ +m_{14}\left(\frac{\partial m_{31}}{\partial x_i}m_{43}+m_{31}\frac{\partial m_{43}}{\partial x_i}-\frac{\partial m_{41}}{\partial x_i}m_{32}-m_{41}\frac{\partial m_{32}}{\partial x_i}\right)\end{array}\tag{1.142}$$

Additionally, it holds that

$$\begin{aligned}\frac{\partial M_{24}}{\partial x_i} &= \frac{\partial m_{11}}{\partial x_i}(m_{32}m_{43} - m_{42}m_{33})+\\ +m_{11}&\left(\frac{\partial m_{32}}{\partial x_i}m_{43} + m_{32}\frac{\partial m_{43}}{\partial x_i} - \frac{\partial m_{42}}{\partial x_i}m_{33} - m_{42}\frac{\partial m_{33}}{\partial x_i}\right) -\\ &-\frac{\partial m_{12}}{\partial x_i}(m_{31}m_{43} - m_{41}m_{33}-)\\ -m_{12}&\left(\frac{\partial m_{31}}{\partial x_i}m_{43} + m_{31}\frac{\partial m_{43}}{\partial x_i} - \frac{\partial m_{41}}{\partial x_i}m_{33} - m_{41}\frac{\partial m_{33}}{\partial x_i}\right) +\\ &+\frac{\partial m_{13}}{\partial x_i}(m_{31}m_{42} - m_{41}m_{22})+\\ +m_{13}&\left(\frac{\partial m_{31}}{\partial x_i}m_{42} + m_{31}\frac{\partial m_{42}}{\partial x_i} - \frac{\partial m_{41}}{\partial x_i}m_{22} - m_{41}\frac{\partial m_{22}}{\partial x_i}\right)\end{aligned} \tag{1.143}$$

In this context, one obtains

$$\begin{aligned}\frac{\partial M_{31}}{\partial x_i} &= \frac{\partial m_{12}}{\partial x_i}(m_{23}m_{44} - m_{43}m_{24})+\\ +m_{12}&\left(\frac{\partial m_{23}}{\partial x_i}m_{44} + m_{23}\frac{\partial m_{44}}{\partial x_i} - \frac{\partial m_{43}}{\partial x_i}m_{24} - m_{43}\frac{\partial m_{24}}{\partial x_i}\right) -\\ &-\frac{\partial m_{13}}{\partial x_i}(m_{22}m_{44} - m_{42}m_{24}-)\\ -m_{13}&\left(\frac{\partial m_{22}}{\partial x_i}m_{44} + m_{22}\frac{\partial m_{44}}{\partial x_i} - \frac{\partial m_{42}}{\partial x_i}m_{24} - m_{42}\frac{\partial m_{24}}{\partial x_i}\right) +\\ &+\frac{\partial m_{14}}{\partial x_i}(m_{22}m_{43} - m_{42}m_{23})+\\ +m_{14}&\left(\frac{\partial m_{32}}{\partial x_i}m_{43} + m_{22}\frac{\partial m_{43}}{\partial x_i} - \frac{\partial m_{42}}{\partial x_i}m_{23} - m_{42}\frac{\partial m_{23}}{\partial x_i}\right)\end{aligned} \tag{1.144}$$

Additionally, one has

$$\begin{aligned}\frac{\partial M_{32}}{\partial x_i} &= \frac{\partial m_{11}}{\partial x_i}(m_{23}m_{44} - m_{42}m_{24})+\\ +m_{11}&\left(\frac{\partial m_{23}}{\partial x_i}m_{44} + m_{23}\frac{\partial m_{44}}{\partial x_i} - \frac{\partial m_{42}}{\partial x_i}m_{24} - m_{42}\frac{\partial m_{24}}{\partial x_i}\right) -\\ &-\frac{\partial m_{13}}{\partial x_i}(m_{12}m_{44} - m_{41}m_{24}-)\\ -m_{13}&\left(\frac{\partial m_{12}}{\partial x_i}m_{44} + m_{12}\frac{\partial m_{44}}{\partial x_i} - \frac{\partial m_{41}}{\partial x_i}m_{24} - m_{41}\frac{\partial m_{24}}{\partial x_i}\right) +\\ &+\frac{\partial m_{14}}{\partial x_i}(m_{12}m_{43} - m_{41}m_{23})+\\ +m_{14}&\left(\frac{\partial m_{12}}{\partial x_i}m_{43} + m_{12}\frac{\partial m_{43}}{\partial x_i} - \frac{\partial m_{41}}{\partial x_i}m_{23} - m_{41}\frac{\partial m_{23}}{\partial x_i}\right)\end{aligned} \tag{1.145}$$

Furthermore, one has

$$\begin{aligned}\frac{\partial M_{33}}{\partial x_i} &= \frac{\partial m_{11}}{\partial x_i}(m_{22}m_{44} - m_{42}m_{24})+\\ +m_{11}&\left(\frac{\partial m_{22}}{\partial x_i}m_{44} + m_{22}\frac{\partial m_{44}}{\partial x_i} - \frac{\partial m_{42}}{\partial x_i}m_{24} - m_{42}\frac{\partial m_{24}}{\partial x_i}\right) -\\ &-\frac{\partial m_{12}}{\partial x_i}(m_{21}m_{44} - m_{41}m_{24}-)\\ -m_{12}&\left(\frac{\partial m_{21}}{\partial x_i}m_{44} + m_{21}\frac{\partial m_{44}}{\partial x_i} - \frac{\partial m_{41}}{\partial x_i}m_{24} - m_{41}\frac{\partial m_{24}}{\partial x_i}\right) +\\ &+\frac{\partial m_{14}}{\partial x_i}(m_{41}m_{22} - m_{21}m_{42})+\\ +m_{14}&\left(\frac{\partial m_{41}}{\partial x_i}m_{22} + m_{41}\frac{\partial m_{22}}{\partial x_i} - \frac{\partial m_{21}}{\partial x_i}m_{42} - m_{21}\frac{\partial m_{42}}{\partial x_i}\right)\end{aligned} \tag{1.146}$$

Continuing in this manner, one gets

$$\begin{aligned}
\frac{\partial M_{34}}{\partial x_i} = & \frac{\partial m_{11}}{\partial x_i}(m_{22}m_{43} - m_{42}m_{23}) + \\
+m_{11}\left(\frac{\partial m_{22}}{\partial x_i}m_{43} + m_{22}\frac{\partial m_{43}}{\partial x_i} - \frac{\partial m_{42}}{\partial x_i}m_{23} - m_{42}\frac{\partial m_{23}}{\partial x_i}\right) - \\
-\frac{\partial m_{12}}{\partial x_i}(m_{21}m_{43} - m_{41}m_{23} -) \\
-m_{12}\left(\frac{\partial m_{21}}{\partial x_i}m_{43} + m_{21}\frac{\partial m_{43}}{\partial x_i} - \frac{\partial m_{41}}{\partial x_i}m_{23} - m_{41}\frac{\partial m_{23}}{\partial x_i}\right) + \\
+\frac{\partial m_{13}}{\partial x_i}(m_{21}m_{42} - m_{41}m_{22}) + \\
+m_{13}\left(\frac{\partial m_{22}}{\partial x_i}m_{42} + m_{22}\frac{\partial m_{42}}{\partial x_i} - \frac{\partial m_{41}}{\partial x_i}m_{22} - m_{41}\frac{\partial m_{22}}{\partial x_i}\right)
\end{aligned} \tag{1.147}$$

Besides, one has

$$\begin{aligned}
\frac{\partial M_{41}}{\partial x_i} = & \frac{\partial m_{12}}{\partial x_i}(m_{22}m_{34} - m_{33}m_{24}) + \\
+m_{12}\left(\frac{\partial m_{22}}{\partial x_i}m_{34} + m_{22}\frac{\partial m_{34}}{\partial x_i} - \frac{\partial m_{33}}{\partial x_i}m_{24} - m_{33}\frac{\partial m_{24}}{\partial x_i}\right) - \\
-\frac{\partial m_{13}}{\partial x_i}(m_{23}m_{34} - m_{32}m_{23} -) \\
-m_{13}\left(\frac{\partial m_{22}}{\partial x_i}m_{34} + m_{22}\frac{\partial m_{34}}{\partial x_i} - \frac{\partial m_{32}}{\partial x_i}m_{23} - m_{32}\frac{\partial m_{23}}{\partial x_i}\right) + \\
+\frac{\partial m_{14}}{\partial x_i}(m_{22}m_{33} - m_{32}m_{23}) + \\
+m_{14}\left(\frac{\partial m_{22}}{\partial x_i}m_{33} + m_{22}\frac{\partial m_{33}}{\partial x_i} - \frac{\partial m_{32}}{\partial x_i}m_{23} - m_{32}\frac{\partial m_{23}}{\partial x_i}\right)
\end{aligned} \tag{1.148}$$

Equivalently, one obtains

$$\begin{aligned}
\frac{\partial M_{42}}{\partial x_i} = & \frac{\partial m_{11}}{\partial x_i}(m_{32}m_{43} - m_{33}m_{42}) + \\
+m_{11}\left(\frac{\partial m_{32}}{\partial x_i}m_{43} + m_{32}\frac{\partial m_{42}}{\partial x_i} - \frac{\partial m_{33}}{\partial x_i}m_{42} - m_{33}\frac{\partial m_{42}}{\partial x_i}\right) - \\
-\frac{\partial m_{31}}{\partial x_i}(m_{12}m_{43} - m_{13}m_{42} -) \\
-m_{31}\left(\frac{\partial m_{12}}{\partial x_i}m_{43} + m_{12}\frac{\partial m_{43}}{\partial x_i} - \frac{\partial m_{13}}{\partial x_i}m_{42} - m_{13}\frac{\partial m_{42}}{\partial x_i}\right) + \\
+\frac{\partial m_{41}}{\partial x_i}(m_{12}m_{33} - m_{13}m_{32}) + \\
+m_{41}\left(\frac{\partial m_{12}}{\partial x_i}m_{33} + m_{12}\frac{\partial m_{33}}{\partial x_i} - \frac{\partial m_{13}}{\partial x_i}m_{32} - m_{13}\frac{\partial m_{32}}{\partial x_i}\right)
\end{aligned} \tag{1.149}$$

In a similar manner, one gets

$$\begin{aligned}
\frac{\partial M_{43}}{\partial x_i} = & \frac{\partial m_{11}}{\partial x_i}(m_{22}m_{34} - m_{32}m_{24}) + \\
+m_{11}\left(\frac{\partial m_{22}}{\partial x_i}m_{34} + m_{22}\frac{\partial m_{34}}{\partial x_i} - \frac{\partial m_{32}}{\partial x_i}m_{24} - m_{32}\frac{\partial m_{24}}{\partial x_i}\right) - \\
-\frac{\partial m_{12}}{\partial x_i}(m_{21}m_{34} - m_{31}m_{24} -) \\
-m_{12}\left(\frac{\partial m_{21}}{\partial x_i}m_{34} + m_{21}\frac{\partial m_{34}}{\partial x_i} - \frac{\partial m_{31}}{\partial x_i}m_{24} - m_{31}\frac{\partial m_{24}}{\partial x_i}\right) + \\
+\frac{\partial m_{14}}{\partial x_i}(m_{21}m_{32} - m_{31}m_{22}) + \\
+m_{14}\left(\frac{\partial m_{21}}{\partial x_i}m_{32} + m_{21}\frac{\partial m_{32}}{\partial x_i} - \frac{\partial m_{31}}{\partial x_i}m_{22} - m_{31}\frac{\partial m_{22}}{\partial x_i}\right)
\end{aligned} \tag{1.150}$$

Finally, one has that

$$\begin{aligned}\frac{\partial M_{44}}{\partial x_i} &= \frac{\partial m_{22}}{\partial x_i}(m_{22}m_{33}-m_{32}m_{23})+\\ +m_{11}&\left(\frac{\partial m_{22}}{\partial x_i}m_{33}+m_{22}\frac{\partial m_{33}}{\partial x_i}-\frac{\partial m_{32}}{\partial x_i}m_{23}-m_{32}\frac{\partial m_{23}}{\partial x_i}\right)-\\ &-\frac{\partial m_{12}}{\partial x_i}(m_{22}m_{33}-m_{31}m_{23}-)\\ -m_{12}&\left(\frac{\partial m_{22}}{\partial x_i}m_{33}+m_{23}\frac{\partial m_{33}}{\partial x_i}-\frac{\partial m_{31}}{\partial x_i}m_{23}-m_{31}\frac{\partial m_{23}}{\partial x_i}\right)+\\ &+\frac{\partial m_{13}}{\partial x_i}(m_{21}m_{32}-m_{31}m_{22})+\\ +m_{13}&\left(\frac{\partial m_{21}}{\partial x_i}m_{32}+m_{21}\frac{\partial m_{32}}{\partial x_i}-\frac{\partial m_{31}}{\partial x_i}m_{22}-m_{31}\frac{\partial m_{22}}{\partial x_i}\right)\end{aligned} \tag{1.151}$$

About the partial derivatives of the determinant $detM$, one has for $i=1,2,\cdots,8$

$$\begin{aligned}\frac{\partial detM}{\partial x_i} &= \frac{\partial m_{11}}{\partial x_i}M_{11}+m_{11}\frac{\partial M_{11}}{\partial x_i}-\frac{\partial m_{12}}{\partial x_i}M_{12}-m_{12}\frac{\partial M_{12}}{\partial x_i}+\\ &+\frac{\partial m_{13}}{\partial x_i}M_{13}+m_{13}\frac{\partial M_{13}}{\partial x_i}-\frac{\partial m_{14}}{\partial x_i}M_{14}-m_{14}\frac{\partial M_{14}}{\partial x_i}\end{aligned} \tag{1.152}$$

Next, the derivatives of the elements of the inertia matrix M are computed.

It holds that $m_{11}=p_1+p_2cos(x_3)$. Thus, one has: $\frac{\partial m_{11}}{\partial x_1}=0$, $\frac{\partial m_{11}}{\partial x_2}=0$, $\frac{\partial m_{11}}{\partial x_3}=-p_2sin(x_3)$, $\frac{\partial m_{11}}{\partial x_4}=0$, $\frac{\partial m_{11}}{\partial x_5}=0$, $\frac{\partial m_{11}}{\partial x_6}=0$, $\frac{\partial m_{11}}{\partial x_7}=0$, $\frac{\partial m_{11}}{\partial x_8}=0$.

Besides, it holds that $m_{12}=m_{21}=p_3+0.5p_2sin(x_3)x_4$, thus $\frac{\partial m_{12}}{\partial x_1}=\frac{\partial m_{21}}{\partial x_1}=0$, $\frac{\partial m_{12}}{\partial x_2}=\frac{\partial m_{21}}{\partial x_2}=0$, $\frac{\partial m_{12}}{\partial x_3}=\frac{\partial m_{21}}{\partial x_3}=0.5p_2cos(x_3)x_4$, $\frac{\partial m_{12}}{\partial x_4}=\frac{\partial m_{21}}{\partial x_4}=0.5p_2sin(x_3)$, $\frac{\partial m_{12}}{\partial x_5}=\frac{\partial m_{21}}{\partial x_5}=0$, $\frac{\partial m_{12}}{\partial x_6}=\frac{\partial m_{21}}{\partial x_6}=0$, $\frac{\partial m_{12}}{\partial x_7}=\frac{\partial m_{21}}{\partial x_7}=0$, $\frac{\partial m_{12}}{\partial x_8}=\frac{\partial m_{21}}{\partial x_8}=0$.

Moreover, it holds that $m_{13}=m_{31}=0$, thus $\frac{\partial m_{13}}{\partial x_1}=\frac{\partial m_{31}}{\partial x_1}=0$, $\frac{\partial m_{13}}{\partial x_2}=\frac{\partial m_{31}}{\partial x_2}=0$, $\frac{\partial m_{13}}{\partial x_3}=\frac{\partial m_{31}}{\partial x_3}=0$, $\frac{\partial m_{13}}{\partial x_4}=\frac{\partial m_{31}}{\partial x_4}=0$, $\frac{\partial m_{13}}{\partial x_5}=\frac{\partial m_{31}}{\partial x_5}=0$, $\frac{\partial m_{13}}{\partial x_6}=\frac{\partial m_{31}}{\partial x_6}=0$, $\frac{\partial m_{13}}{\partial x_7}=\frac{\partial m_{31}}{\partial x_7}=0$, $\frac{\partial m_{13}}{\partial x_8}=\frac{\partial m_{31}}{\partial x_8}=0$.

Additionally, it holds that $m_{14}=m_{41}=-p_5$, thus $\frac{\partial m_{14}}{\partial x_1}=\frac{\partial m_{41}}{\partial x_i}=0$ $\frac{\partial m_{14}}{\partial x_2}=\frac{\partial m_{41}}{\partial x_2}=0$, $\frac{\partial m_{14}}{\partial x_3}=\frac{\partial m_{41}}{\partial x_3}=0$, $\frac{\partial m_{14}}{\partial x_5}=\frac{\partial m_{41}}{\partial x_5}=0$, $\frac{\partial m_{14}}{\partial x_6}=\frac{\partial m_{41}}{\partial x_6}=0$, $\frac{\partial m_{14}}{\partial x_7}=\frac{\partial m_{41}}{\partial x_7}=0$, $\frac{\partial m_{14}}{\partial x_8}=\frac{\partial m_{41}}{\partial x_8}=0$.

Moreover, it holds that $m_{22}=p_2$, thus $\frac{\partial m_{22}}{\partial x_1}=0$, $\frac{\partial m_{22}}{\partial x_2}=0$, $\frac{\partial m_{22}}{\partial x_3}=0$, $\frac{\partial m_{22}}{\partial x_4}=0$, $\frac{\partial m_{22}}{\partial x_5}=0$, $\frac{\partial m_{22}}{\partial x_6}=0$, $\frac{\partial m_{22}}{\partial x_7}=0$, $\frac{\partial m_{22}}{\partial x_8}=0$.

Furthermore, it holds that $m_{23}=m_{32}=0$, thus $\frac{\partial m_{23}}{\partial x_1}=\frac{\partial m_{32}}{\partial x_1}=\frac{\partial m_{23}}{\partial x_2}=\frac{\partial m_{32}}{\partial x_2}=0$, $\frac{\partial m_{23}}{\partial x_3}=\frac{\partial m_{32}}{\partial x_3}=0$, $\frac{\partial m_{23}}{\partial x_4}=\frac{\partial m_{32}}{\partial x_4}=0$, $\frac{\partial m_{23}}{\partial x_5}=\frac{\partial m_{32}}{\partial x_5}=0$, $\frac{\partial m_{23}}{\partial x_6}=\frac{\partial m_{32}}{\partial x_6}=0$, $\frac{\partial m_{23}}{\partial x_7}=\frac{\partial m_{32}}{\partial x_7}=0$, $\frac{\partial m_{23}}{\partial x_8}=\frac{\partial m_{32}}{\partial x_8}=0$.

Besides, it holds that $m_{24}=m_{42}=-p_5$, thus $\frac{\partial m_{24}}{\partial x_1}=\frac{\partial m_{42}}{\partial x_1}=\frac{\partial m_{24}}{\partial x_2}=\frac{\partial m_{42}}{\partial x_2}=0$, $\frac{\partial m_{24}}{\partial x_3}=\frac{\partial m_{42}}{\partial x_3}=0$, $\frac{\partial m_{24}}{\partial x_4}=\frac{\partial m_{42}}{\partial x_4}=0$, $\frac{\partial m_{24}}{\partial x_5}=\frac{\partial m_{42}}{\partial x_5}=0$, $\frac{\partial m_{24}}{\partial x_6}=\frac{\partial m_{42}}{\partial x_6}=0$, $\frac{\partial m_{24}}{\partial x_7}=\frac{\partial m_{42}}{\partial x_7}=0$, $\frac{\partial m_{24}}{\partial x_8}=\frac{\partial m_{42}}{\partial x_8}=0$.

Moreover, it holds that $m_{33}=p_4$, thus $\frac{\partial m_{33}}{\partial x_1}=0$, $\frac{\partial m_{33}}{\partial x_2}=0$, $\frac{\partial m_{33}}{\partial x_3}=0$, $\frac{\partial m_{33}}{\partial x_4}=0$, $\frac{\partial m_{33}}{\partial x_5}=0$, $\frac{\partial m_{33}}{\partial x_6}=0$, $\frac{\partial m_{33}}{\partial x_7}=0$, $\frac{\partial m_{33}}{\partial x_8}=0$.

Additionally, it holds that $m_{34}=m_{43}=$, thus $\frac{\partial m_{34}}{\partial x_1}=\frac{\partial m_{43}}{\partial x_1}=$, $\frac{\partial m_{34}}{\partial x_2}=\frac{\partial m_{43}}{\partial x_2}=$, $\frac{\partial m_{34}}{\partial x_3}=\frac{\partial m_{43}}{\partial x_3}=$, $\frac{\partial m_{34}}{\partial x_4}=\frac{\partial m_{43}}{\partial x_4}=$, $\frac{\partial m_{34}}{\partial x_5}=\frac{\partial m_{43}}{\partial x_5}=$, $\frac{\partial m_{34}}{\partial x_6}=\frac{\partial m_{43}}{\partial x_6}=$, $\frac{\partial m_{34}}{\partial x_7}=\frac{\partial m_{43}}{\partial x_7}=$,. $\frac{\partial m_{34}}{\partial x_8}=\frac{\partial m_{43}}{\partial x_8}=$.

Finally, it holds that $m_{44}=p_5$, thus $\frac{\partial m_{44}}{\partial x_1}=0$, $\frac{\partial m_{44}}{\partial x_2}=0$, $\frac{\partial m_{44}}{\partial x_3}=0$, $\frac{\partial m_{44}}{\partial x_4}=0$, $\frac{\partial m_{44}}{\partial x_5}=0$, $\frac{\partial m_{44}}{\partial x_6}=0$, $\frac{\partial m_{44}}{\partial x_7}=0$,. $\frac{\partial m_{44}}{\partial x_8}=0$.

Finally, about the computation of the partial derivatives of the Coriolis forces vector, one has

$$C(x,\dot{x})\dot{x} = \begin{pmatrix} c_{11}x_2 + c_{12}x_4 + c_{13}x_6 + c_{14}x_8 \\ c_{21}x_2 + c_{22}x_4 + c_{23}x_6 + c_{24}x_8 \\ c_{31}x_2 + c_{22}x_4 + c_{23}x_6 + c_{24}x_8 \\ c_{41}x_2 + c_{42}x_4 + c_{43}x_6 + c_{44}x_8 \end{pmatrix} \tag{1.153}$$

It holds that for $i = 1.3, 5, 7$

$$\frac{\partial c_1}{\partial x_i} = \frac{\partial c_{11}}{\partial x_i}x_2 + \frac{\partial c_{12}}{\partial x_i}x_4 + \frac{\partial c_{13}}{\partial x_i}x_6 + \frac{\partial c_{14}}{\partial x_i}x_8 \tag{1.154}$$

and also

$$\frac{\partial c_1}{\partial x_2} = \frac{\partial c_{11}}{\partial x_2}x_2 + c_{11} + \frac{\partial c_{12}}{\partial x_2}x_4 + \frac{\partial c_{13}}{\partial x_2}x_6 + \frac{\partial c_{14}}{\partial x_2}x_8 \tag{1.155}$$

$$\frac{\partial c_1}{\partial x_4} = \frac{\partial c_{11}}{\partial x_4}x_2 + \frac{\partial c_{12}}{\partial x_4}x_4 + c_{12} + \frac{\partial c_{13}}{\partial x_4}x_6 + \frac{\partial c_{14}}{\partial x_4}x_8 \tag{1.156}$$

$$\frac{\partial c_1}{\partial x_6} = \frac{\partial c_{11}}{\partial x_6}x_2 + \frac{\partial c_{12}}{\partial x_6}x_4 + \frac{\partial c_{13}}{\partial x_6}x_6 + c_{13} + \frac{\partial c_{14}}{\partial x_6}x_8 \tag{1.157}$$

$$\frac{\partial c_1}{\partial x_8} = \frac{\partial c_{11}}{\partial x_8}x_2 + \frac{\partial c_{12}}{\partial x_8}x_4 + \frac{\partial c_{13}}{\partial x_8}x_6 + \frac{\partial c_{14}}{\partial x_8}x_8 + c_{14} \tag{1.158}$$

Equivalently, it holds that for $i = 1.3, 5, 7$

$$\frac{\partial c_2}{\partial x_i} = \frac{\partial c_{21}}{\partial x_i}x_2 + \frac{\partial c_{22}}{\partial x_i}x_4 + \frac{\partial c_{23}}{\partial x_i}x_6 + \frac{\partial c_{24}}{\partial x_i}x_8 \tag{1.159}$$

and also

$$\frac{\partial c_2}{\partial x_2} = \frac{\partial c_{21}}{\partial x_2}x_2 + c_{21} + \frac{\partial c_{22}}{\partial x_2}x_4 + \frac{\partial c_{23}}{\partial x_2}x_6 + \frac{\partial c_{24}}{\partial x_2}x_8 \tag{1.160}$$

$$\frac{\partial c_2}{\partial x_4} = \frac{\partial c_{21}}{\partial x_4}x_2 + \frac{\partial c_{22}}{\partial x_4}x_4 + c_{22} + \frac{\partial c_{23}}{\partial x_4}x_6 + \frac{\partial c_{24}}{\partial x_4}x_8 \tag{1.161}$$

$$\frac{\partial c_2}{\partial x_6} = \frac{\partial c_1}{\partial x_6}x_2 + \frac{\partial c_{22}}{\partial x_6}x_4 + \frac{\partial c_{23}}{\partial x_6}x_6 + c_{23} + \frac{\partial c_{24}}{\partial x_6}x_8 \tag{1.162}$$

$$\frac{\partial c_2}{\partial x_8} = \frac{\partial c_{21}}{\partial x_8}x_2 + \frac{\partial c_{22}}{\partial x_8}x_4 + \frac{\partial c_{23}}{\partial x_8}x_6 + \frac{\partial c_{24}}{\partial x_8}x_8 + c_{24} \tag{1.163}$$

Similarly, it holds that for $i = 1.3, 5, 7$

$$\frac{\partial c_3}{\partial x_i} = \frac{\partial c_{31}}{\partial x_i}x_2 + \frac{\partial c_{32}}{\partial x_i}x_4 + \frac{\partial c_{33}}{\partial x_i}x_6 + \frac{\partial c_{34}}{\partial x_i}x_8 \tag{1.164}$$

and also

$$\frac{\partial c_3}{\partial x_2} = \frac{\partial c_{31}}{\partial x_2}x_2 + c_{31} + \frac{\partial c_{32}}{\partial x_2}x_4 + \frac{\partial c_{33}}{\partial x_2}x_6 + \frac{\partial c_{34}}{\partial x_2}x_8 \tag{1.165}$$

$$\frac{\partial c_3}{\partial x_4} = \frac{\partial c_{31}}{\partial x_4}x_2 + \frac{\partial c_{32}}{\partial x_4}x_4 + c_{32} + \frac{\partial c_{33}}{\partial x_4}x_6 + \frac{\partial c_{34}}{\partial x_4}x_8 \tag{1.166}$$

$$\frac{\partial c_3}{\partial x_6} = \frac{\partial c_3}{\partial x_6}x_2 + \frac{\partial c_{32}}{\partial x_6}x_4 + \frac{\partial c_{33}}{\partial x_6}x_6 + c_{33} + \frac{\partial c_{34}}{\partial x_6}x_8 \tag{1.167}$$

$$\frac{\partial c_3}{\partial x_8} = \frac{\partial c_{31}}{\partial x_8}x_2 + \frac{\partial c_{32}}{\partial x_8}x_4 + \frac{\partial c_{33}}{\partial x_8}x_6 + \frac{\partial c_{34}}{\partial x_8}x_8 + c_{34} \tag{1.168}$$

Finally, it holds that for $i = 1.3, 5, 7$

$$\frac{\partial c_4}{\partial x_i} = \frac{\partial c_{41}}{\partial x_i}x_2 + \frac{\partial c_{42}}{\partial x_i}x_4 + \frac{\partial c_{43}}{\partial x_i}x_6 + \frac{\partial c_{44}}{\partial x_i}x_8 \tag{1.169}$$

and also

$$\frac{\partial c_4}{\partial x_2} = \frac{\partial c_{41}}{\partial x_2}x_2 + c_{41} + \frac{\partial c_{42}}{\partial x_2}x_4 + \frac{\partial c_{43}}{\partial x_2}x_6 + \frac{\partial c_{44}}{\partial x_2}x_8 \tag{1.170}$$

$$\frac{\partial c_4}{\partial x_4} = \frac{\partial c_{41}}{\partial x_4}x_2 + \frac{\partial c_{42}}{\partial x_4}x_4 + c_{42} + \frac{\partial c_{43}}{\partial x_4}x_6 + \frac{\partial c_{44}}{\partial x_4}x_8 \tag{1.171}$$

$$\frac{\partial c_4}{\partial x_6} = \frac{\partial c_4}{\partial x_6}x_2 + \frac{\partial c_{42}}{\partial x_6}x_4 + \frac{\partial c_{43}}{\partial x_6}x_6 + c_{43} + \frac{\partial c_{44}}{\partial x_6}x_8 \tag{1.172}$$

$$\frac{\partial c_4}{\partial x_8} = \frac{\partial c_{41}}{\partial x_8}x_2 + \frac{\partial c_{42}}{\partial x_8}x_4 + \frac{\partial c_{43}}{\partial x_8}x_6 + \frac{\partial c_{44}}{\partial x_8}x_8 + c_{44} \tag{1.173}$$

Next, the following partial derivatives of the elements c_{ij} $i=1,2,3,4$ and $j=1,2,3,4$ of the Coriolis matrix are computed.

It holds $c_{11}=-p_2sin(x_1)x_4$, thus, one has that: $\frac{\partial c_{11}}{\partial x_1}=-p_2cos(x_1)x_4$, $\frac{\partial c_{11}}{\partial x_2}=0$, $\frac{\partial c_{11}}{\partial x_3}=0$, $\frac{\partial c_{11}}{\partial x_4}=-o_2sin(x_1)$, $\frac{\partial c_{11}}{\partial x_5}=0$, $\frac{\partial c_{11}}{\partial x_6}=0$, $\frac{\partial c_{11}}{\partial x_7}=0$, $\frac{\partial c_{11}}{\partial x_8}=0$.

Additionally, it holds that $c_{12}=-0.5p_2sin(x_3)x_4$, thus $\frac{\partial c_{12}}{\partial x_1}=0$, $\frac{\partial c_{11}}{\partial x_2}=0$, $\frac{\partial c_{12}}{\partial x_3}=-0.5p_2cos(x_3)x_4$, $\frac{\partial c_{12}}{\partial x_4}=-0.5p_2sin(x_3)$, $\frac{\partial c_{12}}{\partial x_5}=0$, $\frac{\partial c_{12}}{\partial x_6}=0$, $\frac{\partial c_{12}}{\partial x_7}=0$, $\frac{\partial c_{11}}{\partial x_8}=0$.

Moreover, it holds that $c_{13}=0$; thus, one obtains: $\frac{\partial c_{13}}{\partial x_i}=0$, for $i=1,2,\cdots,8$.

Additionally, it holds that $c_{14}=0$; thus, one obtains: $\frac{\partial c_{14}}{\partial x_i}=$, for $i=1,\cdots,8$.

Additionally, it holds that $c_{21}=0.5p_2sin(x_3)x_2$; thus, one obtains: $\frac{\partial c_{21}}{\partial x_1}=0$, $\frac{\partial c_{21}}{\partial x_2}=0.5p_2sin(x_3)$, $\frac{\partial c_{21}}{\partial x_3}=0.5p_2cos(x_3)x_2$, $\frac{\partial c_{21}}{\partial x_4}=0$, $\frac{\partial c_{21}}{\partial x_5}=0$, $\frac{\partial c_{21}}{\partial x_6}=0$, $\frac{\partial c_{21}}{\partial x_7}=0$, $\frac{\partial c_{21}}{\partial x_8}=0$.

Additionally, it holds that $c_{22}=0$; thus, one obtains: $\frac{\partial c_{22}}{\partial x_i}=$, for $i=1,2,\cdots,8$.

Moreover, it holds that $c_{23}=0$; thus, one obtains: $\frac{\partial c_{23}}{\partial x_i}=0$ for $i=1,2,\cdots,8$.

Furthermore, it holds that $c_{24}=0$; thus, one obtains $\frac{\partial c_{24}}{\partial x_i}=0$ for $i=1,2,\cdots,8$.

Additionally, one has that $c_{31}=0$; thus, one obtains: $\frac{\partial c_{31}}{\partial x_i}=$, for $i=1,2,\cdots,8$.

Furthermore, it holds $c_{32}=0$; thus, one obtains: $\frac{\partial c_{32}}{\partial x_i}=$, for $i=1,2,\cdots,8$.

Moreover, it holds that $c_{33}=0$; thus, one obtains: $\frac{\partial c_{33}}{\partial x_1}=0$, for $i=1,2,\cdots,8$.

Additionally, it holds that $c_{34}=0$; thus, one obtains: $\frac{\partial c_{34}}{\partial x_1}=0$, for $i=1,2,\cdots,8$.

Furthermore, one has that $c_{41}=0$ thus, one obtains: $\frac{\partial c_{41}}{\partial x_i}=$, for $i=1,2,\cdots,8$.

Moreover, it holds that $c_{42}=0$ thus, one obtains: $\frac{\partial c_{42}}{\partial x_i}=$, for $i=1,2,\cdots,8$.

Furthermore, it holds that $c_{43}=0$ thus, one obtains: $\frac{\partial c_{43}}{\partial x_i}=$, for $i=1,2,\cdots,8$.

Finally, it holds that $c_{44}=0$. Thus, one obtains $\frac{\partial c_{44}}{\partial x_i}=0$, for $i=1,\cdots,8$.

In a similar manner, one computes the partial derivatives of the elements of the gravitational forces matrix. It holds that $g_1=0$ that one obtains $\frac{\partial g_1}{\partial x_i}=0$, for $i=1,\cdots,8$.

Additionally, it holds that $g_2=0$ that one obtains $\frac{\partial g_2}{\partial x_i}=0$, for $i=1,\cdots,8$.

Moreover, it holds that $g_3=p_4\bar{g}$, thus, one obtains $\frac{\partial g_3}{\partial x_i}=0$, for $i=1,2,\cdots,8$.

Finally, it holds that $g_4=0$ thus, one obtains $\frac{\partial g_4}{\partial x_i}=0$, for $i=1,2,\cdots,8$.

In a similar manner, one computes the partial derivatives of the elements of the disturbances' vector $\tilde{d}$. It holds that $d_1 = b_1x_2$ that one obtains $\frac{\partial d_1}{\partial x_i} = 0$, for $i = 1, \cdots, 8$ and $i{\neq}2$, while $\frac{\partial d_1}{\partial x_2} = b_1$.

Additionally, it holds that $d_2 = b_2x_4$ that one obtains $\frac{\partial d_2}{\partial x_i} = 0$, for $i = 1, \cdots, 8$ and $i{\neq}4$, while $\frac{\partial d_2}{\partial x_4} = b_2$.

Moreover, it holds that $d_3 = b_3x_6$ that one obtains $\frac{\partial d_3}{\partial x_i} = 0$, for $i = 1, \cdots, 8$ and $i{\neq}6$, while $\frac{\partial d_3}{\partial x_6} = b_3$.

Finally, it holds that $d_4 = b_4x_8$ that one obtains $\frac{\partial d_4}{\partial x_i} = 0$, for $i = 1, \cdots, 8$ and $i{\neq}8$, while $\frac{\partial d_4}{\partial x_8} = b_4$.

1.2.4 Design of an H-infinity nonlinear feedback controller

1.2.4.1 Equivalent linearized dynamics of the 4-DOF SCARA robot

After linearization around its current operating point, the dynamic model for the 4-DOF SCARA robot is written as

$$\dot{x} = Ax + Bu + d_1 \tag{1.174}$$

Parameter d_1 stands for the linearization error in the 4-DOF SCARA robot's model that was given previously in Eq. (1.118). The reference setpoints for the state vector of the aforementioned dynamic model are denoted by $\mathbf{x_d} = [x_1^d, \cdots, x_8^d]$. Tracking of this trajectory is achieved after applying the control input u^*. At every time instant the control input u^* is assumed to differ from the control input u appearing in Eq. (1.174) by an amount equal to Δu, that is $u^* = u + \Delta u$

$$\dot{x}_d = Ax_d + Bu^* + d_2 \tag{1.175}$$

The dynamics of the controlled system described in Eq. (1.174) can be also written as

$$\dot{x} = Ax + Bu + Bu^* - Bu^* + d_1 \tag{1.176}$$

and by denoting $d_3 = -Bu^* + d_1$ as an aggregate disturbance term, one obtains

$$\dot{x} = Ax + Bu + Bu^* + d_3 \tag{1.177}$$

By subtracting Eq. (1.175) from Eq. (1.177), one has

$$\dot{x} - \dot{x}_d = A(x - x_d) + Bu + d_3 - d_2 \tag{1.178}$$

By denoting the tracking error as $e = x - x_d$ and the aggregate disturbance term as $\tilde{d} = d_3 - d_2$, the tracking error dynamics becomes

$$\dot{e} = Ae + Bu + \tilde{d} \tag{1.179}$$

The above linearized form of the 4-DOF SCARA robot's model can be efficiently controlled after applying an H-infinity feedback control scheme.

1.2.5 The nonlinear H-infinity control

The initial nonlinear model of the 4-DOF SCARA robot is in the form

$$\dot{x} = f(x, u) \quad x{\in}R^n, \ u{\in}R^m \tag{1.180}$$

Linearization of the model of the 4-DOF SCARA robot is performed at each iteration of the control algorithm around its present operating point $(x^*, u^*) = (x(t), u(t - T_s))$. The linearized equivalent of the system is described by

$$\dot{x} = Ax + Bu + L\tilde{d} \quad x{\in}R^n, \ u{\in}R^m, \ \tilde{d}{\in}R^q \tag{1.181}$$

where matrices A and B are obtained from the computation of the previously defined Jacobians and vector $\tilde{d}$ denotes disturbance terms due to linearization errors. The problem of disturbance rejection for the linearized model that is described by

$$\begin{aligned} \dot{x} &= Ax + Bu + L\tilde{d} \\ y &= Cx \end{aligned} \tag{1.182}$$

where $x \in R^n$, $u \in R^m$, $\tilde{d} \in R^q$, and $y \in R^p$, cannot be handled efficiently if the classical linear quadratic regulator (LQR) control scheme is applied. This is because of the existence of the perturbation term $\tilde{d}$. The disturbance term $\tilde{d}$ apart from modeling (parametric) uncertainty and external perturbation terms can also represent noise terms of any distribution.

In the H_∞ control approach, a feedback control scheme is designed for trajectory tracking by the system's state vector and simultaneous disturbance rejection, considering that the disturbance affects the system in the worst possible manner. The disturbances' effects are incorporated in the following quadratic cost function:

$$J(t) = \frac{1}{2}\int_0^T [y^T(t)y(t) + ru^T(t)u(t) - \rho^2 \tilde{d}^T(t)\tilde{d}(t)]dt, \quad r, \rho > 0 \tag{1.183}$$

The significance of the negative sign in the cost function's term that is associated with the perturbation variable $\tilde{d}(t)$ is that the disturbance tries to maximize the cost function $J(t)$ while the control signal $u(t)$ tries to minimize it. The physical meaning of the relation given above is that the control signal and the disturbances compete with each other within a min–max differential game. This problem of min–max optimization can be written as $min_u max_{\tilde{d}} J(u, \tilde{d})$.

The objective of the optimization procedure is to compute a control signal $u(t)$ which can compensate for the worst possible disturbance, that is externally imposed to the 4-DOF SCARA robot. However, the solution to the min–max optimization problem is directly related to the value of parameter ρ. This means that there is an upper bound in the disturbances' magnitude that can be annihilated by the control signal.

1.2.5.1 Computation of the feedback control gains

For the linearized system given by Eq. (1.182) the cost function of Eq. (1.183) is defined, where the coefficient r determines the penalization of the control input and the weight coefficient ρ determines the reward of the disturbances' effects. It is assumed that (i) The energy that is transferred from the disturbances' signal $\tilde{d}(t)$ is bounded, that is $\int_0^\infty \tilde{d}^T(t)\tilde{d}(t)dt < \infty$, (ii) matrices $[A, B]$ and $[A, L]$ are stabilizable, (iii) matrix $[A, C]$ is detectable. In the case of a tracking problem, the optimal feedback control law is given by

$$u(t) = -Ke(t) \tag{1.184}$$

with $e = x - x_d$ to be the tracking error, and $K = \frac{1}{r}B^T P$ where P is a positive definite symmetric matrix. As it will be proven in Section 1.2.6, matrix P is obtained from the solution of the Riccati equation:

$$A^T P + PA + Q - P(\frac{2}{r}BB^T - \frac{1}{\rho^2}LL^T)P = 0 \tag{1.185}$$

where Q is a positive semi-definite symmetric matrix. The worst-case disturbance is given by

$$\tilde{d}(t) = \frac{1}{\rho^2}L^T Pe(t) \tag{1.186}$$

The solution of the H-infinity feedback control problem for the 4-DOF SCARA robot and the computation of the worst-case disturbance that the related controller can sustain, comes from superposition of Bellman's optimality principle when considering that the robot is affected by two separate inputs (i) the control input u (ii) the cumulative disturbance input $\tilde{d}(t)$. Solving the optimal control problem for u, that is for the minimum variation (optimal) control input that achieves elimination of the state vector's tracking error, gives $u = -\frac{1}{r}B^T Pe$. Equivalently, solving the optimal control problem for $\tilde{d}$, that is for the worst-case disturbance that the control loop can sustain gives $\tilde{d} = \frac{1}{\rho^2}L^T Pe$.

The diagram of the considered control loop for the 4-DOF SCARA robot is depicted in Figure 1.20.

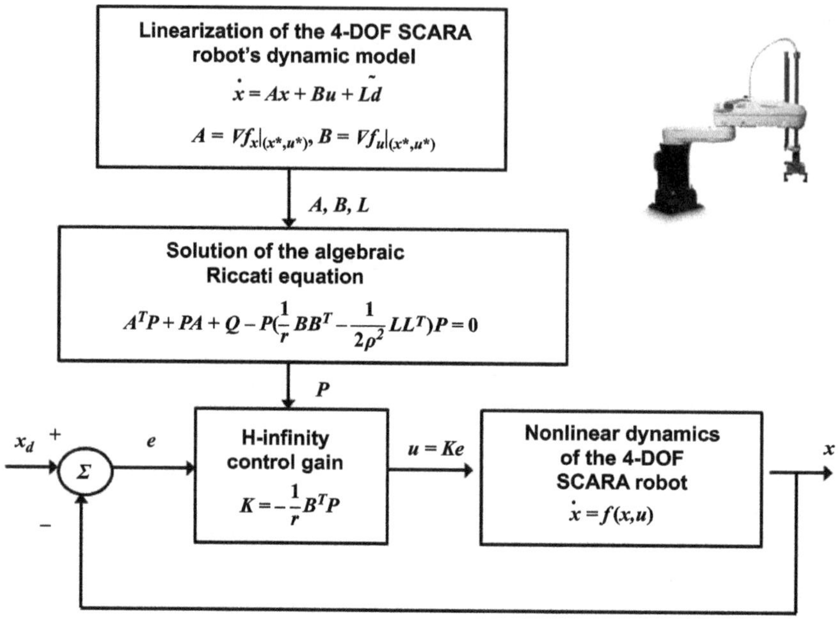

Figure 1.20 Diagram of the control scheme for the 4-DOF SCARA robotic manipulator

1.2.6 Lyapunov stability analysis

Through Lyapunov stability analysis, it will be shown that the proposed nonlinear control scheme assures H_∞ tracking performance for the 4-DOF SCARA robot, and that in case of bounded disturbance terms asymptotic convergence to the reference setpoints is achieved. The tracking error dynamics for the 4-DOF SCARA robot is written in the form

$$\dot{e} = Ae + Bu + L\tilde{d} \tag{1.187}$$

where in the 4-DOF SCARA robot's case $L \in R^{8\times 8}$ is the disturbance inputs gain matrix. Variable $\tilde{d}$ denotes model uncertainties and external disturbances of the 4-DOF SCARA robot's model. The following Lyapunov equation is considered:

$$V = \frac{1}{2}e^TPe \tag{1.188}$$

where $e = x - x_d$ is the tracking error. By differentiating with respect to time, one obtains

$$\dot{V} = \frac{1}{2}\dot{e}^TPe + \frac{1}{2}e^TP\dot{e} \Rightarrow \dot{V} = \frac{1}{2}[Ae + Bu + L\tilde{d}]^TPe + \frac{1}{2}e^TP[Ae + Bu + L\tilde{d}] \Rightarrow \tag{1.189}$$

$$\dot{V} = \frac{1}{2}[e^TA^T + u^TB^T + \tilde{d}^TL^T]Pe + \frac{1}{2}e^TP[Ae + Bu + L\tilde{d}] \Rightarrow \tag{1.190}$$

$$\dot{V} = \frac{1}{2}e^TA^TPe + \frac{1}{2}u^TB^TPe + \frac{1}{2}\tilde{d}^TL^TPe + \frac{1}{2}e^TPAe + \frac{1}{2}e^TPBu + \frac{1}{2}e^TPL\tilde{d} \tag{1.191}$$

The previous equation is rewritten as

$$\dot{V} = \frac{1}{2}e^T(A^TP + PA)e + (\frac{1}{2}u^TB^TPe + \frac{1}{2}e^TPBu) + (\frac{1}{2}\tilde{d}^TL^TPe + \frac{1}{2}e^TPL\tilde{d}) \tag{1.192}$$

Assumption: For given positive definite matrix Q and coefficients r and ρ, there exists a positive definite matrix P, which is the solution of the following matrix equation:

$$A^TP + PA = -Q + P(\frac{2}{r}BB^T - \frac{1}{\rho^2}LL^T)P \tag{1.193}$$

Moreover, the following feedback control law is applied to the system:

$$u = -\frac{1}{r}B^TPe \tag{1.194}$$

By substituting Eq. (1.193) and Eq. (1.194), one obtains

$$\dot{V}=\frac{1}{2}e^T[-Q+P(\frac{2}{r}BB^T-\frac{1}{\rho^2}LL^T)P]e+e^TPB(-\frac{1}{r}B^TPe)+e^TPL\tilde{d}\Rightarrow \tag{1.195}$$

$$\begin{aligned}\dot{V}=-\frac{1}{2}e^TQe+\frac{1}{r}e^TPBB^TPe-\frac{1}{2\rho^2}e^TPLL^TPe\\ -\frac{1}{r}e^TPBB^TPe+e^TPL\tilde{d}\end{aligned} \tag{1.196}$$

which after intermediate operations gives

$$\dot{V}=-\frac{1}{2}e^TQe-\frac{1}{2\rho^2}e^TPLL^TPe+e^TPL\tilde{d} \tag{1.197}$$

or, equivalently,

$$\dot{V}=-\frac{1}{2}e^TQe-\frac{1}{2\rho^2}e^TPLL^TPe+\frac{1}{2}e^TPL\tilde{d}+\frac{1}{2}\tilde{d}^TL^TPe \tag{1.198}$$

Lemma: The following inequality holds:

$$\frac{1}{2}e^TL\tilde{d}+\frac{1}{2}\tilde{d}L^TPe-\frac{1}{2\rho^2}e^TPLL^TPe\le\frac{1}{2}\rho^2\tilde{d}^T\tilde{d} \tag{1.199}$$

Proof: The binomial $(\rho\alpha-\frac{1}{\rho}b)^2$ is considered. Expanding the left part of the above inequality, one gets

$$\begin{aligned}\rho^2a^2+\frac{1}{\rho^2}b^2-2ab\geq 0\Rightarrow\frac{1}{2}\rho^2a^2+\frac{1}{2\rho^2}b^2-ab\geq 0\Rightarrow\\ ab-\frac{1}{2\rho^2}b^2\leq\frac{1}{2}\rho^2a^2\Rightarrow\frac{1}{2}ab+\frac{1}{2}ab-\frac{1}{2\rho^2}b^2\leq\frac{1}{2}\rho^2a^2\end{aligned} \tag{1.200}$$

The following substitutions are carried out: $a=\tilde{d}$ and $b=e^TPL$ and the previous relation becomes

$$\frac{1}{2}\tilde{d}^TL^TPe+\frac{1}{2}e^TPL\tilde{d}-\frac{1}{2\rho^2}e^TPLL^TPe\le\frac{1}{2}\rho^2\tilde{d}^T\tilde{d} \tag{1.201}$$

Eq. (1.201) is substituted in Eq. (1.198) and the inequality is enforced, thus giving

$$\dot{V}\le-\frac{1}{2}e^TQe+\frac{1}{2}\rho^2\tilde{d}^T\tilde{d} \tag{1.202}$$

Eq. (1.202) shows that the H_∞ tracking performance criterion is satisfied. The integration of $\dot{V}$ from 0 to T gives

$$\begin{aligned}\int_0^T\dot{V}(t)dt\le-\frac{1}{2}\int_0^T||e||_Q^2dt+\frac{1}{2}\rho^2\int_0^T||\tilde{d}||^2dt\Rightarrow\\ 2V(T)+\int_0^T||e||_Q^2dt\le 2V(0)+\rho^2\int_0^T||\tilde{d}||^2dt\end{aligned} \tag{1.203}$$

Moreover, if there exists a positive constant $M_d>0$ such that

$$\int_0^\infty||\tilde{d}||^2dt\le M_d \tag{1.204}$$

then, one gets

$$\int_0^\infty||e||_Q^2dt\le 2V(0)+\rho^2M_d \tag{1.205}$$

Thus, the integral $\int_0^\infty||e||_Q^2dt$ is bounded. Moreover, $V(T)$ is bounded and from the definition of the Lyapunov function V in Eq. (1.188) it becomes clear that $e(t)$ will be also bounded since $e(t)\ \in\ \Omega_e=\{e|e^TPe\le 2V(0)+\rho^2M_d\}$. According to the above and with the use of Barbalat's Lemma, one obtains $\lim_{t\to\infty}e(t)=0$.

After following the stages of the stability proof one arrives at Eq. (1.202) which shows that the H-infinity tracking performance criterion holds. By selecting the attenuation coefficient ρ to be sufficiently small and in particular to satisfy $\rho^2<||e||_Q^2/||\tilde{d}||^2$ one has that the first derivative of the Lyapunov function is upper bounded by 0. This condition holds at each sampling instance and consequently global stability for the control loop can be concluded.

1.2.7 Flatness-based control in successive loops for the 4-DOF SCARA robotic manipulator

1.2.7.1 Differential flatness properties of the 4-DOF SCARA robot

The dynamic model of the 4-DOF SCARA robot is differentially flat. The flat outputs' vector of the system is $Y=[y_1,y_2,y_3,y_4]^T=[x_1,x_3,x_5,x_7]^T$. Differential flatness is associated with the following two conditions: (i) all state variables of the system can be expressed as differential functions of its flat outputs, (ii) the flat outputs and their derivatives are differentially independent which signifies that they are not connected through a relation in the form of a linear homogeneous differential equation [721,711].

Obviously, it holds that $x_2 = \dot{x}_1$, $x_4 = \dot{x}_3$, $x_6 = \dot{x}_5$, and $x_8 = \dot{x}_7$. This signifies that state variables x_2, x_4, x_6, and x_8 are differential functions of the system's flat outputs. Besides, using that

$$\begin{pmatrix} \ddot{x}_1 \\ \ddot{x}_3 \\ \ddot{x}_5 \\ \ddot{x}_7 \end{pmatrix} = \begin{pmatrix} f_2(x) \\ f_4(x) \\ f_6(x) \\ f_8(x) \end{pmatrix} + \begin{pmatrix} g_{11}(x) & g_{21}(x) & g_{31}(x) & g_{41}(x) \\ g_{12}(x) & g_{22}(x) & g_{32}(x) & g_{42}(x) \\ g_{13}(x) & g_{23}(x) & g_{33}(x) & g_{43}(x) \\ g_{14}(x) & g_{24}(x) & g_{34}(x) & g_{44}(x) \end{pmatrix} \begin{pmatrix} \tau_1 \\ \tau_2 \\ \tau_3 \\ \tau_4 \end{pmatrix} \tag{1.206}$$

and by solving with respect to the control inputs, one obtains

$$\begin{pmatrix} \tau_1 \\ \tau_2 \\ \tau_3 \\ \tau_4 \end{pmatrix} = \begin{pmatrix} g_{11}(x) & g_{21}(x) & g_{31}(x) & g_{41}(x) \\ g_{12}(x) & g_{22}(x) & g_{32}(x) & g_{42}(x) \\ g_{13}(x) & g_{23}(x) & g_{33}(x) & g_{43}(x) \\ g_{14}(x) & g_{24}(x) & g_{34}(x) & g_{44}(x) \end{pmatrix}^{-1} \left[\begin{pmatrix} \ddot{x}_1 \\ \ddot{x}_3 \\ \ddot{x}_5 \\ \ddot{x}_7 \end{pmatrix} - \begin{pmatrix} f_2(x) \\ f_4(x) \\ f_6(x) \\ f_8(x) \end{pmatrix} \right] \tag{1.207}$$

The above relation signifies that the control inputs τ_1, τ_2, τ_3, and τ_4 are also differential functions of the flat outputs of the system. Consequently, the 4-DOF SCARA robot is differentially flat. The differential flatness property means also [721,711] that (i) the robotic model is input–output linearizable, (ii) setpoints for all state variables of the robot can be defined. Actually, one selects first setpoints for the state variables which coincide with the flat outputs x_1^d, x_3^d, x_5^d, and x_7^d, and next defines setpoints for the rest of the state variables x_2^d, x_4^d, x_6^d, and x_8^d which are associated with the flat outputs through the previously explained differential relations. The differential flatness property is also an implicit proof of the system's controllability.

1.2.7.2 Flatness-based control in successive loops for the SCARA robot

The state-space model for the SCARA robot that was previously defined in Eq. (1.115) and in Eq. (1.116) is used again and the new state vector is defined as $z = [z_1, z_2, z_3, z_4, z_5, z_6, z_7, z_8]^T$ where $z_1 = x_1$, $z_2 = x_3$, $z_3 = x_5$, $z_4 = x_7$, $z_5 = x_2$, $z_6 = x_4$, $z_7 = x_6$, and $z_8 = x_8$. Using the new notation for the robot's state variables the associated state-space model becomes

$$\begin{pmatrix} \dot{z}_1 \\ \dot{z}_2 \\ \dot{z}_3 \\ \dot{z}_4 \\ \dot{z}_5 \\ \dot{z}_6 \\ \dot{z}_7 \\ \dot{z}_8 \end{pmatrix} = \begin{pmatrix} z_5 \\ z_6 \\ z_7 \\ z_8 \\ f_2(z) \\ f_4(z) \\ f_6(z) \\ f_8(x) \end{pmatrix} + \begin{pmatrix} 0 & 0 & 0 & 0 \\ 0 & 0 & 0 & 0 \\ 0 & 0 & 0 & 0 \\ 0 & 0 & 0 & 0 \\ g_{21}(z) & g_{22}(z) & g_{23}(z) & g_{24}(z) \\ g_{41}(z) & g_{42}(z) & g_{43}(z) & g_{44}(z) \\ g_{61}(z) & g_{62}(z) & g_{63}(z) & g_{64}(z) \\ g_{81}(z) & g_{82}(z) & g_{83}(z) & g_{84}(z) \end{pmatrix} \begin{pmatrix} u_1 \\ u_2 \\ u_3 \\ u_4 \end{pmatrix} \tag{1.208}$$

Next, the following subvectors and submatrices are defined:

$$z_{1,4} = [z_1, z_2, z_3, z_4]^T \quad f_{1,4} = 0_{4\times 1} \quad g_{1,4} = I_{4\times 4} \tag{1.209}$$

$$z_{5,8} = [z_5, z_6, z_7, z_8]^T \quad f_{5,8} = [f_2(z), f_4(z), f_6(z), f_8(z)]^T \quad g_{5,8} = \begin{pmatrix} g_{21}(z) & g_{22}(z) & g_{23}(z) & g_{24}(z) \\ g_{41}(z) & g_{42}(z) & g_{43}(z) & g_{44}(z) \\ g_{61}(z) & g_{62}(z) & g_{63}(z) & g_{64}(z) \\ g_{81}(z) & g_{82}(z) & g_{83}(z) & g_{84}(z) \end{pmatrix} \tag{1.210}$$

Thus, the state-space model of the SCARA robot can be decomposed into two subsystems:

$$\dot{z}_{1,4} = f_{1,4}(z_{1,4}) + g_{1,4}(z_{1,4})z_{5,8} \tag{1.211}$$

$$\dot{z}_{5,8} = f_{5,8}(z_{1,4}, z_{5,8}) + g_{5,8}(z_{1,4}, z_{5,8})u \tag{1.212}$$

As it has already been proven, the dynamic model of the SCARA robot is a differentially flat system with flat outputs vector $y = z_{1,4}$. Indeed, from Eq. (1.211), one solves for $z_{5,8}$, giving $z_{5,8} = \dot{z}_{1,4}$ which confirms that $z_{5,8}$ is a differential function of flat outputs vector y. Moreover, from Eq. (1.212), one solves for u which gives $u = g_{5,8}^{-1}(z_{1,4}, z_{5.8})[\dot{z}_{5,8} - f_{5,8}(z_{1,4}, z_{5,8})]$ thus confirming that $z_{5,8}$ is also a differential function of the flat outputs vector y. This completes again the proof of differential flatness properties for the 4-DOF SCARA robot.

Next, it will be proven that each one of the subsystems of Eq. (1.211) and Eq. (1.212), if viewed independently, is also differentially flat. For the subsystem of Eq. (1.211) the flat output is taken to be $y_1 = z_{1,4}$, and the virtual control input is $v_1 = z_{5,8}$. It holds that

$$v_1 = z_{5,8} = \dot{z}_{1,4} \Rightarrow v_1 = h_1(y_1, \dot{y}_1) \tag{1.213}$$

while $f_{1,4}(z_{1,4})$, $g_{1,4}(z_{1,4})$ have constant elements. Therefore, the subsystem of Eq. (1.211) is differentially flat. Additionally, for Eq. (1.212) the flat outputs vector is taken to $y_2 = z_{5,8}$ and the real control input is u, while $z_{1,4}$ is viewed as a vector of coefficients. It holds that

$$\begin{aligned} u = g_{5,8}^{-1}(z_{1,4}, z_{5.8})[\dot{z}_{5,8} - f_{5,8}(z_{1,4}, z_{5,8})] \\ \Rightarrow u = h_1(y_2, \dot{y}_2) \end{aligned} \tag{1.214}$$

Therefore, the subsystem of Eq. (1.212) is also differentially flat. Using the differential flatness property for each one of the subsystems of Eq. (1.211) and Eq. (1.212) one has that each subsystem is input–output linearizable [711]. Thus one can design a stabilizing feedback controller about them by inverting their input–output linearized description.

For the subsystem of Eq. (1.211) and using the virtual control input $v_1 = z_{5,8}$ the stabilizing feedback controller is taken to be

$$v_1 = z_{5,8}^d = g_{1,4}^{-1}(z_{1,4})[\dot{z}_{1,4}^d - f_{1,4}(z_{1,4}) - K_1(z_{1,4} - z_{1,4}^d)] \tag{1.215}$$

where $K_1 \in R^{4\times4}$ is a diagonal gain matrix with diagonal elements $k_{1,ii} > 0, i = 1, \cdots, 4$.

For the subsystem of Eq. (1.212) and using the real control input u the stabilizing feedback controller is taken to be

$$u = g_{5,8}^{-1}(z_{1,4}, z_{5,8})[\dot{z}_{5,8}^d - f_{5,8}(z_{1,4}, z_{5,8}) - K_2(z_{5,8} - z_{5,8}^d)] \tag{1.216}$$

where $K_2 \in R^{4\times4}$ is a diagonal gain matrix with diagonal elements $k_{2,ii} > 0, i = 1, \cdots, 4$. In the above equations $z_{1,4}^d$ and $z_{5,8}^d$ are setpoints for state vectors $z_{1,4}$ and $z_{5,8}$, respectively. Moreover, in Eq. (1.216), one has that the virtual control input v_1 of the subsystem of Eq. (1.211) becomes setpoint $z_{5,8}^d$ for the subsystem of Eq. (1.212).

By substituting the control input of Eq. (1.215) into the subsystem of Eq. (1.211), one obtains

$$\begin{aligned} \dot{z}_{1,4} = f_{1,4}(z_{1,4}) + g_{1,4}(z_{1,4}) \cdot g_{1,4}^{-1}(z_{1,4})[\dot{z}_{1,4}^d - f_{1,4}(z_{1,4}) - K_1(z_{1,4} - z_{1,4}^d)] \\ \Rightarrow (\dot{z}_{1,4} - \dot{z}_{1,4}^d) + K_1(z_{1,4} - z_{1,4}^d) = 0 \end{aligned} \tag{1.217}$$

By substituting the control input of Eq. (1.216) into the subsystem of Eq. (1.212), one obtains

$$\begin{aligned} \dot{z}_{5,8} = f_{5,8}(z_{1,4}, z_{5,8}) + g_{1,4}(z_{1,4}, z_{5.8}) \cdot g_{1,4}^{-1}(z_{1,4}, z_{5,8})[\dot{z}_{5,8}^d - f_{5,8}(z_{1,4}, z_{5,8}) - K_2(z_{5,8} - z_{5,8}^d)] \\ \Rightarrow (\dot{z}_{1,4} - \dot{z}_{1,4}^d) + K_1(z_{1,4} - z_{1,4}^d) = 0 \end{aligned} \tag{1.218}$$

By defining the tracking error variables $e_{1,4} = z_{1,4} - z_{1,4}^d$, $e_{5,8} = z_{5,8} - z_{5,8}^d$, one has the following tracking error dynamics for the two subsystems:

$$\dot{e}_{1,4} + K_1 e_{1,4} = 0 \Rightarrow lim_{t\to\infty} e_{1,4} = 0 \Rightarrow lim_{t\to\infty} z_{1,4} = z_{1,4}^d \tag{1.219}$$

By substituting the control input of Eq. (1.216) into the subsystem of Eq. (1.212), one obtains

$$\dot{e}_{5,8} + K_2 e_{5,8} = 0 \Rightarrow lim_{t\to\infty} e_{5,8} = 0 \Rightarrow lim_{t\to\infty} z_{5,8} = z_{5,8}^d \tag{1.220}$$

Thus all state variables x_i, $i = 1, \cdots, 8$ of the 4-DOF SCARA robotic manipulator converge to the associated setpoints.

A proof of the global stability properties of the proposed flatness-based control scheme in successive loops can be also obtained through Lyapunov stability analysis. To this end, the following Lyapunov function is defined

$$V = \frac{1}{2}[e_{1,4}^T e_{1,4} + e_{5,8}^T e_{5,8}] \tag{1.221}$$

By differentiating in time, one has

$$\dot{V} = \frac{1}{2}[2e_{1,4}^T \dot{e}_{1,4} + 2e_{5,8}^T \dot{e}_{5,8}] \Rightarrow \dot{V} = e_{1,4}^T \dot{e}_{1,4} + e_{5,8}^T \dot{e}_{5,8} \tag{1.222}$$

Moreover, from the equations about the tracking error dynamics of the two subsystems, given in Eq. (1.219) and Eq. (1.220), one has $\dot{e}_{1,4} = -K_1 e_{1,4}$ and $\dot{e}_{5,8} = -K_2 e_{5,8}$. Thus, it holds that

$$\begin{aligned} \dot{V} &= e_{1,4}^T \dot{e}_{1,4} + e_{5,8}^T \dot{e}_{5,8} \Rightarrow \dot{V} = -e_{1,4}^T K_1 e_{1,4} - e_{5,8}^T K_2 e_{5,8} \\ &\Rightarrow \dot{V} < 0 \; \forall \; e_{1,4} \neq 0, \; e_{5,8} \neq 0 \end{aligned} \tag{1.223}$$

while $\dot{V} = 0$ if and only if $e_{1,4} = 0$ and $e_{5,8} = 0$. Consequently, the Lyapunov function of the robotic system converges asymptotically to 0, and finally, one has again that $lim_{t \to \infty} e_{1,4} = 0 \Rightarrow lim_{t \to \infty} z_{1,4} = z_{1,4}^d$ and $lim_{t \to \infty} e_{5,8} = 0 \Rightarrow lim_{t \to \infty} z_{5,8} = z_{8,8}^d$.

1.2.8 Simulation tests

1.2.8.1 Results on nonlinear optimal control of the 4-DOF SCARA robot

The global stability properties of the control method and the elimination of the state vector's tracking error which were previously proven through Lyapunov analysis are further confirmed through simulation experiments. The parameters of the model of the 4-DOF SCARA robot which have been used in the simulation tests have been according to [900]. To compute the stabilizing feedback gains of the controller, the algebraic Riccati equation of Eq. (1.193) had to be repetitively solved at each iteration of the control algorithm. The obtained results are depicted in Figures 1.21–1.36. The real values of the state variables of the 4-DOF SCARA robot are printed in blue, their estimates which are provided by the H-infinity Kalman filter are printed in green color while the associated setpoints are printed in red. The performance of the nonlinear optimal control method was very satisfactory. Actually, through all test cases, it has been confirmed

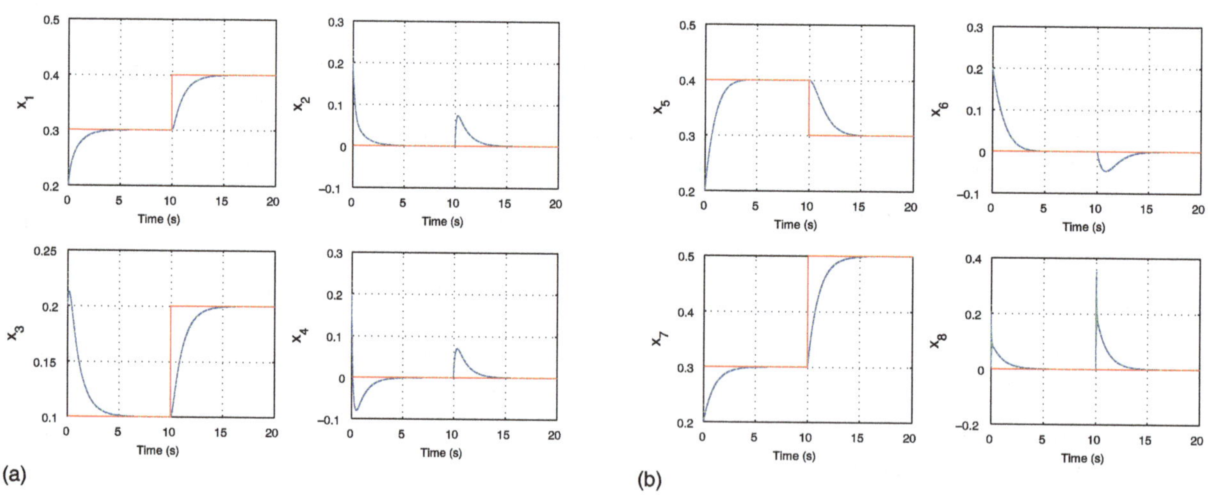

Figure 1.21 Tracking of setpoint 1 for the SCARA robot under nonlinear optimal control: (a) convergence of state variables x_1 to x_4 to their reference setpoints (red line: setpoint, blue line: real value, green line: estimated value); (b) convergence of state variables x_5 to x_8 to their reference setpoints

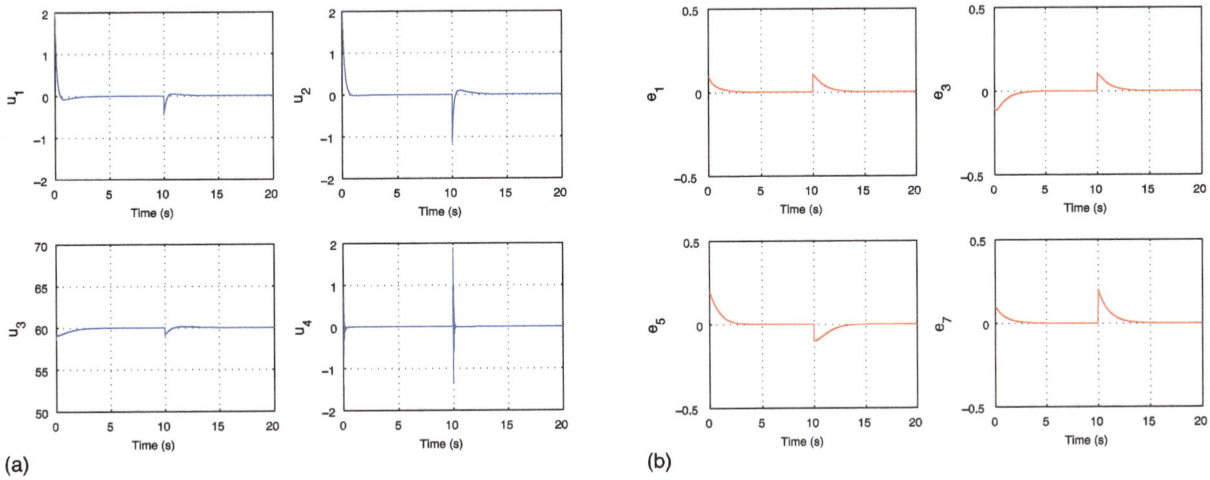

Figure 1.22 Tracking of setpoint 1 for the SCARA robot under nonlinear optimal control: (a) control inputs u_1, u_2 applied to the robot; (b) tracking error variables e_1, e_3, e_5, and e_7 of the SCARA robot

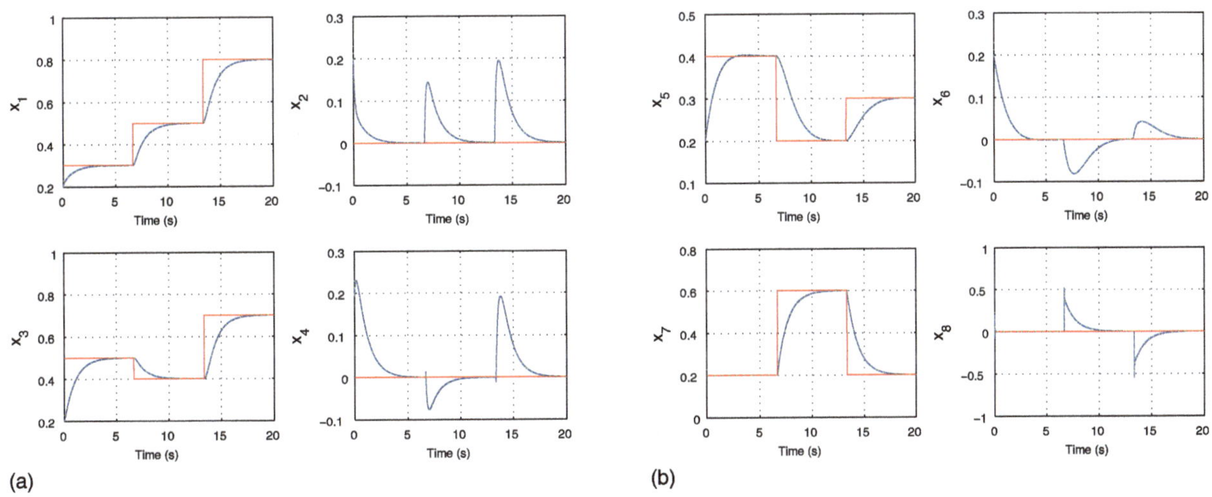

Figure 1.23 Tracking of setpoint 2 for the SCARA robot under nonlinear optimal control: (a) convergence of state variables x_1 to x_4 to their reference setpoints (red line: setpoint, blue line: real value, green line: estimated value); (b) convergence of state variables x_5 to x_8 to their reference setpoints

that the control method can achieve fast and accurate tracking of reference trajectories (setpoints) under moderate variations of the control inputs. The simulation tests come to confirm that the control method has global (and not local) stability properties. Under the nonlinear optimal control method, the state variables of the SCARA robot can track precisely setpoints with fast and abrupt changes. Moreover, the convergence to these setpoints is independent of initial conditions. The variations of the Lyapunov function of the nonlinear optimal control method of the 4-DOF SCARA robotic manipulator are shown in Figure 1.37 to Figure 1.38.

To elaborate on the tracking performance and on the robustness of the proposed nonlinear optimal control method for the SCARA robot, the following tables are given: (i) Table 1.1 which provides information about the accuracy of tracking of the reference setpoints by the state variables of the SCARA robot's state-space model, (ii) Table 1.2 which

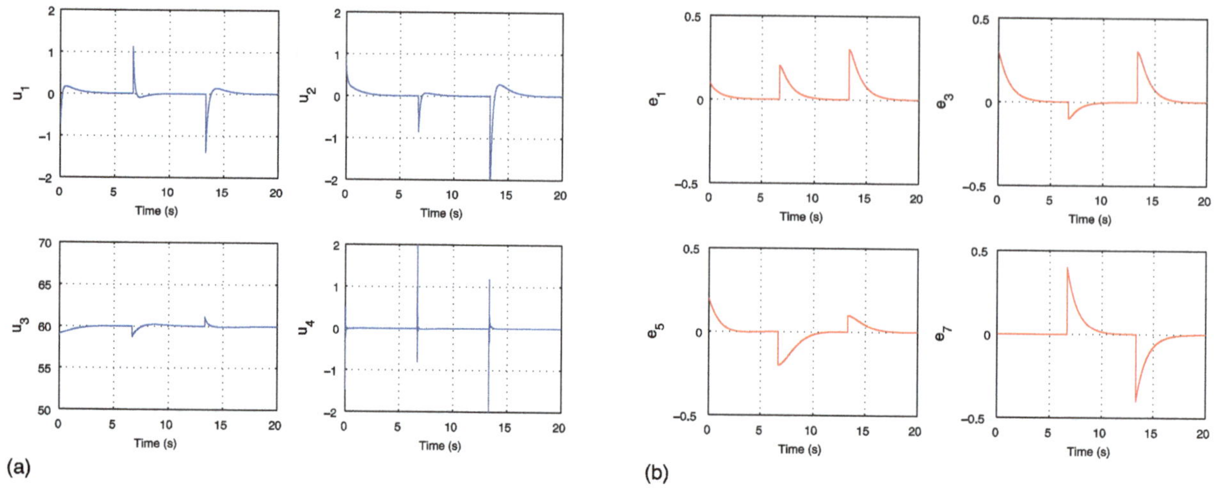

Figure 1.24 Tracking of setpoint 2 for the SCARA robot under nonlinear optimal control: (a) control inputs u_1, u_2 applied to the robot; (b) tracking error variables e_1, e_3, e_5, and e_7 of the SCARA robot

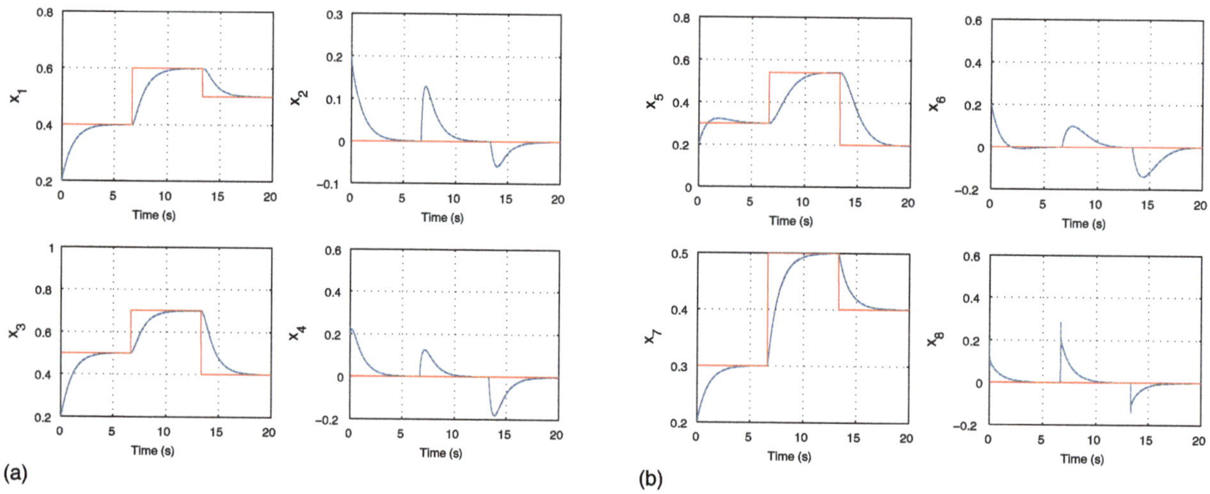

Figure 1.25 Tracking of setpoint 3 for the SCARA robot under nonlinear optimal control: (a) convergence of state variables x_1 to x_4 to their reference setpoints (red line: setpoint, blue line: real value, green line: estimated value); (b) convergence of state variables x_5 to x_8 to their reference setpoints

provides information about the robustness of the control method to parametric changes in the model of the SCARA robot's dynamics (change $\Delta a\%$ in the mass m_2 of the second link), (iii) Table 1.3 which provides information about the precision in state variables' estimation that is achieved by the H-infinity Kalman filter, (iv) Table 1.4 which provides the approximate convergence times of the SCARA robot's state variables to the associated setpoints, (v) Table 1.5 which provides information about the % distribution of the total consumed power in the actuators of the SCARA robot.

1.2.8.2 Results on flatness-based control in successive loops for the 4-DOF SCARA robot

Results about the tracking accuracy and the speed of convergence to setpoints of the successive-loops flatness-based control method, in the case of the 4-DOF SCARA robotic manipulator, are shown in Figures 1.39–1.54. It can be noticed,

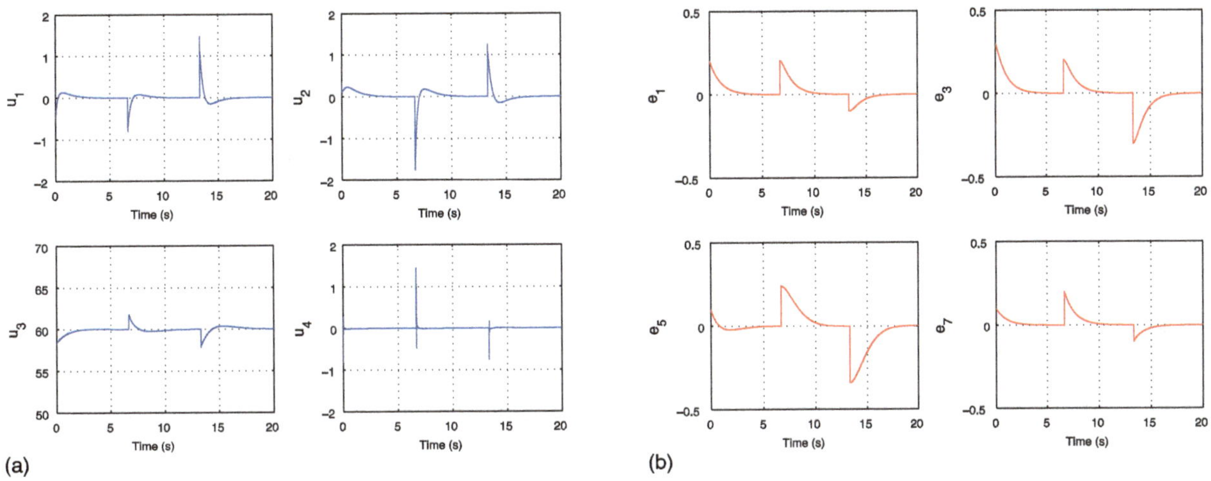

Figure 1.26 Tracking of setpoint 3 for the SCARA robot under nonlinear optimal control: (a) control inputs u_1, u_2 applied to the robot; (b) tracking error variables e_1, e_3, e_5, and e_7 of the SCARA robot

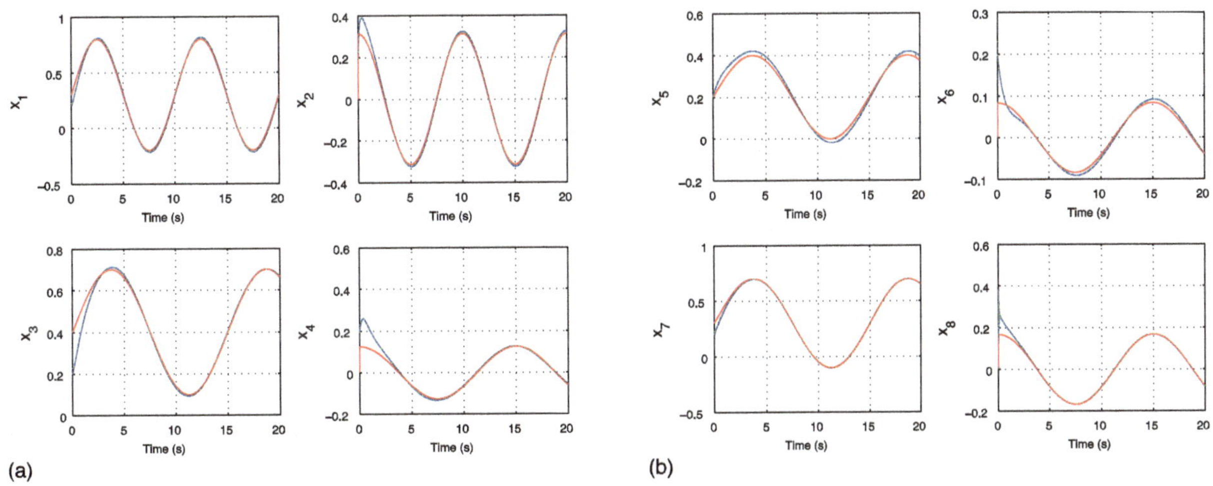

Figure 1.27 Tracking of setpoint 4 for the SCARA robot under nonlinear optimal control: (a) convergence of state variables x_1 to x_4 to their reference setpoints (red line: setpoint, blue line: real value, green line: estimated value); (b) convergence of state variables x_5 to x_8 to their reference setpoints

that under this control scheme, one achieves fast and precise tracking of reference setpoints for all state variables of the robotic system. It is noteworthy, that through the stages of this method one solves also the setpoints definition problem for all state variables of the redundant robotic manipulator. Actually, the selection of setpoints for the state vector $z_{1,4} = [x_1, x_3, x_5, x_7]^T$ is unconstrained. On the other side by defining state vector $z_{5,8} = [x_2, x_4, x_6, x_8]^T$ as virtual control input for the subsystem of $z_{1,4}$ one can find the setpoints for $z_{5,8}$ as functions of the setpoints for $z_{1,4}$. The speed of convergence of the state variables of the robotic system under flatness-based control implemented in successive loops is dependent on the selection of values for the diagonal gain matrices K_1, K_2 of Eq. (1.215) and Eq. (1.216). The variations of the Lyapunov function of the flatness-based control loop of the 4-DOF SCARA robotic manipulator are shown in Figure 1.55 to Figure 1.56.

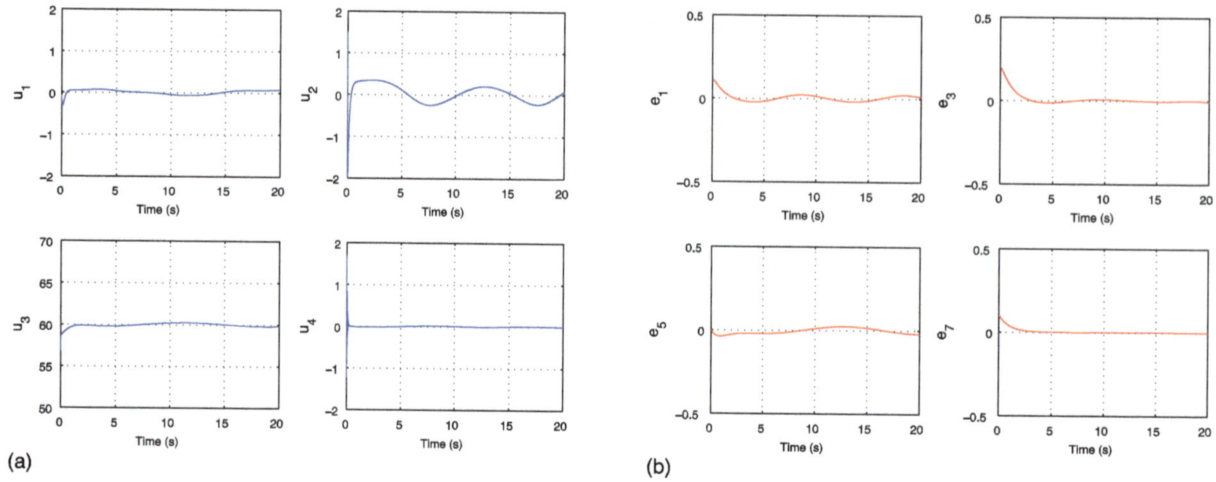

Figure 1.28 Tracking of setpoint 4 for the SCARA robot under nonlinear optimal control: (a) control inputs u_1, u_2 applied to the robot; (b) tracking error variables e_1, e_3, e_5, and e_7 of the SCARA robot

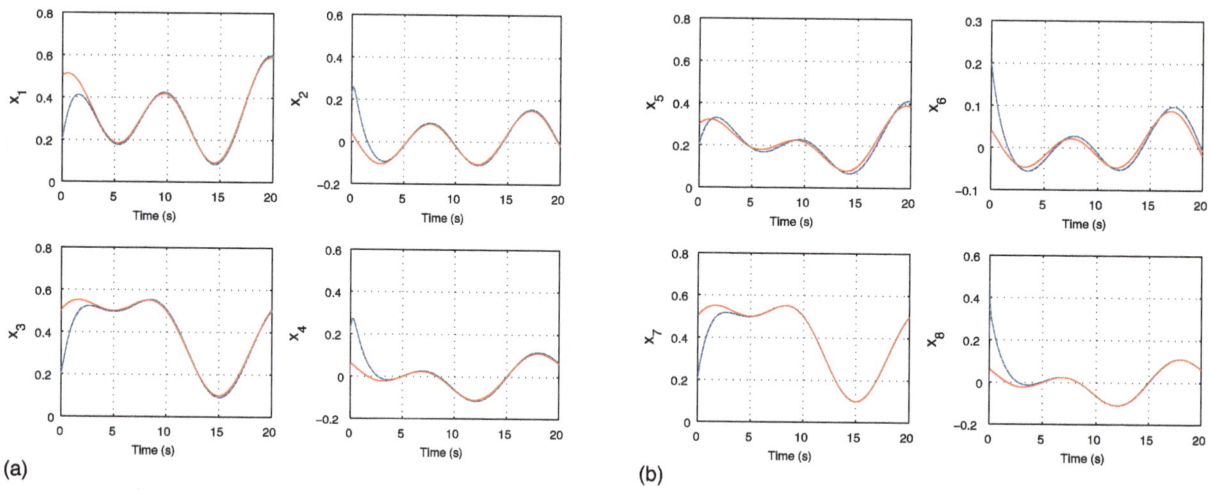

Figure 1.29 Tracking of setpoint 5 for the SCARA robot under nonlinear optimal control: (a) convergence of state variables x_1 to x_4 to their reference setpoints (red line: setpoint, blue line: real value, green line: estimated value); (b) convergence of state variables x_5 to x_8 to their reference setpoints

To elaborate on the tracking performance and on the robustness of the proposed flatness-based control method in successive loops for the SCARA robot, the following Tables are given: (i) Table 1.6 which provides information about the accuracy of tracking of the reference setpoints by the state variables of the SCARA robot's state-space model, (ii) Table 1.7 which provides information about the robustness of the control method to parametric changes in the model of the SCARA robot's dynamics (change $\Delta a\%$ in the mass m_2 of the second link), (iii) Table 1.8 which provides information about the precision in state variables' estimation that is achieved by the H-infinity Kalman filter, (iv) Table 1.9 which provides the approximate convergence times of the SCARA robot's state variables to the associated setpoints, (v) Table 1.10 which provides information about the % distribution of the total consumed power in the actuators of the SCARA robot.

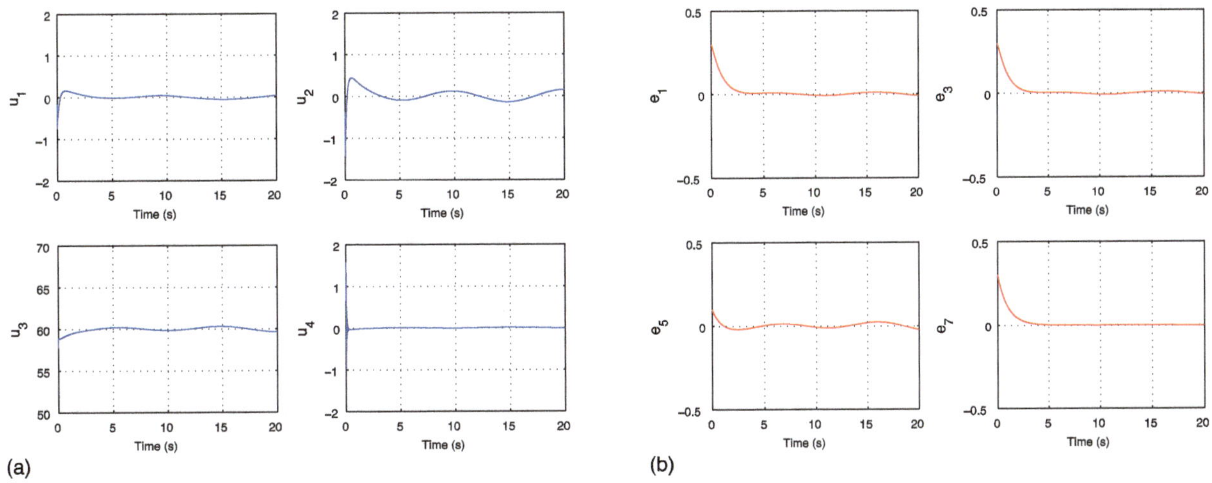

Figure 1.30 *Tracking of setpoint 5 for the SCARA robot under nonlinear optimal control: (a) control inputs u_1, u_2 applied to the robot; (b) tracking error variables e_1, e_3, e_5, and e_7 of the SCARA robot*

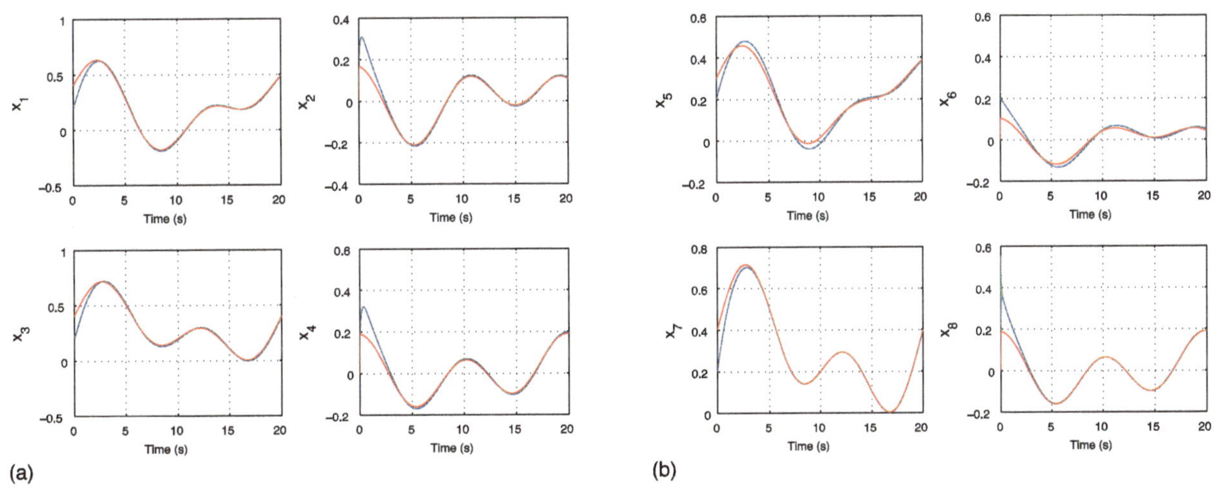

Figure 1.31 *Tracking of setpoint 6 for the SCARA robot under nonlinear optimal control: (a) convergence of state variables x_1 to x_4 to their reference setpoints (red line: setpoint, blue line: real value, green line: estimated value); (b) convergence of state variables x_5 to x_8 to their reference setpoints*

1.3 Nonlinear optimal control of dual-arm robotic manipulators

1.3.1 Outline

Synchronization between robotic manipulators and cooperative execution of tasks by them is an important research area in contemporary robotics [1068]. [1072,220]. In this context, the solution of the nonlinear control and synchronized motion planning problem for dual-arm robots is also a challenging task [23,850,59]. Control for dual-arm robots aims primarily at synchronized and precise positioning of the end-effectors of the two arms [1032,989,364,406]. It also aims at compensating forces exerted on the two robotic arms, which can be due to simultaneous grasping and lifting of objects or due to performing assembling and other tasks that may generate contact and interaction with stiff surfaces

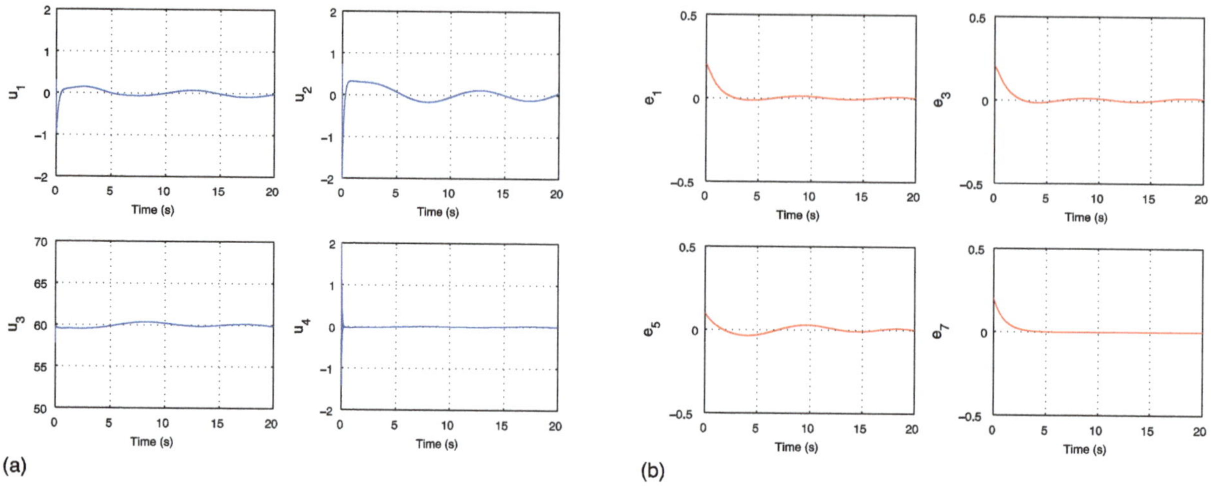

Figure 1.32 Tracking of setpoint 6 for the SCARA robot under nonlinear optimal control: (a) control inputs u_1, u_2 applied to the robot; (b) tracking error variables e_1, e_3, e_5, and e_7 of the SCARA robot

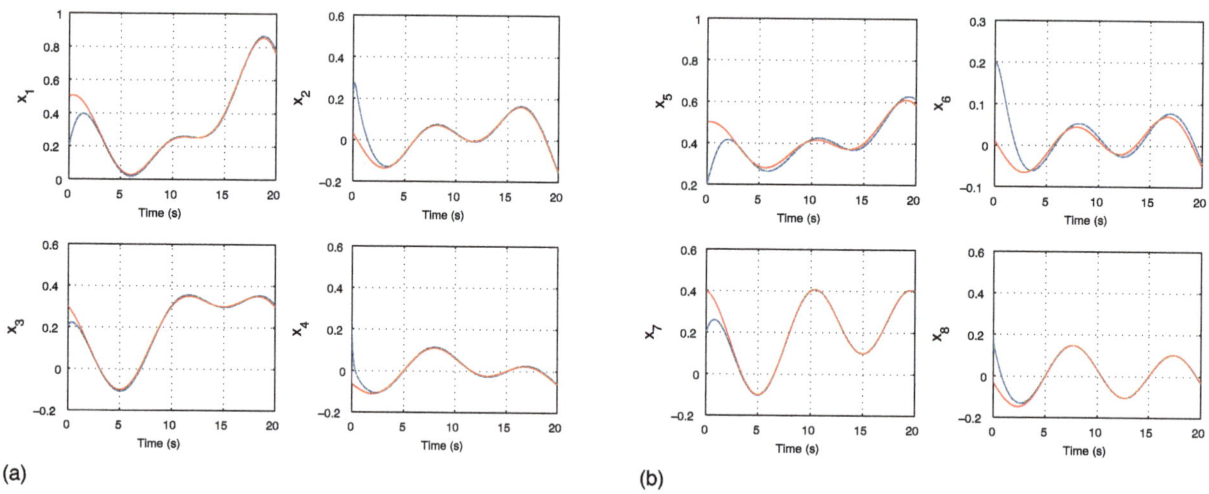

Figure 1.33 Tracking of setpoint 7 for the SCARA robot under nonlinear optimal control: (a) convergence of state variables x_1 to x_4 to their reference setpoints (red line: setpoint, blue line: real value, green line: estimated value); (b) convergence of state variables x_5 to x_8 to their reference setpoints

[111,412,51,641]. Dual-arm robots are widely used in industry, in several manufacturing tasks, in healthcare, in agriculture, and in warehouse tasks [2,196,947,798]. Dual-arm robots appear also in the execution of several space robotics tasks [929,916,788]. Furthermore, they can be used for assistance of the elderly or impaired people, for rehabilitation of the upper limbs, and in robotic surgery [496,948,29]. Among various control schemes that have been proposed for precise positioning, synchronization and position/force control in dual-arm robots, one can distinguish (i) NMPC methods, (ii) sliding-mode control methods, and (iii) backstepping-type control methods. Due to the complexity of the dynamic model and strong nonlinearity of the dynamic model of dual-arm robots the associated nonlinear optimal control problem has been little dealt-with so far [553,879,880,636,44].

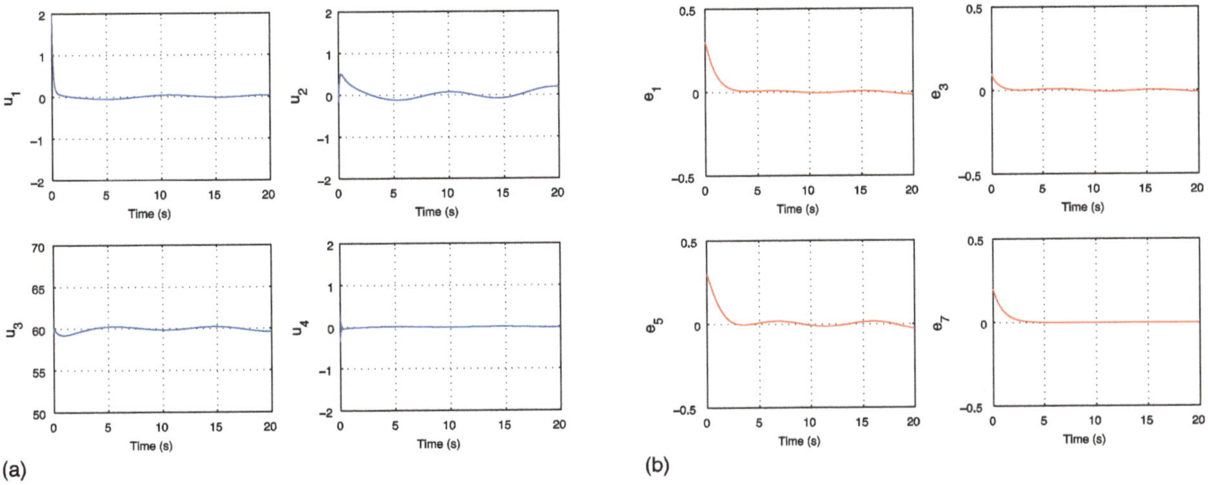

Figure 1.34 Tracking of setpoint 7 for the SCARA robot under nonlinear optimal control: (a) control inputs u_1, u_2 applied to the robot; (b) tracking error variables e_1, e_3, e_5, and e_7 of the SCARA robot

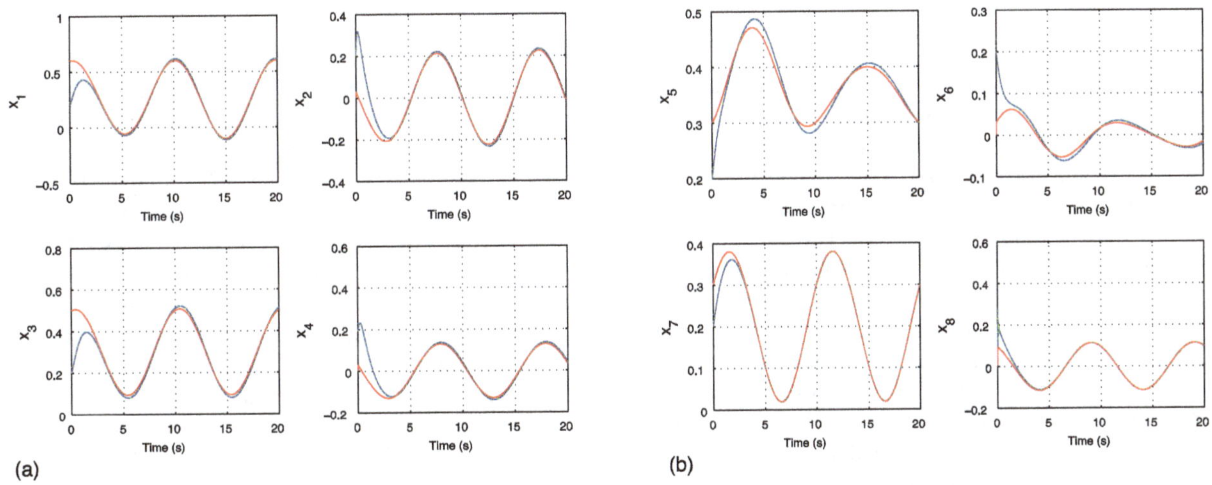

Figure 1.35 Tracking of setpoint 8 for the SCARA robot under nonlinear optimal control: (a) convergence of state variables x_1 to x_4 to their reference setpoints (red line: setpoint, blue line: real value, green line: estimated value); (b) convergence of state variables x_5 to x_8 to their reference setpoints

The present section proposes a novel nonlinear optimal (H-infinity) control method for the dynamic model of dual-arm robotic manipulators [721,711,725,741]. To implement this control scheme, the dynamic model of the dual-arm robotic manipulator undergoes first approximate linearization with the use of first-order Taylor series expansion and through the computation of the associated Jacobian matrices [705,60,706]. The linearization takes place at each sampling instant around a temporary operating point (x^*, u^*) which is defined by the present value of the robot's state vector x^* and the last sampled value of the control inputs vector u^*. The modeling error which is due to the truncation of higher-order terms from the Taylor series expansion is considered to be a perturbation that is asymptotically compensated by the robustness of the control scheme. This controller enables the solution of the optimal control problem for the dual-arm robot after taking into account model imprecision and external disturbances [721,711,725,741]. The

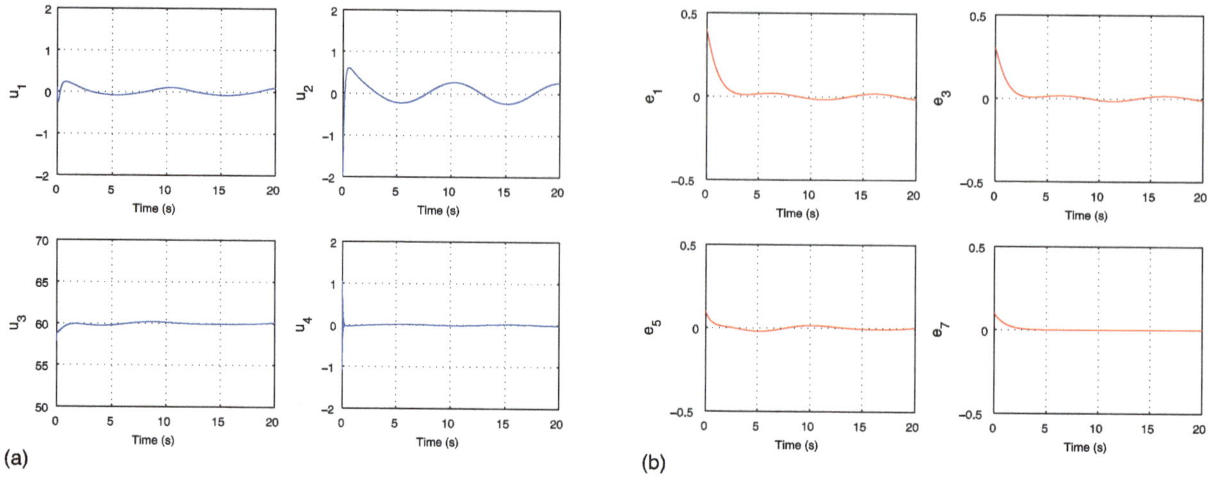

Figure 1.36 Tracking of setpoint 8 for the SCARA robot under nonlinear optimal control: (a) control inputs u_1, u_2 applied to the robot; (b) tracking error variables e_1, e_3, e_5, and e_7 of the SCARA robot

proposed H-infinity controller implements a min–max differential game taking place between (i) the control inputs of the robot which try to minimize an objective function that contains a quadratic term of the state vector's tracking error, (ii) the model uncertainty and disturbance inputs which try to maximize this cost function.

For the linearized model of the robotic system a stabilizing H-infinity controller is designed. To compute the feedback gains of this controller an algebraic Riccati equation is repetitively solved at each time-step of the control algorithm [715,732]. The global stability properties of the control scheme are proven through Lyapunov stability analysis. First, it is demonstrated that the control scheme satisfies an H-infinity tracking performance criterion which signifies elevated robustness against parametric changes in the system's model and exogenous perturbations [719,876]. Moreover, under moderate conditions global asymptotic stability is also proven [715,732]. Additionally, to implement state estimation-based control the H-infinity Kalman filter is considered as a robust state estimator. It is noteworthy that the control inputs that the nonlinear optimal control method computes are applied directly on the initial nonlinear state-space description of the dual-arm robot, while no state-space model transformations or changes of state variables are used. The nonlinear optimal control method achieves fast and accurate tracking of reference setpoints for all state variables of this robotic system under minimal variations of the control inputs.

1.3.2 Dynamic model of the dual-arm robotic manipulator

1.3.2.1 State-space model of the dual-arm robot

The dual-arm robotic manipulator consists of arm A and arm B, as shown in Figure 1.57. Arm A consists of two links, of masses $m_{1,A}$, $m_{2,A}$, and of lengths $l_{1,A}$, $l_{2,A}$, respectively. The turn angles of the joints of these links are denoted as $\theta_{1,A}$ and $\theta_{2,A}$. The position of the first joint of this arm is given by the coordinates (x_0^A, h). The position of the end-effector of this arm is given by the coordinates (x_E^A, y_E^A). Arm B consists also of two links, of masses $m_{1,A}$, $m_{2,A}$, and of lengths $l_{1,A}$, $l_{2,A}$, respectively. The turn angles of the joints of these links are denoted as $\theta_{1,B}$ and $\theta_{2,B}$. The position of the first joint of this arm is given by the coordinates (x_0^B, h). The position of the end-effector of this arm is given by the coordinates (x_E^B, y_E^B). The dual-arm robot is used to grasp and lift-up with its end-effectors an object of weight $B = M{\cdot}g$. The transversal axis of the object forms an angle ϕ with the horizontal axis of the inertial reference frame. The contact force due to grasping which is exerted to each end-effector is denoted by F.

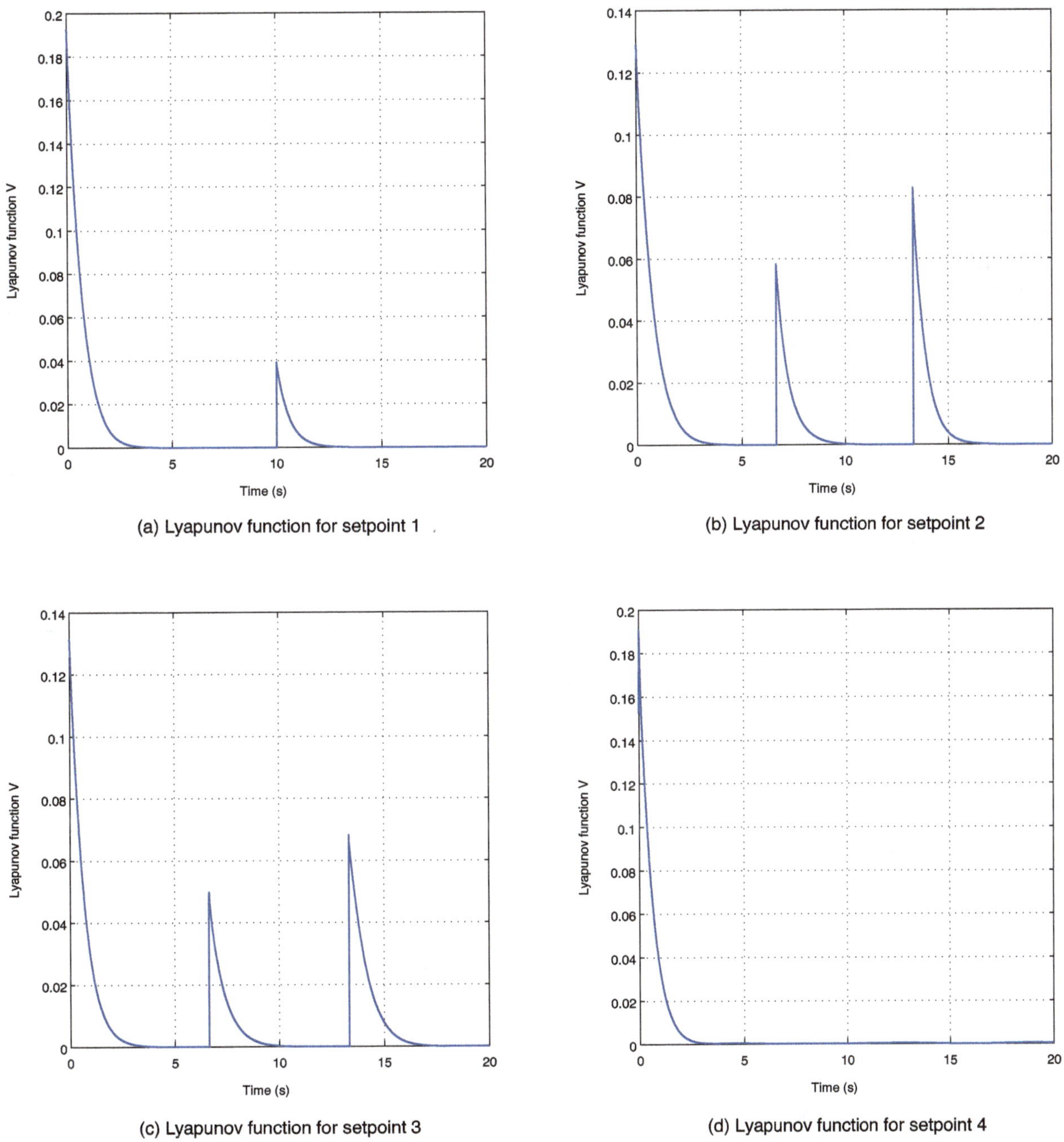

Figure 1.37 Lyapunov functions for nonlinear optimal control of the 4-DOF SCARA robotic manipulator when tracking setpoints 1–4

The dynamic model of arm A is given by [721]

$$M_A(\theta_A)\ddot{\theta}_A + C_A(\theta_A, \dot{\theta}_A) + G_A(\theta_A) = \tau_A - J_A^T B - J_A^T F \tag{1.224}$$

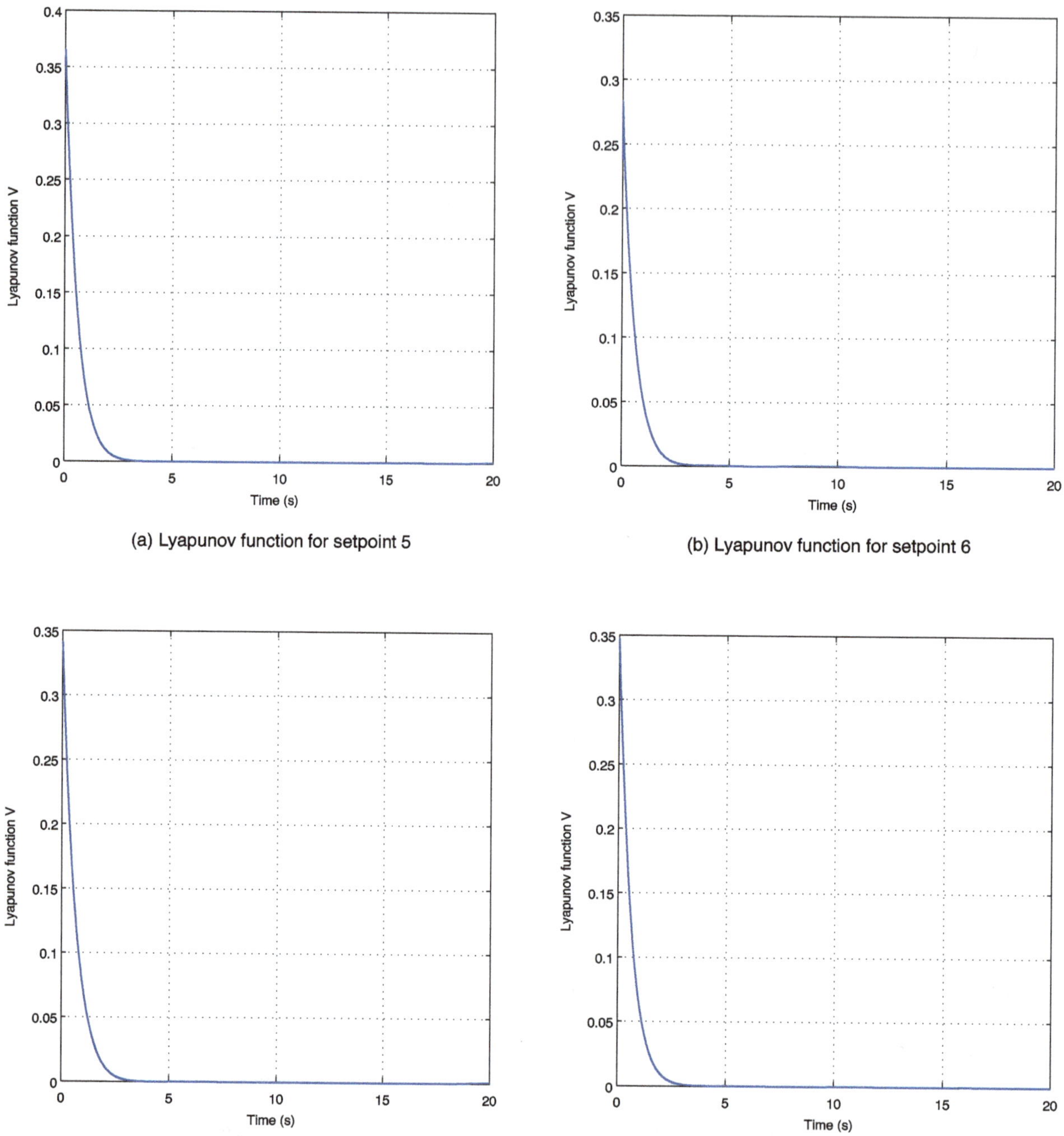

(a) Lyapunov function for setpoint 5

(b) Lyapunov function for setpoint 6

(c) Lyapunov function for setpoint 7

(d) Lyapunov function for setpoint 8

Figure 1.38 Lyapunov functions for nonlinear optimal control of the 4-DOF SCARA robotic manipulator when tracking setpoints 5–8

where $M_A(\theta_A)$ is the inertia matrix of arm A which is given by

$$M_A(\theta_A) = \begin{pmatrix} \bar{m}_{11,A} & \bar{m}_{12,A} \\ \bar{m}_{21,A} & \bar{m}_{22,A} \end{pmatrix} \tag{1.225}$$

Table 1.1 Nonlinear optimal control

Tracking RMSE $\times 10^{-3}$ for the SCARA robot in the disturbance-free case								
	$RMSE_{x_1}$	$RMSE_{x_2}$	$RMSE_{x_3}$	$RMSE_{x_4}$	$RMSE_{x_5}$	$RMSE_{x_6}$	$RMSE_{x_7}$	$RMSE_{x_8}$
$test_1$	0.0009	0.0009	0.0008	0.0008	0.0001	0.0001	0.0017	0.0017
$test_2$	0.0852	0.0875	0.0749	0.0788	0.0201	0.0494	0.0959	0.0955
$test_3$	0.0289	0.0298	0.0865	0.0885	0.0547	0.1660	0.0228	0.0227
$test_4$	0.8982	0.3706	0.0843	0.1459	0.9390	0.2377	0.0107	0.0174
$test_5$	0.2994	0.2798	0.1588	0.2624	0.6638	0.7585	0.0027	0.0069
$test_6$	0.3067	0.1814	0.5778	0.1789	0.3709	0.2655	0.0022	0.0148
$test_7$	0.5771	0.2394	0.3309	0.1792	0.8170	0.6909	0.0050	0.0103
$test_8$	0.4273	0.4844	0.2887	0.4649	0.2049	0.2183	0.0092	0.0056

Table 1.2 Nonlinear optimal control

Tracking RMSE $\times 10^{-3}$ for the SCARA robot in the case of disturbances								
$\Delta a\%$	$RMSE_{x_1}$	$RMSE_{x_2}$	$RMSE_{x_3}$	$RMSE_{x_4}$	$RMSE_{x_5}$	$RMSE_{x_6}$	$RMSE_{x_7}$	$RMSE_{x_8}$
0%	0.0289	0.0298	0.0866	0.0885	0.0617	0.1660	0.0228	0.0227
10%	0.0288	0.0301	0.0870	0.0895	0.0646	0.1657	0.0228	0.0227
20%	0.0288	0.0301	0.0874	0.0901	0.0646	0.1657	0.0228	0.0226
30%	0.0287	0.0302	0.0877	0.0906	0.0646	0.1657	0.0227	0.0226
40%	0.0286	0.0302	0.0880	0.0912	0.0646	0.1657	0.0227	0.0226
50%	0.0295	0.0302	0.0883	0.0917	0.0646	0.1657	0.0227	0.0225
60%	0.0284	0.0303	0.0886	0.0921	0.0646	0.1657	0.0227	0.0225

Table 1.3 Nonlinear optimal control

RMSE $\cdot 10^{-5}$ for the estimation performed by the H-infinity KF								
	$RMSE_{x_1}$	$RMSE_{x_2}$	$RMSE_{x_3}$	$RMSE_{x_4}$	$RMSE_{x_5}$	$RMSE_{x_6}$	$RMSE_{x_7}$	$RMSE_{x_8}$
$test_1$	0.0001	0.0007	0.0001	0.0001	0.0001	0.0001	0.0001	0.0007
$test_2$	0.0004	0.0736	0.0001	0.0045	0.0002	0.0369	0.0003	0.0591
$test_3$	0.0002	0.0371	0.0001	0.0159	0.0007	0.1131	0.0001	0.0039
$test_4$	0.0110	2.2850	0.0010	0.0820	0.0050	0.9320	0.0050	0.9030
$test_5$	0.0110	2.2910	0.0010	0.2840	0.0060	0.2450	0.0030	0.6400
$test_6$	0.0070	1.4090	0.0010	0.1470	0.0010	0.2960	0.0040	0.7460
$test_7$	0.0170	3.4690	0.0020	0.4060	0.0070	0.1320	0.0070	0.1469
$test_8$	0.0190	3.8890	0.0010	0.1870	0.0010	0.1880	0.0030	0.6540

with the elements of the inertia matrix to be given by: $\bar{m}_{11,A} = (m_{1,A} + m_{2,A})l_{1,A}^2 + m_{2,A}l_{2,A}^2 + 2m_{2,A}l_{1,A}l_{2,A}cos(\theta_{2,A})$, $\bar{m}_{12,A} = \bar{m}_{21,A} = m_{2,A}l_{2,A}^2 + m_{2,A}l_{1,A}l_{2,A}cos(\theta_{2,A})$, and $\bar{m}_{22,A} = m_{2,A}l_{2,A}(l_{1,A} + l_{2,A})cos(\theta)_{2,A}$.

The Coriolis and centrifugal forces matrix $C_A(\theta_A, \dot{\theta}_A)$ of arm A is given by

$$C_A(\theta_A, \dot{\theta}_A) = \begin{pmatrix} \bar{C}_{1,A} \\ \bar{C}_{2,A} \end{pmatrix} \tag{1.226}$$

Table 1.4 Nonlinear optimal control

	Convergence time (sec) for the SCARA robot's state variables							
	$T_s\,x_1$	$T_s\,x_2$	$T_s\,x_3$	$T_s\,x_4$	$T_s\,x_5$	$T_s\,x_6$	$T_s\,x_7$	$T_s\,x_8$
$test_1$	3.5	3.0	4.5	3.5	3.5	3.6	3.5	3.0
$test_2$	2.5	3.0	3.5	3.5	2.5	3.5	0.5	0.5
$test_3$	3.0	3.0	3.5	3.5	3.0	2.0	3.5	3.5
$test_4$	0.5	2.0	3.0	3.5	0.5	0.5	0.5	2.5
$test_5$	2.5	3.0	3.0	2.5	1.0	2.0	3.5	3.5
$test_6$	2.0	2.5	2.5	3.0	0.5	0.5	2.0	2.5
$test_7$	2.5	2.5	1.0	1.5	2.5	3.5	2.5	2.5
$test_8$	2.5	2.5	3.0	2.5	1.0	1.5	1.0	2.5

Table 1.5 Nonlinear optimal control

	Distribution of power % in actuators			
	$P\,u_1$	$P\,u_2$	$P\,u_3$	$P\,u_4$
$test_1$	18.41	29.47	43.90	8.22
$test_2$	12.86	32.74	48.14	6.26
$test_3$	6.67	14.06	78.39	0.88
$test_4$	2.55	45.68	50.11	1.76
$test_5$	3.79	19.92	74.56	1.73
$test_6$	8.01	32.65	53.84	3.50
$test_7$	11.78	15.43	72.61	0.18
$test_8$	6.17	52.03	40.77	1.03

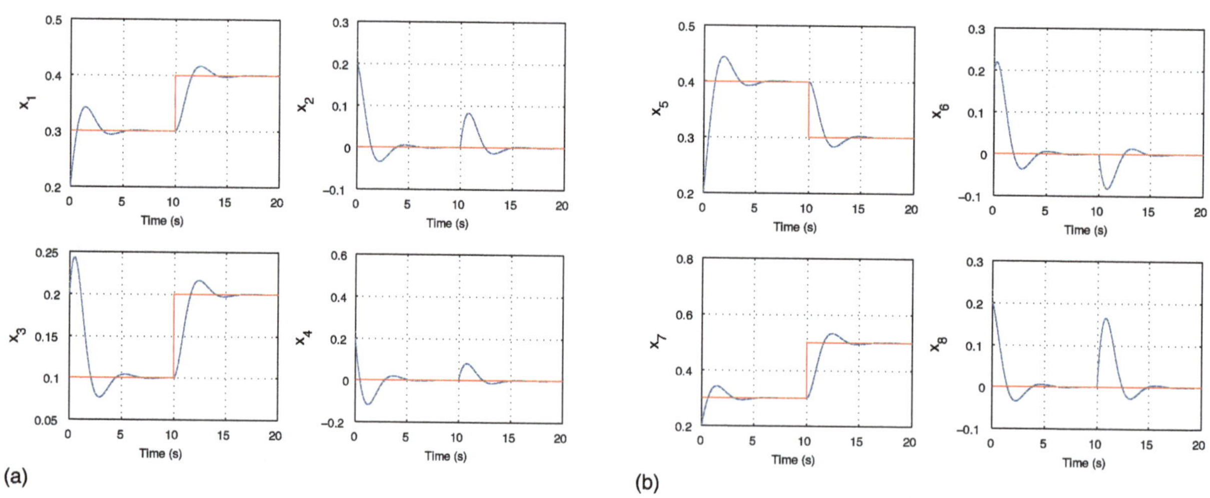

Figure 1.39 Tracking of setpoint 1 for the SCARA robot under flatness-based control in successive loops: (a) convergence of state variables x_1 to x_4 to their reference setpoints (red line: setpoint, blue line: real value, green line: estimated value); (b) convergence of state variables x_5 to x_8 to their reference setpoints

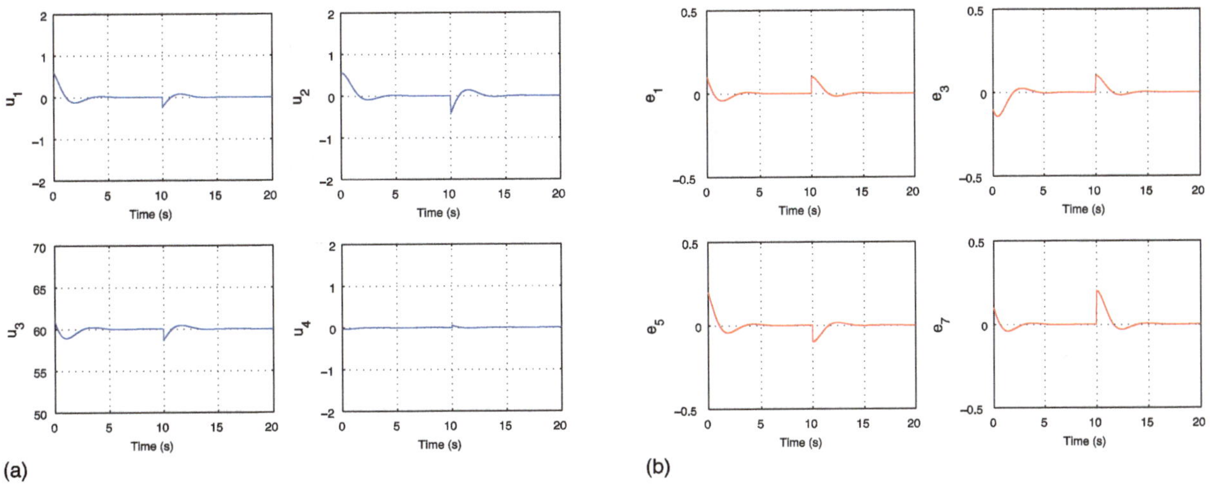

Figure 1.40 Tracking of setpoint 1 for the SCARA robot under flatness-based control in successive loops: (a) control inputs u_1, u_2 applied to the robot; (b) tracking error variables e_1, e_3, e_5, and e_7 of the SCARA robot

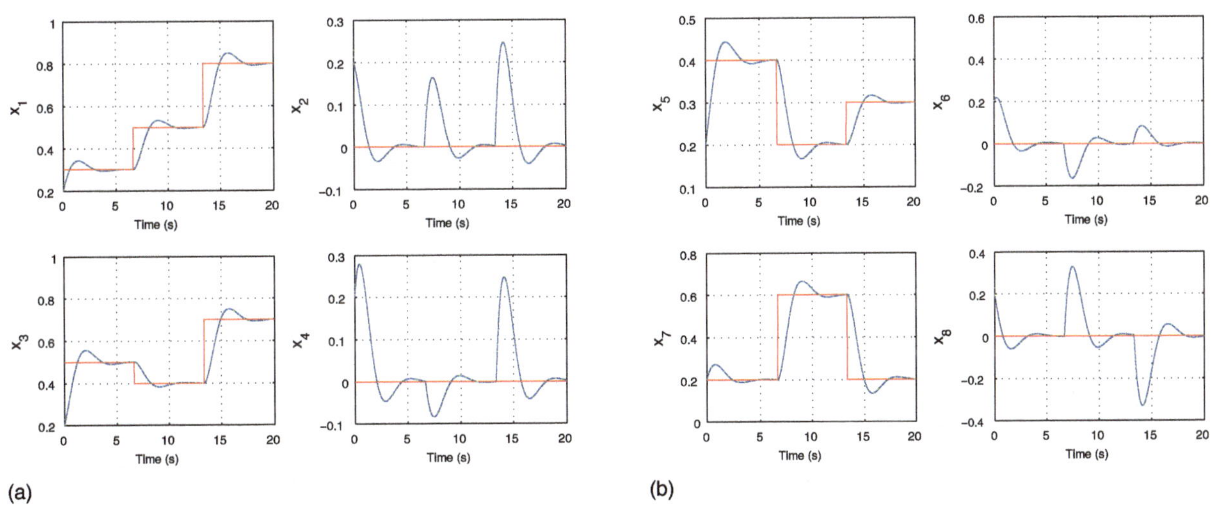

Figure 1.41 Tracking of setpoint 2 for the SCARA robot under flatness-based control in successive loops: (a) convergence of state variables x_1 to x_4 to their reference setpoints (red line: setpoint, blue line: real value, green line: estimated value); (b) convergence of state variables x_5 to x_8 to their reference setpoints

where $\bar{C}_{1,A} = -m_{2,A}l_{1,A}^2 sin(\theta_{2,A})\dot{\theta}_{2,A}^2 - 2m_{2,A}l_{2,A}^2 sin(\theta_{2,A})\dot{\theta}_{1,A}\dot{\theta}_{2,A}$ and $\bar{C}_{2,A} - m_{1,A}l_{1,A}^2 sin(\theta_{2,A})\dot{\theta}_{1,A}^2$.

The gravitational forces vector $G_A(\theta)$ of arm A is given by

$$G_A(\theta_A) = \begin{pmatrix} \bar{G}_{1,A} \\ \bar{G}_{2,A} \end{pmatrix} \tag{1.227}$$

where $\bar{G}_{1,A} = (m_{1,A} + m_{2,A}gl_{1,A})cos(\theta_{1,A}) + m_{2,A}gl_{2,A}cos(\theta_{1,A} + \theta_{2,A})$ and $\bar{G}_{2,A} = m_{2,A}gl_{2,A}cos(\theta_{1,A} + \theta_{2,A})$, with g to be denoting the acceleration of gravity.

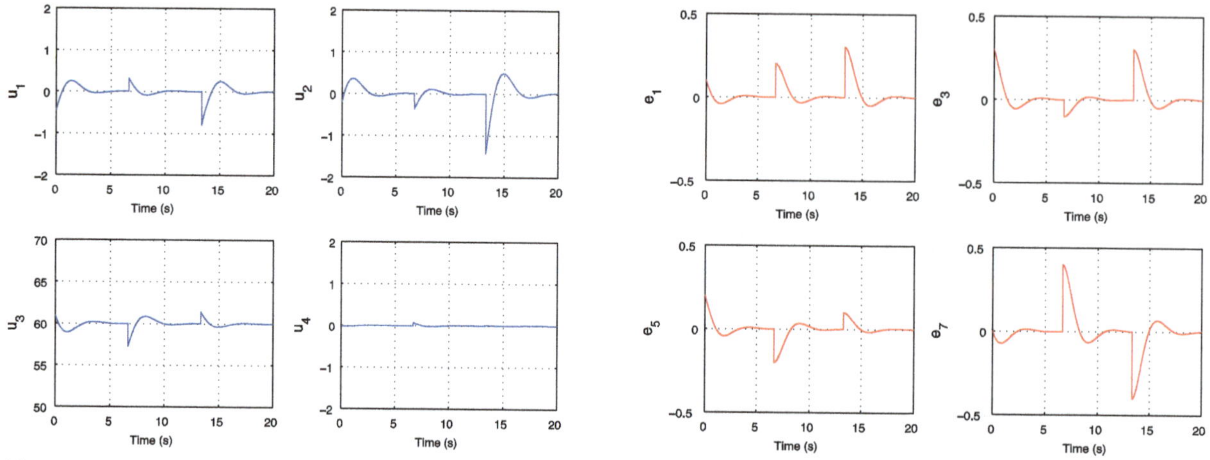

(a) (b)

Figure 1.42 Tracking of setpoint 2 for the SCARA robot under flatness-based control in successive loops: (a) control inputs u_1, u_2 applied to the robot; (b) tracking error variables e_1, e_3, e_5, and e_7 of the SCARA robot

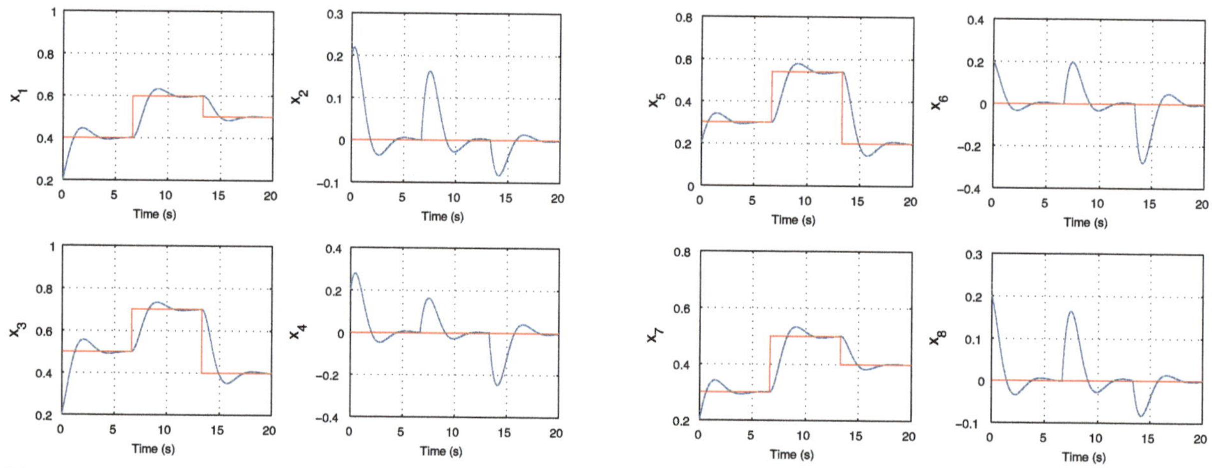

(a) (b)

Figure 1.43 Tracking of setpoint 3 for the SCARA robot under flatness-based control in successive loops: (a) convergence of state variables x_1 to x_4 to their reference setpoints (red line: setpoint, blue line: real value, green line: estimated value); (b) convergence of state variables x_5 to x_8 to their reference setpoints

The torques vector which is provided by the actuators of arm A is given by

$$\tau_A = \begin{pmatrix} \tau_{1,A} \\ \tau_{2,A} \end{pmatrix} \tag{1.228}$$

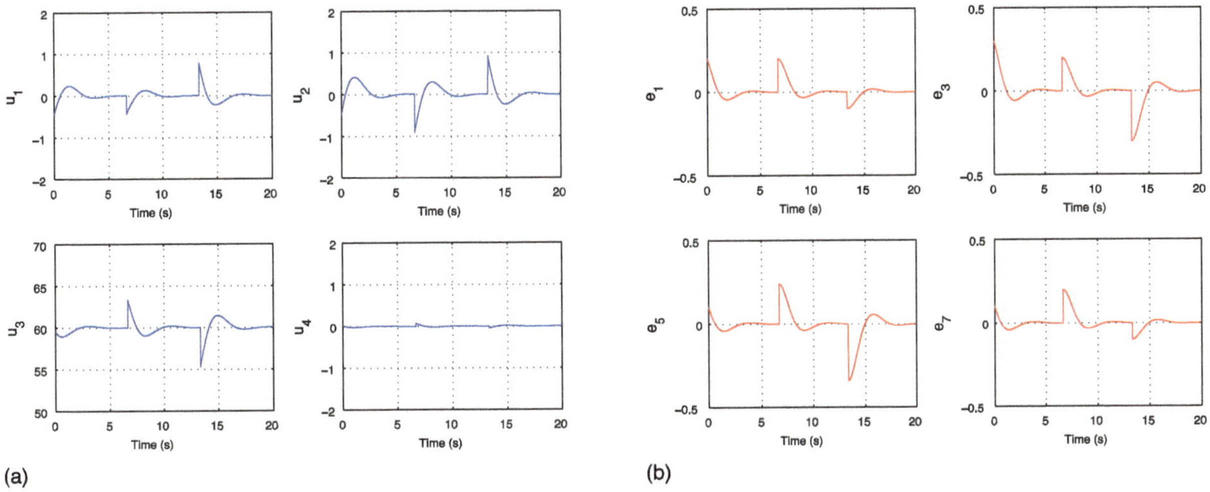

Figure 1.44 Tracking of setpoint 3 for the SCARA robot under flatness-based control in successive loops: (a) control inputs u_1, u_2 applied to the robot; (b) tracking error variables e_1, e_3, e_5, and e_7 of the SCARA robot

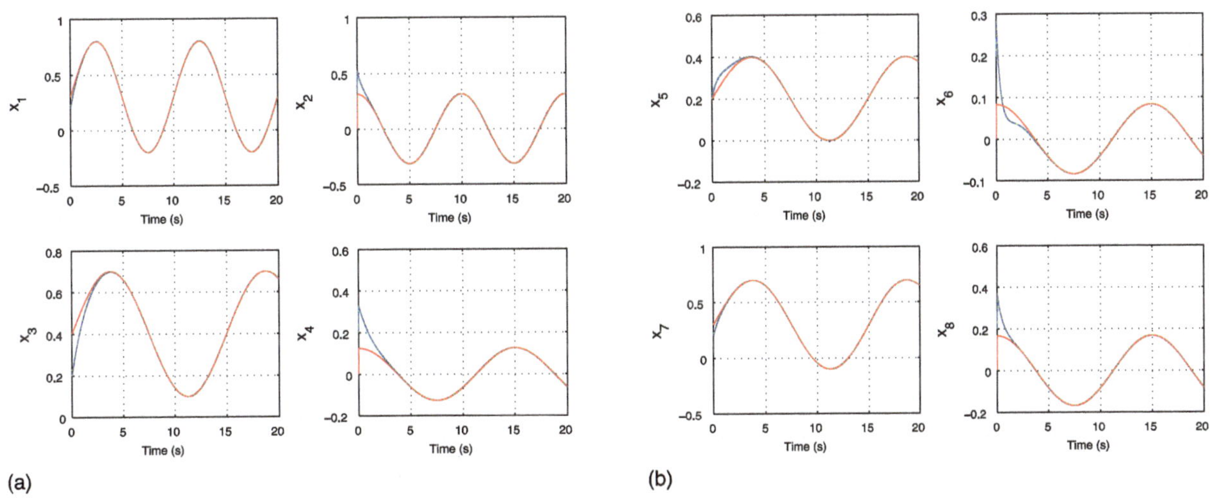

Figure 1.45 Tracking of setpoint 4 for the SCARA robot under flatness-based control in successive loops: (a) convergence of state variables x_1 to x_4 to their reference setpoints (red line: setpoint, blue line: real value, green line: estimated value), (b) convergence of state variables x_5 to x_8 to their reference setpoints

The transpose of the Jacobian matrix J_A^T which transfers torques from the end-effector of arm A to the joints of arm A, due to the load's weight $B = M{\cdot}g$ and due to the grasping forces F, is given by

$$J_A^T = \begin{pmatrix} J_{A,11}^T & J_{A,12}^T \\ J_{A,21}^T & J_{A,22}^T \end{pmatrix} = \begin{pmatrix} -l_{1,A}sin(\theta_{1,A}) - l_{2,A}sin(\theta_{1,A}+\theta_{2,A}) & l_{1,A}cos(\theta_{1,A}) + l_{2,A}cos(\theta_{1,A}+\theta_{2,A}) \\ -l_{2,A}sin(\theta_{1,A}+\theta_{2,A}) & l_{2,A}cos(\theta_{1,A}+\theta_{2,A}) \end{pmatrix} \quad (1.229)$$

The dynamic model of arm B is given by [721]

$$M_B(\theta_B)\ddot{\theta} + C_B(\theta_B, \dot{\theta}_B) + G_B(\theta_B) = \tau_B - J_B^T B + J_B^T F \quad (1.230)$$

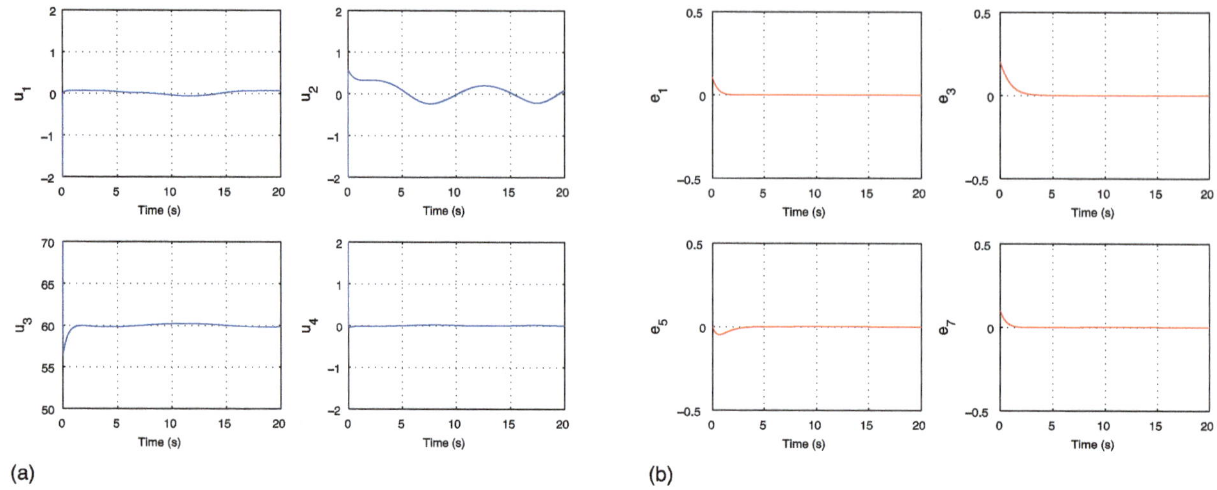

Figure 1.46 Tracking of setpoint 4 for the SCARA robot under flatness-based control in successive loops: (a) control inputs u_1, u_2 applied to the robot; (b) tracking error variables e_1, e_3, e_5, and e_7 of the SCARA robot

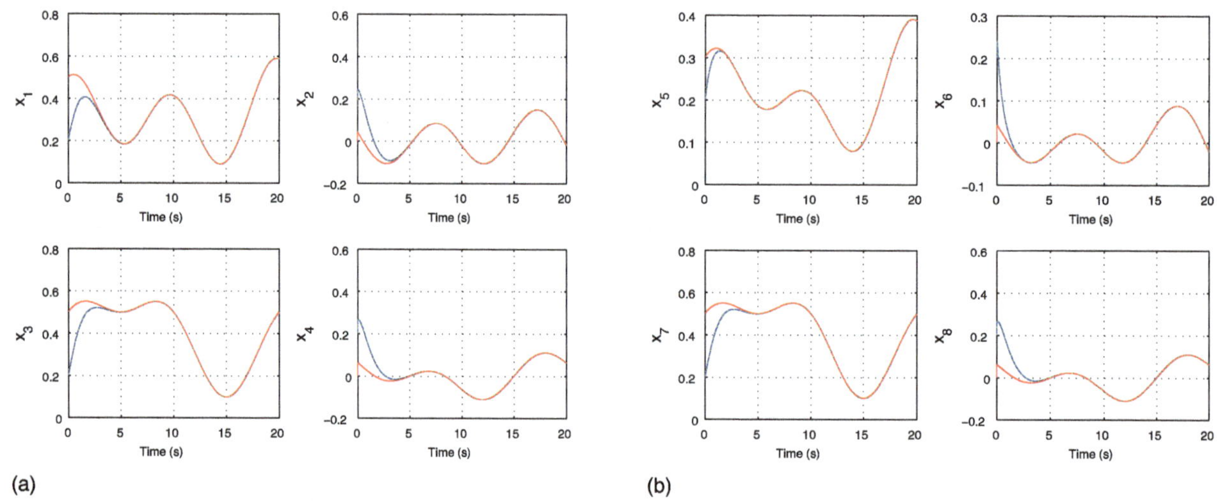

Figure 1.47 Tracking of setpoint 5 for the SCARA robot under flatness-based control in successive loops: (a) convergence of state variables x_1 to x_4 to their reference setpoints (red line: setpoint, blue line: real value, green line: estimated value); (b) convergence of state variables x_5 to x_8 to their reference setpoints

where $M_B(\theta_B)$ is the inertia matrix of arm B which is given by

$$M_B(\theta_B) = \begin{pmatrix} \bar{m}_{11,B} & \bar{m}_{12,B} \\ \bar{m}_{21,B} & \bar{m}_{22,B} \end{pmatrix} \tag{1.231}$$

with the elements of the inertia matrix to be given by: $\bar{m}_{11,B} = (m_{1,B} + m_{2,B})l_{1,B}^2 + m_{2,B}l_{2,B}^2 + 2m_{2,B}l_{1,B}l_{2,B}cos(\theta_{2,B})$, $\bar{m}_{12,B} = \bar{m}_{21,B} = m_{2,B}l_{2,B}^2 + m_{2,B}l_{1,B}l_{2,A}cos(\theta_{2,B})$, and $\bar{m}_{22,B} = m_{2,B}l_{2,B}(l_{1,B} + l_{2,B})cos(\theta)_{2,B}$.

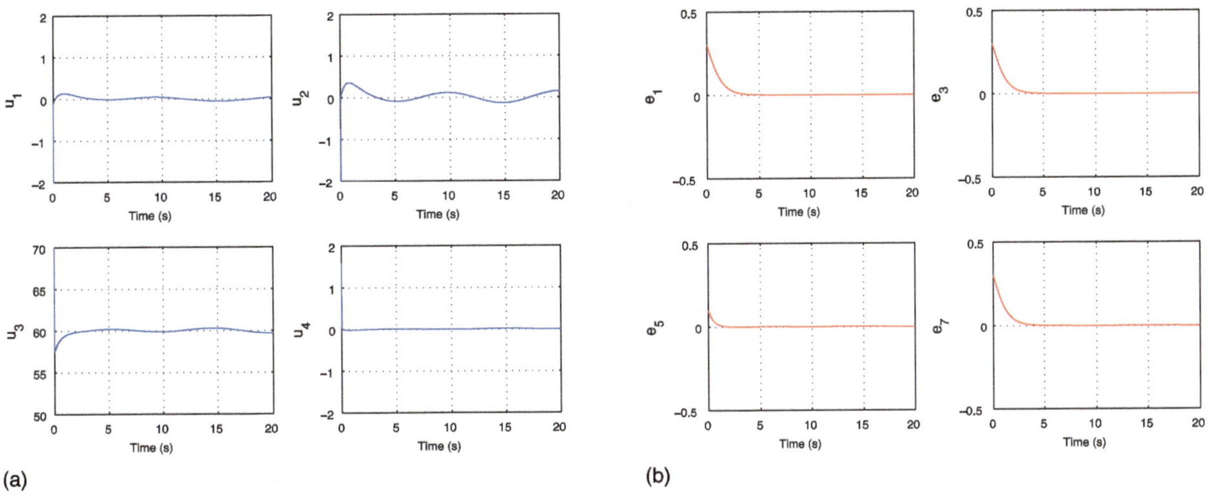

Figure 1.48 *Tracking of setpoint 5 for the SCARA robot under flatness-based control in successive loops: (a) control inputs u_1, u_2 applied to the robot; (b) tracking error variables e_1, e_3, e_5, and e_7 of the SCARA robot*

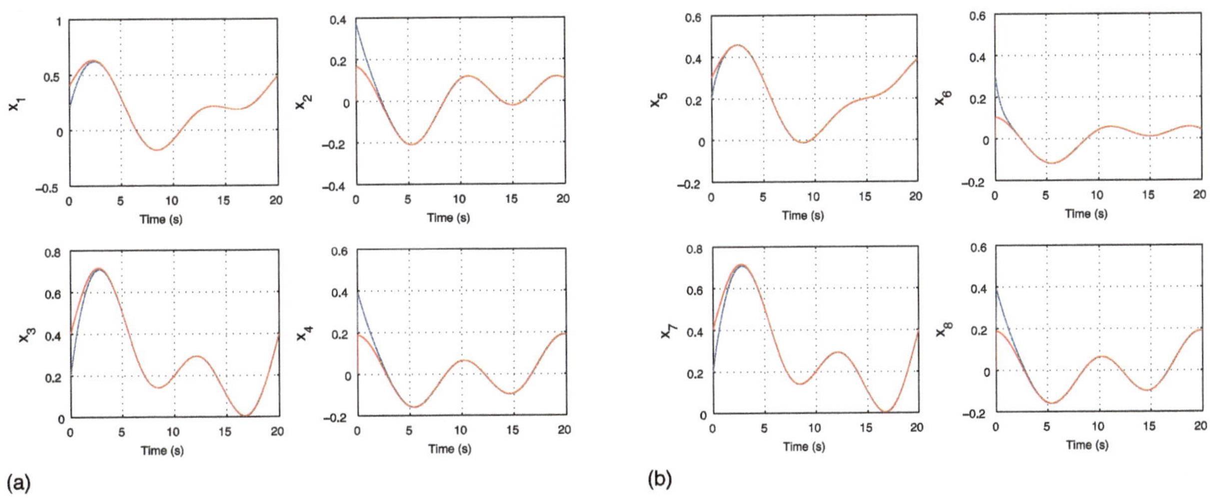

Figure 1.49 *Tracking of setpoint 6 for the SCARA robot under flatness-based control in successive loops: (a) convergence of state variables x_1 to x_4 to their reference setpoints (red line: setpoint, blue line: real value, green line: estimated value); (b) convergence of state variables x_5 to x_8 to their reference setpoints*

The Coriolis and centrifugal forces matrix $C_B(\theta_B, \dot{\theta}_B)$ of arm B is given by

$$C_B(\theta_B, \dot{\theta}_B) = \begin{pmatrix} \bar{C}_{1,B} \\ \bar{C}_{2,B} \end{pmatrix} \tag{1.232}$$

where $\bar{C}_{1,B} = -m_{2,B}l_{1,B}^2 sin(\theta_{2,B})\dot{\theta}_{2,B}^2 - 2m_{2,B}l_{2,B}^2 sin(\theta_{2,B})\dot{\theta}_{1,B}\dot{\theta}_{2,B}$ and $\bar{C}_{2,B} - m_{1,B}l_{1,B}^2 sin(\theta_{2,B})\dot{\theta}_{1,B}^2$.

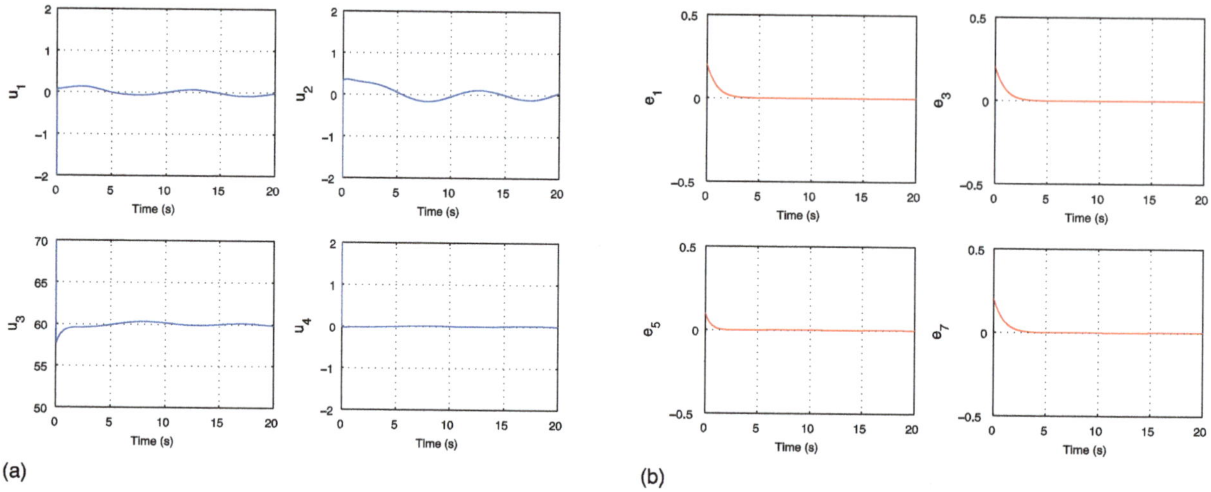

Figure 1.50 *Tracking of setpoint 6 for the SCARA robot under flatness-based control in successive loops: (a) control inputs u_1, u_2 applied to the robot; (b) tracking error variables e_1, e_3, e_5, and e_7 of the SCARA robot*

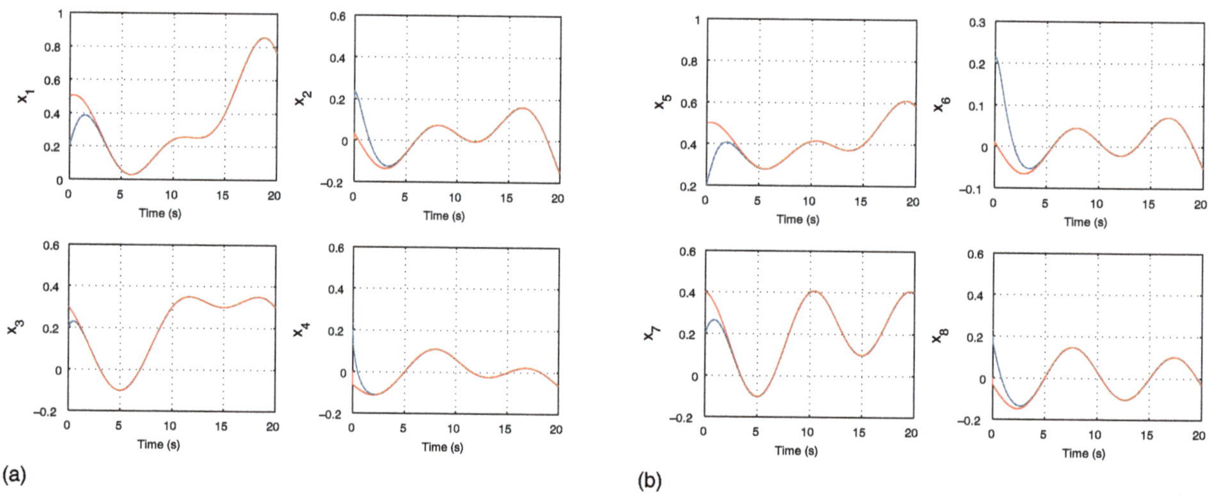

Figure 1.51 *Tracking of setpoint 7 for the SCARA robot under flatness-based control in successive loops: (a) convergence of state variables x_1 to x_4 to their reference setpoints (red line: setpoint, blue line: real value, green line: estimated value); (b) convergence of state variables x_5 to x_8 to their reference setpoints*

The gravitational forces vector $G_B(\theta)$ of arm B is given by

$$G_B(\theta_B) = \begin{pmatrix} \bar{G}_{1,B} \\ \bar{G}_{2,B} \end{pmatrix} \tag{1.233}$$

where $\bar{G}_{1,B} = (m_{1,B} + m_{2,B}gl_{1,B})cos(\theta_{1,B}) + m_{2,B}gl_{2,B}cos(\theta_{1,B} + \theta_{2,B})$ and $\bar{G}_{2,B} = m_{2,B}gl_{2,B}cos(\theta_{1,B} + \theta_{2,B})$, with g to be denoting the acceleration of gravity.

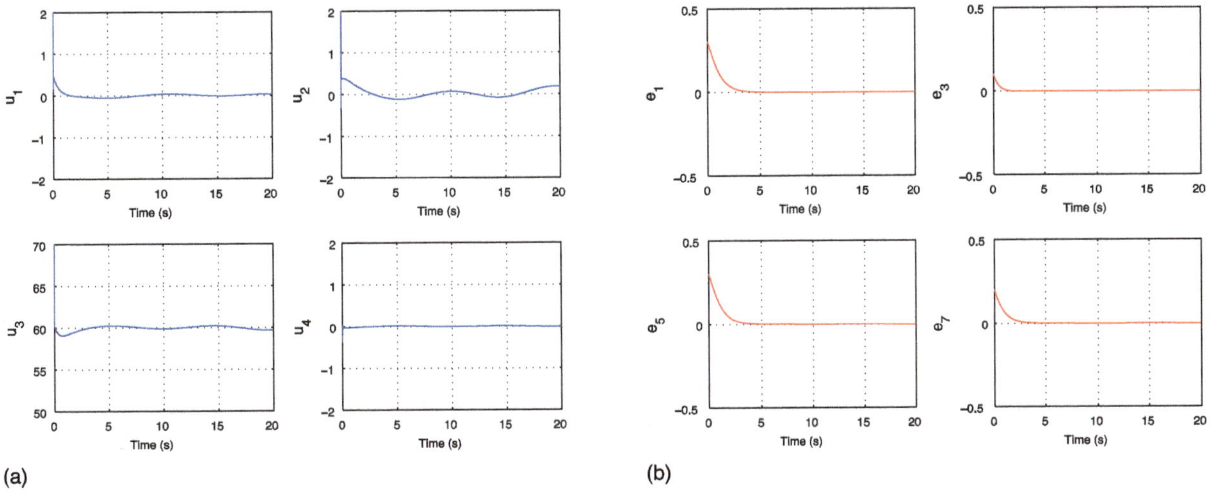

Figure 1.52 *Tracking of setpoint 7 for the SCARA robot under flatness-based control in successive loops: (a) control inputs u_1, u_2 applied to the robot; (b) tracking error variables e_1, e_3, e_5, and e_7 of the SCARA robot*

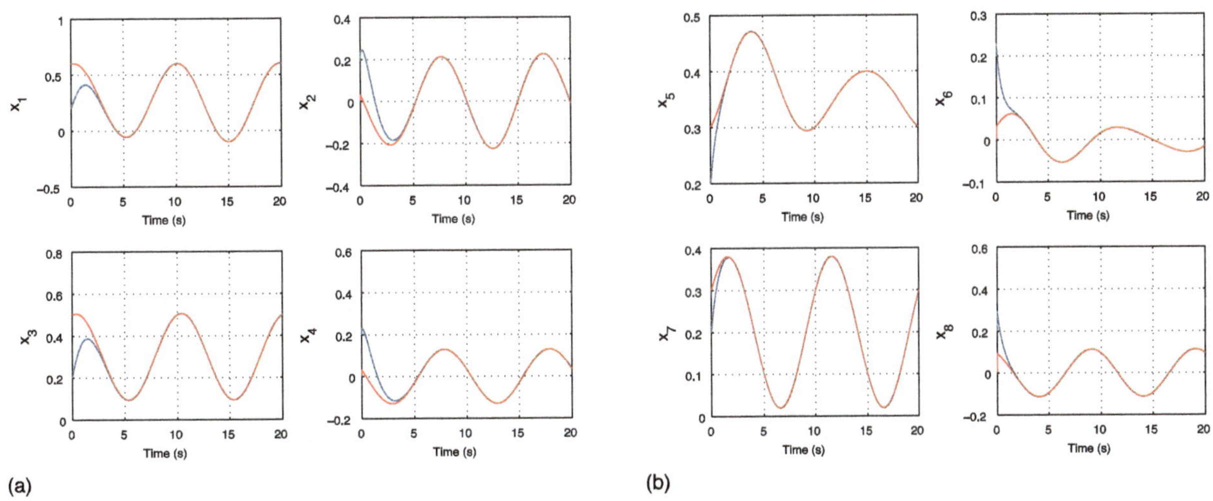

Figure 1.53 *Tracking of setpoint 8 for the SCARA robot under flatness-based control in successive loops: (a) convergence of state variables x_1 to x_4 to their reference setpoints (red line: setpoint, blue line: real value, green line: estimated value); (b) convergence of state variables x_5 to x_8 to their reference setpoints*

The torques vector which is provided by the actuators of arm B is given by

$$\tau_B = \begin{pmatrix} \tau_{1,B} \\ \tau_{2,B} \end{pmatrix} \tag{1.234}$$

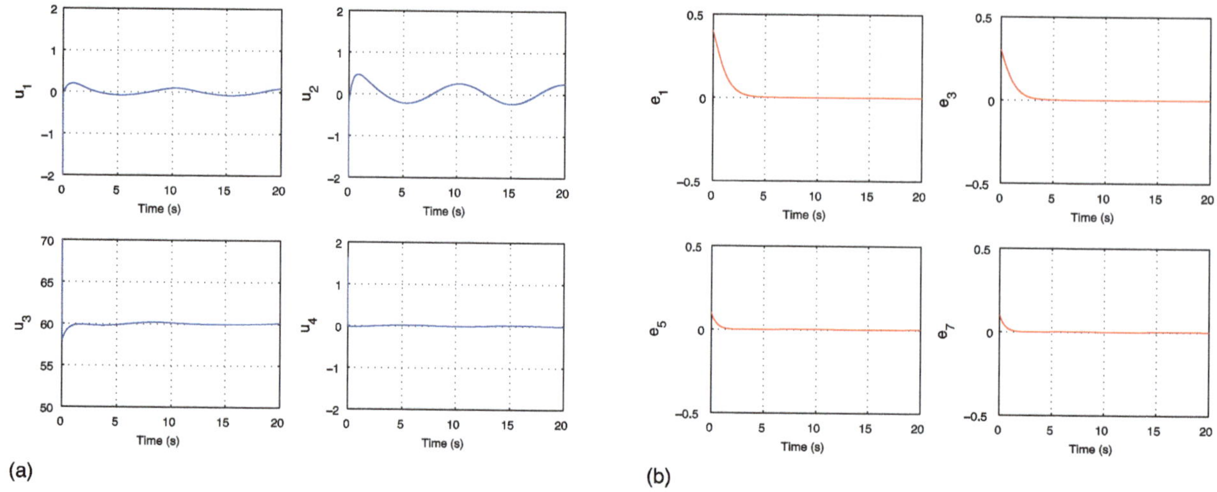

Figure 1.54 Tracking of setpoint 8 for the SCARA robot under flatness-based control in successive loops: (a) control inputs u_1, u_2 applied to the robot; (b) tracking error variables e_1, e_3, e_5, and e_7 of the SCARA robot

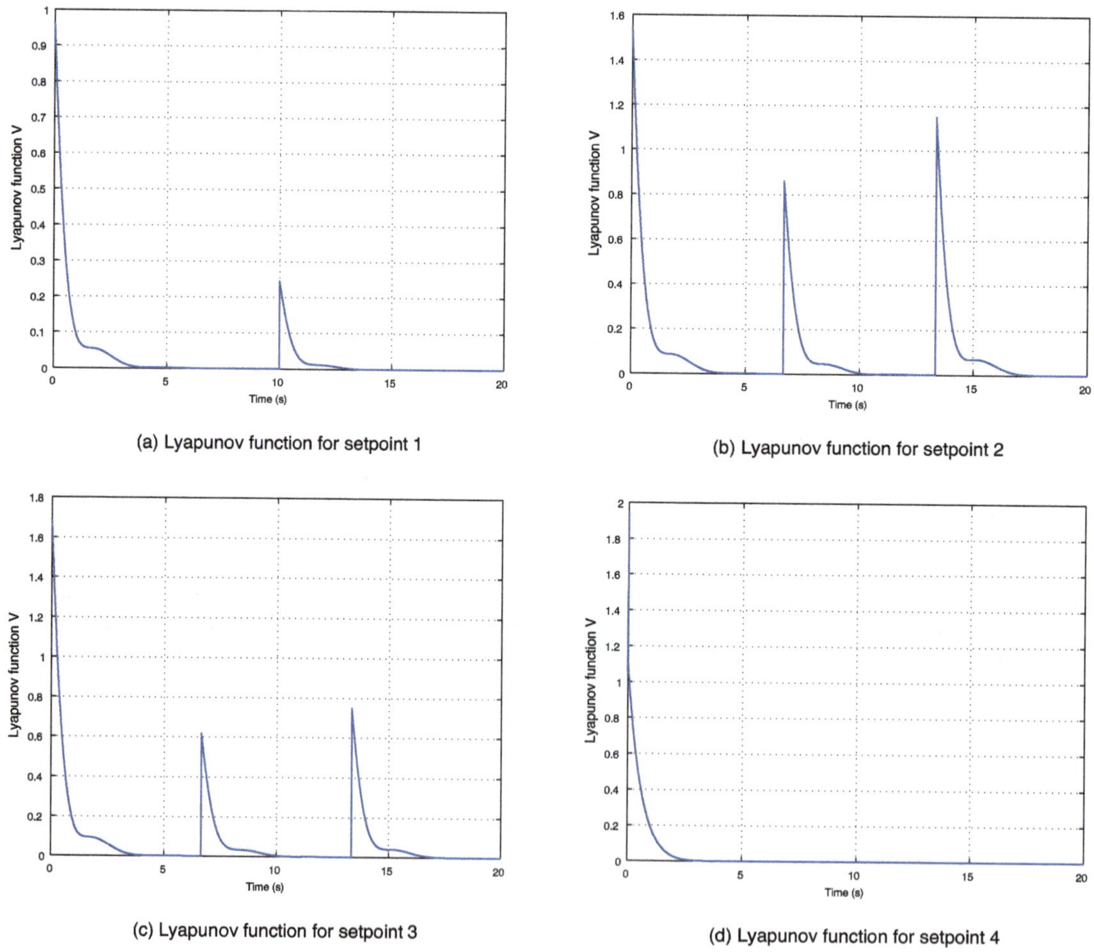

Figure 1.55 Lyapunov functions for flatness-based control in successive loops of the 4-DOF SCARA robotic manipulator when tracking setpoints 1–4

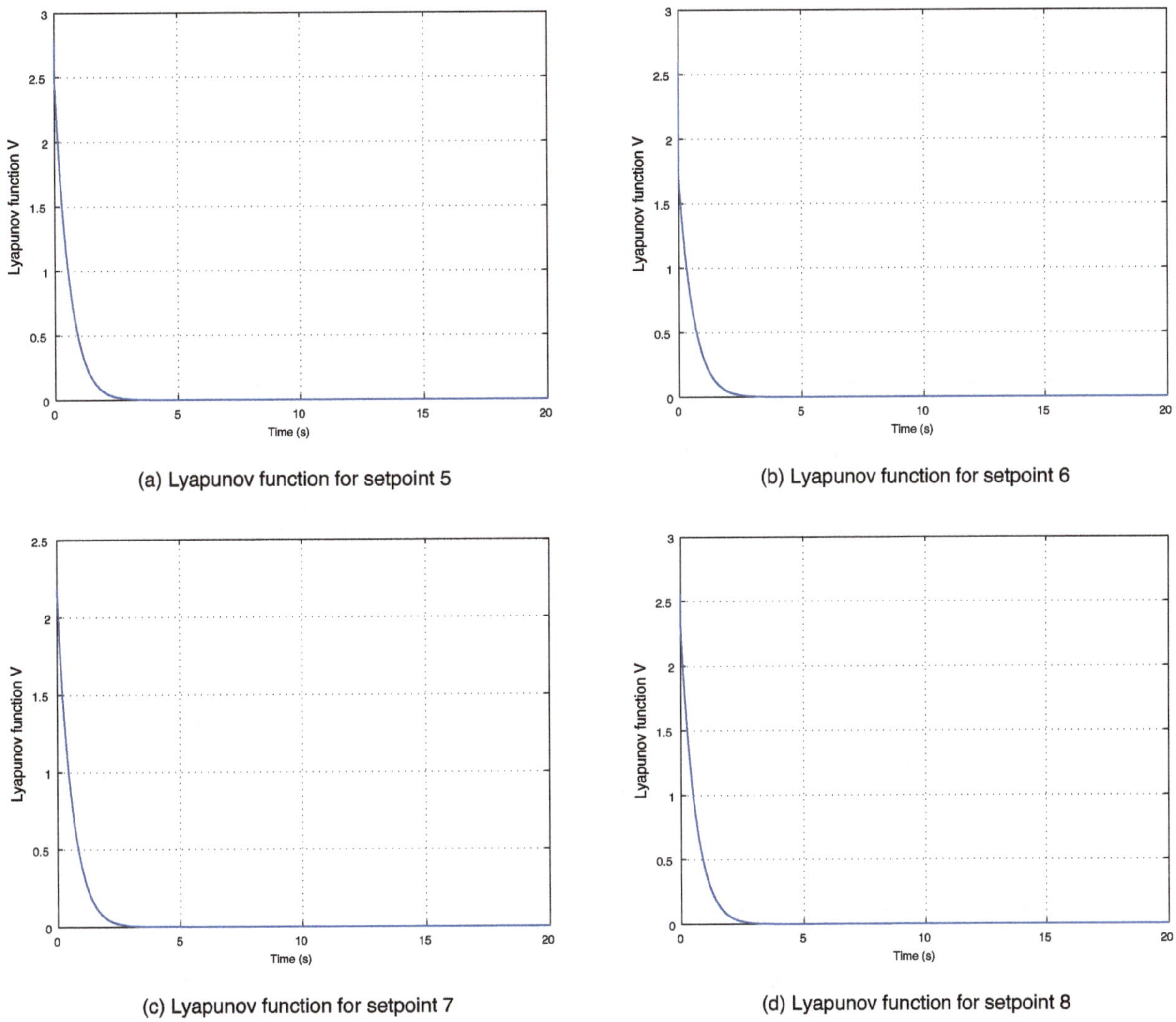

Figure 1.56 Lyapunov functions for flatness-based control in successive loops of the 4-DOF SCARA robotic manipulator when tracking setpoints 5–8

The transpose of the Jacobian matrix J_B^T which transfers torques due to the load's weight $B = M{\cdot}g$ and due to the grasping forces $-F$, from the end-effector of arm B to the joints of arm B is given by

$$J_B^T = \begin{pmatrix} J_{B,11}^T & J_{B,12}^T \\ J_{B,21}^T & J_{B,22}^T \end{pmatrix} = \begin{pmatrix} = -l_{1,B}sin(\theta_{1,B}) - l_{2,B}sin(\theta_{1,B}+\theta_{2,B}) & l_{1,B}cos(\theta_{1,B}) + l_{2,B}cos(\theta_{1,B}+\theta_{2,B}) \\ -l_{2,B}sin(\theta_{1,B}+\theta_{2,B}) & l_{2,B}cos(\theta_{1,B}+\theta_{2,B}) \end{pmatrix} \tag{1.235}$$

Force F exerted on the two end-effectors is due to grasping and is represented as a vector which is collinear to the transversal axis of the lifted object. It is given by

$$F = K_F(L - L_o) \tag{1.236}$$

where L_o is the length of the non-deformed object, L is the length of the lifted object which has been subject to deformation due to the grasping forces applies on its two ends and K_F is an elasticity (spring) coefficient. Considering that

Table 1.6 Flatness-based control in successive loops

Tracking RMSE $\times 10^{-3}$ for the SCARA robot in the disturbance-free case								
	$RMSE_{x_1}$	$RMSE_{x_2}$	$RMSE_{x_3}$	$RMSE_{x_4}$	$RMSE_{x_5}$	$RMSE_{x_6}$	$RMSE_{x_7}$	$RMSE_{x_8}$
$test_1$	0.0068	0.0052	0.0068	0.0052	0.0067	0.0052	0.0135	0.0104
$test_2$	0.1712	0.2543	0.1700	0.2549	0.0560	0.0852	0.2255	0.3403
$test_3$	0.0624	0.0846	0.1683	0.2554	0.1896	0.2893	0.2781	0.0918
$test_4$	0.0317	0.0457	0.0111	0.0128	0.0074	0.0085	0.0148	0.0170
$test_5$	0.0156	0.0139	0.0050	0.0042	0.0096	0.0067	0.0050	0.0042
$test_6$	0.0048	0.0103	0.0126	0.0191	0.0023	0.0093	0.0127	0.0191
$test_7$	0.0249	0.0251	0.0074	0.0069	0.0103	0.0099	0.0127	0.0108
$test_8$	0.0231	0.0196	0.0099	0.0080	0.0014	0.0013	0.0063	0.0112

Table 1.7 Flatness-based control in successive loops

Tracking RMSE $\times 10^{-3}$ for the SCARA robot in the case of disturbances								
$\Delta a\%$	$RMSE_{x_1}$	$RMSE_{x_2}$	$RMSE_{x_3}$	$RMSE_{x_4}$	$RMSE_{x_5}$	$RMSE_{x_6}$	$RMSE_{x_7}$	$RMSE_{x_8}$
0%	0.0624	0.0846	0.1683	0.2354	0.1898	0.2893	0.2381	0.0918
10%	0.0656	0.0915	0.1786	0.2762	0.2015	0.3131	0.2813	0.0994
20%	0.0690	0.0987	0.1891	0.2980	0.2136	0.3379	0.2847	0.1074
30%	0.0723	0.1062	0.1998	0.3208	0.2259	0.3637	0.2982	0.1157
40%	0.0757	0.1139	0.2100	0.3443	0.2383	0.3903	0.2920	0.1243
50%	0.0790	0.1219	0.2215	0.3686	0.2508	0.4179	0.2960	0.1332
60%	0.0824	0.1300	0.2326	0.3937	0.2635	0.4463	0.3003	0.1423

Table 1.8 Flatness-based control in successive loops

RMSE $\cdot 10^{-5}$ for the estimation performed by the H-infinity KF								
	$RMSE_{x_1}$	$RMSE_{x_2}$	$RMSE_{x_3}$	$RMSE_{x_4}$	$RMSE_{x_5}$	$RMSE_{x_6}$	$RMSE_{x_7}$	$RMSE_{x_8}$
$test_1$	0.0001	0.0099	0.0001	0.0005	0.0001	0.0060	0.0001	0.0103
$test_2$	0.0022	0.4332	0.0001	0.0112	0.0004	0.0832	0.0020	0.3930
$test_3$	0.0012	0.2334	0.0002	0.0438	0.0014	0.2858	0.0001	0.0289
$test_4$	0.0100	2.0980	0.0010	0.0820	0.0040	0.8490	0.0050	0.8990
$test_5$	0.0110	2.2510	0.0010	0.2760	0.0060	1.1870	0.0030	0.6440
$test_6$	0.0070	1.3000	0.0010	0.1320	0.0020	0.3020	0.0040	0.7660
$test_7$	0.0170	3.4010	0.0020	0.3970	0.0060	1.2550	0.0070	1.4770
$test_8$	0.0190	3.8110	0.0010	0.1790	0.0010	0.2150	0.0010	0.6560

the position of the end-effector of arm A is (x_E^A, y_E^A) and the position of the end-effector of arm B is (x_E^B, y_E^B), length L is given by

$$L = \sqrt{(x_E^B - x_E^A)^2 + (y_E^B - y_E^A)^2} \Rightarrow$$

$$\begin{aligned} L = \{&[x_0^B + l_{1,B}cos(\theta_{1,B}) + l_{2,B}cos(\theta_{1,B} + \theta_{2,B}) - x_0^A - l_{1,A}cos(\theta_{1,A}) + l_{2,A}cos(\theta_{1,A} + \theta_{2,A})]^2 + \\ &+[h + l_{1,B}sin(\theta_{1,B}) + l_{2,B}sin(\theta_{1,B} + \theta_{2,B}) - h - l_{1,A}sin(\theta_{1,A}) - l_{2,A}sin(\theta_{1,A} + \theta_{2,A})]^2\}^{\frac{1}{2}} \end{aligned} \quad (1.237)$$

Table 1.9 Flatness-based control in successive loops

Convergence time (sec) for the SCARA robot's state variables								
	$T_s\,x_1$	$T_s\,x_2$	$T_s\,x_3$	$T_s\,x_4$	$T_s\,x_5$	$T_s\,x_6$	$T_s\,x_7$	$T_s\,x_8$
$test_1$	4.5	5.0	6.0	4.0	4.0	5.0	4.0	4.5
$test_2$	3.0	4.5	4.5	5.0	4.0	6.0	5.5	4.0
$test_3$	4.0	5.5	4.0	5.0	4.0	4.0	4.0	5.0
$test_4$	0.5	2.0	2.5	3.5	4.0	4.0	0.5	2.0
$test_5$	2.5	4.5	3.5	3.5	2.0	2.5	3.5	3.5
$test_6$	2.0	2.5	2.0	2.5	0.5	2.0	2.0	3.0
$test_7$	2.5	4.0	0.5	2.0	3.5	4.0	2.5	3.5
$test_8$	3.0	4.0	3.0	4.0	1.0	2.0	2.0	2.0

Table 1.10 Flatness-based control in successive loops

Distribution of power % in actuators				
	$P\,u_1$	$P\,u_2$	$P\,u_3$	$P\,u_4$
$test_1$	7.53	13.41	78.99	0.07
$test_2$	5.70	15.41	78.84	0.06
$test_3$	2.42	5.42	92.14	0.02
$test_4$	2.28	43.50	54.01	0.21
$test_5$	4.53	11.97	83.42	0.08
$test_6$	6.03	20.67	73.18	0.12
$test_7$	32.74	9.14	58.06	0.06
$test_8$	2.87	14.16	83.39	0.16

The angle which is formed between the transversal axis of the lifted object and the horizontal axis of the inertial reference frame is denoted as ϕ and is given by

$$\phi = tan^{-1}\left(\frac{(y_E^B - y_E^A)}{(x_E^B - x_E^A)}\right) \tag{1.238}$$

where $(y_E^B - y_E^A) = [h + l_{1,B}sin(\theta_{1,B}) + l_{2,B}sin(\theta_{1,B} + \theta_{2,B}) - h - l_{1,A}sin(\theta_{1,A}) - l_{2,A}sin(\theta_{1,A} + \theta_{2,A})]$ and $(x_E^B - x_E^A) = [x_0^B + l_{1,B}cos(\theta_{1,B}) + l_{2,B}cos(\theta_{1,B} + \theta_{2,B}) - x_0^A - l_{1,A}cos(\theta_{1,A}) + l_{2,A}cos(\theta_{1,A} + \theta_{2,A})]$.

Using that weight B in cartesian coordinates is represented by vector $[0\ B]^T$, then the torque which is transferred to the joints of arm A due to weight B is given by

$$J_A^T B = \begin{pmatrix} J_{A,12}^T Mg \\ J_{A,22}^T Mg \end{pmatrix} \tag{1.239}$$

Using that force F in cartesian coordinates is represented by vector $[Fcos(\phi),\ Fsin(\phi)]^T$, then the torque which is transferred to the joints of arm A due to the grasping force F is given by

$$J_A^T F = \begin{pmatrix} [J_{A,11}^T cos(\phi) + J_{A,12}^T sin(\phi)]F \\ [J_{A,21}^T cos(\phi) + J_{A,22}^T sin(\phi)]F \end{pmatrix} \tag{1.240}$$

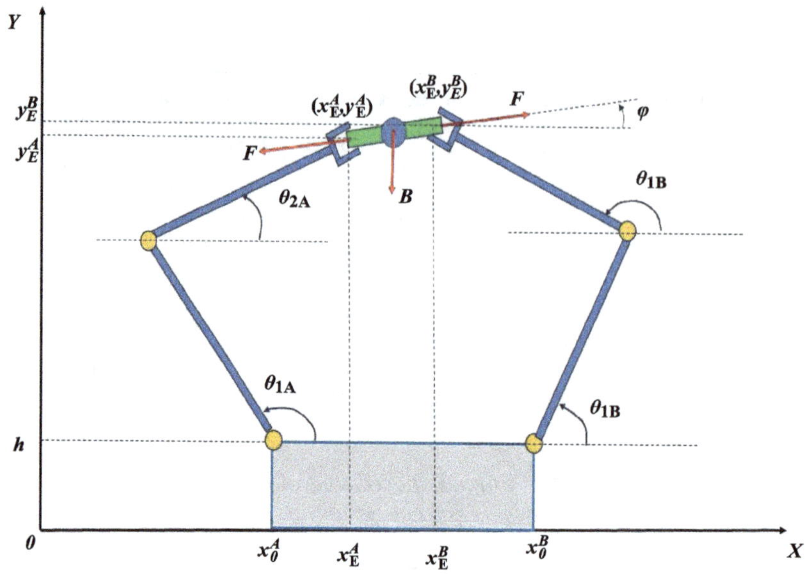

Figure 1.57 Diagram of the dual-arm robotic manipulator in an object-lifting task

Using that weight B in cartesian coordinates is represented by vector $[0\ B]^T$, then the torque which is transferred to the joints of arm B due to weight B is given by

$$J_B^T B = \begin{pmatrix} J_{B,12}^T Mg \\ J_{B,22}^T Mg \end{pmatrix} \tag{1.241}$$

Using that force F in cartesian coordinates is represented by vector $[-Fcos(\phi) - Fsin(\phi)]^T$, then the torque which is transferred to the joints of arm B due to the grasping force F is given by

$$-J_B^T F = \begin{pmatrix} -[J_{B,11}^T cos(\phi) + J_{B,12}^T sin(\phi)]F \\ -[J_{A,21}^T cos(\phi) + J_{B,22}^T sin(\phi)]F \end{pmatrix} \tag{1.242}$$

Consequently, the dynamic model of the dual-arm robotic manipulator is given by the following two equations for matrices A and B, respectively.

$$\begin{aligned} &\begin{pmatrix} \bar{m}_{11,A} & \bar{m}_{12,A} \\ \bar{m}_{21,A} & \bar{m}_{22,A} \end{pmatrix} \begin{pmatrix} \ddot{\theta}_{1,A} \\ \ddot{\theta}_{2,A} \end{pmatrix} + \begin{pmatrix} \bar{C}_{1,A} \\ \bar{C}_{2,A} \end{pmatrix} + \begin{pmatrix} \bar{G}_{1,A} \\ \bar{G}_{2,A} \end{pmatrix} = \begin{pmatrix} \tau_{1,A} \\ \tau_{2,A} \end{pmatrix} - \\ &\quad - \begin{pmatrix} J_{A.12}^T Mg \\ J_{A.22}^T Mg \end{pmatrix} - \begin{pmatrix} [J_{A,11}^T cos(\phi) + J_{A,12}^T sin(\phi)]F \\ [J_{A,21}^T cos(\phi) + J_{A,22}^T sin(\phi)]F \end{pmatrix} \end{aligned} \tag{1.243}$$

$$\begin{aligned} &\begin{pmatrix} \bar{m}_{11,B} & \bar{m}_{12,B} \\ \bar{m}_{21,B} & \bar{m}_{22,B} \end{pmatrix} \begin{pmatrix} \ddot{\theta}_{1,B} \\ \ddot{\theta}_{2,B} \end{pmatrix} + \begin{pmatrix} \bar{C}_{1,B} \\ \bar{C}_{2,B} \end{pmatrix} + \begin{pmatrix} \bar{G}_{1,B} \\ \bar{G}_{2,B} \end{pmatrix} = \begin{pmatrix} \tau_{1,B} \\ \tau_{2,B} \end{pmatrix} - \\ &\quad - \begin{pmatrix} J_{B.12}^T Mg \\ J_{B.22}^T Mg \end{pmatrix} - \begin{pmatrix} -[J_{B,11}^T cos(\phi) + J_{B,12}^T sin(\phi)]F \\ -[J_{B,21}^T cos(\phi) + J_{B,22}^T sin(\phi)]F \end{pmatrix} \end{aligned} \tag{1.244}$$

The inverse of the inertia matrix of arm A is given by

$$M_A^{-1} = \frac{1}{detM_A}\begin{pmatrix} \bar{m}_{22,A} & -\bar{m}_{12,A} \\ -\bar{m}_{21,A} & \bar{m}_{11,A} \end{pmatrix} \tag{1.245}$$

where $detM_A = \bar{m}_{11,A}\bar{m}_{22,A} - \bar{m}_{12,A}\bar{m}_{21,A}$. The inverse of the inertia matrix of arm B is given by

$$M_B^{-1} = \frac{1}{detM_B}\begin{pmatrix} \bar{m}_{22,B} & -\bar{m}_{12,B} \\ -\bar{m}_{21,B} & \bar{m}_{11,B} \end{pmatrix} \tag{1.246}$$

where $detM_B = \bar{m}_{11,B}\bar{m}_{22,B} - \bar{m}_{12,B}\bar{m}_{21,B}$. Thus, after some intermediate operations the dynamic model of arm A and arm B is written as

$$\begin{pmatrix} \ddot{\theta}_{1,A} \\ \ddot{\theta}_{2,A} \end{pmatrix} = -\begin{pmatrix} \frac{\bar{m}_{22,A}}{detM_A} & \frac{-\bar{m}_{12,A}}{detM_A} \\ \frac{-\bar{m}_{21,A}}{detM_A} & \frac{\bar{m}_{11,A}}{detM_A} \end{pmatrix}\begin{pmatrix} \bar{C}_{1,A} + \bar{G}_{1,A} + J_{A,12}^T Mg + [J_{A,11}^T cos(\phi) + J_{A,12}^T sin(\phi)]F \\ \bar{C}_{2,A} + \bar{G}_{2,A} + J_{A,22}^T Mg + [J_{A,21}^T cos(\phi) + J_{A,22}^T sin(\phi)]F \end{pmatrix} + \\ + \begin{pmatrix} \frac{\bar{m}_{22,A}}{detM_A} & \frac{-\bar{m}_{12,A}}{detM_A} \\ \frac{-\bar{m}_{21,A}}{detM_A} & \frac{\bar{m}_{11,A}}{detM_A} \end{pmatrix}\begin{pmatrix} \tau_{1,A} \\ \tau_{2,A} \end{pmatrix} \tag{1.247}$$

$$\begin{pmatrix} \ddot{\theta}_{1,B} \\ \ddot{\theta}_{2,B} \end{pmatrix} = -\begin{pmatrix} \frac{\bar{m}_{22,B}}{detM_B} & \frac{-\bar{m}_{12,B}}{detM_B} \\ \frac{-\bar{m}_{21,B}}{detM_B} & \frac{\bar{m}_{11,B}}{detM_B} \end{pmatrix}\begin{pmatrix} \bar{C}_{1,B} + \bar{G}_{1,B} + J_{B,12}^T Mg - [J_{B,11}^T cos(\phi) + J_{B,12}^T sin(\phi)]F \\ \bar{C}_{2,B} + \bar{G}_{2,B} + J_{B,22}^T Mg - [J_{B,21}^T cos(\phi) + J_{B,22}^T sin(\phi)]F \end{pmatrix} + \\ + \begin{pmatrix} \frac{\bar{m}_{22,B}}{detM_B} & \frac{-\bar{m}_{12,B}}{detM_B} \\ \frac{-\bar{m}_{21,B}}{detM_B} & \frac{\bar{m}_{11,B}}{detM_B} \end{pmatrix}\begin{pmatrix} \tau_{1,B} \\ \tau_{2,B} \end{pmatrix} \tag{1.248}$$

Next, by denoting functions $F_{1,A}$ and $F_{2,A}$

$$\begin{aligned} F_{1,A} &= \bar{C}_{1,A} + \bar{G}_{1,A} + J_{A,12}^T Mg + [J_{A,11}^T cos(\phi) + J_{A,12}^T sin(\phi)]F \\ F_{2,A} &= \bar{C}_{2,A} + \bar{G}_{2,A} + J_{A,22}^T Mg + [J_{A,21}^T cos(\phi) + J_{A,22}^T sin(\phi)]F \end{aligned} \tag{1.249}$$

as well as functions $F_{1,B}$ and $F_{2,B}$

$$\begin{aligned} F_{1,B} &= \bar{C}_{1,B} + \bar{G}_{1,B} + J_{B,12}^T Mg - [J_{B,11}^T cos(\phi) + J_{B,12}^T sin(\phi)]F \\ F_{2,B} &= \bar{C}_{2,B} + \bar{G}_{2,B} + J_{B,22}^T Mg - [J_{B,21}^T cos(\phi) + J_{B,22}^T sin(\phi)]F \end{aligned} \tag{1.250}$$

the dynamic model of arm A and arm B is rewritten as

$$\begin{pmatrix} \ddot{\theta}_{1,A} \\ \ddot{\theta}_{2,A} \end{pmatrix} = \begin{pmatrix} \frac{-\bar{m}_{22,A}\bar{F}_{1,A} + \bar{m}_{12,A}\bar{F}_{2,A}}{detM_A} \\ \frac{\bar{m}_{21,A}\bar{F}_{1,A} - \bar{m}_{11,A}\bar{F}_{2,A}}{detM_A} \end{pmatrix} + \begin{pmatrix} \frac{\bar{m}_{22,A}}{detM_A} & \frac{-\bar{m}_{12,A}}{detM_A} \\ \frac{-\bar{m}_{21,A}}{detM_A} & \frac{\bar{m}_{11,A}}{detM_A} \end{pmatrix}\begin{pmatrix} \tau_{1,A} \\ \tau_{2,A} \end{pmatrix} \tag{1.251}$$

$$\begin{pmatrix} \ddot{\theta}_{1,B} \\ \ddot{\theta}_{2,B} \end{pmatrix} = \begin{pmatrix} \frac{-\bar{m}_{22,B}\bar{F}_{1,B} + \bar{m}_{12,B}\bar{F}_{2,B}}{detM_B} \\ \frac{\bar{m}_{21,B}\bar{F}_{1,B} - \bar{m}_{11,B}\bar{F}_{2,B}}{detM_B} \end{pmatrix} + \begin{pmatrix} \frac{\bar{m}_{22,B}}{detM_B} & \frac{-\bar{m}_{12,B}}{detM_B} \\ \frac{-\bar{m}_{21,B}}{detM_B} & \frac{\bar{m}_{11,B}}{detM_B} \end{pmatrix}\begin{pmatrix} \tau_{1,A} \\ \tau_{2,A} \end{pmatrix} \tag{1.252}$$

By aggregating the dynamics of arm A and arm B into one state-space model, one obtains the following state-space description:

$$\begin{pmatrix} \ddot{\theta}_{1,A} \\ \ddot{\theta}_{2,A} \\ \ddot{\theta}_{1,B} \\ \ddot{\theta}_{2,B} \end{pmatrix} = \begin{pmatrix} \frac{-\bar{m}_{22,A}\bar{F}_{1,A}+\bar{m}_{12,A}\bar{F}_{2,A}}{detM_A} \\ \frac{\bar{m}_{21,A}\bar{F}_{1,A}-\bar{m}_{11,A}\bar{F}_{2,A}}{detM_A} \\ \frac{-\bar{m}_{22,B}\bar{F}_{1,B}+\bar{m}_{12,B}\bar{F}_{2,B}}{detM_B} \\ \frac{\bar{m}_{21,B}\bar{F}_{1,B}-\bar{m}_{11,B}\bar{F}_{2,B}}{detM_B} \end{pmatrix} + \begin{pmatrix} \frac{\bar{m}_{22,A}}{detM_A} & \frac{-\bar{m}_{12,A}}{detM_A} & 0 & 0 \\ \frac{-\bar{m}_{21,A}}{detM_A} & \frac{\bar{m}_{11,A}}{detM_A} & 0 & 0 \\ 0 & 0 & \frac{\bar{m}_{22,B}}{detM_B} & \frac{-\bar{m}_{12,B}}{detM_B} \\ 0 & 0 & \frac{-\bar{m}_{21,B}}{detM_A} & \frac{\bar{m}_{11,B}}{detM_A} \end{pmatrix} \begin{pmatrix} \tau_{1,A} \\ \tau_{2,A} \\ \tau_{1,B} \\ \tau_{2,B} \end{pmatrix} \tag{1.253}$$

Next, the following state vector definition is used:

$$\begin{aligned} x &= [x_1, x_2, x_3, x_4, x_5, x_6, x_7, x_8]^T \Rightarrow \\ x &= [\theta_{1,A}, \dot{\theta}_{1,A}, \theta_{2,A}, \dot{\theta}_{2,A}, \theta_{1,B}, \dot{\theta}_{1,B}, \theta_{2,B}, \dot{\theta}_{2,B}]^T \end{aligned} \tag{1.254}$$

Besides, the following control inputs vector is defined

$$\begin{aligned} u &= [u_1, u_2, u_3, u_4]^T \Rightarrow \\ u &= [\tau_{1,A}, \tau_{2,A}, \tau_{1,B}, \tau_{2,B}]^T \end{aligned} \tag{1.255}$$

Thus, the dynamic model of the dual-arm robotic manipulator can be written in the following state-space description:

$$\begin{pmatrix} \dot{x}_1 \\ \dot{x}_2 \\ \dot{x}_3 \\ \dot{x}_4 \\ \dot{x}_5 \\ \dot{x}_6 \\ \dot{x}_7 \\ \dot{x}_8 \end{pmatrix} = \begin{pmatrix} x_2 \\ \frac{-\bar{m}_{22,A}\bar{F}_{1,A}+\bar{m}_{12,A}\bar{F}_{2,A}}{detM_A} \\ x_4 \\ \frac{\bar{m}_{21,A}\bar{F}_{1,A}-\bar{m}_{11,A}\bar{F}_{2,A}}{detM_A} \\ x_6 \\ \frac{-\bar{m}_{22,B}\bar{F}_{1,B}+\bar{m}_{12,B}\bar{F}_{2,B}}{detM_B} \\ x_8 \\ \frac{\bar{m}_{21,B}\bar{F}_{1,B}-\bar{m}_{11,B}\bar{F}_{2,B}}{detM_B} \end{pmatrix} + \begin{pmatrix} 0 & 0 & 0 & 0 \\ \frac{\bar{m}_{22,A}}{detM_A} & \frac{-\bar{m}_{12,A}}{detM_A} & 0 & 0 \\ 0 & 0 & 0 & 0 \\ \frac{-\bar{m}_{21,A}}{detM_A} & \frac{\bar{m}_{11,A}}{detM_A} & 0 & 0 \\ 0 & 0 & 0 & 0 \\ 0 & 0 & \frac{\bar{m}_{22,B}}{detM_B} & \frac{-\bar{m}_{12,B}}{detM_B} \\ 0 & 0 & 0 & 0 \\ 0 & 0 & \frac{-\bar{m}_{21,B}}{detM_A} & \frac{\bar{m}_{11,B}}{detM_A} \end{pmatrix} \begin{pmatrix} \tau_{1,A} \\ \tau_{2,A} \\ \tau_{1,B} \\ \tau_{2,B} \end{pmatrix} \tag{1.256}$$

The dynamics of the dual-arm robotic manipulator can be also written in the following concise nonlinear affine-in-the-input state-space form:

$$\dot{x} = f(x) + g(x)u \tag{1.257}$$

where $x \in R^{8\times 1}$, $f(x) \in R^{8\times 1}$, $g(x) \in R^{8\times 4}$, and $u \in R^{4\times 1}$.

1.3.2.2 Differential flatness properties of the dynamics of the dual-arm robot

The dual-arm robotic system is differentially flat, with flat outputs vector $Y = [x_1, x_3, x_5, x_7]^T$ or $Y = [\theta_{1,A}, \theta_{2,A}, \theta_{1,B}, \theta_{2,B}]^T$. From the odd-numbered rows of the state-space model of Eq. (1.256), one has

$$\begin{aligned} x_2 = \dot{x}_1 \Rightarrow x_2 = h_2(Y, \dot{Y}) \quad x_4 = \dot{x}_3 \Rightarrow x_4 = h_4(Y, \dot{Y}) \\ x_6 = \dot{x}_5 \Rightarrow x_6 = h_6(Y, \dot{Y}) \quad x_8 = \dot{x}_7 \Rightarrow x_8 = h_8(Y, \dot{Y}) \end{aligned} \tag{1.258}$$

which signifies that state variables x_2, x_4, x_6, and x_8 are differential functions of the flat outputs vector. Besides, from Eq. (1.253), one has that

$$\begin{pmatrix} \ddot{x}_1 \\ \ddot{x}_2 \\ \ddot{x}_3 \\ \ddot{x}_4 \end{pmatrix} = \begin{pmatrix} \frac{-\bar{m}_{22,A}\bar{F}_{1,A}+\bar{m}_{12,A}\bar{F}_{2,A}}{detM_A} \\ \frac{\bar{m}_{21,A}\bar{F}_{1,A}-\bar{m}_{11,A}\bar{F}_{2,A}}{detM_A} \\ \frac{-\bar{m}_{22,B}\bar{F}_{1,B}+\bar{m}_{12,B}\bar{F}_{2,B}}{detM_B} \\ \frac{\bar{m}_{21,B}\bar{F}_{1,B}-\bar{m}_{11,B}\bar{F}_{2,B}}{detM_B} \end{pmatrix} + \begin{pmatrix} \frac{\bar{m}_{22,A}}{detM_A} & \frac{-\bar{m}_{12,A}}{detM_A} & 0 & 0 \\ \frac{-\bar{m}_{21,A}}{detM_A} & \frac{\bar{m}_{11,A}}{detM_A} & 0 & 0 \\ 0 & 0 & \frac{\bar{m}_{22,B}}{detM_B} & \frac{-\bar{m}_{12,B}}{detM_B} \\ 0 & 0 & \frac{-\bar{m}_{21,B}}{detM_A} & \frac{\bar{m}_{11,B}}{detM_A} \end{pmatrix} \begin{pmatrix} u_1 \\ u_2 \\ u_3 \\ u_4 \end{pmatrix} \tag{1.259}$$

One solves Eq. (1.259) for the control inputs vector $u = [u_1, u_2, u_3, u_4]^T$. All terms which appear in the drift vector of the above-noted equation, as well as all terms appearing in the control inputs gain matrix of this equation are differential functions of the flat outputs vector Y.

$$\begin{pmatrix} u_1 \\ u_2 \\ u_3 \\ u_4 \end{pmatrix} = \begin{pmatrix} \frac{\bar{m}_{22,A}}{detM_A} & \frac{-\bar{m}_{12,A}}{detM_A} & 0 & 0 \\ \frac{-\bar{m}_{21,A}}{detM_A} & \frac{\bar{m}_{11,A}}{detM_A} & 0 & 0 \\ 0 & 0 & \frac{\bar{m}_{22,B}}{detM_B} & \frac{-\bar{m}_{12,B}}{detM_B} \\ 0 & 0 & \frac{-\bar{m}_{21,B}}{detM_A} & \frac{\bar{m}_{11,B}}{detM_A} \end{pmatrix}^{-1} \left[\begin{pmatrix} \ddot{x}_1 \\ \ddot{x}_2 \\ \ddot{x}_3 \\ \ddot{x}_4 \end{pmatrix} - \begin{pmatrix} \frac{-\bar{m}_{22,A}\bar{F}_{1,A}+\bar{m}_{12,A}\bar{F}_{2,A}}{detM_A} \\ \frac{\bar{m}_{21,A}\bar{F}_{1,A}-\bar{m}_{11,A}\bar{F}_{2,A}}{detM_A} \\ \frac{-\bar{m}_{22,B}\bar{F}_{1,B}+\bar{m}_{12,B}\bar{F}_{2,B}}{detM_B} \\ \frac{\bar{m}_{21,B}\bar{F}_{1,B}-\bar{m}_{11,B}\bar{F}_{2,B}}{detM_B} \end{pmatrix} \right] \tag{1.260}$$

or equivalently, one has the following concise relation which shows that control input u is a differential function of the flat outputs vector.

$$u = h_u(Y, \dot{Y}) \tag{1.261}$$

By proving differential flatness properties it is confirmed that the dual-arm robot is input–output linearizable. Differential flatness provides also an implicit of the robot's controllability. It also allows to solve the setpoints definition problem. Setpoints are chosen in an unconstrained manner for the state variables x_1, x_3, x_5, x_7 which coincide with the flat outputs of the robot. Next, one chooses setpoints for the rest of the state variables of the robotic system x_2, x_4, x_6, x_8 through the use of the differential relations which connect them to the flat outputs.

1.3.3 Approximate linearization of the dynamic model of the dual-arm robot

The dynamic model of the dual-arm robot undergoes approximate linearization around the temporary operating point (x^*, u^*) at each sampling instant, where x^* is the present value of the robot's state vector and u^* is the last sampled value of the control inputs vector. The linearization is based on first-order Taylor series expansion and on the computation of the system's Jacobian matrices. The dual-arm robot being initially in the state-space form $\dot{x} = f(x) + g(x)u$, is brought after the above-noted linearization process to the state-space form:

$$\dot{x} = Ax + Bu + \tilde{d} \tag{1.262}$$

The Jacobian matrices of the model are computed as

$$\begin{aligned} A &= \nabla_x[f(x) + g(x)u] \mid_{(x^*,u^*)} \Rightarrow \\ A &= \nabla_x f(x) \mid_{(x^*,u^*)} + \nabla_x g_1(x)u_1 \mid_{(x^*,u^*)} + \nabla_x g_2(x)u_2 \mid_{(x^*,u^*)} + \\ &+ \nabla_x g_3(x)u_3 \mid_{(x^*,u^*)} + \nabla_x g_4(x)u_4 \mid_{(x^*,u^*)} \end{aligned} \tag{1.263}$$

$$\begin{aligned} B &= \nabla_u[f(x) + g(x)u] \mid_{(x^*,u^*)} \Rightarrow \\ B &= g(x) \mid_{(x^*,u^*)} \end{aligned} \tag{1.264}$$

The term $\tilde{d}$ is the cumulative disturbance vector which may comprise (i) modeling error due to truncation of higher-order terms from the Taylor series expansion, (ii) exogenous perturbations, (iii) sensor measurement noise of any distribution. Due to linearizing around the present position of the system's state vector (x^*, u^*) and not at points of the desirable trajectory the span of the Taylor series expansion is small and the associated modeling error is negligible.

Computation of the Jacobian matrix $\nabla_x f(x) \mid_{(x^*,u^*)}$:

First row of the Jacobian matrix $\nabla_x f(x) \mid_{(x^*,u^*)}$: $\frac{\nabla_x f_1}{\partial x_1} = 0$, $\frac{\nabla_x f_1}{\partial x_2} = 1$, $\frac{\nabla_x f_1}{\partial x_3} = 0$, $\frac{\nabla_x f_1}{\partial x_4} = 0$, $\frac{\nabla_x f_1}{\partial x_5} = 0$, $\frac{\nabla_x f_1}{\partial x_6} = 0$, $\frac{\nabla_x f_1}{\partial x_7} = 0$, and $\frac{\nabla_x f_1}{\partial x_8} = 0$.

Second row of the Jacobian matrix $\nabla_x f(x) \mid_{(x^*,u^*)}$: It holds that

$$f_2 = \frac{f_{2,num}}{f_{2,den}} \tag{1.265}$$

where $f_{2,num} = -\bar{m}_{22,A}\bar{F}_{1,A} + \bar{m}_{12,A}\bar{F}_{2,A}$ and $f_{2,den} = detM_A$. Thus, one obtains for $i = 1, 2, \cdots, 8$

$$\frac{\partial f_2}{\partial x_i} = \frac{\frac{\partial f_{2,num}}{\partial x_i} detM_A - f_{2,num}\frac{\partial detM_A}{\partial x_i}}{(detM_A)^2} \tag{1.266}$$

where for $i = 1, 2, \cdots, 8$

$$\frac{\partial f_{2,num}}{\partial x_i} = -\frac{\partial \bar{m}_{22,A}}{\partial x_i}\bar{F}_{1,A} - \bar{m}_{22,A}\frac{\partial \bar{F}_{1,A}}{\partial x_i} + \frac{\partial \bar{m}_{12,A}}{\partial x_i}\bar{F}_{2,A} + \bar{m}_{12,A}\frac{\partial \bar{F}_{2,A}}{\partial x_i} \tag{1.267}$$

and also for $i = 1, 2, \cdots, 8$

$$\frac{\partial detM_A}{\partial x_i} = \frac{\partial \bar{m}_{11,A}}{\partial x_i}\bar{m}_{22,A} + \bar{m}_{11,A}\frac{\partial \bar{m}_{22,A}}{\partial x_i} - \frac{\partial \bar{m}_{12,A}}{\partial x_i}\bar{m}_{21,A} - \bar{m}_{12,A}\frac{\partial \bar{m}_{21,A}}{\partial x_i} \tag{1.268}$$

Third row of the Jacobian matrix $\nabla_x f(x) \mid_{(x^*,u^*)}$: $\frac{\nabla_x f_3}{\partial x_1} = 0$, $\frac{\nabla_x f_3}{\partial x_2} = 0$, $\frac{\nabla_x f_3}{\partial x_3} = 0$, $\frac{\nabla_x f_3}{\partial x_4} = 1$, $\frac{\nabla_x f_3}{\partial x_5} = 0$, $\frac{\nabla_x f_3}{\partial x_6} = 0$, $\frac{\nabla_x f_3}{\partial x_7} = 0$, and $\frac{\nabla_x f_3}{\partial x_8} = 0$.

Fourth row of the Jacobian matrix $\nabla_x f(x) \mid_{(x^*,u^*)}$: It holds that

$$f_4 = \frac{f_{4,num}}{f_{4,den}} \tag{1.269}$$

where $f_{4,num} = \bar{m}_{21,A}\bar{F}_{1,A} - \bar{m}_{11,A}\bar{F}_{2,A}$ and $f_{4,den} = detM_A$. Thus, one obtains for $i = 1, 2, \cdots, 8$

$$\frac{\partial f_4}{\partial x_i} = \frac{\frac{\partial f_{4,num}}{\partial x_i} detM_A - f_{4,num}\frac{\partial detM_A}{\partial x_i}}{(detM_A)^2} \tag{1.270}$$

where for $i = 1, 2, \cdots, 8$

$$\frac{\partial f_{4,num}}{\partial x_i} = \frac{\partial \bar{m}_{21,A}}{\partial x_i}\bar{F}_{1,A} + \bar{m}_{21,A}\frac{\partial \bar{F}_{1,A}}{\partial x_i} - \frac{\partial \bar{m}_{11,A}}{\partial x_i}\bar{F}_{2,A} + \bar{m}_{11,A}\frac{\partial \bar{F}_{2,A}}{\partial x_i} \tag{1.271}$$

Fifth row of the Jacobian matrix $\nabla_x f(x) \mid_{(x^*,u^*)}$: $\frac{\nabla_x f_5}{\partial x_1} = 0$, $\frac{\nabla_x f_5}{\partial x_2} = 0$, $\frac{\nabla_x f_5}{\partial x_3} = 0$, $\frac{\nabla_x f_5}{\partial x_4} = 0$, $\frac{\nabla_x f_5}{\partial x_5} = 0$, $\frac{\nabla_x f_5}{\partial x_6} = 1$, $\frac{\nabla_x f_5}{\partial x_7} = 0$, and $\frac{\nabla_x f_5}{\partial x_8} = 0$.

Sixth row of the Jacobian matrix $\nabla_x f(x) \mid_{(x^*,u^*)}$: It holds that

$$f_6 = \frac{f_{6,num}}{f_{6,den}} \tag{1.272}$$

where $f_{6,num} = -\bar{m}_{22,B}\bar{F}_{1,B} + \bar{m}_{12,B}\bar{F}_{2,B}$ and $f_{6,den} = detM_B$. Thus, one obtains for $i = 1, 2, \cdots, 8$:

$$\frac{\partial f_6}{\partial x_i} = \frac{\frac{\partial f_{6,num}}{\partial x_i} detM_B - f_{6,num}\frac{\partial detM_B}{\partial x_i}}{(detM_B)^2} \tag{1.273}$$

where for $i = 1, 2, \cdots, 8$

$$\frac{\partial f_{6,num}}{\partial x_i} = -\frac{\partial \bar{m}_{22,B}}{\partial x_i}\bar{F}_{1,B} - \bar{m}_{22,B}\frac{\partial \bar{F}_{1,B}}{\partial x_i} + \frac{\partial \bar{m}_{12,B}}{\partial x_i}\bar{F}_{2,B} + \bar{m}_{12,B}\frac{\partial \bar{F}_{2,B}}{\partial x_i} \tag{1.274}$$

and also for $i=1,2,\cdots,8$

$$\frac{\partial detM_B}{\partial x_i}=\frac{\partial \bar{m}_{11,B}}{\partial x_i}\bar{m}_{22,B}+\bar{m}_{11,B}\frac{\partial \bar{m}_{22,B}}{\partial x_i}-\frac{\partial \bar{m}_{12,B}}{\partial x_i}\bar{m}_{21,B}-\bar{m}_{12,B}\frac{\partial \bar{m}_{21,B}}{\partial x_i} \quad (1.275)$$

Seventh row of the Jacobian matrix $\nabla_x f(x)\mid_{(x^*,u^*)}$: $\frac{\nabla_x f_7}{\partial x_1}=0$, $\frac{\nabla_x f_7}{\partial x_2}=0$, $\frac{\nabla_x f_7}{\partial x_3}=0$, $\frac{\nabla_x f_7}{\partial x_4}=0$, $\frac{\nabla_x f_7}{\partial x_5}=0$, $\frac{\nabla_x f_7}{\partial x_6}=0$, $\frac{\nabla_x f_7}{\partial x_7}=0$, and $\frac{\nabla_x f_7}{\partial x_8}=1$.

Eighth row of the Jacobian matrix $\nabla_x f(x)\mid_{(x^*,u^*)}$: It holds that

$$f_8=\frac{f_{8,num}}{f_{8,den}} \quad (1.276)$$

where $f_{8,num}=\bar{m}_{21,B}\bar{F}_{1,B}-\bar{m}_{11,B}\bar{F}_{2,B}$ and $f_{8,den}=detM_B$. Thus, one obtains for $i=1,2,\cdots,8$

$$\frac{\partial f_8}{\partial x_i}=\frac{\frac{\partial f_{8,num}}{\partial x_i}detM_B-f_{8,num}\frac{\partial detM_B}{\partial x_i}}{(detM_B)^2} \quad (1.277)$$

where for $i=1,2,\cdots,8$

$$\frac{\partial f_{8,num}}{\partial x_i}=\frac{\partial \bar{m}_{21,B}}{\partial x_i}\bar{F}_{1,B}+\bar{m}_{21,B}\frac{\partial \bar{F}_{1,B}}{\partial x_i}-\frac{\partial \bar{m}_{11,B}}{\partial x_i}\bar{F}_{2,B}+\bar{m}_{11,B}\frac{\partial \bar{F}_{2,B}}{\partial x_i} \quad (1.278)$$

Computation of the Jacobian matrix $\nabla_x g_1(x)$:

First row of the Jacobian matrix $\nabla_x g_1(x)$: It holds that for $i=1,2,\cdots,8$

$$\frac{\partial g_{11}}{\partial x_i}=0 \quad (1.279)$$

Second row of the Jacobian matrix $\nabla_x g_1(x)$: It holds that for $i=1,2,\cdots,8$

$$\frac{\partial g_{12}}{\partial x_i}=\frac{\frac{\partial \bar{m}_{22,A}}{\partial x_i}detM_A-\bar{m}_{22,A}\frac{\partial detM_A}{\partial x_i}}{(detM_A)^2} \quad (1.280)$$

Third row of the Jacobian matrix $\nabla_x g_1(x)$: It holds that for $i=1,2,\cdots,8$

$$\frac{\partial g_{13}}{\partial x_i}=0 \quad (1.281)$$

Fourth row of the Jacobian matrix $\nabla_x g_1(x)$: It holds that for $i=1,2,\cdots,8$

$$\frac{\partial g_{14}}{\partial x_i}=\frac{-\frac{\partial \bar{m}_{21,A}}{\partial x_i}detM_A+\bar{m}_{21,A}\frac{\partial detM_A}{\partial x_i}}{(detM_A)^2} \quad (1.282)$$

Fifth row of the Jacobian matrix $\nabla_x g_1(x)$: It holds that for $i=1,2,\cdots,8$

$$\frac{\partial g_{15}}{\partial x_i}=0 \quad (1.283)$$

Sixth row of the Jacobian matrix $\nabla_x g_1(x)$: It holds that for $i=1,2,\cdots,8$

$$\frac{\partial g_{16}}{\partial x_i}=0 \quad (1.284)$$

Seventh row of the Jacobian matrix $\nabla_x g_1(x)$: It holds that for $i=1,2,\cdots,8$

$$\frac{\partial g_{17}}{\partial x_i}=0 \quad (1.285)$$

Eighth row of the Jacobian matrix $\nabla_x g_1(x)$: It holds that for $i=1,2,\cdots,8$

$$\frac{\partial g_{18}}{\partial x_i}=0 \quad (1.286)$$

Computation of the Jacobian matrix $\nabla_x g_2(x)$:

First row of the Jacobian matrix $\nabla_x g_2(x)$: It holds that for $i=1,2,\cdots,8$

$$\frac{\partial g_{21}}{\partial x_i}=0 \quad (1.287)$$

Second row of the Jacobian matrix $\nabla_x g_2(x)$: It holds that for $i = 1, 2, \cdots, 8$

$$\frac{\partial g_{22}}{\partial x_i} = \frac{\frac{-\partial \bar{m}_{12,A}}{\partial x_i} detM_A + \bar{m}_{12,A} \frac{\partial detM_A}{\partial x_i}}{(detM_A)^2} \quad (1.288)$$

Third row of the Jacobian matrix $\nabla_x g_2(x)$: It holds that for $i = 1, 2, \cdots, 8$

$$\frac{\partial g_{23}}{\partial x_i} = 0 \quad (1.289)$$

Fourth row of the Jacobian matrix $\nabla_x g_2(x)$: It holds that for $i = 1, 2, \cdots, 8$

$$\frac{\partial g_{24}}{\partial x_i} = \frac{\frac{\partial \bar{m}_{11,A}}{\partial x_i} detM_A - \bar{m}_{11,A} \frac{\partial detM_A}{\partial x_i}}{(detM_A)^2} \quad (1.290)$$

Fifth row of the Jacobian matrix $\nabla_x g_2(x)$: It holds that for $i = 1, 2, \cdots, 8$

$$\frac{\partial g_{25}}{\partial x_i} = 0 \quad (1.291)$$

Sixth row of the Jacobian matrix $\nabla_x g_2(x)$: It holds that for $i = 1, 2, \cdots, 8$

$$\frac{\partial g_{26}}{\partial x_i} = 0 \quad (1.292)$$

Seventh row of the Jacobian matrix $\nabla_x g_2(x)$: It holds that for $i = 1, 2, \cdots, 8$

$$\frac{\partial g_{27}}{\partial x_i} = 0 \quad (1.293)$$

Eighth row of the Jacobian matrix $\nabla_x g_2(x)$: It holds that for $i = 1, 2, \cdots, 8$

$$\frac{\partial g_{28}}{\partial x_i} = 0 \quad (1.294)$$

Computation of the Jacobian matrix $\nabla_x g_3(x)$:

First row of the Jacobian matrix $\nabla_x g_3(x)$: It holds that for $i = 1, 2, \cdots, 8$

$$\frac{\partial g_{31}}{\partial x_i} = 0 \quad (1.295)$$

Second row of the Jacobian matrix $\nabla_x g_3(x)$: It holds that for $i = 1, 2, \cdots, 8$

$$\frac{\partial g_{32}}{\partial x_i} = 0 \quad (1.296)$$

Third row of the Jacobian matrix $\nabla_x g_3(x)$: It holds that for $i = 1, 2, \cdots, 8$

$$\frac{\partial g_{33}}{\partial x_i} = 0 \quad (1.297)$$

Fourth row of the Jacobian matrix $\nabla_x g_3(x)$: It holds that for $i = 1, 2, \cdots, 8$

$$\frac{\partial g_{34}}{\partial x_i} = 0 \quad (1.298)$$

Fifth row of the Jacobian matrix $\nabla_x g_3(x)$: It holds that for $i = 1, 2, \cdots, 8$

$$\frac{\partial g_{35}}{\partial x_i} = 0 \quad (1.299)$$

Sixth row of the Jacobian matrix $\nabla_x g_3(x)$: It holds that for $i = 1, 2, \cdots, 8$

$$\frac{\partial g_{36}}{\partial x_i} = \frac{\frac{\partial \bar{m}_{22,B}}{\partial x_i} detM_B - \bar{m}_{22,B} \frac{\partial detM_B}{\partial x_i}}{(detM_B)^2} \quad (1.300)$$

Seventh row of the Jacobian matrix $\nabla_x g_3(x)$: It holds that for $i = 1, 2, \cdots, 8$

$$\frac{\partial g_{37}}{\partial x_i} = 0 \quad (1.301)$$

Eighth row of the Jacobian matrix $\nabla_x g_3(x)$: It holds that for $i = 1, 2, \cdots, 8$

$$\frac{\partial g_{38}}{\partial x_i} = \frac{\frac{-\partial \bar{m}_{21,B}}{\partial x_i} detM_B + \bar{m}_{21,B} \frac{\partial detM_B}{\partial x_i}}{(detM_B)^2} \quad (1.302)$$

Computation of the Jacobian matrix $\nabla_x g_4(x)$:

First row of the Jacobian matrix $\nabla_x g_4(x)$: It holds that for $i = 1, 2, \cdots, 8$

$$\frac{\partial g_{41}}{\partial x_i} = 0 \tag{1.303}$$

Second row of the Jacobian matrix $\nabla_x g_4(x)$: It holds that for $i = 1, 2, \cdots, 8$

$$\frac{\partial g_{42}}{\partial x_i} = 0 \tag{1.304}$$

Third row of the Jacobian matrix $\nabla_x g_4(x)$: It holds that for $i = 1, 2, \cdots, 8$

$$\frac{\partial g_{43}}{\partial x_i} = 0 \tag{1.305}$$

Fourth row of the Jacobian matrix $\nabla_x g_4(x)$: It holds that for $i = 1, 2, \cdots, 8$

$$\frac{\partial g_{44}}{\partial x_i} = 0 \tag{1.306}$$

Fifth row of the Jacobian matrix $\nabla_x g_4(x)$: It holds that for $i = 1, 2, \cdots, 8$

$$\frac{\partial g_{45}}{\partial x_i} = 0 \tag{1.307}$$

Sixth row of the Jacobian matrix $\nabla_x g_4(x)$: It holds that for $i = 1, 2, \cdots, 8$

$$\frac{\partial g_{46}}{\partial x_i} = \frac{-\frac{\partial \bar{m}_{12,B}}{\partial x_i} detM_B + \bar{m}_{12,B} \frac{\partial detM_B}{\partial x_i}}{(detM_B)^2} \tag{1.308}$$

Seventh row of the Jacobian matrix $\nabla_x g_4(x)$: It holds that for $i = 1, 2, \cdots, 8$

$$\frac{\partial g_{47}}{\partial x_i} = 0 \tag{1.309}$$

Eighth row of the Jacobian matrix $\nabla_x g_4(x)$: It holds that for $i = 1, 2, \cdots, 8$

$$\frac{\partial g_{48}}{\partial x_i} = \frac{\frac{\partial \bar{m}_{11,B}}{\partial x_i} detM_B - \bar{m}_{11,B} \frac{\partial detM_B}{\partial x_i}}{(detM_B)^2} \tag{1.310}$$

Computation of the partial derivatives of the elements of the inertia matrix M_A:

It holds that for $i = 1, 2, \cdots, 8$ and $i \neq 3$

$$\frac{\partial \bar{m}_{11,A}}{\partial x_i} = 0 \tag{1.311}$$

while for $i = 3$, one has that

$$\frac{\partial \bar{m}_{11,A}}{\partial x_3} = -2m_{2,A} l_{1,A} l_{2,A} sin(x_3) \tag{1.312}$$

Moreover, for $i = 1, 2, \cdots, 8$ and $i \neq 3$

$$\frac{\partial \bar{m}_{12,A}}{\partial x_i} = \frac{\partial \bar{m}_{21,A}}{\partial x_i} = 0 \tag{1.313}$$

while for $i = 3$, one has that

$$\frac{\partial \bar{m}_{12,A}}{\partial x_3} = \frac{\partial \bar{m}_{21,A}}{\partial x_3} = -m_{2,A} l_{1,A} l_{2,A} sin(x_3) \tag{1.314}$$

Additionally, for $i = 1, 2, \cdots, 8$ and $i \neq 3$

$$\frac{\partial \bar{m}_{22,A}}{\partial x_i} = 0 \tag{1.315}$$

while for $i = 3$, one has that

$$\frac{\partial \bar{m}_{22,A}}{\partial x_3} = -m_{2,A} l_{2,A} (l_{1,A} + l_{2,A}) sin(x_3) \tag{1.316}$$

Computation of the partial derivatives of the elements of the inertia matrix M_B:

It holds that for $i=1,2,\cdots,8$ and $i{\neq}7$

$$\frac{\partial\bar{m}_{11,B}}{\partial x_i}=0 \tag{1.317}$$

while for $i=7$, one has that

$$\frac{\partial\bar{m}_{11,B}}{\partial x_7}=-2m_{2,B}l_{1,B}l_{2,B}sin(x_7) \tag{1.318}$$

Moreover, for $i=1,2,\cdots,8$ and $i{\neq}7$

$$\frac{\partial\bar{m}_{12,B}}{\partial x_i}=\frac{\partial\bar{m}_{21,B}}{\partial x_i}=0 \tag{1.319}$$

while for $i=7$, one has that

$$\frac{\partial\bar{m}_{12,B}}{\partial x_7}=\frac{\partial\bar{m}_{21,B}}{\partial x_7}=-m_{2,B}l_{1,B}l_{2,B}sin(x_7) \tag{1.320}$$

Additionally, for $i=1,2,\cdots,8$ and $i{\neq}7$

$$\frac{\partial\bar{m}_{22,B}}{\partial x_i}=0 \tag{1.321}$$

while for $i=7$, one has that

$$\frac{\partial\bar{m}_{22,B}}{\partial x_7}=-m_{2,B}l_{2,B}(l_{1,B}+l_{2,B})sin(x_7) \tag{1.322}$$

Computation of the partial derivatives of terms $\bar{F}_{1,A}$ and $\bar{F}_{2,A}$:

It holds that for $i=1,2,\cdots,8$

$$\begin{array}{c}\frac{\partial\bar{F}_{1,A}}{\partial x_i}=\frac{\partial\bar{C}_{1,A}}{\partial x_i}+\frac{\partial\bar{G}_{1,A}}{\partial x_i}+\frac{\partial\bar{J}^T_{A,12}}{\partial x_i}Mg+\\ +\left[\frac{\partial\bar{J}^T_{A,11}}{\partial x_i}cos(\phi)+J^T_{A,11}\frac{\partial cos(\phi)}{\partial x_i}+\frac{\partial\bar{J}^T_{A,12}}{\partial x_i}sin(\phi)+J^T_{A,12}\frac{\partial sin(\phi)}{\partial x_i}\right]+\\ +[J^T_{A,11}cos(\phi)+J^T_{A,12}sin(\phi)]\frac{\partial F}{\partial x_i}\end{array} \tag{1.323}$$

$$\begin{array}{c}\frac{\partial\bar{F}_{2,A}}{\partial x_i}=\frac{\partial\bar{C}_{2,A}}{\partial x_i}+\frac{\partial\bar{G}_{2,A}}{\partial x_i}+\frac{\partial\bar{J}^T_{A,22}}{\partial x_i}Mg+\\ +\left[\frac{\partial\bar{J}^T_{A,21}}{\partial x_i}cos(\phi)+J^T_{A,21}\frac{\partial cos(\phi)}{\partial x_i}+\frac{\partial\bar{J}^T_{A,22}}{\partial x_i}sin(\phi)+J^T_{A,22}\frac{\partial sin(\phi)}{\partial x_i}\right]+\\ +[J^T_{A,21}cos(\phi)+J^T_{A,22}sin(\phi)]\frac{\partial F}{\partial x_i}\end{array} \tag{1.324}$$

Computation of the partial derivatives of terms $\bar{F}_{1,B}$ and $\bar{F}_{2,B}$:

It holds that for $i=1,2,\cdots,8$

$$\begin{array}{c}\frac{\partial\bar{F}_{1,B}}{\partial x_i}=\frac{\partial\bar{C}_{1,B}}{\partial x_i}+\frac{\partial\bar{G}_{1,B}}{\partial x_i}+\frac{\partial\bar{J}^T_{B,12}}{\partial x_i}Mg+\\ -\left[\frac{\partial\bar{J}^T_{B,11}}{\partial x_i}cos(\phi)+J^T_{B,11}\frac{\partial cos(\phi)}{\partial x_i}+\frac{\partial\bar{J}^T_{B,12}}{\partial x_i}sin(\phi)+J^T_{B,12}\frac{\partial sin(\phi)}{\partial x_i}\right]-\\ -[J^T_{B,11}cos(\phi)+J^T_{B,12}sin(\phi)]\frac{\partial F}{\partial x_i}\end{array} \tag{1.325}$$

$$\begin{array}{c}\frac{\partial\bar{F}_{2,B}}{\partial x_i}=\frac{\partial\bar{C}_{2,B}}{\partial x_i}+\frac{\partial\bar{G}_{2,B}}{\partial x_i}+\frac{\partial\bar{J}^T_{B,22}}{\partial x_i}Mg+\\ -\left[\frac{\partial\bar{J}^T_{B,21}}{\partial x_i}cos(\phi)+J^T_{B,21}\frac{\partial cos(\phi)}{\partial x_i}+\frac{\partial\bar{J}^T_{B,22}}{\partial x_i}sin(\phi)+J^T_{B,22}\frac{\partial sin(\phi)}{\partial x_i}\right]-\\ -[J^T_{B,21}cos(\phi)+J^T_{B,22}sin(\phi)]\frac{\partial F}{\partial x_i}\end{array} \tag{1.326}$$

Computation of the partial derivatives of the elements of the Coriolis matrix C_A

It holds that for $i = 1, 2, \cdots, 8$:

$$\begin{array}{ll} \frac{\partial \bar{C}_{1,A}}{\partial x_1} = 0 & \frac{\partial \bar{C}_{1,A}}{\partial x_2} = -2m_{2,A}l_{1,A}^2 sin(x_3)x_4 \\ \frac{\partial \bar{C}_{1,A}}{\partial x_3} = -m_{2,A}l_{1.A}^2 cos(x_1)x_4^2 - 2m_{2,A}l_{1,A}^2 cos(x_3)x_2x_4 & \frac{\partial \bar{C}_{1,A}}{\partial x_4} = -2m_{2,A}l_{1,A}^2 sin(x_3)x_4 - 2m_{2,A}l_{1,A}^2 sin(x_3)x_2 \\ \frac{\partial \bar{C}_{1,A}}{\partial x_5} = 0 & \frac{\partial \bar{C}_{1,A}}{\partial x_6} = 0 \\ \frac{\partial \bar{C}_{1,A}}{\partial x_7} = 0 & \frac{\partial \bar{C}_{1,A}}{\partial x_8} = 0 \end{array} \tag{1.327}$$

It holds that for $i = 1, 2, \cdots, 8$:

$$\begin{array}{ll} \frac{\partial \bar{C}_{2,A}}{\partial x_1} = 0 & \frac{\partial \bar{C}_{2,A}}{\partial x_2} = 2m_{2,A}l_{1,A}^2 sin(x_3)x_2 \\ \frac{\partial \bar{C}_{2,A}}{\partial x_3} = m_{2,A}l_{1,A}^2 cos(x_1)x_2^2 & \frac{\partial \bar{C}_{2,A}}{\partial x_4} = 0 \\ \frac{\partial \bar{C}_{2,A}}{\partial x_5} = 0 & \frac{\partial \bar{C}_{2,A}}{\partial x_6} = 0 \\ \frac{\partial \bar{C}_{2,A}}{\partial x_7} = 0 & \frac{\partial \bar{C}_{2,A}}{\partial x_8} = 0 \end{array} \tag{1.328}$$

Computation of the partial derivatives of the elements of the Coriolis matrix C_B

It holds that for $i = 1, 2, \cdots, 8$:

$$\begin{array}{ll} \frac{\partial \bar{C}_{1,B}}{\partial x_1} = 0 & \frac{\partial \bar{C}_{1,B}}{\partial x_2} = 0 \\ \frac{\partial \bar{C}_{1,B}}{\partial x_3} = 0 & \frac{\partial \bar{C}_{1,B}}{\partial x_4} = 0 \\ \frac{\partial \bar{C}_{1,B}}{\partial x_5} = 0 & \frac{\partial \bar{C}_{1,B}}{\partial x_6} = -2m_{2,B}l_{1,B}^2 sin(x_7)x_8 \\ \frac{\partial \bar{C}_{1,B}}{\partial x_7} = -m_{2,B}l_{2,B}^2 cos(x_7)x_8^2 - 2m_{2,B}l_{1,B}^2 cos(x_7)x_6x_8 & \frac{\partial \bar{C}_{1,B}}{\partial x_8} = 2m_{2,B}l_{2,B}^2 sin(x_7)x_8 - 2m_{2,B}l_{1,B}^2 sin(x_7)x_6 \end{array} \tag{1.329}$$

It holds that for $i = 1, 2, \cdots, 8$:

$$\begin{array}{ll} \frac{\partial \bar{C}_{2,B}}{\partial x_1} = 0 & \frac{\partial \bar{C}_{2,B}}{\partial x_2} = 0 \\ \frac{\partial \bar{C}_{2,B}}{\partial x_3} = 0 & \frac{\partial \bar{C}_{2,B}}{\partial x_4} = 0 \\ \frac{\partial \bar{C}_{2,B}}{\partial x_5} = 0 & \frac{\partial \bar{C}_{2,B}}{\partial x_6} = 2m_{2,B}l_{1,B}^2 sin(x_7)x_6 \\ \frac{\partial \bar{C}_{2,B}}{\partial x_7} = m_{2,B}l_{1,B}^2 cos(x_7)x_6^2 & \frac{\partial \bar{C}_{2,B}}{\partial x_8} = 0 \end{array} \tag{1.330}$$

Computation of the elements of the partial derivatives of the gravitational forces vector $\bar{G}_{1,A}$ and $\bar{G}_{2,A}$

It holds that for $i = 1, 2, \cdots, 8$:

$$\begin{array}{ll} \frac{\partial \bar{G}_{1,A}}{\partial x_1} = -(m_{1,A} + m_{2,A})gl_{1,A} sin(x_1) - m_{1,A}gl_{2,A} cos(x_1 + x_3 & \frac{\partial \bar{G}_{1,A}}{\partial x_2} = 0 \\ \frac{\partial \bar{G}_{1,A}}{\partial x_4} = -m_{1,A}gl_{2,A} sin(x_1 + x_3) & \frac{\partial \bar{G}_{1,A}}{\partial x_4} = 0 \\ \frac{\partial \bar{G}_{1,A}}{\partial x_5} = 0 & \frac{\partial \bar{G}_{1,A}}{\partial x_6} = 0 \\ \frac{\partial \bar{G}_{1,A}}{\partial x_7} = 0 & \frac{\partial \bar{G}_{1,A}}{\partial x_8} = 0 \end{array} \tag{1.331}$$

It holds that for $i=1,2,\cdots,8$:

$$\begin{array}{ll}\frac{\partial\bar{G}_{2,A}}{\partial x_1}=-m_{2,A}gl_{2,A}sin(x_1+x_3) & \frac{\partial\bar{G}_{2,A}}{\partial x_2}=0\\ \frac{\partial\bar{G}_{2,A}}{\partial x_3}=-m_{2,A}gl_{2,A}sin(x_1+x_3) & \frac{\partial\bar{G}_{2,A}}{\partial x_4}=0\\ \frac{\partial\bar{G}_{2,A}}{\partial x_5}=0 & \frac{\partial\bar{G}_{2,A}}{\partial x_6}=0\\ \frac{\partial\bar{G}_{2,A}}{\partial x_7}=0 & \frac{\partial\bar{G}_{2,A}}{\partial x_8}=0\end{array} \tag{1.332}$$

Computation of the elements of the partial derivatives of the gravitational forces vector $\bar{G}_{1,B}$ and $\bar{G}_{2,B}$

It holds that for $i=1,2,\cdots,8$:

$$\begin{array}{ll}\frac{\partial\bar{G}_{1,B}}{\partial x_1}=0 & \frac{\partial\bar{G}_{1,B}}{\partial x_2}=0\\ \frac{\partial\bar{G}_{1,B}}{\partial x_4}=0 & \frac{\partial\bar{B}_{1,B}}{\partial x_B}=0\\ \frac{\partial\bar{G}_{1,B}}{\partial x_5}=-(m_{1,B}+m_{2,B})gl_{1,B}sin(x_5)-m_{1,B}gl_{2,B}cos(x_5+x_7) & \frac{\partial\bar{B}_{1,B}}{\partial x_6}=0\\ \frac{\partial\bar{G}_{1,B}}{\partial x_7}=-m_{1,B}gl_{2,B}sin(x_5+x_7) & \frac{\partial\bar{B}_{1,B}}{\partial x_8}=0\end{array} \tag{1.333}$$

It holds that for $i=1,2,\cdots,8$:

$$\begin{array}{ll}\frac{\partial\bar{G}_{2,B}}{\partial x_1}=0 & \frac{\partial\bar{G}_{2,B}}{\partial x_2}=0\\ \frac{\partial\bar{G}_{2,B}}{\partial x_3}=0 & \frac{\partial\bar{G}_{2,B}}{\partial x_4}=0\\ \frac{\partial\bar{G}_{2,B}}{\partial x_5}=-m_{2,B}gl_{2,B}sin(x_5+x_7) & \frac{\partial\bar{G}_{2,B}}{\partial x_6}=0\\ \frac{\partial\bar{G}_{2,B}}{\partial x_7}=-m_{2,B}gl_{2,B}sin(x_5+x_7) & \frac{\partial\bar{G}_{2,B}}{\partial x_8}=0\end{array} \tag{1.334}$$

Computation of the partial derivatives of the elements of the transposed Jacobian matrix J_A^T:
It holds that $\frac{\partial J^T_{A,11}}{\partial x_1}=-l_{1,A}cos(x_1)-l_{2,A}cos(x_1+x_3)$, $\frac{\partial J^T_{A,11}}{\partial x_2}=0$, $\frac{\partial J^T_{A,11}}{\partial x_3}=-l_{2,A}cos(x_1+x_3)$, $\frac{\partial J^T_{A,11}}{\partial x_4}=0$, $\frac{\partial J^T_{A,11}}{\partial x_5}=0$, $\frac{\partial J^T_{A,11}}{\partial x_6}=0$, $\frac{\partial J^T_{A,11}}{\partial x_7}=0$, $\frac{\partial J^T_{A,11}}{\partial x_8}=0$.

Moreover, it holds that $\frac{\partial J^T_{A,12}}{\partial x_1}=-l_{1,A}sin(x_1)-l_{2,A}sin(x_1+x_3)$, $\frac{\partial J^T_{A,12}}{\partial x_2}=0$, $\frac{\partial J^T_{A,12}}{\partial x_3}=-l_{2,A}sin(x_1+x_3)$, $\frac{\partial J^T_{A,12}}{\partial x_4}=0$, $\frac{\partial J^T_{A,12}}{\partial x_5}=0$, $\frac{\partial J^T_{A,12}}{\partial x_6}=0$, $\frac{\partial J^T_{A,12}}{\partial x_7}=0$, $\frac{\partial J^T_{A,12}}{\partial x_8}=0$.

Additionally, it holds that $\frac{\partial J^T_{A,21}}{\partial x_1}=-l_{2,A}cos(x_1+x_3)$, $\frac{\partial J^T_{A,21}}{\partial x_2}=0$, $\frac{\partial J^T_{A,21}}{\partial x_3}=-l_{2,A}cos(x_1+x_3)$, $\frac{\partial J^T_{A,21}}{\partial x_4}=0$, $\frac{\partial J^T_{A,21}}{\partial x_5}=0$, $\frac{\partial J^T_{A,21}}{\partial x_6}=0$, $\frac{\partial J^T_{A,21}}{\partial x_7}=0$, $\frac{\partial J^T_{A,21}}{\partial x_8}=0$.

Finally, it holds that $\frac{\partial J^T_{A,22}}{\partial x_1}=l_{2,A}sin(x_1+x_3)$, $\frac{\partial J^T_{A,22}}{\partial x_2}=0$, $\frac{\partial J^T_{A,22}}{\partial x_3}=l_{2,A}sin(x_1+x_3)$, $\frac{\partial J^T_{A,22}}{\partial x_4}=0$, $\frac{\partial J^T_{A,22}}{\partial x_5}=0$, $\frac{\partial J^T_{A,22}}{\partial x_6}=0$, $\frac{\partial J^T_{A,22}}{\partial x_7}=0$, $\frac{\partial J^T_{A,22}}{\partial x_8}=0$.

Computation of the partial derivatives of the elements of the transposed Jacobian matrix J_B^T:

It holds that $\frac{\partial J^T_{B,11}}{\partial x_1}=$, $\frac{\partial J^T_{B,11}}{\partial x_2}=0$, $\frac{\partial J^T_{B,11}}{\partial x_3}=0$, $\frac{\partial J^T_{B,11}}{\partial x_4}=0$, $\frac{\partial J^T_{B,11}}{\partial x_5}=-l_{1,B}cos(x_5)-l_{2,B}cos(x_5+x_7)$, $\frac{\partial J^T_{B,11}}{\partial x_6}=0$, $\frac{\partial J^T_{B,11}}{\partial x_7}=-l_{2,B}cos(x_5+x_7)$, $\frac{\partial J^T_{B,11}}{\partial x_8}=0$.

Moreover, it holds that $\frac{\partial J_{B,12}^T}{\partial x_1}=0$, $\frac{\partial J_{B,12}^T}{\partial x_2}=0$, $\frac{\partial J_{B,12}^T}{\partial x_3}=0$, $\frac{\partial J_{B,12}^T}{\partial x_4}=0$, $\frac{\partial J_{B,12}^T}{\partial x_5}=-l_{1,B}sin(x_5)-l_{2,B}sin(x_5+x_7)$, $\frac{\partial J_{B,12}^T}{\partial x_6}=0$, $\frac{\partial J_{B,12}^T}{\partial x_7}=-l_{2,B}sin(x_5+x_7)$, $\frac{\partial J_{B,12}^T}{\partial x_8}=0$.

Additionally, it holds that $\frac{\partial J_{B,21}^T}{\partial x_1}=0$, $\frac{\partial J_{B,21}^T}{\partial x_2}=0$, $\frac{\partial J_{B,21}^T}{\partial x_3}=0$, $\frac{\partial J_{B,21}^T}{\partial x_4}=0$, $\frac{\partial J_{B,21}^T}{\partial x_5}=-l_{2,B}cos(x_5+x_7)$, $\frac{\partial J_{B,21}^T}{\partial x_6}=0$, $\frac{\partial J_{B,21}^T}{\partial x_7}=-l_{2,B}cos(x_5+x_7)$, $\frac{\partial J_{B,21}^T}{\partial x_8}=0$.

Finally, it holds that $\frac{\partial J_{B,22}^T}{\partial x_1}=0$, $\frac{\partial J_{B,22}^T}{\partial x_2}=0$, $\frac{\partial J_{B,22}^T}{\partial x_3}=0$, $\frac{\partial J_{B,22}^T}{\partial x_4}=0$, $\frac{\partial J_{B,22}^T}{\partial x_5}=l_{2,B}cos(x_5+x_7)$, $\frac{\partial J_{B,22}^T}{\partial x_6}=0$, $\frac{\partial J_{B,22}^T}{\partial x_7}=l_{2,B}cos(x_5+x_7)$, $\frac{\partial J_{B,22}^T}{\partial x_8}=0$.

Computation of the partial derivatives of force F:

It holds that:

$$\begin{aligned}\frac{\partial L}{\partial x_1}=\frac{1}{2L}\{&2[x_0^B+l_{1,B}cps(x_5)+l_{2,B}cos(x_5+x_7)-x_0^A-l_{1,B}cos(x_1)-l_{2,A}cos(x_1+x_3)]\cdot\\&\cdot[l_{1,A}sin(x_1)+l_{2,A}sin(x_1+x_3)]+\\&2[h+l_{1,B}sin(x_5)+l_{2,B}sin(x_5+x_7)-h-l_{1,A}sin(x_1)-l_{2,A}sin(x_1+x_3)\}\cdot\\&\cdot[-l_{1,A}cos(x_1)-l_{2,A}cos(x_1+x_3)]\end{aligned} \tag{1.335}$$

$$\frac{\partial L}{\partial x_2}=0 \tag{1.336}$$

$$\begin{aligned}\frac{\partial L}{\partial x_3}=\frac{1}{2L}\{&2[x_0^B+l_{1,B}cps(x_5)+l_{2,B}cos(x_5+x_7)-x_0^A-l_{1,B}cos(x_1)-l_{2,A}cos(x_1+x_3)]\cdot\\&\cdot[l_{2,A}sin(x_1+x_3)]+\\&2[h+l_{1,B}sin(x_5)+l_{2,B}sin(x_5+x_7)-h-l_{1,A}sin(x_1)-l_{2,A}sin(x_1+x_3)\}\cdot\\&\cdot[-l_{2,A}cos(x_1+x_3)]\end{aligned} \tag{1.337}$$

$$\frac{\partial L}{\partial x_4}=0 \tag{1.338}$$

$$\begin{aligned}\frac{\partial L}{\partial x_5}=\frac{1}{2L}\{&2[x_0^B+l_{1,B}cos(x_5)+l_{2,B}cos(x_5+x_7)-x_0^A-l_{1,A}cos(x_1)-l_{2,A}cos(x_1+x_4)]\cdot\\&\cdot[-l_{1,B}sin(x_5)-l_{2,B}sin(x_5+x_7)]+\\&+2[h+l_{1,B}sin(x_5)+l_{2,B}sin(x_5+x_7)-l_{1,A}sin(x_1)-l_{1,B}sin(x_5+x_7)]\cdot\\&\cdot[l_{1,B}cos(x_5)+l_{2,B}cos(x_5+x_7)]\end{aligned} \tag{1.339}$$

$$\frac{\partial L}{\partial x_6}=0 \tag{1.340}$$

$$\begin{aligned}\frac{\partial L}{\partial x_7}=\frac{1}{2L}\{&2[x_0^B+l_{1,B}cos(x_5)+l_{2,B}cos(x_5+x_7)-x_0^A-l_{1,A}cos(x_1)-l_{2,A}cos(x_1+x_4)]\cdot\\&\cdot[-l_{2,B}sin(x_5+x_7)]+\\&+2[h+l_{1,B}sin(x_5)+l_{2,B}sin(x_5+x_7)-l_{1,A}sin(x_1)-l_{1,B}sin(x_5+x_7)]\cdot\\&\cdot[l_{2,B}cos(x_5+x_7)]\end{aligned} \tag{1.341}$$

$$\frac{\partial L}{\partial x_8}=0 \tag{1.342}$$

Computation of the partial derivatives of the term $sin(\phi)$:

It holds that

$$sin(\phi)=\frac{y_E^B-y_e^A}{L}\Rightarrow sin(\phi)=\frac{[h+l_{1,B}sin(x_5)+l_{2,B}sin(x_5+x_7)-h-l_{1,A}sin(x_1)-l_{1,B}sin(x_5+x_7)]}{L} \tag{1.343}$$

by defining the numerator of the above fraction as S_{num}, one has

$$\frac{\partial sin(\phi)}{\partial x_i}=\frac{\frac{\partial S_{num}}{\partial x_i}L-S_{num}\frac{\partial L}{\partial x_i}}{L^2} \tag{1.344}$$

where it holds that for $i = 1, 2, \cdots, 8$

$$\begin{array}{ll} \frac{\partial S_{num}}{\partial x_1} = -l_{1,A}cos(x_1) - l_{2,A}cos(x_1+x_3) & \frac{\partial S_{num}}{\partial x_2} = 0 \\ \frac{\partial S_{num}}{\partial x_3} = -l_{2,A}cos(x_1+x_3) & \frac{\partial S_{num}}{\partial x_4} = 0 \\ \frac{\partial S_{num}}{\partial x_5} = l_{1,B}cos(x_5) + l_{2,B}cos(x_5+x_7) & \frac{\partial S_{num}}{\partial x_6} = 0 \\ \frac{\partial S_{num}}{\partial x_7} = l_{2,B}cos(x_5+x_7) & \frac{\partial S_{num}}{\partial x_8} = 0 \end{array} \tag{1.345}$$

Computation of the partial derivatives of the term $cos(\phi)$:

It holds that

$$sin(\phi) = \frac{x_E^B - x_e^A}{L} \Rightarrow sin(\phi) = \frac{[x_0^B + l_{1,B}cos(x_5) + l_{2,B}cos(x_5+x_7) - x_0^A - l_{1,A}cos(x_1) - l_{1,B}cos(x_5+x_7)]}{L} \tag{1.346}$$

by defining the numerator of the above fraction as C_{num}, one has

$$\frac{\partial cos(\phi)}{\partial x_i} = \frac{\frac{\partial C_{num}}{\partial x_i}L - C_{num}\frac{\partial L}{\partial x_i}}{L^2} \tag{1.347}$$

where it holds that for $i = 1, 2, \cdots, 8$

$$\begin{array}{ll} \frac{\partial C_{num}}{\partial x_1} = l_{1,A}sin(x_1) + l_{2,A}sin(x_1+x_3) & \frac{\partial C_{num}}{\partial x_2} = 0 \\ \frac{\partial C_{num}}{\partial x_3} = l_{2,A}sin(x_1+x_3) & \frac{\partial C_{num}}{\partial x_4} = 0 \\ \frac{\partial C_{num}}{\partial x_5} = -l_{1,B}sin(x_5) - l_{2,B}sin(x_5+x_7) & \frac{\partial C_{num}}{\partial x_6} = 0 \\ \frac{\partial C_{num}}{\partial x_7} = -l_{2,B}sin(x_5+x_7) & \frac{\partial C_{num}}{\partial x_8} = 0 \end{array} \tag{1.348}$$

1.3.4 Design of an H-infinity nonlinear feedback controller

1.3.4.1 Equivalent linearized dynamics of the dual-arm robotic manipulator

After linearization around its current operating point, the dynamic model for the dual-arm robotic manipulator is written as

$$\dot{x} = Ax + Bu + d_1 \tag{1.349}$$

Parameter d_1 stands for the linearization error in the dual-arm robotic manipulator's model that was given previously in Eq. (1.349). The reference setpoints for the state vector of the aforementioned dynamic model are denoted by $\mathbf{x_d} = [x_1^d, \cdots, x_8^d]$. Tracking of this trajectory is achieved after applying the control input u^*. At every time instant the control input u^* is assumed to differ from the control input u appearing in Eq. (1.349) by an amount equal to Δu, that is $u^* = u + \Delta u$

$$\dot{x}_d = Ax_d + Bu^* + d_2 \tag{1.350}$$

The dynamics of the controlled system described in Eq. (1.349) can be also written as

$$\dot{x} = Ax + Bu + Bu^* - Bu^* + d_1 \tag{1.351}$$

and by denoting $d_3 = -Bu^* + d_1$ as an aggregate disturbance term, one obtains

$$\dot{x} = Ax + Bu + Bu^* + d_3 \tag{1.352}$$

By subtracting Eq. (1.350) from Eq. (1.352), one has

$$\dot{x} - \dot{x}_d = A(x - x_d) + Bu + d_3 - d_2 \tag{1.353}$$

By denoting the tracking error as $e = x - x_d$ and the aggregate disturbance term as $L\tilde{d} = d_3 - d_2$, the tracking error dynamics becomes

$$\dot{e} = Ae + Bu + L\tilde{d} \tag{1.354}$$

where L is a disturbance inputs gain matrix. The above-linearized form of the dual-arm robotic manipulator's model can be efficiently controlled after applying an H-infinity feedback control scheme.

1.3.4.2 The nonlinear H-infinity control

The initial nonlinear model of the dual-arm robotic manipulator is in the form

$$\dot{x} = f(x, u) \quad x \in R^n, \quad u \in R^m \tag{1.355}$$

Linearization of the model of the dual-arm robotic manipulator is performed at each iteration of the control algorithm around its present operating point $(x^*, u^*) = (x(t), u(t - T_s))$. The linearized equivalent of the system is described by

$$\dot{x} = Ax + Bu + L\tilde{d} \quad x \in R^n, \quad u \in R^m, \quad \tilde{d} \in R^q \tag{1.356}$$

where matrices A and B are obtained from the computation of the previously defined Jacobians and vector $\tilde{d}$ denotes disturbance terms due to linearization errors, while L is a disturbance inputs gain matrix. The problem of disturbance rejection for the linearized model that is described by

$$\begin{aligned} \dot{x} &= Ax + Bu + L\tilde{d} \\ y &= Cx \end{aligned} \tag{1.357}$$

where $x \in R^n$, $u \in R^m$, $\tilde{d} \in R^q$, and $y \in R^p$, cannot be handled efficiently if the classical LQR control scheme is applied. This is because of the existence of the perturbation term $\tilde{d}$. The disturbance term $\tilde{d}$ apart from modeling (parametric) uncertainty and external perturbation terms can also represent noise terms of any distribution.

In the H_∞ control approach, a feedback control scheme is designed for trajectory tracking by the system's state vector and simultaneous disturbance rejection, considering that the disturbance affects the system in the worst possible manner. The disturbances' effects are incorporated in the following quadratic cost function:

$$J(t) = \frac{1}{2}\int_0^T [y^T(t)y(t) + ru^T(t)u(t) - \rho^2\tilde{d}^T(t)\tilde{d}(t)]dt, \quad r, \rho > 0 \tag{1.358}$$

The significance of the negative sign in the cost function's term that is associated with the perturbation variable $\tilde{d}(t)$ is that the disturbance tries to maximize the cost function $J(t)$ while the control signal $u(t)$ tries to minimize it. The physical meaning of the relation given above is that the control signal and the disturbances compete with each other within a min–max differential game. This problem of min–max optimization can be written as $min_u max_{\tilde{d}} J(u, \tilde{d})$.

The objective of the optimization procedure is to compute a control signal $u(t)$ which can compensate for the worst possible disturbance, that is externally imposed on the dual-arm robotic manipulator. However, the solution to the min–max optimization problem is directly related to the value of parameter ρ. This means that there is an upper bound in the disturbances' magnitude that can be annihilated by the control signal.

1.3.4.3 Computation of the feedback control gains

For the linearized system given by Eq. (1.357) the cost function of Eq. (1.358) is defined, where coefficient r determines the penalization of the control input and weight coefficient ρ determines the reward of the disturbances' effects. It is assumed that (i) The energy that is transferred from the disturbances' signal $\tilde{d}(t)$ is bounded, that is $\int_0^\infty \tilde{d}^T(t)\tilde{d}(t)dt < \infty$, (ii) matrices $[A, B]$ and $[A, L]$ are stabilizable, (iii) matrix $[A, C]$ is detectable. In the case of a tracking problem, the optimal feedback control law is given by

$$u(t) = -Ke(t) \tag{1.359}$$

with $e = x - x_d$ to be the tracking error, and $K = \frac{1}{r}B^TP$ where P is a positive definite symmetric matrix. As it will be proven in Section 1.3.5, matrix P is obtained from the solution of the Riccati equation

$$A^TP + PA + Q - P(\tfrac{2}{r}BB^T - \tfrac{1}{\rho^2}LL^T)P = 0 \tag{1.360}$$

where Q is a positive semi-definite symmetric matrix. The worst-case disturbance is given by

$$\tilde{d}(t) = \frac{1}{\rho^2}L^TPe(t) \tag{1.361}$$

The diagram of the considered control loop for the dual-arm robotic manipulator is depicted in Figure 1.58.

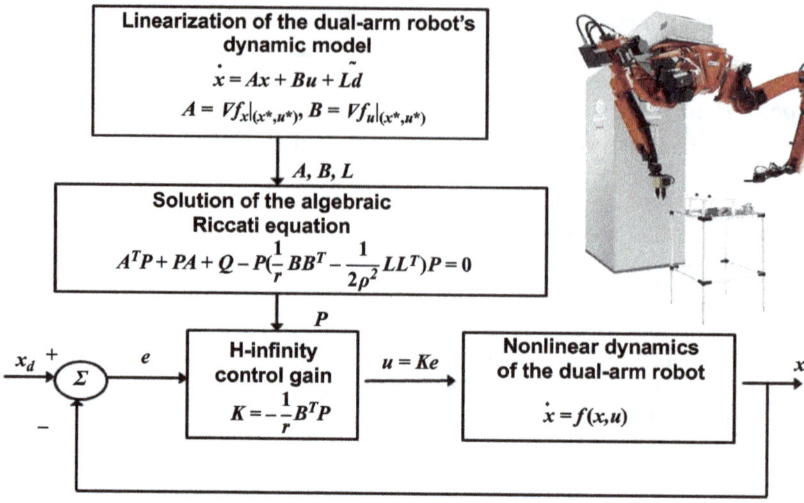

Figure 1.58 Diagram of the control scheme for the dual-arm robotic manipulator

1.3.5 Lyapunov stability analysis

Through Lyapunov stability analysis, it will be shown that the proposed nonlinear control scheme assures H_∞ tracking performance for the dual-arm robotic manipulator and that in case of bounded disturbance terms asymptotic convergence to the reference setpoints is achieved. The tracking error dynamics for the dual-drive H-type gantry crane is written in the form

$$\dot{e} = Ae + Bu + L\tilde{d} \tag{1.362}$$

where in the dual-drive H-type gantry crane's case $L = \in R^{8\times 8}$ to be the disturbance inputs gain matrix. Variable $\tilde{d}$ denotes model uncertainties and external disturbances of the dual-arm robotic manipulator's model. The following Lyapunov equation is considered:

$$V = \tfrac{1}{2}e^T P e \tag{1.363}$$

where $e = x - x_d$ is the tracking error. By differentiating with respect to time, one obtains

$$\begin{aligned} \dot{V} &= \tfrac{1}{2}\dot{e}^T P e + \tfrac{1}{2}e^T P \dot{e} \Rightarrow \\ \dot{V} &= \tfrac{1}{2}[Ae + Bu + L\tilde{d}]^T P e + \tfrac{1}{2}e^T P[Ae + Bu + L\tilde{d}] \Rightarrow \end{aligned} \tag{1.364}$$

$$\begin{aligned} \dot{V} = {} & \tfrac{1}{2}[e^T A^T + u^T B^T + \tilde{d}^T L^T]Pe + \\ & + \tfrac{1}{2}e^T P[Ae + Bu + L\tilde{d}] \Rightarrow \end{aligned} \tag{1.365}$$

$$\begin{aligned} \dot{V} = {} & \tfrac{1}{2}e^T A^T P e + \tfrac{1}{2}u^T B^T P e + \tfrac{1}{2}\tilde{d}^T L^T P e + \\ & \tfrac{1}{2}e^T P A e + \tfrac{1}{2}e^T P B u + \tfrac{1}{2}e^T P L \tilde{d} \end{aligned} \tag{1.366}$$

The previous equation is rewritten as

$$\begin{aligned} \dot{V} = {} & \tfrac{1}{2}e^T(A^T P + PA)e + (\tfrac{1}{2}u^T B^T P e + \tfrac{1}{2}e^T P B u) + \\ & + (\tfrac{1}{2}\tilde{d}^T L^T P e + \tfrac{1}{2}e^T P L \tilde{d}) \end{aligned} \tag{1.367}$$

Assumption: For given positive definite matrix Q and coefficients r and ρ, there exists a positive definite matrix P, which is the solution of the following matrix equation:

$$A^TP + PA = -Q + P(\tfrac{2}{r}BB^T - \tfrac{1}{\rho^2}LL^T)P \tag{1.368}$$

Moreover, the following feedback control law is applied to the system:

$$u = -\tfrac{1}{r}B^TPe \tag{1.369}$$

By substituting Eq. (1.368) and Eq. (1.369), one obtains

$$\begin{aligned}\dot{V} = &\tfrac{1}{2}e^T[-Q + P(\tfrac{2}{r}BB^T - \tfrac{1}{\rho^2}LL^T)P]e+\\ &+e^TPB(-\tfrac{1}{r}B^TPe) + e^TPL\tilde{d} \Rightarrow\end{aligned} \tag{1.370}$$

$$\begin{aligned}\dot{V} = &-\tfrac{1}{2}e^TQe + \tfrac{1}{r}e^TPBB^TPe - \tfrac{1}{2\rho^2}e^TPLL^TPe\\ &-\tfrac{1}{r}e^TPBB^TPe + e^TPL\tilde{d}\end{aligned} \tag{1.371}$$

which after intermediate operations gives

$$\dot{V} = -\tfrac{1}{2}e^TQe - \tfrac{1}{2\rho^2}e^TPLL^TPe + e^TPL\tilde{d} \tag{1.372}$$

or, equivalently,

$$\begin{aligned}\dot{V} = &-\tfrac{1}{2}e^TQe - \tfrac{1}{2\rho^2}e^TPLL^TPe+\\ &+\tfrac{1}{2}e^TPL\tilde{d} + \tfrac{1}{2}\tilde{d}^TL^TPe\end{aligned} \tag{1.373}$$

Lemma: The following inequality holds:

$$\tfrac{1}{2}e^TL\tilde{d} + \tfrac{1}{2}\tilde{d}L^TPe - \tfrac{1}{2\rho^2}e^TPLL^TPe \leq \tfrac{1}{2}\rho^2\tilde{d}^T\tilde{d} \tag{1.374}$$

Proof: The binomial $(\rho\alpha - \frac{1}{\rho}b)^2$ is considered. Expanding the left part of the above inequality, one gets

$$\begin{aligned}&\rho^2a^2 + \tfrac{1}{\rho^2}b^2 - 2ab \geq 0 \Rightarrow \tfrac{1}{2}\rho^2a^2 + \tfrac{1}{2\rho^2}b^2 - ab \geq 0 \Rightarrow\\ &ab - \tfrac{1}{2\rho^2}b^2 \leq \tfrac{1}{2}\rho^2a^2 \Rightarrow \tfrac{1}{2}ab + \tfrac{1}{2}ab - \tfrac{1}{2\rho^2}b^2 \leq \tfrac{1}{2}\rho^2a^2\end{aligned} \tag{1.375}$$

The following substitutions are carried out: $a = \tilde{d}$ and $b = e^TPL$ and the previous relation becomes

$$\tfrac{1}{2}\tilde{d}^TL^TPe + \tfrac{1}{2}e^TPL\tilde{d} - \tfrac{1}{2\rho^2}e^TPLL^TPe \leq \tfrac{1}{2}\rho^2\tilde{d}^T\tilde{d} \tag{1.376}$$

Eq. (1.376) is substituted in Eq. (1.373) and the inequality is enforced, thus giving

$$\dot{V} \leq -\tfrac{1}{2}e^TQe + \tfrac{1}{2}\rho^2\tilde{d}^T\tilde{d} \tag{1.377}$$

Eq. (1.377) shows that the H_∞ tracking performance criterion is satisfied. The integration of $\dot{V}$ from 0 to T gives

$$\begin{aligned}&\int_0^T\dot{V}(t)dt \leq -\tfrac{1}{2}\int_0^T||e||_Q^2dt + \tfrac{1}{2}\rho^2\int_0^T||\tilde{d}||^2dt \Rightarrow\\ &2V(T) + \int_0^T||e||_Q^2dt \leq 2V(0) + \rho^2\int_0^T||\tilde{d}||^2dt\end{aligned} \tag{1.378}$$

Moreover, if there exists a positive constant $M_d > 0$ such that

$$\int_0^\infty ||\tilde{d}||^2dt \leq M_d \tag{1.379}$$

then, one gets

$$\int_0^\infty ||e||_Q^2dt \leq 2V(0) + \rho^2M_d \tag{1.380}$$

Thus, the integral $\int_0^\infty ||e||_Q^2dt$ is bounded. Moreover, $V(T)$ is bounded and from the definition of the Lyapunov function V in Eq. (1.363) it becomes clear that $e(t)$ will be also bounded since $e(t) \in \Omega_e = \{e|e^TPe \leq 2V(0) + \rho^2M_d\}$. According to the above and with the use of Barbalat's Lemma, one obtains $\lim_{t\to\infty} e(t) = 0$.

Through the stages of the stability proof, one arrives at Eq. (1.377) which shows that the H-infinity tracking performance criterion holds. By selecting the attenuation coefficient ρ to be sufficiently small and in particular to satisfy

$\rho^2 < ||e||^2_Q/||\tilde{d}||^2$ one has that the first derivative of the Lyapunov function is upper bounded by 0. This condition holds at each sampling instance and consequently global stability for the control loop can be concluded.

1.3.6 Simulation tests

The global stability properties of the control method and the elimination of the state vector's tracking error which were previously proven through Lyapunov analysis are further confirmed through simulation experiments. To implement the nonlinear optimal (H-infinity) control method in the dynamic model of the dual-arm robotic manipulator the algebraic Riccati equation of this control method which was previously presented in Eq. (1.368) had to be solved at each iteration of the control algorithm. Indicative values for the parameters of the two robotic arms were $m_{1,A} = 2.0kg$, $l_{1,A} = 1.0m$, $m_{2,A} = 1.0kg$, $l_{2,A} = 1.0m$ and also $m_{1,B} = 2.0kg$, $l_{1,B} = 1.0m$, $m_{2,B} = 1.0kg$, $l_{2,b} = 1.0m$. The mass of the lifted load was $M = 1.0kg$ and the elasticity coefficient describing the deformation of the lifted object was $K_L = 10$Nt/m. The sampling period was $T_s = 0.01$ sec. Simulation experiments were concerned with eight different functioning cases of the dual-arm robot resulting in different positions of its end-effector in the robot's workspace. The turn angles of the joints of this robot were measured in rads while the associated angular velocities were measured in rad/sec. The obtained results are depicted in Figures 1.59–1.74. In these diagrams the state variables of the dual-arm robotic system are presented in blue color, their estimated values which are provided by the H-infinity Kalman filter are given in green color, while the associated reference setpoints are plotted in red color. It can be noticed that, in all test cases, the nonlinear optimal control method achieves fast and accurate tracking of reference setpoints by all state variables of the dual-arm robot under moderate variations of the control inputs.

1.4 Flatness-based control in successive loops for electropneumatic robots

1.4.1 Outline

Recent advances in nonlinear control have been benefited much from the theoretical background of differential flatness theory [711,721,732,725,491,252,898]. A system is considered to be differentially flat if all its state variables and its control inputs can be expressed as functions of one single algebraic variable which is the so-called flat output, and also as functions of the flat-output's derivatives [91,594,639,489]. The differential flatness property enables the transformation

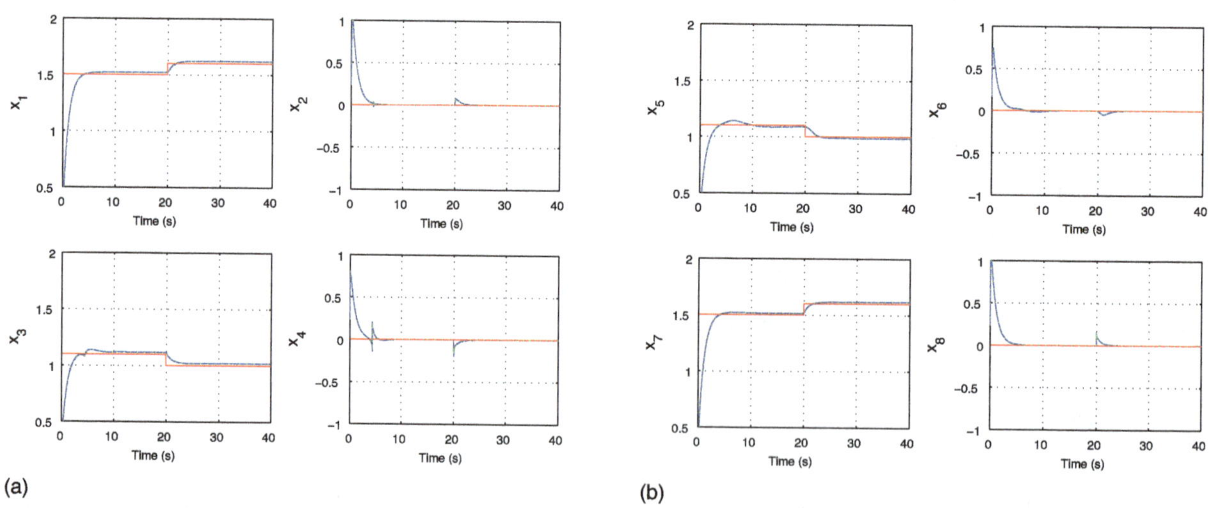

Figure 1.59 Tracking of setpoint 1 by the dual-arm robotic manipulator with nonlinear optimal control: (a) tracking of setpoints (red line) by the state variables of the dual-arm robot x_1 to x_4 (blue line) and estimates provided by the H-infinity Kalman filter; (b) tracking of setpoints (red line) by the state variables of the dual-arm robot x_5 to x_8 (blue line) and estimates provided by the H-infinity Kalman filter

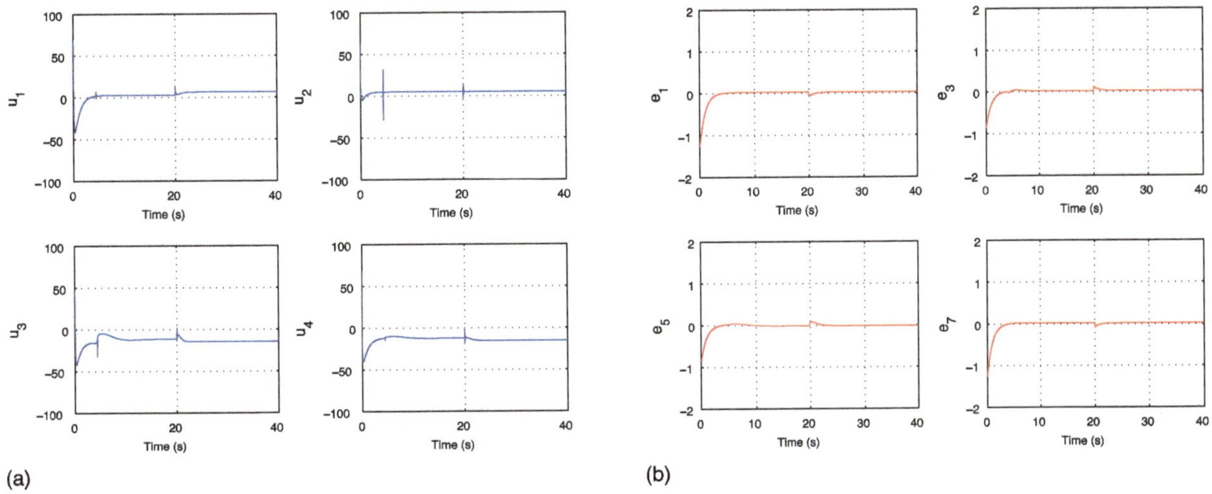

Figure 1.60 *Tracking of setpoint 1 by the dual-arm robotic manipulator with nonlinear optimal control: (a) variations of the control inputs u_1 to u_4 of the dual-arm robot (blue line); (b) tracking error for the turn-angle state variables of the dual-arm robot x_1, x_3, x_5, and x_7 (red line)*

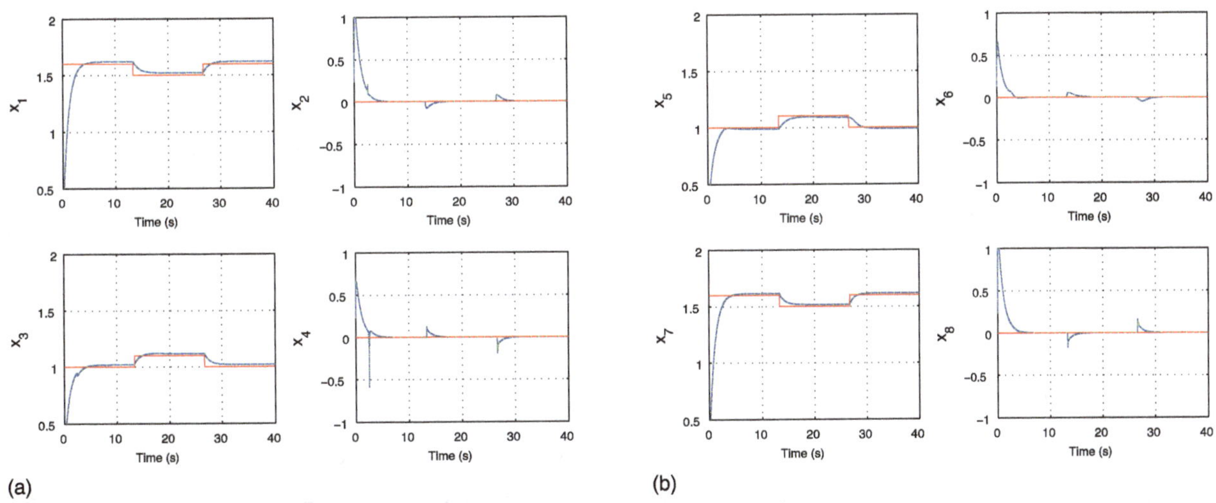

Figure 1.61 *Tracking of setpoint 2 by the dual-arm robotic manipulator with nonlinear optimal control: (a) tracking of setpoints (red line) by the state variables of the dual-arm robot x_1 to x_4 (blue line) and estimates provided by the H-infinity Kalman filter; (b) tracking of setpoints (red line) by the state variables of the dual-arm robot x_5 to x_8 (blue line) and estimates provided by the H-infinity Kalman filter*

of the nonlinear system's dynamics in the linear canonical form [806,492,640,524,54]. The latter description is controllable and observable thus allowing to treat effectively control and estimation problems [439,705,60,706]. In this section, a successive loops approach is developed for controller design in nonlinear dynamical systems which exhibit the differential flatness property. The method makes use of the initial nonlinear model of the system and of its decomposition into a set of nonlinear subsystems for which the differential flatness property holds [712,720,714,716,718].

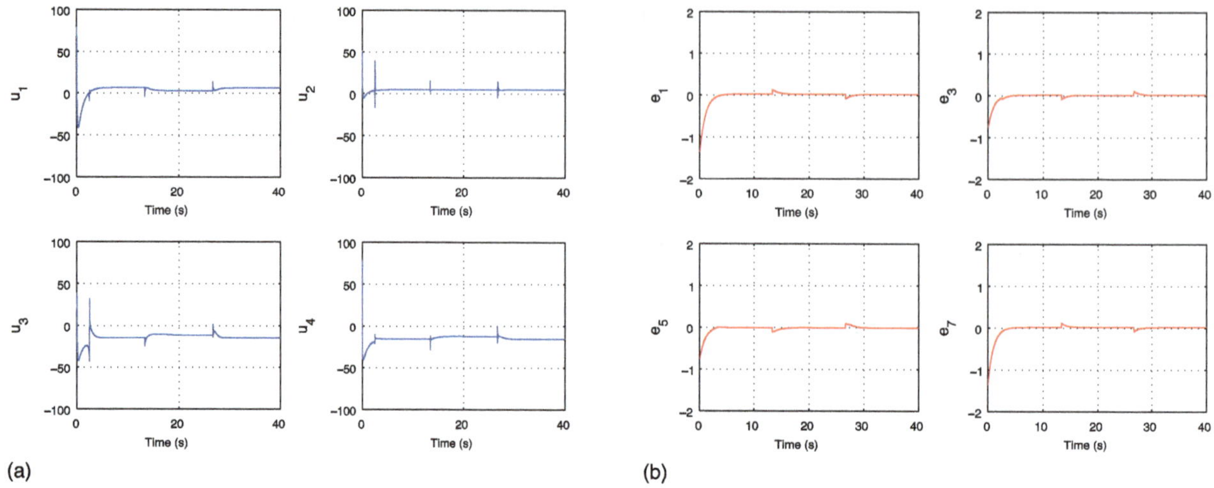

Figure 1.62 Tracking of setpoint 2 by the dual-arm robotic manipulator with nonlinear optimal control: (a) variations of the control inputs u_1 to u_4 of the dual-arm robot (blue line); (b) tracking error for the turn-angle state variables of the dual-arm robot x_1, x_3, x_5, and x_7 (red line)

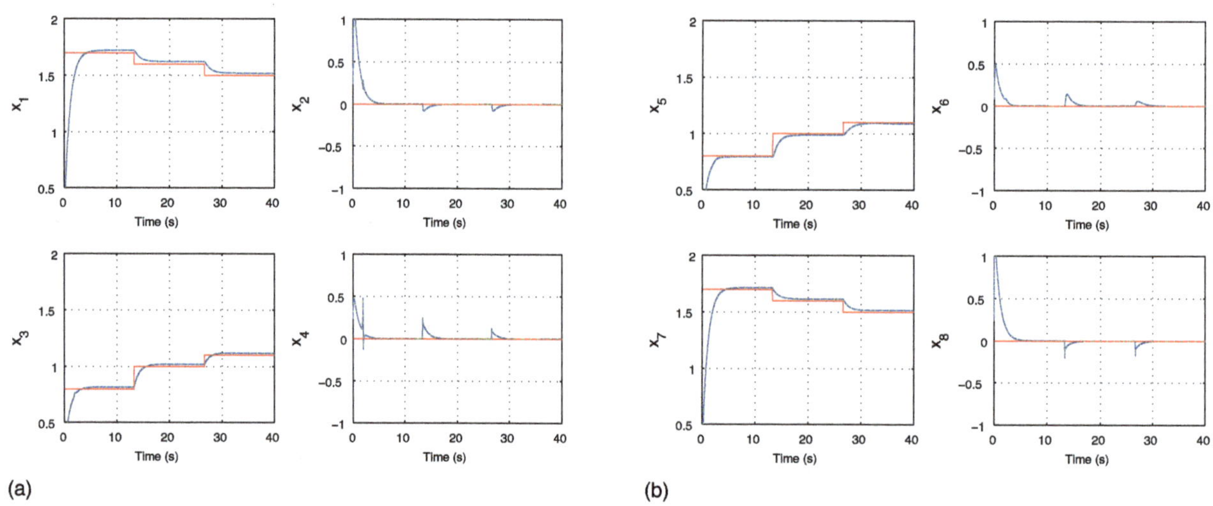

Figure 1.63 Tracking of setpoint 3 by the dual-arm robotic manipulator with nonlinear optimal control: (a) tracking of setpoints (red line) by the state variables of the dual-arm robot x_1 to x_4 (blue line) and estimates provided by the H-infinity Kalman filter; (b) tracking of setpoints (red line) by the state variables of the dual-arm robot x_5 to x_8 (blue line) and estimates provided by the H-infinity Kalman filter

The proposed control method is directly applicable to nonlinear systems of the so-called triangular form, or to systems which can be transformed into such a form [711,721]. The state-space model of the nonlinear system is decomposed into subsystems, which satisfy differential flatness properties. For each subsystem of the state-space model a virtual control input is computed, capable of inverting the subsystem's dynamics and of eliminating the subsystem's tracking error. The control input that is actually applied to the nonlinear system is found from the last rows of the state-space description. This control input incorporates in a recursive manner all virtual control inputs which were computed for the individual subsystems associated with the initial state-space equation. The control input that should be applied to

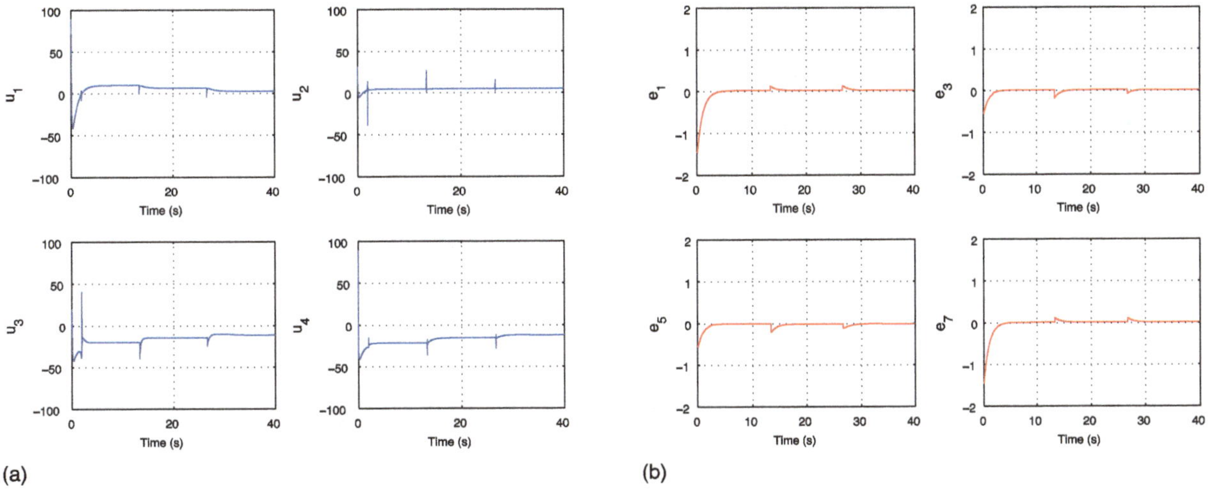

Figure 1.64 Tracking of setpoint 3 by the dual-arm robotic manipulator with nonlinear optimal control: (a) variations of the control inputs u_1 to u_4 of the dual-arm robot (blue line); (b) tracking error for the turn-angle state variables of the dual-arm robot x_1, x_3, x_5, and x_7 (red line)

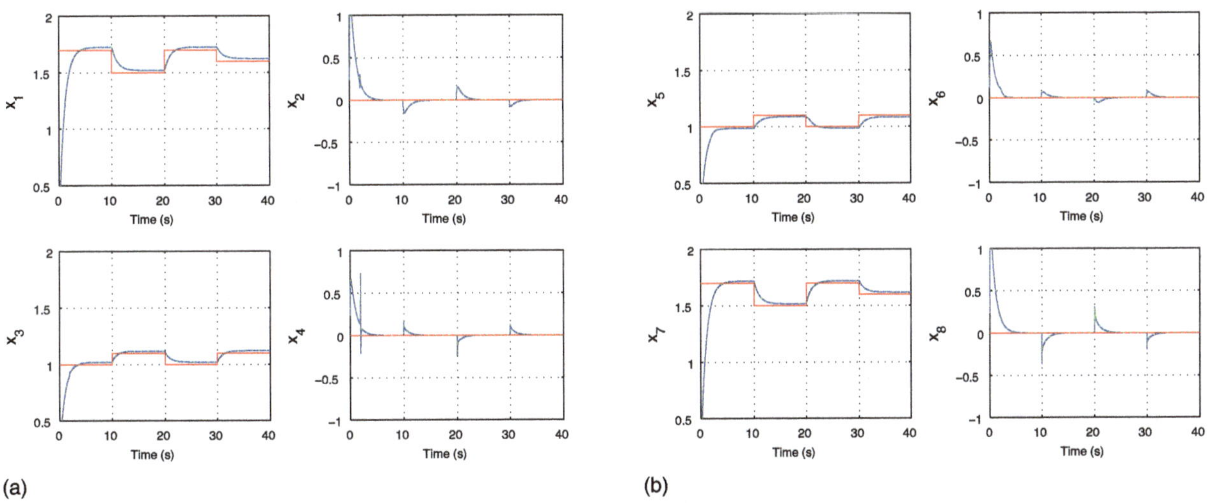

Figure 1.65 Tracking of setpoint 4 by the dual-arm robotic manipulator with nonlinear optimal control: (a) tracking of setpoints (red line) by the state variables of the dual-arm robot x_1 to x_4 (blue line) and estimates provided by the H-infinity Kalman filter; (b) tracking of setpoints (red line) by the state variables of the dual-arm robot x_5 to x_8 (blue line) and estimates provided by the H-infinity Kalman filter

the nonlinear system so as to assure that all its state vector elements will converge to the desirable setpoints, is obtained at each iteration of the control algorithm, by tracing the subsystems of the state-space model backward. The present section gives results on a flatness-based control method which is implemented in successive loops and demonstrates its application to mechatronic and robotic systems with electropneumatic actuation. The proposed control method, being of proven global stability is tested on the dynamic model of a robotic exoskeleton with electropneumatic actuation at its joints.

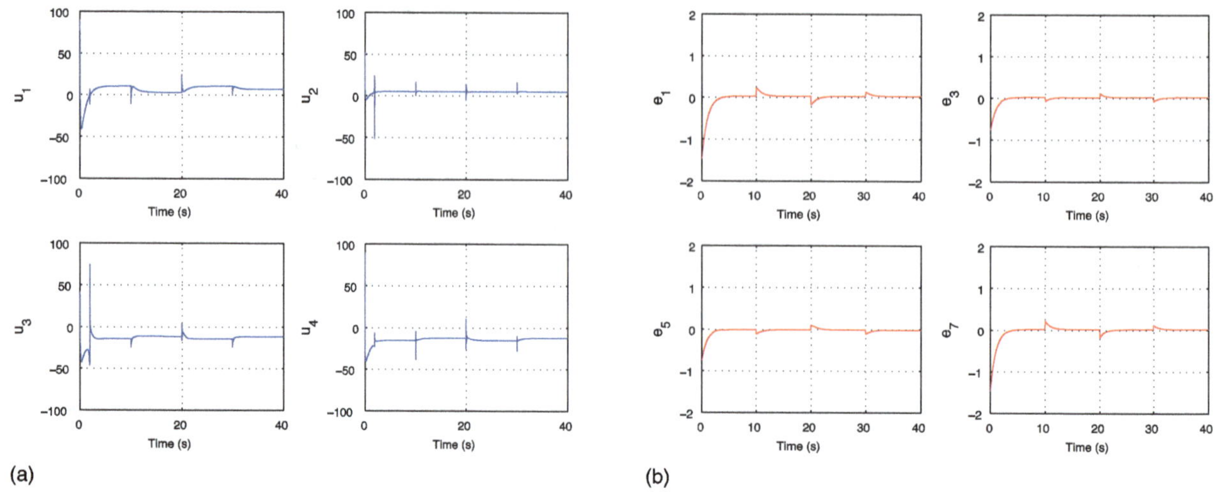

Figure 1.66 *Tracking of setpoint 4 by the dual-arm robotic manipulator with nonlinear optimal control: (a) variations of the control inputs u_1 to u_4 of the dual-arm robot (blue line); (b) tracking error for the turn-angle state variables of the dual-arm robot x_1, x_3, x_5, and x_7 (red line)*

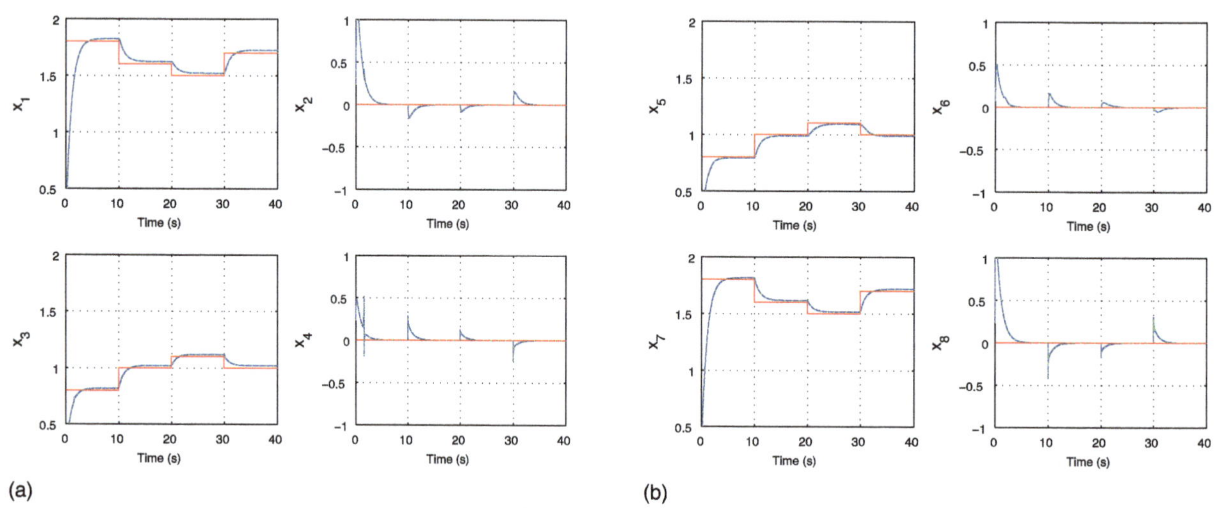

Figure 1.67 *Tracking of setpoint 5 by the dual-arm robotic manipulator with nonlinear optimal control: (a) tracking of setpoints (red line) by the state variables of the dual-arm robot x_1 to x_4 (blue line) and estimates provided by the H-infinity Kalman filter; (b) tracking of setpoints (red line) by the state variables of the dual-arm robot x_5 to x_8 (blue line) and estimates provided by the H-infinity Kalman filter*

This section provides results about flatness-based control in successive loops for a multi-link robotic exoskeleton. To provide high torques needed to move a robot's links, electric actuators are followed by a transmission system with a high transmission rate. For instance, gear ratios of 100:1 are often used in the joints of a robotic manipulator. This results in an actuator with a large mechanical impedance (also known as non-back-drivable actuator). This in turn generates high contact forces when collision of the robotic mechanism occurs and can cause humans injury. Another disadvantage of electric actuators is that they can exhibit overheating when constant torques have to be provided. Compared

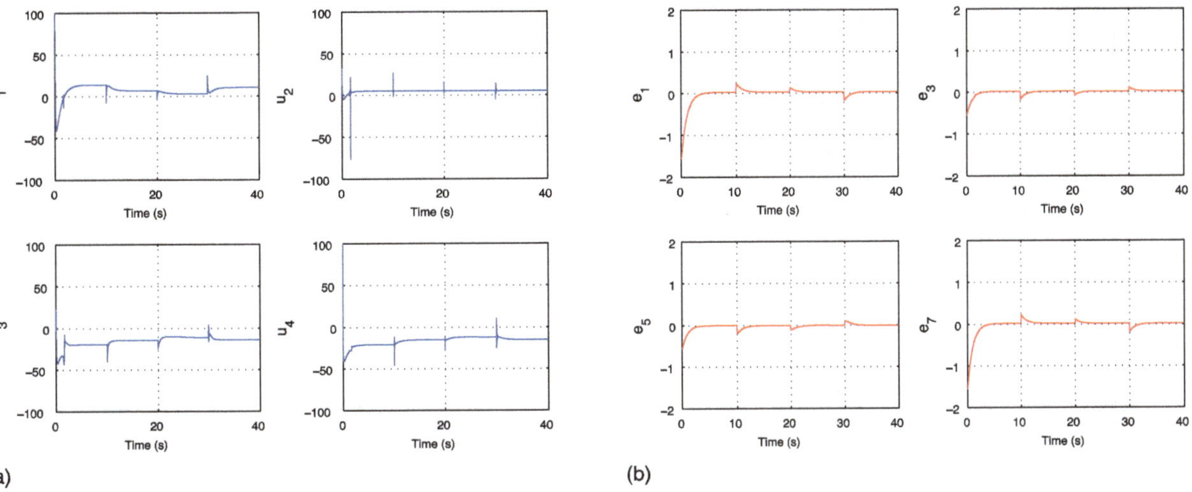

Figure 1.68 Tracking of setpoint 5 by the dual-arm robotic manipulator with nonlinear optimal control: (a) variations of the control inputs u_1 to u_4 of the dual-arm robot (blue line); (b) tracking error for the turn-angle state variables of the dual-arm robot x_1, x_3, x_5, and x_7 (red line)

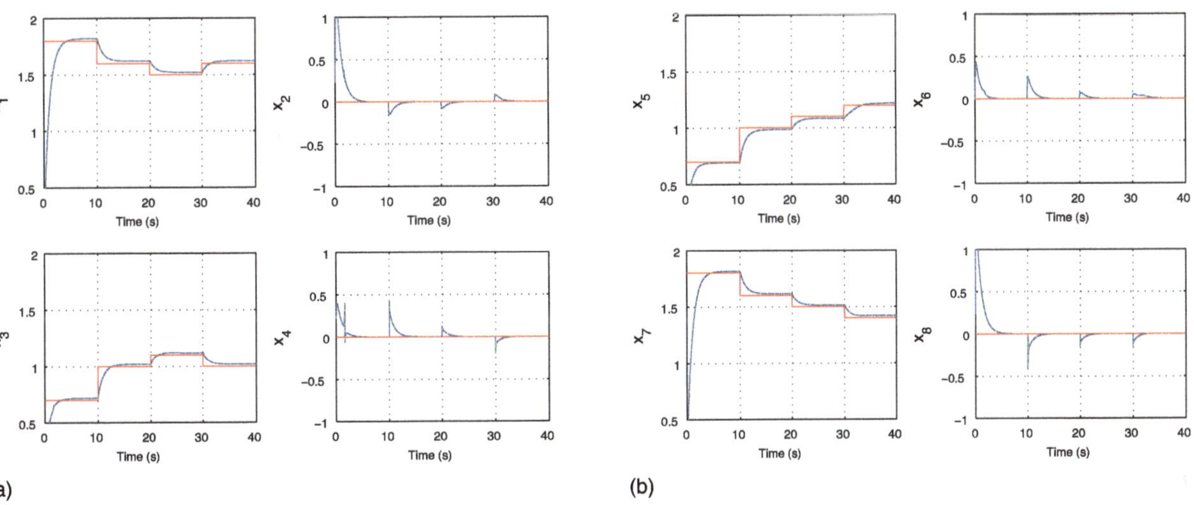

Figure 1.69 Tracking of setpoint 6 by the dual-arm robotic manipulator with nonlinear optimal control: (a) tracking of setpoints (red line) by the state variables of the dual-arm robot x_1 to x_4 (blue line) and estimates provided by the H-infinity Kalman filter; (b) tracking of setpoints (red line) by the state variables of the dual-arm robot x_5 to x_8 (blue line) and estimates provided by the H-infinity Kalman filter

to electric actuators, pneumatic actuators have promising properties for robotic applications, due to their low weight, simple mechanical design, low cost, and good power-to-weight ratio [721,810]. Electropneumatically actuated robots usually have better friction properties [475,65,81,724,894]. Moreover, because of low mechanical impedance, pneumatic robots can provide moderate interaction forces which are important for robotic surgery and rehabilitation tasks [866,433,877]. Pneumatic actuators are also well suited for exoskeleton robots. Actuation in exoskeletons should have a fast and accurate response [723,119,895,120]. While electric motors come against high mechanical impedance and

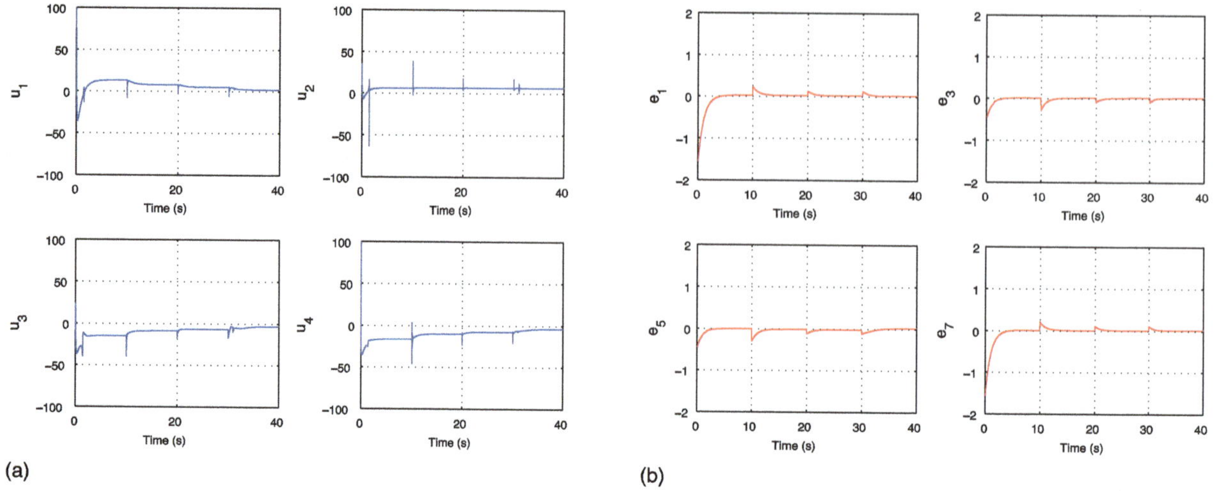

Figure 1.70 Tracking of setpoint 6 by the dual-arm robotic manipulator with nonlinear optimal control: (a) variations of the control inputs u_1 to u_4 of the dual-arm robot (blue line); (b) tracking error for the turn-angle state variables of the dual-arm robot x_1, x_3, x_5, and x_7 (red line)

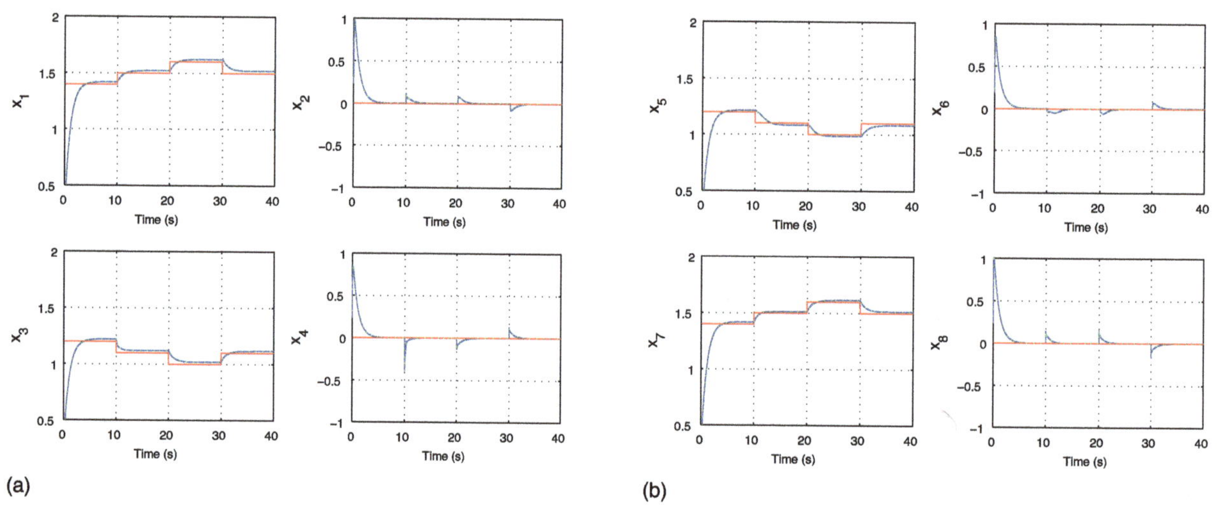

Figure 1.71 Tracking of setpoint 7 by the dual-arm robotic manipulator with nonlinear optimal control: (a) tracking of setpoints (red line) by the state variables of the dual-arm robot x_1 to x_4 (blue line) and estimates provided by the H-infinity Kalman filter; (b) tracking of setpoints (red line) by the state variables of the dual-arm robot x_5 to x_8 (blue line) and estimates provided by the H-infinity Kalman filter

the risk of causing injuries, pneumatic actuators exhibit forces and torques which stay within moderate variation ranges [349,346,530]. Besides, unlike direct current electric motors, pneumatic actuators have an improved weight-to-power ratio and avoid overheating problems [221,769,1091]. However, they are also characterized by high nonlinearities which are due to air compressibility, thermodynamics, and valves behavior, and thus, pneumatic robots require elaborated nonlinear control schemes to ensure their fast and precise positioning [799,84,548]. Among the control methods which have been applied to pneumatic robots one can distinguish differential geometric approaches (Lie algebra-based

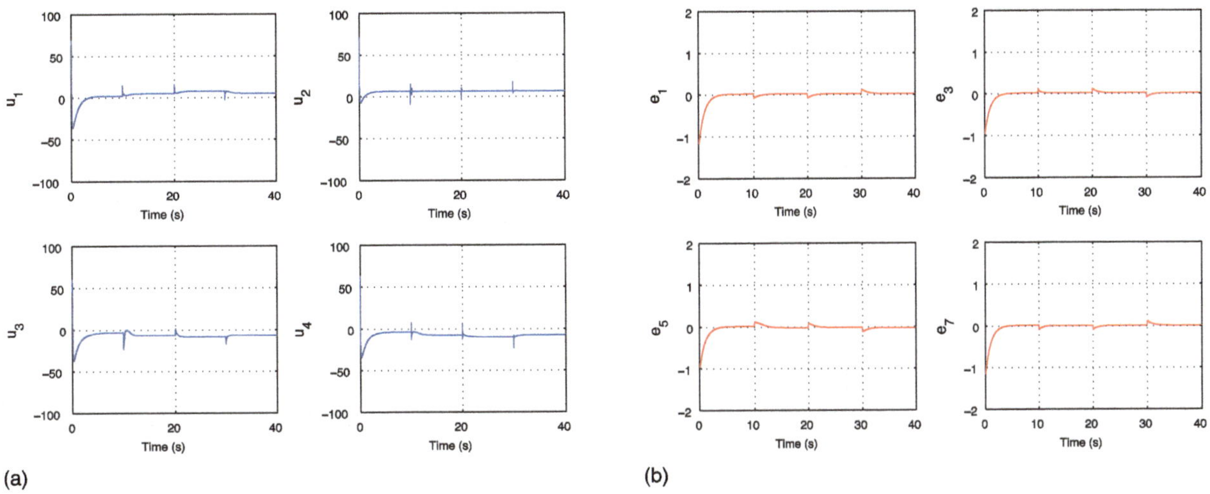

Figure 1.72 Tracking of setpoint 7 by the dual-arm robotic manipulator with nonlinear optimal control: (a) variations of the control inputs u_1 to u_4 of the dual-arm robot (blue line); (b) tracking error for the turn-angle state variables of the dual-arm robot x_1, x_3, x_5, and x_7 (red line)

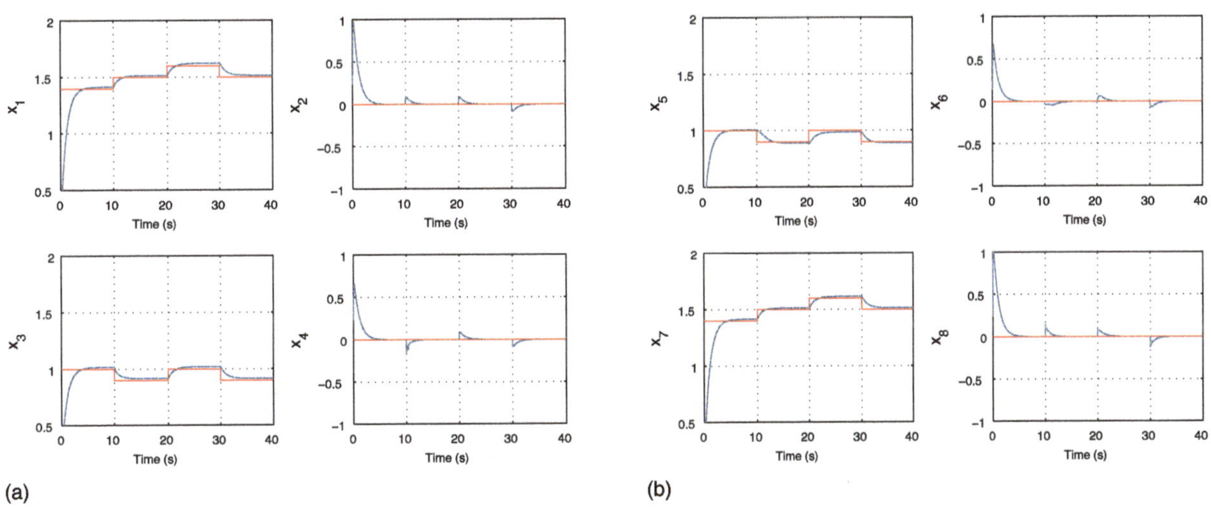

Figure 1.73 Tracking of setpoint 8 by the dual-arm robotic manipulator with nonlinear optimal control: (a) tracking of setpoints (red line) by the state variables of the dual-arm robot x_1 to x_4 (blue line) and estimates provided by the H-infinity Kalman filter; (b) tracking of setpoints (red line) by the state variables of the dual-arm robot x_5 to x_8 (blue line) and estimates provided by the H-infinity Kalman filter

control), differential flatness theory-based control, NMPC, sliding-mode control, backstepping control, and multiple models-based fuzzy control [881,679,72]. The problem of nonlinear fault tolerant control of robotic manipulators with electropneumatic actuators has not been solved yet in optimal and conclusive manner and related recent results can be found in [854,545,347,273,1071,546]. In continuation to the above-noted results, this section's findings come to confirm the fine performance of the flatness-based control method in successive loops in the dynamic model of a multi-link robotic exoskeleton.

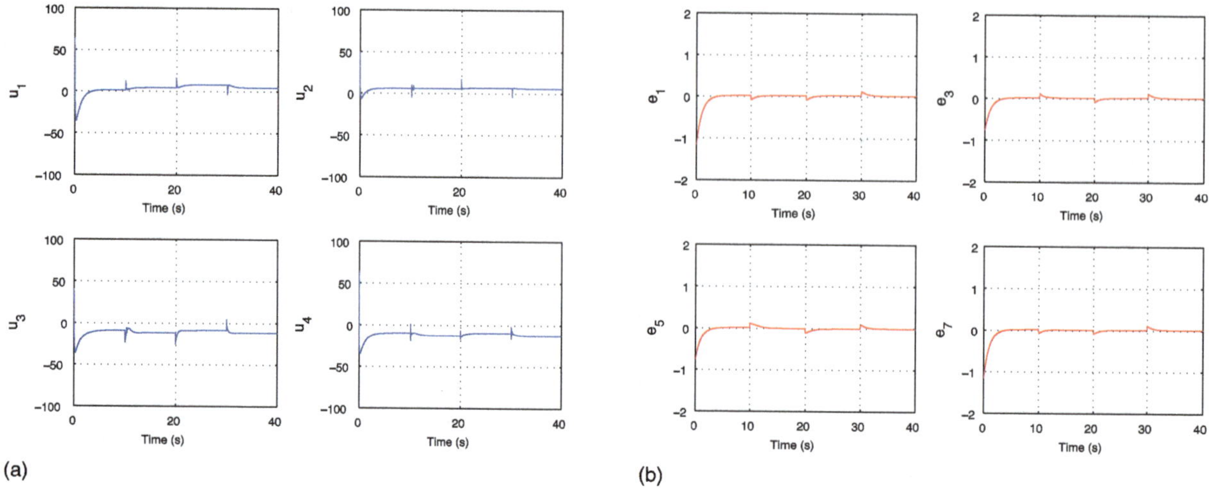

Figure 1.74 Tracking of setpoint 8 by the dual-arm robotic manipulator with nonlinear optimal control: (a) variations of the control inputs u_1 to u_4 of the dual-arm robot (blue line); (b) tracking error for the turn-angle state variables of the dual-arm robot x_1, x_3, x_5, and x_7 (red line)

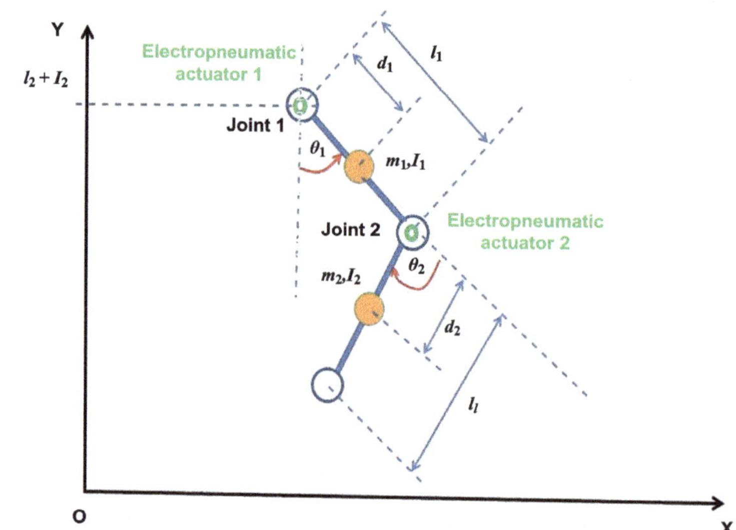

Figure 1.75 Diagram of a robotic exoskeleton with electropneumatic actuators

1.4.2 Dynamic model of the electropneumatic robotic exoskeleton

1.4.2.1 Dynamics of the electropneumatic robotic exoskeleton

The diagram of the two-link robotic system with electropneumatic actuators is shown in Figure 1.75. The dynamic model of the robotic exoskeleton can be also written in the concise form [723]:

$$M(\theta)\ddot{\theta} + C(\theta,\dot{\theta}) + G(\theta) = \tau \tag{1.381}$$

where the state vector is $\theta=[\theta_1,\theta_2]^T$, the joints' torques vector is $\tau=[\tau_1,\tau_2]^T$, while the inertia matrix $M(\theta)$, the Coriolis matrix $C(\theta,\dot{\theta})$, and the gravitational forces matrix $G(\theta)$ are defined as follows [723]: ,

$$M(\theta)=\begin{pmatrix} m_1d_1^2+I_1+m_2l_1^2+m_2d_1^2+2m_2l_1d_2cos(\theta_2)+I_2 & -m_2d_2^2-m_2l_1d_2cos(\theta_2)-I_2 \\ -m_2d_2^2-m_2l_1d_2cos(\theta_2)-I_2 & m_2d_2^2+I_2 \end{pmatrix} \tag{1.382}$$

$$C(\theta,\dot{\theta})=\begin{pmatrix} -m_2l_1d_2sin(\theta_2)\dot{\theta}_2(2\dot{\theta}_1-\dot{\theta}_2) \\ m_2l_1d_2sin(\theta_2)\dot{\theta}_1^2 \end{pmatrix} \quad G(\theta)=\begin{pmatrix} (m_1gd_1+m_2gl_1)sin(\theta_1)+m_2gd_2sin(\theta_1-\theta_2) \\ m_2gd_2sin(\theta_1-\theta_2) \end{pmatrix} \tag{1.383}$$

Using that the symmetric inertia matrix $M(\theta)$ is

$$M=\begin{pmatrix} M_{11} & M_{12} \\ M_{21} & M_{22} \end{pmatrix} \tag{1.384}$$

and that the determinant of $M(\theta)$ is $detM=M_{11}M_{22}-M_{12}^2$ the inverse of $M(\theta)$ is

$$M^{-1}=\begin{pmatrix} \frac{M_{22}}{M_{11}M_{22}-M_{12}^2} & \frac{-M_{12}}{M_{11}M_{22}-M_{12}^2} \\ \frac{-M_{21}}{M_{11}M_{22}-M_{12}^2} & \frac{M_{11}}{M_{11}M_{22}-M_{12}^2} \end{pmatrix} \tag{1.385}$$

Thus the dynamic model of the exoskeleton robot is written as

$$\ddot{\theta}=-M^{-1}(\theta)[C(\theta,\dot{\theta})+G(\theta)]+M^{-1}(\theta)\tau \tag{1.386}$$

By performing operations between the individual matrices, one arrives at the state-space description

$$\begin{pmatrix} \ddot{\theta}_1 \\ \ddot{\theta}_2 \end{pmatrix}=\begin{pmatrix} \frac{-M_{22}(C_1+G_1)+M_{12}(C_2+G_2)}{M_{11}M_{22}-M_{12}^2} \\ \frac{M_{12}(C_1+G_1)-M_{11}(C_2+G_2)}{M_{11}M_{22}-M_{12}^2} \end{pmatrix}+\begin{pmatrix} \frac{M_{22}}{M_{11}M_{22}-M_{12}^2} & \frac{-M_{12}}{M_{11}M_{22}-M_{12}^2} \\ \frac{-M_{21}}{M_{11}M_{22}-M_{12}^2} & \frac{M_{11}}{M_{11}M_{22}-M_{12}^2} \end{pmatrix}\begin{pmatrix} \tau_1 \\ \tau_2 \end{pmatrix} \tag{1.387}$$

Each joint of the robot is connected to an electropneumatic actuator. Transmission of the linear motion x_i of the i-th electromagnetic actuator's free-end to the rotary motion of the i-th joint of the robotic exoskeleton is considered to be given by $\theta_i=k_ix_i$ where k_i is the associated transmission ratio. Equivalently, one has that $x_i=\frac{1}{k_i}\theta_i$ or $x_i=q_i\theta_i$ for $i=1,2$. Using the above, the dynamics of the electropneumatic robotic exoskeleton is given in the following:

First electropneumatic actuator:

$$\begin{aligned} m_1q_1\ddot{\theta}_1 &= [S_{P_1}P_{P_1}-S_{N_1}P_{N_1}-b_1q_1\dot{\theta}_1] \\ \dot{P}_{P_1} &= \frac{K_rT}{[V_{DP_1}+S_{P_1}\frac{l_1}{2}+S_{P_1}q_1\theta_1]}\left[\phi_{P_1}(q_1\theta_1)+\psi_{P_1}(q_1\theta_1)u_{11}-\frac{S_{P_1}}{rT}q_1\dot{\theta}_1P_{P_1}\right] \\ \dot{P}_{N_1} &= \frac{K_rT}{[V_{DN_1}+S_{N_1}\frac{l_1}{2}+S_{N_1}q_1\theta_1]}\left[\phi_{N_1}(q_1\theta_1)+\psi_{N_1}(q_1\theta_1)u_{21}+\frac{S_{N_1}}{rT}q_1\dot{\theta}_1P_{N_1}\right] \end{aligned} \tag{1.388}$$

Second electropneumatic actuator:

$$\begin{aligned} m_2q_2\ddot{\theta}_2 &= [S_{P_2}P_{P_2}-S_{N_2}P_{N_2}-b_2q_2\dot{\theta}_2] \\ \dot{P}_{P_2} &= \frac{K_rT}{[V_{DP_2}+S_{P_2}\frac{l_2}{2}+S_{P_2}q_2\theta_2]}\left[\phi_{P_2}(q_2\theta_2)+\psi_{P_2}(q_2\theta_2)u_{12}-\frac{S_{P_2}}{rT}q_2\dot{\theta}_2P_{P_2}\right] \\ \dot{P}_{N_2} &= \frac{K_rT}{[V_{DN_2}+S_{N_2}\frac{l_2}{2}+S_{N_2}q_12\theta_2]}\left[\phi_{N_2}(q_2\theta_2)+\psi_{N_2}(q_2\theta_2)u_{22}+\frac{S_{N_2}}{rT}q_2\dot{\theta}_2P_{N_2}\right] \end{aligned} \tag{1.389}$$

The torques which are applied from the electropneumatic actuators to the joints of the robotic exoskeleton are

$$\begin{aligned} \tau_1 &= [S_{P_1}P_{P_1}-S_{N_1}P_{N_1}-b_1q_1\dot{\theta}_1] \\ \tau_2 &= [S_{P_2}P_{P_2}-S_{N_2}P_{N_2}-b_2q_2\dot{\theta}_2] \end{aligned} \tag{1.390}$$

Using the above, the aggregate dynamic model of the 2-link robotic exoskeleton with electropneumatic actuators is described by the following set of equations:

$$\ddot{\theta}_1 = \frac{-M_{22}(C_1+G_1)+M_{12}(C_2+G_2)}{M_{11}M_{22}-M_{12}^2} + \frac{M_{22}[S_{P_1}P_{P_1}-S_{N_1}P_{N_1}-b_1q_1\dot{\theta}_1]}{M_{11}M_{22}-M_{12}^2} + \frac{-M_{12}[S_{P_2}P_{P_2}-S_{N_2}P_{N_2}-b_2q_2\dot{\theta}_2]}{M_{11}M_{22}-M_{12}^2} \quad (1.391)$$

$$\ddot{\theta}_2 = \frac{M_{12}(C_1+G_1)-M_{11}(C_2+G_2)}{M_{11}M_{22}-M_{12}^2} + \frac{-M_{21}[S_{P_1}P_{P_1}-S_{N_1}P_{N_1}-b_1q_1\dot{\theta}_1]}{M_{11}M_{22}-M_{12}^2} + \frac{M_{11}[S_{P_2}P_{P_2}-S_{N_2}P_{N_2}-b_2q_2\dot{\theta}_2]}{M_{11}M_{22}-M_{12}^2} \quad (1.392)$$

$$\dot{P}_{P_1} = \frac{K_rT[\phi_{P_1}(q_1\theta_1)-\frac{S_{P_1}}{rT}q_1\dot{\theta}_1P_{P_1}]}{[V_{DP_1}+S_{P_1}\frac{l_1}{2}+S_{P_1}q_1\theta_1]} + \frac{K_rT\psi_{P_1}(q_1\theta_1)}{[V_{DP_1}+S_{N_1}\frac{l_1}{2}+S_{N_1}q_1\theta_1]}u_{11} \quad (1.393)$$

$$\dot{P}_{N_1} = \frac{K_rT[\phi_{N_1}(q_1\theta_1)-\frac{S_{N_1}}{rT}q_1\dot{\theta}_1P_{N_1}]}{[V_{DN_1}+S_{N_1}\frac{l_1}{2}+S_{N_1}q_1\theta_1]} + \frac{K_rT\psi_{N_1}(q_1\theta_1)}{[V_{DN_1}+S_{N_1}\frac{l_1}{2}+S_{N_1}q_1\theta_1]}u_{21} \quad (1.394)$$

$$\dot{P}_{P_2} = \frac{K_rT[\phi_{P_2}(q_2\theta_2)-\frac{S_{P_2}}{rT}q_2\dot{\theta}_2P_{P_2}]}{[V_{DP_2}+S_{P_2}\frac{l_2}{2}+S_{P_2}q_2\theta_2]} + \frac{K_rT\psi_{P_2}(q_2\theta_2)}{[V_{DP_2}+S_{N_2}\frac{l_2}{2}+S_{N_2}q_2\theta_2]}u_{12} \quad (1.395)$$

$$\dot{P}_{N_2} = \frac{K_rT[\phi_{N_2}(q_2\theta_2)-\frac{S_{N_2}}{rT}q_2\dot{\theta}_2P_{N_2}]}{[V_{DN_2}+S_{N_2}\frac{l_2}{2}+S_{N_2}q_2\theta_2]} + \frac{K_rT\psi_{N_2}(q_2\theta_2)}{[V_{DN_2}+S_{N_2}\frac{l_2}{2}+S_{N_2}q_2\theta_2]}u_{22} \quad (1.396)$$

The following state variables are defined for the electropneumatically actuated robotic system: $x_1 = \theta_1$, $x_2 = \dot{\theta}_1$, $x_3 = \theta_2$, $x_4 = \dot{\theta}_4$, $x_5 = P_{P_1}$, $x_6 = P_{N_1}$, $x_7 = P_{P_2}$, and $x_8 = P_{N_2}$. Using this notation, the dynamic model of the robot is written as follows:

$$\dot{x}_1 = x_2 \quad (1.397)$$

$$\dot{x}_2 = \frac{-M_{22}(C_1+G_1)+M_{12}(C_2+G_2)}{M_{11}M_{22}-M_{12}^2} + \frac{M_{22}[S_{P_1}x_5-S_{N_1}x_6-b_1q_1x_2]}{M_{11}M_{22}-M_{12}^2} + \frac{-M_{12}[S_{P_2}x_7-S_{N_2}x_8-b_2q_2x_4]}{M_{11}M_{22}-M_{12}^2} \quad (1.398)$$

$$\dot{x}_3 = x_4 \quad (1.399)$$

$$\dot{x}_4 = \frac{M_{12}(C_1+G_1)-M_{11}(C_2+G_2)}{M_{11}M_{22}-M_{12}^2} + \frac{-M_{21}[S_{P_1}x_5-S_{N_1}x_6-b_1q_1x_2]}{M_{11}M_{22}-M_{12}^2} + \frac{M_{11}[S_{P_2}x_7-S_{N_2}x_8-b_2q_2x_4]}{M_{11}M_{22}-M_{12}^2} \quad (1.400)$$

$$\dot{x}_5 = \frac{K_rT[\phi_{P_1}(q_1x_1)-\frac{S_{P_1}}{rT}q_1x_2x_5]}{[V_{DP_1}+S_{P_1}\frac{l_1}{2}+S_{P_1}q_1x_1]} + \frac{K_rT\psi_{P_1}(q_1x_1))}{[V_{DP_1}+S_{N_1}\frac{l_1}{2}+S_{N_1}q_1x_1]}u_{11} \quad (1.401)$$

$$\dot{x}_6 = \frac{K_rT[\phi_{N_1}(q_1x_1)-\frac{S_{N_1}}{rT}q_1x_2x_6]}{[V_{DN_1}+S_{N_1}\frac{l_1}{2}+S_{N_1}q_1x_1]} + \frac{K_rT\psi_{N_1}(q_1x_1)}{[V_{DN_1}+S_{N_1}\frac{l_1}{2}+S_{N_1}q_1x_1]}u_{21} \quad (1.402)$$

$$\dot{x}_7 = \frac{K_rT[\phi_{P_2}(q_2x_3)-\frac{S_{P_2}}{rT}q_2x_4x_7]}{[V_{DP_2}+S_{P_2}\frac{l_2}{2}+S_{P_2}q_2x_3]} + \frac{K_rT\psi_{P_2}(q_2x_3)}{[V_{DP_2}+S_{N_2}\frac{l_2}{2}+S_{N_2}q_2x_3]}u_{12} \quad (1.403)$$

$$\dot{x}_8 = \frac{K_rT[\phi_{N_2}(q_2x_3)-\frac{S_{N_2}}{rT}q_2x_4x_7]}{[V_{DN_2}+S_{N_2}\frac{l_2}{2}+S_{N_2}q_2x_3]} + \frac{K_rT\psi_{N_2}(q_2x_3)}{[V_{DN_2}+S_{N_2}\frac{l_2}{2}+S_{N_2}q_2x_3]}u_{22} \quad (1.404)$$

The control inputs of this robotic system are generated by the voltages which are applied to the servo-valves of the electropneumatic actuators. Thus, the state-space model of the electropneumatically actuated robotic exoskeleton is written as

$$\dot{x}_1 = f_1(x) + g_{11}u_{11} + g_{21}u_{21} + g_{31}u_{12} + g_{41}u_{22} \quad (1.405)$$

$$\dot{x}_2 = f_2(x) + g_{12}u_{11} + g_{22}u_{21} + g_{32}u_{12} + g_{42}u_{22} \quad (1.406)$$

$$\dot{x}_3 = f_3(x) + g_{13}u_{11} + g_{23}u_{21} + g_{33}u_{12} + g_{43}u_{22} \quad (1.407)$$

$$\dot{x}_4 = f_4(x) + g_{14}u_{11} + g_{24}u_{21} + g_{34}u_{12} + g_{44}u_{22} \quad (1.408)$$

$$\dot{x}_5 = f_5(x) + g_{15}u_{11} + g_{25}u_{21} + g_{35}u_{12} + g_{45}u_{22} \quad (1.409)$$

$$\dot{x}_6 = f_3(x) + g_{16}u_{11} + g_{26}u_{21} + g_{36}u_{12} + g_{46}u_{22} \quad (1.410)$$

$$\dot{x}_7 = f_7(x) + g_{17}u_{11} + g_{27}u_{21} + g_{37}u_{12} + g_{47}u_{22} \quad (1.411)$$

$$\dot{x}_8 = f_7(x) + g_{18}u_{11} + g_{28}u_{21} + g_{38}u_{12} + g_{48}u_{22} \tag{1.412}$$

or equivalently, the state-space model of the electropneumatic robot takes the following matrix form:

$$\begin{pmatrix} \dot{x}_1 \\ \dot{x}_2 \\ \dot{x}_3 \\ \dot{x}_4 \\ \dot{x}_5 \\ \dot{x}_6 \\ \dot{x}_7 \\ \dot{x}_8 \end{pmatrix} = \begin{pmatrix} f_1(x) \\ f_2(x) \\ f_3(x) \\ f_4(x) \\ f_5(x) \\ f_6(x) \\ f_7(x) \\ f_8(x) \end{pmatrix} + \begin{pmatrix} 0 & 0 & 0 & 0 \\ 0 & 0 & 0 & 0 \\ 0 & 0 & 0 & 0 \\ 0 & 0 & 0 & 0 \\ g_{51} & 0 & 0 & 0 \\ 0 & g_{62} & 0 & 0 \\ 0 & 0 & g_{33} & 0 \\ 0 & 0 & 0 & g_{84} \end{pmatrix} \begin{pmatrix} u_{11} \\ u_{21} \\ u_{12} \\ u_{22} \end{pmatrix} \tag{1.413}$$

The model of the robotic exoskeleton with electropneumatic actuation can be also written in the concise form

$$\dot{x} = f(x) + g_1(x)u_{11} + g_2(x)u_{22} + g_3(x)u_{12} + g_4(x)u_{22} \tag{1.414}$$

where $x \in R^{8\times 1}$, $f(x) \in R^{8\times 1}$, $g_1(x) \in R^{8\times 1}$, $g_2(x) \in R^{8\times 1}$, $g_3(x) \in R^{8\times 1}$, and $g_4(x) \in R^{8\times 1}$, while the control inputs vector $u \in R^{4\times 1}$ comprises the control variables $u = [u_{11}, u_{21}, u_{12}, u_{22}]^T$.

It holds that

$$\begin{aligned} f_1(x) &= x_2 \\ f_2(x) &= \frac{-M_{22}(C_1+G_1)+M_{12}(C_2+G_2)}{M_{11}M_{22}-M_{12}^2} + \frac{M_{22}[S_{P_1}x_5-S_{N_1}x_6-b_1q_1x_2]}{M_{11}M_{22}-M_{12}^2} + \frac{-M_{12}[S_{P_2}x_7-S_{N_2}x_8-b_2q_2x_4]}{M_{11}M_{22}-M_{12}^2} \\ f_3(x) &= x_4 \\ f_4(x) &= \frac{M_{12}(C_1+G_1)-M_{11}(C_2+G_2)}{M_{11}M_{22}-M_{12}^2} + \frac{-M_{21}[S_{P_1}x_5-S_{N_1}x_6-b_1q_1x_2]}{M_{11}M_{22}-M_{12}^2} + \frac{M_{11}[S_{P_2}x_7-S_{N_2}x_8-b_2q_2x_4]}{M_{11}M_{22}-M_{12}^2} \\ f_5(x) &= \frac{K_rT[\phi_{P_1}(q_1x_1)-\frac{S_{P_1}}{rT}q_1x_2x_5]}{[V_{DP_1}+S_{P_1}\frac{l_1}{2}+S_{P_1}q_1x_1]} \quad f_6(x) = \frac{K_rT[\phi_{N_1}(q_1x_1)+\frac{S_{N_1}}{rT}q_1x_2x_6]}{[V_{DN_1}+S_{N_1}\frac{l_1}{2}+S_{N_1}q_1x_1]} \\ f_7(x) &= \frac{K_rT[\phi_{P_2}(q_2x_3)-\frac{S_{P_2}}{rT}q_2x_4x_7]}{[V_{DP_2}+S_{P_2}\frac{l_2}{2}+S_{P_2}q_2x_3]} \quad f_8(x) = \frac{K_rT[\phi_{N_2}(q_2x_3)+\frac{S_{N_2}}{rT}q_2x_4x_7]}{[V_{DN_2}+S_{N_2}\frac{l_2}{2}+S_{N_2}q_2x_3]} \end{aligned} \tag{1.415}$$

It also holds that

$$\begin{aligned} g_{51}(x) &= \frac{K_rT\psi_{P_1}(q_1x_1))}{[V_{DP_1}+S_{P_1}\frac{l_1}{2}+S_{P_1}q_1x_1]} \quad g_{62}(x) = \frac{K_rT\psi_{N_1}(q_1x_1)}{[V_{DN_1}+S_{N_1}\frac{l_1}{2}+S_{N_1}q_1x_1]} \\ g_{73}(x) &= \frac{K_rT\psi_{P_2}(q_2x_3)}{[V_{DP_2}+S_{N_2}\frac{l_2}{2}+S_{N_2}q_2x_3]} \quad g_{84}(x) = \frac{K_rT\psi_{N_2}(q_2x_3)}{[V_{DN_2}+S_{N_2}\frac{l_2}{2}+S_{N_2}q_2x_3]} \end{aligned} \tag{1.416}$$

1.4.2.2 Differential flatness of the electropneumatic robotic exoskeleton

It will be proven that the 2-link electropneumatically actuated robotic exoskeleton is a differentially flat system with flat outputs $y = [y_1, y_2, y_3, y_4]$ where

$$y_1 = x_1 = \theta_1 \quad y_2 = x_3 = \theta_2 \quad y_3 = x_5 = P_{P_1} \quad y_4 = x_7 = P_{P_2} \tag{1.417}$$

This signifies that all state variables and the control inputs of the robot can be written as differential functions of the robot's flat outputs. It also signifies that the flat outputs are not connected between them through a relation in the form of an homogeneous differential equation [721,711]. From the first row of the state-space model of the robot, one has

$$x_2 = \dot{x}_1 \Rightarrow x_2 = h_2(y, \dot{y}) \tag{1.418}$$

consequently, state variable x_2 is a differential function of the robot's flat outputs. Equivalently, from the third row of the state-space model, one has

$$x_4 = \dot{x}_3 \Rightarrow x_4 = h_4(y, \dot{y}) \tag{1.419}$$

consequently, state variable x_4 is a differential function of the robot's flat outputs. Next, from the second and fourth rows of the state-space model of the robot, one has a system of equations with respect to state variables x_6 and x_8. Indeed:

$$\begin{aligned}\dot{x}_2 - \frac{-M_{22}(C_1+G_1)+M_{12}(C_2+G2)}{M_{11}M_{22}-M_{12}^2} - \frac{M_{22}[S_{P_1}x_5-b_1q_1x_2]}{M_{11}M_{22}-M_{12}^2} - \frac{M_2[S_{P_1}x_7-b_2q_2x_4]}{M_{11}M_{22}-M_{12}^2} = \\ = \frac{-M_{22}S_{N_1}}{M_{11}M_{22}-M_{12}^2}x_6 + \frac{M_{11}S_{N_2}}{M_{11}M_{22}-M_{12}^2}x_8\end{aligned} \tag{1.420}$$

$$\begin{aligned}\dot{x}_4 - \frac{M_{12}(C_1+G_1)-M_{11}(C_2+G2)}{M_{11}M_{22}-M_{12}^2} - \frac{-M_{21}[S_{P_1}x_5-b_1q_1x_2]}{M_{11}M_{22}-M_{12}^2} - \frac{M_{11}[S_{P_1}x_7-b_2q_2x_4]}{M_{11}M_{22}-M_{12}^2} = \\ = \frac{M_{21}S_{N_1}}{M_{11}M_{22}-M_{12}^2}x_6 - \frac{M_{11}S_{N_2}}{M_{11}M_{22}-M_{12}^2}x_8\end{aligned} \tag{1.421}$$

Equivalently, Eq. (1.420) and Eq. (1.421) can be written as

$$\begin{aligned}\dot{x}_2[M_{11}M_{22}-M_{12}^2] - [-M_{22}(C_1+G_1)+M_{12}(C_2+G_2)] - M_{22}[S_{P_1}x_5-b_1q_1x_2] - M_{12}[S_{P_2}x_7-b_2q_2x_4] = \\ = -M_{22}S_{N_1}x_6 + M_{12}S_{N_2}x_8\end{aligned} \tag{1.422}$$

$$\begin{aligned}\dot{x}_4[M_{11}M_{22}-M_{12}^2] - [M_{12}(C_1+G_1)-M_{11}(C_2+G_2)] + M_{21}[S_{P_1}x_5-b_1q_1x_2] - M_{11}[S_{P_2}x_7-b_2q_2x_4] = \\ = M_{21}S_{N_1}x_6 - M_{11}S_{N_2}x_8\end{aligned} \tag{1.423}$$

From Eq. (1.422) and Eq. (1.423), one can solve with respect to x_6 and x_8. This gives:

$$\begin{aligned}\begin{pmatrix} x_6 \\ x_8 \end{pmatrix} = \begin{pmatrix} -M_{22}S_{N_1} & M_{12}S_{N_2} \\ M_{21}S_{N_1} & -M_{11}S_{N_2} \end{pmatrix}^{-1} \cdot \\ \begin{pmatrix} \dot{x}_2[M_{11}M_{22}-M_{12}^2] - [-M_{22}(C_1+G_1)+M_{12}(C_2+G_2)] - M_{22}[S_{P_1}x_5-b_1q_1x_2] - M_{12}[S_{P_2}x_7-b_2q_2x_4] \\ \dot{x}_4[M_{11}M_{22}-M_{12}^2] - [M_{12}(C_1+G_1)-M_{11}(C_2+G_2)] + M_{21}[S_{P_1}x_5-b_1q_1x_2] - M_{11}[S_{P_2}x_7-b_2q_2x_4] \end{pmatrix}\end{aligned} \tag{1.424}$$

From Eq. (1.424) and taking into account that state variables x_1, x_3, x_5, and x_7 are also the flat outputs of the robot, while state variables x_2 and x_4 have already been expressed in the form of differential functions of the robot's flat outputs, one also obtains that state variables x_6 and x_8 are differential functions of the robotic system's flat outputs. Thus, x_6 and x_8 can be rewritten as

$$x_6 = h_6(y,\dot{y},\ddot{y}) \quad x_8 = h_8(y,\dot{y},\ddot{y}) \tag{1.425}$$

Next, from the fifth row of the state-space model of the robot, one has

$$\begin{aligned}\dot{x}_5[V_{DP_1} + S_{P_1}\tfrac{\bar{l}_1}{2} + S_{P_1}q_1x_1] = K_rT[\phi_{P_1}(q_1x_1) - \tfrac{S_{P_1}}{rT}q_1x_2x_5] + K_rT\psi_{P_1}(q_1x_1)u_{11} \Rightarrow \\ u_{11} = \frac{\dot{x}_5[V_{DP_1}+S_{P_1}\frac{\bar{l}_1}{2}+S_{P_1}q_1x_1] - K_rT[\phi_{P_1}(q_1x_1)-\frac{S_{P_1}}{rT}q_1x_2x_5]}{K_rT\psi_{P_1}(q_1x_1)}\end{aligned} \tag{1.426}$$

Eq. (1.426) denotes that the control input u_{11} is a differential function of the flat outputs of the system, that is

$$u_{11} = h_{u_{11}}(y,\dot{y},\ddot{y}) \tag{1.427}$$

Besides, from the sixth row of the state-space model, one has that

$$\begin{aligned}\dot{x}_6[V_{DN_1} + S_{N_1}\tfrac{\bar{l}_1}{2} + S_{N_1}q_1x_1] = K_rT[\phi_{N_1}(q_1x_1) + \tfrac{S_{N_1}}{rT}q_1x_2x_6] + K_rT\psi_{N_1}(q_1x_1)u_{21} \Rightarrow \\ u_{21} = \frac{\dot{x}_6[V_{DN_1}+S_{N_1}\frac{\bar{l}_1}{2}+S_{N_1}q_1x_1] - K_rT[\phi_{N_1}(q_1x_1)+\frac{S_{N_1}}{rT}q_1x_2x_6]}{K_rT\psi_{N_1}(q_1x_1)}\end{aligned} \tag{1.428}$$

Eq. (1.428) denotes that the control input u_{21} is a differential function of the flat outputs of the system, or

$$u_{21} = h_{u_{21}}(y,\dot{y},\ddot{y}) \tag{1.429}$$

Equivalently, from the seventh row of the state-space model of the robot, one has

$$\begin{aligned}\dot{x}_7[V_{DP_2} + S_{P_2}\tfrac{\bar{l}_2}{2} + S_{P_2}q_2x_3] = K_rT[\phi_{P_2}(q_2x_3) - \tfrac{S_{P_2}}{rT}q_2x_4x_7] + K_rT\psi_{P_2}(q_2x_3)u_{12} \Rightarrow \\ u_{12} = \frac{\dot{x}_7[V_{DP_2}+S_{P_2}\frac{\bar{l}_2}{2}+S_{P_2}q_2x_3] - K_rT[\phi_{P_2}(q_2x_3)-\frac{S_{P_2}}{rT}q_2x_4x_7]}{K_rT\psi_{P_2}(q_2x_3)}\end{aligned} \tag{1.430}$$

Eq. (1.430) denotes that the control input u_{12} is a differential function of the flat outputs of the system, or

$$u_{12} = h_{u_{12}}(y, \dot{y}, \ddot{y}) \tag{1.431}$$

Finally, from the eighth row of the state-space model, one has that

$$\begin{aligned} \dot{x}_8[V_{DN_2} + S_{N_2}\tfrac{\bar{l}_2}{2} + S_{N_2}q_2x_3] &= K_rT[\phi_{N_2}(q_2x_3) + \tfrac{S_{N_2}}{rT}q_2x_4x_8] + K_rT\psi_{N_2}(q_2x_3)u_{22} \Rightarrow \\ u_{22} &= \frac{\dot{x}_8[V_{DN_2} + S_{N_2}\frac{\bar{l}_2}{2} + S_{N_2}q_2x_3] - K_rT[\phi_{N_2}(q_2x_3) + \frac{S_{N_2}}{rT}q_2x_4x_8]}{K_rT\psi_{N_2}(q_2x_3)} \end{aligned} \tag{1.432}$$

Eq. (1.432) denotes that the control input u_{22} is a differential function of the flat outputs of the system, or

$$u_{22} = h_{u_{22}}(y, \dot{y}, \ddot{y}) \tag{1.433}$$

Thus, it has been proven that all state variables and the control inputs of the electropneumatically actuated robot are differential functions of the flat outputs of the system. Consequently, this is a differentially flat robotic system. On the one side, the differential flatness properties allow the transformation of this robotic system into an input–output linearized form. On the other side, by using the relations that connect the state variable of the robot to the flat outputs one can solve the problem of setpoints selection. Actually, setpoints are defined first for the flat outputs of the robot, that is

$$y_1^d = x_1^d \quad y_2^d = x_3^d \quad y_3^d = x_5^d \quad y_4^d = x_7^d \tag{1.434}$$

and using the properties of differential flatness, one defines setpoints also for the rest of the robot's state variables, that is x_2^d, x_4^d, x_6^d, and x_8^d.

1.4.3 Flatness-based control in successive loops for the electropneumatic robot

For the dynamic model of the robotic exoskeleton with electropneumatic actuators that was given in Eq. (1.413), the following subvectors and submatrices are defined:

$$z_{1,2} = \begin{pmatrix} x_1 \\ x_3 \end{pmatrix} \quad f_{1,2}(z_{1,2}) = \begin{pmatrix} 0 \\ 0 \end{pmatrix} \quad g_{1,2}(z_{1,2}) = \begin{pmatrix} 1 & 0 \\ 0 & 1 \end{pmatrix} \tag{1.435}$$

$$\begin{aligned} z_{3,4} &= \begin{pmatrix} x_2 \\ x_4 \end{pmatrix} \quad f_{3,4}(z_{1,2}, z_{3,4}) = \begin{pmatrix} \frac{-M_{22}(C_1+G_1)+M_{12}(C_2+G_2)-M_{22}b_1q_1x_2+M_{12}b_2q_2x_4}{M_{11}M_{22}-M_{12}^2} \\ \frac{M_{12}(C_1+G_1)-M_{11}(C_2+G_2)+M_{21}b_1q_1x_2-M_{11}b_2q_2x_4}{M_{11}M_{22}-M_{12}^2} \end{pmatrix} \\ g_{3,4}(z_{1,2}, z_{3,4}) &= \begin{pmatrix} \frac{M_{22}}{M_{11}M_{22}-M_{12}^2} & \frac{-M_{12}}{M_{11}M_{22}-M_{12}^2} \\ \frac{-M_{12}}{M_{11}M_{22}-M_{12}^2} & \frac{M_{11}}{M_{11}M_{22}-M_{12}^2} \end{pmatrix} \end{aligned} \tag{1.436}$$

$$\begin{aligned} z_{5,8} &= \begin{pmatrix} x_5 \\ x_6 \\ x_7 \\ x_8 \end{pmatrix} \quad f_{5,8}(z_{1,2}, z_{3,4}, z_{5,8}) = \begin{pmatrix} f_5(x) \\ f_6(x) \\ f_7(x) \\ f_8(x) \end{pmatrix} \\ g_{5,8}(z_{1,2}, z_{3,4}, z_{5,8}) &= \begin{pmatrix} g_{51}(x) & 0 & 0 & 0 \\ 0 & g_{62}(x) & 0 & 0 \\ 0 & 0 & g_{73}(x) & 0 \\ 0 & 0 & 0 & g_{74}(x) \end{pmatrix} \end{aligned} \tag{1.437}$$

Thus, using the following vector of virtual control inputs: $v_1 = z_{3,4} = [x_2, x_4]^T$, and $v_2 = [v_{2a}, v_{2b}]^T$ with $v_{2a} = S_{P_1}P_{P_1} - S_{N_1}P_{N_1}$, $v_{2b} = S_{P_2}P_{P_2} - S_{N_2}P_{N_2}$, the dynamic model of the robotic exoskeleton with electropneumatic actuation can be written in the form of three subsystems which are connected in cascading form:

$$\dot{z}_{1,2} = f_{1,2}(z_{1,2}) + g_{1,2}(z_{1,2})v_1 \tag{1.438}$$

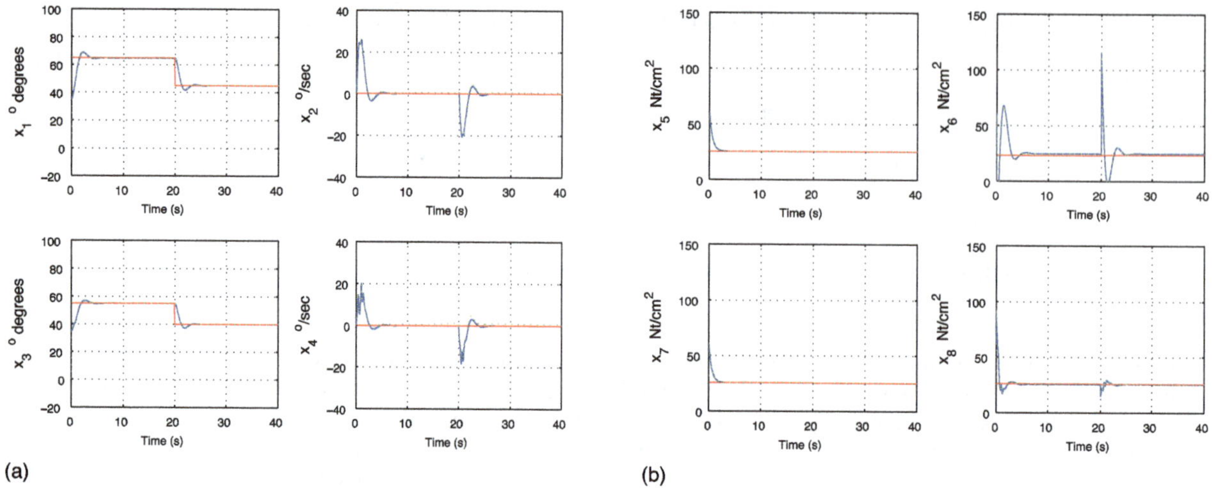

Figure 1.76 Tracking of trajectory 1 by the 2-link electropneumatic robotic manipulator: (a) convergence of state variables x_1 to x_4 to their reference setpoints (red line: setpoint, blue line: real value, green line: estimation by a Kalman filter); (b) convergence of state variables x_5 to x_8 to their reference setpoints (red line: setpoint, blue line: real value, green line: estimation by a Kalman filter)

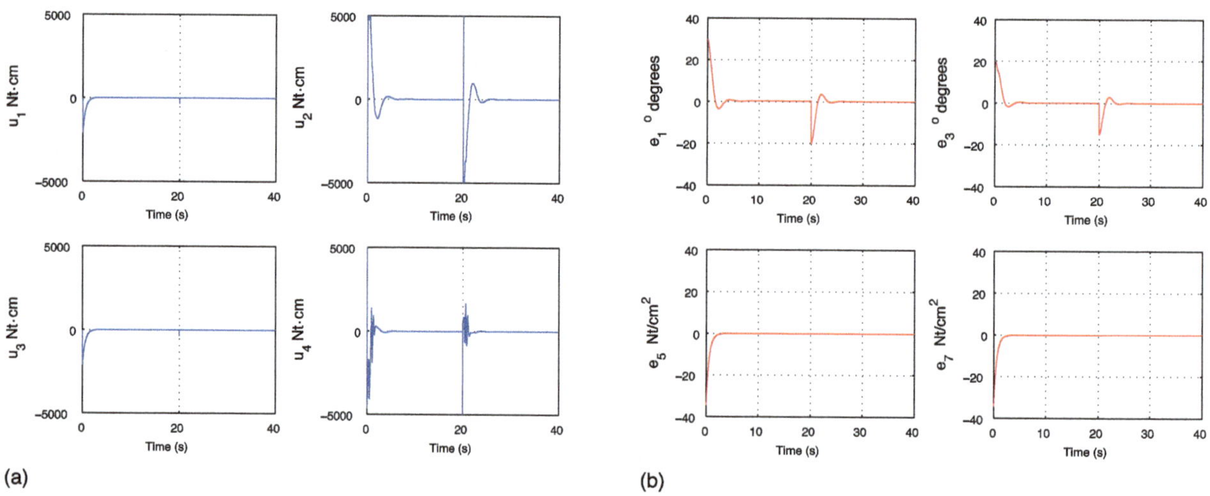

Figure 1.77 Tracking of trajectory 1 by the 2-link electropneumatic robotic manipulator: (a) control inputs u_1 to u_4 generated by the electropneumatic actuators of the robot; (b) tracking error variables

$$\dot{z}_{3,4} = f_{3,4}(z_{1,2}, z_{3,4}) + g_{3,4}(z_{1,2}, z_{3,4})v_2 \tag{1.439}$$

$$\dot{x}_{5,8} = f_{5,8}(z_{1,2}, z_{3,4}, z_{5,8}) + g_{5,8}(z_{1,2}, z_{3,4}, z_{5,8})u \tag{1.440}$$

It will be proven that if each one of Eq. (1.438), Eq. (1.439), and Eq. (1.440) if viewed independently, is a differentially flat subsystem.

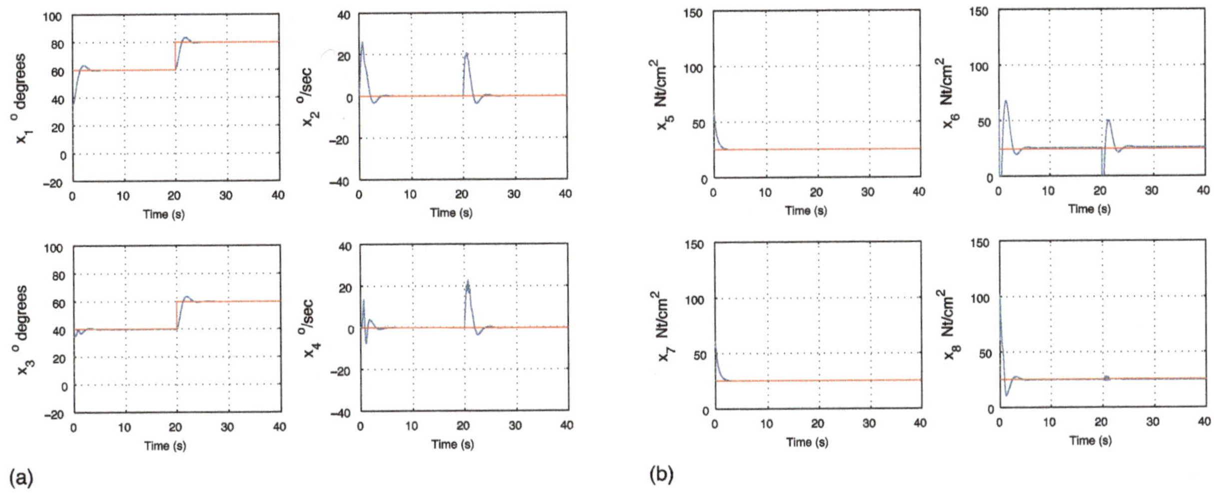

Figure 1.78 Tracking of trajectory 2 by the 2-link electropneumatic robotic manipulator: (a) convergence of state variables x_1 to x_4 to their reference setpoints (red line: setpoint, blue line: real value, green line: estimation by a Kalman filter); (b) convergence of state variables x_5 to x_8 to their reference setpoints (red line: setpoint, blue line: real value, green line: estimation by a Kalman filter)

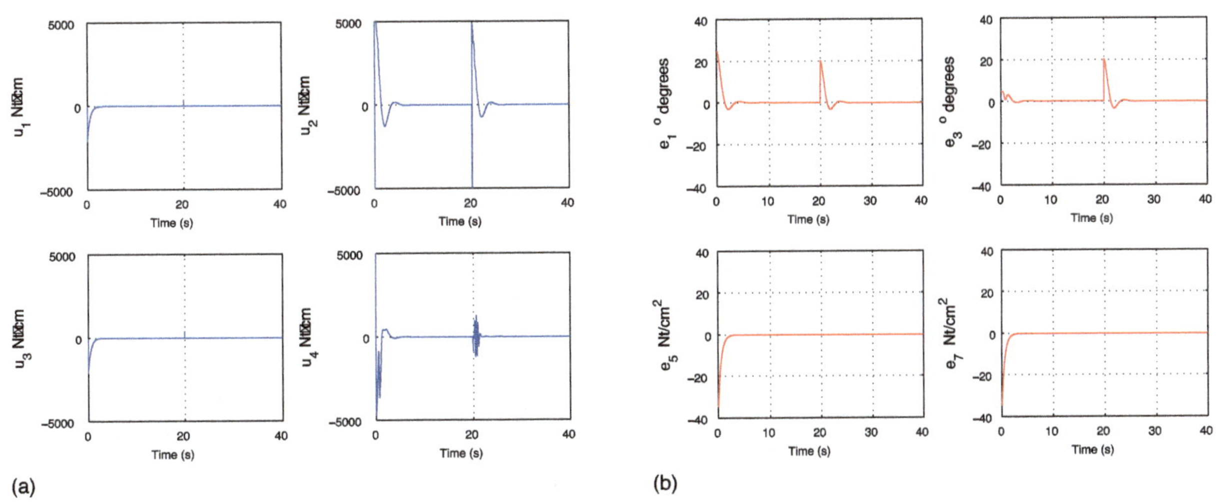

Figure 1.79 Tracking of trajectory 2 by the 2-link electropneumatic robotic manipulator: (a) control inputs u_1 to u_4 generated by the electropneumatic actuators of the robot; (b) tracking error variables

In Eq. (1.438) $y_1 = z_{1,2}$ is taken to be the flat output and $v_1 = z_{3,4}$ is considered to be a virtual control inputs vector. From Eq. (1.438),

$$v_1 = g_{1,2}(z_{1,2})^{-1}[\dot{z}_{1,2} - f_{1,2}(z_{1,2})] \tag{1.441}$$

which signifies that v_1 is a differential function of the flat outputs vector y_1, consequently, Eq. (1.438) is a differentially flat subsystem.

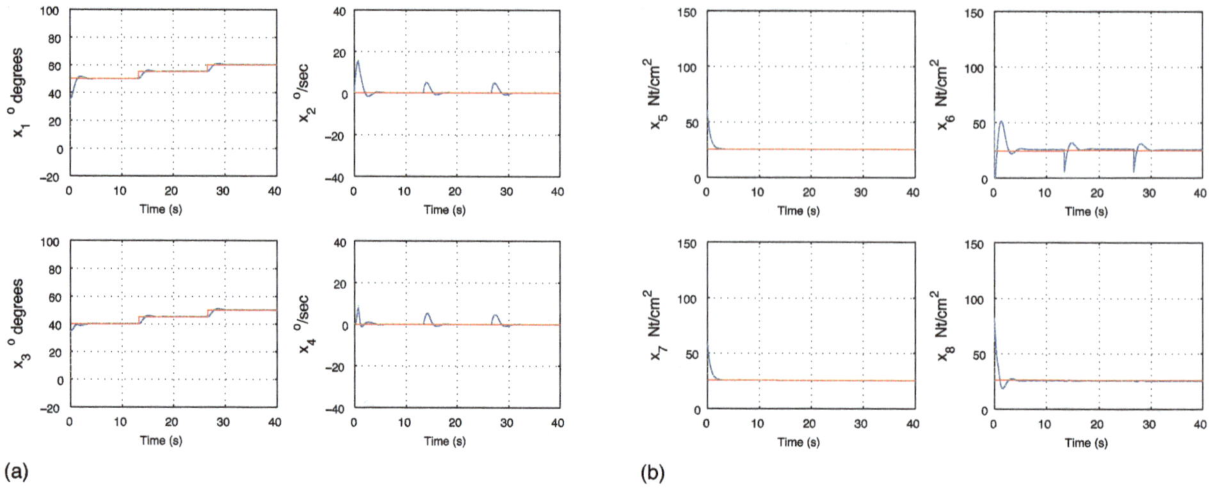

Figure 1.80 Tracking of trajectory 3 by the 2-link electropneumatic robotic manipulator: (a) convergence of state variables x_1 to x_4 to their reference setpoints (red line: setpoint, blue line: real value, green line: estimation by a Kalman filter); (b) convergence of state variables x_5 to x_8 to their reference setpoints (red line: setpoint, blue line: real value, green line: estimation by a Kalman filter)

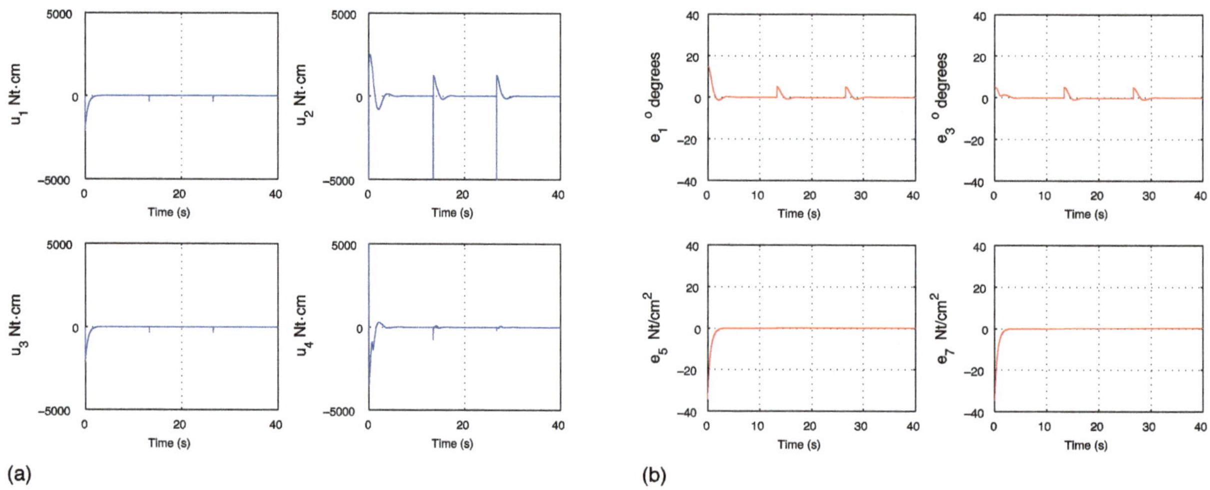

Figure 1.81 Tracking of trajectory 3 by the 2-link electropneumatic robotic manipulator: (a) control inputs u_1 to u_4 generated by the electropneumatic actuators of the robot; (b) tracking error variables

In Eq. (1.439) $y_2 = z_{3,4}$ is taken to be the flat output and v_2 is considered to be a virtual control inputs vector, and $z_{1,2}$ is considered to be a coefficients vector. From Eq. (1.439),

$$v_2 = g_{3,4}(z_{1,2}, z_{3,4})^{-1}[\dot{z}_{3,4} - f_{3,4}(z_{1,2}, z_{3,4})] \tag{1.442}$$

which signifies that v_2 is a differential function of the flat outputs vector y_2, consequently Eq. (1.439) is a differentially flat subsystem.

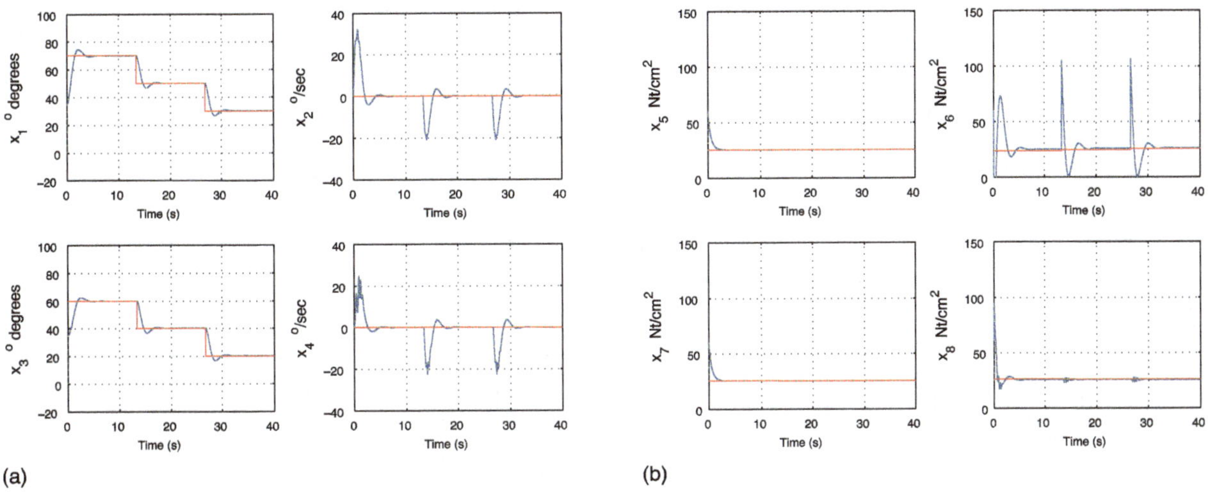

Figure 1.82 Tracking of trajectory 4 by the 2-link electropneumatic robotic manipulator: (a) convergence of state variables x_1 to x_4 to their reference setpoints (red line: setpoint, blue line: real value, green line: estimation by a Kalman filter); (b) convergence of state variables x_5 to x_8 to their reference setpoints (red line: setpoint, blue line: real value, green line: estimation by a Kalman filter)

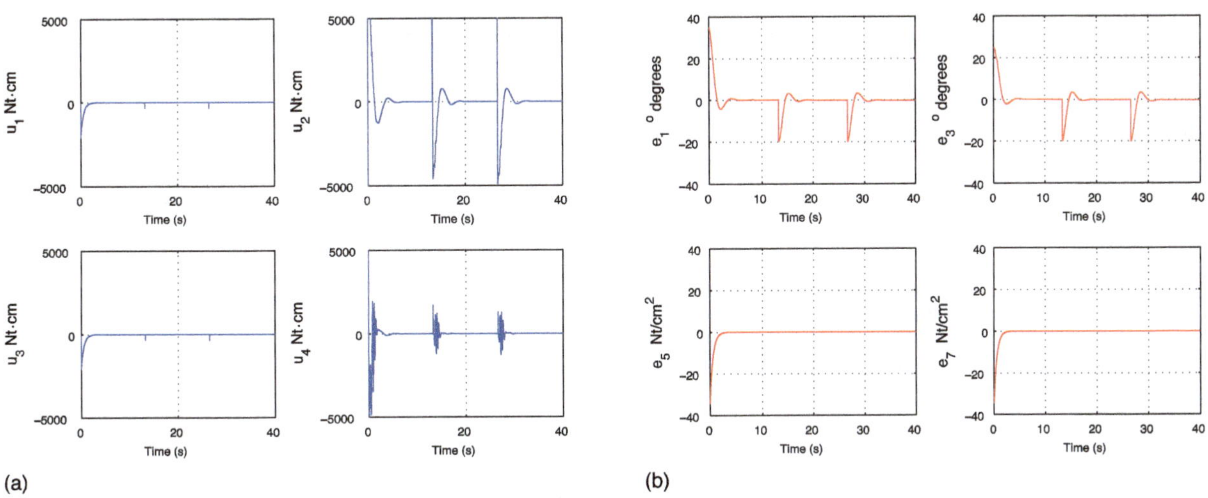

Figure 1.83 Tracking of trajectory 4 by the 2-link electropneumatic robotic manipulator: (a) control inputs u_1 to u_4 generated by the electropneumatic actuators of the robot; (b) tracking error variables

In Eq. (1.440) $y_3 = z_{5,8}$ is taken to be the flat output and u is the real control inputs vector, while $z_{1,2}$, $z_{3,4}$ are considered to be coefficients vectors. From Eq. (1.440),

$$u = g_{5,8}(z_{1,2}, z_{3,4}, z_{5,8})^{-1}[\dot{z}_{5,8} - f_{5,8}(z_{1,2}, z_{3,4}, z_{5,8})] \tag{1.443}$$

which signifies that u is a differential function of the flat outputs vector y_3, consequently Eq. (1.440) is a differentially flat subsystem.

The proof of differential flatness properties for the subsystems of Eq. (1.438), Eq. (1.439) and Eq. (1.440) signifies that each one of them is input–output linearizable and can be controlled and stabilized by inverting its dynamics, as is

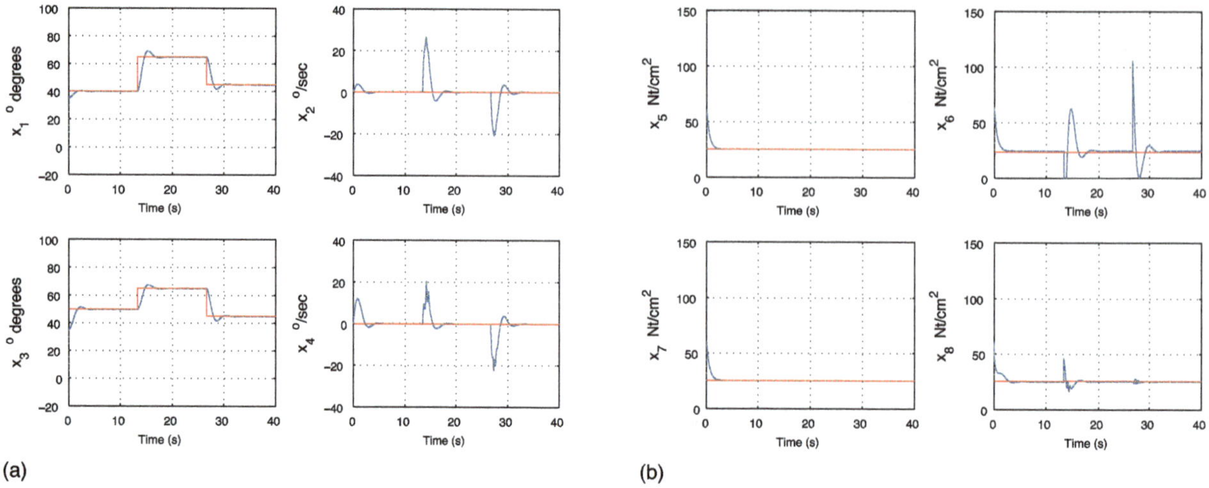

Figure 1.84 *Tracking of trajectory 5 by the 2-link electropneumatic robotic manipulator: (a) convergence of state variables x_1 to x_4 to their reference setpoints (red line: setpoint, blue line: real value, green line: estimation by a Kalman filter); (b) convergence of state variables x_5 to x_8 to their reference setpoints (red line: setpoint, blue line: real value, green line: estimation by a Kalman filter)*

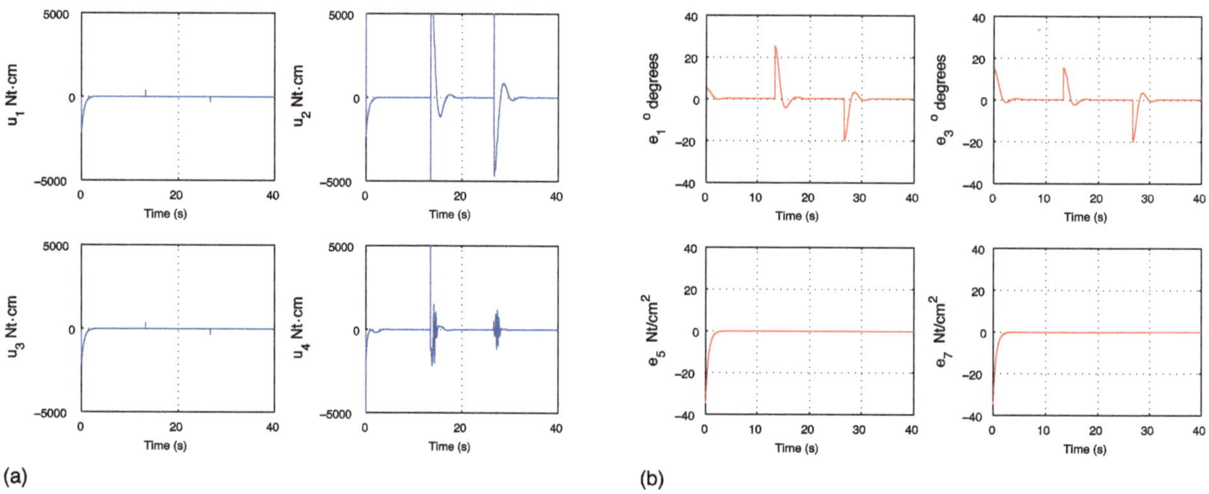

Figure 1.85 *Tracking of trajectory 5 by the 2-link electropneumatic robotic manipulator: (a) control inputs u_1 to u_4 generated by the electropneumatic actuators of the robot; (b) tracking error variables*

done in the case of input–output linearized systems. Thus, for the subsystem of Eq. (1.438) the virtual control input $v_1 = z^d_{3,4}$ is chosen to be

$$v_1 = z^d_{3,4} = g_{1,2}(z_{1,2})^{-1}[\dot{z}^d_{1,2} - f_{1,2}(z_{1,2}) - K_1(z_{1,2} - z^d_{1,2})] \tag{1.444}$$

where $K_1 \in R^{2\times 2}$ is a diagonal matrix with diagonal elements $k_{1,ii} > 0$ for $i = 1, 2$. Eq. (1.444) signifies that the virtual control input $v_1 = z^d_{3,4}$ becomes setpoint for the subsystem of Eq. (1.439).

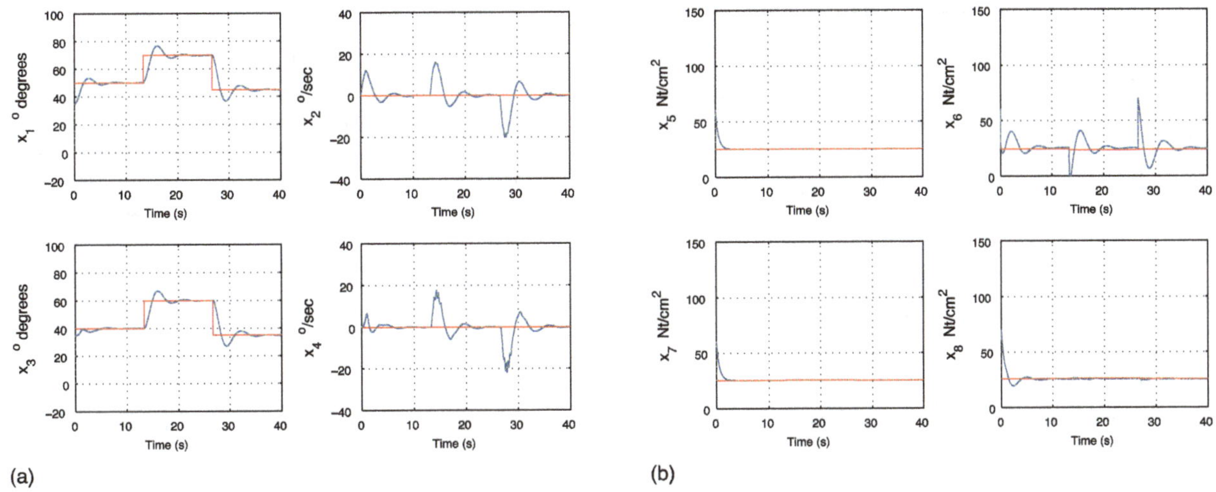

Figure 1.86 Tracking of trajectory 6 by the 2-link electropneumatic robotic manipulator: (a) convergence of state variables x_1 to x_4 to their reference setpoints (red line: setpoint, blue line: real value, green line: estimation by a Kalman filter); (b) convergence of state variables x_5 to x_8 to their reference setpoints (red line: setpoint, blue line: real value, green line: estimation by a Kalman Filter)

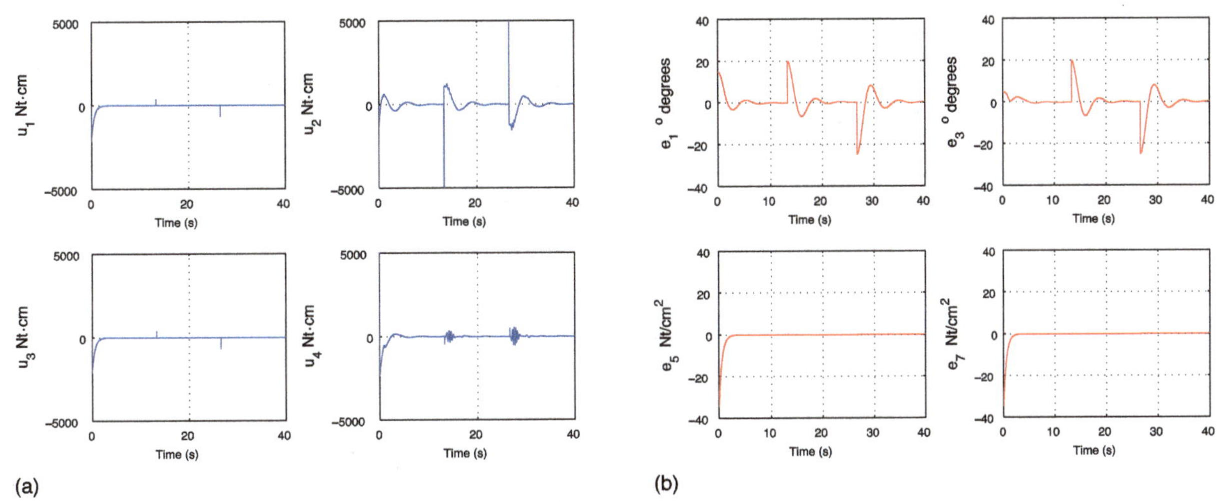

Figure 1.87 Tracking of trajectory 6 by the 2-link electropneumatic robotic manipulator: (a) control inputs u_1 to u_4 generated by the electropneumatic actuators of the robot; (b) tracking error variables

Moreover, for the subsystem of Eq. (1.439) the virtual control input v_2 is chosen to be

$$v_2 = g_{3,4}(z_{1,2}, z_{3,4})^{-1}[\dot{z}^d_{3,4} - f_{3,4}(z_{1,2}, z_{3,4}) - K_2(z_{3,4} - z^d_{3,4})] \tag{1.445}$$

where $K_2 \in R^{2\times 2}$ is a diagonal matrix with diagonal elements $k_{2,ii} > 0$ for $i = 1, 2$. Eq. (1.445) signifies that the virtual control input v_2 provides also the setpoints for the subsystem of Eq. (1.440).

Additionally, for the subsystem of Eq. (1.440) the real control input u is chosen to be

$$u = g_{5,8}(z_{1,2}, z_{3,4}, z_{5,8})^{-1}[\dot{z}^d_{5,8} - f_{5,8}(z_{1,2}, z_{3,4}, z_{5,8}) - K_3(z_{5,8} - z^d_{5,8})] \tag{1.446}$$

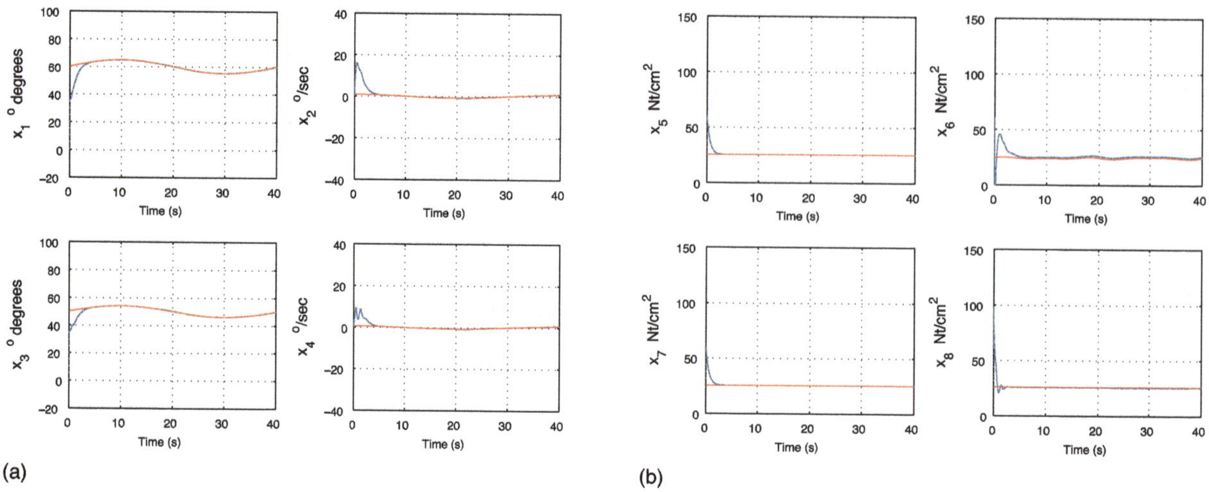

Figure 1.88 Tracking of trajectory 7 by the 2-link electropneumatic robotic manipulator: (a) convergence of state variables x_1 to x_4 to their reference setpoints (red line: setpoint, blue line: real value, green line: estimation by a Kalman filter); (b) convergence of state variables x_5 to x_8 to their reference setpoints (red line: setpoint, blue line: real value, green line: estimation by a Kalman filter)

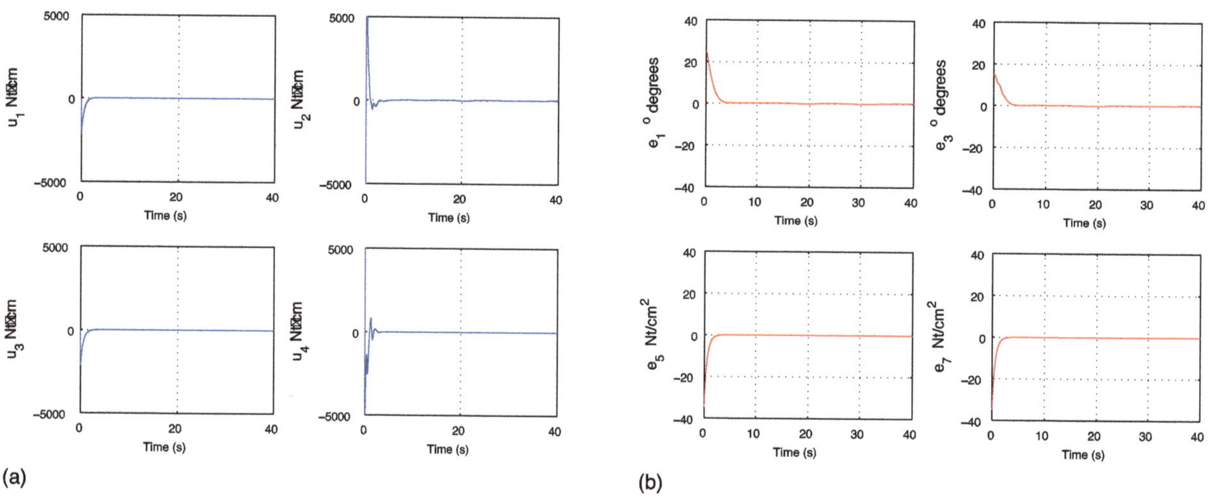

Figure 1.89 Tracking of trajectory 7 by the 2-link electropneumatic robotic manipulator: (a) control inputs u_1 to u_4 generated by the electropneumatic actuators of the robot; (b) tracking error variables

where $K_3{\in}R^{4\times4}$ is a diagonal matrix with diagonal elements $k_{3,ii}>0$ for $i=1,2$.

Next, by substituting the control input of Eq. (1.444) into Eq. (1.438), as well as by substituting the control input of Eq. (1.445) into Eq. (1.439), and finally by substituting the control input of Eq. (1.446) into Eq. (1.440) and by defining the tracking error variables $e_{1,2}=z_{1,2}-z_{1,2}^d$, $e_{3,4}=z_{3,4}-z_{3,4}^d$, and $e_{5,8}=z_{5,8}-z_{5,8}^d$, one obtains

$$\begin{aligned}(\dot{z}_{1,2}-\dot{z}_{1,2}^d)+K_1(z_{1,2}-z_{1,2}^d)=0 &\Rightarrow \dot{e}_{1,2}+K_1e_{1,2}=0\\ &\Rightarrow \lim_{t\to\infty} z_{1,2}=z_{1,2}^d\end{aligned} \tag{1.447}$$

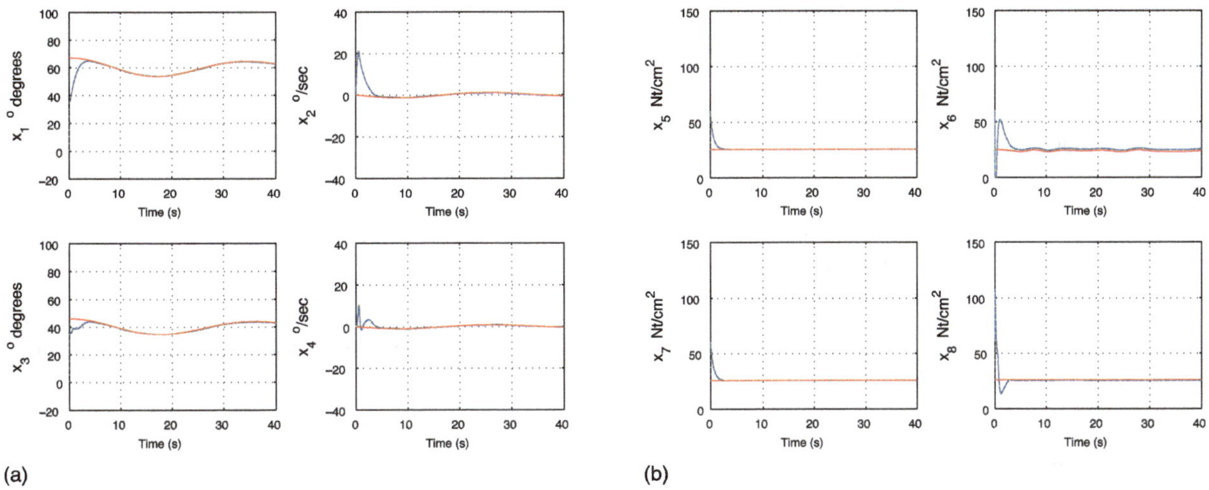

Figure 1.90 Tracking of trajectory 8 by the 2-link electropneumatic robotic manipulator: (a) convergence of state variables x_1 to x_4 to their reference setpoints (red line: setpoint, blue line: real value, green line: estimation by a Kalman filter); (b) convergence of state variables x_5 to x_8 to their reference setpoints (red line: setpoint, blue line: real value, green line: estimation by a Kalman filter)

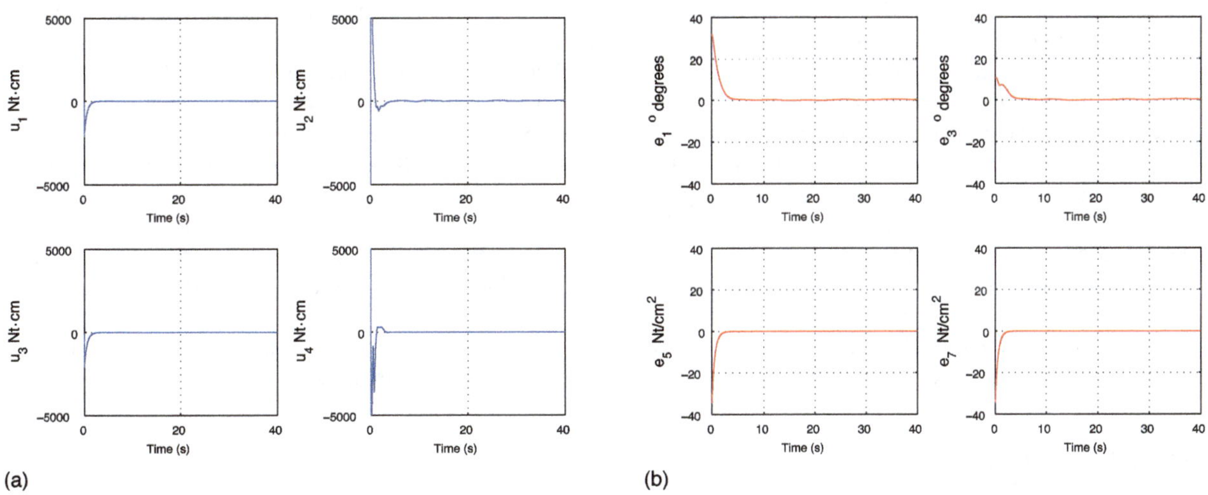

Figure 1.91 Tracking of trajectory 8 by the 2-link electropneumatic robotic manipulator: (a) control inputs u_1 to u_4 generated by the electropneumatic actuators of the robot; (b) tracking error variables

$$(\dot{z}_{3,4} - \dot{z}_{3,4}^d) + K_2(z_{3,4} - z_{3,4}^d) = 0 \Rightarrow \dot{e}_{3,4} + K_2 e_{3,4} = 0 \\ \Rightarrow \lim_{t\to\infty} z_{3,4} = z_{3,4}^d \tag{1.448}$$

$$(\dot{z}_{5,8} - \dot{z}_{5,8}^d) + K_3(z_{5,8} - z_{5,8}^d) = 0 \Rightarrow \dot{e}_{5,8} + K_3 e_{5,8} = 0 \\ \Rightarrow \lim_{t\to\infty} z_{5,8} = z_{5,8}^d \tag{1.449}$$

which confirms that all state variables of the electropneumatic robot converge to the associated setpoints.

The global stability properties of the flatness-based control scheme in successive loops can be also proven through Lyapunov analysis. To this end, the following Lyapunov function is defined:

$$V = \frac{1}{2}[e_{1,2}^T e_{1,2} + e_{3,4}^T e_{3,4} + e_{5,8}^T e_{5,8}] \tag{1.450}$$

By differentiating in time and using the tracking error dynamics of Eq. (1.447), Eq. (1.448) and Eq. (1.449), one obtains

$$\begin{array}{c} \dot{V} = e_{1,2}^T \dot{e}_{1,2} + e_{3,4}^T \dot{e}_{3,4} + e_{5,8}^T \dot{e}_{5,8} \Rightarrow \\ \dot{V} = e_{1,2}^T[-K_1 e_{1,2}] + e_{3,4}^T[-K_2 e_{3,4}] + e_{5,8}^T[-K_3 e_{5,8}] \Rightarrow \\ \dot{V} = -[e_{1,2}^T K_1 e_{1,2} + e_{3,4}^T K_2 e_{3,4} + e_{5,8}^T K_3 e_{5,8}] \\ \Rightarrow \dot{V} < 0 \; \forall \; e_{1,2} \neq 0, e_{3,4} \neq 0, e_{5,8} \neq 0 \end{array} \tag{1.451}$$

and $\dot{V} = 0$ if and only of $e_{1,2} = 0$, $e_{3,4} = 0$, and $e_{5,8} = 0$. Consequently, the closed-loop system of the electropneumatic robot is globally asymptotically stable.

1.4.4 Simulation tests

Indicative values about the parameters of the dynamic model of the electropneumatically actuated robot are as follows: (i) first link and joint of the robotic exoskeleton: $m_1 = 1.0kg$, $l_1 = 1.0m$, $d_1 = 0.25m$, $I_1 = 1.0kg{\cdot}m^2$, $k_1 = 100$, $b_1 = 0.01$ (ii) second link and joint of the robotic exoskeleton: $m_2 = 1.0kg$, $l_2 = 1.0m$, $d_2 = 0.25m$, $I_2 = 1.0kg{\cdot}m^2$, $k_2 = 100$, $b_2 = 0.01$ (iii) first electropneumatic actuator: $S_{P_1} = 5.0cm^2$, $S_{N_1} = 5.0cm^2$, $V_{DP_1} = 35.2cm^3$, $V_{DN_1} = 35.2cm^3$, $K_r = k{\cdot}r$, $k = 1.2$ (polytropic constant), $r = 297J{\cdot}kg^{-1}K^{-1}$ (perfect gas constant) $T = 293^oK = 20^oC$ (temperature) (iv) second electropneumatic actuator: $S_{P_2} = 5.0cm^2$, $S_{N_2} = 5.0cm^2$, $V_{DP_2} = 35.2cm^3$, $V_{DN_2} = 35.2cm^3$, $K_r = k{\cdot}r$, $k = 1.2$ (polytropic constant), $r = 297J{\cdot}kg^{-1}K^{-1}$ (perfect gas constant) $T = 293^oK = 20^oC$ (temperature). The diagonal feedback control matrices K_i had diagonal elements equal to 2.0. Results about the tracking accuracy and the speed of convergence to setpoints of the successive-loops flatness-based control method, in the case of the 2-link electropneumatic robotic exoskeleton, are shown in Figures 1.76–1.91. It can be noticed, that under this control scheme one achieves fast and precise tracking of reference setpoints for all state variables of the robotic system. It is noteworthy, that through the stages of this method one solves also the setpoints definition problem for all state variables of the electropneumatic robotic exoskeleton. Actually, the selection of setpoints for state variables x_1, x_2, x_5, and x_7 is unconstrained. On the other hand, the selection of setpoints for state variables x_3, x_4, x_6, and x_8 is dependent on the setpoints for x_1, x_2, x_5, x_7. The speed of convergence of the state variables of the electropneumatic robotic system under flatness-based control implemented in successive loops is dependent on the selection of values for the diagonal gain matrices K_1, K_2 and K_3 of Eq. (1.444), Eq. (1.445) and Eq. (1.446).

To elaborate on the tracking performance and on the robustness of the proposed flatness-based control method in successive loops for the electropneumatically actuated robot the following tables are given: (i) Table 1.11 which

Table 1.11 electropneumatic robot

	Tracking RMSE in the disturbance-free case							
	$RMSE_{x_1}$	$RMSE_{x_2}$	$RMSE_{x_3}$	$RMSE_{x_4}$	$RMSE_{x_5}$	$RMSE_{x_6}$	$RMSE_{x_7}$	$RMSE_{x_8}$
$test_1$	0.0003	0.0001	0.0001	0.0001	0.0001	0.0526	0.0001	0.0172
$test_2$	0.0004	0.0001	0.0001	0.0001	0.0001	0.0520	0.0001	0.0171
$test_3$	0.0004	0.0001	0.0006	0.0001	0.0001	0.0414	0.0001	0.0187
$test_4$	0.0005	0.0001	0.0011	0.0001	0.0001	0.0404	0.0001	0.0134
$test_5$	0.0007	0.0001	0.0007	0.0001	0.0001	0.0409	0.0001	0.0134
$test_6$	0.0005	0.0001	0.0003	0.0001	0.0001	0.0393	0.0001	0.0129
$test_7$	0.0002	0.0002	0.0002	0.0002	0.0001	0.0531	0.0001	0.0175
$test_8$	0.0005	0.0003	0.0006	0.0001	0.0001	0.0668	0.0001	0.0221

Table 1.12 electropneumatic robot

Tracking RMSE in the case of disturbances								
$\Delta a\%$	$RMSE_{x_1}$	$RMSE_{x_2}$	$RMSE_{x_3}$	$RMSE_{x_4}$	$RMSE_{x_5}$	$RMSE_{x_6}$	$RMSE_{x_7}$	$RMSE_{x_8}$
0%	0.0004	0.0001	0.0001	0.0001	0.0001	0.0520	0.0001	0.0171
10%	0.0004	0.0001	0.0001	0.0001	0.0001	0.0520	0.0001	0.0171
20%	0.0005	0.0001	0.0001	0.0001	0.0001	0.0520	0.0001	0.0171
30%	0.0005	0.0001	0.0001	0.0001	0.0001	0.0520	0.0001	0.0171
40%	0.0006	0.0001	0.0001	0.0001	0.0001	0.0520	0.0001	0.0171
50%	0.0006	0.0001	0.0001	0.0001	0.0001	0.0520	0.0001	0.0171
60%	0.0007	0.0001	0.0001	0.0001	0.0001	0.0520	0.0001	0.0171

Table 1.13 electropneumatic robot

Convergence time (sec) for the pneumatic robot's state variables								
	$T_s\, x_1$	$T_s\, x_2$	$T_s\, x_3$	$T_s\, x_4$	$T_s\, x_5$	$T_s\, x_6$	$T_s\, x_7$	$T_s\, x_8$
$test_1$	3.0	4.0	4.0	4.0	2.0	4.5	2.5	2.5
$test_2$	3.0	4.5	2.0	3.5	3.5	4.5	2.0	4.0
$test_3$	3.0	3.5	2.0	4.5	2.5	4.0	2.5	3.0
$test_4$	3.0	4.5	4.0	4.5	2.5	4.5	2.0	4.0
$test_5$	2.5	2.5	2.5	4.0	2.5	3.0	2.0	2.5
$test_6$	3.5	4.0	2.0	4.5	2.5	4.0	2.0	4.0
$test_7$	3.5	4.0	2.0	4.5	2.5	4.0	2.0	4.0
$test_8$	3.5	4.0	2.0	4.5	2.5	4.0	2.0	4.0

provides information about the accuracy of tracking of the reference setpoints by the state variables of the electropneumatically actuated robot's state-space model, (ii) Table 1.12 which provides information about the robustness of the control method to parametric changes in the model of the electropneumatically actuated robot's dynamics (change $\Delta a\%$ in the moments of inertia of the links), (iii) Table 1.13 which provides the approximate convergence times of the electropneumatically actuated robot's state variables to the associated setpoints.

Chapter 2
Nonlinear optimal and flatness-based control for robotic systems II

Abstract

The chapter analyzes further nonlinear optimal and flatness-based control methods for robotic systems. First, the problem of multi-loop flatness-based control for dual-arm robotic manipulators is solved. Next, a multi-loop flatness-based control method is developed for the dynamic model of the electrohydraulic robotic manipulator. Additionally, a multi-loop flatness-based control approach is proposed for the dynamic model of a multi-DOF flexible-joint robotic manipulator. Moreover, a multi-loop flatness-based control scheme is applied to the model of a three-link redundant robotic manipulator. Finally, the chapter analyzes a method for flatness-based disturbance observer and control with application to robotic mining excavators.

2.1 Flatness-based control in successive loops for dual-arm robotic manipulators

2.1.1 Outline

The monograph continues with additional results on the use of its nonlinear optimal and flatness-based control methods in advanced robotic systems. Differential flatness theory is currently a main research direction in the control of nonlinear dynamic systems with several applications in robotics [711,721,732,725,741,491,252,898]. A dynamic system is considered to be differentially flat if the following two properties hold: (i) all state variables and its control inputs can be written as differential functions of a specific subset of its state variables, the latter standing for the flat outputs of the system, (ii) the flat outputs of the system are differentially independent, which means that they and their time-derivatives are not connected through a relation in the form of a homogeneous linear differential equation [711,91,594,639,489]. Usually, flatness-based control is implemented through transformation of the state-space model of a nonlinear system into the input–output linearized form through successive differentiations of its flat outputs [806,492,640,524,54]. This new state-space description of the system is both controllable and observable [439,705,60,706]. Next, for the linearized state-space description of the system, a stabilizing feedback controller is designed with the eigenvalues assignment technique. To select the stabilizing feedback gains of this controller polynomial equations have to be solved with the requirement that their roots should reside in the left complex semiplane [711]. The aim of the present section is to introduce a new multi-loop flatness-based control method that does not need any changes of state variables or complicated state-space model transformations and which computes control inputs which can be applied directly on the initial nonlinear state-space model of the controlled system. It is also aimed to arrive at a simple procedure for selecting the stabilizing feedback gains of this controller. Actually, in the new multi-loop flatness-based control method, there is no need to solve polynomial equations to select the control gains. Stabilization is achieved through the use of diagonal feedback gain matrices, needing only to assign positive values at their diagonal elements [712,720,714,716,718].

In the new multi-loop flatness-based control method the dynamic model of the controlled system is decomposed into chained subsystems, where each individual subsystem satisfies the differential flatness property [712,720,714,716,718]. The state vector of the subsequent $(i+1)$-th subsystem becomes virtual control input to the preceding i-th subsystem. Besides, the virtual control input of the preceding i-th subsystem becomes setpoints vector for the subsequent $(i+1)$-subsystem. Knowing that each subsystem is differentially flat it is also confirmed that it is input–output linearizable and thus a virtual control input about it can be chosen by inverting its dynamics as it is commonly done for dynamical systems in the input–output linearized form. From the last subsystem, and through inversion of its dynamics, one finds the real control inputs vector which finally stabilizes the entire nonlinear system. The real control inputs incorporate implicitly the virtual control inputs from all preceding subsystems. The stabilization effects are seen by tracing the system backward, that is from the last to its first subsystem. The global stability properties of this control method are demonstrated in two manners: (a) it is proven that the tracking error for the state vectors of the individual subsystems converges asymptotically to zero, and (b) Lyapunov analysis is performed showing that the Lyapunov function of the integrated nonlinear system is a diminishing function which converges asymptotically to zero.

In the present section, the multi-loop flatness-based control method is applied to the dynamic model of dual-arm robotic manipulators. Synchronization between robotic manipulators and cooperative execution of tasks by them is an important research area in contemporary robotics [1068,1072,220]. In this context, the solution of the nonlinear control and synchronized motion planning problem for dual-arm robots is also a challenging task [23,850,59]. Control for dual-arm robots aims primarily at synchronized and precise positioning of the end-effectors of the two arms [1032,989,364,406]. It also aims at compensating forces exerted on the two robotic arms, which can be due to simultaneous grasping and lifting of objects or due to performing assembling and other tasks that may generate contact and interaction with stiff surfaces [111,412,51,641]. Dual-arm robots are widely used in industry, in several manufacturing tasks, in healthcare, in agriculture, and in warehouse tasks [2,196,947,798]. Dual-arm robots appear also in the execution of several space robotics tasks [929,916,788]. Furthermore, they can be used for assistance of the elderly or impaired people, for rehabilitation of the upper limbs and in robotic surgery [496,948,29]. Among various control schemes which have been proposed for precise positioning, synchronization, and position/force control in dual-arm robots, one can distinguish: (i) nonlinear model-predictive control methods, (ii) sliding-mode control methods and (iii) backstepping-type control methods. Besides, there exist some results on attempts to solve the nonlinear optimal control problem for dual-arm robots, mainly with the use of NMPC techniques [553,879,880,636,44].

2.1.2 Dynamic model of the dual-arm robotic manipulator

2.1.2.1 State-space model of the dual-arm robot

The dual-arm robotic manipulator consists of arm A and arm B, as shown in Figure 2.1. Arm A consists of two links, of masses $m_{1,A}$, $m_{2,A}$, and of lengths $l_{1,A}$, $l_{2,A}$, respectively. The turn angles of the joints of these links are denoted as $\theta_{1,A}$ and $\theta_{2,A}$. The position of the first joint of this arm is given by the coordinates (x_0^A, h). The position of the end-effector of this arm is given by the coordinates (x_E^A, y_E^A). Arm B consists also of two links, of masses $m_{1,A}$, $m_{2,A}$, and of lengths $l_{1,A}$, $l_{2,A}$, respectively. The turn angles of the joints of these links are denoted as $\theta_{1,B}$ and $\theta_{2,B}$. The position of the first joint of this arm is given by the coordinates (x_0^B, h). The position of the end-effector of this arm is given by the coordinates (x_E^B, y_E^B). The dual-arm robot is used to grasp and lift-up with its end-effectors an object of weight $B = M{\cdot}g$. The transversal axis of the object forms an angle ϕ with the horizontal axis of the inertial reference frame. The contact force due to grasping which is exerted to each end-effector is denoted by F.
As already explained, the dynamic model of arm A is given by [721]

$$M_A(\theta_A)\ddot{\theta}_A + C_A(\theta_A, \dot{\theta}_A) + G_A(\theta_A) = \tau_A - J_A^T B - J_A^T F \tag{2.1}$$

where $M_A(\theta_A)$ is the inertia matrix of arm A which is given by

$$M_A(\theta_A) = \begin{pmatrix} \bar{m}_{11,A} & \bar{m}_{12,A} \\ \bar{m}_{21,A} & \bar{m}_{22,A} \end{pmatrix} \tag{2.2}$$

with the elements of the inertia matrix to be given by: $\bar{m}_{11,A} = (m_{1,A} + m_{2,A})l_{1,A}^2 + m_{2,A}l_{2,A}^2 + 2m_{2,A}l_{1,A}l_{2,A}cos(\theta_{2,A})$, $\bar{m}_{12,A} = \bar{m}_{21,A} = m_{2,A}l_{2,A}^2 + m_{2,A}l_{1,A}l_{2,A}cos(\theta_{2,A})$ and $\bar{m}_{22,A} = m_{2,A}l_{2,A}(l_{1,A} + l_{2,A})cos(\theta)_{2,A}$.

The Coriolis and centrifugal forces matrix $C_A(\theta_A, \dot{\theta}_A)$ of arm A is given by

$$C_A(\theta_A, \dot{\theta}_A) = \begin{pmatrix} \bar{C}_{1,A} \\ \bar{C}_{2,A} \end{pmatrix} \tag{2.3}$$

where $\bar{C}_{1,A} = -m_{2,A}l_{1,A}^2 sin(\theta_{2,A})\dot{\theta}_{2,A}^2 - 2m_{2,A}l_{2,A}^2 sin(\theta_{2,A})\dot{\theta}_{1,A}\dot{\theta}_{2,A}$ and $\bar{C}_{2,A} - m_{1,A}l_{1,A}^2 sin(\theta_{2,A})\dot{\theta}_{1,A}^2$.

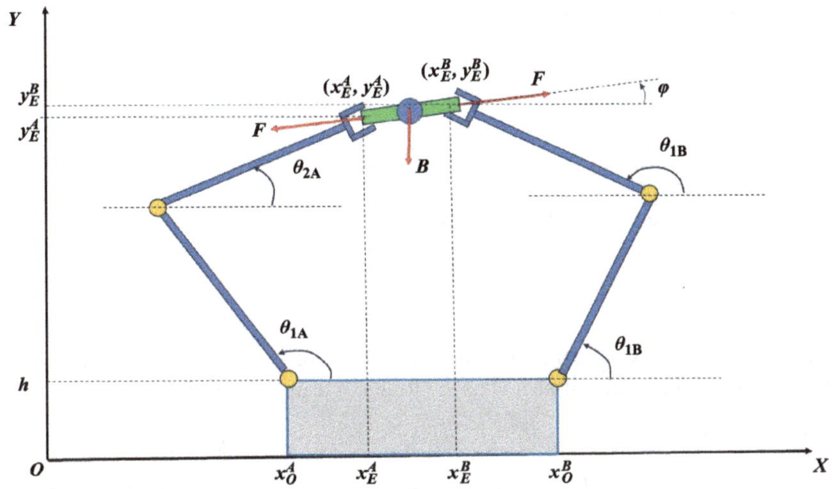

Figure 2.1 Diagram of the dual-arm robotic manipulator in an object-lifting task.

The gravitational forces vector $G_A(\theta)$ of arm A is given by

$$G_A(\theta_A) = \begin{pmatrix} \bar{G}_{1,A} \\ \bar{G}_{2,A} \end{pmatrix} \tag{2.4}$$

where $\bar{G}_{1,A} = (m_{1,A} + m_{2,A}gl_{1,A})cos(\theta_{1,A}) + m_{2,A}gl_{2,A}cos(\theta_{1,A} + \theta_{2,A})$ and $\bar{G}_{2,A} = m_{2,A}gl_{2,A}cos(\theta_{1,A} + \theta_{2,A})$, with g to be denoting the acceleration of gravity.

The torques vector which is provided by the actuators of arm A is given by

$$\tau_A = \begin{pmatrix} \tau_{1,A} \\ \tau_{2,A} \end{pmatrix} \tag{2.5}$$

The transpose of the Jacobian matrix J_A^T which transfers torques from the end-effector of arm A to the joints of arm A, due to the load's weight $B = M{\cdot}g$ and due to the grasping forces F, is given by

$$J_A^T = \begin{pmatrix} J_{A,11}^T & J_{A,12}^T \\ J_{A,21}^T & J_{A,22}^T \end{pmatrix} = \begin{pmatrix} -l_{1,A}sin(\theta_{1,A}) - l_{2,A}sin(\theta_{1,A} + \theta_{2,A}) & l_{1,A}cos(\theta_{1,A}) + l_{2,A}cos(\theta_{1,A} + \theta_{2,A}) \\ -l_{2,A}sin(\theta_{1,A} + \theta_{2,A}) & l_{2,A}cos(\theta_{1,A} + \theta_{2,A}) \end{pmatrix} \tag{2.6}$$

The dynamic model of arm B is given by [721]

$$M_B(\theta_B)\ddot{\theta} + C_B(\theta_B, \dot{\theta}_B) + G_B(\theta_B) = \tau_B - J_B^T B + J_B^T F \tag{2.7}$$

where $M_B(\theta_B)$ is the inertia matrix of arm B which is given by

$$M_B(\theta_B) = \begin{pmatrix} \bar{m}_{11,B} & \bar{m}_{12,B} \\ \bar{m}_{21,B} & \bar{m}_{22,B} \end{pmatrix} \tag{2.8}$$

with the elements of the inertia matrix to be given by: $\bar{m}_{11,B} = (m_{1,B} + m_{2,B})l_{1,B}^2 + m_{2,B}l_{2,B}^2 + 2m_{2,B}l_{1,B}l_{2,B}cos(\theta_{2,B})$, $\bar{m}_{12,B} = \bar{m}_{21,B} = m_{2,B}l_{2,B}^2 + m_{2,B}l_{1,B}l_{2,A}cos(\theta_{2,B})$, and $\bar{m}_{22,B} = m_{2,B}l_{2,B}(l_{1,B} + l_{2,B})cos(\theta)_{2,B}$.

The Coriolis and centrifugal forces matrix $C_B(\theta_B, \dot{\theta}_B)$ of arm B is given by

$$C_B(\theta_B, \dot{\theta}_B) = \begin{pmatrix} \bar{C}_{1,B} \\ \bar{C}_{2,B} \end{pmatrix} \tag{2.9}$$

where $\bar{C}_{1,B} = -m_{2,B}l_{1,B}^2 sin(\theta_{2,B})\dot{\theta}_{2,B}^2 - 2m_{2,B}l_{2,B}^2 sin(\theta_{2,B})\dot{\theta}_{1,B}\dot{\theta}_{2,B}$ and $\bar{C}_{2,B} - m_{1,B}l_{1,B}^2 sin(\theta_{2,B})\dot{\theta}_{1,B}^2$.

The gravitational forces vector $G_B(\theta)$ of arm B is given by

$$G_B(\theta_B) = \begin{pmatrix} \bar{G}_{1,B} \\ \bar{G}_{2,B} \end{pmatrix} \tag{2.10}$$

where $\bar{G}_{1,B} = (m_{1,B} + m_{2,B} g l_{1,B})cos(\theta_{1,B}) + m_{2,B} g l_{2,B} cos(\theta_{1,B} + \theta_{2,B})$ and $\bar{G}_{2,B} = m_{2,B} g l_{2,B} cos(\theta_{1,B} + \theta_{2,B})$, with g to be denoting the acceleration of gravity.

The torques vector which is provided by the actuators of arm B is given by

$$\tau_B = \begin{pmatrix} \tau_{1,B} \\ \tau_{2,B} \end{pmatrix} \tag{2.11}$$

The transpose of the Jacobian matrix J_B^T which transfers torques due to the load's weight $B = M{\cdot}g$ and due to the grasping forces $-F$, from the end-effector of arm B to the joints of arm B is given by

$$J_B^T = \begin{pmatrix} J_{B,11}^T & J_{B,12}^T \\ J_{B,21}^T & J_{B,22}^T \end{pmatrix} = \begin{pmatrix} = -l_{1,B} sin(\theta_{1,B}) - l_{2,B} sin(\theta_{1,B} + \theta_{2,B}) & l_{1,B} cos(\theta_{1,B}) + l_{2,B} cos(\theta_{1,B} + \theta_{2,B}) \\ -l_{2,B} sin(\theta_{1,B} + \theta_{2,B}) & l_{2,B} cos(\theta_{1,B} + \theta_{2,B}) \end{pmatrix} \tag{2.12}$$

Force F exerted on the two end-effectors is due to grasping and is represented as a vector which is collinear to the transversal axis of the lifted object. It is given by

$$F = K_F(L - L_o) \tag{2.13}$$

where L_o is the length of the non-deformed object, L is the length of the lifted object which has been subject to deformation due to the grasping forces applies on its two ends and K_F is an elasticity (spring) coefficient. Considering that the position of the end-effector of arm A is (x_E^A, y_E^A) and the position of the end-effector of arm B is (x_E^B, y_E^B), length L is given by

$$L = \sqrt{(x_E^B - x_E^A)^2 + (y_E^B - y_E^A)^2} \Rightarrow$$

$$\begin{aligned} L = \{&[x_0^B + l_{1,B} cos(\theta_{1,B}) + l_{2,B} cos(\theta_{1,B} + \theta_{2,B}) - x_0^A - l_{1,A} cos(\theta_{1,A}) + l_{2,A} cos(\theta_{1,A} + \theta_{2,A})]^2 + \\ &+[h + l_{1,B} sin(\theta_{1,B}) + l_{2,B} sin(\theta_{1,B} + \theta_{2,B}) - h - l_{1,A} sin(\theta_{1,A}) - l_{2,A} sin(\theta_{1,A} + \theta_{2,A})]^2\}^{\frac{1}{2}} \end{aligned} \tag{2.14}$$

The angle which is formed between the transversal axis of the lifted object and the horizontal axis of the inertial reference frame is denoted as ϕ and is given by

$$\phi = tan^{-1}\left(\frac{(y_E^B - y_E^A)}{(x_E^B - x_E^A)}\right) \tag{2.15}$$

where $(y_E^B - y_E^A) = [h + l_{1,B} sin(\theta_{1,B}) + l_{2,B} sin(\theta_{1,B} + \theta_{2,B}) - h - l_{1,A} sin(\theta_{1,A}) - l_{2,A} sin(\theta_{1,A} + \theta_{2,A})]$ and $(x_E^B - x_E^A) = [x_0^B + l_{1,B} cos(\theta_{1,B}) + l_{2,B} cos(\theta_{1,B} + \theta_{2,B}) - x_0^A - l_{1,A} cos(\theta_{1,A}) + l_{2,A} cos(\theta_{1,A} + \theta_{2,A})]$.

Using that weight B in cartesian coordinates is represented by vector $[0\ B]^T$, then the torque which is transferred to the joints of arm A due to weight B is given by

$$J_A^T B = \begin{pmatrix} J_{A,12}^T Mg \\ J_{A,22}^T Mg \end{pmatrix} \tag{2.16}$$

Using that force F in cartesian coordinates is represented by vector $[Fcos(\phi)Fsin(\phi)]^T$, then the torque which is transferred to the joints of arm A due to the grasping force F is given by

$$J_A^T F = \begin{pmatrix} [J_{A,11}^T cos(\phi) + J_{A,12}^T sin(\phi)]F \\ [J_{A,21}^T cos(\phi) + J_{A,22}^T sin(\phi)]F \end{pmatrix} \tag{2.17}$$

Using that weight B in cartesian coordinates is represented by vector $[0\ B]^T$, then the torque which is transferred to the joints of arm B due to weight B is given by

$$J_B^T B = \begin{pmatrix} J_{B,12}^T Mg \\ J_{B,22}^T Mg \end{pmatrix} \tag{2.18}$$

Using that force F in cartesian coordinates is represented by vector $[-Fcos(\phi) - Fsin(\phi)]^T$, then the torque which is transferred to the joints of arm B due to the grasping force F is given by

$$-J_B^T F = \begin{pmatrix} -[J_{B,11}^T cos(\phi) + J_{B,12}^T sin(\phi)]F \\ -[J_{A,21}^T cos(\phi) + J_{B,22}^T sin(\phi)]F \end{pmatrix} \tag{2.19}$$

Consequently, the dynamic model of the dual-arm robotic manipulator is given by the following two equations for matrices A and B respectively

$$\begin{aligned} \begin{pmatrix} \bar{m}_{11,A} & \bar{m}_{12,A} \\ \bar{m}_{21,A} & \bar{m}_{22,A} \end{pmatrix} \begin{pmatrix} \ddot{\theta}_{1,A} \\ \ddot{\theta}_{2,A} \end{pmatrix} + \begin{pmatrix} \bar{C}_{1,A} \\ \bar{C}_{2,A} \end{pmatrix} + \begin{pmatrix} \bar{G}_{1,A} \\ \bar{G}_{2,A} \end{pmatrix} = \begin{pmatrix} \tau_{1,A} \\ \tau_{2,A} \end{pmatrix} - \\ - \begin{pmatrix} J_{A.12}^T Mg \\ J_{A.22}^T Mg \end{pmatrix} - \begin{pmatrix} [J_{A,11}^T cos(\phi) + J_{A,12}^T sin(\phi)]F \\ [J_{A,21}^T cos(\phi) + J_{A,22}^T sin(\phi)]F \end{pmatrix} \end{aligned} \tag{2.20}$$

$$\begin{aligned} \begin{pmatrix} \bar{m}_{11,B} & \bar{m}_{12,B} \\ \bar{m}_{21,B} & \bar{m}_{22,B} \end{pmatrix} \begin{pmatrix} \ddot{\theta}_{1,B} \\ \ddot{\theta}_{2,B} \end{pmatrix} + \begin{pmatrix} \bar{C}_{1,B} \\ \bar{C}_{2,B} \end{pmatrix} + \begin{pmatrix} \bar{G}_{1,B} \\ \bar{G}_{2,B} \end{pmatrix} = \begin{pmatrix} \tau_{1,B} \\ \tau_{2,B} \end{pmatrix} - \\ - \begin{pmatrix} J_{B.12}^T Mg \\ J_{B.22}^T Mg \end{pmatrix} - \begin{pmatrix} -[J_{B,11}^T cos(\phi) + J_{B,12}^T sin(\phi)]F \\ -[J_{B,21}^T cos(\phi) + J_{B,22}^T sin(\phi)]F \end{pmatrix} \end{aligned} \tag{2.21}$$

The inverse of the inertia matrix of arm A is given by

$$M_A^{-1} = \frac{1}{detM_A} \begin{pmatrix} \bar{m}_{22,A} & -\bar{m}_{12,A} \\ -\bar{m}_{21,A} & \bar{m}_{11,A} \end{pmatrix} \tag{2.22}$$

where $detM_A = \bar{m}_{11,A}\bar{m}_{22,A} - \bar{m}_{12,A}\bar{m}_{21,A}$. The inverse of the inertia matrix of arm B is given by

$$M_B^{-1} = \frac{1}{detM_B} \begin{pmatrix} \bar{m}_{22,B} & -\bar{m}_{12,B} \\ -\bar{m}_{21,B} & \bar{m}_{11,B} \end{pmatrix} \tag{2.23}$$

where $detM_B = \bar{m}_{11,B}\bar{m}_{22,B} - \bar{m}_{12,B}\bar{m}_{21,B}$. Thus, after some intermediate operations the dynamic model of arm A and arm B is written as

$$\begin{aligned} \begin{pmatrix} \ddot{\theta}_{1,A} \\ \ddot{\theta}_{2,A} \end{pmatrix} = - \begin{pmatrix} \frac{\bar{m}_{22,A}}{detM_A} & \frac{-\bar{m}_{12,A}}{detM_A} \\ \frac{-\bar{m}_{21,A}}{detM_A} & \frac{\bar{m}_{11,A}}{detM_A} \end{pmatrix} \begin{pmatrix} \bar{C}_{1,A} + \bar{G}_{1,A} + J_{A,12}^T Mg + [J_{A,11}^T cos(\phi) + J_{A,12}^T sin(\phi)]F \\ \bar{C}_{2,A} + \bar{G}_{2,A} + J_{A,22}^T Mg + [J_{A,21}^T cos(\phi) + J_{A,22}^T sin(\phi)]F \end{pmatrix} + \\ + \begin{pmatrix} \frac{\bar{m}_{22,A}}{detM_A} & \frac{-\bar{m}_{12,A}}{detM_A} \\ \frac{-\bar{m}_{21,A}}{detM_A} & \frac{\bar{m}_{11,A}}{detM_A} \end{pmatrix} \begin{pmatrix} \tau_{1,A} \\ \tau_{2,A} \end{pmatrix} \end{aligned} \tag{2.24}$$

$$\begin{pmatrix} \ddot{\theta}_{1,B} \\ \ddot{\theta}_{2,B} \end{pmatrix} = -\begin{pmatrix} \frac{\bar{m}_{22,B}}{detM_B} & \frac{-\bar{m}_{12,B}}{detM_B} \\ \frac{-\bar{m}_{21,B}}{detM_B} & \frac{\bar{m}_{11,B}}{detM_B} \end{pmatrix} \begin{pmatrix} \bar{C}_{1,B} + \bar{G}_{1,B} + J_{B,12}^T Mg - [J_{B,11}^T cos(\phi) + J_{B,12}^T sin(\phi)]F \\ \bar{C}_{2,B} + \bar{G}_{2,B} + J_{B,22}^T Mg - [J_{B,21}^T cos(\phi) + J_{B,22}^T sin(\phi)]F \end{pmatrix} + \\ + \begin{pmatrix} \frac{\bar{m}_{22,B}}{detM_B} & \frac{-\bar{m}_{12,B}}{detM_B} \\ \frac{-\bar{m}_{21,B}}{detM_B} & \frac{\bar{m}_{11,B}}{detM_B} \end{pmatrix} \begin{pmatrix} \tau_{1,B} \\ \tau_{2,B} \end{pmatrix} \tag{2.25}$$

Next, by denoting functions $F_{1,A}$ and $F_{2,A}$

$$\begin{aligned} F_{1,A} &= \bar{C}_{1,A} + \bar{G}_{1,A} + J_{A,12}^T Mg + [J_{A,11}^T cos(\phi) + J_{A,12}^T sin(\phi)]F \\ F_{2,A} &= \bar{C}_{2,A} + \bar{G}_{2,A} + J_{A,22}^T Mg + [J_{A,21}^T cos(\phi) + J_{A,22}^T sin(\phi)]F \end{aligned} \tag{2.26}$$

as well as functions $F_{1,B}$ and $F_{2,B}$

$$\begin{aligned} F_{1,B} &= \bar{C}_{1,B} + \bar{G}_{1,B} + J_{B,12}^T Mg - [J_{B,11}^T cos(\phi) + J_{B,12}^T sin(\phi)]F \\ F_{2,B} &= \bar{C}_{2,B} + \bar{G}_{2,B} + J_{B,22}^T Mg - [J_{B,21}^T cos(\phi) + J_{B,22}^T sin(\phi)]F \end{aligned} \tag{2.27}$$

the dynamic model of arm A and arm B is rewritten as

$$\begin{pmatrix} \ddot{\theta}_{1,A} \\ \ddot{\theta}_{2,A} \end{pmatrix} = \begin{pmatrix} \frac{-\bar{m}_{22,A}\bar{F}_{1,A} + \bar{m}_{12,A}\bar{F}_{2,A}}{detM_A} \\ \frac{\bar{m}_{21,A}\bar{F}_{1,A} - \bar{m}_{11,A}\bar{F}_{2,A}}{detM_A} \end{pmatrix} + \begin{pmatrix} \frac{\bar{m}_{22,A}}{detM_A} & \frac{-\bar{m}_{12,A}}{detM_A} \\ \frac{-\bar{m}_{21,A}}{detM_A} & \frac{\bar{m}_{11,A}}{detM_A} \end{pmatrix} \begin{pmatrix} \tau_{1,A} \\ \tau_{2,A} \end{pmatrix} \tag{2.28}$$

$$\begin{pmatrix} \ddot{\theta}_{1,B} \\ \ddot{\theta}_{2,B} \end{pmatrix} = \begin{pmatrix} \frac{-\bar{m}_{22,B}\bar{F}_{1,B} + \bar{m}_{12,B}\bar{F}_{2,B}}{detM_B} \\ \frac{\bar{m}_{21,B}\bar{F}_{1,B} - \bar{m}_{11,B}\bar{F}_{2,B}}{detM_B} \end{pmatrix} + \begin{pmatrix} \frac{\bar{m}_{22,B}}{detM_B} & \frac{-\bar{m}_{12,B}}{detM_B} \\ \frac{-\bar{m}_{21,B}}{detM_B} & \frac{\bar{m}_{11,B}}{detM_B} \end{pmatrix} \begin{pmatrix} \tau_{1,A} \\ \tau_{2,A} \end{pmatrix} \tag{2.29}$$

By aggregating the dynamics of arm A and arm B into one state-space model, one obtains the following state-space description:

$$\begin{pmatrix} \ddot{\theta}_{1,A} \\ \ddot{\theta}_{2,A} \\ \ddot{\theta}_{1,B} \\ \ddot{\theta}_{2,B} \end{pmatrix} = \begin{pmatrix} \frac{-\bar{m}_{22,A}\bar{F}_{1,A} + \bar{m}_{12,A}\bar{F}_{2,A}}{detM_A} \\ \frac{\bar{m}_{21,A}\bar{F}_{1,A} - \bar{m}_{11,A}\bar{F}_{2,A}}{detM_A} \\ \frac{-\bar{m}_{22,B}\bar{F}_{1,B} + \bar{m}_{12,B}\bar{F}_{2,B}}{detM_B} \\ \frac{\bar{m}_{21,B}\bar{F}_{1,B} - \bar{m}_{11,B}\bar{F}_{2,B}}{detM_B} \end{pmatrix} + \begin{pmatrix} \frac{\bar{m}_{22,A}}{detM_A} & \frac{-\bar{m}_{12,A}}{detM_A} & 0 & 0 \\ \frac{-\bar{m}_{21,A}}{detM_A} & \frac{\bar{m}_{11,A}}{detM_A} & 0 & 0 \\ 0 & 0 & \frac{\bar{m}_{22,B}}{detM_B} & \frac{-\bar{m}_{12,B}}{detM_B} \\ 0 & 0 & \frac{-\bar{m}_{21,B}}{detM_A} & \frac{\bar{m}_{11,B}}{detM_A} \end{pmatrix} \begin{pmatrix} \tau_{1,A} \\ \tau_{2,A} \\ \tau_{1,B} \\ \tau_{2,B} \end{pmatrix} \tag{2.30}$$

Next, the following state vector definition is used:

$$\begin{aligned} x &= [x_1, x_2, x_3, x_4, x_5, x_6, x_7, x_8]^T \Rightarrow \\ x &= [\theta_{1,A}, \dot{\theta}_{1,A}, \theta_{2,A}, \dot{\theta}_{2,A}, \theta_{1,B}, \dot{\theta}_{1,B}, \theta_{2,B}, \dot{\theta}_{2,B}]^T \end{aligned} \tag{2.31}$$

Besides, the following control inputs vector is defined:

$$\begin{aligned} u &= [u_1, u_2, u_3, u_4]^T \Rightarrow \\ u &= [\tau_{1,A}, \tau_{2,A}, \tau_{1,B}, \tau_{2,B}]^T \end{aligned} \tag{2.32}$$

Thus, the dynamic model of the dual-arm robotic manipulator can be written in the following state-space description:

$$\begin{pmatrix} \dot{x}_1 \\ \dot{x}_2 \\ \dot{x}_3 \\ \dot{x}_4 \\ \dot{x}_5 \\ \dot{x}_6 \\ \dot{x}_7 \\ \dot{x}_8 \end{pmatrix} = \begin{pmatrix} x_2 \\ \frac{-\bar{m}_{22,A}\bar{F}_{1,A}+\bar{m}_{12,A}\bar{F}_{2,A}}{detM_A} \\ x_4 \\ \frac{\bar{m}_{21,A}\bar{F}_{1,A}-\bar{m}_{11,A}\bar{F}_{2,A}}{detM_A} \\ x_6 \\ \frac{-\bar{m}_{22,B}\bar{F}_{1,B}+\bar{m}_{12,B}\bar{F}_{2,B}}{detM_B} \\ x_8 \\ \frac{\bar{m}_{21,B}\bar{F}_{1,B}-\bar{m}_{11,B}\bar{F}_{2,B}}{detM_B} \end{pmatrix} + \begin{pmatrix} 0 & 0 & 0 & 0 \\ \frac{\bar{m}_{22,A}}{detM_A} & \frac{-\bar{m}_{12,A}}{detM_A} & 0 & 0 \\ 0 & 0 & 0 & 0 \\ \frac{-\bar{m}_{21,A}}{detM_A} & \frac{\bar{m}_{11,A}}{detM_A} & 0 & 0 \\ 0 & 0 & 0 & 0 \\ 0 & 0 & \frac{\bar{m}_{22,B}}{detM_B} & \frac{-\bar{m}_{12,B}}{detM_B} \\ 0 & 0 & 0 & 0 \\ 0 & 0 & \frac{-\bar{m}_{21,B}}{detM_A} & \frac{\bar{m}_{11,B}}{detM_A} \end{pmatrix} \begin{pmatrix} \tau_{1,A} \\ \tau_{2,A} \\ \tau_{1,B} \\ \tau_{2,B} \end{pmatrix} \tag{2.33}$$

The dynamics of the dual-arm robotic manipulator can be also written in the following concise nonlinear affine-in-the-input state-space form:

$$\dot{x} = f(x) + g(x)u \tag{2.34}$$

where $x \in R^{8\times 1}, f(x) \in R^{8\times 1}$, $g(x) \in R^{8\times 4}$, and $u \in R^{4\times 1}$.

2.1.2.2 Differential flatness properties of the dynamics of the dual-arm robot

As already analyzed, the dual-arm robotic system is differentially flat, with flat outputs vector $Y = [x_1, x_3, x_5, x_7]^T$ or $Y = [\theta_{1,A}, \theta_{2,A}, \theta_{1,B}, \theta_{2,B}]^T$. From the odd-numbered rows of the state-space model of Eq. (2.33), one has

$$\begin{aligned} x_2 &= \dot{x}_1 \Rightarrow x_2 = h_2(Y, \dot{Y}) \quad x_4 = \dot{x}_3 \Rightarrow x_4 = h_4(Y, \dot{Y}) \\ x_6 &= \dot{x}_5 \Rightarrow x_6 = h_6(Y, \dot{Y}) \quad x_8 = \dot{x}_7 \Rightarrow x_8 = h_8(Y, \dot{Y}) \end{aligned} \tag{2.35}$$

which signifies that state variables x_2, x_4, x_6 and x_8 are differential functions of the flat outputs vector. Besides, from Eq. (2.30), one has that

$$\begin{pmatrix} \ddot{x}_1 \\ \ddot{x}_2 \\ \ddot{x}_3 \\ \ddot{x}_4 \end{pmatrix} = \begin{pmatrix} \frac{-\bar{m}_{22,A}\bar{F}_{1,A}+\bar{m}_{12,A}\bar{F}_{2,A}}{detM_A} \\ \frac{\bar{m}_{21,A}\bar{F}_{1,A}-\bar{m}_{11,A}\bar{F}_{2,A}}{detM_A} \\ \frac{-\bar{m}_{22,B}\bar{F}_{1,B}+\bar{m}_{12,B}\bar{F}_{2,B}}{detM_B} \\ \frac{\bar{m}_{21,B}\bar{F}_{1,B}-\bar{m}_{11,B}\bar{F}_{2,B}}{detM_B} \end{pmatrix} + \begin{pmatrix} \frac{\bar{m}_{22,A}}{detM_A} & \frac{-\bar{m}_{12,A}}{detM_A} & 0 & 0 \\ \frac{-\bar{m}_{21,A}}{detM_A} & \frac{\bar{m}_{11,A}}{detM_A} & 0 & 0 \\ 0 & 0 & \frac{\bar{m}_{22,B}}{detM_B} & \frac{-\bar{m}_{12,B}}{detM_B} \\ 0 & 0 & \frac{-\bar{m}_{21,B}}{detM_A} & \frac{\bar{m}_{11,B}}{detM_A} \end{pmatrix} \begin{pmatrix} u_1 \\ u_2 \\ u_3 \\ u_4 \end{pmatrix} \tag{2.36}$$

One solves Eq. (2.36) for the control inputs vector $u = [u_1, u_2, u_3, u_4]^T$. All terms which appear in the drift vector of the above-noted equation, as well as all terms appearing in the control inputs gain matrix of this equation are differential functions of the flat outputs vector Y.

$$\begin{pmatrix} u_1 \\ u_2 \\ u_3 \\ u_4 \end{pmatrix} = \begin{pmatrix} \frac{\bar{m}_{22,A}}{detM_A} & \frac{-\bar{m}_{12,A}}{detM_A} & 0 & 0 \\ \frac{-\bar{m}_{21,A}}{detM_A} & \frac{\bar{m}_{11,A}}{detM_A} & 0 & 0 \\ 0 & 0 & \frac{\bar{m}_{22,B}}{detM_B} & \frac{-\bar{m}_{12,B}}{detM_B} \\ 0 & 0 & \frac{-\bar{m}_{21,B}}{detM_A} & \frac{\bar{m}_{11,B}}{detM_A} \end{pmatrix}^{-1} \left[\begin{pmatrix} \ddot{x}_1 \\ \ddot{x}_2 \\ \ddot{x}_3 \\ \ddot{x}_4 \end{pmatrix} - \begin{pmatrix} \frac{-\bar{m}_{22,A}\bar{F}_{1,A}+\bar{m}_{12,A}\bar{F}_{2,A}}{detM_A} \\ \frac{\bar{m}_{21,A}\bar{F}_{1,A}-\bar{m}_{11,A}\bar{F}_{2,A}}{detM_A} \\ \frac{-\bar{m}_{22,B}\bar{F}_{1,B}+\bar{m}_{12,B}\bar{F}_{2,B}}{detM_B} \\ \frac{\bar{m}_{21,B}\bar{F}_{1,B}-\bar{m}_{11,B}\bar{F}_{2,B}}{detM_B} \end{pmatrix} \right] \tag{2.37}$$

or equivalently, one has the following concise relation which shows that control input u is a differential function of the flat outputs vector.

$$u = h_u(Y, \dot{Y}) \tag{2.38}$$

By proving differential flatness properties it is confirmed that the dual-arm robot is input–output linearizable. Differential flatness provides also an implicit of the robot's controllability. It also allows to solve the setpoints definition problem. Setpoints are chosen in an unconstrained manner for the state variables x_1, x_3, x_5, x_7 which coincide with the flat outputs of the robot. Next, one chooses setpoints for the rest of the state variables of the robotic system x_2, x_4, x_6, x_8 through the use of the differential relations which connect them to the flat outputs.

2.1.3 A multi-loop state-space model of the dual-arm robot

The following notation of state variables is used next: $z_1 = \theta_{1,A}$, $z_2 = \theta_{2,A}$, $z_3 = \theta_{1,B}$, $z_4 = \theta_{2,B}$, $z_5 = \dot{\theta}_{1,A}$, $z_5 = \dot{\theta}_{2,A}$, $z_7 = \dot{\theta}_{1,B}$, $z_8 = \dot{\theta}_{2,B}$, which also means that $z_1 = x_1$, $z_2 = x_3$, $z_3 = x_5$, $z_4 = x_7$, $z_5 = x_2$, $z_6 = x_4$, $z_7 = x_6$ and $z_8 = x_8$. The control inputs vector remains as before, that is $u = [u_1, u_2, u_3, u_4]^T$ or $u = [\tau_{1,A}, \tau_{1,B}, \tau_{2,A}, \tau_{2,B}]^T$. Thus, the state-space model of the robot is rewritten as

$$\begin{pmatrix} \dot{z}_1 \\ \dot{z}_2 \\ \dot{z}_3 \\ \dot{z}_4 \\ \dot{z}_5 \\ \dot{z}_6 \\ \dot{z}_7 \\ \dot{z}_8 \end{pmatrix} = \begin{pmatrix} z_5 \\ z_6 \\ z_7 \\ z_8 \\ \frac{-\bar{m}_{22,A}\bar{F}_{1,A}+\bar{m}_{12,A}\bar{F}_{2,A}}{detM_A} \\ \frac{\bar{m}_{21,A}\bar{F}_{1,A}-\bar{m}_{11,A}\bar{F}_{2,A}}{detM_A} \\ \frac{-\bar{m}_{22,B}\bar{F}_{1,B}+\bar{m}_{12,B}\bar{F}_{2,B}}{detM_B} \\ \frac{\bar{m}_{21,B}\bar{F}_{1,B}-\bar{m}_{11,B}\bar{F}_{2,B}}{detM_B} \end{pmatrix} + \begin{pmatrix} 0 & 0 & 0 & 0 \\ 0 & 0 & 0 & 0 \\ 0 & 0 & 0 & 0 \\ 0 & 0 & 0 & 0 \\ \frac{\bar{m}_{22,A}}{detM_A} & \frac{-\bar{m}_{12,A}}{detM_A} & 0 & 0 \\ \frac{-\bar{m}_{21,A}}{detM_A} & \frac{\bar{m}_{11,A}}{detM_A} & 0 & 0 \\ 0 & 0 & \frac{\bar{m}_{22,B}}{detM_B} & \frac{-\bar{m}_{12,B}}{detM_B} \\ 0 & 0 & \frac{-\bar{m}_{21,B}}{detM_A} & \frac{\bar{m}_{11,B}}{detM_A} \end{pmatrix} \begin{pmatrix} u_1 \\ u_2 \\ u_3 \\ u_4 \end{pmatrix} \tag{2.39}$$

Next, the dynamic model of the dual-arm robot is decomposed into two subsystems Σ_1 and Σ_2 which are defined by the following subvectors and submatrices:

About subsystem Σ_1, one has

$$z_{1,4} = \begin{pmatrix} z_1 \\ z_2 \\ z_3 \\ z_4 \end{pmatrix} \quad f_{1,4} \begin{pmatrix} 0 \\ 0 \\ 0 \\ 0 \end{pmatrix} \quad g_{1,4} = \begin{pmatrix} 1 & 0 & 0 & 0 \\ 0 & 1 & 0 & 0 \\ 0 & 0 & 1 & 0 \\ 0 & 0 & 0 & 1 \end{pmatrix} \quad v_1 = \begin{pmatrix} z_5 \\ z_6 \\ z_7 \\ z_8 \end{pmatrix} \tag{2.40}$$

where $z_{1,4}$ is the state vector, $f_{1,4}$ is the drift vector, $g_{1,4}$ is the control inputs gain matrix and v_1 is the virtual control inputs vector.

About subsystem Σ_2, one has

$$z_{5,8} = \begin{pmatrix} z_5 \\ z_6 \\ z_7 \\ z_8 \end{pmatrix} \quad f_{5,8} \begin{pmatrix} \frac{-\bar{m}_{22,A}\bar{F}_{1,A}+\bar{m}_{12,A}\bar{F}_{2,A}}{detM_A} \\ \frac{\bar{m}_{21,A}\bar{F}_{1,A}-\bar{m}_{11,A}\bar{F}_{2,A}}{detM_A} \\ \frac{-\bar{m}_{22,B}\bar{F}_{1,B}+\bar{m}_{12,B}\bar{F}_{2,B}}{detM_B} \\ \frac{\bar{m}_{21,B}\bar{F}_{1,B}-\bar{m}_{11,B}\bar{F}_{2,B}}{detM_B} \end{pmatrix} \quad g_{5,8} = \begin{pmatrix} \frac{\bar{m}_{22,A}}{detM_A} & \frac{-\bar{m}_{12,A}}{detM_A} & 0 & 0 \\ \frac{-\bar{m}_{21,A}}{detM_A} & \frac{\bar{m}_{11,A}}{detM_A} & 0 & 0 \\ 0 & 0 & \frac{\bar{m}_{22,B}}{detM_B} & \frac{-\bar{m}_{12,B}}{detM_B} \\ 0 & 0 & \frac{-\bar{m}_{21,B}}{detM_A} & \frac{\bar{m}_{11,B}}{detM_A} \end{pmatrix} \quad u = \begin{pmatrix} u_1 \\ u_2 \\ u_3 \\ u_4 \end{pmatrix} \tag{2.41}$$

where $z_{5,8}$ is the state vector, $f_{5,8}$ is the drift vector, $g_{5,8}$ is the control inputs gain matrix and u is the real control inputs vector.

Consequently, the dynamics of the dual-arm robot can be written in the following chained-subsystems form:

$$\dot{z}_{1,4} = f_{1,4}(z_{1,4}) + g_{1,4}(z_{1,4})v_1 \tag{2.42}$$

$$\dot{z}_{5,8} = f_{5,8}(z_{1,4}, z_{5,8}) + g_{5,8}(z_{1,4}z_{5,8})u \tag{2.43}$$

2.1.4 Design of a multi-loop flatness-based controller

It can be proven that each one of subsystems Σ_1 and Σ_2 is differentially flat. About subsystem Σ_1 given in Eq. (2.42) it is considered that $y_1 = z_{1,4}$ is the flat output. It also holds that $f_{1,4}(z_{1,4})$ and $g_{1,4}(z_{1,4})$ are functions of the flat outputs vector y_1. Thus, by solving Eq. (2.42) for the virtual control input v_1, one obtains

$$v_1 = g_{1,4}(z_{1,4})^{-1}[\dot{z}_{1,4} - f_{1,4}(z_{1,4})] \Rightarrow v_1 = h_{v_1}(y_1, \dot{y}_1) \tag{2.44}$$

which signifies that in subsystem Σ_1 the virtual control input v_1 is a differential function of the flat outputs vector y_1. Thus, subsystem Σ_1 is differentially flat.

About subsystem Σ_2 given in Eq. (2.43) it is considered that $y_2 = z_{5,8}$ is the flat output, while subvector $z_{1,4}$ is viewed as a coefficients vector. It is also holds that $f_{5,8}(z_{1,4}, z_{5,8})$ and $g_{5,8}(z1, 4, z_{5,8})$ are functions of the flat outputs vector y_2. Thus, by solving Eq. (2.43) for the real control input u, one obtains

$$u = g_{5,8}(z_{1,4}, z_{5,8})^{-1}[\dot{z}_{5,8} - f_{5,8}(z_{1,4}, z_{5,9})] \Rightarrow u = h_u(y_2, \dot{y}_3) \tag{2.45}$$

which signifies that in subsystem Σ_2 the real control input u is a differential function of the flat outputs vector y_2. Thus, subsystem Σ_2 is differentially flat.

The proof of differential flatness properties signifies that subsystems Σ_1 and Σ_2 are input–output linearizable and can be controlled by inversion of their dynamics, as it is commonly done for input–output linearized differentially flat systems.

For subsystem Σ_1 the stabilizing feedback control is taken to be

$$v_1 = g_{1,4}(z_{1,4})^{-1}[\dot{z}^d_{1,4} - f_{1,4}(z_{1,4}) - K_1(z_{1,4} - z^d_{1,4})] \tag{2.46}$$

where $z^d_{1,4}$ is the setpoints vector of subsystem Σ_1, while matrix $K_1 \in R^{4\times4}$ is a diagonal matrix with positive elements in its diagonal, that is $k_{1,ii} > 0$ for $i = 1, \cdots, 4$.

For subsystem Σ_2 the stabilizing feedback control is taken to be

$$u = g_{5,8}(z_{1,4}, z_{5,8})^{-1}[\dot{z}^d_{5,8} - f_{5,8}(z_{1,4}, z_{5,8}) - K_2(z_{5,8} - z^d_{5,8})] \tag{2.47}$$

where $z^d_{5,8}$ is the setpoints vector of subsystem Σ_2, while matrix $K_2 \in R^{4\times4}$ is a diagonal matrix with positive elements in its diagonal, that is $k_{2,ii} > 0$ for $i = 1, \cdots, 4$.

By defining the tracking error variable $e_{1,4} = z_{1,4} - z^d_{1,4}$ and by substituting the control input of Eq. (2.46) into subsystem Σ_1 of Eq. (2.40), one has

$$\begin{aligned} \dot{z}_{1,4} &= f_{1,4}(z_{1,4}) + g_{1,4}(z_{1,4}) \cdot g_{1,4}(z_{1,4})^{-1}[\dot{z}^d_{1,4} - f_{1,4}(z_{1,4}) - K_1(z_{1,4} - z^d_{1,4})] \\ &\Rightarrow (\dot{z}_{1,4} - \dot{z}^d_{1,4}) + K_1(z_{1,4} - z^d_{1,4}) \Rightarrow \dot{e}_{1,4} + K_1 e_{1,4} = 0 \\ &\Rightarrow lim_{t\to\infty} e_{1,4}(t) = 0 \Rightarrow lim_{t\to\infty} z_{1,4}(t) = z^d_{1,4}(t) \end{aligned} \tag{2.48}$$

Equivalently, by defining the tracking error variable $e_{5,8} = z_{5,8} - z^d_{5,8}$ and by substituting the control input of Eq. (2.47) into subsystem Σ_2 of Eq. (2.41), one has

$$\begin{aligned} \dot{z}_{5,8} &= f_{5,8}(z_{1,4}, z_{5,8}) + g_{5,8}(z_{1,4}, z_{5,8}) \cdot g_{5,8}(z_{1,4}, z_{5,8})^{-1}[\dot{z}^d_{5,8} - f_{5,8}(z_{1,4}, z_{5.8}) - K_2(z_{5,8} - z^d_{5,8})] \\ &\Rightarrow (\dot{z}_{5,8} - \dot{z}^d_{5,8}) + K_2(z_{5,8} - z^d_{5,8}) \Rightarrow \dot{e}_{5,8} + K_2 e_{5,8} = 0 \\ &\Rightarrow lim_{t\to\infty} e_{5,8}(t) = 0 \Rightarrow lim_{t\to\infty} z_{5,8}(t) = z^d_{5,8}(t) \end{aligned} \tag{2.49}$$

Consequently, all state variables of the dual-arm robot converge to the associated setpoints and this system is globally stable.

Regarding the stabilization of the dynamics of the dual-arm robotic manipulator this is ensured as follows: In Eq. (2.47), there is a real control input variable u which is obtained after inverting the dynamics of the subsystem which is described in Eq. (2.43). Since u is a real control input Eq. (2.47) is exact and the equation between its left and right part definitely holds. In Eq. (2.46), there is a virtual control input variable v_1 which is obtained after inverting the dynamics of the subsystem which is described in Eq. (2.42). Since v_1 is a virtual control input, Eq. (2.46) is true provided that Eq. (2.47) holds. In such a case the equation between the left and right part of Eq. (2.46) holds asymptotically. In terms of rates of convergence the control loop of Eq. (2.47) has to be faster than the control loop of Eq. (2.46). This is achieved by assigning higher values to the positive feedback gains of Eq. (2.47) compared to the feedback gains of Eq. (2.46).

Global asymptotic stability can be also proven through Lyapunov analysis. The following Lyapunov function is defined:

$$V = \frac{1}{2}[e_{1,4}^T e_{1,4} + e_{5,8}^T e_{5,8}] \tag{2.50}$$

By differentiating in time, one gets

$$\begin{aligned} \dot{V} &= \frac{1}{2}2[e_{1,4}^T \dot{e}_{1,4} + e_{5,8}^T \dot{e}_{5,8}] \Rightarrow \\ \dot{V} &= e_{1,4}^T(-K_1 e_{1,4}) + e_{5,8}^T(-K_2 e_{5,8}) \Rightarrow \\ \dot{V} &= -[e_{1,4}^T K_1 e_{1,4} + e_{5,8}^T K_2 e_{5,8}] \Rightarrow \dot{V} < 0 \; \forall \; e_{1,4} \neq 0 \; e_{5,8} \neq 0 \end{aligned} \tag{2.51}$$

$\dot{V} = 0$ if and only if $e_{1,4} = 0$ and $e_{5,8} = 0$. Thus, $\dot{V}$ is always negative and V is a decreasing function which converges asymptotically to the zero equilibrium, that is $lim_{t \Rightarrow \infty} V = 0$.

An explicit demonstration of the exponential stabilization that is achieved by flatness-based control in successive loops is given next. The Lyapunov function of the control loop is written as:

$$V = \frac{1}{2}[\sum_{i=1}^{4} e_i^2 + \sum_{j=5}^{8} e_j^2] \tag{2.52}$$

where e_i $i = 1, \cdots, 4$ are the tracking errors for the state variables of the first subsystem Σ_1 and e_j $j = 7, \cdots, 12$ are the tracking errors for the state variables of the second subsystem Σ_2. Equivalently, the first-order time-derivative of the Lyapunov function is written as

$$\dot{V} = -[\sum_{i=1}^{4} k_{1,i} e_i^2 + \sum_{j=5}^{8} k_{2,j-4} e_j^2] \tag{2.53}$$

where $k_{1,i} > 0$ $i = 1, \cdots, 4$ are the diagonal elements of gain matrix K_1 and $k_{2,j-4} > 0$ $j = 5, \cdots, 8$ are the diagonal elements of gain matrix K_2. By denoting the minimum of the above-noted elements of the feedback gain matrices as k_{min}, that is

$$k_{min} = min\{k_{1,i} : \; i = 1, \cdots, 4 \text{ and } k_{2,j-4} : \; j = 5, \cdots, 8\} \tag{2.54}$$

and using Eq. (2.53), one obtains that

$$\begin{aligned} &\dot{V} \leq -k_{min}[\sum_{i=1}^{4} e_i^2 + \sum_{j=5}^{8} e_j^2] \\ &\Rightarrow \dot{V} \leq -2k_{min} V \Rightarrow \dot{V} + 2k_{min} V \leq 0 \end{aligned} \tag{2.55}$$

From Eq. (2.55), one can demonstrate the exponential convergence of the Lyapunov function V to 0.

Finally, a diagram with the implementation stages of multi-loop flatness-based control method is given in Figure 2.2. Once the state-space model of the robotic system has been separated into chained subsystems which individually satisfy the differential flatness property, flatness-based control is implemented at each time-step through the following steps: (i) the state vector of the subsequent $(i+1)$-th subsystem becomes virtual control input to the preceding i-th subsystem, (ii) Equivalently, the virtual control input of the preceding i-th subsystem becomes setpoint for the subsequent $(i+1)$-th subsystem, (iii) the value of the virtual control input for each subsystem is computed by inverting the dynamics of this subsystem and by selecting feedback gains which allow for eliminating the associated local tracking error, (iv) the real control input is computed from the last subsystem by inverting again the dynamics of this subsystem (v) the real control input makes implicitly use of the virtual control inputs of all preceding subsystems. Its stabilizing effects appear by tracing the subsystems of the state-space model backward from the last to the first one.

2.1.5 Simulation tests

To evaluate the performance of the multi-loop flatness-based control scheme simulation tests have been performed. Indicative values for the parameters of the two robotic arms were $m_{1,A} = 2.0kg$, $l_{1,A} = 1.0m$, $m_{2,A} = 1.0kg$, $l_{2,A} = 1.0m$ and also $m_{1,B} = 2.0kg$, $l_{1,B} = 1.0m$, $m_{2,B} = 1.0kg$, $l_{2,b} = 1.0m$. The mass of the lifted load was $M = 1.0kg$ and

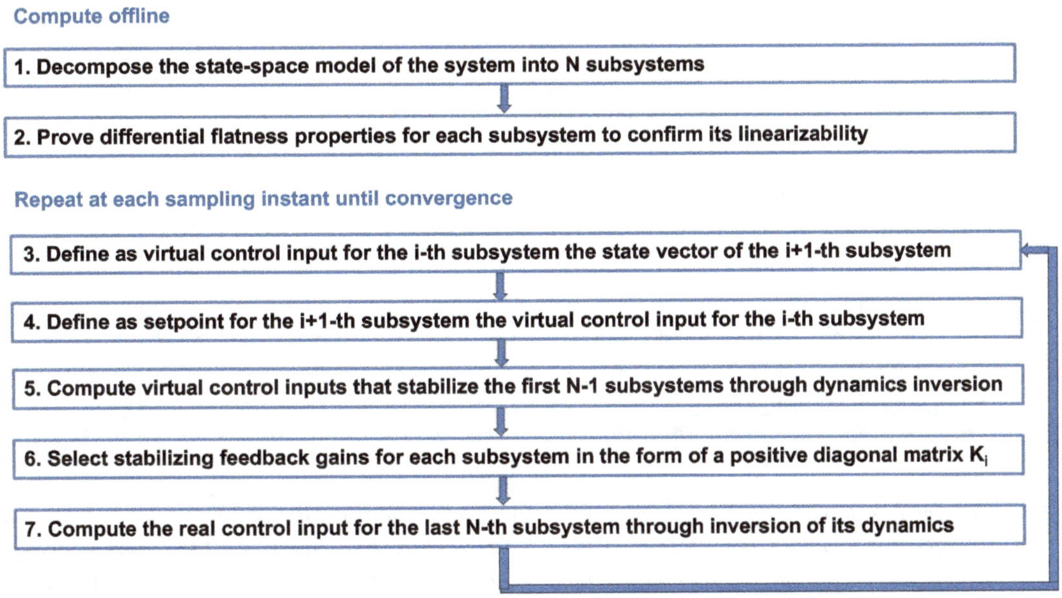

Figure 2.2 Diagram with the implementation stages of multi-loop flatness-based control for the dual-arm robotic manipulator

the elasticity coefficient describing the deformation of the lifted object was $K_L = 10Nt/m$. The sampling period was $T_s = 0.01sec$. Simulation experiments were concerned with eight different functioning cases of the dual-arm robot resulting in different positions of its end-effector in the robot's workspace. The turn angles of the joints of this robot were measured in rads while the associated angular velocities were measured in rad/sec. The obtained results are depicted in Figures 2.3–2.18. In these diagrams the state variables of the dual-arm robotic system are presented in blue color, their estimated values which are provided by the H-infinity Kalman filter are given in green color, while the associated reference setpoints are plotted in red color.

It can be noticed that in all test cases the nonlinear optimal control method achieves fast and accurate tracking of reference setpoints by all state variables of the dual-arm robot under moderate variations of the control inputs. It is noted that the multi-loop flatness-based control method allows also for treating the setpoints definition problem. In the case of the dual-arm robotic manipulator, one defines without constraints setpoints for the first loop which is concerned with the control of the position (turn angles) x_1, x_3, x_5, x_7 of the joints of the two arms. Next, for the second loop which controls the turn speed (angular velocities) x_2, x_4, x_6, x_8 of the joints of the dual-arm robot, setpoints are chosen under the constraint that the vector of joints' angular velocities is also the virtual control inputs vector which stabilizes the first loop of the joints' turn angles. The transient performance of the multi-loop flatness-based control scheme and the speed of convergence of the state variables of the dual-arm robot to their setpoints is determined by the values of the diagonal elements of the gain matrices $K_1 \in R^{4\times 4}$ and $K_2 \in R^{4\times 4}$ which are given in Eq. (2.46) and Eq. (2.47).

2.2 Flatness-based control in successive loops for electrohydraulic robotic manipulators

2.2.1 Outline

Control of electrohydraulic actuators, as well as of electrohydraulically actuated robots is a topic with high importance for industry that has attracted much research interest during the last years [711,721,732]. Electro-hydraulic actuation systems have been widely used in robotic manipulators, in construction machinery, and in aircrafts [436,637]. Nowadays, manipulators and vehicle receiving hydraulic actuation constitute a huge global industrial sector [614,857,216].

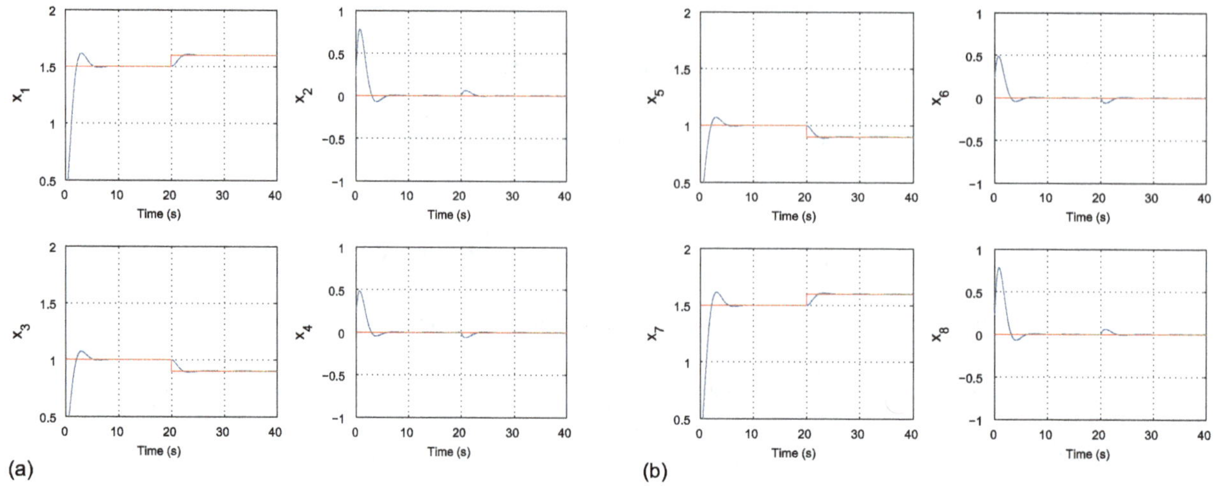

Figure 2.3 *Tracking of setpoint 1 by the dual-arm robot: (a) convergence of state variables x_1 to x_4 (blue lines) to the reference setpoints (red lines) and their KF-based estimation (green line); (b) convergence of state variables x_5 to x_8 (blue lines) to the reference setpoints (red lines) and their KF-based estimation (green line)*

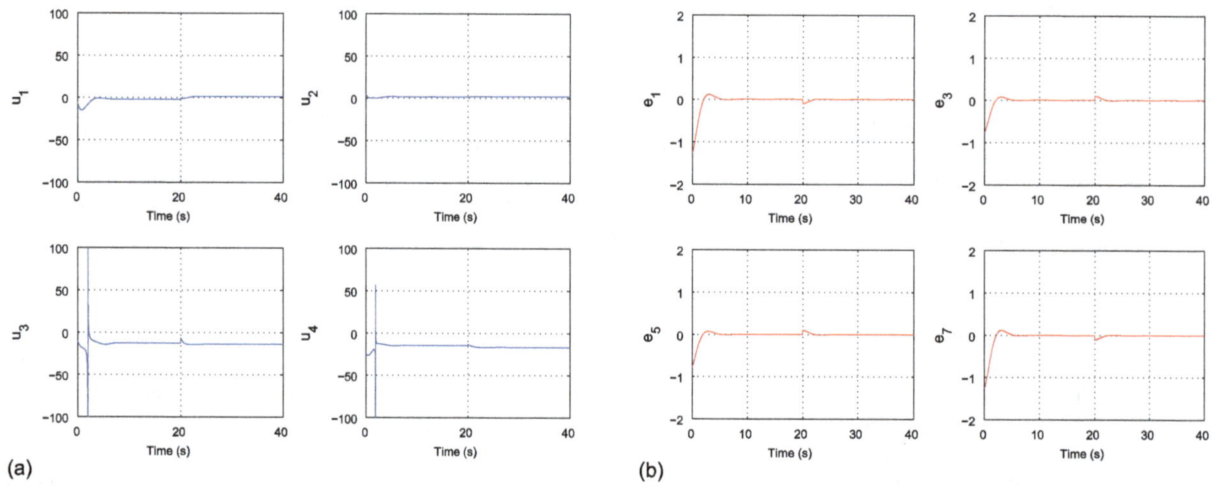

Figure 2.4 *Tracking of setpoint 1 by the dual-arm robot: (a) variations of the control inputs u_1 to u_4; (b) tracking error of state variables $x_1 = \theta_{1,A}$, $x_3 = \theta_{2,A}$, $x_5 = \theta_{1,B}$, and $x_7 = \theta_{2,B}$*

Hydraulic manipulators are favored in industry, mining, agriculture and construction works where large actuation forces are required [514,311,552]. For instance, in 2016, the construction business alone sold 0.7 million units of construction machines, while in 2019 the sales of robotic manipulators, including also hydraulic manipulators, reached the level of 0.4 million units. Hydraulic actuation in robotic manipulators exhibits specific technical advantages, such as large force and torque, fast response, small size-to-power ratio, reliability, and low manufacturing cost [807,615,844]. Hydraulic heavy-duty manipulators are irreplaceable for heavy workpiece handling. They improve the efficiency of fine manipulation of heavy loads, thus ensuring production, safety, and reducing labor costs [447,498,909]. Comparing against electric actuators, hydraulic actuators have the advantage of producing high power while having a smaller size

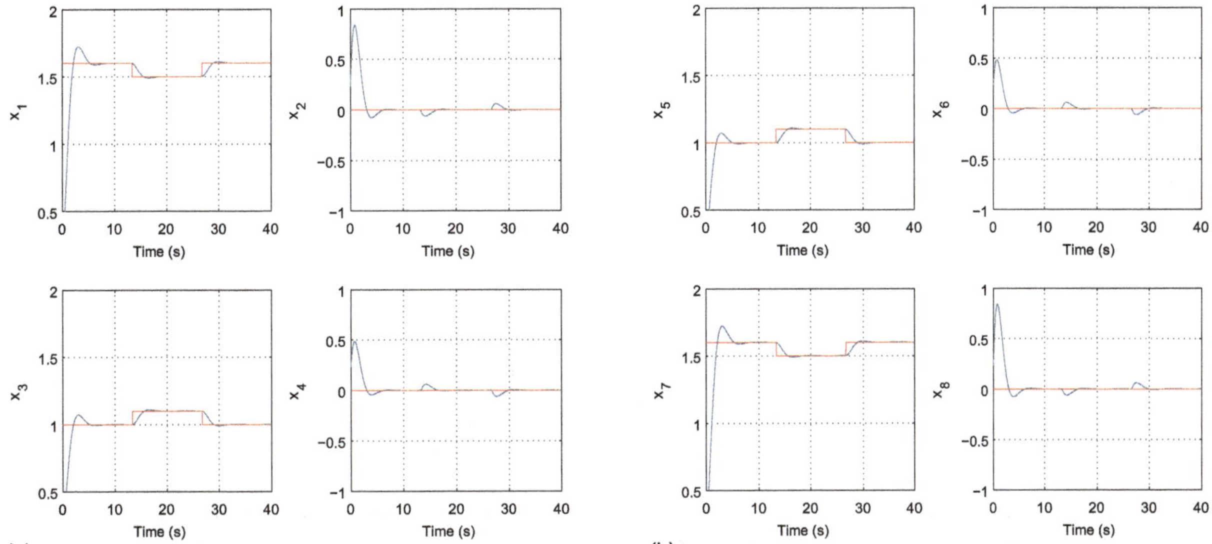

(a) (b)

Figure 2.5 *Tracking of setpoint 2 by the dual-arm robot: (a) convergence of state variables x_1 to x_4 (blue lines) to the reference setpoints (red lines) and their KF-based estimation (green line); (b) convergence of state variables x_5 to x_8 (blue lines) to the reference setpoints (red lines) and their KF-based estimation (green line)*

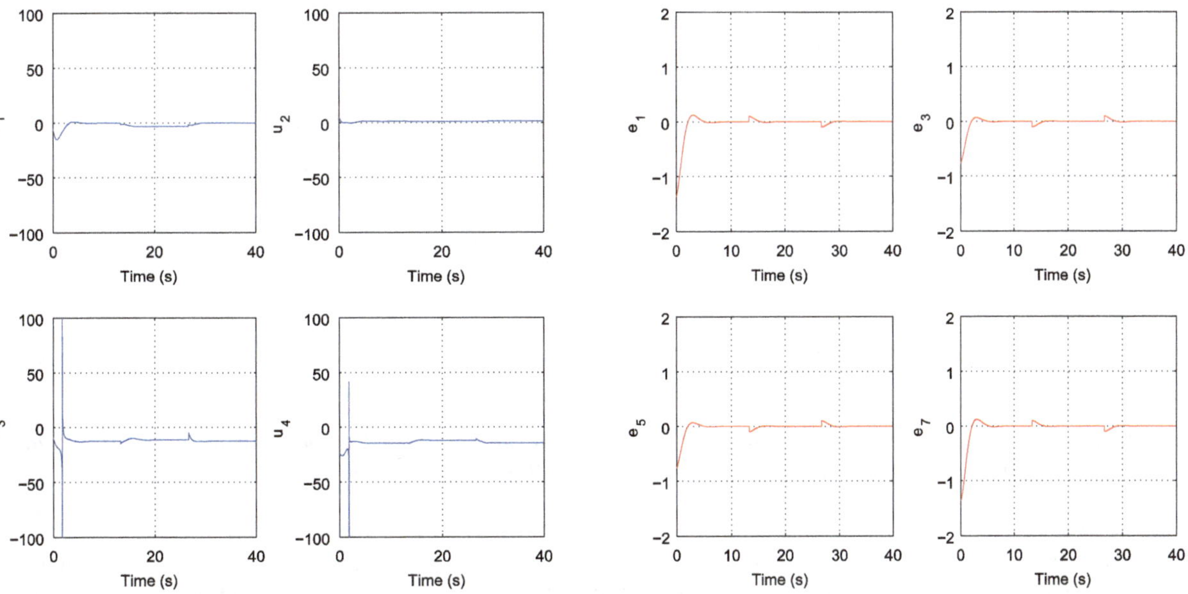

(a) (b)

Figure 2.6 *Tracking of setpoint 2 by the dual-arm robot: (a) variations of the control inputs u_1 to u_4; (b) tracking error of state variables $x_1 = \theta_{1,A}$, $x_3 = \theta_{2,A}$, $x_5 = \theta_{1,B}$, and $x_7 = \theta_{2,B}$*

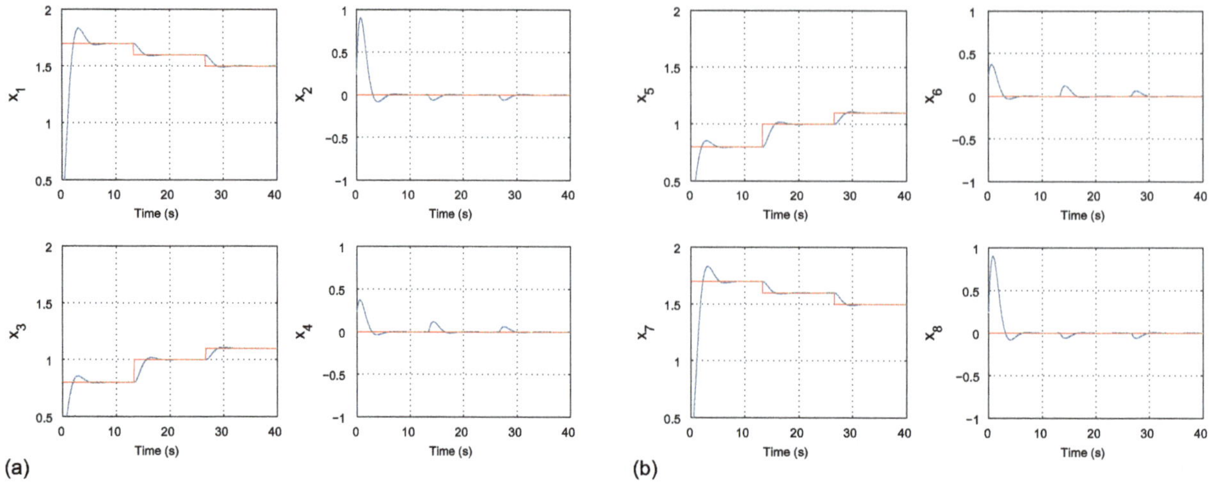

Figure 2.7 Tracking of setpoint 2 by the dual-arm robot: (a) convergence of state variables x_1 to x_4 (blue lines) to the reference setpoints (red lines) and their KF-based estimation (green line); (b) convergence of state variables x_5 to x_8 (blue lines) to the reference setpoints (red lines) and their KF-based estimation (green line)

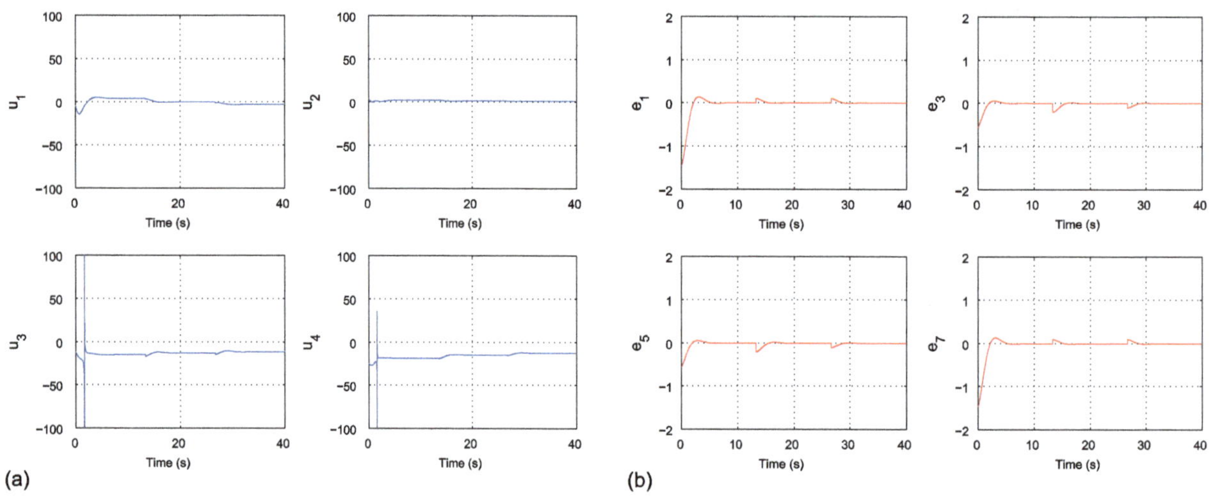

Figure 2.8 Tracking of setpoint 2 by the dual-arm robot: (a) variations of the control inputs u_1 to u_4; (b) tracking error of state variables $x_1 = \theta_{1,A}$, $x_3 = \theta_{2,A}$, $x_5 = \theta_{1,B}$, and $x_7 = \theta_{2,B}$

[343,941]. Minimization of energy consumption is an issue that has to be taken into account in the development of such robotic systems [448,564]. Furthermore, high-precision motion is a vital functionality for electrohydraulic robotic systems [695,446]. Research on nonlinear control for electrohydraulic actuators is growing and several noteworthy results have been obtained during the last years. Control for electrohydraulic actuation in robotic systems can be found in [1087,438,189]. Control for electrohydraulic actuation in mechatronic systems can be found in [677,678,785,312,845]. Moreover, control for electrohydraulic actuation in vehicles can be found in [153,229].

The present section introduces a solution to the problem of nonlinear control of the aforementioned robotic manipulators with electrohydraulic actuation without the need to apply changes of state variables (diffeomorphisms) and

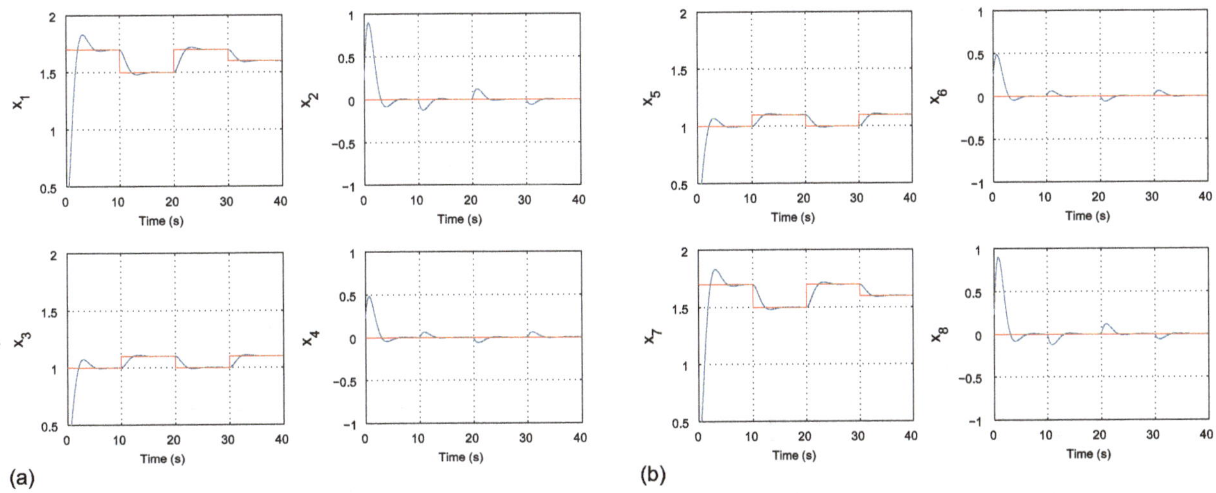

Figure 2.9 *Tracking of setpoint 4 by the dual-arm robot: (a) convergence of state variables x_1 to x_4 (blue lines) to the reference setpoints (red lines) and their KF-based estimation (green line); (b) convergence of state variables x_5 to x_8 (blue lines) to the reference setpoints (red lines) and their KF-based estimation (green line)*

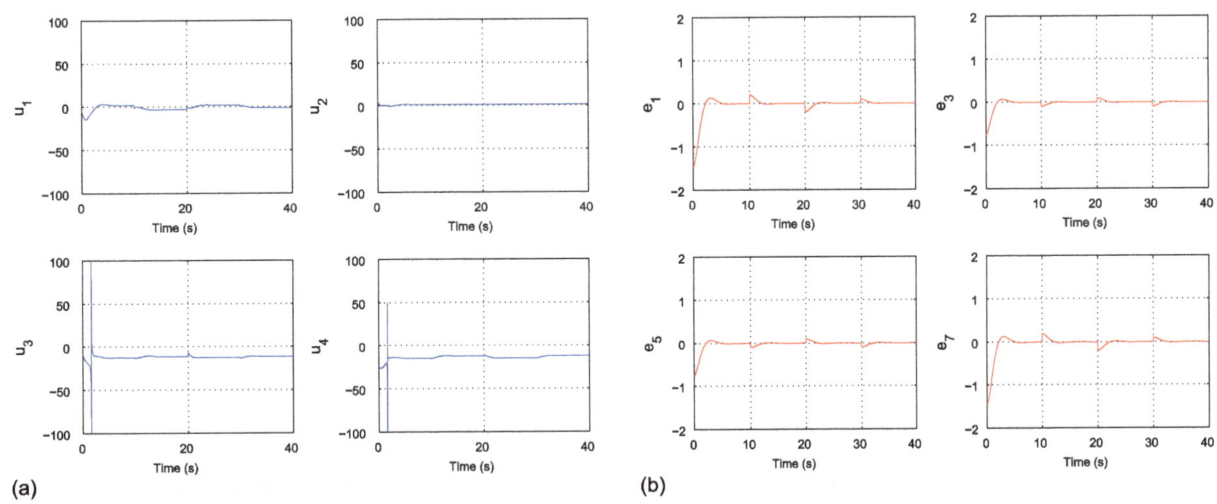

Figure 2.10 *Tracking of setpoint 4 by the dual-arm robot: (a) variations of the control inputs u_1 to u_4; (b) tracking error of state variables $x_1 = \theta_{1,A}$, $x_3 = \theta_{2,A}$, $x_5 = \theta_{1,B}$, and $x_7 = \theta_{2,B}$*

complicated state-space transformations. A system is considered to be differentially flat if all its state variables and its control inputs can be expressed as functions of one single algebraic variable which is the so-called flat output, and also as functions of the flat-output's derivatives [491,252,898,489]. The differential flatness property enables the transformation of the nonlinear system's dynamics in the linear canonical form [91,594,639,489]. The latter description is controllable and observable thus allowing to treat effectively control and estimation problems [806,492,640,524,54]. The demonstration of differential flatness properties for the dynamic model of an electrohydraulically actuated system is an important finding for two reasons (i) It is an implicit proof of the controllability of this nonlinear dynamical system

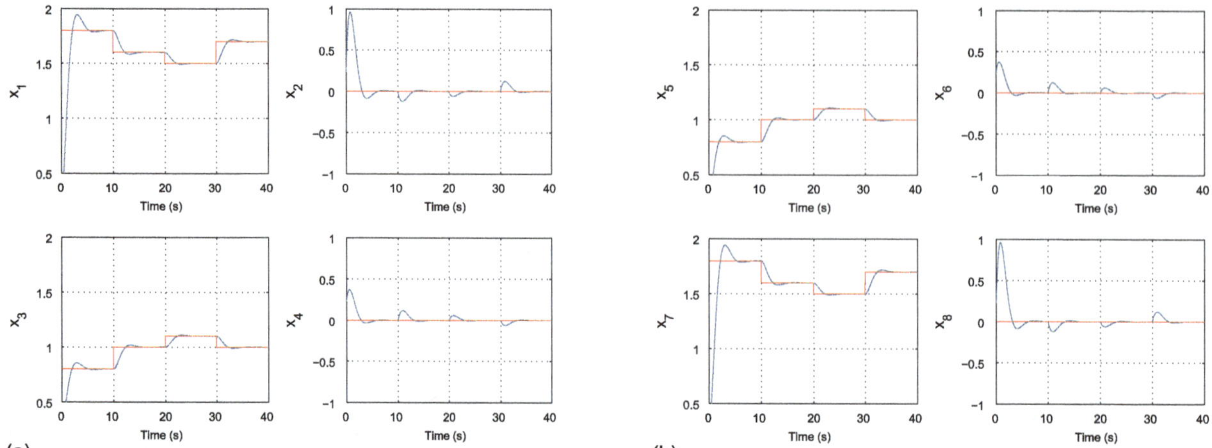

Figure 2.11 *Tracking of setpoint 3 by the dual-arm robot: (a) convergence of state variables x_1 to x_4 (blue lines) to the reference setpoints (red lines) and their KF-based estimation (green line); (b) convergence of state variables x_5 to x_8 (blue lines) to the reference setpoints (red lines) and their KF-based estimation (green line)*

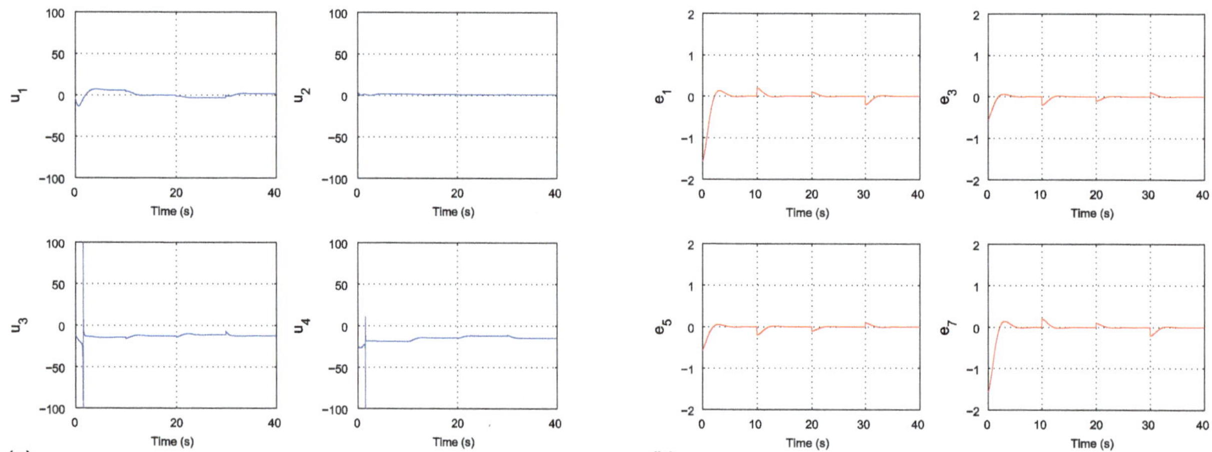

Figure 2.12 *Tracking of setpoint 3 by the dual-arm robot; (a) variations of the control inputs u_1 to u_4; (b) tracking error of state variables $x_1 = \theta_{1,A}$, $x_3 = \theta_{2,A}$, $x_5 = \theta_{1,B}$, and $x_7 = \theta_{2,B}$*

thus signifying that for any targeted setpoints one can compute control inputs which enable the state variables' convergence to these setpoints. Thus, the differential flatness properties confirm that the control problem for this nonlinear dynamical system has a solution (ii) It allows to compute feasible setpoints for all state variables of the system. First, one selects in an unconstraint manner setpoints for the flat outputs of the system [439,705,60,706]. Next, setpoints for the rest of the state variables are chosen subject to the constraint that these state vector elements are differential functions of the flat outputs. By assuring convergence of the flat outputs to their targeted values, one can also ascertain that the rest of the state vector elements of the system will also reach their desirable values.

The present section presents one more example of flatness-based control approach implemented in successive loops. In this method, the dynamic model of the nonlinear system is separated into subsystems which are connected in

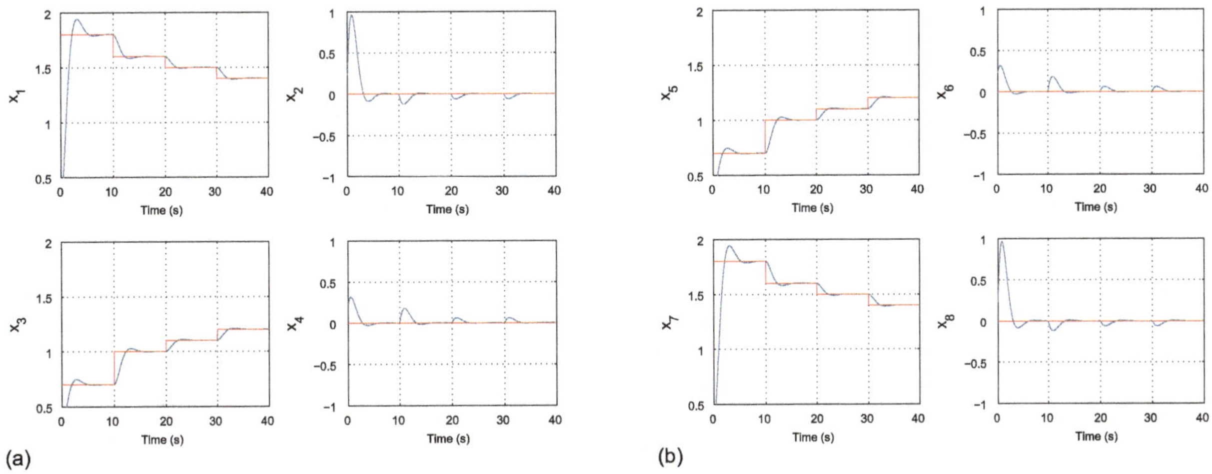

Figure 2.13 Tracking of setpoint 6 by the dual-arm robot: (a) convergence of state variables x_1 to x_4 (blue lines) to the reference setpoints (red lines) and their KF-based estimation (green line); (b) convergence of state variables x_5 to x_8 (blue lines) to the reference setpoints (red lines) and their KF-based estimation (green line)

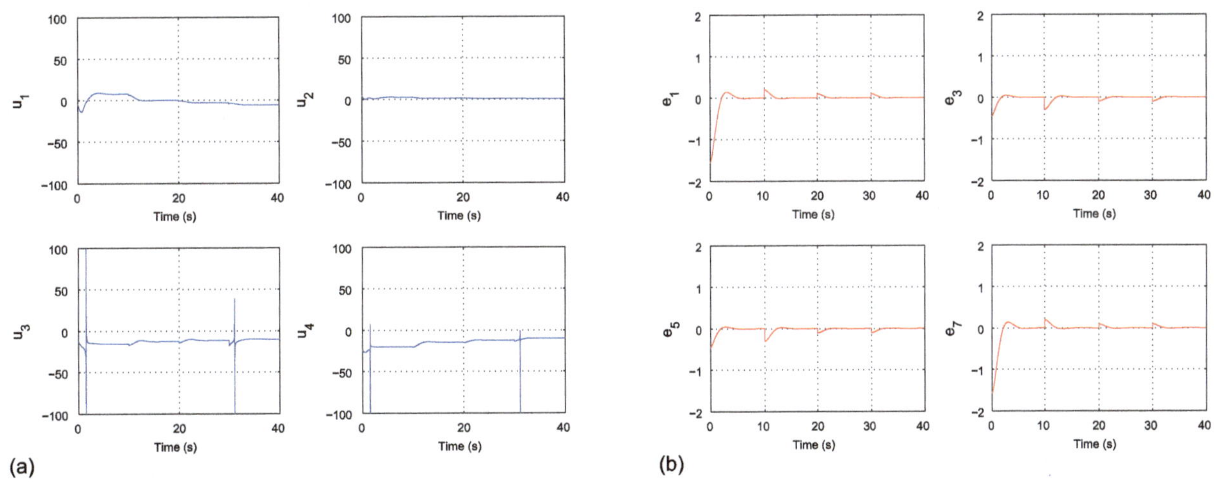

Figure 2.14 Tracking of setpoint 6 by the dual-arm robot: (a) variations of the control inputs u_1 to u_4; (b) tracking error of state variables $x_1 = \theta_{1,A}$, $x_3 = \theta_{2,A}$, $x_5 = \theta_{1,B}$, and $x_7 = \theta_{2,B}$

a cascading manner. This control method is directly applicable (but not limited) to dynamical systems of the triangular form and to nonlinear systems which can be transformed into such a form. The state-space model of the initial nonlinear system is decomposed into cascading subsystems which satisfy differential flatness properties. For each subsystem of the state-space model a virtual control input is computed, capable of inverting the subsystem's dynamics and of eliminating the subsystem's tracking error. The control input which is actually applied to the initial nonlinear system is computed from the last row of the state-space description. This control input incorporates in a recursive manner all virtual control inputs which were computed from the individual subsystems included in the initial state-space equation. The control input that should be applied to the nonlinear system so as to assure that all state vector elements will

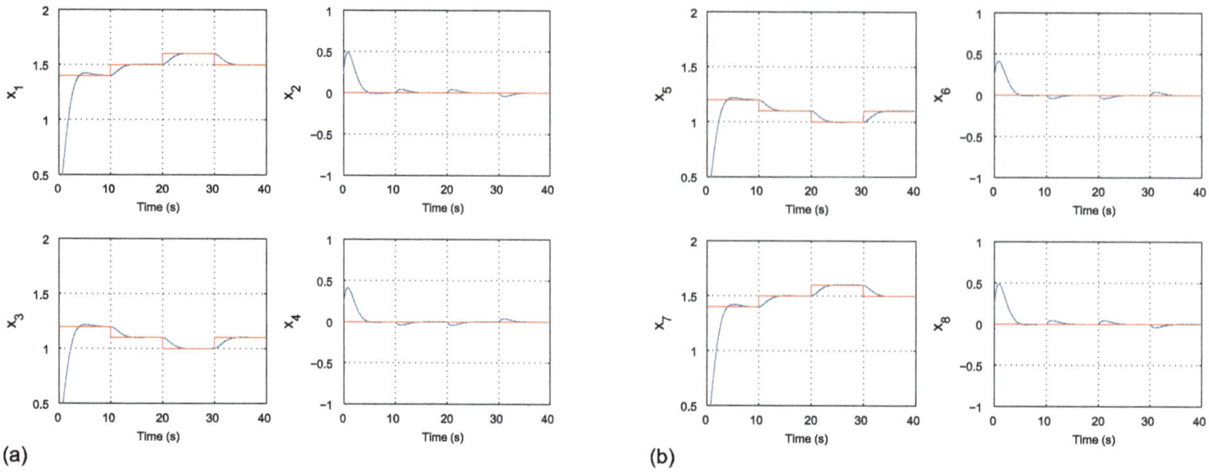

Figure 2.15 Tracking of setpoint 4 by the dual-arm robot: (a) convergence of state variables x_1 to x_4 (blue lines) to the reference setpoints (red lines) and their KF-based estimation (green line); (b) convergence of state variables x_5 to x_8 (blue lines) to the reference setpoints (red lines) and their KF-based estimation (green line)

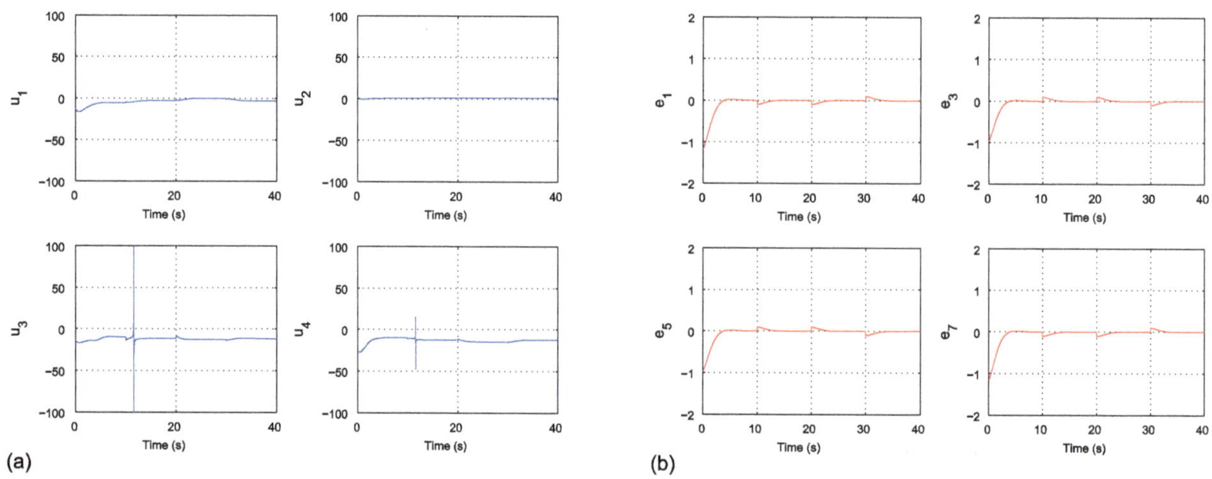

Figure 2.16 Tracking of setpoint 4 by the dual-arm robot: (a) variations of the control inputs u_1 to u_4; (b) tracking error of state variables $x_1 = \theta_{1,A}$, $x_3 = \theta_{2,A}$, $x_5 = \theta_{1,B}$, and $x_7 = \theta_{2,B}$

converge to the desirable setpoints is obtained at each iteration of the control algorithm by tracing backward the subsystems of the state-space model. The flatness-based control method which is implemented in successive loops is being developed during the last years and has been successfully applied and tested in several nonlinear dynamical systems [712,720,714,716,729,731,718].

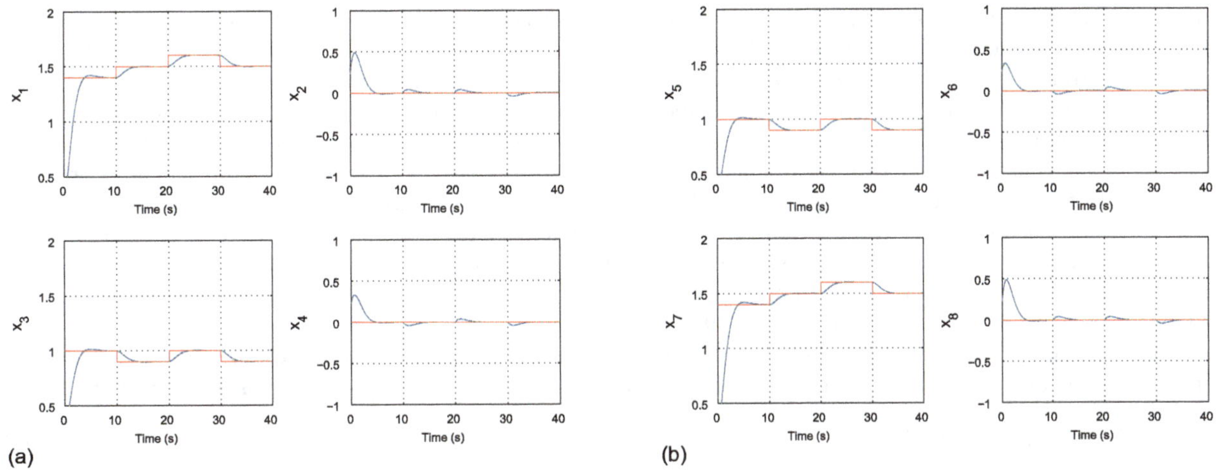

Figure 2.17 Tracking of setpoint 8 by the dual-arm robot: (a) convergence of state variables x_1 to x_4 (blue lines) to the reference setpoints (red lines) and their KF-based estimation (green line); (b) convergence of state variables x_5 to x_8 (blue lines) to the reference setpoints (red lines) and their KF-based estimation (green line)

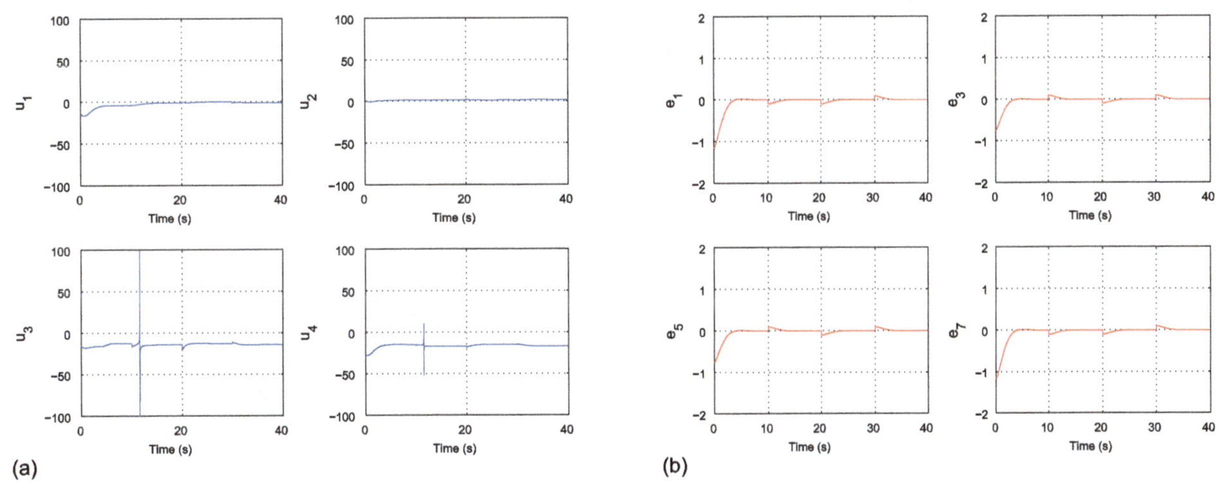

Figure 2.18 Tracking of setpoint 8 by the dual-arm robot: (a) variations of the control inputs u_1 to u_4; (b) tracking error of state variables $x_1 = \theta_{1,A}$, $x_3 = \theta_{2,A}$, $x_5 = \theta_{1,B}$, and $x_7 = \theta_{2,B}$

2.2.2 Dynamic analysis of the electrohydraulic robotic manipulator

2.2.2.1 Dynamic model of the electrohydraulic robot

The test-case in this section is concerned with electrohydraulically actuated robotic manipulators under flatness-based control in successive loops. The control of electrohydraulic robotic manipulators remains a challenging problem that exhibits a higher degree of complexity than the control of electrically actuated manipulators. This is because electrohydraulic manipulators have complicated nonlinear dynamics, and are subject to modeling uncertainties (such as friction, internal leakage, deadbands in valves operation, and control saturation). Besides, the functioning of electrohydraulic manipulators is subject to several parametric variations, such as uncertain or varying loads. Heavy loads lead to significant rigid-flexible coupling characteristics and mechanical deformation in multi-DOF hydraulic manipulators. For this reason, the design of reliable controllers for hydraulic manipulators is acknowledged to be a demanding task. Various control methods have been developed for electrohydraulic robots. Most of the results on the control of these manipulators exclude actuator dynamics from the robots' dynamic behavior. However, actuator dynamics affects also the dynamics of the entire robotic system and the stability properties of the robots' control loop [470,481]. The design of control and actuation schemes for hydraulic manipulators can be improved if rotary hydraulic actuators are used in place of linear hydraulic actuators [342,104,515]. Rotary vane actuators, functioning as rotational drives, provide rotational movements directly because they are constructed as joints and actuators in one. Nevertheless, these actuators may exhibit internal leakage and high friction. Therefore, the controller design for hydraulic manipulators should compensate for such perturbations too. Indicative results about the control of electrohydraulically actuated multi-DOF robotic manipulators can be found in [356,272,498,514,419].

The diagram of the multi-DOF robotic manipulator with electrohydraulic actuators is given in Figure 2.19. Under the assumption that masses are concentrated, the dynamic model of the 2-DOF robotic manipulator is given by

$$M(\theta)\ddot{\theta} + C(\theta,\dot{\theta}) + G(\theta) = \tau \tag{2.56}$$

where $M(\theta)$ is the symmetric inertia matrix of the robot

$$M(\theta) = \begin{pmatrix} M_{11} & M_{12} \\ M_{21} & M_{22} \end{pmatrix} = \begin{pmatrix} (m_1+m_2)l_1^2 + m_1 l_2^2 + 2m_2 l_1 l_2 cos(\theta_2) & m_2 l_2^2 + m_2 l_1 l_2 cos(\theta) \\ m_2 l_2^2 + m_2 l_1 l_2 cos(\theta) & m_2 l_2 (l_1 + l_2) cos(\theta_2) \end{pmatrix} \tag{2.57}$$

The Coriolis forces vector is given by

$$C(\theta,\dot{\theta}) = \begin{pmatrix} -m_2 l_1^2 sin(\theta_2)\dot{\theta}_2^2 - 2m_2 l_1^2 sin(\theta_2)\dot{\theta}_1\dot{\theta}_2 \\ m_2 l_1^2 sin(\theta_2)\dot{\theta}_1^2 \end{pmatrix} \tag{2.58}$$

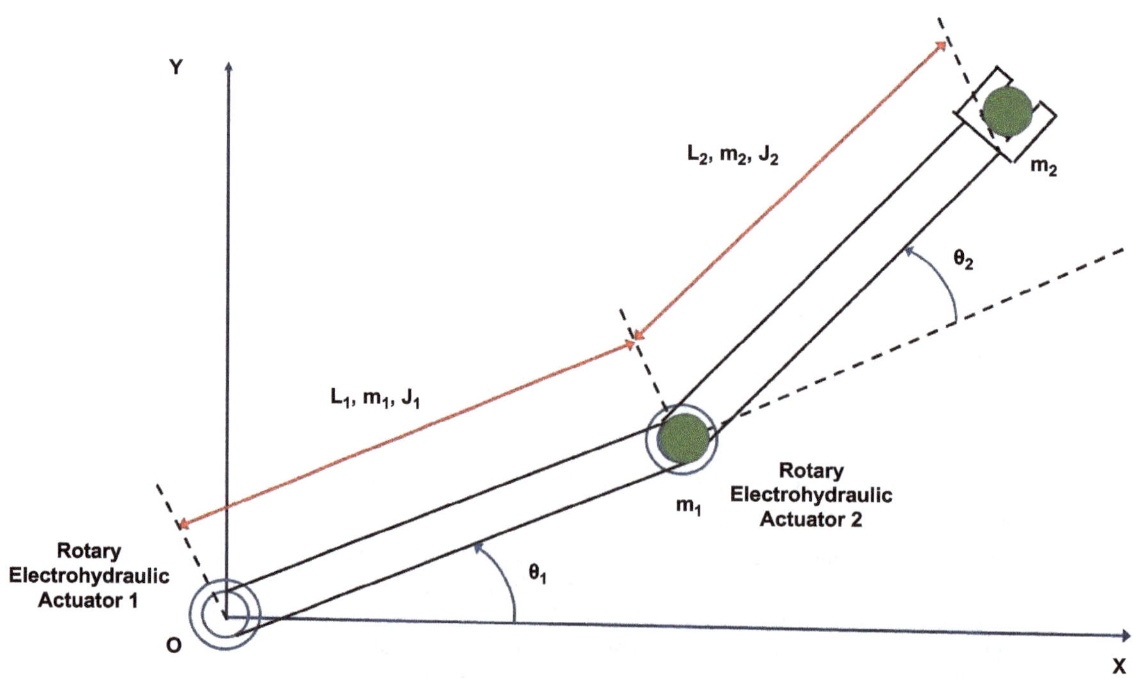

Figure 2.19 Diagram of the multi-DOF electrohydraulic robotic manipulator

The gravitational forces vector is given by

$$G(\theta) = \begin{pmatrix} (m_1+m_2)gl_1cos(\theta_1)+m_2gl_2cos(\theta_1+\theta_2) \\ m_2gl_2cos(\theta_1+\theta_2) \end{pmatrix} \tag{2.59}$$

In Eq. (2.56), $\tau(t)$ is the torques (control inputs) vector which is generated by rotary electrohydraulic actuators mounted on the joints of the robot. It holds that $\tau(t)=[\tau_1,\tau_2]^T$, and by considering elasticity in the second joint, one has $\tau(t)=[\tau_1,\tau_2+k(\theta_1-\theta_2)]^T$.

The determinant of the inertia matrix is $detM = M_{11}M_{22}-M_{12}^2$ and the inverse of the inertial matrix is given by

$$M^{-1} = \frac{1}{detM}\begin{pmatrix} M_{22} & -M_{12} \\ -M_{12} & M_{11} \end{pmatrix} \tag{2.60}$$

and next the dynamics of the two-DOF electrohydraulic robotic manipulator is written as

$$\ddot{\theta} = -M^{-1}(\theta)C(\theta,\dot{\theta}) - M^{-1}(\theta)G(\theta) + M^{-1}(\theta)\tau \tag{2.61}$$

which is also written as

$$\begin{pmatrix} \ddot{\theta}_1 \\ \ddot{\theta}_2 \end{pmatrix} = \begin{pmatrix} \frac{-M_22(C_1+G_1)+M_{12}(C_2+G_2)}{detM} \\ \frac{M_12(C_1+G_1)-M_{11}(C_2+G_2)}{detM} \end{pmatrix} + \begin{pmatrix} \frac{M_22}{detM} & -\frac{M_{12}}{detM} \\ -\frac{M_{12}}{detM} & \frac{M_{11}}{detM} \end{pmatrix}\begin{pmatrix} \tau_1 \\ \tau_2 \end{pmatrix} \tag{2.62}$$

The torques which are generated by the electropneumatic actuators are given by

$$\tau_1 = K_{P_1}A_{P_1}P_{L_1} \quad \tau_2 = K_{P_2}A_{P_2}P_{L_2} \tag{2.63}$$

where K_{P_i}, $i=1,2$ are force to torque conversion coefficients for the i-th actuator, A_{p_i} $i=1,2$ is the efficient area of the cylinder's chamber of the i-th actuator, P_{L_i}, $i=1,2$ is the differential load pressure of the cylinder of the i-th actuator. Moreover, the dynamics of pressure in the two actuators is given by

$$\dot{P}_{L_1} = -\frac{4\beta_{e_1}A_{P_1}}{V_{t_1}}r_1w_1 + \frac{4\beta_{e_1}C_{tl_1}}{V_{t_1}}P_{L_1} + \frac{4\beta_{e_1}C_{d_1}w_1}{V_{t_1}\sqrt{\rho_1}}\sqrt{P_{s_1}-sign(x_{v_1})P_{L_1}}x_{v_1} \tag{2.64}$$

$$\dot{P}_{L_2} = -\frac{4\beta_{e_2}A_{P_2}}{V_{t_2}}r_2w_2 + \frac{4\beta_{e_2}C_{tl_2}}{V_{t_2}}P_{L_2} + \frac{4\beta_{e_2}C_{d_2}w_2}{V_{t_2}\sqrt{\rho_2}}\sqrt{P_{s_2}-sign(x_{v_2})P_{L_2}}x_{v_2} \tag{2.65}$$

Next, by defining the state variables $x_1=\theta_1$, $x_2=\dot{\theta}_1$, $x_3=\theta_2$, $x_4=\dot{\theta}_2$, $x_5=P_{L_1}$, $x_6=x_{v_1}$, $x_7=P_{L_2}$, $x_8=x_{v_2}$, one arrives at the following state-space description:

$$\begin{gathered}
\dot{x}_1 = x_2 \\
\dot{x}_2 = \frac{-M_{22}(C_1+G_1)+M_{12}(C_2+G_2)}{detM} + \frac{M_{22}}{detM}K_{P_1}A_{P_1}x_5 - \frac{M_{12}}{detM}K_{P_2}A_{P_2}x_7 \\
\dot{x}_3 = x_4 \\
\dot{x}_4 = \frac{M_{12}(C_1+G_1)-M_{11}(C_2+G_2)}{detM} - \frac{M_{12}}{detM}K_{P_1}A_{P_1}x_5 + \frac{M_{11}}{detM}K_{P_2}A_{P_2}x_7 \\
\dot{x}_5 = -\frac{4\beta_{e_1}A_{P_1}}{V_{t_1}}r_1x_2 + \frac{4\beta_{e_1}C_{tl_1}}{V_{t_1}}x_5 + \frac{4\beta_{e_1}C_{d_1}x_2}{V_{t_1}\sqrt{\rho_1}}\sqrt{P_{s_1}-sign(x_6)x_5}x_6 \\
\dot{x}_6 = -\frac{1}{T_{sv_1}}x_6 + \frac{K_{sv_1}}{T_{sv_1}}u_1 \\
\dot{x}_7 = -\frac{4\beta_{e_2}A_{P_2}}{V_{t_2}}r_2x_4 + \frac{4\beta_{e_2}C_{tl_2}}{V_{t_2}}x_7 + \frac{4\beta_{e_2}C_{d_2}x_4}{V_{t_2}\sqrt{\rho_2}}\sqrt{P_{s_2}-sign(x_8)x_7}x_8 \\
\dot{x}_8 = -\frac{1}{T_{sv_2}}x_8 + \frac{K_{sv_2}}{T_{sv_2}}u_2
\end{gathered} \tag{2.66}$$

The dynamic model of the electrohydraulic robot can be also written in matrix form

$$\begin{pmatrix} \dot{x}_1 \\ \dot{x}_2 \\ \dot{x}_3 \\ \dot{x}_4 \\ \dot{x}_5 \\ \dot{x}_6 \\ \dot{x}_7 \\ \dot{x}_8 \end{pmatrix} = \begin{pmatrix} x_2 \\ \frac{-M_{22}(C_1+G_1)+M_{12}(C_2+G_2)}{detM} + \frac{M_{22}}{detM}K_{P_1}A_{P_1}x_5 - \frac{M_{12}}{detM}K_{P_2}A_{P_2}x_7 \\ x_4 \\ \frac{M_{12}(C_1+G_1)-M_{11}(C_2+G_2)}{detM} - \frac{M_{12}}{detM}K_{P_1}A_{P_1}x_5 + \frac{M_{11}}{detM}K_{P_2}A_{P_2}x_7 \\ -\frac{4\beta_{e_1}A_{P_1}}{V_{t_1}}r_1x_2 + \frac{4\beta_{e_1}C_{tl_1}}{V_{t_1}}x_5 + \frac{4\beta_{e_1}C_{d_1}x_2}{V_{t_1}\sqrt{\rho_1}}\sqrt{P_{s_1} - sign(x_6)x_5x_6} \\ -\frac{1}{T_{sv_1}}x_6 \\ -\frac{4\beta_{e_2}A_{P_2}}{V_{t_2}}r_2x_4 + \frac{4\beta_{e_2}C_{tl_2}}{V_{t_2}}x_7 + \frac{4\beta_{e_2}C_{d_2}x_4}{V_{t_2}\sqrt{\rho_2}}\sqrt{P_{s_2} - sign(x_8)x_7x_8} \\ -\frac{1}{T_{sv_2}}x_8 \end{pmatrix} + \begin{pmatrix} 0 & 0 \\ 0 & 0 \\ 0 & 0 \\ 0 & 0 \\ 0 & 0 \\ \frac{K_{sv_1}}{T_{sv_1}} & 0 \\ 0 & 0 \\ 0 & \frac{K_{sv_2}}{T_{sv_2}} \end{pmatrix} \begin{pmatrix} u_1 \\ u_2 \end{pmatrix} \tag{2.67}$$

Equivalently, the dynamic model of the electrohydraulic robotic manipulator can be written in the nonlinear affine-in-the-input state-space form:

$$\dot{x} = f(x) + g(x)u \tag{2.68}$$

where $x \in R^{8\times 1}$, $f(x) \in R^{8\times 1}$, $g(x) \in R^{8\times 2}$, and $u \in R$.

2.2.2.2 Proof of differential flatness properties

First, a redefinition of the state variables of the electrohydraulic robot is performed as follows: $z_1 = x_1$, $z_2 = x_3$, $z_2 = x_2$, $x_4 = x_4$, $z_5 = x_5$, $z_6 = x_7$, $z_7 = x_6$ and $z_8 = x_8$. Using the redefinition of the state variables of the robot, its state equations are rewritten as:

$$\begin{aligned} \dot{z}_1 &= z_2 \\ \dot{z}_2 &= z_4 \\ \dot{z}_3 &= \frac{-M_{22}(C_1+G_1)+M_{12}(C_2+G_2)}{detM} + \frac{M_{22}}{detM}K_{P_1}A_{P_1}z_5 - \frac{M_{12}}{detM}K_{P_2}A_{P_2}z_6 \\ \dot{z}_4 &= \frac{M_{12}(C_1+G_1)-M_{11}(C_2+G_2)}{detM} - \frac{M_{12}}{detM}K_{P_1}A_{P_1}z_5 + \frac{M_{11}}{detM}K_{P_2}A_{P_2}z_6 \\ \dot{z}_5 &= -\frac{4\beta_{e_1}A_{P_1}}{V_{t_1}}r_1z_3 + \frac{4\beta_{e_1}C_{tl_1}}{V_{t_1}}z_5 + \frac{4\beta_{e_1}C_{d_1}z_3}{V_{t_1}\sqrt{\rho_1}}\sqrt{P_{s_1} - sign(z_7)z_5z_7} \\ \dot{z}_6 &= -\frac{4\beta_{e_2}A_{P_2}}{V_{t_2}}r_2z_4 + \frac{4\beta_{e_2}C_{tl_2}}{V_{t_2}}z_6 + \frac{4\beta_{e_2}C_{d_2}z_4}{V_{t_2}\sqrt{\rho_2}}\sqrt{P_{s_2} - sign(x_8)z_6z_8} \\ \dot{z}_7 &= -\frac{1}{T_{sv_1}}z_7 + \frac{K_{sv_1}}{T_{sv_1}}u_1 \\ \dot{z}_8 &= -\frac{1}{T_{sv_2}}z_8 + \frac{K_{sv_2}}{T_{sv_2}}u_2 \end{aligned} \tag{2.69}$$

Equivalently, the state-space model of the electrohydraulic robot can be written in the form of four serially interconnected subsystems:

$$\begin{pmatrix} \dot{z}_1 \\ \dot{z}_2 \end{pmatrix} = \begin{pmatrix} 0 \\ 0 \end{pmatrix} + \begin{pmatrix} 1 & 0 \\ 0 & 1 \end{pmatrix} \begin{pmatrix} z_3 \\ z_4 \end{pmatrix} \tag{2.70}$$

$$\begin{pmatrix} \dot{z}_3 \\ \dot{z}_4 \end{pmatrix} = \begin{pmatrix} \frac{-M_{22}(C_1+G_1)+M_{12}(C_2+G_2)}{detM} \\ \frac{M_{12}(C_1+G_1)-M_{11}(C_2+G_2)}{detM} \end{pmatrix} + \begin{pmatrix} \frac{M_{22}}{detM}K_{P_1}A_{P_1} & -\frac{M_{12}}{detM}K_{P_2}A_{P_2} \\ -\frac{M_{12}}{detM}K_{P_1}A_{P_1} & \frac{M_{11}}{detM}K_{P_2}A_{P_2} \end{pmatrix} \begin{pmatrix} z_5 \\ z_6 \end{pmatrix} \tag{2.71}$$

$$\begin{pmatrix} \dot{z}_5 \\ \dot{z}_6 \end{pmatrix} = \begin{pmatrix} -\frac{4\beta_{e_1}A_{P_1}}{V_{t_1}} r_1 z_3 + \frac{4\beta_{e_1}C_{tl_1}}{V_{t_1}} z_5 \\ -\frac{4\beta_{e_2}A_{P_2}}{V_{t_2}} r_2 z_4 + \frac{4\beta_{e_2}C_{tl_2}}{V_{t_2}} z_6 \end{pmatrix} + \\ + \begin{pmatrix} \frac{4\beta_{e_1}C_{d_1}z_3}{V_{t_1}\sqrt{\rho_1}}\sqrt{P_{s_1} - sign(z_7)z_5} & 0 \\ 0 & \frac{4\beta_{e_2}C_{d_2}z_4}{V_{t_2}\sqrt{\rho_2}}\sqrt{P_{s_2} - sign(x_8)z_6} \end{pmatrix} \begin{pmatrix} z_7 \\ z_8 \end{pmatrix} \tag{2.72}$$

$$\begin{pmatrix} \dot{z}_7 \\ \dot{z}_8 \end{pmatrix} = \begin{pmatrix} -\frac{1}{T_{sv_1}} z_7 \\ -\frac{1}{T_{sv_2}} z_8 \end{pmatrix} + \begin{pmatrix} \frac{K_{sv_1}}{T_{sv_1}} & 0 \\ 0 & \frac{K_{sv_2}}{T_{sv_2}} \end{pmatrix} \begin{pmatrix} u_1 \\ u_2 \end{pmatrix} \tag{2.73}$$

In the above-noted subsystems of Eq. (2.70) to Eq. (2.73), the following vectors and matrices are defined:

$$z_{12} = \begin{pmatrix} z_1 \\ z_2 \end{pmatrix} \quad f_{12} = \begin{pmatrix} 0 \\ 0 \end{pmatrix} \quad g_{12} = \begin{pmatrix} 1 & 0 \\ 0 & 1 \end{pmatrix} \tag{2.74}$$

$$z_{34} = \begin{pmatrix} z_2 \\ z_4 \end{pmatrix} \quad f_{34} = \begin{pmatrix} \frac{-M_{22}(C_1+G_1)+M_{12}(C_2+G_2)}{detM} \\ \frac{M_{12}(C_1+G_1)-M_{11}(C_2+G_2)}{detM} \end{pmatrix} \quad g_{34} = \begin{pmatrix} \frac{M_{22}}{detM}K_{P_1}A_{P_1} & -\frac{M_{12}}{detM}K_{P_2}A_{P_2} \\ -\frac{M_{12}}{detM}K_{P_1}A_{P_1} & \frac{M_{11}}{detM}K_{P_2}A_{P_2} \end{pmatrix} \tag{2.75}$$

$$z_{56} = \begin{pmatrix} z_5 \\ z_6 \end{pmatrix} \quad f_{56} = \begin{pmatrix} -\frac{4\beta_{e_1}A_{P_1}}{V_{t_1}} r_1 z_3 + \frac{4\beta_{e_1}C_{tl_1}}{V_{t_1}} z_5 \\ -\frac{4\beta_{e_2}A_{P_2}}{V_{t_2}} r_2 z_4 + \frac{4\beta_{e_2}C_{tl_2}}{V_{t_2}} z_6 \end{pmatrix} \\ g_{56} = \begin{pmatrix} \frac{4\beta_{e_1}C_{d_1}z_3}{V_{t_1}\sqrt{\rho_1}}\sqrt{P_{s_1} - sign(z_7)z_5} & 0 \\ 0 & \frac{4\beta_{e_2}C_{d_2}z_4}{V_{t_2}\sqrt{\rho_2}}\sqrt{P_{s_2} - sign(x_8)z_6} \end{pmatrix} \tag{2.76}$$

$$z_{78} = \begin{pmatrix} z_7 \\ z_8 \end{pmatrix} \quad f_{78} = \begin{pmatrix} -\frac{1}{T_{sv_1}} z_7 \\ -\frac{1}{T_{sv_2}} z_8 \end{pmatrix} \quad g_{78} = \begin{pmatrix} \frac{K_{sv_1}}{T_{sv_1}} & 0 \\ 0 & \frac{K_{sv_2}}{T_{sv_2}} \end{pmatrix} \tag{2.77}$$

Using the previous notation, one arrives at a concise state-space description of the form:

$$\dot{z}_{12} = f_{12} + g_{12}z_{34} \tag{2.78}$$

$$\dot{z}_{34} = f_{34} + g_{34}z_{56} \tag{2.79}$$

$$\dot{z}_{56} = f_{56} + g_{56}z_{78} \tag{2.80}$$

$$\dot{z}_{78} = f_{78} + g_{78}u \tag{2.81}$$

Next, it will be proven that the dynamic model of the electrohydraulic manipulator is differentially flat, with flat output $Y = z_{12}^T = [z_1, z_2]^T = [\theta_1, \theta_2]^T$. From Eq. (2.78), it holds that

$$z_{34} = g_{12}^{-1}(\dot{z}_{12} - f_{12}) \tag{2.82}$$

where f_{12} and g_{12} have constant elements. Thus, z_{34} is a differential function of the flat outputs vector or

$$z_{34} = h_{34}(Y, \dot{Y}) \tag{2.83}$$

From Eq. (2.79), one solves for z_{56} which gives

$$z_{56} = g_{34}^{-1}(\dot{z}_{34} - f_{34}) \tag{2.84}$$

It holds that f_{34}, g_{34} are functions of z_{12} and z_{34}, and consequently they are finally differential functions of the flat outputs vector z_{12}. Thus, z_{56} is a differential function of the flat outputs vector z_{12}. Therefore, z_{56} is a differential function of the flat outputs vector z_{12} or

$$z_{56} = h_{56}(Y, \dot{Y}) \tag{2.85}$$

Additionally, from Eq. (2.80), one solves for z_{78}. Thus, one obtains

$$z_{78} = g_{56}^{-1}(\dot{z}_{56} - f_{56}) \tag{2.86}$$

It holds that f_{56}, g_{56} are functions of z_{12}, z_{34}, z_{56}, and therefore, they are finally differential functions of the flat output z_{12}. Consequently, z_{78} is also a differential function of the flat outputs vector z_{12} or

$$z_{78} = h_{78}(Y, \dot{Y}) \tag{2.87}$$

Finally, from Eq. (2.81), one solves for the control input u. This gives

$$u = g_{78}^{-1}(\dot{z}_{78} - f_{78}) \tag{2.88}$$

where f_{78}, g_{78} are functions of z_{12}, z_{34}, z_{56}, and z_{78}, and thus they are also differential functions of the flat outputs vector z_{12}. Consequently, the control inputs vector u is also a differential function of the flat outputs vector z_{12} or

$$u = h_u(Y, \dot{Y}) \tag{2.89}$$

As a result of the above, all state variables and the control inputs of the dynamic model of the electrohydraulic robotic manipulator are differential functions of the flat outputs vector z_{12}, and the system is differentially flat.

Next, it will be proven that each one of the subsystems of Eq. (2.78) to Eq. (2.81) can be viewed independently as a differentially flat subsystem.

In the subsystem of Eq. (2.78), the flat output is taken to be z_{12} while z_{34} is a virtual control input. Besides, f_{12}, g_{12} have constant elements. Solving for z_{34} gives Eq. (2.83), which demonstrates that z_{34} is a differential function of the flat output z_{12}, and consequently, the subsystem of Eq. (2.78) is differentially flat.

In the subsystem of Eq. (2.79) the flat output is taken to be z_{34} while z_{56} is a virtual control input. Besides, f_{34}, g_{34} are functions of z_{12}, z_{34} where z_{12} is viewed as a coefficients vector and z_{34} is the local flat outputs vector. Solving for z_{56} gives Eq. (2.85), which demonstrates that z_{56} is a differential function of the flat output z_{34}, and consequently, the subsystem of Eq. (2.79) is differentially flat.

In the subsystem of Eq. (2.80), the flat output is taken to be z_{56} while z_{78} is a virtual control input. Besides, f_{56}, g_{56} are functions of z_{12}, z_{34}, z_{56} where z_{12}, z_{34} can be viewed as coefficients vectors and z_{56} is the flat outputs vector. Solving for z_{78} gives Eq. (2.87), which demonstrates that z_{78} is a differential function of the flat output z_{56}, and consequently, the subsystem of Eq. (2.80) is differentially flat.

In the subsystem of Eq. (2.81), the flat output is taken to be z_{78} while u is the real control input. Besides, f_{78}, g_{78} are functions of z_{78} where z_{78} is the flat outputs vector. Solving for u gives Eq. (2.89), which demonstrates that u is a differential function of the flat output z_{78}, and consequently, the subsystem of Eq. (2.80) is differentially flat.

2.2.3 Design of a flatness-based controller in successive loops

Next, for each one of the differentially flat subsystems of Eq. (2.78) to Eq. (2.81), one can design a stabilizing feedback controller following the controller definition stages for input–output linearized state-space descriptions.

For the subsystem of Eq. (2.78), the desirable setpoint is taken to be z_{12}^d. The value of the virtual control input, which achieves convergence to this setpoint, is

$$z_{34}^* = g_{12}^{-1}[\dot{z}_{12}^d - f_{12} - K_1^1(z_{12} - z_{12}^d)] \tag{2.90}$$

where matrix K_1^1 is a diagonal matrix of the form $K_1^1 = diag[K_{1i}^1]$, where $K_{1i}^1 > 0$ for $i = 1, 2$. By substituting Eq. (2.90) into Eq. 2.78, one obtains the closed subsystem dynamics:

$$\begin{aligned}(\dot{z}_{12} - \dot{z}_{12}^d) + K_1^1(z_{12} - z_{12}^d) = 0 \Rightarrow \dot{e}_{12} + K_1^1 e_{12} = 0 \\ \Rightarrow lim_{t\to\infty} e_{12}(t) = 0 \Rightarrow lim_{t\to\infty} z_{12}(t) = z_{12}^d(t)\end{aligned} \tag{2.91}$$

For the subsystem of Eq. (2.78), the control input z_{34}^* becomes the subsystem's setpoint. Thus, $z_{34}^d = z_{34}^*$. The value of the virtual control input which achieves convergence to this setpoint is

$$z_{56}^* = g_{34}^{-1}[\dot{z}_{34}^d - f_{34} - K_1^2(z_{34} - z_{34}^d)] \tag{2.92}$$

where matrix K_1^2 is a diagonal matrix of the form $K_1^2 = diag[K_{1i}^2]$, where $K_{1i}^2 > 0$ for $i = 1, 2$. By substituting Eq. (2.92) into Eq. 2.79, one obtains the closed subsystem dynamics.

$$\begin{aligned} &(\dot{z}_{34} - \dot{z}_{34}^d) + K_1^2(z_{34} - z_{34}^d) = 0 \Rightarrow \dot{e}_{34} + K_1^2 e_{34} = 0 \\ &\quad \Rightarrow lim_{t\to\infty} e_{34}(t) = 0 \Rightarrow lim_{t\to\infty} z_{34}(t) = z_{34}^d(t) \end{aligned} \tag{2.93}$$

For the subsystem of Eq. (2.80) the control input z_{56}^* becomes the subsystem's setpoint. Thus, $z_{56}^d = z_{56}^*$. The value of the virtual control input which achieves convergence to this setpoint is

$$z_{78}^* = g_{56}^{-1}[\dot{z}_{56}^d - f_{56} - K_1^3(z_{56} - z_{56}^d)] \tag{2.94}$$

where matrix K_1^3 is a diagonal matrix of the form $K_1^3 = diag[K_{1i}^3]$, where $K_{1i}^3 > 0$ for $i = 1, 2$. By substituting Eq. (2.94) into Eq. 2.80, one obtains the closed subsystem dynamics:

$$\begin{aligned} &(\dot{z}_{56} - \dot{z}_{56}^d) + K_1^3(z_{56} - z_{56}^d) = 0 \Rightarrow \dot{e}_{56} + K_1^3 e_{56} = 0 \\ &\quad \Rightarrow lim_{t\to\infty} e_{56}(t) = 0 \Rightarrow lim_{t\to\infty} z_{56}(t) = z_{56}^d(t) \end{aligned} \tag{2.95}$$

For the subsystem of Eq. (2.81) the control input z_{78}^* becomes the subsystem's setpoint. Thus, $z_{78}^d = z_{78}^*$. The value of the real control input which achieves convergence to this setpoint is

$$u = g_{78}^{-1}[\dot{z}_{78}^d - f_{78} - K_1^4(z_{78} - z_{78}^d)] \tag{2.96}$$

where matrix K_1^4 is a diagonal matrix of the form $K_1^4 = diag[K_{1i}^4]$, where $K_{1i}^4 > 0$ for $i = 1, 2$. By substituting Eq. (2.96) into Eq. 2.81, one obtains the closed subsystem dynamics:

$$\begin{aligned} &(\dot{z}_{78} - \dot{z}_{78}^d) + K_1^4(z_{78} - z_{78}^d) = 0 \Rightarrow \dot{e}_{78} + K_1^4 e_{78} = 0 \\ &\quad \Rightarrow lim_{t\to\infty} e_{78}(t) = 0 \Rightarrow lim_{t\to\infty} z_{78}(t) = z_{78}^d(t) \end{aligned} \tag{2.97}$$

Therefore, the concept of flatness-based control in successive loops achieves elimination of the tracking error for all state vector elements and the closed-loop system is globally asymptotically stable.

Global asymptotic stability can be also proven through Lyapunov analysis. The following Lyapunov function is defined:

$$V = \frac{1}{2}[e_{12}^T e_{12} + e_{34}^T e_{34} + e_{56}^T e_{56} + e_{78}^T e_{78}] \tag{2.98}$$

By differentiating in time, one obtains

$$\dot{V} = \frac{1}{2}[2e_{12}^T \dot{e}_{12} + 2e_{34}^T \dot{e}_{34} + 2e_{56}^T \dot{e}_{56} + 2e_{78}^T \dot{e}_{78}] \tag{2.99}$$

By substituting in Eq. (2.99) the previously given relations about the tracking error dynamics, one obtains

$$\dot{V} = [e_{12}^T(-K_1^1 e_{12}) + e_{34}^T(-K_1^2 e_{34}) + e_{56}^T(-K_1^3 e_{56}) + e_{78}^T(-K_1^4 e_{78})] \tag{2.100}$$

Thus, one finally gets

$$\dot{V} = -[e_{12}^T K_1^1 e_{12} + e_{34}^T K_1^2 e_{34} + e_{56}^T K_1^3 e_{56} + e_{78}^T K_1^4 e_{78}] \tag{2.101}$$

which signifies that

$$\dot{V} < 0 \text{ for } K_1^1 > 0, K_1^2 > 0, K_1^3 > 0, K_1^4 > 0 \tag{2.102}$$

It holds that $\dot{V}$ is strictly negative $\forall e_{12}, e_{34}, e_{56}, e_{78} \neq 0$. Therefore, $\dot{V}$ becomes 0 when $e_{12} = 0, e_{34} = 0, e_{56} = 0, e_{78} = 0$. Consequently, the Lyapunov function of the system is strictly diminishing and finally converges to 0, no matter what the initial conditions of the electrohydraulic robot are. Thus, the control loop of the electrohydraulic robotic manipulator is globally asymptotically stable.

2.2.4 *Simulation tests*

Results about the tracking accuracy and the speed of convergence to setpoints of the successive-loops flatness-based control method, in the case of the multi-DOF electrohydraulically actuated robot of Section 2.2.2, are shown in

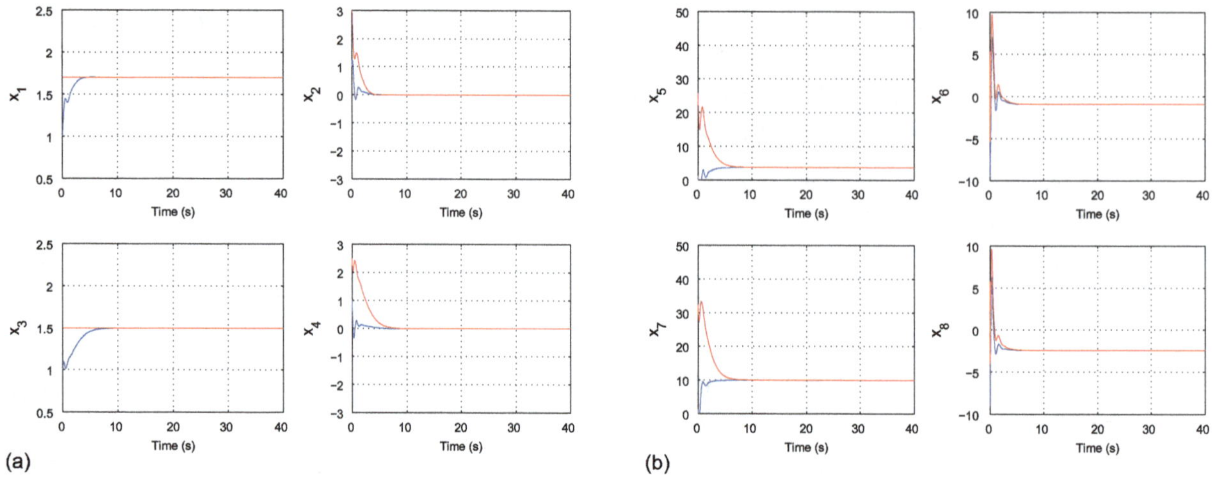

Figure 2.20 *Control for the 2-DOF electrohydraulic robot when tracking setpoint 1: (a) convergence of state variables x_1 to x_4 to their reference setpoints (red line: setpoint, blue line: real value); (b) convergence of state variables x_5 to x_8 to their reference setpoints (red line: setpoint, blue line: real value)*

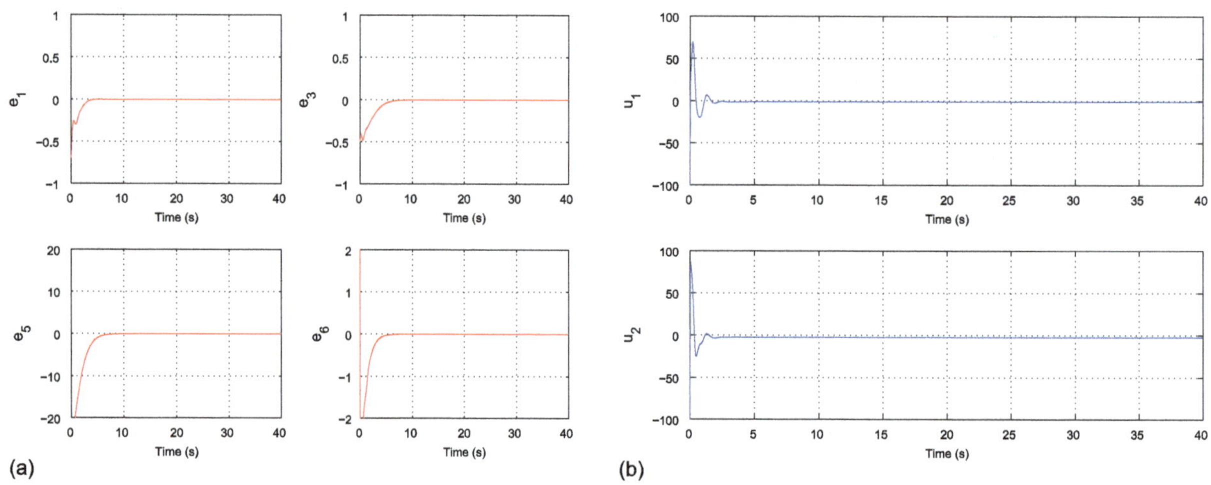

Figure 2.21 *Control for the 2-DOF electrohydraulic robot when tracking setpoint 1: (a) tracking error of state variables x_1, x_2, x_5, x_6 (red line); (b) variation of the control inputs u_1 to u_2 (blue line)*

Figures 2.20–2.35. It can be noticed again that under this control scheme, one achieves fast and precise tracking of reference setpoints for all state variables of the dynamic model of the electrohydraulic robotic manipulator. It is noteworthy that through the stages of this method, one solves also the setpoints definition problem for all state variables of the electrohydraulic robot. Actually, the selection of setpoints for state variables x_1, x_3 is unconstrained. On the other side by defining state variables x_2, x_4, x_5, x_7, and x_6, x_8 as virtual control inputs for the subsystem of x_1, x_3, and for the rest of the subsystems in which the state-space model of the robot is decomposed, one can find the setpoints for x_2, x_4, x_5, x_7, and x_6, x_8 as functions of the setpoints for x_1, x_3. The speed of convergence of the state variables of the electrohydraulic robot under flatness-based control implemented in successive loops is determined by the selection of values for the diagonal gain matrices K_1, K_2, K_3, and K_4.

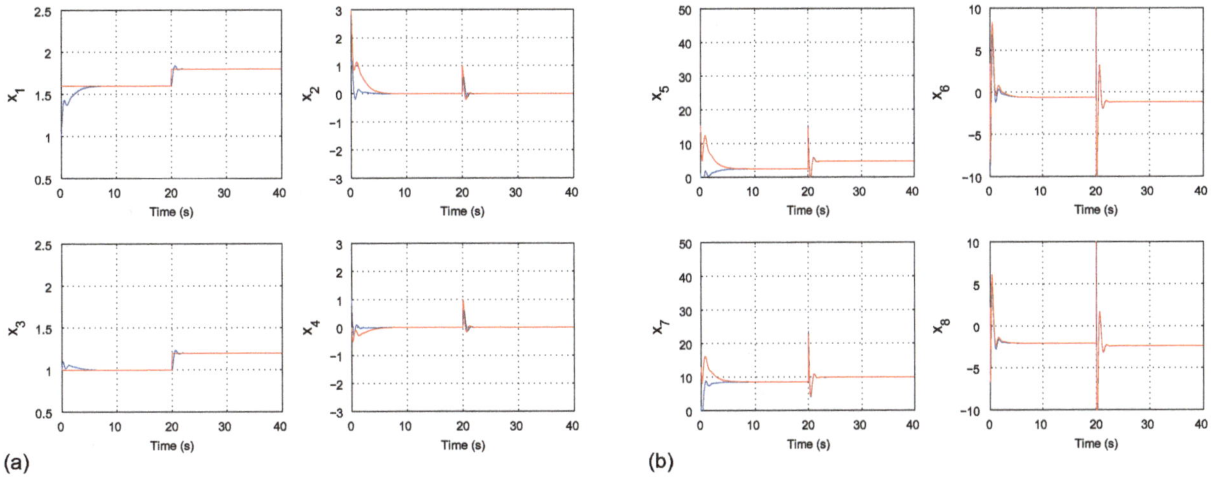

Figure 2.22 *Control for the 2-DOF electrohydraulic robot when tracking setpoint 2: (a) convergence of state variables x_1 to x_4 to their reference setpoints (red line: setpoint, blue line: real value); (b) convergence of state variables x_5 to x_8 to their reference setpoints (red line: setpoint, blue line: real value)*

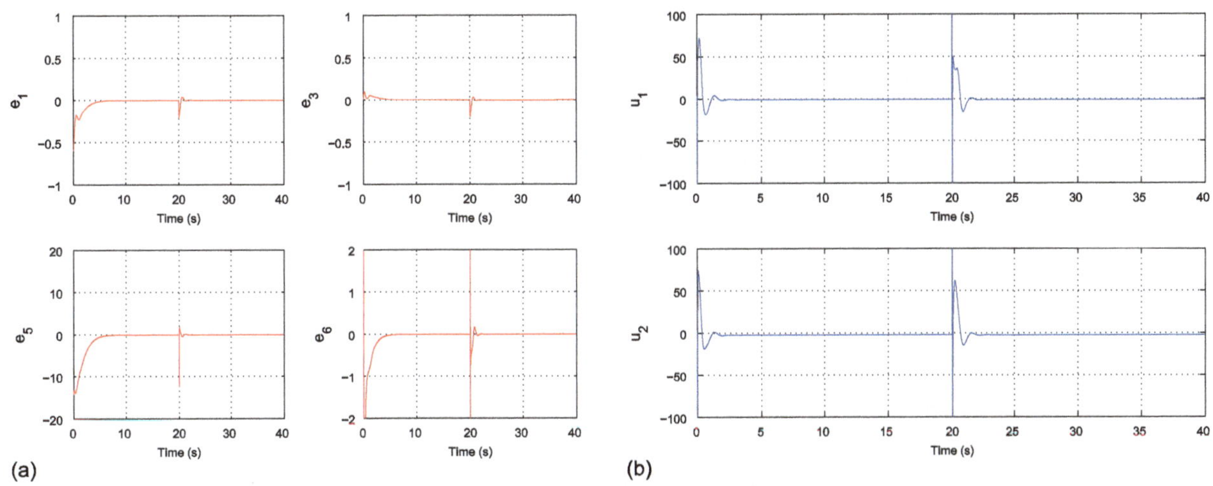

Figure 2.23 *Control for the 2-DOF electrohydraulic robot when tracking setpoint 2: (a) tracking error of state variables x_1, x_2, x_5, x_6 (red line); (b) variation of the control inputs u_1 to u_2 (blue line)*

2.3 Flatness-based control in successive loops of flexible-joint robots

2.3.1 Outline

As previously noted, differential flatness theory under various implementation approaches has enabled the solution of several nonlinear control problems during the last years [711,721,732,725,741,491,252,898]. By demonstrating that all state variables and the control inputs of a dynamical system can be expressed as functions of one single algebraic variable which is the so-called flat output, and also as functions of the flat output's derivatives, one confirms that this system is differentially flat. [91,594,639,489]. By proving differential flatness properties for a nonlinear system it is ensured

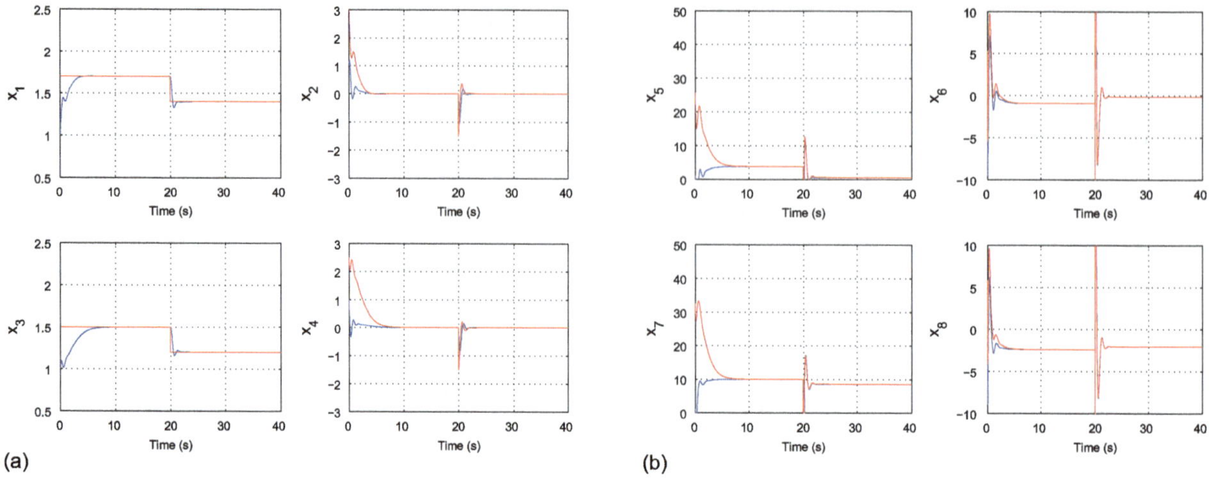

Figure 2.24 Control for the 2-DOF electrohydraulic robot when tracking setpoint 3: (a) convergence of state variables x_1 to x_4 to their reference setpoints (red line: setpoint, blue line: real value); (b) convergence of state variables x_5 to x_8 to their reference setpoints (red line: setpoint, blue line: real value)

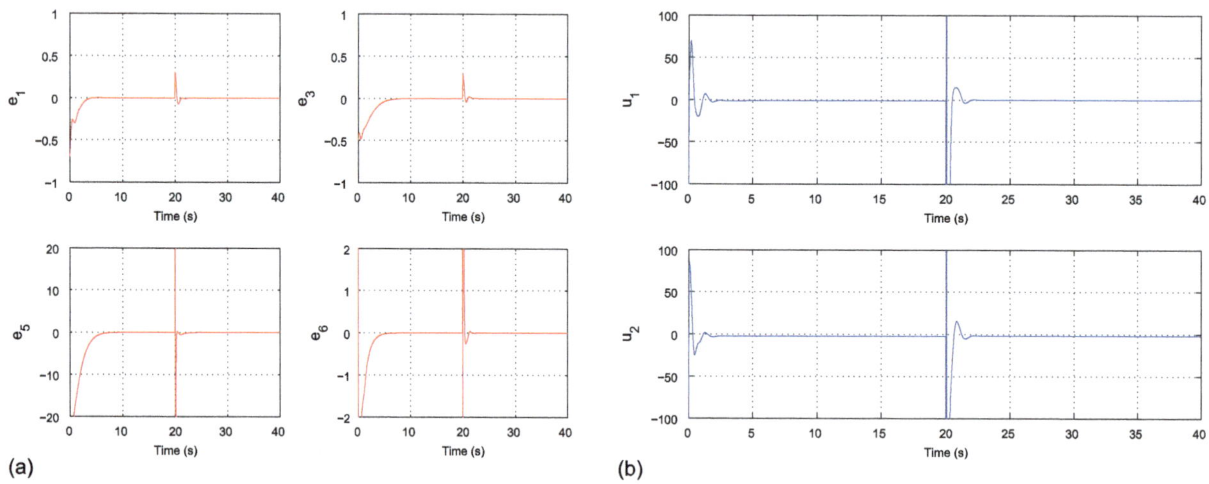

Figure 2.25 Control for the 2-DOF electrohydraulic robot when tracking setpoint 3: (a) tracking error of state variables x_1, x_2, x_5, x_6 (red line); (b) variation of the control inputs u_1 to u_2 (blue line)

that this system can be transformed into the linear canonical (Brunovsky) form [806,492,640,524,54]. The canonical state-space form is controllable and observable thus allowing to treat effectively control and estimation problems [439,705,60,706]. In this section, a successive-loops approach is developed for controller design in mechatronic motion transmission systems which exhibit the differential flatness property. The method makes use of the initial nonlinear model of the system and of its decomposition into a set of nonlinear subsystems for which the differential flatness property holds [712,720,714,716,718].

The concept of the present section has been once again to decompose the state-space model of a nonlinear system into subsystems, which satisfy differential flatness properties. For each subsystem of the state-space model a virtual control input is computed, capable of inverting the subsystem's dynamics and of eliminating the subsystem's tracking error. The control input that is actually applied to the nonlinear system is found from the last row of the state-space description.

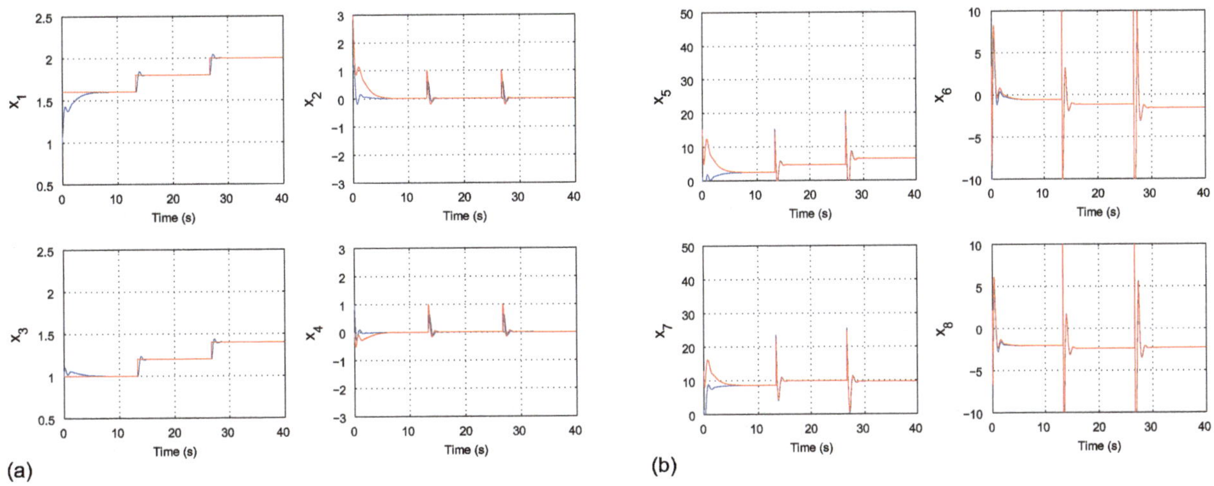

Figure 2.26 Control for the 2-DOF electrohydraulic robot when tracking setpoint 4: (a) convergence of state variables x_1 to x_4 to their reference setpoints (red line: setpoint, blue line: real value); (b) convergence of state variables x_5 to x_8 to their reference setpoints (red line: setpoint, blue line: real value)

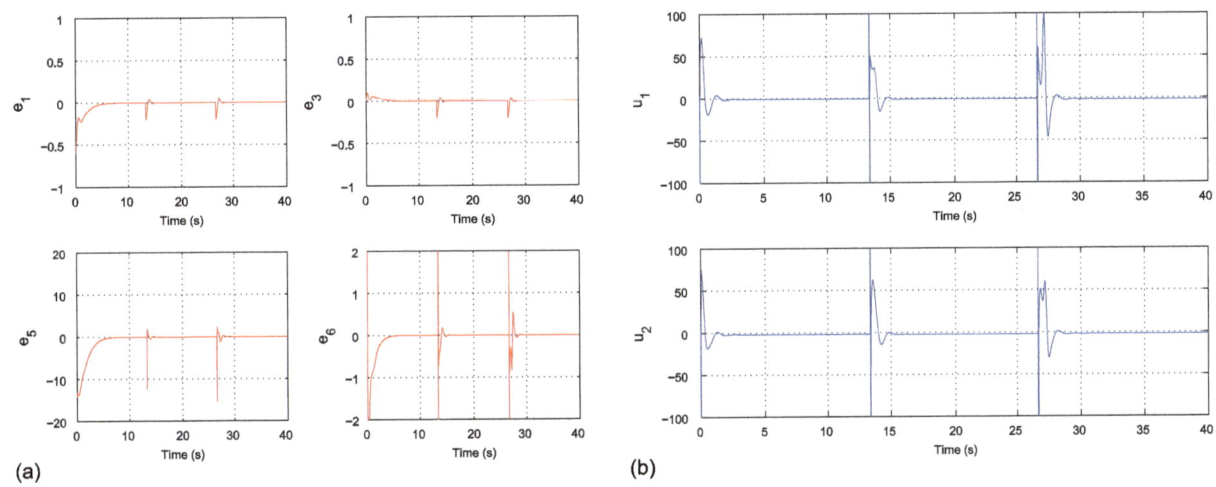

Figure 2.27 Control for the 2-DOF electrohydraulic robot when tracking setpoint 4: (a) tracking error of state variables x_1, x_2, x_5, x_6 (red line); (b) variation of the control inputs u_1 to u_2 (blue line)

This control input incorporates in a recursive manner all virtual control inputs which were computed for the individual subsystems associated with the initial state-space equation. The control input that should be applied to the nonlinear system so as to assure that all its state vector elements will converge to the desirable setpoints, is obtained at each iteration of the control algorithm, by tracing the subsystems of the state-space model backward. Dynamical systems which are inherently found in the strict-feedback form or which can be transformed into such a state-space description are suitable for applying the proposed flatness-based control method in successive loops [711,721]. The present section demonstrates the application of flatness-based control in successive loops in robotic manipulators. Global stability properties are proven for this control scheme and performance tests are given in the case of the dynamic model of a multi-DOF robotic manipulator with flexible joints.

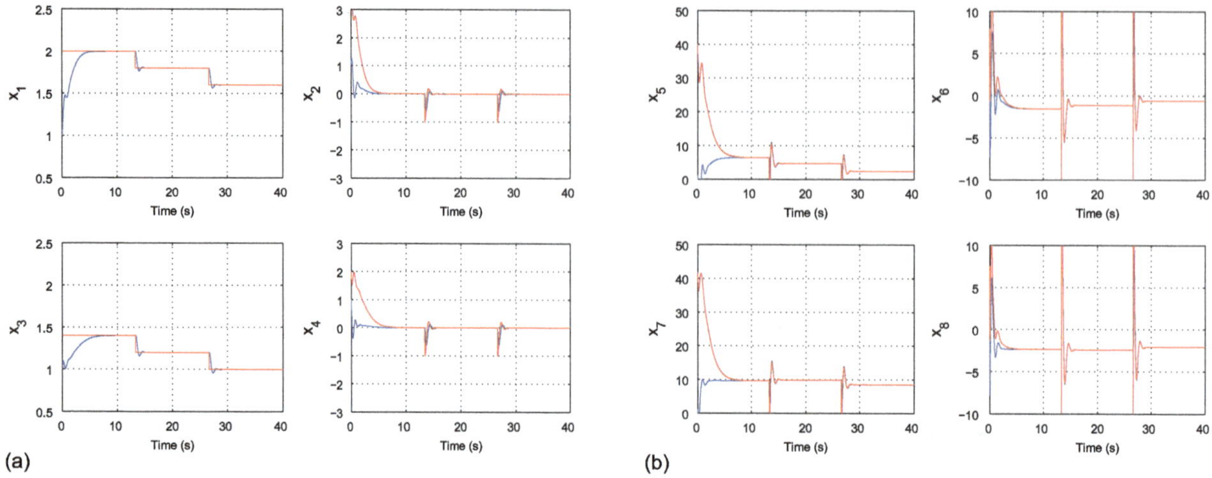

Figure 2.28 Control for the 2-DOF electrohydraulic robot when tracking setpoint 5: (a) convergence of state variables x_1 to x_4 to their reference setpoints (red line: setpoint, blue line: real value); (b) convergence of state variables x_5 to x_8 to their reference setpoints (red line: setpoint, blue line: real value)

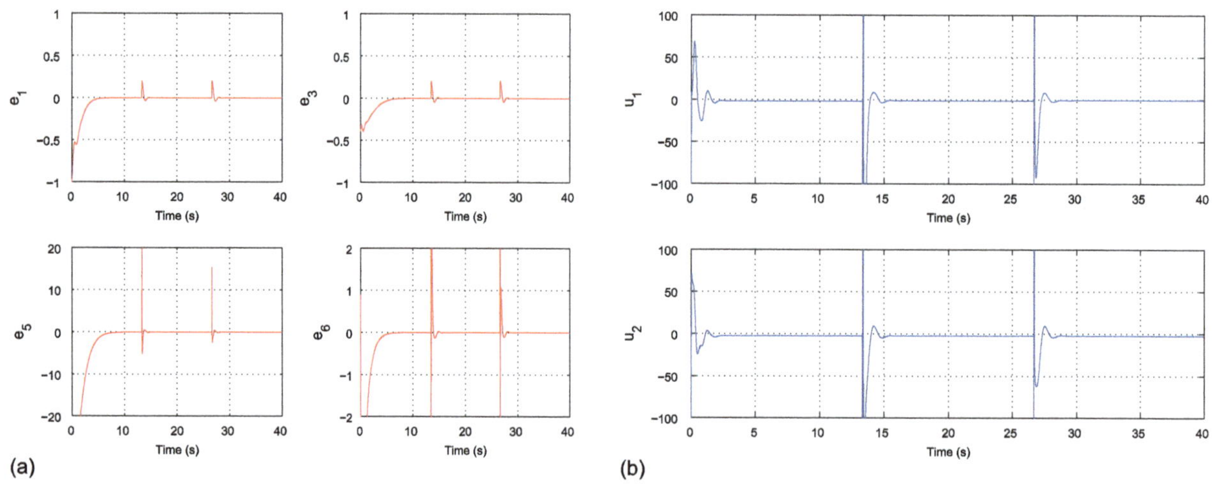

Figure 2.29 Control for the 2-DOF electrohydraulic robot when tracking setpoint 5: (a) tracking error of state variables x_1, x_2, x_5, x_6 (red line); (b) variation of the control inputs u_1 to u_2 (blue line)

The present section analyzes flatness-based control in successive loops for the dynamic model of multi-DOF robotic manipulators with flexible joints. Synchronization between the turn angles of the robots' joints and the turn angles of the motors that provide actuation to these joints is not always direct to achieve [829,917,525,754]. This happens for instance in the case of flexible-joint robots or in the case of robots with compliant actuators [1011,497]. Flexibility in the joints of a robotic manipulator can affect negatively the precise positioning of the robots' end-effector, and consequently can hinder the execution of several robotic tasks in which precise positioning is needed, as for instance robotic-guided radiotherapy, or robotic surgery [643,141]. On the other side, flexibility in the joints of a robotic manipulator or the use of compliant actuators can be useful, for instance when the appearance of abrupt contact forces has to be avoided so as to assure the safe execution of several contact-based tasks [434,99]. So far, several control approaches have been applied to robots with flexible joints and to robots with compliant actuators. One can note for instance

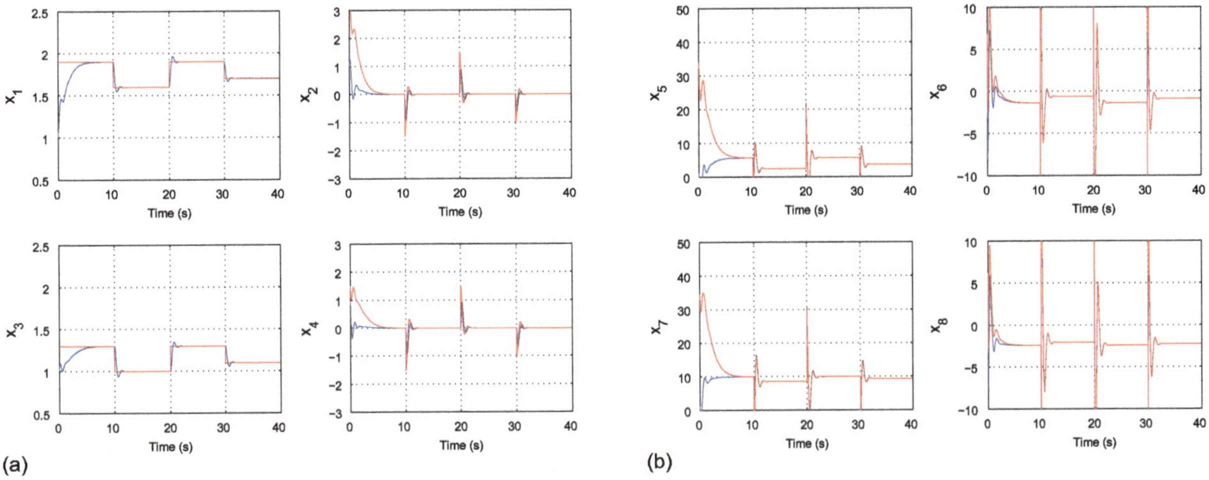

Figure 2.30 Control for the 2-DOF electrohydraulic robot when tracking setpoint 6: (a) convergence of state variables x_1 to x_4 to their reference setpoints (red line: setpoint, blue line: real value); (b) convergence of state variables x_5 to x_8 to their reference setpoints (red line: setpoint, blue line: real value)

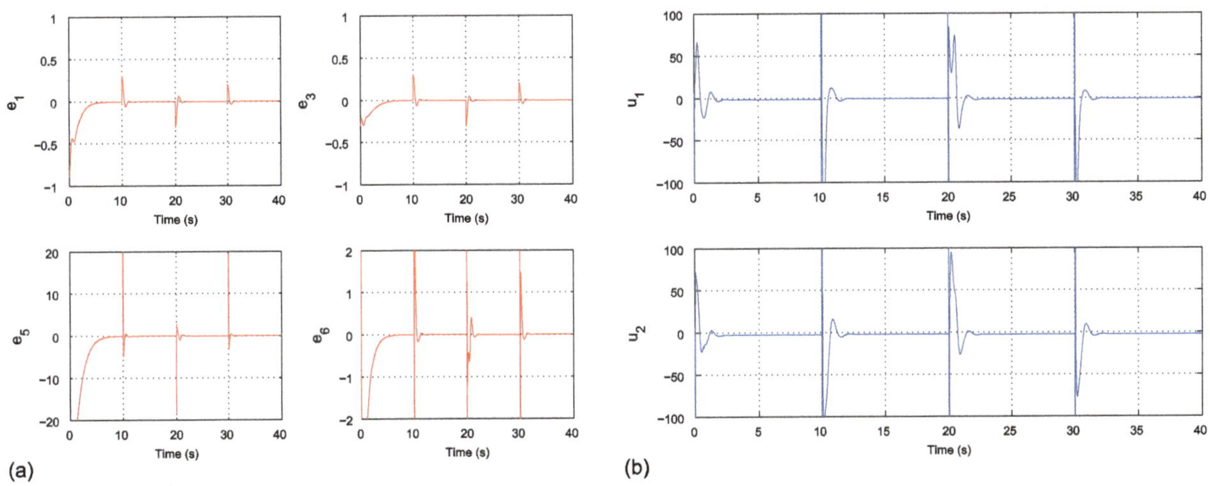

Figure 2.31 Control for the 2-DOF electrohydraulic robot when tracking setpoint 6: (a) tracking error of state variables x_1, x_2, x_5, x_6 (red line); (b) variation of the control inputs u_1 to u_2 (blue line)

sliding-mode control, backstepping control, or model predictive control [978,966,912,977,584,659]. Often, nonlinear state-observers are used in such a type of robotic manipulators, so as to either estimate nonlinear state vector elements or to estimate disturbance forces and torques [753,846,248,1023]. There have been also results on nonlinear optimal control of flexible-joint robots [96,652,653,762,607]. Underactuation in such a type of robots remains an issue that the design of related controllers has to take into account [962,175].

The proposed multi-loop flatness-based control method for flexible-joint robots is based on a widened backstepping control concept. Recent advances in backstepping control of clutch models can be found in [935,792,528,270,868]. Additionally, recent advances in backstepping control of flexible-joint robots can be found in [52,518,506,31,178]. Whereas backstepping control can be applied to dynamical systems which are in the strict-feedback form, the multi-loop flatness-based control method can be applied to all dynamical systems which can be written in the form of chained

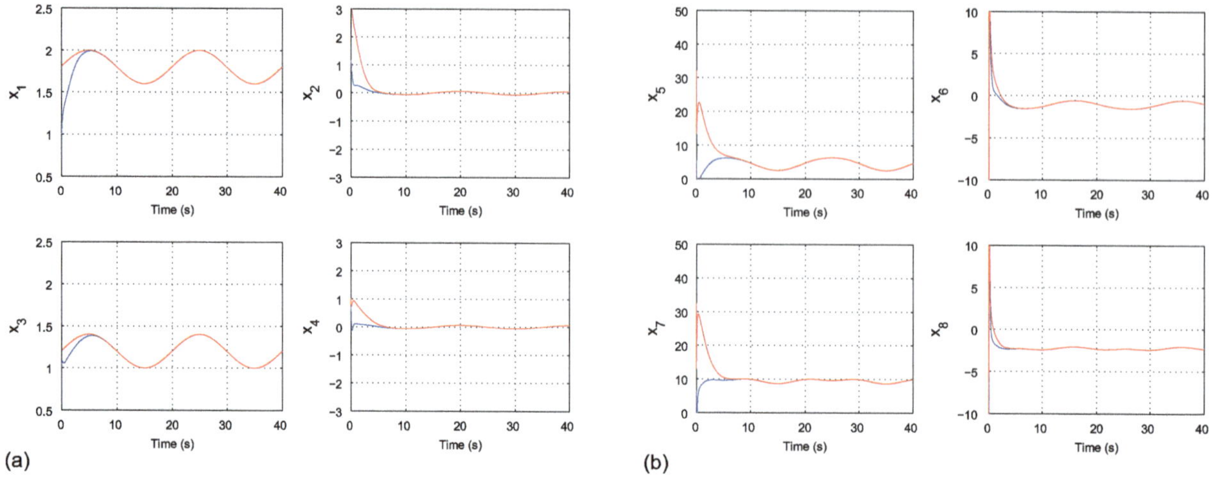

Figure 2.32 Control for the 2-DOF electrohydraulic robot when tracking setpoint 7: (a) convergence of state variables x_1 to x_4 to their reference setpoints (red line: setpoint, blue line: real value); (b) convergence of state variables x_5 to x_8 to their reference setpoints (red line: setpoint, blue line: real value)

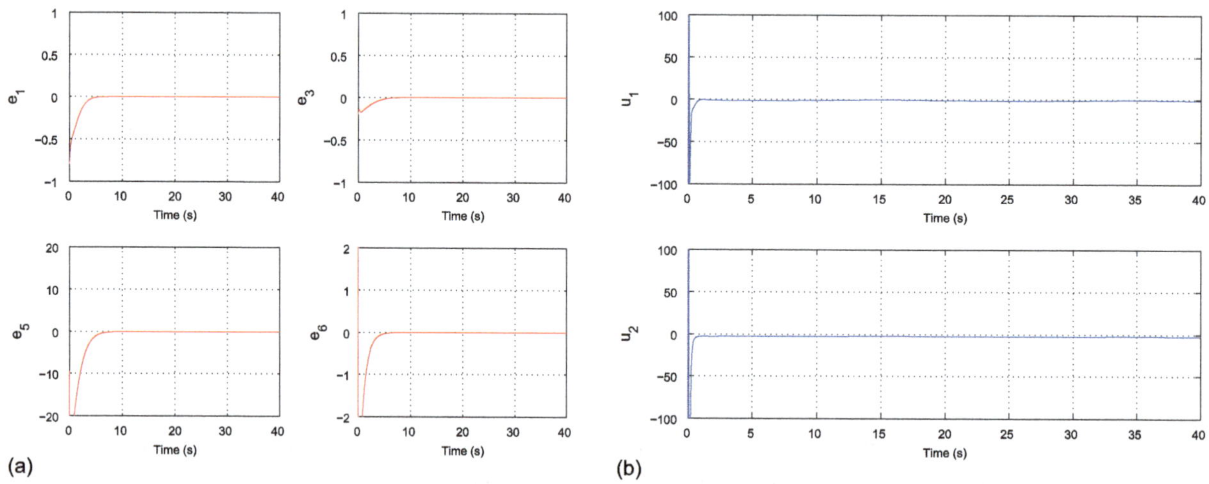

Figure 2.33 Control for the 2-DOF electrohydraulic robot when tracking setpoint 7: (a) tracking error of state variables x_1, x_2, x_5, x_6 (red line); (b) variation of the control inputs u_1 to u_2 (blue line)

subsystems with the individual subsystems to satisfy the differential flatness property. The latter class comprises systems in the strict-feedback (backstepping integral) form. The global stability properties of the proposed multi-loop flatness-based control method are proven with two different manners: (i) by demonstrating that the tracking error converges asymptotically to zero for all state variables of the controlled mechatronic systems, (ii) by providing Lyapunov stability analysis for the associated control loops in which it is shown that the first derivative of the considered Lyapunov functions is always negative and the Lyapunov function converges asymptotically to zero.

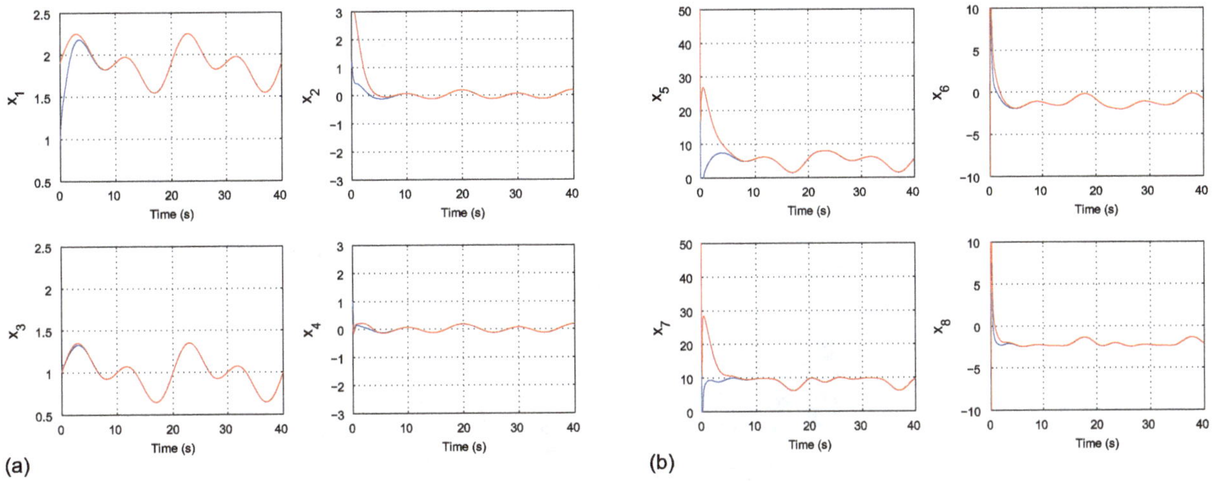

Figure 2.34 Control for the 2-DOF electrohydraulic robot when tracking setpoint 8: (a) convergence of state variables x_1 to x_4 to their reference setpoints (red line: setpoint, blue line: real value); (b) convergence of state variables x_5 to x_8 to their reference setpoints (red line: setpoint, blue line: real value)

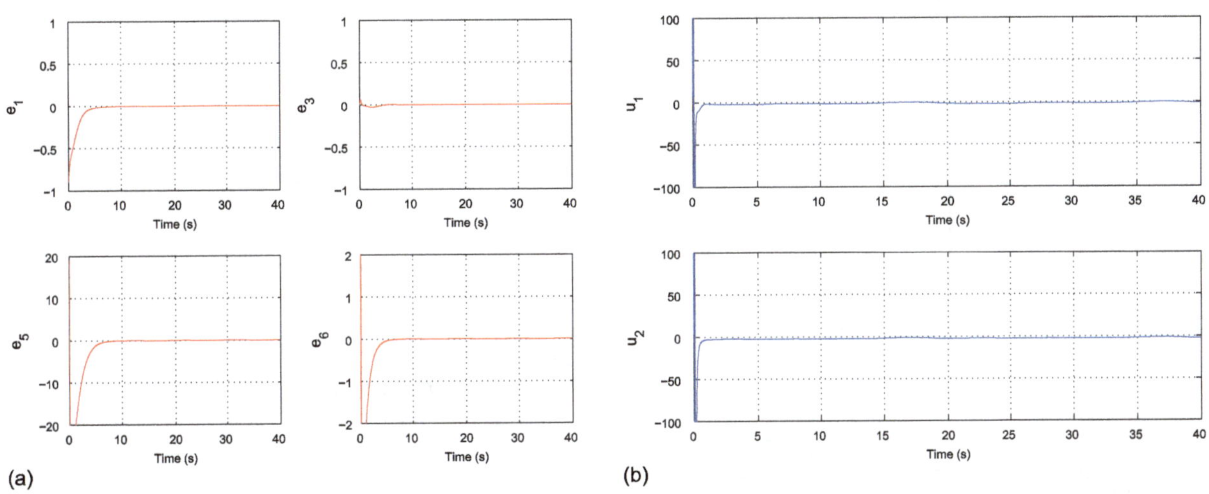

Figure 2.35 Control for the 2-DOF electrohydraulic robot when tracking setpoint 8: (a) tracking error of state variables x_1, x_2, x_5, x_6 (red line); (b) variation of the control inputs u_1 to u_2 (blue line)

2.3.2 *Dynamic model flexible-joint robot*

2.3.2.1 State-space model of the robotic system

Next, the diagram of a robotic link which is actuated by a flexible joint is shown in Figure 2.36. The torque of the electric motor is not applied directly on the link, but the link is actuated by the torque which is generated by the elastic joint, which is proportional to the difference between the turn angle of the link and the turn angle of the motor. The model of a multi-DOF flexible-joint robot is depicted in Figure 2.37. The main parameters of the dynamic model of the flexible-joint robot are outlined in Table 2.1. Under the assumption that the masses of the rigid links are concentrated at the links' end the dynamic model is given by [732,829,917,525,754]:

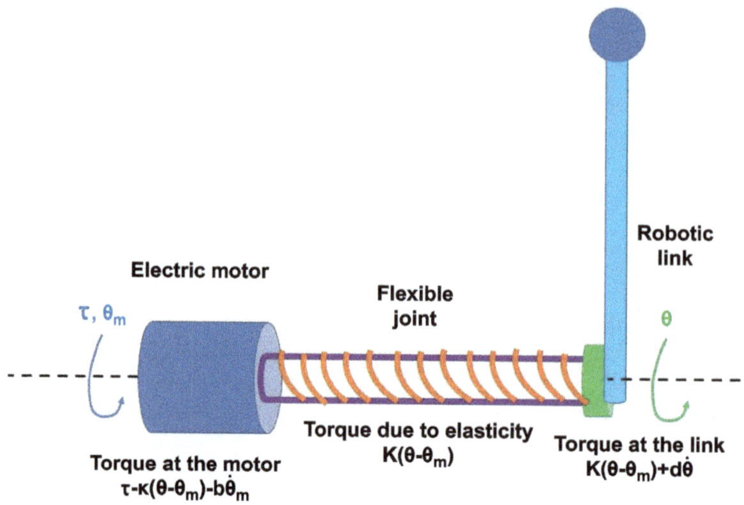

Figure 2.36 Diagram of a robotic link which is actuated by a flexible joint

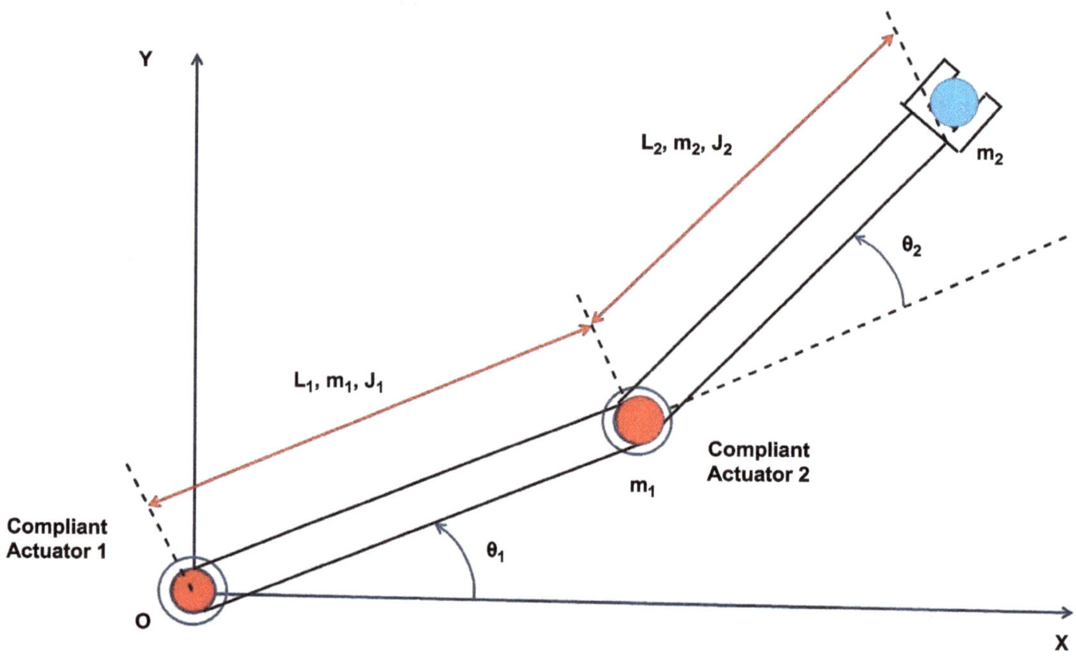

Figure 2.37 Diagram of the 2-link robotic manipulator with flexible joints

$$\begin{aligned}\tilde{M}(\theta)\ddot{\theta}+\tilde{C}(\theta,\dot{\theta})+\tilde{G}(\theta)&=\tilde{K}(\theta-\theta_m)+\tilde{D}\dot{\theta}\\ \tilde{J}\ddot{\theta}_m+\tilde{B}\dot{\theta}_m+\tilde{K}(\theta-\theta_m)&=\tau\end{aligned} \quad (2.103)$$

where $\theta\in R^{2\times 1}$ is the vector of the joints of the robotic manipulator, and $\theta_m\in R^{2\times 1}$ is the vector of the turn angles of the robot's actuators. Besides, $\tilde{M}(\theta)$ is the inertia matrix

$$\tilde{M}(\theta)=\begin{pmatrix}M_{11} & M_{12}\\ M_{21} & M_{22}\end{pmatrix} \quad (2.104)$$

Table 2.1 Parameters of the dynamic model of the flexible-joint robot

Parameter	Definition
θ_1, θ_2	turn angles of the joints of the robot
$\dot{\theta}_1, \dot{\theta}_2$	angular velocities of the joints of the robot
m_1, m_2	masses of the links of the robot
θ_{m1}, θ_{m2}	turn angles of the actuators
$\dot{\theta}_{m1}, \dot{\theta}_{m2}$	angular speeds of the actuators
J_1, J_2	moments of inertia of the actuators
$\tilde{M}$	inertia matrix of the robot
$\tilde{C}$	Coriolis forces vector of the robot
$\tilde{G}$	gravitational forces vector of the robot
$\tilde{D}$	damping coefficients matrix of the robot
$\tilde{K}$	elasticity coefficients matrix of the robot
$\tilde{J}$	inertia matrix of the actuators
B	damping coefficients matrix of the actuators
τ	control torques generated by the actuators

where $M_{11} = (m_1 + m_2)l_1^2 + m_1 l_2^2 + 2m_2 l_1 l_2 cos(\theta_2)$, $M_{12} = m_2 l_2^2 + m_2 l_1 l_2 cos(\theta_2)$, $M_{21} = m_2 l_2^2 + m_2 l_1 l_2 cos(\theta_2)$, and $M_{22} = m_2 l_2(l_1 + l_2) cos(\theta_2)$. Moreover, $C(\theta, \dot{\theta})$ is the Coriolis and centrifugal forces vector

$$\tilde{C}(\theta, \dot{\theta}) = \begin{pmatrix} -m_2 l_1^2 sin(\theta_2)\dot{\theta}_2^2 - 2m_2 l_1^2 sin(\theta_2)\dot{\theta}_1\dot{\theta}_2 \\ m_2 l_1^2 sin(\theta_2)\dot{\theta}_1^2 \end{pmatrix} \tag{2.105}$$

$\tilde{G}(\theta)$ is the gravitational forces vector

$$\tilde{G}(\theta) = \begin{pmatrix} (m_1 + m_2) g l_1 cos(\theta_1) + m_2 g l_2 cos(\theta_1 + \theta_2) \\ m_2 g l_2 cos(\theta_1 + \theta_2) \end{pmatrix} \tag{2.106}$$

$\tilde{K}$ is an elasticity coefficients matrix defining the coupling between the joints and the actuators, $\tilde{D}$ is a damping coefficients matrix for the robots' joints, while $\tilde{J}$ is an inertia matrix of the robots' actuators, and B is a friction coefficients' matrix again for the robots' actuators, which are given by:

$$\tilde{K} = \begin{pmatrix} k_1 & 0 \\ 0 & k_2 \end{pmatrix} \quad \tilde{D} = \begin{pmatrix} d_1 & 0 \\ 0 & d_2 \end{pmatrix} \quad \tilde{J} = \begin{pmatrix} J_1 & 0 \\ 0 & J_2 \end{pmatrix} \quad B = \begin{pmatrix} b_1 & 0 \\ 0 & b_2 \end{pmatrix} \tag{2.107}$$

The control input torques' vector is denoted as $\tau \in R^{2\times 1}$. Writing explicitly the differential equations of the two actuators, one obtains

$$\begin{aligned} J_1\ddot{\theta}_{m_1} + b_1\dot{\theta}_{m_1} + k_1(\theta_1 - \theta_{m_1}) = \tau_1 \\ J_2\ddot{\theta}_{m_2} + b_2\dot{\theta}_{m_2} + k_2(\theta_2 - \theta_{m_2}) = \tau_2 \end{aligned} \tag{2.108}$$

where τ_1 and τ_2 are the torques which drive the turn motion of the two actuators. The above two equations are also written in the following form:

$$\begin{pmatrix} \ddot{\theta}_1 \\ \ddot{\theta}_2 \end{pmatrix} = -\begin{pmatrix} M_{11} & M_{12} \\ M_{21} & M_{22} \end{pmatrix}^{-1} \begin{pmatrix} C_1(\theta, \dot{\theta}) + G_1(\theta) \\ C_2(\theta, \dot{\theta}) + G_2(\theta) \end{pmatrix} + \begin{pmatrix} M_{11} & M_{12} \\ M_{21} & M_{22} \end{pmatrix}^{-1} \begin{pmatrix} k_1(\theta_1 - \theta_{m_1}) + d_1\dot{\theta}_1 \\ k_2(\theta_2 - \theta_{m_2}) + d_2\dot{\theta}_2 \end{pmatrix} \tag{2.109}$$

$$\begin{pmatrix} \ddot{\theta}_{m_1} \\ \ddot{\theta}_{m_2} \end{pmatrix} = \begin{pmatrix} -\frac{b_1}{J_1}\dot{\theta}_{m_1} - \frac{k_1}{J_1}(\theta_{m_1} - \theta_1) \\ -\frac{b_2}{J_2}\dot{\theta}_{m_2} - \frac{k_2}{J_2}(\theta_{m_2} - \theta_2) \end{pmatrix} + \begin{pmatrix} \frac{1}{J_1} & 0 \\ 0 & \frac{1}{J_2} \end{pmatrix} \begin{pmatrix} \tau_1 \\ \tau_2 \end{pmatrix} \tag{2.110}$$

About the inverse inertia matrix of the robot, one has

$$\begin{pmatrix} M_{11} & M_{12} \\ M_{21} & M_{22} \end{pmatrix}^{-1} = \frac{1}{detM}\begin{pmatrix} M_{22} & -M_{12} \\ -M_{21} & M_{11} \end{pmatrix} \tag{2.111}$$

where the determinant is $detM = M_{11}M_{22} - M_{12}M_{21}$. Consequently, the aggregate dynamic model of the 2-link robot with flexible links is written as:

$$\begin{pmatrix} \ddot{\theta}_1 \\ \ddot{\theta}_2 \end{pmatrix} = \begin{pmatrix} -\frac{M_{22}}{detM}(C_1(\theta,\dot{\theta}) + G_1(\theta)) + \frac{M_{12}}{detM}(C_2(\theta,\dot{\theta}) + G_2(\theta)) + \\ \frac{M_{22}}{detM}(k_1(\theta_1 - \theta_{m_1}) + d_1\dot{\theta}_1) - \frac{M_{12}}{detM}(k_2(\theta_2 - \theta_{m_2}) + d_2\dot{\theta}_2) \\ \frac{M_{21}}{detM}(C_1(\theta,\dot{\theta}) + G_1(\theta)) - \frac{M_{11}}{detM}(C_2(\theta,\dot{\theta}) + G_2(\theta)) - \\ \frac{M_{21}}{detM}(k_1(\theta_1 - \theta_{m_1}) + d_1\dot{\theta}_1) + \frac{M_{11}}{detM}(k_2(\theta_2 - \theta_{m_2}) + d_2\dot{\theta}_2) \end{pmatrix} \tag{2.112}$$

$$\begin{pmatrix} \ddot{\theta}_{m_1} \\ \ddot{\theta}_{m_2} \end{pmatrix} = \begin{pmatrix} -\frac{b_1}{J_1\dot{\theta}_{m_1}} - \frac{k_1}{J_1}(\theta_{m_1} - \theta_1) + \frac{1}{J_1}\tau_1 \\ -\frac{b_2}{J_2\dot{\theta}_{m_2}} - \frac{k_2}{J_2}(\theta_{m_2} - \theta_2) + \frac{1}{J_2}\tau_2 \end{pmatrix} \tag{2.113}$$

Next, the state variables of the system are defined as follows: $x_1 = \theta_1$, $x_2 = \dot{\theta}_1$, $x_3 = \theta_2$, $x_4 = \dot{\theta}_2$, $x_5 = \theta_{m_1}$, $x_6 = \dot{\theta}_{m_1}$, $x_7 = \theta_{m_2}$, and $x_8 = \dot{\theta}_{m_2}$. The state vector of the system is $x = [x_1, x_2, x_3, x_4, x_5, x_6, x_7, x_8]^T$ and the control inputs vector is denoted as $u = [u_1, u_2]^T$ or $u = [\tau_1, \tau_2]^T$. Thus, the state-space model of the system is written as:

$$\begin{aligned} \dot{x}_1 &= x_2 \\ \dot{x}_2 &= -\frac{M_{22}}{detM}(C_1(x_1,x_2,x_3,x_4) + G_1(x_1,x_3)) + \frac{M_{12}}{detM}(C_2(x_1,x_2,x_3,x_4) + G_2(x_1,x_3)) + \\ &\quad \frac{M_{22}}{detM}(k_1(x_1 - x_5) + d_1x_2) - \frac{M_{12}}{detM}(k_2(x_3 - x_7) + d_2x_4) \\ \dot{x}_3 &= x_4 \\ \dot{x}_4 &= \frac{M_{21}}{detM}(C_1(x_1,x_2,x_3,x_4) + G_1(x_1,x_3)) - \frac{M_{11}}{detM}(C_2(x_1,x_2,x_3,x_4) + G_2(x_1,x_3)) - \\ &\quad \frac{M_{21}}{detM}(k_1(x_1 - x_5) + d_1x_2) + \frac{M_{11}}{detM}(k_2(x_3 - x_7) + d_2x_4) \\ \dot{x}_5 &= x_6 \\ \dot{x}_6 &= -\frac{b_1}{J_1}x_6 - \frac{k_1}{J_1}(x_5 - x_1) + \frac{1}{J_1}u_1 \\ \dot{x}_7 &= x_8 \\ \dot{x}_8 &= -\frac{b_2}{J_2}x_8 - \frac{k_2}{J_2}(x_7 - x_3) + \frac{1}{J_2}u_2 \end{aligned} \tag{2.114}$$

Besides, the following vector fields $f(x) \in R^{8\times 1}$ and $g(x) \in R^{8\times 2}$ are defined:

$$f(x) = \begin{pmatrix} x_2 \\ -\frac{M_{22}}{detM}(C_1(x_1,x_2,x_3,x_4)+G_1(x_1,x_3))+\frac{M_{12}}{detM}(C_2(x_1,x_2,x_3,x_4)+G_2(x_1,x_3))+ \\ \frac{M_{22}}{detM}(k_1(x_1-x_5)+d_1x_2)-\frac{M_{12}}{detM}(k_2(x_3-x_7)+d_2x_4) \\ x_4 \\ \frac{M_{21}}{detM}(C_1(x_1,x_2,x_3,x_4)+G_1(x_1,x_3))-\frac{M_{11}}{detM}(C_2(x_1,x_2,x_3,x_4)+G_2(x_1,x_3))- \\ \frac{M_{21}}{detM}(k_1(x_1-x_5)+d_1x_2)+\frac{M_{11}}{detM}(k_2(x_3-x_7)+d_2x_4) \\ x_6 \\ -\frac{b_1}{J_1}x_6-\frac{k_1}{J_1}(x_5-x_1) \\ x_8 \\ -\frac{b_2}{J_2}x_8-\frac{k_2}{J_2}(x_7-x_3) \end{pmatrix} \tag{2.115}$$

$$g(x) = \begin{pmatrix} 0 & 0 & 0 & 0 & 0 & \frac{1}{J_1} & 0 & 0 \\ 0 & 0 & 0 & 0 & 0 & 0 & 0 & \frac{1}{J_2} \end{pmatrix}^T$$

and the robotic system with flexible joints is written in the following affine-in-the-input state-space form

$$\dot{x} = f(x) + g(x)u \tag{2.116}$$

where $x \in R^{8\times 1}$, $f(x) \in R^{8\times 1}$, $g(x) \in R^{8\times 2}$, and $u \in R^{2\times 1}$.

2.3.2.2 Differential flatness properties of the multi-DOF robot with flexible joints

It will be proven that the dynamic model of the multi-DOF robot with flexible joints is differentially flat. The flat outputs vector of the system is taken to be $y = [x_1, x_3]^T$. A dynamical system is considered to be differentially flat if the following two properties hold: (i) all state variables and the control inputs of the robotic system can be written as differential functions of a specific subset of its state variables, the latter standing for the flat outputs of the robotic system, (ii) the flat outputs of the robotic system are differentially independent, which means that they are not connected between them through a relation in the form of a homogeneous linear differential equation [711].

From the first row of the state-space model, one has

$$x_2 = \dot{x}_1 \Rightarrow x_2 = h_2(y, \dot{y}) \tag{2.117}$$

Besides, from the third row of the state-space model, one has

$$x_4 = \dot{x}_3 \Rightarrow x_4 = h_4(y, \dot{y}) \tag{2.118}$$

The above two equations signify that state variables x_2 and x_4 are differential functions of the system's flat outputs. Besides, it holds that the terms M_{11}, M_{12}, M_{21}, M_{22}, $detM$, C_1, C_2, G_1 and G_2 are differential functions of the flat outputs of the system $y_1 = x_3$ and $y_2 = x_3$.

Next, from the 2nd and 4th row of the state-space model of the system, one obtains a system of equations for state variables x_5 and x_7:

$$\begin{aligned} &-\frac{M_{22}}{detM}k_1x_5+\frac{M_{12}}{detM}k_2x_7 = \\ &\dot{x}_2+\frac{M_{22}}{detM}(C_1+G_1)-\frac{M_{12}}{detM}(C_2+G_2)-\frac{M_{22}}{detM}(k_1x_1+d_1x_2)+\frac{M_{12}}{detM}(k_2x_3+d_2x_4) \end{aligned} \tag{2.119}$$

$$\begin{aligned} &\frac{M_{21}}{detM}k_1x_5-\frac{M_{11}}{detM}k_2x_7 = \\ &\dot{x}_4-\frac{M_{21}}{detM}(C_1+G_1)+\frac{M_{11}}{detM}(C_2+G_2)+\frac{M_{21}}{detM}(k_1x_1+d_1x_2)-\frac{M_{11}}{detM}(k_2x_3+d_2x_4) \end{aligned} \tag{2.120}$$

The previous two equations are also written in matrix form:

$$\begin{pmatrix} -\frac{M_{22}}{detM}k_1 & \frac{M_{12}}{detM}k_2 \\ \frac{M_{21}}{detM}k_1 & -\frac{M_{11}}{detM}k_2 \end{pmatrix} \begin{pmatrix} x_5 \\ x_7 \end{pmatrix} = \begin{pmatrix} \dot{x}_2 + \frac{M_{22}}{detM}(C_1+G_1) - \frac{M_{12}}{detM}(C_2+G_2) - \frac{M_{22}}{detM}(k_1x_1+d_1x_2) + \frac{M_{12}}{detM}(k_2x_3+d_2x_4) \\ \dot{x}_4 - \frac{M_{21}}{detM}(C_1+G_1) + \frac{M_{11}}{detM}(C_2+G_2) + \frac{M_{21}}{detM}(k_1x_1+d_1x_2) - \frac{M_{11}}{detM}(k_2x_3+d_2x_4) \end{pmatrix} \tag{2.121}$$

or equivalently

$$\begin{pmatrix} x_5 \\ x_7 \end{pmatrix} = \begin{pmatrix} -\frac{M_{22}}{detM}k_1 & \frac{M_{12}}{detM}k_2 \\ \frac{M_{21}}{detM}k_1 & -\frac{M_{11}}{detM}k_2 \end{pmatrix}^{-1} \cdot \begin{pmatrix} \dot{x}_2 + \frac{M_{22}}{detM}(C_1+G_1) - \frac{M_{12}}{detM}(C_2+G_2) - \frac{M_{22}}{detM}(k_1x_1+d_1x_2) + \frac{M_{12}}{detM}(k_2x_3+d_2x_4) \\ \dot{x}_4 - \frac{M_{21}}{detM}(C_1+G_1) + \frac{M_{11}}{detM}(C_2+G_2) + \frac{M_{21}}{detM}(k_1x_1+d_1x_2) - \frac{M_{11}}{detM}(k_2x_3+d_2x_4) \end{pmatrix} \tag{2.122}$$

The above equation demonstrates that state variables x_5 and x_7 can be written as differential functions of the flat outputs of the system. This is also written concisely as

$$x_5 = h_5(y,\dot{y}) \quad x_7 = h_7(y,\dot{y}) \tag{2.123}$$

Additionally, from the fifth row of the state-space model of the flexible joints robot, one has

$$x_6 = \dot{x}_5 \Rightarrow x_6 = h_6(y,\dot{y},\ddot{y}) \tag{2.124}$$

Moreover, from the seventh row of the state-space model of the flexible joints robot, one has

$$x_8 = \dot{x}_7 \Rightarrow x_8 = h_6(y,\dot{y},\ddot{y}) \tag{2.125}$$

The above two equations signify that state variables x_6 and x_8 are also differential functions of the flat outputs of the system. Furthermore, from the sixth row of the state-space model, one has

$$\begin{aligned} u_1 &= J_1[\dot{x}_6 + \tfrac{b_1}{J_1}x_6 + \tfrac{k_1}{J_1}(x_5 - x_1)] \Rightarrow \\ u_1 &= h_a(y,\dot{y},\ddot{y},y^{(3)}) \end{aligned} \tag{2.126}$$

Finally, from the eighth row of the state-space model, one has

$$\begin{aligned} u_2 &= J_2[\dot{x}_8 + \tfrac{b_2}{J_2}x_8 + \tfrac{k_2}{J_2}(x_7 - x_3)] \Rightarrow \\ u_2 &= h_b(y,\dot{y},\ddot{y},y^{(3)}) \end{aligned} \tag{2.127}$$

Consequently, it has been shown that the control inputs of the flexible-joint robot are also differential functions of its flat outputs. Besides, the flat outputs of the flexible-joint robot are differentially independent which means that they are not corrected between them through an ordinary differential equation (see Appendix). As a result of the above, it is confirmed that the dynamic model of the 2-DOF robotic manipulator with flexible joints is a differentially flat system.

2.3.3 *Flatness-based control in successive loops for the flexible-joint robot*

The new state vector of the 2-DOF robotic manipulator with flexible joints is $z = [z_1, z_2, z_3, z_4, z_5, z_6, z_7, z_8]^T$ or $z = [x_1, x_3, x_2, x_4, x_5, x_7, x_6, x_8]^T$. Using this new definition of state variables the dynamic model of the flexible-joints robot comprises the following equations

$$\dot{z}_1 = z_3 \tag{2.128}$$

$$\dot{z}_2 = z_4 \tag{2.129}$$

$$\dot{z}_3 = -\tfrac{M_{22}}{detM}(C_1(z_1,z_2,z_3,z_4)+G_1(z_1,z_2)) + \tfrac{M_{12}}{detM}(C_2(z_1,z_2,z_3,z_4)+G_2(z_1,z_2)) + \tfrac{M_{22}}{detM}(k_1(z_1-z_5)+d_1z_3) - \tfrac{M_{12}}{detM}(k_2(z_2-z_6)+d_2z_4) \tag{2.130}$$

$$\dot{z}_4 = -\tfrac{M_{21}}{detM}(C_1(z_1,z_2,z_3,z_4)+G_1(z_1,z_2)) - \tfrac{M_{11}}{detM}(C_2(z_1,z_2,z_3,z_4)+G_2(z_1,z_2)) - \tfrac{M_{21}}{detM}(k_1(z_1-z_5)+d_1z_3) + \tfrac{M_{11}}{detM}(k_2(z_2-z_6)+d_2z_4) \tag{2.131}$$

$$\dot{z}_5 = z_7 \tag{2.132}$$

$$\dot{z}_6 = z_8 \tag{2.133}$$

$$\dot{z}_7 = -\tfrac{b_1}{J_1}z_7 - \tfrac{k_1}{J_1}(z_5-z_1) + \tfrac{1}{J_1}u_1 \tag{2.134}$$

$$\dot{z}_8 = -\tfrac{b_2}{J_2}z_8 - \tfrac{k_2}{J_2}(z_6-z_2) + \tfrac{1}{J_2}u_2 \tag{2.135}$$

The following subsystems, submatrices, and subvectors are defined in the dynamic model of the flexible-joints robot:

First subsystem consisting of the first and third row of the state-space model, in which $v_1 = [z_3.z_4]^T$ is taken to be the virtual control input

$$z_{1,2} = \begin{pmatrix} z_1 \\ z_2 \end{pmatrix} \quad f_{1,2} = \begin{pmatrix} 0 \\ 0 \end{pmatrix} \quad g_{1,2} = \begin{pmatrix} 1 & 0 \\ 0 & 1 \end{pmatrix} \quad v_1 = \begin{pmatrix} z_3 \\ z_4 \end{pmatrix} \tag{2.136}$$

Second subsystem consisting of the second and fourth row of the state-space model, in which $v_2 = [z_5, z_6]^T$ is taken to be the virtual control input

$$z_{3,4} = \begin{pmatrix} z_3 \\ z_4 \end{pmatrix} \quad g_{3,4} = \begin{pmatrix} -\tfrac{M_{22}}{detM}k_1 & \tfrac{M_{12}}{detM}k_2 \\ \tfrac{M_{21}}{detM}k_1 & -\tfrac{M_{12}}{detM}k_2 \end{pmatrix} \quad v_2 = \begin{pmatrix} z_5 \\ z_6 \end{pmatrix}$$

$$f_{3,4} = \begin{pmatrix} -\tfrac{M_{22}}{detM}(C_1(z_1,z_2,z_3,z_4)+G_1(z_1,z_2)) + \tfrac{M_{12}}{detM}(C_2(z_1,z_2,z_3,z_4)+G_2(z_1,z_2)) + \\ \tfrac{M_{22}}{detM}(k_1z_1+d_1z_3) - \tfrac{M_{12}}{detM}(k_2z_2+d_2z_4) \\ \\ -\tfrac{M_{21}}{detM}(C_1(z_1,z_2,z_3,z_4)+G_1(z_1,z_2)) - \tfrac{M_{11}}{detM}(C_2(z_1,z_2,z_3,z_4)+G_2(z_1,z_2)) - \\ \tfrac{M_{21}}{detM}(k_1z_1+d_1z_3) + \tfrac{M_{11}}{detM}(k_2z_2+d_2z_4) \end{pmatrix} \tag{2.137}$$

Third subsystem consisting of the fifth and seventh row of the state-space model, in which $v_3 = [z_7, z_8]^T$ is defined as a virtual control input

$$z_{5,6} = \begin{pmatrix} z_5 \\ z_6 \end{pmatrix} \quad f_{5,6} = \begin{pmatrix} 0 \\ 0 \end{pmatrix} \quad g_{5,6} = \begin{pmatrix} 1 & 0 \\ 0 & 1 \end{pmatrix} \quad v_3 = \begin{pmatrix} z_7 \\ z_8 \end{pmatrix} \tag{2.138}$$

Fourth subsystem consisting of the sixth and eighth row of the state-space model, $u = [u_1, u_2]^T$ is the real control inputs vector

$$z_{7,8} = \begin{pmatrix} z_7 \\ z_8 \end{pmatrix} \quad f_{7,8} = \begin{pmatrix} -\tfrac{b_1}{J_1}z_7 - \tfrac{k_1}{J_1}(z_5-z_1) \\ -\tfrac{b_2}{J_2}z_8 - \tfrac{k_2}{J_2}(z_6-z_2) \end{pmatrix} \quad g_{7,8} = \begin{pmatrix} \tfrac{1}{J_1} & 0 \\ 0 & \tfrac{1}{J_2} \end{pmatrix} \quad u = \begin{pmatrix} u_1 \\ u_2 \end{pmatrix} \tag{2.139}$$

Thus, the dynamic model of the flexible-joint robot is written as a sequence of subsystems

$$\dot{z}_{1,2} = f_{1,2}(z_{1,2}) + g_{1,2}(z_{1,2})v_1 \tag{2.140}$$

$$\dot{z}_{3,4} = f_{3,4}(z_{1,2}, z_{3,4}) + g_{3,4}(z_{1,2}, z_{3,4})v_2 \tag{2.141}$$

$$\dot{z}_{5,6} = f_{5,6}(z_{1,2}, z_{3,4}, z_{5,6}) + g_{5,6}(z_{1,2}, z_{3,4}, z_{5,6})v_3 \tag{2.142}$$

$$\dot{z}_{7,8} = f_{7,8}(z_{1,2}, z_{3,4}, z_{5,6}, z_{7,8}) + g_{7,8}(z_{1,2}, z_{3,4}, z_{5,6}, z_{7,8})u \tag{2.143}$$

It can be proven that each one of the subsystems of Eq. (2.140) to Eq. (2.143), if viewed independently, is a differentially flat subsystem.

In the subsystem of Eq. (2.140), the flat output is $y_1 = z_{1,2}$, and the virtual control input is v_1. By solving Eq. (2.140) for v_1

$$v_1 = g_{1,2}(z_{1,2})^{-1}[\dot{z}_{1,2} - f_{1,2}(z_{1,2})] \tag{2.144}$$

which signifies that v_1 is a differential function of the flat outputs, therefore the subsystem of Eq. (2.140) is differentially flat. The virtual control input v_1 becomes setpoint to the subsystem of Eq. (2.141) or $v_1 = z^d_{3,4}$.

In the subsystem of Eq. (2.141), the flat output is $y_2 = z_{3,4}$ and the virtual control input is v_2, while $z_{1,2}$ is viewed as a coefficients vector. By solving Eq. (2.141) for v_2,

$$v_2 = g_{3,4}(z_{1,2}, z_{3,4})^{-1}[\dot{z}_{3,4} - f_{3,4}(z_{1,2}, z_{3,4})] \tag{2.145}$$

which signifies that v_2 is a differential function of the flat outputs, therefore the subsystem of Eq. (2.141) is differentially flat. The virtual control input v_2 becomes setpoint to the subsystem of Eq. (2.142) or $v_2 = z^d_{5,6}$.

In the subsystem of Eq. (2.142), the flat output is $y_3 = z_{5,6}$ and the virtual control input is v_3, while $z_{1,2}$, $z_{3,4}$ are viewed as coefficients vectors. By solving Eq. (2.142) for v_3

$$v_3 = g_{5,6}(z_{1,2}, z_{3,4}, z_{5,6})^{-1}[\dot{z}_{5,6} - f_{5,6}(z_{1,2}, z_{3,4}, z_{5,6})] \tag{2.146}$$

which signifies that v_3 is a differential function of the flat outputs, therefore the subsystem of Eq. (2.142) is differentially flat. The virtual control input v_3 becomes setpoint to the subsystem of Eq. (2.143) or $v_3 = z^d_{7,8}$.

In the subsystem of Eq. (2.143), the flat output is $y_4 = z_{7,8}$ and the real control input is u, while $z_{1,2}$, $z_{3,4}$, $z_{5,6}$ are viewed as coefficients vectors. By solving Eq. (2.143) for u

$$u = g_{7,8}(z_{1,2}, z_{3,4}, z_{5,6}, z_{7,8})^{-1}[\dot{z}_{7,8} - f_{7,8}(z_{1,2}, z_{3,4}, z_{5,6}, z_{7,8})] \tag{2.147}$$

which signifies that u is a differential function of the flat outputs, therefore the subsystem of Eq. (2.143) is differentially flat.

Knowing that each one of the subsystems of Eq. (2.140) to Eq. (2.143) is differentially flat, it is concluded that these systems are input–output linearizable. Thus, each one of them can be controlled and stabilized by applying an inverse dynamics controller, as it is usually done for systems in the input–output linearized form.

The stabilizing feedback control for the subsystem of Eq. (2.140) is defined as

$$v_1 = g_{1,2}(z_{1,2})^{-1}[\dot{z}^d_{1,2} - f_{1,2}(z_{1,2}) - K_1(z_{1,2} - z^d_{1,2})] \tag{2.148}$$

where $K_1 \in R^{2\times 2}$, $K_1 > 0$ is a diagonal matrix with diagonal elements $k_{ii,1} > 0$, $i = 1, 2$.

The stabilizing feedback control for the subsystem of Eq. (2.141) is defined as

$$v_2 = g_{3,4}(z_{1,2}, z_{3,4})^{-1}[\dot{z}^d_{3,4} - f_{3,4}(z_{1,2}, z_{3,4}) - K_2(z_{3,4} - z^d_{3,4})] \tag{2.149}$$

where $K_2 \in R^{2\times 2}$, $K_2 > 0$ is a diagonal matrix with diagonal elements $k_{ii,2} > 0$, $i = 1, 2$.

The stabilizing feedback control for the subsystem of Eq. (2.142) is defined as

$$v_3 = g_{5,6}(z_{1,2}, z_{3,4}, z_{5,6})^{-1}[\dot{z}^d_{5,6} - f_{5,6}(z_{1,2}, z_{3,4}, z_{5,6}) - K_3(z_{5,6} - z^d_{5,6})] \tag{2.150}$$

where $K_3 \in R^{2\times 2}$, $K_3 > 0$ is a diagonal matrix with diagonal elements $k_{ii,3} > 0$, $i = 1, 2$.

The stabilizing feedback control for the subsystem of Eq. (2.143) is defined as

$$u = g_{7,8}(z_{1,2}, z_{3,4}, z_{5,6}, z_{7,8})^{-1}[\dot{z}^d_{7,8} - f_{7,8}(z_{1,2}, z_{3,4}, z_{5,6}, z_{7,8}) - K_4(z_{7,8} - z^d_{7,8})] \tag{2.151}$$

where $K_4 \in R^{2\times 2}$, $K_4 > 0$ is a diagonal matrix with diagonal elements $k_{ii,4} > 0$, $i = 1, 2$.

The following tracking error variables are defined next: $e_{1,2} = z_{1,2} - z^d_{1,2}$, $e_{3,4} = z_{3,4} - z^d_{3,4}$, $e_{5,6} = z_{5,6} - z^d_{5,6}$ and $e_{7,8} = z_{7,8} - z^d_{7,6}$. By substituting the control law of Eq. (2.148) into the subsystem of Eq. (2.140), one obtains the following tracking error dynamics:

$$\begin{aligned}&(\dot{z}_{1,2} - \dot{z}^d_{1,2}) + K_1(z_{1,2} - z^d_{1,2}) = 0 \Rightarrow \dot{e}_{1,2} + K_1 e_{1,2} = 0 \Rightarrow \\ &\quad lim_{t\to\infty} e_{1,2}(t) = 0 \Rightarrow lim_{t\to\infty} z_{1,2}(t) = z^d_{1,2}\end{aligned} \tag{2.152}$$

By substituting the control law of Eq. (2.149) into the subsystem of Eq. (2.141), one obtains the following tracking error dynamics:

$$\begin{aligned}&(\dot{z}_{3,4}-\dot{z}^d_{3,4})+K_2(z_{3,4}-z^d_{3,4})=0\Rightarrow\dot{e}_{3,4}+K_2e_{3,4}=0\Rightarrow\\&\qquad lim_{t\to\infty}e_{3,4}(t)=0\Rightarrow lim_{t\to\infty}z_{3,4}(t)=z^d_{3,4}\end{aligned}\tag{2.153}$$

By substituting the control law of Eq. (2.150) into the subsystem of Eq. (2.142), one obtains the following tracking error dynamics:

$$\begin{aligned}&(\dot{z}_{5,6}-\dot{z}^d_{5,6})+K_3(z_{5,6}-z^d_{5,6})=0\Rightarrow\dot{e}_{5,6}+K_3e_{5,6}=0\Rightarrow\\&\qquad lim_{t\to\infty}e_{5,6}(t)=0\Rightarrow lim_{t\to\infty}z_{5,6}(t)=z^d_{5,6}\end{aligned}\tag{2.154}$$

By substituting the control law of Eq. (2.151) into the subsystem of Eq. (2.143), one obtains the following tracking error dynamics:

$$\begin{aligned}&(\dot{z}_{7,8}-\dot{z}^d_{7,8})+K_4(z_{7,8}-z^d_{7,8})=0\Rightarrow\dot{e}_{7,8}+K_4e_{7,8}=0\Rightarrow\\&\qquad lim_{t\to\infty}e_{7,8}(t)=0\Rightarrow lim_{t\to\infty}z_{7,8}(t)=z^d_{7,8}\end{aligned}\tag{2.155}$$

Consequently, all state variables of the flexible joints robot converge to the associated setpoints and this mechatronic system is globally asymptotically stable.

Regarding the stabilization of the dynamics of the flexible-joint robot this is ensured as follows: In Eq. (2.151), there is a real control input variable u which is obtained after inverting the dynamics of the subsystem which is described in Eq. (2.143). Since u is a real control input Eq. (2.151) is exact and the equation between its left and right part definitely holds. In Eq. (2.150), there is a virtual control input variable v_3 which is obtained after inverting the dynamics of the subsystem which is described in Eq. (2.142). Since v_3 is a virtual control input Eq. (2.150) is true provided that Eq. (2.151) holds. In such a case the equation between the left and right part of Eq. (2.150) holds asymptotically. In Eq. (2.149), there is a virtual control input variable v_2 which is obtained after inverting the dynamics of the subsystem which is described in Eq. (2.141). Since v_2 is a virtual control input Eq. (2.149) is true provided that Eq. (2.150) holds. In such a case the equation between the left and right part of Eq. (2.149) holds asymptotically. In Eq. (2.148), there is a virtual control input variable v_1 which is obtained after inverting the dynamics of the subsystem which is described in Eq. (2.140). Since v_1 is a virtual control input Eq. (2.148) is true provided that Eq. (2.149) holds. In such a case the equation between the left and right part of Eq. (2.148) holds asymptotically. Consequently, by ensuring that Eq. (2.151) holds and by tracing the system's dynamics backward from its last subsystem to the first one, one sees that Eq. (2.150) holds, Eq. (2.149) also holds and finally Eq. (2.148) is also true. In terms of rates of convergence the control loop of Eq. (2.151) has to be faster than the control loop of Eq. (2.150), the control loop of Eq. (2.150) should be faster than the control loop of Eq. (2.149) and in turn the control loop of Eq. (2.149) has to be faster than the control loop of Eq. (2.148). This is achieved by assigning higher values to the positive feedback gains of Eq. (2.151) compared to the feedback gains of Eq. (2.150). Equivalently, the feedback gains of Eq. (2.150) should be assigned higher values than the feedback gains for Eq. (2.149). Moreover, the feedback gains of Eq. (2.149) should be assigned higher values than the feedback gains for Eq. (2.148).

Global stability of the flexible joints robot can be also demonstrated through Lyapunov stability analysis. The following Lyapunov function is defined:

$$V=\tfrac{1}{2}[e^T_{1,2}e_{1,2}+e^T_{3,4}e_{3,4}+e^T_{5,6}e_{5,6}+e^T_{7,8}e_{7,8}]\tag{2.156}$$

By differentiating in time, one obtains

$$\dot{V}=e^T_{1,2}\dot{e}_{1,2}+e^T_{3,4}\dot{e}_{3,4}+e^T_{5,6}\dot{e}_{5,6}+e^T_{7,8}\dot{e}_{7,8}\tag{2.157}$$

Next, by substituting the tracking error dynamics of Eq. (2.152) to Eq. (2.155) in Eq. (2.157), one obtains

$$\dot{V}=e^T_{1,2}(-K_1e_{1,2})+e^T_{3,4}(-K_2e_{3,4})+e^T_{5,6}(-K_3e_{5,6})+e^T_{7,8}(-K_4e_{7,8})\tag{2.158}$$

which finally gives

$$\dot{V} = -[e_{1,2}^T K_1 e_{1,2} + e_{3,4}^T K_2 e_{3,4} + e_{5,6}^T K_3 e_{5,6} + e_{7,8}^T K_4 e_{7,8}] \\ \Rightarrow \dot{V} < 0 \, \forall \, e_{1,2} \neq 0, \; e_{3,4} \neq 0, \; e_{5,6} \neq 0, \; e_{7,8} \neq 0, \tag{2.159}$$

while $\dot{V} = 0$ if and only if $e_{1,2} = 0$, $e_{3,4} = 0$, $e_{5,6} = 0$ and $e_{7,8}$. Consequently, the Lyapunov function of the control loop of the flexible-joint robot is strictly diminishing and thus global asymptotic stability can be concluded.

An explicit demonstration of the exponential stabilization that is achieved by flatness-based control in successive loops is given next. The Lyapunov function of the control loop is written as:

$$V = \frac{1}{2}[\sum_{i=1}^{2} e_i^2 + \sum_{j=3}^{4} e_j^2 + \sum_{l=5}^{6} e_l^2 + \sum_{m=7}^{8} e_m^2] \tag{2.160}$$

where $e_i \; i = 1,2$ are the tracking errors for the state variables of the first subsystem of the flexible-joint robot, $e_j \, j = 3,4$ are the tracking errors for the state variables of the second subsystem of the flexible-joint robot, $e_l \; l = 5,6$ are the tracking errors for the state variables of the third subsystem of the flexible-joint robot, and $e_m \; m = 7,8$ are the tracking errors for the state variables of the first subsystem of the flexible-joint robot. Equivalently, the first-order time-derivative of the Lyapunov function is written as

$$\dot{V} = -[\sum_{i=1}^{2} k_{1,i} e_i^2 + \sum_{j=3}^{4} k_{2,j-2} e_j^2 + \sum_{l=5}^{6} k_{3,l-4} e_l^2 + \sum_{m=7}^{8} k_{4,m-6} e_m^2] \tag{2.161}$$

where $k_{1,i} > 0 \; i = 1,2$ are the diagonal elements of gain matrix K_1 and $k_{2,j-2} > 0 \, j = 3,4$ are the diagonal elements of gain matrix K_2, $k_{3,l-4} > 0 \; l = 5,6$ are the diagonal elements of gain matrix K_3 and $k_{1,m-6} > 0 \; m = 7,8$ are the diagonal elements of gain matrix K_4. By denoting the minimum of the above-noted elements of the feedback gain matrices as k_{min}, that is

$$k_{min} = min\{k_{1,i} : \; i = 1,2 \text{ and } k_{2,j-2} : \; j = 3,4 \text{ and } k_{3,l-4} : \; l = 5,6 \text{ and } k_{4,m-6} : \; m = 7,8\} \tag{2.162}$$

and using Eq. (2.161), one obtains that

$$\dot{V} \leq -k_{min}[\sum_{i=1}^{2} e_i^2 + \sum_{j=3}^{4} e_j^2 + \sum_{l=5}^{6} e_l^2 + \sum_{m=7}^{8} e_m^2] \\ \Rightarrow \dot{V} \leq -2k_{min} V \Rightarrow \dot{V} + 2k_{min} V \leq 0 \tag{2.163}$$

From Eq. (2.163), one can demonstrate the exponential convergence of the Lyapunov function V to 0.

2.3.4 Simulation tests

Indicative values about the parameters of the model of the flexible-joint robot are: $m_1 = 1.0kg$, $m_2 = 1.0kg$, $l_1 = 0.5m$, $l_2 = 0.5m$, $g = 10m/sec^2$, $k_1 = 2$, $k_2 = 2$, $d_1 = 0.01$, $d_2 = 0.01$, $b_1 = 0.02$, $b_2 = 0.02$, $J_1 = 0.8kg{\cdot}m^2$, $J_2 = 0.8kg{\cdot}m^2$. The sampling period was $T_s = 0.01sec$. Results about the tracking accuracy and the speed of convergence to setpoints of the successive-loops flatness-based control method, in the case of the 2-DOF flexible-joint robotic manipulator, are shown in Figures 2.38–2.53. It can be noticed, that under this control scheme, one achieves fast and precise tracking of reference setpoints for all state variables of the flexible-joint robot. It is noteworthy that through the stages of this method, one solves also the setpoints definition problem for all state variables of the flexible-joint robot. Actually, the selection of setpoints for state variables x_1, x_3 is unconstrained. On the other side by defining state variables x_2, x_4 as part of the virtual control input v_1, state variables x_5, x_6 as part of the virtual control input v_2, and finally state variables x_7, x_8 as part of the virtual control inputs vector v_3, one can find the setpoints for x_2, x_4, x_5, x_6, as well as x_7 and x_8 as functions of the setpoints for x_1, x_3. The speed of convergence of the state variables of the robot under flatness-based control implemented in successive loops is dependent on the selection of values for the diagonal gain matrices K_i, $i = 1, \cdots, 4$ of Eq. (2.148) to Eq. (2.151).

To elaborate on flatness-based control in successive loops for the flexible-joint robot the following tables are given: (i) Table 2.2 providing results about the accuracy of tracking of setpoints by the state variables of the flexible joint robot under an exact dynamic model, (ii) Table 2.3 providing results about the accuracy of tracking of setpoints by the state variables of the flexible joint robot under a model that is subject to disturbances (for instance change $\Delta\alpha\%$ in the moment of inertia J_2 of the system's dynamic model).

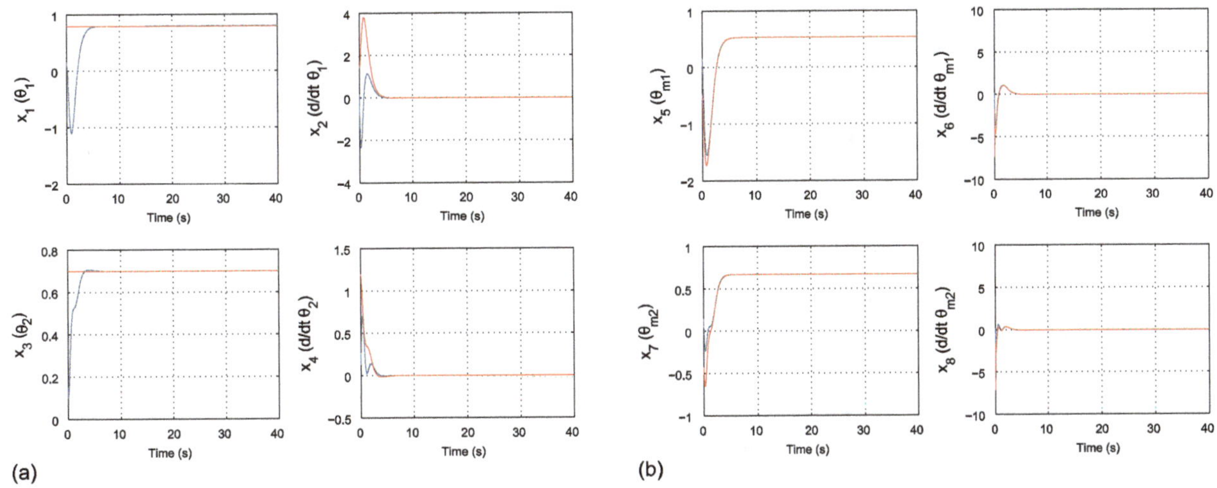

Figure 2.38 Tracking of setpoint 1 by the flexible-joint robot: (a) convergence of state variables x_1 to x_4 (blue lines) to the reference setpoints (red lines) and their KF-based estimation (green line); (b) convergence of state variables x_5 to x_8 (blue lines) to the reference setpoints (red lines) and their KF-based estimation (green line)

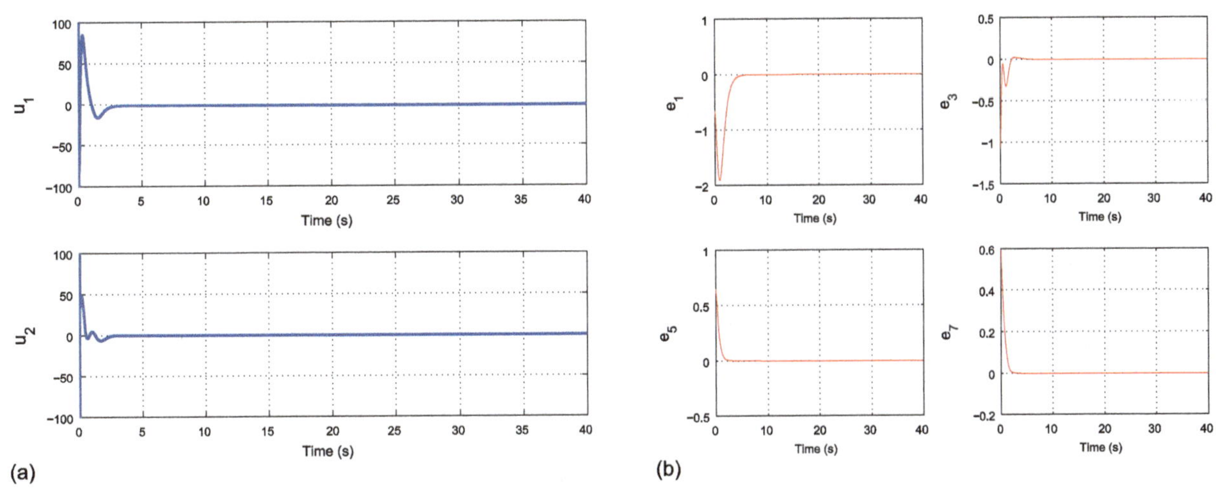

Figure 2.39 Tracking of setpoint 1 by the flexible-joint robot: (a) variations of the control inputs u_1 and u_2; (b) tracking error of state variables $x_1 = \theta_1$, $x_3 = \theta_2$, $x_5 = \theta_{m1}$, and $x_7 = \theta_{m2}$

Finally, the variations of the Lyapunov function of the control scheme for the flexible-joint robot are depicted in Figures 2.54–2.57. These diagrams show clearly convergence of the Lyapunov function of the flexible-joint robot to zero and global stability for the associated control loop.

2.4 Flatness-based control in successive loops for redundant robotic manipulators

2.4.1 Outline

The differential flatness property enables the transformation of the nonlinear system's dynamics in the linear canonical form [806,492,640,524,54]. The latter description is controllable and observable thus allowing to treat effectively

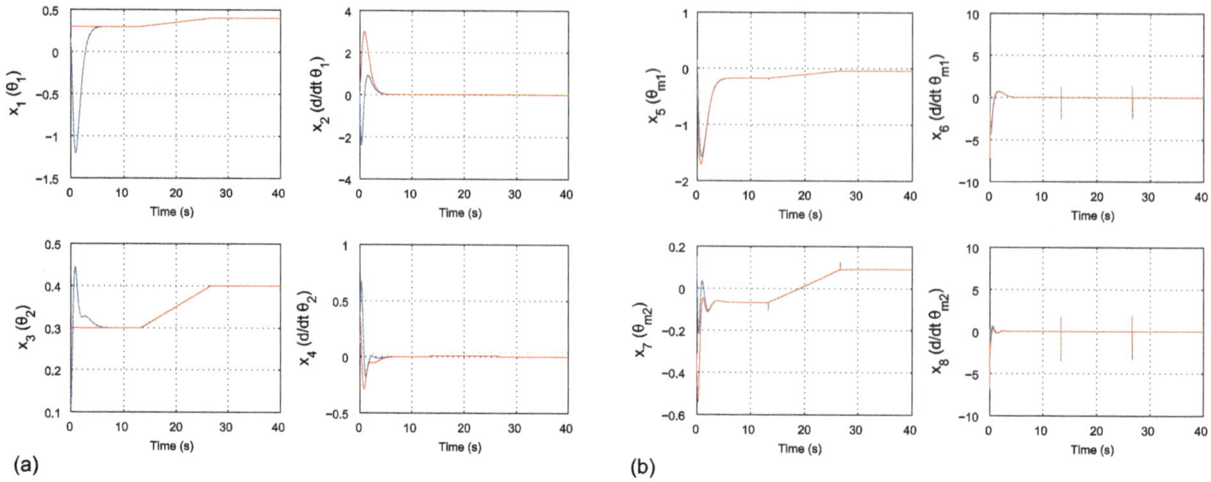

Figure 2.40 *Tracking of setpoint 2 by the flexible-joint robot: (a) convergence of state variables x_1 to x_4 (blue lines) to the reference setpoints (red lines) and their KF-based estimation (green line); (b) convergence of state variables x_5 to x_8 (blue lines) to the reference setpoints (red lines) and their KF-based estimation (green line)*

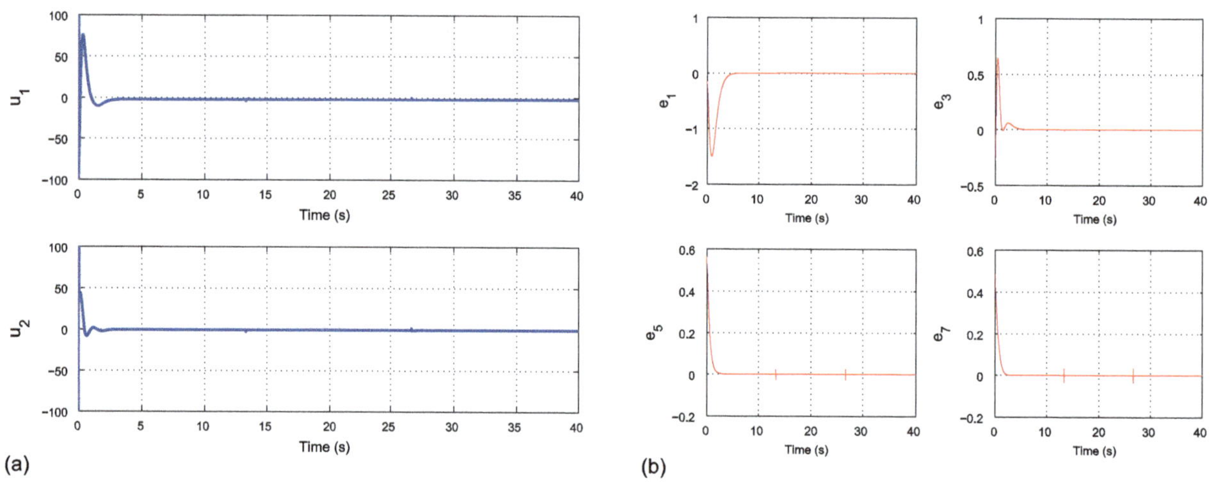

Figure 2.41 *Tracking of setpoint 2 by the flexible-joint robot: (a) variations of the control inputs u_1 and u_2; (b) tracking error of state variables $x_1 = \theta_1$, $x_3 = \theta_2$, $x_5 = \theta_{m1}$, and $x_7 = \theta_{m2}$*

control and estimation problems [439,705,60,706]. In this section, a successive loops approach is developed for controller design in nonlinear dynamical systems which exhibit the differential flatness property. The method makes use of the initial nonlinear model of the system and of its decomposition into a set of nonlinear subsystems for which the differential flatness property holds [712,720,714,716,718]. The proposed control method is directly applicable (but not limited) to nonlinear systems of the so-called triangular form, or to system's which can be transformed into such a form [711,721]. The state-space model of the nonlinear system is decomposed into subsystems, which satisfy differential flatness properties. For each subsystem of the state-space model a virtual control input is computed, capable of inverting the subsystem's dynamics and of eliminating the subsystem's tracking error. The control input that is actually applied to the nonlinear system is found from the last row of the state-space description. This control input incorporates in

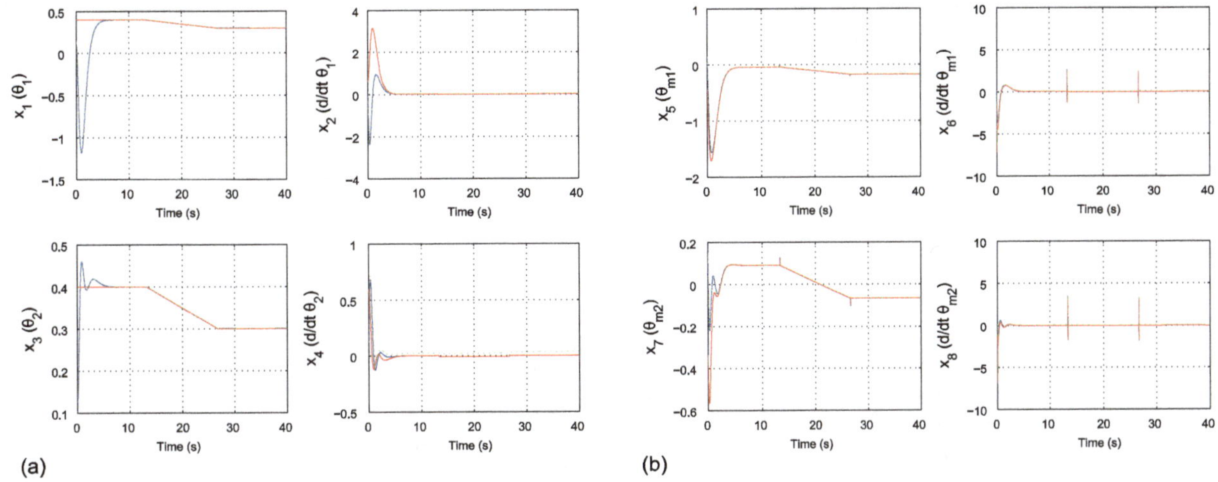

Figure 2.42 Tracking of setpoint 3 by the flexible-joint robot: (a) convergence of state variables x_1 to x_4 (blue lines) to the reference setpoints (red lines) and their KF-based estimation (green line); (b) convergence of state variables x_5 to x_8 (blue lines) to the reference setpoints (red lines) and their KF-based estimation (green line)

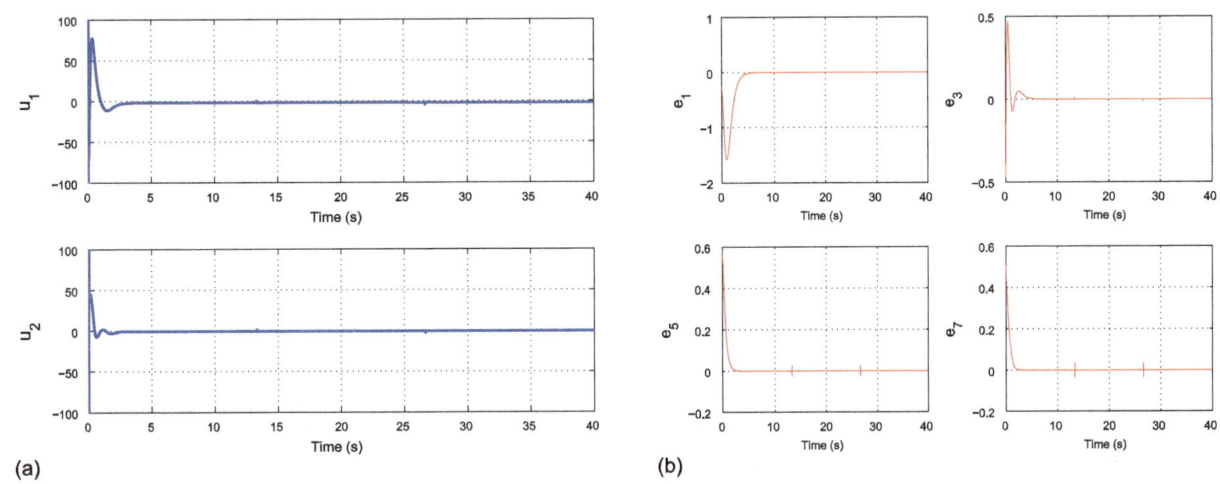

Figure 2.43 Tracking of setpoint 3 by the flexible-joint robot: (a) variations of the control inputs u_1 and u_2; (b) tracking error of state variables $x_1 = \theta_1$, $x_3 = \theta_2$, $x_5 = \theta_{m1}$, and $x_7 = \theta_{m2}$

a recursive manner all virtual control inputs which were computed for the individual subsystems associated with the initial state-space equation. The control input that should be applied to the nonlinear system so as to assure that all its state vector elements will converge to the desirable setpoints, is obtained at each iteration of the control algorithm, by tracing the subsystems of the state-space model backward.

The proposed method of flatness-based control in successive loops is applied to the control problem of the multivariable and to the nonlinear dynamics of multi-DOF redundant robotic manipulators [711,721]. The state-space model of these robotic systems is separated into two subsystems, which are connected between them in cascading loops. Each one of these subsystems can be viewed independently as a differentially flat system and control about it can be performed with inversion of its dynamics as in the case of input–output linearized flat systems. The state variables of the

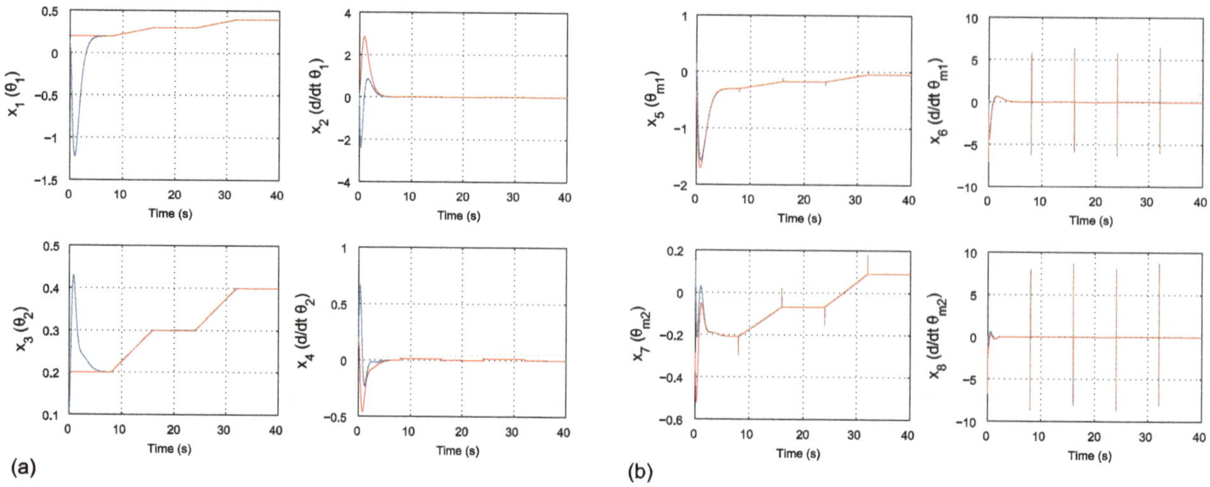

Figure 2.44 *Tracking of setpoint 4 by the flexible-joint robot: (a) convergence of state variables x_1 to x_4 (blue lines) to the reference setpoints (red lines) and their KF-based estimation (green line); (b) convergence of state variables x_5 to x_8 (blue lines) to the reference setpoints (red lines) and their KF-based estimation (green line)*

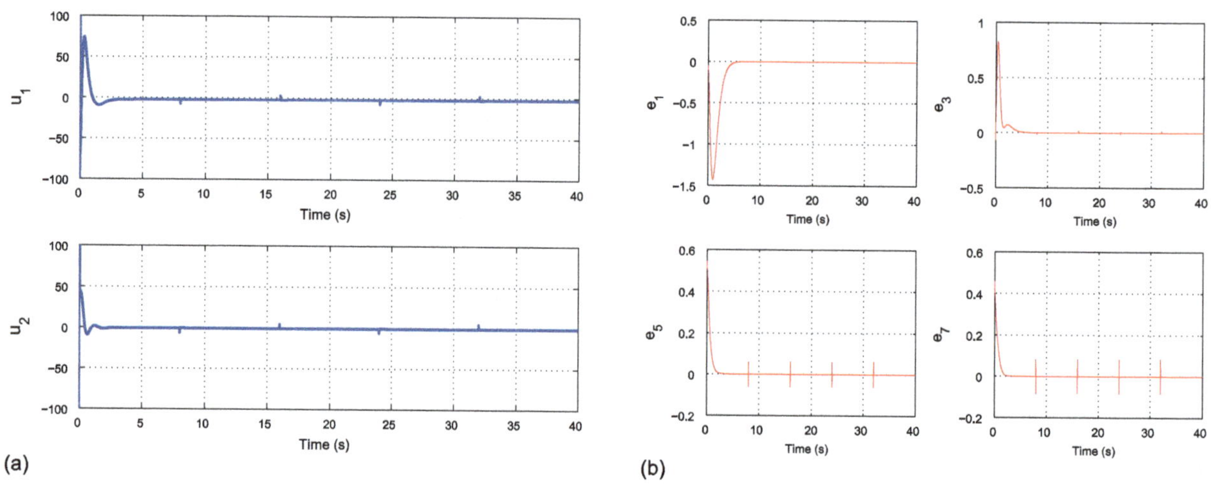

Figure 2.45 *Tracking of setpoint 4 by the flexible-joint robot: (a) variations of the control inputs u_1 and u_2; (b) tracking error of state variables $x_1 = \theta_1$, $x_3 = \theta_2$, $x_5 = \theta_{m1}$, and $x_7 = \theta_{m2}$*

second subsystem become virtual control inputs for the first subsystem. In turn, exogenous control inputs are applied to the first subsystem. The whole control method is implemented in two successive loops and its global stability properties are also proven through Lyapunov stability analysis.

The flatness-based control method in successive loops is advantageous when compared to global linearization-based control methods, such as Lie algebra-based control and flatness-based control through successive differentiation of the flat outputs and transformation of the state-space model of the system into canonical forms. This is because, in the new flatness-based control method in successive loops, there is neither need to perform changes of state variables nor complicated transformations of the state-space model of the controlled system. The control inputs that this method computes can be applied directly on the initial nonlinear state-space model of the system, without the need for inverse

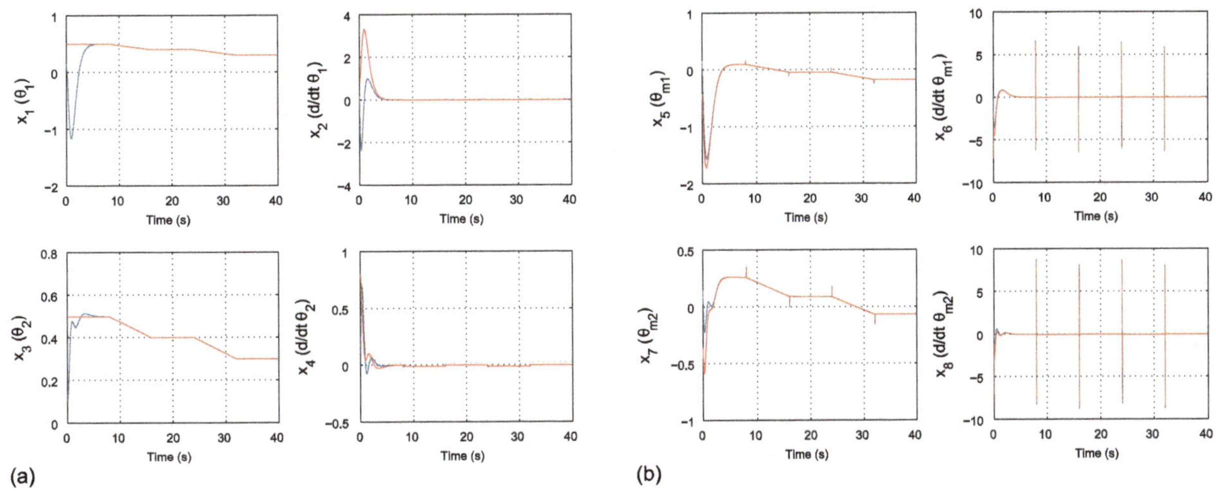

Figure 2.46 *Tracking of setpoint 5 by the flexible-joint robot: (a) convergence of state variables x_1 to x_4 (blue lines) to the reference setpoints (red lines) and their KF-based estimation (green line); (b) convergence of state variables x_5 to x_8 (blue lines) to the reference setpoints (red lines) and their KF-based estimation (green line)*

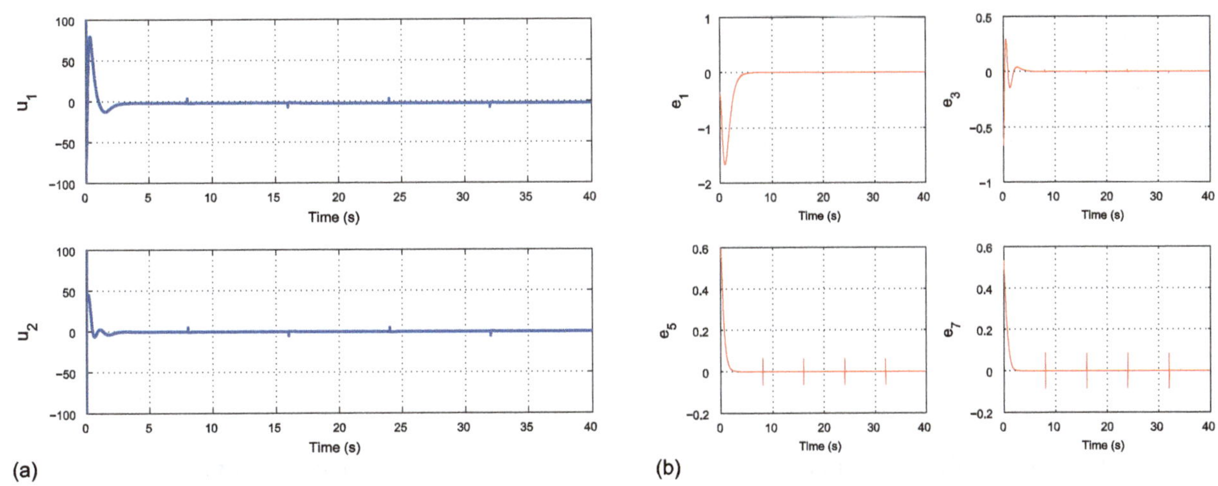

Figure 2.47 *Tracking of setpoint 5 by the flexible-joint robot: (a) variations of the control inputs u_1 and u_2; (b) tracking error of state variables $x_1 = \theta_1$, $x_3 = \theta_2$, $x_5 = \theta_{m1}$, and $x_7 = \theta_{m2}$*

transformations and consequently without the risk for coming against singularity (non-invertibility) cases. Flatness-based control in successive loops is suboptimal in the sense that it does not aim explicitly at the minimization of the amplitude of the control inputs. Despite its simplicity, flatness-based control in successive loops has nice tracking performance and achieves fast and accurate tracking of reference setpoints by the state variables of robotic manipulators and autonomous vehicles under moderate variations of the control inputs.

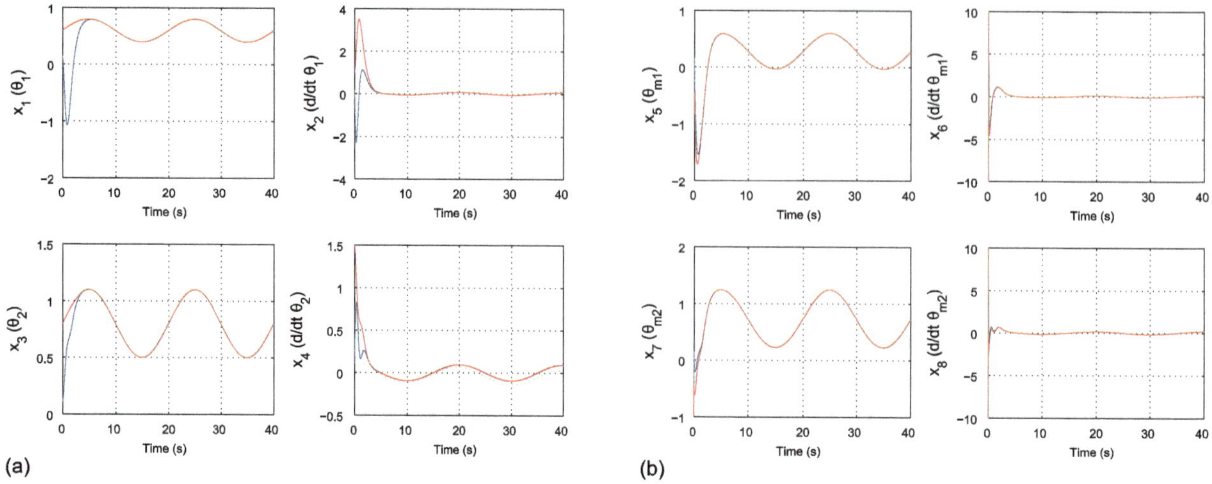

Figure 2.48 *Tracking of setpoint 6 by the flexible-joint robot: (a) convergence of state variables x_1 to x_4 (blue lines) to the reference setpoints (red lines) and their KF-based estimation (green line); (b) convergence of state variables x_5 to x_8 (blue lines) to the reference setpoints (red lines) and their KF-based estimation (green line)*

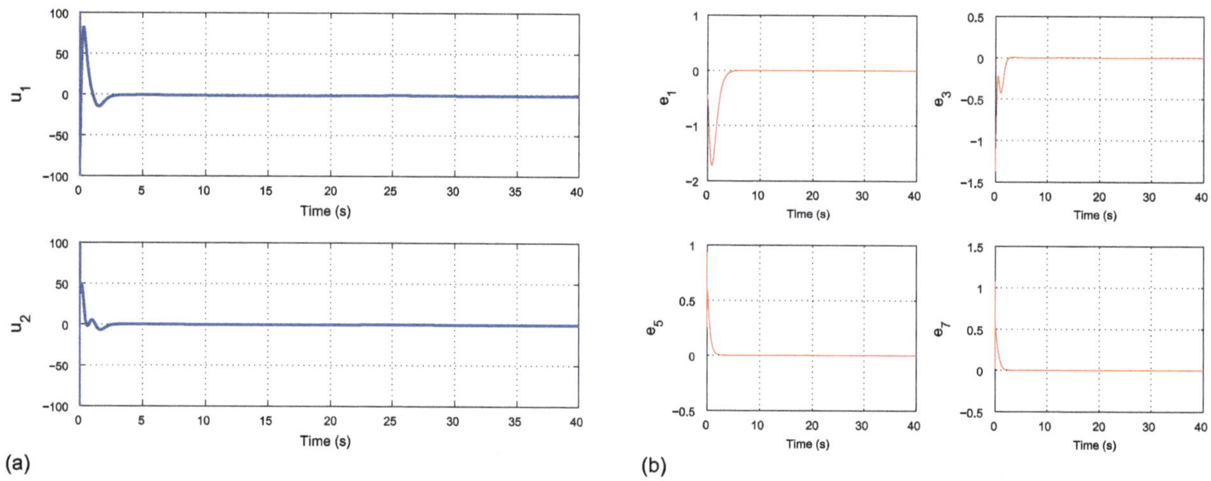

Figure 2.49 *Tracking of setpoint 6 by the flexible-joint robot: (a) variations of the control inputs u_1 and u_2; (b) tracking error of state variables $x_1 = \theta_1$, $x_3 = \theta_2$, $x_5 = \theta_{m1}$, and $x_7 = \theta_{m2}$*

2.4.2 *Dynamic model of the 3-DOF redundant robotic manipulators*

In the present section, the test-case for the flatness-based control method in successive loops is concerned with redundant planar robotic manipulators. Control of robotic manipulators with an end-effector that moves in a tasks space with lower dimension than the robot's degrees of freedom comes against the problem of redundancy [642,410]. Kinematic redundancy occurs when a manipulator has more degrees of freedom than the minimum number required to execute a task. When a manipulator is redundant, it is anticipated that the inverse kinematic problem admits multiple solutions. This means that for a constant position of the end-effector, several configurations of the arm's joints can be obtained [266–409]. Redundant robots achieve a versatile motion in the joints space and dexterity in tasks execution

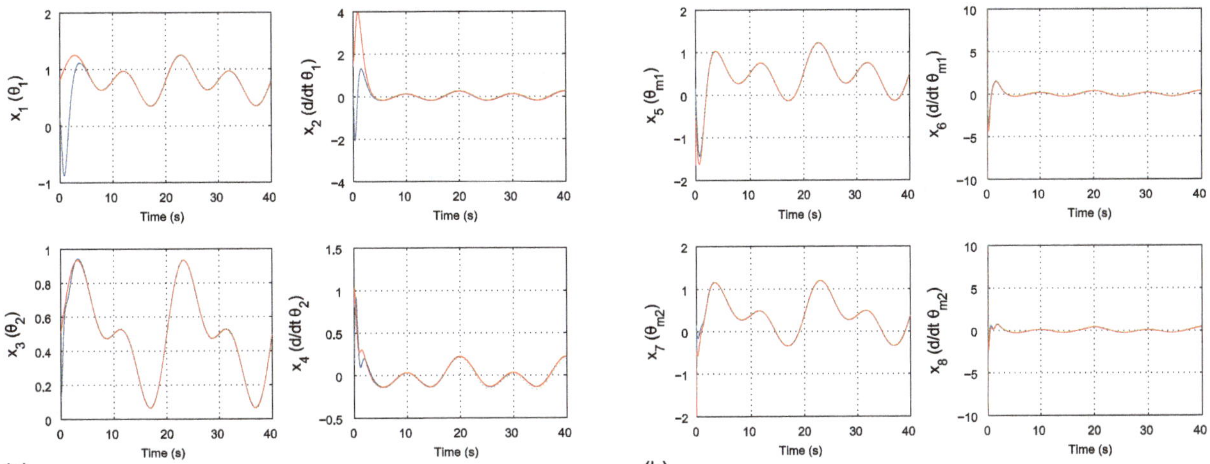

(a) (b)

Figure 2.50 Tracking of setpoint 7 by the flexible-joint robot: (a) convergence of state variables x_1 to x_4 (blue lines) to the reference setpoints (red lines) and their KF-based estimation (green line); (b) convergence of state variables x_5 to x_8 (blue lines) to the reference setpoints (red lines) and their KF-based estimation (green line)

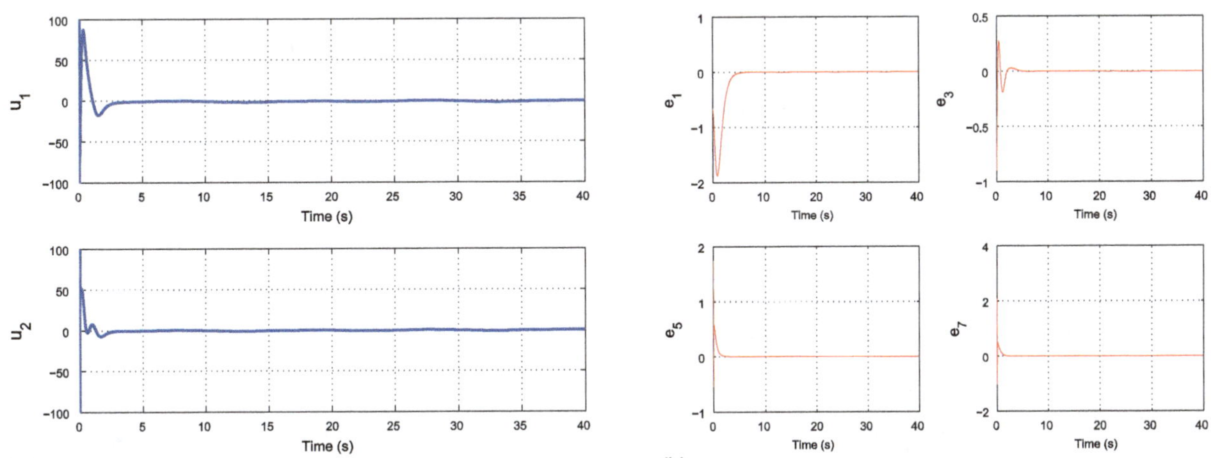

(a) (b)

Figure 2.51 Tracking of setpoint 7 by the flexible-joint robot: (a) variations of the control inputs u_1 and u_2; (b) tracking error of state variables $x_1 = \theta_1$, $x_3 = \theta_2$, $x_5 = \theta_{m1}$, and $x_7 = \theta_{m2}$

[654]. They can find use in robotic surgery as well as on several industrial tasks, because of offering multiple joints configurations that allow the end effector to reach easier the targeted position in the tasks' space [431].

Generally, redundancy specifies a robot's internal movement without affecting the trajectory of the robot's end effector and permits the robot to react in a better way to the environment that surrounds. This in turn signifies that the robot can achieve obstacles' avoidance and can keep moderate the torques developed by its actuators [458,763,683]. Control of redundant manipulators, which is also known as redundancy resolution, is to find a control action in the joints' space of a redundant robotic arm that leads to a desired motion of the end-effector in the tasks' space [908,397,954]. Several approaches have been proposed for the control of redundant robots [152]. The primary objective is to achieve

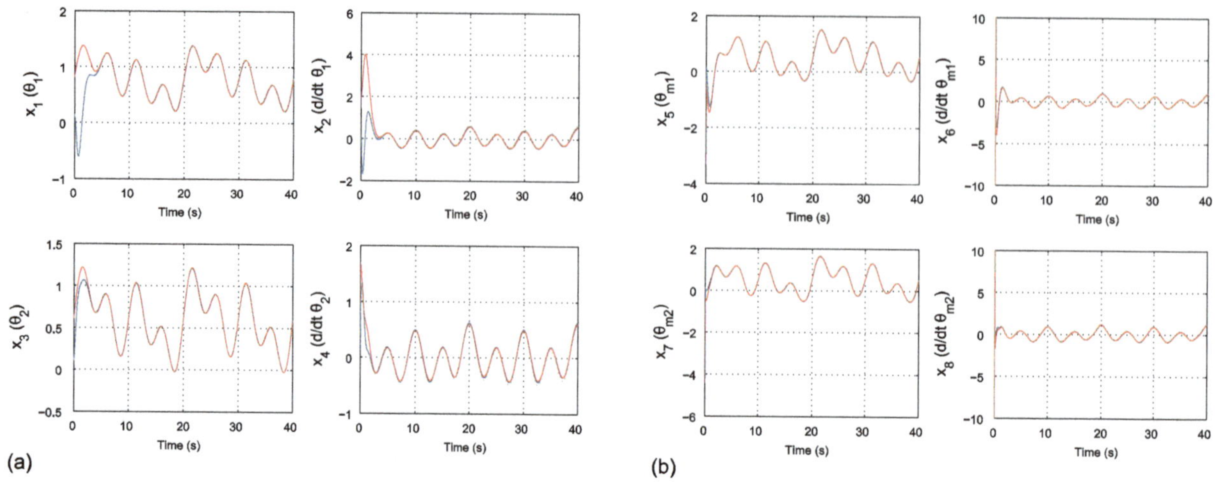

Figure 2.52 Tracking of setpoint 8 by the flexible-joint robot: (a) convergence of state variables x_1 to x_4 (blue lines) to the reference setpoints (red lines) and their KF-based estimation (green line); (b) convergence of state variables x_5 to x_8 (blue lines) to the reference setpoints (red lines) and their KF-based estimation (green line)

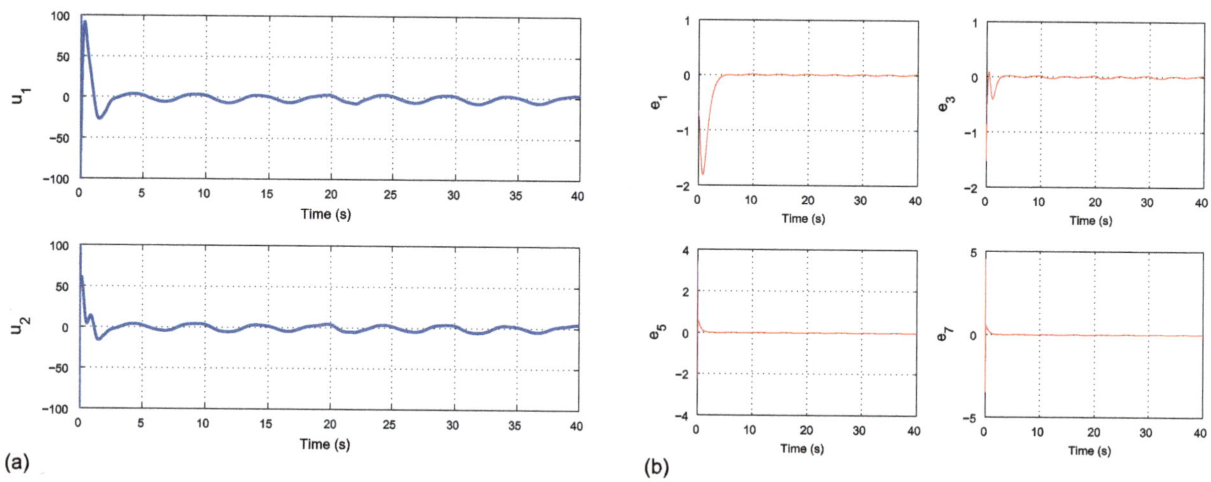

Figure 2.53 Tracking of setpoint 8 by the flexible-joint robot: (a) variations of the control inputs u_1 and u_2; (b) tracking error of state variables $x_1 = \theta_1$, $x_3 = \theta_2$, $x_5 = \theta_{m1}$, and $x_7 = \theta_{m2}$

stability properties for the redundant manipulator A related objective is the avoidance of singularities in the robot's inverse kinematic problem [942]. So far, the nonlinear optimal control problem for redundant manipulators has been little studied.

The diagram of the 3-DOF redundant planar robotic manipulator is given in Figure 2.58 The dynamic model of the 3-DOF redundant robotic manipulator is given by [732]

$$M(\theta)\ddot{\theta} + C(\theta, \dot{\theta}) + G(\theta) = \tau \tag{2.164}$$

Table 2.2 Tracking RMSE $\times 10^{-3}$ for the flexible-joint under exact model

	$RMSE_{x_1}$	$RMSE_{x_2}$	$RMSE_{x_3}$	$RMSE_{x_4}$	$RMSE_{x_5}$	$RMSE_{x_6}$	$RMSE_{x_7}$	$RMSE_{x_8}$
$test_1$	0.0640	0.0030	0.0295	0.0031	0.0002	0.0001	0.0001	0.0001
$test_2$	0.0706	0.0033	0.0313	0.0031	0.0001	0.0001	0.0001	0.0001
$test_3$	0.0584	0.0028	0.0365	0.0032	0.0001	0.0001	0.0001	0.0001
$test_4$	0.0977	0.0066	0.0168	0.0011	0.0002	0.0001	0.0001	0.0001
$test_5$	0.0312	0.0010	0.0500	0.0655	0.0004	0.0001	0.0001	0.0001
$test_6$	0.2380	0.1175	0.5839	0.2763	0.0032	0.0011	0.0052	0.0019
$test_7$	0.3839	0.3684	0.0764	0.0562	0.0249	0.0090	0.0242	0.0087
$test_8$	0.1991	0.7727	0.0718	0.4354	0.0979	0.0807	0.1181	0.0902

Table 2.3 Tracking RMSE $\times 10^{-4}$ for the flexible-joint robot in the case of disturbances

$\Delta a\%$	$RMSE_{x_1}$	$RMSE_{x_2}$	$RMSE_{x_3}$	$RMSE_{x_4}$	$RMSE_{x_5}$	$RMSE_{x_6}$	$RMSE_{x_7}$	$RMSE_{x_8}$
0%	0.0640	0.0030	0.0295	0.0031	0.0002	0.0001	0.0001	0.0001
10%	0.0804	0.0023	0.0785	0.0014	0.0002	0.0001	0.0001	0.0001
20%	0.1158	0.0026	0.1721	0.0103	0.0001	0.0001	0.0002	0.0007
30%	0.0720	0.0085	0.1183	0.0308	0.0001	0.0001	0.0004	0.0014
40%	0.1306	0.0102	0.2823	0.0355	0.0001	0.0001	0.0006	0.0022
50%	0.1734	0.0100	0.4113	0.0332	0.0001	0.0001	0.0007	0.0024
60%	0.2115	0.0216	0.5405	0.0709	0.0001	0.0001	0.0013	0.0045

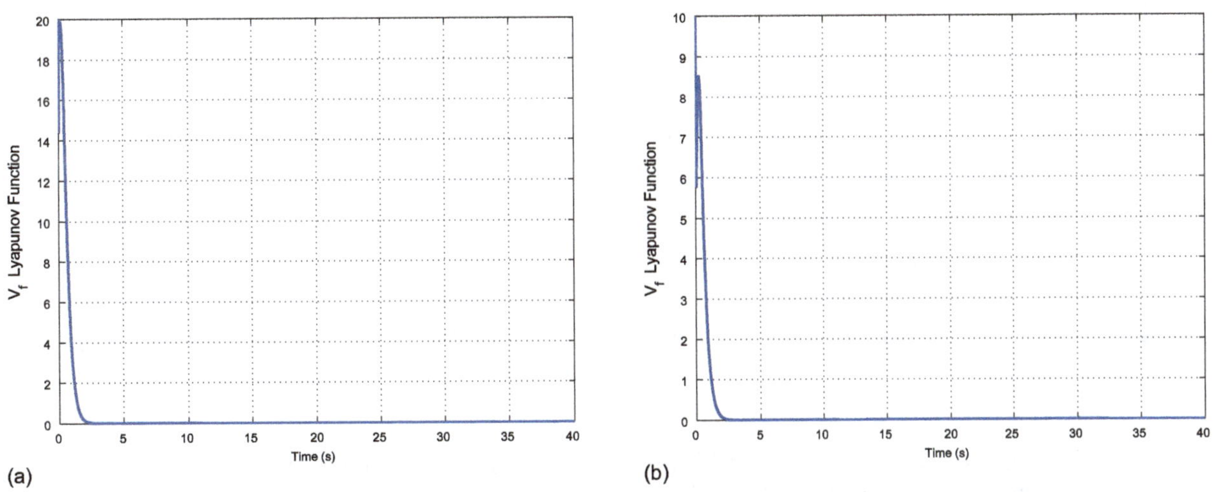

Figure 2.54 Lyapunov functions for multi-loop flatness-based control of the flexible-joint robot when tracking setpoints 1–2

where the robot's inertia matrix is given by

$$M(\theta) = \begin{pmatrix} m_{11} & m_{12} & m_{13} \\ m_{21} & m_{22} & m_{23} \\ m_{31} & m_{32} & m_{33} \end{pmatrix} \tag{2.165}$$

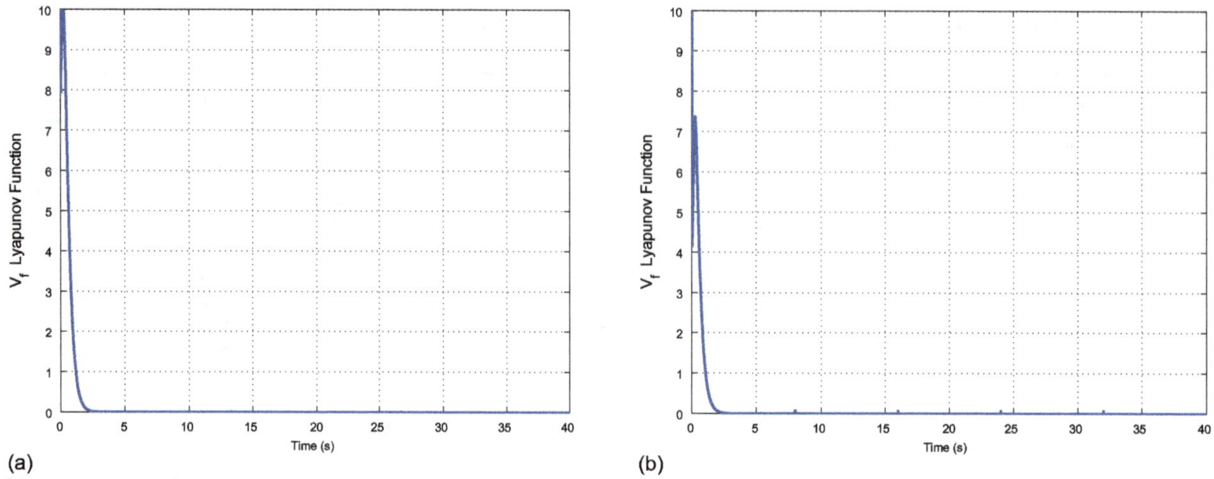

Figure 2.55 Lyapunov functions for multi-loop flatness-based control of the flexible-joint robot when tracking setpoints 3–4

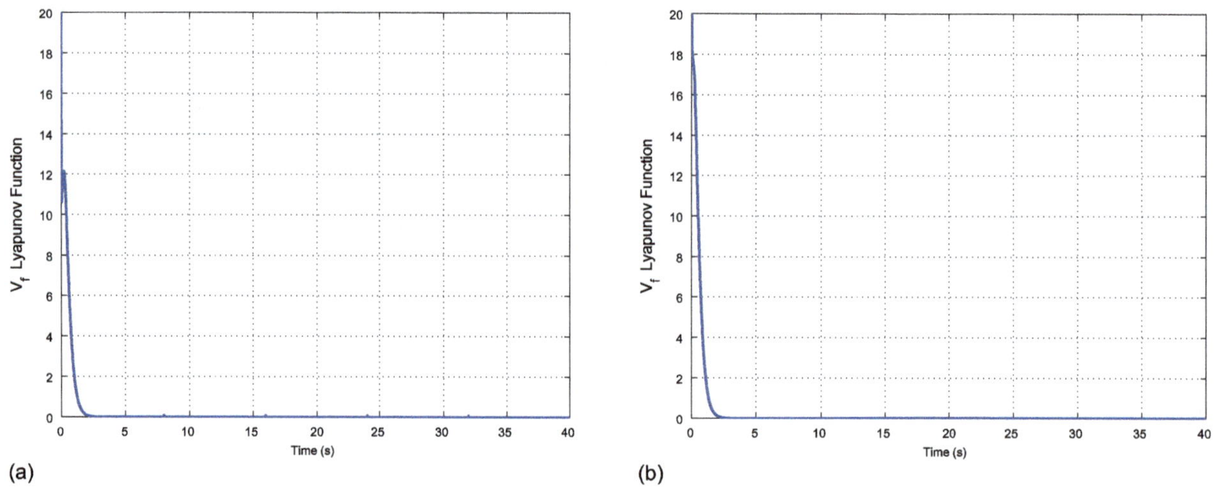

Figure 2.56 Lyapunov functions for multi-loop flatness-based control of the flexible-joint robot when tracking setpoints 5–6

with $m_{11} = a_1 + a_2 + a_4 + 2a_3cos(\theta_3) + 2a_5cos(\theta_2 + \theta_3) + 2a_6cos(\theta_3)$, $m_{12} = m_{21} = a_2 + a_4cos(\theta_2) + a_5cos(\theta_2 + \theta_3) + 2a_6cos(\theta_3)$, $m_{13} = m_{31} = a_4 + a_5cos(\theta_2 + \theta_3) + 2a_6cos(\theta_3)$, $m_{22} = a_2 + a_4 + a_6cos(\theta_3)$, $m_{23} = m_{32} = a_4 + a_6cos(\theta_3)$, $m_{33} = a_4$.

where $a_1 = m_1l_1^2 + I_1 + (m_1 + m_2)l_1^2$, $a_2 = I_2 + m_2l_2^2$, $a_3 = (m_2l_2 + m_3L_2)L_1$, $a_4 = I_3 + m_3l_3^2$, $a_5 = m_3l_3L_1$, $a_6 = m_3L_2l_3$.

with m_i, $i = 1, 2, 3$ to be the masses of the links, I_i, $i = 1, 2, 3$ to be the moments of inertia of the links, $L_,i = 1, 2, 3$ to be the lengths of the links, and l_i, $i = 1, 2, 3$ to be the distances between the centers of the links and their centers of gravity.

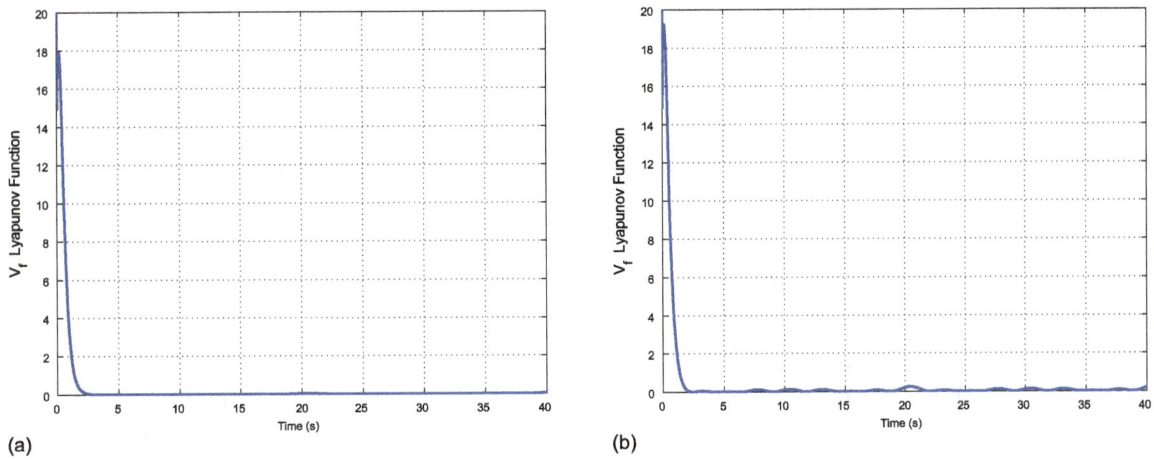

Figure 2.57 Lyapunov functions for multi-loop flatness-based control of the flexible-joint robot when tracking setpoints 7–8

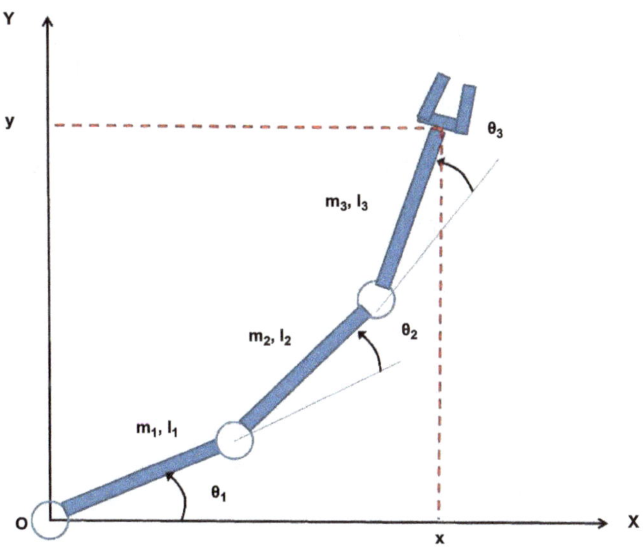

Figure 2.58 Diagram of the 3-DOF redundant planar robotic manipulator

About the Coriolis and centrifugal forces matrix, one has

$$C(\theta,\dot{\theta}) = \begin{pmatrix} -a_3(2\dot{\theta}_1+\dot{\theta}_2)\dot{\theta}_2 sin(\theta)_2 \\ -a_5(2\dot{\theta}_1+\dot{\theta}_2+\dot{\theta}_3)(\dot{\theta}_2+\dot{\theta}_3)sin(\theta_2+\theta_3) \\ -a_6(2\dot{\theta}_1+2\dot{\theta}_2+\dot{\theta}_3)\dot{\theta}_3 sin(\theta)_3 \\ \\ -a_3\dot{\theta}_1^2 sin(\theta_2)+a_5\dot{\theta}_1^2 sin(\theta_2+\theta_3) \\ -a_6(2\dot{\theta}_1+2\dot{\theta}_2+\dot{\theta}_3)\dot{\theta}_3 sin(\theta)_3 \\ \\ a_5\dot{\theta}_1^2 sin(\theta_2+\theta_3)+a_6(\dot{\theta}_1+\dot{\theta}_2)^2 sin(\theta_3) \end{pmatrix} \quad (2.166)$$

The gravitational forces matrix is taken to have zero elements because the manipulator is planar:

$$G(\theta) = \begin{pmatrix} 0 \\ 0 \\ 0 \end{pmatrix} \tag{2.167}$$

Considering next the sub-determinants of the inertia matrix M, the inverse matrix of M is given by

$$M^{-1}(\theta) = \frac{1}{detM} \begin{pmatrix} M_{11} & -M_{21} & M_{31} \\ -M_{12} & M_{22} & -M_{32} \\ M_{13} & -M_{23} & M_{33} \end{pmatrix} \tag{2.168}$$

Thus, about the dynamic model of the redundant robotic manipulator, one has

$$\begin{pmatrix} \ddot{\theta}_1 \\ \ddot{\theta}_2 \\ \ddot{\theta}_3 \end{pmatrix} = \frac{1}{detM} \begin{pmatrix} M_{11} & -M_{21} & M_{31} \\ -M_{12} & M_{22} & -M_{32} \\ M_{13} & -M_{23} & M_{33} \end{pmatrix} \left[\begin{pmatrix} C_1 \\ c_2 \\ C_5 \end{pmatrix} + \begin{pmatrix} G_1 \\ G_2 \\ G_3 \end{pmatrix} \right] + \frac{1}{detM} \begin{pmatrix} M_{11} & -M_{21} & M_{31} \\ -M_{12} & M_{22} & -M_{32} \\ M_{13} & -M_{23} & M_{33} \end{pmatrix} \begin{pmatrix} \tau_1 \\ \tau_2 \\ \tau_3 \end{pmatrix} \tag{2.169}$$

After intermediate operations, the dynamic model of the redundant robotic manipulator is written as

$$\begin{pmatrix} \ddot{\theta}_1 \\ \ddot{\theta}_2 \\ \ddot{\theta}_3 \end{pmatrix} = \begin{pmatrix} \frac{-M_{11}(C_1+G_1)+M_{21}(C_2+G_2)-M_{31}(C_3+G_3)}{detM} \\ \frac{M_{12}(C_1+G_1)-M_{22}(C_2+G_2)+M_{32}(C_3+G_3)}{detM} \\ \frac{-M_{13}(C_1+G_1)+M_{23}(C_2+G_2)-M_{33}(C_3+G_3)}{detM} \end{pmatrix} + \begin{pmatrix} \frac{M_{11}}{detM} & -\frac{M_{21}}{detM} & \frac{M_{31}}{detM} \\ -\frac{M_{12}}{detM} & \frac{M_{22}}{detM} & -\frac{M_{32}}{detM} \\ \frac{M_{13}}{detM} & -\frac{M_{23}}{detM} & \frac{M_{33}}{detM} \end{pmatrix} \begin{pmatrix} \tau_1 \\ \tau_2 \\ \tau_3 \end{pmatrix} \tag{2.170}$$

Next, by defining the state vector $x = [x_1, x_2, x_3, x_4, x_5, x_6]^T = [\theta_1, \theta_2, \theta_3, \dot{\theta}_1, \dot{\theta}_2, \dot{\theta}_3]^T$, one obtains the following state-space description:

$$\begin{aligned} \dot{x}_1 &= x_4 \\ \dot{x}_2 &= x_5 \\ \dot{x}_3 &= x_6 \\ \dot{x}_4 &= \frac{-M_{11}(C_1+G_1)+M_{21}(C_2+G_2)-M_{31}(C_3+G_3)}{detM} + \frac{M_{11}}{detM}\tau_1 - \frac{M_{21}}{setM}\tau_2 + \frac{M_{31}}{detM}\tau_3 \\ \dot{x}_5 &= \frac{M_{12}(C_1+G_1)-M_{22}(C_2+G_2)+M_{33}(C_3+G_3)}{detM} - \frac{M_{12}}{detM}\tau_1 + \frac{M_{22}}{detM}\tau_2 - \frac{M_{32}}{detM}\tau_3 \\ \dot{x}_6 &= \frac{-M_{13}(C_1+G_1)+M_{23}(C_2+G_2)-M_{33}(C_3+G_3)}{detM} + \frac{M_{13}}{detM}\tau_1 - \frac{M_{23}}{setM}\tau_2 + \frac{M_{33}}{detM}\tau_3 \end{aligned} \tag{2.171}$$

Equivalently, one has the following state-space description in matrix form:

$$\begin{pmatrix} \dot{x}_1 \\ \dot{x}_2 \\ \dot{x}_3 \\ \dot{x}_4 \\ \dot{x}_5 \\ \dot{x}_6 \end{pmatrix} = \begin{pmatrix} x_4 \\ x_5 \\ x_6 \\ \frac{-M_{11}(C_1+G_1)+M_{21}(C_2+G_2)-M_{31}(C_3+G_3)}{detM} \\ \frac{M_{12}(C_1+G_1)-M_{22}(C_2+G_2)+M_{33}(C_3+G_3)}{detM} \\ \frac{-M_{13}(C_1+G_1)+M_{23}(C_2+G_2)-M_{33}(C_3+G_3)}{detM} \end{pmatrix} + \begin{pmatrix} 0 & 0 & 0 \\ 0 & 0 & 0 \\ 0 & 0 & 0 \\ \frac{M_{11}}{detM} & -\frac{M_{21}}{setM} & \frac{M_{31}}{detM} \\ -\frac{M_{12}}{detM} & \frac{M_{22}}{detM} & -\frac{M_{32}}{detM} \\ \frac{M_{13}}{detM} & -\frac{M_{23}}{setM} & \frac{M_{33}}{detM} \end{pmatrix} \begin{pmatrix} u_1 \\ u_2 \\ u_3 \end{pmatrix} \tag{2.172}$$

and equivalently, one has the state-space description:

$$\begin{pmatrix} \dot{x}_1 \\ \dot{x}_2 \\ \dot{x}_3 \\ \dot{x}_4 \\ \dot{x}_5 \\ \dot{x}_6 \end{pmatrix} = \begin{pmatrix} x_4 \\ x_5 \\ x_6 \\ f_4(x) \\ f_5(x) \\ f_6(x) \end{pmatrix} + \begin{pmatrix} 0 & 0 & 0 \\ 0 & 0 & 0 \\ 0 & 0 & 0 \\ g_{41}(x) & g_{42}(x) & g_{43}(x) \\ g_{51}(x) & g_{52}(x) & g_{53}(x) \\ g_{61}(x) & g_{62}(x) & g_{63}(x) \end{pmatrix} \begin{pmatrix} u_1 \\ u_2 \\ u_3 \end{pmatrix} \tag{2.173}$$

Next, by denoting the state vectors $x_{13} = [x_1, x_2, x_3]^T$, $x_{46} = [x_4, x_5, x_6]^T$, as well as the vectors $f_{13} = [x_4, x_5, x_6]^T$, $f_{46} = [f_4(x), f_5(x), f_6(x)]^T$ and the matrices $g_{13} = 0_{3\times 3}$

$$g_{46} = \begin{pmatrix} g_{41}(x) & g_{42}(x) & g_{43}(x) \\ g_{51}(x) & g_{52}(x) & g_{53}(x) \\ g_{61}(x) & g_{62}(x) & g_{63}(x) \end{pmatrix} \tag{2.174}$$

the state-space description of the system comes into the following concise (and also triangular) form:

$$\begin{pmatrix} \dot{x}_{13} \\ \dot{x}_{46} \end{pmatrix} = \begin{pmatrix} x_{46} \\ f_{46}(x) \end{pmatrix} + \begin{pmatrix} 0_{3\times 3} \\ g_{46}(x_{46}) \end{pmatrix} u \tag{2.175}$$

Equivalently, one has that the dynamics of the redundant robotic manipulator consists of two subsystems:

$$\dot{x}_{13} = x_{46} \tag{2.176}$$

$$\dot{x}_{46} = f_{46}(x) + g_{46}(x)u \tag{2.177}$$

It can be proven that the redundant robotic manipulator is a differentially flat system with flat output $y = x_{13}$. Indeed, from Eq. (2.176), it holds that

$$x_{46} = \dot{x}_{13} \Rightarrow x_{46} = \dot{y} \tag{2.178}$$

Thus, x_{46} is a differential function of the flat output $y = x_{13}$. Moreover, from Eq. (2.177), and using that $f_{46}(x)$ and $g_{46}(x)$, are functions of x_{31}, x_{46} that is of y, $\dot{y}$, one has

$$u = g_{46}(x)^{-1}[\dot{x}_{46} - f_{46}(x)] \Rightarrow u = g_{46}(y, \dot{y})^{-1}[\ddot{y} - f_{46}(y, \dot{y})] \tag{2.179}$$

Consequently, the control input u is also a differential function of the flat output $y = x_{13}$. As a result of the above, the redundant robotic manipulator which consists of the subsystems of Eq. (2.176) and Eq. (2.177) is a differentially flat system.

2.4.3 Flatness-based controller in successive loops for redundant robotic manipulators

One can also demonstrate that the previously noted subsystems of Eq. (2.176) and Eq. (2.177) stand independently for differentially flat systems.

For the subsystem of Eq. (2.176), x_{13} is the flat output and x_{46} is a virtual control input. As shown before, x_{46} is a differential function of the flat output x_{13}, and thus the subsystem of Eq. (2.176) is differentially flat.

For the subsystem of Eq. (2.177), x_{46} is the flat output. Besides, the state vectors elements of x_{13} are viewed as coefficients, and thus functions $f_{46}(x)$ and $g_{46}(x)$ depend only on the flat output x_{46}. Furthermore, using again Eq. (2.179), one has also that the control input u is also a differential function of the flat output x_{46}. Consequently, the subsystem of Eq. (2.177) is also differentially flat.

The control of the subsystems of Eq. (2.176) and Eq. (2.177) can be carried out following the design process of controllers for input–output linearized differentially flat systems.

For the subsystem of Eq. (2.176), the setpoint is denoted as x_{13}^d and the value of the virtual control input x_{46} which stabilizes the system's dynamics is

$$x_{46}^* = \dot{x}_{13}^d - K_1(x_{13} - x_{13}^d) \tag{2.180}$$

Matrix $K_1 \in R^{3\times3}$ is a diagonal matrix, where its diagonal elements $k_{i1} > 0,\ i = 1, 2, 3$. For the subsystem of Eq. (2.177), the control inputs vector u is selected so as to ensure that x_{46} will converge to the targeted value x_{46}^* which makes the state vector of the subsystem of Eq. (2.176) converge to the associated setpoint. Thus, the setpoint for the subsystem of Eq. (2.177) is $x_{46}^d = x_{46}^*$ and control input u is given by

$$u = g_{46}(x)^{-1}[\dot{x}_{46}^d - f_{46}(x) - K_2(x_{46} - x_{46}^d)] \tag{2.181}$$

Matrix $K_2 \in R^{3\times3}$ is a diagonal matrix, where its diagonal elements $k_{i1} > 0,\ \ i = 1, 2, 3$. By substituting Eq. (2.180) into Eq. (2.176), as well as by substituting Eq. (2.181) into Eq. (2.177), one obtains the following closed-loop system dynamics:

$$(\dot{x}_{13} - \dot{x}_{13}^d) + K_1(x_{13} - x_{13}^d) = 0 \tag{2.182}$$

$$(\dot{x}_{46} - \dot{x}_{46}^d) + K_2(x_{46} - x_{46}^d) = 0 \tag{2.183}$$

Next, by defining the tracking error variables $e_{13} = x_{13} - x_{13}^d$ and $e_{46} = x_{46} - x_{46}^d$, one has that

$$\begin{aligned} &\dot{e}_{13} + K_1 e_{13} = 0 \Rightarrow lim_{t\to\infty} e_{13} = 0 \Rightarrow lim_{t\to\infty} x_{13}(t) = x_{13}^d(t) \\ &\dot{e}_{46} + K_2 e_{46} = 0 \Rightarrow lim_{t\to\infty} e_{46} = 0 \Rightarrow lim_{t\to\infty} x_{46}(t) = x_{46}^d(t) \end{aligned} \tag{2.184}$$

Consequently, all state variables of the redundant robotic manipulator converge to the associated setpoints, or $lim_{t\to\infty} x_i(t) = x_i^d(t)$, for $i = 1, 2, \cdots, 6$.

The global stability properties of the control scheme can be also proven through Lyapunov analysis. The following Lyapunov function is defined:

$$V = \frac{1}{2}[e_{13}^T e_{13} + e_{46}^T e_{46}] \tag{2.185}$$

By differentiating the above Lyapunov function in time,

$$\begin{aligned} \dot{V} = e_{13}^T \dot{e}_{13} + e_{46}^T \dot{e}_{46} &\Rightarrow \dot{V} = e_{13}^T(-K_1 e_{13} + e_{46}^T(-K_2 e_{46}) \\ &\Rightarrow \dot{V} = -K_1 e_{13}^T e_{13} - K_2 e_{46}^T e_{46} \Rightarrow \dot{V} < 0 \end{aligned} \tag{2.186}$$

Thus, it holds that $\dot{V}$ is strictly negative $\forall\ e_{13} \neq 0, e_{46} \neq 0$ while it becomes 0 only when $e_{13} = 0,\ e_{46} = 0$. Consequently, the above given Lyapunov function V is a positive and strictly diminishing function which, no matter what its initial value is, approaches asymptotically the equilibrium $(e_{13}^T, e_{46}^T) = (0_{1\times3}, 0_{1\times3})$. As a result of the above, flatness-based control in successive loops for the 3-DOF redundant robotic manipulator, ensures global asymptotic stability.

2.4.4 Simulation tests

Results about the tracking accuracy and the speed of convergence to setpoints of the successive-loops flatness-based control method, in the case of the 3-DOF redundant robotic manipulator, are shown in Figures 2.59–2.66. The parameters of the model of the 3-link redundant robotic manipulator which have been used in the simulation experiments have been as follows: (i) masses of the links: $m_1 = 1.0kg$, $m_2 = 1.0kg$, $m_3 = 1.0kg$, (ii) lengths of the links: $L_1 = 1.0m$, $L_2 = 1.0m$, $L_3 = 1.0m$, (iii) moments of inertia of the links: $J_1 = 1.0kg{\cdot}m^2$, $J_2 = 1.0kg{\cdot}m^2$, $J_3 = 1.0kg{\cdot}m^2$. The sampling period was $T_s = 0.01sec$. Indicative values about the feedback gain matrices of the control loop are $K_1 = 1.0{\cdot}I_{3\times3}$ and $K_2 = 0.8{\cdot}I_{3\times3}$. It can be noticed that under this control scheme, one achieves fast and precise tracking of reference setpoints for all state variables of the robotic system. It is noteworthy that through the stages of this method, one solves also the setpoints definition problem for all state variables of the redundant robotic manipulator. Actually, the selection of setpoints for state variables x_1, x_2, and x_3 is unconstrained. On the other side by defining state variables x_4, x_5, and x_6 as virtual control inputs for the subsystem of x_1, x_2, and x_3 one can find the setpoints for x_4, x_5, and x_6 as functions of the setpoints for x_1, x_2, and x_3. The speed of convergence of the state variables of the robotic system under flatness-based control implemented in successive loops is dependent on the selection of values for the diagonal gain matrices K_1, K_2 of Eq. (2.180) and Eq. (2.181).

To elaborate on flatness-based control in successive loops for the 3-link redundant robotic manipulator the following tables are given: (i) Table 2.4 providing results about the accuracy of tracking of setpoints by the state variables of

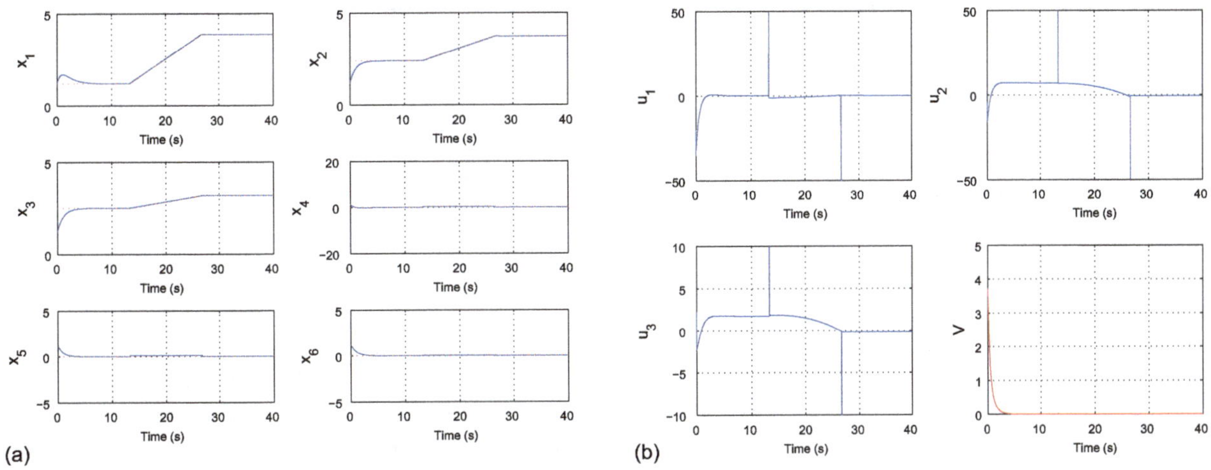

Figure 2.59 Tracking of trajectory 1 by the 3-DOF redundant robotic manipulator: (a) convergence of state variables x_1 to x_6 to their reference setpoints (red line: setpoint, blue line: real value); (b) variation of control inputs u_1 to u_3 (blue lines) and Lyapunov function V (red line)

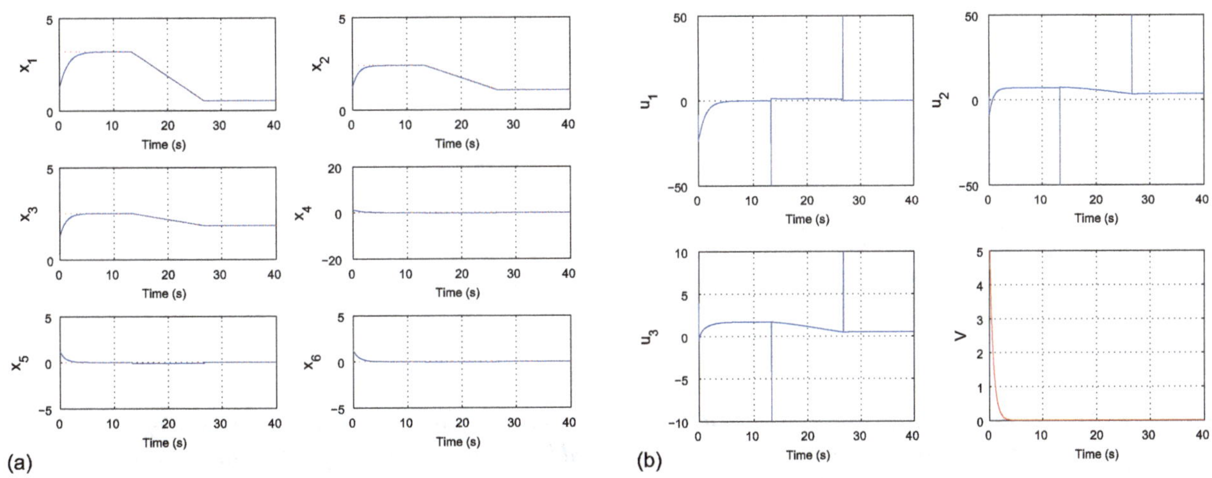

Figure 2.60 Tracking of trajectory 2 by the 3-DOF redundant robotic manipulator: (a) convergence of state variables x_1 to x_6 to their reference setpoints (red line: setpoint, blue line: real value); (b) variation of control inputs u_1 to u_3 (blue lines) and Lyapunov function V (red line) (red line)

the redundant robot under an exact dynamic model, (ii) Table 2.5 providing results about the accuracy of tracking of setpoints by the state variables of the redundant robot under a model that is subject to disturbances (for instance change $\Delta\alpha\%$ in the element M_{33} of the redundant robot's inertia matrix), Table 2.6 providing results about the convergence times of the state variables of the redundant robot's state variables to the associated setpoints.

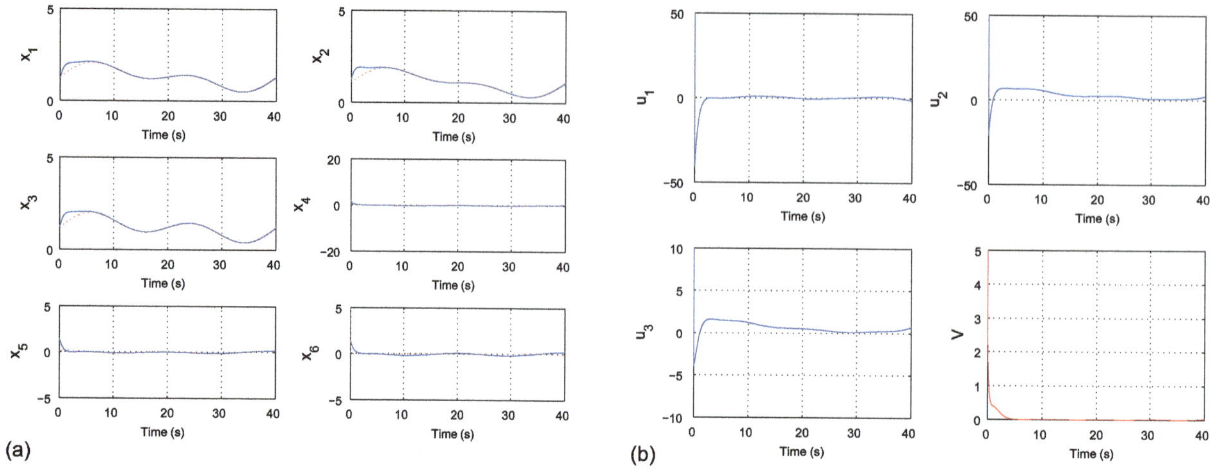

Figure 2.61 *Tracking of trajectory 3 by the 3-DOF redundant robotic manipulator: (a) convergence of state variables x_1 to x_6 to their reference setpoints (red line: setpoint, blue line: real value); (b) variation of control inputs u_1 to u_3 (blue lines) and Lyapunov function V (red line) (red line)*

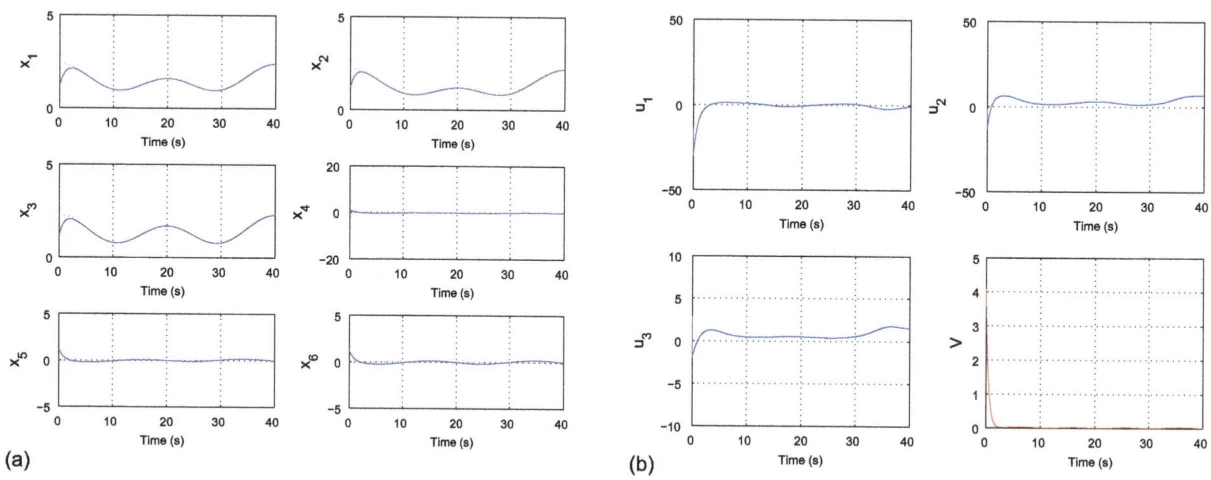

Figure 2.62 *Tracking of trajectory 4 by the 3-DOF redundant robotic manipulator: (a) convergence of state variables x_1 to x_6 to their reference setpoints (red line: setpoint, blue line: real value); (b) variation of control inputs u_1 to u_3 (blue lines) and Lyapunov function V (red line) (red line)*

2.5 Flatness-based disturbance observer and control for a robotic mining excavator

2.5.1 Outline

Surface mining is one of the greatest industries worldwide and electromechanical shovels and excavators play a key role in it [892,893,691]. Mining tasks require often the use of shovels and excavators where their human operators should have several years of experience so as to retain the scheduled duty cycle and to avoid placing undue stress to these machines under difficult digging conditions [1056,246,148,244]. Thus, the robotization of these machines has become a challenging endeavor [483,688,1048]. Precise control of robotic excavators under variable soil conditions can ensure higher mining rates and profitability of the mining business [169,245,232]. A key issue for the accomplishment

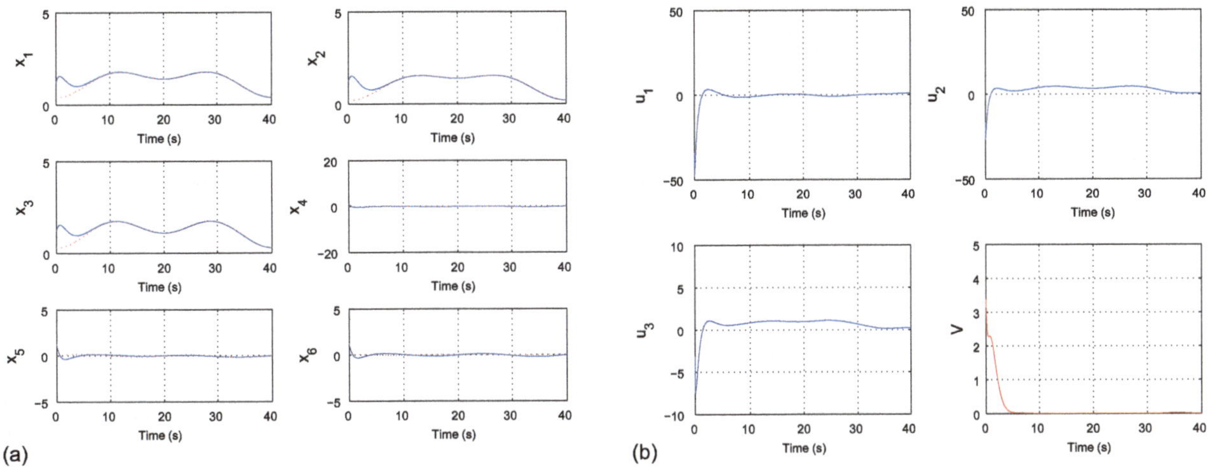

Figure 2.63 Tracking of trajectory 5 by the 3-DOF redundant robotic manipulator: (a) convergence of state variables x_1 to x_6 to their reference setpoints (red line: setpoint, blue line: real value); (b) variation of control inputs u_1 to u_3 (blue lines) and Lyapunov function V (red line) (red line)

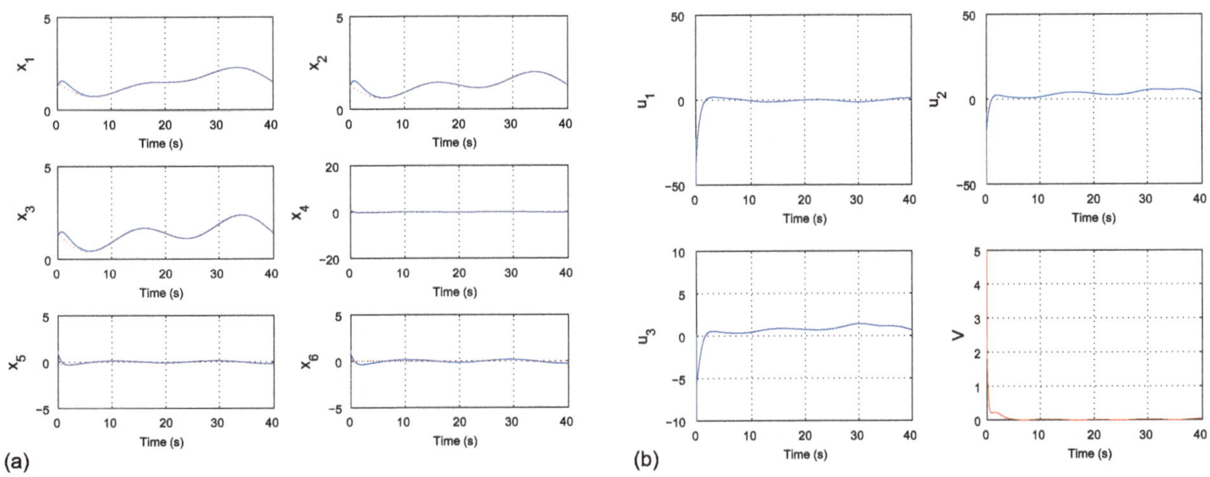

Figure 2.64 Tracking of trajectory 6 by the 3-DOF redundant robotic manipulator: (a) convergence of state variables x_1 to x_6 to their reference setpoints (red line: setpoint, blue line: real value); (b) variation of control inputs u_1 to u_3 (blue lines) and Lyapunov function V (red line) (red line)

of this objective is the estimation in real-time of disturbance inputs which affect the mining robot when digging and loading the ground and ore [764,931,986]. Knowing the external forces and disturbances which are applied to mining shovels and excavators allows for keeping safe the functioning of these machines, while it also eliminates the chances of their components' failure, such as electric motors and electrohydraulic actuators, drives, and power electronics or shears, cables, and hoist ropes [668,701,76]. Besides, it reduces the frequency of equipment repair and maintenance and makes the exploitation of mines more efficient and more profitable [581,582,526]. Moreover, by avoiding abrupt operations, peaks of torques and force variables are also deterred and energy consumption by this mining machinery is kept moderate [657,580,353]. For the control problem of mining shovels and excavators, several nonlinear control methods have been proposed so far, including sliding-mode control and model-predictive control techniques [976,783,578].

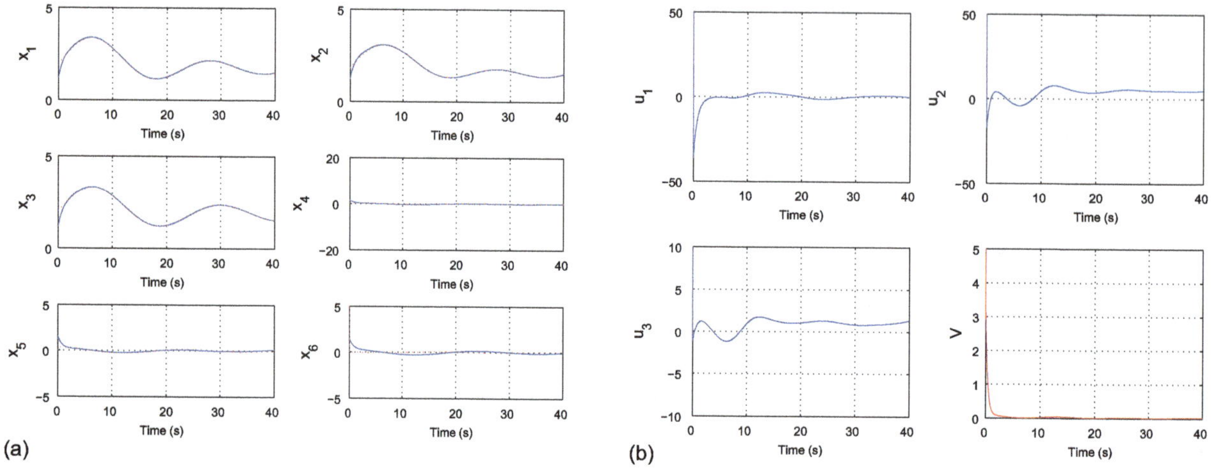

Figure 2.65 *Tracking of trajectory 7 by the 3-DOF redundant robotic manipulator: (a) convergence of state variables x_1 to x_6 to their reference setpoints (red line: setpoint, blue line: real value); (b) variation of control inputs u_1 to u_3 (blue lines) and Lyapunov function V (red line) (red line)*

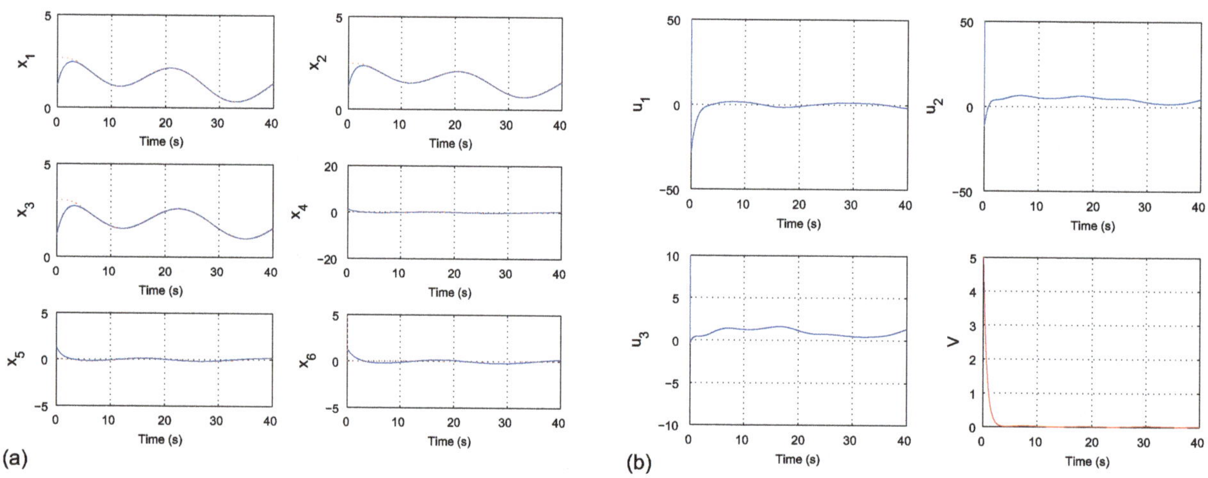

Figure 2.66 *Tracking of trajectory 8 by the 3-DOF redundant robotic manipulator: (a) convergence of state variables x_1 to x_6 to their reference setpoints (red line: setpoint, blue line: real value); (b) variation of control inputs u_1 to u_3 (blue lines) and Lyapunov function V (red line) (red line)*

Energy efficiency and minimization of energy consumption is another important issue about automated shovels and excavators [927,4,870,950]. To address the above-noted issues the section proposes a novel flatness-based disturbance observer and a flatness-based control scheme for robotic excavators [721,711].

First, the dynamic model of the robotic excavator is given in state-space form, and differential flatness properties are proven about it. A system is differentially flat if (i) all its state variables and its control inputs can be expressed as functions of the flat outputs, (ii) the flat outputs and their time-derivatives are differentially independent which means that they are not connected between them through a relation in the form of an homogeneous differential equation [91,594,639,489]. By demonstrating that this system is differentially flat it is also confirmed that it can be transformed to the input–output linearized form and to the canonical Brunvosly form [806,492,640,54]. The canonical form is

Table 2.4 Tracking RMSE for the redundant robot in the disturbance-free case

	$RMSE_{x_1}$	$RMSE_{x_2}$	$RMSE_{x_3}$	$RMSE_{x_4}$	$RMSE_{x_5}$	$RMSE_{x_6}$
$test_1$	$0.1519{\cdot}10^{-4}$	$0.0001{\cdot}10^{-4}$	$0.0299{\cdot}10^{-4}$	$0.0001{\cdot}10^{-4}$	$0.0997{\cdot}10^{-4}$	$0.0001{\cdot}10^{-4}$
$test_2$	$0.1208{\cdot}10^{-4}$	$0.0001{\cdot}10^{-4}$	$0.0538{\cdot}10^{-4}$	$0.0001{\cdot}10^{-4}$	$0.0451{\cdot}10^{-4}$	$0.0001{\cdot}10^{-4}$
$test_3$	$0.1617{\cdot}10^{-4}$	$0.0227{\cdot}10^{-4}$	$0.1331{\cdot}10^{-4}$	$0.0158{\cdot}10^{-4}$	$0.1902{\cdot}10^{-4}$	$0.0267{\cdot}10^{-4}$
$test_4$	$0.1758{\cdot}10^{-4}$	$0.0315{\cdot}10^{-4}$	$0.1571{\cdot}10^{-4}$	$0.0269{\cdot}10^{-4}$	$0.1948{\cdot}10^{-4}$	$0.0370{\cdot}10^{-4}$
$test_5$	$0.1571{\cdot}10^{-4}$	$0.0261{\cdot}10^{-4}$	$0.1388{\cdot}10^{-4}$	$0.0206{\cdot}10^{-4}$	$0.1758{\cdot}10^{-4}$	$0.0315{\cdot}10^{-4}$
$test_6$	$0.1331{\cdot}10^{-4}$	$0.0188{\cdot}10^{-4}$	$0.1564{\cdot}10^{-4}$	$0.0220{\cdot}10^{-4}$	$0.2293{\cdot}10^{-4}$	$0.0322{\cdot}10^{-4}$
$test_7$	$0.1307{\cdot}10^{-4}$	$0.011{\cdot}10^{-4}$	$0.1092{\cdot}10^{-4}$	$0.0077{\cdot}10^{-4}$	$0.0750{\cdot}10^{-4}$	$0.0183{\cdot}10^{-4}$
$test_8$	$0.1401{\cdot}10^{-4}$	$0.0248{\cdot}10^{-4}$	$0.1035{\cdot}10^{-4}$	$0.0193{\cdot}10^{-4}$	$0.1893{\cdot}10^{-4}$	$0.0136{\cdot}10^{-4}$

Table 2.5 Tracking RMSE for the redundant robot in the case of disturbances

$\Delta a\%$	$RMSE_{x_1}$	$RMSE_{x_2}$	$RMSE_{x_3}$	$RMSE_{x_4}$	$RMSE_{x_5}$	$RMSE_{x_6}$
0%	$0.1617{\cdot}10^{-4}$	$0.0227{\cdot}10^{-4}$	$0.1391{\cdot}10^{-4}$	$0.0188{\cdot}10^{-4}$	$0.1902{\cdot}10^{-4}$	$0.0267{\cdot}10^{-4}$
10%	$0.0162{\cdot}10^{-3}$	$0.0023{\cdot}10^{-3}$	$0.0113{\cdot}10^{-3}$	$0.0019{\cdot}10^{-3}$	$0.7700{\cdot}10^{-3}$	$0.7209{\cdot}10^{-3}$
20%	$0.0162{\cdot}10^{-3}$	$0.0023{\cdot}10^{-3}$	$0.0009{\cdot}10^{-3}$	$0.0019{\cdot}10^{-3}$	$1.2907{\cdot}10^{-3}$	$1.2094{\cdot}10^{-3}$
30%	$0.0162{\cdot}10^{-3}$	$0.0023{\cdot}10^{-3}$	$0.0133{\cdot}10^{-3}$	$0.0019{\cdot}10^{-3}$	$1.6594{\cdot}10^{-3}$	$1.5394{\cdot}10^{-3}$
40%	$0.0162{\cdot}10^{-3}$	$0.0023{\cdot}10^{-3}$	$0.0133{\cdot}10^{-3}$	$0.0019{\cdot}10^{-3}$	$1.9328{\cdot}10^{-3}$	$1.7889{\cdot}10^{-3}$
50%	$0.0162{\cdot}10^{-3}$	$0.0023{\cdot}10^{-3}$	$0.0133{\cdot}10^{-3}$	$0.0019{\cdot}10^{-3}$	$2.1458{\cdot}10^{-3}$	$1.9812{\cdot}10^{-3}$
60%	$0.0162{\cdot}10^{-3}$	$0.0023{\cdot}10^{-3}$	$0.0133{\cdot}10^{-3}$	$0.0019{\cdot}10^{-3}$	$2.3145{\cdot}10^{-3}$	$2.1340{\cdot}10^{-3}$

Table 2.6 Convergence times (sec) of states of the redundant robot

No test	$T_s\ x_1$	$T_s\ x_2$	$T_s\ x_3$	$T_s\ x_4$	$T_s\ x_5$	$T_s\ x_6$
$test_1$	5.0	3.5	3.0	0.5	3.0	3.5
$test_2$	5.0	3.0	3.0	3.5	3.0	3.0
$test_3$	5.0	5.0	5.0	1.0	1.5	1.5
$test_4$	3.0	3.0	2.0	1.5	1.5	2.0
$test_5$	5.0	5.0	5.0	0.5	4.0	4.0
$test_6$	4.0	4.5	4.5	0.5	1.5	5.0
$test_7$	2.0	2.0	2.0	3.0	5.0	5.0
$test_8$	2.5	2, 0	2.5	1.0	2.5	5.0

both controllable and observable and the control problem for it can be solved with the eigenvalues assignment technique [898,725,732]. Besides, the state estimation problem can be optimally treated with the use of Kalman filtering [705,60,706]. Under such a transformation, the unknown perturbation terms which affect the system appear as additive disturbance inputs which are superimposed with the excavator's real control inputs. Next, the state-space model of the system is extended by considering as additional state variables the additive disturbance inputs and their time-derivatives.

For the extended state-space model of the robotic excavator a Kalman filter-based disturbance observer is designed [721–711]. Since the extended state-space description satisfies observability conditions it is also confirmed that the above-noted disturbance observer will be stable and convergent and that the estimation error for both the state variables of the system and for the control inputs will be asymptotically eliminated. With the identification of the disturbance inputs of the robotic excavator, one can also update the system's controller by including in it additional terms which annihilate the disturbances' effects. In this manner, the global stability of the excavator's control loop can be ensured and the precise execution of mining tasks can be also ascertained. The use of the Kalman filter-based disturbance observer enables the reliable and precise functioning of the robotic excavator in variable soil and ore conditions.

2.5.2 Dynamic model of the robotic mining excavator

The diagram of the robotic mining excavator is given in Figure 2.67. The hoist rope is made of stainless steel and does not exhibit any deformation. The state variables of the robotic excavator are given in vector $x=[d,\dot{d},\theta,\dot{\theta}]^T$, where θ is the rotation angle of the dipper handle and d is the vertical distance from the center of gravity C_1 to the saddle block [246].
The dynamic model of the robotic excavator is given by [246]

$$\begin{pmatrix} m_d+m_p & 0 \\ 0 & m_dd^2+m_pd_p^2+I_d+I_p \end{pmatrix}\begin{pmatrix}\ddot{d}\\ \ddot{\theta}\end{pmatrix}+\begin{pmatrix} 0 & -(m_dd+m_pd_p)\dot{\theta} \\ 2(m_dd+m_pd_p)\dot{\theta} & 0 \end{pmatrix}\begin{pmatrix}\dot{d}\\ \dot{\theta}\end{pmatrix}$$
$$+\begin{pmatrix} -(m_d+m_p)gcos(\theta) \\ (m_dd+m_pd_pgsin(\theta)) \end{pmatrix}=\begin{pmatrix} F_c-F_hcos(a_3)-F_n \\ F_hl_{OB}sin(a_1+a_2)-F_tl_{OA} \end{pmatrix} \tag{2.187}$$

The parameters of the state-space model are defined as follows [246]:

m_d is the mass of the dipper handle and dipper, m_p is the mass of the payload, θ is the rotation angle of the dipper handle, d is the vertical distance from the center of gravity C_1 of the dipper handle, C_2 is the center of gravity of the payload, d_{offset} is the distance between the center of gravity of the payload and the center of gravity of the dipper, d_p is the sum of the distances d and d_{offset}, I_p is the inertia of the payload, I_d is the inertia of the handler of the dipper, l_{OA} is the distance between the pivot point of the dipper handle O and the end-point of the dipper (below the removed mass of the ground), F_c is the thrust force along the translational axis of the dipper handle (due to an electropneumatic piston or due to a linear AC motor), F_h is the hoist force exerted on the dipper through the hoist rope, F_n is the normal reaction force due to contact of the dipper with the ground, F_t is the tangential reaction force due to contact of the dipper with the ground, and l_{OB} is the distance between the pivot point and the point of suspension of the hoist rope.

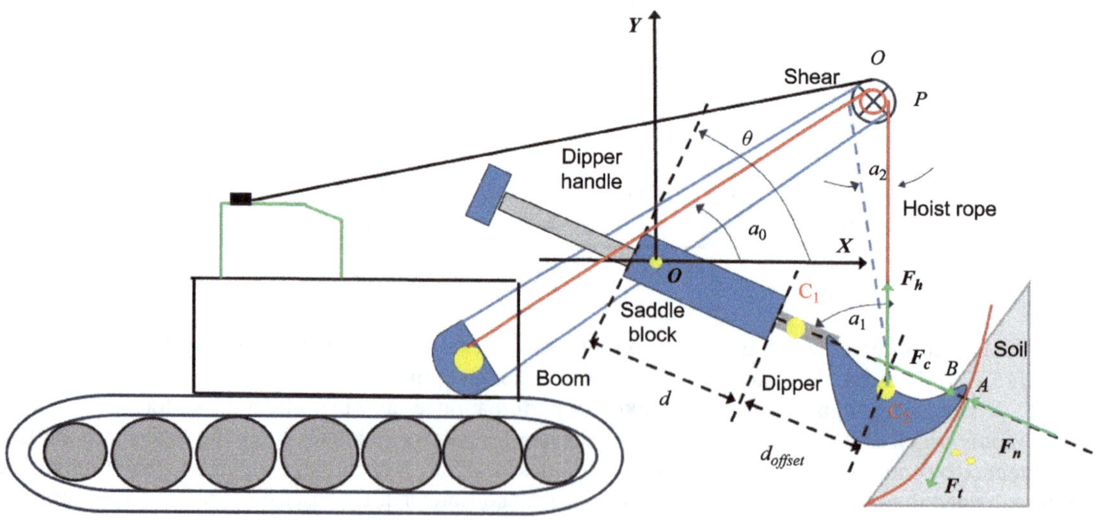

Figure 2.67 Diagram of the robotic mining excavator

In concise form, the dynamic model of the robotic mining excavator is written as

$$M(q)\ddot{q}+C(q,\dot{q})+G(q)=Bu-D\tilde{d} \tag{2.188}$$

In Eq. (2.187), the following matrices and vectors are defined [246]:

$$M(q)=\begin{pmatrix} m_{11} & m_{12} \\ m_{21} & m_{22} \end{pmatrix} \tag{2.189}$$

where $m_{11}=m_d+m_p$, $m_{12}=0$, $m_{21}=0$ and $m_{22}=m_d d^2+m_p d_p^2+I_d+I_p$. The Coriolis and centrifugal forces matrix is

$$C(q,\dot{q})=\begin{pmatrix} C_1(q,\dot{q}) \\ C_2(q,\dot{q}) \end{pmatrix} \tag{2.190}$$

where $C_1(q,\dot{q})=-(m_d d+m_p d_p)\dot{\theta}^2$, $C_2(q,\dot{q})=(2m_d d+m_p d_p)\dot{\theta}\dot{d}$. The gravitational forces matrix is given by

$$G(q)=\begin{pmatrix} G_1(q) \\ G_2(q) \end{pmatrix} \tag{2.191}$$

where $G_1(q)=-(m_d+m_p)g cos(\theta)$, $G_2(q)=(m_d d+m_p d_p)g sin(\theta)$. The control inputs gain matrix is

$$B=\begin{pmatrix} b_{11} & b_{12} \\ b_{21} & b_{22} \end{pmatrix} \tag{2.192}$$

where $b_{11}=1$, $b_{12}=-cos(a_3)$, $b_{21}=0$, $b_{22}=l_{OB} sin(a_1+a_2)$. The disturbances vector $D\tilde{d}$ is

$$D\tilde{d}=\begin{pmatrix} 1 & 0 \\ 0 & l_{OA} \end{pmatrix}\begin{pmatrix} F_n \\ F_t \end{pmatrix} \tag{2.193}$$

Using that the determinant of the inertia matrix is $detM=m_{11}m_{22}-m_{21}m_{12}$, the inverse of the inertia matrix M is

$$M^{-1}=\frac{1}{detM}\begin{pmatrix} m_{22} & -m_{12} \\ -m_{21} & m_{11} \end{pmatrix} \tag{2.194}$$

Using the inverse of the inertia matrix, one obtains

$$\ddot{q}=-M^{-1}(q)[C(q,\dot{q})+G(q)]+M^{-1}(q)Bu-M^{-1}(q)D\tilde{d} \tag{2.195}$$

where the following products of matrices are computed:

$$M^{-1}(q)[C(q,\dot{q})+G(q)]=\begin{pmatrix} \frac{m_{12}(C_1+G_1)-m_{12}(C_2+G_2)}{detM} \\ \frac{-m_{12}(C_1+G_1)+m_{11}(C_2+G_2)}{detM} \end{pmatrix} \tag{2.196}$$

$$M^{-1}(q)B=\begin{pmatrix} \frac{m_{12}}{detM} & \frac{-m_{22}cos(a_3)-m_{12}l_{OB}sin(a_1+a_3)}{detM} \\ -\frac{m_{21}}{detM} & \frac{m_{21}cos(a_3)+m_{11}l_{OB}sin(a_1+a_2)}{detM} \end{pmatrix} \tag{2.197}$$

$$M^{-1}(q)D=\begin{pmatrix} \frac{m_{21}}{detM} & -\frac{m_{12}l_{OA}}{detM} \\ -\frac{m_{21}}{detM} & \frac{m_{11}l_{OA}}{detM} \end{pmatrix} \tag{2.198}$$

Thus, the dynamic model of the robotic mining excavator becomes

$$\begin{pmatrix} \ddot{q}_1 \\ \\ \ddot{q}_3 \end{pmatrix} = -\begin{pmatrix} \frac{m_{12}(C_1+G_1)-m_{12}(C_2+G_2)}{detM} \\ \\ \frac{-m_{12}(C_1+G_1)+m_{11}(C_2+G_2)}{detM} \end{pmatrix} + \begin{pmatrix} \frac{m_{12}}{detM} & \frac{-m_{22}cos(a_3)-m_{12}l_{OB}sin(a_1+a_3)}{detM} \\ \\ -\frac{m_{21}}{detM} & \frac{m_{21}cos(a_3)+m_{11}l_{OB}sin(a_1+a_2)}{detM} \end{pmatrix} \begin{pmatrix} F_c \\ \\ F_h \end{pmatrix} + \\ + \begin{pmatrix} \frac{m_{21}}{detM} & -\frac{m_{12}l_{OA}}{detM} \\ \\ -\frac{m_{21}}{detM} & \frac{m_{11}l_{OA}}{detM} \end{pmatrix} \begin{pmatrix} F_n \\ \\ F_t \end{pmatrix} \tag{2.199}$$

By defining the state vector $x=[x_1,x_2,x_3,x_4]^T=[d,\dot{d},\theta,\dot{\theta}]^T$ and the control inputs vector $u=[u_1,u_2]^T=[F_c,F_h]^T$, as well as the cumulative disturbance vector $d=[d_1,d_2]^T$ where [246]

$$d_1 = \frac{m_{22}}{detM}F_n - \frac{m_{12}l_{OA}}{detM}F_t \quad d_2 = -\frac{m_{21}}{detM}F_n + \frac{m_{11}l_{OA}}{detM}F_t \tag{2.200}$$

Thus, the dynamic model of the robotic excavator is written as

$$\begin{gathered} \dot{x}_1 = x_2 \\ \dot{x}_2 = -\frac{m_{12}(C_1+G_1)-m_{12}(C_2+G_2)}{detM} + \frac{m_{12}}{detM}u_1 + \frac{-m_{22}cos(a_3)-m_{12}l_{OB}sin(a_1+a_3)}{detM}u_2 + d_1 \\ \dot{x}_3 = x_4 \\ \dot{x}_4 = \frac{-m_{12}(C_1+G_1)+m_{11}(C_2+G_2)}{detM} - \frac{m_{21}}{detM}u_1 + \frac{m_{21}cos(a_3)+m_{11}l_{OB}sin(a_1+a_2)}{detM}u_2 + d_2 \end{gathered} \tag{2.201}$$

Additionally, the dynamic model of the robotic excavator is written in the following nonlinear affine-in-the-input state -space form:

$$\begin{pmatrix} \dot{x}_1 \\ \dot{x}_2 \\ \dot{x}_3 \\ \dot{x}_4 \end{pmatrix} = \begin{pmatrix} x_2 \\ -\frac{m_{12}(C_1+G_1)-m_{12}(C_2+G_2)}{detM} \\ x_4 \\ \frac{-m_{12}(C_1+G_1)+m_{11}(C_2+G_2)}{detM} \end{pmatrix} + \begin{pmatrix} 0 & 0 \\ \frac{m_{12}}{detM} & \frac{-m_{22}cos(a_3)-m_{12}l_{OB}sin(a_1+a_3)}{detM} \\ 0 & 0 \\ -\frac{m_{21}}{detM} & \frac{m_{21}cos(a_3)+m_{11}l_{OB}sin(a_1+a_2)}{detM} \end{pmatrix} \begin{pmatrix} u_1 \\ u_2 \end{pmatrix} + \begin{pmatrix} 0 \\ d_1 \\ 0 \\ d_2 \end{pmatrix} \tag{2.202}$$

or equivalently

$$\begin{pmatrix} \dot{x}_1 \\ \dot{x}_2 \\ \dot{x}_3 \\ \dot{x}_4 \end{pmatrix} = \begin{pmatrix} x_2 \\ f_2(x) \\ x_4 \\ f_4(x) \end{pmatrix} + \begin{pmatrix} 0 & 0 \\ g_{21}(x) & g_{22}(x) \\ 0 & 0 \\ g_{41}(x) & g_{42}(x) \end{pmatrix} \begin{pmatrix} u_1 \\ u_2 \end{pmatrix} + \begin{pmatrix} 0 \\ d_1 \\ 0 \\ d_2 \end{pmatrix} \tag{2.203}$$

or finally

$$\ddot{x}_1 = f_2(x) + g_{21}(x)u_1 + g_{22}(x)u_2 + d_1 \tag{2.204}$$

$$\ddot{x}_3 = f_4(x) + g_{41}(x)u_1 + g_{42}(x)u_2 + d_2 \tag{2.205}$$

2.5.3 Canonical form of the robotic mining excavator

2.5.3.1 Definition of differentially flat systems

It will be proven that the robotic excavator is a differentially flat system. The differential flatness property confirms that the system can be transformed into the input–output linearized form and in the canonical Brunovsky form. Differential flatness is a structural property of a class of nonlinear dynamical systems, denoting that all system variables (such as state vector elements and control inputs) can be written in terms of a set of specific variables (the so-called flat outputs and their derivatives). The following nonlinear system is considered [711]:

$$\dot{x}(t) = f(x(t), u(t)) \tag{2.206}$$

The time variable is $t \in R$, the state vector is $x(t) \in R^n$ with initial conditions $x(0) = x_0$, and the input variable is $u(t) \in R^m$. Next, the main principles of differentially flat systems are given:

The finite-dimensional system of Eq. (2.206) can be written in the general form of an ordinary differential equation (ODE), i.e., $S_i(w, \dot{w}, \ddot{w}, \cdots, w^{(i)}), \quad i = 1, 2, \cdots, q$. The entity w is a generic notation for the system variables (these variables are for instance the elements of the system's state vector $x(t)$ and the elements of the control input $u(t)$) while $w^{(i)}$, $i = 1, 2, \cdots, q$ are the associated derivatives. Such a system is said to be differentially flat if there is a collection of m functions $y = (y_1, \cdots, y_m)$ of the system variables and of their time-derivatives, i.e., $y_i = \phi(w, \dot{w}, \ddot{w}, \cdots, w^{(\alpha_i)}), \ i = 1, \cdots, m$ satisfying the following two conditions:

1. There does not exist any differential relation of the form $R(y, \dot{y}, \cdots, y^{(\beta)}) = 0$ which implies that the flat output and its derivatives are not coupled in the form of an ODE, or equivalently it can be said that the flat output is differentially independent.

2. All system variables (i.e., the elements of the system's state vector w and the control input) can be expressed using only the flat output y and its time-derivatives $w_i = \psi_i(y, \dot{y}, \cdots, y^{(\gamma_i)}), \ i = 1, \cdots, s$. An equivalent definition of differentially flat systems is as follows:

Definition: The system $\dot{x} = f(x, u)$, $x \in R^n$, $u \in R^m$ is differentially flat if there exist relations:

$$\begin{aligned} h &: \ R^n \times (R^m)^{r+1} \rightarrow R^m, \\ \phi &: \ (R^m)^r \rightarrow R^n \text{ and} \\ \psi &: \ (R^m)^{r+1} \rightarrow R^m \end{aligned} \tag{2.207}$$

such that

$$\begin{aligned} y &= h(x, u, \dot{u}, \cdots, u^{(r)}), \\ x &= \phi(y, \dot{y}, \cdots, y^{(r-1)}), \text{ and} \\ u &= \psi(y, \dot{y}, \cdots, y^{(r-1)}, y^{(r)}) \end{aligned} \tag{2.208}$$

This means that all system dynamics can be expressed as a function of the flat output and its derivatives, therefore the state vector and the control input can be written as

$$\begin{aligned} x(t) &= \phi(y(t), \dot{y}(t), \cdots, y^{(r-1)}(t)), \text{and} \\ u(t) &= \psi(y(t), \dot{y}(t), \cdots, y^{(r)}(t)) \end{aligned} \tag{2.209}$$

2.5.3.2 Transformation into the canonical form

The robotic excavator is a differentially flat system with flat outputs vector $Y = [y_1, y_2]^T = [x_1, x_3]^T$. Indeed, it holds that

$$x_2 = \dot{x}_1 \quad x_4 = \dot{x}_3 \tag{2.210}$$

while by omitting the unknown disturbance terms d_1, d_2 from the robot's dynamic model, one has

$$\begin{pmatrix} u_1 \\ u_2 \end{pmatrix} = \begin{pmatrix} g_{21}(x) & g_{22}(x) \\ g_{41}(x) & g_{42}(x) \end{pmatrix}^{-1} \begin{pmatrix} \ddot{x}_1 - f_2(x) \\ \ddot{x}_3 - f_4(x) \end{pmatrix} \tag{2.211}$$

Thus, u_1, u_2 are also differential functions of the flat outputs, or

$$u_1 = h_{u_1}(x_1, \dot{x}_1, x_3, \dot{x}_3) \quad u_2 = h_{u_2}(x_1, \dot{x}_1, x_3, \dot{x}_3) \tag{2.212}$$

Consequently, the robotic excavator is a differentially flat system and can be written in the input–output linearized form and in the canonical Brunovsky form. To this end, the following control inputs are defined for the robotic excavator:

$$\begin{aligned} v_1 &= f_2(x) + g_{21}(x)u_1 + g_{22}(x)u_2 \\ v_2 &= f_4(x) + g_{41}(x)u_1 + g_{42}(x)u_2 \end{aligned} \tag{2.213}$$

Thus, the dynamics of the robotic excavator can be written in the following input–output linearized form:

$$\begin{aligned} \ddot{x}_1 &= v_1 + d_1 \\ \ddot{x}_3 &= v_2 + d_2 \end{aligned} \tag{2.214}$$

or equivalently in the canonical form

$$\begin{pmatrix} \dot{x}_1 \\ \dot{x}_2 \\ \dot{x}_3 \\ \dot{x}_4 \end{pmatrix} = \begin{pmatrix} 0 & 1 & 0 & 0 \\ 0 & 0 & 0 & 0 \\ 0 & 0 & 0 & 1 \\ 0 & 0 & 0 & 0 \end{pmatrix} \begin{pmatrix} x_1 \\ x_2 \\ x_3 \\ x_4 \end{pmatrix} + \begin{pmatrix} 0 & 0 \\ 1 & 0 \\ 0 & 0 \\ 0 & 1 \end{pmatrix} \begin{pmatrix} v_1 + d_1 \\ v_2 + d_2 \end{pmatrix} \tag{2.215}$$

By defining the new control inputs $\tilde{v}_1 = v_1 + d_1$ and $\tilde{v}_2 = v_2 + d_2$ and the new control inputs vector $\tilde{v} = [\tilde{v}_1, \tilde{v}_2]^T$, the dynamic model can be further written in the concise form:

$$\dot{x} = Ax + B\tilde{v} \tag{2.216}$$

By defining the desirable setpoints vector as $x_d = [x_{1,d}, x_{2,d}, x_{3,d}, x_{4,d}]^T$ the stabilizing feedback control is

$$\begin{aligned} (v_1 + d_1) &= \ddot{x}_{1,d} - K_{1,1}(\dot{x}_1 - \dot{x}_{1,d}) - K_{2,1}(x_1 - x_{1,d}) \Rightarrow \\ v_1 &= \ddot{x}_{1,d} - K_{1,1}(\dot{x}_1 - \dot{x}_{1,d}) - K_{2,1}(x_1 - x_{1,d}) - d_1 \end{aligned} \tag{2.217}$$

$$\begin{aligned} (v_2 + d_2) &= \ddot{x}_{3,d} - K_{1,2}(\dot{x}_3 - \dot{x}_{3,d}) - K_{3,1}(x_3 - x_{3,d}) \Rightarrow \\ v_2 &= \ddot{x}_{3,d} - K_{3,1}(\dot{x}_3 - \dot{x}_{3,d}) - K_{3,1}(x_3 - x_{3,d}) - d_2 \end{aligned} \tag{2.218}$$

By substituting Eq. (2.217) into Eq. (2.204), as well as by substituting Eq. (2.218) into Eq. (2.205), one obtains the following closed-loop dynamics for the robotic excavator:

$$(\ddot{x}_1 - \ddot{x}_{1,d}) + K_{1,1}(\dot{x}_1 - \dot{x}_{1,d}) + K_{2,1}(x_1 - x_{1,d}) = 0 \tag{2.219}$$

$$(\ddot{x}_3 - \ddot{x}_{3,d}) + K_{1,2}(\dot{x}_3 - \dot{x}_{3,d}) + K_{3,1}(x_3 - x_{3,d}) = 0 \tag{2.220}$$

or equivalently

$$\ddot{e}_1 + K_{1,1}\dot{e}_1 + K_{2,1}e_1 = 0 \tag{2.221}$$

$$\ddot{e}_3 + K_{1,2}\dot{e}_3 + K_{2,2}e_3 = 0 \tag{2.222}$$

By selecting the feedback gains $(K_{1,1}, K_{2,1})$ and $(K_{1,2}, K_{2,2})$ so as the associated characteristic polynomials to be Hurwitz stable, one has that the tracking error converges asymptotically to zero:

$$\begin{aligned} lim_{t\to\infty}e_1(t) = 0 &\Rightarrow lim_{t\to\infty}x_1(t) = x_{1,d}(t) \\ lim_{t\to\infty}e_3(t) = 0 &\Rightarrow lim_{t\to\infty}x_3(t) = x_{3,d}(t) \end{aligned} \tag{2.223}$$

2.5.4 Design of a Kalman filter-based disturbance observer

Since the cumulative disturbance terms d_1, d_2 are unknown and non-measurable, these have to be estimated with the use of a disturbance observer. The dynamic model of the robotic excavator had been written previously in the input–output linearized form:

$$\ddot{x}_1 = v_1 + d_1 \quad \ddot{x}_2 = v_2 + d_2 \tag{2.224}$$

The state vector of the system is extended by considering as additional state variables the disturbance terms d_1, d_2. The extended state-space model is given by the following set of equations:

$$\begin{array}{llll} z_1 = x_1 & z_2 = \dot{x}_1 & z_3 = x_3 & z_4 = \dot{x}_4 \\ z_5 = d_1 & z_6 = \dot{d}_1 & z_7 = d_2 & z_8 = \dot{d}_2 \end{array} \tag{2.225}$$

The disturbance dynamics can be sufficiently described by the time-derivatives of the signals d_1, d_2 and the associated initial conditions. Without loss of generality, it is considered that up to second-order time-derivatives of d_1, d_2, together with initial conditions, describe sufficiently the disturbance signals. This is written as: $\ddot{d}_1 = f_{d_1}, d_1(0), \dot{d}_1(0)$, and $\ddot{d}_2 = f_{d_2}, d_2(0), \dot{d}_2(0)$. Thus, the state-space model of the system comes into the extended state-space form:

$$\begin{pmatrix} \dot{z}_1 \\ \dot{z}_2 \\ \dot{z}_3 \\ \dot{z}_4 \\ \dot{z}_5 \\ \dot{z}_6 \\ \dot{z}_7 \\ \dot{z}_8 \end{pmatrix} = \begin{pmatrix} 0\ 1\ 0\ 0\ 0\ 0\ 0\ 0\ 0\ 0 \\ 0\ 0\ 0\ 0\ 1\ 0\ 0\ 0\ 0\ 0 \\ 0\ 0\ 0\ 1\ 0\ 0\ 0\ 0\ 0\ 0 \\ 0\ 0\ 0\ 0\ 0\ 0\ 0\ 0\ 1\ 0 \\ 0\ 0\ 0\ 0\ 0\ 1\ 0\ 0\ 0\ 0 \\ 0\ 0\ 0\ 0\ 0\ 0\ 0\ 0\ 0\ 0 \\ 0\ 0\ 0\ 0\ 0\ 0\ 0\ 0\ 0\ 1 \\ 0\ 0\ 0\ 0\ 0\ 0\ 0\ 0\ 0\ 0 \end{pmatrix} \begin{pmatrix} z_1 \\ z_2 \\ z_3 \\ z_4 \\ z_5 \\ z_6 \\ z_7 \\ z_8 \end{pmatrix} + \begin{pmatrix} 0\ 0\ 0\ 0 \\ 1\ 0\ 0\ 0 \\ 0\ 0\ 0\ 0 \\ 0\ 1\ 0\ 0 \\ 0\ 0\ 0\ 0 \\ 0\ 0\ 1\ 0 \\ 0\ 0\ 0\ 0 \\ 0\ 0\ 0\ 1 \end{pmatrix} \begin{pmatrix} u_1 \\ u_2 \\ F_{d_1} \\ F_{d_2} \end{pmatrix} \tag{2.226}$$

with measurement equation

$$\begin{pmatrix} z_1^m \\ z_2^m \end{pmatrix} = \begin{pmatrix} 1\ 0\ 0\ 0\ 0\ 0\ 0\ 0 \\ 0\ 0\ 1\ 0\ 0\ 0\ 0\ 0 \end{pmatrix} \begin{pmatrix} z_1 \\ z_2 \\ z_3 \\ z_4 \\ z_5 \\ z_6 \\ z_7 \\ z_8 \end{pmatrix} \tag{2.227}$$

With the use of the extended state-space model, and by defining the control inputs vector $\bar{v} = [u_1, u_2, F_{d_1}, F_{d_2}]^T$, the dynamics of the robotic excavator is written in the concise form:

$$\begin{aligned} \dot{z} &= Az + B\bar{v} \\ z_m &= C{\cdot}z \end{aligned} \tag{2.228}$$

Using Eq. (2.228), one can design an extended state observer (disturbance observer) which can identify dynamically the disturbance terms d_1, d_2 which affect the excavation robot. A Kalman filter-based disturbance observer has the form:

$$\begin{aligned} \dot{\hat{z}} &= A_o\hat{z} + B_o u + K_f(z_m - \hat{z}_m) \\ \hat{z}_m &= C_o z \end{aligned} \tag{2.229}$$

where $A_o = A$, $C_o = C$ and matrix B_o is given by

$$B_o = \begin{pmatrix} 0\ 1\ 0\ 0\ 0\ 0\ 0\ 0 \\ 0\ 0\ 0\ 1\ 0\ 0\ 0\ 0 \end{pmatrix}^T \tag{2.230}$$

The observer's gain K_f is computed through Kalman filter's recursion after expressing Eq. (2.229) in discrete-time form. This state estimation concept is established in the related bibliography are *Derivative-free nonlinear Kalman filter* [721,711]. Matrices A_o, B_o and C_o are substituted by their discrete-time equivalents A_d, B_d, C_d after using common discretization methods. The Kalman filter's recursion consists of a measurement-update stage and of a time-update stage:

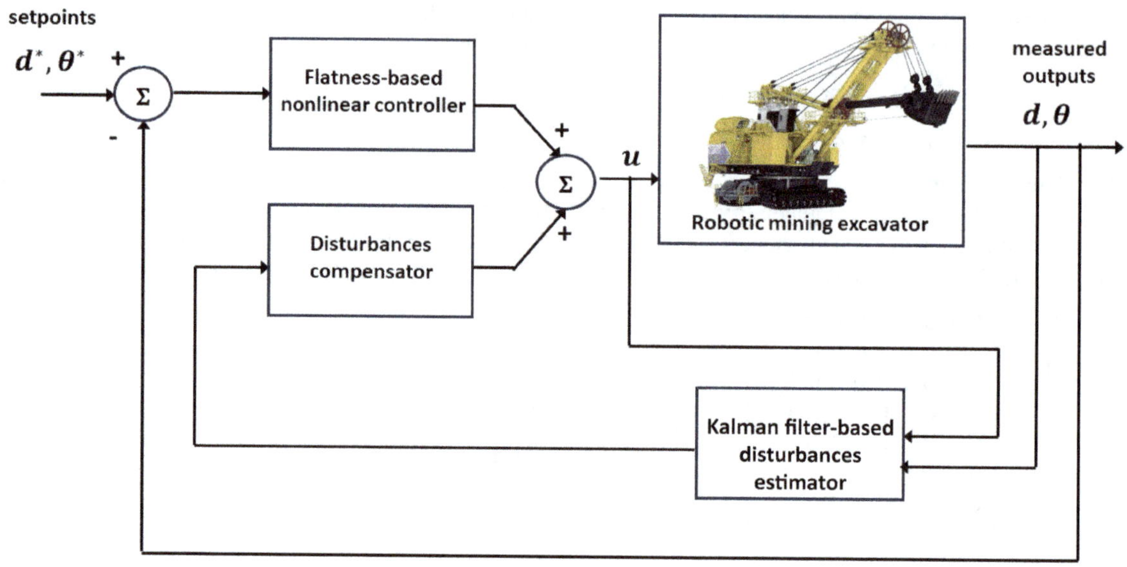

Figure 2.68 Diagram of the flatness-based disturbance observer and feedback controller for the dynamic model of the robotic excavator

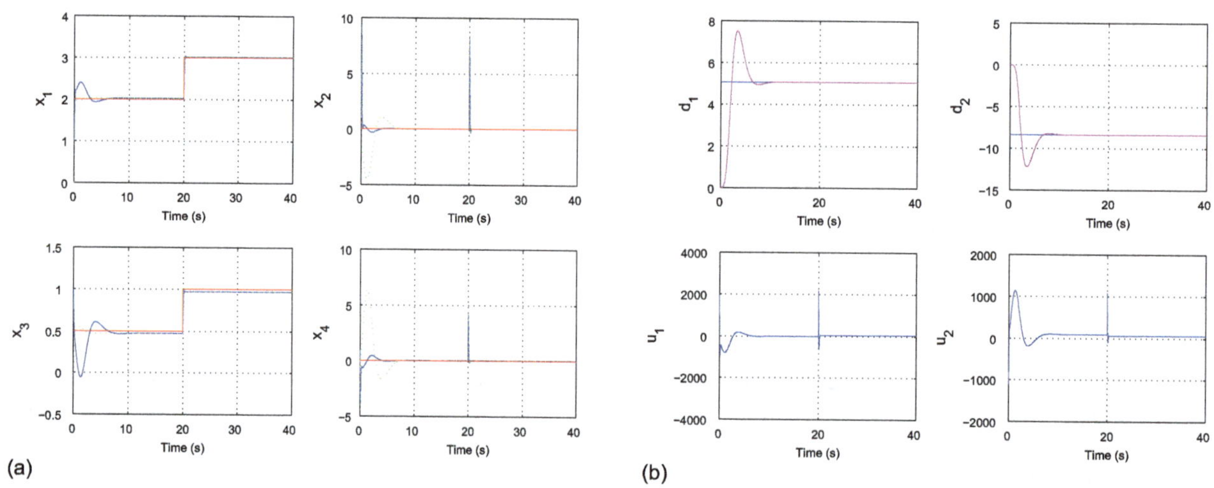

Figure 2.69 Test 1 for the robotic mining excavator: (a) tracking of setpoints (red line) by state variables x_1 to x_4 (blue line) and estimates provided by the Kalman filter (green line); (b) top row: estimated values (magenta line) of additive disturbance inputs d_1 and d_2 (blue line), bottom row: control inputs u_1 and u_2 (blue line)

measurement update:

$$
\begin{aligned}
K_f(k) &= P^-(k)C_d^T[C_dP^-(k)C_d^T + R]^{-1} \\
\hat{z}(k) &= \hat{z}^-(k) + K_f(k)[z_m - \hat{z}_m] \\
P(k) &= P^-(k) - K_f(k)C_dP^-(k)
\end{aligned}
\tag{2.231}
$$

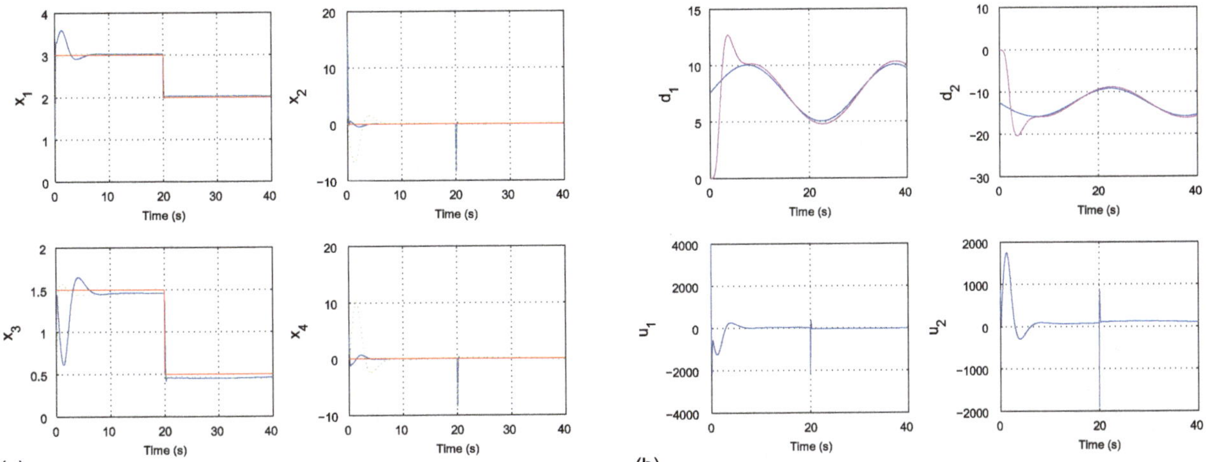

(a) (b)

Figure 2.70 Test 2 for the robotic mining excavator: (a) tracking of setpoints (red line) by state variables x_1 to x_4 (blue line) and estimates provided by the Kalman filter (green line); (b) top row: estimated values (magenta line) of additive disturbance inputs d_1 and d_2 (blue line), bottom row: control inputs u_1 and u_2 (blue line)

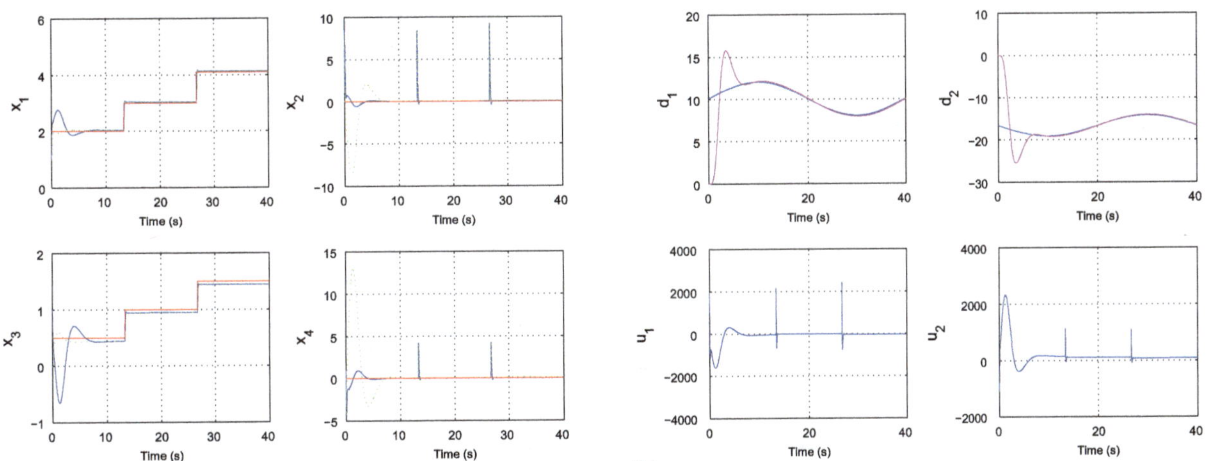

(a) (b)

Figure 2.71 Test 3 for the robotic mining excavator: (a) tracking of setpoints (red line) by state variables x_1 to x_4 (blue line) and estimates provided by the Kalman filter (green line); (b) top row: estimated values (magenta line) of additive disturbance inputs d_1 and d_2 (blue line), bottom row: control inputs u_1 and u_2 (blue line)

time update:

$$\begin{aligned} P^-(k+1) &= A_d P(k) A_d^T + Q \\ \hat{z}^-(k+1) &= A_d \hat{z}(k) + B_d v(k) \end{aligned} \tag{2.232}$$

In the equations of the Kalman filter's recursion P is the a-posteriori state vector's error covariance matrix (after receiving the output measurements at the k-th sampling interval), and P^- is the a priori state vector's error covariance matrix

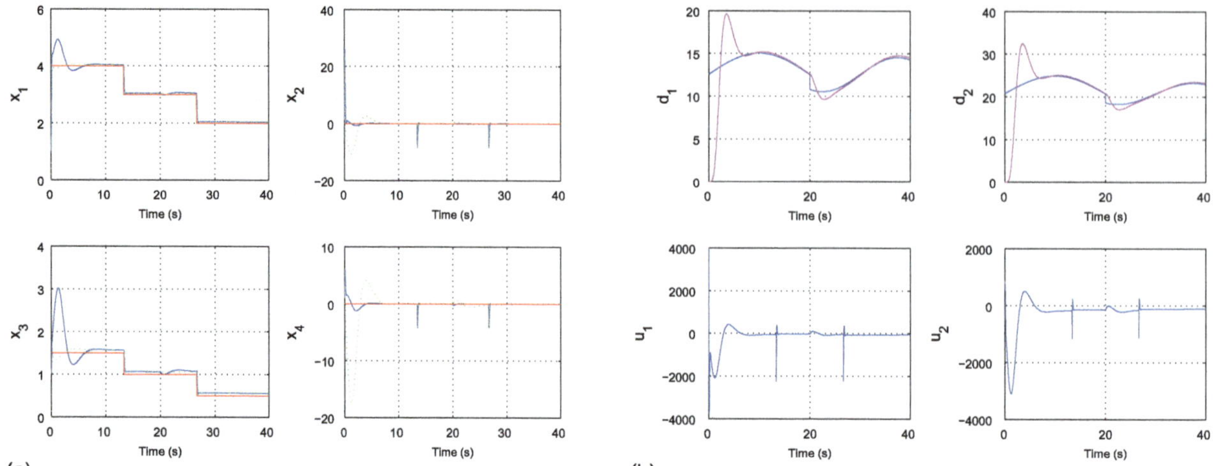

(a) (b)

Figure 2.72 *Test 4 for the robotic mining excavator: (a) tracking of setpoints (red line) by state variables x_1 to x_4 (blue line) and estimates provided by the Kalman filter (green line); (b) top row: estimated values (magenta line) of additive disturbance inputs d_1 and d_2 (blue line), bottom row: control inputs u_1 and u_2 (blue line)*

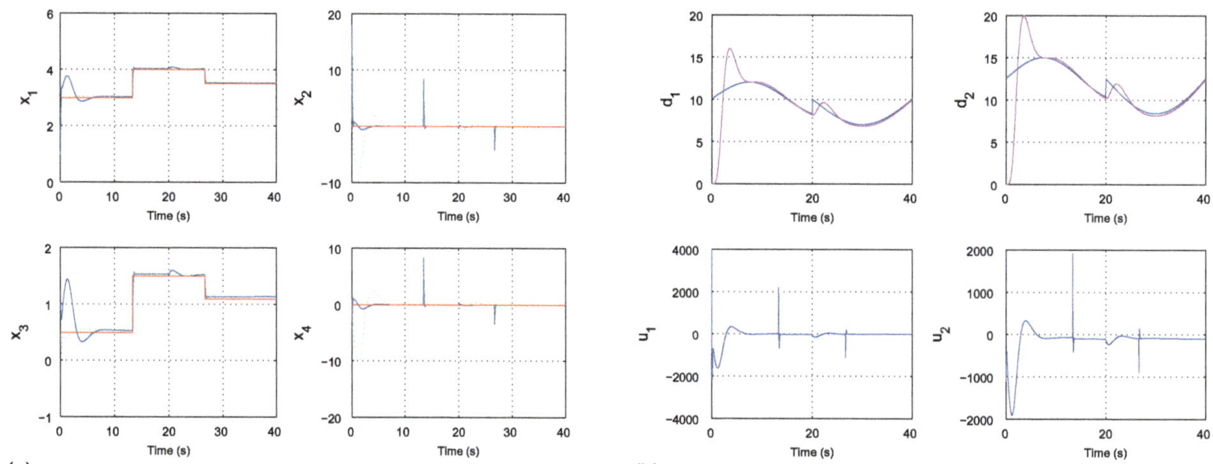

(a) (b)

Figure 2.73 *Test 5 for the robotic mining excavator: (a) tracking of setpoints (red line) by state variables x_1 to x_4 (blue line) and estimates provided by the Kalman filter (green line); (b) top row: estimated values (magenta line) of additive disturbance inputs d_1 and d_2 (blue line), bottom row: control inputs u_1 and u_2 (blue line)*

(before receiving the output measurements at the k-th sampling interval). Besides, Q denotes the process noise covariance matrix, while R is the measurement noise covariance matrix. As confirmed by the related observability matrix the system is observable, thus ensuring that

$$lim_{t\to\infty}\hat{z} = z \tag{2.233}$$

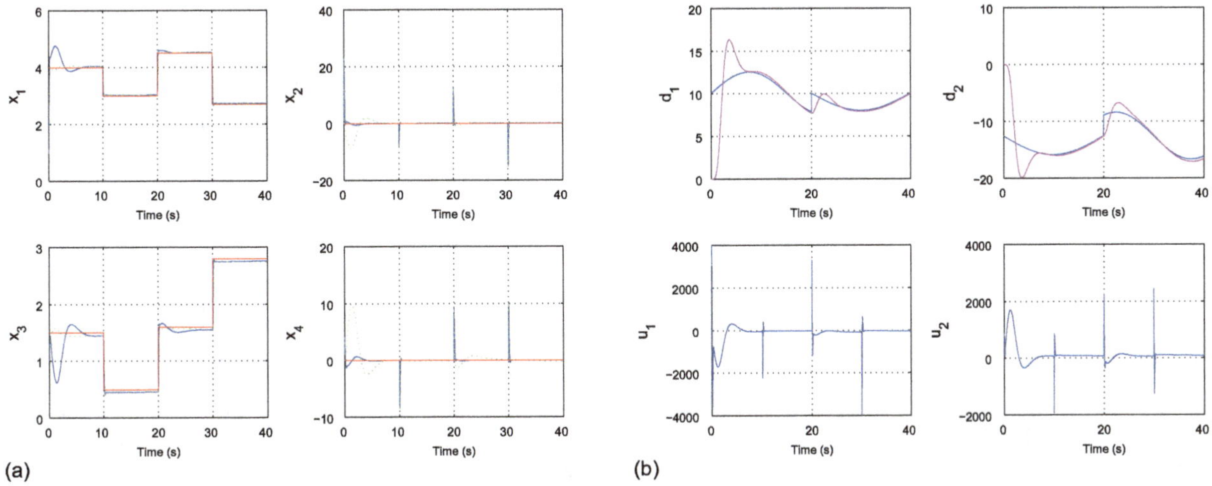

Figure 2.74 Test 6 for the robotic mining excavator: (a) tracking of setpoints (red line) by state variables x_1 to x_4 (blue line) and estimates provided by the Kalman filter (green line); (b) top row: estimated values (magenta line) of additive disturbance inputs d_1 and d_2 (blue line), bottom row: control inputs u_1 and u_2 (blue line)

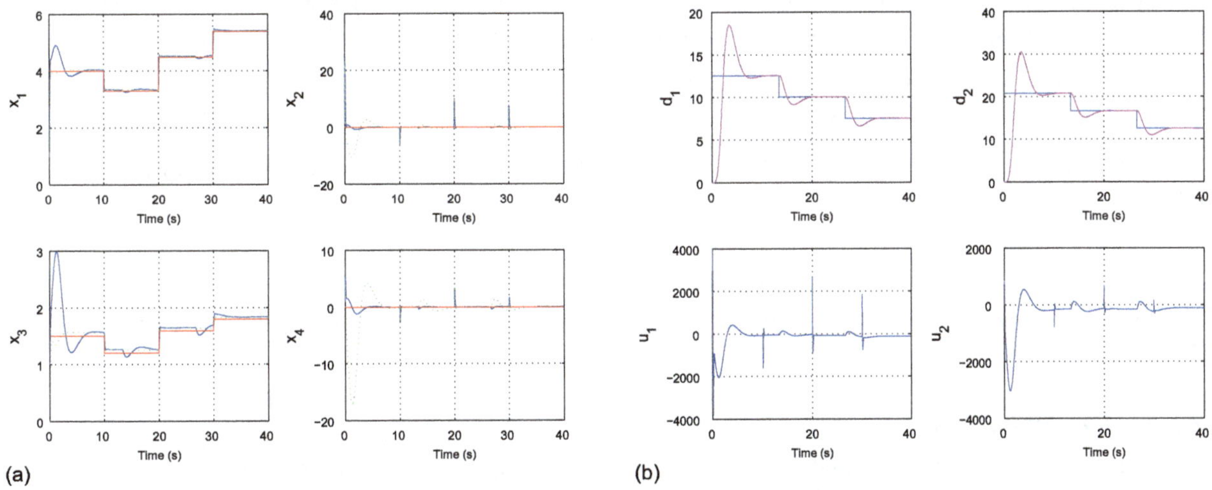

Figure 2.75 Test 7 for the robotic mining excavator: (a) tracking of setpoints (red line) by state variables x_1 to x_4 (blue line) and estimates provided by the Kalman filter (green line); (b) top row: estimated values (magenta line) of additive disturbance inputs d_1 and d_2 (blue line), bottom row: control inputs u_1 and u_2 (blue line)

and particularly that

$$\begin{aligned} &lim_{t\to\infty}\hat{z}_5 = z_5 \Rightarrow lim_{t\to\infty}\hat{d}_1 = d_1 \\ &lim_{t\to\infty}\hat{z}_7 = z_7 \Rightarrow lim_{t\to\infty}\hat{d}_2 = d_2 \end{aligned} \tag{2.234}$$

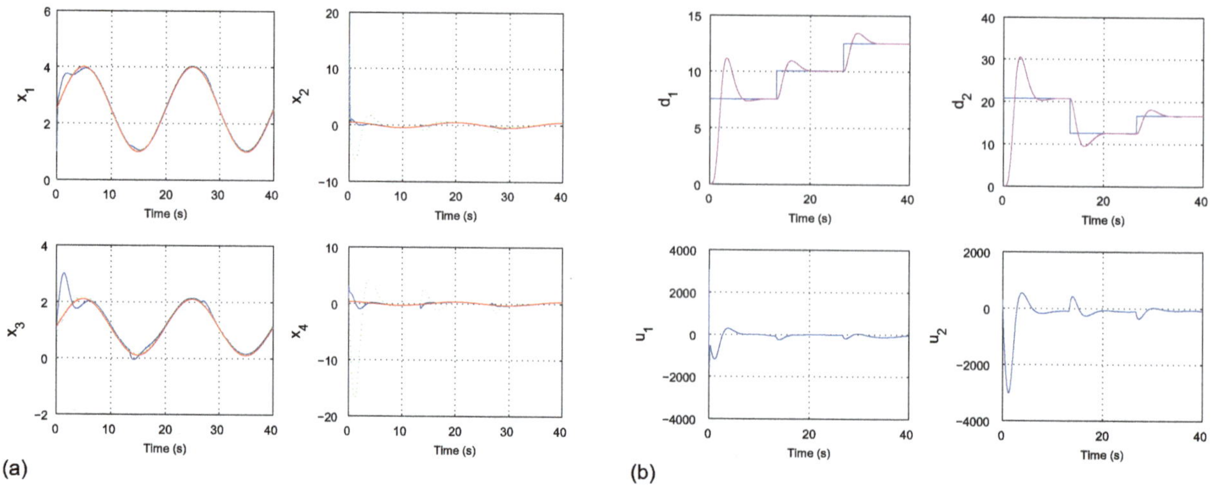

Figure 2.76 Test 8 for the robotic mining excavator: (a) tracking of setpoints (red line) by state variables x_1 to x_4 (blue line) and estimates provided by the Kalman filter (green line); (b) top row: estimated values (magenta line) of additive disturbance inputs d_1 and d_2 (blue line), bottom row: control inputs u_1 and u_2 (blue line)

The dynamic model of the excavation robot has been written in the canonical Brunovsky form which is linear. Thus, the separation principle holds. With the estimates of $\hat{d}_1$ and $\hat{d}_2$ the stabilizing feedback controller becomes:

$$\begin{aligned} v_1 &= \ddot{x}_{1,d} - K_{1,1}(\dot{x}_1 - \dot{x}_{1,d}) - K_{2,1}(x_1 - x_{1,d}) - \hat{d}_1 \\ v_2 &= \ddot{x}_{3,d} - K_{1,2}(\dot{x}_3 - \dot{x}_{3,d}) - K_{2,2}(x_3 - x_{3,d}) - \hat{d}_2 \end{aligned} \tag{2.235}$$

The real control inputs u_1, u_2 which should be applied to the initial nonlinear state-space model of the excavating robot are

$$\begin{pmatrix} u_1 \\ u_2 \end{pmatrix} = \begin{pmatrix} g_{21}(x) & g_{22}(x) \\ g_{41}(x) & g_{42}(x) \end{pmatrix}^{-1} \begin{pmatrix} v_1 - f_2(x) \\ v_2 - f_4(x) \end{pmatrix} \tag{2.236}$$

The cumulative diagram of the flatness-based disturbance observer and feedback controller for the dynamic model of the robotic excavator is given in Figure 2.68.

2.5.5 Simulation tests

The performance of the flatness-based disturbance observer and feedback control scheme for robotic excavators has been further confirmed through simulation experiments. The obtained results are depicted in Figures 2.69–2.76. Eight different simulation experiments have been considered, with different setpoints for the state variables of the robotic excavator and with different disturbance inputs affecting this robotic system. The transient performance of the control loop is determined by the values of the feedback gains $K_{1,1}$, $K_{2,1}$ and $K_{1,2}$, $K_{2,2}$. The following diagrams are given below: (i) diagrams depicting the convergence of the state variables x_1 to x_4 of the robotic mining excavator to the associated setpoints, (ii) diagrams presenting the estimation of the unknown additive disturbance inputs by the disturbance observer for the dynamic model of the robotic excavator, (iii) diagrams showing the variations of the control inputs. The distance along the dipper handle d is measured in m, and the associated linear velocity $\dot{d}$ is measured in m/sec. The turn angle θ of the dipper handle is measured in rad and the associated angular velocity $\dot{\theta}$ is measured in rad/s. Moreover, the control input u_1 which is the torque enabling the rotation of the dipper handle is measured in Nt·m, while the control input u_2 which is the longitudinal force causing the variation of distance d is measured in Nt. It has been confirmed that under the proposed control scheme fast and accurate tracking of setpoints is achieved by all state variables of the robotic excavator. Besides, the method allows for precise estimation of the perturbation terms that are exerted on the mining robot.

Chapter 3
Nonlinear optimal and flatness-based control for cranes and pendulums I

Abstract

The chapter analyzes nonlinear optimal and flatness-based control methods for robotic cranes and underactuated pendulums. First, the problem of nonlinear optimal control of the 4-DOF robotic tower crane is solved. Next, the problem of nonlinear optimal control of the dual PMLSM-driven gantry crane is treated. Moreover, a nonlinear optimal control scheme is developed for the two-cable-driven parallel robot which functions as a crane. Finally, nonlinear optimal control is applied to the dynamic model of the four-cable-driven parallel robot which functions as a crane.

3.1 Nonlinear optimal control of the underactuated 4-DOF robotic tower crane

3.1.1 Outline

The present monograph advances to the analysis of the use of its nonlinear optimal and flatness-based control methods in robotic cranes and pendulums. Tower cranes are complicated mechatronic systems which are widely used for the transportation of payloads in a large workspace and at large height [721,985,833]. Tower cranes can accomplish accurate transport and positioning of heavy items in construction sites, in multistory building erection, in loading and unloading of cargos in ports, and in storage tasks for the logistics industry [1051,233,687]. Tower cranes exhibit the advantages of simple structure, convenient installation, low cost, high payload capacity, and low energy consumption [437,83,945]. Tower cranes usually consist of an actuated jib arm, an actuated translational trolley, and an unactuated payload below the trolley [[946],142,833]. Tower cranes have a complicated and strongly nonlinear dynamics and are also subject to underactuation [531,476,825]. Major issues in the precise and safe operation of tower cranes are the accurate trajectory tracking about the cranes' moving parts, as well as the suppression of the swinging and sway motion of the payload which can emerge when large and fast maneuvers take place [951,725,732]. Under such operating conditions, tower cranes are also exposed to bending and torsional deformations; therefore, their operation is a demanding and delicate task [702,1001]. Even nowadays, most cranes are manually operated by skilled personnel. However, for human operators, it is often difficult to achieve precise positioning and suppression of swinging and swaying phenomena [826,140]. Operational mistakes occur frequently when pursuing highly repetitive tasks under time-pressure. To avoid the deficiencies of the human operation of cranes, automation of such systems has become a pre-requisite, and to this end, elaborated nonlinear control methods have been developed [1055,1052].

The use of robotic tower cranes is rapidly deploying in several constructions, industrial and supply-chain applications, and the treatment of the associated nonlinear control problems has become a necessity. To this end, new results have been recently developed. In [987], a neural adaptive control method is introduced for 4-DOF tower cranes. In [988], a 2-DOF overhead and a 4-DOF rotary crane are controlled through a neural adaptive control scheme. In [996], a control method is presented about vibration suppression in a 4-DOF tower crane with a flexible jib. In [1059], a neural adaptive control scheme is developed for a double-pendulum tower crane. In [1059], a 4-DOF tower crane is controlled by a nonlinear controller that consists of a PD control part and a sliding-mode control part. In [249], an input-shaping control technique is applied to suppress oscillations in the 4-DOF underactuated robotic crane. Finally, in [370], an adaptive control scheme is proposed for an underactuated 2-DOF overhead crane.

In the present section, a nonlinear optimal control method is proposed for the 4-DOF nonlinear and underactuated model of tower cranes [719,713]. The dynamic model of the 4-DOF underactuated tower crane undergoes first approximate linearization around a temporary operating point which is updated at each sampling instance. This operating point is defined by the present value of the crane's state vector and by the last sampled value of the control inputs vector. The linearization process relies on first-order Taylor series expansion and on the computation of the associated Jacobian matrices [705,60,706]. The modeling error, which is due to the truncation of higher-order terms in the Taylor series expansion, is proven to be small and is asymptotically compensated by the robustness of the control algorithm. For the approximately linearized state-space description of the system, a stabilizing H-infinity feedback controller is defined.

The proposed H-infinity controller achieves the solution of the optimal control problem for the tower crane under model uncertainty and external perturbations [725]. Actually, it represents a min–max differential game which takes place between (i) the control inputs of the system that try to minimize a cost function comprising a quadratic term of the state vector's tracking error (2) the model uncertainty and exogenous perturbation terms which try to maximize this cost function. To compute the stabilizing feedback gains of this controller, an algebraic Riccati equation has to be also solved at each time-step of the control method [708,711,715]. The global stability properties of the control scheme are proven through Lyapunov analysis. First, it is proven that the control loop satisfies the H-infinity tracking performance criterion [725,876]. Next, it is proven that under moderate conditions, global asymptotic stability of the control loop is ensured. To implement state estimation-based control without the need to measure the entire state vector of the system, the H-infinity Kalman filter is used as a robust state estimator. The nonlinear optimal control method retains the advantages of linear optimal control, that is fast and accurate tracking of reference setpoints under moderate variations of the control inputs [725].

The present section has its own meaningful contribution to the area of nonlinear control. One can point out the advantages of the nonlinear optimal control method against nonlinear model predictive control (NMPC) [725]. In NMPC, the stability properties of the control scheme remain unproven, and the convergence of the iterative search for an optimum often depends on initialization and parameter values' selection. It is also noteworthy that the nonlinear optimal control method is applicable to a wider class of dynamic systems than approaches based on the solution of state-dependent Riccati equations (SDRE). The SDRE approaches can be applied only to dynamic systems which can be transformed to the linear parameter varying (LPV) form. Besides, the nonlinear optimal control method performs better than nonlinear optimal control schemes which use approximation of the solution of the Hamilton–Jacobi–Bellman equation by Galerkin series expansions. The stability properties of the Galerkin series expansion-based optimal control approaches are still unproven.

3.1.2 Dynamic model of the robotic tower crane

3.1.2.1 State-of-the-art in the control of underactuated robotic cranes

As noted, robotic cranes can find use in construction works, in buildings erection, in loading and unloading of cargo ships in ports, in heavy material lifting for several industrial tasks, and in heavy items pick-and-place tasks in logistics. The solution of the nonlinear optimal control problem for underactuated robotic cranes is important for ensuring the reliable performance of robotic cranes in such applications. In the following, the state-of-the-art in the control of underactuated robotic cranes is outlined. To this end, indicative results are overviewed about (a) the use of established model-based and model-free control approaches in robotics and (b) the use of the proposed nonlinear optimal control methods in specific types of autonomous cranes.

There exist several recent results showing the performance of advanced nonlinear control techniques to the stabilization and trajectory tracking problem of robotic cranes. In [786], Lyapunov stability analysis is used to select the gains of a PID-type controller so as to ensure global stability for an overhead robotic crane. In [1059], a feedback controller which comprises a PID-type control part and an SMC-type control part is proposed for a 4-DOF robotic tower crane. Lyapunov stability analysis is used to select the feedback gains of the controller and to ensure global stability of the control loop. In [1033], an adaptive fuzzy controller jointly with a disturbances estimator is used to stabilize an offshore boom crane. The controller's feedback gains are chosen through Lyapunov stability analysis so as to ensure global stability for the control loop. In [534], an adaptive PID-type controller is developed for dual rotary cranes. The selection of the controller's gains is performed through the system's Lyapunov analysis. Global stability is proven through the Lyapunov technique and LaSalle's invariance principle. In [952], adaptive output feedback control is proposed for a double-pendulum ship-mounted crane while also considering uncertainty about the cargo's mass. The controller's gains and the adaptation law which are used for compensating for the model's uncertainty are chosen through Lyapunov analysis. In [370], a state-variables transformation and the backstepping control technique are used for treating the stabilization of a 2-DOF crane. The stability properties of the control method are proven through Lyapunov analysis. In [304], an energy-based control method is proposed for the dynamic model of a large gantry crane which comprises also elements that are subjected to elastic deformation. Stability properties are demonstrated with the use of Lyapunov analysis. In [883], a knuckle boom crane is controlled by a cascade controller where the inner control loop is designed to damp out the pendulum oscillation, and the crane tip is controlled by the outer loop. An extended Kalman filter is used for the estimation of payload angles and angular velocity. In [1066], modified tracking error variables and Lyapunov stability analysis are used to define a feedback controller that finally stabilizes a 4-DOF overhead crane. The global stability proof makes use of LaSalle's invariance principle. Moreover, in [884], the dynamic model of a knuckle boom crane is considered and a stabilizing feedback controller is designed about it using the nonlinear model predictive control concept and a Lyapunov analysis-based auxiliary controller.

In particular, the proposed nonlinear optimal control method is suitable for use in a variety of autonomous robotic cranes and several related results have been obtained [721,725,732]. Thus, one can note nonlinear optimal control for (i) the 4-DOF overhead crane, (ii) offshore boom cranes, (iii) the double-pendulum overhead crane, (iv) the 4-DOF tower crane, a quay-side cranes. The use of the nonlinear optimal control method in robotic systems is not limited to specific state-space forms (e.g., canonical forms, strict feedback, or pure feedback forms). The computational complexity of the nonlinear optimal control method remains also moderate because (a) it does not require complicated changes of state variables and transformations of the cranes' state-space representation, (b) it does not rely on matrices for computing the robots' control inputs, and consequently, it avoids the risk of singularities, (c) it performs linearization around one single operating point which is easy to define, (d) it does not require the solution of multiple Riccati equations or LMIs, and (e) it avoids the deadlocks of pole-placement techniques which arise in the case of underactuation.

3.1.2.2 State-space model of the tower crane

The diagram of the 4-DOF robotic tower crane is depicted in Figure 3.1. The tower crane consists of an actuated jib arm, an actuated translational trolley, and an unactuated payload below the trolley. The trolley of mass M can perform a one-dimensional motion along the jib. The load mass m is suspended from the vehicle with a rope of length L. The jib can rotate around the vertical axis. The state variables of the tower crane are defined as follows: (i) ϕ is the turn angle of the crane, (ii) $\dot{\phi}$ is the rate of change of the turn angle of the crane, (iii) η is the displacement of the trolley along the jib, (iv) $\dot{\eta}$ is the velocity of the jib, (v) δ_1 is the primary projection angle of the link that connects the trolley with the suspended load (swinging angle), (vi) $\dot{\delta}_1$ is the rate of change of the primary projection angle, (vii) δ_2 is the secondary projection angle of the link that connects the trolley with the suspended load (sway angle), and (viii) $\dot{\delta}_2$ is the rate of change of the secondary projection angle [985,833].

The vector of the main variables of the robotic tower crane is defined as $q = [\phi, \eta, \delta_1, \delta_2]^T$. The dynamic model of the crane is given by [721,985,725]

$$M(q)\ddot{q} + \tilde{C}(q,\dot{q})\dot{q} + G(q) = T - \tilde{d} \tag{3.1}$$

The inertia matrix of the crane $M(q)$ is symmetric and positive definite and is defined as [985,833]

$$M(q) = \begin{pmatrix} m_{11} & m_{12} & m_{13} & m_{14} \\ m_{21} & m_{22} & m_{23} & m_{24} \\ m_{31} & m_{32} & m_{33} & m_{34} \\ m_{41} & m_{42} & m_{43} & m_{44} \end{pmatrix} \tag{3.2}$$

where $m_{11} = m(sin^2(\delta_1)cos^2(\delta_2) + sin^2(\delta_2))L^2 + 2mL\eta cos(\delta_2)sin(\delta_2) + (m+M)\eta^2 + J$, $m_{12} = m_{21} = -mLsin(\delta_2)$, $m_{13} = m_{31} = -mL^2cos(\delta_2)cos(\delta_2)sin(\delta_2)$, $m_{14} = m_{41} = mL(\eta cos(\delta_2) + Lsin(\delta_2))$, $m_{22} = m + M$, $m_{23} = m_{32} = mLcos(\delta_1)cos(\delta_2)$, $m_{24} = m_{42} = -mLsin(\delta_1)sin(\delta_2)$, $m_{33} = mL^2cos(\delta_2)$, $m_{34} = m_{43} = -mLsin(\delta_1)sin(\delta_2)$, $m_{44} = mL^2$.

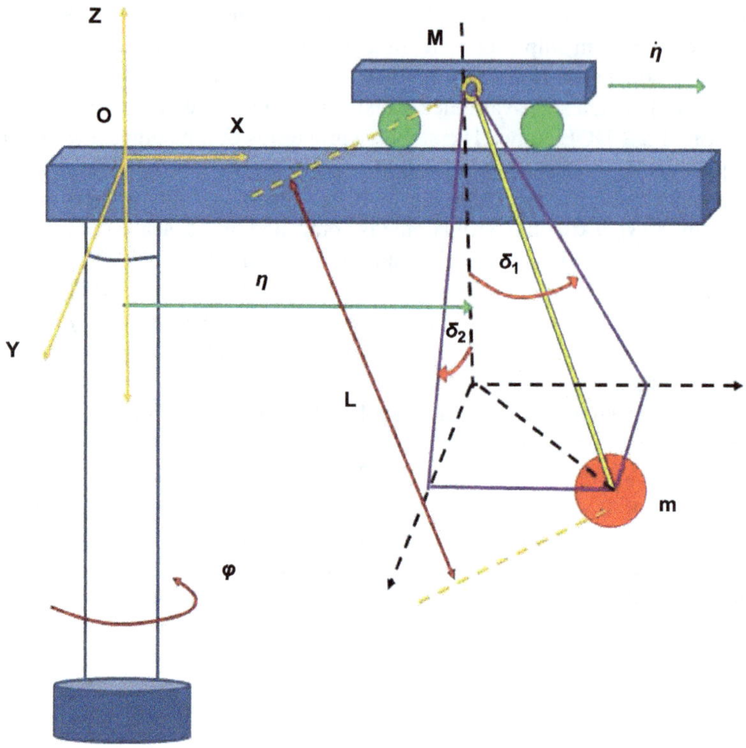

Figure 3.1 Diagram of the 4-DOF robotic tower crane

The Coriolis and centrifugal forces matrix of the crane is [985,833]

$$\tilde{C}(q,\dot{q}) = \begin{pmatrix} c_{11} & c_{12} & c_{13} & c_{14} \\ c_{21} & c_{22} & c_{23} & c_{24} \\ c_{31} & c_{32} & c_{33} & c_{34} \\ c_{41} & c_{42} & c_{43} & c_{44} \end{pmatrix} \tag{3.3}$$

where

$$\begin{gathered} c_{11} = (m+M)\eta\dot{\eta} + mL\eta cos(\delta_1)cos(\delta_2)\dot{\delta}_1 - mL\eta cos(\delta_1)sin(\delta_2)\dot{\delta}_2 - \\ mLsin(\delta_1)sin(\delta_2)\dot{\eta} + mL^2 sin(\delta_1)cos(\delta_1)cos^2(\delta_2)\dot{\delta}_1 + \\ mL^2 cos^2(\delta_1)sin(\delta_2)cos(\delta_2)\dot{\delta}_2. \end{gathered}$$

$$c_{12} = mLsin(\delta_1)cos(\delta_2)\dot{\phi} + (m+M)\eta\dot{\phi}$$

$$\begin{gathered} c_{13} = mL\eta cos(\delta_1)cos(\delta_2)\dot{\phi} + mL^2 sin(\delta_1)sin(\delta_2)cos(\delta_2)\dot{\delta}_1 \\ + mL^2 sin(\delta_1)cos(\delta_1)cos^2(\delta_2)\dot{\phi} + mL^2 cos(\delta_1)sin^2(\delta_2)\dot{\delta}_2 \end{gathered}$$

$$\begin{gathered} c_{14} = -mL\eta sin(\delta_1)sin(\delta_2)\dot{\phi} - mL\eta sin(\delta_2)\dot{\delta}_2 + \\ mL^2 cos^2(\delta_1)sin(\delta_2)cos(\delta_2)\dot{\phi} + mL^2 cos(\delta_1)sin^2(\delta_2)\dot{\delta}_2 \end{gathered}$$

$c_{21} = -(m+M)\eta\dot{\phi} - mLsin(\delta_1)cos(\delta_2)\dot{\phi} - mLcos(\delta_2)\dot{\delta}_2$, $c_{22}=0$, $c_{23} = -mL(sin(\delta_1)cos(\delta_2)\dot{\delta}_1 + cos(\delta_1)sin(\delta_2)\dot{\delta}_2)$,
$c_{24} = -mL(cos(\delta_1)sin(\delta_2)\dot{\delta}_1 + sin(\delta_1)cos(\delta_2)\dot{\delta}_2 + cos(\delta_2)\dot{\phi})$.
$c_{31} = -mLcos(\delta_1)cos(\delta_2)(\eta + Lcos(\delta_1)cos(\delta_2))\dot{\phi} - mL^2cos(\delta_1)cos^2(\delta_2)\dot{\delta}_2$, $c_{32}=0$, $c_{33} = -mL^2 sin(\delta_2)cos(\delta_2)\dot{\delta}_2$,
$c_{34} = -mL^2(cos(\delta_1)cos^2(\delta_2)\dot{\phi} + sin(\delta_2)cos(\delta_2)\dot{\delta}_1)$,

$$\begin{aligned} c_{41} = mL(\eta sin(\delta_1)sin(\delta_2) - Lcos^2(\delta_1)sin(\delta_2)cos(\delta_2))\dot{\phi} + \\ + mL^2cos(\delta_1)cos^2(\delta_2)\dot{\delta}_1 + mLcos(\delta_2)\dot{\eta} \end{aligned}$$

$c_{42} = mLcos(\delta_2)\dot{\phi}$, $c_{43} = mL^2(cos(\delta_1)cos^2(\delta_2)\dot{\phi} + sin(\delta_2)cos(\delta_2)\dot{\delta}_1)$, $c_{44}=0$.

The gravitational vector of the robotic crane is defined as [985,833]

$$G = \begin{pmatrix} g_1 \\ g_2 \\ g_3 \\ g_4 \end{pmatrix} = \begin{pmatrix} 0 \\ 0 \\ gmLsin(\delta_1)sin(\delta_2) \\ gmLcos(\delta_1)sin(\delta_2) \end{pmatrix} \tag{3.4}$$

The disturbances and friction vector of the robotic crane is defined as

$$\tilde{d} = \begin{pmatrix} d_1 \\ d_2 \\ d_3 \\ d_4 \end{pmatrix} = \begin{pmatrix} -w_1\dot{\phi} \\ -w_2\dot{\eta} \\ -w_3\dot{\delta}_1 \\ -w_4\dot{\delta}_2 \end{pmatrix} \tag{3.5}$$

The complexity of the dynamics of the model of the tower crane can be partially alleviated by considering moderate (small angle) swinging or sway motion for the load. In such a case, the associated sinusoidal and cosinusoidal variables $sin(\delta_i)$ and $cos(\delta_i)$, $i=1,2$ can be substituted by the first terms of the related Taylor series expansions. Next, the inverse of the inertia matrix M is defined as

$$N^{-1} = \frac{1}{detM}\begin{pmatrix} M_{11} & -M_{21} & M_{31} & -M_{41} \\ -M_{12} & M_{22} & -M_{32} & M_{42} \\ M_{13} & -M_{23} & M_{33} & -M_{43} \\ -M_{14} & M_{24} & -M_{34} & M_{44} \end{pmatrix} \tag{3.6}$$

where the above noted sub-determinants M_{ij} $i=1,\cdots,4$ and $j=1,\cdots,4$ are defined as

$$\begin{aligned} M_{11} &= m_{22}(m_{33}m_{44} - m_{43}m_{34}) - m_{23}(m_{32}m_{44} - m_{43}m_{34} + m_{24})(m_{32}m_{43} - m_{42}m_{33}) \\ M_{12} &= m_{21}(m_{33}m_{44} - m_{43}m_{34}) - m_{23}(m_{31}m_{44} - m_{41}m_{34} + m_{24})(m_{31}m_{43} - m_{41}m_{33}) \\ M_{13} &= m_{21}(m_{32}m_{44} - m_{42}m_{34}) - m_{22}(m_{31}m_{44} - m_{41}m_{34} + m_{24})(m_{31}m_{43} - m_{41}m_{33}) \\ M_{14} &= m_{21}(m_{32}m_{43} - m_{42}m_{33}) - m_{22}(m_{31}m_{43} - m_{41}m_{33} + m_{23})(m_{31}m_{42} - m_{41}m_{32}) \end{aligned}$$

$$\begin{aligned} M_{21} &= m_{12}(m_{33}m_{44} - m_{43}m_{34}) - m_{13}(m_{32}m_{44} - m_{42}m_{34} + m_{14})(m_{32}m_{43} - m_{42}m_{33}) \\ M_{22} &= m_{11}(m_{33}m_{44} - m_{43}m_{34}) - m_{13}(m_{31}m_{44} - m_{41}m_{34} + m_{14})(m_{31}m_{43} - m_{41}m_{33}) \\ M_{23} &= m_{11}(m_{32}m_{44} - m_{42}m_{34}) - m_{12}(m_{31}m_{44} - m_{41}m_{34} + m_{14})(m_{31}m_{42} - m_{41}m_{32}) \\ M_{24} &= m_{11}(m_{32}m_{43} - m_{42}m_{33}) - m_{12}(m_{31}m_{43} - m_{41}m_{33} + m_{13})(m_{31}m_{42} - m_{41}m_{32}) \end{aligned}$$

$$\begin{aligned} M_{31} &= m_{12}(m_{23}m_{44} - m_{43}m_{24}) - m_{13}(m_{22}m_{44} - m_{42}m_{24} + m_{14})(m_{22}m_{43} - m_{42}m_{23}) \\ M_{32} &= m_{11}(m_{23}m_{44} - m_{43}m_{24}) - m_{13}(m_{21}m_{44} - m_{41}m_{24} + m_{14})(m_{21}m_{43} - m_{41}m_{23}) \\ M_{33} &= m_{11}(m_{22}m_{44} - m_{42}m_{24}) - m_{12}(m_{21}m_{44} - m_{41}m_{24} + m_{14})(m_{41}m_{22} - m_{21}m_{42}) \\ M_{34} &= m_{11}(m_{22}m_{43} - m_{42}m_{23}) - m_{12}(m_{21}m_{43} - m_{41}m_{23} + m_{13})(m_{21}m_{42} - m_{41}m_{22}) \end{aligned}$$

$$M_{41}=m_{12}(m_{23}m_{34}-m_{33}m_{24})-m_{13}(m_{22}m_{34}-m_{32}m_{24}+m_{14})(m_{22}m_{33}-m_{32}m_{23})$$
$$M_{42}=m_{11}(m_{23}m_{34}-m_{33}m_{44})-m_{13}(m_{21}m_{34}-m_{31}m_{24}+m_{14})(m_{21}m_{33}-m_{31}m_{23})$$
$$M_{43}=m_{11}(m_{22}m_{34}-m_{32}m_{24})-m_{12}(m_{21}m_{34}-m_{31}m_{24}+m_{14})(m_{21}m_{32}-m_{31}m_{22})$$
$$M_{44}=m_{11}(m_{22}m_{33}-m_{32}m_{23})-m_{12}(m_{21}m_{33}-m_{31}m_{23}+m_{13})(m_{21}m_{32}-m_{31}m_{22})$$

The determinant of matrix M is

$$detM=m_{11}M_{11}-m_{12}M_{12}+m_{13}M_{13}-m_{14}M_{14}$$

For the dynamic model of the robotic crane that was initially written in the form

$$M(q)\ddot{q}+\tilde{C}(q,\dot{q})\dot{q}+G(q)+\tilde{d}=T \tag{3.7}$$

It holds that

$$\tilde{C}(q,\dot{q})\dot{q}=\begin{pmatrix} c_{11} & c_{12} & c_{13} & c_{14} \\ c_{21} & c_{22} & c_{23} & c_{24} \\ c_{31} & c_{32} & c_{33} & c_{34} \\ c_{41} & c_{42} & c_{43} & c_{44} \end{pmatrix}\begin{pmatrix} \dot{\phi} \\ \dot{\eta} \\ \dot{\delta}_1 \\ \dot{\delta}_2 \end{pmatrix} \tag{3.8}$$

or equivalently

$$C(q,\dot{q})=\begin{pmatrix} c_1 \\ c_2 \\ c_3 \\ c_4 \end{pmatrix}=\begin{pmatrix} c_{11}\dot{\phi} & c_{12}\dot{\eta} & c_{13}\dot{\delta}_1 & c_{14}\dot{\delta}_2 \\ c_{21}\dot{\phi} & c_{22}\dot{\eta} & c_{23}\dot{\delta}_1 & c_{24}\dot{\delta}_2 \\ c_{31}\dot{\phi} & c_{32}\dot{\eta} & c_{33}\dot{\delta}_1 & c_{34}\dot{\delta}_2 \\ c_{41}\dot{\phi} & c_{42}\dot{\eta} & c_{43}\dot{\delta}_1 & c_{44}\dot{\delta}_2 \end{pmatrix} \tag{3.9}$$

Consequently, the dynamic model of the robot can be written as

$$\begin{gathered} M(q)\ddot{q}+C(q,\dot{q})+G(q)+\tilde{d}=T\Rightarrow \\ \ddot{q}+M^{1}(q)C(q,\dot{q})+M^{1}(q)G(q)+M^{1}(q)\tilde{d}=M^{1}(q)T\Rightarrow \\ \ddot{q}=-M^{1}(q)C(q,\dot{q})-M^{1}(q)G(q)-M^{1}(q)\tilde{d}+M^{1}(q)T\Rightarrow \\ \ddot{q}=-M^{1}(q)[C(q,\dot{q})-G(q)+\tilde{d}]+M^{-1}(q)T \end{gathered} \tag{3.10}$$

Next, the dynamic model of the robotic crane is written as

$$\begin{aligned} \begin{pmatrix} \ddot{\phi} \\ \ddot{\eta} \\ \ddot{\delta}_1 \\ \ddot{\delta}_2 \end{pmatrix} = -\frac{1}{detM}\begin{pmatrix} M_{11} & -M_{21} & M_{31} & -M_{41} \\ -M_{12} & M_{22} & -M_{32} & M_{42} \\ M_{13} & -M_{23} & M_{33} & -M_{43} \\ -M_{14} & M_{24} & -M_{34} & M_{44} \end{pmatrix}\begin{pmatrix} c_1+g_1+d_1 \\ c_2+g_2+d_2 \\ c_3+g_3+d_3 \\ c_4+g_4+d_4 \end{pmatrix} + \\ \frac{1}{detM}\begin{pmatrix} M_{11} & -M_{21} & M_{31} & -M_{41} \\ -M_{12} & M_{22} & -M_{32} & M_{42} \\ M_{13} & -M_{23} & M_{33} & -M_{43} \\ -M_{14} & M_{24} & -M_{34} & M_{44} \end{pmatrix}\begin{pmatrix} T_1 \\ T_2 \\ 0 \\ 0 \end{pmatrix} \end{aligned} \tag{3.11}$$

Equivalently, the dynamic model of the robotic crane is written as

$$\begin{pmatrix} \ddot{\phi} \\ \ddot{\eta} \\ \ddot{\delta}_1 \\ \ddot{\delta}_2 \end{pmatrix} = \begin{pmatrix} \frac{-M_{11}(c_1+g_1+d_1)+M_{21}(c_2+g_2+d_2)-M_{31}(c_3+g_3+d_3)+M_{41}(c_4+g_4+d_4)}{detM} \\ \frac{M_{12}(c_1+g_1+d_1)-M_{22}(c_2+g_2+d_2)+M_{32}(c_3+g_3+d_3)-M_{42}(c_4+g_4+d_4)}{detM} \\ \frac{-M_{13}(c_1+g_1+d_1)+M_{23}(c_2+g_2+d_2)-M_{33}(c_3+g_3+d_3)+M_{43}(c_4+g_4+d_4)}{detM} \\ \frac{M_{14}(c_1+g_1+d_1)-M_{24}(c_2+g_2+d_2)+M_{34}(c_3+g_3+d_3)-M_{44}(c_4+g_4+d_4)}{detM} \end{pmatrix} + \begin{pmatrix} \frac{M_{11}}{detM} & -\frac{M_{21}}{detM} \\ -\frac{M_{12}}{detM} & \frac{M_{22}}{detM} \\ \frac{M_{13}}{detM} & -\frac{M_{23}}{detM} \\ -\frac{M_{14}}{detM} & \frac{M_{24}}{detM} \end{pmatrix} \begin{pmatrix} v_1 \\ v_2 \end{pmatrix} \tag{3.12}$$

The state vector of the robotic crane is

$$\begin{aligned} x &= [x_1, x_2, x_3, x_4, x_5, x_6, x_7, x_8]^T \Rightarrow \\ x &= [\phi, \dot{\phi}, \eta, \dot{\eta}, \delta_1, \dot{\delta}_1, \delta_2, \dot{\delta}_2]^T \end{aligned} \tag{3.13}$$

Moreover, the following functions are defined:

$$f_1(x) = x_2 \quad f_2(x) = \frac{-M_{11}(c_1+g_1+d_1)+M_{21}(c_2+g_2+d_2)-M_{31}(c_3+g_3+d_3)+M_{41}(c_4+g_4+d_4)}{detM}$$

$$f_3(x) = x_4 \quad f_4(x) = \frac{M_{12}(c_1+g_1+d_1)-M_{22}(c_2+g_2+d_2)+M_{32}(c_3+g_3+d_3)-M_{42}(c_4+g_4+d_4)}{detM}$$

$$f_5(x) = x_6 \quad f_6(x) = \frac{-M_{13}(c_1+g_1+d_1)+M_{23}(c_2+g_2+d_2)-M_{33}(c_3+g_3+d_3)+M_{43}(c_4+g_4+d_4)}{detM}$$

$$f_7(x) = x_8 \quad f_8(x) = \frac{M_{14}(c_1+g_1+d_1)-M_{24}(c_2+g_2+d_2)+M_{34}(c_3+g_3+d_3)-M_{44}(c_4+g_4+d_4)}{detM}$$

$$\begin{aligned} g_{11}(x) &= 0 & g_{12}(x) &= 0 \\ g_{21}(x) &= \frac{M_{11}}{detM} & g_{22}(x) &= -\frac{M_{21}}{detM} \\ g_{31}(x) &= 0 & g_{32}(x) &= 0 \\ g_{41}(x) &= -\frac{M_{12}}{detM} & g_{42}(x) &= \frac{M_{22}}{detM} \\ g_{51}(x) &= 0 & g_{52}(x) &= 0 \\ g_{61}(x) &= \frac{M_{13}}{detM} & g_{62}(x) &= -\frac{M_{23}}{detM} \\ g_{71}(x) &= 0 & g_{72}(x) &= 0 \\ g_{81}(x) &= -\frac{M_{14}}{detM} & g_{82}(x) &= \frac{M_{24}}{detM} \end{aligned}$$

Thus, the state-space model of the tower crane is written as

$$\begin{pmatrix} \dot{x}_1 \\ \dot{x}_2 \\ \dot{x}_3 \\ \dot{x}_4 \\ \dot{x}_5 \\ \dot{x}_6 \\ \dot{x}_7 \\ \dot{x}_8 \end{pmatrix} = \begin{pmatrix} f_1(x) \\ f_2(x) \\ f_3(x) \\ f_4(x) \\ f_5(x) \\ f_6(x) \\ f_7(x) \\ f_8(x) \end{pmatrix} + \begin{pmatrix} g_{11}(x) & g_{12}(x) \\ g_{21}(x) & g_{22}(x) \\ g_{31}(x) & g_{32}(x) \\ g_{41}(x) & g_{42}(x) \\ g_{51}(x) & g_{52}(x) \\ g_{61}(x) & g_{62}(x) \\ g_{71}(x) & g_{72}(x) \\ g_{81}(x) & g_{82}(x) \end{pmatrix} \begin{pmatrix} \tau_1 \\ \tau_2 \end{pmatrix} \tag{3.14}$$

or in concise form, one has the affine-in-the-input nonlinear state-space model:

$$\dot{x} = f(x) + g(x)u \tag{3.15}$$

where $x \in R^{8\times1}$, $f(x) \in R^{8\times1}$, $g(x) \in R^{8\times2}$, and $u \in R^{2\times1}$. It is noted that the underactuated tower crane is not a 0-flat system. To be brought to a differentially flat form, the dynamic extension approach has to be applied [711].

3.1.3 Approximate linearization of the dynamic model of the robotic crane

3.1.3.1 Approximate linearization with Taylor series expansion

The dynamic model of the tower crane being initially expressed in the state-space form

$$\dot{x} = f(x) + g(x)u \tag{3.16}$$

undergoes approximate linearization at each sampling instance around the temporary operating point (x^*, u^*), where x^* is the present value of the system's state vector and u^* is the last sampled value of the control inputs vector. The linearization process is based on Taylor series expansion and gives

$$\dot{x} = Ax + Bu + \tilde{d} \tag{3.17}$$

where $\tilde{d}$ is the cumulative disturbances vector which may be due to truncation of higher-order terms from (a) the Taylor series expansion, (b) exogenous perturbations, and (c) sensor measurements noise of any distribution. Matrices A and B are Jacobian matrices of the Taylor series expansion which are defined as

$$\begin{gathered} A = \nabla_x[f(x) + g(x)u] \mid_{(x^*,u^*)} \Rightarrow \\ A = \nabla_x f(x) | mid_{(x^*,u^*)} + \nabla_x g_1(x)u \mid_{(x^*,u^*)} + \nabla_x g_2(x)u \mid_{(x^*,u^*)} \end{gathered} \tag{3.18}$$

$$B = \nabla_u[f(x) + g(x)u] \mid_{(x^*,u^*)} \Rightarrow B = g(x) \mid_{(x^*,u^*)} \tag{3.19}$$

This linearization approach which has been followed for implementing the nonlinear optimal control scheme results in a quite accurate model of the system's dynamics. Consider for instance the following affine-in-the-input state-space model:

$$\begin{gathered} \dot{x} = f(x) + g(x)u \Rightarrow \\ \dot{x} = [f(x^*) + \nabla_x f(x) \mid_{x^*} (x - x^*)] + [g(x^*) + \nabla_x g(x) \mid_{x^*} (x - x^*)]u^* + g(x^*)u^* + g(x^*)(u - u^*) + \tilde{d}_1 \Rightarrow \\ \dot{x} = [\nabla_x f(x) \mid_{x^*} + \nabla_x g(x) \mid_{x^*} u^*]x + g(x^*)u - [\nabla_x f(x) \mid_{x^*} + \nabla_x g(x) \mid_{x^*} u^*]x^* + f(x^*) + g(x^*)u^* + \tilde{d}_1 \end{gathered} \tag{3.20}$$

where $\tilde{d}_1$ is the modeling error due to truncation of higher-order terms in the Taylor series expansion of $f(x)$ and $g(x)$. Next, by defining $A = [\nabla_x f(x) \mid_{x^*} + \nabla_x g(x) \mid_{x^*} u^*]$, $B = g(x^*)$, one obtains

$$\dot{x} = Ax + Bu - Ax^* + f(x^*) + g(x^*)u^* + \tilde{d}_1 \tag{3.21}$$

Moreover, by denoting $\tilde{d} = -Ax^* + f(x^*) + g(x^*)u^* + \tilde{d}_1$ about the cumulative modeling error term in the Taylor series expansion procedure, one has

$$\dot{x} = Ax + Bu + \tilde{d} \tag{3.22}$$

which is the approximately linearized model of the dynamics of the system of Eq. (3.17). The term $f(x^*) + g(x^*)u^*$ is the derivative of the state vector at (x^*, u^*) which is almost annihilated by $-Ax^*$.

3.1.3.2 Computation of Jacobian matrices

The computation of the Jacobian matrices A and B proceeds as follows:

Computation of the Jacobian matrix $\nabla_x f(x) \mid_{(x^*,u^*)}$.

First row of the Jacobian matrix $\nabla_x f(x) \mid_{(x^*,u^*)}$: $\frac{\partial f_1}{\partial x_1} = 0$, $\frac{\partial f_1}{\partial x_2} = 1$, $\frac{\partial f_1}{\partial x_3} = 0$, $\frac{\partial f_1}{\partial x_4} = 0$, $\frac{\partial f_1}{\partial x_5} = 0$, $\frac{\partial f_1}{\partial x_6} = 0$, $\frac{\partial f_1}{\partial x_7} = 0$, and $\frac{\partial f_1}{\partial x_8} = 0$.

Second row of the Jacobian matrix $\nabla_x f(x)\mid_{(x^*,u^*)}$: It holds that $f_2(x)=\frac{f_{2,num}}{f_{2,den}}$ with $f_{2,num}=-M_{11}(c_1+g_1+d+1)+M_{21}(c_2+g_2+d_2)-M_{31}(c_3+g_3+d_3)+M_{41}(c_4+g_4+d_4)$ and $f_{2,den}=detM$. Thus, for $i=1,2,\cdots,8$, one has

$$\begin{aligned}\frac{\partial f_{2,num}}{\partial x_1}=&-\frac{\partial M_{11}}{\partial x_i}(c_1+g_1+d_1)-M_{11}(\frac{\partial c_1}{\partial x_i}+\frac{\partial g_1}{\partial x_i}+\frac{\partial d_1}{\partial x_i})+\\&+\frac{\partial M_{21}}{\partial x_i}(c_2+g_2+d_2)+M_{21}(\frac{\partial c_2}{\partial x_i}+\frac{\partial g_2}{\partial x_i}+\frac{\partial d_2}{\partial x_i})-\\&-\frac{\partial M_{31}}{\partial x_i}(c_3+g_3+d_3)-M_{31}(\frac{\partial c_3}{\partial x_i}+\frac{\partial g_3}{\partial x_i}+\frac{\partial d_3}{\partial x_i})+\\&+\frac{\partial M_{41}}{\partial x_i}(c_4+g_4+d_4)+M_{41}(\frac{\partial c_4}{\partial x_i}+\frac{\partial g_4}{\partial x_i}+\frac{\partial d_4}{\partial x_i})\end{aligned}\tag{3.23}$$

and also

$$\frac{\partial f_{2,den}}{\partial x_i}=\frac{\partial detM}{\partial x_i}\tag{3.24}$$

and finally

$$\frac{\partial f_2}{\partial x_i}=\frac{\frac{\partial f_{2,num}}{\partial x_i}f_{2,den}-f_{2,num}\frac{\partial f_{2,den}}{\partial x_i}}{detM^2}\tag{3.25}$$

Third row of the Jacobian matrix $\nabla_x f(x)\mid_{(x^*,u^*)}$: $\frac{\partial f_3}{\partial x_1}=0$, $\frac{\partial f_3}{\partial x_2}=0$, $\frac{\partial f_3}{\partial x_3}=0$, $\frac{\partial f_3}{\partial x_4}=1$, $\frac{\partial f_3}{\partial x_5}=0$, $\frac{\partial f_3}{\partial x_6}=0$, $\frac{\partial f_3}{\partial x_7}=0$, and $\frac{\partial f_3}{\partial x_8}=0$.

Fourth row of the Jacobian matrix $\nabla_x f(x)\mid_{(x^*,u^*)}$: It holds that $f_4(x)=\frac{f_{4,num}}{f_{4,den}}$ with $f_{4,num}=M_{12}(c_1+g_1+d+1)-M_{22}(c_2+g_2+d_2)+M_{32}(c_3+g_3+d_3)-M_{42}(c_4+g_4+d_4)$ and $f_{4,den}=detM$. Thus, for $i=1,2,\cdots,8$, one has

$$\begin{aligned}\frac{\partial f_{4,num}}{\partial x_1}=&\frac{\partial M_{12}}{\partial x_i}(c_1+g_1+d_1)+M_{12}(\frac{\partial c_1}{\partial x_i}+\frac{\partial g_1}{\partial x_i}+\frac{\partial d_1}{\partial x_i})+\\&-\frac{\partial M_{22}}{\partial x_i}(c_2+g_2+d_2)-M_{22}(\frac{\partial c_2}{\partial x_i}+\frac{\partial g_2}{\partial x_i}+\frac{\partial d_2}{\partial x_i})-\\&\frac{\partial M_{32}}{\partial x_i}(c_3+g_3+d_3)+M_{32}(\frac{\partial c_3}{\partial x_i}+\frac{\partial g_3}{\partial x_i}+\frac{\partial d_3}{\partial x_i})-\\&-\frac{\partial M_{42}}{\partial x_i}(c_4+g_4+d_4)-M_{42}(\frac{\partial c_4}{\partial x_i}+\frac{\partial g_4}{\partial x_i}+\frac{\partial d_4}{\partial x_i})\end{aligned}\tag{3.26}$$

and also

$$\frac{\partial f_{4,den}}{\partial x_i}=\frac{\partial detM}{\partial x_i}\tag{3.27}$$

and finally

$$\frac{\partial f_4}{\partial x_i}=\frac{\frac{\partial f_{4,num}}{\partial x_i}f_{4,den}-f_{4,num}\frac{\partial f_{4,den}}{\partial x_i}}{detM^2}\tag{3.28}$$

Fifth row of the Jacobian matrix $\nabla_x f(x)\mid_{(x^*,u^*)}$: $\frac{\partial f_5}{\partial x_1}=0$, $\frac{\partial f_5}{\partial x_2}=0$, $\frac{\partial f_5}{\partial x_3}=0$, $\frac{\partial f_5}{\partial x_4}=0$, $\frac{\partial f_5}{\partial x_5}=0$, $\frac{\partial f_5}{\partial x_6}=1$, $\frac{\partial f_5}{\partial x_7}=0$, and $\frac{\partial f_5}{\partial x_8}=0$.

Sixth row of the Jacobian matrix $\nabla_x f(x)\mid_{(x^*,u^*)}$: It holds that $f_6(x)=\frac{f_{6,num}}{f_{6,den}}$ with $f_{6,num}=-M_{13}(c_1+g_1+d+1)+M_{23}(c_2+g_2+d_2)-M_{33}(c_3+g_3+d_3)+M_{43}(c_4+g_4+d_4)$ and $f_{6,den}=detM$. Thus, for $i=1,2,\cdots,8$, one has

$$\begin{aligned}\frac{\partial f_{6,num}}{\partial x_1}=&-\frac{\partial M_{13}}{\partial x_i}(c_1+g_1+d_1)-M_{13}(\frac{\partial c_1}{\partial x_i}+\frac{\partial g_1}{\partial x_i}+\frac{\partial d_1}{\partial x_i})+\\&+\frac{\partial M_{23}}{\partial x_i}(c_2+g_2+d_2)+M_{23}(\frac{\partial c_2}{\partial x_i}+\frac{\partial g_2}{\partial x_i}+\frac{\partial d_2}{\partial x_i})-\\&-\frac{\partial M_{33}}{\partial x_i}(c_3+g_3+d_3)-M_{33}(\frac{\partial c_3}{\partial x_i}+\frac{\partial g_3}{\partial x_i}+\frac{\partial d_3}{\partial x_i})+\\&+\frac{\partial M_{43}}{\partial x_i}(c_4+g_4+d_4)+M_{43}(\frac{\partial c_4}{\partial x_i}+\frac{\partial g_4}{\partial x_i}+\frac{\partial d_4}{\partial x_i})\end{aligned}\tag{3.29}$$

and also

$$\frac{\partial f_{6,den}}{\partial x_i}=\frac{\partial detM}{\partial x_i}\tag{3.30}$$

and finally

$$\frac{\partial f_6}{\partial x_i} = \frac{\frac{\partial f_{6,num}}{\partial x_i} f_{6,den} - f_{6,num}\frac{\partial f_{6,den}}{\partial x_i}}{detM^2} \tag{3.31}$$

Seventh row of the Jacobian matrix $\nabla_x f(x) \mid_{(x^*,u^*)}$: $\frac{\partial f_7}{\partial x_1}=0$, $\frac{\partial f_7}{\partial x_2}=0$, $\frac{\partial f_7}{\partial x_3}=0$, $\frac{\partial f_7}{\partial x_4}=0$, $\frac{\partial f_7}{\partial x_5}=0$, $\frac{\partial f_7}{\partial x_6}=0$, $\frac{\partial f_7}{\partial x_7}=0$, and $\frac{\partial f_7}{\partial x_8}=0$.

Eighth row of the Jacobian matrix $\nabla_x f(x) \mid_{(x^*,u^*)}$: It holds that $f_4(x) = \frac{f_{8,num}}{f_{8,den}}$ with $f_{8,num} = M_{14}(c_1+g_1+d+1) - M_{24}(c_2+g_2+d_2) + M_{34}(c_3+g_3+d_3) - M_{44}(c_4+g_4+d_4)$ and $f_{8,den} = detM$. Thus, for $i=1,2,\cdots,8$, one has

$$\begin{aligned}\frac{\partial f_{8,num}}{\partial x_1} = & \frac{\partial M_{14}}{\partial x_i}(c_1+g_1+d_1) + M_{14}(\frac{\partial c_1}{\partial x_i} + \frac{\partial g_1}{\partial x_i} + \frac{\partial d_1}{\partial x_i}) + \\ & -\frac{\partial M_{24}}{\partial x_i}(c_2+g_2+d_2) - M_{24}(\frac{\partial c_2}{\partial x_i} + \frac{\partial g_2}{\partial x_i} + \frac{\partial d_2}{\partial x_i}) - \\ & \frac{\partial M_{34}}{\partial x_i}(c_3+g_3+d_3) + M_{34}(\frac{\partial c_3}{\partial x_i} + \frac{\partial g_3}{\partial x_i} + \frac{\partial d_3}{\partial x_i}) - \\ & -\frac{\partial M_{44}}{\partial x_i}(c_4+g_4+d_4) - M_{44}(\frac{\partial c_4}{\partial x_i} + \frac{\partial g_4}{\partial x_i} + \frac{\partial d_4}{\partial x_i})\end{aligned} \tag{3.32}$$

and also

$$\frac{\partial f_{8,den}}{\partial x_i} = \frac{\partial detM}{\partial x_i} \tag{3.33}$$

and finally

$$\frac{\partial f_8}{\partial x_i} = \frac{\frac{\partial f_{8,num}}{\partial x_i} f_{8,den} - f_{8,num}\frac{\partial f_{8,den}}{\partial x_i}}{detM^2} \tag{3.34}$$

Computation of the Jacobian matrix $\nabla_x g_1(x) \mid_{(x^*,u^*)}$.
First row of the Jacobian matrix $\nabla_x g_1(x) \mid_{(x^*,u^*)}$: $\frac{\partial g_{11}(x)}{\partial x_1} = 0$ for $i=1,2,\cdots,8$.

Second row of the Jacobian matrix $\nabla_x g_1(x) \mid_{(x^*,u^*)}$: $\frac{\partial g_{21}(x)}{\partial x_i} = \frac{\frac{\partial M_{11}}{\partial x_i} detM - M_{11}\frac{\partial detM}{\partial x_i}}{detM^2}$, for $i=1,2,\cdots,8$

Third row of the Jacobian matrix $\nabla_x g_1(x) \mid_{(x^*,u^*)}$: $\frac{\partial g_{31}(x)}{\partial x_i} = 0$, for $i=1,2,\cdots,8$.

Fourth row of the Jacobian matrix $\nabla_x g_1(x) \mid_{(x^*,u^*)}$: $\frac{\partial g_{41}(x)}{\partial x_i} = \frac{-\frac{\partial M_{12}}{\partial x_i} detM + M_{12}\frac{\partial detM}{\partial x_i}}{detM^2}$, for $i=1,2,\cdots,8$

Fifth row of the Jacobian matrix $\nabla_x g_1(x) \mid_{(x^*,u^*)}$: $\frac{\partial g_{51}(x)}{\partial x_i} = 0$, for $i=1,2,\cdots,8$.

Sixth row of the Jacobian matrix $\nabla_x g_1(x) \mid_{(x^*,u^*)}$: $\frac{\partial g_{61}(x)}{\partial x_i} = \frac{\frac{\partial M_{13}}{\partial x_i} detM - M_{13}\frac{\partial detM}{\partial x_i}}{detM^2}$, for $i=1,2,\cdots,8$

Seventh row of the Jacobian matrix $\nabla_x g_1(x) \mid_{(x^*,u^*)}$: $\frac{\partial g_{71}(x)}{\partial x_i} = 0$, for $i=1,2,\cdots,8$.

Eighth row of the Jacobian matrix $\nabla_x g_1(x) \mid_{(x^*,u^*)}$: $\frac{\partial g_{81}(x)}{\partial x_i} = \frac{-\frac{\partial M_{14}}{\partial x_i} detM + M_{14}\frac{\partial detM}{\partial x_i}}{detM^2}$, for $i=1,2,\cdots,8$

Computation of the Jacobian matrix $\nabla_x g_2(x) \mid_{(x^*,u^*)}$.

First row of the Jacobian matrix $\nabla_x g_2(x) \mid_{(x^*,u^*)}$: $\frac{\partial g_{12}(x)}{\partial x_1} = 0$ for $i=1,2,\cdots,8$.

Second row of the Jacobian matrix $\nabla_x g_2(x) \mid_{(x^*,u^*)}$: $\frac{\partial g_{22}(x)}{\partial x_i} = \frac{-\frac{\partial M_{21}}{\partial x_i} detM + M_{21}\frac{\partial detM}{\partial x_i}}{detM^2}$, for $i=1,2,\cdots,8$.

Third row of the Jacobian matrix $\nabla_x g_2(x) \mid_{(x^*,u^*)}$: $\frac{\partial g_{32}(x)}{\partial x_i} = 0$, for $i=1,2,\cdots,8$.

Fourth row of the Jacobian matrix $\nabla_x g_2(x)\mid_{(x^*,u^*)}$: $\frac{\partial g_{42}(x)}{\partial x_i}=\frac{\frac{\partial M_{22}}{\partial x_i}detM-M_{22}\frac{\partial detM}{\partial x_i}}{detM^2}$, for $i=1,2,\cdots,8$.

Fifth row of the Jacobian matrix $\nabla_x g_2(x)\mid_{(x^*,u^*)}$: $\frac{\partial g_{52}(x)}{\partial x_i}=0$, for $i=1,2,\cdots,8$.

Sixth row of the Jacobian matrix $\nabla_x g_2(x)\mid_{(x^*,u^*)}$: $\frac{\partial g_{62}(x)}{\partial x_i}=\frac{-\frac{\partial M_{23}}{\partial x_i}detM+M_{23}\frac{\partial detM}{\partial x_i}}{detM^2}$, for $i=1,2,\cdots,8$.

Seventh row of the Jacobian matrix $\nabla_x g_2(x)\mid_{(x^*,u^*)}$: $\frac{\partial g_{72}(x)}{\partial x_i}=0$, for $i=1,2,\cdots,8$.

Eighth row of the Jacobian matrix $\nabla_x g_2(x)\mid_{(x^*,u^*)}$: $\frac{\partial g_{82}(x)}{\partial x_i}=\frac{\frac{\partial M_{24}}{\partial x_i}detM-M_{24}\frac{\partial detM}{\partial x_i}}{detM^2}$, for $i=1,2,\cdots,8$.

Next, one computes the partial derivatives of the sub-determinants M_{ij} and of the determinant $detM$:

$$\begin{array}{c}\frac{\partial M_{11}}{\partial x_i}=\frac{\partial m_{22}}{\partial x_i}(m_{33}m_{44}-m_{43}m_{34})+\\ +m_{22}(\frac{\partial m_{33}}{\partial x_i}m_{44}+m_{33}\frac{\partial m_{44}}{\partial x_i}-\frac{\partial m_{43}}{\partial x_i}m_{34}-m_{43}\frac{\partial m_{34}}{\partial x_i})-\\ -\frac{\partial m_{23}}{\partial x_i}(m_{32}m_{44}-m_{42}m_{34})\\ -m_{23}(\frac{\partial m_{33}}{\partial x_i}m_{44}+m_{32}\frac{\partial m_{44}}{\partial x_i}-\frac{\partial m_{42}}{\partial x_i}m_{34}-m_{42}\frac{\partial m_{34}}{\partial x_i})+\\ +\frac{\partial m_{24}}{\partial x_i}(m_{32}m_{43}-m_{42}m_{33})+\\ +m_{24}(\frac{\partial m_{32}}{\partial x_i}m_{43}+m_{32}\frac{\partial m_{43}}{\partial x_i}-\frac{\partial m_{42}}{\partial x_i}m_{33}-m_{42}\frac{\partial m_{33}}{\partial x_i})\end{array}\tag{3.35}$$

Equivalently, one has

$$\begin{array}{c}\frac{\partial M_{12}}{\partial x_i}=\frac{\partial m_{21}}{\partial x_i}(m_{33}m_{44}-m_{43}m_{34})+\\ +m_{21}(\frac{\partial m_{13}}{\partial x_i}m_{44}+m_{13}\frac{\partial m_{44}}{\partial x_i}-\frac{\partial m_{43}}{\partial x_i}m_{34}-m_{43}\frac{\partial m_{34}}{\partial x_i})-\\ -\frac{\partial m_{23}}{\partial x_i}(m_{31}m_{44}-m_{41}m_{34})\\ -m_{23}(\frac{\partial m_{31}}{\partial x_i}m_{44}+m_{31}\frac{\partial m_{44}}{\partial x_i}-\frac{\partial m_{41}}{\partial x_i}m_{34}-m_{41}\frac{\partial m_{34}}{\partial x_i})+\\ +\frac{\partial m_{24}}{\partial x_i}(m_{31}m_{43}-m_{41}m_{33})+\\ +m_{24}(\frac{\partial m_{31}}{\partial x_i}m_{43}+m_{31}\frac{\partial m_{43}}{\partial x_i}-\frac{\partial m_{41}}{\partial x_i}m_{33}-m_{41}\frac{\partial m_{33}}{\partial x_i})\end{array}\tag{3.36}$$

Moreover, it holds that

$$\begin{array}{c}\frac{\partial M_{13}}{\partial x_i}=\frac{\partial m_{21}}{\partial x_i}(m_{32}m_{44}-m_{42}m_{34})+\\ +m_{21}(\frac{\partial m_{32}}{\partial x_i}m_{44}+m_{32}\frac{\partial m_{44}}{\partial x_i}-\frac{\partial m_{43}}{\partial x_i}m_{34}-m_{43}\frac{\partial m_{34}}{\partial x_i})-\\ -\frac{\partial m_{23}}{\partial x_i}(m_{31}m_{44}-m_{41}m_{34})\\ -m_{23}(\frac{\partial m_{31}}{\partial x_i}m_{44}+m_{31}\frac{\partial m_{44}}{\partial x_i}-\frac{\partial m_{41}}{\partial x_i}m_{34}-m_{41}\frac{\partial m_{34}}{\partial x_i})+\\ +\frac{\partial m_{24}}{\partial x_i}(m_{31}m_{42}-m_{41}m_{32})+\\ +m_{24}(\frac{\partial m_{31}}{\partial x_i}m_{42}+m_{31}\frac{\partial m_{42}}{\partial x_i}-\frac{\partial m_{41}}{\partial x_i}m_{32}-m_{41}\frac{\partial m_{32}}{\partial x_i})\end{array}\tag{3.37}$$

Additionally, it holds that

$$\begin{array}{c}\frac{\partial M_{14}}{\partial x_i}=\frac{\partial m_{21}}{\partial x_i}(m_{32}m_{43}-m_{42}m_{33})+\\ +m_{21}(\frac{\partial m_{32}}{\partial x_i}m_{43}+m_{32}\frac{\partial m_{43}}{\partial x_i}-\frac{\partial m_{44}}{\partial x_i}m_{33}-m_{42}\frac{\partial m_{32}}{\partial x_i})-\\ -\frac{\partial m_{22}}{\partial x_i}(m_{31}m_{43}-m_{41}m_{33})\\ -m_{22}(\frac{\partial m_{31}}{\partial x_i}m_{43}+m_{31}\frac{\partial m_{43}}{\partial x_i}-\frac{\partial m_{33}}{\partial x_i}m_{33}-m_{41}\frac{\partial m_{33}}{\partial x_i})+\\ +\frac{\partial m_{23}}{\partial x_i}(m_{31}m_{42}-m_{41}m_{32})+\\ +m_{23}(\frac{\partial m_{31}}{\partial x_i}m_{42}+m_{31}\frac{\partial m_{42}}{\partial x_i}-\frac{\partial m_{41}}{\partial x_i}m_{32}-m_{41}\frac{\partial m_{32}}{\partial x_i})\end{array}\tag{3.38}$$

In a similar manner, one obtains

$$\begin{array}{c}\frac{\partial M_{21}}{\partial x_i}=\frac{\partial m_{12}}{\partial x_i}(m_{33}m_{44}-m_{43}m_{34})+\\+m_{12}(\frac{\partial m_{33}}{\partial x_i}m_{44}+m_{33}\frac{\partial m_{44}}{\partial x_i}-\frac{\partial m_{43}}{\partial x_i}m_{33}-m_{43}\frac{\partial m_{34}}{\partial x_i})-\\-\frac{\partial m_{13}}{\partial x_i}(m_{32}m_{44}-m_{42}m_{34})\\-m_{13}(\frac{\partial m_{32}}{\partial x_i}m_{44}+m_{32}\frac{\partial m_{44}}{\partial x_i}-\frac{\partial m_{42}}{\partial x_i}m_{34}-m_{42}\frac{\partial m_{34}}{\partial x_i})+\\+\frac{\partial m_{14}}{\partial x_i}(m_{32}m_{43}-m_{42}m_{33})+\\+m_{14}(\frac{\partial m_{32}}{\partial x_i}m_{43}+m_{32}\frac{\partial m_{43}}{\partial x_i}-\frac{\partial m_{42}}{\partial x_i}m_{33}-m_{42}\frac{\partial m_{33}}{\partial x_i})\end{array}\tag{3.39}$$

Equivalently, one has

$$\begin{array}{c}\frac{\partial M_{22}}{\partial x_i}=\frac{\partial m_{11}}{\partial x_i}(m_{33}m_{44}-m_{43}m_{34})+\\+m_{11}(\frac{\partial m_{33}}{\partial x_i}m_{44}+m_{33}\frac{\partial m_{44}}{\partial x_i}-\frac{\partial m_{43}}{\partial x_i}m_{34}-m_{43}\frac{\partial m_{34}}{\partial x_i})-\\-\frac{\partial m_{13}}{\partial x_i}(m_{31}m_{44}-m_{41}m_{24})\\-m_{13}(\frac{\partial m_{31}}{\partial x_i}m_{44}+m_{31}\frac{\partial m_{44}}{\partial x_i}-\frac{\partial m_{41}}{\partial x_i}m_{24}-m_{41}\frac{\partial m_{24}}{\partial x_i})+\\+\frac{\partial m_{14}}{\partial x_i}(m_{31}m_{43}-m_{41}m_{23})+\\+m_{14}(\frac{\partial m_{31}}{\partial x_i}m_{43}+m_{31}\frac{\partial m_{43}}{\partial x_i}-\frac{\partial m_{41}}{\partial x_i}m_{23}-m_{41}\frac{\partial m_{23}}{\partial x_i})\end{array}\tag{3.40}$$

Following this procedure, one gets

$$\begin{array}{c}\frac{\partial M_{23}}{\partial x_i}=\frac{\partial m_{11}}{\partial x_i}(m_{32}m_{44}-m_{42}m_{34})+\\+m_{11}(\frac{\partial m_{32}}{\partial x_i}m_{44}+m_{32}\frac{\partial m_{44}}{\partial x_i}-\frac{\partial m_{42}}{\partial x_i}m_{34}-m_{42}\frac{\partial m_{34}}{\partial x_i})-\\-\frac{\partial m_{12}}{\partial x_i}(m_{31}m_{43}-m_{41}m_{33})\\-m_{12}(\frac{\partial m_{31}}{\partial x_i}m_{43}+m_{31}\frac{\partial m_{43}}{\partial x_i}-\frac{\partial m_{41}}{\partial x_i}m_{33}-m_{41}\frac{\partial m_{33}}{\partial x_i})+\\+\frac{\partial m_{14}}{\partial x_i}(m_{31}m_{42}-m_{41}m_{32})+\\+m_{14}(\frac{\partial m_{31}}{\partial x_i}m_{43}+m_{31}\frac{\partial m_{43}}{\partial x_i}-\frac{\partial m_{41}}{\partial x_i}m_{32}-m_{41}\frac{\partial m_{32}}{\partial x_i})\end{array}\tag{3.41}$$

Additionally, it holds that

$$\begin{array}{c}\frac{\partial M_{24}}{\partial x_i}=\frac{\partial m_{11}}{\partial x_i}(m_{32}m_{43}-m_{42}m_{33})+\\+m_{11}(\frac{\partial m_{32}}{\partial x_i}m_{43}+m_{32}\frac{\partial m_{43}}{\partial x_i}-\frac{\partial m_{42}}{\partial x_i}m_{33}-m_{42}\frac{\partial m_{33}}{\partial x_i})-\\-\frac{\partial m_{12}}{\partial x_i}(m_{31}m_{43}-m_{41}m_{33})\\-m_{12}(\frac{\partial m_{31}}{\partial x_i}m_{43}+m_{31}\frac{\partial m_{43}}{\partial x_i}-\frac{\partial m_{41}}{\partial x_i}m_{33}-m_{41}\frac{\partial m_{33}}{\partial x_i})+\\+\frac{\partial m_{13}}{\partial x_i}(m_{31}m_{42}-m_{41}m_{22})+\\+m_{13}(\frac{\partial m_{31}}{\partial x_i}m_{42}+m_{31}\frac{\partial m_{42}}{\partial x_i}-\frac{\partial m_{41}}{\partial x_i}m_{22}-m_{41}\frac{\partial m_{22}}{\partial x_i})\end{array}\tag{3.42}$$

In this context, one obtains

$$\begin{array}{c}\frac{\partial M_{31}}{\partial x_i}=\frac{\partial m_{12}}{\partial x_i}(m_{23}m_{44}-m_{43}m_{24})+\\+m_{12}(\frac{\partial m_{23}}{\partial x_i}m_{44}+m_{23}\frac{\partial m_{44}}{\partial x_i}-\frac{\partial m_{43}}{\partial x_i}m_{24}-m_{43}\frac{\partial m_{24}}{\partial x_i})-\\-\frac{\partial m_{13}}{\partial x_i}(m_{22}m_{44}-m_{42}m_{24})\\-m_{13}(\frac{\partial m_{22}}{\partial x_i}m_{44}+m_{22}\frac{\partial m_{44}}{\partial x_i}-\frac{\partial m_{42}}{\partial x_i}m_{24}-m_{42}\frac{\partial m_{24}}{\partial x_i})+\\+\frac{\partial m_{14}}{\partial x_i}(m_{22}m_{43}-m_{42}m_{23})+\\+m_{14}(\frac{\partial m_{32}}{\partial x_i}m_{43}+m_{22}\frac{\partial m_{43}}{\partial x_i}-\frac{\partial m_{42}}{\partial x_i}m_{23}-m_{42}\frac{\partial m_{23}}{\partial x_i})\end{array}\tag{3.43}$$

Additionally, one has

$$\begin{array}{c}\frac{\partial M_{32}}{\partial x_i} = \frac{\partial m_{11}}{\partial x_i}(m_{23}m_{44} - m_{43}m_{24}) + \\ +m_{11}(\frac{\partial m_{23}}{\partial x_i}m_{44} + m_{23}\frac{\partial m_{44}}{\partial x_i} - \frac{\partial m_{43}}{\partial x_i}m_{24} - m_{43}\frac{\partial m_{24}}{\partial x_i}) - \\ -\frac{\partial m_{13}}{\partial x_i}(m_{21}m_{44} - m_{41}m_{24}) \\ -m_{13}(\frac{\partial m_{21}}{\partial x_i}m_{44} + m_{21}\frac{\partial m_{44}}{\partial x_i} - \frac{\partial m_{41}}{\partial x_i}m_{24} - m_{41}\frac{\partial m_{24}}{\partial x_i}) + \\ +\frac{\partial m_{14}}{\partial x_i}(m_{12}m_{43} - m_{41}m_{23}) + \\ +m_{14}(\frac{\partial m_{12}}{\partial x_i}m_{43} + m_{12}\frac{\partial m_{43}}{\partial x_i} - \frac{\partial m_{41}}{\partial x_i}m_{23} - m_{41}\frac{\partial m_{23}}{\partial x_i})\end{array} \tag{3.44}$$

Furthermore, one has

$$\begin{array}{c}\frac{\partial M_{33}}{\partial x_i} = \frac{\partial m_{11}}{\partial x_i}(m_{22}m_{44} - m_{42}m_{24}) + \\ +m_{11}(\frac{\partial m_{22}}{\partial x_i}m_{44} + m_{22}\frac{\partial m_{44}}{\partial x_i} - \frac{\partial m_{42}}{\partial x_i}m_{24} - m_{42}\frac{\partial m_{24}}{\partial x_i}) - \\ -\frac{\partial m_{12}}{\partial x_i}(m_{21}m_{44} - m_{41}m_{24}) \\ -m_{12}(\frac{\partial m_{21}}{\partial x_i}m_{44} + m_{21}\frac{\partial m_{44}}{\partial x_i} - \frac{\partial m_{41}}{\partial x_i}m_{24} - m_{41}\frac{\partial m_{24}}{\partial x_i}) + \\ +\frac{\partial m_{14}}{\partial x_i}(m_{41}m_{22} - m_{21}m_{42}) + \\ +m_{14}(\frac{\partial m_{41}}{\partial x_i}m_{22} + m_{41}\frac{\partial m_{22}}{\partial x_i} - \frac{\partial m_{21}}{\partial x_i}m_{42} - m_{21}\frac{\partial m_{42}}{\partial x_i})\end{array} \tag{3.45}$$

Continuing in this manner, one gets

$$\begin{array}{c}\frac{\partial M_{34}}{\partial x_i} = \frac{\partial m_{11}}{\partial x_i}(m_{22}m_{43} - m_{42}m_{23}) + \\ +m_{11}(\frac{\partial m_{22}}{\partial x_i}m_{43} + m_{22}\frac{\partial m_{43}}{\partial x_i} - \frac{\partial m_{42}}{\partial x_i}m_{23} - m_{42}\frac{\partial m_{23}}{\partial x_i}) - \\ -\frac{\partial m_{12}}{\partial x_i}(m_{21}m_{43} - m_{41}m_{23}) \\ -m_{12}(\frac{\partial m_{21}}{\partial x_i}m_{43} + m_{21}\frac{\partial m_{43}}{\partial x_i} - \frac{\partial m_{41}}{\partial x_i}m_{23} - m_{41}\frac{\partial m_{23}}{\partial x_i}) + \\ +\frac{\partial m_{13}}{\partial x_i}(m_{21}m_{42} - m_{41}m_{22}) + \\ +m_{13}(\frac{\partial m_{22}}{\partial x_i}m_{42} + m_{22}\frac{\partial m_{42}}{\partial x_i} - \frac{\partial m_{41}}{\partial x_i}m_{22} - m_{41}\frac{\partial m_{22}}{\partial x_i})\end{array} \tag{3.46}$$

Besides, one has

$$\begin{array}{c}\frac{\partial M_{41}}{\partial x_i} = \frac{\partial m_{12}}{\partial x_i}(m_{22}m_{34} - m_{33}m_{24}) + \\ +m_{12}(\frac{\partial m_{22}}{\partial x_i}m_{34} + m_{22}\frac{\partial m_{34}}{\partial x_i} - \frac{\partial m_{33}}{\partial x_i}m_{24} - m_{33}\frac{\partial m_{24}}{\partial x_i}) - \\ -\frac{\partial m_{13}}{\partial x_i}(m_{22}m_{34} - m_{32}m_{24}) \\ -m_{13}(\frac{\partial m_{22}}{\partial x_i}m_{34} + m_{22}\frac{\partial m_{34}}{\partial x_i} - \frac{\partial m_{32}}{\partial x_i}m_{24} - m_{32}\frac{\partial m_{24}}{\partial x_i}) + \\ +\frac{\partial m_{14}}{\partial x_i}(m_{22}m_{33} - m_{32}m_{23}) + \\ +m_{14}(\frac{\partial m_{22}}{\partial x_i}m_{33} + m_{22}\frac{\partial m_{33}}{\partial x_i} - \frac{\partial m_{32}}{\partial x_i}m_{23} - m_{32}\frac{\partial m_{23}}{\partial x_i})\end{array} \tag{3.47}$$

Equivalently, one obtains

$$\begin{array}{c}\frac{\partial M_{42}}{\partial x_i} = \frac{\partial m_{11}}{\partial x_i}(m_{23}m_{34} - m_{33}m_{24}) + \\ +m_{11}(\frac{\partial m_{23}}{\partial x_i}m_{34} + m_{23}\frac{\partial m_{34}}{\partial x_i} - \frac{\partial m_{33}}{\partial x_i}m_{24} - m_{33}\frac{\partial m_{24}}{\partial x_i}) - \\ -\frac{\partial m_{13}}{\partial x_i}(m_{21}m_{34} - m_{31}m_{24}) \\ -m_{13}(\frac{\partial m_{21}}{\partial x_i}m_{34} + m_{21}\frac{\partial m_{34}}{\partial x_i} - \frac{\partial m_{31}}{\partial x_i}m_{24} - m_{31}\frac{\partial m_{24}}{\partial x_i}) + \\ +\frac{\partial m_{14}}{\partial x_i}(m_{21}m_{33} - m_{31}m_{23}) + \\ +m_{14}(\frac{\partial m_{21}}{\partial x_i}m_{33} + m_{21}\frac{\partial m_{33}}{\partial x_i} - \frac{\partial m_{31}}{\partial x_i}m_{23} - m_{31}\frac{\partial m_{23}}{\partial x_i})\end{array} \tag{3.48}$$

In a similar manner, one gets

$$\begin{array}{c}\frac{\partial M_{43}}{\partial x_i}=\frac{\partial m_{11}}{\partial x_i}(m_{22}m_{34}-m_{32}m_{24})+\\ +m_{11}(\frac{\partial m_{22}}{\partial x_i}m_{34}+m_{22}\frac{\partial m_{34}}{\partial x_i}-\frac{\partial m_{32}}{\partial x_i}m_{24}-m_{32}\frac{\partial m_{24}}{\partial x_i})-\\ -\frac{\partial m_{12}}{\partial x_i}(m_{21}m_{34}-m_{31}m_{24})\\ -m_{12}(\frac{\partial m_{21}}{\partial x_i}m_{34}+m_{21}\frac{\partial m_{34}}{\partial x_i}-\frac{\partial m_{31}}{\partial x_i}m_{24}-m_{31}\frac{\partial m_{24}}{\partial x_i})+\\ +\frac{\partial m_{14}}{\partial x_i}(m_{21}m_{32}-m_{31}m_{22})+\\ +m_{14}(\frac{\partial m_{21}}{\partial x_i}m_{32}+m_{21}\frac{\partial m_{32}}{\partial x_i}-\frac{\partial m_{31}}{\partial x_i}m_{22}-m_{31}\frac{\partial m_{22}}{\partial x_i})\end{array} \tag{3.49}$$

Finally, one has that

$$\begin{array}{c}\frac{\partial M_{44}}{\partial x_i}=\frac{\partial m_{22}}{\partial x_i}(m_{22}m_{33}-m_{32}m_{23})+\\ +m_{11}(\frac{\partial m_{22}}{\partial x_i}m_{33}+m_{22}\frac{\partial m_{33}}{\partial x_i}-\frac{\partial m_{32}}{\partial x_i}m_{23}-m_{32}\frac{\partial m_{23}}{\partial x_i})-\\ -\frac{\partial m_{12}}{\partial x_i}(m_{22}m_{33}-m_{31}m_{23})\\ -m_{12}(\frac{\partial m_{22}}{\partial x_i}m_{33}+m_{23}\frac{\partial m_{33}}{\partial x_i}-\frac{\partial m_{31}}{\partial x_i}m_{23}-m_{31}\frac{\partial m_{23}}{\partial x_i})+\\ +\frac{\partial m_{13}}{\partial x_i}(m_{21}m_{32}-m_{31}m_{22})+\\ +m_{13}(\frac{\partial m_{21}}{\partial x_i}m_{32}+m_{21}\frac{\partial m_{32}}{\partial x_i}-\frac{\partial m_{31}}{\partial x_i}m_{22}-m_{31}\frac{\partial m_{22}}{\partial x_i})\end{array} \tag{3.50}$$

About the partial derivatives of the determinant *detM*, one has for $i=1,2,\cdots,8$

$$\begin{array}{c}\frac{\partial detM}{\partial x_i}=\frac{\partial m_{11}}{\partial x_i}M_{11}+m_{11}\frac{\partial M_{11}}{\partial x_i}-\frac{\partial m_{12}}{\partial x_i}M_{12}-m_{12}\frac{\partial M_{12}}{\partial x_i}+\\ +\frac{\partial m_{13}}{\partial x_i}M_{13}+m_{13}\frac{\partial M_{13}}{\partial x_i}-\frac{\partial m_{14}}{\partial x_i}M_{14}-m_{14}\frac{\partial M_{14}}{\partial x_i}\end{array} \tag{3.51}$$

Next, the derivatives of the elements of the inertia matrix M are computed.

It holds that $m_{11}=m(sin^2(x_3)cos^2(x_7)+sin^2(x_7))L^2+2mLx_3cos(x_7)sin(x_3)+(m+M)x_3^2+J$. Thus, one has:

$\frac{\partial m_{11}}{\partial x_1}=0,\ \frac{\partial m_{11}}{\partial x_2}=0,\ \frac{\partial m_{11}}{\partial x_3}=2mLcos(x_7)sin(x_5)+2(m+M)x_3$

$\frac{\partial m_{11}}{\partial x_4}=0,\ \frac{\partial m_{11}}{\partial x_5}=m(2sin(x_5)cos(x_5)cos^2(x_7))L^2+2mLx_3cos(x_7)cos(x_5)$

$\frac{\partial m_{11}}{\partial x_6}=0,\ \frac{\partial m_{11}}{\partial x_7}=m(-2sin^2(x_5)cos(x_7)sin(x_7)+2sin(x_7)cos(x_7))L^2-2mLx_3sin(x_7)sin(x_5),\ \frac{\partial m_{11}}{\partial x_8}=0.$

Besides, it holds that $m_{12}=m_{21}=-mLsin(x_7)$, thus

$\frac{\partial m_{12}}{\partial x_i}=\frac{\partial m_{21}}{\partial x_i}=0$ if $i=1,2,\cdots,8$ and $i\neq 7$, $\frac{\partial m_{12}}{\partial x_7}=\frac{\partial m_{21}}{\partial x_7}=-mLcos(x_7)$.

Moreover, it holds that $m_{13}=m_{31}=-mLcos(x_5)cos(x_7)sin(x_7)$, thus

$\frac{\partial m_{13}}{\partial x_i}=\frac{\partial m_{31}}{\partial x_i}=0$ if $i=1,2,\cdots,8$ and $i\neq 5,7$, $\frac{\partial m_{13}}{\partial x_5}=\frac{\partial m_{31}}{\partial x_5}=mL^2sin(x_5)cos(x_7)sin(x_7)$,

$\frac{\partial m_{13}}{\partial x_7}=\frac{\partial m_{31}}{\partial x_7}=-mL^2cos(x_5)[-sin^2(x_7)+cos^2(x_7)]$.

Additionally, it holds that $m_{14}=m_{41}=mL(x_3cos(x_7)+Lsin(x_5))$, thus

$\frac{\partial m_{14}}{\partial x_i}=\frac{\partial m_{41}}{\partial x_i}=0$ for $i=1,2,\cdots,8$. and $i\neq 3,5,7$, $\frac{\partial m_{14}}{\partial x_3}=\frac{\partial m_{41}}{\partial x_3}=mLcos(x_7)=mLcos(x_7)$, $\frac{\partial m_{14}}{\partial x_5}=\frac{\partial m_{41}}{\partial x_5}=mLcos(x_7)=mL^2cos(x_5)$, $\frac{\partial m_{14}}{\partial x_7}=\frac{\partial m_{41}}{\partial x_7}=mLcos(x_7)=-mLx_3sin(x_7)$.

Furthermore, it holds that $m_{23}=m_{32}=mLcos(x_5)cos(x_7)$, thus

$\frac{\partial m_{23}}{\partial x_i}=\frac{\partial m_{23}}{\partial x_i}=0$ for $i=1,2,\cdots,8$. and $i\neq 5,7$, $\frac{\partial m_{23}}{\partial x_i}=\frac{\partial m_{32}}{\partial x_i}=-mLsin(x_5)cos(x_7)$, $\frac{\partial m_{23}}{\partial x_7}=\frac{\partial m_{32}}{\partial x_7}=-mLcos(x_5)sin(x_7)$.

Additionally, it holds that $m_{34} = m_{43} = -mLsin(x_5)cos(x_7)$, thus $\frac{\partial m_{34}}{\partial x_i} = \frac{\partial m_{43}}{\partial x_i} = 0$, for $i = 1, 2, \cdots, 8$. and $i \neq 5, 7$, $\frac{\partial m_{34}}{\partial x_5} = \frac{\partial m_{43}}{\partial x_5} = -mLcos(x_5)cos(x_7)$, $\frac{\partial m_{34}}{\partial x_7} = \frac{\partial m_{43}}{\partial x_7} = mLsin(x_5)sin(x_7)$.

Finally, about the computation of the partial derivatives of the Coriolis forces vector, one has

$$C(q, \dot{q}) = \begin{pmatrix} c_{11}x_2 + c_{12}x_4 + c_{13}x_6 + c_{14}x_8 \\ c_{21}x_2 + c_{22}x_4 + c_{23}x_6 + c_{24}x_8 \\ c_{31}x_2 + c_{22}x_4 + c_{23}x_6 + c_{24}x_8 \\ c_{41}x_2 + c_{42}x_4 + c_{43}x_6 + c_{44}x_8 \end{pmatrix} \tag{3.52}$$

It holds that for $i = 1.3, 5, 7$

$$\frac{\partial c_1}{\partial x_i} = \frac{\partial c_{11}}{\partial x_i}x_2 + \frac{\partial c_{12}}{\partial x_i}x_4 + \frac{\partial c_{13}}{\partial x_i}x_6 + \frac{\partial c_{14}}{\partial x_i}x_8 \tag{3.53}$$

and also

$$\frac{\partial c_1}{\partial x_2} = \frac{\partial c_{11}}{\partial x_2}x_2 + c_{11} + \frac{\partial c_{12}}{\partial x_2}x_4 + \frac{\partial c_{13}}{\partial x_2}x_6 + \frac{\partial c_{14}}{\partial x_2}x_8 \tag{3.54}$$

$$\frac{\partial c_1}{\partial x_4} = \frac{\partial c_{11}}{\partial x_4}x_2 + \frac{\partial c_{12}}{\partial x_4}x_4 + c_{12} + \frac{\partial c_{13}}{\partial x_4}x_6 + \frac{\partial c_{14}}{\partial x_4}x_8 \tag{3.55}$$

$$\frac{\partial c_1}{\partial x_6} = \frac{\partial c_{11}}{\partial x_6}x_2 + \frac{\partial c_{12}}{\partial x_6}x_4 + \frac{\partial c_{13}}{\partial x_6}x_6 + c_{13} + \frac{\partial c_{14}}{\partial x_6}x_8 \tag{3.56}$$

$$\frac{\partial c_1}{\partial x_8} = \frac{\partial c_{11}}{\partial x_8}x_2 + \frac{\partial c_{12}}{\partial x_8}x_4 + \frac{\partial c_{13}}{\partial x_8}x_6 + \frac{\partial c_{14}}{\partial x_8}x_8 + c_{14} \tag{3.57}$$

Equivalently, it holds that for $i = 1.3, 5, 7$

$$\frac{\partial c_2}{\partial x_i} = \frac{\partial c_{21}}{\partial x_i}x_2 + \frac{\partial c_{22}}{\partial x_i}x_4 + \frac{\partial c_{23}}{\partial x_i}x_6 + \frac{\partial c_{24}}{\partial x_i}x_8 \tag{3.58}$$

and also

$$\frac{\partial c_2}{\partial x_2} = \frac{\partial c_{21}}{\partial x_2}x_2 + c_{21} + \frac{\partial c_{22}}{\partial x_2}x_4 + \frac{\partial c_{23}}{\partial x_2}x_6 + \frac{\partial c_{24}}{\partial x_2}x_8 \tag{3.59}$$

$$\frac{\partial c_2}{\partial x_4} = \frac{\partial c_{21}}{\partial x_4}x_2 + \frac{\partial c_{22}}{\partial x_4}x_4 + c_{22} + \frac{\partial c_{23}}{\partial x_4}x_6 + \frac{\partial c_{24}}{\partial x_4}x_8 2 \tag{3.60}$$

$$\frac{\partial c_2}{\partial x_6} = \frac{\partial c_1}{\partial x_6}x_2 + \frac{\partial c_{22}}{\partial x_6}x_4 + \frac{\partial c_{23}}{\partial x_6}x_6 + c_{23} + \frac{\partial c_{24}}{\partial x_6}x_8 \tag{3.61}$$

$$\frac{\partial c_2}{\partial x_8} = \frac{\partial c_{21}}{\partial x_8}x_2 + \frac{\partial c_{22}}{\partial x_8}x_4 + \frac{\partial c_{23}}{\partial x_8}x_6 + \frac{\partial c_{24}}{\partial x_8}x_8 + c_{24} \tag{3.62}$$

Similarly, it holds that for $i = 1.3, 5, 7$

$$\frac{\partial c_3}{\partial x_i} = \frac{\partial c_{31}}{\partial x_i}x_2 + \frac{\partial c_{32}}{\partial x_i}x_4 + \frac{\partial c_{33}}{\partial x_i}x_6 + \frac{\partial c_{34}}{\partial x_i}x_8 \tag{3.63}$$

and also

$$\frac{\partial c_3}{\partial x_2} = \frac{\partial c_{31}}{\partial x_2}x_2 + c_{31} + \frac{\partial c_{32}}{\partial x_2}x_4 + \frac{\partial c_{33}}{\partial x_2}x_6 + \frac{\partial c_{34}}{\partial x_2}x_8 \tag{3.64}$$

$$\frac{\partial c_3}{\partial x_4} = \frac{\partial c_{31}}{\partial x_4}x_2 + \frac{\partial c_{32}}{\partial x_4}x_4 + c_{32} + \frac{\partial c_{33}}{\partial x_4}x_6 + \frac{\partial c_{34}}{\partial x_4}x_8 \tag{3.65}$$

$$\frac{\partial c_3}{\partial x_6} = \frac{\partial c_3}{\partial x_6}x_2 + \frac{\partial c_{32}}{\partial x_6}x_4 + \frac{\partial c_{33}}{\partial x_6}x_6 + c_{33} + \frac{\partial c_{34}}{\partial x_6}x_8 \tag{3.66}$$

$$\frac{\partial c_3}{\partial x_8} = \frac{\partial c_{31}}{\partial x_8}x_2 + \frac{\partial c_{32}}{\partial x_8}x_4 + \frac{\partial c_{33}}{\partial x_8}x_6 + \frac{\partial c_{34}}{\partial x_8}x_8 + c_{34} \tag{3.67}$$

Finally, it holds that for $i = 1.3, 5, 7$

$$\frac{\partial c_4}{\partial x_i} = \frac{\partial c_{41}}{\partial x_i}x_2 + \frac{\partial c_{42}}{\partial x_i}x_4 + \frac{\partial c_{43}}{\partial x_i}x_6 + \frac{\partial c_{44}}{\partial x_i}x_8 \tag{3.68}$$

and also

$$\frac{\partial c_4}{\partial x_2} = \frac{\partial c_{41}}{\partial x_2}x_2 + c_{41} + \frac{\partial c_{42}}{\partial x_2}x_4 + \frac{\partial c_{43}}{\partial x_2}x_6 + \frac{\partial c_{44}}{\partial x_2}x_8 \tag{3.69}$$

$$\frac{\partial c_4}{\partial x_4} = \frac{\partial c_{41}}{\partial x_4}x_2 + \frac{\partial c_{42}}{\partial x_4}x_4 + c_{42} + \frac{\partial c_{43}}{\partial x_4}x_6 + \frac{\partial c_{44}}{\partial x_4}x_8 \tag{3.70}$$

$$\frac{\partial c_4}{\partial x_6} = \frac{\partial c_4}{\partial x_6}x_2 + \frac{\partial c_{42}}{\partial x_6}x_4 + \frac{\partial c_{43}}{\partial x_6}x_6 + c_{43} + \frac{\partial c_{44}}{\partial x_6}x_8 \tag{3.71}$$

$$\frac{\partial c_4}{\partial x_8} = \frac{\partial c_{41}}{\partial x_8}x_2 + \frac{\partial c_{42}}{\partial x_8}x_4 + \frac{\partial c_{43}}{\partial x_8}x_6 + \frac{\partial c_{44}}{\partial x_8}x_8 + c_{44} \tag{3.72}$$

Next, the following partial derivatives of the elements c_{ij} $i = 1,2,3,4$ and $j = 1,2,3,4$ of the Coriolis matrix are computed. It holds

$$\begin{array}{c} c_{11} = (m+M)x_3x_4 + mLx_3cos(x_5)cos(x_7)x_6 - mLx_3sin(x_5)sin(x_7)x_8 + \\ +mLsin(x_5)sin(x_7)x_4 + mL^2sin(x_5)cos(x_5)cos^2(x_7)x_6 + \\ +mL^2cos^2(x_5)sin(x_7)cos(x_7)x_8 \end{array} \tag{3.73}$$

Thus, one has that: $\frac{\partial c_{11}}{\partial x_1} = 0$, $\frac{\partial c_{11}}{\partial x_2} = 0$, $\frac{\partial c_{11}}{\partial x_3} = (m+M)x_4 + mLcos(x_5)cos(x_7)x_6 - mLsin(x_5)sin(x_7)x_8$, $\frac{\partial c_{11}}{\partial x_4} = (m+M)x_3 + mLsin(x_5)sin(x_7)$, $\frac{\partial c_{11}}{\partial x_5} = -mLx_3sin(x_5)cos(x_7)x_6 - mLx_3cos(x_5)sin(x_7)x_8 + mLcos(x_5)sin(x_7)x_4 + mL^2[cos^2(x_5) - sin^2(x_5)]cos^2(x_7)x_6 - mL^2[-2cos(x_5)sin(x_5)sin(x_7)cos(x_7)]x_8$, $\frac{\partial c_{11}}{\partial x_6} = mLx_3cos(x_5)cos(x_7) + mL^2sin(x_5)cos(x_5)cos^2(x_7)$, $\frac{\partial c_{11}}{\partial x_7} = -mLx_3cos(x_5)sin(x_7)x_6 - mLx_3sin(x_5)\cdot cos(x_7)x_8 + -mLsin(x_5)cos(x_7)x_4 + mL^2sin(x_5)cos(x_5)[-2cos(x_7)sin(x_7)]x_6 - mL^2cos^2(x_5)[cos^2(x_7) - sin^2(x_7)]x_8$, $\frac{\partial c_{11}}{\partial x_8} = -mLx_3sin(x_5)sin(x_7) + mL^2cos^2(x_5)sin(x_7)cos(x_7)$.

Additionally, it holds that

$$c_{12} = mLsin(x_5)cos(x_7)x_2 + (m+M)x_3x_2 \tag{3.74}$$

Thus: $\frac{\partial c_{12}}{\partial x_7} = 0$, $\frac{\partial c_{11}}{\partial x_2} = mLsin(x_5)cos(x_7) + (m+M)x_3$, $\frac{\partial c_{12}}{\partial x_3} = (m+M)x_2$, $\frac{\partial c_{12}}{\partial x_4} = 0$, $\frac{\partial c_{12}}{\partial x_5} = mLcos(x_5)cos(x_7)x_2$, $\frac{\partial c_{12}}{\partial x_6} = 0$, $\frac{\partial c_{12}}{\partial x_7} = -mLsin(x_5)sin(x_7)x_2$, $\frac{\partial c_{11}}{\partial x_8} = 0$.

Moreover, it holds that

$$\begin{array}{c} c_{13} = mLx_3cos(x_5)cos(x_7)x_2 + mL^2sin(x_5)sin(x_7)cos(x_7)x_6 + \\ mL^2sin(x_5)cos(x_5)cos^2(x_7)x_2 + mL^2cos(x_5)sin^2(x_7)x_8. \end{array} \tag{3.75}$$

Thus, one obtains: $\frac{\partial c_{13}}{\partial x_1} = 0$, $\frac{\partial c_{13}}{\partial x_1} = mLx_3cos(x_5)cos(x_7) + mL^2sin(x_5)cos(x_5)cos^2(x_7)$, $\frac{\partial c_{13}}{\partial x_3} = mLcos(x_5)\cdot cos(x_7)x_2$, $\frac{\partial c_{13}}{\partial x_4} = 0$, $\frac{\partial c_{13}}{\partial x_5} = -mLx_3sin(x_5)cos(x_7)x_2 - mL^2cos(x_5)sin(x_7)cos(x_7)x_6 + mL^2[cos^2(x_5) - sin^2(x_5)]cos^2(x_7)x_2 - mL^2 sin(x_5)sin^2(x_7)x_8$, $\frac{\partial c_{13}}{\partial x_6} = mL^2sin(x_5)sin(x_7)cos(x_7)$, $\frac{\partial c_{13}}{\partial x_7} = -mLx_3cos(x_5)\cdot sin(x_7)x_2 + mL^2sin(x_5)cos(x_5)\cdot[-2cos(x_7)sin(x_7)]x_2 + mL^2cos(x_5)[2sin(x_7)cos(x_7)]x_8$, $\frac{\partial c_{13}}{\partial x_8} = mL^2cos(x_5)\cdot sin^2(x_7)$.

Additionally, it holds that

$$\begin{array}{c} c_{14} = -mLx_3sin(x_5)sin(x_7)x_2 - mLx_3sin(x_7)x_8 + \\ +mL^2cos^2(x_5)sin(x_7)cos(x_7)x_2 + mL^2cos(x_5)sin^2(x_7)x_6 \end{array} \tag{3.76}$$

Thus, one obtains: $\frac{\partial c_{14}}{\partial x_1} = 0$, $\frac{\partial c_{14}}{\partial x_2} = -mLx_3sin(x_5)sin(x_7) + mL^2cos^2(x_5)sin(x_7)cos(x_7)$, $\frac{\partial c_{14}}{\partial x_3} = -mL\cdot sin(x_5)sin(x_7) x_2 - mLsin(x_7)x_8$, $\frac{\partial c_{14}}{\partial x_4} = 0$, $\frac{\partial c_{14}}{\partial x_5} = -mLx_3\cdot cos(x_5)sin(x_7)x_2 + mL^2[-2cos(x_5)sin(x_5)]\cdot sin(x_7)cos(x_7)x_2 - mL^2sin(x_5)sin^2(x_7)x_6$, $\frac{\partial c_{14}}{\partial x_6} = mL^2cos(x_5)sin^2(x_7)$, $\frac{\partial c_{14}}{\partial x_7} = -mLx_3sin(x_5)cos(x_7)x_2 - mLx_3cos(x_7)x_8 - mL^2cos^2(x_5)[cos^2(x_7) - sin^2(x_7)]x_2 + mL^2cos(x_5)[2sin(x_7)cos(x_7)]x_6$, $\frac{\partial c_{14}}{\partial x_8} = -mLx_3sin(x_7)$.

Additionally, it holds that

$$c_{21} = -(m+M)x_2x_3 - mLsin(x_5)cos(x_7)x_2 - mLcos(x_7)x_8 \tag{3.77}$$

Thus, one obtains: $\frac{\partial c_{21}}{\partial x_1}=0$, $\frac{\partial c_{21}}{\partial x_2}=-(m+M)x_3-mLsin(x_7)cos(x_7)$, $\frac{\partial c_{21}}{\partial x_3}=-(m+M)x_2$, $\frac{\partial c_{21}}{\partial x_4}=0$, $\frac{\partial c_{21}}{\partial x_5}=-mLcos(x_5)cos(x_7)x_2$, $\frac{\partial c_{21}}{\partial x_6}=0$, $\frac{\partial c_{21}}{\partial x_7}=mLsin(x_5)sin(x_7)x_2+mLsin(x_7)x_8$, $\frac{\partial c_{21}}{\partial x_8}=-mLcos(x_7)$.

Furthermore, it holds that $c_{22}=0$. Thus, one obtains $\frac{\partial c_{22}}{\partial x_i}=0$, for $i=1,\cdots,8$.

Moreover, it holds that

$$c_{23}=-mL(sin(x_5)cos(x_7)x_6+cos(x_5)sin(x_7)x_8) \tag{3.78}$$

Thus, one obtains: $\frac{\partial c_{23}}{\partial x_1}=0$, $\frac{\partial c_{23}}{\partial x_2}=0$, $\frac{\partial c_{23}}{\partial x_3}=0$, $\frac{\partial c_{23}}{\partial x_4}=0$, $\frac{\partial c_{23}}{\partial x_5}=-mL(cos(x_5)cos(x_7)x_6-sin(x_5)sin(x_7)x_8)$, $\frac{\partial c_{23}}{\partial x_6}==-mLsin(x_5)cos(x_7)$, $\frac{\partial c_{23}}{\partial x_7}=-mL(-sin(x_5)sin(x_7)x_6+cos(x_5)cos(x_7)x_8)$, $\frac{\partial c_{23}}{\partial x_8}=-mLcos(x_5)sin(x_7)$.

Furthermore, it holds that

$$c_{24}=-mL(cos(x_5)sin(x_7)x_6+sin(x_5)cos(x_7)x_8+cos(x_7)x_2) \tag{3.79}$$

Thus, one obtains $\frac{\partial c_{24}}{\partial x_1}=0$, $\frac{\partial c_{24}}{\partial x_2}=-mLcos(x_7)$, $\frac{\partial c_{24}}{\partial x_3}=0$, $\frac{\partial c_{24}}{\partial x_4}=0$, $\frac{\partial c_{24}}{\partial x_5}=-mL(-sin(x_5)sin(x_7)x_6+cos(x_5)cos(x_7)x_8)$, $\frac{\partial c_{24}}{\partial x_6}=-mLcos(x_5)sin(x_7)$, $\frac{\partial c_{24}}{\partial x_7}=-mL(cos(x_5)cos(x_7)x_6-sin(x_5)sin(x_7)x_8)-sin(x_7)x_2$, $\frac{\partial c_{24}}{\partial x_8}=-mLsin(x_5)cos(x_7)$.

Additionally, one has that

$$c_{31}=-mLcos(x_5)cos(x_7)(x_3+Lsin(x_5)cos(x_7))x_2-mL^2cos(x_5)cos^2(x_7)x_8 \tag{3.80}$$

Thus, one obtains: $\frac{\partial c_{31}}{\partial x_1}=0$, $\frac{\partial c_{31}}{\partial x_2}=-mLcos(x_5)cos(x_7)(x_3+Lsin(x_5)cos(x_7))$, $\frac{\partial c_{31}}{\partial x_3}=-mLcos(x_5)cos(x_7)x_2$, $\frac{\partial c_{31}}{\partial x_4}=0$, $\frac{\partial c_{31}}{\partial x_5}=-mL(-sin(x_5))cos(x_7)(x_3+Lsin(x_5)cos(x_7))x_2+mLcos(x_5)cos(x_7)(Lcos(x_5)cos(x_7))x_2+mL^2sin(x_5)cos(x_7)x_8$, $\frac{\partial c_{31}}{\partial x_6}=0$, $\frac{\partial c_{31}}{\partial x_7}=mLcos(x_5)sin(x_7)(x_3+Lsin(x_5)cos(x_7))x_2+mLcos(x_5)cos(x_7)\cdot(Lsin(x_5)sin(x_7))+mL^2cos(x_5)(2cos(x_7)sin(x_7))x_8$, $\frac{\partial c_{31}}{\partial x_8}=-mL^2cos(x_5)cos^2(x_7)$.

Furthermore, it holds that $c_{32}=0$. Thus, one obtains $\frac{\partial c_{32}}{\partial x_i}=0$, for $i=1,\cdots,8$.

Moreover, it holds that

$$c_{33}=-mL^2sin(x_7)cos(x_7)x_8 \tag{3.81}$$

Thus, one obtains: $\frac{\partial c_{33}}{\partial x_1}=0$, $\frac{\partial c_{33}}{\partial x_2}=0$, $\frac{\partial c_{33}}{\partial x_3}=0$, $\frac{\partial c_{33}}{\partial x_4}=0$, $\frac{\partial c_{33}}{\partial x_5}=0$, $\frac{\partial c_{33}}{\partial x_6}=0$, $\frac{\partial c_{33}}{\partial x_7}=-mL^2[cos^2(x_7)-sin^2(x_7)]x_8$, $\frac{\partial c_{33}}{\partial x_8}=-mL^2sin(x_7)cos(x_7)$.

Additionally, it holds that

$$c_{34}=-mL^2(cos(x_5)cos^2(x_7)x_2+sin(x_5)cos(x_7)x_6) \tag{3.82}$$

Thus, one obtains: $\frac{\partial c_{34}}{\partial x_1}=0$, $\frac{\partial c_{34}}{\partial x_2}=-mL^2cos(x_5)cos^2(x_7)$, $\frac{\partial c_{34}}{\partial x_3}=0$, $\frac{\partial c_{34}}{\partial x_4}=0$, $\frac{\partial c_{34}}{\partial x_5}=-mL^2(-sin(x_5)cos^2(x_7)x_2-cos(x_5)cos(x_7)x_6)$, $\frac{\partial c_{34}}{\partial x_6}=-mL^2(sin(x_5)cos(x_7))$, $\frac{\partial c_{37}}{\partial x_7}=-mL^2(-2cos(x_5)cos(x_7)sin(x_7)x_2-sin(x_5)sin(x_7)x_6)$, $\frac{\partial c_{34}}{\partial x_8}=0$.

Furthermore, one has that

$$\begin{aligned}c_{41}=mL(msin(x_5)sin(x_7)-Lcos^2(x_5)sin(x_7)cos(x_7))x_2+\\+mL^2cos(x_5)cos^2(x_7)x_2+mLcos(x_7)x_4\end{aligned} \tag{3.83}$$

Thus, one obtains: $\frac{\partial c_{41}}{\partial x_1}=0$, $\frac{\partial c_{41}}{\partial x_1}=mL(msin(x_5)sin(x_7)-Lcos^2(x_5)sin(x_7)cos(x_7))+mL^2cos(x_5)cos^{(}x_7)$, $\frac{\partial c_{41}}{\partial x_3}=0$, $\frac{\partial c_{41}}{\partial x_4}=mLcos(x_7)$, $\frac{\partial c_{41}}{\partial x_5}=mL(mcos(x_5)sin(x_7)-2Lcos(x_5)sin(x_5)sin(x_7)cos(x_7))x_2-mL^2\cdot sin(x_5)cos^2(x_7)x_2$,

$\frac{\partial c_{41}}{\partial x_6}=0$, $\frac{\partial c_{41}}{\partial x_7}=mL(msin(x_5)cos(x_7)-Lcos^2(x_5))[cos^2(x_7)-sin^2(x_7)]x_2-2mL^2cos(x_5)\cdot cos(x_7)sin(x_7)x_2-mLsin(x_7)x_4$, $\frac{\partial c_{41}}{\partial x_8}=0$.

Moreover, it holds that

$$c_{42}=mLcos(x_7)x_2 \tag{3.84}$$

Thus, one obtains: $\frac{\partial c_{42}}{\partial x_1}=0$, $\frac{\partial c_{42}}{\partial x_2}=mLcos(x_7)$, $\frac{\partial c_{42}}{\partial x_3}=0$, $\frac{\partial c_{42}}{\partial x_4}=0$, $\frac{\partial c_{42}}{\partial x_5}=0$, $\frac{\partial c_{42}}{\partial x_6}=0$, $\frac{\partial c_{42}}{\partial x_7}=-mLsin(x_7)x_2$, $\frac{\partial c_{42}}{\partial x_8}=0$.

Furthermore, it holds that

$$c_{43}=mL^2(cos(x_5)cos^2(x_7)x_2+sin(x_5)cos(x_7)x_6) \tag{3.85}$$

Thus, one obtains: $\frac{\partial c_{43}}{\partial x_1}=0$, $\frac{\partial c_{43}}{\partial x_2}=mL^2cos(x_5)cos^2(x_7)$, $\frac{\partial c_{43}}{\partial x_3}=0$, $\frac{\partial c_{43}}{\partial x_4}=0$, $\frac{\partial c_{43}}{\partial x_5}=mL^2(-sin(x_5)cos^2(x_7)x_2+cos(x_5)cos(x_7)x_6)$, $\frac{\partial c_{43}}{\partial x_6}=mL^2(sin(x_5)cos(x_7))$, $\frac{\partial c_{43}}{\partial x_7}=mL^2(-2cos(x_5)cos(x_7)sin(x_7)-sin(x_5)sin(x_7)x_6)$, $\frac{\partial c_{43}}{\partial x_8}=0$.

Finally, it holds that $c_{44}=0$. Thus, one obtains $\frac{\partial c_{44}}{\partial x_i}=0$, for $i=1,\cdots,8$.

In a similar manner, one computes the partial derivatives of the elements of the gravitational forces matrix. It holds that $g_{11}=0$ that one obtains $\frac{\partial g_{11}}{\partial x_i}=0$, for $i=1,\cdots,8$.

Additionally, it holds that $g_{21}=0$ that one obtains $\frac{\partial g_{21}}{\partial x_i}=0$, for $i=1,\cdots,8$.

Moreover, it holds that

$$g_{31}=gmLsin(x_5)sin(x_7) \tag{3.86}$$

Thus, one obtains $\frac{\partial g_{31}}{\partial x_1}=0$, $\frac{\partial g_{31}}{\partial x_2}=0$, $\frac{\partial g_{31}}{\partial x_3}=0$, $\frac{\partial g_{31}}{\partial x_4}=0$, $\frac{\partial g_{31}}{\partial x_5}=gmLcos(x_5)sin(x_7)$, $\frac{\partial g_{31}}{\partial x_6}=0$, $\frac{\partial g_{31}}{\partial x_7}=gmLsin(x_5)cos(x_7)$, $\frac{\partial g_{31}}{\partial x_8}=0$.

Finally, it holds that

$$g_{41}=gLcos(x_5)sin(x_7) \tag{3.87}$$

Thus, one obtains $\frac{\partial g_{41}}{\partial x_1}=0$, $\frac{\partial g_{41}}{\partial x_2}=0$, $\frac{\partial g_{41}}{\partial x_3}=0$, $\frac{\partial g_{41}}{\partial x_4}=0$, $\frac{\partial g_{41}}{\partial x_5}=-gmLsin(x_5)sin(x_7)$, $\frac{\partial g_{41}}{\partial x_6}=0$, $\frac{\partial g_{41}}{\partial x_7}=gmLcos(x_5)cos(x_7)$, $\frac{\partial g_{41}}{\partial x_8}=0$.

3.1.4 Design of an H-infinity nonlinear feedback controller

3.1.4.1 Equivalent linearized dynamics of the tower crane

After linearization around its current operating point, the dynamic model for the tower crane is written as

$$\dot{x}=Ax+Bu+d_1 \tag{3.88}$$

Parameter d_1 stands for the linearization error in the tower crane's model that was given previously in Eq. (3.17). The reference setpoints for the state vector of the aforementioned dynamic model are denoted by $\mathbf{x_d}=[x_1^d,\cdots,x_8^d]$. Tracking of this trajectory is achieved after applying the control input u^*. At every time instant the control input u^* is assumed to differ from the control input u appearing in Eq. (3.88) by an amount equal to Δu, that is $u^*=u+\Delta u$

$$\dot{x}_d=Ax_d+Bu^*+d_2 \tag{3.89}$$

The dynamics of the controlled system described in Eq. (3.88) can be also written as

$$\dot{x}=Ax+Bu+Bu^*-Bu^*+d_1 \tag{3.90}$$

and by denoting $d_3 = -Bu^* + d_1$ as an aggregate disturbance term, one obtains

$$\dot{x} = Ax + Bu + Bu^* + d_3 \tag{3.91}$$

By subtracting Eq. (3.89) from Eq. (3.91), one has

$$\dot{x} - \dot{x}_d = A(x - x_d) + Bu + d_3 - d_2 \tag{3.92}$$

By denoting the tracking error as $e = x - x_d$ and the aggregate disturbance term as $\tilde{d} = d_3 - d_2$, the tracking error dynamics becomes

$$\dot{e} = Ae + Bu + \tilde{d} \tag{3.93}$$

The above-linearized form of the tower crane's model can be efficiently controlled after applying an H-infinity feedback control scheme.

3.1.5 The nonlinear H-infinity control

The initial nonlinear model of the tower crane is in the form

$$\dot{x} = f(x, u) \;\; x \in R^n, \;\; u \in R^m \tag{3.94}$$

Linearization of the model of the tower crane is performed at each iteration of the control algorithm around its present operating point $(x^*, u^*) = (x(t), u(t - T_s))$. The linearized equivalent of the system is described by

$$\dot{x} = Ax + Bu + L\tilde{d} \;\; x \in R^n, \;\; u \in R^m, \;\; \tilde{d} \in R^q \tag{3.95}$$

where matrices A and B are obtained from the computation of the previously defined Jacobians and vector $\tilde{d}$ denotes disturbance terms due to linearization errors. The problem of disturbance rejection for the linearized model that is described by

$$\begin{aligned} \dot{x} &= Ax + Bu + L\tilde{d} \\ y &= Cx \end{aligned} \tag{3.96}$$

where $x \in R^n$, $u \in R^m$, $\tilde{d} \in R^q$, and $y \in R^p$, cannot be handled efficiently if the classical LQR control scheme is applied. This is because of the existence of the perturbation term $\tilde{d}$. The disturbance term $\tilde{d}$ apart from modeling (parametric) uncertainty and external perturbation terms can also represent noise terms of any distribution.

In the H_∞ control approach, a feedback control scheme is designed for trajectory tracking by the system's state vector and simultaneous disturbance rejection, considering that the disturbance affects the system in the worst possible manner. The disturbances' effects are incorporated in the following quadratic cost function:

$$J(t) = \tfrac{1}{2}\int_0^T [y^T(t)y(t) + ru^T(t)u(t) - \rho^2\tilde{d}^T(t)\tilde{d}(t)]dt, \;\; r, \rho > 0 \tag{3.97}$$

The significance of the negative sign in the cost function's term that is associated with the perturbation variable $\tilde{d}(t)$ is that the disturbance tries to maximize the cost function $J(t)$ while the control signal $u(t)$ tries to minimize it. The physical meaning of the relation given above is that the control signal and the disturbances compete with each other within a min–max differential game. This problem of min–max optimization can be written as $min_u max_{\tilde{d}} J(u, \tilde{d})$.

The objective of the optimization procedure is to compute a control signal $u(t)$ which can compensate for the worst possible disturbance, that is externally imposed on the tower crane. However, the solution to the min–max optimization problem is directly related to the value of the parameter ρ. This means that there is an upper bound in the disturbance magnitude that can be annihilated by the control signal.

3.1.5.1 Computation of the feedback control gains

For the linearized system given by Eq. (3.96) the cost function of Eq. (3.97) is defined, where coefficient r determines the penalization of the control input and weight coefficient ρ determines the reward of the disturbances' effects. It is assumed that (i) The energy that is transferred from the disturbances signal $\tilde{d}(t)$ is bounded, that is $\int_0^\infty \tilde{d}^T(t)\tilde{d}(t)dt < \infty$,

(ii) matrices $[A, B]$ and $[A, L]$ are stabilizable, (iii) matrix $[A, C]$ is detectable. In the case of a tracking problem, the optimal feedback control law is given by

$$u(t) = -Ke(t) \tag{3.98}$$

with $e = x - x_d$ to be the tracking error, and $K = \frac{1}{r}B^TP$ where P is a positive definite symmetric matrix. As it will be proven in Section 3.1.6, matrix P is obtained from the solution of the Riccati equation

$$A^TP + PA + Q - P(\tfrac{2}{r}BB^T - \tfrac{1}{\rho^2}LL^T)P = 0 \tag{3.99}$$

where Q is a positive semi-definite symmetric matrix. The worst-case disturbance is given by

$$\tilde{d}(t) = \tfrac{1}{\rho^2}L^TPe(t) \tag{3.100}$$

The solution of the H-infinity feedback control problem for the tower crane and the computation of the worst-case disturbance that the related controller can sustain, comes from the superposition of Bellman's optimality principle when considering that the robotic crane is affected by two separate inputs (i) the control input u (ii) the cumulative disturbance input $\tilde{d}(t)$. Solving the optimal control problem for u, that is for the minimum variation (optimal) control input that achieves the elimination of the state vector's tracking error, gives $u = -\frac{1}{r}B^TPe$. Equivalently, solving the optimal control problem for $\tilde{d}$, that is for the worst-case disturbance that the control loop can sustain gives $\tilde{d} = \frac{1}{\rho^2}L^TPe$.

The diagram of the considered control loop for the tower crane is depicted in Figure 3.2.

3.1.6 Lyapunov stability analysis

Through Lyapunov stability analysis, it will be shown that the proposed nonlinear control scheme assures H_∞ tracking performance for the underactuated tower crane and that in case of bounded disturbance terms asymptotic convergence to the reference setpoints is achieved. The tracking error dynamics for the tower crane is written in the form

$$\dot{e} = Ae + Bu + L\tilde{d} \tag{3.101}$$

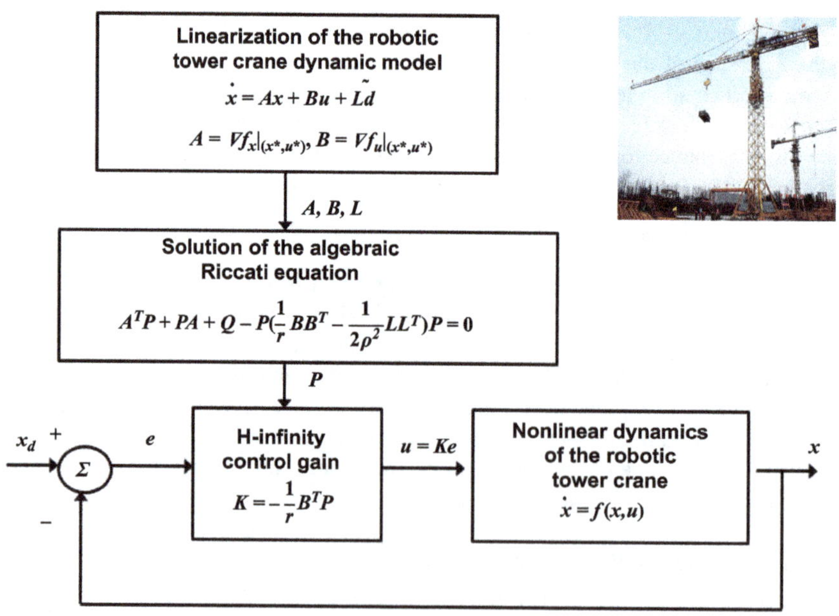

Figure 3.2 Diagram of the control scheme for the 4-DOF underactuated tower crane

where in the tower crane's case $L = \in R^{8\times 8}$ to be the disturbance inputs gain matrix. Variable $\tilde{d}$ denotes model uncertainties and external disturbances of the tower crane's model. The following Lyapunov equation is considered:

$$V = \frac{1}{2}e^T Pe \tag{3.102}$$

where $e = x - x_d$ is the tracking error. By differentiating with respect to time, one obtains

$$\begin{gathered}\dot{V} = \frac{1}{2}\dot{e}^T Pe + \frac{1}{2}eP\dot{e} \Rightarrow \\ \dot{V} = \frac{1}{2}[Ae + Bu + L\tilde{d}]^T Pe + \frac{1}{2}e^T P[Ae + Bu + L\tilde{d}] \Rightarrow\end{gathered} \tag{3.103}$$

$$\begin{aligned}\dot{V} = &\frac{1}{2}[e^T A^T + u^T B^T + \tilde{d}^T L^T]Pe + \\ &+\frac{1}{2}e^T P[Ae + Bu + L\tilde{d}] \Rightarrow\end{aligned} \tag{3.104}$$

$$\begin{aligned}\dot{V} = &\frac{1}{2}e^T A^T Pe + \frac{1}{2}u^T B^T Pe + \frac{1}{2}\tilde{d}^T L^T Pe + \\ &\frac{1}{2}e^T PAe + \frac{1}{2}e^T PBu + \frac{1}{2}e^T PL\tilde{d}\end{aligned} \tag{3.105}$$

The previous equation is rewritten as

$$\begin{aligned}\dot{V} = &\frac{1}{2}e^T(A^T P + PA)e + (\frac{1}{2}u^T B^T Pe + \frac{1}{2}e^T PBu) + \\ &+(\frac{1}{2}\tilde{d}^T L^T Pe + \frac{1}{2}e^T PL\tilde{d})\end{aligned} \tag{3.106}$$

Assumption: For given positive definite matrix Q and coefficients r and ρ there exists a positive definite matrix P, which is the solution of the following matrix equation:

$$A^T P + PA = -Q + P(\frac{2}{r}BB^T - \frac{1}{\rho^2}LL^T)P \tag{3.107}$$

Moreover, the following feedback control law is applied to the system:

$$u = -\frac{1}{r}B^T Pe \tag{3.108}$$

By substituting Eq. (3.107) and Eq. (3.108), one obtains

$$\begin{aligned}\dot{V} = &\frac{1}{2}e^T[-Q + P(\frac{2}{r}BB^T - \frac{1}{\rho^2}LL^T)P]e + \\ &+e^T PB(-\frac{1}{r}B^T Pe) + e^T PL\tilde{d} \Rightarrow\end{aligned} \tag{3.109}$$

$$\begin{aligned}\dot{V} = &-\frac{1}{2}e^T Qe + \frac{1}{r}e^T PBB^T Pe - \frac{1}{2\rho^2}e^T PLL^T Pe \\ &-\frac{1}{r}e^T PBB^T Pe + e^T PL\tilde{d}\end{aligned} \tag{3.110}$$

which after intermediate operations gives

$$\dot{V} = -\frac{1}{2}e^T Qe - \frac{1}{2\rho^2}e^T PLL^T Pe + e^T PL\tilde{d} \tag{3.111}$$

or, equivalently

$$\begin{aligned}\dot{V} = &-\frac{1}{2}e^T Qe - \frac{1}{2\rho^2}e^T PLL^T Pe + \\ &+\frac{1}{2}e^T PL\tilde{d} + \frac{1}{2}\tilde{d}^T L^T Pe\end{aligned} \tag{3.112}$$

Lemma: The following inequality holds

$$\frac{1}{2}e^T L\tilde{d} + \frac{1}{2}\tilde{d}L^T Pe - \frac{1}{2\rho^2}e^T PLL^T Pe \leq \frac{1}{2}\rho^2\tilde{d}^T\tilde{d} \tag{3.113}$$

Proof: The binomial $(\rho\alpha - \frac{1}{\rho}b)^2$ is considered. Expanding the left part of the above inequality, one gets

$$\begin{gathered}\rho^2 a^2 + \frac{1}{\rho^2}b^2 - 2ab \geq 0 \Rightarrow \frac{1}{2}\rho^2 a^2 + \frac{1}{2\rho^2}b^2 - ab \geq 0 \Rightarrow \\ ab - \frac{1}{2\rho^2}b^2 \leq \frac{1}{2}\rho^2 a^2 \Rightarrow \frac{1}{2}ab + \frac{1}{2}ab - \frac{1}{2\rho^2}b^2 \leq \frac{1}{2}\rho^2 a^2\end{gathered} \tag{3.114}$$

The following substitutions are carried out: $a = \tilde{d}$ and $b = e^T PL$ and the previous relation becomes

$$\frac{1}{2}\tilde{d}^T L^T Pe + \frac{1}{2}e^T PL\tilde{d} - \frac{1}{2\rho^2}e^T PLL^T Pe \leq \frac{1}{2}\rho^2\tilde{d}^T\tilde{d} \tag{3.115}$$

Eq. (3.115) is substituted in Eq. (3.112) and the inequality is enforced, thus giving

$$\dot{V} \leq -\frac{1}{2}e^T Q e + \frac{1}{2}\rho^2 \tilde{d}^T \tilde{d} \tag{3.116}$$

Eq. (3.116) shows that the H_∞ tracking performance criterion is satisfied. The integration of $\dot{V}$ from 0 to T gives

$$\begin{aligned} &\int_0^T \dot{V}(t)dt \leq -\frac{1}{2}\int_0^T ||e||_Q^2 dt + \frac{1}{2}\rho^2 \int_0^T ||\tilde{d}||^2 dt \Rightarrow \\ &2V(T) + \int_0^T ||e||_Q^2 dt \leq 2V(0) + \rho^2 \int_0^T ||\tilde{d}||^2 dt \end{aligned} \tag{3.117}$$

Moreover, if there exists a positive constant $M_d > 0$ such that

$$\int_0^\infty ||\tilde{d}||^2 dt \leq M_d \tag{3.118}$$

then, one gets

$$\int_0^\infty ||e||_Q^2 dt \leq 2V(0) + \rho^2 M_d \tag{3.119}$$

Thus, the integral $\int_0^\infty ||e||_Q^2 dt$ is bounded. Moreover, $V(T)$ is bounded and from the definition of the Lyapunov function V in Eq. (3.102) it becomes clear that $e(t)$ will be also bounded since $e(t) \in \Omega_e = \{e | e^T P e \leq 2V(0) + \rho^2 M_d\}$. According to the above and with the use of Barbalat's Lemma, one obtains $\lim_{t \to \infty} e(t) = 0$.

Through the stages of the stability proof, one arrives at Eq. (3.116), which shows that the H-infinity tracking performance criterion holds. By selecting the attenuation coefficient ρ to be sufficiently small and in particular to satisfy $\rho^2 < ||e||_Q^2 / ||\tilde{d}||^2$, one has that the first derivative of the Lyapunov function is upper bounded by 0. This condition holds at each sampling instance, and consequently, global stability for the control loop can be concluded.

3.1.7 Simulation tests

The global stability properties of the control method and the elimination of the state vector's tracking error which were previously proven through Lyapunov analysis are further confirmed through simulation experiments. To compute the stabilizing feedback gains of the controller, the algebraic Riccati equation of Eq. (3.107) had to be repetitively solved at each iteration of the control algorithm. The parameters of the dynamic model of the tower crane which have been used in the simulation experiments were according to [985,833]. All parameters and variables of the crane's model are measured in SI units. The obtained results are depicted in Figures 3.3–3.18. The real values of the state variables of the tower crane are printed in blue, their estimates which are provided by the H-infinity Kalman filter are printed in green color while the associated setpoints are printed in red. The performance of the nonlinear optimal control method was very satisfactory. Actually, through all test cases, it has been shown that the control method can achieve fast and accurate tracking of reference trajectories (setpoints) under moderate variations of the control inputs. The simulation tests come to confirm that the control method has global (and not local) stability properties. Under the nonlinear optimal control method, the state variables of the tower crane can track precisely setpoints with fast and abrupt changes. Moreover, the convergence to these setpoints is independent from initial conditions.

Regarding the selection of values for the controller gains it can be noted that parameters r, ρ, and Q which appear in the method's algebraic Riccati equations are assigned offline constant values, where the gains vector K is updated at each sampling instance, based on the positive definite and symmetric matrix P which is the solution of the method's algebraic Riccati equation. The tracking accuracy and the transient performance of the control scheme depend on the values of coefficients r, ρ and on the values of the elements of the diagonal matrix Q. Actually, for relatively small values of r, one achieves the elimination of the state vectors' tracking error one. Moreover, for relatively high values of the diagonal elements of matrix Q, one achieves fast convergence of the state variables' reference trajectories. Finally, the smallest value of the attenuation coefficient ρ that results in a valid solution of the method's Riccati equation in the form of the positive definite and symmetric matrix P is the one that provides the control loop with maximum robustness.

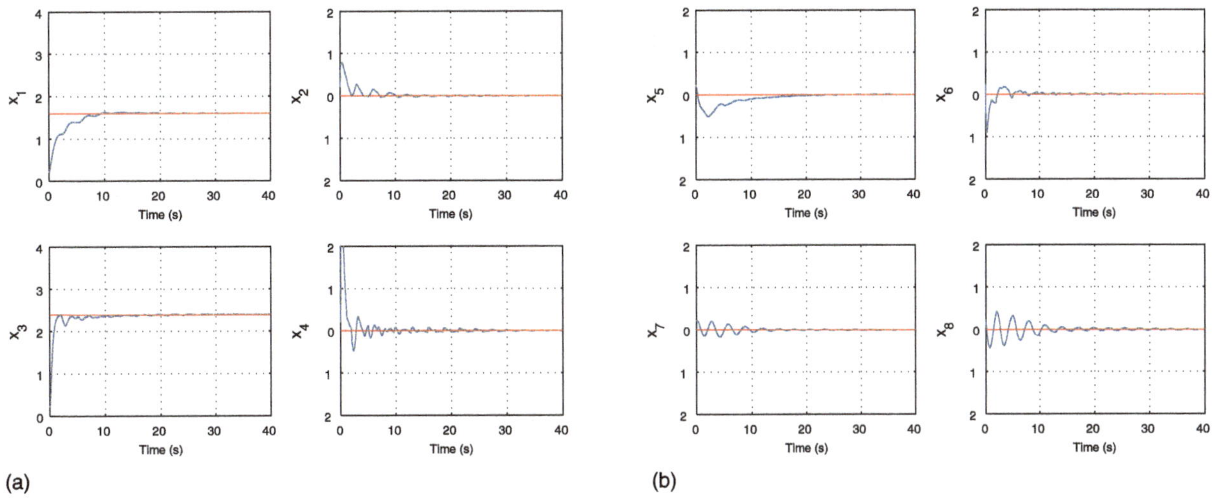

Figure 3.3 *Tracking of setpoint 1 for the tower crane: (a) convergence of state variables x_1 to x_4 to their reference setpoints (red line: setpoint, blue line: real value, green line: estimated value); (b) convergence of state variables x_5 to x_8 to their reference setpoints*

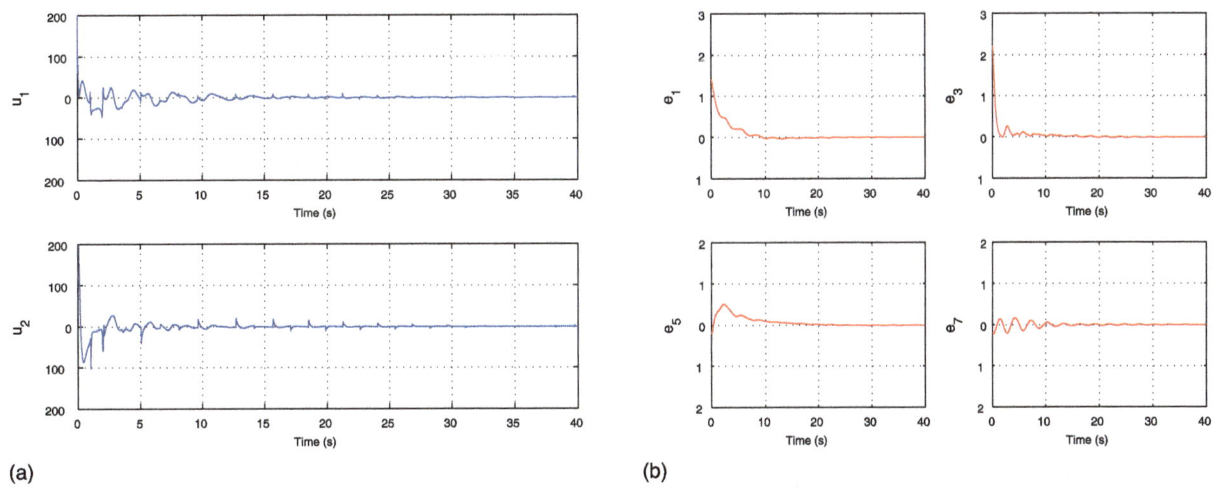

Figure 3.4 *Tracking of setpoint 1 for the tower crane: (a) control inputs u_1, u_2 applied to crane; (b) tracking error variables e_1, e_3, e_5, e_7 of the tower crane*

To elaborate on the tracking performance and on the robustness of the proposed nonlinear optimal control method for the ball and plate system the following tables are given: (i) Table 3.1 which provides information about the accuracy of tracking of the reference setpoints by the state variables of the tower crane's state-space model, (ii) Table 3.2 which provides information about the robustness of the control method to parametric changes in the model of the tower crane's dynamics (change $\Delta a\%$ in the mass M of the trolley), (iii) Table 3.3 which provides information about the precision in state variables' estimation that is achieved by the H-infinity Kalman filter, and (iv) Table 3.4 which provides the approximate convergence times of the tower crane's state variables to the associated setpoints.

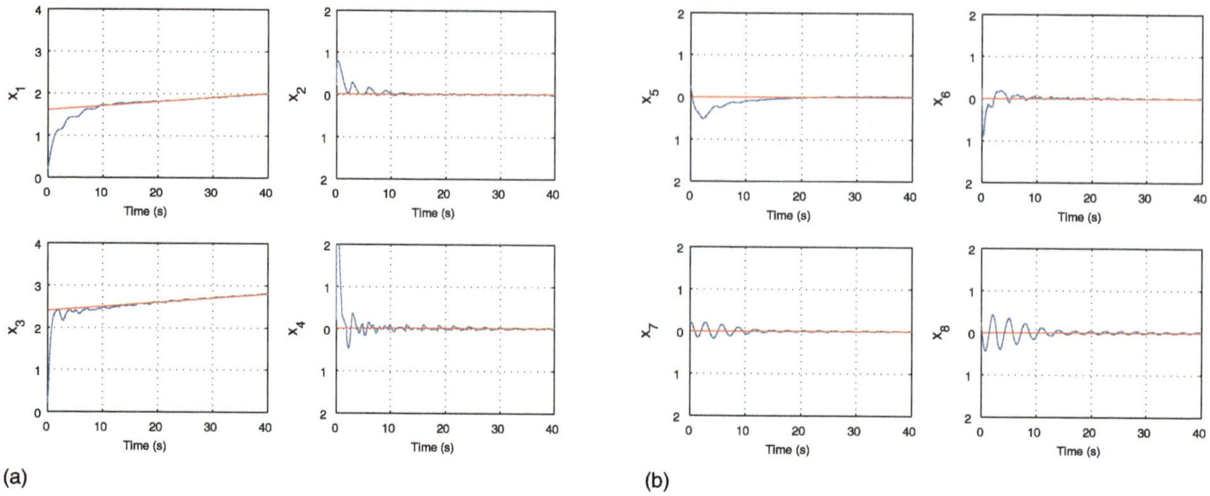

Figure 3.5 *Tracking of setpoint 2 for the tower crane: (a) convergence of state variables x_1 to x_4 to their reference setpoints (red line: setpoint, blue line: real value, green line: estimated value); (b) convergence of state variables x_5 to x_8 to their reference setpoints*

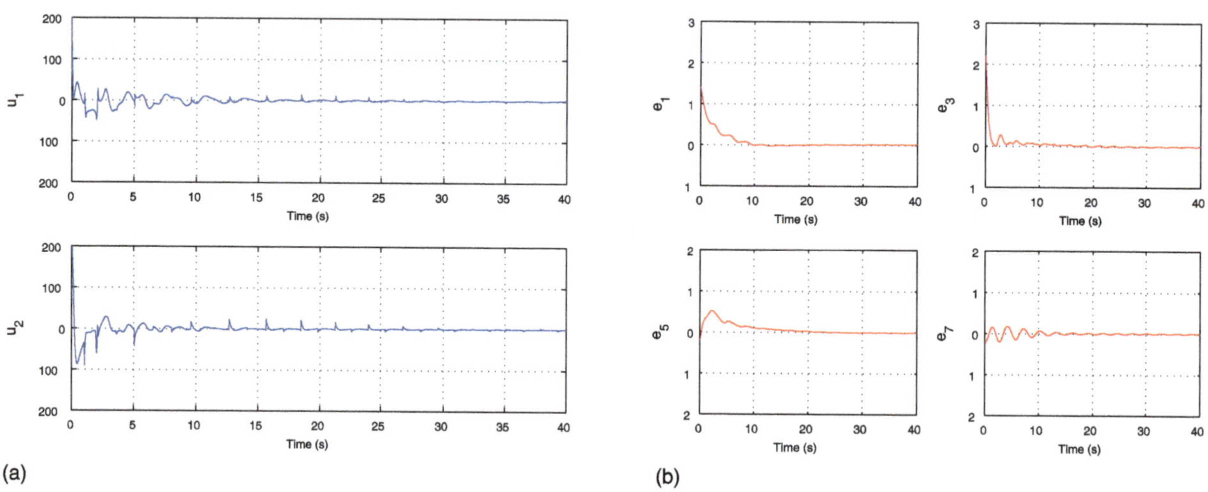

Figure 3.6 *Tracking of setpoint 2 for the tower crane: (a) control inputs u_1, u_2 applied to crane; (b) tracking error variables e_1, e_3, e_5, e_7 of the tower crane*

3.2 Nonlinear optimal control of an H-type gantry crane driven by dual PMLSMs

3.2.1 Outline

Dual-drive H-type gantry cranes are long-stroke high-speed Cartesian robotic systems which can be used in several industrial applications, such as (i) circuit assembly, printing circuit boards, photolithography, (ii) precision machining, computer numerical control (CNC), arc welding and laser or plasma cutting, (iii) as lifting machines and for the loading, unloading, and transfer of cargos, and finally (iv) in biomedical applications such as CT and X-ray scanning [537,535,1050]. The dual-drive H-type gantry crane consists of two linear motors which are arranged in parallel along the vertical axes of an orthogonal board and which are rigidly connected with a cross-beam along the horizontal axis

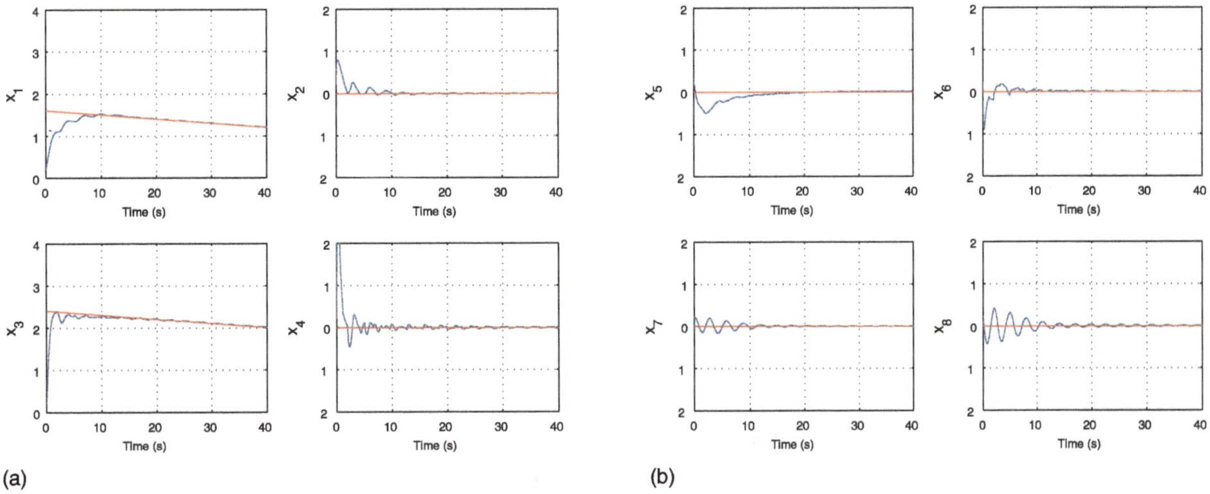

Figure 3.7 Tracking of setpoint 3 for the tower crane: (a) convergence of state variables x_1 to x_4 to their reference setpoints (red line: setpoint, blue line: real value, green line: estimated value); (b) convergence of state variables x_5 to x_8 to their reference setpoints

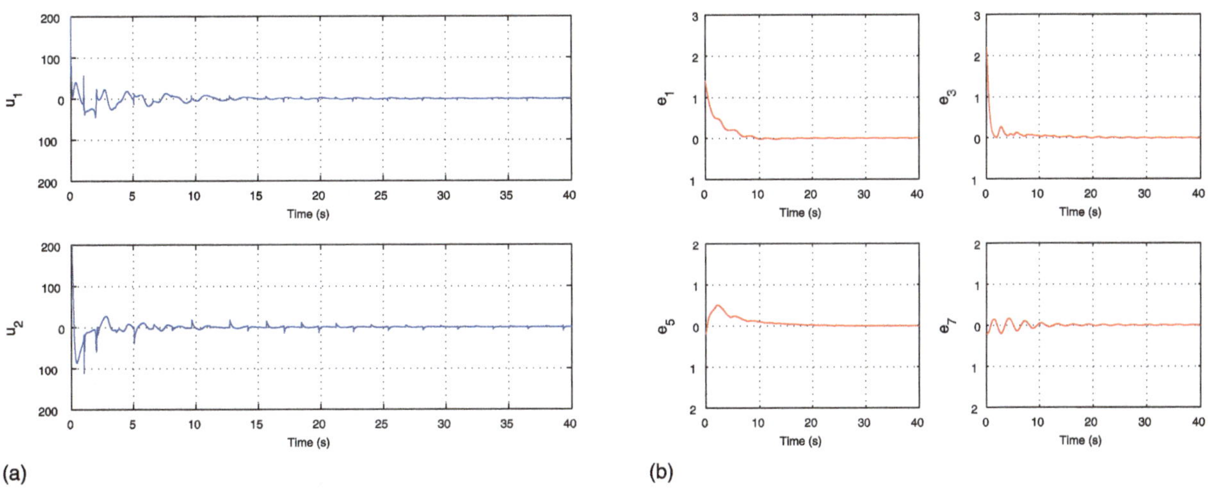

Figure 3.8 Tracking of setpoint 3 for the tower crane: (a) control inputs u_1, u_2 applied to crane; (b) tracking error variables e_1, e_3, e_5, e_7 of the tower crane

[860,923,359]. Dual-drive gantry cranes can achieve high torque and high precision in tasks' execution [860,562,304]. The cross-beam serves also as support to a third linear motor carrying a load [139,791,924,421]. There is a need for precise synchronization between the two parallel motors and for minimization of internal forces [928,277,276,790]. When the two linear motors are not well synchronized the cross-beam will mis-align from the horizontal axis and will rotate by a small angle around its center of gravity [494,133,495]. Synchronization error is often caused by unbalanced forces, mechanical assembly variations, and various disturbances [521,789,493]. Permanent magnet linear synchronous motors (PMLSMs) are a core subsystem in H-type gantry cranes [138,158,1074,167]. PMLSMs can generate the actuation forces which enable precise positioning along the crane's vertical and horizontal axes [906,129,131]. Accurate integration and synchronization of the mechanical part (beams and payload) and of the electrical part (actuators) has

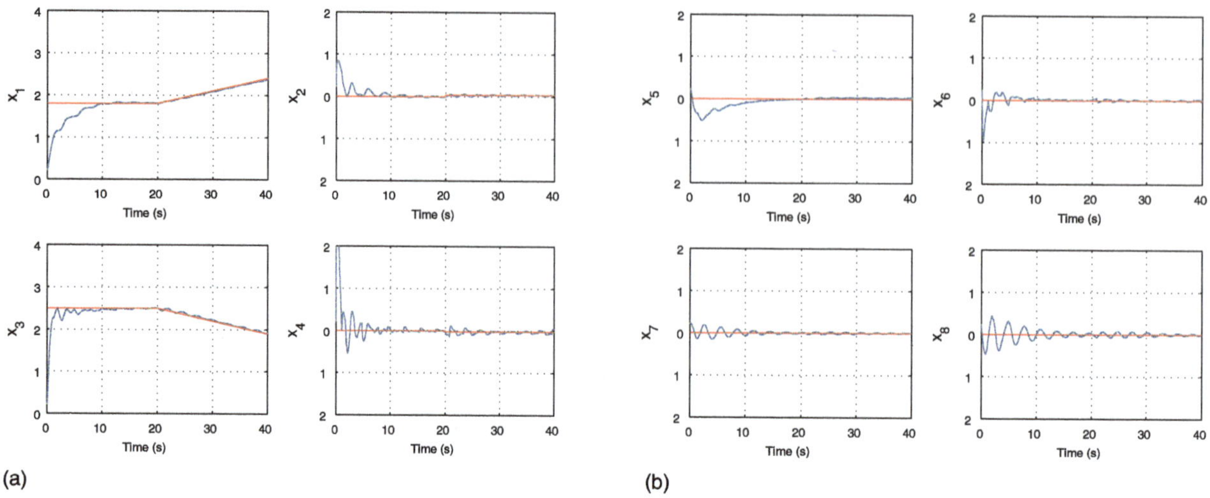

Figure 3.9 Tracking of setpoint 4 for the tower crane: (a) convergence of state variables x_1 to x_4 to their reference setpoints (red line: setpoint, blue line: real value, green line: estimated value); (b) convergence of state variables x_5 to x_8 to their reference setpoints

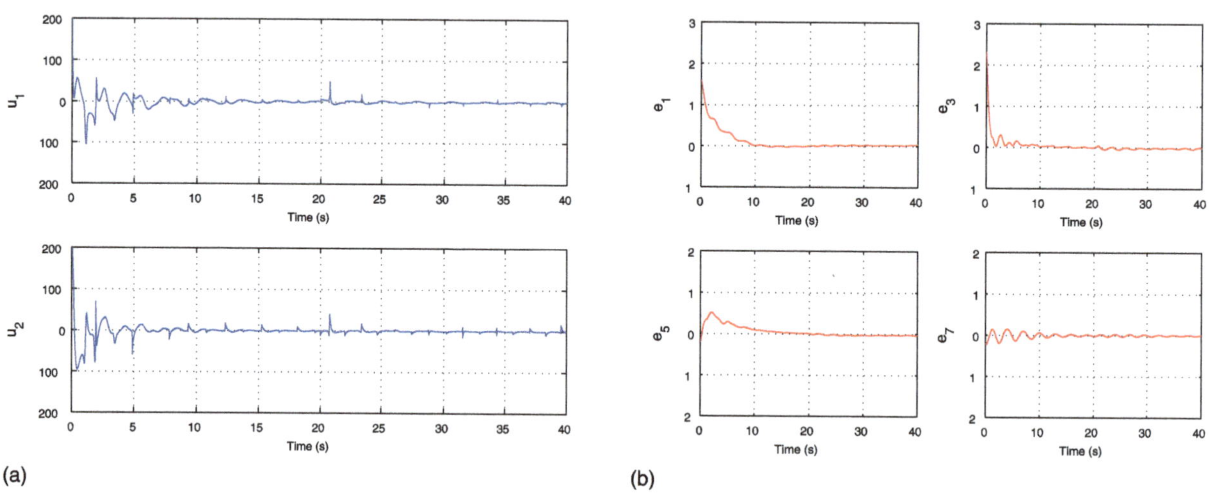

Figure 3.10 Tracking of setpoint 4 for the tower crane: (a) control inputs u_1, u_2 applied to crane; (b) tracking error variables e_1, e_3, e_5, e_7 of the tower crane

been the main objective in the development of control systems for gantry cranes which are driven by dual parallel electric motors [428,1070,991,540,961,930].

In the present section, a nonlinear optimal control method is developed for the dynamic system of the dual PMLSM-driven gantry crane [711,725,741]. By considering the mechanical dynamics of the crane jointly with the electrical dynamics of the electric motors the state-space model of the dual PMLSM-driven gantry crane is obtained. It is proven that this dynamic system is differentially flat, which is also a constrained controllability condition for this system [711,725,741]. Next, a nonlinear optimal control method is proposed for the control and stabilization of this system. In the optimal control approach, the state-space model of the gantry crane undergoes first approximate linearization with the use of first-order Taylor series expansion and through the computation of the associated Jacobian matrices

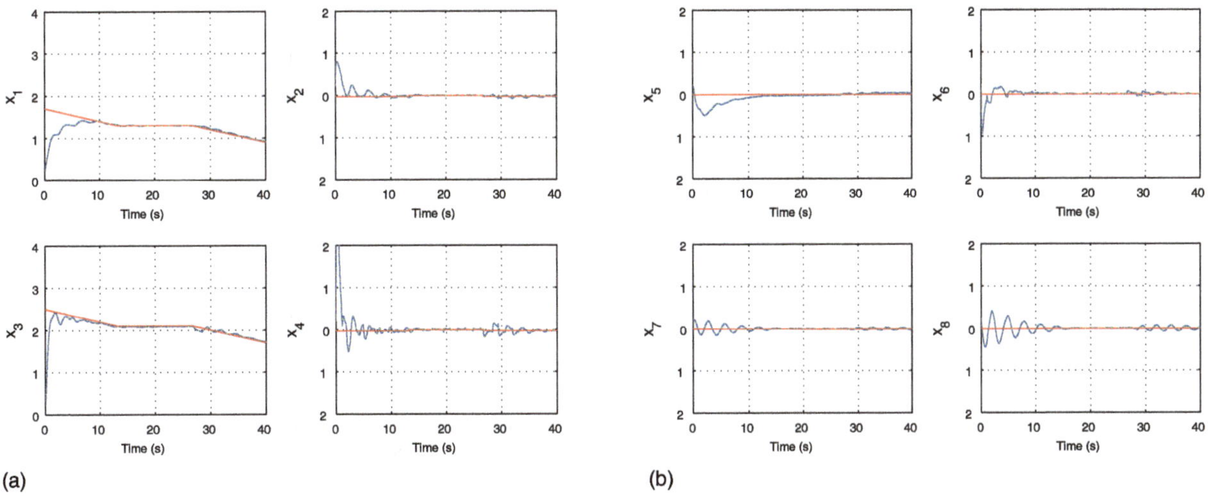

Figure 3.11 Tracking of setpoint 5 for the tower crane: (a) convergence of state variables x_1 to x_4 to their reference setpoints (red line: setpoint, blue line: real value, green line: estimated value); (b) convergence of state variables x_5 to x_8 to their reference setpoints

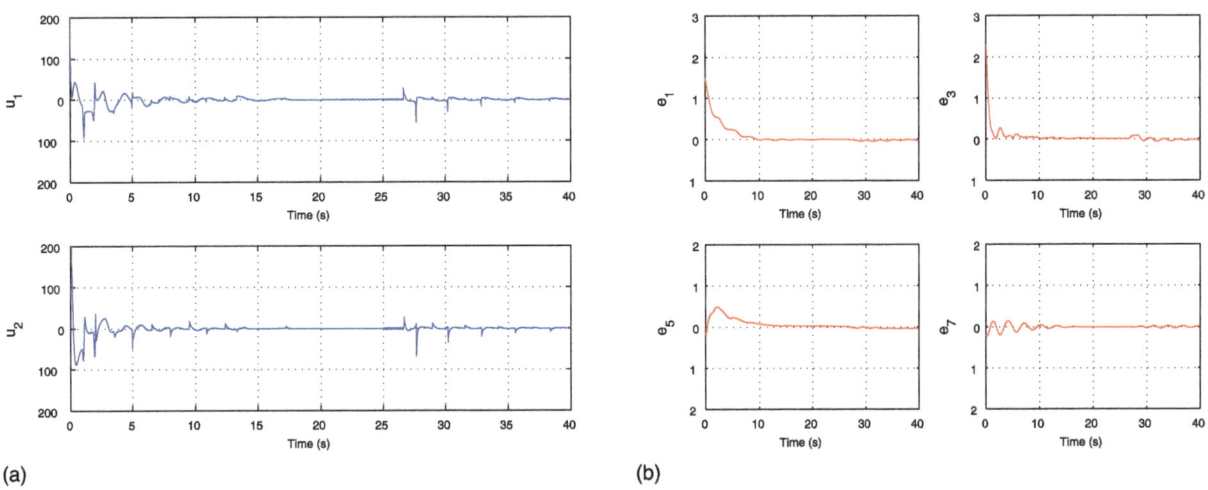

Figure 3.12 Tracking of setpoint 5 for the tower crane: (a) control inputs u_1, u_2 applied to crane; (b) tracking error variables e_1, e_3, e_5, e_7 of the tower crane

[705,60,706]. The linearization process takes place at each sampling instance around a temporary operating point which is defined by the present value of the system's state vector and by the last sampled value of the control inputs vector. The model imprecision which is due to the truncation of higher-order terms in the Taylor series expansion is viewed as a perturbation which is asymptotically compensated by the robustness of the control algorithm. For the approximately linearized model of the system, a stabilizing H-infinity feedback controller is designed.

The control method's H-infinity controller represents a min–max differential game which takes place between: (i) the control inputs of the gantry crane which try to minimize a cost function that contains a quadratic term of the state vector's tracking error, (ii) the model uncertainty and exogenous perturbations which try to maximize this cost

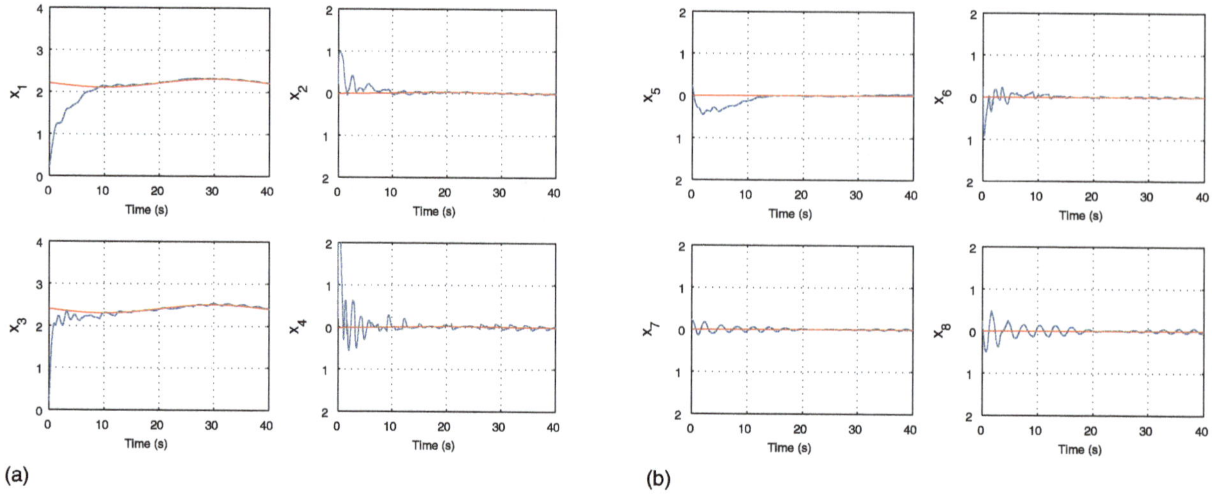

Figure 3.13 *Tracking of setpoint 6 for the tower crane: (a) convergence of state variables x_1 to x_4 to their reference setpoints (red line: setpoint, blue line: real value, green line: estimated value); (b) convergence of state variables x_5 to x_8 to their reference setpoints*

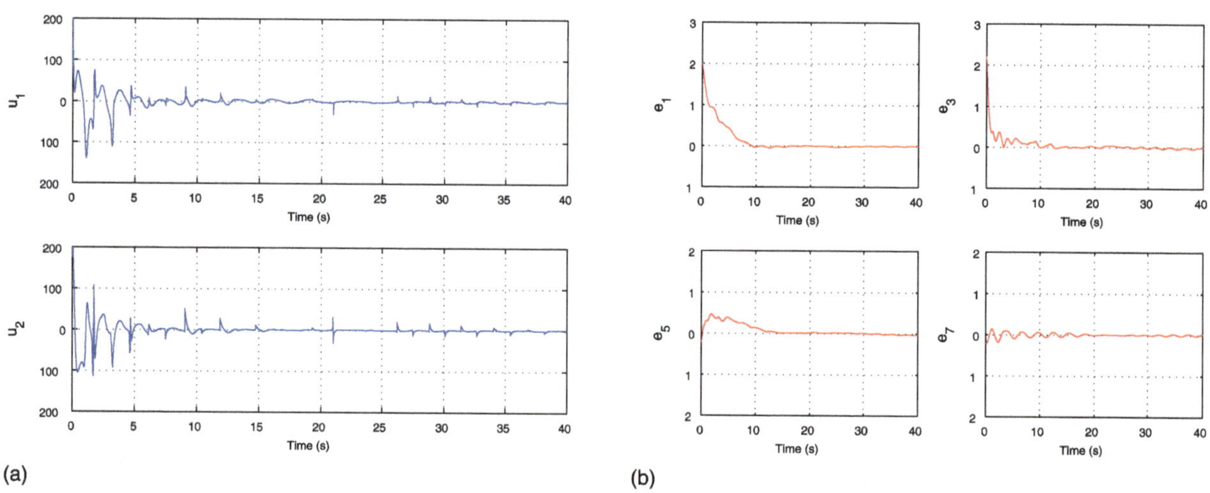

Figure 3.14 *Tracking of setpoint 6 for the tower crane: (a) control inputs u_1, u_2 applied to crane; (b) tracking error variables e_1, e_3, e_5, e_7 of the tower crane*

function. To compute the stabilizing feedback gains of the H-infinity controller an algebraic Riccati equation is repetitively solved at each time-step of the control algorithm [715,732]. The global stability properties of the control scheme are proven through Lyapunov analysis. First, it is demonstrated that the control loop of the gantry crane satisfies the H-infinity tracking performance criterion, which is an indication of the system's robustness to model uncertainty and external disturbances [719,876]. Moreover, it is proven that under moderate conditions the control loop becomes globally asymptotically stable [715,732]. The nonlinear optimal (H-infinity) control method ensures fast and accurate tracking of reference setpoints by the state variables of the dual PMLSM-driven gantry crane under moderate variations of the control inputs. Moreover, to achieve state-estimation-based control without the need of measuring the entire vector of the crane, the H-infinity Kalman filter is used as a robust state estimator.

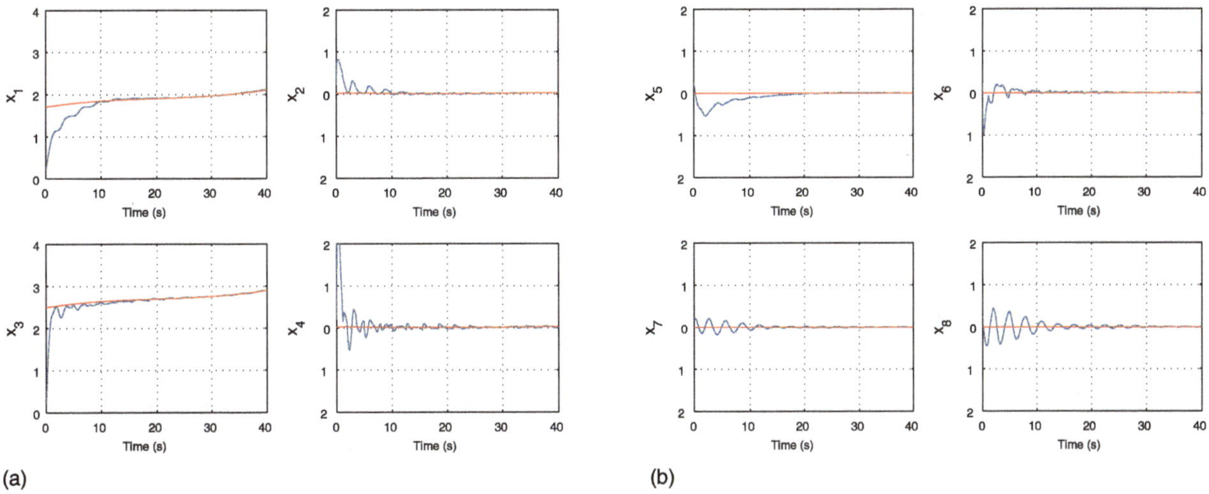

Figure 3.15 Tracking of setpoint 7 for the tower crane: (a) convergence of state variables x_1 to x_4 to their reference setpoints (red line: setpoint, blue line: real value, green line: estimated value); (b) convergence of state variables x_5 to x_8 to their reference setpoints

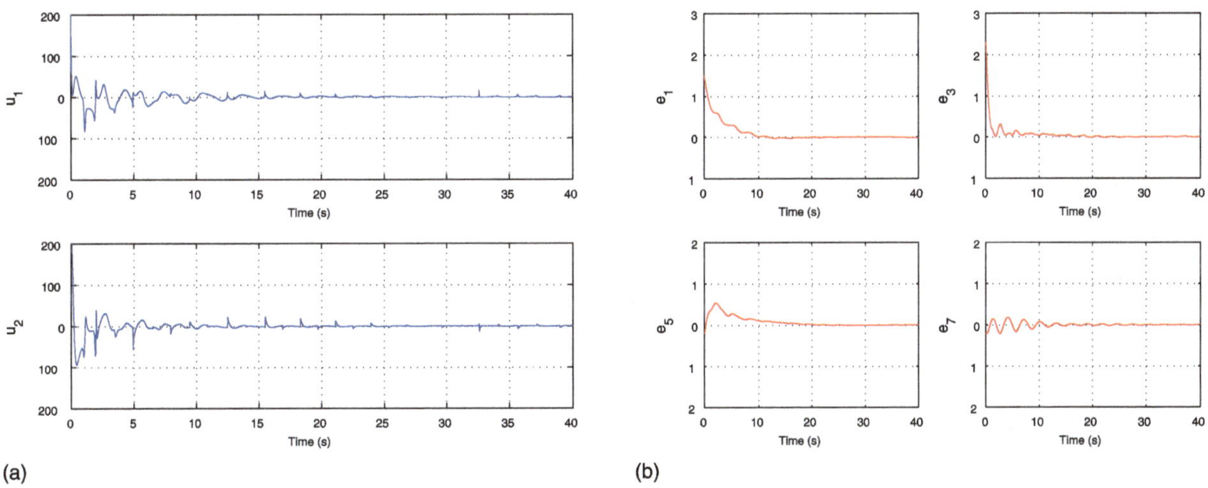

Figure 3.16 Tracking of setpoint 7 for the tower crane: (a) control inputs u_1, u_2 applied to crane; (b) tracking error variables e_1, e_3, e_5, e_7 of the tower crane

3.2.2 Dynamic model of the dual PMLSM-driven H-type gantry crane

The diagram of the dual-drive H-type gantry crane with actuation by two permanent magnet linear synchronous motors is given in Figure 3.19. The model of the gantry crane comprises four masses which are defined as follows: (i) m_b is the mass of the beam, (ii) m_h is the mass of the moving head (load), (iii) m_1 the mass of the first motor, (iv) m_2 the mass of the second motor. The motion of the gantry crane is also defined by four friction coefficients: (i) c_{g_1} is the friction coefficient of the first motor which moves at velocity $\dot{X}_1$, (ii) c_{g_2} is the friction coefficient of the second motor which moves at velocity $\dot{X}_2$, (iii) c_y is the friction coefficient for the actuator that moves the load along the rigid beam, and (iv) $c_b = c_{b_1} + c_{b_2}$ is the friction coefficient of the beam, which is connected to the two motors with two flexible joints that

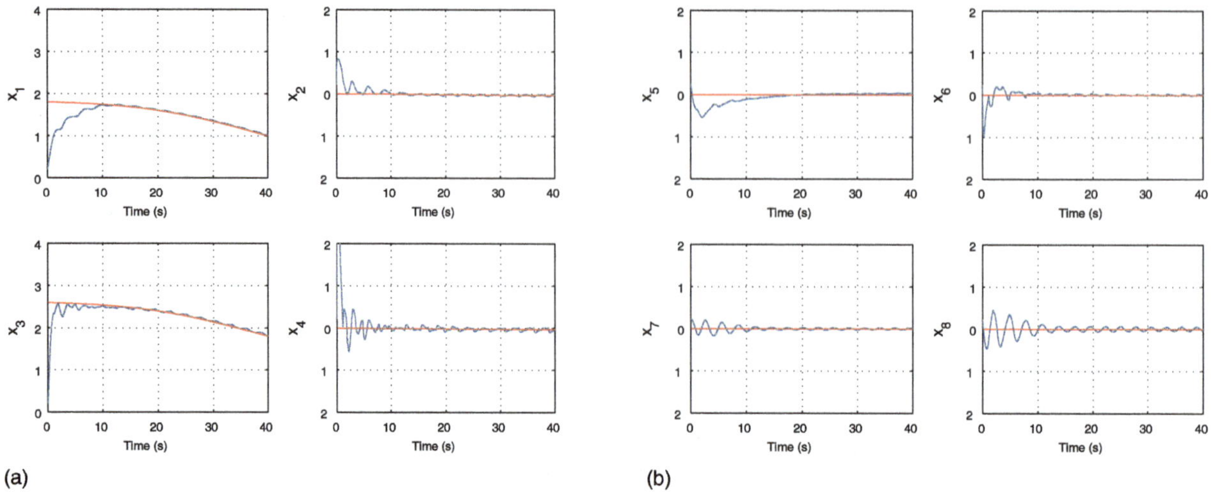

Figure 3.17 Tracking of setpoint 8 for the tower crane: (a) convergence of state variables x_1 to x_4 to their reference setpoints (red line: setpoint, blue line: real value, green line: estimated value); (b) convergence of state variables x_5 to x_8 to their reference setpoints

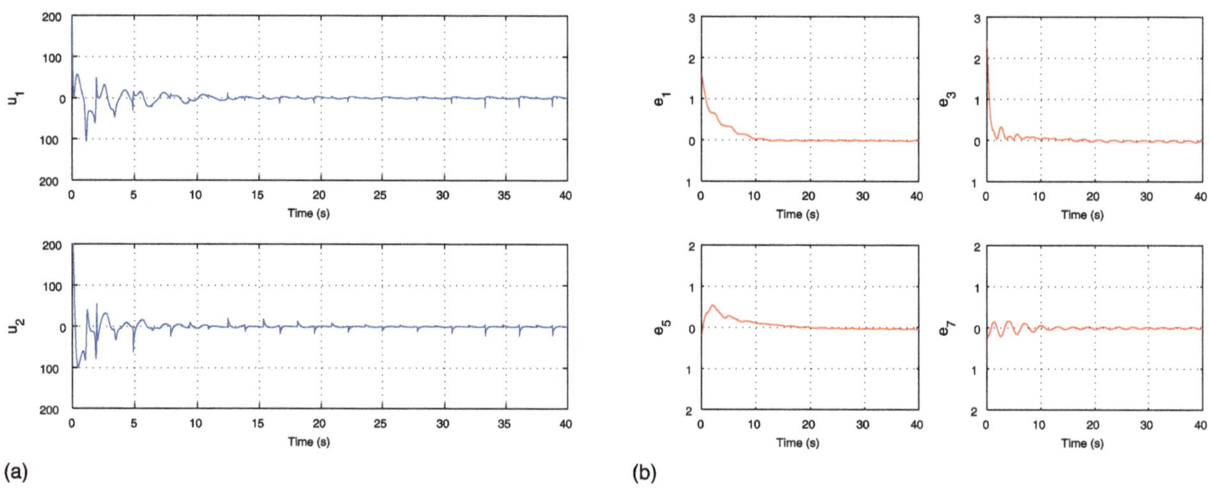

Figure 3.18 Tracking of setpoint 8 for the tower crane: (a) control inputs u_1, u_2 applied to crane; (b) tracking error variables e_1, e_3, e_5, e_7 of the tower crane

exhibit friction coefficients c_{b_1} and c_{b_2}, respectively. There is also a stiffness coefficient for the beam which is defined as $k_b = k_{b_1} + k_{b_2}$. The beam has length l_b. The head (load) has length l_h [537,277].

The position of motor 1 is denoted as X_1, while the position of motor 2 is denoted as X_2. The position of the load with respect to the center of mass of the beam is denoted as Y. Moreover, the following position variables are defined:

$$\begin{pmatrix} X \\ \Theta \\ Y \end{pmatrix} = \begin{pmatrix} \frac{1}{2} & \frac{1}{2} & 0 \\ \frac{1}{L_b} & -\frac{1}{L_b} & 0 \\ 0 & 0 & 1 \end{pmatrix} \begin{pmatrix} X_1 \\ X_2 \\ Y \end{pmatrix} \tag{3.120}$$

Table 3.1 Tracking RMSE $\times 10^{-3}$ *for the robotic tower crane in the disturbance-free case*

	$RMSE_{x_1}$	$RMSE_{x_2}$	$RMSE_{x_3}$	$RMSE_{x_4}$	$RMSE_{x_5}$	$RMSE_{x_6}$	$RMSE_{x_7}$	$RMSE_{x_8}$
$test_1$	0.7736	0.1027	0.6684	0.4821	0.8536	0.1755	0.1200	0.2835
$test_2$	0.9319	0.2061	0.3084	0.8815	0.6038	0.3178	0.2670	0.6294
$test_3$	0.6769	0.2049	0.3693	0.9328	0.4161	0.3288	0.1986	0.4701
$test_4$	0.0022	0.0006	0.0011	0.0030	0.0005	0.0011	0.0002	0.0006
$test_5$	0.4018	0.8466	0.6952	1.3000	1.2000	0.4137	0.7493	1.7000
$test_6$	0.6761	0.6045	0.7306	1.8000	0.4931	0.6634	0.7388	1.7000
$test_7$	0.2752	0.1154	0.9856	0.2283	1.3000	0.1076	0.0545	0.0965
$test_8$	0.3918	0.9022	1.2000	1.9000	1.6000	0.7703	0.7506	1.7000

Table 3.2 Tracking RMSE $\times 10^{-3}$ *for the robotic tower crane in the case of disturbances*

$\Delta a\%$	$RMSE_{x_1}$	$RMSE_{x_2}$	$RMSE_{x_3}$	$RMSE_{x_4}$	$RMSE_{x_5}$	$RMSE_{x_6}$	$RMSE_{x_7}$	$RMSE_{x_8}$
0%	0.4018	0.8466	0.6952	1.3000	1.2000	0.4137	0.7473	1.7000
10%	0.4251	0.8356	0.6932	1.3000	1.0000	0.4039	0.7817	1.8000
20%	0.4697	0.8444	0.6851	1.3000	0.9222	0.4029	0.8072	1.9000
30%	0.5290	0.8559	0.7287	1.5000	0.6807	0.4809	0.8065	1.9000
40%	0.5837	0.8371	0.7505	1.6000	0.6174	0.5169	0.8197	1.7000
50%	0.5841	0.8270	0.6983	1.2000	0.8771	0.3646	0.9044	2.1000
60%	0.5765	1.0000	1.2000	1.4000	1.1000	0.4371	1.1000	2.5000

Table 3.3 RMSE $\times 10^{-3}$ *for the estimation performed by the H-infinity KF*

	$RMSE_{x_1}$	$RMSE_{x_2}$	$RMSE_{x_3}$	$RMSE_{x_4}$	$RMSE_{x_5}$	$RMSE_{x_6}$	$RMSE_{x_7}$	$RMSE_{x_8}$
$test_1$	0.0113	0.3017	0.0107	0.2893	0.0009	0.0010	0.0010	0.0010
$test_2$	0.0115	0.3018	0.0107	0.2891	0.0010	0.0010	0.0009	0.0009
$test_3$	0.0111	0.3015	0.0108	0.2894	0.0011	0.0010	0.0011	0.0009
$test_4$	0.0115	0.3027	0.0110	0.2886	0.0011	0.0010	0.0011	0.0011
$test_5$	0.0115	0.3017	0.0109	0.2892	0.0010	0.0010	0.0009	0.0010
$test_6$	0.0115	0.3036	0.0107	0.2875	0.0010	0.0010	0.0009	0.0010
$test_7$	0.0111	0.3024	0.0104	0.2888	0.0009	0.0010	0.0010	0.0010
$test_8$	0.0114	0.3028	0.0111	0.2887	0.0010	0.0009	0.0009	0.0010

Using the Euler-Lagrange approach, the dynamic model of the dual gantry crane is [277,732]

$$M\ddot{q} + H\dot{q} + C\dot{q} + Kq = F \tag{3.121}$$

where $q = [X, \Theta, Y]^T$ is the state vector of the crane. The inertia matrix $M(q)$ is [277]

$$M = \begin{pmatrix} M_{11} & M_{12} & -m_b sin(\theta) \\ M_{12} & J_t + m_b Y^2 & -m_h d \\ -m_b sin(\Theta) & -m_b d & m_b \end{pmatrix} \tag{3.122}$$

where $J_t = J_b + J_h + 0.25(m_1 + m_2)L_b + m_h d^2$ with J_b to be the moment of inertia of the beam and J_h to be the moment of inertia of the load. Moreover, $M_{11} = m_1 + m_2 + m_b + m_h$ and $M_{12} = m_h d \cdot sin(\Theta) - [m_h Y - 0.5(m_1 - m_2)L_b]cos(\Theta)$.

Table 3.4 Convergence time (sec) for the robotic tower crane's state variables

	$T_s\, x_1$	$T_s\, x_2$	$T_s\, x_3$	$T_s\, x_4$	$T_s\, x_5$	$T_s\, x_6$	$T_s\, x_7$	$T_s\, x_8$
$test_1$	10.0	8.0	7.0	7.0	12.0	8.0	8.0	10.0
$test_2$	10.0	8.0	6.0	7.0	12.0	7.0	8.0	10.0
$test_3$	10.0	10.0	8.0	7.0	10.0	7.0	8.0	10.0
$test_4$	10.0	7.0	8.0	7.0	12.0	7.0	8.0	10.0
$test_5$	8.0	7.0	7.0	6.0	10.0	6.0	6.0	9.0
$test_6$	8.0	8.0	9.0	8.0	12.0	7.0	8.0	12.0
$test_7$	8.0	10.0	8.0	7.0	12.0	7.0	8.0	12.0
$test_8$	8.0	9.0	10.0	7.0	12.0	7.0	8.0	11.0

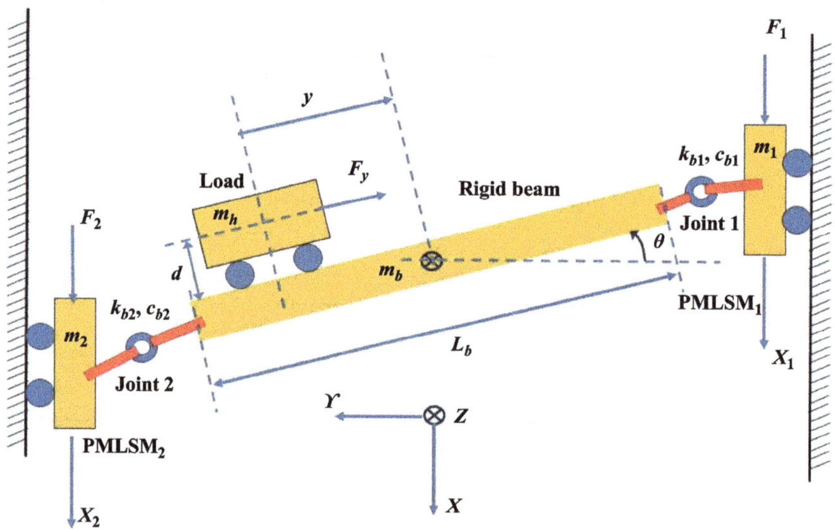

Figure 3.19 Diagram of the dual-drive H-type gantry crane with actuation by two permanent magnet linear synchronous motors

Matrix H is the Coriolis and centrifugal forces matrix: [277]

$$H = \begin{pmatrix} 0 & H_{12}\dot{\Theta} & -2m_b\dot{\Theta}cos(\Theta) \\ 0 & 2m_bY\dot{Y} & 0 \\ 0 & -m_bY\Theta & 0 \end{pmatrix} \tag{3.123}$$

Matrix K is the damping forces matrix which is defined as

$$K = \begin{pmatrix} 0 & 0 & 0 \\ 0 & K_{22} & 0 \\ 0 & 0 & 0 \end{pmatrix} \tag{3.124}$$

where $K_{22} = K_{b_1} + K_{b_2}$. The generalized forces vector is

$$F = \left(F_c \; M_c \; F_y\right)^T \tag{3.125}$$

where F_c is a force causing translation X of the beam, M_c is a torque causing rotation of the beam by angle Θ and F_y is a force causing translation of the load. The relation connecting the generalized forces vector to the forces generated by the PMLSMs of the gantry crane is [277]

$$\begin{pmatrix} F_c \\ M_c \\ F_y \end{pmatrix} = \begin{pmatrix} 1 & 1 & 0 \\ \frac{L_b}{2} & -\frac{L_b}{2} & 0 \\ 0 & 0 & 1 \end{pmatrix} \begin{pmatrix} F_1 \\ F_2 \\ F_y \end{pmatrix} \tag{3.126}$$

The above relation about the forces vectors of the gantry crane can be also written as

$$F = A_F F_1 \tag{3.127}$$

About the moments of inertia of the crane it holds that $J_b = m_b \frac{L_b^2}{12}$ and $J_h = m_h \frac{L_h^2}{12}$.

Using a description of the voltage and current variables of the PMLSM in the dq reference frame, the dynamics of each motor is given by [725]

$$\begin{aligned} \dot{i}_q &= -i_q \frac{R_s}{L_q} - \frac{n_p \pi v}{L_q \tau}(L_d i_d + \lambda_{pm}) + \frac{1}{L_q} v_q \\ \dot{i}_d &= -i_d \frac{R_s}{L_d} - \frac{n_p \pi v}{L_d \tau} L_q i_q + \frac{1}{L_d} v_d \end{aligned} \tag{3.128}$$

The parameters of the electric part of the PMLSM are: R_s is the motor's resistance, λ_{pm} is the magnetic flux due to permanent magnets, n_p is the number of poles of the motor, L_d, L_q are the dq frame inductances. Moreover, i_d, i_q are the dq frame current variables, and v_d, v_q are the dq frame voltage input variables. Additionally, v is the linear velocity of the moving part of each PMLSM. For the first motor, this velocity variable is $\dot{X}_1$, for the second motor it is $\dot{X}_2$ and for the motor moving the load it is $\dot{Y}$. The electromagnetic force for each motor is

$$F_e = \frac{3\pi n_p}{M+M_L} \frac{\lambda_{pm} i_q - (L_d - L_q) i_d i_q}{2\tau} - Dv \tag{3.129}$$

and by considering that $L_d = L_q$ and a small friction coefficient $D = 0$ the electromagnetic force of the motor is given by

$$F_e = \frac{3\pi n_p}{M+M_L} \frac{\lambda_{pm}}{2\tau} i_q \tag{3.130}$$

Next, the dynamic model of the dual-drive H-type gantry crane is written as

$$\begin{aligned} \ddot{q} &= -M^{-1} H \dot{q} - M^{-1} C \dot{q} - M^{-1} K q + M^{-1} A_f F_1 \Rightarrow \\ \ddot{q} &= -M^{-1}(H + C)\dot{q} - M^{-1} K q + M^{-1} A_f F_1 \end{aligned} \tag{3.131}$$

About the inverse of the inertia matrix M of the dual-drive gantry crane it holds that

$$M^{-1} = \frac{1}{detM} \begin{pmatrix} \bar{M}_{11} & -\bar{M}_{21} & \bar{M}_{31} \\ -\bar{M}_{12} & \bar{M}_{22} & -\bar{M}_{32} \\ \bar{M}_{13} & -\bar{M}_{23} & \bar{M}_{33} \end{pmatrix} \tag{3.132}$$

where $\bar{M}_{11} = (J_t + m_b Y^2) m_b - (m_b \cdot d)^2$, $\bar{M}_{12} = M_{12} m_b - (m_b sin(\Theta))(m_b \cdot d)$, $\bar{M}_{13} = -M_{12}(m_b \cdot d) + (m_b sin(\Theta))(J_t + m_b Y^2)$, and also $\bar{M}_{21} = M_{21} \cdot m_b - (m_b sin(\Theta))(m_b \cdot d)$, $\bar{M}_{22} = -M_{11} m_b - (m_b sin(\Theta))^2$, $\bar{M}_{23} = -M_{11} m_b d - M_{12} m_b sin(\Theta)$, $\bar{M}_{31} = M_{12} m_b d + (J_t + m_b Y^2)(m_b sin(\Theta))$, $\bar{M}_{32} = -M_{11} m_b d + M_{12}(m_b sin(\Theta))$, $\bar{M}_{33} = M_{11}(J_t + m_b Y^2) - M_{12}^2$ while the determinant of inertia matrix is $detM = M_{11} \bar{M}_{11} - M_{12} \bar{M}_{12} + M_{13} \bar{M}_{13}$.

About the product $(H + C)\dot{q}$, it holds that

$$(H + C)\dot{q} = \begin{pmatrix} C_{11} & C_{12} + H_{12}\dot{\Theta} & -2m_h \dot{\Theta} cos(\Theta) \\ C_{12} & C_{22} + 2m_h Y \dot{Y} & 0 \\ 0 & -m_h Y \Theta & C_y \end{pmatrix} \begin{pmatrix} \dot{X} \\ \dot{\Theta} \\ \dot{Y} \end{pmatrix} \tag{3.133}$$

which is also written as

$$(H+C)\dot{q} = \begin{pmatrix} C_{11}\dot{X} & (C_{12}+H_{12}\dot{\Theta})\dot{\Theta} & (-2m_h\dot{\Theta}cos(\Theta))\dot{Y} \\ C_{12}\dot{X} & (C_{22}+2m_hY\dot{Y})\dot{\Theta} & 0 \\ 0 & (-m_hY\Theta)\dot{\Theta} & C_y\dot{Y} \end{pmatrix} = \begin{pmatrix} \bar{h}_1 \\ \bar{h}_2 \\ \bar{h}_3 \end{pmatrix} \tag{3.134}$$

Moreover, about the product $-M^{-1}(H+G)\dot{q}$, it holds that

$$M^{-1}(H+C)\dot{q} = \frac{1}{detM}\begin{pmatrix} \bar{M}_{11} & -\bar{M}_{21} & \bar{M}_{31} \\ -\bar{M}_{12} & \bar{M}_{22} & -\bar{M}_{22} \\ \bar{M}_{13} & -\bar{M}_{23} & \bar{M}_{33} \end{pmatrix}\begin{pmatrix} \bar{h}_1 \\ \bar{h}_2 \\ \bar{h}_3 \end{pmatrix} \tag{3.135}$$

which is also written as

$$M^{-1}(H+C)\dot{q} = \frac{1}{detM}\begin{pmatrix} \bar{M}_{11}\bar{h}_1 - \bar{M}_{21}\bar{h}_2 + \bar{M}_{31}\bar{h}_3 \\ -\bar{M}_{12}\bar{h}_1 + \bar{M}_{22}\bar{h}_2 - \bar{M}_{22}\bar{h}_3 \\ \bar{M}_{13}\bar{h}_1 - \bar{M}_{23}\bar{h}_2 + \bar{M}_{33}\bar{h}_3 \end{pmatrix} \tag{3.136}$$

Additionally, about the product $-M^{-1}Kq$, it holds that

$$M^{-1}Kq = \frac{1}{detM}\begin{pmatrix} \bar{M}_{11} & -\bar{M}_{21} & \bar{M}_{31} \\ -\bar{M}_{12} & \bar{M}_{22} & -\bar{M}_{32} \\ \bar{M}_{13} & -\bar{M}_{23} & \bar{M}_{33} \end{pmatrix}\begin{pmatrix} 0 \\ K_{22}\Theta \\ 0 \end{pmatrix} = \frac{1}{detM}\begin{pmatrix} -\bar{M}_{21}K_{22}\Theta \\ \bar{M}_{22}K_{22}\Theta \\ -\bar{M}_{23}K_{22}\Theta \end{pmatrix} = \frac{1}{detM}\begin{pmatrix} -\bar{M}_{21}\bar{k}_1 \\ \bar{M}_{22}\bar{k}_2 \\ -\bar{M}_{23}\bar{k}_3 \end{pmatrix} \tag{3.137}$$

where $\bar{k}_1 = \bar{k}_2 = \bar{k}_3 = K_{22}\Theta$. Consequently, it holds that the aggregate term $-M^{-1}(H+C)\dot{q} - M^{-1}Kq$ takes the form

$$-M^{-1}(H+C)\dot{q} - M^{-1}Kq = \begin{pmatrix} -\bar{M}_{11}\bar{h}_1 + \bar{M}_{21}\bar{h}_2 - \bar{M}_{31}\bar{h}_3 + \bar{M}_{21}\bar{k}_1 \\ \bar{M}_{12}\bar{h}_1 - \bar{M}_{22}\bar{h}_2 + \bar{M}_{32}\bar{h}_3 - \bar{M}_{22}\bar{k}_2 \\ -\bar{M}_{13}\bar{h}_1 + \bar{M}_{23}\bar{h}_2 - \bar{M}_{33}\bar{h}_3 + \bar{M}_{23}\bar{k}_3 \end{pmatrix} \tag{3.138}$$

Additionally, about the control inputs gain matrix of the dual-drive gantry crane, it holds that

$$M^{-1}A_f = \frac{1}{detM}\begin{pmatrix} \bar{M}_{11} & -\bar{M}_{21} & \bar{M}_{31} \\ -\bar{M}_{12} & \bar{M}_22 & -barM_{32} \\ \bar{M}_{13} & -\bar{M}_{23} & \bar{M}_{33} \end{pmatrix}\begin{pmatrix} 1 & 1 & 0 \\ \frac{L_b}{2} & -\frac{L_b}{2} & 0 \\ 0 & 0 & 1 \end{pmatrix} \tag{3.139}$$

and by defining as $L_{b_1} = \frac{L_b}{2}$, the previous equation gives

$$M^{-1}A_f = \frac{1}{detM} = \frac{1}{detM}\begin{pmatrix} \bar{M}_{11} - L_{b_1}\bar{M}_{21} & \bar{M}_{11} + L_{b_1}\bar{M}_{21} & M_{31} \\ -\bar{M}_{12} + L_{b_1}\bar{M}_{22} & -\bar{M}_{12} - L_{b_1}\bar{M}_{22} & -M_{32} \\ \bar{M}_{13} - L_{b_1}\bar{M}_{23} & \bar{M}_{13} + L_{b_1}\bar{M}_{23} & \bar{M}_{33} \end{pmatrix} \tag{3.140}$$

Consequently, the dynamic model of the dual-drive gantry crane is written as

$$\begin{pmatrix} \ddot{X} \\ \ddot{\Theta} \\ \ddot{Y} \end{pmatrix} = \begin{pmatrix} \frac{-\bar{M}_{11}\bar{h}_1 + \bar{M}_{21}\bar{h}_2 - \bar{M}_{31}\bar{h}_3 + \bar{M}_{21}\bar{k}_1}{detM} \\ \frac{\bar{M}_{12}\bar{h}_1 - \bar{M}_{22}\bar{h}_2 + \bar{M}_{32}\bar{h}_3 - \bar{M}_{22}\bar{k}_2}{detM} \\ \frac{-\bar{M}_{13}\bar{h}_1 + \bar{M}_{23}\bar{h}_2 - \bar{M}_{33}\bar{h}_3 - \bar{M}_{23}\bar{k}_3}{detM} \end{pmatrix} + \begin{pmatrix} \frac{\bar{M}_{11} - L_{b_1}\bar{M}_{21}}{detM} & \frac{\bar{M}_{11} + L_{b_1}\bar{M}_{21}}{detM} & \frac{M_{31}}{detM} \\ \frac{-\bar{M}_{12} + L_{b_1}\bar{M}_22}{detM} & \frac{-\bar{M}_{12} - L_{b_1}\bar{M}_{22}}{detM} & \frac{-M_{32}}{detM} \\ \frac{\bar{M}_{13} - L_{b_1}\bar{M}_{23}}{detM} & \frac{\bar{M}_{13} + L_{b_1}\bar{M}_{23}}{detM} & \frac{\bar{M}_{33}}{detM} \end{pmatrix}\begin{pmatrix} F_1 \\ F_2 \\ F_y \end{pmatrix} \tag{3.141}$$

About the forces which are generated by the PMLSM actuators, it holds that

$$F_1 = \frac{3\pi n_{p1}}{m_1 + m_b + m_h}\frac{\lambda_{pm_1}}{2\tau 1} i_{q1} \Rightarrow F_1 = w_1 i_{q1} \tag{3.142}$$

$$F_2 = \frac{3\pi n_{p2}}{m_2+m_b+m_h}\frac{\lambda_{pm_2}}{2\tau 2}i_{q2} \Rightarrow F_2 = w_2 i_{q2} \tag{3.143}$$

$$F_y = \frac{3\pi n_{p3}}{m_h}\frac{\lambda_{pm_3}}{2\tau 3}i_{q3} \Rightarrow F_y = w_3 i_{q3} \tag{3.144}$$

The currents of the PMLSMs are given by the following equations:

First PMLSM of the rigid beam

$$\dot{i}_{q1} = -i_{q1}\frac{R_{s1}}{L_{q1}} - \frac{n_{p1}\pi\dot{X}_1}{L_{q1}\tau_1}(L_{d1}i_{d1} + \lambda_{pm1}) + \frac{1}{L_{q1}}v_{q1} \tag{3.145}$$

$$\dot{i}_{d1} = -i_{d1}\frac{R_{s1}}{L_{d1}} - \frac{n_{p1}\pi\dot{X}_1}{L_{d1}\tau_1}(L_{q1}i_{q1}) + \frac{1}{L_{d1}}v_{d1} \tag{3.146}$$

Second PMLSM of the rigid beam

$$\dot{i}_{q2} = -i_{q2}\frac{R_{s2}}{L_{q2}} - \frac{n_{p2}\pi\dot{X}_2}{L_{q2}\tau_2}(L_{d2}i_{d2} + \lambda_{pm2}) + \frac{1}{L_{q2}}v_{q2} \tag{3.147}$$

$$\dot{i}_{d2} = -i_{d2}\frac{R_{s2}}{L_{d2}} - \frac{n_{p2}\pi\dot{X}_2}{L_{d2}\tau_2}(L_{q2}i_{q2}) + \frac{1}{L_{d2}}v_{d2} \tag{3.148}$$

Third PMLSM of the load

$$\dot{i}_{q3} = -i_{q3}\frac{R_{s3}}{L_{q3}} - \frac{n_{p3}\pi\dot{Y}}{L_{q3}\tau_3}(L_{d3}i_{d3} + \lambda_{pm3}) + \frac{1}{L_{q3}}v_{q3} \tag{3.149}$$

$$\dot{i}_{d3} = -i_{d3}\frac{R_{s3}}{L_{d3}} - \frac{n_{p3}\pi\dot{Y}}{L_{d3}\tau_3}(L_{q3}i_{q3}) + \frac{1}{L_{d3}}v_{d3} \tag{3.150}$$

Besides, from Eq. (3.120), it holds

$$\begin{array}{ll} X_1 = X + L_{b1}\Theta & \dot{X}_1 = \dot{X} + L_{b1}\dot{\Theta} \\ X_2 = X - L_{b2}\Theta & \dot{X}_2 = \dot{X} - L_{b2}\dot{\Theta} \end{array} \tag{3.151}$$

Next, the state vector of the dual-drive gantry crane is defined as

$$\begin{array}{l} x = [x_1, x_2, x_3, x_4, x_5, x_6, x_7, x_8, x_9, x_{10}, x_{11}, x_{12}]^T \Rightarrow \\ \quad x = [X.\Theta, Y, \dot{X}, \dot{\Theta}, \dot{Y}, i_{q1}, i_{d1}, i_{q2}, i_{d2}, i_{q3}, i_{d3}] \end{array} \tag{3.152}$$

while the control inputs vector is defined as

$$\begin{array}{l} u = [u_1, u_2, u_3, u_4, u_5, u_6]^T \Rightarrow \\ u = [v_{d1}, v_{q1}, v_{d2}, v_{q2}, v_{d3}, v_{q3}] \end{array} \tag{3.153}$$

Using this state vector and control inputs vector notation, the dynamic model of the dual-drive gantry crane is written as follows:

$$\dot{x}_1 = x_4 \tag{3.154}$$

$$\dot{x}_2 = x_5 \tag{3.155}$$

$$\dot{x}_3 = x_6 \tag{3.156}$$

$$\dot{x}_4 = \frac{-\bar{M}_{11}\bar{h}_1 + \bar{M}_{21}\bar{h}_2 - \bar{M}_{31}\bar{h}_3 + \bar{M}_{21}\bar{k}_1}{detM} + \frac{\bar{M}_{11} - L_{b1}\bar{M}_{21}}{detM}w_1x_7 + \frac{\bar{M}_{11} + L_{b1}\bar{M}_{21}}{detM}w_2x_9 + \frac{\bar{M}_{31}}{detM}w_3x_{11} \tag{3.157}$$

$$\dot{x}_5 = \frac{\bar{M}_{12}\bar{h}_1 - \bar{M}_{22}\bar{h}_2 + \bar{M}_{32}\bar{h}_3 - \bar{M}_{22}\bar{k}_2}{detM} + \frac{-\bar{M}_{12} + L_{b1}\bar{M}_{22}}{detM}w_1x_7 + \frac{-\bar{M}_{12} + L_{b1}\bar{M}_{22}}{detM}w_2x_9 + \frac{-\bar{M}_{32}}{detM}w_3x_{11} \tag{3.158}$$

$$\dot{x}_6 = \frac{-\bar{M}_{13}\bar{h}_1 + \bar{M}_{23}\bar{h}_2 - \bar{M}_{33}\bar{h}_3 + \bar{M}_{23}\bar{k}_3}{detM} + \frac{\bar{M}_{13} - L_{b1}\bar{M}_{23}}{detM}w_1x_7 + \frac{\bar{M}_{13} + L_{b1}\bar{M}_{23}}{detM}w_2x_9 + \frac{\bar{M}_{33}}{detM}w_3x_{11} \tag{3.159}$$

$$\dot{x}_7 = -x_7\frac{R_{s1}}{L_{q1}} - \frac{n_{p1}\pi(x_4 + L_{b1}x_5)}{L_{q1}\tau_1}(L_{d1}x_8 + \lambda_{pm1}) + \frac{1}{L_{q1}}u_1 \tag{3.160}$$

$$\dot{x}_8 = -x_8\frac{R_{s1}}{L_{d1}} - \frac{n_{p1}\pi(x_4 + L_{b1}x_5)}{L_{d1}\tau_1}(L_{q1}x_7) + \frac{1}{L_{d1}}u_2 \tag{3.161}$$

$$\dot{x}_9 = -x_9\frac{R_{s2}}{L_{q2}} - \frac{n_{p2}\pi(x_4 - L_{b1}x_5)}{L_{q2}\tau_2}(L_{d2}x_{10} + \lambda_{pm2}) + \frac{1}{L_{q2}}u_3 \tag{3.162}$$

$$\dot{x}_{10} = -x_{10}\frac{R_{s2}}{L_{d2}} - \frac{n_{p2}\pi(x_4 - L_{b1}x_5)}{L_{d2}\tau_2}(L_{q2}x_9) + \frac{1}{L_{d2}}u_4 \tag{3.163}$$

$$\dot{x}_{11} = -x_{11}\frac{R_{s3}}{L_{q3}} - \frac{n_{p3}\pi x_6}{L_{q3}\tau_3}(L_{d3}x_{12} + \lambda_{pm3}) + \frac{1}{L_{q3}}u_5 \tag{3.164}$$

$$\dot{x}_{12} = -x_{12}\frac{R_{s3}}{L_{d3}} - \frac{n_{p3}\pi x_6}{L_{d3}\tau_3}(L_{q3}x_{11}) + \frac{1}{L_{d3}}u_6 \tag{3.165}$$

As a result of the above, the state-space model of the dual-drive gantry crane can be also written in the following matrix form:

$$\begin{pmatrix} \dot{x}_1 \\ \dot{x}_2 \\ \dot{x}_3 \\ \dot{x}_4 \\ \dot{x}_5 \\ \dot{x}_6 \\ \dot{x}_7 \\ \dot{x}_8 \\ \dot{x}_9 \\ \dot{x}_{10} \\ \dot{x}_{11} \\ \dot{x}_{12} \end{pmatrix} = \begin{pmatrix} x_4 \\ x_5 \\ x_6 \\ f_{4a} + f_{4b}w_1x_7 + f_{4c}w_2x_9 + f_{4d}w_3x_{11} \\ f_{5a} + f_{5b}w_1x_7 + f_{5c}w_2x_9 + f_{5d}w_3x_{11} \\ f_{6a} + f_{6b}w_1x_7 + f_{6c}w_2x_9 + f_{6d}w_3x_{11} \\ f_7 \\ f_8 \\ f_9 \\ f_{10} \\ f_{11} \\ f_{11} \end{pmatrix} + \begin{pmatrix} 0 & 0 & 0 & 0 & 0 & 0 \\ 0 & 0 & 0 & 0 & 0 & 0 \\ 0 & 0 & 0 & 0 & 0 & 0 \\ 0 & 0 & 0 & 0 & 0 & 0 \\ 0 & 0 & 0 & 0 & 0 & 0 \\ 0 & 0 & 0 & 0 & 0 & 0 \\ g_{7,1} & 0 & 0 & 0 & 0 & 0 \\ 0 & g_{8,2} & 0 & 0 & 0 & 0 \\ 0 & 0 & g_{9,3} & 0 & 0 & 0 \\ 0 & 0 & 0 & g_{10,4} & 0 & 0 \\ 0 & 0 & 0 & 0 & g_{11,5} & 0 \\ 0 & 0 & 0 & 0 & 0 & g_{12,6} \end{pmatrix} \begin{pmatrix} u_1 \\ u_2 \\ u_3 \\ u_4 \\ u_5 \\ u_6 \end{pmatrix} \tag{3.166}$$

where the drift vector of the state-space model comprises the following functions:

$f_{4a} = \frac{-\bar{M}_{11}\bar{h}_1 + \bar{M}_{21}\bar{h}_2 - \bar{M}_{31}\bar{h}_3 + \bar{M}_{21}\bar{k}_1}{detM}$ $f_{4b} = \frac{\bar{M}_{11} - L_{b1}\bar{M}_{21}}{detM}$, $f_{4c} = \frac{\bar{M}_{11} + L_{b1}\bar{M}_{21}}{detM}$, $f_{4d} = \frac{\bar{M}_{31}}{detM}w_3x_{11}$.

$f_{5a} = \frac{\bar{M}_{12}\bar{h}_1 - \bar{M}_{22}\bar{h}_2 + \bar{M}_{32}\bar{h}_3 - \bar{M}_{22}\bar{k}_2}{detM}$, $f_{5b} = \frac{-\bar{M}_{12} + L_{b1}\bar{M}_{22}}{detM}$, $f_{5c} = \frac{-\bar{M}_{12} + L_{b1}\bar{M}_{22}}{detM}$, $f_{5d} = \frac{-\bar{M}_{32}}{detM}$.

$f_6a = \frac{-\bar{M}_{13}\bar{h}_1 + \bar{M}_{23}\bar{h}_2 - \bar{M}_{33}\bar{h}_3 + \bar{M}_{23}\bar{k}_3}{detM}$, $f_{6b} = \frac{\bar{M}_{13} - L_{b1}\bar{M}_{23}}{detM}$, $f_{6c} = \frac{\bar{M}_{13} + L_{b1}\bar{M}_{23}}{detM}$, $f_{6d} = \frac{\bar{M}_{33}}{detM}$

$f_7 = -x_7\frac{R_{s1}}{L_{q1}} - \frac{n_{p1}\pi(x_4 + L_{b1}x_5)}{L_{q1}\tau_1}(L_{d1}x_8 + \lambda_{pm1})$, $f_8 = -x_8\frac{R_{s1}}{L_{d1}} - \frac{n_{p1}\pi(x_4 + L_{b1}x_5)}{L_{d1}\tau_1}(L_{q1}x_7)$, $f_9 = -x_9\frac{R_{s2}}{L_{q2}} - \frac{n_{p2}\pi(x_4 - L_{b1}x_5)}{L_{q2}\tau_2}(L_{d2}x_{10} + \lambda_{pm2})$, $f_{10} = -x_{10}\frac{R_{s2}}{L_{d2}} - \frac{n_{p2}\pi(x_4 - L_{b1}x_5)}{L_{d2}\tau_2}(L_{q2}x_9)$, $f_{11} = -x_{11}\frac{R_{s3}}{L_{q3}} - \frac{n_{p3}\pi x_6}{L_{q3}\tau_3}(L_{d3}x_{12} + \lambda_{pm3})$, and $f_{12} = -x_{12}\frac{R_{s3}}{L_{d3}} - \frac{n_{p3}\pi x_6}{L_{d3}\tau_3}(L_{q3}x_{11})$.

Finally, the state-space model of the dual-drive gantry crane is written in the following concise nonlinear affine-in-the-input state-space form:

$$\dot{x} = f(x) + g(x)u \tag{3.167}$$

where $x \in R^{12\times 1}$, $f(x) \in R^{12\times 1}$, $g(x) \in R^{12\times 6}$, and $u \in R^{6\times 1}$.

3.2.3 Differential flatness properties of the dual-drive H-type gantry crane

The dynamic model of the dual-drive H-type is differentially flat, with flat outputs vector

$$\begin{aligned} Y &= [x_1, x_2, x_3, x_8, x_{10}, x_{12}]^T \Rightarrow \\ Y &= [X, \Theta, Y, i_{d1}, i_{d2}, i_{d3}]^T \end{aligned} \tag{3.168}$$

The state-space model of the dual-drive gantry crane given in Eq. (3.166) comprises 12 state equations. From the first three rows of the state-space model, one has

$$\begin{aligned} \dot{x}_1 &= x_4 \Rightarrow x_4 = \tilde{h}_4(Y,\dot{Y}) \\ \dot{x}_2 &= x_5 \Rightarrow x_5 = \tilde{h}_5(Y,\dot{Y}) \\ \dot{x}_3 &= x_6 \Rightarrow x_6 = \tilde{h}_6(Y,\dot{Y}) \end{aligned} \tag{3.169}$$

which signifies that state variables x_4, x_5, x_6 are differential functions of the flat outputs of the system.

Next, from the fourth, fifth, and sixth rows of the state-space model, one has

$$\begin{pmatrix} \dot{x}_4 \\ \dot{x}_5 \\ \dot{x}_6 \end{pmatrix} = \begin{pmatrix} f_{4a} \\ f_{5a} \\ f_{6a} \end{pmatrix} + \begin{pmatrix} f_{4b}w_1 & f_{4c}w_2 & f_{4d}w_3 \\ f_{5b}w_1 & f_{5c}w_2 & f_{5d}w_3 \\ f_{6b}w_1 & f_{6c}w_2 & f_{6d}w_3 \end{pmatrix} \begin{pmatrix} x_7 \\ x_9 \\ x_{11} \end{pmatrix} \tag{3.170}$$

or equivalently

$$\begin{pmatrix} x_7 \\ x_9 \\ x_{11} \end{pmatrix} = \begin{pmatrix} f_{4b}w_1 & f_{4c}w_2 & f_{4d}w_3 \\ f_{5b}w_1 & f_{5c}w_2 & f_{5d}w_3 \\ f_{6b}w_1 & f_{6c}w_2 & f_{6d}w_3 \end{pmatrix}^{-1} \left[\begin{pmatrix} \dot{x}_4 \\ \dot{x}_5 \\ \dot{x}_6 \end{pmatrix} - \begin{pmatrix} f_{4a} \\ f_{5a} \\ f_{6a} \end{pmatrix} \right] \tag{3.171}$$

where $(f_{4a}, f_{4b}, f_{4c}, f_{4d})$, $(f_{5a}, f_{5b}, f_{5c}, f_{5d})$, and $(f_{6a}, f_{6b}, f_{6c}, f_{6d})$ are functions of state variables x_i $i = 1 \cdots 6$, and thus, they are also differential functions of the flat outputs of the system. Consequently, one obtains about state variables x_7, x_9, x_{11}

$$\begin{aligned} x_7 &= \tilde{h}_7(Y,\dot{Y}) \\ x_9 &= \tilde{h}_9(Y,\dot{Y}) \\ x_{11} &= \tilde{h}_{11}(Y,\dot{Y}) \end{aligned} \tag{3.172}$$

which signifies that state variables x_7, x_9, x_{11} are differential functions of the flat outputs of the system. Moreover, from rows 7 to 12 of the state-space model of Eq. (3.166), one solves for the control inputs u_1 to u_6. This gives:

$$\begin{pmatrix} u_1 \\ u_2 \\ u_3 \\ u_4 \\ u_5 \\ u_6 \end{pmatrix} = \begin{pmatrix} g_{7,1} & 0 & 0 & 0 & 0 & 0 \\ 0 & g_{8,2} & 0 & 0 & 0 & 0 \\ 0 & 0 & g_{9,3} & 0 & 0 & 0 \\ 0 & 0 & 0 & g_{10,4} & 0 & 0 \\ 0 & 0 & 0 & 0 & g_{11,5} & 0 \\ 0 & 0 & 0 & 0 & 0 & g_{12,6} \end{pmatrix}^{-1} \left[\begin{pmatrix} \dot{x}_7 \\ \dot{x}_8 \\ \dot{x}_9 \\ \dot{x}_{10} \\ \dot{x}_{11} \\ \dot{x}_{12} \end{pmatrix} - \begin{pmatrix} f_7 \\ f_8 \\ f_9 \\ f_{10} \\ f_{11} \\ f_{12} \end{pmatrix} \right] \tag{3.173}$$

where x_8, x_{10}, x_{12} are flat outputs of the system, while all other variables and functions which appear in the right part of Eq. (3.173) are differential functions of the flat outputs of the system. Therefore, the control inputs u_i $i = 1 \cdots 6$ are also differential functions of the flat outputs of the system, or

$$u_i = \tilde{h}_{u_i}(Y,\dot{Y}) \ i = 1, 2, \cdots, 6 \tag{3.174}$$

Consequently, the dynamic model of the dual-drive gantry crane is differentially flat. The differential flatness property is also an implicit proof of the system's controllability. It also demonstrates that the system is input–output linearizable through successive differentiations of its flat outputs. Finally, it allows for solving the setpoints definition problem. First, one defines setpoints in an unconstrained manner for the state variables which are also flat outputs of the system, that is for state variables x_1, x_2, x_3, x_8, x_{10}, and x_{12}. For the rest of the state variables of the gantry crane x_4, x_5, x_6, x_7, x_9, x_{11} setpoints are chosen using the differential relations that connect these state variables to the system's flat outputs.

3.2.4 Approximate linearization of the dynamics of the dual-drive gantry crane

The dynamic model of the gantry crane given in Eq. (3.166) undergoes approximate linearization at each sampling instance with the use of first-order Taylor series expansion and through the computation of the associated Jacobian matrices. The linearization is iteratively performed around the temporary operating point (x^*, u^*), where x^* is the present value of the system's state vector and u^* is the last sampled value of the control inputs vector. The linearization procedure results in the state-space model of the form:

$$\dot{x} = Ax + Bu + \tilde{d} \tag{3.175}$$

where A, B are the Jacobian matrices of the system which are given by

$$\begin{gathered} A = \nabla_x[f(x) + g(x)u] \mid_{(x^*,u^*)} \Rightarrow \\ A = \nabla_x f(x) \mid_{(x^*,u^*)} + \nabla_x g(x)u \mid_{(x^*,u^*)} \Rightarrow \\ A = \nabla_x f(x) \mid_{(x^*,u^*)} \end{gathered} \tag{3.176}$$

$$\begin{gathered} B = \nabla_u[f(x) + g(x)u] \mid_{(x^*,u^*)} \Rightarrow \\ B = g(x) \mid_{(x^*,u^*)} \end{gathered} \tag{3.177}$$

The disturbance term $\tilde{d}$ may comprise (i) modeling error due to the truncation of higher-order terms from the Taylor series expansion, (ii) exogenous perturbations, and (iii) sensor measurement noise of any distribution.

The linearization approach which has been followed for implementing the nonlinear optimal control scheme results in a quite accurate model of the system's dynamics. The disturbance $\tilde{d}$ remains small due to linearization around the present position (x^*, u^*) of the system and is asymptotically compensated by the robustness of the control algorithm. Consider for instance the following affine-in-the-input state-space model:

$$\begin{gathered} \dot{x} = f(x) + g(x)u \Rightarrow \\ \dot{x} = [f(x^*) + \nabla_x f(x) \mid_{x^*} (x - x^*)] + [g(x^*) + \nabla_x g(x) \mid_{x^*} (x - x^*)]u^* + g(x^*)u^* + g(x^*)(u - u^*) + \tilde{d}_1 \Rightarrow \\ \dot{x} = [\nabla_x f(x) \mid_{x^*} + \nabla_x g(x) \mid_{x^*} u^*]x + g(x^*)u - [\nabla_x f(x) \mid_{x^*} + \nabla_x g(x) \mid_{x^*} u^*]x^* + f(x^*) + g(x^*)u^* + \tilde{d}_1 \end{gathered} \tag{3.178}$$

where $\tilde{d}_1$ is the modeling error due to truncation of higher-order terms in the Taylor series expansion of $f(x)$ and $g(x)$. Next, by defining $A = [\nabla_x f(x) \mid_{x^*} + \nabla_x g(x) \mid_{x^*} u^*]$, $B = g(x^*)$, one obtains

$$\dot{x} = Ax + Bu - Ax^* + f(x^*) + g(x^*)u^* + \tilde{d}_1 \tag{3.179}$$

Moreover, by denoting $\tilde{d} = -Ax^* + f(x^*) + g(x^*)u^* + \tilde{d}_1$ about the cumulative modeling error term in the Taylor series expansion procedure, one has

$$\dot{x} = Ax + Bu + \tilde{d} \tag{3.180}$$

which is the approximately linearized model of the dynamics of the system of Eq. (3.175). The term $f(x^*) + g(x^*)u^*$ is the derivative of the state vector at (x^*, u^*) which is almost annihilated by $-Ax^*$.

The computation of the Jacobian matrix $\nabla_x f(x) \mid_{(x^*,u^*)}$ is given by

First row of the Jacobian matrix $\nabla_x f(x) \mid_{(x^*,u^*)}$: $\frac{\partial f_1}{\partial x_1} = 0$, $\frac{\partial f_1}{\partial x_2} = 0$, $\frac{\partial f_1}{\partial x_3} = 0$, $\frac{\partial f_1}{\partial x_4} = 1$, $\frac{\partial f_1}{\partial x_5} = 0$, $\frac{\partial f_1}{\partial x_6} = 0$, $\frac{\partial f_1}{\partial x_7} = 0$, $\frac{\partial f_1}{\partial x_8} = 0$, $\frac{\partial f_1}{\partial x_9} = 0$, $\frac{\partial f_1}{\partial x_{10}} = 0$, $\frac{\partial f_1}{\partial x_{11}} = 0$, and $\frac{\partial f_1}{\partial x_{12}} = 0$.

Second row of the Jacobian matrix $\nabla_x f(x) \mid_{(x^*,u^*)}$: $\frac{\partial f_2}{\partial x_1} = 0$, $\frac{\partial f_2}{\partial x_2} = 0$, $\frac{\partial f_2}{\partial x_3} = 0$, $\frac{\partial f_2}{\partial x_4} = 0$, $\frac{\partial f_2}{\partial x_5} = 1$, $\frac{\partial f_2}{\partial x_6} = 0$, $\frac{\partial f_2}{\partial x_7} = 0$, $\frac{\partial f_2}{\partial x_8} = 0$, $\frac{\partial f_2}{\partial x_9} = 0$, $\frac{\partial f_2}{\partial x_{10}} = 0$, $\frac{\partial f_2}{\partial x_{11}} = 0$, and $\frac{\partial f_2}{\partial x_{12}} = 0$.

Third row of the Jacobian matrix $\nabla_x f(x) \mid_{(x^*,u^*)}$: $\frac{\partial f_3}{\partial x_1} = 0$, $\frac{\partial f_3}{\partial x_2} = 0$, $\frac{\partial f_3}{\partial x_3} = 0$, $\frac{\partial f_3}{\partial x_4} = 0$, $\frac{\partial f_3}{\partial x_5} = 0$, $\frac{\partial f_3}{\partial x_6} = 1$, $\frac{\partial f_3}{\partial x_7} = 0$, $\frac{\partial f_3}{\partial x_8} = 0$, $\frac{\partial f_3}{\partial x_9} = 0$, $\frac{\partial f_3}{\partial x_{10}} = 0$, $\frac{\partial f_3}{\partial x_{11}} = 0$, and $\frac{\partial f_3}{\partial x_{12}} = 0$.

Fourth row of the Jacobian matrix $\nabla_x f(x) \mid_{(x^*,u^*)}$: It holds that

$$f_4(x) = f_{4a} + f_{4b}w_1x_7 + f_{4c}w_2x_9 + f_{4d}w_3x_{11} \tag{3.181}$$

where $f_{4a} = \frac{f_{4a,num}}{f_{4a,den}}$ with $f_{4a,num} = -\bar{M}_{11}\bar{h}_1 + \bar{M}_{21}\bar{h}_2 - \bar{M}_{31}\bar{h}_3 + \bar{M}_{21}\bar{k}_1$, and $f_{4a,den} = detM$

additionally, $f_{4b} = \frac{f_{4b,num}}{f_{4b,den}}$ with $f_{4b,num} = \bar{M}_{11} - L_{b1}\bar{M}_{21}$, and $f_{4b,den} = detM$

moreover, $f_{4c} = \frac{f_{4c,num}}{f_{4c,den}}$ with $f_{4c,num} = \bar{M}_{11} + L_{b1}\bar{M}_{21}$, and $f_{4c,den} = detM$

while one has also that $f_{4d} = \frac{f_{4d,num}}{f_{4d,den}}$ with $f_{4d,num} = \bar{M}_{31}$, and $f_{4d,den} = detM$

It holds that for $i = 1, \cdots, 12$ and $i \neq 7, 9, 11$

$$\frac{\partial f_4}{\partial x_i} = \frac{\partial f_{4a}}{\partial x_i} + \frac{\partial f_{4b}}{\partial x_i} w_1 x_7 + \frac{\partial f_{4c}}{\partial x_i} w_2 x_9 + \frac{\partial f_{4d}}{\partial x_i} w_3 x_{11} \tag{3.182}$$

while, for $i = 7$, one has that

$$\frac{\partial f_4}{\partial x_7} = \frac{\partial f_{4a}}{\partial x_7} + [\frac{\partial f_{4b}}{\partial x_7} w_1 x_7 + f_{4b} w_1] + \frac{\partial f_{4c}}{\partial x_7} w_2 x_9 + \frac{\partial f_{4d}}{\partial x_7} w_3 x_{11} \tag{3.183}$$

additionally, for $i = 9$, one has that

$$\frac{\partial f_4}{\partial x_9} = \frac{\partial f_{4a}}{\partial x_9} + \frac{\partial f_{4b}}{\partial x_9} w_1 x_7 + [\frac{\partial f_{4c}}{\partial x_9} w_2 x_9 + f_{4c} w_2] + \frac{\partial f_{4d}}{\partial x_9} w_3 x_{11} \tag{3.184}$$

moreover, for $i = 11$, one has that

$$\frac{\partial f_4}{\partial x_{11}} = \frac{\partial f_{4a}}{\partial x_{11}} + \frac{\partial f_{4b}}{\partial x_{11}} w_1 x_7 + \frac{\partial f_{4c}}{\partial x_{11}} w_2 x_9 + [\frac{\partial f_{4d}}{\partial x_{11}} w_3 x_{11} + f_{4d} w_3] \tag{3.185}$$

Next, one has that for $i = 1, \cdots, 12$

$$\frac{\partial f_{4a}}{\partial x_i} = \frac{\frac{\partial f_{4a,num}}{\partial x_i} detM - f_{4a,num} \frac{\partial detM}{\partial x_i}}{detM^2} \tag{3.186}$$

where it holds that for $i = 1, \cdots, 12$

$$\begin{aligned} \frac{\partial f_{4a,num}}{\partial x_i} = & -\frac{\partial M_{11}}{\partial x_i}\bar{h}_1 - \bar{M}_{11}\frac{\partial \bar{h}_1}{\partial x_i} + \frac{\partial M_{21}}{\partial x_i}\bar{h}_2 + \bar{M}_{21}\frac{\partial \bar{h}_2}{\partial x_i} \\ & -\frac{\partial M_{31}}{\partial x_i}\bar{h}_3 - \bar{M}_{31}\frac{\partial \bar{h}_3}{\partial x_i} + \frac{\partial M_{21}}{\partial x_i}\bar{k}_1 + \bar{M}_{21}\frac{\partial \bar{k}_1}{\partial x_i} \end{aligned} \tag{3.187}$$

and additionally, one has that for $i = 1, \cdots, 12$

$$\frac{\partial detM}{\partial x_i} = \frac{\partial M_{11}}{\partial x_i}\bar{M}_{11} + M_{11}\frac{\partial \bar{M}_{11}}{\partial x_i} - \frac{\partial M_{12}}{\partial x_i}\bar{M}_{12} + M_{12}\frac{\partial \bar{M}_{12}}{\partial x_i} + \frac{\partial M_{13}}{\partial x_i}\bar{M}_{13} + M_{13}\frac{\partial \bar{M}_{13}}{\partial x_i} \tag{3.188}$$

Equivalently, one has that for $i = 1, \cdots, 12$

$$\frac{\partial f_{4b}}{\partial x_i} = \frac{\frac{\partial f_{4b,num}}{\partial x_i} detM - f_{4b,num} \frac{\partial detM}{\partial x_i}}{detM^2} \tag{3.189}$$

where for $i = 1, \cdots, 12$

$$\frac{\partial f_{4b,num}}{\partial x_i} = \frac{\partial \bar{M}_{11}}{\partial x_i} - L_{b1}\frac{\partial \bar{M}_{21}}{\partial x_i} \tag{3.190}$$

Additionally, one has that for $i = 1, \cdots, 12$

$$\frac{\partial f_{4c}}{\partial x_i} = \frac{\frac{\partial f_{4c,num}}{\partial x_i} detM - f_{4c,num} \frac{\partial detM}{\partial x_i}}{detM^2} \tag{3.191}$$

where for $i = 1, \cdots, 12$

$$\frac{\partial f_{4c,num}}{\partial x_i} = \frac{\partial \bar{M}_{11}}{\partial x_i} + L_{b1}\frac{\partial \bar{M}_{21}}{\partial x_i} \tag{3.192}$$

Moreover, one has that for $i = 1, \cdots, 12$

$$\frac{\partial f_{4d}}{\partial x_i} = \frac{\frac{\partial f_{4d,num}}{\partial x_i} detM - f_{4d,num} \frac{\partial detM}{\partial x_i}}{detM^2} \tag{3.193}$$

where for $i = 1, \cdots, 12$

$$\frac{\partial f_{4d,num}}{\partial x_i} = \frac{\partial \bar{M}_{31}}{\partial x_i} \tag{3.194}$$

Fifth row of the Jacobian matrix $\nabla_x f(x) \mid_{(x^*,u^*)}$: It holds that

$$f_5(x) = f_{5a} + f_{5b} w_1 x_7 + f_{5c} w_2 x_9 + f_{5d} w_3 x_{11} \tag{3.195}$$

where $f_{5a} = \frac{f_{4a,num}}{f_{45,den}}$ with $f_{5a,num} = \bar{M}_{12}\bar{h}_1 - \bar{M}_{22}\bar{h}_2 + \bar{M}_{32}\bar{h}_3 - \bar{M}_{22}\bar{k}_2$, and $f_{5a,den} = detM$
additionally, $f_{5b} = \frac{f_{5b,num}}{f_{5b,den}}$ with $f_{5b,num} = -\bar{M}_{12} + L_{b1}\bar{M}_{22}$, and $f_{5b,den} = detM$
moreover, $f_{5c} = \frac{f_{5c,num}}{f_{5c,den}}$ with $f_{5c,num} = -\bar{M}_{12} - L_{b1}\bar{M}_{22}$, and $f_{5b,den} = detM$
while one has also that $f_{5d} = \frac{f_{5d,num}}{f_{5d,den}}$ with $f_{5d,num} = -\bar{M}_{22}$, and $f_{5d,den} = detM$
It holds that for $i = 1, \cdots, 12$ and $i \neq 7, 9, 11$

$$\frac{\partial f_5}{\partial x_i} = \frac{\partial f_{5a}}{\partial x_i} + \frac{\partial f_{5b}}{\partial x_i} w_1 x_7 + \frac{\partial f_{5c}}{\partial x_i} w_2 x_9 + \frac{\partial f_{5d}}{\partial x_i} w_3 x_{11} \tag{3.196}$$

while, for $i = 7$, one has that

$$\frac{\partial f_5}{\partial x_7} = \frac{\partial f_{5a}}{\partial x_7} + [\frac{\partial f_{5b}}{\partial x_7} w_1 x_7 + f_{5b} w_1] + \frac{\partial f_{5c}}{\partial x_7} w_2 x_9 + \frac{\partial f_{5d}}{\partial x_7} w_3 x_{11} \tag{3.197}$$

additionally, for $i = 9$, one has that

$$\frac{\partial f_5}{\partial x_9} = \frac{\partial f_{5a}}{\partial x_9} + \frac{\partial f_{5b}}{\partial x_9} w_1 x_7 + [\frac{\partial f_{5c}}{\partial x_9} w_2 x_9 + f_{5c} w_2] + \frac{\partial f_{5d}}{\partial x_9} w_3 x_{11} \tag{3.198}$$

moreover, for $i = 11$, one has that

$$\frac{\partial f_5}{\partial x_{11}} = \frac{\partial f_{5a}}{\partial x_{11}} + \frac{\partial f_{5b}}{\partial x_{11}} w_1 x_7 + \frac{\partial f_{5c}}{\partial x_{11}} w_2 x_9 + [\frac{\partial f_{5d}}{\partial x_{11}} w_3 x_{11} + f_{5d} w_3] \tag{3.199}$$

Next, one has that for $i = 1, \cdots, 12$

$$\frac{\partial f_{5a}}{\partial x_i} = \frac{\frac{\partial f_{5a,num}}{\partial x_i} detM - f_{5a,num} \frac{\partial detM}{\partial x_i}}{detM^2} \tag{3.200}$$

where it holds that for $i = 1, \cdots, 12$

$$\begin{aligned}\frac{\partial f_{5a,num}}{\partial x_i} = & \frac{\partial M_{12}}{\partial x_i}\bar{h}_1 + \bar{M}_{12}\frac{\partial \bar{h}_1}{\partial x_i} - \frac{\partial M_{22}}{\partial x_i}\bar{h}_2 - \bar{M}_{22}\frac{\partial \bar{h}_2}{\partial x_i} \\ & - \frac{\partial M_{32}}{\partial x_i}\bar{h}_3 - \bar{M}_{32}\frac{\partial \bar{h}_3}{\partial x_i} - \frac{\partial M_{22}}{\partial x_i}\bar{k}_2 - \bar{M}_{22}\frac{\partial \bar{k}_2}{\partial x_i}\end{aligned} \tag{3.201}$$

and additionally, as given before, one has that for $i = 1, \cdots, 12$

$$\frac{\partial detM}{\partial x_i} = \frac{\partial M_{11}}{\partial x_i}\bar{M}_{11} + M_{11}\frac{\partial \bar{M}_{11}}{\partial x_i} - \frac{\partial M_{12}}{\partial x_i}\bar{M}_{12} + M_{12}\frac{\partial \bar{M}_{12}}{\partial x_i} + \frac{\partial M_{13}}{\partial x_i}\bar{M}_{13} + M_{13}\frac{\partial \bar{M}_{13}}{\partial x_i} \tag{3.202}$$

Equivalently, one has that for $i = 1, \cdots, 12$

$$\frac{\partial f_{5b}}{\partial x_i} = \frac{\frac{\partial f_{5b,num}}{\partial x_i} detM - f_{5b,num} \frac{\partial detM}{\partial x_i}}{detM^2} \tag{3.203}$$

where for $i = 1, \cdots, 12$

$$\frac{\partial f_{5b,num}}{\partial x_i} = -\frac{\partial \bar{M}_{12}}{\partial x_i} + L_{b1}\frac{\partial \bar{M}_{22}}{\partial x_i} \tag{3.204}$$

Additionally, one has that for $i = 1, \cdots, 12$

$$\frac{\partial f_{5c}}{\partial x_i} = \frac{\frac{\partial f_{5c,num}}{\partial x_i} detM - f_{5c,num} \frac{\partial detM}{\partial x_i}}{detM^2} \tag{3.205}$$

where for $i = 1, \cdots, 12$

$$\frac{\partial f_{5c,num}}{\partial x_i} = -\frac{\partial \bar{M}_{12}}{\partial x_i} - L_{b1}\frac{\partial \bar{M}_{22}}{\partial x_i} \tag{3.206}$$

Moreover, one has that for $i = 1, \cdots, 12$

$$\frac{\partial f_{5d}}{\partial x_i} = \frac{\frac{\partial f_{5d,num}}{\partial x_i} detM - f_{5d,num} \frac{\partial detM}{\partial x_i}}{detM^2} \tag{3.207}$$

where for $i = 1, \cdots, 12$

$$\frac{\partial f_{5d,num}}{\partial x_i} = -\frac{\partial \bar{M}_{32}}{\partial x_i} \tag{3.208}$$

Sixth row of the Jacobian matrix $\nabla_x f(x) \mid_{(x^*, u^*)}$: It holds that

$$f_6(x) = f_{6a} + f_{6b} w_1 x_7 + f_{6c} w_2 x_9 + f_{6d} w_3 x_{11} \tag{3.209}$$

where $f_{6a} = \frac{f_{6a,num}}{f_{6a,den}}$ with $f_{6a,num} = -\bar{M}_{13}\bar{h}_1 + \bar{M}_{23}\bar{h}_2 - \bar{M}_{33}\bar{h}_3 + \bar{M}_{23}\bar{k}_1$, and $f_{6a,den} = detM$
additionally, $f_{6b} = \frac{f_{6b,num}}{f_{6b,den}}$ with $f_{6b,num} = \bar{M}_{13} - L_{b1}\bar{M}_{23}$, and $f_{6b,den} = detM$
moreover, $f_{6c} = \frac{f_{6c,num}}{f_{6c,den}}$ with $f_{6c,num} = \bar{M}_{13} + L_{b1}\bar{M}_{23}$, and $f_{6c,den} = detM$
while one has also that $f_{6d} = \frac{f_{6d,num}}{f_{6d,den}}$ with $f_{6d,num} = \bar{M}_{33}$, and $f_{6d,den} = detM$
It holds that for $i = 1, \cdots, 12$ and $i \neq 7, 9, 11$

$$\frac{\partial f_6}{\partial x_i} = \frac{\partial f_{6a}}{\partial x_i} + \frac{\partial f_{6b}}{\partial x_i} w_1 x_7 + \frac{\partial f_{6c}}{\partial x_i} w_2 x_9 + \frac{\partial f_{6d}}{\partial x_i} w_3 x_{11} \tag{3.210}$$

while, for $i = 7$, one has that

$$\frac{\partial f_6}{\partial x_7} = \frac{\partial f_{6a}}{\partial x_7} + \left[\frac{\partial f_{6b}}{\partial x_7} w_1 x_7 + f_{6b} w_1\right] + \frac{\partial f_{6c}}{\partial x_7} w_2 x_9 + \frac{\partial f_{6d}}{\partial x_7} w_3 x_{11} \tag{3.211}$$

additionally, for $i = 9$, one has that

$$\frac{\partial f_6}{\partial x_9} = \frac{\partial f_{6a}}{\partial x_9} + \frac{\partial f_{6b}}{\partial x_9} w_1 x_7 + \left[\frac{\partial f_{6c}}{\partial x_9} w_2 x_9 + f_{6c} w_2\right] + \frac{\partial f_{6d}}{\partial x_9} w_3 x_{11} \tag{3.212}$$

moreover, for $i = 11$, one has that

$$\frac{\partial f_6}{\partial x_{11}} = \frac{\partial f_{6a}}{\partial x_{11}} + \frac{\partial f_{6b}}{\partial x_{11}} w_1 x_7 + \frac{\partial f_{6c}}{\partial x_{11}} w_2 x_9 + \left[\frac{\partial f_{6d}}{\partial x_{11}} w_3 x_{11} + f_{6d} w_3\right] \tag{3.213}$$

Next, one has that for $i = 1, \cdots, 12$

$$\frac{\partial f_{6a}}{\partial x_i} = \frac{\frac{\partial f_{6a,num}}{\partial x_i} detM - f_{6a,num}\frac{\partial detM}{\partial x_i}}{detM^2} \tag{3.214}$$

where it holds that for $i = 1, \cdots, 12$

$$\begin{aligned}\frac{\partial f_{6a,num}}{\partial x_i} = & -\frac{\partial M_{13}}{\partial x_i}\bar{h}_1 - \bar{M}_{13}\frac{\partial \bar{h}_1}{\partial x_i} + \frac{\partial M_{23}}{\partial x_i}\bar{h}_2 + \bar{M}_{23}\frac{\partial \bar{h}_2}{\partial x_i} \\ & -\frac{\partial M_{33}}{\partial x_i}\bar{h}_3 - \bar{M}_{33}\frac{\partial \bar{h}_3}{\partial x_i} + \frac{\partial M_{23}}{\partial x_i}\bar{k}_3 + \bar{M}_{23}\frac{\partial \bar{k}_3}{\partial x_i}\end{aligned} \tag{3.215}$$

and additionally, as given before, one has that for $i = 1, \cdots, 12$

$$\frac{\partial detM}{\partial x_i} = \frac{\partial M_{11}}{\partial x_i}\bar{M}_{11} + M_{11}\frac{\partial \bar{M}_{11}}{\partial x_i} - \frac{\partial M_{12}}{\partial x_i}\bar{M}_{12} + M_{12}\frac{\partial \bar{M}_{12}}{\partial x_i} + \frac{\partial M_{13}}{\partial x_i}\bar{M}_{13} + M_{13}\frac{\partial \bar{M}_{13}}{\partial x_i} \tag{3.216}$$

Equivalently, one has that for $i = 1, \cdots, 12$

$$\frac{\partial f_{6b}}{\partial x_i} = \frac{\frac{\partial f_{6b,num}}{\partial x_i} detM - f_{6b,num}\frac{\partial detM}{\partial x_i}}{detM^2} \tag{3.217}$$

where for $i = 1, \cdots, 12$

$$\frac{\partial f_{6b,num}}{\partial x_i} = \frac{\partial \bar{M}_{13}}{\partial x_i} - L_{b1}\frac{\partial \bar{M}_{23}}{\partial x_i} \tag{3.218}$$

Additionally, one has that for $i = 1, \cdots, 12$

$$\frac{\partial f_{6c}}{\partial x_i} = \frac{\frac{\partial f_{6c,num}}{\partial x_i} detM - f_{6c,num}\frac{\partial detM}{\partial x_i}}{detM^2} \tag{3.219}$$

where for $i = 1, \cdots, 12$

$$\frac{\partial f_{6c,num}}{\partial x_i} = \frac{\partial \bar{M}_{13}}{\partial x_i} + L_{b1}\frac{\partial \bar{M}_{23}}{\partial x_i} \tag{3.220}$$

Moreover, one has that for $i = 1, \cdots, 12$

$$\frac{\partial f_{6d}}{\partial x_i} = \frac{\frac{\partial f_{6d,num}}{\partial x_i} detM - f_{6d,num}\frac{\partial detM}{\partial x_i}}{detM^2} \tag{3.221}$$

where for $i = 1, \cdots, 12$

$$\frac{\partial f_{6d,num}}{\partial x_i} = \frac{\partial \bar{M}_{33}}{\partial x_i} \tag{3.222}$$

Seventh row of the Jacobian matrix $\nabla_x f(x) \mid_{(x^*,u^*)}$: $\frac{\partial f_7}{\partial x_1} = 0$, $\frac{\partial f_7}{\partial x_2} = 0$, $\frac{\partial f_7}{\partial x_3} = 0$, $\frac{\partial f_7}{\partial x_4} = -\frac{n_{p1}\pi}{L_{q1}\tau_1}(L_{d1}x_8 + \lambda_{pm1})$, $\frac{\partial f_7}{\partial x_5} = -\frac{n_{p1}\pi L_{b1}}{L_{q1}\tau_1}(L_{d1}x_8 + \lambda_{pm1})$, $\frac{\partial f_7}{\partial x_6} = 0$, $\frac{\partial f_7}{\partial x_7} = -\frac{R_{s1}}{L_{q1}}$, $\frac{\partial f_7}{\partial x_8} = -\frac{n_{p1}\pi(x_4 + L_{b1}x_5)}{L_{q1}\tau_1}L_{d1}$, $\frac{\partial f_7}{\partial x_9} = 0$, $\frac{\partial f_7}{\partial x_{10}} = 0$, $\frac{\partial f_7}{\partial x_{11}} = 0$, and $\frac{\partial f_7}{\partial x_{12}} = 0$.

Eighth row of the Jacobian matrix $\nabla_x f(x)\mid_{(x^*,u^*)}$: $\frac{\partial f_8}{\partial x_1}=0$, $\frac{\partial f_8}{\partial x_2}=0$, $\frac{\partial f_8}{\partial x_3}=0$, $\frac{\partial f_8}{\partial x_4}=-\frac{n_{p1}\pi}{L_{d1}T_1}L_{q1}x_7$, $\frac{\partial f_8}{\partial x_5}=-\frac{n_{p1}\pi L_{b1}}{L_{d1}T_1}L_{q1}x_7$, $\frac{\partial f_8}{\partial x_6}=0$, $\frac{\partial f_8}{\partial x_7}=-\frac{n_{p1}\pi(x_4+L_{b1}x_5)}{L_{d1}T_1}L_{q1}$, $\frac{\partial f_8}{\partial x_8}=-\frac{R_{s1}}{L_{d1}}$, $\frac{\partial f_8}{\partial x_9}=0$, $\frac{\partial f_8}{\partial x_{10}}=0$, $\frac{\partial f_8}{\partial x_{11}}=0$, and $\frac{\partial f_8}{\partial x_{12}}=0$.

Ninth row of the Jacobian matrix $\nabla_x f(x)\mid_{(x^*,u^*)}$: $\frac{\partial f_9}{\partial x_1}=0$, $\frac{\partial f_9}{\partial x_2}=0$, $\frac{\partial f_9}{\partial x_3}=0$, $\frac{\partial f_9}{\partial x_4}=-\frac{n_{p2}\pi}{L_{q2}T_2}(L_{d2}x_{10}+\lambda_{pm2})$, $\frac{\partial f_9}{\partial x_5}=\frac{n_{p2}\pi L_{b1}}{L_{q2}T_2}(L_{d2}x_{10}+\lambda_{pm2})$, $\frac{\partial f_9}{\partial x_6}=0$, $\frac{\partial f_9}{\partial x_7}=0$, $\frac{\partial f_9}{\partial x_8}=0$, $\frac{\partial f_9}{\partial x_9}=-\frac{R_{s2}}{L_{q2}}$, $\frac{\partial f_9}{\partial x_{10}}=-\frac{n_{p2}\pi(x_4-L_{b1}x_5)}{L_{q2}T_2}L_{d2}$, $\frac{\partial f_9}{\partial x_{11}}=0$, and $\frac{\partial f_9}{\partial x_{12}}=0$.

Tenth row of the Jacobian matrix $\nabla_x f(x)\mid_{(x^*,u^*)}$: $\frac{\partial f_{10}}{\partial x_1}=0$, $\frac{\partial f_{10}}{\partial x_2}=0$, $\frac{\partial f_{10}}{\partial x_3}=0$, $\frac{\partial f_{10}}{\partial x_4}=-\frac{n_{p2}\pi}{L_{d2}T_2}L_{q2}x_9$, $\frac{\partial f_{10}}{\partial x_5}=-\frac{n_{p2}\pi L_{b1}}{L_{d2}T_2}L_{q2}x_9$, $\frac{\partial f_{10}}{\partial x_6}=0$, $\frac{\partial f_{10}}{\partial x_7}=0$, $\frac{\partial f_{10}}{\partial x_8}=0$, $\frac{\partial f_{10}}{\partial x_9}=-\frac{n_{p2}\pi(x_4-L_{b1}x_5)}{L_{d2}T_2}L_{q2}$, $\frac{\partial f_{10}}{\partial x_{10}}=-\frac{R_{s2}}{L_{d2}}$, $\frac{\partial f_{10}}{\partial x_{11}}=0$, and $\frac{\partial f_{10}}{\partial x_{12}}=0$.

Eleventh row of the Jacobian matrix $\nabla_x f(x)\mid_{(x^*,u^*)}$: $\frac{\partial f_{11}}{\partial x_1}=0$, $\frac{\partial f_{11}}{\partial x_2}=0$, $\frac{\partial f_{11}}{\partial x_3}=0$, $\frac{\partial f_{11}}{\partial x_4}=0$, $\frac{\partial f_{11}}{\partial x_5}=0$, $\frac{\partial f_{11}}{\partial x_6}=-\frac{n_{p3}\pi}{L_{q3}T_3}(L_{d3}x_{12}+\lambda_{pm3})$, $\frac{\partial f_{11}}{\partial x_7}=0$, $\frac{\partial f_{11}}{\partial x_8}=0$, $\frac{\partial f_{11}}{\partial x_9}=0$, $\frac{\partial f_{11}}{\partial x_{10}}=0$, $\frac{\partial f_{11}}{\partial x_{11}}=-\frac{R_{s3}}{L_{q3}}$, and $\frac{\partial f_{11}}{\partial x_{12}}=-\frac{n_{p3}\pi x_6}{L_{q3}T_3}L_{d3}$.

Twelfth row of the Jacobian matrix $\nabla_x f(x)\mid_{(x^*,u^*)}$: $\frac{\partial f_{12}}{\partial x_1}=0$, $\frac{\partial f_{12}}{\partial x_2}=0$, $\frac{\partial f_{12}}{\partial x_3}=0$, $\frac{\partial f_{12}}{\partial x_4}=0$, $\frac{\partial f_{12}}{\partial x_5}=0$, $\frac{\partial f_{10}}{\partial x_6}=-\frac{n_{p3}\pi}{L_{d3}T_3}L_{q3}x_{11}$, $\frac{\partial f_{10}}{\partial x_7}=0$, $\frac{\partial f_{10}}{\partial x_8}=0$, $\frac{\partial f_{10}}{\partial x_9}=0$, $\frac{\partial f_{10}}{\partial x_{10}}=0$, $\frac{\partial f_{10}}{\partial x_{11}}=\frac{n_{p3}\pi x_6}{L_{d3}T_3}L_{q3}$, and $\frac{\partial f_{10}}{\partial x_{12}}=-\frac{R_{s3}}{L_{d3}}$.

Next, the computation of the partial derivatives of the terms $\bar{M}_{ij}$ $i=1,2,3, j=1,2,3$ is given:

For $\bar{M}_{11}$, it holds that: $\frac{\partial\bar{M}_{11}}{\partial x_i}=0, \forall i=1,2,\cdots,12$ and $i\neq 3$, while $\frac{\partial\bar{M}_{11}}{\partial x_3}=2m_b^2x_3$.

For $\bar{M}_{12}$, it holds that: $\frac{\partial\bar{M}_{12}}{\partial x_i}=\frac{\partial M_{12}}{\partial x_i}, \forall i=1,2,\cdots,12$ and $i\neq 2$

For $\bar{M}_{13}$, it holds that: $\frac{\partial\bar{M}_{13}}{\partial x_i}=-\frac{\partial M_{12}}{\partial x_i}(m_b)d, \forall i=1,2,\cdots,12$ and $i\neq 2,3$, while $\frac{\partial\bar{M}_{13}}{\partial x_2}=-\frac{\partial M_{12}}{\partial x_2}(m_bd)+(m_bcos(x_2))(J_t+m_bx_3^2)$, $\frac{\partial\bar{M}_{13}}{\partial x_3}=-\frac{\partial M_{12}}{\partial x_2}(m_bd)+(m_bcos(x_2))(2m_bx_3)$.

For $\bar{M}_{21}$, it holds that: $\frac{\partial\bar{M}_{21}}{\partial x_i}=\frac{\partial M_{12}}{\partial x_i}m_b, \forall i=1,2,\cdots,12$ and $i\neq 2$, while $\frac{\partial\bar{M}_{21}}{\partial x_2}=\frac{\partial M_{12}}{\partial x_2}m_b-(m_bcos(x_2))(m_bd)$.

For $\bar{M}_{22}$, it holds that: $\frac{\partial\bar{M}_{22}}{\partial x_i}=\frac{\partial M_{11}}{\partial x_i}m_b, \forall i=1,2,\cdots,12$ and $i\neq 2$, while $\frac{\partial\bar{M}_{22}}{\partial x_2}=\frac{\partial M_{11}}{\partial x_2}m_b-2(m_bsin(x_2))(m_bcos(x_2)$.

For $\bar{M}_{23}$, it holds that: $\frac{\partial\bar{M}_{23}}{\partial x_i}=-\frac{\partial M_{11}}{\partial x_i}m_bd-\frac{\partial M_{12}}{\partial x_i}m_bsin(x_2), \forall i=1,2,\cdots,12$ and $i\neq 2$, while $\frac{\partial\bar{M}_{23}}{\partial x_2}=-\frac{\partial M_{11}}{\partial x_2}m_bd-\frac{\partial M_{12}}{\partial x_2}m_bsin(x_2)-M_{12}m_bcos(x_2)$.

For $\bar{M}_{31}$, it holds that: $\frac{\partial\bar{M}_{31}}{\partial x_i}=-\frac{\partial M_{12}}{\partial x_i}m_bd, \forall i=1,2,\cdots,12$ and $i\neq 2,3$, while $\frac{\partial\bar{M}_{31}}{\partial x_2}=-\frac{\partial M_{12}}{\partial x_2}m_bd+(J_t+m_bx_3^2)(m_bcos(x_2))$ and $\frac{\partial\bar{M}_{31}}{\partial x_3}=-\frac{\partial M_{12}}{\partial x_3}m_bd+(2m_bx_3)(m_bcos(x_2))$

For $\bar{M}_{32}$, it holds that: $\frac{\partial\bar{M}_{32}}{\partial x_i}=-\frac{\partial M_{11}}{\partial x_i}m_bd+\frac{\partial M_{12}}{\partial x_i}(m_bsin(x_2)), \forall i=1,2,\cdots,12$ and $i\neq 2$, while $\frac{\partial\bar{M}_{32}}{\partial x_2}=-\frac{\partial M_{11}}{\partial x_2}m_bd+\frac{\partial M_{12}}{\partial x_2}(m_bsin(x_2))+M_{12}(m_bcos(x_2))$.

For $\bar{M}_{33}$, it holds that: $\frac{\partial\bar{M}_{33}}{\partial x_i}=\frac{\partial M_{11}}{\partial x_i}(J_t+m_bx_3^2)-2M_{12}\frac{\partial M_{12}}{\partial x_i}$ $\forall i=1,2,\cdots,12$ and $i\neq 3$, while $\frac{\partial\bar{M}_{33}}{\partial x_3}=\frac{\partial M_{11}}{\partial x_3}(J_t+m_bx_3^2)-M_{11}m_b2x_3-2M_{12}\frac{\partial M_{12}}{\partial x_3}$.

Besides, it holds that: $\frac{\partial M_{11}}{\partial x_i}=0\,\forall i=1,2,\cdots,12$ and $\frac{\partial M_{12}}{\partial x_i}=0\,\forall i=1,2,\cdots,12$ and $i\neq 2,3$ and $\frac{\partial M_{12}}{\partial x_2}=m_hdcos(x_2)+[m_hx_3-0.5(m_1-m_2)L_h]sin(x_2)$, $\frac{\partial M_{12}}{\partial x_3}=-m_hcos(x_2)$.

Next, the computation of the partial derivatives of the terms $\bar{h}_i$, $i=1,2,3$ is given

For term $\bar{h}_1$, one obtains: $\frac{\partial \bar{h}_1}{\partial x_1}=0$, $\frac{\partial \bar{h}_1}{\partial x_2}=\frac{\partial H_{12}}{\partial x_2}x_5^2+2m_hx_5sin(x_2)x_6$, $\frac{\partial \bar{h}_1}{\partial x_3}=\frac{\partial H_{12}}{\partial x_3}x_5^2$, $\frac{\partial \bar{h}_1}{\partial x_4}=C_{11}$, $\frac{\partial \bar{h}_1}{\partial x_5}=\frac{\partial H_{12}}{\partial x_5}x_5^2+H_{12}2x_5-2m_hcos(x_2)x_6$, $\frac{\partial \bar{h}_1}{\partial x_6}=\frac{\partial H_{12}}{\partial x_6}x_5^2-2m_hx_5cos(x_2)$, $\frac{\partial \bar{h}_1}{\partial x_7}=0$, $\frac{\partial \bar{h}_1}{\partial x_8}=0$, $\frac{\partial \bar{h}_1}{\partial x_9}=0$, $\frac{\partial \bar{h}_1}{\partial x_{10}}=0$, $\frac{\partial \bar{h}_1}{\partial x_{11}}=0$, $\frac{\partial \bar{h}_1}{\partial x_{12}}=0$.

For term $\bar{h}_2$, one obtains: $\frac{\partial \bar{h}_2}{\partial x_1}=0$, $\frac{\partial \bar{h}_2}{\partial x_2}=0$, $\frac{\partial \bar{h}_2}{\partial x_3}=2m_hx_6x_5$, $\frac{\partial \bar{h}_2}{\partial x_4}=C_{12}$, $\frac{\partial \bar{h}_2}{\partial x_5}=C_{22}+2m_hx_3x_6$, $\frac{\partial \bar{h}_2}{\partial x_6}=2m_hx_3x_5$, $\frac{\partial \bar{h}_2}{\partial x_7}=0$, $\frac{\partial \bar{h}_2}{\partial x_8}=0$, $\frac{\partial \bar{h}_2}{\partial x_9}=0$, $\frac{\partial \bar{h}_2}{\partial x_{10}}=0$, $\frac{\partial \bar{h}_2}{\partial x_{11}}=0$, $\frac{\partial \bar{h}_2}{\partial x_{12}}=0$.

For term $\bar{h}_3$, one obtains: $\frac{\partial \bar{h}_3}{\partial x_1}=0$, $\frac{\partial \bar{h}_3}{\partial x_2}=-m_hx_3x_4$, $\frac{\partial \bar{h}_3}{\partial x_3}=-m_hx_2x_4$, $\frac{\partial \bar{h}_3}{\partial x_4}=-m_hx_3x_2$, $\frac{\partial \bar{h}_3}{\partial x_5}=0$, $\frac{\partial \bar{h}_3}{\partial x_6}=C_y$, $\frac{\partial \bar{h}_3}{\partial x_7}=0$, $\frac{\partial \bar{h}_3}{\partial x_8}=0$, $\frac{\partial \bar{h}_3}{\partial x_9}=0$, $\frac{\partial \bar{h}_3}{\partial x_{10}}=0$, $\frac{\partial \bar{h}_3}{\partial x_{11}}=0$, $\frac{\partial \bar{h}_3}{\partial x_{12}}=0$.

where it holds that: $\frac{\partial H_{12}}{\partial x_i}=0\ \forall i=1,\cdots,12$ and $i\neq 2,3$, $\frac{\partial H_{12}}{\partial x_2}=-m_hdsin(x_2)+[m_hx_3-0.5(m_1-m_2)L_b]cos(x_2)$ and $\frac{\partial H_{12}}{\partial x_3}=m_hsin(x_2)$.

Finally, the computation of the partial derivatives of the terms $\bar{k}_i$, $i=1,2,3$ is given:

It holds that: $\frac{\partial \bar{k}_1}{\partial x_i}=0$, $\forall i=1,2,\cdots,12$ and $i\neq 2$ and $\frac{\partial \bar{k}_1}{\partial x_2}=k_{22}$.

Additionally: $\frac{\partial \bar{k}_2}{\partial x_i}=0$, $\forall i=1,2,\cdots,12$ and $i\neq 2$ and $\frac{\partial \bar{k}_2}{\partial x_2}=k_{22}$.

Equivalently: $\frac{\partial \bar{k}_3}{\partial x_i}=0$, $\forall i=1,2,\cdots,12$ and $i\neq 2$ and $\frac{\partial \bar{k}_3}{\partial x_2}=k_{22}$.

3.2.5 Design of an H-infinity nonlinear feedback controller

3.2.5.1 Equivalent linearized dynamics of the dual-drive H-type gantry crane

After linearization around its current operating point, the dynamic model for the dual-drive H-type gantry crane is written as

$$\dot{x}=Ax+Bu+d_1 \tag{3.223}$$

Parameter d_1 stands for the linearization error in the dual-drive H-type gantry crane's model that was given previously in Eq. (3.223). The reference setpoints for the state vector of the aforementioned dynamic model are denoted by $\mathbf{x_d}=[x_1^d,\cdots,x_{12}^d]$. Tracking of this trajectory is achieved after applying the control input u^*. At every time instant the control input u^* is assumed to differ from the control input u appearing in Eq. (3.223) by an amount equal to Δu, that is $u^*=u+\Delta u$

$$\dot{x}_d=Ax_d+Bu^*+d_2 \tag{3.224}$$

The dynamics of the controlled system described in Eq. (3.223) can be also written as

$$\dot{x}=Ax+Bu+Bu^*-Bu^*+d_1 \tag{3.225}$$

and by denoting $d_3=-Bu^*+d_1$ as an aggregate disturbance term, one obtains

$$\dot{x}=Ax+Bu+Bu^*+d_3 \tag{3.226}$$

By subtracting Eq. (3.224) from Eq. (3.226), one has

$$\dot{x}-\dot{x}_d=A(x-x_d)+Bu+d_3-d_2 \tag{3.227}$$

By denoting the tracking error as $e=x-x_d$ and the aggregate disturbance term as $\tilde{d}=d_3-d_2$, the tracking error dynamics becomes

$$\dot{e}=Ae+Bu+\tilde{d} \tag{3.228}$$

The above-linearized form of the dual-drive H-type gantry crane's model can be efficiently controlled after applying an H-infinity feedback control scheme.

3.2.5.2 The nonlinear H-infinity control

The initial nonlinear model of the dual-drive H-type gantry crane is in the form

$$\dot{x} = f(x,u) \quad x \in R^n, \ u \in R^m \tag{3.229}$$

Linearization of the model of the dual-drive H-type gantry crane is performed at each iteration of the control algorithm around its present operating point $(x^*, u^*) = (x(t), u(t - T_s))$. The linearized equivalent of the system is described by

$$\dot{x} = Ax + Bu + L\tilde{d} \quad x \in R^n, \ u \in R^m, \ \tilde{d} \in R^q \tag{3.230}$$

where matrices A and B are obtained from the computation of the previously defined Jacobians and vector $\tilde{d}$ denotes disturbance terms due to linearization errors, while L is a disturbance inputs gain matrix. The problem of disturbance rejection for the linearized model that is described by

$$\begin{aligned} \dot{x} &= Ax + Bu + L\tilde{d} \\ y &= Cx \end{aligned} \tag{3.231}$$

where $x \in R^n$, $u \in R^m$, $\tilde{d} \in R^q$, and $y \in R^p$ cannot be handled efficiently if the classical LQR control scheme is applied. This is because of the existence of the perturbation term $\tilde{d}$. The disturbance term $\tilde{d}$ apart from modeling (parametric) uncertainty and external perturbation terms can also represent noise terms of any distribution.

In the H_∞ control approach, a feedback control scheme is designed for trajectory tracking by the system's state vector and simultaneous disturbance rejection, considering that the disturbance affects the system in the worst possible manner. The disturbances' effects are incorporated in the following quadratic cost function:

$$J(t) = \frac{1}{2}\int_0^T [y^T(t)y(t) + ru^T(t)u(t) - \rho^2\tilde{d}^T(t)\tilde{d}(t)]dt, \quad r, \rho > 0 \tag{3.232}$$

The significance of the negative sign in the cost function's term that is associated with the perturbation variable $\tilde{d}(t)$ is that the disturbance tries to maximize the cost function $J(t)$ while the control signal $u(t)$ tries to minimize it. The physical meaning of the relation given above is that the control signal and the disturbances compete with each other within a min–max differential game. This problem of min–max optimization can be written as $min_u max_{\tilde{d}} J(u, \tilde{d})$.

The objective of the optimization procedure is to compute a control signal $u(t)$ which can compensate for the worst possible disturbance, that is externally imposed on the dual-drive H-type gantry crane. However, the solution to the min–max optimization problem is directly related to the value of the parameter ρ. This means that there is an upper bound in the disturbances magnitude that can be annihilated by the control signal.

3.2.5.3 Computation of the feedback control gains

For the linearized system given by Eq. (3.231) the cost function of Eq. (3.232) is defined, where coefficient r determines the penalization of the control input and weight coefficient ρ determines the reward of the disturbances' effects. It is assumed that (i) The energy that is transferred from the disturbances signal $\tilde{d}(t)$ is bounded, that is $\int_0^\infty \tilde{d}^T(t)\tilde{d}(t)dt < \infty$, (ii) matrices $[A, B]$ and $[A, L]$ are stabilizable, (iii) matrix $[A, C]$ is detectable. In the case of a tracking problem, the optimal feedback control law is given by

$$u(t) = -Ke(t) \tag{3.233}$$

with $e = x - x_d$ to be the tracking error, and $K = \frac{1}{r}B^TP$ where P is a positive definite symmetric matrix. As it will be proven in Section 3.2.6, matrix P is obtained from the solution of the Riccati equation

$$A^TP + PA + Q - P\left(\frac{2}{r}BB^T - \frac{1}{\rho^2}LL^T\right)P = 0 \tag{3.234}$$

where Q is a positive semi-definite symmetric matrix. The worst-case disturbance is given by

$$\tilde{d}(t) = \frac{1}{\rho^2}L^TPe(t) \tag{3.235}$$

The solution of the H-infinity feedback control problem for the dual-drive H-type gantry crane and the computation of the worst-case disturbance that the related controller can sustain, comes from the superposition of Bellman's optimality principle when considering that the robotic crane is affected by two separate inputs (i) the control input u (ii) the

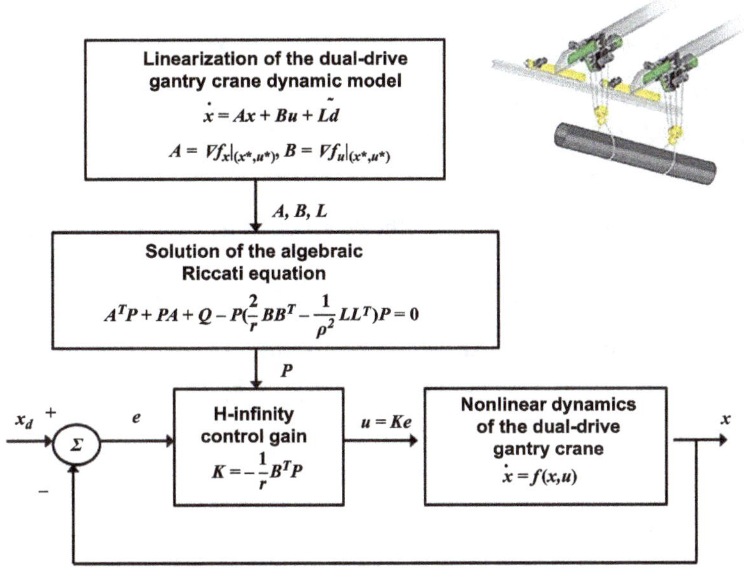

Figure 3.20 Diagram of the control scheme for the dual-drive H-type gantry crane

cumulative disturbance input $\tilde{d}(t)$. Solving the optimal control problem for u, that is for the minimum variation (optimal) control input that achieves the elimination of the state vector's tracking error, gives $u = -\frac{1}{r}B^TPe$. Equivalently, solving the optimal control problem for $\tilde{d}$, that is for the worst-case disturbance that the control loop can sustain gives $\tilde{d} = \frac{1}{\rho^2}L^TPe$.

The diagram of the considered control loop for the dual-drive H-type gantry crane is depicted in Figure 3.20.

3.2.6 Lyapunov stability analysis

Through Lyapunov stability analysis, it will be shown that the proposed nonlinear control scheme assures H_∞ tracking performance for the dual-drive H-type gantry crane and that in case of bounded disturbance terms asymptotic convergence to the reference setpoints is achieved. The tracking error dynamics for the dual-drive H-type gantry crane is written in the form

$$\dot{e} = Ae + Bu + L\tilde{d} \tag{3.236}$$

where in the dual-drive H-type gantry crane's case $L \in R^{12\times 12}$ to be the disturbance inputs gain matrix. Variable $\tilde{d}$ denotes model uncertainties and external disturbances of the dual-drive H-type gantry crane's model. The following Lyapunov equation is considered

$$V = \frac{1}{2}e^TPe \tag{3.237}$$

where $e = x - x_d$ is the tracking error. By differentiating with respect to time, one obtains

$$\begin{gathered}\dot{V} = \frac{1}{2}\dot{e}^TPe + \frac{1}{2}e^TP\dot{e} \Rightarrow \\ \dot{V} = \frac{1}{2}[Ae + Bu + L\tilde{d}]^TPe + \frac{1}{2}e^TP[Ae + Bu + L\tilde{d}] \Rightarrow\end{gathered} \tag{3.238}$$

$$\begin{aligned}\dot{V} = \frac{1}{2}[e^TA^T + u^TB^T + \tilde{d}^TL^T]Pe + \\ + \frac{1}{2}e^TP[Ae + Bu + L\tilde{d}] \Rightarrow\end{aligned} \tag{3.239}$$

$$\begin{array}{c}\dot{V}=\frac{1}{2}e^TA^TPe+\frac{1}{2}u^TB^TPe+\frac{1}{2}\tilde{d}^TL^TPe+\\ \frac{1}{2}e^TPAe+\frac{1}{2}e^TPBu+\frac{1}{2}e^TPL\tilde{d}\end{array} \tag{3.240}$$

The previous equation is rewritten as

$$\begin{array}{c}\dot{V}=\frac{1}{2}e^T(A^TP+PA)e+(\frac{1}{2}u^TB^TPe+\frac{1}{2}e^TPBu)+\\ +(\frac{1}{2}\tilde{d}^TL^TPe+\frac{1}{2}e^TPL\tilde{d})\end{array} \tag{3.241}$$

Assumption: For given positive definite matrix Q and coefficients r and ρ there exists a positive definite matrix P, which is the solution of the following matrix equation

$$A^TP+PA=-Q+P(\frac{2}{r}BB^T-\frac{1}{\rho^2}LL^T)P \tag{3.242}$$

Moreover, the following feedback control law is applied to the system

$$u=-\frac{1}{r}B^TPe \tag{3.243}$$

By substituting Eq. (3.242) and Eq. (3.243), one obtains

$$\begin{array}{c}\dot{V}=\frac{1}{2}e^T[-Q+P(\frac{2}{r}BB^T-\frac{1}{\rho^2}LL^T)P]e+\\ +e^TPB(-\frac{1}{r}B^TPe)+e^TPL\tilde{d}\Rightarrow\end{array} \tag{3.244}$$

$$\begin{array}{c}\dot{V}=-\frac{1}{2}e^TQe+\frac{1}{r}e^TPBB^TPe-\frac{1}{2\rho^2}e^TPLL^TPe\\ -\frac{1}{r}e^TPBB^TPe+e^TPL\tilde{d}\end{array} \tag{3.245}$$

which after intermediate operations gives

$$\dot{V}=-\frac{1}{2}e^TQe-\frac{1}{2\rho^2}e^TPLL^TPe+e^TPL\tilde{d} \tag{3.246}$$

or, equivalently

$$\begin{array}{c}\dot{V}=-\frac{1}{2}e^TQe-\frac{1}{2\rho^2}e^TPLL^TPe+\\ +\frac{1}{2}e^TPL\tilde{d}+\frac{1}{2}\tilde{d}^TL^TPe\end{array} \tag{3.247}$$

Lemma: The following inequality holds

$$\frac{1}{2}e^TL\tilde{d}+\frac{1}{2}\tilde{d}L^TPe-\frac{1}{2\rho^2}e^TPLL^TPe\leq\frac{1}{2}\rho^2\tilde{d}^T\tilde{d} \tag{3.248}$$

Proof: The binomial $(\rho\alpha-\frac{1}{\rho}b)^2$ is considered. Expanding the left part of the above inequality, one gets

$$\begin{array}{c}\rho^2a^2+\frac{1}{\rho^2}b^2-2ab\geq 0\Rightarrow\frac{1}{2}\rho^2a^2+\frac{1}{2\rho^2}b^2-ab\geq 0\Rightarrow\\ ab-\frac{1}{2\rho^2}b^2\leq\frac{1}{2}\rho^2a^2\Rightarrow\frac{1}{2}ab+\frac{1}{2}ab-\frac{1}{2\rho^2}b^2\leq\frac{1}{2}\rho^2a^2\end{array} \tag{3.249}$$

The following substitutions are carried out: $a=\tilde{d}$ and $b=e^TPL$ and the previous relation becomes

$$\frac{1}{2}\tilde{d}^TL^TPe+\frac{1}{2}e^TPL\tilde{d}-\frac{1}{2\rho^2}e^TPLL^TPe\leq\frac{1}{2}\rho^2\tilde{d}^T\tilde{d} \tag{3.250}$$

Eq. (3.250) is substituted in Eq. (3.247) and the inequality is enforced, thus giving

$$\dot{V}\leq-\frac{1}{2}e^TQe+\frac{1}{2}\rho^2\tilde{d}^T\tilde{d} \tag{3.251}$$

Eq. (3.251) shows that the H_∞ tracking performance criterion is satisfied. The integration of $\dot{V}$ from 0 to T gives

$$\begin{array}{c}\int_0^T\dot{V}(t)dt\leq-\frac{1}{2}\int_0^T||e||_Q^2dt+\frac{1}{2}\rho^2\int_0^T||\tilde{d}||^2dt\Rightarrow\\ 2V(T)+\int_0^T||e||_Q^2dt\leq 2V(0)+\rho^2\int_0^T||\tilde{d}||^2dt\end{array} \tag{3.252}$$

Moreover, if there exists a positive constant $M_d>0$ such that

$$\int_0^\infty||\tilde{d}||^2dt\leq M_d \tag{3.253}$$

then, one gets

$$\int_0^\infty ||e||_Q^2 dt \leq 2V(0) + \rho^2 M_d \tag{3.254}$$

Thus, the integral $\int_0^\infty ||e||_Q^2 dt$ is bounded. Moreover, $V(T)$ is bounded and from the definition of the Lyapunov function V in Eq. (3.237) it becomes clear that $e(t)$ will be also bounded since $e(t) \in \Omega_e = \{e|e^TPe \leq 2V(0) + \rho^2 M_d\}$. According to the above and with the use of Barbalat's Lemma, one obtains $\lim_{t\to\infty} e(t) = 0$.

After following the stages of the stability proof, one arrives at Eq. (3.251) which shows that the H-infinity tracking performance criterion holds. By selecting the attenuation coefficient ρ to be sufficiently small and in particular to satisfy $\rho^2 < ||e||_Q^2/||\tilde{d}||^2$, one has that the first derivative of the Lyapunov function is upper bounded by 0. This condition holds at each sampling instance, and consequently, global stability for the control loop can be concluded.

3.2.7 Simulation tests

The global stability properties of the control method and the elimination of the state vector's tracking error which were previously proven through Lyapunov analysis are further confirmed through simulation experiments. To implement the nonlinear optimal (H-infinity) control method in the dynamic model of the dual PMLSM-driven gantry crane, the algebraic Riccati equation of this control method which was previously presented in Eq. (3.242) had to be solved at each iteration of the control algorithm. The obtained results are depicted in Figures 3.21–3.36. The real values of the state vector elements of the dual-PMLSM-driven gantry crane are printed in blue color, the state variables' estimates which have been provided by the H-infinity Kalman filter are plotted in green, while the reference values for each test case are depicted in red. It is confirmed that under the nonlinear optimal control method, one achieves fast and accurate tracking of setpoints under moderate variations of the control inputs. The simulation tests confirm the global stability properties of the control scheme since it is shown that the method does not only enable convergence to fixed points and time-invariant setpoints, but it also ensures precise tracking of time-varying and abruptly changing setpoints. The transient performance of the control method is determined by the selection of the coefficients of the Riccati equation which appears in Eq. (3.242). Thus, for relatively small values of gain r, complete elimination of the state vector's tracking error can be observed. For relatively large values of the elements of the diagonal matrix Q, fast convergence to setpoints is achieved. Finally, the smaller value of the attenuation coefficient ρ for which a valid solution from the

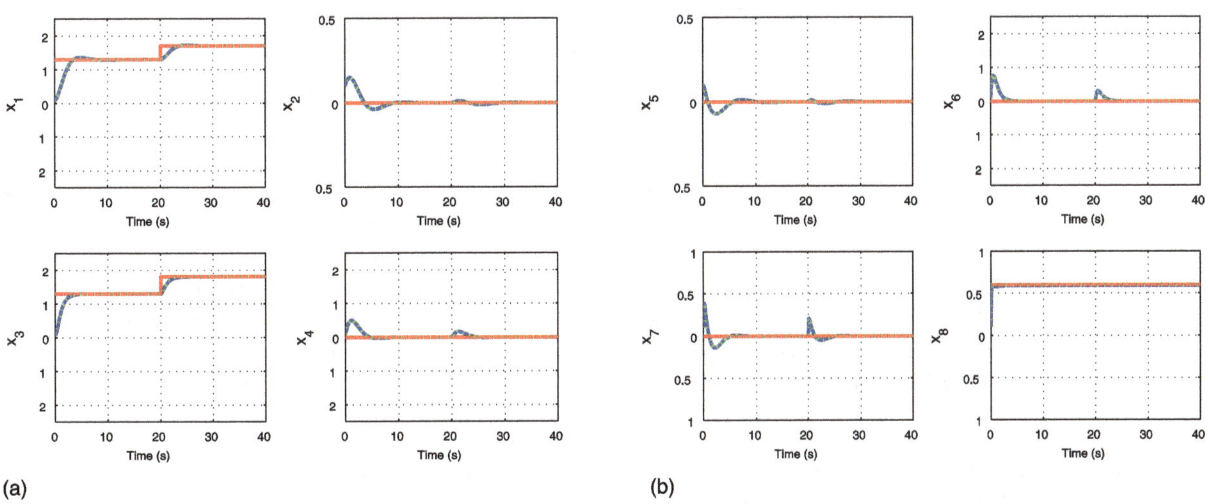

Figure 3.21 Tracking of setpoint 1 by the dual-PMLSM-driven gantry crane with nonlinear optimal control: (a) tracking of setpoints (red line) by the state variables of the gantry crane x_1 to x_4 (blue line) and estimates provided by the H-infinity Kalman filter; (b) tracking of setpoints (red line) by the state variables of the gantry crane x_5 to x_8 (blue line) and estimates provided by the H-infinity Kalman filter

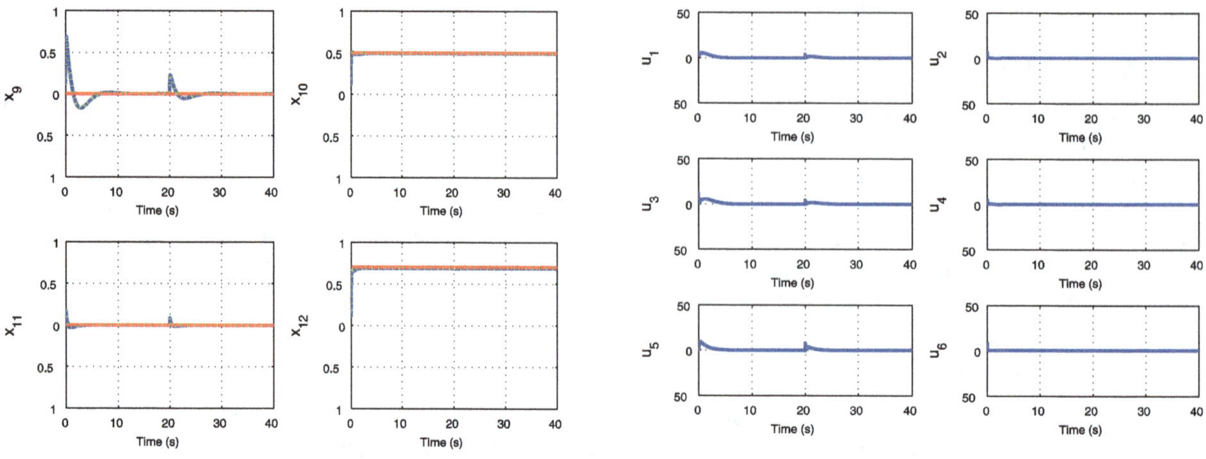

(a) (b)

Figure 3.22 Tracking of setpoint 1 by the dual-PMLSM-driven gantry crane with nonlinear optimal control: (a) tracking of setpoints (red line) by the state variables of the gantry crane x_9 to x_{12} (blue line) and estimates provided by the H-infinity Kalman filter; (b) variations of the control inputs of the gantry crane u_1 to u_6 (blue line)

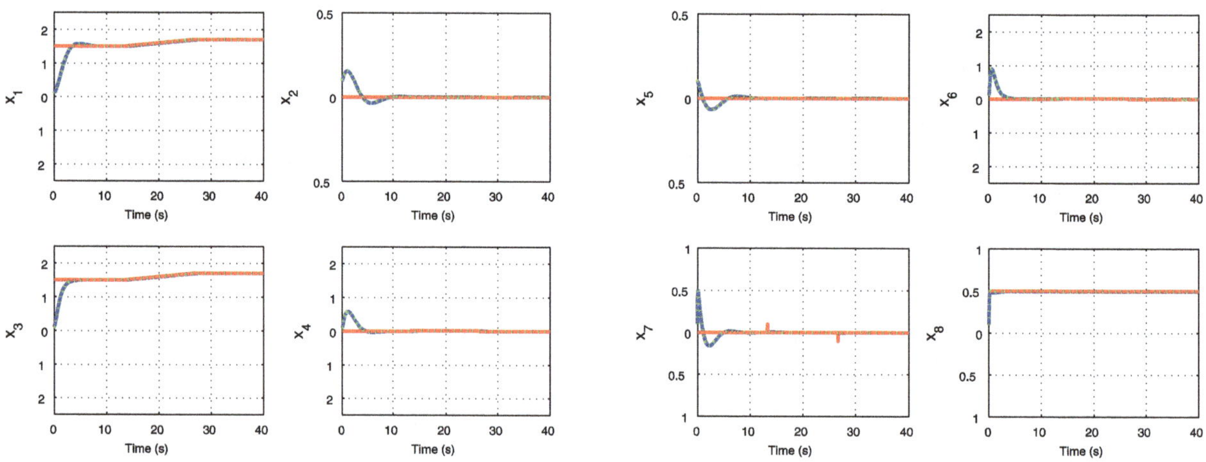

(a) (b)

Figure 3.23 Tracking of setpoint 2 by the dual-PMLSM-driven gantry crane with nonlinear optimal control: (a) tracking of setpoints (red line) by the state variables of the gantry crane x_1 to x_4 (blue line) and estimates provided by the H-infinity Kalman filter; (b) tracking of setpoints (red line) by the state variables of the gantry crane x_5 to x_8 (blue line) and estimates provided by the H-infinity Kalman filter

Riccati equation can be obtained in the form of a positive definite and symmetric matrix P is the one that provides the gantry's control loop with maximum robustness.

To elaborate on nonlinear optimal control for the H-type gantry crane with dual PMLSMs, the following tables are given: (i) Table 3.5 providing results about the accuracy of tracking of setpoints by the state variables of the H-type gantry crane with dual PMLSMs under an exact dynamic model, (ii) Table 3.6 providing results about the accuracy of tracking of setpoints by the state variables of the H-type gantry crane with dual PMLSMs under a model that is subject

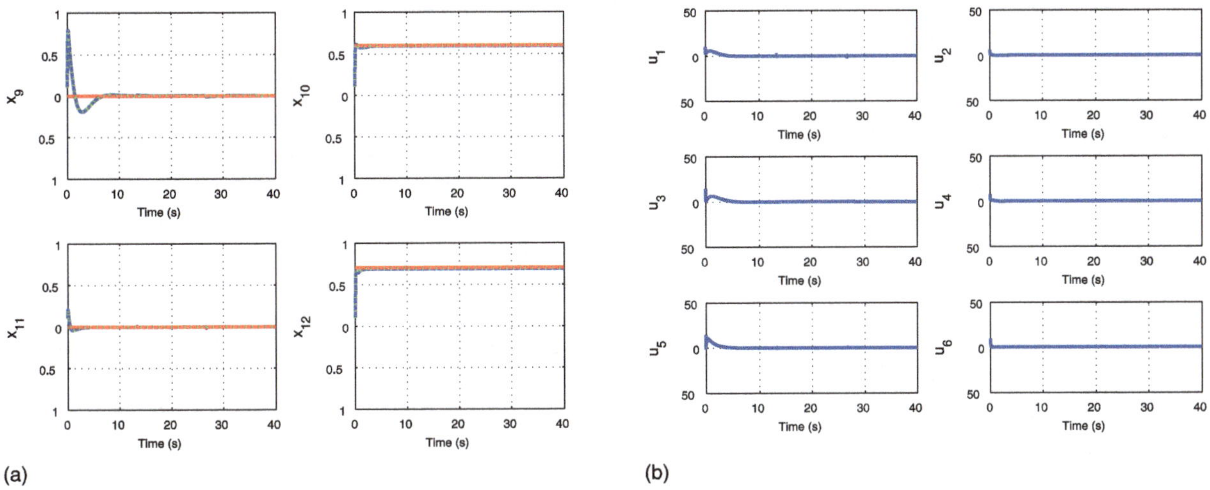

Figure 3.24 Tracking of setpoint 2 by the dual-PMLSM-driven gantry crane with nonlinear optimal control: (a) tracking of setpoints (red line) by the state variables of the gantry crane x_9 to x_{12} (blue line) and estimates provided by the H-infinity Kalman filter; (b) variations of the control inputs of the gantry crane u_1 to u_6 (blue line)

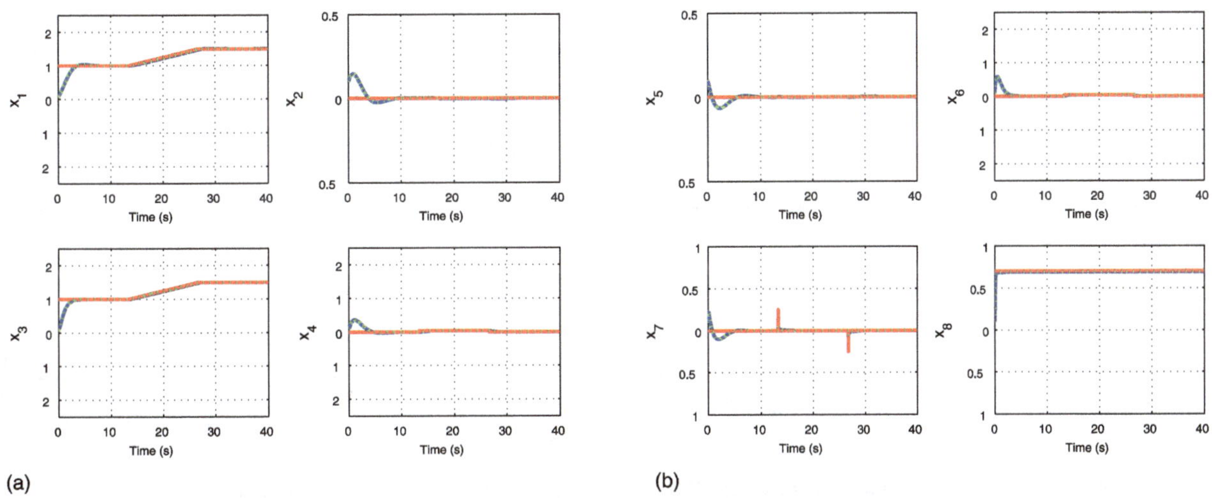

Figure 3.25 Tracking of setpoint 3 by the dual-PMLSM-driven gantry crane with nonlinear optimal control: (a) tracking of setpoints (red line) by the state variables of the gantry crane x_1 to x_4 (blue line) and estimates provided by the H-infinity Kalman filter; (b) tracking of setpoints (red line) by the state variables of the gantry crane x_5 to x_8 (blue line) and estimates provided by the H-infinity Kalman filter

to disturbances (for instance change $\Delta\alpha\%$ in the resistance R_{s1} of the first PMLSM), (iii) Table 3.7 providing results about the accuracy of the state estimation provided by the H-infinity Kalman filter for the state variables of the H-type gantry crane with dual PMLSMs, and (iv) Table 3.8 providing results about the convergence times of the state variables of the H-type gantry crane with dual PMLSMs to the associated setpoints.

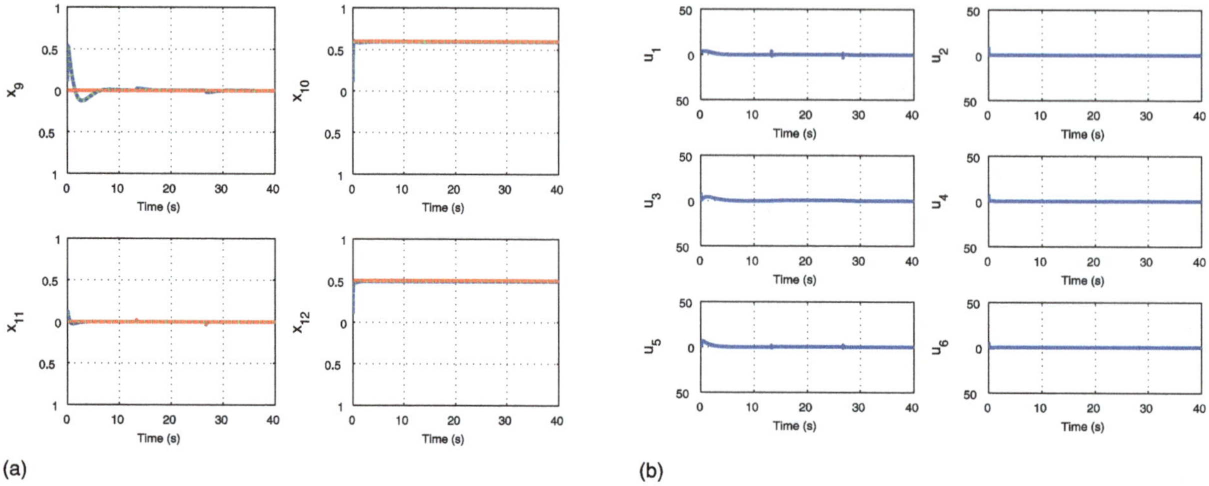

Figure 3.26 *Tracking of setpoint 3 by the dual-PMLSM-driven gantry crane with nonlinear optimal control: (a) tracking of setpoints (red line) by the state variables of the gantry crane x_9 to x_{12} (blue line) and estimates provided by the H-infinity Kalman filter; (b) variations of the control inputs of the gantry crane u_1 to u_6 (blue line)*

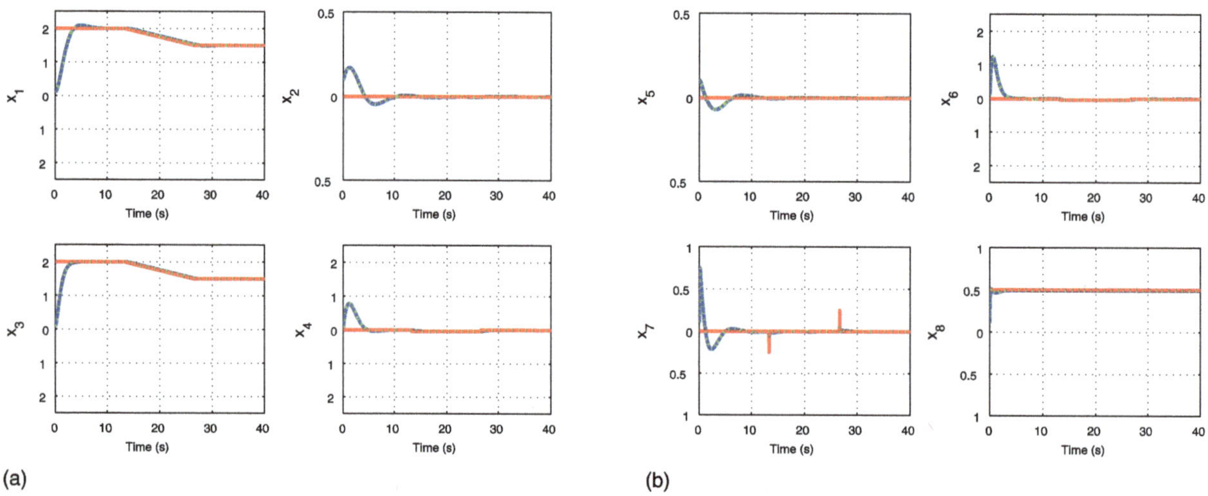

Figure 3.27 *Tracking of setpoint 4 by the dual-PMLSM-driven gantry crane with nonlinear optimal control: (a) tracking of setpoints (red line) by the state variables of the gantry crane x_1 to x_4 (blue line) and estimates provided by the H-infinity Kalman filter; (b) tracking of setpoints (red line) by the state variables of the gantry crane x_5 to x_8 (blue line) and estimates provided by the H-infinity Kalman filter*

3.3 Nonlinear optimal control for a 3-DOF two-cable-driven robotic crane

3.3.1 Outline

Cable-driven cranes are widely used in industry and in applications requiring the crane to operate in a large workspace [617,332,333,1085,351]. Cable-driven robots are light, easy to install and maintain electromechanical systems which are characterized by large payload transfer capability [43,241,305,1092]. However, control of such robots exhibits

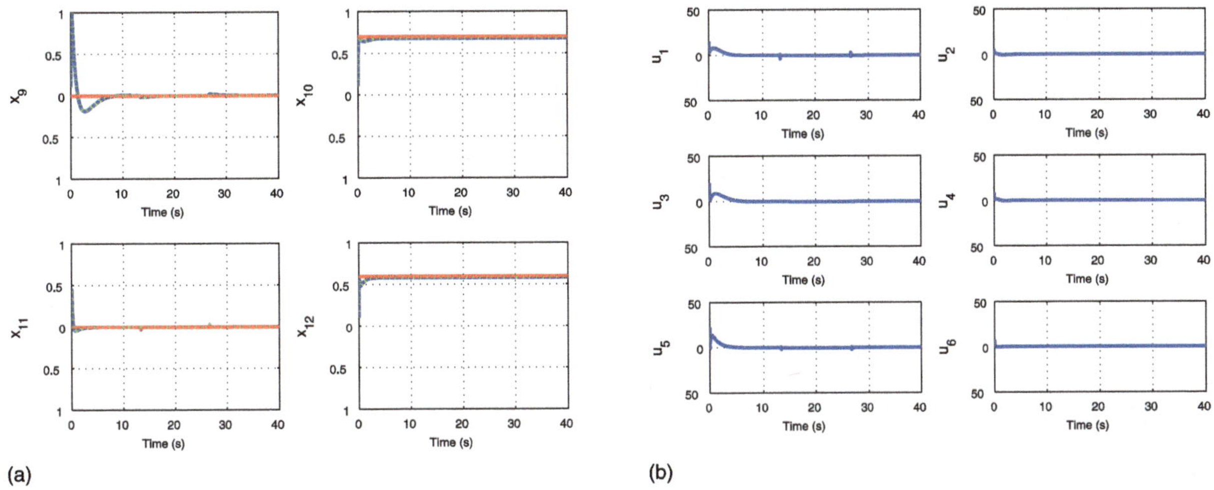

Figure 3.28 Tracking of setpoint 4 by the dual-PMLSM-driven gantry crane with nonlinear optimal control: (a) tracking of setpoints (red line) by the state variables of the gantry crane x_9 to x_{12} (blue line) and estimates provided by the H-infinity Kalman filter; (b) variations of the control inputs of the gantry crane u_1 to u_6 (blue line)

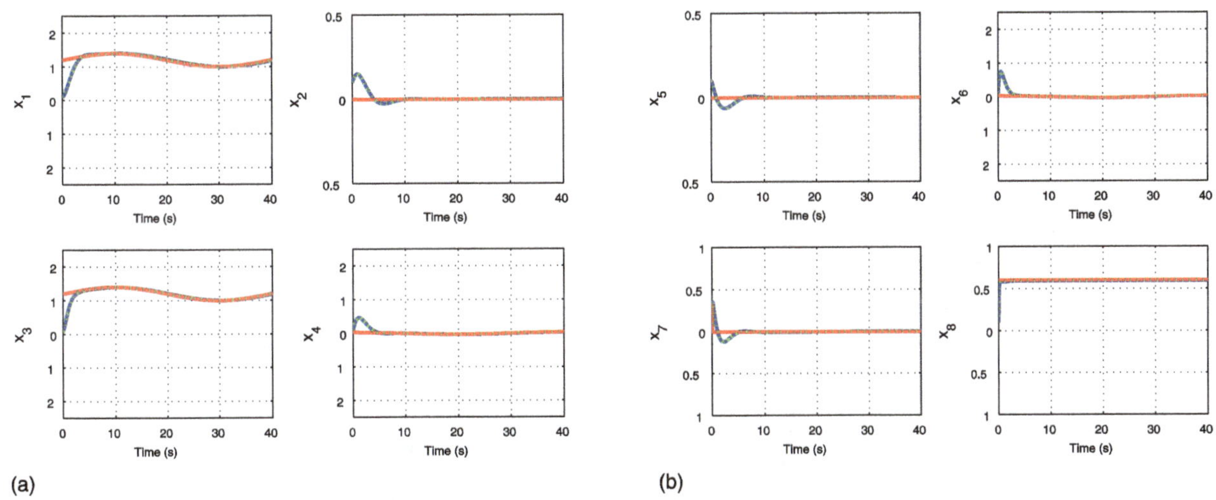

Figure 3.29 Tracking of setpoint 5 by the dual-PMLSM-driven gantry crane with nonlinear optimal control: (a) tracking of setpoints (red line) by the state variables of the gantry crane x_1 to x_4 (blue line) and estimates provided by the H-infinity Kalman filter; (b) tracking of setpoints (red line) by the state variables of the gantry crane x_5 to x_8 (blue line) and estimates provided by the H-infinity Kalman filter

difficulties due to their strongly nonlinear dynamic model and underactuation [649,98,159]. In this section, a specific type of cable-driven robot is examined which is the two-cable-driven 3-DOF robotic crane [100,427]. Such a crane consists of a cart which is suspended in the vertical plane by two cables. Each cable is connected to an electric motor, while by varying the motors' turn speed and torque the length and the tension of the cables are also modified. A payload is attached to the cart with a link of constant length. By changing the motor's torque the position of the cart in the vertical plane changes and the payload is moved to a different position. The cart and payload system performs a

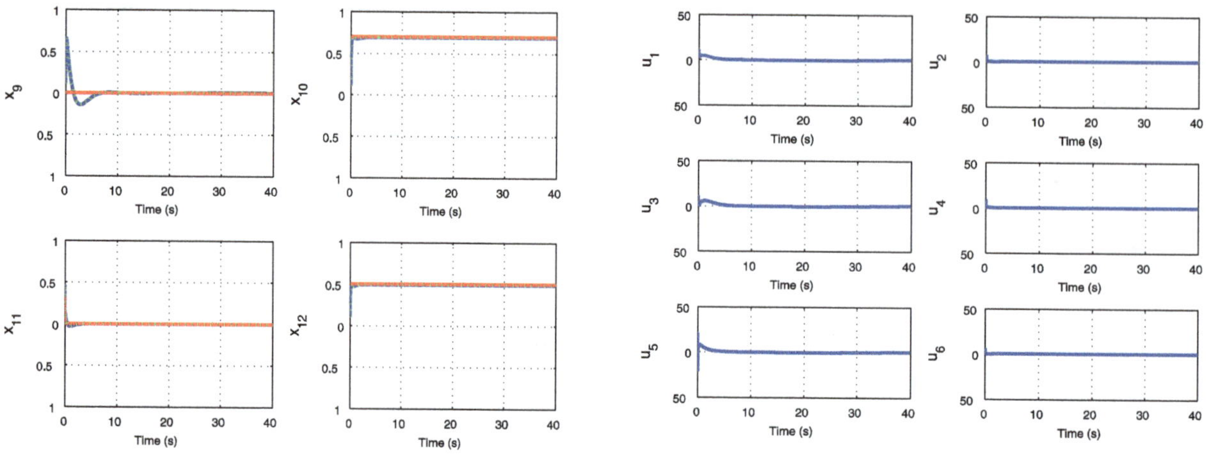

(a) (b)

Figure 3.30 Tracking of setpoint 5 by the dual-PMLSM-driven gantry crane with nonlinear optimal control: (a) tracking of setpoints (red line) by the state variables of the gantry crane x_9 to x_{12} (blue line) and estimates provided by the H-infinity Kalman filter; (b) variations of the control inputs of the gantry crane u_1 to u_6 (blue line)

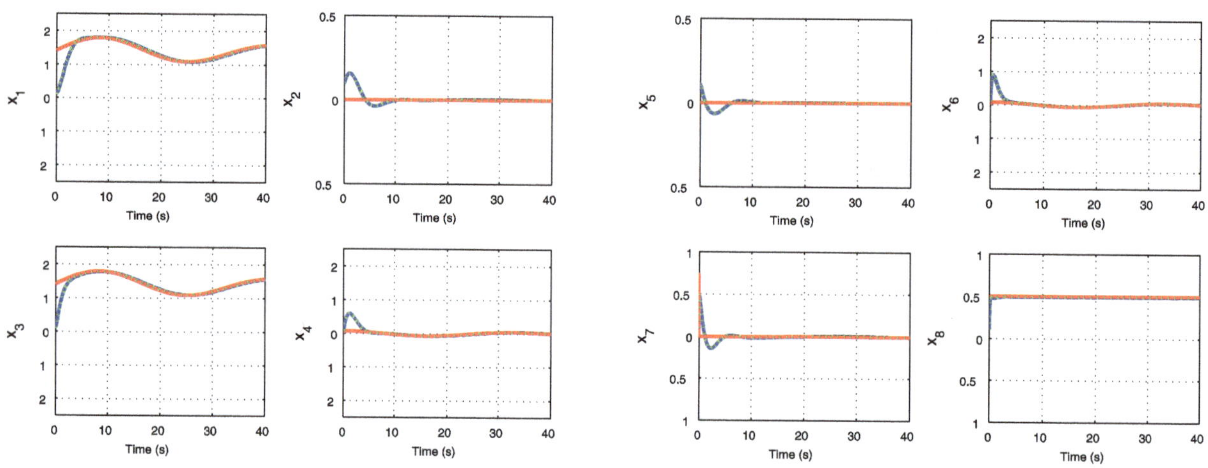

(a) (b)

Figure 3.31 Tracking of setpoint 6 by the dual-PMLSM-driven gantry crane with nonlinear optimal control: (a) tracking of setpoints (red line) by the state variables of the gantry crane x_1 to x_4 (blue line) and estimates provided by the H-infinity Kalman filter; (b) tracking of setpoints (red line) by the state variables of the gantry crane x_5 to x_8 (blue line) and estimates provided by the H-infinity Kalman filter

3-DOF motion which is defined by the cartesian coordinates of the cart and by the swing angle of the payload. So far, there have been several attempts to solve the nonlinear control problem for the two-cable-driven 3-DOF robotic crane [334,750,748,589]. One can distinguish for instance methods of global linearization-based control [101,426]. There exist also results on model predictive control of these robotic cranes [591,422]. One can note also several Lyapunov theory-based and energy-based control schemes [330,749,331,588].

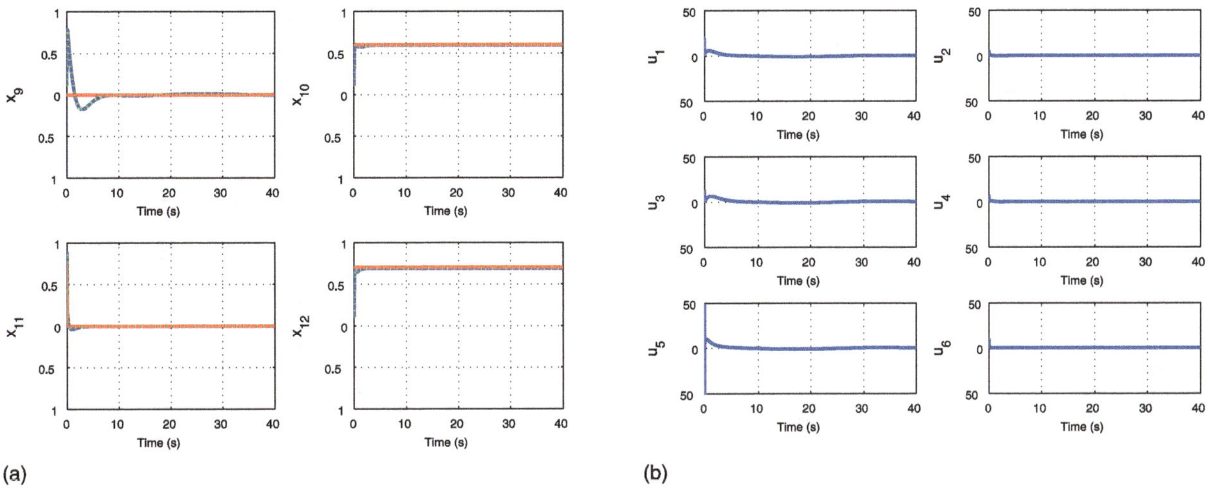

Figure 3.32 *Tracking of setpoint 6 by the dual-PMLSM-driven gantry crane with nonlinear optimal control: (a) tracking of setpoints (red line) by the state variables of the gantry crane x_9 to x_{12} (blue line) and estimates provided by the H-infinity Kalman filter; (b) variations of the control inputs of the gantry crane u_1 to u_6 (blue line)*

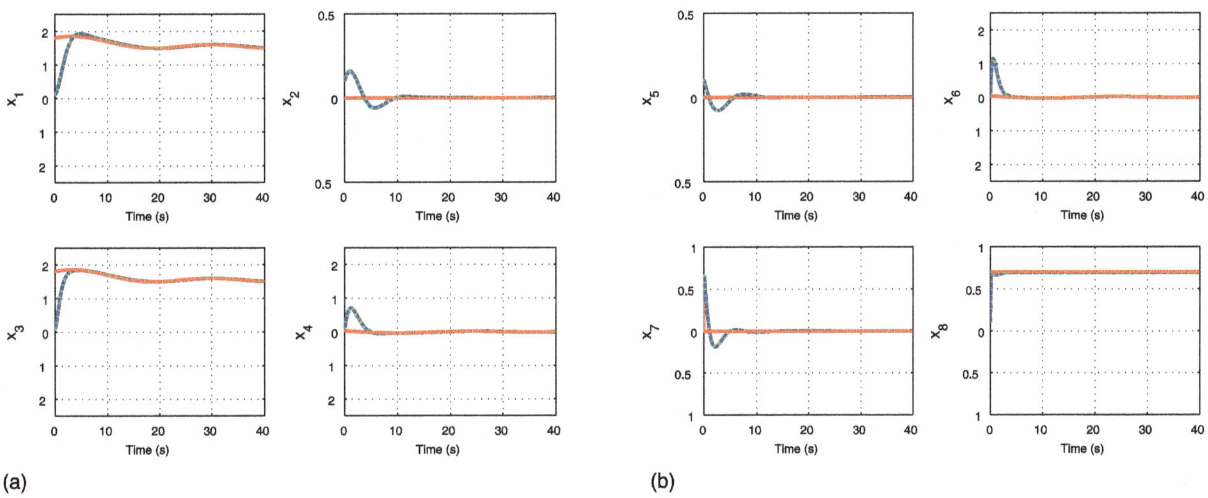

Figure 3.33 *Tracking of setpoint 7 by the dual-PMLSM-driven gantry crane with nonlinear optimal control: (a) tracking of setpoints (red line) by the state variables of the gantry crane x_1 to x_4 (blue line) and estimates provided by the H-infinity Kalman filter; (b) tracking of setpoints (red line) by the state variables of the gantry crane x_5 to x_8 (blue line) and estimates provided by the H-infinity Kalman filter*

In this section, a novel nonlinear optimal control method is proposed for the dynamic model of two-cable-driven 3-DOF robotic cranes [721,732,725,741]. To implement this method, the dynamic model of the robotic crane is formulated first using Euler-Lagrange analysis [721]. Differential flatness properties are also proven for this dynamic system [711]. Next, the model of the robotic crane undergoes approximate linearization with the use of first-order Taylor series expansion and through the computation of the associated Jacobian matrices [705,60,706]. The linearization takes place at each sampling instance around the temporary operating point (x^*, u^*) where x^* is the present value of

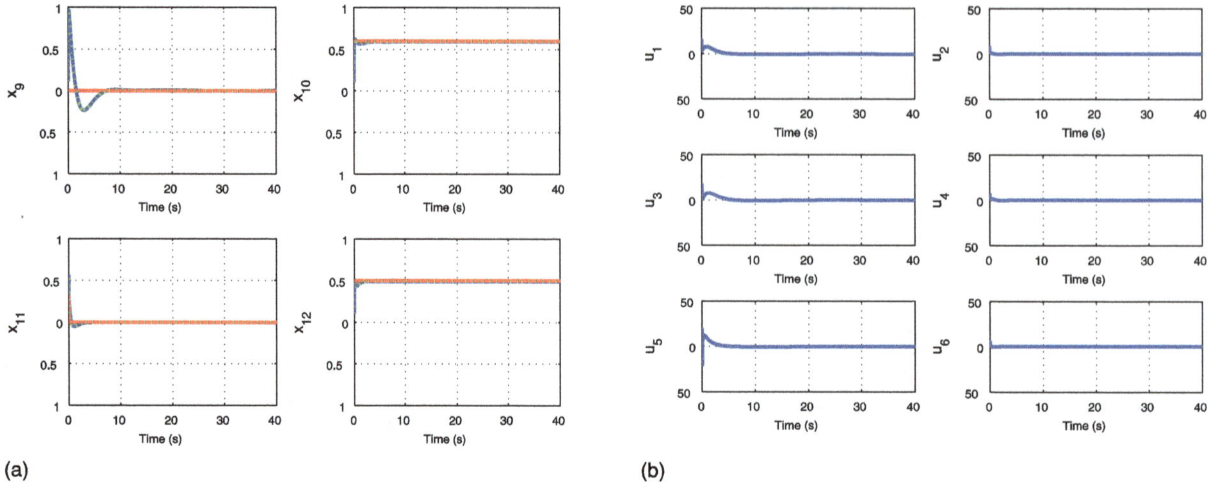

Figure 3.34 Tracking of setpoint 7 by the dual-PMLSM-driven gantry crane with nonlinear optimal control: (a) tracking of setpoints (red line) by the state variables of the gantry crane x_9 to x_{12} (blue line) and estimates provided by the H-infinity Kalman filter; (b) variations of the control inputs of the gantry crane u_1 to u_6 (blue line)

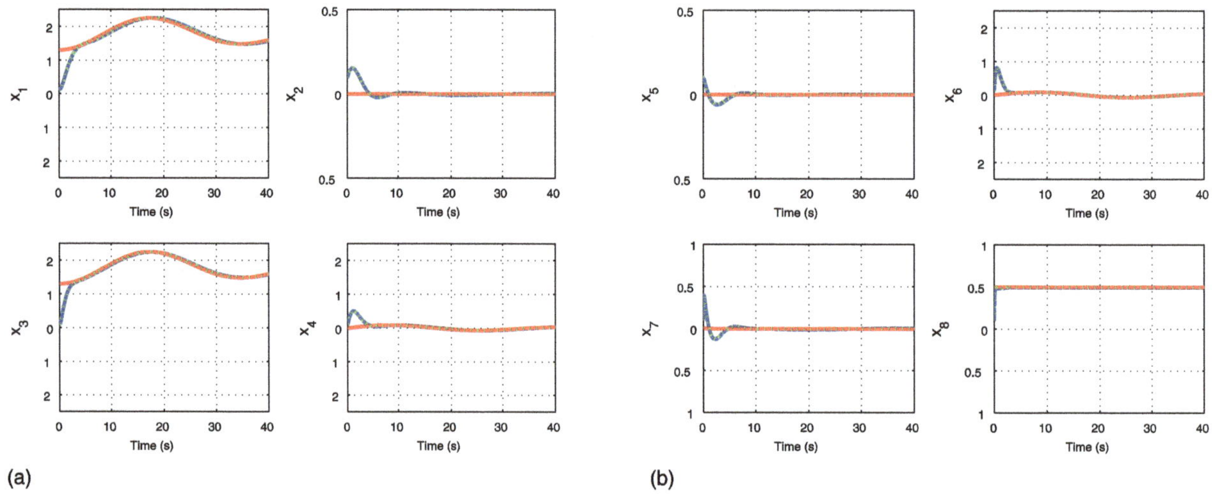

Figure 3.35 Tracking of setpoint 8 by the dual-PMLSM-driven gantry crane with nonlinear optimal control: (a) tracking of setpoints (red line) by the state variables of the gantry crane x_1 to x_4 (blue line) and estimates provided by the H-infinity Kalman filter; (b) tracking of setpoints (red line) by the state variables of the gantry crane x_5 to x_8 (blue line) and estimates provided by the H-infinity Kalman filter

the system's state vector and u^* is the last sampled value of the control inputs vector. The modeling error which is due to the truncation of higher-order terms from the Taylor series is considered to be a perturbation which is asymptotically compensated by the robustness of the control algorithm. For the approximately linearized model of the system, a stabilizing H-infinity feedback controller is designed. This controller represents a min–max differential game taking place between: (i) the control inputs which try to minimize a quadratic term of the state vector's tracking error, (ii) the model uncertainty and external perturbation terms which try to maximize this cost function.

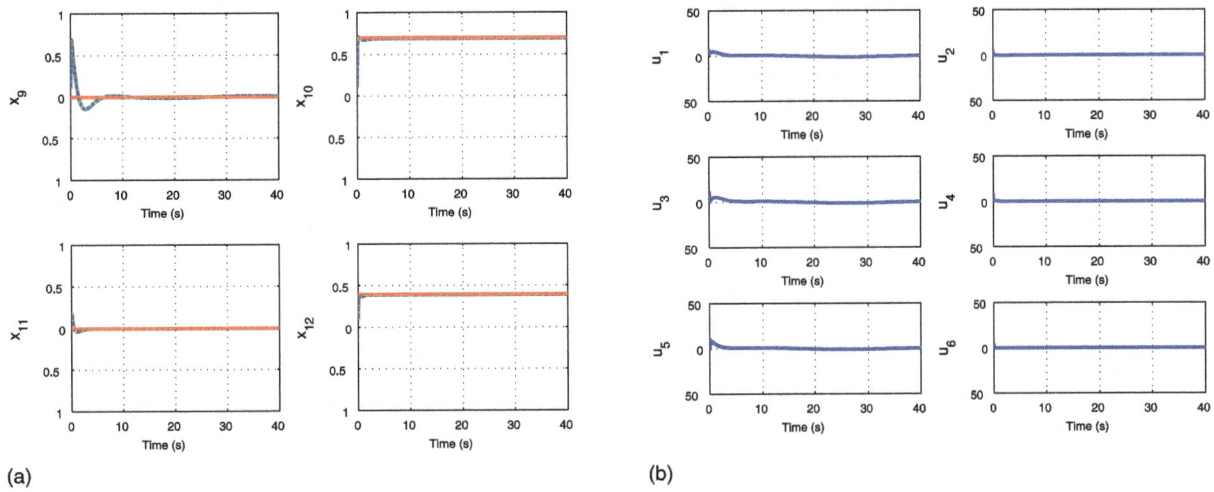

Figure 3.36 *Tracking of setpoint 8 by the dual-PMLSM-driven gantry crane with nonlinear optimal control: (a) tracking of setpoints (red line) by the state variables of the gantry crane x_9 to x_{12} (blue line) and estimates provided by the H-infinity Kalman filter; (b) variations of the control inputs of the gantry crane u_1 to u_6 (blue line)*

Table 3.5 *H-type gantry cranes with dual PMLSMs*

	Tracking RMSE $\times 10^{-3}$ in the disturbance-free case								
	$RMSE_{x_1}$	$RMSE_{x_2}$	$RMSE_{x_3}$	$RMSE_{x_4}$	$RMSE_{x_5}$	$RMSE_{x_6}$	$RMSE_{x_7}$	$RMSE_{x_9}$	$RMSE_{x_{11}}$
$test_1$	0.0001	0.0012	0.0130	0.0002	0.0013	0.0059	0.0009	0.0015	0.0001
$test_2$	0.0002	0.0015	0.0080	0.0003	0.0006	0.0075	0.0011	0.0013	0.0012
$test_3$	0.0005	0.0026	0.0088	0.0008	0.0011	0.0097	0.0014	0.0021	0.0016
$test_4$	0.0002	0.0016	0.0062	0.0004	0.0006	0.0053	0.0012	0.0023	0.0010
$test_5$	0.3932	0.0135	0.2962	0.0385	0.0053	0.0030	0.1329	0.1857	0.2456
$test_6$	0.3042	0.0876	0.5376	0.2107	0.0106	0.1009	0.3705	0.5349	0.3604
$test_7$	0.8039	0.0694	0.8889	0.0911	0.0111	0.0307	0.3045	0.2597	0.2443
$test_8$	0.4846	0.0319	0.2909	0.1519	0.0111	0.0957	0.0629	0.0744	0.1825

Table 3.6 *H-type gantry cranes with dual PMLSMs*

	Tracking RMSE $\times 10^{-3}$ in the case of disturbances								
$\Delta a\%$	$RMSE_{x_1}$	$RMSE_{x_2}$	$RMSE_{x_3}$	$RMSE_{x_4}$	$RMSE_{x_5}$	$RMSE_{x_6}$	$RMSE_{x_7}$	$RMSE_{x_9}$	$RMSE_{x_{11}}$
0%	0.0005	0.0026	0.0088	0.0008	0.0011	0.0097	0.0014	0.0021	0.0016
10%	0.0005	0.0029	0.0083	0.0004	0.0011	0.0080	0.0014	0.0019	0.0013
20%	0.0005	0.0026	0.0169	0.0004	0.0007	0.0080	0.0009	0.0016	0.0011
30%	0.0005	0.0028	0.0072	0.0004	0.0010	0.0089	0.0011	0.0016	0.0010
40%	0.0006	0.0026	0.0059	0.0004	0.0010	0.0087	0.0010	0.0019	0.0014
50%	0.0006	0.0028	0.0088	0.0002	0.0010	0.0008	0.0012	0.0016	0.0012
60%	0.0006	0.0030	0.0133	0.0008	0.0011	0.0104	0.0014	0.0019	0.0012

Table 3.7 H-type gantry cranes with dual PMLSMs

	Tracking RMSE $\times 10^{-4}$ of the estimates provided by the HKF								
	$RMSE_{x_1}$	$RMSE_{x_2}$	$RMSE_{x_3}$	$RMSE_{x_4}$	$RMSE_{x_5}$	$RMSE_{x_6}$	$RMSE_{x_7}$	$RMSE_{x_9}$	$RMSE_{x_{11}}$
$test_1$	0.0443	0.0261	0.1285	0.0097	0.0070	0.1985	0.0932	0.0806	0.0376
$test_2$	0.0343	0.0255	0.1451	0.0088	0.0149	0.1476	0.0902	0.1912	0.0224
$test_3$	0.0230	0.0253	0.1293	0.0119	0.0078	0.1478	0.1348	0.0741	0.0258
$test_4$	0.0281	0.0441	0.0587	0.0066	0.0088	0.1376	0.0645	0.0975	0.0287
$test_5$	0.0235	0.0282	0.0957	0.0154	0.0085	0.1977	0.1590	0.1034	0.0332
$test_6$	0.0337	0.0306	0.1151	0.0112	0.0103	0.1887	0.1072	0.0968	0.0280
$test_7$	0.0517	0.0204	0.2302	0.0100	0.0066	0.3189	0.1167	0.0833	0.0263
$test_8$	0.0275	0.0341	0.1205	0.0075	0.0114	0.1976	0.0757	0.1388	0.0326

Table 3.8 H-type gantry cranes with dual PMLSMs

	Convergence time (sec) of states variables								
No test	$T_s\ x_1$	$T_s\ x_2$	$T_s\ x_3$	$T_s\ x_4$	$T_s\ x_5$	$T_s\ x_6$	$T_s\ x_7$	$T_s\ x_9$	$T_s\ x_{11}$
$test_1$	6.0	8.0	3.0	3.0	8.0	3.0	5.0	8.0	0.5
$test_2$	6.0	8.0	3.0	3.0	8.0	3.0	8.0	8.0	0.5
$test_3$	6.0	8.0	3.0	5.0	8.0	3.0	5.0	6.0	0.5
$test_4$	6.0	8.0	3.0	5.0	9.0	3.0	5.0	6.0	0.5
$test_5$	4.0	8.0	4.0	5.0	8.0	4.0	6.0	6.0	0.5
$test_6$	4.0	8.0	4.0	5.0	8.0	4.0	6.0	8.0	0.5
$test_7$	5.0	9.0	4.0	4.0	9.0	4.0	8.0	8.0	0.5
$test_8$	4.0	8.0	3.0	4.0	8.0	4.0	8.0	8.0	0.5

To select the gains of the H-infinity controller, an algebraic Riccati equation has to be repetitively solved at each time-step of the control algorithm. The global stability properties of the control scheme are proven through Lyapunov analysis. First, it is demonstrated that the control loop satisfies the H-infinity tracking performance criterion, which signifies elevated robustness against model uncertainties and external perturbations [725,876]. Moreover, under moderate conditions, it is proven that the control loop is globally asymptotically stable. To implement state estimation-based control without the need to measure the entire state vector of the controlled system, the H-infinity Kalman filter is used as a robust state estimator [721,725]. The method achieves fast and accurate tracking of reference setpoints by the state variables of the robotic crane. Besides, it achieves all control objectives for this robotic crane without the need for state-space model transformations and without the need for change of state variables.

3.3.2 Dynamic model of the two-cable-driven 3-DOF robotic crane

3.3.2.1 State-space model of the robotic crane

The diagram of the two-cable-driven 3-DOF robotic crane is given in Figure 3.37. The robotic crane consists of a cart which is moved on a 2D xy-plane with the use of two strings. The winding or loosening of the cables is performed by two motors which also vary the tension of the cables. The position of the cart in the xy-plane is denoted as (x_R, y_R). The position of the left motor is denoted as (x_a, y_a), while the position of the right motor is described as (x_b, y_b). The mass of the cart is M. The length of the cable that connects the cart with the left motor is denoted as L_1, while the length of the cable which connects the cart with the right motor is denoted as L_2. Besides, a load of mass m is suspended from

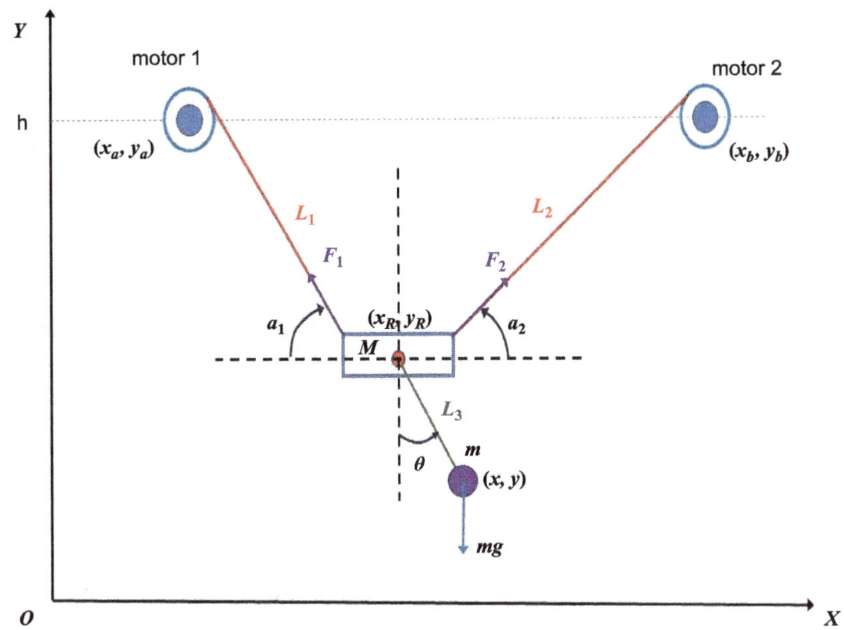

Figure 3.37 Diagram of the two cable-driven 3-DOF robotic crane

the cart using an unactuated link of length L_3. The angle which is formed between the OX horizontal axis and the cable that connects the cart to the left motor is denoted as a_1. The angle which is formed between the OX horizontal axis and the cable that connects the cart to the right motor is denoted as a_2.

The moment of inertia of the left motor is I_a while the moment of inertia of the right motor is I_b. Both motors are placed at a height along the OY axis which is equal to h. The cables are considered to be massless and inelastic. The link which connects the load of mass m to the cart is not subject to deformation either. The parameters which describe the motion of the crane constitute the parameters' vector $[x_R, y_R, \theta]^T$, while the control objective is to make this vector converge fast to the desirable value $[x_{R,d}, y_{R,d}, \theta_d]^T$ under minimal dispersion of energy by the two motors.

Using trigonometric relations the following geometric constraints are given:

$$\begin{array}{l}(x_a - x_R)^2 + (y_a - y_R)^2 = L_1^2 \\ (x_b - x_R)^2 + (y_b - y_R)^2 = L_2^2\end{array} \tag{3.255}$$

To perform Euler-Lagrange analysis for the robotic crane, its kinetic and dynamic energy are computed first. The position of the cart is (x_R, y_R); thus, the associated velocity is $(\dot{x}_R, \dot{y}_R)$. The position of the payload is $(x_R + L_3 sin(\theta), y_R - L_3 cos(\theta))$ and the associated velocity is $(\dot{x}_R + L_3 cos(\theta)\dot{\theta}, \dot{y}_R + L_3 sin(\theta)\dot{\theta})$.

The dynamic model of the system will be defined using Euler-Lagrange analysis [721]. The kinetic energy of the cart is

$$K_c = \tfrac{1}{2}M(\dot{x}_R^2 + \dot{y}_R^2) \tag{3.256}$$

The kinetic energy of the load is

$$K_L = \tfrac{1}{2}m[((\dot{x}_R + L_3 cos(\theta)\dot{\theta})^2 + (\dot{y}_R + L_3 sin(\theta)\dot{\theta}))^2] \tag{3.257}$$

The aggregate kinetic energy of the crane is

$$\begin{array}{c}K = K_c + K_L \Rightarrow K = \tfrac{1}{2}M(\dot{x}_R^2 + \dot{y}_R^2) + \\ + \tfrac{1}{2}m[(\dot{x}_R + L_3 cos(\theta)\dot{\theta})^2 + (\dot{y}_R + L_3 sin(\theta)\dot{\theta})^2]\end{array} \tag{3.258}$$

which after intermediate operations and regrouping of terms gives

$$\begin{aligned}K = \tfrac{1}{2}(M+m)\dot{x}_R^2 + \tfrac{1}{2}(M+m)\dot{y}_R^2+ \\ +m\dot{x}_R L_3 cos(\theta)\dot{\theta} + m\dot{y}_R L_3 sin(\theta)\dot{\theta} + \tfrac{1}{2}mL_3^2\dot{\theta}^2\end{aligned} \quad (3.259)$$

By defining vector $q = [x_R, y_R, \theta]^T$ and vector $\dot{q} = [\dot{x}_R, \dot{y}_R, \dot{\theta}]^T$, the kinetic energy of the crane can be written in matrix form as

$$K = \tfrac{1}{2}\dot{q}\tilde{M}q^T \Rightarrow K = \tfrac{1}{2}\left(\dot{x}_R, \dot{y}_R, \dot{\theta}\right)\begin{pmatrix}(M+m) & 0 & mL_3cos(\theta)\\ 0 & (M+m) & mL_3sin(\theta)\\ mL_3cos(\theta) & mL_3sin(\theta) & mL_3^2\end{pmatrix}\begin{pmatrix}\dot{x}_R\\ \dot{y}_R\\ \dot{\theta}\end{pmatrix} \quad (3.260)$$

Next, the potential energy of the crane, consisting of the potential energy of the cart and the potential energy of the payload, is found to be

$$P = Mgy_R + mg(y_R - L_3cos(\theta)) \quad (3.261)$$

The Lagrangian of the crane is given by

$$\begin{aligned}L = K - P \Rightarrow L = \tfrac{1}{2}(M+m)\dot{x}_R^2 + \tfrac{1}{2}(M+m)\dot{y}_R^2+ \\ +m\dot{x}_R L_3cos(\theta)\dot{\theta} + m\dot{y}_R L_3sin(\theta)\dot{\theta} + \tfrac{1}{2}mL_3^2\dot{\theta}^2- \\ -Mgy_R - mg(y_R - L_3cos(\theta))\end{aligned} \quad (3.262)$$

The forces which are exerted on the robotic crane are the tensions of the two strings F_1 and F_2, while there is no torque. By applying Euler-Lagrange analysis, one obtains:

$$\frac{\partial}{\partial t}\frac{\partial L}{\partial \dot{q}_1} - \frac{\partial L}{\partial q_1} = F_1 \quad (3.263)$$

$$\frac{\partial}{\partial t}\frac{\partial L}{\partial \dot{q}_2} - \frac{\partial L}{\partial q_2} = F_2 \quad (3.264)$$

$$\frac{\partial}{\partial t}\frac{\partial L}{\partial \dot{q}_3} - \frac{\partial L}{\partial q_3} = 0 \quad (3.265)$$

From Eq. (3.263) and after computing the related partial derivatives of the Lagrangian, one gets:

$$(M+m)\ddot{x}_R - mL_3sin(\theta)\dot{\theta}^2 + mL_3cos(\theta)\ddot{\theta} = F_1 \quad (3.266)$$

$$(M+m)\ddot{y}_R + mL_3cos(\theta)\dot{\theta}^2 + mL_3sin(\theta)\ddot{\theta} - (M-m)g = F_2 \quad (3.267)$$

$$\begin{aligned}m\ddot{x}_R L_3cos(\theta) + m\ddot{y}_R L_3sin(\theta) + mL_3^2\ddot{\theta} + mgL_3sin(\theta)- \\ -m\dot{x}_R L_3sin(\theta)\dot{\theta} + m\dot{y}_R L_3cos(\theta)\dot{\theta} = 0\end{aligned} \quad (3.268)$$

In Eq. (3.268), the terms which appear in the last row are zero, that is, it holds $-m\dot{x}_R L_3sin(\theta)\dot{\theta} + m\dot{y}_R L_3cos(\theta)\dot{\theta} = 0$ or $mL_3\dot{\theta}[-\dot{x}_Rsin(\theta) + \dot{y}_Rcos(\theta)] = 0$. Actually, it holds that $-\dot{x}_Rsin(\theta) + \dot{y}_Rcos(\theta) = 0$. This is obtained after considering a body-fixed frame for expressing the velocity of the load and knowing that the velocity of the load along the vertical axis of this frame is 0. Indeed, the transformation of the velocities vector of the load from the inertial reference frame to the body-fixed frame gives

$$\begin{pmatrix}v_1\\ 0\end{pmatrix} = \begin{pmatrix}cos(\theta) & sin(\theta)\\ -\sin(\theta) & cos(\theta)\end{pmatrix}\begin{pmatrix}\dot{x}_R + L_3cos(\theta)\dot{\theta}\\ \dot{y}_R + L_3sin(\theta)\dot{\theta}\end{pmatrix} \quad (3.269)$$

out of which one obtains

$$\begin{aligned}-sin(\theta)[\dot{x}_R + L_3cos(\theta)\dot{\theta}] + cos(\theta)[\dot{y}_R + L_3sin(\theta)\dot{\theta}] = 0 \\ \Rightarrow -\dot{x}_Rsin(\theta) + \dot{y}_Rcos(\theta) = 0\end{aligned} \quad (3.270)$$

Consequently, from Euler-Lagrange analysis, the equations of the system's dynamics become

$$(M+m)\ddot{x}_R + 0\ddot{y}_R - mL_3sin(\theta)\dot{\theta}^2 + mL_3cos(\theta)\ddot{\theta} = F_1 \quad (3.271)$$

$$0\ddot{x}_R + (M+m)\ddot{y}_R + mL_3cos(\theta)\dot{\theta}^2 + mL_3sin(\theta)\ddot{\theta} - (M-m)g = F_2 \quad (3.272)$$

$$m\ddot{x}_R L_3 cos(\theta) + m\ddot{y}_R L_3 sin(\theta) + mL_3^2\ddot{\theta} + mgL_3 sin(\theta) = 0 \tag{3.273}$$

In matrix form, the dynamic model of the two-cable-driven 3-DOF robotic crane is written as

$$\begin{pmatrix} (M+m) & 0 & (mL_3cos(\theta)) \\ 0 & (M+m) & (mL_3sin(\theta)) \\ (mL_3cos(\theta)) & (mL_3sin(\theta)) & mL_3^2 \end{pmatrix}\begin{pmatrix}\ddot{x}\\ \ddot{y}\\ \ddot{\theta}\end{pmatrix} + \begin{pmatrix} -(mL_3sin(\theta))\dot{\theta}^2 \\ (mL_3cos(\theta))\dot{\theta}^2 \\ 0\end{pmatrix} + \begin{pmatrix}0\\0\\ mgL_3cos(\theta)\end{pmatrix} = \begin{pmatrix}F_1\\F_2\\0\end{pmatrix} \tag{3.274}$$

Using that $q = [x_R, y_R, \theta]^T$, this is a dynamic model of a robot with an inertia matrix $M(q)\in R^{3\times 3}$, a Coriolis forces vector $C(q,\dot{q})\in R^{3\times 1}$, gravitational forces vector $G(\theta)\in R^{3\times 1}$, and external (control input) forces/torques vector $\tau\in R^{3\times 3}$. This dynamic model can be written in the concise form:

$$M(q)\ddot{q} + C(q,\dot{q}) + G(q) = \tau \tag{3.275}$$

The inverse of the inertia matrix $M(q)$ is computed as follows:

$$M(q)^{-1} = \frac{1}{detM}\begin{pmatrix} M_{11} & -M_{12} & M_{13} \\ -M_{21} & M_{22} & -M_{23} \\ M_{31} & -M_{32} & M_{33}\end{pmatrix} \tag{3.276}$$

where the sub-determinants of $M(q)$ are: $M_{11} = (M+m)mL_3^2 - (ML_3sin(\theta))^2$, $M_{12} = -(mL_3)^2sin(\theta)cos(\theta)$, $M_{13} = -(M+m)(mL_3cos(\theta))$, $M_{21} = -(mL_3)^2sin(\theta)cos(\theta)$, $M_{22} = (M+m)mL_3^2 - (mL_3cos(\theta)^2)$, $M_{23} = (M+m)(mL_3 cos(\theta))$, $M_{31} = -(M+m)(mL_3cos(\theta))$, $M_{32} = (M+m)(mL_3sin(\theta))$, and $M_{33} = (M+m)^2$ while the determinant of $M(q)$ is $detM = (M+m)M_{11} + (mL_3cos(\theta))M_{13}$.

The Coriolis forces vector is written as $C = [C_1, C_2, 0]^T$ and the gravitational forces vector is written as $G[0, 0, G_3]^T$. Thus, after some intermediate operations, the dynamic model of the robotic crane is written as

$$\begin{pmatrix}\ddot{x}_R\\ \ddot{y}_R\\ \ddot{\theta}\end{pmatrix} = -\frac{1}{detM}\begin{pmatrix} M_{11} & -M_{21} & M_{31} \\ -M_{12} & M_{22} & -M_{32} \\ M_{13} & -M_{23} & M_{33}\end{pmatrix}\begin{pmatrix}C_1\\C_2\\G_3\end{pmatrix} + \frac{1}{detM}\begin{pmatrix} M_{11} & -M_{21} & M_{31} \\ -M_{12} & M_{22} & -M_{32} \\ M_{13} & -M_{23} & M_{33}\end{pmatrix}\begin{pmatrix}F_1\\F_2\\0\end{pmatrix} \tag{3.277}$$

which finally gives

$$\begin{pmatrix}\ddot{x}_R\\ \ddot{y}_R\\ \ddot{\theta}\end{pmatrix} = \begin{pmatrix}\frac{-M_{11}C_1+M_{21}C_2-M_{31}G_3}{detM}\\ \frac{M_{12}C_1-M_{22}C_2+M_{32}G_3}{detM}\\ \frac{-M_{13}C_1+M_{23}C_2-M_{33}G_3}{detM}\end{pmatrix} + \frac{1}{detM}\begin{pmatrix}\frac{M_{11}}{detM} & -\frac{M_{21}}{detM}\\ -\frac{M_{12}}{detM} & \frac{M_{22}}{detM}\\ \frac{M_{13}}{detM} & -\frac{M_{23}}{detM}\end{pmatrix}\begin{pmatrix}F_1\\F_2\end{pmatrix} \tag{3.278}$$

Next, the state vector of the robotic crane is defined as $x = [x_1, x_2, x_3, x_4, x_5, x_6]^T$ or $x = [x_R, \dot{x}_R, y_R, \dot{y}_R, \theta, \dot{\theta}]^T$, and the control inputs vector is defined as $u = [u_1, u_2]^T$ or $u = [F_1, F_2]^T$. Thus, the dynamic model of the crane is written as

$$\begin{pmatrix}\dot{x}_1\\ \dot{x}_2\\ \dot{x}_3\\ \dot{x}_4\\ \dot{x}_5\\ \dot{x}_6\end{pmatrix} = \begin{pmatrix} x_2 \\ \frac{-M_{11}C_1+M_{21}C_2-M_{31}G_3}{detM} \\ x_4 \\ \frac{M_{12}C_1-M_{22}C_2+M_{32}G_3}{detM} \\ x_6 \\ \frac{-M_{13}C_1+M_{23}C_2-M_{33}G_3}{detM}\end{pmatrix} + \begin{pmatrix} 0 & 0 \\ \frac{M_{11}}{detM} & -\frac{M_{21}}{detM} \\ 0 & 0 \\ -\frac{M_{12}}{detM} & \frac{M_{22}}{detM} \\ 0 & 0 \\ \frac{M_{13}}{detM} & -\frac{M_{23}}{detM}\end{pmatrix}\begin{pmatrix}u_1\\u_2\end{pmatrix} \tag{3.279}$$

or equivalently the crane's dynamics can be written in the nonlinear affine-in-the-input state-space form:

$$\dot{x} = f(x) + g(x)u \tag{3.280}$$

$x\in R^{6\times 1}$, $f(x)\in R^{6\times 1}$, $g(x)\in R^{6\times 1}$, and $u\in R^{2\times 1}$. Nonlinear dynamics and underactuation make the solution of the associated control problem be a nontrivial task.

3.3.3 Differential flatness properties of the 3-DOF cable-driven crane

It will be proven that the dynamic model of the two-cable-driven 3-DOF robotic crane is differentially flat, with flat outputs vector $Y=(x_R,y_R)^T$. It holds that

$$\begin{array}{l} x_2=\dot{x}_1 \Rightarrow x_2=h_2(Y,\dot{Y}) \\ x_4=\dot{x}_3 \Rightarrow x_4=h_4(Y,\dot{Y}) \end{array} \tag{3.281}$$

which signifies that state variables x_2, x_4 are differential functions of the flat outputs of the system. Besides, from the condition of Eq. (3.270), it holds that

$$\begin{array}{l} -sin(\theta)\dot{x}_R+cos(\theta)\dot{y}_R=0 \Rightarrow tan(\theta)=\frac{\dot{y}_R}{\dot{x}_R} \Rightarrow \theta=tan^{-1}(\frac{\dot{y}_R}{\dot{x}_R}) \\ \quad \Rightarrow x_5=\tan^{-1}(\frac{\dot{x}_3}{\dot{x}_1}) \Rightarrow x_5=h_5(Y,\dot{Y}) \\ \qquad x_6=\dot{x}_5 \Rightarrow x_6=h_6(Y,\dot{Y}) \end{array} \tag{3.282}$$

which signifies that state variables x_5 and x_6 are also differential functions of the flat outputs of the system. Next, one uses Eq. (3.278) which in concise form is written as

$$\ddot{q}=\tilde{F}(q,\dot{q})+\tilde{G}(q)u \tag{3.283}$$

Using the Moore-Penrose matrix of $\tilde{G}(q)$, one obtains

$$u=[\tilde{G}^T(q)\tilde{G}(q)]^{-1}\tilde{G}^T(q)[\ddot{q}-\tilde{F}(q,\dot{q})] \tag{3.284}$$

where $q=[x_R,y_R,\theta]^T$. Thus, it is confirmed that the control inputs vector $u=[u_1,u_2]^T$ is a differential function of the flat outputs vector, or

$$u_1=h_{u_1}(Y,\dot{Y}) \quad u_2=h_{u_2}(Y,\dot{Y}) \tag{3.285}$$

The proof of differential flatness properties for the robotic crane's model is an implicit proof of the system's controllability. Besides, it allows to treat the setpoints definition problem. One selects setpoints without constraints for state variables $x_1=x_R$ and $x_3=y_R$ which are associated with the flat outputs of the system. For state variable θ, setpoints should be chosen using the differential relation which connects this variable with the flat output $\theta_d=tan^{-1}(\frac{\dot{y}_R^d}{\dot{x}_R^d})$. At steady-state, it holds $\theta_d=0$.

3.3.4 Approximate linearization of the two-cable-driven 3-DOF robotic crane

The dynamic model of the two-cable-driven 3-DOF crane undergoes approximate linearization with the use of first-order Taylor series expansion and through the computation of the associated Jacobian matrices. The linearization process takes place at each sampling instance around the temporary operating point (x^*,u^*) where x^* is the present value of the system's state vector and u^* is the last sampled value of the control inputs vector. The modeling error which is due to the truncation of higher-order terms from the Taylor series expansion is considered to be a perturbation which is asymptotically compensated by the robustness of the control method.

The initial nonlinear state-space model of the crane being in the form $\dot{x}=f(x)+g(x)u$, after linearization comes into the form:

$$\dot{x}=Ax+Bu+\tilde{d} \tag{3.286}$$

where matrices A, B are the Jacobian matrices of the system:

$$\begin{array}{l} A=\nabla_x[f(x)+g(x)u] \mid_{(x^*,u^*)} \Rightarrow A=\nabla_x f(x) \mid_{(x^*,u^*)} + \\ \quad +\nabla_x g_1(x)u_1 \mid_{(x^*,u^*)} +\nabla_x g_2(x)u_2 \mid_{(x^*,u^*)} \end{array} \tag{3.287}$$

$$B=\nabla_u[f(x)+g(x)u] \mid_{(x^*,u^*)} \Rightarrow B=g(x) \mid_{(x^*,u^*)} \tag{3.288}$$

where the cumulative vector of disturbances (x^*,u^*) comprises (i) modeling error due to truncation of higher-order terms from the Taylor series expansion, (ii) external disturbances, and (iii) sensor measurement noise of any distribution.

Because of linearizing at each sampling instance around the present position of the crane's state vector, the span of the Taylor series expansion is small and the modeling error is negligible.

Next, the Jacobian matrix $\nabla_x f(x) \mid_{(x^*,u^*)}$ is computed.

First row of the Jacobian matrix $\nabla_x f(x) \mid_{(x^*,u^*)}$: $\frac{\partial f_1}{\partial x_1}=0$, $\frac{\partial f_1}{\partial x_2}=1$, $\frac{\partial f_1}{\partial x_3}=0$, $\frac{\partial f_1}{\partial x_4}=0$, $\frac{\partial f_1}{\partial x_5}=0$, and $\frac{\partial f_1}{\partial x_6}=0$.

Second row of the Jacobian matrix $\nabla_x f(x) \mid_{(x^*,u^*)}$: It holds that

$$f_2=\frac{f_{2,num}}{f_{2,den}}=\frac{-M_{11}C_1+M_{21}C_2-M_{31}G_3}{detM} \tag{3.289}$$

It holds that for $1,2,\cdots,6$

$$\frac{\partial f_2}{\partial x_i}=\frac{\frac{\partial f_{2,num}}{\partial x_i}detM-f_{2,num}\frac{\partial detM}{\partial x_i}}{detM^2} \tag{3.290}$$

where for $1,2,\cdots,6$

$$\frac{\partial f_{2,num}}{\partial x_i}=-\frac{\partial M_{11}}{\partial x_i}C_1-M_{11}\frac{\partial C_1}{\partial x_i}+\frac{\partial M_{11}}{\partial x_i}C_2+M_{21}\frac{\partial C_2}{\partial x_i}-\frac{\partial M_{31}}{\partial x_i}G_3-M_{31}\frac{\partial G_3}{\partial x_i} \tag{3.291}$$

and also for $1,2,\cdots,6$

$$\frac{\partial detM}{\partial x_i}=(M+m)\frac{\partial M_{11}}{\partial x_i}-(mL_3sin(x_5))M_{13}+mL_3cos(x_5)\frac{\partial M_{13}}{\partial x_i} \tag{3.292}$$

Third row of the Jacobian matrix $\nabla_x f(x) \mid_{(x^*,u^*)}$: $\frac{\partial f_3}{\partial x_1}=0$, $\frac{\partial f_3}{\partial x_2}=0$, $\frac{\partial f_3}{\partial x_3}=0$, $\frac{\partial f_3}{\partial x_4}=1$, $\frac{\partial f_3}{\partial x_5}=0$, and $\frac{\partial f_3}{\partial x_6}=0$.
Fourth row of the Jacobian matrix $\nabla_x f(x) \mid_{(x^*,u^*)}$: It holds that

$$f_4=\frac{f_{4,num}}{f_{4,den}}=\frac{M_{12}C_1-M_{22}C_2+M_{32}G_3}{detM} \tag{3.293}$$

It holds that for $1,2,\cdots,6$

$$\frac{\partial f_4}{\partial x_i}=\frac{\frac{\partial f_{4,num}}{\partial x_i}detM-f_{4,num}\frac{\partial detM}{\partial x_i}}{detM^2} \tag{3.294}$$

where for $1,2,\cdots,6$

$$\frac{\partial f_{4,num}}{\partial x_i}=\frac{\partial M_{12}}{\partial x_i}C_1+M_{12}\frac{\partial C_1}{\partial x_i}-\frac{\partial M_{22}}{\partial x_i}C_2-M_{22}\frac{\partial C_2}{\partial x_i}+\frac{\partial M_{32}}{\partial x_i}G_3+M_{32}\frac{\partial G_3}{\partial x_i} \tag{3.295}$$

Fifth row of the Jacobian matrix $\nabla_x f(x) \mid_{(x^*,u^*)}$: $\frac{\partial f_5|}{\partial x_1}=0$, $\frac{\partial f_5}{\partial x_2}=0$, $\frac{\partial f_5}{\partial x_3}=0$, $\frac{\partial f_5}{\partial x_4}=0$, $\frac{\partial f_3}{\partial x_5}=0$, and $\frac{\partial f_5}{\partial x_6}=1$.
Sixth row of the Jacobian matrix $\nabla_x f(x) \mid_{(x^*,u^*)}$: It holds that

$$f_6=\frac{f_{6,num}}{f_{6,den}}=\frac{-M_{13}C_1+M_{23}C_2-M_{33}G_3}{detM} \tag{3.296}$$

It holds that for $1,2,\cdots,6$

$$\frac{\partial f_6}{\partial x_i}=\frac{\frac{\partial f_{6,num}}{\partial x_i}detM-f_{6,num}\frac{\partial detM}{\partial x_i}}{detM^2} \tag{3.297}$$

where for $1,2,\cdots,6$

$$\frac{\partial f_{6,num}}{\partial x_i}=\frac{-\partial M_{13}}{\partial x_i}C_1-M_{13}\frac{\partial C_1}{\partial x_i}+\frac{\partial M_{23}}{\partial x_i}C_2+M_{23}\frac{\partial C_2}{\partial x_i}-\frac{\partial M_{33}}{\partial x_i}G_3-M_{33}\frac{\partial G_3}{\partial x_i} \tag{3.298}$$

Next, the Jacobian matrix $\nabla_x g_1(x) \mid_{(x^*,u^*)}$ is computed.
First row of the Jacobian matrix $\nabla_x g_1(x) \mid_{(x^*,u^*)}$: $\frac{\partial g_{11}}{\partial x_i}=0$. for $i=1,2,\cdots,6$.
Second row of the Jacobian matrix $\nabla_x g_1(x) \mid_{(x^*,u^*)}$: It holds that for $i=1,2,\cdots,6$

$$\frac{\partial g_{21}}{\partial x_i}=\frac{\frac{\partial M_{11}}{\partial x_i}detM-M_{11}\frac{\partial detM}{\partial x_i}}{detM^2} \tag{3.299}$$

Third row of the Jacobian matrix $\nabla_x g_1(x) \mid_{(x^*,u^*)}$: $\frac{\partial g_{31}}{\partial x_i}=0$. for $i=1,2,\cdots,6$.
Fourth row of the Jacobian matrix $\nabla_x g_1(x) \mid_{(x^*,u^*)}$: It holds that for $i=1,2,\cdots,6$

$$\frac{\partial g_{41}}{\partial x_i}=\frac{-\frac{\partial M_{12}}{\partial x_i}detM+M_{12}\frac{\partial detM}{\partial x_i}}{detM^2} \tag{3.300}$$

Fifth row of the Jacobian matrix $\nabla_x g_1(x) \mid_{(x^*,u^*)}$: $\frac{\partial g_{51}}{\partial x_i}=0$. for $i=1,2,\cdots,6$.
Sixth row of the Jacobian matrix $\nabla_x g_1(x) \mid_{(x^*,u^*)}$: It holds that for $i=1,2,\cdots,6$

$$\frac{\partial g_{61}}{\partial x_i}=\frac{\frac{\partial M_{13}}{\partial x_i}detM-M_{13}\frac{\partial detM}{\partial x_i}}{detM^2} \tag{3.301}$$

Next, the Jacobian matrix $\nabla_x g_2(x) \mid_{(x^*,u^*)}$ is computed.
First row of the Jacobian matrix $\nabla_x g_2(x) \mid_{(x^*,u^*)}$: $\frac{\partial g_{12}}{\partial x_i}=0$. for $i=1,2,\cdots,6$.
Second row of the Jacobian matrix $\nabla_x g_2(x) \mid_{(x^*,u^*)}$: It holds that for $i=1,2,\cdots,6$

$$\frac{\partial g_{22}}{\partial x_i}=\frac{-\frac{\partial M_{12}}{\partial x_i}detM+M_{12}\frac{\partial detM}{\partial x_i}}{detM^2} \tag{3.302}$$

Third row of the Jacobian matrix $\nabla_x g_2(x) \mid_{(x^*,u^*)}$: $\frac{\partial g_{32}}{\partial x_i}=0$. for $i=1,2,\cdots,6$.
Fourth row of the Jacobian matrix $\nabla_x g_2(x) \mid_{(x^*,u^*)}$: It holds that for $i=1,2,\cdots,6$

$$\frac{\partial g_{42}}{\partial x_i}=\frac{\frac{\partial M_{22}}{\partial x_i}detM-M_{22}\frac{\partial detM}{\partial x_i}}{detM^2} \tag{3.303}$$

Fifth row of the Jacobian matrix $\nabla_x g_2(x) \mid_{(x^*,u^*)}$: $\frac{\partial g_{52}}{\partial x_i}=0$. for $i=1,2,\cdots,6$.
Sixth row of the Jacobian matrix $\nabla_x g_2(x) \mid_{(x^*,u^*)}$: It holds that for $i=1,2,\cdots,6$

$$\frac{\partial g_{62}}{\partial x_i}=\frac{-\frac{\partial M_{32}}{\partial x_i}detM+M_{32}\frac{\partial detM}{\partial x_i}}{detM^2} \tag{3.304}$$

Computation of the partial derivatives of the sub-determinants of the inertia matrix:

It holds that $\frac{\partial M_{11}}{\partial x_i}=0$ for $i=1,2,\cdots,6$ and $i{\neq}5$ while $\frac{\partial M_{11}}{\partial x_5}=-(ML_3)^2sin(x_5)cos(x_5)$.

Additionally, it holds that $\frac{\partial M_{12}}{\partial x_i}=0$ for $i=1,2,\cdots,6$ and $i{\neq}5$ while $\frac{\partial M_{12}}{\partial x_5}=-(mL_3)^2[cos(x_5)^2-sin(x_5)^2]$.

Moreover, it holds that $\frac{\partial M_{13}}{\partial x_i}=0$ for $i=1,2,\cdots,6$ and $i{\neq}5$ while $\frac{\partial M_{13}}{\partial x_5}=(M+m)(mL_3)sin(x_5)$.

Furthermore, one has that $\frac{\partial M_{21}}{\partial x_i}=0$ for $i=1,2,\cdots,6$ and $i{\neq}5$ while $\frac{\partial M_{21}}{\partial x_5}=-(mL_3)^2[cos(x_5)^2-sin(x_5)^2]$.

Additionally, it holds that $\frac{\partial M_{22}}{\partial x_i}=0$ for $i=1,2,\cdots,6$ and $i{\neq}5$ while $\frac{\partial M_{22}}{\partial x_5}=2(mL_3)^2sin(x_5)cos(x_5)$.

Moreover, it holds that $\frac{\partial M_{23}}{\partial x_i}=0$ for $i=1,2,\cdots,6$ and $i{\neq}5$ while $\frac{\partial M_{23}}{\partial x_5}=(M+m)(mL_3)cos(x_5)$.

Furthermore, one has that $\frac{\partial M_{31}}{\partial x_i}=0$ for $i=1,2,\cdots,6$ and $i{\neq}5$ while $\frac{\partial M_{31}}{\partial x_5}=(M+m)(mL_3)sin(x_5)$.

Moreover, it holds that $\frac{\partial M_{32}}{\partial x_i}=0$ for $i=1,2,\cdots,6$ and $i{\neq}5$ while $\frac{\partial M_{32}}{\partial x_5}=(M+m)(mL_3)cos(x_5)$.

Finally, it holds that $\frac{\partial M_{33}}{\partial x_i}=0$ for $i=1,2,\cdots,6$.

Computation of the partial derivatives of the non-zero elements of the Coriolis forces vector $C(q,\dot{q})$:

It holds that $\frac{\partial C_1}{\partial x_1}=0$, $\frac{\partial C_1}{\partial x_2}=0$, $\frac{\partial C_1}{\partial x_3}=0$, $\frac{\partial C_1}{\partial x_4}=0$, $\frac{\partial C_1}{\partial x_5}=-(mL_3cos(x_5))x_6^2$, and $\frac{\partial C_1}{\partial x_6}=-2(mL_3sin(x_5))x_6$.

Moreover, it holds that $\frac{\partial C_2}{\partial x_1}=0$, $\frac{\partial C_2}{\partial x_2}=0$, $\frac{\partial C_2}{\partial x_3}=0$, $\frac{\partial C_2}{\partial x_4}=0$, $\frac{\partial C_2}{\partial x_5}=-(mL_3sin(x_5))x_6^2$, and $\frac{\partial C_2}{\partial x_6}=2(mL_3cos(x_5))x_6$.

Computation of the partial derivatives of the non-zero elements of the gravitational forces vector $G(q)$:

It holds that $\frac{\partial G_3}{\partial x_1}=0$, $\frac{\partial G_3}{\partial x_2}=0$, $\frac{\partial G_3}{\partial x_3}=0$, $\frac{\partial G_3}{\partial x_4}=0$, $\frac{\partial G_3}{\partial x_5}=(mgL_3cos(x_5))$, and $\frac{\partial G_3}{\partial x_6}=0$.

As previously explained, this linearization approach which has been followed for implementing the nonlinear optimal control scheme results in a quite accurate model of the system's dynamics. Consider for instance the following affine-in-the-input state-space model:

$$\begin{gathered}\dot{x}=f(x)+g(x)u\Rightarrow\\ \dot{x}=[f(x^*)+\nabla_x f(x)\mid_{x^*}(x-x^*)]+[g(x^*)+\nabla_x g(x)\mid_{x^*}(x-x^*)]u^*+g(x^*)u^*+g(x^*)(u-u^*)+\tilde{d}_1\Rightarrow\\ \dot{x}=[\nabla_x f(x)\mid_{x^*}+\nabla_x g(x)\mid_{x^*}u^*]x+g(x^*)u-[\nabla_x f(x)\mid_{x^*}+\nabla_x g(x)\mid_{x^*}u^*]x^*+f(x^*)+g(x^*)u^*+\tilde{d}_1\end{gathered} \tag{3.305}$$

where $\tilde{d}_1$ is the modeling error due to truncation of higher-order terms in the Taylor series expansion of $f(x)$ and $g(x)$. Next, by defining $A=[\nabla_x f(x)\mid_{x^*}+\nabla_x g(x)\mid_{x^*}u^*]$, $B=g(x^*)$, one obtains

$$\dot{x}=Ax+Bu-Ax^*+f(x^*)+g(x^*)u^*+\tilde{d}_1 \tag{3.306}$$

Moreover, by denoting $\tilde{d}=-Ax^*+f(x^*)+g(x^*)u^*+\tilde{d}_1$ about the cumulative modeling error term in the Taylor series expansion procedure, one has

$$\dot{x}=Ax+Bu+\tilde{d} \tag{3.307}$$

which is the approximately linearized model of the dynamics of the system of Eq. (3.286). The term $f(x^*)+g(x^*)u^*$ is the derivative of the state vector at (x^*,u^*) which is almost annihilated by $-Ax^*$.

3.3.5 Design of an H-infinity nonlinear feedback controller

3.3.5.1 Equivalent linearized dynamics of the two-cable-driven 3-DOF robotic crane

After linearization around its current operating point, the dynamic model for the two-cable-driven 3-DOF robotic crane is written as

$$\dot{x}=Ax+Bu+d_1 \tag{3.308}$$

Parameter d_1 stands for the linearization error in the two-cable-driven 3-DOF robotic crane's model that was given previously in Eq. (3.308). The reference setpoints for the state vector of the aforementioned dynamic model are denoted by $\mathbf{x_d}=[x_1^d,\cdots,x_6^d]$. Tracking of this trajectory is achieved after applying the control input u^*. At every time instant the control input u^* is assumed to differ from the control input u appearing in Eq. (3.308) by an amount equal to Δu, that is $u^*=u+\Delta u$

$$\dot{x}_d=Ax_d+Bu^*+d_2 \tag{3.309}$$

The dynamics of the controlled system described in Eq. (3.308) can be also written as

$$\dot{x}=Ax+Bu+Bu^*-Bu^*+d_1 \tag{3.310}$$

and by denoting $d_3=-Bu^*+d_1$ as an aggregate disturbance term, one obtains

$$\dot{x}=Ax+Bu+Bu^*+d_3 \tag{3.311}$$

By subtracting Eq. (3.309) from Eq. (3.311), one has

$$\dot{x}-\dot{x}_d=A(x-x_d)+Bu+d_3-d_2 \tag{3.312}$$

By denoting the tracking error as $e=x-x_d$ and the aggregate disturbance term as $L\tilde{d}=d_3-d_2$, the tracking error dynamics becomes

$$\dot{e}=Ae+Bu+L\tilde{d} \tag{3.313}$$

where L is a disturbance inputs gain matrix. The above-linearized form of the two-cable-driven 3-DOF robotic crane's model can be efficiently controlled after applying an H-infinity feedback control scheme.

3.3.6 The nonlinear H-infinity control

The initial nonlinear model of the two-cable-driven 3-DOF robotic crane is in the form

$$\dot{x} = f(x, u) \quad x \in R^n, \ u \in R^m \tag{3.314}$$

Linearization of the model of the two-cable-driven 3-DOF robotic crane is performed at each iteration of the control algorithm around its present operating point $(x^*, u^*) = (x(t), u(t - T_s))$. The linearized equivalent of the system is described by

$$\dot{x} = Ax + Bu + L\tilde{d} \quad x \in R^n, \ u \in R^m, \ \tilde{d} \in R^q \tag{3.315}$$

where matrices A and B are obtained from the computation of the previously defined Jacobians and vector $\tilde{d}$ denotes disturbance terms due to linearization errors, while L is a disturbance inputs gain matrix. The problem of disturbance rejection for the linearized model that is described by

$$\begin{aligned} \dot{x} &= Ax + Bu + L\tilde{d} \\ y &= Cx \end{aligned} \tag{3.316}$$

where $x \in R^n$, $u \in R^m$, $\tilde{d} \in R^q$, and $y \in R^p$, cannot be handled efficiently if the classical LQR control scheme is applied. This is because of the existence of the perturbation term $\tilde{d}$. The disturbance term $\tilde{d}$ apart from modeling (parametric) uncertainty and external perturbations can also represent noise terms of any distribution.

In the H_∞ control approach, a feedback control scheme is designed for trajectory tracking by the system's state vector and simultaneous disturbance rejection, considering that the disturbance affects the system in the worst possible manner. The disturbances' effects are incorporated in the following quadratic cost function:

$$J(t) = \frac{1}{2}\int_0^T [y^T(t)y(t) + ru^T(t)u(t) - \rho^2 \tilde{d}^T(t)\tilde{d}(t)]dt, \quad r, \rho > 0 \tag{3.317}$$

The significance of the negative sign in the cost function's term that is associated with the perturbation variable $\tilde{d}(t)$ is that the disturbance tries to maximize the cost function $J(t)$ while the control signal $u(t)$ tries to minimize it. The physical meaning of the relation given above is that the control signal and the disturbances compete with each other within a min–max differential game. This problem of min–max optimization can be written as $min_u max_{\tilde{d}} J(u, \tilde{d})$.

The objective of the optimization procedure is to compute a control signal $u(t)$ which can compensate for the worst possible disturbance, which is externally imposed on the two-cable-driven 3-DOF robotic crane. However, the solution to the min–max optimization problem is directly related to the value of parameter ρ. This means that there is an upper bound in the disturbances magnitude that can be annihilated by the control signal.

3.3.6.1 Computation of the feedback control gains

For the linearized system given by Eq. (3.316), the cost function of Eq. (3.317) is defined, where coefficient r determines the penalization of the control input and weight coefficient ρ determines the reward of the disturbances' effects. It is assumed that (i) The energy that is transferred from the disturbances signal $\tilde{d}(t)$ is bounded, that is $\int_0^\infty \tilde{d}^T(t)\tilde{d}(t)dt < \infty$, (ii) matrices $[A, B]$ and $[A, L]$ are stabilizable, (iii) matrix $[A, C]$ is detectable. In the case of a tracking problem, the optimal feedback control law is given by

$$u(t) = -Ke(t) \tag{3.318}$$

with $e = x - x_d$ to be the tracking error, and $K = \frac{1}{r}B^T P$ where P is a positive definite symmetric matrix. As it will be proven in Section 3.3.7, matrix P is obtained from the solution of the Riccati equation

$$A^T P + PA + Q - P\left(\frac{2}{r}BB^T - \frac{1}{\rho^2}LL^T\right)P = 0 \tag{3.319}$$

where Q is a positive semi-definite symmetric matrix. The worst-case disturbance is given by

$$\tilde{d}(t) = \frac{1}{\rho^2}L^T Pe(t) \tag{3.320}$$

The solution of the H-infinity feedback control problem for the two-cable-driven 3-DOF robotic crane and the computation of the worst-case disturbance that the related controller can sustain comes from superposition of Bellman's optimality principle when considering that the two-cable-driven 3-DOF robotic crane is

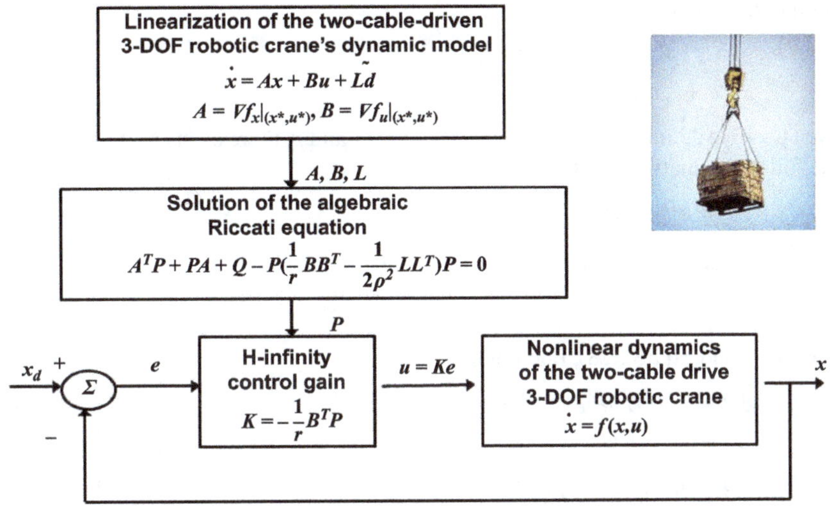

Figure 3.38 Control scheme and implementation stages for the two-cable-driven 3-DOF robotic crane

affected by two separate inputs: (i) the control input u and (ii) the cumulative disturbance input $\tilde{d}(t)$. Solving the optimal control problem for u, that is, for the minimum variation (optimal) control input that achieves the elimination of the state vector's tracking error, gives $u = -\frac{1}{r}B^TPe$. Equivalently, solving the optimal control problem for $\tilde{d}$, that is, for the worst-case disturbance that the control loop can sustain gives $\tilde{d} = \frac{1}{\rho^2}L^TPe$.

The diagram of the considered control loop for the two-cable-driven 3-DOF robotic crane is depicted in Figure 3.38.

3.3.7 Lyapunov stability analysis

Through Lyapunov stability analysis, it will be shown that the proposed nonlinear control scheme assures H_∞ tracking performance for the two-cable-driven 3-DOF robotic crane and that, in case of bounded disturbance terms, asymptotic convergence to the reference setpoints is achieved. The tracking error dynamics for the two-cable-driven 3-DOF robotic crane is written in the form

$$\dot{e} = Ae + Bu + L\tilde{d} \tag{3.321}$$

where in the two-cable-driven 3-DOF robotic crane's case $L = \in R^{6\times 6}$ is the disturbance inputs gain matrix. Variable $\tilde{d}$ denotes model uncertainties and external disturbances of the two-cable-driven 3-DOF robotic crane's model. The following Lyapunov equation is considered:

$$V = \tfrac{1}{2}e^TPe \tag{3.322}$$

where $e = x - x_d$ is the tracking error. By differentiating with respect to time, one obtains

$$\begin{aligned}\dot{V} &= \tfrac{1}{2}\dot{e}^TPe + \tfrac{1}{2}e^TP\dot{e} \Rightarrow \\ \dot{V} &= \tfrac{1}{2}[Ae + Bu + L\tilde{d}]^TPe + \tfrac{1}{2}e^TP[Ae + Bu + L\tilde{d}] \Rightarrow\end{aligned} \tag{3.323}$$

$$\begin{aligned}\dot{V} = &\tfrac{1}{2}[e^TA^T + u^TB^T + \tilde{d}^TL^T]Pe + \\ &+ \tfrac{1}{2}e^TP[Ae + Bu + L\tilde{d}] \Rightarrow\end{aligned} \tag{3.324}$$

$$\begin{aligned}\dot{V} = &\tfrac{1}{2}e^TA^TPe + \tfrac{1}{2}u^TB^TPe + \tfrac{1}{2}\tilde{d}^TL^TPe + \\ &\tfrac{1}{2}e^TPAe + \tfrac{1}{2}e^TPBu + \tfrac{1}{2}e^TPL\tilde{d}\end{aligned} \tag{3.325}$$

The previous equation is rewritten as

$$\dot{V}=\frac{1}{2}e^T(A^TP+PA)e+(\frac{1}{2}u^TB^TPe+\frac{1}{2}e^TPBu)+ \\ +(\frac{1}{2}\tilde{d}^TL^TPe+\frac{1}{2}e^TPL\tilde{d}) \tag{3.326}$$

Assumption: For given positive definite matrix Q and coefficients r and ρ, there exists a positive definite matrix P, which is the solution of the following matrix equation:

$$A^TP+PA=-Q+P(\frac{2}{r}BB^T-\frac{1}{\rho^2}LL^T)P \tag{3.327}$$

Moreover, the following feedback control law is applied to the system:

$$u=-\frac{1}{r}B^TPe \tag{3.328}$$

By substituting Eq. (3.327) and Eq. (3.328), one obtains

$$\dot{V}=\frac{1}{2}e^T[-Q+P(\frac{2}{r}BB^T-\frac{1}{\rho^2}LL^T)P]e+ \\ +e^TPB(-\frac{1}{r}B^TPe)+e^TPL\tilde{d}\Rightarrow \tag{3.329}$$

$$\dot{V}=-\frac{1}{2}e^TQe+\frac{1}{r}e^TPBB^TPe-\frac{1}{2\rho^2}e^TPLL^TPe \\ -\frac{1}{r}e^TPBB^TPe+e^TPL\tilde{d} \tag{3.330}$$

which after intermediate operations gives

$$\dot{V}=-\frac{1}{2}e^TQe-\frac{1}{2\rho^2}e^TPLL^TPe+e^TPL\tilde{d} \tag{3.331}$$

or, equivalently

$$\dot{V}=-\frac{1}{2}e^TQe-\frac{1}{2\rho^2}e^TPLL^TPe+ \\ +\frac{1}{2}e^TPL\tilde{d}+\frac{1}{2}\tilde{d}^TL^TPe \tag{3.332}$$

Lemma: The following inequality holds

$$\frac{1}{2}e^TL\tilde{d}+\frac{1}{2}\tilde{d}L^TPe-\frac{1}{2\rho^2}e^TPLL^TPe\leq\frac{1}{2}\rho^2\tilde{d}^T\tilde{d} \tag{3.333}$$

Proof: The binomial $(\rho\alpha-\frac{1}{\rho}b)^2$ is considered. Expanding the left part of the above inequality, one gets

$$\rho^2a^2+\frac{1}{\rho^2}b^2-2ab\geq 0\Rightarrow\frac{1}{2}\rho^2a^2+\frac{1}{2\rho^2}b^2-ab\geq 0\Rightarrow \\ ab-\frac{1}{2\rho^2}b^2\leq\frac{1}{2}\rho^2a^2\Rightarrow\frac{1}{2}ab+\frac{1}{2}ab-\frac{1}{2\rho^2}b^2\leq\frac{1}{2}\rho^2a^2 \tag{3.334}$$

The following substitutions are carried out: $a=\tilde{d}$ and $b=e^TPL$ and the previous relation becomes

$$\frac{1}{2}\tilde{d}^TL^TPe+\frac{1}{2}e^TPL\tilde{d}-\frac{1}{2\rho^2}e^TPLL^TPe\leq\frac{1}{2}\rho^2\tilde{d}^T\tilde{d} \tag{3.335}$$

Eq. (3.335) is substituted in Eq. (3.332) and the inequality is enforced, thus giving

$$\dot{V}\leq-\frac{1}{2}e^TQe+\frac{1}{2}\rho^2\tilde{d}^T\tilde{d} \tag{3.336}$$

Eq. (3.336) shows that the H_∞ tracking performance criterion is satisfied. The integration of $\dot{V}$ from 0 to T gives

$$\int_0^T\dot{V}(t)dt\leq-\frac{1}{2}\int_0^T||e||_Q^2dt+\frac{1}{2}\rho^2\int_0^T||\tilde{d}||^2dt\Rightarrow \\ 2V(T)+\int_0^T||e||_Q^2dt\leq 2V(0)+\rho^2\int_0^T||\tilde{d}||^2dt \tag{3.337}$$

Moreover, if there exists a positive constant $M_d>0$ such that

$$\int_0^\infty||\tilde{d}||^2dt\leq M_d \tag{3.338}$$

then, one gets

$$\int_0^\infty||e||_Q^2dt\leq 2V(0)+\rho^2M_d \tag{3.339}$$

Thus, the integral $\int_0^{\infty}||e||_Q^2 dt$ is bounded. Moreover, $V(T)$ is bounded and from the definition of the Lyapunov function V in Eq. (3.322) it becomes clear that $e(t)$ will be also bounded since $e(t) \in \Omega_e = \{e|e^TPe \leq 2V(0) + \rho^2 M_d\}$. According to the above and with the use of Barbalat's Lemma, one obtains $\lim_{t\to\infty} e(t) = 0$.

Through the stages of the stability proof, one arrives at Eq. (3.336) which shows that the H-infinity tracking performance criterion holds. By selecting the attenuation coefficient ρ to be sufficiently small and in particular to satisfy $\rho^2 < ||e||_Q^2/||\tilde{d}||^2$, one has that the first derivative of the Lyapunov function is upper bounded by 0. This condition holds at each sampling instance, and consequently, global stability for the control loop can be concluded.

3.3.8 Simulation tests

The global stability properties of the control method and the elimination of the state vector's tracking error which were previously proven through Lyapunov analysis are further confirmed through simulation experiments. Indicative values about the parameters of the crane's model have been: $M = 1.0kg$, $m = 1.0kg$, $L_3 = 0.5m$, and $g = 10m/sec^2$. The sampling period was $T_s = 0.01sec$. To implement the control method, one had to solve at each sampling instance the algebraic Riccati equation of Eq. (3.327). The obtained results are shown in Figures 3.39–3.54. In the related diagrams, the real value of the state vector elements is printed in blue color, the reference setpoints are plotted in red color while the estimated values of the state variables are depicted in blue color.

3.4 Nonlinear optimal control of the 3-DOF four-cable-driven robotic crane

3.4.1 Outline

As noted before, cable-driven parallel robots (CDPR) are used in several industrial applications thanks to their simple structure and actuation [115,20,959,63]. Such robots can be used in construction works and buildings' erection for the lifting of objects, as well as in warehouses for picking and placing of stored products. They can also find use in the construction and repair of building's facades as well as in the painting of the external surfaces of ships [777,55,392,230]. Their end-effector receives actuation by cables of variable length and winding which is performed with the use of electric motors [380,381,770]. Their dynamic model has the form of the model of a robotic manipulator with an inertia

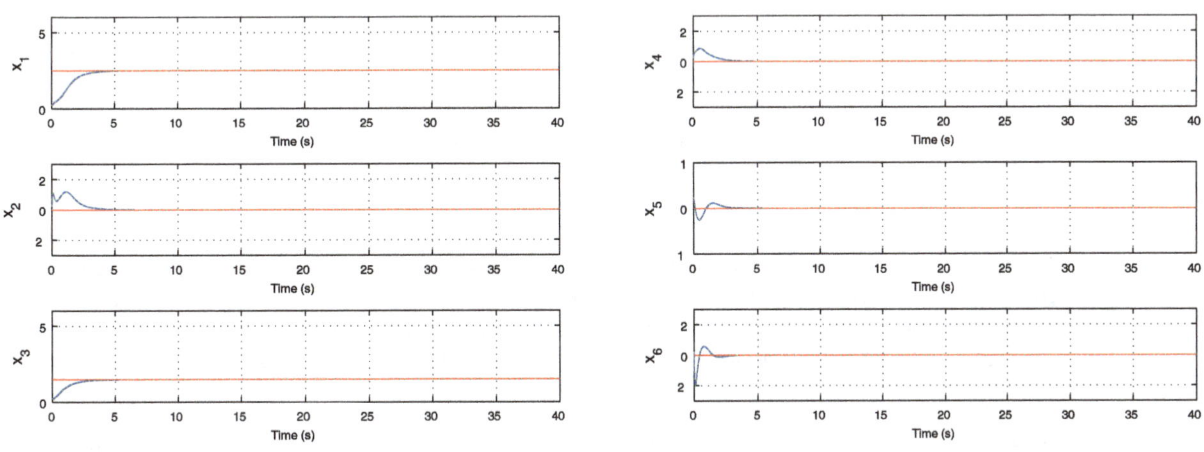

(a) (b)

Figure 3.39 Tracking of setpoint 1 by the two-cable-driven 3-DOF robotic crane with nonlinear optimal control: (a) tracking of setpoints (red line) by the state variables of the crane x_1 to x_3 (blue line) and estimates provided by the H-infinity Kalman filter; (b) tracking of setpoints (red line) by the state variables of the crane x_4 to x_6 (blue line) and estimates provided by the H-infinity Kalman filter

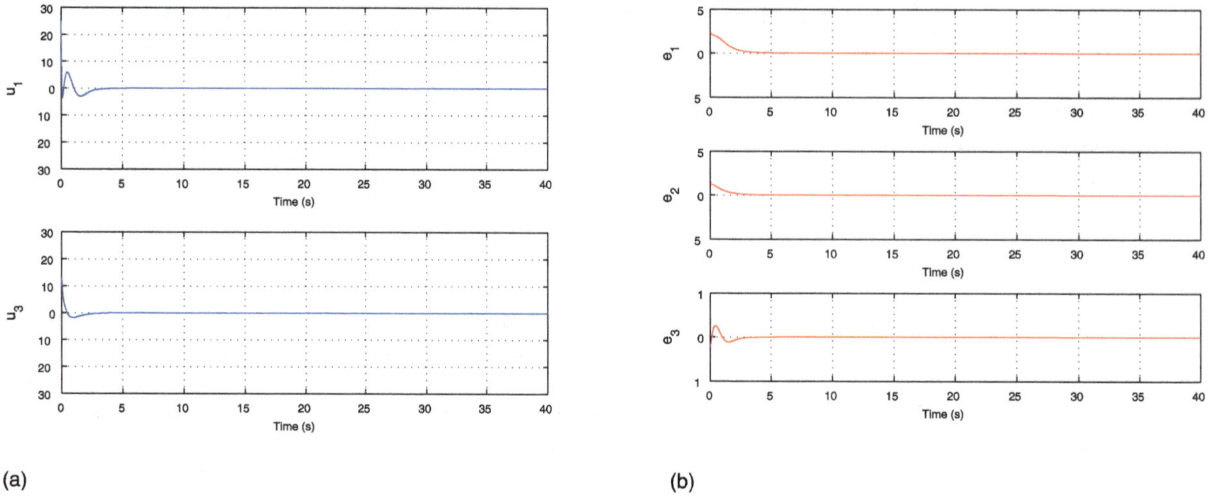

Figure 3.40 Tracking of setpoint 1 by the two-cable-driven 3-DOF robotic crane with nonlinear optimal control: (a) variations of the control inputs of the gantry crane u_1 and u_2 (blue line); (b) convergence to 0 of the tracking error variables e_1, e_3, and e_5 associated with state variables $x_1 = x_R$, $x_3 = y_R$, and $x_5 = \theta$

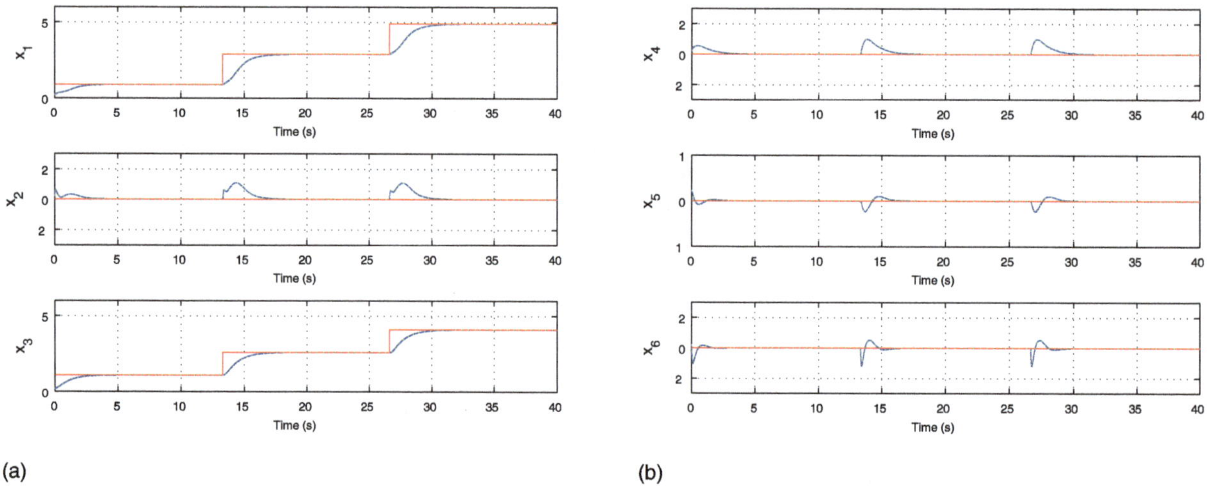

Figure 3.41 Tracking of setpoint 2 by the two-cable-driven 3-DOF robotic crane with nonlinear optimal control: (a) tracking of setpoints (red line) by the state variables of the crane x_1 to x_3 (blue line) and estimates provided by the H-infinity Kalman filter; (b) tracking of setpoints (red line) by the state variables of the crane x_4 to x_6 (blue line) and estimates provided by the H-infinity Kalman filter

matrix, a damping forces matrix, and a gravitational forces matrix [395,574,351]. Moreover, the vector of forces and torques which are applied on the end-effector is computed by multiplying the cable stretching forces generated by the motors with a Jacobian matrix that connects the rate of change of the cables' lengths with the linear and rotational velocities of the end-effector [598,597,464,463]. Several nonlinear control methods have been proposed for the control and stabilization problem of cable-driven parallel robot [47,49,48]. One can distinguish first global linearization-based control schemes, which aim at transforming the cable robot's model into an equivalent input–output linearized form [465,466,778,320]. Besides, one finds LQR, LQG and MPC control schemes which assume linear dynamics of the

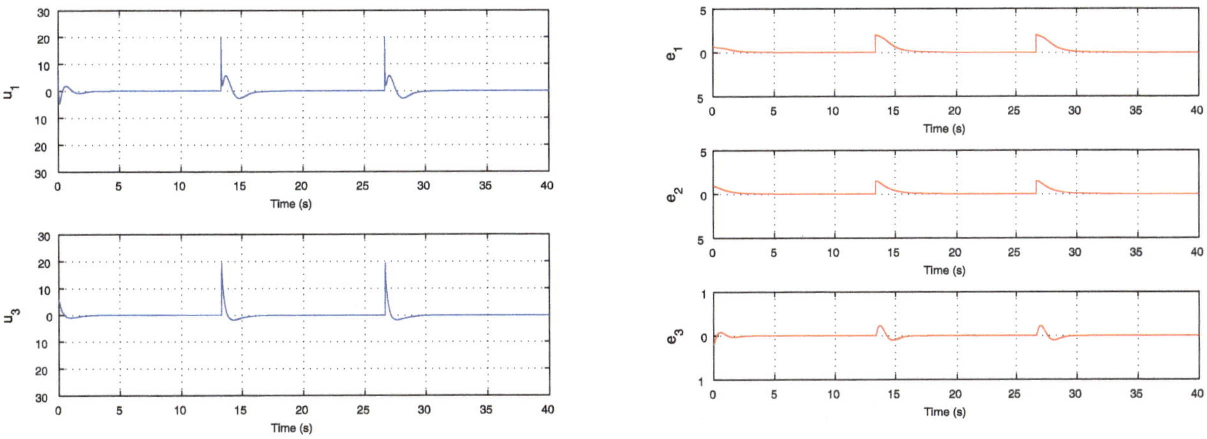

(a) (b)

Figure 3.42 Tracking of setpoint 2 by the two-cable-driven 3-DOF robotic crane with nonlinear optimal control: (a) variations of the control inputs of the gantry crane u_1 and u_2 (blue line); (b) convergence to 0 of the tracking error variables e_1, e_3, and e_5 associated ith state variables $x_1 = x_R$, $x_3 = y_R$, and $x_5 = \theta$

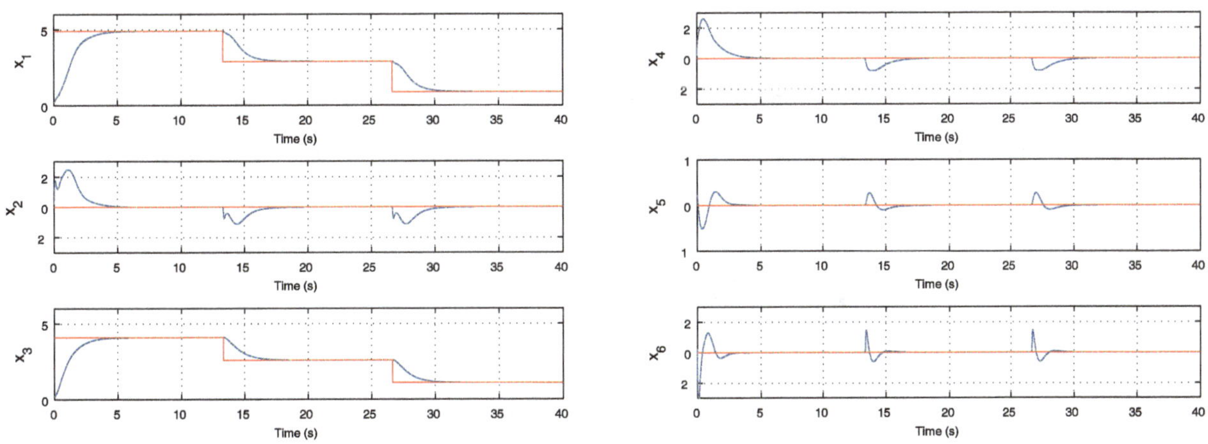

(a) (b)

Figure 3.43 Tracking of setpoint 3 by the two-cable-driven 3-DOF robotic crane with nonlinear optimal control: (a) tracking of setpoints (red line) by the state variables of the crane x_1 to x_3 (blue line) and estimates provided by the H-infinity Kalman filter; (b) tracking of setpoints (red line) by the state variables of the crane x_4 to x_6 (blue line) and estimates provided by the H-infinity Kalman filter

cable-driven parallel robot [143,812,132]. Furthermore, one can note robust control approaches such as sliding-mode control [918,1002,675].

In the present section, a novel nonlinear optimal control method is developed for the dynamic model of 3-DOF four-cable-driven parallel robots. The kinematic analysis of these robots enables to compute the Jacobian matrix that transfers the effect of the forces generated by actuators to the robot's end-effector. Moreover, using Euler-Lagrange analysis, the dynamic model of the 3-DOF four-cable-driven robot is computed which is proven to satisfy differential flatness properties. To apply the proposed nonlinear optimal control method, the state-space model of the robot undergoes

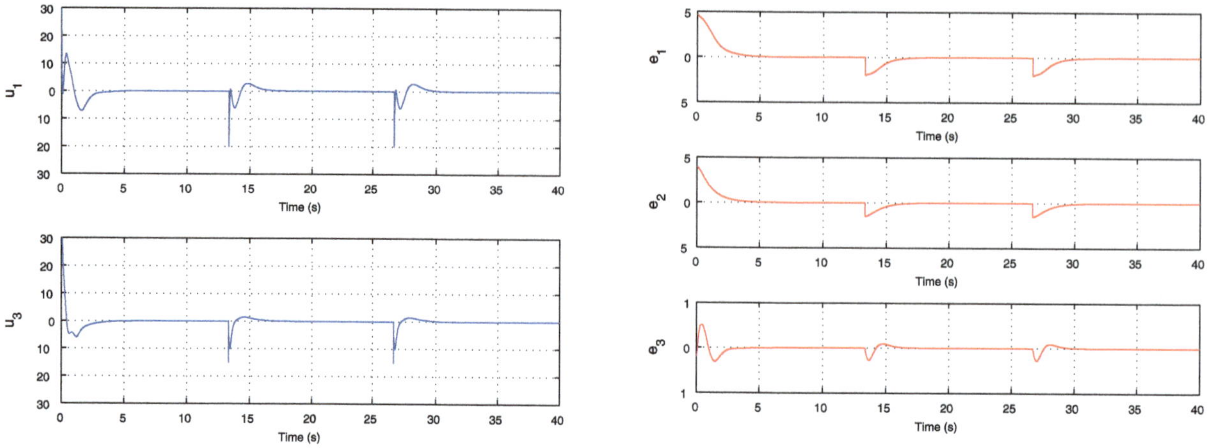

(a) (b)

Figure 3.44 Tracking of setpoint 3 by the two-cable-driven 3-DOF robotic crane with nonlinear optimal control: (a) variations of the control inputs of the gantry crane u_1 and u_2 (blue line); (b) convergence to 0 of the tracking error variables e_1, e_3, and e_5 associated with state variables $x_1 = x_R$, $x_3 = y_R$, and $x_5 = \theta$

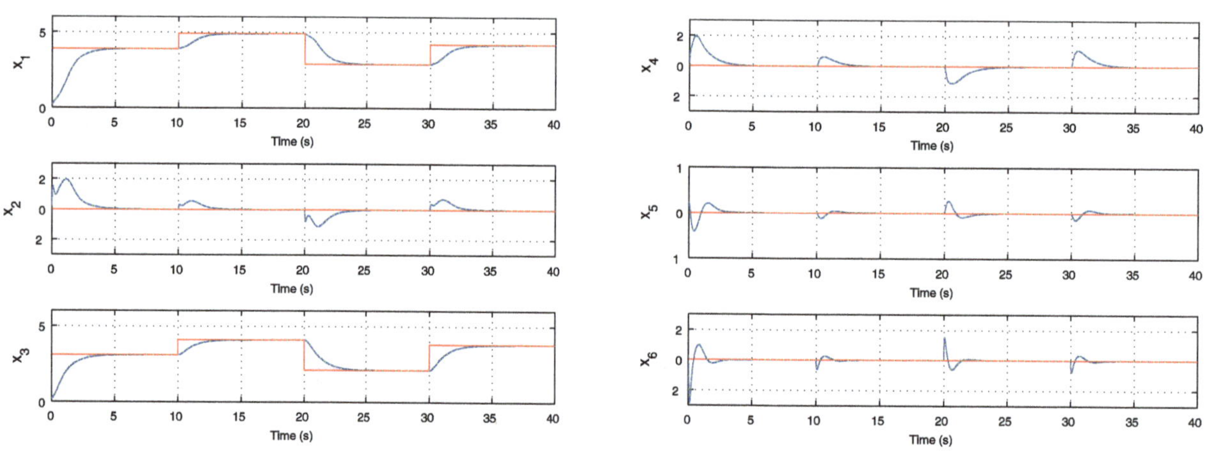

(a) (b)

Figure 3.45 Tracking of setpoint 4 by the two-cable-driven 3-DOF robotic crane with nonlinear optimal control: (a) tracking of setpoints (red line) by the state variables of the crane x_1 to x_3 (blue line) and estimates provided by the H-infinity Kalman filter; (b) tracking of setpoints (red line) by the state variables of the crane x_4 to x_6 (blue line) and estimates provided by the H-infinity Kalman filter

first approximate linearization with the use of first-order Taylor series expansion and through the computation of the associated Jacobian matrices [705,60,706]. The linearization process takes place at each sampling instance around a temporary operating point which is defined by the present value of the system's state vector and by the last sampled of the control inputs vector. The model imprecision which is due to the truncation of higher-order terms from the Taylor series expansion is considered to be a perturbation that is asymptotically compensated by the robustness of the control

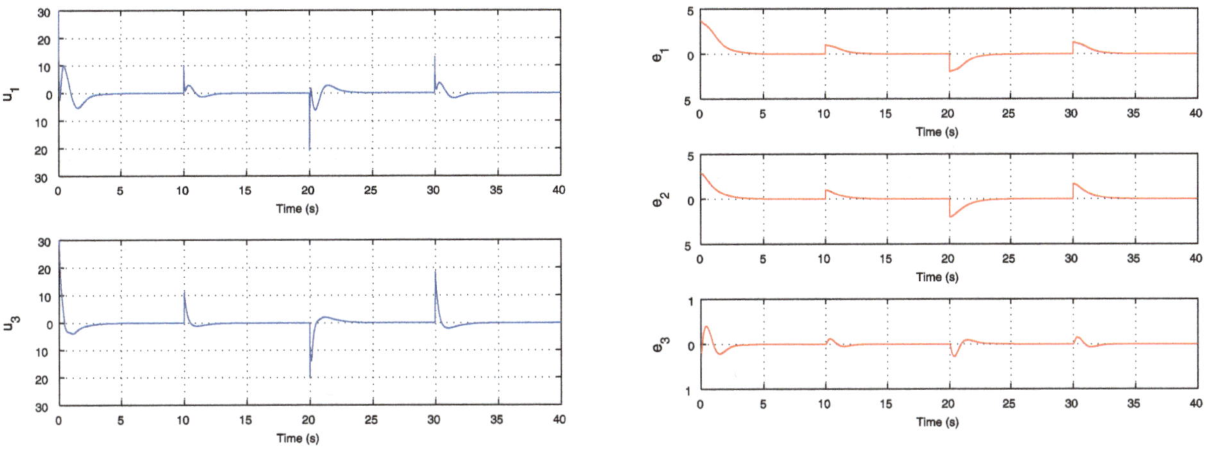

(a) (b)

Figure 3.46 Tracking of setpoint 4 by the two-cable-driven 3-DOF robotic crane with nonlinear optimal control: (a) variations of the control inputs of the gantry crane u_1 and u_2 (blue line); (b) convergence to 0 of the tracking error variables e_1, e_3, and e_5 associated with state variables $x_1 = x_R$, $x_3 = y_R$, and $x_5 = \theta$

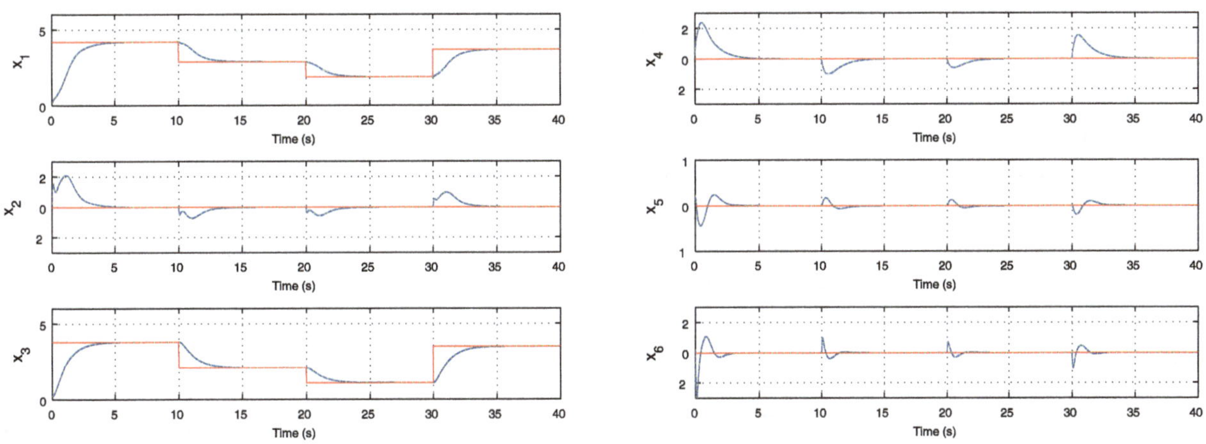

(a) (b)

Figure 3.47 Tracking of setpoint 5 by the two-cable-driven 3-DOF robotic crane with nonlinear optimal control: (a) tracking of setpoints (red line) by the state variables of the crane x_1 to x_3 (blue line) and estimates provided by the H-infinity Kalman filter; (b) tracking of setpoints (red line) by the state variables of the crane x_4 to x_6 (blue line) and estimates provided by the H-infinity Kalman filter

algorithm. For the linearized model of the cable-driven parallel robot, a stabilizing H-infinity feedback controller is designed.

The H-infinity feedback controller for the dynamic model of the 3-DOF cable-driven parallel robot achieves the solution of the associated optimal control problem under model uncertainty and external perturbations. Actually, it represents a min–max differential game taking place between (i) the control inputs which try to minimize a quadratic cost function of the state vector's tracking error and (ii) the model uncertainty and external perturbation terms which try

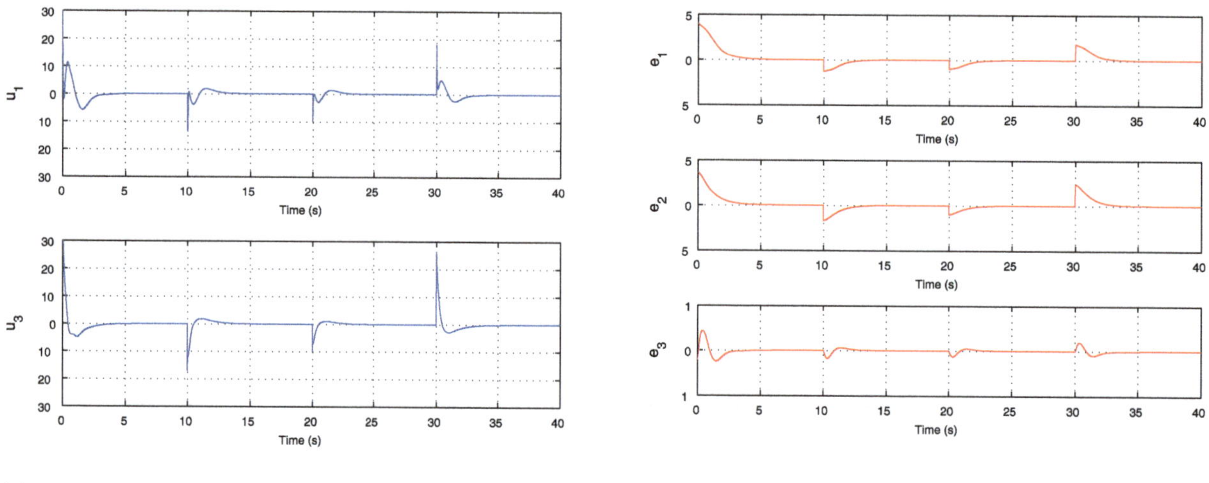

(a) (b)

Figure 3.48 Tracking of setpoint 5 by the two-cable-driven 3-DOF robotic crane with nonlinear optimal control: (a) variations of the control inputs of the gantry crane u_1 and u_2 (blue line); (b) convergence to 0 of the tracking error variables e_1, e_3, and e_5 associated with state variables $x_1 = x_R$, $x_3 = y_R$, and $x_5 = \theta$

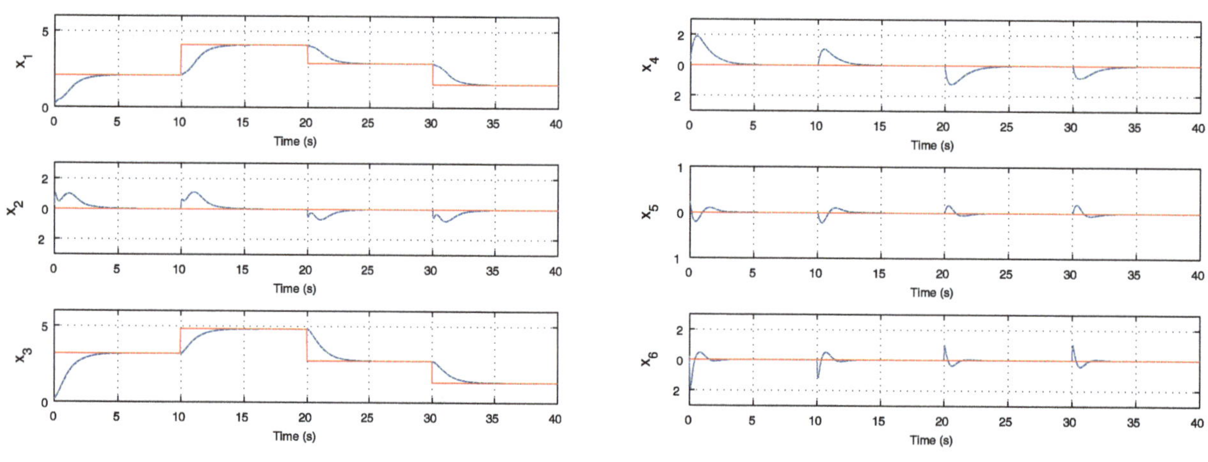

(a) (b)

Figure 3.49 Tracking of setpoint 6 by the two-cable-driven 3-DOF robotic crane with nonlinear optimal control: (a) tracking of setpoints (red line) by the state variables of the crane x_1 to x_3 (blue line) and estimates provided by the H-infinity Kalman filter; (b) tracking of setpoints (red line) by the state variables of the crane x_4 to x_6 (blue line) and estimates provided by the H-infinity Kalman filter

to maximize this cost function. To select the stabilizing feedback gains of the H-infinity controller an algebraic Riccati equation has to be repetitively solved at each time-step of the control algorithm [721,725,732]. The global stability properties of the control scheme are proven through Lyapunov analysis. First, it is demonstrated that the control loop satisfies the H-infinity tracking performance criterion which signifies robustness to model uncertainty and external perturbations [876,711,741]. Moreover, under moderate conditions, the global asymptotic stability is also proven. The

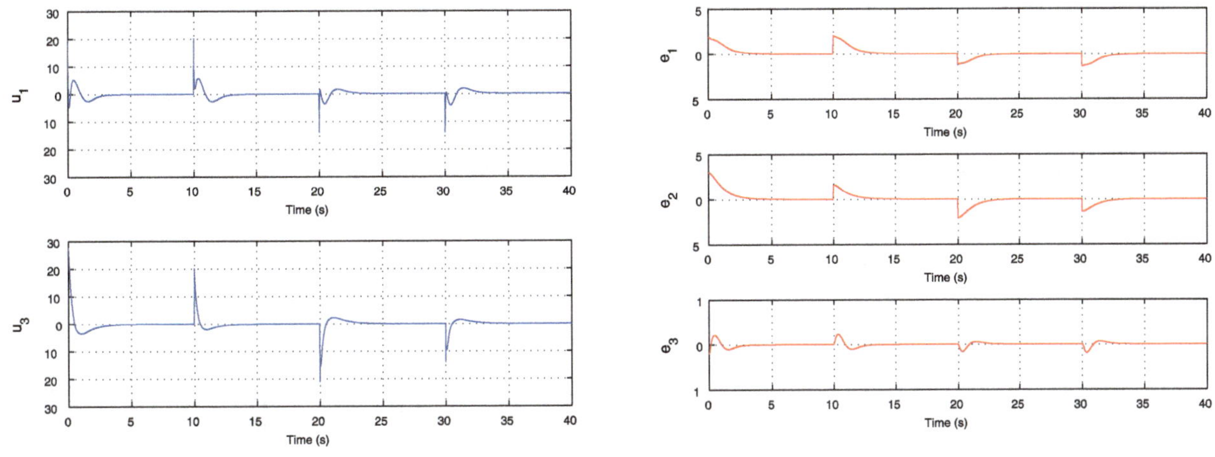

(a) (b)

Figure 3.50 Tracking of setpoint 6 by the two-cable-driven 3-DOF robotic crane with nonlinear optimal control: (a) variations of the control inputs of the gantry crane u_1 and u_2 (blue line); (b) convergence to 0 of the tracking error variables e_1, e_3, and e_5 associated with state variables $x_1 = x_R$, $x_3 = y_R$, and $x_5 = \theta$

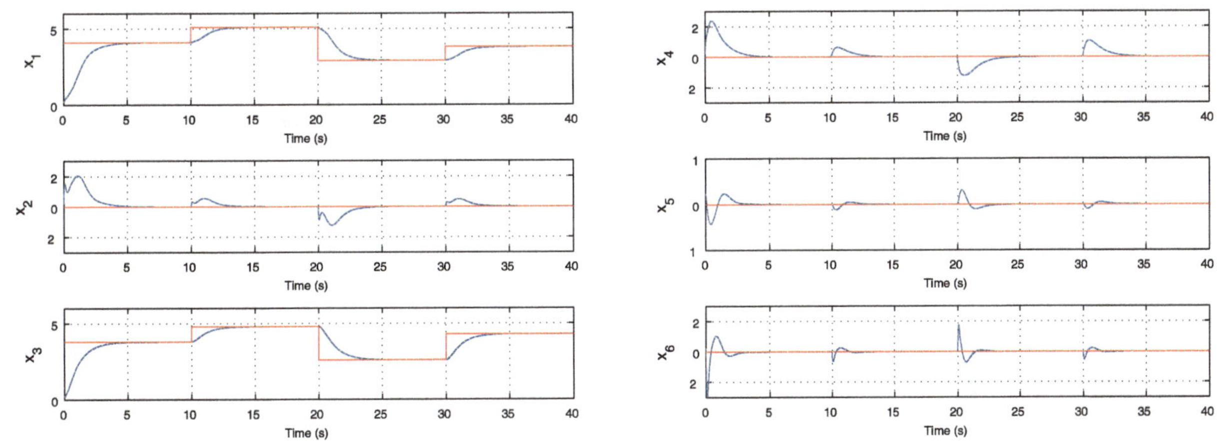

(a) (b)

Figure 3.51 Tracking of setpoint 7 by the two-cable-driven 3-DOF robotic crane with nonlinear optimal control: (a) tracking of setpoints (red line) by the state variables of the crane x_1 to x_3 (blue line) and estimates provided by the H-infinity Kalman filter; (b) tracking of setpoints (red line) by the state variables of the crane x_4 to x_6 (blue line) and estimates provided by the H-infinity Kalman filter

nonlinear optimal control method achieves fast and accurate tracking of reference setpoints under moderate variations of the control inputs.

In this section, a 3-DOF four-cable-driven planar parallel robot is described, which is equivalent in form, dynamics, and kinematics to the ones described in [230,3]. The section's findings can be extended to several types of cable-driven parallel robots and related applications. For instance: In [674], a detailed description of applications of cable-driven parallel robots to industrial tasks and projects is given, including building facades' maintenance and cleaning, ship

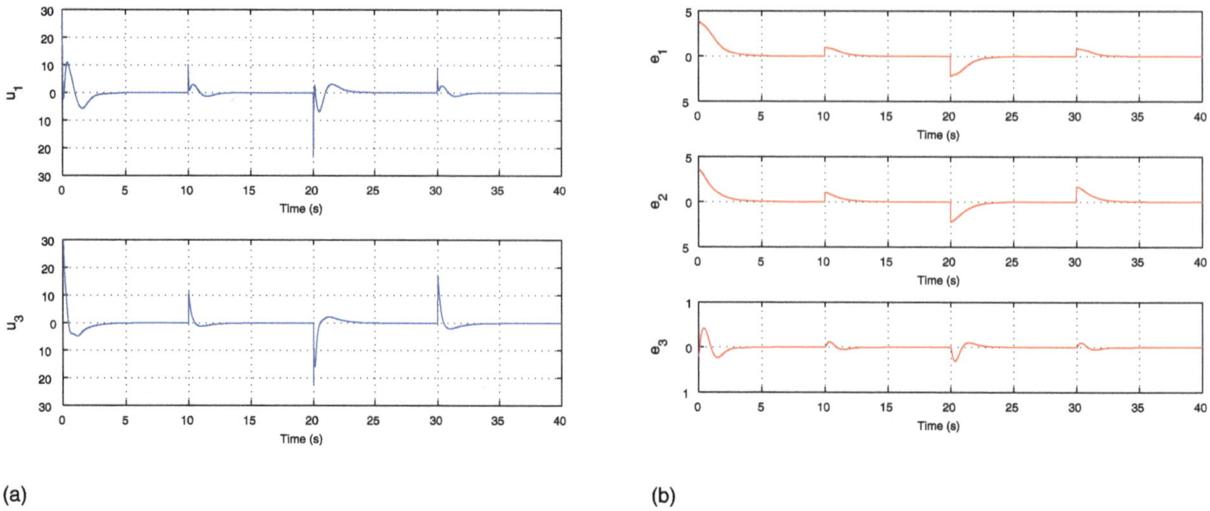

Figure 3.52 *Tracking of setpoint 7 by the two-cable-driven 3-DOF robotic crane with nonlinear optimal control: (a) variations of the control inputs of the gantry crane u_1 and u_2 (blue line); (b) convergence to 0 of the tracking error variables e_1, e_3, and e_5 associated with state variables $x_1 = x_R$, $x_3 = y_R$, and $x_5 = \theta$*

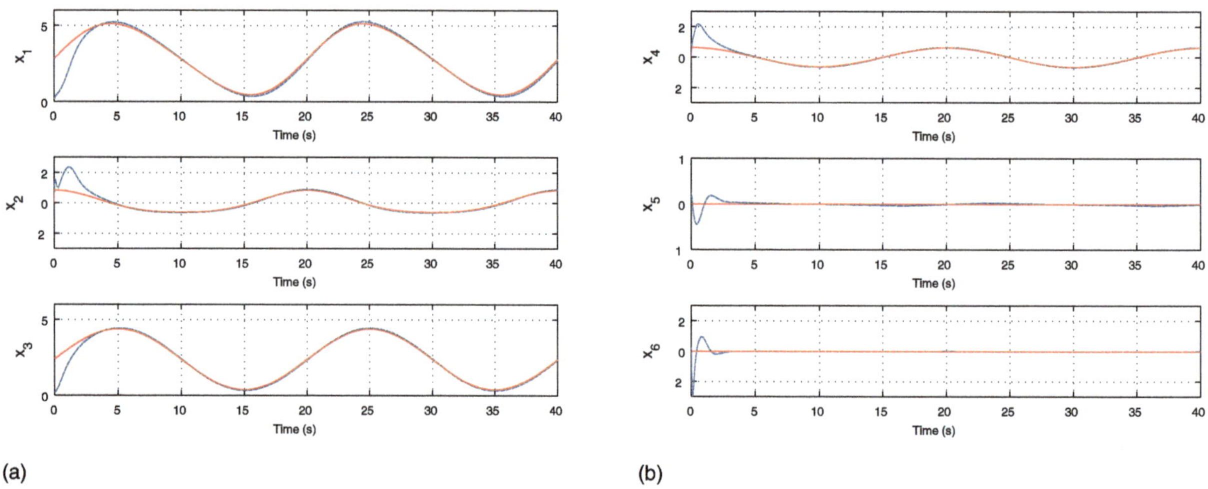

Figure 3.53 *Tracking of setpoint 8 by the two-cable-driven 3-DOF robotic crane with nonlinear optimal control: (a) tracking of setpoints (red line) by the state variables of the crane x_1 to x_3 (blue line) and estimates provided by the H-infinity Kalman filter; (b) tracking of setpoints (red line) by the state variables of the crane x_4 to x_6 (blue line) and estimates provided by the H-infinity Kalman filter*

building with the precise positioning and assembly of metallic plates, as well as painting and cleaning of ships' sides. In [780], the use of cable-driven parallel robots is proposed for cleaning the exterior walls of buildings. In [669], cable-driven parallel robots are used in logistics and for the rapid storage and retrieval of products in warehouses. In [267], cable-driven planar parallel robots are used for sandblasting and painting of large surfaces in buildings. In [858], cable-driven planar parallel robots are proposed for the inspection and scanning of artwork, frescoes, and paintings. Finally, in [218], cable-driven parallel robots are used for home construction based on the precise positioning and assembly of pre-fabricated panels.

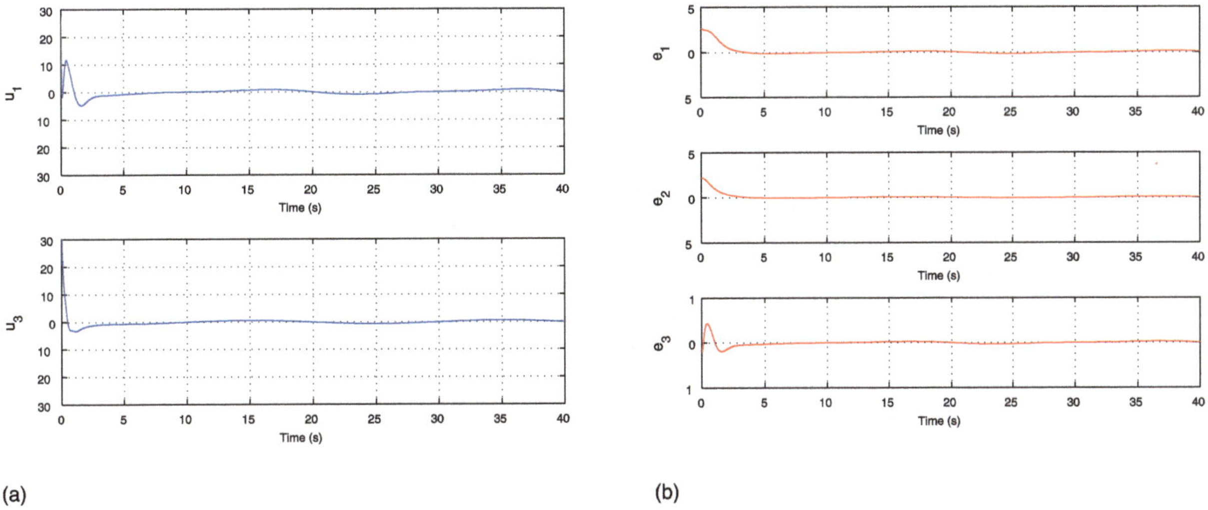

Figure 3.54 Tracking of setpoint 8 by the two-cable-driven 3-DOF robotic crane with nonlinear optimal control: (a) variations of the control inputs of the gantry crane u_1 and u_2 (blue line); (b) convergence to 0 of the tracking error variables e_1, e_3, and e_5 associated with state variables $x_1 = x_R$, $x_3 = y_R$, and $x_5 = \theta$

3.4.2 Dynamic model of the 3-DOF cable-driven parallel robot

3.4.2.1 State-space description of the cable-driven parallel robot

The diagram of the 3-DOF cable-driven parallel robot (CDPR) is given in Figure 3.55. This is a fully constrained CDPR. It consists of a rectangular plate of dimensions a and b which is rotated with the use of four cables attached to its vertices. The initial position of the rectangular plate, with its transversal axis to be parallel to the OX axis of the inertial reference frame is defined by the vertices positions which are denoted as Σ_i, $i = 1, \cdots, 4$. By varying the lengths of the cables, the rectangular plate performs both a translational motion which brings its center of gravity C to position (x, y) and a rotational motion due to which its transversal axis forms an angle θ with the horizontal OX axis. Under the new position and orientation of the plate, the coordinates of its vertices become

$$\begin{array}{ll} \Sigma_1' : (x - \frac{a}{2}, y - \frac{b}{2}) & \Sigma_2' : (x + \frac{a}{2}, y - \frac{b}{2}) \\ \Sigma_3' : (x + \frac{a}{2}, y + \frac{b}{2}) & \Sigma_4' : (x - \frac{a}{2}, y + \frac{b}{2}) \end{array} \tag{3.340}$$

The non-moving ends of the cables are defined by points $A_1 : (a_1, b_1)$, $A_2 : (a_2, b_2)$, $A_3 : (a_3, b_3)$, $A_4 : (a_4, b_4)$. Besides, after some trigonometric calculations, the vectors of the cables are found to be

$$\begin{array}{l} \vec{L}_1 = A_1\vec{\Sigma}_1' : [a_1 - (x - \frac{a}{2})cos(\theta) - (y - \frac{b}{2})sin(\theta),\ b_1 + (x - \frac{a}{2})sin(\theta) - (y - \frac{b}{2})cos(\theta)] \\ \vec{L}_2 = A_2\vec{\Sigma}_2' : [a_2 - (x + \frac{a}{2})cos(\theta) - (y - \frac{b}{2})sin(\theta),\ b_2 + (x + \frac{a}{2})sin(\theta) - (y - \frac{b}{2})cos(\theta)] \\ \vec{L}_3 = A_3\vec{\Sigma}_3' : [a_3 - (x + \frac{a}{2})cos(\theta) - (y + \frac{b}{2})sin(\theta),\ b_3 + (x + \frac{a}{2})sin(\theta) - (y + \frac{b}{2})cos(\theta)] \\ \vec{L}_4 = A_4\vec{\Sigma}_4' : [a_4 - (x - \frac{a}{2})cos(\theta) - (y + \frac{b}{2})sin(\theta),\ b_4 + (x - \frac{a}{2})sin(\theta) - (y + \frac{b}{2})cos(\theta)] \end{array} \tag{3.341}$$

The cables' length is computed as follows:

$$||L_1|| = \{[a_1 - (x - \tfrac{a}{2})cos(\theta) - (y - \tfrac{b}{2})sin(\theta)]^2 + [b_1 - (x - \tfrac{a}{2})sin(\theta) - (y - \tfrac{b}{2})cos(\theta)]^2\}^{\frac{1}{2}} \tag{3.342}$$

$$||L_2|| = \{[a_2 - (x + \tfrac{a}{2})cos(\theta) - (y - \tfrac{b}{2})sin(\theta)]^2 + [b_2 - (x + \tfrac{a}{2})sin(\theta) - (y - \tfrac{b}{2})cos(\theta)]^2\}^{\frac{1}{2}} \tag{3.343}$$

$$||L_3|| = \{[a_3 - (x + \tfrac{a}{2})cos(\theta) - (y + \tfrac{b}{2})sin(\theta)]^2 + [b_3 - (x + \tfrac{a}{2})sin(\theta) - (y + \tfrac{b}{2})cos(\theta)]^2\}^{\frac{1}{2}} \tag{3.344}$$

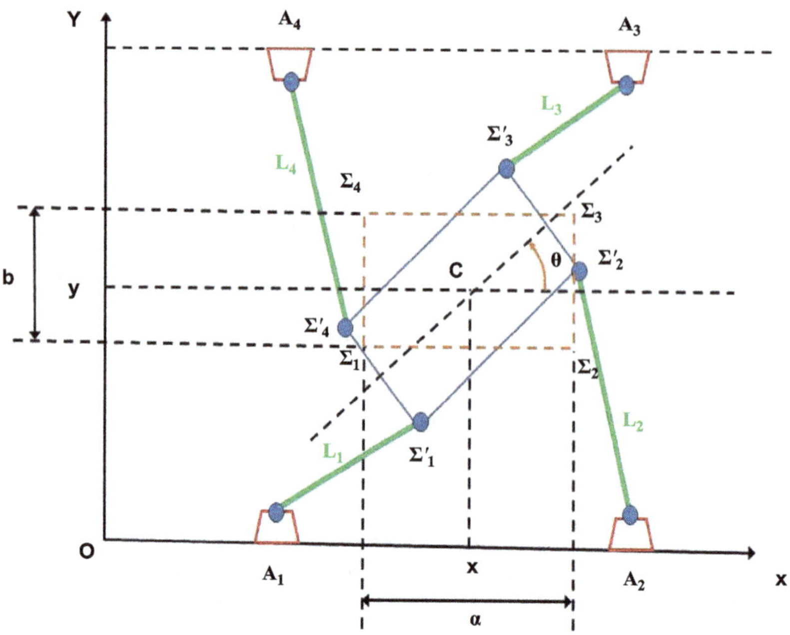

Figure 3.55 Diagram of the 3-DOF four-cable-driven parallel robot

$$||L_4|| = \{[a_4 - (x - \tfrac{a}{2})cos(\theta) - (y + \tfrac{b}{2})sin(\theta)]^2 + [b_4 - (x - \tfrac{a}{2})sin(\theta) - (y + \tfrac{b}{2})cos(\theta)]^2\}^{\frac{1}{2}} \quad (3.345)$$

By differentiating the previous relations in time, one obtains

$$\begin{array}{c} ||\dot{L}_1|| = \frac{1}{||L_1||}\cdot[a_1 - (x - \tfrac{a}{2})cos(\theta) - (y - \tfrac{b}{2})sin(\theta)]\cdot \\ [-\dot{x}cos(\theta) + (x - \tfrac{a}{2}sin(\theta)\dot{\theta} - \dot{y}sin(\theta) - (y - \tfrac{b}{2})cos(\theta)\dot{\theta})] + \\ [b_1 - (x - \tfrac{a}{2})sin(\theta) - (y - \tfrac{b}{2})cos(\theta)]\cdot \\ [\dot{x}sin(\theta) + (x - \tfrac{a}{2}cos(\theta)\dot{\theta} - \dot{y}cos(\theta) + (y - \tfrac{b}{2})sin(\theta)\dot{\theta})] \end{array} \quad (3.346)$$

Next, by defining $X = [\dot{x}\ \dot{y}\ \dot{\theta}]$, the previous relation can be also written as

$$\begin{array}{c} ||\dot{L}_1|| = \frac{1}{||L_1||}\cdot[a_1 - (x - \tfrac{a}{2})cos(\theta) - (y - \tfrac{b}{2})sin(\theta)]\cdot \\ [-cos(\theta)\ sin(\theta)\ (x - \tfrac{a}{2}sin(\theta) + (y - \tfrac{b}{2})cos(\theta)]\dot{X}^T + \\ [b_1 - (x - \tfrac{a}{2})sin(\theta) - (y - \tfrac{b}{2})cos(\theta)]\cdot \\ [[sin(\theta)\ \ -cos(\theta)\ (x - \tfrac{a}{2})cos(\theta) + (y - \tfrac{b}{2})sin(\theta)]]\dot{X}^T \end{array} \quad (3.347)$$

The previous relation can be also written to the following concise form:

$$||\dot{L}_1|| = \frac{1}{||L_1||}[L_{11}\ L_{21}\ L_{31}]\dot{X}^T \quad (3.348)$$

where after intermediate operations one has about terms L_{11}, L_{21}, L_{31}, one has

$$||L_{11}|| = -a_1cos(\theta) + b_1sin(\theta) + (x - \tfrac{a}{2}) \quad (3.349)$$

$$||L_{21}|| = -a_1sin(\theta) - b_1cos(\theta) + (y - \tfrac{b}{2}) \quad (3.350)$$

$$||L_{31}|| = [a_1(x - \tfrac{a}{2}) + b_1(y - \tfrac{b}{2})]sin(\theta) + [-a_1(y - \tfrac{b}{2}) + b_1(x - \tfrac{a}{2})]cos(\theta) \quad (3.351)$$

Similarly, it holds that

$$||\dot{L}_2|| = \frac{1}{||L_2||}[L_{12}\; L_{22}\; L_{32}]\dot{X}^T \tag{3.352}$$

where after intermediate operations one has about terms L_{12}, L_{22}, L_{32}, one has

$$||L_{12}|| = -a_2cos(\theta) + b_2sin(\theta) + (x + \tfrac{a}{2}) \tag{3.353}$$

$$||L_{22}|| = -a_2sin(\theta) - b_2cos(\theta) + (y - \tfrac{b}{2}) \tag{3.354}$$

$$||L_{32}|| = [a_2(x + \tfrac{a}{2}) + b_2(y - \tfrac{b}{2})]sin(\theta) + [-a_2(y - \tfrac{b}{2}) + b_2(x + \tfrac{a}{2})]cos(\theta) \tag{3.355}$$

Additionally, it holds that

$$||\dot{L}_3|| = \frac{1}{||L_3||}[L_{13}\; L_{23}\; L_{33}]\dot{X}^T \tag{3.356}$$

where after intermediate operations one has about terms L_{13}, L_{23}, L_{33}, one has

$$||L_{13}|| = -a_3cos(\theta) + b_3sin(\theta) + (x + \tfrac{a}{2}) \tag{3.357}$$

$$||L_{23}|| = -a_3sin(\theta) - b_3cos(\theta) + (y + \tfrac{b}{2}) \tag{3.358}$$

$$||L_{33}|| = [a_3(x + \tfrac{a}{2}) + b_3(y + \tfrac{b}{2})]sin(\theta) + [-a_3(y + \tfrac{b}{2}) + b_3(x + \tfrac{a}{2})]cos(\theta) \tag{3.359}$$

Moreover, it holds that

$$||\dot{L}_4|| = \frac{1}{||L_4||}[L_{14}\; L_{24}\; L_{34}]\dot{X}^T \tag{3.360}$$

where after intermediate operations one has about terms L_{14}, L_{24}, L_{34}, one has

$$||L_{14}|| = -a_4cos(\theta) + b_4sin(\theta) + (x - \tfrac{a}{2}) \tag{3.361}$$

$$||L_{24}|| = -a_4sin(\theta) - b_4cos(\theta) + (y + \tfrac{b}{2}) \tag{3.362}$$

$$||L_{34}|| = [a_4(x - \tfrac{a}{2}) + b_4(y + \tfrac{b}{2})]sin(\theta) + [-a_4(y + \tfrac{b}{2}) + b_4(x - \tfrac{a}{2})]cos(\theta) \tag{3.363}$$

Using the above, the Jacobian matrix of the robot's kinematic model which connects changes of the length of the links $\dot{L}_i = [||\dot{L}_1||\; ||\dot{L}_2||\; ||\dot{L}_3||\; ||\dot{L}_4||]^T$ to changes of the state variables of the robot $\dot{X} = [\dot{x}\; \dot{y}\; \dot{\theta}]^T$ is given through the relation

$$\begin{pmatrix} \dot{x} \\ \dot{y} \\ \dot{\theta} \end{pmatrix} = \begin{pmatrix} \frac{L_{11}}{||L_1||} & \frac{L_{12}}{||L_2||} & \frac{L_{13}}{||L_3||} & \frac{L_{14}}{||L_4||} \\ \frac{L_{21}}{||L_1||} & \frac{L_{22}}{||L_2||} & \frac{L_{23}}{||L_3||} & \frac{L_{24}}{||L_4||} \\ \frac{L_{31}}{||L_1||} & \frac{L_{32}}{||L_2||} & \frac{L_{33}}{||L_3||} & \frac{L_{34}}{||L_4||} \end{pmatrix} \begin{pmatrix} ||\dot{L}_1|| \\ ||\dot{L}_2|| \\ ||\dot{L}_3|| \\ ||\dot{L}_4|| \end{pmatrix} \tag{3.364}$$

or equivalently it holds that $\dot{X} = J^T\dot{L}_i$, with the transpose of the Jacobian matrix to be given by

$$J^T = \begin{pmatrix} \frac{L_{11}}{||L_1||} & \frac{L_{12}}{||L_2||} & \frac{L_{13}}{||L_3||} & \frac{L_{14}}{||L_4||} \\ \frac{L_{21}}{||L_1||} & \frac{L_{22}}{||L_2||} & \frac{L_{23}}{||L_3||} & \frac{L_{24}}{||L_4||} \\ \frac{L_{31}}{||L_1||} & \frac{L_{32}}{||L_2||} & \frac{L_{33}}{||L_3||} & \frac{L_{34}}{||L_4||} \end{pmatrix} \tag{3.365}$$

The kinetic energy of the robot is $K = \frac{1}{2}(\dot{x}^2 + \dot{y}^2 + \frac{1}{2}J\dot{\theta}^2)$ and the potential energy is $P = mgy$, while the Lagrangian of the system is $L = K - P$. Next, using Euler-Lagrange analysis and by defining as $M(X)$ the inertia matrix, $D(\dot{X})$ the damping forces matrix, $G(X)$ the gravitational forces matrix and $F = [F_1, F_2, F_3, F_4]^T$ the vector of forces which are exerted at the fixed end of the cables, one finds that the dynamic model of the robot is given by [721,732], [465,320]

$$M\ddot{X} + D\dot{X} + G(X) = J^TF \tag{3.366}$$

or equivalently, one arrives at the state-space description of the overactuated system

$$\begin{pmatrix} m & 0 & 0 \\ 0 & m & 0 \\ 0 & 0 & J \end{pmatrix}\begin{pmatrix} \ddot{x} \\ \ddot{y} \\ \ddot{\theta} \end{pmatrix} + \begin{pmatrix} -k_{b1}\dot{x} \\ -k_{b2}\dot{y} \\ -k_{b3}\dot{\theta} \end{pmatrix} + \begin{pmatrix} 0 \\ -mg \\ 0 \end{pmatrix} = \begin{pmatrix} \frac{L_{11}}{||L_1||} & \frac{L_{12}}{||L_2||} & \frac{L_{13}}{||L_3||} & \frac{L_{14}}{||L_4||} \\ \frac{L_{21}}{||L_1||} & \frac{L_{22}}{||L_2||} & \frac{L_{23}}{||L_3||} & \frac{L_{24}}{||L_4||} \\ \frac{L_{31}}{||L_1||} & \frac{L_{32}}{||L_2||} & \frac{L_{33}}{||L_3||} & \frac{L_{34}}{||L_4||} \end{pmatrix}\begin{pmatrix} F_1 \\ F_2 \\ f_3 \\ F_4 \end{pmatrix} \tag{3.367}$$

By inverting the inertia matrix M and after some intermediate operations, one obtains the following state-space description for the robot's dynamics:

$$\begin{pmatrix} \ddot{x} \\ \ddot{y} \\ \ddot{\theta} \end{pmatrix} = \begin{pmatrix} \frac{k_{b1}}{m}\dot{x} \\ \frac{k_{b2}}{m}\dot{y} + g \\ \frac{k_{b3}}{J}\dot{\theta} \end{pmatrix} + \begin{pmatrix} \frac{L_{11}}{m||L_1||} & \frac{L_{12}}{m||L_2||} & \frac{L_{13}}{m||L_3||} & \frac{L_{14}}{m||L_4||} \\ \frac{L_{21}}{m||L_1||} & \frac{L_{22}}{m||L_2||} & \frac{L_{23}}{m||L_3||} & \frac{L_{24}}{m||L_4||} \\ \frac{L_{31}}{J||L_1||} & \frac{L_{32}}{J||L_2||} & \frac{L_{33}}{J||L_3||} & \frac{L_{34}}{J||L_4||} \end{pmatrix}\begin{pmatrix} F_1 \\ F_2 \\ f_3 \\ F_4 \end{pmatrix} \tag{3.368}$$

Next, by defining the state vector of the cable-driven parallel robot as $\tilde{X} = [x, \dot{x}, y, \dot{y}, \theta, \dot{\theta}]^T$, or equivalently $\tilde{X} = [x_1, x_2, x_3, x_4, x_5, x_6]^T$ and the control inputs vector $U = [F_1, F_2, F_3, F_4]^T$ or equivalently $U = [u_1, u_2, u_3, u_4]^T$, one obtains the following state-space description:

$$\begin{pmatrix} \dot{x}_1 \\ \dot{x}_2 \\ \dot{x}_3 \\ \dot{x}_4 \\ \dot{x}_5 \\ \dot{x}_6 \end{pmatrix} = \begin{pmatrix} x_2 \\ \frac{k_{b1}}{m}x_2 \\ x_4 \\ \frac{k_{b2}}{m}x_4 + g \\ x_6 \\ \frac{k_{b3}}{J}x_6 \end{pmatrix} + \begin{pmatrix} 0 & 0 & 0 & 0 \\ \frac{L_{11}}{m||L_1||} & \frac{L_{12}}{m||L_2||} & \frac{L_{13}}{m||L_3||} & \frac{L_{14}}{m||L_4||} \\ 0 & 0 & 0 & 0 \\ \frac{L_{21}}{m||L_1||} & \frac{L_{22}}{m||L_2||} & \frac{L_{23}}{m||L_3||} & \frac{L_{24}}{m||L_4||} \\ 0 & 0 & 0 & 0 \\ \frac{L_{31}}{J||L_1||} & \frac{L_{32}}{J||L_2||} & \frac{L_{33}}{J||L_3||} & \frac{L_{34}}{J||L_4||} \end{pmatrix}\begin{pmatrix} F_1 \\ F_2 \\ F_3 \\ F_4 \end{pmatrix} \tag{3.369}$$

which is a dynamic system in the nonlinear affine-in-the-input state-space form

$$\dot{x} = f(x) + g(x)u \tag{3.370}$$

$x \in R^{6\times 1}$, $f(x) \in R^{6\times 1}$, $g(x) \in R^{6\times 4}$, and $u \in R^{4\times 1}$.

3.4.2.2 Differential flatness properties of the cable-driven parallel robot

It has been shown that the dynamic model of the 3-DOF cable-driven parallel robot can be written in the concise state-space form of Eq. (3.370), that is,

$$\begin{pmatrix} \dot{x}_1 \\ \dot{x}_2 \\ \dot{x}_3 \\ \dot{x}_4 \\ \dot{x}_5 \\ \dot{x}_6 \end{pmatrix} = \begin{pmatrix} x_2 \\ \frac{k_{b1}}{m}x_2 \\ x_4 \\ \frac{k_{b2}}{m}x_4 + g \\ x_6 \\ \frac{k_{b3}}{J}x_6 \end{pmatrix} + \begin{pmatrix} 0 & 0 & 0 & 0 \\ \frac{L_{11}}{m||L_1||} & \frac{L_{12}}{m||L_2||} & \frac{L_{13}}{m||L_3||} & \frac{L_{14}}{m||L_4||} \\ 0 & 0 & 0 & 0 \\ \frac{L_{21}}{m||L_1||} & \frac{L_{22}}{m||L_2||} & \frac{L_{23}}{m||L_3||} & \frac{L_{24}}{m||L_4||} \\ 0 & 0 & 0 & 0 \\ \frac{L_{31}}{J||L_1||} & \frac{L_{32}}{J||L_2||} & \frac{L_{33}}{J||L_3||} & \frac{L_{34}}{J||L_4||} \end{pmatrix}\begin{pmatrix} F_1 \\ F_2 \\ F_3 \\ F_4 \end{pmatrix} \tag{3.371}$$

The flat outputs vector of the system is taken to be $Y = [x_1, x_3, x_5]$. It holds that

$$x_2 = \dot{x}_1 \quad x_4 = \dot{x}_3 \quad x_6 = \dot{x}_5 \tag{3.372}$$

or equivalently

$$x_2 = h_2(Y, \dot{Y}) \quad x_4 = h_4(Y, \dot{Y}) \quad x_6 = h_6(Y, \dot{Y}) \tag{3.373}$$

which signifies that state variables x_2, x_4, x_6 are differential functions of the flat outputs vector of the robot. Besides, Eq. (3.371) can be also written as

$$\begin{pmatrix} \ddot{x}_1 \\ \ddot{x}_3 \\ \ddot{x}_5 \end{pmatrix} = \begin{pmatrix} \frac{k_{b1}}{m} x_2 \\ \frac{k_{b2}}{m} x_4 + g \\ \frac{k_{b3}}{J} x_6 \end{pmatrix} + \begin{pmatrix} \frac{L_{11}}{m||L_1||} & \frac{L_{12}}{m||L_2||} & \frac{L_{13}}{m||L_3||} & \frac{L_{14}}{m||L_4||} \\ \frac{L_{21}}{m||L_1||} & \frac{L_{22}}{m||L_2||} & \frac{L_{23}}{m||L_3||} & \frac{L_{24}}{m||L_4||} \\ \frac{L_{31}}{J||L_1||} & \frac{L_{32}}{J||L_2||} & \frac{L_{33}}{J||L_3||} & \frac{L_{34}}{J||L_4||} \end{pmatrix} \begin{pmatrix} F_1 \\ F_2 \\ F_3 \\ F_4 \end{pmatrix} \tag{3.374}$$

and concisely in the matrix form

$$\ddot{Y} = F_a(Y, \dot{Y}) + G_a(Y, \dot{Y})U \tag{3.375}$$

From Eq. (3.375), one can solve for the control inputs vector U using the Moore-Penrose inverse matrix of $G_a(Y, \dot{Y})$. This gives

$$U = [G_a^T G_a]^{-1} G_a^T [\ddot{Y} - F_a] \tag{3.376}$$

or equivalently

$$U = h_U(Y, \dot{Y}) \tag{3.377}$$

which signifies that the control inputs vector U is also a differential function of the flat outputs of the system. Consequently, all state variables and control inputs of the 3-DOF cable-driven parallel robot are differential functions of the robot's flat outputs and this system is differentially flat. The differential flatness property is also an implicit proof of the controllability of this robotic system.

3.4.3 Approximate linearization of the dynamics of the cable-driven parallel robot

The dynamic model of the cable-driven parallel robot undergoes a linearization process around a temporary operating point (x^*, u^*) which is updated at each sampling instance. The linearization is based on first-order Taylor series expansion and on the computation of the associated Jacobian matrices. The temporary operating point (x^*, u^*) is defined at each sampling interval by the present value of the system's state vector x^* and by the last sampled value of the control inputs vector u^*. The modeling error which is due to the truncation of higher-order terms in the Taylor series expansion is considered to be a perturbation which is asymptotically compensated by the robustness of the control algorithm.

Thus, the robot's dynamic model being initially in the nonlinear state-space form $\dot{x} = f(x) + g(x)u$ is brought through Taylor series-based linearization into the equivalent model

$$\dot{x} = Ax + Bu + \tilde{d} \tag{3.378}$$

where A, B are the Jacobian matrices of the system which are defined as follows:

$$\begin{aligned} A = [\nabla_x f(x) + g(x)u] \mid_{(x^*,u^*)} \Rightarrow \\ A = \nabla_x f(x) \mid_{(x^*,u^*)} + \nabla_x g_1(x) u_1 \mid_{(x^*,u^*)} + \nabla_x g_2(x) u_2 \mid_{(x^*,u^*)} + \\ + \nabla_x g_3(x) u_3 \mid_{(x^*,u^*)} + \nabla_x g_4(x) u_4 \mid_{(x^*,u^*)} \end{aligned} \tag{3.379}$$

$$B = \nabla_x [f(x) + g(x)u] \mid_{(x^*,u^*)} \Rightarrow B = g(x) \mid_{(x^*,u^*)} \tag{3.380}$$

The term $\tilde{d}$ is the cumulative disturbance vector which may comprise: (i) model imprecision due to the truncation of higher-order terms from the Taylor series expansion, (ii) exogenous perturbations, and (iii) sensor measurement noise of any distribution.

Computation of the Jacobian matrix $\nabla_x f(x) \mid_{(x^*,u^*)}$:

$$\nabla_x f(x) \mid_{(x^*,u^*)}= \begin{pmatrix} 0 & 1 & 0 & 0 & 0 & 0 \\ 0 & \frac{k_{b1}}{m} & 0 & 0 & 0 & 0 \\ 0 & 0 & 0 & 1 & 0 & 0 \\ 0 & 0 & 0 & \frac{k_{b2}}{m} & 0 & 0 \\ 0 & 0 & 0 & 0 & 0 & 1 \\ 0 & 0 & 0 & 0 & 0 & \frac{k_{b3}}{J} \end{pmatrix} \tag{3.381}$$

Moreover, the control inputs gain matrix $g(x)$ is decomposed into the following vectors:

$$g_1(x) = \begin{pmatrix} 0 \\ \frac{L_{11}}{m||L_1||} \\ 0 \\ \frac{L_{21}}{m||L_1||} \\ 0 \\ \frac{L_{31}}{J||L_1||} \end{pmatrix} \quad g_2(x) = \begin{pmatrix} 0 \\ \frac{L_{12}}{m||L_2||} \\ 0 \\ \frac{L_{22}}{m||L_2||} \\ 0 \\ \frac{L_{32}}{J||L_2||} \end{pmatrix} \quad g_3(x) = \begin{pmatrix} 0 \\ \frac{L_{13}}{m||L_3||} \\ 0 \\ \frac{L_{23}}{m||L_3||} \\ 0 \\ \frac{L_{33}}{J||L_3||} \end{pmatrix} \quad g_4(x) = \begin{pmatrix} 0 \\ \frac{L_{14}}{m||L_4||} \\ 0 \\ \frac{L_{24}}{m||L_4||} \\ 0 \\ \frac{L_{34}}{J||L_4||} \end{pmatrix} \tag{3.382}$$

Computation of the Jacobian matrix $\nabla_x g_1(x) \mid_{(x^*,u^*)}$:
First row of the Jacobian matrix $\nabla_x g_1(x) \mid_{(x^*,u^*)}$: $\frac{\partial g_{11}}{\partial x_i} = 0$ for $i = 1, 2, \cdots, 6$.
Second row of the Jacobian matrix $\nabla_x g_1(x) \mid_{(x^*,u^*)}$: for $i = 1, 2, \cdots, 6$

$$\frac{\partial g_{12}}{\partial x_i} = \frac{1}{m} \frac{\frac{\partial L_{11}}{\partial x_i}||L_1|| - L_{11}\frac{\partial ||L_1||}{\partial x_i}}{||L_1||^2} \tag{3.383}$$

where it holds that: $\frac{\partial L_{11}}{\partial x_1} = 1$, $\frac{\partial L_{11}}{\partial x_2} = 0$, $\frac{\partial L_{11}}{\partial x_3} = 0$, $\frac{\partial L_{11}}{\partial x_4} = 0$, $\frac{\partial L_{11}}{\partial x_5} = a_1 sin(x_5) + b_1 cos(x_5)$, and $\frac{\partial L_{11}}{\partial x_6} = 0$.
Moreover, about the partial derivative of $||L_1||$ with respect to x_i, $i = 1, 2, \cdots, 6$, it holds that

$$\begin{aligned}\frac{\partial ||L_1||}{\partial x_1} &= \frac{1}{2}\frac{1}{||L_1||}\{2[a_1 - (x_1 - \tfrac{a}{2})cos(x_5) - (x_3 - \tfrac{b}{2})sin(x_5)]cos(x_5) + \\ &\quad 2[b_1 + (x_1 - \tfrac{a}{2})sin(x_5) - (x_3 - \tfrac{b}{2})cos(x_5)]sin(x_5)\}\end{aligned} \tag{3.384}$$

$$\begin{aligned}\frac{\partial ||L_1||}{\partial x_2} &= \frac{1}{2}\frac{1}{||L_1||}\{2[a_1 - (x_1 - \tfrac{a}{2})cos(x_5) - (x_3 - \tfrac{b}{2})sin(x_5)]0 + \\ &\quad 2[b_1 + (x_1 - \tfrac{a}{2})sin(x_5) - (x_3 - \tfrac{b}{2})cos(x_5)]0\} \\ &\Rightarrow \frac{\partial ||L_1||}{\partial x_2} = 0\end{aligned} \tag{3.385}$$

$$\begin{aligned}\frac{\partial ||L_1||}{\partial x_3} &= \frac{1}{2}\frac{1}{||L_1||}\{2[a_1 - (x_1 - \tfrac{a}{2})cos(x_5) - (x_3 - \tfrac{b}{2})sin(x_5)](-sin(x_5)) + \\ &\quad 2[b_1 + (x_1 - \tfrac{a}{2})sin(x_5) - (x_3 - \tfrac{b}{2})cos(x_5)](-cos(x_5))\}\end{aligned} \tag{3.386}$$

$$\begin{aligned}\frac{\partial ||L_1||}{\partial x_4} &= \frac{1}{2}\frac{1}{||L_1||}\{2[a_1 - (x_1 - \tfrac{a}{2})cos(x_5) - (x_3 - \tfrac{b}{2})sin(x_5)]0 + \\ &\quad 2[b_1 + (x_1 - \tfrac{a}{2})sin(x_5) - (x_3 - \tfrac{b}{2})cos(x_5)]0\} \\ &\Rightarrow \frac{\partial ||L_1||}{\partial x_4} = 0\end{aligned} \tag{3.387}$$

$$\begin{aligned}\frac{\partial ||L_1||}{\partial x_5} &= \frac{1}{2}\frac{1}{||L_1||}\{2[a_1 - (x_1 - \tfrac{a}{2})cos(x_5) - (x_3 - \tfrac{b}{2})sin(x_5)][(x_1 - \tfrac{a}{2})(-sin(x_5)) + (x_3 - \tfrac{b}{2})cos(x_5)] + \\ &\quad 2[b_1 + (x_1 - \tfrac{a}{2})sin(x_5) - (x_3 - \tfrac{b}{2})cos(x_5)][(x_1 - \tfrac{a}{2})cos(x_5) - (x_3 - \tfrac{b}{2})](-sin(x_5))\}\end{aligned} \tag{3.388}$$

$$\begin{aligned}\frac{\partial||L_1||}{\partial x_6} &= \frac{1}{2}\frac{1}{||L_1||}\{2[a_1-(x_1-\tfrac{a}{2})cos(x_5)-(x_3-\tfrac{b}{2})sin(x_5)]0+\\ &\quad 2[b_1+(x_1-\tfrac{a}{2})sin(x_5)-(x_3-\tfrac{b}{2})cos(x_5)]0\}\\ &\Rightarrow \frac{\partial||L_1||}{\partial x_6}=0\end{aligned} \tag{3.389}$$

Third row of the Jacobian matrix $\nabla_x g_1(x)\mid_{(x^*,u^*)}$: $\frac{\partial g_{13}}{\partial x_i}=0$ for $i=1,2,\cdots,6$.
Fourth row of the Jacobian matrix $\nabla_x g_1(x)\mid_{(x^*,u^*)}$: for $i=1,2,\cdots,6$

$$\frac{\partial g_{14}}{\partial x_i}=\frac{1}{m}\frac{\frac{\partial L_{21}}{\partial x_i}||L_1||-L_{21}\frac{\partial||L_1||}{\partial x_i}}{||L_1||^2} \tag{3.390}$$

where it holds that: $\frac{\partial L_{21}}{\partial x_1}=0$, $\frac{\partial L_{21}}{\partial x_2}=0$, $\frac{\partial L_{21}}{\partial x_3}=1$, $\frac{\partial L_{21}}{\partial x_4}=0$, $\frac{\partial L_{21}}{\partial x_5}=-a_1cos(x_5)+b_1sin(x_5)$, and $\frac{\partial L_{21}}{\partial x_6}=0$.
Fifth row of the Jacobian matrix $\nabla_x g_1(x)\mid_{(x^*,u^*)}$: $\frac{\partial g_{15}}{\partial x_i}=0$ for $i=1,2,\cdots,6$.
Sixth row of the Jacobian matrix $\nabla_x g_1(x)\mid_{(x^*,u^*)}$: for $i=1,2,\cdots,6$

$$\frac{\partial g_{16}}{\partial x_i}=\frac{1}{J}\frac{\frac{\partial L_{31}}{\partial x_i}||L_1||-L_{31}\frac{\partial||L_1||}{\partial x_i}}{||L_1||^2} \tag{3.391}$$

where it holds that: $\frac{\partial L_{31}}{\partial x_1}=a_1sin(x_5)+b_1cos(x_5)$, $\frac{\partial L_{31}}{\partial x_2}=0$, $\frac{\partial L_{31}}{\partial x_3}=b_1sin(x_5)-a_1sin(x_5)$, $\frac{\partial L_{31}}{\partial x_4}=0$, $\frac{\partial L_{31}}{\partial x_5}=[a_1(x_1-\frac{a}{2})+b_1(x_3-\frac{b}{2})]cos(x_5)+[-a_1(x_3-\frac{b}{2})+b_1(x_1-\frac{a}{2})](-sin(x_5))$, and $\frac{\partial L_{31}}{\partial x_6}=0$.

Thus, about the Jacobian matrix $\nabla_x g_1(x)\mid_{(x^*,u^*)}$, one obtains

$$\nabla_x g_1(x)\mid_{(}x^*,u^*)=\begin{pmatrix}\frac{\partial g_{11}}{\partial x_1}&\frac{\partial g_{11}}{\partial x_2}&\frac{\partial g_{11}}{\partial x_3}&\frac{\partial g_{11}}{\partial x_4}&\frac{\partial g_{11}}{\partial x_5}&\frac{\partial g_{11}}{\partial x_6}\\ \frac{\partial g_{12}}{\partial x_1}&\frac{\partial g_{12}}{\partial x_2}&\frac{\partial g_{12}}{\partial x_3}&\frac{\partial g_{12}}{\partial x_4}&\frac{\partial g_{12}}{\partial x_5}&\frac{\partial g_{12}}{\partial x_6}\\ \frac{\partial g_{13}}{\partial x_1}&\frac{\partial g_{13}}{\partial x_2}&\frac{\partial g_{13}}{\partial x_3}&\frac{\partial g_{13}}{\partial x_4}&\frac{\partial g_{13}}{\partial x_5}&\frac{\partial g_{13}}{\partial x_6}\\ \frac{\partial g_{14}}{\partial x_1}&\frac{\partial g_{14}}{\partial x_2}&\frac{\partial g_{14}}{\partial x_3}&\frac{\partial g_{14}}{\partial x_4}&\frac{\partial g_{14}}{\partial x_5}&\frac{\partial g_{14}}{\partial x_6}\\ \frac{\partial g_{15}}{\partial x_1}&\frac{\partial g_{15}}{\partial x_2}&\frac{\partial g_{15}}{\partial x_3}&\frac{\partial g_{15}}{\partial x_4}&\frac{\partial g_{15}}{\partial x_5}&\frac{\partial g_{15}}{\partial x_6}\\ \frac{\partial g_{16}}{\partial x_1}&\frac{\partial g_{16}}{\partial x_2}&\frac{\partial g_{16}}{\partial x_3}&\frac{\partial g_{16}}{\partial x_4}&\frac{\partial g_{16}}{\partial x_5}&\frac{\partial g_{16}}{\partial x_6}\end{pmatrix} \tag{3.392}$$

Computation of the Jacobian matrix $\nabla_x g_2(x)\mid_{(x^*,u^*)}$:

First row of the Jacobian matrix $\nabla_x g_2(x)\mid_{(x^*,u^*)}$: $\frac{\partial g_{21}}{\partial x_i}=0$ for $i=1,2,\cdots,6$.

Second row of the Jacobian matrix $\nabla_x g_2(x)\mid_{(x^*,u^*)}$: for $i=1,2,\cdots,6$

$$\frac{\partial g_{22}}{\partial x_i}=\frac{1}{m}\frac{\frac{\partial L_{12}}{\partial x_i}||L_2||-L_{12}\frac{\partial||L_2||}{\partial x_i}}{||L_2||^2} \tag{3.393}$$

where it holds that: $\frac{\partial L_{12}}{\partial x_1}=1$, $\frac{\partial L_{12}}{\partial x_2}=0$, $\frac{\partial L_{12}}{\partial x_3}=0$, $\frac{\partial L_{12}}{\partial x_4}=0$, $\frac{\partial L_{12}}{\partial x_5}=a_1sin(x_5)+b_1cos(x_5)$, and $\frac{\partial L_{12}}{\partial x_6}=0$.
Moreover, about the partial derivative of $||L_2||$ with respect to x_i, $i=1,2,\cdots,6$, it holds that

$$\begin{aligned}\frac{\partial||L_2||}{\partial x_1} &= \frac{1}{2}\frac{1}{||L_2||}\{2[a_2-(x_1+\tfrac{a}{2})cos(x_5)-(x_3-\tfrac{b}{2})sin(x_5)](-cos(x_5))+\\ &\quad 2[b_2+(x_1-\tfrac{a}{2})sin(x_5)-(x_3-\tfrac{b}{2})cos(x_5)]sin(x_5)\}\end{aligned} \tag{3.394}$$

$$\begin{aligned}\frac{\partial||L_2||}{\partial x_2} &= \frac{1}{2}\frac{1}{||L_2||}\{2[a_2-(x_1+\tfrac{a}{2})cos(x_5)-(x_3-\tfrac{b}{2})sin(x_5)]0+\\ &\quad 2[b_2+(x_1+\tfrac{a}{2})sin(x_5)-(x_3-\tfrac{b}{2})cos(x_5)]0\}\\ &\Rightarrow \frac{\partial||L_2||}{\partial x_2}=0\end{aligned} \tag{3.395}$$

$$\begin{aligned}\frac{\partial||L_2||}{\partial x_3} &= \frac{1}{2}\frac{1}{||L_2||}\{2[a_2-(x_1+\tfrac{a}{2})cos(x_5)-(x_3-\tfrac{b}{2})sin(x_5)](-sin(x_5))+\\ &\quad 2[b_2+(x_1+\tfrac{a}{2})sin(x_5)-(x_3-\tfrac{b}{2})cos(x_5)](-cos(x_5))\}\end{aligned} \tag{3.396}$$

$$\frac{\partial ||L_2||}{\partial x_4} = \frac{1}{2}\frac{1}{||L_2||}\{2[a_2 - (x_1 + \tfrac{a}{2})cos(x_5) - (x_3 - \tfrac{b}{2})sin(x_5)]0 + 2[b_2 + (x_1 + \tfrac{a}{2})sin(x_5) - (x_3 - \tfrac{b}{2})cos(x_5)]0\} \Rightarrow \frac{\partial ||L_2||}{\partial x_4} = 0 \tag{3.397}$$

$$\frac{\partial ||L_2||}{\partial x_5} = \frac{1}{2}\frac{1}{||L_2||}\{2[a_2 - (x_1 + \tfrac{a}{2})cos(x_5) - (x_3 - \tfrac{b}{2})sin(x_5)][(x_1 + \tfrac{a}{2})sin(x_5) + (x_3 - \tfrac{b}{2})cos(x_5)] + 2[b_2 + (x_1 + \tfrac{a}{2})sin(x_5) - (x_3 - \tfrac{b}{2})cos(x_5)][(x_1 + \tfrac{a}{2})cos(x_5) + (x_3 - \tfrac{b}{2})](sin(x_5))\} \tag{3.398}$$

$$\frac{\partial ||L_2||}{\partial x_6} = \frac{1}{2}\frac{1}{||L_2||}\{2[a_2 - (x_1 + \tfrac{a}{2})cos(x_5) - (x_3 - \tfrac{b}{2})sin(x_5)]0 + 2[b_2 + (x_1 + \tfrac{a}{2})sin(x_5) - (x_3 - \tfrac{b}{2})cos(x_5)]0\} \Rightarrow \frac{\partial ||L_2||}{\partial x_6} = 0 \tag{3.399}$$

Third row of the Jacobian matrix $\nabla_x g_2(x) \mid_{(x^*,u^*)}$: $\frac{\partial g_{23}}{\partial x_i} = 0$ for $i = 1, 2, \cdots, 6$.

Fourth row of the Jacobian matrix $\nabla_x g_2(x) \mid_{(x^*,u^*)}$: for $i = 1, 2, \cdots, 6$

$$\frac{\partial g_{24}}{\partial x_i} = \frac{1}{m}\frac{\frac{\partial L_{22}}{\partial x_i}||L_2|| - L_{22}\frac{\partial ||L_2||}{\partial x_i}}{||L_2||^2} \tag{3.400}$$

where it holds that: $\frac{\partial L_{22}}{\partial x_1} = 0$, $\frac{\partial L_{22}}{\partial x_2} = 0$, $\frac{\partial L_{22}}{\partial x_3} = 1$, $\frac{\partial L_{22}}{\partial x_4} = 0$, $\frac{\partial L_{22}}{\partial x_5} = -a_2cos(x_5) + b_2sin(x_5)$, and $\frac{\partial L_{22}}{\partial x_6} = 0$.

Fifth row of the Jacobian matrix $\nabla_x g_2(x) \mid_{(x^*,u^*)}$: $\frac{\partial g_{25}}{\partial x_i} = 0$ for $i = 1, 2, \cdots, 6$.

Sixth row of the Jacobian matrix $\nabla_x g_2(x) \mid_{(x^*,u^*)}$: for $i = 1, 2, \cdots, 6$

$$\frac{\partial g_{26}}{\partial x_i} = \frac{1}{J}\frac{\frac{\partial L_{32}}{\partial x_i}||L_2|| - L_{32}\frac{\partial ||L_2||}{\partial x_i}}{||L_2||^2} \tag{3.401}$$

where it holds that: $\frac{\partial L_{32}}{\partial x_1} = a_2sin(x_5) + b_2cos(x_5)$, $\frac{\partial L_{32}}{\partial x_2} = 0$, $\frac{\partial L_{32}}{\partial x_3} = b_2sin(x_5) - a_2cos(x_5)$, $\frac{\partial L_{32}}{\partial x_4} = 0$, $\frac{\partial L_{32}}{\partial x_5} = [a_2(x_1 + \frac{a}{2}) + b_2(x_3 - \frac{b}{2})]cos(x_5) - [-a_2(x_3 - \frac{b}{2}) + b_2(x_1 + \frac{a}{2})]sin(x_5)$, and $\frac{\partial L_{32}}{\partial x_6} = 0$.

Thus, about the Jacobian matrix $\nabla_x g_2(x) \mid_{(x^*,u^*)}$, one obtains

$$\nabla_x g_2(x) \mid_{(x^*, u^*)} = \begin{pmatrix} \frac{\partial g_{21}}{\partial x_1} & \frac{\partial g_{21}}{\partial x_2} & \frac{\partial g_{21}}{\partial x_3} & \frac{\partial g_{21}}{\partial x_4} & \frac{\partial g_{21}}{\partial x_5} & \frac{\partial g_{21}}{\partial x_6} \\ \frac{\partial g_{22}}{\partial x_1} & \frac{\partial g_{22}}{\partial x_2} & \frac{\partial g_{22}}{\partial x_3} & \frac{\partial g_{22}}{\partial x_4} & \frac{\partial g_{22}}{\partial x_5} & \frac{\partial g_{22}}{\partial x_6} \\ \frac{\partial g_{23}}{\partial x_1} & \frac{\partial g_{23}}{\partial x_2} & \frac{\partial g_{23}}{\partial x_3} & \frac{\partial g_{23}}{\partial x_4} & \frac{\partial g_{23}}{\partial x_5} & \frac{\partial g_{23}}{\partial x_6} \\ \frac{\partial g_{24}}{\partial x_1} & \frac{\partial g_{24}}{\partial x_2} & \frac{\partial g_{24}}{\partial x_3} & \frac{\partial g_{24}}{\partial x_4} & \frac{\partial g_{24}}{\partial x_5} & \frac{\partial g_{24}}{\partial x_6} \\ \frac{\partial g_{25}}{\partial x_1} & \frac{\partial g_{25}}{\partial x_2} & \frac{\partial g_{25}}{\partial x_3} & \frac{\partial g_{25}}{\partial x_4} & \frac{\partial g_{25}}{\partial x_5} & \frac{\partial g_{25}}{\partial x_6} \\ \frac{\partial g_{26}}{\partial x_1} & \frac{\partial g_{26}}{\partial x_2} & \frac{\partial g_{26}}{\partial x_3} & \frac{\partial g_{26}}{\partial x_4} & \frac{\partial g_{26}}{\partial x_5} & \frac{\partial g_{26}}{\partial x_6} \end{pmatrix} \tag{3.402}$$

Computation of the Jacobian matrix $\nabla_x g_3(x) \mid_{(x^*,u^*)}$:

First row of the Jacobian matrix $\nabla_x g_3(x) \mid_{(x^*,u^*)}$: $\frac{\partial g_{31}}{\partial x_i} = 0$ for $i = 1, 2, \cdots, 6$.

Second row of the Jacobian matrix $\nabla_x g_3(x) \mid_{(x^*,u^*)}$: for $i = 1, 2, \cdots, 6$

$$\frac{\partial g_{32}}{\partial x_i} = \frac{1}{m}\frac{\frac{\partial L_{13}}{\partial x_i}||L_3|| - L_{13}\frac{\partial ||L_3||}{\partial x_i}}{||L_3||^2} \tag{3.403}$$

where it holds that: $\frac{\partial L_{13}}{\partial x_1} = 1$, $\frac{\partial L_{13}}{\partial x_2} = 0$, $\frac{\partial L_{13}}{\partial x_3} = 0$, $\frac{\partial L_{13}}{\partial x_4} = 0$, $\frac{\partial L_{13}}{\partial x_5} = a_3sin(x_5) - b_3cos(x_5)$, and $\frac{\partial L_{13}}{\partial x_6} = 0$.

Moreover, about the partial derivative of $||L_3||$ with respect to x_i, $i=1,2,\cdots,6$, it holds that

$$\begin{aligned}\frac{\partial||L_3||}{\partial x_1}=&\frac{1}{2}\frac{1}{||L_3||}\{2[a_3-(x_1+\tfrac{a}{2})cos(x_5)-(x_3+\tfrac{b}{2})sin(x_5)](-cos(x_5))+\\&2[b_2+(x_1+\tfrac{a}{2})sin(x_5)-(x_3+\tfrac{b}{2})cos(x_5)]sin(x_5)\}\end{aligned}\tag{3.404}$$

$$\begin{aligned}\frac{\partial||L_3||}{\partial x_2}=&\frac{1}{2}\frac{1}{||L_3||}\{2[a_3-(x_1+\tfrac{a}{2})cos(x_5)-(x_3+\tfrac{b}{2})sin(x_5)]0+\\&2[b_3+(x_1+\tfrac{a}{2})sin(x_5)-(x_3+\tfrac{b}{2})cos(x_5)]0\}\\&\Rightarrow\frac{\partial||L_3||}{\partial x_2}=0\end{aligned}\tag{3.405}$$

$$\begin{aligned}\frac{\partial||L_3||}{\partial x_3}=&\frac{1}{2}\frac{1}{||L_3||}\{2[a_3-(x_1+\tfrac{a}{2})cos(x_5)-(x_3+\tfrac{b}{2})sin(x_5)](-sin(x_5))+\\&2[b_3+(x_1+\tfrac{a}{2})sin(x_5)-(x_3+\tfrac{b}{2})cos(x_5)](cos(x_5))\}\end{aligned}\tag{3.406}$$

$$\begin{aligned}\frac{\partial||L_3||}{\partial x_4}=&\frac{1}{2}\frac{1}{||L_3||}\{2[a_3-(x_1+\tfrac{a}{2})cos(x_5)-(x_3+\tfrac{b}{2})sin(x_5)]0+\\&2[b_3+(x_1+\tfrac{a}{2})sin(x_5)-(x_3+\tfrac{b}{2})cos(x_5)]0\}\\&\Rightarrow\frac{\partial||L_3||}{\partial x_4}=0\end{aligned}\tag{3.407}$$

$$\begin{aligned}\frac{\partial||L_3||}{\partial x_5}=&\frac{1}{2}\frac{1}{||L_3||}\{2[a_3-(x_1+\tfrac{a}{2})cos(x_5)-(x_3+\tfrac{b}{2})sin(x_5)][-(x_1+\tfrac{a}{2})sin(x_5)-(x_3+\tfrac{b}{2})cos(x_5)]+\\&2[b_3+(x_1+\tfrac{a}{2})sin(x_5)-(x_3+\tfrac{b}{2})cos(x_5)][(x_1+\tfrac{a}{2})cos(x_5)-(x_3+\tfrac{b}{2})](sin(x_5))\}\end{aligned}\tag{3.408}$$

$$\begin{aligned}\frac{\partial||L_3||}{\partial x_6}=&\frac{1}{2}\frac{1}{||L_3||}\{2[a_3-(x_1+\tfrac{a}{2})cos(x_5)-(x_3+\tfrac{b}{2})sin(x_5)]0+\\&2[b_3+(x_1+\tfrac{a}{2})sin(x_5)-(x_3+\tfrac{b}{2})cos(x_5)]0\}\\&\Rightarrow\frac{\partial||L_3||}{\partial x_6}=0\end{aligned}\tag{3.409}$$

Third row of the Jacobian matrix $\nabla_x g_3(x)\mid_{(x^*,u^*)}$: $\frac{\partial g_{33}}{\partial x_i}=0$ for $i=1,2,\cdots,6$.

Fourth row of the Jacobian matrix $\nabla_x g_3(x)\mid_{(x^*,u^*)}$: for $i=1,2,\cdots,6$

$$\frac{\partial g_{34}}{\partial x_i}=\frac{1}{m}\frac{\frac{\partial L_{23}}{\partial x_i}||L_3||-L_{23}\frac{\partial||L_3||}{\partial x_i}}{||L_3||^2}\tag{3.410}$$

where it holds that: $\frac{\partial L_{23}}{\partial x_1}=0$, $\frac{\partial L_{23}}{\partial x_2}=0$, $\frac{\partial L_{23}}{\partial x_3}=1$, $\frac{\partial L_{23}}{\partial x_4}=0$, $\frac{\partial L_{23}}{\partial x_5}=-a_3cos(x_5)+b_3sin(x_5)$, and $\frac{\partial L_{23}}{\partial x_6}=0$.

Fifth row of the Jacobian matrix $\nabla_x g_3(x)\mid_{(x^*,u^*)}$: $\frac{\partial g_{35}}{\partial x_i}=0$ for $i=1,2,\cdots,6$.

Sixth row of the Jacobian matrix $\nabla_x g_3(x)\mid_{(x^*,u^*)}$: for $i=1,2,\cdots,6$

$$\frac{\partial g_{36}}{\partial x_i}=\frac{1}{J}\frac{\frac{\partial L_{33}}{\partial x_i}||L_3||-L_{33}\frac{\partial||L_3||}{\partial x_i}}{||L_3||^2}\tag{3.411}$$

where it holds that: $\frac{\partial L_{33}}{\partial x_1}=a_3sin(x_5)+b_3cos(x_5)$, $\frac{\partial L_{33}}{\partial x_2}=0$, $\frac{\partial L_{33}}{\partial x_3}=b_3sin(x_5)-a_3cos(x_5)$, $\frac{\partial L_{33}}{\partial x_4}=0$, $\frac{\partial L_{33}}{\partial x_5}=[a_3(x_1+\frac{a}{2})+b_3(x_3+\frac{b}{2})]cos(x_5)-[-a_3(x_3+\frac{b}{2})+b_3(x_1+\frac{a}{2})]sin(x_5)$, and $\frac{\partial L_{33}}{\partial x_6}=0$.

Thus, about the Jacobian matrix $\nabla_x g_3(x) \mid_{(x^*,u^*)}$, one obtains

$$\nabla_x g_3(x) \mid_{(x^*,u^*)} = \begin{pmatrix} \frac{\partial g_{31}}{\partial x_1} & \frac{\partial g_{31}}{\partial x_2} & \frac{\partial g_{31}}{\partial x_3} & \frac{\partial g_{31}}{\partial x_4} & \frac{\partial g_{31}}{\partial x_5} & \frac{\partial g_{31}}{\partial x_6} \\ \frac{\partial g_{32}}{\partial x_1} & \frac{\partial g_{32}}{\partial x_2} & \frac{\partial g_{32}}{\partial x_3} & \frac{\partial g_{32}}{\partial x_4} & \frac{\partial g_{32}}{\partial x_5} & \frac{\partial g_{32}}{\partial x_6} \\ \frac{\partial g_{33}}{\partial x_1} & \frac{\partial g_{33}}{\partial x_2} & \frac{\partial g_{33}}{\partial x_3} & \frac{\partial g_{33}}{\partial x_4} & \frac{\partial g_{33}}{\partial x_5} & \frac{\partial g_{33}}{\partial x_6} \\ \frac{\partial g_{34}}{\partial x_1} & \frac{\partial g_{34}}{\partial x_2} & \frac{\partial g_{34}}{\partial x_3} & \frac{\partial g_{34}}{\partial x_4} & \frac{\partial g_{34}}{\partial x_5} & \frac{\partial g_{34}}{\partial x_6} \\ \frac{\partial g_{35}}{\partial x_1} & \frac{\partial g_{35}}{\partial x_2} & \frac{\partial g_{35}}{\partial x_3} & \frac{\partial g_{35}}{\partial x_4} & \frac{\partial g_{35}}{\partial x_5} & \frac{\partial g_{35}}{\partial x_6} \\ \frac{\partial g_{36}}{\partial x_1} & \frac{\partial g_{36}}{\partial x_2} & \frac{\partial g_{36}}{\partial x_3} & \frac{\partial g_{36}}{\partial x_4} & \frac{\partial g_{36}}{\partial x_5} & \frac{\partial g_{36}}{\partial x_6} \end{pmatrix} \quad (3.412)$$

Computation of the Jacobian matrix $\nabla_x g_4(x) \mid_{(x^*,u^*)}$:
First row of the Jacobian matrix $\nabla_x g_4(x) \mid_{(x^*,u^*)}$: $\frac{\partial g_{41}}{\partial x_i} = 0$ for $i = 1, 2, \cdots, 6$.
Second row of the Jacobian matrix $\nabla_x g_4(x) \mid_{(x^*,u^*)}$: for $i = 1, 2, \cdots, 6$

$$\frac{\partial g_{42}}{\partial x_i} = \frac{1}{m} \frac{\frac{\partial L_{14}}{\partial x_i}||L_4|| - L_{14}\frac{\partial ||L_4||}{\partial x_i}}{||L_4||^2} \quad (3.413)$$

where it holds that: $\frac{\partial L_{14}}{\partial x_1} = 1$, $\frac{\partial L_{14}}{\partial x_2} = 0$, $\frac{\partial L_{14}}{\partial x_3} = 0$, $\frac{\partial L_{14}}{\partial x_4} = 0$, $\frac{\partial L_{14}}{\partial x_5} = a_4 sin(x_5) - b_4 cos(x_5)$, and $\frac{\partial L_{14}}{\partial x_6} = 0$.
Moreover, about the partial derivative of $||L_3||$ with respect to x_i, $i = 1, 2, \cdots, 6$, it holds that

$$\begin{aligned} \frac{\partial ||L_4||}{\partial x_1} = & \frac{1}{2}\frac{1}{||L_4||}\{2[a_4 - (x_1 - \tfrac{a}{2})cos(x_5) - (x_3 + \tfrac{b}{2})sin(x_5)]cos(x_5) + \\ & 2[b_4 + (x_1 - \tfrac{a}{2})sin(x_5) - (x_3 + \tfrac{b}{2})cos(x_5)]sin(x_5)\} \end{aligned} \quad (3.414)$$

$$\begin{aligned} \frac{\partial ||L_4||}{\partial x_2} = & \frac{1}{2}\frac{1}{||L_4||}\{2[a_4 - (x_1 - \tfrac{a}{2})cos(x_5) - (x_3 + \tfrac{b}{2})sin(x_5)]0 + \\ & 2[b_4 + (x_1 + \tfrac{a}{2})sin(x_5) - (x_3 + \tfrac{b}{2})cos(x_5)]0\} \\ & \Rightarrow \frac{\partial ||L_4||}{\partial x_2} = 0 \end{aligned} \quad (3.415)$$

$$\begin{aligned} \frac{\partial ||L_4||}{\partial x_3} = & \frac{1}{2}\frac{1}{||L_4||}\{2[a_4 - (x_1 - \tfrac{a}{2})cos(x_5) - (x_3 + \tfrac{b}{2})sin(x_5)](-sin(x_5)) + \\ & 2[b_4 + (x_1 - \tfrac{a}{2})sin(x_5) - (x_3 + \tfrac{b}{2})cos(x_5)](cos(x_5))\} \end{aligned} \quad (3.416)$$

$$\begin{aligned} \frac{\partial ||L_4||}{\partial x_4} = & \frac{1}{2}\frac{1}{||L_2||}\{2[a_4 - (x_1 - \tfrac{a}{2})cos(x_5) - (x_3 + \tfrac{b}{2})sin(x_5)]0 + \\ & 2[b_4 + (x_1 - \tfrac{a}{2})sin(x_5) - (x_3 + \tfrac{b}{2})cos(x_5)]0\} \\ & \Rightarrow \frac{\partial ||L_3||}{\partial x_4} = 0 \end{aligned} \quad (3.417)$$

$$\begin{aligned} \frac{\partial ||L_4||}{\partial x_5} = & \frac{1}{2}\frac{1}{||L_3||}\{2[a_4 - (x_1 - \tfrac{a}{2})cos(x_5) - (x_3 + \tfrac{b}{2})sin(x_5)][(x_1 - \tfrac{a}{2})sin(x_5) - (x_3 + \tfrac{b}{2})cos(x_5)] + \\ & 2[b_4 + (x_1 - \tfrac{a}{2})sin(x_5) - (x_3 + \tfrac{b}{2})cos(x_5)][(x_1 - \tfrac{a}{2})cos(x_5) - (x_3 + \tfrac{b}{2})](sin(x_5))\} \end{aligned} \quad (3.418)$$

$$\begin{aligned} \frac{\partial ||L_4||}{\partial x_6} = & \frac{1}{2}\frac{1}{||L_4||}\{2[a_4 - (x_1 - \tfrac{a}{2})cos(x_5) - (x_3 + \tfrac{b}{2})sin(x_5)]0 + \\ & 2[b_4 + (x_1 - \tfrac{a}{2})sin(x_5) - (x_3 + \tfrac{b}{2})cos(x_5)]0\} \\ & \Rightarrow \frac{\partial ||L_4||}{\partial x_6} = 0 \end{aligned} \quad (3.419)$$

Third row of the Jacobian matrix $\nabla_x g_4(x) \mid_{(x^*,u^*)}$: $\frac{\partial g_{43}}{\partial x_i} = 0$ for $i = 1, 2, \cdots, 6$.
Fourth row of the Jacobian matrix $\nabla_x g_4(x) \mid_{(x^*,u^*)}$: for $i = 1, 2, \cdots, 6$

$$\frac{\partial g_{44}}{\partial x_i} = \frac{1}{m} \frac{\frac{\partial L_{24}}{\partial x_i}||L_4|| - L_{24}\frac{\partial ||L_4||}{\partial x_i}}{||L_4||^2} \quad (3.420)$$

where it holds that: $\frac{\partial L_{24}}{\partial x_1} = 0$, $\frac{\partial L_{24}}{\partial x_2} = 0$, $\frac{\partial L_{24}}{\partial x_3} = 1$, $\frac{\partial L_{24}}{\partial x_4} = 0$, $\frac{\partial L_{24}}{\partial x_5} = -a_4 cos(x_5) + b_4 sin(x_5)$, and $\frac{\partial L_{24}}{\partial x_6} = 0$.
Fifth row of the Jacobian matrix $\nabla_x g_4(x) \mid_{(x^*,u^*)}$: $\frac{\partial g_{45}}{\partial x_i} = 0$ for $i = 1, 2, \cdots, 6$.

Sixth row of the Jacobian matrix $\nabla_x g_4(x) \mid_{(x^*,u^*)}$: for $i=1,2,\cdots,6$

$$\frac{\partial g_{46}}{\partial x_i} = \frac{1}{J}\frac{\frac{\partial L_{34}}{\partial x_i}||L_4|| - L_{34}\frac{\partial ||L_4||}{\partial x_i}}{||L_4||^2} \tag{3.421}$$

where it holds that: $\frac{\partial L_{34}}{\partial x_1} = a_4 sin(x_5) + b_4 cos(x_5)$, $\frac{\partial L_{34}}{\partial x_2} = 0$, $\frac{\partial L_{34}}{\partial x_3} = b_4 sin(x_5) - a_4 cos(x_5)$, $\frac{\partial L_{34}}{\partial x_4} = 0$, $\frac{\partial L_{34}}{\partial x_5} = [a_4(x_1 - \frac{a}{2}) + b_4(x_3 + \frac{b}{2})]cos(x_5) - [-a_4(x_3 + \frac{b}{2}) + b_4(x_1 - \frac{a}{2})]sin(x_5)$, and $\frac{\partial L_{34}}{\partial x_6} = 0$.

Thus, about the Jacobian matrix $\nabla_x g_4(x) \mid_{(x^*,u^*)}$, one obtains:

$$\nabla_x g_4(x) \mid_{(} x^*, u^*) = \begin{pmatrix} \frac{\partial g_{41}}{\partial x_1} & \frac{\partial g_{41}}{\partial x_2} & \frac{\partial g_{41}}{\partial x_3} & \frac{\partial g_{41}}{\partial x_4} & \frac{\partial g_{41}}{\partial x_5} & \frac{\partial g_{41}}{\partial x_6} \\ \frac{\partial g_{42}}{\partial x_1} & \frac{\partial g_{42}}{\partial x_2} & \frac{\partial g_{42}}{\partial x_3} & \frac{\partial g_{42}}{\partial x_4} & \frac{\partial g_{42}}{\partial x_5} & \frac{\partial g_{42}}{\partial x_6} \\ \frac{\partial g_{43}}{\partial x_1} & \frac{\partial g_{43}}{\partial x_2} & \frac{\partial g_{43}}{\partial x_3} & \frac{\partial g_{43}}{\partial x_4} & \frac{\partial g_{43}}{\partial x_5} & \frac{\partial g_{43}}{\partial x_6} \\ \frac{\partial g_{44}}{\partial x_1} & \frac{\partial g_{44}}{\partial x_2} & \frac{\partial g_{44}}{\partial x_3} & \frac{\partial g_{44}}{\partial x_4} & \frac{\partial g_{44}}{\partial x_5} & \frac{\partial g_{44}}{\partial x_6} \\ \frac{\partial g_{45}}{\partial x_1} & \frac{\partial g_{45}}{\partial x_2} & \frac{\partial g_{45}}{\partial x_3} & \frac{\partial g_{45}}{\partial x_4} & \frac{\partial g_{45}}{\partial x_5} & \frac{\partial g_{45}}{\partial x_6} \\ \frac{\partial g_{46}}{\partial x_1} & \frac{\partial g_{46}}{\partial x_2} & \frac{\partial g_{46}}{\partial x_3} & \frac{\partial g_{46}}{\partial x_4} & \frac{\partial g_{46}}{\partial x_5} & \frac{\partial g_{46}}{\partial x_6} \end{pmatrix} \tag{3.422}$$

3.4.4 Nonlinear optimal control and stability analysis for the cable-driven robot

3.4.4.1 Stabilizing feedback control

After linearization around its current operating point, the 3-DOF cable-driven parallel robot is written in the form [721]

$$\dot{x} = Ax + Bu + d_1 \tag{3.423}$$

Parameter d_1 stands for the linearization error in the 3-DOF cable-driven parallel robot appearing above in Eq. (3.423). The reference setpoints for the cable-driven parallel robot's state vector are denoted by $\mathbf{x_d} = [x_1^d, \cdots, x_6^d]$. Tracking of this trajectory is achieved after applying the control input u_d. At every time instant, the control input u_d is assumed to differ from the control input u appearing in Eq. (3.423) by an amount equal to Δu, that is $u_d = u + \Delta u$

$$\dot{x}_d = Ax_d + Bu_d + d_2 \tag{3.424}$$

The dynamics of the controlled system described in Eq. (3.423) can be also written as

$$\dot{x} = Ax + Bu + Bu_d - Bu_d + d_1 \tag{3.425}$$

and by denoting $d_3 = -Bu_d + d_1$ as an aggregate disturbance term, one obtains

$$\dot{x} = Ax + Bu + Bu_d + d_3 \tag{3.426}$$

By subtracting Eq. (3.424) from Eq. (3.426), one has

$$\dot{x} - \dot{x}_d = A(x - x_d) + Bu + d_3 - d_2 \tag{3.427}$$

By denoting the tracking error as $e = x - x_d$ and the aggregate disturbance term as $\tilde{d} = d_3 - d_2$, the tracking error dynamics becomes

$$\dot{e} = Ae + Bu + \tilde{d} \tag{3.428}$$

For the approximately linearized model of the system, a stabilizing feedback controller is developed. The controller has the form

$$u(t) = -Ke(t) \tag{3.429}$$

with $K = \frac{1}{r}B^T P$ where P is a positive definite symmetric matrix which is obtained from the solution of the Riccati equation [721]

$$A^T P + PA + Q - P(\frac{2}{r}BB^T - \frac{1}{\rho^2}LL^T)P = 0 \tag{3.430}$$

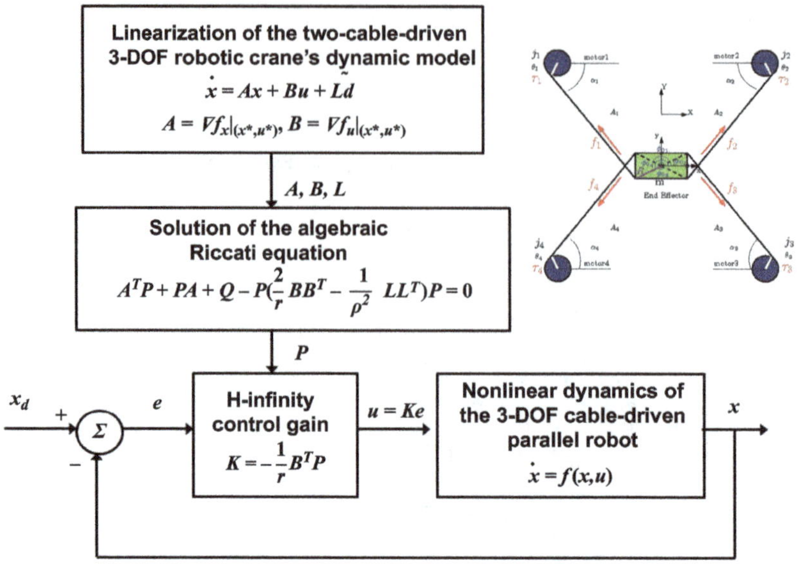

Figure 3.56 Diagram of the nonlinear optimal (H-infinity) control loop for the 3-DOF four-cable-driven parallel robot

where Q is a positive semi-definite symmetric matrix. The diagram of the considered control loop is depicted in Figure 3.56.

It is also noted that the solution of the H-infinity feedback control problem for the 3-DOF four-cable-driven parallel robot and the computation of the worst-case disturbance that this controller can sustain, comes from superposition of Bellman's optimality principle when considering that the robot is affected by two separate inputs (i) the control input u (ii) the cumulative disturbance input $\tilde{d}(t)$. Solving the optimal control problem for u that is for the minimum variation (optimal) control input that achieves the elimination of the state vector's tracking error gives $u = -\frac{1}{r}B^T Pe$. Equivalently, solving the optimal control problem for $\tilde{d}$, that is for the worst-case disturbance that the control loop can sustain gives $\tilde{d} = \frac{1}{\rho^2}L^T Pe$.

3.4.4.2 Lyapunov stability analysis

Through Lyapunov stability analysis, it will be shown that the proposed nonlinear control scheme assures H_∞ tracking performance for the 3-DOF cable-driven parallel robot and that under moderate conditions global stability (asymptotic convergence to the reference setpoints) is achieved. As shown before, the tracking error dynamics for the 3-DOF cable-driven parallel robot is written in the form [721]

$$\dot{e} = Ae + Bu + L\tilde{d} \tag{3.431}$$

where in the 3-DOF cable-driven parallel robot's case $L \in R^{6\times 6}$ is the disturbance inputs gain matrix. Variable $\tilde{d}$ denotes model uncertainties and external disturbances of the robot's model. The following Lyapunov equation is considered:

$$V = \frac{1}{2}e^T Pe \tag{3.432}$$

where $e = x - x_d$ is the tracking error. By differentiating with respect to time, one obtains

$$\begin{gathered}\dot{V} = \frac{1}{2}\dot{e}^T Pe + \frac{1}{2}e^T P\dot{e} \Rightarrow \\ \dot{V} = \frac{1}{2}[Ae + Bu + L\tilde{d}]^T P + \frac{1}{2}e^T P[Ae + Bu + L\tilde{d}] \Rightarrow\end{gathered} \tag{3.433}$$

$$\dot{V}=\frac{1}{2}[e^TA^T+u^TB^T+\tilde{d}^TL^T]Pe+ \\ +\frac{1}{2}e^TP[Ae+Bu+L\tilde{d}]\Rightarrow \tag{3.434}$$

$$\dot{V}=\frac{1}{2}e^TA^TPe+\frac{1}{2}u^TB^TPe+\frac{1}{2}\tilde{d}^TL^TPe+ \\ \frac{1}{2}e^TPAe+\frac{1}{2}e^TPBu+\frac{1}{2}e^TPL\tilde{d} \tag{3.435}$$

The previous equation is rewritten as

$$\dot{V}=\frac{1}{2}e^T(A^TP+PA)e+(\frac{1}{2}u^TB^TPe+\frac{1}{2}e^TPBu)+ \\ +(\frac{1}{2}\tilde{d}^TL^TPe+\frac{1}{2}e^TPL\tilde{d}) \tag{3.436}$$

Assumption: For given positive definite matrix Q and coefficients r and ρ, there exists a positive definite matrix P, which is the solution of the following matrix equation:

$$A^TP+PA=-Q+P(\frac{2}{r}BB^T-\frac{1}{\rho^2}LL^T)P \tag{3.437}$$

Moreover, the following feedback control law is applied to the system:

$$u=-\frac{1}{r}B^TPe \tag{3.438}$$

By substituting Eq. (3.437) and Eq. (3.438), one obtains

$$\dot{V}=\frac{1}{2}e^T[-Q+P(\frac{2}{r}BB^T-\frac{1}{\rho^2}LL^T)P]e+ \\ +e^TPB(-\frac{1}{r}B^TPe)+e^TPL\tilde{d}\Rightarrow \tag{3.439}$$

$$\dot{V}=-\frac{1}{2}e^TQe+(\frac{1}{r}e^TPBB^TPe-\frac{1}{2\rho^2}e^TPLL^TPe \\ -\frac{1}{r}e^TPBB^TPe)+e^TPL\tilde{d} \tag{3.440}$$

which after intermediate operations gives

$$\dot{V}=-\frac{1}{2}e^TQe-\frac{1}{2\rho^2}e^TPLL^TPe+e^TPL\tilde{d} \tag{3.441}$$

or, equivalently

$$\dot{V}=-\frac{1}{2}e^TQe-\frac{1}{2\rho^2}e^TPLL^TPe+ \\ +\frac{1}{2}e^TPL\tilde{d}+\frac{1}{2}\tilde{d}^TL^TPe \tag{3.442}$$

Lemma: The following inequality holds

$$\frac{1}{2}e^TPL\tilde{d}+\frac{1}{2}\tilde{d}L^TPe-\frac{1}{2\rho^2}e^TPLL^TPe\leq\frac{1}{2}\rho^2\tilde{d}^T\tilde{d} \tag{3.443}$$

Proof: The binomial $(\rho\alpha-\frac{1}{\rho}b)^2$ is considered. Expanding the left part of the above inequality, one gets

$$\rho^2a^2+\frac{1}{\rho^2}b^2-2ab\geq 0\Rightarrow\frac{1}{2}\rho^2a^2+\frac{1}{2\rho^2}b^2-ab\geq 0\Rightarrow \\ ab-\frac{1}{2\rho^2}b^2\leq\frac{1}{2}\rho^2a^2\Rightarrow\frac{1}{2}ab+\frac{1}{2}ab-\frac{1}{2\rho^2}b^2\leq\frac{1}{2}\rho^2a^2 \tag{3.444}$$

The following substitutions are carried out: $a=\tilde{d}$ and $b=e^TPL$ and the previous relation becomes

$$\frac{1}{2}\tilde{d}^TL^TPe+\frac{1}{2}e^TPL\tilde{d}-\frac{1}{2\rho^2}e^TPLL^TPe\leq\frac{1}{2}\rho^2\tilde{d}^T\tilde{d} \tag{3.445}$$

Eq. (3.445) is substituted in Eq. (3.442) and the inequality is enforced, thus giving

$$\dot{V}\leq-\frac{1}{2}e^TQe+\frac{1}{2}\rho^2\tilde{d}^T\tilde{d} \tag{3.446}$$

Eq. (3.446) shows that the H_∞ tracking performance criterion is satisfied. The integration of $\dot{V}$ from 0 to T gives

$$\int_0^T\dot{V}(t)dt\leq-\frac{1}{2}\int_0^T||e||_Q^2dt+\frac{1}{2}\rho^2\int_0^T||\tilde{d}||^2dt\Rightarrow \\ 2V(T)+\int_0^T||e||_Q^2dt\leq 2V(0)+\rho^2\int_0^T||\tilde{d}||^2dt \tag{3.447}$$

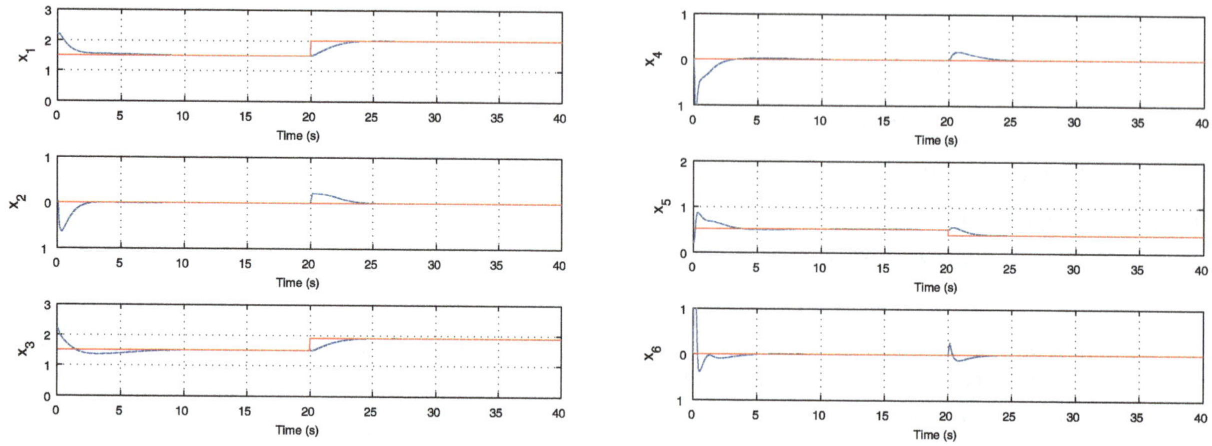

(a) (b)

Figure 3.57 Tracking of setpoint 1 for the 3-DOF cable-driven parallel robot, for motion from initial position (x_A, y_A, θ_A) to final position (x_B, y_B, θ_B): (a) convergence of state variables x_1 to x_3 to their reference setpoints (red line: setpoint, blue line: real value, green line: estimated value); (b) convergence of state variables x_4 to x_6 to their reference setpoints (red line: setpoint, blue line: real value, green line: estimated value)

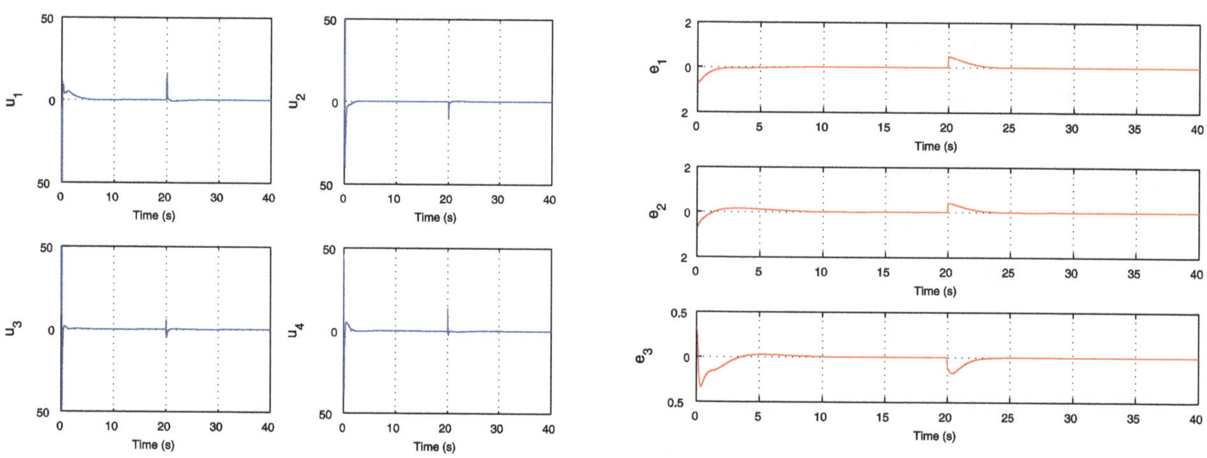

(a) (b)

Figure 3.58 Tracking of setpoint 1 for the 3-DOF cable-driven parallel robot, for motion from initial position (x_A, y_A, θ_A) to final position (x_B, y_B, θ_B): (a) control inputs u_1 to u_4; (b) convergence to zero of the tracking error variables e_1, e_2, e_3 associated with for state variables $x_1 = x$, $x_3 = y$, and $x_5 = \theta$

Moreover, if there exists a positive constant $M_d > 0$ such that

$$\int_0^{\infty} ||\tilde{d}||^2 dt \leq M_d \tag{3.448}$$

then, one gets

$$\int_0^{\infty} ||e||_Q^2 dt \leq 2V(0) + \rho^2 M_d \tag{3.449}$$

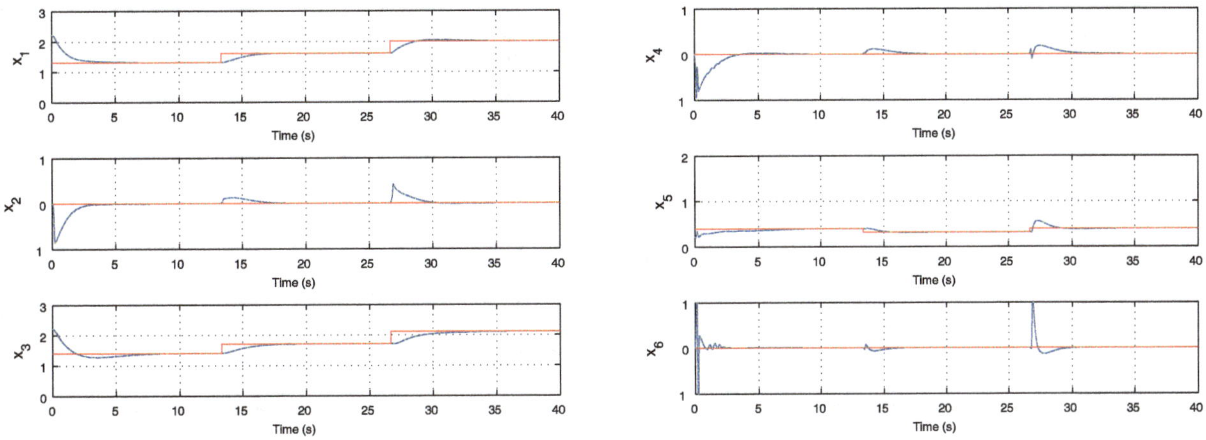

(a) (b)

Figure 3.59 Tracking of setpoint 2 for the 3-DOF cable-driven parallel robot, for motion from initial position (x_A, y_A, θ_A) to final position (x_B, y_B, θ_B): (a) convergence of state variables x_1 to x_3 to their reference setpoints (red line: setpoint, blue line: real value, green line: estimated value); (b) convergence of state variables x_4 to x_6 to their reference setpoints (red line: setpoint, blue line: real value, green line: estimated value)

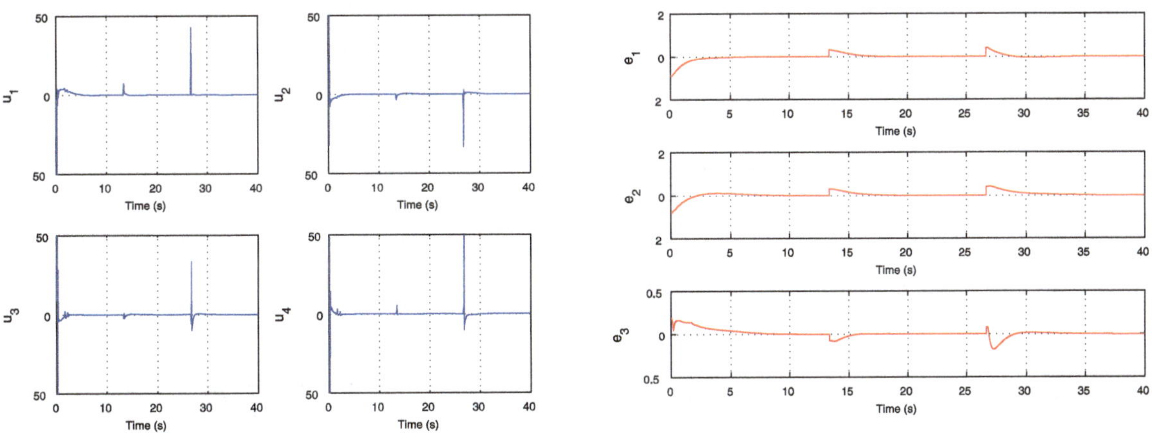

(a) (b)

Figure 3.60 Tracking of setpoint 2 for the 3-DOF cable-driven parallel robot, for motion from initial position (x_A, y_A, θ_A) to final position (x_B, y_B, θ_B): (a) control inputs u_1 to u_4; (b) convergence to zero of the tracking error variables e_1, e_2, e_3 associated with for state variables $x_1 = x$, $x_3 = y$, and $x_5 = \theta$

Thus, the integral $\int_0^{\infty} ||e||_Q^2 dt$ is bounded. Moreover, $V(T)$ is bounded and from the definition of the Lyapunov function V in Eq. (3.432) it becomes clear that $e(t)$ will be also bounded since $e(t) \in \Omega_e = \{e | e^T P e \leq 2V(0) + \rho^2 M_d\}$. According to the above and with the use of Barbalat's Lemma, one obtains $lim_{t \to \infty} e(t) = 0$.

Through the stages of the stability proof, one arrives at Eq. (3.446) which shows that the H-infinity tracking performance criterion holds. By selecting the attenuation coefficient ρ to be sufficiently small and in particular to satisfy $\rho^2 < ||e||_Q^2 / ||\tilde{d}||^2$, one has that the first derivative of the Lyapunov function is upper bounded by 0. This condition holds at each sampling instance, and consequently, global stability for the control loop can be concluded.

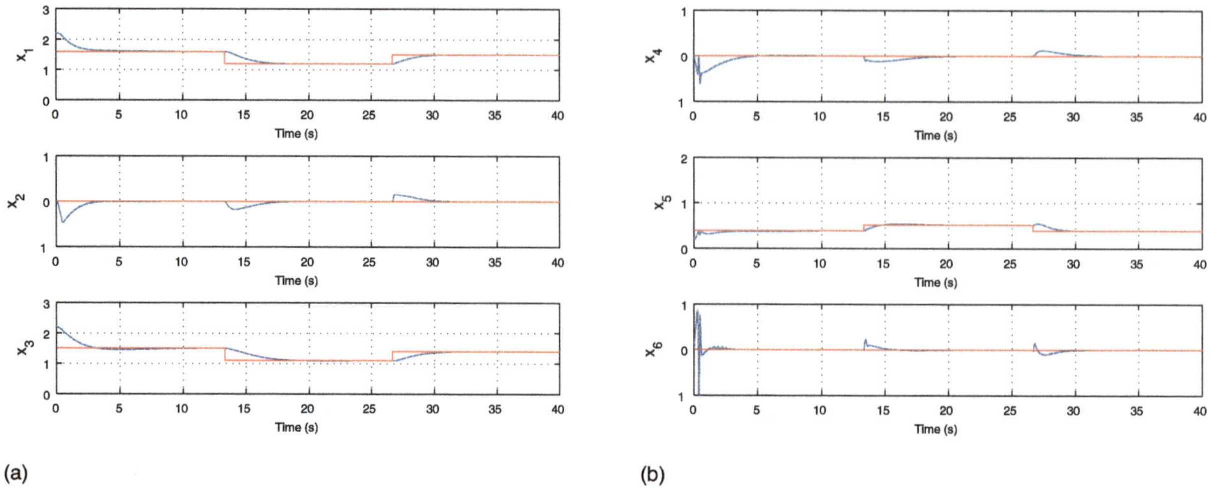

Figure 3.61 *Tracking of setpoint 3 for the 3-DOF cable-driven parallel robot, for motion from initial position (x_A, y_A, θ_A) to final position (x_B, y_B, θ_B): (a) convergence of state variables x_1 to x_3 to their reference setpoints (red line: setpoint, blue line: real value, green line: estimated value); (b) convergence of state variables x_4 to x_6 to their reference setpoints (red line: setpoint, blue line: real value, green line: estimated value)*

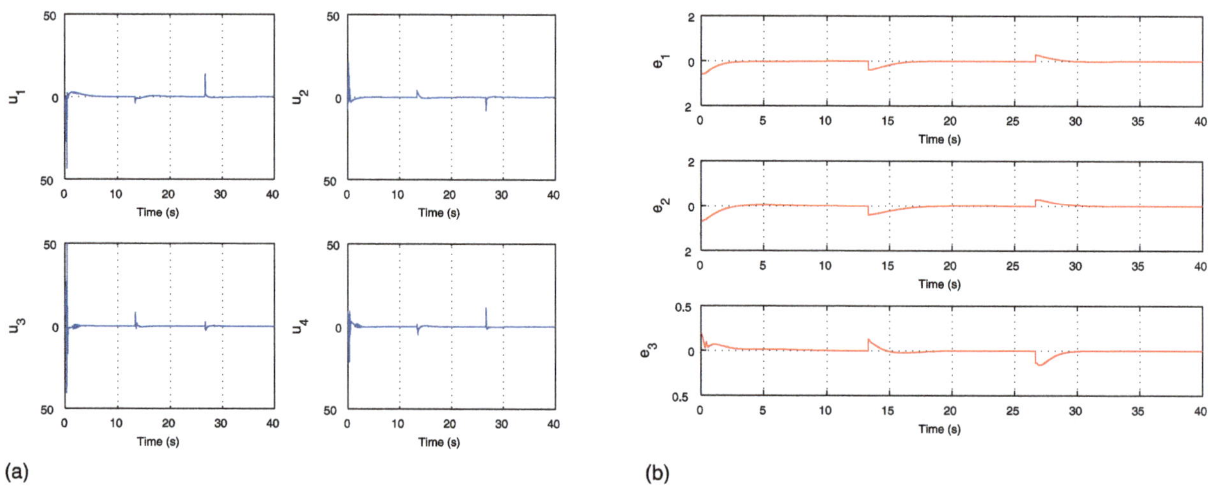

Figure 3.62 *Tracking of setpoint 3 for the 3-DOF cable-driven parallel robot, for motion from initial position (x_A, y_A, θ_A) to final position (x_B, y_B, θ_B): (a) control inputs u_1 to u_4; (b) convergence to zero of the tracking error variables e_1, e_2, e_3 associated with for state variables $x_1 = x$, $x_3 = y$, and $x_5 = \theta$*

3.4.5 Simulation tests

The performance of the nonlinear optimal control loop for the dynamic model of the 3-DOF four-cable-driven parallel robot is further confirmed through simulation experiments. To implement the nonlinear optimal control method the dynamic model of the cable-driven parallel robot had to be linearized at each sampling period through the computation of the Jacobian matrices of its dynamic model, while the method's algebraic Riccati equation of Eq. (3.437) had also to be solved repetitively at each sampling interval. Indicative values about the parameters of the 3-DOF cable-driven parallel robot are as follows: dimensions of the planar robot (rectangular plate) $a = 2.0m$, $b = 1.0m$, mass of the robot

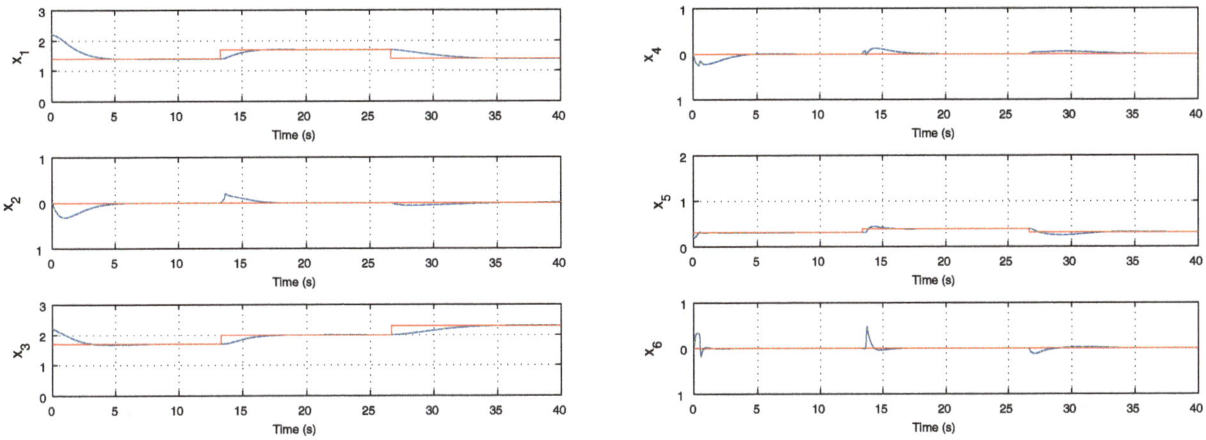

(a) (b)

Figure 3.63 Tracking of setpoint 4 for the 3-DOF cable-driven parallel robot, for motion from initial position (x_A, y_A, θ_A) to final position (x_B, y_B, θ_B): (a) convergence of state variables x_1 to x_3 to their reference setpoints (red line: setpoint, blue line: real value, green line: estimated value); (b) convergence of state variables x_4 to x_6 to their reference setpoints (red line: setpoint, blue line: real value, green line: estimated value)

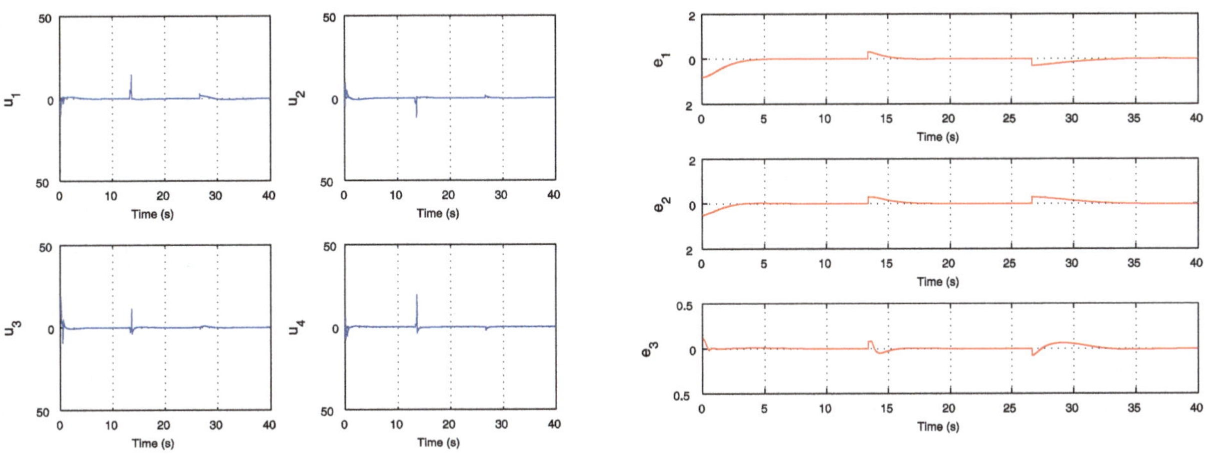

(a) (b)

Figure 3.64 Tracking of setpoint 4 for the 3-DOF cable-driven parallel robot, for motion from initial position (x_A, y_A, θ_A) to final position (x_B, y_B, θ_B): (a) control inputs u_1 to u_4; (b) convergence to zero of the tracking error variables e_1, e_2, e_3 associated with for state variables $x_1 = x$, $x_3 = y$, and $x_5 = \theta$

$m = 20kg$, moment of inertia of the robot $J = 60kg{\cdot}m^2$, friction coefficients $k_{b_i} = 0.1$ for $i = 1, 2, 3$, coordinates of the non-moving ends of the cables A_i, $i = 1, \cdots, 4$: $(a_1, b_1) = (0m, 0m)$, $(a_2, b_2) = (5m, 0m)$, $(a_3, b_3) = (5m, 4m)$, $(a_4, b_4) = (0m, 4m)$. The parameters of the Riccati equation of the control method were $r = 0.005$, $\rho = 0.1$, Q was a diagonal matrix with diagonal elements equal to 0.5 and L was also a diagonal matrix with diagonal elements given the

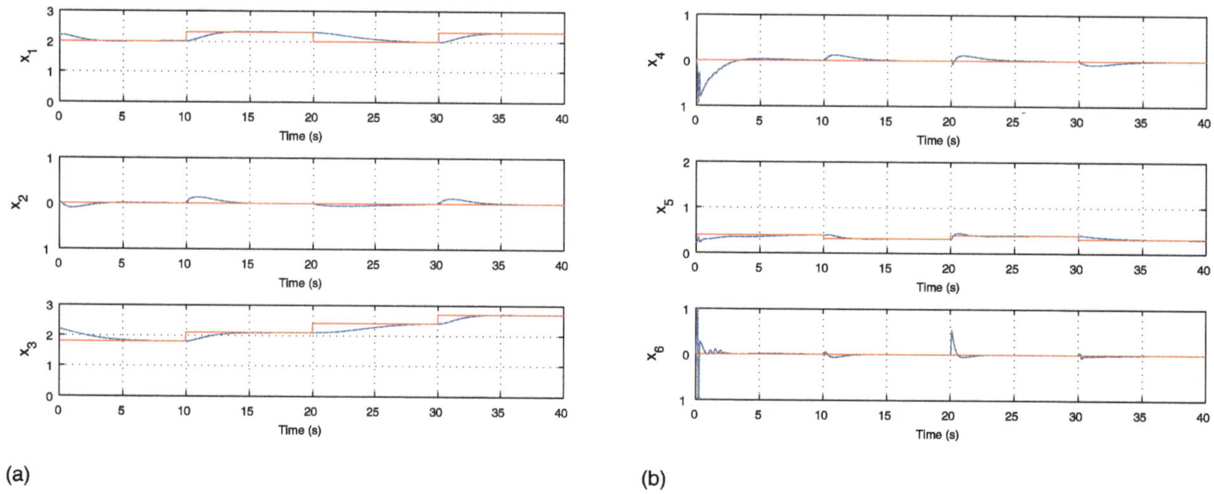

Figure 3.65 Tracking of setpoint 5 for the 3-DOF cable-driven parallel robot, for motion from initial position (x_A,y_A,θ_A) to final position (x_B,y_B,θ_B): (a) convergence of state variables x_1 to x_3 to their reference setpoints (red line: setpoint, blue line: real value, green line: estimated value); (b) convergence of state variables x_4 to x_6 to their reference setpoints (red line: setpoint, blue line: real value, green line: estimated value)

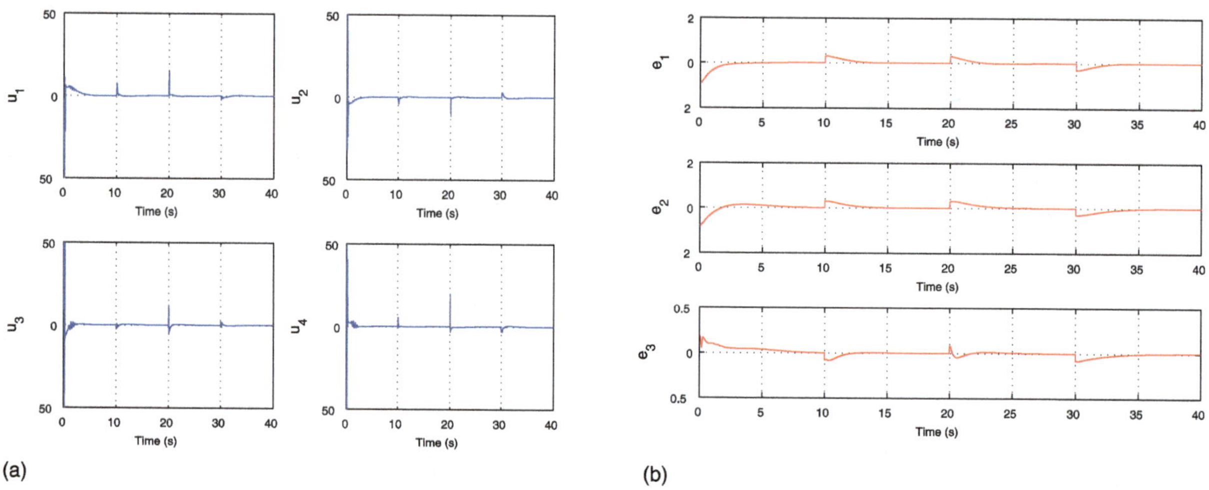

Figure 3.66 Tracking of setpoint 5 for the 3-DOF cable-driven parallel robot, for motion from initial position (x_A,y_A,θ_A) to final position (x_B,y_B,θ_B): (a) control inputs u_1 to u_4; (b) convergence to zero of the tracking error variables e_1, e_2, e_3 associated with for state variables $x_1 = x$, $x_3 = y$, and $x_5 = \theta$

value 10^{-4}. The obtained results are depicted in Figures 3.57–3.72. In all diagrams, it can be noticed that, under the proposed nonlinear optimal control scheme, one achieves fast convergence of the robot's state variables to the associated setpoints under moderate changes in the control inputs and, consequently, under a minimum dispersion of energy by the robot's actuators. It is noted that the transient performance of the control method is determined by the coefficients r, ρ,

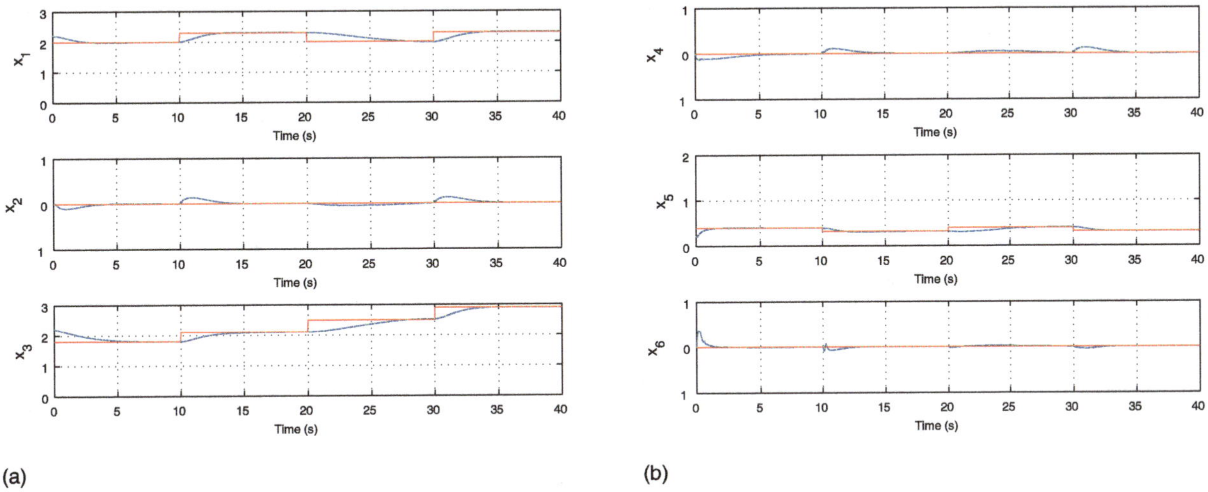

Figure 3.67 Tracking of setpoint 6 for the 3-DOF cable-driven parallel robot, for motion from initial position (x_A, y_A, θ_A) to final position (x_B, y_B, θ_B): (a) convergence of state variables x_1 to x_3 to their reference setpoints (red line: setpoint, blue line: real value, green line: estimated value); (b) convergence of state variables x_4 to x_6 to their reference setpoints (red line: setpoint, blue line: real value, green line: estimated value)

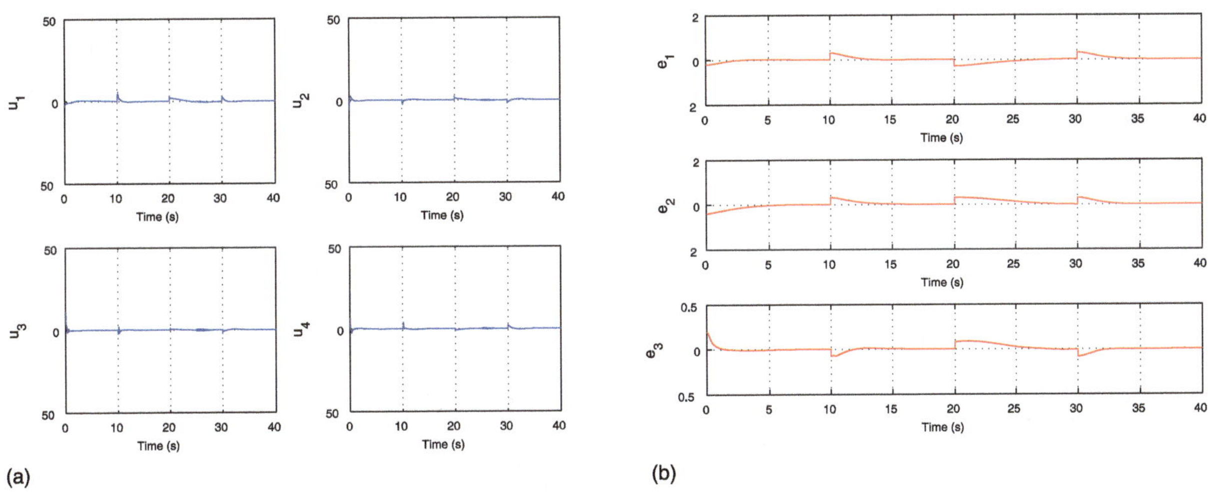

Figure 3.68 Tracking of setpoint 6 for the 3-DOF cable-driven parallel robot, for motion from initial position (x_A, y_A, θ_A) to final position (x_B, y_B, θ_B): (a) control inputs u_1 to u_4; (b) convergence to zero of the tracking error variables e_1, e_2, e_3 associated with for state variables $x_1 = x$, $x_3 = y$, and $x_5 = \theta$

and Q of the method's algebraic Riccati equation. Actually, for small values of r, one achieves complete elimination of the state vector's tracking error, while for relatively large values of the diagonal elements of matrix Q fast convergence of the state variables to their reference values is noticed. Moreover, the smallest value of the attenuation coefficient ρ for which a valid solution of the method's Riccati equation is obtained in the form of a positive definite and symmetric matrix P is the one that results in maximum robustness for the control loop.

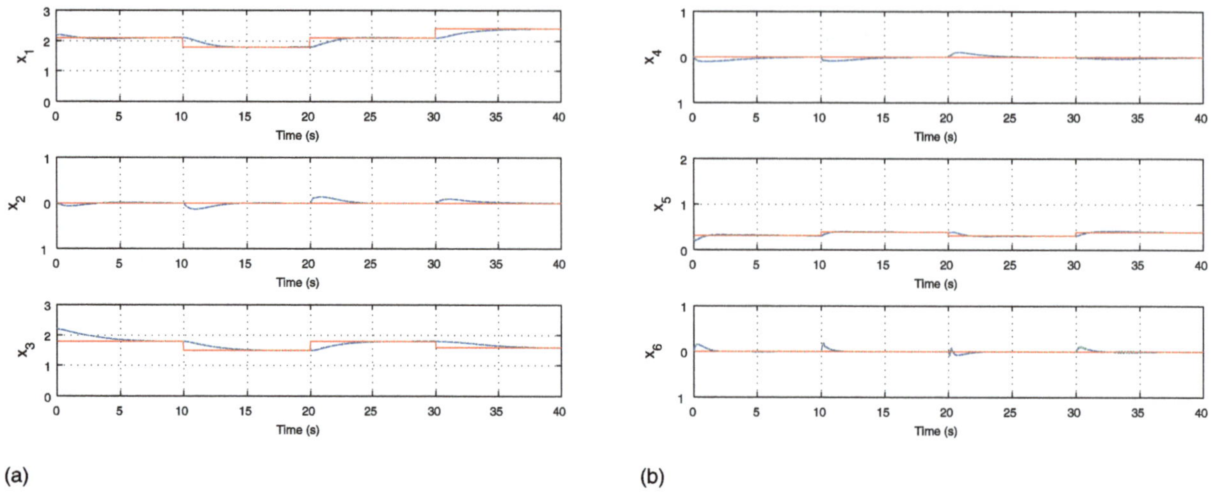

Figure 3.69 Tracking of setpoint 7 for the 3-DOF cable-driven parallel robot, for motion from initial position (x_A, y_A, θ_A) to final position (x_B, y_B, θ_B): (a) convergence of state variables x_1 to x_3 to their reference setpoints (red line: setpoint, blue line: real value, green line: estimated value); (b) convergence of state variables x_4 to x_6 to their reference setpoints (red line: setpoint, blue line: real value, green line: estimated value)

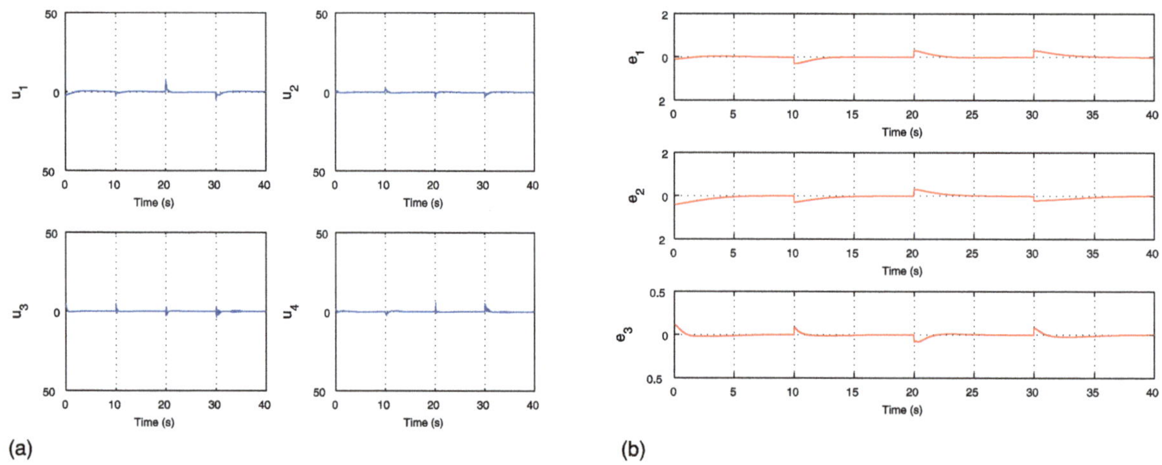

Figure 3.70 Tracking of setpoint 7 for the 3-DOF cable-driven parallel robot, for motion from initial position (x_A, y_A, θ_A) to final position (x_B, y_B, θ_B): (a) control inputs u_1 to u_4; (b) convergence to zero of the tracking error variables e_1, e_2, e_3 associated with for state variables $x_1 = x$, $x_3 = y$, and $x_5 = \theta$

To elaborate on the tracking performance and on the robustness of the proposed nonlinear optimal control method for the dynamics of the 3-DOF cable-driven parallel robot the following tables are given: (i) Table 3.9 which provides information about the accuracy of tracking of the reference setpoints by the state variables of the 3-DOF cable-driven parallel robot under an exact state-space model, (ii) Table 3.10 which provides information about the robustness of

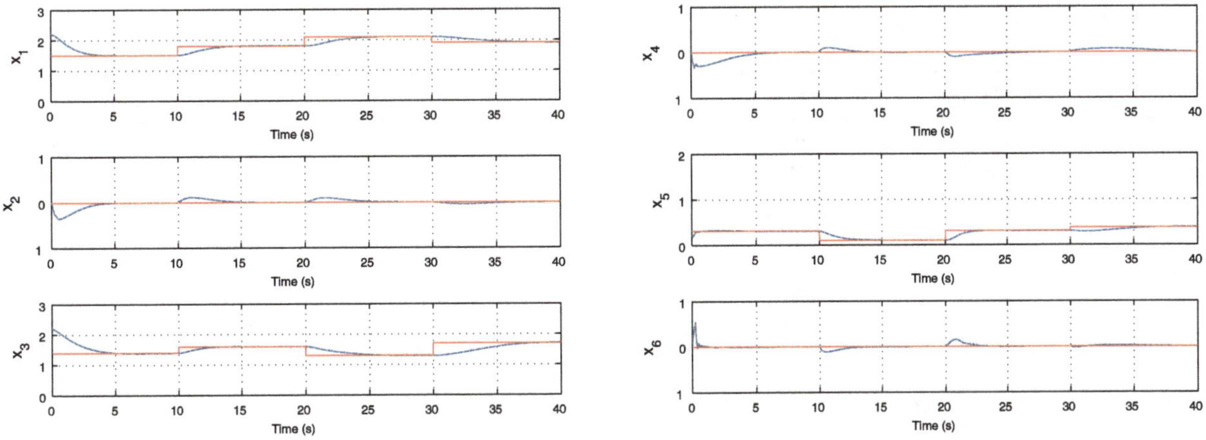

(a) (b)

Figure 3.71 Tracking of setpoint 8 for the 3-DOF cable-driven parallel robot, for motion from initial position (x_A, y_A, θ_A) to final position (x_B, y_B, θ_B): (a) convergence of state variables x_1 to x_3 to their reference setpoints (red line: setpoint, blue line: real value, green line: estimated value); (b) convergence of state variables x_4 to x_6 to their reference setpoints (red line: setpoint, blue line: real value, green line: estimated value)

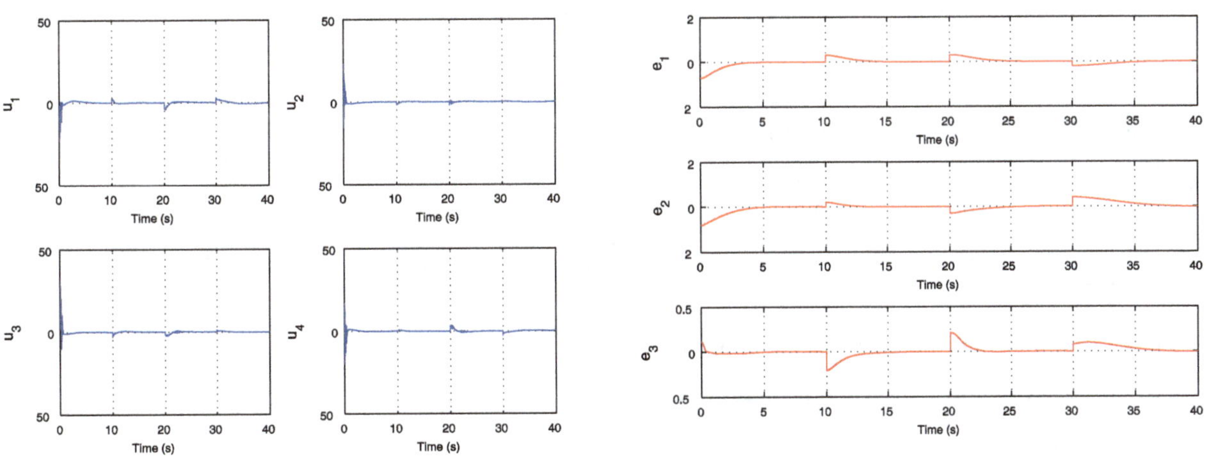

(a) (b)

Figure 3.72 Tracking of setpoint 8 for the 3-DOF cable-driven parallel robot, for motion from initial position (x_A, y_A, θ_A) to final position (x_B, y_B, θ_B): (a) control inputs u_1 to u_4; (b) convergence to zero of the tracking error variables e_1, e_2, e_3 associated with for state variables $x_1 = x$, $x_3 = y$, and $x_5 = \theta$

Table 3.9 Tracking RMSE for the 3-DOF CDPR in the disturbance-free case $\cdot 10^{-3}$

	$RMSE_{x_1}$	$RMSE_{x_2}$	$RMSE_{x_3}$	$RMSE_{x_4}$	$RMSE_{x_5}$	$RMSE_{x_6}$
$setpoint_1$	0.0946	0.0040	0.0947	0.0054	0.3516	0.0063
$setpoint_2$	0.0981	0.0935	0.1275	0.1097	0.1718	0.0219
$setpoint_3$	0.0326	0.0086	0.0931	0.0267	0.1925	0.0123
$setpoint_4$	0.3379	0.1330	0.2634	0.1175	0.3317	0.0683
$setpoint_5$	0.1527	0.0988	0.2437	0.3644	0.1983	0.0909
$setpoint_6$	0.0709	0.1018	0.1856	0.0739	0.1076	0.0370
$setpoint_7$	0.0641	0.0813	0.1578	0.1361	0.0149	0.0221
$setpoint_8$	0.0818	0.1033	0.1466	0.2359	0.1885	0.0472

Table 3.10 Tracking RMSE for the 3-DOF CDPR in the case of disturbances $\cdot 10^{-3}$

$\Delta a\%$	$RMSE_{x_1}$	$RMSE_{x_2}$	$RMSE_{x_3}$	$RMSE_{x_4}$	$RMSE_{x_5}$	$RMSE_{x_6}$
0%	0.0326	0.0086	0.0991	0.0267	0.1914	0.0123
10%	0.0181	0.0110	0.1456	0.0400	0.2118	0.0169
20%	0.0531	0.0180	0.1778	0.0436	0.1851	0.0191
30%	0.0118	0.0205	0.1997	0.0350	0.1681	0.0186
40%	0.1629	0.0157	0.2362	0.0237	0.1618	0.0146
50%	0.1997	0.0252	0.3108	0.0452	0.1626	0.0107
60%	0.2111	0.0333	0.4229	0.0706	0.1796	0.0195

Table 3.11 RMSE for the estimation performed by the H-infinity KF $\cdot 10^{-3}$

	$RMSE_{x_1}$	$RMSE_{x_2}$	$RMSE_{x_3}$	$RMSE_{x_4}$	$RMSE_{x_5}$	$RMSE_{x_6}$
$setpoint_1$	0.0041	0.0181	0.0038	0.0171	0.0584	0.2056
$setpoint_2$	0.0002	0.0007	0.0025	0.0115	0.0261	0.1188
$setpoint_3$	0.0004	0.0016	0.0030	0.0138	0.0271	0.1233
$setpoint_4$	0.0011	0.0049	0.0001	0.0001	0.0138	0.0628
$setpoint_5$	0.0003	0.0012	0.0020	0.0093	0.0240	0.1092
$setpoint_6$	0.0011	0.0049	0.0011	0.0052	0.0081	0.0366
$setpoint_7$	0.0006	0.0028	0.0005	0.0023	0.0074	0.0394
$setpoint_8$	0.0006	0.0028	0.0035	0.0160	0.0287	0.1305

Table 3.12 H-type gantry cranes with dual PMLSMs

	$T_s\ x_1$	$T_s\ x_2$	$T_s\ x_3$	$T_s\ x_4$	$T_s\ x_5$	$T_s\ x_6$
$setpoint_1$	2.5	2.0	6.0	2.5	3.5	3.5
$setpoint_2$	2.5	2.5	6.0	3.0	3.0	2.5
$setpoint_3$	2.0	2.5	2.5	3.5	2.0	2.5
$setpoint_4$	3.5	4.0	2.5	3.5	0.5	1.0
$setpoint_5$	2.0	2.5	5.0	3.5	2.5	2.5
$setpoint_6$	2.5	2.5	5.5	4.5	0.5	1.0
$setpoint_7$	2.0	2.5	5.5	5.0	1.0	1.5
$setpoint_8$	3.5	2.5	2.5	4.0	0.5	1.0

the control method to parametric changes in the model of the 3-DOF cable-driven parallel robot (change in the mass m of the orthogonal metallic plate), (iii) Table 3.11 which provides information about the precision in state variables' estimation that is achieved by the H-infinity Kalman filter, and (iv) Table 3.12 which provides the convergence times of the state variables of the 3-DOF cable-driven parallel robot to the associated setpoints.

Chapter 4
Nonlinear optimal and flatness-based control for cranes and pendulums II

Abstract

The chapter analyzes further nonlinear optimal and flatness-based control methods for robotic cranes and underactuated pendulums. First, a multi-loop flatness-based control method is proposed for the control and stabilization problem of the dual PMLSM-driven gantry crane. Next, the problem of nonlinear optimal control of parallel inverted pendulums is treated. Additionally, a nonlinear optimal control method is proposed for the control and stabilization problem of the rotary double-inverted pendulum (double Furuta pendulum). Moreover, a nonlinear optimal control scheme is developed for the dynamic model of the three-link biped robot.

4.1 Flatness-based control in successive loops of an H-type gantry crane with dual PMLSM

4.1.1 Outline

The monograph analyzes further nonlinear optimal and flatness-based control methods for robotic cranes and underactuated pendulums. In this section, the feedback control and stabilization problem of dual PMLSM-driven H-type gantry cranes is treated with the use of a flatness-based control method which is implemented in successive loops. Differential flatness theory and flatness-based control are currently main research directions in the area of nonlinear dynamic systems [711,725,741,715,732,491,252,898]. As previously analyzed, a system is considered to be differentially flat if (i) all its state variables and its control inputs can be written as differential functions of a subset of its state vector elements which constitute the system's flat outputs and (ii) the flat outputs of the system are differentially independent which means that they are not connected through a relation in the form of a homogenous differential equation [91,594,639,489]. The differential flatness property is a constrained controllability condition, as it demonstrates that there exist control inputs which enable the state vector elements to track precisely the setpoints associated with them. Besides, by proving that a dynamic system is differentially flat, it is also confirmed that it can be transformed into the input–output linearized form and subsequently into the canonical Brunovsky form through successive differentiations of its flat outputs [806,492,640,524,54]. By exploiting this latter property, flatness-based controllers are designed. The controlled system is written into the canonical Brunovsky form through consecutive differentiations of its flat outputs, and this latter state-space form is both controllable and observable [439,705,60,706]. Using this new state-space description, the feedback control and state estimation problems are treated.

However, flatness-based control through description into the input–output linearized form and the canonical Brunovsky form requires also inverse transformation for finding the control inputs which should be applied to the initial nonlinear state-space model of the system. These inverse transformations may come against non-invertibility (singularity) issues. To overcome the need for state-space model transformations and to exclude the occurrence of singularities in the computation of the control inputs, the present section develops a flatness-based control method in successive loops [712,720,714,718]. To this end, the dynamic model of the dual-PMLSM-driven gantry crane is decomposed into a series of subsystems which are connected in chained form. It is proven that, each one of these subsystems, if viewed independently is differentially flat. The state vector of the subsequent $(i+1)$-th subsystem is considered to be a virtual control input for the preceding i-th subsystem. Equivalently, the virtual control inputs vector of the preceding i-th subsystem is considered to be the setpoints vector of the subsequent $(i+1)$-th subsystem. From the last N-th subsystem, the real control inputs of the gantry crane are computed by tracing backward all preceding subsystems $N-1,\cdots,1$. The real control inputs signal contains also recursively the virtual control inputs of all preceding subsystems $1,\cdots,N-1$. Despite its simplicity, the flatness-based control method in successive loops performs remarkably well and achieves also fast and accurate tracking of setpoints by the state variables of the gantry crane. Using the local differential flatness properties of each one of the individual subsystems, the design of a local stabilizing feedback controller becomes an easy procedure. It suffices to invert the dynamics of each subsystem, as it is often done for input–output linearizable systems while it also suffices the control input for each subsystem to use a diagonal feedback gains matrix with positive diagonal elements. The global stability properties of the new control method are also proven through Lyapunov analysis.

The present section proposes flatness-based control in successive loops for dual-PMLSM-driven gantry cranes. As noted, dual-drive H-type gantry cranes are long-stroke high-speed Cartesian robotic systems which can be used in several industrial applications, such as (i) circuit assembly, printing circuit boards, photolithography, (ii) precision machining, computer numerical control (CNC), arc welding and laser or plasma cutting, (iii) as lifting machines and for the loading, unloading and transfer of cargos, and finally (iv) in biomedical applications such as CT and X-ray scanning [537,535,1050]. The dual-drive H-type gantry crane consists of two linear motors which are arranged in parallel along the vertical axes of an orthogonal board and which are rigidly connected with a cross-beam along the horizontal axis [961,923,359]. Dual drive gantry cranes can achieve high torque and high precision in tasks' execution [860,562]. The cross-beam serves also as support to a third linear motor carrying a load [139,791,924,421]. There is a need for precise synchronization between the two parallel motors and for minimization of internal forces [928,277,276,790]. When the two linear motors are not well synchronized, the crossbeam will misalign from the horizontal axis and will rotate by a small angle around its center of gravity [494,133,495]. Synchronization error is often caused by unbalanced forces, mechanical assembly variations and various disturbances [521,789,493]. Permanent magnet linear synchronous motors (PMLSMs) are a core subsystem in H-type gantry cranes [138,158,1074,167]. PMLSMs can generate the actuation forces which enable precise positioning along the crane's vertical and horizontal axes [906,129,131]. Due to the wide expansion of related industrial applications, the need for designing more efficient and optimized electrically actuated gantry cranes has become apparent. Additional results on control of gantry-cranes driven by dual PMLSMs can be found in [540,541,543,1019,1065,275]. Taking into account the above, the use of flatness-based control in successive loops in the dynamic model of the dual-PMLSM-driven gantry crane is a challenging research topic.

4.1.2 Dynamic model of the dual PMLSM-driven H-type gantry crane

The diagram of the dual-drive H-type gantry crane with actuation by two permanent magnet linear synchronous motors is given again in Figure 4.1. The model of the gantry crane comprises four masses which are defined as follows: (i) m_b is the mass of the beam, (ii) m_h is the mass of the moving head (load), (iii) m_1 the mass of the first motor, and (iv) m_2 the mass of the second motor. The motion of the gantry crane is also defined by four friction coefficients: (i) c_{g_1} is the friction coefficient of the first motor which moves at velocity $\dot{X}_1$, (ii) c_{g_2} is the friction coefficient of the second motor which moves at velocity $\dot{X}_2$, (iii) c_y is the friction coefficient for the actuator that moves the load along the rigid beam, and (iv) $c_b=c_{b_1}+c_{b_2}$ is the friction coefficient of the beam, which is connected to the two motors with two flexible joints that exhibit friction coefficients c_{b_1} and c_{b_2}, respectively. There is also a stiffness coefficient for the beam which is defined as $k_b=k_{b_1}+k_{b_2}$. The beam has length l_b. The head (load) has length l_h [537,277].

The position of motor 1 is denoted as X_1, while the position of motor 2 is denoted as X_2. The position of the load with respect to the center of mass of the beam is denoted as Y. Moreover, the following position variables are defined:

$$\begin{pmatrix} X \\ \Theta \\ Y \end{pmatrix} = \begin{pmatrix} \frac{1}{2} & \frac{1}{2} & 0 \\ \frac{1}{L_b} & -\frac{1}{L_b} & 0 \\ 0 & 0 & 1 \end{pmatrix} \begin{pmatrix} X_1 \\ X_2 \\ Y \end{pmatrix} \tag{4.1}$$

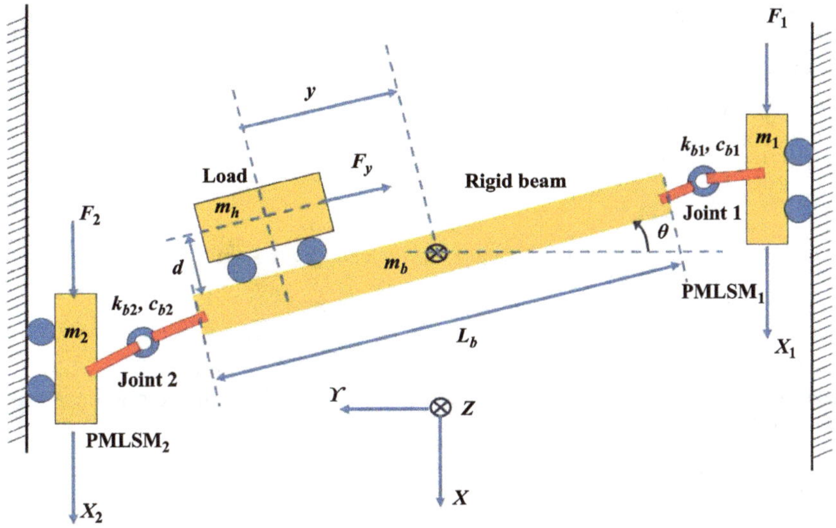

Figure 4.1 Diagram of the dual-drive H-type gantry crane with actuation by two permanent magnet linear synchronous motors

Using the Euler-Lagrange approach, the dynamic model of the dual gantry crane is [277,732]

$$M\ddot{q} + H\dot{q} + C\dot{q} + Kq = F \tag{4.2}$$

where $q = [X, \Theta, Y]^T$ is the state vector of the crane. The inertia matrix $M(q)$ is [277]

$$M = \begin{pmatrix} M_{11} & M_{12} & -m_b sin(\theta) \\ M_{12} & J_t + m_b Y^2 & -m_h d \\ -m_b sin(\Theta) & -m_b d & m_b \end{pmatrix} \tag{4.3}$$

where $J_t = J_b + J_h + 0.25(m_1 + m_2)L_b + m_h d^2$ with J_b to be the moment of inertia of the beam and J_h to be the moment of inertia of the load. Moreover, $M_{11} = m_1 + m_2 + m_b + m_h$ and $M_{12} = m_h d \cdot sin(\Theta) - [m_h Y - 0.5(m_1 - m_2)L_b]cos(\Theta)$.

Matrix H is the Coriolis and centrifugal forces matrix: [277]

$$H = \begin{pmatrix} 0 & H_{12}\dot{\Theta} & -2m_b\dot{\Theta}cos(\Theta) \\ 0 & 2m_b Y\dot{Y} & 0 \\ 0 & -m_b Y\Theta & 0 \end{pmatrix} \tag{4.4}$$

Matrix K is the damping forces matrix which is defined as

$$K = \begin{pmatrix} 0 & 0 & 0 \\ 0 & K_{22} & 0 \\ 0 & 0 & 0 \end{pmatrix} \tag{4.5}$$

where $K_{22} = K_{b_1} + K_{b_2}$. The generalized forces vector is

$$F = \left(F_c \ M_c \ F_y\right)^T \tag{4.6}$$

where F_c is a force causing translation X of the beam, M_c is a torque causing rotation of the beam by angle Θ and F_y is a force causing translation of the load. The relation connecting the generalized forces vector to the forces generated by the PMLSMs of the gantry crane is [277]

$$\begin{pmatrix} F_c \\ M_c \\ F_y \end{pmatrix} = \begin{pmatrix} 1 & 1 & 0 \\ \frac{L_b}{2} & -\frac{L_b}{2} & 0 \\ 0 & 0 & 1 \end{pmatrix} \begin{pmatrix} F_1 \\ F_2 \\ F_y \end{pmatrix} \tag{4.7}$$

The above relation about the forces vectors of the gantry crane can be also written as

$$F = A_f F_1 \tag{4.8}$$

About the moments of inertia of the crane, it holds that $J_b = m_b \frac{L_b^2}{12}$ and $J_h = m_h \frac{L_h^2}{12}$.

Using a description of the voltage and current variables of the PMLSM in the dq reference frame, the dynamics of each motor is given by [725]

$$\begin{aligned} \dot{i}_q &= -i_q \frac{R_s}{L_q} - \frac{n_p \pi v}{L_q \tau}(L_d i_d + \lambda_{pm}) + \frac{1}{L_q} v_q \\ \dot{i}_d &= -i_d \frac{R_s}{L_d} - \frac{n_p \pi v}{L_d \tau} L_q i_q + \frac{1}{L_d} v_d \end{aligned} \tag{4.9}$$

The parameters of the electric part of the PMLSM are: R_s is the motor's resistance, λ_{pm} is the magnetic flux due to permanent magnets, n_p is the number of poles of the motor, L_d, L_q are the dq frame inductances. Moreover, i_d, i_q are the dq frame current variables, and v_d, v_q are the dq frame voltage input variables. Additionally, v is the linear velocity of the moving part of each PMLSM. For the first motor this velocity variable is $\dot{X}_1$, for the second motor it is $\dot{X}_2$ and for the motor moving the load it is $\dot{Y}$. The electromagnetic force for each motor is

$$F_e = \frac{3\pi n_p}{M+M_L} \frac{\lambda_{pm} i_q - (L_d - L_q) i_d i_q}{2\tau} - Dv \tag{4.10}$$

and by considering that $L_d = L_q$ and a small friction coefficient $D = 0$ the electromagnetic force of the motor is given by

$$F_e = \frac{3\pi n_p}{M+M_L} \frac{\lambda_{pm}}{2\tau} i_q \tag{4.11}$$

Next, the dynamic model of the dual-drive H-type gantry crane is written as

$$\begin{aligned} \ddot{q} &= -M^{-1}H\dot{q} - M^{-1}C\dot{q} - M^{-1}Kq + M^{-1}A_f F_1 \Rightarrow \\ \ddot{q} &= -M^{-1}(H+C)\dot{q} - M^{-1}Kq + M^{-1}A_f F_1 \end{aligned} \tag{4.12}$$

About the inverse of the inertia matrix M of the dual-drive gantry crane, it holds that

$$M^{-1} = \frac{1}{detM} \begin{pmatrix} \bar{M}_{11} & -\bar{M}_{21} & \bar{M}_{31} \\ \bar{M}_{12} & \bar{M}_{22} & -\bar{M}_{32} \\ \bar{M}_{13} & -\bar{M}_{23} & \bar{M}_{33} \end{pmatrix} \tag{4.13}$$

where $\bar{M}_{11} = (J_t + m_b Y^2)m_b - (m_b \cdot d)^2$, $\bar{M}_{12} = M_{12}m_b - (m_b sin(\Theta))(m_b \cdot d)$, $\bar{M}_{13} = -M_{12}(m_b \cdot d) + (m_b sin(\Theta))(J_t + m_b Y^2)$, and also $\bar{M}_{21} = M_{21} \cdot m_b - (m_b sin(\Theta))(m_b \cdot d)$, $\bar{M}_{22} = -M_{11}m_b - (m_b sin(\Theta))^2$, $\bar{M}_{23} = -M_{11}m_b d - M_{12}m_b sin(\Theta)$, $\bar{M}_{31} = M_{12}m_b d + (J_t + m_b Y^2)(m_b sin(\Theta))$, $\bar{M}_{32} = -M_{11}m_b d + M_{12}(m_b sin(\Theta))$, $\bar{M}_{33} = M_{11}(J_t + m_b Y^2) - M_{12}^2$ while the determinant of inertia matrix is $detM = M_{11}\bar{M}_{11} - M_{12}\bar{M}_{12} + M_{13}\bar{M}_{13}$.

About the product $(H+C)\dot{q}$, it holds that

$$(H+C)\dot{q} = \begin{pmatrix} C_{11} & C_{12} + H_{12}\dot{\Theta} & -2m_h \dot{\Theta} cos(\Theta) \\ C_{12} & C_{22} + 2m_h Y\dot{Y} & 0 \\ 0 & -m_h Y\Theta & C_y \end{pmatrix} \begin{pmatrix} \dot{X} \\ \dot{\Theta} \\ \dot{Y} \end{pmatrix} \tag{4.14}$$

which is also written as

$$(H+C)\dot{q} = \begin{pmatrix} C_{11}\dot{X} & (C_{12}+H_{12}\dot{\Theta})\dot{\Theta} & (-2m_h\dot{\Theta}cos(\Theta))\dot{Y} \\ C_{12}\dot{X} & (C_{22}+2m_hY\dot{Y})\dot{\Theta} & 0 \\ 0 & (-m_hY\Theta)\dot{\Theta} & C_y\dot{Y} \end{pmatrix} = \begin{pmatrix} \bar{h}_1 \\ \bar{h}_2 \\ \bar{h}_3 \end{pmatrix} \tag{4.15}$$

Moreover, about the product $-M^{-1}(H+G)\dot{q}$, it holds that

$$M^{-1}(H+C)\dot{q} = \frac{1}{detM}\begin{pmatrix} \bar{M}_{11} & -\bar{M}_{21} & \bar{M}_{31} \\ -\bar{M}_{12} & \bar{M}_{22} & -\bar{M}_{22} \\ \bar{M}_{13} & -\bar{M}_{23} & \bar{M}_{33} \end{pmatrix}\begin{pmatrix} \bar{h}_1 \\ \bar{h}_2 \\ \bar{h}_3 \end{pmatrix} \tag{4.16}$$

which is also written as

$$M^{-1}(H+C)\dot{q} = \frac{1}{detM}\begin{pmatrix} \bar{M}_{11}\bar{h}_1 - \bar{M}_{21}\bar{h}_2 + \bar{M}_{31}\bar{h}_3 \\ \bar{M}_{12}\bar{h}_1 + \bar{M}_{22}\bar{h}_2 - \bar{M}_{22}\bar{h}_3 \\ \bar{M}_{13}\bar{h}_1 - \bar{M}_{23}\bar{h}_2 + \bar{M}_{33}\bar{h}_3 \end{pmatrix} \tag{4.17}$$

Additionally, about the product $-M^{-1}Kq$, it holds that

$$M^{-1}Kq = \frac{1}{detM}\begin{pmatrix} \bar{M}_{11} & -\bar{M}_{21} & \bar{M}_{31} \\ -\bar{M}_{12} & \bar{M}_{22} & -\bar{M}_{32} \\ \bar{M}_{13} & -\bar{M}_{23} & \bar{M}_{33} \end{pmatrix}\begin{pmatrix} 0 \\ K_{22}\Theta \\ 0 \end{pmatrix} = \frac{1}{detM}\begin{pmatrix} -\bar{M}_{21}K_{22}\Theta \\ \bar{M}_{22}K_{22}\Theta \\ -\bar{M}_{23}K_{22}\Theta \end{pmatrix} = \frac{1}{detM}\begin{pmatrix} -\bar{M}_{21}\bar{k}_1 \\ \bar{M}_{22}\bar{k}_2 \\ -\bar{M}_{23}\bar{k}_3 \end{pmatrix} \tag{4.18}$$

where $\bar{k}_1 = \bar{k}_2 = \bar{k}_3 = K_{22}\Theta$. Consequently, it holds that the aggregate term $-M^{-1}(H+C)\dot{q} - M^{-1}Kq$ takes the form

$$-M^{-1}(H+C)\dot{q} - M^{-1}Kq = \begin{pmatrix} \bar{M}_{11}\bar{h}_1 + \bar{M}_{21}\bar{h}_2 - \bar{M}_{31}\bar{h}_3 + \bar{M}_{21}\bar{k}_1 \\ \bar{M}_{12}\bar{h}_1 - \bar{M}_{22}\bar{h}_2 + \bar{M}_{32}\bar{h}_3 - \bar{M}_{22}\bar{k}_2 \\ -\bar{M}_{13}\bar{h}_1 + \bar{M}_{23}\bar{h}_2 - \bar{M}_{33}\bar{h}_3 + \bar{M}_{23}\bar{k}_3 \end{pmatrix} \tag{4.19}$$

Additionally, about the control inputs gain matrix of the dual-drive gantry crane, it holds that

$$M^{-1}A_f = \frac{1}{detM}\begin{pmatrix} \bar{M}_{11} & -\bar{M}_{21} & \bar{M}_{31} \\ -\bar{M}_{12} & \bar{M}_{2}2 & -\bar{M}_{32} \\ \bar{M}_{13} & -\bar{M}_{23} & \bar{M}_{33} \end{pmatrix}\begin{pmatrix} 1 & 1 & 0 \\ \frac{L_b}{2} & -\frac{L_b}{2} & 0 \\ 0 & 0 & 1 \end{pmatrix} \tag{4.20}$$

and by defining as $L_{b_1} = \frac{L_b}{2}$, the previous equation gives

$$M^{-1}A_f = \frac{1}{detM}\begin{pmatrix} \bar{M}_{11} - L_{b_1}\bar{M}_{21} & \bar{M}_{11} + L_{b_1}\bar{M}_{21} & M_{31} \\ -\bar{M}_{12} + L_{b_1}\bar{M}_{22} & -\bar{M}_{12} - L_{b_1}\bar{M}_{22} & -M_{32} \\ \bar{M}_{13} - L_{b_1}\bar{M}_{23} & \bar{M}_{13} + L_{b_1}\bar{M}_{23} & \bar{M}_{33} \end{pmatrix} \tag{4.21}$$

Consequently, the dynamic model of the dual-drive gantry crane is written as

$$\begin{pmatrix} \ddot{X} \\ \ddot{\Theta} \\ \ddot{Y} \end{pmatrix} = \begin{pmatrix} \frac{-\bar{M}_{11}\bar{h}_1+\bar{M}_{21}\bar{h}_2-\bar{M}_{31}\bar{h}_3+\bar{M}_{21}\bar{k}_1}{detM} \\ \frac{\bar{M}_{12}\bar{h}_1-\bar{M}_{22}\bar{h}_2+\bar{M}_{32}\bar{h}_3-\bar{M}_{22}\bar{k}_2}{detM} \\ \frac{-\bar{M}_{13}\bar{h}_1+\bar{M}_{23}\bar{h}_2-\bar{M}_{33}\bar{h}_3-\bar{M}_{23}\bar{k}_3}{detM} \end{pmatrix} + \begin{pmatrix} \frac{\bar{M}_{11}-L_{b_1}\bar{M}_{21}}{detM} & \frac{\bar{M}_{11}+L_{b_1}\bar{M}_{21}}{detM} & \frac{M_{31}}{detM} \\ \frac{-\bar{M}_{12}+L_{b_1}\bar{M}_{2}2}{detM} & \frac{-\bar{M}_{12}-L_{b_1}\bar{M}_{22}}{detM} & \frac{-M_{32}}{detM} \\ \frac{\bar{M}_{13}-L_{b_1}\bar{M}_{23}}{detM} & \frac{\bar{M}_{13}+L_{b_1}\bar{M}_{23}}{detM} & \frac{\bar{M}_{33}}{detM} \end{pmatrix}\begin{pmatrix} F_1 \\ F_2 \\ F_y \end{pmatrix} \tag{4.22}$$

About the forces which are generated by the PMLSM actuators, it holds that

$$F_1 = \frac{3\pi n_{p1}}{m_1+m_b+m_h}\frac{\lambda_{pm_1}}{2\tau_1}i_{q1} \Rightarrow F_1 = w_1 i_{q1} \tag{4.23}$$

$$F_2 = \frac{3\pi n_{p2}}{m_2+m_b+m_h}\frac{\lambda_{pm_2}}{2\tau_2}i_{q2} \Rightarrow F_2 = w_2 i_{q2} \tag{4.24}$$

$$F_y = \frac{3\pi n_{p3}}{m_h}\frac{\lambda_{pm_3}}{2\tau_3}i_{q3} \Rightarrow F_y = w_3 i_{q3} \tag{4.25}$$

The currents of the PMLSMs are given by the following equations:

First PMLSM of the rigid beam

$$\dot{i}_{q1} = -i_{q1}\frac{R_{s1}}{L_{q1}} - \frac{n_{p1}\pi\dot{X}_1}{L_{q1}\tau_1}(L_{d1}i_{d1} + \lambda_{pm1}) + \frac{1}{L_{q1}}v_{q1} \tag{4.26}$$

$$\dot{i}_{d1} = -i_{d1}\frac{R_{s1}}{L_{d1}} - \frac{n_{p1}\pi\dot{X}_1}{L_{d1}\tau_1}(L_{q1}i_{q1}) + \frac{1}{L_{d1}}v_{d1} \tag{4.27}$$

Second PMLSM of the rigid beam

$$\dot{i}_{q2} = -i_{q2}\frac{R_{s2}}{L_{q2}} - \frac{n_{p2}\pi\dot{X}_2}{L_{q2}\tau_2}(L_{d2}i_{d2} + \lambda_{pm2}) + \frac{1}{L_{q2}}v_{q2} \tag{4.28}$$

$$\dot{i}_{d2} = -i_{d2}\frac{R_{s2}}{L_{d2}} - \frac{n_{p2}\pi\dot{X}_2}{L_{d2}\tau_2}(L_{q2}i_{q2}) + \frac{1}{L_{d2}}v_{d2} \tag{4.29}$$

Third PMLSM of the load

$$\dot{i}_{q3} = -i_{q3}\frac{R_{s3}}{L_{q3}} - \frac{n_{p3}\pi\dot{Y}}{L_{q3}\tau_3}(L_{d3}i_{d3} + \lambda_{pm3}) + \frac{1}{L_{q3}}v_{q3} \tag{4.30}$$

$$\dot{i}_{d3} = -i_{d3}\frac{R_{s3}}{L_{d3}} - \frac{n_{p3}\pi\dot{Y}}{L_{d3}\tau_3}(L_{q3}i_{q3}) + \frac{1}{L_{d3}}v_{d3} \tag{4.31}$$

Besides, from Eq. (4.1), it holds

$$\begin{aligned} X_1 &= X + L_{b1}\Theta \quad \dot{X}_1 = \dot{X} + L_{b1}\dot{\Theta} \\ X_2 &= X - L_{b2}\Theta \quad \dot{X}_2 = \dot{X} - L_{b2}\dot{\Theta} \end{aligned} \tag{4.32}$$

As explained before in subsection 3.2.2, the state-vector of the dual-drive gantry crane is defined as

$$\begin{aligned} x &= [x_1, x_2, x_3, x_4, x_5, x_6, x_7, x_8, x_9, x_{10}, x_{11}, x_{12}]^T \Rightarrow \\ x &= [X.\Theta, Y, \dot{X}, \dot{\Theta}, \dot{Y}, i_{q1}, i_{d1}, i_{q2}, i_{d2}, i_{q3}, i_{d3}] \end{aligned} \tag{4.33}$$

while the control inputs vector is defined as

$$\begin{aligned} u &= [u_1, u_2, u_3, u_4, u_5, u_6]^T \Rightarrow \\ u &= [v_{d1}, v_{q1}, v_{d2}, v_{q2}, v_{d3}, v_{q3}] \end{aligned} \tag{4.34}$$

Using this state-vector and control inputs vector notation, the dynamic model of the dual-drive gantry crane is written as follows:

$$\dot{x}_1 = x_4 \tag{4.35}$$

$$\dot{x}_2 = x_5 \tag{4.36}$$

$$\dot{x}_3 = x_6 \tag{4.37}$$

$$\dot{x}_4 = \frac{-\bar{M}_{11}\bar{h}_1 + \bar{M}_{21}\bar{h}_2 - \bar{M}_{31}\bar{h}_3 + \bar{M}_{21}\bar{k}_1}{detM} + \frac{\bar{M}_{11} - L_{b1}\bar{M}_{21}}{detM}w_1x_7 + \frac{\bar{M}_{11} + L_{b1}\bar{M}_{21}}{detM}w_2x_9 + \frac{\bar{M}_{31}}{detM}w_3x_{11} \tag{4.38}$$

$$\dot{x}_5 = \frac{\bar{M}_{12}\bar{h}_1 - \bar{M}_{22}\bar{h}_2 + \bar{M}_{32}\bar{h}_3 - \bar{M}_{22}\bar{k}_2}{detM} + \frac{-\bar{M}_{12} + L_{b1}\bar{M}_{22}}{detM}w_1x_7 + \frac{-\bar{M}_{12} + L_{b1}\bar{M}_{22}}{detM}w_2x_9 + \frac{-\bar{M}_{32}}{detM}w_3x_{11} \tag{4.39}$$

$$\dot{x}_6 = \frac{-\bar{M}_{13}\bar{h}_1 + \bar{M}_{23}\bar{h}_2 - \bar{M}_{33}\bar{h}_3 + \bar{M}_{23}\bar{k}_3}{detM} + \frac{\bar{M}_{13} - L_{b1}\bar{M}_{23}}{detM}w_1x_7 + \frac{\bar{M}_{13} + L_{b1}\bar{M}_{23}}{detM}w_2x_9 + \frac{\bar{M}_{33}}{detM}w_3x_{11} \tag{4.40}$$

$$\dot{x}_7 = -x_7\frac{R_{s1}}{L_{q1}} - \frac{n_{p1}\pi(x_4 + L_{b1}x_5)}{L_{q1}\tau_1}(L_{d1}x_8 + \lambda_{pm1}) + \frac{1}{L_{q1}}u_1 \tag{4.41}$$

$$\dot{x}_8 = -x_8\frac{R_{s1}}{L_{d1}} - \frac{n_{p1}\pi(x_4 + L_{b1}x_5)}{L_{d1}\tau_1}(L_{q1}x_7) + \frac{1}{L_{d1}}u_2 \tag{4.42}$$

$$\dot{x}_9 = -x_9\frac{R_{s2}}{L_{q2}} - \frac{n_{p2}\pi(x_4 - L_{b1}x_5)}{L_{q2}\tau_2}(L_{d2}x_{10} + \lambda_{pm2}) + \frac{1}{L_{q2}}u_3 \tag{4.43}$$

$$\dot{x}_{10} = -x_{10}\frac{R_{s2}}{L_{d2}} - \frac{n_{p2}\pi(x_4 - L_{b1}x_5)}{L_{d2}\tau_2}(L_{q2}x_9) + \frac{1}{L_{d2}}u_4 \tag{4.44}$$

$$\dot{x}_{11} = -x_{11}\frac{R_{s3}}{L_{q3}} - \frac{n_{p3}\pi x_6}{L_{q3}\tau_3}(L_{d3}x_{12} + \lambda_{pm3}) + \frac{1}{L_{q3}}u_5 \tag{4.45}$$

$$\dot{x}_{12} = -x_{12}\frac{R_{s3}}{L_{d3}} - \frac{n_{p3}\pi x_6}{L_{d3}\tau_3}(L_{q3}x_{11}) + \frac{1}{L_{d3}}u_6 \tag{4.46}$$

As a result of the above, the state-space model of the dual-drive gantry crane can be also written in the following matrix form:

$$\begin{pmatrix} \dot{x}_1 \\ \dot{x}_2 \\ \dot{x}_3 \\ \dot{x}_4 \\ \dot{x}_5 \\ \dot{x}_6 \\ \dot{x}_7 \\ \dot{x}_8 \\ \dot{x}_9 \\ \dot{x}_{10} \\ \dot{x}_{11} \\ \dot{x}_{12} \end{pmatrix} = \begin{pmatrix} x_4 \\ x_5 \\ x_6 \\ f_{4a} + f_{4b}w_1x_7 + f_{4c}w_2x_9 + f_{4d}w_3x_{11} \\ f_{5a} + f_{5b}w_1x_7 + f_{5c}w_2x_9 + f_{5d}w_3x_{11} \\ f_{6a} + f_{6b}w_1x_7 + f_{6c}w_2x_9 + f_{6d}w_3x_{11} \\ f_7 \\ f_8 \\ f_9 \\ f_{10} \\ f_{11} \\ f_{11} \end{pmatrix} + \begin{pmatrix} 0 & 0 & 0 & 0 & 0 & 0 \\ 0 & 0 & 0 & 0 & 0 & 0 \\ 0 & 0 & 0 & 0 & 0 & 0 \\ 0 & 0 & 0 & 0 & 0 & 0 \\ 0 & 0 & 0 & 0 & 0 & 0 \\ 0 & 0 & 0 & 0 & 0 & 0 \\ g_{7,1} & 0 & 0 & 0 & 0 & 0 \\ 0 & g_{8,2} & 0 & 0 & 0 & 0 \\ 0 & 0 & g_{9,3} & 0 & 0 & 0 \\ 0 & 0 & 0 & g_{10,4} & 0 & 0 \\ 0 & 0 & 0 & 0 & g_{11,5} & 0 \\ 0 & 0 & 0 & 0 & 0 & g_{12,6} \end{pmatrix} \begin{pmatrix} u_1 \\ u_2 \\ u_3 \\ u_4 \\ u_5 \\ u_6 \end{pmatrix} \tag{4.47}$$

where the drift vector of the state-space model comprises the following functions:

$f_{4a} = \frac{-\tilde{M}_{11}\tilde{h}_1 + \tilde{M}_{21}\tilde{h}_2 - \tilde{M}_{31}\tilde{h}_3 + \tilde{M}_{21}\tilde{k}_1}{detM}$ $f_{4b} = \frac{\tilde{M}_{11} - L_{b1}\tilde{M}_{21}}{detM}$, $f_{4c} = \frac{\tilde{M}_{11} + L_{b1}\tilde{M}_{21}}{detM}$, $f_{4d} = \frac{\tilde{M}_{31}}{detM}w_3x_{11}$.

$f_{5a} = \frac{\tilde{M}_{12}\tilde{h}_1 - \tilde{M}_{22}\tilde{h}_2 + \tilde{M}_{32}\tilde{h}_3 - \tilde{M}_{22}\tilde{k}_2}{detM}$, $f_{5b} = \frac{-\tilde{M}_{12} + L_{b1}\tilde{M}_{22}}{detM}$, $f_{5c} = \frac{-\tilde{M}_{12} + L_{b1}\tilde{M}_{22}}{detM}$, $f_{5d} = \frac{-\tilde{M}_{32}}{detM}$.

$f_{6a} = \frac{-\tilde{M}_{13}\tilde{h}_1 + \tilde{M}_{23}\tilde{h}_2 - \tilde{M}_{33}\tilde{h}_3 + \tilde{M}_{23}\tilde{k}_3}{detM}$, $f_{6b} = \frac{\tilde{M}_{13} - L_{b1}\tilde{M}_{23}}{detM}$, $f_{6c} = \frac{\tilde{M}_{13} + L_{b1}\tilde{M}_{23}}{detM}$, $f_{6d} = \frac{\tilde{M}_{33}}{detM}$

$f_7 = -x_7\frac{R_{s1}}{L_{q1}} - \frac{n_{p1}\pi(x_4 + L_{b1}x_5)}{L_{q1}\tau_1}(L_{d1}x_8 + \lambda_{pm1})$, $f_8 = -x_8\frac{R_{s1}}{L_{d1}} - \frac{n_{p1}\pi(x_4 + L_{b1}x_5)}{L_{d1}\tau_1}(L_{q1}x_7)$, $f_9 = -x_9\frac{R_{s2}}{L_{q2}} - \frac{n_{p2}\pi(x_4 - L_{b1}x_5)}{L_{q2}\tau_2}(L_{d2}x_{10} + \lambda_{pm2})$, $f_{10} = -x_{10}\frac{R_{s2}}{L_{d2}} - \frac{n_{p2}\pi(x_4 - L_{b1}x_5)}{L_{d2}\tau_2}(L_{q2}x_9)$, $f_{11} = -x_{11}\frac{R_{s3}}{L_{q3}} - \frac{n_{p3}\pi x_6}{L_{q3}\tau_3}(L_{d3}x_{12} + \lambda_{pm3})$, and $f_{12} = -x_{12}\frac{R_{s3}}{L_{d3}} - \frac{n_{p3}\pi x_6}{L_{d3}\tau_3}(L_{q3}x_{11})$.

Finally, the state-space model of the dual-drive gantry crane is written in the following concise nonlinear affine-in-the-input state-space form:

$$\dot{x} = f(x) + g(x)u \tag{4.48}$$

where $x \in R^{12\times 1}$, $f(x) \in R^{12\times 1}$, $g(x) \in R^{12\times 6}$, and $u \in R^{6\times 1}$. The differential flatness properties of the dual-drive H-type gantry crane were proven before.

4.1.3 Differential flatness properties of the dual-drive H-type gantry crane

It has been previously shown that the dynamic model of the dual-drive H-type is differentially flat, with flat outputs vector

$$\begin{aligned} Y &= [x_1, x_2, x_3, x_8, x_{10}, x_{12}]^T \Rightarrow \\ Y &= [X, \Theta, Y, i_{d1}, i_{d2}, i_{d3}]^T \end{aligned} \tag{4.49}$$

The state-space model of the dual-drive gantry-crane given in Eq. (4.47) comprises 12 state equations. From the first three rows of the state-space model, one has

$$\begin{aligned} \dot{x}_1 &= x_4 \Rightarrow x_4 = \tilde{h}_4(Y, \dot{Y}) \\ \dot{x}_2 &= x_5 \Rightarrow x_5 = \tilde{h}_5(Y, \dot{Y}) \\ \dot{x}_3 &= x_6 \Rightarrow x_6 = \tilde{h}_6(Y, \dot{Y}) \end{aligned} \tag{4.50}$$

which signifies that state variables x_4, x_5, x_6 are differential functions of the flat outputs of the system.

Next, from the fourth, fifth and sixth rows of the state-space model, one has

$$\begin{pmatrix} \dot{x}_4 \\ \dot{x}_5 \\ \dot{x}_6 \end{pmatrix} = \begin{pmatrix} f_{4a} \\ f_{5a} \\ f_{6a} \end{pmatrix} + \begin{pmatrix} f_{4b}w_1 & f_{4c}w_2 & f_{4d}w_3 \\ f_{5b}w_1 & f_{5c}w_2 & f_{5d}w_3 \\ f_{6b}w_1 & f_{6c}w_2 & f_{6d}w_3 \end{pmatrix} \begin{pmatrix} x_7 \\ x_9 \\ x_{11} \end{pmatrix} \tag{4.51}$$

or equivalently

$$\begin{pmatrix} x_7 \\ x_9 \\ x_{11} \end{pmatrix} = \begin{pmatrix} f_{4b}w_1 & f_{4c}w_2 & f_{4d}w_3 \\ f_{5b}w_1 & f_{5c}w_2 & f_{5d}w_3 \\ f_{6b}w_1 & f_{6c}w_2 & f_{6d}w_3 \end{pmatrix}^{-1} \left[\begin{pmatrix} \dot{x}_4 \\ \dot{x}_5 \\ \dot{x}_6 \end{pmatrix} - \begin{pmatrix} f_{4a} \\ f_{5a} \\ f_{6a} \end{pmatrix} \right] \tag{4.52}$$

where $(f_{4a}, f_{4b}, f_{4c}, f_{4d})$, $(f_{5a}, f_{5b}, f_{5c}, f_{5d})$, and $(f_{6a}, f_{6b}, f_{6c}, f_{6d})$ are functions of state variables x_i $i = 1 \cdots 6$ and thus they are also differential functions of the flat outputs of the system. Consequently, one obtains about state variables x_7, x_9, x_{11}

$$\begin{aligned} x_7 &= \tilde{h}_7(Y, \dot{Y}) \\ x_9 &= \tilde{h}_9(Y, \dot{Y}) \\ x_{11} &= \tilde{h}_{11}(Y, \dot{Y}) \end{aligned} \tag{4.53}$$

which signifies that state variables x_7, x_9, x_{11} are differential functions of the flat outputs of the system. Moreover, from rows 7–12 of the state-space model of Eq. (4.47), one solves for the control inputs u_1 to u_6. This gives

$$\begin{pmatrix} u_1 \\ u_2 \\ u_3 \\ u_4 \\ u_5 \\ u_6 \end{pmatrix} = \begin{pmatrix} g_{7,1} & 0 & 0 & 0 & 0 & 0 \\ 0 & g_{8,2} & 0 & 0 & 0 & 0 \\ 0 & 0 & g_{9,3} & 0 & 0 & 0 \\ 0 & 0 & 0 & g_{10,4} & 0 & 0 \\ 0 & 0 & 0 & 0 & g_{11,5} & 0 \\ 0 & 0 & 0 & 0 & 0 & g_{12,6} \end{pmatrix}^{-1} \left[\begin{pmatrix} \dot{x}_7 \\ \dot{x}_8 \\ \dot{x}_9 \\ \dot{x}_{10} \\ \dot{x}_{11} \\ \dot{x}_{12} \end{pmatrix} - \begin{pmatrix} f_7 \\ f_8 \\ f_9 \\ f_{10} \\ f_{11} \\ f_{12} \end{pmatrix} \right] \tag{4.54}$$

where x_8, x_{10}, x_{12} are flat outputs of the system, while all other variables and functions which appear in the right part of Eq. (4.54) are differential functions of the flat outputs of the system. Therefore, the control inputs u_i $i = 1 \cdots 6$ are also differential functions of the flat outputs of the system, or

$$u_i = \tilde{h}_{u_i}(Y, \dot{Y}) \quad i = 1, 2, \cdots, 6 \tag{4.55}$$

Consequently, the dynamic model of the dual-drive gantry crane is differentially flat. The differential flatness property is also an implicit proof of the system's controllability. It also demonstrates that the system is input–output linearizable through successive differentiations of its flat outputs. Finally, it allows for solving the setpoints definition problem. First, one defines setpoints in an unconstrained manner for the state variables which are also flat outputs of the system, that is for state variables x_1, x_2, x_3, x_8, x_{10}, and x_{12}. For the rest of the state variables of the gantry crane x_4, x_5, x_6, x_7, x_9, x_{11} setpoints are chosen using the differential relations that connect these state variables to the system's flat outputs.

4.1.4 Flatness-based control in successive loops for the dual-drive H-type gantry crane

4.1.4.1 Proof of differential flatness properties for the chained subsystems

The state-space model of the dual-drive H-type gantry crane of Eq. (4.47) is decomposed into a series of subsystems which are connected in chained form:

The first subsystem comprises rows 1–3 of the state-space model of Eq. (4.47), and the following sub-vectors and sub-matrices are defined about it:

$$x_{1,3}=\begin{pmatrix}x_1\\x_2\\x_3\end{pmatrix}\quad f_{1,3}=\begin{pmatrix}0\\0\\0\end{pmatrix}\quad g_{1,3}=\begin{pmatrix}1&0&0\\0&1&0\\0&0&1\end{pmatrix}\quad v_1=\begin{pmatrix}x_4\\x_5\\x_6\end{pmatrix}\tag{4.56}$$

where $x_{1,3}$ is the state vector of the subsystem and v_1 is the virtual control inputs vector.

The second subsystem comprises rows 4–6 of the state-space model of Eq. (4.47), and the following sub-vectors and sub-matrices are defined about it:

$$x_{4,6}=\begin{pmatrix}x_4\\x_5\\x_6\end{pmatrix}\quad f_{4,6}=\begin{pmatrix}f_{4a}\\f_{4b}\\f_{4c}\end{pmatrix}\quad g_{4,6}=\begin{pmatrix}f_{4b}w_1&f_{4c}w_2&f_{4d}w_3\\f_{5b}w_1&f_{5c}w_2&f_{5d}w_3\\f_{6b}w_1&f_{6c}w_2&f_{6d}w_3\end{pmatrix}\quad v_2=\begin{pmatrix}x_7\\x_9\\x_{11}\end{pmatrix}\tag{4.57}$$

where $x_{4,6}$ is the state vector of the subsystem and v_2 is the virtual control inputs vector.

The third subsystem comprises rows 7–12 of the state-space model of Eq. (4.47), and the following sub-vectors and sub-matrices are defined about it:

$$x_{7,12}=\begin{pmatrix}x_7\\x_8\\x_9\\x_{10}\\x_{11}\\x_{12}\end{pmatrix}\quad f_{7,12}=\begin{pmatrix}f_7\\f_8\\f_9\\f_{10}\\f_{11}\\f_{12}\end{pmatrix}\quad g_{7,12}=\begin{pmatrix}g_{7,1}&0&0&0&0&0\\0&g_{8,2}&0&0&0&0\\0&0&g_{9,3}&0&0&0\\0&0&0&g_{10,4}&0&0\\0&0&0&0&g_{11,5}&0\\0&0&0&0&0&g_{12,6}\end{pmatrix}\quad u=\begin{pmatrix}u_1\\u_2\\u_3\\u_4\\u_5\\u_6\end{pmatrix}\tag{4.58}$$

where $x_{7,12}$ is the state vector of the third subsystem and u is the virtual control inputs vector.

Consequently, the initial state-space model of Eq. (4.47) is decomposed into the chain of the following three subsystems:

$$\dot{x}_{1,3}=f_{1,3}(x_{1,2})+g_{1,3}(x_{1,3})v_1\tag{4.59}$$

$$\dot{x}_{4,6}=f_{4,6}(x_{1,3},x_{4,6})+g_{4,6}(x_{1,3},x_{4,6})v_2\tag{4.60}$$

$$\dot{x}_{7,12}=f_{7,12}(x_{1,3},x_{4,6},x_{7,12})+g_{7,12}(x_{1,3},x_{4,6},x_{7,12})u\tag{4.61}$$

It will be proven that each one of the subsystems of Eq. (4.59] to Eq. (4.61), if viewed independently, is differentially flat.

For the subsystem of Eq. (4.59), the flat outputs vector is $y_1=x_{1,3}$, the virtual control inputs vector is $v_1=x_{4,6}$ while $f_{1,3}(x_{1,3})$, $g_{1,3}(x_{1,3})$ are functions of $y_1=x_{1,3}$. By solving Eq. (4.59), for v_1, one obtains

$$v_1=g_{1,3}^{-1}(x_{1,3})[\dot{x}_{1,3}-f_{1,3}(x_{1,3})]\tag{4.62}$$

It can be seen that v_1 is also a differential function of the flat output y_1; therefore, the subsystem of Eq. (4.59) is differentially flat.

For the subsystem of Eq. (4.60), the flat outputs vector is $y_2=x_{4,6}$, the virtual control inputs vector is $v_2=[x_7,x_9,x_{11}]^T$, while $x_{1,3}$ is considered to be a coefficients vector. Moreover, $f_{4,6}(x_{1,3},x_{4,6})$ and $g_{4,6}(x_{1,3},x_{4,6})$ are functions of the flat outputs vector $x_{4,6}$ and of the coefficients vector $x_{1,3}$. By solving Eq. (4.60) for v_2, one obtains

$$v_2=g_{4,6}^{-1}(x_{1,3},x_{4,6})[\dot{x}_{4,6}-f_{4,6}(x_{1,3},x_{4,6})] \tag{4.63}$$

It can be seen that v_2 is also a differential function of the flat output y_2; therefore, the subsystem of Eq. (4.60) is differentially flat.

For the subsystem of Eq. (4.61), the flat outputs vector is $y_3=x_{7,12}$, the real control inputs vector is u, while $x_{1,3}$, $x_{4,6}$ are considered to be coefficients' vectors. Moreover, $f_{7,12}(x_{1,3},x_{4,6},x_{7,12})$ and $g_{7,12}(x_{1,3},x_{4,6},x_{7,12})$ are functions of the flat outputs vector $x_{7,12}$ and of the coefficients vector $x_{1,3}$, $x_{4,6}$. By solving Eq. (4.61) for u, one obtains

$$u=g_{7,12}^{-1}(x_{1,3},x_{4,6},x_{7,12})[\dot{x}_{7,12}-f_{7,12}(x_{1,3},x_{4,6},x_{7,12})] \tag{4.64}$$

It can be seen that u is also a differential function of the flat output y_3; therefore, the subsystem of Eq. (4.61) is differentially flat.

4.1.4.2 Design of a multi-loop flatness-based controller

By proving that each one of the subsystems of Eq. (4.59) to Eq. (4.61) is differentially flat, one can infer that it can be brought into the input–output linearized form and that a stabilizing feedback controller can be designed about it through the inversion of its dynamics.

For the subsystem of Eq. (4.59), the stabilizing feedback controller is defined as

$$v_1=g_{1,3}^{-1}(x_{1,3})[\dot{x}_{1,3}^d-f_{1,3}(x_{1,3})-K_1(x_{1,3}-x_{1,3}^d)] \tag{4.65}$$

where setpoint $x_{1,3}^d$ is selected without constraints, while $K_1{\in}R^{3\times 3}$ is a diagonal matrix with positive diagonal elements, that is $K_1>0$ and $K_{1,ii}>0$ for $i=1,2,3$. The virtual control input is $v_1=x_{4,6}$ and becomes setpoint to the subsystem of Eq. (4.60), that is $v_1=x_{4,6}^d$.

For the subsystem of Eq. (4.60), the stabilizing feedback controller is defined to be

$$v_2=g_{4,6}^{-1}(x_{1,3},x_{4,6})[\dot{x}_{4,6}^d-f_{4,6}(x_{1,3},x_{4,6})-K_2(x_{4,6}-x_{4,6}^d)] \tag{4.66}$$

where setpoint $x_{4,6}^d$ is selected without constraints, while $K_3{\in}R^{3\times 3}$ is a diagonal matrix with positive diagonal elements, that is $K_2>0$ and $K_{2,ii}>0$ for $i=1,2,3$. The virtual control input is $v_2=[x_7,x_9,x_{11}]^T$ and jointly with the reference values $[x_8,x_{10},x_{12}]^T$ becomes setpoint to the subsystem of Eq. (4.61), that is $x_{7,12}^d=[x_7^d,x_8^d,x_9^d,x_{10}^d,x_{11}^d,x_{12}^d]^T$.

For the subsystem of Eq. (4.61), the stabilizing feedback controller is defined as

$$\begin{aligned}v_3={}&g_{7,12}^{-1}(x_{1,3},x_{4,6},x_{7,12})\cdot\\ &\cdot[\dot{x}_{7,12}^d-f_{7,12}(x_{1,3},x_{4,6},x_{7,12})-K_3(x_{7,12}-x_{7,12}^d)]\end{aligned} \tag{4.67}$$

where setpoint $x_{7,12}^d=[x_7^d,x_8^d,x_9^d,x_{10}^d,x_{11}^d,x_{12}^d]^T$ has been defined above, while $K_3{\in}R^{6\times 6}$ is a diagonal matrix with positive diagonal elements, that is $K_3>0$ and $K_{3,ii}>0$ for $i=1,\cdots,6$. The real control input is u.

By substituting the control input of Eq. (4.65) into the subsystem of Eq. (4.59) and by defining the tracking error variable $e_{1,3}=x_{1,3}-x_{1,3}^d$, one obtains

$$\begin{aligned}&\dot{x}_{1,3}=f_{1,3}+g_{1,3}{\cdot}g_{1,3}^{-1}[\dot{x}_{1,3}^d-f_{1,3}-K_1(x_{1,3}-x_{1,3}^d)]\Rightarrow\\ &(\dot{x}_{1,3}-\dot{x}_{1,3}^d)+K_1(x_{1,3}-x_{1,3}^d)=0\Rightarrow\dot{e}_{1,3}+K_1e_{1,3}=0\Rightarrow\\ &lim_{t\to\infty}e_{1,3}(t)=0\Rightarrow lim_{t\to\infty}x_{1,3}(t)=x_{1,3}^d(t)\end{aligned} \tag{4.68}$$

By substituting the control input of Eq. (4.66) into the subsystem of Eq. (4.60) and by defining the tracking error variable $e_{4,6} = x_{4,6} - x_{4,6}^d$, one obtains

$$\begin{aligned} \dot{x}_{4,6} &= f_{4,6} + g_{4,6} \cdot g_{4,6}^{-1}[\dot{x}_{4,6}^d - f_{4,6} - K_2(x_{4,6} - x_{4,6}^d)] \Rightarrow \\ (\dot{x}_{4,6} - \dot{x}_{4,6}^d) &+ K_2(x_{4,6} - x_{4,6}^d) = 0 \Rightarrow \dot{e}_{4,6} + K_2 e_{4,6} = 0 \Rightarrow \\ lim_{t\to\infty} e_{4,6}(t) &= 0 \Rightarrow lim_{t\to\infty} x_{4,6}(t) = x_{4,6}^d(t) \end{aligned} \tag{4.69}$$

By substituting the control input of Eq. (4.67) into the subsystem of Eq. (4.61) and by defining the tracking error variable $e_{7,12} = x_{7,12} - x_{7,12}^d$, one obtains

$$\begin{aligned} \dot{x}_{7,12} &= f_{7,12} + g_{7,12} \cdot g_{7,12}^{-1}[\dot{x}_{7,12}^d - f_{7,12} - K_3(x_{7,12} - x_{7,12}^d)] \Rightarrow \\ (\dot{x}_{7,12} - \dot{x}_{7,12}^d) &+ K_3(x_{7,12} - x_{7,12}^d) = 0 \Rightarrow \dot{e}_{7,12} + K_3 e_{7,12} = 0 \Rightarrow \\ lim_{t\to\infty} e_{7,12}(t) &= 0 \Rightarrow lim_{t\to\infty} x_{7,12}(t) = x_{7,12}^d(t) \end{aligned} \tag{4.70}$$

Consequently, all state variables of the dual-drive H-type gantry crane converge to the associated setpoints and the crane's control system is globally asymptotically stable.

Global asymptotic stability for the dual-drive H-type gantry crane can be also proven through Lyapunov analysis. To this end, the following Lyapunov function is defined:

$$V = \tfrac{1}{2}[e_{1,3}^T e_{1,3} + e_{4,6}^T e_{4,6} + e_{7,12}^T e_{7,12}] \tag{4.71}$$

By differentiating in time, one obtains

$$\dot{V} = \tfrac{1}{2} 2[e_{1,3}^T \dot{e}_{1,3} + e_{4,6}^T \dot{e}_{4,6} + e_{7,12}^T \dot{e}_{7,12}] \tag{4.72}$$

Next, by substituting in Eq. (4.72), the relations about the tracking error dynamics of Eq. (4.68), Eq. (4.69), and Eq. (4.70), one obtains

$$\dot{V} = \tfrac{1}{2} 2[e_{1,3}^T(-K_1 e_{1,3}) + e_{4,6}^T(-K_2 e_{4,6}) + e_{7,12}^T(-K_3 e_{7,12})] \tag{4.73}$$

or equivalently

$$\dot{V} = -[e_{1,3}^T K_1 e_{1,3} + e_{4,6}^T K_2 e_{4,6} + e_{7,12}^T K_3 e_{7,12}] \tag{4.74}$$

Thus, finally it is concluded that

$$\dot{V} < 0 \; \forall \; e_{1,3} \neq 0, \; e_{4,6} \neq 0, \; e_{7,12} \neq 0 \tag{4.75}$$

while $V = 0$ iff $e_{1,3} = 0$, $e_{4,6} = 0$, and $e_{7,12} = 0$. Consequently, the first derivative of the system's Lyapunov function is always negative and the Lyapunov function V is a diminishing function which converges asymptotically to 0. Therefore, $lim_{t\to\infty} e_{1,3}(t) = 0$, $lim_{t\to\infty} e_{4,6}(t) = 0$, $lim_{t\to\infty} e_{7,12}(t) = 0$, and the system is globally asymptotically stable.

An explicit demonstration of the exponential stabilization that is achieved by flatness-based control in successive loops is given next. The Lyapunov function of the control loop is written as

$$V = \tfrac{1}{2}[\textstyle\sum_{i=1}^3 e_i^2 + \sum_{j=4}^6 e_j^2 + \sum_{m=7}^{12} e_m^2] \tag{4.76}$$

where e_i $i = 1, \cdots, 3$ are the tracking errors for the position state variables, e_j $j = 4, \cdots, 6$ are the tracking errors for the velocity state variables of the H-type gantry crane with dual PMLSM, and e_m $j = 7, \cdots, 12$ are the tracking errors for the state variables associated with the actuators' currents. Equivalently, the first-order time-derivative of the Lyapunov function is written as

$$\dot{V} = -[\textstyle\sum_{i=1}^3 k_{1,i} e_i^2 + \sum_{j=4}^6 k_{2,j-3} e_j^2 + \sum_{m=7}^{12} k_{3,m-6} e_m^2] \tag{4.77}$$

where $k_{1,i} > 0$ $i = 1, \cdots, 3$ are the diagonal elements of gain matrix K_1, $k_{2,j-3} > 0$ $j = 4, \cdots, 6$ are the diagonal elements of gain matrix K_2 and $k_{3,m-6} > 0$ $m = 7, \cdots, 12$ are the diagonal elements of gain matrix K_3. By denoting the minimum of the above-noted elements of the feedback gain matrices as k_{min}, that is

$$k_{min} = min\{k_{1,i} : \; i = 1, \cdots, 3 \text{ and } k_{2,j-3} : \; j = 4, \cdots, 6 \text{ and } k_{3,m-6} : \; m = 7, \cdots, 12\} \tag{4.78}$$

and using Eq. (4.77), one obtains that

$$\begin{aligned}&\dot{V}\leq -k_{min}[\sum_{i=1}^{3}e_i^2+\sum_{j=4}^{6}e_j^2+\sum_{m=7}^{12}e_m^2]\\&\Rightarrow\dot{V}\leq -2k_{min}V\Rightarrow\dot{V}+2k_{min}V\leq 0\end{aligned}\tag{4.79}$$

From Eq. (4.79), one can demonstrate the exponential convergence of the Lyapunov function V to 0.

The feedback control scheme, which is followed for the cascading subsystems that constitute the dynamic model of the H-type gantry crane with dual PMLSM and which is based on inversion of the subsystems' dynamics of this crane system, is equally robust to sliding-mode control in which the switching control term has been substituted by a saturation function. One can easily confirm this for the first-order i-th subsystem of the form $\dot{x}_i=f_i(x_i)+g_i(x_i)v_i$ by defining the sliding surface $s_i=e_i=x_1-x_i^d$ and the associated sliding mode controller $v_i=\hat{g}_i(x)^{-1}[\dot{x}_i^d-\hat{f}_i(x_i)-K_i sgn(x_i-x_i^d)]$ which after substituting the $sgn(s_i)$ function with the saturation $sat(s_i)$ function becomes $v_i=\hat{g}_i(x)^{-1}[\dot{x}_i^d-\hat{f}_i(x_i)-K_i(x_i-x_i^d)]$. The latter relation coincides with the flatness-based control in successive loops for the i-th subsystem under uncertainty (with use of the estimated functions $\hat{f}_i(x)$ and $\hat{g}_i(x)$) which is computed by this section's control method. Therefore, the proposed flatness-based control method in successive loops provides sufficient robustness margins which enable the reliable and safe functioning of the H-type gantry crane with dual PMLSM under reasonable levels of model uncertainty or external perturbations.

4.1.5 Simulation tests

The performance of the flatness-based control method in successive loops for the dual-PMLSM-driven gantry crane is tested through simulation experiments. Indicative values of the parameters of the H-type gantry crane were as follows: $m_1=3.1kg$, $m_2=3.0kg$, $m_b=2.0kg$, $m_h=1.0kg$, $J_b=8.0kg\cdot m^2$, $J_h=1.0kg\cdot m^2$, $L_b=2.0m$ $d=0.2m$, $c_{g1}=0.3$, $c_{g2}=0.4$, $c_y=0.5$, $c_{b1}=0.6$, $c_{b2}=0.5$, $k_{b1}=0.2$, $k_{b2}=0.3$. Moreover, indicative values about the parameters of the PMLSMs were: $R_{s1}=0.55\Omega$, $L_{d1}=0.060H$, $L_{q1}=0.060H$, $n_{p1}=4$, $\tau_1=1.59$, $R_{s2}=0.58\Omega$, $L_{d2}=0.065H$, $L_{q2}=0.065H$, $n_{p2}=4$, $\tau_2=1.55$, and $R_{s3}=0.51\Omega$, $L_{d3}=0.068H$, $L_{q3}=0.068H$, $n_{p3}=4$, $\tau_3=1.50$. The obtained results are depicted in Figures 4.2–4.17. The variation of the Lyapunov function of the controlled system is shown in Figures 4.18–4.21. They confirm that despite its simplicity the flatness-based control method in successive loops achieves fast and accurate tracking of reference setpoints under moderate variations of the control inputs. The selection of stabilizing values for the feedback gains of the controller is shown to be a straightforward and easy-to-follow procedure. Actually, to stabilize the gantry crane it suffices to select diagonal feedback gain matrices with positive diagonal elements. Thus, for the first subsystem, the feedback gain matrix is $K_1\in R^{3\times 3}>0$ with diagonal elements $k_{1,ii}>0$ for $i=1,\cdots,3$. Equivalently, for the second subsystem, the feedback gain matrix is $K_2\in R^{3\times 3}>0$ with diagonal elements $k_{2,ii}>0$ for $i=1,\cdots,3$. Finally, for the third subsystem the feedback gain matrix is $K_3\in R^{6\times 6}>0$ with diagonal elements $k_{3,ii}>0$ for $i=1,\cdots,3$.

The flatness-based control method in successive loops provides also an automated procedure for selecting setpoints for the dynamic model of the dual-PMLSM-driven gantry crane. For the first subsystem and state variables $[x_1,x_2,x_3]^T$, $i=1,2,3$, setpoints are chosen in an unconstrained manner. Next, the virtual control inputs of the first subsystem, that is state variables $v_1=[x_4,x_5,x_6]^T$, become setpoints for the second subsystem. Additionally, the virtual control input for the second subsystem, that is $v_2=[x_7,x_9,x_{11}]^T$, becomes part of the setpoints vector for the third subsystem. The rest of the setpoints vector for the third subsystem is complemented by setpoint variables $[x_8,x_{10},x_{12}]$ which are selected in an unconstrained manner. Finally, it is noted that the transient performance of the control algorithm and the speed of convergence of the individual state variables to the associated setpoints is determined by the values which are assigned to the diagonal elements of the gain matrices K_i, $i=1,2,3$.

To elaborate on flatness-based control in successive loops for the H-type gantry crane with dual PMLSMs the following tables are given: (i) Table 4.1 providing results about the accuracy of tracking of setpoints by the state variables of the H-type gantry crane with dual PMLSMs under an exact dynamic model, (ii) Table 4.2 providing results about the accuracy of tracking of setpoints by the state variables of the H-type gantry crane with dual PMLSMs under a model that is subject to disturbances (for instance change $\Delta a\%$ in the resistance R_{s1} of the first PMLSM), and (iii)

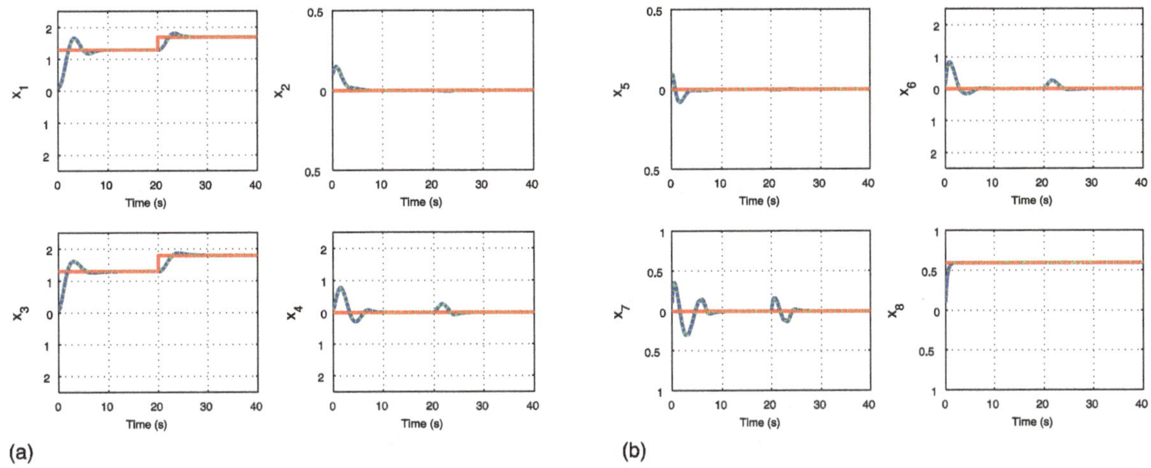

Figure 4.2 *Tracking of setpoint 1 by the dual-PMLSM-driven gantry crane with flatness-based control in successive loops: (a) tracking of setpoints (red line) by the state variables of the gantry crane x_1 to x_4 (blue line) and estimates provided by the H-infinity Kalman filter; (b) tracking of setpoints (red line) by the state variables of the gantry crane x_5 to x_8 (blue line) and estimates provided by the H-infinity Kalman filter*

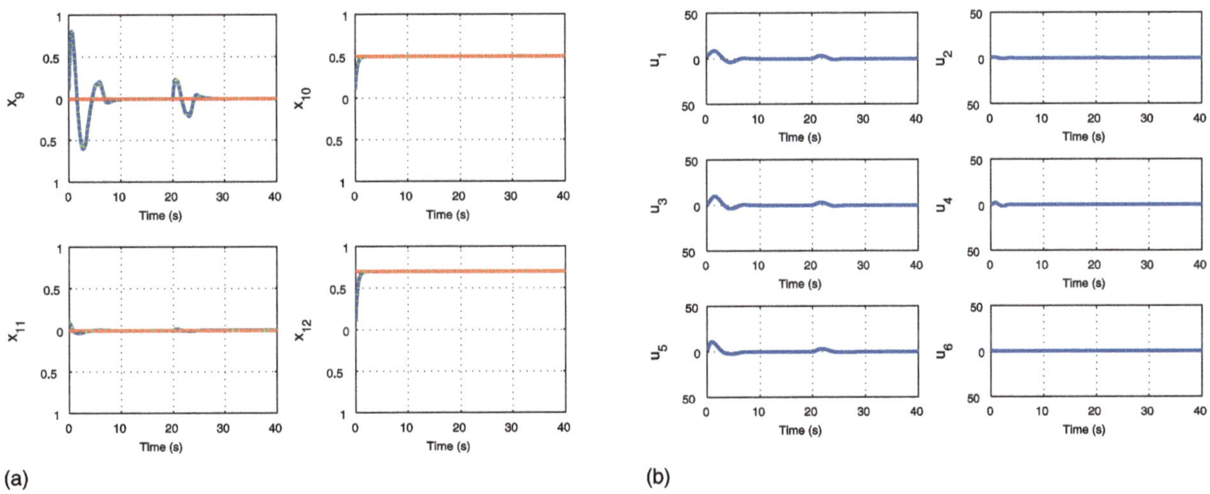

Figure 4.3 *Tracking of setpoint 1 by the dual-PMLSM-driven gantry crane with flatness-based control in successive loops: (a) tracking of setpoints (red line) by the state variables of the gantry crane x_9 to x_{12} (blue line) and estimates provided by the H-infinity Kalman filter; (b) variations of the control inputs of the gantry crane u_1 to u_6 (blue line)*

Table 4.3 providing results about the convergence times of the state variables of the state variables of the H-type gantry crane with dual PMLSMs to the associated setpoints.

4.2 Nonlinear optimal control of the parallel double-inverted pendulum

4.2.1 Outline

The solution of the control problem of coupled nonlinear oscillators is of interest not only for robotics and mechatronics applications but also for electric power systems and models of interconnected power units in the electricity grid [66,213,214]. Among several types of coupled nonlinear oscillators, the parallel double-inverted pendulum is a

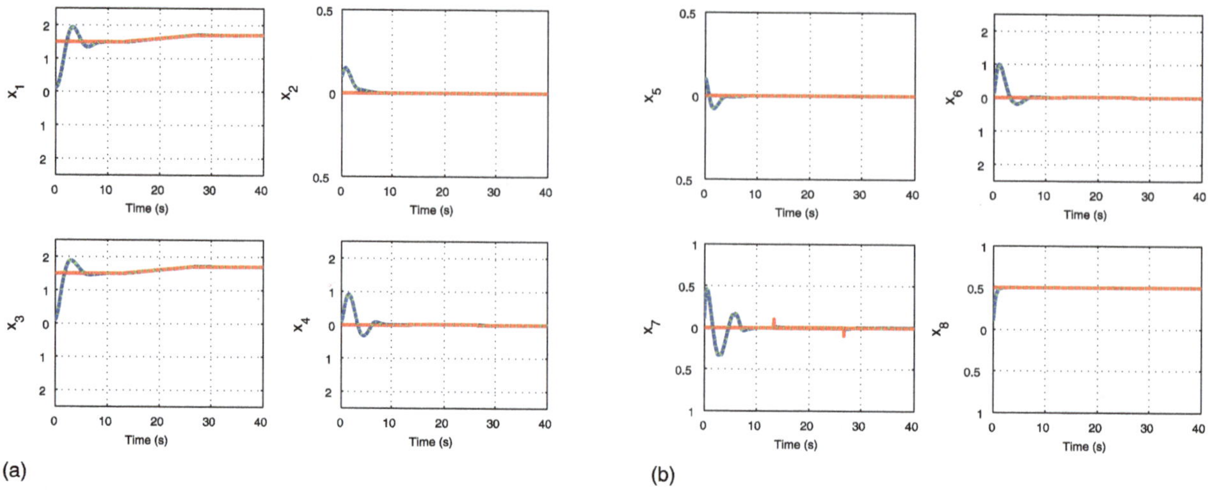

Figure 4.4 Tracking of setpoint 2 by the dual-PMLSM-driven gantry crane with flatness-based control in successive loops: (a) tracking of setpoints (red line) by the state variables of the gantry crane x_1 to x_4 (blue line) and estimates provided by the H-infinity Kalman filter; (b) tracking of setpoints (red line) by the state variables of the gantry crane x_5 to x_8 (blue line) and estimates provided by the H-infinity Kalman filter

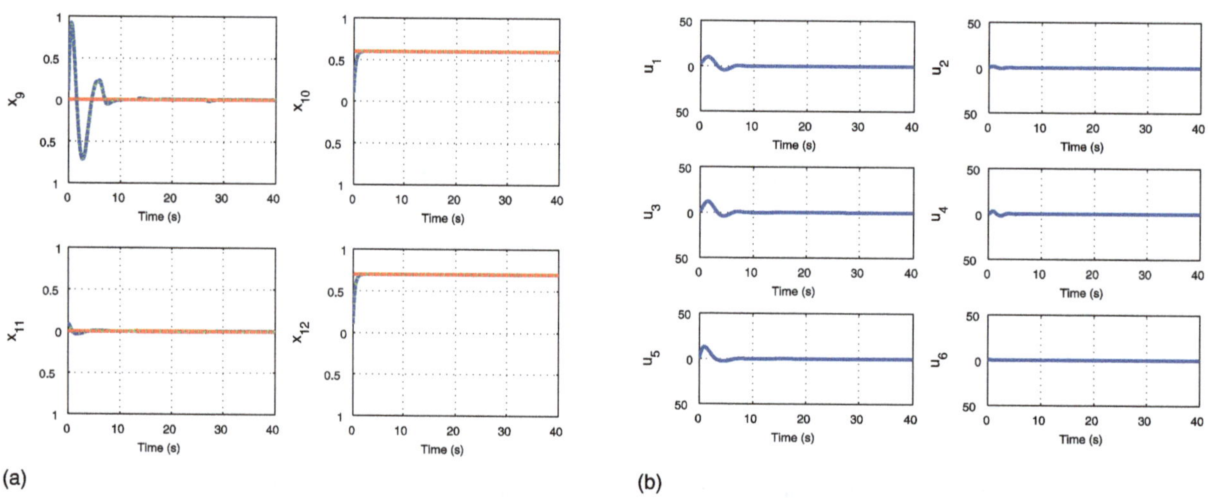

Figure 4.5 Tracking of setpoint 2 by the dual-PMLSM-driven gantry crane with flatness-based control in successive loops: (a) tracking of setpoints (red line) by the state variables of the gantry crane x_9 to x_{12} (blue line) and estimates provided by the H-infinity Kalman filter; (b) variations of the control inputs of the gantry crane u_1 to u_6 (blue line)

strongly nonlinear and underactuated dynamical system [420,473,1004,1003]. The solution of the associated swing-up control and stabilization problem is a nontrivial task [1015,772,680]. There have been so far several attempts to treat the associated nonlinear control problem and to compensate for the highly nonlinear and unstable dynamics of this system [692,257,26]. There have been results on nonlinear model predictive control [306,16]. Moreover, there have been approaches on Lyapunov theory-based and energy-based control [207,1083]. Besides, there have been methods based on fuzzy and sliding-mode control concepts [999,135]. However, little has been done so far for treatment of

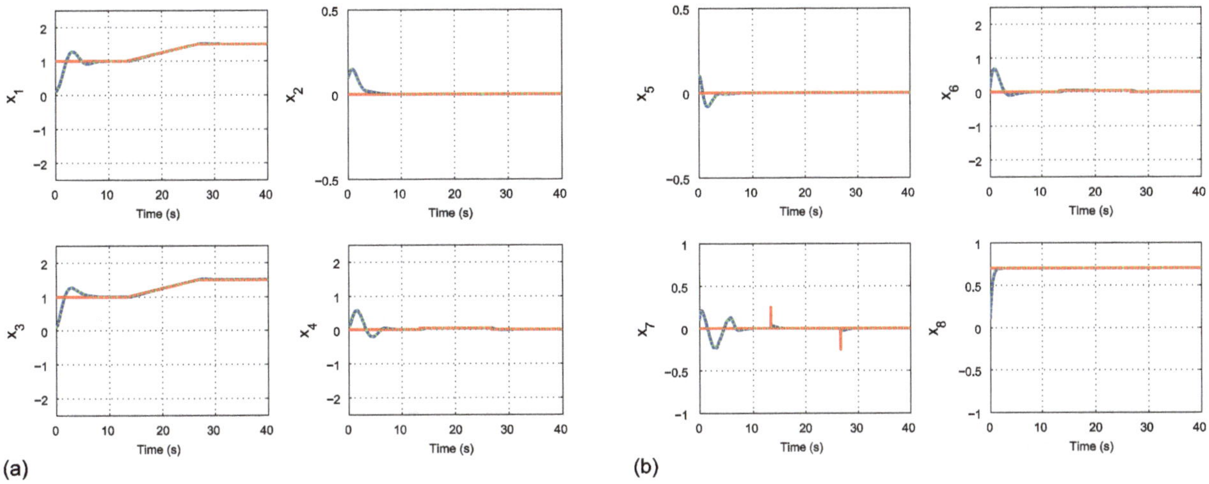

Figure 4.6 *Tracking of setpoint 3 by the dual-PMLSM-driven gantry crane with flatness-based control in successive loops: (a) tracking of setpoints (red line) by the state variables of the gantry crane x_1 to x_4 (blue line) and estimates provided by the H-infinity Kalman filter; (b) tracking of setpoints (red line) by the state variables of the gantry crane x_5 to x_8 (blue line) and estimates provided by the H-infinity Kalman filter*

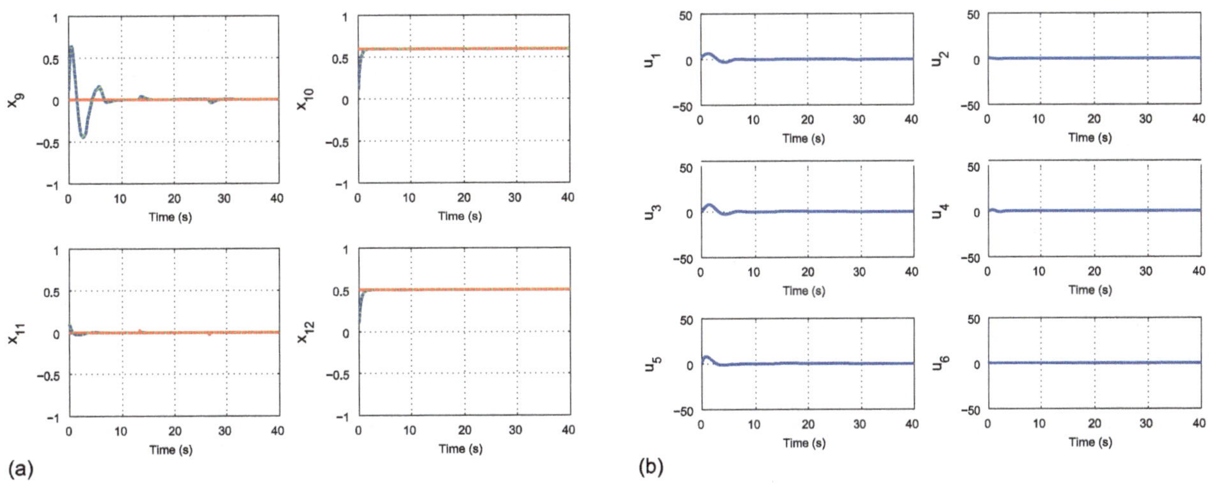

Figure 4.7 *Tracking of setpoint 3 by the dual-PMLSM-driven gantry crane with flatness-based control in successive loops: (a) tracking of setpoints (red line) by the state variables of the gantry crane x_9 to x_{12} (blue line) and estimates provided by the H-infinity Kalman filter; (b) variations of the control inputs of the gantry crane u_1 to u_6 (blue line)*

the associated nonlinear optimal control problem [1016,905,686]. The control problem of coupled and multi-DOF pendulums remains open and additional results about it, presenting methods of confirmed stability, can be found in [533,1010,313,507,215]. In this section, a novel nonlinear optimal (H-infinity) control method is developed for the parallel double-inverted pendulum and is tested in three different forms of this system: (i) a model with four state variables and one control input, describing the dynamics of a cart with double-inverted poles but without including the position of the cart [344], (ii) a model with six state variables and one control input, describing the dynamics of a cart

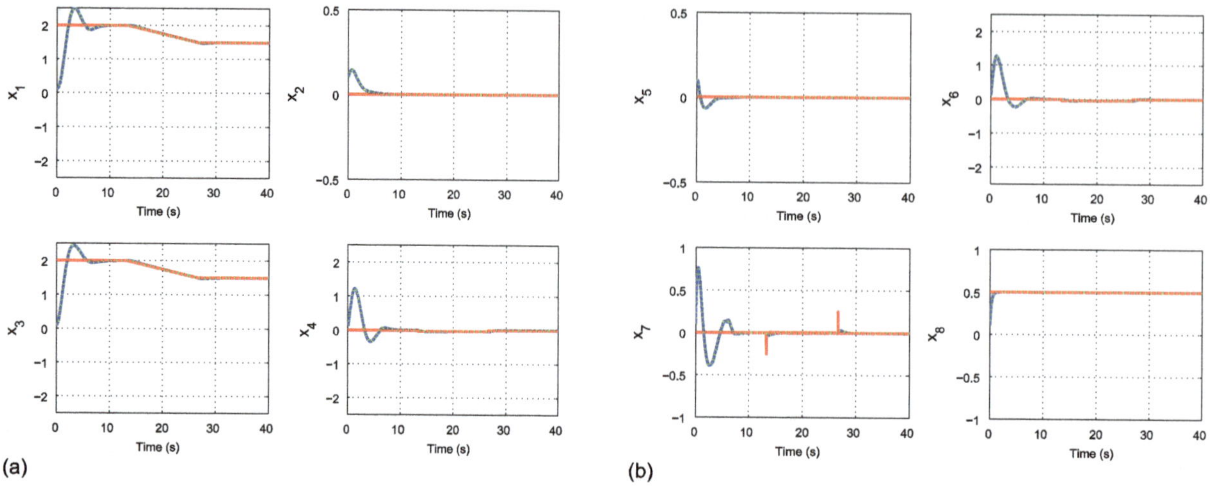

Figure 4.8 *Tracking of setpoint 4 by the dual-PMLSM-driven gantry crane with flatness-based control in successive loops: (a) tracking of setpoints (red line) by the state variables of the gantry crane x_1 to x_4 (blue line) and estimates provided by the H-infinity Kalman filter; (b) tracking of setpoints (red line) by the state variables of the gantry crane x_5 to x_8 (blue line) and estimates provided by the H-infinity Kalman filter*

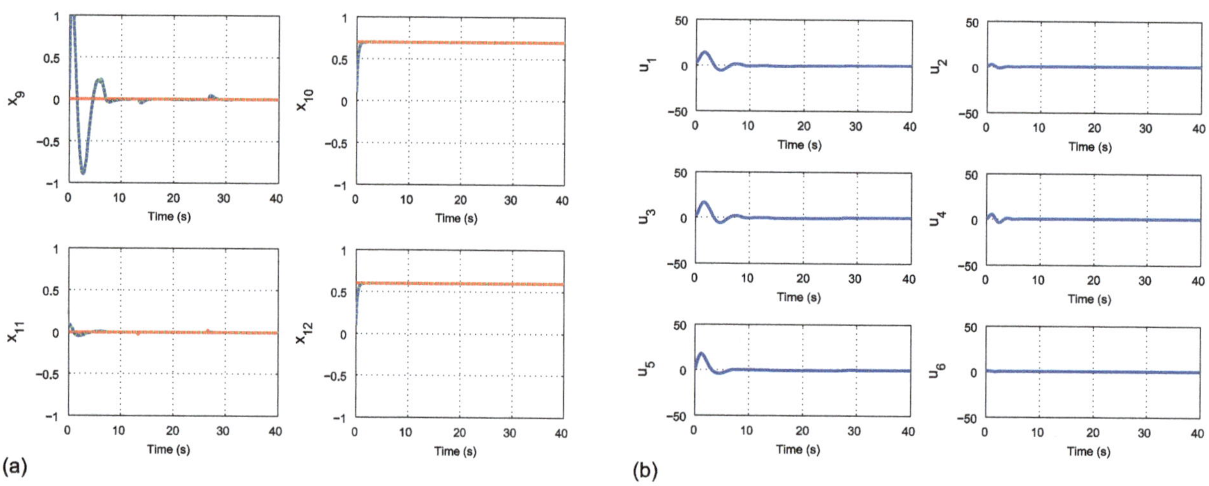

Figure 4.9 *Tracking of setpoint 4 by the dual-PMLSM-driven gantry crane with flatness-based control in successive loops: (a) tracking of setpoints (red line) by the state variables of the gantry crane x_9 to x_{12} (blue line) and estimates provided by the H-infinity Kalman filter; (b) variations of the control inputs of the gantry crane u_1 to u_6 (blue line)*

with double-inverted poles after including the position of the cart [551], (iii) a model with eight state variables and two control inputs, describing the dynamics of two separate cart and inverted pole systems which are connected through an elastic link [666].

In this section a nonlinear optimal (H-infinity) control approach is proposed for the dynamic model of the parallel double-inverted pendulum. The method is based on approximate linearization, through a first-order Taylor series

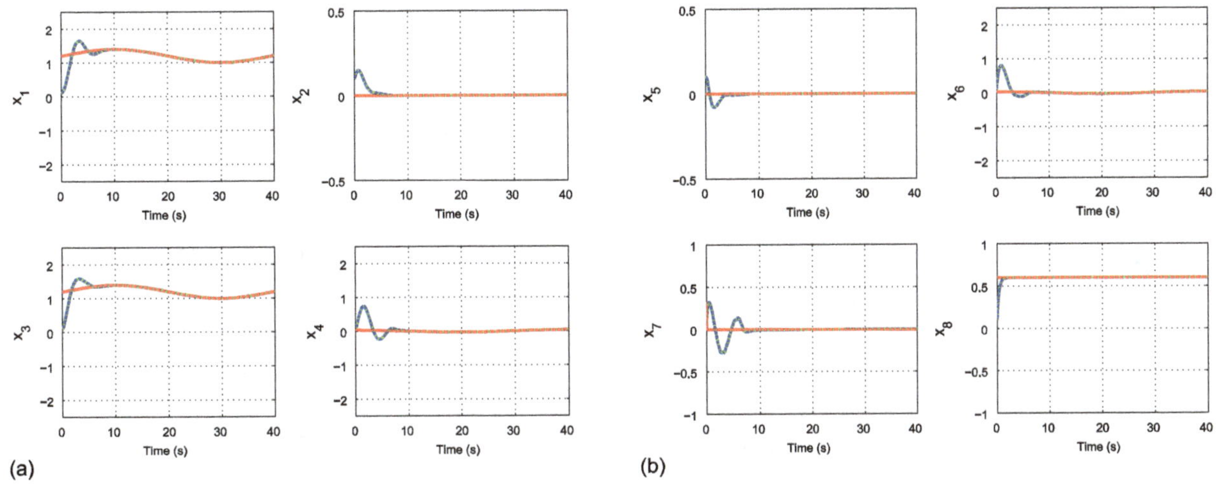

Figure 4.10 Tracking of setpoint 5 by the dual-PMLSM-driven gantry crane with flatness-based control in successive loops: (a) tracking of setpoints (red line) by the state variables of the gantry crane x_1 to x_4 (blue line) and estimates provided by the H-infinity Kalman filter; (b) tracking of setpoints (red line) by the state variables of the gantry crane x_5 to x_8 (blue line) and estimates provided by the H-infinity Kalman filter

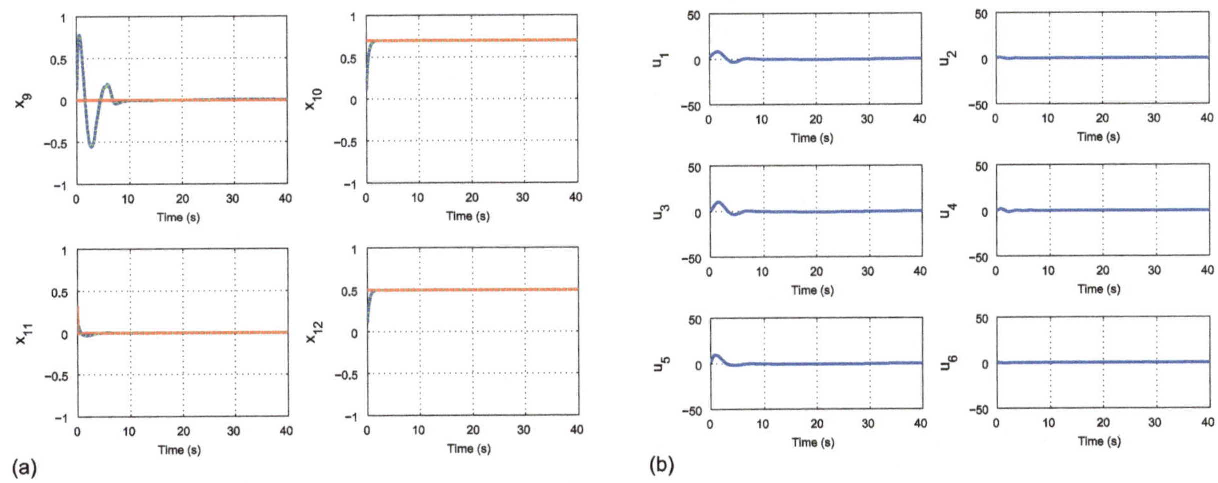

Figure 4.11 Tracking of setpoint 5 by the dual-PMLSM-driven gantry crane with flatness-based control in successive loops: (a) tracking of setpoints (red line) by the state variables of the gantry crane x_9 to x_{12} (blue line) and estimates provided by the H-infinity Kalman filter; (b) variations of the control inputs of the gantry crane u_1 to u_6 (blue line)

expansion process, taking place at each sampling instance [721,719]. The Jacobian matrices of the system are computed around a time-varying operating point [705,60,706]. This linearization point, which is updated at each sampling interval, is defined as (x^*, u^*), where x^* is the present value of state vector of system and u^* is the last sampled value of the control inputs vector. Because of linearizing at the real position of the system and not at points of the desirable trajectory the span of the Taylor series expansion is small and the induced modeling error is also moderate. This model imprecision, due to truncation of higher-order terms from the Taylor-series expansion is finally asymptotically compensated by the robustness of the control algorithm. Next, for the approximately linearized model of the system, a

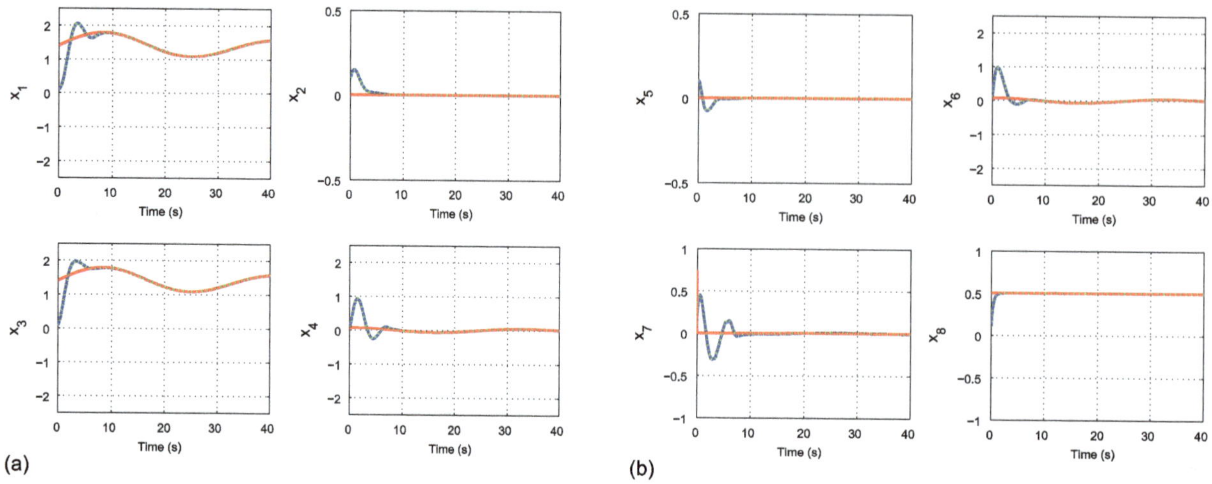

Figure 4.12 *Tracking of setpoint 6 by the dual-PMLSM-driven gantry crane with flatness-based control in successive loops: (a) tracking of setpoints (red line) by the state variables of the gantry crane x_1 to x_4 (blue line) and estimates provided by the H-infinity Kalman filter; (b) tracking of setpoints (red line) by the state variables of the gantry crane x_5 to x_8 (blue line) and estimates provided by the H-infinity Kalman filter*

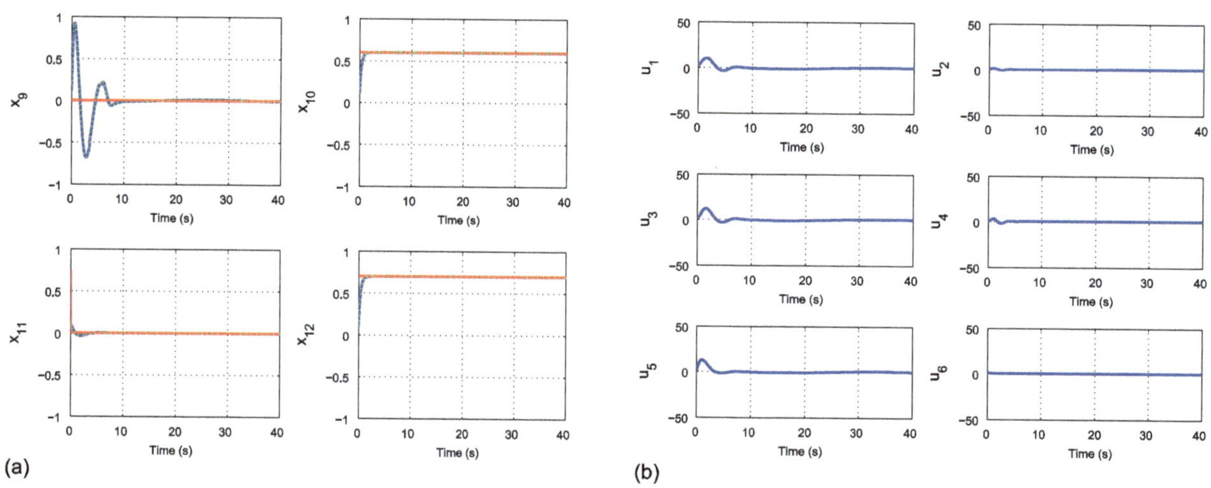

Figure 4.13 *Tracking of setpoint 6 by the dual-PMLSM-driven gantry crane with flatness-based control in successive loops: (a) tracking of setpoints (red line) by the state variables of the gantry crane x_9 to x_{12} (blue line) and estimates provided by the H-infinity Kalman filter; (b) variations of the control inputs of the gantry crane u_1 to u_6 (blue line)*

stabilizing H-infinity feedback controller is designed. The generated control inputs are applied on the initial nonlinear state-space model of the parallel double-inverted pendulum. This controller is a solution for the system's nonlinear optimal control problem under model uncertainty and external perturbations.

The above-noted H-infinity controller represents a min–max differential game taking place between (i) the control inputs of the system which try to minimize a cost function that comprises a quadratic term of the state vector's tracking error, (ii) the model uncertainty and external perturbation terms which try to maximize this cost function. To compute

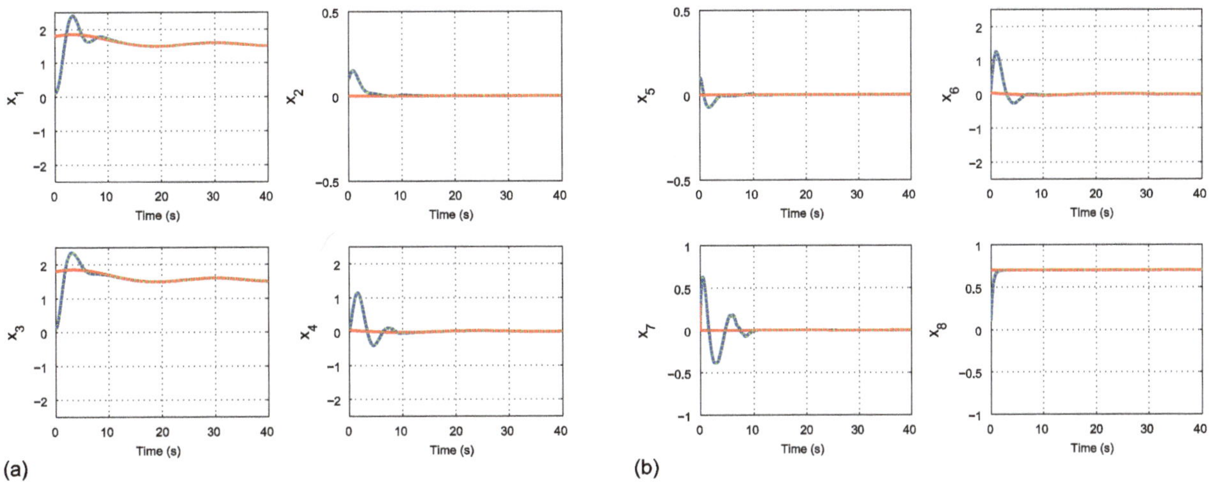

Figure 4.14 *Tracking of setpoint 7 by the dual-PMLSM-driven gantry crane with flatness-based control in successive loops: (a) tracking of setpoints (red line) by the state variables of the gantry crane x_1 to x_4 (blue line) and estimates provided by the H-infinity Kalman filter; (b) tracking of setpoints (red line) by the state variables of the gantry crane x_5 to x_8 (blue line) and estimates provided by the H-infinity Kalman filter*

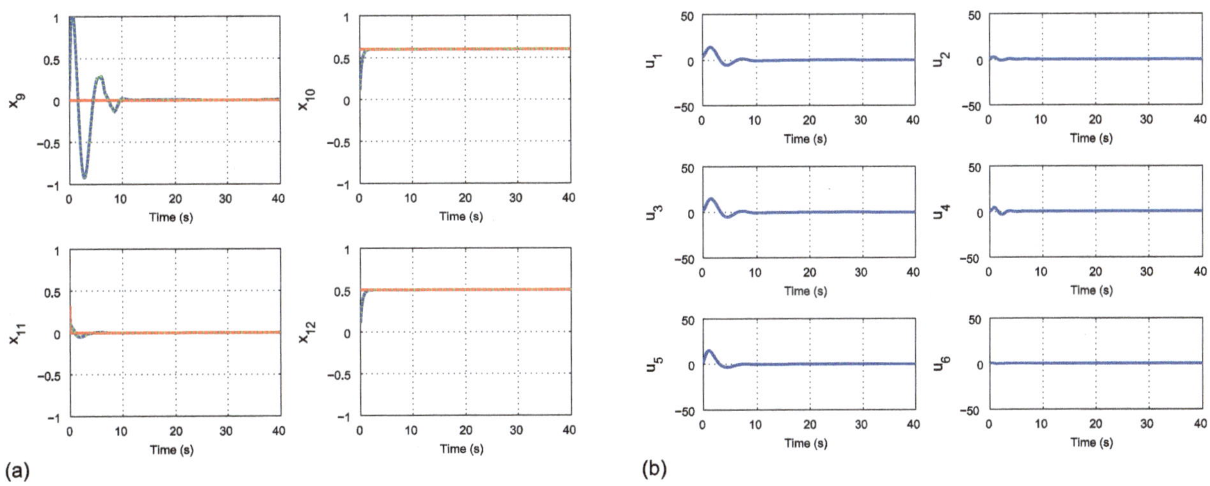

Figure 4.15 *Tracking of setpoint 7 by the dual-PMLSM-driven gantry crane with flatness-based control in successive loops: (a) tracking of setpoints (red line) by the state variables of the gantry crane x_9 to x_{12} (blue line) and estimates provided by the H-infinity Kalman filter; (b) variations of the control inputs of the gantry crane u_1 to u_6 (blue line)*

the feedback gains of the H-infinity controller, an algebraic Riccati equation is repetitively solved at each time-step of the control algorithm [732,741,711]. The global stability properties of the control scheme are proven through Lyapunov analysis. It is proven that the proposed control scheme satisfies the H-infinity tracking performance criterion, which signifies robustness to model uncertainty and external perturbations [725,876]. Moreover, under moderate conditions, it is proven that the control loop is globally asymptotically stable. Furthermore, to enable state estimation-based control without the need to measure the entire state vector of the system, the H-infinity Kalman filter is used as a robust state

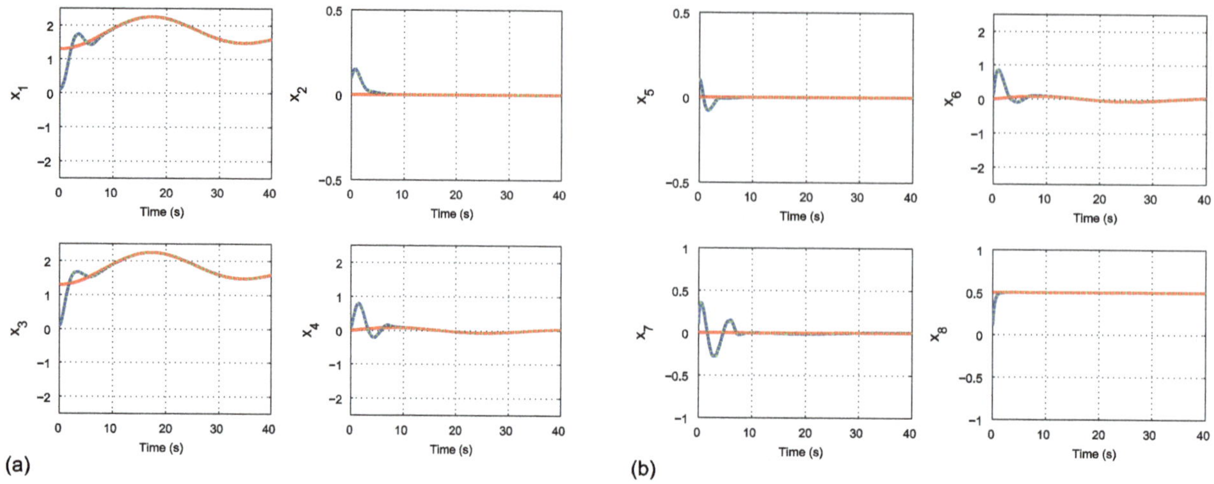

Figure 4.16 Tracking of setpoint 8 by the dual-PMLSM-driven gantry crane with flatness-based control in successive loops: (a) tracking of setpoints (red line) by the state variables of the gantry crane x_1 to x_4 (blue line) and estimates provided by the H-infinity Kalman filter; (b) tracking of setpoints (red line) by the state variables of the gantry crane x_5 to x_8 (blue line) and estimates provided by the H-infinity Kalman filter

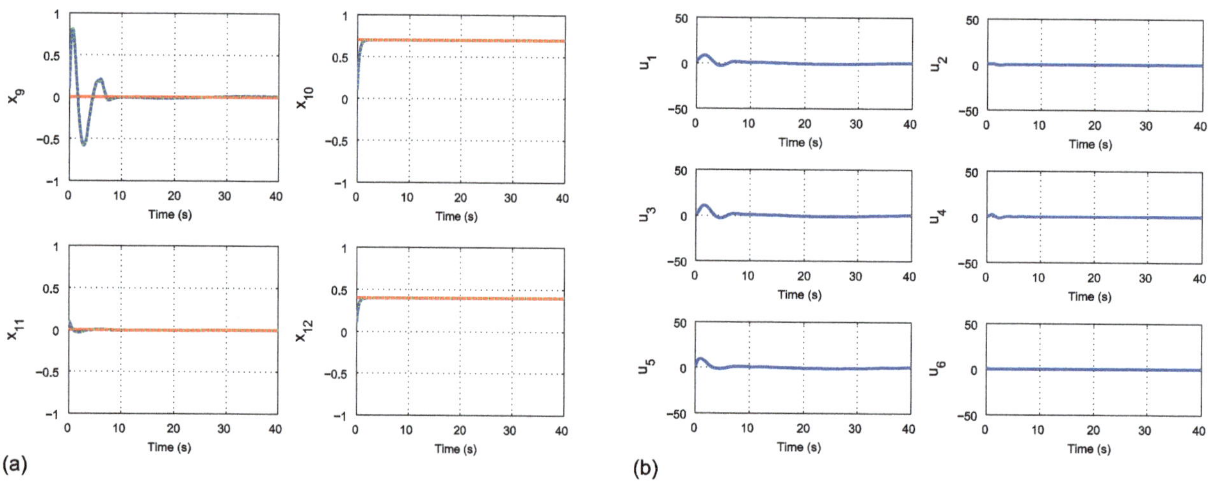

Figure 4.17 Tracking of setpoint 8 by the dual-PMLSM-driven gantry crane with flatness-based control in successive loops: (a) tracking of setpoints (red line) by the state variables of the gantry crane x_9 to x_{12} (blue line) and estimates provided by the H-infinity Kalman filter; (b) variations of the control inputs of the gantry crane u_1 to u_6 (blue line)

estimator. Under the proposed nonlinear control method, one achieves fast and accurate tracking of reference setpoints by all state variables of the parallel double-inverted pendulum under moderate variations of the control inputs.

The present section has a meaningful contribution to the area of nonlinear control. One can point out the advantages of the nonlinear optimal control method against nonlinear model predictive control (NMPC) [725]. In NMPC, the stability properties of the control scheme remain unproven, and the convergence of the iterative search for an optimum often depends on initialization and parameter values' selection. It is also noteworthy that the nonlinear optimal control

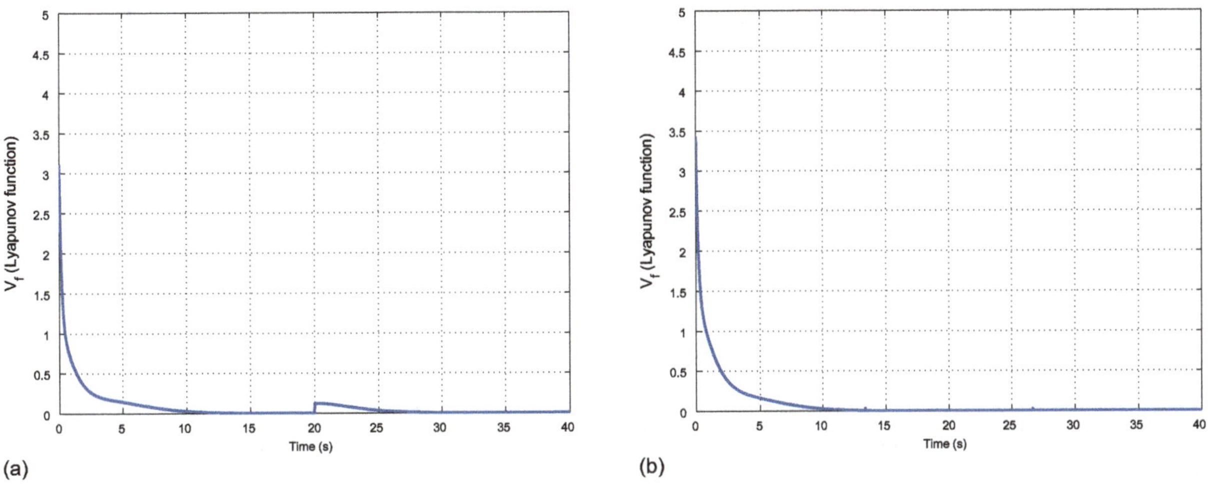

Figure 4.18 *Variation of the Lyapunov function V in the control loop of the H-type gantry crane with dual PMLSM: (a) when tracking setpoint 1; (b) when tracking setpoint 2*

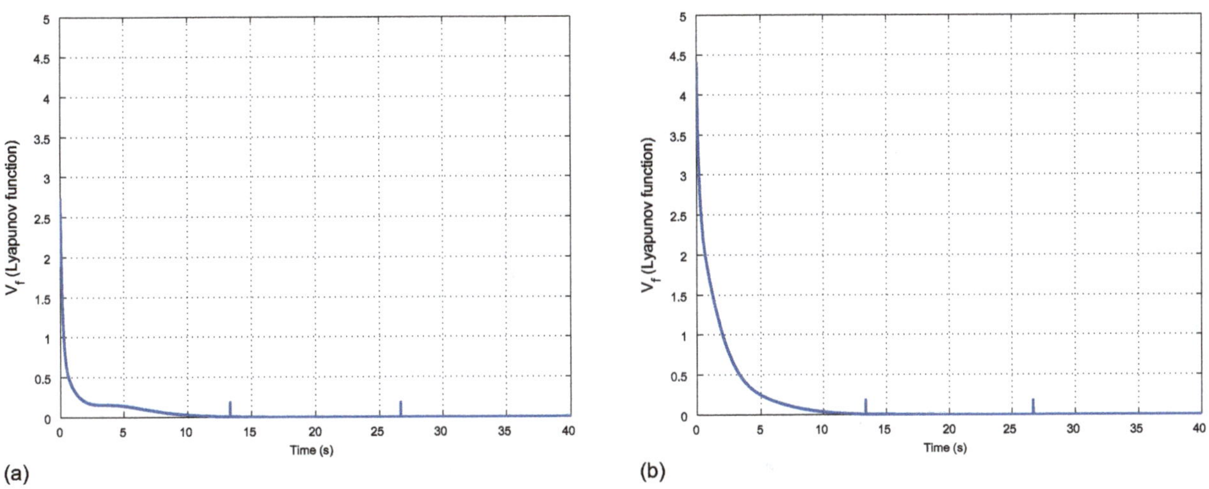

Figure 4.19 *Variation of the Lyapunov function V in the control loop of the H-type gantry crane with dual PMLSM: (a) when tracking setpoint 3; (b) when tracking setpoint 4*

method is applicable to a wider class of dynamical systems than approaches based on the solution of state-dependent Riccati equations (SDRE). The SDRE approaches can be applied only to dynamical systems which can be transformed to the linear parameter varying (LPV) form. Besides, the nonlinear optimal control method performs better than nonlinear optimal control schemes which use approximation of the solution of the Hamilton–Jacobi–Bellman equation by Galerkin series expansions. The stability properties of the Galerkin series expansion -based optimal control approaches are still unproven.

4.2.2 Dynamic model of the double parallel pendulum—System 1

First, a dynamic model of the parallel inverted pendulum with four state variables and one control input is considered. It is the model of a cart with two poles balancing on it, where the position and velocity of the cart are not included among

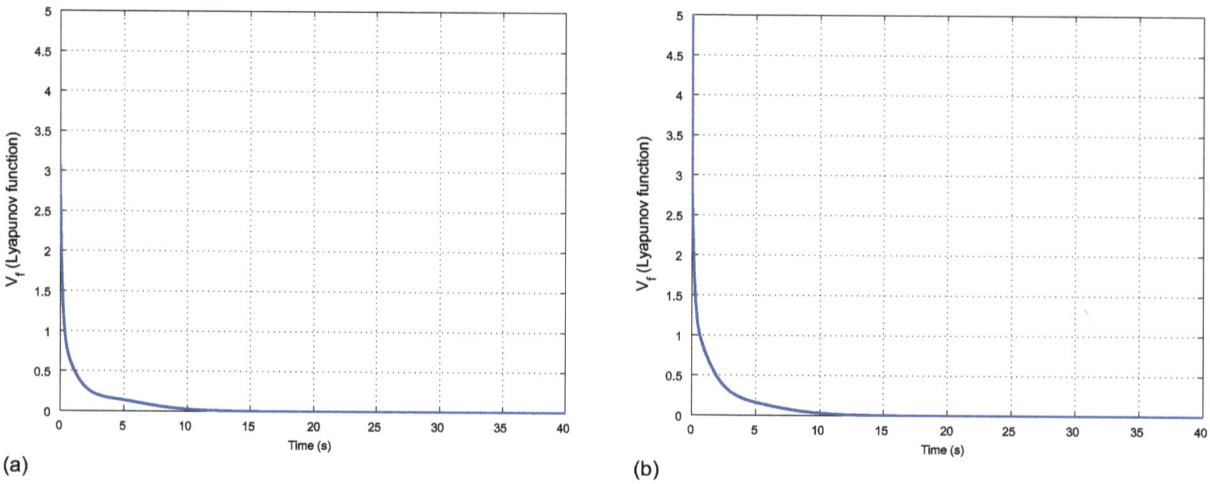

Figure 4.20 Variation of the Lyapunov function V in the control loop of the H-type gantry crane with dual PMLSM: (a) when tracking setpoint 5; (b) when tracking setpoint 6

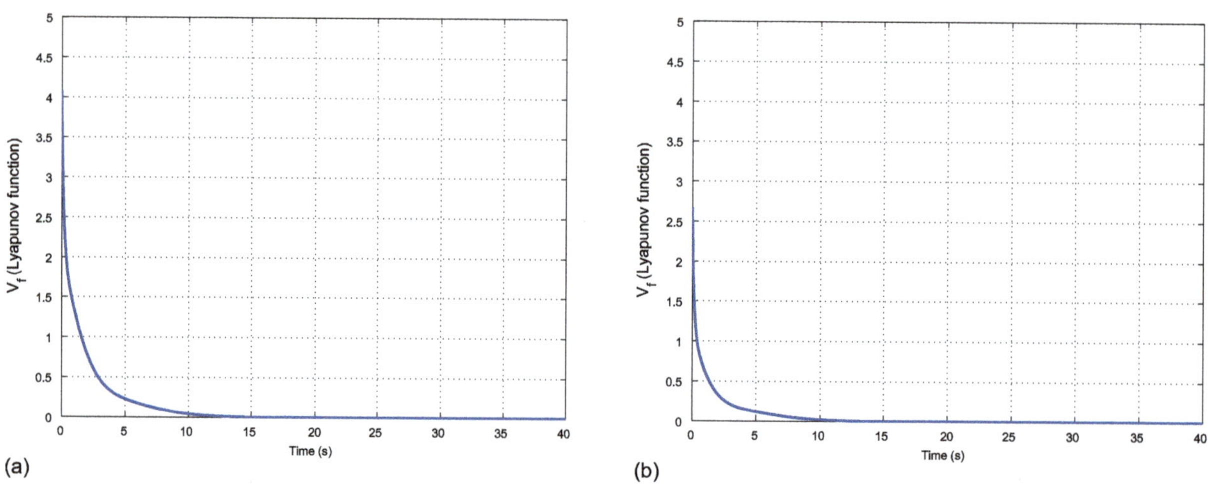

Figure 4.21 Variation of the Lyapunov function V in the control loop of the H-type gantry crane with dual PMLSM: (a) when tracking setpoint 7; (b) when tracking setpoint 8

the system's state variables (System 1). The diagram of the parallel double-inverted pendulum is given in Figure 4.22. The mass of the first pole is m_{p_1}, its length is l_{p_1} and its moment of inertia I_{p_1}. The mass of the second pole is m_{p_2}, its length is l_{p_2} and its moment of inertia I_{p_2}. The position of the cart is denoted as x, while the vehicle's acceleration is $\ddot{x}$, and becomes a control input for this system.

The dynamic model of the parallel double-inverted pendulum is given by [344]

$$\begin{aligned} I_{p_1}\ddot{\theta}_{p_1} + m_{p_1}l_{p_1}\ddot{x}cos(\theta_{p_1}) - m_{p_1}gl_{p_1}sin(\theta_{p_1}) + c_1\dot{\theta}_{p_1} &= 0 \\ I_{p_2}\ddot{\theta}_{p_2} + m_{p_2}l_{p_2}\ddot{x}cos(\theta_{p_2}) - m_{p_2}gl_{p_2}sin(\theta_{p_2}) + c_2\dot{\theta}_{p_2} &= 0 \end{aligned} \tag{4.80}$$

Table 4.1 H-type gantry cranes with dual PMLSMs

	Tracking RMSE $\times 10^{-3}$ in the disturbance-free case								
	$RMSE_{x_1}$	$RMSE_{x_2}$	$RMSE_{x_3}$	$RMSE_{x_4}$	$RMSE_{x_5}$	$RMSE_{x_6}$	$RMSE_{x_7}$	$RMSE_{x_9}$	$RMSE_{x_{11}}$
$test_1$	0.0908	0.0001	0.1134	0.0344	0.0001	0.0430	0.0055	0.0110	0.0009
$test_2$	0.0871	0.0003	0.0640	0.0299	0.0002	0.0222	0.0040	0.0078	0.0004
$test_3$	0.2169	0.0005	0.1927	0.0745	0.0002	0.0662	0.0103	0.0189	0.0012
$test_4$	0.2165	0.0007	0.1924	0.0743	0.0002	0.0661	0.0103	0.0189	0.0112
$test_5$	0.0017	0.0001	0.0016	0.0015	0.0001	0.0015	0.1516	0.1521	0.0018
$test_6$	0.0087	0.0001	0.0086	0.0030	0.0001	0.0030	0.1127	0.2751	0.0152
$test_7$	0.0025	0.0001	0.0024	0.0008	0.0001	0.0008	0.0518	0.1084	0.0058
$test_8$	0.0075	0.0001	0.0072	0.0048	0.0001	0.0046	0.2621	0.3686	0.0108

Table 4.2 H-type gantry cranes with dual PMLSMs

	Tracking RMSE $\times 10^{-3}$ in the case of disturbances								
$\Delta a\%$	$RMSE_{x_1}$	$RMSE_{x_2}$	$RMSE_{x_3}$	$RMSE_{x_4}$	$RMSE_{x_5}$	$RMSE_{x_6}$	$RMSE_{x_7}$	$RMSE_{x_9}$	$RMSE_{x_{11}}$
0%	0.2169	0.0005	0.1927	0.0745	0.0002	0.0662	0.0103	0.0189	0.0012
10%	0.2200	0.0012	0.1925	0.0758	0.0006	0.0661	0.0107	0.0189	0.0012
20%	0.2231	0.0030	0.1921	0.0772	0.0014	0.0660	0.0111	0.0189	0.0012
30%	0.2264	0.0050	0.1919	0.0786	0.0022	0.0659	0.0114	0.0189	0.0012
40%	0.2298	0.0070	0.1916	0.0801	0.0031	0.0658	0.0118	0.0189	0.0011
50%	0.2385	0.0091	0.1913	0.0816	0.0039	0.0656	0.0121	0.0189	0.0011
60%	0.2373	0.0113	0.1910	0.0831	0.0048	0.0655	0.0124	0.0189	0.0011

Table 4.3 H-type gantry cranes with dual PMLSMs

	Convergence time (sec) of states variables								
No test	$T_s\, x_1$	$T_s\, x_2$	$T_s\, x_3$	$T_s\, x_4$	$T_s\, x_5$	$T_s\, x_6$	$T_s\, x_7$	$T_s\, x_9$	$T_s\, x_{11}$
$test_1$	7.5	6.0	5.5	7.5	3.5	6.0	7.5	8.0	3.0
$test_2$	8.0	6.0	6.0	7.5	3.5	7.0	8.0	8.0	3.0
$test_3$	7.5	6.5	7.0	6.0	3.5	6.0	7.0	8.0	1.5
$test_4$	8.0	6.0	8.0	8.0	3.5	6.5	9.0	9.0	3.0
$test_5$	7.5	6.0	6.0	7.5	3.0	6.5	7.0	7.5	1.5
$test_6$	7.5	6.0	6.0	7.5	3.0	6.0	7.5	8.0	3.0
$test_7$	8.0	6.5	6.5	8.0	3.0	7.0	9.0	9.0	3.0
$test_8$	7.5	6.0	6.0	7.5	3.0	7.0	7.5	8.0	2.5

The control input of this dynamic system is taken to be the cart's acceleration $\ddot{x} = u$, thus, one has

$$\begin{aligned} I_{p_1}\ddot{\theta}_{p_1} + m_{p_1}l_{p_1}u{\cdot}cos(\theta_{p_1}) - m_{p_1}gl_{p_1}sin(\theta_{p_1}) + c_1\dot{\theta}_{p_1} = 0 \\ I_{p_2}\ddot{\theta}_{p_2} + m_{p_2}l_{p_2}u{\cdot}cos(\theta_{p_2}) - m_{p_2}gl_{p_2}sin(\theta_{p_2}) + c_2\dot{\theta}_{p_2} = 0 \end{aligned} \tag{4.81}$$

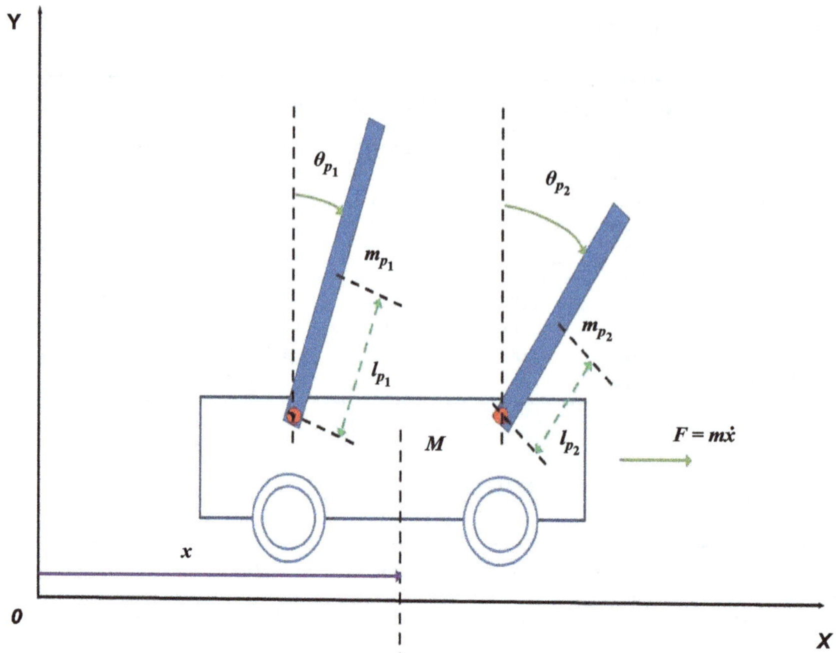

Figure 4.22 Diagram of the parallel double-inverted pendulum—System 1

or equivalently

$$\begin{aligned} \ddot{\theta}_{p_1} &= \frac{m_{p_1}}{I_{p_1}} g l_{p_1} sin(\theta_{p_1}) - \frac{c_1}{I_{p_1}} \dot{\theta}_{p_1} - \frac{m_{p_1}}{I_{p_1}} l_{p_1} \cdot cos(\theta_{p_1}) u \\ \ddot{\theta}_{p_2} &= \frac{m_{p_2}}{I_{p_2}} g l_{p_2} sin(\theta_{p_2}) - \frac{c_2}{I_{p_2}} \dot{\theta}_{p_2} - \frac{m_{p_2}}{I_{p_2}} l_{p_2} \cdot cos(\theta_{p_2}) u \end{aligned} \tag{4.82}$$

The following state vector is defined $x = [x_1, x_2, x_3, x_4]^T = [\theta_{p_1}, \dot{\theta}_{p_1}, \theta_{p_2}, \dot{\theta}_{p_2}]^T$. Thus, the dynamic model of the parallel double-inverted pendulum becomes [344]

$$\begin{aligned} \dot{x}_1 &= x_2 \\ \dot{x}_2 &= \frac{m_{p_1}}{I_{p_1}} g l_{p_1} sin(x_1) - \frac{c_1}{I_{p_1}} x_2 - \frac{m_{p_1}}{I_{p_1}} l_{p_1} \cdot cos(x_1) u \\ \dot{x}_3 &= x_4 \\ \dot{x}_4 &= \frac{m_{p_2}}{I_{p_2}} g l_{p_2} sin(x_3) - \frac{c_2}{I_{p_2}} x_4 - \frac{m_{p_2}}{I_{p_2}} l_{p_2} \cdot cos(x_3) u \end{aligned} \tag{4.83}$$

In matrix form, the state-space model of the parallel double-inverted pendulum becomes

$$\begin{pmatrix} \dot{x}_1 \\ \dot{x}_2 \\ \dot{x}_3 \\ \dot{x}_4 \end{pmatrix} = \begin{pmatrix} x_2 \\ \frac{m_{p_1}}{I_{p_1}} g l_{p_1} sin(x_1) - \frac{c_1}{I_{p_1}} x_2 \\ x_4 \\ \frac{m_{p_2}}{I_{p_2}} g l_{p_2} sin(x_3) - \frac{c_2}{I_{p_2}} x_4 \end{pmatrix} + \begin{pmatrix} 0 \\ -\frac{m_{p_1}}{I_{p_1}} l_{p_1} \cdot cos(x_1) \\ 0 \\ -\frac{m_{p_2}}{I_{p_2}} l_{p_2} \cdot cos(x_3) \end{pmatrix} u \tag{4.84}$$

and finally, the dynamic model of the parallel double-inverted pendulum can be written in the nonlinear affine-in-the-input state-space form

$$\dot{x} = f(x) + g(x)u \tag{4.85}$$

where $x \in R^{4\times 1}$, $f(x) \in R^{4\times 1}$, $g(x) \in R^{4\times 1}$, and $u \in R$. The system is underactuated, and is differentially flat for small turn angles of the poles, having the flat output $y = x_1 - x_3$.

4.2.3 Approximate linearization of the dynamics of the parallel double-inverted pendulum—System 1

The state-space model of the parallel double-inverted pendulum undergoes approximate linearization around the temporary operating point (x^*, u^*), where x^* is the present value of the system's state vector and u^* is the last sampled value of the control inputs vector. The linearization takes place at each sampling instance through the computation of the system's Jacobian matrices. The modeling error which is due to the truncation of higher-order terms in the Taylor series, is considered to be a perturbation which is asymptotically compensated by the robustness of the control algorithm. The linearization process allows for substituting the initial nonlinear state-space model:

$$\dot{x} = f(x) + g(x)u \tag{4.86}$$

with the approximately linearized model

$$\dot{x} = Ax + Bu + \tilde{d} \tag{4.87}$$

where A, B are the Jacobian matrices of the system which are given by

$$\begin{aligned} A &= \nabla_x[f(x) + g(x)u] \mid_{(x^*,u^*)} \Rightarrow \\ A &= \nabla_x[f(x)] \mid_{(x^*,u^*)} + [\nabla_x g(x)]u \mid_{(x^*,u^*)} \end{aligned} \tag{4.88}$$

$$\begin{aligned} B &= \nabla_u[f(x) + g(x)u] \mid_{(x^*,u^*)} \Rightarrow \\ B &= g(x) \mid_{(x^*,u^*)} \end{aligned} \tag{4.89}$$

while $\tilde{d}$ is the cumulative disturbance vector that may comprise (i) exogenous perturbations, (ii) modeling errors due to higher-order terms in the Taylor series, and (iii) sensor measurement noise of any distribution.

This linearization approach which has been followed for implementing the nonlinear optimal control scheme results in a quite accurate model of the system's dynamics. Consider for instance the following affine-in-the-input state-space model:

$$\begin{aligned} \dot{x} &= f(x) + g(x)u \Rightarrow \\ \dot{x} &= [f(x^*) + \nabla_x f(x) \mid_{x^*} (x - x^*)] + [g(x^*) + \nabla_x g(x) \mid_{x^*} (x - x^*)]u^* + g(x^*)u^* + g(x^*)(u - u^*) + \tilde{d}_1 \Rightarrow \\ \dot{x} &= [\nabla_x f(x) \mid_{x^*} + \nabla_x g(x) \mid_{x^*} u^*]x + g(x^*)u - [\nabla_x f(x) \mid_{x^*} + \nabla_x g(x) \mid_{x^*} u^*]x^* + f(x^*) + g(x^*)u^* + \tilde{d}_1 \end{aligned} \tag{4.90}$$

where $\tilde{d}_1$ is the modeling error due to truncation of higher order terms in the Taylor series expansion of $f(x)$ and $g(x)$. Next, by defining $A = [\nabla_x f(x) \mid_{x^*} + \nabla_x g(x) \mid_{x^*} u^*]$, $B = g(x^*)$, one obtains

$$\dot{x} = Ax + Bu - Ax^* + f(x^*) + g(x^*)u^* + \tilde{d}_1 \tag{4.91}$$

Moreover, by denoting $\tilde{d} = -Ax^* + f(x^*) + g(x^*)u^* + \tilde{d}_1$ about the cumulative modeling error term in the Taylor series expansion procedure, one has

$$\dot{x} = Ax + Bu + \tilde{d} \tag{4.92}$$

which is the approximately linearized model of the dynamics of the system of Eq. (4.87). The term $f(x^*) + g(x^*)u^*$ is the derivative of the state vector at (x^*, u^*) which is almost annihilated by $-Ax^*$.

The computation of the Jacobian matrix $\nabla_x f(x) \mid_{(x^*,u^*)}$ is performed as follows:

First row of the Jacobian matrix $\nabla_x f(x) \mid_{(x^*,u^*)}$: $\frac{\partial f_1}{\partial x_1} = 0$, $\frac{\partial f_1}{\partial x_2} = 1$, $\frac{\partial f_1}{\partial x_3} = 0$, $\frac{\partial f_1}{\partial x_4} = 0$.

Second row of the Jacobian matrix $\nabla_x f(x) \mid_{(x^*,u^*)}$: $\frac{\partial f_2}{\partial x_1} = \frac{m_{p_1}}{I_{p_1}} g l_{p_1} cos(x_1)$, $\frac{\partial f_2}{\partial x_2} = -\frac{c_1}{I_{p_1}}$, $\frac{\partial f_2}{\partial x_3} = 0$, $\frac{\partial f_2}{\partial x_4} = 0$.

Third row of the Jacobian matrix $\nabla_x f(x) \mid_{(x^*,u^*)}$: $\frac{\partial f_3}{\partial x_1}=0$, $\frac{\partial f_3}{\partial x_2}=0$, $\frac{\partial f_3}{\partial x_3}=0$, $\frac{\partial f_3}{\partial x_4}=1$.

Fourth row of the Jacobian matrix $\nabla_x f(x) \mid_{(x^*,u^*)}$: $\frac{\partial f_4}{\partial x_1}=0$, $\frac{\partial f_4}{\partial x_2}=0$, $\frac{\partial f_4}{\partial x_3}=\frac{m_{p_2}}{I_{p_2}} g l_{p_2} cos(x_3)$, $\frac{\partial f_4}{\partial x_4}=-\frac{c_2}{I_{p_2}}$.

The computation of the Jacobian matrix $\nabla_x g(x) \mid_{(x^*,u^*)}$ is performed as follows:

First row of the Jacobian matrix $\nabla_x g(x) \mid_{(x^*,u^*)}$: $\frac{\partial g_1}{\partial x_1}=0$, $\frac{\partial g_1}{\partial x_2}=0$, $\frac{\partial g_1}{\partial x_3}=0$, $\frac{\partial g_1}{\partial x_4}=0$.

Second row of the Jacobian matrix $\nabla_x g(x) \mid_{(x^*,u^*)}$: $\frac{\partial g_2}{\partial x_1}=\frac{m_{p_1}}{I_{p_1}} g l_{p_1} sin(x_1)$, $\frac{\partial g_2}{\partial x_2}=0$, $\frac{\partial g_2}{\partial x_3}=0$, $\frac{\partial g_2}{\partial x_4}=0$.

Third row of the Jacobian matrix $\nabla_x g(x) \mid_{(x^*,u^*)}$: $\frac{\partial g_3}{\partial x_1}=0$, $\frac{\partial g_3}{\partial x_2}=0$, $\frac{\partial g_3}{\partial x_3}=0$, $\frac{\partial g_3}{\partial x_4}=0$.

Fourth row of the Jacobian matrix $\nabla_x g(x) \mid_{(x^*,u^*)}$: $\frac{\partial g_4}{\partial x_1}=0$, $\frac{\partial g_4}{\partial x_2}=0$, $\frac{\partial g_4}{\partial x_3}=\frac{m_{p_2}}{I_{p_2}} g l_{p_2} sin(x_3)$, $\frac{\partial g_4}{\partial x_4}=0$.

The Jacobian matrix $B=\nabla_u[f(x)+g(x)u] \mid_{(x^*,u^*)}$ is computed as follows:

$$B=\begin{pmatrix} 0 \\ -\frac{m_{p_1}}{I_{p_1}} l_{p_1} \cdot cos(x_1) \\ 0 \\ -\frac{m_{p_2}}{I_{p_2}} l_{p_2} \cdot cos(x_3) \end{pmatrix} \mid_{(x^*,u^*)} \tag{4.93}$$

4.2.4 Dynamic model of the parallel double-inverted pendulum—System 2

Next, a dynamic model of the parallel inverted pendulum with six state variables and one control input is considered. It is the model of a cart with two poles balancing on it, where the position and velocity of the cart are included among the system's state variables (System 2). The diagram of the parallel double-inverted pendulum is shown in Figure 4.23. The mass of the cart is denoted as m_c, the half-length of the short pole is l_s and the mass of this pole is m_s, while the length of the long pole is l_L and the mass of this pole is m_L. The position of the cart is x. The turn angle of the short pole is θ_s and the turn angle of the long pole is θ_L. The moment of inertia of the short pole is J_s while the moment of inertial of the long pole is J_L [551].

The dynamic model of the parallel double-inverted pendulum is now given by [551]

$$M(\theta)\ddot{\theta}+C(\theta,\dot{\theta})\dot{\theta}+G(\theta)=K{\cdot}u \tag{4.94}$$

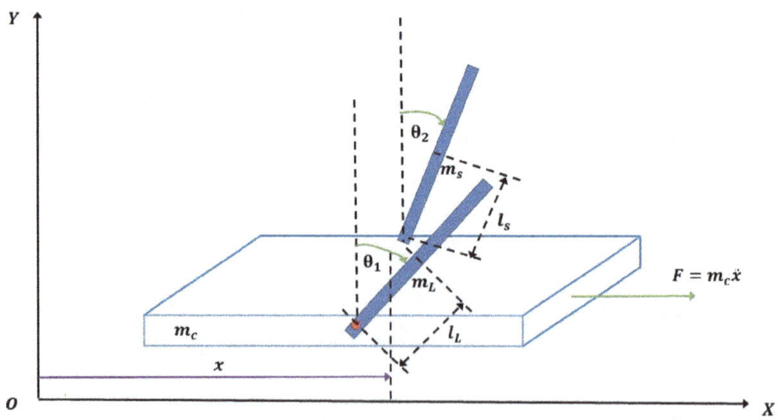

Figure 4.23 Diagram of the parallel double-inverted pendulum—System 2

where the state vector of this robot-type dynamic model is $\theta = [x, \theta_s, \theta_L]$, The control input is u and is the force which is applied on the cart. $M(\theta)$ is the inertia matrix, $C(\theta, \dot{\theta})$ is the Coriolis and centrifugal forces matrix, while $G(\theta)$ is the gravitational forces vector. These elements of the dynamic model are given by

$$M(\theta) = \begin{pmatrix} m_c + m_s + m_L & m_L l_L cos(\theta_L) & m_s l_s cos(\theta_s) \\ m_L l_L cos(\theta_L) & J_L + m_L l_L^2 & 0 \\ m_s l_s cos(\theta_s) & 0 & J_s + m_s l_s^2 \end{pmatrix} \tag{4.95}$$

$$C(\theta, \dot{\theta}) = \begin{pmatrix} C_c & -m_L l_L sin(\theta_L)\dot{\theta}_L & -m_s l_s sin(\theta_s)\dot{\theta}_s \\ 0 & C_L & 0 \\ 0 & 0 & C_s \end{pmatrix} \tag{4.96}$$

while for the gravitational matrix and for the control inputs gain matrix, it holds

$$G(\theta) = \begin{pmatrix} 0 \\ -m_L l_L g \cdot sin(\theta_L) \\ -m_s l_s g \cdot sin(\theta_s) \end{pmatrix} \quad K = \begin{pmatrix} k_v \\ 0 \\ 0 \end{pmatrix} \tag{4.97}$$

The inertia matrix of the parallel double-inverted pendulum can be also written as

$$M = \begin{pmatrix} m_{11} & m_{12} & m_{13} \\ m_{21} & m_{22} & m_{23} \\ m_{31} & m_{32} & m_{33} \end{pmatrix} \tag{4.98}$$

The sub-determinant of matrix M are: $M_{11} = m_{22}m_{33} - m_{32}m_{23}$, $M_{12} = m_{21}m_{33} - m_{31}m_{23}$, $M_{13} = m_{21}m_{32} - m_{31}m_{22}$, $M_{21} = m_{12}m_{33} - m_{32}m_{13}$, $M_{22} = m_{11}m_{33} - m_{31}m_{13}$, $M_{13} = m_{21}m_{32} - m_{31}m_{22}$, and $M_{23} = m_{11}m_{32} - m_{31}m_{12}$.

The determinant of the inertia matrix M is given by

$$detM = m_{11}M_{11} - m_{12}M_{12} + m_{13}M_{13} \tag{4.99}$$

The inverse of the inertia matrix M is given by

$$M^{-1} = \frac{1}{detM} \begin{pmatrix} M_{11} & -M_{21} & M_{31} \\ -M_{12} & M_{22} & -M_{32} \\ M_{13} & -M_{23} & M_{33} \end{pmatrix} \tag{4.100}$$

Thus, the dynamic model of the parallel double-inverted pendulum is given by

$$\ddot{\theta} = -M^{-1}(\theta)[C(\theta, \dot{\theta})\dot{\theta} + G(\theta)] - M(\theta)^{-1}Ku \tag{4.101}$$

or equivalently

$$\begin{gathered} \begin{pmatrix} \ddot{x} \\ \ddot{\theta}_L \\ \ddot{\theta}_s \end{pmatrix} = -\frac{1}{detM} \begin{pmatrix} M_{11} & -M_{21} & M_{31} \\ -M_{12} & M_{22} & -M_{32} \\ M_{13} & -M_{23} & M_{33} \end{pmatrix} \begin{pmatrix} \tilde{C}_{11}\dot{x} + \tilde{C}_{21}\dot{\theta}_L + \tilde{C}_{31}\dot{\theta}_s + G_1 \\ \tilde{C}_{12}\dot{x} + \tilde{C}_{22}\dot{\theta}_L + \tilde{C}_{32}\dot{\theta}_s + G_2 \\ \tilde{C}_{13}\dot{x} + \tilde{C}_{23}\dot{\theta}_L + \tilde{C}_{33}\dot{\theta}_s + G_3 \end{pmatrix} + \\ + \frac{1}{detM} \begin{pmatrix} M_{11} & -M_{21} & M_{31} \\ -M_{12} & M_{22} & -M_{32} \\ M_{13} & -M_{23} & M_{33} \end{pmatrix} \begin{pmatrix} k_v \\ 0 \\ 0 \end{pmatrix} u \end{gathered} \tag{4.102}$$

By denoting $C_1 = \tilde{C}_{11}\dot{x} + \tilde{C}_{21}\dot{\theta}_L + \tilde{C}_{31}\dot{\theta}_s$, $C_2 = \tilde{C}_{12}\dot{x} + \tilde{C}_{22}\dot{\theta}_L + \tilde{C}_{32}\dot{\theta}_s$, and $C_3 = \tilde{C}_{13}\dot{x} + \tilde{C}_{23}\dot{\theta}_L + \tilde{C}_{33}\dot{\theta}_s$, one arrives at the following dynamic model of the parallel double-inverted pendulum:

$$\begin{pmatrix} \ddot{x} \\ \ddot{\theta}_L \\ \ddot{\theta}_s \end{pmatrix} = \begin{pmatrix} \frac{-M_{11}(C_1+G_1)+M_{21}(C_2+G_2)-M_{31}(C_3+G_3)}{detM} \\ \frac{M_{12}(C_1+G_1)-M_{22}(C_2+G_2)+M_{32}(C_3+G_3)}{detM} \\ \frac{-M_{13}(C_1+G_1)+M_{23}(C_2+G_2)-M_{33}(C_3+G_3)}{detM} \end{pmatrix} + \begin{pmatrix} \frac{M_{11}k_v}{detM} \\ -\frac{M_{12}k_v}{detM} \\ \frac{M_{13}k_v}{detM} \end{pmatrix} u \tag{4.103}$$

Next, by defining the state vector $x = [x_1, x_2, x_3, x_4, x_5, x_6]^T$ or $x = [x, \dot{x}, \theta_L, \dot{\theta}_L, \theta_s, \dot{\theta}_s]^T$, one arrives at the following state-space description for the parallel double-inverted pendulum:

$$\begin{pmatrix} \dot{x}_1 \\ \dot{x}_2 \\ \dot{x}_3 \\ \dot{x}_4 \\ \dot{x}_5 \\ \dot{x}_6 \end{pmatrix} = \begin{pmatrix} x_2 \\ \frac{-M_{11}(C_1+G_1)+M_{21}(C_2+G_2)-M_{31}(C_3+G_3)}{detM} \\ x_4 \\ \frac{M_{12}(C_1+G_1)-M_{22}(C_2+G_2)+M_{32}(C_3+G_3)}{detM} \\ x_6 \\ \frac{-M_{13}(C_1+G_1)+M_{23}(C_2+G_2)-M_{33}(C_3+G_3)}{detM} \end{pmatrix} + \begin{pmatrix} 0 \\ \frac{M_{11}k_v}{detM} \\ 0 \\ -\frac{M_{12}k_v}{detM} \\ 0 \\ \frac{M_{13}k_v}{detM} \end{pmatrix} u \tag{4.104}$$

or equivalently, one can write the state-space model in the nonlinear affine-in-the-input state space form:

$$\dot{x} = f(x) + g(x)u \tag{4.105}$$

where $x \in R^{6\times1}$, $f(x) \in R^{6\times1}$, $g(x) \in R^{6\times1}$, and $u \in R$.

4.2.5 Approximate linearization of the dynamics of the parallel double-inverted pendulum—System 2

The dynamic model of the system undergoes approximate linearization through first-order Taylor series expansion and through the computation of the associated Jacobian matrices. The linearization point (x^*, u^*) is updated at each sampling instance and is defined by the present value of the system's state vector and by the last sampled value of the control inputs vector. The modeling error which is due to the truncation of higher-order terms in the Taylor series expansion is viewed as a perturbation which is asymptotically compensated by the robustness of the control algorithm. The system is initially in the form:

$$\dot{x} = f(x) + g(x)u \tag{4.106}$$

while after linearization it becomes

$$\dot{x} = Ax + Bu + \tilde{d} \tag{4.107}$$

where A, B are the Jacobian matrices of the system which are defined as follows:

$$\begin{aligned} A &= \nabla_x[f(x) + g(x)u] \mid_{(x^*,u^*)} \Rightarrow \\ A &= \nabla_x f(x) \mid_{(x^*,u^*)} + \nabla_x g(x)u \mid_{(x^*,u^*)} \end{aligned} \tag{4.108}$$

$$\begin{aligned} B &= \nabla_u[f(x) + g(x)u] \mid_{(x^*,u^*)} \Rightarrow \\ B &= g(x) \mid_{(x^*,u^*)} \end{aligned} \tag{4.109}$$

The modeling error $\tilde{d}$ may comprise: (i) model uncertainty due to truncation of higher-order terms from the Taylor series, (ii) exogenous perturbations, and (iii) sensor measurement noise of any distribution.

The computation of the elements of the Jacobian matrix $\nabla_x f(x) \mid_{x^*,u^*}$ is performed as follows:

First row of the Jacobian matrix $\nabla_x f(x) \mid_{x^*,u^*}$: $\frac{\partial f_1}{\partial x_1}=0$, $\frac{\partial f_1}{\partial x_2}=1$, $\frac{\partial f_1}{\partial x_3}=0$, $\frac{\partial f_1}{\partial x_4}=0$, $\frac{\partial f_1}{\partial x_5}=0$, $\frac{\partial f_1}{\partial x_6}=0$.

Second row of the Jacobian matrix $\nabla_x f(x) \mid_{x^*,u^*}$: It holds that

$$f_2(x)=\frac{f_{2,num}(x)}{f_{2,den}(x)}=\frac{-M_{11}(C_1+G_1)+M_{21}(C_2+G_2)-M_{31}(C_3+G_3)}{detM} \tag{4.110}$$

Therefore, one obtains

$$\frac{\partial f_2(x)}{\partial x_i}=\frac{\frac{\partial f_{2,num}(x)}{\partial x_i}f_{2,den}(x)-f_{2,num}(x)\frac{\partial f_{2,den}(x)}{\partial x_i}}{[f_{2,den}(x)]^2} \tag{4.111}$$

Moreover, for $i=1,2,\cdots .6$, it holds that

$$\begin{aligned}\frac{\partial f_{2,num}(x)}{\partial x_i}=&-\frac{\partial M_{11}}{\partial x_i}(C_1+G_1)-M_{11}\left(\frac{\partial C_1}{\partial x_i}+\frac{\partial G_1}{\partial x_i}\right)\\&\frac{\partial M_{21}}{\partial x_i}(C_2+G_2)+M_{21}\left(\frac{\partial C_2}{\partial x_i}+\frac{\partial G_2}{\partial x_i}\right)\\&-\frac{\partial M_{31}}{\partial x_i}(C_3+G_3)-M_{31}\left(\frac{\partial C_3}{\partial x_i}+\frac{\partial G_3}{\partial x_i}\right)\end{aligned} \tag{4.112}$$

and also for $i=1,2,\cdots .6$

$$\frac{\partial f_{2,den}(x)}{\partial x_i}=\frac{\partial detM}{\partial x_i} \tag{4.113}$$

Third row of the Jacobian matrix $\nabla_x f(x) \mid_{x^*,u^*}$: $\frac{\partial f_3}{\partial x_1}=0$, $\frac{\partial f_3}{\partial x_2}=0$, $\frac{\partial f_3}{\partial x_3}=0$, $\frac{\partial f_3}{\partial x_4}=1$, $\frac{\partial f_3}{\partial x_5}=0$, $\frac{\partial f_3}{\partial x_6}=0$.

Fourth row of the Jacobian matrix $\nabla_x f(x) \mid_{x^*,u^*}$: It holds that

$$f_4(x)=\frac{f_{4,num}(x)}{f_{4,den}(x)}=\frac{M_{12}(C_1+G_1)-M_{21}(C_2+G_2)+M_{31}(C_3+G_3)}{detM} \tag{4.114}$$

Therefore, one obtains

$$\frac{\partial f_4(x)}{\partial x_i}=\frac{\frac{\partial f_{4,num}(x)}{\partial x_i}f_{4,den}(x)-f_{4,num}(x)\frac{\partial f_{4,den}(x)}{\partial x_i}}{[f_{4,den}(x)]^2} \tag{4.115}$$

Moreover, for $i=1,2,\cdots .6$, it holds that

$$\begin{aligned}\frac{\partial f_{4,num}(x)}{\partial x_i}=&\frac{\partial M_{12}}{\partial x_i}(C_1+G_1)+M_{12}\left(\frac{\partial C_1}{\partial x_i}+\frac{\partial G_1}{\partial x_i}\right)\\&-\frac{\partial M_{22}}{\partial x_i}(C_2+G_2)+M_{22}\left(\frac{\partial C_2}{\partial x_i}+\frac{\partial G_2}{\partial x_i}\right)\\&\frac{\partial M_{32}}{\partial x_i}(C_3+G_3)+M_{32}\left(\frac{\partial C_3}{\partial x_i}+\frac{\partial G_3}{\partial x_i}\right)\end{aligned} \tag{4.116}$$

and also for $i=1,2,\cdots .6$

$$\frac{\partial f_{4,den}(x)}{\partial x_i}=\frac{\partial detM}{\partial x_i} \tag{4.117}$$

Fifth row of the Jacobian matrix $\nabla_x f(x) \mid_{x^*,u^*}$: $\frac{\partial f_5}{\partial x_1}=0$, $\frac{\partial f_5}{\partial x_2}=0$, $\frac{\partial f_5}{\partial x_3}=0$, $\frac{\partial f_5}{\partial x_4}=1$, $\frac{\partial f_5}{\partial x_5}=0$, $\frac{\partial f_5}{\partial x_6}=1$.

Sixth row of the Jacobian matrix $\nabla_x f(x) \mid_{x^*,u^*}$: It holds that

$$f_6(x)=\frac{f_{6,num}(x)}{f_{6,den}(x)}=\frac{-M_{13}(C_1+G_1)-M_{23}(C_2+G_2)+M_{33}(C_3+G_3)}{detM} \tag{4.118}$$

Therefore, one obtains

$$\frac{\partial f_6(x)}{\partial x_i}=\frac{\frac{\partial f_{6,num}(x)}{\partial x_i}f_{6,den}(x)-f_{6,num}(x)\frac{\partial f_{6,den}(x)}{\partial x_i}}{[f_{6,den}(x)]^2} \tag{4.119}$$

Moreover, for $i = 1, 2, \cdots .6$, it holds that

$$\begin{aligned}\frac{\partial f_{6,num}(x)}{\partial x_i} &= \frac{\partial M_{13}}{\partial x_i}(C_1+G_1) + M_{13}\left(\frac{\partial C_1}{\partial x_i} + \frac{\partial G_1}{\partial x_i}\right) \\ &-\frac{\partial M_{23}}{\partial x_i}(C_2+G_2) - M_{23}\left(\frac{\partial C_2}{\partial x_i} + \frac{\partial G_2}{\partial x_i}\right) \\ &\frac{\partial M_{33}}{\partial x_i}(C_3+G_3) + M_{33}\left(\frac{\partial C_3}{\partial x_i} + \frac{\partial G_3}{\partial x_i}\right)\end{aligned} \tag{4.120}$$

and also for $i = 1, 2, \cdots .6$

$$\frac{\partial f_{6,den}(x)}{\partial x_i} = \frac{\partial detM}{\partial x_i} \tag{4.121}$$

Additionally, the Jacobian matrix of the control inputs gain matrix $g(x)$, that is $\nabla_x g(x) \mid_{(x^*,u^*)}$ is computed as follows:

First row of the Jacobian matrix $\nabla_x g(x) \mid_{(x^*,u^*)}$: for $i = 1, 2 \cdots, 6$

$$\frac{\partial g_1(x)}{\partial x_i} = 0 \tag{4.122}$$

Second row of the Jacobian matrix $\nabla_x g(x) \mid_{(x^*,u^*)}$: for $i = 1, 2 \cdots, 6$

$$\frac{\partial g_1(x)}{\partial x_i} = \frac{[k_v \frac{\partial M_{11}}{\partial x_i}]detM - k_v M_{11}\frac{\partial detM}{\partial x_i}}{detM^2} \tag{4.123}$$

Third row of the Jacobian matrix $\nabla_x g(x) \mid_{(x^*,u^*)}$: for $i = 1, 2 \cdots, 6$

$$\frac{\partial g_3(x)}{\partial x_i} = 0 \tag{4.124}$$

Fourth row of the Jacobian matrix $\nabla_x g(x) \mid_{(x^*,u^*)}$: for $i = 1, 2 \cdots, 6$

$$\frac{\partial g_4(x)}{\partial x_i} = \frac{[-k_v \frac{\partial M_{12}}{\partial x_i}]detM + k_v M_{12}\frac{\partial detM}{\partial x_i}}{detM^2} \tag{4.125}$$

Fifth row of the Jacobian matrix $\nabla_x g(x) \mid_{(x^*,u^*)}$: for $i = 1, 2 \cdots, 6$

$$\frac{\partial g_5(x)}{\partial x_i} = 0 \tag{4.126}$$

Sixth row of the Jacobian matrix $\nabla_x g(x) \mid_{(x^*,u^*)}$: for $i = 1, 2 \cdots, 6$

$$\frac{\partial g_6(x)}{\partial x_i} = \frac{[k_v \frac{\partial M_{13}}{\partial x_i}]detM - k_v M_{13}\frac{\partial detM}{\partial x_i}}{detM^2} \tag{4.127}$$

Next, one computes the partial derivatives of the subdeterminants of the inertia matrix M, of the elements of the inertia matrix M, of the Coriolis matrix C and of the gravitational forces matrix G.

Partial derivatives of the inertia matrix M: for $i = 1, 2, \cdots, 6$, it holds that:

$$\frac{\partial M_{11}}{\partial x_i} = \frac{\partial m_{22}}{\partial x_i}m_{33} + m_{22}\frac{\partial m_{33}}{\partial x_i} - \frac{\partial m_{32}}{\partial x_i}m_{23} - m_{32}\frac{\partial m_{23}}{\partial x_i} \tag{4.128}$$

$$\frac{\partial M_{12}}{\partial x_i} = \frac{\partial m_{21}}{\partial x_i}m_{33} + m_{21}\frac{\partial m_{33}}{\partial x_i} - \frac{\partial m_{31}}{\partial x_i}m_{23} - m_{31}\frac{\partial m_{23}}{\partial x_i} \tag{4.129}$$

$$\frac{\partial M_{13}}{\partial x_i} = \frac{\partial m_{21}}{\partial x_i}m_{32} + m_{21}\frac{\partial m_{32}}{\partial x_i} - \frac{\partial m_{31}}{\partial x_i}m_{22} - m_{31}\frac{\partial m_{22}}{\partial x_i} \tag{4.130}$$

$$\frac{\partial M_{21}}{\partial x_i} = \frac{\partial m_{12}}{\partial x_i}m_{33} + m_{12}\frac{\partial m_{33}}{\partial x_i} - \frac{\partial m_{32}}{\partial x_i}m_{13} - m_{32}\frac{\partial m_{13}}{\partial x_i} \tag{4.131}$$

$$\frac{\partial M_{22}}{\partial x_i} = \frac{\partial m_{11}}{\partial x_i}m_{33} + m_{11}\frac{\partial m_{33}}{\partial x_i} - \frac{\partial m_{31}}{\partial x_i}m_{13} - m_{31}\frac{\partial m_{13}}{\partial x_i} \tag{4.132}$$

$$\frac{\partial M_{23}}{\partial x_i} = \frac{\partial m_{11}}{\partial x_i}m_{22} + m_{11}\frac{\partial m_{22}}{\partial x_i} - \frac{\partial m_{31}}{\partial x_i}m_{12} - m_{31}\frac{\partial m_{12}}{\partial x_i} \tag{4.133}$$

$$\frac{\partial M_{31}}{\partial x_i} = \frac{\partial m_{12}}{\partial x_i}m_{23} + m_{12}\frac{\partial m_{23}}{\partial x_i} - \frac{\partial m_{22}}{\partial x_i}m_{13} - m_{22}\frac{\partial m_{13}}{\partial x_i} \tag{4.134}$$

$$\frac{\partial M_{32}}{\partial x_i} = \frac{\partial m_{11}}{\partial x_i}m_{23} + m_{11}\frac{\partial m_{23}}{\partial x_i} - \frac{\partial m_{21}}{\partial x_i}m_{13} - m_{21}\frac{\partial m_{13}}{\partial x_i} \tag{4.135}$$

$$\frac{\partial M_{33}}{\partial x_i} = \frac{\partial m_{11}}{\partial x_i} m_{22} + m_{11}\frac{\partial m_{22}}{\partial x_i} - \frac{\partial m_{21}}{\partial x_i} m_{12} - m_{21}\frac{\partial m_{12}}{\partial x_i} \quad (4.136)$$

The partial derivatives of the determinant of the inertia matrix M are given by: for $i = 1, 2, \cdots, 6$

$$\frac{\partial detM}{\partial x_i} = \frac{\partial m_{11}}{\partial x_i} M_{11} - \frac{\partial m_{12}}{\partial x_i} M_{12} + \frac{\partial m_{13}}{\partial x_i} M_{13} \quad (4.137)$$

About the partial derivatives of the elements of the inertia matrix m_{ij}, $i = 1, 2, 3$, $j = 1, 2, 3$, one has: $\frac{\partial m_{11}}{\partial x_i} = 0$ for $i = 1, 2, \cdots, 6$, $\frac{\partial m_{12}}{\partial x_i} = 0$ if $i \neq 3$ and $\frac{\partial m_{12}}{\partial x_i} = -m_L l_L sin(x_3)$ if $i = 3$, $\frac{\partial m_{13}}{\partial x_i} = 0$ if $i \neq 5$ and $\frac{\partial m_{13}}{\partial x_i} = -m_s l_s sin(x_5)$ if $i = 5$. Additionally, $\frac{\partial m_{21}}{\partial x_i} = 0$ if $i \neq 3$ and $\frac{\partial m_{21}}{\partial x_i} = -m_L l_L sin(x_3)$ if $i = 3$, $\frac{\partial m_{22}}{\partial x_i} = 0$ for $i = 1, 2, \cdots, 6$, $\frac{\partial m_{23}}{\partial x_i} = 0$ for $i = 1, 2, \cdots, 6$. Moreover, $\frac{\partial m_{31}}{\partial x_i} = 0$ if $i \neq 5$ and $\frac{\partial m_{31}}{\partial x_i} = -m_s l_s sin(x_5)$ if $i = 5$, $\frac{\partial m_{32}}{\partial x_i} = 0$ for $i = 1, 2, \cdots, 6$ and $\frac{\partial m_{33}}{\partial x_i} = 0$ for $i = 1, 2, \cdots, 6$.

About partial derivatives of the the elements of the Coriolis matrix, one has:

$$\frac{\partial C_1}{\partial x_i} = \frac{\partial \tilde{C}_{11}}{\partial x_i} x_2 + \tilde{C}_{11}\frac{\partial x_2}{\partial x_i} + \frac{\partial \tilde{C}_{21}}{\partial x_i} x_4 + \tilde{C}_{21}\frac{\partial x_4}{\partial x_i} + \frac{\partial \tilde{C}_{31}}{\partial x_i} x_6 + \tilde{C}_{31}\frac{\partial x_6}{\partial x_i} \quad (4.138)$$

$$\frac{\partial C_2}{\partial x_i} = \frac{\partial \tilde{C}_{12}}{\partial x_i} x_2 + \tilde{C}_{12}\frac{\partial x_2}{\partial x_i} + \frac{\partial \tilde{C}_{23}}{\partial x_i} x_4 + \tilde{C}_{23}\frac{\partial x_4}{\partial x_i} + \frac{\partial \tilde{C}_{32}}{\partial x_i} x_6 + \tilde{C}_{32}\frac{\partial x_6}{\partial x_i} \quad (4.139)$$

$$\frac{\partial C_3}{\partial x_i} = \frac{\partial \tilde{C}_{13}}{\partial x_i} x_2 + \tilde{C}_{13}\frac{\partial x_2}{\partial x_i} + \frac{\partial \tilde{C}_{21}}{\partial x_i} x_4 + \tilde{C}_{21}\frac{\partial x_4}{\partial x_i} + \frac{\partial \tilde{C}_{33}}{\partial x_i} x_6 + \tilde{C}_{33}\frac{\partial x_6}{\partial x_i} + \quad (4.140)$$

About the partial derivatives of the elements $\tilde{C}_{ij}$ $i = 1, 2, 3$, $j = 1, 2, 3$ of the Coriolis matrix, one has:

$\frac{\partial \tilde{C}_{11}}{\partial x_i} = 0$ for $i = 1, 2, \cdots, 6$, $\frac{\partial \tilde{C}_{11}}{\partial x_i} = 0$ for $i = 1, 2, 5, 6$, $\frac{\partial \tilde{C}_{12}}{\partial x_3} = -m_L l_L cos(x_3) x_4$, and $\frac{\partial \tilde{C}_{12}}{\partial x_5} = -m_L l_L sin(x_3)$. Additionally, $\frac{\partial \tilde{C}_{13}}{\partial x_i} = 0$ for $i = 1, 2, 3, 4$. Furthermore, $\frac{\partial \tilde{C}_{13}}{\partial x_5} = -m_s l_s cos(x_5) x_6$, and $\frac{\partial \tilde{C}_{13}}{\partial x_6} = -m_s l_s sin(x_5)$. Moreover, it holds that $\frac{\partial \tilde{C}_{21}}{\partial x_i} = 0$ for $i = 1, 2, \cdots, 6$, $\frac{\partial \tilde{C}_{22}}{\partial x_i} = 0$ for $i = 1, 2, \cdots, 6$, $\frac{\partial \tilde{C}_{23}}{\partial x_i} = 0$ for $i = 1, 2, \cdots, 6$, $\frac{\partial \tilde{C}_{31}}{\partial x_i} = 0$ for $i = 1, 2, \cdots, 6$, $\frac{\partial \tilde{C}_{32}}{\partial x_i} = 0$ for $i = 1, 2, \cdots, 6$, $\frac{\partial \tilde{C}_{33}}{\partial x_i} = 0$ for $i = 1, 2, \cdots, 6$.

About the partial derivatives of the elements of the gravitational matrix G_i, $i = 1, 2, 3$, one has: $\frac{\partial G_1}{\partial x_i} = 0$, for $i = 1, 2, \cdots, 6$, $\frac{\partial G_2}{\partial x_i} = 0$, for $i \neq 3$ and $\frac{\partial G_2}{\partial x_3} = -m_L l_L g \cdot cos(x_3)$, $\frac{\partial G_3}{\partial x_i} = 0$, for $i \neq 5$ and $\frac{\partial G_3}{\partial x_5} = -m_s l_s g \cdot cos(x_5)$.

4.2.6 Dynamic model of the parallel double-inverted pendulum—System 3

Another form of the parallel double-inverted pendulum, consisting of two different cart-pole systems which are connected with an elastic link, is shown in the diagram of Figure 4.24. This model of the parallel double-inverted pendulum has eight state variables and two control inputs (System 3).

The parameters of the model of the parallel double-inverted pendulum are: l_1, l_2 are the half-lengths of pendulums 1 and 2, respectively (distance from the pivot point to the center of mass of the pole), m_1, m_2 are the masses of pendulums 1 and 2, M_1, M_2 are the masses of carts 1 and 2, respectively, x_{c_1}, x_{c_2} is the displacement of cart 1 and cart 2 along the OX axis, d is the distance between the slide rails, h is the distance from the spring connection point to the pivot, k is the elasticity of the spring, J_1, J_2 are the moments of inertia of pendulums 1 and 2, μ_1, μ_2 are solid friction coefficients, and ν_1, ν_2 are viscous function coefficients

The dynamic model of the parallel double-inverted pendulum is given by [666]

$$M(\theta)\ddot{\theta} + \tilde{C}(\theta, \dot{\theta})\dot{\theta} + G(\theta) = Hu \quad (4.141)$$

where the state vector of the system is $\theta = [x_{c_1}, \theta_1, x_{c_2}, \theta_2]^T$. The inertia matrix $M(\theta)$ is defined as [666]

$$M(\theta) = \begin{pmatrix} M_1 + m_1 & -m_1 l_1 cos(\theta_1) & 0 & 0 \\ -m_1 l_1 cos(\theta_1) & \frac{7}{3} m_1 l_1^2 & 0 & 0 \\ 0 & 0 & M_2 + m_2 & -m_2 l_2 cos(\theta_2) \\ 0 & 0 & -m_2 l_2 cos(\theta_2) & \frac{7}{3} m_2 l_2^2 \end{pmatrix} \quad (4.142)$$

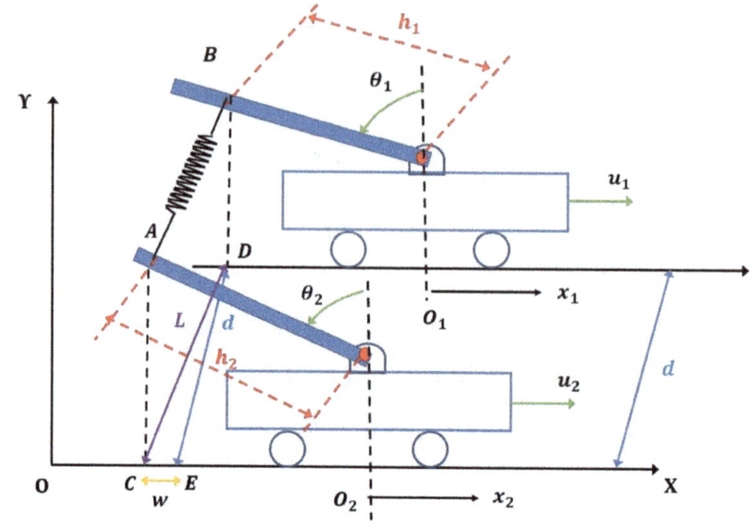

Figure 4.24 Diagram of the parallel double-inverted pendulum—System 3

The Coriolis and centrifugal forces matrix $\tilde{C}(\theta,\dot{\theta})$ is [666]

$$\tilde{C}(\theta,\dot{\theta}) = \begin{pmatrix} \mu_1 & m_1 l_1 \dot{\theta}_1 sin(\theta_1) & 0 & 0 \\ 0 & v_1 & 0 & 0 \\ 0 & 0 & v_2 & m_2 l_2 \dot{\theta}_2 sin(\theta_2) \\ 0 & 0 & 0 & v_2 \end{pmatrix} \tag{4.143}$$

while the gravitational and elastic forces vector and the control inputs gain matrix are [666]

$$G(\theta = \begin{pmatrix} K_1 \\ K_2 - m_1 g l_1 sin(\theta_1) \\ K_3 \\ K_4 - m_2 g l_2 sin(\theta_2) \end{pmatrix} \quad H = \begin{pmatrix} 1 & 0 \\ 0 & 0 \\ 0 & 1 \\ 0 & 0 \end{pmatrix} \tag{4.144}$$

with K_i, $i = 1, 2, 3, 4$ to be nonlinear functions of the angle variables of the parallel double-inverted pendulum which are given by [666]

$$\begin{aligned} K_1 &= -k_s(1 - \tfrac{l_o}{l_d})[x_{c_2} - x_{c_1} - hsin(\theta_2) + hsin(\theta_1)] \\ K_2 &= -hK_1 cos(\theta_1) \\ K_3 &= -K_1 \\ K_4 &= hK_2 cos(\theta_2) \end{aligned} \tag{4.145}$$

In the above equations, k_s is an elasticity coefficient, l_o is the normal length of the elastic link when no forces are applied to its end, l_d is the length of the deformed elastic link, $x_{c_i}, i = 1, 2$ are the positions of carts 1 and 2, θ_i are the turn angles of pendulums 1 and 2, while h_i, $i = 1, 2$ is the distance from the end of the spring to the pivot point of the pendulum. Next, by defining coefficient $\beta = -k_s(1 - \frac{l_o}{l_d})$ and using the notation of the state variables of the system, K_1 in Eq. (4.145) is written as $K_1 = \beta[x_5 - x_1]$. Moreover, by considering small values for the turn angle of each pendulum, the above-noted coefficient K_1 is rewritten as $K_1 = \beta[x_5 - x_1]$. Using the above, the elastic and gravitation forces vector is approximated by

$$G(\theta) = \begin{pmatrix} \beta(x_5 - x_1) \\ -\beta h(x_5 - x_1) - m_1 g l_1 sin(x_3) \\ -\beta(x_5 - x_1) \\ \beta h(x_5 - x_1) - m_2 g l_2 sin(x_7) \end{pmatrix} \tag{4.146}$$

The determinant of the inertia matrix M is $detM = m_{11}m_{22}(m_{33}m_{44} - m_{43}m_{34})$.

The sub-determinants of the inertia matrix M are given by: $M_{11} = m_{22}(m_{33}m_{44}) - m_{43}m_{44}$, $M_{12} = m_{21}(m_{33}m_{44} - m_{43}m_{34})$, $M_{13} = 0$, $M_{14} = 0$, $M_{21} = m_{12}(m_{33}m_{44}) - m_{43}m_{34}$, $M_{22} = m_{11}(m_{33}m_{44} - m_{43}m_{34})$, $M_{23} = 0$, $M_{24} = 0$, $M_{31} = 0$, $M_{32} = 0$, $M_{33} = m_{44}(m_{11}m_{22} - m_{21}m_{12})$, $M_{34} = m_{43}(m_{11}m_{22} - m_{21}m_{12})$, $M_{41} = 0$, $M_{42} = 0$. $M_{43} = m_{22}(m_{11}m_{22} - m_{21}m_{12})$, $M_{44} = m_{33}(m_{11}m_{22} - m_{21}m_{12})$.

The inverse of the inertia matrix M is

$$M^{-1} = \begin{pmatrix} M_{11} & -M_{21} & -M_{31} & M_{41} \\ -M_{12} & M_{22} & M_{32} & -M_{42} \\ M_{13} & -M_{23} & -M_{33} & M_{43} \\ -M_{14} & M_{24} & -M_{34} & M_{44} \end{pmatrix} \tag{4.147}$$

Besides, using that $C(\theta,\dot{\theta}) = \tilde{C}(\theta,\dot{\theta})\dot{\theta}$, one obtains

$$\begin{pmatrix} C_1 \\ C_2 \\ C_3 \\ C_4 \end{pmatrix} = \begin{pmatrix} \mu_1 \dot{x}_{c_1} + m_1 l_1 sin(\theta_1)\dot{\theta}_1^2 \\ v_1 \dot{\theta}_1 \\ \mu_2 \dot{x}_{c_2} + m_2 l_2 sin(\theta_2)\dot{\theta}_2^2 \\ v_2 \dot{\theta}_2 \end{pmatrix} \tag{4.148}$$

Next, the dynamic model of the parallel double-inverted pendulum is written as

$$\ddot{\theta} = -M^{-1}(\theta)[C(\theta,\dot{\theta}) + G(\theta)] + M^{-1}(\theta)Hu \tag{4.149}$$

Using the above equation, the dynamic model of the parallel double-inverted pendulum becomes:

$$\begin{pmatrix} \ddot{x}_{c_1} \\ \ddot{\theta}_1 \\ \ddot{x}_{c_2} \\ \ddot{\theta}_2 \end{pmatrix} = \begin{pmatrix} \frac{-M_{11}(C_1+G_1)+M_{21}(C_2+G_2)-M_{31}(C_3+G_3)+M_{41}(C_4+G_4)}{detM} \\ \frac{M_{12}(C_1+G_1)-M_{22}(C_2+G_2)+M_{31}(C_3+G_3)-M_{41}(C_4+G_4)}{detM} \\ \frac{-M_{13}(C_1+G_1)+M_{23}(C_2+G_2)-M_{33}(C_3+G_3)+M_{43}(C_4+G_4)}{detM} \\ \frac{M_{14}(C_1+G_1)-M_{24}(C_2+G_2)+M_{34}(C_3+G_3)-M_{44}(C_4+G_4)}{detM} \end{pmatrix} + \begin{pmatrix} \frac{M_{11}}{detM} & \frac{M_{31}}{detM} \\ -\frac{M_{12}}{detM} & -\frac{M_{32}}{detM} \\ \frac{M_{13}}{detM} & \frac{M_{33}}{detM} \\ -\frac{M_{14}}{detM} & -\frac{M_{34}}{detM} \end{pmatrix} \begin{pmatrix} u_1 \\ u_2 \end{pmatrix} \tag{4.150}$$

Next, by defining the state vector $x=[x_{c_1},\dot{x}_{c_1},\theta_1,\dot{\theta}_1,x_{c_2},\dot{x}_{c_2},\theta_2,\dot{\theta}_2]^T$ or equivalently $x=[x_1,x_2,x_3,x_4,x_5,x_6,x_7,x_8]^T$, one has

$$\begin{pmatrix}\dot{x}_1\\ \dot{x}_2\\ \dot{x}_3\\ \dot{x}_4\\ \dot{x}_5\\ \dot{x}_6\\ \dot{x}_7\\ \dot{x}_8\end{pmatrix}=\begin{pmatrix}x_2\\ \frac{-M_{11}(C_1+G_1)+M_{21}(C_2+G_2)-M_{31}(C_3+G_3)+M_{41}(C_4+G_4)}{detM}\\ x_4\\ \frac{M_{12}(C_1+G_1)-M_{22}(C_2+G_2)+M_{31}(C_3+G_3)-M_{41}(C_4+G_4)}{detM}\\ x_6\\ \frac{-M_{13}(C_1+G_1)+M_{23}(C_2+G_2)-M_{33}(C_3+G_3)+M_{43}(C_4+G_4)}{detM}\\ x_8\\ \frac{M_{14}(C_1+G_1)-M_{24}(C_2+G_2)+M_{34}(C_3+G_3)-M_{44}(C_4+G_4)}{detM}\end{pmatrix}+\begin{pmatrix}0 & 0\\ \frac{M_{11}}{detM} & \frac{M_{31}}{detM}\\ 0 & 0\\ -\frac{M_{12}}{detM} & -\frac{M_{32}}{detM}\\ 0 & 0\\ \frac{M_{13}}{detM} & \frac{M_{33}}{detM}\\ 0 & 0\\ -\frac{M_{14}}{detM} & -\frac{M_{34}}{detM}\end{pmatrix}\begin{pmatrix}u_1\\ u_2\end{pmatrix} \tag{4.151}$$

Thus, the dynamic model of the double parallel inverted pendulum can be written in the nonlinear affine-in-the-input state-space form:

$$\dot{x}=f(x)+g(x)u \tag{4.152}$$

where $x\in R^{8\times 1}, f(x)\in R^{8\times 1}$, $g(x)\in R^{8\times 2}$, and $u\in R^{2\times 1}$.

4.2.7 Approximate linearization of the dynamics of the parallel double-inverted pendulum—System 3

The dynamic model of the parallel double-inverted pendulum undergoes approximate linearization through first-order Taylor series expansion and the computation of the associated Jacobian matrices. The linearization point is defined at each sampling instance (x^*,u^*), where x^* is the present value of the system's state vector and u^* is the last sampled value of the control inputs vector. The modeling error which is due to the truncation of higher-order terms from the Taylor series expansion is taken to be a perturbation which is asymptotically compensated by the robustness of the control algorithm.

The linearization process allows to express the initial nonlinear system

$$\dot{x}=f(x)+g(x)u \tag{4.153}$$

into the linearized state-space form

$$\dot{x}=Ax+Bu+\tilde{d} \tag{4.154}$$

where A, B are the Jacobian matrices of the system

$$A=\nabla_x[f(x)+g(x)u]\mid_{(x^*,u^*)}\Rightarrow A=\nabla_x f(x)\mid_{(x^*,u^*)}+\nabla_x g(x)u\mid_{(x^*,u^*)} \tag{4.155}$$

$$B=\nabla_u[f(x)+g(x)u]\mid_{(x^*,u^*)}\Rightarrow B=g(x)\mid_{(x^*,u^*)} \tag{4.156}$$

The cumulative disturbance term $\tilde{d}$ may include (i) the modeling error due to the truncation of higher-order terms from the Taylor series expansion, (ii) exogenous perturbations, and (iii) sensor measurement noise of any distribution.

The computation of the Jacobian matrix $A=\nabla_x[f(x)+g(x)u]\mid_{(x^*,u^*)}$ is performed as follows:

First row of the Jacobian matrix $A=\nabla_x[f(x)+g(x)u]\mid_{(x^*,u^*)}$: $\frac{\partial f_1}{\partial x_1}=0$, $\frac{\partial f_1}{\partial x_2}=1$, $\frac{\partial f_1}{\partial x_3}=0$, $\frac{\partial f_1}{\partial x_4}=0$, $\frac{\partial f_1}{\partial x_5}=0$, $\frac{\partial f_1}{\partial x_6}=0$, $\frac{\partial f_1}{\partial x_7}=0$, $\frac{\partial f_1}{\partial x_8}=0$.

Second row of the Jacobian matrix $A=\nabla_x[f(x)+g(x)u]\mid_{(x^*,u^*)}$: It holds that

$$f_2(x)=\frac{f_{2,num}(x)}{f_{2,den}(x)}=\frac{-M_{11}(C_1+G_1)+M_{21}(C_2+G_2)-M_{31}(C_3+G_3)+M_{41}(C_4+G_4)}{detM} \tag{4.157}$$

Thus, for $i = 1, 2, \cdots, 8$, one has

$$\frac{\partial f_2(x)}{\partial x_i} = \frac{\frac{\partial}{\partial x_i}[f_{2,num}(x)]f_{2,den}(x) - f_{2,num}(x)\frac{\partial}{\partial x_i}detM}{detM^2} \tag{4.158}$$

where for $i = 1, 2, \cdots, 8$, it holds that

$$\begin{aligned}\frac{\partial f_{2,num}(x)}{\partial x_i} = & -\frac{\partial M_{11}}{\partial x_i}(C_1 + G_1) - M_{11}\left(\frac{\partial C_1}{\partial x_i} + \frac{\partial G_1}{\partial x_i}\right) \\ & + \frac{\partial M_{21}}{\partial x_i}(C_2 + G_2) + M_{21}\left(\frac{\partial C_2}{\partial x_i} + \frac{\partial G_2}{\partial x_i}\right) \\ & - \frac{\partial M_{31}}{\partial x_i}(C_3 + G_3) - M_{31}\left(\frac{\partial C_3}{\partial x_i} + \frac{\partial G_3}{\partial x_i}\right) \\ & + \frac{\partial M_{41}}{\partial x_i}(C_4 + G_4) + M_{41}\left(\frac{\partial C_4}{\partial x_i} + \frac{\partial G_4}{\partial x_i}\right)\end{aligned} \tag{4.159}$$

and also

$$\frac{\partial f_{2,den}(x)}{\partial x_i} = \frac{\partial detM}{\partial x_i} \tag{4.160}$$

Third row of the Jacobian matrix $A = \nabla_x[f(x) + g(x)u] \mid_{(x^*,u^*)}$: $\frac{\partial f_3}{\partial x_1} = 0$, $\frac{\partial f_3}{\partial x_2} = 0$, $\frac{\partial f_3}{\partial x_3} = 0$, $\frac{\partial f_3}{\partial x_4} = 1$, $\frac{\partial f_3}{\partial x_5} = 0$, $\frac{\partial f_3}{\partial x_6} = 0$, $\frac{\partial f_3}{\partial x_7} = 0$, $\frac{\partial f_3}{\partial x_8} = 0$.

Fourth row of the Jacobian matrix $A = \nabla_x[f(x) + g(x)u] \mid_{(x^*,u^*)}$: It holds that

$$f_4(x) = \frac{f_{4,num}(x)}{f_{4,den}(x)} = \frac{M_{12}(C_1+G_1) - M_{22}(C_2+G_2) + M_{32}(C_3+G_3) + M_{41}(C_4+G_4)}{detM} \tag{4.161}$$

Thus, for $i = 1, 2, \cdots, 8$, one has

$$\frac{\partial f_4(x)}{\partial x_i} = \frac{\frac{\partial}{\partial x_i}[f_{4,num}(x)]f_{4,den}(x) - f_{4,num}(x)\frac{\partial}{\partial x_i}detM}{detM^2} \tag{4.162}$$

where for $i = 1, 2, \cdots, 8$, it holds that

$$\begin{aligned}\frac{\partial f_{4,num}(x)}{\partial x_i} = & \frac{\partial M_{12}}{\partial x_i}(C_1 + G_1) + M_{12}\left(\frac{\partial C_1}{\partial x_i} + \frac{\partial G_1}{\partial x_i}\right) \\ & - \frac{\partial M_{22}}{\partial x_i}(C_2 + G_2) - M_{22}\left(\frac{\partial C_2}{\partial x_i} + \frac{\partial G_2}{\partial x_i}\right) \\ & + \frac{\partial M_{32}}{\partial x_i}(C_3 + G_3) + M_{32}\left(\frac{\partial C_3}{\partial x_i} + \frac{\partial G_3}{\partial x_i}\right) \\ & - \frac{\partial M_{42}}{\partial x_i}(C_4 + G_4) - M_{42}\left(\frac{\partial C_4}{\partial x_i} + \frac{\partial G_4}{\partial x_i}\right)\end{aligned} \tag{4.163}$$

and also

$$\frac{\partial f_{4,den}(x)}{\partial x_i} = \frac{\partial detM}{\partial x_i} \tag{4.164}$$

Fifth row of the Jacobian matrix $A = \nabla_x[f(x) + g(x)u] \mid_{(x^*,u^*)}$: $\frac{\partial f_3}{\partial x_1} = 0$, $\frac{\partial f_5}{\partial x_2} = 0$, $\frac{\partial f_5}{\partial x_3} = 0$, $\frac{\partial f_5}{\partial x_4} = 0$, $\frac{\partial f_5}{\partial x_5} = 0$, $\frac{\partial f_5}{\partial x_6} = 1$, $\frac{\partial f_5}{\partial x_7} = 0$, $\frac{\partial f_5}{\partial x_8} = 0$.

Sixth row of the Jacobian matrix $A = \nabla_x[f(x) + g(x)u] \mid_{(x^*,u^*)}$: It holds that

$$f_6(x) = \frac{f_{6,num}(x)}{f_{6,den}(x)} = \frac{-M_{13}(C_1+G_1) + M_{23}(C_2+G_2) - M_{33}(C_3+G_3) + M_{43}(C_4+G_4)}{detM} \tag{4.165}$$

Thus, for $i = 1, 2, \cdots, 8$, one has

$$\frac{\partial f_6(x)}{\partial x_i} = \frac{\frac{\partial}{\partial x_i}[f_{6,num}(x)]f_{6,den}(x) - f_{6,num}(x)\frac{\partial}{\partial x_i}detM}{detM^2} \tag{4.166}$$

where for $i = 1, 2, \cdots, 8$, it holds that

$$\begin{aligned} \frac{\partial f_{6,num}(x)}{\partial x_i} &= -\frac{\partial M_{13}}{\partial x_i}(C_1+G_1) - M_{13}\left(\frac{\partial C_1}{\partial x_i}+\frac{\partial G_1}{\partial x_i}\right) \\ &+\frac{\partial M_{23}}{\partial x_i}(C_2+G_2) + M_{23}\left(\frac{\partial C_2}{\partial x_i}+\frac{\partial G_2}{\partial x_i}\right) \\ &-\frac{\partial M_{33}}{\partial x_i}(C_3+G_3) - M_{33}\left(\frac{\partial C_3}{\partial x_i}+\frac{\partial G_3}{\partial x_i}\right) \\ &+\frac{\partial M_{43}}{\partial x_i}(C_4+G_4) + M_{43}\left(\frac{\partial C_4}{\partial x_i}+\frac{\partial G_4}{\partial x_i}\right) \end{aligned} \tag{4.167}$$

and also

$$\frac{\partial f_{6,den}(x)}{\partial x_i} = \frac{\partial detM}{\partial x_i} \tag{4.168}$$

Seventh row of the Jacobian matrix $A = \nabla_x[f(x)+g(x)u] \mid_{(x^*,u^*)}$: $\frac{\partial f7}{\partial x_1}=0$, $\frac{\partial f_7}{\partial x_2}=0$, $\frac{\partial f_7}{\partial x_3}=0$, $\frac{\partial f_7}{\partial x_4}=0$, $\frac{\partial f_7}{\partial x_5}=0$, $\frac{\partial f_7}{\partial x_6}=0$, $\frac{\partial f_7}{\partial x_7}=0$, $\frac{\partial f_7}{\partial x_8}=1$.

Eighth row of the Jacobian matrix $A = \nabla_x[f(x)+g(x)u] \mid_{(x^*,u^*)}$: It holds that

$$f_8(x) = \frac{f_{8,num}(x)}{f_{8,den}(x)} = \frac{M_{14}(C_1+G_1)-M_{24}(C_2+G_2)+M_{34}(C_3+G_3)+M_{44}(C_4+G_4)}{detM} \tag{4.169}$$

Thus, for $i = 1, 2, \cdots, 8$, one has

$$\frac{\partial f_8(x)}{\partial x_i} = \frac{\frac{\partial}{\partial x_i}[f_{8,num}(x)]f_{8,den}(x) - f_{8,num}(x)\frac{\partial}{\partial x_i}detM}{detM^2} \tag{4.170}$$

where for $i = 1, 2, \cdots, 8$, it holds that

$$\begin{aligned} \frac{\partial f_{8,num}(x)}{\partial x_i} &= \frac{\partial M_{14}}{\partial x_i}(C_1+G_1) + M_{14}\left(\frac{\partial C_1}{\partial x_i}+\frac{\partial G_1}{\partial x_i}\right) \\ &-\frac{\partial M_{24}}{\partial x_i}(C_2+G_2) - M_{24}\left(\frac{\partial C_2}{\partial x_i}+\frac{\partial G_2}{\partial x_i}\right) \\ &+\frac{\partial M_{34}}{\partial x_i}(C_3+G_3) + M_{34}\left(\frac{\partial C_3}{\partial x_i}+\frac{\partial G_3}{\partial x_i}\right) \\ &-\frac{\partial M_{44}}{\partial x_i}(C_4+G_4) - M_{44}\left(\frac{\partial C_4}{\partial x_i}+\frac{\partial G_4}{\partial x_i}\right) \end{aligned} \tag{4.171}$$

and also

$$\frac{\partial f_{8,den}(x)}{\partial x_i} = \frac{\partial detM}{\partial x_i} \tag{4.172}$$

The computation of the Jacobian matrix $\nabla_x g_1(x) \mid_{(x^*,u^*)}$ is performed as follows:

First row of the Jacobian matrix $\nabla_x g_1(x) \mid_{(x^*,u^*)}$: $\frac{\partial g_{11}}{\partial x_i}=0$, for $i = 1, 2, \cdots, 8$.

Second row of the Jacobian matrix $\nabla_x g_1(x) \mid_{(x^*,u^*)}$: for $i = 1, 2, \cdots, 8$, it holds that

$$\frac{\partial g_{12}}{\partial x_i} = \frac{\frac{\partial M_{11}}{\partial x_i}detM - M_{11}\frac{\partial detM}{\partial x_i}}{detM^2} \tag{4.173}$$

Third row of the Jacobian matrix $\nabla_x g_1(x) \mid_{(x^*,u^*)}$: $\frac{\partial g_{13}}{\partial x_i}=0$, for $i = 1, 2, \cdots, 8$.

Fourth row of the Jacobian matrix $\nabla_x g_1(x) \mid_{(x^*,u^*)}$: for $i = 1, 2, \cdots, 8$, it holds that

$$\frac{\partial g_{14}}{\partial x_i} = \frac{-\frac{\partial M_{12}}{\partial x_i}detM + M_{12}\frac{\partial detM}{\partial x_i}}{detM^2} \tag{4.174}$$

Fifth row of the Jacobian matrix $\nabla_x g_1(x) \mid_{(x^*,u^*)}$: $\frac{\partial g_{13}}{\partial x_i}=0$, for $i = 1, 2, \cdots, 8$.

Sixth row of the Jacobian matrix $\nabla_x g_1(x) \mid_{(x^*,u^*)}$: for $i = 1, 2, \cdots, 8$, it holds that

$$\frac{\partial g_{16}}{\partial x_i} = \frac{\frac{\partial M_{13}}{\partial x_i} detM - M_{13} \frac{\partial detM}{\partial x_i}}{detM^2} \tag{4.175}$$

Seventh row of the Jacobian matrix $\nabla_x g_1(x) \mid_{(x^*,u^*)}$: $\frac{\partial g_{17}}{\partial x_i} = 0$, for $i = 1, 2, \cdots, 8$.

Eighth row of the Jacobian matrix $\nabla_x g_1(x) \mid_{(x^*,u^*)}$: for $i = 1, 2, \cdots, 8$, it holds that

$$\frac{\partial g_{18}}{\partial x_i} = \frac{-\frac{\partial M_{14}}{\partial x_i} detM + M_{14} \frac{\partial detM}{\partial x_i}}{detM^2} \tag{4.176}$$

The computation of the Jacobian matrix $\nabla_x g_2(x) \mid_{(x^*,u^*)}$ is performed as follows:

First row of the Jacobian matrix $\nabla_x g_2(x) \mid_{(x^*,u^*)}$: $\frac{\partial g_{21}}{\partial x_i} = 0$, for $i = 1, 2, \cdots, 8$.

Second row of the Jacobian matrix $\nabla_x g_2(x) \mid_{(x^*,u^*)}$: for $i = 1, 2, \cdots, 8$, it holds that

$$\frac{\partial g_{22}}{\partial x_i} = \frac{\frac{\partial M_{31}}{\partial x_i} detM - M_{31} \frac{\partial detM}{\partial x_i}}{detM^2} \tag{4.177}$$

Third row of the Jacobian matrix $\nabla_x g_2(x) \mid_{(x^*,u^*)}$: $\frac{\partial g_{23}}{\partial x_i} = 0$, for $i = 1, 2, \cdots, 8$.

Fourth row of the Jacobian matrix $\nabla_x g_2(x) \mid_{(x^*,u^*)}$: for $i = 1, 2, \cdots, 8$, it holds that

$$\frac{\partial g_{24}}{\partial x_i} = \frac{-\frac{\partial M_{32}}{\partial x_i} detM + M_{32} \frac{\partial detM}{\partial x_i}}{detM^2} \tag{4.178}$$

Fifth row of the Jacobian matrix $\nabla_x g_2(x) \mid_{(x^*,u^*)}$: $\frac{\partial g_{25}}{\partial x_i} = 0$, for $i = 1, 2, \cdots, 8$.

Sixth row of the Jacobian matrix $\nabla_x g_2(x) \mid_{(x^*,u^*)}$: for $i = 1, 2, \cdots, 8$, it holds that

$$\frac{\partial g_{26}}{\partial x_i} = \frac{\frac{\partial M_{33}}{\partial x_i} detM - M_{33} \frac{\partial detM}{\partial x_i}}{detM^2} \tag{4.179}$$

Seventh row of the Jacobian matrix $\nabla_x g_3(x) \mid_{(x^*,u^*)}$: $\frac{\partial g_{27}}{\partial x_i} = 0$, for $i = 1, 2, \cdots, 8$.

Eighth row of the Jacobian matrix $\nabla_x g_2(x) \mid_{(x^*,u^*)}$: for $i = 1, 2, \cdots, 8$, it holds that

$$\frac{\partial g_{28}}{\partial x_i} = \frac{-\frac{\partial M_{34}}{\partial x_i} detM + M_{34} \frac{\partial detM}{\partial x_i}}{detM^2} \tag{4.180}$$

The partial derivatives of the elements of the inertia matrix M are computed as follows:

$$\frac{\partial M_{11}}{\partial x_i} = \frac{\partial m_{22}}{\partial x_i}(m_{33}m_{44} - m_{43}m_{34}) + m_{22}\left(\frac{\partial m_{33}}{\partial x_i}m_{44} + m_{33}\frac{\partial m_{44}}{\partial x_i} - \frac{\partial m_{43}}{\partial x_i}m_{34} - m_{43}\frac{\partial m_{34}}{\partial x_i}\right) \tag{4.181}$$

$$\frac{\partial M_{12}}{\partial x_i} = \frac{\partial m_{21}}{\partial x_i}(m_{33}m_{44} - m_{43}m_{44}) + m_{21}\left(\frac{\partial m_{33}}{\partial x_i}m_{44} + m_{33}\frac{\partial m_{44}}{\partial x_i} - \frac{\partial m_{43}}{\partial x_i}m_{34} - m_{43}\frac{\partial m_{34}}{\partial x_i}\right) \tag{4.182}$$

$$\frac{\partial M_{13}}{\partial x_i} = 0 \quad \frac{\partial M_{14}}{\partial x_i} = 0. \tag{4.183}$$

$$\frac{\partial M_{21}}{\partial x_i} = \frac{\partial m_{12}}{\partial x_i}(m_{33}m_{44} - m_{43}m_{34}) + m_{12}\left(\frac{\partial m_{33}}{\partial x_i}m_{44} + m_{33}\frac{\partial m_{44}}{\partial x_i} - \frac{\partial m_{43}}{\partial x_i}m_{34} - m_{43}\frac{\partial m_{34}}{\partial x_i}\right) \tag{4.184}$$

$$\frac{\partial M_{22}}{\partial x_i} = \frac{\partial m_{11}}{\partial x_i}(m_{33}m_{44} - m_{43}m_{34}) + m_{11}\left(\frac{\partial m_{33}}{\partial x_i}m_{44} + m_{33}\frac{\partial m_{44}}{\partial x_i} - \frac{\partial m_{43}}{\partial x_i}m_{34} - m_{43}\frac{\partial m_{34}}{\partial x_i}\right) \tag{4.185}$$

$$\frac{\partial M_{23}}{\partial x_i} = 0 \quad \frac{\partial M_{24}}{\partial x_i} = 0. \tag{4.186}$$

$$\frac{\partial M_{31}}{\partial x_i} = 0 \quad \frac{\partial M_{32}}{\partial x_i} = 0. \tag{4.187}$$

$$\frac{\partial M_{33}}{\partial x_i} = \frac{\partial m_{44}}{\partial x_i}(m_{11}m_{22} - m_{21}m_{12}) + m_{44}\left(\frac{\partial m_{11}}{\partial x_i}m_{22} + m_{11}\frac{\partial m_{22}}{\partial x_i} - \frac{\partial m_{21}}{\partial x_i}m_{12} - m_{21}\frac{\partial m_{12}}{\partial x_i}\right) \tag{4.188}$$

$$\frac{\partial M_{34}}{\partial x_i} = \frac{\partial m_{43}}{\partial x_i}(m_{11}m_{22} - m_{21}m_{12}) + m_{43}\left(\frac{\partial m_{11}}{\partial x_i}m_{22} + m_{11}\frac{\partial m_{22}}{\partial x_i} - \frac{\partial m_{21}}{\partial x_i}m_{12} - m_{21}\frac{\partial m_{12}}{\partial x_i}\right) \tag{4.189}$$

$$\frac{\partial M_{41}}{\partial x_i} = 0 \quad \frac{\partial M_{42}}{\partial x_i} = 0. \tag{4.190}$$

$$\frac{\partial M_{43}}{\partial x_i} = \frac{\partial m_{34}}{\partial x_i}(m_{11}m_{22} - m_{21}m_{12}) + m_{34}\left(\frac{\partial m_{11}}{\partial x_i}m_{22} + m_{11}\frac{\partial m_{22}}{\partial x_i} - \frac{\partial m_{21}}{\partial x_i}m_{12} - m_{21}\frac{\partial m_{12}}{\partial x_i}\right) \tag{4.191}$$

$$\frac{\partial M_{44}}{\partial x_i} = \frac{\partial m_{33}}{\partial x_i}(m_{11}m_{22} - m_{21}m_{12}) + m_{33}\left(\frac{\partial m_{11}}{\partial x_i}m_{22} + m_{11}\frac{\partial m_{22}}{\partial x_i} - \frac{\partial m_{21}}{\partial x_i}m_{12} - m_{21}\frac{\partial m_{12}}{\partial x_i}\right) \tag{4.192}$$

The partial derivatives of the determinant $detM$ of the inertia matrix M, with respect to the state variables x_i of the parallel double-inverted pendulum x_i $i = 1, 2, \cdots, 8$, are given by

$$\frac{\partial detM}{\partial x_i} = \frac{\partial m_{11}}{\partial x_i}M_{11} + m_{11}\frac{\partial M_{11}}{\partial x_i} - \frac{\partial m_{12}}{\partial x_i}M_{12} + m_{12}\frac{\partial M_{12}}{\partial x_i} + \frac{\partial m_{13}}{\partial x_i}M_{13} + m_{13}\frac{\partial M_{13}}{\partial x_i} \tag{4.193}$$

The computation of the partial derivatives of the elements of the Coriolis matrix gives:

$\frac{\partial C_1}{\partial x_1} = 0$, $\frac{\partial C_1}{\partial x_2} = \mu_1$, $\frac{\partial C_1}{\partial x_3} = m_1 l_1 cos(x_3)x_4^2$, $\frac{\partial C_1}{\partial x_4} = 2m_1 l_1 x_4 sin(x_3)$, $\frac{\partial C_1}{\partial x_5} = 0$, $\frac{\partial C_1}{\partial x_6} = 0$, $\frac{\partial C_1}{\partial x_7} = 0$, and $\frac{\partial C_1}{\partial x_1} = 8$.

$\frac{\partial C_2}{\partial x_1} = 0$, $\frac{\partial C_2}{\partial x_2} = 0$, $\frac{\partial C_2}{\partial x_3} = 0$, $\frac{\partial C_2}{\partial x_4} = \nu_1$, $\frac{\partial C_2}{\partial x_5} = 0$, $\frac{\partial C_2}{\partial x_6} = 0$, $\frac{\partial C_2}{\partial x_7} = 0$, and $\frac{\partial C_2}{\partial x_8} = 0$.

$\frac{\partial C_3}{\partial x_1} = 0$, $\frac{\partial C_3}{\partial x_2} = 0$, $\frac{\partial C_3}{\partial x_3} = 0$, $\frac{\partial C_3}{\partial x_4} = 0$, $\frac{\partial C_3}{\partial x_5} = 0$, $\frac{\partial C_3}{\partial x_6} = \mu_2$, $\frac{\partial C_3}{\partial x_7} = m_2 l_2 cos(x_7)x_8^2$, and $\frac{\partial C_3}{\partial x_8} = 2m_2 l_2 sin(x_7)x_8$.

$\frac{\partial C_4}{\partial x_1} = 0$, $\frac{\partial C_4}{\partial x_2} = 0$, $\frac{\partial C_4}{\partial x_3} = 0$, $\frac{\partial C_4}{\partial x_4} = 0$, $\frac{\partial C_4}{\partial x_5} = 0$, $\frac{\partial C_4}{\partial x_6} = 0$, $\frac{\partial C_4}{\partial x_7} = 0$, and $\frac{\partial C_4}{\partial x_8} = \nu_2$.

The partial derivatives of the elements of the gravitational and elastic forces matrix are: $\frac{\partial G_1}{\partial x_1} = -\beta$, $\frac{\partial G_1}{\partial x_5} = \beta$, and $\frac{\partial G_1}{\partial x_i} = 0$, for $i \neq 1, 5$. Moreover, $\frac{\partial G_2}{\partial x_i} = 0$ for $i \neq 1, 3, 5$ and $\frac{\partial G_2}{\partial x_1} = \beta h$, $\frac{\partial G_2}{\partial x_3} = -m_1 g l_1 cos(x_3)$, $\frac{\partial G_2}{\partial x_5} = -\beta h$. Additionally, $\frac{\partial G_3}{\partial x_1} = \beta$, $\frac{\partial G_3}{\partial x_5} = -\beta$, and $\frac{\partial G_1}{\partial x_i} = 0$, for $i \neq 1, 5$. Furthermore, $\frac{\partial G_4}{\partial x_i} = 0$ for $i \neq 1, 3, 5$ and $\frac{\partial G_4}{\partial x_1} = -\beta h$, $\frac{\partial G_4}{\partial x_5} = \beta h$, $\frac{\partial G_4}{\partial x_7} = -m_2 g l_2 cos(x_7)$.

Finally, about the partial derivatives of the non-zero elements m_{ij}, $i = 1, \cdots, 4$, $j = 1, \cdots, 4$ of the inertia matrix M, it holds that: $\frac{\partial m_{11}}{\partial x_i} = 0$ for $i = 1, \cdots, 8$, $\frac{\partial m_{12}}{\partial x_i} = 0$ for $i \neq 3$ and $\frac{\partial m_{12}}{\partial x_3} = m_1 l_1 cos(x_3)$, $\frac{\partial m_{21}}{\partial x_i} = 0$ for $i \neq 3$ and $\frac{\partial m_{21}}{\partial x_3} = m_1 l_1 cos(x_3)$, $\frac{\partial m_{22}}{\partial x_i} = 0$ for $i = 1, \cdots, 8$, $\frac{\partial m_{33}}{\partial x_i} = 0$ for $i = 1, \cdots, 8$, $\frac{\partial m_{34}}{\partial x_i} = 0$ for $i \neq 7$ and $\frac{\partial m_{34}}{\partial x_7} = m_2 l_2 cos(x_7)$, $\frac{\partial m_{43}}{\partial x_i} = 0$ for $i \neq 7$ and $\frac{\partial m_{43}}{\partial x_7} = m_2 l_2 cos(x_7)$, $\frac{\partial m_{44}}{\partial x_i} = 0$ for $i = 1, \cdots, 8$.

4.2.8 Design of an H-infinity nonlinear feedback controller

4.2.8.1 Equivalent linearized dynamics of the parallel double-inverted pendulum

After linearization around its current operating point, the dynamic model for the parallel double-inverted pendulum, in all three cases, is written as

$$\dot{x} = Ax + Bu + d_1 \tag{4.194}$$

Parameter d_1 stands for the linearization error in the parallel double-inverted pendulum model that was given previously in Eq. (4.194). The reference setpoints for the state vector of the aforementioned dynamic model are denoted by $\mathbf{x_d} = [x_1^d, \cdots, x_n^d]$ (where in the case of System 1 $n = 4$, in the case of System 2 $n = 6$ and in the case of System 3 $n = 8$). Tracking of this trajectory is achieved after applying the control input u_d. At every time instant, the control input u_d is assumed to differ from the control input u appearing in Eq. (4.194) by an amount equal to Δu, that is $u_d = u + \Delta u$

$$\dot{x}_d = Ax_d + Bu_d + d_2 \tag{4.195}$$

The dynamics of the controlled system described in Eq. (4.194) can be also written as

$$\dot{x} = Ax + Bu + Bu_d - Bu_d + d_1 \tag{4.196}$$

and by denoting $d_3 = -Bu_d + d_1$ as an aggregate disturbance term, one obtains

$$\dot{x} = Ax + Bu + Bu_d + d_3 \tag{4.197}$$

By subtracting Eq. (4.195) from Eq. (4.197), one has

$$\dot{x} - \dot{x}_d = A(x - x_d) + Bu + d_3 - d_2 \tag{4.198}$$

By denoting the tracking error as $e = x - x_d$ and the aggregate disturbance term as $L\tilde{d} = d_3 - d_2$, the tracking error dynamics becomes

$$\dot{e} = Ae + Bu + L\tilde{d} \tag{4.199}$$

with L to be denoting the disturbance inputs gain matrix. The above-linearized form of the model of the parallel double-inverted pendulum can be efficiently controlled after applying an H-infinity feedback control scheme.

4.2.8.2 The nonlinear H-infinity control

The initial nonlinear model of the parallel double-inverted pendulum is in the form

$$\dot{x} = f(x, u) \;\; x \in R^n, \;\; u \in R^m \tag{4.200}$$

Linearization of the model of the parallel double-inverted pendulum is performed at each iteration of the control algorithm round its present operating point $(x^*, u^*) = (x(t), u(t - T_s))$. The linearized equivalent of the system is described by

$$\dot{x} = Ax + Bu + L\tilde{d} \;\; x \in R^n, \;\; u \in R^m, \;\; \tilde{d} \in R^q \tag{4.201}$$

where matrices A and B are obtained from the computation of the previously defined Jacobians and vector $\tilde{d}$ denotes disturbance terms due to linearization errors. The problem of disturbance rejection for the linearized model that is described by

$$\begin{aligned} \dot{x} &= Ax + Bu + L\tilde{d} \\ y &= Cx \end{aligned} \tag{4.202}$$

where $x \in R^n$, $u \in R^m$, $\tilde{d} \in R^q$, and $y \in R^p$, cannot be handled efficiently if the classical LQR control scheme is applied. This is because of the existence of the perturbation term $\tilde{d}$. The disturbance term $\tilde{d}$ apart from modeling (parametric) uncertainty and external perturbation terms can also represent noise terms of any distribution.

In the H_∞ control approach, a feedback control scheme is designed for trajectory tracking by the system's state vector and simultaneous disturbance rejection, considering that the disturbance affects the system in the worst possible manner. The disturbances' effects are incorporated in the following quadratic cost function:

$$J(t) = \frac{1}{2}\int_0^T [y^T(t)y(t) + ru^T(t)u(t) - \rho^2\tilde{d}^T(t)\tilde{d}(t)]dt, \;\; r, \rho > 0 \tag{4.203}$$

The significance of the negative sign in the cost function's term that is associated with the perturbation variable $\tilde{d}(t)$ is that the disturbance tries to maximize the cost function $J(t)$ while the control signal $u(t)$ tries to minimize it. The physical meaning of the relation given above is that the control signal and the disturbances compete with each other within a min–max differential game. This problem of min–max optimization can be written as $min_u max_{\tilde{d}} J(u, \tilde{d})$.

The objective of the optimization procedure is to compute a control signal $u(t)$ which can compensate for the worst possible disturbance, that is externally imposed on the parallel double-inverted pendulum. However, the solution to the min–max optimization problem is directly related to the value of the parameter ρ. This means that there is an upper bound in the disturbances magnitude that can be annihilated by the control signal.

4.2.8.3 Computation of the feedback control gains

For the linearized system given by Eq. (4.202), the cost function of Eq. (4.203) is defined, where coefficient r determines the penalization of the control input and weight coefficient ρ determines the reward of the disturbances' effects. It is assumed that (i) The energy that is transferred from the disturbances signal $\tilde{d}(t)$ is bounded, that is $\int_0^\infty \tilde{d}^T(t)\tilde{d}(t)dt < \infty$,

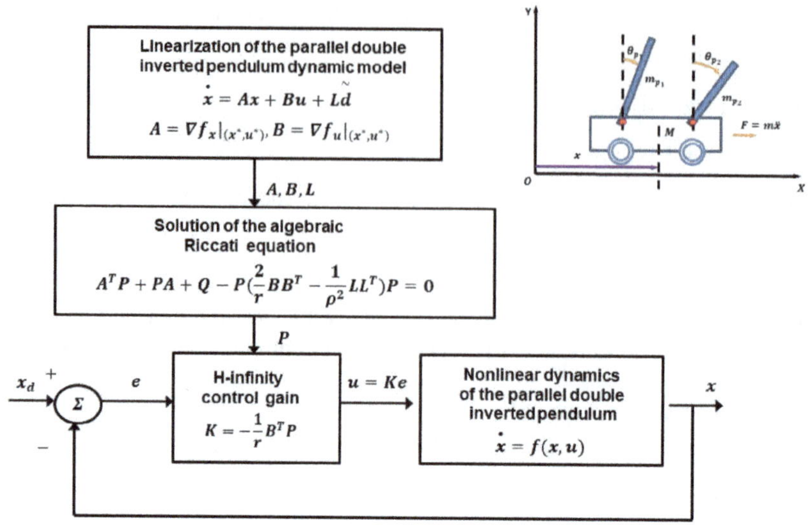

Figure 4.25 Diagram of the control scheme for the parallel double-inverted pendulum

(ii) matrices $[A, B]$ and $[A, L]$ are stabilizable, and (iii) matrix $[A, C]$ is detectable. In the case of a tracking problem, the optimal feedback control law is given by

$$u(t) = -Ke(t) \tag{4.204}$$

with $e = x - x_d$ to be the tracking error, and $K = \frac{1}{r}B^TP$ where P is a positive definite symmetric matrix. As it will be proven in Section 4.2.9, matrix P is obtained from the solution of the Riccati equation

$$A^TP + PA + Q - P(\tfrac{2}{r}BB^T - \tfrac{1}{\rho^2}LL^T)P = 0 \tag{4.205}$$

where Q is a positive semi-definite symmetric matrix. The worst-case disturbance is given by

$$\tilde{d}(t) = \tfrac{1}{\rho^2}L^TPe(t) \tag{4.206}$$

The solution of the H-infinity feedback control problem for the parallel double-inverted pendulum and the computation of the worst-case disturbance that the related controller can sustain, comes from superposition of Bellman's optimality principle when considering that the robot is affected by two separate inputs: (i) the control input u and (ii) the cumulative disturbance input $\tilde{d}(t)$. Solving the optimal control problem for u, that is for the minimum variation (optimal) control input that achieves the elimination of the state vector's tracking error, gives $u = -\frac{1}{r}B^TPe$. Equivalently, solving the optimal control problem for $\tilde{d}$, that is for the worst-case disturbance that the control loop can sustain gives $\tilde{d} = \frac{1}{\rho^2}L^TPe$. The diagram of the considered control loop for the parallel double-inverted pendulum (for all Systems 1–3) is depicted in Figure 4.25.

4.2.9 Lyapunov stability analysis

Through Lyapunov stability analysis, it will be shown that the proposed nonlinear control scheme assures H_∞ tracking performance for the underactuated parallel double-inverted pendulum and that in case of bounded disturbance terms asymptotic convergence to the reference setpoints is achieved. The tracking error dynamics for the parallel double-inverted pendulum is written in the form

$$\dot{e} = Ae + Bu + L\tilde{d} \tag{4.207}$$

where in the parallel double-inverted pendulum's case $L = I \in R^{n\times n}$ with I being the identity matrix (for System 1 $n = 4$, for System 2 $n = 8$, while for System 3 $n = 8$). Variable $\tilde{d}$ denotes model uncertainties and external disturbances of the parallel double-inverted pendulum's model. The following Lyapunov equation is considered:

$$V = \frac{1}{2}e^T P e \tag{4.208}$$

where $e = x - x_d$ is the tracking error. By differentiating with respect to time, one obtains

$$\begin{aligned} \dot{V} &= \tfrac{1}{2}\dot{e}^T P e + \tfrac{1}{2}e^T P \dot{e} \Rightarrow \\ \dot{V} &= \tfrac{1}{2}[Ae + Bu + L\tilde{d}]^T P e + \tfrac{1}{2}e^T P [Ae + Bu + L\tilde{d}] \Rightarrow \end{aligned} \tag{4.209}$$

$$\begin{aligned} \dot{V} &= \tfrac{1}{2}[e^T A^T + u^T B^T + \tilde{d}^T L^T] P e + \\ &+ \tfrac{1}{2}e^T P [Ae + Bu + L\tilde{d}] \Rightarrow \end{aligned} \tag{4.210}$$

$$\begin{aligned} \dot{V} &= \tfrac{1}{2}e^T A^T P e + \tfrac{1}{2}u^T B^T P e + \tfrac{1}{2}\tilde{d}^T L^T P e + \\ &\tfrac{1}{2}e^T P A e + \tfrac{1}{2}e^T P B u + \tfrac{1}{2}e^T P L \tilde{d} \end{aligned} \tag{4.211}$$

The previous equation is rewritten as

$$\begin{aligned} \dot{V} &= \tfrac{1}{2}e^T (A^T P + PA) e + (\tfrac{1}{2}u^T B^T P e + \tfrac{1}{2}e^T P B u) + \\ &+ (\tfrac{1}{2}\tilde{d}^T L^T P e + \tfrac{1}{2}e^T P L \tilde{d}) \end{aligned} \tag{4.212}$$

Assumption: For given positive definite matrix Q and coefficients r and ρ, there exists a positive definite matrix P, which is the solution of the following matrix equation:

$$A^T P + PA = -Q + P(\tfrac{2}{r}BB^T - \tfrac{1}{\rho^2}LL^T)P \tag{4.213}$$

Moreover, the following feedback control law is applied to the system

$$u = -\tfrac{1}{r}B^T P e \tag{4.214}$$

By substituting Eq. (4.213) and Eq. (4.214), one obtains

$$\begin{aligned} \dot{V} &= \tfrac{1}{2}e^T[-Q + P(\tfrac{2}{r}BB^T - \tfrac{1}{\rho^2}LL^T)P]e + \\ &+ e^T PB(-\tfrac{1}{r}B^T P e) + e^T P L \tilde{d} \Rightarrow \end{aligned} \tag{4.215}$$

$$\begin{aligned} \dot{V} &= -\tfrac{1}{2}e^T Q e + \tfrac{1}{r}e^T PBB^T P e - \tfrac{1}{2\rho^2}e^T PLL^T P e \\ &- \tfrac{1}{r}e^T PBB^T P e + e^T P L \tilde{d} \end{aligned} \tag{4.216}$$

which after intermediate operations gives

$$\dot{V} = -\tfrac{1}{2}e^T Q e - \tfrac{1}{2\rho^2}e^T PLL^T P e + e^T P L \tilde{d} \tag{4.217}$$

or, equivalently

$$\begin{aligned} \dot{V} &= -\tfrac{1}{2}e^T Q e - \tfrac{1}{2\rho^2}e^T PLL^T P e + \\ &+ \tfrac{1}{2}e^T P L \tilde{d} + \tfrac{1}{2}\tilde{d}^T L^T P e \end{aligned} \tag{4.218}$$

Lemma: The following inequality holds:

$$\tfrac{1}{2}e^T L \tilde{d} + \tfrac{1}{2}\tilde{d} L^T P e - \tfrac{1}{2\rho^2}e^T PLL^T P e \le \tfrac{1}{2}\rho^2 \tilde{d}^T \tilde{d} \tag{4.219}$$

Proof: The binomial $(\rho\alpha - \frac{1}{\rho}b)^2$ is considered. Expanding the left part of the above inequality, one gets

$$\begin{aligned} &\rho^2 a^2 + \tfrac{1}{\rho^2}b^2 - 2ab \ge 0 \Rightarrow \tfrac{1}{2}\rho^2 a^2 + \tfrac{1}{2\rho^2}b^2 - ab \ge 0 \Rightarrow \\ &ab - \tfrac{1}{2\rho^2}b^2 \le \tfrac{1}{2}\rho^2 a^2 \Rightarrow \tfrac{1}{2}ab + \tfrac{1}{2}ab - \tfrac{1}{2\rho^2}b^2 \le \tfrac{1}{2}\rho^2 a^2 \end{aligned} \tag{4.220}$$

The following substitutions are carried out: $a=\tilde{d}$ and $b=e^TPL$ and the previous relation becomes

$$\frac{1}{2}\tilde{d}^TL^TPe+\frac{1}{2}e^TPL\tilde{d}-\frac{1}{2\rho^2}e^TPLL^TPe\leq\frac{1}{2}\rho^2\tilde{d}^T\tilde{d} \quad (4.221)$$

Eq. (4.221) is substituted in Eq. (4.218) and the inequality is enforced, thus giving

$$\dot{V}\leq-\frac{1}{2}e^TQe+\frac{1}{2}\rho^2\tilde{d}^T\tilde{d} \quad (4.222)$$

Eq. (4.222) shows that the H_∞ tracking performance criterion is satisfied. The integration of $\dot{V}$ from 0 to T gives

$$\begin{array}{l}\int_0^T\dot{V}(t)dt\leq-\frac{1}{2}\int_0^T||e||_Q^2dt+\frac{1}{2}\rho^2\int_0^T||\tilde{d}||^2dt\Rightarrow\\ 2V(T)+\int_0^T||e||_Q^2dt\leq 2V(0)+\rho^2\int_0^T||\tilde{d}||^2dt\end{array} \quad (4.223)$$

Moreover, if there exists a positive constant $M_d>0$ such that

$$\int_0^\infty||\tilde{d}||^2dt\leq M_d \quad (4.224)$$

then, one gets

$$\int_0^\infty||e||_Q^2dt\leq 2V(0)+\rho^2M_d \quad (4.225)$$

Thus, the integral $\int_0^\infty||e||_Q^2dt$ is bounded. Moreover, $V(T)$ is bounded and from the definition of the Lyapunov function V in Eq. (4.208), it becomes clear that $e(t)$ will be also bounded since $e(t)\ \in\ \Omega_e=\{e|e^TPe\leq 2V(0)+\rho^2M_d\}$. According to the above and with the use of Barbalat's Lemma, one obtains $\lim_{t\to\infty}e(t)=0$.

After following the stages of the stability proof, one arrives at Eq. (4.222) which shows that the H-infinity tracking performance criterion holds. By selecting the attenuation coefficient ρ to be sufficiently small and in particular to satisfy $\rho^2<||e||_Q^2/||\tilde{d}||^2$, one has that the first derivative of the Lyapunov function is upper bounded by 0. This condition holds at each sampling instance and consequently global stability for the control loop can be concluded.

4.2.10 Simulation tests

4.2.10.1 Parallel double-inverted pendulum—System 1 with four state variables

The performance of the nonlinear optimal control method for the first type of the parallel double-inverted pendulum (two poles mounted on the same cart resulting in a model with four state variables and one control input) has been further tested and confirmed through simulation experiments. To obtain the feedback gains of the nonlinear optimal (H-infinity), the algebraic Riccati equation of Eq. (4.213) was repetitively being solved at each sampling period. The transient performance of the control method depends on parameters r, ρ, and Q of the above-noted Riccati equation. The smallest value of coefficient ρ, for which a valid solution of the previous Riccati equation is obtained, is the one that provides the control loop with maximum robustness.

Simulation experiments about the performance of the nonlinear optimal control method in the case of the parallel double-inverted pendulum (System 1 with four state variables) are given in Figures 4.26–4.33. The parameters of the dynamic model of the parallel double-inverted pendulum were: $m_{p_1}=0.80kg$, $l_{p_1}=0.70m$, $I_{p_1}=0.40kgm^2$, $c_1=0.12$, $m_{p_2}=0.85kg$, $l_{p_2}=0.85m$, $I_{p_2}=0.54kgm^2$, $c_2=0.15$, and $g=10m/sec^2$. Through the obtained results it can be confirmed that the nonlinear optimal control method achieves fast stabilization of the parallel double-inverted pendulum under moderate variations of the control inputs.

To elaborate on nonlinear optimal control for the dynamic model of the first parallel double-inverted pendulum the following tables are given: (i) Table 4.4 providing results about the accuracy of tracking of setpoints by the state variables of the first pendulum with four state variables and one control input under an exact dynamic model and (ii) Table 4.5 providing results about the accuracy of tracking of setpoints by the state variables of the first pendulum with four state variables and one control input under model uncertainty (parametric change by $\Delta a\%$ in mass m_{p_1} of the first pole).

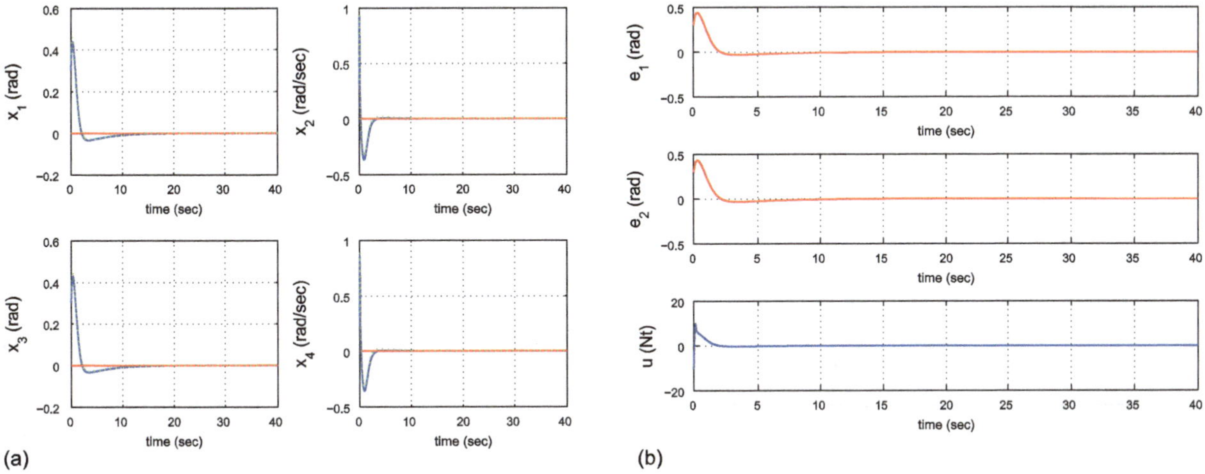

Figure 4.26 *Tracking of setpoint 1 for the parallel double-inverted pendulum (cart with double-inverted pendulum—System 1 with four state variables): (a) convergence of state variable $x_1 = \theta_{p_1}$, $x_2 = \dot{\theta}_{p_1}$, $x_3 = \theta_2$, $x_4 = \dot{\theta}_{p_2}$ to their reference setpoints (red line: setpoint, blue line: real value, green line: estimated value); (b) variation of tracking errors e_1, e_3 and variations of the control input u (acceleration of the cart)*

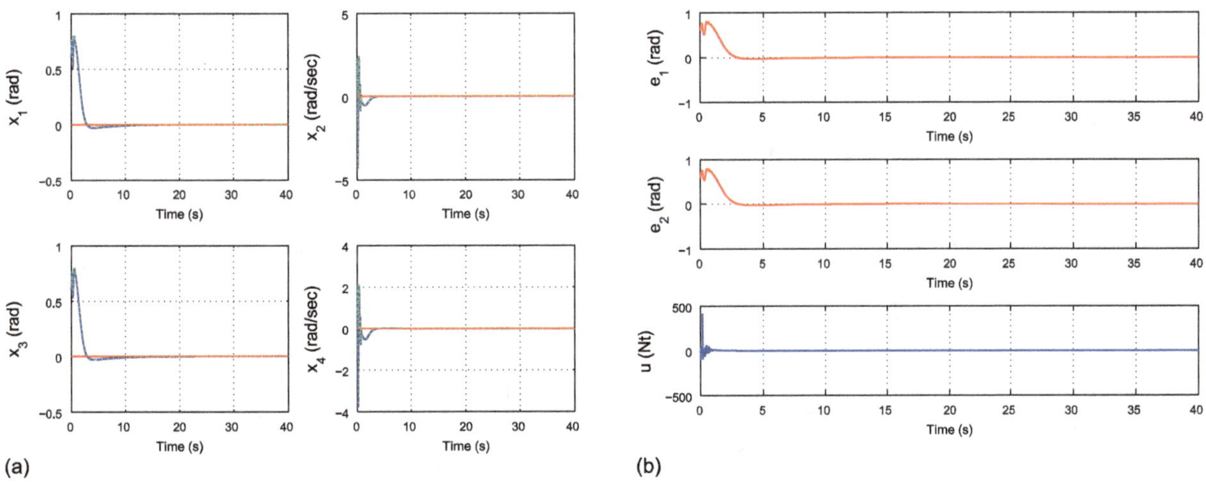

Figure 4.27 *Tracking of setpoint 2 for the parallel double-inverted pendulum (cart with double-inverted pendulum—System 1 with four state variables): (a) convergence of state variable $x_1 = \theta_{p_1}$, $x_2 = \dot{\theta}_{p_1}$, $x_3 = \theta_2$, $x_4 = \dot{\theta}_{p_2}$ to their reference setpoints (red line: setpoint, blue line: real value, green line: estimated value); (b) variation of tracking errors e_1, e_3 and variations of the control input u (acceleration of the cart)*

4.2.10.2 Parallel double-inverted pendulum—System 2 with six state variables

The performance of the nonlinear optimal control method for the second type of the parallel double-inverted pendulum (two poles mounted on the same cart resulting in a model with six state variables and one control input) has been further tested and confirmed through simulation experiments. To obtain the feedback gains of the nonlinear optimal

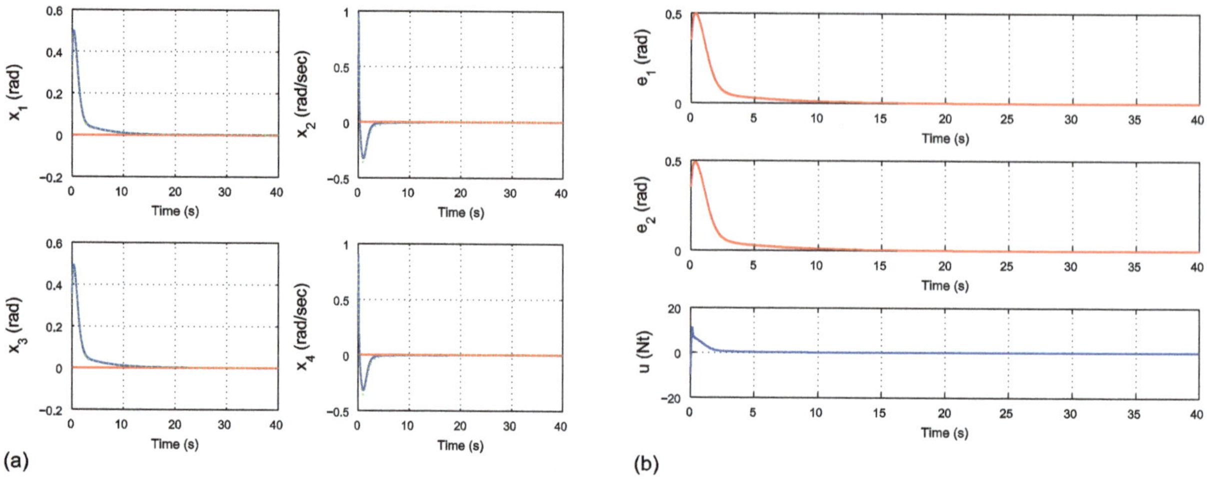

Figure 4.28 Tracking of setpoint 3 for the parallel double-inverted pendulum (cart with double-inverted pendulum—System 1 with four state variables): (a) convergence of state variable $x_1 = \theta_{p_1}$, $x_2 = \dot{\theta}_{p_1}$, $x_3 = \theta_2$, $x_4 = \dot{\theta}_{p_2}$ to their reference setpoints (red line: setpoint, blue line: real value, green line: estimated value); (b) variation of tracking errors e_1, e_3 and variations of the control input u (acceleration of the cart)

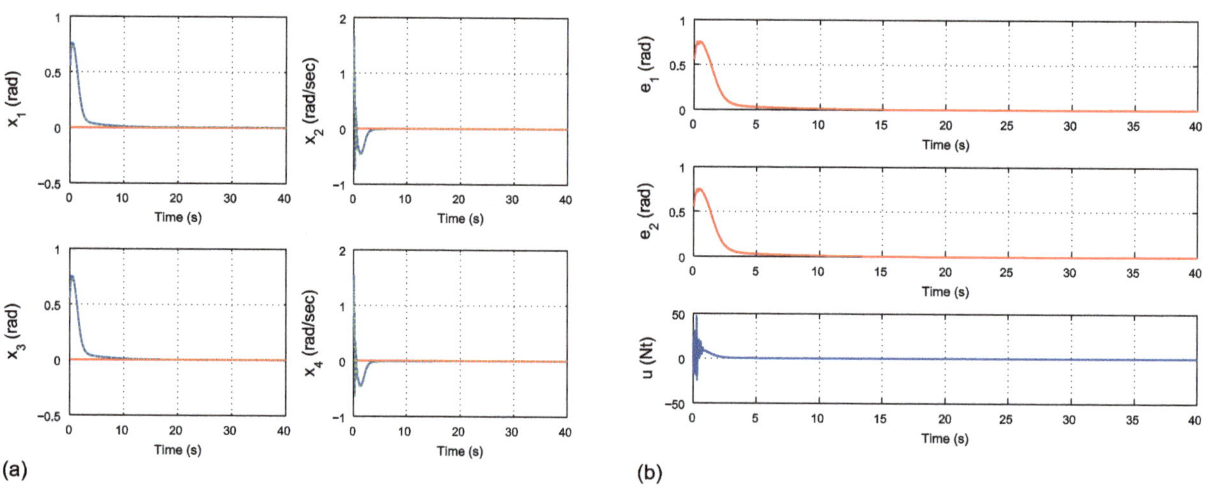

Figure 4.29 Tracking of setpoint 4 for the parallel double-inverted pendulum (cart with double-inverted pendulum—System 1 with four state variables): (a) convergence of state variable $x_1 = \theta_{p_1}$, $x_2 = \dot{\theta}_{p_1}$, $x_3 = \theta_2$, $x_4 = \dot{\theta}_{p_2}$ to their reference setpoints (red line: setpoint, blue line: real value, green line: estimated value); (b) variation of tracking errors e_1, e_3 and variations of the control input u (acceleration of the cart)

(H-infinity) the algebraic Riccati equation of Eq. (A.50) had to be repetitively solved at each sampling period. The transient performance of the control method depends on parameters r, ρ, and Q of the above-noted Riccati equation. The smallest value of coefficient ρ, for which a valid solution of the previous Riccati equation is obtained, is the one that provides the control loop with maximum robustness.

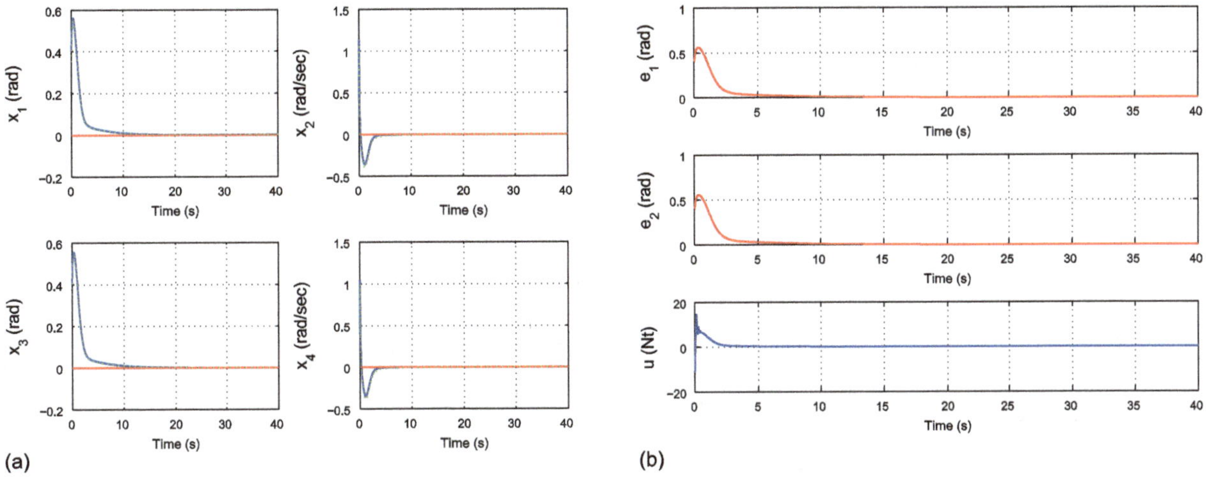

Figure 4.30 Tracking of setpoint 5 for the parallel double-inverted pendulum (cart with double-inverted pendulum—System 1 with four state variables): (a) convergence of state variable $x_1 = \theta_{p_1}$, $x_2 = \dot{\theta}_{p_1}$, $x_3 = \theta_2$, $x_4 = \dot{\theta}_{p_2}$ to their reference setpoints (red line: setpoint, blue line: real value, green line: estimated value); (b) variation of tracking errors e_1, e_3 and variations of the control input u (acceleration of the cart)

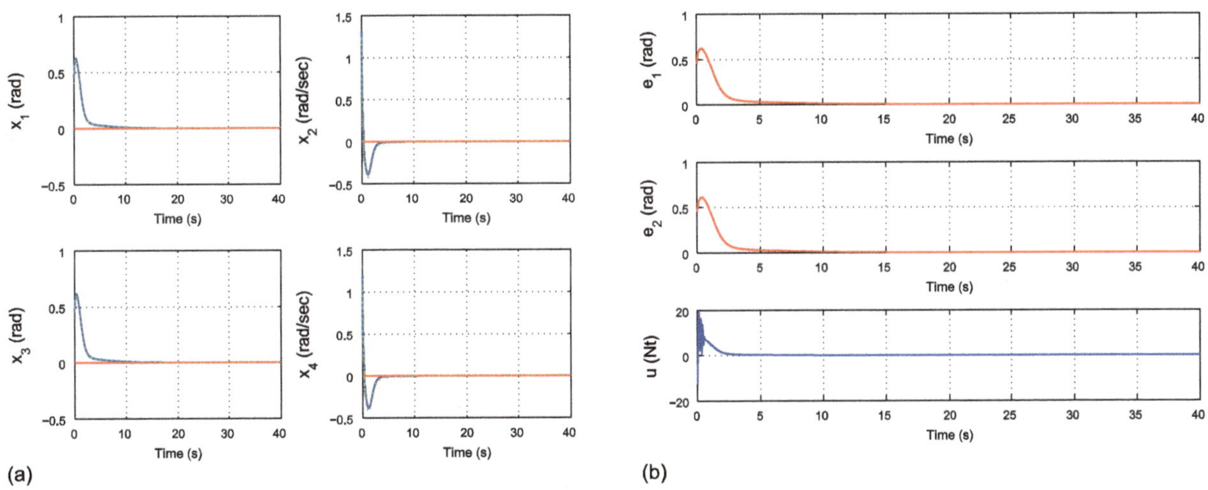

Figure 4.31 Tracking of setpoint 6 for the parallel double-inverted pendulum (cart with double-inverted pendulum—System 1 with four state variables): (a) convergence of state variable $x_1 = \theta_{p_1}$, $x_2 = \dot{\theta}_{p_1}$, $x_3 = \theta_2$, $x_4 = \dot{\theta}_{p_2}$ to their reference setpoints (red line: setpoint, blue line: real value, green line: estimated value); (b) variation of tracking errors e_1, e_3 and variations of the control input u (acceleration of the cart)

Simulation experiments about the performance of the nonlinear optimal control method in the case of the parallel double-inverted pendulum (System 2 with six state variables) are given in Figures 4.34–4.41. The parameters of the dynamic model of the parallel double-inverted pendulum were: parameters of the cart $m_c = 8.0kg$, $C_c = 1.8$, parameters of the first pendulum $m_L = 1.5kg$, $l_L = 1.2m$, $J_L = 1.44kg{\cdot}m^2$, $C_L = 0.10$, parameters of the second pendulum

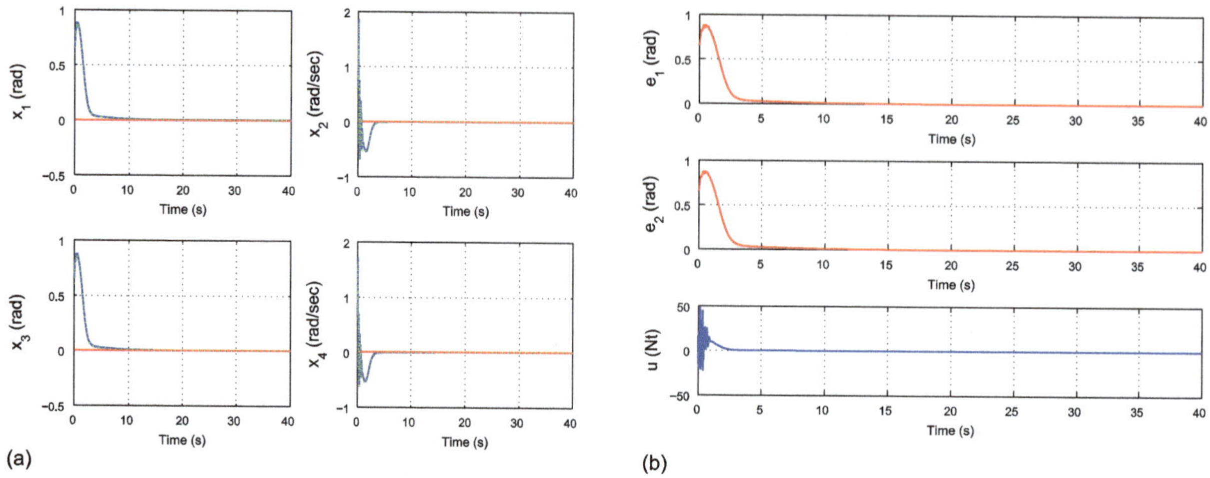

Figure 4.32 Tracking of setpoint 7 for the parallel double-inverted pendulum (cart with double-inverted pendulum—System 1 with four state variables): (a) convergence of state variable $x_1 = \theta_{p_1}$, $x_2 = \dot{\theta}_{p_1}$, $x_3 = \theta_2$, $x_4 = \dot{\theta}_{p_2}$ to their reference setpoints (red line: setpoint, blue line: real value, green line: estimated value); (b) variation of tracking errors e_1, e_3 and variations of the control input u (acceleration of the cart)

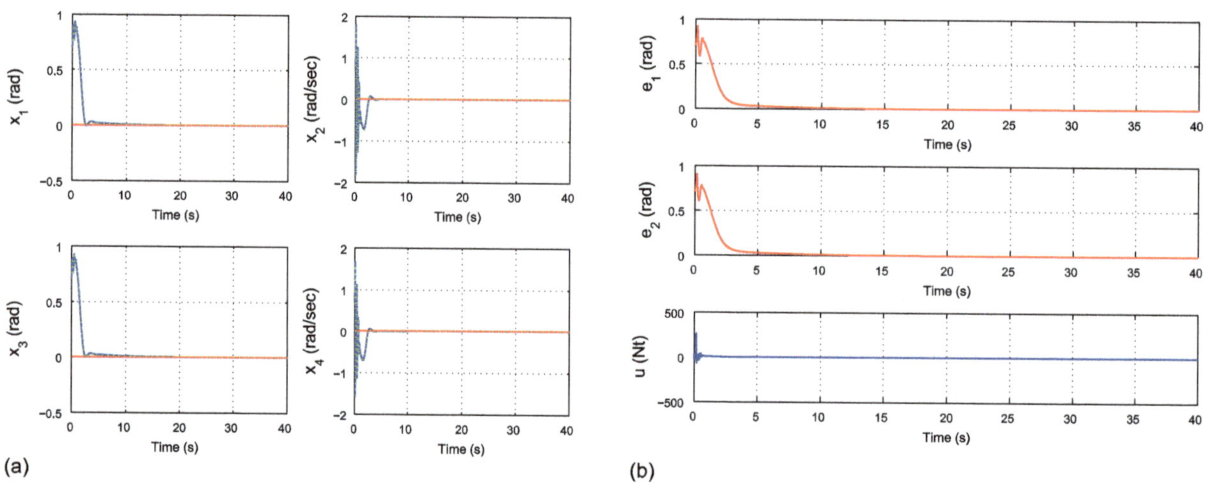

Figure 4.33 Tracking of setpoint 8 for the parallel double-inverted pendulum (cart with double-inverted pendulum—System 1 with four state variables): (a) convergence of state variable $x_1 = \theta_{p_1}$, $x_2 = \dot{\theta}_{p_1}$, $x_3 = \theta_2$, $x_4 = \dot{\theta}_{p_2}$ to their reference setpoints (red line: setpoint, blue line: real value, green line: estimated value); (b) variation of tracking errors e_1, e_3 and variations of the control input u (acceleration of the cart)

$m_s = 1.0kg$, $l_s = 0.9m$, $J_s = 0.81kg{\cdot}m^2$, $C_S = 0.10$, and $g = 10m/sec^2$. It can be noticed that the nonlinear optimal control method achieved fast and accurate tracking of reference setpoints for the parallel double-inverted pendulum with a state vector of dimension 6, under moderate variations of the control inputs.

Table 4.4 Parallel double-inverted pendulum 1

Tracking RMSE $\times 10^{-4}$ in the disturbance-free case				
No test	$RMSE_{x_1}$	$RMSE_{x_2}$	$RMSE_{x_3}$	$RMSE_{x_4}$
$test_1$	0.0043	0.0012	0.0043	0.0012
$test_2$	0.0043	0.0012	0.0043	0.0012
$test_3$	0.0043	0.0012	0.0043	0.0012
$test_4$	0.0043	0.0012	0.0043	0.0012
$test_5$	0.0043	0.0012	0.0043	0.0012
$test_6$	0.0043	0.0012	0.0043	0.0012
$test_7$	0.0049	0.0011	0.0040	0.0011
$test_8$	0.0029	0.0008	0.0029	0.0008

Table 4.5 Parallel double-inverted pendulum 1

Tracking RMSE $\times 10^{-4}$ under disturbances				
$\Delta a\%$	$RMSE_{x_1}$	$RMSE_{x_2}$	$RMSE_{x_3}$	$RMSE_{x_4}$
0%	0.0043	0.0012	0.0043	0.0012
5%	0.0030	0.0007	0.0030	0.0007
10%	0.0053	0.0041	0.0053	0.0041
15%	0.0118	0.0039	0.0118	0.0039
20%	0.0487	0.0137	0.0487	0.0137
25%	0.1397	0.0347	0.1397	0.0347
30%	0.3134	0.0699	0.03134	0.0699

To elaborate on nonlinear optimal control for the dynamic models of the second parallel double-inverted pendulums the following tables are given: (i) Table 4.6 providing results about the accuracy of tracking of setpoints by the state variables of the second pendulum with six state variables and one control input under an exact dynamic model and (ii) Table 4.7 providing results about the accuracy of tracking of setpoints by the state variables of the second pendulum with six state variables and one control input under model uncertainty (parametric change by $\Delta a\%$ in mass m_2 of the second pole.

4.2.10.3 Parallel double-inverted pendulum—System 3 with eight state variables

The performance of the nonlinear optimal control method for the third type of the parallel double-inverted pendulum (two different cart-pole systems connected through an elastic link, resulting in a model with eight state variables and two control inputs) has been further tested and confirmed through simulation experiments. To obtain the feedback gains of the nonlinear optimal (H-infinity) the algebraic Riccati equation of Eq. (A.50) had to be repetitively solved at each sampling interval. The transient performance of the control method depends on parameters r, ρ, and Q of the above-noted Riccati equation. The smallest value of coefficient ρ, for which a valid solution of the previous Riccati equation is obtained, is the one that provides the control loop with maximum robustness.

Simulation experiments about the performance of the nonlinear optimal control method in the case of the parallel double-inverted pendulum (System 3 with eight state variables) are given in Figures 4.42–4.49. The parameters of the dynamic model of the parallel double-inverted pendulum were: mass of the first cart $M_1 = 1.5kg$, mass of the first pendulum $m_1 = 0.2kg$, half-length of the first pendulum $l_1 = 0.15m$, mass of the second cart $M_2 = 1.5kg$, mass of the second pendulum $m_2 = 0.2kg$, half-length of the second pendulum $l_2 = 0.15m$, coefficients of the first pendulum

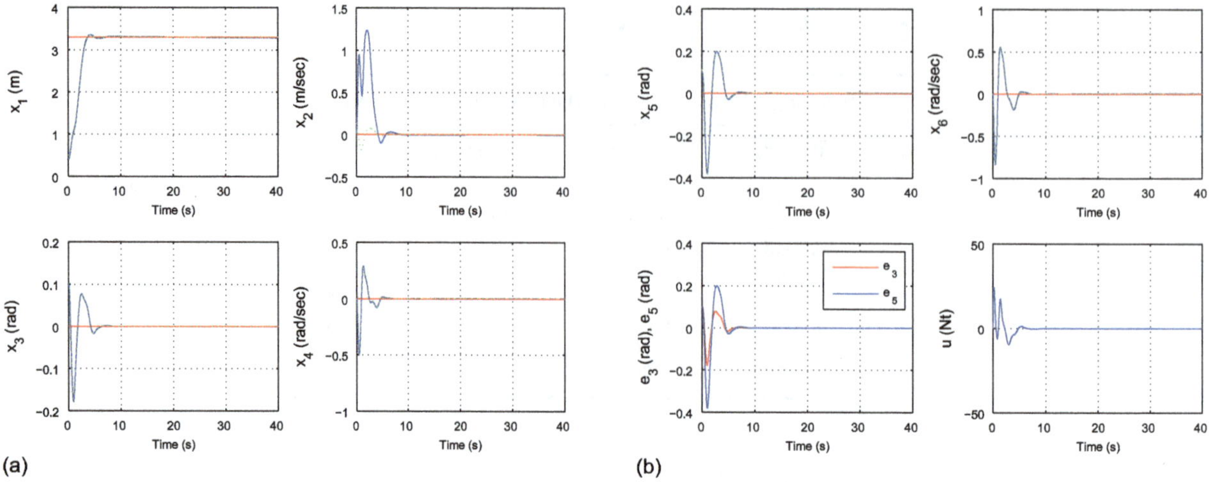

Figure 4.34 Tracking of setpoint 1 for the parallel double-inverted pendulum (cart with double-inverted pendulum—System 2 with six state variables): (a) convergence of state variable $x_1 = x$, $x_2 = \dot{x}$, $x_3 = \theta_1$, $x_4 = \dot{\theta}_1$ to their reference setpoints (red line: setpoint, blue line: real value, green line: estimated value); (b) variation of state variables $x_5 = \theta_2$ and $x_6 = \dot{\theta}_2$, variation of tracking errors e_1, e_3, e_5, and variations of the control input u (traction force on the cart)

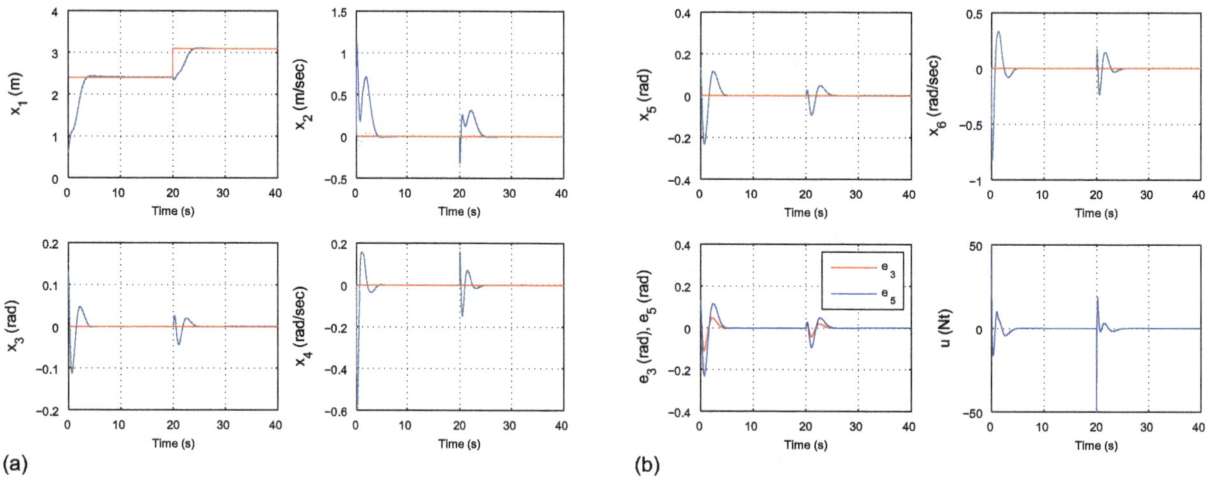

Figure 4.35 Tracking of setpoint 2 for the parallel double-inverted pendulum (cart with double-inverted pendulum—System 2 with six state variables): (a) convergence of state variable $x_1 = x$, $x_2 = \dot{x}$, $x_3 = \theta_1$, $x_4 = \dot{\theta}_1$ to their reference setpoints (red line: setpoint, blue line: real value, green line: estimated value); (b) variation of state variables $x_5 = \theta_2$ and $x_6 = \dot{\theta}_2$, variation of tracking errors e_1, e_3, e_5, and variations of the control input u (traction force on the cart)

$\mu_1 = 0.002$, $v_1 = 0.001$, $\beta_1 = 0.001$, coefficients of the second pendulum $\mu_2 = 0.002$, $v_2 = 0.001$, $\beta_2 = 0.001$, and $g = 10m/sec^2$ while $h = 0.15m$. It can be noticed that the nonlinear optimal control method achieved fast and accurate tracking of reference setpoints for the eight-dimensional state-space model of the double pendulum, under moderate variations of the control inputs.

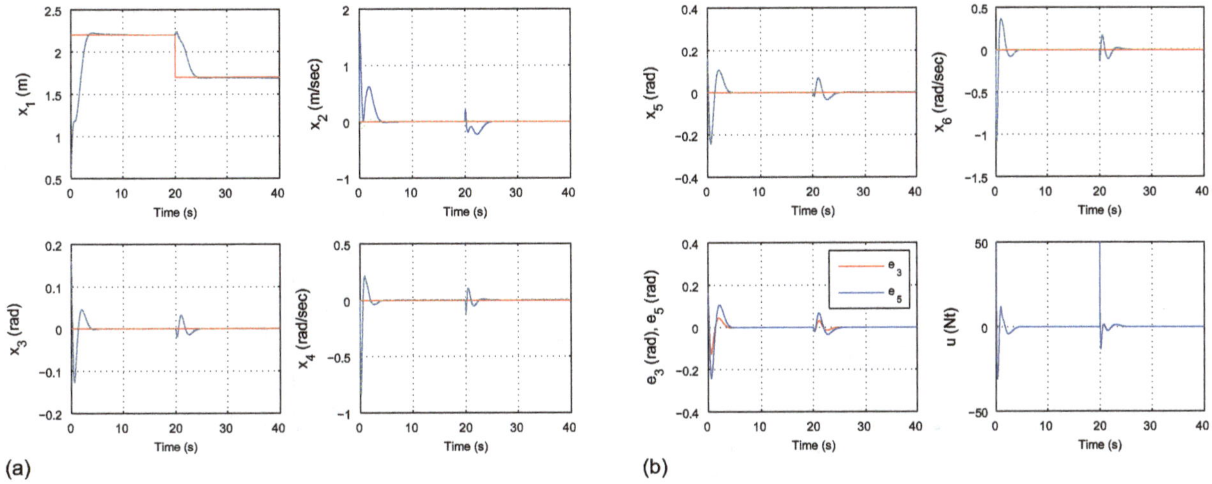

Figure 4.36 Tracking of setpoint 3 for the parallel double-inverted pendulum (cart with double-inverted pendulum—System 2 with six state variables): (a) convergence of state variable $x_1 = x$, $x_2 = \dot{x}$, $x_3 = \theta_1$, $x_4 = \dot{\theta}_1$ to their reference setpoints (red line: setpoint, blue line: real value, green line: estimated value); (b) variation of state variables $x_5 = \theta_2$ and $x_6 = \dot{\theta}_2$, variation of tracking errors e_1, e_3, e_5, and variations of the control input u (traction force on the cart)

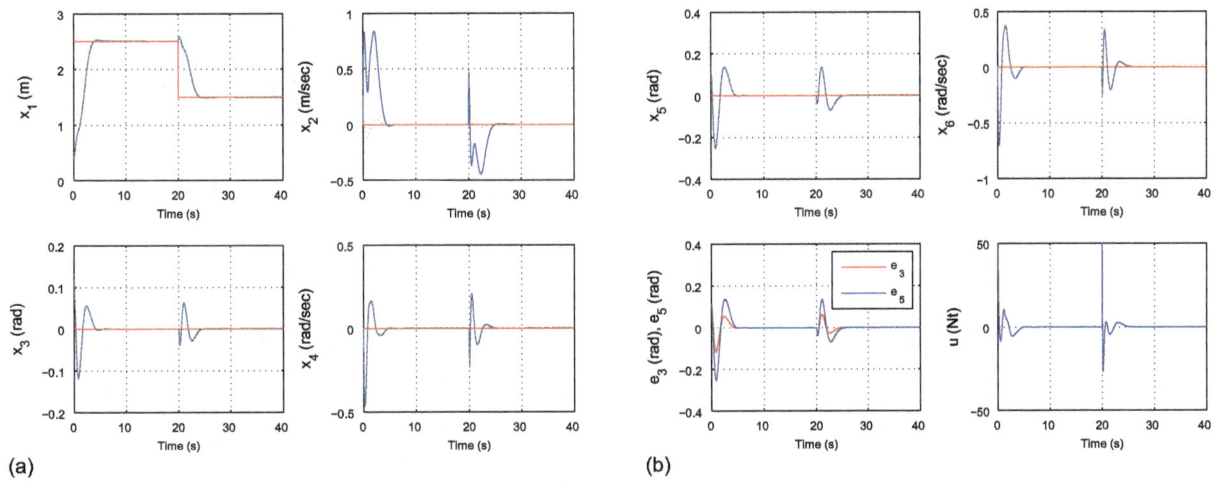

Figure 4.37 Tracking of setpoint 4 for the parallel double-inverted pendulum (cart with double-inverted pendulum—System 2 with six state variables): (a) convergence of state variable $x_1 = x$, $x_2 = \dot{x}$, $x_3 = \theta_1$, $x_4 = \dot{\theta}_1$ to their reference setpoints (red line: setpoint, blue line: real value, green line: estimated value); (b) variation of state variables $x_5 = \theta_2$ and $x_6 = \dot{\theta}_2$, variation of tracking errors e_1, e_3, e_5, and variations of the control input u (traction force on the cart)

The results which are presented in this Simulation tests section confirm the developments of the preceding theoretical analysis about a globally stable and convergent nonlinear optimal control method for the dynamic model of the parallel double-inverted pendulum. As noted, the implementation of the control algorithm has been based on (i) Approximate linearization of the dynamic model of the parallel inverted pendulums at each sampling instance through Taylor

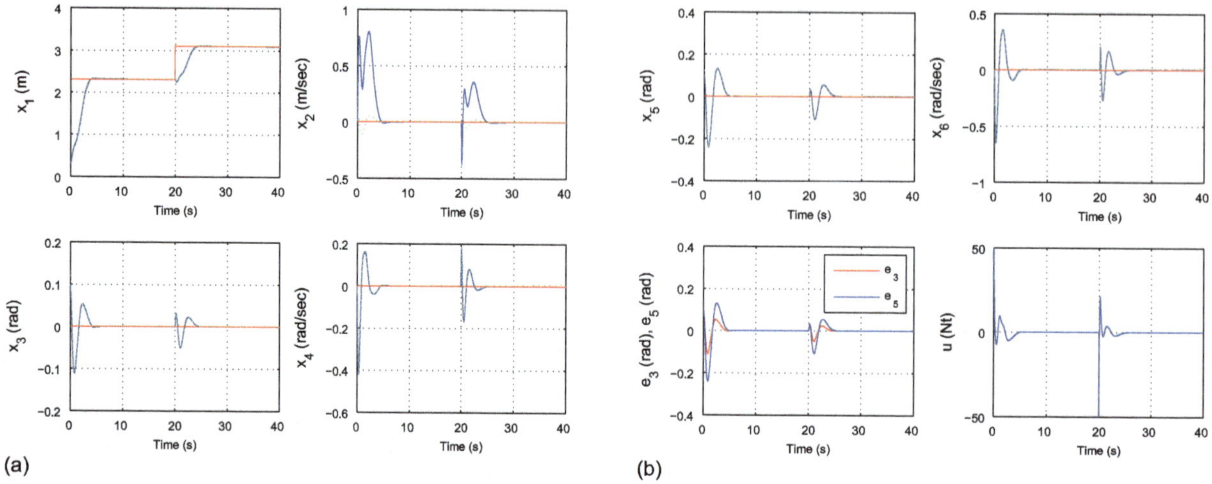

Figure 4.38 Tracking of setpoint 5 for the parallel double-inverted pendulum (cart with double-inverted pendulum—System 2 with six state variables): (a) convergence of state variable $x_1 = x$, $x_2 = \dot{x}$, $x_3 = \theta_1$, $x_4 = \dot{\theta}_1$ to their reference setpoints (red line: setpoint, blue line: real value, green line: estimated value); (b) variation of state variables $x_5 = \theta_2$ and $x_6 = \dot{\theta}_2$, variation of tracking errors e_1, e_3, e_5, and variations of the control input u (traction force on the cart)

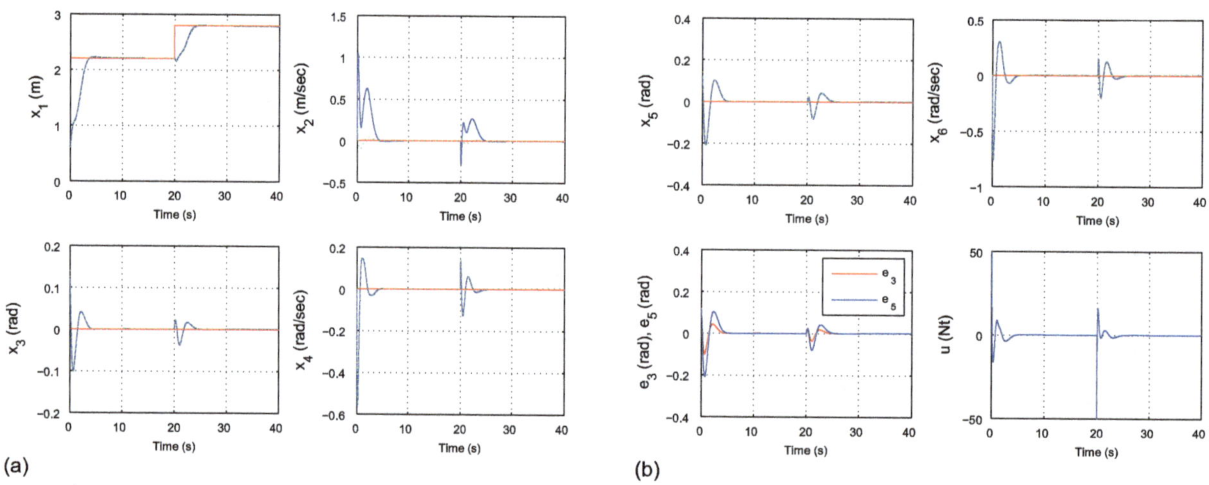

Figure 4.39 Tracking of setpoint 6 for the parallel double-inverted pendulum (cart with double-inverted pendulum—System 2 with six state variables): (a) convergence of state variable $x_1 = x$, $x_2 = \dot{x}$, $x_3 = \theta_1$, $x_4 = \dot{\theta}_1$ to their reference setpoints (red line: setpoint, blue line: real value, green line: estimated value); (b) variation of state variables $x_5 = \theta_2$ and $x_6 = \dot{\theta}_2$, variation of tracking errors e_1, e_3, e_5, and variations of the control input u (traction force on the cart)

series expansion and computation of the associated Jacobian matrices. The accuracy of tracking of setpoints by the state variables of the dynamic models of the pendulums confirms that the modeling error of the linearization process is small. (ii) Repetitive solution of the algebraic Riccati equation at each sampling instance so as to compute the gains

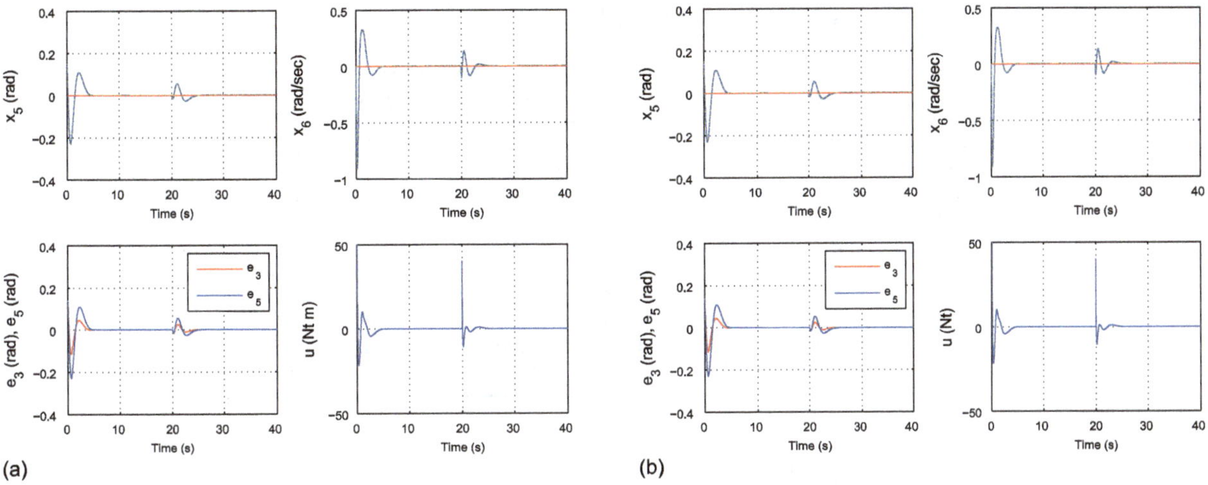

Figure 4.40 Tracking of setpoint 7 for the parallel double-inverted pendulum (cart with double-inverted pendulum—System 2 with six state variables): (a) convergence of state variable $x_1 = x$, $x_2 = \dot{x}$, $x_3 = \theta_1$, $x_4 = \dot{\theta}_1$ to their reference setpoints (red line: setpoint, blue line: real value, green line: estimated value); (b) variation of state variables $x_5 = \theta_2$ and $x_6 = \dot{\theta}_2$, variation of tracking errors e_1, e_3, e_5, and variations of the control input u (traction force on the cart)

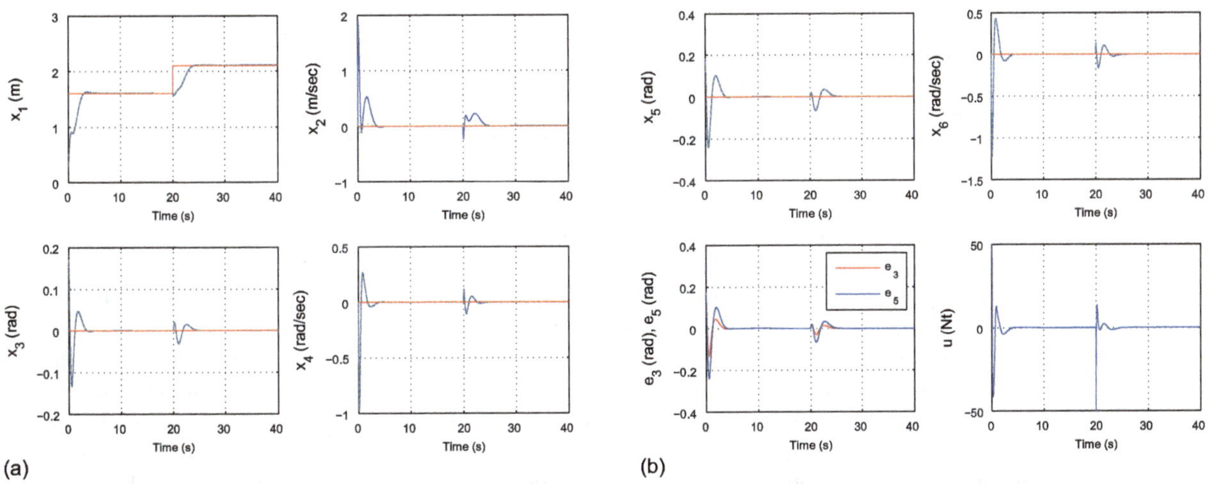

Figure 4.41 Tracking of setpoint 8 for the parallel double-inverted pendulum (cart with double-inverted pendulum—System 2 with six state variables): (a) convergence of state variable $x_1 = x$, $x_2 = \dot{x}$, $x_3 = \theta_1$, $x_4 = \dot{\theta}_1$ to their reference setpoints (red line: setpoint, blue line: real value, green line: estimated value); (b) variation of state variables $x_5 = \theta_2$ and $x_6 = \dot{\theta}_2$, variation of tracking errors e_1, e_3, e_5, and variations of the control input u (traction force on the cart)

of an H-infinity controller that stabilizes the control loop of each pendulum. The solution of the algebraic Riccati equation through the *aresolv()* function in MATLAB® is confirmed to be a computationally efficient procedure since it is completed within a time-interval which is significantly smaller than the sampling period. (iii) Use of the computed control signals on the initial nonlinear model of the pendulums and precise tracking of the targeted setpoints by the state

Table 4.6 Parallel double-inverted pendulum 2

Tracking RMSE $\times 10^{-3}$ in the disturbance-free case						
No test	$RMSE_{x_1}$	$RMSE_{x_2}$	$RMSE_{x_3}$	$RMSE_{x_4}$	$RMSE_{x_5}$	$RMSE_{x_6}$
$test_1$	0.4690	0.0454	0.0001	0.0001	0.0005	0.0001
$test_2$	0.2989	0.0286	0.0001	0.0001	0.0003	0.0001
$test_3$	0.6032	0.0472	0.0002	0.0001	0.0007	0.0001
$test_4$	0.4483	0.0368	0.0001	0.0001	0.0005	0.0001
$test_5$	0.4557	0.0374	0.0001	0.0001	0.0005	0.0001
$test_6$	0.4267	0.0353	0.0001	0.0001	0.0005	0.0001
$test_7$	0.2010	0.0167	0.0001	0.0001	0.0002	0.0001
$test_8$	0.3073	0.0185	0.0001	0.0001	0.0004	0.0001

Table 4.7 Parallel double-inverted pendulum 2

Tracking RMSE $\times 10^{-3}$ under disturbances						
$\Delta a\%$	$RMSE_{x_1}$	$RMSE_{x_2}$	$RMSE_{x_3}$	$RMSE_{x_4}$	$RMSE_{x_5}$	$RMSE_{x_6}$
0%	0.4690	0.0454	0.0001	0.0001	0.0005	0.0001
5%	0.4763	0.0464	0.0001	0.0001	0.0005	0.0001
10%	0.4856	0.0458	0.0001	0.0001	0.0006	0.0001
15%	0.4971	0.0463	0.0001	0.0001	0.0006	0.0001
20%	0.5108	0.0471	0.0002	0.0001	0.0007	0.0001
25%	0.5271	0.0485	0.0001	0.0001	0.0008	0.0001
30%	0.5460	0.0495	0.0001	0.0001	0.0009	0.0001

variables of the pendulums is achieved. The global stability properties of the control scheme are confirmed through the ability of the control method to make the state variables of the pendulums track precisely setpoints which are time-varying and abruptly changing. Thus, what is observed in the simulation experiments is convergence of state vector's tracking error to zero no matter what the initial conditions and the setpoints are.

To elaborate on nonlinear optimal control for the dynamic models of the third parallel double-inverted pendulums the following tables are given: (i) Table 4.8 providing results about the accuracy of tracking of setpoints by the state variables of the third pendulum with eight state variables and two control inputs under an exact dynamic model and (ii) Table 4.9 providing results about the accuracy of tracking of setpoints by the state variables of the third pendulum with eight state variables and two control inputs under model uncertainty (parametric change by $\Delta a\%$ in lengths h_i, $i = 1, 2$ denoting the distance of the end of the spring from the pivot point of each pole.

4.3 Nonlinear optimal control for the rotary double-inverted pendulum

4.3.1 Outline

Nonlinear control for underactuated robotic systems is an elaborated task which has become the subject of much research work during the last years [711,721,725,732]. In this domain, the double rotary pendulum (double Furuta pendulum) is a benchmark nonlinear dynamical system and the solution to the associated nonlinear optimal control problem is challenging [444,631,577,911,577]. Actually, stabilization of the dynamics of the double rotary pendulum is a nontrivial task because of the strong nonlinearities of the associated state-space model and because of the model's

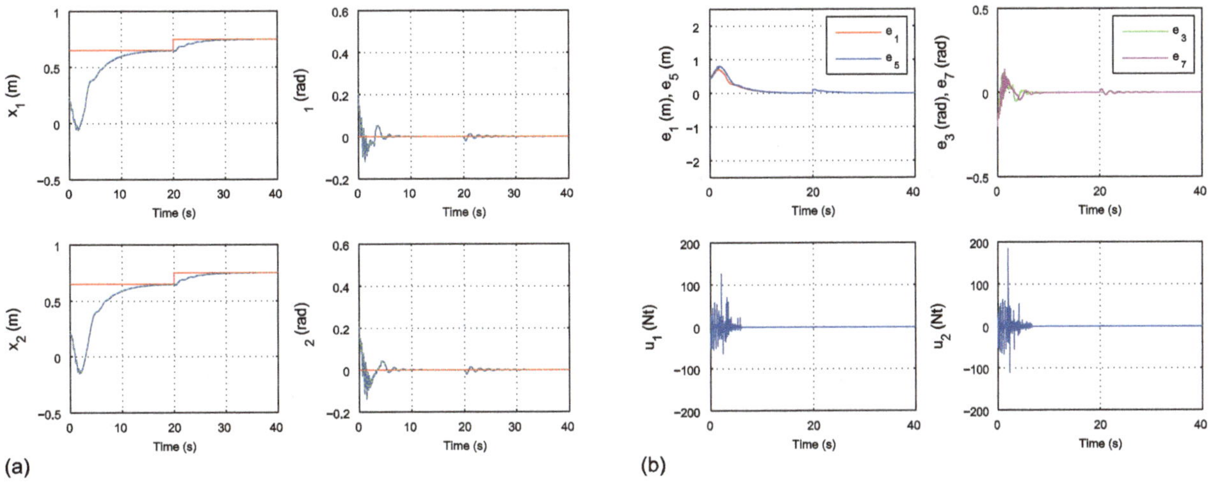

Figure 4.42 Tracking of setpoint 1 for the parallel double-inverted pendulum (System 3 with eight state variables): (a) convergence of state variable x_1 (position of cart 1), x_3 (turn angle of pendulum 1), x_5 (position of cart 2), x_7 (turn angle of pendulum 2) to their reference setpoints (red line: setpoint, blue line: real value, green line: estimated value); (b) variation of tracking errors e_1, e_3, e_5, and e_7 and variations of the control inputs u_1 and u_2 (traction forces on the carts)

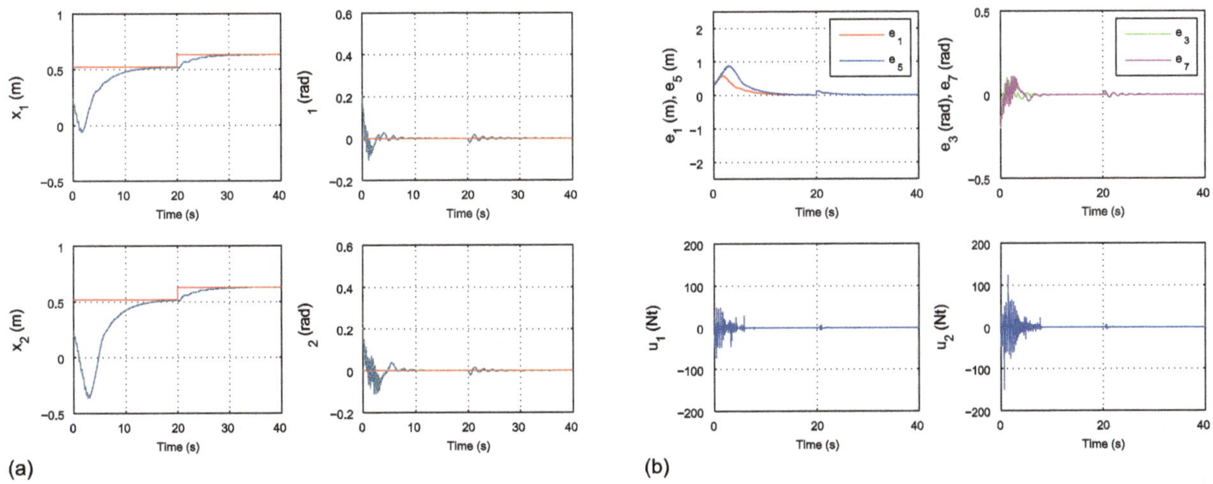

Figure 4.43 Tracking of setpoint 2 for the parallel double-inverted pendulum (System 3 with eight state variables): (a) convergence of state variable x_1 (position of cart 1), x_3 (turn angle of pendulum 1), x_5 (position of cart 2), x_7 (turn angle of pendulum 2) to their reference setpoints (red line: setpoint, blue line: real value, green line: estimated value); (b) variation of tracking errors e_1, e_3, e_5, and e_7 and variations of the control inputs u_1 and u_2 (traction forces on the carts)

underactuation [344,389,97]. Out of the three joints of the pendulum, only one receives mechanical actuation, and this makes significantly difficult the control of the pendulum and its stabilization at the swing-up position [449,1015,21]. So far, several control approaches have been proposed for the model of the double rotary pendulum. These include control based on LQR concepts and approaches based on Linear Parameter Varying formulations of the system's dynamics, as

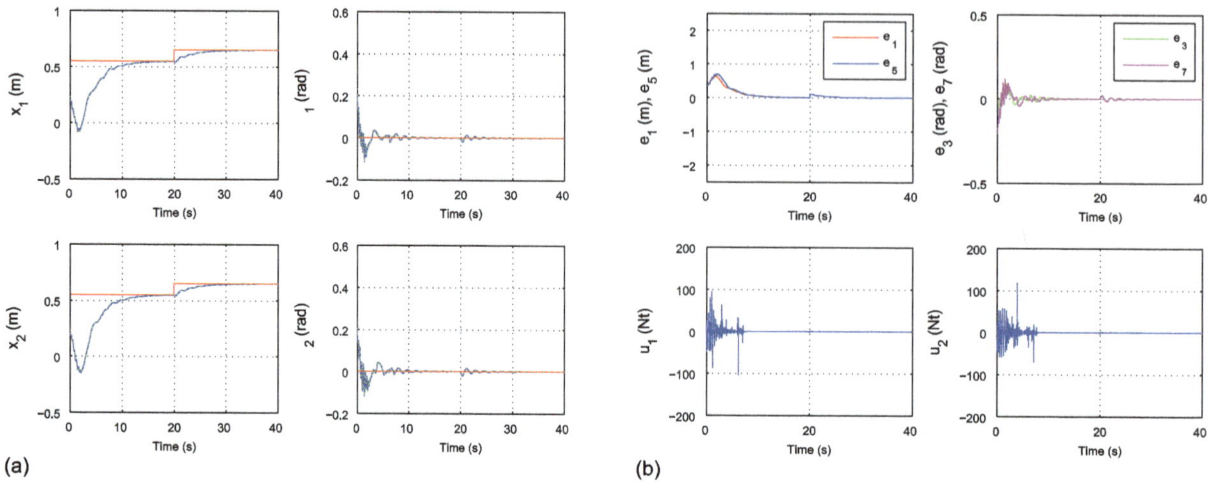

Figure 4.44 *Tracking of setpoint 3 for the parallel double-inverted pendulum (System 3 with eight state variables): (a) convergence of state variable x_1 (position of cart 1), x_3 (turn angle of pendulum 1), x_5 (position of cart 2), x_7 (turn angle of pendulum 2) to their reference setpoints (red line: setpoint, blue line: real value, green line: estimated value); (b) variation of tracking errors e_1, e_3, e_5, and e_7 and variations of the control inputs u_1 and u_2 (traction forces on the carts)*

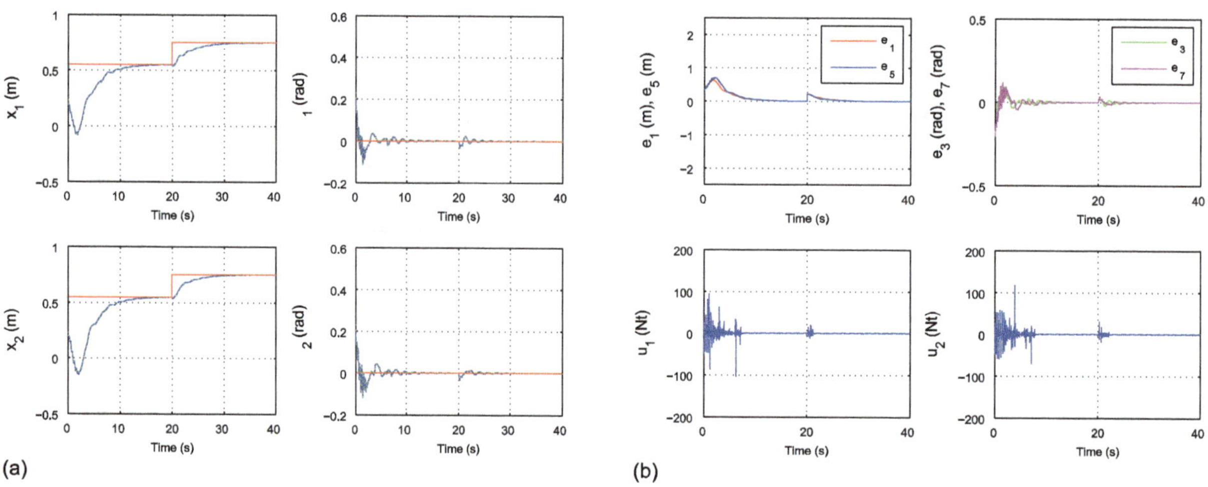

Figure 4.45 *Tracking of setpoint 4 for the parallel double-inverted pendulum (System 3 with eight state variables): (a) convergence of state variable x_1 (position of cart 1), x_3 (turn angle of pendulum 1), x_5 (position of cart 2), x_7 (turn angle of pendulum 2) to their reference setpoints (red line: setpoint, blue line: real value, green line: estimated value); (b) variation of tracking errors e_1, e_3, e_5, and e_7 and variations of the control inputs u_1 and u_2 (traction forces on the carts)*

well as on the solution of state-dependent Riccati equations [838,348,837,186,337]. One also finds sliding-mode and backstepping control approaches. Additionally, there are optimal control concepts which often take the form of Model Predictive Control schemes [664,665,394,586,1015]. It is noted that the solution of the nonlinear control problem for

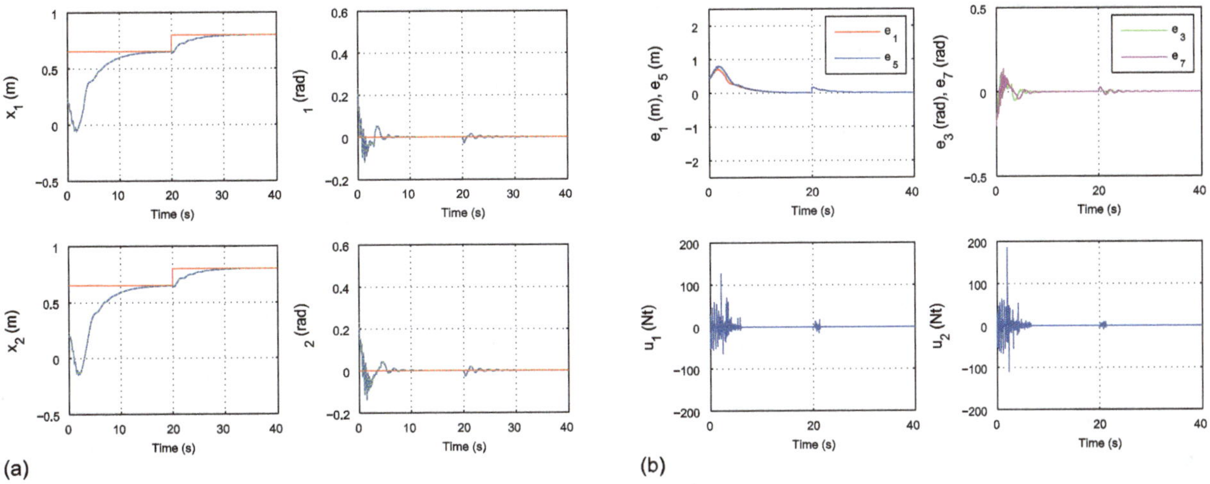

Figure 4.46 Tracking of setpoint 5 for the parallel double-inverted pendulum (System 3 with eight state variables): (a) convergence of state variable x_1 (position of cart 1), x_3 (turn angle of pendulum 1), x_5 (position of cart 2), x_7 (turn angle of pendulum 2) to their reference setpoints (red line: setpoint, blue line: real value, green line: estimated value); (b) variation of tracking errors e_1, e_3, e_5, and e_7 and variations of the control inputs u_1 and u_2 (traction forces on the carts)

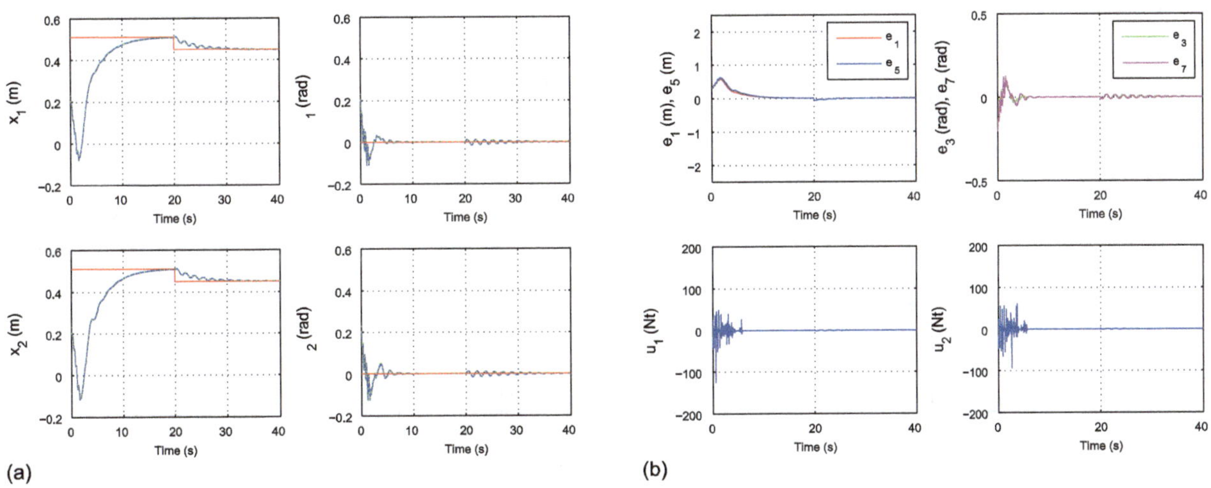

Figure 4.47 Tracking of setpoint 6 for the parallel double-inverted pendulum (System 3 with eight state variables): (a) convergence of state variable x_1 (position of cart 1), x_3 (turn angle of pendulum 1), x_5 (position of cart 2), x_7 (turn angle of pendulum 2) to their reference setpoints (red line: setpoint, blue line: real value, green line: estimated value); (b) variation of tracking errors e_1, e_3, e_5, and e_7 and variations of the control inputs u_1 and u_2 (traction forces on the carts)

the rotary double-inverted pendulum provides also insight for the successful implementation of stabilizing feedback in several highly nonlinear and underactuated robotic systems [257,16,32].

In this section, a novel nonlinear optimal (H-infinity) control method is proposed for the complex dynamics of the rotary double-inverted pendulum [719]. First, the dynamic model of the rotary double-inverted pendulum undergoes

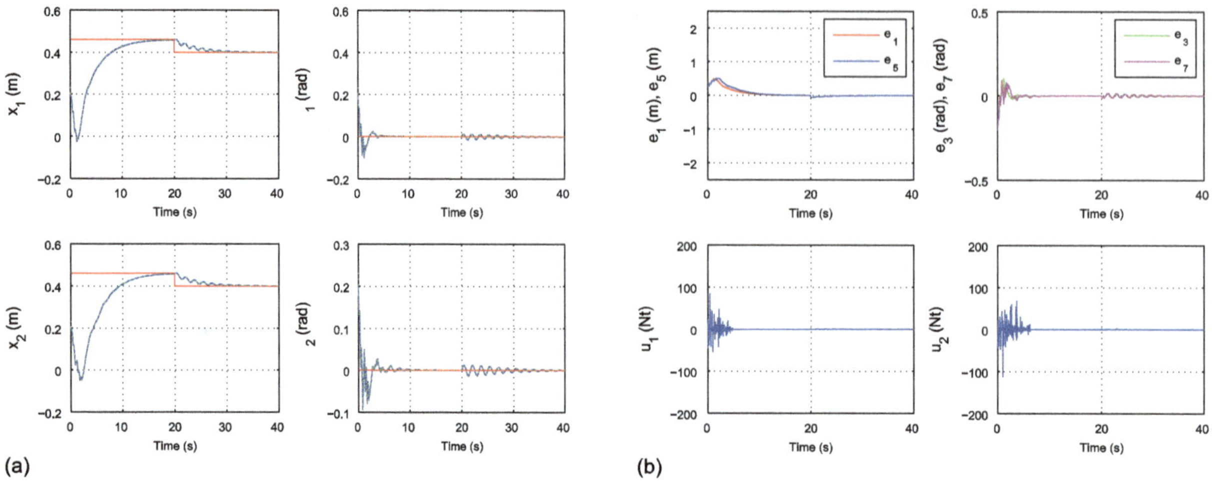

Figure 4.48 *Tracking of setpoint 7 for the parallel double-inverted pendulum (System 3 with eight state variables): (a) convergence of state variable x_1 (position of cart 1), x_3 (turn angle of pendulum 1), x_5 (position of cart 2), x_7 (turn angle of pendulum 2) to their reference setpoints (red line: setpoint, blue line: real value, green line: estimated value); (b) variation of tracking errors e_1, e_3, e_5, and e_7 and variations of the control inputs u_1 and u_2 (traction forces on the carts)*

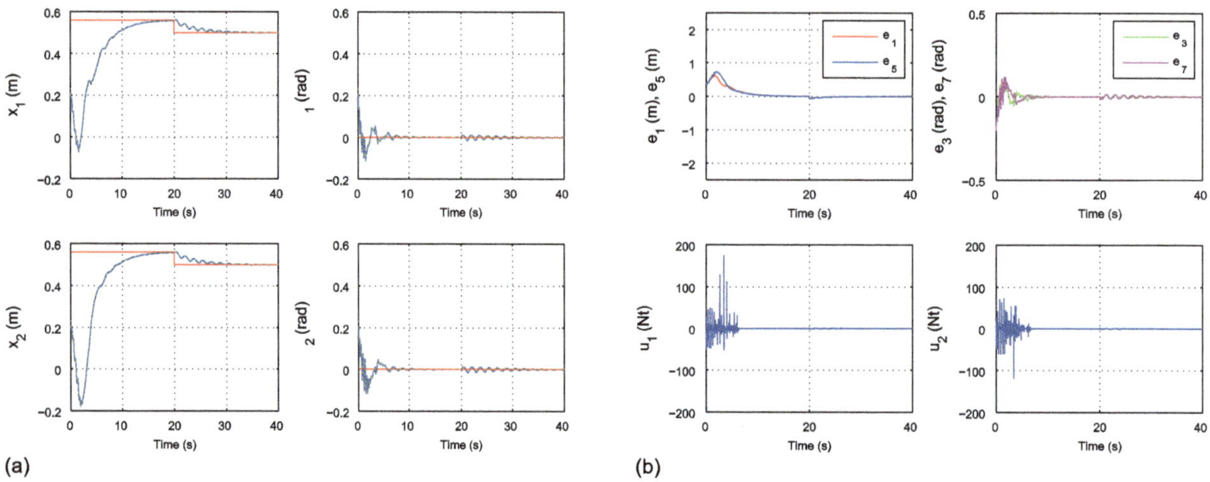

Figure 4.49 *Tracking of setpoint 8 for the parallel double-inverted pendulum (System 3 with eight state variables): (a) convergence of state variable x_1 (position of cart 1), x_3 (turn angle of pendulum 1), x_5 (position of cart 2), x_7 (turn angle of pendulum 2) to their reference setpoints (red line: setpoint, blue line: real value, green line: estimated value); (b) variation of tracking errors e_1, e_3, e_5, and e_7 and variations of the control inputs u_1 and u_2 (traction forces on the carts)*

approximate linearization around a temporary operating point which is updated at each sampling instance. This operating point is defined at each sampling period by the present value of the pendulum's state vector and by the last sampled value of the control inputs vector. The linearization process is based on first-order Taylor series expansion and on the computation of the associated Jacobian matrices [705,60,706]. The model imprecision which is due to the truncation

Table 4.8 Parallel double-inverted pendulum 3

Tracking RMSE $\times 10^{-3}$ in the disturbance-free case								
No test	**RMSE$_{x_1}$**	**RMSE$_{x_2}$**	**RMSE$_{x_3}$**	**RMSE$_{x_4}$**	**RMSE$_{x_5}$**	**RMSE$_{x_6}$**	**RMSE$_{x_7}$**	**RMSE$_{x_8}$**
$test_1$	0.0287	0.0230	0.0087	0.0347	0.0298	0.0289	0.0110	0.0399
$test_2$	0.0299	0.0299	0.0119	0.0452	0.0334	0.0270	0.0097	0.0353
$test_3$	0.0295	0.0227	0.0086	0.0342	0.0295	0.0298	0.0113	0.0419
$test_4$	0.0552	0.0240	0.0084	0.0307	0.0492	0.0240	0.0093	0.0357
$test_5$	0.0383	0.0281	0.0098	0.0368	0.0443	0.0290	0.0092	0.0353
$test_6$	0.0363	0.1127	0.0423	0.1683	0.0350	0.1207	0.0484	0.1774
$test_7$	0.0361	0.1120	0.0421	0.1672	0.0345	0.1187	0.0476	0.1746
$test_8$	0.0358	0.1113	0.0418	0.1661	0.0354	0.1220	0.0485	0.1794

Table 4.9 Parallel double-inverted pendulum 3

Tracking RMSE $\times 10^{-3}$ under disturbances								
Δa%	**RMSE$_{x_1}$**	**RMSE$_{x_2}$**	**RMSE$_{x_3}$**	**RMSE$_{x_4}$**	**RMSE$_{x_5}$**	**RMSE$_{x_6}$**	**RMSE$_{x_7}$**	**RMSE$_{x_8}$**
0%	0.0297	0.0230	0.0087	0.0347	0.0298	0.0289	0.0110	0.0399
5%	0.0287	0.0231	0.0088	0.0348	0.0299	0.0291	0.01100	0.0401
10%	0.0288	0.0228	0.0087	0.0344	0.0296	0.0291	0.0111	0.0402
15%	0.0289	0.0226	0.0086	0.0340	0.0299	0.0292	0.0111	0.0403
20%	0.0288	0.0227	0.0086	0.0342	0.0302	0.0289	0.0110	0.0399
25%	0.0288	0.0227	0.0086	0.0342	0.0300	0.0292	0.0111	0.0404
30%	0.0289	0.0229	0.0085	0.0338	0.0300	0.0294	0.0112	0.0402

of higher-order terms from the Taylor series expansion is treated as a perturbation which is asymptotically compensated by the robustness of the control algorithm. For the approximately linearized model of the rotary double-inverted pendulum a stabilizing H-infinity feedback controller is designed. This controller achieves the solution of the optimal control problem for the nonlinear dynamics of the rotary double-inverted pendulum under model uncertainty and external disturbances [721,732].

Actually, the H-infinity controller represents a min–max differential game which takes place between: (i) the system's control inputs that try to minimize a cost function which contains a quadratic term of the state vector's tracking error and (ii) the system's disturbances and model uncertainty terms which try to maximize this cost function. To select the feedback gains of the H-infinity controller an algebraic Riccati equation has to be repetitively solved at each time-step of the control algorithm [711,732]. The global stability properties of the control scheme are proven through Lyapunov analysis. First, it is demonstrated that the control loop of the rotary double-inverted pendulum satisfies the H-infinity tracking performance criterion [711,732,876]. This signifies elevated robustness against model uncertainty and exogenous perturbations. Moreover, under moderate conditions global asymptotic stability properties are proven [711,721,725,732]. To implement also state estimation-based control without the need to measure the entire state-vector of the rotary double-inverted pendulum, the H-infinity Kalman filter is proposed as a robust state observer. This section's nonlinear optimal (H-infinity) control scheme achieves fast and accurate tracking of reference setpoints under moderate variations of the control inputs. The control problem of rotary single and double-inverted pendulum can be used as a benchmark for testing linear and nonlinear control algorithms, and the results of the present section confirm that the proposed nonlinear optimal control method is a meaningful contribution to the research area of nonlinear control [243,321,638,587,177,264]. The use of this section's nonlinear optimal control method can be extended to more nonlinear underactuated dynamical systems which have been the subject of control systems research during the last years [379,595,596,711,1096].

4.3.2 Dynamic model of the double rotary pendulum

The diagram of the rotary double-inverted pendulum is shown in Figure 4.50. The state vector of the system consists of the turn angles of the rigid links and of the associated angular velocities, that is $x=[\theta_0,\dot{\theta}_0,\theta_1,\dot{\theta}_1,\theta_2,\dot{\theta}_2]^T$. The rotary double-inverted pendulum is underactuated since mechanical torque is applied only to the first link. Thus, the torques vector is $\tau=[T_1,0,0]^T$. Main parameters of the dynamic model of the rotary double-inverted pendulum are: $l_i,\ i=0,1,2$ which are the lengths of the links, $r_i,\ i=0,1,2$ which are the distances between the joint at the basis of the links and the center of gravity of the links and $m_i,\ i=0,1,2$ are the masses of the links [444,631,577,911,577]. The dynamic model of the system is given by

$$M(\theta)\ddot{\theta}+H(\theta,\dot{\theta})\dot{\theta}+G(\theta)=\tau \tag{4.226}$$

The inertia matrix of the rotary double-inverted pendulum is symmetric and positive definite and is given by [444]

$$M(\theta)=\begin{pmatrix} m_{11} & m_{12} & m_{13} \\ m_{21} & m_{22} & m_{33} \\ m_{31} & m_{32} & m_{33} \end{pmatrix} \tag{4.227}$$

where $m_{11}=J_0+(m_1+m_2)l_0^2+(m_1r_1^2+m_2l_1^2)sin^2(\theta_1)+2m_2l_1r_2sin(\theta_1)sin(\theta_1+\theta_2)+m_2r_2^2sin^2(\theta_1+\theta_2)$, $m_{12}=m_{21}=m_1l_0r_1cos(\theta_1)+m_2l_0r_2cos(\theta_1+\theta_2)+m_2l_0l_1cos(\theta_1), m_{13}=m_{31}=m_2l_0r_2cos(\theta_1+\theta_2), m_{22}=J_1+J_2+m_1r_1^2+m_2l_1^2+m_2r_2^2+2m_2l_1r_2^2+2m_2l_1r_2cos(\theta_2)$, $m_{23}=m_{32}=J_2+m_2r_2^2+m_2l_1r_2cos(\theta_2)$, $m_{33}=J_2+m_2r_2^2$.

The Coriolis forces vector of the system is given by $C(\theta,\dot{\theta})=H(\theta,\dot{\theta})\dot{\theta}$ where matrix $H(\theta,\dot{\theta})$ is given by [444]

$$H(\theta.\dot{\theta})=\begin{pmatrix} H_{11} & H_{12} & H_{13} \\ H_{21} & H_{22} & H_{23} \\ H_{31} & H_{32} & \bar{C}_2 \end{pmatrix} \tag{4.228}$$

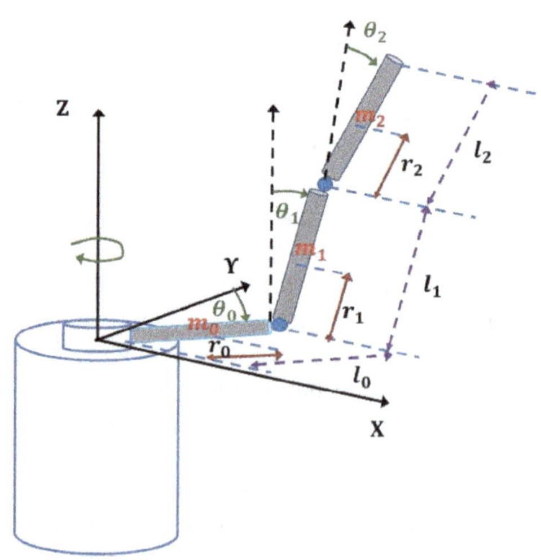

Figure 4.50 Diagram of the rotary double-inverted pendulum

where the elements of matrix $H(\theta,\dot{\theta})$ are [444]

$$\begin{aligned} H_{11} &= m_1 r_1^2 \dot{\theta}_1 sin(2\theta_1) + 2m_2 l_1^2 \dot{\theta}_1 cos(\theta_1) sin(\theta_1) + \\ &+2m_2 r_2^2 \dot{\theta}_1 cos(\theta_1+\theta_2) sin(\theta_1+\theta_2) + \\ &+2m_2 r_2^2 \dot{\theta}_0 cos(\theta_1+\theta_2) sin(\theta_1+\theta_2) + \\ &+2m_2 l_1 r_2 \dot{\theta}_1 cos(\theta_1) sin(\theta_1+\theta_2) + \\ &+2m_2 l_1 r_2 \dot{\theta}_1 cos(\theta_1+\theta_2) sin(\theta_1) + \\ &+2m_2 l_1 r_2 \dot{\theta}_2 cos(\theta_1+\theta_2) sin(\theta_1) + \bar{C}_0 \end{aligned} \tag{4.229}$$

$$H_{12} = -m_1 l_0 r_1 \dot{\theta}_1 sin(\theta_1) - m_2 l_0 l_1 \dot{\theta}_1 sin(\theta_1) - m_2 l_0 r_2 \dot{\theta}_1 sin(\theta_1+\theta_2) \tag{4.230}$$

$$H_{13} = -2m_{12} l_0 r_2 \dot{\theta}_1 sin(\theta_1+\theta_2) - m_2 l_0 r_2 \dot{\theta}_2 sin(\theta_1+\theta_2) \tag{4.231}$$

$$\begin{aligned} H_{21} &= -\tfrac{1}{2} m_1 r_1^2 \dot{\theta}_0 sin(2\theta_1) - 2m_1 r_2^2 \dot{\theta}_0 sin(2(\theta_1+\theta_2)) - \\ &-m_2 l_1^2 \dot{\theta}_0 cos(\theta_1) sin(\theta_1) - m_2 l_1 r_2 \dot{\theta}_0 sin(2\theta_1+\theta_2) \end{aligned} \tag{4.232}$$

$$H_{22} = -2m_2 l_1 r_2 \dot{\theta}_2 sin(\theta_2) + \bar{C}_1 \tag{4.233}$$

$$H_{23} = -m_2 l_1 r_2 \dot{\theta}_2 sin(\theta_2) \tag{4.234}$$

$$H_{31} = -\tfrac{1}{2} m_2 r_2^2 \dot{\theta}_0 sin(2(\theta_1+\theta_2)) - m_2 l_1 r_2 \dot{\theta}_0 sin(\theta_1) cos(\theta_1+\theta_2) \tag{4.235}$$

$$H_{32} = m_2 l_1 r_2 \dot{\theta}_1 sin(\theta_2) \tag{4.236}$$

In particular, about the elements of the Coriolis forces vector, it holds that

$$C(\theta,\dot{\theta}) = \begin{pmatrix} H_{11} & H_{12} & H_{13} \\ H_{21} & H_{22} & H_{23} \\ H_{31} & H_{32} & C_2 \end{pmatrix} \begin{pmatrix} \dot{\theta}_1 \\ \dot{\theta}_2 \\ \dot{\theta}_3 \end{pmatrix} \Rightarrow C(\theta,\dot{\theta}) = \begin{pmatrix} H_{11}\dot{\theta}_1 + H_{12}\dot{\theta}_2 + H_{13}\dot{\theta}_3 \\ H_{21}\dot{\theta}_1 + H_{22}\dot{\theta}_2 + H_{23}\dot{\theta}_3 \\ H_{31}\dot{\theta}_1 + H_{32}\dot{\theta}_2 + H_{33}\dot{\theta}_3 \end{pmatrix} \tag{4.237}$$

In the previous equations m_0, m_1, m_2 are the masses of the links, and J_0, J_1, and J_2 are moments of inertia. Besides, l_0, l_1, l_2 are the lengths of the links and r_0, r_1, and r_2 are the distance between the bases of the links and their centers of gravity. Moreover, $\bar{C}_0$, $\bar{C}_1$, $\bar{C}_2$ are constants. About the gravitational forces vector, one has [444]

$$G(\theta) = \begin{pmatrix} 0 & g_2(\theta) & g_3(\theta) \end{pmatrix}^T \tag{4.238}$$

where the elements of the gravitational forces vector are $g_2(\theta) = -g((m_1 r_1 + m_2 l_1) sin(\theta_1) + m_2 r_2 sin(\theta_1+\theta_2))$ and $g_3(\theta) = -m_2 g r_2 sin(\theta_1+\theta_2)$. Consequently, the dynamic model of the rotary double-inverted pendulum is written as

$$\begin{aligned} &M(\theta)\ddot{\theta} + H(\theta,\dot{\theta})\dot{\theta} + G(\theta) = \tau \Rightarrow \\ &\ddot{\theta} = -M^{-1}(\theta)[C(\theta,\dot{\theta}) + G(\theta)] + M^{-1}(\theta)\tau \end{aligned} \tag{4.239}$$

The inverse of the inertia matrix $M(\theta)$ is denoted as

$$M(\theta)^{-1} = \frac{1}{detM} \begin{pmatrix} M_{11} & -M_{21} & M_{31} \\ -M_{12} & M_{22} & -M_{32} \\ M_{13} & -M_{23} & M_{33} \end{pmatrix} \tag{4.240}$$

where the sub-determinants M_{ij} $i=1,2,3$ $j=1,2,3$ are defined as: $M_{11} = m_{22}m_{33} - m_{32}m_{23}$, $M_{12} = m_{21}m_{33} - m_{31}m_{23}$, $M_{13} = m_{21}m_{32} - m_{31}m_{22}$, $M_{21} = m_{12}m_{33} - m_{32}m_{13}$, $M_{22} = m_{11}m_{33} - m_{31}m_{13}$, $M_{23} = m_{11}m_{32} - m_{31}m_{12}$, $M_{23} = m_{11}m_{32} - m_{31}m_{12}$, $M_{31} = m_{12}m_{23} - m_{22}m_{13}$, $M_{32} = m_{11}m_{23} - m_{21}m_{13}$, $M_{33} = m_{11}m_{22} - m_{21}m_{12}$, and $detM = m_1 M_{11} - m_{12}M_{12} + m_{13}M_{13}$.

Using Eq. (4.239) and Eq. (4.240), the dynamic model of the rotary double-inverted pendulum becomes

$$\begin{pmatrix} \ddot{\theta}_0 \\ \ddot{\theta}_1 \\ \ddot{\theta}_2 \end{pmatrix} = -\frac{1}{detM} \begin{pmatrix} M_{11} & -M_{21} & M_{31} \\ -M_{12} & M_{22} & -M_{32} \\ M_{13} & -M_{23} & M_{33} \end{pmatrix} \begin{pmatrix} C_1(\theta,\dot{\theta}) + G_1(\theta) \\ C_2(\theta,\dot{\theta}) + G_2(\theta) \\ C_3(\theta,\dot{\theta}) + G_3(\theta) \end{pmatrix} + \frac{1}{detM} \begin{pmatrix} M_{11} & -M_{21} & M_{31} \\ -M_{12} & M_{22} & -M_{32} \\ M_{13} & -M_{23} & M_{33} \end{pmatrix} \begin{pmatrix} T_1 \\ 0 \\ 0 \end{pmatrix} \tag{4.241}$$

or equivalently

$$\begin{pmatrix} \ddot{\theta}_0 \\ \ddot{\theta}_1 \\ \ddot{\theta}_2 \end{pmatrix} = \begin{pmatrix} \frac{-M_{11}(C_1(\theta,\dot{\theta})+G_1(\theta))+M_{21}(C_2(\theta,\dot{\theta})+G_2(\theta))-M_{31}(C_3(\theta,\dot{\theta})+G_3(\theta))}{detM} \\ \frac{M_{12}(C_1(\theta,\dot{\theta})+G_1(\theta))-M_{22}(C_2(\theta,\dot{\theta})+G_2(\theta))+M_{32}(C_3(\theta,\dot{\theta})+G_3(\theta))}{detM} \\ \frac{-M_{13}(C_1(\theta,\dot{\theta})+G_1(\theta))+M_{23}(C_2(\theta,\dot{\theta})+G_2(\theta))-M_{33}(C_3(\theta,\dot{\theta})+G_3(\theta))}{detM} \end{pmatrix} + \begin{pmatrix} \frac{M_{11}}{detM} \\ -\frac{M_{12}}{detM} \\ \frac{M_{13}}{detM} \end{pmatrix} T_1 \tag{4.242}$$

Next, by denoting the state vector of the rotary double-inverted pendulum as $x = [x_1, x_2, x_3, x_4, x_5, x_6]^T$, that is $x = [\theta_0, \dot{\theta}_0, \theta_1, \dot{\theta}_1, \theta_2, \dot{\theta}_2]^T$ and the control inputs vector as $u = T_1$, one has the state-space description

$$\begin{pmatrix} \dot{x}_1 \\ \dot{x}_2 \\ \dot{x}_3 \\ \dot{x}_4 \\ \dot{x}_5 \\ \dot{x}_6 \end{pmatrix} = \begin{pmatrix} x_2 \\ f_2(x) \\ x_4 \\ f_4(x) \\ x_6 \\ f_6(x) \end{pmatrix} + \begin{pmatrix} 0 \\ g_2(x) \\ 0 \\ g_4(x) \\ 0 \\ g_6(x) \end{pmatrix} u \tag{4.243}$$

which is a model in the nonlinear affine-in-the-input state-space form

$$\dot{x} = f(x) + g(x)u \tag{4.244}$$

with $x \in R^{6\times 1}$, $f(x) \in R^{6\times 1}$, $g(x) \in R^{6\times 1}$, and $u \in R$. About the elements of vectors $f(x)$ and $g(x)$ which appear in this state-space model, it holds that:

$$\begin{aligned} f_1(x) &= x_2 & g_1(x) &= 0 \\ f_2(x) &= \frac{-M_{11}(C_1(\theta,\dot{\theta})+G_1(\theta))+M_{21}(C_2(\theta,\dot{\theta})+G_2(\theta))-M_{31}(C_3(\theta,\dot{\theta})+G_3(\theta))}{detM} & g_2(x) &= \frac{M_{11}}{detM} \\ f_3(x) &= x_4 & g_3(x) &= 0 \\ f_4(x) &= \frac{M_{12}(C_1(\theta,\dot{\theta})+G_1(\theta))-M_{22}(C_2(\theta,\dot{\theta})+G_2(\theta))+M_{32}(C_3(\theta,\dot{\theta})+G_3(\theta))}{detM} & g_4(x) &= -\frac{M_{12}}{detM} \\ f_5(x) &= x_6 & g_5(x) &= 0 \\ f_6(x) &= \frac{-M_{13}(C_1(\theta,\dot{\theta})+G_1(\theta))+M_{23}(C_2(\theta,\dot{\theta})+G_2(\theta))-M_{33}(C_3(\theta,\dot{\theta})+G_3(\theta))}{detM} & g_6(x) &= \frac{M_{13}}{detM} \end{aligned} \tag{4.245}$$

4.3.3 Approximate linearization of the rotary double-inverted pendulum

4.3.3.1 Linearization process

The dynamic model of the rotary double-inverted pendulum undergoes approximate linearization around the temporary operating point (x^*, u^*) at each sampling instance [721,732]. This operating point is defined by x^* which is the value of the state vector at the present sampling instance and by u^* which is the last sampled value of the control inputs vector. The linearization process is based on first-order Taylor series expansion and on the computation of the associated Jacobian matrices. It brings the system from the initial nonlinear state-space form:

$$\dot{x} = f(x) + g(x)u \tag{4.246}$$

into the equivalent linearized state-space description

$$\dot{x} = Ax + Bu + \tilde{d} \tag{4.247}$$

where $\tilde{d}$ is the cumulative vector of disturbances which may incorporate: (i) modeling errors due to the truncation of higher-order terms from the Taylor series expansion, (ii) exogenous perturbations, and (iii) sensor measurement noise of any distribution. The disturbances' effects will be compensated by the robustness of the nonlinear optimal (H-infinity) control algorithm. The computation of the Jacobian matrices of the system is as follows:

$$A = \nabla_x[f(x) + g(x)u] \mid_{(x^*,u^*)} \Rightarrow A = \nabla_x[f(x)] \mid_{(x^*,u^*)} + \nabla_x[g(x)]u \mid_{(x^*,u^*)} \tag{4.248}$$

$$B = \nabla_u[f(x) + g(x)u] \mid_{(x^*,u^*)} \Rightarrow B = g(x) \mid_{(x^*,u^*)} \tag{4.249}$$

This linearization approach which has been followed for implementing the nonlinear optimal control scheme results in a quite accurate model of the system's dynamics. Consider for instance the following affine-in-the-input state-space model:

$$\begin{gathered} \dot{x} = f(x) + g(x)u \Rightarrow \\ \dot{x} = [f(x^*) + \nabla_x f(x) \mid_{x^*} (x - x^*)] + [g(x^*) + \nabla_x g(x) \mid_{x^*} (x - x^*)]u^* + g(x^*)u^* + g(x^*)(u - u^*) + \tilde{d}_1 \Rightarrow \\ \dot{x} = [\nabla_x f(x) \mid_{x^*} + \nabla_x g(x) \mid_{x^*} u^*]x + g(x^*)u - [\nabla_x f(x) \mid_{x^*} + \nabla_x g(x) \mid_{x^*} u^*]x^* + f(x^*) + g(x^*)u^* + \tilde{d}_1 \end{gathered} \tag{4.250}$$

where $\tilde{d}_1$ is the modeling error due to truncation of higher order terms in the Taylor series expansion of $f(x)$ and $g(x)$. Next, by defining $A = [\nabla_x f(x) \mid_{x^*} + \nabla_x g(x) \mid_{x^*} u^*]$, $B = g(x^*)$, one obtains

$$\dot{x} = Ax + Bu - Ax^* + f(x^*) + g(x^*)u^* + \tilde{d}_1 \tag{4.251}$$

Moreover, by denoting $\tilde{d} = -Ax^* + f(x^*) + g(x^*)u^* + \tilde{d}_1$ about the cumulative modeling error term in the Taylor series expansion procedure, one has

$$\dot{x} = Ax + Bu + \tilde{d} \tag{4.252}$$

which is the approximately linearized model of the dynamics of the system of Eq. (4.247). The term $f(x^*) + g(x^*)u^*$ is the derivative of the state vector at (x^*, u^*) which is almost annihilated by $-Ax^*$.

4.3.3.2 Computation of Jacobian matrices

Computation of the elements of the Jacobian matrix $\nabla_x[f(x)] \mid_{(x^*,u^*)}$ is first carried out.

First row of the Jacobian matrix $\nabla_x[f(x)] \mid_{(x^*,u^*)}$: $\frac{\partial f_1(x)}{\partial x_1} = 0$, $\frac{\partial f_1(x)}{\partial x_2} = 1$, $\frac{\partial f_1(x)}{\partial x_3} = 0$, $\frac{\partial f_1(x)}{\partial x_4} = 0$, $\frac{\partial f_1(x)}{\partial x_5} = 0$, and $\frac{\partial f_1(x)}{\partial x_6} = 0$.

Second row of the Jacobian matrix $\nabla_x[f(x)] \mid_{(x^*,u^*)}$: It holds that $f_2(x) = \frac{f_{2,num}(x)}{f_{2,den}(x)}$ where $f_{2,num}(x) = -M_{11}(C_1(\theta, \dot{\theta}) + G_1(\theta)) + M_{21}(C_2)(\theta, \dot{\theta} + G_2(\theta)) - M_{31}(C_3(\theta, \dot{\theta}) + G_3(\theta))$ and $f_{2,den}(x) = detM$. Moreover, one has that for $i = 1, 2, \cdots, 6$

$$\frac{\partial f_2(x)}{\partial x_i} = \frac{\frac{\partial f_{2,num}(x)}{\partial x_i} f_{2,den}(x) - f_{2,num}(x) \frac{\partial f_{2,den}(x)}{\partial x_i}}{f_{2,den}(x)^2} \tag{4.253}$$

Moreover, it holds that for $i = 1, 2, \cdots, 6$

$$\begin{aligned} \frac{\partial f_{2,num}(x)}{\partial x_i} = & -\frac{\partial M_{11}}{\partial x_i}(C_1(\theta, \dot{\theta}) + G_1(\theta)) - M_{11}\left(\frac{\partial C_1(\theta, \dot{\theta})}{\partial x_i} + \frac{\partial G_1(\theta)}{\partial x_i}\right) + \\ & + \frac{\partial M_{21}}{\partial x_i}(C_2(\theta, \dot{\theta}) + G_2(\theta)) + M_{21}\left(\frac{\partial C_2(\theta, \dot{\theta})}{\partial x_i} + \frac{\partial G_2(\theta)}{\partial x_i}\right) - \\ & - \frac{\partial M_{31}}{\partial x_i}(C_3(\theta, \dot{\theta}) + G_3(\theta)) - M_{31}\left(\frac{\partial C_3(\theta, \dot{\theta})}{\partial x_i} + \frac{\partial G_3(\theta)}{\partial x_i}\right) \end{aligned} \tag{4.254}$$

$$\frac{\partial f_{2,den}(x)}{\partial x_i} = \frac{\partial detM}{\partial x_i} \tag{4.255}$$

Third row of the Jacobian matrix $\nabla_x[f(x)]\mid_{(x^*,u^*)}$: $\frac{\partial f_3(x)}{\partial x_1}=0$, $\frac{\partial f_3(x)}{\partial x_2}=0$, $\frac{\partial f_3(x)}{\partial x_3}=0$, $\frac{\partial f_3(x)}{\partial x_4}=1$, $\frac{\partial f_3(x)}{\partial x_5}=0$, and $\frac{\partial f_3(x)}{\partial x_6}=0$.

Fourth row of the Jacobian matrix $\nabla_x[f(x)]\mid_{(x^*,u^*)}$: It holds that $f_4(x)=\frac{f_{4,num}(x)}{f_{4,den}(x)}$ where $f_{4,num}(x)=M_{12}(C_1(\theta,\dot{\theta})+G_1(\theta))-M_{22}(C_2)(\theta,\dot{\theta}+G_2(\theta))+M_{32}(C_3(\theta,\dot{\theta})+G_3(\theta))$ and $f_{4,den}(x)=detM$. Moreover, one has that for $i=1,2,\cdots,6$

$$\frac{\partial f_4(x)}{\partial x_i}=\frac{\frac{\partial f_{4,num}(x)}{\partial x_i}f_{4,den}(x)-f_{4,num}(x)\frac{\partial f_{4,den}(x)}{\partial x_i}}{f_{4,den}(x)^2} \tag{4.256}$$

Moreover, it holds that for $i=1,2,\cdots,6$

$$\begin{aligned}\frac{\partial f_{4,num}(x)}{\partial x_i}&=\frac{\partial M_{12}}{\partial x_i}(C_1(\theta,\dot{\theta})+G_1(\theta))+M_{13}\left(\frac{\partial C_1(\theta,\dot{\theta})}{\partial x_i}+\frac{\partial G_1(\theta)}{\partial x_i}\right)-\\&-\frac{\partial M_{22}}{\partial x_i}(C_2(\theta,\dot{\theta})+G_2(\theta))-M_{22}\left(\frac{\partial C_2(\theta,\dot{\theta})}{\partial x_i}+\frac{\partial G_2(\theta)}{\partial x_i}\right)+\\&+\frac{\partial M_{32}}{\partial x_i}(C_3(\theta,\dot{\theta})+G_3(\theta))+M_{32}\left(\frac{\partial C_3(\theta,\dot{\theta})}{\partial x_i}+\frac{\partial G_3(\theta)}{\partial x_i}\right)\end{aligned} \tag{4.257}$$

$$\frac{\partial f_{4,den}(x)}{\partial x_i}=\frac{\partial detM}{\partial x_i} \tag{4.258}$$

Fifth row of the Jacobian matrix $\nabla_x[f(x)]\mid_{(x^*,u^*)}$: $\frac{\partial f_5(x)}{\partial x_1}=0$, $\frac{\partial f_5(x)}{\partial x_2}=0$, $\frac{\partial f_5(x)}{\partial x_3}=0$, $\frac{\partial f_5(x)}{\partial x_4}=0$, $\frac{\partial f_5(x)}{\partial x_5}=0$, and $\frac{\partial f_5(x)}{\partial x_6}=0$.

Sixth row of the Jacobian matrix $\nabla_x[f(x)]\mid_{(x^*,u^*)}$: It holds that $f_6(x)=\frac{f_{6,num}(x)}{f_{6,den}(x)}$ where $f_{6,num}(x)=-M_{13}(C_1(\theta,\dot{\theta})+G_1(\theta))+M_{23}(C_2)(\theta,\dot{\theta}+G_2(\theta))-M_{33}(C_3(\theta,\dot{\theta})+G_3(\theta))$ and $f_{6,den}(x)=detM$. Moreover, one has that for $i=1,2,\cdots,6$

$$\frac{\partial f_6(x)}{\partial x_i}=\frac{\frac{\partial f_{6,num}(x)}{\partial x_i}f_{6,den}(x)-f_{6,num}(x)\frac{\partial f_{6,den}(x)}{\partial x_i}}{f_{6,den}(x)^2} \tag{4.259}$$

Moreover, it holds that for $i=1,2,\cdots,6$

$$\begin{aligned}\frac{\partial f_{6,num}(x)}{\partial x_i}&=-\frac{\partial M_{13}}{\partial x_i}(C_1(\theta,\dot{\theta})+G_1(\theta))-M_{13}\left(\frac{\partial C_1(\theta,\dot{\theta})}{\partial x_i}+\frac{\partial G_1(\theta)}{\partial x_i}\right)+\\&+\frac{\partial M_{23}}{\partial x_i}(C_2(\theta,\dot{\theta})+G_2(\theta))+M_{23}\left(\frac{\partial C_2(\theta,\dot{\theta})}{\partial x_i}+\frac{\partial G_2(\theta)}{\partial x_i}\right)-\\&-\frac{\partial M_{33}}{\partial x_i}(C_3(\theta,\dot{\theta})+G_3(\theta))-M_{33}\left(\frac{\partial C_3(\theta,\dot{\theta})}{\partial x_i}+\frac{\partial G_3(\theta)}{\partial x_i}\right)\end{aligned} \tag{4.260}$$

$$\frac{\partial f_{6,den}(x)}{\partial x_i}=\frac{\partial detM}{\partial x_i} \tag{4.261}$$

Computation of the elements of the Jacobian matrix $\nabla_x[g(x)]\mid_{(x^*,u^*)}$ is also carried out.

First row of the Jacobian matrix $\nabla_x[g(x)]\mid_{(x^*,u^*)}$: $\frac{\partial g_1(x)}{\partial x_i}=0$ for $i=1,2,\cdots,6$.

Second row of the Jacobian matrix $\nabla_x[g(x)]\mid_{(x^*,u^*)}$: It holds that for $i=1,2,\cdots,6$

$$\frac{\partial g_{21}}{\partial x_i}=\frac{\frac{dM_{11}}{dx_i}detM-M_{11}\frac{ddetM}{dx_i}}{detM^2} \tag{4.262}$$

Third row of the Jacobian matrix $\nabla_x[g(x)]\mid_{(x^*,u^*)}$: $\frac{\partial g_3(x)}{\partial x_i}=0$ for $i=1,2,\cdots,6$.

Fourth row of the Jacobian matrix $\nabla_x[g(x)]\mid_{(x^*,u^*)}$: It holds that for $i=1,2,\cdots,6$

$$\frac{\partial g_{41}}{\partial x_i}=\frac{-\frac{dM_{12}}{dx_i}detM+M_{11}\frac{ddetM}{dx_i}}{detM^2} \tag{4.263}$$

Fifth row of the Jacobian matrix $\nabla_x[g(x)]\mid_{(x^*,u^*)}$: $\frac{\partial g_3(x)}{\partial x_i}=0$ for $i=1,2,\cdots,6$.

Sixth row of the Jacobian matrix $\nabla_x[g(x)]\mid_{(x^*,u^*)}$: It holds that for $i=1,2,\cdots,6$

$$\frac{\partial g_{61}}{\partial x_i}=\frac{\frac{dM_{13}}{dx_i}detM-M_{13}\frac{ddetM}{dx_i}}{detM^2} \tag{4.264}$$

Besides, the partial derivatives of the sub-determinants of the inertial matrix M are computed as follows: For $i=1,2,\cdots,6$

$$\frac{\partial M_{11}}{\partial x_i}=\frac{\partial m_{22}}{\partial x_i}m_{33}+m_{22}\frac{\partial m_{33}}{\partial x_i}-\frac{\partial m_{32}}{\partial x_i}m_{23}-m_{32}\frac{\partial m_{23}}{\partial x_i} \tag{4.265}$$

$$\frac{\partial M_{12}}{\partial x_i}=\frac{\partial m_{21}}{\partial x_i}m_{33}+m_{21}\frac{\partial m_{33}}{\partial x_i}-\frac{\partial m_{31}}{\partial x_i}m_{23}-m_{31}\frac{\partial m_{23}}{\partial x_i} \tag{4.266}$$

$$\frac{\partial M_{13}}{\partial x_i}=\frac{\partial m_{21}}{\partial x_i}m_{32}+m_{21}\frac{\partial m_{32}}{\partial x_i}-\frac{\partial m_{31}}{\partial x_i}m_{22}-m_{31}\frac{\partial m_{22}}{\partial x_i} \tag{4.267}$$

$$\frac{\partial M_{21}}{\partial x_i}=\frac{\partial m_{12}}{\partial x_i}m_{33}+m_{12}\frac{\partial m_{33}}{\partial x_i}-\frac{\partial m_{32}}{\partial x_i}m_{13}-m_{32}\frac{\partial m_{13}}{\partial x_i} \tag{4.268}$$

$$\frac{\partial M_{22}}{\partial x_i}=\frac{\partial m_{11}}{\partial x_i}m_{33}+m_{11}\frac{\partial m_{33}}{\partial x_i}-\frac{\partial m_{31}}{\partial x_i}m_{13}-m_{31}\frac{\partial m_{13}}{\partial x_i} \tag{4.269}$$

$$\frac{\partial M_{23}}{\partial x_i}=\frac{\partial m_{11}}{\partial x_i}m_{32}+m_{11}\frac{\partial m_{32}}{\partial x_i}-\frac{\partial m_{31}}{\partial x_i}m_{13}-m_{31}\frac{\partial m_{13}}{\partial x_i} \tag{4.270}$$

$$\frac{\partial M_{31}}{\partial x_i}=\frac{\partial m_{12}}{\partial x_i}m_{23}+m_{12}\frac{\partial m_{23}}{\partial x_i}-\frac{\partial m_{22}}{\partial x_i}m_{13}-m_{22}\frac{\partial m_{13}}{\partial x_i} \tag{4.271}$$

$$\frac{\partial M_{32}}{\partial x_i}=\frac{\partial m_{11}}{\partial x_i}m_{23}+m_{11}\frac{\partial m_{23}}{\partial x_i}-\frac{\partial m_{21}}{\partial x_i}m_{13}-m_{21}\frac{\partial m_{13}}{\partial x_i} \tag{4.272}$$

$$\frac{\partial M_{33}}{\partial x_i}=\frac{\partial m_{11}}{\partial x_i}m_{22}+m_{11}\frac{\partial m_{22}}{\partial x_i}-\frac{\partial m_{21}}{\partial x_i}m_{12}-m_{21}\frac{\partial m_{12}}{\partial x_i} \tag{4.273}$$

About the partial derivative of the determinant of the inertia matrix M, it holds that

$$\begin{aligned}\frac{\partial detM}{\partial x_i}&=\frac{\partial m_{11}}{\partial x_i}M_{11}+m_{11}\frac{\partial M_{11}}{\partial x_i}-\\&-\frac{\partial m_{12}}{\partial x_i}M_{12}+m_{12}\frac{\partial M_{12}}{\partial x_i}+\\&+\frac{\partial m_{13}}{\partial x_i}M_{13}+m_{13}\frac{\partial M_{13}}{\partial x_i}\end{aligned} \tag{4.274}$$

Moreover, about the partial derivatives m_{ij}, $i=1,2,3$, and $j=1,2,3$ which appear in the inertia matrix it holds that:

$\frac{\partial m_{11}}{\partial x_1}=0$, $\frac{\partial m_{11}}{\partial x_2}=0$, $\frac{\partial m_{11}}{\partial x_3}=(m_1r_1^2+m_2l_2^2)2sin(x_3)cos(x_3)+2m_2l_1r_2[cos(x_3)sin(x_3+x_5)+sin(x_3)cos(x_3+x_5)]+m_2r_3^22sin(x_3+x_5)cos(x_3+x_5)$, $\frac{\partial m_{11}}{\partial x_4}=0$, $\frac{\partial m_{11}}{\partial x_5}=2m_2l_1r_2sin(x_3)cos(x_3+x_5)+m_2r_2^22sin(x_3+x_5)cos(x_3+x_5)$, $\frac{\partial m_{11}}{\partial x_6}=0$.

$\frac{\partial m_{12}}{\partial x_1}=0$, $\frac{\partial m_{12}}{\partial x_2}=0$, $\frac{\partial m_{12}}{\partial x_3}=-m_1l_0r_1sin(x_3)-m_2l_0r_2sin(x_3+x_5)-m_2l_0l_1sin(x_3)$, $\frac{\partial m_{12}}{\partial x_4}=0$, $\frac{\partial m_{12}}{\partial x_5}=-m_2l_0r_2sin(x_3+x_5)$, $\frac{\partial m_{12}}{\partial x_6}=0$.

$\frac{\partial m_{13}}{\partial x_1}=0$, $\frac{\partial m_{13}}{\partial x_2}=0$, $\frac{\partial m_{13}}{\partial x_3}=-m_21l_0r_2sin(x_3+x_5)$, $\frac{\partial m_{13}}{\partial x_4}=0$, $\frac{\partial m_{13}}{\partial x_5}=-m_2l_0r_2sin(x_3+x_5)$, $\frac{\partial m_{13}}{\partial x_6}=0$.

$\frac{\partial m_{21}}{\partial x_1}=0$, $\frac{\partial m_{21}}{\partial x_2}=0$, $\frac{\partial m_{21}}{\partial x_3}=-m_1l_0r_1sin(x_3)-m_2l_0r_2sin(x_3+x_5)-m_2l_0l_1sin(x_3)$, $\frac{\partial m_{21}}{\partial x_4}=0$, $\frac{\partial m_{21}}{\partial x_5}=-m_2l_0r_2sin(x_3+x_5)$, $\frac{\partial m_{21}}{\partial x_6}=0$.

$\frac{\partial m_{22}}{\partial x_1}=0$, $\frac{\partial m_{22}}{\partial x_2}=0$, $\frac{\partial m_{22}}{\partial x_3}=0$, $\frac{\partial m_{22}}{\partial x_4}=0$, $\frac{\partial m_{22}}{\partial x_5}=-m_2l_1r_2sin(x_5)$, $\frac{\partial m_{22}}{\partial x_6}=0$.

$\frac{\partial m_{23}}{\partial x_1}=0$, $\frac{\partial m_{23}}{\partial x_2}=0$, $\frac{\partial m_{23}}{\partial x_3}=0$, $\frac{\partial m_{23}}{\partial x_4}=0$, $\frac{\partial m_{23}}{\partial x_5}=-m_2l_1r_2sin(x_5)$, $\frac{\partial m_{23}}{\partial x_6}=0$.

$\frac{\partial m_{31}}{\partial x_1}=0$, $\frac{\partial m_{31}}{\partial x_2}=0$, $\frac{\partial m_{31}}{\partial x_3}=-m_21l_0r_2sin(x_3+x_5)$, $\frac{\partial m_{31}}{\partial x_4}=0$, $\frac{\partial m_{31}}{\partial x_5}=-m_2l_0r_2sin(x_3+x_5)$, $\frac{\partial m_{31}}{\partial x_6}=0$.

$\frac{\partial m_{32}}{\partial x_1}=0$, $\frac{\partial m_{32}}{\partial x_2}=0$, $\frac{\partial m_{32}}{\partial x_3}=0$, $\frac{\partial m_{32}}{\partial x_4}=0$, $\frac{\partial m_{32}}{\partial x_5}=-m_2l_1r_2sin(x_5)$, $\frac{\partial m_{32}}{\partial x_6}=0$.

$\frac{\partial m_{33}}{\partial x_1}=0$, $\frac{\partial m_{33}}{\partial x_2}=0$, $\frac{\partial m_{33}}{\partial x_3}=0$, $\frac{\partial m_{33}}{\partial x_4}=0$, $\frac{\partial m_{33}}{\partial x_5}=0$, $\frac{\partial m_{33}}{\partial x_6}=0$.

Next, the partial derivatives of the elements of the Coriolis and centrifugal forces matrix $C(\theta,\dot{\theta})$ are computed. It holds that $C_1(\theta,\dot{\theta})=H_{11}x_2+H_{12}x_4+H_{13}x_6$, $C_2(\theta,\dot{\theta})=H_{21}x_2+H_{22}x_4+H_{23}x_6$, and $C_3(\theta,\dot{\theta})=H_{31}x_2+H_{32}x_4+H_{33}x_6$. Thus:

$\frac{\partial C_1}{\partial x_1}=\frac{\partial H_{11}}{\partial x_1}x_2+\frac{\partial H_{12}}{\partial x_1}x_4+\frac{\partial H_{13}}{\partial x_1}x_6$, $\frac{\partial C_1}{\partial x_2}=[\frac{\partial H_{11}}{\partial x_2}x_2+H_{11}]+\frac{\partial H_{12}}{\partial x_2}x_4+\frac{\partial H_{13}}{\partial x_2}x_6$, $\frac{\partial C_1}{\partial x_3}=\frac{\partial H_{11}}{\partial x_1}x_3+\frac{\partial H_{12}}{\partial x_3}x_4+\frac{\partial H_{13}}{\partial x_3}x_6$, $\frac{\partial C_1}{\partial x_4}=\frac{\partial H_{11}}{\partial x_4}x_2+[\frac{\partial H_{12}}{\partial x_4}x_4+H_{12}]+\frac{\partial H_{13}}{\partial x_4}x_6$, $\frac{\partial C_1}{\partial x_5}=\frac{\partial H_{11}}{\partial x_5}x_2+\frac{\partial H_{12}}{\partial x_5}x_4+\frac{\partial H_{13}}{\partial x_5}x_6$, $\frac{\partial C_1}{\partial x_6}=\frac{\partial H_{11}}{\partial x_6}x_2+\frac{\partial H_{12}}{\partial x_6}x_4+[\frac{\partial H_{13}}{\partial x_1}x_6+H_{13}]$.

$\frac{\partial C_2}{\partial x_1}=\frac{\partial H_{21}}{\partial x_1}x_2+\frac{\partial H_{22}}{\partial x_1}x_4+\frac{\partial H_{23}}{\partial x_1}x_6$, $\frac{\partial C_2}{\partial x_2}=[\frac{\partial H_{21}}{\partial x_2}x_2+H_{21}]+\frac{\partial H_{22}}{\partial x_2}x_4+\frac{\partial H_{23}}{\partial x_2}x_6$, $\frac{\partial C_2}{\partial x_3}=\frac{\partial H_{21}}{\partial x_1}x_3+\frac{\partial H_{22}}{\partial x_3}x_4+\frac{\partial H_{23}}{\partial x_3}x_6$, $\frac{\partial C_2}{\partial x_4}=\frac{\partial H_{21}}{\partial x_4}x_2+[\frac{\partial H_{22}}{\partial x_4}x_4+H_{22}]+\frac{\partial H_{23}}{\partial x_4}x_6$, $\frac{\partial C_2}{\partial x_5}=\frac{\partial H_{21}}{\partial x_5}x_2+\frac{\partial H_{22}}{\partial x_5}x_4+\frac{\partial H_{23}}{\partial x_5}x_6$, $\frac{\partial C_2}{\partial x_6}=\frac{\partial H_{21}}{\partial x_6}x_2+\frac{\partial H_{22}}{\partial x_6}x_4+[\frac{\partial H_{23}}{\partial x_1}x_6+H_{23}]$.

$\frac{\partial C_3}{\partial x_1}=\frac{\partial H_{31}}{\partial x_1}x_2+\frac{\partial H_{32}}{\partial x_1}x_4+\frac{\partial H_{33}}{\partial x_1}x_6$, $\frac{\partial C_3}{\partial x_2}=[\frac{\partial H_{31}}{\partial x_2}x_2+H_{31}]+\frac{\partial H_{32}}{\partial x_2}x_4+\frac{\partial H_{33}}{\partial x_2}x_6$, $\frac{\partial C_3}{\partial x_3}=\frac{\partial H_{31}}{\partial x_1}x_3+\frac{\partial H_{32}}{\partial x_3}x_4+\frac{\partial H_{33}}{\partial x_3}x_6$, $\frac{\partial C_3}{\partial x_4}=\frac{\partial H_{31}}{\partial x_4}x_2+[\frac{\partial H_{32}}{\partial x_4}x_4+H_{32}]+\frac{\partial H_{33}}{\partial x_4}x_6$, $\frac{\partial C_2}{\partial x_5}=\frac{\partial H_{21}}{\partial x_5}x_2+\frac{\partial H_{22}}{\partial x_5}x_4+\frac{\partial H_{23}}{\partial x_5}x_6$, $\frac{\partial C_3}{\partial x_6}=\frac{\partial H_{31}}{\partial x_6}x_2+\frac{\partial H_{32}}{\partial x_6}x_4+[\frac{\partial H_{33}}{\partial x_1}x_6+H_{33}]$.

About the partial derivatives of the terms H_{ij}, $i=1,2,3$, and $j=1,2,3$, one has that $\frac{\partial H_{11}}{\partial x_1}=0$, $\frac{\partial H_{11}}{\partial x_2}=0$

$$\begin{aligned}\frac{\partial H_{11}}{\partial x_3}&=m_1r_1^2x_4^22cos(2x_3)+2m_2l_1^2x_4[cos^2(x_3)-sin^2(x_3)]+\\&+2m_2r_2^2x_4[cos^2(x_3+x_5)-sin^2(x_3+x_5)]+\\&+2m_2r_2^2x_6cos(x_3+x_5)sin(x_3+x_5)+\\&+2m_2l_1r_2x_4[cos(x_3)cos(x_3+x_5)-sin(x_3)sin(x_3+x_5)]+\\&+2m_2l_1r_2x_4[cos(x_3)cos(x_3+x_5)-sin(x_3)sin(x_3+x_5)]+\\&+2m_2l_1r_2x_6[cos(x_3+x_5)cos(x_3)-sin(x_3+x_5)sin(x_3)]\end{aligned}\tag{4.275}$$

$$\begin{aligned}\frac{\partial H_{11}}{\partial x_4}&=m_1r_1^22x_4sin(2x_3)+2m_2l_1^2cos(x_3)sin(x_3)+\\&+2m_2r_2^2cos(x_3+x_5)sin(x_3+x_5)+\\&+2m_2l_1r_2cos(x_3)sin(x_3+x_5)+\\&+2m_2l_1r_2cos(x_3+x_5)sin(x_3)\end{aligned}\tag{4.276}$$

$$\begin{aligned}\frac{\partial H_{11}}{\partial x_5}&=2m_2r_2^2x_4[cos^2(x_3+x_5)-sin^2(x_3+x_5)]+\\&+2m_2r_2^2x_6[cos^2(x_3+x_5)-sin^2(x_3+x_5)]+\\&+2m_3l_1r_2x_4cos(x_3)cos(x_3+x_5)+\\&-2m_2l_1r_2x_4sin(x_3+x_5)sin(x_3)+\\&-2m_2l_1r_2x_6sin(x_3+x_5)sin(x_3)\end{aligned}\tag{4.277}$$

$$\begin{aligned}\frac{\partial H_{11}}{\partial x_6}&=2m_2r_2^2cos(x_3+x_5)sin(x_3+x_5)+\\&+2m_2l_1r_2x_6sin(x_3+x_5)sin(x_3)\end{aligned}\tag{4.278}$$

In an equivalent manner, one finds $\frac{\partial H_{12}}{\partial x_1}=0$, $\frac{\partial H_{12}}{\partial x_2}=0$,

$$\frac{\partial H_{12}}{\partial x_3}=-m_1l_0r_1x_4cos(x_3)-m_2l_0l_1x_4cos(x_3)-m_2l_0r_2x_4cos(x_3+x_5)\tag{4.279}$$

$$\frac{\partial H_{12}}{\partial x_4}=-m_1l_0r_1sin(x_3)-m_2l_0l_1sin(x_3)-m_2l_0r_2sin(x_3++x_5)\tag{4.280}$$

$$\frac{\partial H_{12}}{\partial x_5}=-m_2l_0r_2x_4cos(x_3+x_5)\tag{4.281}$$

and also $\frac{\partial H_{12}}{\partial x_6}=0$. In a similar manner, one obtains $\frac{\partial H_{13}}{\partial x_1}=0$, $\frac{\partial H_{13}}{\partial x_2}=0$ and also

$$\frac{\partial H_{13}}{\partial x_3}=-2m_1l_0r_2x_4cos(x_3+x_5)-m_2l_0r_0x_6cos(x_3+x_5) \tag{4.282}$$

$$\frac{\partial H_{13}}{\partial x_4}=-2m_1l_0r_0sin(x_3+x_5) \tag{4.283}$$

$$\frac{\partial H_{13}}{\partial x_5}=-2m_1l_0r_2x_4cos(x_3+x_5)-m_2l_0r_2x_6cos(x_3+x_5) \tag{4.284}$$

$$\frac{\partial H_{13}}{\partial x_6}=-m_2l_0r_2sin(x_3+x_5) \tag{4.285}$$

Additionally, one computes $\frac{\partial H_{21}}{\partial x_1}=0$ and also

$$\begin{aligned}\frac{\partial H_{21}}{\partial x_2}=&-\tfrac{1}{2}m_1r_1^2sin(2x_3)-\tfrac{1}{2}m_2r_2^2sin(2(x_3+x+5))-\\&-m_2l_1^2cos(x_3)sin(x_3)-m_2l_1r_2sin(2x_3+x_5)\end{aligned} \tag{4.286}$$

$$\begin{aligned}\frac{\partial H_{21}}{\partial x_3}=&-\tfrac{1}{2}m_1r_1^2x_22cos(2x_3)-\tfrac{1}{2}m_2r_2^2x_22cos(2(x_3+x_5))-\\&-m_2l_1^2x_2[cos^2(x_3)-sin^2(x_3)]-m_2l_1r_2x_2^2cos^2(2x_3+x_5)\end{aligned} \tag{4.287}$$

$$\frac{\partial H_{21}}{\partial x_5}=-\tfrac{1}{2}m_2r_2^2x_2^2cos(2(x_3+x_5))-m_2l_1r_2x_2cos(2x_3+x_5) \tag{4.288}$$

while it also holds $\frac{\partial H_{21}}{\partial x_4}=0$ and $\frac{\partial H_{21}}{\partial x_6}=0$. Additionally, one finds $\frac{\partial H_{22}}{\partial x_1}=0$, $\frac{\partial H_{22}}{\partial x_2}=0$, $\frac{\partial H_{22}}{\partial x_3}=0$, $\frac{\partial H_{22}}{\partial x_4}=0$ while it also holds that

$$\frac{\partial H_{22}}{\partial x_4}=-2m_2l_1r_2x_6cos(x_5) \tag{4.289}$$

$$\frac{\partial H_{22}}{\partial x_5}=-2m_2l_1r_2x_6sin(x_5) \tag{4.290}$$

In a similar manner, one obtains $\frac{\partial H_{23}}{\partial x_1}=0$, $\frac{\partial H_{23}}{\partial x_2}=0$, $\frac{\partial H_{23}}{\partial x_3}=0$, $\frac{\partial H_{23}}{\partial x_4}=0$, and also

$$\frac{\partial H_{23}}{\partial x_5}=-m_2l_1r_2x_6cos(x_5) \tag{4.291}$$

$$\frac{\partial H_{23}}{\partial x_6}=-m_2l_1e_2sin(x_5) \tag{4.292}$$

Additionally, one finds $\frac{\partial H_{31}}{\partial x_1}=0$ and also

$$\frac{\partial H_{31}}{\partial x_2}=-\tfrac{1}{2}m_2r_2^2sin(2(x_3+x_5))-m_2l_1r_2sin(x_3)cos(x_3+x_5) \tag{4.293}$$

$$\begin{aligned}\frac{\partial H_{31}}{\partial x_3}=&-\tfrac{1}{2}m_2r_2^2x_22cos(2(x_3+x_5))-\\&-m_2l_1r_2x_2[cos(x_3)cos(x_3+x_5)-sin(x_3)sin(x_3+x_5)]\end{aligned} \tag{4.294}$$

$$\begin{aligned}\frac{\partial H_{31}}{\partial x_5}=&-\tfrac{1}{2}m_2r_2^2x_2-2cos(2(x_3+x_5))+\\&+m_2l_1r_2x_22sin(x_3)sin(x_3+x_5)\end{aligned} \tag{4.295}$$

while it also holds that $\frac{\partial H_{31}}{\partial x_4}=0$ and $\frac{\partial H_{31}}{\partial x_6}=0$. Equivalently, one obtains $\frac{\partial H_{32}}{\partial x_1}=0$, $\frac{\partial H_{32}}{\partial x_2}=0$, $\frac{\partial H_{32}}{\partial x_3}=0$, $\frac{\partial H_{31}}{\partial x_6}=0$ and also

$$\frac{\partial H_{32}}{\partial x_4}=m_2l_1r_2sin(x_5) \tag{4.296}$$

$$\frac{\partial H_{32}}{\partial x_5}=m_2l_1r_2x_4cos(x_5) \tag{4.297}$$

Finally it holds that $\frac{\partial H_{32}}{\partial x_6}=0$ for $i=1,2,\cdots,6$.

Next, the partial derivatives of the elements of the gravitational forces vector are computed. About the first element of the gravitational forces vector, it holds that $\frac{\partial g_1}{\partial x_i}=0$ for $i=1,2,\cdots,6$.

For the second element of the gravitational forces vector, one has $\frac{\partial g_2}{\partial x_1}=0$, $\frac{\partial g_2}{\partial x_2}=0$, $\frac{\partial g_2}{\partial x_4}=0$, $\frac{\partial g_2}{\partial x_6}=0$, and also

$$\frac{\partial g_2}{\partial x_3}=-g(m_1r_1+m_2l_1cos(x_3)+m_2r_2cos(x_3+x_5)) \tag{4.298}$$

For the third element of the gravitational forces vector, one has $\frac{\partial g_3}{\partial x_1}=0$, $\frac{\partial g_3}{\partial x_2}=0$, $\frac{\partial g_3}{\partial x_4}=0$, $\frac{\partial g_3}{\partial x_6}=0$, and also

$$\frac{\partial g_3}{\partial x_3} = -gm_2r_2cos(x_3+x_5) \quad (4.299)$$

$$\frac{\partial g_3}{\partial x_5} = -gm_2r_2cos(x_3+x_5) \quad (4.300)$$

4.3.4 Design of an H-infinity nonlinear feedback controller

4.3.4.1 Equivalent linearized dynamics of the rotary double-inverted pendulum

After linearization around its current operating point, the dynamic model for the rotary double-inverted pendulum is written as [721,732]

$$\dot{x} = Ax + Bu + d_1 \quad (4.301)$$

Parameter d_1 stands for the linearization error in the rotary double-inverted pendulum's model that was given previously in Eq. (4.301). The reference setpoints for the state vector of the aforementioned dynamic model are denoted by $\mathbf{x_d} = [x_1^d, \cdots, x_6^d]$. Tracking of this trajectory is achieved after applying the control input u^*. At every time instant, the control input u^* is assumed to differ from the control input u appearing in Eq. (4.301) by an amount equal to Δu, that is $u^* = u + \Delta u$

$$\dot{x}_d = Ax_d + Bu^* + d_2 \quad (4.302)$$

The dynamics of the controlled system described in Eq. (4.301) can be also written as

$$\dot{x} = Ax + Bu + Bu^* - Bu^* + d_1 \quad (4.303)$$

and by denoting $d_3 = -Bu^* + d_1$ as an aggregate disturbance term, one obtains

$$\dot{x} = Ax + Bu + Bu^* + d_3 \quad (4.304)$$

By subtracting Eq. (4.302) from Eq. (4.304), one has

$$\dot{x} - \dot{x}_d = A(x - x_d) + Bu + d_3 - d_2 \quad (4.305)$$

By denoting the tracking error as $e = x - x_d$ and the aggregate disturbance term as $\tilde{d} = d_3 - d_2$, the tracking error dynamics becomes

$$\dot{e} = Ae + Bu + \tilde{d} \quad (4.306)$$

The above-linearized form of the rotary double-inverted pendulum can be efficiently controlled after applying an H-infinity feedback control scheme.

4.3.4.2 The nonlinear H-infinity control

The initial nonlinear model of the rotary double-inverted pendulum is in the form

$$\dot{x} = f(x,u) \quad x \in R^n, \; u \in R^m \quad (4.307)$$

Linearization of the model of the rotary double-inverted pendulum is performed at each iteration of the control algorithm around its present operating point $(x^*, u^*) = (x(t), u(t - T_s))$. The linearized equivalent of the system is described by

$$\dot{x} = Ax + Bu + L\tilde{d} \quad x \in R^n, \; u \in R^m, \; \tilde{d} \in R^q \quad (4.308)$$

where matrices A and B are obtained from the computation of the previously defined Jacobians and vector $\tilde{d}$ denotes disturbance terms due to linearization errors. The problem of disturbance rejection for the linearized model that is described by

$$\begin{aligned} \dot{x} &= Ax + Bu + L\tilde{d} \\ y &= Cx \end{aligned} \quad (4.309)$$

where $x \in R^n$, $u \in R^m$, $\tilde{d} \in R^q$, and $y \in R^p$, cannot be handled efficiently if the classical LQR control scheme is applied. This is because of the existence of the perturbation term $\tilde{d}$. The disturbance term $\tilde{d}$ apart from modeling (parametric) uncertainty and external perturbation terms can also represent noise terms of any distribution.

In the H_∞ control approach, a feedback control scheme is designed for trajectory tracking by the system's state vector and simultaneous disturbance rejection, considering that the disturbance affects the system in the worst possible manner. The disturbances' effects are incorporated in the following quadratic cost function [721,732]:

$$J(t) = \frac{1}{2}\int_0^T [y^T(t)y(t) + ru^T(t)u(t) - \rho^2\tilde{d}^T(t)\tilde{d}(t)]dt, \quad r, \rho > 0 \tag{4.310}$$

The significance of the negative sign in the cost function's term that is associated with the perturbation variable $\tilde{d}(t)$ is that the disturbance tries to maximize the cost function $J(t)$ while the control signal $u(t)$ tries to minimize it. The physical meaning of the relation given above is that the control signal and the disturbances compete with each other within a min–max differential game. This problem of min–max optimization can be written as $min_u max_{\tilde{d}} J(u, \tilde{d})$.

The objective of the optimization procedure is to compute a control signal $u(t)$ which can compensate for the worst possible disturbance, that is externally imposed on the rotary double-inverted pendulum. However, the solution to the min–max optimization problem is directly related to the value of parameter ρ. This means that there is an upper bound in the disturbances magnitude that can be annihilated by the control signal.

4.3.4.3 Computation of the feedback control gains

For the linearized system given by Eq. (4.309) the cost function of Eq. (4.310) is defined, where coefficient r determines the penalization of the control input and the weight coefficient ρ determines the reward of the disturbances' effects. It is assumed that (i) The energy that is transferred from the disturbances signal $\tilde{d}(t)$ is bounded, that is $\int_0^\infty \tilde{d}^T(t)\tilde{d}(t)dt < \infty$, (ii) matrices $[A, B]$ and $[A, L]$ are stabilizable, and (iii) matrix $[A, C]$ is detectable. In the case of a tracking problem, the optimal feedback control law is given by [721,732]

$$u(t) = -Ke(t) \tag{4.311}$$

with $e = x - x_d$ to be the tracking error, and $K = \frac{1}{r}B^TP$ where P is a positive definite symmetric matrix. As it will be proven in Section 4.3.5, matrix P is obtained from the solution of the Riccati equation [721,732]

$$A^TP + PA + Q - P\left(\frac{2}{r}BB^T - \frac{1}{\rho^2}LL^T\right)P = 0 \tag{4.312}$$

where Q is a positive semi-definite symmetric matrix. The worst-case disturbance is given by

$$\tilde{d}(t) = \frac{1}{\rho^2}L^TPe(t) \tag{4.313}$$

The solution of the H-infinity feedback control problem for the rotary double-inverted pendulum and the computation of the worst-case disturbance that the related controller can sustain, comes from superposition of Bellman's optimality principle when considering that the rotary double-inverted pendulum is affected by two separate inputs: (i) the control input u and (ii) the cumulative disturbance input $\tilde{d}(t)$. Solving the optimal control problem for u, that is for the minimum variation (optimal) control input that achieves the elimination of the state vector's tracking error, gives $u = -\frac{1}{r}B^TPe$. Equivalently, solving the optimal control problem for $\tilde{d}$, that is for the worst-case disturbance that the control loop can sustain gives $\tilde{d} = \frac{1}{\rho^2}L^TPe$.

The diagram of the considered control loop for the rotary double-inverted pendulum is depicted in Figure 4.51.

4.3.5 Lyapunov stability analysis

4.3.5.1 Stability proof

Through Lyapunov stability analysis it will be shown that the proposed nonlinear control scheme assures H_∞ tracking performance for the rotary double-inverted pendulum and that in case of bounded disturbance terms asymptotic convergence to the reference setpoints is achieved. The tracking error dynamics for the rotary double-inverted pendulum is written in the form

$$\dot{e} = Ae + Bu + L\tilde{d} \tag{4.314}$$

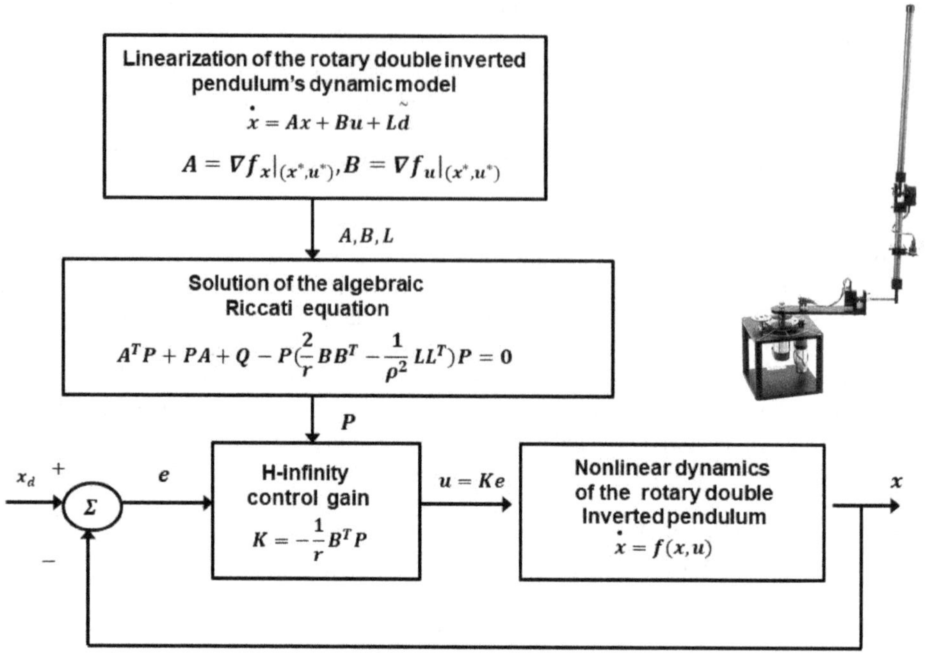

Figure 4.51 Diagram of the control scheme for the rotary double-inverted pendulum

where in the rotary double-inverted pendulum's case $L = \in R^{6\times 6}$ to be the disturbance inputs gain matrix. Variable $\tilde{d}$ denotes model uncertainties and external disturbances of the rotary double-inverted pendulum's model. The following Lyapunov equation is considered [721,732]:

$$V = \frac{1}{2}e^TPe \tag{4.315}$$

where $e = x - x_d$ is the tracking error. By differentiating with respect to time, one obtains

$$\dot{V} = \frac{1}{2}\dot{e}^TPe + \frac{1}{2}e^TP\dot{e} \Rightarrow \dot{V} = \frac{1}{2}[Ae + Bu + L\tilde{d}]^TPe + \frac{1}{2}e^TP[Ae + Bu + L\tilde{d}] \Rightarrow \tag{4.316}$$

$$\dot{V} = \frac{1}{2}[e^TA^T + u^TB^T + \tilde{d}^TL^T]Pe + \frac{1}{2}e^TP[Ae + Bu + L\tilde{d}] \Rightarrow \tag{4.317}$$

$$\dot{V} = \frac{1}{2}e^TA^TPe + \frac{1}{2}u^TB^TPe + \frac{1}{2}\tilde{d}^TL^TPe + \frac{1}{2}e^TPAe + \frac{1}{2}e^TPBu + \frac{1}{2}e^TPL\tilde{d} \tag{4.318}$$

The previous equation is rewritten as

$$\dot{V} = \frac{1}{2}e^T(A^TP + PA)e + (\frac{1}{2}u^TB^TPe + \frac{1}{2}e^TPBu) + (\frac{1}{2}\tilde{d}^TL^TPe + \frac{1}{2}e^TPL\tilde{d}) \tag{4.319}$$

Assumption: For given positive definite matrix Q and coefficients r and ρ, there exists a positive definite matrix P, which is the solution of the following matrix equation:

$$A^TP + PA = -Q + P(\frac{2}{r}BB^T - \frac{1}{\rho^2}LL^T)P \tag{4.320}$$

Moreover, the following feedback control law is applied to the system:

$$u = -\frac{1}{r}B^TPe \tag{4.321}$$

By substituting Eq. (4.320) and Eq. (4.321), one obtains

$$\dot{V} = \frac{1}{2}e^T[-Q + P(\frac{2}{r}BB^T - \frac{1}{\rho^2}LL^T)P]e + e^TPB(-\frac{1}{r}B^TPe) + e^TPL\tilde{d} \Rightarrow \tag{4.322}$$

$$\dot{V} = -\frac{1}{2}e^TQe + \frac{1}{r}e^TPBB^TPe - \frac{1}{2\rho^2}e^TPLL^TPe \\ -\frac{1}{r}e^TPBB^TPe + e^TPL\tilde{d} \tag{4.323}$$

which after intermediate operations gives

$$\dot{V} = -\frac{1}{2}e^TQe - \frac{1}{2\rho^2}e^TPLL^TPe + e^TPL\tilde{d} \tag{4.324}$$

or, equivalently

$$\dot{V} = -\frac{1}{2}e^TQe - \frac{1}{2\rho^2}e^TPLL^TPe + \frac{1}{2}e^TPL\tilde{d} + \frac{1}{2}\tilde{d}^TL^TPe \tag{4.325}$$

Lemma: The following inequality holds:

$$\frac{1}{2}e^TL\tilde{d} + \frac{1}{2}\tilde{d}L^TPe - \frac{1}{2\rho^2}e^TPLL^TPe \leq \frac{1}{2}\rho^2\tilde{d}^T\tilde{d} \tag{4.326}$$

Proof: The binomial $(\rho\alpha - \frac{1}{\rho}b)^2$ is considered. Expanding the left part of the above inequality, one gets

$$\rho^2a^2 + \frac{1}{\rho^2}b^2 - 2ab \geq 0 \Rightarrow \frac{1}{2}\rho^2a^2 + \frac{1}{2\rho^2}b^2 - ab \geq 0 \Rightarrow \\ ab - \frac{1}{2\rho^2}b^2 \leq \frac{1}{2}\rho^2a^2 \Rightarrow \frac{1}{2}ab + \frac{1}{2}ab - \frac{1}{2\rho^2}b^2 \leq \frac{1}{2}\rho^2a^2 \tag{4.327}$$

The following substitutions are carried out: $a = \tilde{d}$ and $b = e^TPL$, and the previous relation becomes

$$\frac{1}{2}\tilde{d}^TL^TPe + \frac{1}{2}e^TPL\tilde{d} - \frac{1}{2\rho^2}e^TPLL^TPe \leq \frac{1}{2}\rho^2\tilde{d}^T\tilde{d} \tag{4.328}$$

Eq. (4.328) is substituted in Eq. (4.325) and the inequality is enforced, thus giving

$$\dot{V} \leq -\frac{1}{2}e^TQe + \frac{1}{2}\rho^2\tilde{d}^T\tilde{d} \tag{4.329}$$

Eq. (4.329) shows that the H_∞ tracking performance criterion is satisfied. The integration of $\dot{V}$ from 0 to T gives

$$\int_0^T\dot{V}(t)dt \leq -\frac{1}{2}\int_0^T||e||_Q^2dt + \frac{1}{2}\rho^2\int_0^T||\tilde{d}||^2dt \Rightarrow \\ 2V(T) + \int_0^T||e||_Q^2dt \leq 2V(0) + \rho^2\int_0^T||\tilde{d}||^2dt \tag{4.330}$$

Moreover, if there exists a positive constant $M_d > 0$ such that

$$\int_0^\infty ||\tilde{d}||^2dt \leq M_d \tag{4.331}$$

then, one gets

$$\int_0^\infty||e||_Q^2dt \leq 2V(0) + \rho^2M_d \tag{4.332}$$

Thus, the integral $\int_0^\infty||e||_Q^2dt$ is bounded. Moreover, $V(T)$ is bounded and from the definition of the Lyapunov function V in Eq. (4.315), it becomes clear that $e(t)$ will be also bounded since $e(t) \in \Omega_e = \{e|e^TPe \leq 2V(0) + \rho^2M_d\}$. According to the above and with the use of Barbalat's Lemma, one obtains $\lim_{t\to\infty} e(t) = 0$.

After following the stages of the stability proof, one arrives at Eq. (4.329), which shows that the H-infinity tracking performance criterion holds. By selecting the attenuation coefficient ρ to be sufficiently small and in particular to satisfy $\rho^2 < ||e||_Q^2/||\tilde{d}||^2$, one has that the first derivative of the Lyapunov function is upper bounded by 0. This condition holds at each sampling instance and consequently global stability for the control loop can be concluded.

4.3.6 Simulation tests

The global stability properties of the control method and the elimination of the state vector's tracking error which were previously proven through Lyapunov analysis are further confirmed through simulation experiments. To implement the nonlinear optimal control scheme, the algebraic Riccati equation of Eq. (4.320) had to be solved at each time-step of the control method. The solution of the Riccati equation was obtained using the *aresolv()* function in MATLAB®. Indicative values for the parameters of the model of the rotary double-inverted pendulum have been given in [444]. In the presented simulation tests the parameters of the rotary double-inverted pendulum have been: (i) lengths of links $l_0 = 1.5m$,

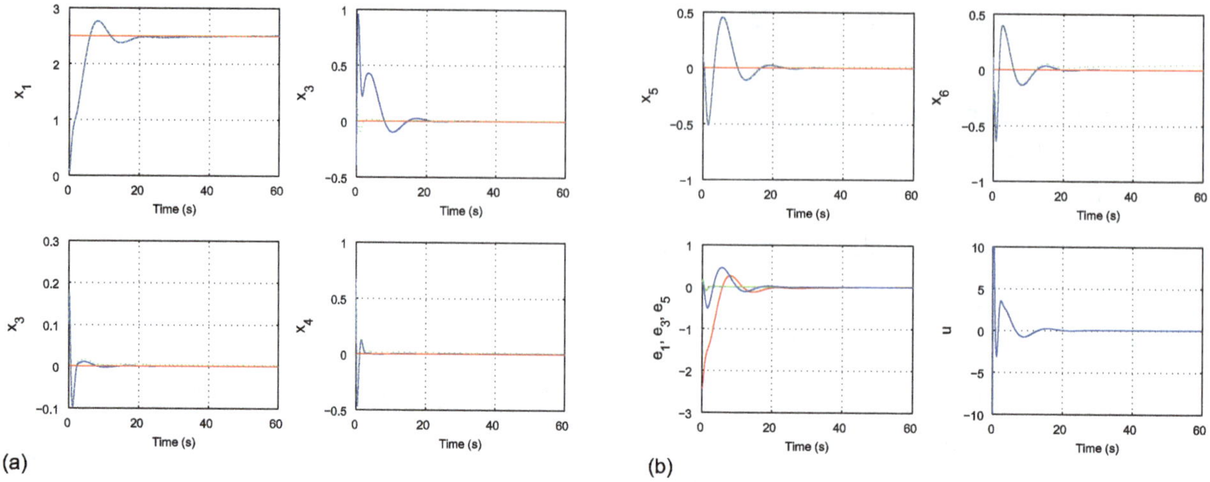

Figure 4.52 Tracking of setpoint 1 for the rotary double-inverted pendulum: (a) convergence of state variable $x_1 = \theta_0$, $x_2 = \dot{\theta}_0$, $x_3 = \theta_1$, $x_4 = \dot{\theta}_1$ to their reference setpoints (red line: setpoint, blue line: real value, green line: estimated value), and variations of the control u; (b) convergence of state variable $x_5 = \theta_2$, $x_6 = \dot{\theta}_2$ to their reference setpoints, variation of tracking errors e_1 (red), e_3 (green), e_5 (blue), and variations of the control u (torque τ) at the first joint)

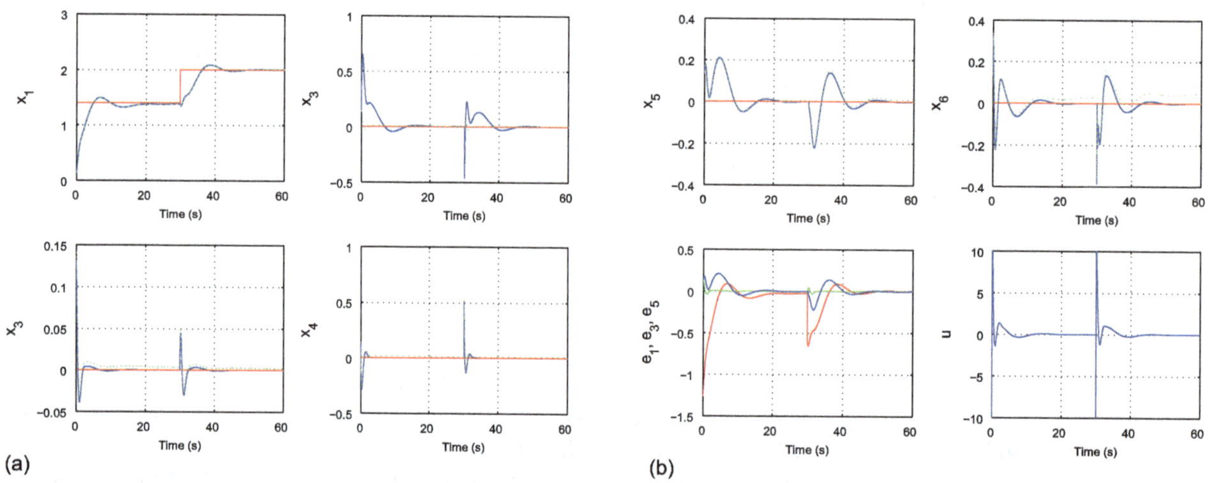

Figure 4.53 Tracking of setpoint 2 for the rotary double-inverted pendulum: (a) convergence of state variable $x_1 = \theta_0$, $x_2 = \dot{\theta}_0$, $x_3 = \theta_1$, $x_4 = \dot{\theta}_1$ to their reference setpoints (red line: setpoint, blue line: real value, green line: estimated value), and variations of the control u; (b) convergence of state variable $x_5 = \theta_2$, $x_6 = \dot{\theta}_2$ to their reference setpoints, variation of tracking errors e_1 (red), e_3 (green), e_5 (blue), and variations of the control u (torque τ) at the first joint)

$l_1 = 0.60m$, $l_2 = 0.65m$, (ii) moments of inertia of the links $J_0 = 1.230kg{\cdot}m^2$, $J_1 = 0.720kg{\cdot}m^2$, $J_2 = 0.590kg{\cdot}m^2$, (iii) distances from the joint to the center of gravity of the links $r_1 = 0.065m$, $r_2 = 0.040$m, (iv) masses of the links $m_1 = 1.60kg$, $m_2 = 1.50kg$, and (v) viscous friction coefficients $\bar{C}_0 = 6.000$, $\bar{C}_1 = 0.0058$, $\bar{C}_2 = 0.060$. The acceleration of gravity was $g = 9.8m/sec^2$. The obtained results are depicted in Figures 4.52–4.59. In these diagrams, the real values of the state variables of the rotary double-inverted pendulum are shown in blue color, the estimated values which

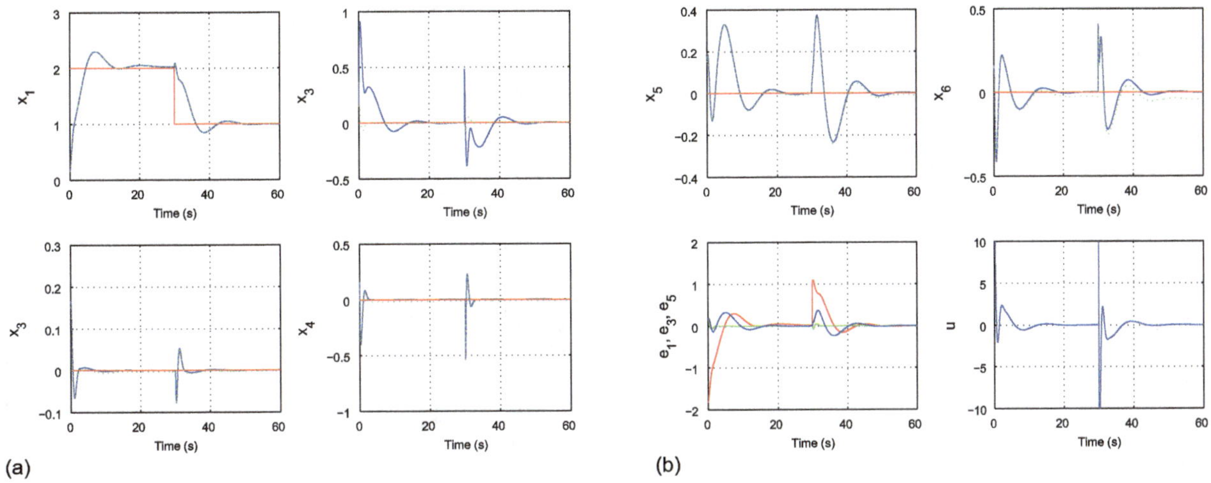

Figure 4.54 *Tracking of setpoint 3 for the rotary double-inverted pendulum: (a) convergence of state variable $x_1 = \theta_0$, $x_2 = \dot{\theta}_0$, $x_3 = \theta_1$, $x_4 = \dot{\theta}_1$ to their reference setpoints (red line: setpoint, blue line: real value, green line: estimated value), and variations of the control u; (b) convergence of state variable $x_5 = \theta_2$, $x_6 = \dot{\theta}_2$ to their reference setpoints, variation of tracking errors e_1 (red), e_3 (green), e_5 (blue), and variations of the control u (torque τ) at the first joint)*

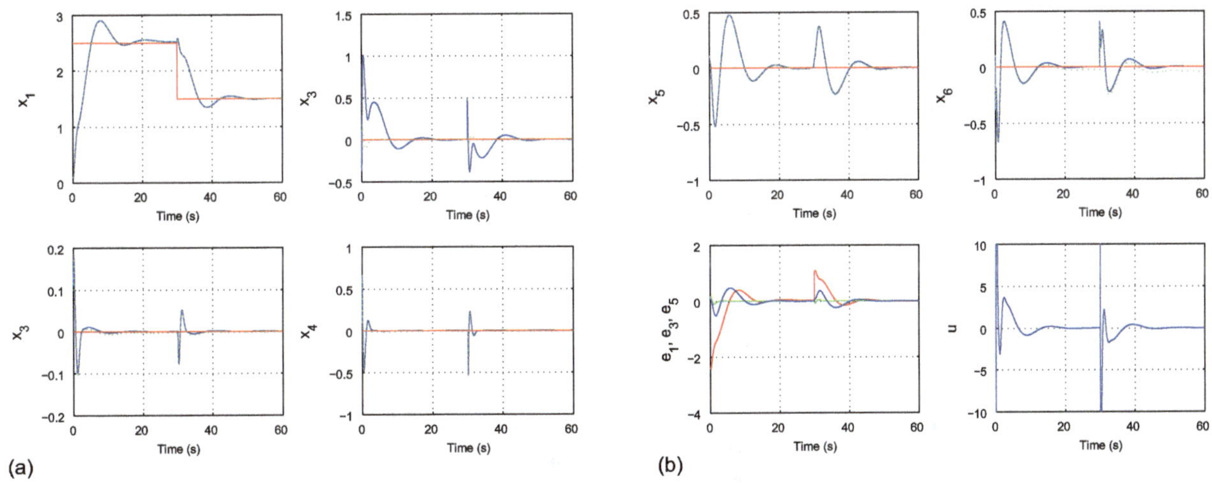

Figure 4.55 *Tracking of setpoint 4 for the rotary double-inverted pendulum: (a) convergence of state variable $x_1 = \theta_0$, $x_2 = \dot{\theta}_0$, $x_3 = \theta_1$, $x_4 = \dot{\theta}_1$ to their reference setpoints (red line: setpoint, blue line: real value, green line: estimated value), and variations of the control u; (b) convergence of state variable $x_5 = \theta_2$, $x_6 = \dot{\theta}_2$ to their reference setpoints, variation of tracking errors e_1 (red), e_3 (green), e_5 (blue), and variations of the control u (torque τ) at the first joint)*

have been obtained with the use of the H-infinity Kalman filter are depicted in green while the associated setpoints are plotted in red.

The simulation experiments have confirmed that the proposed nonlinear optimal control method for the rotary double-inverted pendulum achieves fast and accurate tracking of reference setpoints under moderate variations of the

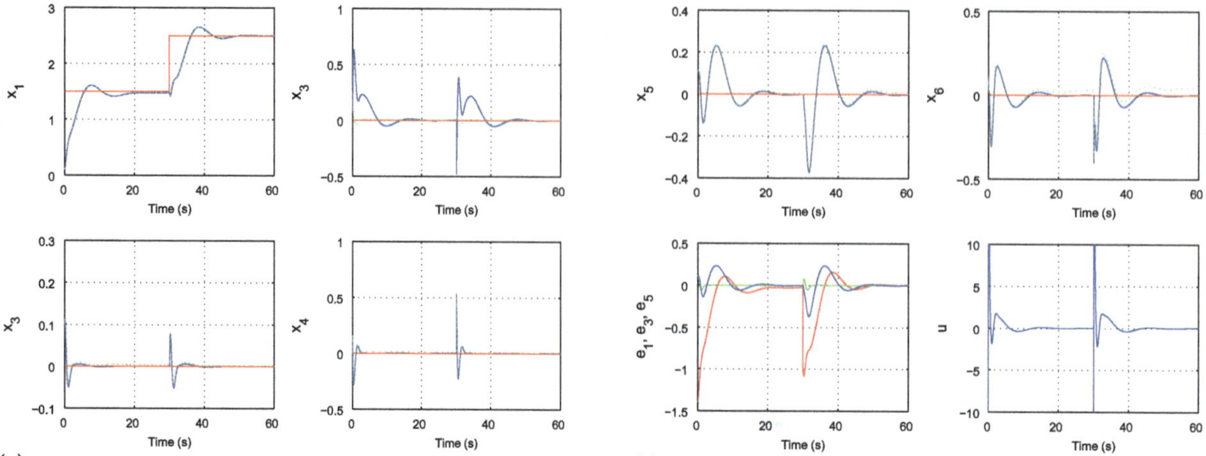

(a) (b)

Figure 4.56 *Tracking of setpoint 5 for the rotary double-inverted pendulum: (a) convergence of state variable* $x_1 = \theta_0$, $x_2 = \dot{\theta}_0$, $x_3 = \theta_1$, $x_4 = \dot{\theta}_1$ *to their reference setpoints (red line: setpoint, blue line: real value, green line: estimated value), and variations of the control u; (b) convergence of state variable* $x_5 = \theta_2$, $x_6 = \dot{\theta}_2$ *to their reference setpoints, variation of tracking errors* e_1 *(red),* e_3 *(green),* e_5 *(blue), and variations of the control u (torque* τ*) at the first joint)*

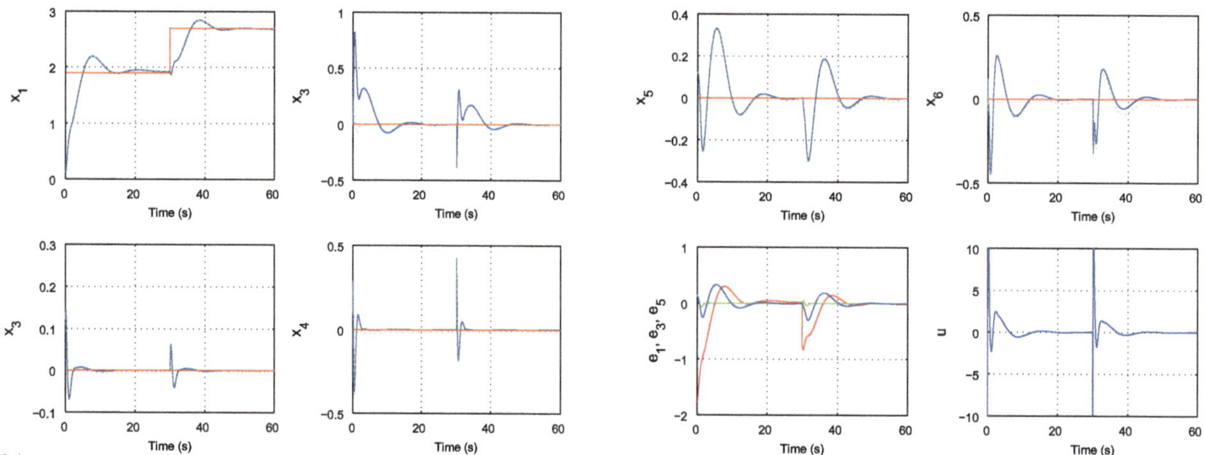

(a) (b)

Figure 4.57 *Tracking of setpoint 6 for the rotary double-inverted pendulum: (a) convergence of state variable* $x_1 = \theta_0$, $x_2 = \dot{\theta}_0$, $x_3 = \theta_1$, $x_4 = \dot{\theta}_1$ *to their reference setpoints (red line: setpoint, blue line: real value, green line: estimated value), and variations of the control u; (b) convergence of state variable* $x_5 = \theta_2$, $x_6 = \dot{\theta}_2$ *to their reference setpoints, variation of tracking errors* e_1 *(red),* e_3 *(green),* e_5 *(blue), and variations of the control u (torque* τ*) at the first joint)*

control inputs. The transient performance of the control method depends on the selection of the parameters of the above-noted Riccati equation r, ρ, and Q. For relatively small values of r the state vectors' tracking error is eliminated. For relatively large values of the diagonal elements of gain matrix Q the speed of convergence of the state variables to the associated setpoints is raised. Moreover, the smallest value of the attenuation coefficient ρ for which a valid solution

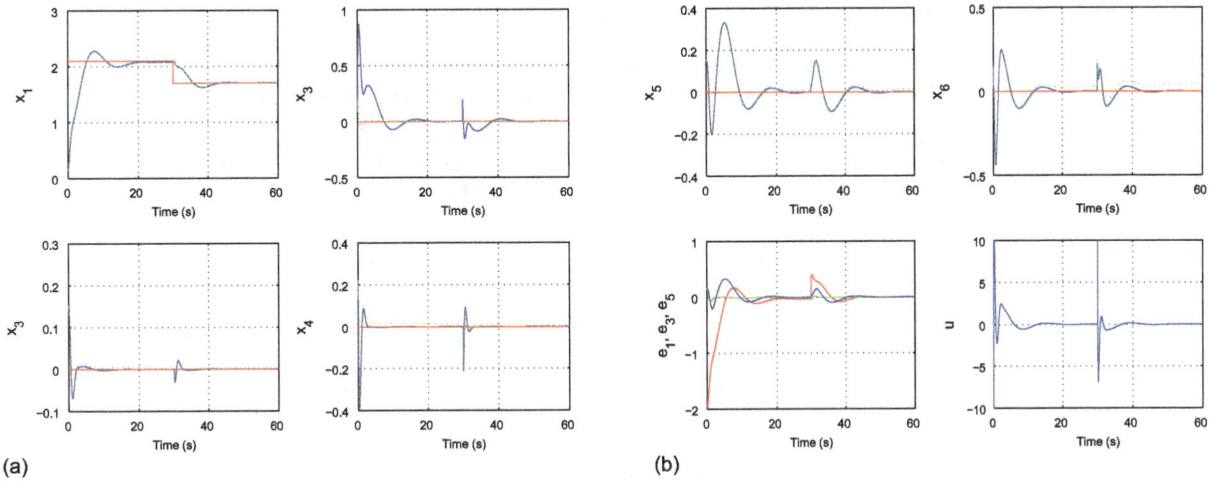

Figure 4.58 *Tracking of setpoint 7 for the rotary double-inverted pendulum: (a) convergence of state variable $x_1 = \theta_0$, $x_2 = \dot{\theta}_0$, $x_3 = \theta_1$, $x_4 = \dot{\theta}_1$ to their reference setpoints (red line: setpoint, blue line: real value, green line: estimated value), and variations of the control u; (b) convergence of state variable $x_5 = \theta_2$, $x_6 = \dot{\theta}_2$ to their reference setpoints, variation of tracking errors e_1 (red), e_3 (green), e_5 (blue), and variations of the control u (torque τ) at the first joint)*

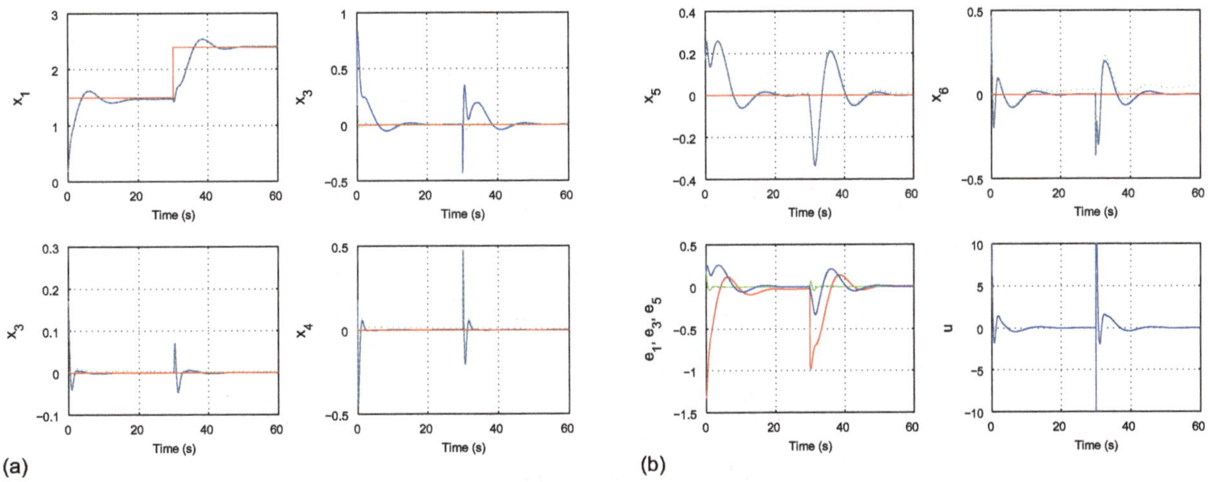

Figure 4.59 *Tracking of setpoint 8 for the rotary double-inverted pendulum: (a) convergence of state variable $x_1 = \theta_0$, $x_2 = \dot{\theta}_0$, $x_3 = \theta_1$, $x_4 = \dot{\theta}_1$ to their reference setpoints (red line: setpoint, blue line: real value, green line: estimated value), and variations of the control u; (b) convergence of state variable $x_5 = \theta_2$, $x_6 = \dot{\theta}_2$ to their reference setpoints, variation of tracking errors e_1 (red), e_3 (green), e_5 (blue), and variations of the control u (torque τ) at the first joint)*

is obtained from the Riccati equation in the form of a positive definite and symmetric matrix P is the one that achieves maximum robustness for the control loop.

To elaborate on the tracking performance and on the robustness of the proposed nonlinear optimal control method for the rotary double-inverted pendulum the following tables are given: (i) Table 4.10 which provides information about

Table 4.10 Rotary double-inverted pendulum

Tracking RMSE for the rotary pendulum in the disturbance-free case $\times 10^{-3}$						
	$RMSE_{x_1}$	$RMSE_{x_2}$	$RMSE_{x_3}$	$RMSE_{x_4}$	$RMSE_{x_5}$	$RMSE_{x_6}$
$test_1$	0.0233	0.0185	0.0001	0.0009	0.0012	0.0007
$test_2$	0.0592	0.0241	0.0008	0.0014	0.0658	0.0223
$test_3$	0.1010	0.0422	0.0014	0.0013	0.1145	0.0402
$test_4$	0.0986	0.0424	0.0014	0.0014	0.1144	0.0400
$test_5$	0.1009	0.0423	0.0014	0.0014	0.1143	0.0400
$test_6$	0.0759	0.0517	0.0011	0.0013	0.0982	0.0320
$test_7$	0.0410	0.0336	0.0006	0.0011	0.0470	0.0161
$test_8$	0.0925	0.0377	0.0012	0.0013	0.1028	0.0359

Table 4.11 Rotary double-inverted pendulum

Tracking RMSE for the rotary pendulum in the case of disturbances $\times 10^{-3}$						
$\Delta a\%$	$RMSE_{x_1}$	$RMSE_{x_2}$	$RMSE_{x_3}$	$RMSE_{x_4}$	$RMSE_{x_5}$	$RMSE_{x_6}$
0%	0.0592	0.0241	0.0008	0.0014	0.0658	0.0223
10%	0.0525	0.0249	0.0008	0.0013	0.0674	0.0239
20%	0.0459	0.0256	0.0009	0.0013	0.0691	0.0258
30%	0.0400	0.0265	0.0009	0.0012	0.0741	0.0277
40%	0.0353	0.0276	0.0009	0.0013	0.0722	0.0296
50%	0.0313	0.0287	0.0008	0.0014	0.0735	0.0352
60%	0.0287	0.0301	0.0009	0.0012	0.0747	0.0337

Table 4.12 Rotary double-inverted pendulum

RMSE for the estimation performed by the H-infinity KF $\times 10^{-3}$						
	$RMSE_{x_1}$	$RMSE_{x_2}$	$RMSE_{x_3}$	$RMSE_{x_4}$	$RMSE_{x_5}$	$RMSE_{x_6}$
$test_1$	0.0200	0.0004	0.0208	0.2192	0.0438	0.9570
$test_2$	0.0189	0.0004	0.0201	0.2194	0.0412	0.9562
$test_3$	0.0191	0.0004	0.0172	0.2193	0.0389	0.9561
$test_4$	0.0203	0.0004	0.0231	0.2190	0.0443	0.9576
$test_5$	0.0214	0.0005	0.0206	0.2193	0.0389	0.9564
$test_6$	0.0189	0.0004	0.0215	0.2193	0.0389	0.9564
$test_7$	0.0189	0.0004	0.0234	0.2195	0.0409	0.9557
$test_8$	0.0189	0.0004	0.0229	0.2195	0.0394	0.9557

the accuracy of tracking of the reference setpoints by the state variables of the rotary double-inverted pendulum's state-space model, (iia) Table 4.11 which provides information about the robustness of the control method to parametric changes in the model of the rotary double-inverted pendulum's dynamics (change $\Delta a\%$ in the moment of inertia J_0 of the first link), and (iii) Table 4.12 which provides information about the precision in state variables' estimation that is achieved by the H-infinity Kalman filter, and (iv) Table 4.13 which provides the approximate convergence times of the rotary double-inverted pendulum's state variables to the associated setpoints.

Table 4.13 Rotary double-inverted pendulum

	Convergence time (sec) for state variables x_1 to x_6					
	$T_s x_1$	$T_s x_2$	$T_s x_3$	$T_s x_4$	$T_s x_5$	$T_s x_6$
$test_1$	18.0	20.0	8.0	4.0	23.0	19.0
$test_2$	15.0	13.0	8.0	3.0	21.0	18.0
$test_3$	12.0	15.0	7.0	4.0	21.0	18.0
$test_4$	16.0	15.0	9.0	4.0	21.0	19.0
$test_5$	18.0	13.0	4.0	4.0	20.0	18.0
$test_6$	12.0	15.0	6.0	4.0	21.0	18.0
$test_7$	10.0	14.0	8.0	4.0	21.0	18.0
$test_8$	16.0	13.0	4.0	4.0	20.0	11.0

4.4 Nonlinear optimal control of the three-link biped robot

4.4.1 Outline

During the last years, there has been much research on biped robots with an aim to design humanoid robots that exhibit precisely the human walking gait [605,310]. Such robots can be used in various service tasks and in rehabilitation tasks [519,794]. Biped walking robots exhibit complex nonlinear dynamics, and in several cases, underactuation and the solution of the associated nonlinear control problem is a nontrivial task [287,759,259]. The walking gait comprises the stance phase and the swing phase. In the stance phase which is almost instantaneous both legs are on the ground. In the swing phase, the stance leg provides support against the robot's weight while the swing leg rotates and moves forward without being in contact with the ground [656,579,795]. There is a swapping of roles between the stance leg and the swing leg each time there is an impact of the swing leg with the ground [208,36,625]. The dynamic model of the biped robot is obtained through Euler-Lagrange analysis. The present section considers the solution of the stabilization and nonlinear optimal control problem for the dynamic model of the three-link biped robot. In this biped robot, the two legs receive actuation by DC motors while the third link which is the torso of the robot remains unactuated. To stabilize this robotic system and to achieve the desirable path following, global linearization control schemes are often applied [309,477,932,467]. These are based on changes of state variables (diffeomorphisms) and in the transformation of the state variables into equivalent linearized state-space descriptions in which both the solution of the control and state estimation problem is enabled [454,121,122,123]. Other control approaches present results on passivity-based control and energy minimization-based control concepts or Lyapunov theory-based design concepts [523,814]. Few results have been given on optimal control of biped robots. One can note related approaches based on MPC control or sequential quadratic programming concepts [442,835,827].

In the present section, a novel nonlinear optimal control method is developed for the dynamic model of the three-link underactuated biped robot [721,725]. The new method avoids changes of state variables and reduces energy consumption by the robot's actuators. The dynamic model of the three-link biped robot is written in an affine-in-the input nonlinear state-space form and differential flatness properties are confirmed about it [711,639,489,640,898,524]. To apply the nonlinear optimal control scheme, the dynamic model of the biped robot undergoes linearization at each sampling instance around a temporary operating point (x^*, u^*) which is defined by the present value of the robot's state vector and by the last sampled value of the control inputs vector. The linearization process is based on first-order Taylor series expansion and on the computation of the associated Jacobian matrices [705,60,706]. The modeling error which is due to the truncation of higher-order terms from the Taylor series is considered to be a perturbation which is asymptotically compensated by the robustness of the control algorithm. For the approximately linearized model of the biped robot, a stabilizing H-infinity feedback controller is designed. The H-infinity controller provides a solution of the associated optimal control problem under model uncertainty and perturbations.

The present section's controller represents a min–max differential game taking place between: (i) the control inputs which try to minimize a cost function that comprises a quadratic term of the state vector's tracking error and (ii) the model uncertainty and exogenous perturbation terms which try to maximize this cost function. To compute the stabilizing gains of the feedback controller an algebraic Riccati equation has to be repetitively solved at each time-step of the control algorithm [732,876,727]. The global stability properties of the control scheme are proven through Lyapunov analysis. First, it is demonstrated that the control loop satisfies the H-infinity tracking performance criterion. Moreover, under moderate conditions, it is also proven that the control loop is globally asymptotically stable [721,725]. To enable state estimation-based control without the need to measure the entire state vector of the biped robot, the H-infinity Kalman filter is used as a robust state observer. The nonlinear optimal control method for the biped robot achieves fast and accurate tracking of setpoints by the state variables of the biped robot under moderate variations of the control inputs.

The deployment of the use of humanoid robots has motivated the development and exploitation of related control methods [255,484,292]. The results of this section can be generalized towards more types of legged and biped robots [940,963,510,1088,544], [423,154]. The dimensionality and the number of degrees of freedom of a robotic system do not impose constraints to the application of the proposed nonlinear optimal control method. This control approach is modular and scalable and can be applied successfully to nonlinear robotic models of high dimensionality. By exploiting the differential flatness properties of the various types of biped and legged robots, one can confirm controllability of these robotic systems and can find feasible setpoints for all state variables in gait or balance mode. Differential flatness confirms that there exist control inputs which make the state variables of biped and legged robots track precisely reference trajectories [725-711,732].

4.4.2 Dynamic model of the three-link underactuated walking robot

4.4.2.1 Formulation of the state-space model

The diagram of the three-link underactuated walking robot is depicted in Figure 4.60. The masses are lumped. The length of the two legs is r and their center of gravity is placed at distance $r/2$ from their end. The distance from the center of gravity of the hips to the center of gravity of the torso is denoted as $l/2$. The mass of each leg is m. The mass of the hips is M_H while the mass of the torso is M_T. The robot receives actuation only at its legs while its hips and torso remain unactuated.

The following assumptions are made about the underactuated three-link biped robot [309]: (a) the impact of the legs with the ground takes place over an infinitesimally small time interval, (b) the external forces during the impact can be represented by impulses, (c) impulse forces result in instantaneous changes of velocities in the generalized coordinates but positions remain continuous, and (d) the torques of the actuators are not impulsive.

The joint angles after the impact are denoted as θ^+, while the joint angles before the impact are denoted as θ^-. The angular velocities after the impact are denoted as $\dot{\theta}^+$ while the angular velocities before the impact are denoted as $\dot{\theta}^-$. The stance leg detaches from the ground without interaction and thus the external forces at the leg's lower end are 0. It is also considered that the swing leg does not rebound or swing upon impact.

When the impact of the swing leg with the ground takes place the dynamic model is re-initialized with a swapping of variables. The stance leg becomes swing leg and inversely. Thus, the variables of joint angles of the stance leg and of the swing leg are also swapped [309]. This is denoted by

$$x^+ = \Delta(x^-) \tag{4.333}$$

where $x^+ = (\theta_1^+, \theta_2^+, \theta_3^+, \omega_1^+, \omega_2^+, \omega_3^+)$ is the vector of the joint angles and angular velocities of the biped robot after the impact and $x^- = (\theta_2^-, \theta_1^-, \theta_3^-, \omega_2^-, \omega_1^-, \omega_3^-)$ is the vector of joint angles and angular velocities of the biped robot before the impact.

To define setpoints for the motion of the biped robot, one may consider that: (i) the angle of the torso has to reach the constant value θ_3^d (ii) the swing leg behaves as a mirror image of the stance leg and inversely that is $\theta_2^d = -\theta_1^d$. At the stance phase, only the stance leg is in contact with the ground while the swing leg is moving forward. The stance leg receives torques from actuation at its hip and ankle joints, while the swing leg receives torque only from actuation at its hip joint.

The main parameters of the dynamic model of the three-link biped robot are outlined in Table 4.14.

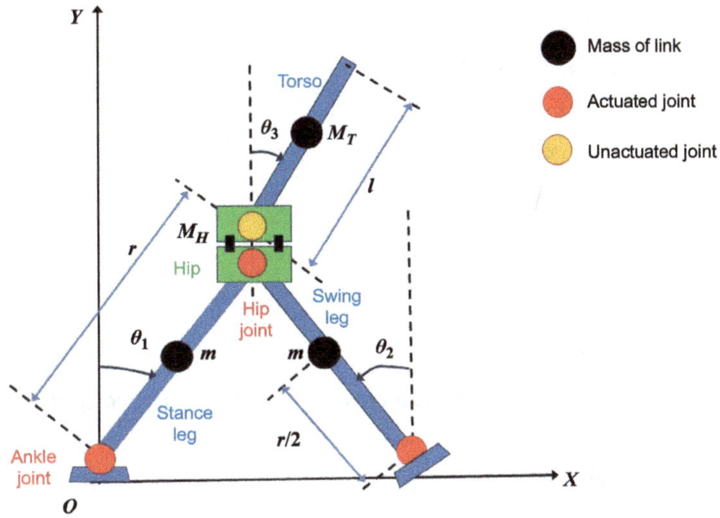

Figure 4.60 Diagram of the three-link underactuated walking robot

Table 4.14 Parameters of the three-link biped robot

Parameter	Definition
m	mass of each leg
r	length of each leg
$r/2$	distance of leg's center of gravity from leg's end
M_H	mass of the hips
M_T	mass of the torso
l	length of the torso
$l/2$	distance of torso's center of gravity from the hips
θ_1	angle between the stance leg and the vertical axis
θ_2	angle between the swing leg and the vertical axis
θ_3	angle between the torso and the vertical axis

The dynamics of the three-link biped robot is obtained through Euler–Lagrange analysis. Considering first actuation only from the hip joints the dynamic model of the biped robot is given by [605,287,309]:

$$D(\theta)\ddot{\theta} + C(\theta,\dot{\theta})\dot{\theta} + G(\theta) = Bu - \tilde{d} \tag{4.334}$$

where $D(\theta)$ is the inertia matrix which is given by

$$D(\theta) = \begin{pmatrix} d_{11} & d_{12} & d_{13} \\ d_{21} & d_{22} & d_{23} \\ d_{31} & d_{32} & d_{33} \end{pmatrix} = \begin{pmatrix} (\frac{5}{4}m + M_H + M_T)r^2 & -\frac{1}{2}mr^2c_{12} & M_Trlc_{13} \\ -\frac{1}{2}mr^2c_{12} & \frac{1}{4}mr^2 & 0 \\ M_Trlc_{13} & 0 & M_Tl^2 \end{pmatrix} \tag{4.335}$$

The Coriolis and centrifugal forces matrix $C(\theta,\dot{\theta})$ is

$$C(\theta,\dot{\theta}) = \begin{pmatrix} C_{11} & C_{12} & C_{13} \\ C_{21} & C_{22} & C_{23} \\ C_{31} & C_{32} & C_{33} \end{pmatrix} = \begin{pmatrix} 0 & -\frac{1}{2}mr^2s_{12}\dot{\theta}_2 & M_Trls_{13}\dot{\theta}_2 \\ \frac{1}{2}mr^2s_{12}\dot{\theta}_1 & 0 & 0 \\ -M_Trls_{12}\dot{\theta}_1 & 0 & 0 \end{pmatrix} \tag{4.336}$$

the gravitational forces vector $G(\theta)$ is

$$G(\theta) = \begin{pmatrix} G_1 \\ G_2 \\ G_3 \end{pmatrix} = \begin{pmatrix} -\frac{1}{2}g(2M_H + 3m + 2M_T)rsin(\theta_1) \\ \frac{1}{2}gmrsin(\theta_2) \\ -gM_T lsin(\theta_3) \end{pmatrix} \tag{4.337}$$

while the control inputs gain matrix B and the disturbances (friction) forces vector $\tilde{d}$ are given by [309]

$$B = \begin{pmatrix} -1 & 0 \\ 0 & -1 \\ 1 & 1 \end{pmatrix} \quad \tilde{d} = \begin{pmatrix} \tilde{d}_1\dot{\theta}_1 \\ \tilde{d}_2\dot{\theta}_2 \\ \tilde{d}_3\dot{\theta}_3 \end{pmatrix} \tag{4.338}$$

The terms s_{ij} and c_{ij}, $i = 1,2,3$, and $j = 1,2,3$ are defined as: $s_{ij} = sin(\theta_i - \theta_j)$ and $c_{ij} = cos(\theta_i - \theta_j)$. The inverse of the inertia matrix D is given by

$$D^{-1} = \frac{1}{detM} \begin{pmatrix} D_{11} & -D_{21} & D_{31} \\ -D_{12} & D_{22} & -D_{32} \\ D_{13} & -D_{23} & D_{33} \end{pmatrix} \tag{4.339}$$

where $D_{11} = d_{22}d_{33} - d_{32}d_{23}$, $D_{12} = d_{21}d_{33} - d_{31}d_{23}$, $D_{13} = d_{21}d_{32} - d_{31}d_{22}$, $D_{21} = d_{12}d_{33} - d_{32}d_{13}$, $D_{22} = d_{11}d_{33} - d_{31}d_{13}$, $D_{23} = d_{11}d_{32} - d_{31}d_{12}$, $D_{31} = d_{12}d_{23} - d_{23}d_{13}$, $D_{32} = d_{11}d_{23} - d_{21}d_{13}$, $D_{33} = d_{11}d_{22} - d_{21}d_{12}$.

Using the inverse of matrix D, the dynamic model of the biped robot is written as

$$\ddot{\theta} = -D(\theta)^{-1}[C(\theta,\dot{\theta})\dot{\theta} + G(\theta) + \tilde{d}] + D(\theta)^{-1}Bu \tag{4.340}$$

Next, the product $C(\theta,\dot{\theta})\dot{\theta}$ is denoted as $\tilde{C}(\theta,\dot{\theta})$ and is given by

$$\tilde{C}(\theta,\dot{\theta}) = \begin{pmatrix} \tilde{C}_1 \\ \tilde{C}_2 \\ \tilde{C}_3 \end{pmatrix} = \begin{pmatrix} C_{11}\dot{\theta}_1 + C_{12}\dot{\theta}_2 + C_{13}\dot{\theta}_3 \\ C_{21}\dot{\theta}_1 + C_{22}\dot{\theta}_2 + C_{23}\dot{\theta}_3 \\ C_{31}\dot{\theta}_1 + C_{32}\dot{\theta}_2 + C_{33}\dot{\theta}_3 \end{pmatrix} \tag{4.341}$$

Using Eq. (4.340), the dynamic model of the biped robot is written as [721]

$$\begin{pmatrix} \ddot{\theta}_1 \\ \ddot{\theta}_2 \\ \ddot{\theta}_3 \end{pmatrix} = -\frac{1}{detM} \begin{pmatrix} D_{11} & -D_{21} & D_{31} \\ -D_{12} & D_{22} & -D_{32} \\ D_{13} & -D_{23} & D_{33} \end{pmatrix} \begin{pmatrix} c_{11}\dot{\theta}_1 + c_{12}\dot{\theta}_2 + c_{13}\dot{\theta}_3 + \tilde{d}_1\dot{\theta}_1 \\ c_{21}\dot{\theta}_1 + c_{22}\dot{\theta}_2 + c_{23}\dot{\theta}_3 + \tilde{d}_2\dot{\theta}_2 \\ c_{31}\dot{\theta}_1 + c_{32}\dot{\theta}_2 + c_{33}\dot{\theta}_3 + \tilde{d}_3\dot{\theta}_3 \end{pmatrix} + \\ +\frac{1}{detM} \begin{pmatrix} D_{11} & -D_{21} & D_{31} \\ -D_{12} & D_{22} & -D_{32} \\ D_{13} & -D_{23} & D_{33} \end{pmatrix} \begin{pmatrix} -1 & 0 \\ 0 & -1 \\ 1 & 1 \end{pmatrix} \begin{pmatrix} u_1 \\ u_2 \end{pmatrix} \tag{4.342}$$

Next, by defining $Q_1 = C_{11}\dot{Q}_1 + C_{12}\dot{Q}_2 + C_{13}\dot{Q}_3 + G_1 + \tilde{d}_1\dot{\theta}_1$, $Q_2 = C_{21}\dot{Q}_1 + C_{22}\dot{Q}_2 + C_{23}\dot{Q}_3 + G_2 + \tilde{d}_2\dot{\theta}_2$, $Q_3 = C_{31}\dot{Q}_1 + C_{32}\dot{Q}_2 + C_{33}\dot{Q}_3 + G_3 + \tilde{d}_3\dot{\theta}_3$, one obtains the following state-space form: for the biped robot

$$\begin{pmatrix} \ddot{\theta}_1 \\ \ddot{\theta}_2 \\ \ddot{\theta}_3 \end{pmatrix} = \begin{pmatrix} \frac{-D_{11}Q_1 + D_{21}Q_2 - D_{31}Q_3}{detM} \\ \frac{D_{12}Q_1 - D_{22}Q_2 + D_{32}Q_3}{detM} \\ \frac{-D_{13}Q_1 + D_{23}Q_2 - D_{33}Q_3}{detM} \end{pmatrix} + \begin{pmatrix} \frac{-D_{11}+D_{31}}{detM} & \frac{D_{21}+D_{31}}{detM} \\ \frac{D_{12}-D_{32}}{detM} & \frac{-D_{22}-D_{32}}{detM} \\ \frac{-D_{13}+D_{33}}{detM} & \frac{D_{23}+D_{33}}{detM} \end{pmatrix} \begin{pmatrix} u_1 \\ u_2 \end{pmatrix} \tag{4.343}$$

Moreover, using the state-vector notation $x=[x_1,x_2,x_3,x_4,x_5,x_6]^T$ or $x=[\theta_1,\dot{\theta}_1,\theta_2,\dot{\theta}_2,\theta_3,\dot{\theta}_3]^T$, one obtains the following state-space description:

$$\begin{pmatrix}\dot{x}_1\\ \dot{x}_2\\ \dot{x}_3\\ \dot{x}_4\\ \dot{x}_5\\ \dot{x}_6\end{pmatrix}=\begin{pmatrix}x_2\\ \frac{-D_{11}Q_1+D_{21}Q_2-D_{31}Q_3}{detM}\\ x_4\\ \frac{D_{12}Q_1-D_{22}Q_2+D_{32}Q_3}{detM}\\ x_6\\ \frac{-D_{13}Q_1+D_{23}Q_2-D_{33}Q_3}{detM}\end{pmatrix}+\begin{pmatrix}0 & 0\\ \frac{-D_{11}+D_{31}}{detM} & \frac{D_{21}+D_{31}}{detM}\\ 0 & 0\\ \frac{D_{12}-D_{32}}{detM} & \frac{-D_{22}-D_{32}}{detM}\\ 0 & 0\\ \frac{-D_{13}+D_{33}}{detM} & \frac{D_{23}+D_{33}}{detM}\end{pmatrix}\begin{pmatrix}u_1\\ u_2\end{pmatrix} \tag{4.344}$$

4.4.2.2 Input-output linearized form of the biped robot

With the use of Lie algebra, one can write the dynamic model of the three-link biped robot in an input–output linearized form. To this end, the following linearizing outputs are defined:

$$y=\begin{pmatrix}y_1\\ y_2\end{pmatrix}=\begin{pmatrix}h_1(\theta)\\ h_2(\theta)\end{pmatrix}=\begin{pmatrix}\theta_3-\theta_3^d\\ \theta_1+\theta_2\end{pmatrix}=\begin{pmatrix}x_5-x_5^d\\ x_1+x_3\end{pmatrix} \tag{4.345}$$

It can be shown that the relative degree of this system is equal to 2. The state-space model of the biped robot was described in the form of Eq. (4.344), that is $\dot{x}=f(x)+g(x)u$ with its state vector $x=[x_1,x_2,x_3,x_4,x_5,x_6]^T$ and vectors $f(x)$, $g_1(x)$, and $g_2(x)$ to be given by

$$f(x)=\begin{pmatrix}x_2\\ \frac{-D_{11}Q_1+D_{21}Q_2-D_{31}Q_3}{detM}\\ x_4\\ \frac{D_{12}Q_1-D_{22}Q_2+D_{32}Q_3}{detM}\\ x_6\\ \frac{-D_{13}Q_1+D_{23}Q_2-D_{33}Q_3}{detM}\end{pmatrix}\quad g_1(x)=\begin{pmatrix}0\\ \frac{-D_{11}+D_{31}}{detM}\\ 0\\ \frac{D_{12}-D_{32}}{detM}\\ 0\\ \frac{-D_{13}+D_{33}}{detM}\end{pmatrix}\quad g_2(x)=\begin{pmatrix}0\\ \frac{D_{21}+D_{31}}{detM}\\ 0\\ \frac{-D_{22}-D_{32}}{detM}\\ 0\\ \frac{D_{23}+D_{33}}{detM}\end{pmatrix} \tag{4.346}$$

The computation of Lie derivatives for the biped robot of Eq. (4.344), using the above defined linearizing outputs h_1, h_2 gives

$$\begin{aligned}L_fh_1&=\tfrac{\partial h_1}{\partial x_1}f_1+\tfrac{\partial h_1}{\partial x_2}f_2+\tfrac{\partial h_1}{\partial x_3}f_3+\tfrac{\partial h_1}{\partial x_4}f_4+\tfrac{\partial h_1}{\partial x_5}f_5+\tfrac{\partial h_1}{\partial x_6}f_6\\ &\Rightarrow L_fh_1=f_5\Rightarrow L_fh_1=x_6\end{aligned} \tag{4.347}$$

$$\begin{aligned}L_{g_1}h_1&=\tfrac{\partial h_1}{\partial x_1}g_{11}+\tfrac{\partial h_1}{\partial x_2}g_{21}+\tfrac{\partial h_1}{\partial x_3}g_{31}+\tfrac{\partial h_1}{\partial x_4}g_{41}+\tfrac{\partial h_1}{\partial x_5}g_{51}+\tfrac{\partial h_1}{\partial g_{61}}f_6\\ &\Rightarrow L_{g_1}h_1=g_{51}\Rightarrow L_{g_1}h_1=0\end{aligned} \tag{4.348}$$

$$\begin{aligned}L_{g_2}h_1&=\tfrac{\partial h_1}{\partial x_1}g_{12}+\tfrac{\partial h_1}{\partial x_2}g_{22}+\tfrac{\partial h_1}{\partial x_3}g_{32}+\tfrac{\partial h_1}{\partial x_4}g_{42}+\tfrac{\partial h_1}{\partial x_5}g_{52}+\tfrac{\partial h_1}{\partial g_{62}}f_6\\ &\Rightarrow L_{g_2}h_1=g_{52}\Rightarrow L_{g_2}h_1=0\end{aligned} \tag{4.349}$$

Similarly, one has

$$\begin{aligned}L_fh_2&=\tfrac{\partial h_2}{\partial x_1}f_1+\tfrac{\partial h_2}{\partial x_2}f_2+\tfrac{\partial h_2}{\partial x_3}f_3+\tfrac{\partial h_2}{\partial x_4}f_4+\tfrac{\partial h_2}{\partial x_5}f_5+\tfrac{\partial h_2}{\partial x_6}f_6\\ &\Rightarrow L_fh_2=f_1+f_3\Rightarrow L_fh_2=x_2+x_4\end{aligned} \tag{4.350}$$

$$\begin{aligned}L_{g_1}h_2&=\tfrac{\partial h_2}{\partial x_1}g_{11}+\tfrac{\partial h_2}{\partial x_2}g_{21}+\tfrac{\partial h_2}{\partial x_3}g_{31}+\tfrac{\partial h_2}{\partial x_4}g_{41}+\tfrac{\partial h_2}{\partial x_5}g_{51}+\tfrac{\partial h_2}{\partial g_{61}}f_6\\ &\Rightarrow L_{g_1}h_2=g_{11}+g_{31}\Rightarrow L_{g_1}h_1=0\end{aligned} \tag{4.351}$$

$$L_{g_2}h_2 = \frac{\partial h_2}{\partial x_1}g_{12} + \frac{\partial h_2}{\partial x_2}g_{22} + \frac{\partial h_2}{\partial x_3}g_{32} + \frac{\partial h_2}{\partial x_4}g_{42} + \frac{\partial h_2}{\partial x_5}g_{52} + \frac{\partial h_2}{\partial g_{62}}f_6$$
$$\Rightarrow L_{g_2}h_2 = g_{12} + g_{32} \Rightarrow L_{g_2}h_2 = 0 \tag{4.352}$$

Additionally, for $L_f h_1 = x_6$, one has

$$L_f^2 h_1 = \frac{\partial L_f h_1}{\partial x_1}f_1 + \frac{\partial L_f h_1}{\partial x_2}f_2 + \frac{\partial L_f h_1}{\partial x_3}f_3 + \frac{\partial L_f h_1}{\partial x_4}f_4 + \frac{\partial L_f h_1}{\partial x_5}f_5 + \frac{\partial L_f h_1}{\partial x_6}f_6$$
$$\Rightarrow L_f^2 h_1 = f_6 \Rightarrow L_f^2 h_1 = \frac{-D_{13}Q_1 + D_{23}Q_2 - D_{33}Q_3}{detM} \tag{4.353}$$

$$L_{g_1}L_f h_1 = \frac{\partial L_f h_1}{\partial x_1}g_{11} + \frac{\partial L_f h_1}{\partial x_2}g_{21} + \frac{\partial L_f h_1}{\partial x_3}g_{31} + \frac{\partial L_f h_1}{\partial x_4}g_{41} + \frac{\partial L_f h_1}{\partial x_5}g_{51} + \frac{\partial L_f h_1}{\partial g_{61}}g_{61}$$
$$\Rightarrow L_{g_1}L_f h_1 = g_{61} \Rightarrow L_{g_1}L_f h_1 = \frac{-D_{13} + D_{33}}{detM} \tag{4.354}$$

$$L_{g_2}L_f h_1 = \frac{\partial L_f h_1}{\partial x_1}g_{12} + \frac{\partial L_f h_1}{\partial x_2}g_{22} + \frac{\partial L_f h_1}{\partial x_3}g_{32} + \frac{\partial L_f h_1}{\partial x_4}g_{42} + \frac{\partial L_f h_1}{\partial x_5}g_{52} + \frac{\partial L_f h_1}{\partial g_{62}}g_{62}$$
$$\Rightarrow L_{g_2}L_f h_1 = g_{62} \Rightarrow L_{g_2}L_f h_1 = \frac{D_{23} + D_{33}}{detM} \tag{4.355}$$

Equivalently, one has

$$L_f^2 h_2 = \frac{\partial L_f h_2}{\partial x_1}f_1 + \frac{\partial L_f h_2}{\partial x_2}f_2 + \frac{\partial L_f h_2}{\partial x_3}f_3 + \frac{\partial L_f h_2}{\partial x_4}f_4 + \frac{\partial L_f h_2}{\partial x_5}f_5 + \frac{\partial L_f h_2}{\partial x_6}f_6$$
$$\Rightarrow L_f^2 h_2 = f_2 + f_4 \Rightarrow L_f^2 h_2 = \frac{-D_{11}Q_1 + D_{21}Q_2 - D_{31}Q_3}{detM} + \frac{D_{12}Q_1 - D_{22}Q_2 + D_{32}Q_3}{detM} \tag{4.356}$$

$$L_{g_1}L_f h_2 = \frac{\partial L_f h_2}{\partial x_1}g_{11} + \frac{\partial L_f h_2}{\partial x_2}g_{21} + \frac{\partial L_f h_2}{\partial x_3}g_{31} + \frac{\partial L_f h_2}{\partial x_4}g_{41} + \frac{\partial L_f h_2}{\partial x_5}g_{51} + \frac{\partial L_f h_2}{\partial g_{61}}g_{61}$$
$$\Rightarrow L_{g_1}L_f h_2 = g_{11} + g_{41} \Rightarrow L_{g_1}L_f h_2 = \frac{-D_{11} + D_{31}}{detM} + \frac{D_{12} - D_{32}}{detM} \tag{4.357}$$

$$L_{g_2}L_f h_2 = \frac{\partial L_f h_2}{\partial x_1}g_{12} + \frac{\partial L_f h_2}{\partial x_2}g_{22} + \frac{\partial L_f h_2}{\partial x_3}g_{32} + \frac{\partial L_f h_2}{\partial x_4}g_{42} + \frac{\partial L_f h_2}{\partial x_5}g_{52} + \frac{\partial L_f h_2}{\partial g_{62}}g_{62}$$
$$\Rightarrow L_{g_2}L_f h_2 = g_{22} + g_{42} \Rightarrow L_{g_2}L_f h_2 = \frac{D_{21} + D_{31}}{detM} + \frac{-D_{22} - D_{32}}{detM} \tag{4.358}$$

Thus, one arrives at the following linearized form of the biped robot's dynamics:

$$\ddot{y}_1 = L_f^2 h_1 + L_{g_1}L_f h_1 u_1 + L_{g_2}L_f h_1 u_2$$
$$\ddot{y}_2 = L_f^2 h_2 + L_{g_1}L_f h_2 u_1 + L_{g_2}L_f h_2 u_2 \tag{4.359}$$

or equivalently

$$\begin{pmatrix}\ddot{y}_1\\ \ddot{y}_2\end{pmatrix} = \begin{pmatrix}L_f^2 h_1\\ L_f^2 h_2\end{pmatrix} + \begin{pmatrix}L_{g_1}L_f h_1 & L_{g_2}L_f h_1\\ L_{g_1}L_f h_2 & L_{g_2}L_f h_2\end{pmatrix}\begin{pmatrix}u_1\\ u_2\end{pmatrix} \tag{4.360}$$

or by defining the transformed control inputs $v_1 = L_f^2 h_1 + L_{g_1}L_f h_1 u_1 + L_{g_2}L_f h_1 u_2$ and $v_2 = L_f^2 h_2 + L_{g_1}L_f h_2 u_1 + L_{g_2}L_f h_2 u_2$, one has the linearized description in the canonical Brunovsky form

$$\ddot{y}_1 = v_1 \quad \ddot{y}_2 = v_2 \tag{4.361}$$

For the dynamic model of the biped robot given in Eq. (4.361), the stabilizing feedback control is

$$v_1 = \ddot{y}_{1,d} - k_{1,1}(\dot{y}_1 - \dot{y}_{1,d}) - k_{2,1}(y_1 - y_{1,d})$$
$$v_2 = \ddot{y}_{2,d} - k_{1,2}(\dot{y}_2 - \dot{y}_{2,d}) - k_{2,2}(y_2 - y_{2,d}) \tag{4.362}$$

By defining the tracking error variables $e_1 = y_1 - y_{1,d}$ and $e_2 = y_2 - y_{2,d}$, the following tracking error dynamics of the closed-loop system is obtained:

$$(\ddot{y}_1 - \ddot{y}_{1,d}) + k_{1,1}(\dot{y}_1 - \dot{y}_{1,d}) + k_{2,1}(y_1 - y_{1,d}) = 0 \Rightarrow \ddot{e}_1 + k_{1,1}\dot{e}_1 + k_{2,1}e_1 = 0$$
$$(\ddot{y}_2 - \ddot{y}_{2,d}) + k_{1,2}(\dot{y}_2 - \dot{y}_{2,d}) + k_{2,2}(y_2 - y_{2,d}) = 0 \Rightarrow \ddot{e}_2 + k_{1,2}\dot{e}_2 + k_{2,2}e_2 = 0 \tag{4.363}$$

Thus, by selecting the feedback gains $k_{1,1}, k_{2,1}$ and $k_{1,2}, k_{2,2}$ so as the associated characteristic polynomials to be Hurwitz stable, one has

$$\begin{aligned} lim_{t\to\infty}e_1(t) = 0 \Rightarrow lim_{t\to\infty}y_1(t) = y_{1,d}(t) \\ lim_{t\to\infty}e_2(t) = 0 \Rightarrow lim_{t\to\infty}y_2(t) = y_{2,d}(t) \end{aligned} \tag{4.364}$$

The real control input which is applied to the initial nonlinear model of the biped robot is given by

$$\begin{pmatrix} u_1 \\ u_2 \end{pmatrix} = \begin{pmatrix} L_{g_1}L_fh_1 & L_{g_2}L_fh_1 \\ L_{g_1}L_fh_2 & L_{g_2}L_fh_2 \end{pmatrix}^{-1} \left[\begin{pmatrix} v_1 \\ v_2 \end{pmatrix} - \begin{pmatrix} L_f^2h_1 \\ L_f^2h_2 \end{pmatrix} \right] \tag{4.365}$$

4.4.2.3 Extended state-space model of the biped robot

Next, the dynamic model of the three-link biped robot which was initially given in Eq. (4.344) is extended with a third control input u_3 to describe also the actuation effects on the stance leg from a torque coming from the ankle joint of this leg. Thus, there is an additional control input u_3 affecting only the turn angle θ_1 of the stance leg, and the dynamic model is rewritten in the following form:

$$\begin{pmatrix} \dot{x}_1 \\ \dot{x}_2 \\ \dot{x}_3 \\ \dot{x}_4 \\ \dot{x}_5 \\ \dot{x}_6 \end{pmatrix} = \begin{pmatrix} x_2 \\ \frac{-D_{11}Q_1 + D_{21}Q_2 - D_{31}Q_3}{detM} \\ x_4 \\ \frac{D_{12}Q_1 - D_{22}Q_2 + D_{32}Q_3}{detM} \\ x_6 \\ \frac{-D_{13}Q_1 + D_{23}Q_2 - D_{33}Q_3}{detM} \end{pmatrix} + \begin{pmatrix} 0 & 0 & 0 \\ \frac{-D_{11}+D_{31}}{detM} & \frac{D_{21}+D_{31}}{detM} & 1 \\ 0 & 0 & 0 \\ \frac{D_{12}-D_{32}}{detM} & \frac{-D_{22}-D_{32}}{detM} & 0 \\ 0 & 0 & 0 \\ \frac{-D_{13}+D_{33}}{detM} & \frac{D_{23}+D_{33}}{detM} & 0 \end{pmatrix} \begin{pmatrix} u_1 \\ u_2 \\ u_3 \end{pmatrix} \tag{4.366}$$

Finally, one has a description in the concise nonlinear affine-in-the-input state-space form

$$\dot{x} = f(x) + g(x)u \tag{4.367}$$

where $x \in R^{6\times1}, f(x) \in R^{6\times1}$, $g(x) \in R^{6\times3}$, and $u \in R^{3\times1}$. It can be directly confirmed that the dynamic model of the three-link biped robot of Eq. (4.366) is differentially flat [711,524]. This means that: (i) all state variables of the system can be written as differential functions of its flat outputs and (ii) the flat outputs are not connected through a relation in the form of an ordinary differential equation. The flat outputs vector is taken to be $Y = [x_1, x_3, x_5]$. Obviously, it holds that $[x_2, x_4.x_6] = [\dot{x}_1, \dot{x}_3, \dot{x}_5]$ thus state variables x_2,x_4,x_6 are differential functions of the flat outputs of the system. Moreover, from Eq. (4.366), one can solve for the control inputs vector u and can express the control inputs as differential functions of the flat outputs. The above demonstrates that the extended state-space model of the biped robot of Eq. (4.366) is a differentially flat system.

4.4.3 Approximate linearization of the dynamics of the biped robot

Next, the extended dynamic model of the three-link biped robot of Eq. (4.366) is considered. The dynamic model of this three-link biped robot undergoes linearization through first-order Taylor series expansion and the computation of the associated Jacobian matrices. The linearization point (x^*, y^*) is defined at each sampling interval and consists of the present position x^* of the system's state vector and by the last sampled value u^* of the control inputs vector. The modeling error which is due to the truncation of higher-order terms from the Taylor series expansion is considered to be a perturbation which is asymptotically compensated by the robustness of the control algorithm. Thus, the initial nonlinear state-space model of the system of Eq. (4.367), given as $\dot{x} = f(x) + g(x)u$, is transformed through the linearization process into the equivalent linearized form

$$\dot{x} = Ax + Bu + \tilde{d} \tag{4.368}$$

where A, B are the Jacobian matrices of the system which are given by

$$A = \nabla_x[f(x) + g(x)u] \mid_{(x^*,u^*)} \Rightarrow$$
$$A = \nabla_x f(x) \mid_{(x^*,u^*)} + \nabla_x g_1(x)u \mid_{(x^*,u^*)} + \nabla_x g_2(x)u \mid_{(x^*,u^*)} \tag{4.369}$$

$$B = \nabla_u[f(x) + g(x)u] \mid_{(x^*,u^*)} \Rightarrow$$
$$B = g(x) \mid_{(x^*,u^*)} \tag{4.370}$$

while the cumulative disturbances vector $\tilde{d}$ may comprise: (i) the modeling error due to truncation of higher-order terms from the Taylor series expansion, (ii) external perturbations, and (iii) sensor measurement noise of any distribution.

Computation of the Jacobian matrix $\nabla_x f(x) \mid_{(x^*,u^*)}$:

First row of the Jacobian matrix $\nabla_x f(x) \mid_{(x^*,u^*)}$: $\frac{\partial f_1}{\partial x_1} = 0$, $\frac{\partial f_1}{\partial x_2} = 1$, $\frac{\partial f_1}{\partial x_3} = 0$, $\frac{\partial f_1}{\partial x_4} = 0$, $\frac{\partial f_1}{\partial x_5} = 0$, $\frac{\partial f_1}{\partial x_6} = 0$.

Second row of the Jacobian matrix $\nabla_x f(x) \mid_{(x^*,u^*)}$: It holds that

$$f_2(x) = \frac{f_{2,num}}{f_{2,den}} = \frac{-D_{11}Q_1 + D_{21}Q_2 - D_{31}Q_3}{detM} \tag{4.371}$$

Moreover, it holds that for $i = 1, 2, \cdots, 6$

$$\frac{\partial f_2}{\partial x_i} = \frac{\frac{\partial f_{2,num}}{\partial x_i} detM - f_{2,num} \frac{\partial detM}{\partial x_i}}{detM^2} \tag{4.372}$$

where for $i = 1, 2, \cdots, 6$

$$\frac{\partial f_2 num}{\partial x_i} = -\frac{\partial D_{11}}{\partial x_i}Q_1 - D_{11}\frac{\partial Q_1}{\partial x_i} + \frac{\partial D_{21}}{\partial x_i}Q_2 + D_{21}\frac{\partial Q_2}{\partial x_i} - \frac{\partial D_{31}}{\partial x_i}Q_3 - D_{31}\frac{\partial Q_3}{\partial x_i} \tag{4.373}$$

Third row of the Jacobian matrix $\nabla_x f(x) \mid_{(x^*,u^*)}$: $\frac{\partial f_3}{\partial x_1} = 0$, $\frac{\partial f_3}{\partial x_2} = 0$, $\frac{\partial f_3}{\partial x_3} = 0$, $\frac{\partial f_3}{\partial x_4} = 1$, $\frac{\partial f_3}{\partial x_5} = 0$, $\frac{\partial f_3}{\partial x_6} = 0$.

Fourth row of the Jacobian matrix $\nabla_x f(x) \mid_{(x^*,u^*)}$: It holds that

$$f_4(x) = \frac{f_{4,num}}{f_{4,den}} = \frac{D_{12}Q_1 - D_{22}Q_2 + D_{32}Q_3}{detM} \tag{4.374}$$

Moreover, it holds that for $i = 1, 2, \cdots, 6$

$$\frac{\partial f_4}{\partial x_i} = \frac{\frac{\partial f_{4,num}}{\partial x_i} detM - f_{4,num} \frac{\partial detM}{\partial x_i}}{detM^2} \tag{4.375}$$

where for $i = 1, 2, \cdots, 6$

$$\frac{\partial f_4 num}{\partial x_i} = \frac{\partial D_{12}}{\partial x_i}Q + D_{12}\frac{\partial Q_1}{\partial x_i} - \frac{\partial D_{22}}{\partial x_i}Q_2 - D_{22}\frac{\partial Q_2}{\partial x_i} + \frac{\partial D_{32}}{\partial x_i}Q_3 + D_{32}\frac{\partial Q_3}{\partial x_i} \tag{4.376}$$

Fifth row of the Jacobian matrix $\nabla_x f(x) \mid_{(x^*,u^*)}$: $\frac{\partial f_5}{\partial x_1} = 0$, $\frac{\partial f_5}{\partial x_2} = 0$, $\frac{\partial f_5}{\partial x_3} = 0$, $\frac{\partial f_5}{\partial x_4} = 0$, $\frac{\partial f_5}{\partial x_5} = 0$, and $\frac{\partial f_5}{\partial x_6} = 1$.

Sixth row of the Jacobian matrix $\nabla_x f(x) \mid_{(x^*,u^*)}$: It holds that

$$f_6(x) = \frac{f_{6,num}}{f_{6,den}} = \frac{-D_{13}Q_1 + D_{23}Q_2 - D_{33}Q_3}{detM} \tag{4.377}$$

Moreover, it holds that for $i = 1, 2, \cdots, 6$

$$\frac{\partial f_6}{\partial x_i} = \frac{\frac{\partial f_{6,num}}{\partial x_i} detM - f_{6,num} \frac{\partial detM}{\partial x_i}}{detM^2} \tag{4.378}$$

where for $i = 1, 2, \cdots, 6$

$$\frac{\partial f_6 num}{\partial x_i} = -\frac{\partial D_{13}}{\partial x_i}Q_1 - D_{13}\frac{\partial Q_1}{\partial x_i} + \frac{\partial D_{23}}{\partial x_i}Q_2 + D_{23}\frac{\partial Q_2}{\partial x_i} - \frac{\partial D_{33}}{\partial x_i}Q_3 - D_{33}\frac{\partial Q_3}{\partial x_i} \tag{4.379}$$

Computation of the Jacobian matrix $\nabla_x g_1(x) \mid_{(x^*,u^*)}$:

First row of the Jacobian matrix $\nabla_x g_1(x)\mid_{(x^*,u^*)}$: $\frac{\partial g_{11}}{\partial x_1}=0$, $\frac{\partial g_{11}}{\partial x_2}=0$, $\frac{\partial g_{11}}{\partial x_3}=0$, $\frac{\partial g_{11}}{\partial x_4}=0$, $\frac{\partial g_{11}}{\partial x_5}=0$, $\frac{\partial g_{11}}{\partial x_6}=0$.

Second row of the Jacobian matrix $\nabla_x g_1(x)\mid_{(x^*,u^*)}$:

$$g_{21}(x)=\frac{D_{11}+D_{31}}{detM} \tag{4.380}$$

Thus, for $i=1,2,\cdots,6$,

$$\frac{\partial g_{21}}{\partial x_i}=\frac{(\frac{-\partial D_{11}}{\partial x_i}+\frac{\partial D_{31}}{\partial x_i})-(-D_{11}+D_{31})\frac{\partial detM}{\partial x_i}}{detM^2} \tag{4.381}$$

Third row of the Jacobian matrix $\nabla_x g_1(x)\mid_{(x^*,u^*)}$: $\frac{\partial g_{31}}{\partial x_1}=0$, $\frac{\partial g_{31}}{\partial x_2}=0$, $\frac{\partial g_{31}}{\partial x_3}=0$, $\frac{\partial g_{31}}{\partial x_4}=0$, $\frac{\partial g_{31}}{\partial x_5}=0$, $\frac{\partial g_{31}}{\partial x_6}=0$.

Fourth row of the Jacobian matrix $\nabla_x g_1(x)\mid_{(x^*,u^*)}$:

$$g_{41}(x)=\frac{-D_{12}-D_{22}}{detM} \tag{4.382}$$

Thus, for $i=1,2,\cdots,6$,

$$\frac{\partial g_{41}}{\partial x_i}=\frac{(\frac{-\partial D_{12}}{\partial x_i}-\frac{\partial D_{22}}{\partial x_i})+(-D_{12}-D_{22})\frac{\partial detM}{\partial x_i}}{detM^2} \tag{4.383}$$

Fifth row of the Jacobian matrix $\nabla_x g_1(x)\mid_{(x^*,u^*)}$: $\frac{\partial g_{51}}{\partial x_1}=0$, $\frac{\partial g_{51}}{\partial x_2}=0$, $\frac{\partial g_{51}}{\partial x_3}=0$, $\frac{\partial g_{51}}{\partial x_4}=0$, $\frac{\partial g_{51}}{\partial x_5}=0$, $\frac{\partial g_{51}}{\partial x_6}=0$.

Sixth row of the Jacobian matrix $\nabla_x g_1(x)\mid_{(x^*,u^*)}$:

$$g_{61}(x)=\frac{-D_{13}+D_{33}}{detM} \tag{4.384}$$

Thus, for $i=1,2,\cdots,6$,

$$\frac{\partial g_{61}}{\partial x_i}=\frac{(\frac{-\partial D_{13}}{\partial x_i}+\frac{\partial D_{33}}{\partial x_i})-(-D_{13}+D_{33})\frac{\partial detM}{\partial x_i}}{detM^2} \tag{4.385}$$

Computation of the Jacobian matrix $\nabla_x g_2(x)\mid_{(x^*,u^*)}$:

First row of the Jacobian matrix $\nabla_x g_2(x)\mid_{(x^*,u^*)}$: $\frac{\partial g_{12}}{\partial x_1}=0$, $\frac{\partial g_{12}}{\partial x_2}=0$, $\frac{\partial g_{12}}{\partial x_3}=0$, $\frac{\partial g_{12}}{\partial x_4}=0$, $\frac{\partial g_{12}}{\partial x_5}=0$, $\frac{\partial g_{12}}{\partial x_6}=0$.

Second row of the Jacobian matrix $\nabla_x g_2(x)\mid_{(x^*,u^*)}$:

$$g_{22}(x)=\frac{D_{21}+D_{31}}{detM} \tag{4.386}$$

Thus, for $i=1,2,\cdots,6$,

$$\frac{\partial g_{22}}{\partial x_i}=\frac{(\frac{\partial D_{21}}{\partial x_i}+\frac{\partial D_{31}}{\partial x_i})-(D_{21}+D_{31})\frac{\partial detM}{\partial x_i}}{detM^2} \tag{4.387}$$

Third row of the Jacobian matrix $\nabla_x g_2(x)\mid_{(x^*,u^*)}$: $\frac{\partial g_{32}}{\partial x_1}=0$, $\frac{\partial g_{32}}{\partial x_2}=0$, $\frac{\partial g_{32}}{\partial x_3}=0$, $\frac{\partial g_{32}}{\partial x_4}=0$, $\frac{\partial g_{32}}{\partial x_5}=0$, $\frac{\partial g_{32}}{\partial x_6}=0$.

Fourth row of the Jacobian matrix $\nabla_x g_2(x)\mid_{(x^*,u^*)}$:

$$g_{42}(x)=\frac{-D_{22}-D_{32}}{detM} \tag{4.388}$$

Thus, for $i=1,2,\cdots,6$,

$$\frac{\partial g_{42}}{\partial x_i}=\frac{(-\frac{\partial D_{22}}{\partial x_i}-\frac{\partial D_{32}}{\partial x_i})+(D_{22}+D_{32})\frac{\partial detM}{\partial x_i}}{detM^2} \tag{4.389}$$

Fifth row of the Jacobian matrix $\nabla_x g_2(x)\mid_{(x^*,u^*)}$: $\frac{\partial g_{52}}{\partial x_1}=0$, $\frac{\partial g_{52}}{\partial x_2}=0$, $\frac{\partial g_{52}}{\partial x_3}=0$, $\frac{\partial g_{52}}{\partial x_4}=0$, $\frac{\partial g_{52}}{\partial x_5}=0$, $\frac{\partial g_{52}}{\partial x_6}=0$.

Sixth row of the Jacobian matrix $\nabla_x g_2(x)\mid_{(x^*,u^*)}$:

$$g_{62}(x)=\frac{D_{23}+D_{33}}{detM} \tag{4.390}$$

Thus, for $i = 1, 2, \cdots, 6$,

$$\frac{\partial g_{62}}{\partial x_i} = \frac{(\frac{\partial D_{23}}{\partial x_i} + \frac{\partial D_{33}}{\partial x_i}) - (D_{23} + D_{33})\frac{\partial detM}{\partial x_i}}{detM^2} \tag{4.391}$$

Computation of the partial derivatives of the terms Q_i, $i = 1, 2, 3$

For $Q_1 = C_{11}x_2 + C_{12}x_4 + C_{13}x_6 + G_1 + \tilde{d}_1 x_2$, it holds that:

$$\frac{\partial Q_1}{\partial x_1} = \frac{\partial C_{11}}{\partial x_1}x_2 + \frac{\partial C_{12}}{\partial x_1}x_4 + \frac{\partial C_{13}}{\partial x_1}x_6 + \frac{\partial G_1}{\partial x_1} \quad \frac{\partial Q_1}{\partial x_2} = \frac{\partial C_{11}}{\partial x_2}x_2 + (C_{11} + \tilde{d}_1) + \frac{\partial C_{12}}{\partial x_2}x_4 + \frac{\partial C_{13}}{\partial x_2}x_6 + \frac{\partial G_1}{\partial x_2}$$

$$\frac{\partial Q_1}{\partial x_3} = \frac{\partial C_{11}}{\partial x_3}x_2 + \frac{\partial C_{12}}{\partial x_3}x_4 + \frac{\partial C_{13}}{\partial x_3}x_6 + \frac{\partial G_1}{\partial x_3} \quad \frac{\partial Q_1}{\partial x_4} = \frac{\partial C_{11}}{\partial x_4}x_2 + \frac{\partial C_{12}}{\partial x_4}x_4 + C_{12} + \frac{\partial C_{13}}{\partial x_4}x_6 + \frac{\partial G_1}{\partial x_4}$$

$$\frac{\partial Q_1}{\partial x_5} = \frac{\partial C_{11}}{\partial x_5}x_2 + \frac{\partial C_{12}}{\partial x_5}x_4 + \frac{\partial C_{13}}{\partial x_5}x_6 + \frac{\partial G_1}{\partial x_5} \quad \frac{\partial Q_1}{\partial x_6} = \frac{\partial C_{11}}{\partial x_6}x_2 + \frac{\partial C_{12}}{\partial x_6}x_4 + \frac{\partial C_{13}}{\partial x_6}x_6 + C_{13} + \frac{\partial G_1}{\partial x_6}$$

For $Q_2 = C_{21}x_2 + C_{22}x_4 + C_{23}x_6 + G_2 + \tilde{d}_2 x_4$, it holds that:

$$\frac{\partial Q_2}{\partial x_1} = \frac{\partial C_{21}}{\partial x_1}x_2 + \frac{\partial C_{22}}{\partial x_1}x_4 + \frac{\partial C_{23}}{\partial x_1}x_6 + \frac{\partial G_2}{\partial x_1} \quad \frac{\partial Q_2}{\partial x_2} = \frac{\partial C_{21}}{\partial x_2}x_2 + C_{21} + \frac{\partial C_{22}}{\partial x_2}x_4 + \frac{\partial C_{23}}{\partial x_2}x_6 + \frac{\partial G_2}{\partial x_2}$$

$$\frac{\partial Q_2}{\partial x_3} = \frac{\partial C_{21}}{\partial x_3}x_2 + \frac{\partial C_{22}}{\partial x_3}x_4 + \frac{\partial C_{23}}{\partial x_3}x_6 + \frac{\partial G_2}{\partial x_3} \quad \frac{\partial Q_2}{\partial x_4} = \frac{\partial C_{21}}{\partial x_4}x_2 + \frac{\partial C_{22}}{\partial x_4}x_4 + (C_{22} + \tilde{d}_2) + \frac{\partial C_{23}}{\partial x_4}x_6 + \frac{\partial G_2}{\partial x_4}$$

$$\frac{\partial Q_2}{\partial x_5} = \frac{\partial C_{21}}{\partial x_5}x_2 + \frac{\partial C_{22}}{\partial x_5}x_4 + \frac{\partial C_{23}}{\partial x_5}x_6 + \frac{\partial G_2}{\partial x_5} \quad \frac{\partial Q_2}{\partial x_6} = \frac{\partial C_{21}}{\partial x_6}x_2 + \frac{\partial C_{22}}{\partial x_6}x_4 + \frac{\partial C_{23}}{\partial x_6}x_6 + C_{23} + \frac{\partial G_2}{\partial x_6}$$

For $Q_3 = C_{31}x_2 + C_{32}x_4 + C_{33}x_6 + G_3 + \tilde{d}_3 x_6$, it holds that:

$$\frac{\partial Q_3}{\partial x_1} = \frac{\partial C_{31}}{\partial x_1}x_2 + \frac{\partial C_{32}}{\partial x_1}x_4 + \frac{\partial C_{33}}{\partial x_1}x_6 + \frac{\partial G_3}{\partial x_1} \quad \frac{\partial Q_3}{\partial x_2} = \frac{\partial C_{31}}{\partial x_2}x_2 + C_{31} + \frac{\partial C_{32}}{\partial x_2}x_4 + \frac{\partial C_{33}}{\partial x_2}x_6 + \frac{\partial G_3}{\partial x_2}$$

$$\frac{\partial Q_3}{\partial x_3} = \frac{\partial C_{31}}{\partial x_3}x_2 + \frac{\partial C_{32}}{\partial x_3}x_4 + \frac{\partial C_{33}}{\partial x_3}x_6 + \frac{\partial G_3}{\partial x_3} \quad \frac{\partial Q_3}{\partial x_4} = \frac{\partial C_{31}}{\partial x_4}x_2 + \frac{\partial C_{32}}{\partial x_4}x_4 + C_{32} + \frac{\partial C_{33}}{\partial x_4}x_6 + \frac{\partial G_3}{\partial x_4}$$

$$\frac{\partial Q_3}{\partial x_5} = \frac{\partial C_{31}}{\partial x_5}x_2 + \frac{\partial C_{32}}{\partial x_5}x_4 + \frac{\partial C_{33}}{\partial x_5}x_6 + \frac{\partial G_3}{\partial x_5} \quad \frac{\partial Q_3}{\partial x_6} = \frac{\partial C_{31}}{\partial x_6}x_2 + \frac{\partial C_{32}}{\partial x_6}x_4 + \frac{\partial C_{33}}{\partial x_6}x_6 + (C_{33} + \tilde{d}_3) + \frac{\partial G_3}{\partial x_6}$$

Computation of the partial derivatives of the sub-determinants D_{ij}, $i = 1, 2, 3, j = 1, 2, 3$ of the inertia matrix D.

For $D_{11} = d_{12}d_{33} - d_{32}d_{23}$, one has that for $i = 1, 2, \cdots, 6$

$$\frac{\partial D_{11}}{\partial x_i} = \frac{\partial d_{12}}{\partial x_i}d_{33} + d_{12}\frac{\partial d_{33}}{\partial x_i} - \frac{\partial d_{32}}{\partial x_i}d_{23} - d_{32}\frac{\partial d_{23}}{\partial x_i} \tag{4.392}$$

For $D_{12} = d_{21}d_{32} - d_{31}d_{23}$, one has that for $i = 1, 2, \cdots, 6$

$$\frac{\partial D_{12}}{\partial x_i} = \frac{\partial d_{21}}{\partial x_i}d_{32} + d_{21}\frac{\partial d_{32}}{\partial x_i} - \frac{\partial d_{31}}{\partial x_i}d_{23} - d_{31}\frac{\partial d_{23}}{\partial x_i} \tag{4.393}$$

For $D_{13} = d_{21}d_{32} - d_{31}d_{22}$, one has that for $i = 1, 2, \cdots, 6$

$$\frac{\partial D_{12}}{\partial x_i} = \frac{\partial d_{21}}{\partial x_i}d_{32} + d_{21}\frac{\partial d_{32}}{\partial x_i} - \frac{\partial d_{31}}{\partial x_i}d_{22} - d_{31}\frac{\partial d_{22}}{\partial x_i} \tag{4.394}$$

For $D_{21} = d_{12}d_{33} - d_{32}d_{13}$, one has that for $i = 1, 2, \cdots, 6$

$$\frac{\partial D_{21}}{\partial x_i} = \frac{\partial d_{12}}{\partial x_i}d_{33} + d_{12}\frac{\partial d_{33}}{\partial x_i} - \frac{\partial d_{32}}{\partial x_i}d_{13} - d_{32}\frac{\partial d_{13}}{\partial x_i} \tag{4.395}$$

For $D_{22} = d_{11}d_{33} - d_{31}d_{13}$, one has that for $i = 1, 2, \cdots, 6$

$$\frac{\partial D_{22}}{\partial x_i} = \frac{\partial d_{11}}{\partial x_i}d_{33} + d_{11}\frac{\partial d_{33}}{\partial x_i} - \frac{\partial d_{31}}{\partial x_i}d_{13} - d_{31}\frac{\partial d_{13}}{\partial x_i} \tag{4.396}$$

For $D_{23} = d_{11}d_{32} - d_{31}d_{12}$, one has that for $i = 1, 2, \cdots, 6$

$$\frac{\partial D_{23}}{\partial x_i} = \frac{\partial d_{11}}{\partial x_i}d_{32} + d_{11}\frac{\partial d_{32}}{\partial x_i} - \frac{\partial d_{31}}{\partial x_i}d_{12} - d_{31}\frac{\partial d_{12}}{\partial x_i} \tag{4.397}$$

For $D_{12} = d_{12}d_{23} - d_{22}d_{13}$, one has that for $i = 1, 2, \cdots, 6$

$$\frac{\partial D_{31}}{\partial x_i} = \frac{\partial d_{12}}{\partial x_i}d_{23} + d_{12}\frac{\partial d_{23}}{\partial x_i} - \frac{\partial d_{22}}{\partial x_i}d_{13} - d_{22}\frac{\partial d_{13}}{\partial x_i} \tag{4.398}$$

For $D_{32}=d_{11}d_{23}-d_{21}d_{13}$, one has that for $i=1,2,\cdots,6$

$$\frac{\partial D_{32}}{\partial x_i}=\frac{\partial d_{11}}{\partial x_i}d_{23}+d_{11}\frac{\partial d_{23}}{\partial x_i}-\frac{\partial d_{21}}{\partial x_i}d_{13}-d_{21}\frac{\partial d_{13}}{\partial x_i} \tag{4.399}$$

For $D_{33}=d_{11}d_{22}-d_{21}d_{12}$, one has that for $i=1,2,\cdots,6$

$$\frac{\partial D_{33}}{\partial x_i}=\frac{\partial d_{11}}{\partial x_i}d_{12}+d_{11}\frac{\partial d_{12}}{\partial x_i}-\frac{\partial d_{21}}{\partial x_i}d_{12}-d_{21}\frac{\partial d_{12}}{\partial x_i} \tag{4.400}$$

Computation of the partial derivatives of the elements d_{ij}, $i=1,2,3$, $j=1,2,3$ of the inertia matrix:

For $d_{11}=(\frac{5}{4}m+M_H+M_T)r^2$: it holds that $\frac{\partial d_{11}}{\partial x_i}=0$ for $i=1,2,\cdots,6$.

For $d_{12}=-\frac{1}{2}mr^2cos(x_1-x_3)$, it holds that: $\frac{\partial d_{12}}{\partial x_1}=\frac{1}{2}mr^2sin(x_1-x_3)$, $\frac{\partial d_{12}}{\partial x_2}=0$, $\frac{\partial d_{12}}{\partial x_3}=-\frac{1}{2}mr^2sin(x_1-x_3)$, $\frac{\partial d_{12}}{\partial x_4}=0$, $\frac{\partial d_{12}}{\partial x_5}=0$, $\frac{\partial d_{12}}{\partial x_6}=0$.

For $d_{13}=M_Trlcos(x_1-x_5)$, it holds that $\frac{\partial d_{13}}{\partial x_1}=-M_Trlsin(x_1-x_5)$, $\frac{\partial d_{13}}{\partial x_2}=0$, $\frac{\partial d_{13}}{\partial x_3}=0$, $\frac{\partial d_{13}}{\partial x_4}=0$, $\frac{\partial d_{13}}{\partial x_5}=M_rrlsin(x_1-x_5)$, $\frac{\partial d_{13}}{\partial x_6}=0$.

For $d_{21}=-\frac{1}{2}mr^2cos(x_1-x_3)$, it holds that $\frac{\partial d_{21}}{\partial x_1}=\frac{1}{2}mr^2sin(x_1-x_3)$, $\frac{\partial d_{21}}{\partial x_2}=0$, $\frac{\partial d_{21}}{\partial x_3}=-\frac{1}{2}mr^2sin(x_1-x_3)$, $\frac{\partial d_{21}}{\partial x_4}=0$, $\frac{\partial d_{21}}{\partial x_5}=0$, $\frac{\partial d_{21}}{\partial x_6}=0$.

For $d_{22}=\frac{1}{4}mr^2$, it holds that $\frac{\partial d_{22}}{\partial x_i}=0$ for $i=1,2,\cdots,6$.

For $d_{23}=0$, it holds that $\frac{\partial d_{23}}{\partial x_i}=0$ for $i=1,2,\cdots,6$.

For $d_{31}=M_Trlcos(x_1-x_5)$, it holds that $\frac{\partial d_{31}}{\partial x_1}=-M_Trlsin(x_1-x_5)$, $\frac{\partial d_{31}}{\partial x_2}=0$, $\frac{\partial d_{31}}{\partial x_3}=0$, $\frac{\partial d_{31}}{\partial x_4}=0$, $\frac{\partial d_{31}}{\partial x_5}=M_rrlsin(x_1-x_5)$, $\frac{\partial d_{31}}{\partial x_6}=0$.

For $d_{32}=0$, it holds that $\frac{\partial d_{32}}{\partial x_i}=0$ for $i=1,2,\cdots,6$.

For $d_{33}=M_Tl^2$, it holds that $\frac{\partial d_{33}}{\partial x_i}=0$ for $i=1,2,\cdots,6$.

It also holds that $detD=d_{11}D_{11}-d_{12}D_{12}+d_{13}D_{13}$ thus for $i=1,2,\cdots,6$

$$\frac{\partial detD}{\partial x_i}=\frac{\partial d_{11}}{\partial x_i}D_{11}+d_{11}\frac{\partial D_{11}}{\partial x_i}-\frac{\partial d_{12}}{\partial x_i}D_{12}-d_{12}\frac{\partial D_{12}}{\partial x_i}+\frac{\partial d_{13}}{\partial x_i}D_{13}+d_{13}\frac{\partial D_{13}}{\partial x_i} \tag{4.401}$$

Computation of the partial derivatives of the elements C_{ij}, $i=1,2,3$, and $j=1,2,3$ of the Coriolis matrix

For $C_{11}=0$, one obtains $\frac{\partial C_{11}}{\partial x_i}=0$ for $i=1,2,\cdots,6$.

For $C_{12}=-\frac{1}{2}mr^2sin(x_1-x_3)x_4$, one obtains $\frac{\partial C_{12}}{\partial x_1}=-\frac{1}{2}mr^2cos(x_1-x_3)x_4$, $\frac{\partial C_{12}}{\partial x_2}=0$, $\frac{\partial C_{12}}{\partial x_3}=\frac{1}{2}mr^2cos(x_1-x_3)x_4$, $\frac{\partial C_{12}}{\partial x_4}=-\frac{1}{2}mr^2sin(x_1-x_3)x_4$, $\frac{\partial C_{12}}{\partial x_5}=0$, $\frac{\partial C_{12}}{\partial x_6}=0$.

For $C_{13}=M_Trlsin(x_11-x_5)x_6$, one obtains $\frac{\partial C_{13}}{\partial x_1}=M_Trlcos(x_1-x_5)x_6$, $\frac{\partial C_{13}}{\partial x_2}=0$, $\frac{\partial C_{13}}{\partial x_3}=0$, $\frac{\partial C_{13}}{\partial x_4}=0$, $\frac{\partial C_{13}}{\partial x_5}=-M_Trlcos(x_1-x_5)x_6$, $\frac{\partial C_{13}}{\partial x_6}=M_Trlsin(x_1-x_5)$.

For $C_{21}=\frac{1}{2}mr^2sin(x_1-x_3)x_2$, one obtains $\frac{\partial C_{21}}{\partial x_1}=\frac{1}{2}mr^2sin(x_1-x_3)x_2$, one obtains $\frac{\partial C_{21}}{\partial x_1}=\frac{1}{2}mr^2cos(x_1-x_3)x_2$, $\frac{\partial C_{13}}{\partial x_2}=\frac{1}{2}mr^2sin(x_1-x_3)$, $\frac{\partial C_{21}}{\partial x_3}=-\frac{1}{2}mr^2cos(x_1-x_3)x_2$, $\frac{\partial C_{21}}{\partial x_4}=0$, $\frac{\partial C_{21}}{\partial x_5}=0$, $\frac{\partial C_{21}}{\partial x_6}=0$.

For $C_{22}=0$, one obtains $\frac{\partial C_{22}}{\partial x_i}=0$ for $i=1,2,\cdots,6$.

For $C_{23}=0$, one obtains $\frac{\partial C_{23}}{\partial x_i}=0$ for $i=1,2,\cdots,6$.

For $C_{31}=-M_T rlcos(x_1-x_3)$, one obtains $\frac{\partial C_{31}}{\partial x_1}=-M_T rlcos(x_1-x_5)x_2$, $\frac{\partial C_{31}}{\partial x_2}=-M_T rlsin(x_1-x_5)$, $\frac{\partial C_{31}}{\partial x_3}=0$, $\frac{\partial C_{31}}{\partial x_4}=0$, $\frac{\partial C_{31}}{\partial x_5}=M_T rlcos(x_1-x_5)x_2$, $\frac{\partial C_{31}}{\partial x_6}=0$.

For $C_{32}=0$, one obtains $\frac{\partial C_{32}}{\partial x_i}=0$ for $i=1,2,\cdots,6$.

For $C_{33}=0$, one obtains $\frac{\partial C_{33}}{\partial x_i}=0$ for $i=1,2,\cdots,6$.

Computation of the partial derivatives of the elements G_i of the gravitational forces vector

For $G_1=-\frac{1}{2}g(2M_H+3m+2M_T)rsin(x_1)$, it holds that $\frac{\partial G_1}{\partial x_1}=-\frac{1}{2}(2M_H+3m+2M_T)rcos(x_1)$ and $\frac{\partial G_1}{\partial x_1}=0$ for $i=2,3,4,5,6$.

For $G_2=\frac{1}{2}gmrsin(x_3)$, it holds that $\frac{\partial G_2}{\partial x_3}=\frac{1}{2}gmrcos(x_3)$ and $\frac{\partial G_2}{\partial x_i}=0$ for $i=1,2,4,5,6$.

For $G_3=-gM_T lsin(x_5)$, it holds that $\frac{\partial G_3}{\partial x_5}=-gM_T lcos(x_5)$ and $\frac{\partial G_3}{\partial x_i}=0$ for $i=1,2,3,4,6$.

The linearization approach which has been followed for implementing the nonlinear optimal control scheme results in a quite accurate model of the system's dynamics. Consider for instance the following affine-in-the-input state-space model:

$$\begin{gathered}\dot{x}=f(x)+g(x)u\Rightarrow\\ \dot{x}=[f(x^*)+\nabla_x f(x)\mid_{x^*}(x-x^*)]+[g(x^*)+\nabla_x g(x)\mid_{x^*}(x-x^*)]u^*+g(x^*)u^*+g(x^*)(u-u^*)+\tilde{d}_1\Rightarrow\\ \dot{x}=[\nabla_x f(x)\mid_{x^*}+\nabla_x g(x)\mid_{x^*}u^*]x+g(x^*)u-[\nabla_x f(x)\mid_{x^*}+\nabla_x g(x)\mid_{x^*}u^*]x^*+f(x^*)+g(x^*)u^*+\tilde{d}_1\end{gathered} \tag{4.402}$$

where $\tilde{d}_1$ is the modeling error due to truncation of higher order terms in the Taylor series expansion of $f(x)$ and $g(x)$. Next, by defining $A=[\nabla_x f(x)\mid_{x^*}+\nabla_x g(x)\mid_{x^*}u^*]$, $B=g(x^*)$, one obtains

$$\dot{x}=Ax+Bu-Ax^*+f(x^*)+g(x^*)u^*+\tilde{d}_1 \tag{4.403}$$

Moreover, by denoting $\tilde{d}=-Ax^*+f(x^*)+g(x^*)u^*+\tilde{d}_1$ about the cumulative modeling error term in the Taylor series expansion, one has

$$\dot{x}=Ax+Bu+\tilde{d} \tag{4.404}$$

which is the approximately linearized model of the dynamics of the system of Eq. (4.368). The term $f(x^*)+g(x^*)u^*$ is the derivative of the state vector at (x^*,u^*) which is almost annihilated by $-Ax^*$.

4.4.4 *Design of an H-infinity controller for the biped robot*

4.4.4.1 Equivalent linearized dynamics of the biped robot

After linearization around its current operating point, the dynamic model for the three-link biped robot of Eq. (4.366), is written as

$$\dot{x}=Ax+Bu+d_1 \tag{4.405}$$

Parameter d_1 stands for the linearization error in the three-link biped robot's model that was given previously in Eq. (4.405). The reference setpoints for the state vector of the aforementioned dynamic model are denoted by $\mathbf{x_d}=[x_1^d,\cdots,x_6^d]$. Tracking of this trajectory is achieved after applying the control input u_d. At every time instant the control input u_d is assumed to differ from the control input u appearing in Eq. (4.405) by an amount equal to Δu, that is $u_d = u + \Delta u$

$$\dot{x}_d = Ax_d + Bu_d + d_2 \tag{4.406}$$

The dynamics of the controlled system described in Eq. (4.405) can be also written as

$$\dot{x} = Ax + Bu + Bu_d - Bu_d + d_1 \tag{4.407}$$

and by denoting $d_3 = -Bu_d + d_1$ as an aggregate disturbance term, one obtains

$$\dot{x} = Ax + Bu + Bu_d + d_3 \tag{4.408}$$

By subtracting Eq. (4.406) from Eq. (4.408), one has

$$\dot{x} - \dot{x}_d = A(x - x_d) + Bu + d_3 - d_2 \tag{4.409}$$

By denoting the tracking error as $e = x - x_d$ and the aggregate disturbance term as $L\tilde{d} = d_3 - d_2$, the tracking error dynamics becomes

$$\dot{e} = Ae + Bu + L\tilde{d} \tag{4.410}$$

with L to be denoting the disturbance inputs gain matrix. The above-linearized form of the three-link biped robot can be efficiently controlled after applying an H-infinity feedback control scheme.

4.4.4.2 The nonlinear H-infinity control

The initial nonlinear model of the three-link biped robot is in the form

$$\dot{x} = f(x, u) \;\; x \in R^n, \; u \in R^m \tag{4.411}$$

Linearization of the model of the three-link biped robot is performed at each iteration of the control algorithm round its present operating point $(x^*, u^*) = (x(t), u(t - T_s))$. The linearized equivalent of the system is described by

$$\dot{x} = Ax + Bu + L\tilde{d} \;\; x \in R^n, \; u \in R^m, \; \tilde{d} \in R^q \tag{4.412}$$

where matrices A and B are obtained from the computation of the previously defined Jacobians and vector $\tilde{d}$ denotes disturbance terms due to linearization errors. The problem of disturbance rejection for the linearized model that is described by

$$\begin{aligned} \dot{x} &= Ax + Bu + L\tilde{d} \\ y &= Cx \end{aligned} \tag{4.413}$$

where $x \in R^n$, $u \in R^m$, $\tilde{d} \in R^q$, and $y \in R^p$, cannot be handled efficiently if the classical LQR control scheme is applied. This is because of the existence of the perturbation term $\tilde{d}$. The disturbance term $\tilde{d}$ apart from modeling (parametric) uncertainty and external perturbation terms can also represent noise terms of any distribution.

In the H_∞ control approach, a feedback control scheme is designed for trajectory tracking by the system's state vector and simultaneous disturbance rejection, considering that the disturbance affects the system in the worst possible manner. The disturbances' effects are incorporated in the following quadratic cost function:

$$J(t) = \frac{1}{2}\int_0^T [y^T(t)y(t) + ru^T(t)u(t) - \rho^2\tilde{d}^T(t)\tilde{d}(t)]dt, \;\; r, \rho > 0 \tag{4.414}$$

The significance of the negative sign in the cost function's term that is associated with the perturbation variable $\tilde{d}(t)$ is that the disturbance tries to maximize the cost function $J(t)$ while the control signal $u(t)$ tries to minimize it. The physical meaning of the relation given above is that the control signal and the disturbances compete with each other within a min–max differential game. This problem of min–max optimization can be written as $min_u max_{\tilde{d}} J(u, \tilde{d})$.

The objective of the optimization procedure is to compute a control signal $u(t)$ which can compensate for the worst possible disturbance, that is externally imposed to the three-link biped robot. However, the solution to the min–max optimization problem is directly related to the value of the parameter ρ. This means that there is an upper bound in the disturbances magnitude that can be annihilated by the control signal.

4.4.4.3 Computation of the feedback control gains

For the linearized system given by Eq. (4.413) the cost function of Eq. (4.414) is defined, where coefficient r determines the penalization of the control input and weight coefficient ρ determines the reward of the disturbances' effects. It is assumed that (i) The energy that is transferred from the disturbances signal $\tilde{d}(t)$ is bounded, that is $\int_0^{\infty}\tilde{d}^T(t)\tilde{d}(t)dt < \infty$, (ii) matrices $[A, B]$ and $[A, L]$ are stabilizable, and (iii) matrix $[A, C]$ is detectable. In the case of a tracking problem, the optimal feedback control law is given by

$$u(t) = -Ke(t) \tag{4.415}$$

with $e = x - x_d$ to be the tracking error, and $K = \frac{1}{r}B^TP$ where P is a positive definite symmetric matrix. As it will be proven in Section 4.4.5, matrix P is obtained from the solution of the Riccati equation

$$A^TP + PA + Q - P(\tfrac{2}{r}BB^T - \tfrac{1}{\rho^2}LL^T)P = 0 \tag{4.416}$$

where Q is a positive semi-definite symmetric matrix. The worst-case disturbance is given by

$$\tilde{d}(t) = \tfrac{1}{\rho^2}L^TPe(t) \tag{4.417}$$

The solution of the H-infinity feedback control problem for the three-link biped robot and the computation of the worst-case disturbance that the related controller can sustain, comes from superposition of Bellman's optimality principle when considering that the robot is affected by two separate inputs: (i) the control input u and (ii) the cumulative disturbance input $\tilde{d}(t)$. Solving the optimal control problem for u, that is for the minimum variation (optimal) control input that achieves the elimination of the state vector's tracking error, gives $u = -\frac{1}{r}B^TPe$. Equivalently, solving the optimal control problem for $\tilde{d}$, that is for the worst-case disturbance that the control loop can sustain gives $\tilde{d} = \frac{1}{\rho^2}L^TPe$. The diagram of the considered control loop for the three-link biped robot is depicted in Figure 4.61.

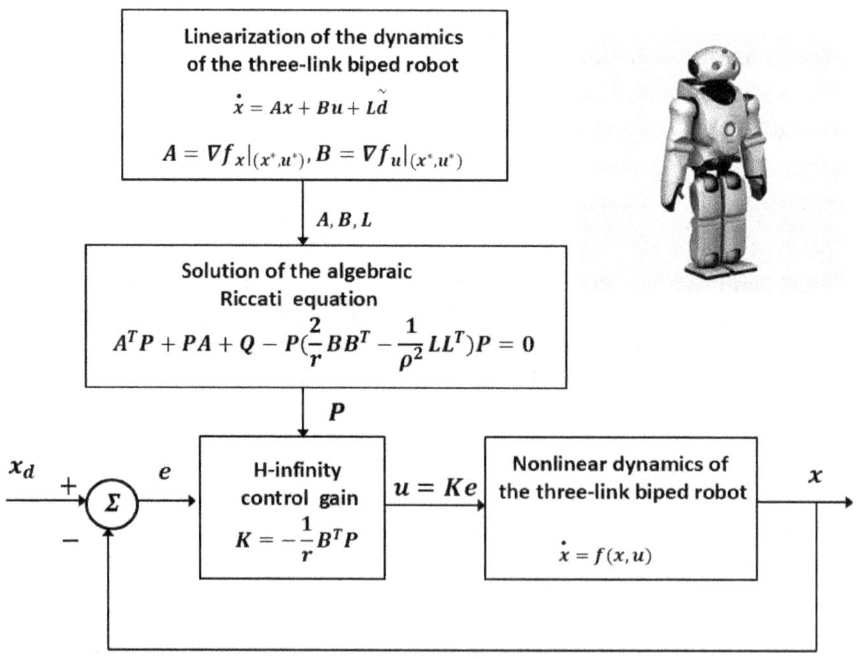

Figure 4.61 Diagram of the control scheme for the three-link biped robot

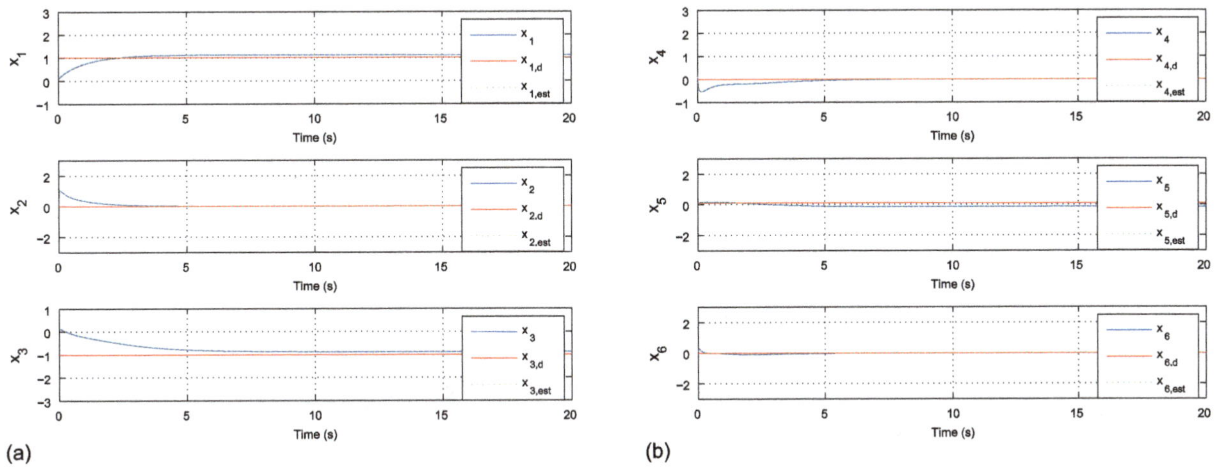

Figure 4.62 Test with setpoint 1 for the three-link biped robot: (a) tracking of setpoints (red line) by state variables x_1 to x_3 (blue line) and estimates provided by the H-infinity Kalman filter (green line); (b) tracking of setpoints (red line) by state variables x_4 to x_6 (blue line) and estimates provided by the H-infinity Kalman filter (green line)

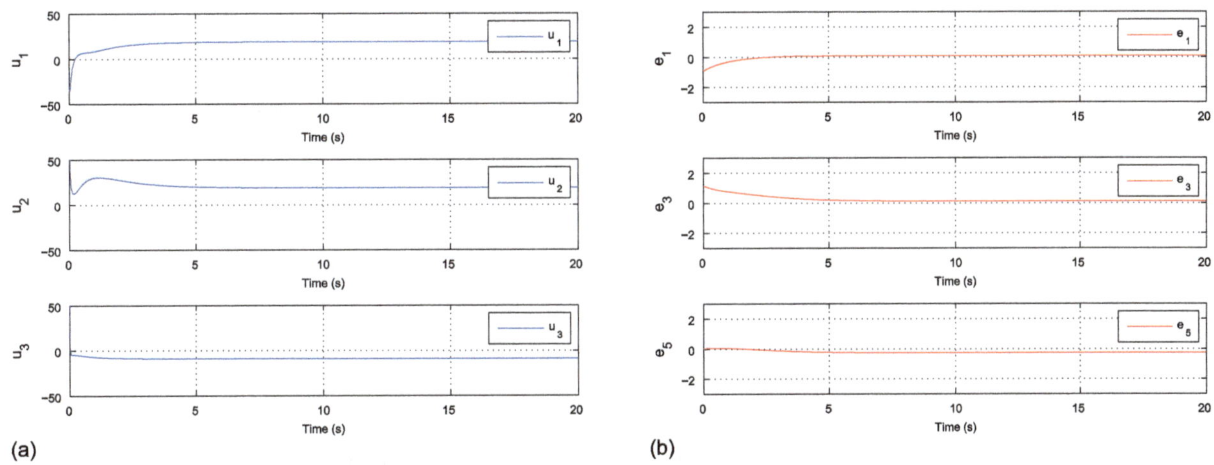

Figure 4.63 Test with setpoint 1 for the three-link biped robot: (a) variation of the control inputs u_1 to u_3 which are applied to the biped robot; (b) tracking error dynamics for state variables x_1, x_3, and x_5

4.4.5 Lyapunov stability analysis

Through Lyapunov stability analysis, it will be shown that the proposed nonlinear control scheme assures H_∞ tracking performance for the three-link biped robot's dynamics, and that under moderate conditions asymptotic convergence to the reference setpoints (global stability) is achieved [721]. As shown before, the tracking error dynamics for the three-link biped robot's dynamics is written in the form [721]

$$\dot{e} = Ae + Bu + L\tilde{d} \tag{4.418}$$

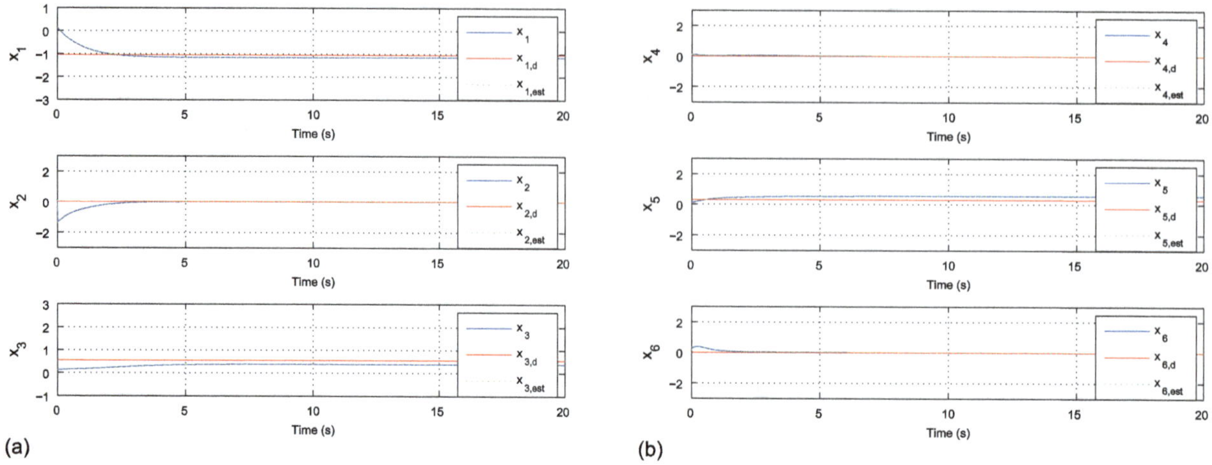

Figure 4.64 Test with setpoint 2 for the three-link biped robot: (a) tracking of setpoints (red line) by state variables x_1 to x_3 (blue line) and estimates provided by the H-infinity Kalman filter (green line); (b) tracking of setpoints (red line) by state variables x_4 to x_6 (blue line) and estimates provided by the H-infinity Kalman filter (green line)

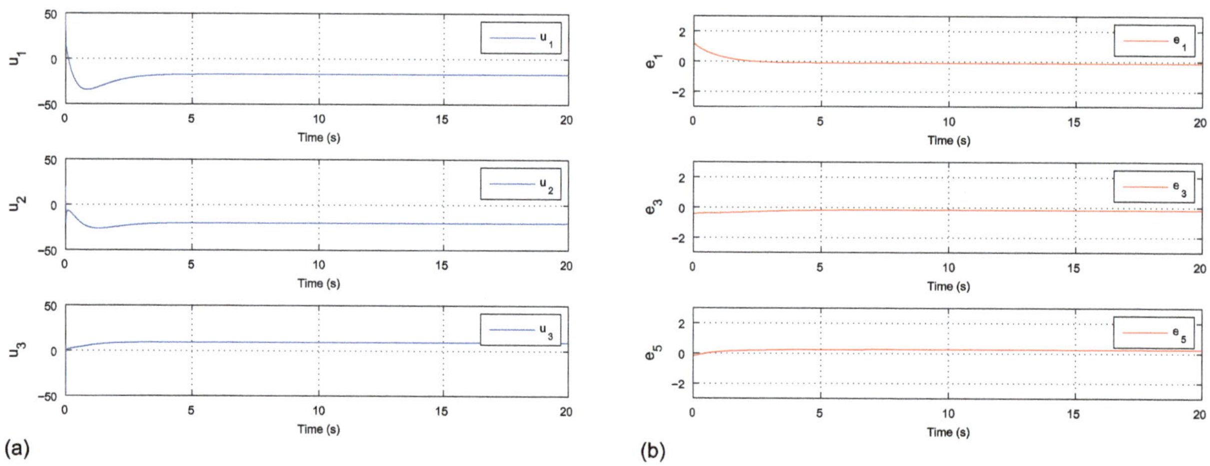

Figure 4.65 Test with setpoint 2 for the three-link biped robot: (a) variation of the control inputs u_1 to u_3 which are applied to the biped robot; (b) tracking error dynamics for state variables x_1, x_3, and x_5

where in the three-link biped robot's case $L\in R^{6\times 6}$ with I being the disturbance inputs gain matrix. Variable $\tilde{d}$ denotes model uncertainties and external disturbances of the biped robot's dynamic model. The following Lyapunov equation is considered

$$V=\frac{1}{2}e^TPe \tag{4.419}$$

where $e=x-x_d$ is the tracking error. By differentiating with respect to time, one obtains

$$\begin{gathered}\dot{V}=\frac{1}{2}\dot{e}^TPe+\frac{1}{2}e^TP\dot{e}\Rightarrow\\ \dot{V}=\frac{1}{2}[Ae+Bu+L\tilde{d}]^TPe+\frac{1}{2}e^TP[Ae+Bu+L\tilde{d}]\Rightarrow\end{gathered} \tag{4.420}$$

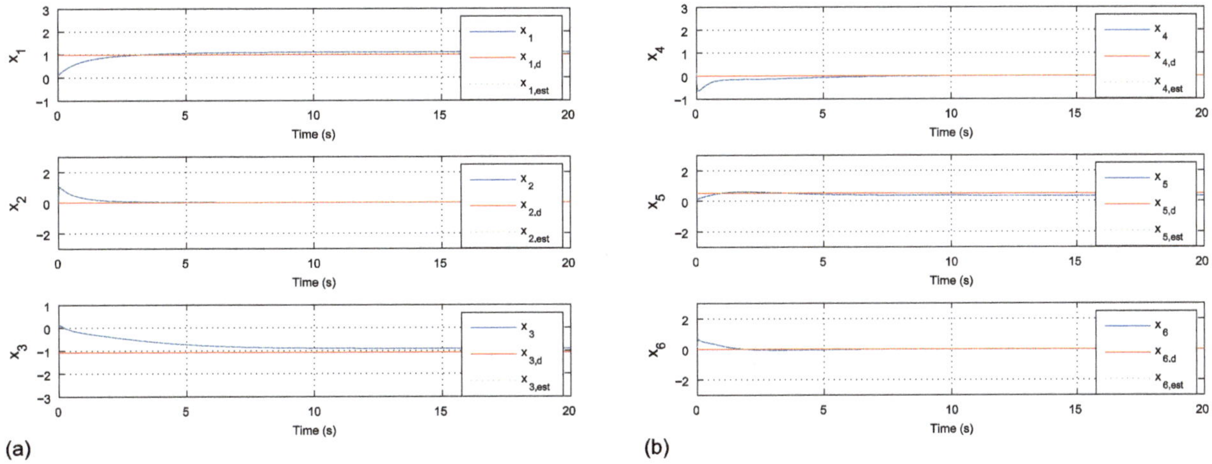

Figure 4.66 *Test with setpoint 3 for the three-link biped robot: (a) tracking of setpoints (red line) by state variables x_1 to x_3 (blue line) and estimates provided by the H-infinity Kalman filter (green line); (b) tracking of setpoints (red line) by state variables x_4 to x_6 (blue line) and estimates provided by the H-infinity Kalman filter (green line)*

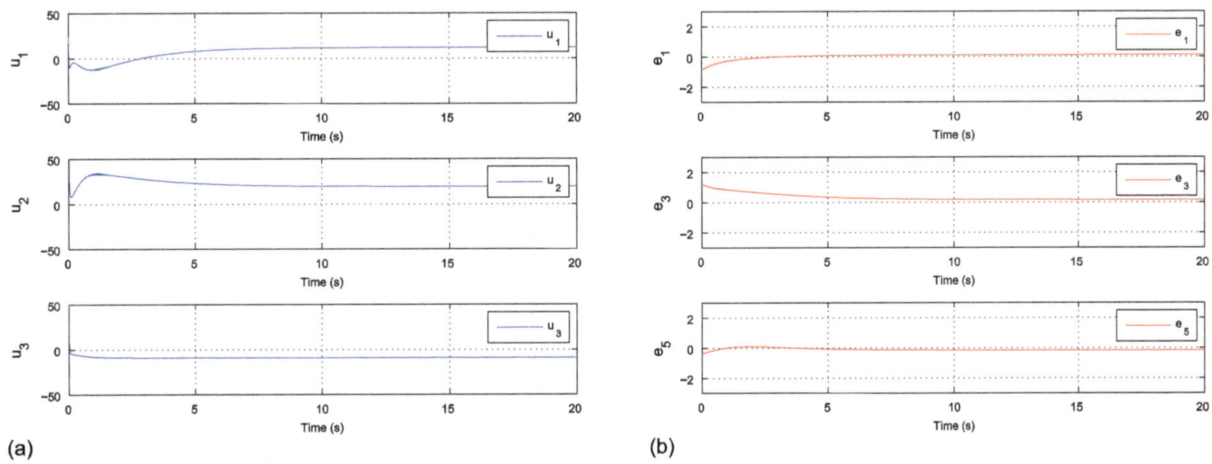

Figure 4.67 *Test with setpoint 3 for the three-link biped robot: (a) variation of the control inputs u_1 to u_3 which are applied to the biped robot; (b) tracking error dynamics for state variables x_1, x_3, and x_5*

$$\begin{aligned}\dot{V} &= \tfrac{1}{2}[e^TA^T + u^TB^T + \tilde{d}^TL^T]Pe \\ &\quad + \tfrac{1}{2}e^TP[Ae + Bu + L\tilde{d}] \Rightarrow\end{aligned} \tag{4.421}$$

$$\begin{aligned}\dot{V} &= \tfrac{1}{2}e^TA^TPe + \tfrac{1}{2}u^TB^TPe + \tfrac{1}{2}\tilde{d}^TL^TPe \\ &\quad + \tfrac{1}{2}e^TPAe + \tfrac{1}{2}e^TPBu + \tfrac{1}{2}e^TPL\tilde{d}\end{aligned} \tag{4.422}$$

The previous equation is rewritten as

$$\begin{aligned}\dot{V} &= \tfrac{1}{2}e^T(A^TP + PA)e + (\tfrac{1}{2}u^TB^TPe + \tfrac{1}{2}e^TPBu) \\ &\quad + (\tfrac{1}{2}\tilde{d}^TL^TPe + \tfrac{1}{2}e^TPL\tilde{d})\end{aligned} \tag{4.423}$$

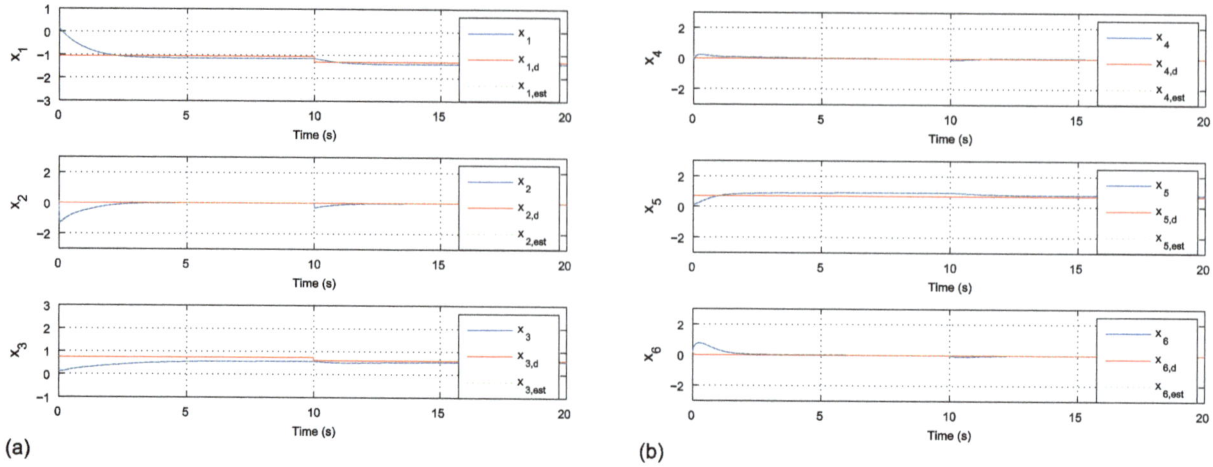

Figure 4.68 *Test with setpoint 4 for the three-link biped robot: (a) tracking of setpoints (red line) by state variables x_1 to x_3 (blue line) and estimates provided by the H-infinity Kalman filter (green line) and (b) tracking of setpoints (red line) by state variables x_4 to x_6 (blue line) and estimates provided by the H-infinity Kalman filter (green line)*

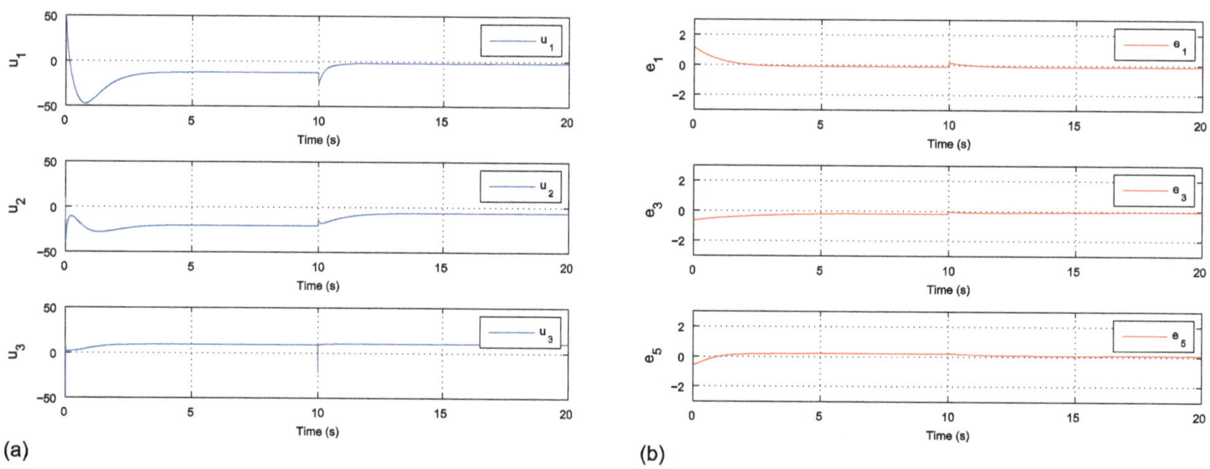

Figure 4.69 *Test with setpoint 4 for the three-link biped robot: (a) variation of the control inputs u_1 to u_3 which are applied to the biped robot; (b) tracking error dynamics for state variables x_1, x_3, and x_5*

Assumption: For given positive definite matrix Q and coefficients r and ρ there exists a positive definite matrix P, which is the solution of the following matrix equation:

$$A^TP + PA = -Q + P(\tfrac{2}{r}BB^T - \tfrac{1}{\rho^2}LL^T)P \tag{4.424}$$

Moreover, the following feedback control law is applied to the system

$$u = -\tfrac{1}{r}B^TPe \tag{4.425}$$

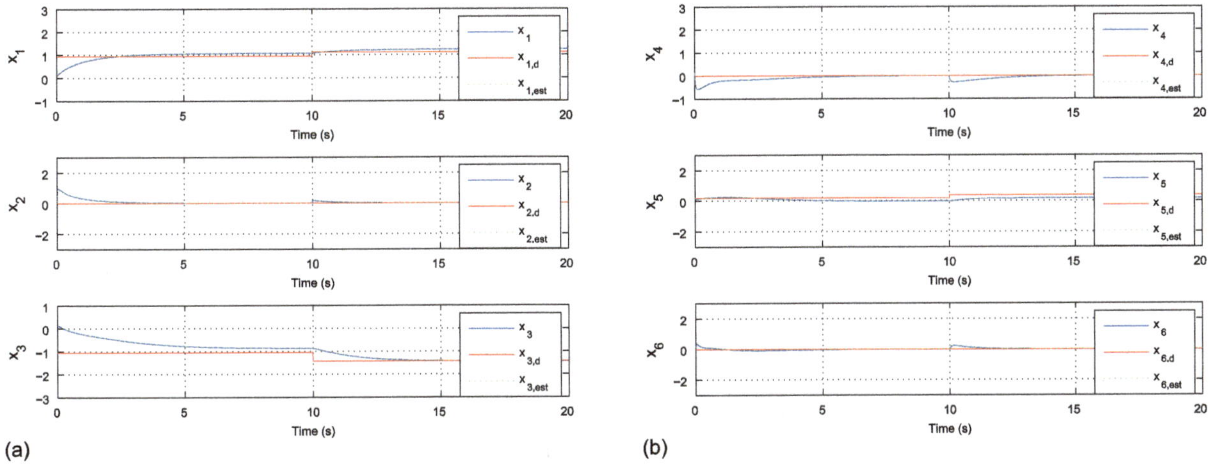

Figure 4.70 *Test with setpoint 5 for the three-link biped robot: (a) tracking of setpoints (red line) by state variables x_1 to x_3 (blue line) and estimates provided by the H-infinity Kalman filter (green line); (b) tracking of setpoints (red line) by state variables x_4 to x_6 (blue line) and estimates provided by the H-infinity Kalman filter (green line)*

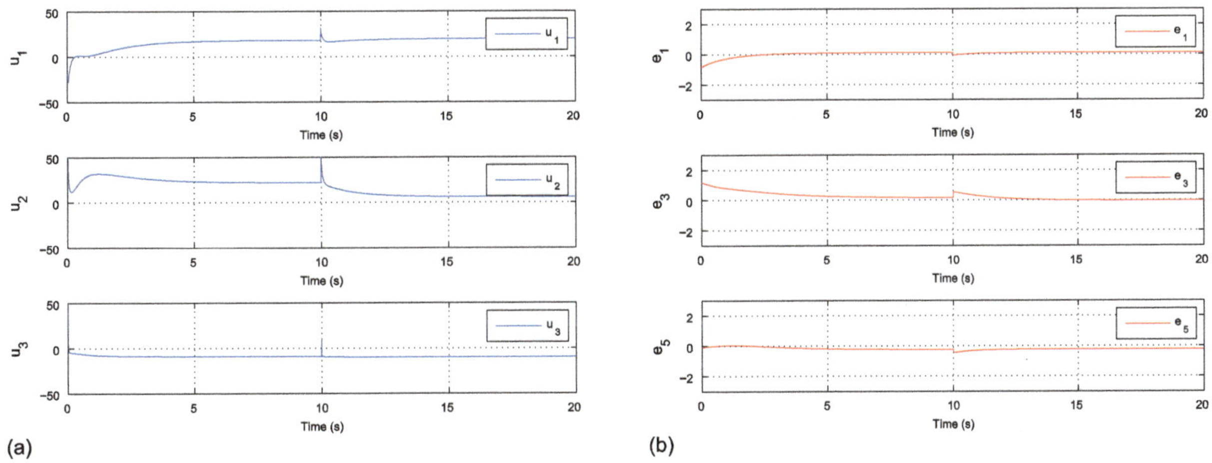

Figure 4.71 *Test with setpoint 5 for the three-link biped robot: (a) variation of the control inputs u_1 to u_3 which are applied to the biped robot; (b) tracking error dynamics for state variables x_1, x_3, and x_5*

By substituting Eq. (4.424) and Eq. (4.425), one obtains

$$\begin{aligned}\dot{V} = &\tfrac{1}{2}e^T[-Q + P(\tfrac{2}{r}BB^T - \tfrac{1}{2\rho^2}LL^T)P]e \\ &+e^TPB(-\tfrac{1}{r}B^TPe) + e^TPL\tilde{d} \Rightarrow\end{aligned} \tag{4.426}$$

$$\begin{aligned}\dot{V} = &-\tfrac{1}{2}e^TQe + (\tfrac{2}{r}e^TPBB^TPe - \tfrac{1}{2\rho^2}e^TPLL^TPe \\ &-\tfrac{1}{r}e^TPBB^TPe) + e^TPL\tilde{d}\end{aligned} \tag{4.427}$$

which after intermediate operations gives

$$\dot{V} = -\tfrac{1}{2}e^TQe - \tfrac{1}{2\rho^2}e^TPLL^TPe + e^TPL\tilde{d} \tag{4.428}$$

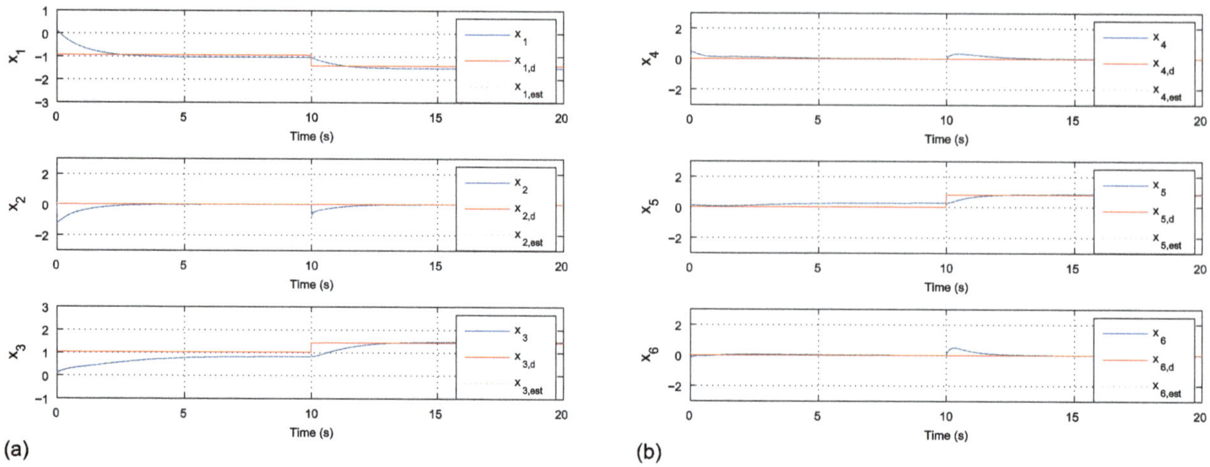

Figure 4.72 *Test with setpoint 6 for the three-link biped robot: (a) tracking of setpoints (red line) by state variables x_1 to x_3 (blue line) and estimates provided by the H-infinity Kalman filter (green line); (b) tracking of setpoints (red line) by state variables x_4 to x_6 (blue line) and estimates provided by the H-infinity Kalman filter (green line)*

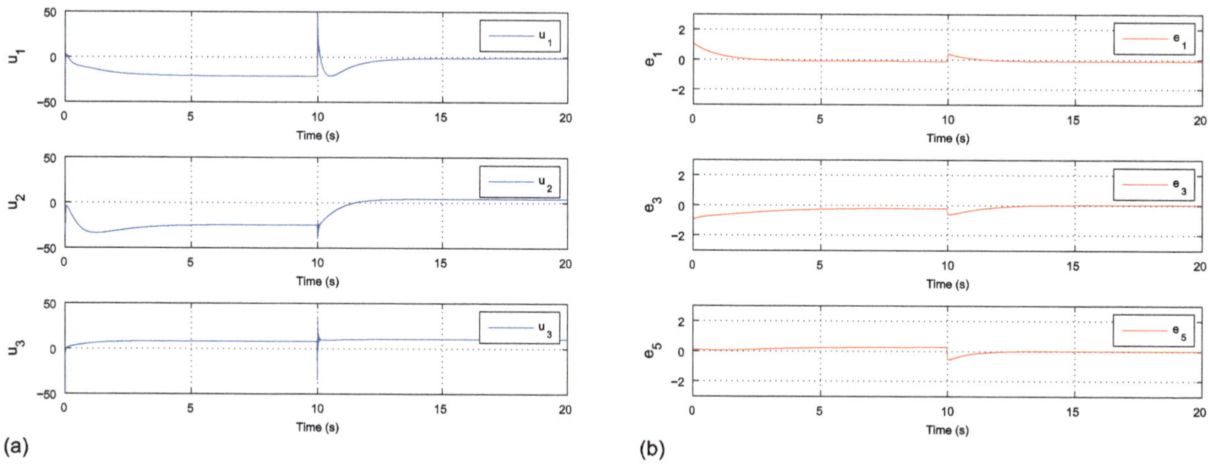

Figure 4.73 *Test with setpoint 6 for the three-link biped robot: (a) variation of the control inputs u_1 to u_3 which are applied to the biped robot; (b) tracking error dynamics for state variables x_1, x_3, and x_5*

or, equivalently

$$\begin{aligned}\dot{V} = & -\tfrac{1}{2}e^TQe - \tfrac{1}{2\rho^2}e^TPLL^TPe \\ & +\tfrac{1}{2}e^TPL\tilde{d} + \tfrac{1}{2}\tilde{d}^TL^TPe\end{aligned} \tag{4.429}$$

Lemma: The following inequality holds:

$$\tfrac{1}{2}e^TPL\tilde{d} + \tfrac{1}{2}\tilde{d}L^TPe - \tfrac{1}{2\rho^2}e^TPLL^TPe \leq \tfrac{1}{2}\rho^2\tilde{d}^T\tilde{d} \tag{4.430}$$

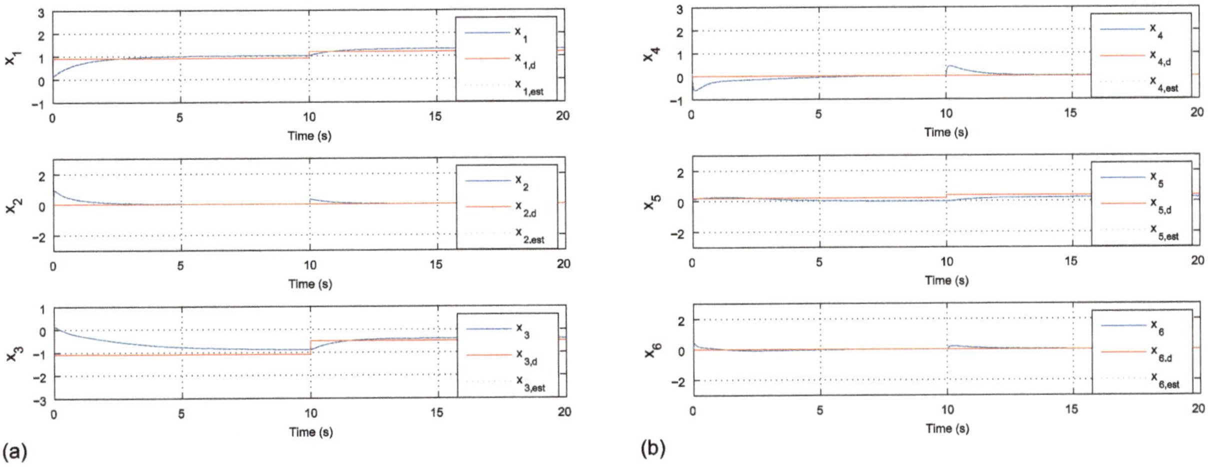

Figure 4.74 Test with setpoint 7 for the three-link biped robot: (a) tracking of setpoints (red line) by state variables x_1 to x_3 (blue line) and estimates provided by the H-infinity Kalman filter (green line); (b) tracking of setpoints (red line) by state variables x_4 to x_6 (blue line) and estimates provided by the H-infinity Kalman filter (green line)

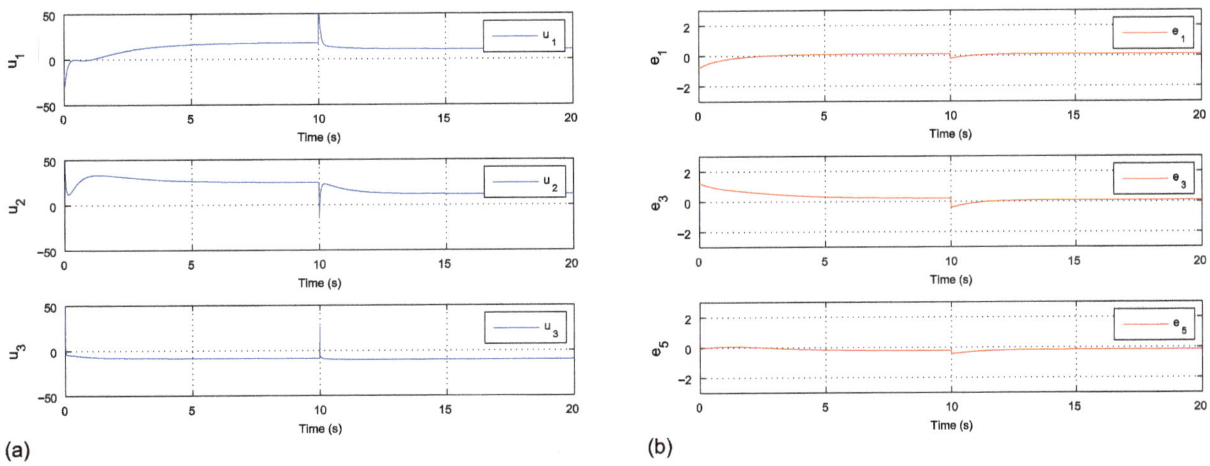

Figure 4.75 Test with setpoint 7 for the three-link biped robot: (a) variation of the control inputs u_1 to u_3 which are applied to the biped robot; (b) tracking error dynamics for state variables x_1, x_3, and x_5

Proof: The binomial $(\rho\alpha - \frac{1}{\rho}b)^2$ is considered. Expanding the left part of the above inequality, one gets

$$\begin{aligned} &\rho^2a^2 + \tfrac{1}{\rho^2}b^2 - 2ab \geq 0 \Rightarrow \tfrac{1}{2}\rho^2a^2 + \tfrac{1}{2\rho^2}b^2 - ab \geq 0 \Rightarrow \\ &ab - \tfrac{1}{2\rho^2}b^2 \leq \tfrac{1}{2}\rho^2a^2 \Rightarrow \tfrac{1}{2}ab + \tfrac{1}{2}ab - \tfrac{1}{2\rho^2}b^2 \leq \tfrac{1}{2}\rho^2a^2 \end{aligned} \tag{4.431}$$

The following substitutions are carried out: $a = \tilde{d}$ and $b = e^TPL$ and the previous relation becomes

$$\tfrac{1}{2}\tilde{d}^TL^TPe + \tfrac{1}{2}e^TPL\tilde{d} - \tfrac{1}{2\rho^2}e^TPLL^TPe \leq \tfrac{1}{2}\rho^2\tilde{d}^T\tilde{d} \tag{4.432}$$

Eq. (4.432) is substituted in Eq. (4.429) and the inequality is enforced, thus giving

$$\dot{V} \leq -\tfrac{1}{2}e^TQe + \tfrac{1}{2}\rho^2\tilde{d}^T\tilde{d} \tag{4.433}$$

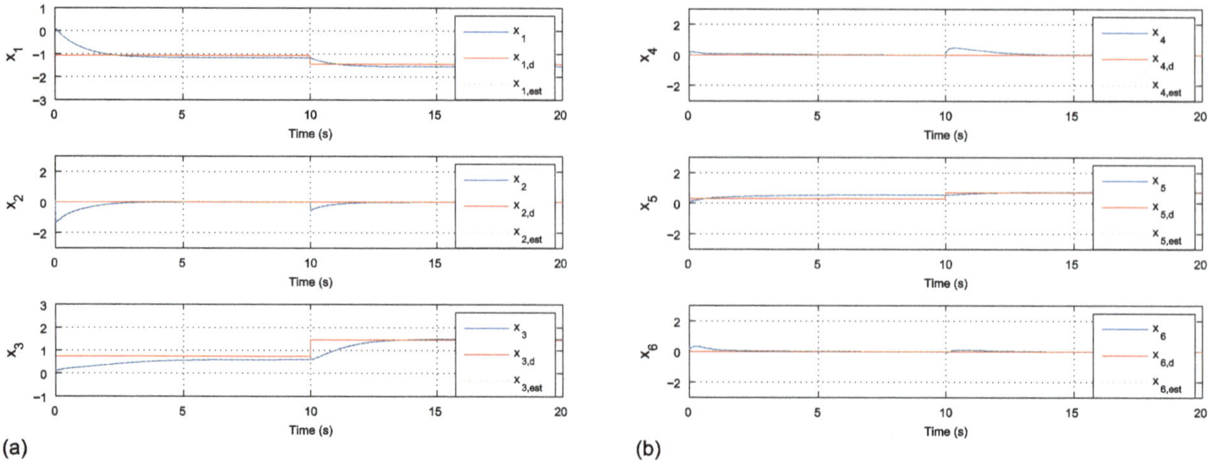

Figure 4.76 *Test with setpoint 8 for the three-link biped robot: (a) tracking of setpoints (red line) by state variables x_1 to x_3 (blue line) and estimates provided by the H-infinity Kalman filter (green line); (b) tracking of setpoints (red line) by state variables x_4 to x_6 (blue line) and estimates provided by the H-infinity Kalman filter (green line)*

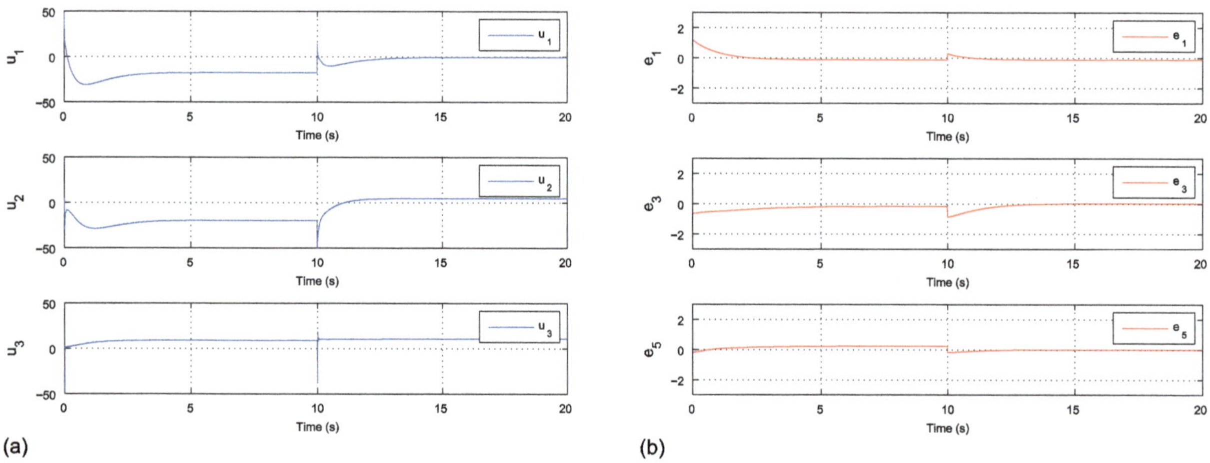

Figure 4.77 *Test with setpoint 8 for the three-link biped robot: (a) variation of the control inputs u_1 to u_3 which are applied to the biped robot; (b) tracking error dynamics for state variables x_1, x_3, and x_5*

Eq. (4.433) shows that the H_∞ tracking performance criterion is satisfied. The integration of $\dot{V}$ from 0 to T gives

$$\begin{aligned} &\int_0^T \dot{V}(t)dt \leq -\tfrac{1}{2}\int_0^T ||e||_Q^2 dt + \tfrac{1}{2}\rho^2 \int_0^T ||\tilde{d}||^2 dt \Rightarrow \\ &2V(T) + \int_0^T ||e||_Q^2 dt \leq 2V(0) + \rho^2 \int_0^T ||\tilde{d}||^2 dt \end{aligned} \tag{4.434}$$

Moreover, if there exists a positive constant $M_d > 0$ such that

$$\int_0^\infty ||\tilde{d}||^2 dt \leq M_d \tag{4.435}$$

then, one gets

$$\int_0^\infty ||e||_Q^2 dt \leq 2V(0) + \rho^2 M_d \tag{4.436}$$

Table 4.15 Tracking RMSE for the biped robot in the disturbance-free case

	$RMSE_{x_1}$	$RMSE_{x_2}$	$RMSE_{x_3}$	$RMSE_{x_4}$	$RMSE_{x_5}$	$RMSE_{x_6}$
$test_1$	0.0028	0.0001	0.0038	0.0003	0.0063	0.0002
$test_2$	0.0027	0.0001	0.0048	0.0001	0.0051	0.0001
$test_3$	0.0028	0.0001	0.0058	0.0005	0.0033	0.0003
$test_4$	0.0023	0.0013	0.0033	0.0005	0.0031	0.0006
$test_5$	0.0028	0.0008	0.0048	0.0021	0.0051	0.0011
$test_6$	0.0029	0.0024	0.0050	0.0026	0.0042	0.0029
$test_7$	0.0027	0.0015	0.0046	0.0022	0.0044	0.0011
$test_8$	0.0026	0.0021	0.0056	0.0036	0.0033	0.0009

Table 4.16 Tracking RMSE for the biped robot in the case of disturbances

$\Delta a\%$	$RMSE_{x_1}$	$RMSE_{x_2}$	$RMSE_{x_3}$	$RMSE_{x_4}$	$RMSE_{x_5}$	$RMSE_{x_6}$
0%	0.0028	0.0001	0.0038	0.0003	0.0063	0.0002
10%	0.0029	0.0001	0.0036	0.0003	0.0066	0.0002
20%	0.0029	0.0001	0.0035	0.0004	0.0070	0.0002
30%	0.0029	0.0001	0.0034	0.0004	0.0074	0.0003
40%	0.0029	0.0001	0.0032	0.0004	0.0078	0.0004
50%	0.0029	0.0001	0.0031	0.0005	0.0084	0.0005
60%	0.0029	0.0001	0.0030	0.0006	0.0090	0.0007

Table 4.17 RMSE for the estimation performed by the H-infinity KF $\times 10^{-5}$

	$RMSE_{x_1}$	$RMSE_{x_2}$	$RMSE_{x_3}$	$RMSE_{x_4}$	$RMSE_{x_5}$	$RMSE_{x_6}$
$test_1$	0.3212	0.0498	0.3153	0.3423	0.5173	0.1580
$test_2$	0.3178	0.0871	0.3235	0.3931	0.3198	0.1698
$test_3$	0.3138	0.2525	0.3226	0.3074	0.3223	0.1256
$test_4$	0.3231	0.0495	0.3237	0.0591	0.3151	0.0691
$test_5$	0.3161	0.2767	0.3137	0.4143	0.3101	0.1557
$test_6$	0.1148	0.0418	0.1117	0.0249	0.1116	0.0230
$test_7$	0.1145	0.0359	0.1159	0.0126	0.1168	0.0064
$test_8$	0.1177	0.0071	0.1143	0.0041	0.1144	0.0133

Thus, the integral $\int_0^\infty ||e||_Q^2 dt$ is bounded. Moreover, $V(T)$ is bounded and from the definition of the Lyapunov function V in Eq. (4.419) it becomes clear that $e(t)$ will be also bounded since $e(t) \in \Omega_e = \{e|e^T Pe \leq 2V(0) + \rho^2 M_d\}$. According to the above and with the use of Barbalat's Lemma, one obtains $lim_{t\to\infty} e(t) = 0$.

After following the stages of the stability proof, one arrives at Eq. (4.433) which shows that the H-infinity tracking performance criterion holds. By selecting the attenuation coefficient ρ to be sufficiently small and in particular to satisfy $\rho^2 < ||e||_Q^2/||\tilde{d}||^2$, one has that the first derivative of the Lyapunov function is upper bounded by 0. This condition holds at each sampling instance, and consequently global stability for the control loop can be concluded.

Table 4.18 Convergence time for the biped robot's state variables

	$T_s\, x_1$	$T_s\, x_2$	$T_s\, x_3$	$T_s\, x_4$	$T_s\, x_5$	$T_s\, x_6$
$test_1$	2.5 sec	3.0 sec	7.0 sec	5.0 sec	0.5 sec	0.5 sec
$test_2$	2.0 sec	2.5 sec	7.0 sec	0.5 sec	0.5 sec	1.0 sec
$test_3$	3.0 sec	3.0 sec	8.0 sec	4.0 sec	3.0 sec	2.0 sec
$test_4$	2.0 sec	3.0 sec	7.0 sec	2.0 sec	3.0 sec	2.5 sec
$test_5$	3.0 sec	3.0 sec	8.0 sec	5.0 sec	5.0 sec	1.5 sec
$test_6$	3.0 sec	2.5 sec	7.0 sec	1.0 sec	1.0 sec	0.5 sec
$test_7$	3.0 sec	3.0 sec	8.0 sec	4.0 sec	0.5 sec	1.0 sec
$test_8$	2.0 sec	3.0 sec	7.0 sec	1.0 sec	1.0 sec	1.0 sec

4.4.6 Simulation tests

The performance of the nonlinear optimal control method has been further confirmed through simulation experiments. Indicative values about the parameters of the three-link biped robot have been as follows: $m = 10kg$, $M_H = 2.0kg$, $M_T = 7.5kg$, $r = 1.0m$, $l = 0.8m$, and $\tilde{d}_i = 0.01$, $i = 1, 2, 3$ and $g = 9.8m/sec^2$. To compute the feedback gains of the H-infinity controller the algebraic Riccati equation of Eq. (4.424) had to be solved at each time-step of the control method using the *aresolv()* function in MATLAB®. The obtained simulation tests are depicted in Figures 4.62–4.77. The presented results describe the motion of the three-link biped robot during the swing phase. The angle of the stance leg and the associated angular velocity are described by state variables x_1, x_2, respectively. The angle of the swing leg and the associated angular velocity are described by state variables x_3 and x_4. The angle of the torso and the associated angular velocity are described by state variables x_5, x_6, respectively. The real values of the robot's state variables are depicted in blue, the estimated values which are provided by the H-infinity Kalman filter are plotted in green and the associated setpoints are printed in red color.

To elaborate on the tracking performance and on the robustness of the proposed nonlinear optimal control method for the three-link biped robot the following tables are given: (i) Table 4.15 which provides information about the accuracy of tracking of the reference setpoints by the state variables of the robot's state-space model, (ii) Table 4.16 which provides information about the robustness of the control method to parametric changes in the model of the robot's dynamics (change $\Delta a\%$ in the mass M_T of the robot's torso), and (iii) Table 4.17 which provides information about the precision in state variables' estimation that is achieved by the H-infinity Kalman filter, and (iv) Table 4.18 which provides the approximate convergence times of the robot's state variables to the associated setpoints.

Chapter 5
Nonlinear optimal and flatness-based control for space robotics I

Abstract

The chapter analyzes nonlinear optimal and flatness-based control for aerospace systems. First, it solves the nonlinear optimal control problem for the 4-DOF dynamic model of a reentry space vehicle. Next, it applies nonlinear optimal control and multi-loop flatness-based control to the 6-DOF attitude dynamics of a reentry space vehicles. Furthermore, it applies multi-loop flatness-based control to the 3-DOF dynamic model of the attitude of microsatellites. Finally, it applies multi-loop flatness-based control to the dynamic model of the 3-DOF free-floating space robotic manipulator.

5.1 Nonlinear optimal control of reentry space vehicles

5.1.1 Outline

The present monograph extends its application domain to the use of its nonlinear optimal and flatness-based control methods in aerospace systems. The control problem of reentry space vehicles has attracted much research interest during the last years because it is closely related with space exploration and exploitation and with the planning and completion of more effective space missions that will allow for deploying satellites and space stations [913,85,242]. Because of the nonlinear dynamics, multivariable structure, and underactuation, the solution of the control problem for reentry space vehicles is a nontrivial task [86,211,212]. One can distinguish between the attitude (orientation) control problem and the path following (gliding) control problem [512,219]. In the latter case, the objective is to control the spacecraft's longitudinal motion along a specific path, thus allowing for either the space vehicle to reach a specific orbit or to reenter into the atmosphere and finally to return safely to earth [620,619,297]. Among related results, one can note the use of various nonlinear control techniques, such as nonlinear model predictive control and sliding-mode control [516,984,982]. There have also been nonlinear optimal control schemes which have been based on Pontryangin's method and iterative optimum search methods [345,634]. One can also find results based on differential flatness theory and inversion of the spacecraft's dynamics, as well as results based on the coupling between differential flatness theory and Linear Parameter Varying systems control [621,1030,1031,40]. Robustness to model uncertainty and perturbations has also been among the objectives in the design of the associated control schemes [784,861]. Recently, interest has also emerged in the topic of cooperation and synchronization between multiple autonomous reentry space vehicles [629,162].

In the present section, a novel nonlinear optimal control method is proposed for the trajectory tracking (gliding) problem of autonomous reentry space vehicles [721,725,732]. First, the dynamic model of the longitudinal motion of such space vehicles is formulated. It is shown that this system is differentially flat and thus a flatness-based controller can be designed about it. The differential flatness properties are also used to select feasible setpoints for the system's state variables [721,725,732]. Next, to apply nonlinear optimal control, the nonlinear model of the reentry space vehicle undergoes approximate linearization with the use of first-order Taylor series expansion and through the computation of the associated Jacobian matrices [705,60,706]. The linearization process takes place at each time-step of the control method around a time-varying operating point which is updated also at each sampling interval and which is defined by the present value of the spacecraft's state vector and by the last sampled value of the control inputs vector. The modeling error which is due to the truncation of higher-order terms from the Taylor series expansion is viewed as a perturbation that is asymptotically compensated by the control loop's robustness. For the linearized state-space description of the system, a stabilizing H-infinity feedback controller is designed.

The proposed H-infinity controller achieves the solution of the optimal control problem for the reentry space vehicle under model uncertainty and external perturbations. It represents a min–max differential game taking place between (i) the control inputs of the system which try to minimize a cost function that contains a quadratic term of the state vector's tracking error (ii) the model uncertainty and external perturbations which try to maximize this cost function. To compute the stabilizing feedback gains of the H-infinity controller, an algebraic Riccati equation is repetitively solved at each iteration of the control algorithm [721,725,732]. The global stability properties of the control scheme are proven through Lyapunov analysis. First, it is shown that the control method satisfies the H-infinity tracking performance criterion which signifies robustness against model uncertainty and exogenous perturbations [711,741,876]. Moreover, it is proven that, under moderate conditions, the control loop is globally asymptotically stable [721,725,732].

5.1.2 Dynamic model of the reentry space vehicle

The dynamic model of the reentry space vehicle is obtained by expressing position variables and orientation angles of the space vehicle in different coordinate frames, namely an earth-fixed, an earth rotating and a body fixed frame. The equations of motion of the re-entry space vehicle are given by [621,1031,784]:

$$\dot{h} = Vsin(\gamma) \tag{5.1}$$

$$\dot{\theta} = \frac{Vcos(\gamma)sin(\psi)}{(h+R_Tcos(\phi))} \tag{5.2}$$

$$\dot{\phi} = \frac{Vcos(\gamma)cos(\psi)}{h+R_T} \tag{5.3}$$

$$\dot{V} = \frac{D(a,M)}{m} - gsin(\gamma) \tag{5.4}$$

$$\dot{\gamma} = \frac{L(a,M)cos(\mu)}{mV} + \left(\frac{V^2}{h+R_T} - g\right)\frac{cos(\gamma)}{V} \tag{5.5}$$

$$\dot{\psi} = \frac{L(a,M)sin(\psi)}{mVcos(\gamma)} + \frac{V}{h+R_T}cos(\gamma)sin(\psi)tan(\phi) \tag{5.6}$$

The earth-fixed, earth -rotating and body-fixed reference frames used to define the dynamic model of reentry space vehicles and the associated position and angular velocity variables and orientation angles are given in Figure 5.1. The main parameters of the model are: h is the altitude of the space vehicle, θ is the angle expressing "longitude" in an earth-fixed reference frame, ϕ is the angle expressing "latitude" in the earth-fixed reference frame, V: is the magnitude of the spacecraft's velocity, γ is the flight path angle, and ψ is the heading angle of the reentry space vehicle. Moreover, g is the acceleration of gravity, which is taken to satisfy the relation $g = \frac{\mu_T}{R^2}$ or $g = \frac{\mu_T}{h+R_T^2}$ with R_T to be earth's radius and μ_T to be earth's gravitational constant. The mass of the spacecraft is denoted by m. Additionally, μ is the bank angle, which is the angle between the drag force and the velocity of the space vehicle. Furthermore, $D(a,M)$ is the drag force function and $L(a,M)$ is the lift-force function, where it holds that [621,1031,784]

$$\begin{aligned} L(a,M) &= \tfrac{1}{2}\rho SV^2C_L(a,M) \\ D(a,M) &= \tfrac{1}{2}\rho SV^2C_D(a,M) \end{aligned} \tag{5.7}$$

where ρ is the atmospheric density and S is the spacecraft's effective area. About lift and drag coefficients $C_L(a,M)$ and $C_D(a,M)$, it has been used that [621,1031,784]

$$\begin{aligned} C_L(a,M_i) &= a_0(M_i) + a_1(M_i)a + a_2(M_i)a^2 + a_3(M_i)a^3 \\ C_D(a,M_i) &= b_0(M_i) + b_1(M_i)a \end{aligned} \tag{5.8}$$

where a is the angle of attack of the space vehicle, that is the pitch angle of the spacecraft in the earth-rotating frame, M_i denotes a Mach number, a_i $(i = 0,1,2,3)$, b_i $(i = 0,1)$ are nonlinear functions obtained for various Mach numbers M_i in the range $[1.5, 40]$ through an interpolation process.

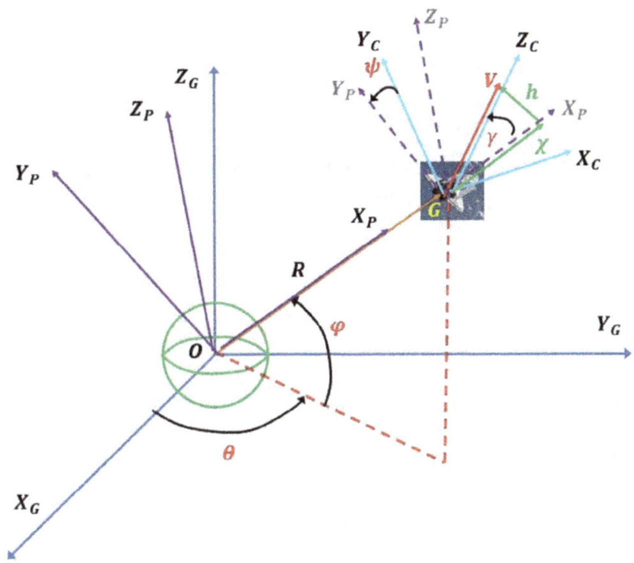

Figure 5.1 Coordinate frames used in the dynamic model of the reentry space vehicle

The earth-fixed frame is denoted as $OX_GY_GZ_G$, while the earth-rotating frame is denoted as $OX_PY_PZ_P$ where G is the center of gravity of the spacecraft. Axis OX_P coincides with the radius connecting the origin of the earth-fixed frame O with the space vehicle's center of gravity G. Transition from the earth-fixed frame $OX_GY_GZ_G$ to the earth rotating frame $OX_PY_PZ_P$ is defined by angles θ and ϕ and the associated rotation matrix. The body-fixed reference frame is denoted as $GX_CY_CZ_C$. In this reference frame, axis GZ_C coincides with the velocity's vector V of the space vehicle. Transition from the earth-rotating reference frame $OX_PY_PZ_P$ to the body-fixed reference frame $GX_CY_CZ_C$ is defined by angles γ and ψ and the associated rotation matrix [621,1031,784].

The transition matrix between the earth-fixed and the earth-rotating frame is given by [620]:

$$R_{GP} = \begin{pmatrix} cos(\phi)cos(\theta) & cos(\phi)sin(\theta) & sin(\phi) \\ -sin(\theta) & cos(\theta) & 0 \\ -sin(\phi)cos(\theta) & -sin(\phi)sin(\theta) & cos(\phi) \end{pmatrix} \tag{5.9}$$

The transition matrix between the earth-rotating frame and the body-fixed frame is given by

$$R_{PC} = \begin{pmatrix} cos(\gamma) & -sin(\gamma)sin(\psi) & -sin(\gamma)cos(\psi) \\ 0 & cos(\psi) & -sin(\psi) \\ sin(\gamma) & cos(\gamma)sin(\psi) & cos(\gamma)cos(\psi) \end{pmatrix} \tag{5.10}$$

It has been proven that the dynamic model of the space-reentry vehicle of Eq. (5.1) is not differentially flat. However, one can examine separately (a) the lateral motion of the spacecraft which is described by state variables h, V, γ plus variable $\chi = Vcos(\gamma)$, from (b) the lateral dynamics which is described by state variables θ, ϕ, ψ. In the latter case, and for the state-space model of the longitudinal motion of the reentry space vehicle which is described by the state vector $x = [h, \chi, V, \gamma]$, one can demonstrate that differential flatness properties hold.

The dynamic model of the longitudinal motion of the reentry space vehicle is given by the following equations [621]:

$$\dot{h} = Vsin(\gamma) \tag{5.11}$$

$$\dot{\chi} = Vcos(\gamma) \tag{5.12}$$

$$\dot{V} = -\frac{1}{2}\frac{\rho SV^2 C_D(a,M)}{m} - gsin(\gamma) \tag{5.13}$$

$$\dot{\gamma} = \frac{1}{2}\frac{\rho SV^2 C_L(a,M)}{mV}cos(\mu) + (\frac{V}{h+R_T} - \frac{g}{V})cos(\gamma) \tag{5.14}$$

The control inputs vector in the state-space model which describes the longitudinal dynamics of the re-entry space vehicle is $U = [C_D(a,M), cos(\mu)]$. Using the state-space notation $x = [x_1, x_2, x_3, x_4]^T = [h, x, V, \gamma]^T$, the state-space model of the dynamics of the spacecraft is written as

$$\begin{aligned} \dot{x}_1 &= x_3 cos(x_4) \\ \dot{x}_2 &= x_3 sin(x_4) \\ \dot{x}_3 &= -\frac{1}{2}\frac{\rho S x_3^2}{m}u_1 - gsin(x_4) \\ \dot{x}_4 &= \frac{1}{2}\frac{\rho S x_3 C_L(a,M)}{m}u_2 + (\frac{x_3}{x_1+R_T} - \frac{g}{x_3})cos(x_4) \end{aligned} \tag{5.15}$$

The state-space model of the re-entry space vehicle is also written in matrix form

$$\begin{pmatrix} \dot{x}_1 \\ \dot{x}_2 \\ \dot{x}_3 \\ \dot{x}_4 \end{pmatrix} = \begin{pmatrix} x_3 sin(x_4) \\ x_3 cos(x_4) \\ -gsin(x_4) \\ (\frac{x_3}{x_1+R_T} - \frac{g}{x_3})cos(x_4) \end{pmatrix} + \begin{pmatrix} 0 & 0 \\ 0 & 0 \\ \frac{\rho S x_3^2}{2m} & 0 \\ 0 & \frac{\rho S x_3}{2m} \end{pmatrix} \begin{pmatrix} u_1 \\ u_2 \end{pmatrix} \tag{5.16}$$

or equivalently, the system is written in the concise nonlinear affine-in-the-input state-space form

$$\dot{x} = f(x) + g(x)u \tag{5.17}$$

where $x \in R^{4\times 1}, f(x) \in R^{4\times 1}$, $g(x) \in R^{4\times 2}$, and $u \in R^{2\times 1}$.

5.1.3 Differential flatness properties of the reentry space vehicle

5.1.3.1 Proof of differential flatness properties

The reentry space-vehicle is a differentially flat system with flat outputs vector $Y = [y_1, y_2]^T = [h, x]^T$. Indeed, all its state variables can be written as differential functions of the flat outputs vector. Besides, the flat outputs are differentially independent, which means that they are not connected through a relation in the form of a linear homogeneous differential equation. From the first two rows of the state-space model, one has:

$$\dot{x}_1^2 + \dot{x}_2^2 = x_3^2 \Rightarrow x_3 = \sqrt{\dot{x}_1^2 + \dot{x}_2^2} \tag{5.18}$$

$$\frac{\dot{x}_1}{\dot{x}_2} = tan(x_4) \Rightarrow x_4 = tan^{-1}(\frac{\dot{x}_1}{\dot{x}_2}) \tag{5.19}$$

which signifies that state variables x_3 and x_4 can be written as differential functions of the flat outputs of the system. Besides, from the third and fourth rows of the state-space model of the system, one can solve for the control inputs vector. This gives:

$$\begin{pmatrix} u_1 \\ u_2 \end{pmatrix} = \begin{pmatrix} \frac{\rho S x_3^2}{2m} & 0 \\ 0 & \frac{\rho S x_3}{2m} \end{pmatrix}^{-1} \left[\begin{pmatrix} \dot{x}_3 \\ \dot{x}_4 \end{pmatrix} - \begin{pmatrix} -gsin(x_4) \\ (\frac{x_3}{x_1+R_T} - \frac{g}{x_3})cos(x_4) \end{pmatrix} \right] \tag{5.20}$$

which signifies that the control inputs vector $u = [u_1, u_2]^T$ is a differential function of the system's flat output. Consequently, the reentry space vehicle's dynamic model is differentially flat.

The differential flatness properties of the system are also an implicit proof about its controllability. Moreover, they can be used for defining acceptable setpoints. One selects first setpoints in an unconstrained manner for state variables x_1, x_2 which coincide with the flat outputs. For the rest of the state variables x_3, x_4, the associated setpoints are computed through the relations which allow for expressing x_3, x_4 as differential functions of x_1, x_2.

5.1.3.2 Design of a flatness-based controller

A flatness-based controller can be designed for the reentry space vehicle by applying an input–output linearization process and the eigenvalues assignment concept. To this end, the flat outputs of the system (x_1, x_2) are successively differentiated until the control inputs reappear. The differentiation of the first flat output x_1 gives

$$\begin{aligned}\ddot{x}_1 = &\{[-gsin(x_4)sin(x_4)] + x_3cos(x_4)[(\tfrac{x_3}{x_1} - \tfrac{g}{x_3})]cos(x_4)\}+ \\ &+[\tfrac{\rho S x_3^2}{2m}sin(x_4)]u_1 + x_3cos(x_4)[\tfrac{\rho S x_3}{2m}C_L(a,M)]u_2\end{aligned} \tag{5.21}$$

or equivalently, in concise form

$$\ddot{x}_1 = f_a(x) + g_{a_1}(x)u_1 + g_{a_2}(x)u_2 \tag{5.22}$$

The differentiation of the second flat output x_2 gives

$$\begin{aligned}\ddot{x}_2 = &\{-gsin(x_4) + x_3sin(x_4)[(\tfrac{x_3}{x_1} - \tfrac{g}{x_3})cos(x_4)]\}+ \\ &+[\tfrac{\rho S x_3^2}{2m}cos(x_4)]u_1 + [-x_3sin(x_4)\tfrac{\rho S x_3}{2m}C_L(a,M)]u_2\end{aligned} \tag{5.23}$$

or equivalently, in concise form

$$\ddot{x}_2 = f_b(x) + g_{b_1}(x)u_1 + g_{b_2}(x)u_2 \tag{5.24}$$

Thus, one arrives at an input–output linearized state-space description of the reentry space vehicle

$$\begin{aligned}\ddot{x}_1 &= f_a(x) + g_{a_1}(x)u_1 + g_{a_2}(x)u_2 \\ \ddot{x}_2 &= f_b(x) + g_{b_1}(x)u_1 + g_{b_2}(x)u_2\end{aligned} \tag{5.25}$$

The following virtual control inputs are defined next:

$$\begin{aligned}v_1 &= f_a(x) + g_{a_1}(x)u_1 + g_{a_2}(x)u_2 \\ v_2 &= f_b(x) + g_{b_1}(x)u_1 + g_{b_2}(x)u_2\end{aligned} \tag{5.26}$$

Consequently, the spacecraft's dynamics can be written as

$$\ddot{x}_1 = v_1 \quad \ddot{x}_2 = v_2 \tag{5.27}$$

The state variables of the system are now defined to be $Z = [z_1, z_2, z_3, z_4]^T = [x_1, \dot{x}_1, x_2, \dot{x}_2]^T$. Thus, the system can be written in the canonical Brunovsky form:

$$\begin{pmatrix}\dot{z}_1 \\ \dot{z}_2 \\ \dot{z}_3 \\ \dot{z}_4\end{pmatrix} = \begin{pmatrix}0 & 1 & 0 & 0 \\ 0 & 0 & 0 & 0 \\ 0 & 0 & 0 & 1 \\ 0 & 0 & 0 & 0\end{pmatrix}\begin{pmatrix}z_1 \\ z_2 \\ z_3 \\ z_4\end{pmatrix} + \begin{pmatrix}0 & 0 \\ 1 & 0 \\ 0 & 0 \\ 0 & 1\end{pmatrix}\begin{pmatrix}v_1 \\ v_2\end{pmatrix} \tag{5.28}$$

The stabilizing feedback controller is based on the eigenvalues assignment concept

$$\begin{aligned}v_1 &= \ddot{x}_{1,d} - K_{1,1}(\dot{x}_1 - \dot{x}_{1,d}) - K_{2,1}(x_1 - x_{1,d}) \\ v_2 &= \ddot{x}_{2,d} - K_{1,2}(\dot{x}_2 - \dot{x}_{2,d}) - K_{2,2}(x_2 - x_{2,d})\end{aligned} \tag{5.29}$$

By substituting the control law of Eq. (5.29) into the dynamic model of Eq. (5.27), one obtains the closed-loop system dynamics

$$\begin{aligned}(\ddot{x}_1 - \ddot{x}_{1,d}) + K_{1,1}(\dot{x}_1 - \dot{x}_{1,d}) + K_{2,1}(x_1 - x_{1,d}) = 0 \\ (\ddot{x}_2 - \ddot{x}_{2,d}) + K_{1,2}(\dot{x}_2 - \dot{x}_{2,d}) + K_{2,2}(x_2 - x_{2,d}) = 0\end{aligned} \tag{5.30}$$

Next, by defining the tracking error variables $e_1 = x_1 - x_{1,d}$, $e_2 = x_2 - x_{2,d}$, one obtains the tracking error dynamics

$$\begin{aligned} &\ddot{e}_1 + K_{1,1}\dot{e}_1 + K_{2,1}e_1 = 0 \\ &\ddot{e}_2 + K_{1,2}\dot{e}_2 + K_{2,2}e_2 = 0 \end{aligned} \tag{5.31}$$

By selecting coefficients $K_{1,1}$, $K_{2,1}$ and $K_{1,2}$, $K_{2,2}$ so as the characteristic polynomials associated with the differential equations of Eq. (5.31) to be Hurwitz stable, one has the global stability conditions

$$\begin{aligned} &lim_{t\to\infty}e_1(t) = 0 \Rightarrow lim_{t\to\infty}x_1(t) = x_{1,d}(t) \\ &lim_{t\to\infty}e_2(t) = 0 \Rightarrow lim_{t\to\infty}x_2(t) = x_{2,d}(t) \end{aligned} \tag{5.32}$$

while from the differential flatness relations for x_3, x_4, one obtains also $lim_{t\to\infty}x_3(t) = x_{3,d}(t)$ and $lim_{t\to\infty}x_4(t) = x_{4,d}(t)$.

Finally, the control inputs which should be applied to the initial nonlinear system are obtained from an inverse transformation

$$\begin{pmatrix} u_1 \\ u_2 \end{pmatrix} = \begin{pmatrix} g_{a_1}(x) & g_{a_2}(x) \\ g_{b_1}(x) & g_{b_2}(x) \end{pmatrix}^{-1} \left[\begin{pmatrix} v_1 \\ v_2 \end{pmatrix} - \begin{pmatrix} f_a(x) \\ f_b(x) \end{pmatrix} \right] \tag{5.33}$$

5.1.4 Design of a nonlinear H-infinity controller for the reentry space vehicle

5.1.4.1 Approximate linearization of the reentry space vehicle

It will be shown that it is possible to control the dynamic model of the reentry space vehicle without the previously analyzed changes of state variables and state-space model transformations. The dynamic model of the reentry space vehicle undergoes approximate linearization around a temporary operating point, with the use of first-order Taylor series expansion and through the computation of the associated Jacobian matrices. The linearization point (x^*, u^*) is defined at each sampling instance by the present value of the system's state vector x^* and by the last sampled value of the control inputs vector u^*. The imprecision of the system's state-space model which is induced by the truncation of higher-order terms from the Taylor series expansion, is considered to be a small perturbation which is asymptotically compensated by the robustness of the control algorithm.

The linearization process turns the nonlinear state-space model of the system $\dot{x} = f(x) + g(x)u$ into the linearized equivalent state-space description

$$\dot{x} = Ax + Bu + \tilde{d} \tag{5.34}$$

where $\tilde{d}$ is the cumulative disturbance comprising exogenous perturbations, modeling errors due to the truncation of higher order terms in the Taylor series expansion and sensor noise of any distribution. About the Jacobian matrices A and B, one has

$$\begin{aligned} A &= \nabla_x[f(x) + g(x)u] \mid_{(x^*,u^*)} \Rightarrow \\ A &= \nabla_x f(x) \mid_{(x^*,u^*)} + \nabla_x g_1(x) \mid_{(x^*,u^*)} + \nabla_x g_2(x) \mid_{(x^*,u^*)} \end{aligned} \tag{5.35}$$

$$\begin{aligned} B &= \nabla_u[f(x) + g(x)u] \mid_{(x^*,u^*)} \Rightarrow \\ B &= g(x) \mid_{(x^*,u^*)} \end{aligned} \tag{5.36}$$

The Jacobian matrix $\nabla_x f(x) \mid_{(x^*,u^*)}$ is as follows:

$$\nabla_x f(x) \mid_{(x^*,u^*)} = \begin{pmatrix} 0 & 0 & sin(x_4) & x_3cos(x_4) \\ 0 & 0 & cos(x_4) & -x_3sin(x_4) \\ 0 & 0 & 0 & -gcos(x_4) \\ -\frac{x_3}{(x_1+R_T)^2}cos(x_4) & 0 & (\frac{1}{x_1+R_T} - \frac{g}{x_3^2})cos(x_4) & -(\frac{x_3}{x_1+R_T} - \frac{g}{x_3})sin(x_4) \end{pmatrix} \mid_{(x^*,u^*)} \tag{5.37}$$

The Jacobian matrix $\nabla_x g_1(x) \mid_{(x^*,u^*)}$ is as follows:

$$\nabla_x g_1(x) \mid_{(x^*,u^*)}= \begin{pmatrix} 0 & 0 & 0 & 0 \\ 0 & 0 & 0 & 0 \\ 0 & 0 & \frac{\rho S x_3}{m} & 0 \\ 0 & 0 & 0 & 0 \end{pmatrix} \mid_{(x^*,u^*)} \tag{5.38}$$

The Jacobian matrix $\nabla_x g_2(x) \mid_{(x^*,u^*)}$ is as follows:

$$\nabla_x g_2(x) \mid_{(x^*,u^*)}= \begin{pmatrix} 0 & 0 & 0 & 0 \\ 0 & 0 & 0 & 0 \\ 0 & 0 & 0 & 0 \\ 0 & 0 & \frac{\rho S}{2m} C_L(a,M) & 0 \end{pmatrix} \mid_{(x^*,u^*)} \tag{5.39}$$

This linearization approach which has been followed for implementing the nonlinear optimal control scheme results in a quite accurate model of the system's dynamics. Consider for instance the following affine-in-the-input state-space model:

$$\begin{gathered} \dot{x} = f(x) + g(x)u \Rightarrow \\ \dot{x} = [f(x^*) + \nabla_x f(x) \mid_{x^*} (x - x^*)] + [g(x^*) + \nabla_x g(x) \mid_{x^*} (x - x^*)]u^* + g(x^*)u^* + g(x^*)(u - u^*) + \tilde{d}_1 \Rightarrow \\ \dot{x} = [\nabla_x f(x) \mid_{x^*} + \nabla_x g(x) \mid_{x^*} u^*]x + g(x^*)u - [\nabla_x f(x) \mid_{x^*} + \nabla_x g(x) \mid_{x^*} u^*]x^* + f(x^*) + g(x^*)u^* + \tilde{d}_1 \end{gathered} \tag{5.40}$$

where $\tilde{d}_1$ is the modeling error due to truncation of higher order terms in the Taylor series expansion of $f(x)$ and $g(x)$. Next, by defining $A = [\nabla_x f(x) \mid_{x^*} + \nabla_x g(x) \mid_{x^*} u^*]$, $B = g(x^*)$, one obtains

$$\dot{x} = Ax + Bu - Ax^* + f(x^*) + g(x^*)u^* + \tilde{d}_1 \tag{5.41}$$

Moreover, by denoting $\tilde{d} = -Ax^* + f(x^*) + g(x^*)u^* + \tilde{d}_1$ about the cumulative modeling error term in the Taylor series expansion procedure, one has

$$\dot{x} = Ax + Bu + \tilde{d} \tag{5.42}$$

which is the approximately linearized model of the dynamics of the system given in Eq. (5.34). The term $f(x^*) + g(x^*)u^*$ is the derivative of the state vector at (x^*, u^*) which is almost annihilated by $-Ax^*$.

5.1.4.2 Min–max control and disturbance rejection

In the H_∞ control approach, a feedback control scheme is designed for setpoint tracking by the reentry space vehicle's state vector and simultaneous disturbance rejection, considering that the disturbance affects the system in the worst possible manner. For the approximately linearized model of Eq. (5.17), with outputs vector $y(t) = Cx(t)$, the disturbances' effects are incorporated in the following quadratic cost function [721]

$$J(t) = \frac{1}{2}\int_0^T [y^T(t)y(t) + ru^T(t)u(t) - \rho^2 \tilde{d}^T(t)\tilde{d}(t)]dt, \quad r, \rho > 0 \tag{5.43}$$

The significance of the negative sign in the cost function's term that is associated with the perturbation variable $\tilde{d}(t)$ is that the disturbance tries to maximize the cost function $J(t)$ while the control signal $u(t)$ tries to minimize it.

The physical meaning of the relation given above is that the control signal and the disturbances compete with each other within a min–max differential game. This problem of min–max optimization can be written as $min_u max_{\tilde{d}} J(u, \tilde{d})$. The objective of the optimization procedure is to compute a control signal $u(t)$ which can compensate for the worst possible disturbance that affects the system. However, the solution to the min–max optimization problem is directly related to the value of parameter ρ. This means that there is an upper bound in the disturbances magnitude that can be annihilated by the control signal.

5.1.4.3 Design of the stabilizing feedback controller

After linearization around its current operating point (x^*, u^*), the dynamic model of the reentry space vehicle is written as

$$\dot{x} = Ax + Bu + d_1 \quad (5.44)$$

Parameter d_1 stands for the linearization error in the reentry space vehicle's model appearing previously in Eq. (5.44. The reference setpoints for the state vector of the reentry space vehicle are denoted by $\mathbf{x_d} = [x_1^d, \cdots, x_4^d]$. Tracking of this trajectory is achieved after applying the control input u_d. At every time instant, the control input u_d is assumed to differ from the control input u appearing in Eq. (5.44) by an amount equal to Δu, that is $u_d = u + \Delta u$

$$\dot{x}_d = Ax_d + Bu_d + d_2 \quad (5.45)$$

The dynamics of the controlled system described in Eq. (5.44) can be also written as

$$\dot{x} = Ax + Bu + Bu_d - Bu_d + d_1 \quad (5.46)$$

and by denoting $d_3 = -Bu_d + d_1$ as an aggregate disturbance term, one obtains

$$\dot{x} = Ax + Bu + Bu_d + d_3 \quad (5.47)$$

By subtracting Eq. (5.45) from Eq. (5.47), one has

$$\dot{x} - \dot{x}_d = A(x - x_d) + Bu + d_3 - d_2 \quad (5.48)$$

By denoting the tracking error as $e = x - x_d$ and the aggregate disturbance term as $L\tilde{d} = d_3 - d_2$, the tracking error dynamics becomes

$$\dot{e} = Ae + Bu + L\tilde{d} \quad (5.49)$$

where L is the disturbance gain matrix. For the linearized system, the cost function of Eq. (5.43) is defined, where coefficient r determines the penalization of the control input and the weight coefficient ρ determines the reward of the disturbances' effects. Then, the optimal feedback control law is given by

$$u(t) = -Ke(t) \quad (5.50)$$

with $K = \frac{1}{r}B^TP$ where P is a positive definite symmetric matrix which is obtained from the solution of the Riccati equation

$$A^TP + PA + Q - P(\tfrac{2}{r}BB^T - \tfrac{1}{\rho^2}LL^T)P = 0 \quad (5.51)$$

where Q is also a semi-positive definite symmetric matrix. The worst-case disturbance is given by $\tilde{d}(t) = \frac{1}{\rho^2}L^TPe(t)$. The computation of the worst-case disturbance that this controller can sustain, comes from superposition of Bellman's optimality principle when considering that the reentry space vehicle is affected by two separate inputs, that is the control input u and the cumulative disturbance input $\tilde{d}$. From the previous relation about the worst-case disturbance, it can be seen that the smallest value of the attenuation coefficient ρ that results in an admissible solution for the methods's algebraic Riccati equation, is the one that provides the control loop with maximum robustness. The diagram of the considered control loop is depicted in Figure 5.2.

5.1.5 Lyapunov stability analysis

Through Lyapunov stability analysis it will be shown that the proposed nonlinear control scheme assures H_∞ tracking performance for the reentry space vehicle, and that in case of bounded disturbance terms asymptotic convergence to the reference setpoints is achieved. The tracking error dynamics for the reentry space vehicle is written in the form [721]

$$\dot{e} = Ae + Bu + L\tilde{d} \quad (5.52)$$

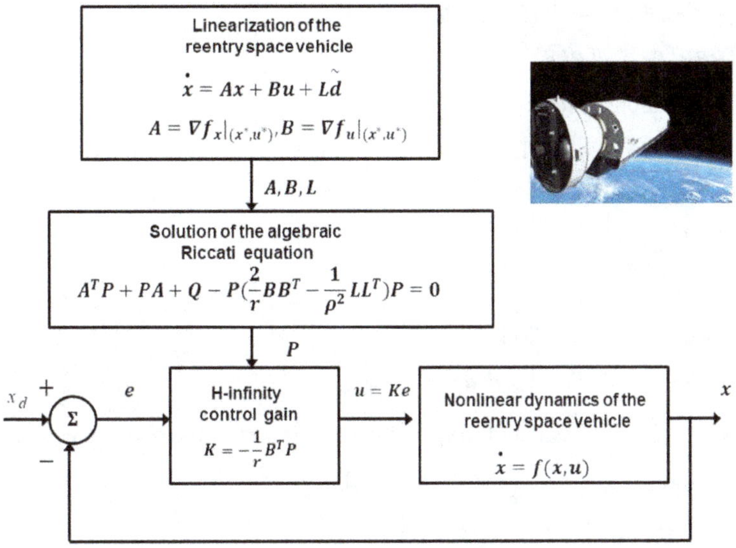

Figure 5.2 Diagram of the Nonlinear Optimal Control loop for reentry space vehicle

where in the reentry space vehicle's case $L \in R^{4\times4}$ is the disturbance inputs gain matrix. Variable $\tilde{d}$ denotes model uncertainties and external disturbances of the reentry space vehicle's model. The following Lyapunov equation is considered

$$V = \tfrac{1}{2}e^TPe \tag{5.53}$$

where $e = x - x_d$ is the tracking error. By differentiating with respect to time, one obtains

$$\begin{aligned} \dot{V} &= \tfrac{1}{2}\dot{e}^TPe + \tfrac{1}{2}e^TP\dot{e} \Rightarrow \\ \dot{V} &= \tfrac{1}{2}[Ae + Bu + L\tilde{d}]^TPe + \tfrac{1}{2}e^TP[Ae + Bu + L\tilde{d}] \Rightarrow \end{aligned} \tag{5.54}$$

$$\begin{aligned} \dot{V} = {} & \tfrac{1}{2}[e^TA^T + u^TB^T + \tilde{d}^TL^T]Pe + \\ & + \tfrac{1}{2}e^TP[Ae + Bu + L\tilde{d}] \Rightarrow \end{aligned} \tag{5.55}$$

$$\begin{aligned} \dot{V} = {} & \tfrac{1}{2}e^TA^TPe + \tfrac{1}{2}u^TB^TPe + \tfrac{1}{2}\tilde{d}^TL^TPe + \\ & \tfrac{1}{2}e^TPAe + \tfrac{1}{2}e^TPBu + \tfrac{1}{2}e^TPL\tilde{d} \end{aligned} \tag{5.56}$$

The previous equation is rewritten as

$$\dot{V} = \tfrac{1}{2}e^T(A^TP + PA)e + (\tfrac{1}{2}u^TB^TPe + \tfrac{1}{2}e^TPBu) + (\tfrac{1}{2}\tilde{d}^TL^TPe + \tfrac{1}{2}e^TPL\tilde{d}) \tag{5.57}$$

Assumption: For given positive definite matrix Q and coefficients r and ρ, there exists a positive definite matrix P, which is the solution of the following matrix equation

$$A^TP + PA = -Q + P(\tfrac{2}{r}BB^T - \tfrac{1}{\rho^2}LL^T)P \tag{5.58}$$

Moreover, the following feedback control law is applied to the system

$$u = -\tfrac{1}{r}B^TPe \tag{5.59}$$

By substituting Eq. (5.58) and Eq. (5.59), one obtains

$$\dot{V} = \tfrac{1}{2}e^T[-Q + P(\tfrac{2}{r}BB^T - \tfrac{1}{\rho^2}LL^T)P]e + e^TPB(-\tfrac{1}{r}B^TPe) + e^TPL\tilde{d} \Rightarrow \tag{5.60}$$

$$\begin{array}{l} \dot{V} = -\frac{1}{2}e^TQe + (\frac{1}{r}e^TPBB^TPe - \frac{1}{2\rho^2}e^TPLL^TPe) \\ \qquad -\frac{1}{r}e^TPBB^TPe + e^TPL\tilde{d} \end{array} \tag{5.61}$$

which after intermediate operations gives

$$\dot{V} = -\frac{1}{2}e^TQe - \frac{1}{2\rho^2}e^TPLL^TPe + e^TPL\tilde{d} \tag{5.62}$$

or, equivalently

$$\begin{array}{l} \dot{V} = -\frac{1}{2}e^TQe - \frac{1}{2\rho^2}e^TPLL^TPe + \\ \qquad +\frac{1}{2}e^TPL\tilde{d} + \frac{1}{2}\tilde{d}^TL^TPe \end{array} \tag{5.63}$$

Lemma: The following inequality holds [721]

$$\frac{1}{2}e^TL\tilde{d} + \frac{1}{2}\tilde{d}L^TPe - \frac{1}{2\rho^2}e^TPLL^TPe \leq \frac{1}{2}\rho^2\tilde{d}^T\tilde{d} \tag{5.64}$$

Proof: The binomial $(\rho\alpha - \frac{1}{\rho}b)^2$ is considered. Expanding the left part of the above inequality, one gets

$$\begin{array}{l} \rho^2a^2 + \frac{1}{\rho^2}b^2 - 2ab \geq 0 \Rightarrow \frac{1}{2}\rho^2a^2 + \frac{1}{2\rho^2}b^2 - ab \geq 0 \Rightarrow \\ ab - \frac{1}{2\rho^2}b^2 \leq \frac{1}{2}\rho^2a^2 \Rightarrow \frac{1}{2}ab + \frac{1}{2}ab - \frac{1}{2\rho^2}b^2 \leq \frac{1}{2}\rho^2a^2 \end{array} \tag{5.65}$$

The following substitutions are carried out: $a = \tilde{d}$ and $b = e^TPL$ and the previous relation becomes

$$\frac{1}{2}\tilde{d}^TL^TPe + \frac{1}{2}e^TPL\tilde{d} - \frac{1}{2\rho^2}e^TPLL^TPe \leq \frac{1}{2}\rho^2\tilde{d}^T\tilde{d} \tag{5.66}$$

Eq. (5.66) is substituted in Eq. (5.63) and the inequality is enforced, thus giving [721]

$$\dot{V} \leq -\frac{1}{2}e^TQe + \frac{1}{2}\rho^2\tilde{d}^T\tilde{d} \tag{5.67}$$

Eq. (5.67) shows that the H_∞ tracking performance criterion is satisfied. The integration of $\dot{V}$ from 0 to T gives

$$\begin{array}{l} \int_0^T \dot{V}(t)dt \leq -\frac{1}{2}\int_0^T ||e||_Q^2 dt + \frac{1}{2}\rho^2 \int_0^T ||\tilde{d}||^2 dt \Rightarrow \\ 2V(T) + \int_0^T ||e||_Q^2 dt \leq 2V(0) + \rho^2 \int_0^T ||\tilde{d}||^2 dt \end{array} \tag{5.68}$$

Moreover, if there exists a positive constant $M_d > 0$ such that

$$\int_0^\infty ||\tilde{d}||^2 dt \leq M_d \tag{5.69}$$

then, one gets

$$\int_0^\infty ||e||_Q^2 dt \leq 2V(0) + \rho^2 M_d \tag{5.70}$$

Thus, the integral $\int_0^\infty ||e||_Q^2 dt$ is bounded. Moreover, $V(T)$ is bounded and from the definition of the Lyapunov function V in Eq. (5.53) it becomes clear that $e(t)$ will also be bounded since $e(t) \in \Omega_e = \{e|e^TPe \leq 2V(0) + \rho^2M_d\}$. According to the above and with the use of Barbalat's Lemma, one obtains $lim_{t\to\infty}e(t) = 0$.

Through the stages of the stability proof, one arrives at Eq. (5.67) which shows that the H-infinity tracking performance criterion holds. By selecting the attenuation coefficient ρ to be sufficiently small and in particular to satisfy $\rho^2 < ||e||_Q^2/||\tilde{d}||^2$, one has that the first derivative of the Lyapunov function is upper bounded by 0. This condition holds at each sampling instance, and consequently global stability for the control loop can be concluded.

5.1.6 Simulation tests

5.1.6.1 Results on nonlinear optimal control

In continuation of the previous global stability proof, the fine performance of the nonlinear optimal control method for autonomous reentry space vehicles is further confirmed through simulation experiments. To implement the nonlinear optimal control method, the algebraic Riccati equation of Eq. (5.58) had to be repetitively solved, at each time-step of the control algorithm using MATLAB® *aresolv()* function. The values of the parameters of the dynamic model of the reentry space vehicle were according to [784]. The obtained results in the case of trajectory tracking by the reentry space vehicle are depicted in Figures 5.3–5.18. The sampling period was $T_s = 0.01$sec. The method is computationally efficient since the time needed to solve the Riccati equation in a PC with an Intel i7 processor at 2.8HGz is much shorter than the sampling interval. As it can be noticed from the simulation experiments, the method achieves fast and accurate tracking of setpoints by all state variables of the reentry space vehicle while the variations of the control inputs remain moderate. The parameters which affect the transient performance of the control scheme are coefficients r, ρ, and Q in the Riccati equation. For relatively small values of r the tracking error of the state variables is eliminated, while for relatively large values of Q fast convergence to the reference values is achieved. Moreover, the smallest value of ρ for which a valid solution of the method's Riccati equation is obtained in the form of a positive definite and symmetric matrix P is the one that provides the control loop with maximum robustness.

The presented simulation tests have been obtained from eight different setpoints about the reentry autonomous vehicle and have been organized as follows: for each simulation experiment, there is (i) a diagram presenting the convergence of the four state vector elements of the RSV to the associated setpoints, (ii) a diagram presenting the variations of the associated tracking error variables and their asymptotic convergence to zero, (iii) a diagram presenting the variations of the two control inputs of the RSV, and (iv) a diagram presenting the path that is followed by the center of gravity of the RSV in the 2D vertical plane of the local (h,x) coordinates frame. The setpoints of the RSV have been chosen using the differential flatness properties of the spacecraft. Reference trajectories have been selected in an unconstrained manner for state variables (h,x) of the RSV which coincide with the flat outputs of the system. For the rest two state variables that is V,γ setpoints are chosen through the relations of Section 5.1.3 which allow to express these state variables as differential functions of the flat outputs of the system. Regarding the height state variable $x_1 = h$ which is also the spacecraft's first flat output this either increases or decreases linearly for certain time-intervals, or remains constant in certain slots of the simulation time. Regarding the position variable $x_2 = x$ of the spacecraft which is also the spacecraft's second flat output, this increases linearly during the simulation experiments, thus implying a forward motion of the spacecraft.

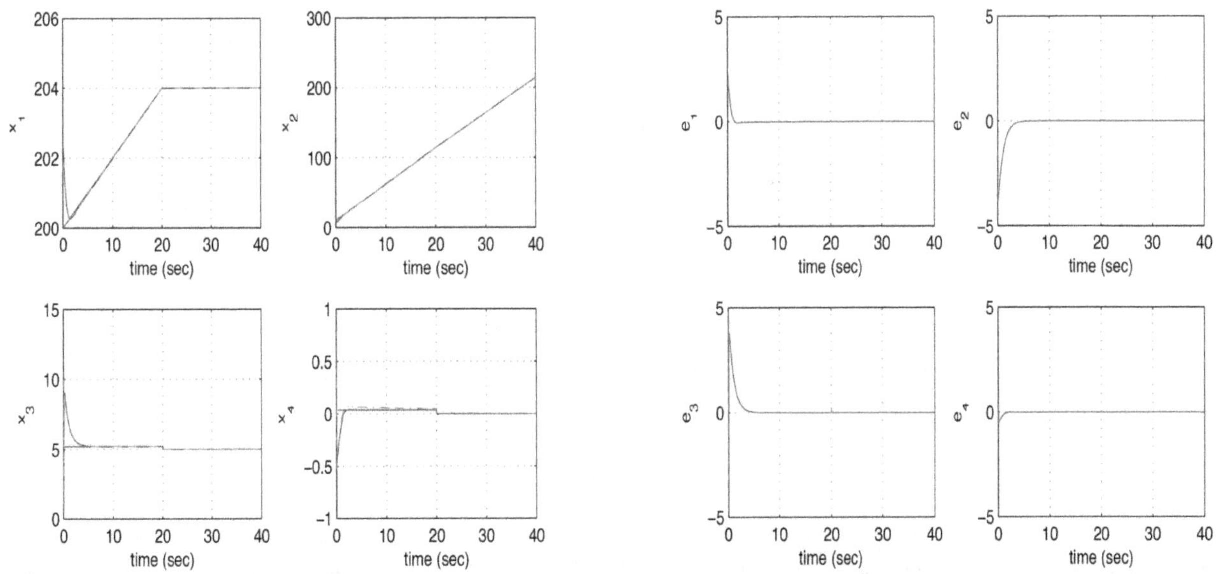

(a) (b)

Figure 5.3 Tracking of setpoint 1 by the reentry space vehicle under nonlinear optimal control: (a) convergence of state variables x_1 to x_4 (blue lines) to the associated setpoints (red lines) and estimated values provided by Kalman filtering; (b) variation of the tracking error variables e_1 to e_4 associated with the state variables of the spacecraft

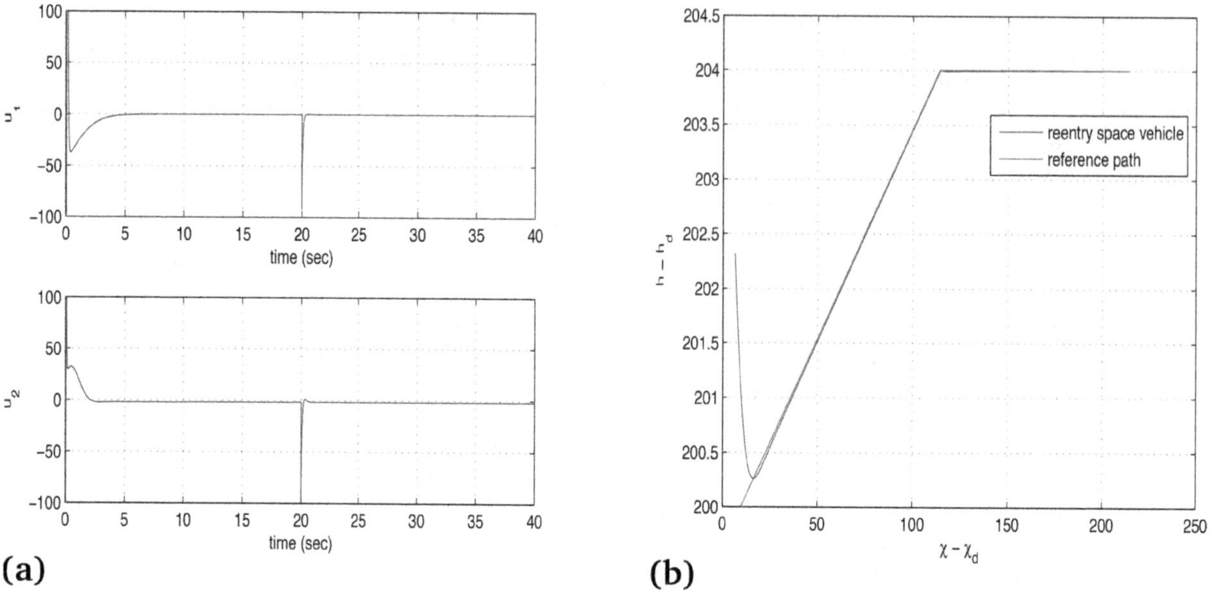

Figure 5.4 *Tracking of setpoint 1 by the reentry space vehicle under nonlinear optimal control: (a) variations of the control inputs u_1 and u_2 of the spacecraft (blue lines); (b) tracking of the reference path (red line) by the spacecraft in the plane of state variables $x_2 = \chi$ and $x_1 = h$*

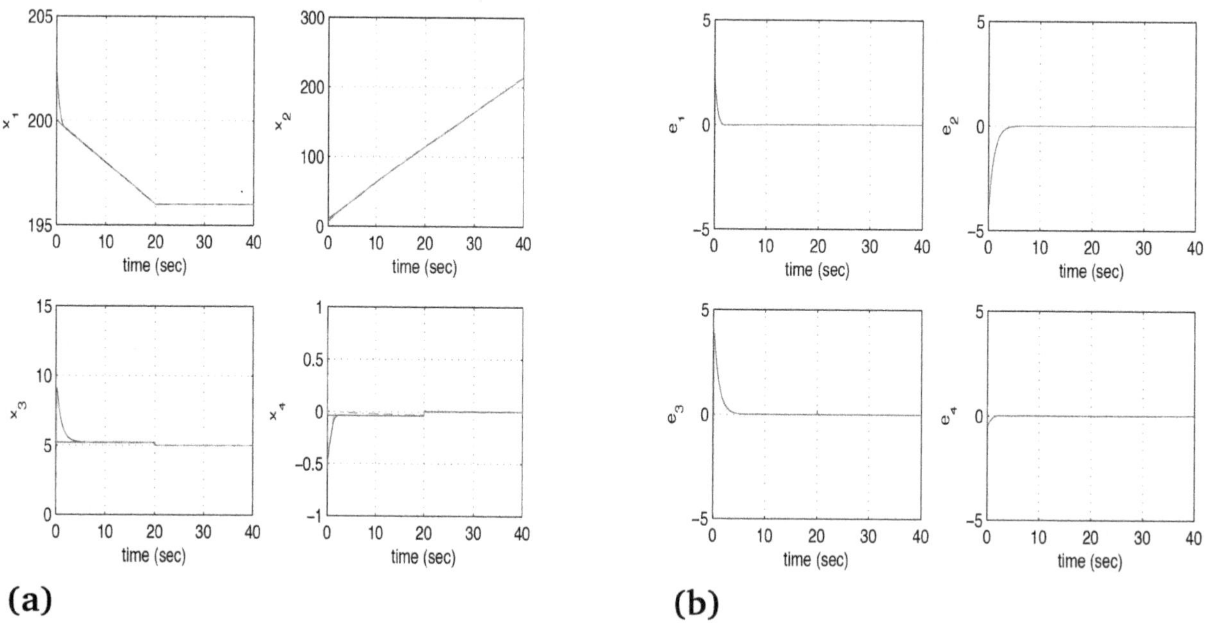

Figure 5.5 *Tracking of setpoint 2 by the reentry space vehicle under nonlinear optimal control: (a) convergence of state variables x_1 to x_4 (blue lines) to the associated setpoints (red lines) and estimated values provided by Kalman filtering; (b) variation of the tracking error variables e_1 to e_4 associated with the state variables of the spacecraft*

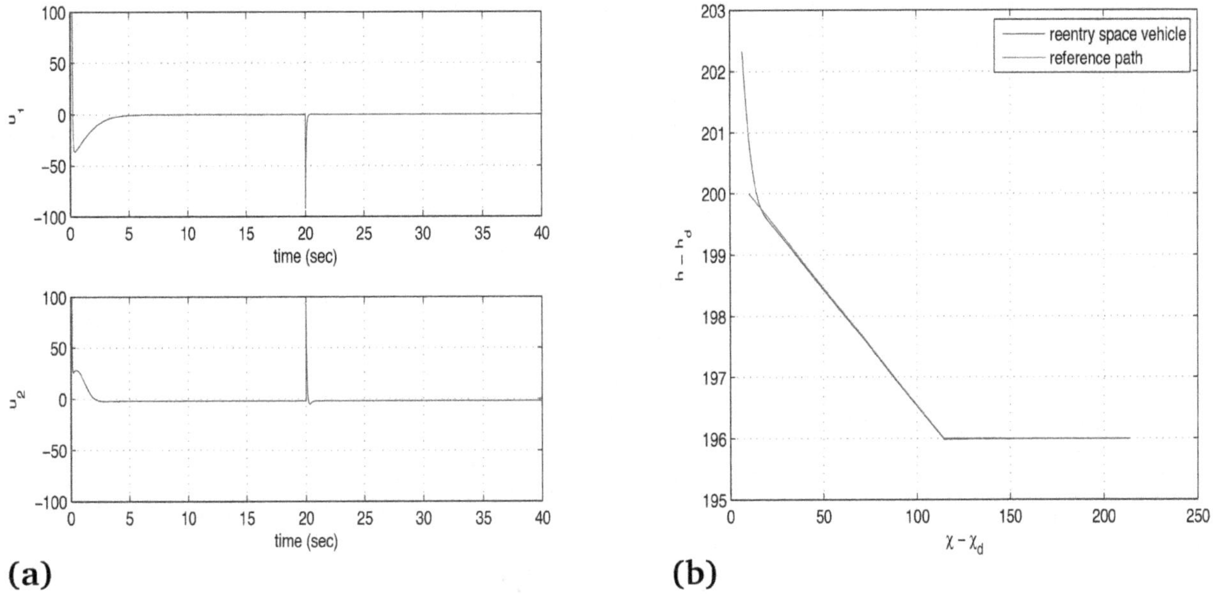

Figure 5.6 *Tracking of setpoint 2 by the reentry space vehicle under nonlinear optimal control: (a) variations of the control inputs u_1 and u_2 of the spacecraft (blue lines); (b) tracking of the reference path (red line) by the spacecraft in the plane of state variables $x_2 = \chi$ and $x_1 = h$*

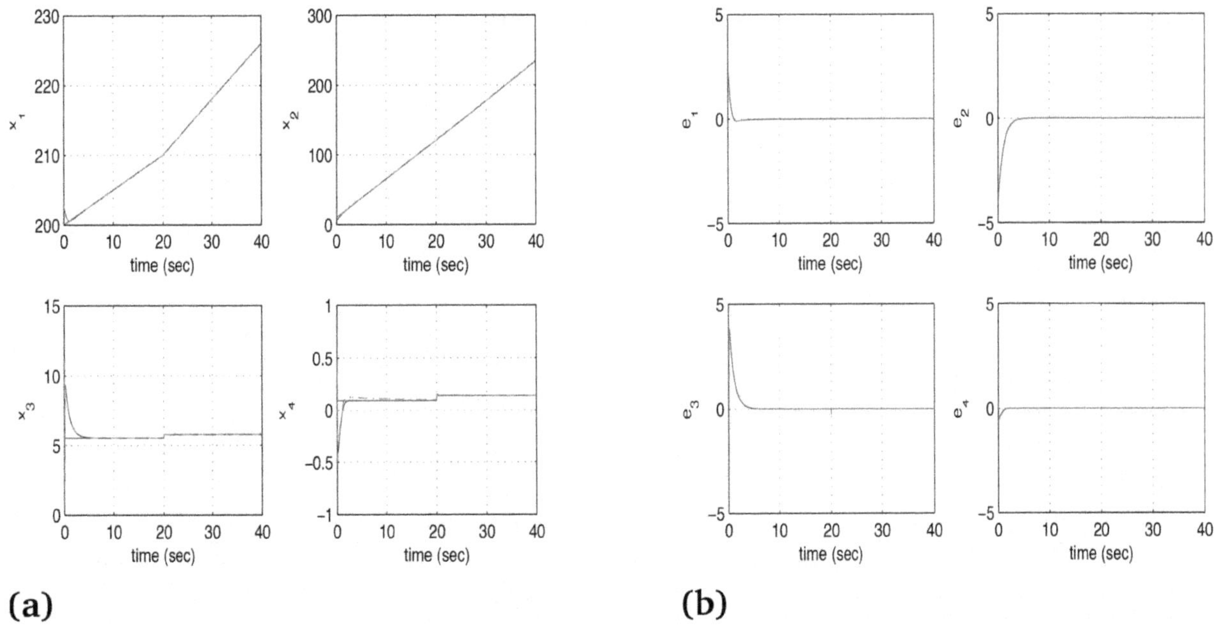

Figure 5.7 *Tracking of setpoint 3 by the reentry space vehicle under nonlinear optimal control: (a) convergence of state variables x_1 to x_4 (blue lines) to the associated setpoints (red lines) and estimated values provided by Kalman filtering; (b) variation of the tracking error variables e_1 to e_4 associated with the state variables of the spacecraft*

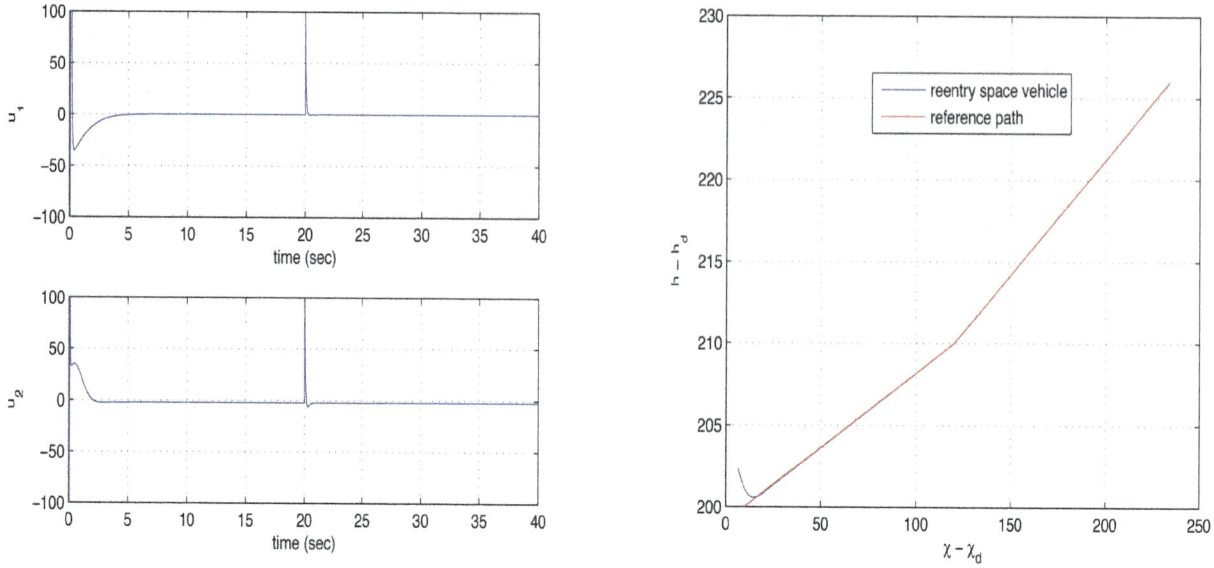

(a) (b)

Figure 5.8 *Tracking of setpoint 3 by the reentry space vehicle under nonlinear optimal control: (a) variations of the control inputs u_1 and u_2 of the spacecraft (blue lines); (b) tracking of the reference path (red line) by the spacecraft in the plane of state variables $x_2 = \chi$ and $x_1 = h$*

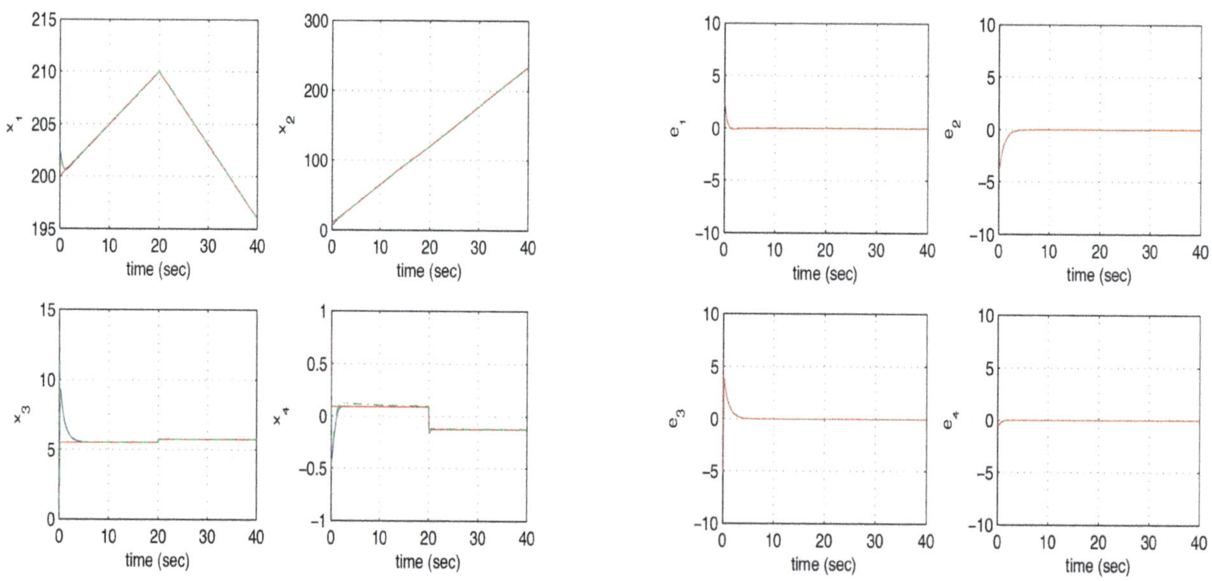

(a) (b)

Figure 5.9 *Tracking of setpoint 4 by the reentry space vehicle under nonlinear optimal control: (a) convergence of state variables x_1 to x_4 (blue lines) to the associated setpoints (red lines) and estimated values provided by Kalman filtering; (b) variation of the tracking error variables e_1 to e_4 associated with the state variables of the spacecraft*

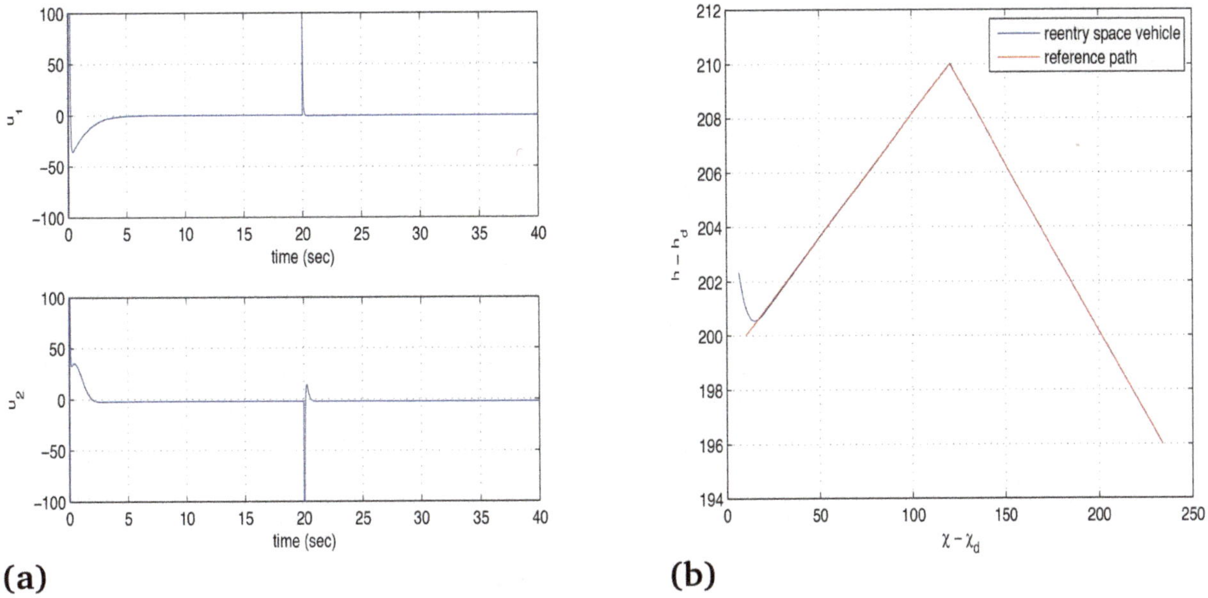

Figure 5.10 *Tracking of setpoint 4 by the reentry space vehicle under nonlinear optimal control: (a) variations of the control inputs u_1 and u_2 of the spacecraft (blue lines); (b) tracking of the reference path (red line) by the spacecraft in the plane of state variables $x_2 = \chi$ and $x_1 = h$*

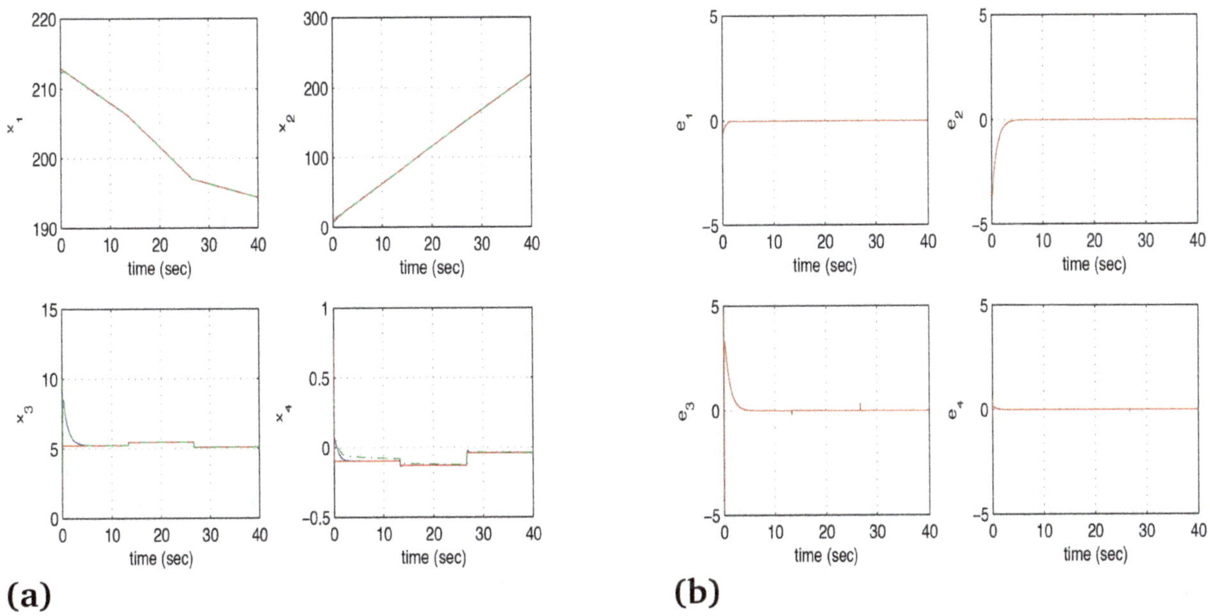

Figure 5.11 *Tracking of setpoint 5 by the reentry space vehicle under nonlinear optimal control: (a) convergence of state variables x_1 to x_4 (blue lines) to the associated setpoints (red lines) and estimated values provided by Kalman filtering; (b) variation of the tracking error variables e_1 to e_4 associated with the state variables of the spacecraft*

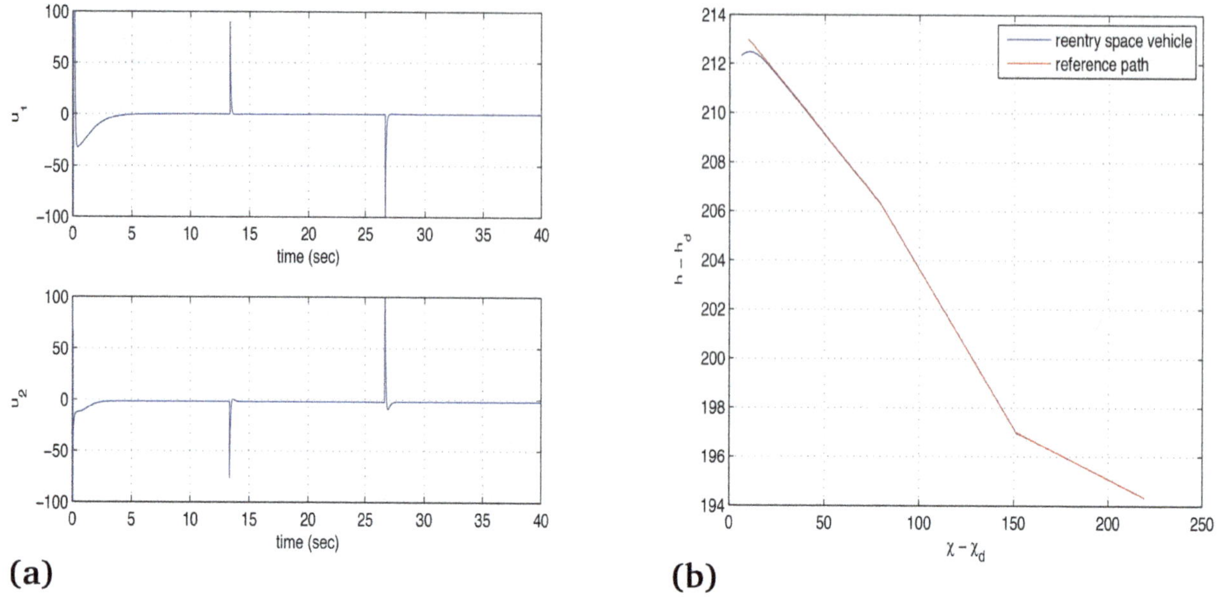

Figure 5.12 *Tracking of setpoint 5 by the reentry space vehicle under nonlinear optimal control: (a) variations of the control inputs u_1 and u_2 of the spacecraft (blue lines); (b) tracking of the reference path (red line) by the spacecraft in the plane of state variables $x_2 = \chi$ and $x_1 = h$*

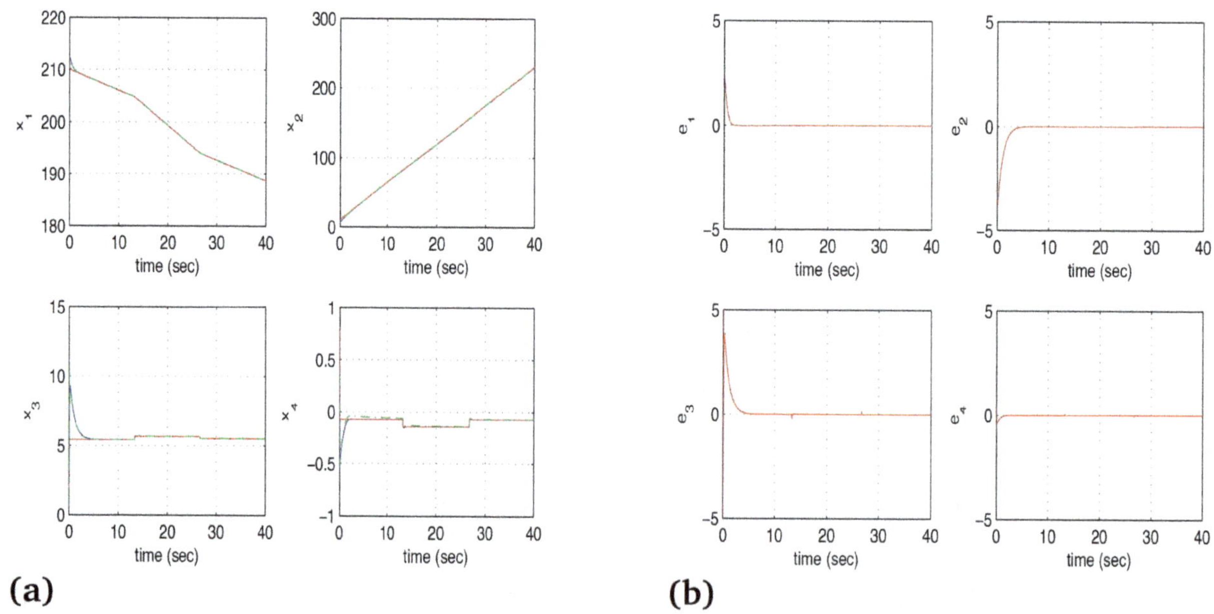

Figure 5.13 *Tracking of setpoint 6 by the reentry space vehicle under nonlinear optimal control: (a) convergence of state variables x_1 to x_4 (blue lines) to the associated setpoints (red lines) and estimated values provided by Kalman filtering; (b) variation of the tracking error variables e_1 to e_4 associated with the state variables of the spacecraft*

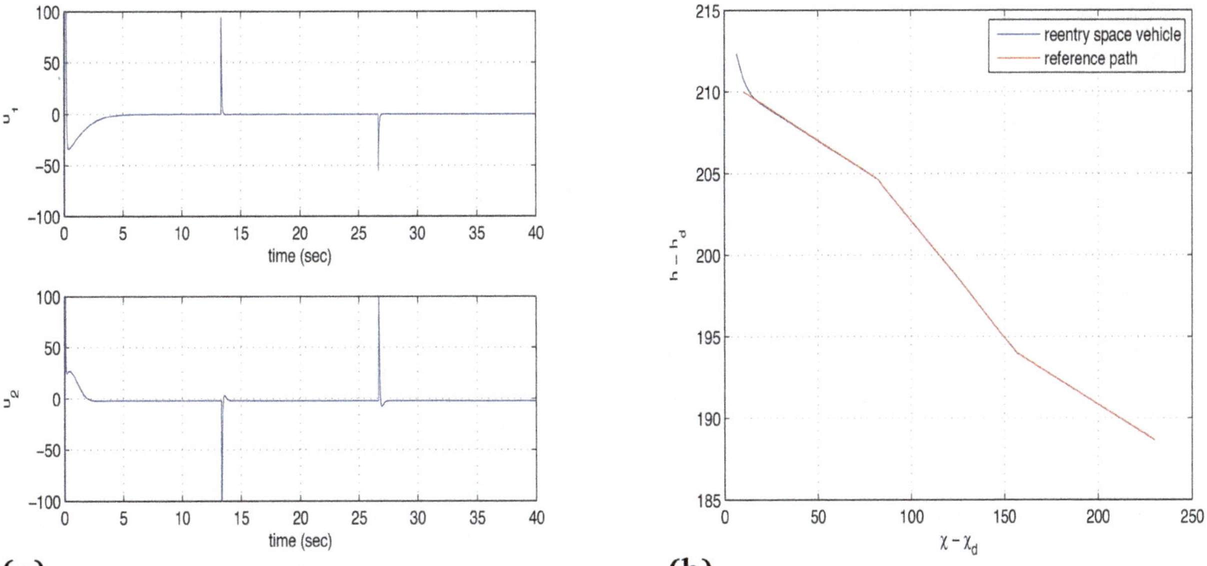

(a) (b)

Figure 5.14 Tracking of setpoint 6 by the reentry space vehicle under nonlinear optimal control: (a) variations of the control inputs u_1 and u_2 of the spacecraft (blue lines); (b) tracking of the reference path (red line) by the spacecraft in the plane of state variables $x_2 = \chi$ and $x_1 = h$

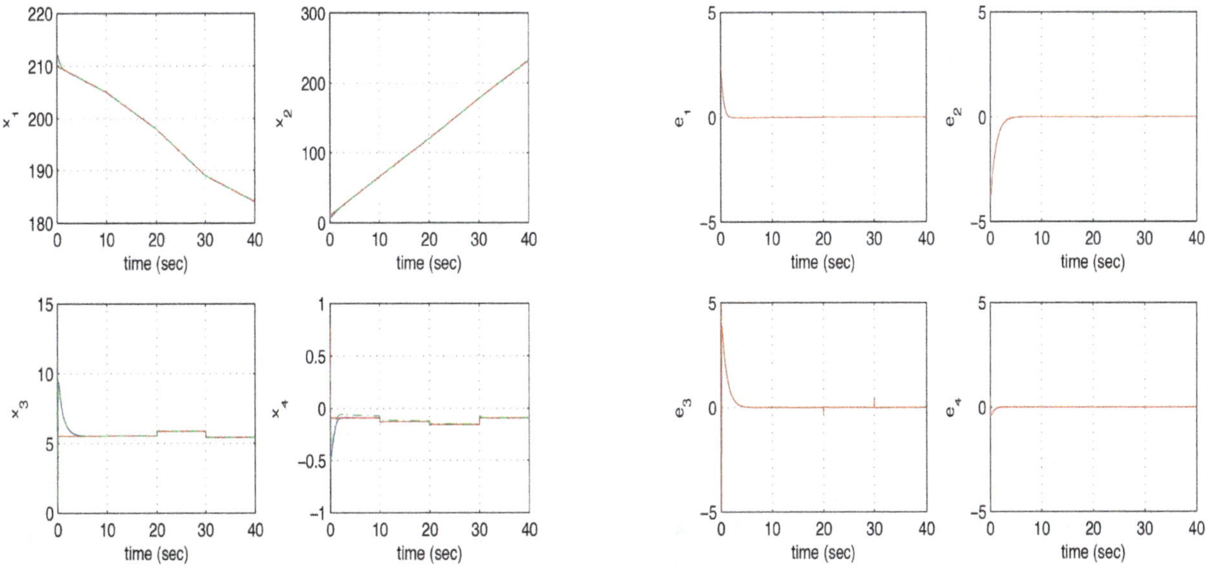

(a) (b)

Figure 5.15 Tracking of setpoint 7 by the reentry space vehicle under nonlinear optimal control: (a) convergence of state variables x_1 to x_4 (blue lines) to the associated setpoints (red lines) and estimated values provided by Kalman filtering; (b) variation of the tracking error variables e_1 to e_4 associated with the state variables of the spacecraft

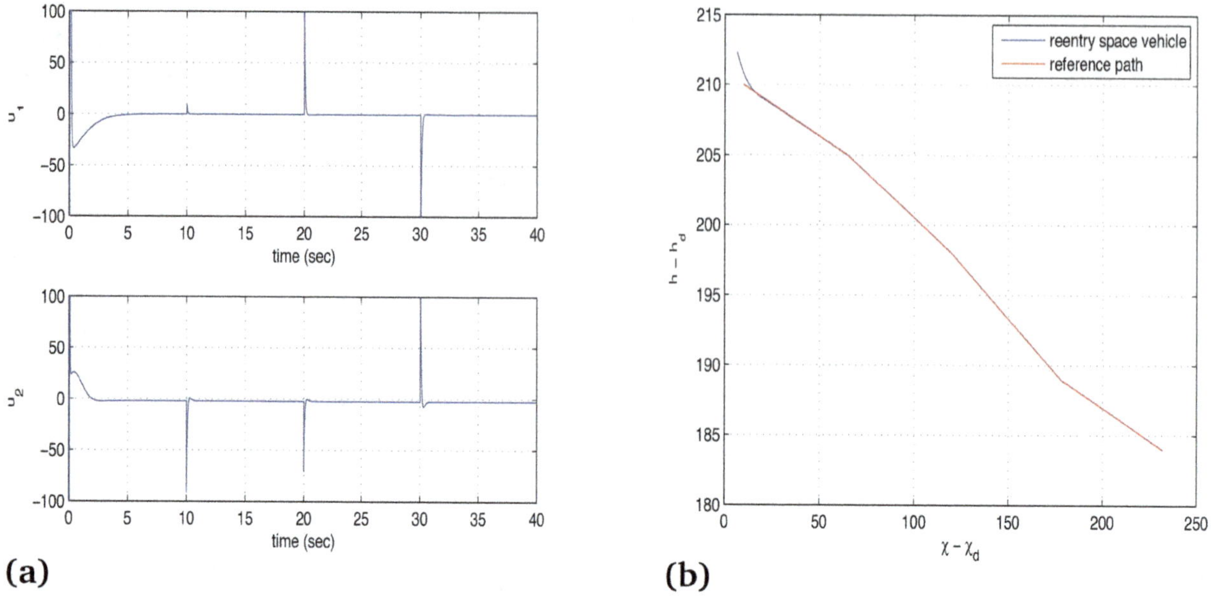

Figure 5.16 *Tracking of setpoint 7 by the reentry space vehicle under nonlinear optimal control: (a) variations of the control inputs u_1 and u_2 of the spacecraft (blue lines); (b) tracking of the reference path (red line) by the spacecraft in the plane of state variables $x_2 = \chi$ and $x_1 = h$*

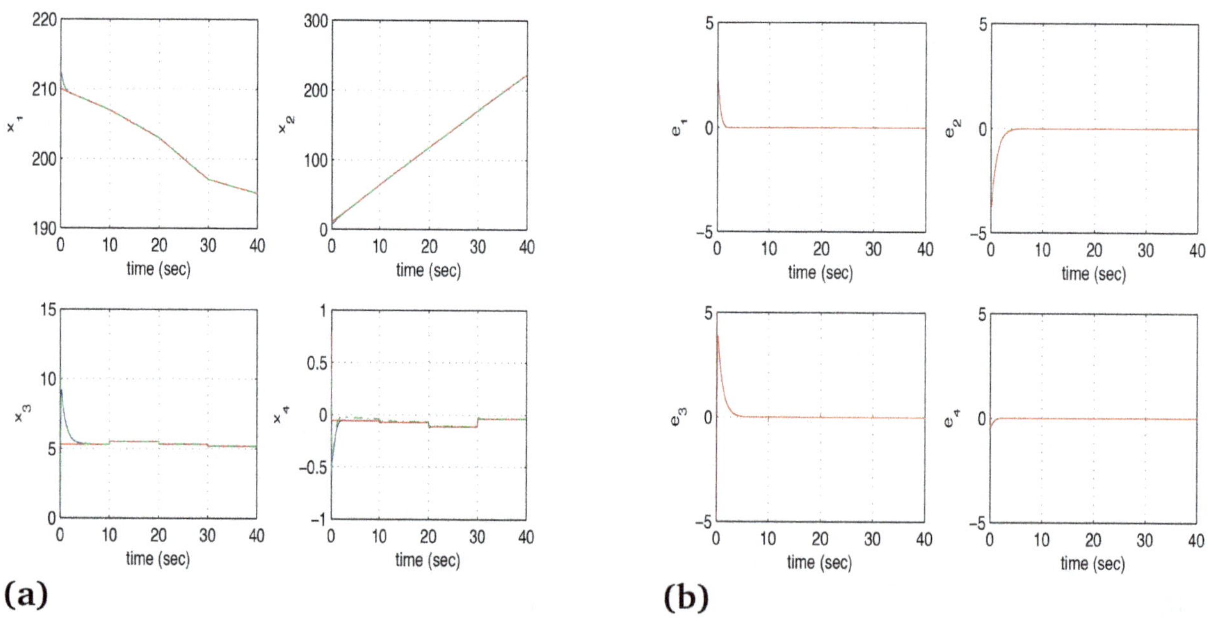

Figure 5.17 *Tracking of setpoint 8 by the reentry space vehicle under nonlinear optimal control: (a) convergence of state variables x_1 to x_4 (blue lines) to the associated setpoints (red lines) and estimated values provided by Kalman filtering; (b) variation of the tracking error variables e_1 to e_4 associated with the state variables of the spacecraft*

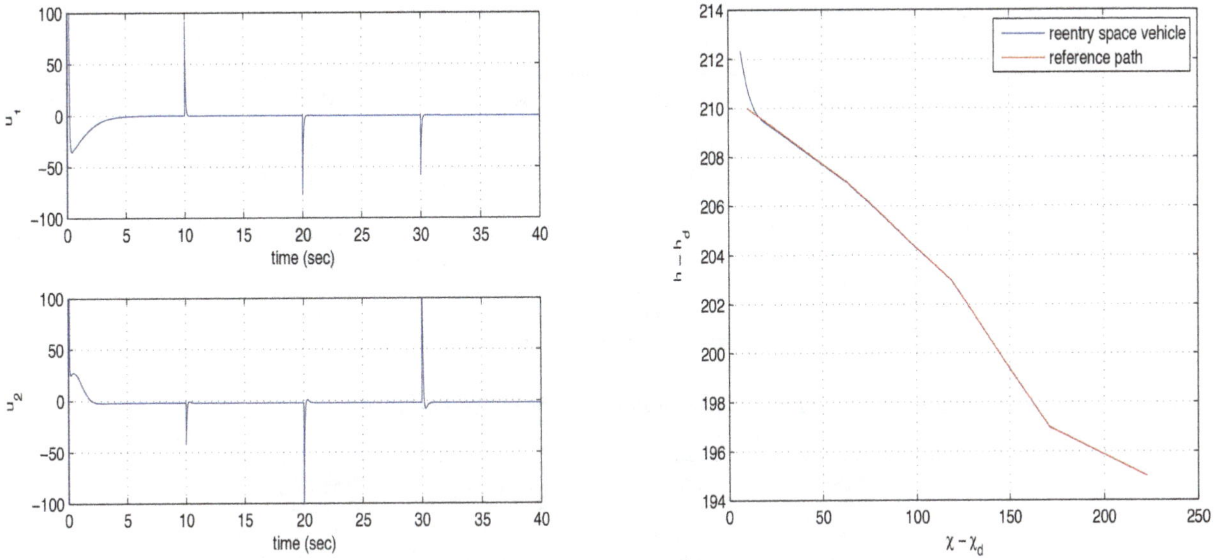

(a) (b)

Figure 5.18 Tracking of setpoint 8 by the reentry space vehicle under nonlinear optimal control: (a) variations of the control inputs u_1 and u_2 of the spacecraft (blue lines); (b) tracking of the reference path (red line) by the spacecraft in the plane of state variables $x_2 = \chi$ and $x_1 = h$

Table 5.1 Tracking RMSE for the RSV in the disturbance-free case

	$\mathbf{RMSE}_{x_1}$	$\mathbf{RMSE}_{x_2}$	$\mathbf{RMSE}_{x_3}$	$\mathbf{RMSE}_{x_4}$
$test_1$	$0.1369 \cdot 10^{-3}$	$0.0004 \cdot 10^{-3}$	$0.0001 \cdot 10^{-3}$	$0.0010 \cdot 10^{-3}$
$test_2$	$0.1359 \cdot 10^{-3}$	$0.0004 \cdot 10^{-3}$	$0.0001 \cdot 10^{-3}$	$0.0011 \cdot 10^{-3}$
$test_3$	$0.0948 \cdot 10^{-3}$	$0.0058 \cdot 10^{-3}$	$0.0001 \cdot 10^{-3}$	$0.0014 \cdot 10^{-3}$
$test_4$	$0.0926 \cdot 10^{-3}$	$0.0057 \cdot 10^{-3}$	$0.0001 \cdot 10^{-3}$	$0.0011 \cdot 10^{-3}$
$test_5$	$0.1300 \cdot 10^{-3}$	$0.0014 \cdot 10^{-3}$	$0.0001 \cdot 10^{-3}$	$0.0011 \cdot 10^{-3}$
$test_6$	$0.1028 \cdot 10^{-3}$	$0.0032 \cdot 10^{-3}$	$0.0001 \cdot 10^{-3}$	$0.0012 \cdot 10^{-3}$
$test_7$	$0.1109 \cdot 10^{-3}$	$0.0044 \cdot 10^{-3}$	$0.0012 \cdot 10^{-3}$	$0.0013 \cdot 10^{-3}$
$test_8$	$0.1284 \cdot 10^{-3}$	$0.0015 \cdot 10^{-3}$	$0.0004 \cdot 10^{-3}$	$0.0015 \cdot 10^{-3}$

To elaborate on the tracking performance and on the robustness of the proposed nonlinear optimal control method for the autonomous reentry space vehicle the following tables are given: (i) Table 5.1 which provides information about the accuracy of tracking of the reference setpoints by the state variables of the reentry space vehicle's state-space model, (ii) Table 5.2 which provides information about the robustness of the control method to parametric changes in the model of the reentry space vehicle's dynamics (change $\Delta a\%$ in gravity's acceleration g for the spacecraft), (iii) Table 5.3 which provides information about the precision in state variables' estimation that is achieved by the H-infinity Kalman filter, and (iv) Table 5.4 which provides the approximate convergence time of the autonomous reentry space vehicle's state variables to the associated setpoints. In particular about the results presented in Table 5.2 it has been considered that a change of $\Delta a\%$ was introduced to the value of gravity's acceleration that the controller of the reentry space vehicle was using, compared to the nominal value of the gravity acceleration for the specific orbit of the spacecraft. Since the reentry space vehicle was considered to be practically out of earth's gravity field the nominal value of gravity's acceleration was $g = 0.06m/sec^2$.

Table 5.2 Tracking RMSE for the RSV in the case of disturbances

$\Delta a\%$	$\mathbf{RMSE}_{x_1}$	$\mathbf{RMSE}_{x_2}$	$\mathbf{RMSE}_{x_3}$	$\mathbf{RMSE}_{x_4}$
0%	$0.1359 \cdot 10^{-3}$	$0.0004 \cdot 10^{-3}$	$0.0001 \cdot 10^{-3}$	$0.0011 \cdot 10^{-3}$
10%	$0.3152 \cdot 10^{-3}$	$0.0019 \cdot 10^{-3}$	$0.0001 \cdot 10^{-3}$	$0.0012 \cdot 10^{-3}$
20%	$0.4918 \cdot 10^{-3}$	$0.0045 \cdot 10^{-3}$	$0.0001 \cdot 10^{-3}$	$0.0011 \cdot 10^{-3}$
30%	$0.6688 \cdot 10^{-3}$	$0.0082 \cdot 10^{-3}$	$0.0001 \cdot 10^{-3}$	$0.0010 \cdot 10^{-3}$
40%	$0.8466 \cdot 10^{-3}$	$0.0130 \cdot 10^{-3}$	$0.0001 \cdot 10^{-3}$	$0.0012 \cdot 10^{-3}$
50%	$1.0240 \cdot 10^{-3}$	$0.0190 \cdot 10^{-3}$	$0.0001 \cdot 10^{-3}$	$0.0010 \cdot 10^{-3}$
60%	$1.2010 \cdot 10^{-3}$	$0.0260 \cdot 10^{-3}$	$0.0001 \cdot 10^{-3}$	$0.0010 \cdot 10^{-3}$

Table 5.3 RMSE for the estimation performed by the H-infinity KF

	$\mathbf{RMSE}_{x_1}$	$\mathbf{RMSE}_{x_2}$	$\mathbf{RMSE}_{x_3}$	$\mathbf{RMSE}_{x_4}$
$test_1$	$0.1000 \cdot 10^{-3}$	$0.1000 \cdot 10^{-3}$	$7.6000 \cdot 10^{-3}$	$0.1000 \cdot 10^{-3}$
$test_2$	$0.1000 \cdot 10^{-3}$	$0.1000 \cdot 10^{-3}$	$7.4000 \cdot 10^{-3}$	$0.1000 \cdot 10^{-3}$
$test_3$	$0.1000 \cdot 10^{-3}$	$0.1000 \cdot 10^{-3}$	$8.9000 \cdot 10^{-3}$	$0.1000 \cdot 10^{-3}$
$test_4$	$0.1000 \cdot 10^{-3}$	$0.1000 \cdot 10^{-3}$	$8.7000 \cdot 10^{-3}$	$0.1000 \cdot 10^{-3}$
$test_5$	$0.1000 \cdot 10^{-3}$	$0.1000 \cdot 10^{-3}$	$7.5000 \cdot 10^{-3}$	$0.1000 \cdot 10^{-3}$
$test_6$	$0.1000 \cdot 10^{-3}$	$0.1000 \cdot 10^{-3}$	$8.1000 \cdot 10^{-3}$	$0.1000 \cdot 10^{-3}$
$test_7$	$0.1000 \cdot 10^{-3}$	$0.1000 \cdot 10^{-3}$	$7.9000 \cdot 10^{-3}$	$0.1000 \cdot 10^{-3}$
$test_8$	$0.1000 \cdot 10^{-3}$	$0.1000 \cdot 10^{-3}$	$7.6000 \cdot 10^{-3}$	$0.1000 \cdot 10^{-3}$

Table 5.4 Convergence time (sec) for the RSV states

	$T_s\ x_1$	$T_s\ x_2$	$T_s\ x_3$	$T_s\ x_4$
$test_1$	2.0	0.2	4.0	2.5
$test_2$	1.5	0.2	4.0	2.5
$test_3$	1.0	0.2	4.0	2.5
$test_4$	1.0	0.2	4.0	2.5
$test_5$	0.5	0.2	3.5	2.0
$test_6$	0.5	0.2	4.5	2.5
$test_7$	0.5	0.2	4.0	2.5
$test_8$	1.0	0.2	4.0	2.5

Finally, the variations of the Lyapunov function of the control scheme for the reentry-space vehicle are depicted in Figures 5.19–5.22. These diagrams show clearly convergence of the Lyapunov function of the RSV to zero and global stability for the associated control loop.

5.1.6.2 Results on flatness-based control through transformation into canonical forms

So far this section has given a theoretical analysis about the proposed nonlinear optimal control method for reentry space vehicles and an experimental part about the evaluation of this control method through simulation tests. Besides, in Section 5.1.3, a theoretical analysis was given about flatness-based control for the reentry space vehicle which is based on the transformation of the state-space model of the spacecraft into the input–output linearized form and subsequently into the canonical Brunovsky form. It was already apparent that the flatness-based control scheme was

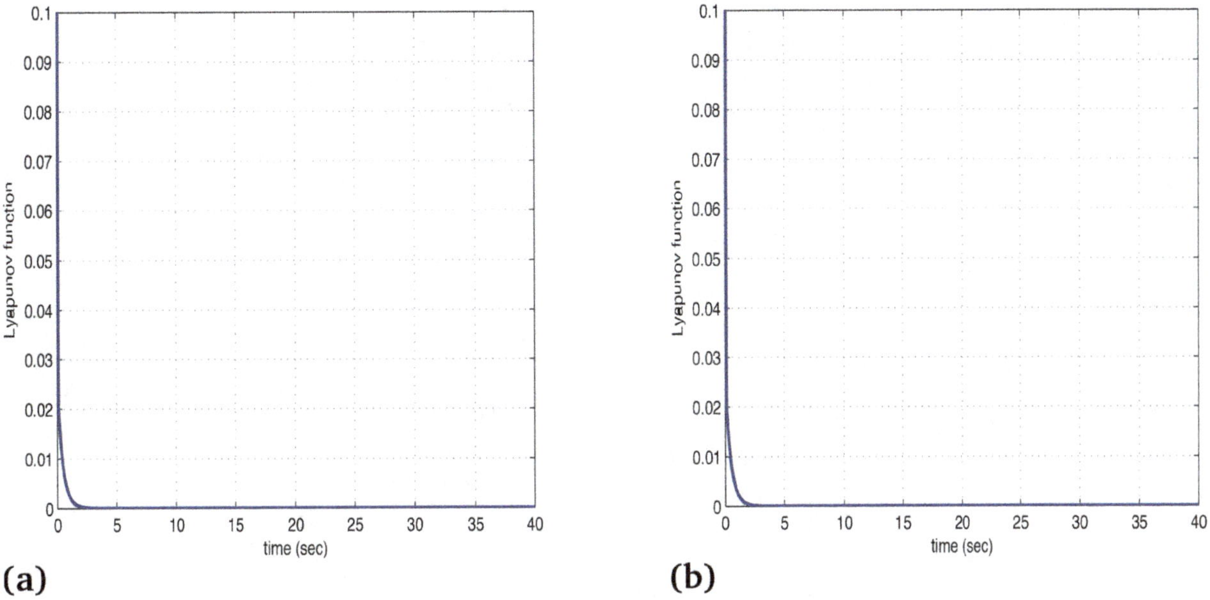

Figure 5.19 Lyapunov functions for nonlinear optimal control of the reentry space vehicle when tracking setpoints 1–2

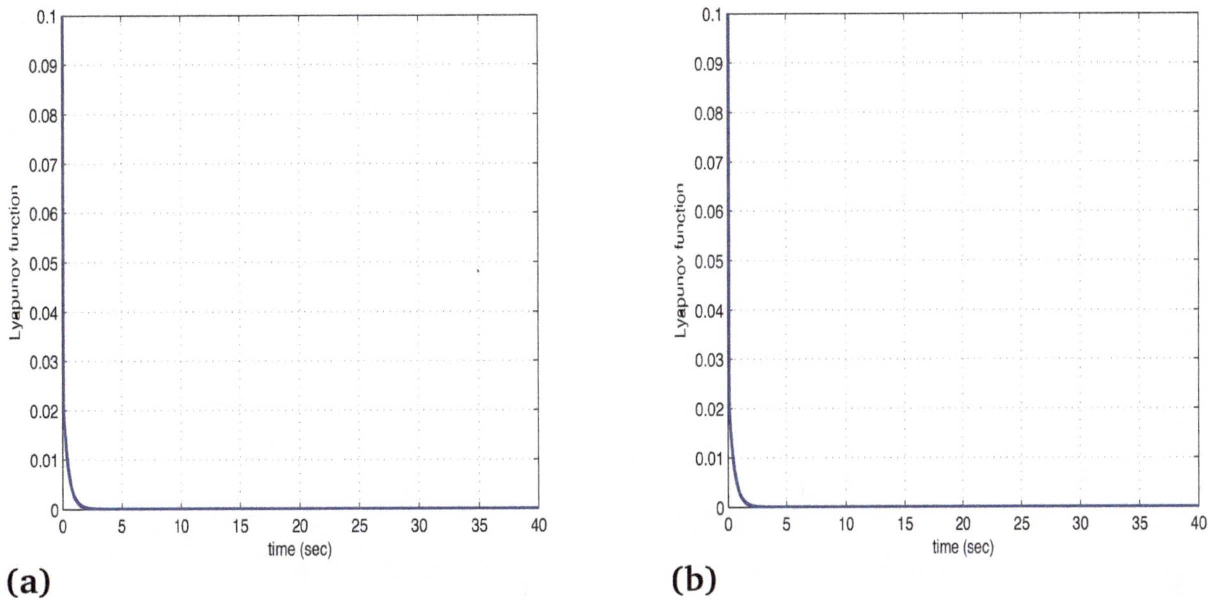

Figure 5.20 Lyapunov functions for nonlinear optimal control of the reentry space vehicle when tracking setpoints 3–4

based on changes of state variables and on transformations of the state-space model of the reentry space vehicles, while the nonlinear optimal control method made no use of diffeomorphisms and changes of state-space descriptions. Besides, the flatness-based control method is suboptimal because it aims only at eliminating the state variables' tracking error and not at the minimization of the control inputs variation. Next, new experimental results are given about

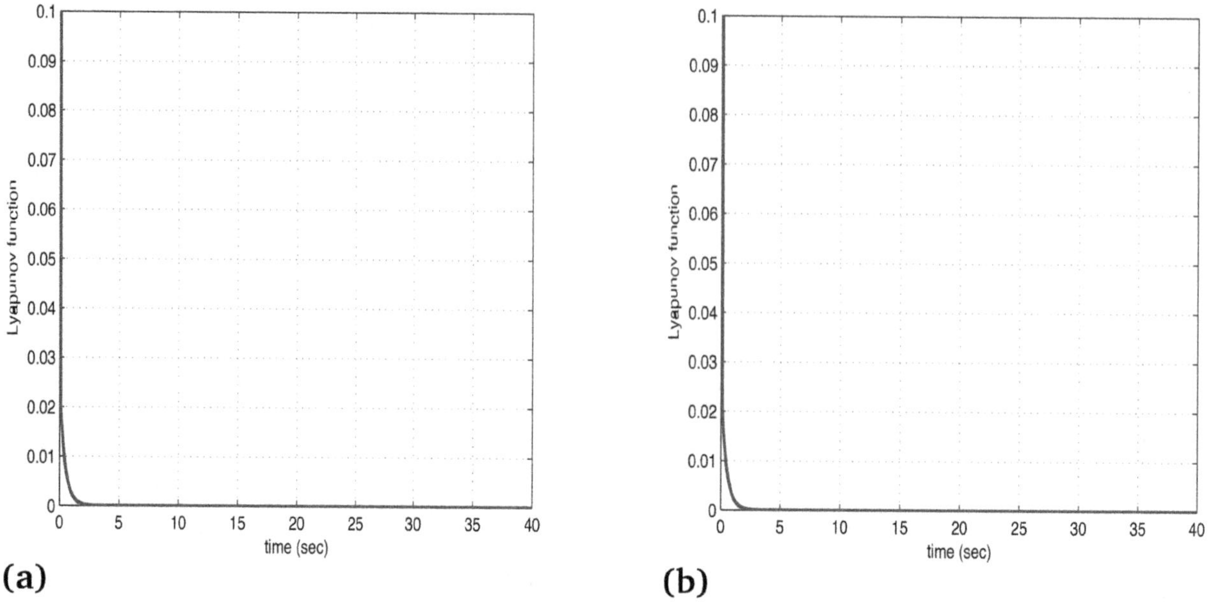

Figure 5.21 Lyapunov functions for nonlinear optimal control of the reentry space vehicle when tracking setpoints 5–6

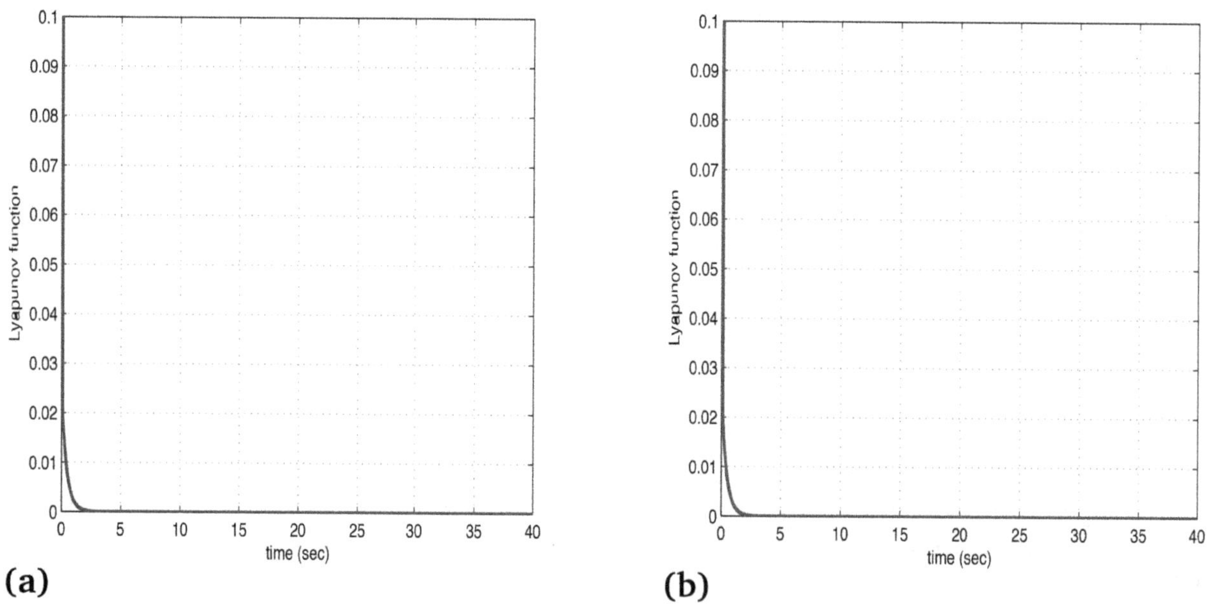

Figure 5.22 Lyapunov functions for nonlinear optimal control of the reentry space vehicle when tracking setpoints 7–8

flatness-based control of the reentry-space vehicle through transformation into canonical forms. These results are depicted in Figures 5.23–5.38 and demonstrate that flatness-based control through transformation into canonical forms achieves also fast and accurate tracking of setpoints by the state variables of the reentry space vehicle under moderate variations of the control inputs.

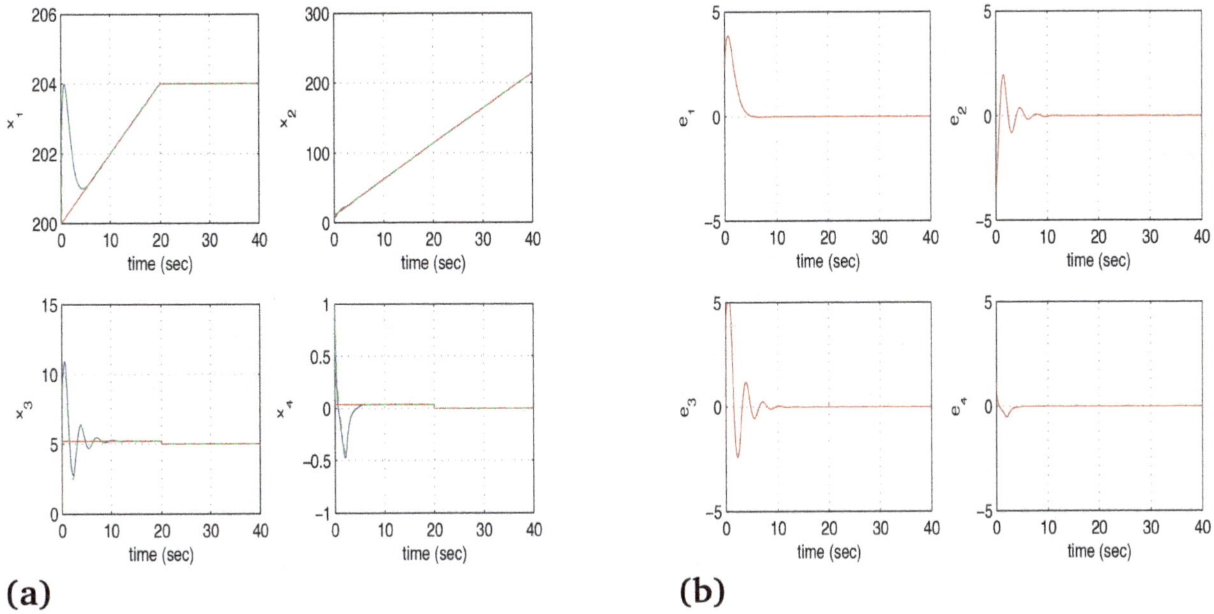

Figure 5.23 *Tracking of setpoint 1 by the reentry space vehicle under flatness-based control: (a) convergence of state variables x_1 to x_4 (blue lines) to the associated setpoints (red lines) and estimated values provided by Kalman filtering; (b) variation of the tracking error variables e_1 to e_4 associated with the state variables of the spacecraft.*

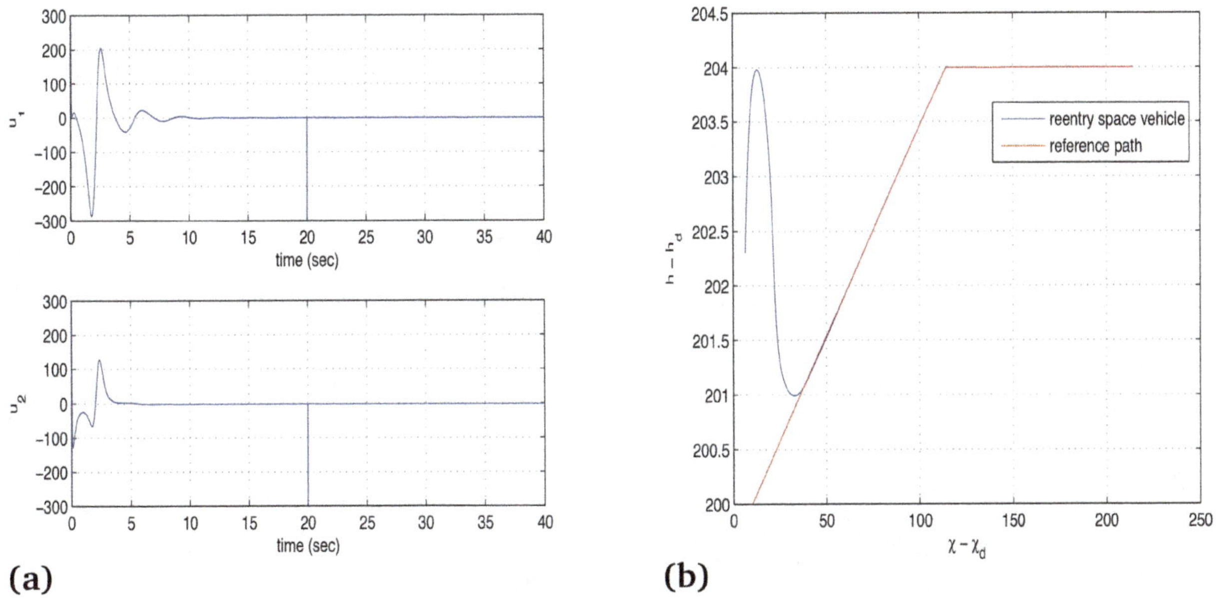

Figure 5.24 *Tracking of setpoint 1 by the reentry space vehicle under flatness-based control: (a) variations of the control inputs u_1 and u_2 of the spacecraft (blue lines); (b) tracking of the reference path (red line) by the spacecraft in the plane of state variables $x_2 = \chi$ and $x_1 = h$*

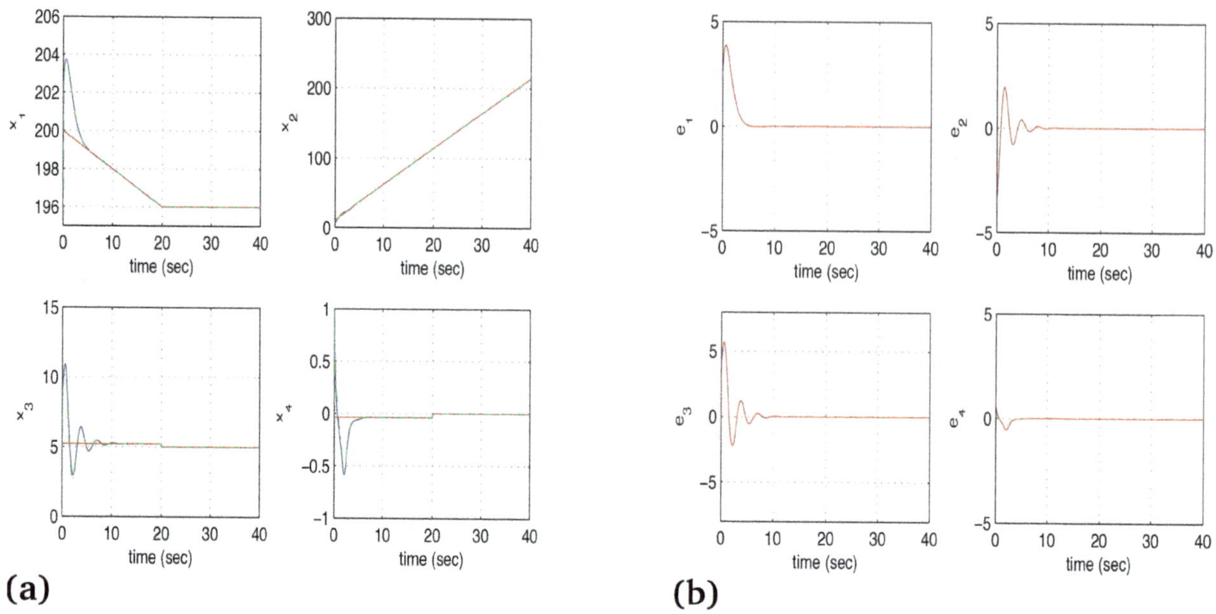

Figure 5.25 Tracking of setpoint 2 by the reentry space vehicle under flatness-based control: (a) convergence of state variables x_1 to x_4 (blue lines) to the associated setpoints (red lines) and estimated values provided by Kalman filtering; (b) variation of the tracking error variables e_1 to e_4 associated with the state variables of the spacecraft

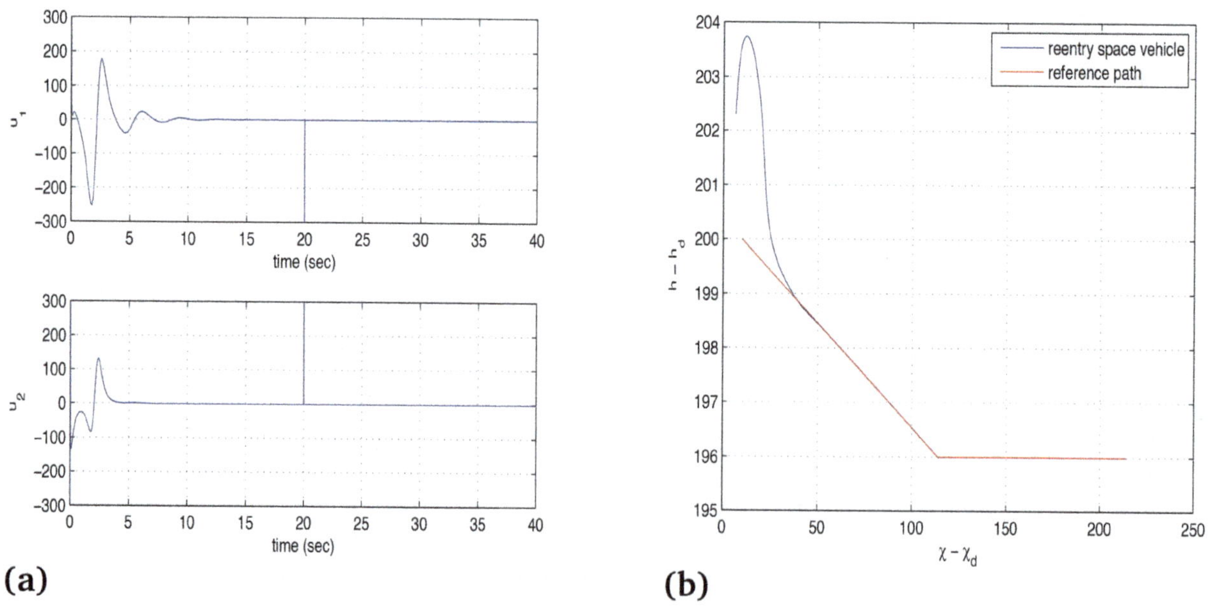

Figure 5.26 Tracking of setpoint 2 by the reentry space vehicle under flatness-based control: (a) variations of the control inputs u_1 and u_2 of the spacecraft (blue lines); (b) tracking of the reference path (red line) by the spacecraft in the plane of state variables $x_2 = \chi$ and $x_1 = h$

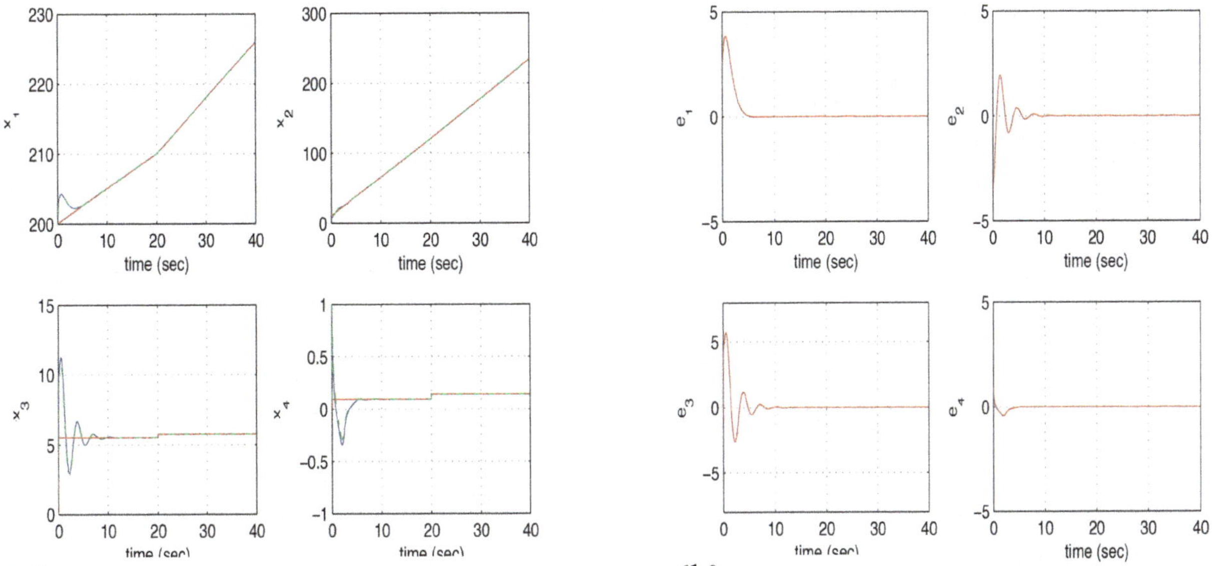

(a) (b)

Figure 5.27 Tracking of setpoint 3 by the reentry space vehicle under flatness-based control: (a) convergence of state variables x_1 to x_4 (blue lines) to the associated setpoints (red lines) and estimated values provided by Kalman filtering; (b) variation of the tracking error variables e_1 to e_4 associated with the state variables of the spacecraft

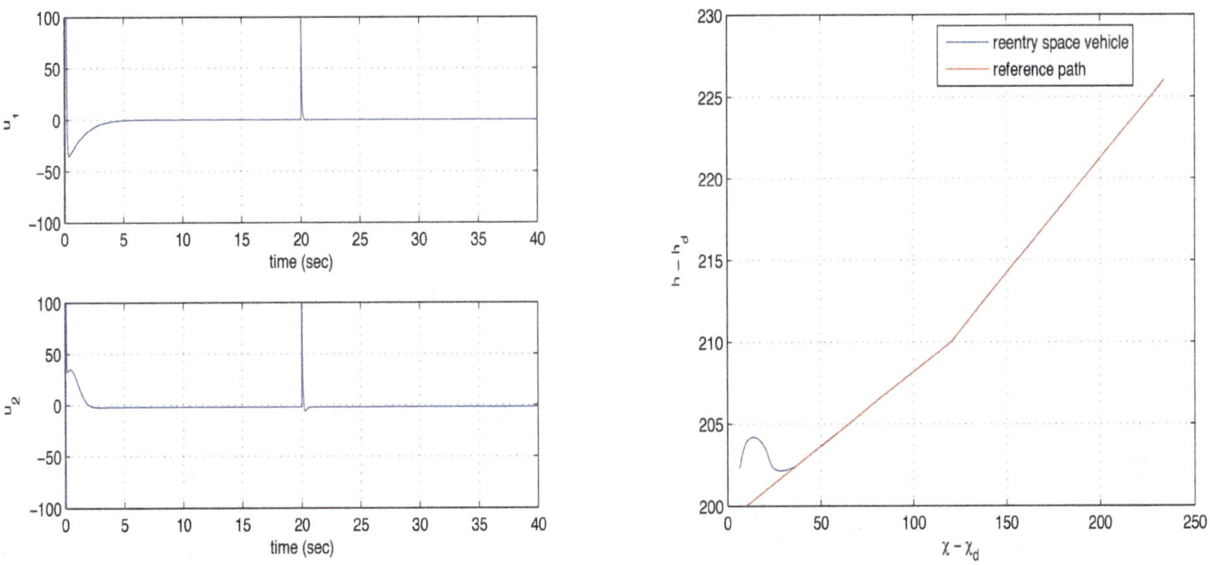

(a) (b)

Figure 5.28 Tracking of setpoint 3 by the reentry space vehicle under flatness-based control: (a) variations of the control inputs u_1 and u_2 of the spacecraft (blue lines); (b) tracking of the reference path (red line) by the spacecraft in the plane of state variables $x_2 = \chi$ and $x_1 = h$

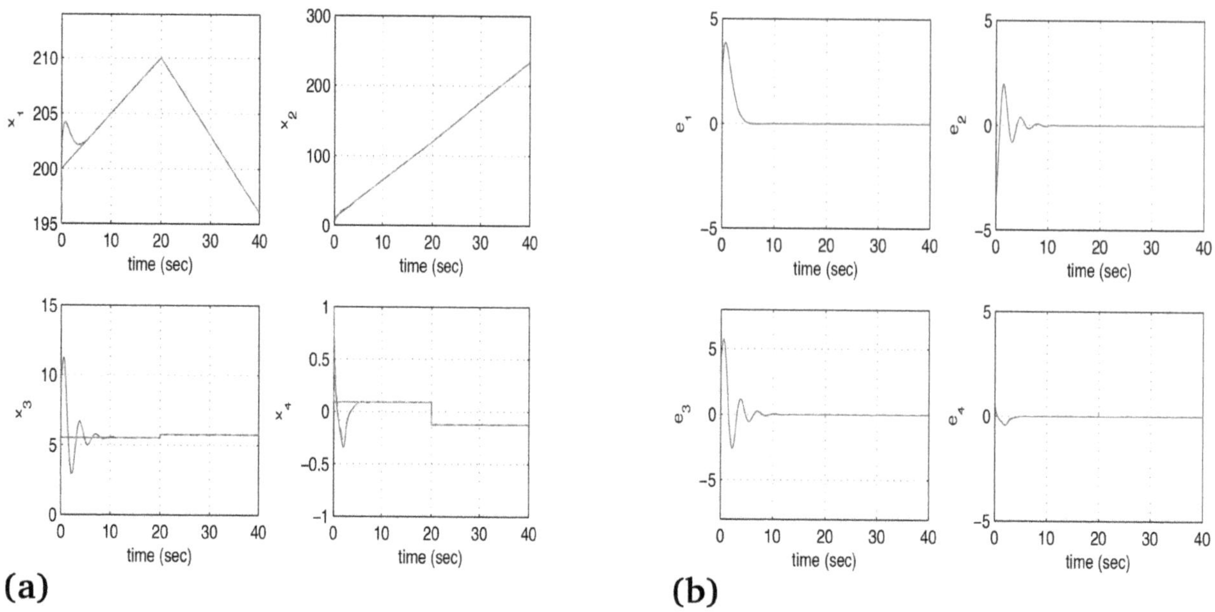

Figure 5.29 *Tracking of setpoint 4 by the reentry space vehicle under flatness-based control: (a) convergence of state variables x_1 to x_4 (blue lines) to the associated setpoints (red lines) and estimated values provided by Kalman filtering; (b) variation of the tracking error variables e_1 to e_4 associated with the state variables of the spacecraft*

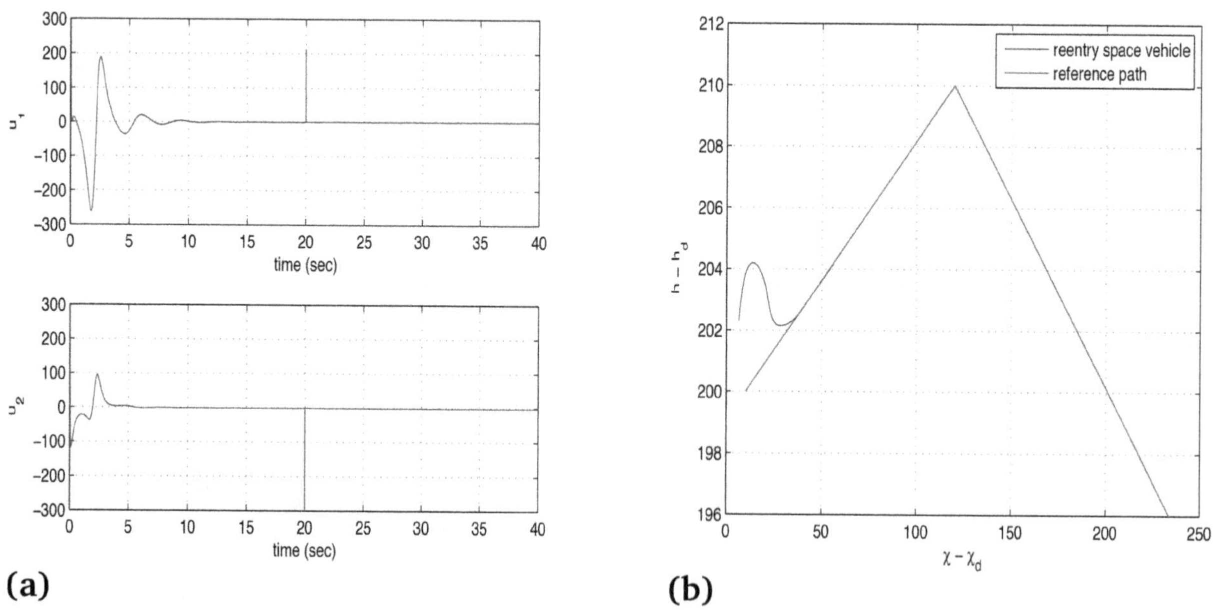

Figure 5.30 *Tracking of setpoint 4 by the reentry space vehicle under flatness-based control: (a) variations of the control inputs u_1 and u_2 of the spacecraft (blue lines); (b) tracking of the reference path (red line) by the spacecraft in the plane of state variables $x_2 = \chi$ and $x_1 = h$*

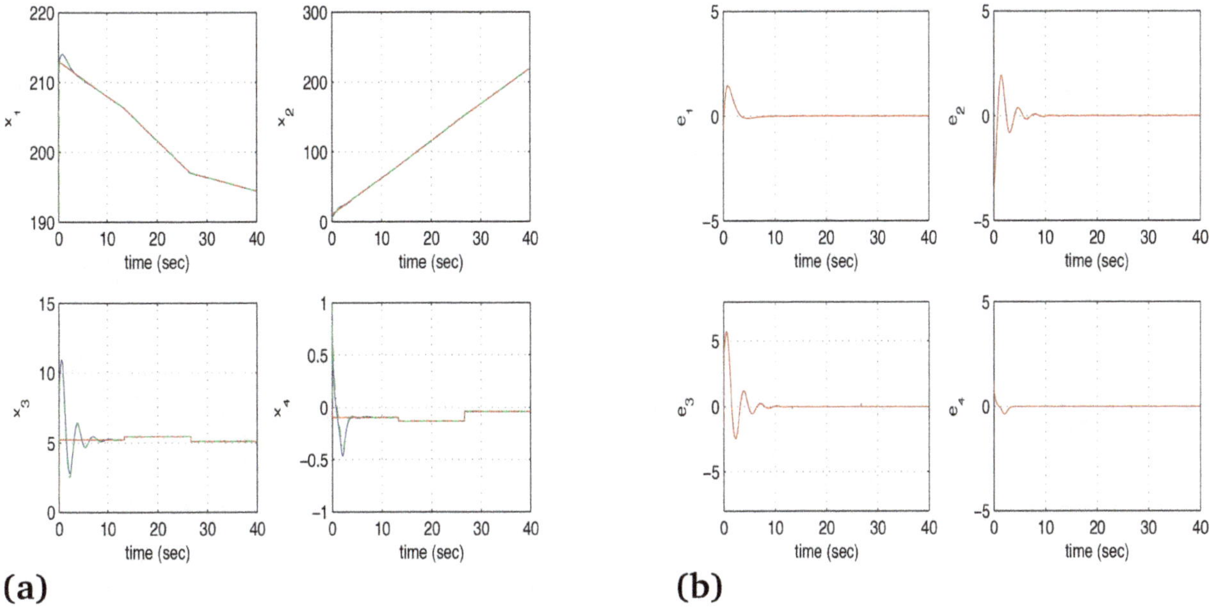

Figure 5.31 Tracking of setpoint 5 by the reentry space vehicle under flatness-based control: (a) convergence of state variables x_1 to x_4 (blue lines) to the associated setpoints (red lines) and estimated values provided by Kalman filtering; (b) variation of the tracking error variables e_1 to e_4 associated with the state variables of the spacecraft

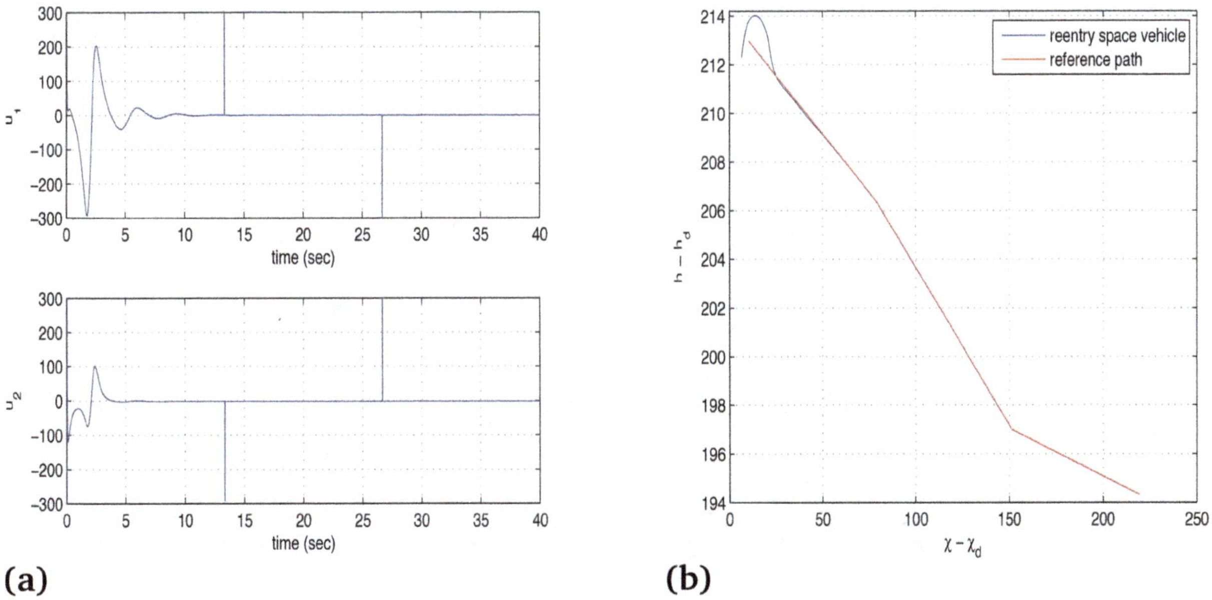

Figure 5.32 Tracking of setpoint 5 by the reentry space vehicle under flatness-based control: (a) variations of the control inputs u_1 and u_2 of the spacecraft (blue lines); (b) tracking of the reference path (red line) by the spacecraft in the plane of state variables $x_2 = \chi$ and $x_1 = h$

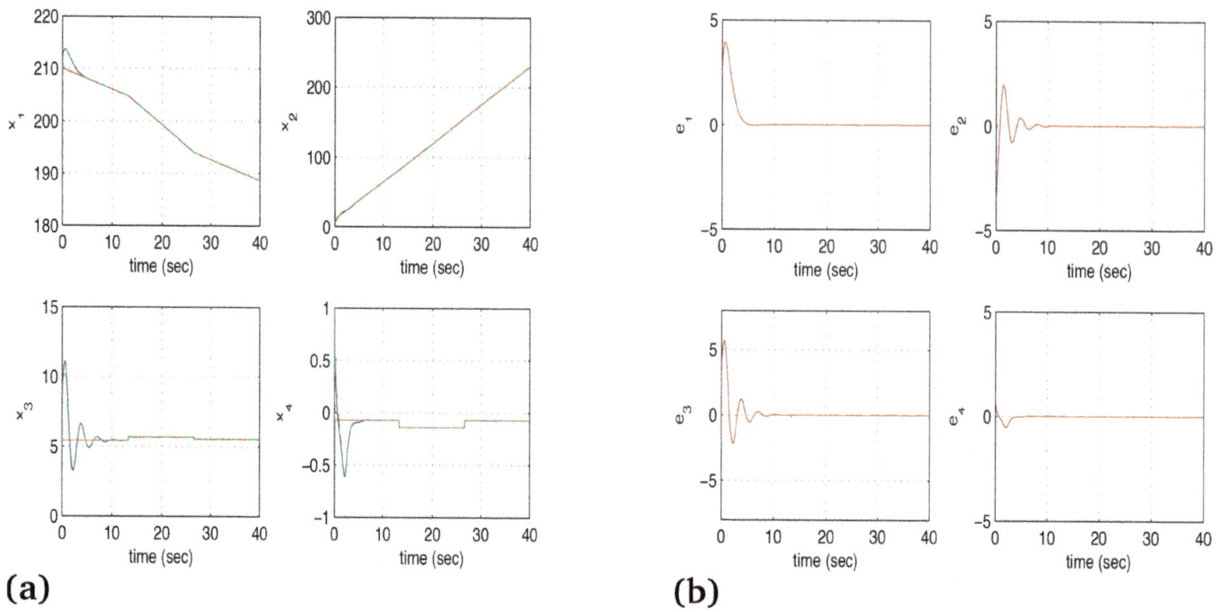

Figure 5.33 Tracking of setpoint 6 by the reentry space vehicle under flatness-based control: (a) convergence of state variables x_1 to x_4 (blue lines) to the associated setpoints (red lines) and estimated values provided by Kalman filtering; (b) variation of the tracking error variables e_1 to e_4 associated with the state variables of the spacecraft

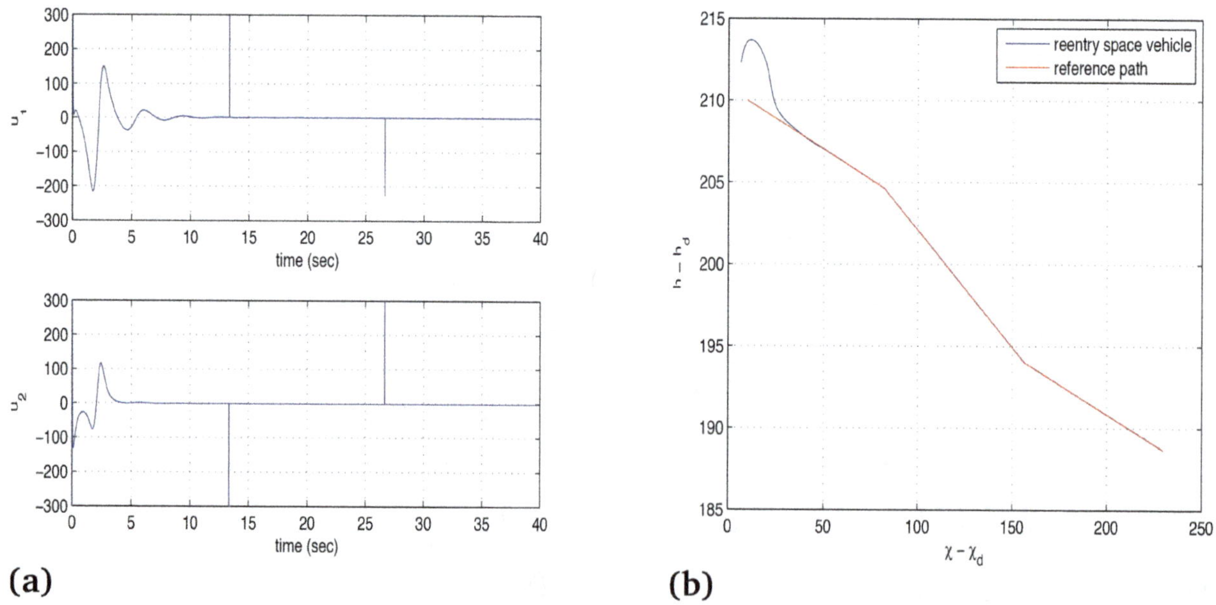

Figure 5.34 Tracking of setpoint 6 by the reentry space vehicle under flatness-based control: (a) variations of the control inputs u_1 and u_2 of the spacecraft (blue lines); (b) tracking of the reference path (red line) by the spacecraft in the plane of state variables $x_2 = \chi$ and $x_1 = h$

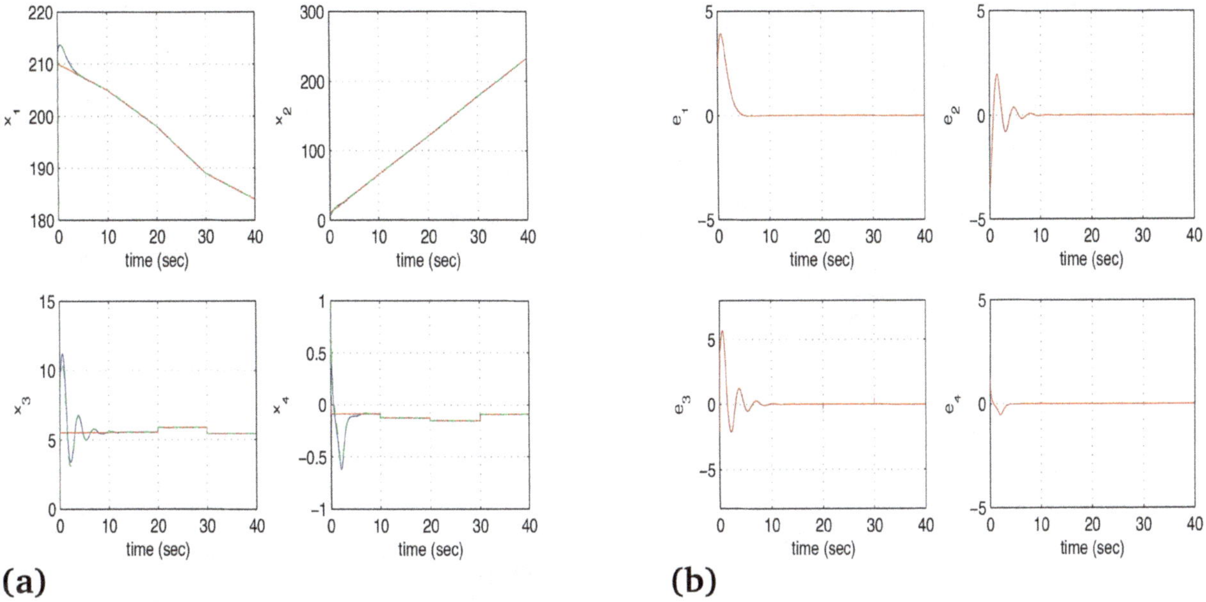

Figure 5.35 *Tracking of setpoint 7 by the reentry space vehicle under flatness-based control: (a) convergence of state variables x_1 to x_4 (blue lines) to the associated setpoints (red lines) and estimated values provided by Kalman filtering; (b) variation of the tracking error variables e_1 to e_4 associated with the state variables of the spacecraft*

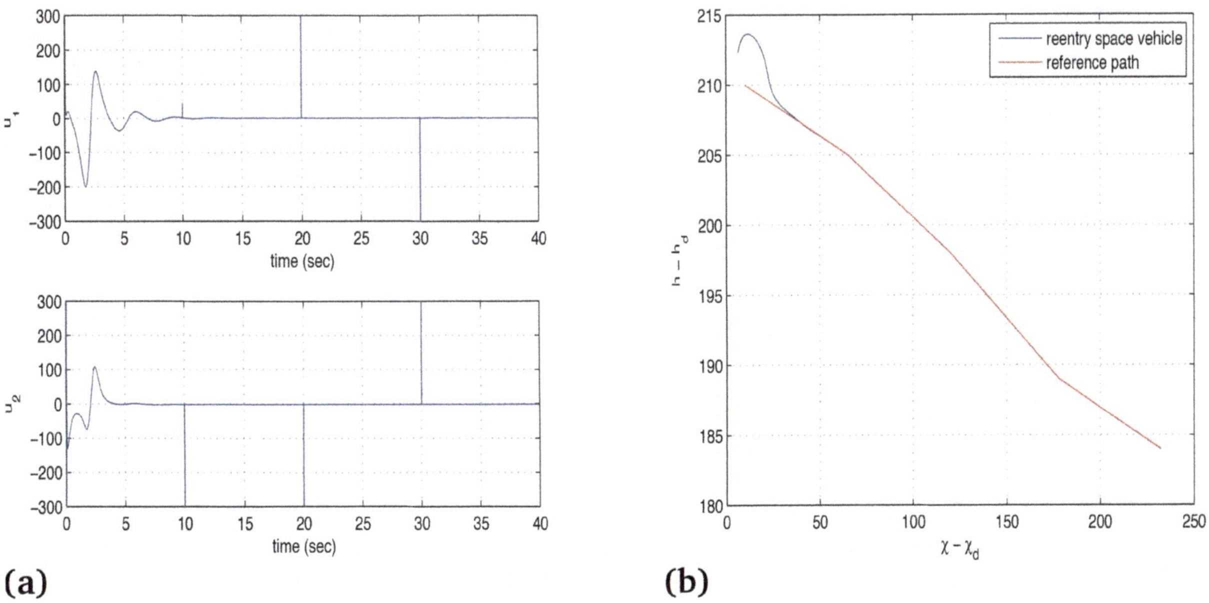

Figure 5.36 *Tracking of setpoint 7 by the reentry space vehicle under flatness-based control: (a) variations of the control inputs u_1 and u_2 of the spacecraft (blue lines); (b) tracking of the reference path (red line) by the spacecraft in the plane of state variables $x_2 = \chi$ and $x_1 = h$*

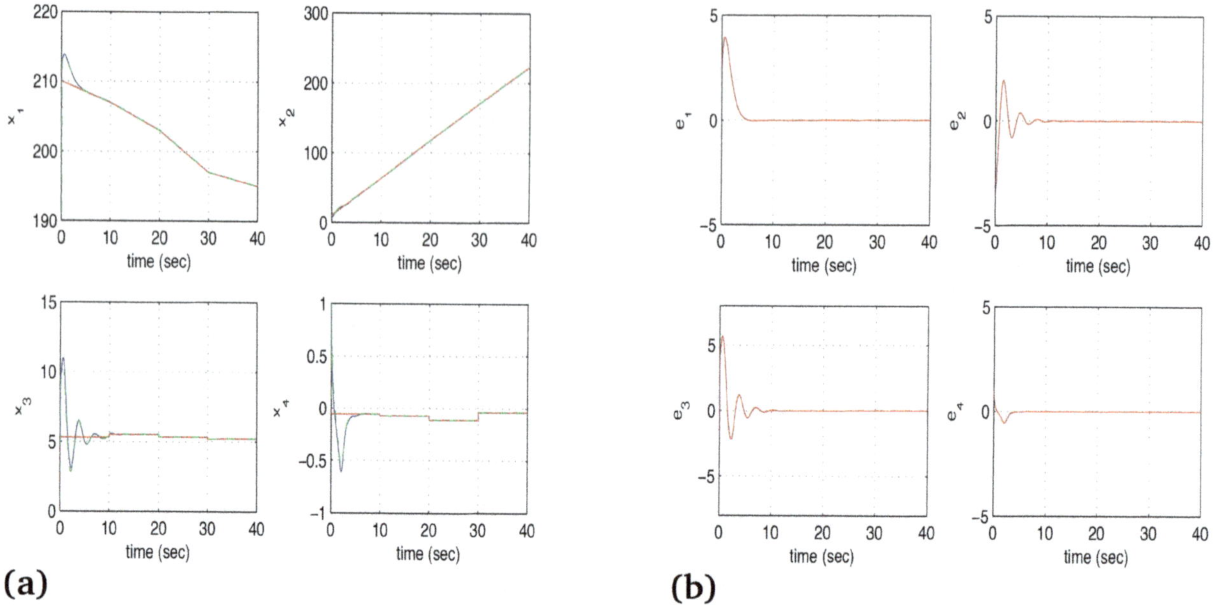

Figure 5.37 Tracking of setpoint 8 by the reentry space vehicle under flatness-based control: (a) convergence of state variables x_1 to x_4 (blue lines) to the associated setpoints (red lines) and estimated values provided by Kalman filtering; (b) variation of the tracking error variables e_1 to e_4 associated with the state variables of the spacecraft

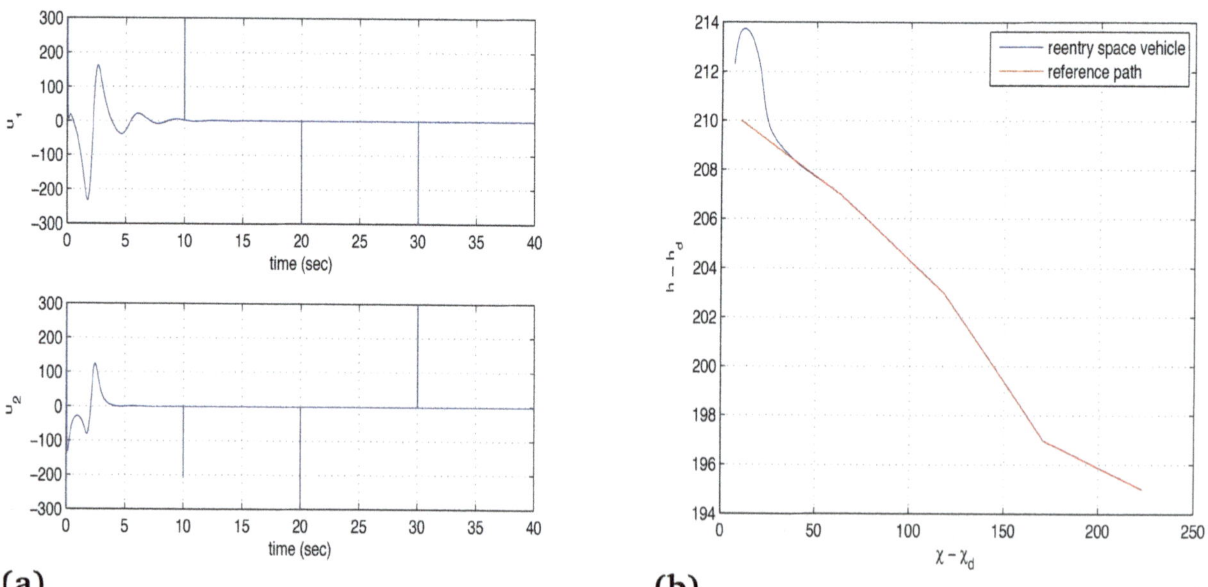

Figure 5.38 Tracking of setpoint 8 by the reentry space vehicle under flatness-based control: (a) variations of the control inputs u_1 and u_2 of the spacecraft (blue lines); (b) tracking of the reference path (red line) by the spacecraft in the plane of state variables $x_2 = \chi$ and $x_1 = h$

5.2 Nonlinear optimal and flatness-based control of the attitude of reentry space vehicles

5.2.1 Outline

Control for autonomous reentry space vehicles has been a nontrivial research topic in the area of aerospace science and technology [1094,958,358]. The associated 6-DOF attitude (orientation) dynamic model is obtained by expressing orientation angles and angular velocities of the space vehicle in different coordinate frames, namely an earth-fixed, an earth rotating and a body fixed frame [128,1093,360]. Among eligible control methods for the spacecraft's attitude dynamics, one can distinguish some sliding-mode control schemes [24,299,188,626,300,965]. Besides, there are several backstepping control approaches [997,828]. Additionally, one can find results on model predictive control concepts and optimal control-type concepts [867,363,613,156,572]. Furthermore, one can distinguish methods which perform disturbance estimation and disturbance compensation thus improving the robustness of the reentry space vehicles' control loop [301,862,863,1006,1078]. Besides, this control problem is often treated with the use of methods which rely on state-space model transformations and on changes of state variables [768,112]. Additional results on nonlinear control of the attitude dynamics of reentry space vehicles are also concerned with flexible spacecrafts and with multiple spacecrafts which pursue orientation in a coordinated manner [765,542,990,836,964]. Solving the attitude control problem for re-entry space vehicles improves also the precision of image captures and video frames in spacecraft-based geo-surveillance tasks [452]. The aim of the present section has been to achieve control and stabilization for the attitude dynamics of reentry space vehicles without forth and back state-space model transformations, changes of state variables (diffeomorphisms), and without the associated singularity issues. To this end, two new nonlinear control methods have been proposed: (i) nonlinear optimal control, (ii) flatness-based control in successive loops.

The first method that this section proposes for the 6-DOF attitude dynamics of reentry space vehicles is a nonlinear optimal (H-infinity) control scheme [721,725]. The dynamic model of the attitude of the reentry space vehicle undergoes approximate linearization around a temporary operating point which is updated at each sampling period. This operating point is defined in each sampling interval by the present value of the system's state vector and by the last sampled value of the control inputs vector. The linearization process is performed through first-order Taylor series expansion and through the computation of the associated Jacobian matrices [705,60,706]. The modeling error which is due to the truncation of higher-order terms in the Taylor series expansion is considered to be a perturbation which is asymptotically compensated by the robustness of the control algorithm. For the approximately linearized state-space description of the system an H-infinity feedback controller is designed. To select the controller's gains an algebraic Riccati equation is repetitively solved at each time-step of the control method [876,741]. The global stability properties of the control scheme are proven through Lyapunov analysis. First, it is demonstrated that the control loop satisfies the H-infinity tracking performance criterion which is a strong indication about the method's robustness to model uncertainty and perturbations. Moreover, under moderate conditions global asymptotic stability is proven.

The second method that this section proposes for the attitude dynamics of reentry space vehicles is flatness-based control implemented in successive loops [711,711,712,720]. The dynamic model of the reentry space vehicle is decomposed into a series of subsystems which are connected in chained form. It is proven that these subsystems are differentially flat. By proving differential flatness properties for the individual subsystems it is confirmed that these are input–output linearizable. It is also confirmed that a stabilizing feedback controller can be designed for each one of them by inverting their dynamics, as it is commonly done for input–output linearized flat dynamical systems [806,492,640,524,54,732]. Next, in the chained subsystems, the state vector of the subsequent ($i+1$-th) subsystem becomes virtual control input to the preceding (i-th) subsystem. From the last (N-th) subsystem, the real control inputs vector is found. This comprises recursively the virtual control inputs associated with the preceding N-1 subsystems. The global stability properties of the flatness-based control method which is implemented in successive loops are also proven through Lyapunov analysis.

The presented nonlinear optimal control method is advantageous because unlike Nonlinear Model Predictive Control (NMPC) it is of ensured global stability and the convergence of its iterative search for the optimum does not depend on empirical parametrization of the controller and on ad hoc initialization. Besides, unlike optimal control based on Linear Parameter Varying (LPV) concepts, in the new nonlinear optimal control method, there is no need to solve state-dependent Riccati equations. Furthermore, the new multi-loop flatness-based control method avoids changes of state variables and transformations of the state-space model of the controlled system into the input–output linearized form. The control inputs it computes are applied directly on the initial nonlinear state-space model of the controlled system without inverse transformations and without the associated singularity issues.

5.2.2 Dynamic model of the attitude of reentry space vehicle

5.2.2.1 State-space model of the attitude of the reentry space vehicle

To describe the motion, position and orientation of the re-entry space vehicle, angle and angular speed variables description is used in an earth-fixed coordinates frame, in an earth-rotating frame and finally in the spacecraft's body-fixed frame. The earth-fixed frame is denoted as $OX_GY_GZ_G$, while the earth-rotating frame is denoted as $GX_PY_PZ_P$ where G is the center of gravity of the spacecraft. Axis OX_P coincides with the radius connecting the origin of the earth-fixed frame O with the space vehicle's center of gravity G. Transition from the earth-fixed frame $OX_GY_GZ_G$ to the earth rotating frame $GX_PY_PZ_P$ is defined by angles θ and ϕ and the associated rotation matrix. The body-fixed reference frame is denoted as $GX_CY_CZ_C$. In this reference frame axis GZ_C coincides with the velocity vector V of the space vehicle. Transition from the earth-rotating reference frame $GX_PY_PZ_P$ to the body-fixed reference frame $GX_CY_CZ_C$ is defined by angles γ and ψ and the associated rotation matrix [621,1031,784].

With reference to the angles defined in Figure 5.39 it is considered that $\dot{\phi} = 0$, $\dot{\theta} = 0$, $\dot{\gamma} = 0$, $\dot{\psi} = 0$, while it is also taken that $\dot{\chi} = 0$. Moreover, the orientation angles (Euler angles) in the body-fixed frame of the spacecraft are defined as $\theta = [\alpha, \beta, \sigma]^T$, where α is the angle of attitude denoting rotation around the GX_c axis, β is the side-slip angle denoting rotation around the GY_c axis and σ is the bank angle denoting rotation around the GZ_c axis. Additionally, the angular velocities vector of the spacecraft around the axes of the earth-rotating frame is defined as $\omega = [p, q, r]^T$, where p denotes angular speed for rotation around the earth-rotating frame axis GX_P, q denotes angular speed for rotation around the earth-rotating frame axis GY_P, and r denotes angular speed for rotation around the earth-rotating frame axis GZ_P [363,862].

The 6-DOF dynamics of the attitude of the spacecraft is given by [862]:

$$\begin{aligned} \dot{\theta} &= R\omega + \Delta_F \\ \dot{\omega} &= -I^{-1}\Omega I\omega + I^{-1}M + \Delta_M \end{aligned} \tag{5.71}$$

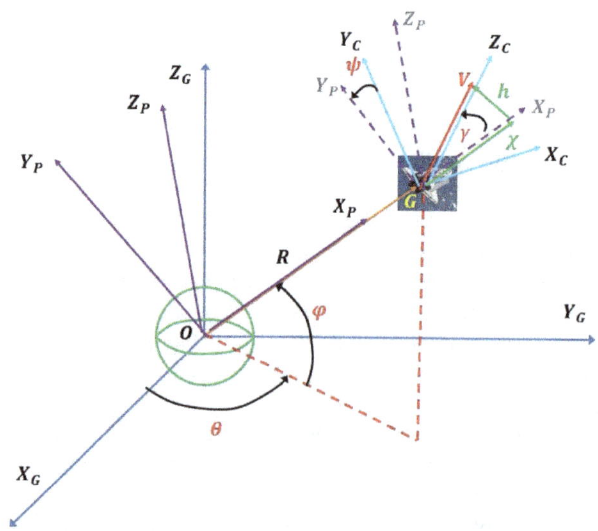

Figure 5.39 Coordinate frames used in the dynamic model of the reentry space vehicle

where $\theta = [\alpha, \beta, \sigma]^T$ is the attitude angular vector and $\omega = [p, q, r]^T$ is the attitude angular speed vector comprising the roll, pitch and yaw angular rates. In Eq. (5.71), Δ_F is the model uncertainty and disturbances term affecting the update of the attitude angles α, β, σ, and Δ_M is the model uncertainty and disturbance term affecting the update of the roll, pitch, and yaw angular rates p, q, and r. Moreover, $M = [M_r, M_y, M_z]^T$ is the control inputs vector representing torques that cause the roll, pitch and yaw motion of the spacecraft, that is rotation around the axes of the $GX_PY_PZ_P$ earth-rotating frame [862]. The representation of the attitude angular velocities of the reentry space vehicle in the body-fixed and earth-related reference frame is given in Figure 5.40.

Additionally, matrices I, Ω, and R are defined as follows [862]:

$$I = \begin{pmatrix} I_{xx} & -I_{xy} & -I_{xz} \\ -I_{xy} & I_{yy} & -I_{yz} \\ -I_{xz} & -I_{yz} & I_{zz} \end{pmatrix} \quad \Omega = \begin{pmatrix} 0 & -r & q \\ r & 0 & p \\ -q & & p & 0 \end{pmatrix} \tag{5.72}$$

$$R = \begin{pmatrix} -cos(a)tan(\beta) & 1 & -sin(a)tan(\beta) \\ sin(a) & 0 & -cos(\alpha) \\ -cos(\alpha)cos(\beta) & -sin(\beta) & -sin(\alpha)cos(\beta) \end{pmatrix} \tag{5.73}$$

where the elements of matrix I represent moments of inertia. Next, the state-vector of the attitude dynamics of the spacecraft is denoted as $x = [x_1, x_2, x_3, x_4, x_5, x_6]^T = [\alpha, \beta, \sigma, p, q, r]^T$. The inverse of matrix I is

$$I^{-1} = \frac{1}{detI} \begin{pmatrix} I_{11} & -I_{21} & I_{31} \\ -I_{12} & I_{22} & -I_{32} \\ I_{13} & -I_{23} & I_{33} \end{pmatrix} \tag{5.74}$$

where $I_{11} = I_{yy}I_{zz} - I_{yz}^2$, $I_{12} = -I_{xy}I_{zz} - I_{xz}I_{yz}$, $I_{13} = I_{xy}I_{yz} + I_{xz}I_{yy}$, $I_{21} = -I_{xy}I_{zz} - I_{yz}I_{xz}$, $I_{22} = I_{xx}I_{zz} - I_{xz}^2$, $I_{23} = -I_{xx}I_{yz} - I_{xz}I_{xy}$, $I_{31} = I_{xy}I_{yz} + I_{yy}I_{xz}$, $I_{32} = -I_{xx}I_{yz} - I_{xy}I_{xz}$, and $I_{33} = I_{xx}I_{yy} - I_{xy}^2$.

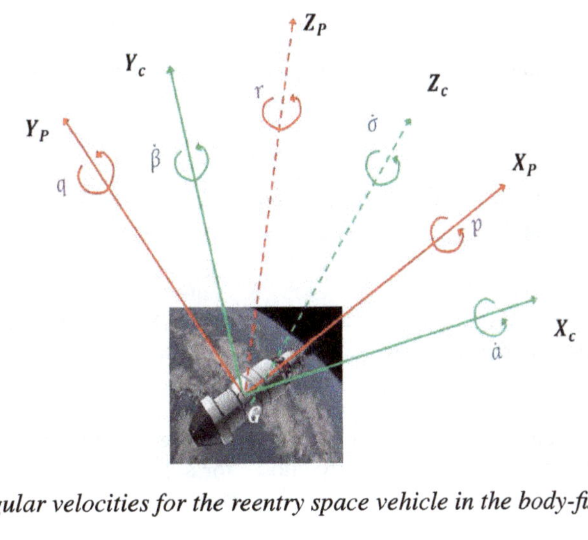

Figure 5.40 Attitude angular velocities for the reentry space vehicle in the body-fixed and earth-related frame

while about the matrices product $I^{-1}\Omega I$, one has

$$I^{-1}\Omega I = \frac{1}{detI}\begin{pmatrix} I_{11} & -I_{21} & I_{31} \\ -I_{12} & I_{22} & -I_{32} \\ I_{13} & -I_{23} & I_{33} \end{pmatrix}\begin{pmatrix} 0 & -r & q \\ r & 0 & p \\ -q & p & 0 \end{pmatrix}\begin{pmatrix} I_{xx} & -I_{xy} & -I_{xz} \\ -I_{xy} & I_{yy} & -I_{yz} \\ -I_{xz} & -I_{yz} & I_{xz} \end{pmatrix} = \begin{pmatrix} N_{11} & N_{12} & N_{13} \\ N_{21} & N_{22} & N_{23} \\ N_{31} & N_{32} & N_{33} \end{pmatrix} \quad (5.75)$$

where $N_{11} = \frac{1}{detI}(I_{11}Q_{11} - I_{12}Q_{21} + I_{31}Q_{31})$, $N_{12} = \frac{1}{detI}(I_{11}Q_{12} - I_{21}Q_{22} + I_{31}Q_{32})$, $N_{13} = \frac{1}{detI}(I_{11}Q_{13} - I_{21}Q_{23} + I_{31}Q_{33})$, and $N_{21} = \frac{1}{detI}(-I_{12}Q_{11} + I_{22}Q_{21} - I_{32}Q_{31})$, $N_{22} = \frac{1}{detI}(-I_{12}Q_{13} + I_{22}Q_{23} - I_{32}Q_{33})$, $N_{23} = \frac{1}{detI}(-I_{12}Q_{13} + I_{22}Q_{23} - I_{32}Q_{33})$, while also $N_{31} = \frac{1}{detI}(I_{13}Q_{11} - I_{23}Q_{21} + I_{33}Q_{31})$, $N_{32} = \frac{1}{detI}(I_{13}Q_{12} - I_{22}Q_{22} + I_{33}Q_{32})$, $N_{33} = \frac{1}{detI}(I_{13}Q_{13} - I_{23}Q_{23} + I_{33}Q_{33})$.

with $Q_{11} = x_6I_{xy} - x_5I_{xz}$, $Q_{12} = -x_6I_{yy} - x_5I_{yz}$, $Q_{13} = x_6I_{yz} + x_5I_{zz}$, $Q_{21} = x_6I_{xx} - x_4I_{xz}$, $Q_{22} = -x_6I_{xy} - x_4I_{yz}$, $Q_{23} = -x_6I_{xz} + x_4I_{zz}$, $Q_{31} - x_5I_{xx} - x_4I_{xy}$, $Q_{32} = x_5I_{xy} + x_4I_{yy}$, $Q_{33} = x_5I_{xz} - x_4I_{yz}$.

Consequently, the dynamic model of the reentry space vehicle becomes:

$$\begin{pmatrix} \dot{x}_1 \\ \dot{x}_2 \\ \dot{x}_3 \\ \dot{x}_4 \\ \dot{x}_5 \\ \dot{x}_6 \end{pmatrix} = \begin{pmatrix} -cos(x_1)tan(x_2)x_4 + x_5 - sin(x_1)tan(x_2)x_6 \\ sin(x_1)x_4 - cos(x_4)x_6 \\ -cos(x_1)cos(x_2)x_4 - sin(x_2)x_5 - sin(x_1)cos(x_2)x_6 \\ -N_{11}x_4 - N_{12}x_5 - N_{13}x_6 \\ -N_{21}x_4 - N_{22}x_5 - N_{23}x_6 \\ -N_{31}x_4 - N_{32}x_5 - N_{33}x_6 \end{pmatrix} + \begin{pmatrix} 0 & 0 & 0 \\ 0 & 0 & 0 \\ 0 & 0 & 0 \\ \frac{I_11}{detM} & -\frac{I_21}{detM} & \frac{I_31}{detM} \\ -\frac{I_12}{detM} & \frac{I_22}{detM} & -\frac{I_32}{detM} \\ \frac{I_13}{detM} & -\frac{I_23}{detM} & \frac{I_33}{detM} \end{pmatrix}\begin{pmatrix} u_1 \\ u_2 \\ u_3 \end{pmatrix} \quad (5.76)$$

or, the attitude dynamics of the reentry space vehicle can be written in the nonlinear affine-in-the-input state-space form

$$\dot{x} = f(x) + g(x)u \quad (5.77)$$

where $x \in R^{6\times 1}$, $f(x) \in R^{6\times 1}$, $g(x) \in R^{6\times 3}$, and $u \in R^{3\times 1}$.

5.2.2.2 Differential flatness of the attitude of the reentry space vehicle

It will be proven that the dynamic model of the attitude of the reentry space vehicle is differentially flat, with flat outputs vector

$$Y = [x_1, x_2, x_3]^T \Rightarrow Y = [\alpha, \beta, \sigma]^T \quad (5.78)$$

For the two equations which constitute the attitude dynamics of the reentry space vehicle, one has

$$\begin{aligned} \dot{x}_{13} &= Rx_{46} \\ \dot{x}_{46} &= -I^{-1}\Omega I x_{46} + I^{-1}u \end{aligned} \quad (5.79)$$

where $x_{13} = [x_1, x_2, x_3]^T$ and $x_{46} = [x_4, x_5, x_6]^T$ and R is a function of $x_{13} = [x_1, x_2, x_3]^T$ which is also written as $R(x_{13})$. By solving the first row of Eq. (5.79) for x_{46}, one obtains

$$x_{46} = R^{-1}(x_{13})\dot{x}_{13} \quad (5.80)$$

which signifies that x_{46} is a differential function of the flat outputs vector $Y = x_{13}$. Moreover, by solving the second row of Eq. (5.79) for the control inputs, one has

$$u = I\dot{x}_{46} + \Omega I x_{46} \quad (5.81)$$

which signifies that u is finally a differential function of $y = x_{13}$. Consequently, the entire attitude dynamics is a differentially flat system. The differential flatness property allows for solving the setpoints definition problem for the system.

5.2.2.3 Approximate linearization of the attitude dynamics of the reentry space vehicle

The attitude dynamics of the reentry space vehicle undergoes approximate linearization around the temporary operating point (x^*, u^*) which is defined at each sampling instant by the present value of the system's state vector and by the last sampled value of the control inputs vector. The linearization process is based of first-order Taylor series expansion and on the computation of the associated Jacobian matrices. The modeling error which is due to the truncation of higher-order terms form the Taylor series expansion is considered to be a perturbation which is asymptotically compensated by the robustness of the control algorithm. The initial state-space model of the attitude dynamics, being in the form $\dot{x} = f(x) + g(x)u$ is brought to the linearized state-space description

$$\dot{x} = Ax + Bu + \tilde{d} \tag{5.82}$$

where A, B are the Jacobian matrices of the system

$$A = \nabla_x[f(x) + g(x)u] \mid_{(x^*,u^*)} \Rightarrow A = \nabla_x f(x) \mid_{(x^*,u^*)} \tag{5.83}$$

$$B = \nabla_u[f(x) + g(x)u] \mid_{(x^*,u^*)} \Rightarrow B = g(x) \mid_{(x^*,u^*)} \tag{5.84}$$

The cumulative disturbance term $\tilde{d}$ may comprise: (i) the modeling error due to the truncation of higher-order terms from the Taylor series (ii) exogenous perturbations, and (iii) sensor measurement noise of any distribution.

The computation of the elements of the Jacobian matrix $\nabla_x f(x) \mid_{(x^*,u^*)}$ proceeds as follows:

First row of the Jacobian matrix $\nabla_x f(x) \mid_{(x^*,u^*)}$: $\frac{\partial f_1}{\partial x_1} = sin(x_1)cos(x_2)x_4 - cos(x_1)tan(x_2)x_6$, $\frac{\partial f_1}{\partial x_2} = -cos(x_1)\frac{1}{cos^2(x_2)}x_4 - sin(x_4)\frac{1}{cos(x_2)}x_6$, $\frac{\partial f_1}{\partial x_3} = 0$, $\frac{\partial f_1}{\partial x_4} = -cos(x_1)tan(x_2)$, $\frac{\partial f_1}{\partial x_5} = 0$, and $\frac{\partial f_1}{\partial x_6} = -sin(x_1)tan(x_2)$.

Second row of the Jacobian matrix $\nabla_x f(x) \mid_{(x^*,u^*)}$: $\frac{\partial f_2}{\partial x_1} = cos(x_1)x_4 + sin(x_1)x_6$, $\frac{\partial f_2}{\partial x_2} = 0$, $\frac{\partial f_2}{\partial x_3} = 0$, $\frac{\partial f_2}{\partial x_4} = sin(x_1)$, $\frac{\partial f_2}{\partial x_5} = 0$, and $\frac{\partial f_2}{\partial x_6} = -cos(x_1)$.

Third row of the Jacobian matrix $\nabla_x f(x) \mid_{(x^*,u^*)}$: $\frac{\partial f_3}{\partial x_1} = sin(x_1)cos(x_2)x_4 - cos(x_1)cos(x_2)x_6$, $\frac{\partial f_3}{\partial x_2} = cos(x_1)sin(x_2)x_4 - cos(x_2)x_5 + sin(x_1)sin(x_2)x_6$, $\frac{\partial f_3}{\partial x_3} = 0$, $\frac{\partial f_3}{\partial x_4} = -cos(x_1)cos(x_4)$, $\frac{\partial f_3}{\partial x_5} = -sin(x_2)$, and $\frac{\partial f_3}{\partial x_6} = -sin(x_1)cos(x_2)$.

Fourth row of the Jacobian matrix $\nabla_x f(x) \mid_{(x^*,u^*)}$: $\frac{\partial f_4}{\partial x_1} = -[\frac{\partial N_{11}}{\partial x_1}x_4 + \frac{\partial N_{12}}{\partial x_1}x_5 + \frac{\partial N_{13}}{\partial x_1}x_6]$, $\frac{\partial f_4}{\partial x_2} = -[\frac{\partial N_{11}}{\partial x_2}x_4 + \frac{\partial N_{12}}{\partial x_2}x_5 + \frac{\partial N_{13}}{\partial x_2}x_6]$, $\frac{\partial f_4}{\partial x_3} = -[\frac{\partial N_{11}}{\partial x_3}x_4 + \frac{\partial N_{12}}{\partial x_3}x_5 + \frac{\partial N_{13}}{\partial x_3}x_6]$, $\frac{\partial f_4}{\partial x_4} = -[\frac{\partial N_{11}}{\partial x_4}x_4 + \frac{\partial N_{12}}{\partial x_4}x_5 + \frac{\partial N_{13}}{\partial x_4}x_6 + N_{11}]$, $\frac{\partial f_4}{\partial x_5} = -[\frac{\partial N_{11}}{\partial x_5}x_4 + \frac{\partial N_{12}}{\partial x_5}x_5 + \frac{\partial N_{13}}{\partial x_5}x_6 + N_{12}]$, and $\frac{\partial f_4}{\partial x_6} = -[\frac{\partial N_{11}}{\partial x_6}x_4 + \frac{\partial N_{12}}{\partial x_6}x_5 + \frac{\partial N_{13}}{\partial x_6}x_6 + N_{13}]$.

Fifth row of the Jacobian matrix $\nabla_x f(x) \mid_{(x^*,u^*)}$: $\frac{\partial f_4}{\partial x_1} = -[\frac{\partial N_{21}}{\partial x_1}x_4 + \frac{\partial N_{22}}{\partial x_1}x_5 + \frac{\partial N_{23}}{\partial x_1}x_6]$, $\frac{\partial f_4}{\partial x_2} = -[\frac{\partial N_{21}}{\partial x_2}x_4 + \frac{\partial N_{22}}{\partial x_2}x_5 + \frac{\partial N_{23}}{\partial x_2}x_6]$, $\frac{\partial f_4}{\partial x_3} = -[\frac{\partial N_{21}}{\partial x_3}x_4 + \frac{\partial N_{22}}{\partial x_3}x_5 + \frac{\partial N_{23}}{\partial x_3}x_6]$, $\frac{\partial f_4}{\partial x_4} = -[\frac{\partial N_{21}}{\partial x_4}x_4 + \frac{\partial N_{22}}{\partial x_4}x_5 + \frac{\partial N_{23}}{\partial x_4}x_6 + N_{21}]$, $\frac{\partial f_4}{\partial x_5} = -[\frac{\partial N_{21}}{\partial x_5}x_4 + \frac{\partial N_{22}}{\partial x_5}x_5 + \frac{\partial N_{23}}{\partial x_5}x_6 + N_{22}]$, and $\frac{\partial f_4}{\partial x_6} = -[\frac{\partial N_{21}}{\partial x_6}x_4 + \frac{\partial N_{22}}{\partial x_6}x_5 + \frac{\partial N_{23}}{\partial x_6}x_6 + N_{23}]$.

Sixth row of the Jacobian matrix $\nabla_x f(x) \mid_{(x^*,u^*)}$: $\frac{\partial f_4}{\partial x_1} - [= \frac{\partial N_{31}}{\partial x_1}x_4 + \frac{\partial N_{32}}{\partial x_1}x_5 + \frac{\partial N_{33}}{\partial x_1}x_6]$, $\frac{\partial f_4}{\partial x_2} = -[\frac{\partial N_{31}}{\partial x_2}x_4 + \frac{\partial N_{32}}{\partial x_2}x_5 + \frac{\partial N_{33}}{\partial x_2}x_6]$, $\frac{\partial f_4}{\partial x_3} = -[\frac{\partial N_{31}}{\partial x_3}x_4 + \frac{\partial N_{32}}{\partial x_3}x_5 + \frac{\partial N_{33}}{\partial x_3}x_6]$, $\frac{\partial f_4}{\partial x_4} = -[\frac{\partial N_{31}}{\partial x_4}x_4 + \frac{\partial N_{32}}{\partial x_4}x_5 + \frac{\partial N_{33}}{\partial x_4}x_6 + N_{31}]$, $\frac{\partial f_4}{\partial x_5} = -[\frac{\partial N_{31}}{\partial x_5}x_4 + \frac{\partial N_{32}}{\partial x_5}x_5 + \frac{\partial N_{33}}{\partial x_5}x_6 + N_{32}]$, and $\frac{\partial f_4}{\partial x_6} = -[\frac{\partial N_{31}}{\partial x_6}x_4 + \frac{\partial N_{32}}{\partial x_6}x_5 + \frac{\partial N_{33}}{\partial x_6}x_6 + N_{33}]$.

Computation of the partial derivatives of the terms N_{ij}, $i = 1, 2, 3$, and $j = 1, 2, 3$ with respect to the state variables x_i, $i = 1, 2, \cdots, 6$.

For $N_{11} = \frac{1}{detI}(I_{11}Q_{11} - I_{21}Q_{21} + I_{31}Q_{31})$, it holds that $\frac{\partial N_{11}}{\partial x_i} = \frac{1}{detI}(I_{11}\frac{\partial Q_{11}}{\partial x_i} - I_{21}\frac{\partial Q_{21}}{\partial x_i} + I_{31}\frac{\partial Q_{31}}{\partial x_i})$.

For $N_{12} = \frac{1}{detI}(I_{11}Q_{12} - I_{21}Q_{22} + I_{31}Q_{32})$, it holds that $\frac{\partial N_{12}}{\partial x_i} = \frac{1}{detI}(I_{11}\frac{\partial Q_{12}}{\partial x_i} - I_{21}\frac{\partial Q_{22}}{\partial x_i} + I_{31}\frac{\partial Q_{32}}{\partial x_i})$.

For $N_{13} = \frac{1}{detI}(I_{13}Q_{13} - I_{21}Q_{23} + I_{31}Q_{33})$, it holds that $\frac{\partial N_{12}}{\partial x_i} = \frac{1}{detI}(I_{11}\frac{\partial Q_{13}}{\partial x_i} - I_{21}\frac{\partial Q_{23}}{\partial x_i} + I_{31}\frac{\partial Q_{33}}{\partial x_i})$.

For $N_{21} = \frac{1}{detI}(-I_{12}Q_{11} + I_{22}Q_{21} - I_{32}Q_{31})$, it holds that $\frac{\partial N_{21}}{\partial x_i} = \frac{1}{detI}(-I_{12}\frac{\partial Q_{11}}{\partial x_i} + I_{22}\frac{\partial Q_{21}}{\partial x_i} - I_{32}\frac{\partial Q_{31}}{\partial x_i})$.

For $N_{22} = \frac{1}{detI}(-I_{12}Q_{12} + I_{22}Q_{22} - I_{32}Q_{32})$, it holds that $\frac{\partial N_{22}}{\partial x_i} = \frac{1}{detI}(-I_{12}\frac{\partial Q_{12}}{\partial x_i} + I_{22}\frac{\partial Q_{22}}{\partial x_i} - I_{32}\frac{\partial Q_{32}}{\partial x_i})$.

For $N_{23} = \frac{1}{detI}(-I_{12}Q_{13} + I_{22}Q_{23} - I_{32}Q_{33})$, it holds that $\frac{\partial N_{23}}{\partial x_i} = \frac{1}{detI}(-I_{12}\frac{\partial Q_{13}}{\partial x_i} + I_{22}\frac{\partial Q_{23}}{\partial x_i} - I_{33}\frac{\partial Q_{32}}{\partial x_i})$.

For $N_{31} = \frac{1}{detI}(I_{13}Q_{11} - I_{23}Q_{21} + I_{33}Q_{31})$, it holds that $\frac{\partial N_{31}}{\partial x_i} = \frac{1}{detI}(I_{13}\frac{\partial Q_{11}}{\partial x_i} - I_{23}\frac{\partial Q_{21}}{\partial x_i} + I_{33}\frac{\partial Q_{31}}{\partial x_i})$.

For $N_{32} = \frac{1}{detI}(I_{13}Q_{12} - I_{23}Q_{22} + I_{33}Q_{32})$, it holds that $\frac{\partial N_{32}}{\partial x_i} = \frac{1}{detI}(I_{13}\frac{\partial Q_{12}}{\partial x_i} - I_{23}\frac{\partial Q_{22}}{\partial x_i} + I_{33}\frac{\partial Q_{32}}{\partial x_i})$.

For $N_{33} = \frac{1}{detI}(I_{13}Q_{13} - I_{23}Q_{23} + I_{33}Q_{33})$, it holds that $\frac{\partial N_{33}}{\partial x_i} = \frac{1}{detI}(I_{13}\frac{\partial Q_{13}}{\partial x_i} - I_{23}\frac{\partial Q_{23}}{\partial x_i} + I_{33}\frac{\partial Q_{33}}{\partial x_i})$.

Computation of the partial derivatives of the terms Q_{ij}, $i = 1, 2, 3$ and $j = 1, 2, 3$.

For $Q_{11} = x_6I_{xy} - x_5I_{xz}$, one has that: $\frac{\partial Q_{11}}{\partial x_1} = 0$, $\frac{\partial Q_{11}}{\partial x_2} = 0$, $\frac{\partial Q_{11}}{\partial x_2} = 0$, $\frac{\partial Q_{11}}{\partial x_4} = 0$, $\frac{\partial Q_{11}}{\partial x_5} = -I_{xz}$, $\frac{\partial Q_{11}}{\partial x_6} = I_{xy}$.

For $Q_{12} = -x_6I_{yy} - x_5I_{yz}$, one has that: $\frac{\partial Q_{12}}{\partial x_1} = 0$, $\frac{\partial Q_{12}}{\partial x_2} = 0$, $\frac{\partial Q_{12}}{\partial x_2} = 0$, $\frac{\partial Q_{12}}{\partial x_4} = 0$, $\frac{\partial Q_{12}}{\partial x_5} = -I_{yz}$, $\frac{\partial Q_{12}}{\partial x_6} = -I_{yy}$.

For $Q_{13} = x_6I_{yz} + x_5I_{zz}$, one has that: $\frac{\partial Q_{13}}{\partial x_1} = 0$, $\frac{\partial Q_{13}}{\partial x_2} = 0$, $\frac{\partial Q_{13}}{\partial x_2} = 0$, $\frac{\partial Q_{13}}{\partial x_4} = 0$, $\frac{\partial Q_{13}}{\partial x_5} = I_{zz}$, $\frac{\partial Q_{13}}{\partial x_6} = I_{yz}$.

For $Q_{21} = -x_6I_{xx} + x_4I_{xz}$, one has that: $\frac{\partial Q_{21}}{\partial x_1} = 0$, $\frac{\partial Q_{21}}{\partial x_2} = 0$, $\frac{\partial Q_{21}}{\partial x_2} = 0$, $\frac{\partial Q_{21}}{\partial x_4} = I_{xz}$, $\frac{\partial Q_{21}}{\partial x_5} = 0$, $\frac{\partial Q_{21}}{\partial x_6} = -I_{xx}$.

For $Q_{22} = -x_6I_{xy} - x_4I_{yz}$, one has that: $\frac{\partial Q_{22}}{\partial x_1} = 0$, $\frac{\partial Q_{22}}{\partial x_2} = 0$, $\frac{\partial Q_{22}}{\partial x_2} = 0$, $\frac{\partial Q_{22}}{\partial x_4} = -I_{yz}$, $\frac{\partial Q_{22}}{\partial x_5} = 0$, $\frac{\partial Q_{22}}{\partial x_6} = -I_{xy}$.

For $Q_{23} = -x_6I_{xz} + x_4I_{zz}$, one has that: $\frac{\partial Q_{23}}{\partial x_1} = 0$, $\frac{\partial Q_{23}}{\partial x_2} = 0$, $\frac{\partial Q_{23}}{\partial x_2} = 0$, $\frac{\partial Q_{23}}{\partial x_4} = I_{zz}$, $\frac{\partial Q_{22}}{\partial x_5} = 0$, $\frac{\partial Q_{23}}{\partial x_6} = -I_{xz}$.

For $Q_{31} = -x_5I_{xx} - x_4I_{xy}$, one has that: $\frac{\partial Q_{31}}{\partial x_1} = 0$, $\frac{\partial Q_{31}}{\partial x_2} = 0$, $\frac{\partial Q_{31}}{\partial x_2} = 0$, $\frac{\partial Q_{31}}{\partial x_4} = -I_{xy}$, $\frac{\partial Q_{31}}{\partial x_5} = -I_{xx}$, $\frac{\partial Q_{31}}{\partial x_6} = 0$.

For $Q_{32} = x_5I_{xy} + x_4I_{yy}$, one has that: $\frac{\partial Q_{32}}{\partial x_1} = 0$, $\frac{\partial Q_{32}}{\partial x_2} = 0$, $\frac{\partial Q_{32}}{\partial x_2} = 0$, $\frac{\partial Q_{32}}{\partial x_4} = I_{yy}$, $\frac{\partial Q_{32}}{\partial x_5} = I_{xy}$, $\frac{\partial Q_{32}}{\partial x_6} = 0$.

For $Q_{33} = x_5I_{xz} - x_4I_{yz}$, one has that: $\frac{\partial Q_{33}}{\partial x_1} = 0$, $\frac{\partial Q_{33}}{\partial x_2} = 0$, $\frac{\partial Q_{33}}{\partial x_2} = 0$, $\frac{\partial Q_{33}}{\partial x_4} = -I_{yz}$, $\frac{\partial Q_{33}}{\partial x_5} = I_{xz}$, $\frac{\partial Q_{33}}{\partial x_6} = 0$.

5.2.3 Design of an H-infinity feedback controller for the attitude's dynamics

5.2.3.1 Stabilizing feedback control

After linearization around its current operating point, the spacecraft's attitude dynamic model is written as [721]

$$\dot{x} = Ax + Bu + d_1 \tag{5.85}$$

Parameter d_1 stands for the linearization error in the spacecraft's attitude dynamic model appearing previously in Eq. (5.85). The reference setpoints for the spacecraft's attitude state vector are denoted by $\mathbf{x_d} = [x_1^d, \cdots, x_6^d]$. Tracking of this trajectory is achieved after applying the control input u_d. At every time instant, the control input u_d is assumed to differ from the control input u appearing in Eq. (5.85) by an amount equal to Δu, that is $u_d = u + \Delta u$

$$\dot{x}_d = Ax_d + Bu_d + d_2 \tag{5.86}$$

The dynamics of the controlled system described in Eq. (5.85) can also be written as

$$\dot{x} = Ax + Bu + Bu_d - Bu_d + d_1 \tag{5.87}$$

and by denoting $d_3 = -Bu_d + d_1$ as an aggregate disturbance term, one obtains

$$\dot{x} = Ax + Bu + Bu_d + d_3 \tag{5.88}$$

By subtracting Eq. (5.86) from Eq. (5.88), one has

$$\dot{x} - \dot{x}_d = A(x - x_d) + Bu + d_3 - d_2 \tag{5.89}$$

By denoting the tracking error as $e = x - x_d$ and the aggregate disturbance term as $L\tilde{d} = d_3 - d_2$, the tracking error dynamics becomes

$$\dot{e} = Ae + Bu + L\tilde{d} \tag{5.90}$$

where L is the disturbance inputs gain matrix. For the approximately linearized model of the system a stabilizing feedback controller is developed. The controller has the form

$$u(t) = -Ke(t) \tag{5.91}$$

with $K = \frac{1}{r}B^T P$ where P is a positive definite symmetric matrix which is obtained from the solution of the Riccati equation [721]

$$A^T P + PA + Q - P(\tfrac{2}{r}BB^T - \tfrac{1}{\rho^2}LL^T)P = 0 \tag{5.92}$$

where Q is a positive semi-definite symmetric matrix. The diagram of the considered control loop is depicted in Figure 5.41. This figure shows that, in the proposed control method: (a) there is linearization at each sampling instance of the state-space model of the reentry space vehicle through Taylor series expansion, (b) there is sequential solution of an algebraic Riccati equation which enables to find the gains of an H-infinity controller that stabilizes the spacecraft, and (c) there is application of the computed optimal control inputs directly on the initial nonlinear state-space model of the spacecraft.

It is also noteworthy that the solution of the H-infinity feedback control problem for the spacecraft's attitude dynamics and the computation of the worst-case disturbance that this controller can sustain, comes from superposition of Bellman's optimality principle when considering that the system is affected by two separate inputs (i) the control input u (ii) the cumulative disturbance input $\tilde{d}(t)$. Solving the optimal control problem for u, that is for the minimum variation (optimal) control input that achieves the elimination of the state vector's tracking error, gives $u = -\frac{1}{r}B^T Pe$. Equivalently, solving the optimal control problem for $\tilde{d}$, that is for the worst-case disturbance that the control loop can sustain gives $\tilde{d} = \frac{1}{\rho^2}L^T Pe$.

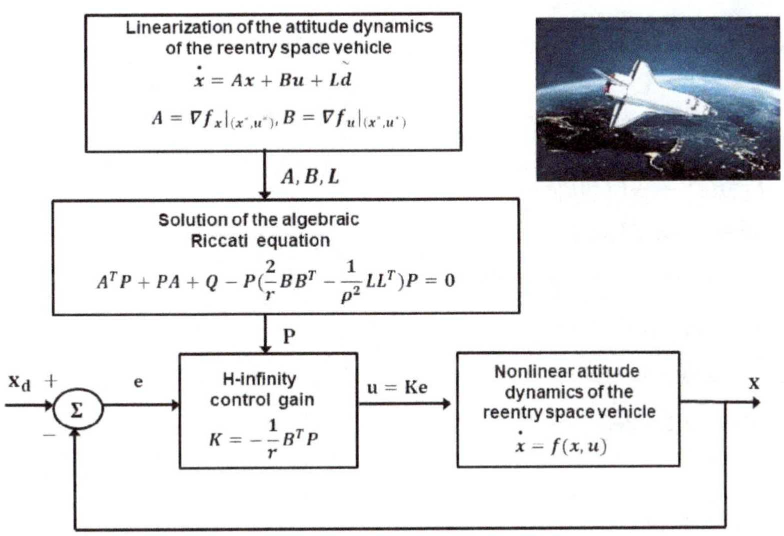

Figure 5.41 Diagram of the spacecraft's attitude nonlinear optimal (H-infinity) control loop

5.2.3.2 Lyapunov stability analysis

Through Lyapunov stability analysis it will be shown that the proposed nonlinear control scheme assures H_∞ tracking performance for the spacecraft's attitude dynamics, and that under moderate conditions asymptotic convergence to the reference setpoints (global stability) is achieved. As shown before, the tracking error dynamics for the spacecraft's attitude model is written in the form [721]

$$\dot{e} = Ae + Bu + L\tilde{d} \tag{5.93}$$

where in the spacecraft's attitude dynamics $L \in R^{6\times 6}$ is the disturbances gain matrix. Variable $\tilde{d}$ denotes model uncertainties and external disturbances of the spacecraft's attitude dynamic model. The following Lyapunov equation is considered

$$V = \frac{1}{2}e^T Pe \tag{5.94}$$

where $e = x - x_d$ is the tracking error. By differentiating with respect to time, one obtains

$$\begin{aligned} \dot{V} &= \tfrac{1}{2}\dot{e}^T Pe + \tfrac{1}{2}e^T P\dot{e} \Rightarrow \\ \dot{V} &= \tfrac{1}{2}[Ae + Bu + L\tilde{d}]^T Pe + \tfrac{1}{2}e^T P[Ae + Bu + L\tilde{d}] \Rightarrow \end{aligned} \tag{5.95}$$

$$\begin{aligned} \dot{V} = &\tfrac{1}{2}[e^T A^T + u^T B^T + \tilde{d}^T L^T]Pe + \\ &+ \tfrac{1}{2}e^T P[Ae + Bu + L\tilde{d}] \Rightarrow \end{aligned} \tag{5.96}$$

$$\begin{aligned} \dot{V} = &\tfrac{1}{2}e^T A^T Pe + \tfrac{1}{2}u^T B^T Pe + \tfrac{1}{2}\tilde{d}^T L^T Pe + \\ &\tfrac{1}{2}e^T PAe + \tfrac{1}{2}e^T PBu + \tfrac{1}{2}e^T PL\tilde{d} \end{aligned} \tag{5.97}$$

The previous equation is rewritten as

$$\begin{aligned} \dot{V} = &\tfrac{1}{2}e^T(A^T P + PA)e + (\tfrac{1}{2}u^T B^T Pe + \tfrac{1}{2}e^T PBu) + \\ &+ (\tfrac{1}{2}\tilde{d}^T L^T Pe + \tfrac{1}{2}e^T PL\tilde{d}) \end{aligned} \tag{5.98}$$

Assumption: For given positive definite matrix Q and coefficients r and ρ, there exists a positive definite matrix P, which is the solution of the following matrix equation

$$A^T P + PA = -Q + P(\tfrac{2}{r}BB^T - \tfrac{1}{\rho^2}LL^T)P \tag{5.99}$$

Moreover, the following feedback control law is applied to the system:

$$u = -\tfrac{1}{r}B^T Pe \tag{5.100}$$

By substituting Eq. (5.99) and Eq. (5.100), one obtains

$$\begin{aligned} \dot{V} = &\tfrac{1}{2}e^T[-Q + P(\tfrac{2}{r}BB^T - \tfrac{1}{\rho^2}LL^T)P]e + \\ &+ e^T PB(-\tfrac{1}{r}B^T Pe) + e^T PL\tilde{d} \Rightarrow \end{aligned} \tag{5.101}$$

$$\begin{aligned} \dot{V} = &-\tfrac{1}{2}e^T Qe + (\tfrac{1}{r}e^T PBB^T Pe - \tfrac{1}{2\rho^2}e^T PLL^T Pe \\ &-\tfrac{1}{r}e^T PBB^T Pe) + e^T PL\tilde{d} \end{aligned} \tag{5.102}$$

which after intermediate operations gives

$$\dot{V} = -\tfrac{1}{2}e^T Qe - \tfrac{1}{2\rho^2}e^T PLL^T Pe + e^T PL\tilde{d} \tag{5.103}$$

or, equivalently

$$\begin{aligned} \dot{V} = &-\tfrac{1}{2}e^T Qe - \tfrac{1}{2\rho^2}e^T PLL^T Pe + \\ &+ \tfrac{1}{2}e^T PL\tilde{d} + \tfrac{1}{2}\tilde{d}^T L^T Pe \end{aligned} \tag{5.104}$$

Lemma: The following inequality holds:

$$\frac{1}{2}e^TPL\tilde{d} + \frac{1}{2}\tilde{d}L^TPe - \frac{1}{2\rho^2}e^TPLL^TPe \leq \frac{1}{2}\rho^2\tilde{d}^T\tilde{d} \tag{5.105}$$

Proof: The binomial $(\rho\alpha - \frac{1}{\rho}b)^2$ is considered. Expanding the left part of the above inequality, one gets

$$\begin{aligned} &\rho^2a^2 + \frac{1}{\rho^2}b^2 - 2ab \geq 0 \Rightarrow \frac{1}{2}\rho^2a^2 + \frac{1}{2\rho^2}b^2 - ab \geq 0 \Rightarrow \\ &ab - \frac{1}{2\rho^2}b^2 \leq \frac{1}{2}\rho^2a^2 \Rightarrow \frac{1}{2}ab + \frac{1}{2}ab - \frac{1}{2\rho^2}b^2 \leq \frac{1}{2}\rho^2a^2 \end{aligned} \tag{5.106}$$

The following substitutions are carried out: $a = \tilde{d}$ and $b = e^TPL$ and the previous relation becomes

$$\frac{1}{2}\tilde{d}^TL^TPe + \frac{1}{2}e^TPL\tilde{d} - \frac{1}{2\rho^2}e^TPLL^TPe \leq \frac{1}{2}\rho^2\tilde{d}^T\tilde{d} \tag{5.107}$$

Eq. (5.107) is substituted in Eq. (5.104) and the inequality is enforced, thus giving

$$\dot{V} \leq -\frac{1}{2}e^TQe + \frac{1}{2}\rho^2\tilde{d}^T\tilde{d} \tag{5.108}$$

Eq. (5.108) shows that the H_∞ tracking performance criterion is satisfied. The integration of $\dot{V}$ from 0 to T gives

$$\begin{aligned} &\int_0^T\dot{V}(t)dt \leq -\frac{1}{2}\int_0^T||e||_Q^2dt + \frac{1}{2}\rho^2\int_0^T||\tilde{d}||^2dt \Rightarrow \\ &2V(T) + \int_0^T||e||_Q^2dt \leq 2V(0) + \rho^2\int_0^T||\tilde{d}||^2dt \end{aligned} \tag{5.109}$$

Moreover, if there exists a positive constant $M_d > 0$ such that

$$\int_0^\infty ||\tilde{d}||^2dt \leq M_d \tag{5.110}$$

then, one gets

$$\int_0^\infty||e||_Q^2dt \leq 2V(0) + \rho^2M_d \tag{5.111}$$

Thus, the integral $\int_0^\infty||e||_Q^2dt$ is bounded. Moreover, $V(T)$ is bounded and from the definition of the Lyapunov function V in Eq. (5.94) it becomes clear that $e(t)$ will also be bounded since $e(t) \in \Omega_e = \{e|e^TPe \leq 2V(0) + \rho^2M_d\}$.

According to the above and with the use of Barbalat's Lemma, one obtains $lim_{t\to\infty}e(t) = 0$.

After following the stages of the stability proof, one arrives at Eq. (5.108) which shows that the H-infinity tracking performance criterion holds. By selecting the attenuation coefficient ρ to be sufficiently small and in particular to satisfy $\rho^2 < ||e||_Q^2/||\tilde{d}||^2$, one has that the first derivative of the Lyapunov function is upper bounded by 0. This condition holds at each sampling instance and consequently global stability for the control loop can be concluded.

5.2.4 Flatness-based control in successive-loops of the spacecraft's attitude

5.2.4.1 Attitude dynamic model in the form of differentially flat chained subsystems

As already shown, the dynamics of the attitude of the reentry space vehicle consists of two subsystems:

$$\dot{\theta} = R\omega \tag{5.112}$$

$$\dot{\omega} = -I^{-1}\Omega I\omega + I^{-1}M \tag{5.113}$$

For the subsystem of Eq. (5.112) the following subvectors and submatrices are defined: $x_{13} = \theta = [\alpha, \beta, \sigma]^T, f_{13}(x_{13}) = 0_{3\times3}, g_{13}(x_{13}) = R$.

For the subsystem of Eq. (5.113) the following subvectors and submatrices are defined: $x_{46} = \omega = [p, q, r]^T$, $f_{46}(x_{13}, x_{46}) = -I^{-1}\Omega I\omega = -I^{-1}\Omega Ix_{46}$, and $g_{46}(x_{13}, x_{46}) = I^{-1}$, while the real control inputs vector is $u = M = [M_x, M_y, M_z]^T$.

Thus, the attitude dynamics of the spacecraft is written in the form of the following chained subsystems:

$$\dot{x}_{13} = f_{13}(x_{13}) + g_{13}(x_{13})x_{46} \tag{5.114}$$

$$\dot{x}_{46} = f_{46}(x_{13}, x_{46}) + g_{46}(x_{13}, x_{46})u \tag{5.115}$$

It will be shown that each one of the subsystems of Eq. (5.114) and Eq. (5.115), if viewed independently, satisfies differential flatness properties.

For the subsystem of Eq. (5.114) the flat outputs vector is taken to be $y_1 = x_{13}$. It holds that $f_{13}(x_{13})$ and $g_{13}(x_{13})$ are functions of the flat outputs vector and $v_1 = x_{46}$ is the virtual control inputs vector. By solving Eq. (5.114) for v_1, one has

$$v_1 = g_{13}(x_{13})^{-1}[\dot{x}_{13} - f_{13}(x_{13})] \tag{5.116}$$

which signifies that v_1 is a differential function of the flat outputs vector $y_1 = x_{13}$. Consequently, in the subsystem of Eq. (5.114) all state variables and the control inputs are differential functions of the flat outputs vector $y_1 = x_{13}$ and this subsystem is differentially flat.

For the subsystem of Eq. (5.115) the flat outputs vector is taken to be $y_2 = x_{46}$, while x_{13} is viewed as a coefficients vector. Thus, $f_{46}(x_{13}, x_{46})$ and $g_{46}(x_{13}, x_{46})$ are differential functions of x_{46}. Moreover, solving for the optimal control inputs vector u gives

$$u = g_{46}(x_{13}, x_{46})^{-1}[\dot{x}_{46} - f_{46}(x_{13}, x_{46})] \tag{5.117}$$

which signifies that u is a differential function of the flat outputs vector $y_2 = x_{46}$. Consequently, in the subsystem of Eq. (5.115), all state variables and the control inputs are differential functions of the flat outputs vector $y_2 = x_{46}$ and this subsystem is differentially flat.

5.2.4.2 Design of a flatness-based controller in successive loops

The proof of differential flatness properties for the subsystems of Eq. (5.114) and Eq. (5.115) signifies that each one of these subsystems is input–output linearizable and a stabilizing feedback controller can be designed about it using the inversion of its dynamics.

For the subsystem of Eq. (5.114), the feedback controller is taken to be

$$v_1 = g_{13}(x_{13})^{-1}[\dot{x}_{13}^d - f_{13}(x_{13}) - K_1(x_{13} - x_{13}^d))] \tag{5.118}$$

where $K_1 > 0$ is the diagonal feedback gains matrix $K_1 \in R^{3\times 3}$ with diagonal elements $k_{1,ii} > 0$ for $i = 1, 2, 3$.

For the subsystem of Eq. (5.115), the feedback controller is taken to be

$$u = g_{46}(x_{13}, x_{46})^{-1}[\dot{x}_{46}^d - f_{46}(x_{13}, x_{46}) - K_2(x_{46} - x_{36}^d))] \tag{5.119}$$

where $K_2 > 0$ is the diagonal feedback gains matrix $K_2 \in R^{3\times 3}$ with diagonal elements $k_{2,ii} > 0$ for $i = 1, 2, 3$.

Control input v_1 of the first control loop becomes setpoint for the second control loop, that is $x_{46}^d = v_1$. Next, by substituting Eq. (5.118) into Eq. (5.114), as well as by substituting Eq. (5.119) into Eq. (5.115), one obtains

$$\dot{x}_{13} = f_{13}(x_{13}) + g_{13}(x_{13})g_{13}(x_{13})^{-1}[\dot{x}_{13}^d - f_{13}(x_{13}) - K_1(x_{13} - x_{13}^d))] \tag{5.120}$$

$$\dot{x}_{46} = f_{46}(x_{13}, x_{46}) + g_{46}(x_{13}, x_{46})g_{46}(x_{13}, x_{46})^{-1}[\dot{x}_{46}^d - f_{46}(x_{13}, x_{46}) - K_2(x_{46} - x_{36}^d))] \tag{5.121}$$

Equivalently, after defining the tracking error variables $e_{13} = x_{13} - x_{13}^d$ and $e_{46} = x_{46} - x_{46}^d$, the closed-loop dynamics of the two subsystems becomes

$$(\dot{x}_{13} - \dot{x}_{13}^d) + K_1(x_{13} - x_{13}^d) \Rightarrow \dot{e}_{13} + K_1 e_{13} = 0 \tag{5.122}$$

$$(\dot{x}_{46} - \dot{x}_{46}^d) + K_2(x_{46} - x_{46}^d) \Rightarrow \dot{e}_{46} + K_2 e_{46} = 0 \tag{5.123}$$

Using that $K_1 > 0$ and $K_2 > 0$, it holds that

$$\lim_{t\to\infty} e_{13}(t) = 0 \Rightarrow \lim_{t\to\infty} x_{13}(t) = x_{13}^d(t) \tag{5.124}$$

$$\lim_{t\to\infty} e_{46}(t) = 0 \Rightarrow \lim_{t\to\infty} x_{46}(t) = x_{46}^d(t) \tag{5.125}$$

Consequently, the state variables of the attitude dynamic model of the reentry space vehicle converge to the associated setpoints and the system is globally asymptotically stable.

The global stability properties of the flatness-based control method in successive loops can also be proven through Lyapunov analysis. The following Lyapunov function is defined

$$V = \frac{1}{2}[e_{13}^T e_{13} + e_{46}^T e_{46}] \tag{5.126}$$

By differentiating in time, one obtains

$$V = \frac{1}{2}2[e_{13}^T \dot{e}_{13} + e_{46}^T \dot{e}_{46}] \tag{5.127}$$

By substituting Eq. (5.124) and Eq. (5.125) into Eq. (5.128), one gets

$$V = \frac{1}{2}2[e_{13}^T(-K_1 e_{13}) + e_{46}^T(-K_2 e_{46})] \tag{5.128}$$

which finally gives

$$V = -[e_{13}^T K_1 e_{13} + e_{46}^T K_2 e_{45}] \tag{5.129}$$

and provides the global stability condition

$$\dot{V} < 0 \, \forall \, e_{13} \neq 0, \; e_{46} \neq 0 \tag{5.130}$$

and $\dot{V} = 0$ iff $e_{13} = 0$ and $e_{46} = 0$. Therefore, the closed-loop system is globally asymptotically stable.

5.2.5 Simulation tests

5.2.5.1 Nonlinear optimal control of the attitude of the reentry space vehicle

The global stability properties and the fine transient performance of nonlinear optimal control for the attitude dynamics of the reentry space vehicle have been tested through simulation experiments. To implement nonlinear optimal control the Jacobian matrices of the dynamic model of the reentry space vehicle had to be computed at each sampling period and the method's Riccati equation had to be solved also at each sampling interval using the *aresolv()* function in MATLAB®. The method is computationally efficient since a solution of this Riccati equation can be achieved within a time interval which is much shorter than the sampling period. The obtained results are depicted in Figures 5.42–5.57. The state variables of the reentry space vehicle are depicted in blue, the associated setpoints are plotted in red while the state estimates which are provided by the H-infinity Kalman filter are printed in green. It has been confirmed that the method achieves fast and accurate tracking of reference setpoints under moderate variations of the control inputs. About the transient performance of the control method this depends on parameters r, ρ, and Q of the method's Riccati equation. Small values of r achieve elimination of the steady-state tracking error while relatively large values of the diagonal elements of matrix Q enable faster convergence of the state variables to their setpoints. Finally, the smallest value of ρ for which a valid solution of the method's Riccati equation can be obtained in the form of a positive definite and symmetric matrix P is the one that gives to the control loop maximum robustness.

To elaborate on nonlinear optimal control for the attitude of the reentry space vehicle the following tables are given (i) Table 5.5 providing results about the accuracy of tracking of setpoints by the state variables of the attitude dynamics of the reentry space vehicle under an exact dynamic model, (ii) Table 5.6 providing results about the accuracy of tracking of setpoints by the state variables of the attitude dynamics of the reentry space vehicle under a model that is subject to disturbances (for instance change $\Delta\alpha\%$ in parameter N_1 of the dynamic model), and (iii) Table 5.7 providing results about the convergence times to setpoints of the state variables of the attitude dynamics of the reentry space vehicles.

5.2.5.2 Flatness-based control in successive loops of the reentry space vehicle

Results about the tracking accuracy and the speed of convergence to setpoints of the flatness-based control method in successive-loops, in the case of the 6-DOF attitude dynamics of the reentry space vehicle, are shown in Figures 5.58–5.73. It can be noticed again that, under this control scheme, one achieves fast and precise tracking of reference setpoints for all state variables of the dynamic model of the attitude of the reentry space vehicle. It is noteworthy, that through the stages of this method, one solves also the setpoints definition problem for all state variables of the reentry space vehicle. Actually, the selection of setpoints for state variables x_1, x_2, and x_3 is unconstrained. On the other side by

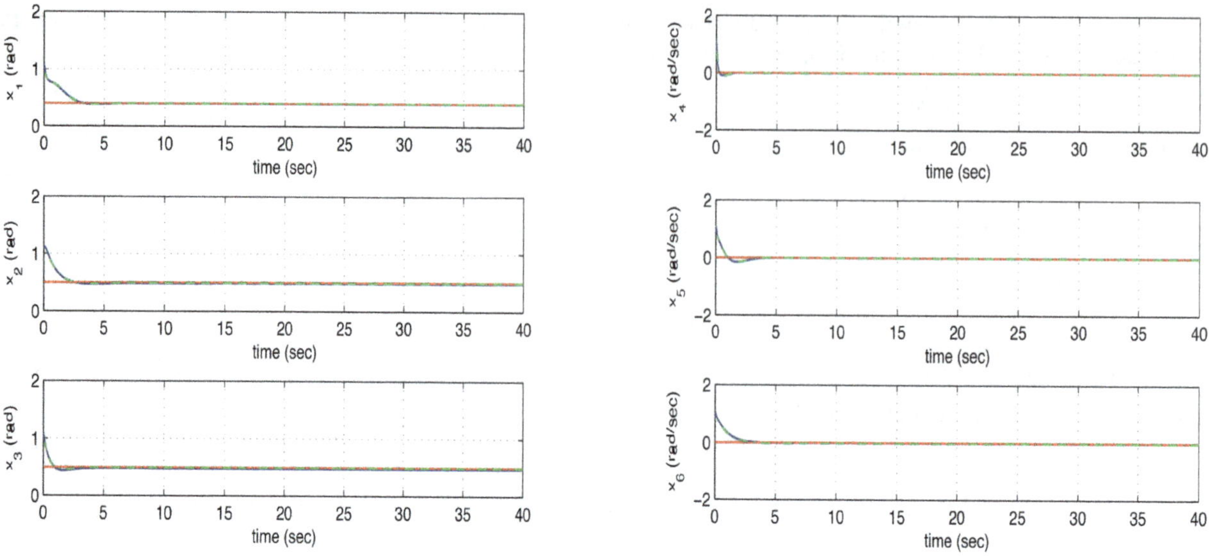

(a) (b)

Figure 5.42 *Tracking of attitude setpoint 1 by the reentry space vehicle with the use of nonlinear optimal control: (a) convergence of state variables x_1 to x_3 (blue lines) to the associated setpoints (red lines) and estimated values provided by Kalman filtering; (b) convergence of state variables x_4 to x_6 (blue lines) to the associated setpoints (red lines) and estimated values provided by Kalman filtering*

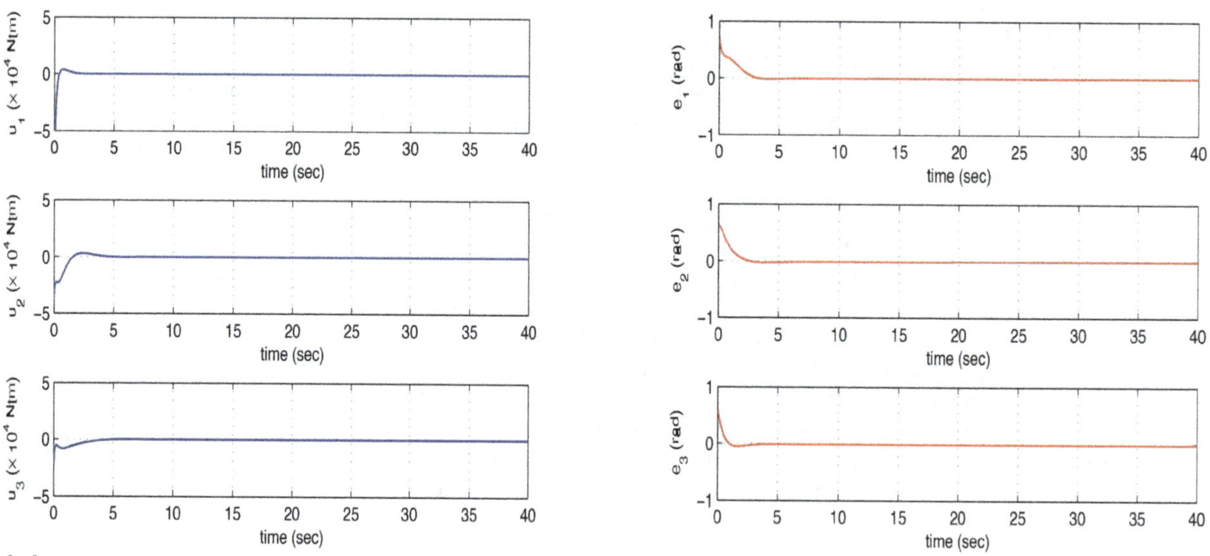

(a) (b)

Figure 5.43 *Tracking of attitude setpoint 1 by the reentry space vehicle: (a) variations of the control inputs u_1 and u_3 of the spacecraft (blue lines); (b) variation of the tracking error variables e_1 to e_3 associated with the state variables x_1 to x_3*

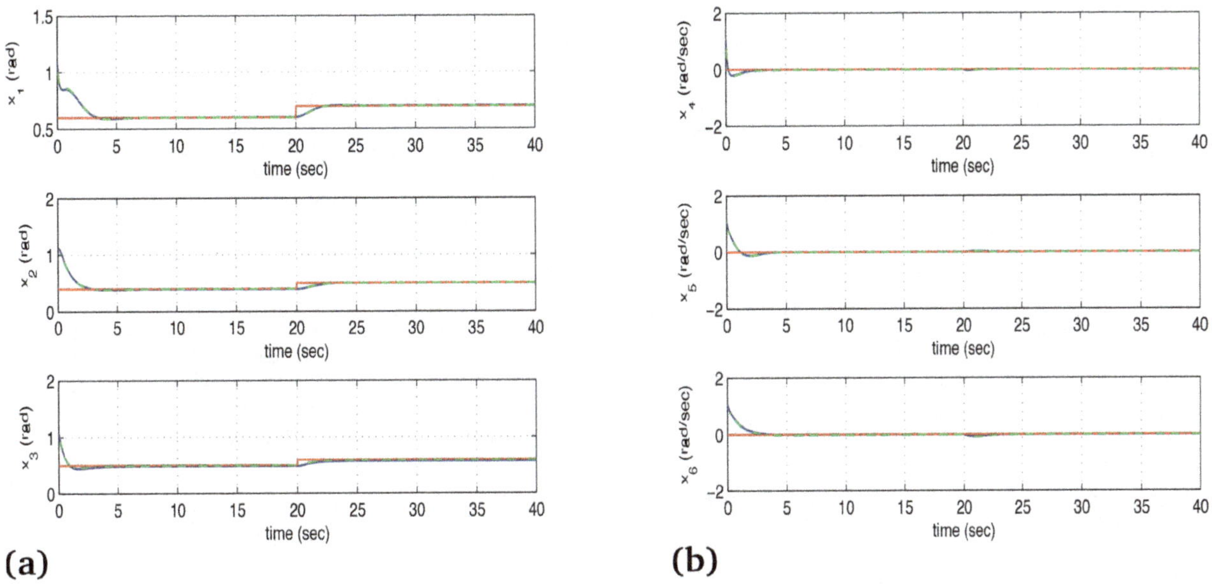

Figure 5.44 *Tracking of attitude setpoint 2 by the reentry space vehicle with the use of nonlinear optimal control: (a) convergence of state variables x_1 to x_3 (blue lines) to the associated setpoints (red lines) and estimated values provided by Kalman filtering; (b) convergence of state variables x_4 to x_6 (blue lines) to the associated setpoints (red lines) and estimated values provided by Kalman filtering*

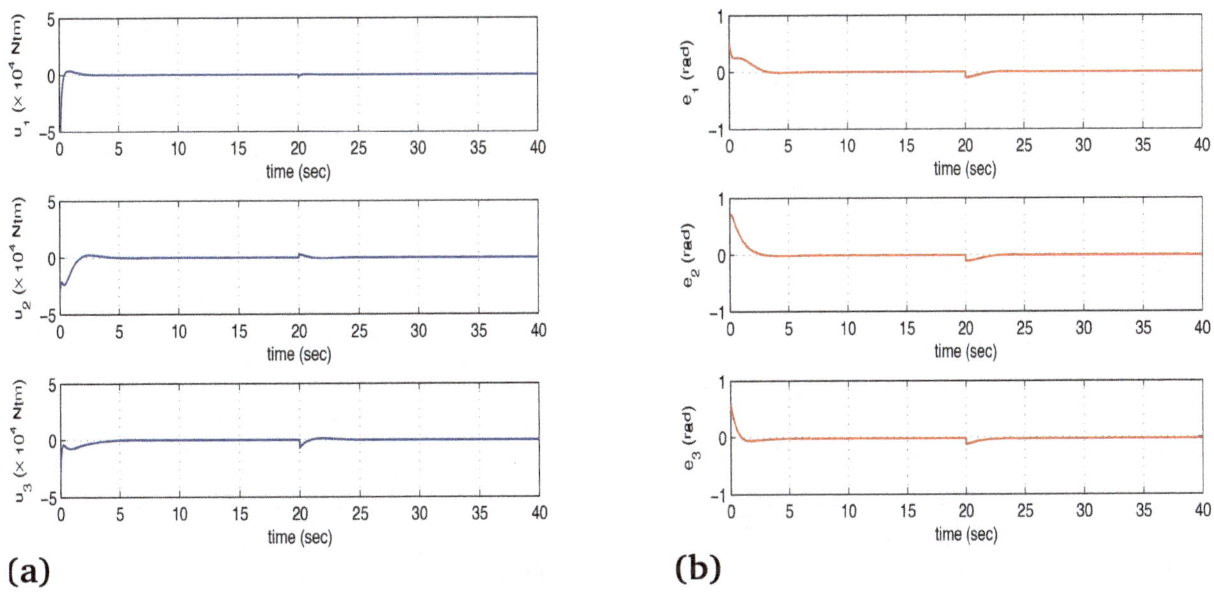

Figure 5.45 *Tracking of attitude setpoint 2 by the reentry space vehicle with the use of nonlinear optimal control: (a) variations of the control inputs u_1 and u_3 of the spacecraft (blue lines); (b) variation of the tracking error variables e_1 to e_3 associated with the state variables x_1 to x_3*

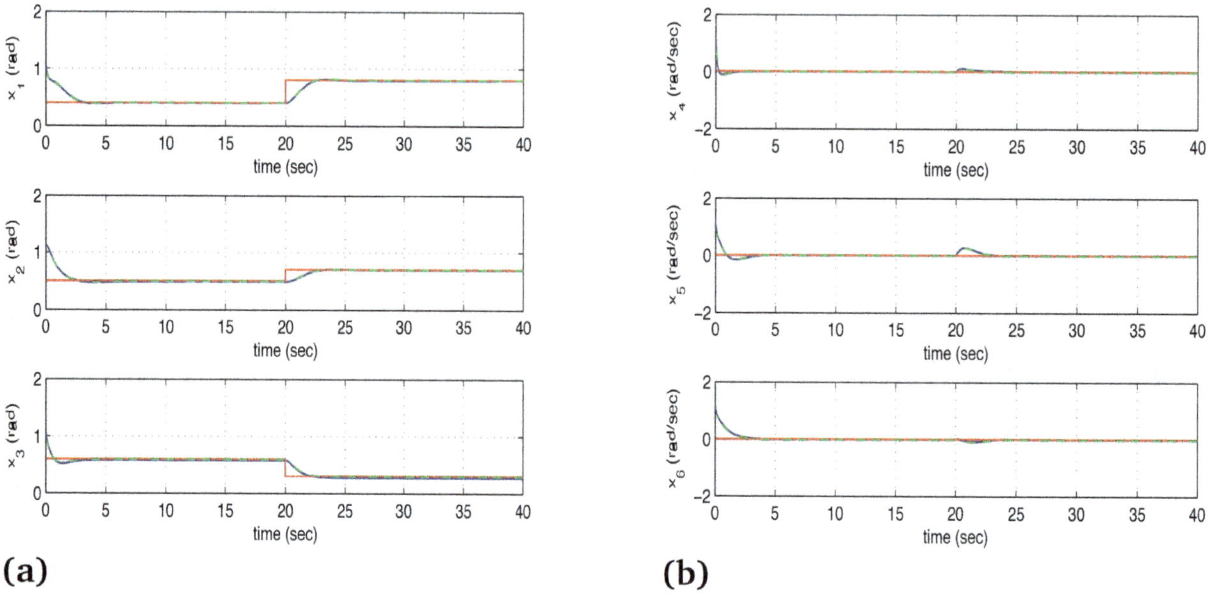

Figure 5.46 Tracking of attitude setpoint 3 by the reentry space vehicle with the use of nonlinear optimal control: (a) convergence of state variables x_1 to x_3 (blue lines) to the associated setpoints (red lines) and estimated values provided by Kalman filtering; (b) convergence of state variables x_4 to x_6 (blue lines) to the associated setpoints (red lines) and estimated values provided by Kalman filtering

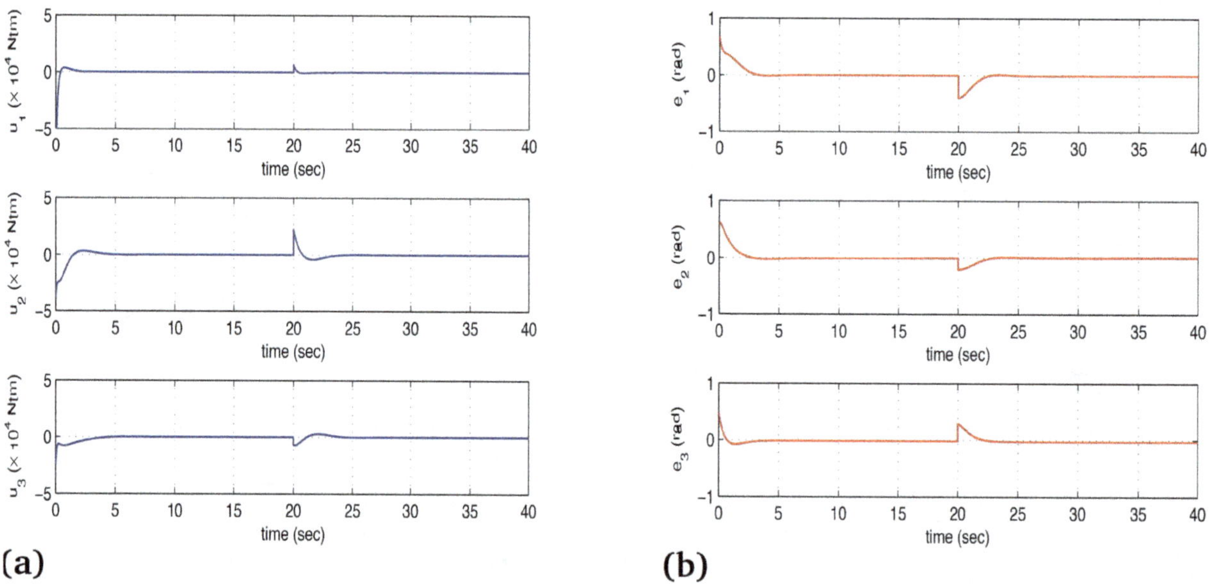

Figure 5.47 Tracking of attitude setpoint 3 by the reentry space vehicle with the use of nonlinear optimal control: (a) variations of the control inputs u_1 and u_3 of the spacecraft (blue lines); (b) variation of the tracking error variables e_1 to e_3 associated with the state variables x_1 to x_3

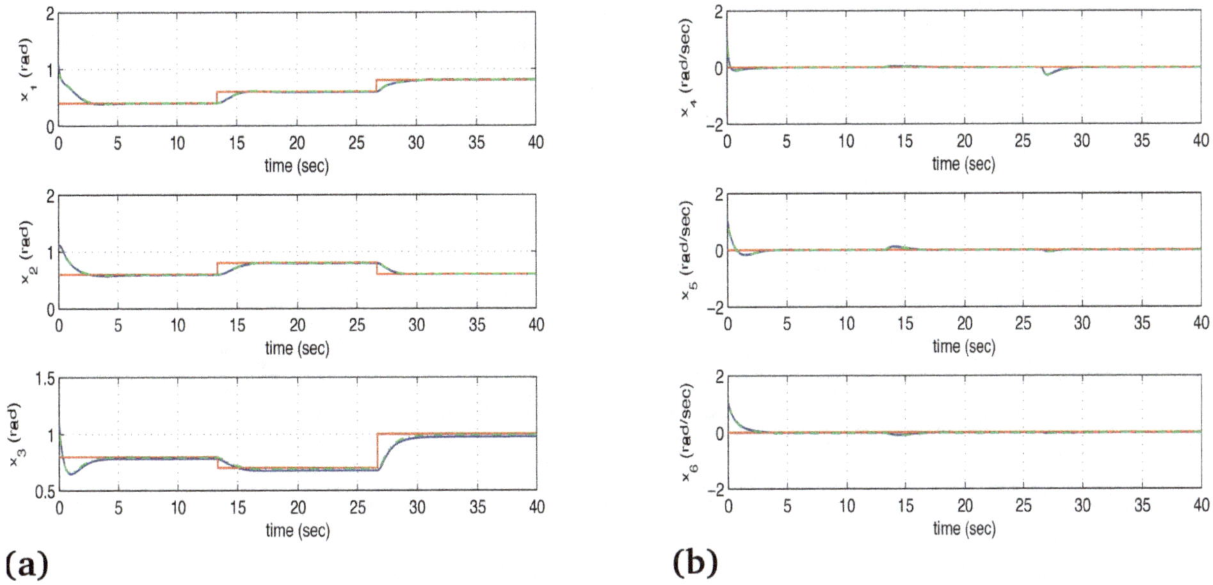

Figure 5.48 *Tracking of attitude setpoint 4 by the reentry space vehicle with the use of nonlinear optimal control: (a) convergence of state variables x_1 to x_3 (blue lines) to the associated setpoints (red lines) and estimated values provided by Kalman filtering; (b) convergence of state variables x_4 to x_6 (blue lines) to the associated setpoints (red lines) and estimated values provided by Kalman filtering*

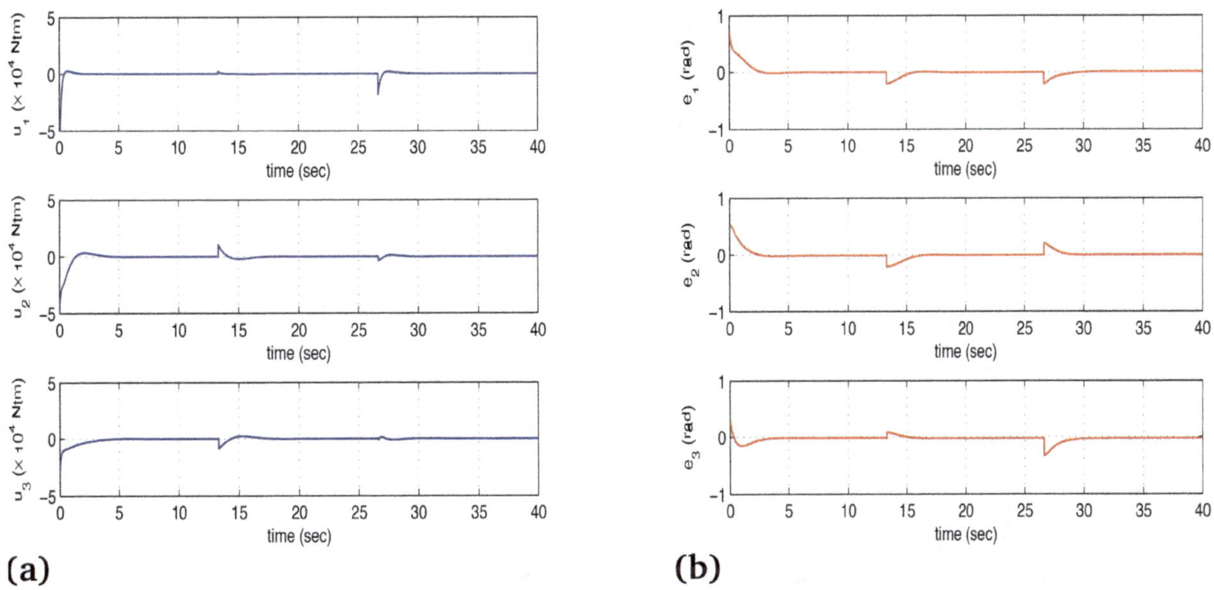

Figure 5.49 *Tracking of attitude setpoint 4 by the reentry space vehicle with the use of nonlinear optimal control: (a) variations of the control inputs u_1 and u_3 of the spacecraft (blue lines); (b) variation of the tracking error variables e_1 to e_3 associated with the state variables x_1 to x_3*

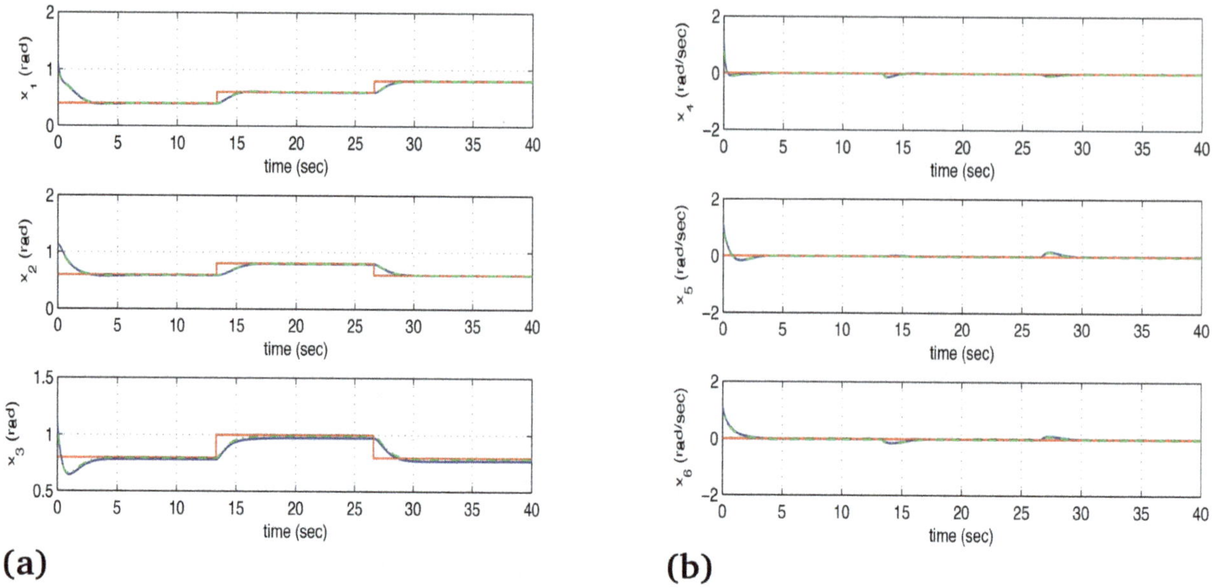

Figure 5.50 *Tracking of attitude setpoint 5 by the reentry space vehicle with the use of nonlinear optimal control: (a) convergence of state variables x_1 to x_3 (blue lines) to the associated setpoints (red lines) and estimated values provided by Kalman filtering; (b) convergence of state variables x_4 to x_6 (blue lines) to the associated setpoints (red lines) and estimated values provided by Kalman filtering*

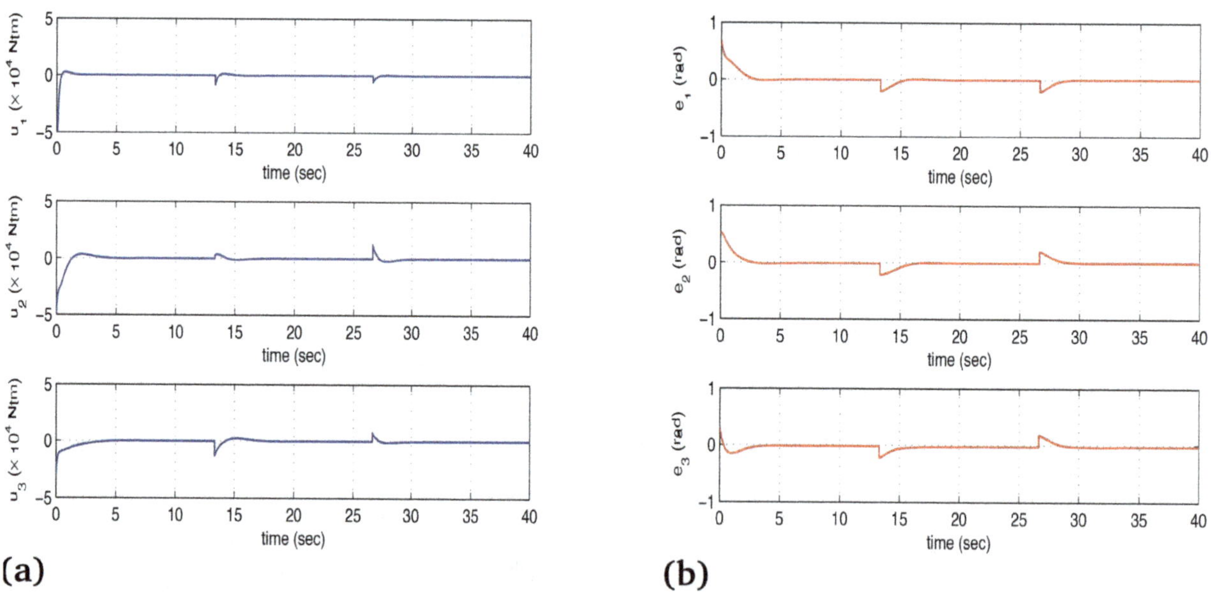

Figure 5.51 *Tracking of attitude setpoint 5 by the reentry space vehicle with the use of nonlinear optimal control: (a) variations of the control inputs u_1 and u_3 of the spacecraft (blue lines); (b) variation of the tracking error variables e_1 to e_3 associated with the state variables x_1 to x_3*

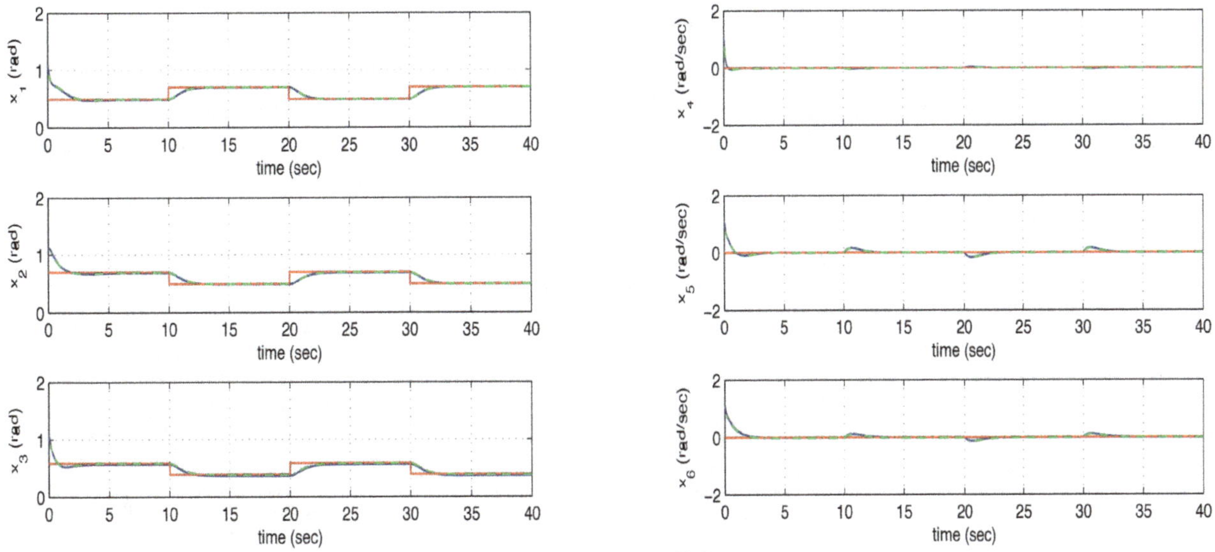

(a) (b)

Figure 5.52 *Tracking of attitude setpoint 6 by the reentry space vehicle with the use of nonlinear optimal control: (a) convergence of state variables x_1 to x_3 (blue lines) to the associated setpoints (red lines) and estimated values provided by Kalman filtering; (b) convergence of state variables x_4 to x_6 (blue lines) to the associated setpoints (red lines) and estimated values provided by Kalman filtering*

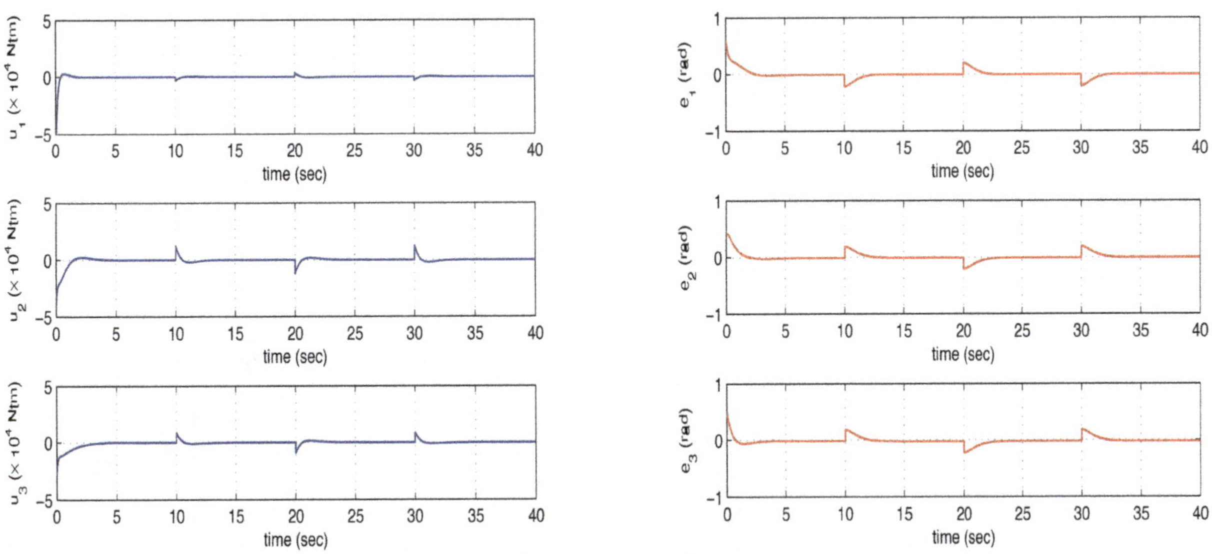

(a) (b)

Figure 5.53 *Tracking of attitude setpoint 6 by the reentry space vehicle with the use of nonlinear optimal control: (a) variations of the control inputs u_1 and u_3 of the spacecraft (blue lines); (b) variation of the tracking error variables e_1 to e_3 associated with the state variables x_1 to x_3*

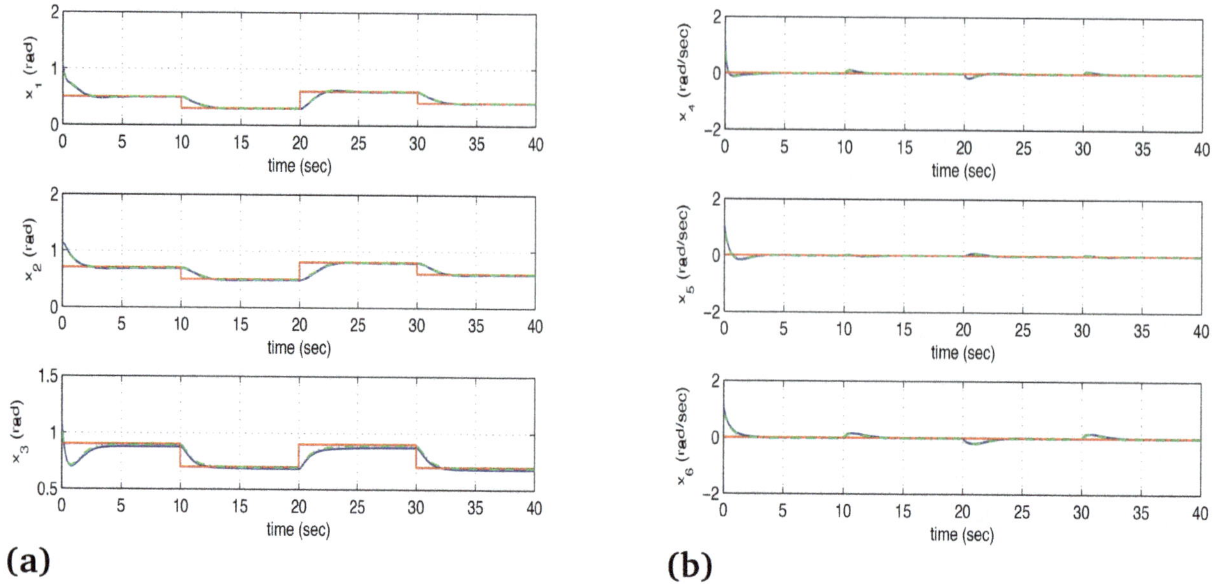

Figure 5.54 *Tracking of attitude setpoint 7 by the reentry space vehicle with the use of nonlinear optimal control: (a) convergence of state variables x_1 to x_3 (blue lines) to the associated setpoints (red lines) and estimated values provided by Kalman filtering; (b) convergence of state variables x_4 to x_6 (blue lines) to the associated setpoints (red lines) and estimated values provided by Kalman filtering*

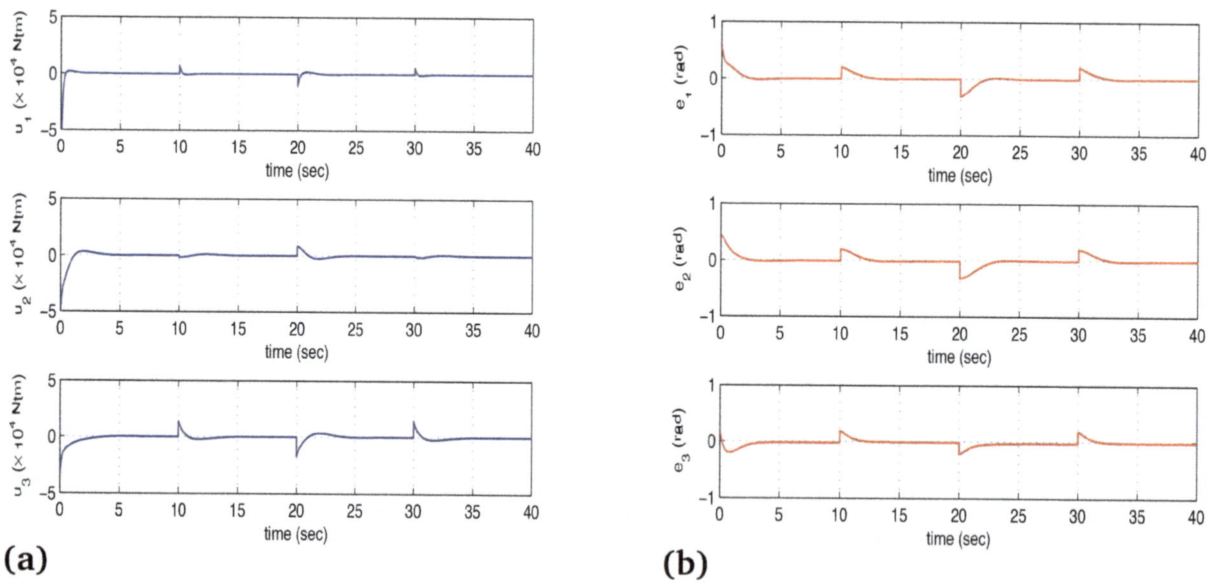

Figure 5.55 *Tracking of attitude setpoint 7 by the reentry space vehicle with the use of nonlinear optimal control: (a) variations of the control inputs u_1 and u_3 of the spacecraft (blue lines); (b) variation of the tracking error variables e_1 to e_3 associated with the state variables x_1 to x_3*

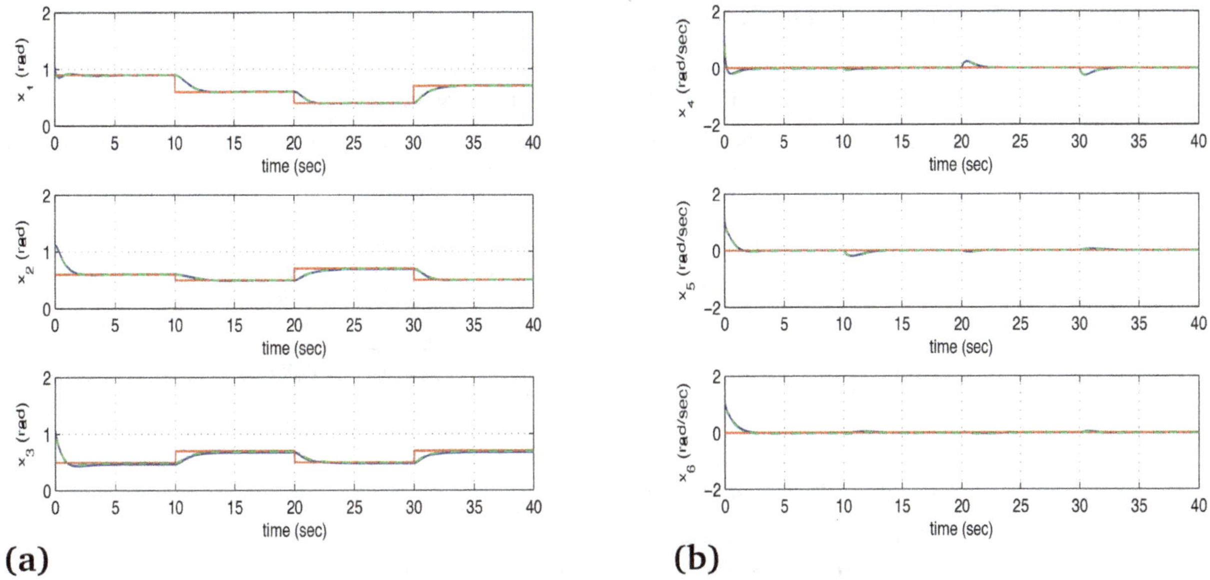

Figure 5.56 *Tracking of attitude setpoint 8 by the reentry space vehicle with the use of nonlinear optimal control: (a) convergence of state variables x_1 to x_3 (blue lines) to the associated setpoints (red lines) and estimated values provided by Kalman filtering; (b) convergence of state variables x_4 to x_6 (blue lines) to the associated setpoints (red lines) and estimated values provided by Kalman filtering*

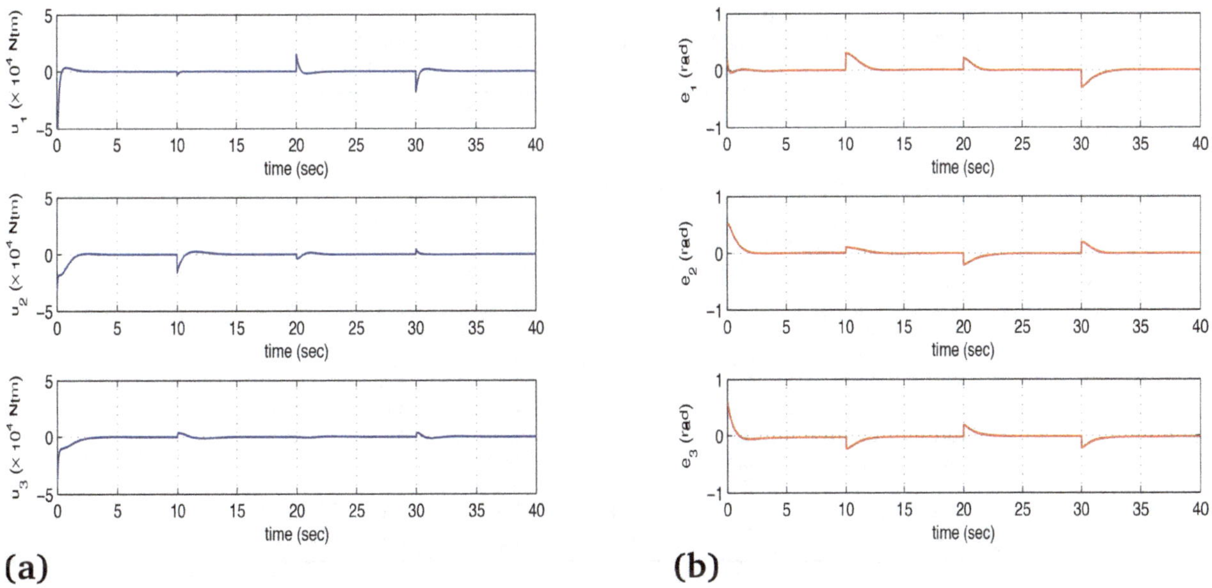

Figure 5.57 *Tracking of attitude setpoint 8 by the reentry space vehicle with the use of nonlinear optimal control: (a) variations of the control inputs u_1 and u_3 of the spacecraft (blue lines); (b) variation of the tracking error variables e_1 to e_3 associated with the state variables x_1 to x_3*

Table 5.5 Nonlinear optimal control

Tracking RMSE $\times 10^{-3}$ in the disturbance-free case						
	RMSE$_{x_1}$	RMSE$_{x_2}$	RMSE$_{x_3}$	RMSE$_{x_4}$	RMSE$_{x_5}$	RMSE$_{x_6}$
test$_1$	0.0351	0.3861	0.6146	0.0040	0.0002	0.0041
test$_2$	0.1076	0.1300	0.8225	0.0040	0.0002	0.0050
test$_3$	0.2453	0.0313	0.9930	0.0031	0.0050	0.0060
test$_4$	0.2500	0.0557	0.9470	0.0082	0.0044	0.0056
test$_5$	0.1135	0.0327	0.9775	0.0065	0.0013	0.0057
test$_6$	0.1135	0.0327	0.9775	0.0065	0.0013	0.0057
test$_7$	0.0127	0.4013	0.6260	0.0073	0.0073	0.0129
test$_8$	0.1989	0.1063	0.8104	0.0247	0.0037	0.0088

Table 5.6 Nonlinear optimal control

Tracking RMSE $\times 10^{-3}$ in the case of disturbances						
$\Delta a\%$	RMSE$_{x_1}$	RMSE$_{x_2}$	RMSE$_{x_3}$	RMSE$_{x_4}$	RMSE$_{x_5}$	RMSE$_{x_6}$
0%	0.0351	0.3841	0.6116	0.0005	0.0025	0.0041
10%	0.0642	0.3823	0.5953	0.0002	0.0029	0.0040
20%	0.0931	0.1812	0.5779	0.0001	0.0033	0.0038
30%	0.1217	0.3805	0.5592	0.0003	0.0037	0.0037
40%	0.1500	0.3804	0.5593	0.0005	0.0041	0.0036
50%	0.1779	0.3804	0.5180	0.0008	0.0045	0.0035
60%	0.2056	0.3816	0.4953	0.0010	0.0048	0.0034

Table 5.7 Nonlinear optimal control

Convergence times T_s (sec) of the state variables						
	$T_s\, x_1$	$T_s\, x_2$	$T_s\, x_3$	$T_s\, x_4$	$T_s\, x_5$	$T_s\, x_6$
test$_1$	3.0	2.5	3.0	2.0	4.0	4.0
test$_2$	2.5	2.5	3.5	2.5	4.0	4.0
test$_3$	2.5	2.5	3.5	2.0	4.0	4.0
test$_4$	3.0	3.5	4.0	2.0	3.5	3.5
test$_5$	2.5	2.5	3.5	2.0	2.5	3.5
test$_6$	2.5	2.0	2.5	1.5	3.5	3.5
test$_7$	2.5	2.5	4.0	2.5	3.5	3.0
test$_8$	1.5	2.5	3.5	2.5	2.5	2.5

defining state variables x_4, x_5, x_6 as virtual control inputs for the subsystem of x_1, x_2, and x_3, one can find the setpoints for x_4 to x_6 as functions of the setpoints for x_1 to x_3. The speed of convergence of the state variables of the attitude dynamics of the reentry space vehicle under flatness-based control implemented in successive loops is determined by the selection of values for the diagonal gain matrices K_1, K_2.

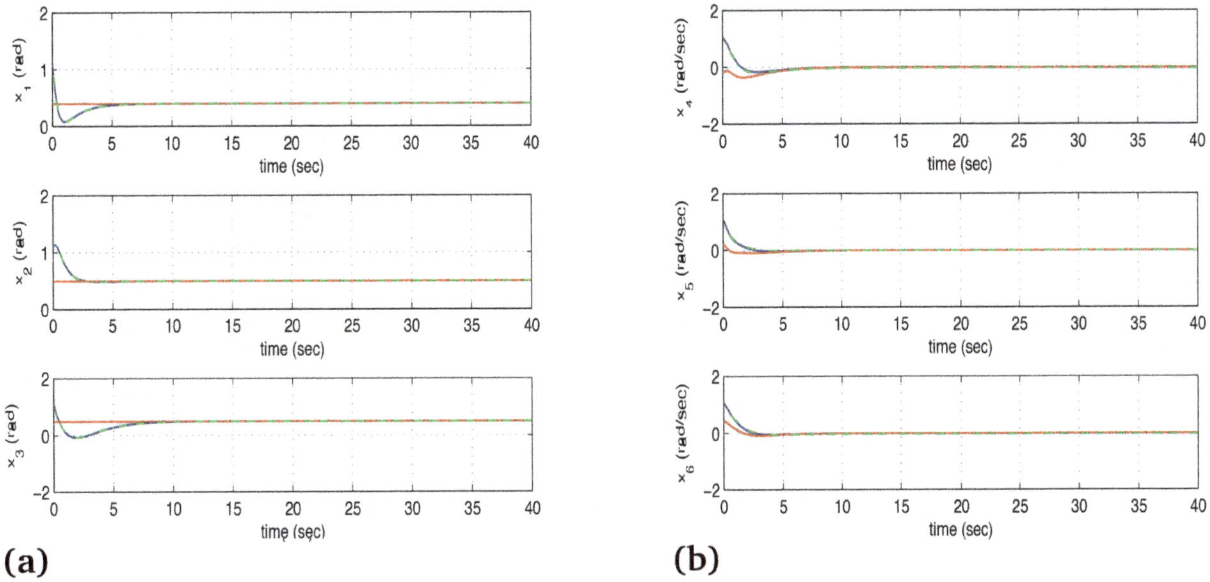

Figure 5.58 *Tracking of attitude setpoint 1 by the reentry space vehicle with the use of flatness-based control in successive loops: (a) convergence of state variables x_1 to x_3 (blue lines) to the associated setpoints (red lines) and estimated values provided by Kalman filtering; (b) convergence of state variables x_4 to x_6 (blue lines) to the associated setpoints (red lines) and estimated values provided by Kalman filtering*

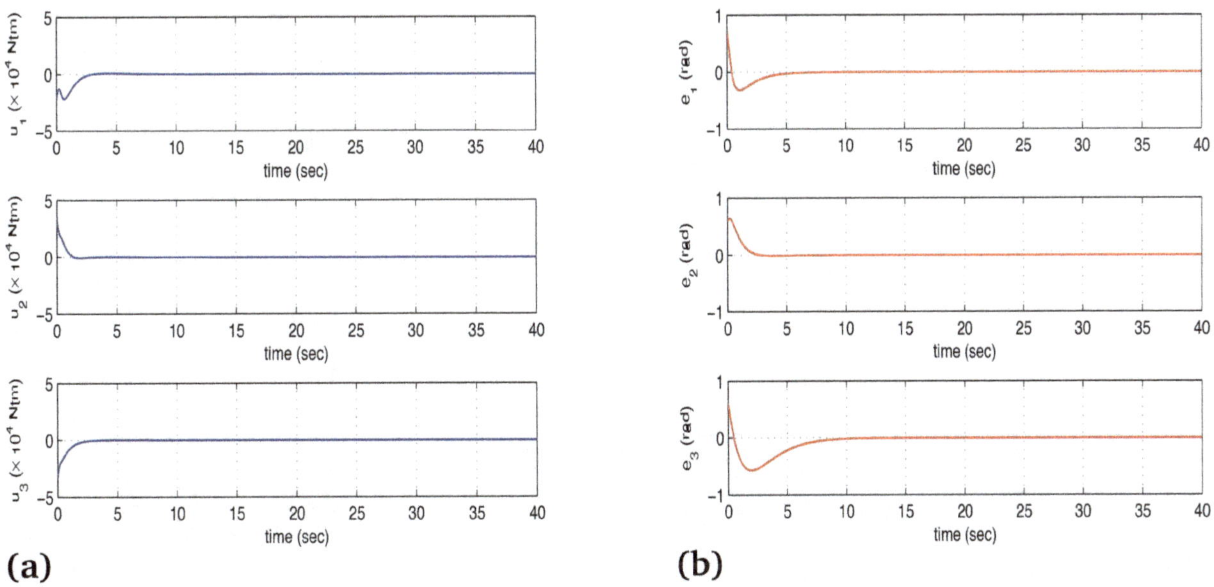

Figure 5.59 *Tracking of attitude setpoint 1 by the reentry space vehicle with the use of flatness-based control in successive loops: (a) variations of the control inputs u_1 and u_3 of the spacecraft (blue lines); (b) variation of the tracking error variables e_1 to e_3 associated with the state variables x_1 to x_3*

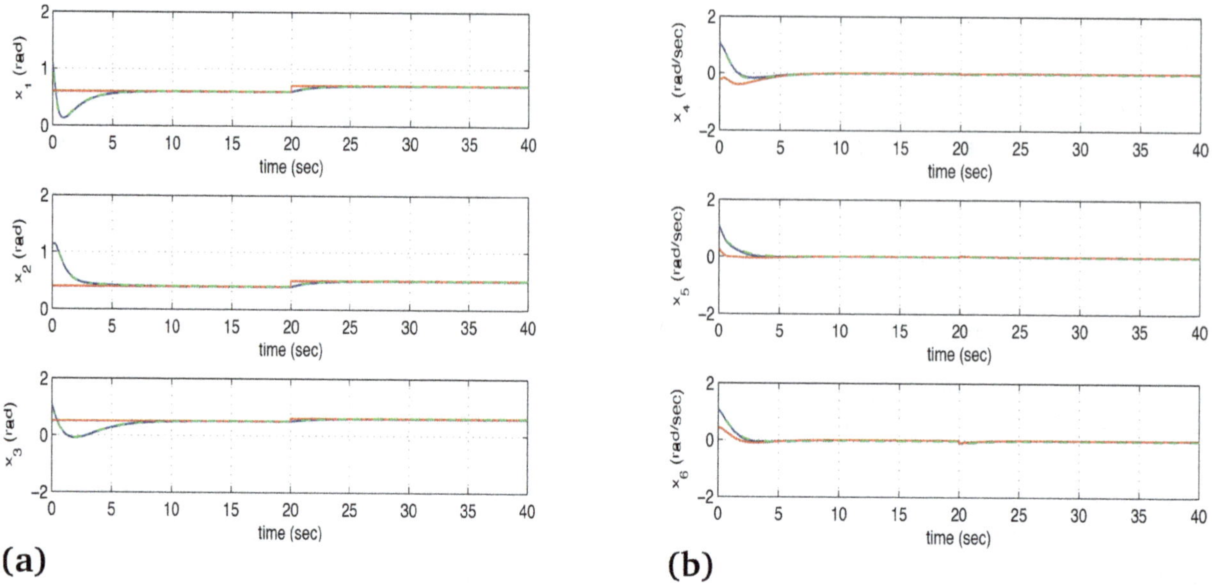

Figure 5.60 *Tracking of attitude setpoint 2 by the reentry space vehicle with the use of flatness-based control in successive loops: (a) convergence of state variables x_1 to x_3 (blue lines) to the associated setpoints (red lines) and estimated values provided by Kalman filtering; (b) convergence of state variables x_4 to x_6 (blue lines) to the associated setpoints (red lines) and estimated values provided by Kalman filtering*

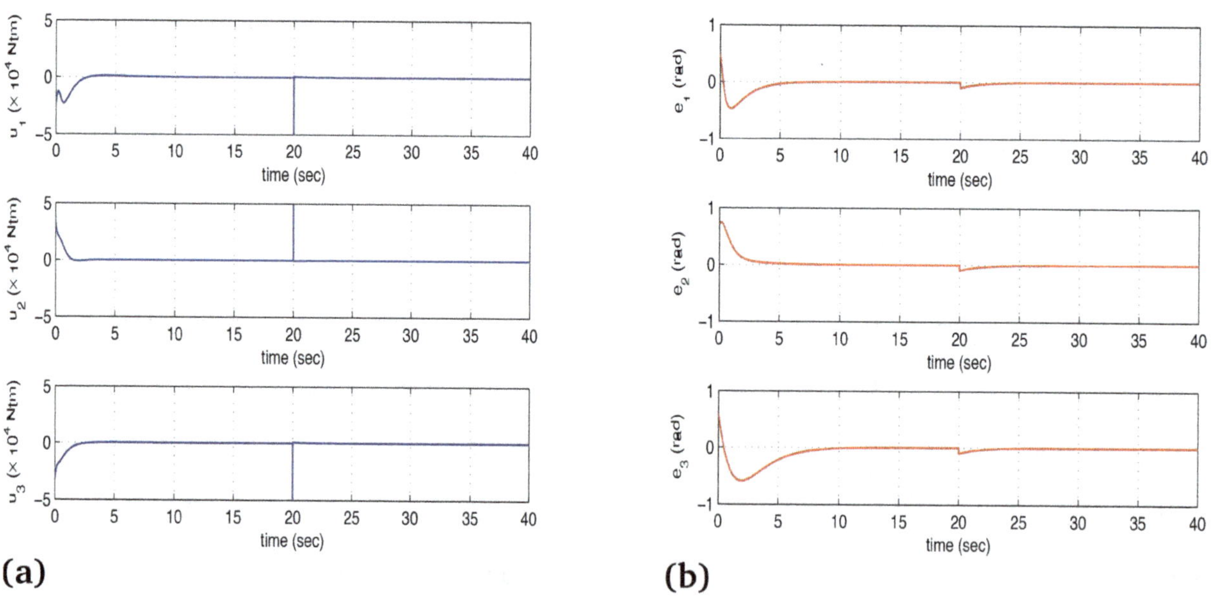

Figure 5.61 *Tracking of attitude setpoint 2 by the reentry space vehicle with the use of flatness-based control in successive loops: (a) variations of the control inputs u_1 and u_3 of the spacecraft (blue lines); (b) variation of the tracking error variables e_1 to e_3 associated with the state variables x_1 to x_3*

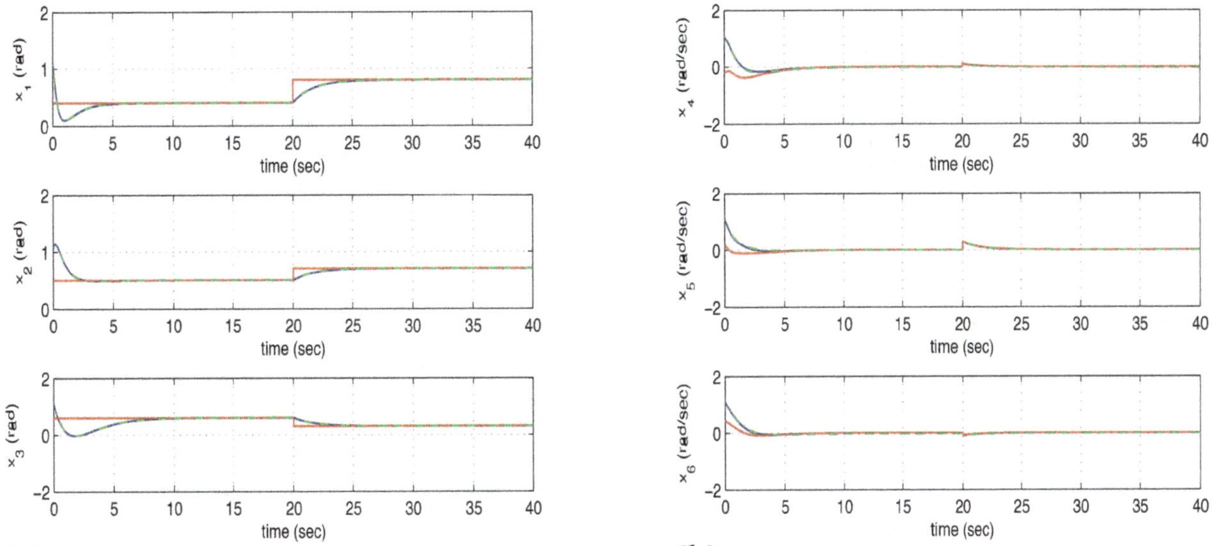

(a) (b)

Figure 5.62 Tracking of attitude setpoint 3 by the reentry space vehicle with the use of flatness-based control in successive loops: (a) convergence of state variables x_1 to x_3 (blue lines) to the associated setpoints (red lines) and estimated values provided by Kalman filtering; (b) convergence of state variables x_4 to x_6 (blue lines) to the associated setpoints (red lines) and estimated values provided by Kalman filtering

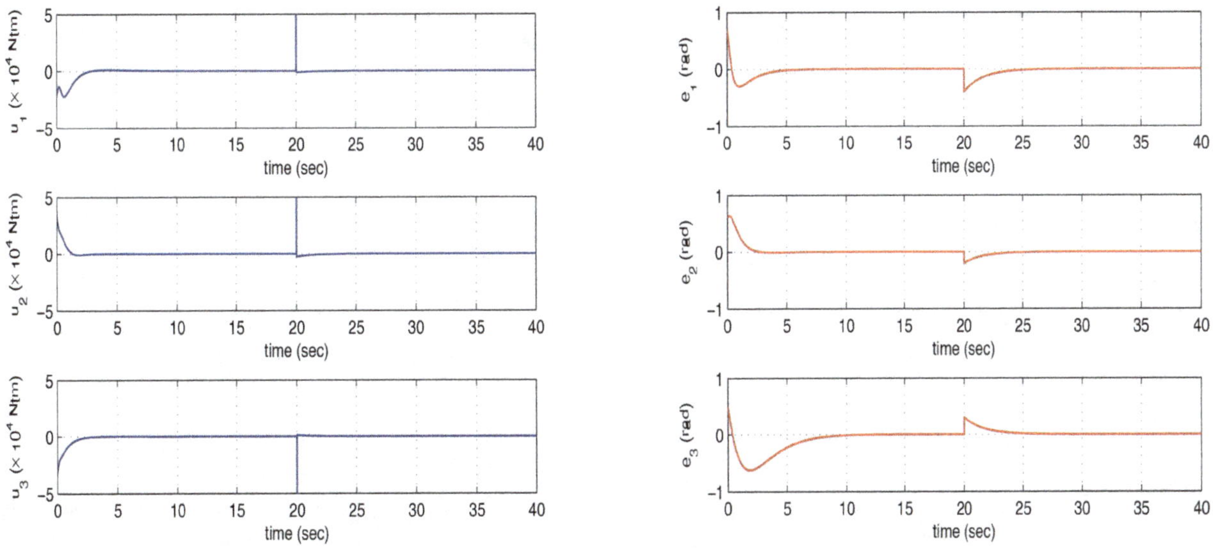

(a) (b)

Figure 5.63 Tracking of attitude setpoint 3 by the reentry space vehicle with the use of flatness-based control in successive loops: (a) variations of the control inputs u_1 and u_3 of the spacecraft (blue lines); (b) variation of the tracking error variables e_1 to e_3 associated with the state variables x_1 to x_3

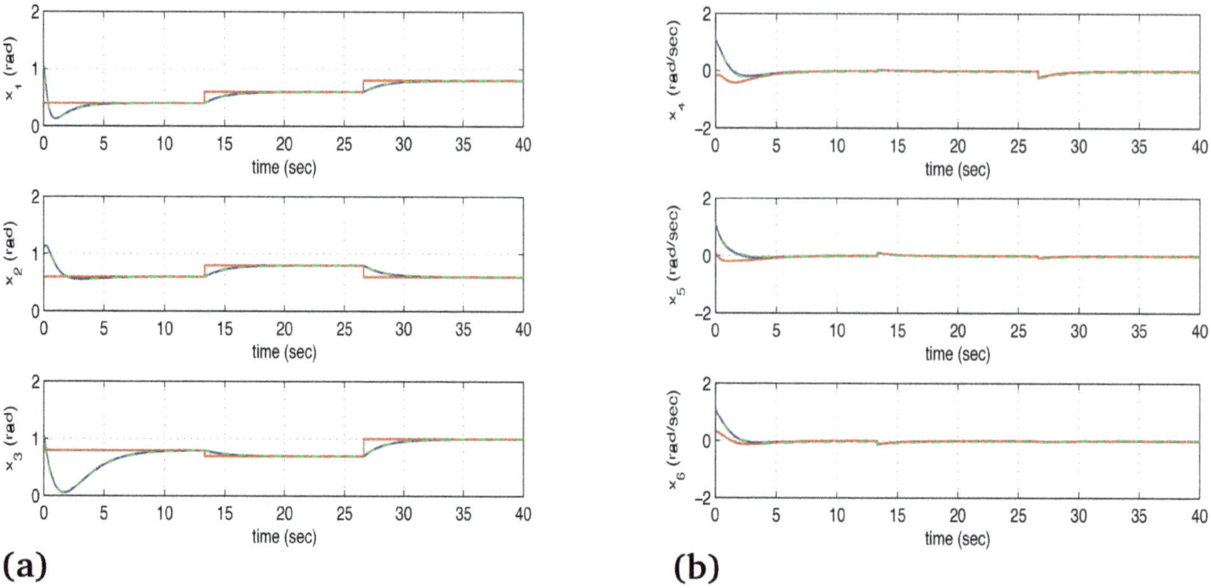

Figure 5.64 *Tracking of attitude setpoint 4 by the reentry space vehicle with the use of flatness-based control in successive loops: (a) convergence of state variables x_1 to x_3 (blue lines) to the associated setpoints (red lines) and estimated values provided by Kalman filtering; (b) convergence of state variables x_4 to x_6 (blue lines) to the associated setpoints (red lines) and estimated values provided by Kalman filtering*

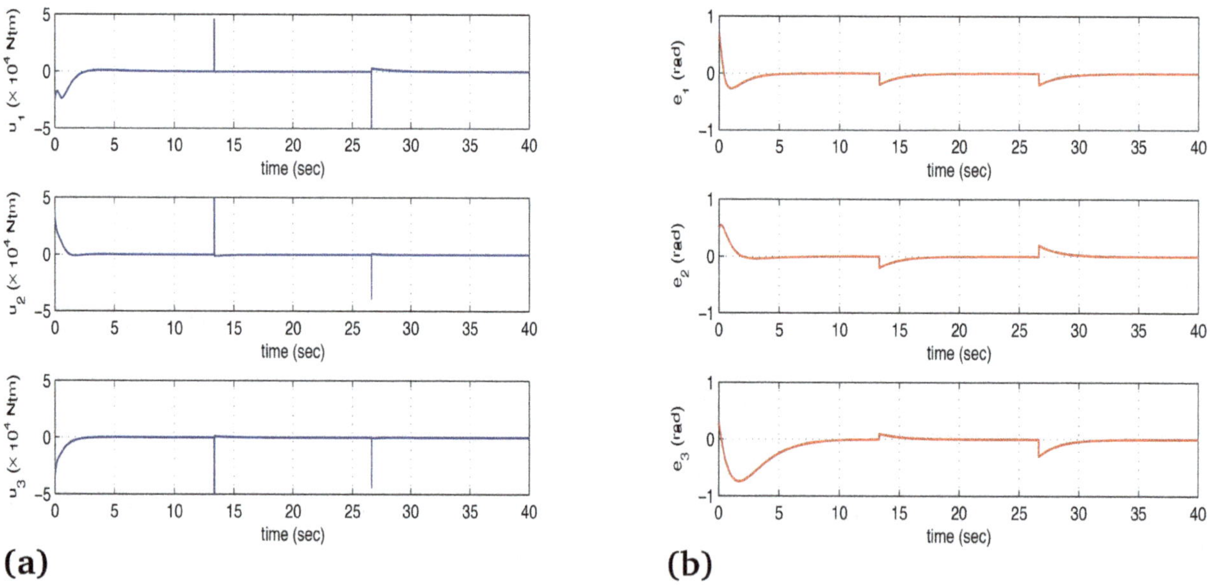

Figure 5.65 *Tracking of attitude setpoint 4 by the reentry space vehicle with the use of flatness-based control in successive loops: (a) variations of the control inputs u_1 and u_3 of the spacecraft (blue lines); (b) variation of the tracking error variables e_1 to e_3 associated with the state variables x_1 to x_3*

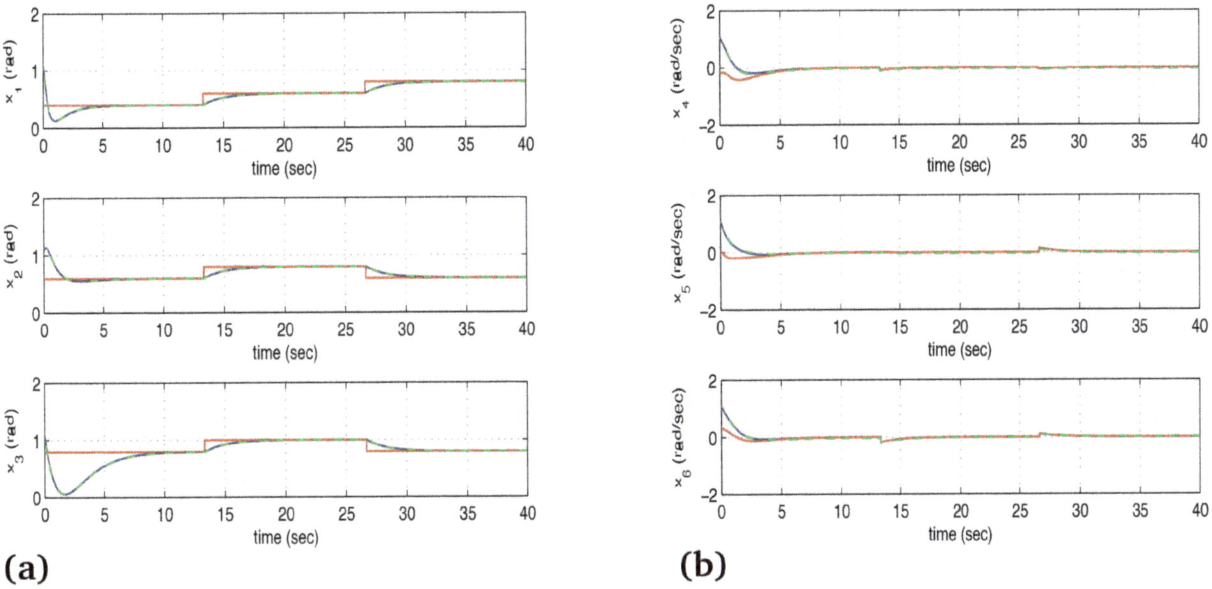

Figure 5.66 *Tracking of attitude setpoint 5 by the reentry space vehicle with the use of flatness-based control in successive loops: (a) convergence of state variables x_1 to x_3 (blue lines) to the associated setpoints (red lines) and estimated values provided by Kalman filtering; (b) convergence of state variables x_4 to x_6 (blue lines) to the associated setpoints (red lines) and estimated values provided by Kalman filtering*

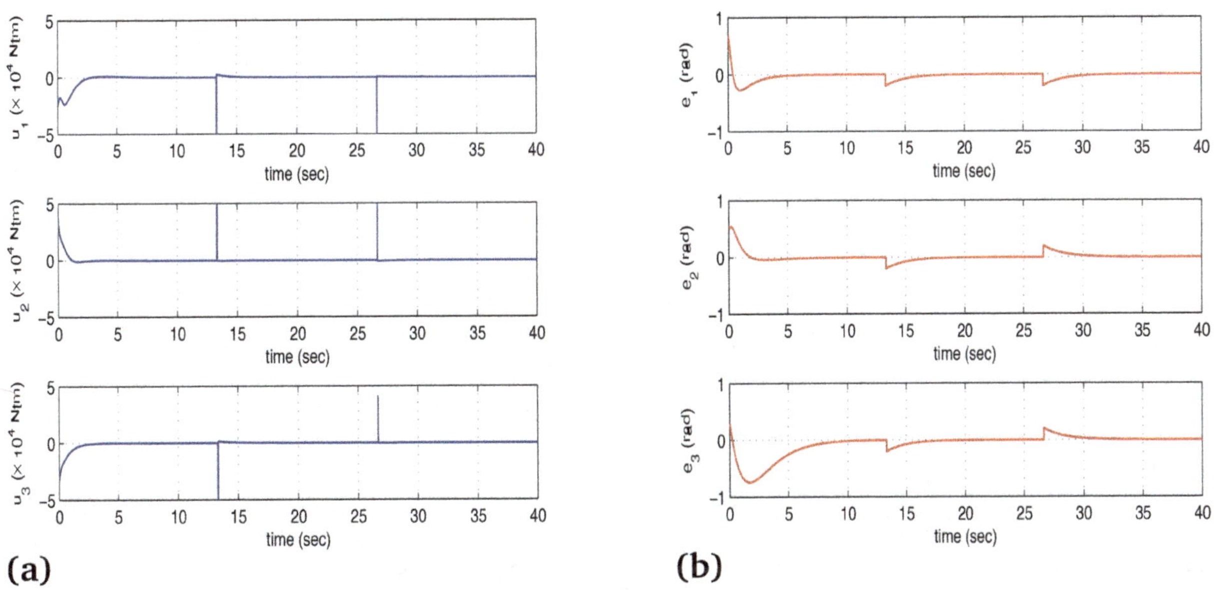

Figure 5.67 *Tracking of attitude setpoint 5 by the reentry space vehicle with the use of flatness-based control in successive loops: (a) variations of the control inputs u_1 and u_3 of the spacecraft (blue lines); (b) variation of the tracking error variables e_1 to e_3 associated with the state variables x_1 to x_3*

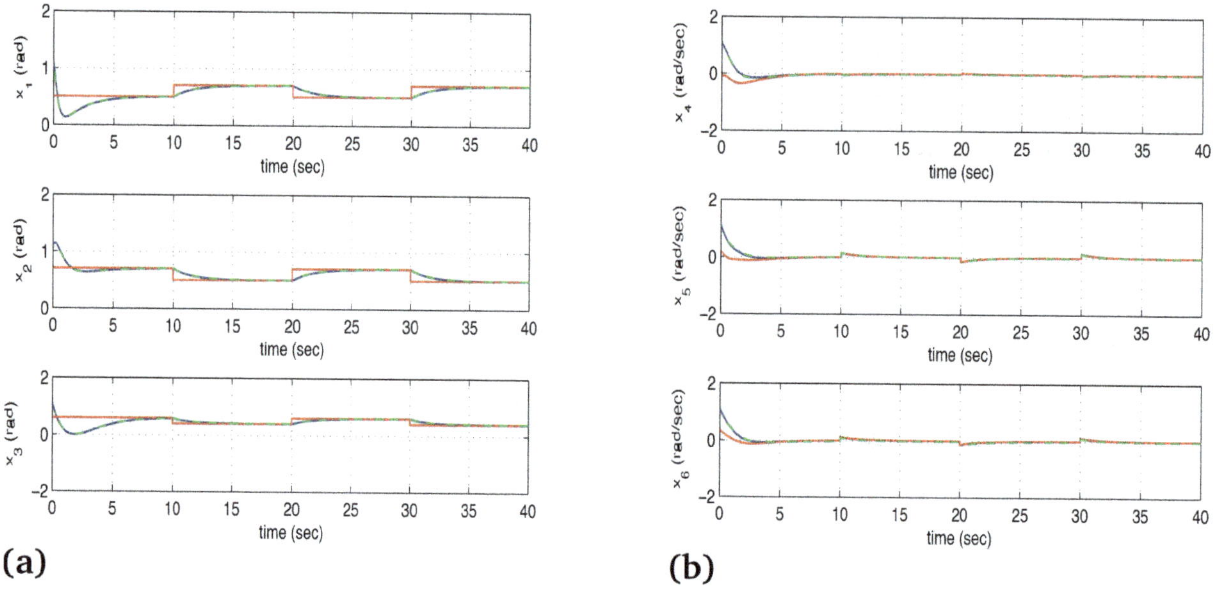

Figure 5.68 *Tracking of attitude setpoint 6 by the reentry space vehicle with the use of flatness-based control in successive loops: (a) convergence of state variables x_1 to x_3 (blue lines) to the associated setpoints (red lines) and estimated values provided by Kalman filtering; (b) convergence of state variables x_4 to x_6 (blue lines) to the associated setpoints (red lines) and estimated values provided by Kalman filtering*

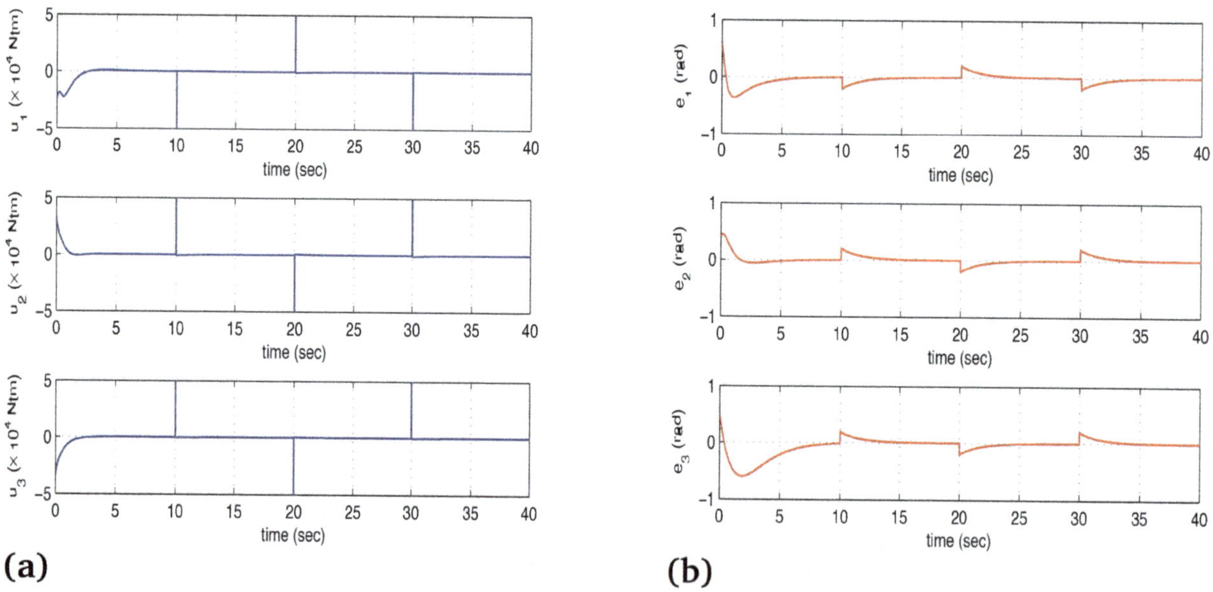

Figure 5.69 *Tracking of attitude setpoint 6 by the reentry space vehicle with the use of flatness-based control in successive loops: (a) variations of the control inputs u_1 and u_3 of the spacecraft (blue lines); (b) variation of the tracking error variables e_1 to e_3 associated with the state variables x_1 to x_3*

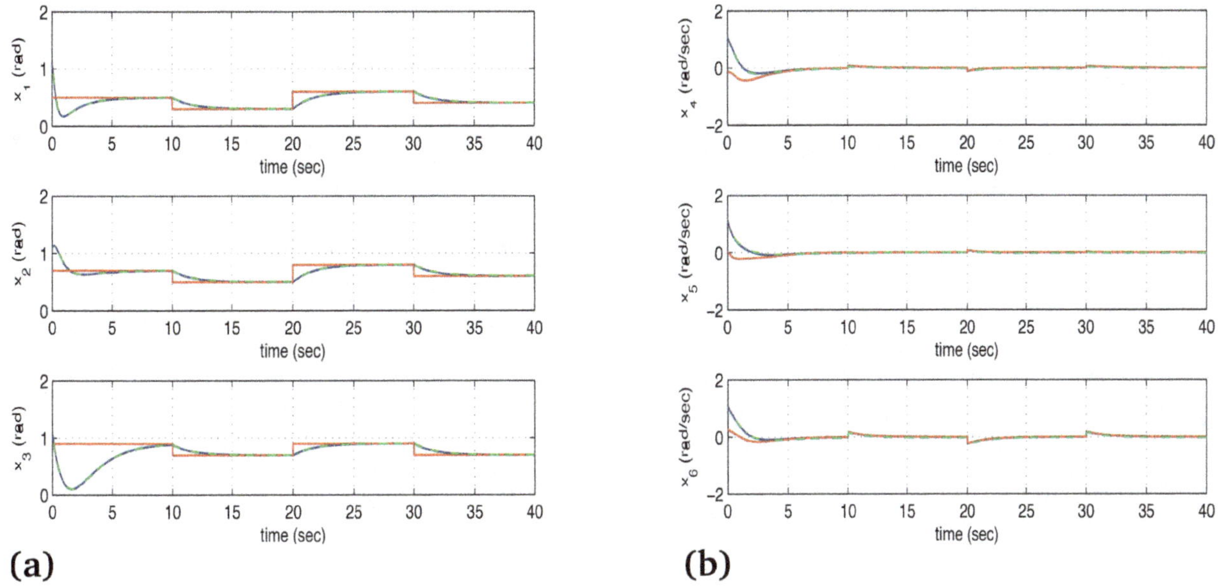

Figure 5.70 Tracking of attitude setpoint 7 by the reentry space vehicle with the use of flatness-based control in successive loops: (a) convergence of state variables x_1 to x_3 (blue lines) to the associated setpoints (red lines) and estimated values provided by Kalman filtering; (b) convergence of state variables x_4 to x_6 (blue lines) to the associated setpoints (red lines) and estimated values provided by Kalman filtering

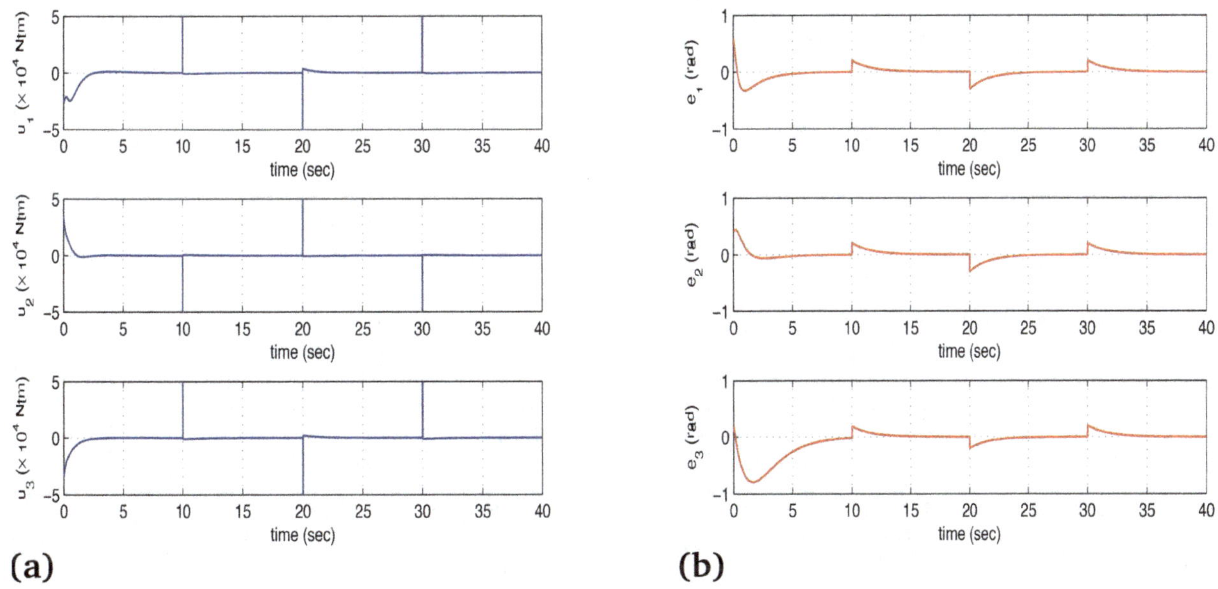

Figure 5.71 Tracking of attitude setpoint 7 by the reentry space vehicle with the use of flatness-based control in successive loops: (a) variations of the control inputs u_1 and u_3 of the spacecraft (blue lines); (b) variation of the tracking error variables e_1 to e_3 associated with the state variables x_1 to x_3

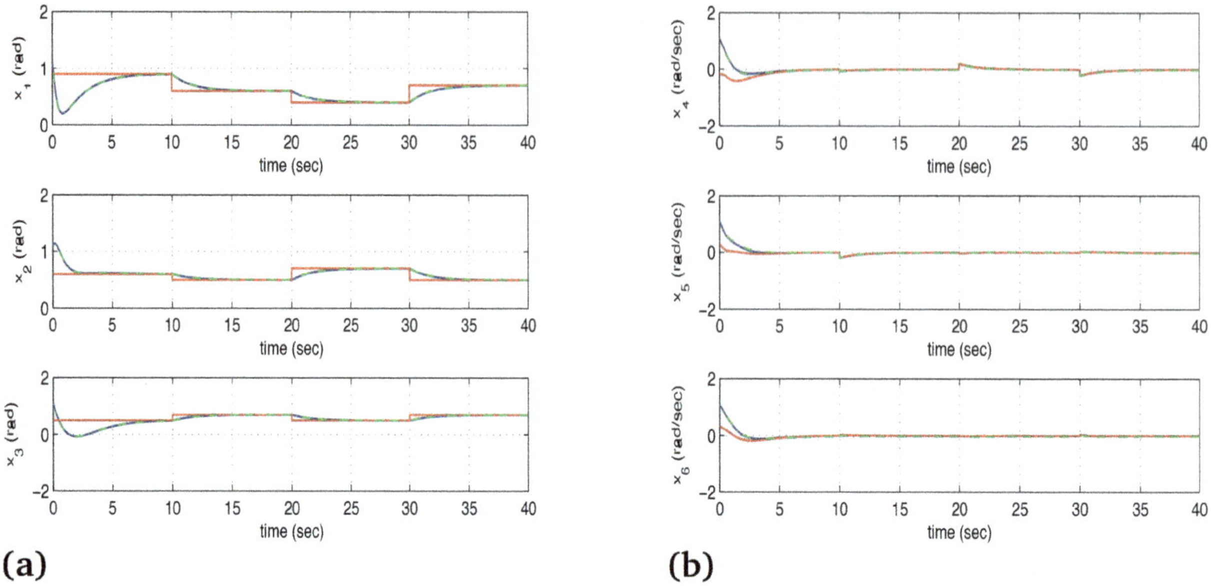

Figure 5.72 Tracking of attitude setpoint 8 by the reentry space vehicle with the use of flatness-based control in successive loops: (a) convergence of state variables x_1 to x_3 (blue lines) to the associated setpoints (red lines) and estimated values provided by Kalman filtering; (b) convergence of state variables x_4 to x_6 (blue lines) to the associated setpoints (red lines) and estimated values provided by Kalman filtering

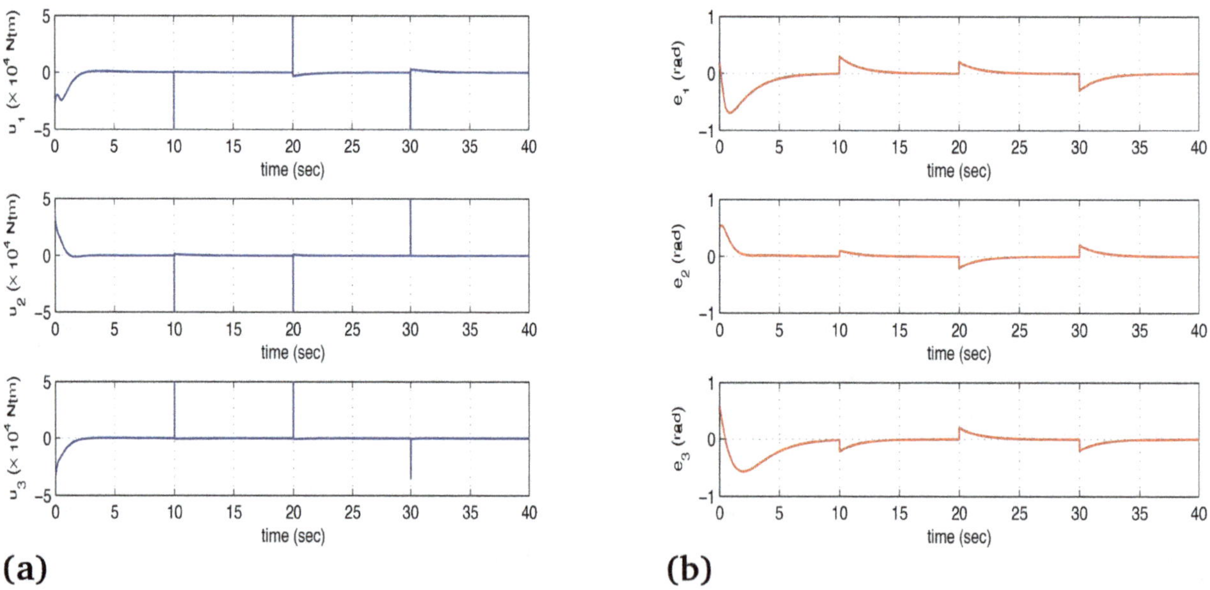

Figure 5.73 Tracking of attitude setpoint 8 by the reentry space vehicle with the use of flatness-based control in successive loops: (a) variations of the control inputs u_1 and u_3 of the spacecraft (blue lines); (b) variation of the tracking error variables e_1 to e_3 associated with the state variables x_1 to x_3

Table 5.8 Flatness-based control in successive loops

Tracking RMSE $\times 10^{-3}$ in the disturbance-free case						
	RMSE_{x_1}	RMSE_{x_2}	RMSE_{x_3}	RMSE_{x_4}	RMSE_{x_5}	RMSE_{x_6}
test_1	0.0612	0.1254	0.9158	0.0001	0.0001	0.0001
test_2	0.0367	0.6178	1.4981	0.0001	0.0001	0.0001
test_3	0.2557	0.1288	0.1807	0.0002	0.0002	0.0002
test_4	0.6988	0.6485	1.0112	0.0003	0.0003	0.0003
test_5	0.6990	0.6484	0.6385	0.0002	0.0002	0.0002
test_6	0.7947	0.7946	0.7943	0.0006	0.0006	0.0006
test_7	0.7821	0.7823	0.7942	0.0007	0.0007	0.0007
test_8	0.8883	0.5892	0.5891	0.0011	0.0011	0.0011

Table 5.9 Flatness-based control in successive loops

Tracking RMSE $\times 10^{-3}$ in the case of disturbances						
$\Delta a\%$	RMSE_{x_1}	RMSE_{x_2}	RMSE_{x_3}	RMSE_{x_4}	RMSE_{x_5}	RMSE_{x_6}
0%	0.0612	0.1254	0.3158	0.0001	0.0001	0.0001
10%	0.0786	0.1319	0.9361	0.0001	0.0001	0.0001
20%	0.0961	0.1393	0.9530	0.0001	0.0001	0.0001
30%	0.1136	0.1477	0.9816	0.0002	0.0002	0.0002
40%	0.1311	0.1572	1.0070	0.0002	0.0002	0.0002
50%	0.1482	0.1679	1.0346	0.0002	0.0002	0.0002
60%	0.1652	0.1798	1.0645	0.0002	0.0002	0.0002

Table 5.10 Flatness-based control in successive loops

Convergence times T_s (sec) of the state variables						
	$T_s\, x_1$	$T_s\, x_2$	$T_s\, x_3$	$T_s\, x_4$	$T_s\, x_5$	$T_s\, x_6$
test_1	5.0	2.5	7.5	7.5	6.5	6.5
test_2	6.5	3.5	7.0	8.0	5.0	6.5
test_3	4.0	2.5	7.5	7.5	7.0	6.0
test_4	4.0	2.0	9.0	7.5	7.5	7.5
test_5	4.0	2.0	9.0	7.5	7.5	7.5
test_6	7.0	4.5	8.0	4.5	4.5	4.5
test_7	5.0	5.0	8.0	4.5	4.5	4.5
test_8	7.0	2.5	8.0	4.5	4.5	4.0

To elaborate on multi-loop flatness-based control for the attitude of the reentry space vehicle the following tables are given (i) Table 5.8 providing results about the accuracy of tracking of setpoints by the state variables of the attitude dynamics of the reentry space vehicle under an exact dynamic model, (ii) Table 5.9 providing results about the accuracy of tracking of setpoints by the state variables of the attitude dynamics of the reentry space vehicle under a model that is subject to disturbances (for instance change $\Delta\alpha\%$ in parameter N_1 of the dynamic model), and (iii) Table 5.10 providing results about the convergence times to setpoints of the state variables of the attitude dynamics of the reentry space vehicles.

5.3 Flatness-based control in successive loops for micro-satellites

5.3.1 Outline

Differential flatness theory is a main research direction in the area of nonlinear dynamical systems [711,721,732,491, 252,898]. A system is considered to be differentially flat if all its state variables and its control inputs can be expressed as functions of one single algebraic variable which is the so-called flat output, and also as functions of the flat-output's derivatives [91,594,639,489]. The differential flatness property enables the transformation of the nonlinear system's dynamics in the linear canonical form [806,492,640,524,54]. The latter description is controllable and observable thus allowing to treat effectively control and estimation problems [439,705,60,706]. In this section, a successive loops approach is developed for controller design in nonlinear dynamical systems which exhibit the differential flatness property. The method makes use of the initial nonlinear model of the system and of its decomposition into a set of nonlinear subsystems for which the differential flatness property holds [712,720,714,716,729,731,718].

This section applies a novel solution to the problem of nonlinear control of complex dynamical systems without the need for changes of state variables (diffeomorphisms) and complicated state-space model transformations. The new solution is a flatness-based control approach implemented in successive loops. In this method, the dynamic model of the nonlinear system is separated into subsystems which are connected in a cascading manner. This control method is directly applicable to dynamical systems of the triangular form and to nonlinear systems which can be transformed into such a form. The state-space model of the initial nonlinear system is decomposed into cascading subsystems which satisfy differential flatness properties. For each subsystem of the state-space model a virtual control input is computed, capable of inverting the subsystem's dynamics and of eliminating the subsystem's tracking error. The control input which is actually applied to the initial nonlinear system is computed from the last row of the state-space description. This control input incorporates in a recursive manner all virtual control inputs which were computed from the individual subsystems included in the initial state-space equation. The control input that should be applied to the nonlinear system so as to assure that all state vector elements will converge to the desirable setpoints is obtained at each iteration of the control algorithm by tracing backward the subsystems of the state-space model.

The application domain for the section's developments of flatness-based control implemented in successive loops is the attitude control problem of micro-satellites. Attitude control for satellites is important for the precise orientation of theirs and consequently for assuring the reliable transmission of communication data or the accomplishment of image acquisition, navigation, geodesic and surveillance tasks [787,408,730,478]. Attitude control is applied to satellites to compensate for various space-environmental disturbances, as well as for inner moment-of-inertia variations [350,662,357,323]. The Attitude Determination and Control Subsystem (ADCS) consists of various sensors and actuators which provide information about attitude and finally stabilize the satellite at the desirable orientation [744,322,11]. Control of the satellite's attitude can be accomplished by using as control inputs electromagnetic actuation, solar-powered actuation, reaction wheels as well as thrusters. The present section is particularly focusing on micro-satellites. These have usually mass that is less than 100 kg and thus their manufacturing is less costly, takes shorter time and has less failure risks. Micro-satellites have a significant role in navigation, earth observation, atmosphere data collection and tele-communications [944,803,980]. Unlike satellites of larger dimensions and weight that comprise flexible parts, micro-satellites can be described by the dynamics of a rigid body. Three-axis attitude control of micro-satellites requires the solution of a multi-variable and highly nonlinear control problem [956,222,1005]. Elimination of the satellite's orientation error is the primary objective of an attitude control system. Besides, another important feature is robustness against the disturbances that may affect the satellite's control loop [284,671,118]. There have been several attempts to solve the satellite's attitude stabilization problem with the use of linear and nonlinear control techniques [800,975,517]. Although significant developments have been witnessed in nonlinear controllers design for the attitude control problem, this research area is still open. The developments of the present section in the satellite and aerospace systems control problem have been preceded by the results given in [730,722].

5.3.2 Dynamic model of the micro-satellite

The diagram of the micro-satellite is shown in Figure 5.74. The dynamic model of the attitude of a micro-satellite is considered. The associated state vector is

$$x=[x_1,x_2,x_3,x_4,x_5,x_6]^T \text{ or } \\ x=[\theta_1,\theta_2,\theta_3,\omega_1,\omega_2,\omega_3]^T \tag{5.131}$$

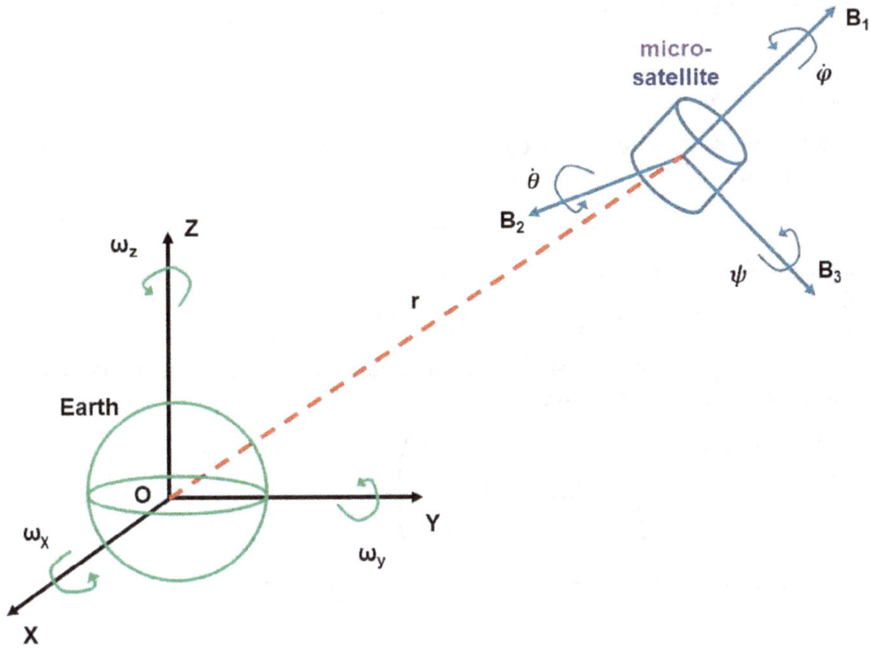

Figure 5.74 Diagram of the micro-satellite

where $\theta_1=\phi,\theta_2=\theta,\theta_3=\psi$ to be the roll, pitch and yaw angles in the body-fixed reference frame, and $\omega_1=\omega_x,\omega_2=\omega_y$, and $\omega_3=\omega_z$ to be the angular velocities of the satellite about the axes of the inertial reference frame. The control inputs vector is $u=[u_1,u_2,u_3]^T=[T_1,T_2,T_3]^T$ are torques which are provided by the satellite's actuation system. The satellite's dynamic model is written in the nonlinear affine-in-the-input state-space form

$$\dot{x}=f(x)+g(x)u+T_d \tag{5.132}$$

with T_d to be the disturbances' vector. About the drift vector $f(x)$ and about the control inputs gain matrix $g(x)$, it holds that

$$f(x)=\begin{pmatrix}\frac{cos(x_3)x_5-sin(x_3)x_6}{cos(x_2)}-\omega_0\\ sin(x_3)x_5+cos(x_3)x_6\\ x_4+\frac{sin(x_2)}{cos(x_2)}[sin(x_3)x_6-cos(x_3)x_5]\\ \frac{I_2-I_3}{I_1}[x_5x_6-3\omega_0^2\zeta_2\zeta_3]\\ \frac{I_3-I_1}{I_2}[x_6x_4-3\omega_0^2\zeta_3\zeta_1]\\ \frac{I_1-I_2}{I_3}[x_4x_5-3\omega_0^2\zeta_1\zeta_2]\end{pmatrix}\quad g(x)=\begin{pmatrix}0&0&0\\0&0&0\\0&0&0\\\frac{1}{I_1}&0&0\\0&\frac{1}{I_2}&0\\0&0&\frac{1}{I_3}\end{pmatrix} \tag{5.133}$$

where vector ζ is given by

$$\zeta=\begin{pmatrix}-sin(x_1)cos(x_2)\\ cos(x_1)sin(x_3)-sin(x_1)sin(x_2)cos(x_3)\\ cos(x_1)cos(x_3)-sin(x_1)sin(x_2)sin(x_3)\end{pmatrix} \tag{5.134}$$

The state-variables of the micro-satellite is now separated into two subvectors:

$$x_a = [x_1, x_2, x_3]^T \Rightarrow x_a = [\theta_1, \theta_2, \theta_3]^T$$
$$x_b = [x_4, x_5, x_6]^T \Rightarrow x_a = [\omega_1, \omega_2, \omega_3]^T \tag{5.135}$$

One also defines the following drift subvector and control inputs gain matrix for subsystem of the micro-satellite which us associated with the state subvector x_a

$$f_a(x) = \begin{pmatrix} -\omega_0 \\ 0 \\ 0 \end{pmatrix} \quad g_a(x) = \begin{pmatrix} 0 & \frac{cos(x_3)}{sin(x_2)} & -\frac{sin(x_3)}{sin(x_2)} \\ 0 & cos(x_3) & sin(x_3) \\ 1 & \frac{sin(x_2)sin(x_3)}{cos(x_2)} & -\frac{sin(x_2)cos(x_3)}{cos(x_2)} \end{pmatrix} \tag{5.136}$$

Additionally, one defines the following drift subvector and control inputs gain matrix for subsystem of the micro-satellite which us associated with the state subvector x_b:

$$f_b(x) = \begin{pmatrix} \frac{I_2-I_3}{I_1}[x_5x_6 - 3\omega_0^2\zeta_2\zeta_3] \\ \frac{I_3-I_1}{I_2}[x_6x_4 - 3\omega_0^2\zeta_3\zeta_1] \\ \frac{I_1-I_2}{I_3}[x_4x_4 - 3\omega_0^2\zeta_1\zeta_2] \end{pmatrix} \quad g_b(x) = \begin{pmatrix} \frac{1}{I_1} & 0 & 0 \\ 0 & \frac{1}{I_2} & 0 \\ 0 & 0 & \frac{1}{I_3} \end{pmatrix} \tag{5.137}$$

Thus, the dynamics of the micro-satellite is decomposed into two-subsystems:

$$\dot{x}_a = f_a(x) + g_a(x)x_b \tag{5.138}$$

$$\dot{x}_b = f_b(x) + g_b(x)u \tag{5.139}$$

It can be proven that the dynamic model of the micro-satellite as described in Eq. (5.138) and Eq. (5.139) is a differentially flat system with flat output $y = x_a$. Indeed, from Eq. (5.138), it holds that the drift vector $f_a(x)$ has constant elements, while the control inputs gain matrix $g_a(x)$ contains elements that depend on the variables which constitute the flat outputs vector x_a. Thus, by solving Eq. (5.138) with respect to vector x_b, one obtains

$$x_b = g_a^{-1}(x)[\dot{x}_a - f_a(x)] \tag{5.140}$$

which signifies that x_b is a differential function of the flat output of the micro-satellite $y = x_a$. Besides, in Eq. (5.139) vector $f_b(x)$ is a function of the flat outputs vector y and matrix $g_b(x)$ is constant. Thus, by solving Eq. (5.139) with respect to the control inputs vector, one obtains

$$u = g_b^{-1}(x)[\dot{x}_b - f_b(x)] \tag{5.141}$$

which signifies that the control inputs vector u is also a differential function of the flat outputs of the system. The above confirm that the micro-satellite's vector is differentially flat. Next, it will be proven that the subsystems of Eq. (5.136) and Eq. (5.139) are independently flat subsystems and that the satellite's dynamics can be controlled in successive flatness-based control loops.

5.3.3 Flatness-based control in successive loops for the micro-satellite's dynamics

First, it will be shown that each one of the subsystems of Eq. (5.138) and Eq. (5.139), if viewed independently, is a differentially flat subsystem. Indeed, Eq. (5.138) is a differentially flat subsystem with flat output x_a. From Eq. (5.138), one solves for x_b which is now viewed as a virtual control input. It holds that

$$x_b = g_a^{-1}(x)[\dot{x}_a - f_a(x)] \tag{5.142}$$

Since the drift vector $f_a(x)$ has constant elements and the control inputs gain matrix $g_a(x)$ has elements that depend on the variables of the flat outputs vector x_a, it holds that x_b is a differential function of the flat output x_a. Therefore, the subsystem of Eq. (5.138) is differentially flat.

Equivalently, in the subsystem of Eq. (5.138) the flat outputs vector is taken to be x_b, while x_a is viewed as a coefficients vector. Thus, $f_b(x)$ is a function of the flat output x_b and $g_b(x)$ has constant elements. Consequently, solving Eq. (5.139) for the control inputs vector u gives

$$u = g_b^{-1}(x)[\dot{x}_b - f_b(x)] \tag{5.143}$$

which signifies that u is also a differential function of the flat outputs vector x_b. Consequently, the subsystem of Eq. (5.139) is also differentially flat.

The control of the subsystems of Eq. (5.138) and Eq. (5.139) can be carried out following the design process of controllers for input–output linearized differentially flat system. For the subsystem of Eq. (5.138) the setpoint is denoted as x_a^d and the value of the virtual control input x_b which stabilizes this subsystem is given by

$$x_b^* = g_a^{-1}(x)[\dot{x}_a^d - f_a(x) - K_1(x_a - x_a^d)] \tag{5.144}$$

Matrix $K_1 \in R^{3\times3}$ is a diagonal matrix with its diagonal elements $k_{ii} > 0$ for $i = 1,2,3$. For the subsystem of Eq. (5.139) the control inputs vector u is selected so as to ensure that x_b will converge to the targeted vector x_b^*, which in turn makes the subsystem of Eq. (5.139) converge to the associated setpoint. Actually, the setpoint of the second subsystem is $x_b^d = x_b^*$ and the control input u is given by

$$u = g_b^{-1}(x)[\dot{x}_b^d - f_b(x) - K_2(x_b - x_b^d)] \tag{5.145}$$

Matrix $K_2 \in R^{3\times3}$ is a diagonal matrix, with its diagonal elements to satisfy $k_{ii} > 0$ for $i = 1,2,3$. By substituting Eq. (5.144) into Eq. (5.138), as well as by substituting Eq. (5.145) into Eq. (5.139), one obtains the dynamics of the closed-loop system

$$\begin{aligned}(\dot{x}_a - \dot{x}_a^d) + K_1(x_a - x_a^d) = 0\\ (\dot{x}_b - \dot{x}_b^d) + K_2(x_b - x_b^d) = 0\end{aligned} \tag{5.146}$$

Next, by defining the tracking error variables $e_a = x_a - x_a^d$ and $e_b = x_b - x_b^d$ the tracking error dynamics of the micro-satellite becomes

$$\begin{aligned}\dot{e}_a + K_1 e_a = 0 \Rightarrow lim_{t\to\infty} e_a(t) = 0 \Rightarrow lim_{t\to\infty} x_a(t) = x_a^d(t)\\ \dot{e}_b + K_2 e_b = 0 \Rightarrow lim_{t\to\infty} e_b(t) = 0 \Rightarrow lim_{t\to\infty} x_b(t) = x_b^d(t)\end{aligned} \tag{5.147}$$

Consequently, all state variables of the micro-satellite's attitude dynamics converge to the associated setpoints or

$$lim_{t\to\infty} x_i(t) = x_i^d(t)\ \forall\, i = 1,2,\cdots,6 \tag{5.148}$$

The global stability properties of the micro-satellite's control loop can also be proven through Lyapunov stability analysis. The following Lyapunov function is defined

$$V = \tfrac{1}{2}[e_a^T e_a + e_b^T e_b] \tag{5.149}$$

By differentiating V in time and with the use of Eq. (5.147), one obtains

$$\begin{aligned}\dot{V} = e_a^T \dot{e}_a + e_b^T \dot{e}_b \Rightarrow \dot{V} = e_a^T(-K_1 e_a) + e_b^T(-K_2 e_b) \Rightarrow\\ \dot{V} = -e_a^T K_1 e_a - e_b^T K_2 e_b \Rightarrow \dot{V} < 0\end{aligned} \tag{5.150}$$

It holds that $\dot{V}$ is strictly negative $\forall e_a \neq 0$ and $\forall e_b \neq 0$. It becomes 0 only when $e_a = 0$ and $e_b = 0$. Therefore, the Lyapunov function of the system is strictly diminishing and finally converges to 0, no matter what the initial conditions of the micro-satellite's tracking error are. Thus, the control loop of the micro-satellite is globally asymptotically stable.

Once again, an explicit demonstration of the exponential stabilization that is achieved by flatness-based control in successive loops is given next. The Lyapunov function of the control loop is written as:

$$V = \frac{1}{2}[\sum_{i=1}^{3} e_i^2 + \sum_{j=4}^{6} e_j^2] \tag{5.151}$$

where e_i $i = 1, \cdots, 3$ are the tracking errors for the state variables of the satellite associated with rotation angles around the axes of the body-fixed frame and e_j $j = 4, \cdots, 6$ are the tracking errors for the state variables of the satellite associated with angular velocities around the axes of the earth-fixed frame. Equivalently, the first-order time-derivative of the Lyapunov function is written as

$$\dot{V} = -[\sum_{i=1}^{3} k_{1,i} e_i^2 + \sum_{j=4}^{6} k_{2,j-3} e_j^2] \tag{5.152}$$

where $k_{1,i} > 0$ $i = 1, \cdots, 3$ are the diagonal elements of gain matrix K_1 and $k_{2,j-3} > 0$ $j = 4, \cdots, 6$ are the diagonal elements of gain matrix K_2. By denoting the minimum of the above-noted elements of the feedback gain matrices as k_{min}, that is

$$k_{min} = min\{k_{1,i} : i = 1, \cdots, 3 \text{ and } k_{2,j-3} : j = 4, \cdots, 6\} \tag{5.153}$$

and using Eq. (5.152), one obtains that

$$\begin{aligned} &\dot{V} \leq -k_{min}[\sum_{i=1}^{3} e_i^2 + \sum_{j=4}^{6} e_j^2] \\ &\Rightarrow \dot{V} \leq -2k_{min} V \Rightarrow \dot{V} + 2k_{min} V \leq 0 \end{aligned} \tag{5.154}$$

From Eq. (5.154), one can demonstrate the exponential convergence of the Lyapunov function V to 0.

5.3.4 Simulation tests

Results about the tracking accuracy and the speed of convergence to setpoints of the successive-loops flatness-based control method, in the case of the 3-DOF micro-satellite's attitude dynamics, are shown in Figures 5.75–5.90. Indicative values about the parameters of the dynamic model of the 3-DOF microsatellite have been as follows: $I_1 = 28.4kg \cdot m^2$, $I_2 = 18.4kg \cdot m^2, I_3 = 16.8kg \cdot m^2$ while the orbital rate was $\omega_0 = 0.0011$. Gain matrix K_1 is diagonal with the elements on its diagonal to be given the value $k_{1,ii} = 0.3$ for $i = 1, \cdots, 3$. Gain matrix K_2 is also diagonal with the elements on its diagonal to be given the value $k_{2,ii} = 0.8$ for $i = 1, \cdots, 3$. Regarding the setpoints which have been used in the 8 tests cases about the micro-satellite's attitude control the following can be noted:

Setpoints 1–8 describe piece-wise constant desirable values θ_1^d, θ_2^d, θ_3^d for the first three state variables of the micro-satellite's attitude dynamic model, that is for the orientation angles in the body-fixed frame. The setpoints for the angular velocities ω_1^d, ω_2^d, ω_3^d of the satellite in the inertial reference frame are computed from the differential relations which connect the angular speed state variables with the turn angle flat outputs of the system. As expected, the setpoints ω_1^d, ω_2^d, ω_3^d converge very fast to the zero value. This signifies that upon convergence of the satellite's state variables to the associated setpoints the satellite gets stabilized at the desirable orientation angles and stops rotating.

It can be noticed again, that under this control scheme, one achieves fast and precise tracking of reference setpoints for all state variables of the dynamic model of the attitude of the micro-satellite. It is noteworthy that, through the stages of this method, one solves also the setpoints definition problem for all state variables of the micro-satellite. Actually, the selection of setpoints for state variables x_1, x_2, and x_3 is unconstrained. On the other side by defining state variables x_4, x_5, and x_6 as virtual control inputs for the subsystem of x_1, x_2, and x_3, one can find the setpoints for x_4 to x_6 as functions of the setpoints for x_1 to x_3. The speed of convergence of the state variables of the micro-satellite's attitude

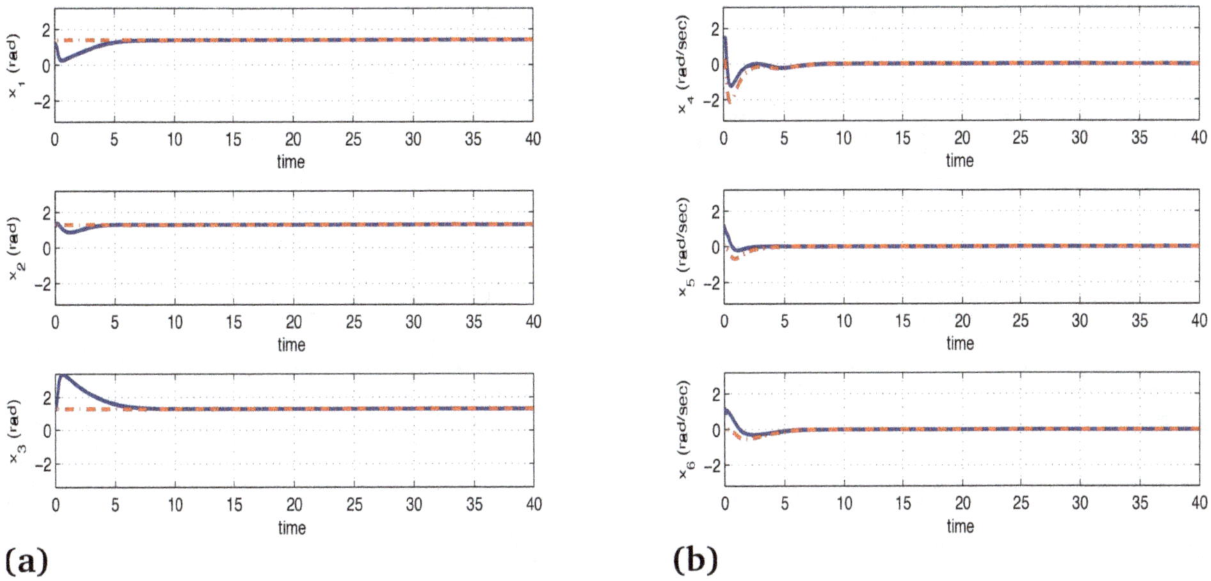

Figure 5.75 *Attitude control for the 3-DOF micro-satellite when tracking setpoint 1: (a) convergence of state variables x_1 to x_3 to their reference setpoints (red line: setpoint, blue line: real value); (b) convergence of state variables x_4 to x_6 to their reference setpoints (red line: setpoint, blue line: real value)*

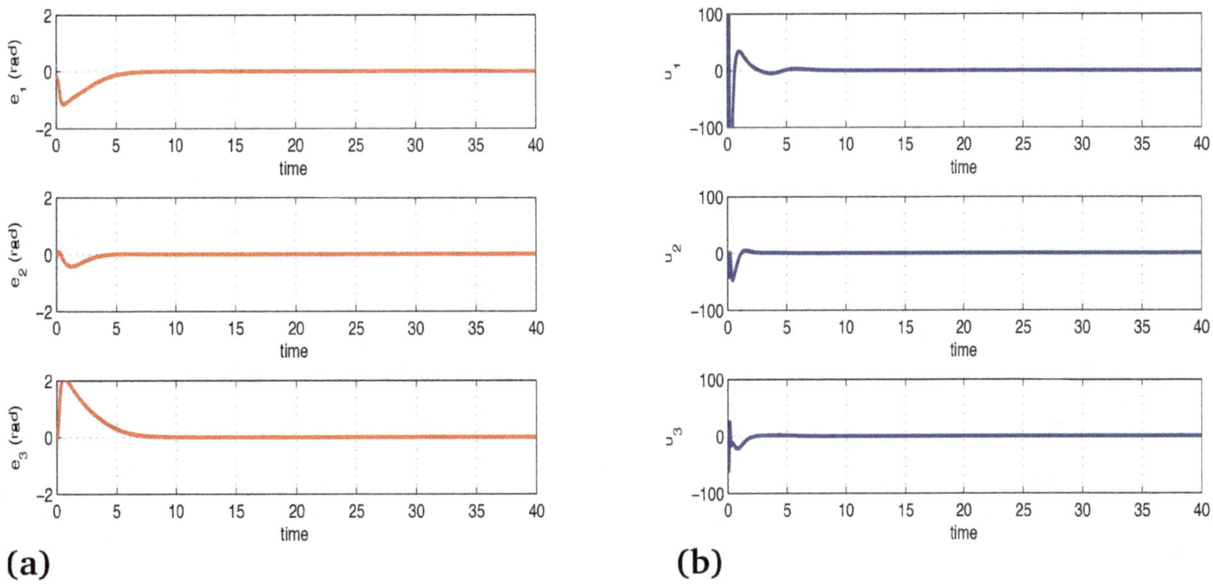

Figure 5.76 *Attitude control for the 3-DOF micro-satellite when tracking setpoint 1: (a) tracking error of state variables x_1 to x_3, that is the orientation angles in the body-fixed frame (red line); (b) variation of the control inputs u_1 to u_3 (blue line)*

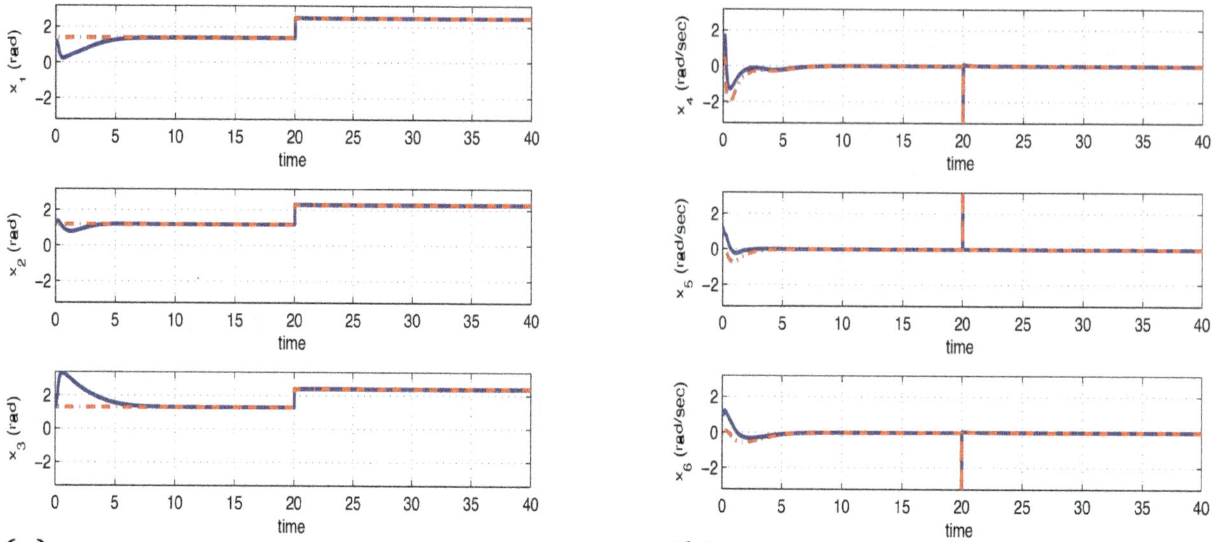

(a) (b)

Figure 5.77 Attitude control for the 3-DOF micro-satellite when tracking setpoint 2: (a) convergence of state variables x_1 to x_3 to their reference setpoints (red line: setpoint, blue line: real value); (b) convergence of state variables x_4 to x_6 to their reference setpoints (red line: setpoint, blue line: real value)

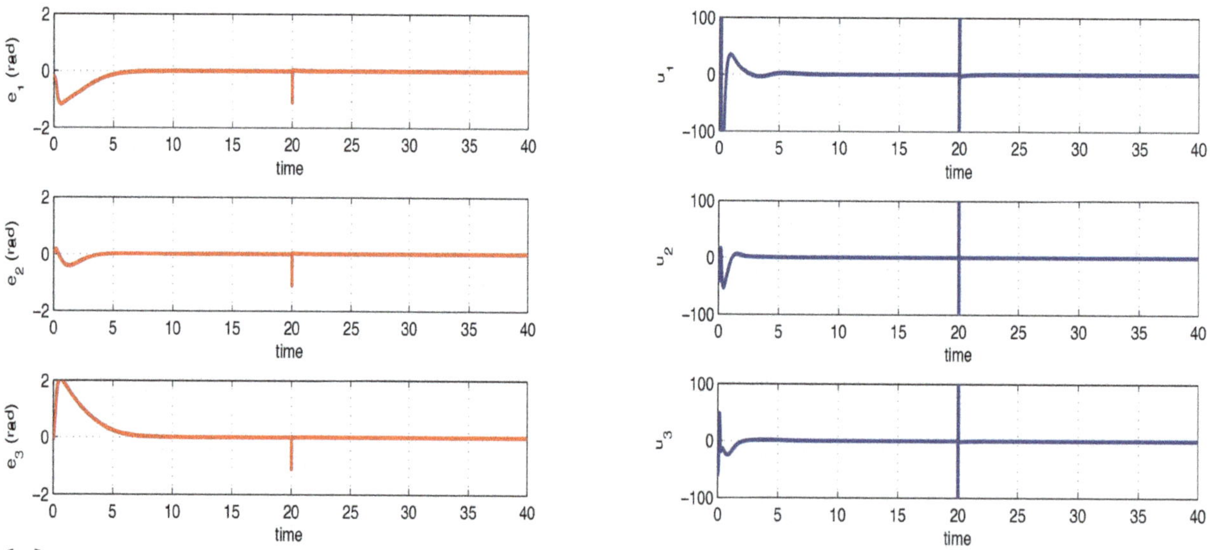

(a) (b)

Figure 5.78 Attitude control for the 3-DOF micro-satellite when tracking setpoint 2: (a) tracking error of state variables x_1 to x_3, that is the orientation angles in the body-fixed frame (red line); (b) variation of the control inputs u_1 to u_3 (blue line)

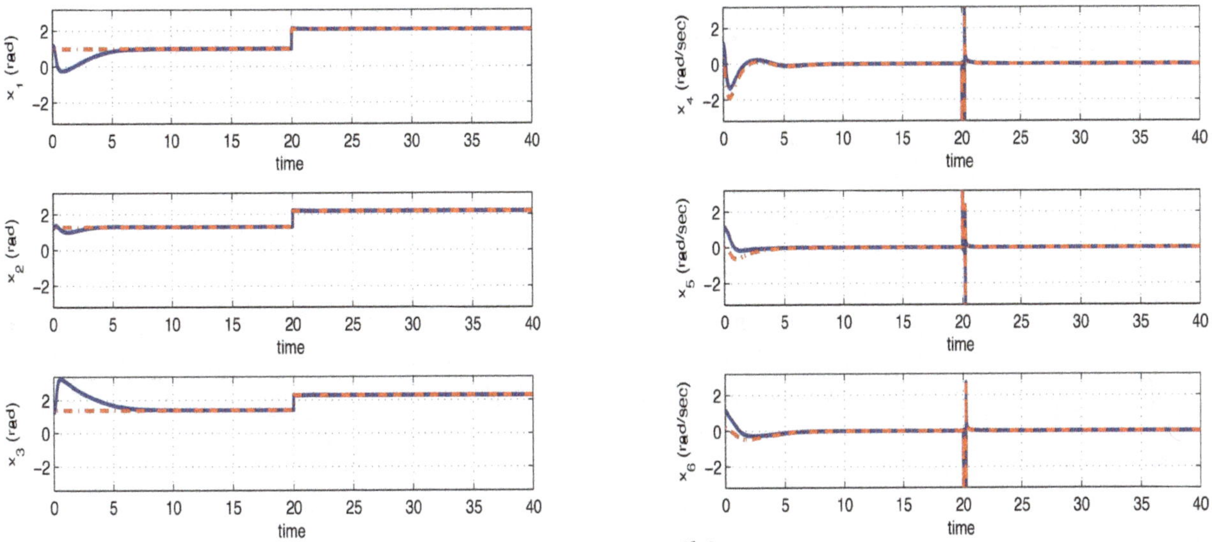

(a) (b)

Figure 5.79 Attitude control for the 3-DOF micro-satellite when tracking setpoint 3: (a) convergence of state variables x_1 to x_3 to their reference setpoints (red line: setpoint, blue line: real value); (b) convergence of state variables x_4 to x_6 to their reference setpoints (red line: setpoint, blue line: real value)

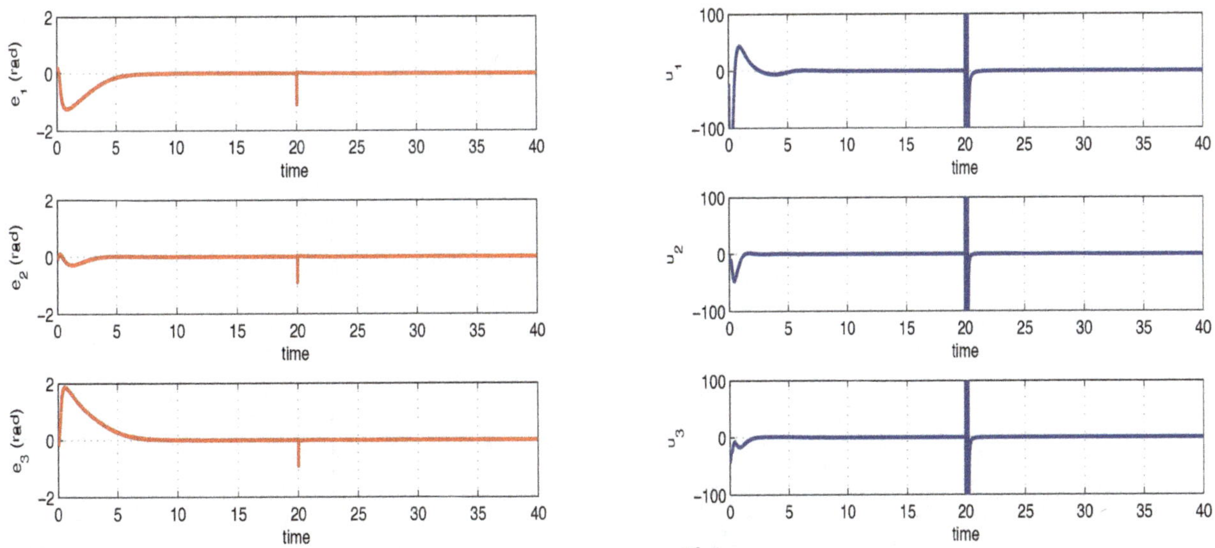

(a) (b)

Figure 5.80 Attitude control for the 3-DOF micro-satellite when tracking setpoint 3: (a) tracking error of state variables x_1 to x_3, that is the orientation angles in the body-fixed frame (red line); (b) variation of the control inputs u_1 to u_3 (blue line)

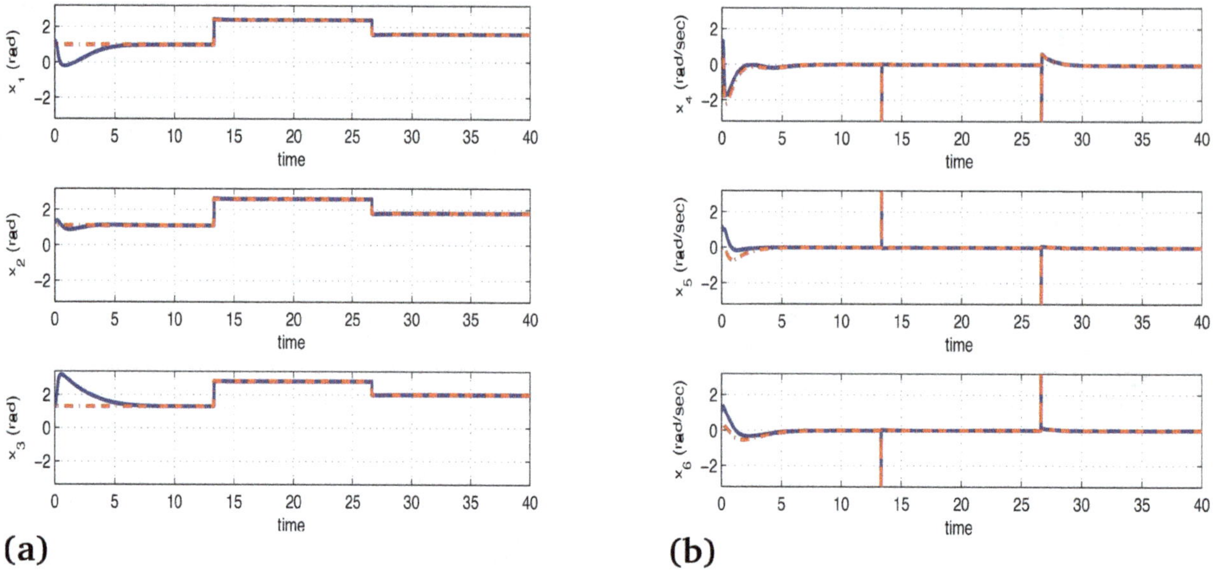

Figure 5.81 *Attitude control for the 3-DOF micro-satellite when tracking setpoint 4: (a) convergence of state variables* x_1 *to* x_3 *to their reference setpoints (red line: setpoint, blue line: real value); (b) convergence of state variables* x_4 *to* x_6 *to their reference setpoints (red line: setpoint, blue line: real value)*

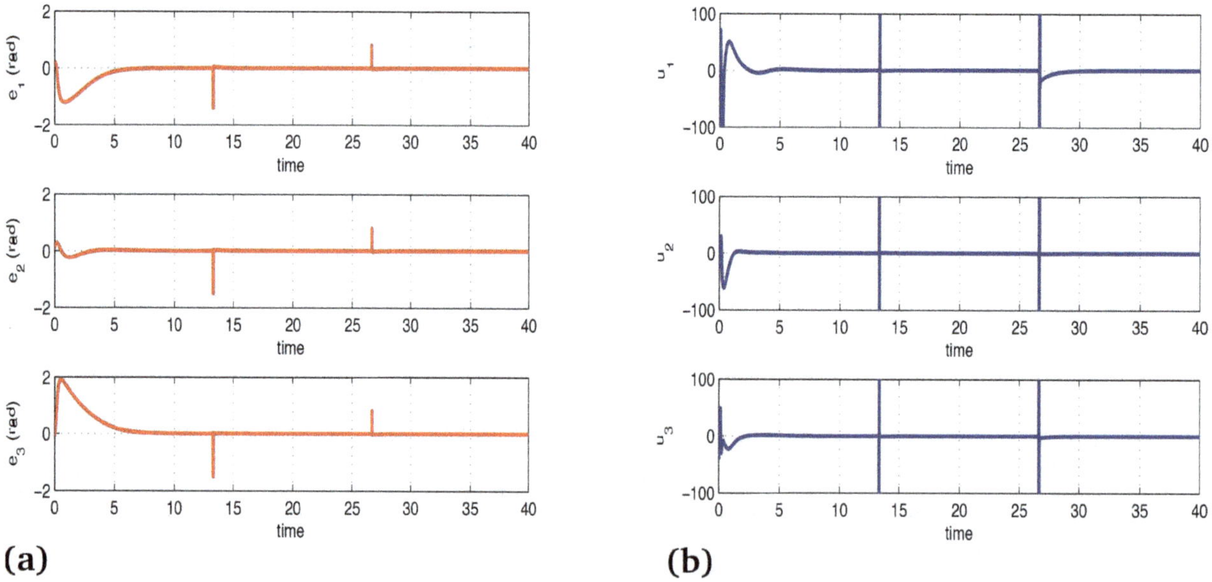

Figure 5.82 *Attitude control for the 3-DOF micro-satellite when tracking setpoint 4: (a) tracking error of state variables* x_1 *to* x_3*, that is the orientation angles in the body-fixed frame (red line); (b) variation of the control inputs* u_1 *to* u_3 *(blue line)*

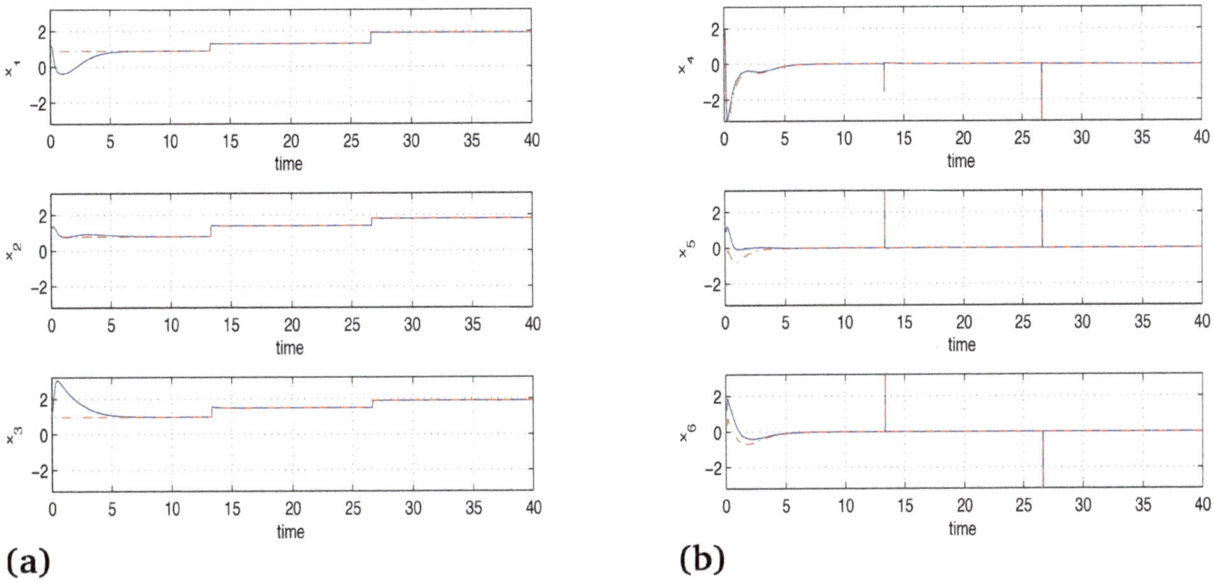

Figure 5.83 *Attitude control for the 3-DOF micro-satellite when tracking setpoint 5: (a) convergence of state variables x_1 to x_3 to their reference setpoints (red line: setpoint, blue line: real value); (b) convergence of state variables x_4 to x_6 to their reference setpoints (red line: setpoint, blue line: real value)*

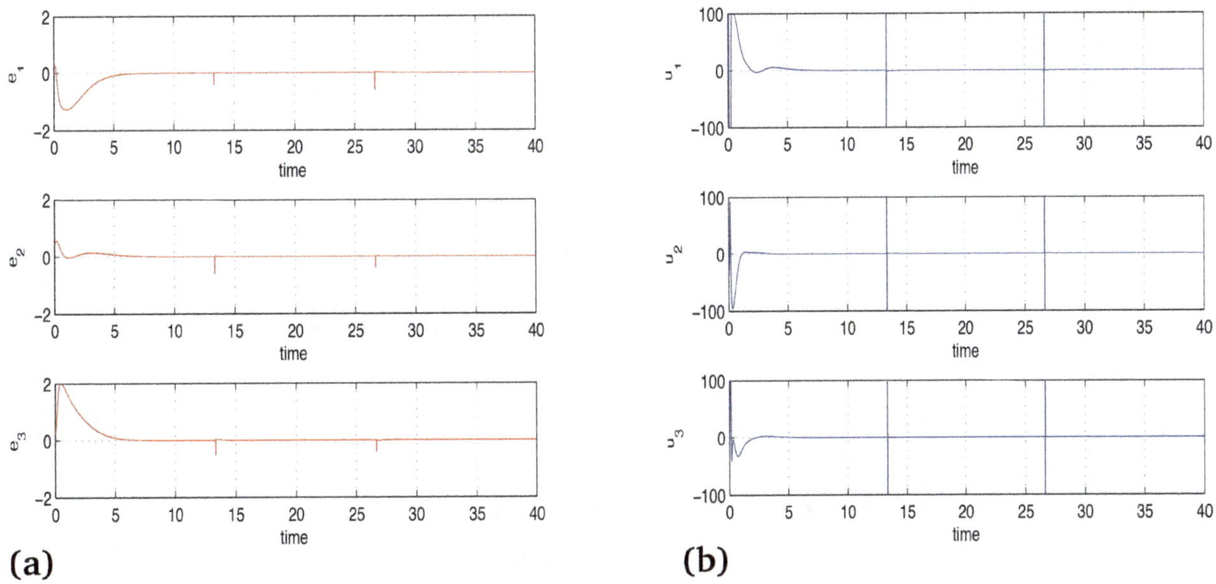

Figure 5.84 *Attitude control for the 3-DOF micro-satellite when tracking setpoint 5: (a) tracking error of state variables x_1 to x_3, that is the orientation angles in the body-fixed frame (red line); (b) variation of the control inputs u_1 to u_3 (blue line)*

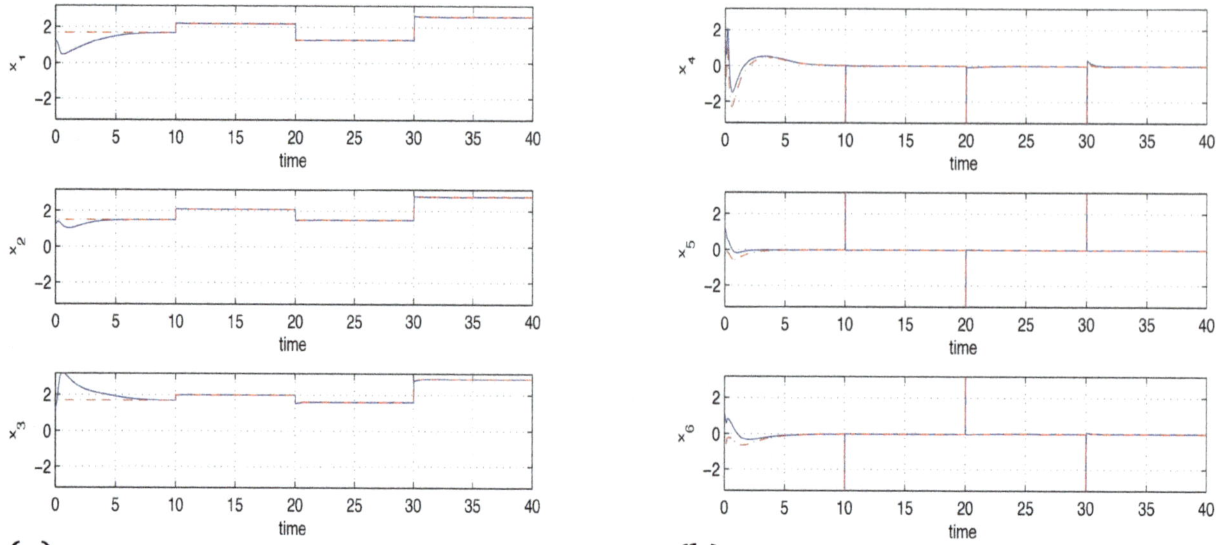

(a) (b)

Figure 5.85 *Attitude control for the 3-DOF micro-satellite when tracking setpoint 6: (a) convergence of state variables x_1 to x_3 to their reference setpoints (red line: setpoint, blue line: real value); (b) convergence of state variables x_4 to x_6 to their reference setpoints (red line: setpoint, blue line: real value)*

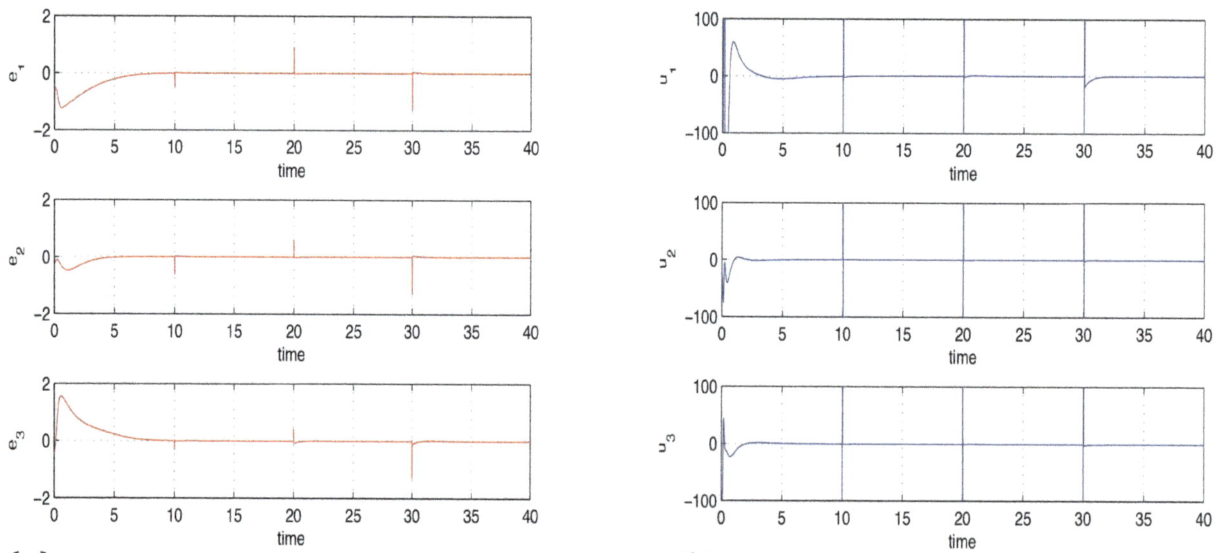

(a) (b)

Figure 5.86 *Attitude control for the 3-DOF micro-satellite when tracking setpoint 6: (a) tracking error of state variables x_1 to x_3, that is the orientation angles in the body-fixed frame (red line); (b) variation of the control inputs u_1 to u_3 (blue line)*

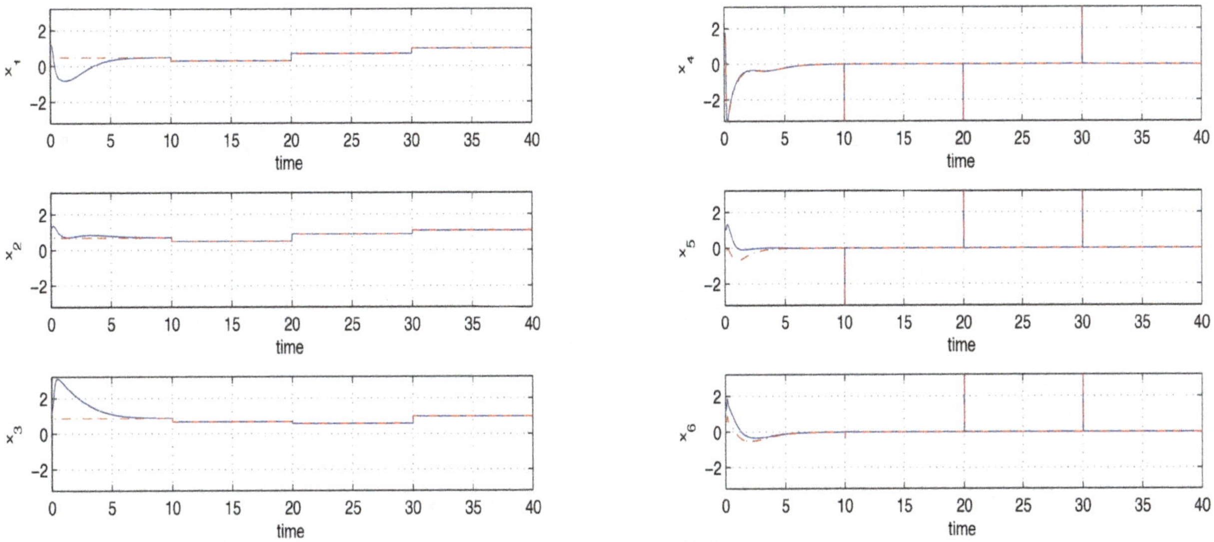

(a) (b)

Figure 5.87 *Attitude control for the 3-DOF micro-satellite when tracking setpoint 7: (a) convergence of state variables x_1 to x_3 to their reference setpoints (red line: setpoint, blue line: real value); (b) convergence of state variables x_4 to x_6 to their reference setpoints (red line: setpoint, blue line: real value)*

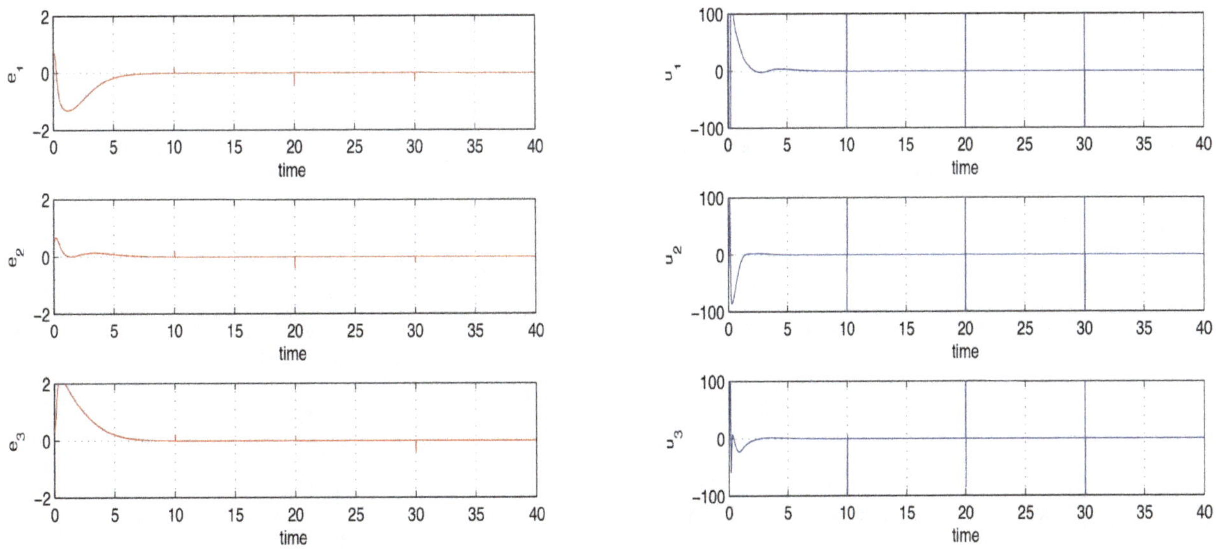

(a) (b)

Figure 5.88 *Attitude control for the 3-DOF micro-satellite when tracking setpoint 7: (a) tracking error of state variables x_1 to x_3, that is the orientation angles in the body-fixed frame (red line); (b) variation of the control inputs u_1 to u_3 (blue line)*

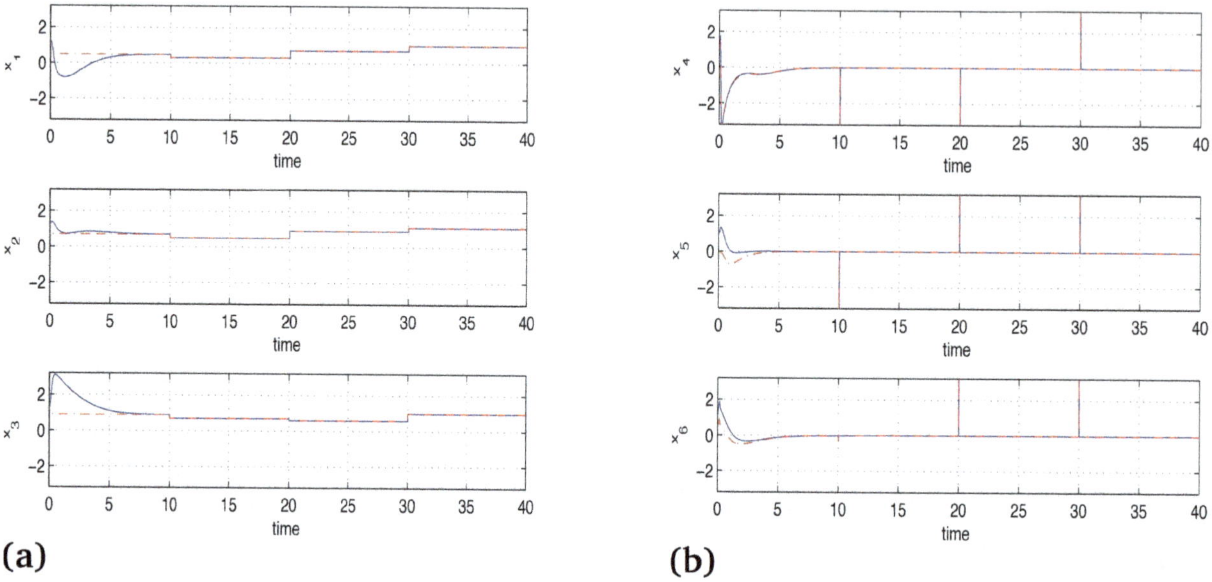

Figure 5.89 *Attitude control for the 3-DOF micro-satellite when tracking setpoint 8: (a) convergence of state variables x_1 to x_3 to their reference setpoints (red line: setpoint, blue line: real value); (b) convergence of state variables x_4 to x_6 to their reference setpoints (red line: setpoint, blue line: real value)*

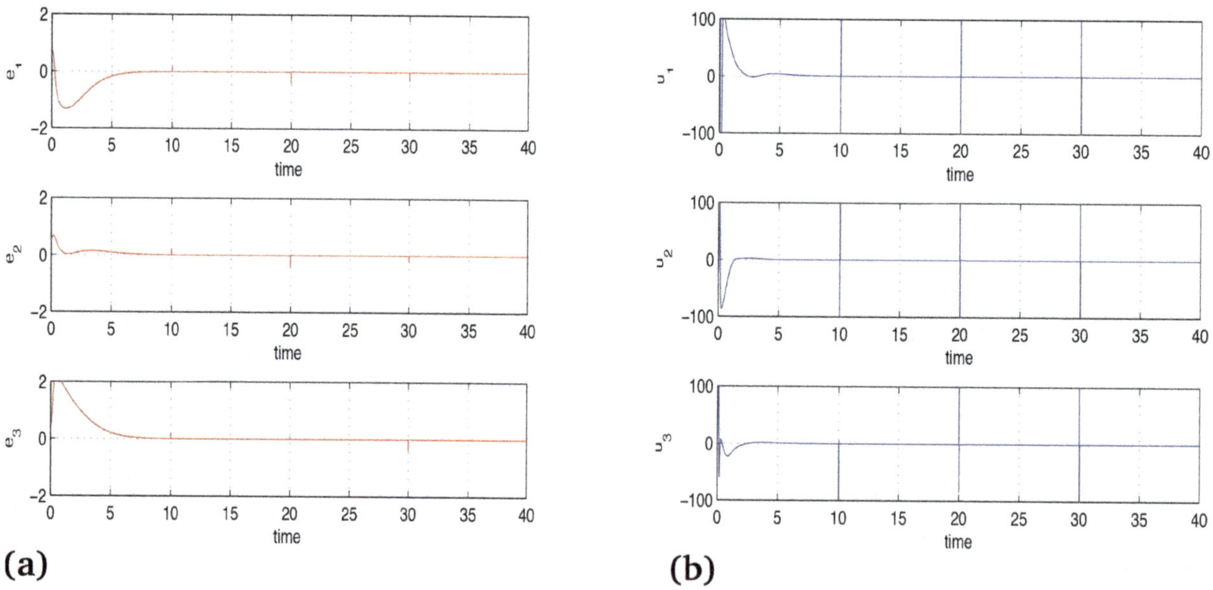

Figure 5.90 *Attitude control for the 3-DOF micro-satellite when tracking setpoint 8: (a) tracking error of state variables x_1 to x_3, that is the orientation angles in the body-fixed frame (red line); (b) variation of the control inputs u_1 to u_3 (blue line)*

dynamics under flatness-based control implemented in successive loops is determined by the selection of values for the diagonal gain matrices K_1, K_2.

To elaborate on flatness-based control in successive loops for the micro-satellite the following tables are given (i) Table 5.12 providing results about the accuracy of tracking of setpoints by the state variables of the micro-satellite under an exact dynamic model, (ii) Table 5.13 providing results about the accuracy of tracking of setpoints by the state variables of the micro-satellite under a model that is subject to disturbances (for instance change $\Delta\alpha\%$ in the orbital rate variable ω_0 of the satellite's dynamic model), and (iv) Table 5.14 providing results about the convergence times of the state variables of the micro-satellite's state variables to the associated setpoints.

Table 5.11 Parameters of the satellite's attitude dynamic model

Parameter	Definition
θ_1	turn angle for rotation around axis B_1 of the satellite-fixed frame
θ_2	turn angle for rotation around axis B_2 of the satellite-fixed frame
θ_3	turn angle for rotation around axis B_3 of the satellite-fixed frame
ω_x	angular speed for rotation around axis Ox of the earth-fixed frame
ω_y	angular speed for rotation around axis Oy of the earth-fixed frame
ω_z	angular speed for rotation around axis Oz of the earth-fixed frame
I_x, I_y, I_z	moments of inertia in rotation around the axes of earth-fixed frame
ω_0	orbital rate of the satellite

Table 5.12 Tracking RMSE $\times 10^{-4}$ for the 3-DOF micro-satellite in the disturbance-free case

	RMSE_{x_1}	RMSE_{x_2}	RMSE_{x_3}	RMSE_{x_4}	RMSE_{x_5}	RMSE_{x_6}
$test_1$	0.0210	0.0146	0.0404	0.0001	0.0001	0.0001
$test_2$	0.0298	0.0073	0.1485	0.0001	0.0001	0.0001
$test_3$	0.0316	0.0155	0.1602	0.0003	0.0001	0.0001
$test_4$	0.0418	0.0777	0.0056	0.0036	0.0003	0.0002
$test_5$	0.0837	0.0267	0.8017	0.0034	0.0005	0.0006
$test_6$	0.1178	0.0885	0.8400	0.2966	0.0237	0.0334
$test_7$	0.0242	0.0117	0.0585	0.0525	0.0486	0.0120
$test_8$	0.0018	0.0263	0.8148	0.1000	0.0097	0.0221

Table 5.13 Tracking RMSE $\times 10^{-4}$ for the 3-DOF micro-satellite in the case of disturbances

$\Delta a\%$	RMSE_{x_1}	RMSE_{x_2}	RMSE_{x_3}	RMSE_{x_4}	RMSE_{x_5}	RMSE_{x_6}
0%	0.0298	0.0073	0.1485	0.0001	0.0001	0.0001
10%	0.0299	0.0073	0.1486	0.0001	0.0001	0.0001
20%	0.0299	0.0073	0.1487	0.0001	0.0001	0.0001
30%	0.0300	0.0074	0.1489	0.0001	0.0001	0.0001
40%	0.0300	0.0074	0.1490	0.0001	0.0001	0.0001
50%	0.0301	0.0074	0.1492	0.0001	0.0002	0.0001
60%	0.0302	0.0074	0.1494	0.0001	0.0003	0.0002

Table 5.14 Convergence times (sec) of states of the 3-DOF micro-satellite

No test	$T_s\, x_1$	$T_s\, x_2$	$T_s\, x_3$	$T_s\, x_4$	$T_s\, x_5$	$T_s\, x_6$
test 1	5.0	2.5	5.5	6.5	2.5	5.0
test 2	5.0	2.5	5.0	6.0	2.5	4.5
test 3	6.0	3.0	6.0	5.0	3.0	5.0
test 4	6.0	3.5	5.0	5.5	2.5	5.0
test 5	5.5	6.0	6.0	6.0	2.5	5.0
test 6	7.0	4.0	7.5	7.5	3.0	4.0
test 7	6.0	5.0	7.0	6.5	2.5	5.0
test 8	7.0	2.5	7.5	7.0	2.5	6.0

5.4 Flatness-based control in successive loops for the 3-DOF free-floating space robot

5.4.1 Outline

Differential flatness theory is among the main research directions in the area of nonlinear control [711,721,491,252,898]. A dynamical system is considered to be differentially flat if the following two properties hold: (i) all state variables and control inputs of it can be written as differential functions of a subset of its state vector elements which constitute the "flat outputs" vector of the system and (ii) the flat output and its time-derivatives are differentially independent which means that they are not connected through a relation in the form of an ordinary (homogeneous) differential equation [91,594,639,489]. Differentially flat systems can be transformed into the input–output linearized form and subsequently in the canonical Brunovsky form through successive differentiations of the flat outputs [806,492,640,524,54]. The latter state-space description is both controllable and observable [732,725]. For the transformed (input–output linearized) state-space model of differentially flat systems, one can design a stabilizing feedback controller with the use of the eigenvalues assignment technique and can also solve the associated state estimation problem through the design of a filter or state observer [439,705,706,60]. However, after this change of variables and state-space model transformation process, inverse transformations should follow which will allow for returning to the initial state-space description and for computing the control inputs of the initial nonlinear dynamics of the systems. These inverse transformations can be complicated and may come against singularity issues.

To overcome the above-noted limitations of global linearization-based flatness-based control methods, the present section proposes a flatness-based control approach with decomposition of the nonlinear dynamics of the controlled system into chained subsystems which satisfy differential flatness properties [712,720,714,737]. For each subsystem of the state-space model, a virtual control input is computed, capable of inverting the subsystem's dynamics and of eliminating the subsystem's tracking error [738,739,740,734]. The control input that is actually applied to the integrated nonlinear system is found from the last row of the state-space description. This control input incorporates in a recursive manner all virtual control inputs which were computed for the individual subsystems associated with the initial state-space equation. Dynamic systems which are inherently found in the strict-feedback form or which can be transformed into such a state-space description admit well the use of the proposed flatness-based control method in successive loops. Global stability properties are proven for this control scheme and performance tests are given through simulation experiments.

In the present section, the new multi-loop flatness-based control scheme is applied to free-floating space robotic manipulators (FSRM). Such space robots consist of a floating platform which can be a spacecraft or a satellite and of a robotic arm which is mounted on the platform [1084,231]. Such a type of robotic systems is widely used for repair tasks in satellites or space stations, as well as for the removal of space debris [413,459,1]. The integrated state-space model of the platform and the robotic manipulator can have many degrees of freedom and can be of high dimensionality [18,693]. In this section, a 3-DOF model of the space robot will be considered, with the first degree of freedom to be coming from the yaw motion of the space platform and the rest two degrees of freedom to be coming from the turn motion of the joints of a two-link robotic manipulator which is mounted on the platform [904,994,992,901]. The associated dynamic model takes the typical form of a multi-DOF robotic arm or mobile manipulator and comprises an inertia matrix, a Coriolis forces vector, a damping forces vector and a control inputs (torques) vector [779,1045,967]. Such a robotic model is feedback linearizable and so far several nonlinear control approaches have been proposed about it [903,902,1039,1042]. These approaches are often based on a transformation of the dynamics of the space robot into a linearized form where the control inputs gain matrix is quadratic and the dimension of the state vector becomes equal to the dimension of the control inputs vector [10,1046]. One can distinguish approaches based on the computed-torque method jointly with disturbance observers for robustifying the control loop, sliding-mode control methods, backstepping control schemes and multi-model feedback control schemes based on linearization of the dynamics of the robotic system at multiple operating points [953,160,1038,339]. Besides, nonlinear model predictive control (NMPC) approaches for space robots follow an optimization concept and do not need state-space model transformations [756,757,907,1064]. Up to now, most approaches for treating the nonlinear optimal control problem for free-floating space robotic manipulators have been based on nonlinear model predictive control (NMPC) [815,453,411]. The new flatness-based control method in successive loops which is analyzed in the present section relies on a generalized backstepping control concept and is suitable for a class of dynamical systems which is wider than systems in the strict feedback (backstepping integral) form.

5.4.2 Dynamic model of the 3-DOF free-floating space robot

5.4.2.1 State-space form of the space robot

The diagram of the 3-DOF free-floating space robotic manipulator (FSRM) is shown in Figure 5.91. The inertial reference frame is defined as OXY. The body-fixed reference frame of the floating platform is defined as $O_A X_A Y_A$, where its origin O_1 coincides with the center of gravity of the floating platform C_0. The rotation angle of the platform with respect to the inertial reference frame is denoted as q_0. The mass of the platform is m_0 and its moment of inertia is I_0 [904,994,992,901].

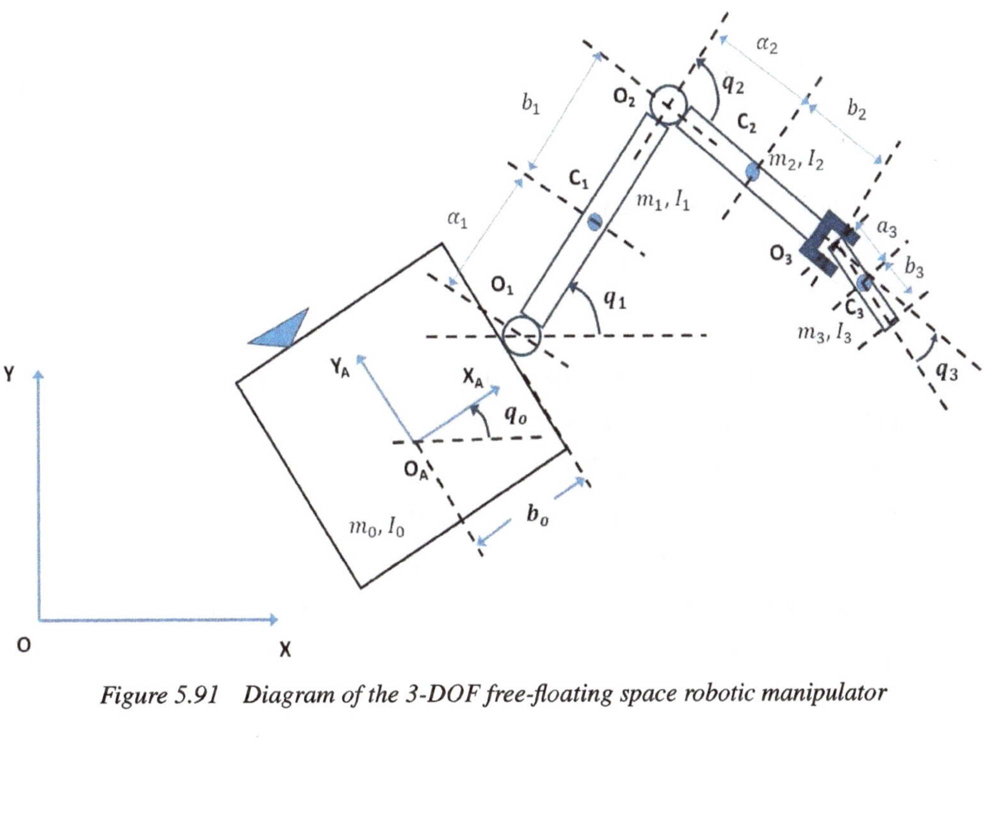

Figure 5.91 Diagram of the 3-DOF free-floating space robotic manipulator

The two joints of the rigid-link manipulator are denoted as O_1, O_2, respectively. The center of gravity of the first link is denoted as C_1 and its distance from O_1 is a_1, while its distance from O_2 is b_1. The turn angle of the first link is denoted as q_1. The mass of the first link is m_1 and its moment of inertia is I_1. The center of gravity of the second link is denoted as C_2 and its distance from O_2 is a_2 while its distance from the end-effector is b_2. The turn angle of the second link is q_2. The mass of the second link is m_2 and its moment of inertia is I_2. The end-effector grasps a non-cooperative target of mass m_3 and of moment of inertia I_3. This object forms an angle q_t with the longitudinal axis which is aligned with the second link. The center of gravity of this load is denoted as C_3 and its distance from the end effector is a_3.

The dynamic model of the 3-DOF planar free-floating space robotic manipulator (FSRM) with two links is given by [904,994,992,901]

$$M(q)\ddot{q} + C(q,\dot{q})\dot{q} = \tau + \tilde{d} \tag{5.155}$$

where $M(q)$ is the inertia matrix, $C(q,\dot{q})$ is the Coriolis and centrifugal forces matrix, τ is the torques vector which is generated by the actuators of this robotic system and $\tilde{d}$ is the disturbances vector. The angles vector of this robot is $q = [q_0, q_1, q_2]^T$. About the symmetric inertia matrix, it holds that

$$M(q) = \begin{pmatrix} M_{11} & M_2 & M_{13} \\ M_{21} & M_{22} & M_{23} \\ M_{31} & M_{32} & M_{33} \end{pmatrix} \tag{5.156}$$

where the terms M_{ij}, $i = 1,2,3$ and $j = 1,2,3$ are defined as follows:

$$\begin{aligned} M_{11} = {} & 2c_1cos(q_1) + 2c_2cos(q_2) + 2c_3cos(q_2 + q_3) + \\ & +2c_4sin(q_2) + 2c_3sin(q_1 + q_2) + c_6 \end{aligned} \tag{5.157}$$

$$\begin{aligned} M_{12} = M_{21} = {} & c_1cos(q_1) + 2c_2cos(q_2) + c_3cos(q_1 + q_2) + \\ & +2c_4sin(q_2) + c_5sin(q_1 + q_2) + c_7 \end{aligned} \tag{5.158}$$

$$\begin{aligned} M_{13} = M_{31} = {} & c_2cos(q_2) + c_3cos(q_1 + q_2) + c_4sin(q_2) + \\ & +c_5sin(q_1 + q_2) + c_8 \end{aligned} \tag{5.159}$$

$$M_{22} = 2c_2cos(q_2) + 2c_4sin(q_2) + c_7 \tag{5.160}$$

$$M_{23} = M_{32} = c_2cos(q_2) + c_4sin(q_2) + c_8 \tag{5.161}$$

$$M_{33} = c_8 \tag{5.162}$$

The Coriolis and centrifugal forces matrix is given by [904,901]

$$C(q,\dot{q}) = \begin{pmatrix} C_{11} & C_{12} & C_{13} \\ C_{21} & C_{22} & C_{23} \\ C_{31} & C_{32} & C_{33} \end{pmatrix} \tag{5.163}$$

The elements C_{ij}, $i = 1,2,3$ and $j = 1,2,3$ of this matrix are given by

$$\begin{aligned} C_{11} = {} & -c_1sin(q_1)\dot{q}_1 - c_2sin(q_2)\dot{q}_2 - c_3sin(q_1 + q_2)(\dot{q}_1 + \dot{q}_2) + \\ & +c_4cos(q_2)\dot{q}_2 + c_5cos(q_1 + q_2)(\dot{q}_1 + \dot{q}_2) \end{aligned} \tag{5.164}$$

$$\begin{aligned} C_{12} = {} & -c_1sin(q_1)(\dot{q}_0 + \dot{q}_1) - c_2sin(q_2)\dot{q}_2 - c_2sin(q_1 + q_2)(\dot{q}_0 + \dot{q}_1 + \dot{q}_2) + \\ & +c_4cos(q_2)\dot{q}_2 + c_5cos(q_1 + q_2)(\dot{q}_1 + \dot{q}_2) \end{aligned} \tag{5.165}$$

$$C_{13} = -[c_2sin(q_2) + c_3sin(q_1 + q_2) - c_4cos(q_2) - c_7cos(q_1 + q_2)](\dot{q}_0 + \dot{q}_1 + \dot{q}_2) \tag{5.166}$$

$$\begin{aligned} C_{21} = {} & c_1cos(q_1)\dot{q}_0 - c_2sin(q_2)\dot{q}_2 + c_3sin(q_1 + q_2)\dot{q}_0 + \\ & +c_4cos(q_2)\dot{q}_2 - c_5cos(q_1 + q_2)\dot{q}_0 \end{aligned} \tag{5.167}$$

$$C_{22} = -c_2 sin(q_2)\dot{q}_2 + c_4 cos(q_2)\dot{q}_2 \tag{5.168}$$

$$C_{23} = [-c_2 sin(q_2) + c_4 cos(q_2)](\dot{q}_0 + \dot{q}_1 + \dot{q}_2) \tag{5.169}$$

$$\begin{aligned} C_{31} = & c_2 sin(q_2)(\dot{q}_0 + \dot{q}_1) + c_3 sin(q_1 + q_2)\dot{q}_0 - \\ & -c_4 cos(q_2)(\dot{q}_0 + \dot{q}_1) - c_5 cos(q_1 + q_2)\dot{q}_0 \end{aligned} \tag{5.170}$$

$$C_{32} = [c_2 sin(q_2) - c_4 cos(q_2)](\dot{q}_0 + \dot{q}_1) \tag{5.171}$$

$$C_{33} = 0 \tag{5.172}$$

Moreover, using that $m_s = m_0 + m_1 + m_2 + m_t$ coefficients $c_i,\ i = 1, 2, \cdots, 8$ are given by [904,994,992,901]

$$c_1 = m_0 b_0 [m_1 a_1 + m_2 a_1 + m_2 b_1 + m_t a_1 + m_t b_1]/m_s \tag{5.173}$$

$$\begin{aligned} c_2 = & [m_0 m_t (a_1 + b_1)(a_2 + b_2 + a_3 cos(q_t)) + m_0 m_2 a_2 (a_1 + b_1) + \\ & +m_1 m_2 b_1 a_2 + m_1 m_t b_1 (a_2 + b_2)]/m_s \end{aligned} \tag{5.174}$$

$$c_3 = m_0 b_0 [m_t (a_2 + b_2 + a_3 cos(q_t)) + m_2 a_2]/m_s \tag{5.175}$$

$$c_4 = -m_t a_3 sin(q_t)(m_0 a_1 + m_0 b_1 + m_1 b1)/m_s \tag{5.176}$$

$$c_5 = -m_t m_0 b_0 a_3 sin(q_t)/m_s \tag{5.177}$$

$$\begin{aligned} c_6 = & [m_0 m_1 (b_0^2 + a_1^2) + m_0 m_2 (b_0^2 + a + 1^2 + b_1^2 + a_2^2 + 2a_1 b_1) + \\ & +m_0 m_t (b_0^2 + a_1^2 + b_1^2 + a_2^2 + b_2^2 + a_3^2 + 2a_1 b_1 + 2a_2 b_2 + 2(a_2 + b_2) a_3 cos(q_t)) + \\ & +m_1 m_t (b_1^2 + a_2^2 + b_2^2 + a_3^2 + 2a_2 b_2 + 2(a_2 + b_2) a_3 cos(q_t)) + \\ & +m_2 m_t (b_2^2 + a_3^2 + 2b_2 a_3 cos(q_t))]/m_s + I_0 + I_1 + I_2 + I_t \end{aligned} \tag{5.178}$$

$$\begin{aligned} c_7 = & [m_0 m_1 a_1^2 + m_0 m_2 (a_1^2 + b_1^2 + a_2^2 + 2a_1 b_1) + m_1 m_2 (b_1^2 + a_2^2) + \\ & m_0 m_1 (a_1^2 + b_1^2 + a_2^2 + b_2^2 + a_3^2 + 2a_1 b_1 + 2a_2 b_2 + 2(a_2 + b_2) a_3 cos(q_t)) + \\ & +m_1 m_t (b_1^2 + a_2^2 + b_2^2 + a_2^2 + 2a_2 b_2 + 2(a_2 + b_2) a_3 cos(q_t)) + \\ & +m_2 m_t (b_2^2 + a_1^2 + 2b_2 a_3 cos(q_t))]/m_s + I_1 + I_2 + I_t \end{aligned} \tag{5.179}$$

$$\begin{aligned} c_8 = & [(m_0 + m_1) m_2 a_2^2 + (m_0 + m_1) m_t (a_2^2 + b_2^2 + a_3^2 + 2a_2 b_2 + 2(a_2 + b_2) a_3 cos(q_t)) + \\ & +m_2 m_1 (b_2^2 + a_3^2 + 2b_2 a_3 cos(q_t))]/m_s + I_2 + I_t \end{aligned} \tag{5.180}$$

The orientation angle of the non-cooperative target q_t is taken to be constant. Thus, coefficients $c_i,\ i = 1, 2, \cdots, 8$ are also constants. The state vector of the floating space robotic manipulator is $x = [q_0, \dot{q}_0, q_1, \dot{q}_1, q_2, \dot{q}_2]^T$ which is also denoted as $x = [x_1, x_2, x_3, x_4, x_5, x_6]^T$.

The inverse of the inertia matrix $M(q)$ is given by:

$$M^{-1}(q) = \frac{1}{detM} \begin{pmatrix} \tilde{M}_{11} & -\tilde{M}_{21} & \tilde{M}_{31} \\ -\tilde{M}_{12} & \tilde{M}_{22} & -\tilde{M}_{32} \\ \tilde{M}_{13} & -\tilde{M}_{23} & \tilde{M}_{33} \end{pmatrix} \tag{5.181}$$

where $detM = M_{11}\tilde{M}_{11} = M_{12}\tilde{M}_{12} + M_{13}\tilde{M}_{13}$ and the subdeterminants $\tilde{M}_{ij},\ i = 1, 2, 3,\ j = 1, 2, 3$ are defined as: $\tilde{M}_{11} = M_{22}M_{33} - M_{32}M_{23}$, $\tilde{M}_{12} = M_{21}M_{33} - M_{31}M_{23}$, $\tilde{M}_{13} = M_{21}M_{32} - M_{31}M_{22}$, $\tilde{M}_{21} = M_{12}M_{33} - M_{32}M_{13}$, $\tilde{M}_{22} = M_{11}M_{33} - M_{21}M_{13}$, $\tilde{M}_{23} = M_{11}M_{32} - M_{31}M_{12}$, $\tilde{M}_{31} = M_{12}M_{23} - M_{22}M_{13}$, $\tilde{M}_{32} = M_{11}M_{23} - M_{21}M_{13}$, $\tilde{M}_{33} = M_{11}M_{22} - M_{21}M_{12}$.

Moreover, the following vector about the Coriolis and centrifugal forces in the space robotic system is computed:

$$\begin{pmatrix}\tilde{C}_1\\ \tilde{C}_2\\ \tilde{C}_3\end{pmatrix} = \begin{pmatrix}C_{11} & C_{12} & C_{13}\\ C_{21} & C_{22} & C_{23}\\ C_{31} & C_{32} & C_{33}\end{pmatrix}\begin{pmatrix}\dot{q}_0\\ \dot{q}_1\\ \dot{q}_2\end{pmatrix} \Rightarrow \begin{pmatrix}\tilde{C}_1\\ \tilde{C}_2\\ \tilde{C}_3\end{pmatrix} = \begin{pmatrix}C_{11}\dot{q}_0 + C_{12}\dot{q}_1 + C_{13}\dot{q}_2\\ C_{21}\dot{q}_0 + C_{22}\dot{q}_1 + C_{23}\dot{q}_2\\ C_{31}\dot{q}_0 + C_{32}\dot{q}_1 + C_{33}\dot{q}_2\end{pmatrix} \tag{5.182}$$

Using this notation, the dynamic model of the space robotic manipulator, without the disturbances vector which is going to be compensated by the robustness of the control scheme, becomes

$$M(q)\ddot{q} + C(q,\dot{q})\dot{q} = \tau \tag{5.183}$$

Next, by defining $\tilde{C}(q,\dot{q}) = C(q,\dot{q})\dot{q}$, the dynamic model of the space robot becomes

$$\ddot{q} = -M^{-1}(q)\tilde{C}(q,\dot{q}) + M^{-1}(q)\tau \tag{5.184}$$

The dynamic model of the space robot of Eq. (5.184) is written analytically in the form:

$$\begin{pmatrix}\ddot{q}_0\\ \ddot{q}_1\\ \ddot{q}_2\end{pmatrix} = \begin{pmatrix}\frac{-\tilde{M}_{11}\tilde{C}_1+\tilde{M}_{21}\tilde{C}_2-\tilde{M}_{31}\tilde{C}_3}{detM}\\ \frac{\tilde{M}_{12}\tilde{C}_1-\tilde{M}_{22}\tilde{C}_2+\tilde{M}_{32}\tilde{C}_3}{detM}\\ \frac{\tilde{M}_{13}\tilde{C}_1-\tilde{M}_{23}\tilde{C}_2+\tilde{M}_{33}\tilde{C}_3}{detM}\end{pmatrix} + \begin{pmatrix}\frac{\tilde{M}_{11}}{detM} & -\frac{\tilde{M}_{21}}{detM} & \frac{\tilde{M}_{31}}{detM}\\ -\frac{\tilde{M}_{12}}{detM} & \frac{\tilde{M}_{22}}{detM} & \frac{\tilde{M}_{32}}{detM}\\ \frac{\tilde{M}_{13}}{detM} & -\frac{\tilde{M}_{23}}{detM} & \frac{\tilde{M}_{33}}{detM}\end{pmatrix}\begin{pmatrix}\tau_1\\ \tau_2\\ \tau_3\end{pmatrix} \tag{5.185}$$

The notation to be used for the state vector of the space robot is $x = [x_1, x_2, x_3, x_4, x_5, x_6]^T = [q_0, \dot{q}_0, q_1, \dot{q}_1, q_2, \dot{q}_2]^T$ while the notation to be used for the control inputs of this robot is $u = [u_1, u_2, u_3]^T = [\tau_1, \tau_2, \tau_2]^T$. Then, the state-space model of the robotic system is written as

$$\begin{pmatrix}\dot{x}_1\\ \dot{x}_2\\ \dot{x}_3\\ \dot{x}_4\\ \dot{x}_5\\ \dot{x}_6\end{pmatrix} = \begin{pmatrix}x_2\\ \frac{-\tilde{M}_{11}\tilde{C}_1+\tilde{M}_{21}\tilde{C}_2-\tilde{M}_{31}\tilde{C}_3}{detM}\\ x_4\\ \frac{\tilde{M}_{12}\tilde{C}_1-\tilde{M}_{22}\tilde{C}_2+\tilde{M}_{32}\tilde{C}_3}{detM}\\ x_6\\ \frac{\tilde{M}_{13}\tilde{C}_1-\tilde{M}_{23}\tilde{C}_2+\tilde{M}_{33}\tilde{C}_3}{detM}\end{pmatrix} + \begin{pmatrix}0 & 0 & 0\\ \frac{\tilde{M}_{11}}{detM} & -\frac{\tilde{M}_{21}}{detM} & \frac{\tilde{M}_{31}}{detM}\\ 0 & 0 & 0\\ -\frac{\tilde{M}_{12}}{detM} & \frac{\tilde{M}_{22}}{detM} & \frac{\tilde{M}_{32}}{detM}\\ 0 & 0 & 0\\ \frac{\tilde{M}_{13}}{detM} & -\frac{\tilde{M}_{23}}{detM} & \frac{\tilde{M}_{33}}{detM}\end{pmatrix}\begin{pmatrix}u_1\\ u_2\\ u_3\end{pmatrix} \tag{5.186}$$

The dynamic model of the space robot can also be written in the concise state-space form

$$\dot{x} = f(x) + g(x)u \tag{5.187}$$

where $x \in R^{6\times 1}$, $f(x) \in R^{6\times 1}$, $g(x) \in R^{6\times 3}$, and $u \in R^{3\times 1}$, or equivalently

$$\dot{x} = f(x) + g_1(x)u_1 + g_2(x)u_2 + g_3(x)u_3 \tag{5.188}$$

where $x \in R^{6\times 1}$, $f(x) \in R^{6\times 1}$, $g_1(x) \in R^{6\times 1}$, $g_2(x) \in R^{6\times 1}$, $g_3(x) \in R^{6\times 1}$, and $u_1, u_2, u_3 \in R$.

5.4.2.2 Differential flatness properties of the dynamics of the space robot

The dynamic model of the free-floating space robotic manipulator is differentially flat, with flat outputs vector $Y = [x_1, x_3, x_5]^T = [q_0, q_1, q_2]^T$. Indeed using that $x_2 = \dot{x}_1$, $x_4 = \dot{x}_3$, and $x_6 = \dot{x}_5$, it holds that x_2, x_4, and x_6 are also differential functions of the flat outputs of system, or

$$x_2 = h_2(Y, \dot{Y}) \quad x_4 = h_4(Y, \dot{Y}) \quad x_6 = h_6(Y, \dot{Y}) \tag{5.189}$$

Besides, from the state-space model of the system given in Eq. (5.185), one solves for the inputs vector, thus obtaining

$$\begin{pmatrix} u_1 \\ u_2 \\ u_3 \end{pmatrix} = \begin{pmatrix} \frac{\tilde{M}_{11}}{detM} & -\frac{\tilde{M}_{21}}{detM} & \frac{\tilde{M}_{31}}{detM} \\ -\frac{\tilde{M}_{12}}{detM} & \frac{\tilde{M}_{22}}{detM} & \frac{\tilde{M}_{32}}{detM} \\ \frac{\tilde{M}_{13}}{detM} & -\frac{\tilde{M}_{13}}{detM} & \frac{\tilde{M}_{33}}{detM} \end{pmatrix}^{-1} \left[\begin{pmatrix} \ddot{q}_0 \\ \ddot{q}_1 \\ \ddot{q}_2 \end{pmatrix} - \begin{pmatrix} \frac{-\tilde{M}_{11}\tilde{C}_1 + \tilde{M}_{21}\tilde{C}_2 - \tilde{M}_{31}\tilde{C}_3}{detM} \\ \frac{\tilde{M}_{12}\tilde{C}_1 - \tilde{M}_{22}\tilde{C}_2 + \tilde{M}_{32}\tilde{C}_3}{detM} \\ \frac{\tilde{M}_{13}\tilde{C}_1 - \tilde{M}_{23}\tilde{C}_2 + \tilde{M}_{31}\tilde{C}_3}{detM} \end{pmatrix} \right] \tag{5.190}$$

Since all terms which appear in the right hand side of the above equation are differential functions of the flat outputs vector, it also holds that the control inputs vector $u = [u_1, u_2, u_3]^T$ is a differential function of the flat outputs vector Y, or

$$u = h_u(Y, \dot{Y}) \tag{5.191}$$

Consequently, the space robot is a differentially flat system. The proof of differential flatness properties for the space robot is also an implicit proof of the system's controllability. Moreover, it demonstrates that the system is input–output linearizable through successive differentiations of its flat outputs. Furthermore, it allows for solving the setpoints definition problem. First, one selects setpoints without constraints, for those state variables which coincide with the flat outputs of the system. For the rest of the state variables, setpoints are chosen using the differential relations which connect these state variables to the flat outputs.

5.4.3 Flatness-based control in successive loops for the floating space robot

5.4.4 Dynamic model in chained subsystems form

The dynamic model of the floating-space robotic manipulator is given by

$$\ddot{q} = -M^{-1}(q)\tilde{C}(q, \dot{q}) + M^{-1}u \tag{5.192}$$

or using the new state vector notation $q_A = [q_0, q_1, q_2]^T$ and $q_B = [\dot{q}_0, \dot{q}_1, \dot{q}_2]^T$, one obtains

$$\begin{aligned} \dot{q}_A &= q_B \\ \dot{q}_B &= -M^{-1}(q_A, q_B)\tilde{C}(q_A, q_B, \dot{q}_A, \dot{q}_B) + M^{-1}(q_A, q_B)u \end{aligned} \tag{5.193}$$

Next, it is considered that control of the position of the end effector (x_{O_3}, y_{O_3}) in the task-space has to be performed. Thus, in place of state vector $q_A = [q_0, q_1, q_2]^T$, the state vector $x_A = [q_0, x_{O_3}, y_{O_3}]^T$ will be used.

Using that $L_1 = a_1 + b_1$, $L_2 = a_2 + b_2$ from the kinematic analysis about the 2-link manipulator, one has that

$$\begin{pmatrix} x_{O_3} \\ y_{O_3} \end{pmatrix} = \begin{pmatrix} b_0 \\ 0 \end{pmatrix} + \begin{pmatrix} cos(q_0) & sin(q_0) \\ -sin(q_0) & cos(q_0) \end{pmatrix} \begin{pmatrix} L_1cos(q_1) + L_2cos(q_1 + q_2) \\ L_1sin(q_1) + L_2sin(q_1 + q_2) \end{pmatrix} \tag{5.194}$$

Analytically, it holds that

$$\begin{aligned} x_{O_3} &= b_0 + cos(q_0)[L_1cos(q_1) + L_2cos(q_1 + q_2)] + sin(q_0)[L_1sin(q_1) + L_2sin(q_1 + q_2)] \\ y_{O_3} &= -sin(q_0)[L_1cos(q_1) + L_2cos(q_1 + q_2)] + cos(q_0)[L_1sin(q_1) + L_2sin(q_1 + q_2)] \end{aligned} \tag{5.195}$$

By differentiating in time, one obtains

$$\begin{aligned} \dot{x}_{O_3} &= (a_{x0} + b_{x0})\dot{q}_0 + (a_{x1} + b_{x1})\dot{q}_1 + (a_{x2} + b_{x2})\dot{q}_2 \\ \dot{y}_{O_3} &= (a_{y0} + b_{y0})\dot{q}_0 + (a_{y1} + b_{y1})\dot{q}_1 + (a_{y2} + b_{y2})\dot{q}_2 \end{aligned} \tag{5.196}$$

where $a_{x0} = -sin(q0)[L_1cos(q_1) + L_2cos(q_1 + q_2)]$, $a_{x1} = cos(q_0)[-L_1sin(q_1) - L_2sin(q_1 + q_2)]$, $a_{x2} = cos(q_0)[-L_2 sin(q_1 + q_2)]$, $b_{x0} = cos(q0)[L_1sin(q_1) + L_2sin(q_1 + q_2)]$, $b_{x1} = sin(q_0)[L_1cos(q_1) + L_2cos(q_1 + q_2)]$, $b_{x2} = sin(q_0) [L_2cos(q_1 + q_2)]$.

and $a_{y0} = -cos(y_0)[L_1cos(q_1) + L_2cos(q_1 + q_2)]$, $a_{y1} = -sin(q_0)[-L_1sin(q_1) - L_2sin(q_1 + q_2)]$, $a_{y2} = -sin(q_2)[-L_2 sin(q_1 + q_2)]$, $b_{y0} = -sin(q_0)[L_1sin(q_1) + L_2sin(q_1 + q2)]$, $b_{y1} = cos(q_0)[L_1cos(q_1) + L_2cos(q_1 + q_2)]$, $b_{y2} = cos(q_0) [L_2cos(q_1 + q_2)]$.

Thus, the following transformation is performed:

$$\begin{pmatrix} \dot{q}_0 \\ \dot{x}_{0_3} \\ \dot{y}_{0_3} \end{pmatrix} = \begin{pmatrix} 1 & 0 & 0 \\ (a_{x0}+b_{x0}) & (a_{x1}+b_{x1}) & (a_{x2}+b_{x2}) \\ (a_{y0}+b_{y0}) & (a_{y1}+b_{y1}) & (a_{y2}+b_{y2}) \end{pmatrix} \begin{pmatrix} \dot{q}_0 \\ \dot{q}_1 \\ \dot{q}_2 \end{pmatrix} \tag{5.197}$$

and the transformation matrix is

$$R = \begin{pmatrix} 1 & 0 & 0 \\ (a_{x0}+b_{x0}) & (a_{x1}+b_{x1}) & (a_{x2}+b_{x2}) \\ (a_{y0}+b_{y0}) & (a_{y1}+b_{y1}) & (a_{y2}+b_{y2}) \end{pmatrix} \tag{5.198}$$

With the use of the above-noted transformation matrix R, the dynamic model of the floating-space robotic manipulator is written as

$$\dot{x}_A = R \cdot q_B \tag{5.199}$$

$$\dot{q}_B = -M^{-1}(x_A, q_B)\tilde{C}(x_A, \dot{x}_A, q_B, \dot{q}_B) + M^{-1}(x_A, q_B)u \tag{5.200}$$

Thus, it can be seen that the dynamic model of the space robotic manipulator is decomposed into two subsystems which are connected into cascading loops. For the subsystem given in Eq. (5.199) $\dot{x}_A = Rq_B$, the following subvectors and submatrices are defined:

$$x_A = \begin{pmatrix} q_0 \\ x_{O_3} \\ y_{O_3} \end{pmatrix} \quad f_A = 0_{3\times 1} \quad g_A = R = \begin{pmatrix} 1 & 0 & 0 \\ (a_{x0}+b_{x0}) & (a_{x1}+b_{x1}) & (a_{x2}+b_{x2}) \\ (a_{y0}+b_{y0}) & (a_{y1}+b_{y1}) & (a_{y2}+b_{y2}) \end{pmatrix} \quad v = q_B = \begin{pmatrix} \dot{q}_0 \\ \dot{q}_1 \\ \dot{q}_2 \end{pmatrix} \tag{5.201}$$

For the subsystem given in Eq. (5.200) $\dot{q}_B = -M^{-1}(x_A, q_B)\tilde{C}(x_A, \dot{x}_A, q_B, \dot{q}_B) + M^{-1}(x_A, q_B)u$, the following subvectors and submatrices are defined:

$$q_B = \begin{pmatrix} \dot{q}_0 \\ \dot{q}_1 \\ \dot{q}_2 \end{pmatrix} \quad f_B = -M^{-1}(x_A, q_B)\tilde{C}(x_A, \dot{x}_A, q_B, \dot{q}_B) \quad g_B = M^{-1}(x_A, q_B) \quad u = \begin{pmatrix} u_1 \\ u_2 \\ u_3 \end{pmatrix} \tag{5.202}$$

Thus, the chained subsystems form of the free-floating space robotic manipulator is written as

$$\dot{x}_A = f_A(x_A) + g_A(x_A)q_B \tag{5.203}$$

$$\dot{q}_B = f_B(x_A, q_B) + g_B(x_A, q_B)u \tag{5.204}$$

As already proven in the second part of Section 5.4.2 the integrated state-space model of the free-floating space robotic manipulator is differentially flat with flat output the angles' vector q_A. Here, differential flatness will be proven once again, using as flat output vector x_A. Indeed, from Eq. (5.203), one solves for q_B thus obtaining

$$q_B = g_A(x_A)^{-1}[\dot{x}_A - f_A(x_A)] \tag{5.205}$$

which signifies that vector q_B is a differential function of the flat outputs vector x_A. Moreover, from Eq. (5.204), one solves for the control input u, thus obtaining

$$u = g_B(x_A, q_B)^{-1}[\dot{q}_B - f_B(x_A, q_B)] \tag{5.206}$$

which signifies that control input u is also a differential function of the flat outputs vector x_A. Consequently, all state variables and the control inputs of the system of Eq. (5.203) and Eq. (5.204) are differential functions of the flat outputs vector x_A. Besides, x_A and its derivatives are differentially independent. Thus, differential flatness is confirmed.

Furthermore, differential flatness properties can also be proven independently for each one of the subsystems of Eq. (5.203) and Eq. (5.204). First, for the subsystem of Eq. (5.203), x_A is taken to be the flat outputs vector and q_B

is the virtual control inputs vector. From Eq. (5.205), one has $q_B = g_A(x_A)^{-1}[\dot{x}_A - f_A(x_A)]$ which shows that q_B is a differential function of x_A; therefore, the subsystem is differentially flat. Next, for the subsystem of Eq. (5.204) q_B is taken to be the flat outputs vector, x_A is taken to be a coefficients vector and u is the control inputs vector. From Eq. (5.206), one has $u = g_B(x_A, q_B)^{-1}[\dot{q}_B - f_B(x_A, q_B)]$ which shows that u is a differential function of q_B; therefore, the subsystem is differentially flat.

5.4.4.1 Design of a multi-loop flatness-based controller

The proof of differential flatness properties for each one of the subsystems of Eq. (5.203) and Eq. (5.204) signifies that these subsystems are input–output linearizable and can be controlled through inversion of their dynamics as it is commonly done for systems in the input–output linearized form.

For the subsystem of Eq. (5.203) $\dot{x}_A = f_A(x_A) + g_A(x_A)q_B$ the virtual control input $v = q_{B.d}$ becomes setpoint for the subsystem of Eq. (5.204) and is given be

$$v = q_{B.d} = g_a(x_A)^{-1}[\dot{x}_{A,d} - f_A(x_A) - K_1(x_A - x_{A,d})] \tag{5.207}$$

where $K_1 \in R^{3\times3} > 0$ is a diagonal matrix with positive elements on its diagonal, or $k_{1,ii} > 0$ for $i = 1, 2, 3$.

For the subsystem of Eq. (5.204) $\dot{x}_B = f_B(x_A, q_B) + g_B(x_A, q_B)u$ the real control input u is given be

$$u = g_b(q_B)^{-1}[\dot{q}_{B,d} - f_B(x_b, q_B) - K_2(q_B - q_{B,d})] \tag{5.208}$$

where $K_2 \in R^{3\times3} > 0$ is a diagonal matrix with positive elements on its diagonal, or $k_{2,ii} > 0$ for $i = 1, 2, 3$.

By substituting the control law of Eq. (5.207) into the subsystem of Eq. (5.203) and by defining the error variable $e_A = (x_A - x_{A,d})$, one obtains

$$\begin{aligned} \dot{x}_A &= f_A(x_A) + g_A(x_A) \cdot g_A(x_A)^{-1}[\dot{x}_{A,d} - f_A(x_A) - K_1(x_A - x_{A,d})] \Rightarrow \\ &(\dot{x}_A - \dot{x}_{A,d}) + K_1(x_A - x_{A,d}) = 0 \Rightarrow \dot{e}_A + K_1 e_A = 0 \Rightarrow \\ &\quad lim_{t\to\infty} e_A(t) = 0 \Rightarrow lim_{t\to\infty} x_A(t) = x_{A,d}(t) \end{aligned} \tag{5.209}$$

By substituting the control law of Eq. (5.208) into the subsystem of Eq. (5.204) and by defining the error variable $e_B = (q_B - q_{B,d})$, one obtains

$$\begin{aligned} \dot{q}_B &= f_B(x_A, q_B) + g_B(x_A, q_B) \cdot g_B(x_A, q_B)^{-1}[\dot{q}_{B,d} - f_B(x_A, q_B) - K_2(q_B - q_{B,d})] \Rightarrow \\ &(\dot{q}_B - \dot{q}_{B,d}) + K_2(q_B - q_{B,d}) = 0 \Rightarrow \dot{e}_B + K_2 e_B = 0 \Rightarrow \\ &\quad lim_{t\to\infty} e_B(t) = 0 \Rightarrow lim_{t\to\infty} q_B(t) = q_{B,d}(t) \end{aligned} \tag{5.210}$$

The global stability properties of the multi-loop flatness-based control method for the free-floating space robotic manipulator can also be proven through Lyapunov analysis. The following Lyapunov function is defined:

$$V = \tfrac{1}{2}[e_A^T e_A + e_B^T e_B] \tag{5.211}$$

By differentiating in time, one gets

$$\dot{V} = \tfrac{1}{2}[2e_A^T \dot{e}_A + 2e_B^T \dot{e}_B] \tag{5.212}$$

Using the tracking error dynamics of Eq. (5.209) and Eq. (5.210), that is $\dot{e}_A + K_1 e_A = 0 \Rightarrow \dot{e}_A = -K_1 e_A$ and $\dot{e}_B + K_2 e_B = 0 \Rightarrow \dot{e}_B = -K_2 e_B$, one obtains

$$\begin{aligned} &\dot{V} = [e_A^T(-K_1 e_A) + e_B^T(-K_2 e_B)] \Rightarrow \\ &\dot{V} = -[e_A^T K_1 e_A + e_B^T K_2 e_B] \Rightarrow \\ &\dot{V} < 0 \; \forall \; e_A \neq 0, \; e_B \neq 0 \end{aligned} \tag{5.213}$$

and $\dot{V}=0$ iff $e_A=0,\ e_B=0$. Consequently, the Lyapunov function of the multi-loop control system is strictly diminishing and converges asymptotically to 0. Thus, one obtains

$$\begin{aligned} &lim_{t\to\infty}e_A(t)\Rightarrow lim_{t\to\infty}x_A(t)=x_{A,d}(t) \\ &lim_{t\to\infty}e_B(t)\Rightarrow lim_{t\to\infty}q_B(t)=q_{B,d}(t) \end{aligned} \tag{5.214}$$

An explicit demonstration of the exponential stabilization that is achieved by flatness-based control in successive loops is given next. The Lyapunov function of the control loop is written as:

$$V=\frac{1}{2}[\sum_{i=1}^{3}e_i^2+\sum_{j=4}^{5}e_j^2] \tag{5.215}$$

where $e_i\ i=1,2,3$ are the tracking errors for the state variables of the first subsystem of the free-floating space robotic manipulator, and $e_j\ j=4,5,6$ are the tracking errors for the state variables of the second subsystem of the space robot. Equivalently, the first-order time-derivative of the Lyapunov function is written as

$$\dot{V}=-[\sum_{i=1}^{3}k_{1,i}e_i^2+\sum_{j=4}^{6}k_{2,j-3}e_j^2] \tag{5.216}$$

where $k_{1,i}>0\ i=1,2,3$ are the diagonal elements of gain matrix K_1 and $k_{2,j-3}>0\ j=4,5,6$ are the diagonal elements of gain matrix K_2. By denoting the minimum of the above-noted elements of the feedback gain matrices as k_{min}, that is

$$k_{min}=min\{k_{1,i}\ :\ i=1,2,3 \text{ and } k_{2,j-3}\ :\ j=4,5,6\} \tag{5.217}$$

and using Eq. (5.216), one obtains that

$$\begin{aligned} &\dot{V}\leq -k_{min}[\sum_{i=1}^{3}e_i^2+\sum_{j=4}^{6}e_j^2] \\ &\Rightarrow\dot{V}\leq -2k_{min}V\Rightarrow\dot{V}+2k_{min}V\leq 0 \end{aligned} \tag{5.218}$$

From Eq. (5.218), one can demonstrate the exponential convergence of the Lyapunov function V to 0.

5.4.5 Simulation tests

The tracking performance of the proposed multi-loop flatness-based control method for free-floating space robotic manipulators has been confirmed through simulation experiments. The obtained results are depicted in Figures 5.92 to 5.107. The state-space model of the system of the robotic manipulator given in Eq. (5.183) has been separated into the chained subsystems of Eq. (5.199) and Eq. (5.200) which individually satisfy the differential flatness property. Flatness-based control is implemented at each time-step through the following steps: (i) the state vector of the subsequent $(i+1)$-th subsystem becomes virtual control input to the preceding i-th subsystem, (ii) Equivalently, the virtual control input of the preceding i-th subsystem becomes setpoint for the subsequent $(i+1)$-th subsystem, (iii) the value of the virtual control input for each subsystem is computed by inverting the dynamics of this subsystem and by selecting feedback gains which allow for eliminating the associated local tracking error, (iv) the real control input is computed from the last subsystem by inverting again the dynamics of this subsystem, and (v) the real control input makes implicitly use of the virtual control inputs of all preceding subsystems. Its stabilizing effects appear by tracing the subsystems of the state-space model backward from the last to the first one.

Indicative values about the parameters of the dynamic model of the 3-DOF space robotic manipulator can be found in [904,901]. These are: $m_0=200$ kg, $m_1=6.0$ kg, $m_2=5.0$ kg, $m_t=2$ kg, $I_0=200$ kg $\cdot m^2$, $I_1=1.125$ kg $\cdot m^2$, $I_2=0.9375$ kg $\cdot m^2$, $I_t=1.0$ kg $\cdot m^2$, $a_0=1.25$m, $b_0=1.25$m, $a_1=0.75$m, $b_1=0.75$m, $a_2=0.75$m, $b_2=0.75$m, $a_3=0.5$m, $b_3=0.5$m and $q_t=\frac{\pi}{6}$ rad. It is noteworthy, that through the stages of this method, one solves also the setpoints definition problem for all state variables of the free-floating space robotic manipulator. Actually, the selection of setpoints for state variables $x_{A,1}, x_{A,2}, x_{A,3}$ is unconstrained. On the other side by defining state variables $q_{B,1}, q_{B,2}, q_{B,3}$ as part of the virtual control input $v=q_B$, one can find the setpoints for $q_{B,1}, q_{B,2}, q_{B,3}$ as functions of the setpoints for $x_{A,1}, x_{A,2}, x_{A,3}$. The speed of convergence of the state variables of the robot under flatness-based control implemented

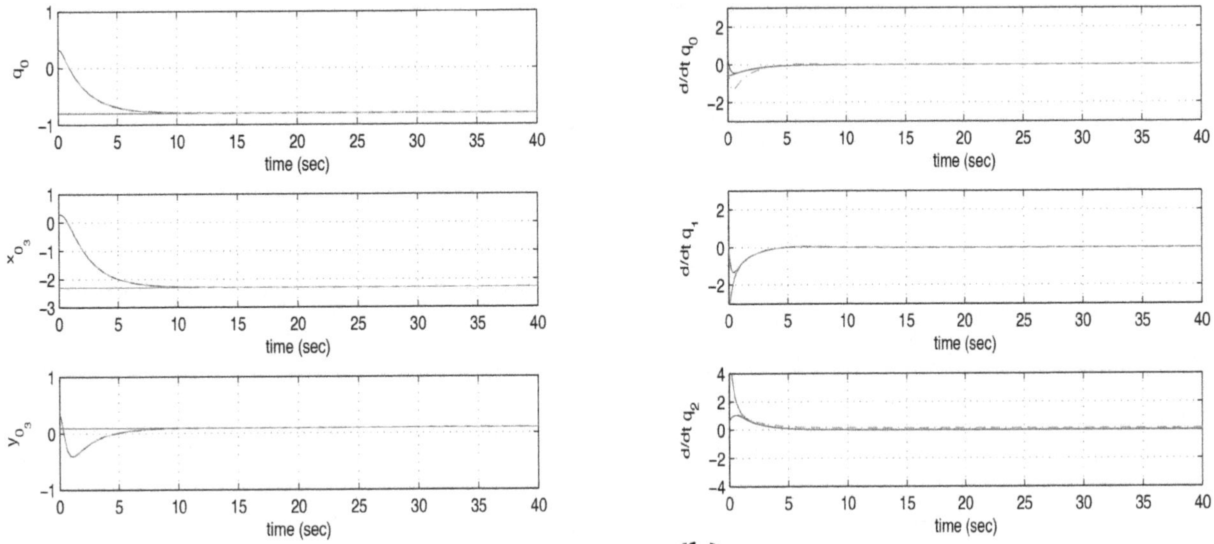

(a) (b)

Figure 5.92 Tracking of setpoint 1 by the free-floating space robotic manipulator: (a) convergence of state variables q_0, $x_{x_{O_3}}$ and y_{O_3} that constitute state subvector x_A (blue lines) to the reference setpoints (red lines) and their KF-based estimation (green line); (b) convergence of state variables $\dot{q}_0$, $\dot{q}_1$ and $\dot{q}_2$ that constitute state subvector q_B (blue lines) to the reference setpoints (red lines) and their KF-based estimation (green line)

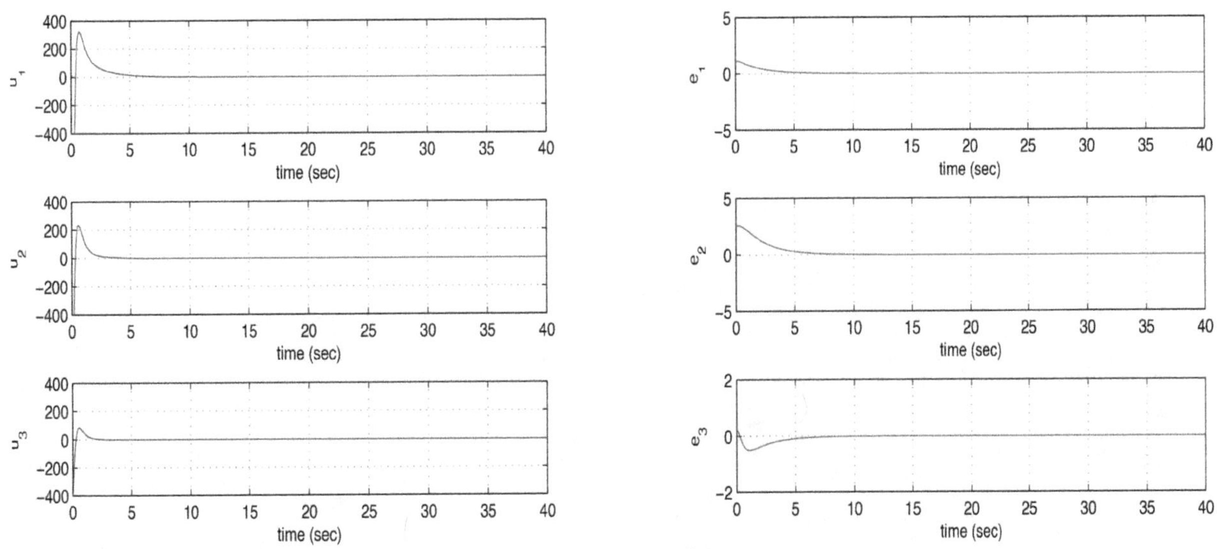

(a) (b)

Figure 5.93 Tracking of setpoint 1 by the free-floating space robotic manipulator: (a) variations of the control inputs u_1, u_2, and u_3; (b) tracking error of state variables q_0, x_{O_3}, and y_{O_3} of state subvector x_A

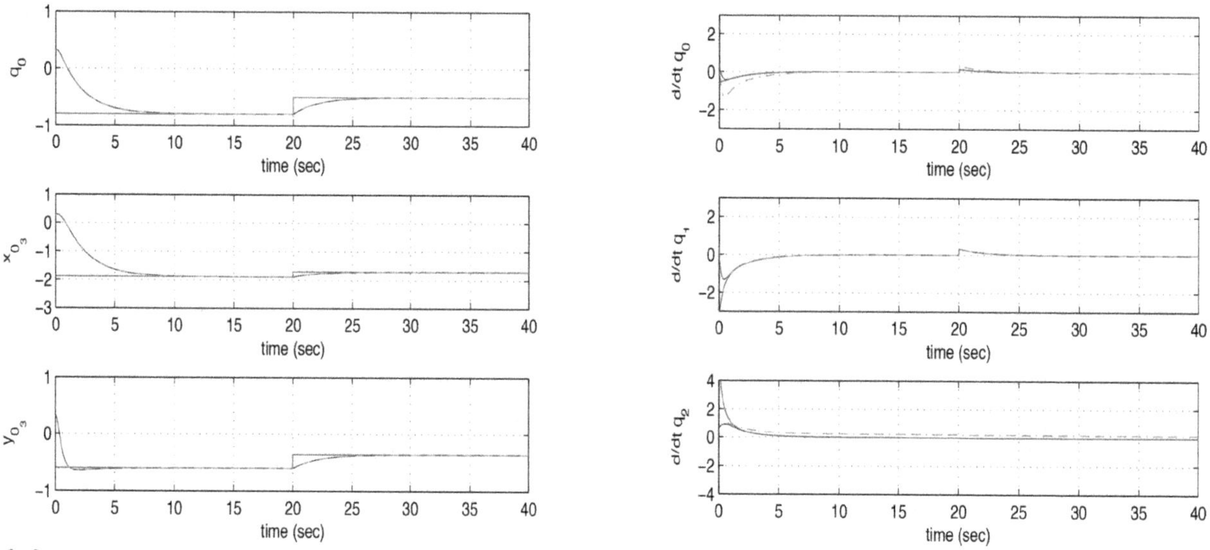

(a) (b)

Figure 5.94 Tracking of setpoint 2 by the free-floating space robotic manipulator: (a) convergence of state variables q_0, $x_{x_{O_3}}$ and y_{O_3} that constitute state subvector x_A (blue lines) to the reference setpoints (red lines) and their KF-based estimation (green line); (b) convergence of state variables $\dot{q}_0$, $\dot{q}_1$ and $\dot{q}_2$ that constitute state subvector q_B (blue lines) to the reference setpoints (red lines) and their KF-based estimation (green line)

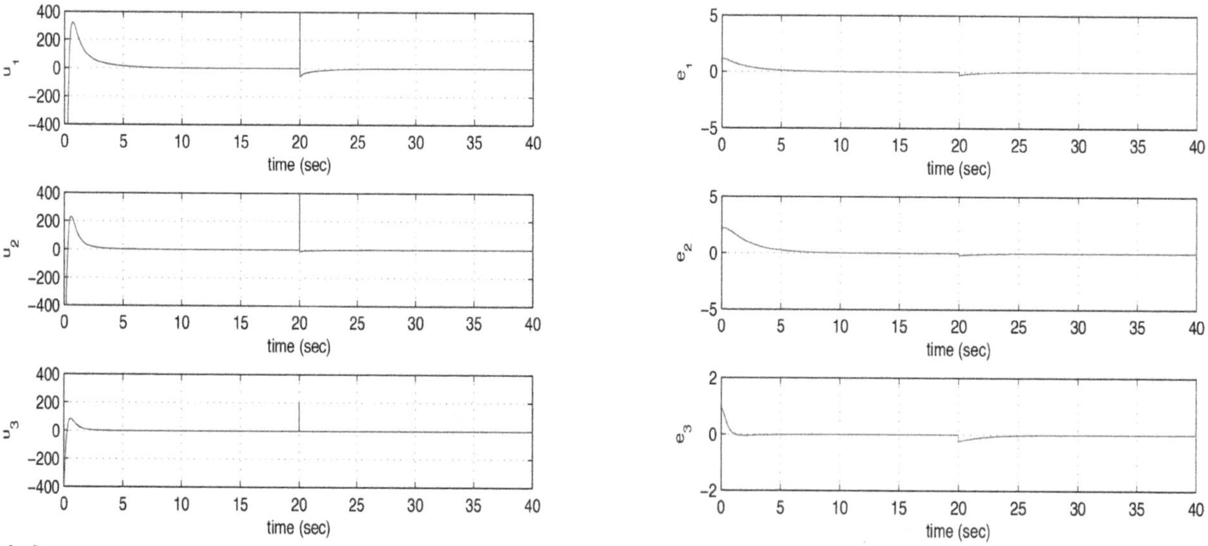

(a) (b)

Figure 5.95 Tracking of setpoint 2 by the free-floating space robotic manipulator: (a) variations of the control inputs u_1, u_2, and u_3; (b) tracking error of state variables q_0, x_{O_3}, and y_{O_3} of state subvector x_A

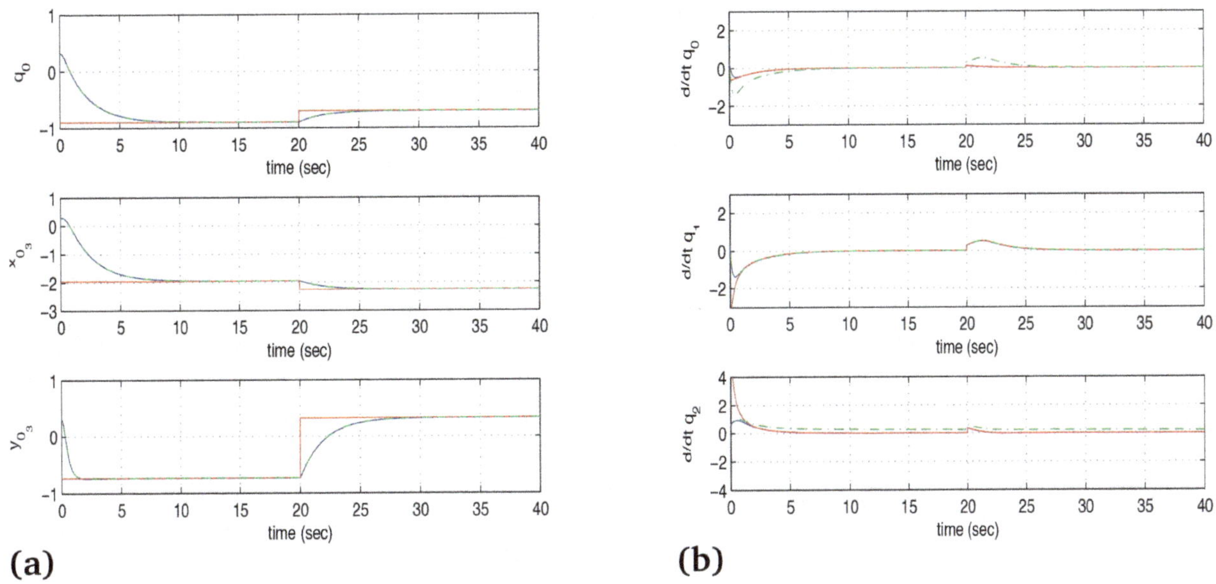

Figure 5.96 Tracking of setpoint 3 by the free-floating space robotic manipulator: (a) convergence of state variables q_0, $x_{x_{O_3}}$ and y_{O_3} that constitute state subvector x_A (blue lines) to the reference setpoints (red lines) and their KF-based estimation (green line); (b) convergence of state variables $\dot{q}_0$, $\dot{q}_1$ and $\dot{q}_2$ that constitute state subvector q_B (blue lines) to the reference setpoints (red lines) and their KF-based estimation (green line)

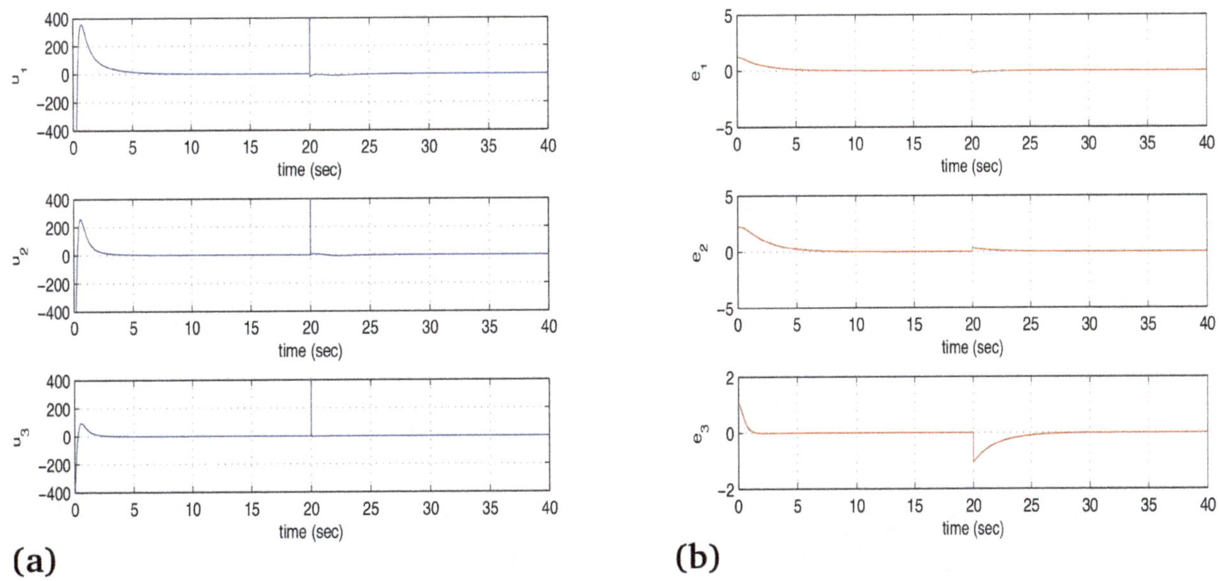

Figure 5.97 Tracking of setpoint 3 by the free-floating space robotic manipulator: (a) variations of the control inputs u_1, u_2, and u_3; (b) tracking error of state variables q_0, x_{O_3}, and y_{O_3} of state subvector x_A

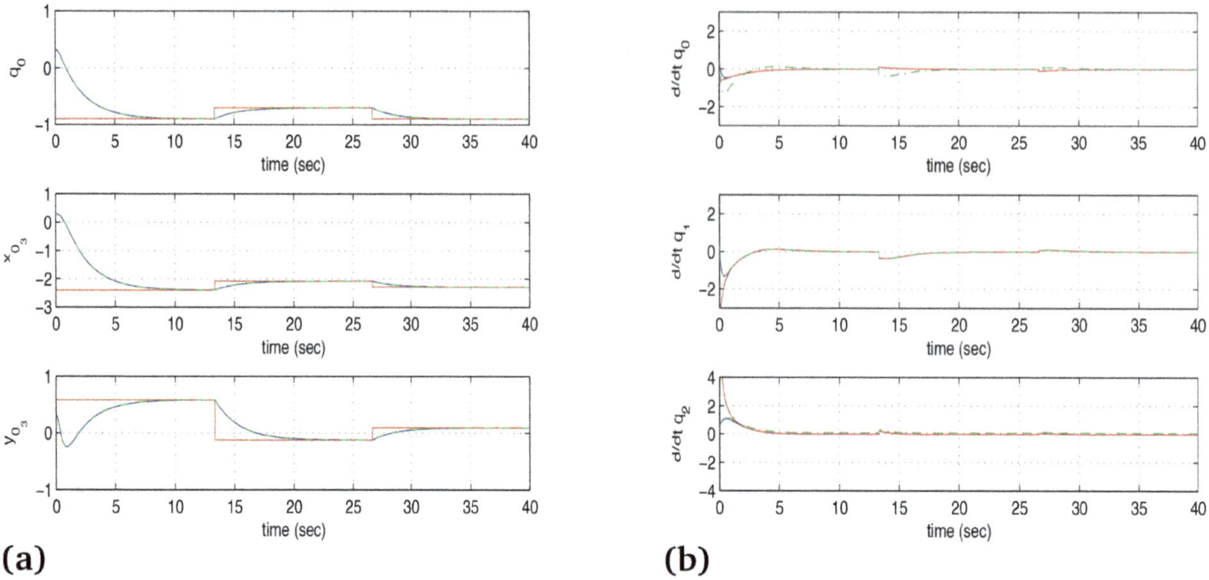

Figure 5.98 Tracking of setpoint 4 by the free-floating space robotic manipulator: (a) convergence of state variables q_0, $x_{x_{O_3}}$ and y_{O_3} that constitute state subvector x_A (blue lines) to the reference setpoints (red lines) and their KF-based estimation (green line); (b) convergence of state variables $\dot{q}_0$, $\dot{q}_1$ and $\dot{q}_2$ that constitute state subvector q_B (blue lines) to the reference setpoints (red lines) and their KF-based estimation (green line)

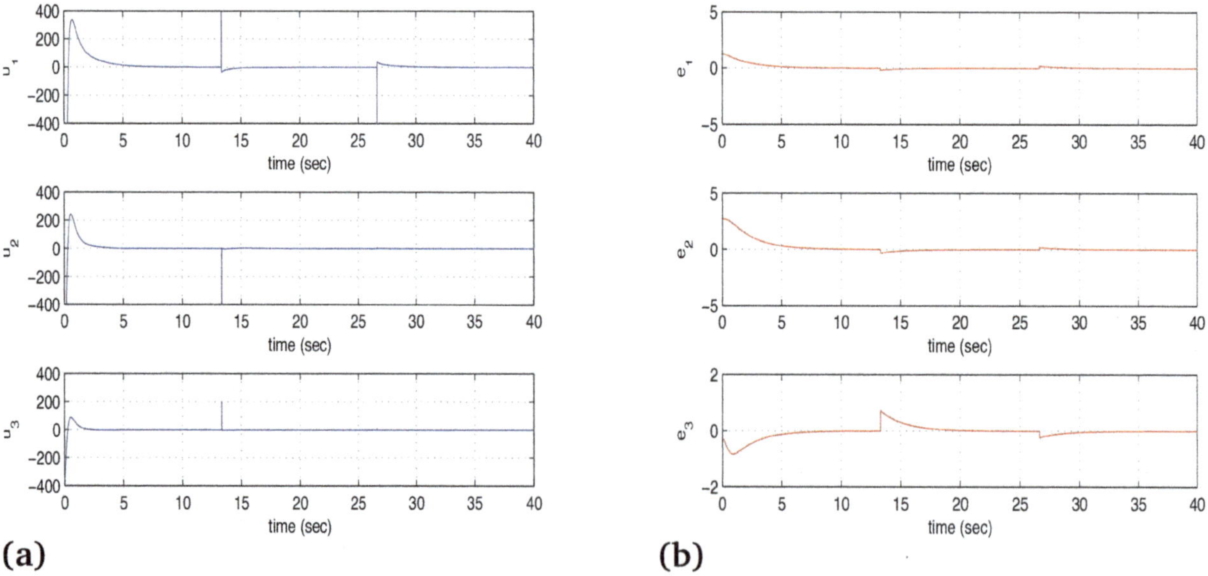

Figure 5.99 Tracking of setpoint 4 by the free-floating space robotic manipulator: (a) variations of the control inputs u_1, u_2, and u_3; (b) tracking error of state variables q_0, x_{O_3}, and y_{O_3} of state subvector x_A

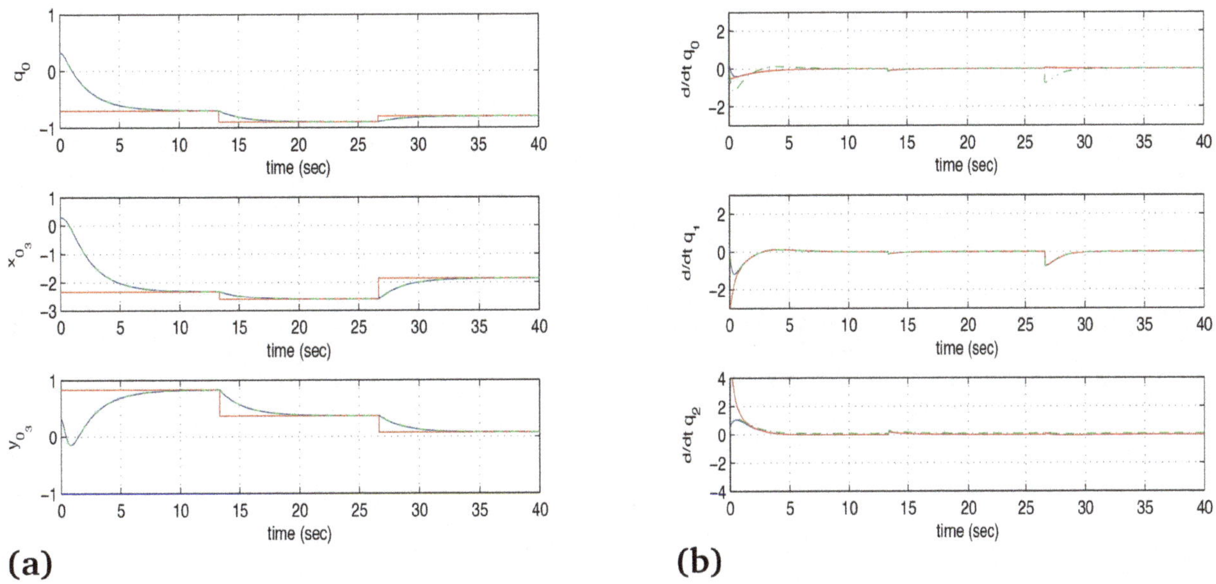

Figure 5.100 *Tracking of setpoint 5 by the free-floating space robotic manipulator: (a) convergence of state variables q_0, $x_{x_{O_3}}$ and y_{O_3} that constitute state subvector x_A (blue lines) to the reference setpoints (red lines) and their KF-based estimation (green line); (b) convergence of state variables $\dot{q}_0$, $\dot{q}_1$ and $\dot{q}_2$ that constitute state subvector q_B (blue lines) to the reference setpoints (red lines) and their KF-based estimation (green line)*

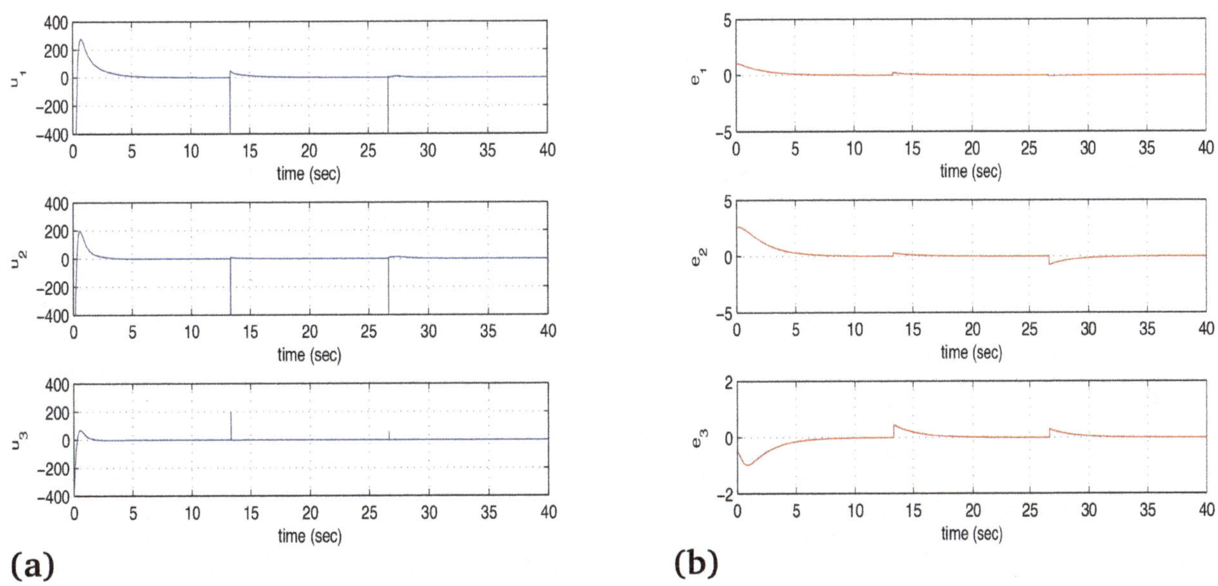

Figure 5.101 *Tracking of setpoint 5 by the free-floating space robotic manipulator: (a) variations of the control inputs u_1, u_2, and u_3; (b) tracking error of state variables q_0, x_{O_3}, and y_{O_3} of state subvector x_A*

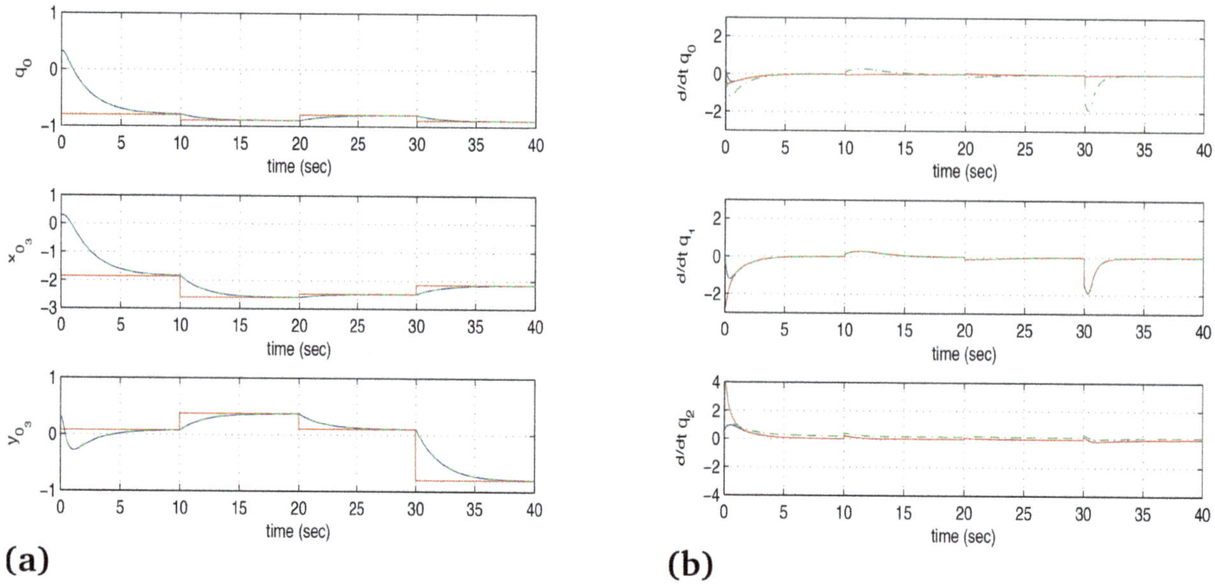

Figure 5.102 *Tracking of setpoint 6 by the free-floating space robotic manipulator: (a) convergence of state variables q_0, $x_{x_{O_3}}$ and y_{O_3} that constitute state subvector x_A (blue lines) to the reference setpoints (red lines) and their KF-based estimation (green line); (b) convergence of state variables $\dot{q}_0$, $\dot{q}_1$ and $\dot{q}_2$ that constitute state subvector q_B (blue lines) to the reference setpoints (red lines) and their KF-based estimation (green line)*

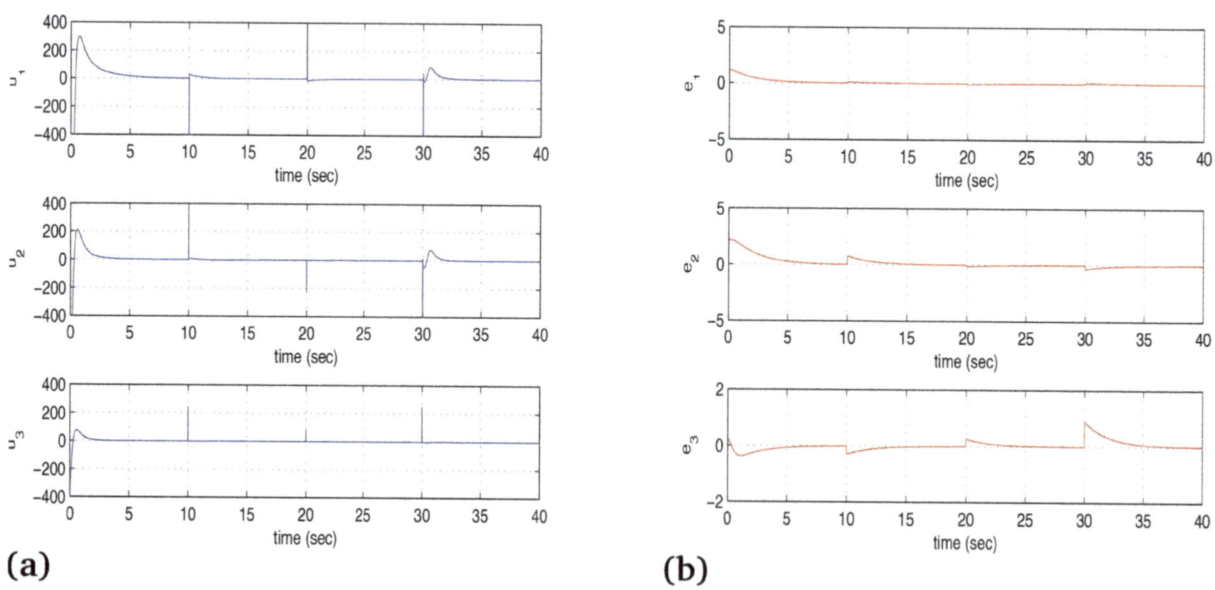

Figure 5.103 *Tracking of setpoint 6 by the free-floating space robotic manipulator: (a) variations of the control inputs u_1, u_2, and u_3; (b) tracking error of state variables q_0, x_{O_3}, and y_{O_3} of state subvector x_A*

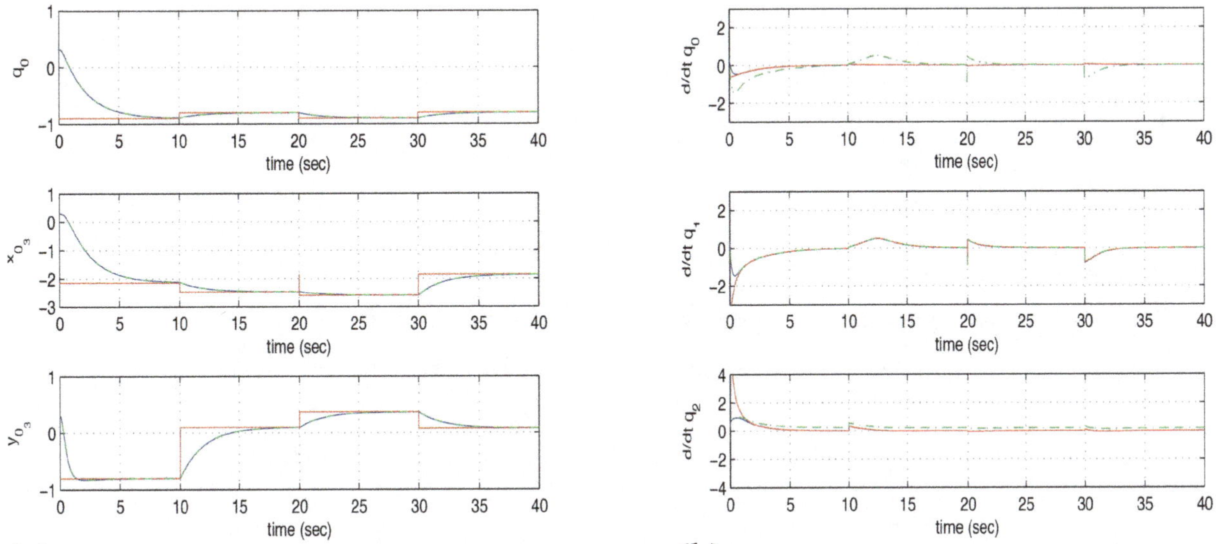

(a) (b)

Figure 5.104 *Tracking of setpoint 7 by the free-floating space robotic manipulator: (a) convergence of state variables q_0, $x_{x_{O_3}}$ and y_{O_3} that constitute state subvector x_A (blue lines) to the reference setpoints (red lines) and their KF-based estimation (green line); (b) convergence of state variables $\dot{q}_0$, $\dot{q}_1$ and $\dot{q}_2$ that constitute state subvector q_B (blue lines) to the reference setpoints (red lines) and their KF-based estimation (green line)*

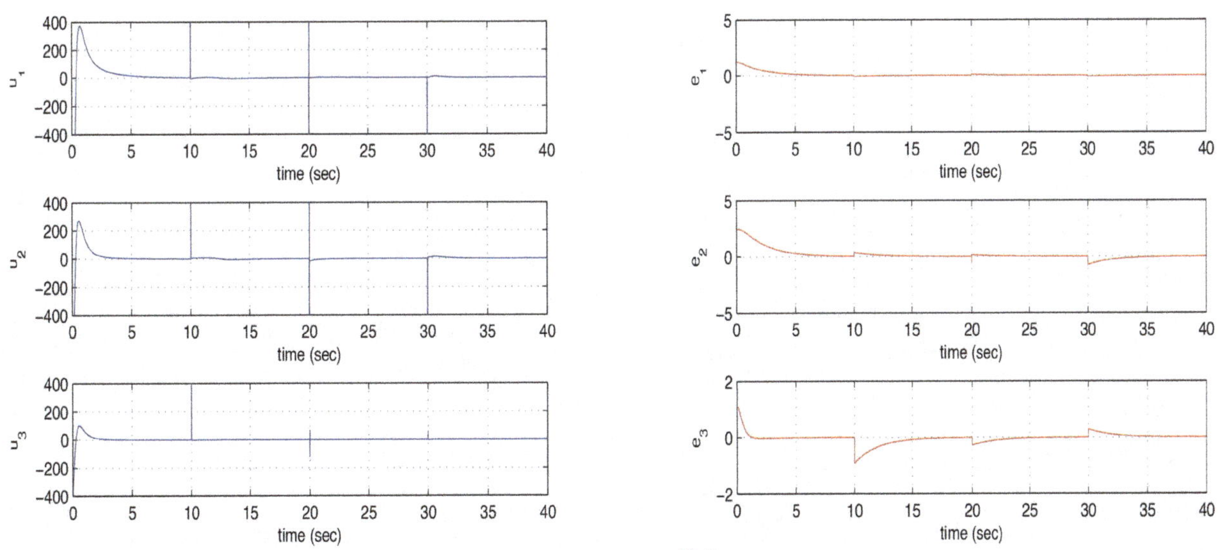

(a) (b)

Figure 5.105 *Tracking of setpoint 7 by the free-floating space robotic manipulator: (a) variations of the control inputs u_1, u_2, and u_3; (b) tracking error of state variables q_0, x_{O_3}, and y_{O_3} of state subvector x_A*

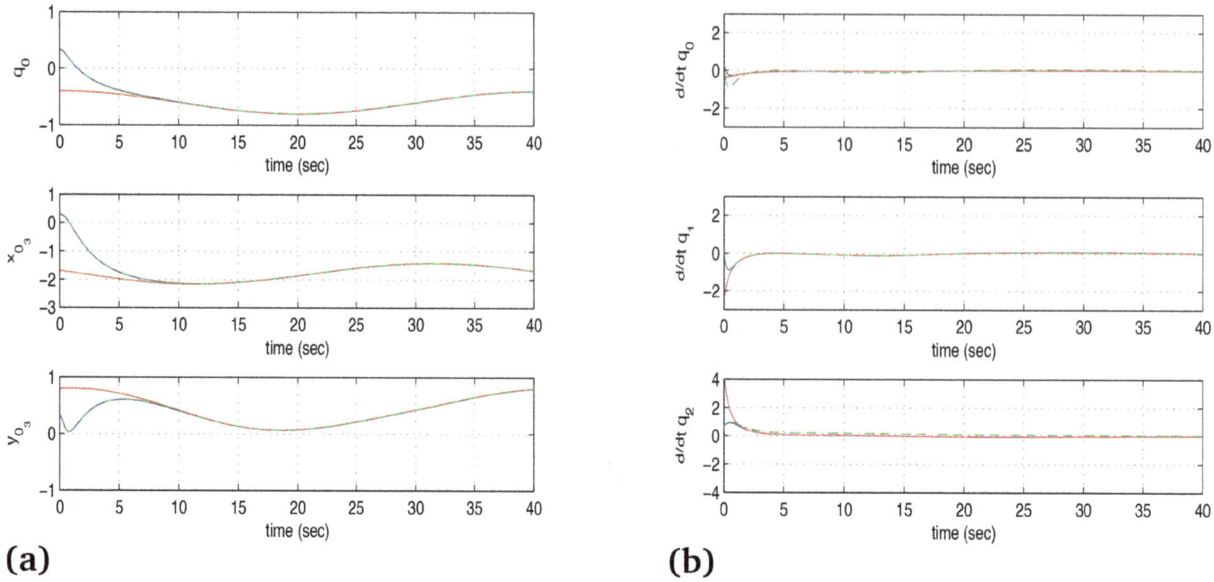

Figure 5.106 *Tracking of setpoint 8 by the free-floating space robotic manipulator: (a) convergence of state variables q_0, $x_{x_{O_3}}$ and y_{O_3} that constitute state subvector x_A (blue lines) to the reference setpoints (red lines) and their KF-based estimation (green line); (b) convergence of state variables $\dot{q}_0$, $\dot{q}_1$ and $\dot{q}_2$ that constitute state subvector q_B (blue lines) to the reference setpoints (red lines) and their KF-based estimation (green line)*

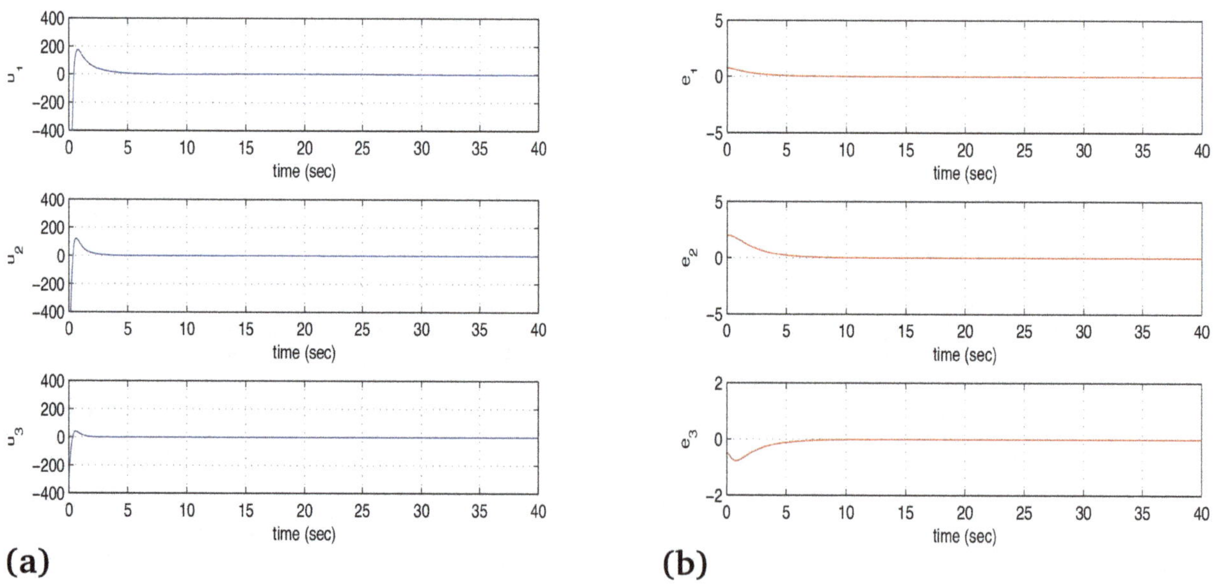

Figure 5.107 *Tracking of setpoint 8 by the free-floating space robotic manipulator: (a) variations of the control inputs u_1, u_2, and u_3; (b) tracking error of state variables q_0, x_{O_3}, and y_{O_3} of state subvector x_A*

in successive loops is dependent on the selection of values for the diagonal gain matrices K_i, $i = 1, 2$ of Eq. (5.207) to Eq. (5.208). The simulation experiments confirm that the proposed multi-loop flatness-based control method achieves fast and accurate tracking of setpoints by all state variables of the space robot under moderate variations of the control inputs. The new control method avoids changes of state variables and state-space model transformations and the control inputs it computes are applied directly on the initial state-space model of the space robot.

Chapter 6

Nonlinear optimal and flatness-based control for space robotics II

Abstract

The chapter analyzes further nonlinear optimal and flatness-based control methods for aerospace systems. First, it solves the nonlinear optimal control problem for the dynamic model of the 3-DOF free-floating space robotic manipulator. Besides, it analyzes nonlinear optimal control for the dynamic model of the underactuated tethered satellite. Next, it generalizes toward nonlinear optimal control for the dynamic model of a tethered multi-satellite system. Finally, it presents results on nonlinear optimal control for the dynamic model of the wing rock effect that appears in aircrafts and reentry space vehicles.

6.1 Nonlinear optimal control for the 3-DOF free-floating space robotic manipulator

6.1.1 Outline

This section elaborates on the use of its nonlinear optimal and flatness-based control methods in aerospace systems. Free-floating space robotic manipulators (FSRM) consist of a floating platform which can be a spacecraft or a satellite and of a robotic arm which is mounted on the platform [1084,994,231]. Such a type of robotic system is widely used for repair tasks in satellites or space stations, as well as for the removal of space debris [413,459,1]. The integrated state-space model of the platform and the robotic manipulator can have many degrees of freedom and can be of high dimensionality [18,693]. In this section, a 3-DOF model of the space robot will be considered again, with the first degree of freedom to be coming from the yaw motion of the space platform and the remaining two degrees of freedom to be coming from the turn motion of the joints of a two-link robotic manipulator which is mounted on the platform [904,994,992,901]. The associated dynamic model takes the typical form of a multi-DOF robotic arm or mobile manipulator and comprises an inertia matrix, a Coriolis forces vector, a damping forces vector, and a control inputs (torques) vector [779,1045,967]. Such a robotic model is feedback linearizable and so far several nonlinear control approaches have been proposed about it [903,902,1039,1042]. These approaches are often based on a transformation of the dynamics of the space robot into a linearized form where the control inputs gain matrix is quadratic and the dimension of the state vector becomes equal to the dimension of the control inputs vector [10,1046]. One can distinguish approaches based on the computed-torque method jointly with disturbance observers for robustifying the control loop, sliding-mode control methods, backstepping control schemes and multi-model feedback control schemes based on linearization of the dynamics of the robotic system at multiple operating points [953,160,1038,339]. Besides, nonlinear model predictive control (NMPC) approaches for space robots follow an optimization concept and do not need state-space model transformations [756,757,907,1064]. Up to now, most approaches for treating the nonlinear optimal control problem for free-floating space robotic manipulators have been based on Nonlinear Model Predictive Control (NMPC) [815,453,411]. In this section, a novel nonlinear optimal control method is proposed for these space robotic systems which is based on sequential linearization of the dynamics of the space robots with Taylor series expansion and on the design of an H-infinity feedback controller [721,725].

It is proven that the dynamic model of the space robot is differentially flat and thus feedback linearizable [711]. To apply the proposed nonlinear optimal control method for the 3-DOF free-floating space robotic manipulator, the dynamic model of this robotic system undergoes first approximate linearization with the use of first-order Taylor series expansion and through the computation of the associated Jacobian matrices [705,60,706]. The modeling error which is due to the truncation of higher-order terms from the Taylor series expansion is considered to be a perturbation which is asymptotically compensated by the robustness of the control algorithm. For the linearized description of the robotic system, an H-infinity (optimal) feedback controller is designed. The H-infinity controller represents a min–max differential game taking place between the control inputs and model uncertainty or disturbance terms. The control inputs try to minimize a cost function which contains a quadratic term of the state vector's tracking error while the model uncertainty and external perturbation terms try to maximize this cost function. To select the controller's feedback gains, an algebraic Riccati equation is repetitively solved at each time step of the control algorithm [732,715]. The global stability properties of the control scheme are proven through Lyapunov analysis. First, it is demonstrated that the control loop satisfies the H-infinity tracking performance criterion. Moreover, under moderate conditions, it is also proven that the control scheme is globally asymptotically stable [721,725]. To implement state estimation-based control without the need to measure the entire state vector of the free-floating space robotic manipulator the H-infinity Kalman filter is used as a robust state estimator. The nonlinear optimal control scheme achieves fast and accurate tracking of the targeted setpoints by the state variables of the controlled system under moderate variations of the control inputs.

6.1.2 Dynamic model of the 3-DOF free-floating space robot

6.1.2.1 State-space form of the space robot

The diagram of the 3-DOF free-floating space robotic manipulator (FSRM) is shown again in Figure 6.1. The inertial reference frame is defined as OXY. The body-fixed reference frame of the floating platform is defined as $O_AX_AY_A$, where its origin O_1 coincides with the center of gravity of the floating platform C_0. The rotation angle of the platform with respect to the inertial reference frame is denoted as q_0. The mass of the platform is m_0, and its moment of inertia is I_0 [904,994,992,901].

The two joints of the rigid-link manipulator are denoted as O_1, O_2, respectively. The center of gravity of the first link is denoted as C_1 and its distance from O_1 is a_1, while its distance from O_2 is b_1. The turn angle of the first link is denoted as q_1. The mass of the first link is m_1, and its moment of inertia is I_1. The center of gravity of the second link is denoted as C_2 and its distance from O_2 is a_2 while its distance from the end-effector is b_2. The turn angle of the second

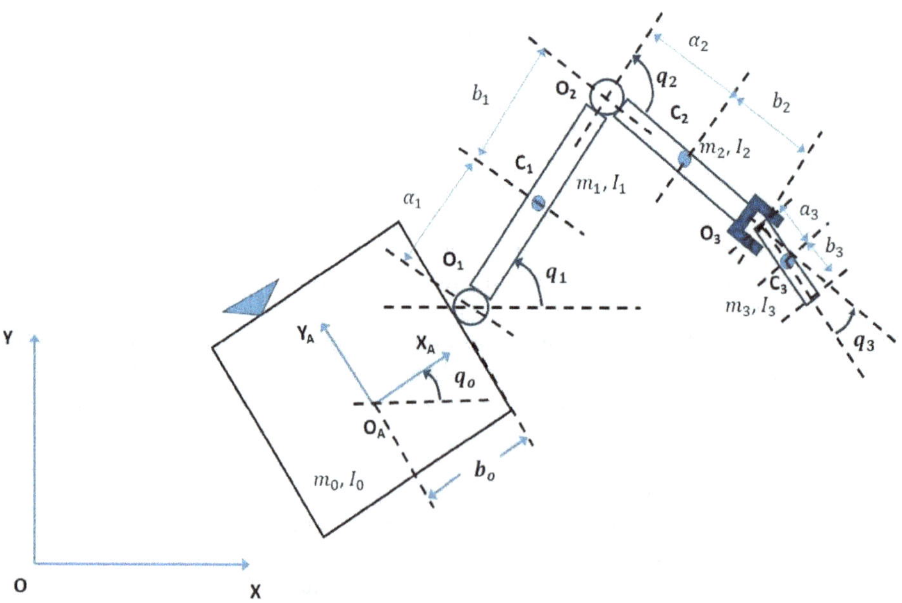

Figure 6.1 Diagram of the 3-DOF free-floating space robotic manipulator

link is q_2. The mass of the second link is m_2, and its moment of inertia is I_2. The end-effector grasps a non-cooperative target of mass m_3 and of moment of inertia I_3. This object forms an angle q_t with the longitudinal axis which is aligned with the second link. The center of gravity of this load is denoted as C_3, and its distance from the end effector is a_3.

The dynamic model of the 3-DOF planar free-floating space robotic manipulator (FSRM) with two links is given by [904,994,992,901]

$$M(q)\ddot{q} + C(q,\dot{q})\dot{q} = \tau + \tilde{d} \tag{6.1}$$

where $M(q)$ is the inertia matrix, $C(q,\dot{q})$ is the Coriolis and centrifugal forces matrix, τ is the torques vector which is generated by the actuators of this robotic system and $\tilde{d}$ is the disturbances vector. The angle vector of this robot is $q = [q_0, q_1, q_2]^T$. About the symmetric inertia matrix it holds that

$$M(q) = \begin{pmatrix} M_{11} & M_{12} & M_{13} \\ M_{21} & M_{22} & M_{23} \\ M_{31} & M_{32} & M_{33} \end{pmatrix} \tag{6.2}$$

where the terms M_{ij}, $i = 1,2,3$ and $j = 1,2,3$ are defined as follows:

$$\begin{aligned} M_{11} = {} & 2c_1cos(q_1) + 2c_2cos(q_2) + 2c_3cos(q_2+q_3) + \\ & +2c_4sin(q_2) + 2c_3sin(q_1+q_2) + c_6 \end{aligned} \tag{6.3}$$

$$\begin{aligned} M_{12} = M_{21} = {} & c_1cos(q_1) + 2c_2cos(q_2) + c_3cos(q_1+q_2) + \\ & +2c_4sin(q_2) + c_5sin(q_1+q_2) + c_7 \end{aligned} \tag{6.4}$$

$$\begin{aligned} M_{13} = M_{31} = {} & c_2cos(q_2) + c_3cos(q_1+q_2) + c_4sin(q_2) + \\ & +c_5sin(q_1+q_2) + c_8 \end{aligned} \tag{6.5}$$

$$M_{22} = 2c_2cos(q_2) + 2c_4sin(q_2) + c_7 \tag{6.6}$$

$$M_{23} = M_{32} = c_2cos(q_2) + c_4sin(q_2) + c_8 \tag{6.7}$$

$$M_{33} = c_8 \tag{6.8}$$

The Coriolis and centrifugal forces matrix is given by [904,994,992,901]

$$C(q,\dot{q}) = \begin{pmatrix} C_{11} & C_{12} & C_{13} \\ C_{21} & C_{22} & C_{23} \\ C_{31} & C_{32} & C_{33} \end{pmatrix} \tag{6.9}$$

The elements C_{ij}, $i = 1,2,3$ and $j = 1,2,3$ of this matrix are given by

$$\begin{aligned} C_{11} = {} & -c_1sin(q_1)\dot{q}_1 - c_2sin(q_2)\dot{q}_2 - c_3sin(q_1+q_2)(\dot{q}_1+\dot{q}_2) + \\ & +c_4cos(q_2)\dot{q}_2 + c_5cos(q_1+q_2)(\dot{q}_1+\dot{q}_2) \end{aligned} \tag{6.10}$$

$$\begin{aligned} C_{12} = {} & -c_1sin(q_1)(\dot{q}_0+\dot{q}_1) - c_2sin(q_2)\dot{q}_2 - c_2sin(q_1+q_2)(\dot{q}_0+\dot{q}_1+\dot{q}_2) + \\ & +c_4cos(q_2)\dot{q}_2 + c_5cos(q_1+q_2)(\dot{q}_1+\dot{q}_2) \end{aligned} \tag{6.11}$$

$$C_{13} = -[c_2sin(q_2) + c_3sin(q_1+q_2) - c_4cos(q_2) - c_7cos(q_1+q_2)](\dot{q}_0+\dot{q}_1+\dot{q}_2) \tag{6.12}$$

$$\begin{aligned} C_{21} = {} & c_1cos(q_1)\dot{q}_0 - c_2sin(q_2)\dot{q}_2 + c_3sin(q_1+q_2)\dot{q}_0 + \\ & +c_4cos(q_2)\dot{q}_2 - c_5cos(q_1+q_2)\dot{q}_0 \end{aligned} \tag{6.13}$$

$$C_{22} = -c_2sin(q_2)\dot{q}_2 + c_4cos(q_2)\dot{q}_2 \tag{6.14}$$

$$C_{23} = [-c_2sin(q_2) + c_4cos(q_2)](\dot{q}_0+\dot{q}_1+\dot{q}_2) \tag{6.15}$$

$$C_{31} = c_2 sin(q_2)(\dot{q}_0 + \dot{q}_1) + c_3 sin(q_1 + q_2)\dot{q}_0 - -c_4 cos(q_2)(\dot{q}_0 + \dot{q}_1) - c_5 cos(q_1 + q_2)\dot{q}_0 \tag{6.16}$$

$$C_{32} = [c_2 sin(q_2) - c_4 cos(q_2)](\dot{q}_0 + \dot{q}_1) \tag{6.17}$$

$$C_{33} = 0 \tag{6.18}$$

Moreover, using that $m_s = m_0 + m_1 + m_2 + m_t$ coefficients c_i, $i = 1, 2, \cdots, 8$ are given by [904,994,992,901]

$$c_1 = m_0 b_0 [m_1 a_1 + m_2 a_1 + m_2 b_1 + m_t a_1 + m_t b_1]/m_s \tag{6.19}$$

$$\begin{aligned} c_2 = [m_0 m_t (a_1 + b_1)(a_2 + b_2 + a_3 cos(q_t)) + m_0 m_2 a_2 (a_1 + b_1) + \\ + m_1 m_2 b_1 a_2 + m_1 m_t b_1 (a_2 + b_2)]/m_s \end{aligned} \tag{6.20}$$

$$c_3 = m_0 b_0 [m_t (a_2 + b_2 + a_3 cos(q_t)) + m_2 a_2]/m_s \tag{6.21}$$

$$c_4 = -m_t a_3 sin(q_t)(m_0 a_1 + m_0 b_1 + m_1 b1)/m_s \tag{6.22}$$

$$c_5 = -m_t m_0 b_0 a_3 sin(q_t)/m_s \tag{6.23}$$

$$\begin{aligned} c_6 = [m_0 m_1 (b_0^2 + a_1^2) + m_0 m_2 (b_0^2 + a + 1^2 + b_1^2 + a_2^2 + 2a_1 b_1) + \\ + m_0 m_t (b_0^2 + a_1^2 + b_1^2 + a_2^2 + b_2^2 + a_3^2 + 2a_1 b_1 + 2a_2 b_2 + 2(a_2 + b_2) a_3 cos(q_t)) + \\ + m_1 m_t (b_1^2 + a_2^2 + b_2^2 + a_3^2 + 2a_2 b_2 + 2(a_2 + b_2) a_3 cos(q_t)) + \\ + m_2 m_t (b_2^2 + a_3^2 + 2b_2 a_3 cos(q_t))]/m_s + I_0 + I_1 + I_2 + I_t \end{aligned} \tag{6.24}$$

$$\begin{aligned} c_7 = [m_0 m_1 a_1^2 + m_0 m_2 (a_1^2 + b_1^2 + a_2^2 + 2a_1 b_1) + m_1 m_2 (b_1^2 + a_2^2) + \\ m_0 m_1 (a_1^2 + b_1^2 + a_2^2 + b_2^2 + a_3^2 + 2a_1 b_1 + 2a_2 b_2 + 2(a_2 + b_2) a_3 cos(q_t)) + \\ + m_1 m_t (b_1^2 + a_2^2 + b_2^2 + a_3^2 + 2a_2 b_2 + 2(a_2 + b_2) a_3 cos(q_t)) + \\ + m_2 m_t (b_2^2 + a_1^2 + 2b_2 a_3 cos(q_t))]/m_s + I_1 + I_2 + I_t \end{aligned} \tag{6.25}$$

$$\begin{aligned} c_8 = [(m_0 + m_1) m_2 a_2^2 + (m_0 + m_1) m_t (a_2^2 + b_2^2 + a_3^2 + 2a_2 b_2 + 2(a_2 + b_2) a_3 cos(q_t)) + \\ + m_2 m_1 (b_2^2 + a_3^2 + 2b_2 a_3 cos(q_t))]/m_s + I_2 + I_t \end{aligned} \tag{6.26}$$

The orientation angle of the non-cooperative target q_t is taken to be constant. Thus, coefficients c_i, $i = 1, 2, \cdots, 8$ are also constants. The state vector of the floating space robotic manipulator is $x = [q_0, \dot{q}_0, q_1, \dot{q}_1, q_2, \dot{q}_2]^T$ which is also denoted as $x = [x_1, x_2, x_3, x_4, x_5, x_6]^T$.

The inverse of the inertia matrix $M(q)$ is given by:

$$M^{-1}(q) = \frac{1}{detM} \begin{pmatrix} \tilde{M}_{11} & -\tilde{M}_{21} & \tilde{M}_{31} \\ -\tilde{M}_{12} & \tilde{M}_{22} & -\tilde{M}_{32} \\ \tilde{M}_{13} & -\tilde{M}_{23} & \tilde{M}_{33} \end{pmatrix} \tag{6.27}$$

where $detM = M_{11}\tilde{M}_{11} = M_{12}\tilde{M}_{12} + M_{13}\tilde{M}_{13}$ and the sub-determinants $\tilde{M}_{ij}$, $i = 1, 2, 3$, $j = 1, 2, 3$ are defined as: $\tilde{M}_{11} = M_{22}M_{33} - M_{32}M_{23}$, $\tilde{M}_{12} = M_{21}M_{33} - M_{31}M_{23}$, $\tilde{M}_{13} = M_{21}M_{32} - M_{31}M_{22}$, $\tilde{M}_{21} = M_{12}M_{33} - M_{32}M_{13}$, $\tilde{M}_{22} = M_{11}M_{33} - M_{21}M_{13}$, $\tilde{M}_{23} = M_{11}M_{32} - M_{31}M_{12}$, $\tilde{M}_{31} = M_{12}M_{23} - M_{22}M_{13}$, $\tilde{M}_{32} = M_{11}M_{23} - M_{21}M_{13}$, $\tilde{M}_{33} = M_{11}M_{22} - M_{21}M_{12}$.

Moreover, the following vector about the Coriolis and centrifugal forces in the space robotic system is computed:

$$\begin{pmatrix} \tilde{C}_1 \\ \tilde{C}_2 \\ \tilde{C}_3 \end{pmatrix} = \begin{pmatrix} C_{11} & C_{12} & C_{13} \\ C_{21} & C_{22} & C_{23} \\ C_{31} & C_{32} & C_{33} \end{pmatrix} \begin{pmatrix} \dot{q}_0 \\ \dot{q}_1 \\ \dot{q}_2 \end{pmatrix} \Rightarrow \begin{pmatrix} \tilde{C}_1 \\ \tilde{C}_2 \\ \tilde{C}_3 \end{pmatrix} = \begin{pmatrix} C_{11}\dot{q}_0 + C_{12}\dot{q}_1 + C_{13}\dot{q}_2 \\ C_{21}\dot{q}_0 + C_{22}\dot{q}_1 + C_{23}\dot{q}_2 \\ C_{31}\dot{q}_0 + C_{32}\dot{q}_1 + C_{33}\dot{q}_2 \end{pmatrix} \tag{6.28}$$

Using this notation, the dynamic model of the space robotic manipulator, without the disturbances vector which is going to be compensated by the robustness of the control scheme, becomes

$$M(q)\ddot{q} + C(q,\dot{q})\dot{q} = \tau \Rightarrow \tag{6.29}$$

Next, by defining $\tilde{C}(q,\dot{q}) = C(q,\dot{q})\dot{q}$, the dynamic model of the space robot becomes

$$\ddot{q} = -M^1(q)\tilde{C}(q,\dot{q}) + M^1(q)\tau \tag{6.30}$$

The dynamic model of the space robot of Eq. (6.30) is written analytically in the form:

$$\begin{pmatrix} \ddot{q}_0 \\ \ddot{q}_1 \\ \ddot{q}_2 \end{pmatrix} = \begin{pmatrix} \frac{-\tilde{M}_{11}\tilde{C}_1+\tilde{M}_{21}\tilde{C}_2-\tilde{M}_{31}\tilde{C}_3}{detM} \\ \frac{\tilde{M}_{12}\tilde{C}_1-\tilde{M}_{22}\tilde{C}_2+\tilde{M}_{32}\tilde{C}_3}{detM} \\ \frac{\tilde{M}_{13}\tilde{C}_1-\tilde{M}_{23}\tilde{C}_2+\tilde{M}_{31}\tilde{C}_3}{detM} \end{pmatrix} + \begin{pmatrix} \frac{\tilde{M}_{11}}{detM} & -\frac{\tilde{M}_{21}}{detM} & \frac{\tilde{M}_{31}}{detM} \\ -\frac{\tilde{M}_{12}}{detM} & \frac{\tilde{M}_{22}}{detM} & \frac{\tilde{M}_{32}}{detM} \\ \frac{\tilde{M}_{13}}{detM} & -\frac{\tilde{M}_{13}}{detM} & \frac{\tilde{M}_{33}}{detM} \end{pmatrix} \begin{pmatrix} \tau_1 \\ \tau_2 \\ \tau_3 \end{pmatrix} \tag{6.31}$$

The notation to be used for the state vector of the space robot is $x = [x_1, x_2, x_3, x_4, x_5, x_6]^T = [q_0, \dot{q}_0, q_1, \dot{q}_1, q_2, \dot{q}_2]^T$ while the notation to be used for the control inputs of this robot is $u = [u_1, u_2, u_3]^T = [\tau_1, \tau_2, \tau_2]^T$. Then the state-space model of the robotic system is written as

$$\begin{pmatrix} \dot{x}_1 \\ \dot{x}_2 \\ \dot{x}_3 \\ \dot{x}_4 \\ \dot{x}_5 \\ \dot{x}_6 \end{pmatrix} = \begin{pmatrix} x_2 \\ \frac{-\tilde{M}_{11}\tilde{C}_1+\tilde{M}_{21}\tilde{C}_2-\tilde{M}_{31}\tilde{C}_3}{detM} \\ x_4 \\ \frac{\tilde{M}_{12}\tilde{C}_1-\tilde{M}_{22}\tilde{C}_2+\tilde{M}_{32}\tilde{C}_3}{detM} \\ x_6 \\ \frac{\tilde{M}_{13}\tilde{C}_1-\tilde{M}_{23}\tilde{C}_2+\tilde{M}_{31}\tilde{C}_3}{detM} \end{pmatrix} + \begin{pmatrix} 0 & 0 & 0 \\ \frac{\tilde{M}_{11}}{detM} & -\frac{\tilde{M}_{21}}{detM} & \frac{\tilde{M}_{31}}{detM} \\ 0 & 0 & 0 \\ -\frac{\tilde{M}_{12}}{detM} & \frac{\tilde{M}_{22}}{detM} & \frac{\tilde{M}_{32}}{detM} \\ 0 & 0 & 0 \\ \frac{\tilde{M}_{13}}{detM} & -\frac{\tilde{M}_{13}}{detM} & \frac{\tilde{M}_{33}}{detM} \end{pmatrix} \begin{pmatrix} u_1 \\ u_2 \\ u_3 \end{pmatrix} \tag{6.32}$$

The dynamic model of the space robot can be also written in the concise state-space form

$$\dot{x} = f(x) + g(x)u \tag{6.33}$$

where $x \in R^{6\times 1}, f(x) \in R^{6\times 1}$, $g(x) \in R^{6\times 3}$, and $u \in R^{3\times 1}$, or equivalently

$$\dot{x} = f(x) + g_1(x)u_1 + g_2(x)u_2 + g_3(x)u_3 \tag{6.34}$$

where $x \in R^{6\times 1}, f(x) \in R^{6\times 1}$, $g_1(x) \in R^{6\times 1}$, $g_2(x) \in R^{6\times 1}$, $g_3(x) \in R^{6\times 1}$ and u_1, u_2, $u_3 \in R$.

6.1.2.2 Differential flatness properties of the dynamics of the space robot

The dynamic model of the free-floating space robotic manipulator is differentially flat, with flat outputs vector $Y = [x_1, x_3, x_5]^T = [q_0, q_1, q_2]^T$. Indeed using that $x_2 = \dot{x}_1$, $x_4 = \dot{x}_3$, and $x_6 = \dot{x}_5$, it holds that x_2, x_4, x_6 are also differential functions of the flat outputs of system, or

$$x_2 = h_2(Y,\dot{Y}) \quad x_4 = h_4(Y,\dot{Y}) \quad x_6 = h_6(Y,\dot{Y}) \tag{6.35}$$

Besides, from the state-space model of the system given in Eq. (6.31), one solves for the inputs vector, thus obtaining

$$\begin{pmatrix} u_1 \\ u_2 \\ u_3 \end{pmatrix} = \begin{pmatrix} \frac{\tilde{M}_{11}}{detM} & -\frac{\tilde{M}_{21}}{detM} & \frac{\tilde{M}_{31}}{detM} \\ -\frac{\tilde{M}_{12}}{detM} & \frac{\tilde{M}_{22}}{detM} & \frac{\tilde{M}_{32}}{detM} \\ \frac{\tilde{M}_{13}}{detM} & -\frac{\tilde{M}_{13}}{detM} & \frac{\tilde{M}_{33}}{detM} \end{pmatrix}^{-1} \left[\begin{pmatrix} \ddot{q}_0 \\ \ddot{q}_1 \\ \ddot{q}_2 \end{pmatrix} - \begin{pmatrix} \frac{-\tilde{M}_{11}\tilde{C}_1+\tilde{M}_{21}\tilde{C}_2-\tilde{M}_{31}\tilde{C}_3}{detM} \\ \frac{\tilde{M}_{12}\tilde{C}_1-\tilde{M}_{22}\tilde{C}_2+\tilde{M}_{32}\tilde{C}_3}{detM} \\ \frac{\tilde{M}_{13}\tilde{C}_1-\tilde{M}_{23}\tilde{C}_2+\tilde{M}_{31}\tilde{C}_3}{detM} \end{pmatrix} \right] \tag{6.36}$$

Since all terms which appear in the right-hand side of the above equation are differential functions of the flat outputs vector, it also holds that the control inputs vector $u=[u_1,u_2,u_3]^T$ is a differential function of the flat outputs vector Y, or

$$u=h_u(Y,\dot{Y}) \tag{6.37}$$

Consequently, the space robot is a differentially flat system. The proof of differential flatness properties for the space robot is also an implicit proof of the system's controllability. Moreover, it demonstrates that the system is input–output linearizable through successive differentiations of its flat outputs. Furthermore, it allows for solving the setpoints definition problem. First, one selects setpoints without constraints, for those state variables which coincide with the flat outputs of the system. For the rest of the state variables, setpoints are chosen using the differential relations which connect these state variables to the flat outputs.

6.1.3 Approximate linearization of the dynamics of the free-floating space robot

The dynamic model of the free-floating space robotic manipulator undergoes approximate linearization around a temporary operating point (x^*,u^*) which is updated at each sampling instance and which is defined by the present value of the system's state vector x^* and by the last sampled value of the control inputs vector u^*. The linearization process is based on first-order Taylor series expansion and on the computation of the associated Jacobian matrices. The modeling error which is due to the truncation of higher-order terms from the Taylor series expansion is considered to be a perturbation which asymptotically compensated by the robustness of the control algorithm.

The initial nonlinear state-space model of the system being in the form $\dot{x}=f(x)+g(x)u$, is written after linearization in the equivalent state-space description

$$\dot{x}=Ax+Bu+\tilde{d} \tag{6.38}$$

where A, B are the Jacobian matrices of the robot which are given by

$$\begin{gathered} A=\nabla_x[f(x)+g(x)u]\mid_{(x^*,u^*)}\Rightarrow \\ A=\nabla_x f(x)\mid_{(x^*,u^*)}+\nabla_x g_1(x)u_1\mid_{(x^*,u^*)}+\nabla_x g_2(x)u_2\mid_{(x^*,u^*)}+\nabla_x g_3(x)u_3\mid_{(x^*,u^*)} \end{gathered} \tag{6.39}$$

$$\begin{gathered} B=\nabla_u[f(x)+g(x)u]\mid_{(x^*,u^*)}\Rightarrow \\ B=g(x)\mid_{(x^*,u^*)} \end{gathered} \tag{6.40}$$

while $\tilde{d}$ is the cumulative disturbances vector which may comprise: (i) the modeling error due to the truncation of higher-order terms from the Taylor series expansion, (ii) exogenous perturbations, and (iii) sensor measurement noise of any distribution.

The Jacobian matrices of the system are computed as follows:

Computation of the Jacobian matrix $\nabla_x f(x)\mid_{(x^*,u^*)}$

First row of the Jacobian matrix $\nabla_x f(x)\mid_{(x^*,u^*)}$: $\frac{\partial f_1}{\partial x_1}=0$, $\frac{\partial f_1}{\partial x_2}=1$, $\frac{\partial f_1}{\partial x_3}=0$, $\frac{\partial f_1}{\partial x_4}=0$, $\frac{\partial f_1}{\partial x_5}=0$, $\frac{\partial f_1}{\partial x_6}=0$.

Second row of the Jacobian matrix $\nabla_x f(x)\mid_{(x^*,u^*)}$: It holds that

$$f_2(x)=\frac{f_{2,num}}{f_{2,den}}=\frac{-\tilde{M}_{11}\tilde{C}_1+\tilde{M}_{21}\tilde{C}_2-\tilde{M}_{31}\tilde{C}_3}{detM} \tag{6.41}$$

Thus, for $i=1,2,\cdots,6$, one gets

$$\frac{\partial f_2}{\partial x_i}=\frac{\frac{\partial f_{2,num}}{\partial x_i}f_{2,den}-f_{2,num}\frac{\partial f_{2,den}}{\partial x_i}}{f_{2,den}^2} \tag{6.42}$$

Moreover, for $i = 1, 2, \cdots, 6$, it holds that

$$\frac{\partial f_{2,num}}{\partial x_i} = -\frac{\partial \tilde{M}_{11}}{\partial x_i}\tilde{C}_1 - \tilde{M}_{11}\frac{\partial \tilde{C}_1}{\partial x_i} + \frac{\partial \tilde{M}_{21}}{\partial x_i}\tilde{C}_2 + \tilde{M}_{21}\frac{\partial \tilde{C}_2}{\partial x_i} - \\ -\frac{\partial \tilde{M}_{31}}{\partial x_i}\tilde{C}_3 - \tilde{M}_{31}\frac{\partial \tilde{C}_3}{\partial x_i} \tag{6.43}$$

Additionally, for $i = 1, 2, \cdots, 6$, one has that

$$\frac{\partial f_{2,den}}{\partial x_i} = \frac{\partial detM}{\partial x_i} \tag{6.44}$$

Third row of the Jacobian matrix $\nabla_x f(x) \mid_{(x^*,u^*)}$: $\frac{\partial f_3}{\partial x_1} = 0$, $\frac{\partial f_3}{\partial x_2} = 0$, $\frac{\partial f_3}{\partial x_3} = 0$, $\frac{\partial f_3}{\partial x_4} = 1$, $\frac{\partial f_3}{\partial x_5} = 0$, $\frac{\partial f_3}{\partial x_6} = 0$.

Fourth row of the Jacobian matrix $\nabla_x f(x) \mid_{(x^*,u^*)}$: It holds that

$$f_4(x) = \frac{f_{4,num}}{f_{4,den}} = \frac{\tilde{M}_{12}\tilde{C}_1 - \tilde{M}_{22}\tilde{C}_2 + \tilde{M}_{32}\tilde{C}_3}{detM} \tag{6.45}$$

Thus, for $i = 1, 2, \cdots, 6$, one gets

$$\frac{\partial f_4}{\partial x_i} = \frac{\frac{\partial f_{4,num}}{\partial x_i} f_{4,den} - f_{4,num}\frac{\partial f_{4,den}}{\partial x_i}}{f_{4,den}^2} \tag{6.46}$$

Moreover, for $i = 1, 2, \cdots, 6$, it holds that

$$\frac{\partial f_{4,num}}{\partial x_i} = \frac{\partial \tilde{M}_{12}}{\partial x_i}\tilde{C}_1 + \tilde{M}_{12}\frac{\partial \tilde{C}_1}{\partial x_i} - \frac{\partial \tilde{M}_{22}}{\partial x_i}\tilde{C}_2 - \tilde{M}_{22}\frac{\partial \tilde{C}_2}{\partial x_i} + \\ +\frac{\partial \tilde{M}_{32}}{\partial x_i}\tilde{C}_3 + \tilde{M}_{32}\frac{\partial \tilde{C}_3}{\partial x_i} \tag{6.47}$$

Additionally, for $i = 1, 2, \cdots, 6$, one has that

$$\frac{\partial f_{4,den}}{\partial x_i} = \frac{\partial detM}{\partial x_i} \tag{6.48}$$

Fifth row of the Jacobian matrix $\nabla_x f(x) \mid_{(x^*,u^*)}$: $\frac{\partial f_5}{\partial x_1} = 0$, $\frac{\partial f_5}{\partial x_2} = 0$, $\frac{\partial f_5}{\partial x_3} = 0$, $\frac{\partial f_5}{\partial x_4} = 0$, $\frac{\partial f_5}{\partial x_5} = 0$, $\frac{\partial f_5}{\partial x_6} = 1$.

Sixth row of the Jacobian matrix $\nabla_x f(x) \mid_{(x^*,u^*)}$: It holds that

$$f_6(x) = \frac{f_{6,num}}{f_{6,den}} = \frac{-\tilde{M}_{13}\tilde{C}_1 + \tilde{M}_{23}\tilde{C}_2 - \tilde{M}_{33}\tilde{C}_3}{detM} \tag{6.49}$$

Thus, for $i = 1, 2, \cdots, 6$, one gets

$$\frac{\partial f_6}{\partial x_i} = \frac{\frac{\partial f_{6,num}}{\partial x_i} f_{6,den} - f_{6,num}\frac{\partial f_{6,den}}{\partial x_i}}{f_{6,den}^2} \tag{6.50}$$

Moreover, for $i = 1, 2, \cdots, 6$, it holds that

$$\frac{\partial f_{6,num}}{\partial x_i} = -\frac{\partial \tilde{M}_{13}}{\partial x_i}\tilde{C}_1 - \tilde{M}_{13}\frac{\partial \tilde{C}_1}{\partial x_i} + \frac{\partial \tilde{M}_{23}}{\partial x_i}\tilde{C}_2 + \tilde{M}_{23}\frac{\partial \tilde{C}_2}{\partial x_i} - \\ -\frac{\partial \tilde{M}_{33}}{\partial x_i}\tilde{C}_3 - \tilde{M}_{33}\frac{\partial \tilde{C}_3}{\partial x_i} \tag{6.51}$$

Additionally, for $i = 1, 2, \cdots, 6$, one has that

$$\frac{\partial f_{6,den}}{\partial x_i} = \frac{\partial detM}{\partial x_i} \tag{6.52}$$

Computation of the Jacobian matrix $\nabla_x g_1(x) \mid_{(x^*,u^*)}$

First row of the Jacobian matrix $\nabla_x g_1(x) \mid_{(x^*,u^*)}$: $\frac{\partial g_{11}}{\partial x_i} = 0$ for $i = 1, 2, \cdots, 6$.

Second row of the Jacobian matrix $\nabla_x g_1(x) \mid_{(x^*,u^*)}$: For $i = 1, 2, \cdots, 6$, it holds that

$$\frac{\partial g_{21}}{\partial x_i} = \frac{\frac{\partial \tilde{M}_{11}}{\partial x_i} detM - \tilde{M}_{11}\frac{\partial detM}{\partial x_i}}{detM^2} \tag{6.53}$$

Third row of the Jacobian matrix $\nabla_x g_1(x)\mid_{(x^*,u^*)}$: $\frac{\partial g_{31}}{\partial x_i}=0$ for $i=1,2,\cdots,6$.

Fourth row of the Jacobian matrix $\nabla_x g_1(x)\mid_{(x^*,u^*)}$: For $i=1,2,\cdots,6$, it holds that

$$\frac{\partial g_{41}}{\partial x_i}=\frac{\frac{-\partial \tilde{M}_{12}}{\partial x_i}detM+\tilde{M}_{12}\frac{\partial detM}{\partial x_i}}{detM^2} \tag{6.54}$$

Fifth row of the Jacobian matrix $\nabla_x g_1(x)\mid_{(x^*,u^*)}$: $\frac{\partial g_{51}}{\partial x_i}=0$ for $i=1,2,\cdots,6$.

Sixth row of the Jacobian matrix $\nabla_x g_1(x)\mid_{(x^*,u^*)}$: For $i=1,2,\cdots,6$, it holds that

$$\frac{\partial g_{61}}{\partial x_i}=\frac{\frac{\partial \tilde{M}_{13}}{\partial x_i}detM-\tilde{M}_{13}\frac{\partial detM}{\partial x_i}}{detM^2} \tag{6.55}$$

Computation of the Jacobian matrix $\nabla_x g_2(x)\mid_{(x^*,u^*)}$

First row of the Jacobian matrix $\nabla_x g_2(x)\mid_{(x^*,u^*)}$: $\frac{\partial g_{12}}{\partial x_i}=0$ for $i=1,2,\cdots,6$.

Second row of the Jacobian matrix $\nabla_x g_2(x)\mid_{(x^*,u^*)}$: For $i=1,2,\cdots,6$, it holds that

$$\frac{\partial g_{22}}{\partial x_i}=\frac{\frac{-\partial \tilde{M}_{21}}{\partial x_i}detM+\tilde{M}_{21}\frac{\partial detM}{\partial x_i}}{detM^2} \tag{6.56}$$

Third row of the Jacobian matrix $\nabla_x g_2(x)\mid_{(x^*,u^*)}$: $\frac{\partial g_{32}}{\partial x_i}=0$ for $i=1,2,\cdots,6$.

Fourth row of the Jacobian matrix $\nabla_x g_2(x)\mid_{(x^*,u^*)}$: For $i=1,2,\cdots,6$, it holds that

$$\frac{\partial g_{42}}{\partial x_i}=\frac{\frac{\partial \tilde{M}_{22}}{\partial x_i}detM-\tilde{M}_{22}\frac{\partial detM}{\partial x_i}}{detM^2} \tag{6.57}$$

Fifth row of the Jacobian matrix $\nabla_x g_2(x)\mid_{(x^*,u^*)}$: $\frac{\partial g_{52}}{\partial x_i}=0$ for $i=1,2,\cdots,6$.

Sixth row of the Jacobian matrix $\nabla_x g_2(x)\mid_{(x^*,u^*)}$: For $i=1,2,\cdots,6$, it holds that

$$\frac{\partial g_{62}}{\partial x_i}=\frac{\frac{-\partial \tilde{M}_{23}}{\partial x_i}detM-\tilde{M}_{23}\frac{\partial detM}{\partial x_i}}{detM^2} \tag{6.58}$$

Computation of the Jacobian matrix $\nabla_x g_3(x)\mid_{(x^*,u^*)}$

First row of the Jacobian matrix $\nabla_x g_3(x)\mid_{(x^*,u^*)}$: $\frac{\partial g_{13}}{\partial x_i}=0$ for $i=1,2,\cdots,6$.

Second row of the Jacobian matrix $\nabla_x g_3(x)\mid_{(x^*,u^*)}$: For $i=1,2,\cdots,6$, it holds that

$$\frac{\partial g_{23}}{\partial x_i}=\frac{\frac{\partial \tilde{M}_{31}}{\partial x_i}detM-\tilde{M}_{31}\frac{\partial detM}{\partial x_i}}{detM^2} \tag{6.59}$$

Third row of the Jacobian matrix $\nabla_x g_3(x)\mid_{(x^*,u^*)}$: $\frac{\partial g_{33}}{\partial x_i}=0$ for $i=1,2,\cdots,6$.

Fourth row of the Jacobian matrix $\nabla_x g_3(x)\mid_{(x^*,u^*)}$: For $i=1,2,\cdots,6$, it holds that

$$\frac{\partial g_{43}}{\partial x_i}=\frac{\frac{-\partial \tilde{M}_{32}}{\partial x_i}detM+\tilde{M}_{32}\frac{\partial detM}{\partial x_i}}{detM^2} \tag{6.60}$$

Fifth row of the Jacobian matrix $\nabla_x g_3(x)\mid_{(x^*,u^*)}$: $\frac{\partial g_{53}}{\partial x_i}=0$ for $i=1,2,\cdots,6$.

Sixth row of the Jacobian matrix $\nabla_x g_3(x)\mid_{(x^*,u^*)}$: For $i=1,2,\cdots,6$, it holds that

$$\frac{\partial g_{63}}{\partial x_i}=\frac{\frac{\partial \tilde{M}_{33}}{\partial x_i}detM-\tilde{M}_{33}\frac{\partial detM}{\partial x_i}}{detM^2} \tag{6.61}$$

Computation of the partial derivatives of the sub-determinants $\tilde{M}_{ij}$, $i=1,2,3$, $j=1,2,3$ of the inertia matrix.

$$\frac{\partial \tilde{M}_{11}}{\partial x_i}=\frac{\partial M_{22}}{\partial x_i}M_{33}+M_{22}\frac{\partial M_{33}}{\partial x_i}-\frac{\partial M_{32}}{\partial x_i}M_{23}-M_{32}\frac{\partial M_{23}}{\partial x_i} \tag{6.62}$$

$$\frac{\partial \tilde{M}_{12}}{\partial x_i}=\frac{\partial M_{21}}{\partial x_i}M_{23}+M_{21}\frac{\partial M_{23}}{\partial x_i}-\frac{\partial M_{31}}{\partial x_i}M_{23}-M_{31}\frac{\partial M_{23}}{\partial x_i} \tag{6.63}$$

$$\frac{\partial \tilde{M}_{13}}{\partial x_i}=\frac{\partial M_{21}}{\partial x_i}M_{32}+M_{21}\frac{\partial M_{32}}{\partial x_i}-\frac{\partial M_{31}}{\partial x_i}M_{22}-M_{31}\frac{\partial M_{22}}{\partial x_i} \tag{6.64}$$

$$\frac{\partial \tilde{M}_{21}}{\partial x_i}=\frac{\partial M_{12}}{\partial x_i}M_{33}+M_{12}\frac{\partial M_{33}}{\partial x_i}-\frac{\partial M_{32}}{\partial x_i}M_{13}-M_{32}\frac{\partial M_{13}}{\partial x_i} \tag{6.65}$$

$$\frac{\partial \tilde{M}_{22}}{\partial x_i}=\frac{\partial M_{11}}{\partial x_i}M_{33}+M_{11}\frac{\partial M_{33}}{\partial x_i}-\frac{\partial M_{31}}{\partial x_i}M_{13}-M_{31}\frac{\partial M_{13}}{\partial x_i} \tag{6.66}$$

$$\frac{\partial \tilde{M}_{23}}{\partial x_i}=\frac{\partial M_{11}}{\partial x_i}M_{32}+M_{11}\frac{\partial M_{32}}{\partial x_i}-\frac{\partial M_{31}}{\partial x_i}M_{12}-M_{31}\frac{\partial M_{12}}{\partial x_i} \tag{6.67}$$

$$\frac{\partial \tilde{M}_{31}}{\partial x_i}=\frac{\partial M_{12}}{\partial x_i}M_{23}+M_{12}\frac{\partial M_{23}}{\partial x_i}-\frac{\partial M_{22}}{\partial x_i}M_{13}-M_{22}\frac{\partial M_{13}}{\partial x_i} \tag{6.68}$$

$$\frac{\partial \tilde{M}_{32}}{\partial x_i}=\frac{\partial M_{11}}{\partial x_i}M_{23}+M_{11}\frac{\partial M_{23}}{\partial x_i}-\frac{\partial M_{21}}{\partial x_i}M_{13}-M_{21}\frac{\partial M_{13}}{\partial x_i} \tag{6.69}$$

$$\frac{\partial \tilde{M}_{33}}{\partial x_i}=\frac{\partial M_{11}}{\partial x_i}M_{22}+M_{11}\frac{\partial M_{22}}{\partial x_i}-\frac{\partial M_{21}}{\partial x_i}M_{12}-M_{21}\frac{\partial M_{12}}{\partial x_i} \tag{6.70}$$

Computation of the partial derivatives of the elements of the Coriolis vector $\tilde{C}_i$, $i=1,2,3$. It holds that

$$\begin{aligned}\tilde{C}_1&=C_{11}\dot{q}_0+C_{12}\dot{q}_1+C_{13}\dot{q}_2 \Rightarrow \tilde{C}_1=C_{11}x_2+C_{12}x_4+C_{13}x_6\\ \tilde{C}_2&=C_{21}\dot{q}_0+C_{22}\dot{q}_1+C_{23}\dot{q}_2 \Rightarrow \tilde{C}_2=C_{21}x_2+C_{22}x_4+C_{23}x_6\\ \tilde{C}_3&=C_{31}\dot{q}_0+C_{32}\dot{q}_1+C_{33}\dot{q}_2 \Rightarrow \tilde{C}_1=C_{31}x_2+C_{32}x_4+C_{33}x_6\end{aligned} \tag{6.71}$$

Thus, one obtains that for $i=1,3,5$

$$\frac{\partial \tilde{C}_1}{\partial x_i}=\frac{\partial C_{11}}{\partial x_i}x_2+\frac{\partial C_{12}}{\partial x_i}x_4+\frac{\partial C_{13}}{\partial x_i}x_6 \tag{6.72}$$

and next

$$\frac{\partial \tilde{C}_1}{\partial x_2}=\frac{\partial C_{11}}{\partial x_2}x_2+C_{11}+\frac{\partial C_{12}}{\partial x_2}x_4+\frac{\partial C_{13}}{\partial x_2}x_6 \tag{6.73}$$

$$\frac{\partial \tilde{C}_1}{\partial x_4}=\frac{\partial C_{11}}{\partial x_4}x_2+\frac{\partial C_{12}}{\partial x_4}x_4+C_{12}+\frac{\partial C_{13}}{\partial x_4}x_6 \tag{6.74}$$

$$\frac{\partial \tilde{C}_1}{\partial x_6}=\frac{\partial C_{11}}{\partial x_6}x_2+\frac{\partial C_{12}}{\partial x_i}x_6+\frac{\partial C_{13}}{\partial x_6}x_6+C_{13} \tag{6.75}$$

Additionally, for $i=1,2,3$, one obtains

$$\frac{\partial \tilde{C}_2}{\partial x_i}=\frac{\partial C_{21}}{\partial x_i}x_2+\frac{\partial C_{22}}{\partial x_i}x_4+\frac{\partial C_{23}}{\partial x_i}x_6 \tag{6.76}$$

and next

$$\frac{\partial \tilde{C}_2}{\partial x_2}=\frac{\partial C_{21}}{\partial x_2}x_2+C_{21}+\frac{\partial C_{22}}{\partial x_2}x_4+\frac{\partial C_{23}}{\partial x_2}x_6 \tag{6.77}$$

$$\frac{\partial \tilde{C}_2}{\partial x_4}=\frac{\partial C_{21}}{\partial x_4}x_2+\frac{\partial C_{22}}{\partial x_4}x_4+C_{22}+\frac{\partial C_{23}}{\partial x_4}x_6 \tag{6.78}$$

$$\frac{\partial \tilde{C}_2}{\partial x_6}=\frac{\partial C_{21}}{\partial x_6}x_2+\frac{\partial C_{22}}{\partial x_i}x_6+\frac{\partial C_{23}}{\partial x_6}x_6+C_{23} \tag{6.79}$$

Moreover, for $i=1,2,3$, one obtains

$$\frac{\partial \tilde{C}_3}{\partial x_i}=\frac{\partial C_{31}}{\partial x_i}x_2+\frac{\partial C_{32}}{\partial x_i}x_4+\frac{\partial C_{33}}{\partial x_i}x_6 \tag{6.80}$$

and next

$$\frac{\partial \tilde{C}_3}{\partial x_2}=\frac{\partial C_{31}}{\partial x_2}x_2+C_{31}+\frac{\partial C_{32}}{\partial x_2}x_4+\frac{\partial C_{33}}{\partial x_2}x_6 \tag{6.81}$$

$$\frac{\partial \tilde{C}_3}{\partial x_4}=\frac{\partial C_{31}}{\partial x_4}x_2+\frac{\partial C_{32}}{\partial x_4}x_4+C_{32}+\frac{\partial C_{33}}{\partial x_4}x_6 \tag{6.82}$$

$$\frac{\partial \tilde{C}_3}{\partial x_6} = \frac{\partial C_{31}}{\partial x_6}x_2 + \frac{\partial C_{32}}{\partial x_i}x_6 + \frac{\partial C_{33}}{\partial x_6}x_6 + C_{33} \tag{6.83}$$

Computation of the partial derivatives of the determinant of the inertia matrix M. It holds that for $i = 1, 2, \cdots, 6$

$$\begin{aligned}\frac{\partial detM}{\partial x_i} = & \frac{\partial M_{11}}{\partial x_i}\tilde{M}_{11} + M_{11}\frac{\partial M_{11}}{\partial x_i} - \frac{\partial M_{12}}{\partial x_i}\tilde{M}_{12} - M_{12}\frac{\partial M_{12}}{\partial x_i} + \\ & + \frac{\partial M_{13}}{\partial x_i}\tilde{M}_{13} + M_{13}\frac{\partial M_{13}}{\partial x_i}\end{aligned} \tag{6.84}$$

Computation of the partial derivatives of the elements M_{ij}, $i = 1, 2, 3, j = 1, 2, 3$ of the inertia matrix M:

$\frac{\partial M_{11}}{\partial x_1} = 0$, $\frac{\partial M_{11}}{\partial x_2} = 0$, $\frac{\partial M_{11}}{\partial x_3} = -2c_1 sin(x_3) - 2c_3 sin(x_3 + x_5) + 2c_3 cos(x_3 + x_5)$, $\frac{\partial M_{11}}{\partial x_4} = 0$, $\frac{\partial M_{11}}{\partial x_5} = -2c_2 sin(x_5) - 2c_3 sin(x_3 + x_5) + 2c_4 cos(x_5) + 2c_3 cos(x_3 + x_5)$, $\frac{\partial M_{11}}{\partial x_6} = 0$.

$\frac{\partial M_{12}}{\partial x_1} = 0$, $\frac{\partial M_{12}}{\partial x_2} = 0$, $\frac{\partial M_{12}}{\partial x_3} = -c_1 sin(x_3) - c_3 sin(x_3 + x_5) + c_5 cos(x_3 + x_5)$, $\frac{\partial M_{12}}{\partial x_4} = 0$, $\frac{\partial M_{12}}{\partial x_5} = -2c_2 sin(x_5) - c_3 sin(x_3 + x_5) + 2c_4 cos(x_5) + c_5 cos(x_3 + x_5)$, $\frac{\partial M_{12}}{\partial x_6} = 0$.

$\frac{\partial M_{13}}{\partial x_1} = 0$, $\frac{\partial M_{13}}{\partial x_2} = 0$, $\frac{\partial M_{13}}{\partial x_3} = -c_3 sin(x_3 + x_5) + c_5 cos(x_3 + x_5)$, $\frac{\partial M_{13}}{\partial x_4} = 0$, $\frac{\partial M_{13}}{\partial x_5} = -c_2 sin(x_5) - c_3 sin(x_3 + x_5) + c_4 cos(x_5) + c_5 cos(x_3 + x_5)$, $\frac{\partial M_{13}}{\partial x_6} = 0$.

$\frac{\partial M_{21}}{\partial x_1} = 0$, $\frac{\partial M_{21}}{\partial x_2} = 0$, $\frac{\partial M_{21}}{\partial x_3} = -c_1 sin(x_3) - c_3 sin(x_3 + x_5) + c_5 cos(x_3 + x_5)$, $\frac{\partial M_{21}}{\partial x_4} = 0$, $\frac{\partial M_{21}}{\partial x_5} = -2c_2 sin(x_5) - c_3 sin(x_3 + x_5) + 2c_4 cos(x_5) + c_5 cos(x_3 + x_5)$, $\frac{\partial M_{21}}{\partial x_6} = 0$.

$\frac{\partial M_{22}}{\partial x_1} = 0$, $\frac{\partial M_{22}}{\partial x_2} = 0$, $\frac{\partial M_{22}}{\partial x_3} = 0$, $\frac{\partial M_{22}}{\partial x_4} = 0$, $\frac{\partial M_{22}}{\partial x_5} = -2c_2 sin(x_5) + 2c_4 cos(x_5)$, $\frac{\partial M_{22}}{\partial x_6} = 0$.
$\frac{\partial M_{23}}{\partial x_1} = 0$, $\frac{\partial M_{23}}{\partial x_2} = 0$, $\frac{\partial M_{23}}{\partial x_3} = 0$, $\frac{\partial M_{23}}{\partial x_4} = 0$, $\frac{\partial M_{23}}{\partial x_5} = -c_2 sin(x_5) + c_4 cos(x_5)$, $\frac{\partial M_{23}}{\partial x_6} = 0$.

$\frac{\partial M_{31}}{\partial x_1} = 0$, $\frac{\partial M_{31}}{\partial x_2} = 0$, $\frac{\partial M_{31}}{\partial x_3} = -c_3 sin(x_3 + x_5) + c_5 cos(x_3 + x_5)$, $\frac{\partial M_{31}}{\partial x_4} = 0$, $\frac{\partial M_{31}}{\partial x_5} = -c_2 sin(x_5) - c_3 sin(x_3 + x_5) + c_4 cos(x_5) + c_5 cos(x_3 + x_5)$, $\frac{\partial M_{31}}{\partial x_6} = 0$.

$\frac{\partial M_{32}}{\partial x_2} = 0$, $\frac{\partial M_{32}}{\partial x_2} = 0$, $\frac{\partial M_{32}}{\partial x_3} = 0$, $\frac{\partial M_{32}}{\partial x_4} = 0$, $\frac{\partial M_{32}}{\partial x_2} = -c_2 sin(x_5)$, $\frac{\partial M_{32}}{\partial x_6} = 0/$
$\frac{\partial M_{31}}{\partial x_i} = 0$ for $i = 1, 2, \cdots, 6$.

Computation of the partial derivatives of the elements C_{ij}, $i = 1, \cdots, 6$ of the Coriolis matrix $\tilde{C}$

$\frac{\partial C_{11}}{\partial x_1} = 0$, $\frac{\partial C_{11}}{\partial x_2} = 0$, $\frac{\partial C_{11}}{\partial x_3} = -c_1 cos(x_3)x_4 - c_3 cos(x_3 + x_5)(x_4 + x_6) - c_5 sin(x_3 + x_5)(x_4 + x_6)$, $\frac{\partial C_{11}}{\partial x_4} = -c_1 sin(x_5) - c_3 sin(x_3 + x_5) + c_5 cos(x_3 + x_5)$, $\frac{\partial C_{11}}{\partial x_5} = -c_2 cos(x_5)x_6 - c_3 cos(x_3 + x_5)(x_4 + x_6) - c_4 sin(x_5)x_6 - c_5 sin(x_3 + x_5)(x_4 + x_6)$, $\frac{\partial C_{11}}{\partial x_6} = -c_2 sin(x_5) - c_3 sin(x_4 + x_5) + c_4 cos(x_5) + c_5 cos(x_3 + x_5)$.

$\frac{\partial C_{12}}{\partial x_1} = 0$, $\frac{\partial C_{12}}{\partial x_2} = -c_1 sin(x_3) - c_2 sin(x_3 + x_5)$, $\frac{\partial C_{12}}{\partial x_3} = -c_1 cos(x_3)(x_2 + x_4) - c_2 cos(x_3 + x_5)(x_2 + x_4 + x_6) + c_5 sin(x_3 + x_5)(x_4 + x_6)$, $\frac{\partial C_{12}}{\partial x_4} = -c_{11} sin(x_3) - c_2 sin(x_3 + x_5) + c_5 cos(x_3 + x_5)$, $\frac{\partial C_{12}}{\partial x_5} = -c_2 cos(x_5)x_6 - c_3 cos(x_3 + x_5)(x_2 + x_4 + x_6) - c_4 sin(x_5)x_6 - c_5 sin(x_3 + x_5)(x_4)$, $\frac{\partial C_{12}}{\partial x_6} = -c_2 sin(x_5) - c_3 sin(x_3 + x_5) + c_4 cos(x_5) + c_5 cos(x_3 + x_5)$.

$\frac{\partial C_{13}}{\partial x_1} = 0$, $\frac{\partial C_{13}}{\partial x_2} = -[c_2 sin(x_5) + c_3 sin(x_3 + x_5) - c_4 cos(x_5) - c_7 cos(x_3 + x_5)]$, $\frac{\partial C_{13}}{\partial x_3} = -[c_3 cos(x_3 + x_5) + c_7 sin(x_3 + x_5)](x_2 + x_4 + x_6)$, $\frac{\partial C_{13}}{\partial x_4} = -[c_2 sin(x_5) + c_3 sin(x_3 + x_5) - c_4 cos(x_5) - c_7 cos(x_3 + x_5)]$, $\frac{\partial C_{13}}{\partial x_5} = -[c_2 cos(x_5) + c_3 cos(x_3 + x_5) + c_4 sin(x_5) + c_7 sin(x_3 + x_5)](x_2 + x_4 + x_6)$, $\frac{\partial C_{13}}{\partial x_6} = -[c_2 sin(x_5) + c_3 sin(x_3 + x_5) - c_4 cos(x_5) - c_7 cos(x_3 + x_5)]$.

$\frac{\partial C_{21}}{\partial x_1} = 0$, $\frac{\partial C_{21}}{\partial x_2} = c_1 sin(x_3) + c_3 sin(x_3 + x_5) - c_5 cos(x_3 + x_5)$, $\frac{\partial C_{21}}{\partial x_3} = c_1 cos(x_3)x_2 + c_3 cos(x_3 + x_5)x_2 + c_5 sin(x_3 + x_5)x_2$, $\frac{\partial C_{21}}{\partial x_4} = 0$, $\frac{\partial C_{21}}{\partial x_5} = -c_2 cos(x_5)x_6 + c_3 cos(x_3 + x_5)x_2 - c_4 sin(x_5)x_6 + c_3 sin(x_3 + x_5)x_2$, $\frac{\partial C_{21}}{\partial x_6} = -c_2 sin(x_5) + c_4 cos(x_5)$.

$\frac{\partial C_{22}}{\partial x_1}=0$, $\frac{\partial C_{22}}{\partial x_2}=0$, $\frac{\partial C_{22}}{\partial x_3}=0$, $\frac{\partial C_{22}}{\partial x_4}=0$, $\frac{\partial C_{22}}{\partial x_5}=-c_2cos(x_5)x_6-c_4sin(x_5)x_6$, $\frac{\partial C_{22}}{\partial x_6}=-c_2sin(x_5)+c_4cos(x_5)$.

$\frac{\partial C_{23}}{\partial x_1}=0$, $\frac{\partial C_{23}}{\partial x_2}=[-c_2sin(x_5)+c_4cos(x_5)]$, $\frac{\partial C_{23}}{\partial x_3}=0$, $\frac{\partial C_{23}}{\partial x_4}=[-c_2sin(x_5)+c_4cos(x_5)]$, $\frac{\partial C_{23}}{\partial x_5}=[-c_2cos(x_5)-c_4sin(x_5)](x_2+x_4+x_6)$, $\frac{\partial C_{23}}{\partial x_6}=[-c_2sin(x_5)+c_4cos(x_5)]$.

$\frac{\partial C_{31}}{\partial x_1}=0$, $\frac{\partial C_{31}}{\partial x_2}=c_2sin(x_5)+c_3sin(x_3+x_5)-c_4cos(x_5)-c_5cos(x_3+x_5)$, $\frac{\partial C_{31}}{\partial x_3}=c_3cos(x_3+x_5)x_2+c_5sin(x_3+x_5)x_2$, $\frac{\partial C_{31}}{\partial x_4}=c_2sin(x_5)-c_4cos(x_5)$, $\frac{\partial C_{31}}{\partial x_5}=c_2cos(x_5)(x_2+x_4)+c_3cos(x_3+x_5)x_2+c_4sin(x_5)(x_2+x_4)+c_5sin(x_3+x_5)x_2$.

$\frac{\partial C_{32}}{\partial x_1}=0$, $\frac{\partial C_{32}}{\partial x_2}=[c_2sin(x_5)-c_4cos(x_5)]$, $\frac{\partial C_{32}}{\partial x_3}=0$, $\frac{\partial C_{32}}{\partial x_4}=[c_2sin(x_5)-c_4cos(x_5)]$, $\frac{\partial C_{32}}{\partial x_5}=[c_2cos(x_5)+c_4sin(x_5)](x_2+x_4)$, $\frac{\partial C_{32}}{\partial x_6}=0$.

$\frac{\partial C_{33}}{\partial x_i}=0$, for $i=1,2,\cdots,6$.

The linearization approach which has been followed for implementing the nonlinear optimal control scheme results in a quite accurate model of the system's dynamics. Consider for instance the following affine-in-the-input state-space model

$$\dot{x}=f(x)+g(x)u\Rightarrow$$
$$\dot{x}=[f(x^*)+\nabla_x f(x)\mid_{x^*}(x-x^*)]+[g(x^*)+\nabla_x g(x)\mid_{x^*}(x-x^*)]u^*+g(x^*)u^*+g(x^*)(u-u^*)+\tilde{d}_1\Rightarrow \tag{6.85}$$
$$\dot{x}=[\nabla_x f(x)\mid_{x^*}+\nabla_x g(x)\mid_{x^*}u^*]x+g(x^*)u-[\nabla_x f(x)\mid_{x^*}+\nabla_x g(x)\mid_{x^*}u^*]x^*+f(x^*)+g(x^*)u^*+\tilde{d}_1$$

where $\tilde{d}_1$ is the modeling error due to truncation of higher order terms in the Taylor series expansion of $f(x)$ and $g(x)$. Next, by defining $A=[\nabla_x f(x)\mid_{x^*}+\nabla_x g(x)\mid_{x^*}u^*]$, $B=g(x^*)$, one obtains

$$\dot{x}=Ax+Bu-Ax^*+f(x^*)+g(x^*)u^*+\tilde{d}_1 \tag{6.86}$$

Moreover, by denoting $\tilde{d}=-Ax^*+f(x^*)+g(x^*)u^*+\tilde{d}_1$ about the cumulative modeling error term in the Taylor series expansion, one has

$$\dot{x}=Ax+Bu+\tilde{d} \tag{6.87}$$

which is the approximately linearized model of the dynamics of the system of Eq. (6.38). The term $f(x^*)+g(x^*)u^*$ is the derivative of the state vector at (x^*,u^*) which is almost annihilated by $-Ax^*$.

6.1.4 The nonlinear H-infinity control

6.1.4.1 Min–max control and disturbance rejection

In the H_∞ control approach, a feedback control scheme is designed for setpoint tracking by the state vector of the space robot and for simultaneous disturbance rejection, considering that the disturbance affects the system in the worst possible manner. For the approximately linearized model of Eq. (6.38), the disturbances' effects are incorporated in the following quadratic cost function [725]

$$J(t)=\frac{1}{2}\int_0^T[y^T(t)y(t)+ru^T(t)u(t)-\rho^2\tilde{d}^T(t)\tilde{d}(t)]dt,\quad r,\rho>0 \tag{6.88}$$

The significance of the negative sign in the cost function's term that is associated with the perturbation variable $\tilde{d}(t)$ is that the disturbance tries to maximize the cost function $J(t)$ while the control signal $u(t)$ tries to minimize it.

The physical meaning of the relation given above is that the control signal and the disturbances compete with each other within a min–max differential game. This problem of min–max optimization can be written as $min_u max_{\tilde{d}} J(u,\tilde{d})$. The objective of the optimization procedure is to compute a control signal $u(t)$ which can compensate for the worst possible disturbance that affects the system. However, the solution to the min–max optimization problem is directly related to the value of parameter ρ. This means that there is an upper bound in the disturbances magnitude that can be annihilated by the control signal.

6.1.4.2 Design of the stabilizing feedback controller

After linearization around its current operating point (x^*, u^*), the dynamic model of the 3-DOF free-floating space robotic manipulator is written as

$$\dot{x} = Ax + Bu + d_1 \tag{6.89}$$

Parameter d_1 stands for the linearization error in the 3-DOF space robot's model appearing previously in Eq. (6.38). The reference setpoints for the state vector of the space robot are denoted by $\mathbf{x_d} = [x_1^d, \cdots, x_6^d]$. Tracking of this trajectory is achieved after applying the control input u_d. At every time instant, the control input u_d is assumed to differ from the control input u appearing in Eq. (6.89) by an amount equal to Δu, that is $u_d = u + \Delta u$

$$\dot{x}_d = Ax_d + Bu_d + d_2 \tag{6.90}$$

The dynamics of the controlled system described in Eq. (6.89) can be also written as

$$\dot{x} = Ax + Bu + Bu_d - Bu_d + d_1 \tag{6.91}$$

and by denoting $d_3 = -Bu_d + d_1$ as an aggregate disturbance term, one obtains

$$\dot{x} = Ax + Bu + Bu_d + d_3 \tag{6.92}$$

By subtracting Eq. (6.90) from Eq. (6.92), one has

$$\dot{x} - \dot{x}_d = A(x - x_d) + Bu + d_3 - d_2 \tag{6.93}$$

By denoting the tracking error as $e = x - x_d$ and the aggregate disturbance term as $L\tilde{d} = d_3 - d_2$, the tracking error dynamics becomes

$$\dot{e} = Ae + Bu + L\tilde{d} \tag{6.94}$$

where L is a disturbance inputs gain matrix. For the linearized system, the cost function of Eq. (6.88) is defined, where coefficient r determines the penalization of the control input and the weight coefficient ρ determines the reward of the disturbances' effects. Then, as it will be also shown in Section 6.1.5, the optimal feedback control law is given by

$$u(t) = -Ke(t) \tag{6.95}$$

with $K = \frac{1}{r}B^TP$ where P is a positive definite symmetric matrix which is obtained from the solution of the Riccati equation

$$A^TP + PA + Q - P(\tfrac{2}{r}BB^T - \tfrac{1}{\rho^2}LL^T)P = 0 \tag{6.96}$$

where Q is also a semi-positive definite symmetric matrix. The worst-case disturbance is given by $\tilde{d}(t) = \frac{1}{\rho^2}L^TPx(t)$. The computation of the worst-case disturbance that this controller can sustain, comes from the superposition of Bellman's optimality principle when considering that the robot is affected by two separate inputs, that is the control input u and the cumulative disturbance input $\tilde{d}$. From the previous relation about the worst-case disturbance, it can be seen that the smallest value of the attenuation coefficient ρ that results in an admissible solution for the method's algebraic Riccati equation, is the one that provides the control loop with maximum robustness. The diagram of the considered control loop is depicted in Figure 6.2.

6.1.5 Lyapunov stability analysis

Through Lyapunov stability analysis it will be shown that the proposed nonlinear control scheme assures H_∞ tracking performance for the free-floating space robotic manipulator, and that in case of bounded disturbance terms asymptotic convergence to the reference setpoints is achieved. The tracking error dynamics for the free-floating space robotic manipulator is written in the form [721,725]:

$$\dot{e} = Ae + Bu + L\tilde{d} \tag{6.97}$$

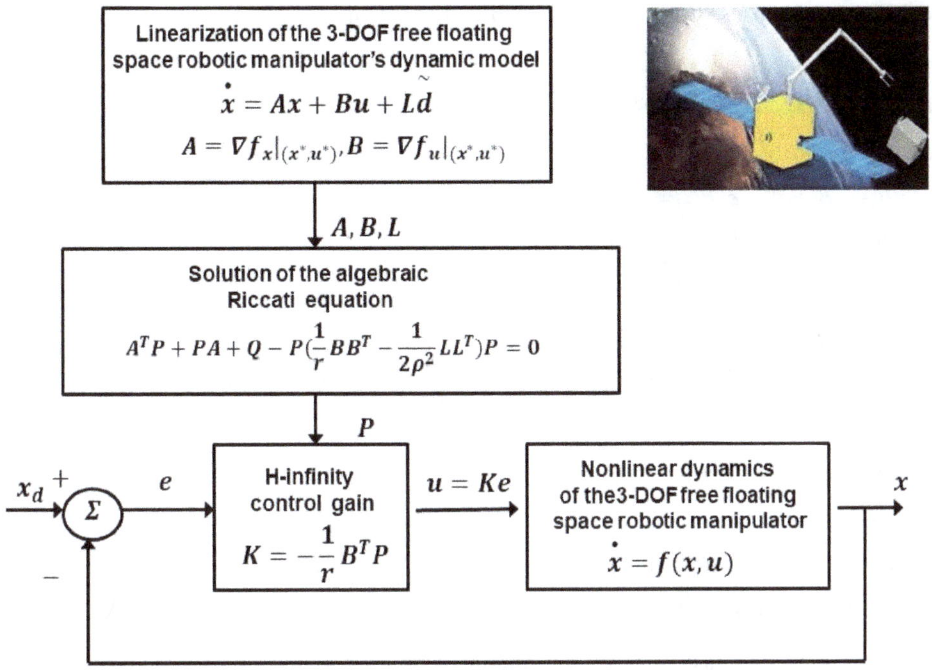

Figure 6.2 Diagram of the nonlinear optimal control loop for 3-DOF the free-floating space robotic manipulator

where in the free-floating space robotic manipulator's case $L \in R^{6\times 6}$ is the disturbance inputs gain matrix. Variable $\tilde{d}$ denotes model uncertainties and external disturbances of the space robot's model. The following Lyapunov equation is considered

$$V = \frac{1}{2}e^TPe \tag{6.98}$$

where $e = x - x_d$ is the tracking error. By differentiating with respect to time, one obtains

$$\begin{gathered}\dot{V} = \frac{1}{2}\dot{e}^TPe + \frac{1}{2}e^TP\dot{e} \Rightarrow \\ \dot{V} = \frac{1}{2}[Ae + Bu + L\tilde{d}]^TPe + \frac{1}{2}e^TP[Ae + Bu + L\tilde{d}] \Rightarrow\end{gathered} \tag{6.99}$$

$$\begin{aligned}\dot{V} = &\frac{1}{2}[e^TA^T + u^TB^T + \tilde{d}^TL^T]Pe + \\ &+\frac{1}{2}e^TP[Ae + Bu + L\tilde{d}] \Rightarrow\end{aligned} \tag{6.100}$$

$$\begin{aligned}\dot{V} = &\frac{1}{2}e^TA^TPe + \frac{1}{2}u^TB^TPe + \frac{1}{2}\tilde{d}^TL^TPe + \\ &\frac{1}{2}e^TPAe + \frac{1}{2}e^TPBu + \frac{1}{2}e^TPL\tilde{d}\end{aligned} \tag{6.101}$$

The previous equation is rewritten as

$$\begin{aligned}\dot{V} = &\frac{1}{2}e^T(A^TP + PA)e + (\frac{1}{2}u^TB^TPe + \frac{1}{2}e^TPBu) + \\ &+(\frac{1}{2}\tilde{d}^TL^TPe + \frac{1}{2}e^TPL\tilde{d})\end{aligned} \tag{6.102}$$

Assumption: For given positive definite matrix Q and coefficients r and ρ, there exists a positive definite matrix P, which is the solution of the following matrix equation:

$$A^TP + PA = -Q + P(\frac{2}{r}BB^T - \frac{1}{\rho^2}LL^T)P \tag{6.103}$$

Moreover, the following feedback control law is applied to the system:

$$u = -\frac{1}{r}B^T Pe \tag{6.104}$$

By substituting Eq. (6.103) and Eq. (6.104), one obtains

$$\begin{aligned}\dot{V} = \frac{1}{2}e^T[-Q + P(\frac{2}{r}BB^T - \frac{1}{\rho^2}LL^T)P]e+ \\ +e^T PB(-\frac{1}{r}B^T Pe) + e^T PL\tilde{d} \Rightarrow\end{aligned} \tag{6.105}$$

$$\begin{aligned}\dot{V} = -\frac{1}{2}e^T Qe + (\frac{1}{r}e^T PBB^T Pe - \frac{1}{2\rho^2}e^T PLL^T Pe) \\ -\frac{1}{r}e^T PBB^T Pe + e^T PL\tilde{d}\end{aligned} \tag{6.106}$$

which after intermediate operations gives

$$\dot{V} = -\frac{1}{2}e^T Qe - \frac{1}{2\rho^2}e^T PLL^T Pe + e^T PL\tilde{d} \tag{6.107}$$

or, equivalently

$$\begin{aligned}\dot{V} = -\frac{1}{2}e^T Qe - \frac{1}{2\rho^2}e^T PLL^T Pe+ \\ +\frac{1}{2}e^T PL\tilde{d} + \frac{1}{2}\tilde{d}^T L^T Pe\end{aligned} \tag{6.108}$$

Lemma: The following inequality holds [721,725]

$$\frac{1}{2}e^T L\tilde{d} + \frac{1}{2}\tilde{d}L^T Pe - \frac{1}{2\rho^2}e^T PLL^T Pe \leq \frac{1}{2}\rho^2\tilde{d}^T\tilde{d} \tag{6.109}$$

Proof: The binomial $(\rho\alpha - \frac{1}{\rho}b)^2$ is considered. Expanding the left part of the above inequality, one gets

$$\begin{aligned}\rho^2 a^2 + \frac{1}{\rho^2}b^2 - 2ab \geq 0 \Rightarrow \frac{1}{2}\rho^2 a^2 + \frac{1}{2\rho^2}b^2 - ab \geq 0 \Rightarrow \\ ab - \frac{1}{2\rho^2}b^2 \leq \frac{1}{2}\rho^2 a^2 \Rightarrow \frac{1}{2}ab + \frac{1}{2}ab - \frac{1}{2\rho^2}b^2 \leq \frac{1}{2}\rho^2 a^2\end{aligned} \tag{6.110}$$

The following substitutions are carried out: $a = \tilde{d}$ and $b = e^T PL$, and the previous relation becomes

$$\frac{1}{2}\tilde{d}^T L^T Pe + \frac{1}{2}e^T PL\tilde{d} - \frac{1}{2\rho^2}e^T PLL^T Pe \leq \frac{1}{2}\rho^2\tilde{d}^T\tilde{d} \tag{6.111}$$

Eq. (6.111) is substituted in Eq. (6.108) and the inequality is enforced, thus giving [721,725]

$$\dot{V} \leq -\frac{1}{2}e^T Qe + \frac{1}{2}\rho^2\tilde{d}^T\tilde{d} \tag{6.112}$$

Eq. (6.112) shows that the H_∞ tracking performance criterion is satisfied. The integration of $\dot{V}$ from 0 to T gives

$$\begin{aligned}\int_0^T \dot{V}(t)dt \leq -\frac{1}{2}\int_0^T ||e||_Q^2 dt + \frac{1}{2}\rho^2\int_0^T ||\tilde{d}||^2 dt \Rightarrow \\ 2V(T) + \int_0^T ||e||_Q^2 dt \leq 2V(0) + \rho^2\int_0^T ||\tilde{d}||^2 dt\end{aligned} \tag{6.113}$$

Moreover, if there exists a positive constant $M_d > 0$ such that

$$\int_0^\infty ||\tilde{d}||^2 dt \leq M_d \tag{6.114}$$

then, one gets

$$\int_0^\infty ||e||_Q^2 dt \leq 2V(0) + \rho^2 M_d \tag{6.115}$$

Thus, the integral $\int_0^\infty ||e||_Q^2 dt$ is bounded. Moreover, $V(T)$ is bounded and from the definition of the Lyapunov function V in Eq. (6.98), it becomes clear that $e(t)$ will be also bounded since $e(t) \in \Omega_e = \{e|e^T Pe \leq 2V(0) + \rho^2 M_d\}$. According to the above and with the use of Barbalat's Lemma, one obtains $lim_{t\to\infty}e(t) = 0$.

Through the stages of the stability proof, one arrives at Eq. (6.112) which shows that the H-infinity tracking performance criterion holds. By selecting the attenuation coefficient ρ to be sufficiently small and in particular to satisfy $\rho^2 < ||e||_Q^2/||\tilde{d}||^2$, one has that the first derivative of the Lyapunov function is upper bounded by 0. This condition holds at each sampling instance, and consequently, global stability for the control loop can be concluded.

6.1.6 Simulation tests

The fine performance of the nonlinear optimal control method for the dynamic model of the 3-DOF free-floating space robotic manipulator has been further confirmed through simulation experiments. To apply the nonlinear optimal control scheme the method's algebraic Riccati equation had to be solved at each iteration of the control algorithm using the *aresolv()* function in MATLAB®. Indicative values about the parameters of the dynamic model of the 3-DOF space robotic manipulator can be found in [904,901]. These are: $m_0 = 200kg$, $m_1 = 6.0kg$, $m_2 = 5.0kg$, $m_t = 2kg$, $I_0 = 200kg{\cdot}m^2$, $I_1 = 1.125kg{\cdot}m^2$, $I_2 = 0.9375kg{\cdot}m^2$, $I_t = 1.0kg{\cdot}m^2$, $a_0 = 1.25m$, $b_0 = 1.25m$, $a_1 = 0.75m$, $b_1 = 0.75m$, $a_2 = 0.75m$, $b_2 = 0.75m$, $a_3 = 0.5m$, $b_3 = 0.5m$, and $q_t = \frac{\pi}{6}$ rad. The obtained results are depicted in a series of simulation tests given in Figures 6.5–6.18, each one using a different set of reference trajectories for the state variables of the space robot. The turn angles of both the floating platform and of the links of the robotic manipulators were measured in radians (rad). In these simulation diagrams the state variables of the space robot are depicted in blue color, the associated setpoints are plotted in red color while the state estimates which are provided by the H-infinity Kalman filter are printed in green color. It can be noticed that the nonlinear optimal control method achieved fast and precise tracking of the reference trajectories under moderate variations of the control inputs, and consequently, under minimal dispersion of energy by the actuators of the robot. The precision of tracking of setpoints by the state variables of this robotic system as well as the transient performance of the control loop depends on the selection of the values of the gains r, ρ and of gain matrix Q of the method's Riccati equation. For relatively small values of gain r, the tracking error becomes negligible, while by raising the values of the diagonal element q_{ii} of the diagonal matrix Q, the speed of convergence of the state variables to the targeted setpoints is also increased. Finally, the smallest value of the attenuation coefficient ρ for which the method's Riccati equation still returns a valid solution in the form of a positive definite and symmetric matrix P is the one that provides the control loop with maximum robustness.

To elaborate on the tracking performance and on the robustness of the proposed nonlinear optimal control method for the 3-DOF floating space robotic manipulator the following tables are given: (i) Table 1 which provides information about the accuracy of tracking of the reference setpoints by the state variables of the space robot's state-space model,

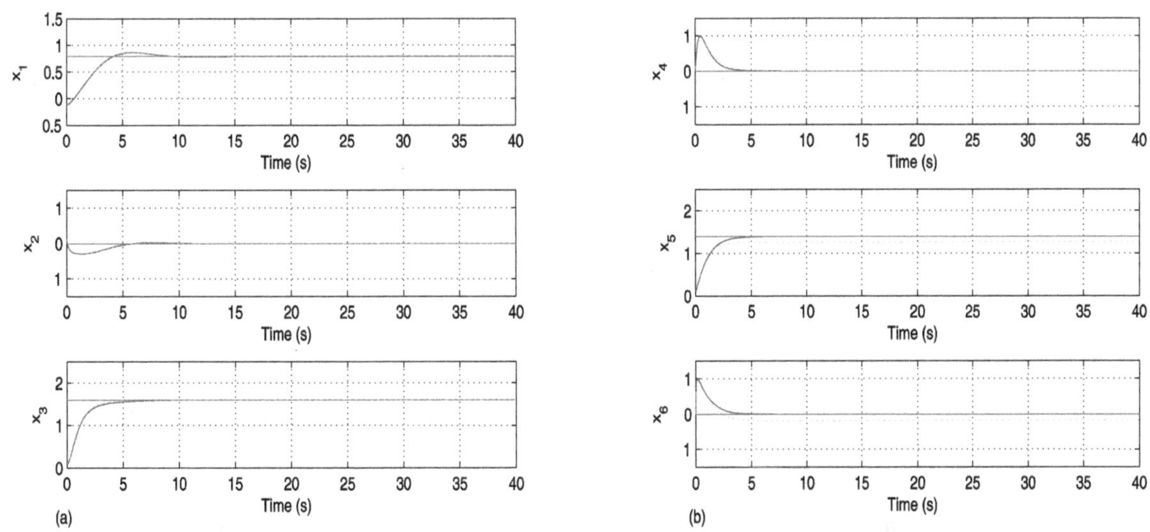

Figure 6.3 Test with setpoint 1 for the free-floating space robotic manipulator: (a) tracking of setpoints (red line) by state variables x_1 to x_3 (blue line) and estimates provided by the H-infinity Kalman filter (green line); (b) tracking of setpoints (red line) by state variables x_4 to x_6 (blue line) and estimates provided by the H-infinity Kalman filter (green line)

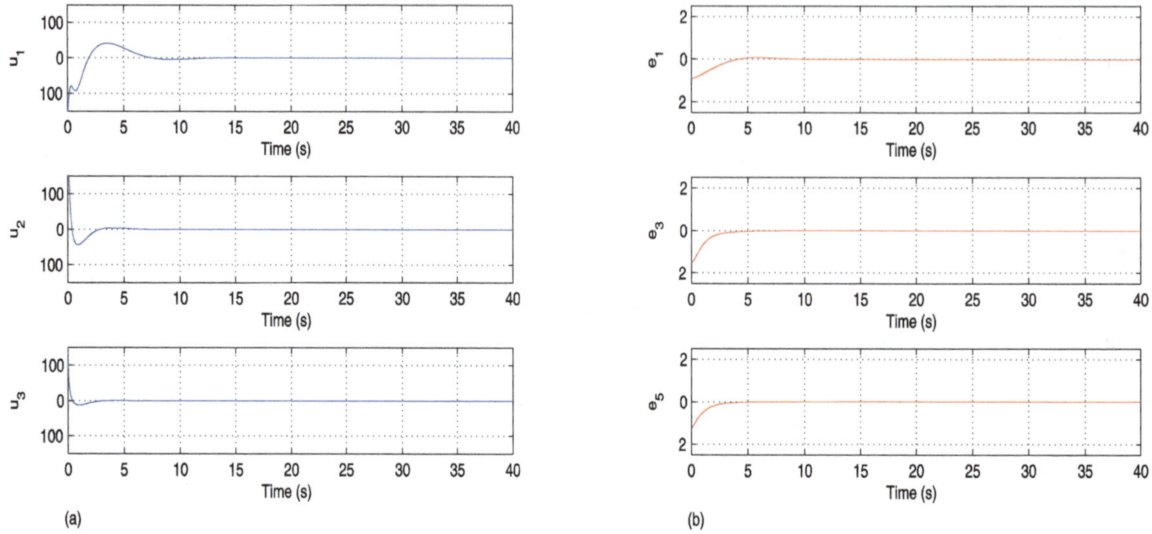

Figure 6.4 *Test with setpoint 1 for the free-floating space robotic manipulator: (a) variation of the control inputs* u_1 *to* u_3 *which are applied to the space robot; (b) tracking error dynamics for state variables* x_1, x_3, *and* x_5

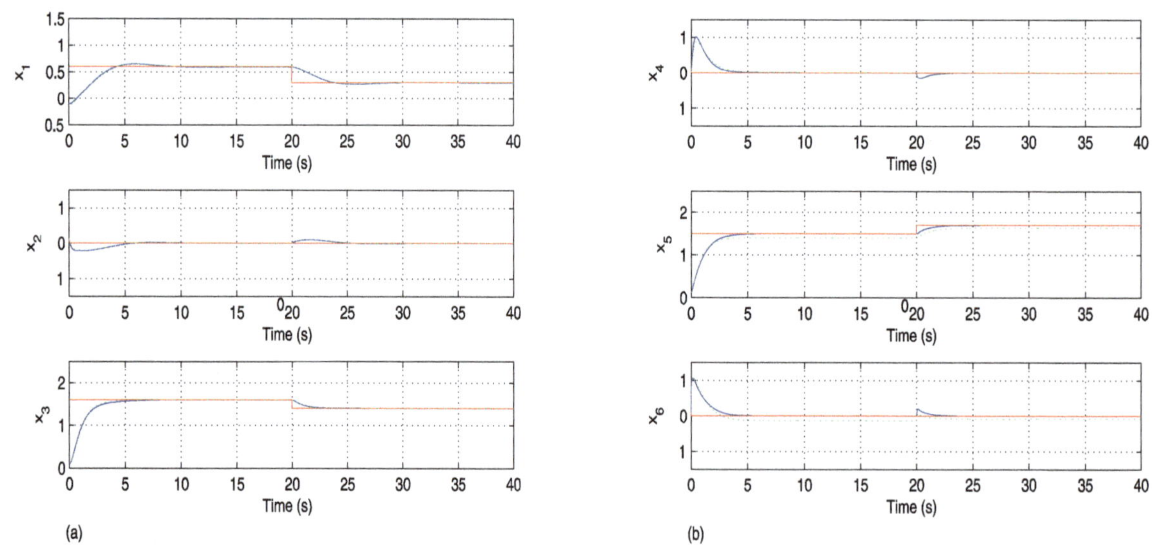

Figure 6.5 *Test with setpoint 2 for the free-floating space robotic manipulator: (a) tracking of setpoints (red line) by state variables* x_1 *to* x_3 *(blue line) and estimates provided by the H-infinity Kalman filter (green line); (b) tracking of setpoints (red line) by state variables* x_4 *to* x_6 *(blue line) and estimates provided by the H-infinity Kalman filter (green line)*

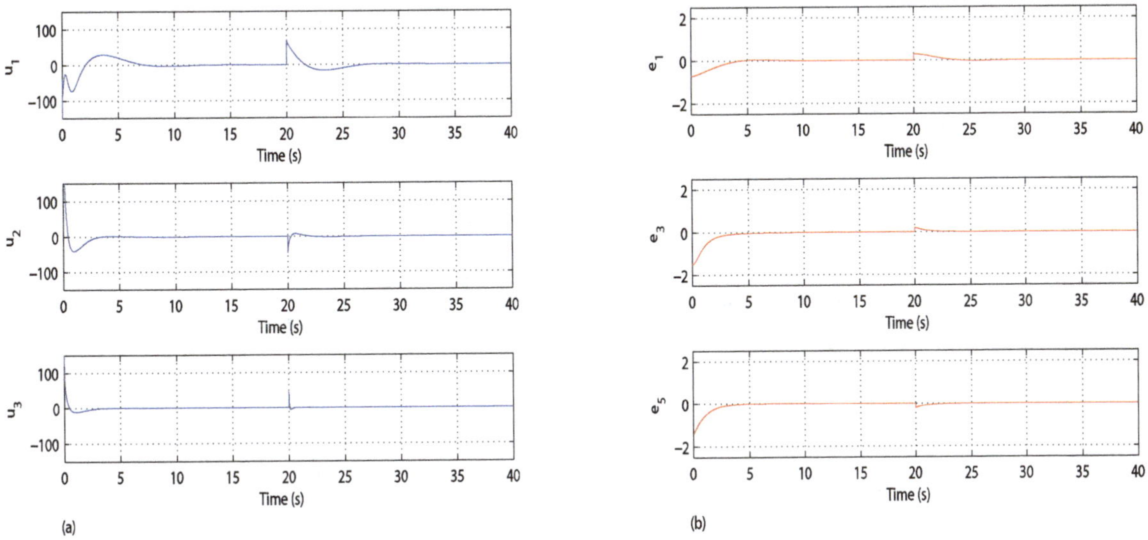

Figure 6.6 *Test with setpoint 2 for the free-floating space robotic manipulator: (a) variation of the control inputs u_1 to u_3 which are applied to the space robot; (b) tracking error dynamics for state variables x_1, x_3, and x_5*

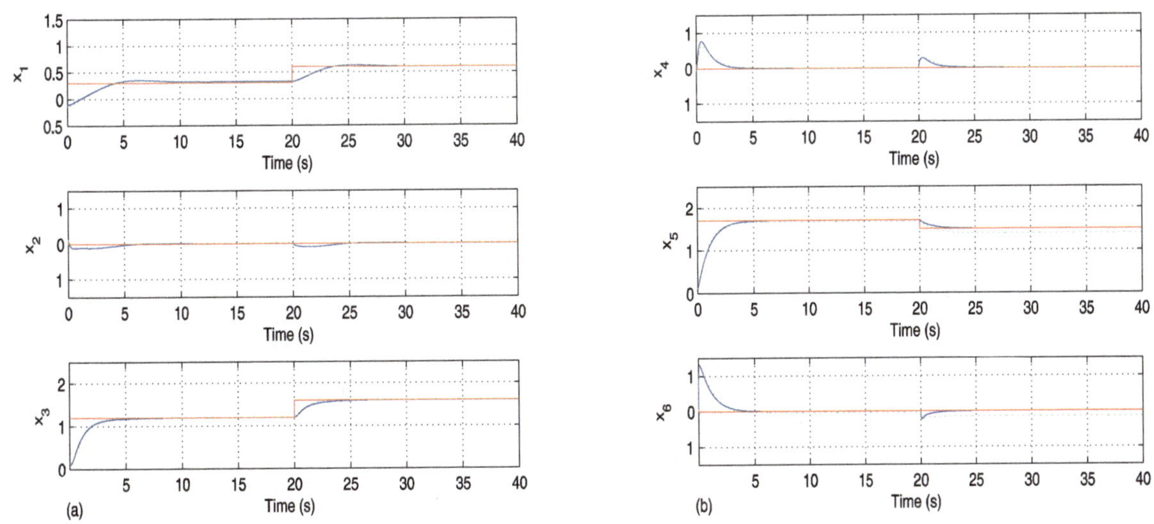

Figure 6.7 *Test with setpoint 3 for the free-floating space robotic manipulator: (a) tracking of setpoints (red line) by state variables x_1 to x_3 (blue line) and estimates provided by the H-infinity Kalman filter (green line); (b) tracking of setpoints (red line) by state variables x_4 to x_6 (blue line) and estimates provided by the H-infinity Kalman filter (green line)*

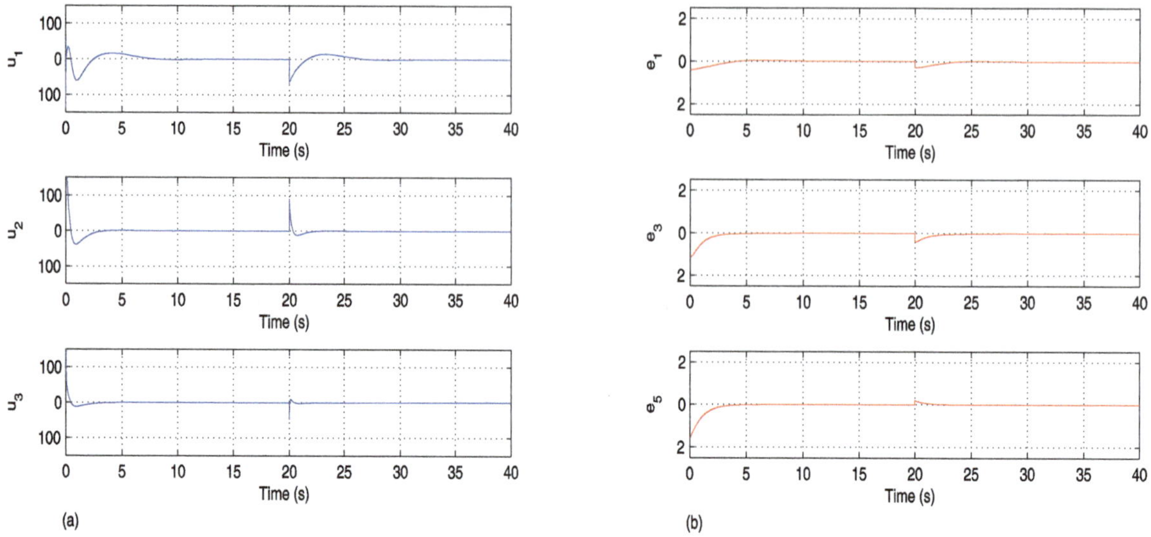

Figure 6.8 Test with setpoint 3 for the free-floating space robotic manipulator: (a) variation of the control inputs u_1 to u_3 which are applied to the space robot; (b) tracking error dynamics for state variables x_1, x_3, and x_5

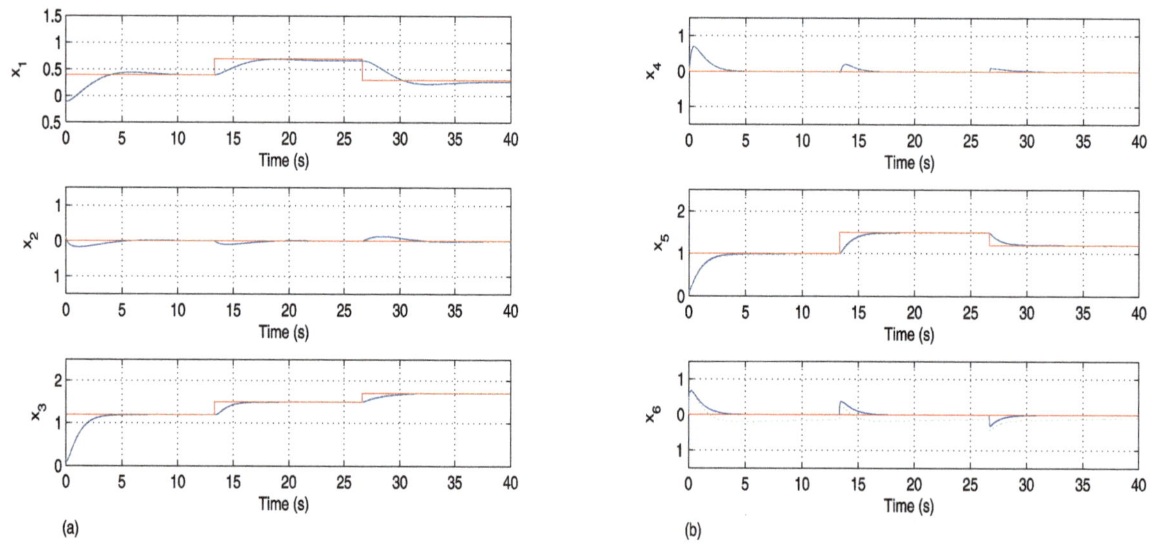

Figure 6.9 Test with setpoint 4 for the free-floating space robotic manipulator: (a) tracking of setpoints (red line) by state variables x_1 to x_3 (blue line) and estimates provided by the H-infinity Kalman filter (green line); (b) tracking of setpoints (red line) by state variables x_4 to x_6 (blue line) and estimates provided by the H-infinity Kalman filter (green line)

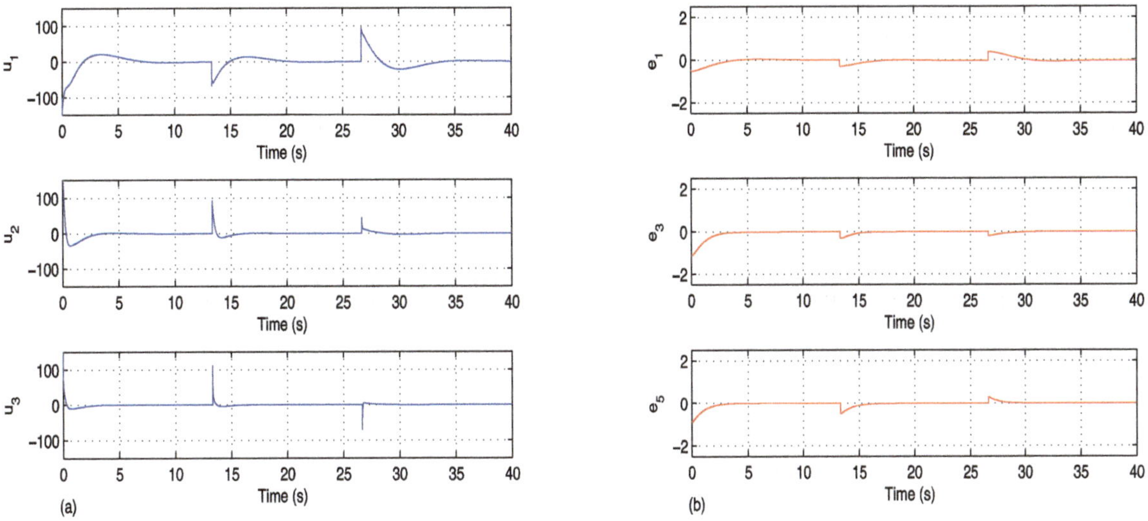

Figure 6.10 Test with setpoint 4 for the free-floating space robotic manipulator: (a) variation of the control inputs u_1 to u_3 which are applied to the space robot; (b) tracking error dynamics for state variables x_1, x_3, and x_5

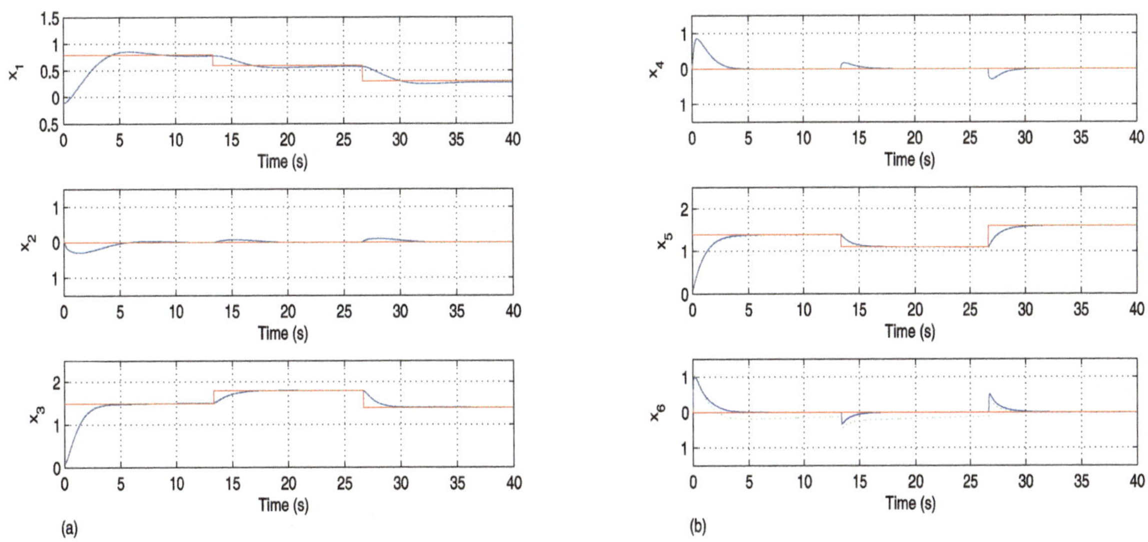

Figure 6.11 Test with setpoint 5 for the free-floating space robotic manipulator: (a) tracking of setpoints (red line) by state variables x_1 to x_3 (blue line) and estimates provided by the H-infinity Kalman filter (green line); (b) tracking of setpoints (red line) by state variables x_4 to x_6 (blue line) and estimates provided by the H-infinity Kalman filter (green line)

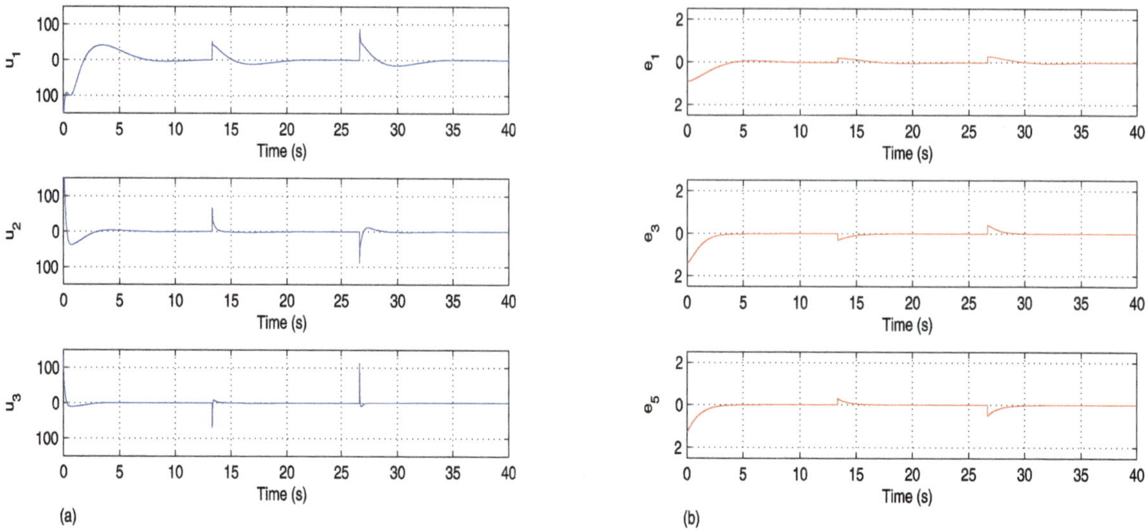

Figure 6.12 *Test with setpoint 5 for the free-floating space robotic manipulator: (a) variation of the control inputs u_1 to u_3 which are applied to the space robot; (b) tracking error dynamics for state variables x_1, x_3, and x_5*

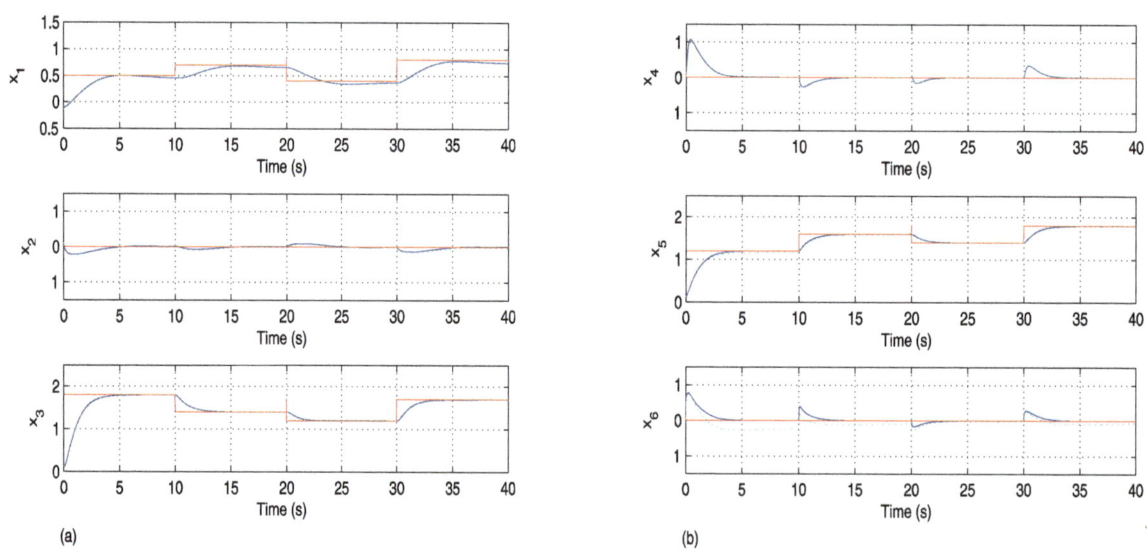

Figure 6.13 *Test with setpoint 6 for the free-floating space robotic manipulator: (a) tracking of setpoints (red line) by state variables x_1 to x_3 (blue line) and estimates provided by the H-infinity Kalman filter (green line); (b) tracking of setpoints (red line) by state variables x_4 to x_6 (blue line) and estimates provided by the H-infinity Kalman filter (green line)*

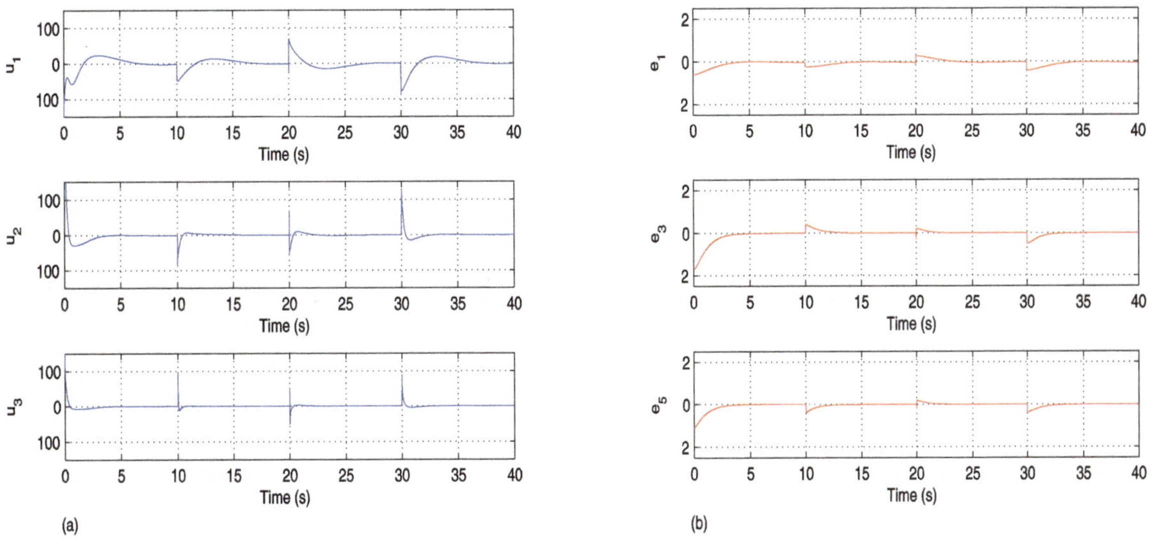

Figure 6.14 *Test with setpoint 6 for the free-floating space robotic manipulator: (a) variation of the control inputs u_1 to u_3 which are applied to the space robot; (b) tracking error dynamics for state variables x_1, x_3, and x_5*

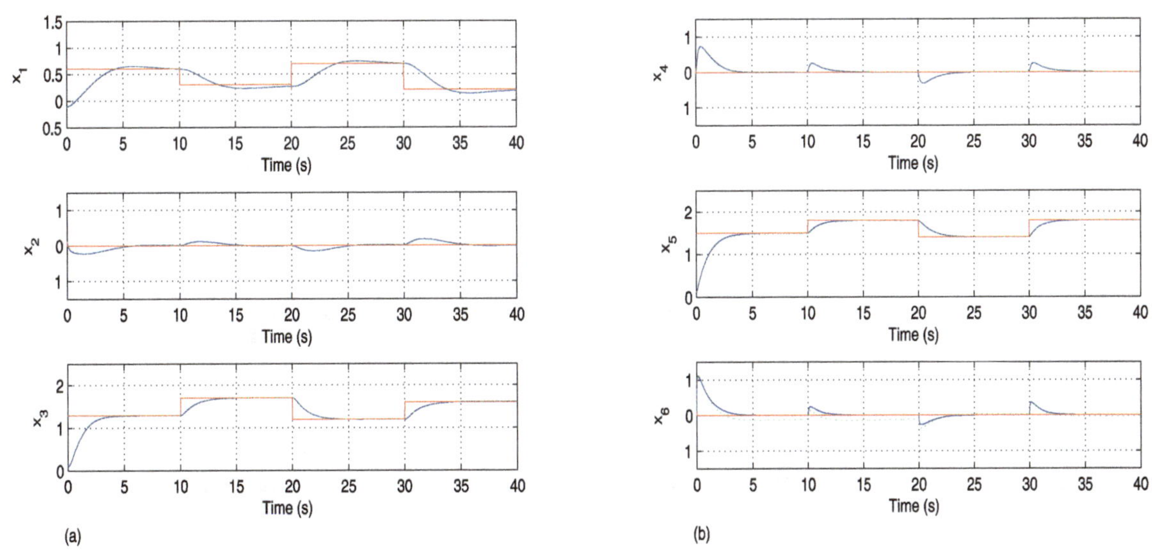

Figure 6.15 *Test with setpoint 7 for the free-floating space robotic manipulator: (a) tracking of setpoints (red line) by state variables x_1 to x_3 (blue line) and estimates provided by the H-infinity Kalman filter (green line); (b) tracking of setpoints (red line) by state variables x_4 to x_6 (blue line) and estimates provided by the H-infinity Kalman filter (green line)*

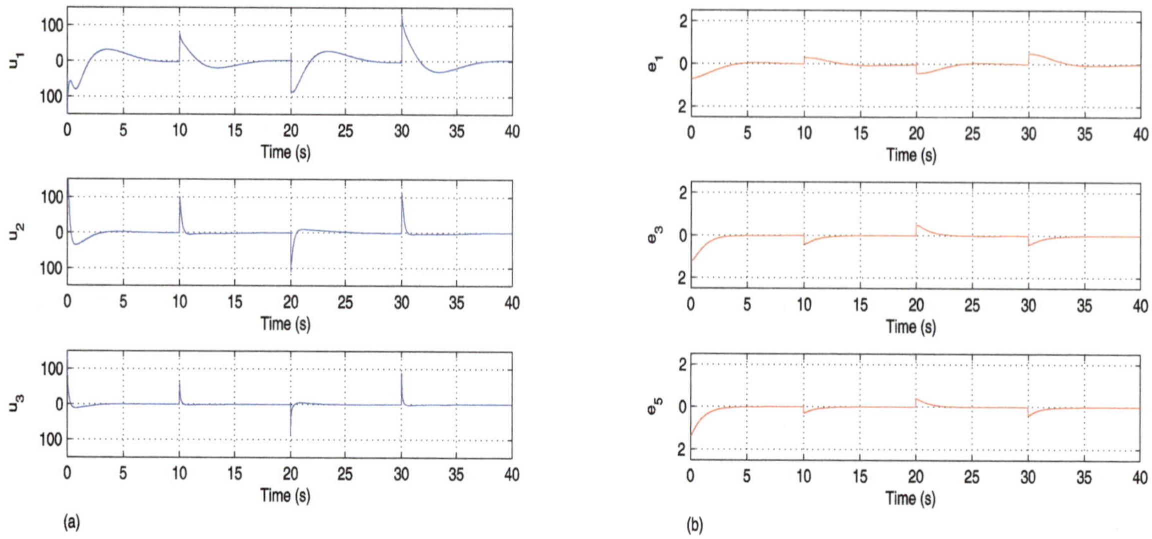

Figure 6.16 *Test with setpoint 7 for the free-floating space robotic manipulator: (a) variation of the control inputs u_1 to u_3 which are applied to the space robot; (b) tracking error dynamics for state variables x_1, x_3, and x_5*

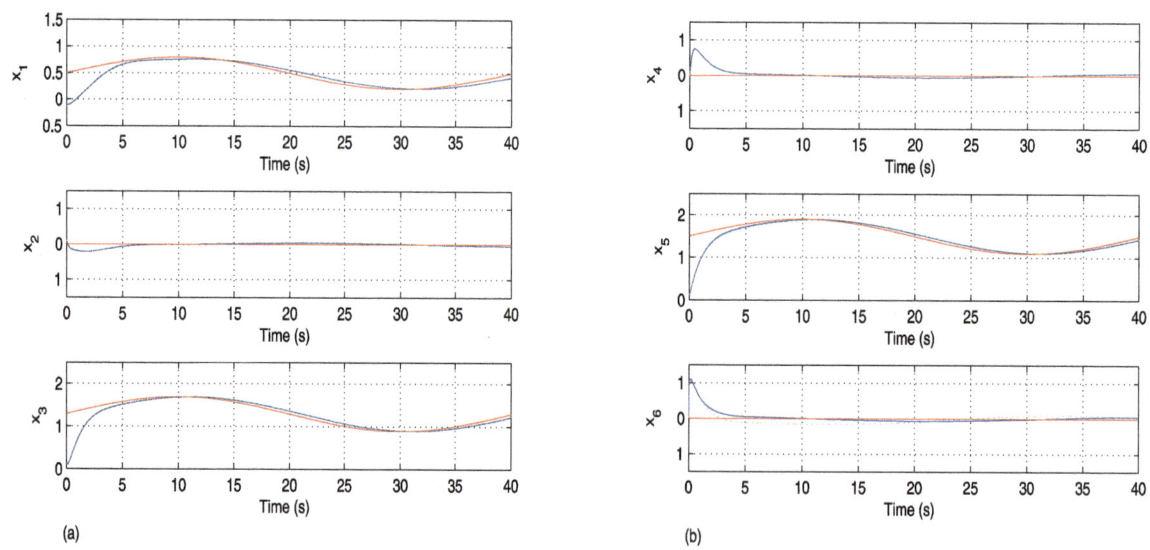

Figure 6.17 *Test with setpoint 8 for the free-floating space robotic manipulator: (a) tracking of setpoints (red line) by state variables x_1 to x_3 (blue line) and estimates provided by the H-infinity Kalman filter (green line); (b) tracking of setpoints (red line) by state variables x_4 to x_6 (blue line) and estimates provided by the H-infinity Kalman filter (green line)*

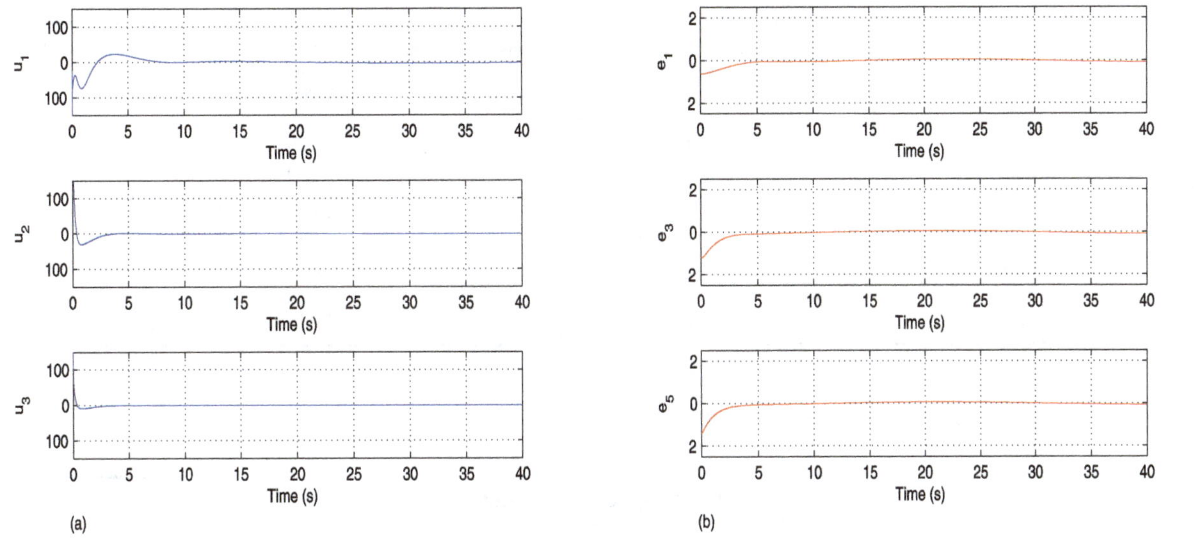

Figure 6.18 Test with setpoint 8 for the free-floating space robotic manipulator: (a) variation of the control inputs u_1 to u_3 which are applied to the space robot; (b) tracking error dynamics for state variables x_1, x_3, and x_5

Table 6.1 Free-floating space-robotic manipulator

	Tracking RMSE for the space robot in the disturbance-free case $\times 10^{-3}$					
	$RMSE_{x_1}$	$RMSE_{x_2}$	$RMSE_{x_3}$	$RMSE_{x_4}$	$RMSE_{x_5}$	$RMSE_{x_6}$
$test_1$	0.2746	0.0047	0.0115	0.0086	0.0010	0.0033
$test_2$	0.0895	0.0077	0.0059	0.0056	0.0002	0.0025
$test_3$	0.3162	0.0099	0.0153	0.0056	0.0009	0.0032
$test_4$	0.7949	0.1513	0.0285	0.0184	0.0053	0.0053
$test_5$	0.1144	0.1295	0.0100	0.0151	0.0015	0.0035
$test_6$	0.5099	0.3401	0.0917	0.0539	0.0381	0.0381
$test_7$	0.9769	0.3775	0.0382	0.0335	0.0338	0.0356
$test_8$	0.4790	0.4960	0.0375	0.7856	0.7766	0.7401

Table 6.2 Free-floating space-robotic manipulator

	Tracking RMSE for the space robot in the case of disturbances $\times 10^{-3}$					
$\Delta a\%$	$RMSE_{x_1}$	$RMSE_{x_2}$	$RMSE_{x_3}$	$RMSE_{x_4}$	$RMSE_{x_5}$	$RMSE_{x_6}$
0%	0.2346	0.0047	0.0115	0.0086	0.0010	0.0033
10%	1.2831	0.0214	0.0538	0.0082	0.0045	0.0036
20%	2.1823	0.0367	0.0913	0.0082	0.0081	0.0031
30%	2.9956	0.0503	0.1272	0.0075	0.0114	0.0034
40%	3.7516	0.0634	0.1595	0.0059	0.0150	0.0029
50%	4.4527	0.0754	0.1890	0.0070	0.0186	0.0032
60%	5.1272	0.0868	0.2208	0.0009	0.0223	0.0033

Table 6.3 Free-floating space-robotic manipulator

	RMSE for the estimation performed by the H-infinity KF $\times 10^{-3}$					
	$RMSE_{x_1}$	$RMSE_{x_2}$	$RMSE_{x_3}$	$RMSE_{x_4}$	$RMSE_{x_5}$	$RMSE_{x_6}$
$test_1$	0.0238	0.2253	0.0231	0.0237	0.0299	0.5886
$test_2$	0.0242	0.0489	0.0227	0.0232	0.0246	0.0141
$test_3$	0.0236	0.1765	0.0232	0.0242	0.0306	0.8718
$test_4$	0.0268	0.4807	0.0215	0.0234	0.0396	0.8718
$test_5$	0.0217	0.0058	0.0235	0.0254	0.0319	0.8647
$test_6$	0.0260	0.4159	0.0228	0.0224	0.0308	0.8992
$test_7$	0.0244	0.2385	0.0227	0.0209	0.0260	0.6162
$test_8$	0.0244	0.2542	0.0238	0.0232	0.0255	0.8763

Table 6.4 Free-floating space-robotic manipulator

	Convergence time for the space robot's state variables					
	$T_s\ x_1$	$T_s\ x_2$	$T_s\ x_3$	$T_s\ x_4$	$T_s\ x_5$	$T_s\ x_6$
$test_1$	8.0 *sec*	6.0 *sec*	3.5 *sec*	4.0 *sec*	4.0 *sec*	4.0 *sec*
$test_2$	4.0 *sec*	5.0 *sec*	3.5 *sec*	3.5 *sec*	4.0 *sec*	3.5 *sec*
$test_3$	4.0 *sec*	4.0 *sec*	3.5 *sec*	3.5 *sec*	3.5 *sec*	3.5 *sec*
$test_4$	3.5 *sec*	4.5 *sec*	3.0 *sec*	3.5 *sec*	3.0 *sec*	3.5 *sec*
$test_5$	8.0 *sec*	6.0 *sec*	4.0 *sec*	3.5 *sec*	3.5 *sec*	4.0 *sec*
$test_6$	4.5 *sec*	4.5 *sec*	4.0 *sec*	4.0 *sec*	3.5 *sec*	4.0 *sec*
$test_7$	4.0 *sec*	4.5 *sec*	4.0 *sec*	4.0 *sec*	3.5 *sec*	4.0 *sec*
$test_8$	5.0 *sec*	6.0 *sec*	4.0 *sec*	4.5 *sec*	4.5 *sec*	6.0 *sec*

(ii) Table 2 which provides information about the robustness of the control method to parametric changes in the model of the space robot's dynamics (change $\Delta a\%$ in the moment of inertia I_0 of the floating platform), (iii) Table 3 which provides information about the precision in state variables' estimation that is achieved by the H-infinity Kalman filter, and (iv) Table 4 which provides the approximate convergence times of the space robot's state variables to the associated setpoints.

6.2 Nonlinear optimal control for tethered underactuated satellites

6.2.1 Outline

Tethered satellite systems are described by the dumbbell satellite model and consist of a primary satellite and a sub-satellite which are connected between them through a tether [937,268]. They find several applications in space missions, such as retrieval of failed satellites, functioning as space tugs, formation of space stations and generation of artificial gravity, assisted re-entrance to the atmosphere of space shuttles, and transfer of power to other satellites [1077,1049]. It is noted that for a failed satellite, capturing and retrieving it allows the repair and re-utilization of its expensive onboard equipment. The dynamics of tethered satellites includes in-plane and off-plane motion (referring to the vertical plane of their body-fixed coordinates frame) and exhibits both longitudinal and lateral oscillations of the tether [1008,352]. Due to the relatively small size of the sub-satellite, the associated attitude motion (rotation around the axes of the body-fixed coordinates frame) is omitted. The control of the deployment of the sub-satellite can be assisted by thrusters. However, incorporating thrusters in satellites to provide full actuation signifies a rise of the satellite's weight, while the use of thrusters results in fuel consumption. On the other side, control of the sub-satellite's deployment through the variations of the tether's tension is a low-cost solution characterized by simplicity and re-usability [1012,969,974].

A main problem for the stabilization of the tethered satellite system is that the sub-satellite, due to the effects of Coriolis acceleration, does not move along the tether's line, which is known as radial equilibrium [1075,608,943]. Instead, an unsteady oscillatory motion may occur and the sub-satellite may deviate significantly from the radial equilibrium [324,938]. This oscillatory motion induces stress forces which may end up at failure of the tether. Controlling the unstable motion of tethered satellites has attracted much research interest. Due to the nonlinear structure and underactuation of the tethered satellites' state-space model, the solution to the related control problem exhibits several difficulties [285,910,939]. These systems are not directly feedback linearizable and the related feedback linearization control concepts are not applicable to them. Attempts to apply sliding-mode control in tethered satellites come against the problem of selection of the sliding surface [561,559,499]. Besides, the use of backstepping control comes against the problem of transforming this system in a strict feedback form. Moreover, linear optimal control methods are inefficient for tethered satellites. On the other side, some nonlinear optimal control methods can be based on Pontryangin's maximum principle or on approximations of the Hamiltonian function [457,816,817,936]. Attempts to solve the nonlinear control problem for tethered satellites through nonlinear programming solvers and Nonlinear Model Predictive Control Techniques (NMPC) are not always of ensured convergence. Additional results on nonlinear control of tethered satellite systems can be found in [849,555,1020,1021,400,402,405,1022]. These techniques try at compensating for underactuation and for suppressing the vibration of multi-satellite systems connected through tethers. The aim is always to achieve precise positioning and orientation of the satellites and at the same time to overcome underactuation and to suppress undesirable oscillations which are induced by the tether's tension. Tethering is modular and re-configurable and this shows its potential for wider use in multi-satellite systems and in satellite maintenance and re-orbiting.

In this section, a novel nonlinear optimal control method is analyzed [721,725,711,741]. This method is shown to be suitable for the dynamic model of a tethered satellite system which consists of a primary satellite connected to a sub-satellite through a massless and inelastic tether. By considering the in-plane motion, a 2-DOF dynamic model of the tethered satellite system has been obtained. This model is characterized by underactuation because of receiving only one control input which is the tether's tension. To implement the nonlinear optimal control scheme the state-space model of the tethered satellite system has undergone approximate linearization with the use of first-order Taylor series expansion and through the computation of the associated Jacobian matrices [705,60,706]. The linearization process is repeated at each sampling instance around a time-varying operating point (x^*, u^*) where x^* describes the present value of the system's state vector and u^* describes the last sampled value of the control inputs vector. The model imprecision which is due to the truncation of higher-order terms from the Taylor series expansion is considered to be a perturbation which could be asymptotically compensated by the robustness of the control algorithm. It is noted that because of linearization at each sampling interval at the present value of the state vector of the tethered satellite system and not at points of the desirable trajectory the span of the Taylor series expansion is small and the modeling error is negligible.

For the approximately linearized model of the tethered satellites, a stabilizing H-infinity controller is designed. The H-infinity controller provides a solution of the associated optimal control problem under model uncertainty and external disturbances [721,741]. Actually, the H-infinity controller represents a min–max differential game taking place between (i) the control inputs of the tethered satellite system which try to minimize a cost function that contains a quadratic term of the state vector's tracking error and (ii) the model imprecision and exogenous perturbations which try to maximize this cost function. To select the feedback gains of this controller an algebraic Riccati equation has to be repetitively solved at each sampling period. The global stability properties of the control method are proven through Lyapunov analysis. First, it was demonstrated that the control loop satisfies the H-infinity tracking performance criterion, which signifies elevated robustness to model uncertainty and perturbations [876,732]. Moreover, under moderate conditions, it is proven that the control loop is globally asymptotically stable [721,741]. To implement state estimation-based control without the need to measure the entire state vector of the tethered satellites the H-infinity Kalman filter has been used as a robust state estimator. The nonlinear optimal control method achieves fast and accurate tracking of all setpoints of the in-plane motion of the tethered satellite system under moderate variations of the control inputs. It is noteworthy that this control method can be applied without any changes of state variables (diffeomorphisms) and without state-space model transformations.

6.2.2 Dynamic model of the tethered satellite system

The considered tethered satellite system consists of the primary satellite and of the sub-satellite which are connected with a tether. The diagram of the tethered satellites is given in Figure 6.19. The angular speed of the primary satellite is Ω. The related cartesian coordinates frame OIJK is fixed on the center of gravity of the primary satellite. Its I axis is tangential to the satellite's orbit. Its J axis is also on the orbit's plane and is perpendicular to the I axis. Its K axis is perpendicular to both the I and J axes and points toward the center of the earth. The mass of the primary satellite is M while the mass of the secondary satellite is m. The tether's length, connecting the two satellites is denoted as L and is time-varying.

The dynamics of the tethered satellite system is described by three variables, namely the tether's length $L(t)$ and the swing angles of the sub-satellite with respect to the primary satellite which are denoted as θ and ϕ, respectively. Angle θ resides on the vertical plane of the body-fixed reference frame and is the in-plane swing angle, while angle ϕ is the off-plane swing angle. The position of the center of gravity the first satellite is given by point O while the position of the center of gravity of the second satellite is given by point A. The projection of point A on the OIK plane in the direction of the OJ axis is given by point B. The angle between line OA that connects the primary satellite with the sub-satellite and line OB is denoted as ϕ. The angle between the vertical axis OK and line OB is denoted as θ. The distance of the primary satellite from the sub-satellite along line OA is the tether's length $L(t)$.

The radius defining the distance between the earth's center of gravity and the center of gravity O of the primary satellite is denoted as r_0. It is considered that $m << M$. Moreover, length $L(t)$ is substituted by the new variable $\xi(t) = \frac{L(t)}{L_d}$ where L_d is the desirable length of the cable. Thus, $\xi(t)$ becomes a normalized and dimensionless distance variable. The dynamic model of the tethered satellite, which is initially computed with the use of Euler–Lagrange analysis comes into the following form [285,936]:

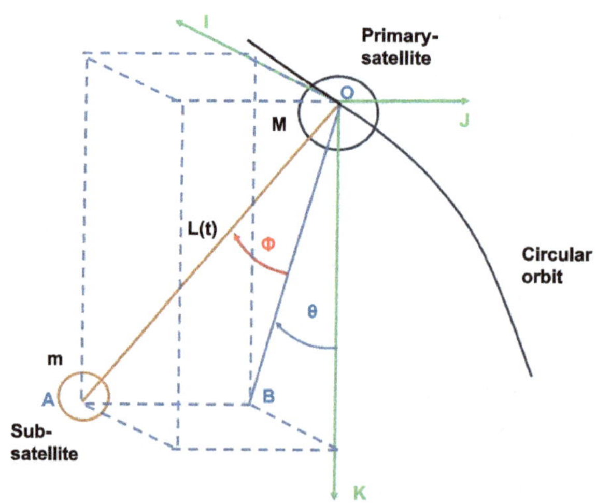

Figure 6.19 Diagram of the tethered satellite system, consisting of the primary satellite with mass M and of the sub-satellite with mass m which are connected with a tether of length L(t).

$$\begin{aligned}
&\ddot{\theta}-2(\dot{\theta}-1)(\dot{\phi}tan(\phi)-\tfrac{\dot{\xi}}{\xi})+3sin(\theta)cos(\theta)=0\\
&\ddot{\phi}+2\tfrac{\dot{\xi}\dot{\phi}}{\xi}+[(\dot{\theta}-1)^2+3cos^2(\theta)]sin(\phi)cos(\phi)=0\\
&\ddot{\xi}-[\dot{\phi}^2+(\dot{\theta}-1)^2cos^2(\phi)+3cos^2(\phi)cos^2(\theta)-1]\xi=-u
\end{aligned}\tag{6.116}$$

In the dynamic model of Eq. (6.116) angle θ defines the pitch motion and angle ϕ defines the roll motion of the sub-satellite. Moreover, the dimensionless length variable ξ defines the longitudinal motion of the subsatellite which is imposed by the tether's force. The tensional force T along the massless tether may vary in the interval $[T_{min}, T_{max}]$. The orbital angular velocity of the primary satellite is $\Omega=\sqrt{\frac{\mu}{r_0^3}}$ where μ is the gravitational parameter of the earth. The earth's angular speed is also Ω. The control input is the normalized tether's tension $u=\frac{T}{m\Omega^2L_d}$, while one can also consider to include in the control input a constant thrust exerted on the sub-satellite.

Next, the dynamic model of the tethered satellite is simplified by considering only in-plane motion for the sub-satellite. This means that the sub-satellite is taken to reside on the vertical OIK plane and thus it holds $\phi=0$ and $\dot{\phi}=0$. It is noted that there is no controllability of the out-of-plane motion when the only control input is the tether's tension. Therefore, the controller design is limited to the in-plane motion. By considering that $\phi=0$ and $\dot{\phi}=0$, the corresponding equations of the dynamic model of the tethered satellite system are rewritten as [285,936]:

$$\begin{aligned}
&\ddot{\theta}+2(\dot{\theta}-1)\tfrac{\dot{\xi}}{\xi}+3sin(\theta)cos(\theta)=0\\
&\ddot{\xi}-[(\dot{\theta}-1)^2+3cos^2(\theta)-1]\xi=-u
\end{aligned}\tag{6.117}$$

The consistency of the dynamic model of Eq. (6.117) can be confirmed by following analytically the stages of Euler–Lagrange analysis [499]. Next, for the dynamic model of the in-plane motion of the tethered satellite system of Eq. (6.117) the state variables notation is $x=[x_1,x_2,x_3,x_4]^T$ or $x=[\theta,\dot{\theta},\xi,\dot{\xi}]^T$, one obtains the following state-space notation for the tethered satellites

$$\begin{aligned}
\dot{x}_1&=x_2\\
\dot{x}_2&=-2(x_2-1)\tfrac{x_4}{x_3}-3sin(x_1)cos(x_1)\\
\dot{x}_3&=x_4\\
\dot{x}_4&=[(x_2-1)^2+3cos^2(x_1)-1]x_3-u
\end{aligned}\tag{6.118}$$

The state-space model of the tethered satellite system can be written in the following nonlinear affine-in-input state-space form

$$\dot{x}=f(x)+g(x)u\tag{6.119}$$

where $x\in R^{4\times1}$, $f(x)\in R^{4\times1}$, $g(x)\in R^{4\times1}$, and $u\in R$, with

$$f(x)=\begin{pmatrix}x_2\\ -2(x_2-1)\frac{x_4}{x_3}-3sin(x_1)cos(x_1)\\ x_4\\ (x_2-1)^2+3cos^2(x_1)-1\end{pmatrix}\quad g(x)=\begin{pmatrix}0\\0\\0\\-1\end{pmatrix}\tag{6.120}$$

6.2.3 Design of an H-infinity feedback controller for the tethered satellites system

6.2.3.1 Approximate linearization of the dynamic model of the tethered satellites

The dynamic model of the tethered satellite system undergoes approximate linearization at each sampling instance around a temporary operating point (x^*, u^*) where x^* is the present value of the system's state vector and u^* is the last sampled value of the control inputs vector. The linearization process is based on first-order Taylor series expansion and on the computation of the associated Jacobian matrices. The modeling error which is due to the truncation of higher-order terms from the Taylor series expansion is considered to be a perturbation which is compensated asymptotically by the robustness of the control algorithm.

By applying Taylor series expansion, the initial nonlinear state-space model of the tethered satellites $\dot{x}=f(x)+g(x)u$ is turned into the equivalent state-space description

$$\dot{x}=Ax+Bu+\tilde{d} \tag{6.121}$$

where matrices A, B are the Jacobian matrices of the tethered satellites system which are given by:

$$A=\nabla_x[f(x)+g(x)u]\mid_{(x^*,u^*)} \Rightarrow A=\nabla_x f(x)\mid_{(x^*,u^*)} \tag{6.122}$$

$$B=\nabla_u[f(x)+g(x)u]\mid_{(x^*,u^*)} \Rightarrow B=g(x)\mid_{(x^*,u^*)} \tag{6.123}$$

and $\tilde{d}$ is the cumulative disturbance vector which may comprise (i) the modeling error due to the truncation of higher-order terms in the Taylor series expansion, (ii) exogenous perturbations, and (iii) sensor measurement noise of any distribution.

The computation of the Jacobian matrix of $A=\nabla_x f(x)\mid_{(x^*,u^*)}$ proceeds as follows:

First row of the Jacobian matrix $A=\nabla_x f(x)\mid_{(x^*,u^*)}$: $\frac{\partial f_1}{\partial x_1}=0$, $\frac{\partial f_1}{\partial x_2}=1$, $\frac{\partial f_1}{\partial x_3}=0$, $\frac{\partial f_1}{\partial x_4}=0$.

Second row of the Jacobian matrix $A=\nabla_x f(x)\mid_{(x^*,u^*)}$: $\frac{\partial f_2}{\partial x_1}=-3[cos^2(x_1)-sin^2(x_1)]$, $\frac{\partial f_2}{\partial x_2}=-2\frac{x_4}{x_3}$, $\frac{\partial f_2}{\partial x_3}=2(x_2-1)\frac{x_4}{x_3^2}$ and $\frac{\partial f_2}{\partial x_4}=-2(x_2-1)\frac{1}{x_3}$

Third row of the Jacobian matrix $A=\nabla_x f(x)\mid_{(x^*,u^*)}$: $\frac{\partial f_3}{\partial x_1}=0$, $\frac{\partial f_3}{\partial x_2}=0$, $\frac{\partial f_3}{\partial x_3}=0$, $\frac{\partial f_3}{\partial x_4}=1$.

Fourth row of the Jacobian matrix $A=\nabla_x f(x)\mid_{(x^*,u^*)}$: $\frac{\partial f_4}{\partial x_1}=-6cos(x_1)sin(x_1)x_3$, $\frac{\partial f_4}{\partial x_2}=2(x_2-1)x_3$, $\frac{\partial f_4}{\partial x_3}=[(x_2-1)^2+3cos^2(x_1)-1]$, $\frac{\partial f_6}{\partial x_4}=0$.

The linearization approach which has been followed for implementing the nonlinear optimal control scheme results in a quite accurate model of the system's dynamics. Consider for instance the following affine-in-the-input state-space model

$$\begin{gathered}\dot{x}=f(x)+g(x)u\Rightarrow\\ \dot{x}=[f(x^*)+\nabla_x f(x)\mid_{x^*}(x-x^*)]+[g(x^*)+\nabla_x g(x)\mid_{x^*}(x-x^*)]u^*+g(x^*)u^*+g(x^*)(u-u^*)+\tilde{d}_1\Rightarrow\\ \dot{x}=[\nabla_x f(x)\mid_{x^*}+\nabla_x g(x)\mid_{x^*}u^*]x+g(x^*)u-[\nabla_x f(x)\mid_{x^*}+\nabla_x g(x)\mid_{x^*}u^*]x^*+f(x^*)+g(x^*)u^*+\tilde{d}_1\end{gathered} \tag{6.124}$$

where $\tilde{d}_1$ is the modeling error due to truncation of higher order terms in the Taylor series expansion of $f(x)$ and $g(x)$. Next, by defining $A=[\nabla_x f(x)\mid_{x^*}+\nabla_x g(x)\mid_{x^*}u^*]$, $B=g(x^*)$, one obtains

$$\dot{x}=Ax+Bu-Ax^*+f(x^*)+g(x^*)u^*+\tilde{d}_1 \tag{6.125}$$

Moreover, by denoting $\tilde{d}=-Ax^*+f(x^*)+g(x^*)u^*+\tilde{d}_1$ about the cumulative modeling error term in the Taylor series expansion, one has

$$\dot{x}=Ax+Bu+\tilde{d} \tag{6.126}$$

which is the approximately linearized model of the dynamics of the system of Eq. (6.121). The term $f(x^*)+g(x^*)u^*$ is the derivative of the state vector at (x^*,u^*) which is almost annihilated by $-Ax^*$.

6.2.3.2 Stabilizing feedback control

After linearization around its current operating point, the tethered satellites dynamic model is written as [721]

$$\dot{x}=Ax+Bu+d_1 \tag{6.127}$$

Parameter d_1 stands for the linearization error in the tethered satellites' dynamic model appearing previously in Eq. (6.127). The reference setpoints for the tethered satellites' state vector are denoted by $\mathbf{x_d} = [x_1^d, \cdots, x_4^d]$. Tracking of this trajectory is achieved after applying the control input u_d. At every time instant, the control input u_d is assumed to differ from the control input u appearing in Eq. (6.127) by an amount equal to Δu, that is $u_d = u + \Delta u$

$$\dot{x}_d = Ax_d + Bu_d + d_2 \tag{6.128}$$

The dynamics of the controlled system described in Eq. (6.127) can be also written as

$$\dot{x} = Ax + Bu + Bu_d - Bu_d + d_1 \tag{6.129}$$

and by denoting $d_3 = -Bu_d + d_1$ as an aggregate disturbance term, one obtains

$$\dot{x} = Ax + Bu + Bu_d + d_3 \tag{6.130}$$

By subtracting Eq. (6.128) from Eq. (6.130), one has

$$\dot{x} - \dot{x}_d = A(x - x_d) + Bu + d_3 - d_2 \tag{6.131}$$

By denoting the tracking error as $e = x - x_d$ and the aggregate disturbance term as $L\tilde{d} = d_3 - d_2$, the tracking error dynamics becomes

$$\dot{e} = Ae + Bu + L\tilde{d} \tag{6.132}$$

where L is the disturbance inputs gain matrix. For the approximately linearized model of the system, a stabilizing feedback controller is developed. The controller has the form

$$u(t) = -Ke(t) \tag{6.133}$$

with $K = \frac{1}{r}B^TP$ where P is a positive definite symmetric matrix which is obtained from the solution of the Riccati equation [721]

$$A^TP + PA + Q - P(\tfrac{2}{r}BB^T - \tfrac{1}{\rho^2}LL^T)P = 0 \tag{6.134}$$

where Q is a positive semi-definite symmetric matrix. The diagram of the considered control loop is depicted in Figure 6.20.

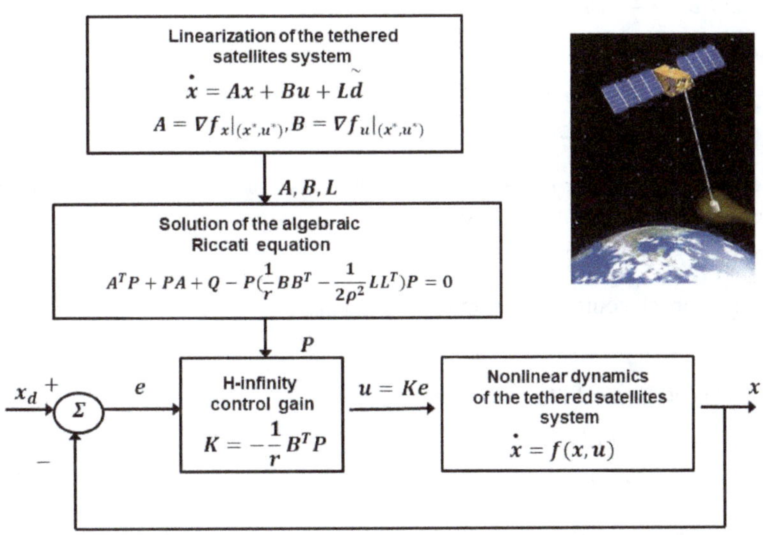

Figure 6.20 Diagram of the tethered satellites' nonlinear optimal (H-infinity) control loop

It is also noteworthy that the solution of the H-infinity feedback control problem for the tethered satellites' dynamics and the computation of the worst-case disturbance that this controller can sustain, comes from the superposition of Bellman's optimality principle when considering that the system is affected by two separate inputs (i) the control input u (ii) the cumulative disturbance input $\tilde{d}(t)$. Solving the optimal control problem for u, that is for the minimum variation (optimal) control input that achieves the elimination of the state vector's tracking error, gives $u = -\frac{1}{r}B^TPe$. Equivalently, solving the optimal control problem for $\tilde{d}$, that is for the worst-case disturbance that the control loop can sustain gives $\tilde{d} = \frac{1}{\rho^2}L^TPe$.

6.2.4 Lyapunov stability analysis

Through Lyapunov stability analysis it will be shown that the proposed nonlinear control scheme assures H_∞ tracking performance for the tethered satellites' attitude dynamics and that under moderate conditions asymptotic convergence to the reference setpoints (global stability) is achieved. As shown before, the tracking error dynamics for the tethered satellites' attitude model is written in the form [721]

$$\dot{e} = Ae + Bu + L\tilde{d} \tag{6.135}$$

where in the tethered satellites' dynamics, and for in-plane motion, $L \in R^{4\times 4}$ is the disturbances gain matrix. Variable $\tilde{d}$ denotes model uncertainties and external disturbances of the tethered satellites' dynamic model. The following Lyapunov equation is considered

$$V = \frac{1}{2}e^TPe \tag{6.136}$$

where $e = x - x_d$ is the tracking error. By differentiating with respect to time, one obtains

$$\begin{gathered} \dot{V} = \frac{1}{2}\dot{e}^TPe + \frac{1}{2}e^TP\dot{e} \Rightarrow \\ \dot{V} = \frac{1}{2}[Ae + Bu + L\tilde{d}]^TPe + \frac{1}{2}e^TP[Ae + Bu + L\tilde{d}] \Rightarrow \end{gathered} \tag{6.137}$$

$$\begin{aligned} \dot{V} = {} & \frac{1}{2}[e^TA^T + u^TB^T + \tilde{d}^TL^T]Pe + \\ & + \frac{1}{2}e^TP[Ae + Bu + L\tilde{d}] \Rightarrow \end{aligned} \tag{6.138}$$

$$\begin{aligned} \dot{V} = {} & \frac{1}{2}e^TA^TPe + \frac{1}{2}u^TB^TPe + \frac{1}{2}\tilde{d}^TL^TPe + \\ & \frac{1}{2}e^TPAe + \frac{1}{2}e^TPBu + \frac{1}{2}e^TPL\tilde{d} \end{aligned} \tag{6.139}$$

The previous equation is rewritten as

$$\begin{aligned} \dot{V} = {} & \frac{1}{2}e^T(A^TP + PA)e + (\frac{1}{2}u^TB^TPe + \frac{1}{2}e^TPBu) + \\ & + (\frac{1}{2}\tilde{d}^TL^TPe + \frac{1}{2}e^TPL\tilde{d}) \end{aligned} \tag{6.140}$$

Assumption: For given positive definite matrix Q and coefficients r and ρ, there exists a positive definite matrix P, which is the solution of the following matrix equation

$$A^TP + PA = -Q + P(\frac{2}{r}BB^T - \frac{1}{\rho^2}LL^T)P \tag{6.141}$$

Moreover, the following feedback control law is applied to the system:

$$u = -\frac{1}{r}B^TPe \tag{6.142}$$

By substituting Eq. (6.141) and Eq. (6.142), one obtains

$$\begin{aligned} \dot{V} = {} & \frac{1}{2}e^T[-Q + P(\frac{2}{r}BB^T - \frac{1}{\rho^2}LL^T)P]e + \\ & + e^TPB(-\frac{1}{r}B^TPe) + e^TPL\tilde{d} \Rightarrow \end{aligned} \tag{6.143}$$

$$\begin{aligned} \dot{V} = {} & -\frac{1}{2}e^TQe + (\frac{1}{r}e^TPBB^TPe - \frac{1}{2\rho^2}e^TPLL^TPe \\ & -\frac{1}{r}e^TPBB^TPe) + e^TPL\tilde{d} \end{aligned} \tag{6.144}$$

which after intermediate operations gives

$$\dot{V} = -\frac{1}{2}e^TQe - \frac{1}{2\rho^2}e^TPLL^TPe + e^TPL\tilde{d} \tag{6.145}$$

or, equivalently

$$\begin{aligned}\dot{V} = -\frac{1}{2}e^TQe - \frac{1}{2\rho^2}e^TPLL^TPe+ \\ +\frac{1}{2}e^TPL\tilde{d} + \frac{1}{2}\tilde{d}^TL^TPe\end{aligned} \tag{6.146}$$

Lemma: The following inequality holds

$$\frac{1}{2}e^TPL\tilde{d} + \frac{1}{2}\tilde{d}L^TPe - \frac{1}{2\rho^2}e^TPLL^TPe \leq \frac{1}{2}\rho^2\tilde{d}^T\tilde{d} \tag{6.147}$$

Proof: The binomial $(\rho\alpha - \frac{1}{\rho}b)^2$ is considered. Expanding the left part of the above inequality, one gets

$$\begin{aligned}\rho^2a^2 + \frac{1}{\rho^2}b^2 - 2ab \geq 0 \Rightarrow \frac{1}{2}\rho^2a^2 + \frac{1}{2\rho^2}b^2 - ab \geq 0 \Rightarrow \\ ab - \frac{1}{2\rho^2}b^2 \leq \frac{1}{2}\rho^2a^2 \Rightarrow \frac{1}{2}ab + \frac{1}{2}ab - \frac{1}{2\rho^2}b^2 \leq \frac{1}{2}\rho^2a^2\end{aligned} \tag{6.148}$$

The following substitutions are carried out: $a = \tilde{d}$ and $b = e^TPL$, and the previous relation becomes

$$\frac{1}{2}\tilde{d}^TL^TPe + \frac{1}{2}e^TPL\tilde{d} - \frac{1}{2\rho^2}e^TPLL^TPe \leq \frac{1}{2}\rho^2\tilde{d}^T\tilde{d} \tag{6.149}$$

Eq. (6.149) is substituted in Eq. (6.146) and the inequality is enforced, thus giving

$$\dot{V} \leq -\frac{1}{2}e^TQe + \frac{1}{2}\rho^2\tilde{d}^T\tilde{d} \tag{6.150}$$

Eq. (6.150) shows that the H_∞ tracking performance criterion is satisfied. The integration of $\dot{V}$ from 0 to T gives

$$\begin{aligned}\int_0^T\dot{V}(t)dt \leq -\frac{1}{2}\int_0^T||e||_Q^2dt + \frac{1}{2}\rho^2\int_0^T||\tilde{d}||^2dt \Rightarrow \\ 2V(T) + \int_0^T||e||_Q^2dt \leq 2V(0) + \rho^2\int_0^T||\tilde{d}||^2dt\end{aligned} \tag{6.151}$$

Moreover, if there exists a positive constant $M_d > 0$ such that

$$\int_0^\infty ||\tilde{d}||^2dt \leq M_d \tag{6.152}$$

then, one gets

$$\int_0^\infty ||e||_Q^2dt \leq 2V(0) + \rho^2M_d \tag{6.153}$$

Thus, the integral $\int_0^\infty ||e||_Q^2dt$ is bounded. Moreover, $V(T)$ is bounded and from the definition of the Lyapunov function V in Eq. (6.136), it becomes clear that $e(t)$ will be also bounded since $e(t) \in \Omega_e = \{e|e^TPe \leq 2V(0) + \rho^2M_d\}$.

According to the above and with the use of Barbalat's Lemma, one obtains $lim_{t\to\infty}e(t) = 0$.

Through the stages of the stability proof, one arrives at Eq. (6.150) which shows that the H-infinity tracking performance criterion holds. By selecting the attenuation coefficient ρ to be sufficiently small and in particular to satisfy $\rho^2 < ||e||_Q^2/||\tilde{d}||^2$, one has that the first derivative of the Lyapunov function is upper bounded by 0. This condition holds at each sampling instance, and consequently, global stability for the control loop can be concluded.

6.2.5 *Simulation tests*

The fine tracking performance and the global stability properties of the nonlinear optimal control scheme for the tethered satellite system are further confirmed through simulation experiments. To implement the nonlinear optimal control scheme the method's algebraic Riccati equation given in Eq. (6.141) had to be solved in each sampling interval using the *aresolv()* function in MATLAB®. Indicative values about the coefficients of the Riccati equation were: $r = 0.0001$, $\rho = 0.1$ while Q was a diagonal matrix with diagonal elements $q_{ii} = 0.2, i = 1, \cdots, 4$. The sampling period was $T_s =$ 0.01 sec. The obtained results are presented in Figures 6.21–6.28. The setpoint for the sub-satellite's in-plane motion was $x_{3,d} = \xi_d = 1.0$, and when $\xi = L/L_d = 1.0$ the tether's length L reaches its targeted value L_d. The variable depicted

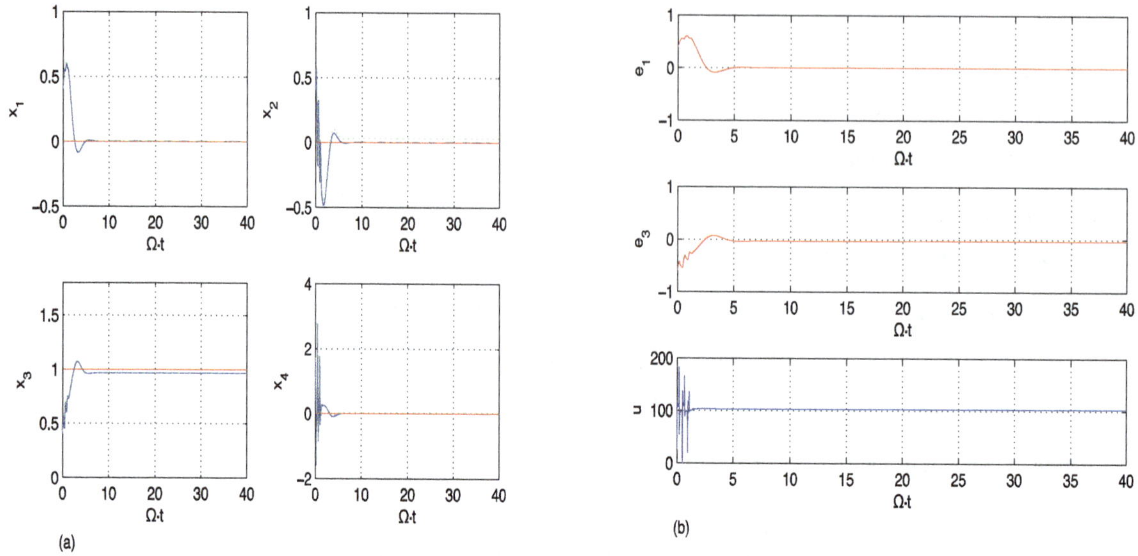

Figure 6.21 *Tracking of setpoint 1 by the tethered satellite system: (a) tracking of setpoints (red line) by state variables x_1 to x_4 (blue line) and estimates provided by the H-infinity Kalman filter; (b) convergence to zero of tracking errors e_1 and e_3 for state variables $x_1 = \theta$ and $x_3 = \xi$ (red line) and control input u (blue line).*

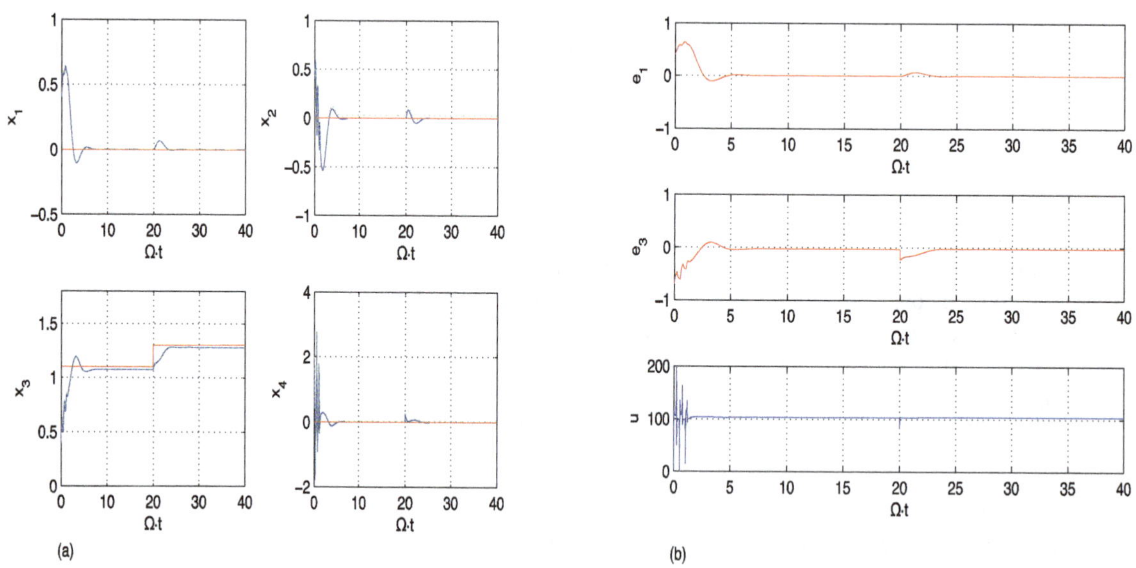

Figure 6.22 *Tracking of setpoint 2 by the tethered satellite system: (a) tracking of setpoints (red line) by state variables x_1 to x_4 (blue line) and estimates provided by the H-infinity Kalman filter; (b) convergence to zero of tracking errors e_1 and e_3 for state variables $x_1 = \theta$ and $x_3 = \xi$ (red line) and control input u (blue line).*

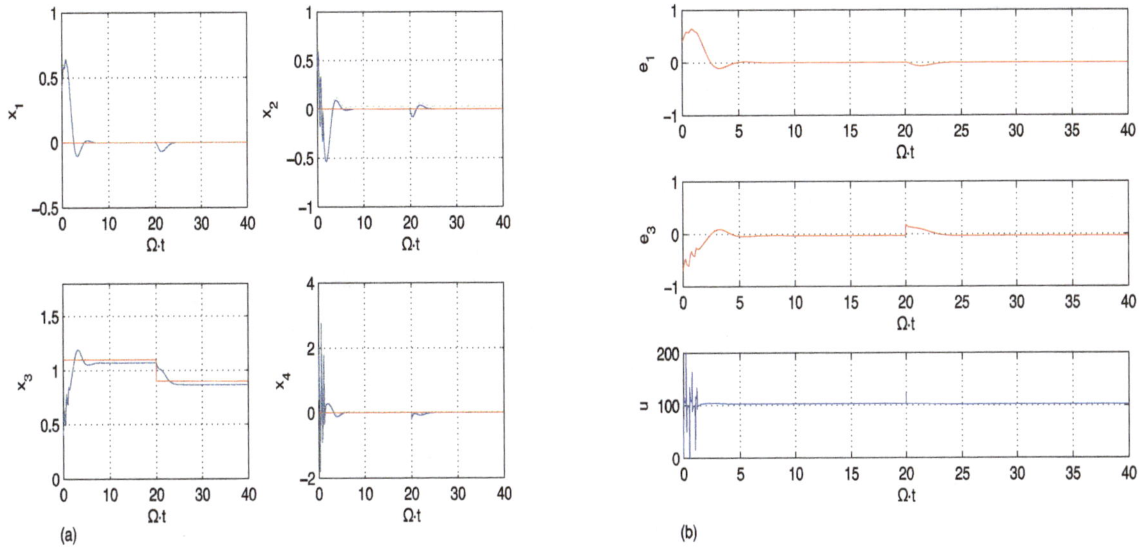

Figure 6.23 *Tracking of setpoint 3 by the tethered satellite system: (a) tracking of setpoints (red line) by state variables x_1 to x_4 (blue line) and estimates provided by the H-infinity Kalman filter; (b) convergence to zero of tracking errors e_1 and e_3 for state variables $x_1=\theta$ and $x_3=\xi$ (red line) and control input u (blue line)*

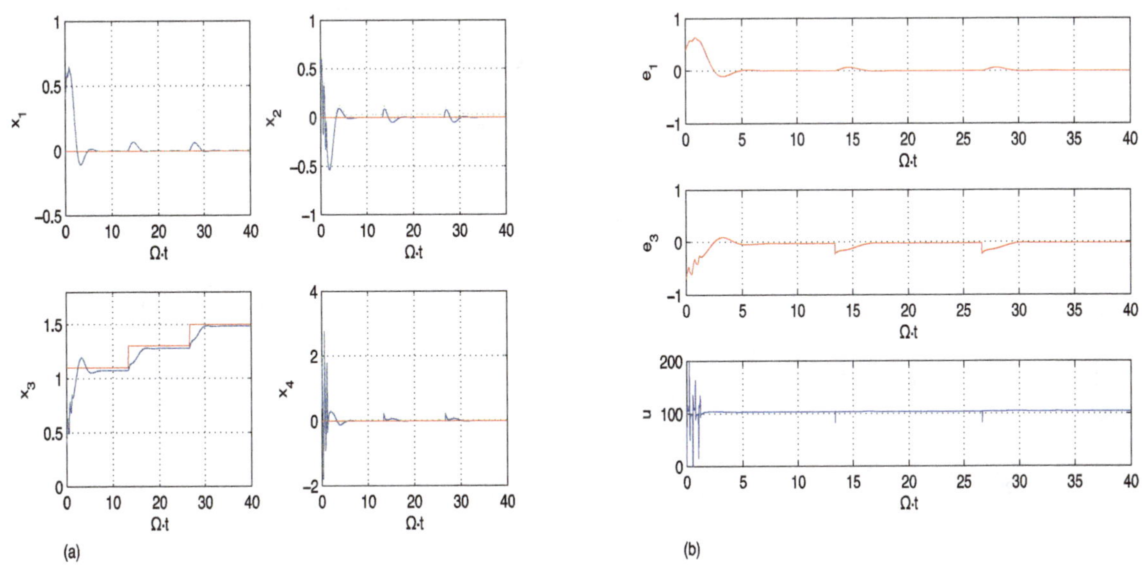

Figure 6.24 *Tracking of setpoint 4 by the tethered satellite system: (a) tracking of setpoints (red line) by state variables x_1 to x_4 (blue line) and estimates provided by the H-infinity Kalman filter; (b) convergence to zero of tracking errors e_1 and e_3 for state variables $x_1=\theta$ and $x_3=\xi$ (red line) and control input u (blue line)*

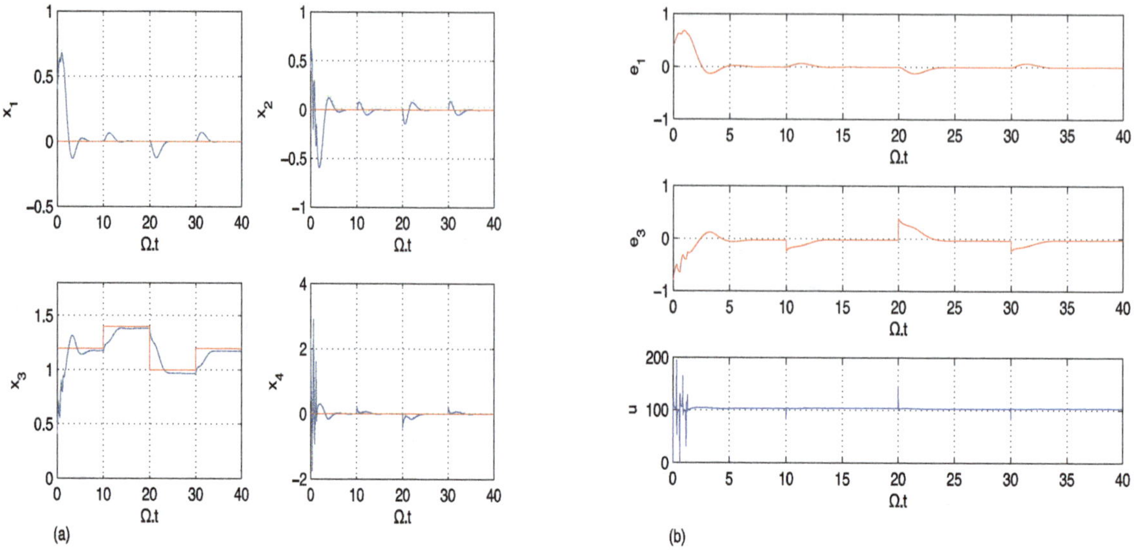

Figure 6.25 *Tracking of setpoint 5 by the tethered satellite system: (a) tracking of setpoints (red line) by state variables x_1 to x_4 (blue line) and estimates provided by the H-infinity Kalman filter; (b) convergence to zero of tracking errors e_1 and e_3 for state variables $x_1 = \theta$ and $x_3 = \xi$ (red line) and control input u (blue line)*

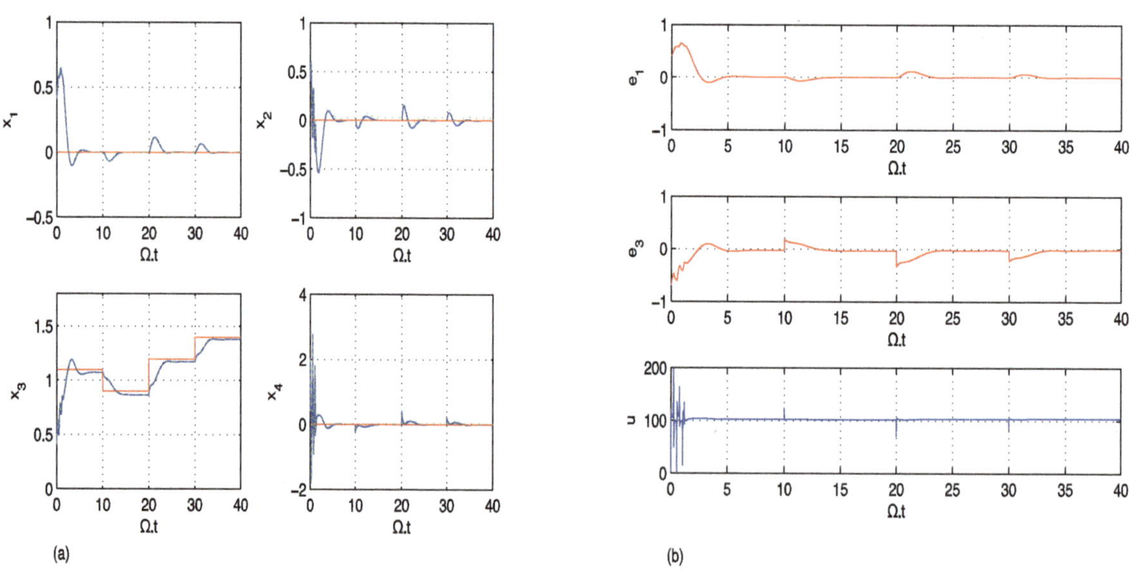

Figure 6.26 *Tracking of setpoint 6 by the tethered satellite system: (a) tracking of setpoints (red line) by state variables x_1 to x_4 (blue line) and estimates provided by the H-infinity Kalman filter; (b) convergence to zero of tracking errors e_1 and e_3 for state variables $x_1 = \theta$ and $x_3 = \xi$ (red line) and control input u (blue line)*

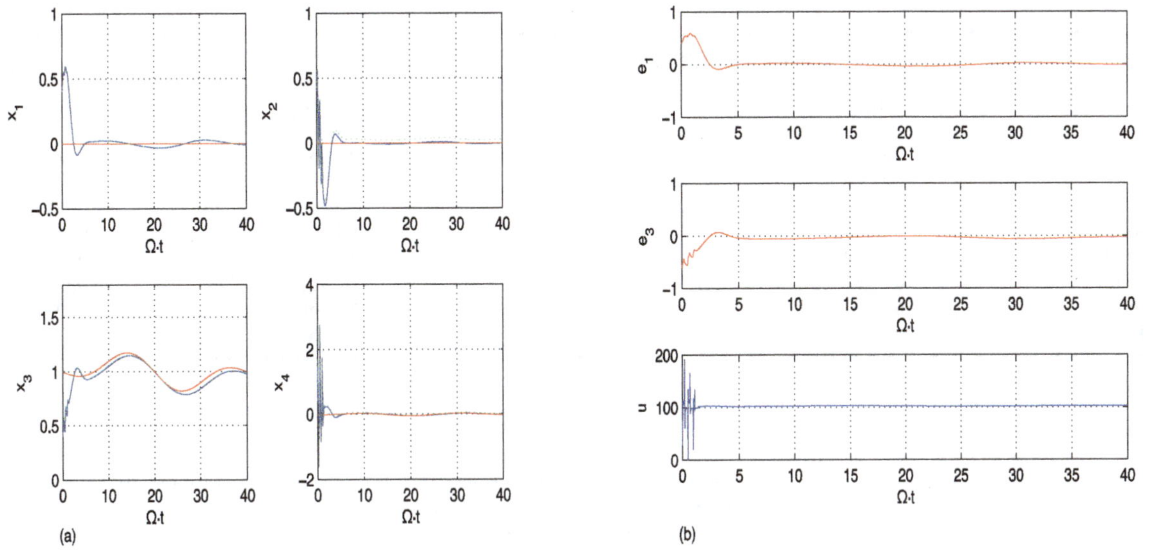

Figure 6.27 Tracking of setpoint 7 by the tethered satellite system: (a) tracking of setpoints (red line) by state variables x_1 to x_4 (blue line) and estimates provided by the H-infinity Kalman filter; (b) convergence to zero of tracking errors e_1 and e_3 for state variables $x_1 = \theta$ and $x_3 = \xi$ (red line) and control input u (blue line)

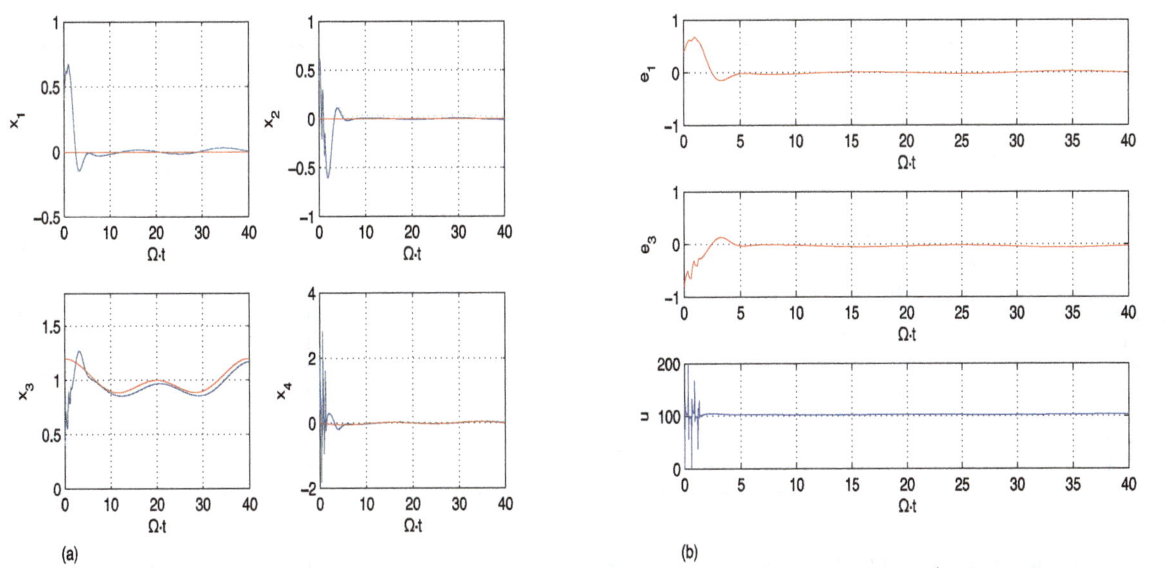

Figure 6.28 Tracking of setpoint 8 by the tethered satellite system: (a) tracking of setpoints (red line) by state variables x_1 to x_4 (blue line) and estimates provided by the H-infinity Kalman filter; (b) convergence to zero of tracking errors e_1 and e_3 for state variables $x_1 = \theta$ and $x_3 = \xi$ (red line) and control input u (blue line)

Table 6.5 Tethered satellites

Tracking RMSE in the disturbance-free case $\times 10^{-3}$				
	$RMSE_{x_1}$	$RMSE_{x_2}$	$RMSE_{x_3}$	$RMSE_{x_4}$
$setpoint_1$	0.0001	0.0001	0.5052	0.0001
$setpoint_2$	0.0001	0.0001	0.9648	0.0001
$setpoint_3$	0.0001	0.0001	0.3512	0.0001
$setpoint_4$	0.0041	0.0053	1.2698	0.0057
$setpoint_5$	0.0024	0.0004	0.7960	0.0396
$setpoint_6$	0.0280	0.0541	1.1061	0.0641
$setpoint_7$	0.3017	0.2084	0.7760	0.2040
$setpoint_8$	0.9158	0.2932	0.3961	0.3247

Table 6.6 Tethered satellites

Tracking RMSE in the disturbance-free case $\times 10^{-3}$				
$\Delta a\%$	$RMSE_{x_1}$	$RMSE_{x_2}$	$RMSE_{x_3}$	$RMSE_{x_4}$
0%	0.0001	0.0001	0.5054	0.0001
10%	0.0001	0.0001	1.3341	0.0001
20%	0.0001	0.0001	2.8676	0.0001
30%	0.0001	0.0001	4.1661	0.0001
40%	0.0001	0.0001	5.2796	0.0001
50%	0.0001	0.0001	6.2453	0.0001
60%	0.0001	0.0001	7.0902	0.0001

Table 6.7 Tethered satellites

RMSE for estimation by the H-infinity KF $\times 10^{-3}$				
	$RMSE_{\hat{x}_1}$	$RMSE_{\hat{x}_2}$	$RMSE_{\hat{x}_3}$	$RMSE_{\hat{x}_4}$
$setpoint_1$	0.0038	0.5083	0.0016	0.4912
$setpoint_2$	0.0024	0.5079	0.0040	0.4922
$setpoint_3$	0.0043	0.5079	0.0017	0.4920
$setpoint_4$	0.0016	0.5080	0.0017	0.4920
$setpoint_5$	0.0023	0.5072	0.0025	0.4930
$setpoint_6$	0.0020	0.5078	0.0017	0.4920
$setpoint_7$	0.0046	0.5083	0.0044	0.4910
$setpoint_8$	0.0027	0.5075	0.0018	0.4927

in the horizontal axis of the presented diagrams is the rotation angle of the earth $\Omega{\cdot}t$. The derivatives of the state space-model $\dot{x}_i$ were defined with respect to this turn angle. The real values of the state variables of the tethered satellite system are printed in blue, their estimated values which were provided by the H-infinity Kalman filter are plotted in green, while the reference setpoints are depicted in red color. The aggregate control input is the sum of the depicted time-varying tether's tension u and of a constant thrust force which is exerted on the sub-satellite and has direction opposite to the tether's tension. It can be noticed that in all test cases, the proposed nonlinear optimal control method achieves fast and accurate tracking of setpoints under moderate variations of the control inputs.

Table 6.8 Tethered satellites

Convergence time of state variables to setpoints				
	$RMSE_{x_1}$	$RMSE_{x_2}$	$RMSE_{x_3}$	$RMSE_{x_4}$
setpoint_1	7.0	6.0	5.0	5.0
setpoint_2	7.0	7.5	5.0	5.0
setpoint_3	7.0	7.5	5.0	5.0
setpoint_4	7.5	7.5	5.0	5.0
setpoint_5	7.5	6.0	5.0	5.0
setpoint_6	7.5	6.0	5.0	5.0
setpoint_7	5.0	5.0	6.0	5.0
setpoint_8	9.0	6.0	5.0	5.0

To elaborate on the tracking performance and on the robustness of the proposed nonlinear optimal control method for tethered satellite systems, the following tables are given: (i) Table 5 which provides information about the accuracy of tracking of the reference setpoints by the state variables of the tethered satellite system, (ii) Table 6 which provides information about the robustness of the control method to parametric changes in the model of the tethered satellite system (change in element f_2 of the drift vector up to 60%), (iii) Table 7 which provides information about the precision in state variables' estimation that is achieved by the H-infinity Kalman filter, and (iv) Table 8 which provides information about the speed of convergence of the state variables of the tethered satellite system to the associated setpoints.

6.3 Nonlinear optimal control for triangular tethered multi-satellite formations

6.3.1 Outline

Tethered satellite systems consist of several satellites which are connected with the use of elastic tethers and which aim at accomplishing a specific mission [1067,820,821]. Tethered satellites have found use in satellite communication networks and in tasks of precise observation of earth and space [822,372]. Applications of tethered satellite systems include the formation of synthetic aperture radars, space interferometry and formation of antenna arrays for fast satellite communication with high channel-capacity between the terrestrial stations and the satellites [371,145]. With the deployment of satellite communication systems, the significance of satellite constellations for enabling fast and reliable data transmission has emerged [401,374,293]. Precise positioning and orientation of the individual satellites within multi-satellite formations is a necessary stage for establishing this new type of satellite communication infrastructure [365,882,205,536]. In the case of a fully-actuated tethered multi-satellite formation, feedback linearization can be performed [981,163]. On this basis, several related control methods have been presented, including mainly decentralized linear and fault-tolerant control schemes [1045,294,824]. Besides, results on adaptive and sliding-mode control techniques can be noted [560,369,823]. One can also mention related results on control of tethered multi-satellite systems with the SDRE technique (State Dependent Riccati Equation) [416,125]. The control problem of tethered multi-satellite systems becomes more complicated in the case of underactuation. For the latter case, implementation of nonlinear optimal control is a feasible solution [166,164,165]. Additional results on control of tethered multi-satellite formations can be found in [1018,1014,1017,375,558].

In this section, a novel nonlinear optimal control method has been developed for the dynamic model of a tethered multi-satellite system [725,721,732]. As a case study the formation of three satellites which are placed on the vertices of a triangle has been considered. The state-space model of the tethered multi-satellite system has been formulated. The system is characterized by strong nonlinearities. To apply the proposed nonlinear optimal control method the dynamic model of the tethered multi-satellite system has undergone approximate linearization with the use of first-order Taylor series expansion and through the computation of the associated Jacobian matrices [705,60,706]. The linearization process takes place at each sampling instance around the temporary operating point (x^*, u^*) where x^* is the present value of the system's state vector and u^* is the last sampled value of the control inputs vector. The modeling error which is due to the truncation of higher-order terms from the Taylor series expansion is considered to be a perturbation which can be asymptotically compensated by the robustness of the control algorithm. For the approximately linearized model of the system, a stabilizing H-infinity feedback controller has been designed [725,721,732].

The proposed nonlinear optimal control method represents a mix–max differential game taking place between (i) the control inputs of the tethered multi-satellite system which try to minimize a cost function that comprises a quadratic term of the state vector's tracking error and (ii) the model uncertainty and external perturbation terms which try to maximize this cost function. The H-infinity control method is actually a solution to the nonlinear optimal control problem for the tethered multi-satellite system under model uncertainty and external perturbations. To compute the controller's feedback gains an algebraic Riccati equation has to be repetitively solved at each time step of the control method [725,721,732]. The global stability properties of the control scheme are proven through Lyapunov analysis. It is demonstrated that the proposed feedback control scheme satisfies the H-infinity tracking performance criterion [711,876]. This confirms the method's robustness to model uncertainty and disturbances. Moreover, under moderate conditions, it is proven that the control loop is globally asymptotically stable. The new control method retains the advantages of optimal control, that is fast and accurate tracking of setpoints under moderate variations of the control inputs and minimal dispersion of energy by the actuators of the tethered multi-satellite system [725,721,732]. Finally, to implement state estimation-based control without the need to measure the entire state vector of the formation of the tethered satellites, the H-infinity Kalman filter has been used as a robust observer.

6.3.2 Dynamic model of the triangular constellation of tethered satellites

The diagram of the triangular constellation of tethered satellites and the associated coordinate frames are shown in Figure 6.29. The dynamic model of the tethered multi-satellite system is obtained with the use of Euler–Lagrange analysis after computing the potential and kinetic energy of the individual satellites in this formation [1067,820,821].

The dynamic model of the tethered satellites formation is given by [1067,820,821]:

$$M(q)\ddot{q} + N(q,\dot{q}) + P(q) = Q_q \tag{6.154}$$

where $q = [L_1.L_2.\theta_1,\theta_2]^T$ is the state vector of the tethered satellites formation, and Q_q is the generalized force vector. The following definitions are used: $\theta = \theta_1 - \theta_2$ and $K = \frac{EA}{m\cdot\Omega^2\cdot L_o}$, where EA is the elastic coefficient of the tether, m is

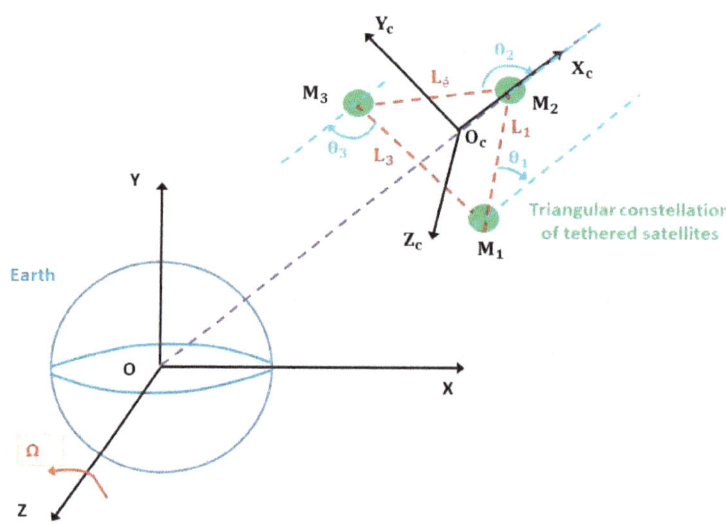

Figure 6.29 Diagram of the triangular constellation of tethered satellites

the total mass of the three satellites, and L_o is the nature-length of the tethers. The matrices of the dynamic model are defined as [1067,820,821]:

$$M = \frac{1}{9}\begin{pmatrix} 2 & cos(\theta) & 0 & L_2 sin(\theta) \\ cos(\theta) & 2 & -L_1 sin(\theta) & 0 \\ 0 & -L_1 sin(\theta) & 2L_1^2 & L_1L_2cos(\theta) \\ L_2 sin(\theta) & 0 & L_1L_2cos(\theta) & 2L_2^2 \end{pmatrix} \tag{6.155}$$

$$N = \frac{1}{9}\begin{pmatrix} 2\dot{L}_2(\dot{\theta}_2+1)sin(\theta) + L_2\dot{\theta}_2(-2-\dot{\theta}_2)cos(\theta) - 2L_1\dot{\theta}_1(\dot{\theta}_1+2) \\ -2\dot{L}_1(\dot{\theta}_1+1)sin(\theta) + L_1\dot{\theta}_1(-2-\dot{\theta}_1)cos(\theta) - 2L_2\dot{\theta}_2(\dot{\theta}_2+2) \\ 4L_1\dot{L}_1(\dot{\theta}_1+1) + 2L_1\dot{L}_2(\dot{\theta}_2+1)cos(\theta) + L_1L_2\dot{\theta}_2(\dot{\theta}_2+2)sin(\theta) \\ 4L_2\dot{L}_2(\dot{\theta}_2+1) + 2L_2\dot{L}_1(\dot{\theta}_1+1)cos(\theta) + L_1L_2\dot{\theta}_1(-\dot{\theta}_1-2)sin(\theta) \end{pmatrix} \tag{6.156}$$

$$P = \frac{1}{9}\begin{pmatrix} \frac{1}{3}(-2L_1cos^2(\theta_1) - L_2cos(\theta_1)cos(\theta_2)) + K[(L_1-1)\xi_1 + \frac{L_3-1}{L_3}(L_1+L_2cos(\theta))]\xi_3] \\ \frac{1}{3}(-2L_2cos^3(\theta_2) - L_1cos(\theta_1)cos(\theta_2)) + K[(L_2-1)\xi_2 + \frac{L_3-1}{L_3}(L_2+L_1cos(\theta))\xi_3] \\ \frac{1}{3}(2L_1^2sin(\theta_1)cos(\theta_1) + L_1L_2sin(\theta_1)cos(\theta_2)) + K\xi_3\frac{L_1L_2}{L_3}(1-L_3)sin(\theta) \\ \frac{1}{3}(2L_2^2sin(\theta_2)cos(\theta_2) + L_1L_2sin(\theta_2)cos(\theta_1)) - K\xi_3\frac{L_1L_2}{L_3}(1-L_3)sin(\theta) \end{pmatrix} \tag{6.157}$$

where about coefficients ξ_i, $i = 1, 2, 3$ and L_3, it holds that

$$\xi_i = \begin{cases} 1 \text{ if } L_i > L_0 \\ 0 \text{ if } L_i \leq L_0 \end{cases} \qquad L_3 = \sqrt{L_1^2 + L_2^2 + 2L_1L_2cos(\theta_1-\theta_2)} \tag{6.158}$$

By denoting as T_i $i = 1, 2, 3$ the tethers' tensions, the generalized forces vector is

$$Q_q = \begin{pmatrix} Q_{L_1} \\ Q_{L_2} \\ Q_{\theta_1} \\ Q_{\theta_2} \end{pmatrix} = \begin{pmatrix} -\frac{T_1}{m\Omega^2 l_o} - \frac{T_3}{m\omega^2 L_3 l_o}[L_1 + L_2cos(\theta_1-\theta_2)] \\ -\frac{T_2}{m\Omega_2 l_o} - \frac{T_3}{m\Omega_2 L_3 l_o}[L_2 + L_1cos(\theta_1-\theta_2)] \\ T_3\frac{L_1L_2sin(\theta_1-\theta_2)}{m\Omega^2 L_3} \\ -T_3\frac{L_1L_2sin(\theta_1-\theta_2)}{m\Omega^2 L_3} \end{pmatrix} \tag{6.159}$$

The generalized forces vector can be also written as

$$Q_q = \begin{pmatrix} -\frac{1}{m\Omega^2 l_o} & 0 & -\frac{1}{m\omega^2 L_3 l_o}[L_1 + L_2cos(\theta_1-\theta_2)] \\ 0 & -\frac{1}{m\Omega_2 l_o} & -\frac{1}{m\Omega_2 L_3 l_o}[L_2 + L_1cos(\theta_1-\theta_2)] \\ 0 & 0 & \frac{L_1L_2sin(\theta_1-\theta_2)}{m\Omega^2 L_3} \\ 0 & 0 & -\frac{L_1L_2sin(\theta_1-\theta_2)}{m\Omega^2 L_3} \end{pmatrix}\begin{pmatrix} T_1 \\ T_2 \\ T_3 \end{pmatrix} \tag{6.160}$$

Next, the inverse of the inertia matrix $M(q)$ is computed as follows:

$$M^{-1}(q) = \frac{9}{detM}\begin{pmatrix} M_{11} & -M_{21} & M_{31} & -M_{41} \\ -M_{12} & M_{22} & -M_{32} & M_{42} \\ M_{13} & -M_{23} & M_{33} & -M_{43} \\ -M_{14} & M_{24} & -M_{34} & M_{44} \end{pmatrix} \tag{6.161}$$

where it holds that $M_{11} = 6L_1^2L_2^2$, $M_{12} = 3L_1^2L_2^2cos(\theta)$, $M_{13} = 0$, $M_{14} = 3L_1^2L_2sin(\theta)$, $M_{21} = 3L_1^2L_2^2cos(\theta)$, $M_{22} = 6x_1^2x_3^2$, $M_{23} = -3L_1L_2^2sin(\theta)$, $M_{24} = 0$, $M_{31} = 0$, $M_{32} = -3L_1L_2^2sin(\theta)$, $M_{33} = 6L_2^2$, $M_{34} = 3L_1L_2cos(\theta)$, $M_{41} =$

$3L_1^2L_2sin(\theta)$, $M_{42}=0$, $M_{43}=3L_1L_2cos(\theta)$, $M_{44}=6L_1^2$. The determinant of matrix M is $detM=2M_{11}-cos(\theta)M_{12}-L_2sin(\theta)M_{14}$.

Moreover, vectors N, P, as well as the control inputs gain matrix B_q are written as

$$N=\frac{1}{9}\begin{pmatrix}N_1\\N_2\\N_3\\N_4\end{pmatrix} \quad \mathbf{P}=\begin{pmatrix}P_1\\P_2\\P_3\\P_4\end{pmatrix} \quad B_q=\begin{pmatrix}B_{q11} & 0 & B_{q13}\\0 & B_{q22} & B_{q23}\\0 & 0 & B_{q33}\\0 & 0 & B_{q43}\end{pmatrix} \tag{6.162}$$

By rewriting the dynamic model of the tethered satellite formation as

$$\ddot{q}=-M^{-1}(q)[N(q,\dot{q})+P(q)]+M^{-1}(q)B_qT \tag{6.163}$$

which also gives

$$\begin{pmatrix}\ddot{q}_1\\\ddot{q}_2\\\ddot{q}_3\\\ddot{q}_4\end{pmatrix}=-\frac{9}{detM}\begin{pmatrix}M_{11} & -M_{21} & M_{31} & -M_{41}\\-M_{12} & M_{22} & -M_{32} & M_{42}\\M_{13} & -M_{23} & M_{33} & -M_{43}\\-M_{14} & M_{24} & -M_{34} & M_{44}\end{pmatrix}\begin{pmatrix}(N_1+P_1)\\(N_2+P_2)\\(N_3+P_3)\\(N_4+P_4)\end{pmatrix}+$$
$$+\frac{9}{detM}\begin{pmatrix}M_{11} & -M_{21} & M_{31} & -M_{41}\\-M_{12} & M_{22} & -M_{32} & M_{42}\\M_{13} & -M_{23} & M_{33} & -M_{43}\\-M_{14} & M_{24} & -M_{34} & M_{44}\end{pmatrix}\begin{pmatrix}B_{q11} & 0 & B_{q13}\\0 & B_{q22} & B_{q23}\\0 & 0 & B_{q33}\\0 & 0 & B_{q43}\end{pmatrix}\begin{pmatrix}T_1\\T_2\\T_3\end{pmatrix} \tag{6.164}$$

The main parameters of the dynamic model of the tethered multi-satellite system are outlined in Table 9.

Table 6.9

Parameters of the tethered multi-satellite model

Parameter	**Definition**
L_i $i=1,2,3$	lengths of the tethers
θ_i $i=1,2,3$	angles between the tethers
T_i $i=1,2,3$	tensions of the tethers
EA	elastic coefficient of the tether
M	inertia matrix of the dynamic model
N	Coriolis-type vector of the dynamic model
P	forces vector that includes elastic coefficient
B_q	control inputs gain matrix
Q_q	generalized forces vector
m	total mass of the three satellites
L_o	natural length of the tethers
Ω	angular speed of the earth

The dynamic model of the multi-satellite tethered formation is also written as

$$\begin{pmatrix} \ddot{q}_1 \\ \ddot{q}_2 \\ \ddot{q}_3 \\ \ddot{q}_4 \end{pmatrix} = \begin{pmatrix} \frac{9[-M_{11}(\frac{N_1}{9}+P_1)+M_{21}(\frac{N_2}{9}+P_2)-M_{31}(\frac{N_3}{9}+P_3)+M_{41}(\frac{N_4}{9}+P_4)]}{detM} \\ \frac{9[M_{12}(\frac{N_1}{9}+P_1)-M_{22}(\frac{N_2}{9}+P_2)+M_{32}(\frac{N_3}{9}+P_3)-M_{42}(\frac{N_4}{9}+P_4)]}{detM} \\ \frac{9[-M_{13}(\frac{N_1}{9}+P_1)+M_{23}(\frac{N_2}{9}+P_2)-M_{33}(\frac{N_3}{9}+P_3)+M_{43}(\frac{N_4}{9}+P_4)]}{detM} \\ \frac{9[M_{14}(\frac{N_1}{9}+P_1)-M_{24}(\frac{N_2}{9}+P_2)+M_{34}(\frac{N_3}{9}+P_3)-M_{44}(\frac{N_4}{9}+P_4)]}{detM} \end{pmatrix} + \tag{6.165}$$

$$+ \begin{pmatrix} \frac{9M_{11}B_{q11}}{detM} & -\frac{9M_{21}B_{q22}}{detM} & \frac{9[M_{11}B_{q13}-M_{21}B_{q23}+M_{31}B_{q33}-M_{41}B_{q43}]}{detM} \\ -\frac{9M_{12}B_{q11}}{detM} & \frac{9M_{22}B_{q22}}{detM} & \frac{9[-M_{12}B_{q13}+M_{22}B_{q23}-M_{32}B_{q33}+M_{42}B_{q43}]}{detM} \\ \frac{9M_{13}B_{q11}}{detM} & -\frac{9M_{23}B_{q22}}{detM} & \frac{9[M_{13}B_{q13}-M_{23}B_{q23}+M_{33}B_{q33}-M_{43}B_{q43}]}{detM} \\ -\frac{9M_{14}B_{q11}}{detM} & \frac{9M_{24}B_{q22}}{detM} & \frac{9[-M_{14}B_{q13}+M_{24}B_{q23}-M_{34}B_{q33}+M_{44}B_{q43}]}{detM} \end{pmatrix} \begin{pmatrix} T_1 \\ T_2 \\ T_3 \end{pmatrix}$$

Next, the state vector of the system is defined as $x=[x_1,x_2,x_3,x_4,x_5,x_6,x_7,x_8]^T$ or $x=[q_1,\dot{q}_1,q_2,\dot{q}_2,q_3,\dot{q}_3,q_4,\dot{q}_4]^T$. The control inputs vector of the system is defined as $u=[u_1,u_2,u_3]^T$ or $u=[T_1,T_2,T_3]^T$.

Moreover, the elements of the drift vector of the dynamic model of the tethered multi-satellite formation are written as

$$\begin{aligned} &f_1(x)=x_2 \quad f_2(x)=\frac{9[-M_{11}(\frac{N_1}{9}+P_1)+M_{21}(\frac{N_2}{9}+P_2)-M_{31}(\frac{N_3}{9}+P_3)+M_{41}(\frac{N_4}{9}+P_4)]}{detM} \\ &f_3(x)=x_4 \quad f_4(x)=\frac{9[M_{12}(\frac{N_1}{9}+P_1)-M_{22}(\frac{N_2}{9}+P_2)+M_{32}(\frac{N_3}{9}+P_3)-M_{42}(\frac{N_4}{9}+P_4)]}{detM} \\ &f_5(x)=x_6 \quad f_6(x)=\frac{9[-M_{13}(\frac{N_1}{9}+P_1)+M_{23}(\frac{N_2}{9}+P_2)-M_{33}(\frac{N_3}{9}+P_3)+M_{43}(\frac{N_4}{9}+P_4)]}{detM} \\ &f_7(x)=x_8 \quad f_8(x)=\frac{9[M_{14}(\frac{N_1}{9}+P_1)-M_{24}(\frac{N_2}{9}+P_2)+M_{34}(\frac{N_3}{9}+P_3)-M_{44}(\frac{N_4}{9}+P_4)]}{detM} \end{aligned} \tag{6.166}$$

Additionally, the elements of the control inputs gain matrix of the dynamic model of the tethered multi-satellite formation are written as

$$\begin{aligned} &g_{11}(x)=0 && g_{12}(x)=0 && g_{13}(x)=0 \\ &g_{21}(x)=\frac{9M_{11}B_{q11}}{detM} && g_{22}(x)=-\frac{9M_{21}B_{q22}}{detM} && g_{32}(x)=\frac{9[M_{11}B_{q13}-M_{21}B_{q23}+M_{31}B_{q33}-M_{41}B_{q43}]}{detM} \\ &g_{31}(x)=0 && g_{32}(x)=0 && g_{33}(x)=0 \\ &g_{41}(x)=-\frac{9M_{12}B_{q11}}{detM} && g_{42}(x)=\frac{9M_{22}B_{q22}}{detM} && g_{43}(x)=\frac{9[-M_{12}B_{q13}+M_{22}B_{q23}-M_{32}B_{q33}+M_{42}B_{q43}]}{detM} \\ &g_{51}(x)=0 && g_{52}(x)=0 && g_{53}(x)=0 \\ &g_{61}(x)=\frac{9M_{13}B_{q11}}{detM} && g_{62}(x)=-\frac{9M_{23}B_{q22}}{detM} && g_{63}(x)=\frac{9[M_{13}B_{q13}-M_{23}B_{q23}+M_{33}B_{q33}-M_{43}B_{q43}]}{detM} \\ &g_{71}(x)=0 && g_{72}(x)=0 && g_{73}(x)=0 \\ &g_{81}(x)=-\frac{9M_{14}B_{q11}}{detM} && g_{82}(x)=\frac{9M_{24}B_{q22}}{detM} && g_{83}(x)=\frac{9[-M_{14}B_{q13}+M_{24}B_{q23}-M_{34}B_{q33}+M_{44}B_{q43}]}{detM} \end{aligned} \tag{6.167}$$

Thus, the dynamic model of the tethered multi-satellite formation can be written as:

$$\begin{pmatrix} \dot{x}_1 \\ \dot{x}_2 \\ \dot{x}_3 \\ \dot{x}_4 \\ \dot{x}_5 \\ \dot{x}_6 \\ \dot{x}_7 \\ \dot{x}_8 \end{pmatrix} = \begin{pmatrix} x_2 \\ f_2(x) \\ x_4 \\ f_4(x) \\ x_6 \\ f_6(x) \\ x_8 \\ f_8(x) \end{pmatrix} + \begin{pmatrix} 0 & 0 & 0 \\ g_{21}(x) & g_{22}(x) & g_{23}(x) \\ 0 & 0 & 0 \\ g_{41}(x) & g_{42}(x) & g_{43}(x) \\ 0 & 0 & 0 \\ g_{61}(x) & g_{62}(x) & g_{63}(x) \\ 0 & 0 & 0 \\ g_{81}(x) & g_{82}(x) & g_{83}(x) \end{pmatrix} \begin{pmatrix} u_1 \\ u_2 \\ u_3 \end{pmatrix} \tag{6.168}$$

Equivalently, the dynamic model of the multi-satellite formation can be written in the following nonlinear affine-in-the-input state-space form

$$\dot{x} = f(x) + g(x)u \tag{6.169}$$

where $x \in R^{8\times1}$, $f(x) \in R^{8\times1}$, $g(x) \in R^{8\times3}$, and $u \in R^{3\times1}$.

This dynamic model of the tethered multi-satellite formation is not directly input–output linearizable and differential flatness properties cannot be proven directly about it. An input–output linearized form of this system can be obtained through dynamic extension. On the contrary, differential flatness properties hold directly for the dynamic model of Eq. (6.154) which uses the generalized force vector Q_q with four control inputs.

6.3.3 Approximate linearization of the tethered multi-satellite formation

The dynamic model of the tethered multi-satellite formation undergoes approximate linearization around the temporary operating point (x^*, u^*). This operating point is updated at each sampling instance and x^* represents the present value of the system's state vector, while u^* denotes the last sampled value of the control inputs vector. The linearization is based on first-order Taylor series expansion and on the computation of the associated Jacobian matrices. The modeling error which is due to the truncation of higher-order terms from the Taylor series expansion is considered to be a perturbation which is asymptotically compensated by the robustness of the control algorithm. By linearizing around the present position of the system x^* and not at points of the desirable trajectory x_d^* the span of the Taylor series expansion is small and the modeling error is also negligible. The initial nonlinear dynamics of the system being in the form $\dot{x} = f(x) + g(x)u$ after linearization becomes

$$\dot{x} = Ax + Bu + \tilde{d} \tag{6.170}$$

where A, B are the Jacobian matrices of the system which are given by

$$A = \nabla_x[f(x) + g(x)u] \mid_{(x^*,u^*)} \Rightarrow A = \nabla_x f(x) \mid_{(x^*,u^*)} + \nabla_x g_1(x)u_1 \mid_{(x^*,u^*)} + \\ + \nabla_x g_2(x)u_2 \mid_{(x^*,u^*)} + \nabla_x g_3(x)u_3 \mid_{(x^*,u^*)} \tag{6.171}$$

$$B = \nabla_u[f(x) + g(x)u] \mid_{(x^*,u^*)} \Rightarrow B = g(x) \mid_{(x^*,u^*)} \tag{6.172}$$

The cumulative disturbance term $\tilde{d}$ may comprise: (i) the modeling error which is due to truncation of higher-order terms from the Taylor series expansion, (ii) exogenous perturbations, and (iii) sensor measurement noise of any distribution.

Computation of the Jacobian matrix $\nabla_x f(x) \mid_{(x^*,u^*)}$:

First row of the Jacobian matrix $\nabla_x f(x)\mid_{(x^*,u^*)}$: $\frac{\partial f_1}{\partial x_1}=0$, $\frac{\partial f_1}{\partial x_2}=1$, $\frac{\partial f_1}{\partial x_3}=0$, $\frac{\partial f_1}{\partial x_4}=0$, $\frac{\partial f_1}{\partial x_5}=0$, $\frac{\partial f_1}{\partial x_6}=0$, $\frac{\partial f_1}{\partial x_7}=0$, and $\frac{\partial f_1}{\partial x_8}=0$.

Second row of the Jacobian matrix $\nabla_x f(x)\mid_{(x^*,u^*)}$: It holds that

$$f_2(x)=\frac{f_{2,num}}{f_{2,den}}=\frac{9[-M_{11}(\frac{N_1}{9}+P_1)+M_{21}(\frac{N_2}{9}+P_2)-M_{31}(\frac{N_3}{9}+P_3)+M_{41}(\frac{N_4}{9}+P_4)]}{detM} \tag{6.173}$$

Thus, one has that for $i=1,2,\cdots,8$

$$\frac{\partial f_2}{\partial x_i}=\frac{\frac{\partial f_{2,num}}{\partial x_i}f_{2,den}-f_{2,num}\frac{\partial f_{2,den}}{\partial x_i}}{f_{2,den}^2} \tag{6.174}$$

where for $i=1,2,\cdots,8$

$$\begin{aligned}\frac{\partial f_2,num}{\partial x_i}&=9[-\frac{\partial M_{11}}{\partial x_i}(\frac{N_1}{9}+P_1)-M_{11}(\frac{1}{9}\frac{\partial N_1}{\partial x_i}+\frac{\partial P_1}{\partial x_i})\\&+\frac{\partial M_{21}}{\partial x_i}(\frac{N_2}{9}+P_2)+M_{21}(\frac{1}{9}\frac{\partial N_2}{\partial x_i}+\frac{\partial P_2}{\partial x_i})\\&-\frac{\partial M_{31}}{\partial x_i}(\frac{N_3}{9}+P_3)-M_{31}(\frac{1}{9}\frac{\partial N_3}{\partial x_i}+\frac{\partial P_3}{\partial x_i})\\&+\frac{\partial M_{41}}{\partial x_i}(\frac{N_4}{9}+P_4)+M_{41}(\frac{1}{9}\frac{\partial N_4}{\partial x_i}+\frac{\partial P_4}{\partial x_i})]\end{aligned} \tag{6.175}$$

while for $i=1,2,\cdots,8$

$$\frac{\partial f_2,den}{\partial x_i}=\frac{\partial detM}{\partial x_i} \tag{6.176}$$

Third row of the Jacobian matrix $\nabla_x f(x)\mid_{(x^*,u^*)}$: $\frac{\partial f_3}{\partial x_1}=0$, $\frac{\partial f_3}{\partial x_2}=0$, $\frac{\partial f_3}{\partial x_3}=0$, $\frac{\partial f_3}{\partial x_4}=1$, $\frac{\partial f_3}{\partial x_5}=0$, $\frac{\partial f_3}{\partial x_6}=0$, $\frac{\partial f_3}{\partial x_7}=0$, and $\frac{\partial f_3}{\partial x_8}=0$.

Fourth row of the Jacobian matrix $\nabla_x f(x)\mid_{(x^*,u^*)}$: It holds that

$$f_4(x)=\frac{f_{4,num}}{f_{4,den}}=\frac{9[M_{12}(\frac{N_1}{9}+P_1)-M_{22}(\frac{N_2}{9}+P_2)+M_{32}(\frac{N_3}{9}+P_3)-M_{42}(\frac{N_4}{9}+P_4)]}{detM} \tag{6.177}$$

Thus, one has that for $i=1,2,\cdots,8$

$$\frac{\partial f_4}{\partial x_i}=\frac{\frac{\partial f_{4,num}}{\partial x_i}f_{4,den}-f_{4,num}\frac{\partial f_{4,den}}{\partial x_i}}{f_{4,den}^2} \tag{6.178}$$

where for $i=1,2,\cdots,8$

$$\begin{aligned}\frac{\partial f_4,num}{\partial x_i}&=9[\frac{\partial M_{12}}{\partial x_i}(\frac{N_1}{9}+P_1)+M_{12}(\frac{1}{9}\frac{\partial N_1}{\partial x_i}+\frac{\partial P_1}{\partial x_i})\\&-\frac{\partial M_{22}}{\partial x_i}(\frac{N_2}{9}+P_2)+M_{22}(\frac{1}{9}\frac{\partial N_2}{\partial x_i}+\frac{\partial P_2}{\partial x_i})\\&+\frac{\partial M_{32}}{\partial x_i}(\frac{N_3}{9}+P_3)+M_{32}(\frac{1}{9}\frac{\partial N_3}{\partial x_i}+\frac{\partial P_3}{\partial x_i})\\&-\frac{\partial M_{42}}{\partial x_i}(\frac{N_4}{9}+P_4)+M_{42}(\frac{1}{9}\frac{\partial N_4}{\partial x_i}+\frac{\partial P_4}{\partial x_i})]\end{aligned} \tag{6.179}$$

while for $i=1,2,\cdots,8$

$$\frac{\partial f_4,den}{\partial x_i}=\frac{\partial detM}{\partial x_i} \tag{6.180}$$

Fifth row of the Jacobian matrix $\nabla_x f(x)\mid_{(x^*,u^*)}$: $\frac{\partial f_5}{\partial x_1}=0$, $\frac{\partial f_5}{\partial x_2}=0$, $\frac{\partial f_5}{\partial x_3}=0$, $\frac{\partial f_5}{\partial x_4}=0$, $\frac{\partial f_5}{\partial x_5}=0$, $\frac{\partial f_5}{\partial x_6}=1$, $\frac{\partial f_5}{\partial x_7}=0$, and $\frac{\partial f_5}{\partial x_8}=0$.

Sixth row of the Jacobian matrix $\nabla_x f(x)\mid_{(x^*,u^*)}$: It holds that

$$f_6(x)=\frac{f_{6,num}}{f_{6,den}}=\frac{9[-M_{13}(\frac{N_1}{9}+P_1)+M_{23}(\frac{N_2}{9}+P_2)-M_{33}(\frac{N_3}{9}+P_3)+M_{43}(\frac{N_4}{9}+P_4)]}{detM} \tag{6.181}$$

Thus, one has that for $i=1,2,\cdots,8$

$$\frac{\partial f_6}{\partial x_i}=\frac{\frac{\partial f_{6,num}}{\partial x_i}f_{6,den}-f_{6,num}\frac{\partial f_{6,den}}{\partial x_i}}{f_{6,den}^2} \tag{6.182}$$

where for $i=1,2,\cdots,8$

$$\begin{aligned}\frac{\partial f_{6,num}}{\partial x_i}&=9[-\frac{\partial M_{13}}{\partial x_i}(\frac{N_1}{9}+P_1)-M_{13}(\frac{1}{9}\frac{\partial N_1}{\partial x_i}+\frac{\partial P_1}{\partial x_i})\\&+\frac{\partial M_{23}}{\partial x_i}(\frac{N_2}{9}+P_2)+M_{23}(\frac{1}{9}\frac{\partial N_2}{\partial x_i}+\frac{\partial P_2}{\partial x_i})\\&-\frac{\partial M_{33}}{\partial x_i}(\frac{N_3}{9}+P_3)-M_{33}(\frac{1}{9}\frac{\partial N_3}{\partial x_i}+\frac{\partial P_3}{\partial x_i})\\&+\frac{\partial M_{43}}{\partial x_i}(\frac{N_4}{9}+P_4)+M_{43}(\frac{1}{9}\frac{\partial N_4}{\partial x_i}+\frac{\partial P_4}{\partial x_i})]\end{aligned} \tag{6.183}$$

while for $i=1,2,\cdots,8$

$$\frac{\partial f_{6,den}}{\partial x_i}=\frac{\partial detM}{\partial x_i} \tag{6.184}$$

Seventh row of the Jacobian matrix $\nabla_x f(x)\mid_{(x^*,u^*)}$: $\frac{\partial f_7}{\partial x_1}=0$, $\frac{\partial f_7}{\partial x_2}=0$, $\frac{\partial f_7}{\partial x_3}=0$, $\frac{\partial f_7}{\partial x_4}=0$, $\frac{\partial f_7}{\partial x_5}=0$, $\frac{\partial f_7}{\partial x_6}=0$, $\frac{\partial f_7}{\partial x_7}=0$, and $\frac{\partial f_7}{\partial x_8}=0$.

Eighth row of the Jacobian matrix $\nabla_x f(x)\mid_{(x^*,u^*)}$: It holds that

$$f_8(x)=\frac{f_{8,num}}{f_{8,den}}=\frac{9[M_{14}(\frac{N_1}{9}+P_1)-M_{24}(\frac{N_2}{9}+P_2)+M_{34}(\frac{N_3}{9}+P_3)-M_{44}(\frac{N_4}{9}+P_4)]}{detM} \tag{6.185}$$

Thus, one has that for $i=1,2,\cdots,8$

$$\frac{\partial f_8}{\partial x_i}=\frac{\frac{\partial f_{8,num}}{\partial x_i}f_{8,den}-f_{8,num}\frac{\partial f_{8,den}}{\partial x_i}}{f_{8,den}^2} \tag{6.186}$$

where for $i=1,2,\cdots,8$

$$\begin{aligned}\frac{\partial f_{8,num}}{\partial x_i}&=9[\frac{\partial M_{14}}{\partial x_i}(\frac{N_1}{9}+P_1)+M_{14}(\frac{1}{9}\frac{\partial N_1}{\partial x_i}+\frac{\partial P_1}{\partial x_i})\\&-\frac{\partial M_{24}}{\partial x_i}(\frac{N_2}{9}+P_2)+M_{24}(\frac{1}{9}\frac{\partial N_2}{\partial x_i}+\frac{\partial P_2}{\partial x_i})\\&+\frac{\partial M_{34}}{\partial x_i}(\frac{N_3}{9}+P_3)+M_{34}(\frac{1}{9}\frac{\partial N_3}{\partial x_i}+\frac{\partial P_3}{\partial x_i})\\&-\frac{\partial M_{44}}{\partial x_i}(\frac{N_4}{9}+P_4)+M_{44}(\frac{1}{9}\frac{\partial N_4}{\partial x_i}+\frac{\partial P_4}{\partial x_i})]\end{aligned} \tag{6.187}$$

while for $i=1,2,\cdots,8$

$$\frac{\partial f_{8,den}}{\partial x_i}=\frac{\partial detM}{\partial x_i} \tag{6.188}$$

Computation of the Jacobian matrix $\nabla_x g_1(x)\mid_{(x^*,u^*)}$:

First row of the Jacobian matrix $\nabla_x g_1(x)\mid_{(x^*,u^*)}$: It holds that $\frac{\partial g_{11}}{\partial x_i}=0$ for $i=1,2,\cdots,8$

Second row of the Jacobian matrix $\nabla_x g_1(x)\mid_{(x^*,u^*)}$: It holds that for $i=1,2,\cdots,8$

$$\frac{\partial g_{21}}{\partial x_i}=\frac{(9\frac{\partial M_{11}}{\partial x_i}B_{q11}+9M_{11}\frac{\partial B_{q11}}{\partial x_i})detM-(9M_{11}B_{q11})\frac{\partial detM}{\partial x_i}}{detM^2} \tag{6.189}$$

Third row of the Jacobian matrix $\nabla_x g_1(x)\mid_{(x^*,u^*)}$: It holds that $\frac{\partial g_{31}}{\partial x_i}=0$ for $i=1,2,\cdots,8$

Fourth row of the Jacobian matrix $\nabla_x g_1(x)\mid_{(x^*,u^*)}$: It holds that for $i=1,2,\cdots,8$

$$\frac{\partial g_{41}}{\partial x_i}=\frac{-(9\frac{\partial M_{12}}{\partial x_i}B_{q11}+9M_{12}\frac{\partial B_{q11}}{\partial x_i})detM+(9M_{12}B_{q11})\frac{\partial detM}{\partial x_i}}{detM^2} \tag{6.190}$$

Fifth row of the Jacobian matrix $\nabla_x g_1(x)\mid_{(x^*,u^*)}$: It holds that $\frac{\partial g_{51}}{\partial x_i}=0$ for $i=1,2,\cdots,8$

Sixth row of the Jacobian matrix $\nabla_x g_1(x) \mid_{(x^*,u^*)}$: It holds that for $i=1,2,\cdots,8$

$$\frac{\partial g_{61}}{\partial x_i} = \frac{(9\frac{\partial M_{13}}{\partial x_i}B_{q11}+9M_{13}\frac{\partial B_{q11}}{\partial x_i})detM-(9M_{13}B_{q11})\frac{\partial detM}{\partial x_i}}{detM^2} \tag{6.191}$$

Seventh row of the Jacobian matrix $\nabla_x g_1(x) \mid_{(x^*,u^*)}$: It holds that $\frac{\partial g_{71}}{\partial x_i}=0$ for $i=1,2,\cdots,8$

Eighth row of the Jacobian matrix $\nabla_x g_1(x) \mid_{(x^*,u^*)}$: It holds that for $i=1,2,\cdots,8$

$$\frac{\partial g_{81}}{\partial x_i} = \frac{-(9\frac{\partial M_{14}}{\partial x_i}B_{q11}+9M_{14}\frac{\partial B_{q11}}{\partial x_i})detM+(9M_{14}B_{q11})\frac{\partial detM}{\partial x_i}}{detM^2} \tag{6.192}$$

Computation of the Jacobian matrix $\nabla_x g_2(x) \mid_{(x^*,u^*)}$:

First row of the Jacobian matrix $\nabla_x g_2(x) \mid_{(x^*,u^*)}$: It holds that $\frac{\partial g_{12}}{\partial x_i}=0$ for $i=1,2,\cdots,8$

Second row of the Jacobian matrix $\nabla_x g_2(x) \mid_{(x^*,u^*)}$: It holds that for $i=1,2,\cdots,8$

$$\frac{\partial g_{22}}{\partial x_i} = \frac{-(9\frac{\partial M_{21}}{\partial x_i}B_{q22}+9M_{21}\frac{\partial B_{q22}}{\partial x_i})detM-(9M_{21}B_{q22})\frac{\partial detM}{\partial x_i}}{detM^2} \tag{6.193}$$

Third row of the Jacobian matrix $\nabla_x g_2(x) \mid_{(x^*,u^*)}$: It holds that $\frac{\partial g_{32}}{\partial x_i}=0$ for $i=1,2,\cdots,8$

Fourth row of the Jacobian matrix $\nabla_x g_2(x) \mid_{(x^*,u^*)}$: It holds that for $i=1,2,\cdots,8$

$$\frac{\partial g_{41}}{\partial x_i} = \frac{(9\frac{\partial M_{22}}{\partial x_i}B_{q22}+9M_{22}\frac{\partial B_{q22}}{\partial x_i})detM-(9M_{22}B_{q22})\frac{\partial detM}{\partial x_i}}{detM^2} \tag{6.194}$$

Fifth row of the Jacobian matrix $\nabla_x g_2(x) \mid_{(x^*,u^*)}$: It holds that $\frac{\partial g_{52}}{\partial x_i}=0$ for $i=1,2,\cdots,8$

Sixth row of the Jacobian matrix $\nabla_x g_2(x) \mid_{(x^*,u^*)}$: It holds that for $i=1,2,\cdots,8$

$$\frac{\partial g_{62}}{\partial x_i} = \frac{-(9\frac{\partial M_{23}}{\partial x_i}B_{q22}+9M_{23}\frac{\partial B_{q22}}{\partial x_i})detM+(9M_{23}B_{q22})\frac{\partial detM}{\partial x_i}}{detM^2} \tag{6.195}$$

Seventh row of the Jacobian matrix $\nabla_x g_2(x) \mid_{(x^*,u^*)}$: It holds that $\frac{\partial g_{72}}{\partial x_i}=0$ for $i=1,2,\cdots,8$

Eighth row of the Jacobian matrix $\nabla_x g_2(x) \mid_{(x^*,u^*)}$: It holds that for $i=1,2,\cdots,8$

$$\frac{\partial g_{82}}{\partial x_i} = \frac{(9\frac{\partial M_{24}}{\partial x_i}B_{q22}+9M_{24}\frac{\partial B_{q22}}{\partial x_i})detM-(9M_{24}B_{q22})\frac{\partial detM}{\partial x_i}}{detM^2} \tag{6.196}$$

Computation of the Jacobian matrix $\nabla_x g_3(x) \mid_{(x^*,u^*)}$:

First row of the Jacobian matrix $\nabla_x g_3(x) \mid_{(x^*,u^*)}$: It holds that $\frac{\partial g_{13}}{\partial x_i}=0$ for $i=1,2,\cdots,8$.

Second row of the Jacobian matrix $\nabla_x g_3(x) \mid_{(x^*,u^*)}$: It holds that for $i=1,2,\cdots,8$

$$g_{23}(x) = \frac{g_{23,num}}{g_{23,den}} = \frac{9[M_{11}B_{q13}-M_{21}B_{q23}+M_{31}B_{q33}-M_{41}B_{q43}]}{detM} \tag{6.197}$$

Thus, one has that

$$\frac{\partial g_{23}}{\partial x_i} = \frac{\frac{\partial g_{23,num}}{\partial x_i}g_{23,den}-g_{23,num}\frac{\partial g_{23,den}}{\partial x_i}}{(g_{23,den})^2} \tag{6.198}$$

where for $i=1,2,\cdots,8$

$$\begin{aligned}\frac{\partial g_{23,num}}{\partial x_i} &= 9[\frac{\partial M_{11}}{\partial x_i}B_{q13}+M_{11}\frac{\partial B_{q13}}{\partial x_i}-\\ &-\frac{\partial M_{21}}{\partial x_i}B_{q23}-M_{21}\frac{\partial B_{q23}}{\partial x_i}+\\ &+\frac{\partial M_{31}}{\partial x_i}B_{q33}+M_{31}\frac{\partial B_{q33}}{\partial x_i}-\\ &-\frac{\partial M_{41}}{\partial x_i}B_{q43}-M_{41}\frac{\partial B_{q43}}{\partial x_i}]\end{aligned} \tag{6.199}$$

and for $i=1,2,\cdots,8$

$$\frac{\partial g_{23,den}}{\partial x_i}=\frac{\partial detM}{\partial x_i} \tag{6.200}$$

Third row of the Jacobian matrix $\nabla_x g_3(x)\mid_{(x^*,u^*)}$: It holds that $\frac{\partial g_{33}}{\partial x_i}=0$ for $i=1,2,\cdots,8$.

Fourth row of the Jacobian matrix $\nabla_x g_3(x)\mid_{(x^*,u^*)}$: It holds that for $i=1,2,\cdots,8$

$$g_{43}(x)=\frac{g_{43,num}}{g_{43,den}}=\frac{9[-M_{12}B_{q13}+M_{22}B_{q23}-M_{32}B_{q33}+M_{42}B_{q43}]}{detM} \tag{6.201}$$

Thus, one has that

$$\frac{\partial g_{43}}{\partial x_i}=\frac{\frac{\partial g_{43,num}}{\partial x_i}g_{43,den}-g_{43,num}\frac{\partial g_{43,den}}{\partial x_i}}{(g_{43,den})^2} \tag{6.202}$$

where for $i=1,2,\cdots,8$

$$\begin{aligned}\frac{\partial g_{43,num}}{\partial x_i} &= 9[-\frac{\partial M_{12}}{\partial x_i}B_{q13}-M_{12}\frac{\partial B_{q13}}{\partial x_i}+\\ &+\frac{\partial M_{22}}{\partial x_i}B_{q23}+M_{22}\frac{\partial B_{q23}}{\partial x_i}-\\ &-\frac{\partial M_{32}}{\partial x_i}B_{q33}-M_{32}\frac{\partial B_{q33}}{\partial x_i}+\\ &+\frac{\partial M_{42}}{\partial x_i}B_{q43}+M_{42}\frac{\partial B_{q43}}{\partial x_i}]\end{aligned} \tag{6.203}$$

and for $i=1,2,\cdots,8$

$$\frac{\partial g_{43,den}}{\partial x_i}=\frac{\partial detM}{\partial x_i} \tag{6.204}$$

Fifth row of the Jacobian matrix $\nabla_x g_3(x)\mid_{(x^*,u^*)}$: It holds that $\frac{\partial g_{53}}{\partial x_i}=0$ for $i=1,2,\cdots,8$.

Sixth row of the Jacobian matrix $\nabla_x g_3(x)\mid_{(x^*,u^*)}$: It holds that for $i=1,2,\cdots,8$

$$g_{63}(x)=\frac{g_{63,num}}{g_{63,den}}=\frac{9[M_{13}B_{q13}-M_{23}B_{q23}+M_{33}B_{q33}-M_{43}B_{q43}]}{detM} \tag{6.205}$$

Thus, one has that

$$\frac{\partial g_{63}}{\partial x_i}=\frac{\frac{\partial g_{63,num}}{\partial x_i}g_{63,den}-g_{63,num}\frac{\partial g_{63,den}}{\partial x_i}}{(g_{63,den})^2} \tag{6.206}$$

where for $i=1,2,\cdots,8$

$$\begin{aligned}\frac{\partial g_{63,num}}{\partial x_i} &= 9[\frac{\partial M_{13}}{\partial x_i}B_{q13}+M_{13}\frac{\partial B_{q13}}{\partial x_i}-\\ &-\frac{\partial M_{23}}{\partial x_i}B_{q23}-M_{23}\frac{\partial B_{q23}}{\partial x_i}+\\ &+\frac{\partial M_{33}}{\partial x_i}B_{q33}+M_{33}\frac{\partial B_{q33}}{\partial x_i}-\\ &-\frac{\partial M_{43}}{\partial x_i}B_{q43}-M_{43}\frac{\partial B_{q43}}{\partial x_i}]\end{aligned} \tag{6.207}$$

and for $i=1,2,\cdots,8$

$$\frac{\partial g_{73,den}}{\partial x_i}=\frac{\partial detM}{\partial x_i} \tag{6.208}$$

Eighth row of the Jacobian matrix $\nabla_x g_3(x) \mid_{(x^*,u^*)}$: It holds that for $i=1,2,\cdots,8$

$$g_{83}(x) = \frac{g_{83,num}}{g_{83,den}} = \frac{9[-M_{14}B_{q13}+M_{24}B_{q24}-M_{34}B_{q34}+M_{44}B_{q44}]}{detM} \tag{6.209}$$

Thus, one has that

$$\frac{\partial g_{83}}{\partial x_i} = \frac{\frac{\partial g_{83,num}}{\partial x_i}g_{83,den}-g_{83,num}\frac{\partial g_{83,den}}{\partial x_i}}{(g_{83,den})^2} \tag{6.210}$$

where for $i=1,2,\cdots,8$

$$\begin{aligned}\frac{\partial g_{83,num}}{\partial x_i} = 9[-\frac{\partial M_{14}}{\partial x_i}B_{q13} - M_{14}\frac{\partial B_{q13}}{\partial x_i} + \\ +\frac{\partial M_{24}}{\partial x_i}B_{q23} + M_{24}\frac{\partial B_{q23}}{\partial x_i} - \\ -\frac{\partial M_{34}}{\partial x_i}B_{q33} - M_{34}\frac{\partial B_{q33}}{\partial x_i} + \\ +\frac{\partial M_{44}}{\partial x_i}B_{q43} + M_{44}\frac{\partial B_{q43}}{\partial x_i}]\end{aligned} \tag{6.211}$$

and for $i=1,2,\cdots,8$

$$\frac{\partial g_{83,den}}{\partial x_i} = \frac{\partial detM}{\partial x_i} \tag{6.212}$$

Next, the partial derivatives of the sub-determinants M_{ij}, $i=1,\cdots,4$, $j=1,\cdots,4$ of the inertia matrix M are computed:

For $M_{11}=6x_1^2x_3^2$: $\frac{\partial M_{11}}{\partial x_1}=12x_1x_3^2$, $\frac{\partial M_{11}}{\partial x_2}=0$, $\frac{\partial M_{11}}{\partial x_3}=12x_1^2x_3$, $\frac{\partial M_{11}}{\partial x_4}=0$, $\frac{\partial M_{11}}{\partial x_5}=0$, $\frac{\partial M_{11}}{\partial x_6}=0$, $\frac{\partial M_{11}}{\partial x_7}=0$, and $\frac{\partial M_{11}}{\partial x_8}=0$.

For $M_{12}=3x_1^2x_3^2cos(x_5-x_7)$: $\frac{\partial M_{12}}{\partial x_1}=6x_1x_3^2cos(x_5-x_7)$, $\frac{\partial M_{12}}{\partial x_2}=0$, $\frac{\partial M_{12}}{\partial x_3}=6x_1^2x_3cos(x_5-x_7)$, $\frac{\partial M_{12}}{\partial x_4}=0$, $\frac{\partial M_{11}}{\partial x_5}=-3x_1^2x_3^2sin(x_5-x_7)$, $\frac{\partial M_{12}}{\partial x_6}=0$, $\frac{\partial M_{12}}{\partial x_7}=3x_1^2x_3^2sin(x_5-x_7)$, $\frac{\partial M_{12}}{\partial x_8}=0$.

For $M_{13}=0$: $\frac{\partial M_{13}}{\partial x_i}=0$ for $i=1,2,\cdots,8$.

For $M_{14}=3x_1^2x_3sin(x_5-x_7)$: $\frac{\partial M_{14}}{\partial x_1}=6x_1x_3sin(x_5-x_7)$, $\frac{\partial M_{14}}{\partial x_2}=0$, $\frac{\partial M_{14}}{\partial x_3}=3x_1^2sin(x_5-x_7)$, $\frac{\partial M_{14}}{\partial x_4}=0$, $\frac{\partial M_{14}}{\partial x_5}=3x_1^2x_3cos(x_5-x_7)$, $\frac{\partial M_{14}}{\partial x_6}=0$, $\frac{\partial M_{14}}{\partial x_7}=-3x_1^2x_3cos(x_5-x_7)$, $\frac{\partial M_{14}}{\partial x_8}=0$.

For $M_{21}=3x_1^2x_3^2cos(x_5-x_7)$: $\frac{\partial M_{11}}{\partial x_1}=6x_1x_3^2cos(x_5-x_7)$, $\frac{\partial M_{21}}{\partial x_2}=0$, $\frac{\partial M_{21}}{\partial x_3}=6x_1^2x_3cos(x_5-x_7)$, $\frac{\partial M_{21}}{\partial x_4}=0$, $\frac{\partial M_{21}}{\partial x_5}=-3x_1^2x_3^2sin(x_5-x_7)$, $\frac{\partial M_{21}}{\partial x_6}=0$, $\frac{\partial M_{21}}{\partial x_7}=3x_1^2x_3^2sin(x_5-x_7)$, $\frac{\partial M_{21}}{\partial x_8}=0$.

For $M_{22}=6x_1^2x_3^2$: $\frac{\partial M_{22}}{\partial x_1}=12x_1x_3^2$, $\frac{\partial M_{22}}{\partial x_2}=0$, $\frac{\partial M_{22}}{\partial x_3}=12x_1^2x_3$, $\frac{\partial M_{22}}{\partial x_4}=0$, $\frac{\partial M_{22}}{\partial x_5}=0$, $\frac{\partial M_{22}}{\partial x_6}=0$, $\frac{\partial M_{22}}{\partial x_7}=0$, and $\frac{\partial M_{22}}{\partial x_8}=0$.

For $M_{23}=-3x_1x_3^2sin(x_5-x_7)$: $\frac{\partial M_{23}}{\partial x_1}=-3x_3^2sin(x_5-x_7)$, $\frac{\partial M_{23}}{\partial x_2}=0$, $\frac{\partial M_{23}}{\partial x_3}=-6x_1x_3sin(x_5-x_7)$, $\frac{\partial M_{23}}{\partial x_4}=0$, $\frac{\partial M_{23}}{\partial x_5}=-3x_1x_3^2cos(x_5-x_7)$, $\frac{\partial M_{23}}{\partial x_6}=0$, $\frac{\partial M_{23}}{\partial x_7}=3x_1x_3^2cos(x_5-x_7)$, $\frac{\partial M_{23}}{\partial x_8}=0$.

For $M_{24}=0$: $\frac{\partial M_{24}}{\partial x_i}=0$ for $i=1,2,\cdots,8$.

For $M_{31}=0$: $\frac{\partial M_{31}}{\partial x_i}=0$ for $i=1,2,\cdots,8$.

For $M_{32}=-3x_1x_3^2sin(x_5-x_7)$: $\frac{\partial M_{32}}{\partial x_1}=-3x_3^2sin(x_5-x_7)$, $\frac{\partial M_{32}}{\partial x_2}=0$, $\frac{\partial M_{32}}{\partial x_3}=-6x_1x_3sin(x_5-x_7)$, $\frac{\partial M_{32}}{\partial x_4}=0$, $\frac{\partial M_{32}}{\partial x_5}=-3x_1x_3^2cos(x_5-x_7)$, $\frac{\partial M_{32}}{\partial x_6}=0$, $\frac{\partial M_{32}}{\partial x_7}=3x_1x_3^2cos(x_5-x_7)$, $\frac{\partial M_{32}}{\partial x_8}=0$.

For $M_{33}=6x_3^2$: $\frac{\partial M_{33}}{\partial x_1}=0$, $\frac{\partial M_{33}}{\partial x_2}=0$, $\frac{\partial M_{33}}{\partial x_3}=12x_3$, $\frac{\partial M_{33}}{\partial x_4}=0$, $\frac{\partial M_{33}}{\partial x_5}=0$, $\frac{\partial M_{33}}{\partial x_6}=0$, $\frac{\partial M_{33}}{\partial x_7}=0$, $\frac{\partial M_{33}}{\partial x_8}=0$.

For $M_{34}=3x_1x_3cos(x_5-x_7)$: $\frac{\partial M_{34}}{\partial x_1}=3x_3cos(x_5-x_7)$, $\frac{\partial M_{34}}{\partial x_2}=0$, $\frac{\partial M_{34}}{\partial x_3}=3x_1cos(x_5-x_7)$, $\frac{\partial M_{34}}{\partial x_4}=0$, $\frac{\partial M_{34}}{\partial x_5}=-3x_1x_3sin(x_5-x_7)$, $\frac{\partial M_{34}}{\partial x_6}=0$, $\frac{\partial M_{34}}{\partial x_7}=3x_1x_3sin(x_5-x_7)$, $\frac{\partial M_{34}}{\partial x_0}=0$.

For $M_{41}=3x_1^2x_3sin(x_5-x_7)$: $\frac{\partial M_{41}}{\partial x_1}=6x_1x_3sin(x_5-x_7)$, $\frac{\partial M_{41}}{\partial x_2}=0$, $\frac{\partial M_{41}}{\partial x_3}=3x_1^2sin(x_5-x_7)$, $\frac{\partial M_{41}}{\partial x_4}=0$, $\frac{\partial M_{41}}{\partial x_5}=3x_1^2x_3cos(x_5-x_7)$, $\frac{\partial M_{41}}{\partial x_6}=0$, $\frac{\partial M_{41}}{\partial x_7}=-3x_1^2x_3cos(x_5-x_7)$, $\frac{\partial M_{41}}{\partial x_8}=0$.

For $M_{42}=0$: $\frac{\partial M_{42}}{\partial x_i}=0$ for $i=1,2,\cdots,8$.

For $M_{43}=3x_1x_3cos(x_5-x_7)$: $\frac{\partial M_{43}}{\partial x_1}=3x_3cos(x_5-x_7)$, $\frac{\partial M_{43}}{\partial x_2}=0$, $\frac{\partial M_{43}}{\partial x_3}=3x_1cos(x_5-x_7)$, $\frac{\partial M_{43}}{\partial x_4}=0$, $\frac{\partial M_{43}}{\partial x_5}=-3x_1x_3sin(x_5-x_7)$, $\frac{\partial M_{43}}{\partial x_6}=0$, $\frac{\partial M_{43}}{\partial x_7}=3x_1x_3sin(x_5-x_7)$, $\frac{\partial M_{43}}{\partial x_0}=0$.

For $M_{44}=6x_1^2$: $\frac{\partial M_{44}}{\partial x_1}=12x_1$, $\frac{\partial M_{44}}{\partial x_2}=0$, $\frac{\partial M_{44}}{\partial x_3}=0$, $\frac{\partial M_{44}}{\partial x_4}=0$, $\frac{\partial M_{44}}{\partial x_5}=0$, $\frac{\partial M_{44}}{\partial x_6}=0$, $\frac{\partial M_{44}}{\partial x_7}=0$, $\frac{\partial M_{44}}{\partial x_8}=0$.

For the determinant $detM=2M_{11}-cos(x_5-x_7)M_{12}-x_3sin(x_5-x_7)M_{13}$, the partial derivatives with respect to x_i, $i=1,2,\cdots,8$ are computed as follows:

$$\begin{gathered}
\frac{\partial detM}{\partial x_1}=2\frac{\partial M_{11}}{\partial x_1}-cos(x_5-x_7)\frac{\partial M_{12}}{\partial x_1}-x_3sin(x_5-x_7)\frac{\partial M_{13}}{\partial x_1}\\
\frac{\partial detM}{\partial x_2}=2\frac{\partial M_{11}}{\partial x_2}-cos(x_5-x_7)\frac{\partial M_{12}}{\partial x_2}-x_3sin(x_5-x_7)\frac{\partial M_{13}}{\partial x_2}\\
\frac{\partial detM}{\partial x_3}=2\frac{\partial M_{11}}{\partial x_3}-cos(x_5-x_7)\frac{\partial M_{12}}{\partial x_3}-sin(x_5-x_7)M_{13}-x_3sin(x_5-x_7)\frac{\partial M_{13}}{\partial x_1}\\
\frac{\partial detM}{\partial x_4}=2\frac{\partial M_{11}}{\partial x_4}-cos(x_5-x_7)\frac{\partial M_{12}}{\partial x_4}-x_3sin(x_5-x_7)\frac{\partial M_{13}}{\partial x_4}\\
\frac{\partial detM}{\partial x_5}=2\frac{\partial M_{11}}{\partial x_5}+sin(x_5-x_7)M_{12}-cos(x_5-x_7)\frac{\partial M_{12}}{\partial x_5}-x_3cos(x_5-x_7)M_{13}-x_3sin(x_5-x_7)\frac{\partial M_{13}}{\partial x_5}\\
\frac{\partial detM}{\partial x_6}=2\frac{\partial M_{11}}{\partial x_6}-cos(x_5-x_7)\frac{\partial M_{12}}{\partial x_6}-x_3sin(x_5-x_7)\frac{\partial M_{13}}{\partial x_6}\\
\frac{\partial detM}{\partial x_7}=2\frac{\partial M_{11}}{\partial x_7}-sin(x_5-x_7)M_{12}-cos(x_5-x_7)\frac{\partial M_{12}}{\partial x_7}+x_3cos(x_5-x_7)M_{13}-x_3sin(x_5-x_7)\frac{\partial M_{13}}{\partial x_7}\\
\frac{\partial detM}{\partial x_8}=2\frac{\partial M_{11}}{\partial x_8}-cos(x_5-x_7)\frac{\partial M_{12}}{\partial x_8}-x_3sin(x_5-x_7)\frac{\partial M_{13}}{\partial x_8}
\end{gathered}\tag{6.213}$$

Computation of the partial derivatives of the elements of matrix N

For $N_1=\frac{1}{9}[2x_4(x_8+1)sin(x_5-x_7)+x_7x_8(-2-x_8)cos(x_5-x_7)-2x_1x_6(x_6+2)]$, it holds that

$\frac{\partial N_1}{\partial x_1}=\frac{1}{9}(-2)x_6(x_6+2)$, $\frac{\partial N_1}{\partial x_2}=0$, $\frac{\partial N_1}{\partial x_3}=0$, $\frac{\partial N_1}{\partial x_4}=\frac{1}{9}2(x_8+1)sin(x_5-x_7)$, $\frac{\partial N_1}{\partial x_5}=\frac{1}{9}[2x_4(x_8+1)cos(x_5-x_7)-x_3x_8(-2-x_8)sin(x_5-x_7)]$, $\frac{\partial N_1}{\partial x_6}=\frac{1}{9}(-2)x_1(2x_6+2)$, $\frac{\partial N_1}{\partial x_7}=\frac{1}{9}[-2x_4(x_8+1)cos(x_5-x_7)+x_3x_8(-2-x_8)sin(x_5-x_7)]$, $\frac{\partial N_1}{\partial x_8}=\frac{1}{9}[2x_4sin(x_5-x_7)+x_3(-2-2x_8)cos(x_5-x_7)]$.

For $N_2=\frac{1}{9}[-2x_2(x_6+1)sin(x_5-x_7)+x_1x_6(-2-x_6)cos(x_5-x_7)-2x_3x_8(x_8+2)]$, it holds that

$\frac{\partial N_2}{\partial x_1}=\frac{1}{9}[x_6(-2-x_6)cos(x_5-x_7)-2x_3x_8(x_8+2)]$, $\frac{\partial N_2}{\partial x_2}=-2(x_6+1)sin(x_5-x_7)+$, $\frac{\partial N_2}{\partial x_3}=\frac{1}{9}[-2x_8(x_8+2)]$, $\frac{\partial N_2}{\partial x_4}=0$, $\frac{\partial N_2}{\partial x_5}=\frac{1}{9}[-2x_1(x_6+1)cos(x_5-x_7)-x_1x_6(-2-x_6)sin(x_5-x_7)]$, $\frac{\partial N_2}{\partial x_6}=\frac{1}{9}[-2x_1sin(x_5-x_7)+x_1(-2-2x_6)sin(x_5-x_7)]$, $\frac{\partial N_2}{\partial x_7}=\frac{1}{9}[2x_1(x_6+1)cos(x_5-x_7)+x_1x_6(-2-x_6)sin(x_5-x_7)]$, $\frac{\partial N_2}{\partial x_8}=\frac{1}{9}[-2x_3(2x_8+2)]$.

For $N_3=\frac{1}{9}[4x_1x_2(x_6+1)+2x_1x_4(x_8+1)cos(x_5-x_7)+x_1x_3x_8(x_8+2)sin(x_5-x_7)]$, it holds that

$\frac{\partial N_3}{\partial x_1} = \frac{1}{9}[4x_2(x_6+1)+2x_4(x_8+1)cos(x_5-x_7)+x_3x_8(x_8+2)sin(x_5-x_7)]$, $\frac{\partial N_3}{\partial x_2} = \frac{1}{9}[4x_1(x_6+1)]$, $\frac{\partial N_3}{\partial x_3} = \frac{1}{9}[x_1x_8(x_8+2)sin(x_5-x_7)]$, $\frac{\partial N_2}{\partial x_4} = \frac{1}{9}[2x_1(x_8+1)cos(x_5-x_7)]$, $\frac{\partial N_3}{\partial x_5} = \frac{1}{9}[-2x_1x_4(x_8+1)sin(x_5-x_7)+x_1x_3x_8(x_8+2)cos(x_5-x_7)]$, $\frac{\partial N_3}{\partial x_6} = \frac{1}{9}[4x_1x_2]$, $\frac{\partial N_3}{\partial x_7} = \frac{1}{9}[2x_1x_4(x_8+1)sin(x_5-x_7)-x_1x_3x_8(x_8+2)cos(x_5-x_7)]$, $\frac{\partial N_3}{\partial x_8} = \frac{1}{9}[2x_1x_4cos(x_5-x_7)+x_1x_3(2x_8+2sin(x_5-x_7))]$.

For $N_4 = \frac{1}{9}[4x_3x_4(x_8+1)+2x_3x_2(x_6+1)cos(x_5-x_)+x_1x_3x_6(-x_6-2)sin(x_5-x_7)]$, it holds that

$\frac{\partial N_4}{\partial x_1} = \frac{1}{9}[x_3x_6(-x_6-2)sin(x_5-x_7)]$, $\frac{\partial N_4}{\partial x_2} = \frac{1}{9}[2x_1(x_6+1)cos(x_5-x_7)]$, $\frac{\partial N_4}{\partial x_3} = \frac{1}{9}[4x_4(x_8+1)+2x_2(x_6+1)cos(x_5-x_7)+x_1x_6(-x_6-2)sin(x_5-x_7)]$, $\frac{\partial N_4}{\partial x_4} = \frac{1}{9}[4x_3(x_8+1)]$, $\frac{\partial N_4}{\partial x_5} = \frac{1}{9}[-2x_3x_2(x_6+1)sin(x_5-x_7)+x_1x_3x_6(-x_6-2)cos(x_5-x_7)]$, $\frac{\partial N_4}{\partial x_6} = \frac{1}{9}[2x_3x_2cos(x_5-x_7)+x_1x_3(-2x_6-2)sin(x_5-x_7)]$, $\frac{\partial N_4}{\partial x_7} = \frac{1}{9}[2x_3x_2(x_6+1)sin(x_5-x_7)-x_1x_5x_6(-x_6-2)cos(x_5-x_7)]$, $\frac{\partial N_4}{\partial x_8} = \frac{1}{9}[4x_3x_4]$.

Computation of the partial derivatives of the elements of matrix P

One uses first the partial derivatives of $L_3 = \sqrt{x_1^2+x_3^2+2x_1x_3cos(x_5-x_7)}$. It holds that

$$\begin{array}{ll}\frac{\partial L_3}{\partial x_1} = \frac{(2x_1+2x_3cos(x_5-x_7))}{\sqrt{x_1^2+x_3^2+2x_1x_3cos(x_5-x_7)}} & \frac{\partial L_3}{\partial x_2} = 0 \\ \frac{\partial L_3}{\partial x_3} = \frac{(2x_3+2x_1cos(x_5-x_7))}{\sqrt{x_1^2+x_3^2+2x_1x_3cos(x_5-x_7)}} & \frac{\partial L_3}{\partial x_4} = 0 \\ \frac{\partial L_3}{\partial x_5} = \frac{(-2x_1x_3sin(x_5-x_7))}{\sqrt{x_1^2+x_3^2+2x_1x_3cos(x_5-x_7)}} & \frac{\partial L_3}{\partial x_6} = 0 \\ \frac{\partial L_3}{\partial x_7} = \frac{(2x_1x_3sin(x_5-x_7))}{\sqrt{x_1^2+x_3^2+2x_1x_3cos(x_5-x_7)}} & \frac{\partial L_3}{\partial x_8} = 0\end{array} \tag{6.214}$$

For $P_1 = \frac{1}{3}(-2x_1cos^2(x_5-x_7)-x_3cos(x_5)cos(x_7))+K[(x_1-1)\xi_1+\frac{L_3-1}{L_3}(x_1+x_3cos(x_5-x_7))\xi_3]$, it holds that

$$\frac{\partial P_1}{\partial x_1} = \frac{1}{3}(-2co^2(x_5-x_7))+K[\xi_1+\frac{\frac{\partial L_3}{\partial x_1}L_3-(L_3-1)\frac{\partial L_3}{\partial x_1}}{L_3^2}(x_1+x_3cos(x_5-x_7))\xi_3+\frac{L_3-1}{L_3}\xi_3] \tag{6.215}$$

$$\frac{\partial P_1}{\partial x_2} = K\frac{\frac{\partial L_3}{\partial x_2}L_3-(L_3-1)\frac{\partial L_3}{\partial x_2}}{L_3^2}(x_1+x_3cos(x_5-x_7))\xi_3 \tag{6.216}$$

$$\frac{\partial P_1}{\partial x_3} = \frac{1}{3}(-cos(x_5)cos(x_7))+K[\frac{\frac{\partial L_3}{\partial x_3}L_3-(L_3-1)\frac{\partial L_3}{\partial x_3}}{L_3^2}(x_1+x_3cos(x_5-x_7))\xi_3+\frac{L_3-1}{L_3}cos(x_5-x_7)\xi_3] \tag{6.217}$$

$$\frac{\partial P_1}{\partial x_4} = K\frac{\frac{\partial L_3}{\partial x_4}L_3-(L_3-1)\frac{\partial L_3}{\partial x_4}}{L_3^2}(x_1+x_3cos(x_5-x_7))\xi_3 \tag{6.218}$$

$$\begin{array}{l}\frac{\partial P_1}{\partial x_5} = \frac{1}{3}(4x_1cos(x_5-x_7)sin(x_5-x_7)+x_3sin(x_5)cos(x_7))+ \\ +K[\frac{\frac{\partial L_3}{\partial x_5}L_3-(L_3-1)\frac{\partial L_3}{\partial x_5}}{L_3^2}(x_1+x_3cos(x_5-x_7))\xi_3+\frac{L_3-1}{L_3}cos(x_5-x_7)\xi_3\end{array} \tag{6.219}$$

$$\frac{\partial P_1}{\partial x_6} = K\frac{\frac{\partial L_3}{\partial x_6}L_3-(L_3-1)\frac{\partial L_3}{\partial x_6}}{L_3^2}(x_1+x_3cos(x_5-x_7))\xi_3 \tag{6.220}$$

$$\begin{array}{l}\frac{\partial P_1}{\partial x_7} = \frac{1}{3}(-4x_1cos(x_5-x_7)sin(x_5-x_7)+x_3cos(x_5)sin(x_7))+ \\ +K[\frac{\frac{\partial L_3}{\partial x_7}L_3-(L_3-1)\frac{\partial L_3}{\partial x_7}}{L_3^2}(x_1+x_3cos(x_5-x_7))\xi_3+\frac{L_3-1}{L_3}(x_3sin(x_5-x_7))\xi_3\end{array} \tag{6.221}$$

$$\frac{\partial P_1}{\partial x_8} = K\frac{\frac{\partial L_3}{\partial x_8}L_3-(L_3-1)\frac{\partial L_3}{\partial x_8}}{L_3^2}(x_1+x_3cos(x_5-x_7))\xi_3 \tag{6.222}$$

For $P_2 = \frac{1}{3}(-2x_3cos^2(x_7) - x_1cos(x_5)cos(x_7)) + K[(x_3-1)\xi_2 + \frac{L_3-1}{L_3}(x_3 + x_1cos(x_5-x_7))\xi_3]$, it holds that

$$\frac{\partial P_2}{\partial x_1} = \frac{1}{3}(-cos(x_5)cos(x_7)) + K[\frac{\frac{\partial L_3}{\partial x_1}L_3 - (L_3-1)\frac{\partial L_3}{\partial x_1}}{L_3^2}(x_3 + x_1cos(x_5-x_7))\xi_3 + \frac{L_3-1}{L_3}cos(x_5-x_7)\xi_3] \tag{6.223}$$

$$\frac{\partial P_2}{\partial x_2} = K[\frac{\frac{\partial L_3}{\partial x_2}L_3 - (L_3-1)\frac{\partial L_3}{\partial x_2}}{L_3^2}(x_3 + x_1cos(x_5-x_7))\xi_3] \tag{6.224}$$

$$\frac{\partial P_2}{\partial x_3} = \frac{1}{3}(-2cos^2(x_7)) + K[\xi_2 + \frac{\frac{\partial L_3}{\partial x_3}L_3 - (L_3-1)\frac{\partial L_3}{\partial x_3}}{L_3^2}(x_3 + x_1cos(x_5-x_7))\xi_3 + \frac{L_3-1}{L_3}\xi_3] \tag{6.225}$$

$$\frac{\partial P_2}{\partial x_4} = K[\frac{\frac{\partial L_3}{\partial x_4}L_3 - (L_3-1)\frac{\partial L_3}{\partial x_4}}{L_3^2}(x_3 + x_1cos(x_5-x_7))\xi_3] \tag{6.226}$$

$$\frac{\partial P_2}{\partial x_5} = \frac{1}{3}(x_1sin(x_5)cos(x_7)) + K[\frac{\frac{\partial L_3}{\partial x_5}L_3 - (L_3-1)\frac{\partial L_3}{\partial x_5}}{L_3^2}(x_3 + x_1cos(x_5-x_7))\xi_3 + \frac{L_3-1}{L_3}(-x_1sin(x_5-x_7))\xi_3] \tag{6.227}$$

$$\frac{\partial P_2}{\partial x_6} = K[\frac{\frac{\partial L_3}{\partial x_6}L_3 - (L_3-1)\frac{\partial L_3}{\partial x_6}}{L_3^2}(x_3 + x_1cos(x_5-x_7))\xi_3] \tag{6.228}$$

$$\frac{\partial P_2}{\partial x_7} = \frac{1}{3}(x_1cos(x_5)sin(x_7)) + K[\frac{\frac{\partial L_3}{\partial x_7}L_3 - (L_3-1)\frac{\partial L_3}{\partial x_7}}{L_3^2}(x_3 + x_1cos(x_5-x_7))\xi_3 + \frac{L_3-1}{L_3}(x_1sin(x_5-x_7))\xi_3] \tag{6.229}$$

$$\frac{\partial P_2}{\partial x_8} = K[\frac{\frac{\partial L_3}{\partial x_8}L_3 - (L_3-1)\frac{\partial L_3}{\partial x_8}}{L_3^2}(x_3 + x_1cos(x_5-x_7))\xi_3] \tag{6.230}$$

For $P_3 = \frac{1}{3}(2x_1^2sin(x_5)cos(x_5) + x_1x_3sin(x_5)cos(x_5)) + K\xi_3\frac{x_1x_3}{L_3}(1-L_3)sin(x_5-x_7)$. it holds that

$$\begin{aligned}\frac{\partial P_3}{\partial x_1} &= \frac{1}{3}(4x_1sin(x_5)cos(x_5) + x_3sin(x_5)cos(x_7)) + \\ &+K[\xi_3\frac{x_3L_3 - x_1x_3\frac{\partial L_3}{\partial x_1}}{L_3^2}(1-L_3)sin(x_5-x_7) + \xi_3\frac{x_1x_3}{L_3}(-\frac{\partial L_3}{\partial x_1})sin(x_5-x_7)]\end{aligned} \tag{6.231}$$

$$\frac{\partial P_3}{\partial x_2} = K[\xi_3x_1x_3\frac{-\frac{\partial L_3}{\partial x_2}L_3 - (1-L_3)\frac{\partial L_3}{\partial x_2}}{L_3^2}sin(x_5-x_7)] \tag{6.232}$$

$$\begin{aligned}\frac{\partial P_3}{\partial x_3} &= \frac{1}{3}(x_1sin(x_5)cos(x_7)) + K[\xi_3x_1\frac{(1-L_3)}{L_3}sin(x_5-x_7)+] \\ &+\xi_3x_1x_3\frac{-\frac{\partial L_3}{\partial x_3}L_3 - (1-L_3)\frac{\partial L_3}{\partial x_3}}{L_3^2}sin(x_5-x_7)]\end{aligned} \tag{6.233}$$

$$\frac{\partial P_3}{\partial x_4} = K[\xi_3x_1x_3\frac{-\frac{\partial L_3}{\partial x_4}L_3 - (1-L_3)\frac{\partial L_3}{\partial x_4}}{L_3^2}sin(x_5-x_7)] \tag{6.234}$$

$$\begin{aligned}\frac{\partial P_3}{\partial x_5} &= \frac{1}{3}(2x_1^2(cos^2(x_5) - sin^2(x_5)) + x_1x_3cos(x_5)cos(x_7)) + \\ &+K[\xi_3x_1x_3\frac{-\frac{\partial L_3}{\partial x_5}L_3 - (1-L_3)\frac{\partial L_3}{\partial x_5}}{L_3^2}sin(x_5-x_7) + \\ &+\xi_3x_1x_3\frac{(1-L_3)}{L_3}cos(x_5-x_7)]\end{aligned} \tag{6.235}$$

$$\frac{\partial P_3}{\partial x_6} = K[\xi_3x_1x_3\frac{-\frac{\partial L_3}{\partial x_6}L_3 - (1-L_3)\frac{\partial L_3}{\partial x_6}}{L_3^2}sin(x_5-x_7)] \tag{6.236}$$

$$\begin{aligned}\frac{\partial P_3}{\partial x_7} &= \frac{1}{3}(-x_1x_3sin(x_5)sin(x_7)) + \\ &+K[\xi_3x_1x_3\frac{-\frac{\partial L_3}{\partial x_7}L_3 - (1-L_3)\frac{\partial L_3}{\partial x_7}}{L_3^2}sin(x_5-x_7) - \\ &-\xi_3x_1x_3\frac{(1-L_3)}{L_3}cos(x_5-x_7)]\end{aligned} \tag{6.237}$$

$$\frac{\partial P_3}{\partial x_8} = K[\xi_3x_1x_3\frac{-\frac{\partial L_3}{\partial x_8}L_3 - (1-L_3)\frac{\partial L_3}{\partial x_8}}{L_3^2}sin(x_5-x_7)] \tag{6.238}$$

For $P_4=\frac{1}{3}(2x_3^2)sin(x_7)cos(x_7)+x_1x_3sin(x_7)cos(x_5)-K\xi_3\frac{x_1x_3}{L_3}(1-L_3)sin(x_5-x_7)$, it holds that

$$\frac{\partial P_4}{\partial x_1}=\frac{1}{3}(x_3sin(x_7)cos(x_5))- \\ -K[\xi_3x_3\frac{(1-L_3)}{L_3}sin(x_5-x_7)+\xi_3x_1x_3\frac{-\frac{\partial L_3}{\partial x_1}L_3-(1-L_3)\frac{\partial L_3}{\partial x_1}}{L_3^2}sin(x_5-x_7)] \tag{6.239}$$

$$\frac{\partial P_4}{\partial x_2}=-K[\xi_3x_1x_3\frac{-\frac{\partial L_3}{\partial x_2}L_3-(1-L_3)\frac{\partial L_3}{\partial x_2}}{L_3^2}sin(x_5-x_7)] \tag{6.240}$$

$$\frac{\partial P_4}{\partial x_3}=\frac{1}{3}(4x_3sin(x_7)cos(x_7)+x_1sin(x_7)cos(x_5))- \\ -K[\xi_3x_1\frac{(1-L_3)}{L_3}sin(x_5-x_7)+\xi_3x_1x_3\frac{-\frac{\partial L_3}{\partial x_3}L_3-(1-L_3)\frac{\partial L_3}{\partial x_3}}{L_3^2}sin(x_5-x_7)] \tag{6.241}$$

$$\frac{\partial P_4}{\partial x_4}=-K[\xi_3x_1x_3\frac{-\frac{\partial L_3}{\partial x_4}L_3-(1-L_3)\frac{\partial L_3}{\partial x_4}}{L_3^2}sin(x_5-x_7)] \tag{6.242}$$

$$\frac{\partial P_4}{\partial x_5}=\frac{1}{3}(-x_1x_3sin(x_7)sin(x_5))- \\ -K[\xi_3x_1x_3\frac{-\frac{\partial L_3}{\partial x_5}L_3-(1-L_3)\frac{\partial L_3}{\partial x_5}}{L_3^2}sin(x_5-x_7)+ \\ +\xi_3x_1x_3\frac{(1-L_3)}{L_3}cos(x_5-x_7)] \tag{6.243}$$

$$\frac{\partial P_4}{\partial x_6}=-K[\xi_3x_1x_3\frac{-\frac{\partial L_3}{\partial x_6}L_3-(1-L_3)\frac{\partial L_3}{\partial x_6}}{L_3^2}sin(x_5-x_7)] \tag{6.244}$$

$$\frac{\partial P_4}{\partial x_7}=\frac{1}{3}(2x_3^2(cos^2(x_7)-sin^2(x_7))+x_1x_3cos(x_7)cos(x_5))- \\ -K[\xi_3x_1x_3\frac{-\frac{\partial L_3}{\partial x_7}L_3-(1-L_3)\frac{\partial L_3}{\partial x_7}}{L_3^2}sin(x_5-x_7)+ \\ +\xi_3x_1x_3\frac{(1-L_3)}{L_3}cos(x_5-x_7)] \tag{6.245}$$

$$\frac{\partial P_4}{\partial x_8}=-K[\xi_3x_1x_3\frac{-\frac{\partial L_3}{\partial x_8}L_3-(1-L_3)\frac{\partial L_3}{\partial x_8}}{L_3^2}sin(x_5-x_7)] \tag{6.246}$$

Computation of the partial derivatives of the elements of matrix B_q:

For $B_{q11}=-\frac{1}{m\Omega^2L_o}$, it holds that $\frac{\partial B_{q11}}{\partial x_i}=0$ for $i=1,2,\cdots,8$.

For $B_{q12}=0$, it holds that $\frac{\partial B_{q12}}{\partial x_i}=0$ for $i=1,2,\cdots,8$.

For B_{q13}, it holds that for $i=1,2,\cdots,8$

$$B_{q13}=\frac{B_{q13,num}}{B_{q13,den}}=-\frac{[x_1+x_3cos(x_5-x_7)]}{m\Omega^2L_3L_o} \tag{6.247}$$

This gives for $i=1,2,\cdots,8$

$$\frac{\partial B_{q13}}{\partial x_i}=\frac{\frac{\partial B_{q13,num}}{\partial x_i}B_{13,den}-B_{q13,num}\frac{\partial B_{q13,den}}{\partial x_i}}{(B_{q13,den})^2} \tag{6.248}$$

where $\frac{\partial B_{q13,num}}{\partial x_1}=-1$, $\frac{\partial B_{q13,num}}{\partial x_2}=0$, $\frac{\partial B_{q13,num}}{\partial x_3}=-cos(x_5-x_7)$, $\frac{\partial B_{q13,num}}{\partial x_4}=0$, $\frac{\partial B_{q13,num}}{\partial x_5}=x_3sin(x_5-x_7)$, $\frac{\partial B_{q13,num}}{\partial x_6}=0$, $\frac{\partial B_{q13,num}}{\partial x_7}=-x_3sin(x_5-x_7)$, $\frac{\partial B_{q13,num}}{\partial x_8}=0$ and also $\frac{\partial B_{q13,den}}{\partial x_i}=m\Omega^2\frac{\partial L_3}{\partial x_i}L_o$ for $i=1,2,\cdots,8$.

For $B_{q21}=0$, it holds that $\frac{\partial B_{q21}}{\partial x_i}=0$ for $i=1,2,\cdots,8$.

For $B_{q22}=-\frac{1}{m\Omega^2L_o}$, it holds that $\frac{\partial B_{q22}}{\partial x_i}=0$ for $i=1,2,\cdots,8$.

For B_{q23}, it holds that for $i=1,2,\cdots,8$

$$B_{q23}=\frac{B_{q23,num}}{B_{q23,den}}=-\frac{[x_3+x_1cos(x_5-x_7)]}{m\Omega^2L_3l_o} \tag{6.249}$$

This gives for $i = 1, 2, \cdots, 8$

$$\frac{\partial B_{q23}}{\partial x_i} = \frac{\frac{\partial B_{q23,num}}{\partial x_i} B_{23,den} - B_{q23,num} \frac{\partial B_{q23,den}}{\partial x_i}}{(B_{q23,den})^2} \tag{6.250}$$

where $\frac{\partial B_{q23,num}}{\partial x_1} = -cos(x_5 - x_7)$, $\frac{\partial B_{q23,num}}{\partial x_2} = 0$, $\frac{\partial B_{q23,num}}{\partial x_3} = -1$, $\frac{\partial B_{q23,num}}{\partial x_4} = 0$, $\frac{\partial B_{q23,num}}{\partial x_5} = x_1 sin(x_5 - x_7)$, $\frac{\partial B_{q23,num}}{\partial x_6} = 0$, $\frac{\partial B_{q23,num}}{\partial x_7} = -x_3 sin(x_5 - x_7)$, $\frac{\partial B_{q23,num}}{\partial x_8} = 0$ and also $\frac{\partial B_{q23,den}}{\partial x_i} = m\Omega^2 \frac{\partial L_3}{\partial x_i} L_o$ for $i = 1, 2, \cdots, 8$.

For $B_{q31} = 0$, it holds that $\frac{\partial B_{q31}}{\partial x_i} = 0$ for $i = 1, 2, \cdots, 8$.

For $B_{q32} = 0$, it holds that $\frac{\partial B_{q32}}{\partial x_i} = 0$ for $i = 1, 2, \cdots, 8$.

For B_{q33}, it holds that for $i = 1, 2, \cdots, 8$

$$B_{q33} = \frac{B_{q33,num}}{B_{q33,den}} = \frac{[x_1 x_3 sin(x_5 - x_7)]}{m\Omega^2 L_3} \tag{6.251}$$

This gives for $i = 1, 2, \cdots, 8$

$$\frac{\partial B_{q33}}{\partial x_i} = \frac{\frac{\partial B_{q33,num}}{\partial x_i} B_{33,den} - B_{q33,num} \frac{\partial B_{q33,den}}{\partial x_i}}{(B_{q33,den})^2} \tag{6.252}$$

where $\frac{\partial B_{q33,num}}{\partial x_1} = x_3 sin(x_5 - x_7)$, $\frac{\partial B_{q33,num}}{\partial x_2} = 0$, $\frac{\partial B_{q33,num}}{\partial x_3} = -x_1 sin(x_5 - x_7)$, $\frac{\partial B_{q33,num}}{\partial x_4} = 0$, $\frac{\partial B_{q33,num}}{\partial x_5} = x_1 x_3 cos(x_5 - x_7)$, $\frac{\partial B_{q33,num}}{\partial x_6} = 0$, $\frac{\partial B_{q33,num}}{\partial x_7} = -x_1 x_3 cos(x_5 - x_7)$, $\frac{\partial B_{q33,num}}{\partial x_8} = 0$ and also $\frac{\partial B_{q33,den}}{\partial x_i} = m\Omega^2 \frac{\partial L_3}{\partial x_i}$ for $i = 1, 2, \cdots, 8$.

For $B_{q41} = 0$, it holds that $\frac{\partial B_{q41}}{\partial x_i} = 0$ for $i = 1, 2, \cdots, 8$.

For $B_{q42} = 0$, it holds that $\frac{\partial B_{q42}}{\partial x_i} = 0$ for $i = 1, 2, \cdots, 8$.

For B_{q43}, it holds that for $i = 1, 2, \cdots, 8$

$$B_{q43} = \frac{B_{q43,num}}{B_{q43,den}} = \frac{[-x_1 x_3 sin(x_5 - x_7)]}{m\Omega^2 L_3} \tag{6.253}$$

This gives for $i = 1, 2, \cdots, 8$

$$\frac{\partial B_{q43}}{\partial x_i} = \frac{\frac{\partial B_{q43,num}}{\partial x_i} B_{43,den} - B_{q43,num} \frac{\partial B_{q43,den}}{\partial x_i}}{(B_{q43,den})^2} \tag{6.254}$$

where $\frac{\partial B_{q43,num}}{\partial x_1} = -x_3 sin(x_5 - x_7)$, $\frac{\partial B_{q43,num}}{\partial x_2} = 0$, $\frac{\partial B_{q43,num}}{\partial x_3} = -x_1 sin(x_5 - x_7)$, $\frac{\partial B_{q43,num}}{\partial x_4} = 0$, $\frac{\partial B_{q43,num}}{\partial x_5} = -x_1 x_3 cos(x_5 - x_7)$, $\frac{\partial B_{q43,num}}{\partial x_6} = 0$, $\frac{\partial B_{q43,num}}{\partial x_7} = x_1 x_3 cos(x_5 - x_7)$, $\frac{\partial B_{q43,num}}{\partial x_8} = 0$ and also $\frac{\partial B_{q43,den}}{\partial x_i} = m\Omega^2 \frac{\partial L_3}{\partial x_i}$ for $i = 1, 2, \cdots, 8$.

With reference to Eq. (6.154) it is noted that often the control inputs vector of the tethered multi-satellite system is taken to be in the form of the generalized forces vector $Q_q \in R^{4\times 1}$. Then, the approximate linearization process is modified by considering that the control inputs gain matrix of Eq. (6.163) to Eq. (6.165) is $M^{-1}(q)$ as described in Eq. (6.161). In such a case, the computation of the Jacobian matrices $\nabla_x g_i(x)$, $i = 1, \cdots, 4$ is updated accordingly.

The linearization approach which has been followed for implementing the nonlinear optimal control scheme results in a quite accurate model of the system's dynamics. Consider for instance the following affine-in-the-input state-space model

$$\begin{gathered} \dot{x} = f(x) + g(x)u \Rightarrow \\ \dot{x} = [f(x^*) + \nabla_x f(x) \mid_{x^*} (x - x^*)] + [g(x^*) + \nabla_x g(x) \mid_{x^*} (x - x^*)]u^* + g(x^*)u^* + g(x^*)(u - u^*) + \tilde{d}_1 \Rightarrow \\ \dot{x} = [\nabla_x f(x) \mid_{x^*} + \nabla_x g(x) \mid_{x^*} u^*]x + g(x^*)u - [\nabla_x f(x) \mid_{x^*} + \nabla_x g(x) \mid_{x^*} u^*]x^* + f(x^*) + g(x^*)u^* + \tilde{d}_1 \end{gathered} \tag{6.255}$$

where $\tilde{d}_1$ is the modeling error due to truncation of higher order terms in the Taylor series expansion of $f(x)$ and $g(x)$. Next, by defining $A = [\nabla_x f(x) \mid_{x^*} + \nabla_x g(x) \mid_{x^*} u^*]$, $B = g(x^*)$, one obtains

$$\dot{x} = Ax + Bu - Ax^* + f(x^*) + g(x^*)u^* + \tilde{d}_1 \tag{6.256}$$

Moreover, by denoting $\tilde{d} = -Ax^* + f(x^*) + g(x^*)u^* + \tilde{d}_1$ about the cumulative modeling error term in the Taylor series expansion, one has

$$\dot{x} = Ax + Bu + \tilde{d} \tag{6.257}$$

which is the approximately linearized model of the dynamics of the system of Eq. (6.170). The term $f(x^*) + g(x^*)u^*$ is the derivative of the state vector at (x^*, u^*) which is almost annihilated by $-Ax^*$.

6.3.3.1 Stabilizing feedback control

After linearization around its current operating point, the tethered multi-satellites dynamic model is written as [721]

$$\dot{x} = Ax + Bu + d_1 \tag{6.258}$$

Parameter d_1 stands for the linearization error in the tethered multi-satellites' dynamic model appearing previously in Eq. (6.258). The reference setpoints for the tethered multi-satellites' state vector are denoted by $\mathbf{x_d} = [x_1^d, \cdots, x_8^d]$. Tracking of this trajectory is achieved after applying the control input u_d. At every time instant, the control input u_d is assumed to differ from the control input u appearing in Eq. (6.258) by an amount equal to Δu, that is $u_d = u + \Delta u$

$$\dot{x}_d = Ax_d + Bu_d + d_2 \tag{6.259}$$

The dynamics of the controlled system described in Eq. (6.258) can be also written as

$$\dot{x} = Ax + Bu + Bu_d - Bu_d + d_1 \tag{6.260}$$

and by denoting $d_3 = -Bu_d + d_1$ as an aggregate disturbance term, one obtains

$$\dot{x} = Ax + Bu + Bu_d + d_3 \tag{6.261}$$

By subtracting Eq. (6.259) from Eq. (6.261), one has

$$\dot{x} - \dot{x}_d = A(x - x_d) + Bu + d_3 - d_2 \tag{6.262}$$

By denoting the tracking error as $e = x - x_d$ and the aggregate disturbance term as $L\tilde{d} = d_3 - d_2$, the tracking error dynamics becomes

$$\dot{e} = Ae + Bu + L\tilde{d} \tag{6.263}$$

where L is the disturbance inputs gain matrix. For the approximately linearized model of the system, a stabilizing feedback controller is developed. The controller has the form

$$u(t) = -Ke(t) \tag{6.264}$$

with $K = \frac{1}{r}B^TP$ where P is a positive definite symmetric matrix which is obtained from the solution of the Riccati equation [721]

$$A^TP + PA + Q - P(\tfrac{2}{r}BB^T - \tfrac{1}{\rho^2}LL^T)P = 0 \tag{6.265}$$

where Q is a positive semi-definite symmetric matrix. The diagram of the considered control loop is depicted in Figure 6.30.

It is also noteworthy that the solution of the H-infinity feedback control problem for the tethered multi satellites' dynamics and the computation of the worst-case disturbance that this controller can sustain, comes from the superposition of Bellman's optimality principle when considering that the system is affected by two separate inputs (i) the control input u (ii) the cumulative disturbance input $\tilde{d}(t)$. Solving the optimal control problem for u, that is for the minimum variation (optimal) control input that achieves the elimination of the state vector's tracking error, gives $u = -\frac{1}{r}B^TPe$. Equivalently, solving the optimal control problem for $\tilde{d}$, that is for the worst-case disturbance that the control loop can sustain gives $\tilde{d} = \frac{1}{\rho^2}L^TPe$.

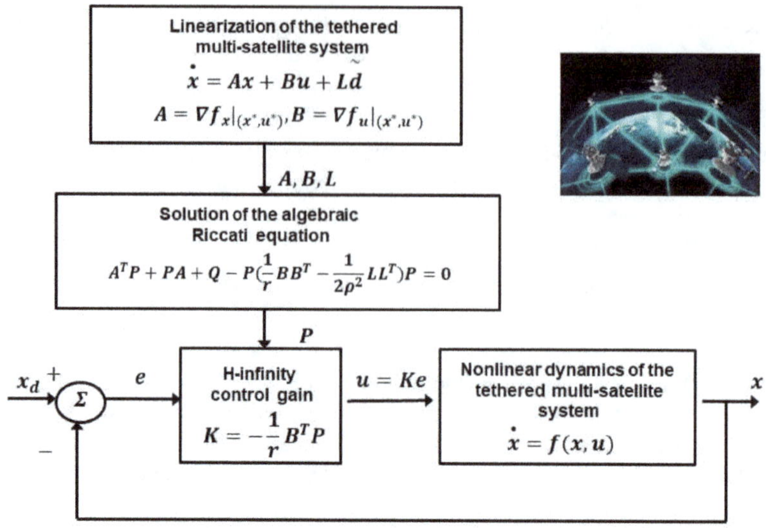

Figure 6.30 Diagram of the tethered multi-satellite nonlinear optimal (H-infinity) control loop

6.3.4 Lyapunov stability analysis

Through Lyapunov stability analysis it will be shown that the proposed nonlinear control scheme assures H_∞ tracking performance for the tethered multi-satellites' attitude dynamics, and that under moderate conditions asymptotic convergence to the reference setpoints (global stability) is achieved. As shown before, the tracking error dynamics for the tethered multi-satellites' attitude model is written in the form [721]

$$\dot{e} = Ae + Bu + L\tilde{d} \tag{6.266}$$

where in the tethered multi-satellites' dynamics, and for in-plane motion, $L \in R^{8\times 8}$ is the disturbances gain matrix. Variable $\tilde{d}$ denotes model uncertainties and external disturbances of the tethered multi-satellites' dynamic model. The following Lyapunov equation is considered

$$V = \tfrac{1}{2}e^TPe \tag{6.267}$$

where $e = x - x_d$ is the tracking error. By differentiating with respect to time, one obtains

$$\begin{aligned} \dot{V} &= \tfrac{1}{2}\dot{e}^TPe + \tfrac{1}{2}e^TP\dot{e} \Rightarrow \\ \dot{V} &= \tfrac{1}{2}[Ae + Bu + L\tilde{d}]^TPe + \tfrac{1}{2}e^TP[Ae + Bu + L\tilde{d}] \Rightarrow \end{aligned} \tag{6.268}$$

$$\begin{aligned} \dot{V} = &\tfrac{1}{2}[e^TA^T + u^TB^T + \tilde{d}^TL^T]Pe + \\ &+ \tfrac{1}{2}e^TP[Ae + Bu + L\tilde{d}] \Rightarrow \end{aligned} \tag{6.269}$$

$$\begin{aligned} \dot{V} = &\tfrac{1}{2}e^TA^TPe + \tfrac{1}{2}u^TB^TPe + \tfrac{1}{2}\tilde{d}^TL^TPe + \\ &\tfrac{1}{2}e^TPAe + \tfrac{1}{2}e^TPBu + \tfrac{1}{2}e^TPL\tilde{d} \end{aligned} \tag{6.270}$$

The previous equation is rewritten as

$$\begin{aligned} \dot{V} = &\tfrac{1}{2}e^T(A^TP + PA)e + (\tfrac{1}{2}u^TB^TPe + \tfrac{1}{2}e^TPBu) + \\ &+ (\tfrac{1}{2}\tilde{d}^TL^TPe + \tfrac{1}{2}e^TPL\tilde{d}) \end{aligned} \tag{6.271}$$

Assumption: For given positive definite matrix Q and coefficients r and ρ, there exists a positive definite matrix P, which is the solution of the following matrix equation

$$A^TP + PA = -Q + P(\frac{2}{r}BB^T - \frac{1}{\rho^2}LL^T)P \tag{6.272}$$

Moreover, the following feedback control law is applied to the system:

$$u = -\frac{1}{r}B^TPe \tag{6.273}$$

By substituting Eq. (6.272) and Eq. (6.273), one obtains

$$\begin{aligned}\dot{V} = &\frac{1}{2}e^T[-Q + P(\frac{2}{r}BB^T - \frac{1}{\rho^2}LL^T)P]e + \\ &+e^TPB(-\frac{1}{r}B^TPe) + e^TPL\tilde{d} \Rightarrow\end{aligned} \tag{6.274}$$

$$\begin{aligned}\dot{V} = &-\frac{1}{2}e^TQe + (\frac{1}{r}e^TPBB^TPe - \frac{1}{2\rho^2}e^TPLL^TPe \\ &-\frac{1}{r}e^TPBB^TPe) + e^TPL\tilde{d}\end{aligned} \tag{6.275}$$

which after intermediate operations gives

$$\dot{V} = -\frac{1}{2}e^TQe - \frac{1}{2\rho^2}e^TPLL^TPe + e^TPL\tilde{d} \tag{6.276}$$

or, equivalently

$$\begin{aligned}\dot{V} = &-\frac{1}{2}e^TQe - \frac{1}{2\rho^2}e^TPLL^TPe + \\ &+\frac{1}{2}e^TPL\tilde{d} + \frac{1}{2}\tilde{d}^TL^TPe\end{aligned} \tag{6.277}$$

Lemma: The following inequality holds

$$\frac{1}{2}e^TPL\tilde{d} + \frac{1}{2}\tilde{d}L^TPe - \frac{1}{2\rho^2}e^TPLL^TPe \leq \frac{1}{2}\rho^2\tilde{d}^T\tilde{d} \tag{6.278}$$

Proof: The binomial $(\rho\alpha - \frac{1}{\rho}b)^2$ is considered. Expanding the left part of the above inequality, one gets

$$\begin{aligned}&\rho^2a^2 + \frac{1}{\rho^2}b^2 - 2ab \geq 0 \Rightarrow \frac{1}{2}\rho^2a^2 + \frac{1}{2\rho^2}b^2 - ab \geq 0 \Rightarrow \\ &ab - \frac{1}{2\rho^2}b^2 \leq \frac{1}{2}\rho^2a^2 \Rightarrow \frac{1}{2}ab + \frac{1}{2}ab - \frac{1}{2\rho^2}b^2 \leq \frac{1}{2}\rho^2a^2\end{aligned} \tag{6.279}$$

The following substitutions are carried out: $a = \tilde{d}$ and $b = e^TPL$, and the previous relation becomes

$$\frac{1}{2}\tilde{d}^TL^TPe + \frac{1}{2}e^TPL\tilde{d} - \frac{1}{2\rho^2}e^TPLL^TPe \leq \frac{1}{2}\rho^2\tilde{d}^T\tilde{d} \tag{6.280}$$

Eq. (6.280) is substituted in Eq. (6.277) and the inequality is enforced, thus giving

$$\dot{V} \leq -\frac{1}{2}e^TQe + \frac{1}{2}\rho^2\tilde{d}^T\tilde{d} \tag{6.281}$$

Eq. (6.281) shows that the H_∞ tracking performance criterion is satisfied. The integration of $\dot{V}$ from 0 to T gives

$$\begin{aligned}&\int_0^T\dot{V}(t)dt \leq -\frac{1}{2}\int_0^T||e||_Q^2dt + \frac{1}{2}\rho^2\int_0^T||\tilde{d}||^2dt \Rightarrow \\ &2V(T) + \int_0^T||e||_Q^2dt \leq 2V(0) + \rho^2\int_0^T||\tilde{d}||^2dt\end{aligned} \tag{6.282}$$

Moreover, if there exists a positive constant $M_d > 0$ such that

$$\int_0^\infty ||\tilde{d}||^2dt \leq M_d \tag{6.283}$$

then, one gets

$$\int_0^\infty||e||_Q^2dt \leq 2V(0) + \rho^2M_d \tag{6.284}$$

Thus, the integral $\int_0^\infty||e||_Q^2dt$ is bounded. Moreover, $V(T)$ is bounded and from the definition of the Lyapunov function V in Eq. (6.267), it becomes clear that $e(t)$ will be also bounded since $e(t) \in \Omega_e = \{e|e^TPe \leq 2V(0) + \rho^2M_d\}$.

According to the above and with the use of Barbalat's Lemma, one obtains $lim_{t\to\infty}e(t) = 0$.

Through the stages of the stability proof, one arrives at Eq. (6.281) which shows that the H-infinity tracking performance criterion holds. By selecting the attenuation coefficient ρ to be sufficiently small and in particular to satisfy $\rho^2 < ||e||_Q^2/||\tilde{d}||^2$, one has that the first derivative of the Lyapunov function is upper bounded by 0. This condition holds at each sampling instance, and consequently, global stability for the control loop can be concluded.

6.3.5 Simulation tests

The fine tracking performance and the global stability properties of the nonlinear optimal control scheme for the tethered multi-satellite system are further confirmed through simulation experiments. The dynamic model of the tethered multi-satellite system of Eq. (6.154) has been used, while the relation between the generalized forces vector Q_q and the control inputs vector u was as described in Eq. (6.160). To implement the nonlinear optimal control scheme, the method's algebraic Riccati equation given in Eq. (6.272) had to be solved in each sampling interval using the *aresolv()* function in MATLAB®. The horizontal axis of the diagrams represents the scaled time variable Ωt where Ω is the angular velocity of the satellites. The obtained results are presented in Figures 6.31–6.46. The real values of the state variables of the tethered multi-satellite system are printed in blue, their estimated values which were provided by the H-infinity Kalman filter are plotted in green, while the reference setpoints are depicted in red color. It can be noticed that in all test cases, the proposed nonlinear optimal control method achieves fast and accurate tracking of setpoints under moderate variations of the control inputs.

To elaborate on the tracking performance and on the robustness of the proposed nonlinear optimal control method for the tethered triangular satellite formation, the following tables are given: (i) Table 10 which provides information about the accuracy of tracking of the reference setpoints by the state variables of the tethered triangular satellite formation's state-space model, (ii) Table 11 which provides information about the robustness of the control method to parametric changes in the model of the tethered triangular satellite formation's dynamics (change $\Delta a\%$ in parameter K of matrix P), (iii) Table 12 which provides information about the precision in state variables' estimation that is achieved

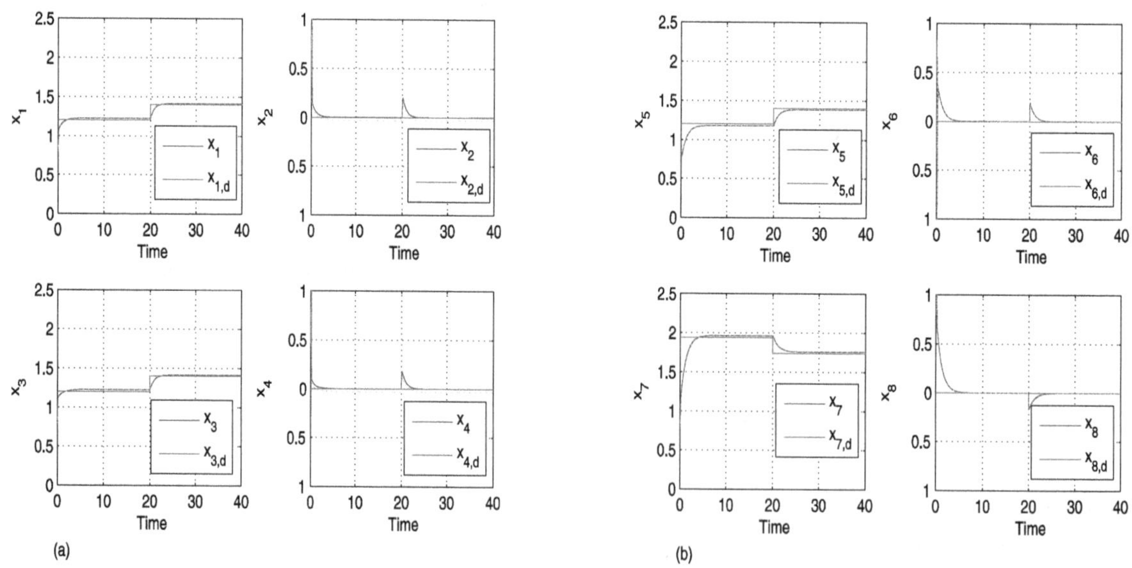

Figure 6.31 *Test-case 1 for the tethered multi-satellite system: (a) tracking of setpoints (red line) by state variables x_1 to x_4 (blue line) and estimates provided by the H-infinity Kalman filter; (b) tracking of setpoints (red line) by state variables line) and estimates provided by the H-infinity Kalman filter*

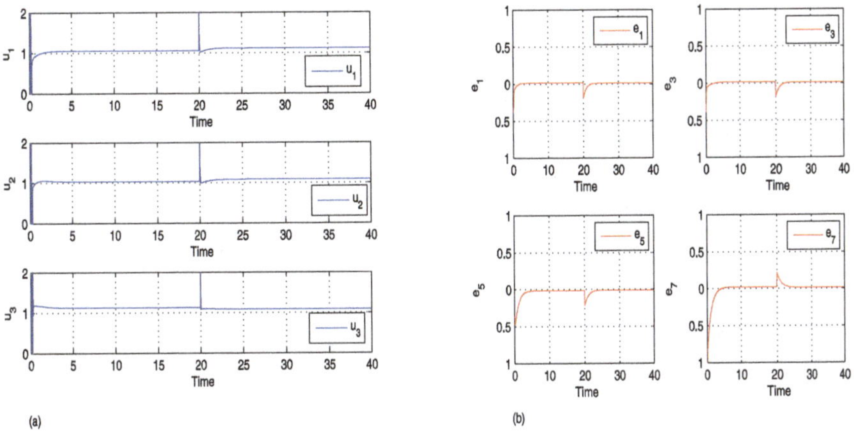

Figure 6.32 *Test-case 1 for the tethered multi-satellite system: (a) variation of the control inputs u_1 to u_3 (tethers' tension) which are applied to the tethered multi-satellite system; (b) tracking error dynamics for state variables x_1, x_3, x_5, and x_7*

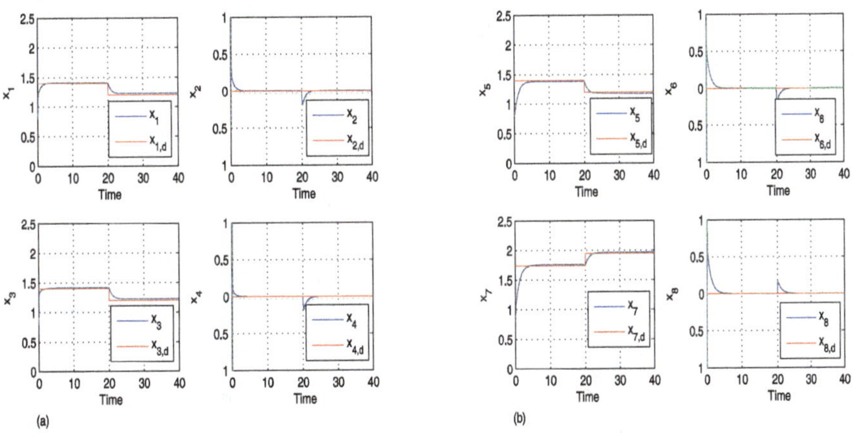

Figure 6.33 *Test-case 2 for the tethered multi-satellite system: (a) tracking of setpoints (red line) by state variables x_1 to x_4 (blue line) and estimates provided by the H-infinity Kalman filter; (b) tracking of setpoints (red line) by state variables x_5 to x_8 (blue line) and estimates provided by the H-infinity Kalman filter*

by the H-infinity Kalman filter, and (iv) Table 13 which provides the approximate convergence times of the tethered triangular satellite formation's state variables to the associated setpoints.

6.4 Nonlinear optimal control of the aircraft's wing-rock effect under a 5-th-order model

6.4.1 Outline

The wing-rock effect is associated with roll-angle oscillations in an aircraft and is determined by the airplane's angle of attack [527,110,296,295]. The oscillations can be persistent and if they are not suppressed they can affect the aiplane's stability and its ability for precise flight-path tracking [366,450,404]. Control of the angles of the aircraft's

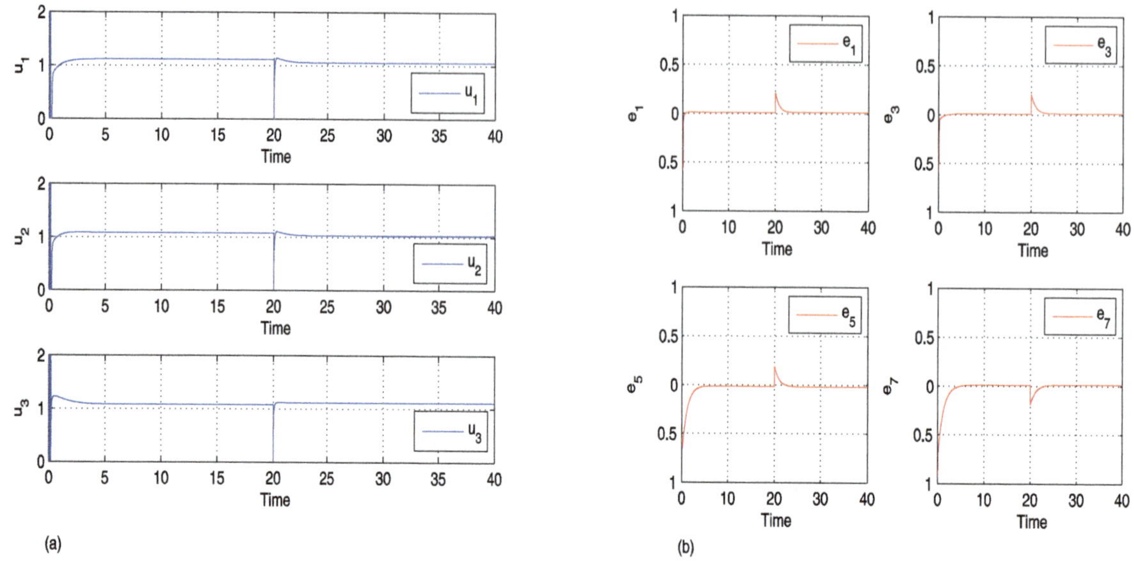

Figure 6.34 *Test-case 2 for the tethered multi-satellite system: (a) variation of the control inputs u_1 to u_3 (tethers' tension) which are applied to the tethered multi-satellite system; (b) tracking error dynamics for state variables x_1, x_3, x_5, and x_7*

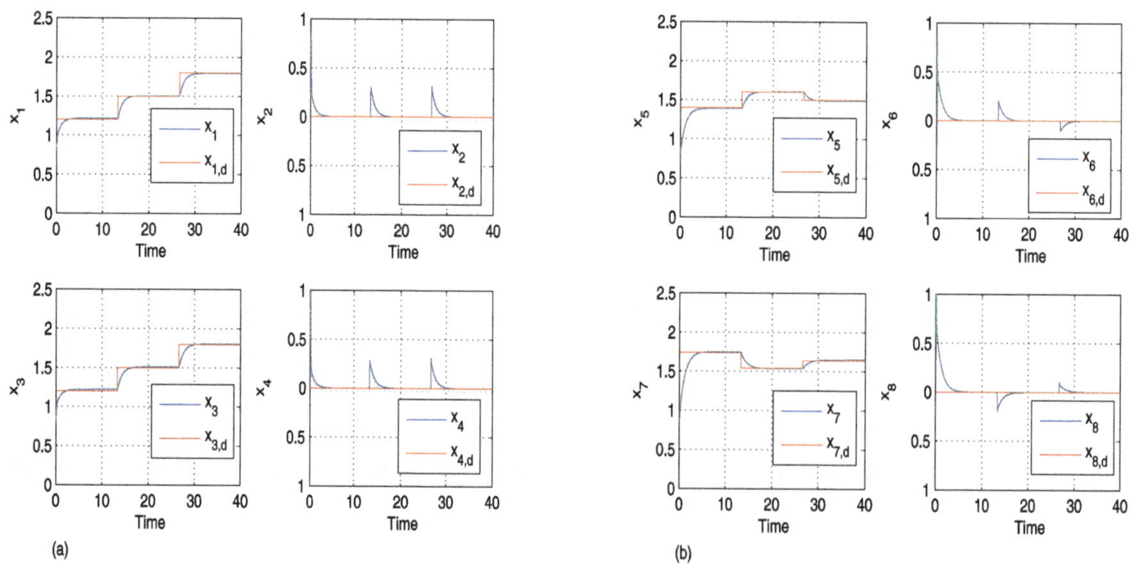

Figure 6.35 *Test-case 3 for the tethered multi-satellite system: (a) tracking of setpoints (red line) by state variables x_1 to x_4 (blue line) and estimates provided by the H-infinity Kalman filter; (b) tracking of setpoints (red line) by state variables x_5 to x_8 (blue line) and estimates provided by the H-infinity Kalman filter*

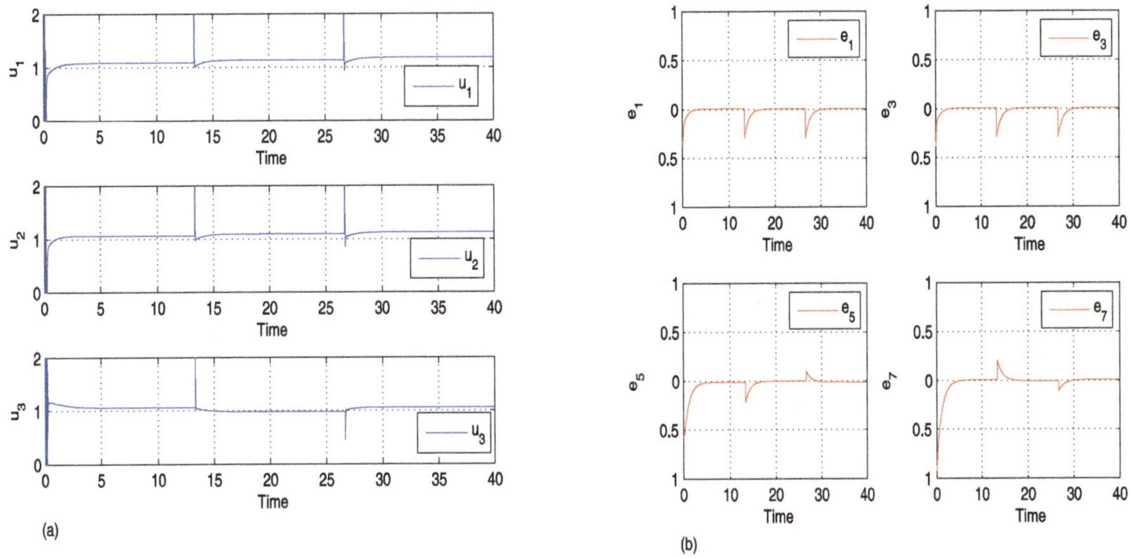

Figure 6.36 *Test-case 3 for the tethered multi-satellite system: (a) variation of the control inputs u_1 to u_3 (tethers' tension) which are applied to the tethered multi-satellite system; (b) tracking error dynamics for state variables x_1, x_3, x_5, and x_7*

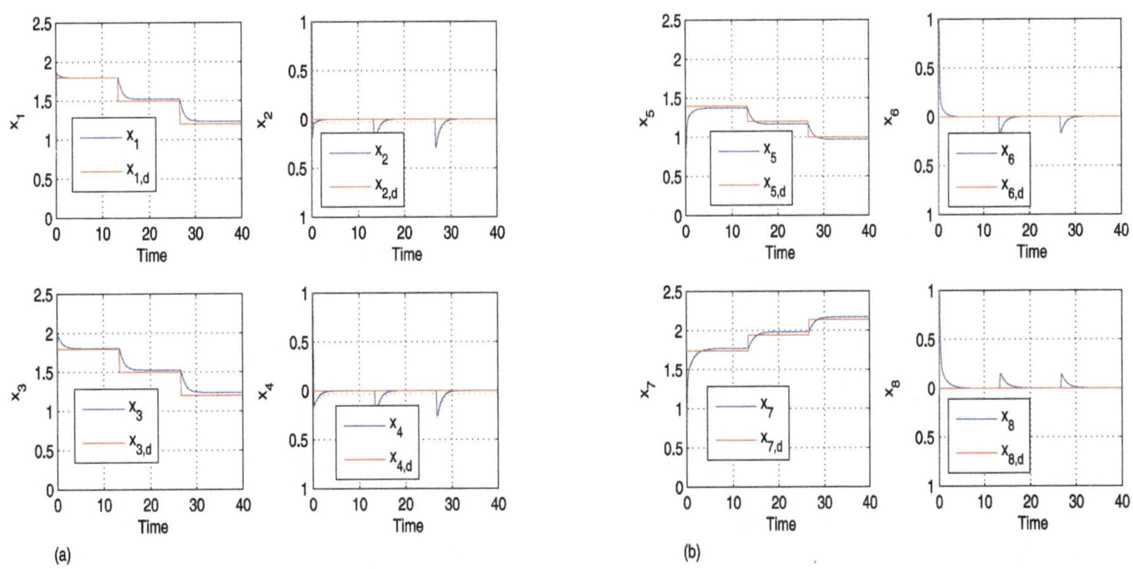

Figure 6.37 *Test-case 4 for the tethered multi-satellite system: (a) tracking of setpoints (red line) by state variables x_1 to x_4 (blue line) and estimates provided by the H-infinity Kalman filter; (b) tracking of setpoints (red line) by state variables x_5 to x_8 (blue line) and estimates provided by the H-infinity Kalman filter*

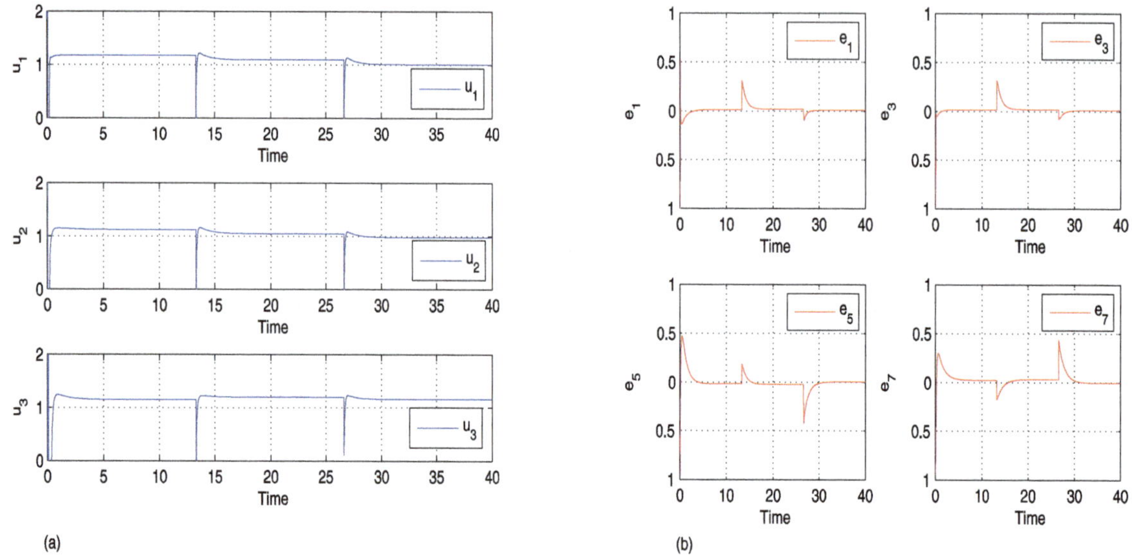

Figure 6.38 *Test-case 4 for the tethered multi-satellite system: (a) variation of the control inputs u_1 to u_3 (tethers' tension) which are applied to the tethered multi-satellite system; (b) tracking error dynamics for state variables x_1, x_3, x_5, and x_7*

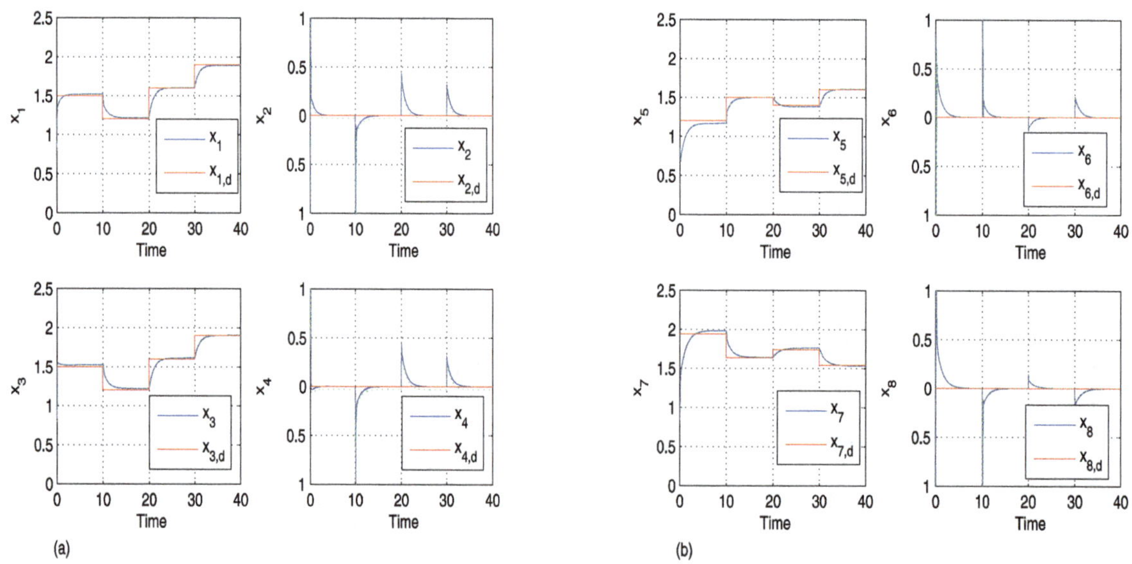

Figure 6.39 *Test-case 5 for the tethered multi-satellite system: (a) tracking of setpoints (red line) by state variables x_1 to x_4 (blue line) and estimates provided by the H-infinity Kalman filter; (b) tracking of setpoints (red line) by state variables x_5 to x_8 (blue line) and estimates provided by the H-infinity Kalman filter*

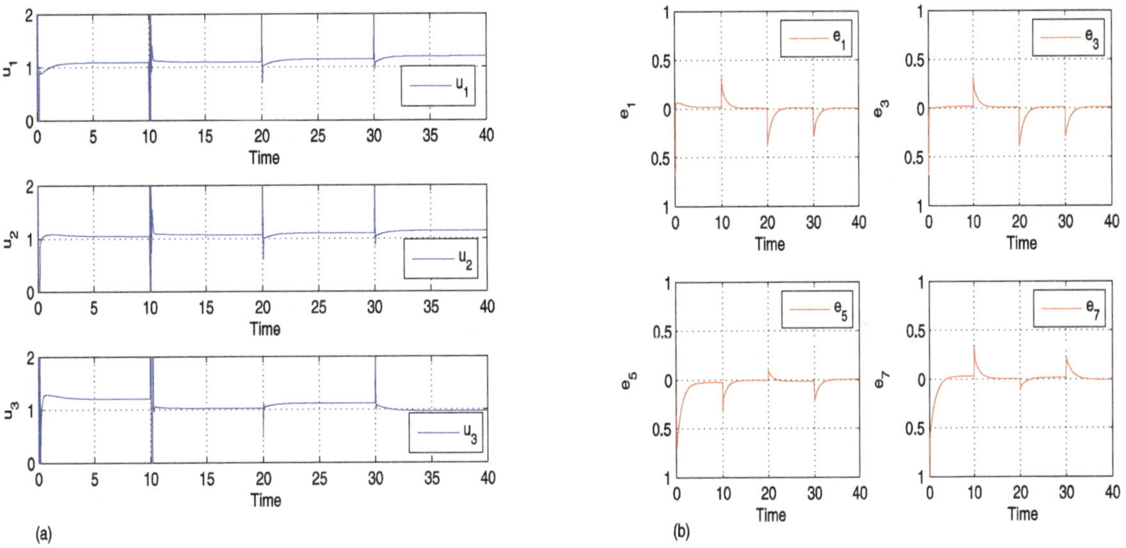

Figure 6.40 *Test-case 5 for the tethered multi-satellite system: (a) variation of the control inputs u_1 to u_3 (tethers' tension) which are applied to the tethered multi-satellite system; (b) tracking error dynamics for state variables x_1, x_3, x_5, and x_7*

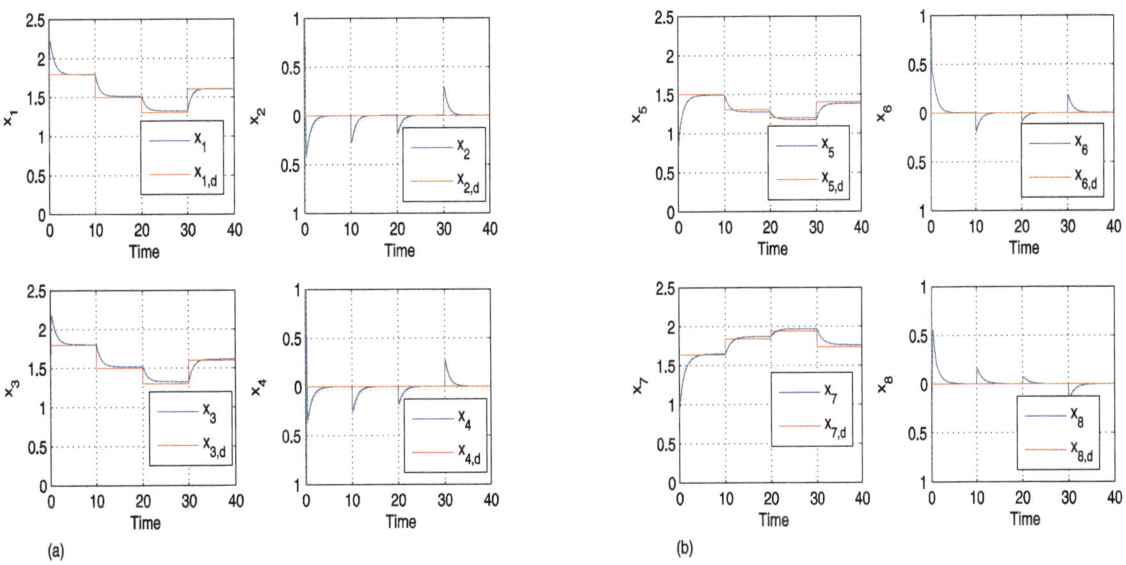

Figure 6.41 *Test-case 6 for the tethered multi-satellite system: (a) tracking of setpoints (red line) by state variables x_1 to x_4 (blue line) and estimates provided by the H-infinity Kalman filter; (b) tracking of setpoints (red line) by state variables x_5 to x_8 (blue line) and estimates provided by the H-infinity Kalman filter*

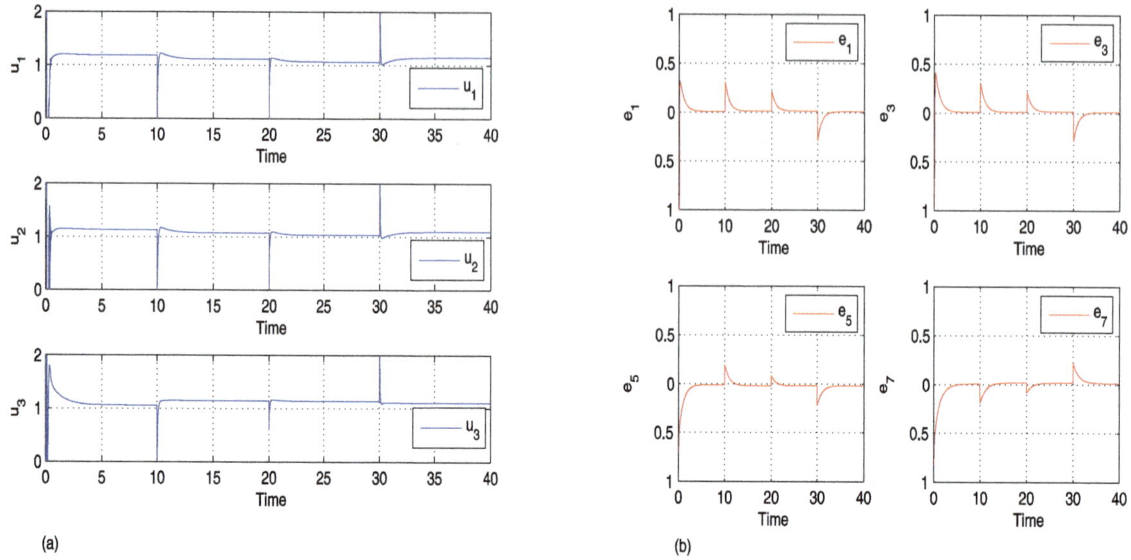

Figure 6.42 *Test-case 6 for the tethered multi-satellite system: (a) variation of the control inputs* u_1 *to* u_3 *(tethers' tension) which are applied to the tethered multi-satellite system; (b) tracking error dynamics for state variables* x_1, x_3, x_5, *and* x_7

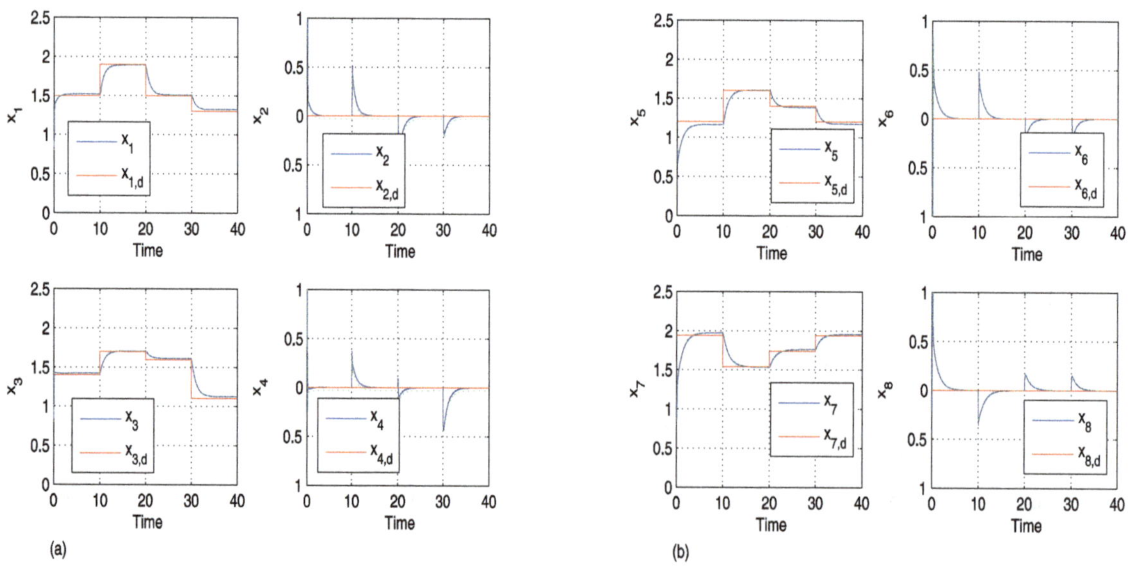

Figure 6.43 *Test-case 7 for the tethered multi-satellite system: (a) tracking of setpoints (red line) by state variables* x_1 *to* x_4 *(blue line) and estimates provided by the H-infinity Kalman filter; (b) tracking of setpoints (red line) by state variables* x_5 *to* x_8 *(blue line) and estimates provided by the H-infinity Kalman filter*

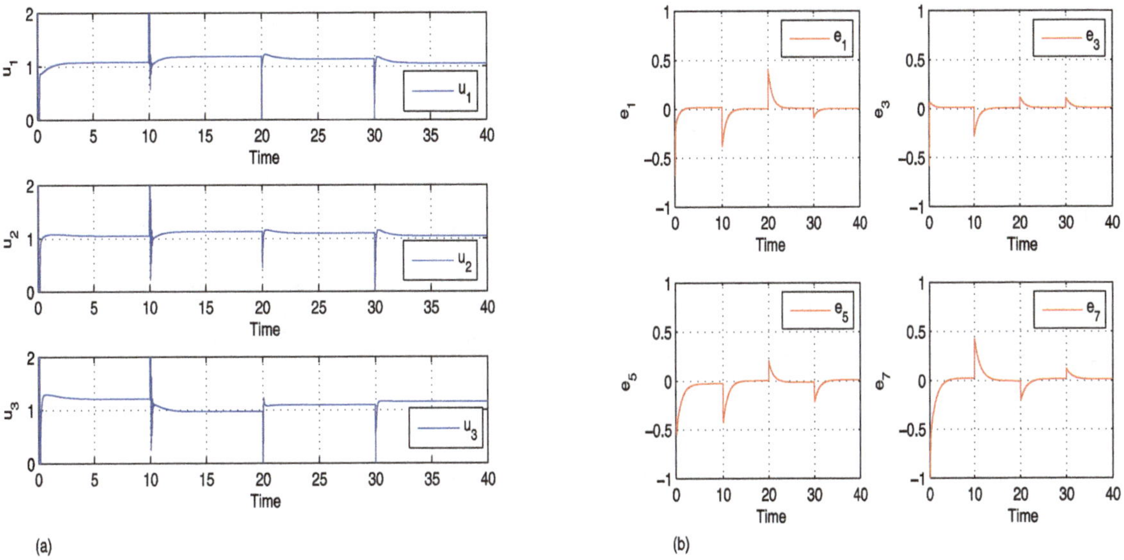

Figure 6.44 Test-case 7 for the tethered multi-satellite system: (a) variation of the control inputs u_1 to u_3 (tethers' tension) which are applied to the tethered multi-satellite system; (b) tracking error dynamics for state variables x_1, x_3, x_5, and x_7

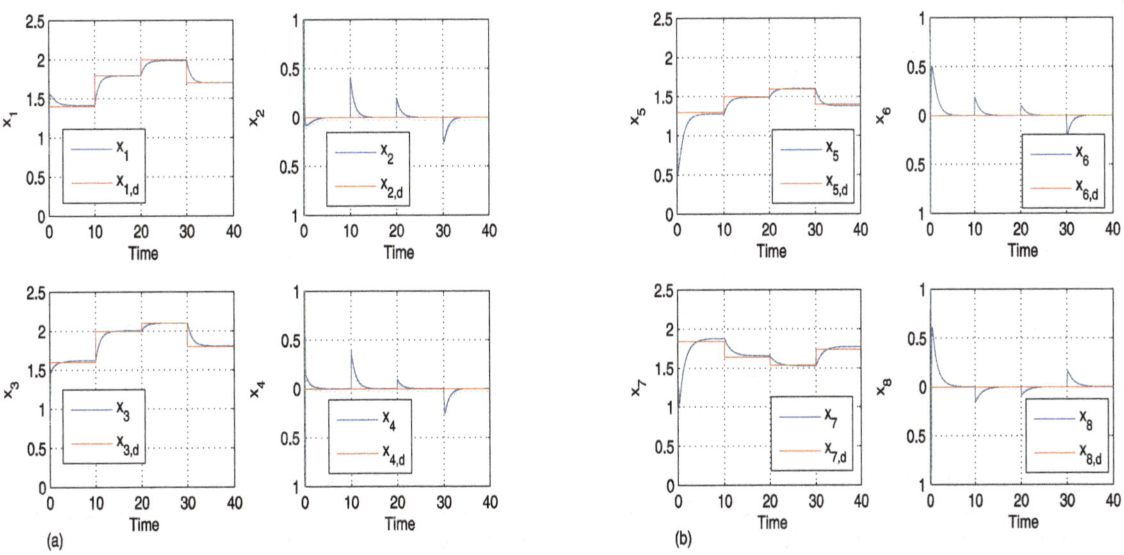

Figure 6.45 Test-case 8 for the tethered multi-satellite system: (a) tracking of setpoints (red line) by state variables x_1 to x_4 (blue line) and estimates provided by the H-infinity Kalman filter; (b) tracking of setpoints (red line) by state variables x_5 to x_8 (blue line) and estimates provided by the H-infinity Kalman filter

ailerons through elaborated control methods enables suppressing the wing-rock effect [53,92,5]. Usually, the wing-rock phenomenon is described by a second-order nonlinear model which may exhibit limit cycles and chaotic behavior

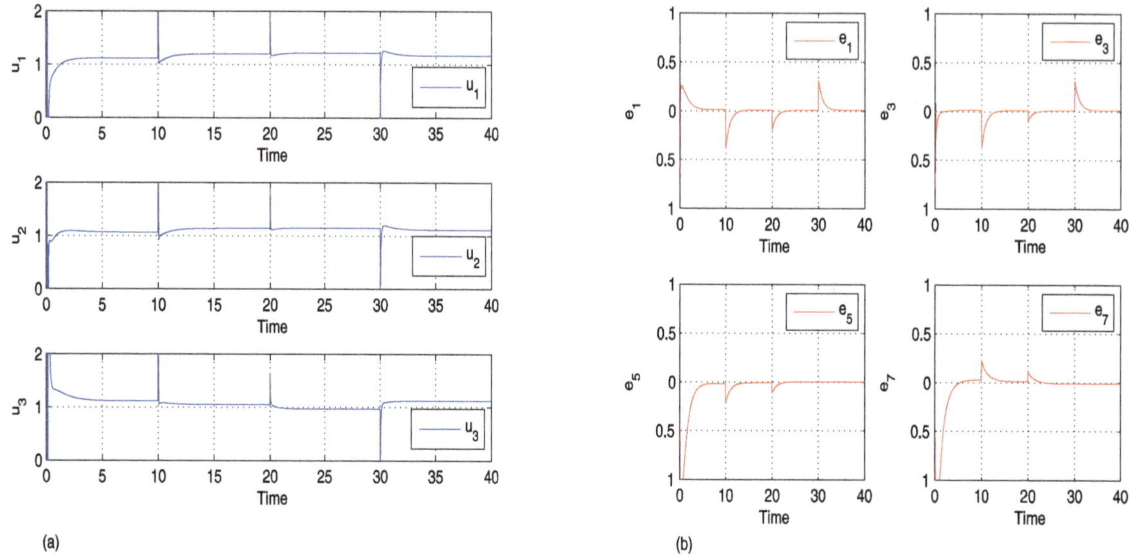

Figure 6.46 Test-case 8 for the tethered multi-satellite system: (a) variation of the control inputs u_1 to u_3 (tethers' tension) which are applied to the tethered multi-satellite system; (b) tracking error dynamics for state variables x_1, x_3, x_5, and x_7

Table 6.10 Nonlinear optimal control

Tracking RMSE $\times 10^{-3}$ for the tethered satellites in the disturbance-free case								
	$RMSE_{x_1}$	$RMSE_{x_2}$	$RMSE_{x_3}$	$RMSE_{x_4}$	$RMSE_{x_5}$	$RMSE_{x_6}$	$RMSE_{x_7}$	$RMSE_{x_8}$
$test_1$	0.5053	0.0001	0.7899	0.0001	0.7388	0.0002	0.8608	0.0001
$test_2$	0.7283	0.0001	0.7825	0.0001	0.8116	0.0002	0.9547	0.0001
$test_3$	0.2530	0.0004	0.4201	0.0006	0.4015	0.0002	0.4821	0.0014
$test_4$	0.2489	0.0002	0.4146	0.0002	0.1178	0.0005	0.1762	0.0032
$test_5$	0.3580	0.0099	0.5017	0.0134	0.1536	0.0088	0.2290	0.0385
$test_6$	0.4179	0.0020	0.5291	0.0089	0.7367	0.0084	0.9276	0.0248
$test_7$	0.2638	0.0023	0.4660	0.0054	0.4915	0.0097	0.7405	0.0106
$test_8$	0.2369	0.0091	0.4094	0.0132	0.1268	0.0002	0.2498	0.0008

[117,286,694]. For the low-order dynamic model of the wing-rock effect, robust and adaptive control methods are often applied [90,480,569]. Besides, there are results on neural or neurofuzzy control which try to treat the problem of control for the wing-rock effect under simultaneous estimation of this system's dynamics [354,658,843,25,520,751]. When the impact of the side-slope angle of the aircraft is taken into account, then a fifth order dynamic model for the wing-rock phenomenon is obtained. This latter dynamic model has as state variables the airplane's roll angle and the rate of change of this angle, the side-slope angle and the rate of change of this angle, as well as the aircraft's ailerons angle [1097,27,568].The control input is the actuation of the ailerons. Additional results on the control problem of the wing-rock dynamics include predictive control, backstepping control and observer–based control [228,318,451]. It is noted that the application of sliding-mode control and backstepping control requires the previous use of diffeomorphisms so as to bring the system into a suitable state-space form that admits such a type of controllers. In this section,

Table 6.11 Nonlinear optimal control

Tracking RMSE $\times 10^{-3}$ for the tethered satellites in the case of disturbances								
$\Delta a\%$	$RMSE_{x_1}$	$RMSE_{x_2}$	$RMSE_{x_3}$	$RMSE_{x_4}$	$RMSE_{x_5}$	$RMSE_{x_6}$	$RMSE_{x_7}$	$RMSE_{x_8}$
0%	0.3580	0.0099	0.5017	0.0134	0.1536	0.0088	0.2280	0.0382
10%	0.8778	0.0429	0.1594	0.0162	0.1713	0.0093	0.2061	0.0376
20%	2.0754	0.0981	0.8138	0.0366	0.1886	0.0099	0.1983	0.0370
30%	3.2477	0.1357	1.4629	0.0580	0.2053	0.0104	0.1832	0.0363
40%	4.3914	0.1798	2.1065	0.0793	0.2216	0.0109	0.1684	0.0357
50%	5.5071	0.2222	2.7444	0.1004	0.2373	0.0144	0.1537	0.0351
60%	6.5956	0.2631	3.3773	0.1212	0.2525	0.0119	0.1394	0.0345

Table 6.12 Nonlinear optimal control

Tracking RMSE $\times 10^{-3}$ for the estimation performed by the H-infinity KF								
	$RMSE_{x_1}$	$RMSE_{x_2}$	$RMSE_{x_3}$	$RMSE_{x_4}$	$RMSE_{x_5}$	$RMSE_{x_6}$	$RMSE_{x_7}$	$RMSE_{x_8}$
$test_1$	0.0004	0.5006	0.0005	0.5006	0.0004	0.0004	0.0004	0.0004
$test_2$	0.0004	0.5006	0.0004	0.5006	0.0004	0.0004	0.0004	0.0004
$test_3$	0.0004	0.5006	0.0004	0.5006	0.0004	0.0004	0.0005	0.0004
$test_4$	0.0005	0.5006	0.0005	0.5006	0.0005	0.0005	0.0005	0.0005
$test_5$	0.0005	0.5006	0.0005	0.5006	0.0005	0.0005	0.0005	0.0005
$test_6$	0.0005	0.5006	0.0005	0.5006	0.0005	0.0005	0.0005	0.0005
$test_7$	0.0005	0.5006	0.0005	0.5006	0.0005	0.0005	0.0005	0.0005
$test_8$	0.0005	0.5006	0.0005	0.5006	0.0005	0.0005	0.0005	0.0005

Table 6.13 Nonlinear optimal control

Convergence time for the tethered satellites' state variables								
	$T_s\ x_1$	$T_s\ x_2$	$T_s\ x_3$	$T_s\ x_4$	$T_s\ x_5$	$T_s\ x_6$	$T_s\ x_7$	$T_s\ x_8$
$test_1$	2.0	3.0	2.0	3.0	4.5	4.0	4.5	5.0
$test_2$	3.0	4.0	1.5	2.5	4.5	4.5	4.0	4.5
$test_3$	3.0	3.5	2.5	3.5	3.0	3.5	4.0	5.0
$test_4$	1.5	2.0	3.5	2.5	3.5	4.0	3.5	4.5
$test_5$	2.5	3.5	2.0	3.5	5.0	3.5	4.0	5.0
$test_6$	3.5	4.0	4.0	3.5	3.5	4.0	4.0	5.0
$test_7$	2.0	3.0	1.0	0.5	5.0	4.0	4.0	5.0
$test_8$	4.0	3.5	2.5	3.0	5.0	4.5	4.5	5.0

the aim has been to treat the nonlinear control problem of the wing-rock effect without the need for changes of state variables and complicated state-space model transformations [725,721,741].

A novel nonlinear optimal control method is proposed for suppressing the wing-rock effects in aircrafts [725–741]. The 5-th order dynamic model of the wing-rock effect has been formulated after considering that oscillatory roll motion can be also induced by the side-slope angle of the airplane. It has been demonstrated that the wing-rock dynamics is a differentially flat system. This confirms that the state-space model of the wing-rock effect is input–output linearizable and controllable. Next, the state-space model of the wing-rock dynamics has undergone approximate linearization with the use of first-order Taylor series expansion and through the computation of the associated Jacobian matrices [705,60,706]. The linearization takes place at each sampling instance around a temporary operating point (x^*, u^*) which is defined by the present value of the system's state vector x^* and by the last sampled value u^* of the control inputs vector. The modeling error which is due to the truncation of higher-order terms from the Taylor series expansion is considered to be a perturbation which is asymptotically compensated by the robustness of the control algorithm. For the approximately linearized wing-rock dynamics a stabilizing H-infinity feedback controller is designed.

The H-infinity feedback controller provides a solution for the optimal control problem of the wing-rock effect under model uncertainties and external perturbations. It represents a min–max differential game which takes place between (i) the control inputs that try to minimize a cost function containing a quadratic term of the state vector's tracking error and (ii) the model uncertainty and external perturbation terms which try to maximize this cost function. To compute the stabilizing feedback gains for the H-infinity controller an algebraic Riccati equation is repetitively solved at each time step of the control algorithm [725,741]. The global stability properties of the control scheme are proven through Lyapunov analysis. First, is is demonstrated that the proposed controller achieves the H-infinity tracking criterion. This means that the system is robust to model imprecision, disturbances and parametric variations [711,732,876]. Moreover, under moderate conditions, it is proven that the control loop is globally asymptotically stable. The control method of this section avoids diffeomorphisms and back or forth transformations for the dynamic model of the wing-rock effect. The proposed control scheme is energy-optimal in the sense that it minimizes the variations of the control inputs. To apply state estimation-based control without the need to measure the entire state vector of the wing-rock dynamics the H-infinity Kalman filter has been used as a robust state estimator.

6.4.2 Dynamic model of the wing-rock effect

The diagram of the wing-rock effect in a delta-wing aircraft and its parameters are shown in Figure 6.47. The main parameters of the dynamic model of the wing-rock effect are: ϕ: is the roll angle of the aircraft (rad), (ii) $\rho = \frac{d\phi}{dt}$ is the

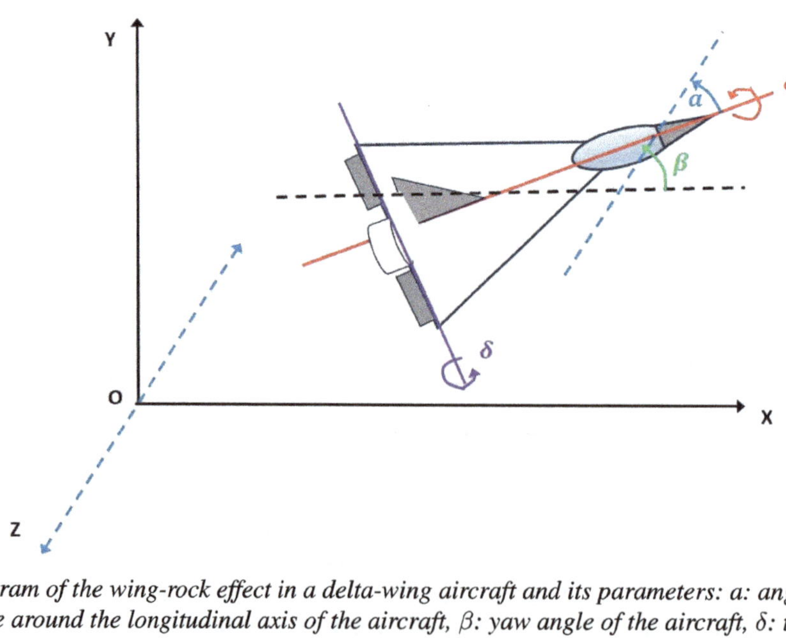

Figure 6.47 Diagram of the wing-rock effect in a delta-wing aircraft and its parameters: a: angle of attack, ϕ: roll angle around the longitudinal axis of the aircraft, β: yaw angle of the aircraft, δ: turn angle of the ailerons

roll rate of the aircraft (rad/sec), (iii) δ is the ailerons angle of the aircraft (rad), (iv) β is the side-slope (yaw) angle of the aircraft (rad), and (v) $\frac{d\beta}{dt}$ is the rate of change of the side-slope angle of the aircraft (rad/sec) [1097,568].
The roll angle dynamics is given by [1097,568]

$$I_{xx}\frac{d\rho}{dt} = \mu\rho + L_{\delta_1}\delta + L_{\beta_1}\beta - L_{r_1}\frac{d\beta}{dt} + f(\phi,\rho) \tag{6.285}$$

where I_{xx} is the moment of inertia of the aircraft for rotation around its longitudinal axis, μ is the constant string-damping coefficient, L_{δ_1} is the coefficient of the ailerons angle, L_{β_1} is the coefficient of the side-slip angle, L_{r_1} is the coefficient of the rate of change of the side-slip angle.
Function $f(\phi,\rho)$ is a nonlinear function of the roll angle ϕ and of the roll angle's change rate ρ and is given by

$$f(\phi,\rho) = a_1\phi + a_2\rho + a_3\rho^3 + a_4\phi^2\rho + a_5\phi\rho^2 \tag{6.286}$$

where coefficients a_i, $i = 1, 2, \cdots, 5$ depend on the angle of attack a of the aircraft. By substituting Eq. (6.286) in Eq. (6.285), one obtains

$$\begin{aligned}\frac{d\rho}{dt} = \frac{d^2\phi}{dt^2} = \frac{1}{I_{xx}}(\mu\rho + a_1\phi + a_2\rho + a_3\rho^3 + a_4\phi^2\rho + a_5\phi\rho^2 + \\ + L_{\delta_1}\delta + L_{\beta_1}\beta - L_{r_1}\frac{d\beta}{dt})\end{aligned} \tag{6.287}$$

$$\begin{aligned}\frac{d^2\phi}{dt^2} = -\omega^2\phi + \mu_1\rho + \mu_2\phi^2\rho + b_1\rho^3 + b_2\phi\rho^2 + \\ L_\delta\delta + L_\beta\beta - L_r\dot{\beta}\end{aligned} \tag{6.288}$$

where $\omega^2 = -\frac{a_1}{I_{xx}}$, $\mu_1 = \frac{a_2+\mu}{I_{xx}}$, $\mu_2 = \frac{a_4}{I_{xx}}$, $b_1 = \frac{a_3}{I_{xx}}$, $b_2 = \frac{a_5}{I_{xx}}$, $L_\delta = \frac{L_{\delta_1}}{I_{xx}}$, $L_\beta = \frac{L_{\beta_1}}{I_{xx}}$, $L_r = \frac{L_{r_1}}{I_{xx}}$.

It is noted that coefficients μ_1, μ_2, b_1, b_2, L_δ, L_β, L_r, and a_i, $i = 1, 2, \cdots, 5$ depend on the aerodynamic properties of the aircraft and on its angle of attack. These parameters are assigned values through identification techniques.

About the dynamics of the yaw angle of the aircraft, one has that

$$\frac{d^2\beta}{dt^2} = -N_p\rho - N_\beta\beta + N_r\dot{\beta} \tag{6.289}$$

where parameters N_p, N_β, and N_r depend on the aerodynamic properties of the aircraft. Finally, the dynamics of the aileron's angle δ is given by

$$\frac{d\delta}{dt} = \frac{1}{\tau}(-\delta + u) \tag{6.290}$$

where δ is the aileron's deflection angle, and u is the actuation from an electric or electro-hydraulic actuator.

The state vector of the dynamic model of the wing-rock effect is given by

$$x = [x_1, x_2, x_3, x_4, x_5]^T = [\phi, \rho, \delta, \beta, \dot{\beta}]^T. \tag{6.291}$$

Next, the state-space equations which describe the dynamics of the wing-rock effect are given by [1097,568]

$$\begin{aligned}\dot{x}_1 &= x_2 \\ \dot{x}_2 &= -\omega^2 x_1 + \mu_1 x_2 + \mu_2 x_1^2 x_2 + b_1 x_2^3 + b_2 x_1 x_2^2 + L_\delta x_3 + L_\beta x_4 - L_r x_5 \\ \dot{x}_3 &= -\frac{1}{\tau}x_3 + \frac{1}{\tau}u \\ \dot{x}_4 &= x_5 \\ \dot{x}_5 &= -N_p x_2 - N_\beta x_4 + N_r x_5\end{aligned} \tag{6.292}$$

To suppress the wing-rock effect, one has to make the roll angle of the aircraft ϕ and the roll rate ρ stabilize at the zero value. The state-space model of the wing-rock effect is also written in the following matrix form:

$$\begin{pmatrix}\dot{x}_1\\ \dot{x}_2\\ \dot{x}_3\\ \dot{x}_4\\ \dot{x}_5\end{pmatrix} = \begin{pmatrix} x_2 \\ -\omega^2 x_1 + \mu_1 x_2 + \mu_2 x_1^2 x_2 + b_1 x_2^3 + b_2 x_1 x_2^2 + L_\delta x_3 + L_\beta x_4 - L_r x_5 \\ -\frac{1}{T} x_3 \\ x_5 \\ -N_p x_2 - N_\beta x_4 + N_r x_5 \end{pmatrix} + \begin{pmatrix}0\\0\\ \frac{1}{T}\\0\\0\end{pmatrix} u \tag{6.293}$$

or, the system's dynamics can be written in the nonlinear affine-in-the-input state-space form

$$\dot{x} = f(x) + g(x)u \tag{6.294}$$

with $x \in R^{5\times1}, f(x) \in R^{5\times1}$, $g(x) \in R^{5\times1}$, and $u \in R$. The system is differentially flat and thus input–output linearizable.

6.4.3 Differential flatness properties of the wing-rock dynamics

6.4.3.1 Proof of differential flatness

A system is considered to be differentially flat if the following two conditions hold: (i) all its state variables and its control inputs can be expressed as differential functions of the system's flat "output," where the "flat output" comprises a subset of the system's state variables, and (ii) the flat output and its time derivatives are "differentially independent" which means that they are not connected through a relation in the form of an ordinary differential equation.

The dynamic model of the wing-rock effect is differentially flat with flat output $y = N_p x_1 - N_r x_4 + x_5$. By differentiating the flat output y in time, one obtains

$$\begin{aligned} &\dot{y} = N_p \dot{x}_1 - N_r \dot{x}_4 + \dot{x}_5 \Rightarrow \dot{y} = -N_\beta x_4 \Rightarrow \\ &\qquad x_4 = -\frac{1}{N_\beta}\dot{y} \Rightarrow x_4 = h_4(y, \dot{y}) \end{aligned} \tag{6.295}$$

which signifies that x_4 is a differential function of the flat output y. By differentiating once again in time, one gets

$$\begin{aligned} &\ddot{y} = -N_\beta \dot{x}_4 \Rightarrow \ddot{y} = -N_\beta x_5 \Rightarrow \\ &\qquad x_5 = -\frac{1}{N_\beta}\ddot{y} \Rightarrow x_5 = h_5(y, \dot{y}) \end{aligned} \tag{6.296}$$

which signifies that x_5 is also a differential function of the flat output y. Next, using the definition of the flat output $y = N_p x_1 - N_r x_4 + x_5$ and after solving for x_1, one has

$$x_1 = \frac{1}{N_p}[y + N_r x_4 - x_5] \Rightarrow x_1 = h_1(y, \dot{y}) \tag{6.297}$$

which signifies that x_1 is a differential function of the flat output y. After one more differentiation in time, one gets

$$\begin{aligned} &y^{(3)} = -N_\beta \dot{x}_5 \Rightarrow y^{(3)} = N_\beta N_p x_2 + N_\beta^2 x_4 - N_\beta N_r x_5 \Rightarrow \\ &\qquad x_2 = \frac{1}{N_\beta N_p}[y^{(3)} - N_\beta^2 x_4 + N_\beta N_r x_5] \Rightarrow x_2 = h_2(y, \dot{y}) \end{aligned} \tag{6.298}$$

which signifies that x_2 is a differential function of the flat output y. By differentiating once again in time, one gets

$$\begin{aligned} &\qquad\qquad y^{(4)} = N_\beta N_p \dot{x}_2 + N_\beta^2 \dot{x}_4 - N_\beta N_r \dot{x}_5 \Rightarrow \\ &y^{(4)} = N_\beta N_p[-\omega^2 x_1 + \mu_1 x_2 + \mu_2 x_1^2 x_2 + b_1 x_2^3 + b_2 x_1 x_2^2 + L_\delta x_3 + L_\beta x_4 - L_r x_5] \\ &\qquad\qquad +N_\beta^2 x_5 - N_\beta N_r[-N_p x_2 - N_\beta x_4 + N_r x_5] \end{aligned} \tag{6.299}$$

Solving for x_3 gives

$$\begin{aligned} x_3 = \frac{1}{N_\beta N_p L_\delta}\{&y^{(4)} - N_\beta N_p[-\omega^2 x_1 + \mu_1 x_2 + \mu_2 x_1^2 x_2 + b_1 x_2^3 + b_2 x_1 x_2^2 + L_\beta x_4 - L_r x_5] \\ &-N_\beta^2 x_5 + N_\beta N_r[-N_p x_2 - N_\beta x_4 + N_r x_5]\} \Rightarrow x_3 = h_3(y, \dot{y}) \end{aligned} \tag{6.300}$$

The above-given relation means that since state variables x_1, x_2, x_4, x_5 are differential functions of the flat output, it also holds that x_3 can be written as a function of the flat output y and its derivatives.

Finally, by differentiating once again in time, one gets

$$\begin{aligned} y^{(5)} = & N_\beta N_p[-\omega^2\dot{x}_1 + \mu_1\dot{x}_2 + \mu_2 2x_1\dot{x}_1x_2 + \mu_2x_1^2\dot{x}_2 + b_1 3x_2^2\dot{x}_2 + \\ & +b_2\dot{x}_1x_2^2 + b_2x_1 2x_2\dot{x}_2 + L_\delta\dot{x}_3 + L_\beta\dot{x}_4 - L_r\dot{x}_5] + \\ & +N_\beta^2\dot{x}_5 - N_\beta N_r[-N_p\dot{x}_2 - N_\beta\dot{x}_4 + N_r\dot{x}_5] \end{aligned} \tag{6.301}$$

By denoting the state-space model of the wing-rock effect as

$$\begin{aligned} &\dot{x}_1 = f_1(x) && f_1(x) = x_2 \\ &\dot{x}_2 = f_2(x) && f_2(x) = -\omega^2x_1 + \mu_1x_2 + \mu_2x_1^2x_2 + b_1x_2^3 + b_2x_1x_2^2 + L_\delta x_3 + L_\beta x_4 - L_rx_5 \\ &\dot{x}_3 = f_3(x) + g_3(x)u && f_3(x) = -\tfrac{1}{\tau}x_3 \quad g_3(x) = \tfrac{1}{\tau} \\ &\dot{x}_4 = f_4(x) && f_4(x) = x_5 \\ &\dot{x}_5 = f_5(x) && f_5(x) = -N_px_2 - N_\beta x_4 + N_rx_5 \end{aligned} \tag{6.302}$$

then Eq. (6.301) becomes

$$\begin{aligned} y^{(5)} = & N_\beta N_p[-\omega^2x_2 + \mu_1f_2 + \mu_2 2x_1x_2^2 + \mu_2x_1^2f_2 + b_1 3x_2^2f_2 + \\ & +b_2x_2^3 + 2b_2x_1x_2f_2 + L_\delta(f_3 + g_3u) + L_\beta x_5 - L_rf_5] + \\ & +N_\beta^2f_5 - N_\beta N_r[-N_pf_2 - N_\beta x_5 + N_rf_5] \end{aligned} \tag{6.303}$$

After re-ordering of terms, one obtains

$$\begin{aligned} y^{(5)} = & \{N_\beta N_p[-\omega^2x_2 + \mu_1f_2 + \mu_2 2x_1x_2^2 + \mu_2x_1^2f_2 + b_1 3x_2^2f_2 + \\ & +b_2x_2^3 + 2b_2x_1x_2f_2 + L_\delta f_3 + L_\beta x_5 - L_rf_5] + \\ & +N_\beta^2f_5 - N_\beta N_r[-N_pf_2 - N_\beta x_5 + N_rf_5]\} + [N_\beta N_pL_\delta g_3]u \end{aligned} \tag{6.304}$$

or a equivalently, one has the description of the system's dynamics in the input–output linearized form

$$y^{(5)} = \bar{f}(x) + \bar{g}(x)u \tag{6.305}$$

with functions $\bar{f}(x)$ and $\bar{g}(x)$ to be defined as

$$\begin{aligned} \bar{f}(x) = & N_\beta N_p[-\omega^2x_2 + \mu_1f_2 + \mu_2 2x_1x_2^2 + \mu_2x_1^2f_2 + b_1 3x_2^2f_2 + \\ & +b_2x_2^3 + 2b_2x_1x_2f_2 + L_\delta f_3 + L_\beta x_5 - L_rf_5] + \\ & +N_\beta^2f_5 - N_\beta N_r[-N_pf_2 - N_\beta x_5 + N_rf_5] \end{aligned} \tag{6.306}$$

$$\bar{g}(x) = N_\beta N_pL_\delta g_3 \tag{6.307}$$

After solving Eq. (6.305) for the control input u, one gets

$$u = \tfrac{1}{\bar{g}(x)}[y^{(5)} - \bar{f}(x)] \tag{6.308}$$

and since all elements of state vector x are differential functions of the flat output y, one concludes also that control input u is a differential function of the system's flat output y.

Therefore, all state variables and the control inputs of the wing-rock effect dynamics are differential functions of the flat output y, and the wing-rock model is a differentially flat system. Moreover, it has been shown above that this system is input–output linearizable and can be written in the form of Eq. (6.305). The differential flatness property is also an implicit proof of the controllability of the wing-rock dynamics.

6.4.3.2 Design of a flatness-based controller for the wing-rock dynamics

For the input–output linearized dynamics of the wing-rock effect, a stabilizing feedback controller can be based on eigenvalues assignment. One defines the control input:

$$u = \frac{1}{\bar{g}(x)}[v - \bar{f}(x)]$$
$$v = y_d^{(5)} - k_1(y^{(4)} - y_d^{(4)}) - k_2(y^{(3)} - y_d^{(3)}) - k_3(\ddot{y} - \ddot{y}_d) - k_4(\dot{y} - \dot{y}_d) - k_5(y - y_d) \quad (6.309)$$

By substituting the control input of Eq. (6.309) into the dynamic model of Eq. (6.305), one obtains the closed-loop system dynamics

$$(y^{(5)} - y_d^{(5)}) + k_1(y^{(4)} - y_d^{(4)}) + k_2(y^{(3)} - y_d^{(3)}) + k_3(\ddot{y} - \ddot{y}_d) + k_4(\dot{y} - \dot{y}_d) + k_5(y - y_d) = 0 \quad (6.310)$$

By denoting the state vector's tracking error as $e = y - y_d$, one has the tracking error dynamics

$$e^{(5)} + k_1 e^{(4)} + k_2 e^{(3)} + k_3 \ddot{e} + k_4 \dot{e} + k_5 e = 0 \quad (6.311)$$

One selects gains k_i, $i = 1, 2, \cdots, 5$ so as the characteristic polynomial of the above-given differential equation to be Hurwitz stable. This ensures that

$$lim_{t \to \infty} e(t) = 0 \Rightarrow lim_{t \to \infty} y(t) = y_d(t) \quad (6.312)$$

It holds that the desirable value of the flat output of the system is associated with the setpoints of the individual state variables x_i, $i = 1, 2, \cdots, 5$ through the following relation

$$y_d = N_p x_{1,d} - N_r x_{4,d} + x_{5,d} \quad (6.313)$$

After selecting $y_d = -N_r x_{4,d}$ with $x_{4,d}$ to be constant, one has $x_{5,d} = \dot{x}_{4,d} \Rightarrow x_{5,d} = 0$. Thus, one obtains

$$-N_r x_{4,d} = N_p x_{1,d} - N_r x_{4,d} + 0 \Rightarrow x_{1,d} = 0 \quad (6.314)$$

It holds that $x_1 = h_1(y, \dot{y})$, that is x_1 is a differential function of the flat output y. Thus, if y converges to its setpoint y_d, then x_1 will also converge to its reference values $x_{1,d}$. This means

$$lim_{t \to \infty} x_1(t) = x_{1,d}(t) = 0 \Rightarrow lim_{t \to \infty} \phi(t) = 0 \quad (6.315)$$

which comes to confirm that the wind-rock effect is completely suppressed and the roll motion of the aircraft is stabilized.

6.4.4 Approximate linearization of the wing-rock dynamics

The dynamic model of the wing-rock dynamics undergoes approximate linearization with the use of first-order Taylor series expansion and through the computation of the associated Jacobian matrices. The linearization takes place at each sampling instance around a temporary operating point (x^*, u^*) which is defined by the present value of the system's state vector x^* and by the last sampled value of the control inputs vector u^*. The linearization point is updated at each sampling period. The modeling error which is due to the truncation of higher-order terms from the Taylor series expansion is considered to be a perturbation which is asymptotically compensated by the robustness of the control algorithm. The wing-rock dynamics, being initial in the nonlinear form of Eq. (6.294) $\dot{x} = f(x) + g(x)u$ is now written into the linearized state-space description

$$\dot{x} = Ax + Bu + \tilde{d} \quad (6.316)$$

where A, B are the Jacobian matrices of the wing-rock dynamics which are given by

$$A = \nabla_x[f(x) + g(x)u] \mid_{(x^*, u^*)} \Rightarrow A = \nabla_x f(x) \mid_{(x^*, u^*)} \quad (6.317)$$
$$B = \nabla_u[f(x) + g(x)u] \mid_{(x^*, u^*)} \Rightarrow B = g(x) \mid_{(x^*, u^*)} \quad (6.318)$$

and $\tilde{d}$ is the cumulative disturbance term which may comprise (i) the modeling error due to the truncation of higher-order terms from the Taylor series expansion, and (ii) exogenous perturbations, and (iii) sensor measurement noise of any distribution.

The computation of the Jacobian matrix $\nabla_x f(x) \mid_{(x^*,u^*)}$ is given below:

First row of the Jacobian matrix $\nabla_x f(x) \mid_{(x^*,u^*)}$: $\frac{\partial f_1}{\partial x_1}=0$, $\frac{\partial f_1}{\partial x_2}=1$, $\frac{\partial f_1}{\partial x_3}=0$, $\frac{\partial f_1}{\partial x_4}=0$, $\frac{\partial f_1}{\partial x_5}=0$.

Second row of the Jacobian matrix $\nabla_x f(x) \mid_{(x^*,u^*)}$: $\frac{\partial f_2}{\partial x_1}=-\omega^2+2\mu_2 x_1 x_2+b_2 x_2^2$, $\frac{\partial f_2}{\partial x_2}=\mu_1+\mu_2 x_1^2+3b_1 x_2^2+2b_2 x_1 x_2$, $\frac{\partial f_2}{\partial x_3}=L_\delta$, $\frac{\partial f_2}{\partial x_4}=L_\beta$, $\frac{\partial f_2}{\partial x_5}=-L_r$.

Third row of the Jacobian matrix $\nabla_x f(x) \mid_{(x^*,u^*)}$: $\frac{\partial f_3}{\partial x_1}=0$, $\frac{\partial f_3}{\partial x_2}=0$, $\frac{\partial f_3}{\partial x_3}=-\frac{1}{\tau}$, $\frac{\partial f_3}{\partial x_4}=0$, $\frac{\partial f_3}{\partial x_5}=0$.

Fourth row of the Jacobian matrix $\nabla_x f(x) \mid_{(x^*,u^*)}$: $\frac{\partial f_4}{\partial x_1}=0$, $\frac{\partial f_4}{\partial x_2}=0$, $\frac{\partial f_4}{\partial x_3}=0$, $\frac{\partial f_4}{\partial x_4}=0$, $\frac{\partial f_4}{\partial x_5}=1$.

Fifth row of he Jacobian matrix $\nabla_x f(x) \mid_{(x^*,u^*)}$: $\frac{\partial f_5}{\partial x_1}=0$, $\frac{\partial f_5}{\partial x_2}=-N_p$, $\frac{\partial f_5}{\partial x_3}=0$, $\frac{\partial f_5}{\partial x_4}=-N_\beta$, $\frac{\partial f_5}{\partial x_5}=N_r$.

This above-given linearization approach which has been followed for implementing the nonlinear optimal control scheme results in a quite accurate model of the system's dynamics. Consider for instance the following affine-in-the-input state-space model:

$$\begin{aligned}\dot{x} &= f(x)+g(x)u \Rightarrow \\ \dot{x} &= [f(x^*)+\nabla_x f(x) \mid_{x^*} (x-x^*)]+[g(x^*)+\nabla_x g(x) \mid_{x^*} (x-x^*)]u^*+g(x^*)u^*+g(x^*)(u-u^*)+\tilde{d}_1 \Rightarrow \\ \dot{x} &= [\nabla_x f(x) \mid_{x^*} +\nabla_x g(x) \mid_{x^*} u^*]x+g(x^*)u-[\nabla_x f(x) \mid_{x^*} +\nabla_x g(x) \mid_{x^*} u^*]x^*+f(x^*)+g(x^*)u^*+\tilde{d}_1\end{aligned} \tag{6.319}$$

where $\tilde{d}_1$ is the modeling error due to truncation of higher order terms in the Taylor series expansion of $f(x)$ and $g(x)$. Next, by defining $A=[\nabla_x f(x) \mid_{x^*} +\nabla_x g(x) \mid_{x^*} u^*]$, $B=g(x^*)$, one obtains

$$\dot{x}=Ax+Bu-Ax^*+f(x^*)+g(x^*)u^*+\tilde{d}_1 \tag{6.320}$$

Moreover, by denoting $\tilde{d}=-Ax^*+f(x^*)+g(x^*)u^*+\tilde{d}_1$ about the cumulative modeling error term in the Taylor series expansion procedure, one has

$$\dot{x}=Ax+Bu+\tilde{d} \tag{6.321}$$

which is the approximately linearized model of the dynamics of the system of Eq. (6.316). The term $f(x^*)+g(x^*)u^*$ is the derivative of the state vector at (x^*,u^*) which is almost annihilated by $-Ax^*$.

6.4.5 An H-infinity controller for the wing-rock dynamics

6.4.5.1 Min–max control and disturbance rejection

In the H_∞ control approach, a feedback control scheme is designed for setpoint tracking by the state vector of the wing-rock dynamics and for simultaneous disturbance rejection, considering that the disturbance affects the system in the worst possible manner. For the approximately linearized model of Eq. (6.316), the disturbances' effects are incorporated in the following quadratic cost function [725]

$$J(t)=\frac{1}{2}\int_0^T[y^T(t)y(t)+ru^T(t)u(t)-\rho^2\tilde{d}^T(t)\tilde{d}(t)]dt, \quad r,\rho>0 \tag{6.322}$$

The significance of the negative sign in the cost function's term which is associated with the perturbation variable $\tilde{d}(t)$ is that the disturbance tries to maximize the cost function $J(t)$ while the control signal $u(t)$ tries to minimize it.

The physical meaning of the relation given above is that the control signal and the disturbances compete with each other within a min–max differential game. This problem of min–max optimization can be written as $min_u max_{\tilde{d}} J(u,\tilde{d})$. The objective of the optimization procedure is to compute a control signal $u(t)$ which can compensate for the worst possible disturbance that affects the system. However, the solution to the min–max optimization problem is directly related to the value of parameter ρ. This means that there is an upper bound in the disturbances magnitude that can be annihilated by the control signal.

6.4.5.2 Design of the stabilizing feedback controller

After linearization around its current operating point (x^*, u^*), the dynamic model of the wing-rock dynamics is written as

$$\dot{x} = Ax + Bu + d_1 \tag{6.323}$$

Parameter d_1 stands for the linearization error in the wing-rock dynamics model appearing above in Eq. (6.323). The reference setpoints for the state vector of the wing-rock dynamics are denoted by $\mathbf{x_d} = [x_1^d, \cdots, x_5^d]$. Tracking of this trajectory is achieved after applying the control input u_d. At every time instant, the control input u_d is assumed to differ from the control input u appearing in Eq. (6.323) by an amount equal to Δu, that is $u_d = u + \Delta u$

$$\dot{x}_d = Ax_d + Bu_d + d_2 \tag{6.324}$$

The dynamics of the controlled system described in Eq. (6.323) can be also written as

$$\dot{x} = Ax + Bu + Bu_d - Bu_d + d_1 \tag{6.325}$$

and by denoting $d_3 = -Bu_d + d_1$ as an aggregate disturbance term, one obtains

$$\dot{x} = Ax + Bu + Bu_d + d_3 \tag{6.326}$$

By subtracting Eq. (6.324) from Eq. (6.326), one has

$$\dot{x} - \dot{x}_d = A(x - x_d) + Bu + d_3 - d_2 \tag{6.327}$$

By denoting the tracking error as $e = x - x_d$ and the aggregate disturbance term as $L\tilde{d} = d_3 - d_2$, the tracking error dynamics becomes

$$\dot{e} = Ae + Bu + L\tilde{d} \tag{6.328}$$

where L is a disturbance inputs gain matrix. For the linearized system the cost function of Eq. (6.322) is defined, where coefficient r determines the penalization of the control input and the weight coefficient ρ determines the reward of the disturbances' effects. Then, as it will be also shown in Section 6.4.6, the optimal feedback control law is given by

$$u(t) = -Ke(t) \tag{6.329}$$

with $K = \frac{1}{r}B^TP$ where P is a positive definite symmetric matrix which is obtained from the solution of the Riccati equation

$$A^TP + PA + Q - P(\tfrac{2}{r}BB^T - \tfrac{1}{\rho^2}LL^T)P = 0 \tag{6.330}$$

where Q is also a semi-positive definite symmetric matrix. The worst-case disturbance is given by $\tilde{d}(t) = \frac{1}{\rho^2}L^TPx(t)$. The computation of the worst-case disturbance that this controller can sustain, comes from the superposition of Bellman's optimality principle when considering that the wing-rock dynamics is affected by two separate inputs, that is the control input u and the cumulative disturbance input $\tilde{d}$. From the previous relation about the worst-case disturbance, it can be seen that the smallest value of the attenuation coefficient ρ which results in an admissible solution for the methods's algebraic Riccati equation, is the one that provides the control loop with maximum robustness. The diagram of the considered control loop is depicted in Figure 6.48.

6.4.6 Lyapunov stability analysis

Through Lyapunov stability analysis it will be shown that the proposed nonlinear control scheme assures H_∞ tracking performance for the wing-rock dynamics, and that in case of bounded disturbance terms asymptotic convergence to the reference setpoints is achieved. The tracking error dynamics for the wing-rock dynamics is written in the form [725,721]:

$$\dot{e} = Ae + Bu + L\tilde{d} \tag{6.331}$$

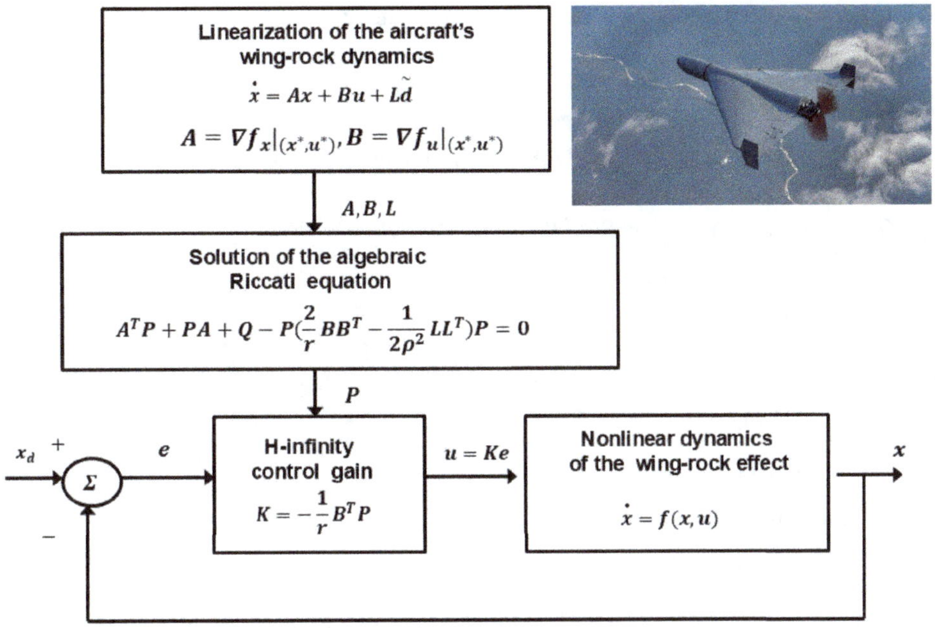

Figure 6.48 Diagram of the nonlinear optimal control loop for the aircraft's roll motion under the wing-rock effect

where in the wing-rock dynamics case $L \in R^{5\times 5}$ is the disturbance inputs gain matrix. Variable $\tilde{d}$ denotes model uncertainties and external disturbances of the wing-rock effect's model. The following Lyapunov equation is considered

$$V = \frac{1}{2}e^TPe \tag{6.332}$$

where $e = x - x_d$ is the tracking error. By differentiating with respect to time, one obtains

$$\begin{gathered}\dot{V} = \frac{1}{2}\dot{e}^TPe + \frac{1}{2}e^TP\dot{e} \Rightarrow \\ \dot{V} = \frac{1}{2}[Ae + Bu + L\tilde{d}]^TPe + \frac{1}{2}e^TP[Ae + Bu + L\tilde{d}] \Rightarrow\end{gathered} \tag{6.333}$$

$$\begin{aligned}\dot{V} = &\frac{1}{2}[e^TA^T + u^TB^T + \tilde{d}^TL^T]Pe + \\ &+\frac{1}{2}e^TP[Ae + Bu + L\tilde{d}] \Rightarrow\end{aligned} \tag{6.334}$$

$$\begin{aligned}\dot{V} = &\frac{1}{2}e^TA^TPe + \frac{1}{2}u^TB^TPe + \frac{1}{2}\tilde{d}^TL^TPe + \\ &\frac{1}{2}e^TPAe + \frac{1}{2}e^TPBu + \frac{1}{2}e^TPL\tilde{d}\end{aligned} \tag{6.335}$$

The previous equation is rewritten as

$$\begin{aligned}\dot{V} = &\frac{1}{2}e^T(A^TP + PA)e + (\frac{1}{2}u^TB^TPe + \frac{1}{2}e^TPBu) + \\ &+(\frac{1}{2}\tilde{d}^TL^TPe + \frac{1}{2}e^TPL\tilde{d})\end{aligned} \tag{6.336}$$

Assumption: For given positive definite matrix Q and coefficients r and ρ, there exists a positive definite matrix P, which is the solution of the following matrix equation:

$$A^TP + PA = -Q + P(\frac{2}{r}BB^T - \frac{1}{\rho^2}LL^T)P \tag{6.337}$$

Moreover, the following feedback control law is applied to the system:

$$u = -\frac{1}{r}B^TPe \tag{6.338}$$

By substituting Eq. (6.337) and Eq. (6.338), one obtains

$$\dot{V} = \frac{1}{2}e^T[-Q + P(\frac{2}{r}BB^T - \frac{1}{\rho^2}LL^T)P]e + e^TPB(-\frac{1}{r}B^TPe) + e^TPL\tilde{d} \Rightarrow \tag{6.339}$$

$$\dot{V} = -\frac{1}{2}e^TQe + (\frac{1}{r}e^TPBB^TPe - \frac{1}{2\rho^2}e^TPLL^TPe) - \frac{1}{r}e^TPBB^TPe + e^TPL\tilde{d} \tag{6.340}$$

which after intermediate operations gives

$$\dot{V} = -\frac{1}{2}e^TQe - \frac{1}{2\rho^2}e^TPLL^TPe + e^TPL\tilde{d} \tag{6.341}$$

or, equivalently

$$\begin{aligned}\dot{V} = &-\frac{1}{2}e^TQe - \frac{1}{2\rho^2}e^TPLL^TPe+ \\ &+\frac{1}{2}e^TPL\tilde{d} + \frac{1}{2}\tilde{d}^TL^TPe\end{aligned} \tag{6.342}$$

Lemma: The following inequality holds [721,725]

$$\frac{1}{2}e^TL\tilde{d} + \frac{1}{2}\tilde{d}L^TPe - \frac{1}{2\rho^2}e^TPLL^TPe \leq \frac{1}{2}\rho^2\tilde{d}^T\tilde{d} \tag{6.343}$$

Proof: The binomial $(\rho\alpha - \frac{1}{\rho}b)^2$ is considered. Expanding the left part of the above inequality, one gets

$$\begin{aligned}&\rho^2a^2 + \frac{1}{\rho^2}b^2 - 2ab \geq 0 \Rightarrow \frac{1}{2}\rho^2a^2 + \frac{1}{2\rho^2}b^2 - ab \geq 0 \Rightarrow \\ &ab - \frac{1}{2\rho^2}b^2 \leq \frac{1}{2}\rho^2a^2 \Rightarrow \frac{1}{2}ab + \frac{1}{2}ab - \frac{1}{2\rho^2}b^2 \leq \frac{1}{2}\rho^2a^2\end{aligned} \tag{6.344}$$

The following substitutions are carried out: $a = \tilde{d}$ and $b = e^TPL$, and the previous relation becomes

$$\frac{1}{2}\tilde{d}^TL^TPe + \frac{1}{2}e^TPL\tilde{d} - \frac{1}{2\rho^2}e^TPLL^TPe \leq \frac{1}{2}\rho^2\tilde{d}^T\tilde{d} \tag{6.345}$$

Eq. (6.345) is substituted in Eq. (6.108) and the inequality is enforced, thus giving [721,725]

$$\dot{V} \leq -\frac{1}{2}e^TQe + \frac{1}{2}\rho^2\tilde{d}^T\tilde{d} \tag{6.346}$$

Eq. (6.346) shows that the H_∞ tracking performance criterion is satisfied. The integration of $\dot{V}$ from 0 to T gives

$$\begin{aligned}&\int_0^T\dot{V}(t)dt \leq -\frac{1}{2}\int_0^T||e||_Q^2dt + \frac{1}{2}\rho^2\int_0^T||\tilde{d}||^2dt \Rightarrow \\ &2V(T) + \int_0^T||e||_Q^2dt \leq 2V(0) + \rho^2\int_0^T||\tilde{d}||^2dt\end{aligned} \tag{6.347}$$

Moreover, if there exists a positive constant $M_d > 0$ such that

$$\int_0^\infty ||\tilde{d}||^2dt \leq M_d \tag{6.348}$$

then, one gets

$$\int_0^\infty ||e||_Q^2dt \leq 2V(0) + \rho^2M_d \tag{6.349}$$

Thus, the integral $\int_0^\infty ||e||_Q^2dt$ is bounded. Moreover, $V(T)$ is bounded and from the definition of the Lyapunov function V in Eq. (6.332), it becomes clear that $e(t)$ will be also bounded since $e(t) \in \Omega_e = \{e|e^TPe \leq 2V(0) + \rho^2M_d\}$. According to the above and with the use of Barbalat's Lemma, one obtains $lim_{t\to\infty}e(t) = 0$.

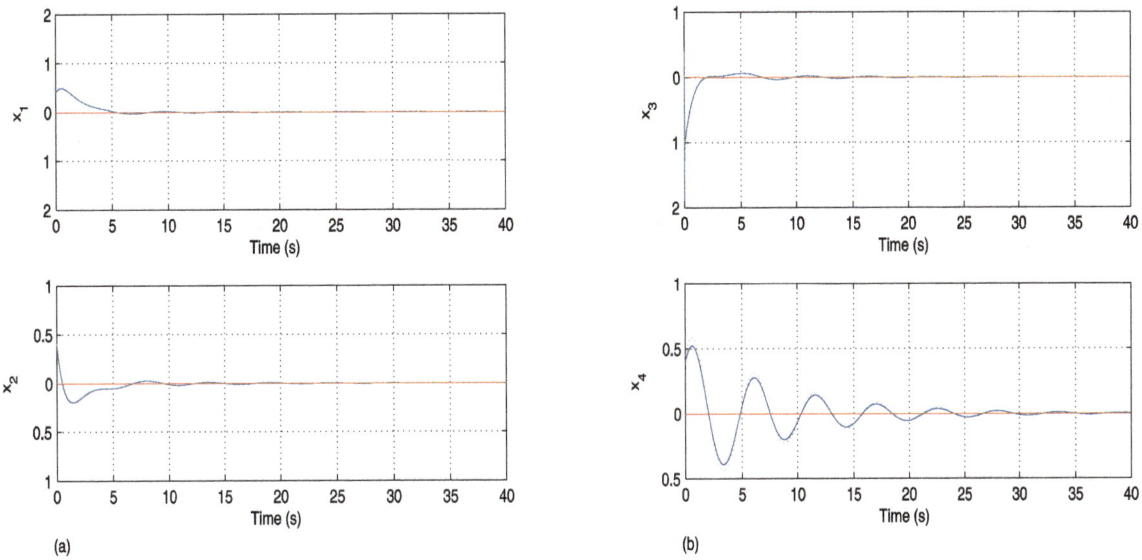

Figure 6.49 *Test 1 about the stabilization of the wing-rock dynamics with nonlinear optimal control: (a) tracking of setpoints (red line) by the state variables of the wing-rock model x_1 to x_2 (blue line) and estimates provided by the H-infinity Kalman filter; (b) tracking of setpoints (red line) by the state variables of the wing-rock model x_3 to x_4 (blue line) and estimates provided by the H-infinity Kalman filter*

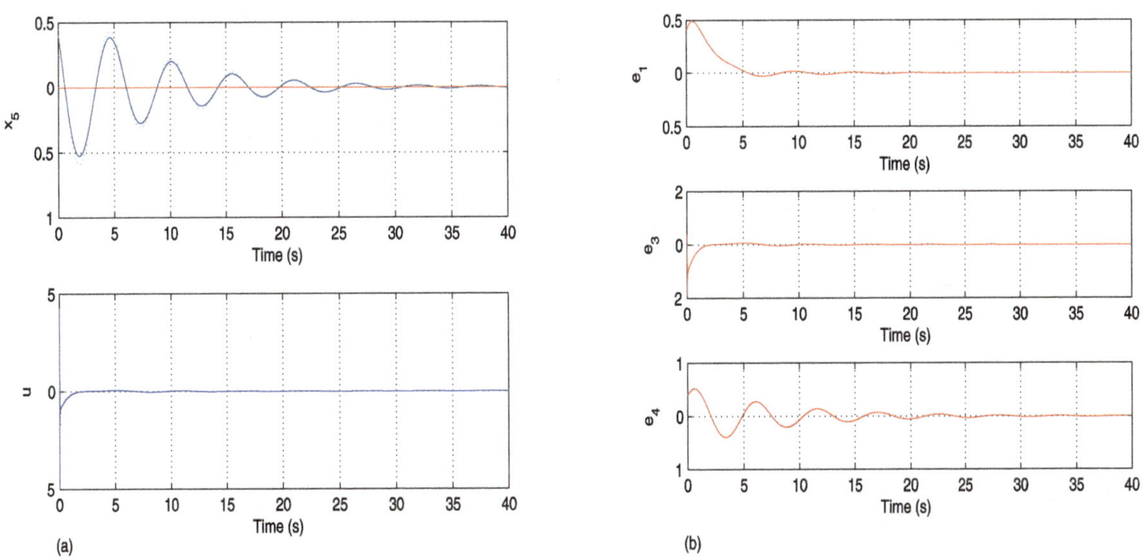

Figure 6.50 *Test 1 about the stabilization of the wing-rock dynamics with nonlinear optimal control: (a) tracking of setpoint (red line) by the state variable of the wing-rock model x_5 (blue line) and estimate provided by the H-infinity Kalman filter; (b) tracking error for the state variables of the wing-rock model x_1, x_3, and x_4 (red line)*

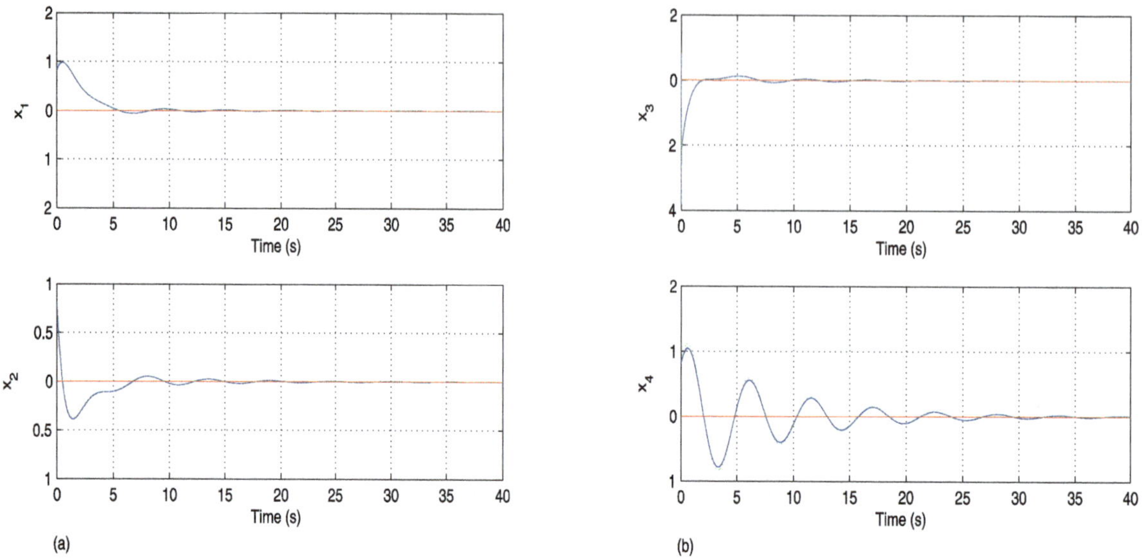

Figure 6.51 *Test 2 about the stabilization of the wing-rock dynamics with nonlinear optimal control: (a) tracking of setpoints (red line) by the state variables of the wing-rock model x_1 to x_2 (blue line) and estimates provided by the H-infinity Kalman filter; (b) tracking of setpoints (red line) by the state variables of the wing-rock model x_3 to x_4 (blue line) and estimates provided by the H-infinity Kalman filter*

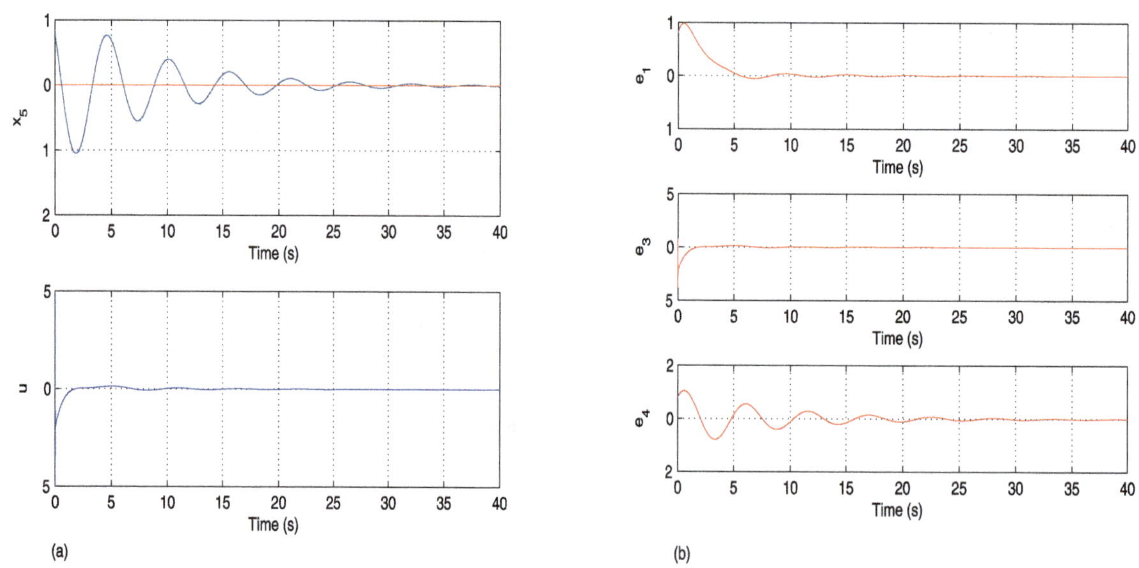

Figure 6.52 *Test 2 about the stabilization of the wing-rock dynamics with nonlinear optimal control: (a) tracking of setpoint (red line) by the state variable of the wing-rock model x_5 (blue line) and estimate provided by the H-infinity Kalman filter; (b) tracking error for the state variables of the wing-rock model x_1, x_3, and x_4 (red line)*

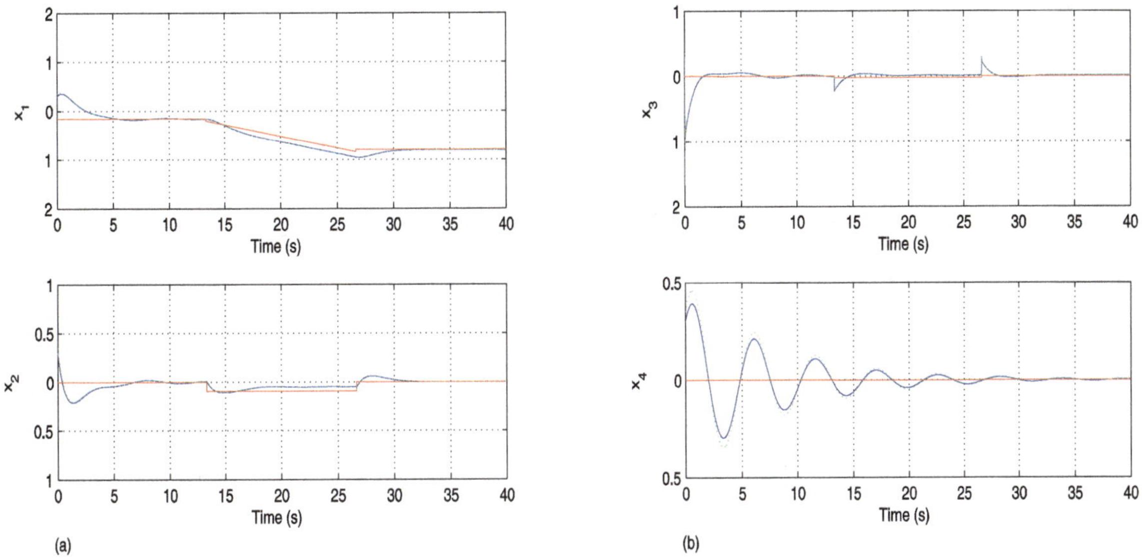

Figure 6.53 *Test 3 about the stabilization of the wing-rock dynamics with nonlinear optimal control: (a) tracking of setpoints (red line) by the state variables of the wing-rock model x_1 to x_2 (blue line) and estimates provided by the H-infinity Kalman filter; (b) tracking of setpoints (red line) by the state variables of the wing-rock model x_3 to x_4 (blue line) and estimates provided by the H-infinity Kalman filter*

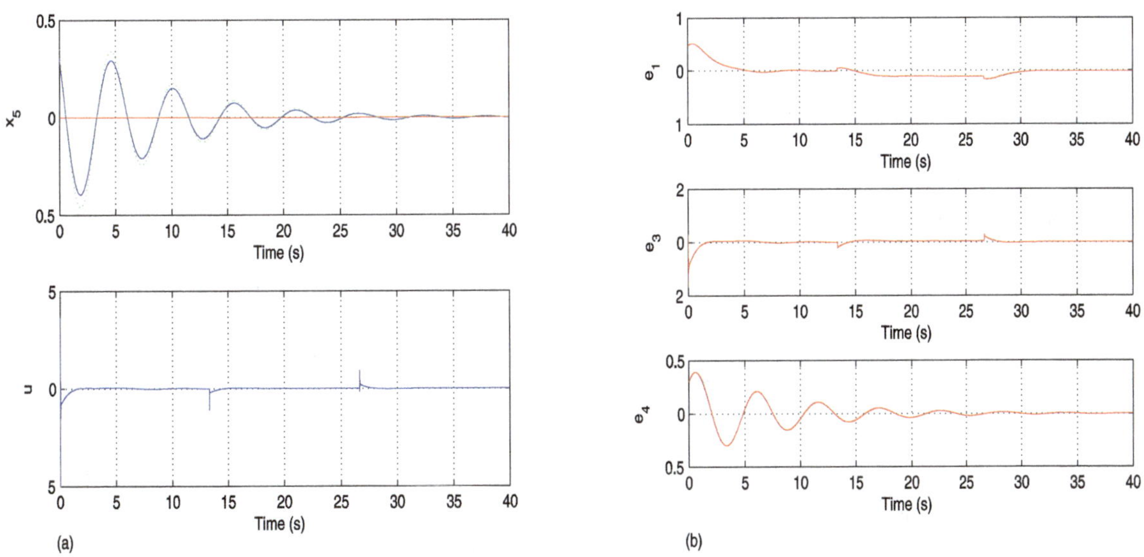

Figure 6.54 *Test 3 about the stabilization of the wing-rock dynamics with nonlinear optimal control: (a) tracking of setpoint (red line) by the state variable of the wing-rock model x_5 (blue line) and estimate provided by the H-infinity Kalman filter; (b) tracking error for the state variables of the wing-rock model x_1, x_3, and x_4 (red line)*

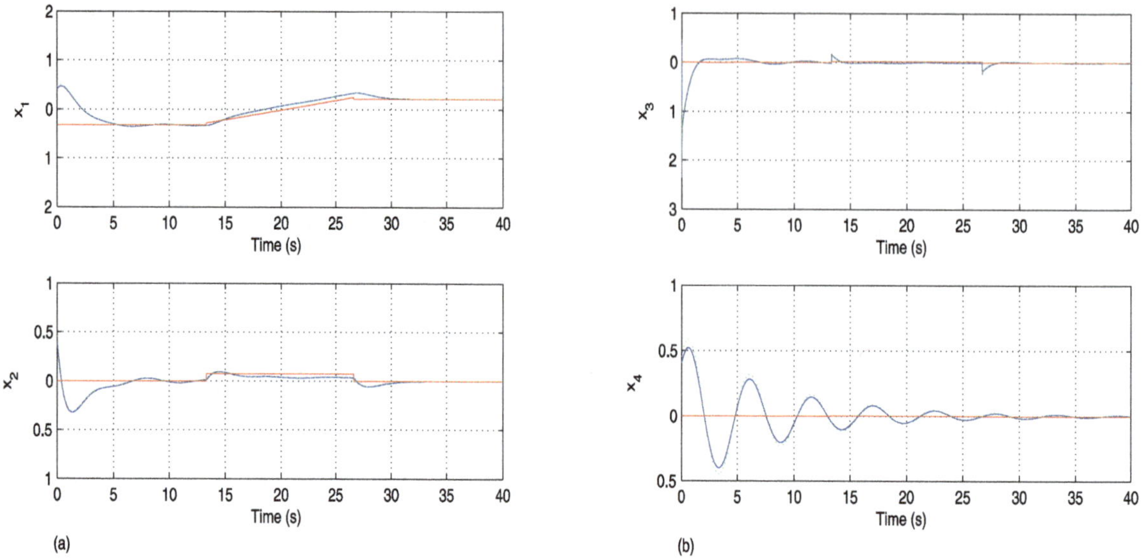

Figure 6.55 *Test 4 about the stabilization of the wing-rock dynamics with nonlinear optimal control: (a) tracking of setpoints (red line) by the state variables of the wing-rock model x_1 to x_2 (blue line) and estimates provided by the H-infinity Kalman filter; (b) tracking of setpoints (red line) by the state variables of the wing-rock model x_3 to x_4 (blue line) and estimates provided by the H-infinity Kalman filter*

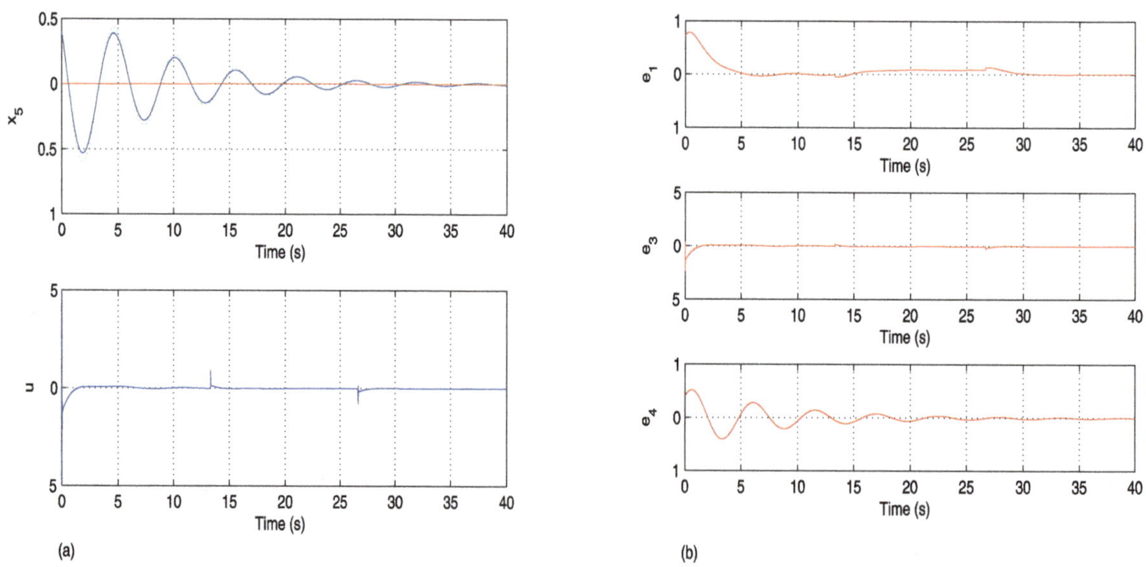

Figure 6.56 *Test 4 about the stabilization of the wing-rock dynamics with nonlinear optimal control: (a) tracking of setpoint (red line) by the state variable of the wing-rock model x_5 (blue line) and estimate provided by the H-infinity Kalman filter; (b) tracking error for the state variables of the wing-rock model x_1, x_3, and x_4 (red line)*

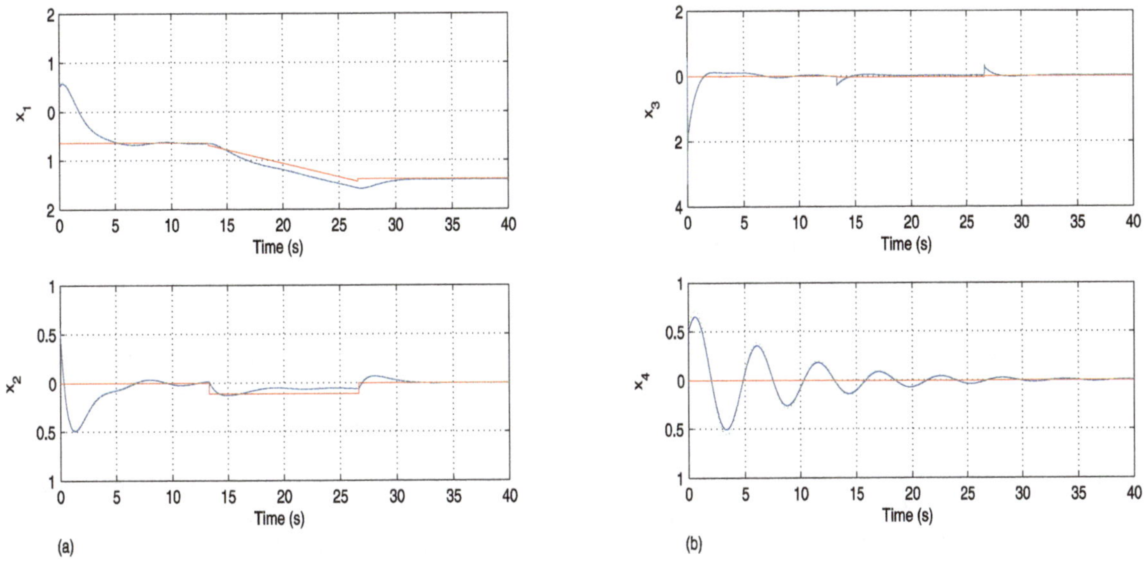

Figure 6.57 Test 5 about the stabilization of the wing-rock dynamics with nonlinear optimal control: (a) tracking of setpoints (red line) by the state variables of the wing-rock model x_1 to x_2 (blue line) and estimates provided by the H-infinity Kalman filter; (b) tracking of setpoints (red line) by the state variables of the wing-rock model x_3 to x_4 (blue line) and estimates provided by the H-infinity Kalman filter

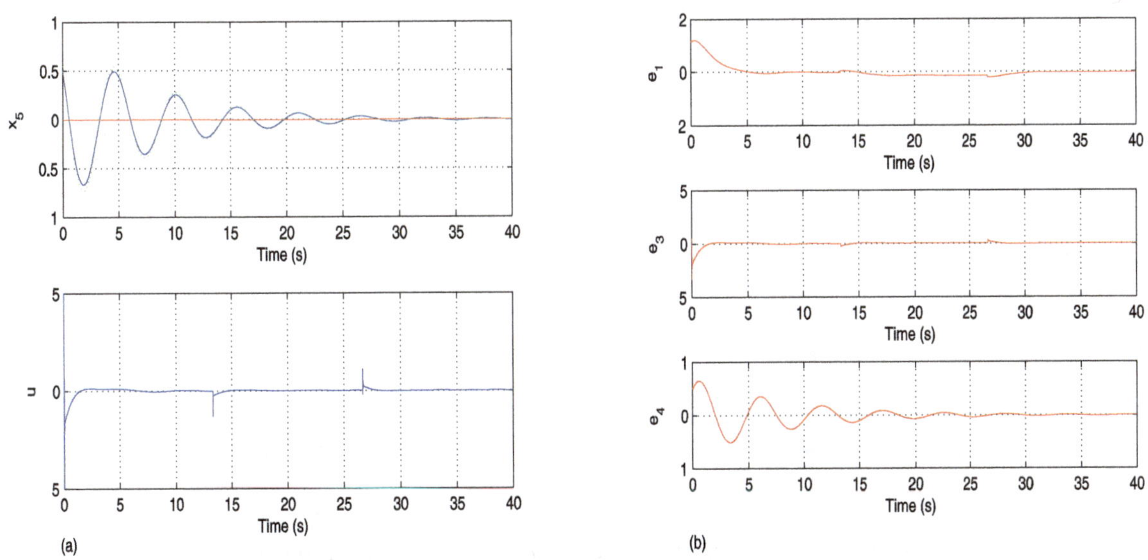

Figure 6.58 Test 5 about the stabilization of the wing-rock dynamics with nonlinear optimal control: (a) tracking of setpoint (red line) by the state variable of the wing-rock model x_5 (blue line) and estimate provided by the H-infinity Kalman filter; (b) tracking error for the state variables of the wing-rock model x_1, x_3, and x_4 (red line)

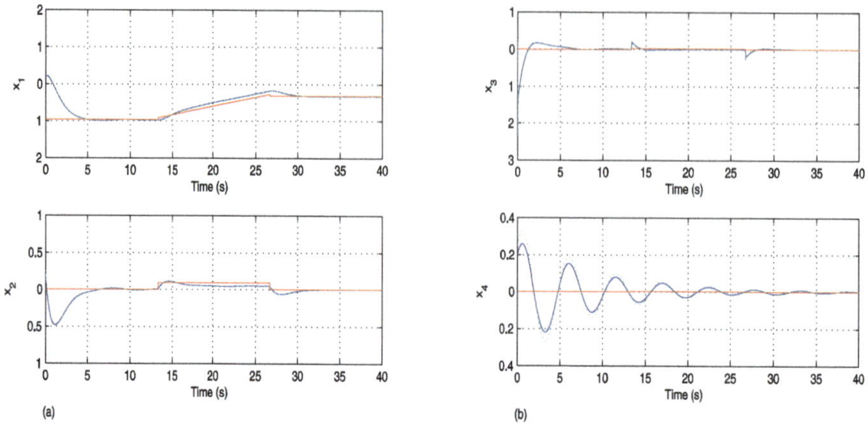

Figure 6.59 *Test 6 about the stabilization of the wing-rock dynamics with nonlinear optimal control: (a) tracking of setpoints (red line) by the state variables of the wing-rock model x_1 to x_2 (blue line) and estimates provided by the H-infinity Kalman filter; (b) tracking of setpoints (red line) by the state variables of the wing-rock model x_3 to x_4 (blue line) and estimates provided by the H-infinity Kalman filter*

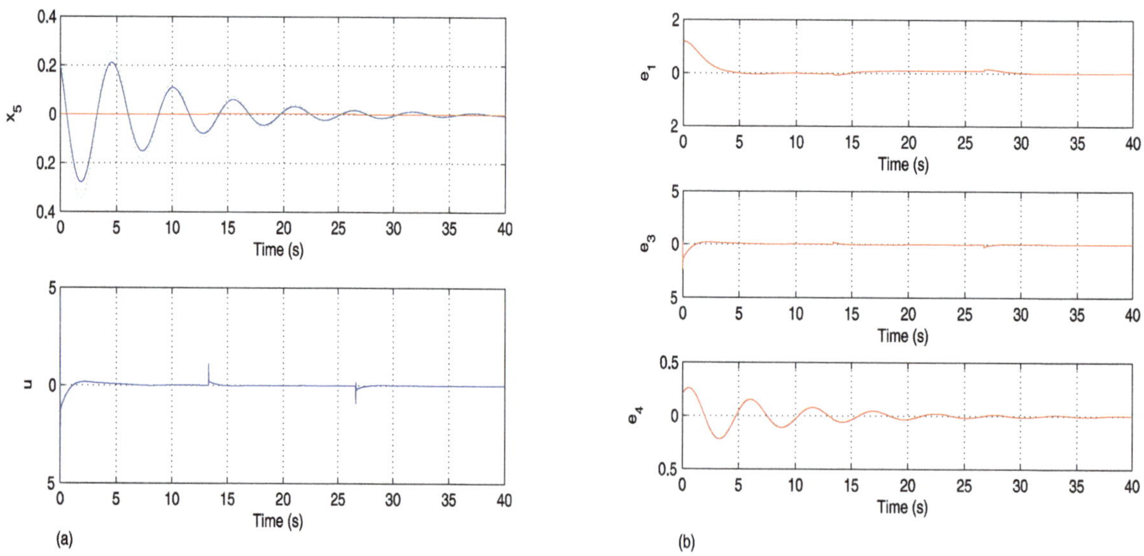

Figure 6.60 *Test 6 about the stabilization of the wing-rock dynamics with nonlinear optimal control: (a) tracking of setpoint (red line) by the state variable of the wing-rock model x_5 (blue line) and estimate provided by the H-infinity Kalman filter; (b) tracking error for the state variables of the wing-rock model x_1, x_3, and x_4 (red line)*

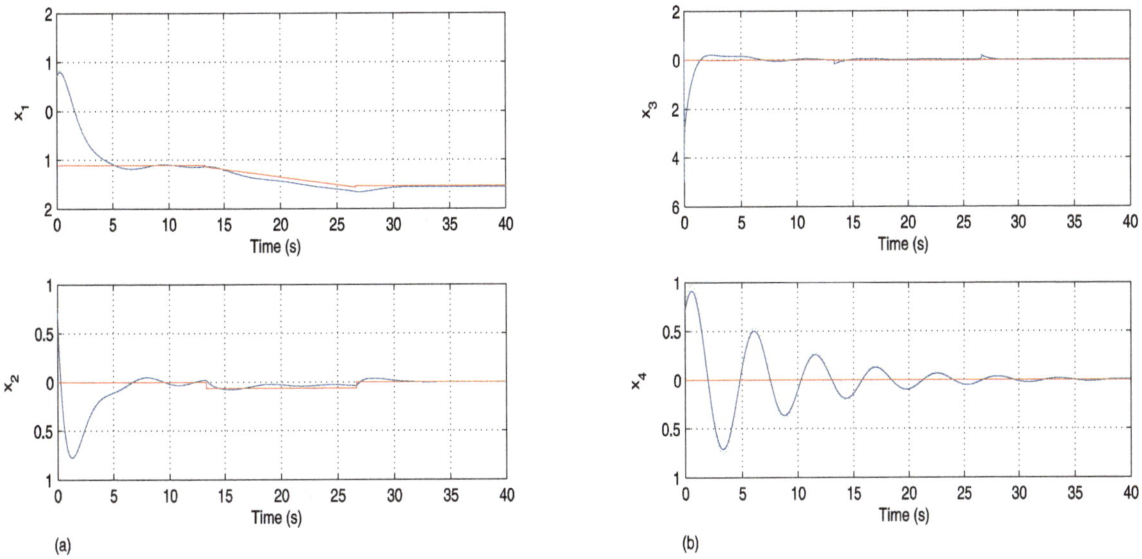

Figure 6.61 Test 7 about the stabilization of the wing-rock dynamics with nonlinear optimal control: (a) tracking of setpoints (red line) by the state variables of the wing-rock model x_1 to x_2 (blue line) and estimates provided by the H-infinity Kalman filter; (b) tracking of setpoints (red line) by the state variables of the wing-rock model x_3 to x_4 (blue line) and estimates provided by the H-infinity Kalman filter

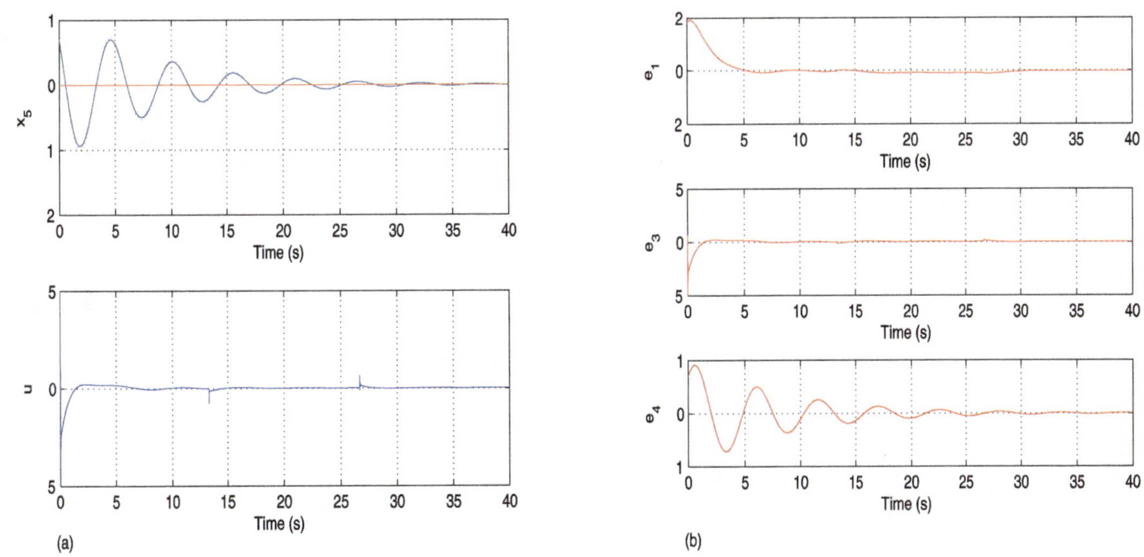

Figure 6.62 Test 7 about the stabilization of the wing-rock dynamics with nonlinear optimal control: (a) tracking of setpoint (red line) by the state variable of the wing-rock model x_5 (blue line) and estimate provided by the H-infinity Kalman filter; (b) tracking error for the state variables of the wing-rock model x_1, x_3, and x_4 (red line)

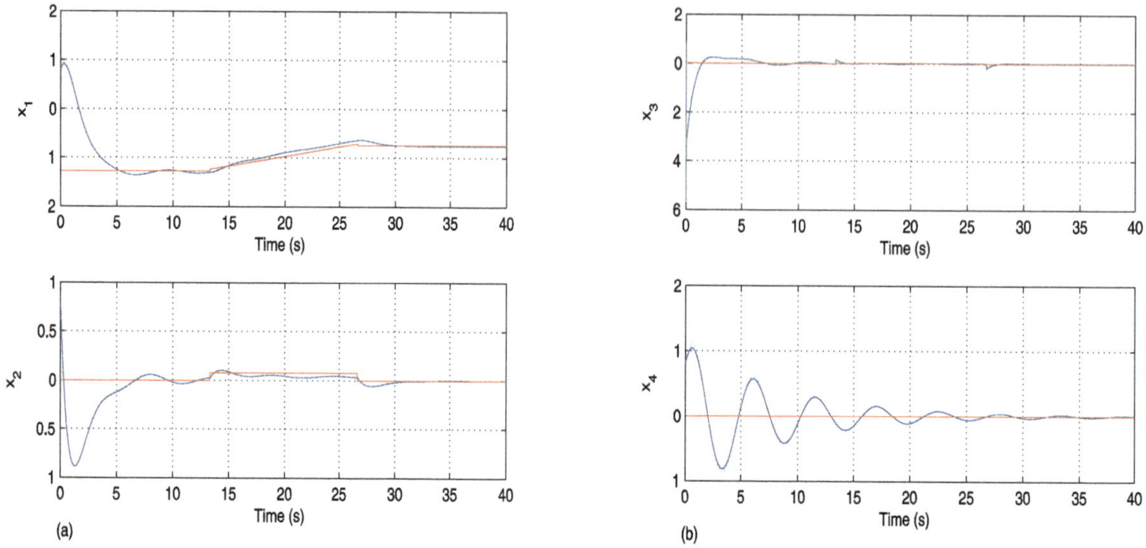

Figure 6.63 Test 8 about the stabilization of the wing-rock dynamics with nonlinear optimal control: (a) tracking of setpoints (red line) by the state variables of the wing-rock model x_1 to x_2 (blue line) and estimates provided by the H-infinity Kalman filter; (b) tracking of setpoints (red line) by the state variables of the wing-rock model x_3 to x_4 (blue line) and estimates provided by the H-infinity Kalman filter

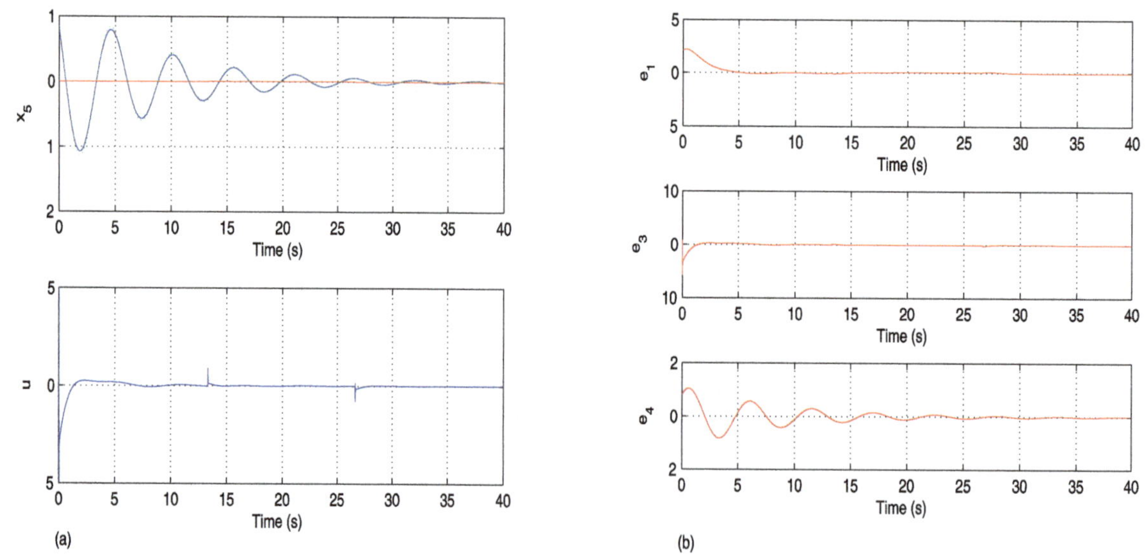

Figure 6.64 Test 8 about the stabilization of the wing-rock dynamics with nonlinear optimal control: (a) tracking of setpoint (red line) by the state variable of the wing-rock model x_5 (blue line) and estimate provided by the H-infinity Kalman filter; (b) tracking error for the state variables of the wing-rock model x_1, x_3, and x_4 (red line)

Through the stages of the stability proof, one arrives at Eq. (6.346) which shows that the H-infinity tracking performance criterion holds. By selecting the attenuation coefficient ρ to be sufficiently small and in particular to satisfy $\rho^2 < \|e\|^2_Q/\|\tilde{d}\|^2$, one has that the first derivative of the Lyapunov function is upper bounded by 0. This condition holds at each sampling instance, and consequently, global stability for the control loop can be concluded.

6.4.7 *Simulation tests*

The performance of the nonlinear optimal control scheme for the wing-rock dynamics is also confirmed through simulation experiments. To implement the nonlinear optimal control method, the dynamic model of the wing-rock motion had to be successively linearized at each time step of the control algorithm and the Jacobian matrices of this model had to be computed. Moreover, the algebraic Riccati equation of Eq. (6.337) had to be solved successively at each sampling period using the *aresolv()* function in MATLAB®. The values of the parameters of the wing-rock dynamic model have been taken from [1097]. Using the notation a_i, $i = 1, \cdots, 5$ about the aerodynamic coefficients which change with the aircraft's angle of attack the following parameter values have been used: $\mu_1 :\ 0.354a_2 - 0.001$, $\mu_2 :\ 0.354a_4$, $b_1 :\ 0.354a_3$, $b_2 :\ 0.354a_5$, $\omega^2 :\ -0.354a_1$, $L_\delta :\ 1$, $L_\beta :\ -0.02822$, $L_r :\ 0.1517$, $\tau : 0.0495$, $N_p :\ -0.0629$, $N_\beta : 1.3214$, $N_r : -0.2491$. The obtained results are depicted in Figures 6.49–6.64. The real values of the state variables of the wing-rock dynamics are printed in blue color, the associated setpoints are plotted in red color while the estimates of the state variables which were provided by the H-infinity Kalman filter are depicted in green color. It can be noticed that in all simulation tests the proposed nonlinear optimal control method achieved the elimination of the wing-rock oscillations while the variations of the control inputs remained moderate.

Chapter 7
Nonlinear optimal and flatness-based control for mechatronics and power electronics

Abstract

The chapter analyzes nonlinear optimal control and multi-loop flatness-based control for mechatronic systems and for power electronics. First, it solves the nonlinear optimal and flatness-based control problem for the dynamic model of induction motor-driven desalination units. Moreover, it develops a multi-loop flatness-based control method for centrifugal gas compressors driven by induction motors and permanent magnet synchronous motors. Besides, it introduces nonlinear optimal control for wave energy conversion units which consist of a permanent magnet linear synchronous generator connected to a three-phase AC/DC converter. Additionally, it analyzes nonlinear optimal control and multi-loop flatness-based control for the dynamic model of 5-DOF active magnetic bearings. Next, a differential flatness and Kalman filter-based disturbance observer is developed for the fault diagnosis problem of 5-DOF active magnetic bearings. Finally, a differential flatness and Kalman filter-bases disturbance observer is proposed for the fault diagnosis problem of the VSC-HVDC transmission system in the electric power grid.

7.1 Nonlinear optimal and multi-loop flatness-based control for IM-driven desalination

7.1.1 Outline

The present section shows also the use of its nonlinear optimal and flatness-based control methods in mechatronic systems and in power electronics. The scarcity of water resources worldwide and also in the Mediterranean region has been the reason for the wide development of several new desalination plants of large processing capacity. Several littoral zones of the Mediterranean cover the water supply needs of their population with large quantities of desalinated water which are produced on a daily basis by reverse-osmosis desalination plants. Electrically actuated centrifugal pumps are used to feed with sea water the desalination process [747,635,703,161,441]. The main stages of the process are described as follows: (i) sea water is pumped into the desalination units with the use of variable-speed electrically actuated centrifugal pumps (ii) the sea water undergoes a first level of crude processing in which unclear water is returned to sea, while transparent and phenomenally clear water is transferred to reverse-osmosis membranes, and (iii) Water passing through membranes is desalinated and thus becomes potable, while the quantity of water that is retained by the membranes is still of elevated salinity and thus unsuitable for drinking. This unsuitable for drinking quantity of water (brine) is either rejected to the sea or is forwarded to subsequent processing stages through more membranes until it is sufficiently purified [1029,58,57,811]. The desalination process with electrically actuated pumps and valves is controlled through the voltage inputs to the pumps' electric motors and to the actuators of valves [609,660,12,983]. The process is energy consuming, and if motors and actuators are tuned to function sub-optimally, there can be significant energy dispersion which also raises the cost of production of desalinated water [571,171,108,550]. There exist several results on control of electric motors and actuators in desalination plants of the type of PID control and MPC control with the confirmation of the associated and stability properties to be a challenging research topic [425,610,570,435,479]. Power supply from microgrids and optimized power management in desalination units can reduce the cost of the desalination process and can make affordable the large-scale water supply from desalination plants [752,38,201,69]. Furthermore, the use of renewable energy sources for power supply in desalination units is a promising solution for achieving the functioning of desalination plants at zero carbon emissions and at no environmental cost [611,685,856,696,775].

The present section aims at developing a nonlinear optimal control method for induction motor-driven reverse-osmosis desalination plants which will achieve minimization of energy dispersion and lowering of the functioning cost of desalination units [721,725]. The method will also guarantee global stability of the related control loops, thus ensuring the reliable and uninterrupted functioning of desalination units. It is proven that the dynamic model of an electrically actuated desalination unit is differentially flat [711,492,806]. Next, to solve the associated nonlinear optimal control problem, the dynamic model of the induction motor-driven reverse-osmosis desalination unit undergoes approximate linearization with the use of first-order Taylor series expansion and through the computation of the associated Jacobian matrices [705,60,706]. The linearization point is updated at each sampling instance and is defined by the present value of the system's state vector and by the last sampled value of the control inputs vector. The modeling error which is due to the truncation of higher-order terms from the Taylor-series expansion is considered to be a perturbation which is asymptotically compensated by the robustness of the control method. For the approximately linearized model of the system, a stabilizing H-infinity controller is defined. To compute the feedback gains of this controller an algebraic Riccati equation has to be repetitively solved at each time step of the control algorithm [876,741]. The global stability properties of the control scheme are proven through Lyapunov analysis. First, it is demonstrated that the control loop satisfies the H-infinity tracking performance criterion. Moreover, under moderate conditions, it is shown that the control method is globally asymptotically stable. Additionally, to implement state estimation-based control without the need to measure the entire state vector of the desalination process, the H-infinity Kalman filter is used as a robust state estimator.

Additionally, this section proposes a second solution about the nonlinear control problem of the induction motor-driven reverse-osmosis process, which is a flatness-based control method implemented in successive loops [712,715]. The state-space model of the electrically actuated desalination plant is decomposed into three subsystems which are connected in chained form. It is proven that each one of these subsystems, if viewed independently, is a differentially flat system which can be written in the input–output linearized form. As a result, stabilizing feedback control for each one of these subsystems can be designed by dynamics inversion [640,524,54,732]. In this method, the state vector of the subsequent ($i+1$-th) subsystem becomes a virtual control input to the preceding (i-th) subsystem. Equivalently, it can be considered that the virtual control input of the preceding (i-th) subsystem becomes setpoint for the subsequent ($i+1$-th) subsystem. The real control inputs of the integrated system (that is the input voltages of valves and the input voltages of the induction motor) are computed by tracing backward the virtual control inputs of the individual subsystems. The global stability properties of the flatness-based control method in successive loops are also confirmed through Lyapunov analysis. The method is characterized by simplicity because, in the control loop of each subsystem, it suffices to assign positive values to the diagonal elements of the associated diagonal feedback gain matrices. These values determine also the transient performance of the individual control loops.

7.1.2 Dynamics of the induction motor-driven reverse osmosis desalination unit

7.1.2.1 State-space model of the induction motor-driven desalination unit

In this section's dynamic model of the desalination unit, key state variables are (i) the turn speed of the of induction motor which is proportional to the feed rate of sea water into the desalination process, (ii) the preliminary rejection rate of unclear sea water, (iii) the rate of accumulation of brine water by the membranes, and (iv) the rotor magnetic flux and stator current variables of the induction motor. The difference between the sea-water feed rate and the rejection rates of unclear and unsuitable for drinking water defines the volume of the produced potable water. The control inputs to this system are (i) the aperture of the valve which allows rejection of the brine water that has been retained by the membranes and (ii) the stator voltage inputs of the induction motor. The state-space model of the desalination process is multi-variable and strongly nonlinear and the solution of the associated control problem is a non-trivial task. The diagram of the induction motor-driven reverse osmosis desalination unit is given in Figure 7.1.

The dynamic model of the reverse osmosis desalination process consists of two differential equations [635,57]:

$$\frac{dv_b}{dt} = \frac{A_P^2}{A_m K_m V}[v_f - v_b - v_r] + \frac{A_P}{\rho V}\Delta\pi - \frac{1}{2}\frac{A_P e_{vb} v_b^2}{V} \tag{7.1}$$

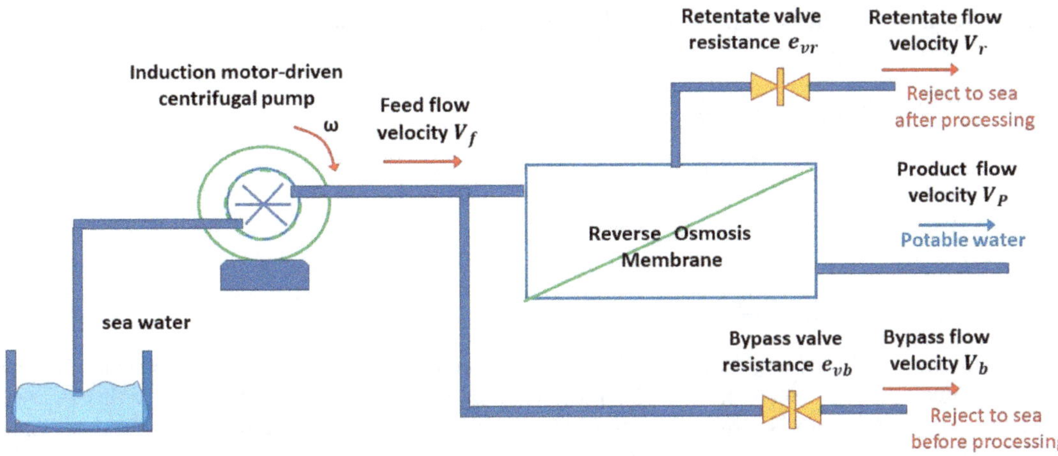

Figure 7.1 Diagram of the induction motor-driven reverse osmosis desalination unit

$$\frac{dv_r}{dt} = \frac{A_p^2}{A_m K_m V}[v_f - v_b - v_r] + \frac{A_p}{\rho V}\Delta\pi - \frac{1}{2}\frac{A_P e_{vr} v_r^2}{V} \tag{7.2}$$

Moreover, the model of the reverse osmosis is completed by the following algebraic equations [635,57]:

$$\Delta\pi = \delta C_{eff}(T + 273) \Rightarrow \tag{7.3}$$

$$C_{eff} = C_{feed}[a + (1 - a)\frac{(1-R)+R(v_f - v_b)}{v_r}] \tag{7.4}$$

$$v_p = \frac{A_m K_m}{\rho A_p}(P_{sys} - \Delta_\pi) \tag{7.5}$$

$$P_{sys} = \frac{\rho A_p}{A_m K_m}(v_f - v_b - v_r) + \Delta\pi \tag{7.6}$$

$$\Delta\pi = \delta C_{feed}[a(T + 273)] + \delta C_{feed}(1 - a) \cdot \frac{(1-R)+R(v_f - v_b)}{v_r} \tag{7.7}$$

and by defining coefficients $c_1 = \delta C_{feed}[a(T + 273)]$ and $d_1 = \delta C_{feed}(1 - a)$ Eq. (7.7) is written as

$$\Delta\pi = c_1 + d_1 \frac{(1-R)+R(v_f - v_b)}{v_r} \tag{7.8}$$

The parameters of the reverse-osmosis dynamic model are defined as follows: ρ is the fluid density, V is the system's volume, v_f is the feed velocity, A_p is the tube's cross-sectional area, A_m is the membrane's area, K_m is the membrane mass transfer coefficient, C_{feed} is the concentration of total dissolved solids in the feed, a is the effective concentration weighting factor, δ is a constant relating effective concentration to osmotic pressure, v_b is the bypass flow velocity, v_r is the flow velocity of the water that has been retained by the membrane (brine), P_{sys} is the system's pressure, R is the fractional salt rejection of the membrane, e_{vb} is the bypass valve resistance, e_{vr} is the retentate valve resistance.

The values of e_{vb} and e_{vr} denote the opening or closure of the two valves. Small value of the valve resistance signifies that the tube is open. Large value of the valve resistance means that the tube is obstructed.

Under the assumption of field orientation, the dynamic model of the induction motor is given by [711,715]:

$$\begin{aligned}
\dot{\theta} &= \omega \\
\dot{\omega} &= \frac{\mu}{J}\psi_{rd}i_{sq} - \frac{T_L}{J} - \frac{T_b}{J} \\
\dot{\psi}_{rd} &= -a\psi_{rd} + aMi_{s_d} \\
\dot{i}_{s_d} &= -\gamma i_{s_d} + \alpha\beta\psi_{r_d} + n_p\omega i_{s_q} + \frac{aMi_{s_q}^2}{\psi_{rd}} + \frac{1}{\sigma L_s}v_{sd} \\
\dot{i}_{s_q} &= -\gamma i_{s_q} - \beta n_p\omega\psi_{r_d} - n_p\omega i_{sd} - \frac{\alpha M i_{s_q} i_{s_d}}{\psi_{rd}} + \frac{1}{\sigma L_s}v_{sq} \\
\dot{\rho} &= n_p\omega + \frac{\alpha M i_{s_q}}{\psi_{rd}}
\end{aligned} \tag{7.9}$$

where θ, ω are the turn angle and angular speed of the rotor, respectively, ψ_{r_d} is the d-axis component of the magnetic flux of the rotor (due to field orientation, about the q-axis component of the magnetic flux it holds that $\psi_{r_q} = 0$), i_{s_d} is the d-axis component of the stator's currents vector, i_{s_q} is the q-axis component of the stator's currents vector, and ρ is the orientation angle of the magnetic field of the motor which changes at an asynchronous rate. In the dynamic model of the induction motor parameters α, β, γ, μ. L_s. M, σ are associated with the coefficients of the electric circuit of the stator and rotor (resistance, inductance, mutual inductance, magnetic permeability, and number of poles).

Next, using that the feed rate of the reverse-osmosis process is proportional to the turn speed of the induction motor (and also proportional to the turn speed of the centrifugal pump which is actuated by the motor), one has the relation $v_f = k_f\omega$. Moreover, regarding the orientation angle ρ of the magnetic field of the induction motor, this is determined by the controlled state variables of the motor ω, ψ_{r_d}, and i_{s_q} without appearing in the state equations of these state variables. Therefore, the last row of the dynamic model of the induction motor of Eq. (7.9) can be omitted from the integrated model of the induction motor-driven reverse-osmosis process without affecting the solution of the associated control problem.

Using the dynamics of the reverse-osmosis process of Eq. (7.1) and Eq. (7.2) jointly with the dynamics of the induction motor which is given in Eq. (7.9), and by substituting the friction torque of the induction motor by $T_b = k_b\omega$ and by considering that T_L is an additional load torque that may affect the rotation of the motor, one obtains the state equations of the integrated dynamical system of the induction motor-driven reverse osmosis process.

$$\dot{v}_b = \frac{A_p^2}{A_m K_m V}(k_f\omega - v_b - v_r) + \frac{A_P}{\rho V}[c_1 + d_1\frac{(1-R)+R(k_f\omega - v_b)}{v_r}] - \frac{1}{2}\frac{A_p e_{vb} v_b^2}{V} \tag{7.10}$$

$$\dot{v}_r = \frac{A_p^2}{A_m K_m V}(k_f\omega - v_b - v_r) + \frac{A_P}{\rho V}[c_1 + d_1\frac{(1-R)+R(k_f\omega - v_b)}{v_r}] - \frac{1}{2}\frac{A_p e_{vr} v_r^2}{V} \tag{7.11}$$

$$\dot{\omega} = \frac{\mu}{J}\psi_{r_d}i_{sq} - \frac{T_L}{J} - \frac{k_b\omega}{J} \tag{7.12}$$

$$\dot{\psi}_{rd} = -a\psi_{rd} + aMi_{s_d} \tag{7.13}$$

$$\dot{i}_{s_d} = -\gamma i_{s_d} + \alpha\beta\psi_{r_d} + n_p\omega i_{s_q} + \frac{aMi_{s_q}^2}{\psi_{rd}} + \frac{1}{\sigma L_s}v_{sd} \tag{7.14}$$

$$\dot{i}_{s_q} = -\gamma i_{s_q} - \beta n_p\omega\psi_{r_d} - n_p\omega i_{sd} - \frac{\alpha M i_{s_q} i_{s_d}}{\psi_{rd}} + \frac{1}{\sigma L_s}v_{sq} \tag{7.15}$$

Next, the following state variables and control inputs are defined

$$x = [x_1, x_2, x_3, x_4, x_5, x_6]^T = [v_b, v_r, \omega, \psi_{r_d}, i_{s_d}, i_{s_q}]^T \tag{7.16}$$

$$u = [u_1, u_2, u_3]^T = [e_{vr}, v_{s_d}, v_{s_q}]^T \tag{7.17}$$

Thus, the dynamic model of the induction motor-driven reverse osmosis process is written in the following state-space form:

$$\dot{x}_1 = \frac{A_p^2}{A_m K_m V}(k_f x_3 - x_1 - x_2) + \frac{A_P}{\rho V}[c_1 + d_1\frac{(1-R)+R(k_f x_3 - x_1)}{x_2}] - \frac{1}{2}\frac{A_p e_{vb} x_1^2}{V} \tag{7.18}$$

$$\dot{x}_2 = \frac{A_p^2}{A_m K_m V}(k_f x_3 - x_1 - x_2) + \frac{A_P}{\rho V}[c_1 + d_1\frac{(1-R)+R(k_f x_3 - x_1)}{x_2}] - \frac{1}{2}\frac{A_p x_2^2 u_1}{V} \tag{7.19}$$

$$\dot{x}_3 = \frac{\mu}{J}x_4x_6 - \frac{T_L}{J} - \frac{k_bx_3}{J} \tag{7.20}$$

$$\dot{x}_4 = -ax_4 + aMx_5 \tag{7.21}$$

$$\dot{x}_5 = -\gamma x_5 + \alpha\beta x_4 + n_px_3x_6 + \frac{aMx_6^2}{x_4} + \frac{1}{\sigma L_s}u_2 \tag{7.22}$$

$$\dot{x}_6 = -\gamma x_6 - \beta n_px_3x_4 - n_px_3x_5 - \frac{\alpha Mx_6x_5}{x_4} + \frac{1}{\sigma L_s}u_3 \tag{7.23}$$

In matrix form the dynamic model of the induction motor-driven reverse osmosis process is written as

$$\begin{pmatrix}\dot{x}_1\\ \dot{x}_2\\ \dot{x}_3\\ \dot{x}_4\\ \dot{x}_5\\ \dot{x}_6\end{pmatrix} = \begin{pmatrix}\frac{A_P^2}{A_mK_mV}(k_fx_3 - x_1 - x_2) + \frac{A_P}{\rho V}[c_1 + d_1\frac{(1-R)+R(k_fx_3-x_1)}{x_2}] - \frac{1}{2}\frac{A_pe_{vb}x_1^2}{V}\\ \frac{A_P^2}{A_mK_mV}(k_fx_3 - x_1 - x_2) + \frac{A_P}{\rho V}[c_1 + d_1\frac{(1-R)+R(k_fx_3-x_1)}{x_2}]\\ \frac{\mu}{J}x_4x_6 - \frac{T_L}{J} - \frac{k_bx_3}{J}\\ -ax_4 + aMx_5\\ -\gamma x_5 + \alpha\beta x_4 + n_px_3x_6 + \frac{aMx_6^2}{x_4}\\ -\gamma x_6 - \beta n_px_3x_4 - n_px_3x_5 - \frac{\alpha Mx_6x_5}{x_4}\end{pmatrix} + \begin{pmatrix}0 & 0 & 0\\ -\frac{1}{2}\frac{A_px_2^2}{V} & 0 & 0\\ 0 & 0 & 0\\ 0 & 0 & 0\\ 0 & \frac{1}{\sigma L_s} & 0\\ 0 & 0 & \frac{1}{\sigma L_s}\end{pmatrix}\begin{pmatrix}u_1\\ u_2\\ u_3\end{pmatrix} \tag{7.24}$$

Equivalently, the dynamic model of the desalination unit is written in the nonlinear affine-in-the-input state-space form

$$\dot{x} = f(x) + g(x)u \tag{7.25}$$

where $x\in R^{6\times 1}$, $f(x)\in R^{6\times 1}$, $g(x)\in R^{6\times 3}$, and $u\in R^{3\times 1}$.

7.1.2.2 Differential flatness properties of the induction motor-driven desalination unit

The dynamic model of the induction motor-driven reverse osmosis process given in Eq. (7.25) is differentially flat [711,715]. Its flat outputs vector is

$$Y = [y_1, y_2, y_3]^T = [x_1, x_3, x_4]^T = [v_b, \omega, \psi_{r_d}]^T \tag{7.26}$$

From Eq. (7.18) of the state-space model, one solves for x_2. This gives

$$\dot{x}_1 = \frac{A_P^2}{A_mK_mV}(k_fx_3 - x_1 - x_2) + \frac{A_P}{\rho V}[c_1 + d_1\frac{(1-R)+R(k_fx_3-x_1)}{x_2}] - \frac{1}{2}\frac{A_pe_{vb}x_1^2}{V} \tag{7.27}$$

By regrouping terms, one obtains

$$\dot{x}_1 - \frac{A_P^2}{A_mK_mV}(k_fx_3 - x_1) - \frac{A_P}{\rho V}c_1 + \frac{1}{2}\frac{A_pe_{vb}x_1^2}{V} = -\frac{A_P^2}{A_mK_mV}x_2 + \frac{A_P}{\rho V}d_1\frac{(1-R)+R(k_fx_3-x_1)}{x_2} \tag{7.28}$$

Next, one defines the following functions and $f_a(x_1, x_3)$, $f_b(x_1, x_3)$, and coefficient q

$$f_a(x_1, x_3) = \dot{x}_1 - \frac{A_P^2}{A_mK_mV}(k_fx_3 - x_1) - \frac{A_P}{\rho V}c_1 + \frac{1}{2}\frac{A_pe_{vb}x_1^2}{V}$$

$$f_b(x_1, x_3) = \frac{A_P}{\rho V}d_1(1 - R) + R(k_fx_3 - x_1) \tag{7.29}$$

$$q = -\frac{A_P^2}{A_mK_mV}$$

This gives a binomial relation for x_2

$$f_a(x_1,x_3) = qx_2 + f_b(x_1,x_3)\frac{1}{x_2} \Rightarrow$$
$$qx_2^2 - f_a(x_1,x_3)x_2 + f_b(x_1,x_3) = 0 \tag{7.30}$$

Since x_1, x_3 have already been defined as flat outputs of the system they are allowed to vary in the entire positive real numbers axis, including those regions where an admissible solution x_2 of the above-noted binomial exists. In conclusion, one has that

$$x_2 = h_2(Y, \dot{Y}) \tag{7.31}$$

which signifies that state variable x_2 is a differential function of the flat outputs vector of the system. Next, from the second row of the state-space model, one solves for control input u_1

$$u_1 = -\frac{\dot{x}_2 - \frac{A_p^2}{A_m K_m V}(k_f x_3 - x_1 - x_2) - \frac{A_p}{\rho V}[c_1 + \frac{d_1(1-R)+Rx_1}{x_2}]}{\frac{A_p x_2^2}{2V}} \Rightarrow u_1 = h_{u_1}(Y, \dot{Y}) \tag{7.32}$$

which signifies that u_1 is a differential function of the flat outputs of the system. Moreover, from Eq. (7.20), one solves for x_6. This gives

$$x_6 = \frac{J\dot{x}_3 + T_L + k_b x_3}{\mu x_4} \Rightarrow x_6 = h_6(Y, \dot{Y}) \tag{7.33}$$

which signifies that x_6 is a differential function of the flat outputs of the system. Moreover, from Eq. (7.21), one solves for x_5. This gives

$$x_5 = \frac{\dot{x}_4 + ax_4}{aM} \Rightarrow x_5 = h_5(Y, \dot{Y}) \tag{7.34}$$

which signifies that x_5 is a differential function of the flat outputs of the system. Moreover, from Eq. (7.22), one solves for u_2. This gives

$$u_2 = \sigma L_s[\dot{x}_5 + \gamma x_5 - \alpha\beta x_4 - n_p x_3 x_6 - \frac{\alpha M x_6^2}{x_4}] \Rightarrow u_2 = h_{u_2}(Y, \dot{Y}) \tag{7.35}$$

which signifies that u_2 is a differential function of the flat outputs of the system. Moreover, from Eq. (7.23), one solves for u_3. This gives

$$u_3 = \sigma L_s[\dot{x}_6 + \gamma x_6 - \beta n_p x_3 x_4 - n_p x_3 x_5 - \frac{\alpha M x_5 x_6}{x_4}] \Rightarrow u_3 = h_{u_3}(Y, \dot{Y}) \tag{7.36}$$

which signifies that u_3 is a differential function of the flat outputs of the system. Consequently, by proving that all state variables and the control inputs are differential functions of the flat outputs vector Y, it can be concluded that the integrated induction motor-driven reverse-osmosis process is a differentially flat system.

The differential flatness properties of the desalination unit mean that (i) the system can be transformed into the input–output linearized form through successive differentiations of its flat outputs and (ii) one can solve the associated setpoints definition problem. First, one defines in an unconstrained manner setpoints for the flat outputs. Next, the setpoints of the rest of the state variables are chosen so as to satisfy the differential relations which connect them to the flat outputs, or $x_i = h_i(Y, \dot{Y})$ for $i = 2, 5, 6$.

7.1.3 *Design of a nonlinear optimal controller for the induction motor-driven desalination process*

7.1.3.1 Approximate linearization of the induction motor-driven desalination unit

The dynamic model of the induction motor-driven inverse-osmosis undergoes approximate linearization through first-order Taylor series expansion and through the computation of the associated Jacobian matrices. The linearization process takes place at each sampling instance around the temporary operating point (x^*, u^*) where x^* is the present value of the system's state vector and u^* is the last sampled value of the control inputs vector. The modeling error which is due to the truncation of higher-order terms from the Taylor series expansion is considered to be a perturbation which is asymptotically compensated by the robustness of the control algorithm. This procedure allows for bringing the initial nonlinear model of the system $\dot{x}=f(x)+g(x)u$ into the equivalent linearized form

$$\dot{x}=Ax+Bu+\tilde{d} \tag{7.37}$$

where the Jacobian matrices of the system are defined as

$$\begin{gathered}A=\nabla_x[f(x)+g(x)u]\,|_{(x^*,u^*)}\Rightarrow\\ A=\nabla_x[f(x)]\,|_{(x^*,u^*)}+\nabla_x[g_1(x)u_1]\,|_{(x^*,u^*)}+\nabla_x[g_2(x)u_2]\,|_{(x^*,u^*)}+\nabla_x[g_3(x)u_3]\,|_{(x^*,u^*)}\end{gathered} \tag{7.38}$$

$$\begin{gathered}B=\nabla_u[f(x)+g(x)u]\,|_{(x^*,u^*)}\Rightarrow\\ B=g(x)\,|_{(x^*,u^*)}\end{gathered} \tag{7.39}$$

while the cumulative disturbance term $\tilde{d}$ may comprise (a) modeling error due to the truncation of higher-order term from the Taylor series expansion (b) exogenous perturbations (c) sensor measurement noise of any distribution. Because of linearizing at the present position of the system and not at points of the desirable trajectory the span of the Taylor series expansion is small and finally the associated modeling error is negligible.

The computation of the elements of the Jacobian matrix $\nabla_x[f(x)]\,|_{(x^*,u^*)}$ gives:

First row of the Jacobian matrix $\nabla_x[f(x)]\,|_{(x^*,u^*)}$: $\frac{\partial f_1}{\partial x_1}=-\frac{A_p^2}{A_mK_mV}-\frac{A_p}{\rho V}d_1\frac{R}{x_2}-\frac{A_pe_{vb}}{V}x_1$, $\frac{\partial f_1}{\partial x_2}=-\frac{A_p^2}{A_mK_mV}-\frac{A_p}{\rho V}[d_1\frac{(1-R)+R(k_fx_3-x_1)}{x_2^2}]$, $\frac{\partial f_1}{\partial x_3}=\frac{A_p^2}{A_mK_mV}k_f+\frac{A_p}{\rho V}(\frac{d_1Rk_f}{x_2})$, $\frac{\partial f_1}{\partial x_4}=0$, $\frac{\partial f_1}{\partial x_5}=0$, $\frac{\partial f_1}{\partial x_6}=0$.

Second row of the Jacobian matrix $\nabla_x[f(x)]\,|_{(x^*,u^*)}$: $\frac{\partial f_2}{\partial x_1}=-\frac{A_p^2}{A_mK_mV}-\frac{A_p}{\rho V}d_1\frac{R}{x_2}$, $\frac{\partial f_2}{\partial x_2}=-\frac{A_p^2}{A_mK_mV}-\frac{A_p}{\rho V}[d_1\frac{(1-R)+R(k_fx_3-x_1)}{x_2^2}]$, $\frac{\partial f_2}{\partial x_3}=\frac{A_p^2}{A_mK_mV}k_f+\frac{A_p}{\rho V}(\frac{d_1Rk_f}{x_2})$, $\frac{\partial f_2}{\partial x_4}=0$, $\frac{\partial f_2}{\partial x_5}=0$, $\frac{\partial f_2}{\partial x_6}=0$.

Third row of the Jacobian matrix $\nabla_x[f(x)]\,|_{(x^*,u^*)}$: $\frac{\partial f_3}{\partial x_1}=0$, $\frac{\partial f_3}{\partial x_2}=0$, $\frac{\partial f_3}{\partial x_3}=-\frac{k_b}{J}$, $\frac{\partial f_3}{\partial x_4}=\frac{\mu}{J}x_6$, $\frac{\partial f_3}{\partial x_5}=0$, $\frac{\partial f_3}{\partial x_6}=\frac{\mu}{J}x_4$.

Fourth row of the Jacobian matrix $\nabla_x[f(x)]\,|_{(x^*,u^*)}$: $\frac{\partial f_4}{\partial x_1}=0$, $\frac{\partial f_4}{\partial x_2}=0$, $\frac{\partial f_4}{\partial x_3}=0$, $\frac{\partial f_4}{\partial x_4}=-\alpha$, $\frac{\partial f_4}{\partial x_5}=\alpha M$, $\frac{\partial f_4}{\partial x_6}=0$.

Fifth row of the Jacobian matrix $\nabla_x[f(x)]\,|_{(x^*,u^*)}$: $\frac{\partial f_5}{\partial x_1}=0$, $\frac{\partial f_5}{\partial x_2}=0$, $\frac{\partial f_5}{\partial x_3}=n_px_6$, $\frac{\partial f_5}{\partial x_4}=\alpha\beta-\frac{\alpha Mx_6^2}{x_4^2}$, $\frac{\partial f_5}{\partial x_5}=-\gamma$, $\frac{\partial f_5}{\partial x_6}=n_px_3+\frac{2\alpha Mx_6}{x_4}$.

Sixth row of the Jacobian matrix $\nabla_x[f(x)]\,|_{(x^*,u^*)}$: $\frac{\partial f_6}{\partial x_1}=0$, $\frac{\partial f_6}{\partial x_2}=0$, $\frac{\partial f_6}{\partial x_3}=-\beta n_px_4-n_px_5$, $\frac{\partial f_6}{\partial x_4}=-\beta n_px_3+\frac{\alpha Mx_5x_6}{z_4^2}$, $\frac{\partial f_6}{\partial x_5}=-n_px_3-\frac{\alpha Mx_6}{x_4}$, $\frac{\partial f_6}{\partial x_6}=-\gamma-\frac{aMx_5}{x_4}$.

The computation of the elements of the Jacobian matrix $\nabla_x[g_1(x)]\,|_{(x^*,u^*)}$ gives:

$$\nabla_xg_1(x)\,|_{(x^*,u^*)}=\begin{pmatrix}0&0&0\;0\;0\;0\\0&-\frac{A_px_2}{V}&0\;0\;0\;0\\0&0&0\;0\;0\;0\\0&0&0\;0\;0\;0\\0&0&0\;0\;0\;0\\0&0&0\;0\;0\;0\end{pmatrix}|_{(x^*,u^*)} \tag{7.40}$$

Moreover, it holds that

$$\nabla_xg_2(x)\,|_{(x^*,u^*)}=0_{6\times6}\quad\nabla_xg_3(x)\,|_{(x^*,u^*)}=0_{6\times6} \tag{7.41}$$

7.1.4 Stabilizing feedback control

After linearization around its current operating point at each sampling interval, the model of the induction motor-driven desalination unit is written in the form [725,711,732]

$$\dot{x} = Ax + Bu + d_1 \tag{7.42}$$

Parameter d_1 stands for the linearization error in the desalination unit's model appearing above in Eq. (7.42). The reference setpoints for the state vector of the desalination unit's dynamics are denoted by $\mathbf{x_d} = [x_1^d, \cdots, x_6^d]$. Tracking of this trajectory is achieved after applying the control input u_d. At every time instant, the control input u_d is assumed to differ from the control input u appearing in Eq. (7.42) by an amount equal to Δu, that is $u_d = u + \Delta u$

$$\dot{x}_d = Ax_d + Bu_d + d_2 \tag{7.43}$$

The dynamics of the controlled system described in Eq. (7.42) can be also written as

$$\dot{x} = Ax + Bu + Bu_d - Bu_d + d_1 \tag{7.44}$$

and by denoting $d_3 = -Bu_d + d_1$ as an aggregate disturbance term, one obtains

$$\dot{x} = Ax + Bu + Bu_d + d_3 \tag{7.45}$$

By subtracting Eq. (7.43) from Eq. (7.45), one has

$$\dot{x} - \dot{x}_d = A(x - x_d) + Bu + d_3 - d_2 \tag{7.46}$$

By denoting the tracking error as $e = x - x_d$ and the aggregate disturbance term as $L\tilde{d} = d_3 - d_2$, the tracking error dynamics becomes

$$\dot{e} = Ae + Bu + L\tilde{d} \tag{7.47}$$

where $L \in R^{6\times 6}$ is the disturbance inputs gain matrix. For the approximately linearized model of the system, a stabilizing feedback controller is developed. The controller has the form

$$u(t) = -Ke(t) \tag{7.48}$$

with $K = \frac{1}{r}B^TP$ where P is a positive definite symmetric matrix which is obtained from the solution of the Riccati equation [725,711,732]

$$A^TP + PA + Q - P\left(\frac{2}{r}BB^T - \frac{1}{\rho^2}LL^T\right)P = 0 \tag{7.49}$$

where Q is a positive semi-definite symmetric matrix. The diagram of the considered control loop is depicted in Figure 7.2

It is also noted that the solution of the H-infinity feedback control problem for the dynamics of the induction motor-driven desalination unit and the computation of the worst-case disturbance that this controller can sustain, comes from the superposition of Bellman's optimality principle when considering that the desalination process is affected by two separate inputs (i) the control input u (ii) the cumulative disturbance input $\tilde{d}(t)$. Solving the optimal control problem for u that is for the minimum variation (optimal) control input that achieves the elimination of the state vector's tracking error gives $u = -\frac{1}{r}B^TPe$. Equivalently, solving the optimal control problem for $\tilde{d}$, that is for the worst-case disturbance that the control loop can sustain, gives $\tilde{d} = \frac{1}{\rho^2}L^TPe$.

7.1.4.1 Lyapunov stability analysis

Through Lyapunov stability analysis it will be shown that the proposed nonlinear control scheme assures H_∞ tracking performance for the dynamics of the induction motor-driven desalination unit and that under moderate conditions global stability (asymptotic convergence to the reference setpoints) is achieved. As shown before, the tracking error dynamics for the desalination unit's dynamics is written in the form [725,711,732]

$$\dot{e} = Ae + Bu + L\tilde{d} \tag{7.50}$$

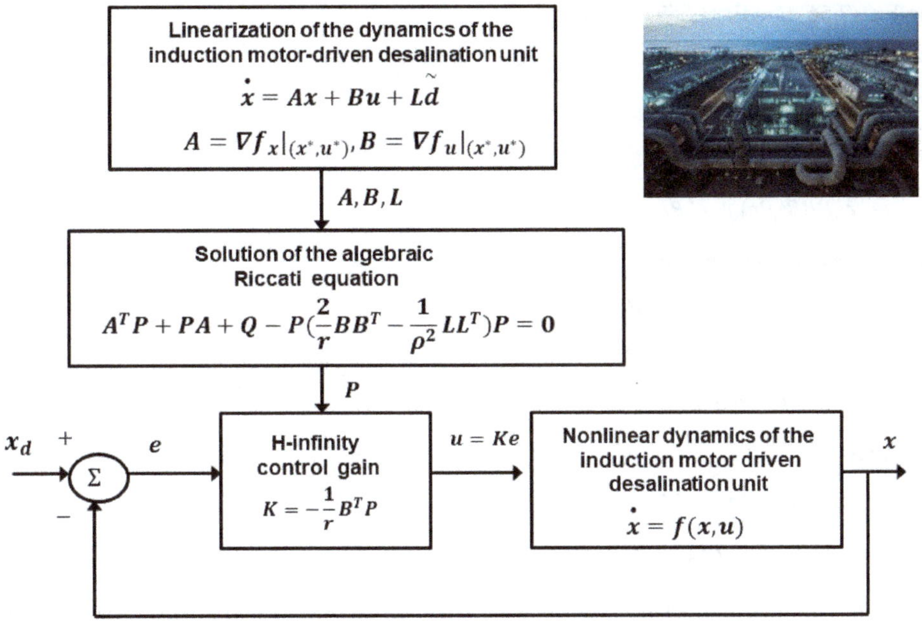

Figure 7.2 Diagram of nonlinear optimal (H-infinity) control loop for the induction motor-driven desalination unit

where in the desalination unit's case $L = R^{6\times 6}$ is the disturbance inputs gain matrix. Variable $\tilde{d}$ denotes model uncertainties and external disturbances of the desalination unit's dynamic model. The following Lyapunov equation is considered:

$$V = \frac{1}{2}e^TPe \tag{7.51}$$

where $e = x - x_d$ is the tracking error. By differentiating with respect to time, one obtains

$$\begin{aligned} \dot{V} &= \frac{1}{2}\dot{e}^TPe + \frac{1}{2}e^TP\dot{e} \Rightarrow \\ \dot{V} &= \frac{1}{2}[Ae + Bu + L\tilde{d}]^TP + \frac{1}{2}e^TP[Ae + Bu + L\tilde{d}] \Rightarrow \end{aligned} \tag{7.52}$$

$$\begin{aligned} \dot{V} &= \frac{1}{2}[e^TA^T + u^TB^T + \tilde{d}^TL^T]Pe + \\ &+ \frac{1}{2}e^TP[Ae + Bu + L\tilde{d}] \Rightarrow \end{aligned} \tag{7.53}$$

$$\begin{aligned} \dot{V} &= \frac{1}{2}e^TA^TPe + \frac{1}{2}u^TB^TPe + \frac{1}{2}\tilde{d}^TL^TPe + \\ &\frac{1}{2}e^TPAe + \frac{1}{2}e^TPBu + \frac{1}{2}e^TPL\tilde{d} \end{aligned} \tag{7.54}$$

The previous equation is rewritten as

$$\begin{aligned} \dot{V} &= \frac{1}{2}e^T(A^TP + PA)e + (\frac{1}{2}u^TB^TPe + \frac{1}{2}e^TPBu) + \\ &+ (\frac{1}{2}\tilde{d}^TL^TPe + \frac{1}{2}e^TPL\tilde{d}) \end{aligned} \tag{7.55}$$

Assumption: For given positive definite matrix Q and coefficients r and ρ, there exists a positive definite matrix P, which is the solution of the following matrix equation:

$$A^TP + PA = -Q + P(\frac{2}{r}BB^T - \frac{1}{\rho^2}LL^T)P \tag{7.56}$$

Moreover, the following feedback control law is applied to the system:

$$u = -\frac{1}{r}B^TPe \tag{7.57}$$

By substituting Eq. (7.56) and Eq. (7.57), one obtains

$$\dot{V} = \frac{1}{2}e^T[-Q + P(\frac{2}{r}BB^T - \frac{1}{\rho^2}LL^T)P]e+ \\ +e^TPB(-\frac{1}{r}B^TPe) + e^TPL\tilde{d} \Rightarrow \tag{7.58}$$

$$\dot{V} = -\frac{1}{2}e^TQe + (\frac{1}{r}e^TPBB^TPe - \frac{1}{2\rho^2}e^TPLL^TPe \\ -\frac{1}{r}e^TPBB^TPe) + e^TPL\tilde{d} \tag{7.59}$$

which after intermediate operations gives

$$\dot{V} = -\frac{1}{2}e^TQe - \frac{1}{2\rho^2}e^TPLL^TPe + e^TPL\tilde{d} \tag{7.60}$$

or, equivalently

$$\dot{V} = -\frac{1}{2}e^TQe - \frac{1}{2\rho^2}e^TPLL^TPe+ \\ +\frac{1}{2}e^TPL\tilde{d} + \frac{1}{2}\tilde{d}^TL^TPe \tag{7.61}$$

Lemma: The following inequality holds

$$\frac{1}{2}e^TPL\tilde{d} + \frac{1}{2}\tilde{d}L^TPe - \frac{1}{2\rho^2}e^TPLL^TPe \leq \frac{1}{2}\rho^2\tilde{d}^T\tilde{d} \tag{7.62}$$

Proof: The binomial $(\rho\alpha - \frac{1}{\rho}b)^2$ is considered. Expanding the left part of the above inequality, one gets

$$\rho^2a^2 + \frac{1}{\rho^2}b^2 - 2ab \geq 0 \Rightarrow \frac{1}{2}\rho^2a^2 + \frac{1}{2\rho^2}b^2 - ab \geq 0 \Rightarrow \\ ab - \frac{1}{2\rho^2}b^2 \leq \frac{1}{2}\rho^2a^2 \Rightarrow \frac{1}{2}ab + \frac{1}{2}ab - \frac{1}{2\rho^2}b^2 \leq \frac{1}{2}\rho^2a^2 \tag{7.63}$$

The following substitutions are carried out: $a = \tilde{d}$ and $b = e^TPL$, and the previous relation becomes

$$\frac{1}{2}\tilde{d}^TL^TPe + \frac{1}{2}e^TPL\tilde{d} - \frac{1}{2\rho^2}e^TPLL^TPe \leq \frac{1}{2}\rho^2\tilde{d}^T\tilde{d} \tag{7.64}$$

Eq. (7.64) is substituted in Eq. (7.61) and the inequality is enforced, thus giving

$$\dot{V} \leq -\frac{1}{2}e^TQe + \frac{1}{2}\rho^2\tilde{d}^T\tilde{d} \tag{7.65}$$

Eq. (7.65) shows that the H_∞ tracking performance criterion is satisfied. The integration of $\dot{V}$ from 0 to T gives

$$\int_0^T\dot{V}(t)dt \leq -\frac{1}{2}\int_0^T||e||_Q^2dt + \frac{1}{2}\rho^2\int_0^T||\tilde{d}||^2dt \Rightarrow \\ 2V(T) + \int_0^T||e||_Q^2dt \leq 2V(0) + \rho^2\int_0^T||\tilde{d}||^2dt \tag{7.66}$$

Moreover, if there exists a positive constant $M_d > 0$ such that

$$\int_0^\infty ||\tilde{d}||^2dt \leq M_d \tag{7.67}$$

then, one gets

$$\int_0^\infty||e||_Q^2dt \leq 2V(0) + \rho^2M_d \tag{7.68}$$

Thus, the integral $\int_0^\infty||e||_Q^2dt$ is bounded. Moreover, $V(T)$ is bounded and from the definition of the Lyapunov function V in Eq. (7.51) it becomes clear that $e(t)$ will be also bounded since $e(t) \in \Omega_e = \{e|e^TPe \leq 2V(0) + \rho^2M_d\}$. According to the above and with the use of Barbalat's Lemma, one obtains $lim_{t\to\infty}e(t) = 0$.

After following the stages of the stability proof, one arrives at Eq. (7.65) which shows that the H-infinity tracking performance criterion holds. By selecting the attenuation coefficient ρ to be sufficiently small and in particular to satisfy $\rho^2 < ||e||_Q^2/||\tilde{d}||^2$, one has that the first derivative of the Lyapunov function is upper bounded by 0. This condition holds at each sampling instant, and consequently, global stability for the control loop can be concluded.

7.1.5 Flatness-based control in successive loops for the induction motor-driven desalination unit

7.1.5.1 State-space model in chained subsystems form

Next, the dynamic model of the induction motor-driven reverse osmosis process is separated into three chained subsystems. The first one consists of Eq. (7.10) and Eq. (7.11), the second one of Eq. (7.12) and Eq. (7.13), and the third one of Eq. (7.14) and Eq. (7.15).

The state equations of the first subsystem are rewritten as:

$$\dot{x}_1 = \frac{A_p^2}{A_m K_m V}(k_f x_3 - x_1 - x_2) + \frac{A_p}{\rho V}[c_1 + d_1\frac{(1-R)+R(k_f x_3 - x_1)}{x_2}] - \frac{1}{2}\frac{A_p e_{vb} x_1^2}{V} + 0u_1 \tag{7.69}$$

$$\dot{x}_2 = \frac{A_p^2}{A_m K_m V}(k_f x_3 - x_1 - x_2) + \frac{A_p}{\rho V}[c_1 + d_1\frac{(1-R)+R(k_f x_3 - x_1)}{x_2}] - \frac{1}{2}\frac{A_p x_2^2 u_1}{V} \tag{7.70}$$

Next, subvectors $x_{12} = [x_1, x_2]^T$ and $v_1 = [x_3, u_1]^T$ are defined while the following submatrices $f_{12}(x_{12})$ and $g_{12}(x_{12})$ are also considered

$$f_{12} = \begin{pmatrix} \frac{A_p^2}{A_m K_m V}(-x_1 - x_2) + \frac{A_p}{\rho V}[c_1 + d_1\frac{(1-R)+R(-x_1)}{x_2}] - \frac{1}{2}\frac{A_p e_{vb} x_1^2}{V} \\ \frac{A_p^2}{A_m K_m V}(-x_1 - x_2) + \frac{A_p}{\rho V}[c_1 + d_1\frac{(1-R)+R(-x_1)}{x_2}] \end{pmatrix}$$

$$g_{12} = \begin{pmatrix} \frac{A_p^2}{A_m K_m V}k_f + \frac{A_p}{\rho V}d_1\frac{Rk_f}{x_2} & 0 \\ \frac{A_p^2}{A_m K_m V}k_f + \frac{A_p}{\rho V}d_1\frac{Rk_f}{x_2} & -\frac{1}{2}\frac{A_p x_1^2}{V} \end{pmatrix} \tag{7.71}$$

Thus, the subsystem of Eq. (7.69) and Eq. (7.70) is also written as

$$\dot{x}_{12} = f_{12}(x_{12}) + g_{12}(x_{22})v_1 \tag{7.72}$$

The state equations of the second subsystem are rewritten as:

$$\dot{x}_3 = -\frac{T_L}{J} - \frac{k_b x_3}{J} + 0x_5 + \frac{\mu}{J}x_6 \tag{7.73}$$

$$\dot{x}_4 = -\alpha x_4 + aMx_5 + 0x_6 \tag{7.74}$$

Next, subvectors $x_{34} = [x_3, x_4]^T$ and $v_2 = [x_5, x_6]^T$ are defined while the following submatrices $f_{34}(x_{12}, x_{34})$ and $g_{34}(x_{12}, x_{34})$ are also considered:

$$f_{34} = \begin{pmatrix} -\frac{T_L}{J} - \frac{k_b x_3}{J} \\ -\alpha x_4 \end{pmatrix} \quad g_{34} = \begin{pmatrix} 0 & \frac{\mu}{J}x_4 \\ \alpha M & 0 \end{pmatrix} \tag{7.75}$$

Thus, the subsystem of Eq. (7.73) and Eq. (7.74) is also written as

$$\dot{x}_{34} = f_{34}(x_{12}, x_{34}) + g_{34}(x_{12}, x_{34})v_2 \tag{7.76}$$

The state equations of the third subsystem are rewritten as:

$$\dot{x}_5 = -\gamma x_5 + \alpha\beta x_4 + n_p x_3 x_6 + \frac{aMx_6^2}{x_4} + \frac{1}{\sigma L_s}u_2 + 0u_3 \tag{7.77}$$

$$\dot{x}_6 = -\gamma x_6 - \beta n_p x_3 x_4 - n_p x_3 x_5 - \frac{\alpha M x_6 x_5}{x_4} + 0u_2 + \frac{1}{\sigma L_s}u_3 \tag{7.78}$$

Next, subvectors $x_{56} = [x_5, x_6]^T$ and $v_3 = [u_2, u_3]^T$ are defined while the following submatrices $f_{56}(x_{12}, x_{34}, x_{56})$ and $g_{56}(x_{12}, x_{34}, x_{56})$ are also considered:

$$f_{56} = \begin{pmatrix} -\gamma x_5 + \alpha\beta x_4 + n_p x_3 x_6 + \frac{aMx_6^2}{x_4} \\ -\gamma x_5 + \alpha\beta x_4 + n_p x_3 x_6 + \frac{aMx_6^2}{x_4} \end{pmatrix} \quad g_{56} = \begin{pmatrix} \frac{1}{\sigma L_s} & 0 \\ 0 & \frac{1}{\sigma L_s} \end{pmatrix} \tag{7.79}$$

Thus, the subsystem of Eq. (7.77) and Eq. (7.78) is also written as

$$\dot{x}_{56} = f_{56}(x_{12}, x_{34}, x_{56}) + g_{56}(x_{12}, x_{34}, x_{56})v_3 \tag{7.80}$$

Thus, the integrated induction motor-driven desalination unit is written in the form of three chained subsystems

$$\begin{aligned}\dot{x}_{12} &= f_{12}(x_{12}) + g_{12}(x_{12})v_1 \\ \dot{x}_{34} &= f_{34}(x_{12}, x_{34}) + g_{34}(x_{12}, x_{34})v_2 \\ \dot{x}_{56} &= f_{56}(x_{12}, x_{34}, x_{56}) + g_{56}(x_{12}, x_{34}, x_{56})v_3\end{aligned} \tag{7.81}$$

7.1.6 Proof of differential flatness properties

It will be proven that each one of the subsystems of Eq. (7.72), Eq. (7.76), and Eq. (7.80), viewed independently, is differentially flat.

For the subsystem of Eq. (7.72) the flat output vector is taken to be x_{12} and v_1 is considered as a virtual control input. Eq. (7.72) is solved for v_1, thus giving

$$v_1 = g_{12}(x_{12})^{-1}[\dot{x}_{12} - f_{12}(x_{12})] \tag{7.82}$$

This signifies that v_1 is a differential function of the flat output $y_1 = x_{12}$ and thus the subsystem of Eq. (7.72) is differentially flat.

For the subsystem of Eq. (7.76), the flat output vector is taken to be $y_2 = x_{34}$ and v_2 is considered as a virtual control input, while x_{12} is considered to be a coefficient vector. Eq. (7.76) is solved for the virtual control input v_2 thus giving

$$v_2 = g_{34}(x_{12}, x_{34})^{-1}[\dot{x}_{34} - f_{34}(x_{12}, x_{34})] \tag{7.83}$$

which signifies that v_2 is a differential function of x_{34}, therefore the subsystem of Eq. (7.76) is differentially flat.

For the subsystem of Eq. (7.80), the flat output vector is taken to be $y_3 = x_{56}$ and v_3 is considered as a virtual control input, while x_{12}, x_{34} are considered to be coefficient vectors. Eq. (7.80) is solved for the real control input v_3 thus giving

$$v_3 = g_{56}(x_{12}, x_{34}, x_{56})^{-1}[\dot{x}_{56} - f_{56}(x_{12}, x_{34}, x_{56})] \tag{7.84}$$

which signifies that v_3 is a differential function of x_{56}, therefore the subsystem of Eq. (7.80) is differentially flat.

By proving that the subsystems of Eq. (7.72), Eq. (7.76) and Eq. (7.80) are differentially flat it can be concluded that (i) each one of these subsystems is input–output linearizable and (ii) a stabilizing feedback controller can be designed about them by inverting their dynamics as in the case of input–output linearized flat systems.

7.1.7 Design of a flatness-based controller in successive loops

For each one of the subsystems of Eq. (7.72), Eq. (7.76), and Eq. (7.80), the virtual control inputs of the preceding (i-th) subsystem become setpoints to the subsequent $(i+1)$-th subsystem,

For the subsystem of Eq. (7.72), the stabilizing feedback control is

$$v_1 = g_{12}(x_{12})^{-1}[\dot{x}_{12}^d - f_{12}(x_{12}) - K_1(x_{12} - x_{12}^d)] \tag{7.85}$$

where K_1 is a diagonal matrix $K_1 \in R^{2\times 2}$ with diagonal elements $K_{1,ii} > 0$, $i = 1, 2$. Control input v_1 becomes setpoint for the subsequent subsystem of Eq. (7.76), that is used to define x_3^d, while x_4^d can be assigned any desirable value.

For the subsystem of Eq. (7.76), the stabilizing feedback control is

$$v_2 = g_{34}(x_{12}, x_{34})^{-1}[\dot{x}_{34}^d - f_{34}(x_{12}, x_{34}) - K_2(x_{34} - x_{34}^d)] \tag{7.86}$$

where K_2 is a diagonal matrix $K_2 \in R^{2\times 2}$ with diagonal elements $K_{2,ii} > 0$, $i = 1, 2$. Control input v_2 becomes setpoint for the subsequent subsystem of Eq. (7.80) that is $x_{56}^d = v_2$.

For the subsystem of Eq. (7.80), the stabilizing feedback control is

$$v_3 = g_{56}(x_{12}, x_{34}, x_{56})^{-1}[\dot{x}_{56}^d - f_{56}(x_{12}, x_{34}, x_{56}) - K_3(x_{56} - x_{56}^d)] \tag{7.87}$$

where K_3 is a diagonal matrix $K_3 \in R^{2\times 2}$ with diagonal elements $K_{3,ii} > 0$, $i = 1, 2$. The real control input v_3 is computed by tracing backward the virtual control inputs v_2, v_1 of the preceding subsystems.

By substituting the control input of Eq. (7.85) into Eq. (7.69) and by defining the tracking error variable $e_{12} = x_{12} - x_{12}^d$, one obtains the following tracking error dynamics:

$$\begin{aligned} \dot{x}_{12} &= f_{12}(x_{12}) + g_{12}(x_{12})g_{12}(x_{12})^{-1}[\dot{x}_{12}^d - f_{12}(x_{12}) - K_1(x_{12} - x_{12}^d)] \\ &\Rightarrow (\dot{x}_{12} - \dot{x}_{12}^d) + K_1(x_{12} - x_{12}^d) = 0 \Rightarrow \dot{e}_{12} + K_1 e_{12} = 0 \\ &\Rightarrow \lim_{t\to\infty} e_{12}(t) = 0 \Rightarrow \lim_{t\to\infty} x_{12}(t) = x_{12}^d(t) \end{aligned} \tag{7.88}$$

By substituting the control input of Eq. (7.86) into Eq. (7.70) and by defining the tracking error variable $e_{34} = x_{34} - x_{34}^d$, one obtains the following tracking error dynamics:

$$\begin{aligned} \dot{x}_{34} &= f_{34}(x_{12}, x_{34}) + g_{34}(x_{12}, x_{34})g_{34}(x_{12}, x_{34})^{-1}[\dot{x}_{34}^d - f_{34}(x_{12}, x_{34}) - K_2(x_{34} - x_{34}^d)] \\ &\Rightarrow (\dot{x}_{34} - \dot{x}_{34}^d) + K_2(x_{34} - x_{34}^d) = 0 \Rightarrow \dot{e}_{34} + K_2 e_{34} = 0 \\ &\Rightarrow \lim_{t\to\infty} e_{34}(t) = 0 \Rightarrow \lim_{t\to\infty} x_{34}(t) = x_{34}^d(t) \end{aligned} \tag{7.89}$$

By substituting the control input of Eq. (7.87) into Eq. (7.73) and by defining the tracking error variable $e_{56} = x_{56} - x_{56}^d$, one obtains the following tracking error dynamics:

$$\begin{aligned} \dot{x}_{56} &= f_{56}(x_{12}, x_{34}, x_{56}) + g_{56}(x_{12}, x_{34}, x_{56})g_{56}(x_{12}, x_{34}, x_{56})^{-1}[\dot{x}_{56}^d - f_{56}(x_{12}, x_{34}, x_{56}) - K_3(x_{56} - x_{56}^d)] \\ &\Rightarrow (\dot{x}_{56} - \dot{x}_{56}^d) + K_3(x_{56} - x_{56}^d) = 0 \Rightarrow \dot{e}_{56} + K_3 e_{56} = 0 \\ &\Rightarrow \lim_{t\to\infty} e_{56}(t) = 0 \Rightarrow \lim_{t\to\infty} x_{56}(t) = x_{56}^d(t) \end{aligned} \tag{7.90}$$

Consequently, all state variables $x_i(t)$ $i = 1, 2 \cdots, 6$ of the induction motor-driven desalination unit converge to the associated setpoints and this control system is globally asymptotically stable.

Global stability for the induction motor-driven reverse-osmosis process can be also proven with the use of Lyapunov analysis. The following Lyapunov function is defined:

$$V = \tfrac{1}{2}[e_{12}^T e_{12} + e_{34}^T e_{34} + e_{56}^T e_{56}] \tag{7.91}$$

By differentiating the Lyapunov function of Eq. (7.91) in time, one obtains

$$\dot{V} = \tfrac{1}{2}2[e_{12}^T \dot{e}_{12} + e_{34}^T \dot{e}_{34} + e_{56}^T \dot{e}_{56}] \tag{7.92}$$

By substituting in Eq. (7.92), the tracking error dynamics which has been given in Eq. (7.88), Eq. (7.89), and Eq. (7.90), one obtains

$$\begin{aligned} \dot{V} &= e_{12}^T(-K_1 e_{12}) + e_{34}^T(-K_2 e_{34}) + e_{56}^T(-K_3 e_{56}) \Rightarrow \\ \dot{V} &= -[e_{12}^T K_1 e_{12} + e_{34}^T K_2 e_{34} + e_{56}^T K_3 e_{56}] \end{aligned} \tag{7.93}$$

The above equation signifies that

$$\dot{V} < 0 \, \forall \, e_{12} \neq 0, \; e_{34} \neq 0, \; e_{56} \neq 0 \tag{7.94}$$

while $\dot{V} = 0$ if and only if $e_{12} = 0$, $e_{34} = 0$, and $e_{56} = 0$. Consequently, $\dot{V}$ is always negative and the Lyapunov function V is a strictly diminishing function which converges asymptotically to 0. Thus, the closed-loop system is globally asymptotically stable.

An explicit demonstration of the exponential stabilization that is achieved by flatness-based control in successive loops is given next. The Lyapunov function of the control loop is written as:

$$V = \tfrac{1}{2}[\textstyle\sum_{i=1}^{2} e_i^2 + \sum_{j=3}^{4} e_j^2 + \sum_{m=5}^{6} e_m^2] \tag{7.95}$$

where e_i $i = 1, 2$ are the tracking errors for the state variables of the first subsystem, e_j $j = 3, 4$ are the tracking errors for the state variables of the second subsystem, and e_m $m = 5, 6$ are the tracking errors for the state variables of the third subsystem of the desalination unit. Equivalently, the first-order time derivative of the Lyapunov function is written as

$$\dot{V} = -[\textstyle\sum_{i=1}^{2} k_{1,i} e_i^2 + \sum_{j=3}^{4} k_{2,j} e_j^2 + \sum_{m=5}^{6} k_{3,m} e_m^2] \tag{7.96}$$

where $k_{1,i} > 0$ $i = 1,2$ are the diagonal elements of gain matrix K_1, $k_{2,j} > 0$ $j = 3,4$ are the diagonal elements of gain matrix K_2, and $k_{3,m} > 0$ $m = 5,6$ are the diagonal elements of gain matrix K_3. By denoting the minimum of the above-noted elements of the feedback gain matrices as k_{min}, that is

$$k_{min} = min\{k_{1,i} : i = 1,2 \text{ and } k_{2,j} : j = 3,4 \text{ and } k_{3,m} : m = 5,6\} \tag{7.97}$$

and using Eq. (7.96), one obtains that

$$\begin{aligned}&\dot{V} \leq -k_{min}[\textstyle\sum_{i=1}^{2} e_i^2 + \sum_{j=3}^{4} e_j^2 + \sum_{m=5}^{6} e_m^2]\\&\Rightarrow \dot{V} \leq -2k_{min}V \Rightarrow \dot{V} + 2k_{min}V \leq 0\end{aligned} \tag{7.98}$$

From Eq. (7.98), one can demonstrate the exponential convergence of the Lyapunov function V to 0.

It is noted that the feedback control scheme, which is followed for the cascading subsystems that constitute the dynamic model of induction motor-driven desalination unit and which is based on the inversion of the subsystems' dynamics of the desalination process, is equally robust to sliding-mode control in which the switching control term has been substituted by a saturation function. One can easily confirm this for the first-order i-th subsystem of the form $\dot{x}_i = f_i(x_i) + g_i(x_i)v_i$ by defining the sliding surface $s_i = e_i = x_1 - x_i^d$ and the associated sliding mode controller $v_i = \hat{g}_i(x)^{-1}[\dot{x}_i^d - \hat{f}_i(x_i) - K_i sgn(x_i - x_i^d)]$ which after substituting the $sgn(s_i)$ function with the saturation $sat(s_i)$ function becomes $v_i = \hat{g}_i(x)^{-1}[\dot{x}_i^d - \hat{f}_i(x_i) - K_i(x_i - x_i^d)]$. The latter relation coincides with the flatness-based control in successive loops for the i-th subsystem under uncertainty (with the use of the estimated functions $\hat{f}_i(x)$ and $\hat{g}_i(x)$) which is computed by this section's control method. Therefore, the proposed flatness-based control method in successive loops provides sufficient robustness margins which enable the reliable and safe functioning of the induction motor-driven desalination unit under reasonable levels of model uncertainty or external perturbations.

7.1.8 Simulation tests

7.1.8.1 Results on nonlinear optimal control

Results about the use of the nonlinear optimal control method in the dynamic model of the induction motor-driven desalination unit are shown in the following diagrams, given in Figures 7.3–7.18. Eight different setpoints have been considered which allow to test the control loop under variable operating conditions. It can be observed that the nonlinear optimal control method achieves fast and accurate tracking of reference setpoints by the state variables of the induction motor-driven reverse-osmosis process, while keeping also moderate the variations of the control inputs. It is noted that to implement this nonlinear control method, one does not have to measure the magnetic flux $x_4 = \psi_{r_d}$ of the induction motor. One can measure at each sampling instance state variables x_1, x_2, x_3, x_5, x_6 while state variable x_4 can be estimated with the use of a nonlinear observer such as the nonlinear H-infinity Kalman filter [725]. It is meaningful to aim at minimization of the control inputs amplitude for the desalination process, because this also signifies a reduction of the cost and of energy dispersion for the functioning of desalination units. In the case of the analyzed desalination unit, the minimization of the variations of the control inputs signifies that minimal electric power has to be used to achieve the targeted clean water production.

To elaborate on nonlinear optimal control for the desalination unit the following tables are given (i) Table 7.1 providing results about the accuracy of tracking of setpoints by the state variables of the desalination unit under an exact dynamic model, (ii) Table 7.2 providing results about the accuracy of tracking of setpoints by the state variables of the desalination unit under a model that is subject to disturbances (for instance, change $\Delta a\%$ in parameter μ of the induction motor's dynamic model), and (iii) Table 7.3 providing results about the convergence times of the state variables of the induction motor's state variables to the associated setpoints.

7.1.8.2 Results on flatness-based control in successive loops

Results about the use of flatness-based control in successive loops in the dynamic model of the induction motor-driven desalination unit are shown in the following diagrams, given in Figures 7.19–7.34. Eight different setpoints have been considered, the same as in the case of nonlinear optimal control. It can be noticed, that under this control scheme, one achieves again fast and precise tracking of reference setpoints for all state variables of the induction motor-driven

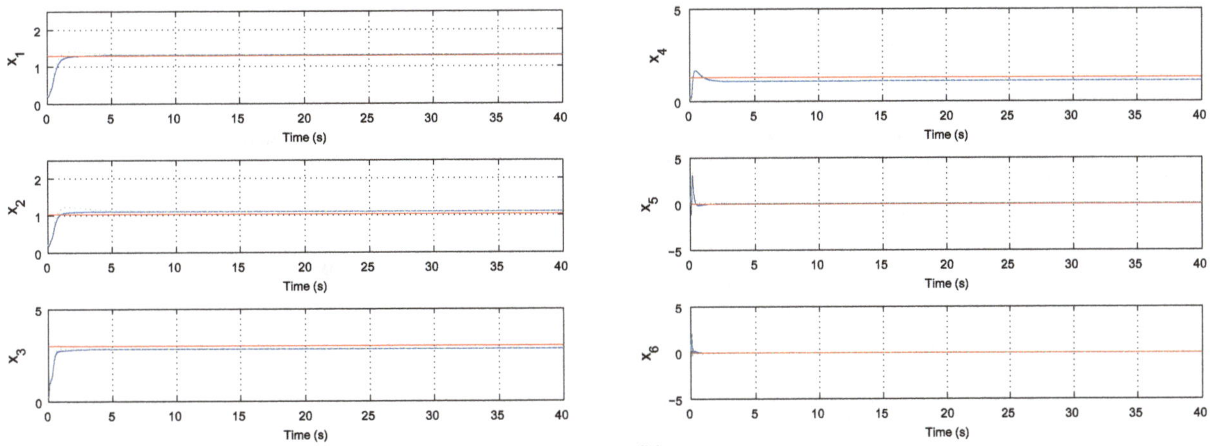

(a) (b)

Figure 7.3 Tracking of trajectory 1 by the induction motor-driven reverse-osmosis process using nonlinear optimal control: (a) convergence of state variables x_1 to x_3 to their reference setpoints (red line: setpoint, blue line: real value), green line: estimation by the HKF; (b) convergence of state variables x_4 to x_6 to their reference setpoints (red line: setpoint, blue line: real value, green line: estimation by the HKF)

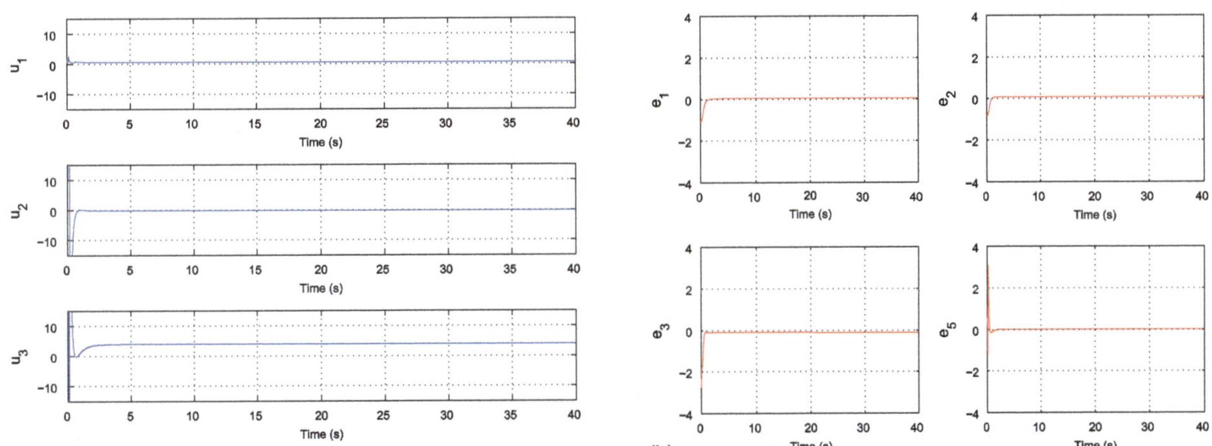

(a) (b)

Figure 7.4 Tracking of trajectory 1 by the induction motor-driven reverse-osmosis process using nonlinear optimal control: (a) variation of the control inputs u_1 to u_3; (b) variation of the tracking error variables e_i, $i = 1,2,3,5$ (associated with state variables x_i of the desalination process)

reverse osmosis process. It is noteworthy, that through the stages of this method, one solves also the setpoints definition problem for all state variables of the controlled system. Actually, the selection of setpoints for state variables x_1 and x_2 is unconstrained. On the other side by defining state variable x_3 as part of the virtual control input for the subsystem of x_1, x_2, one can find the setpoints for x_3 as a function of the setpoints for x_1, x_2. The selection of the setpoint of x_4 is also unconstrained. Equivalently, by defining state variables x_5, x_6 as virtual control inputs for the subsystem of x_3, x_4, one can find the setpoints for x_5, x_5 as functions of the setpoints for x_3, x_4. The speed of convergence of the state variables of the induction motor-driven reverse-osmosis process under flatness-based control implemented in successive loops is dependent on the selection of values for the diagonal gain matrices K_1, K_2, K_3 of Eq. (7.85), Eq. (7.86), and Eq. (7.87).

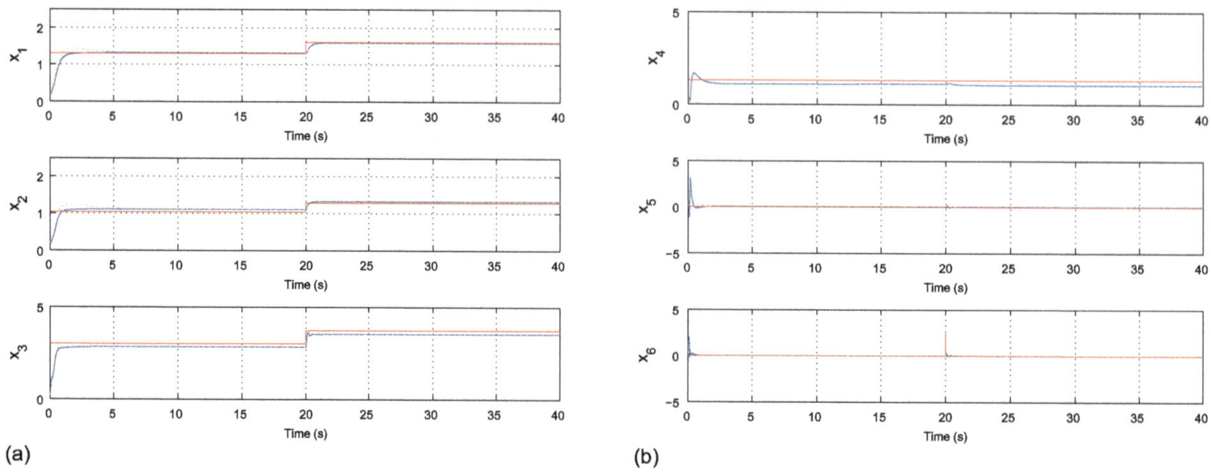

Figure 7.5 Tracking of trajectory 2 by the induction motor-driven reverse-osmosis process using nonlinear optimal control: (a) convergence of state variables x_1 to x_3 to their reference setpoints (red line: setpoint, blue line: real value), green line: estimation by the HKF; (b) convergence of state variables x_4 to x_6 to their reference setpoints (red line: setpoint, blue line: real value, green line: estimation by the HKF)

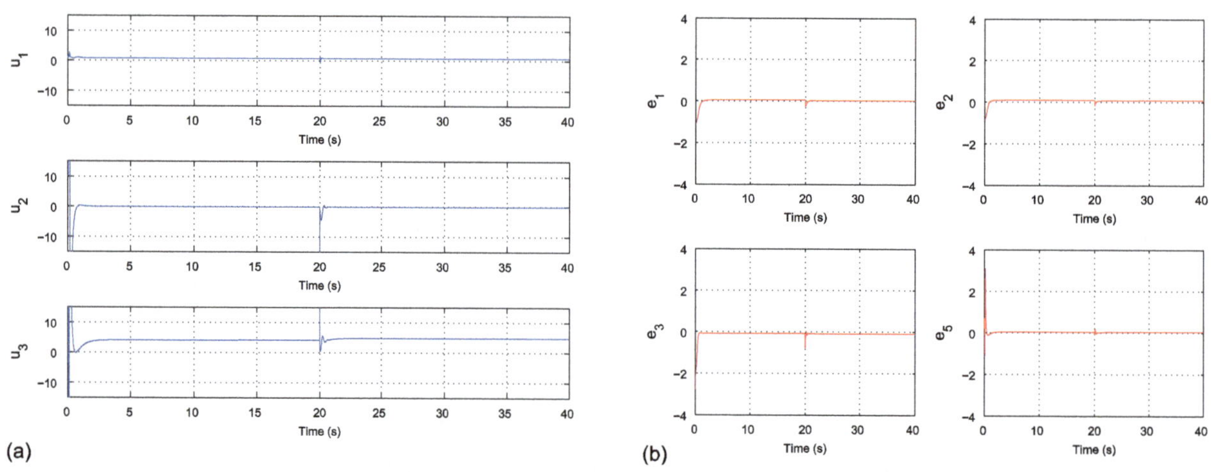

Figure 7.6 Tracking of trajectory 2 by the induction motor-driven reverse-osmosis process using nonlinear optimal control: (a) variation of the control inputs u_1 to u_3; (b) variation of the tracking error variables e_i, $i = 1,2,3,5$ (associated with state variables x_i of the desalination process)

To elaborate on flatness-based control in successive loops for the desalination unit the following tables are given (i) Table 7.4 providing results about the accuracy of tracking of setpoints by the state variables of the desalination unit under an exact dynamic model, (ii) Table 7.5 providing results about the accuracy of tracking of setpoints by the state variables of the desalination unit under a model that is subject to disturbances (for instance, change $\Delta\alpha\%$ in parameter μ of the induction motor's dynamic model), and (iii) Table 7.6 providing results about the convergence times of the state variables of the desalination's state variables to the associated setpoints.

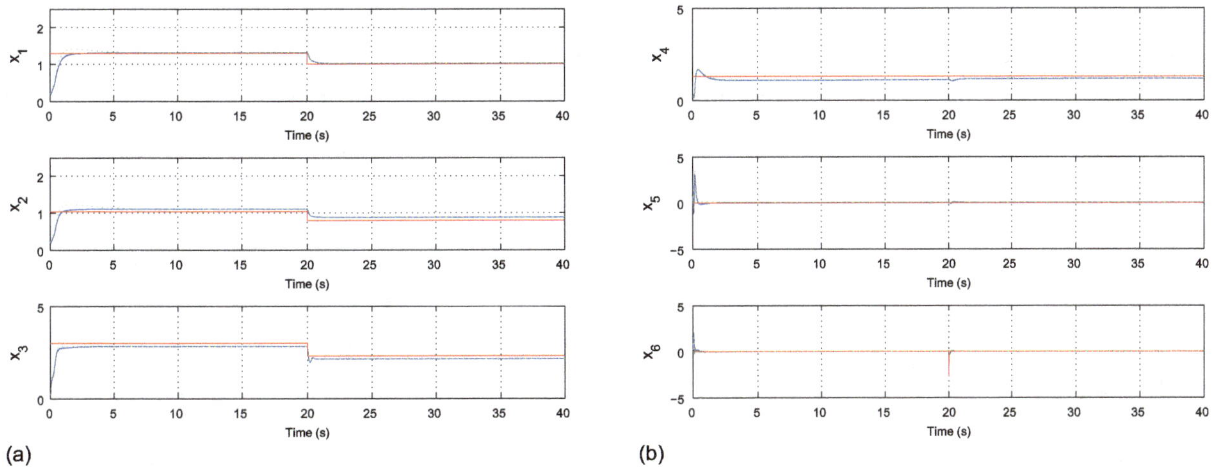

Figure 7.7 *Tracking of trajectory 3 by the induction motor-driven reverse-osmosis process using nonlinear optimal control: (a) convergence of state variables x_1 to x_3 to their reference setpoints (red line: setpoint, blue line: real value), green line: estimation by the HKF; (b) convergence of state variables x_4 to x_6 to their reference setpoints (red line: setpoint, blue line: real value, green line: estimation by the HKF)*

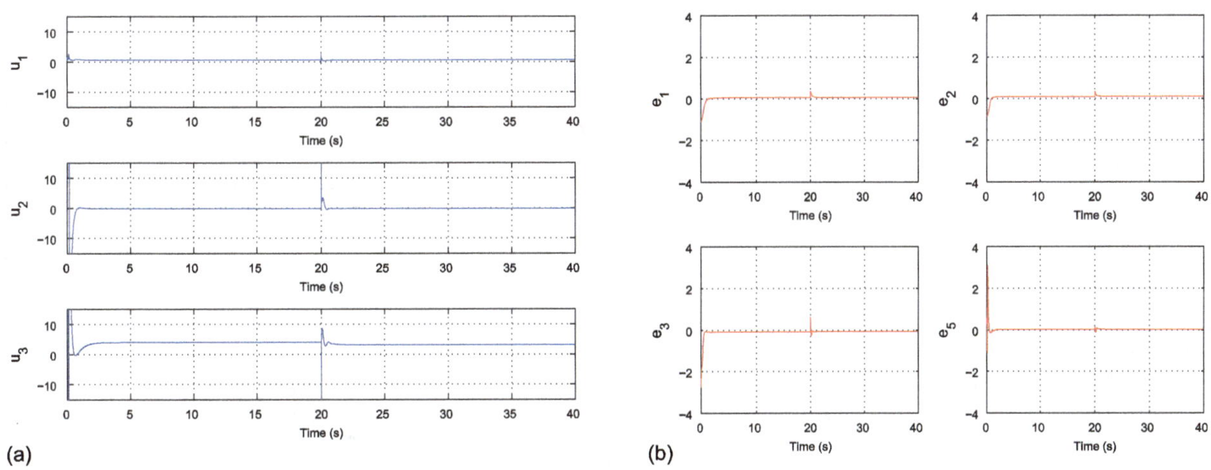

Figure 7.8 *Tracking of trajectory 3 by the induction motor-driven reverse-osmosis process using nonlinear optimal control: (a) variation of the control inputs u_1 to u_3; (b) variation of the tracking error variables e_i, $i = 1,2,3,5$ (associated with state variables x_i of the desalination process)*

7.2 Flatness-based control in successive-loops for gas compressors driven by IM and PMSM

7.2.1 Outline

As previously noted, flatness-based control has been one of the main research directions on nonlinear control systems during the last years [711,732]. The method allows for ensuring controllability and designing stabilizing feedback controllers for complex nonlinear dynamical systems [715,721,725]. A system is considered to be differentially flat if the following two conditions hold: (i) all its state variables and its control inputs can be written as differential functions

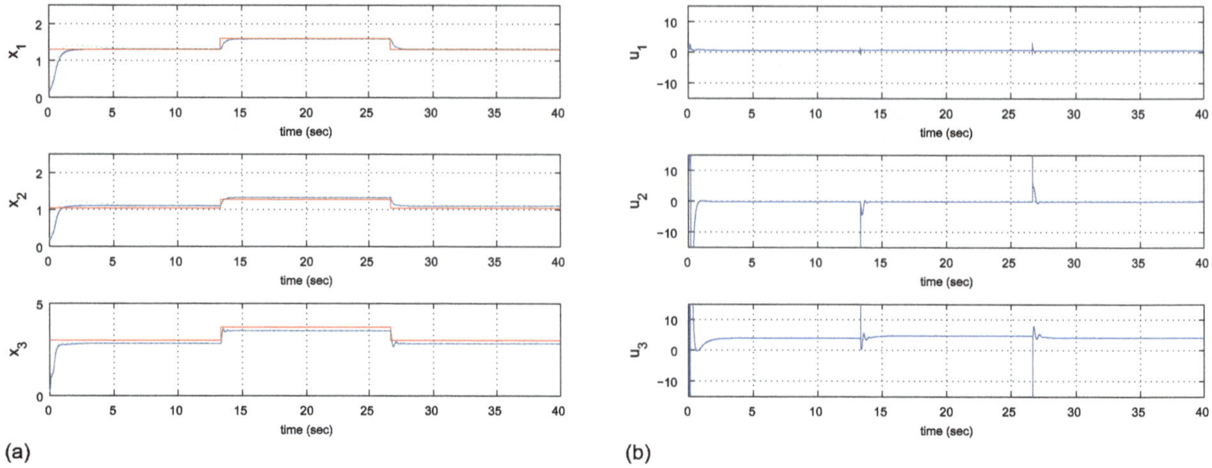

Figure 7.9 Tracking of trajectory 4 by the induction motor-driven reverse-osmosis process using nonlinear optimal control: (a) convergence of state variables x_1 to x_3 to their reference setpoints (red line: setpoint, blue line: real value), green line: estimation by the HKF; (b) convergence of state variables x_4 to x_6 to their reference setpoints (red line: setpoint, blue line: real value, green line: estimation by the HKF)

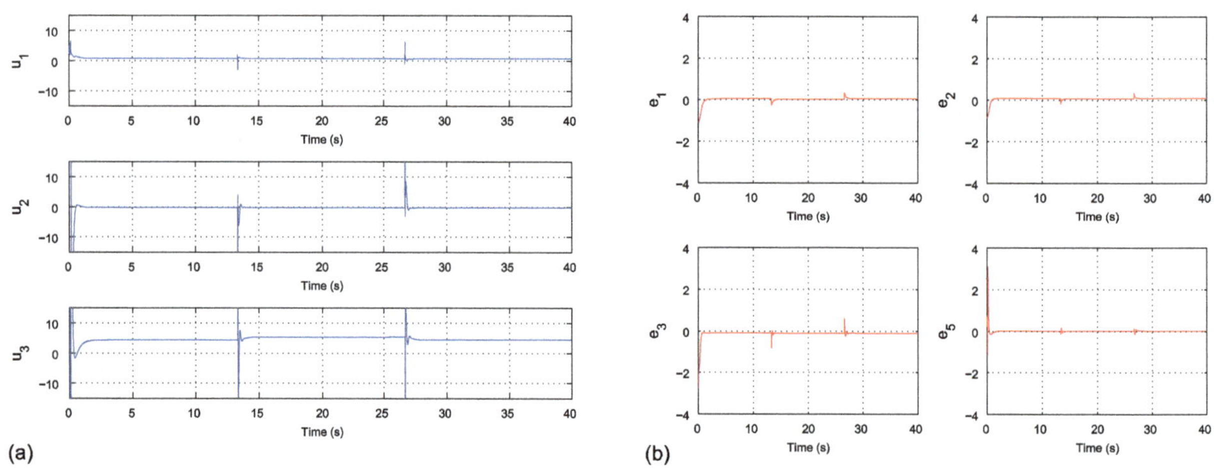

Figure 7.10 Tracking of trajectory 4 by the induction motor-driven reverse-osmosis process using nonlinear optimal control: (a) variation of the control inputs u_1 to u_3; (b) variation of the tracking error variables e_i, $i = 1,2,3,5$ (associated with state variables x_i of the desalination process)

of a subset of its state variables which define the flat outputs of the system (ii) the flat outputs of the system are differentially independent, which means that they and their time derivatives are not connected through a relation in the form of a homogeneous differential equation [91,594,639,489]. The proof of differential flatness properties for a system signifies that through successive differentiations of its flat outputs and a change of state variables, the system can be transformed into the input–output linearized form [491,704,252,898]. Equivalently, differentially flat systems can be written into the canonical Brunovsky form [806,492,640,524,54]. The latter description of the system's state-space model is both controllable and observable and this enables the design of a stabilizing feedback controller with the use of the eigenvalues assignment technique as well as the design of a convergent state estimator, for instance, in

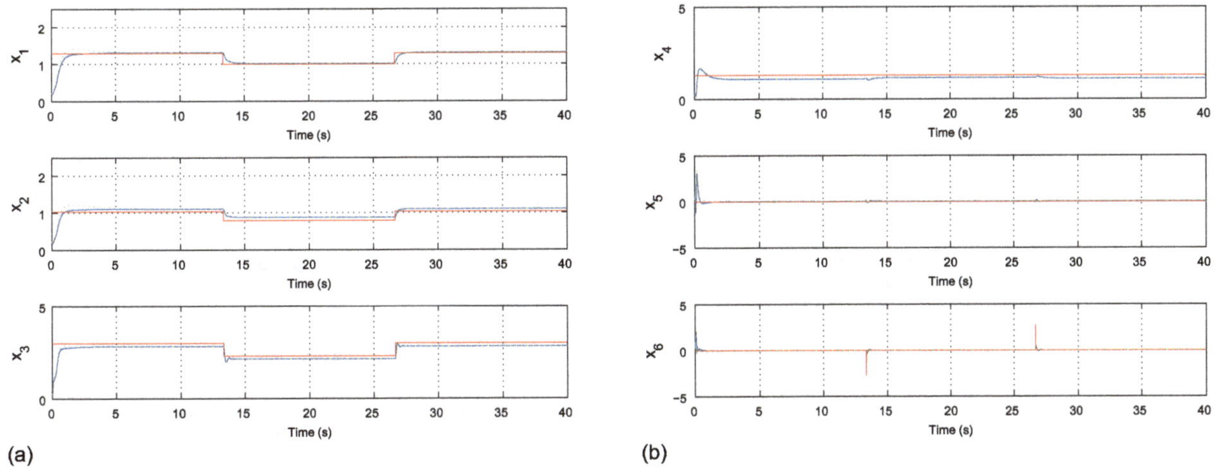

Figure 7.11 Tracking of trajectory 5 by the induction motor-driven reverse-osmosis process using nonlinear optimal control: (a) convergence of state variables x_1 to x_3 to their reference setpoints (red line: setpoint, blue line: real value), green line: estimation by the HKF; (b) convergence of state variables x_4 to x_6 to their reference setpoints (red line: setpoint, blue line: real value, green line: estimation by the HKF)

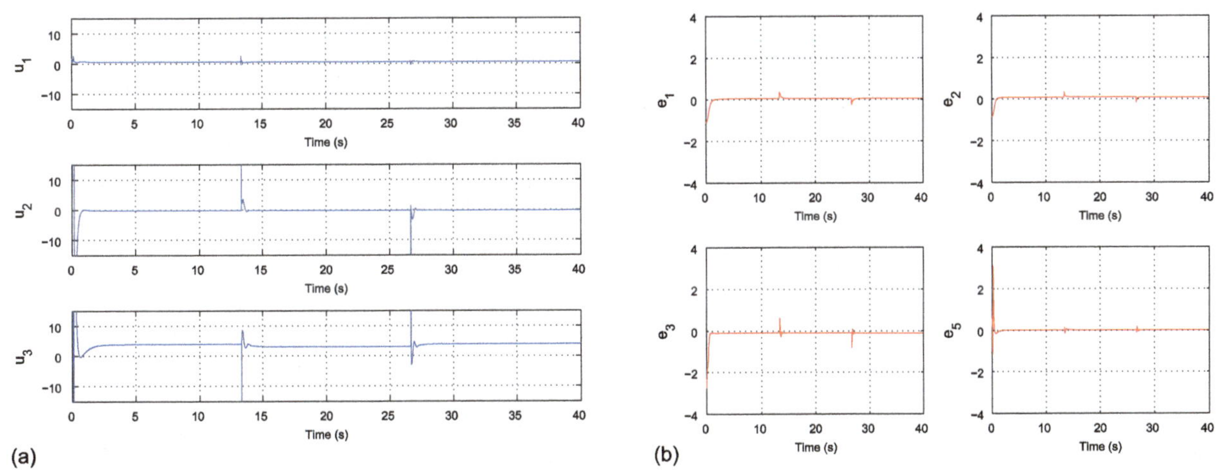

Figure 7.12 Tracking of trajectory 5 by the induction motor-driven reverse-osmosis process using nonlinear optimal control: (a) variation of the control inputs u_1 to u_3; (b) variation of the tracking error variables e_i, $i = 1,2,3,5$ (associated with state variables x_i of the desalination process)

the form of the Kalman filter [439,705,60,706]. However, to compute the control inputs that should be applied on the initial nonlinear state-space model of the system, one should perform inverse transformations, without excluding the appearance of singularities. To avoid such a case and to design a stabilizing feedback controller based on the differential flatness properties of the system, but without the need for changes of state variables, successive differentiations of the flat outputs and state-space model transformations, the present section proposes once again the concept of flatness-based control in successive loops [712,720,714,716].

In the flatness-based control approach in successive loops, the initial nonlinear state-space model of the system is written in the form of a chain of subsystems, where for each one of them it is shown that differential flatness properties

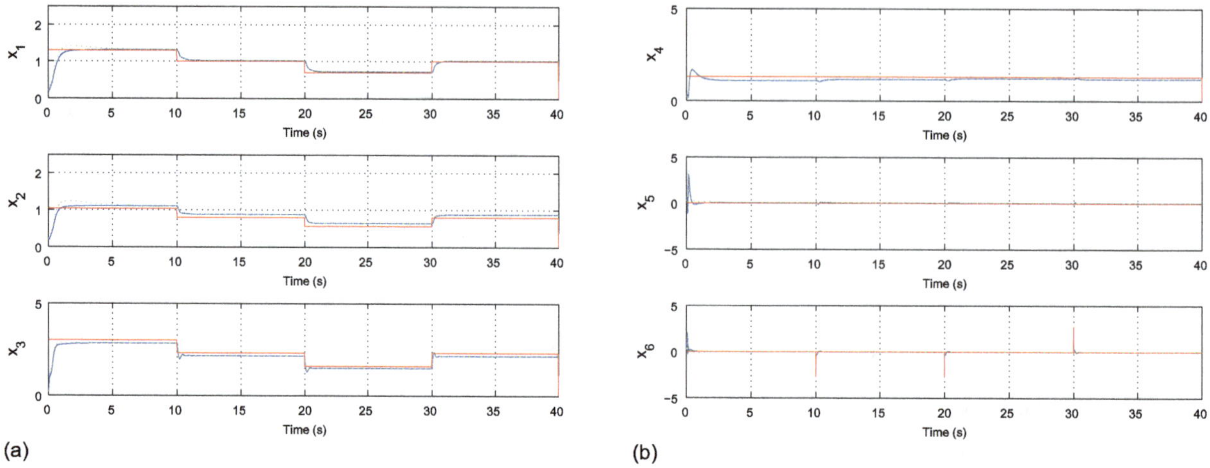

Figure 7.13 *Tracking of trajectory 6 by the induction motor-driven reverse-osmosis process using nonlinear optimal control: (a) convergence of state variables x_1 to x_3 to their reference setpoints (red line: setpoint, blue line: real value), green line: estimation by the HKF; (b) convergence of state variables x_4 to x_6 to their reference setpoints (red line: setpoint, blue line: real value, green line: estimation by the HKF)*

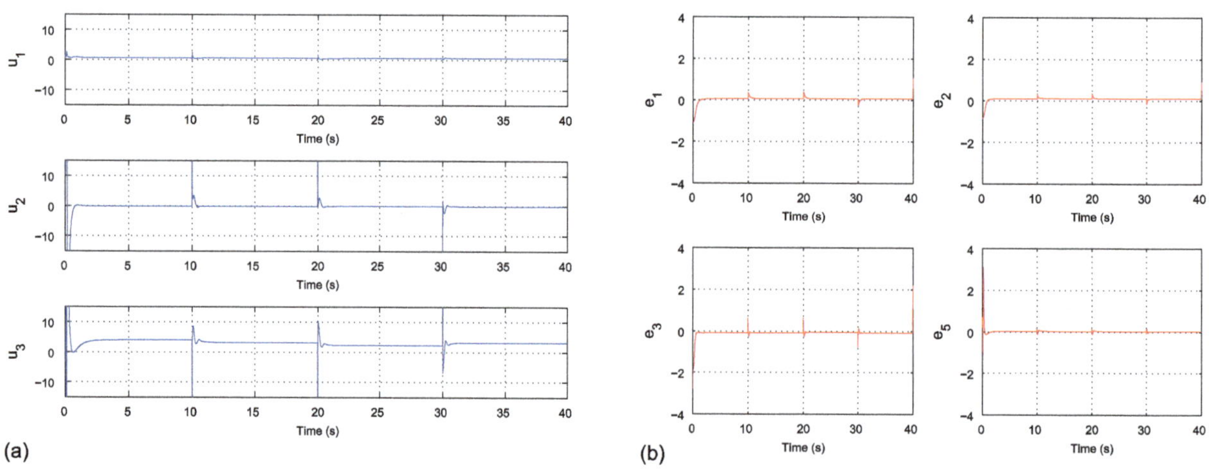

Figure 7.14 *Tracking of trajectory 6 by the induction motor-driven reverse-osmosis process using nonlinear optimal control: (a) variation of the control inputs u_1 to u_3; (b) variation of the tracking error variables e_i, $i = 1,2,3,5$ (associated with state variables x_i of the desalination process)*

hold [712,720,714,716]. The state vector of the (i-th) subsystem becomes also a flat outputs vector about it while the state vector of the $(i+1)$-th subsystem forms a virtual control inputs vector for the i-th subsystem. The proof of differential flatness properties for these subsystems signifies that each one of them can be written in the input–output linearized form and that a stabilizing feedback controller can be designed for each one of them by inverting its dynamics. In this concept, the virtual control inputs vector that stabilizes the (i-th) subsystem should be also a setpoints vector for the $(i+1)$-th subsystem. Finally, the N-th (last) subsystem in this chain of state-space models contains the real control inputs. The inversion of its dynamics gives the stabilizing feedback controller for the initial nonlinear state-space model. The control signal which achieves stabilization for the initial extended nonlinear state-space model uses

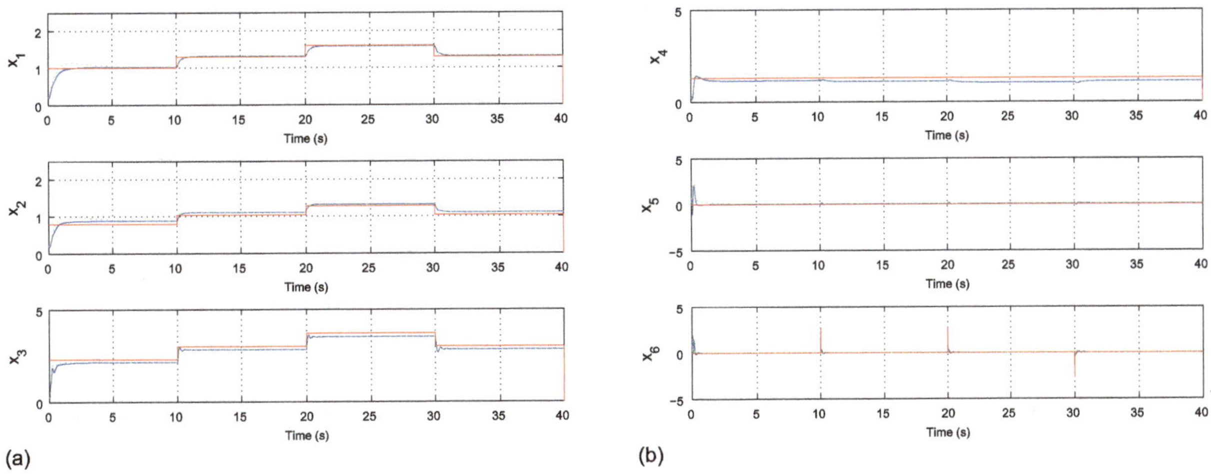

Figure 7.15 Tracking of trajectory 7 by the induction motor-driven reverse-osmosis process using nonlinear optimal control: (a) convergence of state variables x_1 to x_3 to their reference setpoints (red line: setpoint, blue line: real value), green line: estimation by the HKF; (b) convergence of state variables x_4 to x_6 to their reference setpoints (red line: setpoint, blue line: real value, green line: estimation by the HKF)

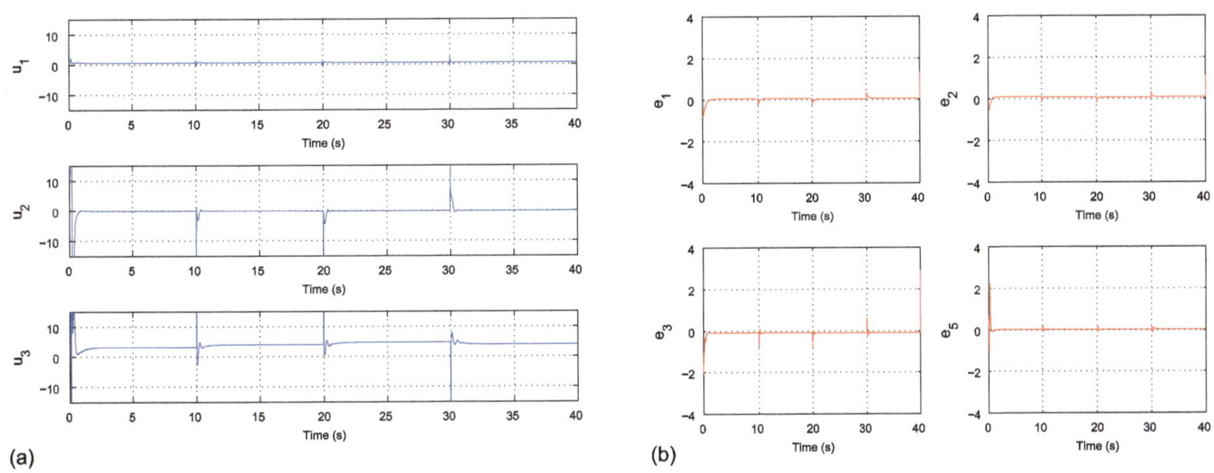

Figure 7.16 Tracking of trajectory 7 by the induction motor-driven reverse-osmosis process using nonlinear optimal control: (a) variation of the control inputs u_1 to u_3; (b) variation of the tracking error variables e_i, $i = 1,2,3,5$ (associated with state variables x_i of the desalination process)

recursively the virtual control inputs of the preceding $N-1$ subsystems, by tracing these subsystems backwards that is from the $(N-1)$-th subsystem to the first one. The global stability of the flatness-based control method in successive loops is proven analytically by showing the asymptotic elimination of the tracking error of each one of the subsystems that constitute the integrated system's dynamics. A global stability proof can be also obtained using Lyapunov analysis [729,731,718].

In the present section, the flatness-based control method in successive loops is tested in two nonlinear industrial systems: (i) centrifugal gas compressors driven by induction motors (ii) centrifugal gas compressors driven by permanent magnet synchronous motors. Gas compressors are of paramount importance for natural gas processing stations and

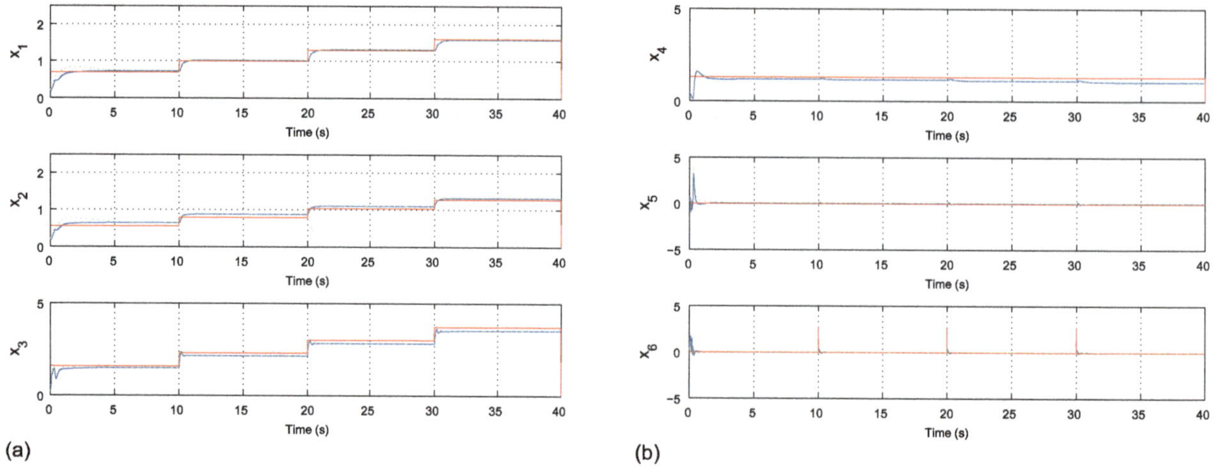

Figure 7.17 *Tracking of trajectory 8 by the induction motor-driven reverse-osmosis process using nonlinear optimal control: (a) convergence of state variables x_1 to x_3 to their reference setpoints (red line: setpoint, blue line: real value), green line: estimation by the HKF; (b) convergence of state variables x_4 to x_6 to their reference setpoints (red line: setpoint, blue line: real value, green line: estimation by the HKF)*

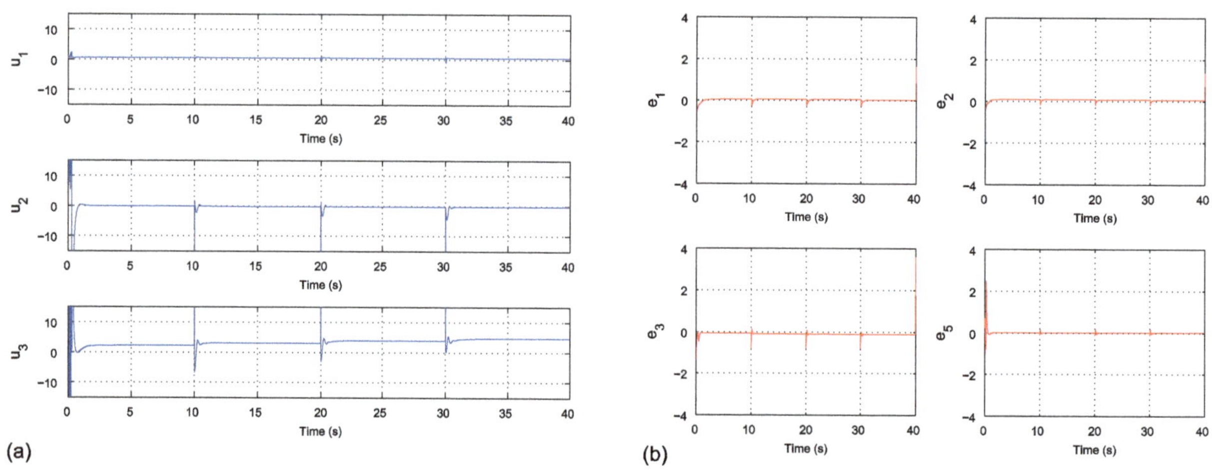

Figure 7.18 *Tracking of trajectory 8 by the induction motor-driven reverse-osmosis process using nonlinear optimal control: (a) variation of the control inputs u_1 to u_3; (b) variation of the tracking error variables e_i, $i = 1,2,3,5$ (associated with state variables x_i, $i = 1,2,3,5$ of the desalination process)*

have progressively become the backbone of the gas distribution network [326,563,873]. They perform liquidization of natural gas thus facilitating its storage and transport [455,1076]. Besides, they enable the transfer of natural gas through pipelines under agreed pressure specifications [256,235]. The growing interest in the optimized functioning of gas distribution networks and gas terminals has motivated much-related research during the last years. Recent advancements on optimization techniques for the gas distribution network are given in [144,187,424,236]. Recent advancements in the control of gas compressors and pressure levels in the gas network are outlined in [590,279,283].

Centrifugal gas compressors exhibit multivariable and nonlinear dynamics [64,872,289]. If inefficiently controlled gas compressors may exhibit surge, which is associated with unsteady mass flow through the compressor and sustained

Table 7.1 Nonlinear optimal control

Tracking RMSE for the desalination unit in the disturbance-free case						
	$RMSE_{x_1}$	$RMSE_{x_2}$	$RMSE_{x_3}$	$RMSE_{x_4}$	$RMSE_{x_5}$	$RMSE_{x_6}$
$test_1$	0.0005	0.0030	0.0078	0.0089	0.0012	0.0001
$test_2$	0.0001	0.0020	0.0079	0.0091	0.0012	0.0001
$test_3$	0.0010	0.0036	0.0067	0.0063	0.0013	0.0001
$test_4$	0.0005	0.0030	0.0070	0.0089	0.0012	0.0001
$test_5$	0.0005	0.0030	0.0078	0.0089	0.0011	0.0001
$test_6$	0.0010	0.0036	0.0067	0.0063	0.0010	0.0001
$test_7$	0.0005	0.0030	0.0078	0.0089	0.0017	0.0001
$test_8$	0.0006	0.0018	0.0065	0.0073	0.0073	0.0001

Table 7.2 Nonlinear optimal control

Tracking RMSE for the desalination unit in the case of disturbances						
$\Delta a\%$	$RMSE_{x_1}$	$RMSE_{x_2}$	$RMSE_{x_3}$	$RMSE_{x_4}$	$RMSE_{x_5}$	$RMSE_{x_6}$
0%	0.0010	0.0036	0.0067	0.0063	0.0013	0.0001
10%	0.0011	0.0037	0.0064	0.0063	0.0013	0.0001
20%	0.0012	0.0038	0.0062	0.0063	0.0012	0.0001
30%	0.0013	0.0038	0.0060	0.0064	0.0013	0.0001
40%	0.0014	0.0039	0.0058	0.0064	0.0013	0.0001
50%	0.0015	0.0039	0.0057	0.0064	0.0013	0.0001
60%	0.0015	0.0040	0.0056	0.0064	0.0013	0.0001

Table 7.3 Nonlinear optimal control

Convergence times (sec) of states of the desalination unit						
No test	$T_s\, x_1$	$T_s\, x_2$	$T_s\, x_3$	$T_s\, x_4$	$T_s\, x_5$	$T_s\, x_6$
$test_1$	2.5	2.5	4.0	2.5	1.0	0.5
$test_2$	2.5	2.0	3.5	2.0	1.0	0.5
$test_3$	2.5	2.0	4.0	2.0	1.0	0.5
$test_4$	2.5	2.0	3.5	2.5	0.5	0.5
$test_5$	2.5	2.0	4.0	2.5	0.5	0.5
$test_6$	2.5	2.0	3.5	2.5	0.5	0.5
$test_7$	2.0	2.0	3.0	2.0	0.5	0.5
$test_8$	2.0	2.0	3.0	1.5	0.5	0.5

oscillations that may finally damage the compressor's blades [33,137,130]. To treat the complicated nonlinear and multi-variable dynamics of gas compressors, elaborated nonlinear control algorithms have been proposed [73,328,414]. Fault tolerance, precision, and robustness are prerequisites for the actuation systems of gas compressors [960,602,382]. One can distinguish for instance several results on NMPC (nonlinear model predictive control), as well as on MPC which is applied to linearized models of the compressors [172,173]. Besides, one can note results on the use of global linearization-based control and state estimation-based control [217,103,41]. Additionally, one can find results based

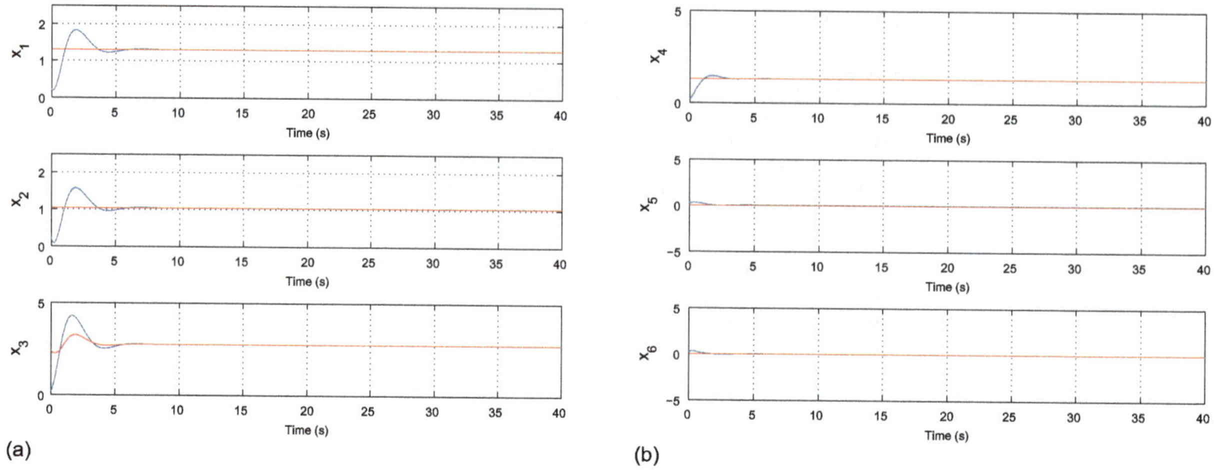

Figure 7.19 *Tracking of trajectory 1 by the induction motor-driven reverse-osmosis process using flatness-based control in successive loops: (a) convergence of state variables x_1 to x_3 to their reference setpoints (red line: setpoint, blue line: real value), green line: estimation by the HKF; (b) convergence of state variables x_4 to x_6 to their reference setpoints (red line: setpoint, blue line: real value, green line: estimation by the HKF)*

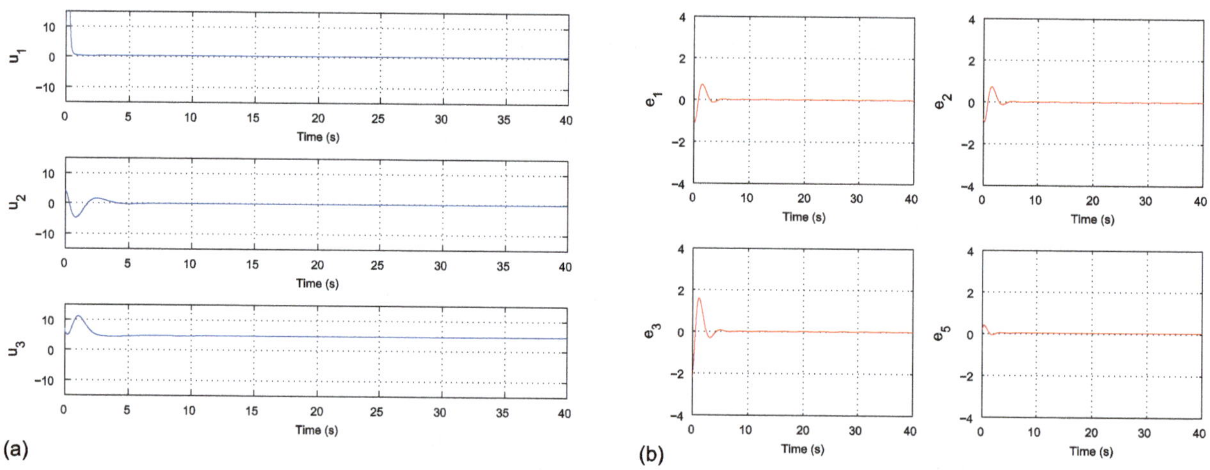

Figure 7.20 *Tracking of trajectory 1 by the induction motor-driven reverse-osmosis process using flatness-based control in successive loops: (a) variation of the control inputs u_1 to u_3; (b) variation of the tracking error variables e_i, $i = 1,2,3,5$ (associated with state variables x_i of the desalination process)*

on linearization around multiple equilibria and subsequent use of LMI techniques [281,848]. Usually, the opening or closure of the compressor's inlet, outlet, and retrofitting valves function as control inputs [262,263]. Moreover, when the compressor receives actuation from an electric motor, the motor's torque becomes also a control input that determines significantly the stability properties and the optimal function of the compressor [45,80,174].

Several types of electric motors have been considered as actuators of centrifugal gas compressors, and these are mainly synchronous three-phase motors, asynchronous (induction) motors and multi-phase motors [647,891,532]. To stabilize the functioning of the compressor and to make its input and output pressure as well as its mass flow reach

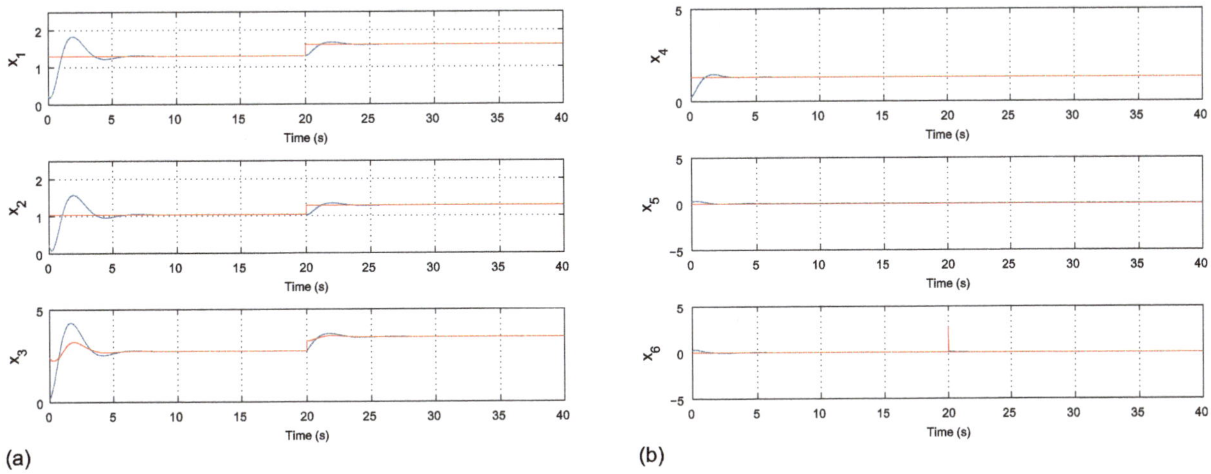

Figure 7.21 *Tracking of trajectory 2 by the induction motor-driven reverse-osmosis process using flatness-based control in successive loops: (a) convergence of state variables x_1 to x_3 to their reference setpoints (red line: setpoint, blue line: real value), green line: estimation by the HKF; (b) convergence of state variables x_4 to x_6 to their reference setpoints (red line: setpoint, blue line: real value, green line: estimation by the HKF)*

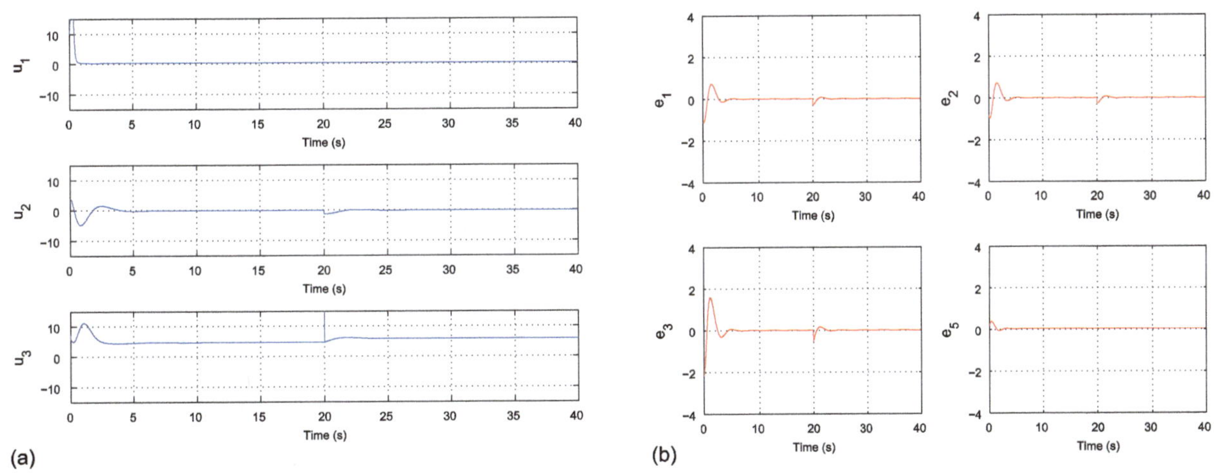

Figure 7.22 *Tracking of trajectory 2 by the induction motor-driven reverse-osmosis process using flatness-based control in successive loops: (a) variation of the control inputs u_1 to u_3; (b) variation of the tracking error variables e_i, $i = 1,2,3,5$ (associated with state variables x_i of the desalination process)*

the targeted setpoints, elaborated nonlinear control methods have been proposed [102,399]. The control problem of distributed gas compressors is primarily found in gas-processing stations where the aim is to mix volumes of gas, arriving through different gas-transmission pipelines under different mass rate and pressure characteristics [1089,399]. Different configurations of gas compressors, being connected serially or in parallel, can result into gas processing units of nonlinear, complicated and high-dimensional state-space models [1026,601,600,462]. The development of nonlinear control and optimization methods that can assure the reliable functioning of such gas-processing stations is obviously a non-trivial task [418,673,603]. Similar are the control problems in the case of electrically driven air compressors

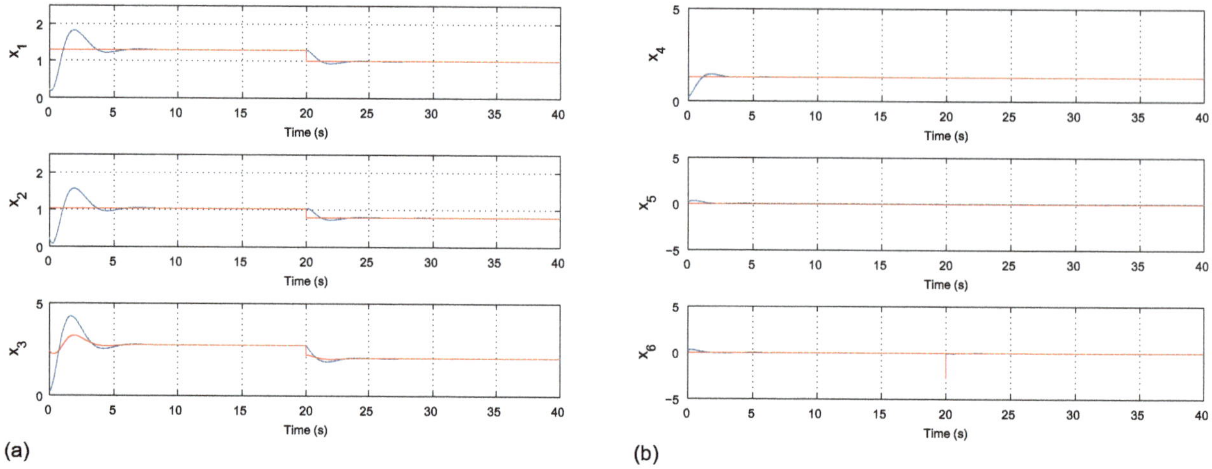

Figure 7.23 Tracking of trajectory 3 by the induction motor-driven reverse-osmosis process using flatness-based control in successive loops: (a) convergence of state variables x_1 to x_3 to their reference setpoints (red line: setpoint, blue line: real value), green line: estimation by the HKF; (b) convergence of state variables x_4 to x_6 to their reference setpoints (red line: setpoint, blue line: real value, green line: estimation by the HKF)

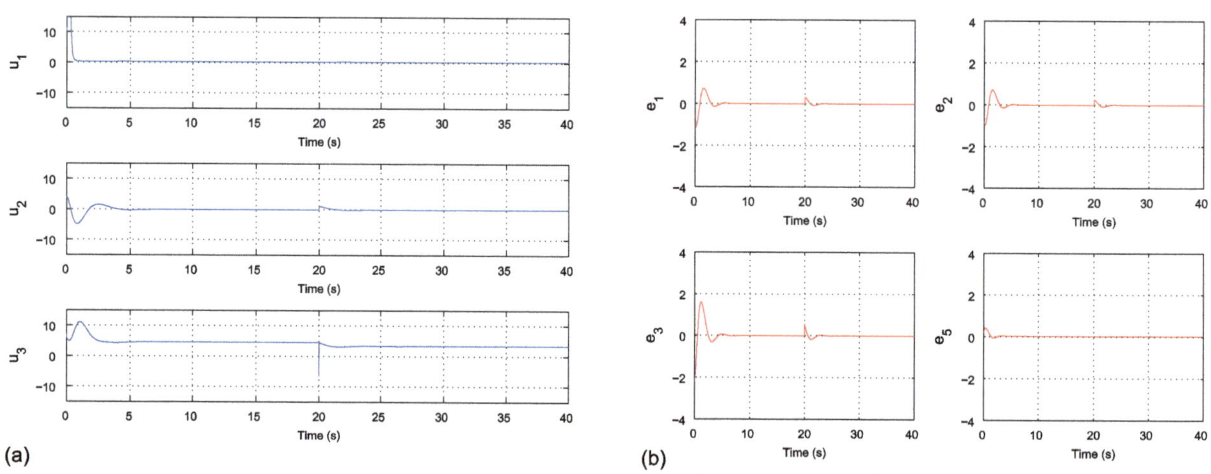

Figure 7.24 Tracking of trajectory 3 by the induction motor-driven reverse-osmosis process using flatness-based control in successive loops: (a) variation of the control inputs u_1 to u_3; (b) variation of the tracking error variables e_i, $i = 1,2,3,5$ (associated with state variables x_i of the desalination process)

and this is another industrial domain where this section's results potentially can be used [329,1086,361,661,733]. The present section considers models of gas compressors with electric actuation coming from either three-phase induction (asynchronous) motors or from three-phase permanent magnet synchronous motors. The global stability properties of the flatness-based control method in successive loops are analytically proven and experimentally confirmed through simulation tests.

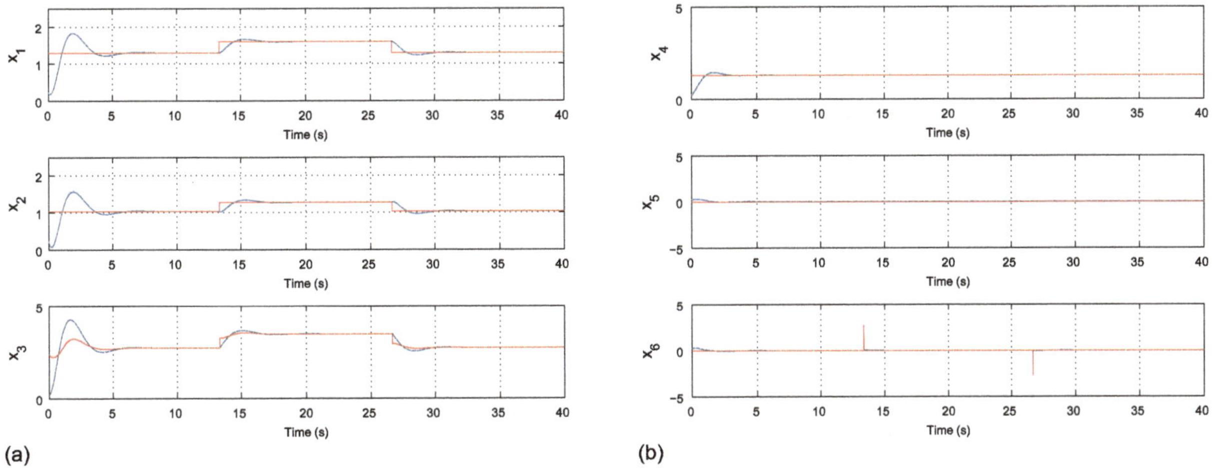

Figure 7.25 Tracking of trajectory 4 by the induction motor-driven reverse-osmosis process using flatness-based control in successive loops: (a) convergence of state variables x_1 to x_3 to their reference setpoints (red line: setpoint, blue line: real value), green line: estimation by the HKF; (b) convergence of state variables x_4 to x_6 to their reference setpoints (red line: setpoint, blue line: real value, green line: estimation by the HKF)

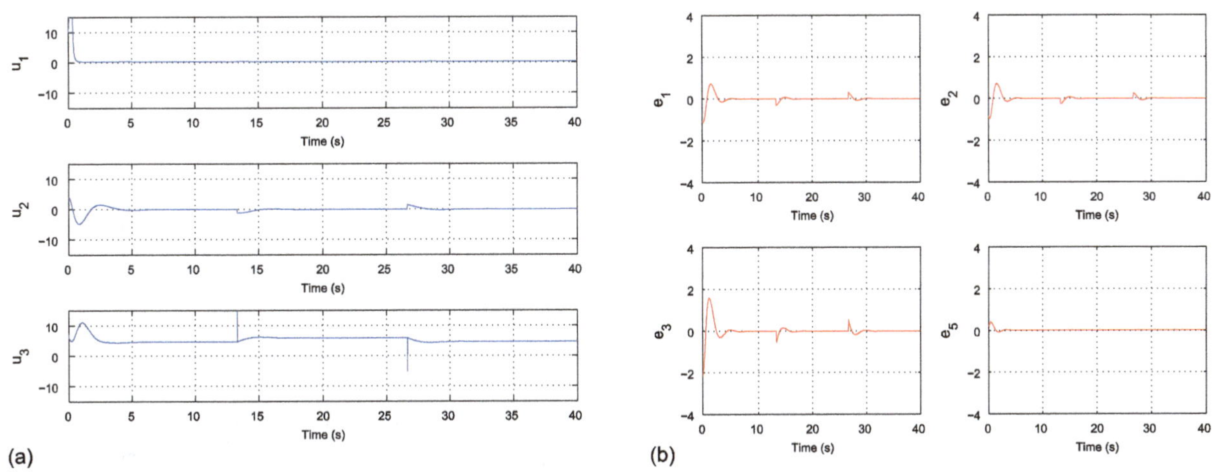

Figure 7.26 Tracking of trajectory 4 by the induction motor-driven reverse-osmosis process using flatness-based control in successive loops: (a) variation of the control inputs u_1 to u_3; (b) variation of the tracking error variables e_i, $i = 1,2,3,5$ (associated with state variables x_i of the desalination process)

7.2.2 Dynamic model of the induction motor-driven gas-compressor

7.2.2.1 Dynamics of the IM-driven gas compressor

The diagram of the integrated system which includes a gas compressor that is actuated by an Induction Motor is given in Figure 7.35. Two tanks are considered of volumes V_i and V_o, respectively. These represent the plenum before and after compression. The inlet tank stores uncompressed gas arriving to it through the inlet valve. The outlet tank stores compressed gas arriving to it through the outlet valve. The pressure of the inlet valve is denoted as p_i. The pressure of the outlet valve is denoted as p_o. The mass flow through the compressor is denoted as m. The mass flow is dependent

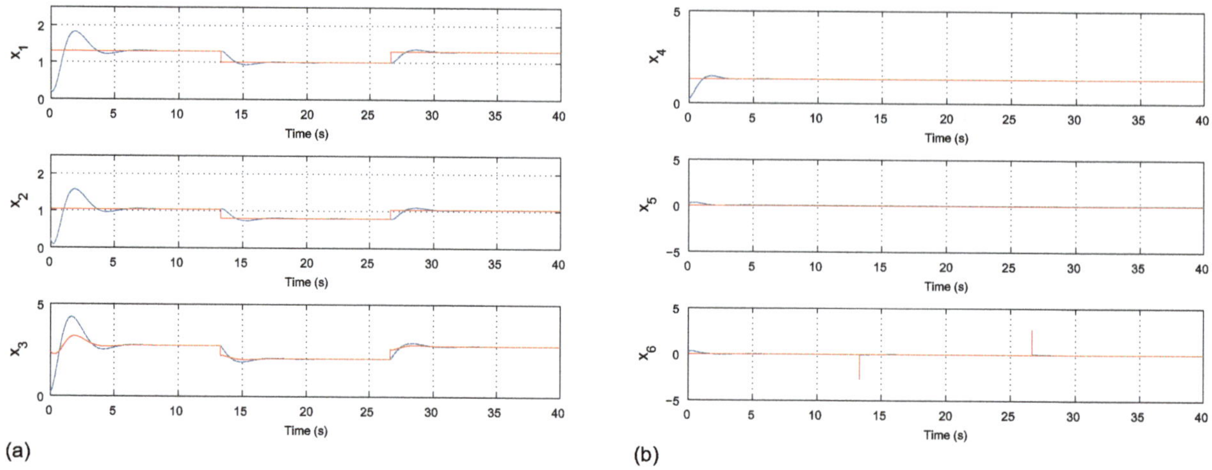

Figure 7.27 Tracking of trajectory 5 by the induction motor-driven reverse-osmosis process using flatness-based control in successive loops: (a) convergence of state variables x_1 to x_3 to their reference setpoints (red line: setpoint, blue line: real value), green line: estimation by the HKF; (b) convergence of state variables x_4 to x_6 to their reference setpoints (red line: setpoint, blue line: real value, green line: estimation by the HKF)

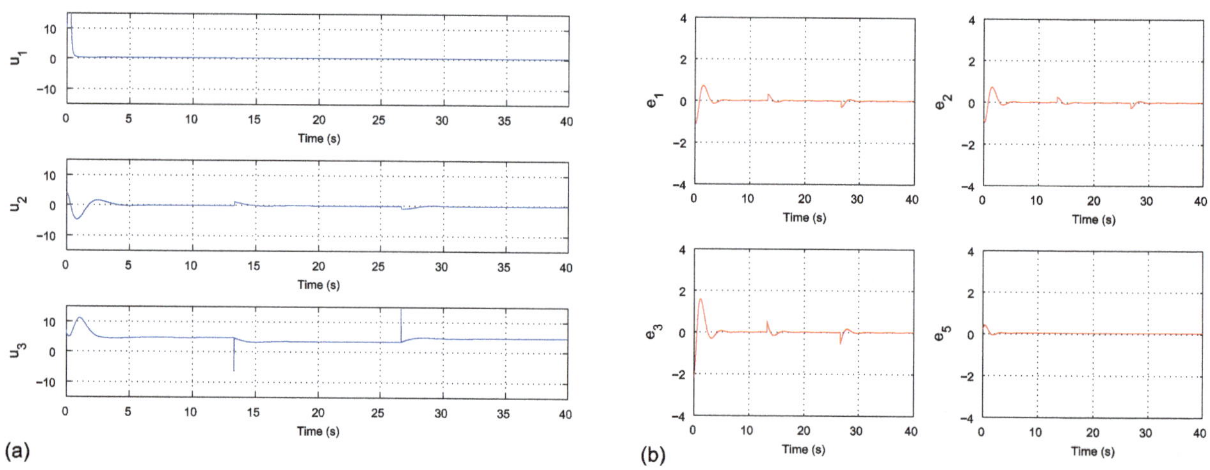

Figure 7.28 Tracking of trajectory 5 by the induction motor-driven reverse-osmosis process using flatness-based control in successive loops: (a) variation of the control inputs u_1 to u_3; (b) variation of the tracking error variables e_i, $i = 1,2,3,5$ (associated with state variables x_i of the desalination process)

on the tank volumes and gas velocities, as well as on inlet and outlet tank conditions, denoted as a_i and a_o, respectively. The compressor's turn speed, which is also the motor's turn speed is denoted as ω. The compressor's torque is denoted as τ_c. The motor's torque is denoted as τ_d and is a control input for the compressor. The moment of inertia of the turning part of the compressor is denoted as J [873,45–174,733].

The mass flow through the recycle valve is described by a first-order differential equation with a time constant which is denoted as T_r. The steady-state recycle mass flow is denoted as $m_{r,s}$. The compressor's characteristics which define the mass flow rate through it are denoted by function $\pi(m, \omega)$. The compressor's torque τ_c is given by function

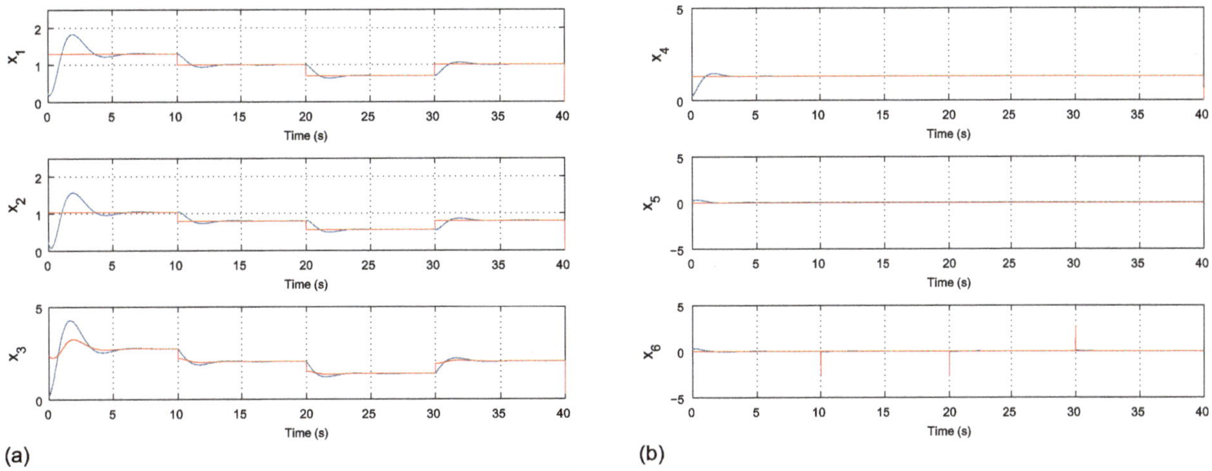

Figure 7.29 Tracking of trajectory 6 by the induction motor-driven reverse-osmosis process using flatness-based control in successive loops: (a) convergence of state variables x_1 to x_3 to their reference setpoints (red line: setpoint, blue line: real value), green line: estimation by the HKF; (b) convergence of state variables x_4 to x_6 to their reference setpoints (red line: setpoint, blue line: real value, green line: estimation by the HKF)

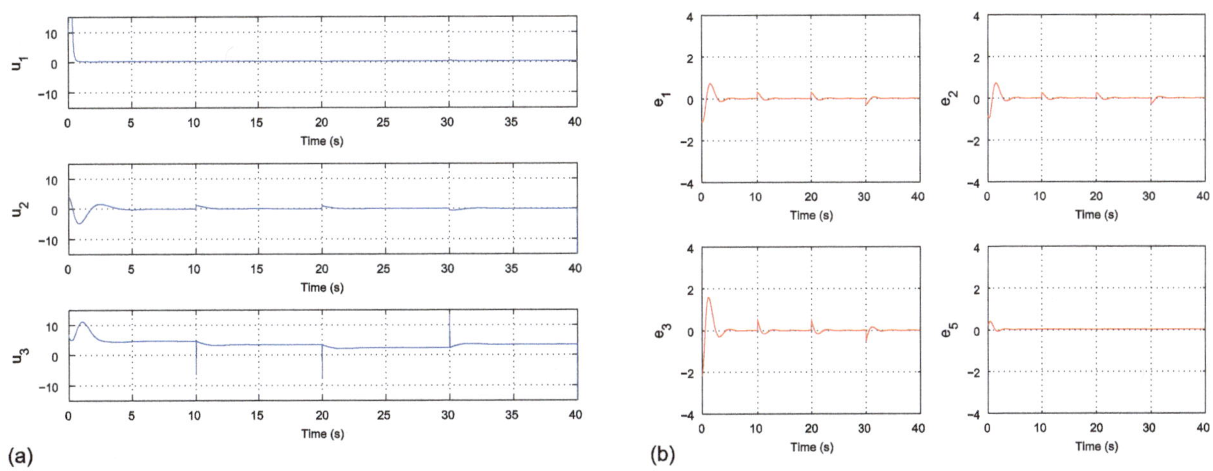

Figure 7.30 Tracking of trajectory 6 by the induction motor-driven reverse-osmosis process using flatness-based control in successive loops: (a) variation of the control inputs u_1 to u_3; (b) variation of the tracking error variables e_i, $i = 1,2,3,5$ (associated with state variables x_i of the desalination process)

$\tau_c(m,\omega)$ which signifies that the torque which is developed by the compressor depends on its turn speed and on the mass flow rate through it. In aggregate, the dynamic model of the compressor is given by the following equations [873]:

$$\dot{p}_i = \tfrac{a_i^2}{V_i}[m_i(p_i) - m + m_p] \tag{7.99}$$

$$\dot{p}_o = \tfrac{a_o^2}{V_o}[-m_o(p_o) + m - m_r] \tag{7.100}$$

$$\dot{m} = \tfrac{1}{L}[\pi(\omega,m)p_i - p_o] \tag{7.101}$$

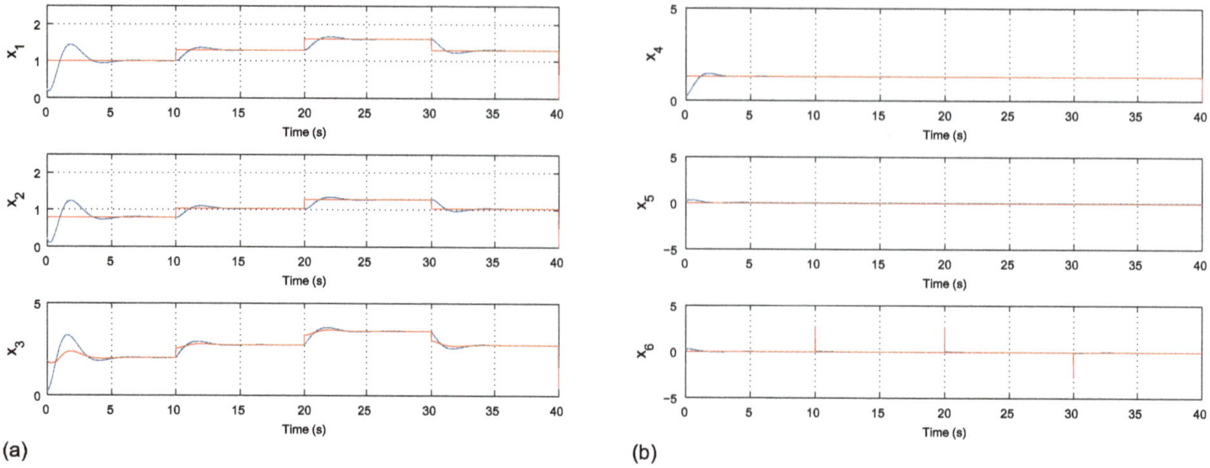

Figure 7.31 Tracking of trajectory 7 by the induction motor-driven reverse-osmosis process using flatness-based control in successive loops: (a) convergence of state variables x_1 to x_3 to their reference setpoints (red line: setpoint, blue line: real value), green line: estimation by the HKF; (b) convergence of state variables x_4 to x_6 to their reference setpoints (red line: setpoint, blue line: real value, green line: estimation by the HKF)

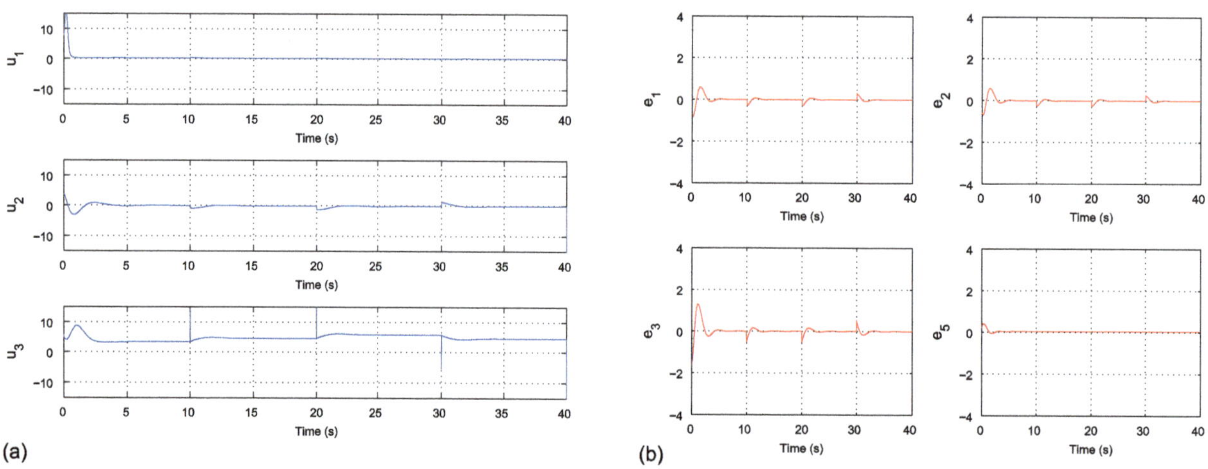

Figure 7.32 Tracking of trajectory 7 by the induction motor-driven reverse-osmosis process using flatness-based control in successive loops: (a) variation of the control inputs u_1 to u_3; (b) variation of the tracking error variables e_i, $i = 1,2,3,5$ (associated with state variables x_i of the desalination process)

$$\dot{\omega} = \frac{1}{J}[\tau_d - \tau_c(\omega, m) - \nu\omega] \tag{7.102}$$

$$\dot{m}_r = \frac{1}{T_r}[m_{r,ss}(p_i, p_o, K_i) - m_i] \tag{7.103}$$

where ν is a mechanical friction coefficient. Besides, about the inlet, outlet, and reverse mass flow in the gas compressor, it holds that [873]

$$m_i(p_i) = K_i\sqrt{p_{atm} - p_i} \tag{7.104}$$

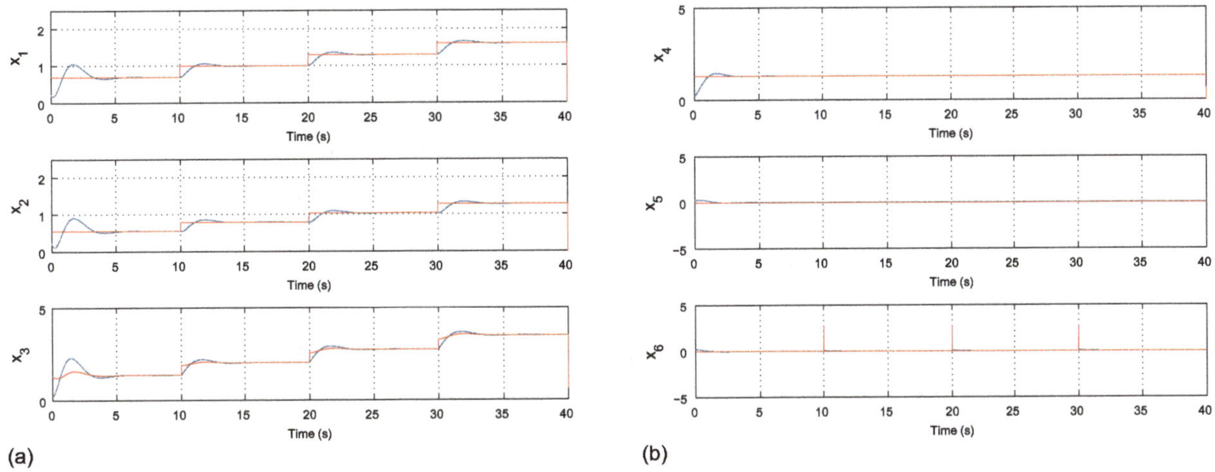

Figure 7.33 *Tracking of trajectory 1 by the induction motor-driven reverse-osmosis process using flatness-based control in successive loops: (a) convergence of state variables x_1 to x_3 to their reference setpoints (red line: setpoint, blue line: real value), green line: estimation by the HKF; (b) convergence of state variables x_4 to x_6 to their reference setpoints (red line: setpoint, blue line: real value, green line: estimation by the HKF)*

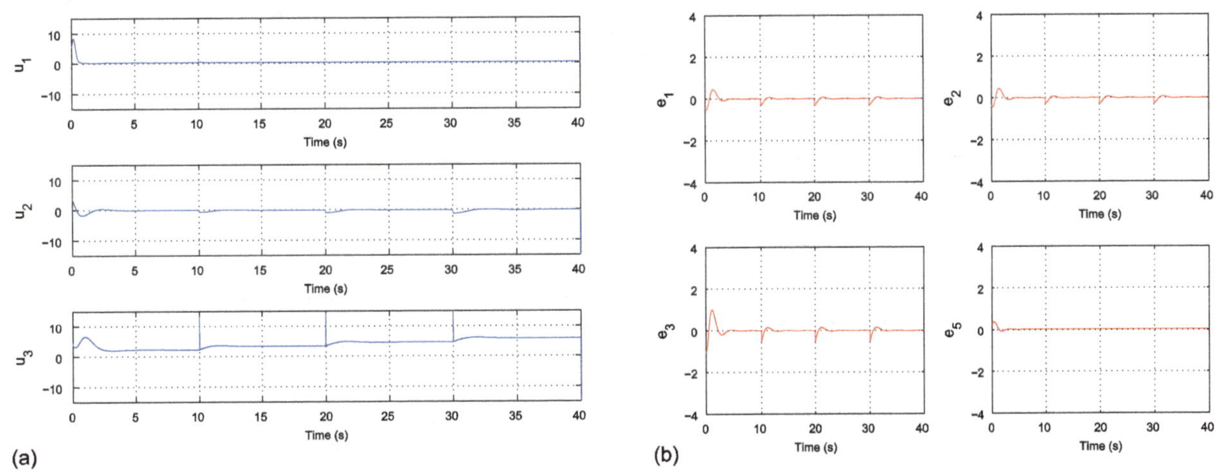

Figure 7.34 *Tracking of trajectory 8 by the induction motor-driven reverse-osmosis process using flatness-based control in successive loops: (a) variation of the control inputs u_1 to u_3; (b) variation of the tracking error variables e_i, $i = 1,2,3,5$ (associated with state variables x_i of the desalination process)*

$$m_o(p_o) = K_o\sqrt{p_o - p_{atm}} \tag{7.105}$$

$$m_{r,ss}(p_i, p_o, K_r) = K_r\sqrt{p_2 - p_1} \tag{7.106}$$

where K_i is the inlet valve constant which is used as a control input, K_o is the outlet valve constant, p_{atm} is the atmospheric pressure, and K_r is the recycle valve constant which also serves as a control input. The three valves' constants depend on each valve's throttle position. Non-return values ensure also that mass flow rates m_i, m_o, and m_r cannot change direction.

Table 7.4 Flatness-based control in successive loops

Tracking RMSE for the desalination unit in the disturbance-free case						
	$RMSE_{x_1}$	$RMSE_{x_2}$	$RMSE_{x_3}$	$RMSE_{x_4}$	$RMSE_{x_5}$	$RMSE_{x_6}$
$test_1$	0.0004	0.0004	0.0122	0.0001	0.0014	0.0001
$test_2$	0.0009	0.0009	0.0068	0.0001	0.0014	0.0001
$test_3$	0.0014	0.0015	0.0151	0.0001	0.0014	0.0001
$test_4$	0.0006	0.0006	0.0123	0.0001	0.0014	0.0001
$test_5$	0.0013	0.0012	0.0125	0.0001	0.0014	0.0001
$test_6$	0.0041	0.0038	0.0142	0.0001	0.0014	0.0001
$test_7$	0.0040	0.0036	0.0048	0.0001	0.0014	0.0001
$test_8$	0.0021	0.0018	0.0129	0.0001	0.0014	0.0001

Table 7.5 Flatness-based control in successive loops

Tracking RMSE for the desalination unit in the case of disturbances						
$\Delta a\%$	$RMSE_{x_1}$	$RMSE_{x_2}$	$RMSE_{x_3}$	$RMSE_{x_4}$	$RMSE_{x_5}$	$RMSE_{x_6}$
0%	0.0014	0.0015	0.0151	0.0001	0.0014	0.0001
10%	0.0014	0.0015	0.0151	0.0001	0.0014	0.0001
20%	0.0014	0.0015	0.0150	0.0001	0.0014	0.0001
30%	0.0014	0.0015	0.0150	0.0001	0.0014	0.0001
40%	0.0015	0.0015	0.0149	0.0001	0.0014	0.0001
50%	0.0015	0.0015	0.0149	0.0001	0.0014	0, 0001
60%	0.0015	0.0015	0.0149	0.0001	0, 0014	0.0001

Table 7.6 Flatness-based control in successive loops

Convergence times (sec) of states of the desalination unit						
No test	$T_s\, x_1$	$T_s\, x_2$	$T_s\, x_3$	$T_s\, x_4$	$T_s\, x_5$	$T_s\, x_6$
$test_1$	6.0	6.0	6.0	3.5	2.5	2.0
$test_2$	6.0	6.0	6.0	3.5	2.5	1.5
$test_3$	6.0	6.0	6.0	3.0	2.5	2.0
$test_4$	6.0	6.0	6.0	3.0	2.0	1.5
$test_5$	6.0	6.0	6.0	3.5	2.5	2.0
$test_6$	6.0	6.0	6.0	3.5	2.5	1.5
$test_7$	5.5	5.5	5.0	3.5	2.5	1.5
$test_8$	5.0	5.0	5.0	3.5	2.5	1.5

By substituting Eq. (7.104), Eq. (7.105), and Eq. (7.106) into Eq. (7.99) to Eq. (7.103), one obtains [873]

$$\dot{p}_i = \frac{a_i^2}{V_i}[K_i\sqrt{p_{atm} - p_i} - m + m_r] \tag{7.107}$$

$$\dot{p}_o = \frac{a_o^2}{V_o}[-K_o\sqrt{p_o - p_{atm}} + m - m_r] \tag{7.108}$$

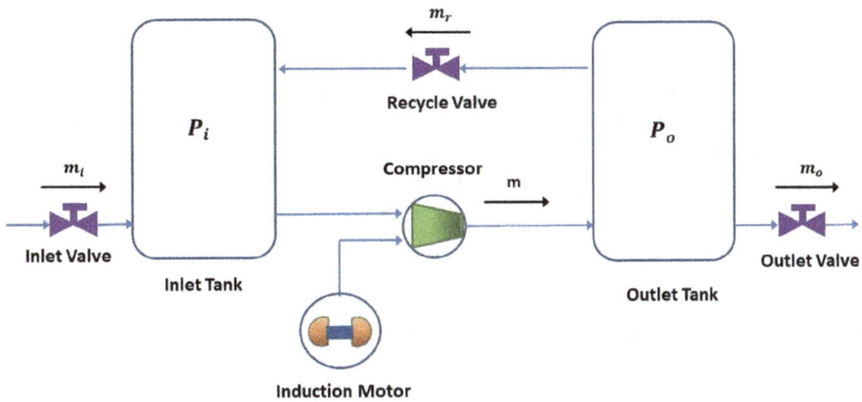

Figure 7.35 Diagram of the integrated system that includes a gas compressor being actuated by an Induction Motor

$$\dot{m} = \frac{1}{L}[\pi(\omega, m)p_i - p_o] \tag{7.109}$$

$$\dot{\omega} = \frac{1}{J}[\tau_d - \tau_c(m, \omega) - v\omega] \tag{7.110}$$

$$\dot{m}_r = \frac{1}{T_r}[K_r\sqrt{p_o - p_i} - K_i\sqrt{p_{atm} - p_i}] \tag{7.111}$$

By defining the state vector $x = [x_1, x_2, x_3, x_4, x_5]^T = [p_i, p_o, m, \omega, m_r]^T$ and the control inputs vector K_i, K_r, τ_d the dynamic model of the compressor is written as

$$\dot{x}_1 = \frac{a_i^2}{V_i}[K_i\sqrt{p_{atm} - x_1} - x_3 + x_5] \tag{7.112}$$

$$\dot{x}_2 = \frac{a_o^2}{V_o}[-K_o\sqrt{x_2 - p_{atm}} + x_3 - x_5] \tag{7.113}$$

$$\dot{x}_3 = \frac{1}{L}[\pi(x_3, x_4)x_1 - x_2] \tag{7.114}$$

$$\dot{x}_4 = \frac{1}{J}[\tau_d - \tau_c(x_3, x_4) - vx_4] \tag{7.115}$$

$$\dot{x}_5 = \frac{1}{T_r}[K_r\sqrt{x_2 - x_1} - K_i\sqrt{p_{atm} - x_1}] \tag{7.116}$$

Under the assumption of field orientation, the magnetic field of the induction motor rotates asynchronously at a rate which is defined by the first-order time derivative of the field-orientation angle [715,711–725]

$$\rho = tan^{-1}(\frac{\psi_{rb}}{\psi_{ra}}) \tag{7.117}$$

In the dq asynchronously rotating reference frame only the d-axis constituent of the magnetic flux remains while the q-axis constituent of the magnetic flux vanishes, that in $\psi = \psi_{rd}$ and $\psi_{rq} = 0$. In the new coordinates frame, the induction motor model is written as: [715,711,725]

$$\dot{\theta} = \omega \tag{7.118}$$

$$\dot{\omega} = \frac{\mu}{J}\psi_{rd}i_{sq} - \frac{T_L}{J} - \frac{T_b}{J} \tag{7.119}$$

$$\dot{\psi}_{rd} = -\alpha\psi_{rd} + \alpha M i_{sd} \tag{7.120}$$

$$\dot{i}_{sd} = -\gamma i_{sd} + \alpha\beta\psi_{rd} + n_p\omega i_{sq} + \frac{\alpha M {i_{sq}}^2}{\psi_{rd}} + \frac{1}{\sigma L_s}v_{sd} \tag{7.121}$$

$$\dot{i}_{sq} = -\gamma i_{sq} - \beta n_p\omega\psi_{rd} - n_p\omega i_{sd} - \frac{\alpha M i_{sq} i_{sd}}{\psi_d} + \frac{1}{\sigma L_s}v_{sq} \tag{7.122}$$

$$\dot{\rho} = n_p\omega + \frac{\alpha M i_{sq}}{\psi_{rd}} \tag{7.123}$$

In Eq. (7.119), T_L is the load torque and in the case of the compressor $T_L = \tau_c(m, \omega)$, while T_b is the torque due to mechanical friction with $T_b = v\omega$. M denotes mutual inductance, and n_p is the number of poles. Parameters σ, γ, μ, a, and β are associated with the coefficients of the electric circuit of the stator and the rotor (resistance, inductance, mutual inductance and number of poles) [711,725,733]. Next, the following state variables are defined: $x = [x_6, x_7, x_8, x_9]^T = [\psi_{rd}, i_{sd}, i_{sq}, \rho]^T$ while the control inputs vector is $u = [v_{sd}, v_{sq}]^T$. Thus, in the dynamic model of the induction motor, the driving torque is rewritten as

$$\tau_d = \mu x_6 x_8 \tag{7.124}$$

while about the dynamics of the electric part of the induction motor, one has

$$\dot{x}_6 = -\alpha x_6 + \alpha M x_7 \tag{7.125}$$

$$\dot{x}_7 = -\gamma x_7 + \alpha\beta x_6 + n_p x_4 x_8 + \frac{\alpha M x_8^2}{x_6} + \frac{1}{\sigma L_s} v_{sd} \tag{7.126}$$

$$\dot{x}_8 = -\gamma x_8 - \beta n_p x_4 x_6 - n_p x_4 x_7 - \frac{\alpha M x_8 x_7}{x_6} + \frac{1}{\sigma L_s} v_{sq} \tag{7.127}$$

$$\dot{x}_9 = n_p x_4 + \frac{\alpha M x_8}{x_6} \tag{7.128}$$

7.2.3 State-space model of the electrically actuated gas compressor

The state vector of the integrated gas compression system which consists of a centrifugal gas-compressor and a three-phase induction motor is given by:

$$\begin{aligned} x &= [x_1, x_2, x_3, x_4, x_5, x_6, x_7, x_8, x_9]^T \Rightarrow \\ x &= [P_i, P_o, m, \omega, m_r, \psi_{r_d}, i_{s_d}, i_{s_q}, \rho]^T \end{aligned} \tag{7.129}$$

The dynamic model of the induction motor-driven gas-compressor has the following state-space description:

$$\begin{pmatrix} \dot{x}_1 \\ \dot{x}_2 \\ \dot{x}_3 \\ \dot{x}_4 \\ \dot{x}_5 \\ \dot{x}_6 \\ \dot{x}_7 \\ \dot{x}_8 \\ \dot{x}_9 \end{pmatrix} = \begin{pmatrix} \frac{a_i^2}{V_i}(-x_3 + x_5) \\ \frac{a_o^2}{V_o}[-K_o\sqrt{x_2 - p_{atm}} + x_3 - x_5] \\ \frac{1}{L}[\pi(x_3, x_4)x_1 - x_2] \\ \frac{1}{J}[\mu x_6 x_8 - \tau_c(x_3, x_4) - v x_4] \\ 0 \\ -a x_6 + a M x_7 \\ -\gamma x_7 + a\beta x_6 + n_p x_4 x_6 + \frac{a M x_8^2}{x_6} \\ -\gamma x_6 - \beta n_p x_4 x_6 - n_p x_4 x_7 - \frac{a M x_8 x_7}{x_6} \\ n_p x_4 + \frac{a M x_8}{x_6} \end{pmatrix} + \begin{pmatrix} \frac{a_i^2}{V_i}\sqrt{P_{atm} - x_1} & 0 & 0 & 0 \\ 0 & 0 & 0 & 0 \\ 0 & 0 & 0 & 0 \\ 0 & 0 & 0 & 0 \\ 0 & 0 & 0 & 0 \\ -\frac{1}{T_r}\sqrt{P_{atm} - x_1} & \frac{1}{T_r}\sqrt{x_2 - x_1} & 0 & 0 \\ 0 & 0 & 0 & 0 \\ 0 & 0 & \frac{1}{\sigma L_s} & 0 \\ 0 & 0 & 0 & \frac{1}{\sigma L_s} \\ 0 & 0 & 0 & 0 \end{pmatrix} \begin{pmatrix} u_1 \\ u_2 \\ u_3 \\ u_4 \end{pmatrix} \tag{7.130}$$

First, the rows of the state-space model are reorganized as follows:

$$\dot{x}_2 = \frac{a_o^2}{V_o}[-K_o\sqrt{x_2 - p_{atm}} + x_3 - x_5] \tag{7.131}$$

$$\dot{x}_3 = \frac{1}{L}[\pi(x_3, x_4)x_1 - x_2] \tag{7.132}$$

$$\dot{x}_4 = \frac{1}{J}[\mu x_6 x_8 - \tau_c(x_3, x_4) - v x_4] \tag{7.133}$$

$$\dot{x}_6 = -ax_6 + aMx_7 \tag{7.134}$$

$$\dot{x}_1 = \frac{a_i^2}{V_i}(-x_3 + x_5) + \frac{a_i^2}{V_i}\sqrt{p_{atm} - x_1}u_1 \tag{7.135}$$

$$\dot{x}_5 = -\frac{1}{T_r}\sqrt{p_{atm} - x_1}u_1 + \frac{1}{T_r}\sqrt{x_2 - x_1}u_2 \tag{7.136}$$

$$\dot{x}_7 = -\gamma x_7 + a\beta x_6 + n_p x_4 x_6 + \frac{aMx_8^2}{x_6} + \frac{1}{\sigma L_s}u_3 \tag{7.137}$$

$$\dot{x}_8 = -\gamma x_6 - \beta n_p x_4 x_6 - n_p x_4 x_7 - \frac{aMx_8 x_7}{x_6} + \frac{1}{\sigma L_s}u_4 \tag{7.138}$$

The integrated system can be also written in the concise nonlinear affine-in-the-input state-space form

$$\dot{x} = f(x) + g(x)u \tag{7.139}$$

where $x \in R^{9\times 1}$, $f(x) \in R^{9\times 1}$, $g(x) \in R^{9\times 4}$ and $u \in R^{4\times 1}$.

Next, the state variables of the electrically actuated system are re-defined as follows: $z_1 = x_2$, $z_2 = x_3$, $z_3 = x_4$, $z_4 = x_6$, $z_5 = x_1$, $z_6 = x_5$, $z_7 = x_7$, and $z_8 = x_8$. By substituting state variables x_i by state variables z_i, one obtains

$$\dot{z}_1 = \frac{a_o^2}{V_o}[-K_o\sqrt{z_1 - p_{atm}} + z_2 - z_6] \tag{7.140}$$

$$\dot{z}_2 = \frac{1}{L}[\pi(z_2, z_3)z_5 - z_1] \tag{7.141}$$

$$\dot{z}_3 = \frac{1}{J}[\mu z_4 z_8 - \tau_c(z_2, z_3) - \nu z_3] \tag{7.142}$$

$$\dot{z}_4 = -az_4 + aMz_7 \tag{7.143}$$

$$\dot{z}_5 = \frac{a_i^2}{V_i}(-z_2 + z_6) + \frac{a_i^2}{V_i}\sqrt{p_{atm} - z_5}u_1 \tag{7.144}$$

$$\dot{z}_6 = -\frac{1}{T_r}\sqrt{p_{atm} - z_5}u_1 + \frac{1}{T_r}\sqrt{z_1 - z_5}u_2 \tag{7.145}$$

$$\dot{z}_7 = -\gamma z_7 + a\beta z_4 + n_p z_3 z_4 + \frac{aMz_2^2}{z_4} + \frac{1}{\sigma L_s}u_3 \tag{7.146}$$

$$\dot{z}_8 = -\gamma z_4 - \beta n_p z_3 z_4 - n_p z_3 z_7 - \frac{aMz_8 z_7}{z_4} + \frac{1}{\sigma L_s}u_4 \tag{7.147}$$

Thus, the state-space model of the induction motor-driven gas compressor takes the following form:

$$\begin{pmatrix} \dot{z}_1 \\ \dot{z}_2 \\ \dot{z}_3 \\ \dot{z}_4 \\ \dot{z}_5 \\ \dot{z}_6 \\ \dot{z}_7 \\ \dot{z}_8 \end{pmatrix} = \begin{pmatrix} \frac{a_o^2}{V_o}[-K_o\sqrt{z_1 - p_{atm}} + z_2 - z_6] \\ \frac{1}{L}[\pi(z_2, z_3)z_5 - z_1] \\ \frac{1}{J}[\mu z_4 z_8 - \tau_c(z_2, z_3) - \nu z_3] \\ -az_4 + aMz_7 \\ \frac{a_i^2}{V_i}(-z_2 + z_6) \\ 0 \\ -\gamma z_7 + a\beta z_4 + n_p z_3 z_4 + \frac{aMz_2^2}{z_4} \\ -\gamma z_4 - \beta n_p z_3 z_4 - n_p z_3 z_7 - \frac{aMz_8 z_7}{z_4} \end{pmatrix} + \begin{pmatrix} 0 & 0 & 0 & 0 \\ 0 & 0 & 0 & 0 \\ 0 & 0 & 0 & 0 \\ 0 & 0 & 0 & 0 \\ \frac{a_i^2}{V_i}\sqrt{p_{atm} - z_5} & 0 & 0 & 0 \\ -\frac{1}{T_r}\sqrt{p_{atm} - z_5} & \frac{1}{T_r}\sqrt{z_1 - z_5} & 0 & 0 \\ 0 & 0 & \frac{1}{\sigma L_s} & 0 \\ 0 & 0 & 0 & \frac{1}{\sigma L_s} \end{pmatrix} \begin{pmatrix} u_1 \\ u_2 \\ u_3 \\ u_4 \end{pmatrix} \tag{7.148}$$

Next, subsystem Σ_1 is defined with the following subvectors and submatrices: $z_{1,4} = [z_1, z_2, z_3, z_4]^T$, $f_{1,4}(z_{1,4})$ and $g_{1,4}(z_{1,4})$

$$f_{1,4} = \begin{pmatrix} \frac{a_o^2}{V_o}[-K_o\sqrt{z_1 - p_{atm}} + z_2] \\ -\frac{1}{L}z_1 \\ \frac{1}{J}[-\tau_c(z_2, z_3) - vz_3] \\ -az_4 + aMz_7 \end{pmatrix} \quad g_{1,4} = \begin{pmatrix} 0 & -\frac{a_o^2}{V_o} & 0 & 0 \\ \frac{1}{L}[\pi(z_2, z_3) & 0 & 0 & 0 \\ 0 & 0 & 0 & \frac{1}{J}[\mu z_4 \\ 0 & 0 & aM & 0 \end{pmatrix} \tag{7.149}$$

Moreover, subsystem Σ_2 is defined with the following subvectors and submatrices: $z_{5,8} = [z_5, z_6, z_7, z_7]^T, f_{5,8}(z_{1,4}, z_{5,8})$, and $g_{5,8}(z_{1,4}, z_{5,8})$

$$f_{5,8} = \begin{pmatrix} \frac{a_i^2}{V_i}(-z_2 + z_6) \\ 0 \\ -\gamma z_7 + a\beta z_4 + n_p z_3 z_4 + \frac{aMz_2^2}{z_4} \\ -\gamma z_4 - \beta n_p z_3 z_4 - n_p z_3 z_7 - \frac{aMz_8 z_7}{z_4} \end{pmatrix} \quad g_{5,8} = \begin{pmatrix} \frac{a_i^2}{V_i}\sqrt{p_{atm} - z_5} & 0 & 0 & 0 \\ -\frac{1}{T_r}\sqrt{p_{atm} - z_5} & \frac{1}{T_r}\sqrt{z_1 - z_5} & 0 & 0 \\ 0 & 0 & \frac{1}{\sigma L_s} & 0 \\ 0 & 0 & 0 & \frac{1}{\sigma L_s} \end{pmatrix} \tag{7.150}$$

Thus, the dynamic model of the induction motor-driven gas-compressor can be now written in the form of the two chained subsystems Σ_1 and Σ_2

$$\dot{z}_{1,4} = f_{1,4(z_{1,4})} + g_{1,4}(z_{1,4})z_{5,8} \tag{7.151}$$

$$\dot{z}_{5,8} = f_{5,8}(z_{1,4}, z_{5,8}) + g_{5,8}(z_{1,4}, z_{5,8})u \tag{7.152}$$

7.2.3.1 Differential flatness of the IM-driven gas compressor

It can be proven that the state-space model of the IM-driven gas compressor which consists of the cascading subsystems of Eq. (7.151) and Eq. (7.152) is differentially flat, with flat outputs vector $y_1 = z_{1,4}$. From Eq.(7.151), one solves for subvector $z_{5,8}$ which gives

$$z_{5,8} = g_{1,4}(z_{1,4})^{-1}[\dot{z}_{1,4} - f_{1,4}(z_{1,4})] \Rightarrow z_{5,8} = h_a(z_{1,4}, \dot{z}_{1,4}) \tag{7.153}$$

which signifies that $z_{5,8}$ is a differential function of the flat outputs vector of the system. Moreover, from Eq. (7.152), one solves for the control input u. This gives

$$u = g_{5,8}(z_{1,4}, z_{5,8})^{-1}[\dot{z}_{5,8} - f_{5,8}(z_{1,4}, z_{5,8})] \Rightarrow u = h_b(z_{1,4}, \dot{z}_{1,4}) \tag{7.154}$$

which signifies that u is a differential function of the flat outputs vector of the system. This confirms that the integrated gas-compression system is differentially flat.

7.2.4 *Flatness-based controller in successive loops for the IM-driven gas-compressor*

It will be proven that each one of the subsystems Σ_1 and Σ_2 of Eq. (7.151) and Eq. (7.152) which constitute the dynamics of the induction motor-driven gas compressor is a differentially flat subsystem. Indeed, in the subsystem of Eq. (7.151) the flat output is taken to be $y_1 = z_{1,4}$ and the virtual control input is $v_1 = z_{5,8}$. Then, solving Eq. (7.151) for the virtual control input v_1, one has

$$v_1 = g_{1,4}(z_{1,4})^{-1}[\dot{z}_{1,4} - f_{1,4}(z_{1,4})] \tag{7.155}$$

which signifies that v_1 is a differential function of the flat output y_1. Consequently, Eq. (7.151) represents a differentially flat subsystem. Next, in the subsystem of Eq. (7.152), the flat output is taken to be $y_2 = z_{5,8}$, and $z_{1,4}$ is considered to be a coefficients vector and u is the real control input. Solving for u gives

$$u = g_{5,8}(z_{1,4}, z_{5,8})^{-1}[\dot{z}_{1,4} - f_{5,8}(z_{1,4}, z_{5,8})] \tag{7.156}$$

Consequently, u is a differential function of the flat output y_2, and the subsystem of Eq. (7.152) is also differentially flat.

Knowing that the subsystems of Eq. (7.151) and Eq. (7.152) are differentially flat, it is also inferred that they can be written in the input–output linearized form. Next, a stabilizing feedback controller can be designed for each one of them by inverting their dynamics, as it is commonly done for input–output linearized systems.

For the subsystem of Eq. (7.151), the stabilizing feedback control input which is provided by $v_1 = z_{5,8}$ is

$$v_1 = g_{1,4}(z_{1,4})^{-1}[\dot{z}^d_{1,4} - f_{1,4}(z_{1,4}) - K_1(z_{1,4} - z^d_{1,4})] \tag{7.157}$$

Next, the value of the virtual control input $v_1 = z^d_{5,8}$ becomes setpoint $z^d_{5,8}$ for the subsystem of Eq. (7.152). The stabilizing feedback control for the latter subsystem, coming from the real control input u is given by

$$u = g_{5,8}(z_{1,4}, z_{5.8})^{-1}[\dot{z}^d_{5,8} - f_{5,8}(z_{1,4}, z_{5.8}) - K_2(z_{5,8} - z^d_{5,8})] \tag{7.158}$$

By substituting Eq. (7.157) into Eq. (7.151) and by defining the tracking error variable $e_{1,4} = z_{1,4} - z^d_{1,4}$, as well as the diagonal gain matrix $K_1 \in R^{4\times4} > 0$ with diagonal elements $k_{1,ii} > 0$ for $i = 1, \cdots, 4$, one obtains the tracking error dynamics for the first subsystem. This is given by:

$$\begin{aligned} \dot{z}_{1,4} = & f_{1,4}(z_{1,4}) + g_{1,4}(z_{1,4})g_{1,4}(z_{1,4})^{-1}[\dot{z}^d_{1,4} - f_{1,4}(z_{1,4}) - K_1(z_{1,4} - z^d_{1,4})] \Rightarrow \\ & (\dot{z}_{1,4} - \dot{z}^d_{1,4}) + K_1(z_{1,4} - z^d_{1,4}) = 0 \Rightarrow \dot{e}_{1,4} + K_1 e_{1,4} = 0 \Rightarrow \\ & \lim_{t\to\infty} e_{1,4}(t) = 0 \Rightarrow z_{1,4} = z^d_{1,4} \end{aligned} \tag{7.159}$$

By substituting Eq. (7.158) into Eq. (7.152) and by defining the tracking error variable $e_{5,8} = z_{5,8} - z^d_{5,8}$, as well as the diagonal gain matrix $K_2 \in R^{4\times4} > 0$ with diagonal elements $k_{2,ii} > 0$ for $i = 1, \cdots, 4$, one obtains the tracking error dynamics for the first subsystem. This is given by:

$$\begin{aligned} \dot{z}_{5,8} = & f_{5,8}(z_{1,4}, z_{5,8}) + g_{5,8}(z_{1,4}, z_{5,8})g_{5,8}(z_{1,4}, z_{5,8})^{-1}[\dot{z}^d_{5,8} - f_{5,8}(z_{1,4}, z_{5,8}) - K_2(z_{5,8} - z^d_{5,8})] \Rightarrow \\ & (\dot{z}_{5,8} - \dot{z}^d_{5,8}) + K_2(z_{5,8} - z^d_{5,8}) = 0 \Rightarrow \dot{e}_{5,8} + K_2 e_{5,8} = 0 \Rightarrow \\ & \lim_{t\to\infty} e_{5,8}(t) = 0 \Rightarrow z_{5,8} = z^d_{5,8} \end{aligned} \tag{7.160}$$

Consequently, all state vector elements z_i $i = 1. \cdots, 8$ converge to the associated setpoints z^d_i, and through them, the state variables of the initial nonlinear system x_i, $i = 1, \cdots, 8$ converge to the associated setpoints x^d_i. Therefore, the control loop of the induction motor-driven gas-compressor is globally asymptotically stable.

Global stability for the control loop of the IM-driven gas compressor can be also proven through Lyapunov analysis. To this end, the following Lyapunov function is defined:

$$V = \tfrac{1}{2}[e^T_{1,4}e_{1,4} + e^T_{5,8}e_{5,8}] \tag{7.161}$$

By differentiating in time, one obtains

$$\dot{V} = \tfrac{1}{2}2[e^T_{1,4}\dot{e}_{1,4} + e^T_{5,8}\dot{e}_{5,8}] \tag{7.162}$$

Next, using the relations about the tracking error dynamics of the system's subvectors which were given in Eq. 7.159 and Eq. (7.160), one obtains

$$\begin{aligned} \dot{V} = & [e^T_{1,4}(-K_1 e_{1,4}) + e^T_{5,8}(-K_2 e_{5,8})] \Rightarrow \\ & \dot{V} = -[e^T_{1,4}K_1 e_{1,4} - e^T_{5,8}K_2 e_{5,8}] \Rightarrow \\ & \quad \dot{V} < 0 \, \forall \, e_{1,4} \neq 0, \ e_{5,8} \neq 0 \end{aligned} \tag{7.163}$$

Moreover, $\dot{V} = 0$ if and only if $e_{1,4} = 0$ and $e_{5,8} = 0$. Consequently, V is a strictly diminishing function which converges asymptotically to the equilibrium $[e_{1,4} = 0, \ e_{5,8} = 0]$. Thus, once again global asymptotic stability is proven for the control loop of the induction motor-driven gas-compressor.

7.2.5 Dynamic model of the PMSM-driven gas-compressor

7.2.5.1 Dynamics of the PMSM-driven gas compressor

The diagram of the PMSM electrically driven gas compression system is given in Figure 7.36. The pressure of the inlet valve is denoted again as P_i. The pressure of the outlet valve is denoted as P_o. The mass flow through the compressor is denoted as m. The mass flow depends on the tank volumes and gas velocities, as well as on inlet and outlet tank conditions. The compressor's turn speed is denoted as ω. The compressor's torque is denoted as τ_c. The motor's torque is denoted as τ_d and is a control input for the compressor. The moment of inertia of the turning part of the compressor is denoted as J. By defining the state vector of the electrically driven gas compressor as $x_a = [x_1, x_2, x_3, x_4, x_5]^T = [P_i, P_o, m, \omega, m_r]^T$ the associated state-space model is given by [873,733]

$$\begin{aligned}
\dot{x}_1 &= \frac{a_i^2}{V_i}[K_i\sqrt{p_{atm} - x_1} - x_3 + x_5] \\
\dot{x}_2 &= \frac{a_o^2}{V_o}[-K_o\sqrt{x_2 - p_{atm}} + x_3 - x_5] \\
\dot{x}_3 &= \frac{1}{L}[\pi(x_3, x_4)x_1 - x_2] \\
\dot{x}_4 &= \frac{1}{J}[\tau_d - \tau_c(x_3, x_4) - vx_4] \\
\dot{x}_5 &= \frac{1}{T_r}[K_r\sqrt{x_2 - x_1} - K_i\sqrt{p_{atm} - x_1}]
\end{aligned} \tag{7.164}$$

By defining the state vector of the permanent magnet synchronous motor as $x_b = [x_6, x_7, x_8, x_9]^T = [\theta, \omega, i_d, i_q]^T$ the state-space model of this electric motor is given by [715,725]

$$\begin{aligned}
\dot{x}_6 &= x_7 \\
\dot{x}_7 &= \frac{1}{J}[px_9(\Psi_m - (L_d - L_q)x_8) - Bx_7 - T_L] \\
\dot{x}_8 &= \frac{1}{L_d}(v_d - Rx_8 - px_7L_qx_9) \\
\dot{x}_9 &= \frac{1}{L_q}(v_q - Rx_9 - px_7L_dx_8 - x_7\Psi_m)
\end{aligned} \tag{7.165}$$

where J is the moment of inertia of the PMSM, p is the number of poles, Ψ_m is the magnetic flux due to the machine's permanent magnets, L_d, L_q are the inductance coefficients in the dq reference frame, R is the stator's resistance, and B is the mechanical friction coefficient.

Using that the torque which provides the rotational motion of the compressor is the electromagnetic torque of the PMSM, that is

$$\tau_d = px_9(\Psi_m - (L_d - L_q)x_8) \tag{7.166}$$

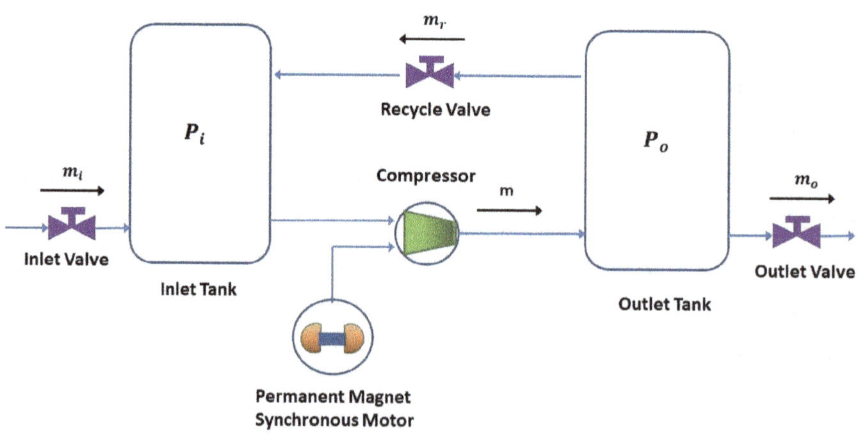

Figure 7.36 Diagram of the PMSM electrically driven gas compressor

and applying Eq. (7.164) about the dynamics of the compressor, as well as Eq. (7.165) about the dynamics of the PMSM, and by considering that the complete state vector of the integrated system is

$$x = [x_1, x_2, x_3, x_4, x_5, x_6, x_7]^T = [P_i, P_o, m, \omega, m_r, i_d, i_q]^T \tag{7.167}$$

and that the control inputs vector of the integrated system is

$$u = [u_1, u_2, u_3, u_4]^T = [K_i, K_r, v_d, v_q]^T \tag{7.168}$$

the integrated state-space model of the compressor-PMSM system comprises the following equations:

$$\dot{x}_1 = \frac{a_i^2}{V_i}[u_1\sqrt{p_{atm} - x_1} - x_3 + x_5] \tag{7.169}$$

$$\dot{x}_2 = \frac{a_o^2}{V_o}[-K_o\sqrt{x_2 - p_{atm}} + x_3 - x_5] \tag{7.170}$$

$$\dot{x}_3 = \frac{1}{L}[\pi(x_3, x_4)x_1 - x_2] \tag{7.171}$$

$$\dot{x}_4 = \frac{1}{J}\{[px_7(\Psi_m - (L_d - L_q)x_6)] - \tau_c(x_3, x_4) - vx_4\} \tag{7.172}$$

$$\dot{x}_5 = \frac{1}{T_r}[u_2\sqrt{x_2 - x_1} - u_1\sqrt{p_{atm} - x_1}] \tag{7.173}$$

$$\dot{x}_6 = \frac{1}{L_d}(u_3 - Rx_6 - px_4L_qx_7) \tag{7.174}$$

$$\dot{x}_7 = \frac{1}{L_q}(u_4 - Rx_7 - px_4L_dx_6 - x_4\Psi_m) \tag{7.175}$$

The previous state-space model of Eq. (7.169) to Eq. (7.175) can be also written in the nonlinear affine-in-the-input state-space form

$$\dot{x} = f(x) + g(x)u \tag{7.176}$$

where $x \in R^{7\times1}$, $f(x) \in R^{7\times1}$, $g(x) \in R^{7\times4}$, and $u \in R^{4\times1}$. The associated matrix equations are given next

$$f(x) = \begin{pmatrix} \frac{a_i^2}{V_i}[-x_3 + x_5] \\ \frac{a_o^2}{V_o}[-K_o\sqrt{x_2 - p_{atm}} + x_3 - x_5] \\ \frac{1}{L}[\pi(x_3, x_4)x_1 - x_2] \\ \frac{1}{J}\{[px_7(\Psi_m) - (L_d - L_q)x_6] - \tau_c(x_3, x_4) - vx_4\} \\ 0 \\ \frac{1}{L_d}(-Rx_6 - px_4L_qx_7) \\ \frac{1}{L_q}(-Rx_7 - px_4L_dx_6 - x_4\Psi_m) \end{pmatrix} \tag{7.177}$$

$$g(x) = \begin{pmatrix} \frac{a_i^2}{V_i}\sqrt{p_{atm} - x_1} & 0 & 0 & 0 \\ 0 & 0 & 0 & 0 \\ 0 & 0 & 0 & 0 \\ 0 & 0 & 0 & 0 \\ -\frac{1}{T_r}\sqrt{p_{atm} - x_1} & \frac{1}{T_r}\sqrt{x_2 - x_1} & 0 & 0 \\ 0 & 0 & \frac{1}{L_d} & 0 \\ 0 & 0 & 0 & \frac{1}{L_q} \end{pmatrix} \tag{7.178}$$

7.2.6 State-space model of the PMSM-driven gas-compressor

The state vector of the PMSM-driven gas-compressor is given by:

$$x = [x_1, x_2, x_3, x_4, x_5, x_6, x_7]^T = [P_1, P_2, m, \omega, m_r, i_d, i_q]^T \tag{7.179}$$

and the control inputs vector is given by

$$u = [u_1, u_2, u_3, u_4]^T = [K_i, K_r, v_d, v_q]^T \tag{7.180}$$

The dynamic model of the PMSM-driven gas compressor is given by

$$\begin{pmatrix} \dot{x}_1 \\ \dot{x}_2 \\ \dot{x}_3 \\ \dot{x}_4 \\ \dot{x}_5 \\ \dot{x}_6 \\ \dot{x}_7 \end{pmatrix} = \begin{pmatrix} \frac{a_i^2}{V_i}(-x_3 + x_5) \\ \frac{a_o^2}{V_o}(-K_o\sqrt{x_2 - p_{atm}} + x_3 - x_5) \\ \frac{1}{L}[\pi(x_3, x_4)x_1 - x_2] \\ \frac{1}{J}\{[px_7\Psi_m - (L_d - L_q)x_6] - \tau_c(x_3, x_4) - vx_4\} \\ 0 \\ \frac{1}{L_d}(-Rx_6 - px_4L_qx_7) \\ \frac{1}{L_q}(-Rx_7 - px_4L_dx_6 - x_4\Psi_m) \end{pmatrix} + \begin{pmatrix} \frac{a_i^2}{V_i}\sqrt{P_{atm-x_1}} & 0 & 0 & 0 \\ 0 & 0 & 0 & 0 \\ 0 & 0 & 0 & 0 \\ 0 & 0 & 0 & 0 \\ -\frac{1}{T_r}\sqrt{P_{atm} - x_1} & \frac{1}{T_r}\sqrt{x_2 - x_1} & 0 & 0 \\ 0 & 0 & \frac{1}{L_d} & 0 \\ 0 & 0 & 0 & \frac{1}{L_q} \end{pmatrix} \begin{pmatrix} u_1 \\ u_2 \\ u_3 \\ u_4 \end{pmatrix} \tag{7.181}$$

Next, the system's state-space model is reorganized as follows:

$$\dot{x}_2 = \frac{a_o^2}{V_o}(-K_o\sqrt{x_2 - p_{atm}} + x_3 - x_5) \tag{7.182}$$

$$\dot{x}_3 = \frac{1}{L}[\pi(x_3, x_4)z_1 - x_2] \tag{7.183}$$

$$\dot{x}_1 = \frac{a_i^2}{V_i}(-x_3 + x_5) + \frac{a_i^2}{V_i}\sqrt{P_{atm-x_1}}u_1 \tag{7.184}$$

$$\dot{x}_5 = -\frac{1}{T_r}\sqrt{P_{atm} - x_1}u_1 + \frac{1}{T_r}\sqrt{x_2 - x_1}u_2 \tag{7.185}$$

$$\dot{x}_4 = \frac{1}{J}\{[px_7\Psi_m - (L_d - L_q)x_6] - \tau_c(x_3, x_4) - vx_4\} \tag{7.186}$$

$$\dot{x}_6 = \frac{1}{L_d}(-Rx_6 - px_4L_qx_7) + \frac{1}{L_d}u_3 \tag{7.187}$$

$$\dot{x}_7 = \frac{1}{L_q}(-Rx_7 - px_4L_dx_6 - x_4\Psi_m) + \frac{1}{L_q}u_4 \tag{7.188}$$

Next, the state variables of the system are re-defined as: $z_1 = x_2$, $z_2 = x_3$, $z_3 = x_1$, $z_4 = x_5$, $z_5 = x_4$, $z_6 = x_6$, and $z_7 = x_7$. Using this state variables notation, the system's state-space model is rewritten as::

$$\dot{z}_1 = \frac{a_o^2}{V_o}(-K_o\sqrt{z_3 - p_{atm}} + z_2 - z_4) \tag{7.189}$$

$$\dot{z}_2 = \frac{1}{L}[\pi(z_2, z_5)z_3 - z_1] \tag{7.190}$$

$$\dot{z}_3 = \frac{a_i^2}{V_i}(-z_2 + z_4) + \frac{a_i^2}{V_i}\sqrt{P_{atm-z_3}}u_1 \tag{7.191}$$

$$\dot{z}_4 = -\frac{1}{T_r}\sqrt{P_{atm} - z_3}u_1 + \frac{1}{T_r}\sqrt{z_1 - z_3}u_2 \tag{7.192}$$

$$\dot{z}_5 = \frac{1}{J}\{[pz_7\Psi_m - (L_d - L_q)z_6] - \tau_c(z_2, z_5) - vz_5\} \tag{7.193}$$

$$\dot{z}_6 = \frac{1}{L_d}(-Rz_6 - pz_5L_qz_7) + \frac{1}{L_d}u_3 \tag{7.194}$$

$$\dot{z}_7 = \frac{1}{L_q}(-Rz_7 - pz_5L_dz_6 - z_5\Psi_m) + \frac{1}{L_q}u_4 \tag{7.195}$$

Next, the dynamics of the PMSM-driven gas-compressor is written in the form of the following chained subsystems:

$$\begin{pmatrix} \dot{z}_1 \\ \dot{z}_2 \end{pmatrix} = \begin{pmatrix} \frac{a_o^2}{V_o}(-K_o\sqrt{z_3 - p_{atm}} + z_2) \\ -\frac{1}{L}z_1 \end{pmatrix} + \begin{pmatrix} 0 & -\frac{a_o^2}{V_o} \\ -\frac{1}{L}\pi(z_2, z_5) & 0 \end{pmatrix} \begin{pmatrix} z_3 \\ z_4 \end{pmatrix} \tag{7.196}$$

$$\begin{pmatrix}\dot{z}_3\\ \dot{z}_4\end{pmatrix}=\begin{pmatrix}\frac{a_i^2}{V_i}(-z_2+z_4)\\ 0\end{pmatrix}+\begin{pmatrix}-\frac{a_i^2}{V_i}\sqrt{p_{atm}-z_1} & 0\\ -\frac{1}{T_r}\sqrt{p_{atm}-z_3} & \frac{1}{T_r}\sqrt{z_1-z_3}\end{pmatrix}\begin{pmatrix}u_1\\ u_2\end{pmatrix} \tag{7.197}$$

$$\dot{z}_5=[-\tfrac{1}{J}\tau_c(z_2,z_5)-\tfrac{\nu}{J}z_5]+[\tfrac{L_d-L_q}{J}z_6+\tfrac{p\Psi_m}{J}z_7] \tag{7.198}$$

$$\begin{pmatrix}z_6\\ z_7\end{pmatrix}=\begin{pmatrix}\frac{1}{L_d}(-Rz_6-pz_5L_qz_7)\\ \frac{1}{L_q}(-Rz_7-pz_5L_dz_6-z_5\Psi_m)\end{pmatrix}+\begin{pmatrix}\frac{1}{L_d} & 0\\ 0 & \frac{1}{L_q}\end{pmatrix}\begin{pmatrix}u_3\\ u_4\end{pmatrix} \tag{7.199}$$

7.2.6.1 Differential flatness properties of the PMSM-driven gas-compressor

The dynamic system of the PMSM-driven gas compressors is differentially flat with flat outputs vector

$$Y=[y_1,y_2,y_3,y_4]^T=[x_1,x_2,x_4,x_6]^T=[z_3,z_1,z_2,z_6]^T \tag{7.200}$$

From the first row of the re-arranged state-space model, one has:

$$\dot{z}_1=\tfrac{a_o^2}{V_o}(-K_o\sqrt{z_1-p_{atm}}+z_2-z_4) \tag{7.201}$$

and by solving for z_4, it holds that

$$z_4=\tfrac{1}{a_o^2}[-V_o\dot{z}_1-a_o^2K_o\sqrt{z_1-p_{atm}}+a_o^2z_2] \tag{7.202}$$

which signifies that z_4 is a differential function of the flat output of the system or $z_4=h_4(Y,\dot{Y})$. From the second row of the re-arranged state-space model, one has

$$\dot{z}_2=\tfrac{1}{L}[\pi(z_2,z_5)z_3-z_1] \tag{7.203}$$

thus, by solving for z_5, it holds that

$$\pi(z_2,z_5)=\tfrac{1}{z_3}[L\dot{z}_2+z_1] \tag{7.204}$$

which allows for expressing z_5 in the form of a differential function of the flat output vector, or $z_5=h_5(Y,\dot{Y})$. From the fifth row of the state-space model, one has

$$\dot{z}_5=\tfrac{1}{J}[pz_7\Psi_m-(L_d-L_q)z_6]-\tau_c(z_2,z_5)-\nu z_6 \tag{7.205}$$

thus by solving for z_7, one has

$$z_7=\tfrac{1}{p\Psi_m}[J\dot{z}_5+(L_d-L_q)z_6+\tau_c(z_3,z_4)+\nu z_5] \tag{7.206}$$

which signifies that z_7 is a differential function of the flat outputs vector Y or $z_7=h_7(Y,\dot{Y})$. Next, from the 3rd, 4th, 6th and 7th rows of the state-space model, one has the following system of equations:

$$\begin{pmatrix}\dot{z}_3\\ \dot{z}_4\\ \dot{z}_6\\ \dot{z}_7\end{pmatrix}=\begin{pmatrix}\frac{a_i^2}{V_i}(-z_2+z_4)\\ 0\\ \frac{1}{L_d}(-Rz_6-pz_5L_qz_7)\\ \frac{1}{L_q}(-Rz_7-pz_5L_dz_6-z_5\Psi_m)\end{pmatrix}+\begin{pmatrix}\frac{a_i^2}{V_i} & 0 & 0 & 0\\ -\frac{1}{T_r}\sqrt{p_{atm}-z_3} & \frac{1}{T_r}\sqrt{z_1-z_3} & 0 & 0\\ 0 & 0 & \frac{1}{L_d} & 0\\ 0 & 0 & 0 & \frac{1}{L_q}\end{pmatrix}\begin{pmatrix}u_1\\ u_2\\ u_3\\ u_4\end{pmatrix} \tag{7.207}$$

or, in concise form the previous matrix equation is written as

$$\dot{\tilde{z}}=\tilde{f}+\tilde{g}u\Rightarrow u=\tilde{g}^1[\dot{\tilde{z}}-\tilde{F}] \tag{7.208}$$

where $\tilde{z}$, $\tilde{f}$, $\tilde{g}$, are differential functions of the flat outputs vector. Thus, it is also inferred that u is also a differential function of the flat outputs vector Y, or equivalently $u=h_u(Y,\dot{Y})$. Consequently, the dynamic model of the PMSM-driven gas compressor is differentially flat. The differential flatness properties of this system confirm its linearizability and implicitly also its controllability.

7.2.7 Flatness-based control in successive loops for the PMSM-driven gas-compressor

It will be proven that each one of the subsystems of Eq. (7.196) to Eq. (7.199) is differentially flat and a flatness-based controller can be designed about it. The following sub-vectors and sub-matrices are defined:

For subsystem Σ_1, one has subvector $z_{1,2}$ and submatrices $f_{1,2}(z_{1,2})$, $g_{1,2}(z_{1,2})$ and virtual control input $v_1 = [v_3, v_4]^T$:

$$\dot{z}_{1,2} = \begin{pmatrix} z_1 \\ z_2 \end{pmatrix} \quad f_{1,2} = \begin{pmatrix} \frac{a_o^2}{V_o}(-K_o\sqrt{z_1 - p_{atm}} + z_2) \\ -\frac{1}{L}z_1 \end{pmatrix} \quad g_{1,2} = \begin{pmatrix} 0 & -\frac{a_o^2}{V_o} \\ -\frac{1}{L}\tau(z_2, z_5) & 0 \end{pmatrix} \tag{7.209}$$

For subsystem Σ_2, one has subvector $z_{3,4}$ and submatrices $f_{3,4}(z_{1,2}, z_{3,4})$, $g_{3,4}(z_{1,2}, z_{3,4})$, and real control input $u_a = [u_1, u_2]^T$:

$$z_{3,4} = \begin{pmatrix} z_3 \\ z_4 \end{pmatrix} \quad f_{3,4} = \begin{pmatrix} \frac{a_i^2}{V_i}(-z_2 + z_4) \\ 0 \end{pmatrix} \quad g_{3,4} = \begin{pmatrix} -\frac{a_i^2}{V_i}\sqrt{p_{atm} - z_1} & 0 \\ -\frac{1}{T_r}\sqrt{p_{atm} - z_3} & \frac{1}{T_r}\sqrt{z_1 - z_3} \end{pmatrix} \tag{7.210}$$

For subsystem Σ_3, one has subvector z_5 and submatrices $f_5(z_{1,2}, z_{3,4}, z_5)$, $g_5(z_{1,2}, z_{3,4}, z_5)$ and virtual control input $v_2 = -\frac{(L_d - L_q)}{J}z_6 + \frac{\mu\Psi_m}{J}z_7$:

$$z_5 \quad f_5 = -\frac{1}{J}\tau_c(z_2, z_5) - \frac{v}{J}z_5 \quad g_5 = 1 \tag{7.211}$$

For subsystem Σ_4, one has subvector $z_{6,7}$ and submatrices $f_{6,7}(z_{1,2}, z_{3,4}, z_5, z_{6,7})$, $g_{6,7}(z_{1,2}, z_{3,4}, z_5, z_{6,7})$ and real control input $u_b = [u_3, u_4]^T$:

$$z_{6,7} = \begin{pmatrix} z_6 \\ z_7 \end{pmatrix} \quad f_{6,7} = \begin{pmatrix} \frac{1}{L_d}(-Rz_6 - pz_5L_qz_7) \\ \frac{1}{L_q}(-Rz_7 - pz_5L_dz_6 - z_5\Psi_m) \end{pmatrix} \quad g_{6,7} = \begin{pmatrix} \frac{1}{L_d} & 0 \\ 0 & \frac{1}{L_q} \end{pmatrix} \tag{7.212}$$

Thus, the dynamics of the PMSM-driven gas-compressor can be written in the form of the following subsystems Σ_1 to Σ_4:

$$\dot{z}_{1,2} = f_{1,2}(z_{1,2}) + g_{1,2}(z_{1,2})v_1 \tag{7.213}$$

$$\dot{z}_{3,4} = f_{3,4}(z_{1,2}, z_{3,4}) + g_{3,4}(z_{1,2}, z_{3,4})u_a \tag{7.214}$$

$$\dot{z}_5 = f_5(z_{1,2}, z_{3,4}, z_5) + g_5(z_{1,2}, z_{3,4}, z_5)v_2 \tag{7.215}$$

$$\dot{z}_{6.7} = f_{6,7}(z_{1,2}, z_{3,4}, z_5, z_{6,7}) + g_{6,7}(z_{1,2}, z_{3,4}, z_5, z_{6,7})u_b \tag{7.216}$$

For the subsystem Σ_1 of Eq. (7.213) the flat output is defined as $y_1 = z_{1,2}$ and the virtual control input is $v_1 = z_{3,4}$. By solving Eq. (7.213) for v_1, one has

$$v_1 = g_{1,2}(z_{1,2})^{-1}[\dot{z}_{1,2} - f_{1,2}(z_{1,2})] \tag{7.217}$$

which signifies that v_1 is a differential function of the flat output y_1 and that the subsystem of Eq. (7.213) is differentially flat.

For the subsystem Σ_2 of Eq. (7.213) the flat output is defined as $y_2 = z_{3,4}$, the real control input is u_a, while $z_{1,2}$ is viewed as a coefficients vector. By solving for u_a, one has

$$u_a = g_{3,4}(z_{1,2}, z_{3,4})^{-1}[\dot{z}_{3,4} - f_{3,4}(z_{1,2}, z_{3,4})] \tag{7.218}$$

which signifies that u_a is a differential function of the flat output y_2, and thus the subsystem of Eq. (7.214) is differentially flat.

For the subsystem Σ_3 of Eq. (7.215) the flat outputs vector is $y_3 = z_5$, the virtual control input is v_2 while $z_{1,2}$ and $z_{3,4}$ are viewed as coefficients. Solving Eq. (7.215) for v_2 gives

$$v_2 = g_5(z_{1,2}, z_{3,4}, z_5)^{-1}[\dot{z}_5 - f_5(z_{1,2}, z_{3,4}, z_5)] \tag{7.219}$$

which signifies that v_2 is a differential function of the flat output y_3 and that the subsystem of Eq. (7.215) is differentially flat.

For the subsystem Σ_4 of Eq. (7.216) the flat output vector is $y_4 = z_{6,7}$, the real control input is u_b, while $z_{1,2}, z_{3,4}, z_5$ are viewed as coefficients. By solving Eq. (7.216) for v_b, one obtains

$$u_b = g_{6,7}(z_{1,2}, z_{3,4}, z_5, z_{6,7})^{-1}[\dot{z}_{6,7} - f_{6,7}(z_{1,2}, z_{3,4}, z_5, z_{6,7})] \tag{7.220}$$

which signifies that u_b is a differential function of the flat output y_4 and that the subsystem of Eq. (7.216) is differentially flat.

Knowing that each one of the subsystems Σ_1 to Σ_4 of Eq. (7.213) to Eq. (7.216) is differentially flat means also that each one of them can be written in the input–output linearized form. Besides, it means that a stabilizing feedback controller can be designed for each one of these subsystems by applying inversion of its dynamics, as it is usually done for input–output linearized systems.

For the subsystem Σ_1 of Eq. (7.213), the stabilizing feedback control is taken to be $v_1 = z^d_{1.2}$

$$v_1 = g_{1,2}(z_{1,2})[\dot{z}^d_{1,2} - f_{1,2}(z_{1,2}) - K_1(z_{1,2} - z^d_{1,2})] \tag{7.221}$$

where $K_1 > 0$ is a diagonal matrix with diagonal elements $k_{1,ii} > 0$ for $i = 1, 2$. The virtual control input $v_1 = z^d_{1,2}$ becomes setpoint to the subsystem of Eq. (7.214). For the subsystem Σ_2 of Eq. (7.214), the stabilizing feedback control is taken to be

$$u_a = g_{3,4}(z_{1,2}, z_{3,4})[\dot{z}^d_{3,4} - f_{3,4}(z_{1,2}, z_{3,4}) - K_2(z_{3,4} - z^d_{3,4})] \tag{7.222}$$

where $K_2 > 0$ is a diagonal matrix with diagonal elements $k_{2,ii} > 0$ for $i = 1, 2$. For the subsystem Σ_3 of Eq. (7.215), the stabilizing feedback control is taken to be

$$v_2 = g_5(z_{1,2}, z_{3,4}, z_5)[\dot{z}^d_5 - f_5(z_{1,2}, z_{3,4}, z_5) - K_3(z_5 - z^d_5)] \tag{7.223}$$

where $K_3 > 0$ is a positive scalar feedback gain. Knowing that the desirable value for control input v_2 is $v_2^* = -\frac{(L_d - L_q)}{J} z_6 + \frac{\mu\Psi_m}{J} z_7$ and by assigning a desirable setpoint value to z_6 which is denoted as z^d_6, one can also find the desirable (setpoint) value for z_7 which is denoted as z^d_7. Actually, it holds that

$$z^d_7 = \frac{1}{p\Psi_m}[Jv_2^* + (L_d - L_q)z^d_6] \tag{7.224}$$

For the subsystem Σ_4 of Eq. (7.216), the stabilizing feedback control is taken to be

$$u_b = g_{6,7}(z_{1,2}, z_{3,4}, z_{5,6}, z_7)[\dot{z}^d_{6,7} - f_{6,7}(z_{1,2}, z_{3,4}, z_{5,6}, z_7) - K_4(z_{6,7} - z^d_{6,7})] \tag{7.225}$$

where $K_4 > 0$ is a diagonal matrix with diagonal elements $k_{4,ii} > 0$ for $i = 1, 2$. The setpoints vector $z^d_{6,7} = [z^d_6, z^d_7]$ has been defined before using the value of z^d_6 and the value of z^d_7 that was computed from Eq. (7.224).

Next, the following tracking error variables are defined: $e_{1,2} = z_{1,2} - z^d_{1,2}$, $e_{3,4} = z_{3,4} - z^d_{3,4}$, $e_5 = z_5 - z^d_5$, $e_{6,7} = z_{6,7} - z^d_{6,7}$. By substituting Eq. (7.221) into Eq. (7.213), the following tracking error dynamics is obtained:

$$\begin{aligned}\dot{z}_{1,2} = &f_{1,2}(z_{1,2}) + g_{1,2}(z_{1,2})g_{1,2}(z_{1,2})^{-1}[\dot{z}^d_{1,2} - f_{1,2}(z_{1,2}) - K_1(z_{1,2} - z^d_{1,2})] \Rightarrow \\ &(\dot{z}_{1,2} - \dot{z}^d_{1,2}) + K_1(z_{1,2} - z^d_{1,2}) = 0 \Rightarrow \dot{e}_{1,2} + K_1 e_{1,2} = 0 \Rightarrow \\ &lim_{t\to\infty} e_{1,2}(t) = 0 \Rightarrow lim_{t\to\infty} z_{1,2}(t) = z^d_{1,2}(t)\end{aligned} \tag{7.226}$$

By substituting Eq. (7.222) into Eq. (7.214), the following tracking error dynamics is obtained:

$$\begin{aligned}\dot{z}_{3,4} = &f_{3,4}(z_{1,2}, z_{3,4}) + g_{3,4}(z_{1,2}, z_{3,4})g_{1,2}(z_{1,2}, z_{3,4})^{-1}[\dot{z}^d_{3,4} - f_{3,4}(z_{1,2}, z_{3,4}) - K_2(z_{3,4} - z^d_{3,4})] \Rightarrow \\ &(\dot{z}_{3,4} - \dot{z}^d_{3,4}) + K_2(z_{3,4} - z^d_{3,4}) = 0 \Rightarrow \dot{e}_{3,4} + K_2 e_{3,4} = 0 \Rightarrow \\ &lim_{t\to\infty} e_{3,4}(t) = 0 \Rightarrow lim_{t\to\infty} z_{3,4}(t) = z^d_{3,4}(t)\end{aligned} \tag{7.227}$$

By substituting Eq. (7.223) into Eq. (7.215), the following tracking error dynamics is obtained:

$$\begin{aligned}\dot{z}_5 = &f_5(z_{1,2}, z_{3,4}, z_5) + g_5(z_{1,2}, z_{3,4}, z_5)g_5(z_{1,2}, z_{3,4}, z_5)^{-1}[\dot{z}^d_5 - f_5(z_{1,2}, z_{3,4}, z_5) - K_3(z_5 - z^d_5)] \Rightarrow \\ &(\dot{z}_5 - \dot{z}^d_5) + K_3(z_5 - z^d_5) = 0 \Rightarrow \dot{e}_5 + K_3 e_5 = 0 \Rightarrow \\ &lim_{t\to\infty} e_5(t) = 0 \Rightarrow lim_{t\to\infty} z_5(t) = z^d_5(t)\end{aligned} \tag{7.228}$$

Finally, by substituting Eq. (7.225) into Eq. (7.216), the following tracking error dynamics is obtained:

$$\begin{aligned}\dot{z}_{6,7} =& f_{6,7}(z_{1,2},z_{3,4},z_5,z_{6,7}) + g_{6,7}(z_{1,2},z_{3,4},z_5,z_{6,7})g_{6,7}(z_{1,2},z_{3,4},z_5,z_{6,7})^{-1} \\ &\cdot[\dot{z}^d_{6,7} - f_{6,7}(z_{1,2},z_{3,4},z_5,z_{6,7}) - K_4(z_{6,7}-z^d_{6,7})]\Rightarrow \\ &(\dot{z}_{6,7}-\dot{z}^d_{6,7}) + K_4(z_{6,7}-z^d_{6,7}) = 0 \Rightarrow \dot{e}_{6,7} + K_4 e_{6,7} = 0 \Rightarrow \\ &lim_{t\to\infty} e_{6,7}(t) = 0 \Rightarrow lim_{t\to\infty} z_{6,7}(t) = z^d_{6,7}(t)\end{aligned} \tag{7.229}$$

By proving that state vector elements z_i converge to the associated setpoints z_i^d, it is also proven that the state variables of the initial nonlinear system x_i converge to their setpoints x_i^d. Thus, the control loop of the PMSM-driven gas compressor is globally asymptotically stable.

Global stability can be also proven through Lyapunov analysis. To this end, the following Lyapunov function is defined:

$$V = \frac{1}{2}[e^T_{1,2}e_{1,2} + e^T_{3,4}e_{3,4} + e_5^2 + e^T_{6,7}e_{6,7}] \tag{7.230}$$

By differentiating the above equation in time, one obtains:

$$\dot{V} = \frac{1}{2}2[e^T_{1,2}\dot{e}_{1,2} + e^T_{3,4}\dot{e}_{3,4} + e_5\dot{e}_5 + e^T_{6,7}\dot{e}_{6,7}] \tag{7.231}$$

Next, by substituting in Eq. (7.231) the relations about the tracking error dynamics which are provided by Eq. (7.226) to Eq. (7.229), one gets

$$\begin{aligned}\dot{V} =& [e^T_{1,2}(-K_1e_{1,2}) + e^T_{3,4}(-K_2e_{3,4}) + e_5(-K_3e_5) + e^T_{6,7}(-K_4e_{6,7})]\Rightarrow \\ &\dot{V} = -[e^T_{1,2}(K_1e_{1,2}) + e^T_{3,4}(K_2e_{3,4}) + (K_3e_5^2) + e^T_{6,7}(K_4e_{6,7})]\Rightarrow \\ &\dot{V} < 0 \ \forall \ e_{1,2}\neq 0,\ e_{3,4}\neq 0,\ e_5\neq 0,\ e_{6,7}\neq 0,\end{aligned} \tag{7.232}$$

and $\dot{V}=0$ if and only if $e_{1,2}=0$, $e_{3,4}=0$, $e_5=0$, and $e_{6,7}=0$. Consequently, Lyapunov function V is a strictly diminishing function which converges asymptotically to the fixed point $[e^T_{1,2} = 0_{1\times 2}, e^T_{3,4} = 0_{1\times 2}, e_5 = 0, e^T_{6,7} = 0_{1\times 2}]$. In this manner, it is proven once again that the dynamic model of the PMSM-driven gas compressor is globally asymptotically stable.

7.2.8 Simulation tests

7.2.8.1 Results on control of IM-driven gas compressors

Results about the tracking accuracy and the speed of convergence to setpoints of the successive-loops flatness-based control method, in the case of the induction motor-driven gas compressor, are shown in Figures 7.37–7.52. A per-unit (p.u.) state variables notation has been used. It can be noticed, that under this control scheme, one achieves fast and precise tracking of reference setpoints for all state variables of the induction motor-driven gas compressor. It is noteworthy, that through the stages of this method, one solves also the setpoints definition problem for all state variables of the induction motor-driven gas compressor. Actually, the selection of setpoints for state variables x_2, x_3, x_4, and x_6 is unconstrained. On the other side by defining state variables x_1, x_5, x_6, x_7 as virtual control inputs to the subsystem of x_2, x_3, x_4, x_6, one can find the setpoints for x_1, x_5, x_6, x_7 as functions of the setpoints for x_2, x_3, x_4, x_6. The speed of convergence of the state variables of the induction motor-driven gas compressor under flatness-based control implemented in successive loops is dependent on the selection of values for the diagonal gain matrices K_i, $i=1,2$ of Eq. (7.157) to Eq. (7.158).

7.2.8.2 Results on control of PMSM-driven gas compressor

Results about the tracking accuracy and the speed of convergence to setpoints of the successive-loops flatness-based control method, in the case of the PMSM-driven gas compressor, are shown in Figures 7.53–7.68. A per-unit (p.u.) state variables notation has been used. It can be noticed, that under this control scheme, one achieves fast and precise tracking of reference setpoints for all state variables of the PMSM-driven gas compressor. It is noteworthy, that through

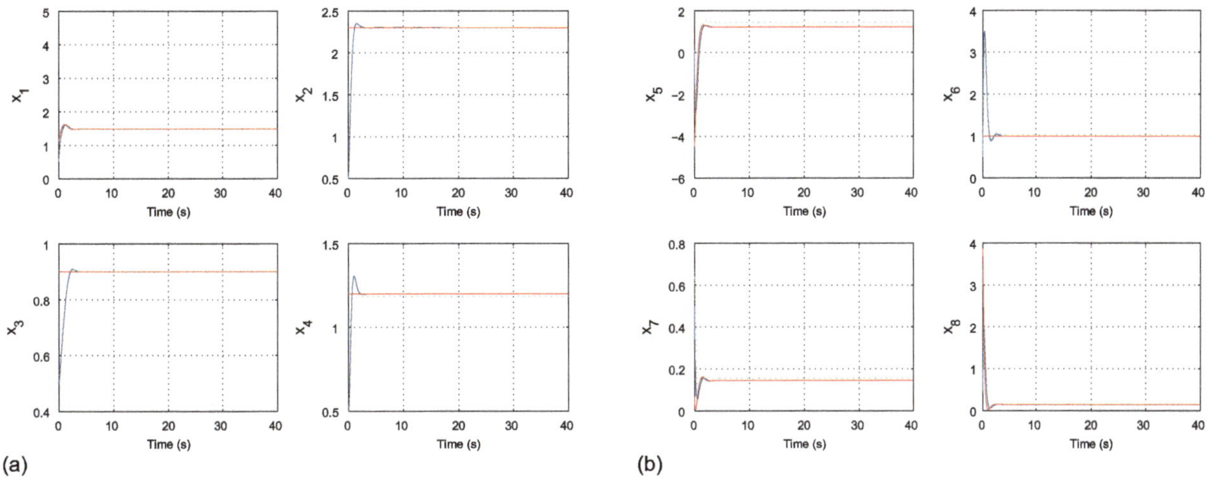

Figure 7.37 Tracking of setpoint 1 by the induction motor-driven gas compressor: (a) convergence of state variables x_1 to x_4 (blue lines) to the reference setpoints (red lines) and their KF-based estimation (green line); (b) convergence of state variables x_5 to x_8 (blue lines) to the reference setpoints (red lines) and their KF-based estimation (green line)

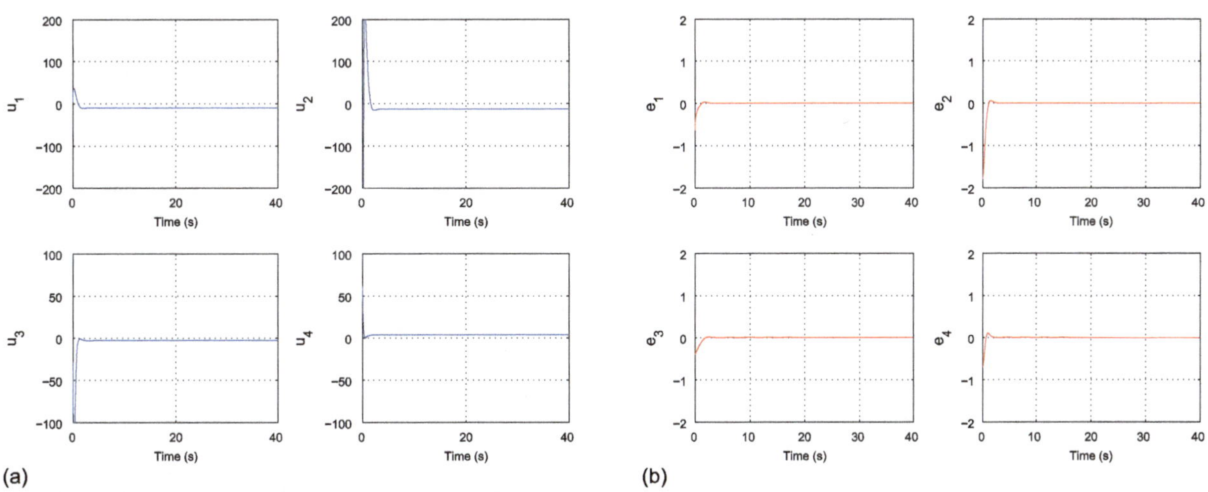

Figure 7.38 Tracking of setpoint 1 by the induction motor-driven gas compressor: (a) variations of the control inputs u_1 to u_4; (b) tracking error of state variables x_1 to x_4

the stages of this method, one solves also the setpoints definition problem for all state variables of the PMSM-driven gas compressor. Actually, the selection of setpoints for state variables x_2 and x_3 is unconstrained. On the other side by defining state variables x_1 and x_5 as virtual control inputs for the subsystem of x_2, x_3, one can find the setpoints for x_1, x_5 as functions of the setpoints for x_2, x_3. In a similar manner, the selection of setpoints for the turn speed x_4 of the compressor is unconstrained. On the other side by defining state variables x_6 and x_7 as parts of the virtual control input for the subsystem of x_2, and by assigning without constraints a value to the setpoint of x_6, one can also find the setpoint for x_7 as a function of the setpoints of x_4, x_6. The speed of convergence of the state variables of the PMSM-driven gas

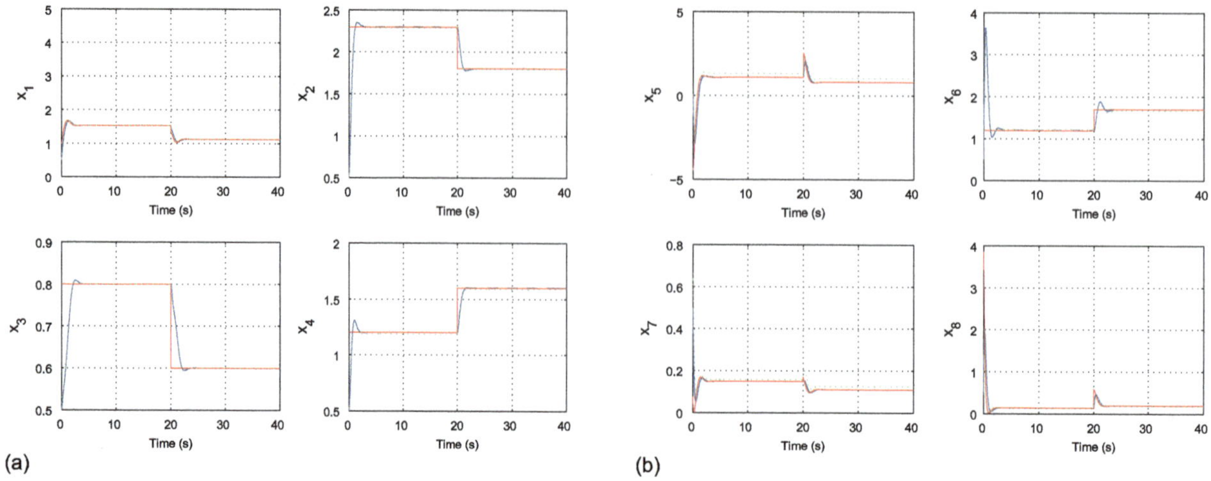

Figure 7.39 *Tracking of setpoint 2 by the induction motor-driven gas compressor: (a) convergence of state variables x_1 to x_4 (blue lines) to the reference setpoints (red lines) and their KF-based estimation (green line); (b) convergence of state variables x_5 to x_8 (blue lines) to the reference setpoints (red lines) and their KF-based estimation (green line)*

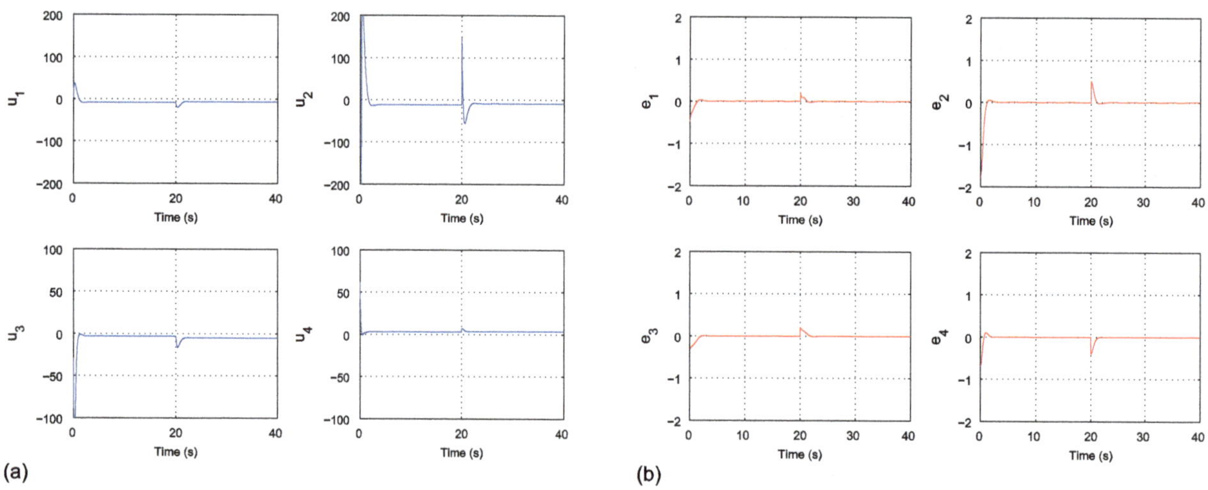

Figure 7.40 *Tracking of setpoint 2 by the induction motor-driven gas compressor: (a) variations of the control inputs u_1 to u_4; (b) tracking error of state variables x_1 to x_4*

compressor under flatness-based control implemented in successive loops is dependent on the selection of values for the diagonal gain matrices K_i, $i = 1, \cdots, 4$ of Eq. (7.221) to Eq. (7.221).

7.3 Nonlinear optimal control for a PMLSG-VSC wave energy conversion unit

7.3.1 Outline

Electric power generation from wave power is a promising solution for covering the continuously growing energy needs in an environmental friendly and without-pollution manner [373,1069,599,1079]. However, forces due to waves are

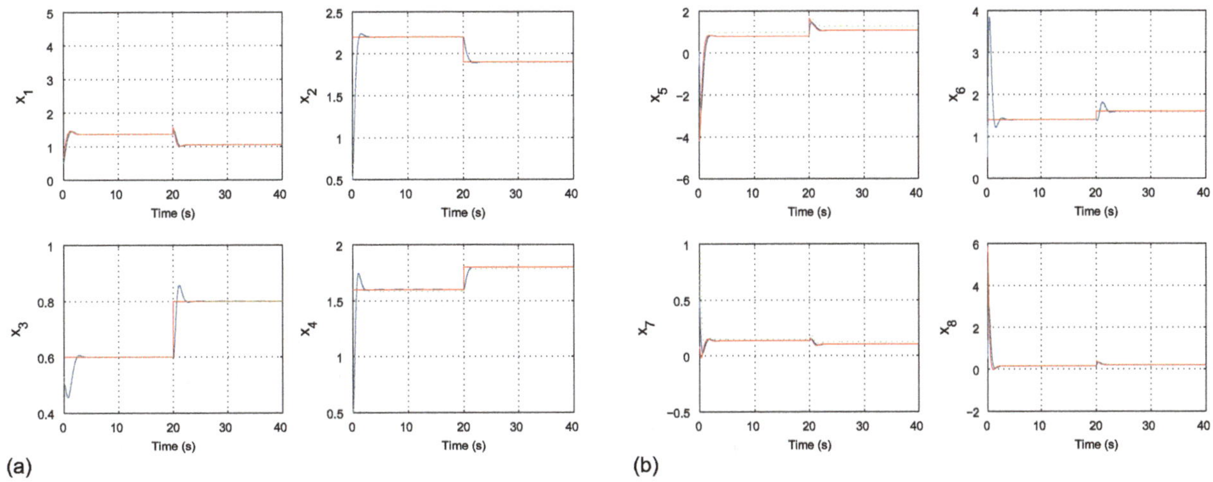

Figure 7.41 Tracking of setpoint 3 by the induction motor-driven gas compressor: (a) convergence of state variables x_1 to x_4 (blue lines) to the reference setpoints (red lines) and their KF-based estimation (green line); (b) convergence of state variables x_5 to x_8 (blue lines) to the reference setpoints (red lines) and their KF-based estimation (green line)

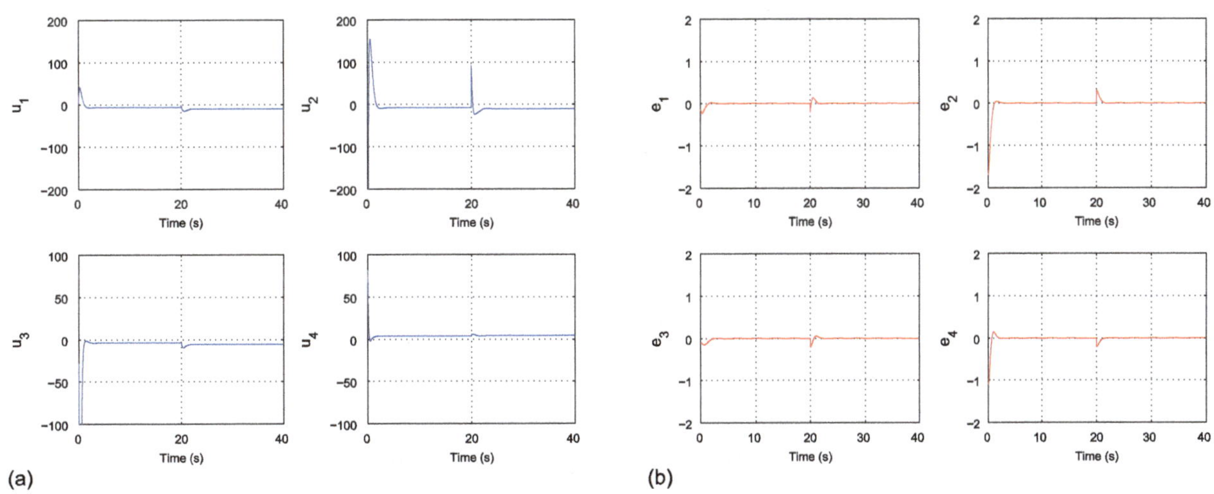

Figure 7.42 Tracking of setpoint 3 by the induction motor-driven gas compressor: (a) variations of the control inputs u_1 to u_4; (b) tracking error of state variables x_1 to x_4

of variable amplitude and frequency and the voltage outputs which are produced by power generators receiving such a kind of actuation are not directly exploitable by the main electricity grid [565,46,474,227]. To harmonize the voltage outputs of power generators with the power grid Voltage Source Converters (VSCs) have to be connected to them [202,368,396,335]. Wave energy conversion based on Permanent Magnet Linear Synchronous Generators (PMLSGs) and AC/DC Voltage Source Converters (VSCs) allows for exploiting efficiently the energy of waves and for producing significant amounts of electric power which can be fed in the electricity network [760,995,508,864]. The control of the motion of such type of generators under wave-induced excitation is a non-trivial task which should be based on elaborated nonlinear control algorithms [7,8,915,1063]. Besides, control of AC/DC voltage source converters so

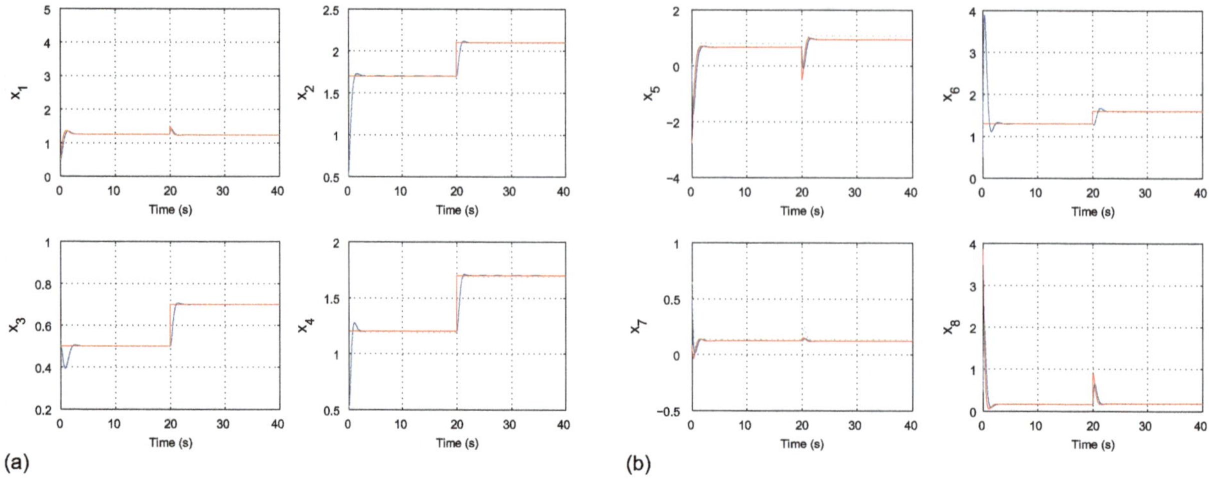

Figure 7.43 *Tracking of setpoint 4 by the induction motor-driven gas compressor: (a) convergence of state variables x_1 to x_4 (blue lines) to the reference setpoints (red lines) and their KF-based estimation (green line); (b) convergence of state variables x_5 to x_8 (blue lines) to the reference setpoints (red lines) and their KF-based estimation (green line)*

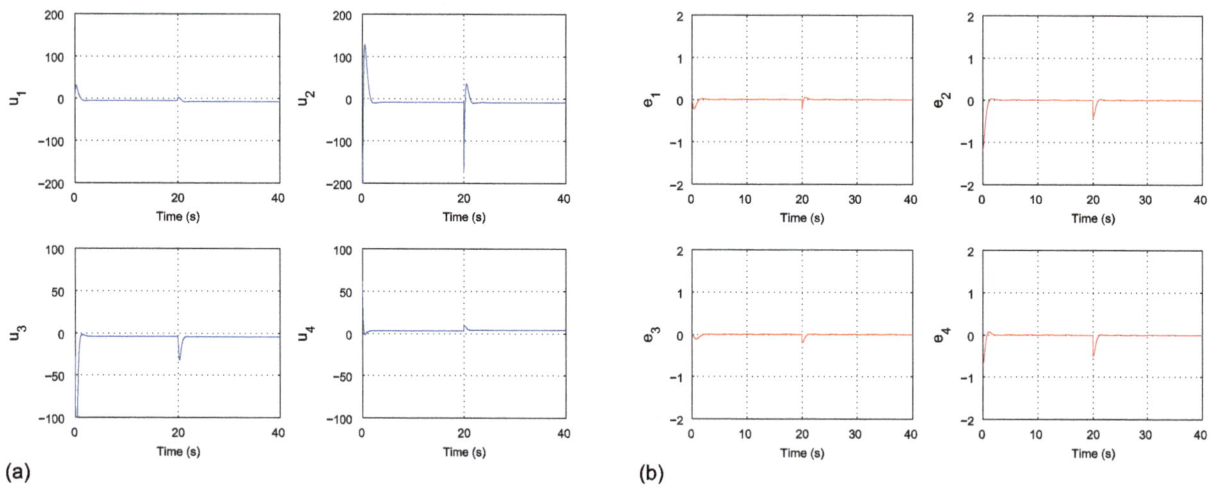

Figure 7.44 *Tracking of setpoint 4 by the induction motor-driven gas compressor: (a) variations of the control inputs u_1 to u_4; (b) tracking error of state variables x_1 to x_4*

as to synchronize wave power units with the main electricity grid is a delicate procedure which also requires the use of advanced nonlinear control methods [847,468,469]. Additionally, control of the integrated wave power generation systems which comprise PMLSGs serially connected to VSCs is a subtle task that should compensate for the complicated nonlinear dynamics and multivariable structure of these energy conversion units [700,697,699,698]. Among recent results in the area of PMLSG and VSC-based wave power conversion systems, one can distinguish methods for maximizing the amount of wave power that is finally converted to exploitable electric power [265,278,522,970,834]. There exist also several other results on the stabilization of WECs and on the VSC-based integration of these power units to the electricity grid [505,1060,566,645,758].

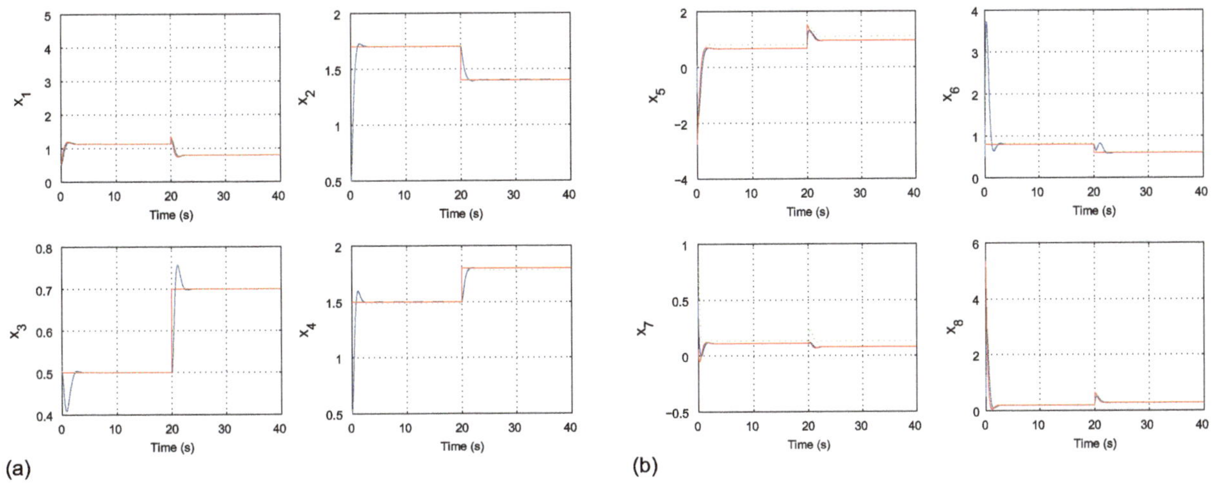

Figure 7.45 Tracking of setpoint 5 by the induction motor-driven gas compressor: (a) convergence of state variables x_1 to x_4 (blue lines) to the reference setpoints (red lines) and their KF-based estimation (green line); (b) convergence of state variables x_5 to x_8 (blue lines) to the reference setpoints (red lines) and their KF-based estimation (green line)

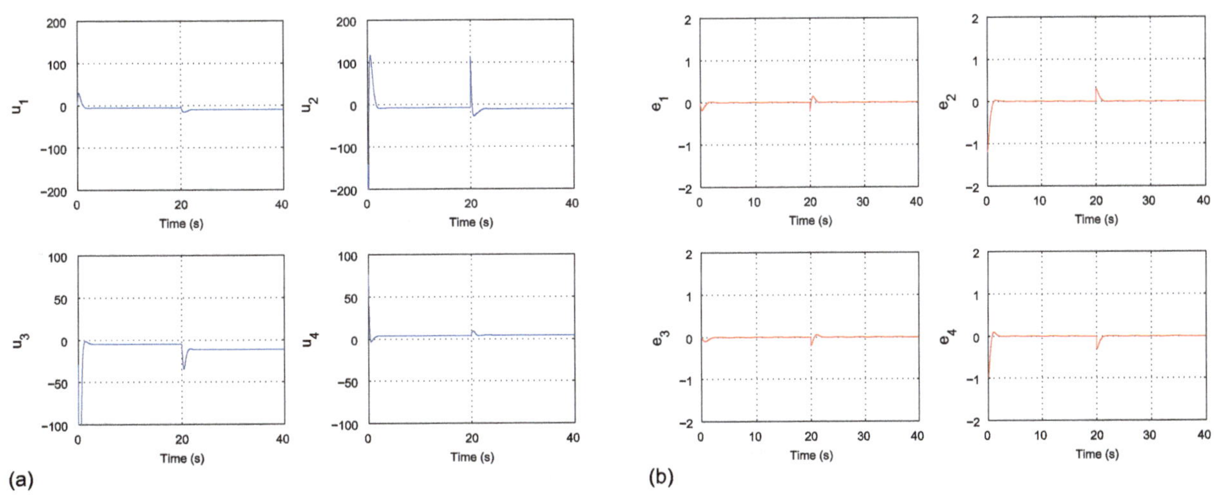

Figure 7.46 Tracking of setpoint 5 by the induction motor-driven gas compressor: (a) variations of the control inputs u_1 to u_4; (b) tracking error of state variables x_1 to x_4

In this section, a novel nonlinear optimal control method is developed for the dynamic model of a wave energy conversion unit which comprises a Permanent Magnet Linear Synchronous Generator (PMLSG) which is serially connected to a three-phase AC/DC Voltage Source Converter (VSC) [715,711,725]. The dynamic model of the integrated PMLSG-VSC-based power unit is given in state-space form and differential flatness properties are proven about it. Differential flatness is also an implicit proof of the system's controllability and of its input–output linearizability [711]. Indeed, through successive differentiations of the power unit's flat outputs, formed into the input–output linearized form and a stabilizing feedback controller is developed about it using the eigenvalues assignment technique. This control method is characterized by complicated and elongated state-space model transformations which can be avoided by

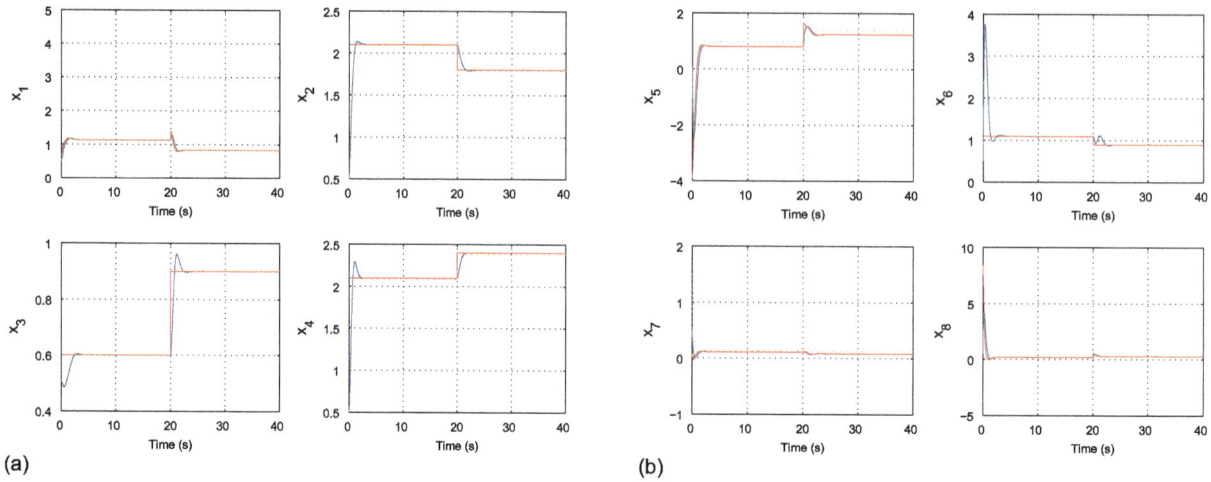

Figure 7.47 *Tracking of setpoint 6 by the induction motor-driven gas compressor: (a) convergence of state variables x_1 to x_4 (blue lines) to the reference setpoints (red lines) and their KF-based estimation (green line); (b) convergence of state variables x_5 to x_8 (blue lines) to the reference setpoints (red lines) and their KF-based estimation (green line)*

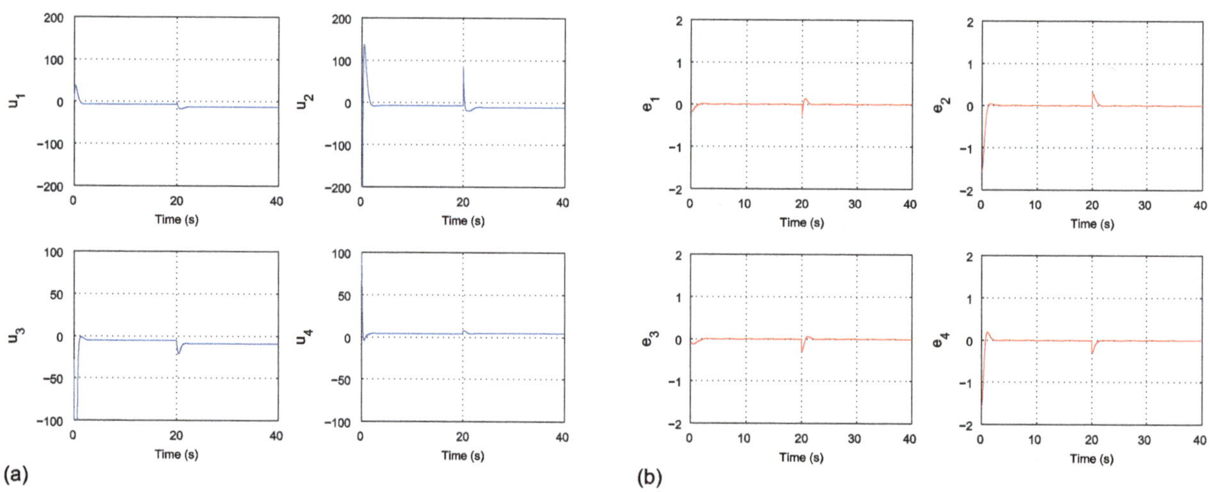

Figure 7.48 *Tracking of setpoint 6 by the induction motor-driven gas compressor: (a) variations of the control inputs u_1 to u_4; (b) tracking error of state variables x_1 to x_4*

applying the aforementioned nonlinear optimal control scheme. In this section's nonlinear optimal control approach, the dynamic model of the PMSLG-VSC-based wave energy conversion system undergoes first approximate linearization through Taylor series expansion [711,705,60,706]. This approximate linearization process takes place at each sampling instance around a time-varying operating point which is defined by the present value of the system's state vector and the last sampled value of the control inputs vector. The linearization is mainly based on the computation of the system's Jacobian matrices and their update at each sampling interval. The modeling error which is due to the truncation of higher-order terms from the Taylor series is considered to be a perturbation which is asymptotically compensated by the robustness of the control algorithm.

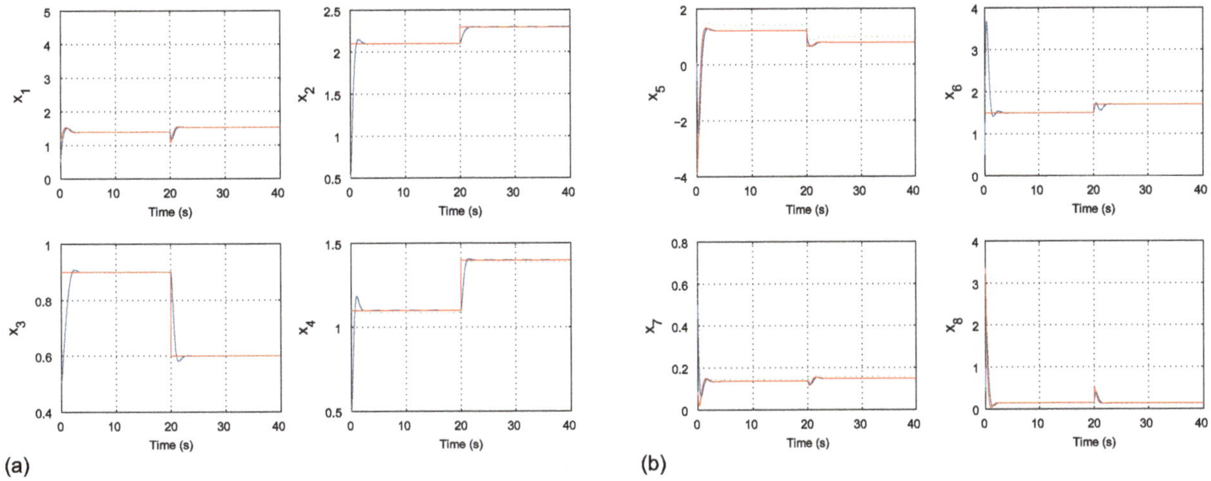

Figure 7.49 Tracking of setpoint 7 by the induction motor-driven gas compressor: (a) convergence of state variables x_1 to x_4 (blue lines) to the reference setpoints (red lines) and their KF-based estimation (green line); (b) convergence of state variables x_5 to x_8 (blue lines) to the reference setpoints (red lines) and their KF-based estimation (green line)

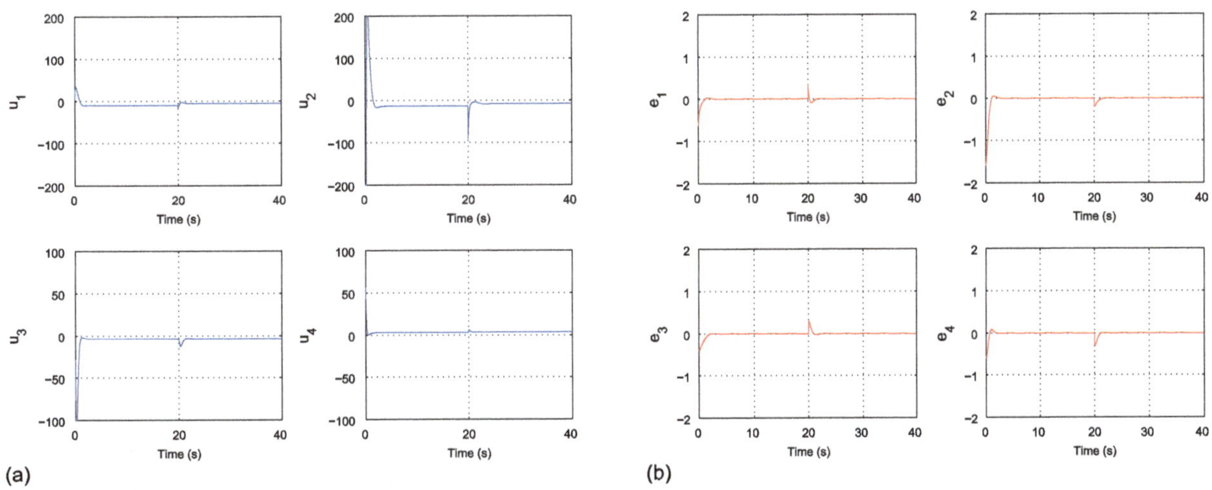

Figure 7.50 Tracking of setpoint 7 by the induction motor-driven gas compressor: (a) variations of the control inputs u_1 to u_4; (b) tracking error of state variables x_1 to x_4

For the approximately linearized model of the PMSLG-VSC-based wave energy conversion unit a stabilizing H-infinity feedback controller is designed. This controller achieves the solution of the optimal control problem for the wave energy conversion system under model uncertainty and external perturbations. It represents a min–max differential game taking place between (i) the control inputs which try to minimize a quadratic function of the state vector's tracking error and (ii) the model uncertainty or perturbation terms which try to maximize this cost function. To compute the controller's gains an algebraic Riccati equation is solved repetitively at each time step of the control method [708,741,721]. The global stability properties of this control scheme are proven through Lyapunov analysis. First, it is proven that the control loop satisfies the H-infinity tracking performance criterion which signifies that the control

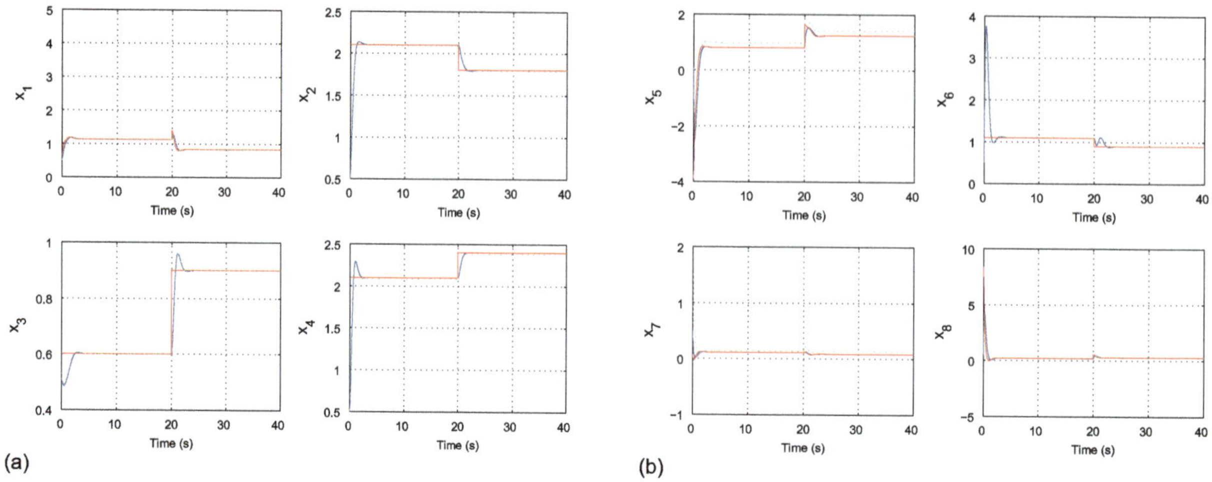

Figure 7.51 Tracking of setpoint 8 by the induction motor-driven gas compressor: (a) convergence of state variables x_1 to x_4 (blue lines) to the reference setpoints (red lines) and their KF-based estimation (green line); (b) convergence of state variables x_5 to x_8 (blue lines) to the reference setpoints (red lines) and their KF-based estimation (green line)

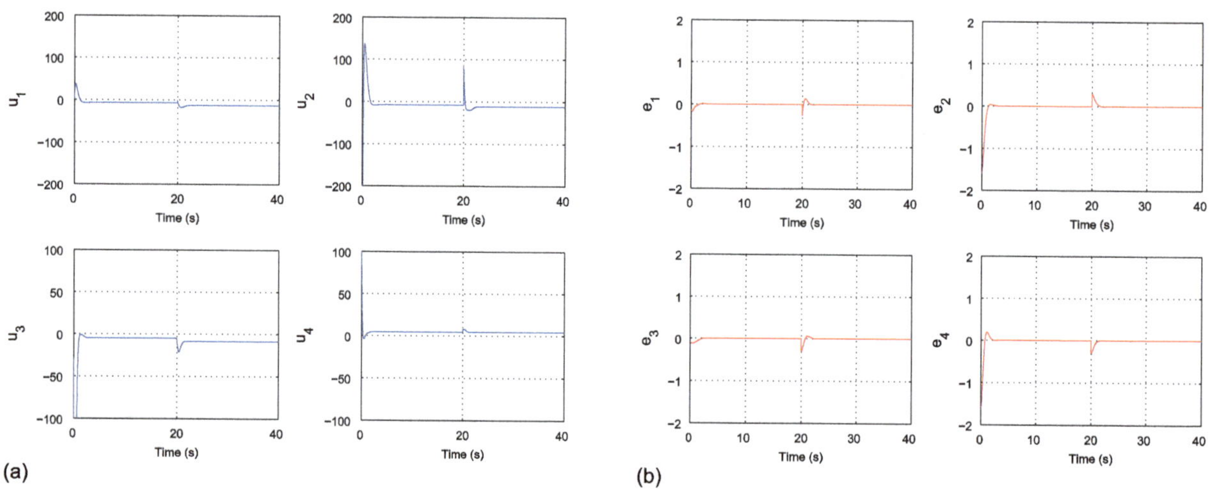

Figure 7.52 Tracking of setpoint 8 by the induction motor-driven gas compressor: (a) variations of the control inputs u_1 to u_4; (b) tracking error of state variables x_1 to x_4

method is significantly robust to model uncertainties and external perturbations [741,876,732]. Moreover, under moderate conditions, global asymptotic stability for this control approach is shown to hold. Thus, the nonlinear optimal control method achieves fast and accurate tracking of reference setpoints by the state variables of the wave energy conversion system under moderate variations of the control inputs [715,725]. Finally, to implement state estimation-based control without the need to measure the entire state vector of the power unit, the H-infinity Kalman filter is used as a robust state estimator.

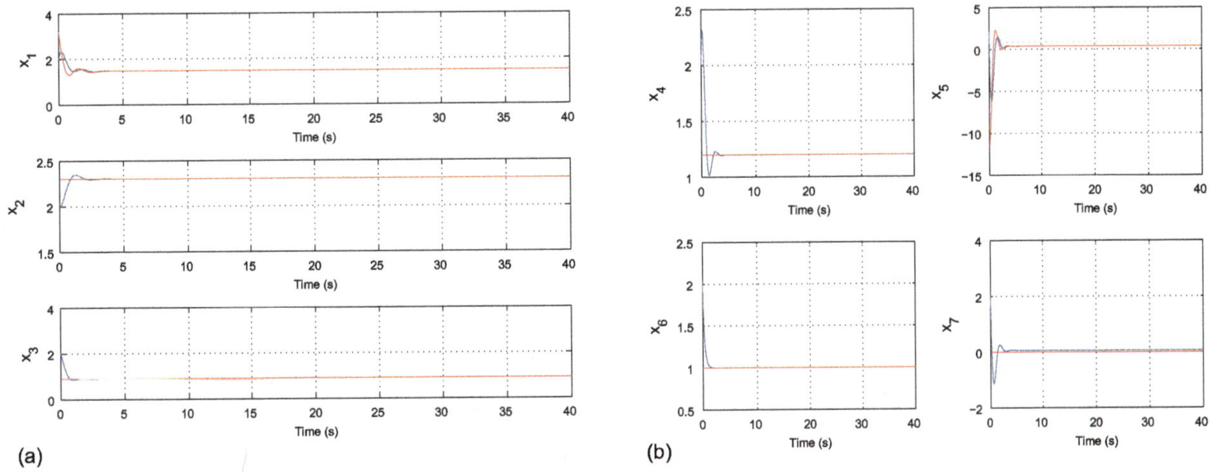

Figure 7.53 *Tracking of setpoint 1 by the PM synchronous motor-driven gas compressor: (a) convergence of state variables x_1 to x_3 (blue lines) to the reference setpoints (red lines) and their KF-based estimation (green line); (b) convergence of state variables x_4 to x_7 (blue lines) to the reference setpoints (red lines) and their KF-based estimation (green line)*

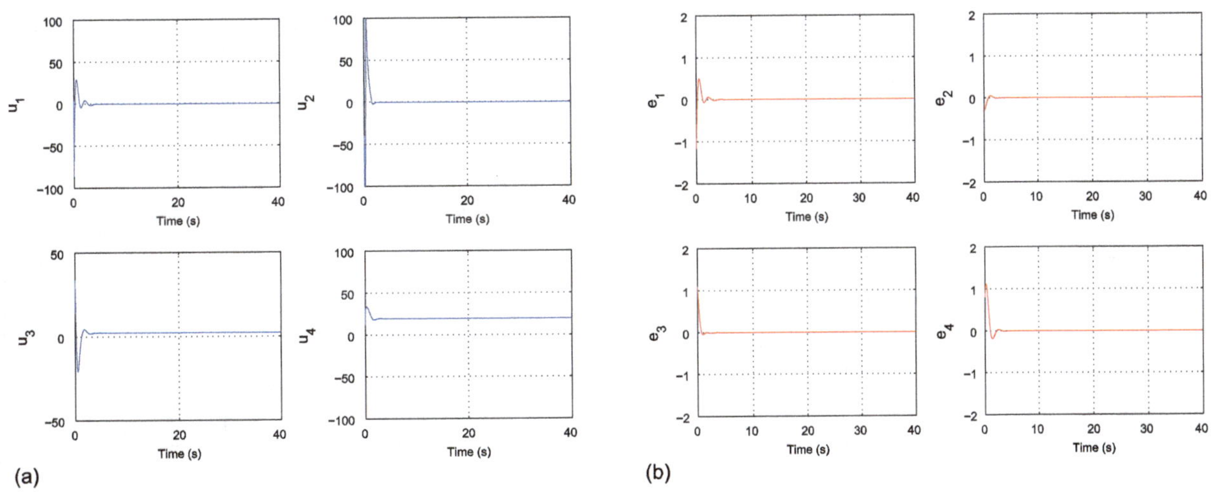

Figure 7.54 *Tracking of setpoint 1 by the PM induction motor-driven gas compressor: (a) variations of the control inputs u_1 to u_4; (b) tracking error of state variables x_1 to x_4*

7.3.2 Dynamic model of the PMLSG-VSC wave energy conversion unit

The diagram of the wave energy conversion unit that comprises a permanent magnet linear synchronous generator (PMLSG) serially connected to an AC/DC three-phase voltage source converter (VSC) is shown in Figure 7.69. The motion of the moving part (mover) of the PMSLG is given by

$$M\ddot{z} = -b_g\dot{z} - b_w\dot{z} - k_s z + F_e + F_{br} \tag{7.233}$$

where $b_g\dot{z}$ is the friction force that resists to the motion of the moving part of the tubular PMLSG, $b_w\dot{z}$ is the hydrodynamic resistance force that resists to the motion of the buoy, $k_s z$ is a spring-type force exerted on the buoy, F_e is the

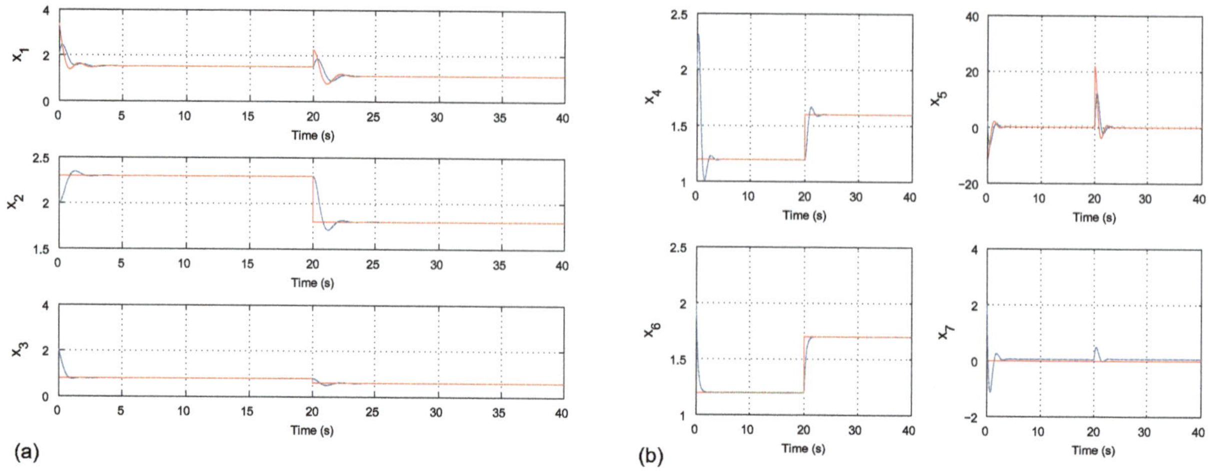

Figure 7.55 Tracking of setpoint 2 by the PM synchronous motor-driven gas compressor: (a) convergence of state variables x_1 to x_3 (blue lines) to the reference setpoints (red lines) and their KF-based estimation (green line); (b) convergence of state variables x_4 to x_7 (blue lines) to the reference setpoints (red lines) and their KF-based estimation (green line)

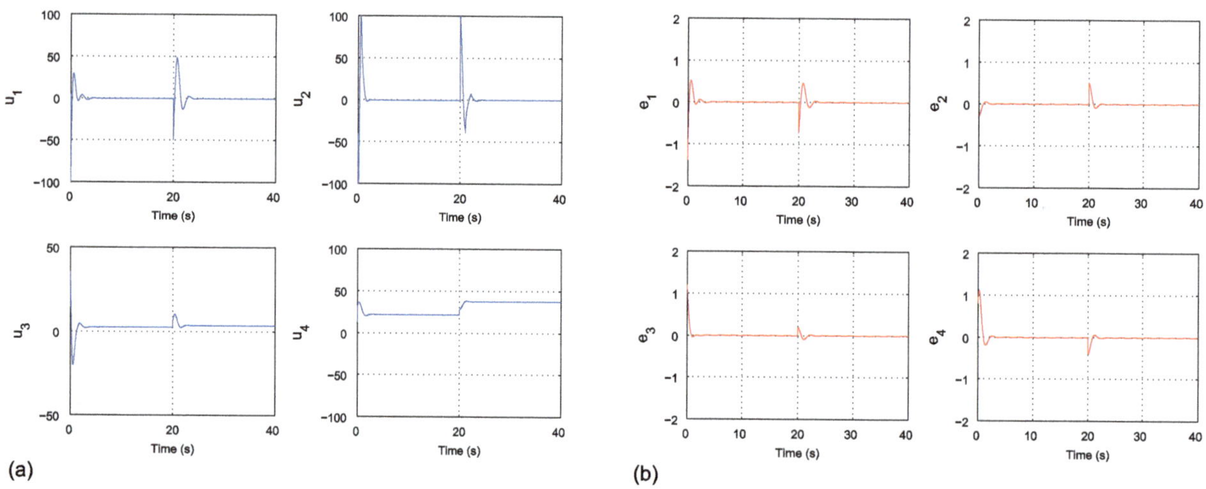

Figure 7.56 Tracking of setpoint 2 by the PM induction motor-driven gas compressor: (a) variations of the control inputs u_1 to u_4; (b) tracking error of state variables x_1 to x_4

excitation force due to the incident wave on the buoy, and F_{br} is a brake force which can decelerate the mover. Additional forces that one can include in the dynamic model of the motion of the PMLSG's mover are: F_a is an Archimedean (lift) force which is applied to the partially submerged buoy, F_r are radiated forces acting on the buoy because of the wave that is radiated as a result of the buoy's oscillations [732].

The electromagnetic force F_{em} which is applied to the moving part of the PMLSG because of the generator's magnetic field is given by

$$F_{em} = \frac{3}{2}\frac{\pi}{\tau_p}[i_d i_q(L_d - L_q) - \Psi_{PM} i_q] \tag{7.234}$$

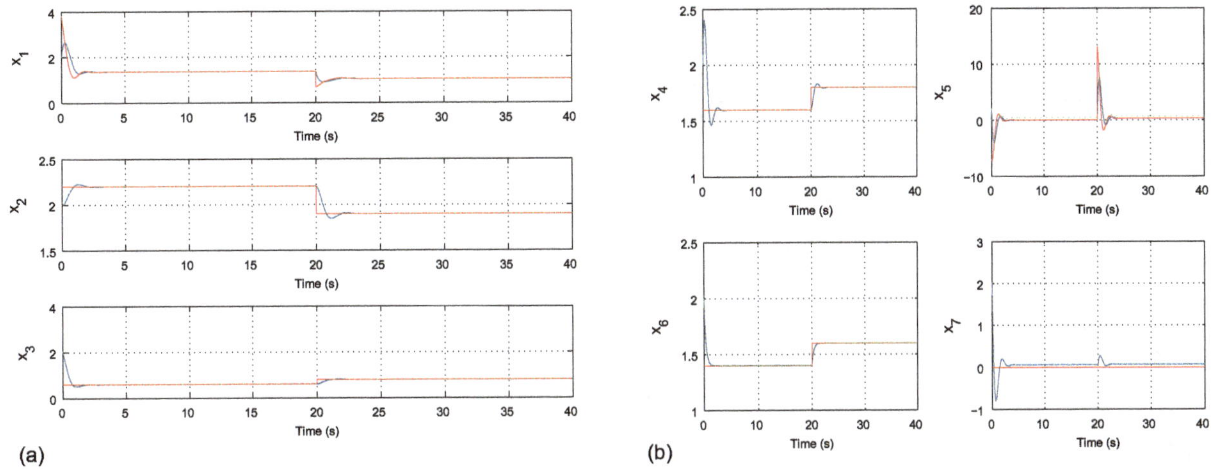

Figure 7.57 *Tracking of setpoint 3 by the PM synchronous motor-driven gas compressor: (a) convergence of state variables x_1 to x_3 (blue lines) to the reference setpoints (red lines) and their KF-based estimation (green line); (b) convergence of state variables x_4 to x_7 (blue lines) to the reference setpoints (red lines) and their KF-based estimation (green line)*

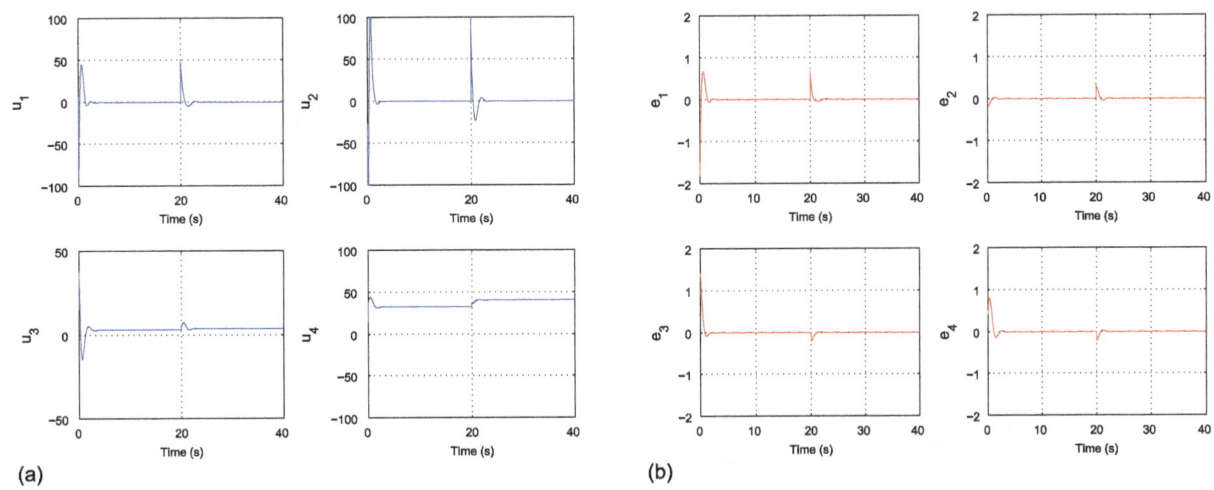

Figure 7.58 *Tracking of setpoint 3 by the PM induction motor-driven gas compressor: (a) variations of the control inputs u_1 to u_4; (b) tracking error of state variables x_1 to x_4*

where τ_p is the poles' pitch, i_d is the d-axis component of the current in the primary (mover) part of the generator, i_q is the q-axis component of the current in primary (mover) part of the generator, L_d is the d-axis inductance at the primary (mover) part of the generator, L_q is the q-axis inductance at the primary (mover) part of the generator, Ψ_{PM} is the magnetic flux due to permanent magnets.

The dynamics of the electrical part of the permanent magnet linear synchronous generator is obtained by decomposing the electric circuit of the generator in the dq rotating frame and is given by the following differential equations: [202,732]:

$$v_d = -Ri_d - \frac{d\psi_d}{dt} + v_t\frac{\pi}{\tau_p}\psi_q \tag{7.235}$$

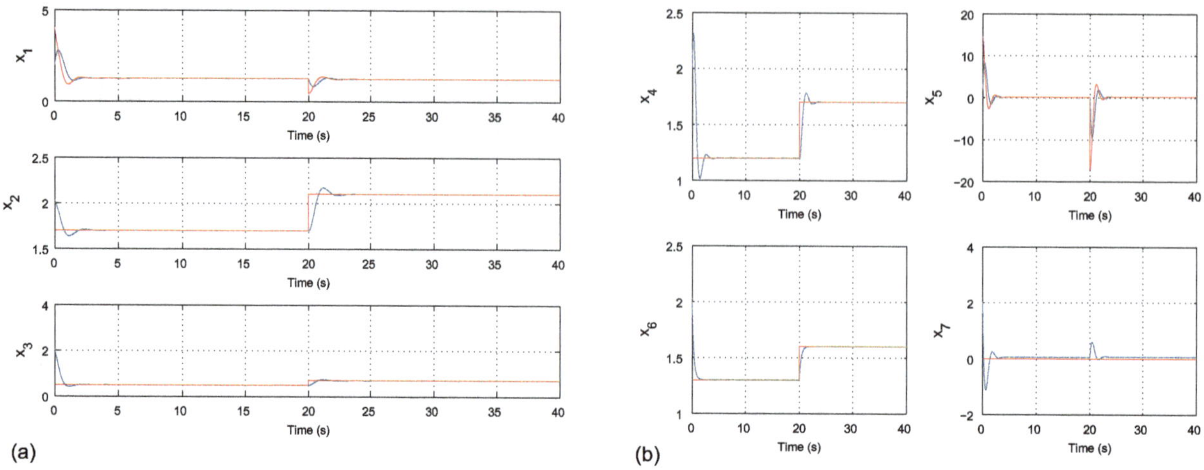

Figure 7.59 Tracking of setpoint 4 by the PM synchronous motor-driven gas compressor: (a) convergence of state variables x_1 to x_3 (blue lines) to the reference setpoints (red lines) and their KF-based estimation (green line); (b) convergence of state variables x_4 to x_7 (blue lines) to the reference setpoints (red lines) and their KF-based estimation (green line)

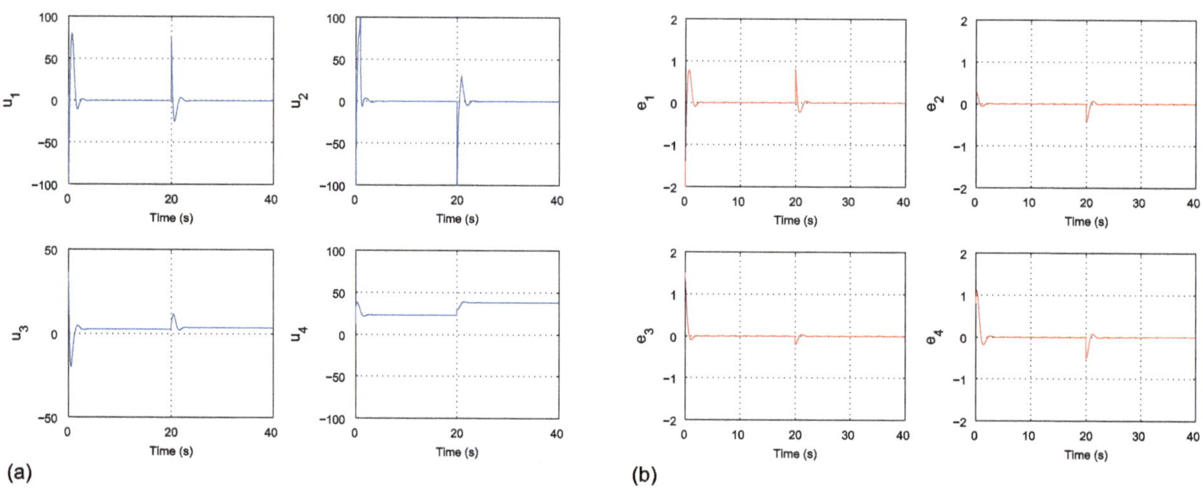

Figure 7.60 Tracking of setpoint 4 by the PM induction motor-driven gas compressor: (a) variations of the control inputs u_1 to u_4; (b) tracking error of state variables x_1 to x_4

$$v_q = -Ri_q - \frac{d\psi_q}{dt} - v_t\frac{\pi}{T_p}\psi_d \tag{7.236}$$

where $v_t = \dot{z}$ is the linear velocity of the primary part (mover) of the PMSLG along the z axis, while about the magnetic flux, one has the following equations:

$$\psi_d = L_d i_d - \psi_{PM} \tag{7.237}$$

$$\psi_q = L_q i_q \tag{7.238}$$

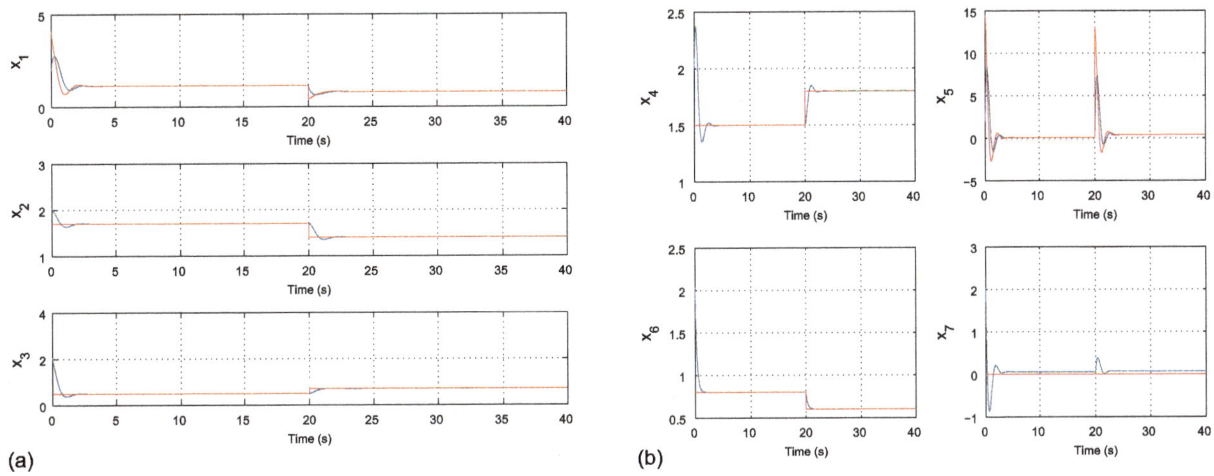

Figure 7.61 *Tracking of setpoint 5 by the PM synchronous motor-driven gas compressor: (a) convergence of state variables x_1 to x_3 (blue lines) to the reference setpoints (red lines) and their KF-based estimation (green line); (b) convergence of state variables x_4 to x_7 (blue lines) to the reference setpoints (red lines) and their KF-based estimation (green line)*

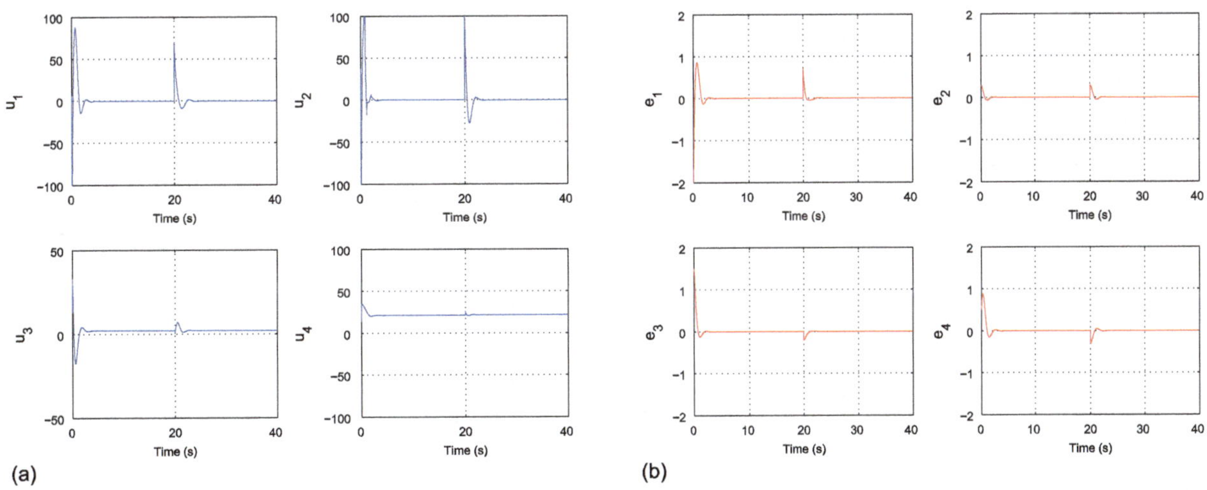

Figure 7.62 *Tracking of setpoint 5 by the PM induction motor-driven gas compressor: (a) variations of the control inputs u_1 to u_4; (b) tracking error of state variables x_1 to x_4*

Considering that the external forces which are applied to the PMLSG's mover are (i) the excitation force F_e that is due to the incidence of the waves on the buoy and (ii) the brake force F_{br} which can be exerted on the mover by an electrohydraulic actuator, one has that the dynamics of the PMLSG are given by the following equations [202,732]:

$$\ddot{z} = -\frac{(b_g+b_w)}{M}\dot{z} - \frac{k_v}{M}z + \frac{3}{2M}\frac{\pi}{\tau_p}[i_d i_q(L_d - L_q) - \Psi_{PM} i_q] + \frac{F_e}{M} + \frac{F_{br}}{M} \quad (7.239)$$

$$\frac{di_d}{dt} = -\frac{R}{L_d}i_d + \frac{\dot{z}}{L_d}\frac{\pi}{\tau_p}(L_q i_q) - \frac{v_d}{L_d} \quad (7.240)$$

$$\frac{di_q}{dt} = -\frac{R}{L_q}i_q - \frac{\dot{z}}{L_q}\frac{\pi}{\tau_p}(L_d i_d - \Psi_{PM}) - \frac{v_q}{L_d} \quad (7.241)$$

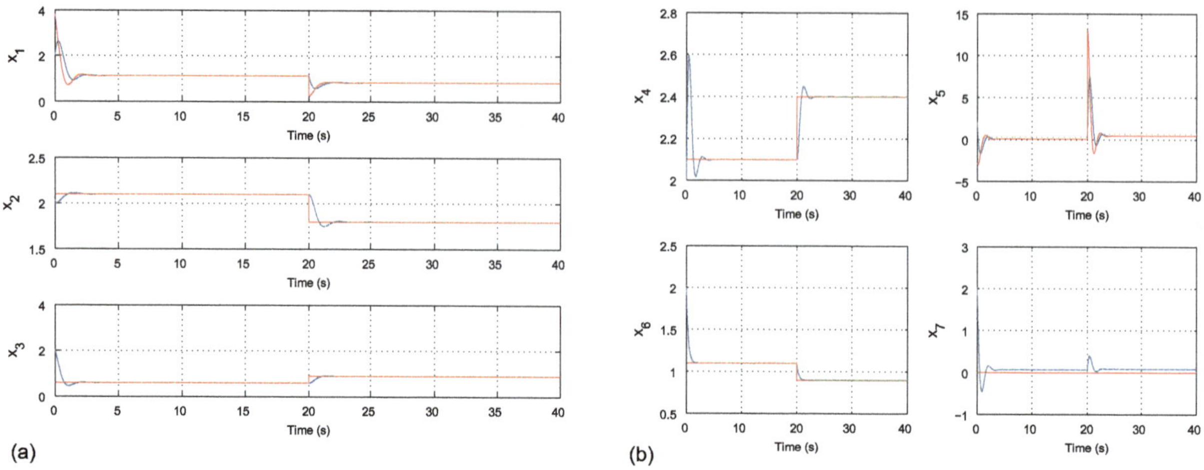

Figure 7.63 Tracking of setpoint 6 by the PM synchronous motor-driven gas compressor: (a) convergence of state variables x_1 to x_3 (blue lines) to the reference setpoints (red lines) and their KF-based estimation (green line); (b) convergence of state variables x_4 to x_7 (blue lines) to the reference setpoints (red lines) and their KF-based estimation (green line)

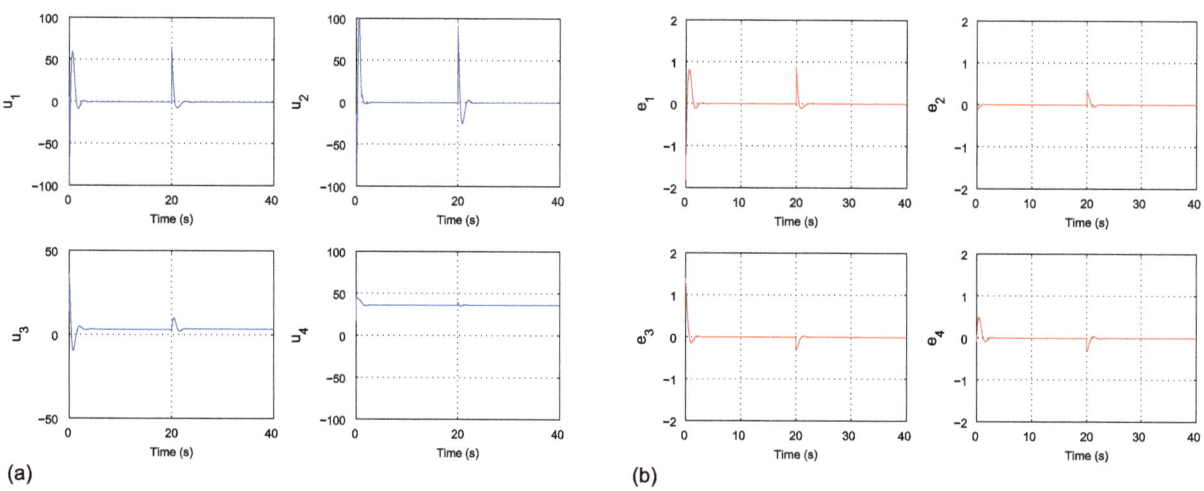

Figure 7.64 Tracking of setpoint 6 by the PM induction motor-driven gas compressor: (a) variations of the control inputs u_1 to u_4; (b) tracking error of state variables x_1 to x_4

Next, by defining the state variables $x_1 = z$, $x_2 = \dot{z}$, $x_3 = i_d$, $x_4 = i_q$ and the control input variables $u_1 = F_{br}$, $u_2 = v_d$, $u_3 = v_q$, the dynamic model of the PMLSG becomes

$$\begin{aligned}\dot{x}_1 &= x_2\\ \dot{x}_2 &= -\frac{(b_g+b_w)}{M}x_2 - \frac{k_v}{M}x_1 + \frac{3}{2M}\frac{\pi}{\tau_p}[x_3x_4(L_d - L_q) - \Psi_{PM}x_4] + \frac{F_e}{M} + \frac{u_1}{M}\\ \dot{x}_3 &= -\frac{R}{L_d}x_3 + \frac{x_2}{L_d}\frac{\pi}{\tau_p}(L_qx_4) - \frac{u_2}{L_d}\\ \dot{x}_4 &= -\frac{R}{L_q}x_4 - \frac{x_2}{L_q}\frac{\pi}{\tau_p}(L_dx_3 - \Psi_{PM}) - \frac{u_3}{L_d}\end{aligned} \tag{7.242}$$

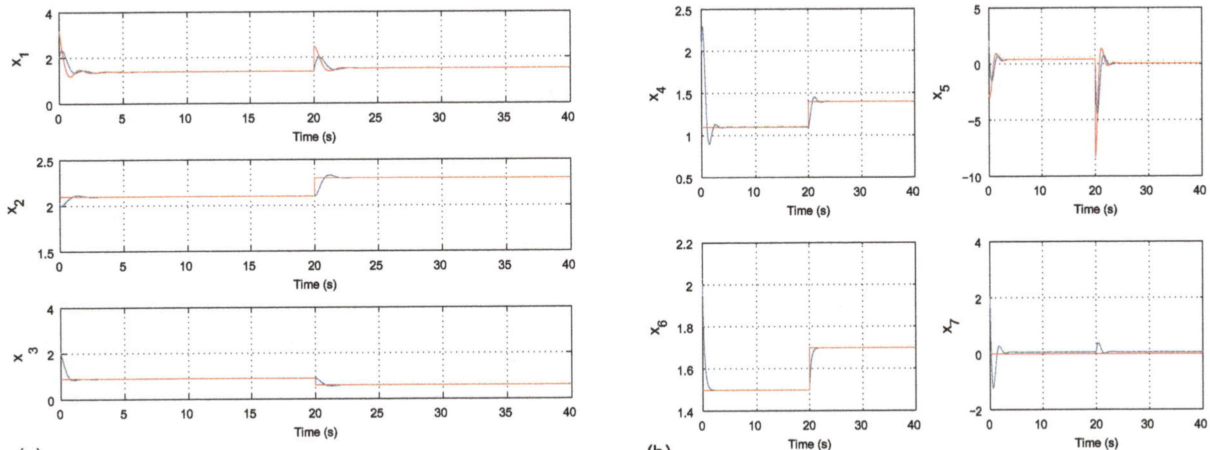

(a) (b)

Figure 7.65 Tracking of setpoint 7 by the PM synchronous motor-driven gas compressor: (a) convergence of state variables x_1 to x_3 (blue lines) to the reference setpoints (red lines) and their KF-based estimation (green line); (b) convergence of state variables x_4 to x_7 (blue lines) to the reference setpoints (red lines) and their KF-based estimation (green line)

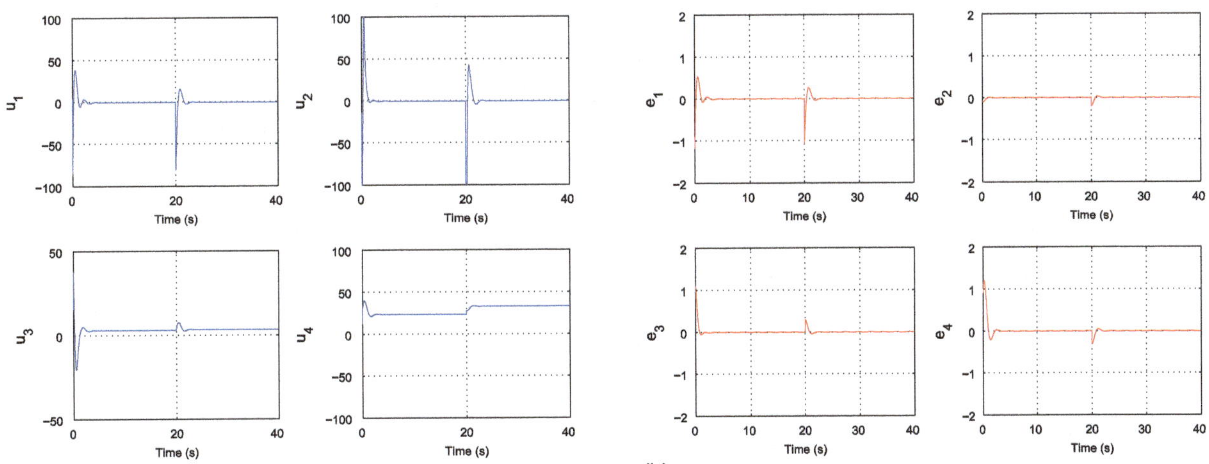

(a) (b)

Figure 7.66 Tracking of setpoint 7 by the PM induction motor-driven gas compressor: (a) variations of the control inputs u_1 to u_4; (b) tracking error of state variables x_1 to x_4

The diagram of the AC/DC three-phase voltage source converter is given in Figure 7.70. The dynamics of the AC/DC converter is given by

$$L'\dot{i}_d' = -R'i_d' + L'\omega_{dq}i_q' + v_d' - \frac{V_{dc}}{2}\eta_1 \tag{7.243}$$

$$L'\dot{i}_q' = -L'\omega_{dq}i_d' - R'i_q' + v_q' - \frac{V_{dc}}{2}\eta_2 \tag{7.244}$$

$$C_{dc}\dot{V}_{dc} = -\frac{1}{R_c}V_{dc} + \frac{3}{4}i_d'\eta_1 + \frac{3}{4}i_q'\eta_2 \tag{7.245}$$

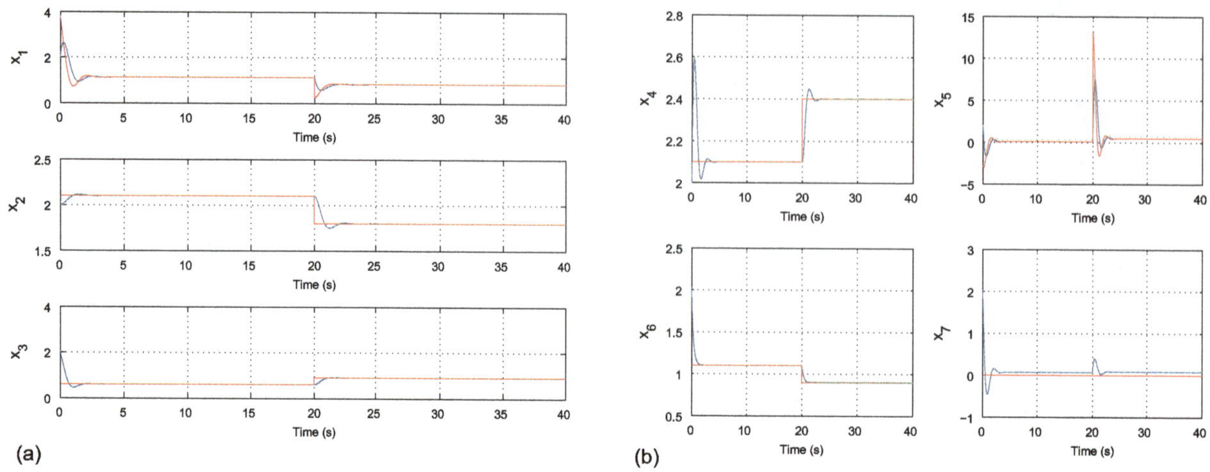

Figure 7.67 Tracking of setpoint 8 by the PM synchronous motor-driven gas compressor: (a) convergence of state variables x_1 to x_3 (blue lines) to the reference setpoints (red lines) and their KF-based estimation (green line); (b) convergence of state variables x_4 to x_7 (blue lines) to the reference setpoints (red lines) and their KF-based estimation (green line)

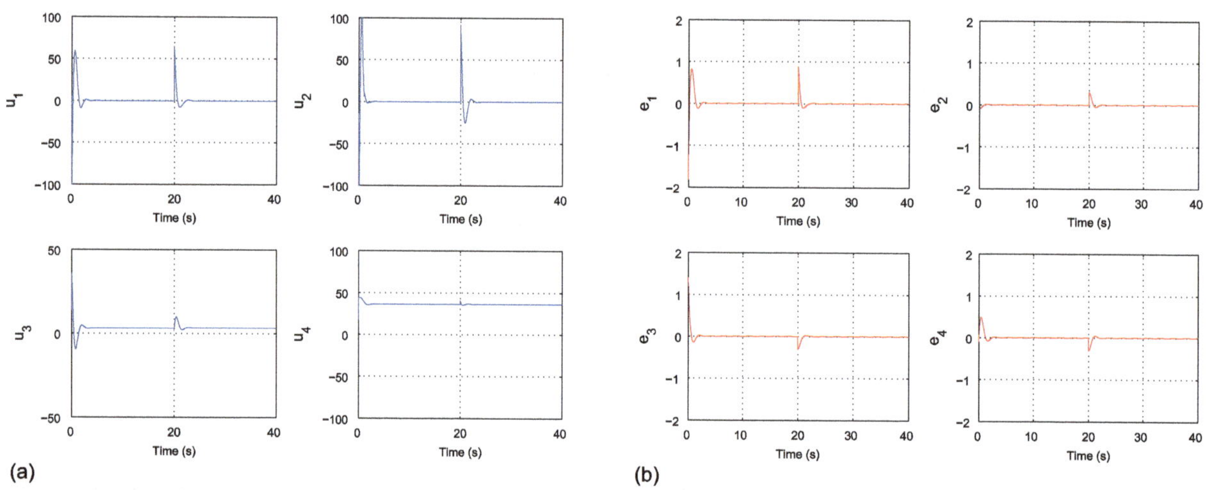

Figure 7.68 Tracking of setpoint 8 by the PM induction motor-driven gas compressor: (a) variations of the control inputs u_1 to u_4; (b) tracking error of state variables x_1 to x_4

where $i_d^{'}$, $i_q^{'}$ are the line currents i_a, i_b, i_c after transformation in the dq reference frame, $v_d^{'}$, $v_q^{'}$ are the line voltages v_a, v_b, v_c after transformation in the dq reference frame, V_{dc} is the voltage output of the converter, η_1 and η_2 stand for external control inputs. The line losses are modeled by resistance $R^{'}$, the converter losses are modeled by resistance R_c, parameter $L^{'}$ denotes inductance, and coefficient ω_{dq} is the grid's frequency.

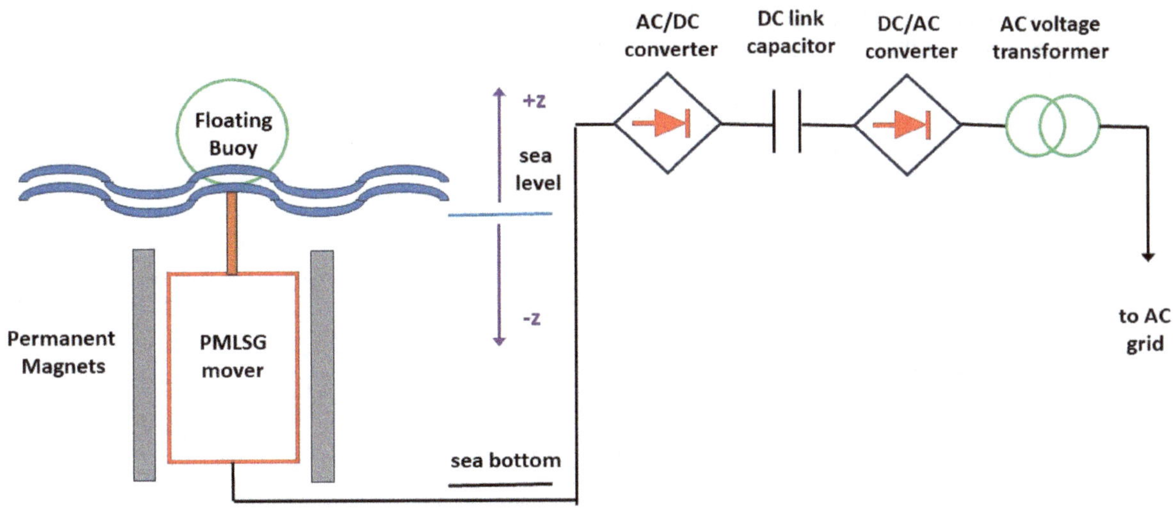

Figure 7.69 Diagram of the wave energy conversion unit that comprises a permanent magnet linear synchronous generator (PMLSG) serially connected to an AC/DC three-phase voltage source converter (VSC)

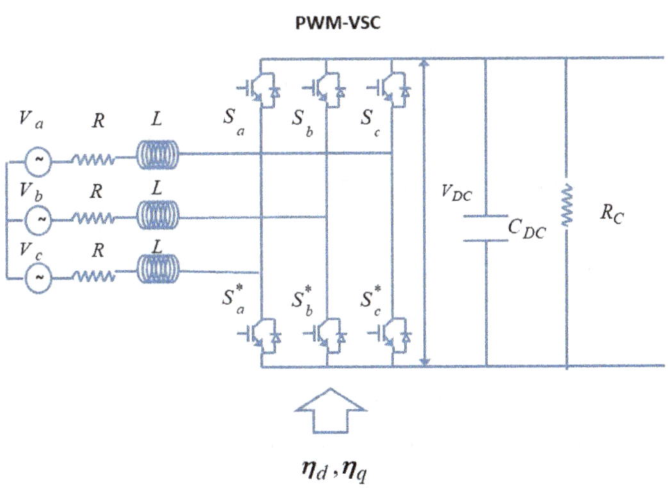

Figure 7.70 Diagram of the three-phase AC/DC voltage source converter

Because of connecting the AC/DC converter to the PMLSG it holds that $\eta_1 = k_1 i_d^{'}$, $\eta_2 = k_2 i_q^{'}$, where k_1, k_2 are constants. By defining the state variables $x_5 = i_d^{'}$, $x_6 = i_q^{'}$, $x_7 = V_{dc}$ the dynamic model of the AC/DC converter becomes [715]

$$\begin{aligned}
\dot{x}_5 &= -\frac{R^{'}}{L^{'}}x_5 + \omega_{dq}x_6 + \frac{v_d^{'}}{L^{'}} - \frac{k_1}{2L^{'}}x_3x_7 \\
\dot{x}_6 &= -\omega_{dq}x_5 - \frac{R^{'}}{L^{'}}x_6 + \frac{v_q^{'}}{L^{'}} - \frac{k_2}{2L^{'}}x_4x_7 \\
\dot{x}_7 &= -\frac{1}{R_cC_{dc}}x_7 + \frac{3k_1}{4C_{dc}}x_3x_5 + \frac{3k_2}{4C_{dc}}x_4x_6
\end{aligned} \tag{7.246}$$

By combining the dynamic model of the PMLSG given in Eq. (7.242) with the dynamic model of the VSC given in Eq. (7.246), one obtains the following state-space model for the wave energy conversion unit:

$$\begin{aligned}
\dot{x}_1 &= x_2 \\
\dot{x}_2 &= -\frac{(b_g+b_w)}{M}x_2 - \frac{k_v}{M}x_1 + \frac{3}{2M}\frac{\pi}{\tau_p}[x_3x_4(L_d-L_q) - \Psi_{PM}x_4] + \frac{F_e}{M} + \frac{1}{M}u_1 \\
\dot{x}_3 &= -\frac{R}{L_d}x_3 + \frac{x_2}{L_d}\frac{\pi}{\tau_p}(L_qx_4) - \frac{1}{L_d}u_2 \\
\dot{x}_4 &= -\frac{R}{L_q}x_4 - \frac{x_2}{L_q}\frac{\pi}{\tau_p}(L_dx_3 - \Psi_{PM}) - \frac{1}{L_d}u_3 \\
\dot{x}_5 &= -\frac{R'}{L'}x_5 + \omega_{dq}x_6 + \frac{v_d'}{L'} - \frac{k_1}{2L'}x_3x_7 \\
\dot{x}_6 &= -\omega_{dq}x_5 - \frac{R'}{L'}x_6 + \frac{v_q'}{L'} - \frac{k_2}{2L'}x_4x_7 \\
\dot{x}_7 &= -\frac{1}{R_cC_{dc}}x_7 + \frac{3k_1}{4C_{dc}}x_3x_5 + \frac{3k_2}{4C_{dc}}x_4x_6
\end{aligned} \tag{7.247}$$

Thus, the state-space model of the wave energy conversion system can be written in the following matrix form:

$$\begin{pmatrix}\dot{x}_1\\ \dot{x}_2\\ \dot{x}_3\\ \dot{x}_4\\ \dot{x}_5\\ \dot{x}_6\\ \dot{x}_7\end{pmatrix} = \begin{pmatrix} x_2 \\ -\frac{(b_g+b_w)}{M}x_2 - \frac{k_v}{M}x_1 + \frac{3}{2M}\frac{\pi}{\tau_p}[x_3x_4(L_d-L_q) - \Psi_{PM}x_4] + \frac{F_e}{M} \\ -\frac{R}{L_d}x_3 + \frac{x_2}{L_d}\frac{\pi}{\tau_p}(L_qx_4) \\ -\frac{R}{L_q}x_4 - \frac{x_2}{L_q}\frac{\pi}{\tau_p}(L_dx_3 - \Psi_{PM}) \\ -\frac{R'}{L'}x_5 + \omega_{dq}x_6 + \frac{v_d'}{L'} - \frac{k_1}{2L'}x_3x_7 \\ -\omega_{dq}x_5 - \frac{R'}{L'}x_6 + \frac{v_q'}{L'} - \frac{k_2}{2L'}x_4x_7 \\ -\frac{1}{R_cC_{dc}}x_7 + \frac{3k_1}{4C_{dc}}x_3x_5 + \frac{3k_2}{4C_{dc}}x_4x_6 \end{pmatrix} + \begin{pmatrix} 0 & 0 & 0 \\ \frac{1}{M} & 0 & 0 \\ 0 & -\frac{1}{L_d} & 0 \\ 0 & 0 & -\frac{1}{L_q} \\ 0 & 0 & 0 \\ 0 & 0 & 0 \\ 0 & 0 & 0 \end{pmatrix}\begin{pmatrix}u_1\\ u_2\\ u_3\end{pmatrix} \tag{7.248}$$

Finally, the dynamic model of the wave energy conversion system can be also written in the concise nonlinear affine-in-the-input state-space form:

$$\dot{x} = f(x) + g(x)u \tag{7.249}$$

where $x \in R^{7\times 1}$, $f(x) \in R^{7\times 1}$, $g(x) \in R^{7\times 3}$, and $u \in R^{3\times 1}$.

7.3.3 *Differential flatness of the PMLSG-VSC wave energy conversion unit*

7.3.3.1 Proof of differential flatness of the wave energy conversion system

Differential flatness for a dynamical system is confirmed if (i) all its state variables can be written as differential functions of a subset of selected state vector elements of it which constitute the flat outputs vector and (ii) the flat outputs are differentially independent which means that they and their time derivatives are not connected through a relation in the form of a homogeneous differential equation [711]. It will be proven that the dynamic model of the PMLSG-VSC wave energy conversion unit is differentially flat with flat outputs vector:

$$Y = [y_1, y_2, y_3]^T = [x_1,\ \frac{3L'}{4}(x_5^2 + x_6^2) + \frac{C_{dc}}{2}x_7^2,\ x_6]^T \tag{7.250}$$

By differentiating the flat output y_2 and after some intermediate operations, one obtains

$$\dot{y}_2 = -\frac{3R'}{2}(x_5^2 + x_6^2) - \frac{1}{R_c'}x_7^2 + \frac{3L'}{2}v_d x_5 + \frac{3L'}{2}v_q x_6 \tag{7.251}$$

Besides, from the definition of the flat output $y_2 = \frac{3L'}{4}(x_5^2 + x_6^2) + \frac{C_{dc}}{2}x_7^2$ and after solving for x_7^2, one has

$$x_7^2 = \frac{2y_2 - \frac{3L'}{2}(x_5^2 + x_6^2)}{C_{dc}} \tag{7.252}$$

By substituting Eq. (7.252) into Eq. (7.251) and using that the third flat output is $y_3 = x_2$, one obtains

$$\dot{y}_2 = -\frac{3R'}{2}x_5^2 - \frac{3R'}{2}y_3^2 - \frac{2}{R_c C_{dc}}y_2 + \frac{3L'}{2R_c C_{dc}}x_5^2 + \frac{3L'}{2R_c C_{dc}}y_2^2 + \frac{3L'}{2}v_d x_5 + \frac{3L'}{2}v_q y_3 \tag{7.253}$$

By regrouping terms in Eq. (7.253), one obtains a binomial for x_5

$$[-\frac{3R'}{2} + \frac{3L'}{2R_c C_{dc}}]x_5^2 + \frac{3L'}{2}v_d x_5 + [-\frac{3R'}{2}y_3^2 - \frac{2}{R_c C_{dc}}y_2 + \frac{3L'}{2R_c C_{dc}}y_3^2 + \frac{3L'}{2}v_q y_3] = 0 \tag{7.254}$$

By defining coefficients $a = [-\frac{3R'}{2} + \frac{3L'}{2R_c C_{dc}}]$, $b = \frac{3L'}{2}v_d$ and $c = [-\frac{3R'}{2}y_3^2 - \frac{2}{R_c C_{dc}}y_2 + \frac{3L'}{2R_c C_{dc}}y_3^2 + \frac{3L'}{2}v_q y_3]$ the binomial of Eq. (7.254) is written as

$$ax_5^2 + bx_5 + c = 0 \tag{7.255}$$

By solving Eq. (7.255) for x_5 and by retaining the maximum of the two solutions, one obtains

$$x_5 = \frac{-b + \sqrt{|b^2 - 4ac|}}{2a} \Rightarrow x_5 = h_5(Y, \dot{Y}) \tag{7.256}$$

which signifies that state variable x_5 is a differential function of the flat outputs vector. Next, returning to Eq. (7.252)

$$x_7 = \sqrt{\frac{2y_2 - \frac{3L'}{2}(h_5(Y,\dot{Y})^2 + y_3^2)}{C_{dc}}} \Rightarrow x_7 = h_7(Y, \dot{Y}) \tag{7.257}$$

which signifies that state variable x_7 is also a differential function of the flat outputs vector. From the 5th row of the state-space model, one solves for x_3. It holds that

$$x_3 = -2L' \cdot \frac{\dot{x}_5 + \frac{R'}{L'}x_5 - \omega_{dq}x_6 - v_d}{K_1 x_2} \tag{7.258}$$

where x_5, x_6, x_7 are differential functions of the flat outputs vector Y. This gives

$$x_3 = h_3(Y, \dot{Y}) \tag{7.259}$$

which signifies that x_3 is a differential function of the flat outputs of the system. Additionally, from the 6th row of the state-space model, one solves for x_4. This gives

$$x_4 = -2L' \cdot \frac{\dot{x}_6 + \omega_{dq}x_5 + \frac{R'}{L'}x_6 - v_q}{K_2 v_q} \tag{7.260}$$

where x_5, x_6, x_7 are differential functions of the flat outputs vector Y. This gives

$$x_4 = h_4(Y, \dot{Y}) \tag{7.261}$$

which signifies that x_4 is a differential function of the flat outputs of the system. Furthermore, from the 1st equation of the state-space model, one has

$$x_2 = \dot{x}_1 \Rightarrow x_2 = h_2(Y, \dot{Y}) \tag{7.262}$$

which signifies that x_2 is a differential function of the flat outputs vector. Consequently, all state variables x_i, $i = 1, 2, \cdots, 7$ are differential functions of the flat outputs vector. Moreover, from the 2nd row of the state-space model, one solves for control input u_1. This gives

$$u_1 = M\dot{x}_2 + K_3 x_1 + bx_2 + \frac{3}{2}\frac{\pi}{\tau_p}[x_3 x_4(L_d - L_q) - \Psi_{PM}x_4] - F_e \tag{7.263}$$

where the excitation force F_e is considered to be independent of x_i. Therefore, u_1 is a differential function of the flat outputs vector, or

$$u_1 = h_{u_1}(Y, \dot{Y}) \tag{7.264}$$

Equivalently, from the 3rd row of the state-space model, one solves for control input u_2. It holds that

$$u_2 = -[L_d\dot{x}_3 + Rx_3 + x_2\frac{\pi}{\tau_p}(L_qx_4)] \Rightarrow u_2 = h_{u_2}(Y, \dot{Y}) \tag{7.265}$$

Finally, from the 4th row of the state-space model, one solves for control input u_3. It holds that

$$u_3 = -[L_q\dot{x}_4 + Rx_4 + x_2\frac{\pi}{\tau_p}(L_dx_3)] \Rightarrow u_3 = h_{u_2}(Y, \dot{Y}) \tag{7.266}$$

which signifies that u_3 is a differential function of the flat outputs vector. As a result of the above all state variables and the control inputs are differential functions of the flat outputs vector and the wave energy conversion system is differentially flat.

The differential flatness of the system is an implicit proof of the system's controllability. It also demonstrates that the system is input–output linearizable through successive differentiations of its flat outputs. Finally, it allows for solving the setpoints definition problem. First, one selects setpoints in an unconstrained manner for the flat outputs of the system. Based on these setpoints, one can also obtain reference values for the rest of the state vector elements of the system using the differential relations which connect them to the flat outputs.

7.3.3.2 Input–output linearization and eigenvalues assignment-based control

By proving differential flatness properties for the dynamic model of the wave energy conversion system, one can transform the associated state-space model into the input–output linearized form and subsequently into the canonical Brunovsky form. Linearization is performed through successive differentiations of the flat outputs. The state-space model of the system given in Eq. (7.247) was shown to be comprising the following state equations

$$\begin{aligned}
\dot{x}_1 &= f_1(x) \\
\dot{x}_2 &= f_2(x) + g_2(x)u_1 \\
\dot{x}_3 &= f_3(x) + g_3(x)u_2 \\
\dot{x}_4 &= f_4(x) + g_4(x)u_3 \\
\dot{x}_5 &= f_5(x) \\
\dot{x}_6 &= f_6(x) \\
\dot{x}_7 &= f_7(x)
\end{aligned} \tag{7.267}$$

where $f_1(x) = x_2$, $g_1(x) = 0$, $f_2(x) = -\frac{(b_g+b_w)}{M}x_2 - \frac{k_v}{M}x_1 + \frac{3}{2M}\frac{\pi}{\tau_p}[x_3x_4(L_d - L_q) - \Psi_{PM}x_4] + \frac{F_e}{M}$, $g_2(x) = \frac{1}{M}$, $f_3(x) = -\frac{R}{L_d}x_3 + \frac{x_2}{L_d}\frac{\pi}{\tau_p}(L_qx_4)$, $g_3(x) = -\frac{1}{L_d}$, $f_4(x) = -\frac{R}{L_q}x_4 - \frac{x_2}{L_q}\frac{\pi}{\tau_p}(L_dx_3 - \Psi_{PM})$, $g_4(x) = -\frac{1}{L_d}$, $f_5(x) = -\frac{R'}{L'}x_5 + \omega_{dq}x_6 + \frac{v_d'}{L'} - \frac{k_1}{2L'}x_3x_7$, $g_5(x) = 0$, $f_6(x) = -\omega_{dq}x_5 - \frac{R'}{L'}x_6 + \frac{v_q'}{L'} - \frac{k_2}{2L'}x_4x_7$, $g_6(x) = 0$, $f_7(x) = -\frac{1}{R_cC_{dc}}x_7 + \frac{3k_1}{4C_{dc}}x_3x_5 + \frac{3k_2}{4C_{dc}}x_4x_6$, $g_7(x) = 0$.

By differentiating twice the first flat output of the system $y_1 = x_1$, one obtains

$$\ddot{y}_1 = f_a(x) + g_{a1}(x)u_1 + g_{a2}(x)u_2 + g_{a3}(x)u_3 \tag{7.268}$$

where about functions $f_a(x)$, $g_{a1}(x)$, $g_{a2}(x)$, $g_{a3}(x)$, one has that

$$f_{a1}(x) = -\frac{(b_g+b_w)}{M}x_2 - \frac{k_v}{M}x_1 + \frac{3}{2M}\frac{\pi}{\tau_p}[x_3x_4(L_d - L_q) - \Psi_{PM}x_4] + \frac{F_e}{M} \tag{7.269}$$

$$g_{a1}(x) = \frac{1}{M} \quad g_{a2}(x) = 0 \; g_{a3}(x) = 0 \tag{7.270}$$

By differentiating three times the flat output $y_2 = \frac{3L'}{4}(x_5^2 + x_6^2) + \frac{C_{dc}}{2}x_7^2$, one obtains

$$y_2^{(3)} = f_b(x) + g_{b1}(x)u_1 + g_{b2}(x)u_2 + g_{b3}(x)u_3 \tag{7.271}$$

where about functions $f_b(x)$, $g_{b1}(x)$, $g_{b2}(x)$, $g_{b3}(x)$, one has that

$$f_b(x) = f_{b1}(x) + f_{b2}(x) + f_{b3}(x) \tag{7.272}$$

with functions $f_{b1}(x), f_{b2}(x), f_{b3}(x)$ to be given by

$$\begin{aligned}
f_{b1}(x) = &-3R'\dot{x}_5[-\tfrac{R'}{L'}x_5 + \omega_d q x_6 - \tfrac{k_1}{2L'}x_3x_7 + v_d] \\
&-3R'x_5[-\tfrac{R'}{L'}\dot{x}_5 + \omega_d q\dot{x}_6 - \tfrac{k_1}{2L'}x_3\dot{x}_7] \\
&-3R'\dot{x}_6[-\omega_d q x_5 - \tfrac{R'}{L'}x_6 - \tfrac{k_2}{2L'}x_4x_7 + v_q] \\
&-3R'\dot{x}_6[-\omega_d q\dot{x}_5 - \tfrac{R'}{L'}\dot{x}_6 - \tfrac{k_2}{2L'}x_4\dot{x}_7] \\
&-\tfrac{1}{R_c}\dot{x}_7[-\tfrac{1}{R_cC_{dc}}x_7 + \tfrac{3k_1}{4C_{dc}}x_3x_5 + \tfrac{3k_2}{4C_{dc}}x_4x_6] \\
&-\tfrac{1}{R_c}x_7[-\tfrac{1}{R_cC_{dc}}\dot{x}_7 + \tfrac{3k_1}{4C_{dc}}x_3\dot{x}_5 + \tfrac{3k_2}{4C_{dc}}x_4\dot{x}_6] \\
&\tfrac{3L'}{2}v_d[-\tfrac{R'}{L'}\dot{x}_5 + \omega_{dq}\dot{x}_6 - \tfrac{k_1}{2L'}\dot{x}_7x_3] \\
&\tfrac{3L'}{2}v_q[-\omega_{dq}\dot{x}_5 - \tfrac{R'}{L'}\dot{x}_6 - \tfrac{k_2}{2L'}\dot{x}_7x_4]
\end{aligned} \tag{7.273}$$

$$f_{b2}(x) = [3R'x_5\tfrac{k_1}{2L'}x_7 - \tfrac{1}{R_c}x_7\tfrac{3k_1}{4C_{dc}}x_5 - \tfrac{3L'}{2}v_d\tfrac{k_1}{2L'}x_7]f_3(x) \tag{7.274}$$

$$f_{b3}(x) = [3R'x_6\tfrac{k_2}{2L'}x_7 - \tfrac{1}{R_c}x_7\tfrac{3k_2}{4C_{dc}}x_6 - \tfrac{3L'}{2}v_q\tfrac{k_2}{2L'}x_7]f_4(x) \tag{7.275}$$

while about functions $g_{b1}(x)$, $g_{b2}(x)$, $g_{b3}(x)$, one has that

$$\begin{aligned}
g_{b1}(x) &= 0 \\
g_{b2}(x) &= [3R'x_5\tfrac{k_1}{2L'}x_7 - \tfrac{1}{R_c}x_7\tfrac{3k_1}{4C_{dc}}x_5 - \tfrac{3L'}{2}v_d\tfrac{k_1}{2L'}x_7]g_3(x) \\
g_{b3}(x) &= [3R'x_6\tfrac{k_2}{2L'}x_7 - \tfrac{1}{R_c}x_7\tfrac{3k_2}{4C_{dc}}x_6 - \tfrac{3L'}{2}v_q\tfrac{k_2}{2L'}x_7]g_4(x)
\end{aligned} \tag{7.276}$$

Moreover, by differentiating twice flat output y_3, one obtains

$$\ddot{y}_3 = f_c(x) + g_{c1}(x)u_1 + g_{c2}(x)u_2 + g_{c3}(x)u_3 \tag{7.277}$$

where about function $f_c(x)$, it holds that

$$\begin{aligned}
f_c(x) = &-\omega_{dq}\dot{x}_5 - \tfrac{R'}{L'}\dot{x}_6 - \tfrac{k_1}{2L'}\dot{x}_7x_4 \\
&-\tfrac{k_1}{2L'}x_7[-\tfrac{R}{L_q}x_4 - \tfrac{x_2}{L_q}\tfrac{\pi}{\tau_p}(L_dx_3)] + \tfrac{x_2}{L_q}\tfrac{\pi}{\tau_p}\Psi_{PM}
\end{aligned} \tag{7.278}$$

while about functions $g_{c1}(x)$, $g_{c2}(x)$, $g_{c3}(x)$, one has that

$$\begin{aligned}
g_{c1}(x) &= 0 \\
g_{c2}(x) &= 0 \\
g_{c3}(x) &= \tfrac{k_1}{2L'}x_7\tfrac{1}{L_q}
\end{aligned} \tag{7.279}$$

In aggregate the input–output linearized dynamics of the wave energy conversion system which comprises a PMLSG and a VSC is given by:

$$\begin{pmatrix} \ddot{y}_1 \\ y_2^{(3)} \\ \ddot{y}_3 \end{pmatrix} = \begin{pmatrix} f_a(x) \\ f_b(x) \\ f_c(x) \end{pmatrix} + \begin{pmatrix} g_{a1}(x) & g_{a2}(x) & g_{a3}(x) \\ g_{b1}(x) & g_{b2}(x) & g_{b3}(x) \\ g_{c1}(x) & g_{c2}(x) & g_{c3}(x) \end{pmatrix} \begin{pmatrix} u_1 \\ u_2 \\ u_3 \end{pmatrix} \tag{7.280}$$

The previous state-space model in matrix form is also outlined in the following equation:

$$\dot{\tilde{Y}} = \tilde{F} + \tilde{G}\tilde{u} \tag{7.281}$$

Moreover, by defining the virtual control inputs

$$\begin{aligned} v_1 &= g_{a1}(x)u_1 + g_{a2}(x)u_2 + g_{a3}(x)u_3 \\ v_2 &= g_{b1}(x)u_1 + g_{b2}(x)u_2 + g_{b3}(x)u_3 \\ v_3 &= g_{c1}(x)u_1 + g_{c2}(x)u_2 + g_{c3}(x)u_3 \end{aligned} \tag{7.282}$$

the input–output linearized model of the wave energy conversion system is given by the following equations:

$$\ddot{y}_1 = v_1 \quad y_2^{(3)} = v_2 \quad \ddot{y}_3 = v_3 \tag{7.283}$$

By defining the state variables $z_1 = y_1$, $z_2 = \dot{y}_1$, $z_3 = y_2$, $z_4 = \dot{y}_2$, $z_5 = \ddot{y}_2$, $z_6 = y_3$, $z_7 = \dot{y}_3$, one arrives also at a canonical Brunovsky form of the wave energy conversion system which comprises the following equations: $\dot{z}_1 = z_2$, $\dot{z}_2 = v_1$, $\dot{z}_3 = z_4$, $\dot{z}_4 = z_5$, $\dot{z}_5 = v_2$, $\dot{z}_6 = z_7$, and $\dot{z}_7 = v_3$. In matrix form, the canonical Brunovsky state-space description of the system becomes

$$\begin{pmatrix} \dot{z}_1 \\ \dot{z}_2 \\ \dot{z}_3 \\ \dot{z}_4 \\ \dot{z}_5 \\ \dot{z}_6 \\ \dot{z}_7 \end{pmatrix} = \begin{pmatrix} 0&1&0&0&0&0&0 \\ 0&0&0&0&0&0&0 \\ 0&0&0&1&0&0&0 \\ 0&0&0&0&1&0&0 \\ 0&0&0&0&0&0&0 \\ 0&1&0&0&0&0&1 \\ 0&0&0&0&0&0&0 \end{pmatrix} \begin{pmatrix} z_1 \\ z_2 \\ z_3 \\ z_4 \\ z_5 \\ z_6 \\ z_7 \end{pmatrix} + \begin{pmatrix} 0&0&0 \\ 1&0&0 \\ 0&0&0 \\ 0&0&0 \\ 0&1&0 \\ 0&0&0 \\ 0&0&1 \end{pmatrix} \begin{pmatrix} u_1 \\ u_2 \\ u_3 \end{pmatrix} \tag{7.284}$$

with measurement equation

$$\begin{pmatrix} z_1^m \\ z_2^m \\ z_3^m \end{pmatrix} = \begin{pmatrix} 1&0&0&0&0&0&0 \\ 0&0&1&0&0&0&0 \\ 0&0&0&0&0&1&0 \end{pmatrix} \begin{pmatrix} z_1 \\ z_2 \\ z_3 \\ z_4 \\ z_5 \\ z_6 \\ z_7 \end{pmatrix} \tag{7.285}$$

The stabilizing feedback control for the wave energy conversion system is

$$\begin{aligned} v_1 &= \ddot{z}_{1,d} - K_{1,1}(\dot{z}_1 - \dot{z}_{1,d}) - K_{2,1}(z_1 - z_{1,d}) \\ v_2 &= z_{3,d}^{(3)} - K_{1,2}(\ddot{z}_3 - \ddot{z}_{3,d}) - K_{2,2}(\dot{z}_3 - \dot{z}_{3,d}) - K_{3,2}(z_3 - z_{3,d}) \\ v_3 &= \ddot{z}_{6,d} - K_{1,3}(\dot{z}_6 - \dot{z}_{6,d}) - K_{2,3}(z_6 - z_{6,d}) \end{aligned} \tag{7.286}$$

By substituting the feedback control law of Eq. (7.286) into the state-space model of Eq. (7.284), one obtains the following closed-loop system dynamics

$$\begin{aligned}(\ddot{z}_1-\ddot{z}_{1,d})+K_{1,1}(\dot{z}_1-\dot{z}_{1,d})+K_{2,1}(z_1-z_{1,d})&=0\\ (z_3^{(3)}-z_{3,d}^{(3)})+K_{1,2}(\ddot{z}_3-\ddot{z}_{3,d})+K_{2,2}(\dot{z}_3-\dot{z}_{3,d})+K_{3,2}(z_3-z_{3,d})&=0\\ (\ddot{z}_6-\ddot{z}_{6,d})+K_{1,3}(\dot{z}_6-\dot{z}_{6,d})+K_{2,3}(z_6-z_{6,d})&=0\end{aligned} \tag{7.287}$$

Next, by defining the tracking error variables $e_1=z_1-z_{1,d}$, $e_3=z_3-z_{3,d}$, $e_6=z_6-z_{6,d}$, one obtains that the tracking error dynamics of the closed-loop system is given by

$$\begin{aligned}\ddot{e}_1+K_{1,1}\dot{e}_1+K_{2,1}e_1&=0\\ e_3^{(3)}+K_{1,2}\ddot{e}_3+K_{2,2}\dot{e}_3+K_{3,2}e_3&=0\\ \ddot{e}_6+K_{1,3}\dot{e}_6+K_{2,3}e_6&=0\end{aligned} \tag{7.288}$$

Besides, by selecting the feedback gains $(K_{1,1},K_{2,1})$, $(K_{1,2},K_{2,2},K_{3,2})$ and $(K_{1,3},K_{2,3})$ so as the characteristic polynomials which are associated with the differential equations of Eq. (7.288) to be Hurwitz stable, one has that

$$\begin{aligned}&lim_{t\to\infty}e_1=0\Rightarrow lim_{t\to\infty}z_1=z_{1,d}\Rightarrow lim_{t\to\infty}y_1=y_{1,d}\\ &lim_{t\to\infty}e_3=0\Rightarrow lim_{t\to\infty}z_3=z_{3,d}\Rightarrow lim_{t\to\infty}y_2=y_{2,d}\\ &lim_{t\to\infty}e_6=0\Rightarrow lim_{t\to\infty}z_6=z_{6,d}\Rightarrow lim_{t\to\infty}y_3=y_{3,d}\end{aligned} \tag{7.289}$$

Through the computation the control inputs vector $\tilde{v}=[v_1,v_2,v_3]^T$, one can also find the real control inputs vector $\tilde{u}$ of Eq. (7.281) which should be applied to the initial nonlinear state-space model of the wave energy conversion system. This is given by the inverse transformation

$$\tilde{u}=\tilde{G}^{-1}(\tilde{v}-\tilde{F}) \tag{7.290}$$

By proving that the flat outputs of the system converge to the associated setpoints it is also ensured that all state variables $x_i(t)$ converge to their reference values. However, as shown above, control of the PMLSG and VSC-based wave energy conversion system through input–output linearization and eigenvalues assignment ends to be a complicated procedure which requires complex forward and inverse transformations. To overcome this complexity a nonlinear optimal control approach is applied next.

7.3.4 Nonlinear optimal control for the wave energy conversion system

7.3.4.1 Approximate linearization of the dynamics of the wave energy conversion system

The dynamic model of the wave energy conversion system undergoes approximate linearization around the temporary operating point (x^*,u^*) which is updated at each sampling instant and is defined by the present value of the system's state vector x^* and by the last sampled value of the control inputs vector u^*. The linearization process is based on first-order Taylor series expansion and on the computation of the associated Jacobian matrices. The modeling error which is due to the truncation of higher-order terms from the Taylor series expansion is considered to be a perturbation which is asymptotically compensated by the robustness of the control algorithm [715,725].

Through the proposed linearization process, the wave energy conversion system, being initially in the nonlinear state-space form $\dot{x}=f(x)+g(x)u$, is turned into the equivalent linear state-space description

$$\dot{x}=Ax+Bu+\tilde{d} \tag{7.291}$$

where A, B are the Jacobian matrices of the system, and $\tilde{d}$ is the disturbances vector which may comprise (i) the modeling error due to the truncation of higher-order terms in the Taylor series expansion, (ii) exogenous perturbations, and (iii) sensor measurement noise of any distribution. The Jacobian matrices of the system are given by

$$A=\nabla_x[f(x)+g(x)u]\,|_{(x^*,u^*)}\Rightarrow A=\nabla_x f(x)\,|_{(x^*,u^*)} \tag{7.292}$$

$$B = \nabla_u[f(x) + g(x)u] \mid_{(x^*,u^*)} \Rightarrow B = g(x) \mid_{(x^*,u^*)} \tag{7.293}$$

The linearization approach which has been followed for implementing the nonlinear optimal control scheme results in a quite accurate model of the system's dynamics. Consider for instance the following affine-in-the-input state-space model

$$\begin{gathered}\dot{x} = f(x) + g(x)u \Rightarrow \\ \dot{x} = [f(x^*) + \nabla_x f(x) \mid_{x^*} (x - x^*)] + [g(x^*) + \nabla_x g(x) \mid_{x^*} (x - x^*)]u^* + g(x^*)u^* + g(x^*)(u - u^*) + \tilde{d}_1 \Rightarrow \\ \dot{x} = [\nabla_x f(x) \mid_{x^*} + \nabla_x g(x) \mid_{x^*} u^*]x + g(x^*)u - [\nabla_x f(x) \mid_{x^*} + \nabla_x g(x) \mid_{x^*} u^*]x^* + f(x^*) + g(x^*)u^* + \tilde{d}_1\end{gathered} \tag{7.294}$$

where $\tilde{d}_1$ is the modeling error due to truncation of higher order terms in the Taylor series expansion of $f(x)$ and $g(x)$. Next, by defining $A = [\nabla_x f(x) \mid_{x^*} + \nabla_x g(x) \mid_{x^*} u^*]$, $B = g(x^*)$, one obtains

$$\dot{x} = Ax + Bu - Ax^* + f(x^*) + g(x^*)u^* + \tilde{d}_1 \tag{7.295}$$

Moreover, by denoting $\tilde{d} = -Ax^* + f(x^*) + g(x^*)u^* + \tilde{d}_1$ about the cumulative modeling error term in the Taylor series expansion procedure, one has

$$\dot{x} = Ax + Bu + \tilde{d} \tag{7.296}$$

which is the approximately linearized model of the dynamics of the system of Eq. (7.291). The term $f(x^*) + g(x^*)u^*$ is the derivative of the state vector at (x^*, u^*) which is almost annihilated by $-Ax^*$.

Next, the computation of the Jacobian matrix $\nabla_x f(x) \mid_{(x^*,u^*)}$ proceeds as follows:

First row of the Jacobian matrix $\nabla_x f(x) \mid_{(x^*,u^*)}$: $\frac{\partial f_1}{\partial x_1} = 0$, $\frac{\partial f_1}{\partial x_2} = 1$, $\frac{\partial f_1}{\partial x_3} = 0$, $\frac{\partial f_1}{\partial x_4} = 0$, $\frac{\partial f_1}{\partial x_5} = 0$, $\frac{\partial f_1}{\partial x_6} = 0$, $\frac{\partial f_1}{\partial x_7} = 0$.

Second row of the Jacobian matrix $\nabla_x f(x) \mid_{(x^*,u^*)}$: $\frac{\partial f_2}{\partial x_1} = -\frac{k_3}{M}$, $\frac{\partial f_2}{\partial x_2} = -\frac{b}{M}$, $\frac{\partial f_2}{\partial x_3} = \frac{3}{2M}\frac{\pi}{\tau_p}x_4(L_d - L_q)$, $\frac{\partial f_2}{\partial x_4} = \frac{3}{2M}\frac{\pi}{\tau_p}[x_3(L_d - L_q) - \Psi_{PM}]$, $\frac{\partial f_2}{\partial x_5} = 0$, $\frac{\partial f_2}{\partial x_6} = 0$, $\frac{\partial f_2}{\partial x_7} = 0$.

Third row of the Jacobian matrix $\nabla_x f(x) \mid_{(x^*,u^*)}$: $\frac{\partial f_3}{\partial x_1} = 0$, $\frac{\partial f_3}{\partial x_2} = -\frac{1}{L_d}\frac{\pi}{\tau_p}L_q$, $\frac{\partial f_3}{\partial x_3} = -\frac{R}{L_d}$, $\frac{\partial f_3}{\partial x_4} = \frac{x_2}{L_d}\frac{\pi}{\tau_p}L_q$, $\frac{\partial f_3}{\partial x_5} = 0$, $\frac{\partial f_3}{\partial x_6} = 0$, $\frac{\partial f_3}{\partial x_7} = 0$.

Fourth row of the Jacobian matrix $\nabla_x f(x) \mid_{(x^*,u^*)}$: $\frac{\partial f_4}{\partial x_1} = 0$, $\frac{\partial f_4}{\partial x_2} = -\frac{1}{L_q}\frac{\pi}{\tau_p}(L_d x_3 - \Psi_{PM})$, $\frac{\partial f_4}{\partial x_3} = -\frac{x_2}{L_q}\frac{\pi}{\tau_p}L_d$, $\frac{\partial f_4}{\partial x_4} = -\frac{R}{L_q}$, $\frac{\partial f_4}{\partial x_5} = 0$, $\frac{\partial f_4}{\partial x_6} = 0$, $\frac{\partial f_4}{\partial x_7} = 0$.

Fifth row of the Jacobian matrix $\nabla_x f(x) \mid_{(x^*,u^*)}$: $\frac{\partial f_5}{\partial x_1} = 0$, $\frac{\partial f_5}{\partial x_2} = 0$, $\frac{\partial f_5}{\partial x_3} = -\frac{k_1 x_7}{2L'}$, $\frac{\partial f_5}{\partial x_4} = 0$, $\frac{\partial f_5}{\partial x_5} = -\frac{R'}{L'}$, $\frac{\partial f_5}{\partial x_6} = \omega_{dq}$, $\frac{\partial f_5}{\partial x_7} = -\frac{k_1 x_3}{2L'}$.

Sixth row of the Jacobian matrix $\nabla_x f(x) \mid_{(x^*,u^*)}$: $\frac{\partial f_6}{\partial x_1} = 0$, $\frac{\partial f_6}{\partial x_2} = 0$, $\frac{\partial f_6}{\partial x_3} = 0$, $\frac{\partial f_6}{\partial x_4} = -\frac{k_2 x_7}{2L'}$, $\frac{\partial f_6}{\partial x_5} = -\omega_{dq}$, $\frac{\partial f_6}{\partial x_6} = -\frac{R'}{L'}$, $\frac{\partial f_6}{\partial x_7} = -\frac{k_2 x_4}{2L'}$.

Seventh row of the Jacobian matrix $\nabla_x f(x) \mid_{(x^*,u^*)}$: $\frac{\partial f_7}{\partial x_1} = 0$, $\frac{\partial f_7}{\partial x_2} = 0$, $\frac{\partial f_7}{\partial x_3} = \frac{3}{4C_{dc}}k_1 x_5$, $\frac{\partial f_7}{\partial x_4} = \frac{3}{4C_{dc}}k_2 x_6$, $\frac{\partial f_7}{\partial x_5} = \frac{3}{4C_{dc}}k_1 x_3$, $\frac{\partial f_7}{\partial x_6} = \frac{3}{4C_{dc}}k_2 x_4$, $\frac{\partial f_7}{\partial x_7} = -\frac{1}{R_c C_{dc}}$.

7.3.4.2 Equivalent linearized dynamics of the wave energy conversion system

After linearization around its current operating point, the dynamic model for the PMLSG-VSC-based wave energy conversion system is written as

$$\dot{x} = Ax + Bu + d_1 \tag{7.297}$$

Parameter d_1 stands for the linearization error in the PMLSG-VSC-based wave energy conversion system that was given previously in Eq. (7.291). The reference setpoints for the state vector of the aforementioned dynamic model are denoted by $\mathbf{x_d} = [x_1^d, \cdots, x_7^d]$. Tracking of this setpoint vector is achieved after applying the control input u^*. At every time instant, the control input u^* is assumed to differ from the control input u appearing in Eq. (7.297) by an amount equal to Δu, that is $u^* = u + \Delta u$

$$\dot{x}_d = Ax_d + Bu^* + d_2 \tag{7.298}$$

The dynamics of the controlled system described in Eq. (7.297) can be also written as

$$\dot{x} = Ax + Bu + Bu^* - Bu^* + d_1 \tag{7.299}$$

and by denoting $d_3 = -Bu^* + d_1$ as an aggregate disturbance term, one obtains

$$\dot{x} = Ax + Bu + Bu^* + d_3 \tag{7.300}$$

By subtracting Eq. (7.298) from Eq. (7.300), one has

$$\dot{x} - \dot{x}_d = A(x - x_d) + Bu + d_3 - d_2 \tag{7.301}$$

By denoting the tracking error as $e = x - x_d$ and the aggregate disturbance term as $L\tilde{d} = d_3 - d_2$, the tracking error dynamics becomes

$$\dot{e} = Ae + Bu + L\tilde{d} \tag{7.302}$$

where L is the disturbance inputs gain matrix. The above linearized form of the PMLSG-VSC-based wave energy conversion system can be efficiently controlled after applying an H-infinity feedback control scheme.

7.3.4.3 The nonlinear H-infinity control

The initial nonlinear model of the PMLSG-VSC-based wave energy conversion system is in the form

$$\dot{x} = f(x, u) \quad x \in R^n, \; u \in R^m \tag{7.303}$$

Linearization of the model of the PMLSG-VSC-based wave energy conversion system is performed at each iteration of the control algorithm around its present operating point $(x^*, u^*) = (x(t), u(t - T_s))$. The linearized equivalent of the system is described by

$$\dot{x} = Ax + Bu + L\tilde{d} \quad x \in R^n, \; u \in R^m, \; \tilde{d} \in R^q \tag{7.304}$$

where matrices A and B are obtained from the computation of the previously defined Jacobians and vector $\tilde{d}$ denotes disturbance terms due to linearization errors, while L is a disturbance inputs gain matrix. The problem of disturbance rejection for the linearized model that is described by

$$\begin{aligned} \dot{x} &= Ax + Bu + L\tilde{d} \\ y &= Cx \end{aligned} \tag{7.305}$$

where $x \in R^n$, $u \in R^m$, $\tilde{d} \in R^q$, and $y \in R^p$ cannot be handled efficiently if the classical LQR control scheme is applied. This is because of the existence of the perturbation term $\tilde{d}$. The disturbance term $\tilde{d}$ apart from modeling (parametric) uncertainty and external perturbation terms can also represent noise terms of any distribution.

In the H_∞ control approach, a feedback control scheme is designed for setpoints tracking by the system's state vector and simultaneous disturbance rejection, considering that the disturbance affects the system in the worst possible manner. The disturbances' effects are incorporated in the following quadratic cost function:

$$J(t) = \frac{1}{2}\int_0^T [y^T(t)y(t) + ru^T(t)u(t) - \rho^2\tilde{d}^T(t)\tilde{d}(t)]dt, \quad r, \rho > 0 \tag{7.306}$$

The significance of the negative sign in the cost function's term that is associated with the perturbation variable $\tilde{d}(t)$ is that the disturbance tries to maximize the cost function $J(t)$ while the control signal $u(t)$ tries to minimize it. The physical meaning of the relation given above is that the control signal and the disturbances compete with each other within a min–max differential game. This problem of min–max optimization can be written as $min_u max_{\tilde{d}} J(u, \tilde{d})$.

The objective of the optimization procedure is to compute a control signal $u(t)$ which can compensate for the worst possible disturbance, that is externally imposed to the PMLSG-VSC-based wave energy conversion system. It can be seen that the solution to the min–max optimization problem is directly related to the value of parameter ρ. This means that there is an upper bound in the disturbances magnitude that can be annihilated by the control signal.

7.3.4.4 Computation of the feedback control gains

For the linearized system given by Eq. (7.305) the cost function of Eq. (7.306) is defined, where coefficient r determines the penalization of the control input and weight coefficient ρ determines the reward of the disturbances' effects. It is assumed that (i) The energy that is transferred from the disturbances signal $\tilde{d}(t)$ is bounded, that is $\int_0^\infty \tilde{d}^T(t)\tilde{d}(t)dt < \infty$, (ii) matrices $[A, B]$ and $[A, L]$ are stabilizable, and (iii) matrix $[A, C]$ is detectable. In the case of a tracking problem, the optimal feedback control law is given by

$$u(t) = -Ke(t) \tag{7.307}$$

with $e = x - x_d$ to be the tracking error, and $K = \frac{1}{r}B^T P$ where P is a positive definite symmetric matrix. As it will be proven in Section 7.3.5, matrix P is obtained from the solution of the Riccati equation

$$A^T P + PA + Q - P(\tfrac{2}{r}BB^T - \tfrac{1}{\rho^2}LL^T)P = 0 \tag{7.308}$$

where Q is a positive semi-definite symmetric matrix. The worst-case disturbance is given by

$$\tilde{d}(t) = \tfrac{1}{\rho^2}L^T Pe(t) \tag{7.309}$$

The solution of the H-infinity feedback control problem for the PMLSG-VSC-based wave energy conversion system and the computation of the worst-case disturbance that the related controller can sustain, comes from the superposition of Bellman's optimality principle when considering that this renewable energy unit is affected by two separate inputs (i) the control input u (ii) the cumulative disturbance input $\tilde{d}(t)$. Solving the optimal control problem for u, that is for the minimum variation (optimal) control input that achieves the elimination of the state vector's tracking error, gives $u = -\frac{1}{r}B^T Pe$. Equivalently, solving the optimal control problem for $\tilde{d}$, that is for the worst-case disturbance that the control loop can sustain gives $\tilde{d} = \frac{1}{\rho^2}L^T Pe$.

The diagram of the considered control loop for the PMLSG-VSC-based wave energy conversion system is depicted in Figure 7.71.

7.3.5 Lyapunov stability analysis

Through Lyapunov stability analysis it will be shown that the proposed nonlinear control scheme assures H_∞ tracking performance for the PMLSG-VSC-based wave energy conversion system and that in case of bounded disturbance terms asymptotic convergence to the reference setpoints is achieved. The tracking error dynamics for the PMLSG-VSC-based wave energy conversion system is written in the form [715,725]

$$\dot{e} = Ae + Bu + L\tilde{d} \tag{7.310}$$

where in the PMLSG-VSC-based wave energy conversion unit's case $L \in R^{7\times 7}$ is the disturbance inputs gain matrix. Variable $\tilde{d}$ denotes model uncertainties and external disturbances of the wave energy conversion system's model. The following Lyapunov equation is considered:

$$V = \tfrac{1}{2}e^T Pe \tag{7.311}$$

where $e = x - x_d$ is the tracking error. By differentiating with respect to time, one obtains

$$\begin{aligned} \dot{V} &= \tfrac{1}{2}\dot{e}^T Pe + \tfrac{1}{2}e^T P\dot{e} \Rightarrow \\ \dot{V} &= \tfrac{1}{2}[Ae + Bu + L\tilde{d}]^T Pe + \tfrac{1}{2}e^T P[Ae + Bu + L\tilde{d}] \Rightarrow \end{aligned} \tag{7.312}$$

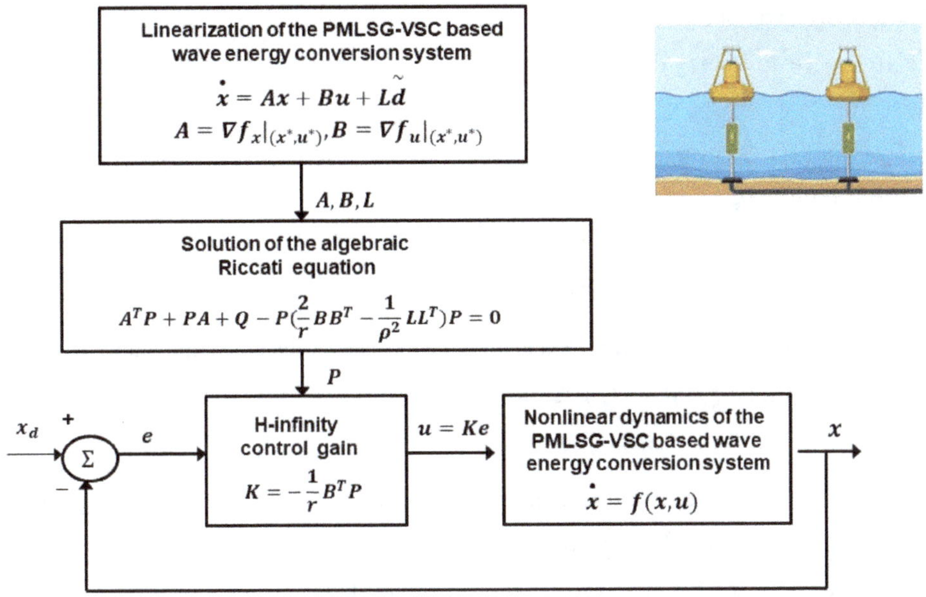

Figure 7.71 Diagram of the control scheme for the PMLSG-VSC-based wave energy conversion system

$$\begin{aligned}\dot{V} = &\tfrac{1}{2}[e^TA^T + u^TB^T + \tilde{d}^TL^T]Pe+ \\ &+\tfrac{1}{2}e^TP[Ae + Bu + L\tilde{d}]\Rightarrow\end{aligned} \tag{7.313}$$

$$\begin{aligned}\dot{V} = &\tfrac{1}{2}e^TA^TPe + \tfrac{1}{2}u^TB^TPe + \tfrac{1}{2}\tilde{d}^TL^TPe+ \\ &\tfrac{1}{2}e^TPAe + \tfrac{1}{2}e^TPBu + \tfrac{1}{2}e^TPL\tilde{d}\end{aligned} \tag{7.314}$$

The previous equation is rewritten as

$$\begin{aligned}\dot{V} = &\tfrac{1}{2}e^T(A^TP + PA)e + (\tfrac{1}{2}u^TB^TPe + \tfrac{1}{2}e^TPBu)+ \\ &+(\tfrac{1}{2}\tilde{d}^TL^TPe + \tfrac{1}{2}e^TPL\tilde{d})\end{aligned} \tag{7.315}$$

Assumption: For given positive definite matrix Q and coefficients r and ρ, there exists a positive definite matrix P, which is the solution of the following matrix equation:

$$A^TP + PA = -Q + P(\tfrac{2}{r}BB^T - \tfrac{1}{\rho^2}LL^T)P \tag{7.316}$$

Moreover, the following feedback control law is applied to the system:

$$u = -\tfrac{1}{r}B^TPe \tag{7.317}$$

By substituting Eq. (7.316) and Eq. (7.317) in $\dot{V}$, one obtains

$$\begin{aligned}\dot{V} = &\tfrac{1}{2}e^T[-Q + P(\tfrac{2}{r}BB^T - \tfrac{1}{\rho^2}LL^T)P]e+ \\ &+e^TPB(-\tfrac{1}{r}B^TPe) + e^TPL\tilde{d}\Rightarrow\end{aligned} \tag{7.318}$$

$$\begin{aligned}\dot{V} = &-\tfrac{1}{2}e^TQe + (\tfrac{1}{r}e^TPBB^TPe - \tfrac{1}{2\rho^2}e^TPLL^TPe) \\ &-\tfrac{1}{r}e^TPBB^TPe + e^TPL\tilde{d}\end{aligned} \tag{7.319}$$

which after intermediate operations gives

$$\dot{V} = -\frac{1}{2}e^TQe - \frac{1}{2\rho^2}e^TPLL^TPe + e^TPL\tilde{d} \tag{7.320}$$

or, equivalently

$$\begin{aligned}\dot{V} = -\frac{1}{2}e^TQe - \frac{1}{2\rho^2}e^TPLL^TPe+ \\ +\frac{1}{2}e^TPL\tilde{d} + \frac{1}{2}\tilde{d}^TL^TPe\end{aligned} \tag{7.321}$$

Lemma: The following inequality holds [715,721]

$$\frac{1}{2}e^TL\tilde{d} + \frac{1}{2}\tilde{d}L^TPe - \frac{1}{2\rho^2}e^TPLL^TPe \leq \frac{1}{2}\rho^2\tilde{d}^T\tilde{d} \tag{7.322}$$

Proof: The binomial $(\rho\alpha - \frac{1}{\rho}b)^2$ is considered. Expanding the left part of the above inequality, one gets

$$\begin{aligned}\rho^2a^2 + \frac{1}{\rho^2}b^2 - 2ab \geq 0 \Rightarrow \frac{1}{2}\rho^2a^2 + \frac{1}{2\rho^2}b^2 - ab \geq 0 \Rightarrow \\ ab - \frac{1}{2\rho^2}b^2 \leq \frac{1}{2}\rho^2a^2 \Rightarrow \frac{1}{2}ab + \frac{1}{2}ab - \frac{1}{2\rho^2}b^2 \leq \frac{1}{2}\rho^2a^2\end{aligned} \tag{7.323}$$

The following substitutions are carried out: $a = \tilde{d}$ and $b = e^TPL$, and the previous relation becomes

$$\frac{1}{2}\tilde{d}^TL^TPe + \frac{1}{2}e^TPL\tilde{d} - \frac{1}{2\rho^2}e^TPLL^TPe \leq \frac{1}{2}\rho^2\tilde{d}^T\tilde{d} \tag{7.324}$$

Eq. (7.324) is substituted in Eq. (7.321) and the inequality is enforced, thus giving [715,721]

$$\dot{V} \leq -\frac{1}{2}e^TQe + \frac{1}{2}\rho^2\tilde{d}^T\tilde{d} \tag{7.325}$$

Eq. (7.325) shows that the H_∞ tracking performance criterion is satisfied. The integration of $\dot{V}$ from 0 to T gives

$$\begin{aligned}\int_0^T \dot{V}(t)dt \leq -\frac{1}{2}\int_0^T ||e||_Q^2 dt + \frac{1}{2}\rho^2\int_0^T ||\tilde{d}||^2 dt \Rightarrow \\ 2V(T) + \int_0^T ||e||_Q^2 dt \leq 2V(0) + \rho^2\int_0^T ||\tilde{d}||^2 dt\end{aligned} \tag{7.326}$$

Moreover, if there exists a positive constant $M_d > 0$ such that

$$\int_0^\infty ||\tilde{d}||^2 dt \leq M_d \tag{7.327}$$

then, one gets

$$\int_0^\infty ||e||_Q^2 dt \leq 2V(0) + \rho^2 M_d \tag{7.328}$$

Thus, the integral $\int_0^\infty ||e||_Q^2 dt$ is bounded. Moreover, $V(T)$ is bounded and from the definition of the Lyapunov function V in Eq. (7.311), it becomes clear that $e(t)$ will be also bounded since $e(t) \in \Omega_e = \{e|e^TPe \leq 2V(0) + \rho^2M_d\}$. According to the above and with the use of Barbalat's Lemma, one obtains $lim_{t\to\infty}e(t) = 0$.

After following the stages of the stability proof, one arrives at Eq. (7.325) which shows that the H-infinity tracking performance criterion holds. By selecting the attenuation coefficient ρ to be sufficiently small and in particular to satisfy $\rho^2 < ||e||_Q^2/||\tilde{d}||^2$, one has that the first derivative of the Lyapunov function is upper bounded by 0. This condition holds at each sampling instance, and consequently, global stability for the control loop can be concluded.

7.3.6 Simulation tests

The global stability properties of the proposed nonlinear optimal control method for the PMLSG-VSC-based wave energy conversion system and the precise tracking of setpoints that this method achieves for the state variables of the wave energy conversion system are further confirmed through simulation experiments. To implement this section's nonlinear optimal control approach the algebraic Riccati equation of Eq. (7.316) has to be repetitively solved at each time step of the control algorithm. The sampling period has been chosen to be $T_s = 0.01$sec. The parameters of this Riccati equation are gains r, ρ, and matrix Q. The method is computationally efficient. For instance, at a PC with moderate processing power (multi-core i7 Intel processor at 2.8 GHz), one achieves a solution of the Riccati equation

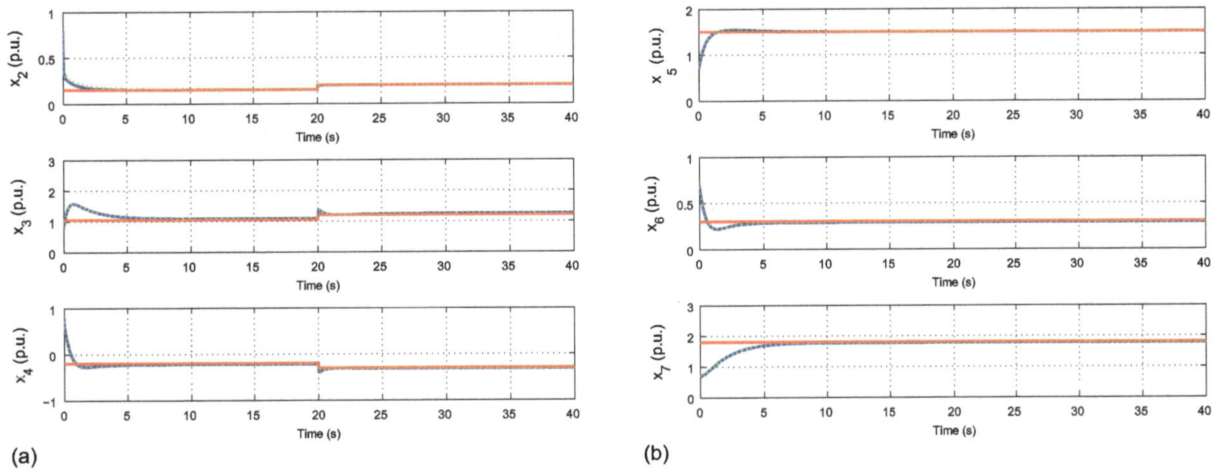

Figure 7.72 Tracking of setpoint 1 for the PMLSG-VSC-based wave energy conversion system under nonlinear optimal control: (a) convergence of state variables $x_2 = \dot{z}$, $x_3 = i_d$, $x_4 = i_d$ of the PMLSG to their reference setpoints (red line: setpoint, blue line: real value, green line: estimated value); (b) convergence of state variables $x_5 = i_d^{'}$, $x_6 = i_q^{'}$, $x_7 = V_{dc}$ of the VSC to their reference setpoints

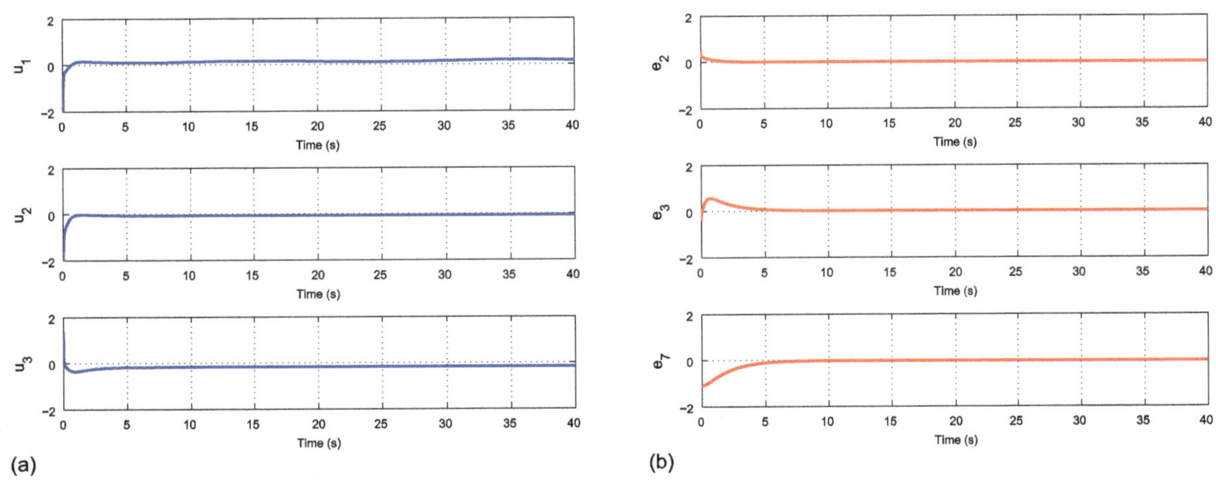

Figure 7.73 Tracking of setpoint 1 for the PMLSG-VSC-based wave energy conversion system under nonlinear optimal control: (a) control inputs u_1 to u_3 applied to the wave energy conversion unit; (b) tracking error e_1, e_2, e_3, for state variables $x_2 = \dot{z}$, $x_3 = i_d$, $x_7 = V_{dc}$ of the wave energy conversion unit

within a time interval which is much shorter than the sampling period. The obtained results are depicted in Figures 7.72–7.87. The state variables of the wave energy conversion unit are depicted in blue color, the associated setpoints are printed in red color while the estimates which are produced by the H-infinity Kalman filter are plotted in green color. It can be noticed that the proposed nonlinear optimal control scheme achieves fast and accurate tracking of setpoints by the state variables of the PMLSG-VSC-based wave energy conversion system under moderate variations of the control inputs. The parameters which define the transient performance of the nonlinear optimal control method are gains r, ρ,

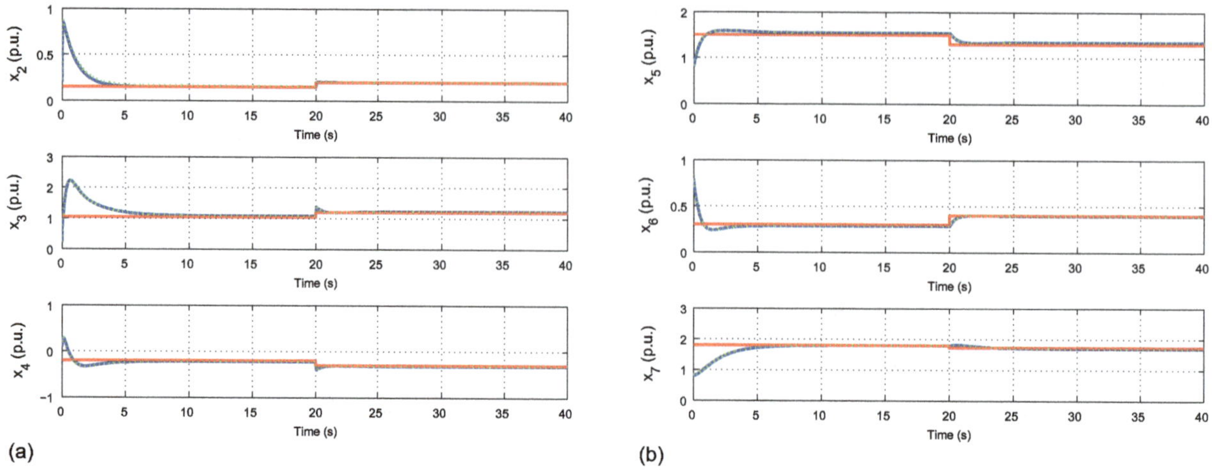

Figure 7.74 Tracking of setpoint 2 for the PMLSG-VSC-based wave energy conversion system under nonlinear optimal control: (a) convergence of state variables $x_2 = \dot{z}$, $x_3 = i_d$, $x_4 = i_d$ of the PMLSG to their reference setpoints (red line: setpoint, blue line: real value, green line: estimated value); (b) convergence of state variables $x_5 = i_d^{'}$, $x_6 = i_q^{'}$, $x_7 = V_{dc}$ of the VSC to their reference setpoints

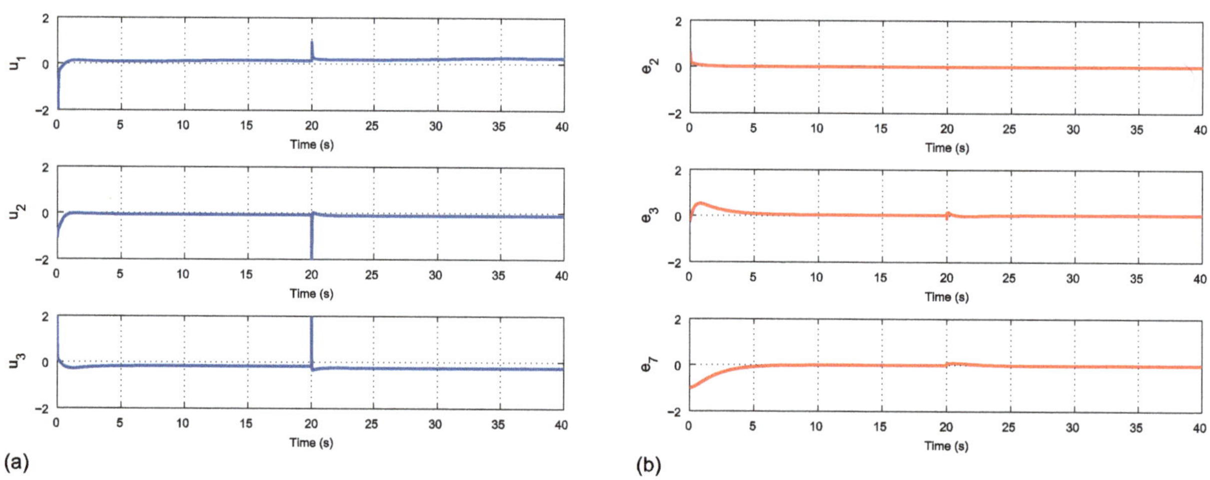

Figure 7.75 Tracking of setpoint 2 for the PMLSG-VSC-based wave energy conversion system under nonlinear optimal control: (a) control inputs u_1 to u_3 applied to the wave energy conversion unit; (b) tracking error e_1, e_2, e_3, for state variables $x_2 = \dot{z}$, $x_3 = i_d$, $x_7 = V_{dc}$ of the wave energy conversion unit

and Q. For relatively small values of r the steady-state tracking error of the state variables is eliminated. Additionally, for relatively large values of the elements of the diagonal matrix Q faster convergence of the state variables to their reference values is achieved. Moreover, the smallest value of the attenuation coefficient ρ for which one still obtains a valid solution of the method's Riccati equation is the one that provides the control loop with maximum robustness.

To elaborate on the tracking performance and on the robustness of the proposed nonlinear optimal control method for the PMLSG and VSC-based wave energy conversion unit the following tables are given: (i) Table 7.7 which provides

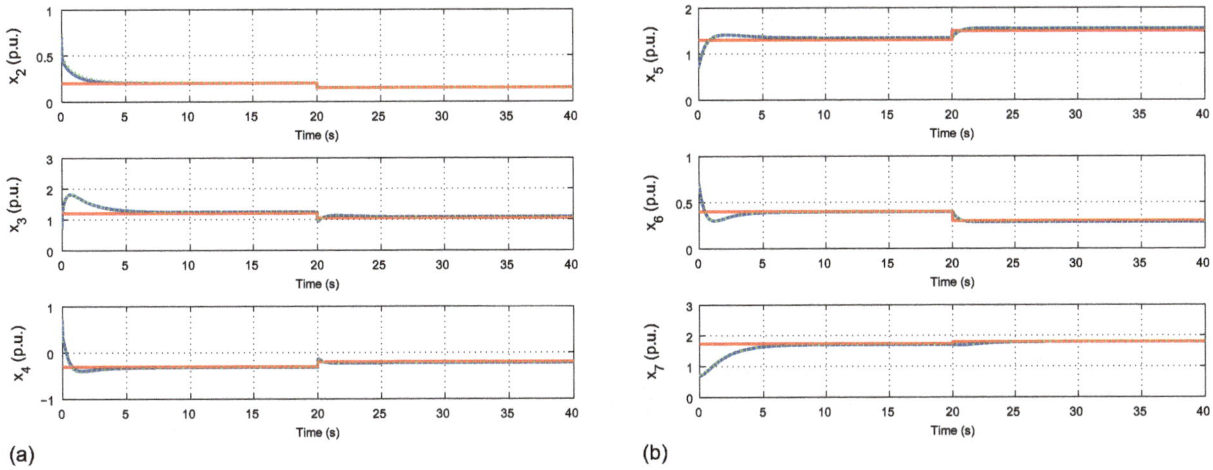

Figure 7.76 Tracking of setpoint 3 for the PMLSG-VSC-based wave energy conversion system under nonlinear optimal control: (a) convergence of state variables $x_2 = \dot{z}$, $x_3 = i_d$, $x_4 = i_d$ of the PMLSG to their reference setpoints (red line: setpoint, blue line: real value, green line: estimated value); (b) convergence of state variables $x_5 = i_d^{'}$, $x_6 = i_q^{'}$, $x_7 = V_{dc}$ of the VSC to their reference setpoints

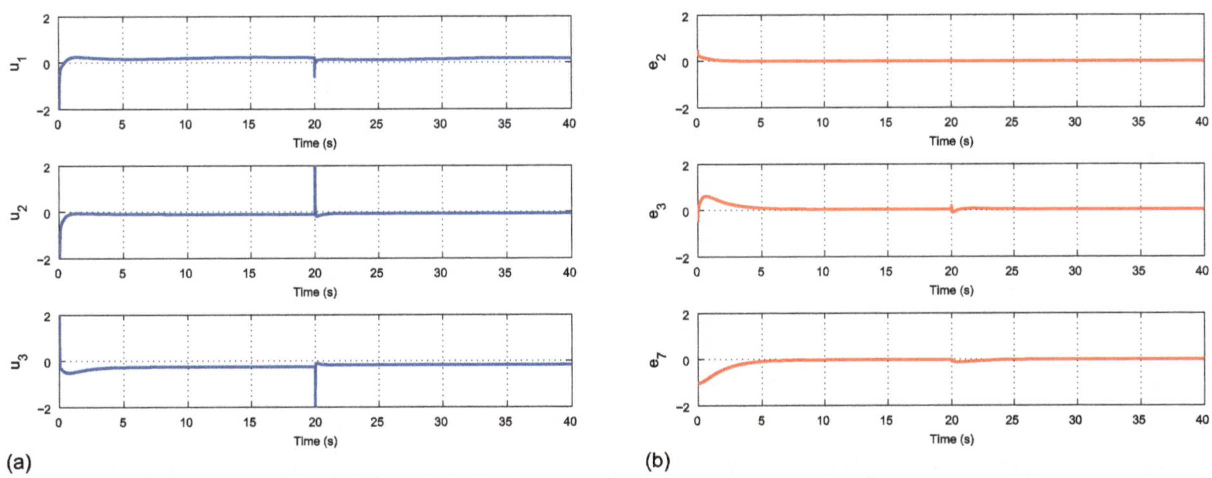

Figure 7.77 Tracking of setpoint 3 for the PMLSG-VSC-based wave energy conversion system under nonlinear optimal control: (a) control inputs u_1 to u_3 applied to the wave energy conversion unit; (b) tracking error e_1, e_2, e_3, for state variables $x_2 = \dot{z}$, $x_3 = i_d$, $x_7 = V_{dc}$ of the wave energy conversion unit

information about the accuracy of tracking of the reference setpoints by the state variables of the wave power conversion unit's state-space model, (ii) Table 7.8 which provides information about the robustness of the control method to parametric changes in the model of the electrically actuated wave power conversion unit's dynamics (change $\Delta a\%$ in the mass M of the moving part of the PMLSG), (iii) Table 7.9 which provides information about the precision in state variables' estimation that is achieved by the H-infinity Kalman filter, and (iv) Table 7.10 which provides the convergence times of the wave power conversion unit's state variables to the associated setpoints.

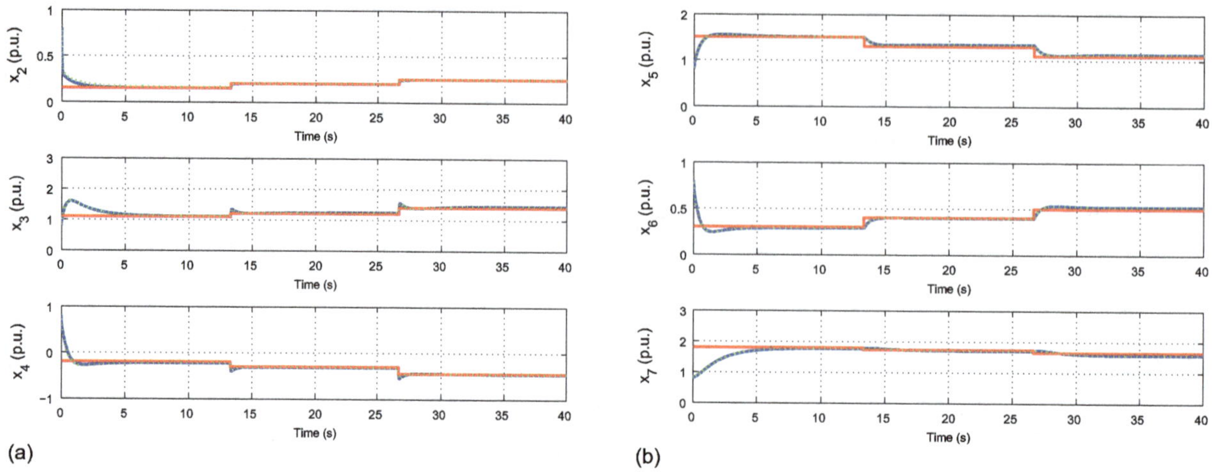

Figure 7.78 *Tracking of setpoint 4 for the PMLSG-VSC-based wave energy conversion system under nonlinear optimal control: (a) convergence of state variables $x_2 = \dot{z}$, $x_3 = i_d$, $x_4 = i_d$ of the PMLSG to their reference setpoints (red line: setpoint, blue line: real value, green line: estimated value); (b) convergence of state variables $x_5 = i_d^{'}$, $x_6 = i_q^{'}$, $x_7 = V_{dc}$ of the VSC to their reference setpoints*

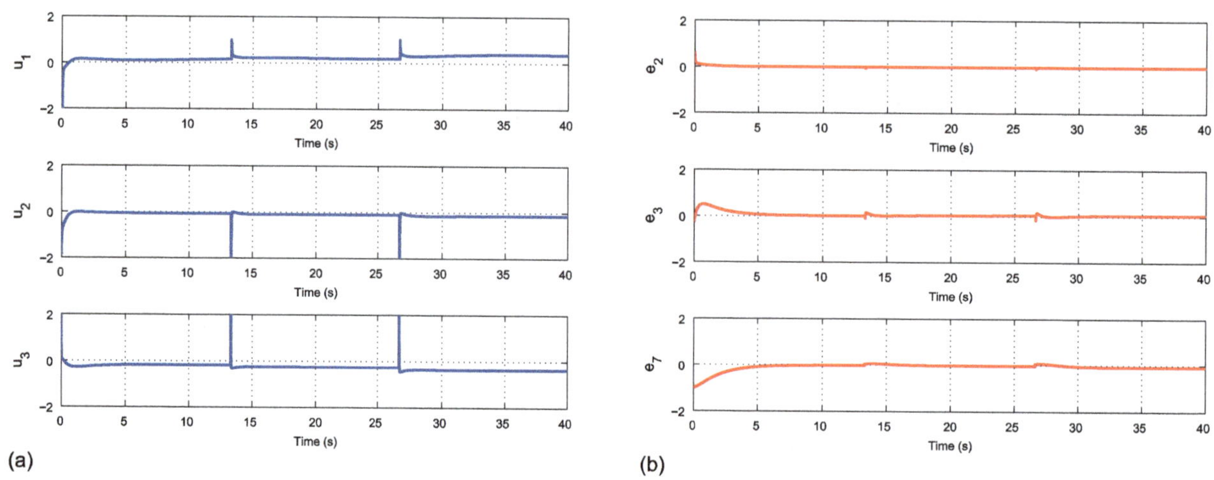

Figure 7.79 *Tracking of setpoint 4 for the PMLSG-VSC-based wave energy conversion system under nonlinear optimal control: (a) control inputs u_1 to u_3 applied to the wave energy conversion unit; (b) tracking error e_1, e_2, e_3, for state variables $x_2 = \dot{z}$, $x_3 = i_d$, $x_7 = V_{dc}$ of the wave energy conversion unit*

7.4 Nonlinear optimal and flatness-based control for the 5-DOF rotor and AMB system

7.4.1 Outline

Active magnetic bearings (AMBs) are widely used in high-speed rotating machines [839,269,841]. They are met in several industrial applications as for instance in motion transmission systems based on rotational or linear electric motors, in gas compressors, in wind turbines, and in flywheel-based energy storage systems [504,554,949]. Active magnetic bearings find also use in the actuation and orientation system of satellites and in biomedical systems such as blood pumps [869,109,195]. Besides, they are a key component in aircraft engines and in gas-turbine power generation units

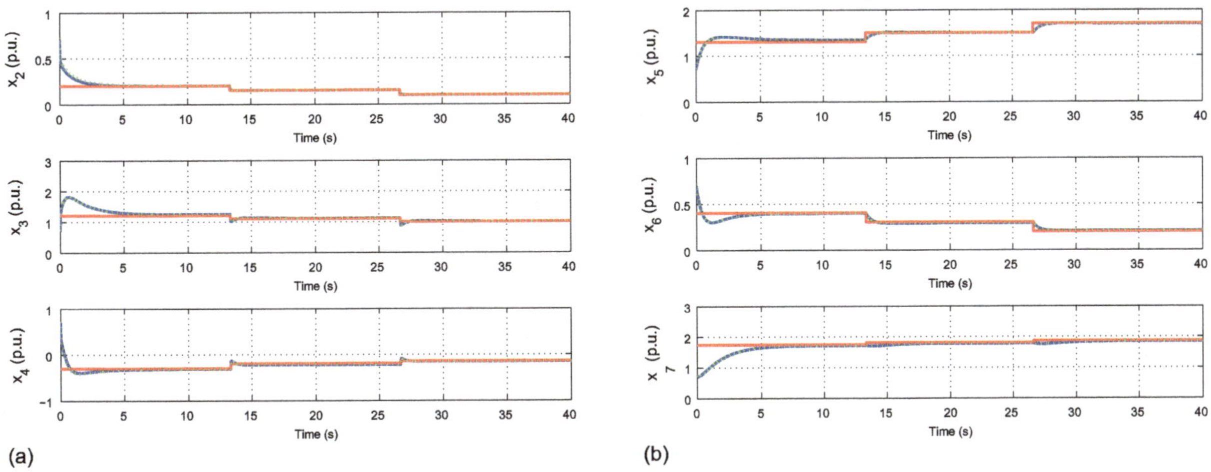

Figure 7.80 Tracking of setpoint 5 for the PMLSG-VSC-based wave energy conversion system under nonlinear optimal control: (a) convergence of state variables $x_2 = \dot{z}$, $x_3 = i_d$, $x_4 = i_d$ of the PMLSG to their reference setpoints (red line: setpoint, blue line: real value, green line: estimated value); (b) convergence of state variables $x_5 = i_d^{'}$, $x_6 = i_q^{'}$, $x_7 = V_{dc}$ of the VSC to their reference setpoints

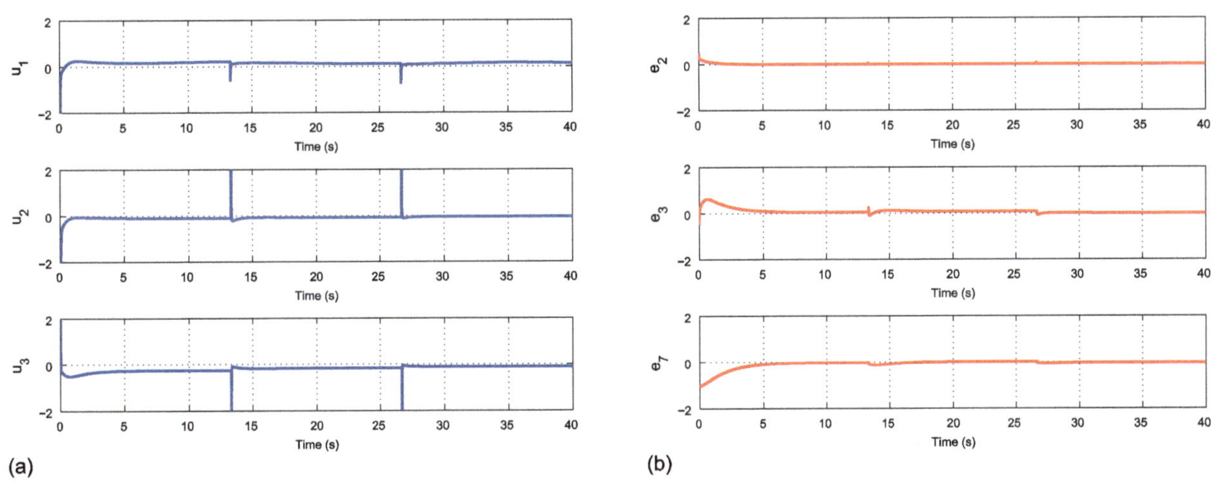

Figure 7.81 Tracking of setpoint 5 for the PMLSG-VSC-based wave energy conversion system under nonlinear optimal control: (a) control inputs u_1 to u_3 applied to the wave energy conversion unit; (b) tracking error e_1, e_2, e_3, for state variables $x_2 = \dot{z}$, $x_3 = i_d$, $x_7 = V_{dc}$ of the wave energy conversion unit

[204,1000]. AMBs' industrial use includes also the weapon, metal industry and transportation industries [290,1082]. Magnetic bearing-supported rotating machinery, whether based on a vertical or a horizontal motor, needs several subsystems to adjust the radial and axial levitations [274,136,1013]. Typically, two radial and one axial AMB are used for a 5-DOF levitation [622,511,766]. By achieving contactless operation of the rotor and its spindle AMBs eliminate mechanical friction and enable high rotational speeds. Compared to mechanical bearings AMBs exhibit the following benefits: (i) they work efficiently under higher rotational speed, (ii) they have reduced friction losses, (iii) they do not need lubrication, (iv) they consume less energy, and (v) they are more resilient to failures and have extended useful life [22,501]. AMBs are prone to instability because of the relation of their electromagnetic force and the length of the air

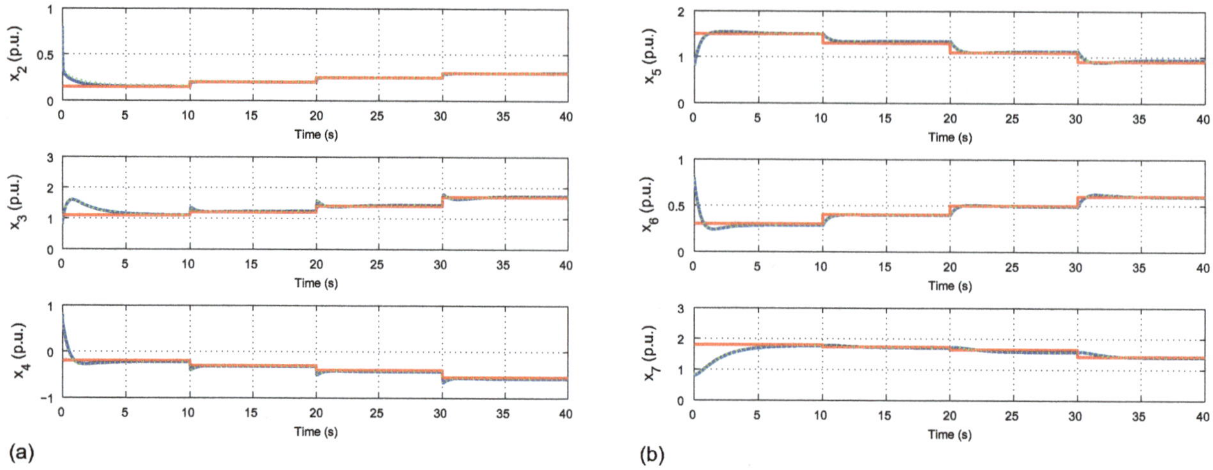

Figure 7.82 Tracking of setpoint 6 for the PMLSG-VSC-based wave energy conversion system under nonlinear optimal control: (a) convergence of state variables $x_2 = \dot{z}$, $x_3 = i_d$, $x_4 = i_d$ of the PMLSG to their reference setpoints (red line: setpoint, blue line: real value, green line: estimated value); (b) convergence of state variables $x_5 = i_d^{'}$, $x_6 = i_q^{'}$, $x_7 = V_{dc}$ of the VSC to their reference setpoints

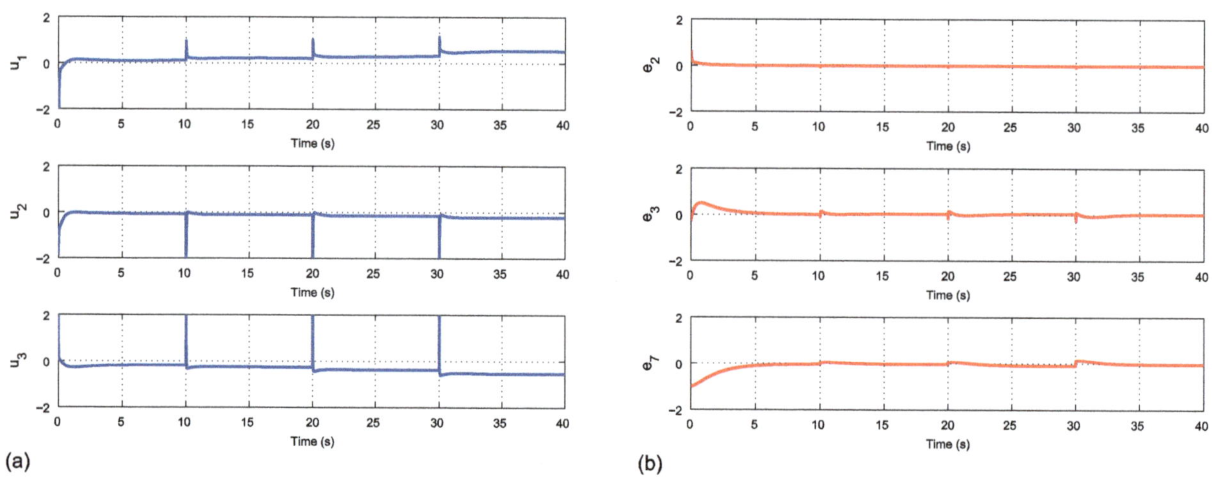

Figure 7.83 Tracking of setpoint 6 for the PMLSG-VSC-based wave energy conversion system under nonlinear optimal control: (a) control inputs u_1 to u_3 applied to the wave energy conversion unit; (b) tracking error e_1, e_2, e_3, for state variables $x_2 = \dot{z}$, $x_3 = i_d$, $x_7 = V_{dc}$ of the wave energy conversion unit

gap. Through the control system of AMBs the appearance of mechanical vibrations can be eliminated [529,151,1082]. The control problem of AMBs exhibits difficulties due to nonlinearities and a multi-variable structure of the associated state-space model [538,258,1007]. So far, there have been attempts to solve the control and stabilization problem for AMBs using model-predictive control, sliding-mode control, and backstepping control [490,150,247].

In this section, two new control methods are applied to the dynamic model of the 5-DOF rotor and active magnetic bearing system. These are (i) nonlinear optimal control implemented through the sequential linearization of the state-space model of the system and (ii) flatness-based control implemented in successive loops [711,721,732]. Both control methods are applied directly on the initial nonlinear state-space model of the 5-DOF rotor and AMB system

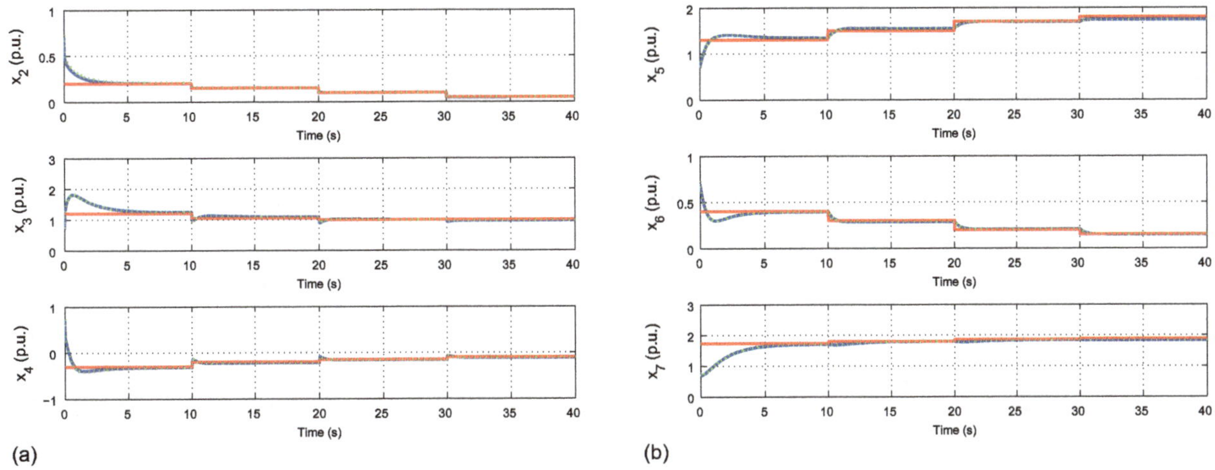

Figure 7.84 Tracking of setpoint 7 for the PMLSG-VSC-based wave energy conversion system under nonlinear optimal control: (a) convergence of state variables $x_2 = \dot{z}$, $x_3 = i_d$, $x_4 = i_d$ of the PMLSG to their reference setpoints (red line: setpoint, blue line: real value, green line: estimated value); (b) convergence of state variables $x_5 = i_d^{'}$, $x_6 = i_q^{'}$, $x_7 = V_{dc}$ of the VSC to their reference setpoints

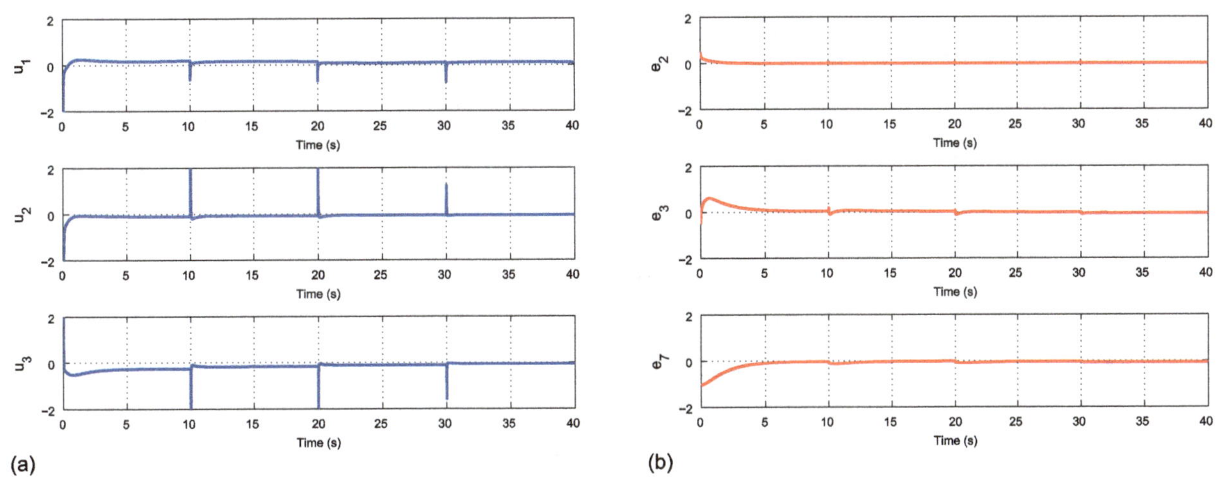

Figure 7.85 Tracking of setpoint 7 for the PMLSG-VSC-based wave energy conversion system under nonlinear optimal control: (a) control inputs u_1 to u_3 applied to the wave energy conversion unit; (b) tracking error e_1, e_2, e_3, for state variables $x_2 = \dot{z}$, $x_3 = i_d$, $x_7 = V_{dc}$ of the wave energy conversion unit

without the need for changes of state variables or complicated state-space model transformations. In the first approach, that is nonlinear optimal control with sequential linearization, the dynamic model of the 5-DOF rotor and AMB system undergoes approximate linearization through Taylor series expansion and the computation of the associated Jacobian matrices [705,60,706]. The linearization process takes place at each sampling instance around a temporary operating point which is defined by the present value of the system's state vector and by the last sampled value of the control inputs vector. The modeling error which is due to truncation of higher-order terms in the Taylor series expansion is considered to be a perturbation which is asymptotically compensated by the robustness of the control algorithm. For the approximately linearized model of the system, an H-infinity feedback controller is defined [711,721,732]. This

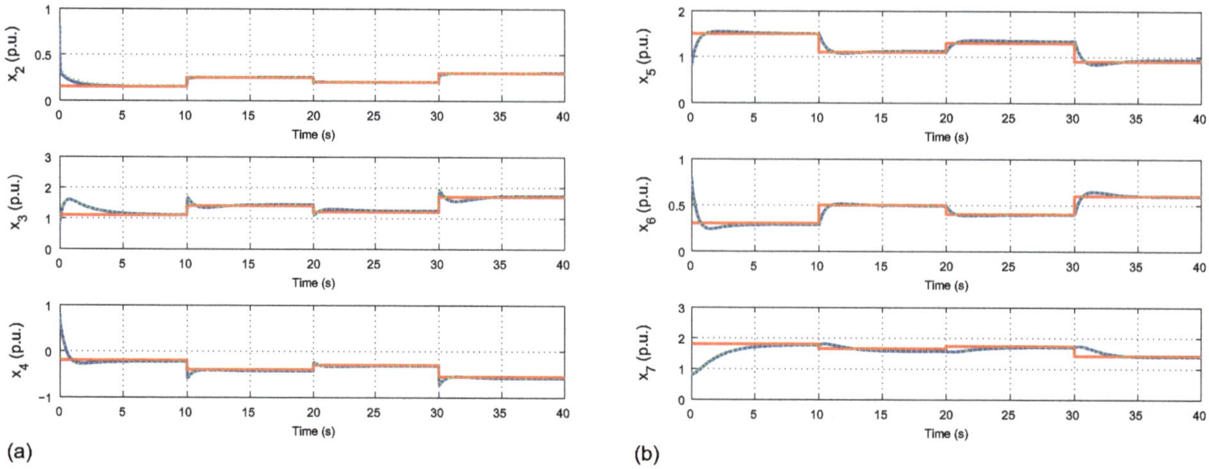

Figure 7.86 Tracking of setpoint 8 for the PMLSG-VSC-based wave energy conversion system under nonlinear optimal control: (a) convergence of state variables $x_2 = \dot{z}$, $x_3 = i_d$, $x_4 = i_d$ of the PMLSG to their reference setpoints (red line: setpoint, blue line: real value, green line: estimated value); (b) convergence of state variables $x_5 = i_d^{'}$, $x_6 = i_q^{'}$, $x_7 = V_{dc}$ of the VSC to their reference setpoints

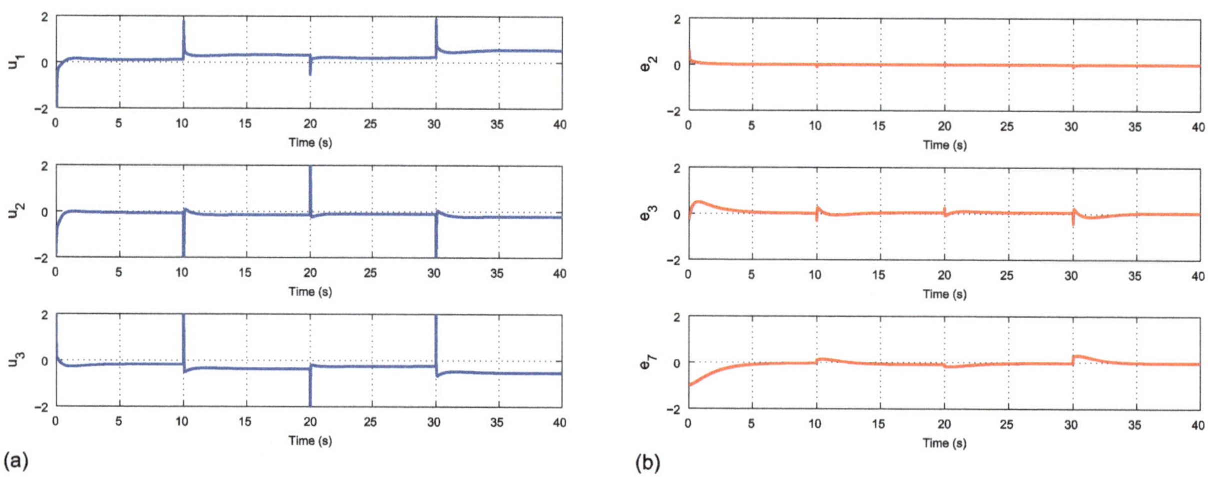

Figure 7.87 Tracking of setpoint 8 for the PMLSG-VSC-based wave energy conversion system under nonlinear optimal control: (a) control inputs u_1 to u_3 applied to the wave energy conversion unit; (b) tracking error e_1, e_2, e_3, for state variables $x_2 = \dot{z}$, $x_3 = i_d$, $x_7 = V_{dc}$ of the wave energy conversion unit

controller provides a solution to the associated optimal control problem under model uncertainties and external perturbations. Actually, it represents a min–max differential game taking place between the controller and the disturbance or model uncertainty terms. To compute the feedback gains of the H-infinity controller an algebraic Riccati equation has to be solved repetitively at each time step of the control method. The global stability properties of the control scheme are proven through Lyapunov analysis. First, it is demonstrated that the H-infinity tracking performance criterion holds while under moderate conditions global asymptotic stability is also proven.

The second control method of this section makes use of differential flatness theory. Differentially flat systems are currently a main direction in nonlinear control systems analysis and synthesis [755,491,704,252,898]. A system

Table 7.7 Tracking RMSE for the WEC system in the disturbance-free case

	$RMSE_{x_1}$	$RMSE_{x_2}$	$RMSE_{x_3}$	$RMSE_{x_4}$	$RMSE_{x_5}$	$RMSE_{x_6}$	$RMSE_{x_7}$
$test_1$	0.0017	0.0001	0.0018	0.0027	0.0007	0.0013	0.0039
$test_2$	0.0020	0.0002	0.0053	0.0024	0.0061	0.0014	0.0031
$test_3$	0.0018	0.0002	0.0057	0.0024	0.0059	0.0013	0.0040
$test_4$	0.0024	0.0002	0.0051	0.0023	0.0052	0.0023	0.0055
$test_5$	0.0017	0.0002	0.0038	0.0022	0.0038	0.0015	0.0054
$test_6$	0.0029	0.0003	0.0045	0.0033	0.0052	0.0019	0.0063
$test_7$	0.0015	0.0003	0.0051	0.0019	0.0059	0.0016	0.0055
$test_8$	0.0029	0.0005	0.0068	0.0037	0.0071	0.0030	0.0107

Table 7.8 Tracking RMSE for the WEC system in the case of disturbances

$\Delta a\%$	$RMSE_{x_1}$	$RMSE_{x_2}$	$RMSE_{x_3}$	$RMSE_{x_4}$	$RMSE_{x_5}$	$RMSE_{x_6}$	$RMSE_{x_7}$
0%	0.0020	0.0002	0.0053	0.0024	0.0061	0.0014	0.0031
10%	0.0034	0.0003	0.0053	0.0023	0.0060	0.0014	0.0032
20%	0.0049	0.0004	0.0053	0.0023	0.0060	0.0014	0.0032
30%	0.0065	0.0006	0.0053	0.0023	0.0060	0.0015	0.0032
40%	0.0080	0.0008	0.0053	0.0023	0.0059	0.0015	0.0032
50%	0.0094	0.0010	0.0053	0.0023	0.0059	0.0015	0.0032
60%	0.0109	0.0012	0.0053	0.0023	0.0058	0.0015	0.0032

Table 7.9 RMSE $\times$ 10^{-3} for the estimation performed by the H-infinity KF

	$RMSE_{\hat{x}_1}$	$RMSE_{\hat{x}_2}$	$RMSE_{\hat{x}_3}$	$RMSE_{\hat{x}_4}$	$RMSE_{\hat{x}_5}$	$RMSE_{\hat{x}_6}$	$RMSE_{\hat{x}_7}$
$test_1$	0.0825	0.6456	0.0390	0.3652	0.0003	0.0025	0.0003
$test_2$	0.0830	0.6890	0.0417	0.3843	0.0004	0.0027	0.0003
$test_3$	0.0631	0.4922	0.0422	0.3114	0.0004	0.0021	0.0003
$test_4$	0.0841	0.6576	0.0427	0.3781	0.0004	0.0026	0.0003
$test_5$	0.0638	0.4980	0.0414	0.3114	0.0004	0.0021	0.0003
$test_6$	0.0824	0.6442	0.0452	0.3781	0.0004	0.0026	0.0003
$test_7$	0.0653	0.5112	0.0398	0.3112	0.0009	0.0021	0.0003
$test_8$	0.0798	0.6290	0.0486	0.3784	0.0004	0.0026	0.0003

is differentially flat if all its state variables and its control inputs are functions of the flat output [91,594,639,489]. The differential flatness property enables transformation of the system's dynamics in the linear canonical form [806,492,640,524,54]. The latter description is controllable and observable thus allowing for treating control and estimation problems [705,60,706,439]. Alternatively, a successive loops control scheme has been also developed for dynamical systems which exhibit the differential flatness property [712,718,720,714]. In the second approach, that is flatness-based control in successive loops, the dynamic model of the 5-DOF rotor and AMB system is separated into subsystems which are connected in a chained form. Actually, the state vector of the successive $(i+1)$-th subsystem becomes virtual control input for the preceding i-th subsystem and inversely [734,736,718]. The chained form constitutes a wider class of systems than the strict-feedback form. It is proven that each one of the subsystems in which the initial nonlinear state-space model is decomposed is differentially flat. This signifies that it can be written in the

Table 7.10 Convergence time (sec) for the WEC system's state variables

	$T_s\,x_1$	$T_s\,x_2$	$T_s\,x_3$	$T_s\,x_4$	$T_s\,x_5$	$T_s\,x_6$	$T_s\,x_7$
$test_1$	0.5	3.0	5.0	4.0	2.0	3.5	7.0
$test_2$	0.5	3.0	7.0	7.0	6.0	3.0	5.0
$test_3$	0.5	3.0	7.0	4.0	7.0	3.5	5.0
$test_4$	0.5	2.5	5.0	4.0	2.0	3.5	7.0
$test_5$	0.5	3.0	5.0	3.5	7.0	4.0	5.0
$test_6$	0.5	2.5	5.0	4.0	2.0	3.5	6.0
$test_7$	0.5	2.5	7.0	3.5	7.0	3.5	5.0
$test_8$	0.5	2.5	5.0	3.5	2.0	3.5	6.0

input–output linearized form and it can be controlled by inverting its dynamics, as it is commonly done for input–output linearized differentially flat systems. The real control input which is finally applied to the nonlinear state-space model is computed from the last N-th subsystem. It retrieves recursively, the virtual control inputs of the preceding $N-1$ subsystems, after tracing them at each sampling instance backward, that is from subsystem $N-1$ up to subsystem 1. The global stability properties of this multi-loop flatness-based control method are also proven through Lyapunov stability analysis.

The presented nonlinear optimal control method is advantageous because, unlike nonlinear model predictive control (NMPC), it is of ensured global stability and the convergence of its iterative search for the optimum does not depend on empirical parametrization of the controller and on ad hoc initialization. Besides, unlike optimal control based on linear parameter varying (LPV) concepts, in the new nonlinear optimal control method, there is no need to solve state-dependent Riccati equations. Furthermore, the new multi-loop flatness-based control method avoids changes of state variables and transformations of the state-space model of the controlled system into canonical forms. The control inputs it computes are applied directly on the initial nonlinear state-space model of the controlled system without inverse transformations and without the associated singularity issues.

7.4.2 Dynamic model of the 5-DOF rotor and active magnetic bearing system

The 5-DOF rotor and active magnetic bearing system comprises a rotor and front, rear, and axial active magnetic bearings (AMBs) which enable the rotor and its spindle remain suspended and reduce significantly the friction effects in this motion transmission system. The diagram of the 5-DOF rotor and active magnetic bearing system is given in Figure 7.88.

The state vector of the 5-DOF rotor and active magnetic bearing system is given by [150]

$$\begin{aligned} x &= [x_1, x_2, x_3, x_4, x_5, x_6, x_7, x_8, x_9, x_{10}]^T \text{ or} \\ x &= [x_a, x_b, y_a, y_b, z_g, \dot{x}_a, \dot{x}_b, \dot{y}_a, \dot{y}_b, \dot{z}_g]^T \end{aligned} \tag{7.329}$$

where x_a is the position of radial bearing AMB_a along the x axis, x_b is the position of radial bearing AMB_b along the x axis, y_a is the position of AMB_a along the y axis, y_b is the position of AMB_b along the y axis, and z_g is the position of the axial bearing along the z-axis. Besides, $\dot{x}_a$ is the speed of AMB_a along the x axis, $\dot{x}_b$ is the speed of AMB_b along the x axis, $\dot{y}_a$ is the speed of AMB_a along the y axis, $\dot{y}_b$ is the speed of AMB_b along the y axis, and $\dot{z}_g$ is the speed of the axial bearing along the z-axis.

The control inputs vector is [150]

$$\begin{aligned} U &= [u_1, u_2, u_3, u_4, u_5]^T \; textor \\ U &= [i_{ax}, i_{bx}, i_{ay}, i_{by, i_{gz}}]^T \end{aligned} \tag{7.330}$$

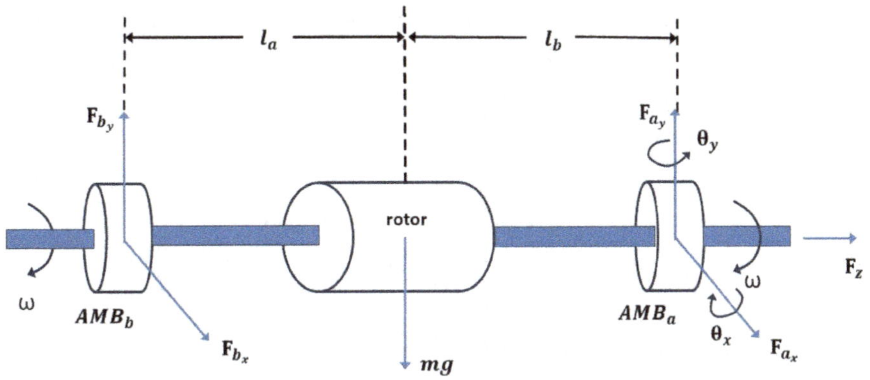

Figure 7.88 Diagram of the 5-DOF rotor and active magnetic bearing system

and consists of current variables which control the position of the front, rear, and axial magnetic bearings. The state-space model of the system is given by [150]:

$$\begin{aligned}
\dot{x}_1 &= x_6\\
\dot{x}_2 &= x_7\\
\dot{x}_3 &= x_8\\
\dot{x}_4 &= x_9\\
\dot{x}_5 &= x_{10}\\
\dot{x}_6 &= R_a^x G_1 x_1 + R_b^x G_3 x_2 - M_a(x_8 - x_9) + R_1^a G_1 u_1 + R_1^b G_3 u_2\\
\dot{x}_7 &= R_a^x G_3 x_1 + R_b^x G_2 x_2 + M_b(x_8 - x_9) + R_1^a G_3 u_1 + R_1^b G_2 u_2\\
\dot{x}_8 &= R_a^x G_1 x_3 + R_b^x G_3 x_4 + M_a(x_6 - x_7) + R_1^a G_1 u_3 + R_1^b G_3 u_4 - g\\
\dot{x}_9 &= R_a^x G_3 x_3 + R_b^x G_2 x_4 - M_b(x_6 - x_7) + R_1^a G_3 u_3 + R_1^b G_3 u_4 - g\\
\dot{x}_{10} &= \frac{K_{2x}}{m} x_5 + \frac{K_{zl}}{m} u_5
\end{aligned} \tag{7.331}$$

where $R_a^x = \frac{K_x^a}{m}$, $R_b^x = \frac{K_x^b}{m}$, $R_i^a = \frac{K_i^a}{m}$, $R_i^b = \frac{K_i^b}{m}$, $G_1 = 1 + \frac{ml_a^2}{J_{xy}}$, $G_2 = 1 + \frac{ml_b^2}{J_{xy}}$, $G_3 = 1 - \frac{ml_a l_b}{J_{xy}}$, $M_a = \omega\frac{J_z l_a}{J_{xy} l}$, $M_b = \omega\frac{J_z l_b}{J_{xy} l}$.

Furthermore, K_x^a, K_x^b are the displacement stiffness coefficients, and K_i^a, K_i^b are the current stiffness coefficients for the front and rear radial bearings. Additionally, K_x^z, K_i^z are the displacement stiffness and the current stiffness coefficients of the axial bearing.

J_{xy} is the moment of inertia for rotation around the x or y axis. J_z is the moment of inertia for rotation around the z axis. Besides, l_a, l_b are distances measured from the front and rear bearing up to the center of the rotor. Their sum is defined as $l = l_a + l_b$. The mass of the rotor is defined as m. The angular speed of the rotor and the spindle is defined as ω.

By defining the new control inputs vector $\tilde{u} = [u_1, u_2, v_3, v_4, u_5]^T$ with the new control inputs v_3, v_4 to be given by $v_3 = R_i^a G_1 u_3 + R_1^b G_3 u_4 - g$, and $v_4 = R_i^a G_1 u_3 - R_1^b G_3 u_4 - g$, or equivalently

$$\begin{pmatrix} R_i^a G_1 & R_1^b G_3 \\ R_i^a G_1 & R_1^b G_3 \end{pmatrix} \begin{pmatrix} u_3 \\ u_4 \end{pmatrix} = \begin{pmatrix} v_1 - g \\ v_2 - g \end{pmatrix} \tag{7.332}$$

the dynamic model of the 5-DOF rotor and active magnetic bearing system can be written in the following matrix form:

$$\dot{x} = f(x) + g(x)\tilde{u} \tag{7.333}$$

where $x \in R^{10\times 1}$, $f(x) \in R^{10\times 1}$, $g(x) \in R^{10\times 5}$, and $u \in R^{5\times 1}$. Analytically, the matrix-form representation of the 5-DOF rotor and active magnetic bearings system is given by:

$$\begin{pmatrix} \dot{x}_1 \\ \dot{x}_2 \\ \dot{x}_3 \\ \dot{x}_4 \\ \dot{x}_5 \\ \dot{x}_6 \\ \dot{x}_7 \\ \dot{x}_8 \\ \dot{x}_9 \\ \dot{x}_{10} \end{pmatrix} = \begin{pmatrix} x_6 \\ x_7 \\ x_8 \\ x_9 \\ x_{10} \\ R_a^x G_1 x_1 + R_b^x G_3 x_2 - M_a(x_8 - x_9) \\ R_a^x G_3 x_1 + R_b^x G_2 x_2 + M_b(x_8 - x_9) \\ R_a^x G_1 x_3 + R_b^x G_3 x_4 + M_a(x_6 - x_7) \\ R_a^x G_3 x_3 + R_b^x G_2 x_4 - M_b(x_6 - x_7) \\ \frac{K_{2x}}{m} x_5 \end{pmatrix} + \begin{pmatrix} 0 & 0 & 0 & 0 & 0 \\ 0 & 0 & 0 & 0 & 0 \\ 0 & 0 & 0 & 0 & 0 \\ 0 & 0 & 0 & 0 & 0 \\ 0 & 0 & 0 & 0 & 0 \\ R_1^a G_1 & R_1^b G_3 & 0 & 0 & 0 \\ R_1^a G_3 & R_1^b G_2 & 0 & 0 & 0 \\ 0 & 0 & 1 & 0 & 0 \\ 0 & 0 & 0 & 1 & 0 \\ 0 & 0 & 0 & 0 & \frac{K_{zl}}{m} \end{pmatrix} \begin{pmatrix} u_! \\ u_2 \\ v_3 \\ v_4 \\ u_5 \end{pmatrix} \tag{7.334}$$

7.4.3 Differential flatness properties of the 5-DOF rotor and AMB system

The dynamic model of the 5-DOF rotor and AMB system is differentially flat, with flat outputs vector $Y = [x_1, x_2, x_3, x_4, x_5]^T$. Indeed, for the first five rows of the state-space model, one has:

$$\dot{x}_6 = x_1 \quad \dot{x}_7 = x_2 \quad \dot{x}_8 = x_3 \quad \dot{x}_9 = x_4 \quad \dot{x}_{10} = x_5 \tag{7.335}$$

which signifies that all state variables of the system are differential functions of the flat outputs. From the last five lines of the state-space model, one has

$$\begin{aligned} \dot{x}_6 &= f_6(x) + g_{6a}(x)u_1 + g_{6b}(x)u_2 \\ \dot{x}_7 &= f_7(x) + g_{7a}(x)u_1 + g_{7b}(x)u_2 \\ \dot{x}_8 &= f_8(x) + g_{8a}(x)u_3 + g_{8b}(x)u_4 - g \\ \dot{x}_9 &= f_9(x) + g_{9a}(x)u_3 + g_{9b}(x)u_4 - g \\ \dot{x}_{10} &= f_{10}(x) + g_{10}(x)u_5 \end{aligned} \tag{7.336}$$

Solving the above set of equations for the control inputs gives:

$$\begin{pmatrix} u_1 \\ u_2 \\ u_3 \\ u_4 \\ u_5 \end{pmatrix} = \begin{pmatrix} g_{6a} & g_{6b} & 0 & 0 & 0 \\ g_{7a} & g_{7b} & 0 & 0 & 0 \\ 0 & 0 & g_{8a} & g_{8b} & 0 \\ 0 & 0 & g_{9a} & g_{9b} & 0 \\ 0 & 0 & g_{8a} & g_{8b} & 0 \\ 0 & 0 & 0 & 0 & g_{10} \end{pmatrix} \left[\begin{pmatrix} \dot{x}_6 \\ \dot{x}_7 \\ \dot{x}_8 \\ \dot{x}_9 \\ \dot{x}_{10} \end{pmatrix} - \begin{pmatrix} f_6(x) \\ f_7(x) \\ f_8(x) - g \\ f_9(x) - g \\ f_{10}(x) \end{pmatrix} \right] \tag{7.337}$$

which signifies that the control inputs vector u of the system is a differential function of the flat outputs vector Y, or

$$u = h_u(Y, \dot{Y}) \tag{7.338}$$

Consequently, all state variables of the rotor and AMB system can be written as differential functions of the flat outputs and the 5-DOF rotor and AMB system is differentially flat. The differential flatness property is an implicit proof of the system's controllability, Besides, it allows for treating the setpoints definition problem. First, one defines setpoints, in an unconstrained manner, for those state variables which coincide with the flat outputs. Next, for the rest of the state variables, setpoints are chosen using the differential relations which connect them to the flat outputs.

7.4.4 Approximate linearization of the 5-DOF rotor and AMB system

The state-space model of the 5-DOF rotor and AMB system undergoes approximate linearization with the use of first-order Taylor series expansion and through the computation of the associated Jacobian matrices. The linearization takes place at each sampling instance around the temporary operating point (x^*, u^*), where x^* is the present value of the system's state vector and u^* is the last sampled value of the control inputs vector. The modeling error which is due to the truncation of higher-order terms from the Taylor series expansion ia considered to be a perturbation which is asymptotically compensated by the robustness of the control algorithm.
The system, being initially in the nonlinear state-space form $\dot{x} = f(x, u)$ after linearization is transformed into the equivalent linear form:

$$\dot{x} = Ax + Bu + \tilde{d} \tag{7.339}$$

where A, B are the Jacobian matrices of the system

$$A = \nabla_x[f(x) + g(x)u] \mid_{(x^*,u^*)} \Rightarrow A = \nabla_x f(x) \mid_{(x^*,u^*)} \tag{7.340}$$

$$B = \nabla_u[f(x) + g(x)u] \mid_{(x^*,u^*)} \Rightarrow B = g(x) \mid_{(x^*,u^*)} \tag{7.341}$$

The term $\tilde{d}$ denotes the cumulative disturbance which may comprise: (i) modeling error due to truncation of terms from the Taylor series, (ii) exogenous perturbations, and (iii) sensor measurement noise of any distribution.
Computation of the Jacobian matrix $A = \nabla_x f(x) \mid_{(x^*,u^*)}$:

First row of the Jacobian matrix $A = \nabla_x f(x) \mid_{(x^*,u^*)}$: $\frac{\partial f_1}{\partial x_i} = 0$ for $i = 1, \cdots, 10$ and $i \neq 6$, while $\frac{\partial f_1}{\partial x_6} = 1$.

Second row of the Jacobian matrix $A = \nabla_x f(x) \mid_{(x^*,u^*)}$: $\frac{\partial f_2}{\partial x_i} = 0$ for $i = 1, \cdots, 10$ and $i \neq 7$, while $\frac{\partial f_2}{\partial x_7} = 1$.

Third row of the Jacobian matrix $A = \nabla_x f(x) \mid_{(x^*,u^*)}$: $\frac{\partial f_3}{\partial x_i} = 0$ for $i = 1, \cdots, 10$ and $i \neq 8$, while $\frac{\partial f_3}{\partial x_8} = 1$.

Fourth row of the Jacobian matrix $A = \nabla_x f(x) \mid_{(x^*,u^*)}$: $\frac{\partial f_4}{\partial x_i} = 0$ for $i = 1, \cdots, 10$ and $i \neq 9$, while $\frac{\partial f_4}{\partial x_9} = 1$.

Fifth row of the Jacobian matrix $A = \nabla_x f(x) \mid_{(x^*,u^*)}$: $\frac{\partial f_5}{\partial x_i} = 0$ for $i = 1, \cdots, 9$ and $i \neq 10$, while $\frac{\partial f_5}{\partial x_{10}} = 1$.

Sixth row of the Jacobian matrix $A = \nabla_x f(x) \mid_{(x^*,u^*)}$: $\frac{\partial f_6}{\partial x_1} = R_a^x G_1$, $\frac{\partial f_6}{\partial x_2} = R_b^x G_2$, $\frac{\partial f_6}{\partial x_3} = 0$, $\frac{\partial f_6}{\partial x_4} = 0$, $\frac{\partial f_6}{\partial x_5} = 0$, $\frac{\partial f_6}{\partial x_6} = 0$, $\frac{\partial f_6}{\partial x_7} = 0$, $\frac{\partial f_6}{\partial x_8} = -M_a$, $\frac{\partial f_6}{\partial x_9} = M_a$, $\frac{\partial f_6}{\partial x_{10}} = 0$.

Seventh row of the Jacobian matrix $A = \nabla_x f(x) \mid_{(x^*,u^*)}$: $\frac{\partial f_7}{\partial x_1} = R_a^x G_3$, $\frac{\partial f_7}{\partial x_2} = R_b^x G_2$, $\frac{\partial f_7}{\partial x_3} = 0$, $\frac{\partial f_7}{\partial x_4} = 0$, $\frac{\partial f_7}{\partial x_5} = 0$, $\frac{\partial f_7}{\partial x_6} = 0$, $\frac{\partial f_7}{\partial x_7} = 0$, $\frac{\partial f_7}{\partial x_8} = M_b$, $\frac{\partial f_7}{\partial x_9} = -M_b$, $\frac{\partial f_7}{\partial x_{10}} = 0$.

Eighth row of the Jacobian matrix $A = \nabla_x f(x) \mid_{(x^*,u^*)}$: $\frac{\partial f_8}{\partial x_1} = 0$, $\frac{\partial f_8}{\partial x_2} = 0$, $\frac{\partial f_8}{\partial x_3} = R_a^x G_1$, $\frac{\partial f_8}{\partial x_4} = R_b^x G_3$, $\frac{\partial f_8}{\partial x_5} = 0$, $\frac{\partial f_8}{\partial x_6} = M_a$, $\frac{\partial f_8}{\partial x_7} = -M_a$, $\frac{\partial f_8}{\partial x_8} = 0$, $\frac{\partial f_8}{\partial x_9} = 0$, $\frac{\partial f_8}{\partial x_{10}} = 0$.

Ninth row of the Jacobian matrix $A = \nabla_x f(x) \mid_{(x^*,u^*)}$: $\frac{\partial f_9}{\partial x_1} = 0$, $\frac{\partial f_9}{\partial x_2} = 0$, $\frac{\partial f_9}{\partial x_3} = R_a^x G_3$, $\frac{\partial f_9}{\partial x_4} = R_b^x G_2$, $\frac{\partial f_9}{\partial x_5} = 0$, $\frac{\partial f_9}{\partial x_6} = -M_b$, $\frac{\partial f_9}{\partial x_7} = M_b$, $\frac{\partial f_9}{\partial x_8} = 0$, $\frac{\partial f_9}{\partial x_9} = 0$, $\frac{\partial f_9}{\partial x_{10}} = 0$.

Tenth row of the Jacobian matrix $A = \nabla_x f(x) \mid_{(x^*,u^*)}$: $\frac{\partial f_{10}}{\partial x_i} = 0$, for $i = 1, \cdots, 10$ and $i \neq 5$, while $\frac{\partial f_{10}}{\partial x_5} = \frac{K_{zx}}{m}$.

This linearization approach which has been followed for implementing the nonlinear optimal control scheme results in a quite accurate model of the system's dynamics. Consider for instance the following affine-in-the-input state-space model

$$\dot{x}=f(x)+g(x)u\Rightarrow$$
$$\dot{x}=[f(x^*)+\nabla_x f(x)\mid_{x^*}(x-x^*)]+[g(x^*)+\nabla_x g(x)\mid_{x^*}(x-x^*)]u^*+g(x^*)u^*+g(x^*)(u-u^*)+\tilde{d}_1\Rightarrow \quad (7.342)$$
$$\dot{x}=[\nabla_x f(x)\mid_{x^*}+\nabla_x g(x)\mid_{x^*}u^*]x+g(x^*)u-[\nabla_x f(x)\mid_{x^*}+\nabla_x g(x)\mid_{x^*}u^*]x^*+f(x^*)+g(x^*)u^*+\tilde{d}_1$$

where $\tilde{d}_1$ is the modeling error due to truncation of higher order terms in the Taylor series expansion of $f(x)$ and $g(x)$. Next, by defining $A=[\nabla_x f(x)\mid_{x^*}+\nabla_x g(x)\mid_{x^*}u^*]$, $B=g(x^*)$, one obtains

$$\dot{x}=Ax+Bu-Ax^*+f(x^*)+g(x^*)u^*+\tilde{d}_1 \quad (7.343)$$

Moreover, by denoting $\tilde{d}=-Ax^*+f(x^*)+g(x^*)u^*+\tilde{d}_1$ about the cumulative modeling error term in the Taylor series expansion, one has

$$\dot{x}=Ax+Bu+\tilde{d} \quad (7.344)$$

which is the approximately linearized model of the dynamics of the system of Eq. (7.339). The term $f(x^*)+g(x^*)u^*$ is the derivative of the state vector at (x^*,u^*) which is almost annihilated by $-Ax^*$.

7.4.5 Design of an H-infinity nonlinear feedback controller

7.4.5.1 Equivalent linearized dynamics of the 5-DOF rotor and AMB system

After linearization around its current operating point, the dynamic model for the 5-DOF rotor and AMB system is written as

$$\dot{x}=Ax+Bu+d_1 \quad (7.345)$$

Parameter d_1 stands for the linearization error in the 5-DOF rotor and AMB system's model that was given previously in Eq. (7.345). The reference setpoints for the state vector of the aforementioned dynamic model are denoted by $\mathbf{x_d}=[x_1^d,\cdots,x_{10}^d]$. Tracking of this trajectory is achieved after applying the control input u^*. At every time instant, the control input u^* is assumed to differ from the control input u appearing in Eq. (7.345) by an amount equal to Δu, that is $u^*=u+\Delta u$

$$\dot{x}_d=Ax_d+Bu^*+d_2 \quad (7.346)$$

The dynamics of the controlled system described in Eq. (7.345) can be also written as

$$\dot{x}=Ax+Bu+Bu^*-Bu^*+d_1 \quad (7.347)$$

and by denoting $d_3=-Bu^*+d_1$ as an aggregate disturbance term, one obtains

$$\dot{x}=Ax+Bu+Bu^*+d_3 \quad (7.348)$$

By subtracting Eq. (7.346) from Eq. (7.348), one has

$$\dot{x}-\dot{x}_d=A(x-x_d)+Bu+d_3-d_2 \quad (7.349)$$

By denoting the tracking error as $e=x-x_d$ and the aggregate disturbance term as $L\tilde{d}=d_3-d_2$, the tracking error dynamics becomes

$$\dot{e}=Ae+Bu+L\tilde{d} \quad (7.350)$$

where L is a disturbance inputs gain matrix. The above linearized form of the 5-DOF rotor and AMB system's model can be efficiently controlled after applying an H-infinity feedback control scheme.

7.4.5.2 The nonlinear H-infinity control

The initial nonlinear model of the 5-DOF rotor and AMB system is in the form

$$\dot{x} = f(x, u) \quad x \in R^n, \ u \in R^m \tag{7.351}$$

Linearization of the model of the 5-DOF rotor and AMB system is performed at each iteration of the control algorithm around its present operating point $(x^*, u^*) = (x(t), u(t - T_s))$. The linearized equivalent of the system is described by

$$\dot{x} = Ax + Bu + L\tilde{d} \quad x \in R^n, \ u \in R^m, \ \tilde{d} \in R^q \tag{7.352}$$

where matrices A and B are obtained from the computation of the previously defined Jacobians and vector $\tilde{d}$ denotes disturbance terms due to linearization errors, while L is a disturbance inputs gain matrix. The problem of disturbance rejection for the linearized model that is described by

$$\begin{aligned} \dot{x} &= Ax + Bu + L\tilde{d} \\ y &= Cx \end{aligned} \tag{7.353}$$

where $x \in R^n$, $u \in R^m$, $\tilde{d} \in R^q$, and $y \in R^p$ cannot be handled efficiently if the classical LQR control scheme is applied. This is because of the existence of the perturbation term $\tilde{d}$. The disturbance term $\tilde{d}$ apart from modeling (parametric) uncertainty and external perturbations can also represent noise terms of any distribution.

In the H_∞ control approach, a feedback control scheme is designed for trajectory tracking by the system's state vector and simultaneous disturbance rejection, considering that the disturbance affects the system in the worst possible manner. The disturbances' effects are incorporated in the following quadratic cost function:

$$J(t) = \frac{1}{2}\int_0^T [y^T(t)y(t) + ru^T(t)u(t) - \rho^2\tilde{d}^T(t)\tilde{d}(t)]dt, \quad r, \rho > 0 \tag{7.354}$$

The significance of the negative sign in the cost function's term which is associated with the perturbation variable $\tilde{d}(t)$ is that the disturbance tries to maximize the cost function $J(t)$ while the control signal $u(t)$ tries to minimize it. The physical meaning of the relation given above is that the control signal and the disturbances compete with each other within a min–max differential game. This problem of min–max optimization can be written as $min_u max_{\tilde{d}} J(u, \tilde{d})$.

The objective of the optimization procedure is to compute a control signal $u(t)$ which can compensate for the worst possible disturbance, that is externally imposed to the 5-DOF rotor and AMB system. However, the solution to the min–max optimization problem is directly related to the value of parameter ρ. This means that there is an upper bound in the disturbances magnitude that can be annihilated by the control signal.

7.4.5.3 Computation of the feedback control gains

For the linearized system given by Eq. (7.353) the cost function of Eq. (7.354) is defined, where coefficient r determines the penalization of the control input and weight coefficient ρ determines the reward of the disturbances' effects. It is assumed that (i) The energy that is transferred from the disturbances signal $\tilde{d}(t)$ is bounded, that is $\int_0^\infty \tilde{d}^T(t)\tilde{d}(t)dt < \infty$, (ii) matrices $[A, B]$ and $[A, L]$ are stabilizable, and (iii) matrix $[A, C]$ is detectable. In the case of a tracking problem, the optimal feedback control law is given by

$$u(t) = -Ke(t) \tag{7.355}$$

with $e = x - x_d$ to be the tracking error, and $K = \frac{1}{r}B^TP$ where P is a positive definite symmetric matrix. As it will be proven in Section 7.4.6, matrix P is obtained from the solution of the Riccati equation

$$A^TP + PA + Q - P(\tfrac{2}{r}BB^T - \tfrac{1}{\rho^2}LL^T)P = 0 \tag{7.356}$$

where Q is a positive semi-definite symmetric matrix. The worst-case disturbance is given by

$$\tilde{d}(t) = \frac{1}{\rho^2}L^TPe(t) \tag{7.357}$$

The solution of the H-infinity feedback control problem for the 5-DOF rotor and AMB system and the computation of the worst-case disturbance that the related controller can sustain, comes from the superposition of Bellman's optimality principle when considering that the 5-DOF rotor and AMB system is affected by two separate inputs (i) the control

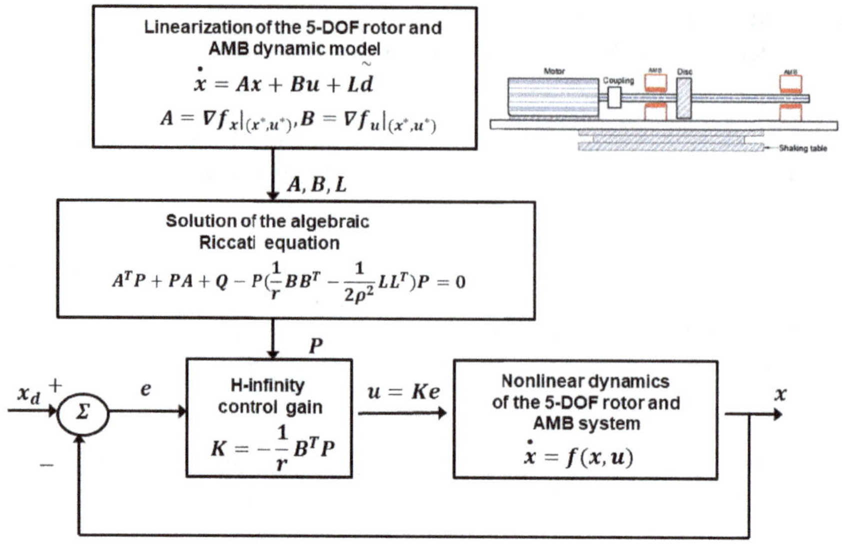

Figure 7.89 Control scheme and implementation stages for the 5-DOF rotor and AMB system

input u (ii) the cumulative disturbance input $\tilde{d}(t)$. Solving the optimal control problem for u, that is for the minimum variation (optimal) control input that achieves the elimination of the state vector's tracking error, gives $u = -\frac{1}{r}B^TPe$. Equivalently, solving the optimal control problem for $\tilde{d}$, that is for the worst-case disturbance that the control loop can sustain gives $\tilde{d} = \frac{1}{\rho^2}L^TPe$.

The diagram of the considered control loop for the 5-DOF rotor and AMB system is depicted in Figure 7.89.

7.4.6 Lyapunov stability analysis

Through Lyapunov stability analysis it will be shown that the proposed nonlinear control scheme assures H_∞ tracking performance for the 5-DOF rotor and AMB system and that in case of bounded disturbance terms asymptotic convergence to the reference setpoints is achieved. The tracking error dynamics for the 5-DOF rotor and AMB system is written in the form

$$\dot{e} = Ae + Bu + L\tilde{d} \tag{7.358}$$

where in the 5-DOF rotor and AMB system's case $L = \in R^{10\times 10}$ is the disturbance inputs gain matrix. Variable $\tilde{d}$ denotes model uncertainties and external disturbances of the 5-DOF rotor and AMB system's model. The following Lyapunov equation is considered:

$$V = \frac{1}{2}e^TPe \tag{7.359}$$

where $e = x - x_d$ is the tracking error. By differentiating with respect to time, one obtains

$$\begin{gathered}\dot{V} = \frac{1}{2}\dot{e}^TPe + \frac{1}{2}e^TP\dot{e} \Rightarrow \\ \dot{V} = \frac{1}{2}[Ae + Bu + L\tilde{d}]^TPe + \frac{1}{2}e^TP[Ae + Bu + L\tilde{d}] \Rightarrow\end{gathered} \tag{7.360}$$

$$\begin{aligned}\dot{V} = &\frac{1}{2}[e^TA^T + u^TB^T + \tilde{d}^TL^T]Pe + \\ &+\frac{1}{2}e^TP[Ae + Bu + L\tilde{d}] \Rightarrow\end{aligned} \tag{7.361}$$

$$\begin{aligned}\dot{V} = &\frac{1}{2}e^TA^TPe + \frac{1}{2}u^TB^TPe + \frac{1}{2}\tilde{d}^TL^TPe + \\ &\frac{1}{2}e^TPAe + \frac{1}{2}e^TPBu + \frac{1}{2}e^TPL\tilde{d}\end{aligned} \tag{7.362}$$

The previous equation is rewritten as

$$\dot{V} = \frac{1}{2}e^T(A^TP + PA)e + (\frac{1}{2}u^TB^TPe + \frac{1}{2}e^TPBu) + \\ +(\frac{1}{2}\tilde{d}^TL^TPe + \frac{1}{2}e^TPL\tilde{d}) \tag{7.363}$$

Assumption: For given positive definite matrix Q and coefficients r and ρ, there exists a positive definite matrix P, which is the solution of the following matrix equation:

$$A^TP + PA = -Q + P(\frac{2}{r}BB^T - \frac{1}{\rho^2}LL^T)P \tag{7.364}$$

Moreover, the following feedback control law is applied to the system:

$$u = -\frac{1}{r}B^TPe \tag{7.365}$$

By substituting Eq. (7.364) and Eq. (7.365), one obtains

$$\dot{V} = \frac{1}{2}e^T[-Q + P(\frac{2}{r}BB^T - \frac{1}{\rho^2}LL^T)P]e + \\ +e^TPB(-\frac{1}{r}B^TPe) + e^TPL\tilde{d} \Rightarrow \tag{7.366}$$

$$\dot{V} = -\frac{1}{2}e^TQe + \frac{1}{r}e^TPBB^TPe - \frac{1}{2\rho^2}e^TPLL^TPe \\ -\frac{1}{r}e^TPBB^TPe + e^TPL\tilde{d} \tag{7.367}$$

which after intermediate operations gives

$$\dot{V} = -\frac{1}{2}e^TQe - \frac{1}{2\rho^2}e^TPLL^TPe + e^TPL\tilde{d} \tag{7.368}$$

or, equivalently

$$\dot{V} = -\frac{1}{2}e^TQe - \frac{1}{2\rho^2}e^TPLL^TPe + \\ +\frac{1}{2}e^TPL\tilde{d} + \frac{1}{2}\tilde{d}^TL^TPe \tag{7.369}$$

Lemma: The following inequality holds

$$\frac{1}{2}e^TL\tilde{d} + \frac{1}{2}\tilde{d}L^TPe - \frac{1}{2\rho^2}e^TPLL^TPe \leq \frac{1}{2}\rho^2\tilde{d}^T\tilde{d} \tag{7.370}$$

Proof: The binomial $(\rho\alpha - \frac{1}{\rho}b)^2$ is considered. Expanding the left part of the above inequality, one gets

$$\rho^2a^2 + \frac{1}{\rho^2}b^2 - 2ab \geq 0 \Rightarrow \frac{1}{2}\rho^2a^2 + \frac{1}{2\rho^2}b^2 - ab \geq 0 \Rightarrow \\ ab - \frac{1}{2\rho^2}b^2 \leq \frac{1}{2}\rho^2a^2 \Rightarrow \frac{1}{2}ab + \frac{1}{2}ab - \frac{1}{2\rho^2}b^2 \leq \frac{1}{2}\rho^2a^2 \tag{7.371}$$

The following substitutions are carried out: $a = \tilde{d}$ and $b = e^TPL$ and the previous relation becomes

$$\frac{1}{2}\tilde{d}^TL^TPe + \frac{1}{2}e^TPL\tilde{d} - \frac{1}{2\rho^2}e^TPLL^TPe \leq \frac{1}{2}\rho^2\tilde{d}^T\tilde{d} \tag{7.372}$$

Eq. (7.372) is substituted in Eq. (7.369) and the inequality is enforced, thus giving

$$\dot{V} \leq -\frac{1}{2}e^TQe + \frac{1}{2}\rho^2\tilde{d}^T\tilde{d} \tag{7.373}$$

Eq. (7.373) shows that the H_∞ tracking performance criterion is satisfied. The integration of $\dot{V}$ from 0 to T gives

$$\int_0^T\dot{V}(t)dt \leq -\frac{1}{2}\int_0^T||e||_Q^2dt + \frac{1}{2}\rho^2\int_0^T||\tilde{d}||^2dt \Rightarrow \\ 2V(T) + \int_0^T||e||_Q^2dt \leq 2V(0) + \rho^2\int_0^T||\tilde{d}||^2dt \tag{7.374}$$

Moreover, if there exists a positive constant $M_d > 0$ such that

$$\int_0^\infty||\tilde{d}||^2dt \leq M_d \tag{7.375}$$

then, one gets

$$\int_0^\infty||e||_Q^2dt \leq 2V(0) + \rho^2M_d \tag{7.376}$$

Thus, the integral $\int_0^{\infty}||e||_Q^2 dt$ is bounded. Moreover, $V(T)$ is bounded and from the definition of the Lyapunov function V in Eq. (7.359) it becomes clear that $e(t)$ will be also bounded since $e(t) \in \Omega_e = \{e|e^TPe \leq 2V(0) + \rho^2 M_{\tilde{d}}\}$. According to the above and with the use of Barbalat's Lemma, one obtains $\lim_{t\to\infty} e(t) = 0$.

Through the stages of the stability proof, one arrives at Eq. (7.373) which shows that the H-infinity tracking performance criterion holds. By selecting the attenuation coefficient ρ to be sufficiently small and in particular to satisfy $\rho^2 < ||e||_Q^2/||\tilde{d}||^2$, one has that the first derivative of the Lyapunov function is upper bounded by 0. This condition holds at each sampling instance, and consequently, global stability for the control loop can be concluded.

7.4.7 Flatness-based control in successive loops for the 5-DOF rotor with AMB

As noted before, a dynamic system is considered to be differentially flat if the following two conditions hold: (i) all its state variables and its control inputs can be written as differential functions of a subset of variables which constitute the flat outputs of the system (usually these some of the state variables of the system) and (ii) the flat outputs are differentially independent which means that they are not connected through a relation in the form of an homogenous ordinary differential equation. For the dynamic model of the 5-DOF rotor and AMB system, the initial nonlinear state-space form is

$$\dot{x} = f(x,u) \ x \in R^{10\times 1}, \ f(x) \in R^{10\times 1}, \ g(x) \in R^{10\times 5}, \ u \in R^{5\times 1}, \tag{7.377}$$

two subsystems Σ_1 and Σ_2 are defined consisting of the following sub-vectors and sub-matrices:

subsystem Σ_1: with state sub-vector $x_{1,5} = [x_1, x_2, x_3, x_4, x_5]^T$, drift vector $f_{1,5}(x_{1,5}) = 0_{5\times 1}$, control inputs gain matrix $g_{1,5}(x_{1,5}) = I_{5\times 5}$ and virtual control input $v_1 = [x_6, x_7, x_8, x_9, x_{10}]^T$

subsystem Σ_2: with state sub-vector $x_{6,10} = [x_6, x_7, x_8, x_9, x_{10}]^T$, drift vector $f_{6,10}(x_{1,5}, x_{6,10}) \in R^{5\times 1}$, control inputs gain matrix $g_{6,10}(x_{1,5}, x_{6,10}) \in R^{5\times 5}$ and real control input $u = [u_1, u_2, v_3, v_4, u_5]^T$ where

$$f_{6,10} = \begin{pmatrix} R_a^x G_1 x_1 + R_b^x G_3 x_2 - M_a(x_8 - x_9) \\ R_a^x G_3 x_1 + R_b^x G_2 x_2 + M_b(x_8 - x_9) \\ R_a^x G_1 x_3 + R_b^x G_3 x_4 + M_a(x_6 - x_7) \\ R_a^x G_3 x_3 + R_b^x G_2 x_4 - M_b(x_6 - x_7) \\ \frac{K_{2x}}{m} x_5 \end{pmatrix} \quad g_{6,10} = \begin{pmatrix} R_1^a G_1 & R_1^b G_3 & 0 & 0 & 0 \\ R_1^a G_3 & R_1^b G_2 & 0 & 0 & 0 \\ 0 & 0 & 1 & 0 & 0 \\ 0 & 0 & 0 & 1 & 0 \\ 0 & 0 & 0 & 0 & \frac{K_{zl}}{m} \end{pmatrix} \tag{7.378}$$

Thus, the state-space model of the 5-DOF rotor and AMB system is written in the following chained subsystems form:

$$\dot{x}_{1,5} = f_{1,5}(x_{1,5}) + g_{1,5}(x_{1,5})v_1 \tag{7.379}$$

$$\dot{x}_{6,10} = f_{6,10}(x_{1,5}, x_{6,10}) + g_{6,10}(x_{1,5}, x_{6,10})u \tag{7.380}$$

It will be proven that each one of these subsystems, if viewed independently, is differentially flat. For subsystem Σ_1 the flat outputs vector is taken to be $y_1 = x_{1,5}$ and the virtual control input about it is defined as $v_1 = x_{6,10}$. It holds that

$$v_1 = g_{1,5}(x_{1,5})^{-1}[\dot{x}_{1,5} - f_{1,5}(x_{1,5})] \tag{7.381}$$

which signifies that v_1 is a differential function of the flat outputs vector. Consequently, in subsystem Σ_1, all state variables and the control inputs are differential functions of the flat outputs vector $y_1 = x_{1,5}$ and this subsystem is differentially flat.

For subsystem Σ_2 the flat outputs vector is taken to be $y_2 = x_{6,10}$, u is the real control input, while sub-vector $x_{1,5}$ is viewed as a coefficients vector. It holds that

$$u = g_{6,10}(x_{1,5}, x_{6,10})^{-1}[\dot{x}_{6,10} - f_{6,10}(x_{1,5}, x_{6,10})] \tag{7.382}$$

which signifies that control input u is a differential function of the flat outputs vector. Consequently, in subsystem Σ_2, all state variables and the control inputs are differential functions of the flat outputs vector y_2 and this subsystem is also differentially flat.

The proof of differential flatness for the individual subsystems demonstrated that each one of them is input–output linearizable and a stabilizing feedback controller about it can be designed as it is commonly done for state-space models in the input–output linearized form.

7.4.8 Design of a multi-loop flatness-based controller

For subsystem Σ_1 a stabilizing feedback controller has the form

$$v_1 = g_{1,5}(x_{1,5})^{-1}[\dot{x}_{1,5}^d - f_{1,5}(x_{1,5}) - K_1(x_{1,5} - x_{1,5}^d)] \tag{7.383}$$

where $x_{1,5}^d$ is the setpoints vector of subsystem Σ_1 and $K_1 \in R^{5\times 5}$ is a diagonal matrix with diagonal elements $K_{1,ii} > 0$ for $i = 1, \cdots, 5$. The virtual control input v_1 becomes setpoints vector for the second subsystem Σ_2, that is $v_1 = x_{6,10}^d$. Then, for subsystem Σ_2, a stabilizing feedback controller has the form

$$u = g_{6,10}(x_{1,5}, x_{6,10})^{-1}[\dot{x}_{6,10}^d - f_{6,10}(x_{1,5}, x_{6,10}) - K_2(x_{6,10} - x_{6,10}^d)] \tag{7.384}$$

where the control inputs gain matrix $K_2 \in R^{5\times 5}$ is a diagonal matrix with positive elements $K_{2,ii} > 0$ for $i = 1, \cdots, 5$.

By substituting Eq. (7.383) into Eq. (7.379) and by defining the tracking error vector $e_{1,5} = x_{1,5} - x_{1,5}^d$, one obtains

$$\begin{aligned}
\dot{x}_{1,5} = & f_{1,5}(x_{1,5}) + g_{1,5}(x_{1,5})g_{1,5}(x_{1,5})^{-1} \cdot \\
& \cdot [\dot{x}_{1,5}^d - f_{1,5}(x_{1,5}) - K_1(x_{1,5} - x_{1,5}^d)] \\
\Rightarrow & (\dot{x}_{1,5} - \dot{x}_{1,5}^d) + K_1(x_{1,5} - x_{1,5}^d) = 0 \Rightarrow \dot{e}_{1,5} + K_1 e_{1,5} = 0 \\
& \Rightarrow lim_{t\to\infty} e_{1,5} = 0 \Rightarrow lim_{t\to\infty} x_{1,5} = x_{1,5}^d
\end{aligned} \tag{7.385}$$

Besides, by substituting Eq. (7.384) into Eq. (7.380) and by defining the tracking error vector $e_{6,10} = x_{6,10} - x_{6,10}^d$, one obtains

$$\begin{aligned}
\dot{x}_{6,10} = & f_{6,10}(x_{1,5}, x_{6,10}) + g_{6,10}(x_{1,5}, x_{6,10})g_{6,10}(x_{1,5}, x_{6,10})^{-1} \cdot \\
& \cdot [\dot{x}_{6,10}^d - f_{6,10}(x_{1,5}, x_{6,10}) - K_2(x_{6,10} - x_{6,10}^d)] \\
\Rightarrow & (\dot{x}_{1,5} - \dot{x}_{1,5}^d) + K_1(x_{1,5} - x_{1,5}^d) = 0 \Rightarrow \dot{e}_{1,5} + K_1 e_{1,5} = 0 \\
& \Rightarrow lim_{t\to\infty} e_{6,10} = 0 \Rightarrow lim_{t\to\infty} x_{6,10} = x_{6,10}^d
\end{aligned} \tag{7.386}$$

Consequently, the state variables of the 5-DOF rotor and AMB system converge to the targeted setpoints. Global stability can be also concluded through Lyapunov analysis. The following Lyapunov functions are defined:

$$V = \frac{1}{2}[e_{1,5}^T e_{1,5} + e_{6,10}^T e_{6,10}] \tag{7.387}$$

By differentiating in time, one obtains:

$$\dot{V} = \frac{1}{2}2[e_{1,5}^T \dot{e}_{1,5} + e_{6,10}^T \dot{e}_{6,10}] \tag{7.388}$$

Using the previous relations about the tracking error dynamics of $e_{1,5}$ and $e_{6.10}$, one gets

$$\begin{aligned}
& \dot{V} = \frac{1}{2}2[e_{1,5}^T(-K_1 e_{1,5}) + e_{6,10}^T(-K_2 e_{6,10})] \Rightarrow \\
& \quad \dot{V} = -[e_{1,5}^T K_1 e_{1,5} + e_{6,10}^T K_2 e_{6,10}] \Rightarrow \\
& \quad \dot{V} < 0 \, \forall \, e_{1,5} \neq 0, e_{6,10} \neq 0 \text{ and} \\
& \quad \dot{V} = 0 \text{ iff } e_{1,5} = 0, e_{6,10} = 0
\end{aligned} \tag{7.389}$$

Therefore, V is a continuously decreasing function which converges asymptotically to 0. This confirms once again global asymptotic stability for the 5-DOF rotor and AMB system.

7.4.9 Simulation tests

7.4.9.1 Nonlinear optimal control of the 5-DOF rotor and AMB system

The global stability properties and the fine transient performance of nonlinear optimal control for the 5-DOF and active magnetic bearing system has been tested through simulation experiments. To implement nonlinear optimal control the Jacobian matrices of the dynamic model of the 5-DOF and AMB system had to be computed at each sampling period and the method's Riccati equation had to be solved also at sampling interval using the *aresolv()* function in MATLAB®. The method is computationally efficient since a solution of this Riccati equation can be achieved within a time interval which is much shorter than the sampling period. The obtained results are depicted in Figures 7.90–7.105. The state

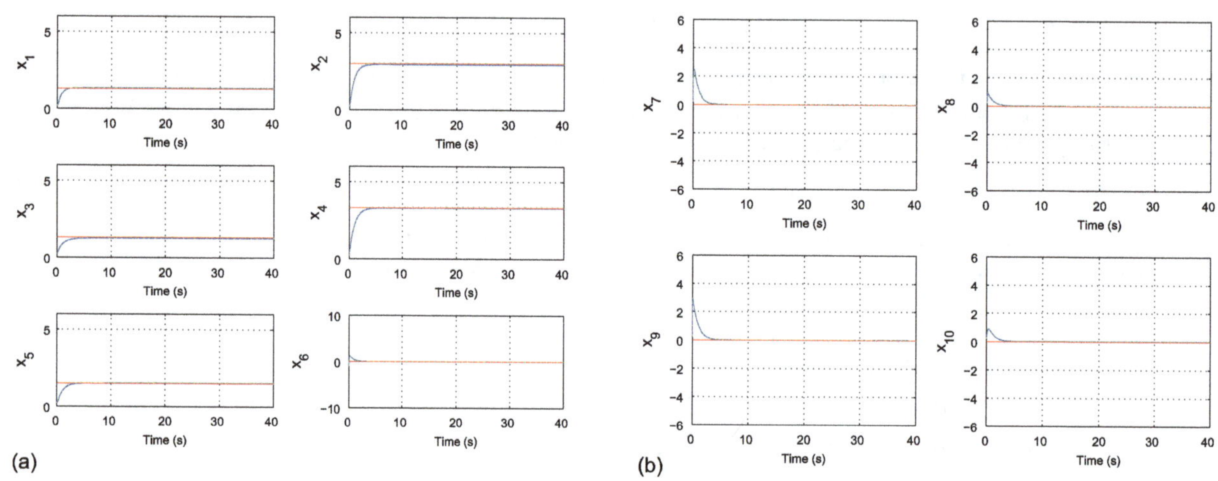

Figure 7.90 Tracking of setpoint 1 by the 5-DOF rotor and AMB system with the use of nonlinear optimal control: (a) convergence of state variables x_1 to x_6 (blue lines) to the associated setpoints (red lines) and estimated values provided by Kalman filtering; (b) convergence of state variables x_7 to x_{10} (blue lines) to the associated setpoints (red lines) and estimated values provided by Kalman filtering

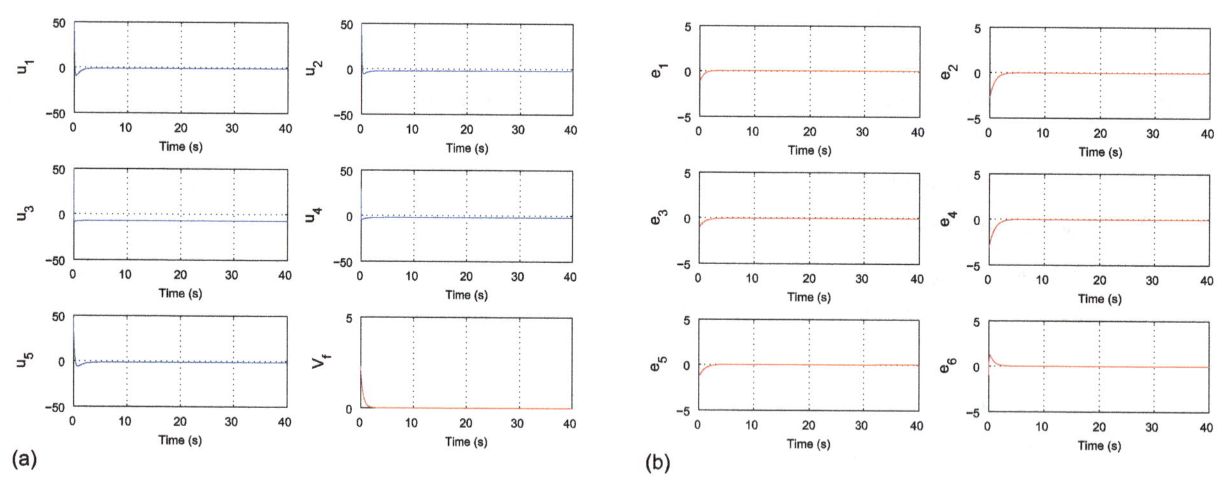

Figure 7.91 Tracking of setpoint 1 by the 5-DOF rotor and AMB system with the use of nonlinear optimal control: (a) variations of the control inputs u_1 and u_5 (blue lines) and Lyapunov function of the system (red line); (b) variation of the tracking error variables e_1 to e_6 associated with the state variables x_1 to x_6

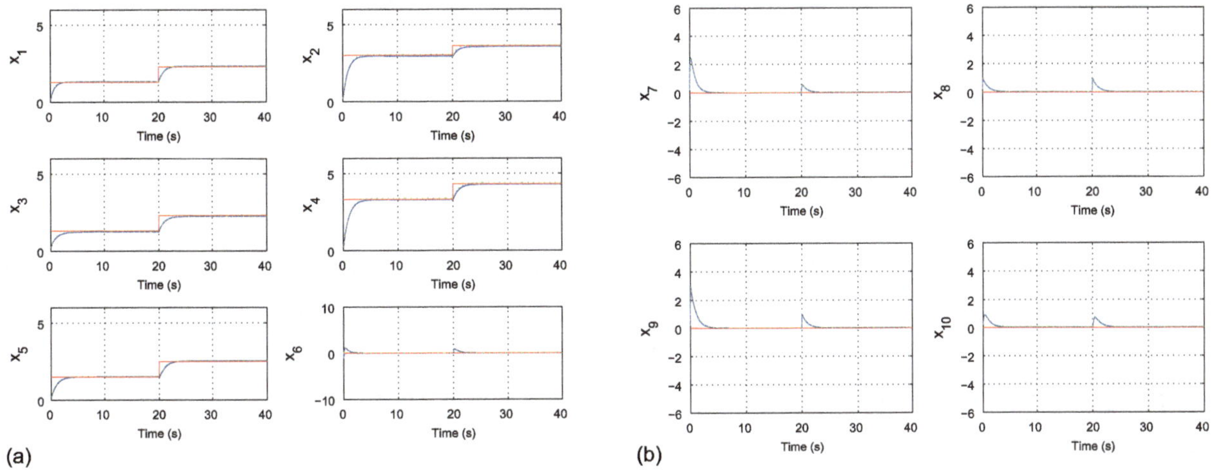

Figure 7.92 *Tracking of setpoint 2 by the 5-DOF rotor and AMB system with the use of nonlinear optimal control: (a) convergence of state variables x_1 to x_6 (blue lines) to the associated setpoints (red lines) and estimated values provided by Kalman filtering; (b) convergence of state variables x_7 to x_{10} (blue lines) to the associated setpoints (red lines) and estimated values provided by Kalman filtering*

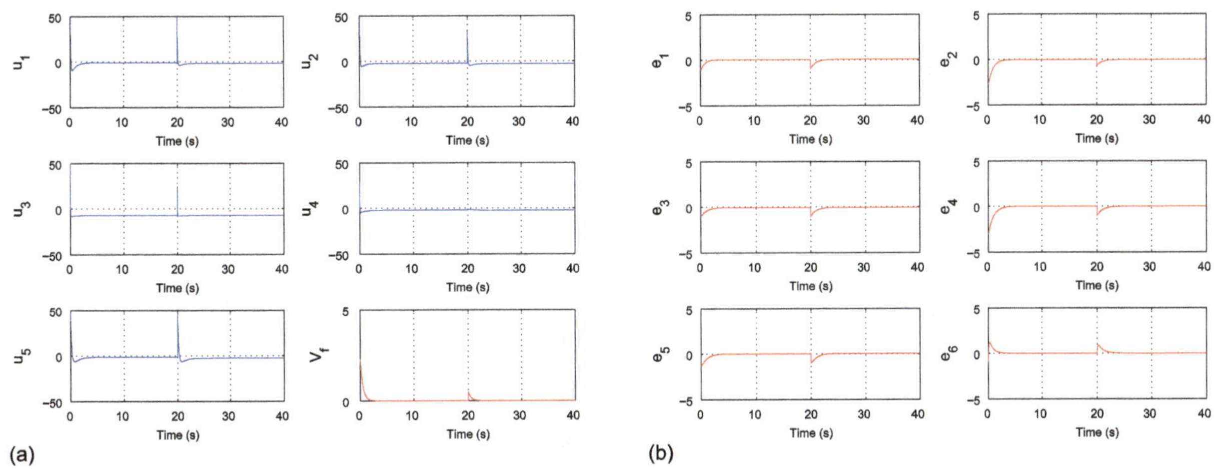

Figure 7.93 *Tracking of setpoint 2 by the 5-DOF rotor and AMB system with the use of nonlinear optimal control: (a) variations of the control inputs u_1 and u_5 (blue lines) and Lyapunov function of the system (red line); (b) variation of the tracking error variables e_1 to e_6 associated with the state variables x_1 to x_6*

variables of the 5-DOF rotor and AMB system are depicted in blue, the associated setpoints are plotted in red, while the state estimates which are provided by the H-infinity Kalman filter are printed in green. It has been confirmed that the method achieves fast and accurate tracking of setpoints under moderate variations of the control inputs. About the transient performance of the control method, this depends on parameters r, ρ, and Q of the associated Riccati equation. Small values of r achieve the elimination of the steady-state tracking error while relatively large values of the diagonal elements of matrix Q enable faster convergence of the state variables to their setpoints. Finally, the smallest value of

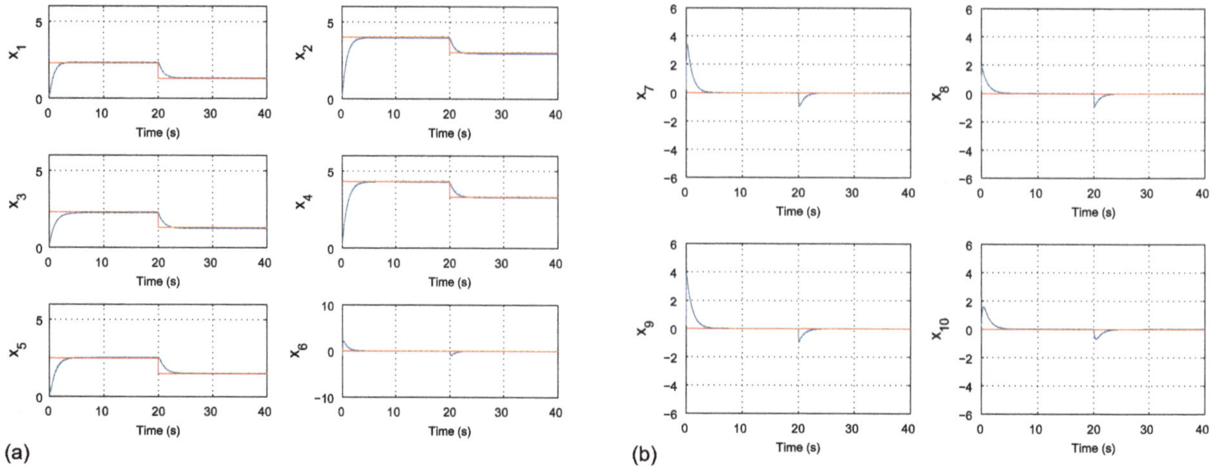

Figure 7.94 *Tracking of setpoint 3 by the 5-DOF rotor and AMB system with the use of nonlinear optimal control: (a) convergence of state variables x_1 to x_6 (blue lines) to the associated setpoints (red lines) and estimated values provided by Kalman filtering; (b) convergence of state variables x_7 to x_{10} (blue lines) to the associated setpoints (red lines) and estimated values provided by Kalman filtering*

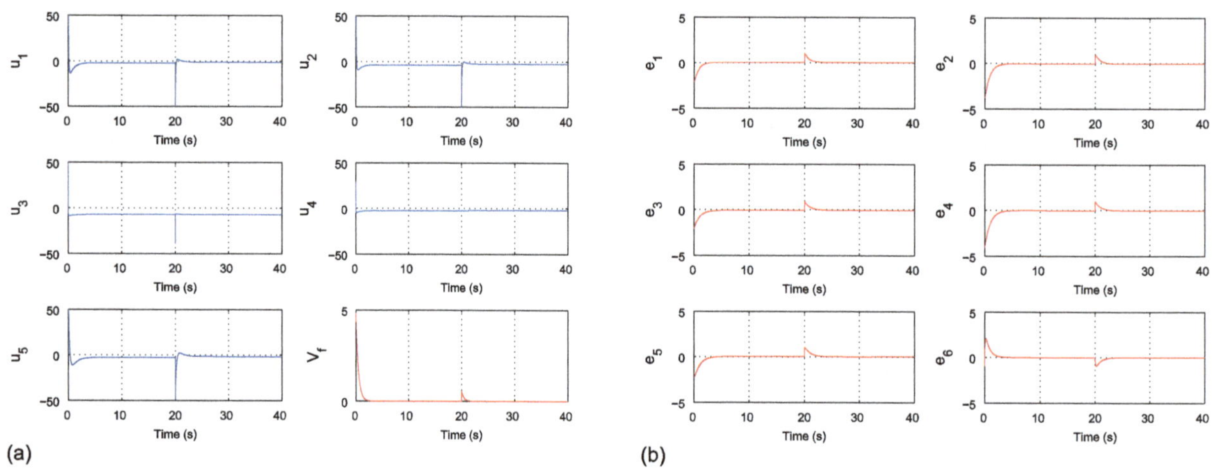

Figure 7.95 *Tracking of setpoint 3 by the 5-DOF rotor and AMB system with the use of nonlinear optimal control: (a) variations of the control inputs u_1 and u_5 (blue lines) and Lyapunov function of the system (red line); (b) variation of the tracking error variables e_1 to e_6 associated with the state variables x_1 to x_6*

ρ for which a valid solution of the method's Riccati equation can be obtained in the form of a positive definite and symmetric matrix P is the one that gives to the control loop maximum robustness.

7.4.9.2 Flatness-based control in successive loops of the 5-DOF rotor and AMB system

Results about the tracking accuracy and the speed of convergence to setpoints of the flatness-based control method in successive-loops, in the case of the 5-DOF rotor and active magnetic bearings system, are shown in Figures 7.106–7.121. It can be noticed again, that under this control scheme, one achieves fast and precise tracking of reference setpoints for

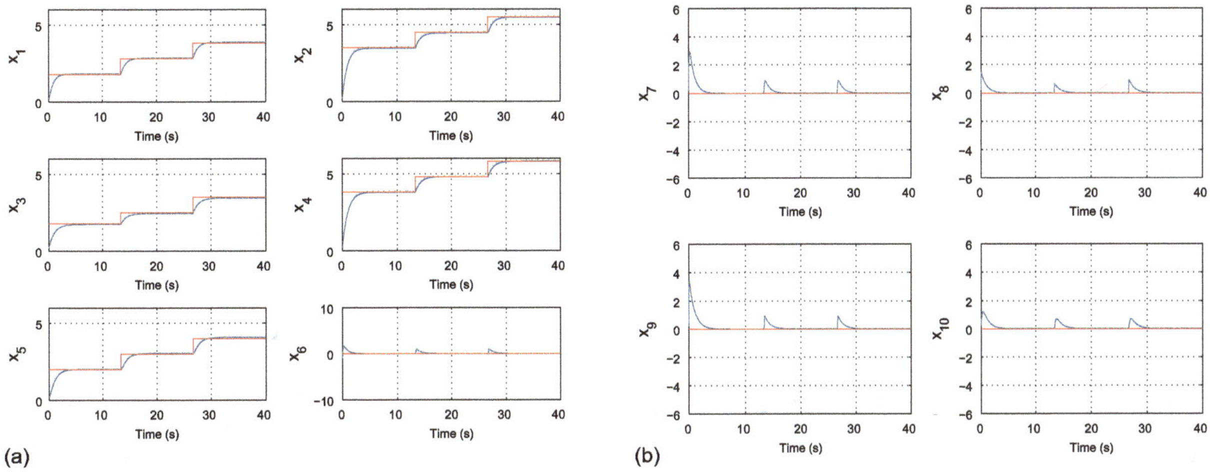

Figure 7.96 Tracking of setpoint 4 by the 5-DOF rotor and AMB system with the use of nonlinear optimal control: (a) convergence of state variables x_1 to x_6 (blue lines) to the associated setpoints (red lines) and estimated values provided by Kalman filtering; (b) convergence of state variables x_7 to x_{10} (blue lines) to the associated setpoints (red lines) and estimated values provided by Kalman filtering

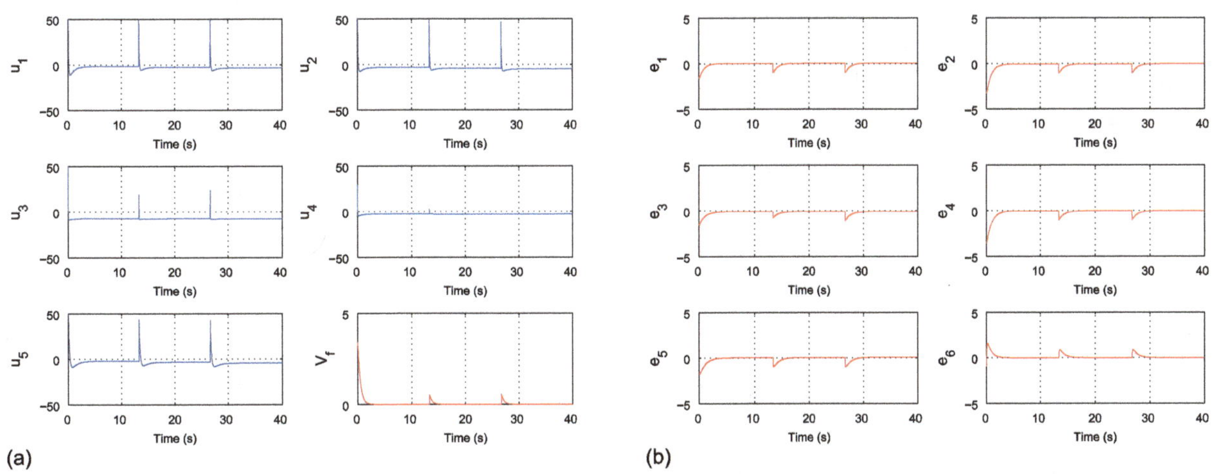

Figure 7.97 Tracking of setpoint 4 by the 5-DOF rotor and AMB system with the use of nonlinear optimal control: (a) variations of the control inputs u_1 and u_5 (blue lines) and Lyapunov function of the system (red line); (b) variation of the tracking error variables e_1 to e_6 associated with the state variables x_1 to x_6

all state variables of the dynamic model of the 5-DOF rotor and active magnetic bearings system. It is noteworthy, that through the stages of this method, one solves also the setpoints definition problem for all state variables of the 5-DOF rotor and AMB system. Actually, the selection of setpoints for state variables x_1 to x_5 is unconstrained. On the other side by defining state variables x_6 to x_{10} as virtual control inputs for the subsystem of x_1 to x_5, one can find the setpoints for x_6 to x_{10} as functions of the setpoints for x_1 to x_5. The speed of convergence of the state variables of the 5-DOF rotor and AMB system under flatness-based control implemented in successive loops is determined by the selection of values for the diagonal gain matrices K_1, K_2.

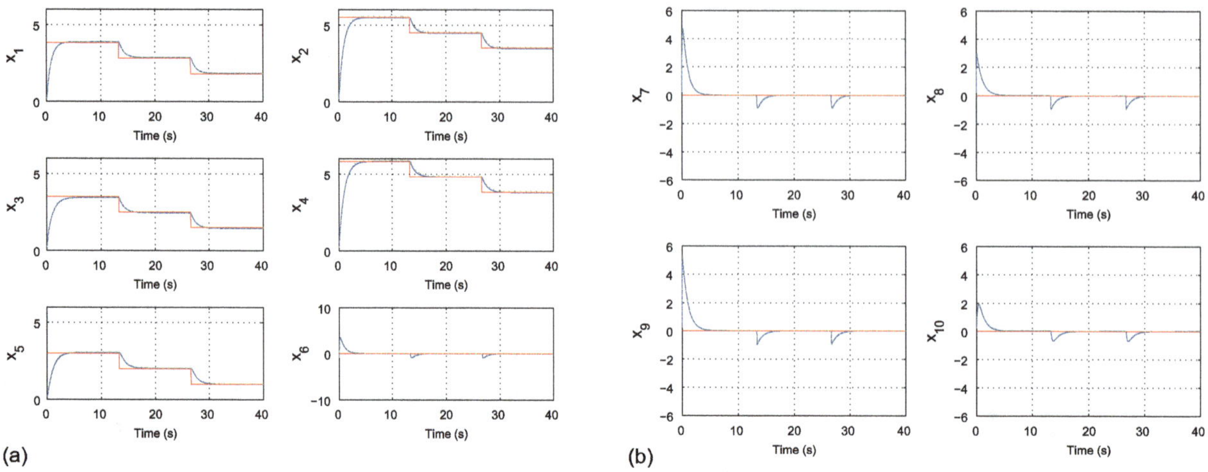

Figure 7.98 *Tracking of setpoint 5 by the 5-DOF rotor and AMB system with the use of nonlinear optimal control: (a) convergence of state variables x_1 to x_6 (blue lines) to the associated setpoints (red lines) and estimated values provided by Kalman filtering; (b) convergence of state variables x_7 to x_{10} (blue lines) to the associated setpoints (red lines) and estimated values provided by Kalman filtering*

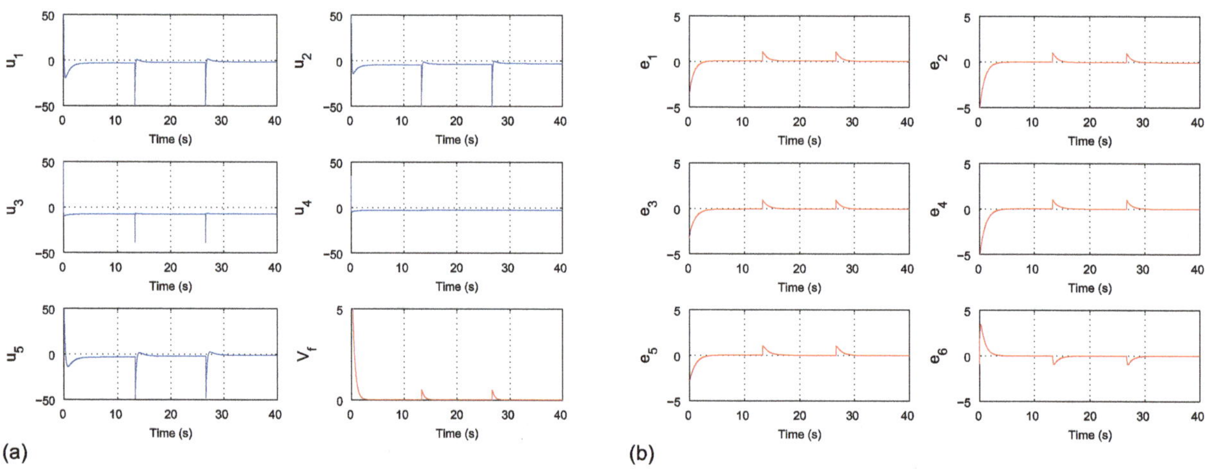

Figure 7.99 *Tracking of setpoint 5 by the 5-DOF rotor and AMB system with the use of nonlinear optimal control: (a) variations of the control inputs u_1 and u_5 (blue lines) and Lyapunov function of the system (red line); (b) variation of the tracking error variables e_1 to e_6 associated with the state variables x_1 to x_6*

7.5 Fault diagnosis for 5-DOF rotor and AMB with KF-based disturbance observers

7.5.1 Outline

As previously noted, active magnetic bearings (AMBs) are an essential component of high-speed rotating machines [839,269,841]. An overview has been given about industrial applications of AMBs, as for instance in motion transmission systems based on rotational or linear electric motors, in gas compressors, in wind turbines, and in flywheel-based energy storage systems [504,554,949]. Active magnetic bearings find also use in the actuation and orientation system of satellites and in biomedical systems such as blood pumps [869,109,195]. Besides, they are a key component in aircraft

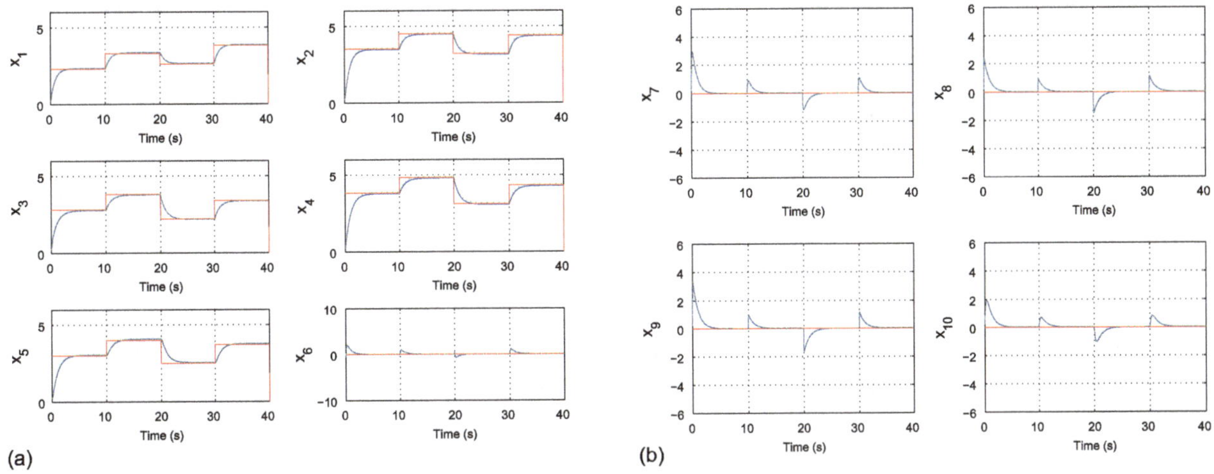

Figure 7.100 *Tracking of setpoint 6 by the 5-DOF rotor and AMB system with the use of nonlinear optimal control: (a) convergence of state variables x_1 to x_6 (blue lines) to the associated setpoints (red lines) and estimated values provided by Kalman filtering; (b) convergence of state variables x_7 to x_{10} (blue lines) to the associated setpoints (red lines) and estimated values provided by Kalman filtering*

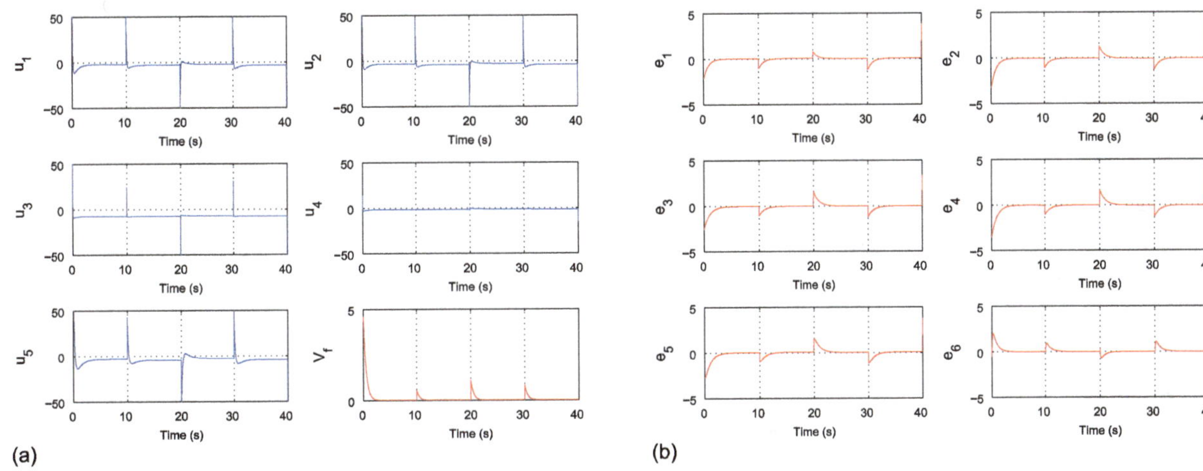

Figure 7.101 *Tracking of setpoint 6 by the 5-DOF rotor and AMB system with the use of nonlinear optimal control: (a) variations of the control inputs u_1 and u_5 (blue lines) and Lyapunov function of the system (red line); (b) variation of the tracking error variables e_1 to e_6 associated with the state variables x_1 to x_6*

engines and in gas-turbine power generation units [204,1000]. AMBs' industrial use includes also the weapon, metal industry, and transportation industries [290,1082]. Magnetic bearing-supported rotating machinery, whether based on a vertical or a horizontal motor, needs several subsystems to adjust the radial and axial levitations [274,136,1013]. Typically, two radial and one axial AMB are used for a 5-DOF levitation [622,511,766]. By achieving contactless operation of the rotor and its spindle AMBs eliminate mechanical friction and enable high rotational speeds. Compared to mechanical bearings AMBs exhibit the following benefits: (i) they work efficiently under higher rotational speed, (ii) they have reduced friction losses, (iii) they do not need lubrication, (iv) they consume less energy, and (v) they are more resilient to failures and have extended useful life [22,501]. AMBs are prone to instability because of the relation

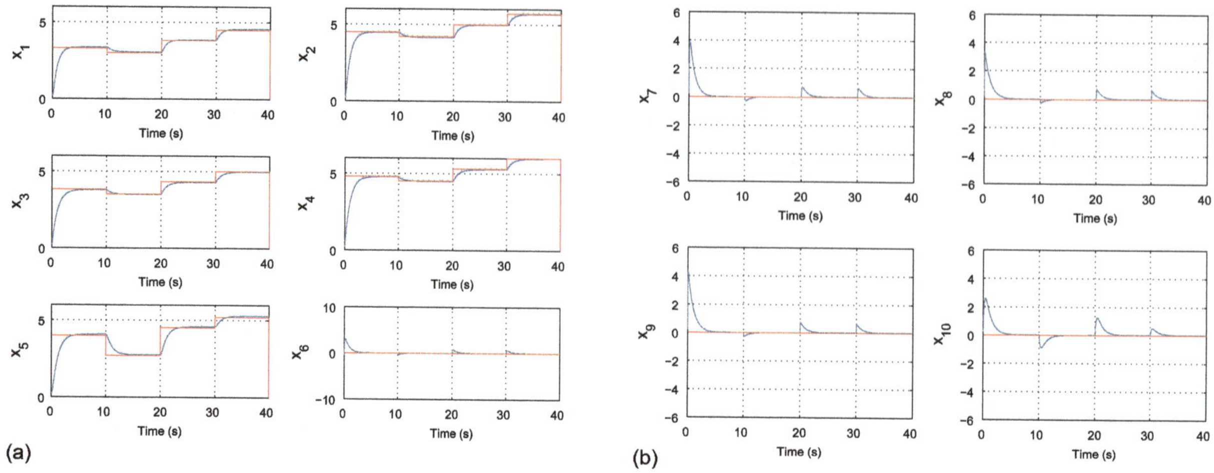

Figure 7.102 *Tracking of setpoint 7 by the 5-DOF rotor and AMB system with the use of nonlinear optimal control: (a) convergence of state variables x_1 to x_6 (blue lines) to the associated setpoints (red lines) and estimated values provided by Kalman filtering; (b) convergence of state variables x_7 to x_{10} (blue lines) to the associated setpoints (red lines) and estimated values provided by Kalman filtering*

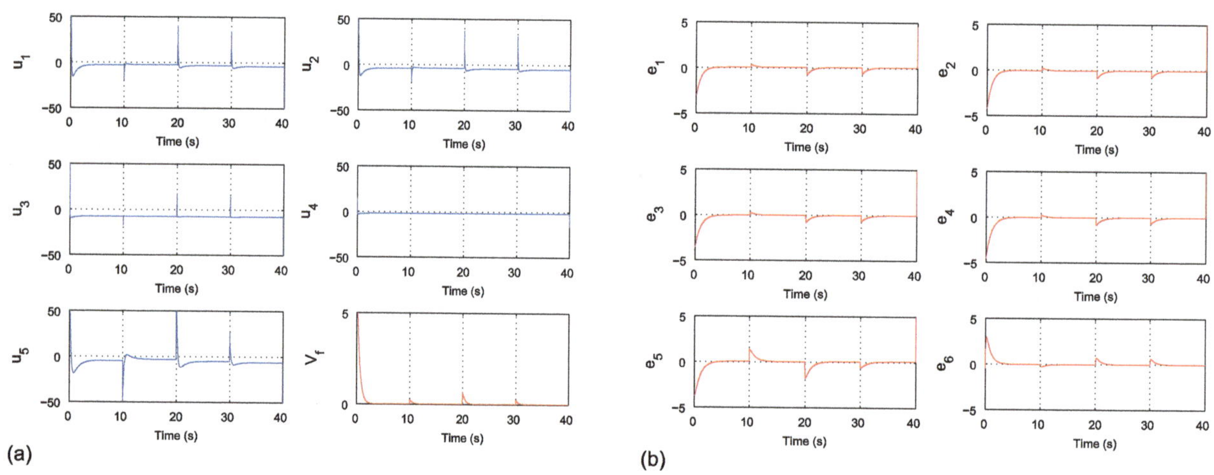

Figure 7.103 *Tracking of setpoint 7 by the 5-DOF rotor and AMB system with the use of nonlinear optimal control: (a) variations of the control inputs u_1 and u_5 (blue lines) and Lyapunov function of the system (red line); (b) variation of the tracking error variables e_1 to e_6 associated with the state variables x_1 to x_6*

of their electromagnetic force and the length of the air gap. Through the control system of AMBs the appearance of mechanical vibrations can be eliminated [529,151,1082]. The control problem of AMBs exhibits difficulties due to nonlinearities and a multi-variable structure of the associated state-space model [538,258,1007]. The control and stabilization problem for AMBs can be treated using model nonlinear optimal control, -predictive control, sliding-mode control and backstepping control [490,150,247].

Defects on bearings are the most common cause of machinery failures, especially for small or medium rotating machines. Bearing faults can be local or distributed. In the case of bearing faults in electric motors, common reasons are the improper power supply, the electric structure degradation, and the out-of-range operating conditions [606,260].

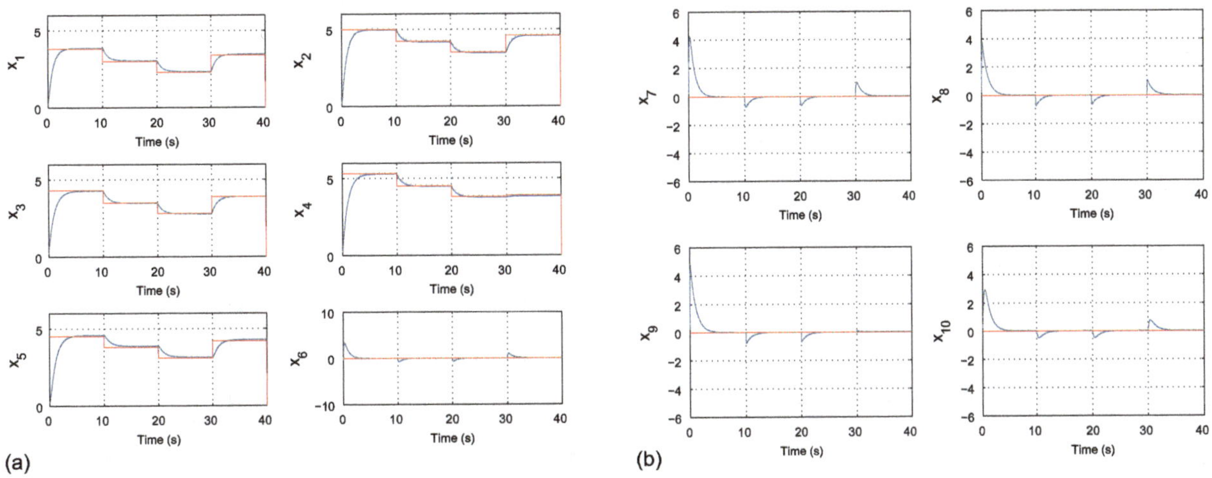

Figure 7.104 Tracking of setpoint 8 by the 5-DOF rotor and AMB system with the use of nonlinear optimal control: (a) convergence of state variables x_1 to x_6 (blue lines) to the associated setpoints (red lines) and estimated values provided by Kalman filtering; (b) convergence of state variables x_7 to x_{10} (blue lines) to the associated setpoints (red lines) and estimated values provided by Kalman filtering

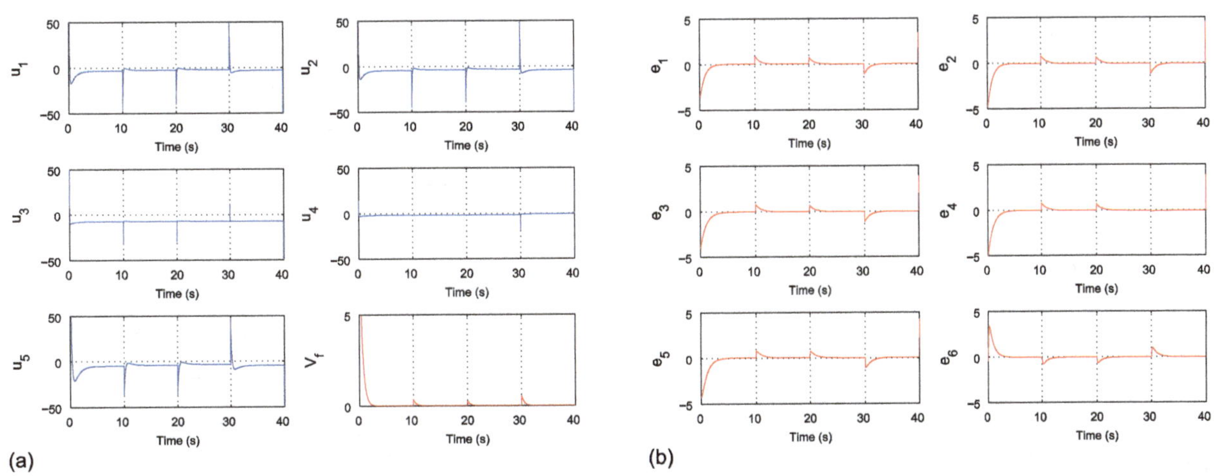

Figure 7.105 Tracking of setpoint 8 by the 5-DOF rotor and AMB system with the use of nonlinear optimal control: (a) variations of the control inputs u_1 and u_5 (blue lines) and Lyapunov function of the system (red line); (b) variation of the tracking error variables e_1 to e_6 associated with the state variables x_1 to x_6

Bearing fault detection is important for preventing machinery performance degradation and for improving production quality [655,878]. Bearing fault diagnosis can be performed in several manners, for instance, through acoustic information, measurement of electric currents, displacement estimation, and vibration analysis [146,37]. Bearing fault diagnosis can take place in the time domain, in the frequency domain, and in the time–frequency domain. In time-domain fault diagnosis, the accumulated measurements undergo statistical processing and generate a statistical index which is compared against fault thresholds [1061,325,403]. In the frequency domain, there is an assessment of the frequency information which is contained in the measurements collected from the bearings [919,68]. Fourier transform and its variants are used to extract this information. In the time–frequency domain, there is use of techniques such

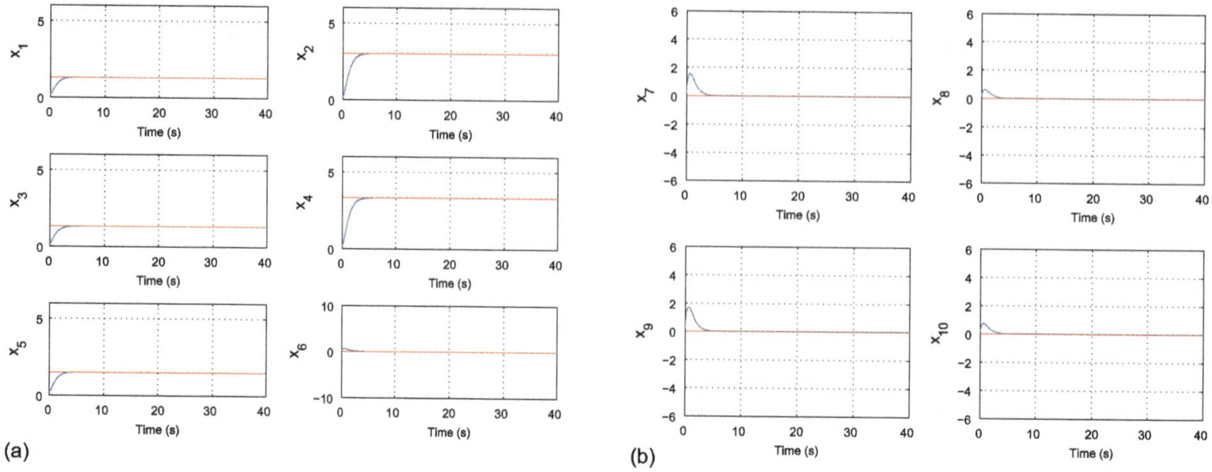

Figure 7.106 *Tracking of setpoint 1 by the 5-DOF rotor and AMB system with the use of multi-loop flatness-based control: (a) convergence of state variables x_1 to x_6 (blue lines) to the associated setpoints (red lines) and estimated values provided by Kalman filtering; (b) convergence of state variables x_7 to x_{10} (blue lines) to the associated setpoints (red lines) and estimated values provided by Kalman filtering*

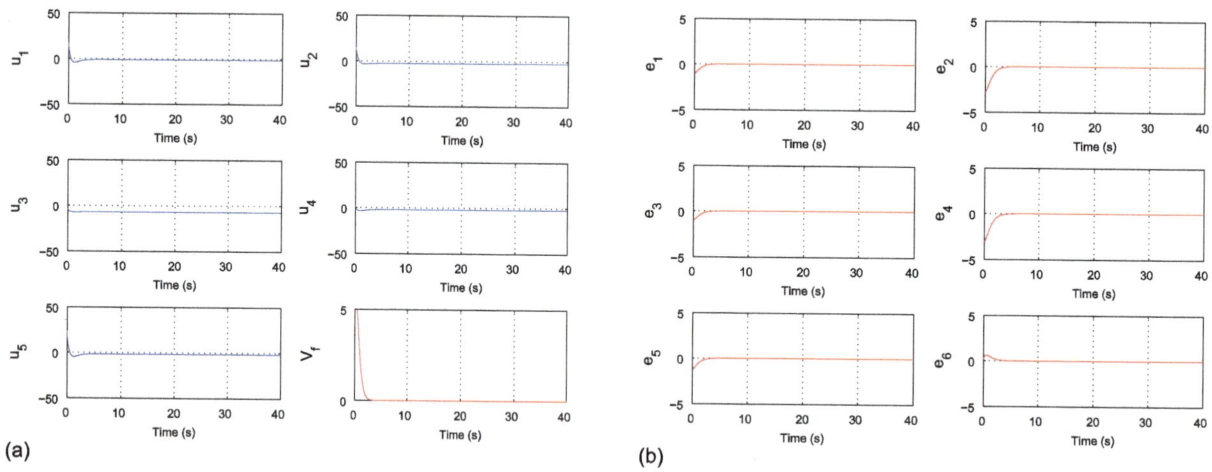

Figure 7.107 *Tracking of setpoint 1 by the 5-DOF rotor and AMB system with the use of multi-loops flatness-based control: (a) variations of the control inputs u_1 and u_5 (blue lines) and Lyapunov function of the system (red line); (b) variation of the tracking error variables e_1 to e_6 associated with the state variables x_1 to x_6*

as the wavelet transform and the Hilbert-Huang transform to extract information that reveals the existence of failures [955,107,832]. In time-domain fault diagnosis, one can distinguish model-based and model-free techniques, while fault thresholds definition is a significant issue [1037,1040,1041,1047]. In model-based fault diagnosis knowledge about the dynamics of the active magnetic bearings and the rotor is used to develop mathematical models that emulate the fault-free status of the AMB and rotor system [859,1081,797,62]. In model-free fault diagnosis input–output data from the active magnetic bearings and the rotor are used to train nonlinear regressors such as neural networks that finally emulate

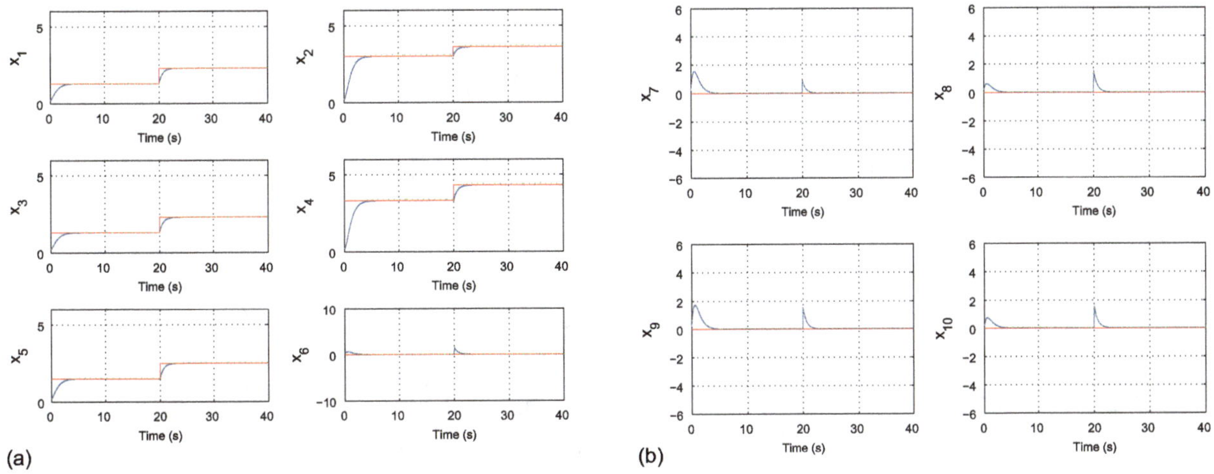

Figure 7.108 *Tracking of setpoint 2 by the 5-DOF rotor and AMB system with the use of multi-loop flatness-based control: (a) convergence of state variables x_1 to x_6 (blue lines) to the associated setpoints (red lines) and estimated values provided by Kalman filtering; (b) convergence of state variables x_7 to x_{10} (blue lines) to the associated setpoints (red lines) and estimated values provided by Kalman filtering*

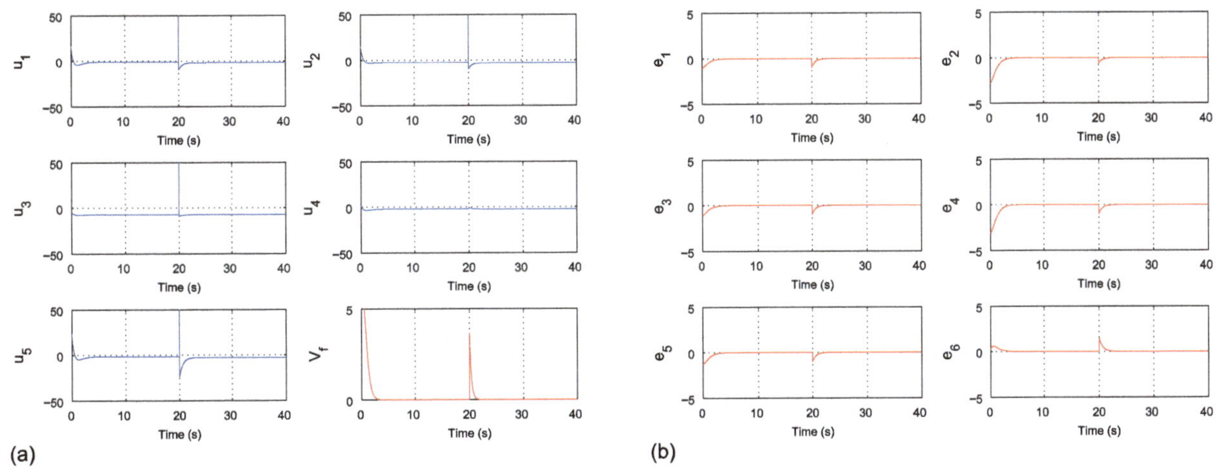

Figure 7.109 *Tracking of setpoint 2 by the 5-DOF rotor and AMB system with the use of multi-loops flatness-based control: (a) variations of the control inputs u_1 and u_5 (blue lines) and Lyapunov function of the system (red line); (b) variation of the tracking error variables e_1 to e_6 associated with the state variables x_1 to x_6*

the fault-free functioning of the AMB and rotor system [796,706]. The measured signals can be the vibrations of the AMB and rotor system or currents in their power supply circuits [922,842].

In the present section a disturbance observer is used jointly with statistical change detection criteria to perform fault detection and isolation in the 5-DOF rotor and AMB system. It is proven that the dynamic model of the 5-DOF rotor and AMB system is differentially flat and this property confirms that the system can be transformed into the input–output linearized form and into the canonical Brunovsky form [91,594,639]. The effects of model uncertainty

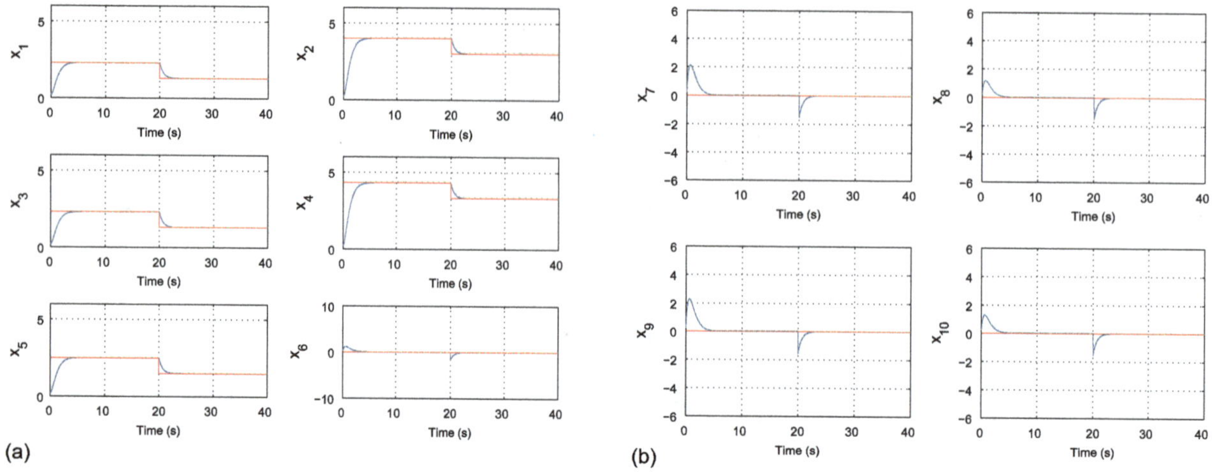

Figure 7.110 *Tracking of setpoint 3 by the 5-DOF rotor and AMB system with the use of multi-loop flatness-based control: (a) convergence of state variables x_1 to x_6 (blue lines) to the associated setpoints (red lines) and estimated values provided by Kalman filtering; (b) convergence of state variables x_7 to x_{10} (blue lines) to the associated setpoints (red lines) and estimated values provided by Kalman filtering*

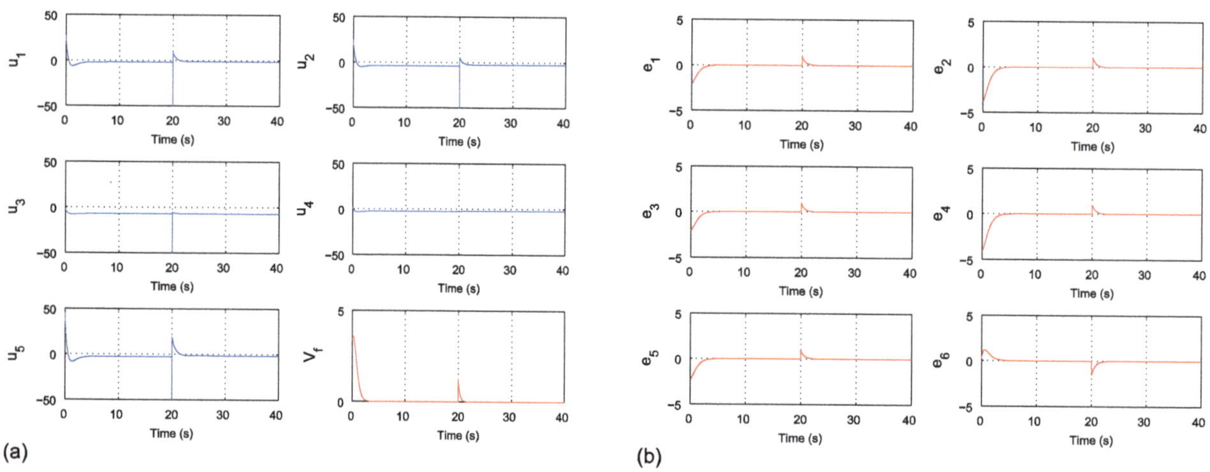

Figure 7.111 *Tracking of setpoint 3 by the 5-DOF rotor and AMB system with the use of multi-loops flatness-based control: (a) variations of the control inputs u_1 and u_5 (blue lines) and Lyapunov function of the system (red line); (b) variation of the tracking error variables e_1 to e_6 associated with the state variables x_1 to x_6*

and external perturbations are also included in this model as additive disturbance inputs. Next, by considering the additive disturbance terms as additional state variables an extended state space model in canonical form is obtained and observability properties are proven about it. For the latter state-space description a Kalman filter-based disturbance observer is designed, enabling simultaneous estimation of the non-measured state variables and of the disturbance additive inputs [711,721,732,741]. The residuals of the Kalman filter-based disturbance observer, are the differences between the measured outputs vector and the estimated outputs that the disturbance observer generates [705]. The

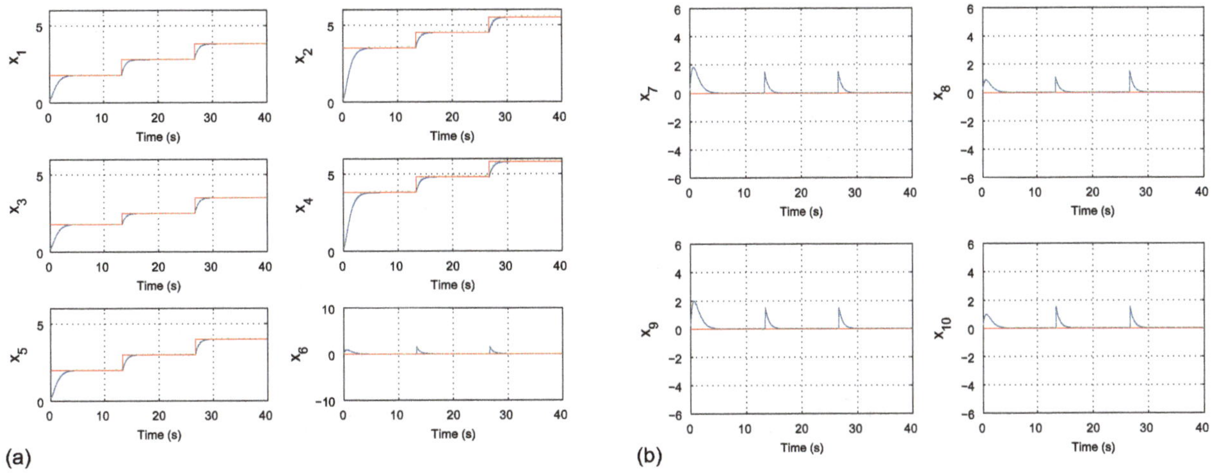

Figure 7.112 Tracking of setpoint 4 by the 5-DOF rotor and AMB system with the use of multi-loop flatness-based control: (a) convergence of state variables x_1 to x_6 (blue lines) to the associated setpoints (red lines) and estimated values provided by Kalman filtering; (b) convergence of state variables x_7 to x_{10} (blue lines) to the associated setpoints (red lines) and estimated values provided by Kalman filtering

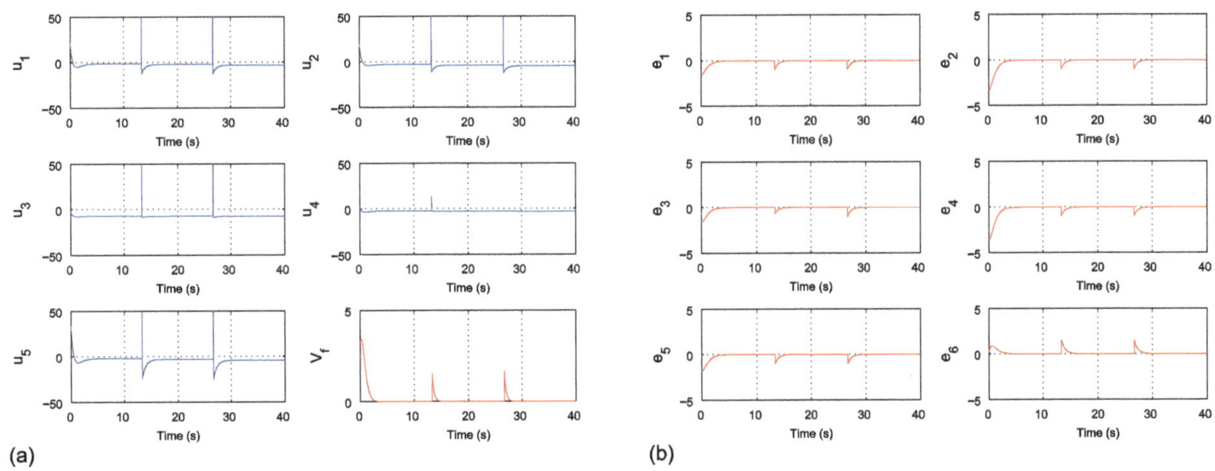

Figure 7.113 Tracking of setpoint 4 by the 5-DOF rotor and AMB system with the use of multi-loops flatness-based control: (a) variations of the control inputs u_1 and u_5 (blue lines) and Lyapunov function of the system (red line); (b) variation of the tracking error variables e_1 to e_6 associated with the state variables x_1 to x_6

statistical processing of the sequence of the residuals enables to define fault thresholds and based on them to treat the fault detection and isolation problem for the 5-DOF rotor and AMB system [60].

Actually, it holds that the sum of the squares of the residuals vector being weighted by the inverse of the associated covariance matrix forms a stochastic variable which follows the χ^2 distribution and which is also called "statistical test." It also holds that in the fault-free case the mean value of the "statistical test" should approximate closely the mean value of the χ^2 distribution with degrees of freedom which are equal to the dimension of the residuals vector. On the other

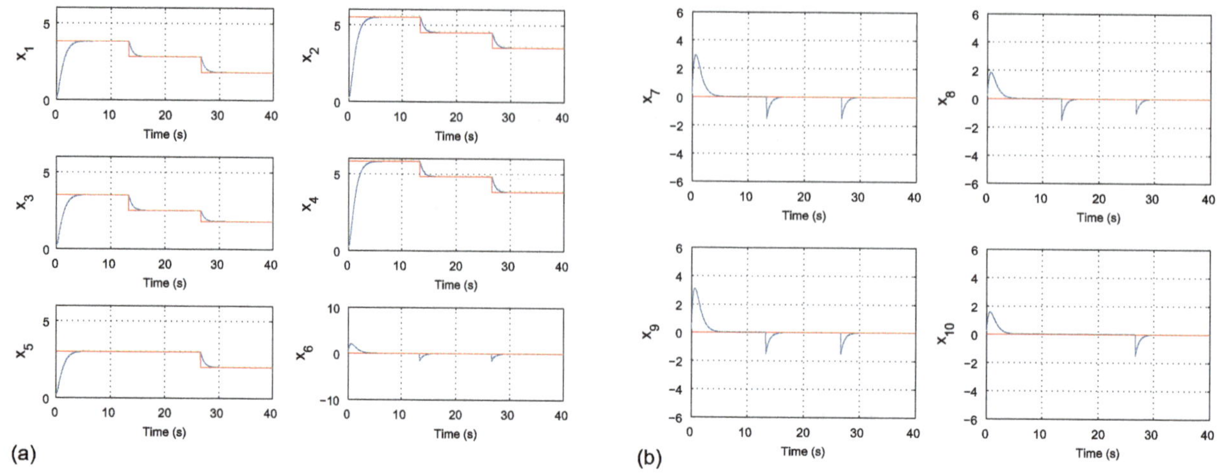

Figure 7.114 Tracking of setpoint 5 by the 5-DOF rotor and AMB system with the use of multi-loop flatness-based control: (a) convergence of state variables x_1 to x_6 (blue lines) to the associated setpoints (red lines) and estimated values provided by Kalman filtering; (b) convergence of state variables x_7 to x_{10} (blue lines) to the associated setpoints (red lines) and estimated values provided by Kalman filtering

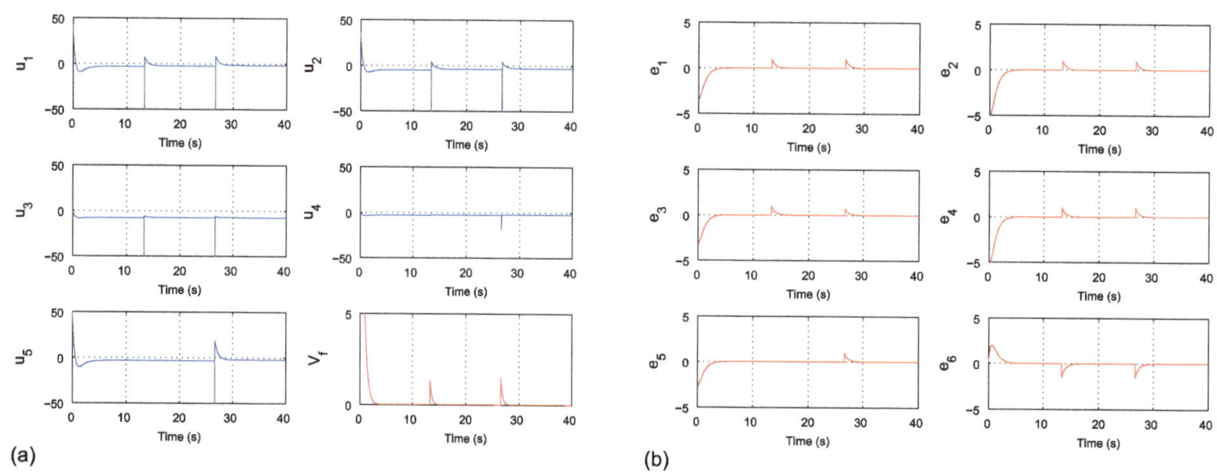

Figure 7.115 Tracking of setpoint 5 by the 5-DOF rotor and AMB system with the use of multi-loops flatness-based control: (a) variations of the control inputs u_1 and u_5 (blue lines) and Lyapunov function of the system (red line); (b) variation of the tracking error variables e_1 to e_6 associated with the state variables x_1 to x_6

hand, in the case of faults, the "statistical test" deviates from the above-noted mean value. Furthermore, the confidence intervals of the χ^2 distribution define fault thresholds which enable to conclude with a high confidence level (of the order of 95% and beyond) the occurrence of a failure. As long as the "statistical test" falls within the range defined by the previously described confidence interval the fault-free condition of the 5-DOF rotor and AMB system can be inferred. On the contrary when the "statistical tests" exceed persistently the aforementioned confidence interval, then there is almost certainty that the monitored system has been subjected to a failure. The disturbance estimates which are provided by the Kalman filter-based disturbance observer allow also for performing fault isolation.

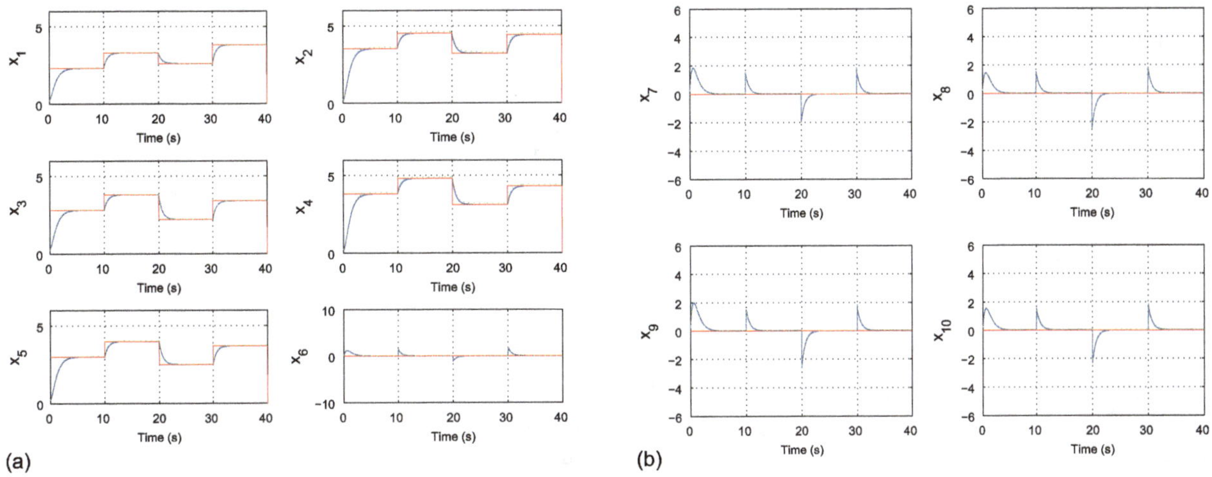

Figure 7.116 *Tracking of setpoint 6 by the 5-DOF rotor and AMB system with the use of multi-loop flatness-based control: (a) convergence of state variables x_1 to x_6 (blue lines) to the associated setpoints (red lines) and estimated values provided by Kalman filtering; (b) convergence of state variables x_7 to x_{10} (blue lines) to the associated setpoints (red lines) and estimated values provided by Kalman filtering*

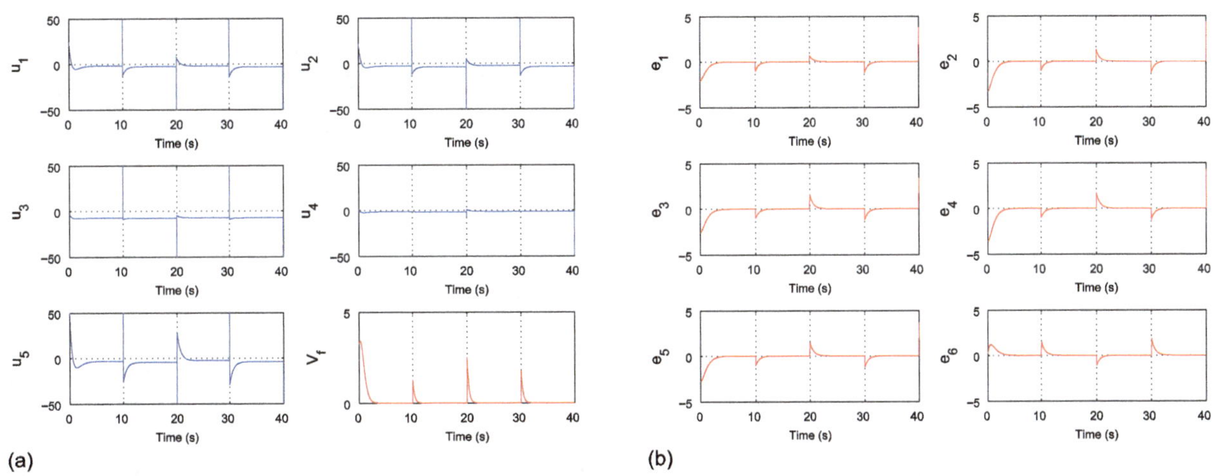

Figure 7.117 *Tracking of setpoint 6 by the 5-DOF rotor and AMB system with the use of multi-loops flatness-based control: (a) variations of the control inputs u_1 and u_5 (blue lines) and Lyapunov function of the system (red line); (b) variation of the tracking error variables e_1 to e_6 associated with the state variables x_1 to x_6*

7.5.2 Dynamic model of the 5-DOF rotor and active magnetic bearing system

As explained before, the 5-DOF rotor and active magnetic bearing system comprises a rotor and front, rear, and axial active magnetic bearings (AMBs) which enable the rotor and its spindle remain suspended and reduce significantly the friction effects in this motion transmission system. The diagram of the 5-DOF rotor and active magnetic bearing system is given in Figure 7.122.

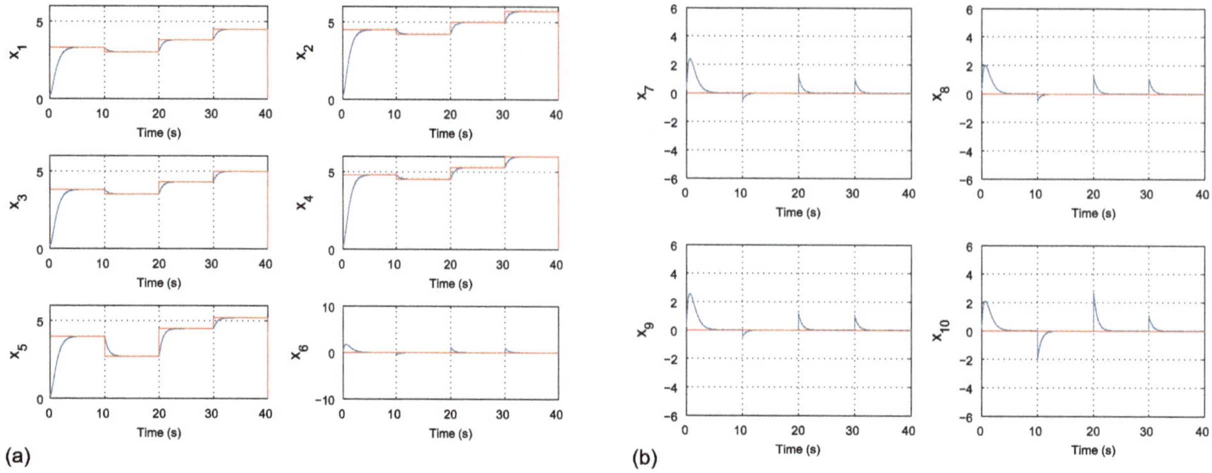

Figure 7.118 Tracking of setpoint 7 by the 5-DOF rotor and AMB system with the use of multi-loop flatness-based control: (a) convergence of state variables x_1 to x_6 (blue lines) to the associated setpoints (red lines) and estimated values provided by Kalman filtering; (b) convergence of state variables x_7 to x_{10} (blue lines) to the associated setpoints (red lines) and estimated values provided by Kalman filtering

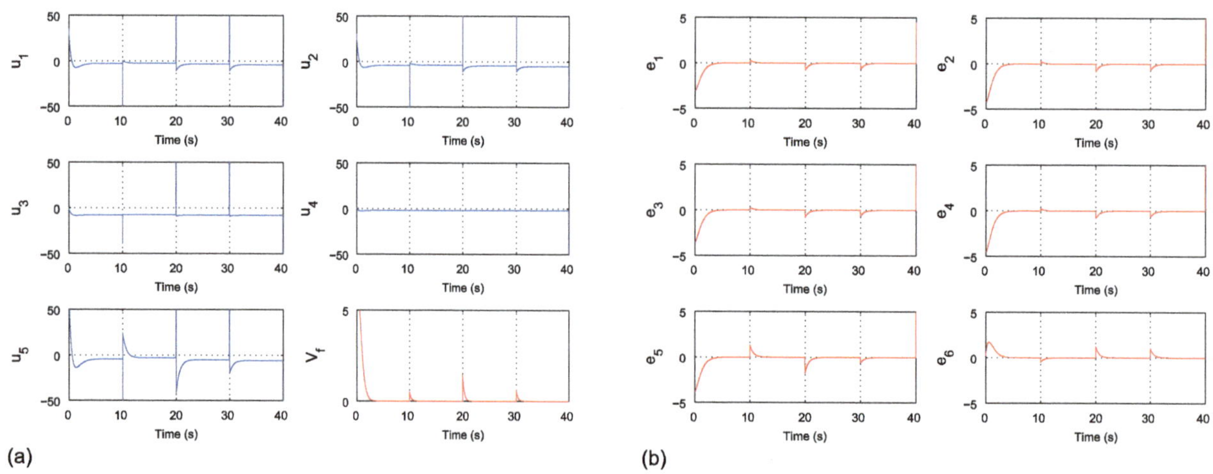

Figure 7.119 Tracking of setpoint 7 by the 5-DOF rotor and AMB system with the use of multi-loops flatness-based control: (a) variations of the control inputs u_1 and u_5 (blue lines) and Lyapunov function of the system (red line); (b) variation of the tracking error variables e_1 to e_6 associated with the state variables x_1 to x_6

The state vector of the 5-DOF rotor and active magnetic bearing system is given by [150]

$$x = [x_1, x_2, x_3, x_4, x_5, x_6, x_7, x_8, x_9, x_{10}]^T \text{ or}$$
$$x = [x_a, x_b, y_a, y_b, z_g, \dot{x}_a, \dot{x}_b, \dot{y}_a, \dot{y}_b, \dot{z}_g]^T \tag{7.390}$$

where x_a is the position of radial bearing AMB_a along the x axis, x_b is the position of radial bearing AMB_b along the x axis, y_a is the position of AMB_a along the y axis, y_b is the position of AMB_b along the y axis, and z_g is the position of the axial bearing along the z-axis. Besides, $\dot{x}_a$ is the speed of AMB_a along the x axis, $\dot{x}_b$ is the speed of AMB_b along the

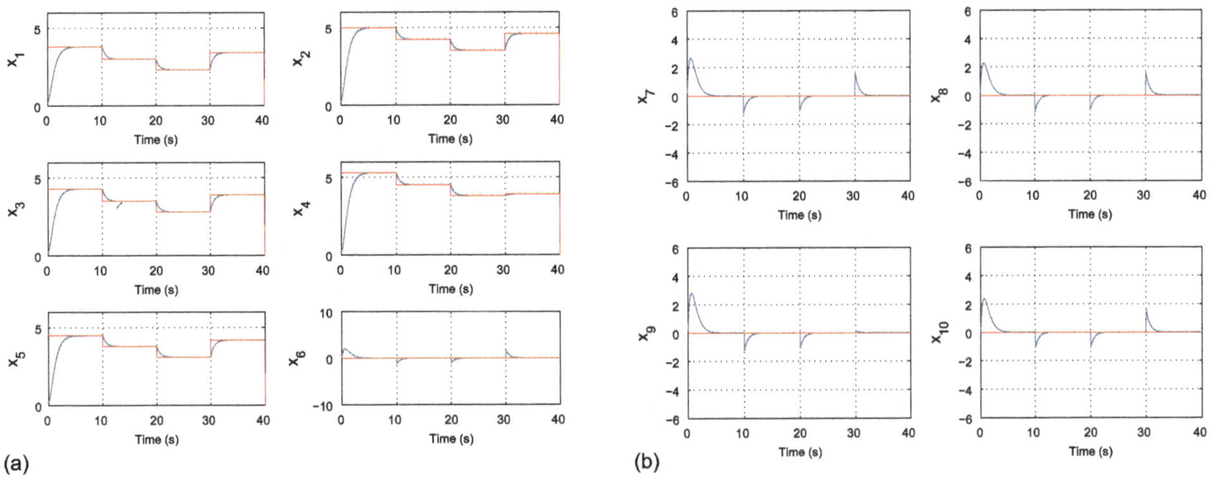

Figure 7.120 Tracking of setpoint 8 by the 5-DOF rotor and AMB system with the use of multi-loop flatness-based control: (a) convergence of state variables x_1 to x_6 (blue lines) to the associated setpoints (red lines) and estimated values provided by Kalman filtering; (b) convergence of state variables x_7 to x_{10} (blue lines) to the associated setpoints (red lines) and estimated values provided by Kalman filtering

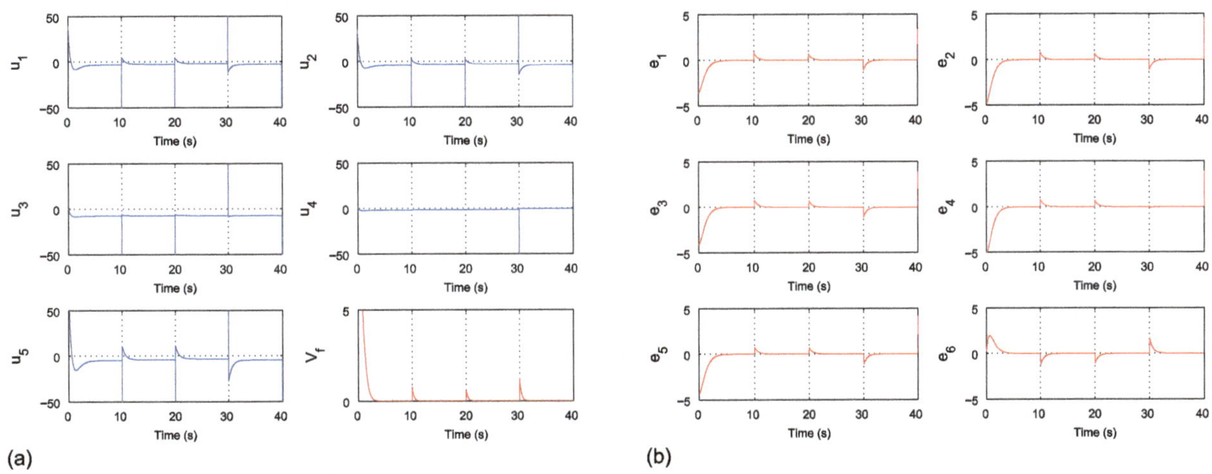

Figure 7.121 Tracking of setpoint 8 by the 5-DOF rotor and AMB system with the use of multi-loops flatness-based control: (a) variations of the control inputs u_1 and u_5 (blue lines) and Lyapunov function of the system (red line); (b) variation of the tracking error variables e_1 to e_6 associated with the state variables x_1 to x_6

x axis, $\dot{y}_a$ is the speed of AMB_a along the y axis, $\dot{y}_b$ is the speed of AMB_b along the y axis, and $\dot{z}_g$ is the speed of the axial bearing along the z-axis.

The control inputs vector is [150]

$$U = [u_1, u_2, u_3, u_4, u_5]^T \Rightarrow \\ U = [i_{ax}, i_{bx}, i_{ay}, i_{by,i_{gz}}]^T \tag{7.391}$$

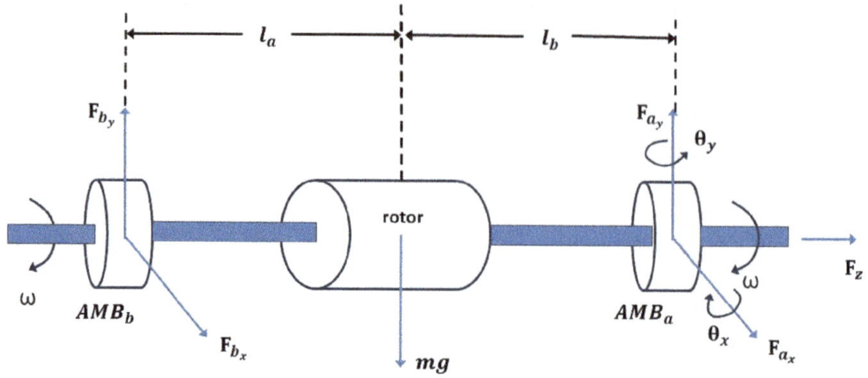

Figure 7.122 Diagram of the 5-DOF rotor and active magnetic bearing system

and consists of current variables which control the position of the front, rear, and axial magnetic bearings. The state-space model of the system is given by [150]:

$$
\begin{aligned}
\dot{x}_1 &= x_6 \\
\dot{x}_2 &= x_7 \\
\dot{x}_3 &= x_8 \\
\dot{x}_4 &= x_9 \\
\dot{x}_5 &= x_{10} \\
\dot{x}_6 &= R_a^x G_1 x_1 + R_b^x G_3 x_2 - M_a(x_8 - x_9) + R_1^a G_1 u_1 + R_1^b G_3 u_2 \\
\dot{x}_7 &= R_a^x G_3 x_1 + R_b^x G_2 x_2 + M_b(x_8 - x_9) + R_1^a G_3 u_1 + R_1^b G_2 u_2 \\
\dot{x}_8 &= R_a^x G_1 x_3 + R_b^x G_3 x_4 + M_a(x_6 - x_7) + R_1^a G_1 u_3 + R_1^b G_3 u_4 - g \\
\dot{x}_9 &= R_a^x G_3 x_3 + R_b^x G_2 x_4 - M_b(x_6 - x_7) + R_1^a G_3 u_3 + R_1^b G_3 u_4 - g \\
\dot{x}_{10} &= \tfrac{K_{2x}}{m} x_5 + \tfrac{K_{zl}}{m} u_5
\end{aligned}
\tag{7.392}
$$

where $R_a^x = \frac{K_x^a}{m}$, $R_b^x = \frac{K_x^b}{m}$, $R_i^a = \frac{K_i^a}{m}$, $R_i^b = \frac{K_i^b}{m}$, $G_1 = 1 + \frac{ml_a^2}{J_{xy}}$, $G_2 = 1 + \frac{ml_b^2}{J_{xy}}$, $G_3 = 1 - \frac{ml_a l_b}{J_{xy}}$, $M_a = \omega\frac{J_z l_a}{J_{xy} l}$, $M_b = \omega\frac{J_z l_b}{J_{xy} l}$.

Furthermore, K_x^a, K_x^b are the displacement stiffness coefficients, and K_i^a, K_i^b are the current stiffness coefficients for the front and rear radial bearings. Additionally, K_x^z, K_i^z are the displacement stiffness and the current stiffness coefficients of the axial bearing.

J_{xy} is the moment of inertia for rotation around the x or y axis. J_z is the moment of inertia for rotation around the z axis. Besides, l_a, l_b are distances measured from the front and rear bearing up to the center of the rotor. Their sum is defined as $l = l_a + l_b$. The mass of the rotor is defined as m. The angular speed of the rotor and the spindle is defined as ω.

The dynamic model of the 5-DOF rotor and active magnetic bearing system can be written in the following matrix form:

$$\dot{x} = f(x) + g(x)u \tag{7.393}$$

where $x \in R^{10\times1}$, $f(x) \in R^{10\times1}$, $g(x) \in R^{10\times5}$, and $u \in R^{5\times1}$. Analytically, the matrix-form representation of the 5-DOF rotor and active magnetic bearings system is given by

$$\begin{pmatrix} \dot{x}_1 \\ \dot{x}_2 \\ \dot{x}_3 \\ \dot{x}_4 \\ \dot{x}_5 \\ \dot{x}_6 \\ \dot{x}_7 \\ \dot{x}_8 \\ \dot{x}_9 \\ \dot{x}_{10} \end{pmatrix} = \begin{pmatrix} x_6 \\ x_7 \\ x_8 \\ x_9 \\ x_{10} \\ R_a^x G_1 x_1 + R_b^x G_3 x_2 - M_a(x_8 - x_9) \\ R_a^x G_3 x_1 + R_b^x G_2 x_2 + M_b(x_8 - x_9) \\ R_a^x G_1 x_3 + R_b^x G_3 x_4 + M_a(x_6 - x_7) - g \\ R_a^x G_3 x_3 + R_b^x G_2 x_4 - M_b(x_6 - x_7) - g \\ \frac{K_{2x}}{m} x_5 \end{pmatrix} + \begin{pmatrix} 0 & 0 & 0 & 0 & 0 \\ 0 & 0 & 0 & 0 & 0 \\ 0 & 0 & 0 & 0 & 0 \\ 0 & 0 & 0 & 0 & 0 \\ 0 & 0 & 0 & 0 & 0 \\ R_1^a G_1 & R_1^b G_3 & 0 & 0 & 0 \\ R_1^a G_3 & R_1^b G_2 & 0 & 0 & 0 \\ 0 & 0 & 1 & 0 & 0 \\ 0 & 0 & 0 & 1 & 0 \\ 0 & 0 & 0 & 0 & \frac{K_{zi}}{m} \end{pmatrix} \begin{pmatrix} u_! \\ u_2 \\ u_3 \\ u_4 \\ u_5 \end{pmatrix} \tag{7.394}$$

As shown before the dynamic model of the 5-DOF rotor and AMB system is differentially flat, with flat outputs vector $Y = [x_1, x_2, x_3, x_4, x_5]^T$.

7.5.3 Disturbance observers for the 5-DOF rotor and AMB system

The dynamic model of the 5-DOF rotor and AMB system given in Eq. (7.393) is re-considered. When this model is subjected to disturbances it can be also written in the following form:

$$\begin{aligned} \ddot{x}_1 &= f_6(x) + g_{6a}(x)u_1 + g_{6b}(x)u_2 + d_1 \\ \ddot{x}_2 &= f_7(x) + g_{7a}(x)u_1 + g_{7b}(x)u_2 + d_2 \\ \ddot{x}_3 &= f_8(x) + g_{8a}(x)u_3 + g_{8b}(x)u_4 + d_3 \\ \ddot{x}_4 &= f_9(x) + g_{9a}(x)u_3 + g_{9b}(x)u_4 + d_4 \\ \ddot{x}_5 &= f_{10}(x) + g_{10}(x)u_5 + d_5 \end{aligned} \tag{7.395}$$

where $d_i, \ i = 1, 2, \cdots, 5$ are unknown disturbance terms which have to be dynamically identified. The disturbance terms d_i can be equivalently described by their (n-th) order time-derivative and the associated initial conditions. In case of the second-order derivative, that is when $n = 2$ the equivalent description of the disturbances' dynamics is

$$\ddot{d}_i = f_{d_i}(t), \ \text{with } d_i(0), \ \dot{d}_i(0) \tag{7.396}$$

The state vector of the system is extended by including as additional state variables the disturbance terms and their time derivatives. The new state vector comprises the following state variables: $z_1 = x_1$, $z_2 = \dot{x}_1$, $z_3 = x_2$, $z_4 = \dot{x}_2$, $z_5 = x_3$, $z_6 = \dot{x}_3$, $z_7 = x_4$, $z_8 = \dot{x}_4$, $z_9 = x_5$, $z_{10} = \dot{x}_5$, $z_{11} = d_1$, $z_{12} = \dot{d}_1$, $z_{13} = d_2$, $z_{14} = \dot{d}_2$, $z_{15} = d_3$, $z_{16} = \dot{d}_3$, $z_{17} = d_6$, $z_{18} = \dot{d}_4$, $z_{19} = d_5$, $z_{20} = \dot{d}_5$.

Thus, the extended state-space model of the 5-DOF rotor and AMB system becomes

$$
\begin{aligned}
\dot{z}_1 &= z_2 & \dot{z}_{11} &= z_{12} \\
\dot{z}_2 &= f_6(z) + g_{6a}(z)u_1 + g_{6b}(z)u_2 + z_{11} & \dot{z}_{12} &= f_{d_1} \\
\dot{z}_3 &= z_4 & \dot{z}_{13} &= z_{14} \\
\dot{z}_4 &= f_7(z) + g_{7a}(z) + g_{7b}(z)u_2 + z_{13} & \dot{z}_{14} &= f_{d_2} \\
\dot{z}_5 &= z_6 & \dot{z}_{15} &= z_{16} \\
\dot{z}_6 &= f_8(z) + g_{8a}(z)u_3 + g_{8b}(z)u_4 + z_{15} & \dot{z}_{16} &= f_{d_3} \\
\dot{z}_7 &= z_8 & \dot{z}_{17} &= z_{18} \\
\dot{z}_8 &= f_9(z) + g_{9a}(z)u_3 + g_{9b}(z)u_4 + z_{17} & \dot{z}_{18} &= f_{d_4} \\
\dot{z}_9 &= z_{10} & \dot{z}_{19} &= z_{20} \\
\dot{z}_{10} &= f_{10}(z) + g_{10}(z)u_5 + z_{19} & \dot{z}_{20} &= f_{d_5}
\end{aligned}
\tag{7.397}
$$

Besides, the following cumulative control inputs are defined:

$$
\begin{aligned}
\tilde{v}_1 &= f_6(z) + g_{6a}(z)u_1 + g_{6b}(z)u_2 \\
\tilde{v}_2 &= f_7(z) + g_{7a}(z)u_1 + g_{7b}(z)u_2 \\
\tilde{v}_3 &= f_8(z) + g_{8a}(z)u_3 + g_{8b}(z)u_4 \\
\tilde{v}_4 &= f_9(z) + g_{9a}(z)u_3 + g_{9b}(z)u_4 \\
\tilde{v}_5 &= f_{10}(z) + g_{10}(z)u_5
\end{aligned}
\tag{7.398}
$$

The extended state-space model of the 5-DOF rotor and AMB system, which was previously given in Eq. (7.397) is now written as

$$
\begin{aligned}
\dot{z}_1 &= z_2 & \dot{z}_{11} &= z_{12} \\
\dot{z}_2 &= \tilde{v}_1 + z_{11} & \dot{z}_{12} &= f_{d_1} \\
\dot{z}_3 &= z_4 & \dot{z}_{13} &= z_{14} \\
\dot{z}_4 &= \tilde{v}_2 + z_{13} & \dot{z}_{14} &= f_{d_2} \\
\dot{z}_5 &= z_6 & \dot{z}_{15} &= z_{16} \\
\dot{z}_6 &= \tilde{v}_3 + z_{15} & \dot{z}_{16} &= f_{d_3} \\
\dot{z}_7 &= z_8 & \dot{z}_{17} &= z_{18} \\
\dot{z}_8 &= \tilde{v}_4 + z_{17} & \dot{z}_{18} &= f_{d_4} \\
\dot{z}_9 &= z_{10} & \dot{z}_{19} &= z_{20} \\
\dot{z}_{10} &= \tilde{v}_5 + z_{19} & \dot{z}_{20} &= f_{d_5}
\end{aligned}
\tag{7.399}
$$

Additionally, the extended state-space model of the system can be written in the following matrix form:

$$\begin{pmatrix} \dot{z}_1 \\ \dot{z}_2 \\ \dot{z}_3 \\ \dot{z}_4 \\ \dot{z}_5 \\ \dot{z}_6 \\ \dot{z}_7 \\ \dot{z}_8 \\ \dot{z}_9 \\ \dot{z}_{10} \\ \dot{z}_{11} \\ \dot{z}_{12} \\ \dot{z}_{13} \\ \dot{z}_{14} \\ \dot{z}_{15} \\ \dot{z}_{16} \\ \dot{z}_{17} \\ \dot{z}_{18} \\ \dot{z}_{19} \\ \dot{z}_{20} \end{pmatrix} = \begin{pmatrix} 0 & 1 & 0_{1\times 18} \\ 0_{1\times 10} & 1 & 0_{1\times 9} \\ 0_{1\times 3} & 1 & 0_{1\times 16} \\ 0_{1\times 12} & 1 & 0_{1\times 7} \\ 0_{1\times 5} & 1 & 0_{1\times 14} \\ 0_{1\times 14} & 1 & 0_{1\times 5} \\ 0_{1\times 7} & 1 & 0_{1\times 12} \\ 0_{1\times 16} & 1 & 0_{1\times 3} \\ 0_{1\times 9} & 1 & 0_{1\times 10} \\ 0_{1\times 18} & 1 & 0_{1\times 1} \\ 0_{1\times 11} & 1 & 0_{1\times 8} \\ & 0_{1\times 20} & \\ 0_{1\times 13} & 1 & 0_{1\times 6} \\ & 0_{1\times 20} & \\ 0_{1\times 15} & 1 & 0_{1\times 4} \\ & 0_{1\times 20} & \\ 0_{1\times 17} & 1 & 0_{1\times 2} \\ & 0_{1\times 20} & \\ 0_{1\times 18} & 0 & 1 \\ & 0_{1\times 20} & \end{pmatrix} \begin{pmatrix} z_1 \\ z_2 \\ z_3 \\ z_4 \\ z_5 \\ z_6 \\ z_7 \\ z_8 \\ z_9 \\ z_{10} \\ z_{11} \\ z_{12} \\ z_{13} \\ z_{14} \\ z_{15} \\ z_{16} \\ z_{17} \\ z_{18} \\ z_{19} \\ z_{20} \end{pmatrix} + \begin{pmatrix} 0&0&0&0&0&0&0&0&0&0 \\ 1&0&0&0&0&0&0&0&0&0 \\ 0&0&0&0&0&0&0&0&0&0 \\ 0&1&0&0&0&0&0&0&0&0 \\ 0&0&0&0&0&0&0&0&0&0 \\ 0&0&1&0&0&0&0&0&0&0 \\ 0&0&0&0&0&0&0&0&0&0 \\ 0&0&0&1&0&0&0&0&0&0 \\ 0&0&0&0&0&0&0&0&0&0 \\ 0&0&0&0&1&0&0&0&0&0 \\ 0&0&0&0&0&0&0&0&0&0 \\ 0&0&0&0&0&1&0&0&0&0 \\ 0&0&0&0&0&0&0&0&0&0 \\ 0&0&0&0&0&0&1&0&0&0 \\ 0&0&0&0&0&0&0&0&0&0 \\ 0&0&0&0&0&0&0&1&0&0 \\ 0&0&0&0&0&0&0&0&0&0 \\ 0&0&0&0&0&0&0&0&1&0 \\ 0&0&0&0&0&0&0&0&0&0 \\ 0&0&0&0&0&0&0&0&0&1 \end{pmatrix} \begin{pmatrix} \tilde{v}_1 \\ \tilde{v}_2 \\ \tilde{v}_3 \\ \tilde{v}_4 \\ \tilde{v}_5 \\ f_{d_1} \\ f_{d_2} \\ f_{d_3} \\ f_{d_4} \\ f_{d_5} \end{pmatrix} \tag{7.400}$$

The associated measurement equation is

$$\begin{pmatrix} z_1^m \\ z_3^m \\ z_5^m \\ z_7^m \\ z_9^m \end{pmatrix} = \begin{pmatrix} 1&0&0&0&0&0&0&0&0&0_{1\times 11} \\ 0&0&1&0&0&0&0&0&0&0_{1\times 11} \\ 0&0&0&0&1&0&0&0&0&0_{1\times 11} \\ 0&0&0&0&0&0&1&0&0&0_{1\times 11} \\ 0&0&0&0&0&0&0&0&1&0_{1\times 11} \end{pmatrix} \begin{pmatrix} z_1 \\ z_2 \\ z_3 \\ z_4 \\ z_5 \\ z_6 \\ z_7 \\ z_8 \\ z_9 \\ [z_{10} \cdots z_{20}]^T \end{pmatrix} \tag{7.401}$$

The concise state-space description of the 5-DOF rotor and extended AMB system is

$$\begin{aligned} \dot{z} &= A_e z + B_e \tilde{v}_e \\ z_m &= C_e z \end{aligned} \tag{7.402}$$

where $z \in R^{20\times 1}$, $A_e \in R^{20\times 20}$, $B_e \in R^{20\times 10}$, $C_e \in R^{5\times 20}$ and $v_e \in R^{10\times 1}$. The associated state estimation problem for the extended state-space model can be solved with the design of a state estimator (disturbance observer) in the form:

$$\begin{aligned} \dot{\hat{z}} &= A_o \hat{z} + B_o \tilde{v}_o + K_f (z_m - \hat{z}_m) \\ \hat{z}_m &= C_o \hat{z} \end{aligned} \tag{7.403}$$

where matrix $A_o = A_e$ coincides with matrix A_e, while matrix B_o consists of the first five columns of matrix B and matrix C_o coincides with matrix C_e. The observer's gain matrix K_f can be computed sequentially using the Kalman filter recursion [711,705,60].

The input vector of the disturbance observer $\tilde{v}_o$ comprises the first five rows of the control inputs vector $\tilde{v}_e$ of the extended state-space model of the system, that is $\tilde{v}_o = [u_1, u_2, v_3, v_4, u_5]^T$. The control inputs f_{d_i} $i = 1, 2, \cdots, 5$ of the extended state-space model are unknown and are not used by the disturbance observer. Besides, the initial values of the disturbance terms $d_i,\ i=1,2,\cdots,5$ are also unknown and are not used by the disturbance observer either.

The state-space model which is considered by the disturbance observer in the joint state variables and disturbances estimation problem is defined by matrices A_o, B_o, and C_o and is observable. To implement the disturbance observer, matrices A_o, B_o, C_o undergo first discretization with the use of common discretization methods, for instance, the Tustin or the exact method. The discrete-time equivalents of these matrices are denoted as $A_{o,d}$, $B_{o,d}$, and $C_{o,d}$, respectively.

The recursion of the Kalman filter which computes the observer's gain K_f consists of the measurement-update part (after receiving the measurements of the k-th sampling instance) and of the time-update part (prior to receiving the measurements of the (k+1)-th sampling instance). These are outlined below

measurement update:

$$\begin{aligned} K_f(k) &= P^-(k)C_{o,d}^T[C_{o,d}P^-(k)C_{o,d}^T+R(k)]^{-1} \\ \hat{x}(k) &= \hat{x}^-(k)+K_f(k)[z_m(k)-\hat{z}_m(k)] \\ P(k) &= P^-(k)-K_f(k)C_{o.d}P^-(k) \end{aligned} \tag{7.404}$$

time update:

$$\begin{aligned} &P^-(k+1)A_{o,d}P(k)A_{o,d}^T+Q(k) \\ &\hat{x}^-(k+1)=A_{o,d}\hat{x}(k)+B_{o,d}\tilde{v}_o(k) \end{aligned} \tag{7.405}$$

The process and measurement noises are considered to be white Gaussian signals. $P^-(k)$ is the state estimation error covariance matrix before receiving measurements at the k-th sampling instance, $P(k)$ is the state estimation error covariance matrix after having received measurements at the k-th sampling instance, $P^-(k+1)$ is the state estimation error covariance matrix before receiving measurements at the $(k+1)$-th sampling instance, $Q(k)$ is the process noise covariance matrix, and $R(k)$ is the measurement noise covariance matrix.

Unlike the Extended Kalman filter (EKF), the previously defined procedure for estimating state and disturbance variables in the dynamic model of the 5-DOF rotor and AMB system does not depend on linearization of the system's model with the computation of Jacobian matrices and the associated partial derivatives. For this reason, it is also known as *Derivative-free nonlinear Kalman filter* (DKF) [711,741,742].

7.5.4 Statistical criteria for fault detection and isolation

7.5.4.1 Fault detection

The residuals' sequence, that is the differences between (i) the zero-valued disturbance inputs $d_i\ i=1,\cdots,5$ of the 5-DOF rotor and AMB system in the fault-free case and (ii) the estimated values of the disturbance inputs $\hat{d}_i\ i=1,\cdots,5$ which are provided by the Kalman filter-based disturbance observer (Figure 7.139) is a discrete error process e_k with dimension $m\times 1$ (here $m=N$ is the dimension of the output measurements vector). Actually, it is a zero-mean Gaussian white-noise process with covariance given by E_k.

A conclusion can be stated based on a measure of certainty that the 5-DOF rotor and AMB system has not been subjected to a fault or parametric change (being included in the cumulative signals of additive disturbance inputs). To this end, the following *normalized error square* (NES) is defined [706,711]

$$\epsilon_k = e_k^T E_k^{-1} e_k \tag{7.406}$$

The sum of this normalized residuals' square follows a χ^2 distribution, with a number of degrees of freedom that is equal to the dimension of the residuals' vector. The form of the χ^2 distribution for various degrees of freedom is shown in Figure 7.124. An appropriate test for the normalized error sum is to numerically show that the following condition is met within a level of confidence (according to the properties of the χ^2 distribution)

$$E\{\epsilon_k\} = m \tag{7.407}$$

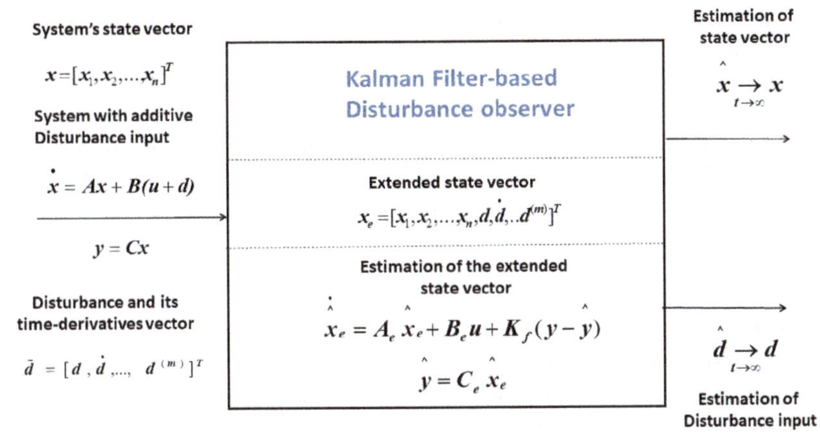

Figure 7.123 Residuals' generation for the 5-DOF rotor and AMB system, with a Kalman filter-based disturbance estimator and through the processing of the estimated disturbance inputs

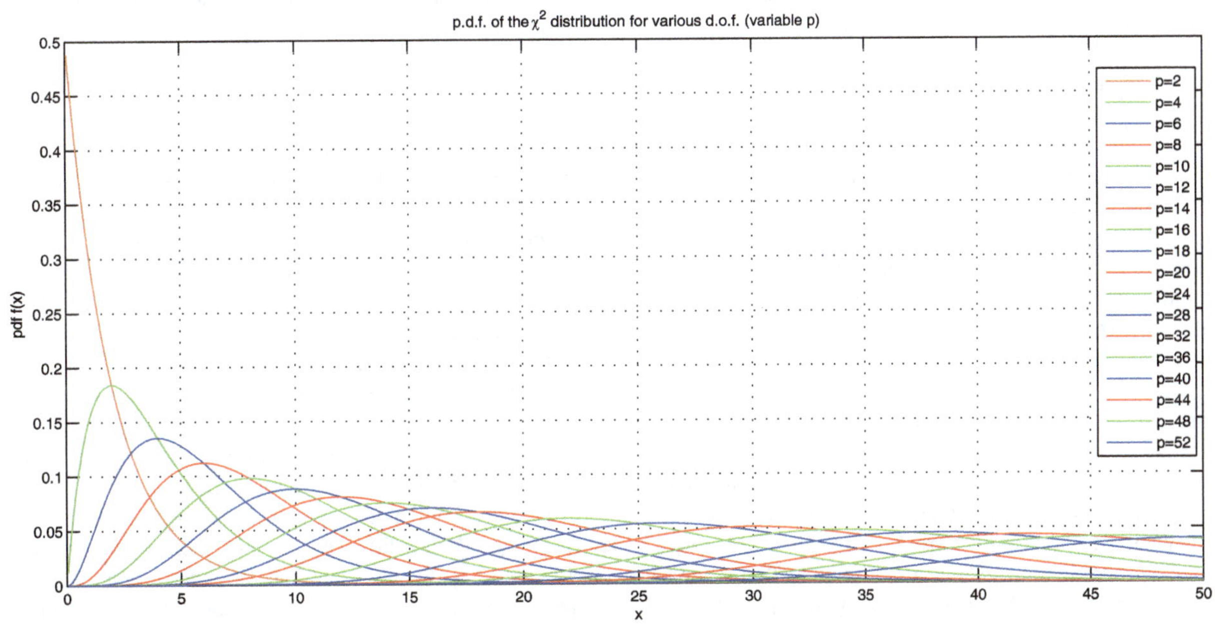

Figure 7.124 Probability density function of the χ^2 distribution for several values of the degrees of freedom (variable p)

This can be achieved using statistical hypothesis tests, which is associated with confidence intervals. For instance, a 95% confidence interval is frequently applied, which is specified using $100(1-a)$ with $a=0.05$. Actually, a two-sided probability region is considered cutting-off two end tails of 2.5% each. For M runs the normalized error square that is obtained is given by

$$\bar{\epsilon}_k = \frac{1}{M}\sum_{i=1}^{M}\epsilon_k(i) = \frac{1}{M}\sum_{i=1}^{M}e_k^T(i)E_k^{-1}(i)e_k(i) \tag{7.408}$$

where ϵ_i stands for the i-th run at time t_k. Then $M\bar{\epsilon}_k$ will follow a χ^2 density with Mm degrees of freedom. This condition can be checked using a χ^2 test. The hypothesis holds if the following condition is satisfied

$$\bar{\epsilon}_k \in [\zeta_1, \zeta_2] \tag{7.409}$$

where ζ_1 and ζ_2 are derived from the tail probabilities of the χ^2 density. For example, for $m = 20$ (dimension of the measurements vector) and $M = 100$ (total number of the output vector's samples), one has $\chi^2_{Mm}(0.025) = 1878$ and $\chi^2_{Mm}(0.975) = 2126$. Using that $M = 100$, one obtains $\zeta_1 = \chi^2_{Mm}(0.025)/M = 18.78$ and $\zeta_2 = \chi^2_{Mm}(0.975)/M = 21.26$.

7.5.4.2 Fault isolation

By applying the statistical test in the individual components of the 5-DOF rotor and AMB system, it is also possible to find out the specific component that has been subjected to a fault [711,741]. For a 5-DOF rotor and AMB system of n parameters suspected for change, one has to carry out n χ^2 statistical change detection tests, where each test is applied to the subset that comprises parameters $i-1$, i, and $i+1$, $i = 1, 2, \cdots, n$. Actually, out of the n χ^2 statistical change detection tests, the one that exhibits the highest score are those that identify the parameter that has been subject to change.

In the case of multiple faults, one can identify the subset of parameters that has been subjected to change by applying the χ^2 statistical change detection test according to a combinatorial sequence. This means:

$$\binom{n}{k} = \frac{n}{k!(n-k)!} \tag{7.410}$$

tests have to take place, for all clusters in the 5-DOF rotor and AMB system, that finally comprise $n, n-1, n-2, \cdots$, 2, 1 parameters. Again the χ^2 tests that give the highest scores indicate the parameters which are most likely to have been subjected to change.

As a whole, the concept of the proposed statistical fault detection isolation test is simple. The residuals' vector is formed by taking at each sampling instance the differences between the estimated values of the additive disturbance inputs $\hat{d}_i$, $i = 1, 2, \cdots, 5$ and the zero-valued vector $d_i = 0$, $i = 1, 2, \cdots, 5$ that is the values of the disturbance inputs in the fault-free case. The sum of the squares of the residuals' vector, weighted by the inverse of the residuals' covariance matrix, stands for a stochastic variable which follows the χ^2 distribution. Actually, this is a multi-dimensional χ^2 distribution and the number of its degrees of freedom is equal to the dimension of the residuals' vector. Since there are five estimated disturbance inputs at the 5-DOF rotor AMB system, the residuals' vector is of dimension 5 and the number of degrees of freedom is also 5. Next, from the properties of the χ^2 distribution, the mean value of the aforementioned stochastic variable in the fault-free case should be also 5. However, due to small fluctuations in the estimated variables, the value of the statistical test in the fault-free case will not be precisely equal to 5 but it may vary within a small range around this value. This range is determined by the confidence intervals of the χ^2 distribution. For a probability of 98% to get a value of the stochastic variable about 5, the associated confidence interval is given by the lower bound $L = 4.84$ and by the upper bound $U = 5.16$. Consequently, as long as the statistical test provides an indication that the aforementioned stochastic variable is in the interval $[L, U]$, the functioning of the power unit can be concluded to be free of faults. On the other side, when the bounds of the previously given confidence interval are exceeded it can be concluded that the power unit has been subject to a fault. Finally, by performing the statistical test to subspaces of the 5-DOF rotor and AMB's state-space model, where each subspace is associated with different state subvectors, one can also achieve fault isolation. This signifies that the part of the system that has caused the malfunctioning can be identified.

7.5.5 Simulation tests

First, results about flatness-based control with transformation of the model of the 5-DOF rotor and AMB system in the canonical (Brunovsky) form of Eq. (7.399), are given [711]. These are presented in Figure 7.125–7.127. The real values of the disturbance signals are depicted in blue color while their estimates, which have been provided by the Kalman

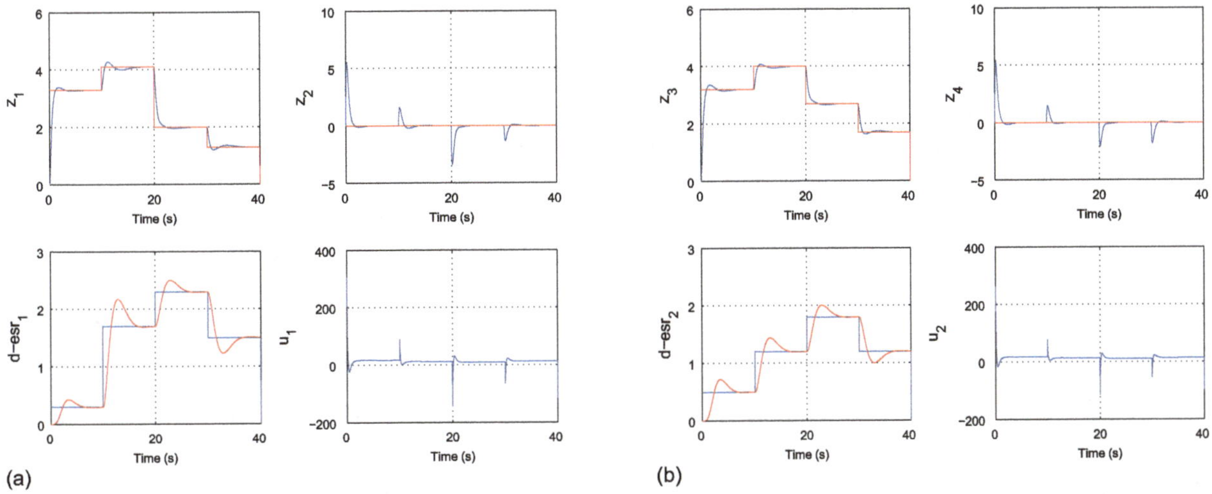

Figure 7.125 *Control of the 5-DOF rotor and AMB system with the use of the Kalman filter-based disturbance observer: (a) top row: convergence of state variables z_1 and z_2 (blue line) to setpoints (red line), bottom row: estimation of the additive disturbance term $\hat{d}_1$ (red line) vs real disturbance value (blue line) and control input u_1 (blue line); (b) top row: convergence of state variables z_3 and z_4 (blue line) to setpoints (red line), bottom row: estimation of the additive disturbance term $\hat{d}_2$ (red line) vs real disturbance value (blue line) and control input u_2 (blue line)*

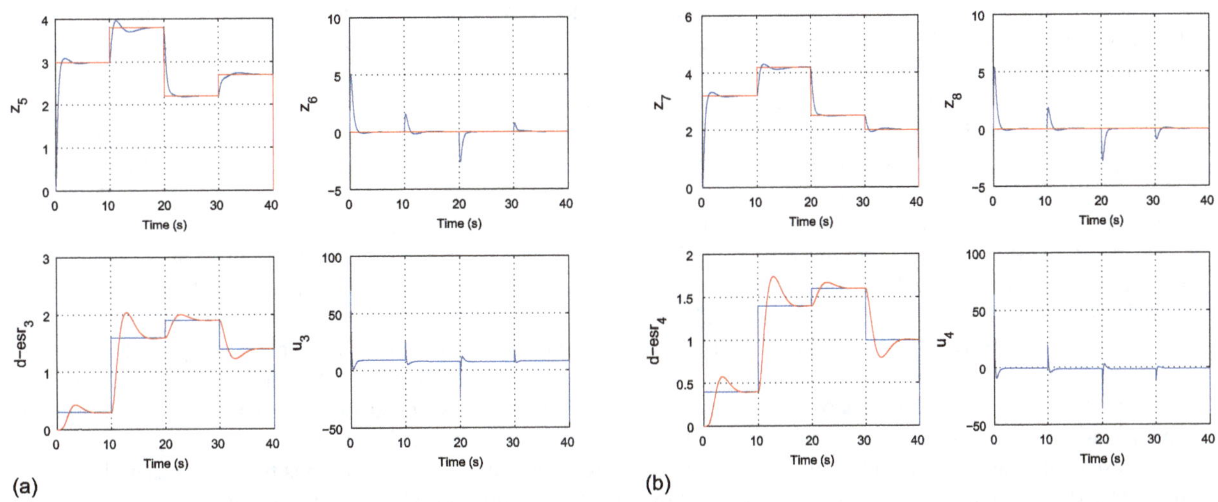

Figure 7.126 *Control of the 5-DOF rotor and AMB system with the use of the Kalman filter-based disturbance observer: (a) top row: convergence of state variables z_5 and z_6 (blue line) to the setpoints (red line), bottom row: estimation of the additive disturbance term $\hat{d}_3$ (red line) vs real disturbance value (blue line) and control input u_3 (blue line); (b) top row: convergence of state variables z_7 and z_8 (blue line) to the setpoints (red line), bottom row: estimation of the additive disturbance term $\hat{d}_4$ (red line) vs real disturbance value (blue line) and control input u_4 (blue line)*

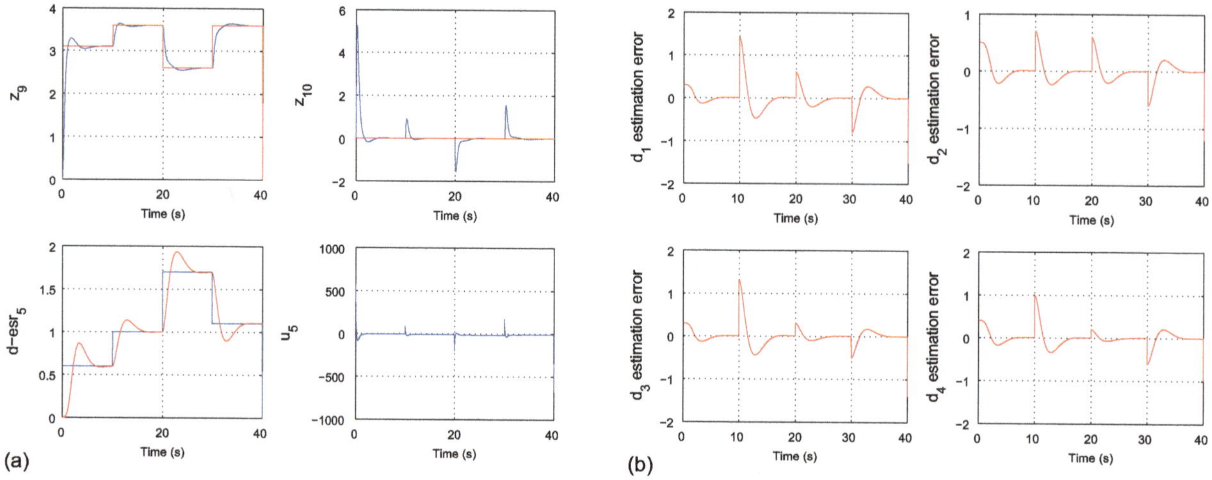

Figure 7.127 Control of the 5-DOF rotor and AMB system with the use of the Kalman filter-based disturbance observer: (a) top row: convergence of state variables z_9 and z_{10} (blue line) to the associated setpoint (red line), bottom row: estimation of the additive disturbance term $\hat{d}_5$ (red line) vs real disturbance value (blue line) and control input u_5 (blue line); (b) top row: convergence to zero of the estimation error for disturbance terms $\hat{d}_1$ (red line) and $\hat{d}_2$ (red line), bottom row: convergence to zero of the estimation error for disturbance terms $\hat{d}_3$ (red line) and $\hat{d}_4$ (red line)

filter, are shown in red color. The simulation tests demonstrate that the disturbance observer allows for simultaneous estimation of the non-measurable state variables and of the additive disturbance inputs that were affecting the 5-DOF rotor and AMB system. Although additive disturbance faults d_1 to d_5 affect the dynamics of the system, these cannot be distinguished in the system's outputs z_1, z_3, z_5, z_7, and z_9. The robustness of the control loop "hides" the faults and enables the system's outputs reach the targeted reference values,

Next, the statistical fault diagnosis test is performed in the disturbance inputs d_1 to d_5. Considering $n = 5$ disturbance inputs for the 5-DOF rotor and AMB system, to conclude the normal functioning of the system, the previously analyzed statistical test should take a value that is very close to the mean value of χ^2 distribution, that is 5. The residuals' vector is formed by taking at each sampling instance the differences between the estimated values of the additive disturbance inputs $[\hat{d}_1, \cdots, \hat{d}_5]$ and the zero-valued vector $[d_1 = 0, \cdots, d_5 = 0]$ that is the values of the disturbance inputs in the fault-free case. Thus, the dimension of the residuals' vector is $m = 5$. Considering that the number of residual vector samples is $M = 2000$ and using a 98% confidence interval for the χ^2 distribution, the fault thresholds can be given the values $L = 4.84$ and $U = 5.16$. In an equivalent manner, when the statistical test is applied exclusively to each one of the five subsystems that constitute the 5-DOF rotor and AMB system, the fault thresholds are defined as $L = 0.89$ and $U = 1.11$. The proposed fault diagnosis approach gives early warnings even about the appearance of incipient failures, for instance, disturbance terms of the order of $d_i = 0.002$ or $d_i = 0.003$.

In case of fault detection with χ^2 tests, the Derivative-free nonlinear Kalman filter has been used as a disturbance estimator and the statistical χ^2 test has been applied to the 5-DOF rotor and AMB system. In particular, the following cases have been examined:

- case (0): functioning of the 5-DOF rotor and AMB system in the fault-free condition.$d_i = 0$ for $i = 1, \cdots, 5$. The associated results are depicted in Figure 7.128.
- case (1) functioning of the 5-DOF rotor and AMB system under additive input disturbance d_1 at its first subsystem. The associated results are depicted in Figure 7.129
- case (2): functioning of the 5-DOF rotor and AMB system under additive input disturbance d_2 at its second subsystem. The associated results are depicted in Figure 7.130

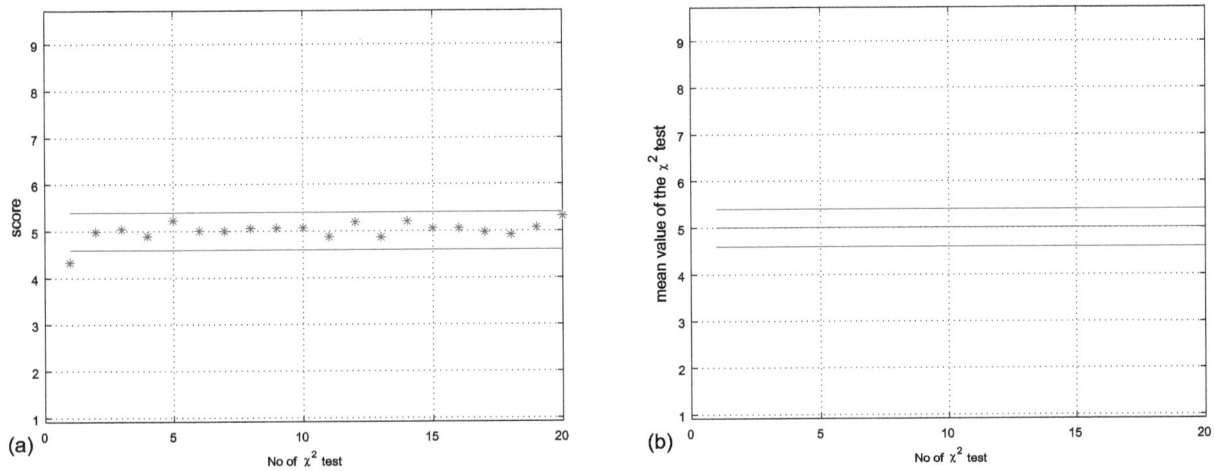

Figure 7.128 Fault diagnosis of the 5-DOF rotor and AMB system with the use of the Kalman filter-based disturbance observer in case of no fault in the system: (a) individual χ^2 tests (blue marking) vs the fault thresholds (red lines); (b) mean value of the χ^2 tests (blue line) vs the fault thresholds (red lines)

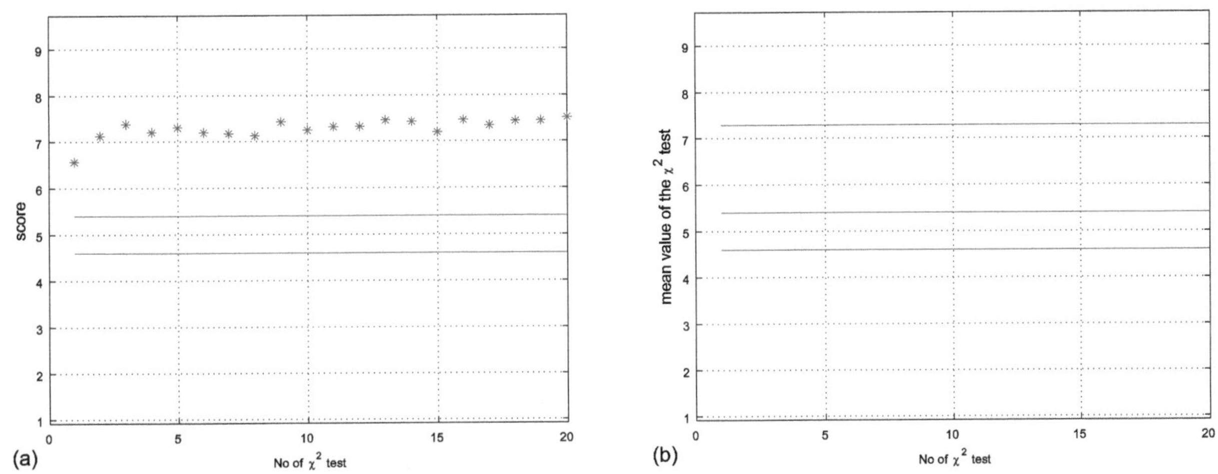

Figure 7.129 Fault diagnosis of the 5-DOF rotor and AMB system with the use of the Kalman filter-based disturbance observer in case of fault associated with disturbance term d_1 (fault's value $d_1 = 0.0020$): (a) individual χ^2 tests (blue marking) vs the fault thresholds (red lines); (b) mean value of the χ^2 tests (blue line) vs the fault thresholds (red lines)

- case (3) functioning of the 5-DOF rotor and AMB system under additive input disturbance d_3 at its third subsystem. The associated results are depicted in Figure 7.131
- case (4) functioning of the 5-DOF rotor and AMB system under additive input disturbance d_4 at its fourth subsystem. The associated results are depicted in Figure 7.132
- case (5) functioning of the 5-DOF rotor and AMB system under additive input disturbance d_5 at its fifth subsystem. The associated results are depicted in Figure 7.133
- case (6) functioning of the 5-DOF rotor and AMB system under additive input disturbances d_1 and d_2 at its first and second subsystem. The associated results are depicted in Figure 7.134

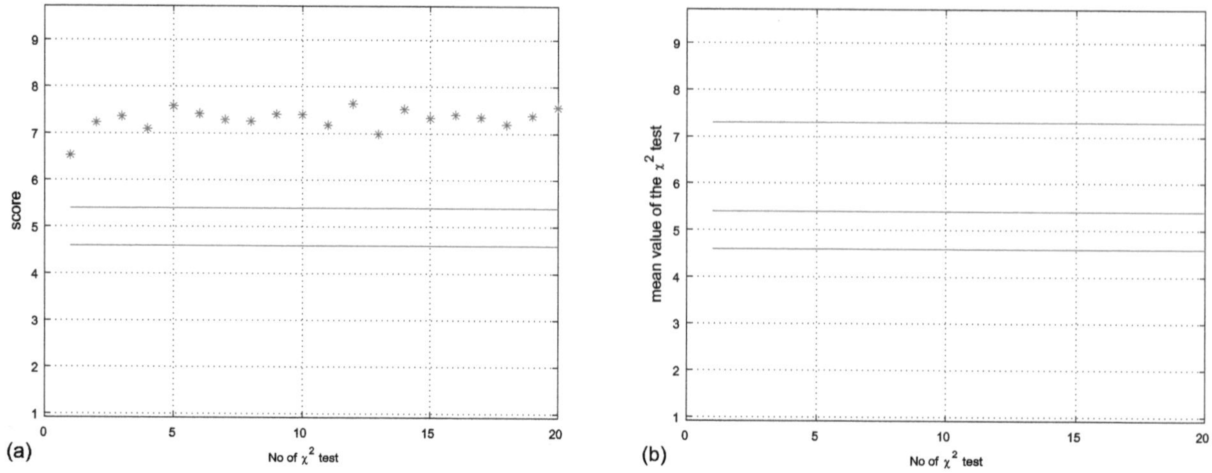

Figure 7.130 Fault diagnosis of the 5-DOF rotor and AMB system with the use of the Kalman filter-based disturbance observer in case of fault associated with disturbance term d_2 (fault's value $d_2 = 0.0030$): (a) individual χ^2 tests (blue marking) vs the fault thresholds (red lines); (b) mean value of the χ^2 tests (blue line) vs the fault thresholds (red lines)

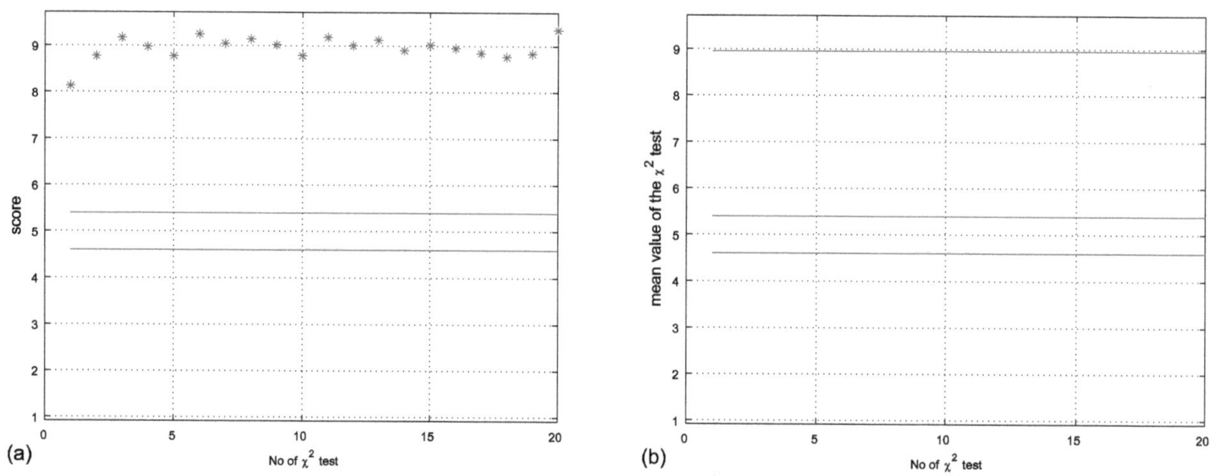

Figure 7.131 Fault diagnosis of the 5-DOF rotor and AMB system with the use of the Kalman filter-based disturbance observer in case of fault associated with disturbance term d_3 (fault's value $d_3 = 0.0040$): (a) individual χ^2 tests (blue marking) vs the fault thresholds (red lines); (b) mean value of the χ^2 tests (blue line) vs the fault thresholds (red lines)

- case (7) functioning of the 5-DOF rotor and AMB system under additive input disturbances d_1 and d_3 at its first and third subsystem. The associated results are depicted in Figure 7.135
- case (8) functioning of the 5-DOF rotor and AMB system under additive input disturbance d_2 and d_3 at its second and third subsystem. The associated results are depicted in Figure 7.136

It has been confirmed that, in all aforementioned cases, the proposed condition monitoring method was capable of detecting incipient faults appearing in components of the 5-DOF rotor and AMB system.

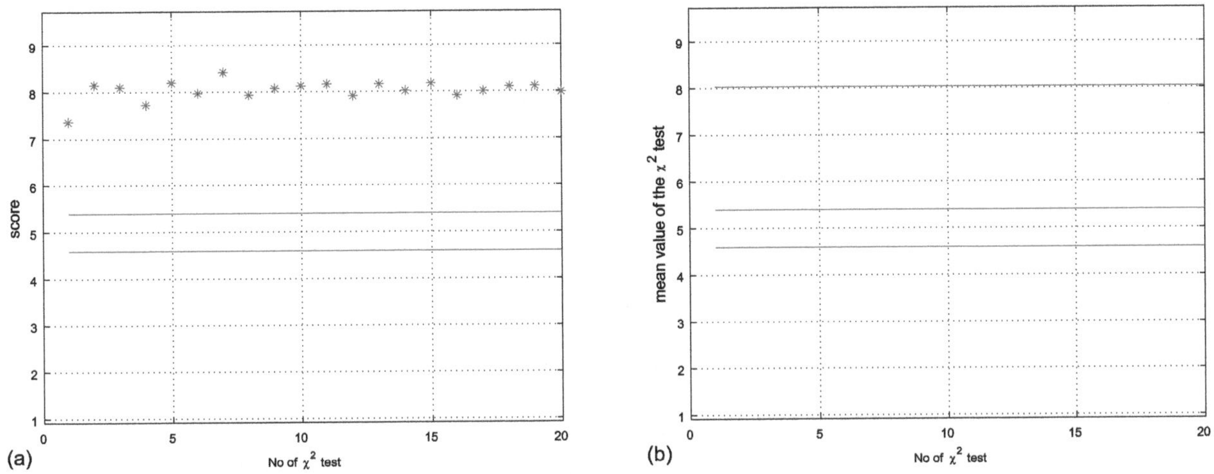

Figure 7.132 Fault diagnosis of the 5-DOF rotor and AMB system with the use of the Kalman filter-based disturbance observer in case of fault associated with disturbance term d_4 (fault's value $d_4 = 0.0035$): (a) individual χ^2 tests (blue marking) vs the fault thresholds (red lines); (b) mean value of the χ^2 tests (blue line) vs the fault thresholds (red lines)

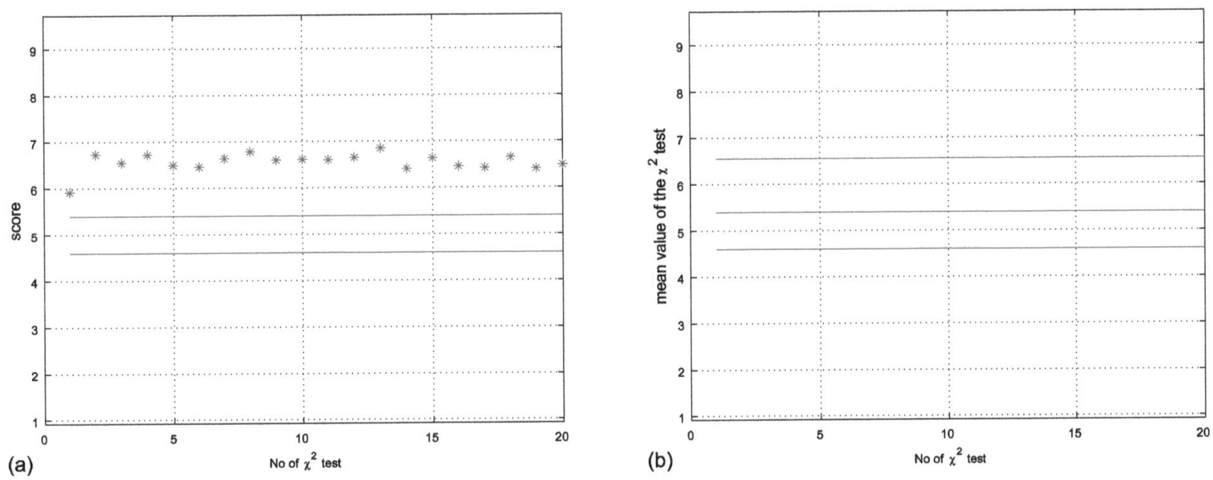

Figure 7.133 Fault diagnosis of the 5-DOF rotor and AMB system with the use of the Kalman filter-based disturbance observer in case of fault associated with disturbance term d_5 (fault's value $d_5 = 0.0025$): (a) individual χ^2 tests (blue marking) vs the fault thresholds (red lines); (b) mean value of the χ^2 tests (blue line) vs the fault thresholds (red lines)

7.6 Fault diagnosis for VSC-HVDC transmission with KF-based disturbance observers

7.6.1 Outline

VSC-HVDC transmission systems are a significant part of the smart grid and of distributed power generation systems based on renewable energy sources [576,616]. Wind farms or marine power generation units are often connected to the main electricity grid through VSC-HVDC transmission lines [126,612]. VSC-HVDC transmission lines exhibit specific advantages: (i) they improve the stability and controllability of an electricity grid that integrates distributed

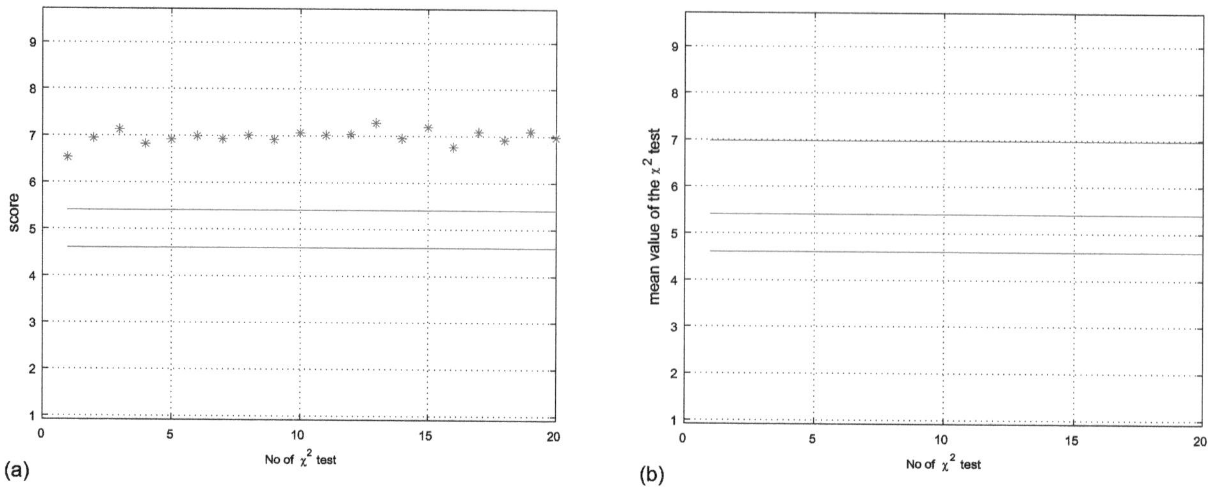

Figure 7.134 *Fault diagnosis of the 5-DOF rotor and AMB system with the use of the Kalman filter-based disturbance observer in case of fault associated with disturbance terms d_1 (fault's value $d_1 = 0.0020$) and d_2 (fault's value $d_2 = 0.0020$): (a) individual χ^2 tests (blue marking) vs the fault thresholds (red lines); (b) mean value of the χ^2 tests (blue line) vs the fault thresholds (red lines)*

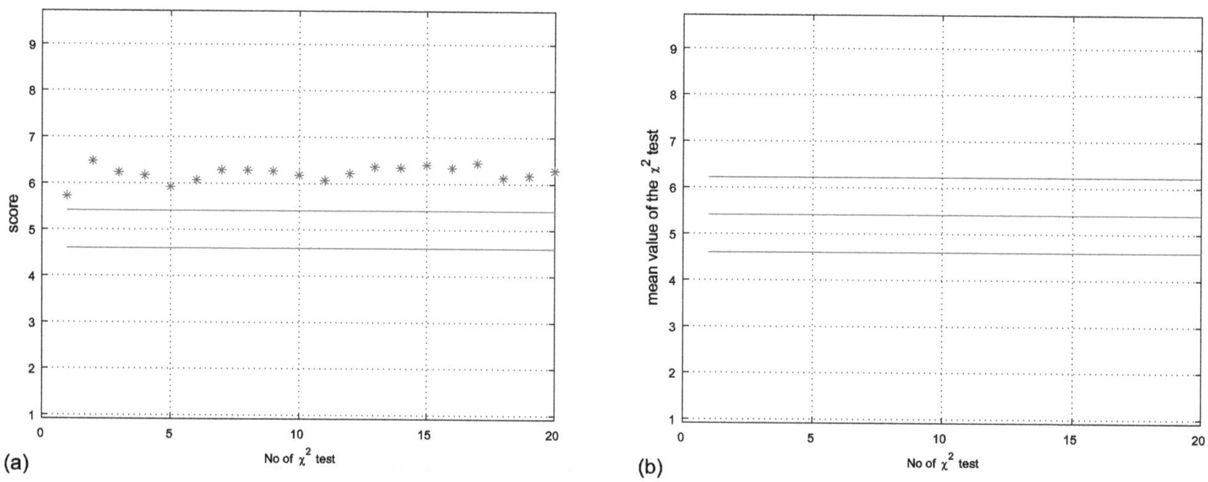

Figure 7.135 *Fault diagnosis of the 5-DOF rotor and AMB system with the use of the Kalman filter-based disturbance observer in case of fault associated with disturbance terms d_1 (fault's value $d_1 = 0.0020$) and d_1 (fault's value $d_3 = 0.0010$): (a) individual χ^2 tests (blue marking) vs the fault thresholds (red lines); (b) mean value of the χ^2 tests (blue line) vs the fault thresholds (red lines)*

and heterogenous power sources, (ii) they facilitate AC asynchronous connection, and (iii) they allow for more efficient management of power compared to long-distance AC transmission lines [746,871]. VSC-HVDC transmission links comprise an AC/DC converter (rectifier), a high voltage DC line, and a DC/AC inverter [682,628]. Due to harsh and variable operating conditions the above-noted components of the VSC-HVDC transmission system can be subject to faults [239,627]. Failures can affect the transistors of the rectifier and the inverter as well as the LC filters

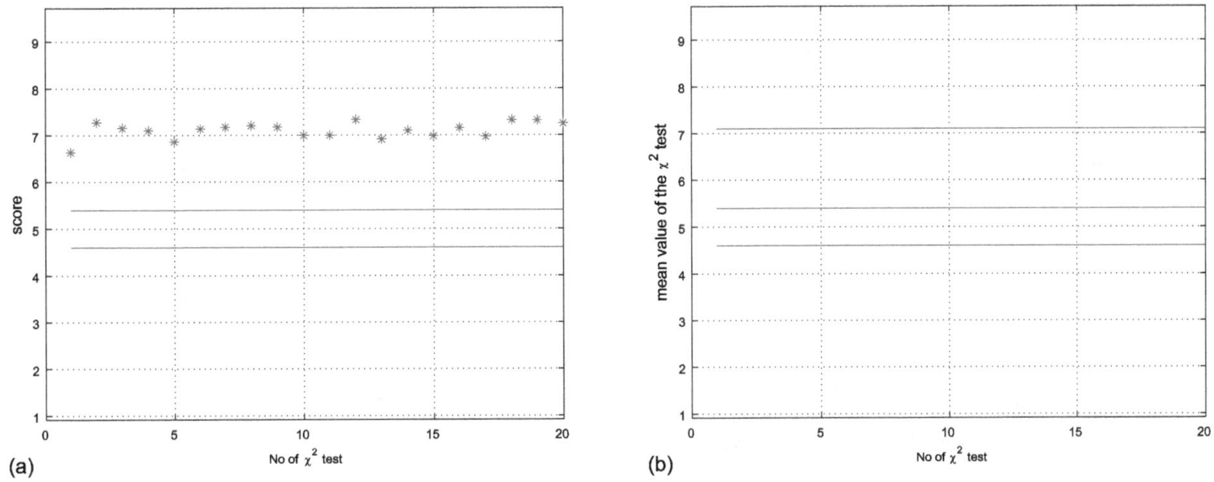

Figure 7.136 Fault diagnosis of the 5-DOF rotor and AMB system with the use of the Kalman filter-based disturbance observer in case of fault associated with disturbance terms d_2 (fault's value $d_2 = 0.0015$) and d_3 (fault's value $d_3 = 0.0025$): (a) individual χ^2 tests (blue marking) vs the fault thresholds (red lines); (b) mean value of the χ^2 tests (blue line) vs the fault thresholds (red lines)

connected to them [865,1062,921,539]. Moreover, faults can appear in the DC transmission lines. Early fault diagnosis is important for achieving the secure and reliable operation of this power transmission system and for avoiding power losses [896,897]. Fault diagnosis for VSC-HVDC lines can take place in the time domain, in the frequency domain and in the time–frequency domain [920,650]. In time-domain fault diagnosis, the accumulated measurements undergo statistical processing and generate a statistical index which is compared against fault thresholds [383,15]. In the frequency domain, there is an assessment of the frequency information which is contained in the measurements collected from the VSC-HVDC transmission line. Fourier transform and its variants are used to extract this information [503,35,926]. In the time–frequency domain, there is the use of techniques such as the wavelet transform and the Hilbert-Huang transform to extract information that reveals the existence of failures [684,384]. In time-domain fault diagnosis, one can distinguish model-based and model-free techniques, while fault thresholds definition is a significant issue [1037,1040,1041,1047,706].

In this section, a novel fault detection and isolation method is proposed for the VSC-HVDC transmission system using Kalman filter-based disturbance observers and statistical criteria for the definition of fault thresholds [711,715,741,742,743]. The dynamic model of the VSC-HVDC transmission line is separated in two subsystems, where the first one comprises the rectifier and the DC-line while the second one consists of the inverter at the end of the transmission system. It is proven that each one of these subsystems is differentially flat. This also demonstrates that each one of these subsystems can be transformed into the input–output linearized form and in the canonical Brunovsky form through successive differentiations of its flat outputs [91,594,639]. For the linearized description of these subsystems, optimal state estimation can be performed with the use of Kalman filtering [705,60]. Faults and the effects of parametric changes appear as additive disturbances to the inputs of the transformed state-space model of these subsystems. Next, the state vector of the two subsystems is extended by including as additional state variables the additive disturbance inputs and their time derivatives. For the extended state-space model of the two subsystems, two Kalman filter-based disturbance observers are designed. These disturbance estimators can estimate the perturbation inputs that affect the two subsystems. The residuals of the disturbance inputs, that is the differences between the estimated values of the perturbation terms and the zero value which signifies the fault-free case are used for performing fault diagnosis in the rectified and DC-line system, as well as in the inverter.

The proposed fault detection and isolation concept uses that the sum of the squares of the residuals vector being weighted by the inverse of the associated covariance matrix forms a stochastic variable which follows the χ^2 distribution and which is also called "statistical test." It holds that in the fault-free case the mean value of the "statistical test" should approximate closely the mean value of the χ^2 distribution with degrees of freedom which are equal to the dimension of the residuals vector. On the other hand, in the case of faults, the "statistical test" deviates from the above-noted mean value. Furthermore, the confidence intervals of the χ^2 distribution define fault thresholds which enable to conclude with a high confidence level (of the order of 95% and beyond) the occurrence of a failure. As long as the "statistical test" falls within the range defined by the previously described confidence interval the fault-free condition of the rectifier and DC-line subsystem as well as of the inverter subsystem can be inferred. On the contrary, when the "statistical tests" exceed persistently the aforementioned confidence interval, then there is almost certainty that the monitored subsystem has been subjected to failure. The disturbance estimates which are provided by the Kalman filter-based disturbance observer allow also to perform fault isolation and to identify which specific component of the VSC-HVDC line has been subjected to failure.

7.6.2 Dynamic model of the VSC rectifier and of the DC transmission line

7.6.2.1 State-space model of the rectifier and DC line

The diagram of the 2-area VSC-HVDC transmission system, comprising an AC/DC converter (rectifier), the DC transmission line and a DC/AC inverter is shown in Figure 7.137. The detailed diagram of the VSCs is shown in Figure 7.138. Each VSC has 3 arms and each arm comprises two transistors which function as switches. The diagram of the HVDC transmission line is also shown in Figure 7.138.

The dynamics of VSC-1 (rectifier) jointly with the dynamics of the HVDC transmission line is given by the following set of equations [715]:

$$L\dot{i}_d = -Ri_d + L\omega i_q + v_d - \tfrac{V_{DC_1}}{2}u_1 \qquad (7.411)$$

$$L\dot{i}_q = -L\omega i_d - Ri_q + v_q - \tfrac{V_{DC_2}}{2}u_2 \qquad (7.412)$$

$$C_{DC_2}\dot{V}_{DC_2} = \tfrac{1}{Z_L}V_{DC_1} - \tfrac{1}{Z_L}V_{DC_2} \qquad (7.413)$$

$$C_{DC_1}\dot{V}_{DC_1} = \tfrac{1}{Z_L}V_{DC_1} + \tfrac{1}{Z_L}V_{DC_2} + \tfrac{3}{4}i_du_1 + \tfrac{3}{4}i_qu_2 \qquad (7.414)$$

where (i_d, i_q) are the rectifier's currents in the dq-frame, (v_d, v_q) are the grid voltages in the dq frame, (V_{DC_1}, V_{DC_2}) are voltages after the rectifier and before the inverter, (u_1, u_2) are the control inputs of the rectifier and ω is the grid's frequency.

The state variables of the rectifier and DC-line's dynamic model are: $x = [x_1, x_2, x_3, x_4]^T = [i_d, i_q, V_{DC_2}, V_{DC_1}]^T$. Thus, the dynamic model of the rectifier and the DC-link is written in the following state-space form:

$$L\dot{x}_1 = -Rx_1 + L\omega x_2 + v_d - \tfrac{x_4}{2}u_1 \qquad (7.415)$$

$$L\dot{x}_2 = -L\omega x_1 - Rx_2 + v_q - \tfrac{x_4}{2}u_2 \qquad (7.416)$$

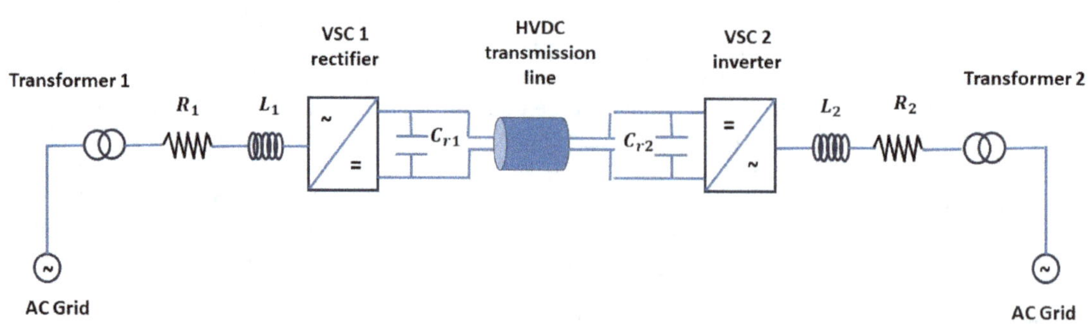

Figure 7.137 Diagram of the 2-area VSC-HVDC transmission line

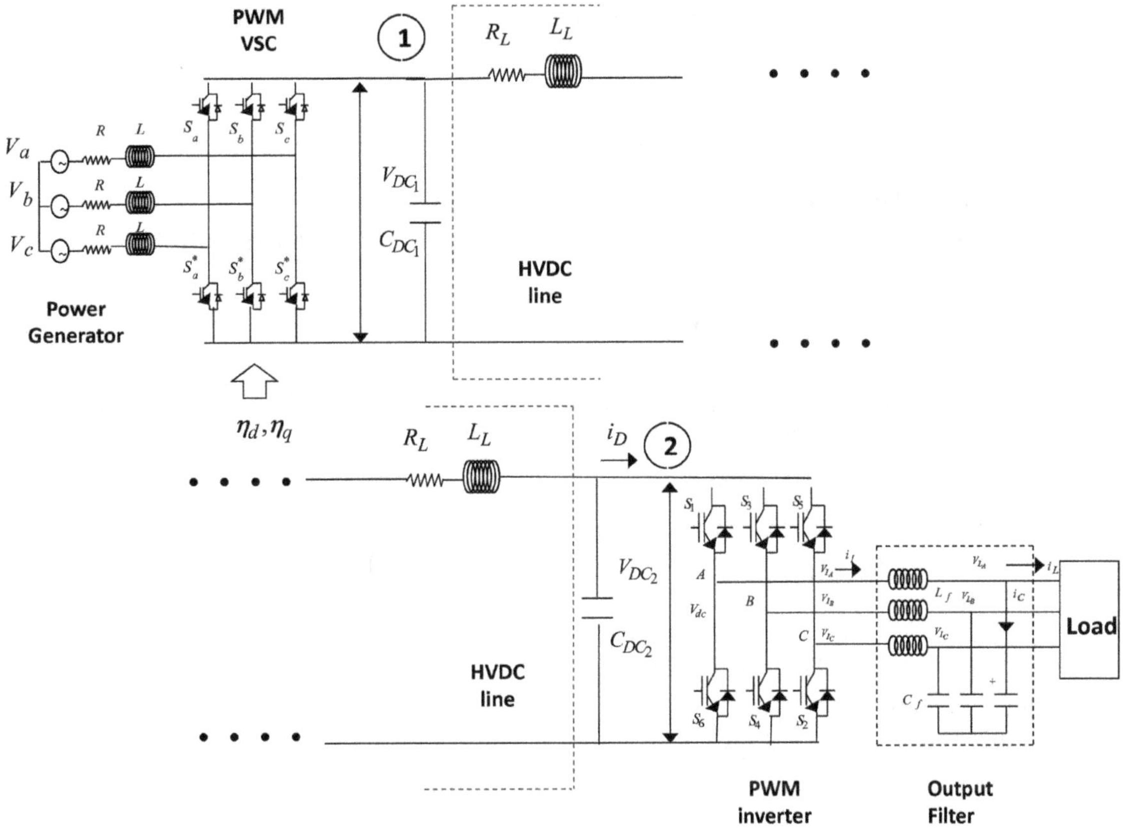

Figure 7.138 VSC-HVDC transmission system comprising an AC/DC converter, a DC line, and an inverter

$$C_{DC_2}\dot{x}_3 = \frac{1}{Z_L}x_4 - \frac{1}{Z_L}x_3 \tag{7.417}$$

$$C_{DC_1}\dot{x}_4 = \frac{1}{Z_L}x_4 + \frac{1}{Z_L}x_3 + \frac{3}{4}x_1u_1 + \frac{3}{4}x_2u_2 \tag{7.418}$$

The dynamic model of the rectifier and HVDC line can be also written in matrix form

$$\begin{pmatrix}\dot{x}_1\\ \dot{x}_2\\ \dot{x}_3\\ \dot{x}_4\end{pmatrix} = \begin{pmatrix}-\frac{R}{L}x_1 + \omega x_2 + \frac{v_d}{L}\\ -\omega x_1 - \frac{R}{L}x_2 + \frac{v_q}{L}\\ \frac{1}{Z_LC_{DC_2}}x_4 - \frac{1}{Z_LC_{DC_2}}x_3\\ \frac{1}{Z_LC_{DC_1}}x_4 + \frac{1}{Z_LC_{DC_1}}x_3\end{pmatrix} + \begin{pmatrix}-\frac{x_4}{2L} & 0\\ 0 & -\frac{x_4}{2L}\\ 0 & 0\\ \frac{3x_1}{4C_{DC_1}} & \frac{3x_2}{4C_{DC_1}}\end{pmatrix}\begin{pmatrix}u_1\\ u_2\end{pmatrix} \tag{7.419}$$

The measured outputs of this system are taken to be its linearizing outputs and are given by

$$h = \begin{pmatrix}h_1\\ h_2\end{pmatrix} = \begin{pmatrix}\frac{3}{4}L(i_d^2 + i_q^2) + \frac{C_{DC_1}}{2}V_{DC_1}^2\\ V_{DC_2}\end{pmatrix} \tag{7.420}$$

or equivalently

$$h = \begin{pmatrix}h_1\\ h_2\end{pmatrix} = \begin{pmatrix}\frac{3}{4}L(x_1^2 + x_2^2) + \frac{C_{DC_1}}{2}x_4^2\\ x_3\end{pmatrix} \tag{7.421}$$

The rectifier and DC-line system is also written in the nonlinear affine-in-the-input state-space form

$$\begin{aligned}\dot{x} &= f(x) + g(x)u \\ y &= h(x)\end{aligned} \tag{7.422}$$

where $x \in R^{4\times1}, f(x) \in R^{4\times1}$, $g(x) \in R^{4\times2}$, and $u \in R^{2\times1}$.

7.6.2.2 Differential flatness of the rectifier and DC line dynamics

The dynamic model of the rectifier in connection with the DC transmission line is differentially flat, with flat outputs vector [711]

$$y_f = \begin{pmatrix} y_{f_1} \\ y_{f_2} \end{pmatrix} = \begin{pmatrix} \frac{3}{4}L(x_1^2 + x_2^2) + \frac{C_{DC_1}}{2}x_4^2 \\ x_3 \end{pmatrix} \tag{7.423}$$

It is considered that v_d is constant while v_q is 0. After the regrouping of terms and intermediate computations, the differentiation of the flat output y_{f_1} gives

$$\dot{y}_{f_1} = -\frac{3R}{2}(x_1^2 + x_2^2) + \frac{3}{2}x_1 v_d - \frac{x_4^2}{Z_L} + \frac{1}{Z_L}x_3 x_4 \tag{7.424}$$

Using again the definition of the first flat output, one has

$$\begin{aligned}y_{f_1} &= \frac{3L}{4}(x_1^2 + x_2^2) + \frac{C_{DC}}{2}x_4^2 \Rightarrow \\ (x_1^2 + x_2^2) &= \frac{4}{3L}(y_{f_1} - \frac{C_{DC}}{2}y_{f_2})^2\end{aligned} \tag{7.425}$$

which means that the term $(x_1^2 + x_2^2)$ is a differential function of the flat outputs vector, or

$$(x_1^2 + x_2^2) = f_a(y_f, \dot{y}_f) \tag{7.426}$$

Besides, from the third row of the state-space model, one has

$$\begin{aligned}C_{DC_2}\dot{x}_3 = \frac{1}{Z_L}x_4 - \frac{1}{Z_L}x_3 \Rightarrow x_4 = Z_L(C_{DC_2}\dot{y}_{f_2} + \frac{1}{Z_L}y_{f_2}) \\ \Rightarrow x_4 = f_b(y_f, \dot{y}_f)\end{aligned} \tag{7.427}$$

which means that x_4 is a differential function of the flat outputs vector of the system. By substituting Eq. (7.426) and Eq. (7.427) into Eq. (7.424), one gets

$$\dot{y}_{f_1} = -\frac{3R}{2}f_a(y_f, \dot{y}_f) + \frac{3v_d}{2}x_1 - \frac{f_b(y_f, \dot{y}_f)^2}{Z_L} - \frac{1}{Z_L}y_{f_2}f_b(y_f, \dot{y}_f) \tag{7.428}$$

By solving for x_1, one has

$$\begin{aligned}x_1 = \frac{2}{3v_d}[\dot{y}_{f_1} + \frac{3R}{2}f_a(y_f, \dot{y}_f) + \frac{f_b(y_f, \dot{y}_f)^2}{Z_L} + \frac{1}{Z_L}y_{f_2}f_b(y_f, \dot{y}_f)] \\ \Rightarrow x_1 = f_c(y_f, \dot{y}_f)\end{aligned} \tag{7.429}$$

which signifies that x_1 is a differential function of the flat outputs vector. Next using Eq. (7.426) and Eq. (7.429), one has

$$\begin{aligned}x_2^2 = f_a(y_f, \dot{y}_f) - f_c(y_f, \dot{y}_f)^2 \Rightarrow x_2 = \sqrt{f_a(y_f, \dot{y}_f) - f_c(y_f, \dot{y}_f)^2} \\ \Rightarrow x_2 = f_d(y_f, \dot{y}_f)\end{aligned} \tag{7.430}$$

Thus, it has been proven that all state variables of the system that comprises the rectifier and the DC-line are differential functions of the associated flat outputs vector y_F. Next, using Eq. (7.415), one solves for the control input u_1, which gives

$$u_1 = \frac{2L}{x_4}[Rx_1 + L\omega x_2 + v_d - L\dot{x}_1] \Rightarrow u_1 = f_e(y_f, \dot{y}_f) \tag{7.431}$$

This signifies that u_2 is a differential function of the flat outputs vector. Additionally, from Eq. (7.416), one solves for control input u_2 which gives

$$u_2 = \frac{2L}{x_4}[-\omega x_1 - Rx_2 + v_q - L\dot{x}_2] \Rightarrow u_2 = f_g(y_f, \dot{y}_f) \tag{7.432}$$

which also signifies that u_2 is a differential function of the flat outputs vector. Consequently, all state variables and the control inputs of the dynamic model of the rectifier and DC-line subsystem are differential functions of the system's flat outputs vector and the subsystem is differentially flat.

Knowing that the VSC (rectifier) and DC-line subsystem is differentially flat, it is also concluded that this subsystem can be written in the input–output linearized form through successive differentiations of its flat outputs. First, about flat output y_{f_1}, it has been shown that

$$\dot{y}_{f_1} = -\frac{3R}{2}(x_1^2+x_2^2)+\frac{3}{2}x_1v_d-\frac{x_4^2}{Z_L}+\frac{1}{Z_L}x_3x_4 \tag{7.433}$$

By differentiating once again in time, and after intermediate operations and regrouping of terms, one gets

$$\ddot{y}_{f_1} = f_a(x)+g_{a_1}(x)u_1+g_{a_2}(x)u_2 \tag{7.434}$$

where functions $f_a(x)$, $g_{a_1}(x)$, and $g_{a_2}(x)$ are given by

$$\begin{aligned} f_a(x) = \{(-3Rx_1+\tfrac{3v_d}{2})[-\tfrac{R}{L}x_1+\omega x_2+\tfrac{v_d}{L}-\tfrac{x_4}{2}u_1]- \\ -3Rx_2[-\omega x_1+\tfrac{R}{L}x_2]+(\tfrac{2x_4}{Z_L})+\tfrac{x_3}{Z_L}[-\tfrac{1}{Z_LC_{DC_1}}x_4+\tfrac{1}{Z_LC_{DC_1}}x_3]+ \\ +\tfrac{x_4}{Z_L}[\tfrac{1}{Z_LC_{DC_1}}x_4-\tfrac{1}{Z_LC_{DC_1}}x_3]\} \end{aligned} \tag{7.435}$$

$$g_{a_1}(x) = \{(-3Rx_1+\tfrac{3v_d}{2})(-\tfrac{x_4}{2L})+(\tfrac{2x_4}{Z_L}+\tfrac{x_3}{Z_L})(\tfrac{3x_1}{4C_{DC_2}})\} \tag{7.436}$$

$$g_{a_2}(x) = \{(-3Rx_2)(-\tfrac{x_4}{2L})+(\tfrac{2x_4}{Z_L}+\tfrac{x_3}{Z_L})\tfrac{3x_2}{4C_{DC_1}})\} \tag{7.437}$$

About the second flat output $y_{f_2}=x_3$, differentiation in time gives

$$\dot{y}_{f_2} = \dot{x_3}\Rightarrow\dot{y}_{f_2} = \frac{1}{Z_LC_{DC_2}}x_4-\frac{1}{Z_LC_{DC_2}}x_3 \tag{7.438}$$

By differentiating once again in time and through intermediate operations and the regrouping of terms, one gets

$$\ddot{y}_{f_2} = f_b(x)+g_{b_1}(x)u_1+g_{b_2}(x)u_2 \tag{7.439}$$

where functions $f_b(x)$, $g_{b_1}(x)$, and $g_{b_2}(x)$ are given by

$$f_b(x) = \{\tfrac{1}{Z_LC_{DC_2}}[-\tfrac{1}{Z_LC_{DC_1}}x_4+\tfrac{1}{Z_LC_{DC_1}}x_3]-\tfrac{1}{Z_LC_{DC_2}}[\tfrac{1}{Z_LC_{DC_2}}x_4+\tfrac{1}{Z_LC_{DC_2}}x_3]\} \tag{7.440}$$

$$g_{b_1}(x) = [\tfrac{1}{Z_LC_{DC_2}}\tfrac{3x_1}{4C_{DC_1}}] \tag{7.441}$$

$$g_{b_2}(x) = [\tfrac{1}{Z_LC_{DC_2}}\tfrac{3x_2}{4C_{DC_1}}] \tag{7.442}$$

Thus, finally, the dynamic model of the VSC (rectifier) and of the DC-line is written in the input–output linearized form

$$\begin{aligned} \ddot{y}_{f_1} &= f_a(x)+g_{a_1}(x)u_1+g_{a_2}(x)u_2 \\ \ddot{y}_{f_2} &= f_b(x)+g_{b_1}(x)u_1+g_{b_2}(x)u_2 \end{aligned} \tag{7.443}$$

or in matrix form

$$\begin{pmatrix}\ddot{y}_{f_1}\\ \ddot{y}_{f_2}\end{pmatrix} = \begin{pmatrix}f_a(x)\\ f_b(x)\end{pmatrix}+\begin{pmatrix}g_{a_1}(x) & g_{a_2}(x)\\ g_{b_1}(x) & g_{b_2}(x)\end{pmatrix}\begin{pmatrix}u_1\\ u_2\end{pmatrix} \tag{7.444}$$

Next, the following state variables are defined: $z_1=y_{f_1}$, $z_2=\dot{y}_{f_1}$, $z_3=y_{f_2}$, $z_4=\dot{y}_{f_2}$. Besides, the following virtual control inputs are defined:

$$\begin{aligned} v_1 &= f_a(x)+g_{a_1}(x)u_1+g_{a_2}(x)u_2 \\ v_2 &= f_b(x)+g_{b_1}(x)u_1+g_{b_2}(x)u_2 \end{aligned} \tag{7.445}$$

Thus, one arrives at the concise input–output state-space form:

$$\ddot{z}_1 = v_1 \quad \ddot{z}_3 = v_2 \tag{7.446}$$

which is also written in the following matrix form

$$\begin{pmatrix} \dot{z}_1 \\ \dot{z}_2 \\ \dot{z}_3 \\ \dot{z}_4 \end{pmatrix} = \begin{pmatrix} 0 & 1 & 0 & 0 \\ 0 & 0 & 0 & 0 \\ 0 & 0 & 0 & 1 \\ 0 & 0 & 0 & 0 \end{pmatrix} \begin{pmatrix} z_1 \\ z_2 \\ z_3 \\ z_4 \end{pmatrix} + \begin{pmatrix} 0 & 0 \\ 1 & 0 \\ 0 & 0 \\ 0 & 1 \end{pmatrix} \begin{pmatrix} v_1 \\ v_2 \end{pmatrix} \tag{7.447}$$

For the input–output linearized dynamic model of the rectifier and DC-line a stabilizing feedback controller has the form:

$$\begin{aligned} v_1 &= \ddot{z}_{1,d} - k_{1,1}(\dot{z}_1 - \dot{z}_{1,d}) - k_{2,1}(z_1 - z_1, d) \\ v_2 &= \ddot{z}_{2,d} - k_{1,2}(\dot{z}_2 - \dot{z}_{2,d}) - k_{2,2}(z_2 - z_2, d) \end{aligned} \tag{7.448}$$

This results in the tracking error dynamics

$$\begin{aligned} &(\ddot{z}_1 - \ddot{z}_{1,d}) + k_{1,1}(\dot{z}_1 - \dot{z}_{1,d}) + k_{2,1}(z_1 - z_1, d) = 0 \\ &(\ddot{z}_2 - \ddot{z}_{2,d}) + k_{1,2}(\dot{z}_2 - \dot{z}_{2,d}) + k_{2,2}(z_2 - z_2, d) = 0 \end{aligned} \tag{7.449}$$

and through suitable selection of the feedback gains $(k_{1,1}, k_{2,1})$ and $(k_{1,2}, k_{2,2})$ so as the associated characteristic polynomials to be Hurwitz stable, one has

$$lim_{t\to\infty} z_1 = z_{1,d} \quad lim_{t\to\infty} z_3 = z_{3,d} \tag{7.450}$$

The above condition signifies that the entire state vector of the rectifier and DC-line converges to the associated setpoints vector.

7.6.3 Design of a disturbance observer for the rectifier and DC-line

The input–output linearized dynamic model of the rectifier and DC-line is extended with the inclusion of additive disturbance terms [711]. These terms describe both parametric changes and external perturbations. In case of the perturbed input–output linearized model, one has

$$\ddot{z}_1 = v_1 + d_1 \quad \ddot{z}_3 = v_2 + d_2 \tag{7.451}$$

The dynamics of the disturbance terms d_i $i = 1, 2$ is taken to be described the disturbances' derivatives up to order 2 and the associated initial conditions. The related conditions are: $\ddot{d}_1 = f_{d_1}, d_1(0), \dot{d}_1(0)$ and $\ddot{d}_2 = f_{d_2}, d_2(0), \dot{d}_2(0)$.

The state-space model of the disturbed input–output linearized model of the system is extended by considering as additional state variables the disturbance inputs and their time derivatives. Thus, one defines the extended state vector of the system: $z_1 = x_1$, $z_2 = \dot{x}_1$, $z_3 = x_2$, $z_4 = \dot{x}_2$, $z_5 = d_1$, $z_6 = \dot{d}_1$, $z_7 = d_2$, $z_8 = \dot{d}_2$.

This gives the extended state-space model

$$\begin{aligned} &\dot{z}_1 = z_2 \quad \dot{z}_2 = v_1 + z_5 \quad \dot{z}_3 = z_4 \quad \dot{z}_4 = v_2 + z_7 \\ &\dot{z}_5 = z_6 \quad \dot{z}_6 = f_{d_1} \quad \dot{z}_7 = z_8 \quad \dot{z}_8 = f_{d_2} \end{aligned} \tag{7.452}$$

or in matrix form, one has

$$\begin{pmatrix} \dot{z}_1 \\ \dot{z}_2 \\ \dot{z}_3 \\ \dot{z}_4 \\ \dot{z}_5 \\ \dot{z}_6 \\ \dot{z}_7 \\ \dot{z}_8 \end{pmatrix} = \begin{pmatrix} 0 & 1 & 0 & 0 & 0 & 0 & 0 & 0 \\ 0 & 0 & 0 & 0 & 1 & 0 & 0 & 0 \\ 0 & 0 & 0 & 1 & 0 & 0 & 0 & 0 \\ 0 & 0 & 0 & 0 & 0 & 0 & 1 & 0 \\ 0 & 0 & 0 & 0 & 0 & 1 & 0 & 0 \\ 0 & 0 & 0 & 0 & 0 & 0 & 0 & 0 \\ 0 & 0 & 0 & 0 & 0 & 0 & 0 & 1 \\ 0 & 0 & 0 & 0 & 0 & 0 & 0 & 0 \end{pmatrix} \begin{pmatrix} z_1 \\ z_2 \\ z_3 \\ z_4 \\ z_5 \\ z_6 \\ z_7 \\ z_8 \end{pmatrix} + \begin{pmatrix} 0 & 0 & 0 & 0 \\ 1 & 0 & 0 & 0 \\ 0 & 0 & 0 & 0 \\ 0 & 1 & 0 & 0 \\ 0 & 0 & 0 & 0 \\ 0 & 0 & 1 & 0 \\ 0 & 0 & 0 & 0 \\ 0 & 0 & 0 & 1 \end{pmatrix} \begin{pmatrix} v_! \\ v_2 \\ f_{d_1} \\ f_{d_2} \end{pmatrix} \tag{7.453}$$

The measurement equation for the rectifier and DC-line subsystem is

$$\begin{pmatrix} z_1^m \\ z_2^m \end{pmatrix} = \begin{pmatrix} 1 & 0 & 0 & 0 & 0 & 0 & 0 & 0 \\ 0 & 0 & 1 & 0 & 0 & 0 & 0 & 0 \end{pmatrix} \begin{pmatrix} z_1 \\ z_2 \\ z_3 \\ z_4 \\ z_5 \\ z_6 \\ z_7 \\ z_9 \end{pmatrix} \tag{7.454}$$

For the extended state-space model of the rectifier and DC line given in Eq. (7.453) and Eq. (7.454) a disturbance observer is designed. While the extended state-space model of the rectifier and DC line has the form

$$\begin{aligned} \dot{z}_e &= A_e z_e + B_e u_e \\ z_e^m &= C_e z_e \end{aligned} \tag{7.455}$$

the disturbance observer has the form

$$\begin{aligned} \dot{\hat{z}}_o &= A_o z_o + B_o u_o + K_f(z_e^m - \hat{z}_o^m) \\ \hat{z}_o^m &= C_o \hat{z}_o \end{aligned} \tag{7.456}$$

In Eq. (7.456) matrix A_o is equal to the transition matrix A_e of the extended state-space model, while matrix B_o is equal to the first two columns of the control inputs gain matrix B_e of the extended system. Moreover, matrix C_o is equal to the measurement matrix C_e of the extended state-space model.

The disturbance observer's gain K_f is computed sequentially using the Kalman filter's recursion. The control inputs vector of the disturbance observer comprises the first two variables of the control inputs vector of the extended system, that is $v_o = [v_1, v_2]^T$. Control inputs f_{d_i} $i = 1, 2$ are unknown and are not used by the disturbances estimator. Initial conditions about the disturbance terms and their derivatives are also unknown and are not used by the disturbance observer. The extended state-space model of the rectifier and DC line which is defined by matrices A_o, B_o, C_o is observable and the convergence of the disturbance observer does not depend on initial conditions.

Matrices A_o, B_o, C_o undergo discretization and their discrete-time equivalents are denoted as $A_{o,d}$, $B_{o,d}$, and $C_{o,d}$. The recursion of the Kalman filter which computes the observer's gain K_f consists of a time-update and of a measurement-update part.

measurement update:

$$\begin{aligned} K_f(k) &= P^-(k)C_{o,d}^T[C_{o,d}P^-(k)C_{o,d}^T + R(k)]^{-1} \\ \hat{x}(k) &= \hat{x}^-(k) + K_f(k)[z_m(k) - \hat{z}_m(k)] \\ P(k) &= P^-(k) - K_f(k)C_{o.d}P^-(k) \end{aligned} \tag{7.457}$$

time update:

$$\begin{aligned} &P^-(k+1)A_{o,d}P(k)A_{o,d}^T + Q(k) \\ &\hat{x}^-(k+1) = A_{o,d}\hat{x}(k) + B_{o,d}\tilde{v}_o(k) \end{aligned} \tag{7.458}$$

The process and measurement noises are considered to be white Gaussian signals. $P^-(k)$ is the state estimation error covariance matrix before receiving measurements at the k-th sampling instance, $P(k)$ is the state estimation error covariance matrix after having received measurements at the k-th sampling instance, $P^-(k+1)$ is the state estimation error covariance matrix before receiving measurements at the $(k+1)$-th sampling instance, $Q(k)$ is the process noise covariance matrix, and $R(k)$ is the measurement noise covariance matrix.

Unlike the Extended Kalman filter (EKF), the previously defined procedure for estimating state and disturbance variables in the dynamic model of the rectifier and DC line subsystem does not depend on linearization of the subsystem's model with the computation of Jacobian matrices and the associated partial derivatives. For this reason, it is also known as *Derivative-free nonlinear Kalman filter* (DKF) [711,715].

7.6.4 Dynamic model of the inverter

7.6.4.1 State-space model of the inverter

The diagram of the inverter at the end of the VSC-HVDC transmission system is shown in Figure 7.138. The state vector of the inverter is taken to be $x = [V_{Ld}, V_{Lq}, I_{Id}, I_{Iq}]^T$. Variables V_{Ld}, V_{Lq} denote voltages at the output of the LC filter, and variables i_{Ld}, i_{Lq} denote currents again at the output of the LC filter, while variables I_{Id}, I_{Iq} denote currents at the output of the transistors. The active and reactive power of the inverter are]715]

$$p_f = V_{Ld}i_{Ld} + V_{Lq}i_{Lq} \tag{7.459}$$

$$q_f = V_{Lq}i_{Ld} - V_{Ld}i_{Lq} - \omega C_f(V_{Ld}^2 + V_{Lq}^2)^2 + \omega L_f(i_{Ld}^2 + i_{Lq}^2) \tag{7.460}$$

The current variables at the output of the LC filter i_{Ld}, i_{Lq} are expressed as functions of the active and reactive power, thus giving

$$i_{Ld} = \frac{p_fV_{Ld}+q_fV_{Lq}}{V_{Ld}^2+V_{Lq}^2} + \omega C_fV_{Lq} - \frac{\omega L_fV_{Lq}(i_{Id}^2+i_{Iq}^2)}{(V_{Ld}^2+V_{Lq}^2)} \tag{7.461}$$

$$i_{Lq} = \frac{p_fV_{Lq}-q_fV_{Ld}}{V_{Ld}^2+V_{Lq}^2} - \omega C_fV_{Ld} + \frac{\omega L_fV_{Ld}(i_{Id}^2+i_{Iq}^2)}{(V_{Ld}^2+V_{Lq}^2)} \tag{7.462}$$

Then the dynamic model of the inverter is written in the matrix form [715]

$$\begin{pmatrix} \dot{V}_{Ld} \\ \dot{V}_{Lq} \\ \dot{i}_{Id} \\ \dot{i}_{Lq} \end{pmatrix} = \begin{pmatrix} \omega V_{Lq} + \frac{1}{C_f}i_{Id} - \frac{1}{C_f}\left[\frac{p_fV_{Ld}+q_fV_{Lq}}{V_{Ld}^2+V_{Lq}^2} + \omega C_fV_{Lq} - \frac{\omega L_fV_{Lq}(i_{Id}^2+i_{Iq}^2)}{(V_{Ld}^2+V_{Lq}^2)}\right] \\ -\omega V_{Ld} + \frac{1}{C_f}i_{Iq} - \frac{1}{C_f}\left[\frac{p_fV_{Lq}-q_fV_{Ld}}{V_{Ld}^2+V_{Lq}^2} - \omega C_fV_{Ld} + \frac{\omega L_fV_{Ld}(i_{Id}^2+i_{Iq}^2)}{(V_{Ld}^2+V_{Lq}^2)}\right] \\ \omega i_{Iq} - \frac{1}{L_f}V_{Ld} \\ -\omega i_{Id} - \frac{1}{L_f}V_{Lq} \end{pmatrix} + \begin{pmatrix} 0 & 0 \\ 0 & 0 \\ \frac{1}{L_d} & 0 \\ 0 & \frac{1}{L_q} \end{pmatrix} \begin{pmatrix} V_{Id} \\ V_{Iq} \end{pmatrix} \tag{7.463}$$

Using the state vector's notation $x=[V_{Ld},V_{Lq},i_{Id},i_{Iq}]^T$ and the control inputs vector notation $u=[V_{Id},V_{Iq}]^T$ the state-space model of the inverter is written as

$$\begin{pmatrix}\dot{x}_1\\ \dot{x}_2\\ \dot{x}_3\\ \dot{x}_4\end{pmatrix}=\begin{pmatrix}\omega x_2+\frac{1}{C_f}x_3-\frac{1}{C_f}\left[\frac{p_fx_1+q_fx_2}{x_1^2+x_2^2}+\omega C_fx_2-\frac{\omega L_fx_2(x_3^2+x_4^2)}{(x_1^2+x_2^2)}\right]\\ -\omega x_1+\frac{1}{C_f}x_4-\frac{1}{C_f}\left[\frac{p_fx_2-q_fx_1}{x_1^2+x_2^2}-\omega C_fx_1+\frac{\omega L_fx_1(x_3^2+x_4^2)}{(x_1^2+x_2^2)}\right]\\ \omega x_4-\frac{1}{L_f}x_1\\ -\omega x_3-\frac{1}{L_f}x_2\end{pmatrix}+\begin{pmatrix}0&0\\0&0\\ \frac{1}{L_d}&0\\0&\frac{1}{L_q}\end{pmatrix}\begin{pmatrix}u_1\\u_2\end{pmatrix}\quad(7.464)$$

Finally, the inverter's model is written in the nonlinear affine-in-the-input state-space form:

$$\dot{x}=f(x)+g(x)u\quad(7.465)$$

where $x\in R^{4\times1}, f(x)\in R^{4\times1}$, $g(x)\in R^{4\times2}$, and $u\in R^{2\times1}$.

7.6.4.2 Differential flatness properties of the inverter's model

The inverter's model is differentially flat, with flat outputs vector $y=[y_{f_1},y_{f_2}]^T=[V_{L_d},V_{L_q}]^T$ [711]. From the first two rows of the inverter's state-space model, one has

$$\dot{x}_1=\omega x_2+\frac{1}{C_f}x_3-\frac{1}{C_f}\left[\frac{p_fx_1+q_fx_2}{x_1^2+x_2^2}+\omega C_fx_2-\frac{\omega L_fx_2(x_3^2+x_4^2)}{(x_1^2+x_2^2)}\right]\quad(7.466)$$

$$\dot{x}_2=-\omega x_1+\frac{1}{C_f}x_4-\frac{1}{C_f}\left[\frac{p_fx_2-q_fx_1}{x_1^2+x_2^2}-\omega C_fx_1+\frac{\omega L_fx_1(x_3^2+x_4^2)}{(x_1^2+x_2^2)}\right]\quad(7.467)$$

By solving these equations with respect to their last term, one gets:

$$-\frac{1}{C_f}\frac{\omega L_fx_2(x_3^2+x_4^2)}{(x_1^2+x_2^2)}=\dot{x}_1-\omega x_2-\frac{1}{C_f}x_3+\frac{1}{C_f}\left[\frac{p_fx_1-q_fx_2}{x_1^2+x_2^2}\right]+\frac{1}{C_f}\omega x_2\quad(7.468)$$

$$\frac{1}{C_f}\frac{\omega L_fx_1(x_3^2+x_4^2)}{(x_1^2+x_2^2)}=\dot{x}_2+\omega x_1-\frac{1}{C_f}x_4+\frac{1}{C_f}\left[\frac{p_fx_2+q_fx_1}{x_1^2+x_2^2}\right]+\frac{1}{C_f}\omega x_1\quad(7.469)$$

By dividing the above two equations, one gets

$$-\frac{x_2}{x_1}=\frac{\dot{x}_1-\omega x_2-\frac{1}{C_f}x_3+\frac{1}{C_f}\left[\frac{p_fx_1-q_fx_2}{x_1^2+x_2^2}\right]+\frac{1}{C_f}\omega x_2}{\dot{x}_2+\omega x_1-\frac{1}{C_f}x_4+\frac{1}{C_f}\left[\frac{p_fx_2+q_fx_1}{x_1^2+x_2^2}\right]+\frac{1}{C_f}\omega x_1}\quad(7.470)$$

Using that the flat outputs of the system are $y_{f_1}=x_1$ and $y_{f_2}=x_2$ from Eq. (7.470), one has

$$x_3=C_f\{-\frac{y_{f_2}}{y_{f_1}}\dot{y}_{f_2}-\omega y_{f_2}+\frac{1}{C_f}\frac{y_{f_2}}{y_{f_1}}\frac{p_fy_{f_2}-q_fy_{f_1}}{y_{f_1}^2+y_{f_2}^2}-\dot{y}_{f_1}+\omega y_{f_2}+\frac{1}{C_f}\frac{p_fy_{f_1}+q_fy_{f_2}}{y_{f_1}^2+y_{f_2}^2}\}-\frac{y_{f_2}}{y_{f_1}}x_4\quad(7.471)$$

The above equation can be also written in the concise form

$$x_3=f_a(y_{f_1},\dot{y}_{f_1},y_{f_2},\dot{y}_{f_2})-\frac{y_{f_2}}{y_{f_1}}x_4\quad(7.472)$$

Next, Eq. (7.472) is substituted in the second row of the state-space model. This gives

$$\dot{y}_{f_2}=\omega y_{f_1}+\frac{1}{C_f}x_4-\frac{1}{C_f}\frac{p_fy_{f_1}+q_fy_{f_2}}{y_{f_1}^2+y_{f_2}^2}-\omega y_{f_2}--\frac{1}{C_f}\omega L_fy_{f_2}\frac{\{[f_a(y_{f_1},\dot{y}_{f_1},y_{f_2},\dot{y}_{f_2})-\frac{y_{f_2}}{y_{f_1}}x_4]^2+x_4^2\}}{y_{f_1}^2+y_{f_2}^2}\quad(7.473)$$

Eq. (7.473) is a binomial which can be solved for x_4 thus giving

$$x_4=f_b(y_{f_1},\dot{y}_{f_1},y_{f_2},\dot{y}_{f_2})\quad(7.474)$$

which signifies that x_4 is a differential function of the flat outputs vector. By substituting Eq. (7.474) in Eq. (7.472), one confirms also that

$$x_3=f_a(y_{f_1},\dot{y}_{f_1},y_{f_2},\dot{y}_{f_2})-\frac{y_{f_2}}{y_{f_1}}f_b(y_{f_1},\dot{y}_{f_1},y_{f_2},\dot{y}_{f_2})\quad(7.475)$$

From Eq. (7.475), one concludes that x_3 is also a differential function of the flat outputs vector. Next, from the third row of the state-space model of the inverter, one solves for the control input u_1. This gives

$$u_1 = L_f\{\dot{x}_3 - \omega x_4 + \tfrac{1}{L_f}x_1\} \Rightarrow u_1 = f_c(y_{f_1}, \dot{y}_{f_1}, y_2, \dot{y}_{f_2}) \tag{7.476}$$

which signifies that u_1 is a differential function of the flat outputs vector. Moreover, from the fourth row of the state-space model, one gets

$$u_2 = L_f\{\dot{x}_4 - \omega x_3 + \tfrac{1}{L_f}x_2\} \Rightarrow u_2 = f_d(y_{f_1}, \dot{y}_{f_1}, y_2, \dot{y}_{f_2}) \tag{7.477}$$

which signifies that u_2 is also a differential function of the flat outputs vector. Consequently, all state variables and the control input variables of the inverter are differential functions of the flat outputs vector and the system is differentially flat.

The inverter's dynamic model is also input–output linearizable. To obtain the system's input–output linearized form, one differentiates its flat outputs until the control inputs re-appear. It is used that the inverter's dynamics can be written as

$$\begin{aligned} \dot{x}_1 &= f_1(x) \\ \dot{x}_2 &= f_2(x) \\ \dot{x}_3 &= f_3(x) + g_3(x)u_1 \\ \dot{x}_4 &= f_4(x) + g_4(x)u_2 \end{aligned} \tag{7.478}$$

From the first row of the state-space model, differentiation with respect to time gives

$$\ddot{x}_1 = f_a(x) + g_{a_1}(x)u_1 + g_{a_2}(x)u_2 \tag{7.479}$$

where functions $f_a(x)$, $g_{a_1}(x)$, $g_{a_2}(x)$ are given by

$$\begin{aligned} f_a(x) = {} & \omega f_2 + \tfrac{1}{C_f}f_3 - \tfrac{1}{C_f}\tfrac{(p_f f_1 + q_f f_2)(x_1^2+x_2^2) - (p_f x_1 + q_f x_2)(2x_1 f_1 + 2x_2 f_2)}{(x_1^2+x_2^2)^2} + \\ & + \omega C_f f_2 - \tfrac{\omega L_f[f_2(x_3^2+x_4^2) + 2x_2x_3f_3 + 2x_2x_4f_4](x_1^2+x_2^2)}{(x_1^2+x_2^2)^2} + \\ & + \tfrac{\omega L_f x_2(x_3^2+x_4^2)(2x_1f_1+2x_2f_2)}{(x_1^2+x_2^2)^2} \end{aligned} \tag{7.480}$$

$$g_{a_1}(x) = \tfrac{g_3}{C_f} - \tfrac{(\omega L_f 2x_2x_3g_3)(x_1^2+x_2^2)}{(x_1^2+x_2^2)^2} \tag{7.481}$$

$$g_{a_2}(x) = \tfrac{(\omega L_f 2x_2x_4g_4)(x_1^2+x_2^2)}{(x_1^2+x_2^2)^2} \tag{7.482}$$

From the second row of the state-space model, differentiation with respect to time gives

$$\ddot{x}_2 = f_b(x) + g_{b_1}(x)u_1 + g_{b_2}(x)u_2 \tag{7.483}$$

where functions $f_b(x)$, $g_{b_1}(x)$, $g_{b_2}(x)$ are given by

$$\begin{aligned} f_b(x) = {} & -\omega f_1 + \tfrac{1}{C_f}f_4 - \tfrac{1}{C_f}\tfrac{(p_f f_2 + q_f f_1)(x_1^2+x_2^2) - (p_f x_2 - q_f x_1)(2x_1 f_1 + 2x_2 f_2)}{(x_1^2+x_2^2)^2} - \\ & - \omega C_f f_1 + \tfrac{\omega L_f[f_1(x_3^2+x_4^2) + 2x_1x_3f_3 + 2x_1x_4f_4](x_1^2+x_2^2)}{(x_1^2+x_2^2)^2} - \\ & - \tfrac{\omega L_f x_1(x_3^2+x_4^2)(2x_1f_1+2x_2f_2)}{(x_1^2+x_2^2)^2} \end{aligned} \tag{7.484}$$

$$g_{b_1}(x) = \tfrac{(\omega L_f 2x_1x_3g_3)(x_1^2+x_2^2)}{(x_1^2+x_2^2)^2} \tag{7.485}$$

$$g_{b_2}(x) = \tfrac{g_4}{C_f} + \tfrac{(\omega L_f 2x_1x_4g_4)(x_1^2+x_2^2)}{(x_1^2+x_2^2)^2} \tag{7.486}$$

Thus, finally, the dynamic model of the inverter is written in the input–output linearized form

$$\ddot{y}_{f_1} = f_a(x) + g_{a_1}(x)u_1 + g_{a_2}(x)u_2$$
$$\ddot{y}_{f_2} = f_b(x) + g_{b_1}(x)u_1 + g_{b_2}(x)u_2 \tag{7.487}$$

or in matrix form

$$\begin{pmatrix}\ddot{y}_{f_1}\\ \ddot{y}_{f_2}\end{pmatrix} = \begin{pmatrix}f_a(x)\\ f_b(x)\end{pmatrix} + \begin{pmatrix}g_{a_1}(x) & g_{a_2}(x)\\ g_{b_1}(x) & g_{b_2}(x)\end{pmatrix}\begin{pmatrix}u_1\\ u_2\end{pmatrix} \tag{7.488}$$

Next, the following state variables are defined: $z_1 = y_{f_1}, z_2 = \dot{y}_{f_1}, z_3 = y_{f_2}, z_4 = \dot{y}_{f_2}$. Besides, the following virtual control inputs are defined:

$$v_1 = f_a(x) + g_{a_1}(x)u_1 + g_{a_2}(x)u_2$$
$$v_2 = f_b(x) + g_{b_1}(x)u_1 + g_{b_2}(x)u_2 \tag{7.489}$$

Thus, one arrives at the concise input–output state-space form:

$$\ddot{z}_1 = v_1 \quad \ddot{z}_3 = v_2 \tag{7.490}$$

which is also written in the following matrix form

$$\begin{pmatrix}\dot{z}_1\\ \dot{z}_2\\ \dot{z}_3\\ \dot{z}_4\end{pmatrix} = \begin{pmatrix}0&1&0&0\\0&0&0&0\\0&0&0&1\\0&0&0&0\end{pmatrix}\begin{pmatrix}z_1\\ z_2\\ z_3\\ z_4\end{pmatrix} + \begin{pmatrix}0&0\\1&0\\0&0\\0&1\end{pmatrix}\begin{pmatrix}v_1\\ v_2\end{pmatrix} \tag{7.491}$$

For the input–output linearized dynamic model of the inverter a stabilizing feedback controller has the form:

$$v_1 = \ddot{z}_{1,d} - k_{1,1}(\dot{z}_1 - \dot{z}_{1,d}) - k_{2,1}(z_1 - z_1, d)$$
$$v_2 = \ddot{z}_{2,d} - k_{1,2}(\dot{z}_2 - \dot{z}_{2,d}) - k_{2,2}(z_2 - z_2, d) \tag{7.492}$$

This results in the tracking error dynamics

$$(\ddot{z}_1 - \ddot{z}_{1,d}) + k_{1,1}(\dot{z}_1 - \dot{z}_{1,d}) + k_{2,1}(z_1 - z_1, d) = 0$$
$$(\ddot{z}_2 - \ddot{z}_{2,d}) + k_{1,2}(\dot{z}_2 - \dot{z}_{2,d}) + k_{2,2}(z_2 - z_2, d) = 0 \tag{7.493}$$

and through suitable selection of the feedback gains $(k_{1,1}, k_{2,1})$ and $(k_{1,2}, k_{2,2})$ so as the associated characteristic polynomials to be Hurwitz stable, one has

$$lim_{t\to\infty} z_1 = z_{1,d} \quad lim_{t\to\infty} z_3 = z_{3,d} \tag{7.494}$$

The above condition signifies that the entire state vector of the inverter converges to the associated setpoints vector.

7.6.5 Design of a disturbance observer for the inverter

The input–output linearized dynamic model of the inverter is extended with the inclusion of additive disturbance terms. These terms describe both parametric changes and external perturbations. In case of the perturbed input–output linearized model, one has

$$\ddot{z}_1 = v_1 + d_1 \quad \ddot{z}_3 = v_2 + d_2 \tag{7.495}$$

The dynamics of the disturbance terms d_i $i = 1, 2$ is taken to be described the disturbances' derivatives up to order 2 and the associated initial conditions. The related conditions are: $\ddot{d}_1 = f_{d_1}, d_1(0), \dot{d}_1(0)$ and $\ddot{d}_2 = f_{d_2}, d_2(0), \dot{d}_2(0)$.

The state-space model of the disturbed input–output linearized model of the system is extended by considering as additional state variables the disturbance inputs and their time derivatives. Thus, one defines the extended state vector of the system: $z_1 = x_1, z_2 = \dot{x}_1, z_3 = x_2, z_4 = \dot{x}_2, z_5 = d_1, z_6 = \dot{d}_1, z_7 = d_2, z_8 = \dot{d}_2$.

This gives the extended state-space model

$$\begin{matrix} \dot{z}_1 = z_2 & \dot{z}_2 = v_1 + z_5 & \dot{z}_3 = z_4 & \dot{z}_4 = v_2 + z_7 \\ \dot{z}_5 = z_6 & \dot{z}_6 = f_{d_1} & \dot{z}_7 = z_8 & \dot{z}_8 = f_{d_2} \end{matrix} \tag{7.496}$$

or in matrix form one has

$$\begin{pmatrix} \dot{z}_1 \\ \dot{z}_2 \\ \dot{z}_3 \\ \dot{z}_4 \\ \dot{z}_5 \\ \dot{z}_6 \\ \dot{z}_7 \\ \dot{z}_8 \end{pmatrix} = \begin{pmatrix} 0 & 1 & 0 & 0 & 0 & 0 & 0 & 0 \\ 0 & 0 & 0 & 0 & 1 & 0 & 0 & 0 \\ 0 & 0 & 0 & 1 & 0 & 0 & 0 & 0 \\ 0 & 0 & 0 & 0 & 0 & 0 & 1 & 0 \\ 0 & 0 & 0 & 0 & 0 & 1 & 0 & 0 \\ 0 & 0 & 0 & 0 & 0 & 0 & 0 & 0 \\ 0 & 0 & 0 & 0 & 0 & 0 & 0 & 1 \\ 0 & 0 & 0 & 0 & 0 & 0 & 0 & 0 \end{pmatrix} \begin{pmatrix} z_1 \\ z_2 \\ z_3 \\ z_4 \\ z_5 \\ z_6 \\ z_7 \\ z_8 \end{pmatrix} + \begin{pmatrix} 0 & 0 & 0 & 0 \\ 1 & 0 & 0 & 0 \\ 0 & 0 & 0 & 0 \\ 0 & 1 & 0 & 0 \\ 0 & 0 & 0 & 0 \\ 0 & 0 & 1 & 0 \\ 0 & 0 & 0 & 0 \\ 0 & 0 & 0 & 1 \end{pmatrix} \begin{pmatrix} v! \\ v_2 \\ f_{d_1} \\ f_{d_2} \end{pmatrix} \tag{7.497}$$

The measurement equation for the inverter is

$$\begin{pmatrix} z_1^m \\ z_2^m \end{pmatrix} = \begin{pmatrix} 1 & 0 & 0 & 0 & 0 & 0 & 0 & 0 \\ 0 & 0 & 1 & 0 & 0 & 0 & 0 & 0 \end{pmatrix} \begin{pmatrix} z_1 \\ z_2 \\ z_3 \\ z_4 \\ z_5 \\ z_6 \\ z_7 \\ z_9 \end{pmatrix} \tag{7.498}$$

For the extended state-space model of the inverter given in Eq. (7.497) and Eq. (7.498) a disturbance observer is designed [711]. While the extended state-space model of the inverter has the form

$$\begin{aligned} \dot{z}_e &= A_e z_e + B_e u_e \\ z_e^m &= C_e z_e \end{aligned} \tag{7.499}$$

the disturbance observer has the form

$$\begin{aligned} \dot{\hat{z}}_o &= A_o z_o + B_o u_o + K_f(z_e^m - \hat{z}_o^m) \\ \hat{z}_o^m &= C_o \hat{z}_o \end{aligned} \tag{7.500}$$

In Eq. (7.500) matrix A_o is equal to the transition matrix A_e of the extended state-space model, while matrix B_o is equal to the first two columns of the control inputs gain matrix B_e of the extended system. Moreover, matrix C_o is equal to the measurement matrix C_e of the extended state-space model.

The disturbance observer's gain K_f is computed sequentially using the Kalman filter's recursion. The control inputs vector of the disturbance observer comprises the first two variables of the control inputs vector of the extended system, that is $v_o = [v_1, v_2]^T$. Control inputs f_{d_i} $i = 1, 2$ are unknown and are not used by the disturbances estimator. Initial conditions about the disturbance terms and their derivatives are also unknown and are not used by the disturbance observer. The extended state-space model of the inverter is which is defined by matrices A_o, B_o, C_o is observable and the convergence of the disturbance observer does not depend on initial conditions.

Matrices A_o, B_o, C_o undergo discretization and their discrete-time equivalents are denoted as $A_{o,d}$, $B_{o,d}$, and $C_{o,d}$. The recursion of the Kalman filter which computes the observer's gain K_f consists of a time-update and of a measurement-update part.

measurement update:

$$\begin{aligned} K_f(k) &= P^-(k)C_{o,d}^T[C_{o,d}P^-(k)C_{o,d}^T + R(k)]^{-1} \\ \hat{x}(k) &= \hat{x}^-(k) + K_f(k)[z_m(k) - \hat{z}_m(k)] \\ P(k) &= P^-(k) - K_f(k)C_{o.d}P^-(k) \end{aligned} \tag{7.501}$$

time update:

$$\begin{aligned} &P^-(k+1)A_{o,d}P(k)A_{o,d}^T + Q(k) \\ &\hat{x}^-(k+1) = A_{o,d}\hat{x}(k) + B_{o,d}\tilde{v}_o(k) \end{aligned} \tag{7.502}$$

Once again the process and measurement noises are considered to be white Gaussian signals. $P^-(k)$ is the state estimation error covariance matrix before receiving measurements at the k-th sampling instance, $P(k)$ is the state estimation error covariance matrix after having received measurements at the k-th sampling instance, $P^-(k+1)$ is the state estimation error covariance matrix before receiving measurements at the $(k+1)$-th sampling instance, $Q(k)$ is the process noise covariance matrix, and $R(k)$ is the measurement noise covariance matrix.

Unlike the Extended Kalman filter (EKF), the previously defined procedure for estimating state and disturbance variables in the dynamic model of the inverter's subsystem does not depend on linearization of this subsystem's model with the computation of Jacobian matrices and the associated partial derivatives. For this reason, it is also known as *Derivative-free nonlinear Kalman filter* (DKF) [711,715].

7.6.6 Statistical criteria for fault detection and isolation

7.6.6.1 Fault detection

The residuals' sequence, that is the differences between (i) the zero-valued disturbance inputs d_i $i = 1, \cdots, 2$ of the monitored subsystem, and either the rectifier and DC-line or the inverter in the fault-free case and (ii) the estimated values of the disturbance inputs $\hat{d}_i$ $i = 1, \cdots, 2$ which are provided by the Kalman filter-based disturbance observer (Figure 7.139) is a discrete error process e_k with dimension $m \times 1$ (here $m = N$ is the dimension of the output measurements vector). Actually, it is a zero-mean Gaussian white-noise process with covariance given by E_k.

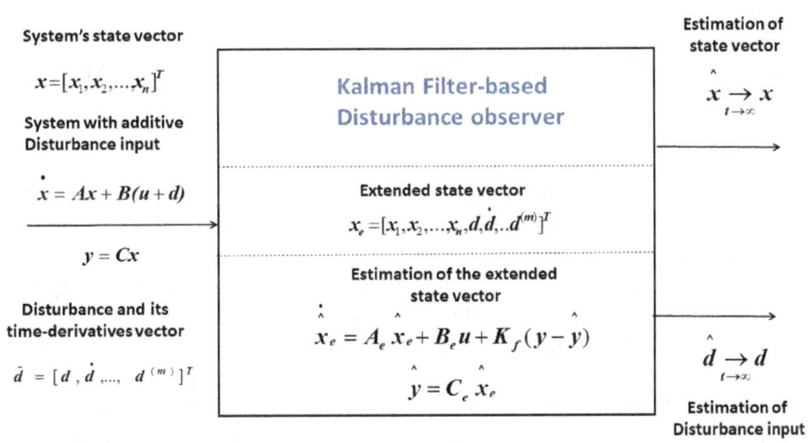

Figure 7.139 Residuals' generation for the rectifier and HVDC-line system and for the inverter, with a Kalman filter-based disturbance estimator and through the processing of the estimated disturbance inputs

A conclusion can be stated based on a measure of certainty that the rectifier and DC-line system or the inverter has not been subjected to a fault or parametric change (being included in the cumulative signals of additive disturbance inputs). To this end, the following *normalized error square* (NES) is defined [706,711]

$$\epsilon_k = e_k^T E_k^{-1} e_k \tag{7.503}$$

The sum of this normalized residuals' square follows a χ^2 distribution, with a number of degrees of freedom that is equal to the dimension of the residuals' vector. An appropriate test for the normalized error sum is to numerically show that the following condition is met within a level of confidence (according to the properties of the χ^2 distribution)

$$E\{\epsilon_k\} = m \tag{7.504}$$

This can be achieved using statistical hypothesis tests, which are associated with confidence intervals. For instance, a 95% confidence interval is frequently applied, which is specified with the notation $100(1-a)$ with $a=0.05$. Actually, a two-sided probability region is considered cutting-off two end tails of 2.5% each. For M runs the normalized error square that is obtained is given by

$$\bar{\epsilon}_k = \frac{1}{M}\sum_{i=1}^{M}\epsilon_k(i) = \frac{1}{M}\sum_{i=1}^{M}e_k^T(i)E_k^{-1}(i)e_k(i) \tag{7.505}$$

where ϵ_i stands for the i-th run at time t_k. Then $M\bar{\epsilon}_k$ will follow a χ^2 density with Mm degrees of freedom. This condition can be checked using a χ^2 test. The hypothesis holds, if the following condition is satisfied

$$\bar{\epsilon}_k \in [\zeta_1, \zeta_2] \tag{7.506}$$

where ζ_1 and ζ_2 are derived from the tail probabilities of the χ^2 density. For example, for $m=20$ (dimension of the measurements vector) and $M=100$ (total number of the output vector's samples), one has $\chi^2_{Mm}(0.025)=1878$ and $\chi^2_{Mm}(0.975)=2126$. Using that $M=100$, one obtains $\zeta_1=\chi^2_{Mm}(0.025)/M=18.78$ and $\zeta_2=\chi^2_{Mm}(0.975)/M=21.26$.

7.6.6.2 Fault isolation

By applying the statistical test into the individual components of the rectifier and DC-line subsystem and of the inverter, it is also possible to find out the specific component that has been subjected to a fault [711,715]. For a rectifier and DC-line subsystem as well as for an inverter of n parameters suspected for change, one can carry out n χ^2 statistical change detection tests. For example, each test is applied to the subset that comprises parameters $i-1$, i, and $i+1$, $i=1,2,\cdots,n$. Actually, out of the n χ^2 statistical change detection tests, the ones that exhibit the highest score are those that identify the parameter that has been subject to change.

In the case of multiple faults, one can identify the subset of parameters that has been subjected to change by applying the χ^2 statistical change detection test according to a combinatorial sequence. This means:

$$\binom{n}{k} = \frac{n!}{k!(n-k)!} \tag{7.507}$$

tests have to take place, for all clusters in the rectifier and DC-line system and for all clusters in the inverter, that finally comprise n, $n-1$, $n-2$, $\cdots$, 2, 1 parameters. Again the χ^2 tests that give the highest scores indicate the parameters which are most likely to have been subjected to change.

As a whole, the concept of the proposed statistical fault detection and isolation test is simple. The residuals' vector is formed by taking at each sampling instance the differences between the estimated values of the additive disturbance inputs $\hat{d}_i$, $i=1,2$ and the zero-valued vector $d_i=0$, $i=1,2$ that is the values of the disturbance inputs in the fault-free case. The sum of the squares of the residuals' vector, weighted by the inverse of the residuals' covariance matrix, stands for a stochastic variable which follows the χ^2 distribution. Actually, this is a multi-dimensional χ^2 distribution and the number of its degrees of freedom is equal to the dimension of the residuals' vector. Since, there are two estimated disturbance inputs in the case of the rectifier and DC-line system, as well as two disturbance inputs in the case of the inverter, the residuals' vector is of dimension 2 and the number of degrees of freedom is also 2. Next, from the properties of the χ^2 distribution, the mean value of the aforementioned stochastic variable in the fault-free case should be also 2. However, due to small fluctuations of the estimated variables, the value of the statistical test in the fault-free case will not be precisely equal to 2 but it may vary within a small range around this value. This range is determined by the confidence intervals of the χ^2 distribution. For a probability of 98% to get a value of the stochastic variable about 2, the associated confidence interval is given by the lower bound $L = 1.84$ and by the upper bound $U = 2.16$. Consequently, as long as the statistical test provides an indication that the aforementioned stochastic variable is in the interval $[L, U]$, the functioning of the rectifier and DC-line system and equivalently the functioning of the inverter can be concluded to be free of faults. On the other side, when the bounds of the previously given confidence interval are exceeded it can be concluded that the rectifier and DC-line system and equivalently the inverter has been subject to a fault. Finally, by performing the statistical test into subspaces of the rectifier and DC-line system's state-space model and equivalently in the inverter's state-space model, where each subspace is associated with different state sub-vectors, one can also achieve fault isolation. This signifies that the part of the VSC-HVDC transmission line that has caused the malfunctioning can be identified.

7.6.7 *Simulation tests*

7.6.7.1 Fault diagnosis for the rectifier and DC-line

Results about fault diagnosis of the model of the rectifier and DC-line subsystem, based on the transformation of this subsystem into the canonical (Brunovsky) form of Eq. (7.446), are given. These are presented in Figures 7.140–7.144. First, in Figure 7.140, the simulation tests demonstrate that the disturbance observer allows for simultaneous estimation of the non-measurable state variables and of the additive disturbance inputs that were affecting the rectifier and DC-line subsystem. The real values of the disturbance signals are depicted in blue color while their estimates, which have been provided by the Kalman filter, are shown in red color. Although additive disturbance faults d_1, d_2 affect the dynamics of the system, these cannot be distinguished in the system's outputs z_1, z_3. The robustness of the control loop "hides" the faults and enables the subsystem's outputs reach the targeted reference values,

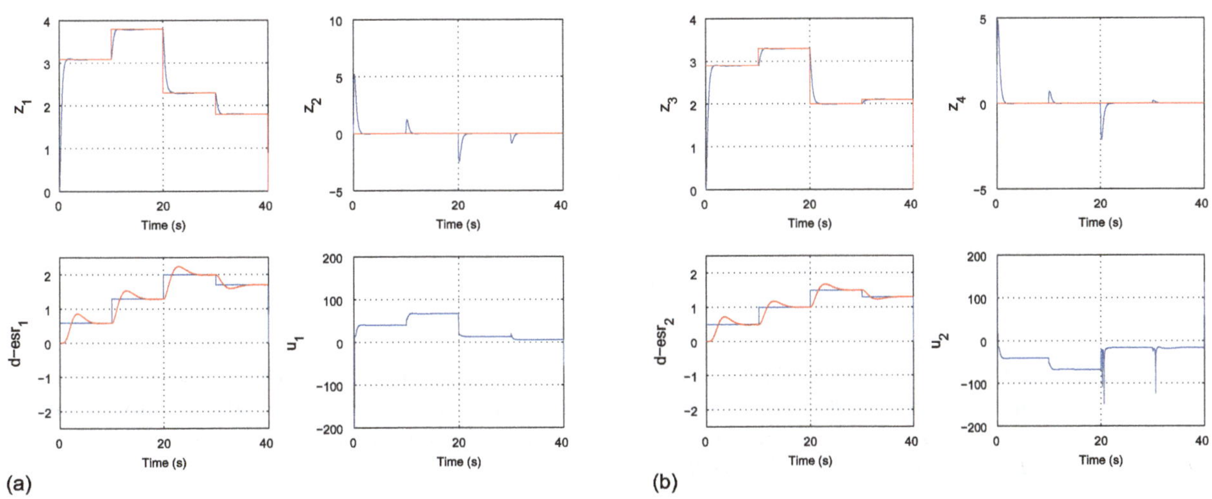

Figure 7.140 Control of the VSC-HVDC transmission system at the rectifier and DC-line with the use of the Kalman filter-based disturbance observer: (a) top row: convergence of state variables z_1 and z_2 (blue line) to the associated setpoint (red line); bottom row: estimation of the additive disturbance term $\hat{d}_1$ (red line) vs real disturbance value (blue line) and control input u_1 (blue line); (b) top row: convergence of state variables z_3 and z_4 (blue line) to the associated setpoint (red line), bottom row: estimation of the additive disturbance term $\hat{d}_2$ (red line) vs real disturbance value (blue line) and control input u_2 (blue line)

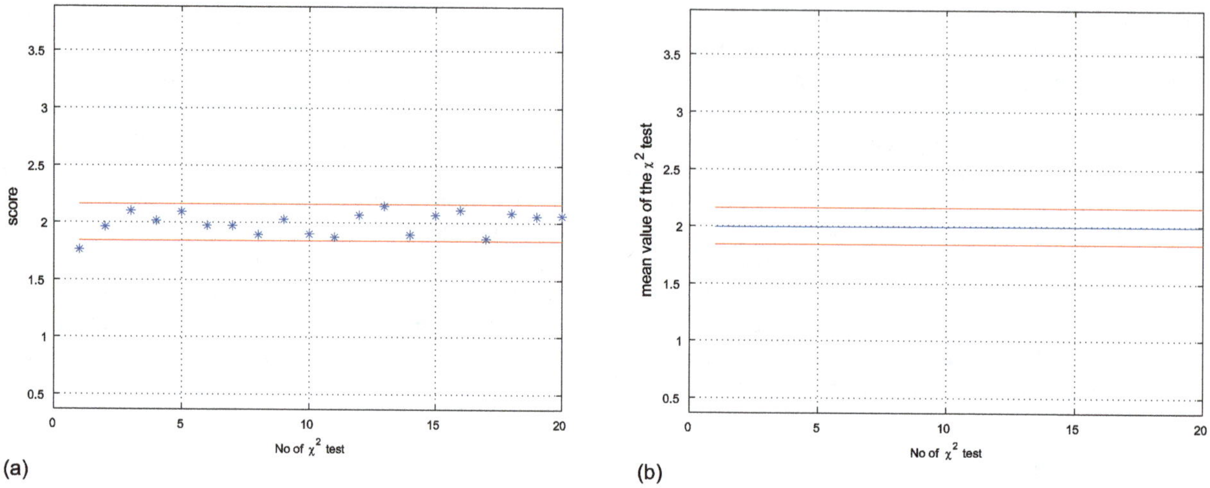

Figure 7.141 Fault diagnosis of the VSC-HVDC transmission system at the rectifier and DC-line with the use of the Kalman filter-based disturbance observer in case of no fault in this subsystem: (a) individual χ^2 tests (blue marking) vs the fault thresholds (red lines); (b) mean value of the χ^2 tests (blue line) vs the fault thresholds (red lines)

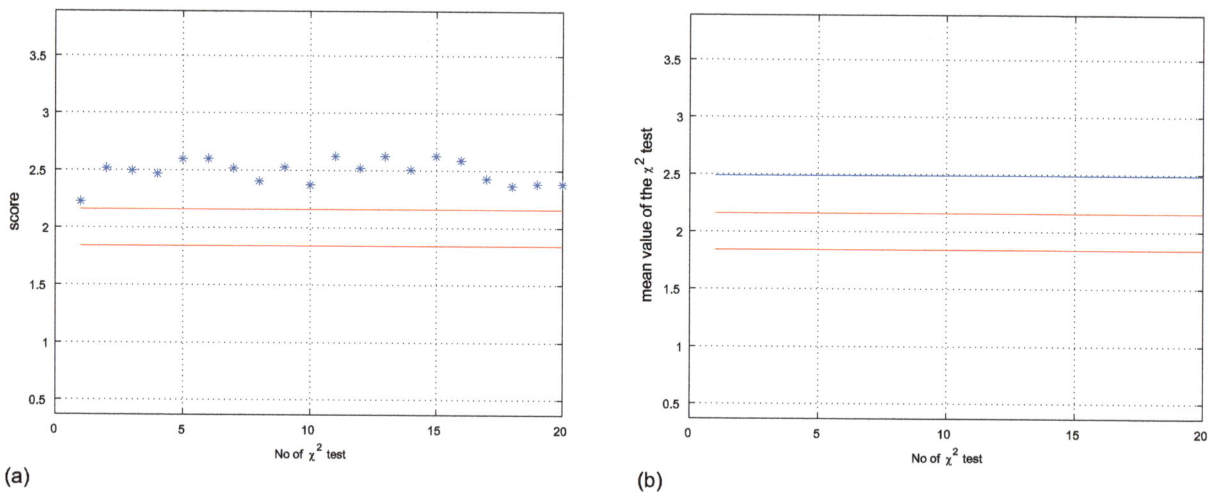

Figure 7.142 Fault diagnosis of the VSC-HVDC transmission system at the rectifier and DC-line with the use of the Kalman filter-based disturbance observer in case of fault associated with disturbance term d_1 (fault's value $d_1 = 0.0010$): (a) individual χ^2 tests (blue marking) vs the fault thresholds (red lines); (b) mean value of the χ^2 tests (blue line) vs the fault thresholds (red lines)

Next, the statistical fault diagnosis test is performed in the disturbance inputs d_1, d_2. Considering $n = 2$ disturbance inputs for the rectifier and DC-line subsystem, to conclude the normal functioning of the subsystem, the previously analyzed statistical test should take a value that is very close to the mean value of χ^2 distribution, that is 2. The residuals' vector is formed by taking at each sampling instance the differences between the estimated values of the additive disturbance inputs $[\hat{d}_1, \hat{d}_2]$ and the zero-valued vector $[d_i = 0, d_2 = 0]$ that is the values of the disturbance inputs in the fault-free case. Thus, the dimension of the residuals' vector is $m = 2$. Considering that the number of residual vector

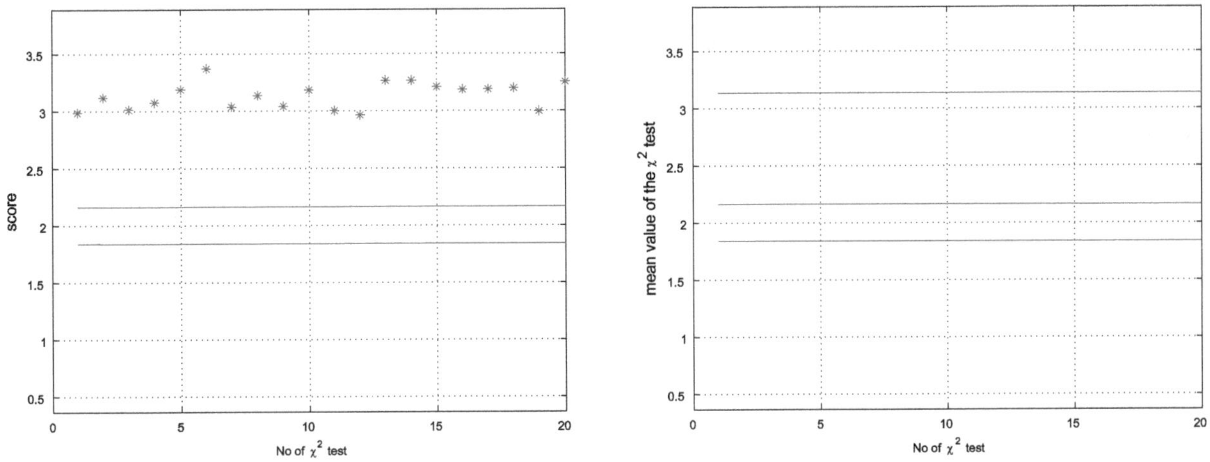

(a) (b)

Figure 7.143 Fault diagnosis of the VSC-HVDC transmission system at the rectifier and DC-line with the use of the Kalman filter-based disturbance observer in case of fault associated with disturbance term d_2 (fault's value $d_2 = 0.0015$): (a) individual χ^2 tests (blue marking) vs the fault thresholds (red lines); (b) mean value of the χ^2 tests (blue line) vs the fault thresholds (red lines)

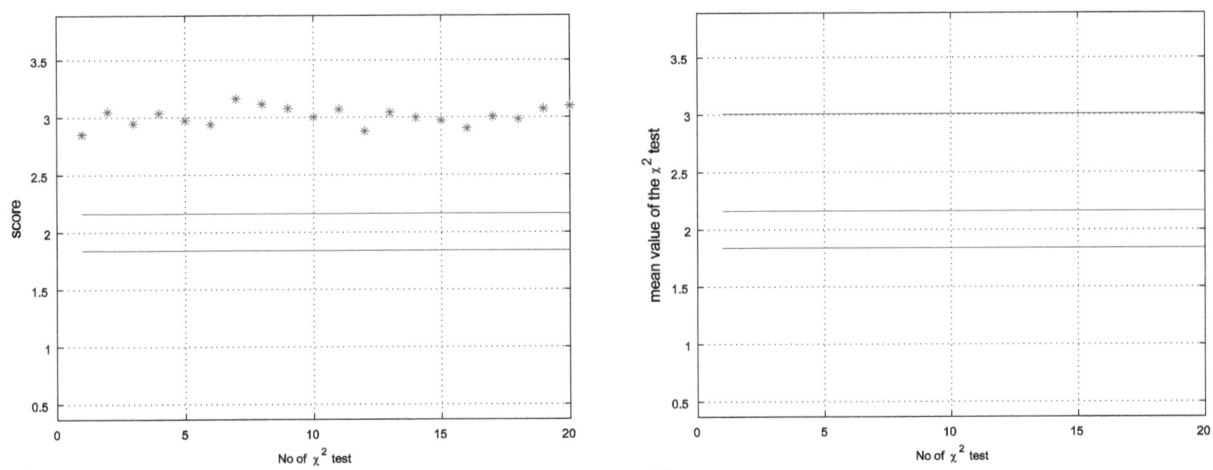

(a) (b)

Figure 7.144 Fault diagnosis of the VSC-HVDC system at the rectifier and DC-line with the use of the Kalman filter-based disturbance observer in case of fault associated with disturbance term d_1 (fault's value $d_1 = 0.0005$) and d_2 (fault's value $d_2 = 0.0010$): (a) individual χ^2 tests (blue marking) vs the fault thresholds (red lines); (b) mean value of the χ^2 tests (blue line) vs the fault thresholds (red lines)

samples is $M = 2000$ and using a 98% confidence interval for the χ^2 distribution the fault thresholds can be given the values $L = 1.84$ and $U = 2.16$. In an equivalent manner, when the statistical test is applied exclusively to each one of the 2 subsystems that constitute the rectifier and DC-line system the fault thresholds are defined as $L = 0.89$ and $U = 1.11$. The proposed fault diagnosis approach gives early warnings even about the appearance of incipient failures, for instance, disturbance terms of the order of $d_i = 0.001$ or $d_i = 0.002$.

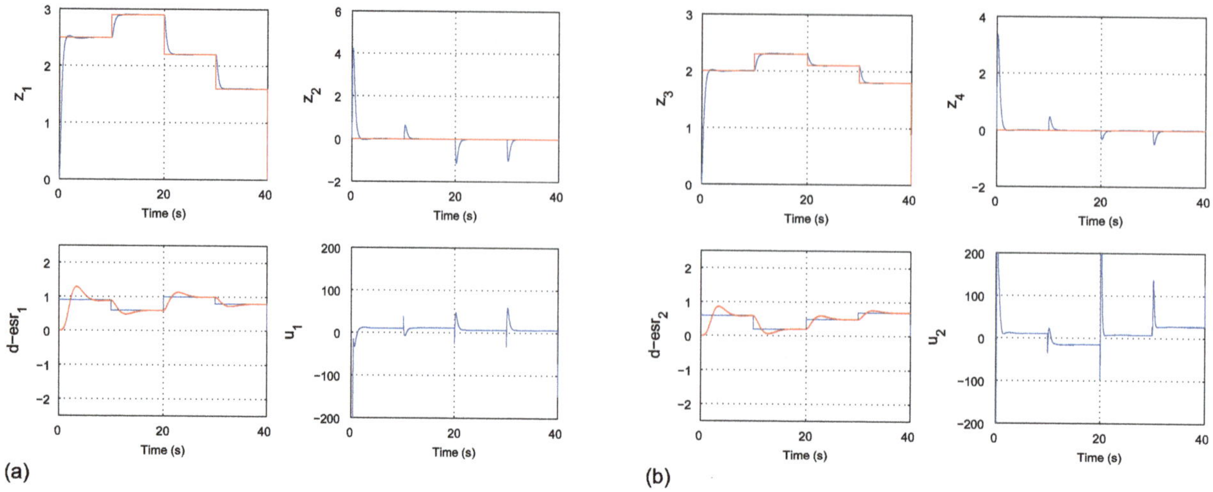

Figure 7.145 Control of the VSC-HVDC transmission system at the inverter with the use of the Kalman filter-based disturbance observer: (a) top row: convergence of state variables z_1 and z_2 (blue line) to the associated setpoint (red line), bottom row: estimation of the additive disturbance term $\hat{d}_1$ (red line) vs real disturbance value (blue line) and control input u_1 (blue line); (b) top row: convergence of state variables z_3 and z_4 (blue line) to the associated setpoint (red line), bottom row: estimation of the additive disturbance term $\hat{d}_2$ (red line) vs real disturbance value (blue line) and control input u_2 (blue line)

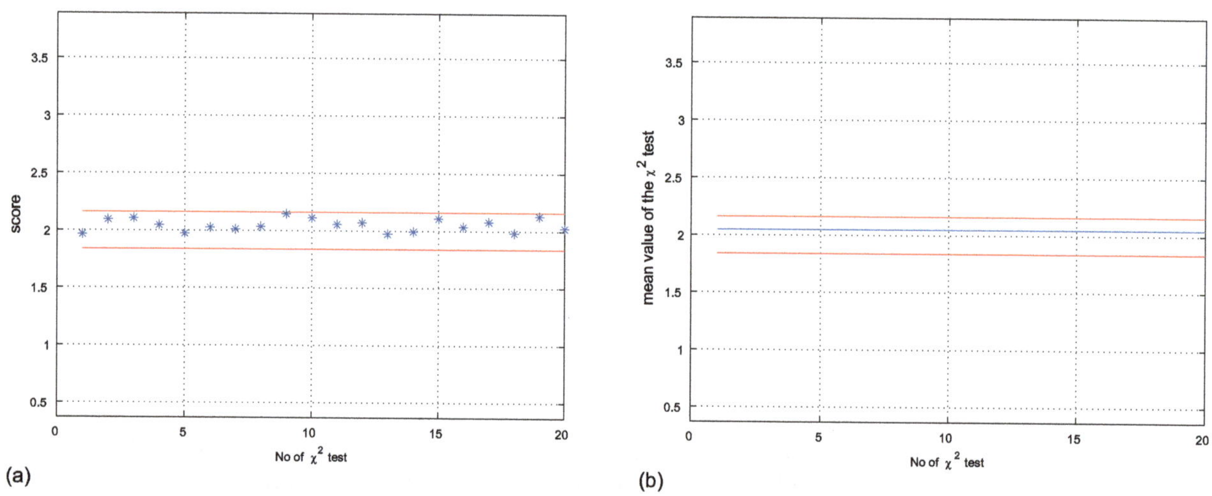

Figure 7.146 Fault diagnosis of the VSC-HVDC transmission system at the inverter with the use of the Kalman filter-based disturbance observer in case of no fault in this subsystem: (a) individual χ^2 tests (blue marking) vs the fault thresholds (red lines); (b) mean value of the χ^2 tests (blue line) vs the fault thresholds (red lines)

In case of fault detection with χ^2 tests, the Derivative-free nonlinear Kalman filter has been used as a disturbance estimator and the statistical χ^2 test has been applied to the rectifier and DC-line subsystem. In particular, the following cases have been examined:

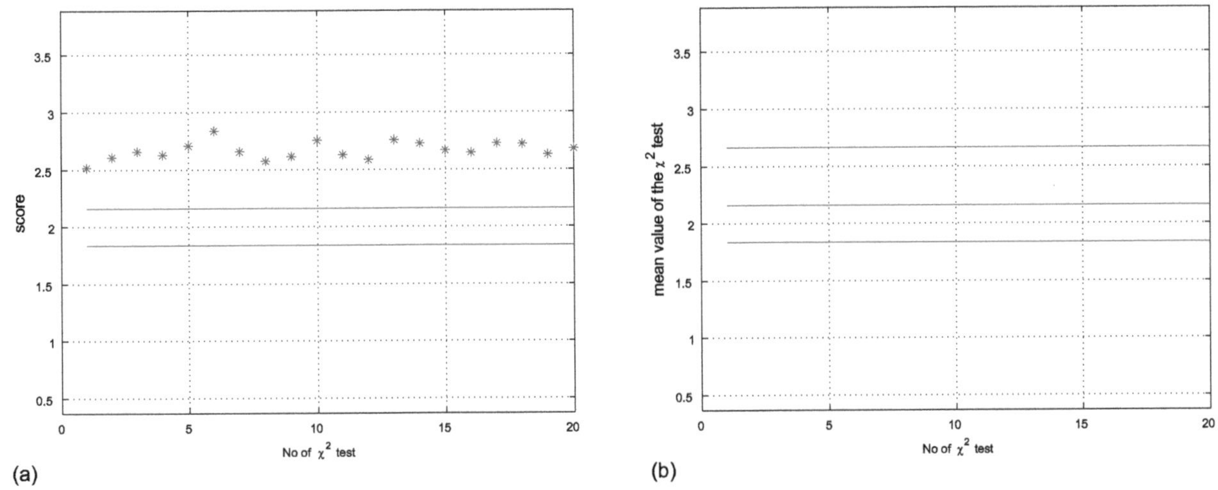

Figure 7.147 Fault diagnosis of the VSC-HVDC transmission system at the inverter with the use of the Kalman filter-based disturbance observer in case of fault associated with disturbance term d_1 (fault's value $d_1 = 0.0015$): (a) individual χ^2 tests (blue marking) vs the fault thresholds (red lines); (b) mean value of the χ^2 tests (blue line) vs the fault thresholds (red lines)

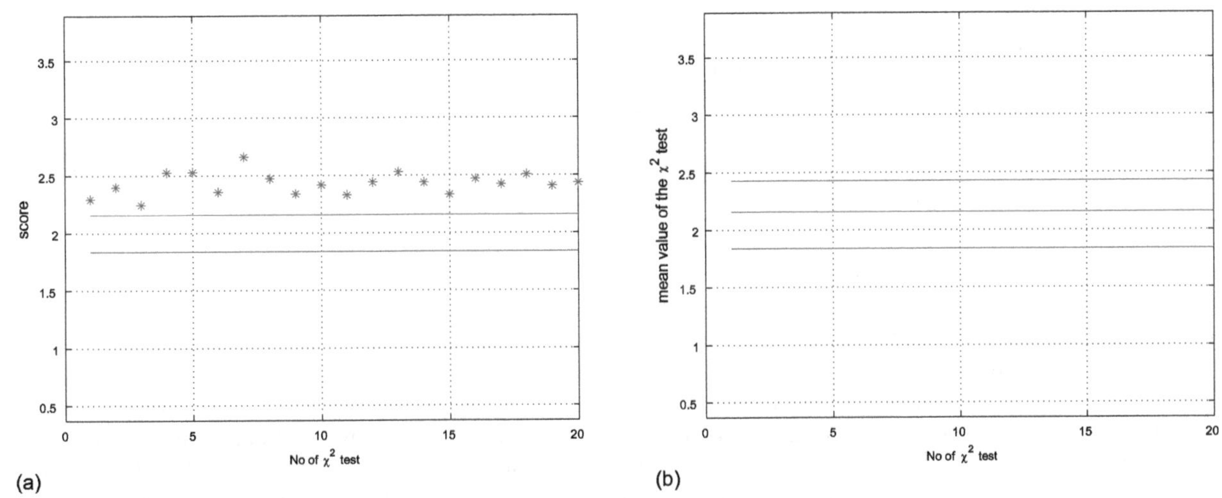

Figure 7.148 Fault diagnosis of the VSC-HVDC transmission system at the inverter with the use of the Kalman filter-based disturbance observer in case of fault associated with disturbance term d_2 (fault's value $d_2 = 0.0010$): (a) individual χ^2 tests (blue marking) vs the fault thresholds (red lines); (b) mean value of the χ^2 tests (blue line) vs the fault thresholds (red lines)

- case (0): functioning of the rectifier and DC-line subsystem in the fault-free condition $d_i = 0$ for $i = 1, 2$. The associated results are depicted in Figure 7.141.
- case (1) functioning of the rectifier and DC-line subsystem under additive input disturbance d_1 at its first control input. The associated results are depicted in Figure 7.142
- case (2): functioning of the rectifier and DC-line subsystem under additive input disturbance d_2 at its second control input. The associated results are depicted in Figure 7.143

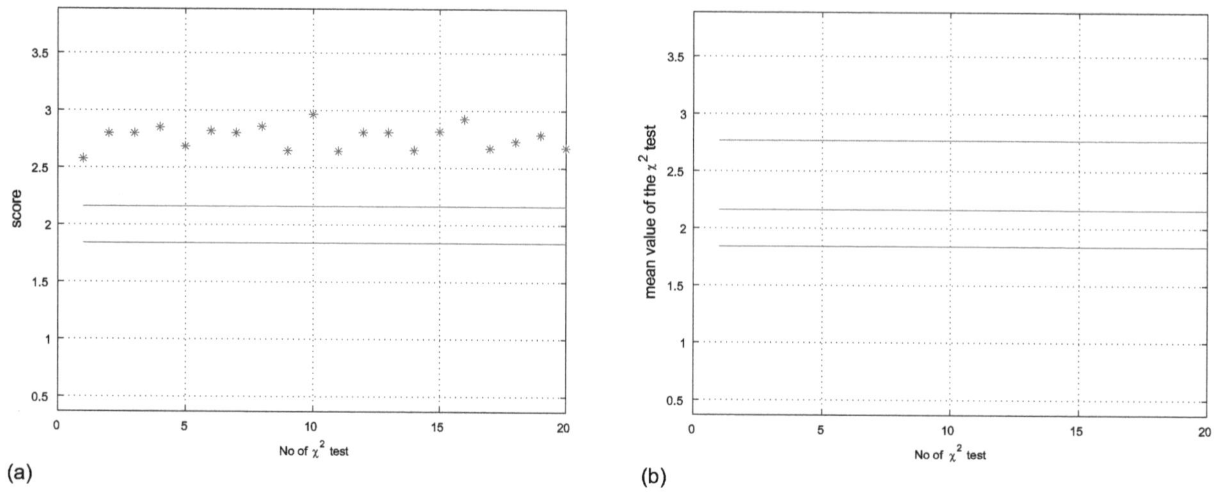

Figure 7.149 Fault diagnosis of the VSC-HVDC system at the inverter with the use of the Kalman filter-based disturbance observer in case of fault associated with disturbance term d_1 (fault's value $d_1 = 0.0010$) and d_2 (fault's value $d_2 = 0.0005$): (a) individual χ^2 tests (blue marking) vs the fault thresholds (red lines); (b) mean value of the χ^2 tests (blue line) vs the fault thresholds (red lines)

- case (3) functioning of the rectifier and DC-line subsystem under additive input disturbances d_1 and d_2 at both its first and second control input. The associated results are depicted in Figure 7.144

It has been confirmed that in all aforementioned cases the proposed condition monitoring method was capable of detecting incipient faults appearing in components of the rectifier and DC-line system.

7.6.7.2 Fault diagnosis for the inverter

Results about fault diagnosis of the model of inverter, based on the transformation of this subsystem into the canonical (Brunovsky) form of Eq. (7.490), are given. These are presented in Figures 7.145–7.149. First, in Figure 7.145. The simulation tests demonstrate that the disturbance observer allows for simultaneous estimation of the non-measurable state variables and of the additive disturbance inputs that were affecting the inverter. The real values of the disturbance signals are depicted in blue color while their estimates, which have been provided by the Kalman filter, are shown in red color. Although additive disturbance faults d_1, d_2 affect the dynamics of the system, these cannot be distinguished in the system's outputs z_1, z_3. The robustness of the control loop "hides" the faults and enables the system's outputs to reach the targeted reference values,

Next, the statistical fault diagnosis test is performed in the disturbance inputs d_1, d_2. Considering $n = 2$ disturbance inputs for the inverter, to conclude the normal functioning of this subsystem, the previously analyzed statistical test should take a value that is very close to the mean value of χ^2 distribution, that is 2. The residuals' vector is formed by taking at each sampling instance the differences between the estimated values of the additive disturbance inputs $[\hat{d}_1, \hat{d}_2]$ and the zero-valued vector $[d_i = 0, d_2 = 0]$ that is the values of the disturbance inputs in the fault-free case. Thus, the dimension of the residuals' vector is $m = 2$. Considering that the number of residual vector samples is $M = 2000$ and using a 98% confidence interval for the χ^2 distribution the fault thresholds can be given the values $L = 1.84$ and $U = 2.16$. In an equivalent manner, when the statistical test is applied exclusively to each one of the two subsystems that constitute the inverter the fault thresholds are defined as $L = 0.89$ and $U = 1.11$. The proposed fault diagnosis approach gives early warnings even about the appearance of incipient failures, for instance, disturbance terms of the order of $d_i = 0.001$ or $d_i = 0.002$.

In case of fault detection with χ^2 tests, the Derivative-free nonlinear Kalman filter has been used as a disturbance estimator and the statistical χ^2 test has been applied to the inverter. In particular, the following cases have been examined:

- case (0): functioning of the inverter in the fault-free condition $d_i = 0$ for $i = 1, 2$. The associated results are depicted in Figure 7.146
- case (1) functioning of the inverter under additive input disturbance d_1 at its first control input. The associated results are depicted in Figure 7.147
- case (2): functioning of the inverter under additive input disturbance d_2 at its second control input. The associated results are depicted in Figure 7.148
- case (3) functioning of the inverter under additive input disturbances d_1 and d_2 at both its first and second control input. The associated results are depicted in Figure 7.149

It has been confirmed that, in all aforementioned cases, the proposed condition monitoring method was capable of detecting incipient faults appearing in components of the inverter.

Chapter 8
Nonlinear optimal and flatness-based control for biological and financial systems

Abstract

The chapter analyzes nonlinear optimal control and flatness-based control for dynamical models of biological systems and financial systems. First, nonlinear optimal control is applied for the treatment of bacterial infections with resistance to antibiotics. Additionally, nonlinear optimal control is proposed for the optimized infusion of chemotherapy and the treatment of leukemia. In the same context, one could also consider nonlinear optimal control and flatness-based control for the treatment of tumor growth. In general, nonlinear optimal control and multi-loop flatness-based control can be used for the stabilization of competing population dynamics, among which cell populations are a specific case. Finally, in this chapter, nonlinear optimal control is used for the optimized management of multi-echelon supply chain networks under time delays.

8.1 Nonlinear optimal control for bacterial infections with resistance to antibiotics

8.1.1 Introduction

Finally, the monograph extends its application domain to the use of its nonlinear optimal and flatness-based control methods in biosystems and in financial systems. Antibiotics are secondary metabolites of certain micro-organisms which can selectively inhibit or eliminate other micro-organisms without bringing serious toxicity to the host. The overuse and misuse of antibiotics have become a major problem for public health since it has resulted in raising the resistance of bacterial infections to antibiotics treatment [377,663]. Under antibiotics, sensitive bacteria are inhibited or killed. However, resistant bacteria change also their metabolic pathways and survive. Bacterial resistance caused by prolonged administration of the same antibiotics, exacerbates the threat of bacterial infections [42,79]. Resistance to antibiotics can be natural having a genetic substrate or can be acquired by biochemical mechanisms [979,327]. Natural resistance is an innate feature of certain bacteria and is not caused by high doses of an antibiotic. Acquired resistance is due to change of the DNA of bacteria, or due to contact of their genetic material with external sources, such as plasmids [194,1035]. There exist mathematical models that describe the resistance of bacteria to antibiotics either at the host's level, or at the level of interaction between different hosts, as for instance in the case of in-hospital bacterial infections [592,575].

It is estimated that nearly 1 million people die every year because of bacterial infections that cannot be treated with common antibiotics. In-hospital bacterial infections have become more severe, hospitalization lasts for longer periods and the cost for recuperation of patients is significantly raised. The apparent ease at which resistance to antibiotics spreads, is primarily due to the ability of bacteria to modify their DNA through the attachment of extrachromosomal material which is transferred to them by plasmids [28,500]. The use of antibiotics must be carried out in a manner that minimizes the risk of growth of the population of resistant bacteria [430,670]. To this end, the nonlinear optimal control problem for the treatment with antibiotics has been considered [1090,271]. On the one side, it has been attempted to solve the nonlinear optimal control problem with the use of Bellman's optimality principle [417,87,191]. On the other side the use of Pontryagin's maximum principle has been proposed [670,1090,417], [93,87]. The latter approach requires the definition of a new Hamiltonian function for the controlled system so as to describe the optimization objectives [592,500,378,367]. Besides, it needs the repetitive solution of a set of differential equations at each time step of the control algorithm so as to compute "adjoint coefficients" [1035,271,191], [155]. The control inputs it generates are often of the bang-bang type and switch between two extreme values. Additional results on the control problem of cells and microbial populations can be found in [488,314,557,203,88,6,390]. Administered provision of anti-infection and anti-bacterial treatment is a key factor for the successful abatement of the spreading of such infections.

In this section it is proven that the dynamic model of the bacterial infection with resistance to antibiotics is differentially flat [711,709]. This provides an implicit proof of the controllability of this system and allows for defining feasible setpoints for all of its state variables. Besides, in this section, a novel solution is proposed for the nonlinear optimal control problem of bacterial infections with resistance to antibiotics [732,725]. The method is based on approximate linearization of the dynamic model of bacterial infection with the use of first-order Taylor series expansion and through the computation of the associated Jacobian matrices [705,60,706]. The linearization process takes place at each sampling instance, around a temporary operating point (x^*, u^*) which is defined by the present value of the system's state vector x^* and by the last sampled value of the control inputs vector u^*. The modeling error, which is due to the truncation of higher-order terms from the Taylor series expansion, is considered to be a perturbation that is asymptotically compensated by the robustness of the control algorithm. For the approximately linearized model of the bacterial infection with resistance to antibiotics an H-infinity feedback controller is designed.

The H-infinity controller enables the solution of the optimal control problem and the optimized supply of the patient with antibiotics under model uncertainty about the infection's dynamics and under the effects of exogenous perturbations. Actually, the H-infinity controller represents a min–max differential game taking place between: (i) the control inputs which try to minimize a cost function that contains a quadratic term of the state vector's tracking error and (ii) the model uncertainty and external disturbances which try to maximize this cost function. To compute the stabilizing gains of the H-infinity controller an algebraic Riccati equation is also repetitively solved at each time step of the control algorithm. The global stability properties of the control scheme are proven through Lyapunov analysis [726,735]. First, it is demonstrated that the control scheme satisfies the H-infinity tracking performance criterion which signifies robustness to model uncertainty and perturbations. Moreover, under moderate conditions, it is proven that the control scheme is globally asymptotically stable [725,876]. The nonlinear optimal control method achieves fast and accurate tracking of reference values by the state variables that represent bacteria populations in the host, and this is done under moderate variations of the control inputs. Moreover, to implement state estimation-based control without the need to measure the entire state vector of the bacteria populations, the H-infinity Kalman filter is used as a robust state observer.

The proposed nonlinear optimal control method for bacterial infection dynamics exhibits advantages compared to input–output linearization-based control schemes for this system. Unlike the input–output linearization-based control approach, the nonlinear optimal control method does not need changes of state variables and complicated state-space model transformations. The control inputs it computes are applied directly on the initial nonlinear model of the bacterial infection dynamics without the need for inverse transformations and without coming against singularity issues. Besides, it is only the nonlinear optimal control method that ensures minimization of the variations of the control inputs, and thus it enables reaching the targeted therapeutic effects without excessive consumption of antibiotics.

8.1.2 Dynamic model of bacterial infection with resistance to antibiotics

8.1.2.1 Formulation of the state-space model

The considered model of bacterial infection comprises two populations: namely $S(t)$ is the bacterial population which is sensitive to antibiotics, and $R(t)$ is the bacteria population which has developed resistance to antibiotics [378]. Both populations exist in the same host (infected patient). Moreover, there exists the plasmids population $P(t)$ in the host which is the genetic material that leaves the bacteria with resistance to antibiotics $R(t)$ and which is finally attached to the bacteria of population $S(t)$, thus making progressively the latter population be also resistant to the use of antibiotics. The reproduction rate of bacteria which are sensitive to antibiotics is denoted as β_s, and the associated decay rate is described as μ_s. The reproduction rate of bacteria which are resistant to antibiotics is denoted as β_r, and the associated decay rate is described as μ_r.

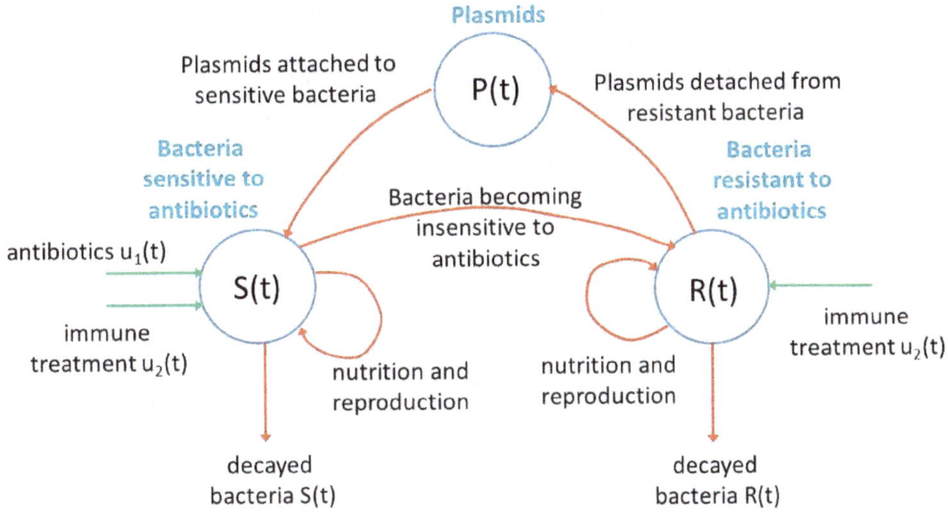

Figure 8.1 Diagram of the dynamics of the bacterial infection with resistance to antibiotics: S(t) population of bacteria with sensitivity to antibiotics, R(t) population of bacteria with resistance to antibiotics, P(t) plasmids

Plasmids increase proportionally to the population of the resistant bacteria, at a replication rate which is denoted as σ_p, and they degrade at a constant rate which is denoted as μ_p. In the related dynamic model, the term K is the carrying capacity of the bacteria and is a coefficient that determines the sufficiency of their nutrition in the host. Sensitive-to-antibiotics bacteria get mutated at a rate defined by coefficient q and at an amount equal to $qS(t)$. The population of the sensitive-to-antibiotics bacteria is decreased at each time instance due to the effect of plasmids, at a rate which is equal to δPS. The population of the resistant to antibiotics bacteria is increased at each time instance due to the effect of plasmids, at a rate which is again equal to δPS.

Combining the above into one dynamic model, without the effect of exogenous control (that is without considering the use of antibiotics or immune treatment), the evolution of the bacterial infection is given in the following [378]:

$$
\begin{aligned}
\frac{dS}{dt} &= \beta_s S\left(1-\frac{S+R}{K}\right) - qS - \delta PS - \mu_s S \\
\frac{dR}{dt} &= \beta_r R\left(1-\frac{S+R}{K}\right) + qS + \delta PS - \mu_r R \\
\frac{dP}{dt} &= \sigma_p R - \mu_p P
\end{aligned}
\tag{8.1}
$$

One can confirm the existence of equilibria if no control inputs are applied to the dynamic model of the bacterial infection: (i) first equilibrium $[x_1^*, x_2^*, x_3^*] = [0, 0, 0]$, (ii) second equilibrium $[x_1^*, x_2^*, x_3^*] = [0, K(1-\frac{\mu_r}{\beta_r}), \frac{\sigma_p}{\mu_p}K(1-\frac{\mu_r}{\beta_r})]$ for $\frac{\mu_r}{\beta_r} < 1$, and (iii) third equilibrium $[x_1^*, x_2^*, x_3^*]$ with x_3^* to be the solution of a binomial in the form $x_3^* = \frac{-b+\sqrt{b^2-4ac}}{2a}$ with a, b, c to be constants, while $x_1^* = -\frac{K}{\beta_s}[q + \delta x_3^* + \mu_s - \beta_s + \frac{\beta_s}{K}\frac{\mu_p}{\sigma_p}x_3^*]$ and $x_2^* = \frac{\mu_p}{\sigma_p}x_3^*$.

The stability properties of these equilibria can be found by substituting them to the Jacobian matrix of the system and by computing the associated eigenvalues. The Jacobian matrix of the autonomous system is:

$$
\nabla_x f(x)\mid_{(x_1^*, x_2^*, x_3^*)} = \begin{pmatrix} \beta_s\left(1-\frac{x_1^*+x_2^*}{K}\right) + \beta_s x_1^*\left(-\frac{1}{K}\right) - q - \delta x_3^* - \mu_s & 0 & \beta_s x_1^*\left(-\frac{1}{K}\right) - \delta x_1^* \\ \beta_r\left(1-\frac{x_1^*+x_2^*}{K}\right) + \beta_r x_1^*\left(-\frac{1}{K}\right) + q + \delta x_3^* - \mu_r & 0 & \beta_r x_1^*\left(-\frac{1}{K}\right) + \delta x_1^* \\ 0 & \sigma_p & -\mu_p \end{pmatrix}
\tag{8.2}
$$

Next, the dynamic model of the bacterial infection is modified with the inclusion of exogenous control inputs. These control inputs are: (i) the use of antibiotics which affects only the population of the sensitive bacteria by an amount equal to u_1S and (ii) the use of immune treatment (e.g. vaccination or monoclonal antibodies) that results in decay of the population of the sensitive bacteria by an amount equal to u_2S and in decay of the population of resistant bacteria by an amount equal to u_2R.

Under the effect of exogenous control inputs the dynamic model of the bacterial infection is given by [378]

$$\begin{aligned}\frac{dS}{dt} &= \beta_s S(1-\tfrac{S+R}{K}) - (q+u_1(t))S - \delta PS - u_2(t)S - \mu_s S\\ \frac{dR}{dt} &= \beta_r R(1-\tfrac{S+R}{K}) + qS + \delta PS - u_2(t)R - \mu_r R\\ \frac{dP}{dt} &= \sigma_p R - \mu_p P\end{aligned} \tag{8.3}$$

Next, the state vector in the bacterial infection dynamics is defined as $x=[x_1,x_2,x_3]^T$ or $x=[S,R,P]^T$, while the control inputs vector of the system is $u=[u_1,u_2]^T$. Thus, the state-space description of the bacterial infection is written as

$$\begin{aligned}\dot{x}_1 &= \beta_s x_1(1-\tfrac{x_1+x_2}{K}) - qx_1 - \delta x_3x_1 - \mu_s x_1 - qx_1u_1(t) - x_1u_2(t)\\ \dot{x}_2 &= \beta_r x_2(1-\tfrac{x_1+x_2}{K}) + qx_1 + \delta x_3x_1 - \mu_r x_2 - x_2u_2(t)\\ \dot{x}_3 &= \sigma_p x_2 - \mu_p x_3\end{aligned} \tag{8.4}$$

In matrix form, the dynamic model of the bacterial infection is written as

$$\begin{pmatrix}\dot{x}_1\\ \dot{x}_2\\ \dot{x}_3\end{pmatrix} = \begin{pmatrix}\beta_s x_1(1-\frac{x_1+x_2}{K}) - qx_1 - \delta x_3x_1 - \mu_s x_1\\ \beta_r x_2(1-\frac{x_1+x_2}{K}) + qx_1 + \delta x_3x_1 - \mu_r x_2\\ \sigma_p x_2 - \mu_p x_3\end{pmatrix} + \begin{pmatrix}-qx_1 & -x_1\\ 0 & -x_2\\ 0 & 0\end{pmatrix}\begin{pmatrix}u_1\\ u_2\end{pmatrix} \tag{8.5}$$

which can be also written in the concise nonlinear affine-in-the-input state-space form

$$\dot{x} = f(x) + g(x)u \tag{8.6}$$

with $x \in R^{3\times1}$, $f(x) \in R^{3\times1}$, $g(x) \in R^{3\times1}$, and $u \in R^{2\times1}$. This system, in its present form, is not directly feedback linearizable.

8.1.2.2 Differential flatness properties of the bacterial infection dynamics

It will be proven that the dynamic model of the bacterial infection is differentially flat, with flat outputs vector $Y=[x_1,x_3]^T=[S,P]^T$. From the third row of the state-space model, one has [711,709]

$$\begin{aligned}\dot{x}_3 = \sigma_p x_2 - \mu_p x_3 &\Rightarrow x_2 = \frac{1}{\sigma_p}(\dot{x}_3 + \mu_p x_3)\\ &\Rightarrow x_2 = h_2(Y,\dot{Y})\end{aligned} \tag{8.7}$$

which signifies that x_2 is a differential function of the flat outputs vector. Next, from the first two rows of the state-space model, one has

$$\begin{pmatrix}\dot{x}_1\\ \dot{x}_2\end{pmatrix} = \begin{pmatrix}\beta_s x_1(1-\frac{x_1+x_2}{K}) - qx_1 - \delta x_3x_1 - \mu_s x_1\\ \beta_r x_2(1-\frac{x_1+x_2}{K}) + qx_1 + \delta x_3x_1 - \mu_r x_2\end{pmatrix} + \begin{pmatrix}-qx_1 & -x_1\\ 0 & -x_2\end{pmatrix}\begin{pmatrix}u_1\\ u_2\end{pmatrix} \tag{8.8}$$

and under the condition that $x_1 \neq 0$, $x_2 \neq 0$, one can solve for the control inputs vector u thus obtaining

$$\begin{pmatrix}u_1\\ u_2\end{pmatrix} = \begin{pmatrix}-qx_1 & -x_1\\ 0 & -x_2\end{pmatrix}^{-1}\left\{\begin{pmatrix}\dot{x}_1\\ \dot{x}_2\end{pmatrix} - \begin{pmatrix}\beta_s x_1(1-\frac{x_1+x_2}{K}) - qx_1 - \delta x_3x_1 - \mu_s x_1\\ \beta_r x_2(1-\frac{x_1+x_2}{K}) + qx_1 + \delta x_3x_1 - \mu_r x_2\end{pmatrix}\right\} \tag{8.9}$$

Knowing that x_1, x_2, x_3 are differential functions of the flat outputs vector, then from Eq. (8.9), one has that the control inputs u_1, u_2 are also differential functions of the flat outputs, or

$$u_1 = h_{u_1}(Y,\dot{Y}) \quad u_2 = h_{u_2}(Y,\dot{Y}) \tag{8.10}$$

Consequently, all state variables and the control inputs of the model of the bacterial infection under resistance to antibiotics are differential functions of the flat outputs vector and the system is differentially flat. Differential flatness is an implicit proof of the controllability of this system. Besides, it allows for computing feasible setpoints for all state variables of the system. First, one selects setpoints in an unconstrained manner for state variables x_1, x_3 which coincide with the flat outputs of the system. Next, for the rest of the state variables, that is for x_2, state variables are chosen using the differential relations which connect them to the flat outputs.

8.1.2.3 Input–output linearization-based control for the bacterial infection dynamics

It has been shown that the dynamic model of the bacterial infection with resistance to antibiotics comprises the following state equations:

$$\dot{x}_1 = \beta_s x_1\left(1 - \tfrac{x_1+x_2}{K}\right) - qx_1 - \delta x_1 - \delta x_3 x_1 - \mu_s x_1 - qx_1u_1 - x_1u_2 \tag{8.11}$$

$$\dot{x}_2 = \beta_r x_2\left(1 - \tfrac{x_1+x_2}{K}\right) + qx_1 + \delta x_3 x_1 - \mu_r x_2 - x_2u_2 \tag{8.12}$$

$$\dot{x}_3 = \sigma_p x_2 - \mu_r x_3 \tag{8.13}$$

It has been shown that the system is differentially flat, with flat outputs vector $Y = [x_1, x_3]^T$. This also confirms that the system is input–output linearizable. To obtain the input–output linearized description, the flat outputs of the system are differentiated until the control inputs re-appear.

The first state equation of the system about $\dot{x}_1$ given in Eq. (8.11) contains already the control inputs u_1 and u_2 and does not need additional differentiation.

The third state equation of the system about $\dot{x}_3$ given in Eq. (8.13) after differentiation gives

$$\begin{aligned} \ddot{x}_3 &= \sigma_p\dot{x}_2 - \mu_p\dot{x}_3 \Rightarrow \\ \ddot{x}_3 &= \sigma_p\left[\beta_r x_2\left(1 - \tfrac{x_1+x_2}{K}\right) + qx_1 + \delta x_3x_1 - \mu_r x_2 - x_2u_2\right] \\ &\quad - \mu_p[\sigma_p x_2 - \mu_r x_3] \end{aligned} \tag{8.14}$$

By regrouping terms in Eq. (8.14), one gets

$$\begin{aligned} \ddot{x}_3 = &\left\{\sigma_p\left[\beta_r x_2\left(1 - \tfrac{x_1+x_2}{K}\right) + qx_1 + \delta x_3x_1 - \mu_r x_2\right] - \mu_p[\sigma_p x_2 - \mu_r x_3]\right\} + \\ &-[\sigma_p x_2]u_2 \end{aligned} \tag{8.15}$$

The new notation of state variables is used: $z_1 = x_1$, $z_2 = x_3$, $z_3 = \dot{x}_3$. It also holds that x_2 is a differential function of the flat outputs x_1, x_3, thus x_2 is also a differential function of z_1, z_2, z_3.

Besides, in the right part of Eq. (8.11), the following functions are defined:

$$\begin{aligned} f_a(z_1, z_2, z_3) &= \beta_s x_1\left(1 - \tfrac{x_1+x_2}{K}\right) - qx_1 - \delta x_1 - \delta x_3x_1 - \mu_s x_1 \\ g_{a_1}(z_1, z_2, z_3) &= -qx_1 \quad g_{a_2}(z_1, z_2, z_3) = -x_1 \end{aligned} \tag{8.16}$$

Moreover, in the right part of Eq. (8.15), the following equations are defined:

$$\begin{aligned} f_b(z_1, z_2, z_3) &= \left\{\sigma_p\left[\beta_r x_2\left(1 - \tfrac{x_1+x_2}{K}\right) + qx_1 + \delta x_3x_1 - \mu_r x_2\right] - \mu_p[\sigma_p x_2 - \mu_r x_3]\right\} \\ g_{b_1}(z_1, z_2, z_3) &= 0 \quad g_{b_2}(z_1, z_2, z_3) = -\sigma_p x_2 \end{aligned} \tag{8.17}$$

Thus, using the state vector notation $z = [z_1, z_2, z_3]^T$, the state-space description of the input–output linearized system becomes

$$\begin{aligned} \dot{z}_1 &= f_a(z) + g_{a_1}(z)u_1 + g_{a_2}(z)u_2 \\ \ddot{z}_2 &= f_b(z) + g_{b_1}(z)u_1 + g_{b_2}(z)u_2 \end{aligned} \tag{8.18}$$

The following virtual control input variables are defined

$$\begin{aligned} v_1 &= f_a(z) + g_{a_1}(z)u_1 + g_{a_2}(z)u_2 \\ v_2 &= f_b(z) + g_{b_1}(z)u_1 + g_{b_2}(z)u_2 \end{aligned} \tag{8.19}$$

one gets the concise description of the bacteria infection dynamics

$$\dot{z}_1 = v_1 \quad \ddot{z}_2 = v_2 \tag{8.20}$$

or equivalently in matrix form

$$\begin{pmatrix} \dot{z}_1 \\ \dot{z}_2 \\ \dot{z}_3 \end{pmatrix} = \begin{pmatrix} 0 & 0 & 0 \\ 0 & 0 & 1 \\ 0 & 0 & 0 \end{pmatrix} \begin{pmatrix} z_1 \\ z_2 \\ z_3 \end{pmatrix} + \begin{pmatrix} 1 & 0 \\ 0 & 0 \\ 0 & 1 \end{pmatrix} \begin{pmatrix} v_1 \\ v_2 \end{pmatrix} \tag{8.21}$$

which is a canonical Brunovsky form. Additionally, the stabilizing feedback control for the input–output linearized system is taken to be

$$\begin{aligned} v_1 &= \dot{z}_{1,d} - k_{1,1}(z_1 - z_{1,d}) \\ v_2 &= \ddot{z}_{2,d} - k_{1,2}(\dot{z}_2 - \dot{z}_{2,d}) - k_{2,2}(z_2 - z_{2,d}) \end{aligned} \tag{8.22}$$

where $z_{1,d}$ and $z_{2,d}$ are setpoint values. By defining the tracking error variables $e_1 = z_1 - z_{1,d}$, $e_2 = z_2 - z_{2,d}$ and by substituting the control law of Eq. (8.22) in the concise description of the system's dynamics of Eq. (8.20), one obtains the closed-loop system dynamics:

$$\begin{aligned} (\dot{z}_1 - \dot{z}_{1,d}) + k_{1,1}(z_1 - z_{1,d}) &= 0 \\ (\ddot{z}_2 - \ddot{z}_{2,d}) + k_{1,2}(\dot{z}_2 - \dot{z}_{2,d}) + k_{2,2}(z_2 - z_{2,d}) &= 0 \end{aligned} \tag{8.23}$$

or, equivalently, one gets the tracking error dynamics:

$$\begin{aligned} \dot{e}_1 + k_{1,1}e_1 &= 0 \\ \ddot{e}_2 + k_{1,2}\dot{e}_2 + k_{2,2}e_2 &= 0 \end{aligned} \tag{8.24}$$

Next, by selecting the feedback gains $k_{1,1}$ and $(k_{1,1}, k_{2.2})$ so as the characteristic polynomials of the differential equations of the first and second row of Eq. (8.24) to be Hurwitz stable, one has the global stability condition:

$$\begin{aligned} lim_{t\to\infty}e_1(t) = 0 &\Rightarrow lim_{t\to\infty}z_1(t) = z_{1,d}(t) \\ lim_{t\to\infty}e_2(t) = 0 &\Rightarrow lim_{t\to\infty}z_2(t) = z_{2,d}(t) \end{aligned} \tag{8.25}$$

As a result of the above it also holds that $lim_{t\to\infty}z_3(t) = z_{3,d}(t)$, Since the flat outputs z_1 $i = 1, 2, 3$ of the system converge to setpoints, the state variables of the initial nonlinear system x_i $i = 1, 2, 3$ also converge to the associated setpoints.

Finally, the control inputs that should be applied to the initial nonlinear system are computed using the definition of the virtual control inputs v_i $i = 1, 2$ which was given in Eq. (8.19). It holds that

$$\begin{pmatrix} v_1 \\ v_2 \end{pmatrix} = \begin{pmatrix} f_a(x) \\ f_b(x) \end{pmatrix} + \begin{pmatrix} g_{a_1}(x) & g_{a_2}(x) \\ g_{b_1}(x) & g_{b_2}(x) \end{pmatrix} \begin{pmatrix} u_1 \\ u_2 \end{pmatrix} \tag{8.26}$$

Thus, by solving for the real control inputs vector $u = [u_1, u_2]^T$, one obtains

$$\begin{pmatrix} u_1 \\ u_2 \end{pmatrix} = \begin{pmatrix} g_{a_1}(x) & g_{a_2}(x) \\ g_{b_1}(x) & g_{b_2}(x) \end{pmatrix}^{-1} \left[\begin{pmatrix} v_1 \\ v_2 \end{pmatrix} - \begin{pmatrix} f_a(x) \\ f_b(x) \end{pmatrix} \right] \tag{8.27}$$

Through the previous analysis it can be seen that the input–output linearization-based control approach requires changes of state variables and complicated state-space model transformations. Besides, because of the appearance of the inverse matrix in Eq. (8.27), singularities cannot be excluded.

8.1.3 Design of an H-infinity feedback controller

8.1.3.1 Approximate linearization of the dynamic model of the bacterial infection

The dynamic model of the bacterial infection undergoes approximate linearization at each sampling instance (x^*, u^*), around a temporary operating point which is defined by the present value x^* of the system's state vector and by last sampled value u^* of the control inputs vector. The linearization process is based on first-order Taylor series expansion and on the computation of the associated Jacobian matrices. The system, being initially in the nonlinear form $\dot{x} = f(x) + g(x)u$, after linearization is turned into the equivalent state-space form

$$\dot{x} = Ax + Bu + \tilde{d} \tag{8.28}$$

where A, B are the Jacobian matrices of the system which are given by

$$\begin{aligned} A = \nabla_x[f(x) + g(x)u] \mid_{(x^*,u^*)} \Rightarrow A = \nabla_x f(x) \mid_{(x^*,u^*)} + \\ +\nabla_x g_1(x)u_1 \mid_{(x^*,u^*)} + \nabla_x g_2(x)u_2 \mid_{(x^*,u^*)} \end{aligned} \tag{8.29}$$

$$B = \nabla_u[f(x) + g(x)u] \mid_{(x^*,u^*)} \Rightarrow B = g(x) \mid_{(x^*,u^*)} \tag{8.30}$$

and $\tilde{d}$ is the cumulative disturbance vector which may comprise: (i) modeling error due to the truncation of higher-order terms from the Taylor series expansion, (ii) exogenous perturbations, and (iii) sensor measurement noise of any distribution.

Computation of the Jacobian matrix $\nabla_x f(x) \mid_{(x^*,u^*)}$:

First row of the Jacobian matrix $\nabla_x f(x) \mid_{(x^*,u^*)}$: $\frac{\partial f_1}{\partial x_1} = \beta_s(1 - \frac{x_1+x_2}{K}) + \beta_s x_1(-\frac{1}{K}) - q - \delta x_3 - \mu_s$, $\frac{\partial f_1}{\partial x_2} = \beta_s x_1(-\frac{1}{K})$, $\frac{\partial f_1}{\partial x_3} = -\delta x_1$.

Second row of the Jacobian matrix $\nabla_x f(x) \mid_{(x^*,u^*)}$: $\frac{\partial f_2}{\partial x_1} = \beta_r x_2(-\frac{1}{K}) + q + \delta x_3$, $\frac{\partial f_2}{\partial x_2} = \beta_r(1 - \frac{x_1+x_2}{K}) + \beta_r x_2(-\frac{1}{K}) - \mu_r$, $\frac{\partial f_2}{\partial x_3} = \delta x_1$.

Third row of the Jacobian matrix $\nabla_x f(x) \mid_{(x^*,u^*)}$: $\frac{\partial f_3}{\partial x_1} = 0$, $\frac{\partial f_3}{\partial x_2} = \sigma_p$, and $\frac{\partial f_3}{\partial x_3} = -\mu_p$.

Thus, one obtains the Jacobian matrix:

$$\nabla_x f(x) \mid_{(x^*,u^*)} = \begin{pmatrix} \beta_s(1 - \frac{x_1+x_2}{K}) + \beta_s x_1(-\frac{1}{K}) - q - \delta x_3 - \mu_s & 0 & \beta_s x_1(-\frac{1}{K}) - \delta x_1 \\ \beta_r(1 - \frac{x_1+x_2}{K}) + \beta_r x_1(-\frac{1}{K}) + q + \delta x_3 - \mu_r & 0 & \beta_r x_1(-\frac{1}{K}) + \delta x_1 \\ 0 & \sigma_p & -\mu_p \end{pmatrix} \tag{8.31}$$

Computation of the Jacobian matrices $\nabla_x g_1(x) \mid_{(x^*,u^*)}$ and $\nabla_x g_2(x) \mid_{(x^*,u^*)}$:

$$\nabla_x g_1(x) \mid_{(x^*,u^*)} = \begin{pmatrix} -q & 0 & 0 \\ 0 & 0 & 0 \\ 0 & 0 & 0 \end{pmatrix} \quad \nabla_x g_2(x) \mid_{(x^*,u^*)} = \begin{pmatrix} -1 & 0 & 0 \\ 0 & -1 & 0 \\ 0 & 0 & 0 \end{pmatrix} \tag{8.32}$$

This linearization approach which has been followed for implementing the nonlinear optimal control scheme results in a quite accurate model of the system's dynamics. Consider for instance the following affine-in-the-input state-space model:

$$\begin{aligned} \dot{x} &= f(x) + g(x)u \Rightarrow \\ \dot{x} &= [f(x^*) + \nabla_x f(x)\mid_{x^*}(x - x^*)] + [g(x^*) + \\ &+ \nabla_x g(x)\mid_{x^*}(x - x^*)]u^* + g(x^*)u^* + g(x^*)(u - u^*) + \tilde{d}_1 \Rightarrow \\ \dot{x} &= [\nabla_x f(x)\mid_{x^*} + \nabla_x g(x)\mid_{x^*} u^*]x + g(x^*)u - \\ &- [\nabla_x f(x)\mid_{x^*} + \nabla_x g(x)\mid_{x^*} u^*]x^* + f(x^*) + g(x^*)u^* + \tilde{d}_1 \end{aligned} \tag{8.33}$$

where $\tilde{d}_1$ is the modeling error due to truncation of higher order terms in the Taylor series expansion of $f(x)$ and $g(x)$. Next, by defining $A = [\nabla_x f(x)\mid_{x^*} + \nabla_x g(x)\mid_{x^*} u^*]$, $B = g(x^*)$, one obtains

$$\dot{x} = Ax + Bu - Ax^* + f(x^*) + g(x^*)u^* + \tilde{d}_1 \tag{8.34}$$

Moreover, by denoting $\tilde{d} = -Ax^* + f(x^*) + g(x^*)u^* + \tilde{d}_1$ about the cumulative modeling error term in the Taylor series expansion procedure, one has

$$\dot{x} = Ax + Bu + \tilde{d} \tag{8.35}$$

which is the approximately linearized model of the dynamics of the system of Eq. (8.28). The term $f(x^*) + g(x^*)u^*$ is the derivative of the state vector at (x^*, u^*) which is almost annihilated by $-Ax^*$.

8.1.3.2 Stabilizing feedback control

After linearization around its current operating point, the dynamic model of the bacterial infection with resistance to antibiotics is written as [725]

$$\dot{x} = Ax + Bu + d_1 \tag{8.36}$$

Parameter d_1 stands for the linearization error in the dynamics of bacterial infection with resistance to antibiotics appearing previously in Eq. (8.28). The reference setpoints for the bacterial infection's state vector are denoted by $\mathbf{x_d} = [x_1^d, x_2^d, x_3^d]$. Tracking of this trajectory is achieved after applying the control input u^*. At every time instant the control input u^* is assumed to differ from the control input u appearing in Eq. (8.36) by an amount equal to Δu, that is $u^* = u + \Delta u$

$$\dot{x}_d = Ax_d + Bu^* + d_2 \tag{8.37}$$

The dynamics of the controlled system described in Eq. (8.36) can be also written as

$$\dot{x} = Ax + Bu + Bu^* - Bu^* + d_1 \tag{8.38}$$

and by denoting $d_3 = -Bu^* + d_1$ as an aggregate disturbance term, one obtains

$$\dot{x} = Ax + Bu + Bu^* + d_3 \tag{8.39}$$

By subtracting Eq. (8.37) from Eq. (8.39), one has

$$\dot{x} - \dot{x}_d = A(x - x_d) + Bu + d_3 - d_2 \tag{8.40}$$

By denoting the tracking error as $e = x - x_d$ and the aggregate disturbance term as $L\tilde{d} = d_3 - d_2$, the tracking error dynamics becomes

$$\dot{e} = Ae + Bu + L\tilde{d} \tag{8.41}$$

where L is a disturbance inputs gain matrix. For the approximately linearized model of the system, a stabilizing feedback controller is developed. The controller has the form

$$u(t) = -Ke(t) \tag{8.42}$$

with $K = \frac{1}{r}B^TP$ where P is a positive definite symmetric matrix which is obtained from the solution of the Riccati equation [725]

$$A^TP + PA + Q - P(\tfrac{2}{r}BB^T - \tfrac{1}{\rho^2}LL^T)P = 0 \tag{8.43}$$

where Q is a positive semi-definite symmetric matrix. The diagram of the considered control loop is depicted in Figure 8.2.

From Figure 8.2 it can be seen that the nonlinear optimal control approach takes place through the following steps: (i) at each sampling instance the nonlinear dynamic model of the bacterial infection with resistance to antibiotics undergoes approximate linearization with use of first-order Taylor series expansion and through the computation of the associated Jacobian matrices. Because of linearizing at the present value of the system's state vector and not at points of the desirable trajectory, the span of the Taylor-series expansion is kept small, and modeling error is negligible (ii) for the approximately linearized model of the system, an H-infinity feedback controller is designed. The stabilizing gains of this controller are computed by solving at each sampling period the method's algebraic Riccati equation after using the dynamically updated matrices of the linearized state-space model of the system (iii) the computed control inputs are applied at each sampling interval to the initial nonlinear dynamics of the bacteria population. The global stability properties of the control scheme which has considered the original nonlinear state-space description of the system, are proven through Lyapunov analysis.

It is also noted that the solution of the H-infinity feedback control problem for the dynamics of the bacterial infection with resistance to antibiotics and the computation of the worst-case disturbance that this control scheme can sustain, comes from the superposition of Bellman's optimality principle when considering that the bacterial infection's dynamics is affected by two separate inputs (i) the control input u (ii) the cumulative disturbance input $\tilde{d}(t)$. Solving the optimal control problem for u that is for the minimum variation (optimal) control input that achieves the elimination of the state vector's tracking error gives $u = -\frac{1}{r}B^TPe$. Equivalently, solving the optimal control problem for $\tilde{d}$, that is for the worst-case disturbance that the control loop can sustain, gives $\tilde{d} = \frac{1}{\rho^2}L^TPe$.

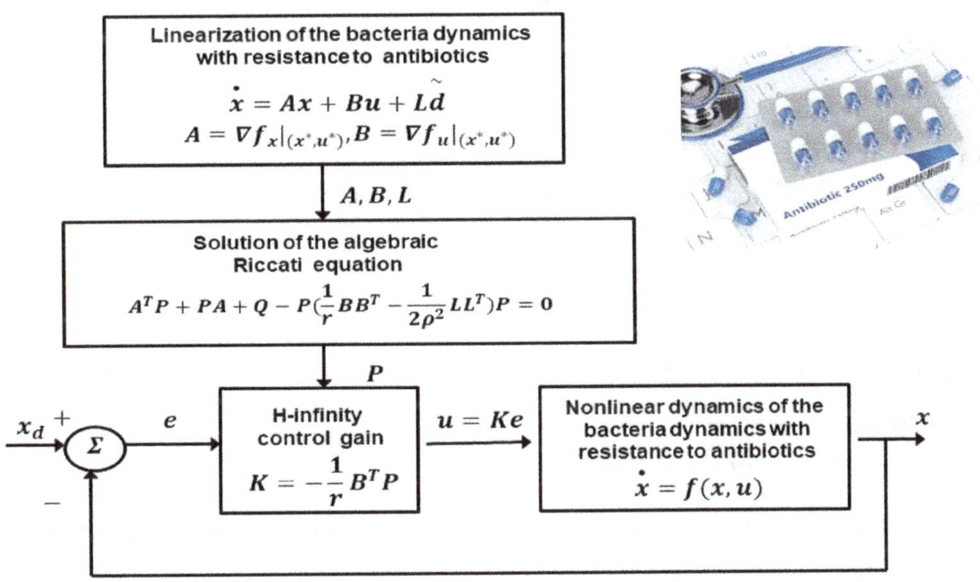

Figure 8.2 Diagram of the optimal (H-infinity) control loop of the bacterial infection with resistance to antibiotics

8.1.4 Lyapunov stability analysis

Through Lyapunov stability analysis it will be shown that the proposed nonlinear control scheme assures H-infinity (H_∞) tracking performance for the control loop of the bacterial infection with resistance to antibiotics, and that under moderate conditions about the disturbance terms, asymptotic convergence to the reference setpoints (global stability) is achieved. As shown before, the tracking error dynamics for the bacterial infection with resistance to antibiotics is written in the form

$$\dot{e} = Ae + Bu + L\tilde{d} \tag{8.44}$$

where in the bacterial infection's case $L \in R^{3\times 3}$ to be the disturbance inputs gain matrix. Variable $\tilde{d}$ denotes model uncertainties and external disturbances of the bacterial infection's dynamic model. The following Lyapunov equation is considered

$$V = \frac{1}{2}e^T P e \tag{8.45}$$

where $e = x - x_d$ is the tracking error. By differentiating with respect to time, one obtains

$$\begin{aligned}\dot{V} &= \frac{1}{2}\dot{e}^T P e + \frac{1}{2}e^T P \dot{e} \Rightarrow \\ \dot{V} &= \frac{1}{2}[Ae + Bu + L\tilde{d}]^T P + \frac{1}{2}e^T P[Ae + Bu + L\tilde{d}] \Rightarrow\end{aligned} \tag{8.46}$$

$$\dot{V} = \frac{1}{2}[e^T A^T + u^T B^T + \tilde{d}^T L^T]Pe + \frac{1}{2}e^T P[Ae + Bu + L\tilde{d}] \Rightarrow \tag{8.47}$$

$$\dot{V} = \frac{1}{2}e^T A^T Pe + \frac{1}{2}u^T B^T Pe + \frac{1}{2}\tilde{d}^T L^T Pe + \frac{1}{2}e^T PAe + \frac{1}{2}e^T PBu + \frac{1}{2}e^T PL\tilde{d} \tag{8.48}$$

The previous equation is rewritten as

$$\dot{V} = \frac{1}{2}e^T(A^T P + PA)e + \left(\frac{1}{2}u^T B^T Pe + \frac{1}{2}e^T PBu\right) + \left(\frac{1}{2}\tilde{d}^T L^T Pe + \frac{1}{2}e^T PL\tilde{d}\right) \tag{8.49}$$

Assumption: For given positive definite matrix Q and coefficients r and ρ, there exists a positive definite matrix P, which is the solution of the following matrix equation:

$$A^T P + PA = -Q + P\left(\frac{2}{r}BB^T - \frac{1}{\rho^2}LL^T\right)P \tag{8.50}$$

Moreover, the following feedback control law is applied to the system:

$$u = -\frac{1}{r}B^T Pe \tag{8.51}$$

By substituting Eq. (8.50) and Eq. (8.51), one obtains

$$\dot{V} = \frac{1}{2}e^T\left[-Q + P\left(\frac{2}{r}BB^T - \frac{1}{\rho^2}LL^T\right)P\right]e + e^T PB\left(-\frac{1}{r}B^T Pe\right) + e^T PL\tilde{d} \Rightarrow \tag{8.52}$$

$$\dot{V} = -\frac{1}{2}e^T Qe + \left(\frac{1}{r}e^T PBB^T Pe - \frac{1}{2\rho^2}e^T PLL^T Pe - \frac{1}{r}e^T PBB^T Pe\right) + e^T PL\tilde{d} \tag{8.53}$$

which after intermediate operations gives

$$\dot{V} = -\frac{1}{2}e^T Qe - \frac{1}{2\rho^2}e^T PLL^T Pe + e^T PL\tilde{d} \tag{8.54}$$

or, equivalently

$$\dot{V} = -\frac{1}{2}e^T Qe - \frac{1}{2\rho^2}e^T PLL^T Pe + \frac{1}{2}e^T PL\tilde{d} + \frac{1}{2}\tilde{d}^T L^T Pe \tag{8.55}$$

Lemma: The following inequality holds

$$\frac{1}{2}e^T PL\tilde{d} + \frac{1}{2}\tilde{d}L^T Pe - \frac{1}{2\rho^2}e^T PLL^T Pe \leq \frac{1}{2}\rho^2 \tilde{d}^T \tilde{d} \tag{8.56}$$

Proof: The binomial $(\rho\alpha - \frac{1}{\rho}b)^2$ is considered. Expanding the left part of the above inequality, one gets

$$\begin{aligned}&\rho^2 a^2 + \frac{1}{\rho^2}b^2 - 2ab \geq 0 \Rightarrow \frac{1}{2}\rho^2 a^2 + \frac{1}{2\rho^2}b^2 - ab \geq 0 \Rightarrow \\ &ab - \frac{1}{2\rho^2}b^2 \leq \frac{1}{2}\rho^2 a^2 \Rightarrow \frac{1}{2}ab + \frac{1}{2}ab - \frac{1}{2\rho^2}b^2 \leq \frac{1}{2}\rho^2 a^2\end{aligned} \tag{8.57}$$

The following substitutions are carried out: $a = \tilde{d}$ and $b = e^T PL$, and the previous relation becomes

$$\frac{1}{2}\tilde{d}^T L^T Pe + \frac{1}{2}e^T PL\tilde{d} - \frac{1}{2\rho^2}e^T PLL^T Pe \leq \frac{1}{2}\rho^2 \tilde{d}^T \tilde{d} \tag{8.58}$$

Eq. (8.58) is substituted in Eq. (8.55) and the inequality is enforced, thus giving

$$\dot{V} \leq -\frac{1}{2}e^T Qe + \frac{1}{2}\rho^2 \tilde{d}^T \tilde{d} \tag{8.59}$$

Eq. (8.59) shows that the H-infinity (H_∞) tracking performance criterion is satisfied. The integration of $\dot{V}$ from 0 to T gives

$$\begin{aligned}&\int_0^T \dot{V}(t)dt \leq -\frac{1}{2}\int_0^T ||e||_Q^2 dt + \frac{1}{2}\rho^2 \int_0^T ||\tilde{d}||^2 dt \Rightarrow \\ &2V(T) + \int_0^T ||e||_Q^2 dt \leq 2V(0) + \rho^2 \int_0^T ||\tilde{d}||^2 dt\end{aligned} \tag{8.60}$$

Moreover, if there exists a positive constant $M_d > 0$ such that

$$\int_0^\infty ||\tilde{d}||^2 dt \leq M_d \tag{8.61}$$

then, one gets

$$\int_0^\infty ||e||_Q^2 dt \leq 2V(0) + \rho^2 M_d \tag{8.62}$$

Thus, the integral $\int_0^\infty ||e||_Q^2 dt$ is bounded. Moreover, $V(T)$ is bounded and from the definition of the Lyapunov function V in Eq. (8.45), it becomes clear that $e(t)$ will be also bounded since $e(t) \in \Omega_e = \{e | e^T Pe \leq 2V(0) + \rho^2 M_d\}$.

According to the above and with the use of Barbalat's Lemma, one obtains $lim_{t \to \infty} e(t) = 0$.

After following the stages of the stability proof, one arrives at Eq. (8.59) which shows that the H-infinity tracking performance criterion holds. By selecting the attenuation coefficient ρ to be sufficiently small and in particular to satisfy $\rho^2 < ||e||_Q^2 / ||\tilde{d}||^2$, one has that the first derivative of the Lyapunov function is upper bounded by 0. This condition holds at each sampling instant, and consequently, global stability for the control loop can be concluded.

8.1.5 Simulation tests

The performance of the nonlinear optimal control scheme for the bacterial infection with resistance to antibiotics has been further confirmed through simulation experiments. The simulation code was developed in MATLAB® and to solve the control method's algebraic Riccati equation function *aresolv()* has been used. Indicative values of the parameters of the model come from [378] and are outlined as follows: $\beta_s = 0.30$, $\mu_s = 0.20$, $\beta_r = 0.25$, $\mu_r = 0.20$, $K = 2{\cdot}10^3$, $q = 6.4{\cdot}10^{-4}$, $\delta = 0.001$, $\sigma_p = 0.40$, and $\mu_p = 0.40$. The obtained results are depicted in Figures 8.3–8.10. In the related diagrams, the real values of the state variables of the bacterial infection under resistance to antibiotics are depicted in blue color, the associated setpoints are printed in red while the estimated values which have been provided by the H-infinity Kalman filter are plotted in green. It can be noticed that in all cases the nonlinear optimal control method achieved fast and accurate tracking of setpoints by the populations of the bacteria, and this was done with moderate variations of the control inputs. The transient characteristics of the control method depend on the selection of the parameters of the aforementioned Riccati equation. For small values of gain r the steady-state tracking error is eliminated, while for relatively large values of the diagonal elements of matrix Q faster convergence of the state variables to their setpoints is achieved. Moreover, the smallest value of the attenuation coefficient ρ for which one can still obtain a valid solution for the method's Riccati equation, in the form of a positive definite and symmetric matrix P, is the one which provides the control loop with maximum robustness.

To elaborate on the tracking performance and on the robustness of the proposed nonlinear optimal control method for the dynamics of the bacterial infection with the use of antibiotics the following tables are given: (i) Table 8.1 which provides information about the accuracy of tracking of the reference setpoints by the state variables of the bacterial infection with the use of antibiotics under an exact state-space model, (ii) Table 8.2 which provides information about the robustness of the control method to parametric changes in the model of the bacterial infection with the use of antibiotics (change in parameter β_s of Eq. (8.1)), (iii) Table 8.3 which provides information about the precision in state variables' estimation that is achieved by the H-infinity Kalman filter, and (iv) Table 8.4 which provides the convergence times of the state variables of the bacterial infection with the use of antibiotics to the associated setpoints.

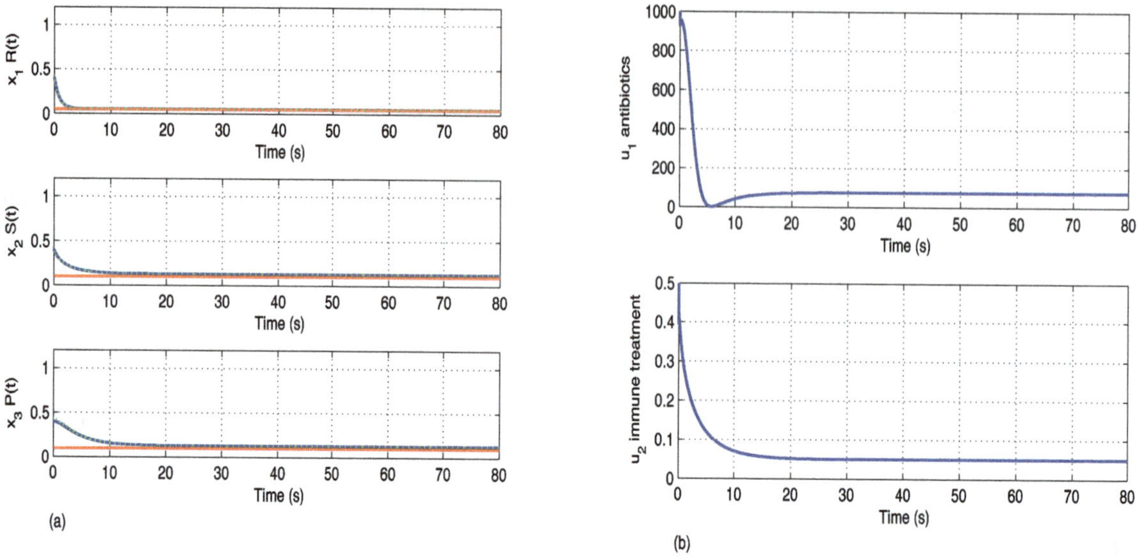

Figure 8.3 Test case 1 on nonlinear optimal control of the bacterial infection with resistance to antibiotics: (a) convergence of the state variables x_1 to x_3 of the bacteria population dynamics model $\times 10^4$ to the reference setpoints (red lines) and state variables estimates (green lines); (b) optimal control inputs (antibiotics u_1 and immune treatment u_2)

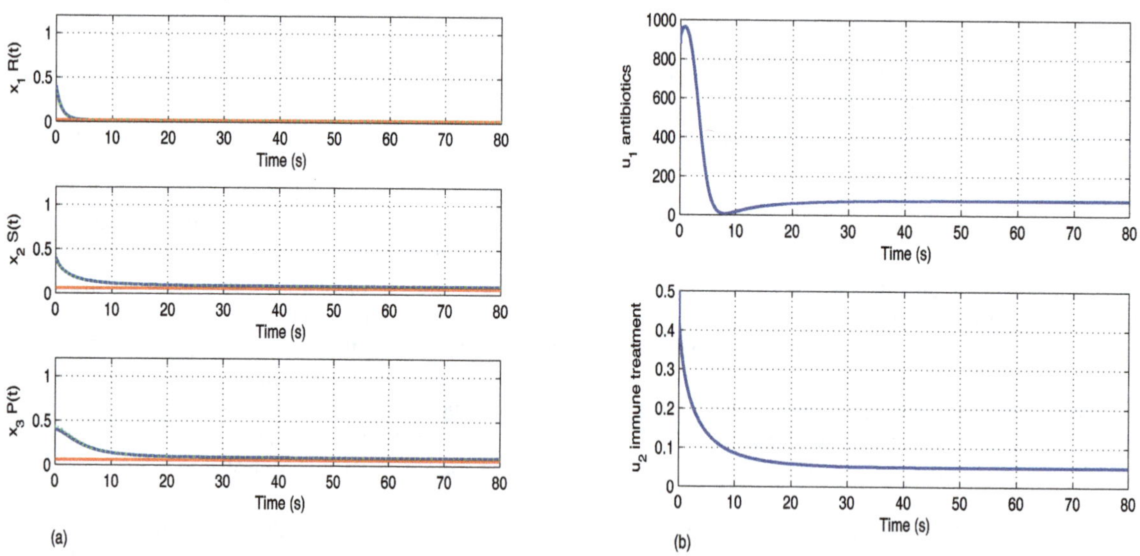

Figure 8.4 Test case 2 on nonlinear optimal control of the bacterial infection with resistance to antibiotics: (a) convergence of the state variables x_1 to x_3 of the bacteria population dynamics model $\times 10^4$ to the reference setpoints (red lines) and state variables estimates (green lines); (b) optimal control inputs (antibiotics u_1 and immune treatment u_2)

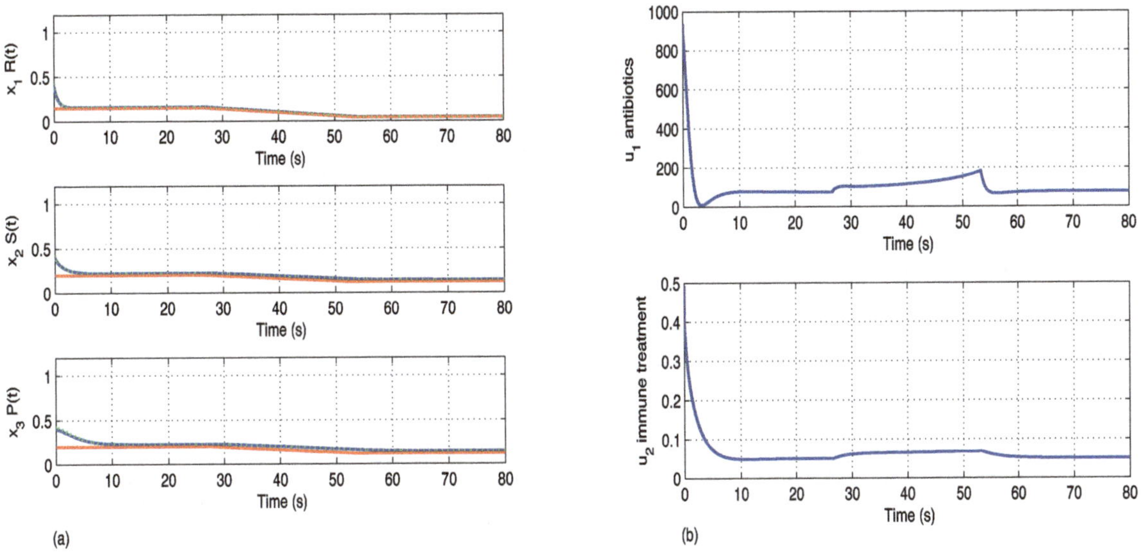

Figure 8.5 *Test case 3 on nonlinear optimal control of the bacterial infection with resistance to antibiotics: (a) convergence of the state variables x_1 to x_3 of the bacteria population dynamics model $\times 10^4$ to the reference setpoints (red lines) and state variables estimates (green lines); (b) optimal control inputs (antibiotics u_1 and immune treatment u_2)*

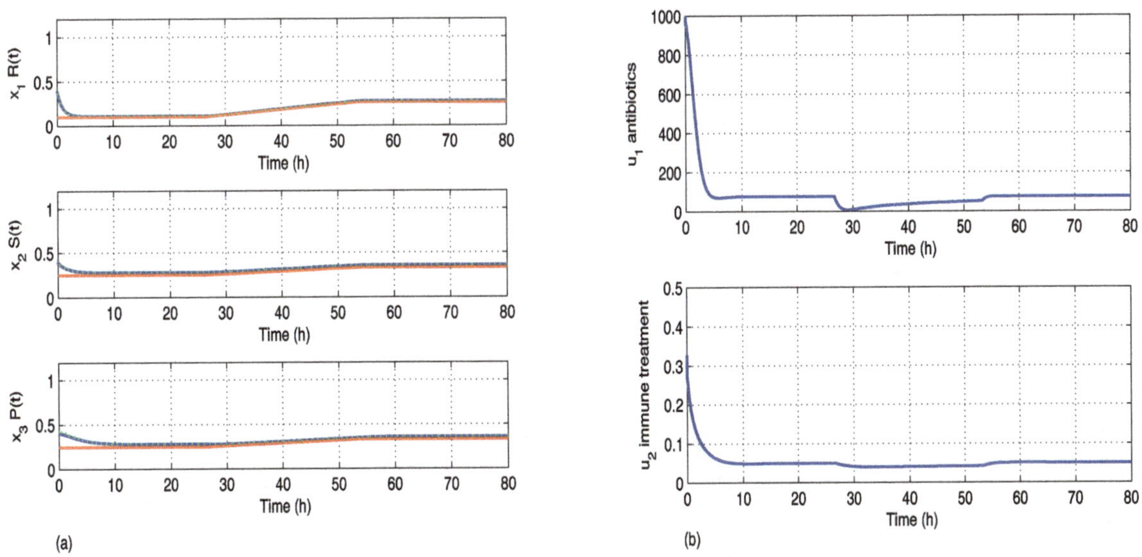

Figure 8.6 *Test case 14 on nonlinear optimal control of the bacterial infection with resistance to antibiotics: (a) convergence of the state variables x_1 to x_3 of the bacteria population dynamics model $\times 10^4$ to the reference setpoints (red lines) and state variables estimates (green lines); (b) optimal control inputs (antibiotics u_1 and immune treatment u_2)*

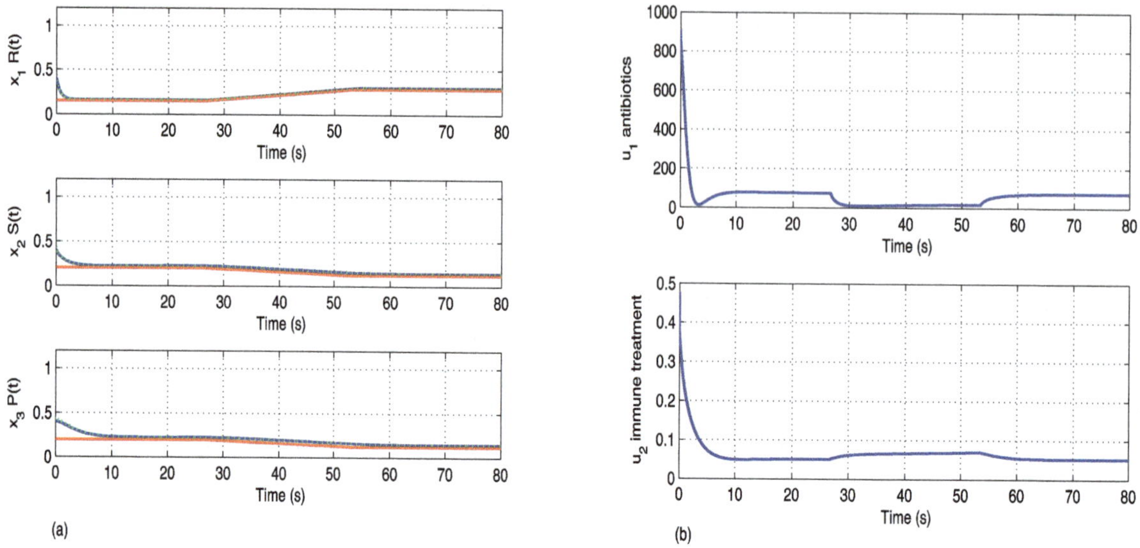

Figure 8.7 Test case 5 on nonlinear optimal control of the bacterial infection with resistance to antibiotics: (a) convergence of the state variables x_1 to x_3 of the bacteria population dynamics model $\times 10^4$ to the reference setpoints (red lines) and state variables estimates (green lines); (b) optimal control inputs (antibiotics u_1 and immune treatment u_2)

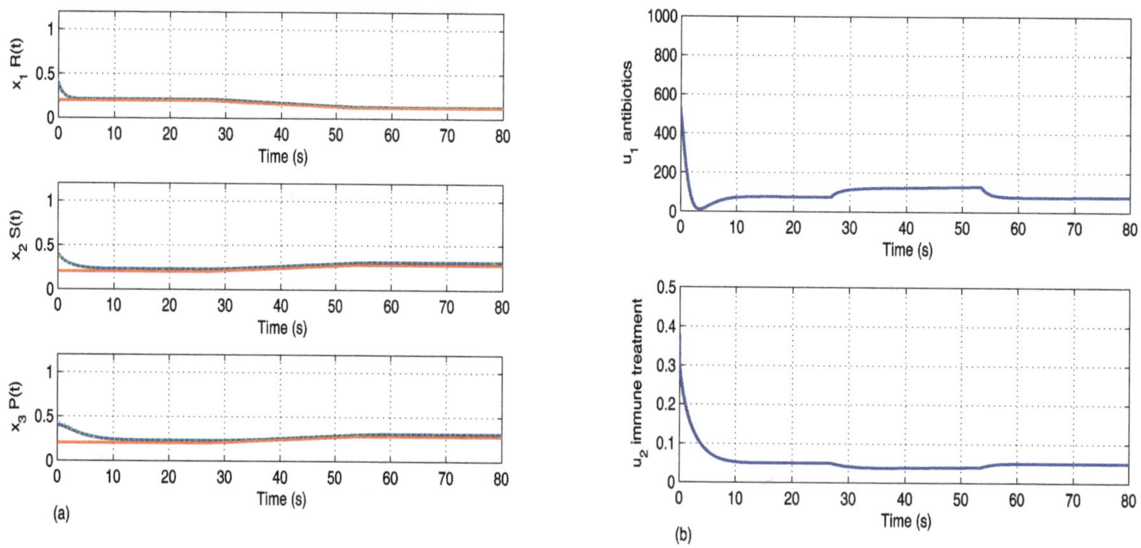

Figure 8.8 Test case 6 on nonlinear optimal control of the bacterial infection with resistance to antibiotics: (a) convergence of the state variables x_1 to x_3 of the bacteria population dynamics model $\times 10^4$ to the reference setpoints (red lines) and state variables estimates (green lines); (b) optimal control inputs (antibiotics u_1 and immune treatment u_2)

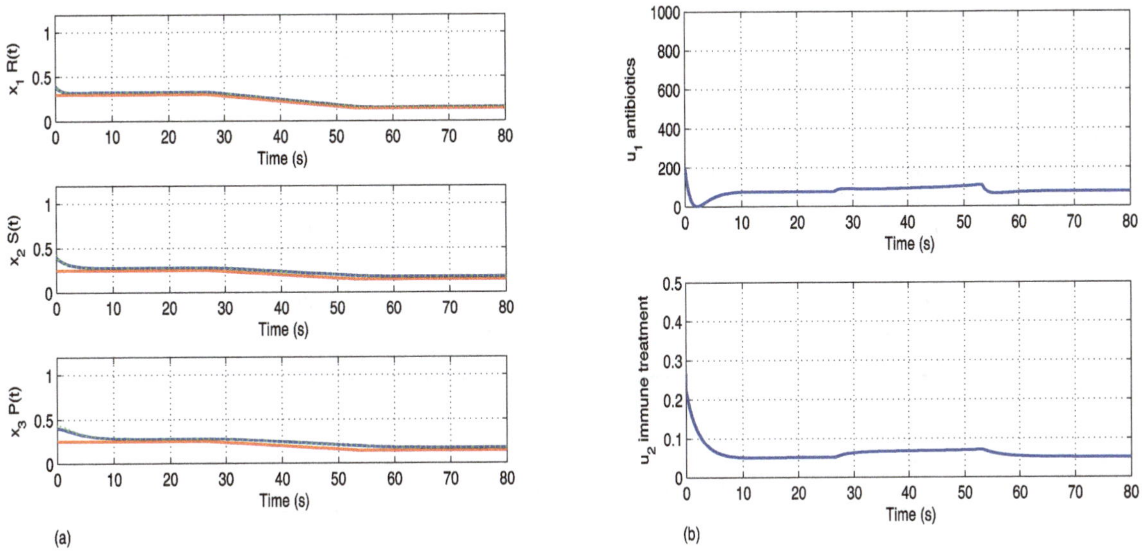

Figure 8.9 *Test case 7 on nonlinear optimal control of the bacterial infection with resistance to antibiotics: (a) convergence of the state variables x_1 to x_3 of the bacteria population dynamics model $\times\ 10^4$ to the reference setpoints (red lines) and state variables estimates (green lines); (b) optimal control inputs (antibiotics u_1 and immune treatment u_2)*

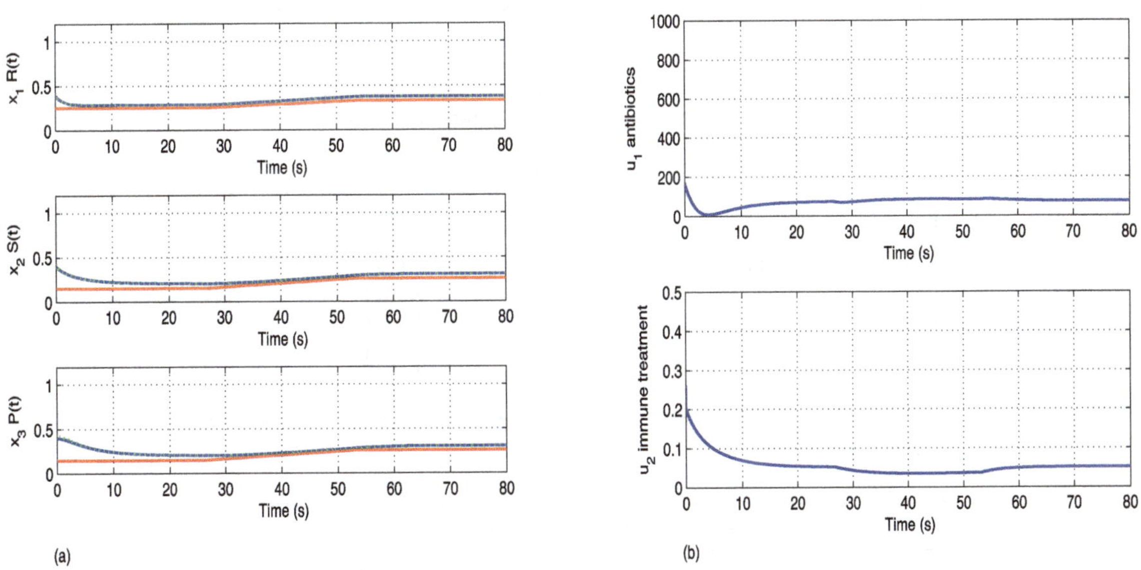

Figure 8.10 *Test case 8 on nonlinear optimal control of the bacterial infection with resistance to antibiotics: (a) convergence of the state variables x_1 to x_3 of the bacteria population dynamics model $\times\ 10^4$ to the reference setpoints (red lines) and state variables estimates (green lines); (b) optimal control inputs (antibiotics u_1 and immune treatment u_2)*

Table 8.1

Tracking RMSE in the disturbance-free case			
	$RMSE_{x_1}$	$RMSE_{x_2}$	$RMSE_{x_3}$
$setpoint_1$	0.0002	0.0002	0.0013
$setpoint_2$	0.0001	0.0016	0.0016
$setpoint_3$	0.0001	0.0011	0.0011
$setpoint_4$	0.0009	0.0013	0.0013
$setpoint_5$	0.0009	0.0011	0.0011
$setpoint_6$	0.0004	0.0014	0.0014
$setpoint_7$	0.0005	0.0014	0.0014
$setpoint_8$	0.0021	0.0022	0.0022

Table 8.2

Tracking RMSE in the case of disturbances			
$\Delta a\%$	$RMSE_{x_1}$	$RMSE_{x_2}$	$RMSE_{x_3}$
0%	0.0001	0.0016	0.0016
10%	0.0001	0.0016	0.0016
20%	0.0002	0.0016	0.0016
30%	0.0002	0.0016	0.0016
40%	0.0002	0.0016	0.0016
50%	0.0003	0.0016	0.0016
60%	0.0003	0.0016	0.0016

Table 8.3

RMSE for the estimation of the H-infinity KF $\times 10^{-4}$			
	$RMSE_{x_1}$	$RMSE_{x_2}$	$RMSE_{x_3}$
$setpoint_1$	0.1331	0.1326	0.1254
$setpoint_2$	0.1022	0.1000	0.0944
$setpoint_3$	0.1442	0.1409	0.1326
$setpoint_4$	0.3607	0.3599	0.3994
$setpoint_5$	0.1343	0.1350	0.1262
$setpoint_6$	0.2926	0.2904	0.2732
$setpoint_7$	0.1505	0.1511	0.1428
$setpoint_8$	0.2646	0.2608	0.2641

Table 8.4

Convergence times (h) for state variables			
	$T_s\, x_1$	$T_s\, x_2$	$T_s\, x_3$
$setpoint_1$	4.0	16.0	16.0
$setpoint_2$	4.0	20.0	20.0
$setpoint_3$	3.0	6.0	10.0
$setpoint_4$	3.0	6.0	10.0
$setpoint_5$	3.0	6.0	12.0
$setpoint_6$	3.0	6.0	12.0
$setpoint_7$	3.0	5.0	9.0
$setpoint_8$	4.0	20.0	20.0

8.2 Nonlinear optimal control for the treatment of leukemia

8.2.1 Outline

Leukemia is caused by the uncontrolled growth of immature white blood cells (hemopoietic stem cells) in the blood [998,94,818]. In leukemia, abnormal blood cells dominate healthy blood cells in the bone marrow. As a result, normal blood cells progressively die out from the blood [415,14]. The treatment of leukemia is usually based on the simultaneous supply of chemotherapy and immunotherapy drugs [234,70]. By viewing hematopoiesis as a dynamical system and disease as a perturbation to this system, control and systems theory can contribute to better modeling and understanding of pathological states such as leukemia [74,781]. Different dynamic models consisting of nonlinear differential equations have been developed to describe hematopoiesis [690,745]. Optimization of leukemia's treatment is beneficial and necessary for achieving better therapeutic results and for minimizing negative side effects for the patient [632,689]. Established dynamic models of leukemia are strongly nonlinear while they are not directly feedback linearizable [482,302]. Thus, the solution of the associated nonlinear control problem exhibits significant difficulty [1028,50]. Sliding mode control has been proposed for treating the nonlinear control problem of acute leukemia [440,773,387,388]. Besides, nonlinear optimal control methods for this system have been mainly based on Pontryagin's maximum principle where the associated control inputs are often of the bang-bang type and oscillate between two extreme values [250,95]. Additional control methods which have been proposed for the dynamic model of leukemia include fractional order feedback control and neural adaptive control [667,303,13]. Optimal control of leukemia has been used to improve the effectiveness of drugs and to reduce side effects as well as to avoid the excessive dosage of medication [782,398,648,14].

This section considers a dynamic model of acute leukemia with three state variables, namely the number of normal cells, the number of leukemic cells, and the quantity of a chemotherapy (cytotoxic) drug which is diffused in blood. This model has two control inputs, where the first one is an agent that modifies the replication rate of normal cells and the second drug is the infusion rate of the cytotoxic drug. It is proven that this system is differentially flat [711,709]. Next, to apply the proposed nonlinear optimal control method the dynamic model of acute leukemia undergoes approximate linearization at each sampling instance around a temporary operating point (x^*, u^*), where x^* is the present value of the system's state vector, and u^* is the last sampled value of the control inputs vector [725,728]. The linearization point is updated at each sampling instance and the linearization takes place through first-order Taylor series expansion and through the computation of the associated Jacobian matrices [705,60,706]. The modeling error which is due to the truncation of higher-order terms from the Taylor series expansion is considered to be a perturbation which is asymptotically compensated by the robustness of the control algorithm. For the approximately linearized model of the system, an H-infinity feedback controller is designed.

The proposed H-infinity control scheme provides a solution to the optimal control problem under model uncertainty and external perturbations. Actually, it represents a min–max differential game taking place between: (i) the control inputs which try to minimize a cost function that contains a quadratic term of the state vector's tracking error and (ii) the model uncertainty and external perturbations which try to maximize this cost function [725,728]. To compute the stabilizing feedback gains of the controller an algebraic Riccati equation has to be repetitively solved at each time step of the control method [726,735]. The stability properties of the control scheme are proven through Lyapunov analysis. First it is demonstrated that the control scheme satisfies the H-infinity tracking performance criterion. This signifies robustness to changes in the model parameters as well as to exogenous disturbances [876,732]. Moreover, under mild conditions, it is proven that the control scheme is globally asymptotically stable. To perform state estimation-based control without measuring the entire state vector of leukemia's dynamics, the H-infinity Kalman filter is used as a robust state estimator. Simulation tests confirm that method achieves fast and accurate tracking of setpoints by the state variables of leukemia's model under moderate variations of the control inputs.

8.2.2 Dynamic model of the acute leukemia

Leukemia is the result of uncontrolled growth of immature white blood cells (hemopoietic stem cells) in the blood. The disease affects the blood and the bone marrow. It has two possible manifestations: (a) acute myeloid leukemia (AML) and (b) chronic myeloid leukemia (CML). If the disease is acute then immature lymphoid cells grow rapidly in the blood while if the disease is chronic immature lymphoid cells grow slowly. Leukemia has been further classified into lymphocytic and myeloid on the basis of the white blood cells which are affected by the disease. Myeloid leukemia is associated with myeloid stem cells that give rise to red blood cells, some types of white blood cells and platelets. Lymphocytic leukemia is associated with lymphoid stem cells that give rise to some types of white blood cells. In leukemia, abnormal blood cells dominate healthy blood cells in the bone marrow. As a result, the three types of blood cells (white blood cells/ T cells, red blood cells and platelets) progressively die out from the blood. The treatment of leukemia is usually based on the simultaneous supply of chemotherapy and immunotherapy drugs.

The diagram of leukemia's dynamic model is given in Figure 8.11. The considered dynamic model of the acute myelogenic leukemia consists of the following differential equations [440,773,387,388], [250,95]:

$$\dot{x}_1 = r_n x_1 ln(\frac{N_a}{x_1}) - \gamma_n x_1 - c x_1 x_2 - x_1 f_n(x_3) \tag{8.63}$$

$$\dot{x}_2 = r_l x_2 ln(\frac{L_a}{x_2}) - \gamma_1 x_2 - x_2 f_l(x_3) \tag{8.64}$$

$$\dot{x}_3 = -\gamma_h x_3 + u_2 \tag{8.65}$$

where x_1 is the number of normal cells, x_2 is the number of leukemic cells, and x_3 is the quantity of the chemotherapeutic agent, r_n is the replication rate of normal cells, r_l is the replication rate of leukemic cells, L_a is the bound of the leukemic cells population, N_a is the bound of the normal cells population, γ_n is the mortality rate of normal cells, γ_l is the mortality rate of leukemic cells, γ_h is the dissipation rate of the therapeutic agent, and c is the infection rate of cells by leukemia.

In Eq. (8.63) the term $r_n x_1 ln(\frac{N_a}{x_1})$ represents the growth of normal cells based on Gombertz law. The second term $\gamma_n x_1$ represents the killing or reduction of normal cells with a decay rate to be given by γ_n. The third term $c x_1 x_2$ represents the infection of normal cells due to leukemic cells with infection rate c

In Eq. (8.64), the first term $r_1 x_2 ln(\frac{L_a}{x_2})$ represents again the growth of the leukemic cells according to Gombertz law. The second term $\gamma_1 x_2$ represents the killing or reduction of leukemic cells with mortality rate γ_l.

In Eq. (8.65), the term $\gamma_h x_3$ represents the reduction of the concentration of the chemotherapeutic agent due to its dissipation by the human body. Besides, u_2 is the control input which represents the quantity of the chemotherapeutic agent being released to the human body.

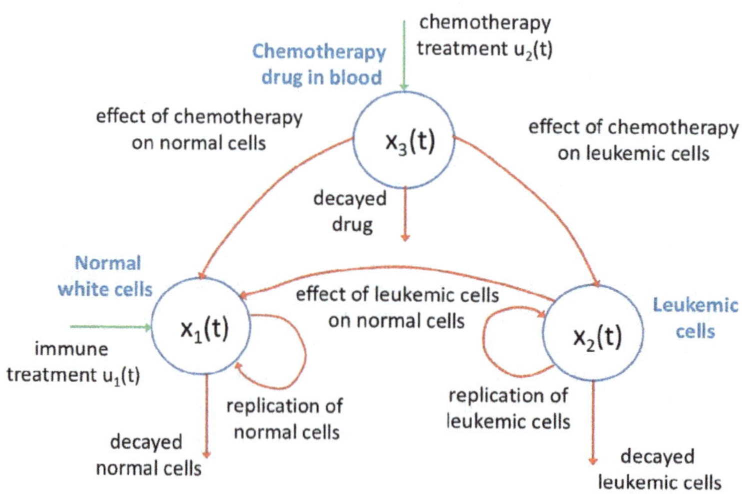

Figure 8.11 Diagram of the dynamics of the acute myelogenic leukemia: $x_1(t)$ population of normal white cells, $x_2(t)$ population leukemic cells, $x_3(t)$ chemotherapy drug in blood

Function $f_n(x)$ represents the effect of the chemotherapeutic drug on normal cells, while function $f_l(x_n)$ represents the effect of the chemotherapeutic drug on leukemic cells. It is considered that $f_l(x_3) > f_n(x_3)$ for $x_3 > 0$ which means that the chemotherapeutic drug affects more leukemic cells.

Function $f(x)$ can be considered to be monotonous in accordance to the Michaelis-Mendes law, that is $f(x) = \frac{\lambda x}{1+x}$. In such a case the mathematical model of the acute leukemia of Eq. (8.63) to Eq. (8.65) becomes

$$\dot{x}_1 = r_n x_1 ln(\tfrac{N_a}{x_1}) - \gamma_n x_1 - c x_1 x_2 - \tfrac{\lambda_n x_1 x_3}{1+x_3} \tag{8.66}$$

$$\dot{x}_2 = r_l x_2 ln(\tfrac{L_a}{x_2}) - \gamma_1 x_2 - \tfrac{\lambda_l x_2 x_3}{1+x_3} \tag{8.67}$$

$$\dot{x}_3 = -\gamma_h x_3 + u_2 \tag{8.68}$$

Besides, function $f(x)$ can be considered to be non-monotonous and to be given by $f(x) = axe^{-bx}$ or $f(x) = \frac{x}{Ax^2+Bx+C}$. In the latter case, the dynamic model of the acute leukemia becomes

$$\dot{x}_1 = r_n x_1 ln(\tfrac{N_a}{x_1}) - \gamma_n x_1 - c x_1 x_2 - \tfrac{x_1 x_3}{a_1 x_3^2 + b_1 x_3 + c_1} \tag{8.69}$$

$$\dot{x}_2 = r_l x_2 ln(\tfrac{L_a}{x_2}) - \gamma_1 x_2 - \tfrac{x_2 x_3}{a_2 x_3^2 + b_2 x_3 + c_2} \tag{8.70}$$

$$\dot{x}_3 = -\gamma_h x_3 + u_2 \tag{8.71}$$

Control input u_2 is considered to be a cytotoxic drug. Besides, the replication rate r_n of normal cells is also taken to be a control input regulated by another drug. In such a case, the control input vector of acute leukemia is $u = [u_1, u_2]^T$ with $u_1 = r_n$ and u_2 as before, and the dynamic model is written as

$$\dot{x}_1 = u_1 x_1 ln(\tfrac{N_a}{x_1}) - \gamma_n x_1 - c x_1 x_2 - \tfrac{x_1 x_3}{a_1 x_3^2 + b_1 x_3 + c_1} \tag{8.72}$$

$$\dot{x}_2 = r_l x_2 ln(\tfrac{L_a}{x_2}) - \gamma_1 x_2 - \tfrac{x_2 x_3}{a_2 x_3^2 + b_2 x_3 + c_2} \tag{8.73}$$

$$\dot{x}_3 = -\gamma_h x_3 + u_2 \tag{8.74}$$

The dynamic model of acute leukemia is also written in the following matrix form

$$\begin{pmatrix} \dot{x}_1 \\ \dot{x}_2 \\ \dot{x}_3 \end{pmatrix} = \begin{pmatrix} -\gamma_n x_1 - c x_1 x_2 - \frac{x_1 x_3}{a_1 x_3^2 + b_1 x_3 + c_1} \\ r_l x_2 ln(\frac{L_a}{x_2}) - \gamma_1 x_2 - \frac{x_2 x_3}{a_2 x_3^2 + b_2 x_3 + c_2} \\ -\gamma_h x_3 \end{pmatrix} + \begin{pmatrix} x_1 ln(\frac{N_a}{x_1}) & 0 \\ 0 & 0 \\ 0 & 1 \end{pmatrix} \begin{pmatrix} u_1 \\ u_2 \end{pmatrix} \tag{8.75}$$

Equivalently, the dynamic model of acute leukemia can be written in the nonlinear affine-in-the-input state-space form

$$\dot{x} = f(x) + g(x)u \tag{8.76}$$

where $x \in R^{3\times1}$, $f(x) \in R^{3\times1}$, $g(x) \in R^{3\times2}$, and $u \in R^{2\times1}$.

The parameters and state variables of the dynamic model of leukemia are outlined in the following Table:

8.2.3 Differential flatness of the acute leukemia dynamics

The dynamic model of acute leukemia is differentially flat with flat outputs vector $Y = [y_1, y_2]^T = [x_1, x_2]^T$. From the second row of the state-space model, one has

$$\tfrac{1}{x_2}[\dot{x}_2 - r_l x_2 ln(\tfrac{L_a}{x_2}) - \gamma_1 x_2] = \tfrac{-x_3}{a_2 x_3^2 + b_2 x_3 + c_2} \tag{8.77}$$

By defining $q(x_2, \dot{x}_2) = \frac{1}{x_2}[\dot{x}_2 - r_l x_2 ln(\frac{L_a}{x_2}) - \gamma_1 x_2]$, one obtains the following binomial for x_3

$$a_3 q(x_2, \dot{x}_2) x_3^2 + [b_3 q(x_2, \dot{x}_2) + 1] x_3 + c_2 q(x_2, \dot{x}_2) = 0 \tag{8.78}$$

Solving this binomial for x_3, one obtains that x_3 is a differential function of the flat output $y_2 = x_2$, and consequently, a differential function of the flat outputs vector Y, or

$$x_3 = h_3(Y, \dot{Y}) \tag{8.79}$$

Table 8.5

Parameters of the model of leukemia	
Parameter	*Definition*
x_1	population of normal white cells
x_2	population of leukemic cells
x_3	chemotherapy drug in blood
u_1	agent that modifies cells' replication
u_2	infusion of chemotherapy drug
r_n	replication rate of normal cells
r_l	replication rate of leukemic cells
N_a	bound of normal cells' population
L_a	bound of leukemic cells' population
γ_n	mortality rate of normal cells
γ_l	mortality rate of leukemic cells
γ_h	dissipation rate of chemotherapy drug
c	infection rate of cells by leukemia

From the third row of the state-space model, one solves for control input u_2 which gives

$$u_2 = \dot{x}_3 + \gamma_h x_3 \Rightarrow u_2 = h_{u_2}(Y, \dot{Y}) \tag{8.80}$$

which signifies that control input u_2 is a differential function of the flat outputs vector. Moreover, from the first row of the state-space model, one solves for control input u_1 which gives

$$u_1 = \frac{1}{x_1 ln(\frac{N_a}{x_1})}(\dot{x}_1 + \gamma_n x_2 + c x_1 x_2 + \frac{x_1 x_3}{a_1 x_3^2 + b_1 x_3 + c_1}) \tag{8.81}$$

which signifies that control input u_1 is a differential function of the flat outputs vector Y, or

$$u_1 = h_{u_1}(Y, \dot{Y}) \tag{8.82}$$

Consequently, all state variables x_i, $i = 1, 2, 3$ and all control inputs u_i, $i = 1, 2$ of leukemia's dynamic model are differential functions of the flat outputs vector Y, and the system is differentially flat.

The differential flatness property is an implicit proof of the system's controllability. Besides, by demonstrating differential flatness it is confirmed that the system can be transformed into the input–output linearized form and into the canonical Brunovsky form. Furthermore, it allows to solve the setpoint definition problem. First, one defines setpoints in an unconstraint manner for those state variables which coincide with the flat outputs, that is x_1 and x_2. Next, for the rest of the state variables, setpoints are chosen using the differential relations which connect them to the flat outputs In leukemia's model, state x_1 (number of normal cells) and state x_2 (number of leukemic cells) have setpoints which are allowed to be given any value. For state x_3 (concentration of chemotherapy drug) setpoints are chosen using the relation $x_3 = h_3(Y, \dot{Y})$.

8.2.4 Approximate linearization of leukemia's dynamic model

The nonlinear dynamic model of leukemia, being in the form $\dot{x} = f(x) + g(x)u$, undergoes approximate linearization through first-order Taylor series expansion and through the computation of the associated Jacobian matrices. The linearization point (x^*, u^*) is defined at each sampling instance by the present value of the system's state vector x^* and by the last sampled value of the control inputs vector u^*. The modeling error which is due to the truncation of higher-order terms in the Taylor series expansion, is considered to be a perturbation which is asymptotically compensated by the robustness of the control algorithm.

Through the linearization process, the dynamic model of leukemia is transformed into the linear form:

$$\dot{x} = Ax + Bu + \tilde{d} \tag{8.83}$$

where A, B are the Jacobian matrices of the system which are given by

$$\begin{gathered} A = \nabla_x[f(x) + g(x)u] \mid_{(x^*,u^*)} \Rightarrow \\ A = \nabla_x f(x) \mid_{(x^*,u^*)} + \nabla_x g_1(x)u \mid_{(x^*,u^*)} + \nabla_x g_2(x)u \mid_{(x^*,u^*)} \end{gathered} \tag{8.84}$$

$$B = \nabla_u[f(x) + g(x)u] \mid_{(x^*,u^*)} \Rightarrow B = g(x) \mid_{(x^*,u^*)} \tag{8.85}$$

where $g(x) = [g_1(x) g_2(x)]$ is the control inputs gain matrix consisting of two column vectors and $\tilde{d}$ is the cumulative disturbance vector which may comprise: (a) modeling error due to truncation of higher-order terms from the Taylor series expansion, (b) exogenous perturbations, and (c) sensor measurement noise of any distribution.

Computation of the Jacobian matrix $\nabla_x f(x) \mid_{(x^*,u^*)}$:

First row of the Jacobian matrix $\nabla_x f(x) \mid_{(x^*,u^*)}$: $\frac{\partial f_1}{\partial x_1} = -\gamma_n - cx_2 - \frac{x_3}{a_1 x_3^2 + b_1 x_3 + c_1}$, $\frac{\partial f_1}{\partial x_2} = -cx_2$, $\frac{\partial f_1}{\partial x_3} = -\frac{x_1(a_1 x_3^2 + b_1 x_3 + c_1) - x_1 x_3(2a_1 x_3 + b_1)}{(a_1 x_3^2 + b_1 x_3 + c_1)^2}$.

Second row of the Jacobian matrix $\nabla_x f(x) \mid_{(x^*,u^*)}$: $\frac{\partial f_2}{\partial x_1} = 0$, $\frac{\partial f_2}{\partial x_2} = r_l ln(\frac{L_a}{x_2}) + r_l x_2(\frac{x_2}{L_a})(\frac{-L_a}{x_2^2}) - \gamma_1 - \frac{x_3}{a_2 x_3^2 + b_2 x_3 + c_2}$, $\frac{\partial f_2}{\partial x_3} = -\frac{x_2(a_2 x_3^2 + b_2 x_3 + c_2) - x_2 x_3(2a_2 x_3 + b_3)}{(a_2 {x_3}^2 + b_2 x_3 + c_3)^2}$.

Third row of the Jacobian matrix $\nabla_x f(x) \mid_{(x^*,u^*)}$: $\frac{\partial f_3}{\partial x_1} = 0$, $\frac{\partial f_3}{\partial x_2} = 0$, $\frac{\partial f_3}{\partial x_3} = \gamma_n$.

Computation of the Jacobian matrix $\nabla_x g_1(x) \mid_{(x^*,u^*)}$:

$$\nabla_x g_1(x) \mid_{(x^*,u^*)} = \begin{pmatrix} ln(\frac{N_a}{x_1}) + x_1(\frac{x_1}{N_a})(\frac{-N_a}{x_1^2}) & 0 & 0 \\ 0 & 0 & 0 \\ 0 & 0 & 0 \end{pmatrix} \mid_{(x^*,u^*)} \tag{8.86}$$

Computation of the Jacobian matrix $\nabla_x g_2(x) \mid_{(x^*,u^*)}$: It holds that the Jacobian matrix $\nabla_x g_1(x) \mid_{(x^*,u^*)} = 0_{3\times 3}$:

This linearization approach which has been followed for implementing the nonlinear optimal control scheme results in a quite accurate model of the system's dynamics. Consider for instance the following affine-in-the-input state-space model

$$\begin{gathered} \dot{x} = f(x) + g(x)u \Rightarrow \\ \dot{x} = [f(x^*) + \nabla_x f(x) \mid_{x^*} (x - x^*)] + [g(x^*) + \nabla_x g(x) \mid_{x^*} (x - x^*)]u^* + g(x^*)u^* + g(x^*)(u - u^*) + \tilde{d}_1 \Rightarrow \\ \dot{x} = [\nabla_x f(x) \mid_{x^*} + \nabla_x g(x) \mid_{x^*} u^*]x + g(x^*)u - [\nabla_x f(x) \mid_{x^*} + \nabla_x g(x) \mid_{x^*} u^*]x^* + f(x^*) + g(x^*)u^* + \tilde{d}_1 \end{gathered} \tag{8.87}$$

where $\tilde{d}_1$ is the modeling error due to truncation of higher order terms in the Taylor series expansion of $f(x)$ and $g(x)$. Next, by defining $A = [\nabla_x f(x) \mid_{x^*} + \nabla_x g(x) \mid_{x^*} u^*]$, $B = g(x^*)$, one obtains

$$\dot{x} = Ax + Bu - Ax^* + f(x^*) + g(x^*)u^* + \tilde{d}_1 \tag{8.88}$$

Moreover, by denoting $\tilde{d} = -Ax^* + f(x^*) + g(x^*)u^* + \tilde{d}_1$ about the cumulative modeling error term in the Taylor series expansion procedure, one has

$$\dot{x} = Ax + Bu + \tilde{d} \tag{8.89}$$

which is the approximately linearized model of the dynamics of the system of Eq. (8.83). The term $f(x^*) + g(x^*)u^*$ is the derivative of the state vector at (x^*, u^*) which is almost annihilated by $-Ax^*$.

8.2.5 Min–max control and disturbance rejection

In the H_∞ control approach, a feedback control scheme is designed for setpoint tracking by the dynamic model of leukemia and for simultaneous disturbance rejection, considering that the disturbance affects the system in the worst possible manner. For the approximately linearized model of Eq. (8.89), one considers also measurable outputs $y = Cx$. Then, the disturbances' effects are incorporated in the following quadratic cost function [725]

$$J(t) = \frac{1}{2}\int_0^T [y^T(t)y(t) + ru^T(t)u(t) - \rho^2\tilde{d}^T(t)\tilde{d}(t)]dt, \quad r, \rho > 0 \tag{8.90}$$

The significance of the negative sign in the cost function's term that is associated with the perturbation variable $\tilde{d}(t)$ is that the disturbance tries to maximize the cost function $J(t)$ while the control signal $u(t)$ tries to minimize it.

The physical meaning of the relation given above is that the control signal and the disturbances compete with each other within a min–max differential game. This problem of min–max optimization can be written as $min_u max_{\tilde{d}} J(u, \tilde{d})$. The objective of the optimization procedure is to compute a control signal $u(t)$ which can compensate for the worst possible disturbance that affects the system. However, the solution to the min–max optimization problem is directly related to the value of parameter ρ. This means that there is an upper bound in the disturbances magnitude that can be annihilated by the control signal.

8.2.5.1 Stabilizing feedback control

After linearization around its current operating point, the dynamic model of leukemia is written as [725]

$$\dot{x} = Ax + Bu + d_1 \tag{8.91}$$

Parameter d_1 stands for the linearization error in the dynamics of leukemia appearing previously in Eq. (8.83). The reference setpoints for leukemia's dynamic model are denoted by $\mathbf{x_d} = [x_1^d, x_2^d, x_3^d]$. Tracking of this trajectory is achieved after applying the control input u^*. At every time instant the control input u^* is assumed to differ from the control input u appearing in Eq. (8.91) by an amount equal to Δu, that is $u^* = u + \Delta u$

$$\dot{x}_d = Ax_d + Bu^* + d_2 \tag{8.92}$$

The dynamics of the controlled system described in Eq. (8.91) can be also written as

$$\dot{x} = Ax + Bu + Bu^* - Bu^* + d_1 \tag{8.93}$$

and by denoting $d_3 = -Bu^* + d_1$ as an aggregate disturbance term, one obtains

$$\dot{x} = Ax + Bu + Bu^* + d_3 \tag{8.94}$$

By subtracting Eq. (8.92) from Eq. (8.94), one has

$$\dot{x} - \dot{x}_d = A(x - x_d) + Bu + d_3 - d_2 \tag{8.95}$$

By denoting the tracking error as $e = x - x_d$ and the aggregate disturbance term as $L\tilde{d} = d_3 - d_2$, the tracking error dynamics becomes

$$\dot{e} = Ae + Bu + L\tilde{d} \tag{8.96}$$

where L is a disturbance inputs gain matrix. For the approximately linearized model of the system, a stabilizing feedback controller is developed. The controller has the form

$$u(t) = -Ke(t) \tag{8.97}$$

with $K = \frac{1}{r}B^TP$ where P is a positive definite symmetric matrix which is obtained from the solution of the Riccati equation [725]

$$A^TP + PA + Q - P(\frac{2}{r}BB^T - \frac{1}{\rho^2}LL^T)P = 0 \tag{8.98}$$

where Q is a positive semi-definite symmetric matrix. The diagram of the considered control loop is depicted in Figure 8.12.

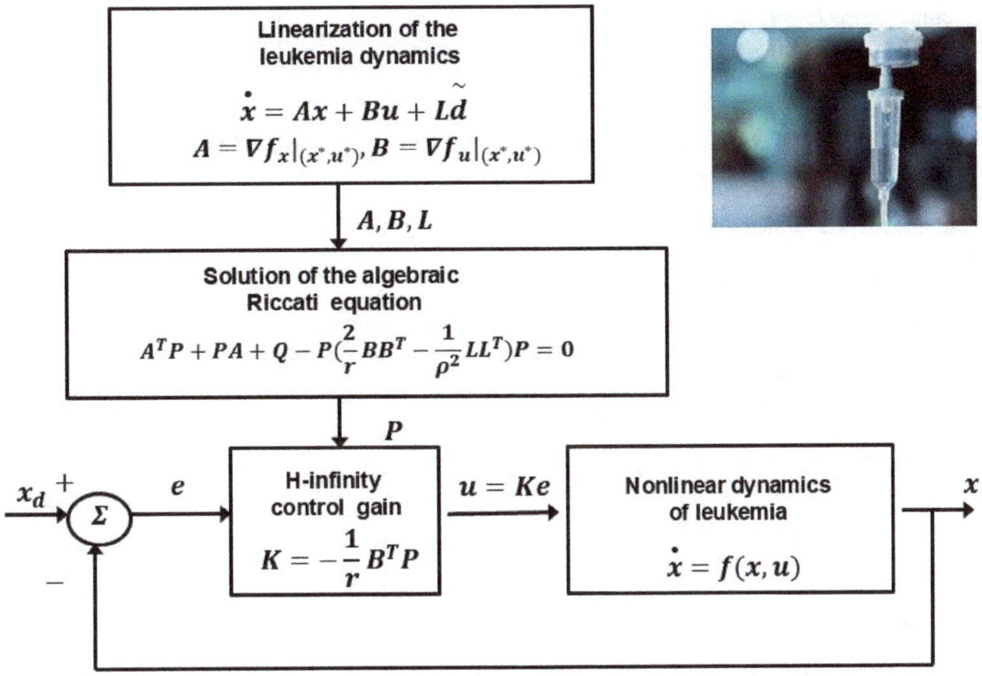

Figure 8.12 Diagram of the optimal (H-infinity) control loop for the treatment of leukemia

From Figure 8.12 it can be seen that the nonlinear optimal control approach takes place through the following steps: (i) at each sampling instance the nonlinear dynamic model of leukemia undergoes approximate linearization with use of first-order Taylor series expansion and through the computation of the associated Jacobian matrices. Because of linearizing at the present value of the system's state vector and not at points of the desirable trajectory, the span of the Taylor-series expansion is kept small and the modeling error is negligible (ii) for the approximately linearized model of the system, an H-infinity feedback controller is designed. The stabilizing gains of this controller are computed by solving at each sampling period the method's algebraic Riccati equation after using the dynamically updated matrices of the linearized state-space model of the system (iii) the computed control inputs are applied at each sampling interval to the initial nonlinear dynamics of leukemia. The global stability properties of the control scheme which has considered the original nonlinear state-space description of the system, are proven through Lyapunov analysis.

It is also noted that the solution of the H-infinity feedback control problem for leukemia's dynamics and the computation of the worst-case disturbance that this control scheme can sustain, comes from the superposition of Bellman's optimality principle when considering that leukemia's dynamics is affected by two separate inputs (i) the control input u (ii) the cumulative disturbance input $\tilde{d}(t)$. Solving the optimal control problem for u, that is for the minimum variation (optimal) control input which achieves the elimination of the state vector's tracking error, gives $u = -\frac{1}{r}B^T Pe$. Equivalently, solving the optimal control problem for $\tilde{d}$, that is for the worst-case disturbance that the control loop can sustain, gives $\tilde{d} = \frac{1}{\rho^2}L^T Pe$.

8.2.6 Lyapunov stability analysis

Through Lyapunov stability analysis it will be shown that the proposed nonlinear control scheme assures H_∞ tracking performance for the control loop of leukemia's dynamics and that under moderate conditions about the disturbance terms, asymptotic convergence to the reference setpoints (global stability) is achieved. As shown before, the tracking error dynamics of leukemia is written in the form

$$\dot{e} = Ae + Bu + L\tilde{d} \tag{8.99}$$

where in leukemia's case $L = I \in R^{3\times 3}$ with I being the identity matrix. Variable $\tilde{d}$ denotes model uncertainties and external disturbances of leukemia's dynamic model. The following Lyapunov equation is considered

$$V = \frac{1}{2}e^T P e \tag{8.100}$$

where $e = x - x_d$ is the tracking error. By differentiating with respect to time, one obtains

$$\begin{aligned} \dot{V} &= \frac{1}{2}\dot{e}^T P e + \frac{1}{2}e^T P \dot{e} \Rightarrow \\ \dot{V} &= \frac{1}{2}[Ae + Bu + L\tilde{d}]^T P + \frac{1}{2}e^T P [Ae + Bu + L\tilde{d}] \Rightarrow \end{aligned} \tag{8.101}$$

$$\dot{V} = \frac{1}{2}[e^T A^T + u^T B^T + \tilde{d}^T L^T]Pe + \frac{1}{2}e^T P[Ae + Bu + L\tilde{d}] \Rightarrow \tag{8.102}$$

$$\dot{V} = \frac{1}{2}e^T A^T P e + \frac{1}{2}u^T B^T P e + \frac{1}{2}\tilde{d}^T L^T P e + \frac{1}{2}e^T P A e + \frac{1}{2}e^T P B u + \frac{1}{2}e^T P L \tilde{d} \tag{8.103}$$

The previous equation is rewritten as

$$\dot{V} = \frac{1}{2}e^T(A^T P + PA)e + \left(\frac{1}{2}u^T B^T P e + \frac{1}{2}e^T P B u\right) + \left(\frac{1}{2}\tilde{d}^T L^T P e + \frac{1}{2}e^T P L \tilde{d}\right) \tag{8.104}$$

Assumption: For given positive definite matrix Q and coefficients r and ρ, there exists a positive definite matrix P, which is the solution of the following matrix equation:

$$A^T P + PA = -Q + P\left(\frac{2}{r}BB^T - \frac{1}{\rho^2}LL^T\right)P \tag{8.105}$$

Moreover, the following feedback control law is applied to the system:

$$u = -\frac{1}{r}B^T P e \tag{8.106}$$

By substituting Eq. (8.105) and Eq. (8.106), one obtains

$$\dot{V} = \frac{1}{2}e^T\left[-Q + P\left(\frac{2}{r}BB^T - \frac{1}{\rho^2}LL^T\right)P\right]e + e^T PB\left(-\frac{1}{r}B^T P e\right) + e^T P L \tilde{d} \Rightarrow \tag{8.107}$$

$$\dot{V} = -\frac{1}{2}e^T Q e + \left(\frac{1}{r}e^T PBB^T P e - \frac{1}{2\rho^2}e^T PLL^T P e - \frac{1}{r}e^T PBB^T P e\right) + e^T P L \tilde{d} \tag{8.108}$$

which after intermediate operations gives

$$\dot{V} = -\frac{1}{2}e^T Q e - \frac{1}{2\rho^2}e^T PLL^T P e + e^T P L \tilde{d} \tag{8.109}$$

or, equivalently,

$$\dot{V} = -\frac{1}{2}e^T Q e - \frac{1}{2\rho^2}e^T PLL^T P e + \frac{1}{2}e^T P L \tilde{d} + \frac{1}{2}\tilde{d}^T L^T P e \tag{8.110}$$

Lemma: The following inequality holds

$$\frac{1}{2}e^T P L \tilde{d} + \frac{1}{2}\tilde{d}L^T P e - \frac{1}{2\rho^2}e^T PLL^T P e \leq \frac{1}{2}\rho^2 \tilde{d}^T \tilde{d} \tag{8.111}$$

Proof: The binomial $(\rho\alpha - \frac{1}{\rho}b)^2$ is considered. Expanding the left part of the above inequality, one gets

$$\begin{aligned} &\rho^2 a^2 + \frac{1}{\rho^2}b^2 - 2ab \geq 0 \Rightarrow \frac{1}{2}\rho^2 a^2 + \frac{1}{2\rho^2}b^2 - ab \geq 0 \Rightarrow \\ &ab - \frac{1}{2\rho^2}b^2 \leq \frac{1}{2}\rho^2 a^2 \Rightarrow \frac{1}{2}ab + \frac{1}{2}ab - \frac{1}{2\rho^2}b^2 \leq \frac{1}{2}\rho^2 a^2 \end{aligned} \tag{8.112}$$

The following substitutions are carried out: $a = \tilde{d}$ and $b = e^T PL$, and the previous relation becomes

$$\frac{1}{2}\tilde{d}^T L^T P e + \frac{1}{2}e^T P L \tilde{d} - \frac{1}{2\rho^2}e^T PLL^T P e \leq \frac{1}{2}\rho^2 \tilde{d}^T \tilde{d} \tag{8.113}$$

Eq. (8.113) is substituted in Eq. (8.110) and the inequality is enforced, thus giving

$$\dot{V} \leq -\frac{1}{2}e^T Q e + \frac{1}{2}\rho^2 \tilde{d}^T \tilde{d} \tag{8.114}$$

Eq. (8.114) shows that the H_∞ tracking performance criterion is satisfied. The integration of $\dot{V}$ from 0 to T gives

$$\begin{aligned} &\int_0^T \dot{V}(t)dt \leq -\frac{1}{2}\int_0^T \|e\|_Q^2 dt + \frac{1}{2}\rho^2 \int_0^T \|\tilde{d}\|^2 dt \Rightarrow \\ &2V(T) + \int_0^T \|e\|_Q^2 dt \leq 2V(0) + \rho^2 \int_0^T \|\tilde{d}\|^2 dt \end{aligned} \tag{8.115}$$

Moreover, if there exists a positive constant $M_d > 0$ such that

$$\int_0^\infty ||\tilde{d}||^2 dt \leq M_d \tag{8.116}$$

then, one gets

$$\int_0^\infty ||e||_Q^2 dt \leq 2V(0) + \rho^2 M_d \tag{8.117}$$

Thus, the integral $\int_0^\infty ||e||_Q^2 dt$ is bounded. Moreover, $V(T)$ is bounded and from the definition of the Lyapunov function V in Eq. (8.100), it becomes clear that $e(t)$ will be also bounded since $e(t) \in \Omega_e = \{e|e^T Pe \leq 2V(0) + \rho^2 M_d\}$.

According to the above and with the use of Barbalat's Lemma, one obtains $lim_{t\to\infty} e(t) = 0$.

Through the stages of the stability proof, one arrives at Eq. (8.114) which shows that the H-infinity tracking performance criterion holds. By selecting the attenuation coefficient ρ to be sufficiently small and in particular to satisfy $\rho^2 < ||e||_Q^2/||\tilde{d}||^2$, one has that the first derivative of the Lyapunov function is upper bounded by 0. This condition holds at each sampling instant, and consequently, global stability for the control loop can be concluded.

8.2.7 Input–output linearization of the leukemia dynamics

The dynamic model of leukemia $\dot{x} = f(x) + g(x)u$ can be also transformed into the input–output linearized form and next to the canonical Brunovsky form through successive differentiations of its flat outputs. It has been shown that the flat outputs vector is $Y = [x_1, x_2]^T$. By differentiating in time $y_1 = x_1$, one obtains the first row of the leukemia's dynamic model that was given in Eq. (8.75)

$$\dot{x}_1 = -\gamma_n x_1 - c x_1 x_2 - \frac{x_1 x_3}{a_1 x_3^2 + b_1 x_3 + c_1} + x_1 ln(\frac{N_a}{x_1}) u_1 \tag{8.118}$$

By defining functions

$$\begin{aligned} \bar{f}_1(x) &= -\gamma_n x_1 - c x_1 x_2 - \frac{x_1 x_3}{a_1 x_3^2 + b_1 x_3 + c_1} \\ \bar{g}_1(x) &= x_1 ln(\frac{N_a}{x_1}) \end{aligned} \tag{8.119}$$

one obtains the concise description

$$\dot{x}_1 = \bar{f}_1(x) + \bar{g}_1(x) u_1 \tag{8.120}$$

By differentiating in time the flat output $y_2 = x_2$, one obtains the second row of the state-space model, that is

$$\dot{x}_2 = r_l x_2 ln(\frac{L_a}{x_2}) - \gamma_1 x_2 - \frac{x_2 x_3}{a_2 x_3^2 + b_2 x_3 + c_3} \tag{8.121}$$

By differentiating once again in time the second flat output, one gets

$$\ddot{x}_2 = r_l ln(\frac{L_a}{x_2}) + r_l x_2 \frac{x_2}{L_a}(\frac{-L_a}{x_2^2})\dot{x}_2 - \gamma_1 \dot{x}_2 - \frac{(\dot{x}_2 x_3 + x_2 \dot{x}_3)(a_2 x_3^2 + b_2 x_3 + c_2) - x_2 x_3 (2a_2 x_3 + b_2)\dot{x}_3}{(a_2 x_3^2 + b_2 x_3 + c_2)^2} \tag{8.122}$$

Using that $\dot{x}_2 = f_2(x)$ the previous relation becomes

$$\ddot{x}_2 = r_l ln(\frac{L_a}{x_2}) + r_l x_2 \frac{x_2}{L_a}(\frac{-L_a}{x_2^2}) f_2 - \gamma_1 f_2 - \frac{(f_2 x_3 + x_2 \dot{x}_3)(a_2 x_3^2 + b_2 x_3 + c_2) - x_2 x_3 (2a_2 x_3 + b_2)\dot{x}_3}{(a_2 x_3^2 + b_2 x_3 + c_2)^2} \tag{8.123}$$

By grouping together the terms which contain the derivative $\dot{x}_3$, one has

$$\begin{aligned} \ddot{x}_2 = & [r_l ln(\frac{L_a}{x_2}) + r_l x_2 \frac{x_2}{L_a}(\frac{-L_a}{x_2^2}) f_2 - \gamma_1 f_2 - \frac{(f_2 x_3)(a_2 x_3^2 + b_2 x_3 + c_2)}{(a_2 x_3^2 + b_2 x_3 + c_2)^2}] \\ & + [\frac{-x_2 (a_2 x_3^2 + b_2 x_3 + c_2) - x_2 x_3 (2a_2 x_3 + b_2)}{(a_2 x_3^2 + b_2 x_3 + c_2)^2}] \dot{x}_3 \end{aligned} \tag{8.124}$$

Next, from the third row of the state-space model, one uses that $\dot{x}_3 = -\gamma_h x_3 + u_2$ which gives

$$\ddot{x}_2 = [r_l ln(\frac{L_a}{x_2}) + r_l x_2 \frac{x_2}{L_a}(\frac{-L_a}{x_2^2})f_2 - \gamma_1 f_2 - \frac{(f_2x_3)(a_2x_3^2+b_2x_3+c_2)}{(a_2x_3^2+b_2x_3+c_2)^2}]$$
$$+[\frac{-x_2(a_2x_3^2+b_2x_3+c_2)-x_2x_3(2a_2x_3+b_2)}{(a_2x_3^2+b_2x_3+c_2)^2}](-\gamma_h x_3) \quad (8.125)$$
$$+[\frac{-x_2(a_2x_3^2+b_2x_3+c_2)-x_2x_3(2a_2x_3+b_2)}{(a_2x_3^2+b_2x_3+c_2)^2}]u_2$$

By defining the following functions

$$\bar{f}_2(x) = [r_l ln(\frac{L_a}{x_2}) + r_l x_2 \frac{x_2}{L_a}(\frac{-L_a}{x_2^2})f_2 - \gamma_1 f_2 - \frac{(f_2x_3)(a_2x_3^2+b_2x_3+c_2)}{(a_2x_3^2+b_2x_3+c_2)^2}]$$
$$+[\frac{-x_2(a_2x_3^2+b_2x_3+c_2)-x_2x_3(2a_2x_3+b_2)}{(a_2x_3^2+b_2x_3+c_2)^2}](-\gamma_h x_3) \quad (8.126)$$

$$\bar{g}_2(x) = [\frac{-x_2(a_2x_3^2+b_2x_3+c_2)-x_2x_3(2a_2x_3+b_2)}{(a_2x_3^2+b_2x_3+c_2)^2}] \quad (8.127)$$

one obtains a concise form about the second-order time-derivative of the second flat output x_2

$$\ddot{x}_2 = \bar{f}_2(x) + \bar{g}_2(x)u_2 \quad (8.128)$$

Thus, one arrives at the input–output linearized form:

$$\dot{x}_1 = \bar{f}_1(x) + \bar{g}_1(x)u_1$$
$$\ddot{x}_2 = \bar{f}_2(x) + \bar{g}_2(x)u_2 \quad (8.129)$$

or equivalently in matrix form

$$\begin{pmatrix} \dot{x}_1 \\ \ddot{x}_2 \end{pmatrix} = \begin{pmatrix} \bar{f}_1(x) \\ \bar{f}_2(x) \end{pmatrix} + \begin{pmatrix} \bar{g}_1(x) & 0 \\ 0 & \bar{g}_2(x) \end{pmatrix} \begin{pmatrix} u_1 \\ u_2 \end{pmatrix} \quad (8.130)$$

Next, by defining the cumulative control inputs

$$v_1 = \bar{f}_1(x) + \bar{g}_1(x)u_1 \quad v_2 = \bar{f}_2(x) + \bar{g}_2(x)u_2 \quad (8.131)$$

one arrives at the following input–output linearized form

$$\dot{x}_1 = v_1 \quad \ddot{x}_2 = v_2 \quad (8.132)$$

By defining the new state variables $z_1 = x_1$, $z_2 = x_2$, $z_3 = \dot{x}_3$, one obtains a canonical Brunovsky form for the dynamics of acute leukemia

$$\begin{pmatrix} \dot{z}_1 \\ \dot{z}_2 \\ \dot{z}_3 \end{pmatrix} = \begin{pmatrix} 0 & 0 & 0 \\ 0 & 0 & 1 \\ 0 & 0 & 0 \end{pmatrix} \begin{pmatrix} z_1 \\ z_2 \\ z_3 \end{pmatrix} + \begin{pmatrix} 1 & 0 \\ 0 & 0 \\ 0 & 1 \end{pmatrix} \begin{pmatrix} v_1 \\ v_2 \end{pmatrix} \quad (8.133)$$

Next, the design of a stabilizing feedback controller is performed using the eigenvalues assignment technique. The targeted setpoint for z_1 is denoted as $z_{1,d}$ while the targeted setpoint for z_2 is $z_{2,d}$. The feedback control inputs are taken to be:

$$v_1 = \dot{z}_{1,d} + k_{1,1}(z_1 - z_{1,d})$$
$$v_2 = \ddot{z}_{2,d} + k_{1,2}(\dot{z}_2 - \dot{z}_{2,d}) + k_{2,2}(z_2 - z_{2,d}) \quad (8.134)$$

By stabilizing the above-noted feedback control inputs of Eqs. (8.134) to the input–output linearized model of Eq. (8.133), one obtains the following closed-loop system dynamics

$$\begin{aligned}(\dot{z}_1-\dot{z}_{1,d})+k_{1,1}(z_1-z_{1,d})&=0\\(\ddot{z}_2-\ddot{z}_{2,d})+k_{1,2}(\dot{z}_1-\dot{z}_{1,d})+k_{2,2}(z_1-z_{1,d})&=0\end{aligned}\tag{8.135}$$

By defining the tracking error variables $e_1=z_1-z_{1,d}$ and $e_2=z_2-z_{2,d}$, one has the tracking error dynamics of the closed-loop system:

$$\begin{aligned}\dot{e}_1+k_{1,1}e_1&=0\\\ddot{e}_2+k_{1,2}\dot{e}_1+k_{2,2}e_2&=0\end{aligned}\tag{8.136}$$

By selecting the feedback gains $k_{1,1}$ and $k_{1,2}$, $k_{2,2}$ so as the characteristic polynomials which are associated with the above differential equations to be Hurwitz stable, one has

$$\begin{aligned}lim_{t\to\infty}e_1=0&\Rightarrow lim_{t\to\infty}z_1=z_{1,d}\Rightarrow lim_{t\to\infty}x_1=x_{1,d}\\lim_{t\to\infty}e_2=0&\Rightarrow lim_{t\to\infty}z_2=z_{2,d}\Rightarrow lim_{t\to\infty}x_2=x_{2,d}\end{aligned}\tag{8.137}$$

The third state variable x_3 was chosen to be a differential function of the flat outputs $y_1=x_1$ and $y_2=x_2$. The set-point $x_{3,d}$ of the third state variable is also a differential function of the targeted values of the flat outputs $y_{1,d}=x_{1,d}$ and $y_{2,d}=x_{2,d}$. Thus, by proving that $lim_{t\to\infty}x_1=x_{1,d}$ and $lim_{t\to\infty}x_2=x_{2,d}$, one also ascertains that $lim_{t\to\infty}x_3=x_{3,d}$.

The control input which should be applied to the initial nonlinear system is

$$\begin{pmatrix}u_1\\u_2\end{pmatrix}=\begin{pmatrix}\bar{g}_1(x)&0\\0&\bar{g}_2(x)\end{pmatrix}^{-1}\left[\begin{pmatrix}\dot{x}_1\\\ddot{x}_2\end{pmatrix}-\begin{pmatrix}\bar{f}_1(x)\\\bar{f}_2(x)\end{pmatrix}\right]\tag{8.138}$$

Through the previous analysis it has been shown that unlike the proposed nonlinear optimal control method, control which is based on input–output linearization of leukemia's dynamics (i) requires transformations of the associated state-space model and changes of state variables and (ii) requires inverse transformations (based on the inversion of matrices) so as to compute the control inputs which should be applied to leukemia's initial nonlinear model and this can finally result in the appearance of singularities.

8.2.8 Simulation tests

The previous results about convergence and global stability of the nonlinear optimal control method for leukemia's dynamics are further confirmed through simulation experiments. To compute the optimal control inputs the method's algebraic Riccati equation was being solved repetitively at each time step of the control algorithm using the *aresolv()* function in MATLAB®. Time was measured in days and the sampling period was $T_s=0.01$ days (that is 14.4 minutes). The values of the model's parameters were in accordance with [440,773,387,388]. Such values are: $\gamma_n=0.2$, $c=0.2$, $N_a=15$, $L_a=14$, $\gamma_l=0.4$, $r_l=0.1$, $\gamma_h=1.3$. The coefficients of the model's binomials were assigned the values $a_1=0.2$, $b_1=3.0$, $c_1=0.2$, $a_2=0.5$, $b_2=1.0$, $c_2=0.1$. The obtained results can be seen in Figures 8.13–8.28. The real values of the system's state variables are printed in blue, the associated setpoints are marked in red while the state estimates which are provided by the H-infinity Kalman filter are depicted in green color. The values of the populations of the normal cells x_1 and of the leukemic cells x_2 are multiplied by 10^8. It can be noticed that, in all test cases, fast and accurate tracing of setpoints by the state variables of leukemia's dynamic model is achieved under moderate variations of the control inputs.

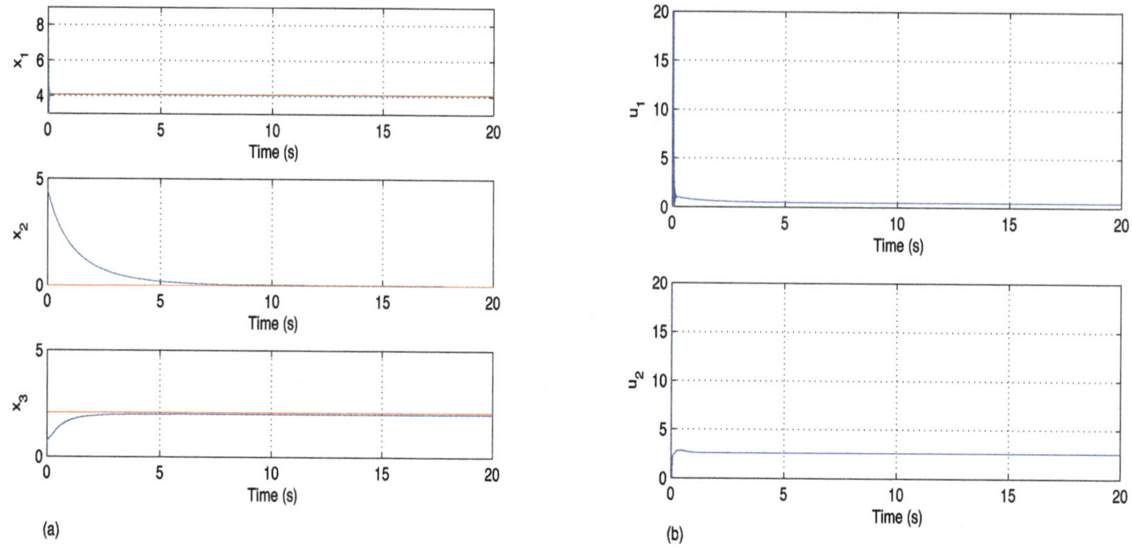

Figure 8.13 *Test case 1 on nonlinear optimal control of leukemia: (a) convergence of the state variables x_1 to x_3 to the reference setpoints (red lines) and state variables estimates (green lines); (b) optimal control inputs u_1 and u_2*

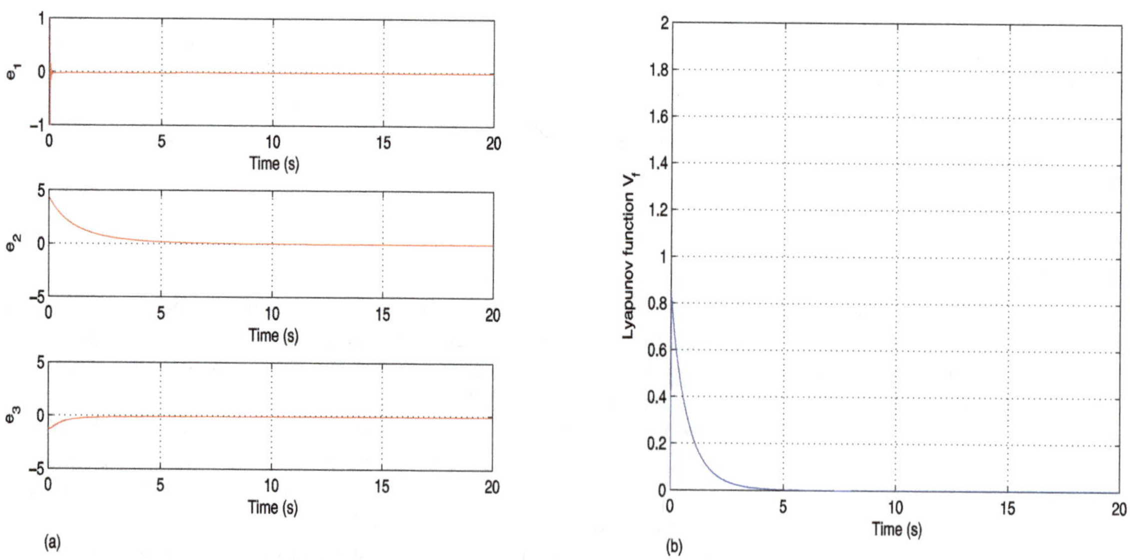

Figure 8.14 *Test case 1 on nonlinear optimal control of leukemia: (a) convergence of the tracking error variables e_1 to e_3 to zero (red lines); (b) Lyapunov function V_f of the control loop*

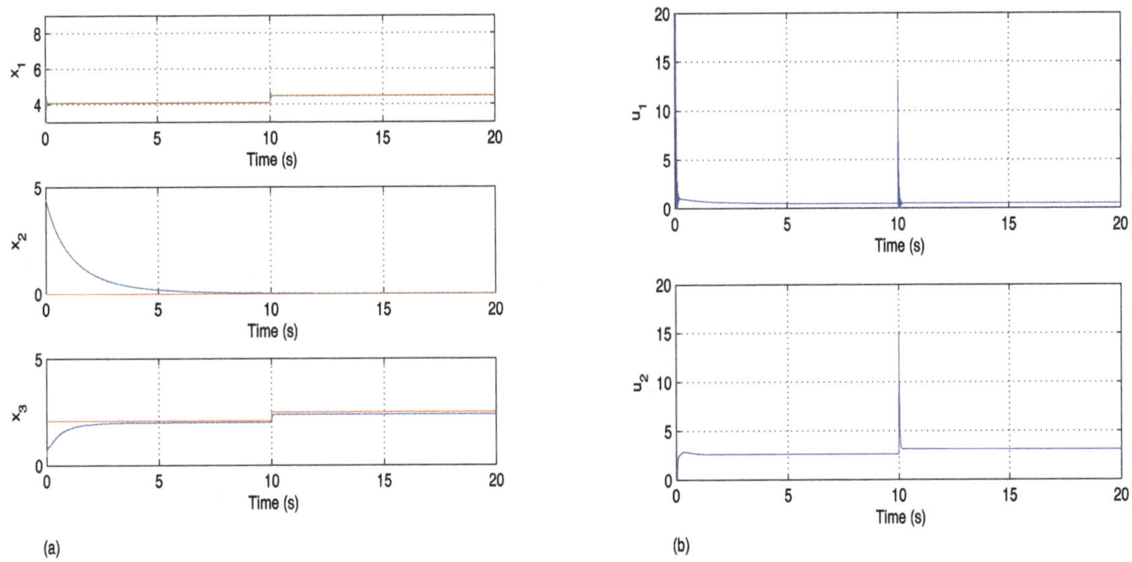

Figure 8.15 Test case 2 on nonlinear optimal control of leukemia: (a) convergence of the state variables x_1 to x_3 to the reference setpoints (red lines) and state variables estimates (green lines); (b) optimal control inputs u_1 and u_2

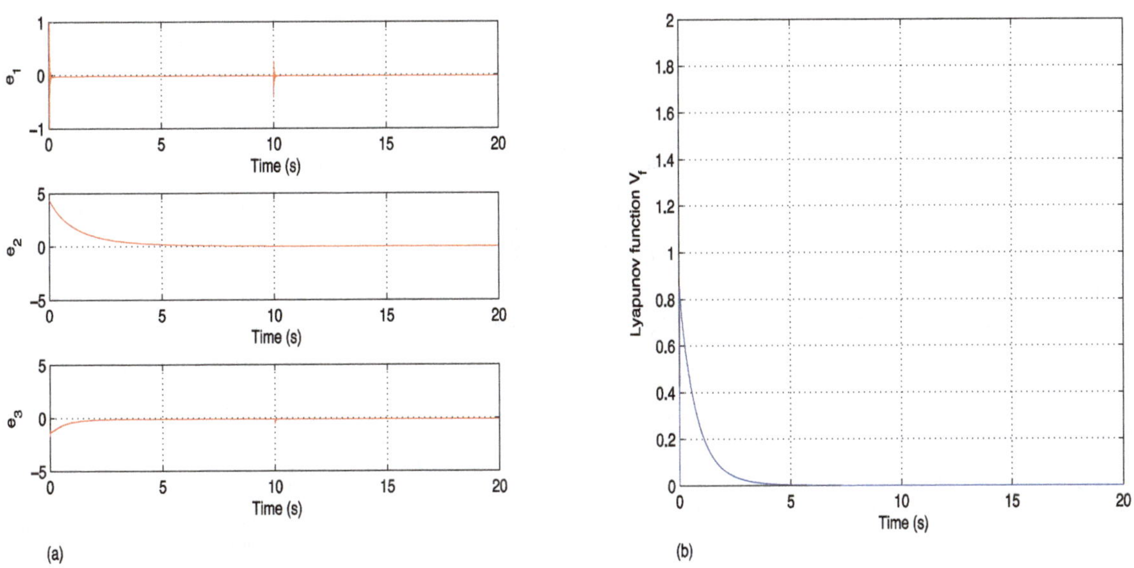

Figure 8.16 Test case 2 on nonlinear optimal control of leukemia: (a) convergence of the tracking error variables e_1 to e_3 to zero (red lines); (b) Lyapunov function V_f of the control loop

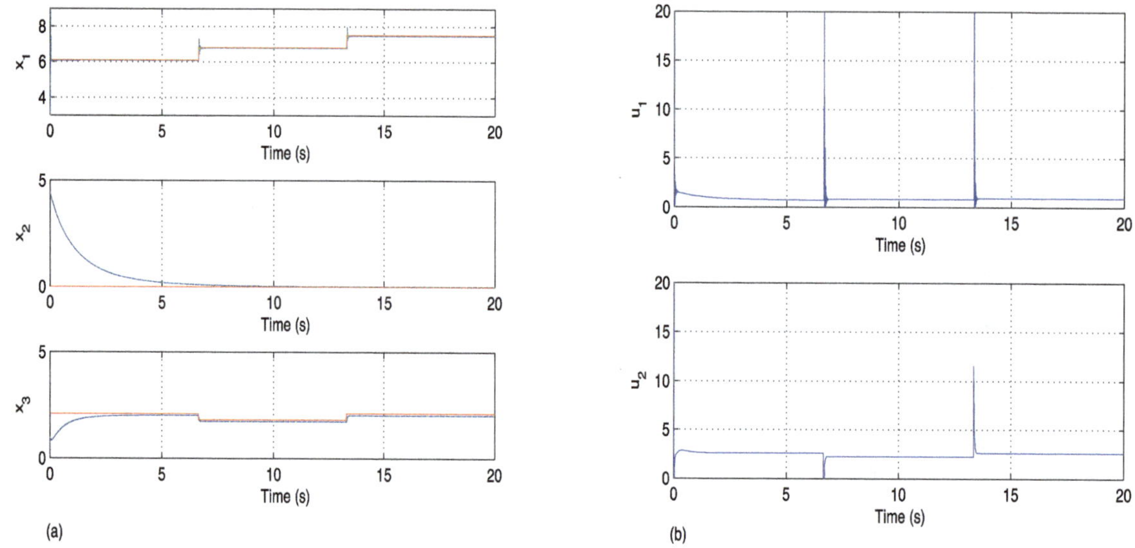

Figure 8.17 *Test case 3 on nonlinear optimal control of leukemia: (a) convergence of the state variables x_1 to x_3 to the reference setpoints (red lines) and state variables estimates (green lines); (b) optimal control inputs u_1 and u_2*

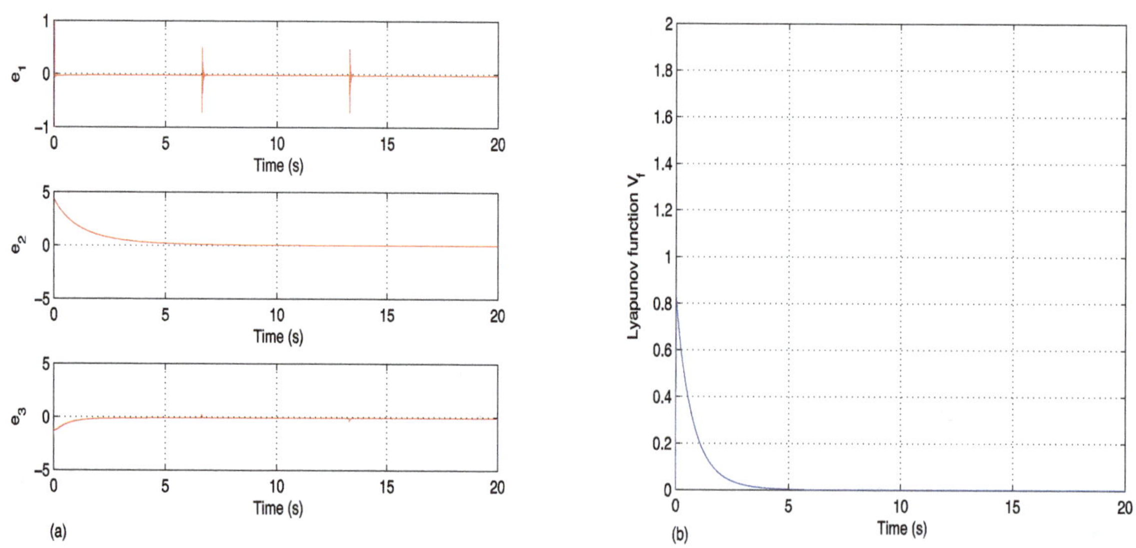

Figure 8.18 *Test case 3 on nonlinear optimal control of leukemia: (a) convergence of the tracking error variables e_1 to e_3 to zero (red lines); (b) Lyapunov function V_f of the control loop*

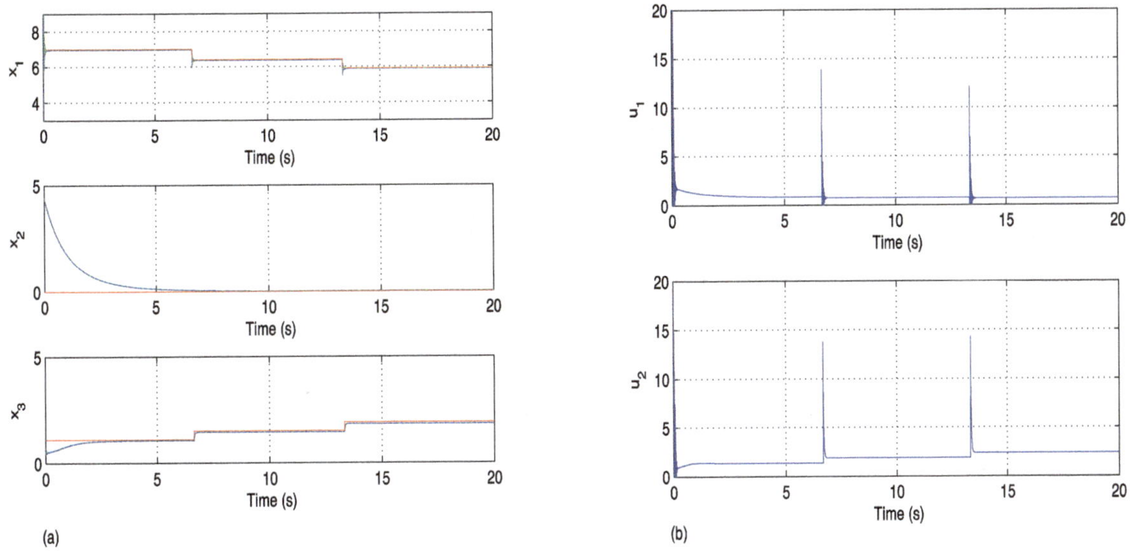

Figure 8.19 *Test case 4 on nonlinear optimal control of leukemia: (a) convergence of the state variables x_1 to x_3 to the reference setpoints (red lines) and state variables estimates (green lines); (b) optimal control inputs u_1 and u_2*

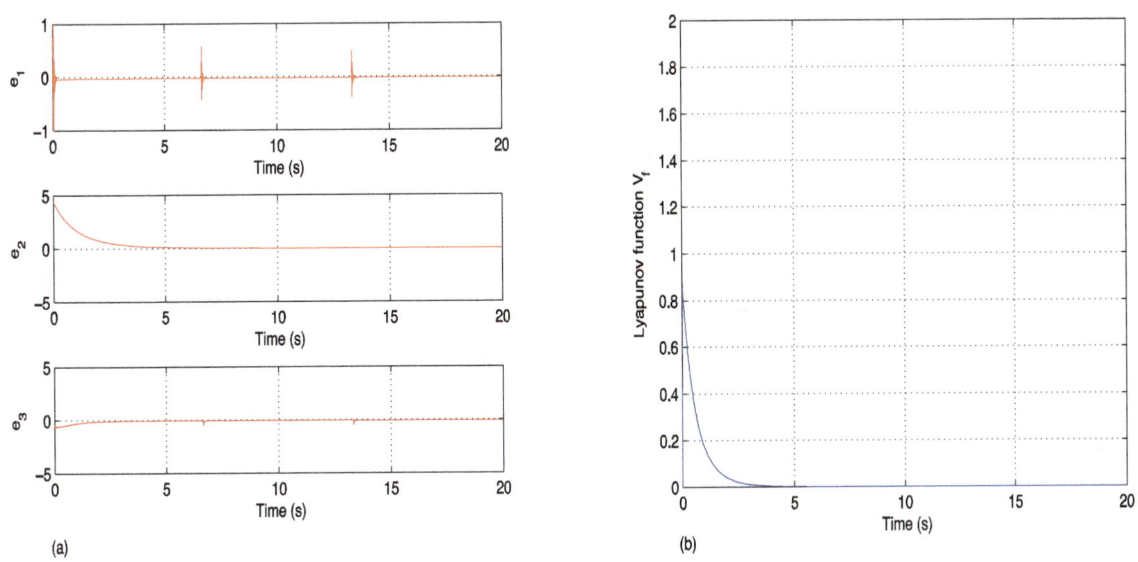

Figure 8.20 *Test case 4 on nonlinear optimal control of leukemia: (a) convergence of the tracking error variables e_1 to e_3 to zero (red lines); (b) Lyapunov function V_f of the control loop*

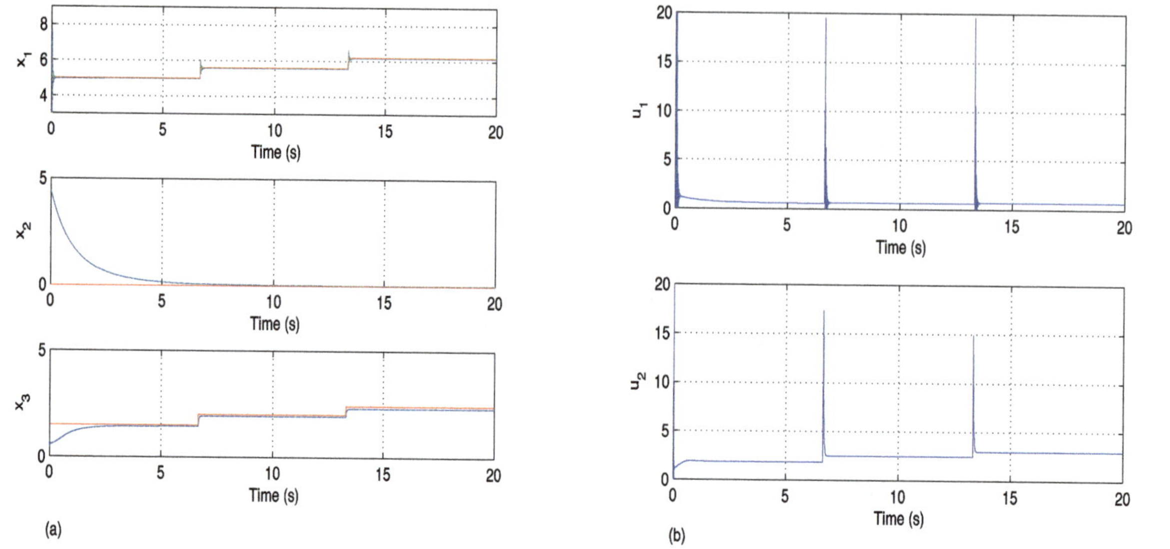

Figure 8.21 *Test case 5 on nonlinear optimal control of leukemia: (a) convergence of the state variables x_1 to x_3 to the reference setpoints (red lines) and state variables estimates (green lines); (b) optimal control inputs u_1 and u_2*

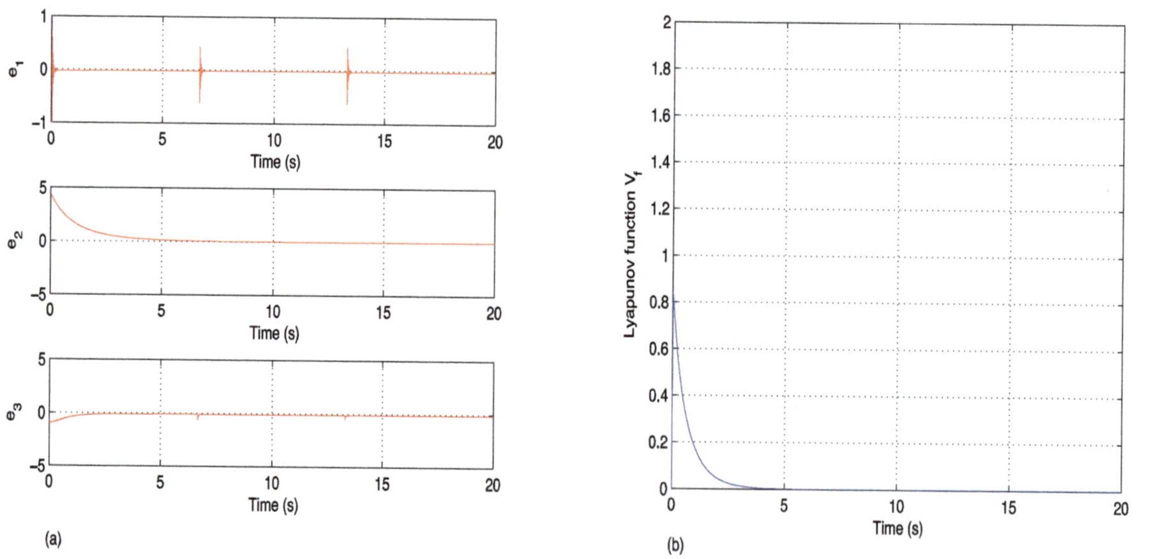

Figure 8.22 *Test case 5 on nonlinear optimal control of leukemia: (a) convergence of the tracking error variables e_1 to e_3 to zero (red lines); (b) Lyapunov function V_f of the control loop*

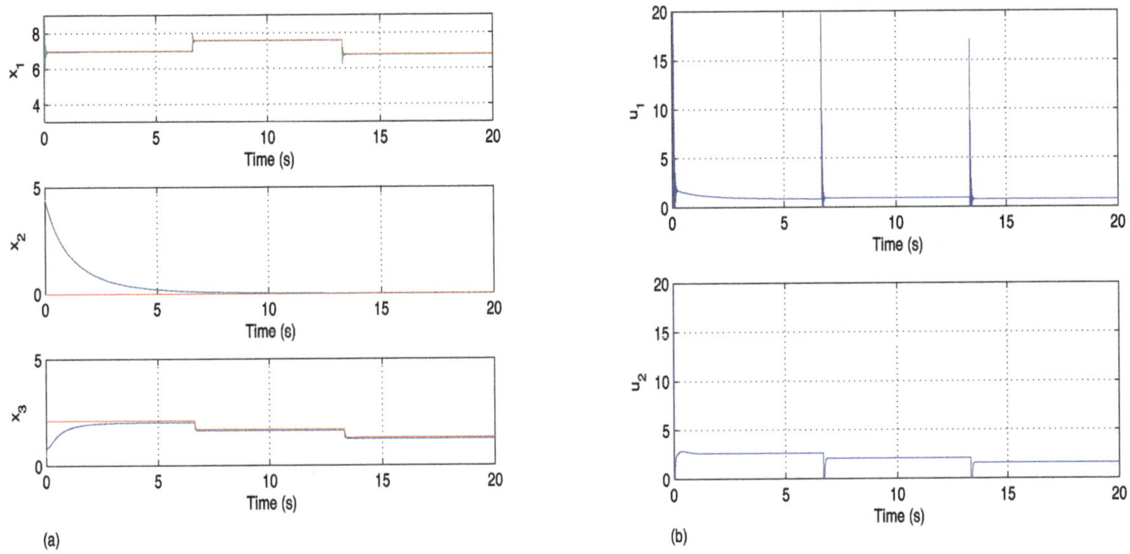

Figure 8.23 *Test case 6 on nonlinear optimal control of leukemia: (a) convergence of the state variables x_1 to x_3 to the reference setpoints (red lines) and state variables estimates (green lines); (b) optimal control inputs u_1 and u_2*

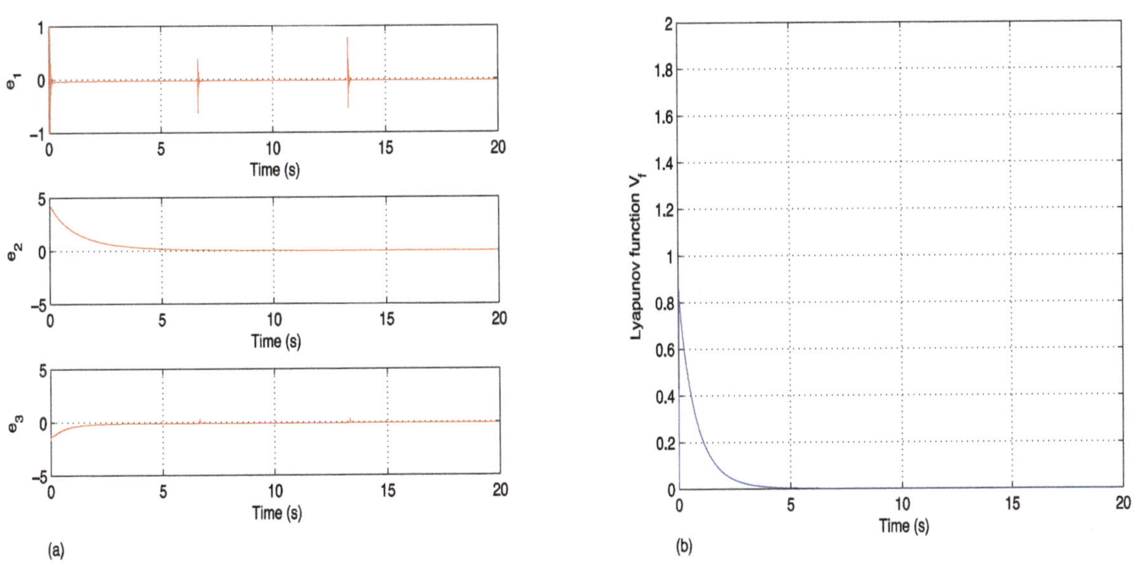

Figure 8.24 *Test case 6 on nonlinear optimal control of leukemia: (a) convergence of the tracking error variables e_1 to e_3 to zero (red lines); (b) Lyapunov function V_f of the control loop*

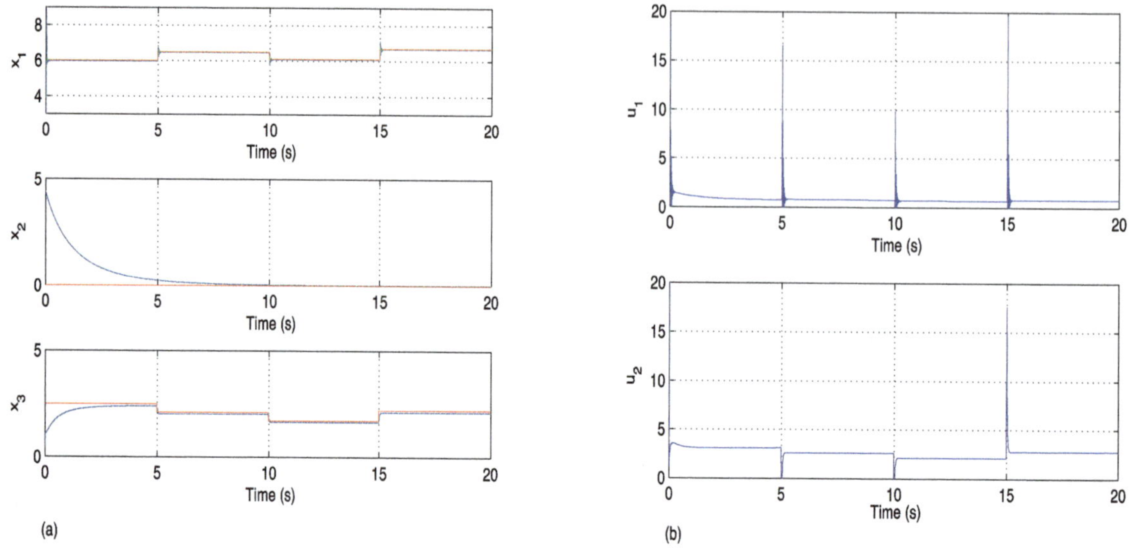

Figure 8.25 Test case 7 on nonlinear optimal control of leukemia: (a) convergence of the state variables x_1 to x_3 to the reference setpoints (red lines) and state variables estimates (green lines); (b) optimal control inputs u_1 and u_2

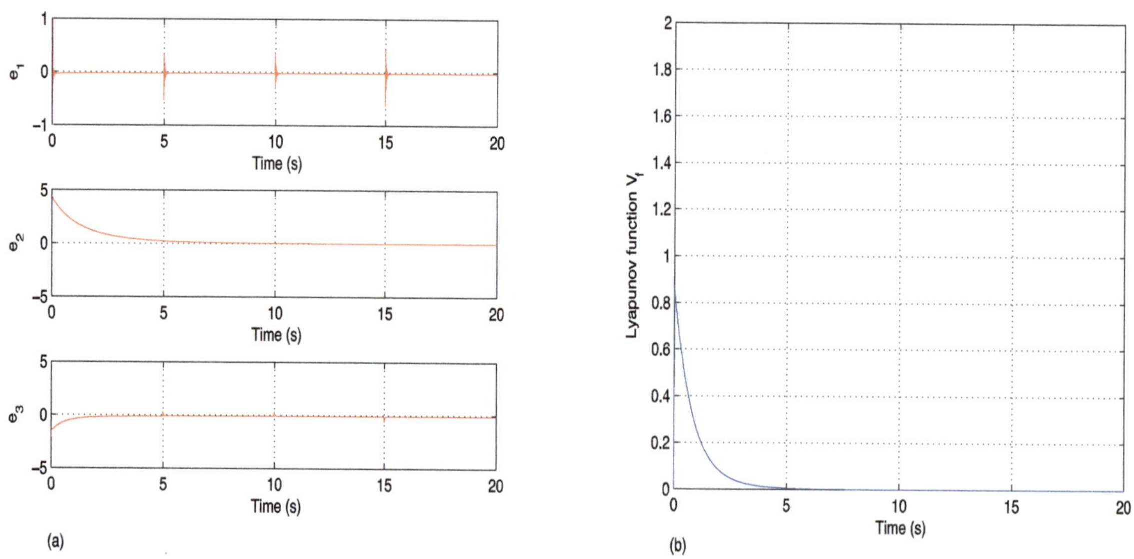

Figure 8.26 Test case 7 on nonlinear optimal control of leukemia: (a) convergence of the tracking error variables e_1 to e_3 to zero (red lines); (b) Lyapunov function V_f of the control loop

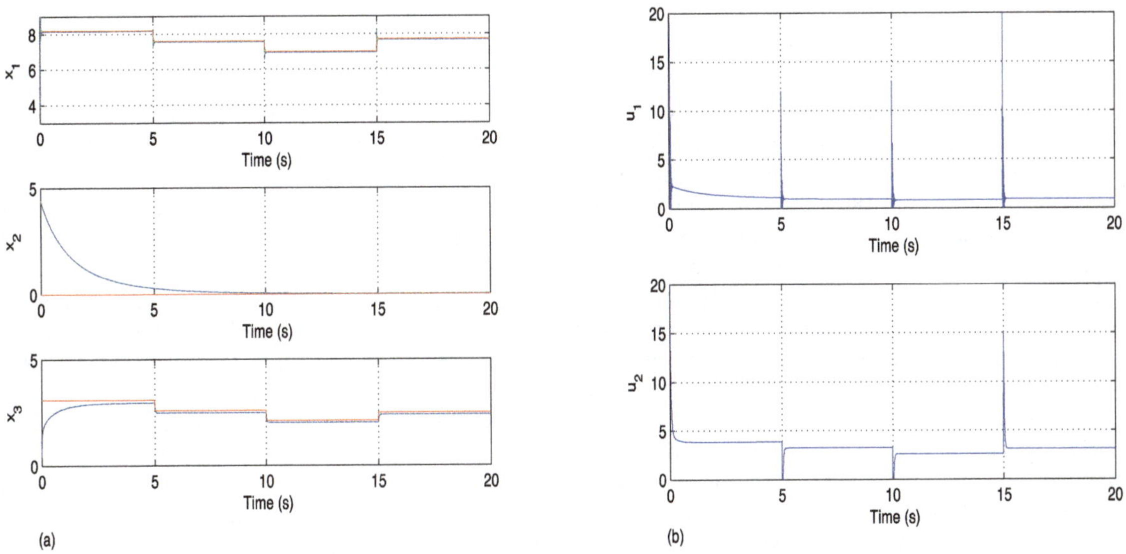

Figure 8.27 *Test case 8 on nonlinear optimal control of leukemia: (a) convergence of the state variables x_1 to x_3 to the reference setpoints (red lines) and state variables estimates (green lines); (b) optimal control inputs u_1 and u_2*

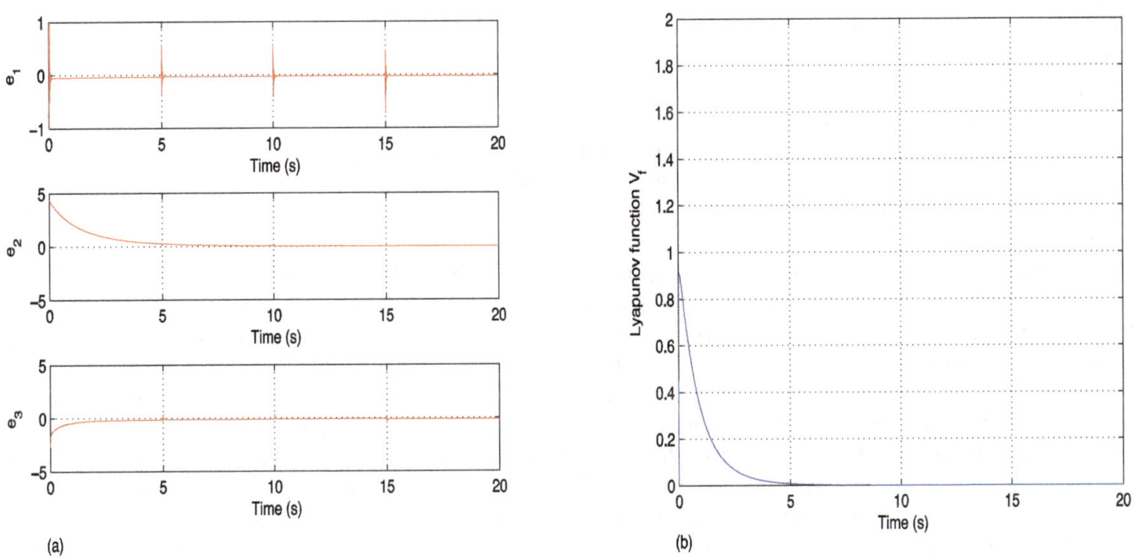

Figure 8.28 *Test case 8 on nonlinear optimal control of leukemia: (a) convergence of the tracking error variables e_1 to e_3 to zero (red lines); (b) Lyapunov function V_f of the control loop*

8.3 Nonlinear optimal control of supply chain networks under time-delays

8.3.1 Outline

A supply chain network is a complex dynamical system that connects the activities of enterprises from upstream manufacturing and supply to downstream product distribution and sales [771,78]. There is a forward flow of materials and products and a backward flow of information and orders [487,1043]. A general form of a supply chain network consists of four echelons, associated with customers' demand and inventory levels at the side of the retailers, the distributors, and the manufacturers [549,180]. The efficiency of a supply chain depends on the maintenance of targeted inventory levels at its individual stages [179,972]. Low inventory levels are the cause for the scarcity of final products while high inventory levels signify a raise of storage costs [445,181]. Supply chain networks are often subjected to disruptions such as time-delays and the "bullwhip effect" [82,261]. Time-delays are mainly due to the lack of synchronization between the individual units and due to transport of materials and products [147,319]. The "bullwhip effect" takes the form of amplification of demand from a downstream site to an upstream site and can provoke redundant inventory, excessive production and financial losses [355,1034]. Disruptions in the supply chain network may also end up at a chaotic behavior in which the levels of local inventories exhibit oscillations [288,1024]. Thus, synchronization and control between the individual units which constitute a supply chain is essential for high system performance, reliability, maximization of profits and minimization of risks [19,973]. Often the management of supply chains is performed with decentralized decision taking procedures which can be conflicting. On the other side, centralized control schemes can ensure stability of supply chains, synchronization and optimized inventory management [618,646]. Among various control approaches for supply chain networks, one can distinguish model-free control, model predictive control and sliding-mode control [432,971]. Optimal control addressed to the nonlinear dynamics of supply chains is a topic that has been little investigated [774,393,681].

In this section a four-echelon supply chain network has been considered. A nonlinear optimal control method is proposed for the dynamics of the supply chain and for inventory levels control [717,725]. The state variables of the model are the customers' demand and the inventories at the retailers, the distributors, and the manufacturers. The control inputs of the system are the orders' rate at the side of the retailers and the production rate at the side of the manufacturer. It has been proven that the dynamic model of the supply chain network is differentially flat. This proof is also an implicit confirmation of the system's controllability. To apply the new nonlinear optimal control scheme the dynamic model of the supply chain network undergoes approximate linearization with the use of first-order Taylor series expansion and through the computation of the associated Jacobian matrices [705,60,706]. The linearization point (x^*, y^*) is updated at each sampling instance and is defined by the present value of the system's state vector x^* and by the last sampled value of the control inputs vector u^*. The modeling error which is due to the truncation of higher-order terms from the Taylor series is considered to be a perturbation which is asymptotically compensated by the robustness of the control method. For the approximately linearized model of the supply chain, an H-infinity feedback controller is designed.

The H-infinity controller is a solution to the optimal control problem under model uncertainty and disturbances. The H-infinity controller represents a min–max differential game which takes place between (i) the control inputs that try to minimize a cost function which contains a quadratic term of the state vector's tracking error and (ii) the model uncertainty and external perturbation terms which try to maximize this cost function [717,725]. To compute the feedback gains of the H-infinity controller an algebraic Riccati equation is repetitively solved at each time step of the control algorithm [711,876,732]. The global stability properties of the control scheme are proven through Lyapunov analysis. First, it is demonstrated that the control loop satisfies the H-infinity tracking performance criterion which signifies robustness against model uncertainties and perturbations. Moreover, under moderate conditions, global asymptotic stability is proven [717,725]. The nonlinear optimal control method achieves fast and accurate tracking of setpoints by all state variables of the supply chain network under moderate variations of the control inputs. To apply state estimation-based control without the need to measure the entire state vector of the system at each sampling interval the H-infinity Kalman filter is proposed as a robust state estimator.

8.3.2 Dynamic model of the supply chain network

8.3.2.1 State-space model of the supply chain network

A four-echelon supply chain network is considered [549,180]. The diagram of the supply chain network and of inventory control under time-delays is given in Figure 8.29. The state variables of the supply chain network are: x_1 is the demand of the customers, x_2 is the inventory at the retailers, x_3 is the inventory at the distributors, and x_4 is the inventory at the factory. The control inputs of the supply chain network are: u_1 is the rate of orders at the retailers, and u_2 is the production rate at the factory. Continuous-time multi-echelon models can be used to describe industrial production planning and product distribution in several cases.

In discrete-time form the dynamics of the supply chain network is given by the following difference equations:

$$\begin{aligned}
x_1(k+1) &= -x_2(k-\theta_2) + nx_1(k) \\
x_2(k+1) &= x_3(k-\theta_3) - \eta x_2(k) + cx_1(k-\theta_1) + u_1(k) \\
x_3(k+1) &= rx_2(k) - x_2(k)x_4(k-\theta_4) \\
x_4(k+1) &= x_2(k-\theta_2)x_3(k) + sx_4(k) + u_2(k)
\end{aligned} \tag{8.139}$$

where assuming a sampling period of T_s, $x_i(k)$, $i = 1,2,3,4$ denotes the i-th state variable $x_i(t)$ at time instant $t = k{\cdot}T_s$. Parameters θ_i, $i = 1,2,3,4$ are time-delays equal to $\theta_i T_s$. Using the approximation about the time derivatives of the state variables $\dot{x}_i(t) \simeq \frac{x_i(k+1)-x_i(k)}{T_s}$, and the approximation about the state variables which are subjected to time-delays

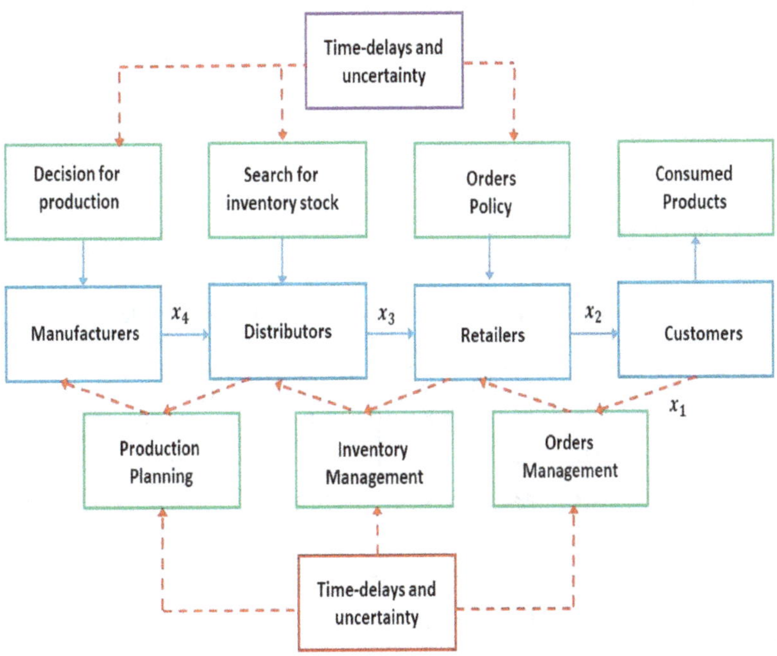

Figure 8.29 Diagram of the supply chain network and of inventory levels control under time-delays

$x_i(k-\theta_i) \simeq x_i(t) - \theta_i T_s \dot{x}_i(t)$, one obtains also the following continuous-time description of the state-space model of the supply chain network:

$$\begin{aligned}\dot{x}_1 &= -\tfrac{1}{T_s}x_1 - \tfrac{1}{T_s}x_2 + \theta_2\dot{x}_2 + \tfrac{n}{T_s}x_1 \\ \dot{x}_2 &= -\tfrac{1}{T_s}x_2 - \tfrac{1}{T_s}x_3 + \theta_3\dot{x}_3 - \tfrac{\eta}{T_s}x_2 + \tfrac{c}{T_s}x_1 - c\theta_1\dot{x}_1 + \tfrac{1}{T_s}u_1 \\ \dot{x}_3 &= -\tfrac{1}{T_s}x_3 + \tfrac{r}{T_s}x_2 - \tfrac{1}{T_s}x_2x_4 + \theta_4 x_2\dot{x}_4 \\ \dot{x}_4 &= -\tfrac{1}{T_s}x_4 + \tfrac{1}{T_s}x_2x_3 - \theta_2 x_3\dot{x}_2 + \tfrac{s}{T_s}x_4 + \tfrac{1}{T_s}u_2\end{aligned} \tag{8.140}$$

By transferring the time-derivatives of the state variables to the left part of the above-given equations, one obtains the equivalent state-space description:

$$\begin{pmatrix} 1 & -\theta_2 & 0 & 0 \\ c\theta_1 & 1 & -\theta_3 & 0 \\ 0 & 0 & 1 & -\theta_4 x_2 \\ 0 & \theta_2 x_3 & 0 & 1 \end{pmatrix}\begin{pmatrix}\dot{x}_1\\ \dot{x}_2\\ \dot{x}_3\\ \dot{x}_4\end{pmatrix} = \begin{pmatrix} -\frac{(1-\eta)}{T_s}x_1 - \frac{1}{T_s}x_2 \\ -\frac{c}{T_s}x_1 - \frac{(1+\eta)}{T_s}x_2 - \frac{1}{T_s}x_3 \\ -\frac{1}{T_s}x_3 + \frac{n}{T_s}x_2 - \frac{1}{T_s}x_2x_4 \\ \frac{1}{T_s}x_2x_3 + \frac{(s-1)}{T_s}x_4 \end{pmatrix} + \begin{pmatrix} 0 & 0 \\ \frac{1}{T_s} & 0 \\ 0 & 0 \\ 0 & \frac{1}{T_s}\end{pmatrix}\begin{pmatrix}u_1\\ u_2\end{pmatrix} \tag{8.141}$$

The following notation of matrices and vectors is used:

$$M = \begin{pmatrix} 1 & -\theta_2 & 0 & 0 \\ c\theta_1 & 1 & -\theta_3 & 0 \\ 0 & 0 & 1 & -\theta_4 x_2 \\ 0 & \theta_2 x_3 & 0 & 1 \end{pmatrix} \quad F(x) = \begin{pmatrix} -\frac{(1-\eta)}{T_s}x_1 - \frac{1}{T_s}x_2 \\ -\frac{c}{T_s}x_1 - \frac{(1+\eta)}{T_s}x_2 - \frac{1}{T_s}x_3 \\ -\frac{1}{T_s}x_3 + \frac{n}{T_s}x_2 - \frac{1}{T_s}x_2x_4 \\ \frac{1}{T_s}x_2x_3 + \frac{(s-1)}{T_s}x_4 \end{pmatrix} \quad G(x) = \begin{pmatrix} 0 & 0 \\ \frac{1}{T_s} & 0 \\ 0 & 0 \\ 0 & \frac{1}{T_s}\end{pmatrix} \tag{8.142}$$

The inverse of matrix M is given by

$$M = \frac{1}{detM}\begin{pmatrix} M_{11} & -M_{21} & M_{31} & -M_{41} \\ -M_{12} & M_{22} & -M_{32} & M_{42} \\ M_{13} & -M_{23} & M_{33} & -M_{43} \\ -M_{14} & M_{24} & -M_{34} & M_{44}\end{pmatrix} \tag{8.143}$$

where $M_{11} = 1 + \theta_2\theta_3\theta_4 x_2 x_3$, $M_{12} = c\theta_1$, $M_{13} = c\theta_1\theta_2\theta_4 x_2 x_3$, $M_{14} = -c\theta_1\theta_2 x_3$, $M_{21} = -\theta_2$, $M_{22} = 1$, $M_{23} = \theta_2\theta_4 x_2 x_3$, $M_{24} = -\theta_2 x_3$, $M_{31} = -\theta_2\theta_3$, $M_{32} = -\theta_3$, $M_{33} = 1 + \theta_1\theta_2$, $M_{34} = \theta_2\theta_3 x_3$, $M_{41} = -\theta_2\theta_3\theta_4 x_2$, $M_{42} = \theta_3\theta_4 x_2$, $M_{43} = -\theta_4(1 + c\theta_1\theta_2)x_2$, $M_{44} = 1 - \theta_1\theta_2$, and $detM = M_{11} + \theta_2 M_{12}$.

Using the previous notation about the inverse of matrix M, the dynamic model of the supply chain network is written as

$$\begin{pmatrix}\dot{x}_1\\ \dot{x}_2\\ \dot{x}_3\\ \dot{x}_4\end{pmatrix} = \frac{1}{detM}\begin{pmatrix} M_{11} & -M_{21} & M_{31} & -M_{41} \\ -M_{12} & M_{22} & -M_{32} & M_{42} \\ M_{13} & -M_{23} & M_{33} & -M_{43} \\ -M_{14} & M_{24} & -M_{34} & M_{44}\end{pmatrix}\left[\begin{pmatrix}F_1\\ F_2\\ F_3\\ F_4\end{pmatrix} + \begin{pmatrix} G_{11} & G_{21} \\ G_{21} & G_{22} \\ G_{31} & G_{32} \\ G_{41} & G_{42}\end{pmatrix}\begin{pmatrix}u_1\\ u_2\end{pmatrix}\right] \tag{8.144}$$

or equivalently

$$\begin{pmatrix}\dot{x}_1\\ \dot{x}_2\\ \dot{x}_3\\ \dot{x}_4\end{pmatrix} = \begin{pmatrix} \frac{M_{11}F_1 - M_{21}F_2 + M_{31}F_3 - M_{41}F_4}{detM} \\ \frac{-M_{12}F_1 + M_{22}F_2 - M_{32}F_3 + M_{42}F_4}{detM} \\ \frac{M_{13}F_1 - M_{23}F_2 + M_{33}F_3 - M_{43}F_4}{detM} \\ \frac{-M_{14}F_1 + M_{24}F_2 - M_{34}F_3 + M_{44}F_4}{detM}\end{pmatrix} + \begin{pmatrix} \frac{-M_{21}G_{21}}{detM} & \frac{-M_{41}G_{42}}{detM} \\ \frac{M_{22}G_{21}}{detM} & \frac{M_{42}G_{42}}{detM} \\ \frac{-M_{23}G_{21}}{detM} & \frac{-M_{43}G_{42}}{detM} \\ \frac{M_{24}G_{21}}{detM} & \frac{M_{44}G_{42}}{detM}\end{pmatrix}\begin{pmatrix}u_1\\ u_2\end{pmatrix} \tag{8.145}$$

Next, the dynamic model of the supply chain network is written in the nonlinear affine in-the-input state-space form

$$\dot{x} = f(x) + g(x)u \tag{8.146}$$

where $x \in R^{4\times1}$, $f(x) \in R^{4\times1}$, $g(x) \in R^{4\times2}$, and $u \in R^{2\times1}$. Using that $g(x) = [g_1(x)\ g_2(x)]$, one has the following definition about vectors $f(x)$, $g_1(x)$, $g_2(x)$:

$$f(x) = \begin{pmatrix} \frac{M_{11}F_1 - M_{21}F_2 + M_{31}F_3 - M_{41}F_4}{detM} \\ \frac{-M_{12}F_1 + M_{22}F_2 - M_{32}F_3 + M_{42}F_4}{detM} \\ \frac{M_{13}F_1 - M_{23}F_2 + M_{33}F_3 - M_{43}F_4}{detM} \\ \frac{-M_{14}F_1 + M_{24}F_2 - M_{34}F_3 + M_{44}F_4}{detM} \end{pmatrix} \quad g_1(x) = \begin{pmatrix} \frac{-M_{21}G_{21}}{detM} \\ \frac{M_{22}G_{21}}{detM} \\ \frac{-M_{23}G_{21}}{detM} \\ \frac{M_{24}G_{21}}{detM} \end{pmatrix} \quad g_2(x) = \begin{pmatrix} \frac{-M_{41}G_{42}}{detM} \\ \frac{M_{42}G_{42}}{detM} \\ \frac{-M_{43}G_{42}}{detM} \\ \frac{M_{44}G_{42}}{detM} \end{pmatrix} \tag{8.147}$$

8.3.2.2 Differential flatness properties of the supply chain network

If $\theta_2 = 0$ and $\theta_4 = 0$ then $M_{21} = 0$, $M_{41} = 0$, $M_{23} = 0$ and $M_{43} = 0$. In such a case, the dynamic model of the supply chain network given in Eq. (8.145) and Eq. (8.146) is differentially flat, with flat outputs vector $Y = [x_1, x_3]^T$ [711]. Indeed from the first and third rows of the state-space model, one has

$$\begin{aligned} \dot{x}_1 &= \frac{M_{11}F_1 - M_{21}F_2 + M_{31}F_3 - M_{41}F_4}{detM} \Rightarrow \dot{x}_1 = f_1(x_1, x_2, x_3, x_4) \\ \dot{x}_3 &= \frac{M_{13}F_1 - M_{23}F_2 + M_{33}F_3 - M_{43}F_4}{detM} \Rightarrow \dot{x}_3 = f_3(x_1, x_2, x_3, x_4) \end{aligned} \tag{8.148}$$

Therefore, one has two polynomial equations with x_1, $\dot{x}_1$, x_3, $\dot{x}_3$ to be the coefficients and x_2, x_4 to be the unknowns. These can be solved for x_2, x_4 thus giving

$$\begin{aligned} x_2 &= h_2(x_1, \dot{x}_1, x_3, \dot{x}_3) \Rightarrow x_2 = h_2(Y, \dot{Y}) \\ x_4 &= h_4(x_1, \dot{x}_1, x_3, \dot{x}_3) \Rightarrow x_4 = h_4(Y, \dot{Y}) \end{aligned} \tag{8.149}$$

Therefore, it is demonstrated that x_2, x_4 are differential functions of the flat outputs. Moreover, from the second and fourth row of the state-space model, one has

$$\begin{aligned} \dot{x}_2 &= f_2(x) + g_{21}(x)u_1 + g_{22}(x)u_2 \Rightarrow \dot{h}_2(Y, \dot{Y}) = f_2(Y) + g_{21}(Y)u_1 + g_{22}(Y)u_2 \\ \dot{x}_4 &= f_4(x) + g_{41}(x)u_1 + g_{42}(x)u_2 \Rightarrow \dot{h}_4(Y, \dot{Y}) = f_4(Y) + g_{41}(Y)u_1 + g_{42}(Y)u_2 \end{aligned} \tag{8.150}$$

Solving the above relation for u_1, u_2 gives

$$\begin{pmatrix} u_1 \\ u_2 \end{pmatrix} = \begin{pmatrix} g_{21}(Y) & g_{22}(Y) \\ g_{41}(Y) & g_{42}(Y) \end{pmatrix}^{-1} \begin{pmatrix} \dot{h}_2(Y, \dot{Y}) - f_2(Y) \\ \dot{h}_4(Y, \dot{Y}) - f_4(Y) \end{pmatrix} \tag{8.151}$$

The above relation demonstrates that control inputs u_1, u_2 are also differential functions of the system, or equivalently

$$\begin{pmatrix} u_1 \\ u_2 \end{pmatrix} = \begin{pmatrix} h_{u_1}(Y, \dot{Y}) \\ h_{u_2}(Y, \dot{Y}) \end{pmatrix} \tag{8.152}$$

Since (i) all state variables and the control inputs of the dynamic model of the supply chain network are differential functions of the flat outputs and (ii) the flat outputs are differentially independent (that is they are not connected through a relation in the form of an ordinary differential equation), it can be concluded that the supply chain network is a differentially flat system. The proof of differential flatness is also an implicit demonstration of the system's controllability, considering that the control inputs are the rate of orders at the retailers and the rate of industrial production at the manufacturer's side.

8.3.3 *Approximate linearization of the dynamics of the supply chain network*

The dynamic model of the supply chain network, being initially in the nonlinear state-space form of Eq. (8.146) $\dot{x} = f(x) + g(x)u$, undergoes approximate linearization through first-order Taylor series expansion and the computation of the associated Jacobian matrices. The linearization point (x^*, u^*) is updated at each sampling instance and is defined by the present value of the system's state vector x^* and by the last sampled value of the control inputs vector u^*. The modeling error which is due to the truncation of higher-order terms from the Taylor series expansion is considered to be a perturbation which is asymptotically compensated by the robustness of the control algorithm. The linearization process gives

$$\dot{x} = Ax + Bu + \tilde{d} \tag{8.153}$$

A, B are the Jacobian matrices of the dynamic model of the supply chain network which are given by

$$\begin{aligned} A &= \nabla_x[f(x) + g(x)u] \mid_{(x^*,u^*)} \Rightarrow \\ A &= \nabla_x f(x) \mid_{(x^*,u^*)} + \nabla_x g_1(x)u_1 \mid_{(x^*,u^*)} + \nabla_x g_2(x)u_2 \mid_{(x^*,u^*)} \end{aligned} \tag{8.154}$$

$$B = \nabla_u[f(x) + g(x)u] \mid_{(x^*,u^*)} \Rightarrow B = g(x) \mid_{(x^*,u^*)} \tag{8.155}$$

and $\tilde{d}$ is the cumulative disturbances vector which may comprise (i) modeling error due to the truncation of higher-order terms from the Taylor series expansion, (ii) exogenous perturbation, and (iii) measurement uncertainty and noise of any distribution.

Computation of the Jacobian matrix $\nabla_x f(x) \mid_{(x^*,u^*)}$:

First row of the Jacobian matrix $\nabla_x f(x) \mid_{(x^*,u^*)}$: It holds that

$$f_1 = \frac{f_{1,num}}{f_{1,den}} = \frac{M_{11}F_1 - M_{21}F_2 + M_{31}F_3 - M_{41}F_4}{detM} \tag{8.156}$$

Thus, for $i = 1, 2, 3, 4$, one has

$$\frac{\partial f_1}{\partial x_i} = \frac{\frac{\partial f_{1,num}}{\partial x_i} detM - f_{1,num}\frac{\partial detM}{\partial x_i}}{detM^2} \tag{8.157}$$

where

$$\begin{aligned} \frac{\partial f_{1,num}}{\partial x_i} &= \frac{\partial M_{11}}{\partial x_i}F_1 + M_{11}\frac{\partial F_1}{\partial x_i} - \frac{\partial M_{21}}{\partial x_i}F_2 - M_{21}\frac{\partial F_2}{\partial x_i} + \\ &+ \frac{\partial M_{31}}{\partial x_i}F_3 + M_{31}\frac{\partial F_3}{\partial x_i} - -\frac{\partial M_{41}}{\partial x_i}F_4 - M_{41}\frac{\partial F_4}{\partial x_i} \end{aligned} \tag{8.158}$$

Second row of the Jacobian matrix $\nabla_x f(x) \mid_{(x^*,u^*)}$: It holds that

$$f_2 = \frac{f_{2,num}}{f_{2,den}} = \frac{-M_{12}F_1 + M_{22}F_2 - M_{32}F_3 + M_{42}F_4}{detM} \tag{8.159}$$

Thus, for $i = 1, 2, 3, 4$, one has

$$\frac{\partial f_2}{\partial x_i} = \frac{\frac{\partial f_{2,num}}{\partial x_i} detM - f_{2,num}\frac{\partial detM}{\partial x_i}}{detM^2} \tag{8.160}$$

where

$$\begin{aligned} \frac{\partial f_{2,num}}{\partial x_i} &= \frac{-\partial M_{12}}{\partial x_i}F_1 - M_{12}\frac{\partial F_1}{\partial x_i} + \frac{\partial M_{22}}{\partial x_i}F_2 + M_{22}\frac{\partial F_2}{\partial x_i} - \\ &- \frac{\partial M_{32}}{\partial x_i}F_3 - M_{32}\frac{\partial F_3}{\partial x_i} + +\frac{\partial M_{42}}{\partial x_i}F_4 + M_{42}\frac{\partial F_4}{\partial x_i} \end{aligned} \tag{8.161}$$

Third row of the Jacobian matrix $\nabla_x f(x) \mid_{(x^*,u^*)}$: It holds that

$$f_3 = \frac{f_{3,num}}{f_{3,den}} = \frac{M_{13}F_1 - M_{23}F_2 + M_{33}F_3 - M_{43}F_4}{detM} \tag{8.162}$$

Thus, for $i = 1, 2, 3, 4$, one has

$$\frac{\partial f_3}{\partial x_i} = \frac{\frac{\partial f_{3,num}}{\partial x_i} detM - f_{3,num}\frac{\partial detM}{\partial x_i}}{detM^2} \tag{8.163}$$

where

$$\begin{aligned} \frac{\partial f_{3,num}}{\partial x_i} &= \frac{\partial M_{13}}{\partial x_i}F_1 + M_{13}\frac{\partial F_1}{\partial x_i} - \frac{\partial M_{23}}{\partial x_i}F_2 - M_{23}\frac{\partial F_2}{\partial x_i} + \\ &+ \frac{\partial M_{33}}{\partial x_i}F_3 + M_{33}\frac{\partial F_3}{\partial x_i} - -\frac{\partial M_{43}}{\partial x_i}F_4 - M_{43}\frac{\partial F_4}{\partial x_i} \end{aligned} \tag{8.164}$$

Fourth row of the Jacobian matrix $\nabla_x f(x)\mid_{(x^*,u^*)}$: It holds that

$$f_4 = \frac{f_{4,num}}{f_{4,den}} = \frac{-M_{14}F_1+M_{24}F_2-M_{34}F_3+M_{44}F_4}{detM} \quad (8.165)$$

Thus, for $i=1,2,3,4$, one has

$$\frac{\partial f_4}{\partial x_i} = \frac{\frac{\partial f_{4,num}}{\partial x_i}detM - f_{4,num}\frac{\partial detM}{\partial x_i}}{detM^2} \quad (8.166)$$

where

$$\frac{\partial f_{4,num}}{\partial x_i} = \frac{-\partial M_{14}}{\partial x_i}F_1 - M_{14}\frac{\partial F_1}{\partial x_i} + \frac{\partial M_{24}}{\partial x_i}F_2 + M_{24}\frac{\partial F_2}{\partial x_i} - \\ -\frac{\partial M_{34}}{\partial x_i}F_3 - M_{34}\frac{\partial F_3}{\partial x_i} + +\frac{\partial M_{44}}{\partial x_i}F_4 + M_{44}\frac{\partial F_4}{\partial x_i} \quad (8.167)$$

Computation of the Jacobian matrix $\nabla_x g_1(x)\mid_{(x^*,u^*)}$:

First row of the Jacobian matrix $\nabla_x g_1(x)\mid_{(x^*,u^*)}$: It holds that $g_{11} = -\frac{M_{21}G_{21}}{detM}$ where $G_{21} = \frac{1}{T_s}$. Thus, for $i=1,2,3,4$,

$$\frac{\partial g_{11}}{\partial x_i} = \frac{(-\frac{\partial M_{21}}{\partial x_i}detM + M_{21}\frac{\partial detM}{\partial x_i})\frac{1}{T_s}}{detM^2} \quad (8.168)$$

Second row of the Jacobian matrix $\nabla_x g_1(x)\mid_{(x^*,u^*)}$: It holds that $g_{21} = \frac{M_{22}G_{21}}{detM}$ where $G_{21} = \frac{1}{T_s}$. Thus, for $i=1,2,3,4$,

$$\frac{\partial g_{21}}{\partial x_i} = \frac{(\frac{\partial M_{22}}{\partial x_i}detM - M_{22}\frac{\partial detM}{\partial x_i})\frac{1}{T_s}}{detM^2} \quad (8.169)$$

Third row of the Jacobian matrix $\nabla_x g_1(x)\mid_{(x^*,u^*)}$: It holds that $g_{31} = -\frac{M_{23}G_{21}}{detM}$ where $G_{21} = \frac{1}{T_s}$. Thus, for $i=1,2,3,4$,

$$\frac{\partial g_{31}}{\partial x_i} = \frac{(-\frac{\partial M_{23}}{\partial x_i}detM + M_{23}\frac{\partial detM}{\partial x_i})\frac{1}{T_s}}{detM^2} \quad (8.170)$$

Fourth row of the Jacobian matrix $\nabla_x g_1(x)\mid_{(x^*,u^*)}$: It holds that $g_{41} = \frac{M_{24}G_{21}}{detM}$ where $G_{21} = \frac{1}{T_s}$. Thus, for $i=1,2,3,4$,

$$\frac{\partial g_{41}}{\partial x_i} = \frac{(\frac{\partial M_{24}}{\partial x_i}detM - M_{24}\frac{\partial detM}{\partial x_i})\frac{1}{T_s}}{detM^2} \quad (8.171)$$

Computation of the Jacobian matrix $\nabla_x g_2(x)\mid_{(x^*,u^*)}$:

First row of the Jacobian matrix $\nabla_x g_2(x)\mid_{(x^*,u^*)}$: It holds that $g_{12} = -\frac{M_{41}G_{42}}{detM}$ where $G_{42} = \frac{1}{T_s}$. Thus, for $i=1,2,3,4$,

$$\frac{\partial g_{12}}{\partial x_i} = \frac{(-\frac{\partial M_{41}}{\partial x_i}detM + M_{41}\frac{\partial detM}{\partial x_i})\frac{1}{T_s}}{detM^2} \quad (8.172)$$

Second row of the Jacobian matrix $\nabla_x g_2(x)\mid_{(x^*,u^*)}$: It holds that $g_{22} = \frac{M_{42}G_{42}}{detM}$ where $G_{42} = \frac{1}{T_s}$. Thus, for $i=1,2,3,4$,

$$\frac{\partial g_{22}}{\partial x_i} = \frac{(\frac{\partial M_{42}}{\partial x_i}detM - M_{42}\frac{\partial detM}{\partial x_i})\frac{1}{T_s}}{detM^2} \quad (8.173)$$

Third row of the Jacobian matrix $\nabla_x g_2(x)\mid_{(x^*,u^*)}$: It holds that $g_{32} = -\frac{M_{43}G_{42}}{detM}$ where $G_{42} = \frac{1}{T_s}$. Thus, for $i=1,2,3,4$,

$$\frac{\partial g_{32}}{\partial x_i} = \frac{(-\frac{\partial M_{43}}{\partial x_i}detM + M_{43}\frac{\partial detM}{\partial x_i})\frac{1}{T_s}}{detM^2} \quad (8.174)$$

Fourth row of the Jacobian matrix $\nabla_x g_1(x)\mid_{(x^*,u^*)}$: It holds that $g_{42} = \frac{M_{44}G_{42}}{detM}$ where $G_{42} = \frac{1}{T_s}$. Thus, for $i=1,2,3,4$,

$$\frac{\partial g_{42}}{\partial x_i} = \frac{(\frac{\partial M_{44}}{\partial x_i}detM - M_{44}\frac{\partial detM}{\partial x_i})\frac{1}{T_s}}{detM^2} \quad (8.175)$$

Computation of the partial derivatives of the elements M_{ij}, $i=1,2,3,4, j=1,2,3,4$ of matrix M:

It holds that $\frac{\partial M_{11}}{\partial x_1}=0$, $\frac{\partial M_{11}}{\partial x_2}=\theta_2\theta_3\theta_4x_3$, $\frac{\partial M_{11}}{\partial x_3}=\theta_2\theta_3\theta_4x_2$, $\frac{\partial M_{11}}{\partial x_4}=0$. Additionally, $\frac{\partial M_{12}}{\partial x_1}=0$, $\frac{\partial M_{12}}{\partial x_2}=0$, $\frac{\partial M_{12}}{\partial x_3}=0$, $\frac{\partial M_{12}}{\partial x_4}=0$. Moreover, $\frac{\partial M_{13}}{\partial x_1}=0$, $\frac{\partial M_{13}}{\partial x_2}=\theta_1\theta_2\theta_4x_3$, $\frac{\partial M_{13}}{\partial x_3}=\theta_1\theta_2\theta_4x_2$, $\frac{\partial M_{13}}{\partial x_4}=0$. Furthermore, $\frac{\partial M_{14}}{\partial x_1}=0$, $\frac{\partial M_{14}}{\partial x_2}=0$,

$\frac{\partial M_{14}}{\partial x_3}=-c\theta_1\theta_2$, $\frac{\partial M_{14}}{\partial x_4}=0$.

It holds that $\frac{\partial M_{21}}{\partial x_1}=0$, $\frac{\partial M_{21}}{\partial x_2}=0$, $\frac{\partial M_{21}}{\partial x_3}=0$, $\frac{\partial M_{21}}{\partial x_4}=0$. Additionally, $\frac{\partial M_{22}}{\partial x_1}=0$, $\frac{\partial M_{22}}{\partial x_2}=0$, $\frac{\partial M_{22}}{\partial x_3}=0$, $\frac{\partial M_{22}}{\partial x_4}=0$. Moreover, $\frac{\partial M_{23}}{\partial x_1}=0$, $\frac{\partial M_{23}}{\partial x_2}=0$, $\frac{\partial M_{23}}{\partial x_3}=\theta_2\theta_4x_2$, $\frac{\partial M_{23}}{\partial x_4}=0$. Furthermore, $\frac{\partial M_{24}}{\partial x_1}=0$, $\frac{\partial M_{24}}{\partial x_2}=0$, $\frac{\partial M_{24}}{\partial x_3}=-\theta_2$, $\frac{\partial M_{24}}{\partial x_4}=0$.

It holds that $\frac{\partial M_{31}}{\partial x_1}=0$, $\frac{\partial M_{31}}{\partial x_2}=0$, $\frac{\partial M_{31}}{\partial x_3}=0$, $\frac{\partial M_{31}}{\partial x_4}=0$. Additionally, $\frac{\partial M_{32}}{\partial x_1}=0$, $\frac{\partial M_{32}}{\partial x_2}=0$, $\frac{\partial M_{32}}{\partial x_3}=0$, $\frac{\partial M_{32}}{\partial x_4}=0$. Moreover, $\frac{\partial M_{33}}{\partial x_1}=0$, $\frac{\partial M_{33}}{\partial x_2}=0$, $\frac{\partial M_{33}}{\partial x_3}=0$, $\frac{\partial M_{33}}{\partial x_4}=0$. Furthermore, $\frac{\partial M_{34}}{\partial x_1}=0$, $\frac{\partial M_{34}}{\partial x_2}=0$, $\frac{\partial M_{34}}{\partial x_3}=\theta_2\theta_3$, $\frac{\partial M_{34}}{\partial x_4}=0$.

It holds that $\frac{\partial M_{41}}{\partial x_1}=0$, $\frac{\partial M_{41}}{\partial x_2}=-\theta_2\theta_3\theta_4$, $\frac{\partial M_{41}}{\partial x_3}=0$, $\frac{\partial M_{41}}{\partial x_4}=0$. Additionally, $\frac{\partial M_{42}}{\partial x_1}=0$, $\frac{\partial M_{42}}{\partial x_2}=\theta_2\theta_4$, $\frac{\partial M_{42}}{\partial x_3}=0$, $\frac{\partial M_{42}}{\partial x_4}=0$. Moreover, $\frac{\partial M_{43}}{\partial x_1}=0$, $\frac{\partial M_{43}}{\partial x_2}=-\theta_4(1+c\theta_1\theta_2)$, $\frac{\partial M_{43}}{\partial x_3}=0$, $\frac{\partial M_{43}}{\partial x_4}=0$. Furthermore, $\frac{\partial M_{44}}{\partial x_1}=0$, $\frac{\partial M_{44}}{\partial x_2}=0$, $\frac{\partial M_{44}}{\partial x_3}=0$, $\frac{\partial M_{44}}{\partial x_4}=0$.

One also obtains that $\frac{\partial detM}{\partial x_i}=\frac{\partial M_{11}}{\partial x_i}+\theta_2\frac{\partial M_{12}}{\partial x_i}$ for $i=1,2,3,4$.

Computation of the partial derivatives of the elements F_i, $i=1,2,3,4$ of vector F:

It holds that $\frac{\partial F_1}{\partial x_1}=-\frac{(1-n)}{T_s}$, $\frac{\partial F_1}{\partial x_2}=-\frac{1}{T_s}$, $\frac{\partial F_1}{\partial x_3}=0$, $\frac{\partial F_1}{\partial x_4}=0$. Additionally, $\frac{\partial F_2}{\partial x_1}=-\frac{c}{T_s}$, $\frac{\partial F_2}{\partial x_2}=-\frac{(1+n)}{T_s}$, $\frac{\partial F_2}{\partial x_3}=-\frac{1}{T_s}$, $\frac{\partial F_2}{\partial x_4}=\frac{1}{T_s}$. Moreover, $\frac{\partial F_3}{\partial x_1}=0$, $\frac{\partial F_3}{\partial x_2}=\frac{n}{T_s}-\frac{x_4}{T_s}$, $\frac{\partial F_3}{\partial x_3}=-\frac{1}{T_s}$, $\frac{\partial F_3}{\partial x_4}=-\frac{1}{T_s}x_2$. Furthermore, $\frac{\partial F_4}{\partial x_1}=0$, $\frac{\partial F_4}{\partial x_2}=\frac{1}{T_s}x_3$, $\frac{\partial F_4}{\partial x_3}=\frac{1}{T_s}x_3$, $\frac{\partial F_4}{\partial x_4}=\frac{(s-1)}{T_s}$.

The linearization approach which has been followed for implementing the nonlinear optimal control scheme results in a quite accurate model of the system's dynamics. Consider for instance the following affine-in-the-input state-space model

$$\begin{aligned}\dot{x}&=f(x)+g(x)u\Rightarrow\\ \dot{x}&=[f(x^*)+\nabla_xf(x)\mid_{x^*}(x-x^*)]+[g(x^*)+\nabla_xg(x)\mid_{x^*}(x-x^*)]u^*+g(x^*)u^*+g(x^*)(u-u^*)+\tilde{d}_1\Rightarrow\\ \dot{x}&=[\nabla_xf(x)\mid_{x^*}+\nabla_xg(x)\mid_{x^*}u^*]x+g(x^*)u-[\nabla_xf(x)\mid_{x^*}+\nabla_xg(x)\mid_{x^*}u^*]x^*+f(x^*)+g(x^*)u^*+\tilde{d}_1\end{aligned} \tag{8.176}$$

where $\tilde{d}_1$ is the modeling error due to truncation of higher order terms in the Taylor series expansion of $f(x)$ and $g(x)$. Next, by defining $A=[\nabla_xf(x)\mid_{x^*}+\nabla_xg(x)\mid_{x^*}u^*]$, $B=g(x^*)$, one obtains

$$\dot{x}=Ax+Bu-Ax^*+f(x^*)+g(x^*)u^*+\tilde{d}_1 \tag{8.177}$$

Moreover, by denoting $\tilde{d}=-Ax^*+f(x^*)+g(x^*)u^*+\tilde{d}_1$ about the cumulative modeling error term in the Taylor series expansion, one has

$$\dot{x}=Ax+Bu+\tilde{d} \tag{8.178}$$

which is the approximately linearized model of the dynamics of the system of Eq. (8.153). The term $f(x^*)+g(x^*)u^*$ is the derivative of the state vector at (x^*,u^*) which is almost annihilated by $-Ax^*$.

8.3.4 The nonlinear H-infinity control

8.3.4.1 Min–max control and disturbance rejection

In the H_∞ control approach, a feedback control scheme is designed for setpoint tracking by the state vector of the supply chain network and for simultaneous disturbance rejection, considering that the disturbance affects the system in the worst possible manner. For the approximately linearized model of Eq. (8.153), the disturbances' effects are incorporated in the following quadratic cost function [725]

$$J(t)=\frac{1}{2}\int_0^T[y^T(t)y(t)+ru^T(t)u(t)-\rho^2\tilde{d}^T(t)\tilde{d}(t)]dt,\quad r,\rho>0 \tag{8.179}$$

The significance of the negative sign in the cost function's term that is associated with the perturbation variable $\tilde{d}(t)$ is that the disturbance tries to maximize the cost function $J(t)$ while the control signal $u(t)$ tries to minimize it.

The physical meaning of the relation given above is that the control signal and the disturbances compete with each other within a min–max differential game. This problem of min–max optimization can be written as $min_u max_{\tilde{d}} J(u,\tilde{d})$. The objective of the optimization procedure is to compute a control signal $u(t)$ which can compensate for the worst possible disturbance that affects the system. The solution to the min–max optimization problem is directly related to the value of parameter ρ. This means that there is an upper bound in the disturbances magnitude that can be annihilated by the control signal.

8.3.4.2 Design of the stabilizing feedback controller

After linearization around its current operating point (x^*, u^*), the dynamic model of the supply chain network is written as

$$\dot{x} = Ax + Bu + d_1 \tag{8.180}$$

Parameter d_1 stands for the linearization error in the supply chain network's model appearing previously in Eq. (8.153). The reference setpoints for the state vector of the supply chain network are denoted by $\mathbf{x_d} = [x_1^d, \cdots, x_4^d]$. Tracking of this trajectory is achieved after applying the control input u_d. At every time instant the control input u_d is assumed to differ from the control input u appearing in Eq. (8.180) by an amount equal to Δu, that is $u_d = u + \Delta u$

$$\dot{x}_d = Ax_d + Bu_d + d_2 \tag{8.181}$$

The dynamics of the controlled system described in Eq. (8.180) can be also written as

$$\dot{x} = Ax + Bu + Bu_d - Bu_d + d_1 \tag{8.182}$$

and by denoting $d_3 = -Bu_d + d_1$ as an aggregate disturbance term, one obtains

$$\dot{x} = Ax + Bu + Bu_d + d_3 \tag{8.183}$$

By subtracting Eq. (8.181) from Eq. (8.183), one has

$$\dot{x} - \dot{x}_d = A(x - x_d) + Bu + d_3 - d_2 \tag{8.184}$$

By denoting the tracking error as $e = x - x_d$ and the aggregate disturbance term as $L\tilde{d} = d_3 - d_2$, the tracking error dynamics becomes

$$\dot{e} = Ae + Bu + L\tilde{d} \tag{8.185}$$

where L is a disturbance inputs gain matrix. For the linearized system, the cost function of Eq. (8.179) is defined, where coefficient r determines the penalization of the control input and the weight coefficient ρ determines the reward of the disturbances' effects. Then, as it will be also shown in Section 8.3.5, the optimal feedback control law is given by

$$u(t) = -Ke(t) \tag{8.186}$$

with $K = \frac{1}{r}B^TP$ where P is a positive definite symmetric matrix which is obtained from the solution of the Riccati equation

$$A^TP + PA + Q - P(\tfrac{2}{r}BB^T - \tfrac{1}{\rho^2}LL^T)P = 0 \tag{8.187}$$

where Q is also a semi-positive definite symmetric matrix. The worst-case disturbance is given by $\tilde{d}(t) = \frac{1}{\rho^2}L^TPx(t)$. The computation of the worst-case disturbance that this controller can sustain, comes from the superposition of Bellman's optimality principle when considering that the supply chain network is affected by two separate inputs, that is the control input u and the cumulative disturbance input $\tilde{d}$. From the previous relation about the worst-case disturbance, it can be seen that the smallest value of the attenuation coefficient ρ that results in an admissible solution for the methods's algebraic Riccati equation, is the one that provides the control loop with maximum robustness. The diagram of the considered control loop is depicted in Figure 8.30.

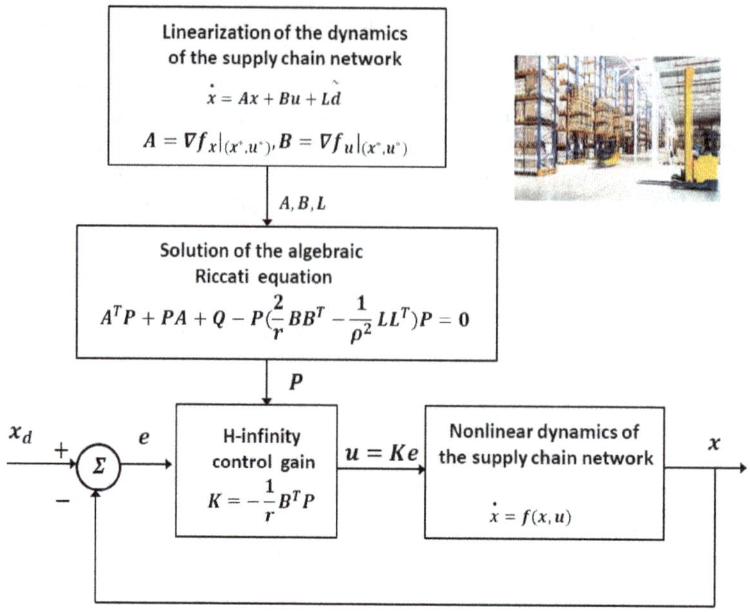

Figure 8.30 Diagram of the nonlinear optimal control loop for the supply chain network

8.3.5 Lyapunov stability analysis

Through Lyapunov stability analysis it will be shown that the proposed nonlinear control scheme assures H_∞ tracking performance for the supply chain network and that in case of bounded disturbance terms asymptotic convergence to the reference setpoints is achieved. The tracking error dynamics for the supply chain network is written in the form [717,725]:

$$\dot{e} = Ae + Bu + L\tilde{d} \tag{8.188}$$

where in the supply chain network's case $L \in R^{4\times4}$ is the disturbance inputs gain matrix. Variable $\tilde{d}$ denotes model uncertainties and external disturbances of the supply chain network's model. The following Lyapunov equation is considered

$$V = \frac{1}{2}e^TPe \tag{8.189}$$

where $e = x - x_d$ is the tracking error. By differentiating with respect to time, one obtains

$$\begin{gathered}\dot{V} = \frac{1}{2}\dot{e}^TPe + \frac{1}{2}e^TP\dot{e} \Rightarrow \\ \dot{V} = \frac{1}{2}[Ae + Bu + L\tilde{d}]^TPe + \frac{1}{2}e^TP[Ae + Bu + L\tilde{d}] \Rightarrow\end{gathered} \tag{8.190}$$

$$\begin{aligned}\dot{V} = &\frac{1}{2}[e^TA^T + u^TB^T + \tilde{d}^TL^T]Pe + \\ &+\frac{1}{2}e^TP[Ae + Bu + L\tilde{d}] \Rightarrow\end{aligned} \tag{8.191}$$

$$\begin{aligned}\dot{V} = &\frac{1}{2}e^TA^TPe + \frac{1}{2}u^TB^TPe + \frac{1}{2}\tilde{d}^TL^TPe + \\ &\frac{1}{2}e^TPAe + \frac{1}{2}e^TPBu + \frac{1}{2}e^TPL\tilde{d}\end{aligned} \tag{8.192}$$

The previous equation is rewritten as

$$\begin{aligned}\dot{V} = &\frac{1}{2}e^T(A^TP + PA)e + (\frac{1}{2}u^TB^TPe + \frac{1}{2}e^TPBu) + \\ &+(\frac{1}{2}\tilde{d}^TL^TPe + \frac{1}{2}e^TPL\tilde{d})\end{aligned} \tag{8.193}$$

Assumption: For given positive definite matrix Q and coefficients r and ρ, there exists a positive definite matrix P, which is the solution of the following matrix equation:

$$A^TP + PA = -Q + P(\tfrac{2}{r}BB^T - \tfrac{1}{\rho^2}LL^T)P \tag{8.194}$$

Moreover, the following feedback control law is applied to the system:

$$u = -\tfrac{1}{r}B^TPe \tag{8.195}$$

By substituting Eq. (8.194) and Eq. (8.195), one obtains

$$\begin{aligned}\dot{V} = &\tfrac{1}{2}e^T[-Q + P(\tfrac{2}{r}BB^T - \tfrac{1}{\rho^2}LL^T)P]e+\\ &+e^TPB(-\tfrac{1}{r}B^TPe) + e^TPL\tilde{d}\Rightarrow\end{aligned} \tag{8.196}$$

$$\begin{aligned}\dot{V} = &-\tfrac{1}{2}e^TQe + (\tfrac{1}{r}e^TPBB^TPe - \tfrac{1}{2\rho^2}e^TPLL^TPe)\\ &-\tfrac{1}{r}e^TPBB^TPe + e^TPL\tilde{d}\end{aligned} \tag{8.197}$$

which after intermediate operations gives

$$\dot{V} = -\tfrac{1}{2}e^TQe - \tfrac{1}{2\rho^2}e^TPLL^TPe + e^TPL\tilde{d} \tag{8.198}$$

or, equivalently,

$$\begin{aligned}\dot{V} = &-\tfrac{1}{2}e^TQe - \tfrac{1}{2\rho^2}e^TPLL^TPe+\\ &+\tfrac{1}{2}e^TPL\tilde{d} + \tfrac{1}{2}\tilde{d}^TL^TPe\end{aligned} \tag{8.199}$$

Lemma: The following inequality holds [717,725]

$$\tfrac{1}{2}e^TL\tilde{d} + \tfrac{1}{2}\tilde{d}L^TPe - \tfrac{1}{2\rho^2}e^TPLL^TPe \leq \tfrac{1}{2}\rho^2\tilde{d}^T\tilde{d} \tag{8.200}$$

Proof: The binomial $(\rho\alpha - \frac{1}{\rho}b)^2$ is considered. Expanding the left part of the above inequality, one gets

$$\begin{aligned}&\rho^2a^2 + \tfrac{1}{\rho^2}b^2 - 2ab \geq 0 \Rightarrow \tfrac{1}{2}\rho^2a^2 + \tfrac{1}{2\rho^2}b^2 - ab \geq 0 \Rightarrow\\ &ab - \tfrac{1}{2\rho^2}b^2 \leq \tfrac{1}{2}\rho^2a^2 \Rightarrow \tfrac{1}{2}ab + \tfrac{1}{2}ab - \tfrac{1}{2\rho^2}b^2 \leq \tfrac{1}{2}\rho^2a^2\end{aligned} \tag{8.201}$$

The following substitutions are carried out: $a = \tilde{d}$ and $b = e^TPL$, and the previous relation becomes

$$\tfrac{1}{2}\tilde{d}^TL^TPe + \tfrac{1}{2}e^TPL\tilde{d} - \tfrac{1}{2\rho^2}e^TPLL^TPe \leq \tfrac{1}{2}\rho^2\tilde{d}^T\tilde{d} \tag{8.202}$$

Eq. (8.202) is substituted in Eq. (8.199) and the inequality is enforced, thus giving [717,725]

$$\dot{V} \leq -\tfrac{1}{2}e^TQe + \tfrac{1}{2}\rho^2\tilde{d}^T\tilde{d} \tag{8.203}$$

Eq. (8.203) shows that the H_∞ tracking performance criterion is satisfied. The integration of $\dot{V}$ from 0 to T gives

$$\begin{aligned}&\int_0^T\dot{V}(t)dt \leq -\tfrac{1}{2}\int_0^T||e||_Q^2dt + \tfrac{1}{2}\rho^2\int_0^T||\tilde{d}||^2dt \Rightarrow\\ &2V(T) + \int_0^T||e||_Q^2dt \leq 2V(0) + \rho^2\int_0^T||\tilde{d}||^2dt\end{aligned} \tag{8.204}$$

Moreover, if there exists a positive constant $M_d > 0$ such that

$$\int_0^\infty ||\tilde{d}||^2dt \leq M_d \tag{8.205}$$

then, one gets

$$\int_0^\infty ||e||_Q^2dt \leq 2V(0) + \rho^2M_d \tag{8.206}$$

Thus, the integral $\int_0^\infty ||e||_Q^2dt$ is bounded. Moreover, $V(T)$ is bounded and from the definition of the Lyapunov function V in Eq. (8.189), it becomes clear that $e(t)$ will be also bounded since $e(t) \in \Omega_e = \{e|e^TPe \leq 2V(0) + \rho^2M_d\}$. According to the above and with the use of Barbalat's Lemma, one obtains $lim_{t\rightarrow\infty}e(t) = 0$.

Through the stages of the stability proof, one arrives at Eq. (8.203) which shows that the H-infinity tracking performance criterion holds. By selecting the attenuation coefficient ρ to be sufficiently small and in particular to satisfy $\rho^2 < \|e\|_Q^2/\|\tilde{d}\|^2$, one has that the first derivative of the Lyapunov function is upper bounded by 0. This condition holds at each sampling instance, and consequently, global stability for the control loop can be concluded.

8.3.6 Simulation tests

The fine performance of the nonlinear optimal control method for the supply chain network has been also tested and confirmed through simulation experiments. To compute the controller's feedback gains the algebraic Riccati equation of Eq. (8.194) had to be solved at each time step of the control method using the *aresolv()* function in MATLAB®. To perform state estimation-based control without the need to measure all state variables of the supply chain network the H-infinity Kalman filter has been used as a state observer. The total simulation time was 100h and the sampling period was $T_s = 0.125$h, that is the database of the supply chain network was being updated every 7.5 min. The time-delays at the individual stages of the supply chain were of the order of a few sampling periods. The obtained results are depicted in Figures 8.31–8.46. The real values of the state vector of the supply chain network are depicted in blue, the estimated values which are provided by the H-infinity Kalman filter are printed in green while the setpoints for the individual state variables are plotted in red. In all simulation experiments, it is confirmed that the nonlinear optimal control scheme achieves fast and accurate tracking of setpoints under moderate variations of the control inputs. The transient performance of the control method depends on the selection of values for the coefficients of the above-noted Riccati equation, namely gain r, the attenuation coefficient ρ and matrix Q. For relatively small values of gain r

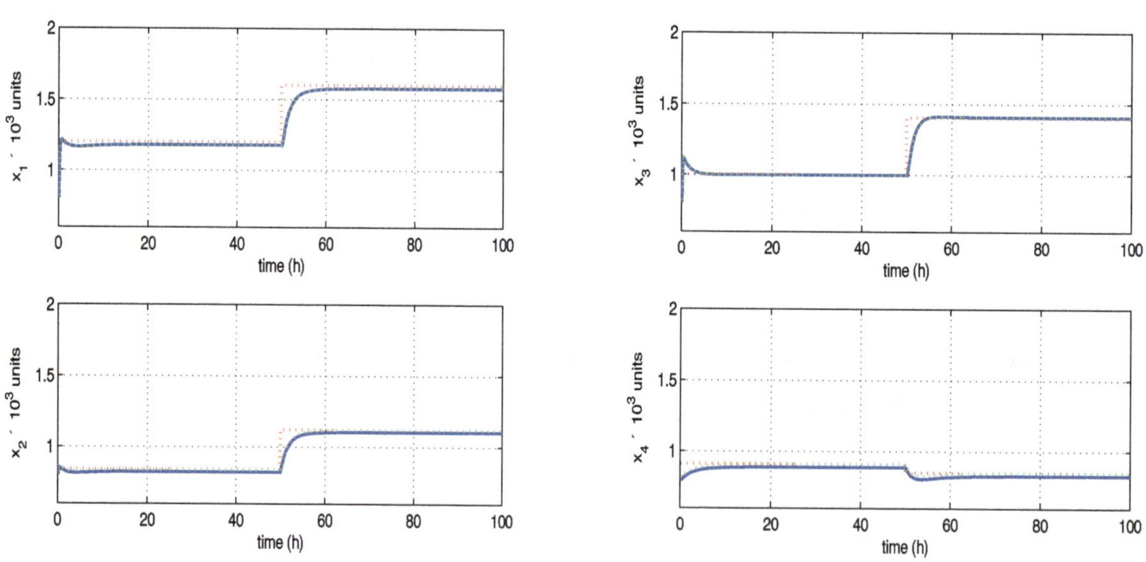

Figure 8.31 Tracking of setpoint 1 by the supply chain network: (a) convergence of state variables x_1 to x_2 (blue lines) to the reference setpoints (red lines) and their KF-based estimation (green line); (b) convergence of state variables x_3 to x_4 (blue lines) to the reference setpoints (red lines) and their KF-based estimation (green line)

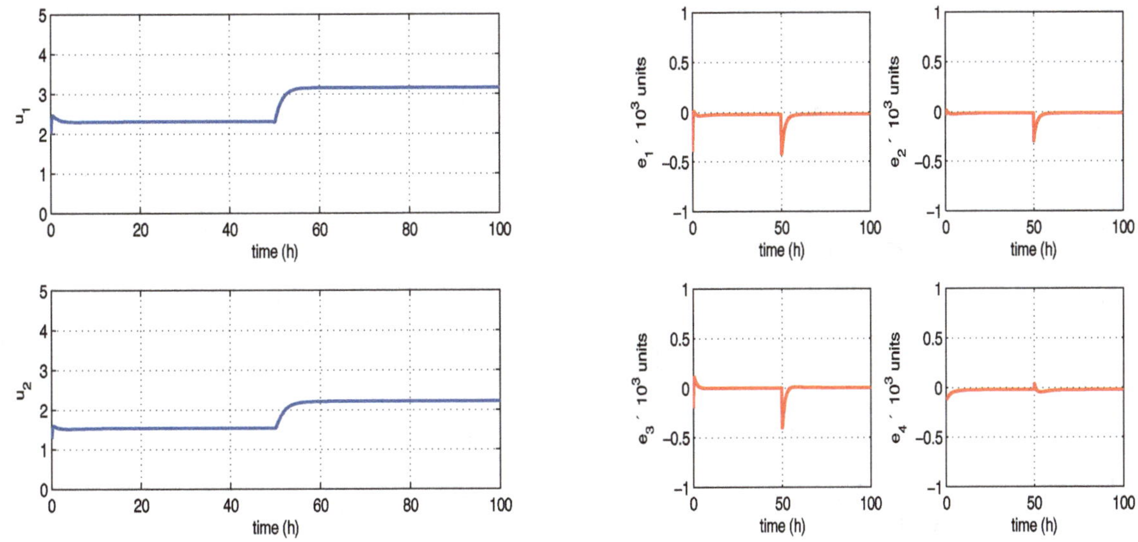

Figure 8.32 Tracking of setpoint 1 by the supply chain network: (a) variations of the control inputs u_1 and u_2; (b) tracking error of state variables x_1 to x_4

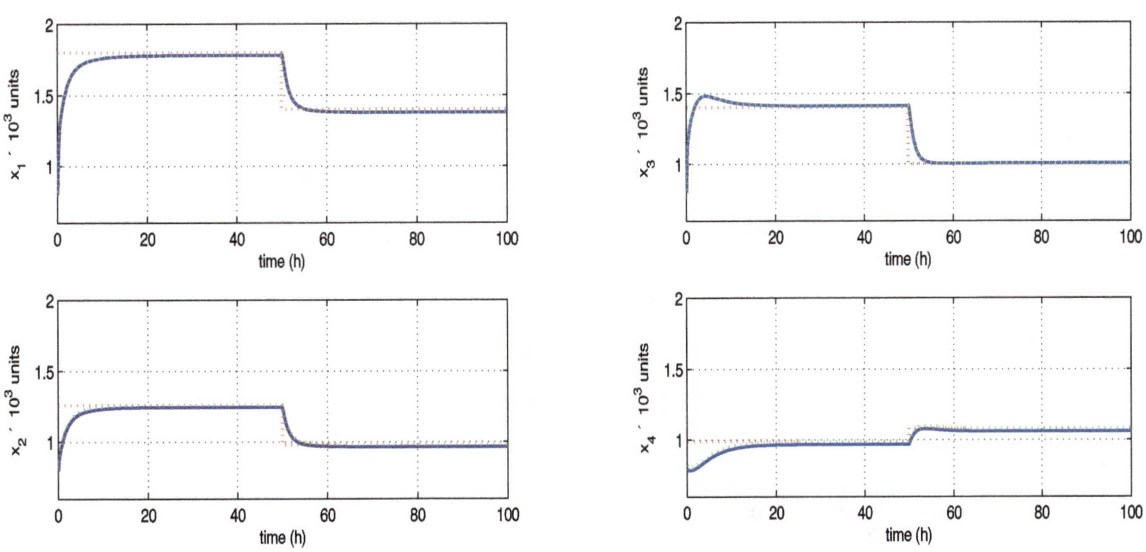

Figure 8.33 Tracking of setpoint 2 by the supply chain network: (a) convergence of state variables x_1 to x_2 (blue lines) to the reference setpoints (red lines) and their KF-based estimation (green line); (b) convergence of state variables x_3 to x_4 (blue lines) to the reference setpoints (red lines) and their KF-based estimation (green line)

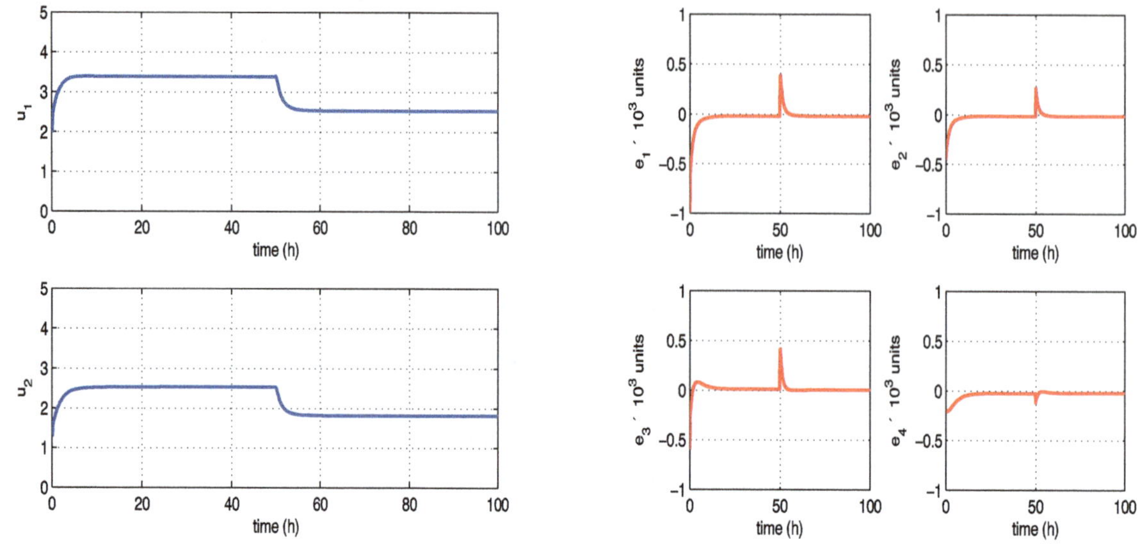

Figure 8.34 Tracking of setpoint 2 by the supply chain network: (a) variations of the control inputs u_1 and u_2; (b) tracking error of state variables x_1 to x_4

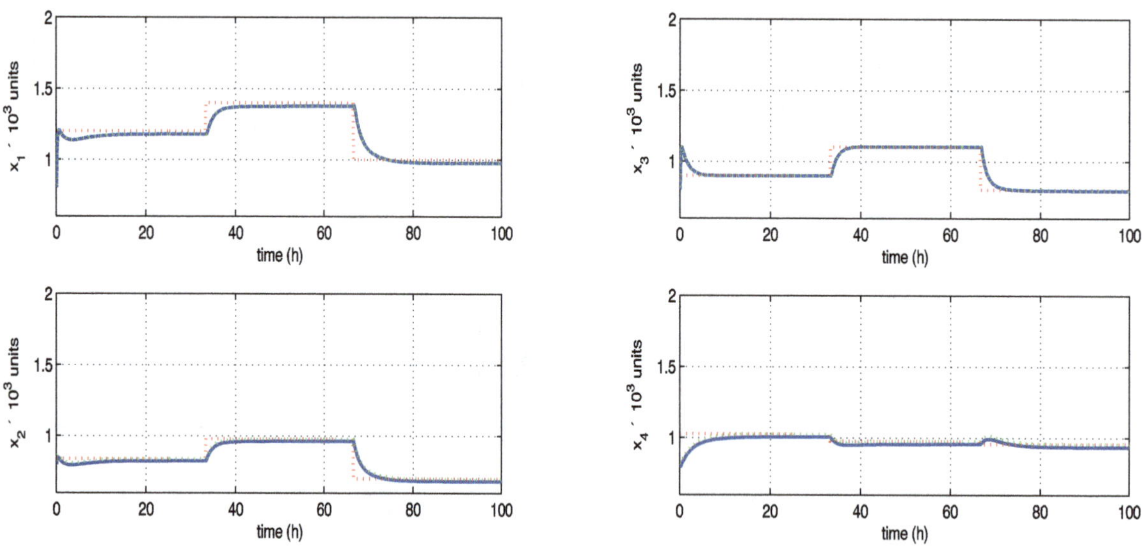

Figure 8.35 Tracking of setpoint 3 by the supply chain network: (a) convergence of state variables x_1 to x_2 (blue lines) to the reference setpoints (red lines) and their KF-based estimation (green line); (b) convergence of state variables x_3 to x_4 (blue lines) to the reference setpoints (red lines) and their KF-based estimation (green line)

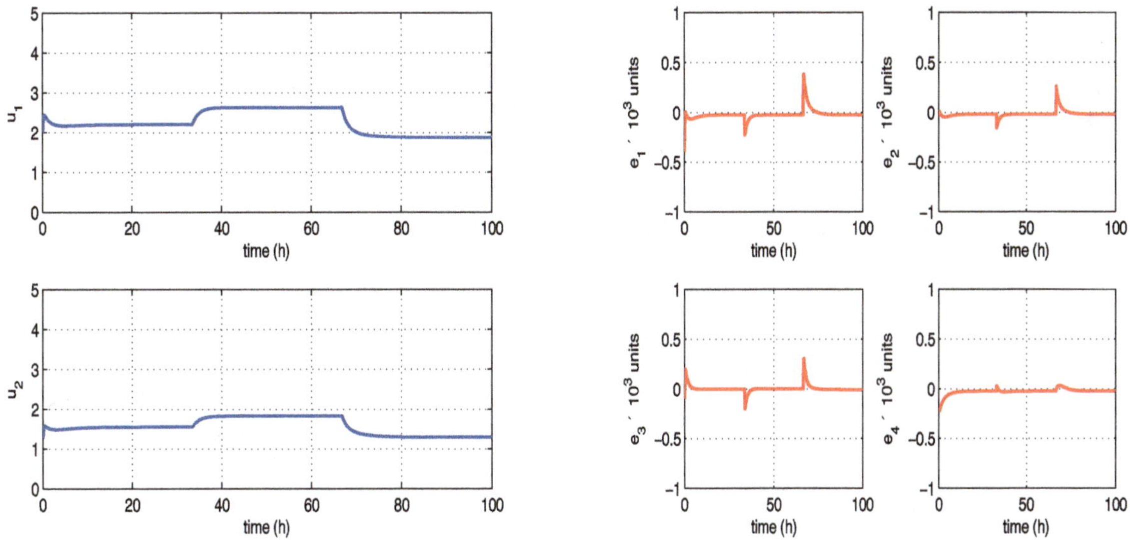

Figure 8.36 *Tracking of setpoint 3 by the supply chain network: (a) variations of the control inputs u_1 and u_2; (b) tracking error of state variables x_1 to x_4*

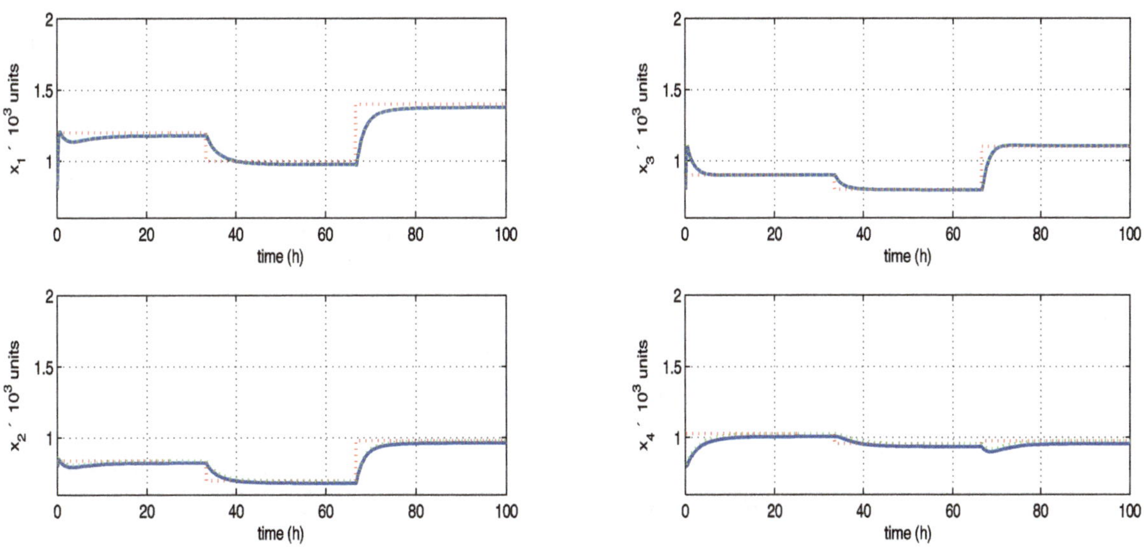

Figure 8.37 *Tracking of setpoint 4 by the supply chain network: (a) convergence of state variables x_1 to x_2 (blue lines) to the reference setpoints (red lines) and their KF-based estimation (green line); (b) convergence of state variables x_3 to x_4 (blue lines) to the reference setpoints (red lines) and their KF-based estimation (green line)*

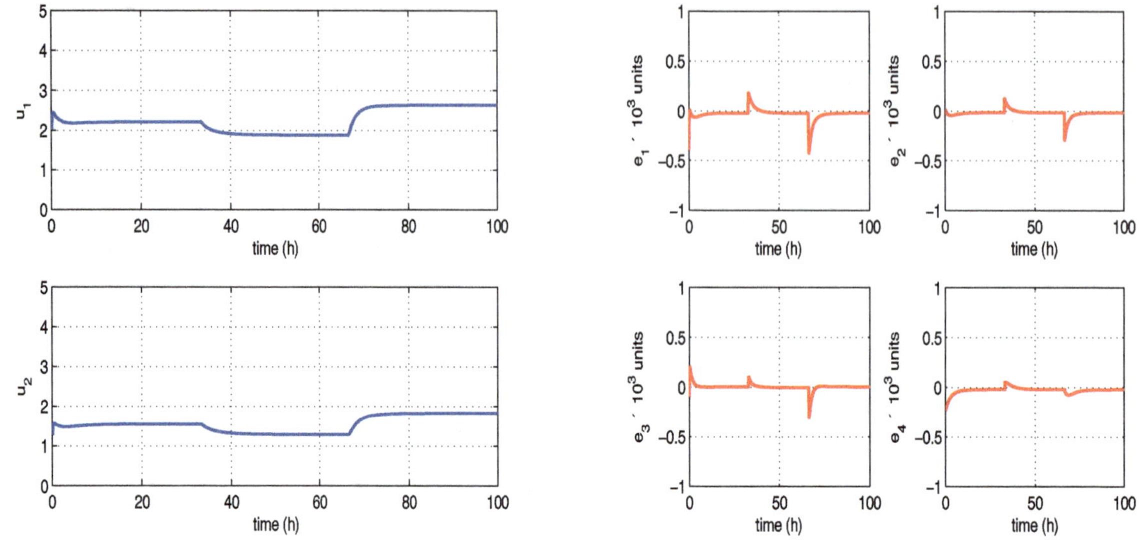

Figure 8.38 *Tracking of setpoint 4 by the supply chain network: (a) variations of the control inputs u_1 and u_2; (b) tracking error of state variables x_1 to x_4*

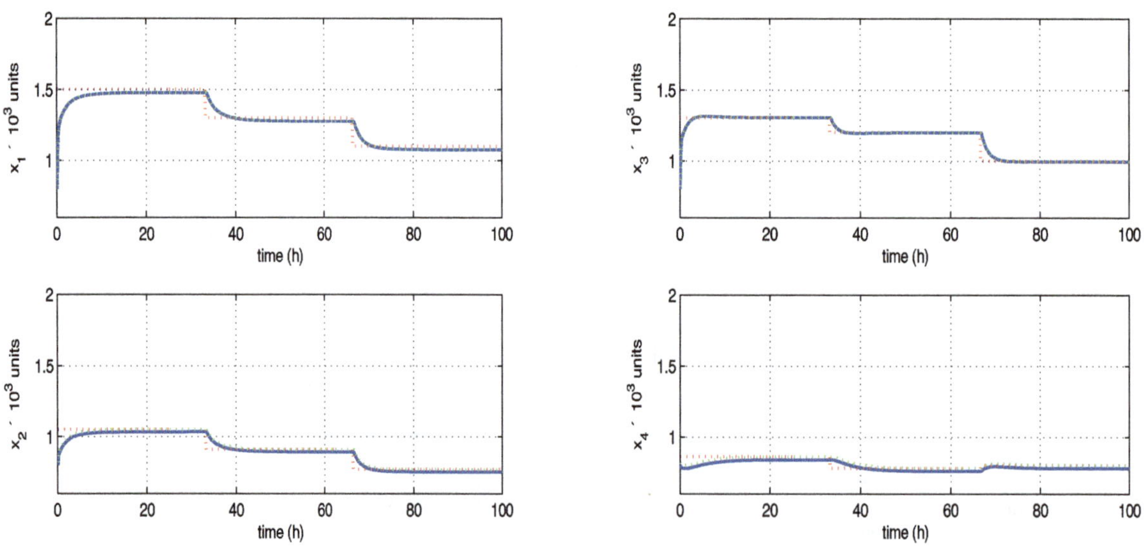

Figure 8.39 *Tracking of setpoint 5 by the supply chain network: (a) convergence of state variables x_1 to x_2 (blue lines) to the reference setpoints (red lines) and their KF-based estimation (green line); (b) convergence of state variables x_3 to x_4 (blue lines) to the reference setpoints (red lines) and their KF-based estimation (green line)*

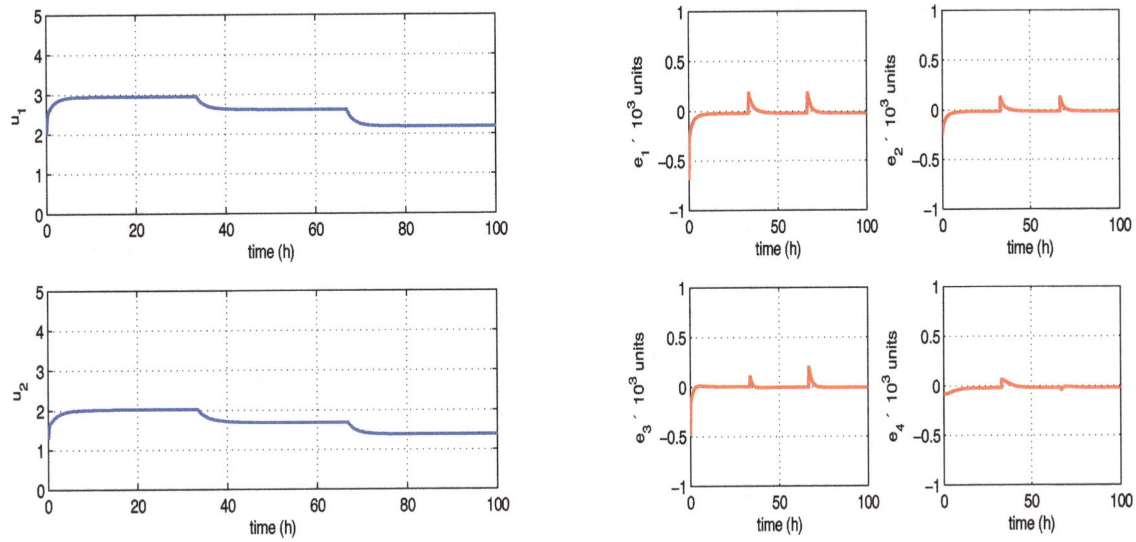

Figure 8.40 Tracking of setpoint 5 by the supply chain network: (a) variations of the control inputs u_1 and u_2; (b) tracking error of state variables x_1 to x_4

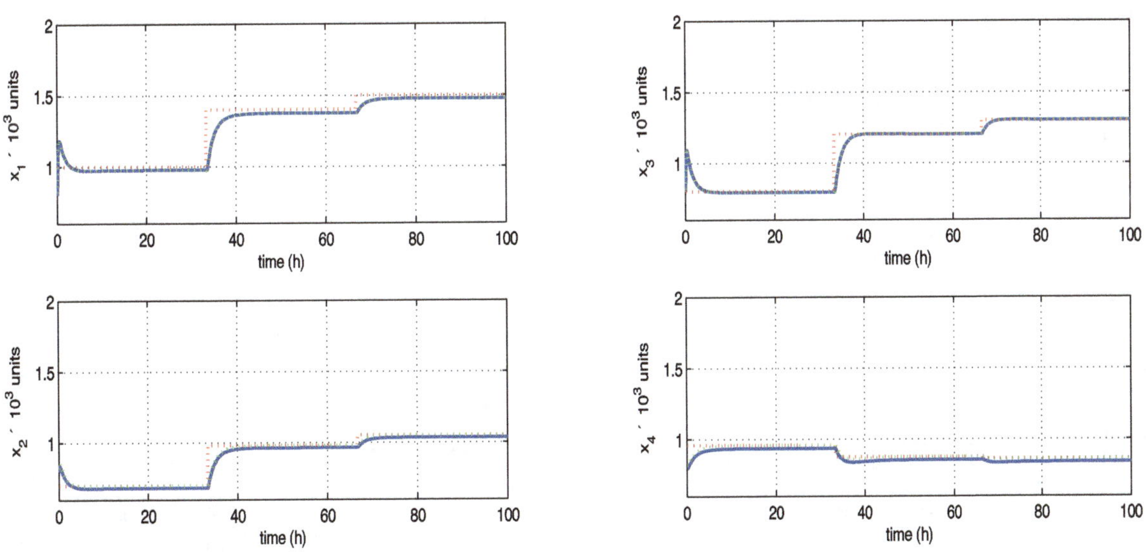

Figure 8.41 Tracking of setpoint 6 by the supply chain network: (a) convergence of state variables x_1 to x_2 (blue lines) to the reference setpoints (red lines) and their KF-based estimation (green line); (b) convergence of state variables x_3 to x_4 (blue lines) to the reference setpoints (red lines) and their KF-based estimation (green line)

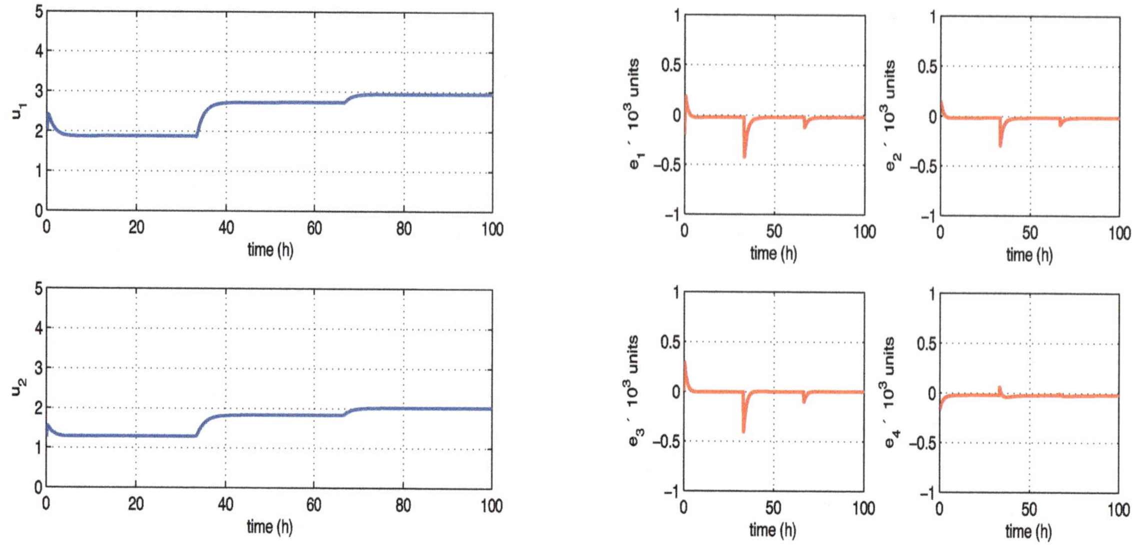

Figure 8.42 Tracking of setpoint 6 by the supply chain network: (a) variations of the control inputs u_1 and u_2; (b) tracking error of state variables x_1 to x_4

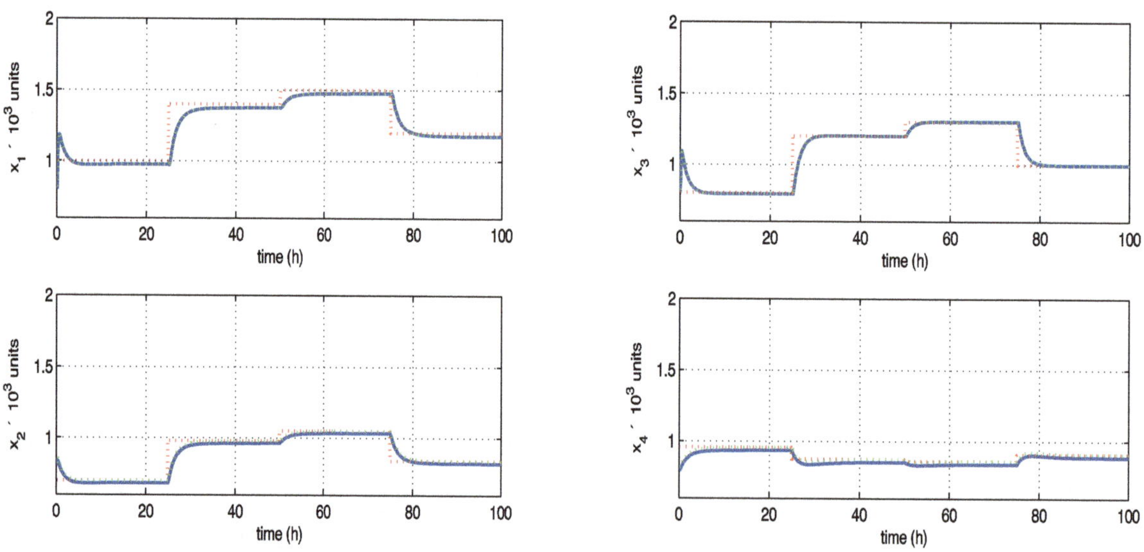

Figure 8.43 Tracking of setpoint 7 by the supply chain network: (a) convergence of state variables x_1 to x_2 (blue lines) to the reference setpoints (red lines) and their KF-based estimation (green line); (b) convergence of state variables x_3 to x_4 (blue lines) to the reference setpoints (red lines) and their KF-based estimation (green line)

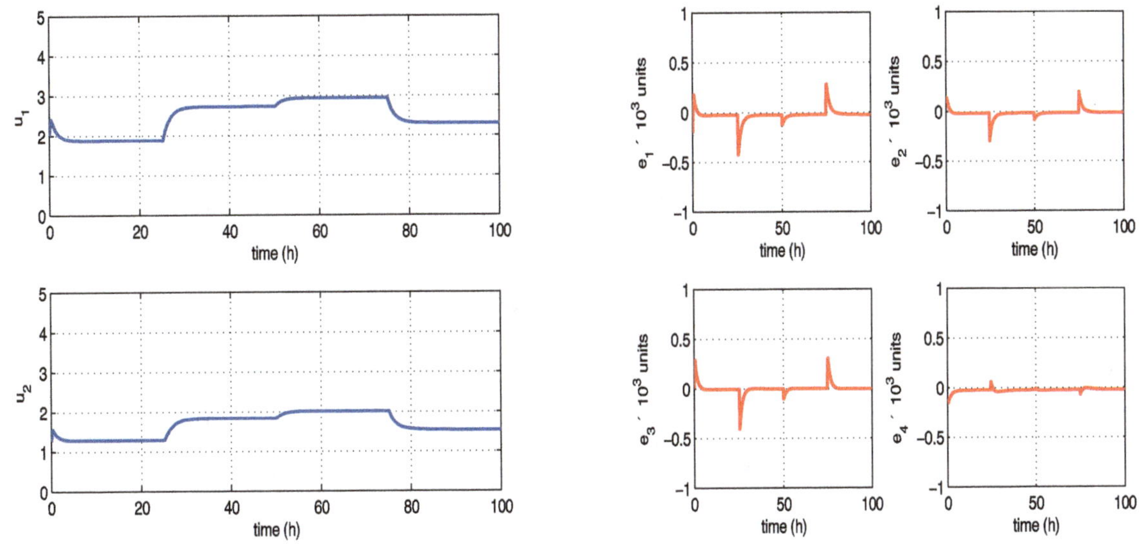

Figure 8.44 Tracking of setpoint 7 by the supply chain network: (a) variations of the control inputs u_1 and u_2; (b) tracking error of state variables x_1 to x_4

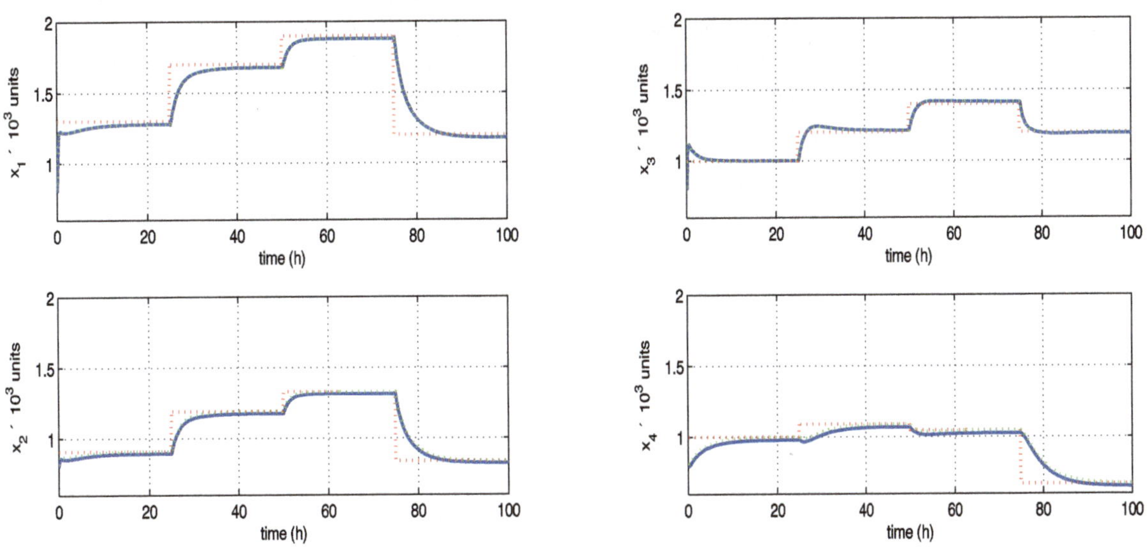

Figure 8.45 Tracking of setpoint 8 by the supply chain network: (a) convergence of state variables x_1 to x_2 (blue lines) to the reference setpoints (red lines) and their KF-based estimation (green line); (b) convergence of state variables x_3 to x_4 (blue lines) to the reference setpoints (red lines) and their KF-based estimation (green line)

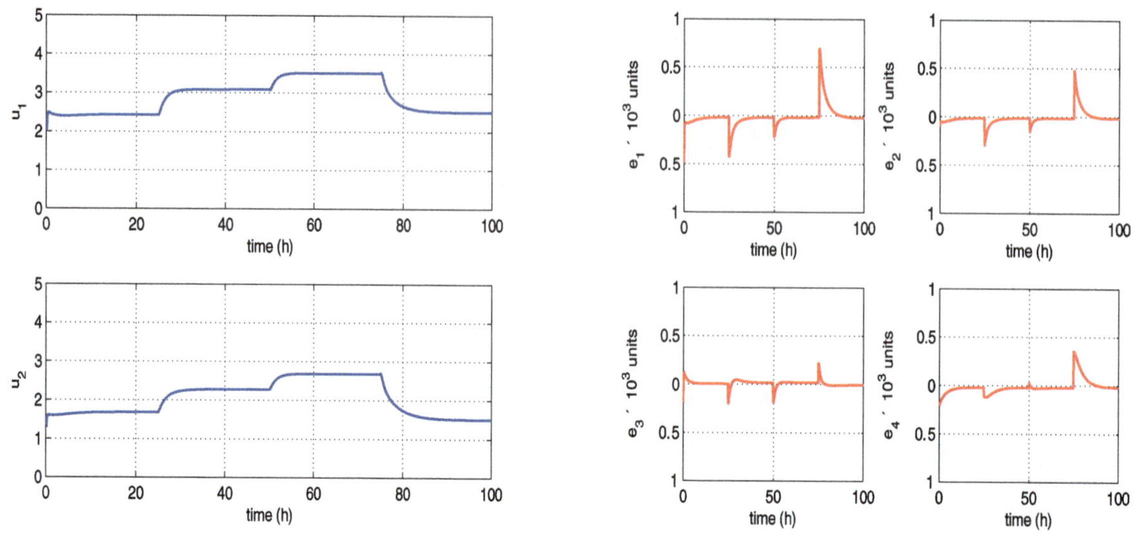

Figure 8.46 Tracking of setpoint 8 by the supply chain network: (a) variations of the control inputs u_1 and u_2; (b) tracking error of state variables x_1 to x_4

elimination of the state vectors' tracking error is achieved. For relatively large values of the elements of the diagonal matrix Q fast convergence of the state variables to setpoints is achieved. Finally, the smallest value of ρ for which one still obtains an acceptable solution of the method's Riccati equation in the form of a positive definite and symmetric matrix P is the one that provides the control loop with maximum robustness.

Appendix A
Nonlinear optimal control and Lie algebra-based control

Abstract

Appendix A analyzes nonlinear optimal control and Lie algebra-based control. In the first section of this appendix, control and state estimation based of approximate linearization of the dynamic models of nonlinear systems is explained. Linearization is performed sequentially through Taylor series expansion and the solution of a Riccati equation. The H-infinity Kalman filter is developed as a robust state estimator. In the second section of this appendix, global linearization-based control and state estimation is developed for nonlinear systems. To this end, input–output and input–state linearization conditions are analyzed. Results from Lie algebra-based control and Lie algebra-based state observers are explained. The concept of dynamic extension is also outlined.

A.1 Control based on approximate linearization

A.1.1 Overview of the optimal control concept

A.1.1.1 Optimal control for continuous-time systems with Bellman's principle

The monograph presents a novel solution to the nonlinear optimal control problem for dynamical systems, including electric power systems and electric traction and propulsion systems. The method is of proven stability and remains computationally tractable. It retains the advantages of linear optimal control, that is fast and accurate tracking of reference setpoints under moderate variations of the control inputs. One can also point out the advantages of this nonlinear optimal control method against nonlinear model predictive control (NMPC) (in NMPC, the stability properties of the control scheme remain unproven and the convergence of the iterative search for an optimum often depends on initialization and parameter values' selection). It is also noteworthy that the new nonlinear optimal control method is applicable to a wider class of dynamical systems than approaches based on the solution of state-dependent Riccati equations (SDRE) (the SDRE approaches can be applied only to dynamical systems which can be transformed to the linear parameter varying—LPV—form). Besides, the nonlinear optimal control method performs better than nonlinear optimal control schemes which use an approximation of the solution of the Hamilton–Jacobi–Bellman equation by Galerkin series expansions (actually, the stability properties of the Galerkin series expansion-based optimal control approaches are still unproven).

Optimal control and its applications exhibit a rapid development in the last years [933,513,547,89]. However, nonlinear optimal control problems remain difficult to handle and a conclusive solution about them has not been achieved so far [291,34,75,761]. In the area of optimal control, Bellman's optimality principle stands for one of the most efficient approaches [176]. The following continuous-time dynamical system is considered:

$$\dot{x} = f(x, u, t), \quad x(t_0) = x_0 \tag{A.1}$$

The objective is to select the control input $u(t)$, $t \in [t_0, t_f]$ so as to achieve minimization of the following cost function

$$J(u) = \int_{t_0}^{t_f} L(x, u, t)dt \tag{A.2}$$

To this end, one defines

$$J^0(x,t) = \min_{u\in[t,\,t_f]} \int_t^{t_f} L(x,u,t)dt \tag{A.3}$$

where $J^0(x,t_0) = J^0(x_0)$. According to the principle of optimality, optimization can take place in two stages: (i) from t to $t+\Delta t$ and (ii) from $t+\Delta t$ to t_f. Consequently, the previous cost function is decomposed as follows:

$$\begin{aligned} J^0(x,t) &= \min_{u\in[t,\,t_f]} \left\{\int_t^{t+\Delta t} L(x,u,t)dt + \int_{t+\Delta t}^{t_f} L(x,u,t)dt\right\} \\ &= \min_{u\in[t,\,t_f]} \left\{\int_t^{t+\Delta t} L(x,u,t)dt + J^0(x+\Delta x, t+\Delta t)\right\} \end{aligned} \tag{A.4}$$

After taking the Taylor series expansion of $J^0(x+\Delta x, t+\Delta t)$, one obtains

$$J^0(x+\Delta x, t+\Delta t) = J^0(x,t) + \frac{\partial J^0}{\partial x}\Delta x + \frac{\partial J^0}{\partial t}\Delta t + O(\Delta) \tag{A.5}$$

where $O(\Delta)$ stands for higher-order terms to be omitted from the Taylor series expansion. By substituting Eq. (A.5) in Eq. (A.4), one obtains

$$0 = \min_{u\in[t,\,t_f]} \left\{L\Delta t + \frac{\partial J^0}{\partial x}\Delta x + \frac{\partial J^0}{\partial t}\Delta t + O(\Delta)\right\} \tag{A.6}$$

and after dividing by Δt and taking the limit $\Delta t \to 0$, one arrives at the relation

$$-\frac{\partial J^0}{\partial t} = \min_{u\in[t,\,t_f]} \mathrm{H} \tag{A.7}$$

$$H = L(x,u,t) + \frac{\partial J^0}{\partial x}\frac{dx}{dt} = L(x,u,t) + \frac{\partial J^0}{\partial x} f(x,u,t) \tag{A.8}$$

In the above relations, Eq. (A.7) is known as the Hamilton–Jacobi–Bellman equation, whereas Eq. (A.8) is the Hamiltonian function of the optimal control problem. By solving the Hamilton–Jacobi–Bellman equation, one obtains both the control input $u(t)$ $t\in[t,t_f]$ that minimizes the objective function $J^0(x,t)$, as well as the minimized value of $J^0(x,t)$.

The solution of the Hamilton–Jacobi–Bellman equation is a non-trivial problem and in the case of nonlinear dynamical systems it may become a complicated or even non-tractable problem. If the system is a linear one, and the cost function comprises quadratic terms, then the solution of the Hamilton–Jacobi–Bellman equation is substituted by the solution of a Riccati differential equation.

A.1.1.2 Solution of the optimal control problem in closed form

The solution of the optimal control problem in closed form can be achieved in the case of linear continuous-time systems with a quadratic cost function. For a linear system of the form:

$$\dot{x}(t) = A(t)x(t) + B(t)u(t) \tag{A.9}$$

and for a cost function that comprises quadratic terms in the form [176]:

$$J = \int_{t_0}^{t_f} L dt \tag{A.10}$$

with

$$L = \tfrac{1}{2}x^T Q(t)x + \tfrac{1}{2}u^T R u \tag{A.11}$$

then the optimal value of the cost function is

$$J^0(x,t) = \tfrac{1}{2}x^T(t)P(t)x(t) \tag{A.12}$$

with $P(t)$ being a symmetric positive definite matrix. For such a system, the Hamiltonian function is [176]

$$H = \tfrac{1}{2}x^T Q x + \tfrac{1}{2}u^T R u + \tfrac{1}{2}x^T P(Ax+Bu) + \tfrac{1}{2}(Ax+Bu)^T P x \tag{A.13}$$

By deriving the Hamiltonian function with respect to the control input u and by taking its extremum condition, one gets

$$\frac{\partial H}{\partial u} = Ru + B^T Px = 0 \tag{A.14}$$

By solving the above equation with respect to $u(t)$, one finds the optimal control input that minimizes the quadratic cost function of the system. This is given by

$$u^0(t) = -R^1(t)B^T(t)P(t)x(t) \tag{A.15}$$

By substituting the optimal control input to the Hamiltonian of Eq. (A.13), one obtains

$$\begin{aligned} H^0 = \min_{u\in[t,t_f]} H &= \tfrac{1}{2}x^T Qx + \tfrac{1}{2}x^T PBR^{-1}B^T Px + \tfrac{1}{2}x^T A^T Px \\ &+ \tfrac{1}{2}x^T PAx - x^T PBR^{-1}B^T Px \\ &= \tfrac{1}{2}x^T(Q + A^T P + PA - PBR^{-1}B^T P)x \end{aligned} \tag{A.16}$$

Next, by differentiating the cost function of Eq. (A.12), one obtains

$$\frac{\partial J^0(x,t)}{\partial t} = \tfrac{1}{2}x^T(t)\frac{dP(t)}{dt}x(t) \tag{A.17}$$

Next, by substituting Eq. (A.16) and Eq. (A.17) in Eq. (A.7), one has the following formulation of the Hamilton–Jacobi–Bellman equation

$$\begin{aligned} -\tfrac{1}{2}x^T(t)\frac{dP(t)}{dt}x(t) &= \\ \min_{u\in[t,t_f]} H &= \tfrac{1}{2}x^T Qx + \tfrac{1}{2}x^T PBR^{-1}B^T Px + \tfrac{1}{2}x^T A^T Px \\ &+ \tfrac{1}{2}x^T PAx - x^T PBR^{-1}B^T Px \\ &= \tfrac{1}{2}x^T(Q + A^T P + PA - PBR^{-1}B^T P)x \end{aligned} \tag{A.18}$$

Through intermediate computations, one arrives at the following differential Riccati equation [176]:

$$-\frac{dP(t)}{dt} = A^T P + PA + Q - PBR^{-1}B^T P \tag{A.19}$$

The boundary condition for the solution of this differential equation comes from the relation

$$J^0(x, t_f) = \tfrac{1}{2}x(t_f)^T P(t_f)x(t_f) = 0 \tag{A.20}$$

having also $P(t_f) = 0$. By integrating Eq. (A.19) in inverse time from t_f to t_0, one can find matrix $P(t) :\ t\in[t_0, t_f]$. Besides, using Eq. (A.15), one can find the optimal control input to be applied to the system. Moreover, using Eq. (A.12), one can find the minimum value of the system's cost function $J^0(x, t)$.

The above procedure for finding a solution to the optimal control problem of a linear dynamical system arrives at the design of a linear quadratic regulator. If matrices A and B which appear in the state-space model of the system of Eq. (A.9) are time invariant and the weight matrices Q and R which appear in the quadratic cost function of Eq. (A.10) and Eq. (A.11) are time-invariant too, and the finite-time $t_f = \infty$ (infinite time horizon) then the solution of the differential Riccati equation $P(t)$ arrives at a steady-state value, that is $\frac{dP(t)}{dt} = 0$ for $t\to\infty$. In such a case, the differential Riccati equation of Eq. (A.19) becomes an algebraic Riccati equation of the form:

$$A^T P + PA + Q - PBR^{-1}B^T P = 0 \tag{A.21}$$

In such a case, the optimal feedback control becomes

$$u = -R^{-1}B^T Px(t) \tag{A.22}$$

where P now is the solution of the algebraic Riccati equation of Eq. (A.21). Thus, the optimized cost function becomes $J^0 = \frac{1}{2}x^T(0)Px(0)$.

A.1.1.3 Stability conditions for optimal control

The stability of a linear optimal control loop is related with the stabilizability properties of such a system. The linear system $\dot{x}(t) = Ax(t) + Bu(t)$ $x{\in}R^n$ $u{\in}R^m$ is stabilizable if the control input $u(t)$ can be chosen such that all closed-loop poles are stable. The stabilizability of a system is a more generic property than controllability. It actually means that some of the poles of the system can be placed at the left complex semi-plane through state feedback, while the rest of the system poles which cannot be relocated with feedback are inherently found in the left complex semi-plane. The conditions that assure global stability for the closed-loop of an optimal linear quadratic control scheme are given in the following theorem:

Theorem: Assume C to be a $n{\times}n$ square root of matrix Q where Q is the weight matrix appearing in the quadratic cost function of Eq. (A.10) and Eq. (A.11). This signifies that $C^TC = Q$. Assume next that the matrices' pair (A, B) is stabilizable, and the matrices' pair (A, C) is observable. Then, the following holds:

(i) There exists a unique symmetric positive-definite steady-state solution P of the differential Riccati equation given in Eq. (A.19), which is not dependent on the final condition $P(t_f)$. (ii) The aforementioned matrix P is also the unique solution of the algebraic Riccati equation given in Eq. (A.21). (iii) The closed-loop system is globally asymptotically stable.

According to this theorem, the stability properties of an optimal linear quadratic control loop depend on the structure of the open loop system, as defined by matrices A and B, and on the structure of the weight matrix Q which appear as design parameters in the quadratic cost function. The stabilizability of the pair (A, B) assures that through state-feedback all poles of the closed-loop system can be finally found in the left complex semi-plane. The observability of the pair $(A, \sqrt{Q})$ signifies that all state variables of the system appear finally in the cost function J which is subject to minimization. The poles of the closed-loop system, and consequently the transient characteristics of the optimal control loop are determined by the selection of matrices $Q{\geq}0$ and $R > 0$.

A.1.2 Design of an H-infinity nonlinear optimal controller

A.1.2.1 Equivalent linearized dynamics of the system

The proposed nonlinear optimal control method is addressed to nonlinear dynamical systems in the form:

$$\dot{x} = f(x) + g(x)u \tag{A.23}$$

where $x{\in}R^n$, $f(x){\in}R^n$, $g(x){\in}R^n$, and $u{\in}R^m$. It can be also applied to nonlinear systems having an even more generic state-space description in the form $\dot{x} = \tilde{f}(x, u)$. After linearization of Eq. (A.23) around its current operating point, the system's dynamic model is also written as

$$\dot{x} = Ax + Bu + d_1 \tag{A.24}$$

where A, B are the Jacobian matrices of the system which are given by $A = [\nabla_x f(x) \mid_{x^*} + \nabla_x g(x) \mid_{x^*} u^*]$, $B = g(x^*)$. Parameter d_1 stands for the linearization error in the system's dynamic model appearing in Eq. (A.24).

The linearization approach which has been followed for implementing the nonlinear optimal control scheme is based on Taylor series expansion and on the computation of Jacobian matrices for the state-space model of the system. The linearization takes place at each sampling instant around the time-varying operating point (x^*, u^*) where x^* is the present value of the system's state vector and u^* is the last sampled value of the control inputs vector. Consider for instance the following affine-in-the-input state-space model:

$$\begin{aligned} &\dot{x} = f(x) + g(x)u \Rightarrow \\ &\dot{x} = [f(x^*) + \nabla_x f(x) \mid_{x^*} (x - x^*)] + [g(x^*) + \nabla_x g(x) \mid_{x^*} (x - x^*)]u^* + \\ &\quad + g(x^*)u^* + g(x^*)(u - u^*) + \tilde{d}_1 \Rightarrow \\ &\dot{x} = [\nabla_x f(x) \mid_{x^*} + \nabla_x g(x) \mid_{x^*} u^*]x + g(x^*)u - [\nabla_x f(x) \mid_{x^*} + \nabla_x g(x) \mid_{x^*} u^*]x^* + \\ &\quad + f(x^*) + g(x^*)u^* + \tilde{d}_1 \end{aligned} \tag{A.25}$$

where $\tilde{d}_1$ is the modeling error due to truncation of higher order terms in the Taylor series expansion of $f(x)$ and $g(x)$. Next, by defining $A=[\nabla_x f(x)\mid_{x^*} + \nabla_x g(x)\mid_{x^*} u^*]$, $B=g(x^*)$, one obtains

$$\dot{x}=Ax+Bu-Ax^*+f(x^*)+g(x^*)u^*+\tilde{d}_1 \tag{A.26}$$

Moreover by denoting $d_1=-Ax^*+f(x^*)+g(x^*)u^*+\tilde{d}_1$ about the cumulative modeling error term in the Taylor series expansion procedure, one has

$$\dot{x}=Ax+Bu+d_1 \tag{A.27}$$

which is the approximately linearized model of the dynamics of the system of Eq. (A.24). The term $f(x^*)+g(x^*)u^*$ is the derivative of the state vector at (x^*,u^*) which is almost annihilated by $-Ax^*$. The Jacobian matrices A and B are indeed updated at each sampling instance by substituting the state vector elements by their values at the linearization point. As proven by this chapter's Lyapunov stability analysis, the proposed optimal (H-infinity) controller achieves the compensation of the above-noted modeling error and the precise tracking of reference setpoints by the system's state variables.

The reference setpoints for the system's state vector are denoted by $\mathbf{x_d}=[x_1^d,\cdots,x_n^d]$. Tracking of this trajectory is achieved after applying the control input u^*. At every time instant, the control input u_d is assumed to differ from the control input u appearing in Eq. (A.24) by an amount equal to Δu, that is $u_d=u+\Delta u$

$$\dot{x}_d=Ax_d+Bu^*+d_2 \tag{A.28}$$

The dynamics of the controlled system described in Eq. (A.24) can be also written as

$$\dot{x}=Ax+Bu+Bu_d-Bu_d+d_1 \tag{A.29}$$

and by denoting $d_3=-Bu_d+d_1$ as an aggregate disturbance term, one obtains

$$\dot{x}=Ax+Bu+Bu_d+d_3 \tag{A.30}$$

By subtracting Eq. (A.28) from Eq. (A.30), one has

$$\dot{x}-\dot{x}_d=A(x-x_d)+Bu+d_3-d_2 \tag{A.31}$$

By denoting the tracking error as $e=x-x_d$ and the aggregate disturbance term as $L\tilde{d}=d_3-d_2$, the tracking error dynamics becomes

$$\dot{e}=Ae+Bu+L\tilde{d} \tag{A.32}$$

where L is the disturbance inputs gain matrix. The above linearized form of the system's model can be efficiently controlled after applying an H-infinity feedback control scheme.

A.1.2.2 The nonlinear H-infinity control

Consider the case in which the nonlinear model of the system is written in the form

$$\dot{x}=\tilde{f}(x,u) \quad x\in R^n,\ u\in R^m \tag{A.33}$$

Linearization of the system is performed at each iteration of the control algorithm around its present operating point $(x^*,u^*)=(x(t),u(t-T_s))$, where T_s is the sampling period. The linearized equivalent model of the system is described by

$$\dot{x}=Ax+Bu+L\tilde{d} \quad x\in R^n,\ u\in R^m,\ \tilde{d}\in R^q \tag{A.34}$$

where matrices A and B are obtained from the computation of the Jacobians

$$A=\begin{pmatrix} \frac{\partial\tilde{f}_1}{\partial x_1} & \frac{\partial\tilde{f}_1}{\partial x_2} & \cdots & \frac{\partial\tilde{f}_1}{\partial x_n} \\ \frac{\partial\tilde{f}_2}{\partial x_1} & \frac{\partial\tilde{f}_2}{\partial x_2} & \cdots & \frac{\partial\tilde{f}_2}{\partial x_n} \\ \cdots & \cdots & \cdots & \cdots \\ \frac{\partial\tilde{f}_n}{\partial x_1} & \frac{\partial\tilde{f}_n}{\partial x_2} & \cdots & \frac{\partial\tilde{f}_n}{\partial x_n} \end{pmatrix}\Bigg|_{(x^*,u^*)} \quad B=\begin{pmatrix} \frac{\partial\tilde{f}_1}{\partial u_1} & \frac{\partial\tilde{f}_1}{\partial u_2} & \cdots & \frac{\partial\tilde{f}_1}{\partial u_m} \\ \frac{\partial\tilde{f}_2}{\partial u_1} & \frac{\partial\tilde{f}_2}{\partial u_2} & \cdots & \frac{\partial\tilde{f}_2}{\partial u_m} \\ \cdots & \cdots & \cdots & \cdots \\ \frac{\partial\tilde{f}_n}{\partial u_1} & \frac{\partial\tilde{f}_n}{\partial u_2} & \cdots & \frac{\partial\tilde{f}_n}{\partial u_m} \end{pmatrix}\Bigg|_{(x^*,u^*)} \tag{A.35}$$

and vector $\tilde{d}$ denotes disturbance terms due to linearization errors. The problem of disturbance rejection for the linearized model that is described by

$$\dot{x} = Ax + Bu + L\tilde{d}$$
$$y = Cx \tag{A.36}$$

where $x \in R^n$, $u \in R^m$, $\tilde{d} \in R^q$, and $y \in R^p$ cannot be treated efficiently if the classical LQR control scheme is applied. This is because of the existence of the perturbation term $\tilde{d}$. The disturbance term $\tilde{d}$ apart from modeling (parametric) uncertainty and external perturbation terms, can also represent noise terms of any distribution.

In the H_∞ control approach, a feedback control scheme is designed for trajectory tracking by the system's state vector and simultaneous disturbance rejection, considering that the disturbance affects the system in the worst possible manner. The disturbances' effects are incorporated in the following quadratic cost function:

$$J(t) = \frac{1}{2}\int_0^T [y^T(t)y(t) + ru^T(t)u(t) - \rho^2\tilde{d}^T(t)\tilde{d}(t)]dt, \quad r, \rho > 0 \tag{A.37}$$

The significance of the negative sign in the cost function's term that is associated with the perturbation variable $\tilde{d}(t)$ is that the disturbance tries to maximize the cost function $J(t)$ while the control signal $u(t)$ tries to minimize it. The physical meaning of the relation given above is that the control signal and the disturbances compete with each other within a min–max differential game. This problem of min–max optimization can be written as

$$min_u max_{\tilde{d}} J(u, \tilde{d}) \tag{A.38}$$

The objective of the optimization procedure is to compute a control signal $u(t)$ which can compensate for the worst possible disturbance, that is externally imposed to the system. However, the solution to the min–max optimization problem is directly related to the value of parameter ρ. This means that there is an upper bound in the disturbances magnitude that can be annihilated by the control signal.

A.1.2.3 Computation of the feedback control gains

For the linearized system given by Eq. (A.36) the cost function of Eq. (A.37) is defined, where coefficient r determines the penalization of the control input and weight coefficient ρ determines the reward of the disturbances' effects.

It is assumed that (i) The energy that is transferred from the disturbances signal $\tilde{d}(t)$ to the dynamical system is bounded, that is $\int_0^\infty \tilde{d}^T(t)\tilde{d}(t)dt < \infty$, (ii) matrices $[A, B]$ and $[A, L]$ are stabilizable, and (iii) matrix $[A, C]$ is detectable. Then, the optimal feedback control law is given by

$$u(t) = -Ke(t) \tag{A.39}$$

with

$$K = \frac{1}{r}B^T P \tag{A.40}$$

where P is a positive definite symmetric matrix, which is obtained from the solution of the Riccati equation:

$$A^T P + PA + Q - P(\frac{2}{r}BB^T - \frac{1}{\rho^2}LL^T)P = 0 \tag{A.41}$$

where Q is also a positive semi-definite symmetric matrix. The worst-case disturbance is given by

$$\tilde{d}(t) = \frac{1}{\rho^2}L^T Pe(t) \tag{A.42}$$

The computation of the worst-case disturbance that this controller can sustain comes from the superposition of Bellman's optimality principle when considering that the system is affected by two separate inputs: (i) the control input u; (ii) the cumulative disturbance input $\tilde{d}(t)$. Solving the optimal control problem for u that is for the minimum variation (optimal) control input that achieves the elimination of the state vector's tracking error gives $u = -\frac{1}{r}B^T Pe$. Equivalently, solving the optimal control problem for $\tilde{d}$, that is for the worst-case disturbance that the control loop can sustain gives $\tilde{d} = \frac{1}{\rho^2}L^T Pe$.

A.1.2.4 Riccati equation coefficients in H-infinity control robustness

Parameter ρ in Eq. (A.37) is an indication of the closed-loop system robustness. If the values of $\rho > 0$ are excessively decreased with respect to r, then the solution of the Riccati equation is no longer a positive definite matrix. Consequently, there is a lower bound ρ_{min} of ρ for which the H_∞ control problem has a solution. The acceptable values of ρ lie in the interval $[\rho_{min}, \infty)$. If ρ_{min} is found and used in the design of the H_∞ controller, then the closed-loop system will have elevated robustness. Unlike this, if a value $\rho > \rho_{min}$ is used, then an admissible stabilizing H_∞ controller will be derived but it will be a suboptimal one. The Hamiltonian matrix

$$H = \begin{pmatrix} A & -(\frac{1}{r}BB^T - \frac{1}{\rho^2}LL^T) \\ -Q & -A^T \end{pmatrix} \tag{A.43}$$

provides a criterion for the existence of a solution of the Riccati equation Eq. (A.41). A necessary condition for the solution of the algebraic Riccati equation to be a positive definite symmetric matrix P is that H has no imaginary eigenvalues [708,556].

A.1.2.5 Lyapunov stability analysis

Through Lyapunov stability analysis, it will be shown that the proposed nonlinear control scheme assures H_∞ tracking performance for the nonlinear system and that in case of bounded disturbance terms asymptotic convergence to the reference setpoints is achieved.

The tracking error dynamics for the nonlinear system is written in the form

$$\dot{e} = Ae + Bu + L\tilde{d} \tag{A.44}$$

where in the case of a nonlinear system with a state vector of dimension equal to n, one can define $L \in R^{n \times n}$ to be the disturbance inputs gain matrix. Variable $\tilde{d}$ denotes model uncertainties and external disturbances of the system's model. The following Lyapunov equation is considered:

$$V = \frac{1}{2}e^TPe \tag{A.45}$$

where $e = x - x_d$ is the tracking error. By differentiating with respect to time, one obtains

$$\begin{aligned} \dot{V} &= \tfrac{1}{2}\dot{e}^TPe + \tfrac{1}{2}e^TP\dot{e} \Rightarrow \\ \dot{V} &= \tfrac{1}{2}[Ae + Bu + L\tilde{d}]^TPe + \tfrac{1}{2}e^TP[Ae + Bu + L\tilde{d}] \Rightarrow \end{aligned} \tag{A.46}$$

$$\begin{aligned} \dot{V} &= \tfrac{1}{2}[e^TA^T + u^TB^T + \tilde{d}^TL^T]Pe + \\ &+ \tfrac{1}{2}e^TP[Ae + Bu + L\tilde{d}] \Rightarrow \end{aligned} \tag{A.47}$$

$$\begin{aligned} \dot{V} &= \tfrac{1}{2}e^TA^TPe + \tfrac{1}{2}u^TB^TPe + \tfrac{1}{2}\tilde{d}^TL^TPe + \\ &\tfrac{1}{2}e^TPAe + \tfrac{1}{2}e^TPBu + \tfrac{1}{2}e^TPL\tilde{d} \end{aligned} \tag{A.48}$$

The previous equation is rewritten as

$$\begin{aligned} \dot{V} &= \tfrac{1}{2}e^T(A^TP + PA)e + (\tfrac{1}{2}u^TB^TPe + \tfrac{1}{2}e^TPBu) + \\ &+ (\tfrac{1}{2}\tilde{d}^TL^TPe + \tfrac{1}{2}e^TPL\tilde{d}) \end{aligned} \tag{A.49}$$

Assumption: For given positive definite matrix Q and coefficients r and ρ, there exists a positive definite matrix P, which is the solution of the following matrix equation:

$$A^TP + PA = -Q + P(\tfrac{2}{r}BB^T - \tfrac{1}{\rho^2}LL^T)P \tag{A.50}$$

Moreover, the following feedback control law is applied to the system:

$$u = -\tfrac{1}{r}B^TPe \tag{A.51}$$

By substituting Eq. (A.50) and Eq. (A.51), one obtains

$$\dot{V} = \frac{1}{2}e^T[-Q + P(\frac{2}{r}BB^T - \frac{1}{\rho^2}LL^T)P]e+ \\ +e^T PB(-\frac{1}{r}B^T Pe) + e^T PL\tilde{d} \Rightarrow \tag{A.52}$$

$$\dot{V} = -\frac{1}{2}e^T Qe + \frac{1}{r}e^T PBB^T Pe - \frac{1}{2\rho^2}e^T PLL^T Pe \\ -\frac{1}{r}e^T PBB^T Pe + e^T PL\tilde{d} \tag{A.53}$$

which after intermediate operations gives

$$\dot{V} = -\frac{1}{2}e^T Qe - \frac{1}{2\rho^2}e^T PLL^T Pe + e^T PL\tilde{d} \tag{A.54}$$

or, equivalently

$$\dot{V} = -\frac{1}{2}e^T Qe - \frac{1}{2\rho^2}e^T PLL^T Pe+ \\ +\frac{1}{2}e^T PL\tilde{d} + \frac{1}{2}\tilde{d}^T L^T Pe \tag{A.55}$$

Lemma: The following inequality holds:

$$\frac{1}{2}e^T PL\tilde{d} + \frac{1}{2}\tilde{d}L^T Pe - \frac{1}{2\rho^2}e^T PLL^T Pe \leq \frac{1}{2}\rho^2 \tilde{d}^T \tilde{d} \tag{A.56}$$

Proof: The binomial $(\rho\alpha - \frac{1}{\rho}b)^2$ is considered. Expanding the left part of the above inequality, one gets

$$\rho^2 a^2 + \frac{1}{\rho^2}b^2 - 2ab \geq 0 \Rightarrow \frac{1}{2}\rho^2 a^2 + \frac{1}{2\rho^2}b^2 - ab \geq 0 \Rightarrow \\ ab - \frac{1}{2\rho^2}b^2 \leq \frac{1}{2}\rho^2 a^2 \Rightarrow \frac{1}{2}ab + \frac{1}{2}ab - \frac{1}{2\rho^2}b^2 \leq \frac{1}{2}\rho^2 a^2 \tag{A.57}$$

The following substitutions are carried out: $a = \tilde{d}$ and $b = e^T PL$ and the previous relation becomes

$$\frac{1}{2}\tilde{d}^T L^T Pe + \frac{1}{2}e^T PL\tilde{d} - \frac{1}{2\rho^2}e^T PLL^T Pe \leq \frac{1}{2}\rho^2 \tilde{d}^T \tilde{d} \tag{A.58}$$

Eq. (A.58) is substituted in Eq. (A.55) and the inequality is enforced, thus giving

$$\dot{V} \leq -\frac{1}{2}e^T Qe + \frac{1}{2}\rho^2 \tilde{d}^T \tilde{d} \tag{A.59}$$

Eq. (A.59) shows that the H_∞ tracking performance criterion is satisfied. The integration of $\dot{V}$ from 0 to T gives

$$\int_0^T \dot{V}(t)dt \leq -\frac{1}{2}\int_0^T ||e||_Q^2 dt + \frac{1}{2}\rho^2 \int_0^T ||\tilde{d}||^2 dt \Rightarrow \\ 2V(T) + \int_0^T ||e||_Q^2 dt \leq 2V(0) + \rho^2 \int_0^T ||\tilde{d}||^2 dt \tag{A.60}$$

Moreover, if there exists a positive constant $M_d > 0$ such that

$$\int_0^\infty ||\tilde{d}||^2 dt \leq M_d \tag{A.61}$$

then, one gets

$$\int_0^\infty ||e||_Q^2 dt \leq 2V(0) + \rho^2 M_d \tag{A.62}$$

Thus, the integral $\int_0^\infty ||e||_Q^2 dt$ is bounded. Moreover, $V(T)$ is bounded and from the definition of the Lyapunov function V in Eq. (A.45) it becomes clear that $e(t)$ will be also bounded since $e(t) \in \Omega_e = \{e|e^T Pe \leq 2V(0) + \rho^2 M_d\}$. According to the above and with the use of Barbalat's Lemma, one obtains $lim_{t\to\infty} e(t) = 0$.

Elaborating on the above, it can be noted that the proof of global asymptotic stability for the control loop of the nonlinear system is based on Eq. (A.59) and on the application of Barbalat's Lemma. It uses the condition of Eq. (A.61) about the boundedness of the square of the aggregate disturbance and modeling error term $\tilde{d}$ that affects the model. However, as explained above the proof of global asymptotic stability is not restricted by this condition. By selecting the attenuation coefficient ρ to be sufficiently small and in particular to satisfy $\rho^2 < ||e||_Q^2/||\tilde{d}||^2$, one has that the first derivative of the Lyapunov function is upper bounded by 0. Therefore, for the i-th time interval, it is proven that the Lyapunov function defined in Eq (A.45) is a decreasing one. This also assures that the Lyapunov function of the system defined in Eq. (A.45) will always have a negative first-order derivative.

A.1.2.6 Robust state estimation with the use of the H_∞ Kalman filter

The control loop has to be implemented with the use of information provided by a small number of sensors and by processing only a small number of state variables. To reconstruct the missing information about the state vector of the nonlinear system, it is proposed to use a filtering scheme and based on it to apply state estimation-based control [711,282,802]. By denoting as $A(k)$, $B(k)$, and $C(k)$, the discrete-time equivalents of matrices A, B, and C of the linearized state-space model of the system, the recursion of the H_∞ Kalman filter, for the model of the nonlinear system, can be formulated in terms of a *Measurement update* and a *Time update* part

Measurement update:

$$\begin{aligned} D(k) &= [I - \theta W(k)P^-(k) + C^T(k)R(k)^{-1}C(k)P^-(k)]^{-1} \\ K(k) &= P^-(k)D(k)C^T(k)R(k)^{-1} \\ \hat{x}(k) &= \hat{x}^-(k) + K(k)[y(k) - C\hat{x}^-(k)] \end{aligned} \tag{A.63}$$

Time update:

$$\begin{aligned} \hat{x}^-(k+1) &= A(k)x(k) + B(k)u(k) \\ P^-(k+1) &= A(k)P^-(k)D(k)A^T(k) + Q(k) \end{aligned} \tag{A.64}$$

where it is assumed that parameter θ is sufficiently small to assure that the covariance matrix ${P^-(k)}^{-1} - \theta W(k) + C^T(k)R(k)^{-1}C(k)$ will be positive definite. When $\theta = 0$ the H_∞ Kalman filter becomes equivalent to the standard Kalman filter. One can measure only a part of the state vector of the nonlinear system and can estimate through filtering the rest of the state vector elements.

A.1.3 Optimal state estimation with the H-infinity Kalman filter

A.1.3.1 Outline of the filter

Throughout this monograph, the H-infinity Kalman filter is often used as a robust optimal state estimator for complex dynamical systems [282,802]. This is a variant of the typical Kalman filter. This filter is the result of an optimization procedure which maximizes the robustness of estimation against noise and against errors about the initial value of the system's state vector. The functioning and the performance of the filter are assessed as follows:

(i) Due to the nonlinearity of the systems under control, state estimation should be normally performed with a nonlinear filter or state-observer. This problem cannot be treated with the use of the linear Kalman filter or with the use of linear state observers. However, the monograph's approach allows the use of the H-infinity Kalman filter, although this is a linear filter too. This is achieved by applying approximate linearization around a time-varying operating point (equilibrium). The linearization is based on Taylor series expansion and on the computation of the associated Jacobian matrices.

(ii) Usually, the problem of state estimation for nonlinear systems is treated with nonlinear filtering techniques. However, the related extended Kalman filtering methods are not particularly robust against linearization modeling errors and against measurement noise. Furthermore, Sigma Point Kalman filtering methods are not of proven convergence and stability. Additionally, Particle Filtering techniques demand high computation power and their slow convergence shows that they are not the recommended approach for real-time state estimation purposes [715,711]. Comparing to the aforementioned nonlinear filtering schemes, the H-infinity Kalman filtering approach exhibits elevated robustness against noise and is of proven convergence and stability.

(iii) To elaborate on the matrices which appear in the *Measurement update* part and in the *Time update part* of the H-infinity Kalman filter, the following can be noted: Matrix $R(k)$ is the measurement noise covariance matrix, that is the covariance matrix of the measurement error vector of the system. Matrix $P^-(k)$ is the a-priori state vector estimation error covariance matrix of the system, that is the covariance matrix of the state vector estimation error prior to receiving the updated measurement of the system's outputs. Matrix $W(k)$ is a weight matrix which defines the significance to be attributed by the H-infinity Kalman filter for minimizing the state vector's estimation error, relatively to the effects that the noise affecting the system may have. Finally, matrix $D(k)$ stands for a modified a-posteriori state vector estimation error covariance matrix, that is the covariance matrix of the state vector estimation error after receiving the updated measurement of the system's outputs. Conclusively, the H-infinity Kalman filter retains the structure of the typical Kalman filter, that is a recursion in discrete time comprising a Time update part (computation of variables prior to receiving measurements) and a Measurement update part (computation of variables after measurements have been received). There is a modified a-posteriori state vector estimation error covariance matrix, which in turn takes into account a weight matrix that defines the accuracy of the state estimation under the effects of elevated noise.

(iv) The stability of the H-infinity Kalman filter and the convergence of the state estimation procedure that is performed by it relies on the detectability/observability properties of the nonlinear systems treated in the monograph and not on initialization of the matrices that appear in the filter's recursion. The monograph's concept of performing nonlinear state estimation with the use of the H-infinity Kalman filter is clear and widely applicable [715,711]. The nonlinear state-space model of the monitored system undergoes approximate linearization around a temporary operating point, which is recomputed at each time-step of the estimation algorithm. For the approximately linearized model, state estimation is performed with the use of the H-infinity Kalman filter. The effects of measurement noise, as well as the modeling error which is due to truncation of higher-order terms in the Taylor series expansion, is considered to be a perturbation which is asymptotically compensated by the robustness of the H-infinity Kalman filter.

A.1.3.2 Implementation stages of the filter

The H-infinity Kalman filter is often used in this monograph for the state estimation-based (sensorless) implementation of the proposed optimal control scheme. Filters designed to minimize a weighted norm of state errors are called H_∞ or minimax filters [282,802]. The discrete time H_∞ filter uses the same state-space model as the Kalman filter, which has the form

$$\begin{aligned} x(k+1) &= A(k)x(k) + B(k)u(k) + w(k) \\ z(k) &= C(k)x(k) + v(k) \end{aligned} \tag{A.65}$$

$E[w(k)] = 0$, $E[w(k)w(k)^T] = Q(k)\delta ij$, $E[v(k)] = 0$, $E[v(k)v(k)^T] = R(k)\delta_{ij}$, and $E(w(k)v(k)^T) = 0$. The update of the state estimate is again given by

$$\hat{x}(k) = \hat{x}^-(k) + K(k)(z(k) - C(k)\hat{x}^-(k)) \tag{A.66}$$

and minimizes the trace of the covariance matrix of the state vector estimation error

$$J = \tfrac{1}{2}E\{\tilde{x}(k)^T \cdot \tilde{x}(k)\} = \tfrac{1}{2}tr(P^-(k)) \tag{A.67}$$

where $\tilde{x}^-(k) = x(k) - \hat{x}^-(k)$ and $P^-(k) = E[\tilde{x}^-(k)^T \cdot \tilde{x}^-(k)]$. The H_∞ filtering approach defines first a transformation $d(k) = L(k)x(k)$, where $L(k) \in R^{n \times n}$ is a full rank matrix. The use of the aforementioned transformation allows certain combinations of states to be given more weight than others. Next, by defining the estimation error variable $\tilde{d}_1(i) = d(i) - \hat{d}(i)$, the cost function of the H_∞ filter is initially formulated as [282,802]

$$\begin{aligned} J(k) &= \textstyle\sum_{i=0}^{k-1}\tilde{d}(i+1)^T S(i)\tilde{d}(i+1)/b \\ b &= \tilde{x}^-(0)^T P^-(0)^{-1}\tilde{x}^-(0) + \textstyle\sum_{i=0}^{k-1} w^T(i+1)Q(i+1)^{-1}w(i+1) + \\ &\quad + \textstyle\sum_{i=0}^{k-1} v^T(i)R(i)^{-1}v(i) \end{aligned} \tag{A.68}$$

where S_i is a positive-definite symmetric weighting matrix. It can be observed that both matrices $S(k)$ and $L(k)$ appear in the cost function and thus affect the solution $\hat{x}^-(k+1)$ of the optimization problem. The objective is to find state vector estimates $\hat{x}^-(k)$ and $\hat{x}(k)$ that keep the cost function below a given value $1/\theta$ for worst-case conditions, i.e.,

$$J(k) < \tfrac{1}{\theta} \tag{A.69}$$

By rewriting Eq. (A.68) and substituting Eq. (A.65), a modified cost function is obtained:

$$\begin{aligned} J_a(k) &= -\tfrac{1}{\theta}\tilde{x}^-(0)^T P^-(0)\tilde{x}^-(0) + \textstyle\sum_{i=0}^{k-1}\Gamma(i) \\ \Gamma(i) &= (x(i+1) - \hat{x}^-(i+1))^T W_i (x(i+1) - \hat{x}^-(i+1)) - \\ &\quad -\tfrac{1}{\theta}(w^T(i+1)Q(i+1)^{-1}w(i+1) + \\ &\quad +(y(i) - C(i)x^-(i))^T R(i)^{-1}(y(i) - C(i)x^-(i))) \end{aligned} \tag{A.70}$$

and

$$W(i) = L(i)^T S(i) L(i) \tag{A.71}$$

This cost function does not include the dynamic model of the system given in Eq. (A.65), and this cost is added by using a vector of Lagrange multipliers $\lambda(i+1)$. This gives

$$\begin{aligned} J(k) &= -\tfrac{1}{\theta}\tilde{x}^-(0)^T P^-(0)\tilde{x}^-(0) + \\ &+\textstyle\sum_{i=0}^{k-1}(\Gamma_i + 2\frac{\lambda(i+1)^T}{\theta})(A(i)\hat{x}(i) + B(i)u(i) + \\ &+w(i) - x(i+1)) + \tfrac{2\lambda(0)^T}{\theta}x(0) - \tfrac{2\lambda(0)^T}{\theta}x(0) \end{aligned} \tag{A.72}$$

The cost function of the filter given in Eq. (A.72) can be used as the basis for the solution. It is aimed to find equations defining $\hat{x}^-(k+1)$, or equivalently a measurement weighting matrix (similar to the Kalman gain matrix), that minimize the cost for worst-case assumptions about $x(0)$, $w(i)$, and $y(i)$. Thus, the optimization objective is formulated as

$$J^*(k) = \min_{x_i} \max_{x(0), w(i), y(i)} J(k) \tag{A.73}$$

It is noted that the estimation algorithm uses knowledge of the output measurement $y(i)$ but receives no information about the initial conditions of the system $x(0)$ and the process noise $w(i)$. Under this assumption, the estimation should be able to compensate for worst-case values for the unknown parameters. This is a game theoretic problem, that is solved in two steps [282,802].

In the first step of optimization, partial derivatives of $J(k)$ with respect to $x(0)$, $w(i)$, and $\lambda(i)$ are set to zero so as to maximize the cost function of Eq. (A.72), now being dependent only on the terms $\hat{x}^-(k+1)$ and $y(k)$ which are included in Γ_i. In the second step of optimization, the partial derivatives of $J(k)$ with respect to $\hat{x}^-(k+1)$ and $y(k)$ are set to zero, to obtain a condition for the filter's gain matrix that minimizes this cost function. From the optimization conditions, $\partial J(k)/\partial x_0 = 0^T$, $\partial J(k)/\partial w(i) = 0^T$, $\partial J(k)/\partial \lambda(i) = 0^T$, one obtains an expression of $J(k)$ as function of $\hat{x}^-(k+1)$ and $y(k)$. Next, from the optimization conditions, $\partial J(k)/\partial \hat{x}^-(i+1) = 0^T$, and $\partial J(k)/\partial y(i) = 0^T$, one obtains the filter's equations.

The recursion of the H-infinity Kalman filter can be formulated, as it was previously noted, in terms of a *Measurement update* and a *Time update* part [282,802]

Measurement update:

$$\begin{aligned} D(k) &= [I - \theta W(k)P^-(k) + C^T(k)R(k)^{-1}C(k)P^-(k)]^{-1} \\ K(k) &= P^-(k)D(k)C^T(k)R(k)^{-1} \\ \hat{x}(k) &= \hat{x}^-(k) + K(k)[y(k) - C\hat{x}^-(k)] \end{aligned} \tag{A.74}$$

Time update:

$$\begin{aligned} \hat{x}^-(k+1) &= A(k)x(k) + B(k)u(k) \\ P^-(k+1) &= A(k)P^-(k)D(k)A^T(k) + Q(k) \end{aligned} \tag{A.75}$$

where it is assumed that parameter θ is sufficiently small to maintain

$$P^{-}(k)^{-1} - \theta W(k) + C^{T}(k)R(k)^{-1}C(k) \tag{A.76}$$

positive definite. When $\theta = 0$ the H_{∞} Kalman filter becomes equivalent to the standard Kalman filter. It is noted that apart from the process noise covariance matrix $Q(k)$ and the measurement noise covariance matrix $R(k)$ the H_{∞} Kalman filter requires tuning of the weight matrices L and S, as well as of parameter θ.

A.2 Global linearization-based control concepts

A.2.1 Foundations of global linearization-based control

A.2.1.1 The differential geometric approach

Global linearization-based control is also widely used in this monograph for complex nonlinear dynamical systems. In the area of global linearization-based control, one distinguishes primarily differential geometric approaches (Lie algebra-based control) and differential flatness theory approaches. Before proceeding to the analysis of the differential geometric approach for the control of nonlinear systems the following definitions are given [386,439]:

Vector Field: A mapping $f: D \rightarrow R^n$ where $D \subset R^n$ is a domain, is said to be a vector field on D. A vector field is a n-dimensional column.

Covector Field: The transpose of a vector field is said to be a covector field. A covector field is a n-dimensional row.

Inner Product: If f and w are, respectively, a vector field and a covector field on D then, the inner product $< w,f >$ is defined as

$$< w,f >= w(x)f(x) = \sum_{i=1}^{n} w_i(x)f_i(x) \tag{A.77}$$

Differential: Let $h : D \rightarrow R$. The differential of h is a covector field defined as

$$dh = \frac{\partial h}{\partial x} = [\frac{\partial h}{\partial x_1}, \frac{\partial h}{\partial x_2}, \cdots, \frac{\partial h}{\partial x_n}] \tag{A.78}$$

Lie Derivatives: Let $h : D \rightarrow R$ and $f: D \rightarrow R^n$. The Lie derivative of h with respect to f or along f, is denoted as $L_f h$ and is defined as

$$L_f h = \frac{\partial h}{\partial x} f(x) \tag{A.79}$$

This is the meaning of computing the derivatives of h along the trajectories of the system $\dot{x} = f(x)$. The previous notation can be generalized toward the computation of Lie derivatives of higher order with respect to the same vector field, or with respect to a new vector field. Thus, one has

$$L_g L_f h(x) = \frac{\partial L_f h}{\partial x} g(x) \tag{A.80}$$

while it also holds

$$\begin{aligned} &L_f^0 h(x) = h(x) \\ &L_f h(x) = \frac{\partial h_1}{\partial x_1} f_1 + \frac{\partial h_2}{\partial x_2} f_2 + \cdots + \frac{\partial h_n}{\partial x_n} f_n \\ &\cdots \\ &L_f^k h(x) = L_f L_f^{k-1} = \frac{\partial L_f^{k-1}}{\partial x} f(x) \end{aligned} \tag{A.81}$$

Using the Lie derivative notation, one can arrive at the definition of the relative degree for nonlinear dynamical systems. Consider for example the following single-input single-output nonlinear system

$$\begin{aligned} \dot{x} &= f(x) + g(x)u \\ y &= h(x) \end{aligned} \tag{A.82}$$

This has relative degree r in a domain D_0 if

$$L_g L_f^i h(x) = 0 \text{ for } 0 \leq i \leq r-2 \text{ and } L_g L_f^{r-1} h(x) \neq 0 \ \forall \ x \in D_0 \tag{A.83}$$

Lie Bracket: Let f and g be two vector fields on $D \subset R^n$. The Lie Bracket of f and g is written as $[f, g]$ and is a third vector field which is defined as

$$[f, g](x) = \frac{\partial g}{\partial x} f - \frac{\partial f}{\partial x} g \tag{A.84}$$

where $\frac{\partial g}{\partial x}$ and $\frac{\partial f}{\partial x}$ are Jacobian matrices. The definition of the Lie Brackets can be repetitively used as follows:

$$\begin{aligned} ad_f^0 g(x) &= g(x) \\ ad_f g(x) &= [f, g](x) \\ &\cdots \\ ad_f^k g(x) &= [f, ad_f^{k-1} g](x) \end{aligned} \tag{A.85}$$

For Lie Brackets the following properties hold

Bilinear: Let f_1, f_2 and g_1, g_2 be vector fields and r_1, r_2 be real numbers. Then,

$$\begin{aligned} [r_1 f_1 + r_2 f_2, g_1] &= r_1 [f_1, g_1] + r_2 [f_2, g_2] \\ [f_1, r_1 g_1 + r_2 g_2] &= r_1 [f_1, g_1] + r_2 [f_1, g_2] \end{aligned} \tag{A.86}$$

Skew commutative: $[f, g] = -[g, f]$

Jacobi identity: If f and g are vector fields and h is a real-valued function, then

$$L_{[f,g]} h(x) = L_f L_g h(x) = L_g L_f h(x) \tag{A.87}$$

Diffeomorphism: A mapping $T: \ D \to R^n$ is a diffeomorphism on D if (i) it is invertible on D, that is there exists a function $T^{-1}(x)$ such that $T^{-1}(T(x)) = x$ for all $x \in D$ and (ii) both $T(x)$ and $T^{-1}(x)$ are continuously differentiable. If the Jacobian matrix $[\frac{\partial T}{\partial x}]$ is nonsingular at a point $x_0 \in D$, then from the inverse function theorem it is concluded that there is a neighborhood N of x_0, such that T restricted to N is a diffeomorphism on N, T is said to be a global diffeomorphism if it is a diffeomorphism on R^n and $T(R^n) = R^n$. T is a global diffeomorphism if and only if

(i) $[\partial T / \partial x]$ is nonsingular for all $x \in R^n$
(ii) T is proper, that is $lim_{||x|| \to \infty} ||T(x)|| = \infty$

Distribution: Let $f_1, f_2, \cdots, f_k$ be vector fields on $D \subset R^n$. At any fixed point $x \in D$, $f_1(x), f_2(x), \cdots, f_k(x)$ are vectors in R^n and

$$\Delta(x) = span\{f_1(x), f_2(x), \cdots, f_n(x)\} \tag{A.88}$$

is a subspace on R^n. To each point $x \in R^n$ a subspace $\Delta(x)$ is assigned. One can refer to this assignment using

$$\Delta = span(f_1, f_2, \cdots, f_n) \tag{A.89}$$

and this is called a distribution. This implies that, Δ is the collection of all vector spaces $\Delta(x)$ for $x \in D$. The dimension of $\Delta(x)$ which is denoted as

$$dim(\Delta(x)) = rank[f_1(x), f_2(x), \cdots, f_k(x)] \tag{A.90}$$

may vary with x. However, when $\Delta = span\{f_1, f_2, \cdots, f_r\}$, where $\{f_1(x), \cdots, f_r(x)\}$ are linearly independent for all $x \in D$, then $dim(\Delta(x)) = r$ for all $x \in D$. In that case, Δ is said to be a non-singular distribution on D, generated by $f_1, f_2, \cdots, f_r$. It follows from the smoothness of the vector fields that every $y \in \Delta$ can be expressed as

$$g(x) = \sum_{i=1}^{r} c_i(x) f_i(x) \tag{A.91}$$

where $c_i(x)$ are smooth functions defined on D.

Involutive distribution: A distribution Δ is involutive if

$$g_1 \in \Delta \text{ and } g_2 \in \Delta \Rightarrow [g_1, g_2] \in \Delta \tag{A.92}$$

If Δ is a non-singular distribution on Δ, generated by $f_1, f_2, \cdots, f_r$, then it can be verified that Δ is an involutive distribution if and only if

$$[f_i, f_j] \in \Delta \ \forall \ 1 \leq i, j \leq r \tag{A.93}$$

or equivalently

$$[f_i, f_j](x) = \sum_{k=1}^{r} \alpha_{ijk} f_k(x) \ \ \forall \ i, j \tag{A.94}$$

Equivalently, involutivity means that if one forms the Lie bracket of any pairs of vector fields from the set $[f_1, f_2, \cdots, f_m]$, then the produced vector field can be expressed as a linear combination of the original set of vector fields. It is noted that checking if a set of vector fields $[f_1, \cdots, f_m]$ is involutive is equivalent to checking if

$$rank(f_1(x), \cdots, f_m(x)) = rank(f_1(x), \cdots, f_m(x), \ [f_i, f_j](x)) \tag{A.95}$$

for all x and all i, j.

Codistribution: Codistributions are dual objects to distributions. They are defined using covector fields and possess properties in analogy to those of distributions. A particular codistribution is the one that annihilates a distribution Δ, which is denoted as $\Delta^{\perp}$, and which is defined as

$$\Delta^{\perp}(x) = \{w \in (R^n)^* | < w, v >= 0, \forall \ v \in \Delta(x)\} \tag{A.96}$$

where $(R^n)^*$ is the n-dimensional space of new vectors.

Complete integrability: Let Δ be a non-singular distribution on D, generated by $f_1, \cdots, f_n$. Then Δ is completely integrable, if for each $x_0 \in D$, there exists a neighborhood N of x_0 and $n - r$ real-valued smooth functions $h_1(x), \cdots, h_{n-r}(x)$, such that $h_1(x), \cdots, h_{n-r}(x)$ satisfy the partial differential equations

$$\frac{\partial h_j}{\partial x} f_i(x) = 0, \ \forall \ 1 \leq i \leq r \text{ and } 1 \leq j \leq n - r \tag{A.97}$$

and the covector fields $dh_j(x)$ are linearly independent for all $x \in D$. Equivalently, one can write

$$\Delta^{\perp} = span\{dh_1, \cdots, dh_{n-r}\} \tag{A.98}$$

Finally, a key result in differential geometry is *Frobenius' theorem* which states that: a nonsingular distribution is completely integrable if it is involutive.

A.2.1.2 Elaboration on the Frobenius theorem

The Frobenius theorem is important for the formal treatment of feedback linearization of n-order nonlinear systems. It provides a necessary and sufficient condition for the solvability of a special class of partial differential equations. Consider for example the set of first-order partial differential equations:

$$\begin{array}{l} \frac{\partial h}{\partial x_1} f_1 + \frac{\partial h}{\partial x_2} f_2 + \frac{\partial h}{\partial x_3} f_3 = 0 \\ \frac{\partial h}{\partial g_1} g_1 + \frac{\partial h}{\partial x_2} g_2 + \frac{\partial h}{\partial x_3} g_3 = 0 \end{array} \tag{A.99}$$

where $f_i(x_1, x_2, x_3)$ and $g_i(x_1, x_2, x_3)$, $i = 1, 2, 3$ are scalar functions of x_1, x_2, x_3 and $h(x_1, x_2, x_3)$ is an unknown function. If a solution $h(x_1, x_2, x_3)$ exists for the above set of partial differential equations, then it is said that the set of vector fields $[f, g]$ where $f = (f_1, f_2, f_3)^T$ and $g = (g_1, g_2, g_3)^T$ is completely integrable.

It has to be determined when this set of partial differential equations is solvable. The Frobenius theorem provides a relatively simple condition about this. Eq. (A.99) has a solution $h(x_1, x_2, x_3)$ if and only if there exist scalar functions $\alpha_1(x_1, x_2, x_3)$ and $\alpha_2(x_1, x_2, x_3)$, such that

$$[f, g] = \alpha_1 f + \alpha_2 g \tag{A.100}$$

which means that the Lie bracket of f and g can be expressed as a linear combination of f and g. This is the involutivity condition on the vector fields $[f, g]$. Thus, the Frobenius theorem states that the set of vector fields $[f, g]$ is completely integrable, if and only if, it is involutive.

Next, the Frobenius theorem is re-stated, according to the previous definitions of *complete integrability* and *involutivity*.

Frobenius theorem: Let $f_1, f_2, \cdots, f_m$ be a set of linearly independent vector fields. The set is completely integrable, if and only if it is involutive.

A.2.1.3 Input–output linearization

Input–output linearization is analyzed for nonlinear dynamical systems of the form

$$\begin{aligned} \dot{x} &= f(x) + g(x)u \\ y &= h(x) \end{aligned} \tag{A.101}$$

where $f : D \rightarrow R^n$, $g : D \rightarrow R^n$ and $h : D \rightarrow R$ are smooth, $D \subset R^n$ is a domain. Moreover, y is the system's output. By input–output linearization, one means the generation of the linear differential relation between the output y and a new equivalent input v [808]. The basic approach of input–output linearization is to differentiate the output function y repeatedly until the input u appears.

The single-input single-output system of Eq. (A.101) is considered [439]

$$\begin{aligned} \dot{x} &= f(x) + g(x)u \\ y &= h(x) \end{aligned} \tag{A.102}$$

The system has relative degree r on D, which means

$$L_g L_f^i h(x) = 0 \text{ for } 0 \leq i \leq r-2, \text{ and } L_g L_f^{r-1} h(x) \neq 0, \ \forall x \in D \tag{A.103}$$

The following two lemmas are applied to the system [439]:

Lemma 1: For all $x \in D$ and all integers k and j such that $k \geq 0$ and $0 \leq r-1$, one has

$$L_{ad_f^j g} L_f^k h(k) = \begin{cases} 0 \ \text{ if } 0 \leq j+k < r-1 \\ (-1)^j L_g L_f^{r-1} h(x) \neq 0 \ \text{ if } j+k = r-1 \end{cases} \tag{A.104}$$

Proof: The proof is based on induction. For $j = 0$ Eq. (A.104) holds because it becomes equivalent to the definition of the relative degree of the system of Eq. (A.102). Next, it is assumed that Eq. (A.104) holds for some j and it is proven for $j+1$. Using the Jacobi identity, one has

$$L_{[f,\beta]} \lambda(x) = L_f L_\beta \lambda(x) - L_\beta L_f \lambda(x) \tag{A.105}$$

for any real-valued function λ and any vector fields f and β. Taking $\lambda = L_f^k$, $\beta = ad_f^j g$, one obtains

$$L_{ad_f^{j+1} g} L_f^k h(k) = L_{[f, ad_f^j g]} L_f^k h(x) = L_f L_{ad_f^j g} L_f^k h(x) - L_{ad_f^j g} L_f^{k+1} h(x) \tag{A.106}$$

In the previous equation, the first term of the right hand side is equal to 0, that is $L_f L_{ad_f^j g} L_f^k h(x) = 0$ because $j+k+1 \leq r-1$ or $j+k < r-1$.

Moreover, using that Eq. (A.104) holds for j, one has also

$$L_{ad_f^j g}L_f^{k+1}h(k) = \begin{cases} 0 \text{ if } 0 \le j+k+1 < r-1 \\ (-1)^j L_g L_f^{r-1}h(x) \neq 0 \text{ if } j+k+1 = r-1 \end{cases} \qquad \text{(A.107)}$$

Consequently, one obtains

$$L_{ad_f^{j+1} g}L_f^{k}h(k) = \begin{cases} 0 \text{ if } 0 \le j+k+1 < r-1 \\ (-1)^{j+1} L_g L_f^{r-1}h(x) \neq 0 \text{ if } j+k+1 = r-1 \end{cases} \qquad \text{(A.108)}$$

which completes the proof of the Lemma.

Lemma 2: For all $x \in D$

- the row vectors $dh(x), d_{L_f}h(x), \cdots, dL_f^{r-1}(x)$ are linearly independent
- the column vectors $g(x), ad_f g(x), \cdots, ad_f^{r-1}g(x)$ are linearly independent

Proof: One has

$$\begin{pmatrix} dh(x) \\ \cdots \\ dL_f^{r-1}h(x) \end{pmatrix} \begin{pmatrix} g(x) & \cdots & ad_f^{r-1}g(x) \end{pmatrix} =$$

$$\begin{pmatrix} L_g h(x) & L_{ad_f g}h(x) & \cdots & \cdots & L_{ad_f^{r-1}g}h(x) \\ L_g L_f h(x) & \cdots & \cdots & L_{ad_f^{r-2}g}h(x) & * \\ \cdots & \cdots & \cdots & \cdots & \cdots \\ L_g L_f^{r-1}h(x) & * & \cdots & \cdots & * \end{pmatrix} \qquad \text{(A.109)}$$

From the previous Lemma, the right-hand side matrix is written as

$$\begin{pmatrix} 0 & \cdots & \cdots & 0 & \diamond \\ 0 & \cdots & \cdots & \diamond & * \\ \cdots & \cdots & \cdots & \cdots & \cdots \\ 0 & \diamond & \cdots & \cdots & * \\ \diamond & * & \cdots & \cdots & * \end{pmatrix} \qquad \text{(A.110)}$$

where $\diamond$ denotes a non-zero elements. Thus, the matrix is non-singular. This provides a proof for the Lemma, because if any of the two matrices appearing in the left side of the equation had rank less than n, their product should be a singular matrix.

The previous lemma proves that $r \le n$. Next, the main result of input–output linearization will be formulated: a single-input single-output system of relative degree $r < n$ can be locally transformed into the canonical (normal) form. This is given in the following theorem:

Theorem: The system of Eq. (A.102) is considered and it is assumed that it has relative degree $r < n$ in D. Then, for every $x_0 \in D$, there exist a neighborhood N of x_0 and smooth functions $\phi_1(x), \cdots, \phi_{n-r}(x)$, such that

$$L_g \phi_i(x) = 0, \; \forall \, 1 \le i \le n-r, \; \forall \, x \in N \qquad \text{(A.111)}$$

and the mapping

$$T(x) = \begin{pmatrix} \phi_1(x) \\ \cdots \\ \phi_{n-r}(x) \\ h(x) \\ \cdots \\ L_f^{r-1}h(x) \end{pmatrix} \tag{A.112}$$

restricted to N, is a diffeomorphism on N.

Proof: The distribution $\Delta = span\{g\}$ is non-singular, involutive and has dimension 1. By Frobenius' theorem Δ is completely integrable. Consequently, for every $x_0 \in D$, there exist a neighborhood N_1 of x_0 and $n-1$ smooth functions $\phi_1(x), \cdots, \phi_{n-1}(x)$ with linearly independent differentials such that

$$L_g\phi_i(x) = 0 \text{ for } 1 \leq i \leq n-1 \ \forall \ x \in N_1 \tag{A.113}$$

Equivalently,

$$\Delta^{\perp} = span\{d\phi_1, \cdots, d\phi_{n-1}\} \tag{A.114}$$

on N_1. Since

$$L_gL_f^ih(x) = 0, \ \text{for } 0 \leq i \leq r-2 \tag{A.115}$$

and $dh_1, \cdots, dL_f^{r-2}h(x)$ are linearly independent, one can use $h, \cdots, L_f^{r-2}h$ as part of these $n-1$ functions. In particular, they can be denoted as $\phi_{n-r+1}, \cdots, \phi_{n-1}$. Since $L_gL_f^{r-1}h(x) \neq 0$, one has

$$\begin{aligned} dL_f^{r-1}h(x_0) \notin \Delta^{\perp}(x_0) \Rightarrow rank[\tfrac{\partial T}{\partial x}(x_0)] = n \Rightarrow \\ \tfrac{\partial T}{\partial x}(x_0) \text{ is non-singular} \end{aligned} \tag{A.116}$$

This shows that there is a neighborhood N_2 of x_0 such that $T(x)$, restricted to N_2, is a diffeomorphism on N_2. Taking $N = N_1 \bigcap N_2$ completes the theorem's proof.

A.2.2 Elaborating on input–output linearization

A.2.2.1 The case of a well-defined relative degree

First a region Ω_x of the state-space is considered. Using the notations of differential geometry, the process of repeated differentiation gives

$$\dot{y} = \nabla h(f + gu) = L_fh(x) + L_gh(x)u = v \tag{A.117}$$

If $L_gh(x) \neq 0$ for some $x = x_0$ in Ω_x, then by continuity this relation is verified in a finite neighborhood Ω of x_0. In Ω, the control input that is finally applied on the system is given by

$$u = \tfrac{1}{L_gh}(-L_fh + v) \tag{A.118}$$

results in a linear relation between u and v, which is $\dot{y} = v$.

If however, $L_gh(x) = 0$ for all $x \in \Omega_x$, one can attempt to differentiate $\dot{y}$ once more with respect to time, and this gives

$$\ddot{y} = L_f^2h(x) + L_gL_fh(x)u \tag{A.119}$$

If $L_gL_fh(x)$ is again zero for all $x \in \Omega_x$, additional differentiations with respect to time can be attempted

$$y^{(i)} = L_f^ih(x) + L_gL_f^{i-1}h(x)u \tag{A.120}$$

until for some integer r, one arrives at

$$L_gL_f^{i-1}h(x){\neq}0 \tag{A.121}$$

for some $x=x_0$ in Ω_x. If the above relation indeed holds, then one can compute the stabilizing control input that is finally applied to the system

$$u=\frac{1}{L_gL_f^{i-1}h(x)}(-L_f^rh+v) \tag{A.122}$$

By applying the control input of Eq. (A.122) to the equivalent description of the system's dynamics, one gets

$$y^{(r)}=L_f^rh(x)+L_gL_f^{r-1}h(x)u \tag{A.123}$$

one arrives at the input–output linearized description of the system

$$y^{(r)}=v \tag{A.124}$$

The number r of the differentiations required for the input u to appear is the relative-degree of the system. One has $r{\leq}n$, where n is the system's order. If $r=n$ then input–output linearization becomes equivalent to the input–state linearization of the system.

A.2.2.2 Input–output linearization and normal forms

When the relative degree r is defined and $r<n$, the nonlinear system of Eq. (A.101) can be transformed to the normal form using the new state vector $[y,\dot{y},\cdots,y^{(r-1)}]$. Let

$$\mu=[\mu_1,\mu_2,\cdots,\mu_r]^T=[y,\dot{y},\cdots,y^{(r-1)}]^T \tag{A.125}$$

In a neighborhood Ω of a point x_0, the normal form of the system can be written as

$$\begin{pmatrix}\dot{\mu}_1\\ \dot{\mu}_2\\ \dot{\mu}_3\\ \cdots\\ \cdots\\ \dot{\mu}_{r-1}\\ \dot{\mu}_r\end{pmatrix}=\begin{pmatrix}0&1&0&\cdots&\cdots&0&0\\ 0&0&1&\cdots&\cdots&0&0\\ 0&0&0&\cdots&\cdots&0&0\\ \cdots&\cdots&\cdots&\cdots&\cdots&\cdots&\cdots\\ \cdots&\cdots&\cdots&\cdots&\cdots&\cdots&\cdots\\ 0&0&0&\cdots&\cdots&0&1\\ 0&0&0&\cdots&\cdots&0&0\end{pmatrix}\begin{pmatrix}\mu_1\\ \mu_2\\ \mu_3\\ \cdots\\ \cdots\\ \mu_{r-1}\\ \mu_r\end{pmatrix}+\begin{pmatrix}0\\ 0\\ 0\\ \cdots\\ \cdots\\ 0\\ 1\end{pmatrix}(\alpha(\mu,\psi)+(\mu,\psi)u) \tag{A.126}$$

with internal dynamics defined by

$$\dot{\psi}=w(\mu,\psi) \tag{A.127}$$

while the output function is defined by

$$y=\mu_1 \tag{A.128}$$

Variables μ_i and ψ_j are referred to as *normal coordinates* or *normal states* in Ω (or at x_0). It is noted that the internal dynamics of the system do not depend on the control input u.

To demonstrate that the nonlinear system of Eq. (A.101) can indeed be transformed into the normal form of Eq. (A.126), it has to be shown not only that a suitable transformation exists. Equivalently, a diffeomorphism has to be defined

$$\phi(x)=[\mu_1,\cdots,\mu_r,\psi_1,\cdots,\psi_{n-r}]^T \tag{A.129}$$

such that Eq. (A.126) and Eq. (A.127) are verified. To show that a transformation ϕ is a diffeomorphism it suffices to show that its Jacobian is invertible, i.e., that the gradients $\nabla\mu_i$ and $\nabla\psi_j$ are linearly independent.

First it will be shown that the gradients $\nabla\mu_i$ are linearly independent which means that the components of μ can serve as state variables. This result states that the system's output and its first $r-1$ derivatives can serve as new state

variables. Next, it will be shown that other $n-r$ vector fields ψ_j can be found to complete the new state vector, which means that the gradients $\nabla\mu_i$ and $\nabla\psi_j$ are also linearly independent.
The following Lemma holds about the linear independence of the gradients $\nabla\mu_i$ [808]:

Lemma: If the relative degree of the system of Eq. (A.101) is r in the region Ω, then the gradients $\nabla\mu_1, \nabla\mu_2, \cdots, \nabla\mu_r$ are also linearly independent in Ω.

Proof: It is first noted, that in terms of μ Eq. (A.121) can be written as

$$\begin{aligned} &\nabla\mu_i g = 0 \quad 1 \leq i < r \\ &\nabla\mu_r g \neq 0 \end{aligned} \tag{A.130}$$

Next, the proof is provided for the case that $r=3$, however it can be generalized to higher values of r. It is assumed that there exists smooth functions $\alpha_i(x)$, such that everywhere in Ω

$$a_1\nabla\mu_1 + a_2\nabla\mu_2 + a_3\nabla\mu_3 = 0 \tag{A.131}$$

Multiplying Eq. (A.131) by g yields

$$[a_1\nabla\mu_1 + a_2\nabla\mu_2 + a_3\nabla\mu_3]g = 0 \tag{A.132}$$

which from Eq. (A.131) implies that $\alpha_3 = 0$ (everywhere in Ω). Replacing $a_3 = 0$ in Eq. (A.132) gives

$$[a_1\nabla\mu_1 + a_2\nabla\mu_2] = 0 \tag{A.133}$$

Multiplying Eq. (A.133) by the Lie bracket $ad_f g$, and using Jacobi's identity, as well as the definition of the relative degree, one has

$$\begin{aligned} 0 &= a_1 L_{ad_f g}\mu_1 + a_2 L_{ad_f g}\mu_2 = \\ &= \alpha_1[L_f L_g h - L_g L_f h] + \alpha_2[L_f L_g - L_g L_f]L_f h = \\ &\quad -a_2 L_g L_f^2 h \end{aligned} \tag{A.134}$$

which from the definition of the relative degree implies that $a_2 = 0$, everywhere in Ω. Replacing $a_2 = 0$ in Eq. (A.133) implies that

$$a_1\nabla\mu_1 = 0 \tag{A.135}$$

Multiplying Eq. (A.135) by $ad_f^2 g$ and using the definition of the relative degree, as well as the following result from Jacobi's identity

$$\begin{aligned} L_{ad_f^2 g}h &= L_{ad_f ad_f g}h = L_f L_{ad_f g}h - L_{ad_f g}L_f h = \\ &= L_f[L_f L_g h - L_g L_f h] - [L_f L_g - L_g L_f]L_f h = \\ &= L_f^2 L_g h - 2L_f L_g L_f h + L_g L_f^2 h \end{aligned} \tag{A.136}$$

one gets

$$0 = a_1 L_{ad_f^2 g}\mu_1 = a_1[L_f^2 L_g h - 2L_f L_g L_f h + L_g L_f^2 h] = a_1 L_g L_f^2 h \tag{A.137}$$

which again from the definition of the relative degree implies that $a_1 = 0$, everywhere in Ω. This finally confirms that the gradients $\nabla\mu_i$ are linearly independent. This completes the Lemma's proof ⋄.

Next, it is shown that there exist $n-r$ more functions ψ_j to complete the coordinates transformation. The proof about this is as follows:

It is needed to show that the gradient vectors $\nabla\psi_j$ can be found such that the $\nabla\mu_i$ and $\nabla\psi_j$ are all linearly independent. At x_0 Eq. (A.130) indicates that the first $r-1$ vectors $\nabla\mu_i$, $(i = 1, 2, \cdots, r-1)$ which have been shown to be linearly independent, are all within an hyperplane orthogonal to g. Since the dimension of such hyperplane is $n-1$, one can find

$$n - r = (n-1) - (r-1) \tag{A.138}$$

vectors in that hyperplane that are linearly independent of the $\nabla\mu_i$ and linearly independent of each other. These vectors can be called $\nabla\psi_j$ $(j=1,2,\cdots,n-r)$ and by definition they verify

$$\nabla\psi_j g = 0 \quad 1 \leq j \leq n-r \tag{A.139}$$

Through Frobenius theorem, it can be also assured that there exist $n-1$ linearly independent gradient functions ∇h_k $k=1,\cdots,n-1$ which satisfy $\nabla h_k g = 0$. Furthermore, from Eq. (A.130), it is assured that μ_r is not in the hyperplane orthogonal to g. Thus, the gradients $\nabla\mu_i$ $(i=1,\cdots,r)$ and $\nabla\psi_j$ $(j=1,\cdots,n-r)$ are all linearly independent. Consequently, the Jacobian matrix of the transformation given in Eq. (A.129) is invertible.

By continuity, the Jacobian remains invertible in a neighborhood Ω_1 of x_0. Redefining Ω as the intersection of Ω_1 and Ω, the transformation ϕ defines a diffeomorphism in Ω. Hence, in Ω this transformation is a state transformation which brings the nonlinear system in the form of Eq. (A.126), with

$$\begin{aligned} \alpha(\mu,\psi) &= L_f^r h(x) = L_f^r h[\psi^{-1}(\mu,\psi)] \\ \beta(\mu,\psi) &= L_g L_f^{r-1} h(x) = L_g L_f^{r-1} h[\psi^{-1}(\mu,\psi)] \end{aligned} \tag{A.140}$$

while in Eq. (A.127) the input u does not appear, since the ψ_j verify

$$L_g\psi_j(x) = 0 \ \forall x \in \Omega \tag{A.141}$$

From a practical point of view, to find the vector field ψ that completes the transformation into a normal form requires the non-trivial step of solving the set of partial differential equations given in Eq. (A.139).

A.2.3 Input–state linearization

A different approach for obtaining a linearized equivalent description of a nonlinear dynamical system is input–state linearization.

Definition: A nonlinear system

$$\dot{x} = f(x) + g(x)u \tag{A.142}$$

where $f: D_r \rightarrow R^n$ and $G: D_r \rightarrow R^{n\times p}$ are sufficiently smooth on a domain $D_x \subset R^n$, is said to be input–state linearizable if there exists a diffeomorphism $T: D_x \Rightarrow R^n$ such that $D_x = T(D_x)$ contains the origin, and the change of variables $z = T(x)$ transforms the system of Eq. (A.142) into the form

$$\dot{z} = Az + B\beta^{-1}(x)[u - \alpha(x)] \tag{A.143}$$

The following, single-input system is considered

$$\dot{x} = f(x) + g(x)u \tag{A.144}$$

where $f: D \rightarrow R^n$ and $g: D \rightarrow R^n$ are smooth functions on a domain $D \subset R^n$. The following Theorem gives necessary and sufficient conditions for the system of Eq. (A.144) to be input-to-state linearizable near a point $x_0 \in D$ [439].

Theorem: The system of Eq. (A.144) is input–state linearizable if and only if there exists a domain $D_x \subset D$ such that

1. the matrix $G(x) = [g(x), ad_f g(x), \cdots, ad_f^{n-1} g(x)]$ has rank n for all $x \in D_x$
2. the distribution $D = span\{g, ad_f g, \cdots, ad_f^{n-2} g\}$ is involutive in D_x

Proof: The system of Eq. (A.144) is input-to-state linearizable if and only if there is a real-valued function $h(x)$ such that the system

$$\begin{aligned} \dot{x} &= f(x) + g(x)u \\ y &= h(x) \end{aligned} \tag{A.145}$$

has relative degree n in D_x. That is $h(x)$ satisfies

$$L_gL_f^ih(x) = 0 \text{ for } 0 \leq i \leq n-2 \text{ and } L_gL_f^{n-1}h(x) \neq 0 \ \forall x \in D_x \tag{A.146}$$

The "only if" part follows from the definition of input-to-state linearization while the "if part" follows from applying the change of variables which is given by $z = T(x) = [h(x), L_fh(x), \cdots, L_f^{n-1}h(x)]^T$. The previously given Lemma 2 confirms that the Jacobian matrix $[\frac{\partial T}{\partial x}](x)$ is non-singular for all $x \in D_x$. Hence, for each $x_0 \in D_x$, there is a neighborhood N of x_0 such that $T(x)$ is restricted to N, is a diffeomorphism. For the theorem's proof it is also necessary to show that the existence of $h(x)$ satisfying Eq. (A.146) is equivalent to the previous stated conditions 1 and 2.

Necessity: Suppose there is $h(x)$ satisfying Eq. (A.146) shows that $rank(G) = n$. Then D is non-singular and has dimension $n-1$. From Eq. (A.104), with $k = 0$ and $r = n$, one has

$$L_gh(x) = L_{ad_fg}h(x) = \cdots = L_{ad_f^{(n-2)}g}h(x) = 0 \tag{A.147}$$

which can be also written as

$$dh(x)[g(x), ad_fg(x), \cdots, ad_f^{n-2}g(x)] = 0 \tag{A.148}$$

The above equation implies that $D^{\perp} = span(dh)$. Hence, D is completely integrable and it follows from Frobenius theorem that D is involutive.

Sufficiency: It is assumed that conditions 1 and 2 hold. Then D is non-singular and has dimension $n-1$. By Frobenius' theorem, there exists $h(x)$ satisfying

$$L_gh(x) = L_{ad_fg}h(x) = \cdots = L_{ad_f^{n-2}g}h(x) = 0 \tag{A.149}$$

Using the Jacobi identity, it can be verified that

$$L_gh(x) = L_gL_fh(x) = \cdots = L_gL_f^{n-2}h(x) = 0 \tag{A.150}$$

Furthermore,

$$dh(x)G(x) = dh(x)[g(x), ad_fg(x), \cdots, ad_f^{n-1}g(x)] = [0, \cdots, 0, L_{ad_f^{n-1}g}h(x)] \tag{A.151}$$

Since $rank(G) = n$ and $dh(x) \neq 0$, it must be true that $L_{ad_f^{n-1}g}h(x) \neq 0$. Using the Jacobi identity, it can be verified that $L_gL_f^{n-1}h(x) \neq 0$, which completes the proof of the theorem.

A.2.3.1 Relation between input–state and input–output linearization

The formal definition of input-to-state-linearization is as follows:

Using the transformed state-space description of the system of Eq. (A.143) and by defining the new control input $v = \beta^{-1}(x)[u - \alpha(x)]$, one obtains

$$\dot{z} = Az + Bv \tag{A.152}$$

where

$$A = \begin{pmatrix} 0 & 1 & 0 & \cdots & 0 \\ 0 & 0 & 1 & \cdots & 0 \\ \cdots & \cdots & \cdots & \cdots & \cdots \\ \cdots & \cdots & \cdots & \cdots & \cdots \\ 0 & 0 & 0 & \cdots & 1 \\ 0 & 0 & 0 & \cdots & 0 \end{pmatrix} \quad B = \begin{pmatrix} 0 \\ 0 \\ \cdots \\ \cdots \\ 0 \\ 1 \end{pmatrix} \tag{A.153}$$

The new state vector z is called linearizing state and the control input of Eq. (A.152) is the linearizing control law. The transformed system dynamics given in Eq. (A.152) and Eq. (A.153) is in the so-called linear canonical form, which is both controllable and observable.

The canonical form of Eq. (A.152) stands for a special case of the input–output linearization, where the output function leads to a relative degree n. This means that if a system is input–output linearizable with relative degree n, it is also input–state linearizable. On the other hand, is a system is input–state linearizable with a first new state z_1 representing the output, the system is also input–output linearizable with relative degree n.

Consequently, the relation between input–state linearization and input–output linearization can be summarized as follows:

Lemma: An n-th order nonlinear system is input–state linearizable, if and only if, there exists a scalar function $z_1(x)$ such that the system's input–output linearization with $z_1(x)$ as output has relative degree n.

However, the above lemma provides no guidance about how to find the desirable output function $z_1(x)$.

A.2.3.2 Conditions for input–state linearization

It should be specified which are the nonlinear systems of the form of Eq. (A.154)

$$\dot{x} = f_0(x) + \sum_{j=1}^{n-1} g_j(x)u_j \;\; x \in R^n \tag{A.154}$$

which can be subjected to input–state linearization. The above question stands for one of the most important issues in feedback linearization theory. The necessary and sufficient conditions for input–output linearization of a dynamical system have been stated in the theorem given in subsection A.2.3. Next, the proof of this theorem is revisited [808].

Theorem: The nonlinear system of Eq. (A.154) with $f(x)$ and $g(x)$ being smooth vector fields, is input–state linearizable, if and only if, there exists a region Ω such that the following conditions hold:

- the vector fields $[g, ad_f g, \cdots, ad_f^{n-1} g]$ are linearly independent in Ω
- the set $[g, ad_f g, \cdots, ad_f^{n-2} g]$ is involutive in Ω

Prior to proving the theorem some remarks about the aforementioned conditions are given:

The first condition can be interpreted as a controllability condition for the nonlinear system of Eq. (A.145). For linear systems, the vector fields $[g, ad_f g, \cdots, ad_f^{n-1} g]$ become $[b, Ab, \cdots, A^{n-1}b]$ and independence implies that the associated matrix is a full-rank one which is also invertible. It can be also shown that if a system's linear approximations in a closed connected region $\Omega \in R$ are all controllable, then under some mild smoothness assumptions, the system can be driven from any point in Ω to any point in Ω (the inverse does not hold because, a nonlinear system can be controllable whereas its linear approximation may not be).

The involutivity condition is less intuitive. It is trivially satisfied for linear systems (which have constant vector fields) but does not hold in general in the nonlinear case.

Next, to proceed to the Theorem's proof, the following Lemma is analyzed:

Lemma: Let $z(x)$ be a smooth function in a region Ω. Then, in Ω, the set of equations

$$L_g z = L_g L_f z = \cdots = L_g L_f^k z = 0 \tag{A.155}$$

is equivalent to

$$L_g z = L_g L_{ad_f g} z = \cdots = L_g L_{ad_f^k g} z = 0 \tag{A.156}$$

for any positive integer k.

Proof: The Lemma's proof is based on induction. When $k=0$ the result is obvious. When $k=1$, then from Jacobi's identity, one obtains

$$L_{ad_f g}z = L_f L_g z - L_g L_f z = 0 - 0 = 0 \tag{A.157}$$

When $k=2$, the using again Jacobi's identity, then one obtains

$$L_{ad_f^2 g}z = L_f^2 L_g z - 2L_f L_g L_f z + L_g L_f^2 z = 0 - 0 + 0 = 0 \tag{A.158}$$

Repeating the procedure, it can be shown by induction that Eq. (A.157) implies Eq. (A.158) for any k. Finally, using Jacobi's identity in an inverse manner it can be shown that from Eq. (A.158), one can arrive at Eq. (A.157).

Proof of Theorem: First the necessity condition is proven. It is assumed that there exists a state transformation $z=z(x)$ and an input transformation $u=\alpha(x)+\beta(x)v$, such that z and v satisfy Eq. (A.152). The first row of Eq. (A.152) can be rewritten as

$$\dot{z}_1 = \tfrac{\partial z_1}{\partial x}(f+gu) = z_2 \tag{A.159}$$

Continuing, in a similar manner with the rest of the elements of the state vector, one has the following differential equations:

$$\tfrac{\partial z_1}{\partial x}f + \tfrac{\partial z_1}{\partial x}gu = z_2 \tag{A.160}$$

$$\tfrac{\partial z_2}{\partial x}f + \tfrac{\partial z_2}{\partial x}gu = z_3 \tag{A.161}$$

$$\cdots\ \cdots\ \cdots$$

$$\tfrac{\partial z_n}{\partial x}f + \tfrac{\partial z_n}{\partial x}gu = v \tag{A.162}$$

Since $z_1,\cdots,z_n$ are independent of u, while v is not, from the above equations, it can be concluded that

$$L_g z_1 = L_g z_2 = \cdots = L_g z_{n-1} = 0 \text{ and } L_g z_n \neq 0 \tag{A.163}$$

$$L_f z_i = z_{i+1} \quad i=1,2,\cdots,n-1 \tag{A.164}$$

The above equations on z_i can be written concisely into one single equation concerning only z_1. Using the Lemma proven above and Eq. (A.163), one gets

$$\nabla z_1 ad_f^k g = 0 \quad k=0,1,2,\cdots,n-2 \tag{A.165}$$

Moreover, by continuing as in the proof of the previous Lemma it can be shown that

$$\nabla z_1 ad_f^{n-1} g = (-1)^{n-1} L_g z_n \tag{A.166}$$

which implies that

$$\nabla z_1 ad_f^{n-1} g \neq 0 \tag{A.167}$$

From Eq. (A.165), Eq. (A.166) and Eq. (A.167), one can conclude that the state vector fields $[g, ad_f g, \cdots, ad_f^{n-1} g]$ are linearly independent. Indeed, if for some number $i (i \leq n-1)$ there existed scalar functions $\alpha_1(x),\cdots,\alpha_{i-1}(x)$, such that

$$ad_f^i g = \textstyle\sum_{k=0}^{i-1} \alpha_k ad_f^k g \tag{A.168}$$

one would also have

$$ad_f^{n-1} g = \textstyle\sum_{k=0}^{n-2} \alpha_k ad_f^k g \tag{A.169}$$

Eq. (A.169) together with Eq. (A.166) would imply that

$$\nabla z_1 ad_f^{n-1} g = \textstyle\sum_{k=0}^{n-2} \alpha_k ad_f^k g = 0 \tag{A.170}$$

which is in contradiction to Eq. (A.167).

From Eqs. (A.165), (A.166), and (A.167), it is also possible to conclude that the vector fields are involutive. This follows from the existence of a scalar function z_1 satisfying the $n-1$ partial differential equations of Eq. (A.165) and from the necessity part of the Frobenius theorem. This completes the necessity part of the present Theorem proof.

Next, it will be proven that the two conditions in the present Theorem are also sufficient for the state linearizability of the system of Eq. (A.145), which means that it is possible to find a state transformation and an input transformation such that Eq. (A.152) is satisfied. The stages of the proof are as follows: Since the involutivity condition is satisfied, from the Frobenius theorem there exists a non-zero scalar function $z_1(x)$ satisfying

$$L_g z_1 = L_{ad_f g} z_1 = \cdots = L_{ad_f^{n-2} g} z_1 = 0 \tag{A.171}$$

According to the previous Lemma, the above equation can be written as

$$L_g z_1 = L_g L_f z_1 = \cdots = L_g L_f^{n-2} z_1 = 0 \tag{A.172}$$

Consequently, if one uses as new state variables the elements of the vector $z = [z_1, L_f z_1, \cdots, L_f^{n-1} z_1]^T$, the first $n-1$ state equations give

$$\dot{z}_k = z_{k+1} \quad k = 1, \cdots, n-1 \tag{A.173}$$

while the last equation becomes

$$\dot{z}_n = L_f^n z_1 + L_g L_f^{n-1} z_1 u \tag{A.174}$$

A next question is if gain function $L_g L_f^{n-1} z_1$ can be equal to zero. Since the vector fields $[g, ad_f g, \cdots, ad_f^{n-1} g]$ are linearly independent in Ω, and using Eq. (A.172), one obtains in accordance to the proof of the previous Lemma that

$$L_g L_f^{n-1} z_1 = (-1)^{n-1} L_{ad_f^{n-1} g} z_1 \tag{A.175}$$

and consequently it should hold

$$L_{ad_f^{n-1} g} z_1(x) \neq 0 \quad \forall x \in \Omega \tag{A.176}$$

Otherwise, the non-zero vector ∇z_1 should satisfy

$$\nabla z_1 [g, ad_f g, \cdots, ad_f^{n-1} g] = 0 \tag{A.177}$$

which implies that ∇z_1 would be orthogonal to n linearly independent vectors, which is a contradiction. For the system's dynamics described in Eq. (A.174), and by denoting as new control input $v = L_f^n z_1 + L_g L_f^{n-1} z_1 u$, the stabilizing feedback control law that is finally applied to the system is

$$u = (L_g L_f^{n-1} z_1)^{-1} (-L_f^n z_1 + v) \tag{A.178}$$

A.2.4 Stages in the implementation of input–state linearization

The stages for implementing input–state linearization of dynamical systems of the form of Eq. (A.145) and for designing the associated feedback control law, are as follows:

(i) the vector fields $[g, ad_f g, \cdots, ad_f^{n-1} g]$ for the given system is constructed. (ii) it is checked if the controllability and involutivity conditions are satisfied.

If both conditions are satisfied, then one should find the first state z_1 (which is the output function that leads to input–output linearization of relative degree n) from Eq. (A.171) and Eq. (A.172), that is

$$\begin{aligned} &\nabla z_1 ad_f^i g = 0 \quad i = 0, 1, \cdots, n-2 \\ &\nabla z_1 ad_f^{n-1} g \neq 0 \end{aligned} \tag{A.179}$$

Next, the state transformation $z(x) = [z_1, L_f z_1, \cdots, L_f^{n-1} z_1]^T$ and the input transformation $u = \alpha(x) + \beta(x)v$ are computed where

$$\begin{aligned} \alpha(x) &= -\frac{L_f^n z_1}{L_g L_f^{n-1} z_1} \\ \beta(x) &= \frac{1}{L_g L_f^{n-1} z_1} \end{aligned} \tag{A.180}$$

The transformed control input that makes the linearized system $z_1^{(n)} = v$ be a Hurwitz stable one, i.e., to have all its poles in the left complex semi-plane is given by

$$\begin{aligned} v = z_{1,d}^{(n)} - K_1(z_1^{(n-1)} - z_{1,d}^{(n-1)}) - K_2(z_1^{(n-2)} - z_{1,d}^{(n-2)}) - \cdots - \\ -K_{n-1}(\dot{z}_1 - \dot{z}_{1,d}) - K_n(z_1 - z_{1,d}) \end{aligned} \tag{A.181}$$

A.2.5 *Input–output and input–state linearization for MIMO systems*

The previously analyzed approaches to input–output and input–state linearization which were formulated for SISO systems can be also extended to MIMO systems. Input–output linearization of MIMO systems is obtained similarly to the SISO case, by differentiating the selected outputs y_i until the inputs re-appear. It is assumed that r_i stands for the smallest order of output derivations, such that at least one of the inputs appears in $y_i^{(r_i)}$, that is

$$y_i^{(r_i)} = L_f^{r_i} h_i + \sum_{j=1}^m L_{g_j} L_f^{r_i-1} h_i u_j \tag{A.182}$$

with $L_{g_j} L_f^{r_i-1} h_i(x) \neq 0$, for at least one j, in a neighborhood Ω of point x_0. Following this procedure of successive derivations for all chosen outputs of the system, one obtains the input–output linearized description of the system's dynamics in matrix form

$$\begin{pmatrix} y_1^{r_1} \\ \cdots \\ \cdots \\ y_m^{r_m} \end{pmatrix} = \begin{pmatrix} L_f^{r_1} h_1(x) \\ \cdots \\ \cdots \\ L_f^{r_m} h_m(x) \end{pmatrix} + \begin{pmatrix} L_{g_1^1} L_f^{r_1-1} h_1(x) & \cdots & L_{g_1^m} L_f^{r_m-1} h_m(x) \\ \cdots & \cdots & \cdots \\ \cdots & \cdots & \cdots \\ L_{g_1^m} L_f^{r_1-1} h_1(x) & \cdots & L_{g_m^m} L_f^{r_m-1} h_m(x) \end{pmatrix} \begin{pmatrix} u_1 \\ \cdots \\ \cdots \\ u_m \end{pmatrix} \tag{A.183}$$

In the above description the linearized dynamical system is a square one, that is the number of its outputs is equal to the number of control inputs. Therefore, through the followed procedure, one achieves at the same time linearization and decoupling of the system's dynamics. Thus, finally the linearized system is written in the form

$$\begin{aligned} \tilde{Y} = \tilde{F} + \tilde{G}u \Rightarrow \\ \tilde{Y} = v \end{aligned} \tag{A.184}$$

and the feedback control law that makes the elements of the output vector $\tilde{Y}$ converge to the desirable setpoints is $u = \tilde{G}^{-1}(v - \tilde{F})$. The above system is said to have relative degree $(r_1, \cdots, r_m)$ and the scalar parameter $r = r_1 + \cdots + r_m$ is the total relative degree.

A.2.6 *Dynamic extension*

In certain cases, and despite successive derivations of the system's outputs y_i, $i = 1, 2, \cdots, m$ the control inputs may not re-appear and thus matrix $\tilde{G}$ given in Eq. (A.184) may be singular. To handle this case, the so-called dynamic extension or dynamic feedback linearization is performed. This means that the system's state vector is extended by considering as additional state vector elements some of the control inputs and their derivatives. For the extended state-space description it becomes possible to perform input–output linearization and to obtain an equivalent linearized and decoupled form of it. This control approach has as a result to generate a control input that contains integral terms of the initial state vector of the system.

Without loss of generality, one can consider dynamic extension in the case of a system with two inputs and two outputs and that matrix $\tilde{G}$ given in Eq. (A.184) is of rank 1. Assuming that one column of matrix $\tilde{G}$ becomes equal to zero, we get a description of the system (after successive differentiations) in form

$$\begin{pmatrix} y_1^{(r_1)} \\ y_2^{(r_2)} \end{pmatrix} = \begin{pmatrix} L_f^{r_1} h_1 \\ L_f^{r_2} h_2 \end{pmatrix} + \tilde{g}_1 u_1 \tag{A.185}$$

By considering the control input as additional state variable, and by differentiating Eq. (A.185) once more with respect to time while also using the initial system's dynamics, one arrives at the description

$$\begin{pmatrix} y_1^{(r_1+1)} \\ y_2^{(r_2+1)} \end{pmatrix} = \begin{pmatrix} b_1(x,u_1) \\ b_2(x,u_1) \end{pmatrix} + \begin{pmatrix} \tilde{e}_{1,1}(x,u_1) & \tilde{e}_{1,2}(x,u_1) \\ \tilde{e}_{2,1}(x,u_1) & \tilde{e}_{2,2}(x,u_2) \end{pmatrix} \begin{pmatrix} \dot{u}_1 \\ u_2 \end{pmatrix} \tag{A.186}$$

which is also written as

$$\begin{pmatrix} y_1^{(r_1+1)} \\ y_2^{(r_2+1)} \end{pmatrix} = \tilde{B}(x,u_1) + \tilde{E}(x,u_1) \begin{pmatrix} \dot{u}_1 \\ u_2 \end{pmatrix} = \tilde{v} \tag{A.187}$$

The control input $\tilde{v}$ is chosen to place the poles of the input–output linearized system to the left complex semiplane. The control input $[\dot{u}_1, u_2]^T$ is given by

$$\begin{pmatrix} \dot{u}_1 \\ u_2 \end{pmatrix} = \tilde{E}^{-1}(\tilde{v} - \tilde{B}(x,u_1)) \tag{A.188}$$

After computing $\dot{u}_1$ from the above relation, the control input that is finally applied to the dynamical system is given by $[\int \dot{u}_1, u_2]^T$. The term $\int \dot{u}_1$ implies that the integral of the initial state vector of the system is used for implementing the feedback control.

Appendix B

Differential flatness theory and flatness-based control methods

Abstract

Appendix B analyzes differential flatness theory and flatness-based control methods. In the first section of the Appendix, global linearization-based control with the use of differential flatness theory is analyzed, comprising issues such as differential flatness for finite dimensional systems, equivalence and differential flatness, feedback control and equivalence flatness-based control, and state feedback for systems with model uncertainties and classification of types of differentially flat systems. In the second section of this chapter, the concept of flatness-based control of nonlinear dynamical systems in cascading loops is analyzed. This comprises issues such as decomposition of the state-space model into cascading differentially flat subsystems, tracking error dynamics for flatness-based control in successive loops, and comparison to backstepping control for nonlinear systems.

B.1 Global linearization-based control with the use of differential flatness theory

B.1.1 The background of differential flatness theory

Flatness-based control is widely used in this monograph for electric power systems and for electric traction and propulsion systems. Differential flatness theory and flatness-based control were introduced in the late 1980s by Michel Fliess and co-researchers, and since then they keep on being developed and on providing efficient solutions to advanced control and state estimation problems [251].

The definition of a differentially flat system is as follows: a system $\dot{x}=f(x,u)$ with state vector $x \in R^n$, input vector $u \in R^m$ where f is a smooth vector field, is differentially flat if there exists a vector $y \in R^m$ in the form

$$y = h(x, u, \dot{u}, \cdots, u^{(r)}) \tag{B.1}$$

such that

$$\begin{aligned} x &= \phi(y, \dot{y}, \cdots, y^{(q-1)}) \\ u &= \alpha(y, \dot{y}, \cdots, y^{(q)}) \end{aligned} \tag{B.2}$$

where h, ϕ, and α are smooth functions. This means that the new system's description is given by the m algebraic variables y_i, $i = 1, 2, \cdots, m$. The definition of the flat output given above is $y = h(x, u, \dot{u}, \cdots, u^{(r)})$. If the flat output is exclusively a function of the state vector x then the system is a 0-flat one. However, there may be need to express the flat output as a function of not only the state vector x but also as a function of the control u and of its derivatives. For instance, the latter holds in the case of dynamic feedback linearization and in the application of the so-called dynamic extension. This means that the state vector of the system is extended by considering as additional state variables the control inputs and its derivatives.

Eq. (B.2) shows that the state vector of the differentially flat system and its control inputs can be expressed as function of the flat output and of the flat output's derivatives. The basic question that arises in the study of differential flatness is whether, given the differential equations that describe the nonlinear system dynamics $\dot{x}=f(x,u)$, there exists a function $y = h()$ given by $y = h(x, u, \dot{u}, \cdots, u^{(r)})$, such that the state vector of the system x and the control input u can be expressed as functions of y and of its derivatives, as in Eq. (B.2).

This problem was initially set by D. Hilbert in 1912, for the second order Monge's equation

$$\frac{d^2y}{dx^2} = F(x, y, z, \frac{dy}{dx}, \frac{dz}{dx}) \tag{B.3}$$

which is an under-determined differential-equation system (there is one differential equation with respect to x, connecting two functions of x, namely $y(x)$ and $z(x)$). Hilbert speaks about a solution that is not based on the computation of integrals. He has shown that this problem is related to the classification of under-determined differential-equation systems through a group of transformations called "invertible without integral."

Soon afterward, Èlie Cartan reworks on the question set by Hilbert and shows how the calculations on the Pfaff systems permit to classify second-order Monge equations which admit a solution without integral. He deduced an explicit description for all curves in the Euclidean space R^3 for which the curvature ratio and the torsion are constant. The results of Èlie Cartan are complete for Pfaff systems of codimension 2, which are under-determined differential-equation systems of co-dimension 1, i.e., systems in which only one arbitrary function intervenes (otherwise stated, systems having only one control input). Èlie Cartan also suggested the notion of absolute equivalence; however, he did not define it with precision. He also noted that for Pfaff systems of codimension equal or greater than 2 (which means systems having at least two control inputs), the solution to the above problem becomes extremely complicated.

Next, a formal definition will be given about the analogy in terms of the control theory, of what was described above as under-determined differential-equation systems which can be solved without integration.

B.1.2 Differential flatness for finite dimensional systems

As noted in Eq. (B.1) and Eq. (B.2), differential flatness is a structural property of a class of nonlinear dynamical systems, denoting that all system variables (such as state vector elements and control inputs) can be written in terms of a set of specific variables (the so-called flat outputs) and their derivatives. The following nonlinear system is considered:

$$\dot{x}(t) = f(x(t), u(t)) \tag{B.4}$$

The time variable is $t \in R$, the state vector is $x(t) \in R^n$ with initial conditions $x(0) = x_0$, and the input variable is $u(t) \in R^m$. Next, the main principles of differentially flat systems are given [755,898]:

The finite dimensional system of Eq. (B.4) can be written in the general form of an ordinary differential equation (ODE), i.e., $S_i(w, \dot{w}, \ddot{w}, \cdots, w^{(i)}), \quad i = 1, 2, \cdots, q$. The entity w is a generic notation for the system variables (these variables are for instance the elements of the system's state vector $x(t)$ and the elements of the control input $u(t)$) while $w^{(i)}$, $i = 1, 2, \cdots, q$ are the associated derivatives. Such a system is said to be differentially flat if there is a collection of m functions $y = (y_1, \cdots, y_m)$ of the system variables and of their time-derivatives, i.e., $y_i = \phi(w, \dot{w}, \ddot{w}, \cdots, w^{(\alpha_i)})$, $i = 1, \cdots, m$ satisfying the following two conditions [252,573,623,624,707]:

1. There does not exist any differential relation of the form $R(y, \dot{y}, \cdots, y^{(\beta)}) = 0$ which implies that the derivatives of the flat output are not coupled in the sense of an ODE, or equivalently it can be said that the flat output is differentially independent.

2. All system variables (i.e., the elements of the system's state vector w and the control input) can be expressed using only the flat output y and its time derivatives $w_i = \psi_i(y, \dot{y}, \cdots, y^{(\gamma_i)})$, $i = 1, \cdots, s$. An equivalent definition of differentially flat systems is as follows:

Definition: The system $\dot{x} = f(x, u)$, $x \in R^n$, $u \in R^m$ is differentially flat if there exist relations

$$\begin{aligned} h &: R^n \times (R^m)^{r+1} \rightarrow R^m, \\ \phi &: (R^m)^r \rightarrow R^n \text{ and} \\ \psi &: (R^m)^{r+1} \rightarrow R^m \end{aligned} \tag{B.5}$$

such that

$$
\begin{aligned}
&y = h(x, u, \dot{u}, \cdots, u^{(r)}),\\
&x = \phi(y, \dot{y}, \cdots, y^{(r-1)}), \text{ and}\\
&u = \psi(y, \dot{y}, \cdots, y^{(r-1)}, y^{(r)}).
\end{aligned}
\tag{B.6}
$$

This means that all system dynamics can be expressed as a function of the flat output and its derivatives; therefore, the state vector and the control input can be written as

$$
\begin{aligned}
&x(t) = \phi(y(t), \dot{y}(t), \cdots, y^{(r-1)}(t)), \text{and}\\
&u(t) = \psi(y(t), \dot{y}(t), \cdots, y^{(r)}(t))
\end{aligned}
\tag{B.7}
$$

B.1.3 Equivalence and differential flatness

B.1.3.1 Representation of system dynamics as vector fields of infinite dimension

A basic property of differentially flat systems is that through a change of variables (diffeomorphism), they can be transformed to an equivalent description, which is the linear canonical (Brunovsky) form. This is analyzed next and stands for the Lie-Backlünd isomorphism to equivalence and differential flatness [251].

First, a dynamical system of the form

$$\dot{x} = f(x) \; x \in X \subset R^n \tag{B.8}$$

is considered. This is described by the couple (X, f), where X is defined in R^n and f is a vector field on X. A *trajectory* is considered to be the function $t \rightarrow x(t)$ such that $\dot{x}(t) = f(x(t)) \; \forall t \geq 0$. One can consider also the output mapping $x \rightarrow h(x)$ for which holds

$$\tfrac{d}{dt} h(x(t)) = \tfrac{dh}{dt}(x(t)) \cdot \dot{x}(t) = \tfrac{dh}{dt}(x(t)) \cdot f(x(t)) \; \forall \, t \geq 0 \tag{B.9}$$

Eq. (B.9) gives the total derivative, i.e., $\frac{dh}{dt}(x(t)) \cdot f(x(t))$ is also called *time derivative* of the function h, and is noted by $\dot{h}$.

Next, the notions of the total derivative and of the time derivative of function h can be generalized to the case of a nonlinear system with control input:

$$\dot{x} = f(x, u) \tag{B.10}$$

where $X \times U \in R^n \times R^m$. Here, f is no longer a vector field but is an infinite collection of vector fields parameterized by u. Actually, for every u, the function $x \rightarrow f_u(x) = f(x, u)$ is a vector field on X. One can also consider the case of dynamic feedback in which the system's state vector is extended by defining as state vector element the control input (and its derivatives), while the equations describing the system's dynamics are extended as follows:

$$
\begin{aligned}
\dot{x} &= f(x, u)\\
\dot{u} &= v
\end{aligned}
\tag{B.11}
$$

In the latter case, in place of the state-space X, one has the state-space $X \times U$. Assume now that the solution of $\dot{x} = f(x, u)$ is the function $t \rightarrow (x(t, u(t)))$ which has values in $X \times U$, such that $\dot{x}(t) = f(x(t), u(t)) \; \forall \, t \geq 0$. Moreover, one can consider the infinite function

$$t \rightarrow \xi(t) = (x(t), u(t), \dot{u}(t), \cdots) \tag{B.12}$$

which takes values in $X \times U \times R_m^\infty$, where $R_m^\infty = R^m \times R^m \times \ldots$ represents an infinite product formed by vectors defined in R^m. A point of R_m^∞ is of the form $u^{(1)}, u^{(2)}, \cdots$ with $u^i \in R^m$, using that $\dot{x} = f(x, u)$. This function satisfies also the relation

$$\dot{\xi}(t) = (f(x(t), u(t)), \dot{u}(t), \ddot{u}(t), \cdots) \; \forall \, t \geq 0 \tag{B.13}$$

and this can be interpreted as a trajectory of an infinite vectors field

$$(x,u,u^{(1)},\cdots)\rightarrow F(x,u,u^{(1)},\cdots)=(f(x,u),u^{(1)},u^{(2)},\cdots,\dots) \tag{B.14}$$

According to the above, the dynamical system $\dot{x}=f(x,u)$ is defined by the space $X\times U\times R_m^{\infty}$ and an infinite number of vector fields F on that space. As in the case of an autonomous dynamical system $\dot{x}=f(x)$, for the non-autonomous dynamical system $\dot{x}=f(x,u)$ one can also define the time derivative of the smooth output function $(x,u,u^{(1)},\cdots)\rightarrow h(x,u,u^{(1)},\cdots,u^{(k)})$ which is written as

$$\begin{aligned}&\dot{h}(x,u,u^{(1)},\cdots,u^{(k+1)}):=Dh{\cdot}F=\\&=\tfrac{\partial h}{\partial x}f(x,u)+\tfrac{\partial h}{\partial u}u^{(1)}+\tfrac{\partial h}{\partial u^{(1)}}u^{(2)}+\cdots\end{aligned} \tag{B.15}$$

If $h()$ depends on a finite number of variables then the above sum becomes finite too. For the above type of functions, one can define the so-called Fréchet topology which implies computations on smooth functions which are defined in k copies of R^m, for k being sufficiently large.

After a change of variables, a differentially flat system is written in the canonical Brunovsky form. Thus, it becomes equivalent to the trivial system R_m^{∞}, F_m with coordinates $y, y^{(1)}, y^{(2)}, \cdots$, and the vector fields

$$(y^{(1)},y^{(2)},y^{(3)},\cdots)=F_m(y,y^{(1)},y^{(2)},\cdots) \tag{B.16}$$

which describe systems composed by m chains of integrators (this is the canonical Brunovsky form for linear controllable systems).

B.1.3.2 Equivalence of systems

Two dynamical systems are considered to be equivalent if there exists an invertible relationship which exchanges their trajectories. The definition of equivalent systems is as follows:

Definition: Two systems (M,F) and (N,G) (notation referring to state vector and vector field, respectively) are equivalent in $(p,q)\in(M,N)$, if and only if, there exists an endogenous transformation from a neighborhood of p to a neighborhood of q. The two systems (M,F) and (N,G) are equivalent if the equivalence holds for all pairs of points (p,q) of the space $M\times N$.

Using coordinates, the previous notions are expressed as follows:

Considering the two systems $(X\times U\times R_m^{\infty},F)$ and $(Y\times V\times R_s^{\infty},G)$ describing the initial system dynamics

$$\dot{x}=f(x,u) \text{ with } (x,u)\in X\times U\subset R^n\times R^m \tag{B.17}$$

and the equivalent system dynamics

$$\dot{y}=g(y,v) \text{ with } (y,v)\in Y\times V\subset R^r\times R^s \tag{B.18}$$

The vector fields F, G are defined as

$$\begin{aligned}F(x,u,u^{(1)},\cdots)&=(f(x,u),u^{(1)},u^{(2)},\cdots)\\G(y,v,v^{(1)},\cdots)&=(g(y,v),v^{(1)},v^{(2)},\cdots)\end{aligned} \tag{B.19}$$

If the two systems are equivalent, the endogenous transformation takes the form

$$\Psi(x,u,u^{(1)},\cdots)=(\psi(x,\bar{u}),\beta(x,\bar{u}),\dot{\beta}(x,\bar{u}),\cdots) \tag{B.20}$$

where $\bar{u}$ is the abbreviated notation $\bar{u}=(u,u^{(1)},\cdots,u^{(k)})$. Similarly, one can define the inverse endogenous transformation

$$\Phi(y,v,v^{(1)},\cdots)=(\phi(y,\bar{v}),\alpha(y,\bar{v}),\dot{\alpha}(y,\bar{v}),\cdots) \tag{B.21}$$

Since Φ and Ψ are inverse to each other, one has

$$\begin{aligned}&\psi(\phi(y,\bar{v}),\bar{\alpha}(y,\bar{v}))=y \quad\text{and}\quad \phi(\psi(x,\bar{u}),\bar{\beta}(x,\bar{u}))=x\\&\beta(\phi(y,\bar{v}),\bar{\alpha}(y,\bar{v}))=v \quad\text{and}\quad \alpha(\psi(x,\bar{u}),\bar{\beta}(x,\bar{u}))=u\end{aligned} \tag{B.22}$$

About vector fields F and G which are related to the derivatives of x and y, respectively, one has

$$f(\phi(y,\bar{v}),\alpha(y,\bar{v})) = D\phi(y,\bar{v})\cdot\bar{g}(y,\bar{v}) \tag{B.23}$$

where $\bar{g}$ corresponds to $(g, v^{(1)}, \cdots, v^{(k)})$

$$g(\psi(x,\bar{u}),\beta(x,\bar{u})) = D\psi(x,\bar{u})\bar{f}(x,\bar{u}) \tag{B.24}$$

where $\bar{f}$ corresponds to $(f, u^{(1)}, \cdots, u^{(k)})$. Additionally, it can be stated that if the trajectory of the first system is denoted as

$$t\rightarrow(x(t),u(t)) \tag{B.25}$$

then the trajectory of the second system becomes

$$t\rightarrow(y(t),v(t)) = (\psi(x(t),\bar{u}(t)),\beta(x(t),\bar{u}(t))) \tag{B.26}$$

An important property of the endogenous transformations is that the number of control inputs in the initial and the final description of the system remains the same.

Theorem: If the two systems $(X\times U\times R_m^{\infty}, F)$ and $Y\times V\times R_s^{\infty}, G$ are equivalent, then they will have the same number of control inputs, that is $m = s$.

Proof: Consider the truncation Φ_m of Φ on $X\times U\times(R^m)^{\mu}$

$$\begin{gathered}\Phi_{\mu}:\ X\times U\times(R^{m+k})^{\mu}\rightarrow Y\times V\times(R^s)^{\mu}\\(x,u,u^{(1)},\cdots,u^{(k+\mu)})\rightarrow(\phi,\alpha,\dot{\alpha},\cdots,\alpha^{(\mu)})\end{gathered} \tag{B.27}$$

which means the first $\mu+2$ blocks of the components of Ψ, while μ is a sufficiently large entity. Since Ψ is invertible, Ψ_{mu} is a submersion for all μ. Thus, the dimension of the space from which Ψ has emanated is superior or equal to the dimension of the space of arrival. This is written as

$$n + m(k+\mu+1)\geq s(\mu+1)\ \forall\mu > 0 \tag{B.28}$$

This implies $m\geq s$. In a similar manner, it is shown that with the mapping Ψ it holds $s\geq m$. Consequently, one finally has $m = s$ which means that the number of control inputs in the two descriptions remains the same.

This definition of the equivalence is adapted to the equivalence between two so-called *diffieties*. The concept of a diffiety is as follows: variable (X,τ_x) comprising the state vector of the nonlinear system and its successive derivatives with respect to time, as well as the Lie derivatives of the mapping transformation along a vector field defined by the state vector elements and their derivatives is called a *manifold of jets* or a *diffiety* .

Given two diffieties (M, CTM) and (N, CTN) it is said that a function Ψ from M to N is Lie-Backlünd or a C-morphism, if the tangent (projection) function $T\Psi$ satisfies $T\Phi(CTM)\subset CTN$. Moreover, if Ψ has an inverse function Φ, such that $T\Psi(CTN)\subset CTM$, then Ψ is a Lie-Backlünd isomorphism or a C isomorphism. When such an isomorphism exists, the diffieties are called equivalent. Therefore, an endogenous transformation is a special type of a Lie-Backlünd transformation which preserves the parametrization in time of the integral curves. It is also possible to introduce a more general concept which is the orbital equivalence, about the Lie-Backlünd isomorphisms which preserve only the geometric locus of the integral curves.

B.1.3.3 Differentially flat systems and equivalence to the trivial system

Definition: The dynamical system (M,F) where M: is the state vector and F: is a vector field, is differentially flat in p (or respectively flat), if and only if it is equivalent in p (respectively equivalent) with the trivial system (therefore it can be described in the Brunovsky canonical form, as shown in the examples of subsection B.1.2).

The definition of equivalent systems can be used, and one can consider a differentially flat system $(X\times U\times R_m^{\infty}, F)$ associated with

$$\dot{x} = f(x,u) \text{ where } (x,u)\in X\times U\subset R^n\times R^m \tag{B.29}$$

By definition, the above system is equivalent to the trivial system R_n^{∞}, F_s, where the endogenous transformation Ψ takes the form

$$\Psi(x,u,u^{(1)},\cdots) = (h(x,u),\dot{h}(x,u),\ddot{h}(x,u),\cdots) \tag{B.30}$$

In other terms Ψ is the infinite extension of function $h()$. The inverse transformation of Ψ is denoted as Φ and holds

$$\Psi(\bar{y}) = (\psi(\bar{\psi}),\beta(\bar{\psi}),\dot{\beta}(\bar{\psi}).\cdots) \tag{B.31}$$

Since Φ and Ψ are inverse applications, it holds that

$$\phi(\bar{h}(x,\bar{u})) = x \quad \text{and} \quad \alpha(\bar{h}(x,\bar{u})) = u \tag{B.32}$$

Moreover, $t{\rightarrow}y(t)$ is a trajectory of $y = v$ and

$$t{\rightarrow}(x(t),u(t)) = (\psi(\bar{\psi}(t)),\beta(\bar{\psi}(t))) \tag{B.33}$$

is a trajectory of $\dot{x} = f(x,u)$.

Definition: Assume that $\{M,F\}$ is a system and that Ψ is the transformation that brings it to the canonical form. The first component of Ψ is considered to be the flat output.

Another important property of differentially flat systems is that the dimension of the flat output is equal to the number of the control inputs (and thus linearization and decoupling is achieved). However, it is noted that the flat output is not unique. If y is a flat output, then $\gamma(y)$ is also a flat output, provided that γ is a smooth diffeomorphism. For single-input systems, the flat output is a scalar $s = m = 1$. It is easy to show that one can pass from one flat output to another by using only a static change of variables. The multi-variable case is clearly more complicated. For example, for $s = m = 2$, it is easy to see that if y_1, y_2 is a flat output then $(y_1, y_2 + \gamma(y_1^{(r)}))$ is also a flat output, provided that $\gamma()$ is smooth function and $r{\geq}0$. Since the flat output of a system is not unique, it is preferable to select the flat output which leads to simple computation for the linearization of the system and the design of the feedback controller.

B.1.4 Differential flatness and trajectory planning

B.1.4.1 Applications of differential flatness in trajectory planning

Next, it will be shown how differential flatness can be used to solve the problem of trajectory planning (setpoints definition) for dynamical systems' control. Assume the following nonlinear system in which the control input u does not implement feedback control

$$\dot{x} = f(x,u), \quad x{\in}R^n \ \ u{\in}R^m \tag{B.34}$$

with flat output $y = h(x,u,\dot{u},\cdots,u^{(r)})$. In such a manner, the system's trajectories $(x(t),u(t))$ are written as a function of the flat output y and its derivatives $x = \phi(y,\dot{y},\cdots,y^{(q)})$ and $u = \alpha(y,\dot{y},\cdots,y^{(q)})$. The trajectory planning problem contains a start and an arrival state. It is assumed that the flat output is a vector with elements y_i which are written as

$$y_i(t) = \sum_j A_{ij}\lambda_j(t) \tag{B.35}$$

where $\lambda_j, j = 1,\cdots,N$ are basis functions. The problem of definition of the flat output components y_i becomes equivalent to finding its projections in the space spanned by the basis functions λ_i.

Next, it is assumed that the initial state of the system is x_0 at t_0, while the final state of the system is x_f at t_f. The coefficients A_{ij} should satisfy the following:

$$\begin{array}{ccc} y_i(t_0) = \sum_j A_{ij}\lambda_j(t_0) & & y_i(t_f) = \sum_j A_{ij}\lambda_j(t_f) \\ \cdots & \cdots & \cdots \\ y_i^{(q)}(t_0) = \sum_j A_{ij}\lambda_j^{(q)}(t_0) & & y_i^{(q)}(t_f) = \sum_j A_{ij}\lambda_j^{(q)}(t_f) \end{array} \tag{B.36}$$

Without loss of generality, it assumed that the vector of the flat output is of dimension $i = 1$, that is $y = [y_1]$ (these results can be also generalized to the multi-dimensional case). The following vectors are defined:

$$\begin{aligned}\tilde{y}_0 &= (y_1(\tau_0), \cdots, y_1^{(q)}(\tau_0))\\ \tilde{y}_f &= (y_1(\tau_f), \cdots, y_1^{(q)}(\tau_f))\\ \tilde{y} &= (\tilde{y}_0, \tilde{y}_f)\end{aligned} \tag{B.37}$$

Using Eq. (B.37) and Eq. (B.36), one has that

$$\tilde{y} = \begin{pmatrix}\Lambda(\tau_0)\\ \Lambda(\tau_j)\end{pmatrix} A = \Lambda A \tag{B.38}$$

The elements of A should satisfy Eq. (B.38). The only condition for Eq. (B.38) to have a solution, is that matrix Λ is a full rank one. This means that the space of functions λ_j has to be sufficiently rich. It can be concluded that the path planning problem for the case of differentially flat systems ends up to linear algebra theory.

B.1.4.2 Planning under constraints

In this case, the objective is to find a trajectory going from point a to point b, which will satisfy the constraints $K(x, u, \cdots, u^{(r)}) \leq 0$ at each time instant. In the flat coordinates, this consists in finding $T > 0$ a function $[0, T] \ni t \rightarrow y(t)$ with $(y, \cdots, y^{(q)})$ to be given at $t = 0$ and T and to verify that $\forall t \in [0, T]$ the constraint about function $K(y, \cdots, y^{(v)})(t) \leq 0$ will hold, for a certain v and a specific function K, obtained from k, ϕ, and α, where k is the constraint condition and ϕ, α are the previously defined mappings $\dot{x} = \phi(y, \bar{v})$ and $u = \alpha(y, \bar{v})$. The difficulty of this problem increases when $q = v = 0$.

Next, it is simply assumed that the initial state Y_0 and the final state Y_f are equilibria. It is also assumed that the motion from the initial to the final state satisfies the following constraint: there exists a path $[0, 1] \ni \sigma \rightarrow Y(\sigma)$ such that $Y(0)$ and $Y(1)$ correspond to start and arrival equilibria, for all $\sigma \in [0, 1]$, $K(Y(\sigma), 0, \cdots, 0) < 0$. Therefore, there exists $T > 0$ and $[0, T] \ni t \rightarrow y(t)$ which is a solution to the problem. It suffices to take $Y(\eta(t/T))$ where T is sufficiently large, and with η to be a smooth increasing function $[0, 1] \ni s \rightarrow \eta(s) \in [0, 1]$ with $\eta(0) = 0$, $\eta(1) = 1$ and $\frac{d^i \eta}{ds^i} = 0$ for $i = 1, \cdots, max(q, v)$.

The problem becomes more complicated in the case of trajectory planning for systems subjected to constraints.

B.1.4.3 Planning of trajectories with singularities

With the previous analysis, it has been shown that the endogenous transformation

$$\Psi(x, u, u^{(1)}, \cdots) = (h(x, \bar{u}), \dot{h}(x, \bar{u}), \ddot{h}(x, \bar{u}), \cdots) \tag{B.39}$$

associated with the flat output $y = h(x, \bar{u})$ is defined everywhere, is smooth and invertible in a manner that always enables state variable x and control input u to be expressed as functions of the flat output and its derivatives

$$(y, \dot{y}, \cdots, y^{(q)}) \rightarrow (x, u) = \phi(y, \dot{y}, \cdots, y^{(q)}) \tag{B.40}$$

However, a point of singularity may exist in the area of the state-space into which the control tries to bring the system's state vector (y is no longer invertible and Eq. (B.40) cannot be solved). As function ϕ cannot be defined in such a point, the previous computations become meaningless. A manner to circumvent this problem and to avoid the singularity is to choose a trajectory for the system

$$t \rightarrow \phi(y(t), \dot{y}(t), \cdots, y^{(q)}(t)) \tag{B.41}$$

such that the point of potential singularity is bypassed. This procedure is depicted in the following example:

Example: The following system with differentially flat dynamics is considered:

$$\dot{x}_1 = u_1 \quad \dot{x}_2 = u_2 u_1 \quad \dot{x}_3 = x_2 u_1 \tag{B.42}$$

where the flat output is defined as $y=(x_1,x_3)$. In case that $u_1=0$, it holds that $\dot{x}_1=\dot{y}_1=0$ and a singularity arises because it holds

$$(y,\dot{y},\ddot{y})\overset{\phi}{\rightarrow}(x_1,x_2,x_3,u_1,u_2)=(y_1,\tfrac{\dot{y}_2}{\dot{y}_1},y_2,\dot{y}_1,\tfrac{\ddot{y}_2\dot{y}_1-\ddot{y}_1\dot{y}_2}{\dot{y}_1^3}) \quad (B.43)$$

and the inverse transformation ϕ is not defined at the points of singularity. However, if one considers trajectories of the following form $t{\rightarrow}y(t):=(\sigma(t),p(\sigma(t)))$, with σ and π being smooth functions, he obtains that

$$\frac{y_2(t)}{y_1(t)}=\frac{p(\sigma(t))}{\sigma(t)} \quad (B.44)$$

$$\frac{\dot{y}_2(t)}{\dot{y}_1(t)}=\frac{\frac{dp}{d\sigma}(\sigma(t)\cdot)\dot{\sigma}(t)}{\dot{\sigma}(t)} \quad (B.45)$$

and

$$\frac{\ddot{y}_2\dot{y}_1-\ddot{y}_1\dot{y}_2}{\dot{y}_1^3}=\frac{\frac{d^2p}{d\sigma^2}(\sigma(t))\cdot\dot{\sigma}^3(t)}{\dot{\sigma}^3(t)} \quad (B.46)$$

Therefore, one can extend $t{\rightarrow}\phi(y(t),\dot{y}(t),\ddot{y}(t))$ everywhere in time using

$$t{\rightarrow}(\sigma(t),\tfrac{dp}{d\sigma}(\sigma(t)),p(\sigma(t)),\dot{\sigma}(t),\tfrac{d^2p}{d\sigma^2}(\sigma(t))) \quad (B.47)$$

Trajectory planning is performed in a similar manner as before: actually, functions σ and p and their derivatives are constrained at the start and at the arrival point of the trajectory, while elsewhere they are free. In this example, the avoidance of the singularity is internally related with a symmetry of the system. Equations

$$\dot{x}_1=u_1,\quad \dot{x}_2=u_2u_1,\quad \dot{x}_3=x_2u_1 \quad (B.48)$$

are linear in u_1. The transformation, $t{\rightarrow}\tilde{t}=\sigma(t)$ and $u_1{\rightarrow}\tilde{u}_1=\frac{u_1}{\dot{\sigma}(t)}$, where only the time t and the control input u change, leaves the equations invariants.

B.1.5 Differential flatness, feedback control and equivalence

B.1.5.1 Closed-loop systems under state feedback control and equivalence

The analysis of subsection B.1.4 referred to a system's dynamics $\dot{x}=f(x,u)$ in which u was not specifically implementing a feedback control action. Next, the notion of equivalence will be extended to the case of closed-loop control. To this end, the initial system dynamics $\dot{x}=f(x,u)$ and its equivalent description $\dot{y}=g(y,v)$ are considered:

$$\begin{aligned}&\dot{x}=f(x,u),\ (x,u){\in}X{\times}U{\subset}R^n{\times}R^m\\&\dot{y}=g(y,v),\ (y,v){\in}Y{\times}V{\subset}R^r{\times}R^p\end{aligned} \quad (B.49)$$

These correspond to the systems $(X{\times}U{\times}R_m^\infty,F)$ and $(Y{\times}V{\times}R_s^\infty,G)$, where F and G are defined by

$$\begin{aligned}&F(x,u,u^{(1)},\cdots):=(f(x,u),u^{(1)},u^{(2)},\cdots)\\&G(y,v,v^{(1)},\cdots):=(g(y,v),v^{(1)},v^{(2)},\cdots)\end{aligned} \quad (B.50)$$

Considering that these systems are equivalent, they have the same trajectories. A question that arises is if a transition is possible from the system $\dot{x}=f(x,u)$ to the system $\dot{y}=g(y,v)$ by a feedback loop (primarily dynamic or alternatively static), or inversely. For the transformed state vectors z and control inputs v, one has

$$\begin{aligned}&\dot{z}=\alpha(x,z,v)\ \ z{\in}Z{\subset}R^q\\&u=\kappa(x,z,v)\end{aligned} \quad (B.51)$$

Next, the following theorem is introduced [573]:

Theorem: Assume that the systems $\dot{x}=f(x,u)$ and $\dot{y}=g(y,v)$ are equivalent. Thus, $\dot{x}=f(x,u)$ can be transformed by a dynamic loop and change of coordinates (dynamic feedback linearization) into

$$\dot{y}=g(y,v),\ \dot{v}=v^{(1)},\ \dot{v}^{(1)}=v^{(2)},\cdots,\dot{v}^{(\mu)}=w \tag{B.52}$$

for an entity μ being sufficiently large. Inversely, $\dot{y}=g(y,v)$ can be transformed by a feedback loop (dynamic) and a change of coordinates into

$$\dot{x}=g(x,u),\ \dot{u}=u^{(1)},\ \dot{u}^{(1)}=u^{(2)},\cdots,\dot{u}^{(\mu)}=w \tag{B.53}$$

for an entity v being sufficiently large.

Proof: The notation F and G is used for the infinite dimensional vector fields which represent the two systems. The equivalence shows that there exists an invertible function

$$\Phi(y,\bar{v})=(\phi(y,\bar{v}),\alpha(y,\bar{v}),\dot{\alpha}(y,\bar{v}),\cdots) \tag{B.54}$$

such that

$$F(\Phi(y,\bar{v}))=D\Phi(y,\bar{v})\cdot G(y,\bar{v}) \tag{B.55}$$

Next, one sets $\bar{y}=(y,v,v^{(1)},\cdots,v^{(\mu)})$ and $w=v^{(\mu+1)}$. For μ sufficiently large, it holds that ϕ (respectively α) depends exclusively on $\bar{y}$ (respectively on $\bar{y}$ and w). With these notations Φ is written as

$$\phi(\tilde{y},w)=(\phi(\tilde{y}),\alpha(\tilde{y},\bar{w}),\dot{\alpha}(\tilde{y},\bar{w}),\cdots) \tag{B.56}$$

and Eq. (B.55) implies in particular that

$$f(\phi(\tilde{y},\alpha(\tilde{y},w)))=D\phi(\tilde{y})\bar{g}(\tilde{y},w) \tag{B.57}$$

where $\bar{g}:=(g,v^{(1)},\cdots,v^{(k)})$. Since Φ is invertible, and ϕ is of full rank it can thus be completed by a function π to obtain a change of coordinates

$$\tilde{y}\rightarrow\phi(\tilde{\psi})=(\phi(\tilde{y}),\pi(\tilde{\psi})) \tag{B.58}$$

Considering now the dynamic feedback, one has

$$\begin{aligned} u&=\alpha(\phi^{-1}(x,z),w)\\ \dot{z}&=D\pi(\phi^{-1}(x,z))\cdot\tilde{g}(\phi^{-1}(x,z),w)\end{aligned} \tag{B.59}$$

which transforms $\dot{x}=f(x,u)$ into

$$\begin{pmatrix}\dot{x}\\ z\end{pmatrix}=\tilde{f}(x,z,w):=\begin{pmatrix}f(x,\alpha(\phi^{-1}(x,z),w))\\ D\pi(\phi^{-1}(x,z))\cdot\tilde{g}(\phi^{-1}(x,z),w)\end{pmatrix} \tag{B.60}$$

Using the previous example, one has

$$\tilde{f}(\phi(\tilde{y},w))=\begin{pmatrix}f(\phi(\tilde{y},\alpha(\tilde{y},w)))\\ D\pi(\tilde{y})\cdot\tilde{g}(\tilde{y},w)\end{pmatrix}=\begin{pmatrix}D\phi(\tilde{y}\\ D\pi(\tilde{y}))\end{pmatrix}\tilde{g}(\tilde{y},w)=D\phi(\tilde{y})\cdot\tilde{g}(\tilde{y},w) \tag{B.61}$$

Thus, $\tilde{f}$ and $\tilde{g}$ are ϕ-conjugates which completes the proof.

A generic remark is that a differentially flat system is equivalent to a system in the linear canonical (Brunovsky) form, and thus one arrives immediately at the following conclusion: a flat dynamics is linearizable by static or dynamic feedback and change of coordinates.

B.1.5.2 Closed-loop systems and equivalence under endogenous feedback

The dynamical system $\dot{x}=f(x,u)$ is considered. The feedback loop

$$\begin{aligned} u&=\kappa(x,z,w)\\ \dot{z}&=\alpha(x,z,w)\end{aligned} \tag{B.62}$$

is called endogenous since the system's dynamics in open loop $\dot{x}=f(x,u)$ is equivalent to the dynamics of the system in closed loop, the latter given by

$$\begin{aligned}\dot{x}&=f(x,\kappa(x,z,w))\\ \dot{z}&=\alpha(x,z,w)\end{aligned} \tag{B.63}$$

The term endogenous has the meaning that variables z and w are not included in the state vector of the system (case also known as "zero dynamics"). Using the above, one arrives at a more refined definition of equivalence between systems and differential flatness.

Theorem: Two dynamics $\dot{x}=f(x,u)$ and $\dot{y}=g(y,v)$ are equivalent, if and only if, $\dot{x}=f(x,u)$ can be transformed by endogenous feedback and change of coordinates.

Corollary 1: A dynamics is flat, if and only if it is linearizable by dynamic endogenous feedback and change of coordinates.

Corollary 2: Consider the system description

$$\begin{aligned}\dot{x}&=f(x,\kappa(x,z,w))\\ \dot{z}&=\alpha(x,z,w)\end{aligned} \tag{B.64}$$

where the constituents of the dynamic endogenous feedback are

$$\begin{aligned}\dot{u}&=\kappa(x,z,w)\\ \dot{z}&=\alpha(x,z,w)\end{aligned} \tag{B.65}$$

This system can be transformed by endogenous feedback into

$$\dot{x}=f(x,u),\ \text{where}\ \dot{u}=u^{(1)},\cdots,\dot{u}^{(\mu)}=w \tag{B.66}$$

for μ sufficiently large (thus finally the system's control input contains multiple integral terms). This shows clearly the properties which are preserved by equivalence. The property which is maintained by introducing integral terms and change of coordinates is controllability. An endogenous feedback is actually reversible to the effect of multiple integrators.

B.1.5.3 Trajectory tracking in feedback control of differentially flat systems

The problem of trajectory tracking for the system $\dot{x}=f(x,u)$ consists in finding a controller u which will permit to the system state variables to track the reference setpoint $t\rightarrow(x_r(t),u_r(t))$. It is convenient to add to the open-loop control a correction term Δu, which is a function of the state vector error $\Delta x=x-x_r$. For differentially flat systems, there is a systematic method to calculate Δu from the tracking error Δx.

If the dynamics admits as flat output $y=h(x,\bar{u})$, one can use the corollary that a differentially flat system is linearizable by (dynamic) feedback and change of coordinates, into the linear canonical form $y^{(\mu+1)}=w$. For the linearized equivalent model, the control input is defined as $y_r^{(\mu+1)}-K\Delta\tilde{y}$, where K is a gains vector and Δy is a vector having as elements the flat output tracking error and its derivatives up to order μ. By applying such a feedback control input, the tracking error dynamics becomes

$$\Delta y^{(\mu+1)}=-K\Delta\tilde{y} \tag{B.67}$$

where $y_r(t):=(x_r(t),\bar{u}_r(t))$ and $\tilde{y}:=(y,\dot{y},\cdots,y^{\mu})$, and Δy represents the vector of the flat output tracking error. Such a feedback control law assures asymptotic tracking. There exists an invertible transformation

$$\Phi(\bar{y})=(\phi(\bar{y}),\alpha(\bar{y}),\dot{\alpha}(\bar{y}),\cdots) \tag{B.68}$$

which connects the infinite vector fields $F(x,\bar{u}) := (f(x,u),u,u^{(1)},\cdots)$ and $G(\bar{y}) := (y,y^{(1)},\cdots)$. From the proof of the theorem on the equivalence of two systems (theorem given in subsubsection B.1.5.1), the above means that for the state vector, one has

$$\begin{aligned} x &= \phi(\tilde{y}_r(t)+\Delta\tilde{y}) = \\ &= \phi(\tilde{y}_r(t)) + R_\phi(y_r(t),\Delta\tilde{y})\Delta\tilde{y} = \\ &= x_r(t) + R_\phi(y_r(t),\Delta\tilde{y})\Delta\tilde{y} \end{aligned} \tag{B.69}$$

and for the control input, one has

$$\begin{aligned} u &= \alpha(\tilde{y}_r(t)+\Delta\tilde{y},-K\Delta\tilde{y}) = \\ &= \alpha(\tilde{y}_r(t)) + R_\alpha(y_r^{(\mu+1)}(t),\Delta\tilde{y})\begin{pmatrix}\Delta\tilde{y}\\ -K\Delta\tilde{y}\end{pmatrix} = \\ &= u_r(t) + R_\alpha(\tilde{y}_r(t),y_r^{(\mu+1)}(t),\Delta\tilde{y},\Delta w)\begin{pmatrix}\Delta\tilde{y}\\ -K\Delta\tilde{y}\end{pmatrix} \end{aligned} \tag{B.70}$$

where the following classical result of factorization has been used:

$$\begin{aligned} R_\phi(Y,\Delta Y) &:= \textstyle\int_0^1 D\phi(Y+t\Delta Y)dt \\ R_\alpha(Y,w,\Delta Y,\Delta w) &:= \textstyle\int_0^1 D\alpha(Y+t\Delta Y,w+t\Delta w)dt \end{aligned} \tag{B.71}$$

Since $\Delta y \to 0$ for $t\to\infty$, this means that $x\to x_r(t)$ and $u\to u_r(t)$.

B.1.6 Flatness-based control and state feedback for systems with model uncertainties

Up to now, for the systems used for analyzing differential flatness properties and flatness-based control it has been assumed that an exact dynamic model was available. Flatness-based control can be also applied to systems characterized by model uncertainties and exogenous disturbances. Denoting the coefficients vector of the system as $\theta\in R^p$, the generalized system dynamics of Eq. (B.72) is written as

$$\dot{x}(t) = f(\theta,x(t),u(t)) \tag{B.72}$$

In the generic case, the system's coefficients θ are subject to uncertainty, i.e.,

$$\theta = \theta_0 + \tilde{\theta},\ \tilde{\theta}_i \in [\underline{\theta}_i,\bar{\theta}_i],\ i=1,\cdots,p \tag{B.73}$$

where θ_0 is the nominal value of the parameter vector. In Eq. (B.72), the vector field $f: R^p \times R^n \times R^m \to R^n$ is smooth. Denoting the flat output $y(t)\in R$ and parameters vector θ, the definition of differential flatness given in subsection B.1.2 is rewritten as [315,317]

$$\begin{aligned} y &= h(\theta,x,u,\dot{u},\cdots,u^{(r)}) \\ x &= \phi(\theta,y,\dot{y},\cdots,y^{(r-1)}) \\ u &= \psi(\theta,y,\dot{y},\cdots,y^{(r)}) \end{aligned} \tag{B.74}$$

where h, ϕ, and ψ are smooth functions defining mappings $h: R^p\times R^n\times(R^m)^{r+1} \to R^m$, $\phi: R^p\times(R^m)^r\to R^n$ and $\psi: R^p\times(R^m)^{r+1}\to R^m$, respectively. It is also assumed that the flat output $y(t)$ is independent of the coefficients (parameters) vector θ. Eq. (B.74) shows that for every given trajectory of the flat output $t\to y(t)$ the evolution of all other variables of the system $t\to x(t)$ and $t\to u(t)$ are given without integration of any differential equation. Moreover, for a sufficiently smooth desired trajectory of the flat output $t\to y^*(t)$, Eq. (B.74) can be used to design the corresponding feed-forward control $u^*(t)$ directly for the nominal system parameters θ_0. The trajectory y^* is called the nominal trajectory, while the trajectory u^* is called the nominal control. Thus, knowing the desirable system's output, one can also find the associated flat output y^* and subsequently the control input that makes the system track the desirable trajectory is given by

$$u^*(t) = \psi(\theta_0, y^*(t), \dot{y}^*(t), \cdots, y^{*(r)}(t)) \tag{B.75}$$

which means that for each admissible nominal trajectory $y^*(t)$, there is a nominal control input u^*. The initial condition of the desired trajectory of the flat output $t{\to}y^*(t)$ is defined as the vector $y_0^* = [y^*(0), \dot{y}^*(0), \cdots, y^{*(r-1)}(0)]^T$. The system is *consistent* with respect to the initial conditions x_0 if $x_0 = \phi(\theta, y_0^*)$. The following theorem has been proven [315]:

Theorem: If the desired trajectory of the flat output is consistent with the initial condition x_0 and $\theta = \theta_0$, then when applying the nominal control input of Eq. (B.75) the system becomes equivalent, by a change of coordinates, to a linear system in Brunovsky (canonical) form .

Without loss of generality, a single-input single-output flat system is considered. It is easy to show that such a flat system can be represented as follows: Setting $y = [y, \dot{y}, \cdots, y^{(n-1)}]^T = [y_1, y_2, \cdots, y_n]^T$, the system can be transformed via the diffeomorphism

$$y = h(\theta, x) \tag{B.76}$$

where $h = \phi^{-1}$ with ϕ defined in Eq. (B.74), into the control normal (canonical) form

$$\begin{aligned} \dot{y}_i(t) &= y_{i+1}(t) \;\; i{\in}\{1, \cdots, n-1\} \\ \dot{y}_n(t) &= \alpha(\theta, y(t), u(t)) \end{aligned} \tag{B.77}$$

where $y^{(i)}$ is the $i-1$-th order derivative of the flat output. From the second row of Eq. (B.77), one can see that the n-th order derivative of the flat output is finally written as a nonlinear function of the parameters' vector θ, of the flat output's vector $y(t)$ and of the control input $u(t)$. The system in canonical form described by Eq. (B.77) and the system of Eq. (B.72), Eq. (B.74), and Eq. (B.75) have the same solution $t{\to}y(t)$.

Defining the new control input $v(t) = \alpha(\theta, y(t), u(t))$, the Brunovsky form of the transformed initial system is written as

$$\begin{aligned} \dot{y}_i(t) &= y_{i+1}(t) \;\; i{\in}\{1, \cdots, n-1\} \\ \dot{y}_n(t) &= v(t) \end{aligned} \tag{B.78}$$

Once the system has been written in the Brunovsky (canonical) form of Eq. (B.78), a control input that can assure tracking of any desirable trajectory $\dot{y}_n^*$ is given by

$$v = \dot{y}_n^* + K^T\bar{e} \tag{B.79}$$

where the term $K^T\bar{e}$ corresponds to an error feedback controller with an integral term and thus processes the augmented error vector $\bar{e} = [\int_0^t e_1(\tau)d\tau, e_1, e_2, \cdots, e_n]$. The feedback control term can be also of the form of a proportional-derivative controller and in the latter case processes the error vector $e = [e_1, e_2, \cdots, e_n]$, thus taking the form

$$K^Te = -\sum_{i=1}^n k_{n+1-i}e_i(t), \;\; k_{n+1-i} > 0 \tag{B.80}$$

Therefore, in case that the exact values of the model's parameters θ are known and these are equal to the nominal values θ_0, and no integration of the flat output error is used in the feedback loop, i.e., the feedback control signal is of the form $K^Te = -\sum_{i=1}^n k_{n+1-i}e_i(t)$, the closed-loop system error vector dynamics become

$$\begin{aligned} \dot{y}_n = v {\Rightarrow} \dot{y}_n &= \dot{y}_n^* + K^Te{\Rightarrow} \\ \dot{y}_n &= \dot{y}_n^* - k_ne_1 - k_{n-1}e_2 + \cdots - k_1e_n{\Rightarrow} \\ \dot{y}_n - \dot{y}_n^* &= -k_ne_1 - k_{n-1}e_2 + \cdots - k_1e_n{\Rightarrow} \\ e^{(n)} + k_1e^{(n-1)} &+ k_2e^{(n-2)} + \cdots + k_{n-1}\dot{e} + k_ne = 0 \end{aligned} \tag{B.81}$$

where $e_i = y_i - y_i^* = y^{(i-1)} - y^{(i-1)*}$. The dynamics of the flat output tracking error can be also written in a matrix form

$$\begin{pmatrix} \dot{e}_1 \\ \dot{e}_2 \\ \cdots \\ \cdots \\ \dot{e}_n \end{pmatrix} = \begin{pmatrix} 0 & 1 & 0 & \cdots & 0 \\ 0 & 0 & 1 & \cdots & 0 \\ \cdots & \cdots & \cdots & \cdots & \cdots \\ 0 & 0 & 0 & \cdots & 1 \\ -k_n & -k_{n-1} & -k_{n-2} & \cdots & -k_1 \end{pmatrix} \begin{pmatrix} e_1 \\ e_2 \\ \cdots \\ \cdots \\ e_n \end{pmatrix} \tag{B.82}$$

In case of the error dynamics of Eq. (B.82), to achieve the elimination of the tracking error, it suffices to select the controller gains $k_1, k_2, \cdots, k_n$ so as the polynomial $e^{(n)} + k_1 e^{(n-1)} + k_2 e^{(n-2)} + \cdots + k_{n-1}\dot{e} + k_n e = 0$ to be a Hurwitz one.

When the model's parameters θ are not precisely known, or have deviated from their nominal values θ_0, one can apply again a control input of the form described in Eq. (B.79) and Eq. (B.80), and for specific magnitude of the exogenous disturbances or parametric uncertainties can expect again satisfactory tracking of the reference trajectory. In such a case, the robustness features of the closed loop can be analyzed with the use of results from interval polynomial stability theory (e.g. Kharitonov's theory) or other results on interval polynomial analysis (e.g., see [708]). The problem of robustness of a flatness-based controller, in case of parametric uncertainties or exogenous perturbations, has been studied in [315,317] and also in [708,711].

B.1.7 Classification of types of differentially flat systems

B.1.7.1 Criteria about the differential flatness of a system

Definition of 0-flat systems

The existence of a criterion on whether the dynamical system $\dot{x} = f(x, u)$, $x \in R^n$, $u \in R^m$ is differentially flat remains open. This means that there is no systematic method to construct a flat output. This situation is somehow equivalent to the definition of Lyapunov functions for nonlinear dynamical systems. Even though Lyapunov functions are both theoretically and practically important for the study of the stability features of nonlinear dynamical systems, there is no systematic manner to define them.

The main difficulty in the computation of the flat output $y = h(x, u, \cdots, u^{(r)})$ remains in its dependence to the derivatives of the control input u, up to order r which can be arbitrarily large. To know if the order of r admits an upper bound (which can be dependent on n and m) remains a question. It is not known if for a certain dimension of the state vector and for a certain dimension of the control inputs vector such a finite boundary exists. In the following it will be noted that a dynamical system is r-flat if it admits a flat output depending up to the r-th order derivative of the control input. Equivalently, in a 0-flat system, its flat output depends exclusively on the elements of the system's state vector.

The next example shows that such an upper bound should depend, in the best case, linearly to the dimension of the state vector. Consider the dynamical system

$$x_1^{(a_1)} = u_1 \quad x_1^{(a_2)} = u_2 \quad \dot{x}_3 = u_1 u_2 \tag{B.83}$$

The flat output is taken to be

$$y_1 = x_3 + \sum_{i=1}^{\alpha_1} (-1)^i x_1^{(\alpha_1 - i)} u_2^{(i-1)}, \qquad y_2 = x_2 \tag{B.84}$$

This system is r-flat, with $r = min(\alpha_1, \alpha_2) - 1$. There is no flat output depending on the derivatives of u which has an order equal or less than $r - 1$.

Systems which can be linearized by static or dynamic state feedback

It has been shown that dynamical systems which can be linearized with static state feedback can be also transformed to the linear canonical (Brunovsky) form. Thus, such systems are differentially flat, with the flat outputs to be exclusively functions of the state vector x (0-flat system). It is noted that a differentially flat system is not always a system which can be linearized with static state feedback. There exist systems which can be linearized through dynamic feedback linearization, that is their state vector is extended by including as additional state variables the control inputs and their derivatives (see examples on the linearization of underactuated electromechanical systems). Such systems can be also written in the linear canonical form, and thus are also differentially flat.

Systems with only one control input
Single input dynamical systems, that is

$$\dot{x}=f(x,u)\ x\in R^n,\ u\in R \tag{B.85}$$

can be linearized with static state feedback linearization and can be finally written in the linear canonical form. Such systems are differentially flat.

Systems affine-in-the-input of co-dimension 1
A system of the form

$$\dot{x}=f_0(x)+\sum_{j=1}^{n-1}g_j(x)u_j\ \ x\in R^n \tag{B.86}$$

that is systems in which the number of control inputs is by one less than the dimension of the system's state vector, are 0-flat, as long as they are controllable (strongly accessible in x). The problem becomes more complicated when the system is not an affine in the input one. However, the latter systems can be finally transformed into affine-in-the-input ones too.

Systems without drift term
These are systems of the form $\dot{x}=\sum_{i=1}^{m}f_i(x)u_i$. Within this class of systems, one can distinguish several cases. The following theorem holds: [573]

Theorem 1: Systems without drift and with two control inputs

$$\dot{x}=f_1(x)u_1+f_2(x)u_2 \tag{B.87}$$

are flat if and only if, the generic rank of E_k is equal to $k+2$, for $k=0,\cdots,n-2$, where $E_0=\{f_1,f_2\}$ and $E_{k+1}=\{E_k,[E_k,E_k]\},\ \ k>0$. A dynamical system without drift and with two control inputs is always 0-flat. Such a system can be written in the chained form (which is a precursor of the linear canonical form), through static state feedback and a change of variables. For example, for a system of the form

$$\dot{x}_1=u_1\quad \dot{x}_2=u_2\quad \dot{x}_3=x_2u_1\quad \cdots\quad \dot{x}_n=x_{n-1}u_1 \tag{B.88}$$

is a differentially flat one and the flat output is $y=(x_1,x_n)$.

Systems in triangular form
The state-space description of systems in triangular form is as follows:

$$\begin{aligned}
\dot{x}_1&=f_1(x_1)+g_1(x_1)x_2\\
\dot{x}_2&=f_2(x_1,x_2)+g_2(x_1,x_2)x_3\\
\dot{x}_3&=f_3(x_1,x_2,x_3)+g_3(x_1,x_2,x_3)x_4\\
&\cdots\\
\dot{x}_i&=f_i(x_1,x_2,\cdots,x_i)+g_i(x_1,x_2,\cdots,x_i)x_{i+1}\\
&\cdots\\
\dot{x}_{n-1}&=f_{n-1}(x_1,x_2,\cdots,x_{n-1})+g_{n-1}(x_1,x_2,\cdots,x_{n-1})x_n\\
\dot{x}_n&=f_n(x_1,x_2,\cdots,x_n)+g_n(x_1,x_2,\cdots,x_n)u
\end{aligned} \tag{B.89}$$

The flat output for such systems is $y=x_1$.

Theorem 2: Driftless systems with a state vector of dimension n and $n-2$ control inputs. The system of the form

$$\dot{x}=\sum_{i=1}^{n-2}f_i(x)u_i \tag{B.90}$$

is a differentially flat one if it is controllable (strongly accessible for almost all x). More precisely, such a system is 0-flat if n is an odd number and is a 1-flat one if n is an even number.

B.1.7.2 A sufficient condition for showing that a system is not differentially flat

The following theorem provides a sufficient condition for showing that dynamical systems are not differentially flat [573]:

Theorem: The criterion of the regulated variety. Assume that the dynamical system $\dot{x}=f(x,u)$ is a differentially flat one. Then, the projection to the p-dimensional space of its state-space equation $p=f(x,u)$ is a subvariety regulated for all x.

This criterion signifies that the elimination of the control input u from the n equations $\dot{x}=f(x,u)$ leads to $n-m$ equations of the form $F(x,\dot{x})=0$ with the following property: for all (x,p) such that $F(x,p)=0$, there exists $\alpha \in R^n$, $\alpha \neq 0$ such that

$$\forall \lambda \in R \;\; F(x,p+\lambda a)=0 \tag{B.91}$$

Proof: Consider $(\bar{x},\bar{u},\bar{p})$ such that $\bar{p}=f(\bar{x},\bar{u})$. Generally, f is of the order $m=dim(u)$ with respect to u. If x is removed from the state space description then one arrives at $n-m$ equations which are based exclusively on x and p, that is $F(x,p)=0$. Thus, one has the constraint about the trajectory around $(\bar{x},\bar{u}):\;\; F(x,\dot{x})=0$. If the system is differentially flat, it holds that

$$x=\phi(y,\dot{y},\cdots,y^{(q)}) \tag{B.92}$$

where variable y is the flat output. Thus, there exists $(\bar{y},\dot{\bar{y}},\cdots,\bar{y}^{(q)})$ such that $\bar{x}=\phi(\bar{y},\dot{\bar{y}},\cdots,\bar{y}^{(q)})$. Then, the following identity holds:

$$F(\phi(\bar{y},\dot{\bar{y}},\cdots,\bar{y}^{(q)}),\phi_y\dot{y}+\phi_{\dot{y}}\ddot{y}+\cdots+\phi_{y^{(q)}}y^{(q+1)})=0 \tag{B.93}$$

Taking now that the derivatives of function y at $t=0$ up to order q are given by $y^{(r)}(0)=\bar{y}^{(r)}\;\; r=0.\cdots,q$ and $y^{\alpha}(0)=\bar{y}^{q}+\xi$ where ξ is an arbitrary vector in R^n, one has the following identity for all $\xi \in R^n$:

$$F(\bar{x},\bar{x}+\phi_{y^{\alpha}}\xi)=0 \tag{B.94}$$

with $\phi_y^{(q)}$ to be evaluated at $\bar{y},\dot{\bar{y}},\cdots,\bar{y}^{(q)}$. From the point $(\bar{x},\bar{p})$ of $F(x,p)=0$ passes the affine space which is parallel to the image of $\phi_{y^{(q)}}$, one has a space of dimension at most equal to m and at least equal to 1.

Therefore, by showing that there does not exist any flat output with derivatives which are not coupled in the sense of an ODE (or equivalently it can be said that there does not exist a flat output being differentially independent), one can prove that a dynamical system is not differentially flat. This result is particularly simple. It stands however for an efficient method to show that certain multi-input systems are not differentially flat.

Example 1: The system

$$\dot{x}_1=u_1,\quad \dot{x}_2=u_2,\quad \dot{x}_3=(u_1)^2+(u_2)^3 \tag{B.95}$$

is not a differentially flat one. This is because the sub-variety $p_3=p_1^2+p_2^3$ is not a regulated one. There does not exist an $\alpha \in R^3$, $\alpha \neq 0$ such that

$$\forall \lambda \in R \;\; p_3+\lambda\alpha_3=(p_1+\lambda\alpha_1)^2+(p_2+\lambda\alpha_2)^3 \tag{B.96}$$

Actually, the cubic term implies that $\alpha_2=0$ and the quadratic term implies that $\alpha_1=0$ and thus one has $\alpha_3=0$.

B.1.7.3 Liouvillian and non-differentially flat systems

There are some non-flat systems which do not satisfy differential flatness properties. Such systems cannot be linearized by state feedback (neither by static feedback linearization nor by dynamic feedback linearization). For such cases, the notion of Liouvillian systems has been introduced, which can be seen as an extension of differentially flat systems.

Non-flat systems are algebraically characterized by an integer called the *defect*. This integer is well defined by introducing the notion of maximal linearizing output. Using an informal definition, a set of variables $y=(y_1,\cdots,y_m)$

is a maximal linearizing output, if the number of system variables which can be expressed as a differential function of y is maximal. Therefore, the maximal linearizing output characterizes the largest subsystem which satisfies the differential flatness property. This largest subsystem is called flat subsystem.

The defect represents the number of the system's variables which do not belong to the flat subsystem. A nonlinear system is said to be *Liouvillian* or integrable by quadratures if the elements which do not belong to the flat subsystem can be obtained by elementary integrations of the elements of the flat subsystem. In the following examples, the notion of integrability by quadratures and that of Liouvillian systems is explained [573].

Example 1:

$$\dot{x}_1 = x_2 + x_3^2, \quad \dot{x}_2 = x_3, \quad \dot{x}_3 = u \tag{B.97}$$

This system is not linearizable by static or dynamic state feedback. The system is of defect equal to 1 and is found to be Liouvillian (there is a state variable which does not belong to the flat subsystem). A maximum linearizing output is given by $y = x_2$, and x_1 is the only variable that does not belong to the flat subsystem. However, x_1 can be obtained from the integration $x_1 = \int y + \dot{y}^2$, that is x_1 is given by a pure integral of a differential function of y.

Example 2: The following dynamic model is considered

$$\dot{x}_1 = u_1, \quad \dot{x}_2 = u_2, \quad \dot{x}_3 = x_1u_2 - x_2u_1, \quad \dot{x}_4 = x_3u_1 \quad \dot{x}_5 = x_3u_2 \tag{B.98}$$

This system is nilpotent and completely controllable since it is driftless and satisfies the Lie algebra rank condition. However, the system is not flat and cannot be transformed into the canonical form. Nevertheless, there is a subsystem within it defined by the first four equations, which is differentially flat, and consequently, the system's defect is equal to 1. The maximum linearizing output of the system is given by $y = (y_1, y_2)$, where $y_1 = x_1$ and $y_2 = (zx_1 - x_4)/3$, with z to be defined by $z = (x_1x_2 - x_3)/2$. Indeed, $z = \dot{y}_2/\dot{y}_1$, thus $x_4 = y_1z - 3y_2 = \alpha(y_1, \dot{y}_1, y_2, \dot{y}_2)$, $x_3 = \dot{x}_4/\dot{y}_1 = b(y_1, \dot{y}_1, \ddot{y}_1, y_2, \dot{y}_2, \ddot{y}_2)$, $x_2 = (2z + x_3)/y_1 = c(y_1, \dot{y}_1, \ddot{y}_1, y_2, \dot{y}_2, \ddot{y}_2)$, $u_2 = \dot{x}_2 = q(y_1, \cdots, y_1^{(3)}, y_2, \cdots, y_2^{(3)})$. Since $x_5 = \int x_3u_2 = \int b(y_1, \cdots, \ddot{y}_2)q(y_1, \cdots, y_2^{(3)})$, one gets that the above system of Eq. (B.98) is a Liouvillian one.

B.2 Flatness-based control of nonlinear dynamical systems in cascading loops

B.2.1 Decomposition of the state-space model into cascading differentially flat subsystems

The following nonlinear dynamical system is now examined:

$$\begin{matrix} \dot{x} = f(x) + g(x)u \; x \in R^m \;\; u \in R^q \\ y = h(x) \end{matrix} \tag{B.99}$$

Moreover, it is considered that the system can be decomposed into n subsystems which have the so-called triangular form [712,718]:

$$\begin{aligned} \dot{x}_1 &= f_1(x_1) + g_1(x_1)x_2 \\ \dot{x}_2 &= f_2(x_1, x_2) + g_2(x_1, x_2)x_3 \\ \dot{x}_3 &= f_3(x_1, x_2, x_3) + g_3(x_1, x_2, x_3)x_4 \\ &\cdots \\ \dot{x}_i &= f_i(x_1, x_2, \cdots, x_i) + g_i(x_1, x_2, \cdots, x_i)x_{i+1} \\ &\cdots \\ \dot{x}_{n-1} &= f_{n-1}(x_1, x_2, \cdots, x_{n-1}) + g_{n-1}(x_1, x_2, \cdots, x_{n-1})x_n \\ \dot{x}_n &= f_n(x_1, x_2, \cdots, x_n) + g_n(x_1, x_2, \cdots, x_n)u \end{aligned} \tag{B.100}$$

The following virtual control inputs $\alpha_i = x_{i+1}$ are defined for the i-th subsystem of the state-space model of Eq. (B.100)

$$\begin{aligned}
\dot{x}_1 &= f_1(x_1) + g_1(x_1)\alpha_1 \\
\dot{x}_2 &= f_2(x_1, x_2) + g_2(x_1, x_2)\alpha_2 \\
\dot{x}_3 &= f_3(x_1, x_2, x_3) + g_3(x_1, x_2, x_3)\alpha_3 \\
&\cdots \\
\dot{x}_i &= f_i(x_1, x_2, \cdots, x_i) + g_i(x_1, x_2, \cdots, x_i)\alpha_i \\
&\cdots \\
\dot{x}_{n-1} &= f_{n-1}(x_1, x_2, \cdots, x_{n-1}) + g_{n-1}(x_1, x_2, \cdots, x_{n-1})\alpha_{n-1} \\
\dot{x}_n &= f_n(x_1, x_2, \cdots, x_n) + g_n(x_1, x_2, \cdots, x_n)u
\end{aligned} \tag{B.101}$$

The system of Eq. (B.101) is a differentially flat one. It is considered that $y = x_1$ is the flat output of this system. It can be easily shown that each virtual control input $\alpha_i = x_{i+1}$, $i = 1, 2, \cdots$ can be expressed as a function of the flat output and its derivatives, since it holds

$$\alpha_i = \frac{1}{g_i(x_1, x_2, \cdots, x_i)}(\dot{x}_i - f(x_1, x_2, \cdots, x_i)) \tag{B.102}$$

For $i = 1$, one has

$$\alpha_1 = \frac{1}{g_1(x_1)}(\dot{x}_1 - f(x_1)) \tag{B.103}$$

which means that α_1 is a function of the flat output and its derivative. For $i = 2$, one has

$$\alpha_2 = \frac{1}{g_2(x_1, x_2)}(\dot{x}_2 - f_2(x_1, x_2)) \tag{B.104}$$

which means that $\alpha_2 = x_3$ is a function of the flat output $y = x_1$ and its derivatives. Continuing in a similar manner, one has that $\alpha_{n-1} = x_n$, and consequently $\alpha_n = u$ is also a function of the flat output $y = x_1$ and its derivatives. According to the above, one has a nonlinear dynamical system in which, all its state variables and the control input can be written as functions of the flat output and its derivatives. Therefore, such a system is differentially flat.

Additionally, by considering each subsystem of the model of Eq. (B.101), one has a set of n subsystems of the form

$$\dot{x}_i = f_i(x_1, x_2, \cdots, x_i) + g_i(x_1, x_2, \cdots, x_i)\alpha_i \tag{B.105}$$

where each subsystem describes the dynamics of the single state variable x_i. For each one of these subsystems, one can consider state variable x_i as the flat output. Obviously, the virtual control input α_i is a function of this flat output and its derivatives. Therefore, each local subsystem is also differentially flat.

Next, one can compute the virtual inputs which are applied to each subsystem. For the first subsystem, which is associated with the first row of Eq. (B.100), and by defining $z_i = x_i - x_i^* = x_1 - \alpha_{i-1}$, the virtual control input is given by

$$\begin{aligned}
\alpha_1 = x_2^* &= \frac{1}{g_1(x_1)}(\dot{x}_1^* - f(x_1) - K_1^1(x_1 - x_1^*)) \Rightarrow \\
\alpha_1 = x_2^* &= \frac{1}{g_1(x_1)}(\dot{x}_1^* - f(x_1) - K_1^1 z_1)
\end{aligned} \tag{B.106}$$

From the second row of Eq. (B.100), and using that $z_2 = x_2 - x_2^* = x_2 - \alpha_1$, one has

$$\begin{aligned}
\alpha_2 = x_3^* &= \frac{1}{g_2(x_1, x_2)}(\dot{x}_2^* - f_2(x_1, x_2) - K_1^2(x_2 - x_2^*)) \Rightarrow \\
\alpha_2 = x_3^* &= \frac{1}{g_2(x_1, x_2)}(\dot{\alpha}_1 - f_2(x_1, x_2) - K_1^2 z_2)
\end{aligned} \tag{B.107}$$

From the third row of Eq. (B.100), and using that $z_3 = x_3 - x_3^* = x_3 - \alpha_2$, one has

$$\alpha_3 = x_4^* = \frac{1}{g_3(x_1,x_2,x_3)}(\dot{x}_3^* - f_3(x_1,x_2,x_3) - K_1^3(x_3 - x_3^*)) \Rightarrow$$
$$\alpha_3 = x_4^* = \frac{1}{g_3(x_1,x_2,x_3)}(\dot{\alpha}_2 - f_3(x_1,x_2,x_3) - K_1^3 z_3) \tag{B.108}$$

Continuing in a similar manner and from the i-th row of the state-space description of the system given in Eq. (B.100), and while also using that $z_i = x_i - x_i^* = x_i - \alpha_{i-1}$, one obtains

$$\alpha_i = x_{i+1}^* = \frac{1}{g_i(x_1,x_2,\cdots,x_i)}(\dot{x}_i^* - f_i(x_1,x_2,\cdots,x_i) - K_1^i(x_i - x_i^*)) \Rightarrow$$
$$\alpha_i = x_{i+1}^* = \frac{1}{g_i(x_1,x_2,\cdots,x_i)}(\dot{\alpha}_{i-1} - f_i(x_1,x_2,\cdots,x_3) - K_1^i z_i) \tag{B.109}$$

Equivalently, from the $(n-1)$-th row of the state-space model of Eq. (B.100) and using that $z_{n-1} = x_{n-1} - x_{n-1}^* = x_{n-1} - \alpha_{n-2}$ one has

$$\alpha_{n-1} = x_n^* = \frac{1}{g_{n-1}(x_1,x_2,\cdots,x_{n-1})}(\dot{x}_{n-1}^* - f_{n-1}(x_1,x_2,\cdots,x_{n-1}) - K_1^{n-1}(x_{n-1} - x_{n-1}^*)) \Rightarrow$$
$$\alpha_{n-1} = x_n^* = \frac{1}{g_{n-1}(x_1,x_2,\cdots,x_{n-1})}(\dot{\alpha}_{n-2} - f_{n-1}(x_1,x_2,\cdots,x_{n-1}) - K_1^{n-1} z_{n-1}) \tag{B.110}$$

Finally, from the n-th row of the state-space model of Eq. (B.100) and using that $z_n = x_n - x_n^* = x_n - \alpha_{n-1}$, one has

$$\alpha_n = u = \frac{1}{g_n(x_1,x_2,\cdots,x_n)}(\dot{x}_n^* - f_n(x_1,x_2,\cdots,x_n) - K_1^n(x_n - x_n^*)) \Rightarrow$$
$$\alpha_n = u = \frac{1}{g_n(x_1,x_2,\cdots,x_n)}(\dot{\alpha}_{n-1} - f_n(x_1,x_2,\cdots,x_n) - K_1^n z_n) \tag{B.111}$$

The computation of the control input u that should be actually applied to the nonlinear system is performed in a recursive manner by processing backwards the virtual control inputs described in Eq. (B.106) to Eq. (B.111).

Thus, from the last subsystem of the state-space description, the control input that is actually applied to the nonlinear system is found. This control input contains recursively all virtual control inputs which were computed for the individual subsystems associated with the state-space equation. Consequently, by tracing the rows of the state-space model backwards, at each iteration of the control algorithm, one can finally obtain the control input that should be applied to the nonlinear system so as to assure that all its state vector elements will converge to the desirable setpoints.

B.2.2 Tracking error dynamics for flatness-based control in successive loops

By substituting Eq. (B.111) into the last row of the state space model of Eq. (B.100), and using the definition $x_n - a_{n-1} = z_n$, one obtains:

$$\begin{aligned}\dot{x}_n &= \dot{a}_{n-1} - K_1^n(x_n - \alpha_{n-1}) \Rightarrow\\ (\dot{x}_n - \dot{a}_{n-1}) &+ K_1^n(x_n - \alpha_{n-1}) = 0 \Rightarrow\\ \dot{z}_n &+ K_1^n z_n = 0\end{aligned} \tag{B.112}$$

By substituting Eq. (B.110) into the last row of the state space model of Eq. (B.100), and using the definition $x_{n-1} - a_{n-2} = z_{n-1}$, one obtains:

$$\begin{aligned}\dot{x}_{n-1} &= \dot{a}_{n-2} - K_1^{n-1}(x_{n-1} - \alpha_{n-2}) \Rightarrow\\ (\dot{x}_{n-1} - \dot{a}_{n-2}) &+ K_1^{n-1}(x_{n-1} - \alpha_{n-2}) = 0 \Rightarrow\\ \dot{z}_{n-1} &+ K_1^{n-1} z_{n-1} = 0\end{aligned} \tag{B.113}$$

By substituting Eq. (B.109) into the last row of the state space model of Eq. (B.100), and using the definition $x_i - a_{i-1} = z_i$, one obtains

$$\begin{aligned} \dot{x}_i &= \dot{a}_{i-1} - K_1^{n-1}(x_i - \alpha_{i-1}) \Rightarrow \\ (\dot{x}_i - \dot{a}_{i-1}) &+ K_1^i(x_i - \alpha_{i-1}) = 0 \Rightarrow \\ \dot{z}_i &+ K_1^i z_i = 0 \end{aligned} \tag{B.114}$$

while continuing backwards and by substituting Eq. (B.107) into the second row of the state space model of Eq. (B.100), and using the definition $x_2 - a_1 = z_2$, one gets

$$\begin{aligned} \dot{x}_2 &= \dot{a}_1 - K_1^2(x_2 - \alpha_1) \Rightarrow \\ (\dot{x}_2 - \dot{a}_1) &+ K_1^2(x_2 - \alpha_1) = 0 \Rightarrow \\ \dot{z}_2 &+ K_1^2 z_2 = 0 \end{aligned} \tag{B.115}$$

Finally, by substituting Eq. (B.106) into the first row of the state space model of Eq. (B.100), one has

$$\begin{aligned} \dot{x}_1 &= \dot{x}_1^d - K_1^1(x_1 - x_1^d) \Rightarrow \\ (\dot{x}_1 - \dot{x}_1^d) &+ K_1^1(x_1 - x_1^d) = 0 \Rightarrow \\ \dot{z}_1 &+ K_1^1 z_1 = 0 \end{aligned} \tag{B.116}$$

Therefore, after the application of the feedback control law, the closed-loop dynamics becomes $\dot{z}_1 + K_1^1 z_1 = 0$, $\dot{z}_2 + K_1^2 z_2 = 0$, $\dot{z}_3 + K_1^3 z_3 = 0, \cdots, \dot{z}_i + K_1^i z_i = 0, \cdots, \dot{z}_{n-1} + K_1^{n-1} z_{n-1} = 0$, $\dot{z}_n + K_1^n z_n = 0$.

In matrix form, the closed-loop dynamics is written as

$$\begin{pmatrix} \dot{z}_1 \\ \dot{z}_2 \\ \dot{z}_3 \\ \cdots \\ \dot{z}_i \\ \cdots \\ \dot{z}_{n-1} \\ \dot{z}_n \end{pmatrix} = \begin{pmatrix} -K_1^1 & 0 & 0 & \cdots & 0 & \cdots & 0 & 0 \\ 0 & -K_1^2 & 0 & \cdots & 0 & \cdots & 0 & 0 \\ 0 & 0 & -K_1^3 & \cdots & 0 & \cdots & 0 & 0 \\ \cdots & \cdots & \cdots & \cdots & \cdots & & & \\ 0 & 0 & 0 & \cdots & -K_1^i & \cdots & 0 & 0 \\ \cdots & \cdots & \cdots & \cdots & \cdots & & & \\ 0 & 0 & 0 & \cdots & 0 & \cdots & -K_1^{n-1} & 0 \\ 0 & 0 & 0 & \cdots & 0 & \cdots & 0 & -K_1^n \end{pmatrix} \begin{pmatrix} z_1 \\ z_2 \\ z_3 \\ \cdots \\ z_i \\ \cdots \\ z_{n-1} \\ z_n \end{pmatrix} \tag{B.117}$$

or equivalently

$$\dot{Z} = KZ \tag{B.118}$$

By selecting the eigenvalues of matrix K to be in the left complex semiplane, one has that

$$lim_{t \to \infty} Z = 0_{n \times 1} \tag{B.119}$$

which also implies that $lim_{t \to \infty} x_1 = x_1^d$, $lim_{t \to \infty} x_2 = \alpha_1 = x_2^d$, $lim_{t \to \infty} x_3 = \alpha_2 = x_3^d, \cdots, lim_{t \to \infty} x_i = \alpha_{i-1} = x_i^d, \cdots$, $lim_{t \to \infty} x_{n-1} = \alpha_{n-2} = x_{n-1}^d$, and $lim_{t \to \infty} x_n = \alpha_{n-1} = x_n^d$.

To prove asymptotic stability for the proposed control scheme, the following Lyapunov function can be defined:

$$V = \sum_{i=1}^N \frac{1}{2} z_i^2 \tag{B.120}$$

The time derivative of the aforementioned Lyapunov function is

$$\dot{V} = \sum_{i=1}^N z_i \dot{z}_i \Rightarrow \dot{V} = -\sum_{i=1}^N K_1^i z_i^2 \Rightarrow \dot{V} < 0 \tag{B.121}$$

By selecting the feedback control gains K_1^i, $i = 1, \cdots, n$ to be $K_1^i > 0$, the asymptotic stability of the control loop is assured.

B.2.3 Comparison to backstepping control for nonlinear systems

It can be also noted that the so-called backstepping control which is based on the recursive computation of the control signal of the system after applying virtual control inputs to the individual rows of the state-space model, is a subcase of the proposed flatness-based control method. By showing that each row of the state-space model stands for a subsystem that satisfies differential flatness properties, one can apply effectively, to each subsystem, the controller design stages found in input–output linearizing flatness-based control methods. A backstepping control law can be derived for systems of the form [439,711]:

$$\begin{aligned} \dot{x} &= f(x) + g(x)u \\ y &= h(x) \end{aligned} \tag{B.122}$$

for which holds

$$\dot{y} = \frac{\partial h(x)}{\partial x}\dot{x} = \frac{\partial h(x)}{\partial x}[f(x) + g(x)u] = L_f h(x) + L_g h(x)u \tag{B.123}$$

where the Lie derivatives are defined as

$$L_f h(x) = \frac{\partial}{\partial x}h(x)f(x),\ L_g h(x) = \frac{\partial}{\partial x}h(x)g(x) \tag{B.124}$$

The system of Eq. (B.122) can be written in cascading form

$$\begin{aligned} \dot{x}_1 &= f_1(x_1) + g_1(x_1)x_2 \\ \dot{x}_2 &= f_2(x_1, x_2) + g_2(x_1, x_2)x_3 \\ \dot{x}_3 &= f_3(x_1, x_2, x_3) + g_3(x_1, x_2, x_3)x_4 \\ &\cdots\ \cdots \\ &\cdots\ \cdots \\ \dot{x}_{n_1} &= f_{n-1}(x_1, x_2, \cdots, x_{n-1}) + g_{n-1}(x_1, x_2, \cdots, x_{n-1})x_n \\ \dot{x}_n &= f_n(x_1, x_2, \cdots, x_n) + g_{n-1}(x_1, x_2, \cdots, x_n)u \\ y &= h(x_1) \end{aligned} \tag{B.125}$$

Then, the n-th order backstepping SISO controller is given by the recursive relation

$$\alpha_1 = \frac{1}{L_{g_1}h(x_1)}[\dot{y}_d - L_{f_1}h(x_1) - k_1 z_1 - n_1(z_1)z_1]$$

$$\alpha_2 = \frac{1}{L_{g_2}h(x_1, x_2)}[\dot{\alpha}_1 - f_2(x_1, x_2) - L_{g_1}h(x_1)z_1 - k_2 z_2 - n_2(z_2)z_2] \cdots$$

$$\cdots\ \cdots\ \cdots$$

$$\alpha_i = \frac{1}{L_{g_i}h(x_1, x_2, \cdots, x_i)}[\dot{\alpha}_{i-1} - f_i(x_1, \cdots, x_i) - g_{i-1}(x_1, x_2, \cdots, x_{i-1})z_{i-1} - k_i z_1 - n_i(z_i)z_i] \cdots$$

$$\cdots\ \cdots\ \cdots$$

$$\alpha_n = \frac{1}{L_{g_n}h(x_1, x_2, \cdots, x_n)}[\dot{\alpha}_{n-1} - f_n(x_1, \cdots, x_n) - g_{n-1}(x_1, x_2, \cdots, x_{n-2})z_{n-1} - k_n z_n - n_n(z_n)z_n]$$

$$u = \alpha_n$$

with $z_1 = h(x_1) - y_d$ and $z_i = x_i - \alpha_{i-1}$. Such a backstepping controller results in closed-loop dynamics given by $\dot{z} = -K(z)z + S(x)z$, with

$$K(z) = \begin{pmatrix} k_1 + n_1(z_1) & 0 & \cdots & 0 \\ 0 & k_2 + n_2(z_2) & \cdots & 0 \\ \cdots & \cdots & \cdots & \cdots \\ 0 & 0 & \cdots & k_n + n_n(z_n) \end{pmatrix} \tag{B.126}$$

$$S(x) = \begin{pmatrix} 0 & L_{g_1}(x_1) & 0 & \cdots & 0 & 0 & 0 \\ -L_{g_1}(x_1) & 0 & g_2(x_1, x_2) & 0 & 0 & 0 & \\ 0 & -g_2(x_1, x_2) & 0 & \cdots & 0 & 0 & 0 \\ \cdots & \cdots & \cdots & \cdots & \cdots & \cdots & \cdots \\ 0 & 0 & 0 & \cdots & 0 & g_{n-1}(x_1, \cdots, x_{n-1}) & 0 \\ 0 & 0 & 0 & \cdots & -g_{n-1}(x_1, \cdots, x_{n-1}) & 0 & 0 \\ 0 & 0 & 0 & \cdots & 0 & -g_n(x_1, \cdots, x_n & 0) \end{pmatrix} \tag{B.127}$$

Epilogue

The monograph has presented advances on nonlinear control and multi-loop flatness-based control for complex dynamical systems. The new control methods are free of shortcomings met in control schemes which are based diffeomorphisms and global linearization (complicated changes of state variables, forward and backward state-space transformations, singularities). It is shown that such methods can be used in a wide class of nonlinear dynamical systems without needing to transform the systems' state-space model into equivalent linearized forms. It is also shown that the new control methods can be implemented in a computationally simple manner and are also followed by global stability proofs. Application domains of the nonlinear optimal and flatness-based control concepts for complex dynamical systems have been: (i) robotic systems; (ii) robotic cranes and pendulums; (iii) space robots; aerospace systems and satellites; (iv) mechatronic systems; (v) power electronics; (vi) biosystems; (vii) financial systems. One can also distinguish the potential for using the aforementioned control approaches not only in areas of high technological interest but also in domains of wider societal impact. The research domain is indeed wide, and there is ongoing work which is expected to give additional important results in the forthcoming years. Research on the above topics will continue in the following years, and significant findings are expected to be obtained.

Bibliography

[1] A.F. Abad, L. Zhang, Z. Wei and C. Ma, Optimal capture of a tumbling object in orbit using a space manipulator, *Journal of Intelligent and Robotic Systems*, vol. 86, pp. 199–211, 2017.

[2] M. Abbas, J. Narayan and S.K. Dwivedy, A systematic review on cooperative dual-arm manipulators: modelling, planning, control and vision strategies, *Journal of Intelligent Robotics and Applications*, pp. 1–25, 2023.

[3] G. Abbasnejad, J. Yoon and H. Lee, Optimum kinematics design of a planar cable-driven parallel robot with wrench closure gait trajectory, *Mechanism and Machine Theory*, vol. 99, pp. 1–18, 2016.

[4] O.J. Abdelbaqi, A. Nasir and P.J. Miller, Energy management for a 8000hp hybrid hydraulic mining shovel, *IEEE Transactions on Industry Applications*, vol. 52, no. 6, pp. 5041–5050, 2016.

[5] E.N. Abdulwahab and C. Hangquan, Periodic motion suppression based on control of wing-rock in aircraft lateral dynamics, *Aerospace Science and Technology*, vol. 12, pp. 205–301, 2008.

[6] S. Achanya, B. Mondal, R.K. Upadhyaya and P. Das, Exploring noise-induced dynamics and optimal control strategy of iSIR cholera transmission model, *Nonlinear Dynamics*, vol. 112, pp. 3951–3975, 2024.

[7] M.R. Adaryani, S.A. Taher and J.M. Guerrero, Model predictive control of direct-drive wave power generation system connected to DC microgrid through DC cable, *International Transactions of Electrical Energy Systems*, vol. 30, article no. e12484, pp. 1–18, 2020.

[8] M.R. Adaryani, S.A. Taher and J.M. Guerrero, Improved direct model predictive control for variable magnitude variable frequency wave energy converter connected to constant power load, *Journal of Energy Storage*, vol. 43, pp. 103175–103183, 2021.

[9] A. Aghaeeyand and M. Yazdanpanah, Self-triggered impulsive control of a nonlinear time-delay system: application to chemotherapeutic dose regimen design, *Nonlinear Analysis: Hybrid Systems*, vol. 41, pp. 1–16, 2021.

[10] S.K. Agrawal, K. Pathak, J. Franch, R. Lampariello and G. Hirzinger, A differentially-flat open-chain space robot with arbitrarily oriented joint axes and two momentum wheels at the base, *IEEE Transactions on Automatic Control*, vol. 54, no. 9, pp. 2185–2192, 2009.

[11] S. Ahmed and E.C. Kerrigan, Suboptimal predictive control for satellite detumbling, *Journal of Guidance, Control and Dynamics*, vol. 37, no. 3, pp. 850–859, 2014.

[12] A.A. Ahmed, B.A. Moharam and E.E. Rashad, Improving energy efficiency and economies of motor-pump system using electric variable speed drives for automatic transition of working points, *Computers and Electrical Engineering*, vol. 97, pp. 107607–107629, 2022.

[13] S. Ahmed, A.T. Azar, M. Abdel-Ary, H. Khan and J Alcabut, A nonlinear system of hybrid fractional differential equations with application to fixed time sliding-mode control for leukemia therapy, *Aim Shams Engineering Journal*, vol. 15, pp. 102566–102578, 2022.

[14] B.E Aiseba and C. Benosman, CML dynamics: optimal control of age-structured stem cells population, *Mathematics and Computers in Simulation*, vol. 81, pp. 1962–1977, 2011.

[15] A. Al-Huge Ali and B. Buchstaller, A multiple-model approach to fault detection and localization in multi-terminal HVDC systems, In: *IEEE IECON 2015, 41st Intl. Conference of the Industrial Electronics Society*, Yokohama, Japan, Nov. 2015.

[16] M. Alamir and A. Murilo, Swing-up and stabilization of a twin-pendulum under state and control constraints by a fast NMPC scheme, *Automatica*, vol. 44, no. 5, pp. 1319–1324, 2008.

[17] M. Alamir, Robust feedback design for combined therapy of cancer, *Optimal Control Applications and Methods*, vol. 35, no. 1, pp. 77–88, 2014.

[18] J.P. Alepuz, M.R. Emami and J. Pomares, Direct image-based visual servoing of free-floating space manipulators, *Journal of Aerospace Science and Technology*, vol. 55, pp. 1–9, 2016.

[19] K.R. Anne, J.C. Chejdou and K. Kyamakya, Bifurcation analysis and synchronization issues in a three-echelon supply chain, *International Journal of Logistics Research and Applications*, vol. 12, no. 5, pp. 347–362, 2007.

[20] A. Amari, A. Molaei, M.A. Khosravi and M. Hossani, Control-based tension distribution for fully-constrained cable-driven robots, *IEEE Transactions on Industrial Electronics*, vol. 69, no. 11, pp. 11383–11393, 2022.

[21] T.S. Amer, M.A. Bek, M.S. Nael, M.A. Sirwah and A. Arab, Stability of the dynamical motion of a damped 3-DOF autoparametric pendulum system, *Journal of Vibration Engineering & Technologies*, pp. 1–21, 2022.

[22] S.M. Amrr, A. Alturki and K. Zafar, Advanced control design for the support of fast rotating shaft using active magnetic bearings, *Transportation Engineering*, vol. 10, pp. 100139–100148, 2022.

[23] Q. An, Y. Zhang, Q. Hu, M. Li, J. Li and A. Mao, Time-optimal path tracking for dual-arm free-floating space manipulator system using convex programming, *IEEE Transactions on Aerospace and Electronic Systems*, vol. 59, no. 5, pp. 6670–6682, 2023.

[24] B. An, B. Wang, Y. Wang and L. Liu, Adaptive terminal sliding-mode control for reentry vehicle based on nonlinear disturbance observer, *IEEE Access*, vol. 7, pp. 154502–154514, 2019.

[25] S.G. Anavatti, J.Y. Choi and P.P. Wang, Design and implementation of fuzzy logic controller for wing-rock, *International Journal of Control, Automation and Systems*, vol. 2, no. 4, pp. 494–500, 2004.

[26] B. Andrievsky and K.B. Boykov, Numerical and laboratory experiments with controlled coupled pendulum, *5th IFAC Symposium on Nonlinear Control Systems, IFAC Proceedings*, vol. 14, no. 6, pp. 901–806, 2001.

[27] B. Andrievsky, E.V. Kudryashova, N.V. Kuznetsov and O.A. Kuznetsova, Aircraft wing-rock oscillations suppression by simple adaptive control, *Aerospace Science and Technology*, vol. 105, article no. 100049, pp. 1–10, 2020.

[28] P. Ankomah and B.R. Levin, Exploring the collaboration between antibiotics and the immune response in the treatment of acute self-limiting infections, *Proceedings of the National Academy of Sciences (PNAS)*, vol. 111, no. 23, pp. 9331–8338, 2014.

[29] E. D' Antonio, G. Galofaro, F. Patané, M. Casadia and L. Masia, A dual-arm haptic exoskeleton for dynamical coupled manipulation, In: *IEEE/ASME 2021 Conference on Advanced Intelligent Mechatronics*, Delft, Netherlands, July 2020.

[30] V. Arakelian and S. Briot, Dynamic balancing of the SCARA robot, In: *RmanSys 2008, 17th CISM-IFToMM Symposium on Robot Design*, Tokyo, Japan, July 2008.

[31] M.A. Arefi, N. Vafamand, B. Hamayoun and M. Davoudi, Command filtered backstepping control of constrained flexible joint robotic manipulators, *IET Control Theory and Applications*, pp. 1–13, 2023.

[32] E. Arondo-Escolastico, M. Guinaldo, M. Santos and S. Dormido, Control of a chain pendulum: a fuzzy logic approach, *International Journal of Computational Intelligence Systems*, vol. 9, no. 2, pp. 281–295, 2016.

[33] K.M. Arthur, M. Basu and S.Y. Yoon, Stabilization of compressor surge: systems with uncertain equilibrium flow, *ISA Transactions*, vol. 93, pp. 115–134, 2019.

[34] L.T. Aschepkov, D.V. Dolgy, Taekyun Kim, and Ravi P Agarwal, *Optimal Control*, Springer, 2017.

[35] M. Ashouri, F. Faria da Silva and C.L. Bak, Analysis and validation of mathematical morphology filters for single-ended fault localization in VSC-HVDC links, *Electrical Engineering*, vol. 103, pp. 1583–1596, 2021.

[36] F. Asuno, Stealth walking of 3-link planar underactuated biped, In: *IEEE IROS 2017, IEEE Intl. Conference on Intelligent Robots and Systems*, Vancouver, Canada, Sep. 2020.

[37] V. Avina-Corral, J. Rangel-Magdalena, C. Moralez-Perez and J. Hernamdez, Bearing fault detection in adjustable speed-drive powered induction machine by using motor current signature analysis and goodness-of-fit tests, *IEEE Transactions on Industrial Informatics*, vol. 17, no. 12, pp. 8265–8274, 2021.

[38] A.M. Ba-Alawi, H.T. Nguyen and C.K. Yoo, Sustainable design of a solar/wind powered reverse osmosis with cooperation demand-side water management: a coordinated sizing approach with a fuzzy decision-making model, *Energy Conversion and Management*, vol. 295, pp. 117624–117841, 2023.

[39] N. Babaei and M.U. Salamci, State-dependent Riccati equation based model reference adaptive stabilization of nonlinear systems with application to cancer treatment, In: *19th IFAC World Congress*, Cape Town, South Africa, Aug. 2014.

[40] R.S. Babu, U.P. Rajeev and R. Lethakumori, Hybrid guidance algorithm using flatness and dynamic inversion for RLV, In: *IFAC 3rd Intl. Conf. on Control and Optimization of Dynamical Systems*, Kanpur, India, March 2014.

[41] C.J. Backl, J.T. Gravdahl and S. Skogestad, Robust control of a two-state Greitzer compressor model by state-feedback linearization, In: *IEEE CCA 2016, IEEE 2016 Conference on Control Applications*, Buenos Aires, Argentina, Sep. 2016.

[42] M. Badzioch, P. Bajger and U. Forys, Competition between populations: preventing domination of resistant population with optimal control, *Applied Mathematical Modelling*, vol. 114, pp. 671–693, 2023.

[43] S. Bahadure, C. Venkatesh and F. Kazi, Structure preserving optimal control of 2D SpiderCrane, In: *IEEE SysCon 2012, Intl. Conf. on Systems and Control*, Vancouver, Canada, 2012.

[44] B. Baigzadehnoe, Z. Rahmani, A. Khosravvi and B. Rezaie, On position/force tracking control problem of cooperative robot manipulators using adaptive fuzzy backstepping approach, *ISA Transactions*, vol. 70, pp. 432–446, 2017.

[45] K.O. Bainov, E.A. Lomonova, A.J.A. Vandenput and A. Tyagunov, Surge control of the electrically driven centrifugal compressors, *IEEE Transactions on Industry Applications*, vol. 42, no. 6, pp. 1523–1532, 2006.

[46] N.J. Baker, M.A.H. Raihar and A.A. Almeraya, A cylindrical linear permanent magnet Vernier hybrid machine for wave energy, *IEEE Transactions on Energy Conversion*, vol. 34, no. 2, pp. 691–700, 2018.

[47] S. Baklouti, E. Courtelle, P. Lemoine and S. Caro, Vibration reduction of cable-driven parallel robots through elasto-dynamic model-based control, *Mechanism and Machine Theory*, vol. 139, pp. 329–345, 2019.

[48] S. Baklouti, Vibration analysis and reduction of cable-driven parallel robots, Thèse de doctorat, INSA de Rennes, 2018, Thèse No 18ISAR 35/D18-35.

[49] S. Baklouti, E. Courtelle, P. Lemoine and S. Caro, Input-shaping for feedforward control of cable-driven parallel robots, *Journal of Dynamical Systems, Measurement and Control*, vol. 143, no. 2, 10.1115/1.4048354, hal-02932088v2, 2021.

[50] S. Balea, A. Halanay, D. Jordan, M. Neamtu and C.A. Safta, Stability analysis of a feedback model for the action of the immune system in leukemia, *Mathematical Modelling of Natural Phenomena*, vol. 9, no. 1, pp. 108–132, 2014.

[51] H. Bambili and A. Billard, Dual-arm control for coordinated fast grabbing and tossing of an object, *IEEE Robotics and Automation Magazine*, vol. 29, no. 3, pp. 127–138, 2022.

[52] J.S. Bang, H. Shin, S.K. Park and J.M. Seo, Robust tracking and vibration suppression for a two-inertia system by combining backstepping approach and disturbance observer, *IEEE Transactions on Industrial Electronics*, vol. 57, no. 9, pp. 1197–1206, 2010.

[53] N. Baofeng, D. Xueying and W. Bing, Effects of wing locations on wing-rock induced by forebody vortices, *Chinese Journal of Aeronautics*, vol. 29, no. 5, pp. 1226–1236, 2016.

[54] J.P. Barbot, M. Fliess and T. Floquet, An algebraic framework for the design of nonlinear observers with unknown inputs, In: *IEEE CDC 2007, IEEE 46th Intl. Conference on Decision and Control*, New Orleans, USA, Dec. 2007.

[55] L. Barbozza, D. Zanotto, G. Rosati and S.K. Agrawal, Design and optimal control of an underactuated cable-driven micro-macro robot, *IEEE Robotics and Automation Letters*, vol. 2, no. 2, pp. 896–903, 2017.

[56] A. Barreo and M.E. Hernandez, Optimal control of a delayed breast cancer stem cells nonlinear model, *Optimal Control Applications and Methods*, vol. 37, no. 2, pp. 248–258, 2016.

[57] A.R. Bartman, C.W. McFall, P.D. Christofides and Y. Cohen, Model-predictive control of feed flow reversal in a reverse osmosis desalination process, *Journal of Process Control*, vol. 19, pp. 413–442, 2009.

[58] A.R. Bartman, A. Zhu, P.D. Christofides and Y. Cohen, Minimizing energy consumption in reverse osmosis membrane desalination using optimization-based control, *Journal of Process Control*, vol. 20, pp. 1261–1269, 2010.

[59] F. Basile, F. Caccavale, P. Chiacchio, J. Coppola and A. Marino, A decentralized kinematic control architecture for collaborative and cooperative multi-arm systems, *Mechatronics*, vol. 23, pp. 1100–1112, 2013.

[60] M. Basseville and I. Nikiforov, *Detection of Abrupt Changes: Theory and Applications*, Prentice-Hall, 1993.

[61] Y. Batmani and H. Khalazadeh, Optimal chemotherapy in cancer treatment: state dependent Riccati equation control and extended Kalman filter, *Optimal Control Applications and Methods*, vol. 34, no. 5, pp. 562–577, 2013.

[62] P. Beckerle, H. Shaede, N. Butzek and S. Rinderknecht, Balancing filters: an approach to improve model-based fault diagnosis based on parity equations, *Mechanical Systems and Signal Processing*, vol. 29, pp. 137–147, 2012.

[63] J. Begey, L. Cuvillon, M. Lesellier M. Gouttefarde and J. Gangloff, Dynamic control of parallel robots driven by flexible cables and actuated by position controlled winches, *IEEE Transactions on Robotics*, vol. 55, no. 1, pp. 286–293, 2019.

[64] B. Behegen and T.M. Gravdahl, Active surge control of compression system using drive torque, *Automatica*, vol. 44, pp. 1135–1140, 2008.

[65] R. Behrad, G.M. Bone, G. Ashby and E. Li, Design, implementation and control of an improved hybrid pneumatic-electric actuator for robot arms, *IEEE Access*, vol. 7, pp. 14699–14714, 2019.

[66] J.P. Belfo, A.J. Aguiar and J.M. Lemos, Distributed inverse optimal control for discrete-time nonlinear multi-agent systems, *IEEE Control Systems Letters*, vol. 5, no. 6, pp. 2096–2103, 2021.

[67] Z. Belkhotr, M. Pavon, J.C. Mathews, *et al.*, Stochastic Norton-Simon-Messanguéé tumor growth modelling: controlled and mixed effect, uncontrolled analysis, *IEEE Transactions on Control Systems Technology*, vol. 29, no. 2, pp. 704–717, 2021.

[68] A. Bellini, F. Filippeti, C. Tassoni and G.A. Capolino, Advances in Diagnostic Techniques for induction machines, *IEEE Transactions on Industrial Electronics*, vol. 55, no. 13, pp. 4103–4124, 2003.

[69] I. Ben Ali, M. Turki, J. Belhadj, and X. Roboam, System's design and energy management of a stand-alone battery-less PV/wind driven brackish water reverse osmosis desalination system, *Sustainable Energy Technologies and Assessments*, vol. 42, pp. 100884–100898, 2020.

[70] C. Benosman, B.E, Ainseba and A. Ducrat, Optimization of cytostatic leukemia therapy in an Advection-Reaction-Diffusion model, *Journal of Optimization Theory and Applications*, vol. 167, pp. 296–325, 2015.

[71] Y. Bensafia, S. Ladaci, K. Khettab and A. Chemoni, Fractional order model reference adaptive control for SCARA robot trajectory tracking, *International Journal of Industrial and Systems Engineering, Inderscience*, vol 30, no. 2, pp. 138–156, 2018.

[72] C.M. Best, M.T. Gillespie, P. Hyatt, L. Rapert, V. Sherrad and M.D. Killpack, A new soft robot control method: using model predictive control for a pneumatically actuated humanoid, *IEEE Robotics and Automation Magazine*, vol. 23, no. 3, pp. 75–84, 2016.

[73] T. Bentaleb, A. Cacitti, S. De Franciscis and A. Garulli, Model predictive control for pressure regulation and surge prevention in centrifugal compressors, In: *IEEE ECC 2015 Linz*, Austria, July 2015.

[74] I. Berezansky, S. Bunimmovitch-Mendrazitsky and B. Sklyar, Stability and controllability issues in mathematical modelling of the intensive treatment of leukemia, *Journal of Optimization Theory and Applications*, vol. 167, pp. 326–341, 2015.

[75] L.D. Berkovitz and N.G. Medhin, *Nonlinear Optimal Control Theory*, Chapman and Hall/CRC Press, 2012.

[76] Q. Bi, G. Wang, R. Yang, Y. Liu, Y. Lu and S. Xing, Study on theory and methods of payload online estimation for cable shovels, *Journal of the Brazilian Society of Mechanical Sciences and Engineering*, vol. 41, no. 53, pp. 52–64, 2019.

[77] M. Bianchi, A. van der Maas, E. Maljaari and W.P.M.H. Heemels, Offset-free MPC for resource sharing on a nonlinear SCARA robot, In: *IFAC NMPC 2018, 6th IFAC Conference on Nonlinear Model Predictive Control*, Madison, Wisconsin, USA, Aug. 2018.

[78] I. Bicer, *Supply Chain Analytics: An Uncertainty Modelling Approach*, Springer, 2023.

[79] F. Blanquart, Evolutionary epidemiology models to predict the dynamics of antibiotic resistance, *Journal of Evolutionary applications*, vol. 12, pp. 365–383, 2010.

[80] B. Blunier, M. Pucci, G. Cirrincione and A. Miraoui, A stroll compressor with a high-performance induction motor drive for the air management of a PEMFC system for automotive applications, *IEEE Transactions on Industry Applications*, vol. 44, no. 6, pp. 1966–1976, 2008.

[81] J.E. Bobrow and B.W. McDonell, Modelling, identification and control of a pneumatically actuated, force controllable robot, *IEEE Transactions on Robotics and Automation*, vol. 14, no. 5, pp. 732–742, 1998.

[82] M. Boccadoro, F. Martinelli and P. Valigi, Supply chain management by H-infinity control, *IEEE Transactions on Automation Science and Engineering*, vol. 5, no. 4, pp. 703–708, 2008.

[83] M. Boek and A. Kugi, Real-time nonlinear model-predictive path-following control of a laboratory tower crane, *IEEE Transactions on Control Systems Technology*, vol. 22, no. 4, pp. 1461–1473, 2014.

[84] G.M. Bone, M. Xitad and J. Flet, Position control of hybrid pneumatic-electric actuator using discrete-valued model-predictive control, *Mechatronics*, vol. 25, pp. 1–10, 2015.

[85] R. Bonalli, B. Hérissé and E. Trélat, Optimal control of endo-atmospheric launch vehicle systems: geometric and computational issues, *IEEE Transactions on Automatic Control*, vol. 65, no. 6, pp. 2418–2493, 2020.

[86] B. Bonnard, L. Faoubourg and E. Trélat, *Mécanique céleste et contrôle des véhicules spatiaux*, Springer, 2005.

[87] B. Bonnard and J. Rouot, Optimal control of the controlled Lotka-Volterra equations with applications – the permanent case, *SIAM Journal of Applied Dynamical Systems*, vol. 22, no. 4, pp. 2671–2691, 2023.

[88] B. Bonnard, J. Rouat and C.J. Silva, Geometric optimal control of the generalized Lotka-Volterra model of the intestinal microbioma, *Optimal Control Applications and Methods*, vol. 45, pp. 544–574, 2024.

[89] F. Borrelli, A. Bemporad and M. Morari, *Predictive Control for Linear and Hybrid Systems*, Cambridge University Press, 2017.

[90] H. Bouchaita and S. Ladaci, Fractional adaptive SMC fault tolerant control against actuator failures for wing-rock supervision, *Aerospace Science and Technology*, vol. 114, article no. 106745, pp. 1–9, 2021.

[91] S. Bououden, D. Boutat, G. Zheng, J.P. Barbot and F. Kratz, A triangular canonical form for a class of 0-flat nonlinear systems, *International Journal of Control*, vol. 84, no. 2, pp. 261–269, 2011.

[92] S. Bouloum, S. Labiod and H. Boubertakh, Linear adaptive actuator failure compensation for wing-rock motion control, *Aerospace Science and Technology*, vol. 67, pp. 155–168, 2017.

[93] S. Bowong, Optimal control of the transmission dynamics of tuberculosis, *Nonlinear Dynamics*, vol. 61, pp. 729–748, 2010.

[94] A.S. Bratus, E. Fimmel, Y. Todorov, Y.S. Semenov and F. Nurnberg, On strategies on a mathematical model for leukemia therapy, *Nonlinear analysis: real world applications*, vol. 13, pp. 1044–1059, 2012.

[95] A. Bratus, I. Yegorov and D. Yurchenko, Dynamic mathematical models of therapy processes against glioma and leukemia under stochastic uncertainties, *Meccanica di Materialli e delle Strutture*, vol. 6, no. 1, pp. 131–138, 2016.

[96] D.J. Braun, F. Petit, F. Huber, *et al.*, Robots driven by compliant actuators: optimal control under actuation constraints, *IEEE Transactions on Robotics*, vol. 29, no. 5, pp. 1085–1101, 2013.

[97] R.W. Brockett and M. Li, A light weight rotary double pendulum: maximizing the domain of attraction, In: *IEEE CDC 2007, IEEE 42nd Conference on Decision and Control*, Maui, Hawaii, USA, Dec. 2003.

[98] B. Brogliato, Modelling, analysis and control of robot-object nonsmooth underactuated Lagrangian systems: a tutorial overview and perspectives, *Annual Reviews in Control*, vol. 55, pp. 297–337, 2023.

[99] G. Buandonno and A. de Luca, Efficient computation of inverse dynamics and feedback linearization for VSA-based robots, *IEEE Robotics and Automation Letters*, vol. 1, no. 2, pp. 908–915, 2018.

[100] D. Buccieri, P. Mullhaupt and D. Bonvin, SpiderCrane: model and properties of a fast weight handling equipment, In: *IFAC 16th Triennial World Congress*, Prague, Czech Republic, 2005.

[101] D. Buccieri, J. Sanchez, S. Dormido, R. Mullhaupt and D. Bonvin, Interactive 3D Simulation of Flat Systems: the SpiderCrane as a case study, In: *IEEE CDC 2005, IEEE 44th Intl. Conference on Decision and Control*, Seville, Spain, Dec. 2005.

[102] S. Budini and N.F. Thornhill, Control of centrifugal compression via model predictive control for enhanced oil recovery applications, In: *2nd IFAC Workshop on Automatic Control in Offshore Oil and Gas Production*, Florianopolis, Brazil, May 2015.

[103] B.H. Bue, S.K. Sul, J.H. Kwon and J.S. Bycon, Implementation of sensorless vector control for super-high-speed PMSM turbo-compressors, *IEEE Transactions on Industry Applications*, vol. 39, no. 3, pp. 811–818, 2003.

[104] M. Brucker and P. Heidtmann, Flatness-based control of a rotary vane actuator, In: *Proc of 16th Intl. Conf. on Mathematical Theory of Networks and Systems, IEEE MTNS 2004*, Leuven, Belgium, July 2004.

[105] F. Cacace, V. Cusimano, A. Germani, P. Palumbo and F. Pepe, Closed-loop control of tumor growth by means of anti-angiogenic administration, *Mathematical Biosciences and Engineering*, vol. 15, no. 4, pp. 827–839, 2018.

[106] F. Cacace, V. Cusimano and P. Palumbo, Optimal impulsive control with application to anti-angiogenic tumor therapy, *IEEE Transactions on Control Systems Technology*, vol. 28, no. 1, pp. 106–117, 2020.

[107] I.S. Cade, P.T. Keogh and M.N. Sahinkaya, Fault identification of rotor/magnetic bearing systems using discrete-time wavelet coefficients, *IEEE/ASME Transactions on Mechatronics*, vol. 10, pp. 648–657, 2005.

[108] F. Calise, F. Liberati Cappiello, R. Vaneli and M. Vicidomini, Economic assessment of renewable energy systems integrating photovoltaic panels, seawater desalination and water storage, *Applied Energy*, vol. 253, pp. 113575–113591, 2019.

[109] Z. Cao, J. Dong, F. Wani, *et al.*, Sliding-mode control with neural network for active suspension bearing system, In: *IEEE IECON 2019, IEEE 45th Annual Conference of the Industrial Electronics Society*, Lisbon, Portugal, Oct. 2019.

[110] E. Capello, G. Guglieri and D. Sartori, Performance evaluation of a L_1 adaptive controller for wing-body rock suppression, *Journal of Guidance, Control and Dynamics*, vol. 35, no. 6, pp. 1702–1708, 2012.

[111] F. Caccavale, P. Chiacchio, A. Marino and L. Villani, Six-DOF impedance control of dual-arm cooperative manipulators, *IEEE/ASME Transactions on Mechatronics*, vol. 13, no. 5, pp. 576–586, 2008.

[112] W.W. Cai, L.P. Yang and Y.W. Zhu, Optimal reorientation of asymmetric underactuated spacecraft using differential flatness and receding horizon control, *Advances in Space Research*, vol. 55, pp. 343–353, 2015.

[113] Y. Cai, S. Zheng, W. Liu, Z. Qu, J. Zhu and J. Han, Sliding-mode control of ship-mounted Stewart platforms for wave compensation using velocity feed-forward, *Ocean Engineering*, vol. 236, pp. 1–11, 2021.

[114] Y. Cai, S. Zhang, W. Liu, Z. Qu and J. Han, Model analysis and modified control method of ship-mounted Stewart platforms for wave compensation, *IEEE Access*, vol. 9, pp. 4505–4517, 2021.

[115] J. Calcavanti-Santos, M. Gouttefard and A. Chemori, A nonlinear model predictive control for the position tracking of cable-driven parallel robots, *IEEE Transactions on Robotics*, vol. 38, no. 4, pp. 2597–2616, 2022.

[116] C. Carrère, Optimization of an in-vitro chemotherapy to avoid resistant tumors, *Journal of Theoretical Biology*, vol. 413, pp. 24–33, 2017.

[117] M.L. de Carvalho Moura Moreira, T.R. Oliveira and K.I. Pinhero de Melo Queiroz, Adaptive backstepping control based on global HOSM differentiators with dynamic gains, *IEEE Access*, vol. 12, pp. 84498–84514, 2024.

[118] L. Cao, X.Q. Chen, and T. Sheng, Fault tolerant small satellite attitude control using adaptive non-singular terminal sliding modes, *Advances in Space Research*, vol. 51, pp. 2374–2393, 2013.

[119] J. Cao, S.G. Xie and R. Das, MIMO sliding-mode controller for gait exoskeleton driven by pneumatic muscles, *IEEE Transactions on Control Systems Technology*, vol. 26, no. 1, pp. 274–283, 2018.

[120] Y. Cao and J. Huang Neural-network-based nonlinear model predictive tracking control of a pneumatic muscle actuator-driven exoskeleton, *IEEE/CAA Journal of Automatica Sinica*, vol. 7, no. 6, pp. 1478–1488, 2020.

[121] S. Celikovsky and M. Anderle, Stable walking gaits for a three-link planar biped robot with two actuators based on the collocated virtual impedance constraints and the cyclic unactuated variables, In: *12th IFAC Symposium on Robot Control, IFAC Syroco 2018*, Budapest, Hungary, Aug. 2018.

[122] S. Celikovsky and M. Anderle, Exact feedback linearization of the collocated constrained dynamics of the three-link with adjustable torso and its application in the underactuated planar walking, In: *IEEE ICCA 2019, IEEE 2019 15th Intl. Conference on Control and Automation*, Edinburgh, Scotland, July 2019.

[123] S. Celikovsky and M. Anderle, Chain of four integrators as a possible essence of the underactuated planar walking, In: *IEEE MICNON 2021, 3rd IFAC Conference on Modelling, Identification and Control of Nonlinear Systems*, Tokyo, Japan, Sep. 2021.

[124] T. Chang, P. Jaroonshiriphan, M. Bernhardt and P. Ludden, Web-based command shaping of Cobra 600 Robot with a swinging load, *IEEE Transactions on Industrial Informatics*, vol. 2,, no. 1, pp. 59–69, 2006.

[125] I. Chang, S.Y. Park and K.M. Choi, Nonlinear attitude control of tethered-connected multi-satellite in three-dimensional space, *IEEE Transactions on Aerospace and Electronic Systems*, vol. 46, no 4, pp. 1950–1968, 2010.

[126] B. Chang, O. Cwikowski, M. Barnes, R. Shuttleworth, A. Beddard and P. Coventry, Review of different fault detection methods and their impact on pre-emptive HSC-HVDC dc protection performance, *IET High Voltage*, vol. 2, no. 2, pp. 211–219, 2017.

[127] T.L. Chen, C.C. Chen and C.I. Huang, Feedback linearization control and its application to MIMO cancer immunotherapy, *IEEE Transactions on Control Systems Technology*, vol. 18, no. 4, pp. 953–960, 2010.

[128] D. Chao, R. Qi and B. Jiang, Adaptive fault-tolerant attitude control for hypersonic reentry vehicle subject to complex uncertainties, *Journal of the Franklin Institute*, vol. 359, pp. 5458–5487, 2022.

[129] B. Chappui, S. Gavin, L. Rigazzi and M. Carpita, Speed control of a multi-phase active way linear motor based on back EMF estimator, *IEEE Transactions on Industrial Electronics*, vol. 62, no. 12, pp. 7299–7308, 2015.

[130] N.A. Chaturvedi and S.P. Bhat, Output feedback semiglobal stabilization of stall dynamics for preventing hysteresis and surge in axial-flow compressors, *IEEE Transactions on Control Systems Technology*, vol. 14, no. 2, pp. 301–307, 2006.

[131] M.A.M. Cheema, J.E. Fletcher, M. Forshadmin, D. Xiao and M.F. Rahman, Combined speed and direct thrust force control of linear permanent-magnet synchronous motors with sensorless speed estimation using a sliding-mode control with integral action, *IEEE Transactions on Industrial Electronics*, vol. 64, no. 5, pp. 3489–3501, 2017.

[132] R. Chellal, E. Laroche and L. Cuvillon, An H_∞ methodology for position control of cable-driven parallel robots, *IEEE ECC 2014, European Control Conference*, 2014.

[133] Z. Chem, C. Li, B. Yao, M. Yuan and C. Yang, Integrated coordinated/synchronized contouring control of a dual-linear-motor-driven gantry, *IEEE Transactions on Industrial Electronics*, vol. 67, no. 5, pp. 3944–3954, 2020.

[134] A. Chemori, R. Kouki and F. Bouani, A new fast nonlinear model predictive control of parallel manipulators, design and experiments, *Control Engineering Practice*, vol. 130, pp. 1–13, 2023.

[135] Y.P. Chen, J.L. Chang and S.R. Chu, PC-based sliding-mode control applied to parallel-type double inverted pendulum system, *Mechatronics*, vol. 9, pp. 553–564, 1999.

[136] X Chen, C.Y. Su and T. Fukuda, A nonlinear disturbance observer for multi-variable systems and its application to magnetic bearing systems, *IEEE Transactions on Control Systems Technology*, vol. 12, no. 4, pp. 569–577, 2004.

[137] T.W. Chen, J.R. Ahn, H.H. Lee, H.G. Kim and E.C Nho, A novel strategy of efficiency control for a linear compressor system driven by a PWM inverter, *IEEE Transactions on Industrial Electronics*, vol. 55, no. 1, pp. 296–30, 2008.

[138] M.Y. Chen and J.S. Lu, High-precision motion control for a linear permanent magnet iron-core synchronous motor drive in position platform, *IEEE Transactions on Industrial Informatics*, vol. 10, no.1, pp. 99–108, 2014.

[139] R. Chen, L. Yao, Z. Jiao and Y. Shang, Dynamic modeling and analysis of flexible H-type gantry stage, *Journal of Sound and Vibration*, vol. 439, pp. 144–155, 2019.

[140] H. Chen, Y. Fang and N. Sun, An adaptive tracking control method with swing suppression for 4-DOF tower crane systems, *Mechanical Systems and Signal Processing*, vol. 123, pp. 426–442, 2019.

[141] Y. Chen and B. Guo, Sliding-mode fault-tolerant tracking control for a single-link flexible-joint manipulator system, *IEEE Access*, vol. 7, pp. 83046–83055, 2019.

[142] H. Chen and N. Sun, Nonlinear control of underactuated systems subject to both actuated and unactuated state constraints with experimental verification, *IEEE Transactions on Industrial Electronics*, vol. 67, no. 9, pp. 7702–7714, 2020.

[143] G. Chen, S. Hutchinson and F. Dellaert, Locally optimal estimation and control of cable-driven parallel robots using time-varying linear quadratic Gaussian control, *IEEE IROS 2022, IEEE 2022 International Conference on Intelligent Robots and Systems*, Kyoto, Japan, Oct. 2022.

[144] Q. Chen, L. Zuo, C. Wu, *et al.*, Optimization of compressor standby schemes for gas transmission pipeline systems based on gas delivery reliability, *Reliability Engineering and System Safety*, vol. 221, pp. 108351–108363, 2022.

[145] S. Chen and Y.M. Zabolotnov, Model of forming a regular triangular tethered constellation of microsatellites with considering their motion relative to the center of mass, *Journal of Computers and Systems Science*, no. 2, pp. 44–59, 2023.

[146] K. Chen, Y. Lu, R. Zhang and H. Wang, The adaptive bearing fault diagnosis based on optimal regulation of generalized SR behavior in fluctuating-damping induced harmonic oscillator, *Mechanical Systems and Signal Processing*, vol. 189, pp. 110078–110197, 2023.

[147] D. Chen, H.Y. Fang, Y. Huang, M. Tan, Q.Y. Chen, and X.S. Wei, Robust control of bullwhip effect for supply chain system with time-varying delay on basis of discrete-time approach, *IEEE Access*, vol. 11, pp. 61049–61058, 2023.

[148] Z. Chen, W. Guan, J Guo, D. Xue, Z. Liu, G. Wang and L. Quan, Electric shovel trajectory tracking with inversion sliding mode based on Lyapunov function, *Automation in Construction*, vol. 162, pp. 1056340106651, 2024.

[149] S.H. Chena and L.C. Fu, Output feedback sliding-mode control for a Stewart platform with a nonlinear observer-based forward kinematic solution, *IEEE Transactions on Control Systems Technology*, vol. 21, no. 1, pp. 176–185, 2013.

[150] X. Cheng, B.X. Chang, S. Deng, R.G. Zhou, M.Q. Lu and B. Wang, State-feedback decoupling control of 5-DOF magnetic bearings based on a-order inverse system, *Mechatronics*, vol. 68, pp. 102358–102367, 2020.

[151] X. Cheng, S. Deng, B.X. Cheng, Y.F. Yu, H.C. Wu and R.G. Zhou, Design and implementation of a fault-tolerant magnetic bearing control system combined with a novel fault diagnosis of actuators, *IEEE Access*, vol. 9, pp. 2454–2465, 2021.

[152] G. Chiaverini, G. Oriolo and A.A. Maciejewski, Redundant robots, In: *Springer Handbooks on Robotics*, Springer, 2nd Ed., pp. 221–242, 2016.

[153] H.D. Cho and S.H. You, Fuzzy finite memory state estimation for electrohydraulic active suspension systems, *IEEE Access*, vol. 9, pp. 99364–99373, 2021.

[154] Z. Choe, J.H. Kim, S. Hong, J. Lee and H.W. Park, Seamless reaction strategy for bipedal locomotion exploiting real-time nonlinear model predictive control, *IEEE Robotics and Automation Letters*, vol. 8, no. 8, pp. 5031–5038, 2023.

[155] S. Choi and E Jung, Optimal tuberculosis prevention and control strategy from a mathematical model based on real data, *Bulletin of Mathematical Biology*, vol. 76, pp. 1566–1589, 2014.

[156] R. Choi, A. Tsourdos, H. Gao, Y. Xia and S. Choi, Dual-loop tube-based robust model predictive attitude tracking control for spacecraft with system constraints and additive disturbances, *IEEE Transactions on Industrial Electronics*, vol. 69, no. 4, pp. 4022–4033, 2022.

[157] H. Chonal, J.B. Guyon, A. Koessler, *et al.*, Geometrical defect identification of a SCARA robot from a vector modelling of kinematic joints invariant, *Mechanism and Machine Theory*, vol. 162, pp. 104333–104362, 2021.

[158] P.H. Chou, C.S. Chen and F.J. Lin, DSP-based synchronous control of dual linear motors via Sugeno-type fuzzy neural network compensator, *Journal of the Franklin Institute*, vol. 349, pp. 792–812, 2012.

[159] A. Choukchou-Braham, B. Cherki, M. Djemai and K. Busawon, *Analysis and Control of Underactuated Mechanical Systems*, Springer, 2014.

[160] Z.Y. Chu, J. Cai and F.C. Sun, Fuzzy adaptive disturbance observer-based robust tracking control of electrically driven free-floating space manipulator, *IEEE Systems Journal*, vol. 8, no. 2, pp. 343–353, 2014.

[161] S. Chu, S. Zhang, X. Ma, *et al.*, Experimental study of the influence of flexible control on key parameters in reverse osmosis desalination, *IEEE Access*, vol. 10, pp. 4844–4860, 2022.

[162] N.A. Chudinov, R.F. Murtazin and V.A. Soloviev, Counteracting contingencies in fast rendezvous with ISS, *Journal of Space Safety Engineering*, 2023.

[163] S.J. Chung, E.M. Kong and D.W. Miller, Dynamics and control of tethered formation flight spacecraft using the spheres testbed, *AIAA Guidance, Navigation and Control Conference*, San Francisco, California, Aug. 2005.

[164] S.J. Chung, Nonlinear model reduction and decentralized control of tethered formation flight by oscillation synchronisation, *AIAA 2006 Conference on Guidance, Navigation and Control*, Keystone, Colorado, USA, Aug. 2006.

[165] S.J. Chung, J.J.E. Slotine and D.W. Miller, Nonlinear model reduction and decentralized control of tethered formation flight, *Journal of Guidance, Control and Dynamics*, vol. 30, no. 2, pp. 1–11, 2007.

[166] S.J. Chung, J.J.E. Slotine and D.W. Miller, Propellant-free control of tethered formation flight: Part 2, Nonlinear Underactuated Control, *Journal of Guidance, Control and Dynamics*, vol. 31, no. 5, pp. 1–10, 2008.

[167] S.U. Chung, J.W. Kim, B.C. Woo, *et al.*, Design and experimental validation of doubly salient permanent magnet linear synchronous motor for precision position control, *Mechatronics*, vol. 23, pp. 172–181, 2013.

[168] M.K. Ciliz and M.O. Tuncay, Comparative experiments with a multiple model-based adaptive controller for a SCARA type direct-drive manipulator, *Robotica*, vol. 23, pp. 721–729, 2005.

[169] I. Cires and V.M. Nani, Stability control for a huge excavator for surface excavation, *Applied Mathematical Modelling*, vol. 40, pp. 388–397, 2016.

[170] J.P. Corriou, *Numerical Methods and Optimization: Theory and Practice for Engineers*, Springer, 2021.

[171] J.A. Corte and P. Cabrera, Optimal sizing of stand-alone wind-powered seawater reverse osmosis plant without use of massive energy storage, *Applied Energy*, vol. 304, pp. 117888–117910, 2021.

[172] A. Cortinovis, D. Pareschi, M. Mercangoz and T. Besselmann, Model-predictive anti-surge control of centrifugal compressors and variable speed dries, In: *Proc. 2012 IFAC Workshop on Automatic Control in Offshore Oil and Gas Production*, Trondheim, Norway, June 2012.

[173] A. Cortinovis, H.J. Ferreau, D. Lewandowski and M. Mercangoz, Safe and efficient operation of centrifugal compression using linearized MPC, In: *53rd IEEE Conference on Decision and Control*, Los Angeles, USA, Dec. 2005.

[174] A. Cortinovis, M. Mercengoz, T. Besselmann, *et al.*, Enabling voltage drop ride-throughs of large electric driven gas compressors, *Journal of Process Control*, vol. 74, pp. 76–87, 2019.

[175] R.J. Coverly, D.E. Zlotnik and J.R. Forbes, Saturated control of flexible-joint manipulator using a Hammerstein strictly positive real compensator, *Robotica*, vol. 34, no. 6, pp. 1367–1382, 2016.

[176] J.L. Crassidis and J.L. Junkims, *Optimal Estimation of Dynamical Systems*, CRC Press, 2012.

[177] D. Cruz-Ortiz, M. Ballesteros, A. Polyakov, D. Efimov, I. Chairez and A. Poznyak, Practical realization of implicit homogeneous controllers for linearized systems, *IEEE Transactions on Industrial Electronics*, vol. 69, no. 5, pp. 5142–5151, 2021.

[178] M. Cui and Z. Wu, Trajectory tracking of flexible-joint manipulators actuated by DC-motors under random disturbances, *Journal of the Franklin Institute*, vol. 356, pp. 9330–9343, 2019.

[179] T.N. Cuong, X. Xu, S.D. Lee and S.S. You, Dynamic analysis and management optimization for maritime supply chains using nonlinear control theory, *International Journal of Maritime Safety, Environmental Affairs and Shipping*, vol. 4, no. 2, pp. 48–55, 2020.

[180] T.N. Cuong, H.S. Kim, D.A. Nguyen and S.S. You, Nonlinear analysis and active management of production – distribution in nonlinear supply chain model using sliding-mode control theory, *Applied Mathematical Modelling*, vol. 97, pp. 418–437, 2021.

[181] T.N. Cuong, H.S. Kim, L.N.B Long and S.S. You, Decision support system for managing multi-echelon supply chain network against disruptions, *RAIRO Operations Research*, vol. 57, pp. 787–815, 2023.

[182] V. Cusimano, P. Palumbo and F. Pepe, Closed-loop control of tumor growth by means of anti-angiogenic administration, In: *IEEE CDC 2015, IEEE 2015 54th Conference on Decision and Control*, Osaka, Japan, Dec. 2015.

[183] B.G. Czako, D.A. Drexler and L. Kovacs, Impulsive control of tumor growth via nonlinear model predictive control using direct multiple shooting, In: *IEEE ECC 2020*, San Petersburg, Russia, May 2020.

[184] B. Czako, M. Siket, D.A. Drexler and L. Kovacs, Chemotherapy optimization using moving horizon estimation-based nonlinear model predictive control, In: *IFAC BMS 2021, 11th IFAC Symposium on Biological and Medical Systems*, Ghent, Belgium, Sep. 2022.

[185] X. Dai, S. Song, W. Xu, Z. Huang, and D. Gong, Modal space neural network compensation control for Gough-Stewart robot with uncertain load, *Neurocomputing*, vol. 490, pp. 245–257, 2023.

[186] N. Dakovic and M. Radulovich, Flatness and LQR control of Furuta pendulum, *ETF Journal of Electrical Engineering*, vol. 21, no. 1, pp. 138–146, 2015.

[187] P. Damschke, O. Kahl and J. Lang, Fast and reliable transient simulation and continuous optimization of large-scale gas-networks, *Mathematical Methods of Operations Research*, vol. 95, pp. 475–501, 2022.

[188] Q. Dang, Q. Zong, B. Tian and F. Wang, Adaptive-gain multivariable super-twisting sliding-mode control for reentry RLV with target perturbation, *International Journal of Robust and Nonlinear Control*, vol. 27, pp. 620–638, 2017.

[189] H.V. Dao and K.K. Ahn, Extended sliding-mode observer-based admittance control for hydraulic robots, *IEEE Robotics and Automation Letters*, vol. 7, no. 2, pp. 3992–3999, 2022.

[190] M.T. Das and L.C. Dulger, Mathematical modelling simulation and experimental verification of a SCARA robot, *Simulation Modelling: Practice and Theory*, vol. 13, pp. 257–271, 2005.

[191] S. Das, P. Das and P. Das, Dynamics and control of multi-drug resistant bacterial infection in hospitals with multiple delays, *Communications in Nonlinear Sciences and Numerical Simulation*, vol. 89, pp. 105279–105905, 2020.

[192] A. Das, K. Dehingia, H. Kr. Sarmah and K. Hosseini, An optimally controlled chemotherapy treatment for cancer eradication, *International Journal of Modelling and Simulation*, pp. 1–17, 2022, doi: 10.1080/02286203.2022.2155601.

[193] A. Das, H. Kr. Sarmah, D. Bhattacharya, K. Dehingia and K. Hosseini, Combination of virotherapy and chemotherapy with optimal control for combating cancer, *Mathematics and Computer in Simulation*, vol. 194, pp. 460–488, 2022.

[194] P. De Leenheer and N.G. Cogan, Failure of antibiotic treatment in microbial populations, *Mathematical Biology*, vol. 59, pp. 563–579, 2009.

[195] J. De Miras, C. Join, M. Fliess, S. Riachy and S. Bonnet, Active magnetic bearing: a new step for model-free control, In: *IEEE CDC 2012, IEEE 32nd Intl. Conference on Decision and Control*, Florence, Italy, Dec. 2013.

[196] N. Dehia, Y. Wang and A. Kheddar, Dual-arm box-grabbing with impact-aware MPC utilizing soft-deformable end-effector pads, *IEEE Robotics and Automation Letters*, vol. 7, no. 2, pp. 5647–5654, 2022.

[197] K. Dehingia, H.K. Sarmah, K. Hosseini, K. Sadni, S. Sulashour and C. Park, An optimal control problem of immuno-chemotherapy in presence of gene therapy, *AIMS Mathematics*, vol 6. no. 10, pp. 11530–11549, 2021.

[198] K. Dehingia, S.W. Yao, K. Sadri, *et al.*, A study on cancer-obesity-treatment model with quadratic optimal control approach for better outcomes, *Results in Physics*, vol. 42, pp. 105963–105977, 2022.

[199] K. Dehingia, P. Das, R.K. Upadhyay, A.K. Misra, F.A. Rihan and K. Hosseini, Modelling and analysis of delayed tumor-immune system with hunting T-cells, *Mathematics and Computers in Simulation*, vol. 203, pp. 669–684, 2023.

[200] A. Dehuk, A.T. Nguyen, A. Dequidt, L. Vermeiren and M. Dambrine, Disturbance observer-based tracking control for industrial SCARA robot manipulators, *IEEE IECON 2019, IEEE 45th Annual Conference of the Industrial Electronics Society*, Lisbon, Portugal, Oct. 2019.

[201] A.M. Delgado-Torres and L. Garcia-Rodriguez, Off-grid seawater reverse osmosis (SWRO) desalination driven by hybrid tidal rang/solar PV systems: sensitivity analysis and criteria for preliminary design, *Sustainable Energy Technologies*, vol. 53, pp. 102425–102497, 2022.

[202] V. Delli Colli, P. Cancelliere, F. Marignetti, R. Di Stefano and M. Scarano, A tubular-generator drive for wave energy conversion, *IEEE Transactions on Industrial Electronics*, vol. 53, no. 4, pp. 1153–1160, 2006.

[203] A. Dey, V. Bokka and S. Sen, Dependence of bacterial growth rate on dynamic temperature changes, *IET Systems Biology*, vol. 14, no. 2, pp. 68–74, 2020.

[204] G. Donati, M. Basso, M. Mugnains and M.O. Neri, Automated tuning of augmented PIDs for active magnetic bearings supporting turbomachinery, *IEEE/ASME Transactions on Mecchatronics*, pp. 1–11, 2024.

[205] R. Deng, B. Di and L. Song, Ultra-dense LEO satellite based formation flying, *IEEE Transactions on Communications*, vol. 69, pp. 3091–3105, 2021.

[206] M. Deng, A. Inoue, S. Nakagawa, T. Henmi, N. Ueki, and Y. Hirashima, Experimental study on simultaneous swing-up of parallel double inverted pendulum, In: *Proc. of the 23rd Chinese Control Conference*, pp. 1627–1631, 2004.

[207] M. Deng, A. Inoue, T. Henmi and N. Ueki, Analysis and experiment on simultaneous swing-up of a parallel cart-type double inverted pendulum, *Asian Journal of Control*, vol. 10, no. 1 pp. 121–128, 2008.

[208] K. Deng, M. Zhao and W. Xu, Level-ground walking for a bipedal robot with a torso via hip series elastic actuators and its gain bifurcation control, *Robotics and Autonomous Systems*, vol. 79, pp. 58–71, 2016.

[209] R. Deng, B. Di and L. Song, Ultra-dense LEO satellite based formation flying, *IEEE Transactions on Communications*, vol. 69, pp. 3091–3105, 2021.

[210] L.A. Dessaint, M. Saad and K. Al-Haddad, An adaptive controller of a direct drive robot, *IEEE Transactions on Industrial Electronics*, vol. 39, no. 2, pp. 105–111, 1992.

[211] D. Desiderio and M. Lovera, Guidance and control for planetary landing: flatness-based approach, *IEEE Transactions on Control Systems Technology*, vol. 21, no. 4, pp. 1280–1294, 2013.

[212] D. Desiderio and M. Lovera, Flatness-based guidance for planetary landing, In: *IEEE ACC 2010, IEEE 2010 American Control, Conference*, Baltimore, USA, June 2010.

[213] W. Dib, R. Ortega, and D. Hill, Transient stability enhancement of multi-machine power systems: synchronisation via immersion of a pendular system, *Asian Journal od Control*, vol. 16, no. 1, pp. 50–58, 2014.

[214] W. Dib, E. Ortega, A. Astolfi and D. Hill, Improving transient stability of multi-machine power systems: synchronization via immersion of a pendular system, In: *IEEE ACC 2011, IEEE 2011 American Control Conference*, Sam Francisco, California, July 2011.

[215] Y. Ding, X.G. Yan, Z. Ma, S.K. Spurgeon and B. Jiang, System structure-based decentralized sliding-mode output tracking control for nonlinear interconnected systems, *Journal of Robust and Nonlinear Control*, vol. 33, no. 3, pp. 1704–1719, 2023.

[216] T.X. Dinh, T.D. Thien, T.H.V. Anh and K.K. Ahn, Disturbance observer-based finite-time trajectory tracking control for a 3-DOF hydraulic manipulator including actuator dynamics, *IEEE Access*, vol. 6, pp. 36798–36809, 2018.

[217] S. Dominic, Y. Lohr, A. Schwung and S.X. Ding, PLC-based real-time realization of flatness-based feedforward control for industrial compressor systems, *IEEE Transactions on Industrial Electronics*, vol. 64, no. 2, pp. 1323–1331, 2017.

[218] E.M. Donald, M. Arsenault and S. Beites, Design of a suspended cable-driven parallel robot for the semi-automation of construction tasks, In: *Canadian Committee for the Theory of Machines and Mechanisms*, Quebéc, Canada, June 2023.

[219] O. Dong, Q. Zong, B. Tian, C. Zhang and W. Liu, Adaptive disturbance observer-based finite-time continuous fault-tolerant control for reentry RLV, *International Journal of Robust and Nonlinear Control*, vol. 27, pp. 4275–4295, 2017.

[220] Y. Dong, T. Nuchkrua and T. Shen, Asymptotical stability contouring control of dual-arm robot with holonomic constraints: modified distributed control framework, *IET Control Theory and Applications*, vol. 13, no. 17, pp. 2877–2885, 2019.

[221] J. Dong, J. Shi, C. Liu and T. Yu, Research on pneumatic polishing force control system based on a high-speed on/off with PWM controlling, *Robotics and Computer Integrated Manufacturing*, vol. 70, pp. 102133–102147, 2021.

[222] A.V. Doroshin, Attitude dynamics of gyrostat-satellites under control by magnetic actuators at small perturbations, *Communications in Nonlinear Science and Numerical Simulation*, vol. 49, pp. 159–175, 2017.

[223] D.A. Drexler, J. Sapi and L. Kovacs, H-infinity control of nonlinear systems with positive input with application to antiangiogenic therapy, In: *IFAC ROCOND 2018, 3rd IFAC Symposium on Robust Control Design*, Florianopolis, Brazil, Sep. 2018.

[224] D. Drexler, J. Sapi, A. Szeler, I. Harmati, A. Kovacs and L. Kovacs, Flat control of tumor growth with angiogenic inhibitor, In: *7th IEEE Intl. Symposium on Applied Computational Intelligence and Informatics*, Timisoara, Romania, May 2020.

[225] D.A. Drexler, T. Ferrenci and L. Kovacs, Extended tumor growth model for combined therapy, In: *IEEE SMC 2019 Conference on Systems, Man and Cybernetics*, Bari, Italy, Oct. 2019.

[226] D.A. Drexler and L. Kovacs, Optimization of low dose metronomic therapy based on pharmacological parameters, In: *11th IFAC Symposium on biological and modular systems, IFAC BMC 2021*, Ghent, Belgium, Sep. 2021.

[227] Y. Du, M. Chang, K. Tong Chau, X. Liu, F. Xiao and W. Zhao, Linear primary permanent magnet Vernier machine for wave energy conversion, *IET Electric Power Applications*, vol. 9, no. 3, pp. 203–214, 2015.

[228] X. Du, J. Huang and S. Zhang, An analytical nonlinear predictive controller for wing rock suppression, In: *Proceedings of the 38th Chinese Control Conference*, Guangzhou, China, July 2019.

[229] H. Du, J. Shi, J. Chou, Z. Zhang and X. Feng, High-gain observer-based integral sliding-mode tracking control for heavy vehicle electrohydraulic steering system, *Mechatronics*, vol. 74, pp. 102484–102498, 2021.

[230] J. Duan, Z. Shao, H. Liiu, Z. Zhang, Y. Wang and H. Zhao, Design analysis of a cable-driven parallel robot with parallel cables for ship-side painting, In: *IEEE CACRE 2023, 8th Intl Conference on Automation, Control and Robotic Engineering*, Hong Kong, China, July 2023.

[231] V. Dubunchet, D. Saussie, D. Alazard, C. Bèrard and C. Le Peurodie, Modelling and control of a space robot for active debris removal, *CEAS Space*, vol. 7, pp. 203–218, 2015.

[232] M. Dunbabin and R. Corke, Autonomous excavation using a rope shovel, *Journal of Field Robotics*, vol. 23, no. 6–7, pp. 379–394, 2006.

[233] S.C. Duong, E. Uezato, H. Kinjo and T. Yamamoto, A hybrid evolutionary algorithm for recurrent neural network control of a three-dimensional tower crane, *Automation in Construction*, vol. 49, pp. 1318–1325, 2013.

[234] X. Dupuis, Optimal control of leukemia cell population dynamics, *Mathematical Modelling of Natural Phenomena*, vol. 9, no. 1, pp. 4–26, 2014.

[235] L. Durantay and P. Laboube, Technology trends for LNG compression using innovative electric solutions, In: *IOP Conference Series on Material Sciences and Engineering*, vol. 643, no. 012149, pp. 118, 2019.

[236] V. Dvornik, D. Mallapragada, A. Batterud, J. Kazempour and P. Pinson, Multi-stage linear decision rules for stochastic control of natural gas networks with linepack, *Electric Power Systems Research*, vol. 212, pp. 108388–108394, 2022.

[237] G. Eigner, D.A. Drexler and L. Kovacs, Tumor-growth control by TP-LPV-LMI based controller, In: *IEEE 2018 Intl. Conf. on Systems, Man and Cybernetics*, Miyazaki, Japan, Oct. 2018.

[238] G. Eigner, M. Siket, B. Czako, *et al.*, Model Predictive tumor volume control using nonlinear optimization, In: P. Shi (Ed.) *Complex Systems, Spanning Control and Computational Cybernetics: Applications, Studies in Systems, Decision and Control*, Springer, vol. 415, pp. 235–250, 2022.

[239] M.M. Elgamassy, M.A. Izzaalarab and X.P. Zhang, Single-end based fault location method for VSC-HVDC transmission systems, *IEEE Access*, vol. 10, pp. 43129–43142, 2022.

[240] N.M.J. Er, C.B. Low, K.H. Nah, M.H. Lim and S.Y. Ng, Real-time implementation of a dynamic fuzzy neural network for a SCARA, *Microprocessors and Microsystems*, vol. 26, pp. 449–461, 2002.

[241] L. Étienne, P. Cardou, M. Métillon and S. Caro, Design of a planar cable-driven parallel crane without parasitic tilt, *ASME Journal of Mechanisms and Robotics*, vol. 14, no. 4, 041006 pp. 1–11, 2022.

[242] A. Falcoz, D. Henry and A. Zolghadri, Robust fault diagnosis for atmospheric reentry vehicles: a case study, *IEEE Transactions on Systems, Man and Cybernetics – Part A: Systems and Humans*, vol. 40, no. 5, pp. 886–899, 2010.

[243] L. Fan, A. Zhang, G. Pan, Y. Du and J. Qiu, Swing-up and fixed-time stabilization control of underactuated cart-double pendulum system, *IET Control Theory and Applications*, pp. 1–10, 2022.

[244] R. Fan, Y. Li and L. Yang, Multi-objective trajectory optimization of intelligent electrohydraulic shovel, *Frontiers in Mechanical Engineering*, vol. 17, no. 4, pp. 1–24, 2022.

[245] R. Fan and Y. Li, An adaptive fuzzy trajectory tracking control via improved cerebellar model articulation controller for electrohydraulic shovel, *IEEE/ASME Transactions on Mechatronics*, vol. 26, no. 6, pp. 2870–2880, 2022.

[246] R. Fan, Y. Li, L. Yang, T. Qin, Y. Zhang and Z. Wang, Coordinated trajectory tracking control for mining excavator with oscillating external disturbance, *Journal of Field Robotics*, vol. 41, pp. 258–272, 2024.

[247] J. Fang and Y. Ren, Decoupling control of magnetically suspended rotor system in control moment gyros based on an inverse system method, *IEEE/ASME Transactions on Mechatronics*, vol. 17, no. 6, pp. 1133–1144, 2012.

[248] M. Farza, M.M. Saad, M.L. Fall, O. Gehan and K. Busawon, Continuous-discrete time observers for a class of MIMO nonlinear systems, *IEEE Transactions on Automatic Control*, vol. 59, no.4, pp. 1060–1065, 2013.

[249] S.M. Fasih ur Rahman and M.A. Abbasi, Input shaping with an adaptive scheme for swing control of an underactuated tower crane under payload hoisting and mass variation, *Mechanical Systems and Signal Processing*, vol. 175, pp. 109106–109151, 2022.

[250] E. Fimmel, Y. Semenov and A. Bratus, On optimal and suboptimal treatment strategies for a mathematical model of leukemia, *Mathematical Biosciences and Engineering*, vol. 10, no. 1, pp. 151–166, 2013.

[251] M.Fliess, J. Levine, P. Martin and P. Rouchon, A Lie-Backlünd approach to equivalence and flatness of nonlinear systems, *IEEE Transactions on Automatic Control*, vol. 44, no. 5, pp. 922–937, 1999.

[252] M. Fliess and H. Mounier, Tracking control and π-freeness of infinite dimensional linear systems, In: G. Picci and D.S. Gilliam (Eds.), *Dynamical Systems, Control, Coding and Computer Vision*, vol. 258, pp. 41–68, *Birkhaüser*, 1999.

[253] M. Fliess, C. Join, K. Moussa, S.M. Djouadi and M.W. Alsager, Toward simple in silico experiments for drugs administration in some cancer treatments, In: *11th IFAC Symposium on Biological and Medical Systems, IFAC BMS 2021*, Ghent, Belgium, Sep. 2021.

[254] J.A. Florian Jr, J.L. Esseman and R.S. Parker, Nonlinear model predictive control law for closing daily anticancer agents using a novel saturating-rate cell-cycle model, *Computers in Biology and Medicine*, vol. 38, pp. 339–347, 2008.

[255] M. Folgheraiter and B. Aubakir, Design and modeling of a lightweight and low power consumption full-scale biped robot, *Journal of Humanoid robotics*, vol. 15, no. 5, article no. 1850022, pp. 1–32, 2018.

[256] D. Fontaine, S. Liao, J. Padueiro and P.V. Kokotovic, Nonlinear control experiments on axial flow compressor, *IEEE Transactions on Control Systems Technology*, vol. 12, no. 5, pp. 683–693, 2004.

[257] A. Fradkov, B. Andrievsky and K. Boykov, Control of the coupled double pendulum system, *Mechatronics*, vol. 15, pp. 1289–1303, 2005.

[258] A.L.D. Franco, H. Barlés, E.R. de Pieri and H Guillald, Robust nonlinear control associating robust feedback linearization and H_∞ control, *IEEE Transactions on Automatic Control*, vol. 51, no. 7, pp. 1200–1207, 2006.

[259] L.B. Freidovich, A.S. Shiriaev and I.R. Manchester, Stability analysis and control design for an underactuated walking robot via computation of a transverse linearization, In: *IFAC WC 2008, Prod. of the 17th World Congress of the Intl. Federation of Automatic Control*, Seoul, Korea, July 2008.

[260] L. Frosini, C. Garlesca and L. Stabo, Induction machine bearing fault detection by means of statistical processing of the stray flux measurement, *IEEE Transactions on Industrial Electronics*, vol. 62, no. 3, pp. 1945–1954, 2015

[261] D. Fu, H.J. Zhang, Y. Yu, C.M. Ionescu, E.H. Aghezzal and R. de Keyser, A distributed model predictive control strategy for the bullwhip reducing inventory management policy, *IEEE Transactions on Industrial Informatics*, vol. 15, no. 2, pp. 932–941, 2019.

[262] X. Fu, L. Fu and H. Nalekizadeh, Finite-time action fuzzy sliding mode approach for deep surge control in nonlinear distorted compressor system with uncertainty in characteristic curve, *Automatika*, vol. 62, no. 3–4, pp. 530–540, 2021.

[263] L. Fu, X. Fu, M.T. Ziabari, Finite-time adaptive sliding-mode control for compressor surge with uncertain characteristics in the presence of disturbances, *Systems Science and Control Engineering*, vol. 9, no. 1, pp. 369–379, 2021.

[264] Y. Fujita, M. Izutsu and S. Hatakeyama, Swing-up and stabilization control of Twin Furuta pendulums by energy control, In: *IEEE IECON, 2014, IEEE 40th Annual Conference of the IEEE Industrial Electronics Society*, Dallas, Texas, Oct. 2014.

[265] F. Fusco and J.V. Ringwood, Hierarchical robust control of oscillating wave energy conversion with uncertain dynamics, *IEEE Transactions on Sustainable Energy*, vol. 5, no. 3, pp. 958–967, 2014.

[266] M. Galicki, Path-constrained control of a redundant manipulator in a task space, *Robotics and Autonomous Systems*, vol. 34, pp. 234–243, 2006.

[267] L. Gagliardini, S. Caro, M. Guttefarde, P. Wagner and A. Girin, A reconfigurable cable-driven parallel robot for sandblasting and painting of large structures, In: *Proc. CableCon 2014L Cable-Driven Parallel Robots*, Duisburg, Germany, Aug. 2014.

[268] X.W. Gan, A. J. Li, Y.T. Ge and C.Q. Wang, Fractional-order attitude stability control for sub-satellite of tethered satellite system during deployment, *Applied Mathematical Modelling*, vol. 62, pp. 271–286, 2018.

[269] L. Gang and C. Zhu, Four factor polarity switching control for synchronous vibration suppression of active magnetic bearings rigid rotors system in the full rotational speed range, *IEEE Transactions on Industrial Electronics*, vol. 68, no. 12, pp. 11870–11880, 2021.

[270] B.Z. Gao, H. Chen, K. Sanada and Y. Hu, Design of clutch-slip controller for automatic transmission using backstepping, *IEEE/ASME Transactions on Mechatronics*, vol. 16, no. 3, pp. 498–508, 2011.

[271] D.P. Gao and N.J. Huang, Optimal control analysis of a tuberculosis model, *Applied Mathematical Modelling*, vol. 58, pp. 57–64, 2018.

[272] Q. Gao, Y. Liu and D. Jiang, Nonlinear control method of electro-hydraulic system driving two-DOF robot arm with output constraint, In: *IOP Conf. Series on Materials Science and Engineering*, vol. 715, pp. 012090, 2020.

[273] X. Gao, J.X. Zhaang and L. Hao, Fault-tolerant control of pneumatic continuum manipulators under actuator faults, *IEEE Transactions on Industrial Informatics*, vol. 17, no. 12, pp. 8299–8307, 2021.

[274] X. Gao, E. Cui, D. Yang, Z. Tan and J. Sun, Adaptive displacement constraint control with predefined performance for active magnetic bearings, *IEEE Transactions on Automation Science and Engineering*, pp. 1–12, 2024.

[275] H. Gao, Y. Liu, W. Sun and X. Yu, Adaptive wavelet tracking control of dual linear-motor-driven gantry stage with suppression of crossbeam rotation, *IEEE/ASME Transactions on Mechatronics*, pp. 1–9, 2024.

[276] I. Garcia-Herreros, X. Kestelyn, J. Gomard and P.J. Barre, Model-based control of a dual-drive H-type gantry stage on a decoupling base, In: *IEEE ICIT 2010, IEEE 2010 Intl. Conference on Industrial Technology*, Via del Mar, Chiel, March 2010.

[277] I. Garcia-Herreros, X. Kestelyn, J. Gomard, R. Coleman and P.J. Barre, Model-based decoupling control method for dual-drive gantry-stages: a case-study with experimental validation, *Control Engineering Practice*, vol. 21, pp. 298–307, 2013.

[278] R. Genest and G.V. Ringwood, Receding horizon pseudospectral control for energy maximization with application to wave energy devices, *IEEE Transactions on Control Systems Technology*, vol. 25, no. 1, pp. 29–38, 2017.

[279] R.S. Gesser, R. Sartori, T.P. Damo, *et al.*, Advanced control applied to a gas compression system of an offshore platform: from modelling to related system infrastructure, *Journal of Petroleum Science and Engineering*, vol. 208, pp. 109428–109447, 2022.

[280] S. Gholami, H. Salarieh and A. Alamy, Observer-based feedback control of a biodynamical model of tumor growth with sampled measurements, In: *IEEE ECC 2016*, Aalborg, Denmark, June 2016.

[281] L. Giarré, D. Bauso, P. Falugi and B. Bamie, LPV model identification for gain scheduling control: an application to rotating stall and surge control problem, *Control Engineering Practice*, vol. 14, pp. 351–361, 2006.

[282] B.P. Gibbs, *Advanced Kalman Filtering, Least Squares and Modelling: A Practical Handbook*, J. Wiley, 2011.

[283] S.A.C. Girardo, R.C. Supelano, T.C. d' Avila, *et al.*, Model-predictive control with dead-time compensation applied to a gas-compression system, *Journal of Petroleum Science and Engineering*, vol. 203, pp. 108580–108588, 2021.

[284] A. Giuseppi, A. Pietrabissa, S. Cilione and L. Galvani, Feedback linearization-based satellite attitude control with a life-support device without communications, *Control Engineering Practice*, vol. 90, pp. 221–230, 2019.

[285] H. Glassel, F. Ziemmermann, S. Bruckner, U.M. Schattle and S. Rudolph, Adaptive neural control of the deployment procedure for tethered-assisted re-entry, *Aerospace Science and Technology*, vol. 8, pp. 73–81, 2004.

[286] A. Glushchenko, V. Petrov and K. Lastochkin, Regression filtration with resetting to provide exponential convergence of MRAC for planes with jump change or unknown parameters, *IEEE Transactions on Automatic Control*, vol. 68, no. 8, pp. 5127–5134, 2023.

[287] M. Gnucci and R. Marino, Adaptive tracking control of underactuated mechanical systems with relative degree two, *Automatica*, vol. 129, pp. 109633–109637, 2021.

[288] A. Goksu, U.E. Kocamaz and Y. Uyaroglou, Synchronization and control of chaos in supply chain management, *Computers and Industrial Engineering*, vol. 86, pp. 107–115, 2016.

[289] J.T. Gravdahl, O. Egeland and S.O. Vatland. Drive torque actuation in active surge control of centrifugal compressors, *Automatica*, vol. 38, pp. 1881–1893, 2002.

[290] T.R. Grochmal and A.R. Lynch, Precision tracking of a rotating shaft with magnetic bearing by nonlinear decoupled disturbance observers, *IEEE Transactions on Control Systems Technology*, vol. 16, no. 6, pp. 1112–1121, 2007.

[291] L. Grüne and Jü Pannek, *Nonlinear Model Predictive Control: Theory and Algorithms*, Springer, 2017.

[292] J.R. Guadarrama Olivera, E.D. Leon, F. Bergner, and G. Cheng, Plantar tactile feedback for biped balance and locomotion on unknown terrain, *Journal of Humanoid Robotics*, vol. 17, no. 1, article no. 1950036, pp. 1–21, 2020.

[293] Z. Guang, B. Xingzi and L. Bin, Optimal deployment of spin-stabilized tethered formations with continuous thrusters, *Nonlinear dynamics*, vol. 95, pp. 2143–2162, 2014.

[294] Z. Guang, L. Yuyang and B. Liang, Observer enhanced control for spin-stabilized tethered formation in earth orbit, *Acta Astronautica*, vol. 145, pp. 216–228, 2018.

[295] G. Guglieri and B. Guagliotti, Analytical and experimental analysis of wing-rock, *Nonlinear Dynamics*, vol. 24, pp. 129–146, 2001.

[296] G. Guglieri, A comprehensive analysis of wing-rock dynamics for slender delta wings configurations, *Nonlinear Dynamics*, vol. 69, pp. 1559–1575, 2012.

[297] J. Guo, X. Wu and S. Tang, Autonomous gliding entry guidance with geographic constraints. *Chinese Journal of Aeronautics*, vol. 28, no. 5, pp. 1343–1354, 2015.

[298] L. Guo, W. Liu, X. Li, *et al.*, Energy management system for stand-alone wind-powered desalination microgrid, *IEEE Transactions on Smart Grid*, vol. 7, no. 2, pp. 1079–1088, 2016.

[299] Z. Guo, J. Chang, J. Guo and J. Zhou, Adaptive twisting sliding-mode algorithm for hypersonic reentry vehicle attitude control based on finite-time observer, *ISA Transactions*, vol. 771, pp. 20–29, 2018.

[300] Z. Guo, Q. Ma, J. Guo, B. Zhao and J. Zhou, Performance-involved coupling effect-triggered scheme for robust attitude control of HRV, *IEEE/ASME Transactions on Mechatronics*, vol. 25, no. 3, pp. 1288–1298, 2020.

[301] Z. Guo, J. Guo, J. Zhou and J. Chang, Robust tracking for hypersonic reentry vehicles via disturbance estimation-triggered control, *IEEE Transactions on Aerospace and Electronics*, vol. 56, no. 2, pp. 1279–1289, 2020.

[302] P.J. Gutierez-Diaz and J. Russo, Design of personalized cancer treatment by use of optimal control problems: the case of chronic myeloid leukemia, *Mathematical Biosciences and Engineering*, vol. 17, no. 5, pp. 4773–4800, 2020.

[303] P.J. Gutierez-Diaz, M.A. Lopez Marcos, J. Martinez-Rodriguez and J. Russo, The effects of time valuation in cancer optimal therapies: a study of chronic myeloid leukemia, *Journal of Theoretical Biology and Medical Modelling*, vol. 16, no. 10, pp. 1–20, 2019.

[304] I. Golovin, A. Maksakov, M. Shysh and S. Palis, Discrepancy-based control for positioning of large gantry crane, *Mechanical Systems and Signal Processing*, vol. 163, pp. 108199–108116, 2022.

[305] G. Gogte, C. Venkatesh, F. Kazi, N.M. Singh and R. Pasumarthy, Passivity-based control of underactuated 2D SpiderCrane manipulator, In: *MTNS 2012, 20th Intl. Symp. on Mathematical Theory, Networks and Systems*, Melbourne, Australia, July 2012.

[306] O. Gonzalez and A. Rossiter, Fast hybrid dual mode NMPC for a parallel double inverted pendulum with experimental validation, *IET Control Theory and Applications*, vol. 14, no. 16, pp. 2239–2338, 2020.

[307] K.S. Grewal, R. Dixon and J. Pearson, LQG controller design applied to a prismatic Stewart-Gough platform, *International Journal of Automation and Computing*, vol. 9, no. 1, pp. 45–53, 2012.

[308] M.J. Grimble and P. Majeski, *Nonlinear Industrial Control Systems: Optimal Polynomial Systems and State-Space Approach*, Springer, 2020.

[309] J.W. Grizzle, G. Abbu and F. Plestan, Asymptotically stable walking for biped robots: analysis via systems with impulse effects, *IEEE Transactions on Automatic Control*, vol. 46, no. 1, pp. 51–64, 2001.

[310] J.W. Grizzle, C. Chevallereau, R.W. Sinnet, and A.D. Ames, Models, feedback control, and open problems of 3D bipedal robotic walking, *Automatica*, vol. 50, pp. 1955–1988, 2014.

[311] Q. Guo, P. Sun, J.M. Yin, T. Yu and D. Jiang, Parametric adaptive estimation and backstepping control of electro-hydraulic actuator with delayed memory filter, *ISA Transactions*, vol. 62, pp. 202–214, 2016.

[312] Q. Guo, Z. Chen, Y. Shi, *et al.*, Synchronous control for multiple electrohydraulic actuators with feedback linearization, *Mechanical Systems and Signal Processing*, vol. 178, pp. 109280–109295, 2022.

[313] Q. Guo, L. Chai and H. Liu, Anti-swing sliding-mode control of three-dimensional double pendulum overhead cranes based on extended state observer, *Nonlinear Dynamics*, vol. 111, pp. 391–410, 2023.

[314] R.I. Gwerryina, M.Y. Kura and J.T. Ashezuo, Optimal control model for the infectiology of *Staphylococcus aureus* with dual transmission pathways, *Results in Control and Optimization*, vol. 14, pp. 100364, 2024.

[315] V. Hagenmeyer and E. Delaleau, Robustness analysis of exact feedforward linearization based on differential flatness, *Automatica*, vol. 39, pp. 1941–1946, 2003.

[316] V. Hagenmeyer, S. Streif and M. Zeitz, Flatness-based feedforward and feedback linearization of the ball and plate lab experiment, In: *IFAC NOLCOS 2004, 6th IFAC Conference on Nonlinear Control Systems*, Stuttgart, Germany, Sep. 2004.

[317] V. Hagenmeyer and E. Delaleau, Robustness analysis with respect to exogenous perturbations for flatness-based exact feedforward linearization, *IEEE Transactions on Automatic Control*, vol. 55, no.3, pp. 727–731, 2010.

[318] A. Hamadai, A.H. Hammed and I.K. Ibraheem, Design and performance study of two sliding mode backstepping control schemes for roll channel of Delta Wing Aircraft 2019. In: *6th International Conference on Control, Decision and Information Technologies (CoDIT'19)*, Paris, France, April 2019.

[319] K. Hamiche, M. Fliess, C. Join and H. Abouaissa, Bullwhip effect attenuation to supply chain management via control-theoretic tools and short-term forecasts: a preliminary study with an application to perishable inventories, In: *IEEE CoDIT 2019, IEEE 2019 6th Intl. Conference on Control, Decision and Information Technology*, Paris, France, April 2019.

[320] M. Hamman and C. Ament, Model-based control of a planar 3-DOF cable robot using exact linearization, In: M. Gouttefarde et al. (Eds.), *Proc. CableCon 2021, 5th Intl Conference on Cable-driven parallel robots*, Springer, pp. 260–270, 2021.

[321] M.F. Hamza, H.J. Yap, I.A. Choudhury, A.I. Isa, A.Y. Zimit and T. Kumbasar, Current development on using rotary inverted pendulum as a benchmark for testing linear and nonlinear control algorithms, *Mechanical Systems and Signal Processing*, vol. 116, pp. 347–369, 2019.

[322] C. Han and A.N. Pechev, Underactuated satellite attitude control with two parallel CMGs, In: *IEEE 2007 Intl. Conference on Control and Automation*, Guangzhou, China, May 2007.

[323] C. Han, J. Guo and A. Pechev, Nonlinear H_∞-based underactuated attitude control for small satellites with two reaction wheels, *Acta Astronautica*, vol. 194, pp. 159–172, 2014.

[324] Y. Han and J. Hong, Retrieved strategy for failed satellite on tether's optimal balance swing angle, *Journal of Systems Engineering and Electronics*, vol. 30, no. 4, pp. 749–759, 2019.

[325] Q. Han, X. Shenm, B. Wu, R. Zhu, D. Wang and B. Ruan, Fault diagnosis of the rolling bearings vibration signals based on Kalman Filter method, In: *IEEE 2021 Global Reliability and Prognostics and Health Management*, Nanjing, China, Oct. 2021.

[326] X. Han, C. Liu, B. Chen and S. Zhang, Surge disturbance suppression of AMB-rotor systems in magnetically suspended centrifugal compressors, *IEEE Transactions on Control Systems Technology*, 2022.

[327] A. Handels, E. Margolis and B.R. Levin, Exploring the role of immune response in preventing antibiotic resistance, *Journal of Theoretical Biology*, vol. 256, pp. 655–662, 2009.

[328] J. Hang, W.J. Lee, S. Park, Y. Kim, K.J. Lee and C. Han, Improved control strategy for fixed-speed compressors in parallel system, *Journal of Process Control*, vol. 53, pp. 57–69, 2017.

[329] Z. Hao, Y. Yang, Y. Tian, *et al.*, Phase-lead compensation of a phase-locked loop in an IPMSM for air conditioner compressors, *IEEE Transactions on Energy Conversion*, vol. 35, no. 2, pp. 1090–1100, 2021.

[330] M.R.J. Harandi and H.D. Taghirad, Stabilization of a class of underactuated parallel robots via energy shaping: application to cable-driven manipulators, *Automatica*, vol. 156, pp. 111201–111205, 2023.

[331] M.R.J. Harandi and H.D. Taghirad, On the matching equations of kinetic energy shaping in IDA-PBC, *Journal of the Franklin Institute*, vol. 358, pp. 8639–8655, 2021.

[332] M.R.J. Harandi, S.A. Khalilpour and H.D. Taghirad, Adaptive energy-shaping control of a 3-DOF underactuated cable-driven parallel robot, *IEEE Transactions on Industrial Informatics*, vol. 19, no. 6, pp. 7552–7560, 2023.

[333] M.R.J. Harandi and H.D. Taghirad, Adaptive interconnection and damping assignment passivity-based control for an underactuated cable-driven robot, *International Journal of Adaptive Control and Signal Processing*, vol. 35, pp. 2487–2498, 2021.

[334] M.R.J. Harandi, M. Namvar and H.D. Taghirad, Stabilization of robots with actuator constraints via interconnection and damping assignment, *IEEE Transactions on Control Systems Technology*, pp. 1–8, 2023.

[335] H.M. Hasanien, Transient stability augmentation for a wave energy conversion system using a water cycle algorithm-based multi-objective optimal control strategy, *IEEE Transactions on Industrial Informatics*, vol. 15, no. 6, pp. 3411–3419, 2019.

[336] U. Hasvci, T.C. Burg and R.E. Graff, Closed-loop nonlinear adaptive control of anti-angiogenic tumor therapy, In: *IEEE ASCC 2013 9th Asian Control Conference*, Istanbul, Turkey, June 2019.

[337] Z.B. Hazem, M.J. Fotuhi and Z. Bingul, A study of anti-swing fuzzy LQR control for a double serial link rotary pendulum, *IETE Journal of Research*, pp. 1–13, 2021.

[338] Y. He, X. Mai, C. Cui, *et al.*, Dynamic modeling simulation and experimental verification of a wafer handling SCARA Robot with decoupling servo control, *IEEE Access*, vol. 7, pp. 47143–47153, 2019.

[339] J. He, F. Xu, X. Wang, J. Yang and B. Lang, Modelling and control of free-floating space manipulator using the TS fuzzy descriptor system approach, In: *IEEE SMC 2019, IEEE 2019 Intl Conf. on Systems, Man and Cybernetics*, Bari, Italy, Oct. 2019.

[340] Y. He, X. Li, Z. Xu, X. Zhou and S. Li, Collaboration of multiple SCARA robots with guaranteed safety using recurrent neural networks, *Neurocomputing*, vol. 456, pp. 1–10, 2021.

[341] Y. He, Y. Wu and W. Li, Sliding-mode control for offshore parallel antenna platform with large orientation workspace, *ISA Transactions*, vol. 125, pp. 90–108, 2022.

[342] P. Heidtmann and M. Brucker, Nonlinear modelling and tracking control of a hydraulic rotary vane actuator, *PAMM Proceeding on Applied Mathematics and Mechanics*, vol. 5, pp. 161–162, 2005.

[343] J. Henikl, W. Kemmetmuller, T. Meurer and A. Kugi, Infinite dimensional decentralized damping control of large-scale manipulators with hydraulic actuation, *Automatica*, vol. 63, pp. 101–115, 2016.

[344] T. Henmi, M. Deng and A. Inoue, Unified method for swing-up control of double inverted pendulum systems, In: *IEEE AMS 2014, Proc. of the 2014 Intl. Conference on Advanced Mechatronic Systems*, Kumamoto, Japan, Aug. 2014.

[345] A. Hermant, Optimal control of the atmospheric reentry pf a space shuttle by a homotopy method, *Rapport de Recherche INRIA*, no. 6627, Sep. 2008.

[346] A. Hilderbrandt, R. Neumann and O. Sawodny, Optimal system design of SISO-servopneumatic position drives, *IEEE Transactions on Control Systems Technology*, vol. 18, no. 1, pp. 35–44, 2010.

[347] C.M. Ho, D.T. Tran, C.H. Nguyen and K.K. Ahn, Adaptive neural command filtered control for pneumatic active suspension with prescribed performance and input saturation, *IEEE Access*, vol. 2, pp. 56855–56868, 2021.

[348] H. Hoang and M. Wengsaisuwan, Robust controller design for a rotary double inverted pendulum using linear matrix inequalities, *IEEE SICE 2007 Conferences*, Kagawa University, Japan, Sep. 2007.

[349] Q.D. Hoang, V.A. Rosas-Cervantes, S.G. Lee, J.S. Weon, J.H. Choi and Y.H. Kwon, Robust finite-time convergence control mechanism for high-precision tracking in a hybrid fluid power actuator, *IEEE Access*, vol. 8, pp. 196775–196790, 2020.

[350] N.M. Horri, P. Palmer and M. Roberts, Gain-scheduled inverse optimal satellite attitude control, *IEEE Transactions of Aerospace and Electronic Systems*, vol. 48, no. 3, pp. 2437–2457, 2012.

[351] H.J. Hosseini, S.A. Khalilpour and H.D. Taghirad, Practical robust nonlinear PD controller for cable-driven parallel manipulators, *Nonlinear dynamics*, vol. 106, pp. 405–424, 2021.

[352] Y. Hou and J. Hon, Retrieval strategy for failed satellite on tether's optimal balance swing angle, *Journal of Systems Engineering and Electronics*, vol. 30, no. 4, pp. 749–759, 2019.

[353] L. Houley and S. Alahakoon, Impact assessment of AC and DC electric rope shovel on coal mine power distribution system, In: *IEEE AUPEC 2012, IEEE 2012 22nd Australasian Universities Power Engineering Conference*, Bali, Indonesia, Sep. 2012.

[354] C.F. Hsu, C.M. Lin and T.T. Lee, Wavelet adaptive backstepping control for a class of nonlinear systems, *IEEE Transactions on Neural Networks*, vol. 17, no. 5, pp. 1175–1183, 2006.

[355] X.Y. Huang, N.N. You and M.F. Guo, An H-infinity control method of the bullwhip effect for a class of supply chain systems, *International Journal of Production Research*, vol. 45, no. 1, pp. 207–226, 2007.

[356] S.H. Hyan, Y. Taniai, K. Hianum, K. Yosunaga and H. Mizui, Overpressure compensation for hydraulic hybrid servo booster applied to hydraulic manipulators, *IEEE Robotics and Automation Letters*, vol. 4, no. 2, pp. 942–949, 2020.

[357] Q. Hu, L. Xiao and C. Wang, Adaptive fault-tolerant attitude tracking control for spacecraft with time-varying inertia uncertainties, *Chinese Journal of Aeronautics*, vol. 32, no. 3, pp. 674–687, 2019.

[358] H. Hu, L. Liu, Y. Wang, Z. Cheng and Q. Luo, Active fault-tolerant attitude tracking control with adaptive gain for spacecrafts, *Aerospace Science and Technology*, vol. 98, pp. 105706–105719, 2020.

[359] C. Hu, B. Yao and Q. Wang, Coordinated adaptive robust contouring controller design for an industrial biaxial precision gantry, *IEEE Transactions on Mechatronics*, vol. 15, no. 5, pp. 728–736, 2020.

[360] H. Hu, B. Wang, Z. Cheng, L. Liu, Y. Wang and X. Luo, A novel active fault-tolerant control for spacecrafts with full state constraints and input saturation, *Aerospace Science and Technology*, vol. 108, pp. 106368–106381, 2021.

[361] J. Hu, N. Zhou, R. Guo, G. Wang, G. Zhang and D. Xu, Torque ripple compensation with anti-over-voltage for electrolytic capacitorless PMSM compression drives, *IEEE Journal on Emerging and Selected Topics in Power Electronics*, vol. 10, no. 5, pp. 6148–6159, 2022.

[362] Y. Hu, Y. Wu and W. Li, Nonlinear extended state observer-based adaptive higher-order sliding-mode control for parallel antenna platform with input saturation, *Nonlinear Dynamics*, vol. 11, pp. 16111–16132, 2023.

[363] G. Hu, J. Guo, Z. Guo J. Cieslak and D. Henry, ADP-based intelligent tracking algorithm for reentry vehicles subjected to model and state uncertainty, *IEEE Transactions on Industrial Informatics*, vol. 19, no. 4, pp. 6047–6054, 2023.

[364] H. Hu, J. Cao and Y. Cao, Prescribed time tracking control window velocity measurements for dual-arm robots, *Information Sciences*, vol. 629, pp. 533–550, 2023.

[365] P. Huang, B. Liu and F. Zhang, Configuration maintaining control of three-body ring tethered system based on thrust compensation, *Acta Astronautica*, vol. 123, pp. 37–50, 2016.

[366] R. Huang, J. Zhang and X. Zhang, Adaptive tracking control of uncertain switched nonlinear systems with application to aircraft wing-rock, *IET Control Theory and Applications*, vol. 10, no. 15, pp. 1755–1762, 2016.

[367] O. Hunang, X. Hua and R. Ruan, Optimal control of environmental cleaning and antibiotic prescription in an epidemiological model of methicillin-resistant *Staphylococcus aureus* infections in hospitals, *Mathematical Biosciences*, vol. 311, pp. 13–30, 2019.

[368] W. Huang and J. Yang, A novel piecewise velocity control method using passivity-based controller for wave energy conversion, *IEEE Access*, vol. 5, pp. 59029–59043, 2020.

[369] B. Huang, F. Zhang and P. Huang, Stable spinning deployment of space triangle tethered formation system using event-triggered sliding mode control, *Advances in Space Research*, vol. 70, pp. 1222–1233, 2022

[370] J. Huang, W. Wang and J. Zhou, Adaptive control design for underactuated cranes with guaranteed transient performance: theoretical design and experimental verification, *IEEE Transactions on Industrial Electronics*, vol. 69, no. 3, pp. 2822–2832, 2022.

[371] B. Huang, F. Zhang, M. Song and P. Huang, Adaptive dynamic event-triggered deployment control for space triangle tethered formation system with external disturbances, *Acta Astronautica*, vol. 200, pp. 1–10, 2022.

[372] B. Huang, F. Zhang, M. Song and P. Huang, Event-model-based fuzzy sliding-mode deployment control for triangular tethered satellite systems suffering external disturbances, *Journal of Aerospace Science and Technology*, vol. 139, pp. 108399–108410, 2023.

[373] L. Huang, L. Wei, J. Liu, J. Yang and X. Zhang, Research on an improved model predictive current control for direct-drive wave energy converters with linear generator, *IET Renewable Power Generation*, vol. 17, pp. 1670–1699, 2023.

[374] B. Huang, F. Zhang, M. Song and P. Huang, Event-based predefined time-deployment control for space triangular tethered satellite system with input optimization, *IEEE Transactions on Aerospace and Electronic Systems*, vol. 59, no. 5, pp. 4936–4946, 2023.

[375] B. Huang, F. Zhang and P. Huang, Predefined-time disturbance observer based predefined-time sliding-mode control for space triangle tethered formation system, In: L. Yan, H. Duan and Y. Deng (Eds.), *Advance in Guidance, Navigation and Control, Lecture Notes in Electrical Engineering*, vol. 845, pp. 1163–1173, 2023.

[376] S. Ibaraki and R. Usui, A novel error mapping of bi-directional angular positioning deviation of rotary axes in a SCARA-type robot by "open-loop" tracking interferometer measurement, *Precision Engineering*, vol. 74, pp. 60–68, 2022.

[377] E. Ibarguen-Modragan, S. Mosquera, M. Ceron, *et al.*, Mathematical Modelling on bacterial resistance to multiple antibiotics causes by spontaneous mutations, *Biosystems*, vol. 117, pp. 60–67, 2014.

[378] E. Ibarguen-Mongragan, L. Esteva and M. Ceron-Gomez, An optimal control problem applied to plasmid-mediated antibiotic resistance, *Journal of Applied Mathematics and Computing*, vol. 68, pp. 1635–1667, 2022.

[379] A. Ibeas, A. Esmaeili, J. Herrera and F. Zouari, Discrete-time observer-based state feedback control of heart rate during treadmill exercise, In: *IEEE ICSTCC 2016, IEEE 20th International Conference on System Theory, Control and Computing*, Sinaia, Romania, Oct. 2016.

[380] E. Ida, S. Briot and M. Carricato, Neural oscillations of underactuated cable-driven parallel robots, *IEEE Access*, vol. 9, pp. 71660–71672, 2021.

[381] E. Ida, S. Briot and M. Carricato, Robust trajectory planning of underactuated cable-driven parallel robot with 3-cables, In: J. Lenarcic and B. Siciliano (Eds.), *Proc. ARL 2020*, pp. 65572, 2021.

[382] H. Imani, M.R. Jahed-Matlagh, K. Salashoor, A. Ramezani and A. Moarefianpur, Robust decentralized model predictive control approach or a multi-compressor system surge instability including piping acoustic, *Cogent Engineering*, vol. 5, no. 1, pp. 148381, 2018.

[383] A. Imani, Z. Morawej and M. Pazoki, A novel time-domain method for fault detection and classification in VDC-HVDC transmission lines, *International Journal of Electrical Power and Energy Systems*, vol. 140, article no. 109056, 2022.

[384] A. Imani, Z. Moravej and M. Pazoki, A novel MODWT-based fault detection and classification scheme in VSC-HVDC transmission line, *Electric Power Systems Research*, vol. 221, article no. 109434, 2023.

[385] M. Indri, G. Calafiore, G. Legnani, F. Juttu and A. Visioli, Optimized dynamic calibration of a SCARA Robot, *IFAC 2002 15th Triennial World Congress*, Barcelona, Spain, July 2002.

[386] A. Isidori, *Nonlinear Control Systems*, New York: Springer, 3rd ed., 1995.

[387] Y. Islam, I. Ahmed, M. Zabar and K. Shahzad, Double integral sliding-mode control of leukemia therapy, *Biomedical Signal Processing and Control*, vol. 61, pp. 102046, 2020.

[388] Y. Islam, I. Ahmed, M. Zabair and A. Islam, Adaptive terminal and super-twisting sliding-mode controlled for acute leukemia therapy, *Biomedcal Signal Processing and Control*, vol. 21, pp. 103121–103132, 2022.

[389] J. Ismail and S. Liu, Efficient planning of optimal trajectory for a Furuta double pendulum using discrete mechanics and optimal control, In: *20th IFAC World Conference*, Toulouse, France, July 2017.

[390] S. Issam, K. Bouchaib, A. Lobzai, G. Hicham and B. Mohamed, Mathematical modeling and optimal control strategy for a discrete-time cholear model, *Communications in Mathematical Biological Neurosciences*, vol. 135, pp. 1–16, 2023.

[391] M. Itik, M.U. Salamci and S.P. Banks, Optimal control of drug therapy in cancer treatment, *Nonlinear analysis*, vol. 71, pp. 1473–1486, 2009.

[392] J.B. Izard, M. Gouttefarde, C. Baradat, D. Cullo and D. Sullé, Integration of a parallel cable-driven robot on an existing building facade, In: *1st Intl. Conference on Cable-Driven Parallel Robots*, Stuttgart, Germany, 2012.

[393] D. Ivanov, A. Dolgui and B. Sokolov, On applicability of optimal control theory to adaptive supply chain planning and scheduling, *Annual Reviews in Control*, vol. 36, no. 1, pp. 73–84, 2012.

[394] A. Jabbari, F.M. Malik and S.A. Sheikh, Nonlinear stabilizing control of a rotary double inverted pendulum: a modified backstepping approach, *Transactions of the Institute of Measurement and Control*, vol. 38, no. 11, pp. 1721–1734, 2017.

[395] H. Jabbari-Asl and J. Yoon, Robust trajectory tracking control of cable-driven parallel robots, *Nonlinear Dynamics*, vol. 89, pp. 2769–2784, 2017.

[396] M. Jama, B.F. Mon and A. Wahyudie, Maximum capturing energy approach for heaving wave energy converters using an estimator-based finite control set model predictive control, *IEEE Access*, vol. 9, pp. 67648–67659, 2021.

[397] A.M. Jassour and M. Farrokhi, Adaptive neuropredictive control for redundant robot manipulators in presence of static and dynamic obstacles: a Lyapunov-based approach, *International Journal of Adaptive Control and Signal Processing*, vol. 28, nos. 3–5, pp. 386–411, 2014.

[398] J. Jayachandran, A.E. Rundell, R.E. Hannemann, T.A. Vik and D. Ramkrishna, Optimal chemotherapy for leukemia: a model-based strategy for individualized treatment, *Plos One*, vol. 9–10, e109623, 2014.

[399] F. Jeng Lin, S.Y. Lee, H.C. Chsang and Z.Y. Kuo, Design and implementation of sensorless DC inverter-fed compressor drive system, *Journal of the Chinese Institute of Engineers*, vol. 35, no. 6, pp. 655–673, 2012.

[400] Z. Ji and G. Shi, Adaptive neural dynamic-based hybrid control strategy for stable retrieval of tethered satellite systems *Astrodynamics*, vol. 8, no. 8, pp. 261–269, 2024.

[401] C. Jia, Z. Meng and B. Wang, Deployment of three-body chain-type tethered satellites in low eccentricity orbits using only tethers, *Space: Science and Technology*, pp. 1–9, 2023.

[402] C. Jia, Z. Meng and X. Guo, Dynamic surface control for satellite attitude of the chained three-body tethered system during deployment, *Advances in Space Research*, pp. 1–29, 2024.

[403] H Jiang, Z. Su, D. Wang and C. Wu, Multi-parameter identification for active magnetic bearing with uncertainties based on coupled nonlinear model, *IEEE Transactions on Industrial Electronics*, vol. 70, no. 10, pp. 10431–10441, 2013.

[404] Y. Jiang, D. Li, Z. Kan, B. Dong, C. Chen and J. Xiang, Bifurcation analysis of wing-rock and routes to chaos of a low aspect ratio flying wing, *Nonlinear Dynamics*, vol. 4, pp. 1–14, 2014.

[405] X. Jiang and Z. Bai, Dynamic analysis of the tethered satellite system considering uncertain but bounded parameters, *Journal of Defence Technology*, pp. 1–128, DT 1471, 2024

[406] C. Jiao, X. Wei, H. Zhao and X. Su, Adaptive hybrid impedance control for a dual-arm robot manipulating an unknown object, In: *IEEE IECON 2020, IEEE 46th Annual Conference of the Industrial Electronics Society*, Singapore, Oct. 2020.

[407] H. Jiao, Q. Shen, Y. Shi and P. Shi, Adaptive tracking control for uncertain cancer-tumor immune systems, *IEEE ACM Transactions on Computational Biology and Bioinformatics*, vol. 18, no. 6, pp. 2753–2758, 2021.

[408] T. Jiang, K. Khorasani and S. Tafazoli, Parameter estimation-based fault detection, isolation and recovery for nonlinear satellite model, *IEEE Transactions on Control Systems Technology*, vol. 16, no. 4, pp. 799–808, 2008.

[409] P. Jiang, S. Huang, J. Xiang and M.Z.Q. Chen, Iteratively successive projection: a novel continuous approach for the task-based control of redundant robots, *IEEE Access*, vol. 7, pp. 25347–25358, 2019.

[410] J. Jin and Y. Zhang, G2-type SRMPC scheme for synchronous manipulation of two redundant robot arms, *IEEE Transactions on Cybernetics*, vol. 45, no. 2, pp. 153–164, 2015.

[411] R. Jin, P. Rocco, X. Chen and Y. Geng, LPV-based offline model predictive control for free-floating space-robots, *IEEE Transactions on Aerospace and Electronic Systems*, vol. 57, no. 6, pp. 3896–3905, 2021.

[412] D. Jinjan, G. Yahui, C. Ming and D. Xanzhang, Symmetrical adaptive variable admittance control for position/force tracking of dual-arm cooperative manipulator with unknown trajectory deviations, *Robotics and Computer Integrated Manufacturing*, vol. 57, pp. 357–369. 2019.

[413] S. Jiu, Y. Jiu, S. Xu and Q. Hu, Maneuver and active vibration suppression of free-flying space robot, *IEEE Transactions on Aerospace and Electronic Systems*, vol. 54, no. 3, pp. 1115–1134, 2018

[414] K. Jones, A. Cortinovis, M. Mercengoz and H.J. Ferreau, Distributed model predictive control of centrifugal compressor systems, In: *20th IFAC World Congress*, Toulouse, France, July, 2017.

[415] F. Jost, E. Schalk, D. Weber, H. Dohner, T. Fischer and S. Sagem, Model-based optimal AML consolidation treatment, *IEEE Transactions on Biomedical Engineering*, vol. 67, no. 12, pp. 3298–3308, 2020.

[416] A. Jugat and A.J. Sinclair, Nonlinear control for spacecraft pursuit-evasion game using State-Dependent Riccati equation method, *IEEE Transactions on Aerospace and Electronic Systems*, vol. 53, no. 6, pp. 3032–3042, 2017.

[417] E. Jung, S. Lenhart and E. Jung, Optimal control of treatment in a two-strain tuberculosis model, *Discrete and Continuous Dynamical Systems: Series B*, vol. 2, no. 4, pp. 473–482, 2002.

[418] J. Jung, W.J. Lee, S. Park, Y. Kim, C.J. Lee, and C. Han, Improved control strategy for fixed-speed compressors in parallel system, *Journal of Process Control*, vol. 53, pp. 57–69, 2017.

[419] J. Kalmari, J. Backman and A. Visala, Coordinated motion of a hydraulic forestry crane and vehicle using nonlinear model predictive control, *Computers and Electronics in Agriculture*, vol. 133, pp. 119–127, 2017.

[420] P. Kalud, P. Perlikowski, K. Czalczyanski and T. Kapitaniak, Synchronization configurations of two coupled double pendulums, *Communications in Nonlinear Science and Numerical Simulation*, vol. 19, pp. 977–990. 2014.

[421] N. Kamaldin, S.L. Chen, C.S. Teo, W. Lin, A. Tay and K.K. Tan, A novel adaptive jerk control with application to large work-piece tracking on a flexure-linked dual drive gantry, *IEEE Transactions on Industrial Electronics*, vol. 66, no. 7, pp. 5353–5363, 2019.

[422] A.K. Kamath, F. Kazi and N.M. Singh, Dynamics and control of 2D SpiderCrane: a RHC Approach, In: *IFAC MTNS 2010, Proc. of the 19th Intl. Symposium on Mathematical Theory of Networks and Systems*, Budapest, Hungary, July 2010.

[423] D. Kang, J. Chen, M. Zamara, F. Zangerbashi and S. Coros, RL+Model-based control using on-demand optimal control to learn versatile legged locomotion, *IEEE Robotics and Automation Letters*, vol. 8, no. 13, pp. 6619–6626, 2023.

[424] B.K. Kanth Hari, K. Sundar, S. Srinivashan, A. Zlotnik and R. Bent, Operation of natural gas pipeline networks with storage under transient flow conditions, *IEEE Transactions on Control Systems Technology*, vol. 30, no. 2, pp. 667–679, 2022.

[425] S.M. Kargar and R. Mehral, Robust model predictive control for a small reverse osmosis desalination unit subject to uncertainty and actuator fault, *Journal of Water Supply*, vol. 20, no. 4, pp. 1229–1240, 2020.

[426] F. Kazi, P. Mullhaupt, R.M. Banavar and D. Bonvin, Dynamics and control of 2D SpiderCrane: a Lie-Backlund approach, In: *Proc. ECC 2009, European Control Conference*, Budapest, Hungary, Aug. 2009.

[427] F. Kazi, R.N. Banavar, P. Mullhaupt and D. Bonvin, Stabilization of a 2D SpiderCrane Mechanism using damping assignment passivity-based control, In: *IFAC WC 2008, The 17th IFAC World Congress*, Seoul, Korea, July 2008.

[428] Z. Kang, W. Lin, Z. Liu and R. Xu, Accurate Contour Error Estimation-based robust contour control for dual-linear-motor-driven gantry stages, *Mechatronics*, vol. 100, pp. 103174–103182, 2024.

[429] A. Karami, H. Sadeghian, M. Keshmiri and G. Oriolo, Force orientation and position control in redundant manipulators in prioritized scheme with null space compliance, *Control Engineering Practice*, vol. 85, pp. 23–33, 2019.

[430] G. Katriel, Optimizing antimicrobial treatment schedules: some fundamental analysis results, *Bulletin of Mathematical Biology*, vol. 86, no. 1, pp. 1–37, 2023.

[431] S. Kayastha, L. Shi, J. Katapitiya and G. Pearce, Nonlinear model predictive control of a planar three-link space manipulator, In: *IEEE ASCC 2017, 11th Asian Control Conference*, Gold Coast, Australia, Dec. 2017.

[432] S.A. Khiani, H. Khaloozadeh and F. Soltanian, Suboptimal, sliding manifold for nonlinear supply chain with time-delay, *Journal of Combinatorial optimization*, vol. 43, pp. 151–154, 2021.

[433] C. Kim, J.H. Chang and D. Hong, Coordination control of an active deburring tool, *Robotics and Computer Integrated Manufacturing*, vol. 24, pp. 462–471, 2008.

[434] B.S. Kim and J.B. Song, Design and control of a variable stiffness actuator, based on adjustable moment arm, *IEEE Transactions on Robotics*, vol. 28, no. 5, pp. 1145–1151, 2012.

[435] J.S. Kim, J. Chen and H.E. Garcia, Modeling, control and dynamic performance analysis of a reverse osmosis desalination plant integrated within hybrid energy system, *Energy*, vol. 112, pp. 52–66, 2016.

[436] J. Kim, n, W. Choi and J. Lee, Discrete-time delay control for hydraulic excavator motion control with terminal sliding-mode control, *Mechatronics*, vol. 60, pp. 15–25, 2019.

[437] J. Kim, D. Lee, B. Kiss and D. Kim, An adaptive unscented Kalman filter with selective scaling (AUKF-SS) for overhead cranes, *IEEE Transactions on Industrial Electronics*, vol. 68, no. 7, pp. 6131–6140, 2021.

[438] S. Kittissares, Y. Hirota, H. Nabae, G. Endo and K. Suzumori, Alternating pressure control system for hydraulic robots, *Mechatronics*, vol. 85, pp. 102822–102831, 2022.

[439] H. Khalil, Nonlinear Systems, Prentice Hall, 1996.

[440] H. Khan, S. Ahmed, J. Alzabut and A.T. Azar, A generalized coupled system of fractional differential equations with application to finite-time sliding-mode control of leukemia therapy, *Chaos, Solitons and Fractals*, vol. 174, pp. 113901–113910, 2023.

[441] W. Khitur, M. Turki and J. Belhadj, Power control strategy for PV/wind reverse osmosis desalination without battery, *Control Engineering Practice*, vol. 89, pp. 169–179, 2019.

[442] M.T. Khorsardi, B.M. Fard and A. Bagheri, Optimal tracking control of a biped robot walking in the lateral plane, In: *IEEE 2011 Intl. Symposium on Innovation in Intelligent Systems and Applications*, Istanbul, Turkey, June 2011.

[443] E. Khoshbin, K. Youssef, R. Meziane and M.J.D. Otis, Reconfigurable fully constrained cable-driven parallel mechanism for avoiding collision between cables with human, *Robotica*, vol. 40, pp. 4405–4430, 2022.

[444] T. Kobayashi, M. Iwase, S. Suzuki and K. Furuta, Swinging-up and balancing control of double Furuta pendulum by using State Dependent Riccati Equation, In: *IFAC 2004 Conference on Mechatronic Systems*, Sydney, Australia, 2004.

[445] U.E. Kocamaz, H. Taksin, Y. Uyaroglu and A. Goksu, Control and synchronization of chaotic supply chains using intelligent approaches, *Computers and Industrial Engineering*, vol. 102, pp. 476–487, 2016.

[446] J. Koivumaki and J. Mattila, High performance nonlinear motion/force controller design for redundant hydraulic construction crane automation, *Automation in Construction*, vol. 51, pp. 59–97, 2015.

[447] J. Koivumaki and J. Mattila, Stability-generated force-sensorless contact force-motion control of heavy duty hydraulic manipulators, *IEEE Transactions on Robotixs*, vol. 31, no. 4, pp 916–936, 2015.

[448] J. Koivumaki, W.H. Zhu and J. Mattila, Energy-efficient and high-precision control of hydraulic robots, *Control Engineering Practice*, vol. 85, pp. 176–193, 2019.

[449] T. Komine, M. Iwase, S. Suzuki and K. Furuta, Rotational control of double pendulum, In: *IFAC 2004 on Mechatronic Systems*, Sydney, Australia, 2004.

[450] P. Konstadinopoulos, D.T. Mook and A. H. Nayfeh, Subsonic wing rock of slender delta wings, *Journal of Aircraft*, vol. 22, pp. 223–228, 1985.

[451] D.K. Kori, J.P. Kolhe and S.E. Talole, Extended state observer-based robust control of wing-rock motion, *Aerospace Science and Technology*, vol. 39, pp. 107–117, 2014.

[452] A. Kosari, A. Sarifi, A. Ahmadi and M. Khashima, Remote sensing satellite's attitude control system: rapid performance sizing for passive scan imaging mode, *Journal of Aircraft Engineering and Aerospace Technology*, vol. 92, no. 7, pp. 1073–1083, 2020.

[453] S. Koyostha, L Shi, J. Katupitiya and G. Pearce, Nonlinear model predictive control of a planar three-link space manipulator, In: *IEEE ASCC 2017, IEEE 2017 Asian Control Conference*, Gold Coast, Australia, Dec. 2017.

[454] M. Krause, J. Rudolph and F. Woittennek, Motion planning for planar bipeds using differential flatness: the single support phase, *Proceedings of Applied Mathematics and Mechanics*, vol. 9, pp. 627–630, 2009.

[455] T.T. Kristoffersen and C. Holden, Modeling and control of a wet-gas centrifugal compressor, *IEEE Transactions on Control Systems Technology*, vol. 29, mo. 3, pp. 1175–1190, 2021.

[456] L. Krönig, P. Hörler, S. Caseiro, *et al.*, Precision control of miniature SCARA robots for multi-object spectrographs, *Journal of Optomechatronics*, vol. 14, no. 1, pp. 53–77, 2020.

[457] G.K.D. Kumar and D. Tan, Nonlinear optimal control of tethered satellite systems using tether offset in the presence of tether failure, *Acta Astronautica*, vol. 66, pp. 1434–1448, 2010.

[458] N. Kumar, V. Panwar, N. Sukovanam, S.P. Sharma and J.H. Borm, Neural network-based nonlinear tracking control of kinematically redundant robot manipulator, *Mathematics and Computer Modelling*, vol. 57, pp. 1889–1901, 2011.

[459] N. Kumar, V. Panwar, J.W. Borm, J. Choi and J. Woon, Adaptive neural controller for space robot system with an attitude controlled base, *Computing and Applications*, vol. 23, pp. 2333–2340, 2013.

[460] R. Kumar, A. Chalanga and B. Bandyopadhyay, Smooth integral sliding-mode controller for the position control of Stewart platform, *ISA Transactions*, vol. 58, pp. 543–554, 2015.

[461] P.R. Kumar, A.K. Behera and B. Bandyopadhyay, Robust finite-time tracking of Stewart platform: a super-twisting like observer-based forward kinematic solution, *IEEE Transactions on Industrial Electronics*, vol. 64, no. 5, pp. 3776–3785, 2017.

[462] S. Kumar and A. Cortinovis, Load sharing optimization for parallel and serial compressor systems, In: *IEEE CCTA 2018, IEEE 2018 Conf. on Control Technology and Applications*, Hawai, USA, Aug. 2017.

[463] A.A. Kumar, J.F. Antoine and G. Abba, Linéarisation à rétroaction partielle d' un robot parallèle commande par câble, 24éme Congrés Français de Mécanique, Brest, France, Aug. 2019.

[464] A.A. Kumar, Conception et commande d' un robot â câbles pour la manipulation dextre de pièces sur des chaines de production, Thèse de doctorat, Université de Lorraine, 2020.

[465] A.A. Kumar, J.F. Antoine and G. Abba, Input-output feedback linearization for the control of 4-cable driven parallel robot, In: *IFAC MIM 2019, IFAC 2019 Conference on Manufacturing, Modelling, Management and Control*, Berlin, Germany, 2019.

[466] A.A. Kumar, J.F. Antoine and G. Abba, Control of an underactuated 4 cable-driven parallel robot using modified input-output feedback linearization, In: *21st IFAC World Congress 2020*, Berlin, Germany, July 2020.

[467] A. Kumar and V. Sangwan, Planar bipedal robot with differentially flat dynamics, In: *Mathmod 2022, 10th Vienna Conference on Mathematical Modelling*, Vienna, Austria, July 2022.

[468] A. Kumar Maurya and S.P. Singh, Analysis of cascaded buck-boost inverter for PMLSG-based ocean wave energy converter, *IETE Journal of Research*, vol. 68, no. 6, pp. 4109–4119, 2022.

[469] A. Kumar Maurya and S.P. Singh, Analysis of buck-boost inverter installed in permanent magnet linear synchronous generator-based ocean wave energy converter, *IETE Journal of Research*, doi: 10.1080/03772064, pp. 1–14, 2022.

[470] P.M. La Hera, U. Mettin, S. Westenberg and A.S. Shiriaev, Modelling and control of hydraulic rotary actuators used in forestry cranes, In: *IEEE ICRA 2009, IEEE 2009 Intl. Conference on Robotics and Automation*, Kobe, Japan, May 2009.

[471] A.S. Lafmiejani, M.T. Masouleh and A. Kolhar, Trajectory tracking control of a pneumatically actuated 6-DOF Gough-Stewart parallel robot using backstepping sliding-mode controller and geometry-based quasi-forward kinematic method, *Robotics and Computer Integrated Manufacturing*, vol. 54, pp. 96–114, 2018.

[472] G. Lahet, K. Liu and F.L. Lewis, Dynamic analysis and control of a Stewart platform manipulator, *Journal of Robotic Systems*, vol. 10, no. 5, pp. 629–655, 1993.

[473] Y. Lan and M. Fei, Design of state-feedback controller by pole placement for a coupled set of inverted pendulums, In: *IEEE ICEM 2014, 10th International Conference on Electronic Measurement and Investments*, Chengdu, China, Aug. 2011.

[474] M. Lawali Ali, Etude des structures des génerateurs linéaires pour la conversion de l' énergie de la houle, Thèse de Doctorat, Université de Normandie, Le Havre, 2021

[475] O.M. Le, M.T. Pham, M. Tavakoli, R. Moreau, J.P. Simon and T. Redarce, Bilateral control of nonlinear pneumatic teleoperation system with solenoid valves, *IEEE Transactions on Control Systems Technology*, vol. 21, no. 4, pp. 1463–1471, 2013.

[476] A.T. Le and S.G. Lee, 3D comparative control of tower cranes using robust adaptive techniques, *Journal of the Franklin Institute*, vol. 354, pp. 8333–8357, 2017.

[477] V. Lebastard, Y. Aoustin and F. Plestan, Step-by-step sliding-mode observer for control of a walking biped robot by using only actuated variable measurements, In: *IEEE IROS 2005, Proc. IEEE Intl. Conf. on Intelligent Robots and Systems*, Edmonton, Canada, Aug. 2005.

[478] K.W. Lee and S.N. Singh, Attractive manifold-based adaptive solar attitude control of satellites in elliptic orbits, *Acta Astronautica*, vol. 68, pp. 185–196, 2011.

[479] S. Lee, J. Hong and D. Har, Jointly optimized control for reverse osmosis desalination process with different types of energy resources, *Energy*, vol. 117, pp. 116–130, 2016.

[480] K.W. Lee and S.N. Singh, Immersion and invariance-based adaptive wing-rock control with nonlinear terminal manifold, *Nonlinear Dynamics*, vol. 88, pp. 955–972, 2017.

[481] W. Lee and W.K. Chung, Disturbance observer-based compliance control of electo-hydraulic actuators with backdrivability, *IEEE Robotics and Automation Letters*, vol. 4, no. 2, pp. 1722–1730, 2019.

[482] J. Lee, D. Lee and Y. Kim, Mathematical model of STAT signaling pathways in cancer development and optimal control approaches, *Royal Society Open Sources*, vol. 8, 210694, 2021.

[483] M. Lee, H. Choi, C.U. Kim, J. Moon, D. Kim and D. Lee, Precision motion control of robotized industrial robotic excavators via data-driven model inversion, *IEEE Robotics and Automation Letters*, vol. 7, no. 2, pp. 1912–1919, 2022.

[484] A.C.-H. Lee, H.-K. Hsu, H.-P. Huang, Optimized system identification of humanoid robots with physical consistency constraints and floating-based exciting motions, *Journal of Humanoid Robotics*, vol. 19, no. 5, article no., 2250015, pp. 1–21, 2022.

[485] J.M. Lemos and D.V. Caiado, Receding horizon control of tumor growth based on optimal control, In: *IEEE MED 2015, IEEE 2015 Mediterranean Conference on Control and Automation*, Torremolinos, Spain, June 2015.

[486] J.M. Lemos, D.V. Calado, R. Coehlo and S. Vinga, Optimal and receding horizon control of tumor growth in myeloma bone disease, *Biomedical Signal Processing and Control*, vol. 24, pp. 128–134, 2016.

[487] P. Lesniewski and A. Bartoszewicz, Optimal model-reference sliding-mode control of perishable inventory system, *IEEE Transactions on Automation Science and Engineering*, vol. 17, no. 3, pp. 1647–1656, 2020.

[488] M. Leszczynski, U. Ledzewicz, and H. Schattler, Optimal control for a mathematical model for chemotherapy with pharmacometrics, *Mathematical Modelling of Natural Phenomena*, vol. 15, no. 69, pp. 1–23, 2020.

[489] C. Letelier and J.P. Barbot, Optimal flatness placement of sensors and actuators for controlling chaotic systems, *Chaos*, vol. 31, no. 10, article no. 103114, 2021.

[490] J. Lévine, J. Lottin and J.G. Pousart, A nonlinear approach to the control of magnetic bearings, *IEEE Transactions on Control Systems Technology*, vol. 4, no. 5, pp. 524–544, 1996.

[491] J. Lévine, *Analysis and Control of Nonlinear Systems: A Flatness-Based Approach*, Springer, 2009.

[492] J. Lévine, On necessary and sufficient conditions for differential flatness, *Applicable Algebra in Engineering, Communications and Computing*, vol. 22, no. 1, pp. 47–90, 2011.

[493] C. Li, B. Yu and Q. Wang, Modelling and synchronization control of a dual-drive industrial gantry stage, *IEEE/ASME Transactions on Mechatronics*, vol. 23, no. 6, pp. 2340–2351, 2018.

[494] C. Li, Z. Chen and B. Yao, Adaptive robust synchronization control of a dual-linear-motor-driven gantry with rotational dynamics and accurate online parameter estimation, *IEEE Transactions on Industrial Informatics*, vol. 14, no. 7, pp. 2013–3022, 2018.

[495] C. Li, C. Li, Z. Chou and B. Yao, Advanced synchronization control of a dual-linear-motor-driven gantry with rotational dynamics, *IEEE Transactions on Industrial Electronics*, vol. 65, no. 9, pp. 7526–7535, 2018.

[496] Z. Li, B. Huang, A. Ajoudani, C. Yanh, C.Y. Su and A. Bicchi, Asymmetric binomial control of dual-arm exoskeletons for human cooperative manipulators, *IEEE Transactions on Robotics*, vol. 34, no. 1, pp. 264–271, 2018.

[497] X. Li, Y. Pan, G. Chen and H. Yu, Continuous tracking control for a compliant actuator with two-stage stiffness, *IEEE Transactions on Automation Science and Engineering*, vol. 15, no. 1, pp. 57–66, 2018.

[498] L. Li, L. Xie, X. Luo and Z. Wang, Compliance control using hydraulic heavy-duty manipulator, *IEEE Transactions on Industrial Informatics*, vol. 15, no. 2, pp. 1193–1204, 2019.

[499] X. Li, G. Sun and X. Shao, Discrete-time pure-tension sliding-mode predictive control for the deployment of space-tethered satellite with input saturation, *Acta Astronautica*, vol. 170, pp. 521–529, 2020.

[500] G. Li, C.Y. Leung, Y. Wardi, L. Debarbieux and J.S. Weitz, Optimizing the timing and composition of therapeutic phage cocktails: a control theoretic approach, *Bulletin on Mathematical Biology*, vol. 82, no. 75, pp. 1–29, 2020.

[501] X. Li, A. Palazzolo and Z. Wang, A combination 5-DOF active magnetic bearing for energy storage flywheels, *IEEE Transactions on Transportation Electrification*, vol. 7, no. 4, pp. 2344–2355, 2021.

[502] X. Li, T. Cui, K. Huan and X. Ma, Optimization of load sharing for parallel compressors using a novel hybrid intelligent algorithm, *Energy Science and Engineering*, vol. 9, pp. 330–342, 2021.

[503] D. Li and A. Ukil, Fault detection method using high-pass filtering in VSC-based multi-terminal DC system, *International Journal of Electrical Power and Energy Systems*, vol. 132, article no. 107207, 2021.

[504] K. Li, C. Peng, Z. Deng, W. Huang and Z. Zhang, Field dynamic balancing for active magnetic bearings supporting rigid rotor shaft based on extended state observer, *Mechanical Systems and Signal Processing*, vol. 158, pp. 107801–107814, 2021.

[505] Y. Li, L. Huang, M. Chen, P. Tan and M. Hu, Maximum power-point tracking control based on inertia force for underwater direct-drive wave-energy converter, *Renewable Energy*, vol. 215. pp. 118964–118976, 2023.

[506] T. Li, J. Zhang, S. Li, P. Zhou and D. Lv, Neural-based adaptive fixed-time prescribed performance control for the flexible-joint robot with actuator failures, *Nonlinear Dynamics*, vol. 111, pp. 16187–16214, 2023.

[507] G. Li, X. Ma, Z. Li, P. Guo and Y. Li, Kinematic coupling-based trajectory planning for rotary crane system with double pendulum effects and output constraints, *Journal of Field Robotics*, vol. 40, pp. 289–305, 2023.

[508] Y. Li, N. Wang, X. Hui, H.R. Karimi and S. Liu, Finite-time cascade-like tracking control of direct-drive wave energy converters, *Ocean Engineering*, vol. 266, pp. 112622–112630, 2022.

[509] W. Li, Y. Zou, X. Ma, B. Qiu and D. Guo, Novel neural controller for kinematic redundancy resolution of joint-constrained Gough-Stewart robot, *IEEE Transactions on Industrial Informatics*, vol. 20, no. 3, pp. 4559–4570, 2024.

[510] B. Li, L. Zhu and J. Huang, Stable locomotion of biped robots with gaits of sinusoidal harmonics, *IEEE Transactions on Control Systems Technology*, vol. 32, no. 3, pp. 805–817, 2024.

[511] L. Li, H. Ouyang, H. Qu and L. Gao, Sliding-mode control of active magnetic bearing motor system based on state observer, *Advances in Mechanical Engineering*, vol. 16, no. 5, pp. 1–12, 2024.

[512] Z. Liang, Z. Ren and X. Shao, Decoupling trajectory tracking for gliding reentry vehicles, *IEEE/CAA Journal of Automatica Sinica*, vol. 2, no. 1, pp. 115–120, 2015.

[513] D. Liu, Q. Wei, D. Wang and X. Yang, Hongliang Li, *Adaptive Dynamic Programming with Applications in Optimal Control*, Springer, 2017.

[514] X. Liang, Y. Wan, C. Zhang, Y. Kim, O. Xin and W. Yi, Robust position control of hydraulic manipulators using time-delay estimation and non-singular fast terminal sliding-mode, In: *Proceedings of the IMehE, Journal of Systems and Control Engineering*, vol. 232, no. 1, pp. 50–61, 2018.

[515] L.H. Liang, L.Y. Wang and J.F. Wang, Energy-saving nonlinear friction compensation method for direct drive volume control flange-type rotary-vane steering gear, *IEEE Access*, vol. 7, pp. 148351–148362, 2019.

[516] Y. Liang, Y. Jin, C. Wanchun and L. Hao, Entry guidance with no-fly zone avoidance using linear pseudospectral model predictive control, *IEEE Access*, vol. 7, pp. 98589–98602, 2019.

[517] X. Liang, G. Wang, C. Hu and C. Dong, Observer-based H-infinity fault tolerant attitude control for satellite with actuator sensor faults, *Journal of Aerospace Science and Technology*, vol. 95, 2019.

[518] S. Liang, H. Wang and P.X. Liu, Adaptive fuzzy tracking control of flexible-joint robots based on command filtering, *IEEE Transactions on Industrial Electronics*, vol. 67, no. 5, pp. 4046–4055, 2020.

[519] F. Liao, Y. Zhou and Q. Zhang, Gait transition and orbital stability analysis for a biped robot based on the V-DSLIP model with torso and swing leg dynamics, *Nonlinear Dynamics*, vol. 108, pp. 3053–3073, 2022.

[520] C.H. Lin, Senior and C.F. Hsu, Supervisory recurrent fuzzy neural network control of wing rock for slender delta wings, *IEEE Transactions on Fuzzy Systems*, vol. 12, no. 5, pp. 733–742, 2004.

[521] F.J. Lin, P.H. Chen, C.J. Chen and Y.S. Lin, DSP-based cross-coupled synchronous control for dual linear motors via intelligent complementary sliding-mode control, *IEEE Transactions on Industrial Electronics*, vol. 59, no. 2, pp. 1061–1073, 2012.

[522] Y. Lin, N. Wang, X. Hui, H.R. Karimi and S. Liu, Finite-time cascade-like tracking control of direct-drive wave energy converters, *Ocean Engineering*, vol. 266, pp. 112622–112630, 2022.

[523] C. Liu, J. Yang, K. An, M Liu and Q. Chen, Robust control of semi-positive biped dynamic locomotion based on a discrete control Lyapunov function, *Robotica*, vol. 38, no. 8, pp. 1345–1358, 2013.

[524] J.O. Limaverde Filho, E.C.R. Fortaleza and M.C.M. Campos, A derivative-free nonlinear Kalman Filtering approach using flat inputs, *International Journal of Control*, 2021.

[525] C.A. Lightcap and S.A. Banks, An extended Kalman filter for real-time estimation and control of a rigid-link flexible-joint manipulator, *IEEE Transactions on Control Systems Technology*, vol. 18, no. 1, pp. 91–101, 2010.

[526] T. Lin, Y. Lin, H. Ren, H Chen, Z. Li and Q. Chen, A double variable control load sensing system for electric hydraulic excavators, *Energy*, vol. 223. pp. 119999–120012, 2021.

[527] L. Lion, Y.J. Liu, S. Tong and Z. Gao, Relative threshold-based event-triggered control for nonlinear constrained systems with application to aircraft wing-rock motion, *IEEE Transactions on Industrial Informatics*, vol. 18, no. 2, pp. 1175–1183, 2022.

[528] Z. Liu, J. Gao and Q. Zhang, Antishudder gearshift controller design for automatic transmission, *IEEE Transactions on Vehicular Technology*, vol. 60, no. 9, pp. 4261–4275, 2011.

[529] C. Liu, G. Lu and J. Fang, Feedback linearization and extended observer-based control for rotor-AMBs system with mismatched uncertainties, *IEEE Transactions on Industrial Electronics*, vol. 64. no. 2, pp. 1313–1322, 2017.

[530] W. Liu, R. Guon, L. Yuan, Z. Li and M. Tong, Position feedback dynamic surface control for pneumatic actuator position servo system, *Systems Science and Control Engineering*, vol. 6, no. 1, pp. 388–397, 2018.

[531] Z. Liu, T. Yang, N. Sun and Y. Fang, An antiswing trajectory planning method with state constraints for 4-DOF tower cranes: design and experiments, *IEEE Access*, vol. 7, pp. 62142–62151, 2019.

[532] E. Liu, L. Lv, Y.Yi and P. Xie, Research on the steady operation optimization model of natural gas pipeline considering the combined operation of air coolers and compressors, *IEEE Access*, vol. 7, pp. 83521–83535, 2019.

[533] T. Liu, J.P. Jiang and P. Zhang, Decentralized event-triggered control of large-scale nonlinear systems, *Journal of Robust and Nonlinear Control*, vol. 30, pp. 1451–1466, 2020.

[534] Z. Liu, Y. Fu, N. Sun, T. Yang and Y. Fang, Collaborative antiswing hoisting for dual rotary cranes with motion constraints, *IEEE Transactions on Industrial Informatics*, vol. 18, no. 9, pp. 6120–6130, 2022.

[535] Z. Liu, W. Lu, X. Yu, J.J. Rodriguez-Adina and H. Gao, Approximation-free robust synchronization control for dual-linear-motors-driven system with uncertainties and disturbances, *IEEE Transactions on Industrial Electronics*, vol. 69, no. 10, pp. 10500–10509, 2022.

[536] C. Liu, W. Wang, J. Kong and Z.H. Zhou, Spin-deployment of Hub-spoke tethered satellite formation with sliding-mode tether tension control, *Advances in Space Research*, vol. 71, pp. 2509–2520, 2023.

[537] Y. Liu, W. Su, C. Buccella and C. Cecati, Robust control of a dual-linear-motor-driven gantry stage for coordinated contouring tasks based on feed velocity, *IEEE Transactions on Industrial Electronics*, vol. 70, no. 6, pp. 6229–6238, 2023.

[538] X. Liu, X. Ma, S. Zhang, J. Zhou and Y. Chen, Feedback linearization and robust control for whirl mode with operating point deviation in active magnetic bearing-rotor system, *IEEE Transactions on Industrial Electronics*, vol. 70, no. 8, pp. 7673–7682, 2023.

[539] T. Liu, Y. Zhang, S. Wang, X. Li, H.B. Goui and A.M.M. Ghias, Fault identification and fault location methods for VSC-HVDC transmission lines based on the travelling waveform difference, *International Journal of Electrical Power and Energy Systems*, vol. 147, article no. 108867, 2023.

[540] Z. Liu, H. Gao, X. Yu, *et al.*, B-spline wavelet neural-network-based adaptive control for linear-motor-driven systems via a novel gradient descent algorithm, *IEEE Transactions on Industrial Electronics*, vol. 71, no. 2, pp. 1896–1905, 2024.

[541] Z. Liu, W. Lin, X. Yu, J.J. Rodriguez-Andina and H. Gao, Approximation-free robust synchronization control for dual linear-motor-driven systems with uncertainties and disturbances, *IEEE Transactions on Industrial Eletronics*, vol. 69, no. 10, pp. 10500–10509, 2023.

[542] Q.Z. Liu, L. Zhang, B. Sun, Y. Xiao and G.W. Fun, Fixed-time disturbance observer-based attitude prescribed performance predictive control for flexible spacecraft, *IEEE Transactions on Aerospace and Electronic Systems*, vol. 60, no. 3, pp. 3209–3220, 2024.

[543] Y. Liu and W. Sun, High-performance position control for repetitive tasks of motor-drive servo systems based on periodic disturbance observers, *IEEE/ASME Transactions on Mechatronics*, vol. 28, no. 5, pp. 2461–2470, 2023.

[544] X. Liu, H. Rong, F. Neri, *et al.*, Human-simulated intelligent walking control for biped robots, *IEEE Transactions on Automation Science and Engineering*, pp. 1–13, 2024.

[545] D. Liang, Y. Wu, Y. Chen, Y. Fang and L. Liu, Energy-based motion control for pneumatic artificial muscle-actuated robots with experiments. *IEEE Transactions on Industrial Electronics*, 2022.

[546] Q. Liu, Y. Liu, C.Zhu, X. Guo, W. Meng, Q. AI and J. Hu, Design and control of a reconfigurable upper limb rehabilitation exoskeleton with soft modular joints, *IEEE Access*, vol. 9, pp. 166815–166824, 2021.

[547] J. Löber, *Optimal Trajectory Tracking of Nonlinear Dynamical Systems*, Springer, 2017.

[548] A.S. Lofimejani, M.T. Masouleh and A. Kalhor, Trajectory tracking control of a pneumatically actuated 6-DOF Gough-Stewart parallel robot using backstepping sliding-mode controller and geometry-based quasi-forward kinematic model, *Robotics and Computer Integrated Manufacturing*, vol. 54, pp. 96–114, 2016.

[549] L.N.B Long, T.C. Cuong, H.S. Kim and S.S. You, Sustainability and robust decision support strategy for multi-echelon supply chain system against disruptions, *International Journal of Logistics Research and Applications*, pp. 1–12, 2023.

[550] Y.Y. Lu, H.D. Hu, X.L. Zhang, L.Y. Wu and Q.Z. Liu, Optimum design of reverse osmosis system under different feed concentration and product specification, *Journal of Membrane Science*, vol. 287, pp. 219–229, 2007.

[551] C.N. Lu, C.C. Tsai, M.C. Tsai, K.V. Ling and W.S. Yuo, Application of model predictive control to parallel-type double-inverted pendulum driven by a linear motor, In: *IEEE IECON 2007, IEEE 33rd Annual Conference of the Industrial Electronics Society*, Taipei, Taiwan, Nov. 2007.

[552] L. Lu and B. Yao, Energy-saving adaptive robust control of a hydraulic manipulator using five cartridge valves with an accumulator, *IEEE Transactions on Industrial Electronics*, vol. 61, no. 12, pp. 7046–7055, 2014.

[553] M. Lu, M. Salvaggio, T. Wang, *et al.*, Dual-arm object transportation via model predictive control and external disturbance estimation, In: *IEEE CASE 2022, IEEE 18th Intl. Conference on Automation Science and Engineering*, Mexico City, Aug. 2022.

[554] J. Lu, C. Ye, S. Wan, K. Liu and C. Ye, Improved active disturbance rejection control for AMB flywheel system based on inverse system, *International Journal of Dynamics and Control*, vol. 12, pp. 1086–1097, 2024.

[555] H. Lu, C. Wang, A. Li, Y.M. Zabolotnov and Y. Guo, Sliding mode control strategy of spinning electrodynamic tether formation during its spin-up process, *IEEE Transaction on Aerospace and Electronic Systems*, vol. 60, no. 1, pp. 449–462, 2024.

[556] L. Lublin and M. Athans, An experimental comparison of and designs for interferometer testbed, In: B. Francis and A. Tannenbaum (Eds.) *Lectures Notes in Control and Information Sciences: Feedback Control, Nonlinear Systems and Complexity*, Springer, pp. 150–172, 1995.

[557] C. Lunz and J.F. Bonnans, Modelling and optimal control of a two-species bioproducing microbial consortium, *SIAM Journal of Applied Mathematics*, vol. 83, no. 1, pp. 10.1137, 2023.

[558] C. Luo, H. Wan, D. Jin and J. Sun, Retargeting control fo a multi-tethered satellite formation at sun-earth libration point, *Advances in Space Research*, vol. 70, pp. 268–285, 2022.

[559] Z. Ma and G. Sun, Adaptive sliding-mode control of tethered satellite deployment with input limitation, *Acta Astronautica*, vol. 127, pp. 67–75, 2016.

[560] Z. Ma and G. Sun, Adaptive hierarchical sliding-mode control with input saturation for attitude regulation of multi-satellite tethered system, *Journal of Astronautical Sciences*, vol. 64, pp. 207–230, 2017.

[561] Z. Ma, G. Sun, Z. Cheng, and Z. Li, Pure-tension non-linear sliding mode control for deployment of tethered satellite system, *Proceedings of the Institute of Mechanical Engineers, Part G, Journal of Aerospace Engineering*, vol. 232, no. 13, pp. 2541–2551, 2018.

[562] J. Ma, S.L. Chen, N. Kamaldin, *et al.*, Integrated mechatronic design in the flexure-linked dual-drive gantry by constrained linear-quadratic optimization, *IEEE Transactions on Industrial Electronics*, vol. 65, no. 3, pp. 2408–2418, 2018.

[563] X. Ma, S. Zhang and K. Wang, Active surge control for magnetically suspended centrifugal compressors using a variable equilibrium point approach, *IEEE Transactions on Industrial Electronics*, vol. 66, no. 12, pp. 9363–9393, 2019.

[564] Y. Maddahi, S. Liao, W.K. Fung and N. Sepehri, Position referenced force augmentation in teleoperated hydraulic manipulators operating under delayed and lossy networks: a pilot survey, *Robotics and Autonomus Systems*, vol. 83, pp. 231–242, 2016.

[565] A. Mahdy, H.M. Hasanian, W. Holmy, R.A. Turky and S.H.E. Abdel Aleem, Transient stability improvement of wave energy conversion system connected to power grid using anti-windup-coot optimization strategy, *Energy*, vol. 245, pp. 123321–123336, 2022.

[566] A. Mahdy, H.M. Hasanien, R.A. Turky and S.H. Abdel Aleem, Modeling and optimal operation of hybrid wave energy and PV system feeding supercharging stations based on golden jackal optimal control strategy, *Energy*, vol. 263, pp. 125932–125950, 2023.

[567] K. Maissu, M. Fiaschini and A. Alamir, Robust optimal control-based design of a combined chemo and immunotherapy delivery profiles, In: *IEEE FOSBE 2019, 8th IFAC Conference on Foundations of Systems Biology in Engineering*, Valencia, Spain, Oct. 2019.

[568] N. Malekzadeh, A. Khosrani, H. Rasouli and A.R. Noei, Wing-rock suppression via backstepping controller, In: *IEEE KBEI, 2015 IEEE 2nd Intl. Conf. on Knowledge-Based Engineering and Innovation*, Tehran, Iran, Nov. 2015.

[569] M. Malekzadeh, J. Sadati and N. Alizadeh, Adaptive PID controller design for wing rock suppression using self-recurrent wavelet neural network identifier, *Nonlinear Dynamics*, vol. 7, pp. 267–275, 2016.

[570] R. Mahadeva, M. Kumar, S.P. Putole and G. Manik, PID control design using AGPSO technique and its application in TITO reverse osmosis desalination plant, *IEEE Access*, vol. 10, pp. 125881–125890, 2022.

[571] T.M. Mansour, T.M. Ismail, K. Ramzy and M. Abd El-Salam, Energy recovery system in small reverse osmosis desalination plant: experimental and theoretical investigations, *Alexandria Journal of Engineering*, vol. 59, pp. 3741–3753, 2020.

[572] Q. Mao, L. Dou, Q. Zong and Z. Ding, Attitude controller design for reusable launch vehicles during reentry phase via compound adaptive fuzzy H-infinity control, *Aerospace Science and Technology*, vol. 72, pp. 36–48, 2018.

[573] Ph. Martin and P. Rouchon, Systèmes plats: planification et suivi des trajectoires, Journées X-UPS, École des Mines de Paris, Centre Automatique et Systèmes, Mai, 1999.

[574] M. Marufkhani and M.A. Khosravi, Robust optimal constrained control of fully-constrained cable-driven parallel robots based on GSDRE, *Nonlinear Dynamics*, pp. 1–16, 2023.

[575] E. Massad, M. Nasciemento-Burratini and F.A. Bezerra-Coutinho, An optimization model for antibiotic use, *Applied Mathematics and Computation*, vol. 201, pp. 161–167, 2008.

[576] J.V. Matos Farias, L.A. Gregoire, A.F. Cupertino, M.A. Pereira, S.I. Seleme and M. Fadil, A sliding-mode observer for MMC-HVDC systems fault tolerant scheme with reduced number of sensors, *IEEE Transactions on Power Delivery*, vol. 38, no. 2, pp. 867–876, 2023.

[577] M. Matsuda, M. Izursu and K. Furuta, Simultaneous swinging-up and stabilization of double Furuta pendulum, In: *IEEE SICE 2007*, Kagawa University, Japan, Sep. 2007.

[578] P.R. McAree and K.J. Austin, Real-time friction estimation using the Kalman inverse filter, In: *12th IFToMM 2007, 12th Intl. Congress on Mechanism and Machine Sciences*, Besancon, France, June 2007.

[579] T.G. McGee and M.W. Pang, Trajectory planning and control of a novel walking biped, In: *IEEE CCA 2001, IEEE 2001 Intl. Conference on Control Applications*, Mexico City, Mexico, Sep. 2001.

[580] G.J. Maeda and D.C. Rye, Learning disturbances in automatic excavators, In: *IEEE 2012 Intl. Conference on Intelligent Robots and Systems*, Algarve, Portugal, Oct. 2012.

[581] G.J. Maeda, I.R. Manchester and D.C. Rye, Combined ILC and disturbance observer for the rejection of near-repetitive disturbances with application to excavation, *IEEE Transactions on Control Systems Technology*, vol. 29, no. 5, pp. 1754–1769, 2015.

[582] G. Maeda, D. Rye and S. Singh, Iterative autonomous excavation, In: *International Conference on Field and Service Robots, Springer Tracts in Advanced Robotics*, vol. 32, pp. 369–382, 2012.

[583] H. Makino, Development of the SCARA, *Journal of Robotics and Mechatronics*, vol. 26, no. 1, pp. 1–14, 2014.

[584] I. El Makrini, C. Rodriguez-Guerrero, D. Lefeber and B. Vanderborght, The variable boundary layer sliding-mode control: a safe and performant control for compliant joint manipulators, *IEEE Robotics and Automation Letters*, vol. 2, no. 1, pp. 187–192, 2017.

[585] A. van der Maars, Y.F. Steinbuch, A. Boverhof and W.P Heemel, Switched control of a SCARA robot with shared actuation resources, In: *IFAC WC 2017, IFAC 2017 World Congress*, Toulouse, France, July 2017.

[586] I.M. Mehedi, U.M. Al-Saggaf, R. Mansouri and M. Bettayeb, Stabilization of a double inverted rotary pendulum through fractional-order integral control scheme, *International Journal of Advanced Robotic Systems*, pp. 1–9, July–Aug. 2019.

[587] I. Mehedi, U, Ansari, A. Bajodah, U.M. AL-Saggaf, B. Kada and M.J. Rawa, Underactuated rotary inverted pendulum control using robust generalized dynamic inversion, *Journal of Vibration and Control*, vol. 26 (23–24) 2210–2220, 2020.

[588] R. Mehra, S. Satpute, F. Kazi and N.M. Singh, Passivity-based control of stochastic mechanical systems, In: *IEEE ICARCV 2012, IEEE 2012 12th Intl Conf. on Control, Automation, Robotics and Vision*, Guangzhou, China, Dec. 2012.

[589] R. Mehra, S.G. Satputo, F. Kazi and N.M. Singh, Control of a class of underactuated mechanical systems obviating matching conditions, *Automatica*, vol. 86, pp. 98–103, 2017.

[590] R.L. Meira, M.A.F. Martins, R.A. Kalid and G.M.N. Costa, Implementable MPC-based surge avoidance nonlinear control strategies for non-ideally modelled natural gas compression systems, *Journal of Natural Gas Science and Engineering*, vol. 102, pp. 104573–104586, 2022.

[591] M. Mejari, A. Gupta, V. Ketkar, M. Datar, N. Singh and F. Kazi, Stabilization of 2D SpiderCrane with non-convex state constraints using MPC, In: *IFAC Intl. Conf. on Advances in Control and Optimization of Dynamical Systems*, Kanpur, India, March 2014.

[592] H. Mena, L.M. Pfurtscheller and J.R. Romero-Leiton, Random perturbations in a mathematical model of bacterial resistance: analysis and optimal control, *Mathematical Biosciences and Engineering*, vol. 17, no. 5, pp. 4477–4499, 2020.

[593] Q. Meng, T. Zhang, X. Gao and J.Y. Song, Adaptive sliding-mode fault-tolerant control of the uncertain Stewart platform based on offline multibody dynamics, *IEEE/ASME Transactions on Mechatronics*, vol. 19, no. 3, pp. 882–894, 2014.

[594] L. Menhour, B. d'Andre'a-Novel, M. Fliess and H. Mounier, Coupled nonlinear vehicle control: flatness-based setting with algebraic estimation techniques, *Control Engineering Practice*, vol. 22, pp. 135–146, 2014.

[595] L. Merazka, F. Zouari and A. Boulkroune, Fuzzy state-feedback control of uncertain nonlinear MIMO systems, In: *IEEE ICSC 2017, IEEE 2017 6th International Conference on Systems and Control*, Batna, Algeria, May 2017.

[596] L. Merazka, F. Zouari; A. Boulkroune, High-gain observer-based adaptive fuzzy control for a class of multivariable nonlinear systems, In: *IEEE ICSC 2017, IEEE 2017 6th International Conference on Systems and Control*, Batna, Algeria, May 2017.

[597] J.P. Merlet, Direct kinematics of CDPR with extra cable orientation sensors: the 2 and 3 cables case with perfect measurement and sagging cables, In: *IEEE IROS 2017, IEEE 2017 Intl Conference on Intelligent Robots and Systems*, Vancouver, Canada, Sep. 2017.

[598] J.P. Merlet, On the redundancy of cable-driven parallel robots, *Mechanism and Machine Science*, vol. 26, pp. 31-39–2019.

[599] P.E. Meunier, Contrôle collaboratif d' une ferme de génératrices houlomotrices, Thèse de Doctorat, Ecole Centrale de Nantes, 2018.

[600] P. Milosavljevic, A. Cortinovis, A.G. Marchetti, T. Faulwasser, M. Mercangoz and D. Bonvin, Optimal load sharing of parallel compressor via modifier adaptation, In: *IEEE CCA 2016, IEEE 2016 Conference on Control Applications*, Buenos Aires, Argentina, Sep. 2016.

[601] P. Milosavljevic, R. Schneider, A. Cortinovis, T. Faulwasser and D. Bonvin, A distributed feasible-side convergent modifier adaptation scheme for interconnected systems with application to gas compressor station, *Computers and Chemical Engineering*, vol. 115, pp. 474–486, 2018.

[602] P. Miloslavljevic, A. Cortinovis, R. Schneider, T. Faulwasser, M. Mercangoz and D. Bonvin, Optimal load sharing for serial compressors with modifier adaptation, In: *IEEE ECC 2018, European Control Conference 2018*, Limassol, Cyprus, June 2018.

[603] P. Milosavljevic, R. Schneider, A. Cortinovis, T. Faulwasser and D. Bonvins, A distributed feasible-side convergent modifier-adaptation scheme for interconnected systems, with application to gas-compressor stations, *Computers and Chemical Engineering*, vol. 115, pp. 474–486, 2018.

[604] P. Milosavljevic, A.G. Marchetti, A. Cortinovis, T. Faulwasser, M. Mercangoz and D. Bonvin, Real-time optimization for load sharing for gas compressors in the presence of uncertainty, *Applied Energy*, vol. 272, pp. 114853, 2020.

[605] P.X. Miranda La Hera, A.S. Shiriaev, B. Freidovich, U. Mettin and S.V. Gusev, Stable walking gaits for a three-link planar biped robot with one actuator, *IEEE Transactions on Robotics*, vol. 29, no. 3, pp. 589–601, 2013.

[606] S.M. Mirimani, A. Vahedi, P. Marignetti and R. Di Stefano, An online method for static eccentricity fault detection in axial flux machines, *IEEE Transactions on Industrial Electronics*, vol. 62, no. 1, pp. 1931–1962, 2015.

[607] B.J.E. Misgeld, L. Hewing, L. Liu and S. Leanhardt, Closed-loop positive real optimal control of variable stiffness actuators, *Control Engineering Practice*, vol. 82, pp. 142–150, 2019.

[608] A.K. Misra, M.S. Nixon and V.J. Modi, Nonlinear dynamics of two-body tethered satellite systems: constant length case, *The Journal of Astronautical Sciences*, vol. 49, no. 2, pp. 219–238, 2001.

[609] M.T. Mita, X. Ma, H. Albuflasa and P.A. Davies, Reverse osmosis (RO) membrane desalination driven by wind and solar photovoltaic (PV) energy: state of the art and challenges for large scale implementation, *Renewable and Sustainable Energy Review*, vol. 112, pp. 6669–685, 2019.

[610] M.T. Mita, X. Ma, H. Albuflasa and P.A. Davies, Variable operation of a renewable energy-driven reverse osmosis system using model predictive control and variable recovery: towards large-scale implementation, *Journal of Desalination*, vol. 532, pp. 115715–115734, 2022.

[611] M.T. Mito, X. Ma, H. Albuflassa and P.A. Davies, Modular operation of renewable energy-driven reverse-osmosis using neural networks for wind-speed prediction and scheduling, *Desalination*, vol. 567, pp. 116950–116971, 2023.

[612] B. Mitra, B. Chowdhury and M. Manjekar, HVDC transmission for access to offshore renewable energy: a review of technology and fault detection techniques, *IET Renewable Power Generation*, vol. 12, no. 13, pp. 1563–1571, 2018.

[613] L. Mohammadi, D. Gucik-Derigny and D. Henry, Improvement of navigation unit in space applications using H_∞ control. In: *IEEE SMC 2022, IEEE Intl. Conference on Systems, Man and Cybernetics*, Prague, Czech Republic, Oct. 2022.

[614] A. Mohanty and B. Yao, Indirect adaptive robust control of hydraulic manipulators with accurate parameter estimates, *IEEE Transactions on Control Systems Technology*, vol. 19, no. 3, pp. 567–576, 2011.

[615] A. Mohanty and B. Yao, Integrated direct/indirect adaptive robust control of hydraulic manipulators with valve deadband, *IEEE Transactions on Mechatronics*, vol. 16, no. 6, pp. 707–716, 2011.

[616] A. Moharana and P.K. Dash, Input-output linearization and robust sliding-mode controller for the VSC-HVDC transmission link, *IEEE Transactions on Power Delivery*, vol. 25, no. 3, pp. 1952–1961, 2010.

[617] M.R. Mojallizadeh, B. Brogliato and C. Prieur, Modelling and control of overhead cranes: a tutorial overview and perspectives, *Annual Reviews in Control*, pp. 100877–100910, 2023.

[618] S. Mondal, A new supply chain model and its synchronization behavior, *Chaos, Solitons and Fractals*, vol. 123, pp. 140–148, 2019.

[619] V. Morio, P. Vemis and F. Cazaurang, Hypersonic reentry and flatness theory application to medium L/D entry vehicle, In: *Proc. of the 1st Intl. ARA days Conference*, Arachon, France, July 2006.

[620] V. Morio, Contribution au développement d'une loi de guidage autonome par platitude. Application à une mission de rentrée atmosphérique. Automatique/Robotique. Université Sciences et Technologies – Bordeaux I, 2009.

[621] V. Morio, F. Cazaurang and P. Vernis, Flatness-based hypersonic reentry guidance of a lifting body vehicle, *Control Engineering Practice*, vol. 17, pp. 588–596, 2009.

[622] A. Morsi, M.C. Abbasi, S.M. Agmed and A.M. Mohamed, Model Predictive Control based on linear parameter varying models of active magnetic bearing systems, *IEEE Access*, vol. 9, pp. 23633–23642, 2022.

[623] H. Mounier and J. Rudolph, Trajectory tracking for π-flat nonlinear delay systems with a motor example, In: A. Isidori, F. Lamnabhi-Lagarrigue and W. Respondek (Eds.), *Nonlinear control in the year 2000, vol. 1, Lecture Notes in Control and Inform. Sci.*, Springer, vol. 258, pp. 339–352, 2001.

[624] H. Mounier, Document de synthèse pour l' habilitation à diriger des recherches, Université de Paris XI, Juillet 2005.

[625] W.L. Mu, Y. Or and A.D. Ames, Dynamic walking on slippery surfaces: demonstrating stable bipedal gaits with planned ground slippage, In: *IEEE ICRA 2019, IEEE 2019 Intl. Conference on Robotics and Automation*, Montreal, Canada, May 2019.

[626] R. Mu, J. Chen, K. Peng, X. Zhang, Y. Deng and N. Cui, Finite-time super-twisting controller based on SESO design for RLV reentry phase, *IEEE Access*, vol. 7, pp. 37371–37380, 2019.

[627] M. Munnapian, A comprehensive review of DC fault protection methods in HVDC transmission systems, *Protection and Control of Modern Power Systems*, vol. 6, no. 1, pp. 1–20, 2021.

[628] R. Murdani and M.V. Ehtheshami, Efficient model-based DC-fault detection and location scheme for multi-terminal HVDC systems with voltage source converter, *Electrical Engineering*, vol. 104, pp. 1553–1564, 2022.

[629] R. Murtazin, N. Sevastiyanov and N. Chudinov, Fast rendezvous profile evaluation: from ISS to lunar station, *Acta Astronautica*, vol. 173, pp. 139–144, 2020.

[630] E.S. Nadini, T.Bak and R. Izadi-Zamanabadi, Model predictive controller combined with LQG controller and velocity feedback to control the Stewart platform, In: IEEE AMC 2006, 9th IEEE Intl. Workshop on Advanced Motion Control, Istanbul, Turkey, May 2006.

[631] K. Nagachi, M. Izatsu, N. Kamamichi. T. Shiotsuki, J. Ishikawa and K Furuta, Swinging-up and balancing control of double Furuta pendulum by using State Dependant Riccati Equation, In: *IEEE/RSJ 2009 Intl. Conf. on Intelligent Robots and Systems*, St. Louis, USA, Oct. 2009.

[632] S. Nanda, H. Moore and S. Lenhart, Optimal control of treatment in a mathematical model of chronic myelogenous leukemia, *Mathematical Biosciences*, vol. 210, pp. 143–156, 2007.

[633] H. Nasri and A. Akbarzadeh Kalat, Adaptive fuzzy backstepping control of drug dosage regimen in cancer treatment, *Biomedical Signal Processing and Control*, vol. 42, pp. 267–276, 2018.

[634] T. Neckel, C. Talbot and N. Petit, Collocation and inversion for a reentry optimal control problem, In: *Proc. of the 5th International Conference on Launch Technology*, 2003.

[635] H.D. Nejad, M. Nazari, M. Nazari, *et al.*, Fuzzy state-dependent Riccati equation (FSDRE) control of the reverse osmosis desalination system with photovoltaic power supply, *IEEE Access*, vol. 10, pp. 95595–95607, 2022.

[636] J.K. Nguyen, N.H. Thai, H.T. Pham, *et al.*, Adaptive neural network-based backstepping sliding-mode control approach for dual-arm robots, *Journal of Control, Automation and Electrical Systems*, vol. 30, pp. 512–521, 2019.

[637] M.N. Nguyen, D.T. Tran and K.K. Ahn, Robust position and vibration control of an electrohydraulic series elastic manipulator against disturbance generated by a variable stiffness actuator, *Mechatronics*, vol. 52, pp. 22–35, 2018.

[638] N.P. Nguyen, H. Oh, Y. Kim, J. Moon, J. Yang and W.H. Chen, Fuzzy-based super-twisting sliding mode stabilization control for under-actuated rotary inverted pendulum systems, *IEEE Access*, vol. 8, pp. 185079–185093, 2020.

[639] F. Nicolau, W. Respondek and J.P. Barbot, How to minimally modify a dynamical system when constructing flat inputs, *International Journal of Robust and Nonlinear Control*, 2022.

[640] F. Nicolau, W. Respondek and J.P. Barbot, Construction of flat inputs for mechanical systems, In: 7th IFAC Workshop on Lagrangian and Hamiltonian methods for nonlinear control, Berlin, Germany, Oct. 2021.

[641] D. Nicolis, A.M. Zanchettin and P. Rocco, Constraint-based and sensorless force control with an application to a lightweight dual-arm robot, *IEEE Robotics and Automation Letters*, vol. 1, no. 1, pp. 340–348, 2016.

[642] D. Nicolis, F. Allewi and P. Rocco, Operational space model predictive sliding-mode control for redundant manipulators, *IEEE Transactions on Robotics*, pp. 1–10, 2020.

[643] C. Niu, H. Shen, J. Li and Y. Yi, Disturbance observer-based L_1 robust tracking control for single-link flexible-joint manipulator systems, *IET Journal of Engineering*, vol. 2019, no. 22, pp. 8390–8394, 2019.

[644] S.L. Noble, E. Sheren, R.E. Hannemann, D. Ramkrishna, T. Vik and A.E. Rundell, Using adaptive model predictive control to customize maintenance therapy chemotherapeutic dosing for childhood acute lymphoblastic leukemia, *Journal of Theoretical Biology*, vol. 364, pp. 990–1000, 2010.

[645] M. Noman, G Li, K. Wang and B. Hun, Electrical control strategy for an ocean energy conversion system, *Protection and Control of Modern Power Systems*, vol. 6, no. 12, pp. 1–18, 2021.

[646] H. Norouzi Nov, M.R. Jahed Motlagh and A. Makui, Robust controlling of chaotic behavior in supply chain networks, *Journal of the Operational Research Society*, vol. 68, no. 6, pp. 711–724, 2017.

[647] Y. Notohara, D. Li, Y. Iwagi, M. Tomura and K. Tsukei, Study on vibration suppression control for rotary compressors, *IEEJ Journal on Industry Applications*, vol. 10, no. 4, pp. 481–486, 2021.

[648] A. Novak, A nonlinear optimal control problem with an application to optimal dosing of cytotoxic drugs, *Chaos, Solitons and Fractals*, vol. 177, pp. 104301–104309, 2023.

[649] A. Nurkanovic, S. Albrecht, B. Brogliato and M. Diehl, The time-freezing reformulation for numerical optimal control of complementary Lagrangian systems with state jumps, *Automatica*, vol. 158, pp. 111295–111308, 2023.

[650] J.A. de Oliveira Neto, C.A. Franca Sartori and G. Manassero, Fault location in overhead transmission lines based on magnetic signatures and on the extended Kalman filter, *IEEE Access*. vol. 9, pp. 15259–15270, 2021.

[651] A. Oral, Patterning automation of square mosaics using computer assisted SCARA robot, *Robotica*, vol. 27, pp. 897–903, 2009.

[652] P. Ordaz and A. Poznyak, The Furuta's pendulum stabilization without the use of a mathematical model: attractive ellipsoid method with KL-adaptation, In: *IEEE CDC 2012, IEEE 2012 51st IEEE Conference on Decision and Control*, Maui, Hawaii, USA, Dec. 2012.

[653] P. Ordaz and A. Poznyak, Adaptive-robust stabilization of the Furuta's pendulum via attractive ellipsoid method, *Journal of Dynamic System, Measurement and Control*, vol. 138, no. 2, pp. 021005–021013, 2015.

[654] G. Oriolo, M. Cefalo and M. Venditelli, Repeatable motion planning for redundant robots over cyclic tasks, *IEEE Transactions on Robotics*, vol. 33, no. 5, pp. 1170–1187, 2017.

[655] T. Orlowska-Kowalska, M. Wolkrewicz, P. Petrzak, *et al.*, Fault diagnosis and fault tolerant control of PMSM drives: state of the art and future challenges, vol. 10, pp. 59979–60024, 2022.

[656] D. Ortiz Morales and P.X. Miranda La Hera, Design of energy efficient walking gaits for a three-link planar biped walker with two unactuated degrees of freedom, In: *IEEE 2012 Conf. on Robotics and Automation*, Saint-Paul, Minnesota, May 2012.

[657] I. Palumbo, D. Richiedei, A. Trevisani, E. Sanjurgo, A. Luaces and J. Cuadrado, Estimation of the digging and payload forces in excavators by means of state observers, *Mechanical Systems and Signal Processing*, vol. 134, pp. 106356–106371, 2019.

[658] Y. Pan, Y. Zhou, T. Sub and M.J. Er, Composite adaptive fuzzy H_∞ tracking control of uncertain nonlinear systems, *Neurocomputing*, vol. 99, pp. 15–24, 2014.

[659] Y. Pan, H. Wang, X. Li and H. Yu, Adaptive command-filtered backstepping control of robot arms with compliant actuators, *IEEE Transactions on Control Systems Technology*, vol. 26, no. 3, pp. 1149–1157, 2018.

[660] G.L. Park, A.Y. Schafer and B.S. Richard, Renewable energy-powered membrane technology: supercapacitors for buffering resource fluctuations in a wind-powered membrane system for brackish water desalination, *Renewable Energy*, vol. 50, pp. 126–135, 2013.

[661] C.S. Park, J.H. Kim, S.H. Park, Y.D. Yoon and M.S. Lim, Multi-physics characteristics of PMSM for compressor according to driving mode considering PWM frequency, *IEEE Access*, vol. 10, pp. 114490–1114500, 2022.

[662] T.R. Patel, K.D. Kumar and K. Behdinan, Variable structure control for satellite attitude stabilization on elliptic orbits suing solar radiation pressure, *Acta Astronautica*, vol. 64, pp. 359–373, 2009.

[663] I.K. Paterson, A. Hayle, G. Ochan, C. Baker-Austin and N.G. Taylor, Optimizing antibiotic usage to treat bacterial infections, *Journal of Scientific Reports, Nature*, vol. 6, pp. 37853–37862, 2016.

[664] M. Patil and S. Kurode, Stabilization of rotary double inverted pendulum using high-order sliding-modes, In: *IEEE ASCC 2017. IEEE 2017 11th Asian Control Conference*, Gold Coast Convention Center, Australia, Dec. 2017.

[665] H. Patil and S. Kurode, Stabilization of RDIP using sliding modes, In: *IEEE ICC 2018, IEEE 2018 Indian Control Conference*, Kanpur, India, Jan. 2018.

[666] A. Paul, I. Kucukdemiral and G. Bevon, Modelling and offset-free predictive control of the parallel-type double inverted pendulum, In: *IEEE ICSC 2022, Proc. of the 2022 10th International Conference on Systems and Control*, Marseille, France, Nov. 2022.

[667] R Padhi and M. Kothari, A neuro-adaptive augmented optimal dynamic inversion approach for effective and efficient treatment of chronic myelogenous leukemia, In: *IEEE ACC 2006, Proc. of the 2006 American Control Conference*, Minneapolis, Minnesota, USA, June 2006.

[668] B. Parkhideh, S. Bhattacharya, J. Mazumdar and W. Koellner, Modelling and control of large shovel converter systems integrated with supercapacitor, In: *IEEE 2008 Annual Meeting of the Industry Application Society*, Edmonton, Canada, Oct. 2008.

[669] N. Pedemonte, T. Rashid, D. Marquez-Gamez, *et al.*, Fastkit: a mobile cable-driven parallel robot for logistics, In: *Advances in Robotic Research: From Lab to Market, Springer Tracts in Advanced Robotics*, vol. 132, pp. 141–163, 2020.

[670] R. Pena-Miller, D. Lahnemann, H. Schulenburg, M. Ackermann and R. Beardmore, Selecting against antibiotic resistant pathogens: optimal treatments in the presence of commensal bacteria, *Bulletin of Mathematical Biology*, Springer, vol. 14, pp. 908–934, 2012.

[671] C.D. Petersen, F. Leve and J. Kolmanovsky, Model Predictive Control of an underactuated spacecraft with two reaction wheels, *Journal of Guidance, Control and Dynamics*, vol. 40, no. 2, pp. 320–333, 2017.

[672] P.B. Petrovic and N.A. Lukic, Configuration-based compliance control of kinematically redundant robot arm, In: *Part I: Theoretical Framework, Transactions of the Faculty of Mechanical Engineering*, Univ. of Belgrade, Feb. 2017.

[673] M. Peyrl and A. Cortinovis, Computationally efficient solution of a compressor load sharing problem using the alternating direction method of multiplies, In: *IEEE ECC 2015, European Control Conference 2015*, Linz, Austria, July 2015.

[674] E. Picard, Modelling and Robust Control of cable-driven parallel robots for industrial applications, Thése de Doctorat, Ecole Centrale de Nantes, 2019.

[675] E. Picard, E. Tahoumi, F. Plestan, S. Caro and F. Claveau, A new control scheme of cable-driven parallel robot balancing between sliding-mode and linear feedback, In: *IFAC 21st World Congress*, Berlin, Germany, July 2020.

[676] S. Pietrosanti, W. Holderbaum and V.M. Beccera, Modelling power-flow in a hoist motor of a rubber-tired gantry crane, *IEEE Transactions on Industry Applications*, vol. 52, no. 3, pp. 2028–2014, 2016.

[677] V.D. Pham, C.P. Vo, H.V. Dao and K.K. Ahn, Actuator fault-tolerant control for an electrohydraulic actuator using time-delay estimation and feedback linearization, *IEEE Access*, vol. 9, pp. 107111–107122, 2021.

[678] V.D. Pham, C.P. Vo, H.V. Dao and K.K. Ahn, Robust fault tolerant control of an electrohydraulic actuator with a novel nonlinear Unknown Input Observer, *IEEE Access*, vol. 9, pp. 30750–30760, 2021.

[679] A. Pfeffer, T. Glück, F. Schausberger and A. Kugi, Control and estimation strategies for pneumatic drives with partial position information, *Mechatronics*, vol. 50, pp. 259–270, 2018.

[680] A.Y. Pogvansky, V.N., Belykh and H. Nijmejer, Controlled synchronization of pendula, In: *Proc. of the 42nd IEEE Conference on Decision and Control*, Maui, Hawaii, USA, Dec. 2001.

[681] L. Popescu, Applications of optimal control to production planning, *Journal of Information Technology and Control*, vol. 49, no. 1, pp. 89–99, 2020.

[682] M. Pourmirasghariyan, S.F. Zarei, M. Hamizah and F. Blaabjerg, An interior-angle analysis for fault detection in VSC-based HVDC-grids, In: *International Journal of Electrical Power and Energy Systems*, vol. 159, article no. 110058, 2024.

[683] V. Potkonjak, M. Popovic, M. Lazarevic and J. Sinanovic, Redundancy problem in writing: from human to anthropomorphic robot arm, *IEEE Transactions on Systems, Man and Cybernetics - Part B: Cybernetics*, vol. 28, no. 6, pp. 790–805, 1998.

[684] A. Pragati, D.A. Gadanayak and M. Mishra, Fault detection and fault place identification on a VSC-MT-HVDC link using Stockwell transform Teager energy operator, *Electric Power Systems Research*, vol. 236, article no. 110937, 2024.

[685] D.R. Prathapaneni and K. Detroja, Optimal design of energy sources and reverse osmosis desalination plant with demand side management for cost-effective freshwater production, *Desalination*, vol. 496, pp. 114741–114755, 2020.

[686] V. Purba, X. Wu, M. Sinha, S.V. Dhople and M.R. Jovanovic, Design of optimal coupling gains for synchronization of nonlinear oscillators, In: *IEEE CDC 2015, IEEE 2015 Annual Conference on Decision and Control*, Osaka, Japan, Dec. 2015.

[687] Y. Qian and Y. Fang, Switching logic-based nonlinear feedback control of offshore ship-mounted tower cranes: a disturbance observer-based approach, *IEEE Transactions on Automation Science and Engineering*, vol. 16, no. 3, pp. 1125–1136, 2019.

[688] T. Qin, Y. Li, L. Quan and L. Yang, An adaptive robust impedance control considering energy saving of hydraulic excavator boom and stick systems, *IEEE Transactions on Mechatronics*, vol. 27, no. 4, pp. 1928–1936, 2022.

[689] I.R. Radulescu, D. Candea and A. Halanay, Optimal control analysis of a leukemia model under imanitib treatment, *Mathematics and Computers in Simulation*, vol. 121, pp. 1–11, 2016.

[690] I.R. Radulescu. D. Candea and A. Halanay, A complex mathematical model with competition in leukemia with immune response: an optimal control approach, In: *IFIP Advances on Information and Communication Technology, IFIPICT*, vol. 494, pp. 430–441, 2017.

[691] M. Ramezani and S. Tafazoli, Using artificial intelligence in mining excavators, *IEEE Industrial Electronics Magazine*, vol. 15, no. 1, pp. 6–11, 2021.

[692] J.P. Ramirez, R.C. Garcia and J. Alvarez, Synchronization of two inverted pendula via dynamic coupling, *IEEE 2016 Australian Control Conference*, Newcastle, Australia, Nov. 2016.

[693] J.L. Ramon, J. Pomares and L. Felicetti, Direct visual servoing and interaction control for a two-arms on-orbit servicing spacecraft, *Acta Astronautica*, vol. 192, pp. 368–378, 2022.

[694] M. Ran, J. Li and L. Xie, A new extended state observer for uncertain nonlinear systems, *Automatica*, vol. 131, article no. 109772, pp. 1–7, 2021.

[695] P. Ranjan, G. Wrat, M. Bhala, S.K. Mishara and J. Das, A novel approach for the energy recovery and position control of a hybrid hydraulic excavator, *ISA Transactions*, 2020.

[696] M.A. Rashidi, I. Macariq, N. Murshid, S. Wangwises, O. Mohian and M.A. Nazari, Applying with energy as a clean source for reverse osmosis desalination: a comprehensive review, *Alexandria Journal of Engineering*, vol. 61, pp. 12977–12989, 2022.

[697] S. Rasool, F. Islam, K.M. Muttaqi and D. Suttanto, Coupled modelling and advanced control for smooth operation of a grid-connected linear electric generator based wave-to-wire system, *IEEE Transactions on Industry Applications*, vol. 56, no.3, pp. 5575–5584, 2020.

[698] S. Rasool, K.M. Muttaqi and D. Suttanto, Modelling of a wave-to-wire system for a wave farm and its response analysis against power quality and grid codes, *Renewable Energy*, vol. 162, pp. 2021–2055, 2020.

[699] S. Rasool, K.M. Muttaqi and D. Suttanto, A multi-filter based dynamic power sharing control for a hybrid energy storage system integrated to a wave energy converter for output power smoothing, *IEEE Transactions on Sustainable Energy*, vol. 15, no. 9, pp. 1693–1706, 2022.

[700] S. Rasool, K.M. Muttaqi and D. Sutanto, An investigation on the integration of a hybrid off-shore wind-wave energy conversion system with the distribution network, *IEEE Transactions on Industry Applications*, vol. 59, no. 4, pp. 4562–4571, 2023.

[701] A. Rasuli, S. Tafazoli and W.G. Dunford, Dynamic modeling, parameter identification and payload estimation of mining cable shovels, In: *IEEE 2014 Industry Application Society Annual Meeting*, Vancouver, Canada, Oct. 2014.

[702] F. Rauscher and O. Sawodny, Modeling and control of tower cranes with elastic structure, *IEEE Transactions on Control Systems Technology*, vol. 25, no. 1, pp. 64–79, 2021.

[703] H. Rezk, M.A. Dhoifallah, Y.B. Hassan and H.A. Ziedan, Optimization and energy management of hybrid photovoltaic-diesel-battery system to pump and desalinate water at isolated regions, *IEEE Access*, vol. 8, pp. 102512–102529, 2020.

[704] S. Riachy, M. Fliess, C. Join and J.P. Barbot. Vers une simplification de la commande non linéaire : l'exemple d'un avion à décollage vertical, *Sixième Conférence Internationale Francophone d'Automatique, CIFA 2010*, Nancy, France, Jun 2010.

[705] G.G. Rigatos and S.G. Tzafestas, Extended Kalman filtering for fuzzy modelling and multi-sensor fusion, *Mathematical and Computer Modelling of Dynamical Systems*, vol. 13, pp. 251–266, 2007.

[706] G. Rigatos and Q. Zhang, Fuzzy model validation using the local statistical approach, *Fuzzy Sets and Systems*, vol. 60, no. 7, pp. 882–904, 2009.

[707] G.G. Rigatos, Extended Kalman and particle filtering for sensor fusion in motion control of mobile robots, *Mathematics and Computers in Simulation*, vol. 81, no. 3, pp. 590–607, 2010.

[708] G. Rigatos, *Modelling and Control for Intelligent Industrial Systems: Adaptive Algorithms in Robotics and Industrial Engineering*, Springer, 2011.

[709] G. Rigatos, *Advanced Models of Neural Networks: Nonlinear Dynamics and Stochasticity in Biological Neurons*, Springer, 2013.

[710] G. Rigatos, A differential flatness theory approach to observer-based adaptive fuzzy control of MIMO nonlinear dynamical systems, *Nonlinear Dynamics*, vol. 76, no. 2, pp. 1335–1354, 2014.

[711] G. Rigatos, *Nonlinear Control and Filtering using Differential Flatness Theory Approaches: Applications to Electromechanical systems*, Springer, 2015.

[712] G. Rigatos, P. Siano and N. Zervos, A new concept of flatness-based control of nonlinear dynamical systems, *IEEE INDIN 2015, 13th IEEE Intl. Conf. on Industrial Informatics*, Cambridge, UK, July 2015.

[713] G. Rigatos, P. Siano, P. Wira and F. Profumo, Nonlinear H-infinity feedback control for asynchronous motors of electric trains, *Journal of Intelligent Industrial Systems*, 2015.

[714] G. Rigatos, Flatness-based embedded control in successive loops for spark ignited engines, *Journal of Physics Conference Series*, vol. 659, 012019, *IFAC ACD 2015, Proc. of the 12th European Workshop on Advanced Control and Diagnosis*, 2015.

[715] G. Rigatos, *Intelligent Renewable Energy Systems: Modelling and Control*, Springer, 2016.

[716] G. Rigatos and A. Melkikh, Flatness-based control in successive loops for stabilization of heart's electrical activity, In: *Proc. ICNAAM 2016, AIP Conf. Proceedings No 1790 060004, Intl. Conf. on Numerical Analysis and Applied Mathematics*, 2016.

[717] G. Rigatos, *State-Space Approaches for Modelling and Control in Financial Engineering*, Springer, 2017.

[718] G. Rigatos and P. Siano, Flatness-based control in successive loops for business cycles of finance agents, *Journal of Intelligent Industrial Systems*, vol. 3, pp. 77–89, 2017.

[719] G. Rigatos, P. Siano and M. Abbaszadeh, Nonlinear H-infinity control for 4-DOF underactuated overhead cranes, *Transactions of the Institute of Measurement and Control*, vol. 40, no. 7, pp. 2364–2377, 2018.

[720] G. Rigatos, P. Siano, S. Ademi and P. Wira, Flatness-based control of DC-DC converters implemented in successive loops, *Electric Power Components and Systems*, vol. 46, no. 6, pp. 673–687, 2018.

[721] G. Rigatos and K. Busawon, *Robotic Manipulators and Vehicles: Control, Estimation and Filtering*, Springer, 2018.

[722] G. Rigatos, P. Wira, M. Abbaszadeh, K. Busawon and L. Dala, Nonlinear optimal control autonomous hypersonic vehicles, *Aerospace Systems*, vol. 2, pp. 197–213, 2019.

[723] G. Rigatos, M. Abbaszadeh, J. Pomares, and P. Wira, A nonlinear optimal control approach for a lower-limb robotic exoskeleton, *World Scientific*, vol. 17, no. 5, pp. 2050018, 2020.

[724] G. Rigatos, N. Zervos, M. Abbaszadeh, J. Pomares and P. Wira, Nonlinear optimal control for multi-DOF electro-hydraulic robotic manipulators, *IET Cyber-Systems and Robotics*, 2020.

[725] G. Rigatos and E. Karapanou, *Advances in Applied Nonlinear Optimal Control*, Cambridge Scholars Publishing, 2020.

[726] G. Rigatos, K. Busawon, and M. Abbaszadeh, Nonlinear optimal control of the acute inflammatory response, *Biomedical Signal Processing and Control*, vol. 55, 2020.

[727] G. Rigatos, M. Abbaszadeh, K. Busawon, Z. Gao, and J. Pomares, A nonlinear optimal control approach for multi-DOF brachiation robots, *International Journal of Humanoid Robotics*, vol. 18, no. 05, 2150015, 2021.

[728] G. Rigatos, M. Abbazadeh and G. Cuccurullo, A nonlinear optimal control method against the spreading of epidemics, *International Journal of Biomathematics*, vol. 15, no. 05, 2250026, 2022.

[729] G. Rigatos, P. Wira, M. Abbaszadeh and J. Pomares, Flatness-based control in successive loops for industrial and mobile robots, In: *IEEE IECON 2022, IEEE 48th Annual Conference of the Industrial Electronics Society*, Brussels, Belgium, Oct. 2022.

[730] G. Rigatos, M. Abbaszadeh, K. Busawon, and L. Dala, A nonlinear optimal control method for attitude stabilization of micro-satellites, *Journal of Guidance, Navigation and Control*, vol. 2, no. 3, 2250016, 2022.

[731] G. Rigatos, P. Wira, M. Abbaszadeh and P. Siano, A nonlinear optimal control approach for the Lotka-Volterra dynamical system, In: *IEEE IECON 2022, IEEE 48th Annual Conference of the Industrial Electronics Society*, Brussels, Belgium, Oct. 2022.

[732] G. Rigatos, M. Abbaszadeh, P. Siano, *Control and Estimation of Dynamical Nonlinear and Partial Differential Equation Systems: Theory and Applications*, IET Publications, 2022.

[733] G. Rigatos, M. Abbaszadeh, B. Sari, P. Siano, G. Cuccurullo and F. Zouari, Nonlinear optimal control for a gas compressor driven by an induction motor, *Results in Control and Optimization*, vol. 11, pp. 1–24, 2023.

[734] G. Rigatos, M. Abbaszadeh and J. Pomares, Flatness-based control in successive loops for electropneumatic actuators and robots, *IFAC Journal of Systems and Control*, vol. 25, p. 10222, 2023.

[735] G. Rigatos, Nonlinear optimal control for the multi-variable tumor-growth dynamics, *Journal of Computer Methods in Biomechanics and Biomedical Engineering*, 2023.

[736] G. Rigatos, M. Abbaszadeh, K. Busawon, L. Dala and J. Pomares, Flatness-based control in successive loops for autonomous quadrotors, *ASME Journal of Dynamic Systems, Measurement and Control*, pp. 1–19, 2023.

[737] G. Rigatos, M. Abbaszadeh, J. Pomares, P. Wira and G. Cuccurullo, Flatness-based control in successive loops for robotic manipulators and autonomous vehicles, *International Journal of Systems Science*, vol. 55, no. 5, pp. 954–979, 2024.

[738] G. Rigatos, M. Abbaszadeh, K. Busawon, L. Dala, J. Pomares and F Zouari, Flatness-based control in successive loops of autonomous quadrotors, *ASME Journal of Dynamic Systems, Measurement and Control*, vol. 146, pp. 024501(1–10), 2024

[739] G. Rigatos, M. Abbaszadeh, K. Busawon and L. Dala, Flatness-based control in successive loops for unmanned aerial vehicles and micro-satellites, *Journal of Guidance, Navigation and Control*, vol. 3, no. 4, pp. 2350026, 2024.

[740] G. Rigatos, P. Siano, M. Al-Numay, M. Abbaszadeh, J. Pomares and G. Cuccurullo, Flatness-based control in successive loops of an H-type gantry crane with dual PMLSM, *Journal of Vibration Engineering & Technologies*, pp. 1–24, 2024.

[741] G. Rigatos, N. Abbaszadeh, M. Hamida and P. Siano, *Intelligent Control for Electric Power Systems and Electric Vehicles*, CRC Publications, 2024.

[742] G. Rigatos, M. Abbaszadeh, M. Hamida and P, Siano, *Fault Diagnosis for Electric Power Systems and Electric Vehicles*, CRC Publications, 2024.

[743] G. Rigatos, M. Abbaszadeh, P. Siano and P. Wira, *Autonomous Electric Vehicles: Nonlinear Control, Traction and Propulsion*, Elsevier, 2025.

[744] R. Rodrigues, A. Murillo, R. Vilela-Lopes and L.C. Cadelh de Souza, Hardware in the loop simulation for model predictive control applied to satellite attitude control, *IEEE Access*, vol. 7, pp. 157401–157416, 2019.

[745] D.S. Rodriguez, P.F.A. Mancera, T. Carvalho and L.F. Goncalves, A mathematical model of chemoimmunotherapy of chronic lymphocytic leukemia, *Applied Mathematics and Computation*, vol. 349, pp. 118–133, 2019.

[746] R. Rohani and A. Koochaki, A hybrid method based on optimized neuro-fuzzy system and effective features for fault location in VSC-HVDC systems, *IEEE Access*, vol. 8, pp. 70861–70869, 2020.

[747] R. Roi, S. Shakla and B. Singh, Sensorless field-oriented SMCC based integral sliding-mode for solar PV based induction motor drive for water pumping, *IEEE Transactions on Industry Applications*, vol. 56, no. 5, pp. 3036–3064, 2022.

[748] J.G. Romero and R. Ortega, Two globally convergent adaptive speed observers for mechanical systems, *Automatica*, vol. 60, pp, 7–11, 2015.

[749] J.C. Romero, A. Donaire, R. Ortega and P. Borja, Global stabilization of underactuated mechanical systems via PID passivity-based control, *Automatica*, vol. 98, pp. 178–185, 2018.

[750] J.D Romero, I. Gandarilla, V. Sanibanez and B. Yi, A constructive procedure for orbital stabilization of a class of underactuated mechanical systems, *IEEE Transactions on Control Systems Technology*, vol. 30, no. 6, pp. 2698–2706, 2022.

[751] M.J. Rong, S. Han, J.M. Bai and Y.Q. Liang, Improved adaptive control for wing-rock via fuzzy neural network with randomly assigned fuzzy membership function parameters, *Aerospace Science and Technology*, vol. 39, pp. 614–627, 2014.

[752] R. Rosales-Asensio, F. Jose Garcia-Maya, D. Borge-Diaz and A. Colmenar-Santos, *Sea-water desalination in microgrids*, Springer, 2022.

[753] K. Rsetam, Z. Cao and Z. Man, Cascaded Extended State Observer-based sliding-mode control for underactuated flexible-joint robot, *IEEE Transactions on Industrial Electronics*, 2020.

[754] M. Ruderman, Compensation of nonlinear torsion in flexible-joint robots: comparison of two approaches, *IEEE Transactions on Industrial Electronics*, vol. 83, no. 9, pp. 5744–5751, 2016.

[755] J. Rudolph, *Flatness Based Control of Distributed Parameter Systems, Examples and Computer Exercises from Various Technological Domains*, Shaker Verlag, Aachen, 2003.

[756] T. Rybus, K. Seweryn and J.Z. Sasiadek, Control system for free-floating space robotic manipulator based on nonlinear model predictive control (NMPC), *Journal of Intelligent and Robotic Systems*, vol. 85, pp. 491–509, 2017.

[757] T. Rybus, K. Seweryn and J.Z. Sasiadek, Nonlinear model predictive control (NMPC) for free-floating space manipulator, *Aerospace Robotics III: GeoPlanet - Earth and Planetary Sciences*, pp. 17–29, 2019.

[758] A. Saenz-Aguirre, A. Ulazin, G. Ibarra-Beeastegui and J. Saenz, Extension and improvement of synchronous linear generator-based point absorber operation of high-wave excitation scenarios, *Ocean Engineering*, vol. 233, pp. 109844–109856, 2021.

[759] C.F. Saetre, A. Shiriaev and T. Astensrad, Trajectory optimization and orbital stabilization of underactuated Euler-Lagrange systems with impacts, In: *IEEE 2019 European Control Conference*, Naples, Italy, June 2019.

[760] H.A. Said, D. Garcia-Violini and J.V. Ringwood, Wave to grid (W2G) control of a wave energy converter, *Energy Conversion and Management: X*, vol. 14, pp. 100190–100209, 2022.

[761] E.N. Sanchez and F. Ornelas-Tellez, *Discrete-Time Inverse Optimal Control for Nonlinear Systems*, CRC Press, 2017.

[762] B. Sanchez, C. Cuvas, P. Ordaz, O. Santos-Sanchez and A. Poznyak, Observer for a class of nonlinear systems with unmatched uncertainties: joint attractive ellipsoid and sliding mode concepts, *IEEE Transactions on Industrial Electronics*, vol. 67, no. 7, pp. 5677–5686, 2020.

[763] J. Sandeval, P. Vieyes and G. Poisson. Generalized framework for control of redundant manipulators in robot-assisted minimal invasive surgery, *IRBM Journal (Innovation and Research in Biomedical Engineering)*, vol. 39, pp. 160–166, 2018.

[764] R.J. Sandzimer and H.H. Asada, A data-driven approach to prediction and optimal bucket-filling control for autonomous excavators, *IEEE Robotics and Automation Letters*, vol. 5, no. 2, pp. 2662–2609, 2020.

[765] P. dos Santos and P. Pliveira, Integrated architecture for navigation and attitude control of low-cost sub-orbital launch vehicle, *Acta Astronautica*, vol. 222, pp. 52–68, 2024.

[766] T. Santra, D. Roy, A.B. Choudhury and S. Yamada, Vibration control of a hybrid magnetic bearing using an adaptive sliding-mode technique, *Journal of Vibration and Control*, vol. 24, no. 10, pp. 1948–1960, 2019.

[767] J. Sapi, D.A. Drexler, J. Harmati, Z. Sapi and L. Kovacs, Qualitative analysis of tumor-growth model under anti-angiogenic therapy choosing the effective operating point and design parameters for controller design, *Optimal Control Applications and Methods*, vol. 37, pp. 848–866, 2016.

[768] R. Sarkar, T. Mukherjee, D. Patel and I.N. Kar, Reentry trajectory tracking of reusable launch vehicle using artificial delay based robust guidance law, *Advances in Space Research*, vol. 67, 557–579, 2021.

[769] T. Sasaki and K. Kawashima, Remote control of backhoe at construction site with a pneumatic robot system, *Automation in Construction*, vol. 17, pp. 907–914, 2008.

[770] L. Scalera, A. Gasparetto and D. Zanotto, Design and experimental validation of a 3-DOF underactuated pendulum-like robot, *IEEE/ASME Transactions on Mechatronics*, vol. 28, no. 1, pp. 217–228, 2021.

[771] P. Schönsleben, *Handbook Integral Logistics Management: Operations and Supply Chains Management Within and Across Companies*, Springer, 2016.

[772] M.L. Semenov, A.M. Solovyov and P.A. Meleshenko, Stabilization of coupled inverted pendula: from discrete to continuous case, *Journal of Vibration and Control*, vol. 27, no. 1–2, pp. 43–58, 2021.

[773] A. Sengupta and D.K. Das, Adaptive boundary layer double integral sliding-mode controller design for leukemia therapy, *International Journal of Adaptive Control and Signal Processing*, vol. 77, pp. 2796–2811, 2023.

[774] S.P. Sethi, *Optimal Control Theory: Applications to Management Systems and Economics*, Springer, 2021.

[775] S.M. Shalaby, F.A. Hammad and M.E. Zayed, Current progress in integrated solar desalination systems: prospects from coupling configuration to energy conversion and desalination processes, *Process Safety and Environmental Protection*, vol. 178, pp. 494–510, 2023.

[776] X. Shan, Y. Li, H. Liu and T. Huang, Residual vibration reduction of high-speed pick-and-place parallel robot using input shaping, *Chinese Journal of Mechanical Engineering*, vol. 35, no. 16, pp. 1–8, 2022.

[777] W. Shang, B. Zhang, B. Zhang, F. Zhang, and S. Cong, Synchronization control in the cable space for cable-driven parallel robots, *IEEE Transactions on Industrial Electronics*, vol. 66, no. 6, pp. 4544–4554, 2019.

[778] W. Shang, F. Xie, B. Zhang, S. Cong and Z. Li, Adaptive cross-coupled control of cable-driven parallel robots with model uncertainties, *IEEE Robotics and Automation Letters*, vol. 5, no. 3, pp. 4110–4117, 2020.

[779] X. Shao, G. Sun, G. Xue and X. Li, Nonsingular terminal sliding-mode control for free-floating space manipulator with disturbance, *Acta Astronautica*, vol. 181, pp. 396–404, 2021.

[780] Z. Shao, G. Xie, Z. Zhang and L. Wong, Design and analysis of the cable-driven parallel robot for cleaning exterior wall building, *International Journal of Advanced Robotic Systems*, pp. 1–11, Jan.–Feb. 2021.

[781] J.A, Sharo, A.P. Browing, Mapder, K. Burrage and M.J. Simpson, Optimal control of acute myeloid leukemia, *Journal of Theoretical Biology*, vol. 470, pp. 30–42, 2019.

[782] M. Shanta Chatun and M. Maider Ali Biswas, Mathematical analysis and optimal control applied to the treatment of leukemia, *Journal of Applied Mathematics and Computing*, vol. 64, pp. 331–353, 2020.

[783] R.C. Shekhar and J.M. Maciejowski, Surface excavation with model predictive control, In: *IEEE CDC 2010, IEEE 49th Conf. on Decision and Control*, Atlanta, Georgia, USA, Dec. 2010.

[784] G. Shen, Y. Xiu, J. Zhang and B. Cui, Finite-time trajectory tracking control for entry guidance, *International Journal of Robust and Nonlinear Control*, vol. 28, pp. 5855–5914, 2018.

[785] W. Shen, H. Zhao, G. Zhang and Y.J. Lin, An event triggered robust control for electrohydraulic servo machines with active disturbances rejection, *Journal of the Franklin Institute*, vol. 359, pp. 2857–2885, 2022.

[786] Z. Shengzheng, H. Xiongxiong, Z. Haiyut, L. Xiaocong and L. Xingguo, PID-like coupling control of underactuated overhead cranes with input constraints, *Mechanical Systems and Signal Processing*, vol. 178, pp. 109274–109283, 2022.

[787] L. Shi, N. Kinkaid and J. Katupitiya, Robust control for satellite attitude regulation during on-orbit assembly, *IEEE Transactions on Aerospace Science and Electronic Systems*, vol. 52, no. 1, pp. 49–59, 2016.

[788] L. Shi, S. Kayastha and J. Katupitiya, Robust coordinated control of a dual-arm space robot, *Acta Astronautica*, vol. 138, pp. 475–489, 2019.

[789] P. Shi, X. Yu, X. Yang, J.J. Rodriguez-Andina, W. Sun and H. Gao, Composite adaptive synchronous control of dual-drive gantry stage with load measurement, *IEEE Open Journal of the Industrial Electronics Society*, vol. 4, pp. 63–74, 2023.

[790] P. Shi, W. Sun and X. Yang, RBF neural network-based adaptive robust synchronization control of dual drive gantry stage with rotational coupling dynamics, *IEEE Transactions on Automation Science and Engineering*, vol. 20, no. 2, pp. 1059–1068, 2023.

[791] P. Shi, W. Sun, X. Yang, I.J. Rudas and H. Gao, Master-slave synchronization control of dual-drive gantry stage with cogging force computation, *IEEE Transactions on Systems, Man and Cybernetics*, vol. 53, no. 1, pp. 216–226, 2023.

[792] J. Shi, L. Li, X. Wang and C. Liu, Robust output feedback controller with high-gain observer for automatic clutch, *Mechanical Systems and Signal Processing*, vol. 132, pp. 806–822, 2019.

[793] Y. Shi, W. Sheng, J. Wang, L. Jin, B. Li, and X. Sun, Real-time tracking control and efficiency analyses for Stewart platform based on discrete-time recurrent neural network, *IEEE Transactions on Systems, Man and Cybernetics: Systems*, vol. 54, no. 7, pp. 5099–5111, 2024.

[794] H. Shibata and F. Asano, Strict stealth walking generation for 3-link underactuated biped robots, *Artificial Life and Robotics*, vol. 25, pp. 603–611, 2020.

[795] C.L. Shih, J.W. Grizzle and C. Chevallereau, Asymptotically stable walking of a simple underactuated 3D bipedal robot, In: *IEEE IECON 2007, IEEE 33rd Annual Conference of the IEEE Industrial Electronics Society*, Taipei, Taiwan, March 2008.

[796] G.M. da Silva and R. Pederiva, Fault diagnosis of active magnetic bearings, *Mechatronics*, vol. 84, pp. 102801–102815, 2022.

[797] R.K. Singleton, E.G. Strangas and S. Avinyente, Extended Kalman filtering for remaining useful life estimation of bearings, *IEEE Transactions on Industrial Electronics*, vol. 62, no. 3, pp. 1781–1789, 2015.

[798] S.Y. Shin and C. Kim, Human-like motion generation and control for humanoid's dual arm object manipulation, *IEEE Transactions on Industrial Electronics*, vol. 62, no. 4, pp. 2265–2276, 2015.

[799] H. Schulte and K. Guelton, Descriptor modelling control of two-link pneumatic robot manipulator: a T-S multimodel approach, *Nonlinear Analysis: Hybrid Systems*, vol. 3, no. 2, pp. 124–132, 2009.

[800] L.L. Show, J.C. Juang and Y.W. Jan, An LMI-based nonlinear attitude control approach, *IEEE Transactions on Control Systems Technology*, vol. 11, no. 1, pp. 73–83, 2014.

[801] H.S. Sim, Y.M. Kao, S.H. Jang, D.K. Ahn, B.N. Cho and S.H. Han, A study on visual feedback control of SCARA Robot arm, In: *IEEE ICCAS 2015, IEEE 2015 15th Intl. Conference on Control, Automation and Systems*, Busan, Korea, Oct. 2018.

[802] D. Simon, A game theory approach to constrained minimax state estimation, *IEEE Transactions on Signal Processing*, vol. 54, no. 2, pp. 405–412, 2006.

[803] S. Singh and A Iyer, Nonlinear decoupling sliding-mode control and attitude control of spacecraft, *IEEE Transactions on Aerospace and Electronic Letters*, vol. 25, no. 5, pp. 621–633, 1989.

[804] A. Singh and P. Deolio, Dynamical analysis and chaos control in discrete-time prey-predator model, *Communications in Nonlinear Science and Numerical Simulation*, vol. 90, pp. 105313–105335, 2020.

[805] C. Singh and G. Liu, Energy-aware redundant actuation for safe spring-assisted modular and reconfigurable robot, *Robotica*, vol. 40, pp. 4498–4511, 2022.

[806] H. Sira-Ramirez and S. Agrawal, *Differentially Flat Systems*, Marcel Dekker, New York, 2004.

[807] M.R. Sirouspour and S.E. Salcudean, Nonlinear control of hydraulic robots, *IEEE Transactions on Robotics and Automation*, vol. 17, no. 2, pp. 173–182, 2001.

[808] J.J. Slotine, *Applied Nonlinear Control*, Prentice-Hall, 1991.

[809] M. Smaoui, X. Brun and D. Thomasset, Systematic control of an electropneumatic system: integrator backstepping and sliding-mode control, *IEEE Transactions on Control Systems Technology*, vol. 14, no. 5, pp. 905–913, 2006.

[810] M. Smaoui, X. Brun and D. Thomasset, A study on tracking position control of an electropneumatic system using backstepping design, *Control Engineering Practice*, vol. 14, pp. 923–933, 2006.

[811] M. Smaoui and L. Krichen, Control, energy management and performance evaluation of desalination unit based renewable energies using a graphical user interface, *Energy*, vol. 116, pp. 1187–1206, 2016.

[812] C. Song and D. Lau, Workspace-based model predictive control for cable-driven robots, *IEEE Transactions on Robotics*, vol. 38, no. 4, pp. 2577–2596, 2022.

[813] H. Soulay Ali, L. Boutat-Baddus, Y. Bacis-Aubny and M. Darouach, H_∞ control of a SCARA robot using polytopic LPV approach, In: *IEEE MED 2006, IEEE 2006 Mediterranean Conference on Control and Automation*, Ancona, Italy, June 2006.

[814] M.W. Spong, J.K. Holm and D. Lee, Passivity-based control of a bipedal locomotion: regulating walking by exploiting passive gaits in 2D and 3D bipeds, *IEEE Robotics and Automation Magazine*, pp. 31–40, 2007.

[815] R. Srivastava, R. Sah and K. Das, Nonlinear model predictive control of rotation flying space robots for autonomous active debris removal, In: *IFAC ACA 2022, IFAC 22nd Symposium on Automatic Control in Aerospace*, Mumbai, India, Nov. 2022.

[816] A. Steindl, Time-optimal control for the deployment of a tethered satellite allowing for a massive tether, *Meccanica*, vol. 51, pp. 2741–2751, 2016.

[817] A. Steindl, Optimal deployment of a tethered satellite allowing for a massive tether, *Meccanica*, vol. 51, pp. 2741–2751, 2016.

[818] T. Stiehl and A. Marciniak-Czochra, Mathematical model of leukemogenesis and cancer stem cells dynamics, *Mathematical Modelling of Natural Phenomena, EDP Sciences*, vol. 7, no. 1, pp. 106–202, 2012.

[819] Y.X. Su, B.Y. Duan, C.M. Zheng, Y.F. Zhang, G.D. Chen and J.W. Mi, Disturbance-rejection high-precision motion control of a Stewart platform, *IEEE Transactions on Control Systems Technology*, vol. 12, no. 3, pp. 364–374, 2004.

[820] B. Su, F. Zhang and P. Huang, Robust control of triangular tethered satellite formation with unmeasured velocities, *Acta Astronautica*, vol. 186, pp. 190–202, 2021.

[821] B. Su, F. Zhang and P. Huang, Nonlinear state observer and control design for triangular tethered satellite formation, *IEEE Transactions on Aerospace and Electronic Systems*, vol. 58, no. 5, pp. 4718–4728, 2022.

[822] B. Su, F. Zha and P. Huang, Stability research of a triangular tethered satellite formation: dynamics, filtering and control, *IEEE Transactions on Aerospace and Electronic Systems*, vol. 58, no. 6, pp. 5240–5255, 2022.

[823] B. Su, F. Zhang and P. Huang, Observer-based adaptive backstepping control and the application for a class of multi-input multi-output nonlinear systems with structural uncertainties and perturbations, *International Journal of Robust and Nonlinear Control*, vol. 33, pp. 6211–6222, 2023.

[824] B. Su, F. Zhang and P. Huang, Proportional derivative control of second-order tethered satellites system based on extended state observer and feedforward compensation, *Journal of Guidance, Navigation and Control*, pp. 1–24, 2023.

[825] N. Sun, Y. Fang and X. Zhang, Energy coupling output feedback control of 4-DOF underactuated cranes with saturated inputs, *Automatica*, vol. 49, pp. 1318–1325, 2013.

[826] N. Sun, Y. Fang, H. Chen, B. Lu and Y. Fu, Slow/translational positioning and swing suppression for 4-DOF tower cranes with parametric uncertainties: design and hardware experimentation, *IEEE Transactions on Industrial Electronics*, vol. 63, no. 10, pp. 6407–6418, 2016.

[827] Z. Sun, Y. Tian, M. Lu and J. Wang, A superlinear convergence feasible sequential quadratic programming algorithm for bipedal dynamic walking robot via discrete mechanics and optimal control, *Optimal Control Applications and Methods*, vol. 37, pp. 1139–1161, 2016.

[828] L. Sun, W. Huo, and Z. Jiao, Adaptive backstepping control of spacecraft rendezvous and proximity operations with input saturation and full-state constraint, *IEEE Transactions on Industrial Electronics*, vol. 64, no. 1, pp. 480–492, 2017.

[829] L. Sun, W. Yu, M. Yang and J. Liu, Position control for flexible-joint robot based on online gravity compensation with vibration suppression, *IEEE Transactions on Industrial Electronics*, vol. 65, no. 6, pp. 4840–4848, 2018.

[830] N. Sun, Y. Wu, Y Fang and H. Chen, Non-linear anti-swing control for crane systems with double-pendulum swing effects and uncertain parameters: design and experiment, *IEEE Transactions on Automation Science and Engineering*, vol. 15, no. 3, pp. 1413–1422, 2018.

[831] N. Sun, Y. Wu, X. Liang and Y. Fang, Nonlinear state transportation control for double-pendulum shipboard cranes with ship-motion-induced disturbances, *IEEE Transactions on Industrial Electronics*, vol. 66, mo. 12, pp. 96467–9479, 2019.

[832] X. Sun, Z, Sun, J. Shao, Z. Shi and C.A. Zhang, Fault diagnosis of rotating machinery equipped with multiple sensors using space-time fragments, *Journal of Sound and Vibration*, vol. 456, pp. 49–64, 2019.

[833] N. Sun, Y. Wu, H. Chen and Y. Fang, Antiswing cargo transportation of underactuated tower crane systems by a nonlinear controller embedded with an integral term, *IEEE Transactions on Automation Science and Engineering*, vol. 16, no. 3, pp. 1387–1398, 2019.

[834] X. Sun, M. Wu, Z. Yang, G. Lei and Y. Guo, High-performance control for a permanent magnet linear synchronous generator using state feedback control scheme plus grey wolf optimization, *IET Electric Power Applications*, vol. 14, no. 5, pp. 771–780, 2020.

[835] Z. Sun, B. Zhang, Y. Sun, Z. Peng and C. Chang, A novel superlinearly convergent trust region-requested quadratic programming approach for optimal gait of bipedal robots via nonlinear model predictive control, *Journal of Intelligent and Robotic Systems*, vol. 100, pp. 401–416, 2020.

[836] Q. Sun and Y. Jia, Attitude-orbit integrated control for spacecraft fly-around mission based on Chebyshev neural networks, *IEEE Transactions on Aerospace and Electronic Systems*, vol. 60, no. 3, pp. 3177–3181, 2024.

[837] S.S. Sunjeewa and M. Parnichkun, Control of rotary double inverted pendulum system using mixed sensitivity H_∞ control, *International Journal of Advanced Robotic Systems*, pp. 1–17, 2019.

[838] S.D.A Sunjeewa and M. Parnichkun, Control of rotary double inverted pendulum system using LQR sliding surface based sliding-mode controller, *Journal of Control and Decision*, vol. 9, no. 1, pp. 89–101, 2022.

[839] G. Schweitzer and E.H. Maslen, *Magnetic Bearings: Theory, Design, and Application to Rotating Machinery*, Springer, 2009.

[840] A. Szeles, D.A. Drexler, J. Sapi, I. Harmati and L. Kovacs, Model-based angiogenic inhibition of tumor growth using feedback linearization, In: *IEEE CDC 2013, IEEE 2013 Conference on Decision and Control*, Firenze, Italy, Dec. 2013.

[841] J. Tang, K. Wang and B. Xiang, Stable control of high-speed rotor suspended by superconducting magnetic bearings and active magnetic bearings, *IEEE Transactions on Industrial Electronics*, vol. 64, no. 4, pp. 3319–3329, 2012.

[842] H. Tang, H.L. Dai and Y. Du, Bearing fault detection for doubly-fed induction generator based on stator currents, *IEEE Transactions on Industrial Electronics*, vol. 69, no. 3, pp. 5267–5276, 2022.

[843] C.W. Tao, J.S. Taur, C.W. Chang and Y.H. Chang, Simplified type-2 fuzzy sliding controller for wing-rock system, *Fuzzy Sets and Systems*, vol. 207, pp. 111–129, 2012.

[844] S. Tafazoli, S.F. Salcudean, K. Hashtradi-Zaad and P.D. Lawrence, Impedance control of a teleoperated excavator, *IEEE Transactions on Control Systems Technology*, vol. 9, no. 5, pp. 355–369, 2002.

[845] A. Taheri, P. Gustafsson, M. Rösth, R. Ghabeheloo and J. Pajarinen, Nonlinear model learning for compensation and feedforward control of real-world hydraulic actuators using Gaussian processes, *IEEE Robotics and Automation Letters*, vol. 7, no. 4, pp. 3525–3592, 2022.

[846] S.E. Talole, J.P. Kolhe and S.N. Phadke, Extended-state-observer-based control of flexible-joint system with experimental validation, *IEEE Transactions on Industrial Electronics*, vol. 57, no. 4, pp. 1411–1419, 2010.

[847] P. Tan, L. Huang, M. Chen, Y. Li, L. Wei and R. Ma, A robust faster joint control of a direct-drive wave energy converter connected with supercapacitor and battery energy storage, *IEEE Journal of Emerging and Selected Topics in Power Electronics*, pp. 1–11, 2023.

[848] N. Tangesdal, T.T. Kristofferesen and C. Holden, Applied nonlinear compressor control with gain scheduling and state estimation, In: *3rd IFAC Workshop on Automatic Control in Offshore, Oil and Gas Production*, Esbjerg, Denmark, June 2011.

[849] X. Tao, F. Zhang, B. Huang, G. Shen and P. Huang, Fixed-time sliding mode coordinated deployment control for space triangle tethered formation system, *IEEE Transaction on Aerospace and Electronic Systems*, vol. 60, no. 2, pp. 1351–1363, 2024.

[850] S. Tarbouriech, B. Navarra, P. Fraisse, A. Crasnier, A. Cherubini and D. Sallé, An admittance-based hierarchical control framework for dual-arm robots, *Mechatronics*, vol. 86, pp. 102814–102825, 2022.

[851] S. Takagi and N. Uchiyama, Robust control system design for SCARA Robots using adaptive pole placement, *IEEE Transactions on Industrial Electronics*, vol. 52, no. 3, pp. 315–321, 2005.

[852] F. Tajdari and N.E. Toulkani, Implementation and intelligent gain-tuning feedback-based optimal torque control of a rotary parallel robot, *Journal of Vibration and Control*, vol. 28, no. 19–20, pp. 2678–2695, 2022.

[853] T.S. Tamor, G. Xiang, X. Dong, Q. Fang, S. Liu, E. Ladhi, Z. Shen and F.Y. Wang, Design and optimization of a control framework for robot-assisted additive manufacturing based on Stewart platform, *International Journal of Control, Automation and Systems*, vol. 20, no. 3, pp. 968–982, 2022.

[854] K. Tanaka, S. Nishikawa, R. Niiyama and Y. Kaniyoshi, Immediate generation of jump-and-hit motions by a pneumatic humanoid robot using a lookup table of learned dynamics, *IEEE Robotics and Automation Letters*, vol. 6, no. 3, pp. 5557–5565, 2021.

[855] K.L. Tao, B. Li and Y. Rehbock, *Applied and Computational Optimal Control: A Control Parametrization Approach*, Springer, 2021.

[856] D. Tashtoush, W. Alyahy, M.A. Ghadi, J.A. Oman and T. Morosuk, Renewable energy integration in water desalination: state-of-the-art review and cooperative analysis, *Applied Energy*, vol. 352, pp. 121950–121993, 2023.

[857] C.J. Taylor and D. Robertson, State-dependent control of a hydraulically actuated nuclear decommissioning robot, *Control Engineering Practice*, vol. 21, pp. 1716–1725, 2013.

[858] P. Tempel, M. Alfeld and V. van der Wijk, Design and analysis of cable-driven parallel robot CoRISA: a cable robot for inspecting and scanning artwork, In: *Proc. Romansy 23 Conf. on Robot Design, Dynamics and Control - CISM Intl. Center for Mechanical Sciences*, pp. 136–144, Sapporo, Japan, Sep. 2020.

[859] W. Teng, C. Han, Y. Hu, X. Cheng, L. Song and Y. Liu, A robust model-based approach for bearing remaining useful life progress in wind turbines, *IEEE Access*, vol. 8, pp. 47133–47143, 2020.

[860] C.S. Teo, K.K. Tan, S.Y. Lim, S. Huang and E.B. Tay, Dynamic modelling and adaptive control of an H-type gantry stage, *Mechatronics*, vol. 17. pp. 361–367, 2007.

[861] B. Tian, W. Fan, Q. Zong, J. Wang and F. Wang, Nonlinear robust control for reusable launch vehicle in reentry phase based on time-varying high-order sliding-mode, *Journal of the Franklin Institute*, vol. 350, pp. 1787–1807, 2013.

[862] B. Tian, L. Yin and H. Wang, Finite-time reentry attitude control based on adaptive multivariable disturbance compensation, *IEEE Transactions on Industrial Electronics*, vol. 60, no. 9, pp. 5889–5898, 2015.

[863] B. Tian, H. Lu, Z. Zuo and Q. Zong, Multi-variable uniform finite-time output feedback reentry attitude control for RLV with mismatched disturbances, *Journal of the Franklin Institute*, vol. 355, pp. 3470–3487, 2018.

[864] J. Tian, H. Polinder, A. Jarquin Laguna and S. Miedema, A wave-to-wire analysis of the adjustable draft point absorber wave energy converter coupled with a linear permanent magnet generator, *Ocean Engineering*, vol. 276, pp. 114195–114215, 2023.

[865] R.S. Tiwari, R. Kumar and O.M. Gupta, Detection and classification of faults in VSC-HVDC system based on single-end grounded pole differential current components, In: *Recent Advances in Power Systems, Lecture Notes in Electrical Engineering*, vol. 960, pp. 218–227, 2023.

[866] A. Toedtheide, T. Lilge and S. Haddadin, Antagonistic impedance control for pneumatically actuated robot joints, *IEEE Robotics and Automation Letters*, vol. 1, no. 1, pp. 161–169,2016.

[867] B. Tian, W. Fan, R. Su and Q. Zong, Real-time trajectory and attitude coordination control for reusable launch vehicle in reentry phases, *IEEE Transactions on Industrial Electronics*, vol. 62, no. 3, pp. 1639–1650, 2015.

[868] C.S. Ting, Y. Chang and Y.T. Cheng, Backstepping direct drive thrust force control for sensorless PMLSM drive, *IET Electric Power Applications*, vol. 19, no. 3, pp. 322–331, 2019.

[869] W. Tong and P. Jiancheng, A feedback linearization control for the nonlinear 5-DOF flywheel suspended by the permanent magnet biased hybrid magnetic bearings, *Acta Astronautica*, vol. 79, pp. 131–139, 2012.

[870] S.A. Topno, L.K. Kumar Sahoo and B.S. Umre, Energy efficiency assessment of electric shovel operating in opencase mine, *Energy*, vol. 230, pp. 120703–120712, 2021.

[871] J.A. Torres, R.C. dos Santos, Q. Yang and J. Li, Analysis of different approaches for detecting, classifying and locating faults in a three-terminal VSC-HVDC system, *International Journal of Electrical Power and Energy Systems*, vol. 135, article no. 107514, 2022.

[872] G. Torrisi, V. Jaramillo, J.R. Ottewill, M. Mariethoz, M. Morari and R.S. Smith, Active surge control of electrically driven centrifugal compressors, In: *2015 European Control Conference*, Linz, Austria, July 2015.

[873] G. Torrissi, S. Grammatico, A. Cortinovis, M. Mercangöz, M. Morari and R.S. Smith, Model predictive approaches for active surge control in centrifugal compressors, *IEEE Transactions on Control Systems Technology*, vol. 29, no. 6, pp. 1947–1960, 2019.

[874] M. Tourajizadeh, M. Yousefzadeh and A. Tajik, Closed-loop optimal control for a Stewart platform using an optimal feedback linearization method, *Journal of Advanced Robotic Systems*, vol. 13, pp. 134–142, 2016.

[875] M. Tourajizadeh and S. Manteghi, Design and optimal control of dual-stage Stewart platform using feedback linearized quadratic regulation, *Advanced Robotics*, vol. 30, no. 20, pp. 1305–1323, 2016.

[876] G.J. Toussaint, T. Basar and F. Bullo, H_∞ optimal tracking control techniques for nonlinear underactuated systems, In: *Proc. IEEE CDC 2000, 39th IEEE Conference on Decision and Control*, Sydney, Australia, 2000.

[877] M. Trumic, K. Jovanovic and A. Fagiolini, Decoupled nonlinear adaptive control of position and stiffness for pneumatic soft robots, *International Journal of Robotics Research*, vol. 40, no. 1, pp. 277–295, 2020.

[878] N.C. Tsai, Y.H. King and R.M. Lee, Fault diagnosis for magnetic bearing systems, *Mechanical Systems and Signal Processing*, vol. 23, pp. 1139–1151, 2005.

[879] L.A. Tuan, Y.H. Joo, P.X. Duong and L.O. Tian, Parameter estimator integrated sliding-mode control of dual-arm robots, *Journal of Control, Automation and Systems*, vol. 15, no. 6, pp. 2754–2763, 2017.

[880] L.A. Tuan, Y.H. Joo, L.Q. Tien and P.X. Duang, Adaptive neural-network second-order sliding-mode control of dual-arm robots, *Journal of Control, Automation and Systems*, vol. 15, no. 6, pp. 2883–2891, 2017.

[881] W. Tuvayonond and M. Parnichkun, Position control of a pneumatic surgical robot using PSO-based 2-DOF H_∞ loop shaping structural controller, *Mechatronics*, vol. 43, pp. 40–55, 2017.

[882] D. Tuzi, T. Delamotte and A. Knopp, Satellite swarm-based antenna arrays for 6G direct-to-cell connectivity, *IEEE Access*, vol. 11. pp. 36907–36228, 2023.

[883] G.O. Tysse, A. Cibicik and O. Egeland, Vision-based control of a Knuckle boom crane with online cable length estimation, *IEEE/ASME Transactions on Mechatronics*, vol. 26, no 1. pp. 416–426, 2021.

[884] G.O. Tysse, A. Cibicik, L. Tingelstad and O. Egeland, Lyapunov-based damping controller with nonlinear MPC control of payload position for a knuckle boom crane, *Automatica*, vol. 140, pp. 110219–110229, 2022.

[885] S. Tzafestas, M. Kotsis and T. Pimenides, Observer-based optimal control of flexible Stewart parallel robots, *Journal of Intelligent Robotic Systems*, vol. 24, pp. 489–503, 2002.

[886] C.S. Ukidre, J.B. McInroy and F. Jafari, Using redundancy to optimize manipulability of Stewart platforms, *IEEE/ASME Transactions on Mechatronics*, vol. 13, no. 4, pp. 475–479, 2008.

[887] C. Urea and J. Kern, Modelling and control of a redundant SCARA-type manipulator robot, *International Journal of Advanced Robotic Systems*, vol. 9, no. 58, pp. 1–14, 2012.

[888] C. Urea, J. Cortes and J. Pascal, Design, construction and control of a SCARA manipulator with 6 degrees of freedom, *Journal of Applied Research and Technology*, vol. 14, pp. 396–404, 2014.

[889] C. Urea and J. Kern, Trajectory tracking control of a real redundant manipulator of the SCARA type, *Journal of Electrical Engineering and Technology*, vol. 11, no. 1, pp. 215–226, 2016.

[890] C. Urea, I. Kerr and E. Alvarez, Design and implementation of fault-tolerant control strategies for a real underactuated manipulator robot, *Journal of Complex and Intelligent Systems*, 2022.

[891] N. Uzhegov, A. Smirnov, C.H. Park, J.H. Ahn, J. Heikinnen and J. Pyrhonnen, Design aspects of high-speed electrical machines with active magnetic bearings for compressor applications, *IEEE Transactions on Industrial Electronics*, vol. 64, no. 11, pp. 8427–8436, 2019.

[892] M. Valenzuela Guzman and M. Anibal Valenzuela, Integrated mechanical-electrical modeling of an AC electric mining shovel and evaluation of power requirements, during a truck loading cycle, *IEEE Transactions on Industry Applications*, vol. 51, no. 3, pp. 2580–2599, 2015.

[893] M. Valenzuela and M. Anibal Valenzuela, Payload estimation in AC electric mining shovels using drive signals, *IEEE Transactions on Industry Applications*, vol. 52, no. 5, pp. 4420–4429, 2016.

[894] M. Van Damme, B. Vanderborght, R. Van Ham, B. Verrelst, F. Daersden and B. Lefeber, Sliding-mode control of a 2-DOF planar pneumatic manipulator, *Journal of Dynamical Systems, Measurement and Control*, vol. 131, no. 2, 021013 (12 pages), 2009.

[895] J. Varghese, V.M. Akhil, P.K Rajendrakumar and K.S. Sirandan, A rotary pneumatic actuator for the actuation of the exoskeleton knee joint, *Theoretical and Applied Mechanics Letters*, vol. 7, pp. 222–230, 2012.

[896] P. Verrax, A. Bertinato, M. Kieffer and B. Roison, Transient-based fault identification algorithm using parametric models for meshed HVDC grids, *Electric Power Systems Research*, vol. 185, article no. 196387, 2020.

[897] P. Verrax, A. Bertinato, M. Kieffer and B. Roison, Fault identification in HVDC grids using a transient parametric model, In: *21st IFAC World Congress*, Berlin, Germany, July 2020.

[898] J. Villagra, B. d'Andrea-Novel, H. Mounier and M. Pengov, Flatness-based vehicle steering control strategy with SDRE feedback gains tuned via a sensitivity approach, *IEEE Transactions on Control Systems Technology*, vol. 15, pp. 554–565, 2007.

[899] A. Visioli and G. Legnani, On the trajectory tracking control of industrial SCARA robot manipulators, *IEEE Transactions on Industrial Electronics*, vol. 49, no. 1, pp. 224–233, 2002.

[900] P. Voglewede, A.P.C. Smith and A. Monti, Dynamic performance of a SCARA robot manipulator with uncertainty using polynomial chaos theory, *IEEE Transactions on Robotics*, vol 25, no. 1, pp. 2-6-219, 2009.

[901] H. Wang and Y. Xie, Adaptive Jacobian position/force tracking control of free-floating manipulators, *Robotics and Autonomous Systems*, vol. 57, pp. 173–181, 2009.

[902] H. Wang and Y. Xie, Passivity-based adaptive Jacobian tracking for free-floating space manipulators without using spacecraft acceleration *Automatica*, vol. 45, pp. 1510–1517, 2009.

[903] H. Wang, On adaptive inverse dynamics for free-floating space manipulators, *Robotics and Autonomous Systems*, vol. 59, pp. 782–788, 2011.

[904] H. Wang and Y. Xie, Prediction error-based adaptive Jacobian tracking for free-floating space manipulators, *IEEE Transactions on Aerospace and Electronic Systems*, vol. 48, no. 4, pp. 1207–1226, 2012.

[905] L. Wang, H. Ni, W. Zhou, P.M. Pardalos, J. Fang and M. Fei, MBPOA-based LQR controller and its application tp the double-parallel inverted pendulum system, *Engineering Applications of Artificial Intelligence*, vol. 36, pp. 262–269, 2014.

[906] M. Wang, L. Li and D. Pan, Detent force compensation for PMLSM systems based on structural design and control method combination, *IEEE Transactions on Industrial Electronics*, vol. 62, no. 11, pp. 6845–6854, 2015.

[907] M. Wang, J. Luo and U. Walter, A Nonlinear Model Predictive Controller with obstacle avoidance for a space robot, *Advances in Space Research*, vol. 57, pp. 1737–1746, 2016.

[908] Z. Wang, B. Wang, L. Xu and Q. Xie, Feedback-added pseudoinverse type balanced minimization scheme for kinematic control of redundant robot manipulators, *IEEE Access*, vol. 7, pp. 23806–23815, 2017.

[909] W. Wang, H. Chi, S. Zhao and Z. Du, A control method for hydraulic manipulators in automatic emulsion filling, *Automation in Construction*, vol. 91, pp. 92–99, 2018.

[910] C. Wang and F. Zhang, Finite-time stability of an underactuated tethered satellite system, *Acta Astronautica*, vol. 159, pp. 199–202, 2019.

[911] Z. Wang, L.B. Freidovich and M. Zhang, Periodic motion planning and control for double rotary pendulum via virtual balancing constrains, *IEEE/CAA Journal of Automatica Sinica*, vol. 6, no. 1, pp. 291–298, 2019.

[912] H. Wang, Y. Pan, S. Li and H. Yu, Robust sliding-mode control for robots driven by compliant actuators, *IEEE Transactions on Control Systems Technology*, vol. 27, no. 3, pp. 1259–1266, 2019.

[913] R. Wang, S. Tang and D. Zhang, Short-range reentry guidance with impact angle and impact velocity constraints for hypersonic gliding reentry vehicle. *IEEE Access*, vol. 7, pp. 47437–47450, 2019.

[914] S. Wang, Optimal control for cancer chemotherapy under tumor heterogeneity, In: *IEEE CDC 2019, IEEE 2019 58th Conference on Decision and Control*, Nice, France, Dec. 2019.

[915] L. Wang, M. Liu, E. Tedeschi, J. Engstrom and J. Insberg, Improving electric power generation of a stand-alone wave energy converter via optimal electric load control, *Energy*, vol. 211, pp. 118945–118953, 2020.

[916] X. Wang, L. Shi and J. Katupitiya, Coordinated control of a dual-arm space robot to approach and synchronize with the motion of a spinning target in 3D space, *Acta Astronautica*, vol. 176, pp. 99–110, 2020.

[917] L. Wang, Q. Shi, J.Liu and D. Zhang, Backstepping control of flexible joint manipulator based on hyperbolic tangent function with control input and rate constraints, *Asian Journal of Control*, vol. 22, no. 3, pp. 1268–1279, 2020.

[918] Y. Wang, Q. Liu, J. Huang, *et al.*, Sliding-mode robust control of wire-driven parallel robot based on HJI theory and a disturbance observer, *IEEE Access*, vol. 8, pp. 215235–215245, 2020.

[919] W. Wang, Analysis of fault detection in rolling element bearings, *IEEE Magazine on Instrumentation and Measurement*, vol. 24, no. 23, pp. 42-49, 2021.

[920] B. Wang, Y. Liu, K. Yue, D. Lu and J. Zhao, Improved dynamic state estimation-based protection on transmission line in MMC-HVDC grids, *IEEE Transactions on Power Delivery*, vol. 37, no. 5, pp. 3567–3581, 2021.

[921] D. Wang, J. Fu, M. Hao, F. Qiao and M. Gao, Novel travelling wave fault location principle for VSC-HVDC transmission line, *Electric Power Systems Research*, vol. 196, article no. 107226, 2021.

[922] X. Wang, S. Lu, K. Chen, O. Wang and S. Zhang, Bearing fault diagnosis of switched reluctance motor in electric vehicle powertrain via multisensor data fusion, *IEEE Transactions on Industrial Informatics*, vol. 18, no. 4, pp. 2452–2464, 2021.

[923] Z. Wang, C. Hu, Y. Zhu and L. Zhu, Prediction model-based-contouring error iterative precompensation scheme for position multiaxis motion systems, *IEEE/ASME Transactions on Mechatronics*, vol. 26, no. 5, pp. 2274–2284, 2021.

[924] W. Wang, J. Ma, Z. Chang, X. Li, A. Al Mamoun and T.M. Lee, Generalized iterative super-twisting sliding-mode control: a case study on flexure-joint, dual-drive H-gantry stage, In: *IEEE ICIT 2021, IEEE 2021 Intl. Conference on Industrial Technology*, Valencia, Spain, March 2021.

[925] L. Wang, J. Jiang and M. Wang, Research on MPC control of a 6-DOF Stewart platform, In: *IEEE 3M-Nano, IEEE, 2022 Intl. Conf. on Manipulation, Manufacturing and Measurement on the Nanoscale*, Tianjin, China, Aug. 2022.

[926] B. Wang, Y. Liu, D. Lu, R. Pen and K. Yue, Multi-section frequency-dependent line model enabled underground cable fault location in MMC-HVDC grids using dynamic state estimation, *International Journal of Electrical Power and Energy Systems*, vol. 153, article no. 109322, 2023.

[927] X. Wang, L. Ge, Y. Li and L. Quan, Research on energy efficiency characteristics of mining shovel hoisting and slewing system driven by hydraulic-electric hybrid system, *Chinese Journal of Mechanical Engineering*, vol. 36, no. 156, pp. 1–15, 2023.

[928] W. Wang, J. Ma, Z. Chang, X. li, C. da Silva and T.H. Lee, Global iterative sliding-mode control of an industrial biaxial gantry system for contouring motion tasks, *IEEE/ASME Transactions on Mechatronics*, vol. 27, no. 3, pp. 1617–1628, 2022.

[929] X. Wang and J. Katupitiya, Robust control of a dual-arm space robot to capture a non-cooperative target in 3D space, *Aerospace Systems and Technology*, vol. 141, pp. 108538–108551, 2023.

[930] T. Wang, C. Lin, R. Li, *et al.*, Nonlinear enhanced coupled feedback control for bridge crane with uncertain disturbances: theoretical and experimental investigations, *Nonlinear Dynamics*, vol. 111, pp. 19021–19032, 2023.

[931] X. Wang, L. Xia, L. Ge, L. Quan, J. Huang and J. Cui, Electro-hydraulic active passive hybrid drive slewing system for electric shovel, *IEEE Access*, pp. 1–10, 2024.

[932] W.M. Weerakoon, T.W.U. Madhushani, D.H.S. Maithrapula and J.M. Berg, Feedback regularization and geometric PID control for robust stabilization of a planar three-link hybrid walking robot, In: *IEEE ACC 2018, IEEE 2018 American Control Conference*, Wisconsin, Milwaukee, June 2018.

[933] Q. Wei, R. Song, B. Li, and X. Lin, *Self-Learning Optimal Control of Nonlinear Systems: Adaptive Dynamic Programming Approach*, Springer, 2016.

[934] B. Wei and D. Zhang, A review of dynamic balancing for robotic mechanisms, *Robotica*, vol. 39, pp. 55–71, 2021.

[935] W. Wei, H. Dourra, and G. Zhu, Integrated clutch torque control and touch point estimation using deadbeat adaptive backstepping, *IEEE Transactions on Control Systems Technology*, vol. 30, no. 1, pp. 768–775, 2022.

[936] H. Wen, D.P. Jin and H.Y. Hu, Optimal feedback control of the deployment of a tethered sub-satellite subject to perturbations, *Nonlinear Dynamics*, vol. 51, pp. 501–514, 2008.

[937] H. Wen, Z.H. Zhu, D. Jin and H. Hu, Tension control of space tether via online quasi-linearization structures, *Advances in Space Research*, vol. 57, pp. 754–763, 2016.

[938] H. Wen, Z. Zhu, D. Jin, H. Hu, Constrained tension control of a tethered space-tug system with only length measurement, *Acta Astronautica*, vol. 199, pp. 110–117, 2016.

[939] H. Wen, Z.H. Zhu, D. Jin and H. Hu, Space tether deployment control and explicit tension constraint and saturation function, *Journal of Guidance, Control and Dynamics*, vol. 29, no. 4, pp. 1–6, 2016.

[940] P.W. Wensing, M. Posu, Y. Hu, A. Escande, N. Mansard and A. del Preto, Optimization-based control for dynamic legged robots, *IEEE Transactions on Robotics*, vol. 40, pp. 43–63, 2024.

[941] H. Wind, A. Renner, S. Schmit, S. Albrecht and O. Sawodny, Comparison of joint angle, velocity and acceleration estimates for hydraulically actuated manipulators to a novel dynamical approach, *Control Engineering Practice*, vol. 91, pp. 10411, 2019.

[942] J. Woolfrey, W. Lu and D. Liu, A control method for joint-torque minimization of redundant manipulators handling large external forces, *Journal of Intelligent and Robotic Systems*, vol. 93, pp. 3–16, 2019.

[943] S. Wu, Tethered satellite system analysis (1) - Two dimensional case and regular dynamics, *Acta Astronautica*, vol. 47, no. 12. pp. 849–858, 2000.

[944] Q. Wu and M. Saif, Neural adaptive observer-based fault detection and identification for satellite attitude control systems, In: *IEEE ACC 2005, IEEE 2005 American Control Conference*, Portland, Oregon, USA, July 2005.

[945] T.S. Wu, M. Karkoub, W.S. Yu, C.T. Chen, M.G. Henand and K.W. Wu, Anti-sway tracking control of tower cranes with delayed uncertainty using a robust adaptive fuzzy control, *Fuzzy Sets and Systems*, vol. 290, pp. 118–137, 2016.

[946] T.S. Wu, M. Karhoub, H. Wang, H.S. Chen and T.H. Chen, Robust tracking control of MIMO underactuated nonlinear systems with dead-zone band and delayed uncertainty using an adaptive fuzzy control, *IEEE Transactions on Fuzzy Systems*, vol. 25, no. 4, pp. 905–918, 2017.

[947] Q. Wu, M. Li, X. Qi, Y. Hu, B. Li and J. Zhang, Coordinated control of a dual-arm robot for surgical instrument sorting tasks, *Robotics and Autonomous Systems*, vol. 112, pp. 1–12, 2019.

[948] X. Wu and Z. Li, Cooperative manipulation of wearable dual-arm exoskeleton using force communication between partners, *IEEE Transactions on Industrial Electronics*, vol. 67, no. 8, pp. 6629–6638, 2020.

[949] T. Wu, B. Xiang and S. Zhang, Optimal control for hybrid magnetically suspended flywheel rotor based on state feedback exact linearization model, *Science Progress*, vol. 103, no.3, pp. 1–23, 2020.

[950] J. Wu, G. Wang, Q. Bi and R. Hull, Digging force and power consumption during robotic excavation of cable shovel: experimental study and DEM simulation, *International Journal of Mining, Reclamation and Environment*, vol. 35, no. 1, pp. 12–33, 2021.

[951] Y. Wu, N. Sun, H. Chen and Y. Fang, Adaptive output feedback control for 5-DOF varying-cable-length tower cranes with cargo mass estimation, *IEEE Transactions on Industrial Informatics*, vol. 17, no. 4, pp. 2453–2464, 2021.

[952] Y. Wu, N. Sun, H. Chen and Y. Fang, New adaptive dynamic output feedback control of double-pendulum ship-mounted cranes with accurate gravitational compensation and constrained inputs, *IEEE Transactions on Industrial Electronics*, vol. 69, no. 9, pp. 9196–9205, 2022.

[953] Z. Xia, T. Sun, T.H. Kwon, Z. Ma and X. Wu, A new reinforcement learning-based adaptive sliding-mode control scheme for free-floating space robotic manipulator, *IEEE Access*, vol. 8, pp. 127048–127064, 2020.

[954] B. Xian, M.S. de Queiroz, D. Dawson and I. Walker, Task-based tracking control of redundant robot manipulators via quaternion feedback, *IEEE Transactions on Robotics and Automation*, vol. 20, no. 1, pp. 160–167, 2004.

[955] H. Xiang, J. Xiao, D. Yang, B. Zhao and X. Gao, Displacement estimation of self-sending magnetic bearings based on biorthogonal spline wavelets, *IEEE Access*, vol. 9, pp. 76213–76223, 2021.

[956] B. Xiao, L. Cao and D. Ram, Attitude exponential stabilization control of rigid bodies via disturbance observer, *IEEE Transactions on Systems, Man and Cybernetics: Systems*, 2019.

[957] B. Xie and S. Dai, Optimal trajectory generation for Stewart platform using discrete mechanics and optimal control, In: *IEEE ICIEA 2018, 13th 2018 IEEE Conference on Industrial Electronics and Applications*, Wuhan, China, June 2018.

[958] Y. Xie, Y. Lei, J. Guo and B. Meng, Spacecraft dynamics and control, In: *Springer Series in Space Science and Technology*, Springer, 2020.

[959] F. Xie, W. Shang, B. Zhang, S. Cong and Z. Li, High-precision trajectory tracking control of cable-driven parallel robots using robust synchronization, *IEEE Transactions on Industrial Informatics*, vol. 17, no. 4, pp. 2488–2499, 2021.

[960] C. Xie and H. Imani Marrani, Design of a robust LMI-based model predictive control method for surge instability in interconnected compressor systems, *Systems Science and Control Engineering*, vol. 9, no. 1, pp. 358–368, 2021.

[961] B. Xie, L. Lei, Y. Yao, and Y. Li, Advanced synchronous control of dual parallel motion systems, *IEEE Transactions on Industrial Electronics*, vol. 70, no. 2, pp. 1270–1281, 2023.

[962] Y. Xie, Q. Ma, J. Gu, and G. Zhou. Event-triggered fixed-time practical tracking control for flexible-joint robot, *IEEE Transactions of Fuzzy Systems*, vol. 11, no. 31, pp. 67–76, 2023.

[963] Z. Xie, Y. Wang, X. Luo, P. Arpenti, F. Ruggiero and B. Siciliano, Three-dimensional variable center of mass height biped walking using a new model and nonlinear model predictive control, *Mechanism and Machine Theory*, vol. 197, article no. 105651, pp. 1–22, 2024.

[964] X. Xie, T. Sheng and X. Chen, Dynamic event-triggered and self-triggered fault-tolerant attitude control for multiple space systems with uncertainties and input saturation, *IEEE Transactions on Aerospace and Electronic Systems*, vol. 60, no. 3, pp. 2922–2933, 2024.

[965] S. Xingling and W. Honglun, Sliding-mode based trajectory linearization control for hypersonic reeentry vehicle via extended disturbance observer, *ISA Transactions*, vol. 53, pp. 1771–1786, 2014.

[966] K. Xiu, H. Xing, L. Ding, H. Gao, G. Liu and Z. Deng, Virtual decomposition-based modeling for multi-DOF manipulator with flexible joints, *IEEE Access*, vol. 7, pp. 91582–91592, 2019.

[967] S. Xu, H. Wang, B. Zhang and B. Yang, Adaptive reactionless motion control for free-floating space robotic manipulators with uncertain kinematics and dynamics, In: *3rd IFAC Intl. Conf. on Intelligent Control and Autonomous Science*, Chengdu, China, Sep. 2013.

[968] Y. Xu, H. Liao, L. Liu and Y. Wang, Modeling and robust H-infinity control of a novel non-contact ultra-quiet Stewart spacecraft, *Acta Astronautica*, vol. 107, pp. 274–289, 2015.

[969] S. Xu, H. Wen, Z. Huang and D. Jin, A fuzzy control scheme for deployment of space tethered system with tension constraint, *Aerospace Science and Technology*, vol. 106, pp. 106143–106151, 2020.

[970] J. Xu, Y. Zhuo, Y. Zhan and Y. Yang, Maximum power point tracking control for mechanical rectification wave energy converter, *IET Renewable Power Generation*, vol. 15, pp. 3138–3150, 2021.

[971] X. Xu, S.D. Lee, H.S. Kim and S.S. You, Management and optimization of chaotic supply chain system using adaptive sliding-mode control algorithm, *International Journal of Production Research*, vol. 59, no.9, pp. 2571–2587, 2021.

[972] X. Xu, T.N. Cuong, H.S. Kim and S.S. You, Observer-based decision making strategy of supply chain management under business disruptions, *Journal of Control and Decision*, vol. 10, no. 4, pp. 504–513, 2022.

[973] X. Xu, H.S. Kim, S.S. You and S.D. Lee, Active management strategy for supply chain system using nonlinear control synthesis, *Journal of Dynamics and Control*, vol. 19, pp. 1981–1995, 2022.

[974] S. Xu, T. Chen, H. Wen and D. Jin, Digital deployment control of a tethered satellite system with constrained tension, *Advances in Space Research*, vol. 72, pp. 4184–4196, 2023.

[975] C. Xue, S. Tan and H. Liu, An improved iterative sliding-mode control for tethered satellite, In: *Proc. of the 36th Chinese Control Conference*, Dalian, China, July 2012.

[976] H. Yan, Y. Tang, G. Shen and S. Wang, Excavation trajectory research for mining electric shovel using sliding-mode control strategy based on disturbance observer, In: *IEEE ICICIP 2024, 12th International Conference on Intelligent Control and Information Processing*, Nanjing, China, March 2014.

[977] Y. Yan, C. Zhang, A. Narayan, J. Yang, S. Li and H. Yu, Generalized dynamic predictive control for non-parametric uncertain systems with application to series elastic actuators, *IEEE Transactions on Industrial Informatics*, vol. 14, no. 11, pp. 4829–4840, 2018.

[978] Z. Yan, X. Lai, O. Meng and M. Wu, A novel robust control method for motion control of uncertain single-link flexible-joint manipulator, *IEEE Transactions on Systems, Man and Cybernetics*, 2020.

[979] X. Yan, Z. Zhap, Y. Hu and J. Yang, Dynamic analysis of a bacterial resistance model with impulsive state feedback control, *Mathematical Biosciences and Engineering*, vol. 29, no. 12, pp. 20422–20436, 2023.

[980] L.D. Yang an Y.P. Sun, Mixed H_2/H_∞ state feedback design for micro-satellite attitude control, *Control Engineering Practice*, vol. 10, pp. 951–970, 2002.

[981] H. Yang, B. Jiang and V. Cocquempot, Decentralized fault-tolerant formation control for a class of tethered spacecraft, In: *IFAC Safeprocess 2015, 9th IFAC Symposium of Fault Detection, Supervision and Safety*, Paris, France, Sep. 2015.

[982] L. Yang, W. Chen, X. Liu and H. Zhou, Robust entry guidance using multi-segment linear pseudospectral model predictive control, *Journal of Systems Engineering and Electronics*, vol. 28, no. 1, pp. 103–125, 2017.

[983] X. Yang, G. Gong, H. Yang, L. Jiu and J. Zhou, An investigation to performance of a variable-speed displacement pump controlled motor system, *IEEE/ASME Transactions on Mechatronics*, vol. 22, no. 2, pp. 647–656, 2017.

[984] L. Yang, X. Liu, W. Chen and H. Zhou, Autonomous entry guidance using linear pseudospectral model predictive control, *Aerospace Science and Technology*, vol. 80, pp. 38–55, 2018.

[985] T. Yang, N. Sun, H. Chen and Y. Fang, Observer-based nonlinear control for tower cranes suffering from uncertain function and actuator constraints with experimental verification, *IEEE Transactions on Industrial Electronics*, vol. 68, no. 7, pp. 6192–6204, 2021.

[986] Y. Yang, P. Long, X. Sang, J. Pan and L. Zhang, Optimization-based framework for excavation trajectory generation, *IEEE Robotics and Automation Letters*, vol. 6, no. 2, pp. 1479–1487, 2021.

[987] T. Yang, N. Sun and Y. Fang, Neuroadaptive control for complicated underactuated systems with simultaneous output and velocity constraints exerted on both actuated and unactuated states, *IEEE Transactions on Neural Networks and Learning Systems*, 2022.

[988] T. Yang, H. Chen, N. Sun and Y. Fang, Adaptive neural network output feedback control of uncertain underactuated systems with actuated and unactuated state constraints, *IEEE Transactions on Systems, Man and Cybernetics*, 2022.

[989] S. Yang, Y. Zhang, H. Wen and D. Jin, Coordinated control of dual-arm robot on space structure for capturing space targets, *Advances in Space Research*, vol. 71, pp. 2437–2448, 2023.

[990] X.R. Yang, X. Lin, Y. Tang and A.M. Zou, Finite-time attitude tracking control of rigid spacecraft with multiple constraints, *IEEE Transactions on Aerospace and Electronic Systems*, vol. 60, no. 3, pp. 3688–3697, 2024.

[991] Y. Yang, B. Jin, J. Li, and C. Hua, Adaptive PID-like fault-tolerant control for uncertain H-type gantry stage with actuator failure, *IEEE Transactions on Industrial Electronics*, vol. 71, no. 11, pp. 14440–14448, 2024.

[992] Q. Yao, Adaptive trajectory tracking control of a free-flying space robot subject to input nonlinearities, *Journal of the Brazilian Society of Mechanical Sciences and Engineering*, vol. 42, no. 574, pp. 1–14, 2020.

[993] O. Yao, Adaptive trajectory tracking control of a free-flying space manipulator with guaranteed prescribed performance and actuator saturation, *Acta Astronautica*, vol. 185, pp. 283–298, 2021.

[994] Q. Yao, Adaptive fuzzy neural network control for a space manipulator in the presence of output constraints and input nonlinearities, *Advances in Space Research*, vol. 67, pp. 1830–1843, 2021.

[995] G. Yao, Z. Luo, Z. Li, M. Wang, J. Shang and J.M. Guerrero, Principle and control strategy of a novel wave-to-wire system embedded ocean energy optimization, *Ocean Engineering*, vol. 271, pp. 113762–113775, 2023.

[996] J. Ye and J. Huang, Control of beam-pendulum dynamics in a tower crane with a slender jib transporting a distributed mass-load *IEEE Transactions on Industrial Electronics*, 2022.

[997] L. Ye, B. Tian, H. Liu, Q. Zong, B. Liung and B. Yuan, Anti-windup robust backstepping control for an underactuated reusable launch vehicle, *IEEE Transactions on Systems, Man and Cybernetics: Systems*, vol. 52, no. 9, pp. 1492–1502, 2022.

[998] I. Yegorov and Y. Todorov, Synthesis of optimal control in a mathematical model of tumor-immune dynamics, *Optimal Control Applications and Methods*, vol. 96, pp. 93–108, 2015.

[999] J. Yi, N. Yubazaki and K. Hirota, A new fuzzy controller for stabilization of parallel-type double inverted pendulum system, *Fuzzy Sets and Systems*, vol. 126, pp. 105–119, 2002.

[1000] S.Y. Yoon, Z. Lin and P.E. Allaire, *Control of Surge in Centrifugal Compressors by Active Magnetic Bearings: Theory and Implementation*, Springer, 2013.

[1001] J. Yoon, S. Nution, W. Singhose and J.E. Vaugham, Control of crane payloads that bounce during hoisting, *IEEE Transactions on Control Systems Technology*, vol. 22, pp. 1233–1238, 2014.

[1002] J. Yoon, S.W. Hwang, J.H. Bak and J.H. Park, Vibration suppression of CDPRs based on differential flatness, In: *IEEE CCTA 2918, IEEE 2018 Conference on Control Theory and Applications*.

[1003] K. Yoshida and H. Ohta, Coupled inverted pendulum model of competition and cooperation, *Journal of Systems Design and Dynamics*, vol. 2, no. 3, pp. 727–737, 2008.

[1004] K. Yoshida, S. Matsu-Mata and Y. Matsue, Artificial wrestling: a dynamical formulation of autonomous agents fighting in a coupled inverted pendulum framework, *Mechanical Engineering Journal*, vol. 2, no. 5, pp. 1–11, 2015.

[1005] Y. Yoshimoto, T. Matsuno and S. Hokamoto, Position and attitude control of an underactuated satellite with constant thrust, In: *AIAA Guidance, Navigation, and Control Conference*, Portland, Oregon, USA, Aug. 2011.

[1006] M. You, Q. Zong, B. Tian, X. Zhou and F. Zong, Comprehensive decision of uniform robust exact disturbance observer and fixed-time controller for reusable launch vehicles, *IET Control Theory and Applications*, vol. 12, no. 5, pp. 638–648, 2018.

[1007] Y. You, G. Ren and S. Yu, Division linearization zero-bias current control for AMBs-rotor system with uncertainties and saturation, *IEEE Transactions on Industrial Electronics*, vol. 70, no. 10, pp. 10557–10566, 2023.

[1008] P. Yousefian and H. Salarieh, Nonlinear control of sway in a tethered satellite system via attitude control of the main satellite, *Aerospace Science and Technology*, vol. 63, pp. 317–327, 2017.

[1009] K. Yovchev and L. Miteva, Payload adaptive iterative learning control for robotic manipulators, *Mechanical Sciences*, vol. 13, pp. 427–326, 2022.

[1010] J. Yu, L. Nan and L. Huang, Controller design of flexible double-indented pendulum with uncertainties based on the T-S fuzzy inference system, In: *Lecture Notes in Electrical Engineering*, vol. 338, pp. 233–243, 2015.

[1011] H. Yu, S. Huang, G. Chen and Z. Gao, Human-robot interaction control of rehabilitation robots with series elastic actuators, *IEEE Transactions on Robotics*, vol. 31, no. 5, pp. 1089–1100, 2015.

[1012] B.S. Yu, D.P. Jin and H. Wu, An analytical control law of length rate of tethered satellite system, *Meccanica*, vol. 52, pp. 2035–2046, 2017.

[1013] Y. Yu, Z. Yung, C. Hun and H. Liu, Active vibration control of magnetically suspended wheel using active shaft deflection, *IEEE Transactions on Industrial Electronics*, vol. 64, no. 8, pp. 6329–6337, 2017.

[1014] B.S. Yu, Z.H. Huang, L.L. Geng and D.P. Jin, Stability and ground experiments of a spinning triangular tethered satellite formation on a low earth orbit, *Aerospace Science and Technology*, vol. 32, pp. 595–604, 2019.

[1015] J. Yu and X. Zhang, The global control of first-order rotary parallel double inverted pendulum systems, In: *IEEE CCC 2021, 40th Chinese Control Conference*, Shanghai, China, July 2021.

[1016] Z. Yu, L. Liu, J. Yu, X. Zhang, Y. Guo and S. Wang, EFC/H_∞ based dual mode switching global control of the first-order parallel rotating double inverted pendulum system, In: *IEEE IECON 2022, IEEE 49th Annual Conference of the IEEE Industrial Electronics Society*, Brussels, Belgium, Dec. 2022.

[1017] B.S. Yu, K. Ji, Z.T. Wei and D.P. Jin, In-plane global dynamics and ground experiment of a linear tethered formation with three satellites, *Nonlinear Dynamics*, vol. 108, pp. 3247–3278, 2022.

[1018] B.S. Yu and Z.H. Zhu, Global dynamics of a ring-tethered three-satellite system in any plane, *Nonlinear Dynamics*, vol. 111, pp. 11811–11821, 2023.

[1019] C. Yu, J. Ma, H. Pan and M.V. Basin, Adaptive iterative learning constrained control for linear-motor-driven gantry stage, *IEEE/ASME Transactions on Mechatronics*, vol. 29, no. 3, pp. 1647–1658, 2024.

[1020] P. Yu, S. Tan, J. Guo, and Y. Song, Data-driven optimal controller design for sub-satellite deployment of tethered satellite system, *Electronic Research Archive*, vol. 32, no. 1, pp. 505–522, 2024.

[1021] B.S. Yu and Z.H. Zhu, Characterization of out-of-plane libration of tethered satellite systems in elliptical orbits. *Acta Astronautica*, vol. 219, pp. 108–114, 2024.

[1022] P. Yu, J.F, Zhang, S. Tan, J. Guo and Y. Zhao, On the optimality of uniform velocity–deceleration separation scheme for tethered satellite systems, *Journal of Control Theory and Technology*, vol. 22, pp. 503–516, 2024.

[1023] J.N. Yun and J.B. Sun, Design of a disturbance observer for a two-link manipulator with flexible-joints, *IEEE Transactions on Control Systems Technology*, vol. 22, no. 2, pp. 809–815, 2014.

[1024] L. Yun, J. Liu, F. Xu, K.L. Teo and M. Lai, Control and synchronization of hyperchaos in digital manufacturing supply chain, *Applied Mathematics and Computation*, vol. 331, pp. 125646–125654, 2021.

[1025] M. Zagorowska, N. Thornhill, T. Haugen and Skourup, Load sharing strategy taking account of compressor degradation, In: *IEEE CCTA 2018, IEEE 2018 Conference on Control Technology and Applications*, Copenhagen, Denmark, Aug. 2018.

[1026] M. Zagorowska, C Skourup and N.T. Thornhill, Influence of compressor degradation on optimal operation of a compressor system, *Computers and Chemical Engineering*, vol. 143, pp. 107104, 2020.

[1027] M. Zagorowska and N.F. Thornhill, Optimal operation of compressors in terms of power and efficiency, In: *IEEE 2021 Mediterranean Conf. on Control and Automation*, Bari, Italy, June 2021.

[1028] A. Zanati, M. Choker and M. Tadjine, Global stability analysis and optimal control therapy of blood cell production process (hematopoiesis) in acute myeloid leukemia, *Journal of Theoretical Biology*, vol. 458, pp. 15–30, 2018.

[1029] A. Zein, S. Karaki, M. Al-Hindi, Analysis of variable reverse osmosis operation powered by solar energy, *Renewable Energy*, vol. 208, pp. 385–398, 2023.

[1030] M. Zerar, F. Cazaurang and A. Zolghadri, LPV modelling of atmospheric reentry demonstrator for guidance reentry problem, In: *IEEE CDC 2006, IEEE 44th Conference on Decision and Control and the European Control Conference*, Seville, Spain, Dec. 2005.

[1031] M. Zerar, F. Cazaurang and A. Zolghadri, Coupled LPV and flatness-based approach for space reentry vehicles guidance, *IET Control Theory and Applications*, vol. 3, no. 8, pp. 1081–1092, 2009.

[1032] A. Zhai, J. Wang, H. Zhang, G. Lu and H. Li, Adaptive robust synchronized control for cooperative manipulators with uncertain base coordinate system, *ISA Transactions*, vol. 126, pp. 134–143, 2022.

[1033] M. Zhai, T. Yang, N. Sun and Y. Fang, Observer-based adaptive fuzzy control of underactuated offshore cranes for cargo stabilization with respect to ship decks, *Mechanism and Machine Theory*, vol. 175, pp. 104927–104942, 2022.

[1034] S. Zhang, X. Li and C. Zhang, A fuzzy control model for restraint of bullwhip effect in uncertain closed-loop supply chain with hybrid recycling channels, *IEEE Transactions on Fuzzy Systems*, vol. 25, no. 2, pp. 475–482, 2017.

[1035] Y. Zhao, J. Jia, Y. Dong and K. Meng, Optimal dosing strategy and sensitivity analysis of a within-host drug resistance model with continuous and impulsive drug treatment, *Journal of Applied Mathematics and Computing*, vol. 69, pp. 2277–2293, 2023.

[1036] Q. Zhang, M. Basseville, and A. Benveniste, Fault detection and isolation in nonlinear dynamic systems: a combined input-output and local approach, *Automatica*, vol. 34, no. 11, pp. 1359–1373, 1998.

[1037] Q. Zhang, Adaptive observer for multiple input-multiple-output (MIMO) linear time-varying systems. *IEEE Transactions on Automatic Control*, vol. 47, no. 3, pp. 525–529, 2002.

[1038] C. Zhang, S. Fuchun and C. Jing, Disturbance observer-based robust control of free-floating space manipulators, *IEEE Systems Journal*, vol. 2, no. 1, pp. 114–122, 2008.

[1039] W.H. Zhang, X. Yu, L. Jiang, Y. Zhu, X. Ji and X. Hu, Output feedback control for free-floating space robotic manipulators based on adaptive fuzzy neural network, *Aerospace Science and Technology*, vol. 29, pp. 135–143, 2013.

[1040] Q. Zhang and M. Basseville, Statistical detection and isolation of additive faults in linear time-varying systems, *Automatica*, vol. 50, no. 10, pp. 2577–2538, 2014.

[1041] Q. Zhang, Stochastic Hybrid System Actuator Fault Diagnosis by Adaptive Estimation. In: *IFAC Safeprocess 2015. 9th IFAC Symposium on Fault Detection, Supervision and Safety for Technical Processes*, Paris, France, 2015.

[1042] F. Zhang, Y. Fu, J. Qu and S. Wang, Robust adaptive control of a free-floating space robot system in Cartesian space, *Journal of Advanced Robotic Systems*, vol. 12, pp. 1–11, 2015.

[1043] S. Zhang and X. Zhao, Fuzzy robust control for uncertain switched dual-channel closed-loop supply chain model, *IEEE Transactions on Fuzzy Systems*, vol. 23, no. 3, pp. 485–500, 2015.

[1044] Z. Zhang, H. Huang, B. Jiang, V. Cocquempot and Y. Chen, Decentralized fault-tolerant control of tethered spacecraft formation: an interconnected system approach, *IET Control Theory and Applications*, vol. 11, no. 17, pp. 3047–3055, 2017.

[1045] Z. Wenhui, L. Hongsheng, Y. Xiaoping, H. Jiacai and H. Mingying. Adaptive robust control for free-floating space robot with unknown uncertainty based on neural network, *International Journal of Advanced Robotic Systems*, vol. 15, no. 6, pp. 1–11, 2018.

[1046] W.H. Zhang, H. Li, X. Ye, J. Huang and M. Huo, Adaptive robust control for free-floating space robot with unknown uncertainty based on neural network, *Journal of Advanced Robotic Systems*, vol. 15, no. 6, pp. 1–11, 2018.

[1047] Q. Zhang, Adaptive Kalman filter for actuator fault diagnosis, *Automatica*, vol. 93, pp. 333–342, 2018.

[1048] X. Zhang, S. Qiao, L. Quan and L. Ge, Velocity and position hybrid control for excavator based on impedance metering system, *IEEE Access*, vol. 7, pp. 71999–72011, 2019.

[1049] F. Zhang and P. Huang, A novel underactuated control scheme for deployment/retrieval of space-tethered system, *Nonlinear Dynamics*, vol. 95, pp. 3465–3476, 2019.

[1050] G. Zhang, H. Yi and W. Dou, Design of dual-drive vertical lift servo system and synchronous control performance analysis, *IEEE/ASME Transactions on Mechatronics*, vol. 25, no. 6, pp. 2927–2937, 2020.

[1051] M. Zhang, Y. Zhang, B.J.C. Ma and X. Cheng, Adaptive sway reduction for tower crane systems with variable cable lengths, *Automation in Construction*, vol. 119, pp. 10342–10351, 2020.

[1052] M. Zhang, Y. Zhang, B.J.C Ma and X. Cheng, Modeling and energy-based sway reduction control for tower-crane systems with double-pendulum and spherical-pendulum effects, *Measurement and Control*, vol. 53, no. 1–2. pp. 141–150, 2020.

[1053] S. Zhang, X. He, Q. Chen and Y. Feng, Improved energy dissipation control of overhead cranes, *Asian Journal of Control*, vol. 20, no. 4, pp. 1729–1735, 2020.

[1054] Y. Zhang, P. Zhang, G. Fang and P. Huang, A state estimation scheme with minimal sensor configuration for three-body tethered satellite formations, *Advances in Space Research*, vol. 66, pp. 2667–2680, 2020.

[1055] M. Zhang, X. Jing and Z. Zhu, Disturbance observer-based sliding-mode control for 4-DOF tower crane systems, *Mechanical Systems and Signal Processing*, vol. 161, pp. 107946–107954, 2021.

[1056] T. Zhang, T. Fu, X. Song and F. Qu, Multi-objective excavation trajectory/optimization for unmanned electric shovels based on pseudo-spectral models, *Automation in Construction*, vol. 136, pp. 104176–104191, 2022.

[1057] Q. Zhang and S. Tang, Bifurcation analysis of an ecological model with nonlinear state-dependent feedback control by Poincaré map defined in phase set, *Communications in Nonlinear Science and Numerical Simulation*, vol. 108, pp. 106212–106236, 2022.

[1058] M. Zhang and X. Jing, Model-free saturated PD-SMC method for 4-DOF tower crane systems, *IEEE Transactions on Industrial Electronics*, vol. 69, no. 10, pp. 10270–10280, 2022.

[1059] M. Zhang and X. Jing, Adaptive neural network tracking control for double-pendulum tower crane systems with nonideal inputs, *IEEE Transactions on Systems, Man and Cybernetics*, vol. 52, no. 4, pp. 254–2530, 2022.

[1060] J. Zhang, H. Hu and M. Chen, Direct-drive wave energy conversion with linear generator: a review of research status and challenges, *IET Renewable Power Generation*, vol. 1, pp. 1020–1034, 2023.

[1061] Q. Zhang, R. Yuan, Y. Lv, Z. Li and H. Wu, Multivariate direct mode decomposition and its application to bearing fault diagnosis, *IEEE Sensors Journal*, vol. 23, no. 7, pp. 7514–7524, 2023.

[1062] Y. Zhang, S. Wang and T. Liu, A waveform similarity-based protection scheme for the VSC-HVDC transmission lines, *International Journal of Electrical Power and Energy Systems*, vol. 144, article no. 108571, 2023.

[1063] Y. Zhang and G. Li, Robust tube-based model predictive control for wave energy converters, *IEEE Transactions on Sustainable Energy*, vol. 14, no. 1, pp. 65–74, 2023

[1064] T. Zhang, P. Shi, W. Li and X. Yue, EKF enhanced MPC for rapid attitude stabilization of space robots with bounded control torque in postcapture, *Journal of the Franklin Institute*, vol. 360, pp. 7105–7127, 2023.

[1065] C. Zhang, X. Yan, H. Pan, H. Gao and J.J. Rodriguez-Andina, Integrated control of linear motors for coordination and synchronization, *IEEE Transactions on Transportation Electrification*, pp. 1–11, 2023.

[1066] S. Zhang, H. Zhou, X. He, Y. Feng and C.K. Pang, Passivity-based coupling control for underactuated three-dimensional overhead cranes, *ISA Transactions*, vol. 126, pp. 352–360, 2022.

[1067] F. Zhang, H. Zhou and P. Huang, Stable spinning deployment control of triangle tethered formation system, *IEEE Transactions on Cybernetics*, vol. 52, no. 11, pp. 11442–11452, 2022.

[1068] Y. Zhang, X. Zhao, B. Tao and H. Ding, Multi-objective synchronization control for dual-robot interactive cooperation using nonlinear model predictive policy, *IEEE Transactions on Industrial Electronics*, vol. 70 no. 1, pp. 582–592, 2023.

[1069] J. Zhang, H. Yu and M. Chen, Direct drive wave energy conversion with linear generator: a review of research status and challenges, *IET Renewable Power Generation*, vol. 17, pp. 1020–1034, 2023.

[1070] C. Zhang, J. Wang and H. Pan, Safety control of precision motion system with gantry structure based on fault-tolerant gradient descent B-spline wavelet neural network, *Control Engineering Practice*, vol. 149, pp. 105971–105989, 2024.

[1071] L. Zhao, B. Zhang, H. Yang and Y. Wang, Observer-based integral sliding-mode tracking control for a pneumatic cylinder with varying loads, *IEEE Transactions on Systems, Man and Cybernetics*, vol. 50, no. 7, pp. 2650–2658, 2020.

[1072] X. Zhao, Y. Zhang, W. Ding, B. Tao and H. Ding, A dual-arm robot cooperation framework based on nonlinear model predictive cooperative control, *IEEE/ASME Transactions on Mechatronics*, pp. 1–13, 2023.

[1073] S. Zhen, Z. Zhou, X. Liu, F. Chen, H. Zhao and Y.H. Chen, A novel practical robust control inheriting PID for SCARA robot, *IEEE Access*, vol. 8, pp. 227407–227419, 2020.

[1074] J. Zheng, H. Wang, Z. Man, J. Lin and M. Fu, Robust motion control of a linear motor positioner using fast nonsingular terminal sliding-mode, *IEEE/ASME Transactions on Mechatronics*, vol. 20, no. 4, pp. 1743–1752, 2015.

[1075] M. Zheng, Y. Zhang, L. Fan, Y. Xu, Retrieval dynamics and control for approach of tethered on-orbit service satellite, *Advances in Space Research*, vol. 71, pp. 4995–5006, 2023.

[1076] Y. Zhou, X. Xiu, M. Qadrdan and J. Wu, Optimal operation of compressor units in gas networks to provided flexibility to power systems, *Applied Energy*, vol. 290, pp. 116740–116750, 2022.

[1077] G. Zhou, L. Sun and H. Huang, Optimal attitude control of a tethered system for noncoplanar orbital transfer under a constant thrust, *IEEE Transactions on Aerospace and Electronic Systems*, vol. 56, no. 3, pp. 1814–1855, 2020.

[1078] C. Zhou, G. Zhang and Q. Dong, Fixed-time disturbance observer-based nearly optimal control for reusable launch vehicle with input constraints, *ISA Transactions*, vol. 122, pp. 182–197, 2022.

[1079] X. Zhou, S. Zou, W.W. Weaver and O. Abdelhalik, Assessment of electrical power generation of wave energy converters with wave-to-wire modeling, *IEEE Transactions on Sustainable Energy*, vol. 13, no. 3, pp. 1654, 1665, 2022.

[1080] Y. Zhou, J. She, F. Wang and M. Iwasaki, Disturbance rejection for Stewart platform based on integration of equivalent-input-disturbance and sliding-mode control methods, *IEEE/ASME Transactions on Mechatronics*, vol. 29, no. 4, pp. 2364–2374, 2023.

[1081] H. Zhou, B. Wang, E. Zio, *et al.*, Hybrid system response model for condition monitoring on bearings under time-varying operating conditions, *Reliability Engineering and Systems Safety*, vol. 234, pp. 109528–109560, 2023.

[1082] T. Zhou and C. Zhi, Output feedback dynamical compensation via eigenstructure assignment for active magnetic bearings-rigid rotor systems, *IEEE Transactions on Industrial Electronics*, vol. 71, no.2, pp. 1328–1340, 2024.

[1083] C.L. Zhu and Y.Q. Wu, Energy-based controller for parallel-type double inverted pendulum with restricted cart rail length, In: *IEEE CCC 2015, Proc. of the 34th Chinese Control Conference*, Hangzhou, China, July 2015.

[1084] Y. Zhu, J. Qiao and L. Guo, Adaptive sliding-mode disturbance observer-based composite control with prescribed performance of space manipulators for target capturing, *IEEE Transactions on Industrial Electronics*, vol. 66, no. 5, pp. 1873–1882, 2019.

[1085] B. Zhu, T. Zhao, Z. Tang, S. Ding and E. Li, Hierarchical coupling control of cable-driven multi-loop crane for underactuated positioning, *International Journal of Mechanical Sciences*, vol. 259, pp. 108620–108645, 2023.

[1086] Z. Zhu, J. Deng, H. Ouyama and X. Dou, Optimized sampling mechanism for full-state current control of LCL-equipped high-speed PMSM for fuel-cell air compressor, *IEEE Transactions on Transportation Electrification*, pp. 1–12, 2023.

[1087] Q. Zhu, D. Huang, B. Yu, K. Ba, X. Kang and S. Wang, An improved method combined SMC and MLESO for impedance control of legged robot's electrohydraulic servo system, *ISA Transactions*, pp. 1–12, 2022.

[1088] Z. Zhu, G. Zhang, Y. Li, *et al.*, Observer-based state feedback model predictive control framework for legged robots, *IEEE/ASME Transactions on Mechatronics*, pp. 1–11, 2024.

[1089] M.T. Ziabari, M.R. Jahed-Motlagh, K. Salashoor, A. Ramazani and A. Mourefianpur, Robust adaptive control of surge instability in constant speed centrifugal compressors using tube-MPC, *Cogent Engineering*, vol. 4, no. 1, pp. 1–21, 2017.

[1090] E.M. Zilanova and A.S. Bratus, Optimal strategies in antibiotic treatment of microbial populations, *Journal of Applicable analysis*, vol. 95, no. 7, pp. 1534–1547, 2016.

[1091] T. Zilic, D. Pavkovic and D. Zorc, Modelling and control for a pneumatically actuated inverted pendulum, *ISA Transactions*, vol. 48, pp. 327–345, 2009.

[1092] N. Zoso and C. Gosselin, Point-to-point motion planning of a parallel 3-DOF underactuated cable-suspended robot, In: *IEEE Intl. Conf. on Robotics and Automation*, Saint Paul, Minnesota, May 2012.

[1093] A.M. Zou, A.H.J. de Ruiter, and K. Dev Kumar, Finite-time attitude tracking control for rigid spacecraft with control input constraints, *IET Control Theory and Applications*, vol. 11, no. 7, pp. 931–940, 2016.

[1094] A.M. Zou and Z. Fan, Fixed-time attitude tracking control for rigid spacecraft without angular velocity measurements, *IEEE Transactions on Industrial Electronics*, vol. 67, no. 8, pp. 6795–6805, 2020.

[1095] Q. Zou, D. Zhang, and G. Huang, Kinematic joint matrix and block diagram for a group of parallel manipulators, *Robotica*, vol. 41, pp. 939–961, 2023.

[1096] F. Zouari, K, Ben-Saad and M. Benrejeb, Adaptive backstepping control for a single-link flexible robot manipulator driven DC motor, In: *IEEE CoDIT 2013, IEEE 2013 International Conference on Control, Decision and Information Technologies*, Hammamet, Tunisia, May 2013.

[1097] M. Zribi, S. Alshamali and M. Al-Kendari, Suppression of the wing-rock phenomenon using nonlinear controllers, *Nonlinear Dynamics*, vol. 71, pp. 313–322, 2013.

[1098] F. Zouari, Neural network-based adaptive backstepping dynamic surface control of drug dosage regime in cancer treatment, *Neurocomputing*, vol. 366, pp. 249–263, 2019.

Index